T0417891

CONCISE ENCYCLOPEDIA OF

BIOMEDICAL POLYMERS AND POLYMERIC BIOMATERIALS

VOLUME II

Melt Extrusion — Zwitterionic Polymeric Materials

Encyclopedias from the Taylor & Francis Group

Print	Online

Agriculture

Encyclopedia of Agricultural, Food, and Biological Engineering,
2nd Ed., 2 Vols. **Pub'd. 10/21/10**
K10554 (978-1-4398-1111-5) K11382 (978-1-4398-2806-9)

Encyclopedia of Animal Science, 2nd Ed., 2 Vols. **Pub'd. 2/1/11**
K10463 (978-1-4398-0932-7) K10528 (978-0-415-80286-4)

Encyclopedia of Biotechnology in Agriculture and Food
Pub'd. 7/16/10
DK271X (978-0-8493-5027-6) DKE5044 (978-0-8493-5044-3)

Encyclopedia of Pest Management **Pub'd. 5/9/02**
DK6323 (978-0-8247-0632-6) DKE517X (978-0-8247-0517-6)

Encyclopedia of Plant and Crop Science **Pub'd. 2/27/04**
DK1190 (978-0-8247-0944-0) DKE9438 (978-0-8247-0943-3)

Encyclopedia of Soil Science, 3rd Ed., 3 Vols. **Pub'd 11/1/ 2016**
K26612 (978-1-4987-3890-3) K26614 (978-1-4987-3893-4)

Encyclopedia of Water Science, 2nd Ed., 2 Vols. **Pub'd. 12/26/07**
DK9627 (978-0-8493-9627-4) DKE9619 (978-0-8493-9619-9)

Business and Computer Science

Encyclopedia of Computer Sci. & Tech., 2nd Ed., 2 Vols.
Pub'd 12/21/2016
K21573 (978-1-4822-0819-1) K21578 (978-1-4822-0822-1)

Encyclopedia of Information Assurance, 4 Vols. **Pub'd. 12/21/10**
AU6620 (978-1-4200-6620-3) AUE6620 (978-1-4200-6622-7)

Encyclopedia of Information Systems and Technology, 2 Vols.
Pub'd. 12/29/15
K15911 (978-1-4665-6077-2) K21745 (978-1-4822-1432-1)

Encyclopedia of Library and Information Sciences, 4th Ed.
Publishing 2017
K15223 (978-1-4665-5259-3) K15224 (978-1-4665-5260-9)

Encyclopedia of Software Engineering, 2 Vols. **Pub'd. 11/24/10**
AU5977 (978-1-4200-5977-9) AUE5977 (978-1-4200-5978-6)

Encyclopedia of Supply Chain Management, 2 Vols. **Pub'd. 12/21/11**
K12842 (978-1-4398-6148-6) K12843 (978-1-4398-6152-3)

Encyclopedia of U.S. Intelligence, 2 Vols. **Pub'd. 12/19/14**
AU8957 (978-1-4200-8957-8) AUE8957 (978-1-4200-8958-5)

Encyclopedia of Wireless and Mobile Communications,
2nd Ed., 3 Vols. **Pub'd. 12/18/12**
K14731 (978-1-4665-0956-6) KE16352 (978-1-4665-0969-6)

Chemistry, Materials and Chemical Engineering

Encyclopedia of Chemical Processing, 5 Vols. **Pub'd. 11/1/05**
DK2243 (978-0-8247-5563-8) DKE499X (978-0-8247-5499-0)

Encyclopedia of Chromatography, 3rd Ed. **Pub'd. 10/12/09**
84593 (978-1-4200-8459-7) 84836 (978-1-4200-8483-2)

Encyclopedia of Iron, Steel, and Their Alloys, 5 Vols. **Pub'd. 1/6/16**
K14814 (978-1-4665-1104-0) K14815 (978-1-4665-1105-7)

Encyclopedia of Plasma Technology, 2 Vols. **Pub'd 12/12/2016**
K14378 (978-1-4665-0059-4) K21744 (978-1-4822-1431-4)

Encyclopedia of Supramolecular Chemistry, 2 Vols. **Pub'd. 5/5/04**
DK056X (978-0-8247-5056-5) DKE7259 (978-0-8247-4725-1)

Encyclopedia of Surface & Colloid Science, 3rd Ed., 10 Vols.
Pub'd. 8/27/15
K20465 (978-1-4665-9045-8) K20478 (978-1-4665-9061-8)

Engineering

Dekker Encyclopedia of Nanoscience and Nanotechnology,
3rd Ed., 7 Vols. **Pub'd. 3/20/14**
K14119 (978-1-4398-9134-6) K14120 (978-1-4398-9135-3)

Encyclopedia of Energy Engineering and Technology, 2nd Ed.,
4 Vols. **Pub'd. 12/1/14**
K14633 (978-1-4665-0673-2) KE16142 (978-1-4665-0674-9)

Encyclopedia of Optical and Photonic Engineering, 2nd Ed.,
5 Vols. **Pub'd. 9/22/15**
K12323 (978-1-4398-5097-8) K12325 (978-1-4398-5099-2)

Environment

Encyclopedia of Environmental Management, 4 Vols.
Pub'd. 12/13/12
K11434 (978-1-4398-2927-1) K11440 (978-1-4398-2933-2)

Encyclopedia of Environmental Science and Engineering,
6th Ed., 2 Vols. **Pub'd. 6/25/12**
K10243 (978-1-4398-0442-1) KE0278 (978-1-4398-0517-6)

Encyclopedia of Natural Resources, 2 Vols. **Pub'd. 7/23/14**
K12418 (978-1-4398-5258-3) K12420 (978-1-4398-5260-6)

Medicine

Encyclopedia of Biomaterials and Biomedical Engineering,
2nd Ed. **Pub'd. 5/28/08**
H7802 (978-1-4200-7802-2) HE7803 (978-1-4200-7803-9)

Encyclopedia of Biomedical Polymers and Polymeric Biomaterials,
11 Vols. **Pub'd. 4/2/15**
K14324 (978-1-4398-9879-6) K14404 (978-1-4665-0179-9)

Concise Encyclopedia of Biomedical Polymers and Polymeric
Biomaterials, 2 Vols. **Publishing 2017**
K14313 (978-1-4398-9855-0) KE42253 (978-1-315-11644-0)

Encyclopedia of Biopharmaceutical Statistics, 3rd Ed.
Pub'd. 5/20/10
H100102 (978-1-4398-2245-6) HE10326 (978-1-4398-2246-3)

Encyclopedia of Clinical Pharmacy **Pub'd. 11/14/02**
DK7524 (978-0-8247-0752-1) DKE6080 (978-0-8247-0608-1)

Encyclopedia of Dietary Supplements, 2nd Ed. **Pub'd. 6/25/10**
H100094 (978-1-4398-1928-9) HE10315 (978-1-4398-1929-6)

Encyclopedia of Medical Genomics and Proteomics, 2 Vols.
Pub'd. 12/29/04
DK2208 (978-0-8247-5564-5) DK501X (978-0-8247-5501-0)

Encyclopedia of Pharmaceutical Science and Technology,
4th Ed., 6 Vols. **Pub'd. 7/1/13**
H100233 (978-1-84184-819-8) HE10420 (978-1-84184-820-4)

Routledge Encyclopedias

Encyclopedia of Public Administration and Public Policy,
3rd Ed., 5 Vols. **Pub'd. 11/6/15**
K16418 (978-1-4665-6909-6) K16434 (978-1-4665-6936-2)

Routledge Encyclopedia of Modernism **Pub'd. 5/11/16**
Y137844 (978-1-135-00035-6)

Routledge Encyclopedia of Philosophy Online **Pub'd. 11/1/00**
RU22334 (978-0-415-24909-6)

Routledge Performance Archive **Pub'd. 11/12/12**
Y148405 (978-0-203-77466-3)

CONCISE ENCYCLOPEDIA OF

BIOMEDICAL POLYMERS AND POLYMERIC BIOMATERIALS

VOLUME II

Melt Extrusion — Zwitterionic Polymeric Materials

edited by

Munmaya Mishra

CRC Press
Taylor & Francis Group
Boca Raton London New York

CRC Press is an imprint of the
Taylor & Francis Group, an **informa** business

CRC Press
Taylor & Francis Group
6000 Broken Sound Parkway NW, Suite 300
Boca Raton, FL 33487-2742

© 2018 by Taylor & Francis Group, LLC
CRC Press is an imprint of Taylor & Francis Group, an Informa business

No claim to original U.S. Government works

Printed in Canada on acid-free paper

International Standard Book Number-13: 978-1-4398-9855-0 (Set) 978-1-138-56314-8 (Volume I) 978-1-138-56318-6 (Volume II)

Visit the Taylor & Francis Web site at
http://www.taylorandfrancis.com

and the CRC Press Web site at
http://www.crcpress.com

Volume I

Volume II

To my family

Also, to those who made and will make a difference through polymer research for improving the quality of life and for the betterment of human health.

Concise Encyclopedia of Biomedical Polymers and Polymeric Biomaterials

Editor-in-Chief
Munmaya K. Mishra, Ph. D.
c/o Altria Research Center, Richmond, Virginia U.S.A.

Contributors

Carlos A. Aguilar / *Department of Mechanical Engineering, University of Texas at Austin, Austin, Texas, U.S.A.*

M.R. Aguilar / *Spanish National Research Council Institute of Polymer Science and Technology, Madrid, Spain*

Arti Ahluwalia / *E. Piaggio Interdepartmental Research Center, Department of Chemical Engineering, University of Pisa, Pisa, Italy*

Nasir M. Ahmad / *Polymer and Surface Engineering Lab, Department of Materials Engineering, School of Chemical and Materials Engineering (SCME), National University of Sciences and Technology (NUST), Islamabad, Pakistan*

Naveed Ahmed / *Department of Pharmacy, Quaid-i-Azam University, Islamabad, Pakistan*

Tamer A. E. Ahmed / *Medical Biotechnology Department, Genetic Engineering and Biotechnology Research Institute, City of Scientific Research and Technology Applications (SRTA-City), Alexandria, Egypt Department of Cellular and Molecular Medicine, Faculty of Medicine, University of Ottawa, Ottawa, Ontario, Canada*

A.K.M. Moshiul Alam / *Institute of Radiation and Polymer Technology, Bangladesh Atomic Energy Commission, Dhaka, Bangladesh*

A.-C. Albertsson / *Department of Polymer Technology, Royal Institute of Technology (KTH), Stockholm, Sweden*

Carmen Alvarez-Lorenzo / *Department of Pharmacy and Pharmaceutical Technology, University of Santiago de Compostela (USC), Santiago de Compostela, Spain*

Garima Ameta / *Sonochemistry Laboratory, Department of Chemistry, University College of Science, Mohanlal Sukhadia University, Udaipur, India*

Rakshit Ameta / *Department of Chemistry, Pacific College of Basic and Applied Sciences, Pacific Academy of Higher Education and Research Society (PAHER) University, Udaipur, India*

Suresh C. Ameta / *Department of Chemistry, PAHER University, Udaipur, India*

Sambandam Anandan / *Nanomaterials and Solar Energy Conversion Lab, Department of Chemistry, National Institute of Technology, Tiruchirappalli, India*

S. Anandhakumar / *Department of Physics and Nanotechnology, Sri Ramaswamy Memorial (SRM) University, Kattankulathur, Chennai, India*

Fernanda Andrade / *Laboratory of Pharmaceutical Technology (LTFCICF), Faculty of Pharmacy, University of Porto, Porto, Portugal*

Kirk P. Andriano / *APS Research Institute, Redwood City, California, U.S.A.*

Bruce L. Anneaux / *Poly-Med, Incorporated, Anderson, South Carolina, U.S.A.*

Ilaria Armentano / *Materials Science and Technology Center, University of Perugia, Terni, Italy*

Alexandra Arranja / *Research Institute for Medicine and Pharmaceutical Sciences (iMed. UL), School of Pharmacy, University of Lisbon, Lisbon, Portugal*

Hamed Asadi / *Polymer Chemistry Department, School of Science, University of Tehran, Tehran, Iran*

Murat Ates / *Department of Chemistry, Faculty of Arts and Sciences, Namik Kemal University, Degirmenalti Campus, Tekirdag, Turkey*

A.A. Attama / *Department of Pharmaceutics, Faculty of Pharmaceutical Sciences, University of Nigeria, Nsukka, Nigeria*

Vimalkumar Balasubramanian / *Division of Pharmaceutical Technology, University of Basel, Basel, Switzerland*

Florence Bally / *Institute of Materials Science of Mulhouse (IS2M), University of Upper Alsace, Mulhouse, France, and Institute of Functional Interfaces (IFG), Karlsruhe Institute of Technology (KIT), Eggenstein-Leopoldshafen, Germany*

Indranil Banerjee / *Department of Biotechnology and Medical Engineering, National Institute of Technology, Rourkela, India*

Clark E. Barrett / *Department of Materials Science and Metallurgy, University of Cambridge, Cambridge, U. K.*

Beauty Behera / *Department of Biotechnology and Medical Engineering, National Institute of Technology, Rourkela, India*

Andreas Bernkop-Schnürch / *Institute of Pharmacy, University of Innsbruck, Innsbruck, Austria*

Serena M. Best / *Department of Materials Science and Metallurgy, University of Cambridge, Cambridge, U. K.*

Kumudini Bhanat / *Department of Chemistry, University College of Science, Mohanlal Sukhadia University, Udaipur, India*

Maria Cristina Bonferoni / *Department of Drug Sciences, University of Pavia, Pavia, Italy*

Kellye D. Branch / *Biomedical Engineering Program, Agricultural and Biological Engineering Department, Mississippi State University, Mississippi State, Mississippi, U.S.A.*

Chad Brown / *Merck & Co., Inc., West Point, Pennsylvania, U.S.A.*

Toby D. Brown / *Institute for Health and Biomedical Innovation, Queensland University of Technology, Brisbane, Queensland, Australia*

Joel D. Bumgardner / *Biomedical Engineering Program, Agricultural and Biological Engineering Department, Mississippi State University, Mississippi State, Mississippi, U.S.A.*

K. J. L. Burg / *Department of Bioengineering, Clemson University, Clemson, South Carolina, U.S.A.*

Marine Camblin / *Division of Pharmaceutical Technology, University of Basel, Basel, Switzerland*

Ruth E. Cameron / *Department of Materials Science and Metallurgy, University of Cambridge, Cambridge, U. K.*

Carla Caramella / *Department of Drug Sciences, University of Pavia, Pavia, Italy*

Yu Chang / *Department of Obstetrics and Gynecology, Kaohsiung Medical University Hospital, Kaohsiung Medical University, Kaohsiung, Taiwan*

Robert Chapman / *Key Center for Polymers and Colloids, School of Chemistry, University of Sydney, Sydney, New South Wales, Australia*

Jhunu Chatterjee / *Florida A&M University and College of Engineering, Florida State University, Tallahassee, Florida, U.S.A.*

Jyoti Chaudhary / *Department of Polymer Science, University College of Science, Mohanlal Sukhadia University, Udaipur, India*

Narendra Pal Singh Chauhan / *Department of Chemistry, Bhupal Nobles' Post-Graduate (B.N.P.G.) College, Udaipur, India*

Guoping Chen / *Tissue Regeneration Materials Unit, International Center for Materials Nanoarchitectonics, National Institute for Materials Science, Tsukuba, Japan*

Jianjun Chen / *Department of Materials Science and Engineering, University of Illinois at Urbana-Champaign, Urbana, Illinois, U.S.A.*

Shaochen Chen / *Department of Mechanical Engineering, University of Texas at Austin, Austin, Texas, U.S.A.*

Yung Hung Chen / *Department of Obstetrics and Gynecology, Kaohsiung Medical University Hospital, Kaohsiung Medical University, Kaohsiung, Taiwan*

S.A. Chime / *Department of Pharmaceutics, Faculty of Pharmaceutical Sciences, University of Nigeria, Nsukka, Nigeria*

Joon Sig Choi / *Department of Biochemistry, Chungnam National University, Daejeon, South Korea*

Soonmo Choi / *Department of Nano, Medical and Polymer Materials, Yeungnam University, Gyeongsan, South Korea*

Seth I. Christian / *Biomedical Engineering Program, Agricultural and Biological Engineering Department, Mississippi State University, Mississippi State, Mississippi, U.S.A.*

Chih-Chang Chu / *Department of Textiles and Apparel and Biomedical Engineering Program, Cornell University, Ithaca, New York, U.S.A.*

Jeffrey Chuang / *Boston College, Chestnut Hill, Massachusetts, U.S.A.*

Giuseppe Cirillo / *Department of Pharmacy, Health, and Nutritional Sciences, University of Calabria, Rende, Italy*

Angel Concheiro / *Department of Pharmacy and Pharmaceutical Technology, University of Santiago de Compostela (USC), Santiago de Compostela, Spain*

Patrick Crowley / *Callum Consultancy, Devon, Pennsylvania, U.S.A.*

Paul D. Dalton / *Institute for Health and Biomedical Innovation, Queensland University of Technology, Brisbane, Queensland, Australia*

Vinod B. Damodaran / *New Jersey Center for Biomaterials, Rutgers University, Piscataway, New Jersey, U.S.A.*

Martin Dauner / *Denkendorf Forschungsbereich Blomedizintechnik, Institute of Textile and Process Engineering (ITV) Denkendorf*

Sabrina Dehn / *Key Center for Polymers and Colloids, School of Chemistry, University of Sydney, Sydney, New South Wales, Australia*

Claudia Del Fante / *Immunohaematology and Transfusion Service and Cell Therapy Unit, San Matteo University Hospital, Pavia, Italy*

Prashant K. Deshmukh / *Post Graduate Department of Pharmaceutics and Quality Assurance, H. R. Patel Institute of Pharmaceutical Education and Research, Shirpur, India*

James DiNunzio / *Merck & Co., Inc., Summit, New Jersey, U.S.A.*

Ryan F. Donnelly / *School of Pharmacy, Medical Biology Center, Queen's University Belfast, Belfast, U.K.*

Garry P. Duffy / *Tissue Engineering Research Group, Royal College of Surgeons in Ireland, Dublin, Ireland, Trinity Center for Bioengineering, Trinity College, Dublin, Ireland, Advanced Materials and BioEngineering Research Center (AMBER), Dublin, Ireland*

Abdelhamid Elaissari / *University of Lyon, Lyon, France*

Monzer Fanun / *Colloids and Surfaces Research Center, Al-Quds University, East Jerusalem, Palestine*

Miguel Faria / *Pharmacology Laboratory, Multidisciplinary Biomedical Research Unit, Abel Salazar Institute of Biomedical Science, University of Porto (ICBAS-UP), Oporto, Portugal*

Zahoor H. Farooqi / *Institute of Chemistry, University of the Punjab, New Campus, Lahore, Pakistan*

George Fercana / *Biocompatibility and Tissue Regeneration Laboratories, Department of Bioengineering, Clemson University, Clemson, and Laboratory of Regenerative Medicine, Patewood/ CU Bioengineering Translational Research Center, Greenville Hospital System, Greenville, South Carolina, U.S.A.*

M.M. Fernández / *Spanish National Research Council Institute of Polymer Science and Technology, Madrid, Spain*

Franca Ferrari / *Department of Drug Sciences, University of Pavia, Pavia, Italy*

Seth Forster / *Merck & Co., Inc., West Point, Pennsylvania, U.S.A.*

Keertik S. Fulzele / *Biomedical Engineering Program, Agricultural and Biological Engineering Department, Mississippi State University, Mississippi State, Mississippi, U.S.A.*

Bernard R. Gallot / *Laboratoire des Materiaux Organiques, Proprietes Specifi ques, National Center for Scientific Research (CNRS), France*

Miguel Gama / *Center of Biological Engineering (CBE), University of Minho, Braga, Portugal*

L. García-Fernández / *Networking Biomedical Research Center on Bioengineering, Biomaterials, and Nanomedicine (CIBER-BBN), Madrid, Spain*

Surendra G. Gattani / *School of Pharmacy, Swami Ramanand Teerth Marathwada University, Nanded, India*

Trent M. Gause / *Department of Plastic Surgery, University of Pittsburgh, Pittsburgh, Pennsylvania, U.S.A.*

Shyamasree Ghosh / *School of Biological Sciences, National Institute of Science, Education, and Research, Bhubaneswar, India*

Rory L.D. Gibney / *Departments of Materials Engineering, Chemistry, and Mechanical Engineering, KU Leuven, Leuven, Belgium*

Abhijit Gokhale / *Product Development Services, Patheon Pharmaceuticals Inc., Cincinnati, Ohio, U.S.A.*

Aleksander Góra / *Center for Nanofibers and Nanotechnology, Nanoscience and Nanotechnology Initiative, Faculty of Engineering and Department of Mechanical Engineering, Faculty of Engineering, National University of Singapore, Singapore*

Erin Grassl / *University of Minnesota, Minneapolis, Minnesota, U.S.A.*

Alexander Yu. Grosberg / *Department of Physics, University of Minnesota, Minneapolis, Minnesota, U.S.A.*

E. Haimer / *Division of Chemistry of Renewable Resources, University of Natural Resources and Life Sciences Vienna, Tulln, Austria*

Dalia A.M. Hamza / *Department of Biochemistry, Faculty of Science, Ain Shams University, Cairo, Egypt*

Dong Keun Han / *Polymer Chemistry Laboratory, Korea Institute of Science and Technology, Seoul, South Korea*

Sung Soo Han / *Department of Nano, Medical, and Polymer Materials, Polymer Gel Cluster Research Center, Yeungnam University, Daedong, South Korea*

Takao Hanawa / *Institute of Biomaterials and Bioengineering, Tokyo Medical and Dental University, Tokyo, Japan*

Ali Hashemi / *Chemical and Petroleum Engineering Department, Sharif University of Technology, Tehran, Iran*

Jorge Heller / *APS Research Institute, Redwood City, California, U.S.A.*

Akon Higuchi / *Department of Chemical and Materials Engineering, National Central University, Jhongli, Taiwan, and Department of Botany and Microbiology, King Saud University, Riyadh, Saudi Arabia*

Georgios T. Hilas / *Poly-Med, Incorporated, Anderson, South Carolina, U.S.A.*

Maxwell T. Hincke / *Department of Cellular and Molecular Medicine and Department of Innovation in Medical Education, Faculty of Medicine, University of Ottawa, Ottawa, Ontario, Canada*

Allan S. Hoffman / *Center for Bioengineering, University of Washington, Seattle, Washington, U.S.A.*

J.M. Hook / *Mark Wainwright Analytical Centre, University of New South Wales, Sydney, Australia*

Dietmar W. Hutmacher / *Division of Bioengineering, Department of Orthopaedic Surgery, National University of Singapore, Singapore*

Jörg Huwyler / *Division of Pharmaceutical Technology, University of Basel, Basel, Switzerland*

Francesca Iemma / *Department of Pharmacy, Health, and Nutritional Sciences, University of Calabria, Rende, Italy*

Yoshita Ikada / *Department of Clinical Engineering, Faculty of Medical Engineering, Suzuka University of Medical Science, Mie, Japan*

Kazuhiko Ishihara / *Department of Materials Engineering, School of Engineering, University of Tokyo, Tokyo, Japan*

Saeed Jafarirad / *Research Institute for Fundamental Sciences (RIFS), University of Tabriz, Tabriz, Iran*

B. Jansen / *Institute of Medical Microbiology and Hygiene, University of Cologne, Cologne, Germany*

Katrina A. Jolliffe / *School of Chemistry, University of Sydney, Sydney, New South Wales, Australia*

Priya Juneja / *Jubilant Life Sciences, Noida, India*

Rajesh Kalia / *Department of Physics, Maharishi Markandeshwar University, Mullana, India*

Sapna Kalia / *Department of Physics, Maharishi Markandeshwar University, Mullana, India*

Ta-Chun Kao / *Department of Chemical and Materials Engineering, National Central University, Jhongli, Taiwan*

Paridhi Kataria / *Department of Chemistry, University College of Science, Mohanlal Sukhadia University, Udaipur, India*

Harmeet Kaur / *Prabhu Dayal Memorial (PDM) College of Pharmacy, Bahadurgarh, India*

F.C. Kenechukwu / *Department of Pharmaceutics, Faculty of Pharmaceutical Sciences, University of Nigeria, Nsukka, Nigeria*

Josè Maria Kenny / *Materials Science and Technology Center, University of Perugia, Terni, Italy*

Asad Ullah Khan / *Department of Chemical Engineering, Commission on Science and Technology for Sustainable Development in the South (COMSATS) Institute of Information Technology, Lahore, Pakistan*

Sepideh Khoee / *Polymer Chemistry Department, School of Science, University of Tehran, Tehran, Iran*

Arezoo Khosravi / *Institute for Nanoscience and Nanotechnology, Sharif University of Technology, Tehran, Iran*

Kwang J. Kim / *College of Engineering, University of Nevada–Reno, Reno, Nevada, U.S.A.*

Sanghyo Kim / *Department of Bionanotechnology, Gachon University, Gyeonggi-do, South Korea*

Young Ha Kim / *Polymer Chemistry Laboratory, Korea Institute of Science and Technology, Seoul, South Korea*

Yoshihiro Kiritoshi / *Department of Materials Engineering, School of Engineering, University of Tokyo, Tokyo, Japan*

W. Kohnen / *Institute of Medical Microbiology and Hygiene, University of Cologne, Cologne, Germany*

Soheila S. Kordestani / *ChitoTech Inc., Tehran, Iran*

Abhijith Kundadka Kudva / *Departments of Materials Engineering, Chemistry, and Mechanical Engineering, KU Leuven, Leuven, Belgium*

Ashok Kumar / *Department of Biological Sciences and Bioengineering, Indian Institute of Technology, Kanpur, India*

S. Suresh Kumar / *Department of Medical Microbiology and Parasitology, Putra University, Slangor, Malaysia*

Jay F. Künzler / *Department of Chemistry and Polymer Development, Bausch and Lomb Incorporated*

A. Lauto / *Biomedical Engineering and Neuroscience Research Group, The MARCS Institute, Western Sydney University, Penrith, Australia*

L. James Lee / *Department of Chemical Engineering, Ohio State University, Columbus, Ohio, U.S.A.*

Alexandre F. Leitão / *Center of Biological Engineering (CBE), University of Minho, Braga, Portugal*

Kimon Alexandros Leonidakis / *Departments of Materials Engineering, Chemistry, and Mechanical Engineering, KU Leuven, Leuven, Belgium*

Victoria Leszczak / *Department of Mechanical Engineering, Colorado State University, Fort Collins, Colorado, U.S.A.*

Pei-Tsz Li / *Department of Chemical and Materials Engineering, National Central University, Jhongli, Taiwan*

F. Liebner / *Division of Chemistry of Renewable Resources, University of Natural Resources and Life Sciences Vienna, Tulln, Austria*

Cai Lloyd-Griffith / *Tissue Engineering Research Group, Royal College of Surgeons in Ireland, Dublin, Ireland, Trinity Center for Bioengineering, Trinity College, Dublin, Ireland, Advanced Materials and BioEngineering Research Center (AMBER), Dublin, Ireland*

A. Löfgren / *Department of Polymer Technology, Royal Institute of Technology (KTH), Stockholm, Sweden*

M.L. López-Donaire / *Spanish National Research Council Institute of Polymer Science and Technology, Madrid, Spain*

Yi Lu / *Department of Mechanical Engineering, University of Texas at Austin, Austin, Texas, U.S.A.*

Sofia Luís / *Research Institute for Medicine and Pharmaceutical Sciences (iMed.UL), School of Pharmacy, University of Lisbon, Lisbon, Portugal*

Mizuo Maeda / *Graduate School of Engineering, Kyushu University, Fukuoka, Japan*

Bernard Mandrand / *bioMérieux, National Center for Scientific Research (CNRS), Lyon, France*

João F. Mano / *Biomaterials, Biodegradables and Biomimetics (3Bs) Research Group, Department of Polymer Engineering, School of Engineering, European Institute of Excellence on Tissue Engineering and Regenerative Medicine, University of Minho, Braga and Life and Health Sciences Research Institute (ICVS)Portuguese Government Associate Laboratory, Guimaraes, Portugal*

Alexandra A. P. Mansur / *Center of Nanoscience, Nanotechnology, and Innovation (CeNano), Department of Metallurgical and Materials Engineering, Federal University of Minas Gerais, Belo Horizonte, Minas Gerais, Brazil*

Herman S. Mansur / *Center of Nanoscience, Nanotechnology, and Innovation (CeNano), Department of Metallurgical and Materials Engineering, Federal University of Minas Gerais, Belo Horizonte, Minas Gerais, Brazil*

Asghari Maqsood / *Department of Physics, Center for Emerging Sciences, Engineering and Technology (CESET), Islamabad, Pakistan*

Mohana Marimuthu / *Department of Bionanotechnology, Gachon University, Gyeonggi-do, South Korea*

Kacey G. Marra / *Departments of Plastic Surgery and Bioengineering, McGowan Institute for Regenerative Medicine, University of Pittsburgh, Pittsburgh, Pennsylvania, U.S.A.*

Luigi G. Martini / *Institute of Pharmaceutical Sciences, King's College, London, U.K.*

Shojiro Matsuda / *Research and Development Department, Gunze Limited, Ayabe, Japan*

Mohammad Mazinani / *Department of Materials Engineering, Faculty of Engineering, Ferdowsi University of Mashhad, Mashhad, Iran*

Tara M. McFadden / *Tissue Engineering Research Group, Royal College of Surgeons in Ireland, Dublin, Ireland, Trinity Center for Bioengineering, Trinity College, Dublin, Ireland, Advanced Materials and BioEngineering Research Center (AMBER), Dublin, Ireland*

James W. McGinity / *Division of Pharmaceutics, College of Pharmacy, University of Texas at Austin, Austin, Texas, U.S.A.*

Kiran Meghwal / *Department of Chemistry, University College of Science, Mohanlal Sukhadia University, Udaipur, India*

Shahram Mehdipour-Ataei / *Iran Polymer and Petrochemical Institute, Tehran, Iran*

Taíla O. Meiga / *Departments of Materials Engineering, Chemistry, and Mechanical Engineering, KU Leuven, Leuven, Belgium*

Meymanant Sadat Mohsenzadeh / *Department of Materials Engineering, Faculty of Engineering, Ferdowsi University of Mashhad, Mashhad, Iran*

Mehran Mojarrad / *Pharmaceutical Delivery Systems, Eli Lilly and Co., Indianapolis, Indiana, U.S.A.*

Kwang Woo Nam / *Department of Materials Engineering, School of Engineering, University of Tokyo, Tokyo, Japan*

Robert M. Nerem / *Parker H. Petit Institute for Bioengineering and Bioscience, Georgia Institute of Technology, Atlanta, Georgia, U.S.A.*

Fergal J. O'Brien / *Tissue Engineering Research Group, Royal College of Surgeons in Ireland, Dublin, Ireland, Trinity Center for Bioengineering, Trinity College, Dublin, Ireland, Advanced Materials and BioEngineering Research Center (AMBER), Dublin, Ireland*

J.D.N. Ogbonna / *Department of Pharmaceutics, Faculty of Pharmaceutical Sciences, University of Nigeria, Nsukka, Nigeria*

Maryam Oroujzadeh / *Iran Polymer and Petrochemical Institute, Tehran, Iran*

Kunal Pal / *Department of Biotechnology and Medical Engineering, National Institute of Technology, Rourkela, India*

Abhijeet P. Pandey / *Post Graduate Department of Pharmaceutics and Quality Assurance, H. R. Patel Institute of Pharmaceutical Education and Research, Shirpur, India*

Arpita Pandey / *Department of Chemistry, University College of Science, Mohanlal Sukhadia*

Jong-Sang Park / *Department of Chemistry, Seoul National University, Seoul, South Korea*

Ki Dong Park / *Polymer Chemistry Laboratory, Korea Institute of Science and Technology, Seoul, South Korea*

F. Parra / *Spanish National Research Council Institute of Polymer Science and Technology, Madrid, Spain*

Arpit Kumar Pathak / *Department of Chemistry, University College of Science, Mohanlal Sukhadia University, Udaipur, India*

Pravin O. Patil / *Post Graduate Department of Pharmaceutics and Quality Assurance, H. R. Patel Institute of Pharmaceutical Education and Research, Shirpur, India*

Jennifer Patterson / *Departments of Materials Engineering, Chemistry, and Mechanical Engineering, KU Leuven, Leuven, Belgium*

Shawn J. Peniston / *Poly-Med, Incorporated, Anderson, South Carolina, U.S.A.*

Cesare Perotti / *Immunohaematology and Transfusion Service and Cell Therapy Unit, San Matteo University Hospital, Pavia, Italy*

Sébastien Perrier / *Key Center for Polymers and Colloids, School of Chemistry, University of Sydney, Sydney, New South Wales, Australia*

Susanna Piluso / *Departments of Materials Engineering, Chemistry, and Mechanical Engineering, KU Leuven, Leuven, Belgium*

N. Pircher / *Division of Chemistry of Renewable Resources, University of Natural Resources and Life Sciences Vienna, Tulln, Austria*

Stergios Pispas / *Theoretical and Physical Chemistry Institute of the National Hellenic Research Foundation (TPCI-NHRF), Athens, Greece*

Heinrich Planck / *Denkendorf Forschungsbereich Blomedizintechnik, Institute of Textile and Process Engineering (ITV) Denkendorf*

Ketul C. Popat / *Department of Mechanical Engineering and School of Biomedical Engineering, Colorado State University, Fort Collins, Colorado, U.S.A.*

Krishna Pramanik / *Department of Biotechnology and Medical Engineering, National Institute of Technology, Rourkela, India*

Pinki B. Punjabi / *Department of Chemistry, University College of Science, Mohanlal Sukhadia University, Udaipur, India*

Diana Rafael / *Research Institute for Medicine and Pharmaceutical Sciences (iMed.UL), School of Pharmacy, University of Lisbon, Lisbon, Portugal*

Seeram Ramakrishna / *Center for Nanofibers and Nanotechnology, Nanoscience and Nanotechnology Initiative, Faculty of Engineering and Department of Mechanical Engineering, Faculty of Engineering, National University of Singapore, Singapore*

Stanislav Rangelov / *Laboratory of Polymerization Processes, Scientific Council of the Institute of Polymers, Bulgarian Academy of Sciences, Sofia, Bulgaria*

Manish Kumar Rawal / *Department of Chemistry, Vidya Bhawan Rural Institute, Udaipur, India*

Melissa M. Reynolds / *Department of Chemistry and School of Biomedical Engineering, Colorado State University, Fort Collins, Colorado, U.S.A.*

G. Rodríguez / *Spanish National Research Council Institute of Polymer Science and Technology, Madrid, Spain*

L. Rojo / *Department of Materials and Institute of Bioengineering, Imperial College London, London, U.K.*

T. Rosenau / *Division of Chemistry of Renewable Resources, University of Natural Resources and Life Sciences Vienna, Tulln, Austria*

Aftin M. Ross / *Institute of Functional Interfaces (IFG), Karlsruhe Institute of Technology (KIT), Eggenstein-Leopoldshafen, Germany, and Center for Devices and Radiological Health, Food and Drug Administration, Silver Spring, Maryland, U.S.A.*

Silvia Rossi / *Department of Drug Sciences, University of Pavia, Pavia, Italy*

Shaneen L. Rowe / *Department of Biomedical Engineering, Rensselaer Polytechnic Institute, Troy, New York, U.S.A.*

H. Ruprai / *School of Science and Health, Western Sydney University, Penrith, Australia.*

Elizabeth Ryan / *School of Pharmacy, Medical Biology Center, Queen's University of Belfast, Belfast, U.K.*

Sai Sateesh Sagiri / *Department of Biotechnology and Medical Engineering, National Institute of Technology, Rourkela, India*

Soham Saha / *School of Biological Sciences, National Institute of Science, Education, and Research, Bhubaneswar, India*

Rahul Sahay / *Fluid Division, Department of Mechanical Engineering, National University of Singapore, Singapore*

J. San Román / *Institute for Health and Biomedical Innovation, Queensland University of Technology, Kelvin Grove, Queensland, Australia*

Giuseppina Sandri / *Department of Drug Sciences, University of Pavia, Pavia, Italy*

Smritimala Sarmah / *Department of Physics, Girijananda Chowdhury Institute of Management and Technology, Guwahati, India and Raktim Pratim Tamuli, Department of Forensic Medicine, Guwahati Medical College and Hospital, Guwahati, India*

M. Sasidharan / *Department of Physics and Nanotechnology, Sri Ramaswamy Memorial (SRM) University, Kattankulathur, Chennai, India*

C. Schimper / *Division of Chemistry of Renewable Resources, University of Natural Resources and Life Sciences Vienna, Tulln, Austria*

Deepak Kumar Semwal / *Department of Chemistry, Panjab University, Chandigarh, India*

Ravindra Semwal / *Faculty of Pharmacy, Dehradun Institute of Technology, Dehradun, India*

Ruchi Badoni Semwal / *Department of Chemistry, Panjab University, Chandigarh, India*

Mohsen Shahinpoor / *Artificial Muscle Research Institute (AMRI), School of Engineering and School of Medicine, University of New Mexico, Albuquerque, New Mexico, U.S.A.*

Haseeb Shaikh / *Polymer and Surface Engineering Lab, Department of Materials Engineering, School of Chemical and Materials Engineering (SCME), National University of Sciences and Technology (NUST), Islamabad, Pakistan*

Waleed S. W. Shalaby / *Poly-Med Incorporated, Anderson, South Carolina, U.S.A.*

V. Prasad Shastri / *Vanderbilt University, Nashville, Tennessee, U.S.A.*

Mikhail I. Shtilman / *Biomaterials Scientific and Teaching Center, D. I. Mendeleyev University of Chemical Technology, Moscow, Russia*

Quazi T. H. Shubhra / *Doctoral School of Molecular and Nanotechnologies, Faculty of Information Technology, University of Pannonia, Veszprém, Hungary*

Mohammad Siddiq / *Department of Chemistry, Quaid-I-Azam University, Islamabad, Pakistan*

Ivone Silva / *Department of Angiology and Vascular Surgery, Hospital of Porto, Porto, Portugal*

Dan Simionescu / *Department of Bioengineering, Clemson University, Clemson, and Patewood/CU Bioengineering Translational Research Center, Greenville Hospital System, Greenville, South Carolina, U.S.A., and Department of Anatomy, University of Medicine and Pharmacy, Tirgu Mures, Romania*

Deepti Singh / *Department of Nano, Medical, and Polymer Materials, Polymer Gel Cluster Research Center, Yeungnam University, Daedong, South Korea*

Jasbir Singh / *Department of Pharmacy, University of Health Sciences, Rohtak, India*

Thakur Raghu Raj Singh / *School of Pharmacy, Medical Biology Center, Queen's University of Belfast, Belfast, U.K.*

Vinay K. Singh / *Department of Biotechnology and Medical Engineering, National Institute of Technology, Rourkela, India*

Wesley N. Sivak / *Department of Plastic Surgery, University of Pittsburgh, Pittsburgh, Pennsylvania, U.S.A.*

Radhakrishnan Sivakumar / *Nanomaterials and Solar Energy Conversion Lab, Department of Chemistry, National Institute of Technology, Tiruchirappalli, India*

Daniel H. Smith / *Biomedical Engineering Program, Agricultural and Biological Engineering Department, Mississippi State University, Mississippi State, Mississippi, U.S.A.*

Al Halifa Soultan / *Departments of Materials Engineering, Chemistry, and Mechanical Engineering, KU Leuven, Leuven, Belgium*

Umile Gianfranco Spizzirri / *Department of Pharmacy, Health, and Nutritional Sciences, University of Calabria, Rende, Italy*

Jan P. Stegeman / *Department of Biomedical Engineering, Rensselaer Polytechnic Institute, Troy, New York, U.S.A.*

Kaliappa Gounder Subramanian / *Department of Biotechnology, Bannari Amman Institute of Technology, Sathyamangalam, India*

Abhinav Sur / *School of Biological Sciences, National Institute of Science, Education, and Research, Bhubaneswar, India*

Michael Szycher / *Sterling Biomedical, Incorporated, Lynnfield, Massachusetts, U.S.A.*

G. Lawrence Thatcher / *TESco Associates Incorporated, Tyngsborough, Massachusetts, U.S.A.*

Velmurugan Thavasi / *Elam Pte Ltd., Singapore*

Rong Tong / *Department of Materials Science and Engineering, University of Illinois at Urbana-Champaign, Urbana, Illinois, U.S.A.*

Luigi Torre / *Materials Science and Technology Center, University of Perugia, Terni, Italy*

Robert T. Tranquillo / *Department of Chemical Engineering and Materials Science, University of Minnesota, Minneapolis, Minnesota, U.S.A.*

Yasuhiko Tsuchitani / *Faculty of Dentistry, Osaka University, Osaka, Japan Tohru Wada, Kuraray Company Ltd., Tokyo, Japan*

Laurien Van den Broeck / *Departments of Materials Engineering, Chemistry, and Mechanical Engineering, KU Leuven, Leuven, Belgium*

Jan C. M. van Hest / *Organic Chemistry, Institute for Molecules and Materials, Radboud University Nijmegen, Nijmegen, the Netherlands*

Cedryck Vaquette / *Institute for Health and Biomedical Innovation, Queensland University of Technology, Brisbane, Queensland, Australia*

Marcia Vasquez-Lee / *Biomedical Engineering Program, Agricultural and Biological Engineering Department, Mississippi State University, Mississippi State, Mississippi, U.S.A.*

Jason Vaughn / *Product Development Services, Patheon Pharmaceuticals Inc., Cincinnati, Ohio, U.S.A.*

Raphael Veyret / *bioMérieux, National Center for Scientific Research (CNRS), Lyon, France*

Mafalda Videira / *Research Institute for Medicine and Pharmaceutical Sciences (iMed. UL), School of Pharmacy, University of Lisbon, Lisbon, Portugal*

Vediappan Vijayakumar / *Department of Biotechnology, Bannari Amman Institute of Technology, Sathyamangalam, India*

Tanushree Vishnoi / *Department of Biological Sciences and Bioengineering, Indian Institute of Technology, Kanpur, India*

S. Vivekananthan / *Department of Physics and Nanotechnology, Sri Ramaswamy Memorial (SRM) University, Kattankulathur, Chennai, India*

Giovanni Vozzi / *E. Piaggio Interdepartmental Research Center, Department of Chemical Engineering, University of Pisa, Pisa, Italy*

Ernst Wagner / *Department of Pharmacy, Ludwig Maximilians University, Munich, Germany*

Yadong Wang / *Wallace H. Coulter School of Biomedical Engineering and School of Chemistry and Biochemistry, Georgia Institute of Technology, Atlanta, Georgia, U.S.A.*

Junji Watanabe / *Department of Materials Engineering, School of Engineering, University of Tokyo, Tokyo, Japan*

Jennifer L. West / *Department of Bioengineering, Rice University, Houston, Texas, U.S.A.*

David L. Williams / *Biomedical Engineering Program, Agricultural and Biological Engineering Department, Mississippi State University, Mississippi State, Mississippi, U.S.A.*

Thomas Williams / *Product Development Services, Patheon Pharmaceuticals Inc., Cincinnati, Ohio, U.S.A.*

Dominik Witzigmann / *Division of Pharmaceutical Technology, University of Basel, Basel, Switzerland*

A. David Woolfson / *School of Pharmacy, Medical Biology Center, Queen's University of Belfast, Belfast, U.K.*

Tetsuji Yamaoka / *Department of Polymer Science and Engineering, Kyoto Institute of Technology, Kyoto, Japan*

Can Yang / *College of Engineering, Zhejiang Normal University, Zhejiang, China*

Siou-Ting Yang / *Department of Chemical and Materials Engineering, National Central University, Jhongli, Taiwan*

Xiao-Hong Yin / *College of Engineering, Zhejiang Normal University, Zhejiang, China*

Ali Zarrabi / *Department of Biotechnology, Faculty of Advanced Sciences and Technologies, University of Isfahan, Isfahan, Iran*

Seyed Mojtaba Zebarjad / *Department of Materials Engineering, Faculty of Engineering, Ferdowsi University of Mashhad, Mashhad, Iran*

Contents

Volume I

Volume II

Topical Table of Contents

Body Systems (cont'd.)

Chemistry, Properties, and Polymer Architectures

Devices, Implants, Grafts, and Prosthetics

Drugs and Drug Delivery

Electroactive and Ionic Materials

Fabrication Techniques and Processing

Gels

Gels (cont'd.)

Gene Delivery and Gene Therapy

Immobilization and Sensors

Micro- and Nanotechnology

Natural Sources

Natural Sources (cont'd.)

Synthetic and Custom-Designed Materials

Tissue Engineering

Wound Healing Applications

Preface

This compact desk reference includes carefully selected topics from the print and online versions of the multivolume *Encyclopedia of Biomedical Polymers and Polymeric Biomaterials*. This distillation was skillfully assembled in broad subject area of polymer applications in the medical field, which will enable readers to have an enriching experience in general as well as to gain targeted knowledge in this evolving area. This concise encyclopedia is designed to serve as a ready-reference guide for the researchers in the field, and is expected to provide quick access to some recent advances in the area.

This handy reference work is designed for novices to experienced researchers; it caters to engineers and scientists (polymer and materials scientists, biomedical engineers, biochemists, molecular biologists, and macromolecular chemists), pharmacists, doctors, cardiovascular and plastic surgeons, and students, as well as general readers in academia, industry, and research institutions.

I feel honored to undertake the important and challenging endeavor of developing the *Concise Encyclopedia of Biomedical Polymers and Polymeric Biomaterials* that will cater to the needs of many who are working in the field. I would like to express my sincere gratitude and appreciation to the authors for their excellent professionalism and dedicated work. I would like to thank the entire management of encyclopedia program of Taylor & Francis Group (T&F - CRC Press) and particularly to Ms. Megan Hilands who made this possible. Her hard work and professionalism made this a wonderful experience for all parties involved.

I take the opportunity to express my appreciation to my wife Bidu Mishra, Ph.D. for her encouragement, sacrifice, and support, during weekends, early mornings, and holidays spent on this project. Without their help and support, this project would have never been started nor completed.

Munmaya K. Mishra
Editor-in-Chief

About the Editor-in-Chief

Munmaya Mishra, Ph.D. is a polymer scientist who has worked in the industry for more than 30 years. He has been engaged in research, management, technology innovations, and product development. He has contributed immensely to multiple aspects of polymer applications, including biomedical polymers, encapsulation and controlled release technologies, etc. He has authored and coauthored hundreds of scientific articles is the author or editor of multiple books, and hold over 50 U.S. patents, over 50 U.S. patent-pending applications, and over 150 world patents. He has received numerous recognitions and awards. He is the editor-in-chief of three renowned polymer journals and the editor-in-chief of the recently published multivolume *Encyclopedia of Biomedical Polymers and Polymeric Biomaterials*. He is also the founder of new scientific organization "International Society of Biomedical Polymers and Polymeric Biomaterials." In 1995 he founded and established a scientific meeting titled "Advanced Polymers via Macromolecular Engineering," which has gained international recognition and is still being held under the sponsorship of the IUPAC organization.

Concise Encyclopedia of Biomedical Polymers and Polymeric Biomaterials

Volume II
Melt Extrusion through Zwitterionic
Pages 827–1672

Melt—Microgels

Molecular—Nanogels

Nanomaterials—Nanoparticles

Nanoparticles—Nerve

Neural—Rapid Prototyping

Scaffolds—Smart

Stem Cell—Ultrasound

Vascular—Zwitterionic

Melt Extrusion: Pharmaceutical Applications

James DiNunzio
Merck & Co., Inc., Summit, New Jersey, U.S.A.

Seth Forster
Chad Brown
Merck & Co., Inc., West Point, Pennsylvania, U.S.A.

Abstract

Melt extrusion technology plays a vital role in the health care sector, supporting the production of a range of products, including stents, tablets, and injectable depot systems. The design of such systems using extrusion is achieved through the combined engineering of composition and extrudate geometry, allowing for unique release profiles not obtainable with conventional drug product manufacturing technologies. This entry reviews the basic formulation design concepts for the production of solid dispersions using melt extrusion, with a particular focus on oral drug delivery applications. Also covered within the text are application case studies using melt extrusion technology for the production of enabled formulations, controlled release products, foamed materials, and directly shaped products.

INTRODUCTION

Drug development pipelines have changed dramatically over the last 40 years, yielding a number of candidate molecules with challenging properties that limit drugability.[1–4] These properties range from physicochemical stability to permeability, with the most significant limitation for oral molecules being poor aqueous solubility. Driven by the advent of high throughput screening and the need to enhance molecular structure to facilitate target binding efficiency, compounds have evolved lower solubility, making many of the lead compounds in modern pipelines more insoluble than marble.[5] This has led to the characterization of many modern active pharmaceutical ingredients as "brick dust." These compounds are often physically delineated by a higher melting temperature and larger melting enthalpy.[6] Another defining characteristic of many modern new chemical entities has been a transition in lipophilicity,[7] increasing to high log P values thereby reducing the ability to wet and subsequently dissolve. Additionally, the higher degree of molecular complexity challenges the ability to generate a stable crystalline drug substance. Other non-ideal properties such as chemical instability, the need for non-traditional release profiles, and/or alternative delivery routes makes solid dispersion technology an ideal approach for enabling next-generation therapies.

Extrusion technology traces its origins back nearly 2000 years to the first uses of screw conveyors to move water in ancient times.[8] Since then, the technology has changed significantly and found utility in a range of applications including food manufacturing, industrial ceramics, and electronic components.[9,10] Today, modern extruders are available in a variety of different forms based on the number of screws in the system and direction in which the screws rotate. Broadly speaking, extruders are classified as single screw, twin screw, and multiple screw extruders. For the pharmaceutical industry, twin screw extruders are the most prevalent system, with the co-rotating version the most commonly leveraged sub-type within the class. It is also important to note that to date there are no reported applications of multi-screw extruders in the pharmaceutical manufacturing of drug products. Twin screw extruders provide several key advantages when compared to the single screw class and the multi-screw class. First, because of the twin screw configuration it is possible to achieve a higher degree of mixing than that of a single screw extruder. Secondly, twin screw extruders are available in an intermeshing screw design which allows for self-wiping.[11] This concept, initially introduced by Erdmenger,[12] allows one screw to remove material from the surface of the other during the rotation. This leads to more effective conveyance along the length of the processing section, which translates into shorter and more uniform residence time distributions. These attributes are key to the success of dispersion manufacture, particularly in the case of pharmaceutical systems where a high degree of uniformity and purity are required. There is one limitation associated with co-rotating twin screw extruders when compared to single screw or counter-rotating extruders that should be noted. When compared to these systems, co-rotating extruders provide less consistent pumping action, which can impact performance in the production of directly shaped products. To compensate for this gear pumps can be added onto the

Concise Encyclopedia of Biomedical Polymers and Polymeric Biomaterials DOI: 10.1081/E-EBPPC-120050541

end of the co-rotating extruder to allow for consistent pumping and enhanced flow behavior. A summary of the key attributes of commonly available extruder classes and subclasses is shown in Table 1. Although a detailed summary of the equipment is outside of the scope of this work, a number of useful references are cited here for the interested reader.[13,14]

To date, melt extrusion has served as the backbone for 14 publically disclosed marketed products, with numerous more in development. As shown in Table 2, these products represent a range of applications, many of which provide some form of novel delivery behavior not available with traditional technologies. In cases where extrusion technology was not applied for a drug delivery benefit, it was leveraged for continuous processing that can more easily support high volume products while still being operated in the batch mode paradigm required by regulatory agencies.

This section is structured to provide the reader an overview of pharmaceutical applications of melt extrusion to enhance the delivery of pharmaceutical products. The first section on formulation design covers aspects related to the composition of pharmaceutical dispersions. It highlights the important properties of drugs and carriers to achieve the desired critical quality attributes while also describing strategies for the structured design of extruded products. The second section covers applications of melt extrusion for controlled release and targeted release, highlighting the use of this technology for dissolution enhancement, controlled release, foaming, and direct shaping. Finally, a concluding section summarizes the current state of melt extrusion technology and the future outlook for pharmaceutical manufacturing.

FORMULATION DESIGN AND CHARACTERIZATION

Pharmaceutical melt extrusion technology is primarily used in the production of solid dispersions.[15–19] As summarized in Fig. 1, depending on the nature of the product being produced, the material can exist as one of several different types of solid dispersions based on the nature of the drug substance and carrier. In the case of crystalline solid dispersions, the drug substance is dispersed in the carrier phase as discrete crystalline particulates. For amorphous solid dispersions, they encompass amorphous drug dispersed in the carrier phase, which can include drug molecularly dispersed, as is the case of an amorphous solid solution, and also systems containing discrete domains of amorphous drug in the larger carrier phase. To denote systems where the drug is molecularly dispersed in the carrier, the term "solid solution" was developed. In practice however, because of the drive to develop a homogenous amorphous system with the drug ideally molecularly dispersed, the terms "solid dispersion" and "solid solution" can be used interchangeably. Within this text, the term "solid dispersion" refers to a homogeneous system where ideally the drug is dispersed uniformly at a molecular level.

Table 1 Comparison of equipment attributes for classes and sub-classes of extruders for pharmaceutical applications

| Class | | Twin screw | | |
Subclass	Single screw	Co-rotating	Counter rotating	Multi-screw
Schematic				
Advantages	High pumping efficiency	Narrow residence time distribution	Very narrow residence time distribution	Excellent surface renewal for devolatilization
	Simple equipment design	High dispersive and distributive mixing	High pumping efficiency	
		Very high feeding efficiency	High feeding efficiency	
Disadvantages	Broad residence time distribution	Low pumping efficiency	Lower output than co-rotating extruder	Higher cost & complexity
	No self-wiping	Reduced degassing efficiency	Not as effective for distributive mixing when compared to co-rotating extruder	Not commonly used in Pharma
	Lower dispersive mixing			
	Reduced feeding efficiency			

Table 2 Marketed pharmaceutical products using melt extrusion technology

Product	Indication	Application	Company
Eudragit E PO (methacrylic acid copolymer)	Excipient	Excipient manufacturing	Evonik
Gris-PEG (Griseofulvin)[a]	Anti-fungal	Dissolution enhanced	Pedinol Pharmacal
Rezulin (Troglitazone)	Diabetes	Dissolution enhanced	Wyeth
Onmel (itraconazole)	Anti-fungal	Dissolution enhanced	Merz Pharmaceuticals
Norvir (ritonavir)	HIV therapy	Dissolution enhanced	Abbott Laboratories
Kaletra (ritonavir/lopinavir)	HIV therapy	Dissolution enhanced	Abbott Laboratories
Isoptin SR (verapamil)	Angina	CR/direct shape tab.	Abbott Laboratories
Opana ER (oxymorphone HCl)	Pain	CR/direct shape tab.	Endo Pharmaceuticals
Palladone (Hydromorphone)	Pain	CR/direct shape pellet	Purdue Pharma
Ozurdex (Dexamethasone)	Macular edema	CR/shaped device	Allergan
Zoladex (Goserelin Acetate Implant)	Prostate cancer	CR/shaped device	AstraZeneca
Nuvaring (Etonogestrel/Ethinyl Estradiol)	Contraceptive	CR/shaped device	Merck & Co., Inc.
Implanon (Etonogestrel)	Contraceptive	CR/shaped device	Merck & Co., Inc.
Lacrisert (HPC Rod)	Dry eye syndrome	Shaped device	Merck & Co., Inc.
Eucreas (Vildagliptin/Metformin HCl)	Diabetes	Melt granulation	Novartis
Zithromax (Azythromycin)[b]	Antibiotic	Taste masking	Pfizer

[a]Undisclosed melt quench process.
[b]Melt spray congeal technology.

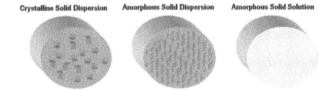

Fig. 1 Types of solid dispersions.

It is also important to note that each of these dispersions can be exploited to enhance the delivery of the drug substance. For example, the majority of dissolution-enhanced products are prepared in the amorphous state to leverage the higher solubility of the material and increase release rates from the drug product. Furthermore, the homogeneous nature of the dispersion provides a kinetic barrier for recrystallization of the amorphous form that contributes to improved physical stability. However, in the case of controlled release and directly shaped products where the systems may be prepared to have specific structural and mechanical attributes, the use of solid dispersions wherein both the drug and polymer exist as crystalline materials within the system may be beneficial.

The resulting solid dispersion is a strong function of formulation and process, both of which are contoured to the drug product application. Composition considerations for each application can be broken down based on the primary function of each of the excipients, as shown in Table 3. For successful manufacturing using extrusion, it is necessary that one or more of the components be a thermoplastic material capable of facilitating molten flow within

the process section. In the case of an amorphous dispersion for bioavailability enhancement applications, the polymeric carrier is amorphous and thermoplastic in nature. Materials such as copovidone,[20] polyvinyl caprolactam—polyvinyl acetate—polyethylene glycol graft copolymer,[21] and hypromellose acetate succinate[22] exhibit rheological properties necessary for extrusion within pharmaceutical formulations and can even be extruded independently. However, other materials do not exhibit ideal rheological characteristics. Amongst these are high molecular weight povidone and methacrylic acid copolymer,[23] which exhibit a narrow difference between glass transition temperature (T_g) and degradation temperature (T_{deg}). This ratio, T_g:T_{deg}, should ideally be less than 0.85 to provide a sufficient operational window. One strategy to address this is to include a plasticizer, which can be an additional component, such as triethyl citrate or dibutyl sebacate,[24,25] that provides molecular lubrication and lowers the T_g of the system. In other select cases, the drug substance may also function as a plasticizer; however, this requires specific properties of the active ingredient. In general, the addition of a plasticizer presents unique challenges, particularly for the development of amorphous dispersions where the reduced T_g can negatively impact physical stability. Additional risks for the use of small molecule plasticizers include concerns for uniformity, chemical stability, and volatility. A viable alternative to using plasticizers is the use of polymer blends to enhance the processability of formulations while maintaining higher T_gs.[26]

Many solid dispersions are prepared as multicomponent systems to engineer specific properties within the product.[27–30]

Table 3 Functional excipient roles for different types of dosage forms prepared by melt extrusion

Polymer	T_g or T_m (°C)	Grades	Notes
Hypromellose	170–180 (Tg)	Methocel® E5	Non-thermoplastic
			API must plasticize
			Excellent nucleation inhibition
			Difficult to mill
Vinylpyrrolidone	168 (Tg)	Povidone® K30	API must plasticize
			Potential for H-bonding
			Hygroscopic
			Residual peroxides
			Easily milled
Vinylpyrrolidone-vinylacetate copolymer	106 (Tg)	Kollidon® VA 64	Easily processed by melt extrusion
			No API plasticization required
			More hydrophobic than vinylpyrrolidone
			Processed around 130°C
Polyethylene glycol, vinyl acetate, vinyl caprolactam graft co-polymer	70 (Tg)	Soluplus®	Newest excipient for melt extruded dispersions
			Easily process by melt extrusion
			Low T_g can limit stability
			Not of compendial status
			Stable up to >180°C
Polymethacrylates	130 (Tg)	Eudragit® L100-55	Not easily extruded without plasticizer
		Eudragit® L100	Degradation onset is 155°C
			Ionic polymer soluble above pH 5.5
Hypromellose acetate succinate	120–135 (Tg)	AQOAT®-L	Easily extruded without plasticizer
		AQOAT®-M	Process temperatures >140°C
		AQOAT®-H	Ionic polymer soluble above pH 5.5 depending on grade
			Excellent concentration enhancing polymer
			Stable to 190°C depending on processing conditions
Amino methacrylate copolymer	56 (Tg)	Eudragit® E PO	Processing at ~100°C
			Degradation onset is >200°C
			Low T_g can limit stability
Methacrylic acid ester	65–70 (Tg)	Eudragit® RS	Extrudable at moderate temperatures, >100°C
		Eudragit® RL	Excellent CR polymer
Poly(ethylene vinylacetate)	35–205 (Tm)	Elvax®	Extrudable at low temperatures, >60°C
			Excellent controlled release polymer but non-biodegradable
Poly(ethylene oxide)	<25–80 (Tm)	Polyox®	Mechanical properties ideal for abuse-deterrent applications and CR
			Process temperatures >70°C
			Excellent CR polymer
Poly(lactic-co-glycolic acid)	40–60 (Tm)	RESOMER®	Low melt viscosity for certain grades is challenging to process
			Biodegradation rate controlled by polymer chemistry
			Excellent for implantable systems

For amorphous systems where the drug substance exhibits a strong lipophilicity, wettability can be limited. To compensate for this, many formulations include surfactants to aid in surface wetting and dissolution of the formulation. Wettability can also be an important consideration in the preparation of controlled release and directly shaped products, contributing to the overall performance of the drug product. Multicomponent dispersions are also particularly important for controlled release and directly shaped products, where extended drug release and mechanical integrity are critical attributes.[27,28] Controlled release products and many directly shaped products are designed to incorporate crystalline drug substance where they may also utilize semi-crystalline polymers. These systems also have engineered structures with limited internal porosity.[31,32] Through the addition of pore formers, it is possible to facilitate drug release. Similarly, controlled release polymers can be engineered into these systems to design release rates from the drug product. Interplay with the shaping process may also influence the formulation. For injection molding and calendering operations, it is necessary to have a finished product that can be removed from the shaping cavity. Adhesion between the melt and surface during the shaping and cooling process can lead to surface defects and product failure. To prevent this, lubricants are often included in the formulation. These materials mitigate surface interactions and lead to more effective discharge from the die.

Formulation design is also influenced by the properties of the drug substance and the operating conditions during extrusion. Specifically, the mode of manufacturing is defined based on the melting temperature of the drug substance in relation to the processing temperature wherein two distinct regimes can be defined: the miscibility regime and the solubilization regime.[33]

In the miscibility regime, the processing temperature is greater than the melting temperature of the drug substance leading to the situation where lower viscosity molten drug substance must be mixed within the molten polymer. In this case, the ability to form a stable dispersion is a function of miscibility between the components and the ability of the extrusion process to homogeneously distribute the materials. When materials are selected with appropriate miscibility, the formation of an amorphous solid dispersion can be easily achieved. A major advantage for processing in this mode is the limited number of critical kinetic processes that occur during manufacturing, specifically the distributive mixing process during extrusion. In other processing variants in this regime, it is possible to produce bottom-up engineered particles by selecting immiscible drug:carrier combinations. In this application, the molten drug and carrier are intimately mixed in the extruder before the drug recrystallizes out during the subsequent cooling and mixing process.

Processing in the solubilization regime occurs when the operating temperatures are below the melting temperature of the drug substance. In this mode, the formation of an amorphous form is achieved by dissolution of the drug substance into the molten carrier. Careful control must be provided for the process and coupled with judicious selection of formulation additives. During operation in this regime, the molten polymer functions as a solvent, allowing the drug substance to dissolve within the carrier at a temperature below the melting temperature of the compound. The selection of the formulation becomes driven not only by the ability of the formulation to provide the necessary delivery characteristics but also the ability to solubilize the drug substance within the molten carrier. The diffusivity of drug in molten polymer can be related to key descriptors of the process and composition through the Stokes–Einstein equation. Changing formulation additives to enhance molten solubility or reduce viscosity leads to increased dissolution rates. These attributes can be screened in early development to aid in the selection of solubilizing material, with differential scanning calorimetry and hot stage microscopy being well-utilized techniques.[20] The application of these methods are shown in Fig. 2 and Fig. 3, respectively. Similarly, control of starting particle size governs specific surface area, which can be manipulated to control dissolution rates. With respect to process, temperature influences viscosity and solubility, where higher temperatures increase dissolution rates. Higher shear rates achieved by more aggressive screw design and higher screw speeds can also be shown to reduce the viscosity of most pharmaceutical systems and shorten characteristic transport path lengths leading to improved mass transfer. Noting that the preceding discussion focused on the application of solubilizing a drug within a carrier system, most crystalline systems are also produced in this processing mode; however, the formulation and process are designed to minimize dissolution rates. In this case, one seeks a formulation with limited molten solubility for drug within the carrier and is also operated at lower temperatures

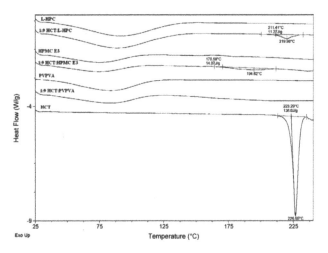

Fig. 2 Application of differential scanning calorimetry for assessment of melt solubilization.
Source: From DiNunzio et al.[20]

with minimal dispersive mechanical energy input where the free energy of drug dissolution in the carrier is greater than zero. This allows the drug substance to be distributed with the carrier while retaining the intrinsic crystalline structure of the feedstock.

Characterization of these dispersions is also a necessary step to establish manufacturing control of the product while also yielding valuable insight into the nature of the system. In general, the determination of the type of solid dispersion product is achieved through the use of several concurrent technologies aimed at identifying crystallinity, distribution, and chemical interactions of the components. For the assessment of crystallinity, the most commonly used technique is X-ray diffraction (XRD),[34–36] which elucidates the long range order of crystalline material. In the case of amorphous materials, a characteristic amorphous halo is observed. Similarly, different crystalline materials exhibit different diffraction patterns allowing for identification of the specific crystalline phase. Crystallinity confirmation and distribution within the dispersion can be assessed using differential scanning calorimetry (DSC) to determine crystalline material based on melting endotherm and modulated differential scanning calorimetry (mDSC) to identify glass transition temperature of the individual components or the resulting dispersion.[37–42] With this technique, it may be possible to identify the number of phases within the solid dispersion, where a homogeneous amorphous dispersion is indicated by a single glass transition temperature and a heterogeneous system is identifiable by multiple thermal events associated with glass transitions or melting. The application of thermal techniques should be coupled with secondary isothermal techniques, in particular XRD, because of the possibility of solubilizing the drug substance within the polymer system during preparation and/or analysis. The ability to form and stabilize solid dispersions is also a function of the intermolecular interactions between the components, detectable by a variety of spectroscopic techniques[43–45] including Fourier transform infrared spectroscopy (FTIR), solid state nuclear magnetic resonance (NMR), Raman, and near infrared spectroscopy (NIR). With these techniques, it is possible to determine the transition from crystalline to amorphous and the resulting formation of drug–polymer interactions by the change of wavelengths associated with different chemical interactions. It also becomes possible using these techniques to quantitate the level of crystalline material within a formulation depending on the properties of the native crystalline material.

Discrimination between products as it relates to bioperformance is also a necessary consideration when developing a melt extruded dosage form, particularly when the technology is leveraged for solubility enhancement and controlled release. Specific factors must be considered both in the formulation design and characterization of the system. In the case of poorly soluble compounds, carrier selection is identified to both produce the amorphous form and maximize the apparent solubility.[38,40,46–53] Noting the considerations of polymer selection for processing as discussed previously, the aim of producing an amorphous dispersion is to exploit the free energy benefit associated with the absence of a crystal structure, which leads to rapid dissolution and is often termed "the spring." Without the aid of stabilizing materials, the excess drug in solution will often precipitate until reaching the solubility of the crystalline material. In order to reduce precipitation rates, the stabilizing carrier material used during extrusion should also interact molecularly with the drug substance to enhance its solubility. This behavior is commonly referred to as "the parachute" effect. Conceptually, this combined "spring and parachute" effect[54] is illustrated in Fig. 4.

Conveniently, many of the polymers that most effectively stabilize supersaturation also provide suitable thermoplastic characteristics for extrusion. Two of the most commonly used materials are copovidone and

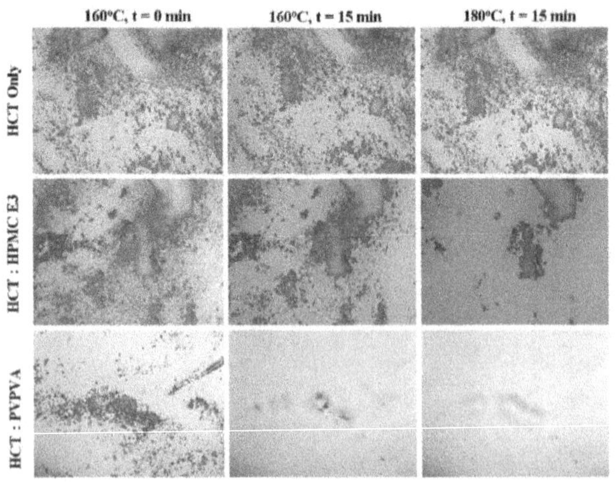

Fig. 3 Application of hot stage microscopy for the assessment of melt solubilization.
Source: From DiNunzio et al.[20]

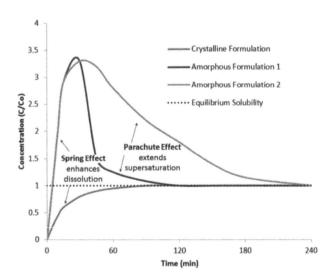

Fig. 4 Spring and parachute performance of amorphous dispersions in comparison to crystalline material.

hypromellose acetate succinate, which are non-ionic and enteric materials, respectively.

In order to optimally exploit the spring and parachute behavior, it is necessary to understand the origin of the phenomena. Numerous researchers have reported on the nature of these interactions, ranging from hydrogen bonding, hydrophobic interactions, ionic interaction of the carrier, and steric hindrance.[54–59] In reality, it is likely a combination of drug and polymer-specific interactions as well as contributions of drug substance properties such as glass transition temperature and recrystallization potential[60] that drive stabilization. Most importantly, a wealth of academic and industrial information has highlighted the use of these materials in solid dispersions for enhancing the oral bioavailability of poorly soluble compounds.

For controlled release products, the considerations are very different. For these systems, it is necessary that extrudable formulations lead to controlled release characteristics of the delivery system. In many of these cases, the drug substance is intended to retain its native form, i.e., crystalline structure, which requires processing of the formulation in a mode where the drug remains insoluble in the melt. Challenges further emerge in that many of the materials needed for controlled release lack desirable characteristics for extrusion. For example, hypromellose is a commonly used polymer for controlled oral delivery but is non-thermoplastic. This mandates the need for additional materials to aid in processability, with plasticizer and low melting point surfactants frequently meeting this need.[22,26,27,61,62] The presence of plasticizers may be leveraged to regulate the ability of the polymeric phase to control the release rate of the active due to changes in diffusivity, wettability, and water penetration into the system. In the case of other systems, which exhibit the necessary thermoplastic characteristics for processing, particularly polyethylene oxide (discussed in greater detail in the controlled release section), the semi-crystalline nature of the polymer can present additional challenges.[63] The level of crystallinity in a material impacts the microstructure of the system, leading to transport variations. Additionally, as changes occur on stability resulting in crystallization, a volume contraction occurs due to the ordering within the system that can impact the stability and distribution of other components in the formulations. These changes, both of formulation and physical stability, can also influence mechanical properties of the system, which is a key factor in the production of directly shaped products.[30] For these reasons, in the case of crystalline and semi-crystalline polymers it is necessary that they reach a similar endpoint to enable batch-to-batch consistency and achieve requisite stability. Another application for controlled release products prepared by extrusion is the case of amorphous controlled release products. Exploiting the increased solubility of the amorphous form, it becomes possible to tailor release properties from the dosage form. Careful design of these systems is necessary however, because of the complex interactions that occur within the dosage form and use environment.

Numerous publications have covered the design and implementation of extrusion operations, with several seminal references available for the interested reader.[11,64–67] The subsequent sections of this entry discuss the applications of melt extrusion for controlled delivery, focusing on the interplay between formulation design, characterization, and the desired drug product attributes for the delivery routes.

APPLICATIONS

Broadly speaking, melt extrusion is used to produce pharmaceutical products that can be broken down into four application classes: dissolution-enhanced products, controlled released products, foamed products, and shaped delivery systems. The subsequent sections describe the design of drug products to achieve the desired controlled release characteristics.

DISSOLUTION-ENHANCED PRODUCTS

In the absence of solubility, drug substances are poorly absorbed *in vivo* leading to ineffective therapies. One strategy to address this is the formation of an amorphous dispersion that leads to higher solubility and faster dissolution rates, whereby the drug substance becomes more bioavailable. In other cases where crystalline interactions are favorable for dissolution, crystalline dispersions can be manufactured to provide enhanced wetting. This section describes examples of both types of dispersions, with a focus on the influence of formulation to achieve the desired attributes.

Applications of Dissolution-Enhanced Dispersions

Lopinavir and ritonavir are both antiviral compounds that exhibit low solubility, metabolic conversion, and efflux leading to low oral exposure, which are currently formulated as melt extruded solid dispersions.[68] Chronologically, ritonavir was developed prior to lopinavir utilizing a metastable polymorph, Form II. After initial agency approval and marketing, stability issues emerged and the product was pulled from the market and reformulated.[69] A soft gelatin capsule product was subsequently launched. As lopinavir was developed, liabilities associated with metabolic conversion and efflux were noted that limited oral bioavailability. Given the need for a fixed dose combination product with other antivirals to enhance treatment, lopinavir was combined with ritonavir, a powerful cytochrome P450 (CYP) enzyme and P-glycoprotein (PGP) efflux inhibitor. This product was also formulated as a soft-gelatin capsule; however, due to the storage restrictions and undesirable manufacturing process, researchers at Abbott looked to reformulate the product.

Noting that both compounds exhibit molecular characteristics appropriate for use in melt extrusion, specifically low melting points that enable processing at moderate temperatures, a combination product of lopinavir and ritonavir was developed (Kaletra®) as well as a standalone drug product of ritonavir alone (Norvir®). In the design of the formulation researchers looked to design a system capable of forming a colloidal dispersion on exposure with an aqueous environment.[68] Additionally, the need for physical stability at room temperature storage conditions required the development of a formulation, where the drug loading did not exceed the solubility of API in the carrier matrix. Despite this constraint, the total number of dosage units administered compared with the soft-gel formulation was reduced from 6 units to 4. Another consideration in the design of the formulation was the utilization of formulation and processing conditions to minimize impurity formation in the solid dispersion. The incorporation of inviscid polymer, copovidone, and low levels of the surfactant, sorbitan monolaurate, enabled production at lower temperatures while also aiding in the release of intermediate from the calendering process. These changes also minimized impurities, which was further supplemented through the incorporation of customized screw elements,[70] shown in Fig. 5, which minimized local excess energy build-up during extrusion that can lead to decomposition.

Crystalline solid dispersions are also effective in enhancing dissolution rates by creating solid dispersions with intimate contact between the drug and polymer. Historically, one of the earliest applications of thermally produced solid dispersions for dissolution enhancement was the development of Gris-PEG (griseofulvin ultramicrosize), a crystalline solid dispersion of griseofulvin prepared in polyethylene glycol.[71] Due to the rapid dissolution, illustrated in Fig. 6, even greater exposures were possible when compared to conventionally formulated higher dose products. In this system, the particle size of the drug substance is set by the properties of the material prior to the extrusion operation. Achieving a high degree of dispersive mixing was key to uniform spatial distribution of the API in the carrier.

In another novel application of melt extrusion for the production of dissolution enhanced drug products, the extrusion process was used for the production of bottom-up nanoparticles as part of the development of dalcetrapib.[72] Outlined in a patent application, the inventors described the process of melting the drug within an immiscible carrier matrix. After intimate mixing, nucleation and growth of the drug substance were driven in the later stages of the process section via temperature reduction while the continued agitation from the screws aided in the control of the embedded crystalline drug particle size. As a result of the dispersion structure enhanced dissolution was observed while maintaining enhanced stability characteristics when compared to amorphous concepts. Similar process concepts have also been reported for the direct extrusion of engineered nanoparticles. Miller et al.[49] demonstrated the use of non-solubilizing carriers to densify particles by incorporation into melt-extruded matrices. In this particular process variant, carrier selection was critical to both dissolution enhancement and stabilization of the particles. As shown in Fig. 7,[58] when processed within a non-solubilizing formulation using a single screw extruder to improve pumping, the particles can be maintained largely intact, preserving the dissolution benefits of the engineered materials. In other variants, researchers have prepared top-down nanoparticles in aqueous media and then used the combined devolatilization and compounding capabilities to generate nanoparticle-based solid dispersions. For such cases where it is necessary to preserve the particle properties, processing is concluded in the solubilization regime using a design intended to minimize dissolution within the molten carrier. Conversely, the production of bottom-up engineered particles relies on processing the miscibility regime where the drug and carrier are immiscible and processing conditions aid in phase separation and crystal growth.

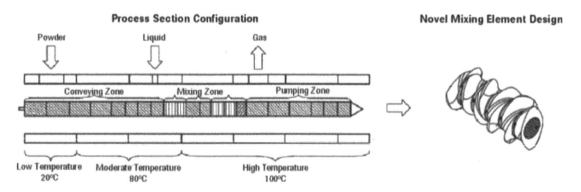

Fig. 5   Screw elements and designs used for minimization of impurities during extrusion processing of ritonavir and lopinavir.
Source: From Kessler et al.[70]

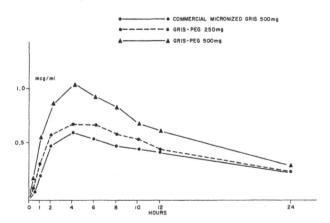

Fig. 6 *In vitro* dissolution rates of griseofulvin:polyethylene glycol solid dispersions.
Source: From Riegelman & Chiou.[71]

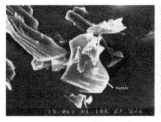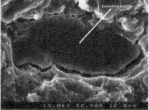

Fig. 7 Embedded engineered particles in solid dispersions prepared by melt extrusion.
Source: Adapted from Miller.[58]

CONTROLLED RELEASE PRODUCTS

Controlled release products, including those with delayed, sustained, or extended release, are designed for reduced frequency of dosing, reduced absorption variability, improved efficacy, or improved safety over immediate release formulations. In 2008, the controlled release market was estimated at US$21 billion globally and is anticipated to increase to US$29.5 billion in 2017. In the past, controlled release products have been developed to protect innovators from generic erosion of immediate release products while improving patient compliance or product performance. Innovators are now seeking earlier entry of controlled release products to increase patient benefit and sales.[73]

One area where melt extruded controlled release products have made a major impact is in the abuse deterrent space. By creating a system with intimate contact between the drug and polymer, as well as the use of highly deformable carrier materials, it is possible to engineer abuse deterrence. An epidemic of opioid dependency and abuse prompted the Food & Drug Administration (FDA) to issue a draft guidance describing the desired attributes for abuse-deterrent formulations, including limited ability to inhale or inject the active ingredient via crushing, grinding, melting, or dissolving of the drug product.[31] These recommendations apply to both immediate release and controlled release products; however, many new opioid-based products have been recently developed as controlled release systems to improve patient compliance. In April 2013, the FDA approved an abuse-deterrent formulation of OxyContin as the previous formulation was withdrawn from the market for safety concerns.[32] The FDA explicitly will not accept Abbreviated New Drug Applications (ANDAs) for the previous formulation.

Controlled release (CR) products can be broadly classified into two major categories: extended release and delayed release. While a striking range of innovative techniques exist, drug release from the dosage form is regulated by a number of methods based on diffusion, swelling, erosion, and osmotic pressure. Extended release is desired to maintain a therapeutic level of drug in the body over a longer time, reduce peak plasma concentrations, or minimize exposure variation. For these systems, drug products contain polymers that do not rapidly dissolve in the use environment, leading to a slower rate of erosion or swelling that creates a gel layer. Both erosion of the exterior layer and diffusion through the gel layer govern the release of the drug from the device. Other variants of the extended release technology can also rely on osmotic pressure to "push" the drug from the drug product in a controlled fashion. Similarly, a delayed release product has a drug release profile that does not begin immediately after administration and is desired if the drug substance irritates the upper gastrointestinal (GI) tract, degrades in the presence of low pH, or exhibits a large change in solubility with minor changes in gastric pH. Delayed release often exploits the pH-solubility of a polymer with a pKa at or near the target physiological pH. On contact with an environment where the pH is greater than the pKa the polymers ionize and dissolve to release the payload.

Compared to "conventional" CR formulation technologies, Hot-Melt Extrusion (HME) offers several potential technical and business advantages. HME can be used to simultaneously generate an amorphous solid dispersion (ASD) and CR/ASD behavior, providing unique release profiles for low solubility molecules. It is also a convenient trend that a number of polymers used to generate amorphous solid dispersions with HME have inherent delayed or extended release functionality, allowing for a healthy selection of available materials. Crystalline solid dispersions produced with HME also tend to have high density, low porosity, and low specific surface area, which allows for a higher drug loading and lower polymer loading while maintaining comparable release characteristics from a smaller dosage form. Amorphous systems, by nature of the reduced surface area, may also benefit from improved physical stability. From a mechanical properties perspective, many polymeric excipients used in HME tend to deform plastically, frustrating attempts to grind or mill it for abuse. Further engineering of the drug product is also possible through the production of multi-layer systems and directly manufactured products via calendering and injection molding. From a business perspective, the technology can be adapted as a continuous process.

As discussed in the earlier sections, the desired solid state properties of the extrudate are critical to consider and monitor throughout the formulation and process development. This is particularly critical in the case of melt extruded dispersions for CR applications, where these systems can be intentionally designed as crystalline or amorphous dispersions. If a crystalline solid suspension is desired, the extrusion process must impart sufficient thermal and shear energy to melt the excipients, but not melt the API. To achieve this goal, the API may need to be added after melting and mixing the excipients, late in the extrusion process to limit time at high temperatures and exposure to high shear zones. The melting points and/or glass transition temperatures of excipients will be crucial to understand. For an amorphous solid dispersion, the HME process parameters must be appropriate to achieve a single, continuous phase by imparting sufficient thermal and shear energy, and then the system must be monitored for physical stability since the thermodynamically unstable amorphous state could result in API crystallization or phase separation could occur on stability.

It is often challenging to achieve immediate release with amorphous solid formulations due to the physical nature of HME particles and the polymers generally used for this purpose. Many of the commonly used excipients tend to be enteric polymers (e.g., HPMCAS) or binders (PVP, PVP/PVAc). Furthermore, the high density and low porosity nature of the dispersion results in lower specific surface area that drives slower release rates. Coupled with the lower disintegration rates for the monolithic HME products, release rates will tend to remain slower after introduction to the aqueous environment. However, for CR applications these attributes present unique advantages that can be exploited in the development process.

In design of CR products using melt extrusion, a number of commonly used oral controlled release excipients could be successfully processed as is or with the addition of plasticizers. Desired drug release profiles can be achieved by manipulating the polymer chemistry, grade and level of polymers, and geometric considerations of the drug product. Specific polymers used in CR HME applications include the following:

- Cellulose esters or ethers such as methylcellulose (MC), ethylcellulose (EC), hydroxyethyl cellulose (HEC), hydroxypropyl cellulose (HPC), hydroxypropyl methylcellulose/hypromellose (HPMC), hypromellose phthalate (HPMCP), hypromellose acetate phthalate (HPMCAP), hypromellose acetate succinate (HPMCAS), cellulose acetate (CA) and derivates
- Poly (methyl) methacrylates (Eudragit® L, S, E, RS/RL)
- Polyethylene oxide (PEO)
- Polyvinyl pyrrolidone (PVP), polyvinyl acetate (PVAc), or combination as Kollidon SR®

- Cross-linked polyacrylic anhydride (PAA), e.g., Carbopol®
- Starches or modified starches
- Long-chain fatty acids, mono- or di-glycerides
- Natural products such as gelatin, gums, chitosan, carnauba wax, and alginates

Some of these polymers have a narrow T_g:T_{deg} ratios or have such high viscosities that require the addition of plasticizers such as phthalate or citrate esters, low molecular weight polyethylene glycol (PEG), and/or propylene glycol to process robustly or improve flow during extrusion. Often a surfactant is required for bioavailability enhancement and can conveniently provide process enhancement as well due to a reduction of melt viscosity.[27,74] Similarly, drug substance in the formulation can also provide autoplasticization. In more complex manufacturing design, fugitive plasticizers can be used to lower viscosity during extrusion. One unique advantage of fugitive plasticizers is that they are not retained in the drug product, which allows the formulation to maintain an elevated T_g. In the case of non-fugitive plasticizers, changes to the stability and release rate of a formulation are possible so it is important to understand the impact to performance across different levels.[75] Another strategy to improve performance is to develop formulations of multiple polymers, exploiting the beneficial aspects of the individual materials. This leads to improved processing as well as release and bioavailability characteristics.[76]

Beyond judicious polymer selection, the controlled release profile can be further modified by the addition of disintegrants or soluble components such as saccharides, salts, or pH-modifying excipients. For a selected formulation, the morphology and surface area of the HME particle or product (i.e., direct shaped pellet) can also be manipulated to change the drug release profile and likely will be important to control for consistent performance. Exemplary control of the pellet or tablet surface area and morphology can be achieved through die-face or strand pelletization. Pellets can then be spheronized to remove asperities and improve the consistency of release.[77]

Controlled Release Excipients and Applications

Hypromellose (HPMC) and polyethylene oxide (PEO) are commonly used as eroding hydrophilic matrices for sustained release. Ethylcellulose (EC) is commonly used as an insoluble matrix polymer. These polymers are often used in combination, e.g., HPMC/EC or PEO/EC, in order to tailor release profiles. All have been used to enhance solubility by producing solid dispersions.[31,32,38,63,78–80] Of these materials, PEO and EC are readily amenable to HME based on the physical properties, but the narrow processing window for HPMC and very high melt viscosity requires the use of high levels of plasticizer. In many cases the plasticizer level can be as much as 30% by weight.[81]

While evaluating different formulations, HME process parameters should be considered as well. Of note is the study of HME process on guaifenesin release from EC matrices, which demonstrated a statistically significant effect due to changes in the matrix porosity induced by higher temperatures.[32] Soluble components in an insoluble matrix can be used to promote release by leaving voids upon dissolution that increase surface exposed for diffusion.[80]

Poly(methylmethacrylate) (PMMA) was originally developed and marketed in 1933 as a glass replacement called Plexiglas®. This polymer and derivatives were first used to coat oral solid dosage forms in the 1950s and were extruded for pharmaceutical applications in the late 1990s. A wide range of Eudragit polymers cover the physiological pH range and act as insoluble sustained release polymers, with the specific pH of release determined by the side group chemistry of the material grade. However, many grades have relatively low degradation temperatures and high melt viscosities, which complicates their use in extrusion applications. For example, Eudragit S100 and L100 functional groups degrade near the polymers' T_gs, so a plasticizer is required.[23] Several recommended plasticizers for Eudragit L100 and S100 are triethyl citrate (TEC), PEG 6000, or propylene glycol at concentrations ranging from 5% to 25% polymer weight. In specific cases the drug may also provide plasticizer functionality, concomitantly reducing viscosity, and providing therapeutic efficacy through improved release performance. Eudragit E and L 100-55 have precedence of use for solubility enhancement, and HME has been used to make crystalline suspensions for theophylline enteric pellets (L100, S100, L100-55), indomethacin controlled release (L100, S100), and sustained release of diltiazem hydrochloride and chlorpheniramine maleate (RSPO).[82–85] Although most CR HME systems are crystalline dispersions, CR amorphous solid solutions may also be achievable with the right stabilizing formulation.

Beyond the abovementioned polymers, many other excipients have been used for HME and CR applications. Notably, Kollidon SR is a physical blend of PVP and PVAc, which generates a matrix of insoluble (PVAc) and soluble (PVP) domains in the extrudate and has been used for combined CR and amorphous solid solution of ibuprofen[86] for sustained release. The addition of long fatty acids, glycerides, or waxes that slow drug release may be of interest since they act as polymer plasticizers and have demonstrated a reduction of *in vitro* food effect risk.

Abuse-deterrent products have additional requirements for demonstrating reduced ability to crush, cut, grate, or grind with common household implements like spoons, cutters, and coffee grinders, under hot or cold conditions. The particle size before and after these treatments should be compared to assess risk of inhaled use. Lab assessments of the extractability of the drug in common household solvents,

especially ethanol, and syringeability/injectability are also required.[31] Several abuse-deterrent systems take advantage of PEO's significant plastic strain properties that make milling or crushing difficult. Further, the hydrogel nature makes dissolving or injecting the formulation challenging. These products usually provide controlled release characteristics to support once a day dosing and may also be formulated as such due to the matrix polymer viscosity. These PEO/drug formulations can be calendered using equipment such as Abbott/Soliqs Meltrex calendering technology (e.g., Isoptin SR)[87] or melt extruded/compression molded as in Grünenthal's Intac® technology (e.g., Opana ER)[88,89] to produce a low porosity tablet that releases slowly. Other variants have been developed to provide immediate release characteristics through pelletization, illustrating that these systems can be used for immediate release or controlled release applications. Figure 8 highlights the Grünenthal Intac technology that provides improved abuse deterrence through greater crushing resistance.

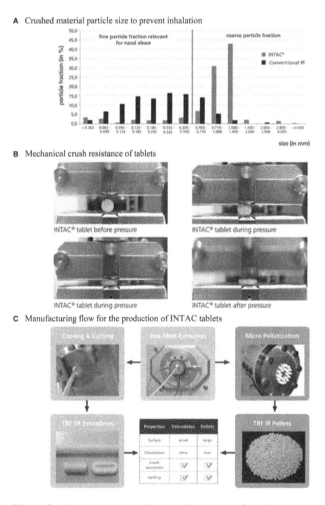

Fig. 8 Abuse-resistant formulation manufacturing and characterization.
Source: Used with permission from Grünenthal GmbH, © 2013.

FOAMED PRODUCTS

In addition to traditional extrusion compounding to produce solid dispersion or solid solutions for immediate and controlled release applications, foam extrusion has emerged as a means to provide additional product and process value in the pharmaceutical industry.[90–92] In general, foam extrusion can be utilized to produce extrudate formulations with improved milling properties,[93–95] increased dissolution rates,[93,96,97] and potentially improved *in vivo* performance. In addition, this technology can be employed to generate low density, buoyant dosage forms with or without entrapped gas that allow for targeted release of drug through gastric retention.[98–100] Finally, unique dosage forms, such as orally disintegrating tablets, can be generated to allow for the targeted delivery of actives.[101]

Foam extrusion, represented schematically in Fig. 9, is achieved through (1) the uniform dissolution of gas into the polymer melt at high pressure, (2) the introduction of thermodynamic instability via a decrease in pressure or an increase in temperature at the die resulting in cell nucleation, (3) followed by cell growth and expansion, and finally (4) stabilization of the resulting foamed structure by cooling below the T_g of the material. Foaming of hot melt extrudates can be achieved using chemical agents such as sodium bicarbonate and citric acid that decompose with temperature to form carbon dioxide[9] or physical blowing agents such as carbon dioxide, nitrogen, or water.[90,92,100,102] Physical blowing agents have the advantage of being removed during processing and as such may have minimal impact on the physical and chemical stability or dissolution performance of the resulting foamed extrudate. In addition, because of the plasticizing property of physical blowing agents, they can be utilized as fugitive plasticizers allowing for processing at lower product temperatures in cases where the active ingredients or excipients are thermally labile, or to lower the viscosity of the extruded formulation

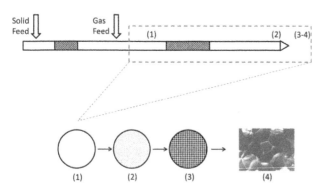

Fig. 9 Schematic representation of foaming process: (1) dissolution of gas into polymer, (2) cell nucleation, (3) cell growth and expansion, and (4) SEM of hydroxypropyl methyl cellulose acetate succinate (HPMCAS) extrudate foamed with nitrogen gas.

potentially leading to increased productivity.[93–96,102] Unlike chemical blowing agents that are easily added to the pre-extruded feed at a desired ratio, physical blowing agents require additional equipment to process in a robust and safe manner. Typically, high pressure pumps capable of metering supercritical fluids are utilized to maintain pressure and control the delivery of known amounts of gas to the extruder. Importantly, a screw design must be considered to allow for the correct pressure profile and solubilization of gas into the polymer melt. A sufficient melt seal must be formed upstream of the injection site to prevent gas escaping through the feed zone. In addition, a filled section must be present downstream of gas injection to allow adequate pressure and time for gas dissolution. Finally, one must utilize the extruder to cool the polymer melt to achieve a melt strength that is supportive of foaming.[103] If the melt strength is too low, the foam will collapse. If the melt viscosity is too high, there is the potential the gas will phase separate or the polymer will not achieve full expansion.

Application of Foam Extrusion to Modified Release Formulations

The most straightforward application of foam extrusion to control the release of active ingredients is the fabrication of low density floating dosage forms that can achieve gastric retention allowing for enhanced delivery of drugs that act locally (in stomach) or to improve the absorption of drugs with an absorption window limited to the stomach or upper gastrointestinal tract. Fukuda et al. prepared foam matrices of acetohydroxamic acid and chlorpheniramine maleate with Eudragit RS PO using 5–10 weight percent sodium bicarbonate as a chemical foaming agent.[98] Sodium bicarbonate decomposes thermally into carbon dioxide and water as the foaming agents leaving behind sodium carbonate in the formulation matrix. Extrudate tablets were produced on a single screw Randcastle extruder and manually cut into tablets. Foamed tablets with apparent densities ranging from 0.6 to 1.0 g/cm³ demonstrated the ability to remain buoyant in dissolution media (0.1N HCl) for 24 hr. In addition, both compounds in their work showed sustained release over this period of time, as shown in Fig. 10. The release rate of chlorpheniramine maleate could be increased by adding Eudragit E PO (0–65%) to the formulation or by decreasing the diameter of the extruder die (from 3 mm to 1 mm). The formulations were found to be physically stable at 40°C and 75% relative humidity with no changes in buoyancy or dissolution properties; however, unknown is the long-term impact of residual sodium carbonate in the formulation. Nakamichi et al. utilized water as a physical blowing agent to produce floating dosage forms of hydroxypropyl methylcellulose acetate succinate (HPMCAS, AQOAT MF grade) and nicardipine hydrochloride.[104] Formulations

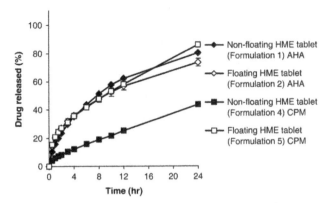

Fig. 10 Controlled release profiles for conventional CR tablets produced by hot-melt extrusion and porous floating tablets produced using hot-melt extrusion. Acetohydroxamic acid (AHA) tested in 0.01N HCl, chlorpheniramine maleate (CPM) tested in 0.1N HCl.
Source: From Fukuda et al.[98] © 2006, with permission from Elsevier.

containing 10–30 weight percent nicardipine hydrochloride were extruded in a co-rotating twin screw extruder and cut to desired lengths for testing. In addition, varying amounts of calcium phosphate dihydrate was added to the formulation as a nucleating agent. Nucleating agents are utilized to increase the number of sites for gas nucleation leading to an overall increase in cell number and a decrease in foam density. As the amount of calcium phosphate dihydrate increased from 0 to 12 wt. percent, the porosity of the tablets increased from 5% to 72%. Foamed tablets retained their buoyancy for 4–6 hr in pH 1.2 media. The foamed tablets showed low rates of API release at pH 1.2 and rapid and complete release at pH 6.8 due to the enteric nature of the HPMCAS polymer. In addition to gastric retentive dosage forms, extrusion foaming can be utilized to create fast disintegrating oral tablets to target drug delivery in the buccal or sublingual space. Clarke describes the use of extrusion compounding coupled with injection molding and an inert physical blowing agent to form foamed dosage forms with rapid disintegration.[101] Clarke explored the ability to extrude, injection mold, and foam formulations of several polymers, sweeteners, flavorants, and disintegrants. The foaming process developed would allow for the continuous manufacture of orally disintegrating tablets as well as gastric retentive dosage forms.

DIRECTLY SHAPED PRODUCTS

Although the majority of pharmaceutical formulations are delivered orally, a large number of products have been developed for non-oral administration based on compound properties, delivery frequency, patient compliance, or

specific site targeting. In order to prepare such systems, it is important that in addition to the desired release characteristics consideration must be given for geometry, mechanical behavior, and immunogenicity. Utilizing extrusion it is possible to prepare monolithic drug products for advanced non-oral applications, where the unique advantages of a continuous non-solvent process are used to more rapidly and effectively prepare products when compared to similar solvent-based approaches. To date products including subcutaneous filaments, intravitreal implants, buccal fibers, and vaginal rings have been developed using melt extrusion. Release mechanisms from directly shaped systems rely on similar principles to those outlined in the controlled release section, where delivery into the bulk is driven by diffusion, swelling, erosion, and/or pressure differentials coupled with the geometric design of the device.

Rod-shaped implants or vaginal rings can be made using hot-melt extrusion, formulated as drug in polymer matrix or membrane-controlled reservoir systems. Several polymer options have been demonstrated for commercial and development purposes. For flexible rings, low Tg polymers are preferred like ethylene vinyl acetate (EVA), polyurethane elastomers, and silicones produced by reactive injection molding. Poly (lactide) (PLA),[105] poly (lactide-co-glycolide) (PLGA),[105] and polycaprolactone (PCL)[106] have been extruded and cut into strands or injection molded to produce bioabsorbable implants (e.g., with dexamethasone,[107] gentamicin sulfate,[108] and lysozyme).[109]

Applications of Directly Shaped Products by HME

One of the most successful implant systems developed to date is the Nuvaring®, which is a vaginal implant containing hormonal contraceptives that are delivered over a 21 day duration.[110] During the use of the product, insertion and removal are performed by the patient, which allows for the use of non-degradable polymers without compromising patient compliance. The circular ring is comprised of an inner core of EVA embedded with etonogestrel and ethinyl estradiol. Release rates of 120 and 15 μg/day, respectively, are achieved by diffusion through an exterior EVA layer with a thickness of approximately 110 μm.[110] During the manufacturing of this product, a dual extrusion setup into a coaxial die is used to prepare the drug-loaded core while a second layer of pure EVA is extruded at a product matching feed rate to achieve a rod intermediate. The rod is cut to length before the ends of the rod are welded together to form the final ring. The finished product exhibits the necessary mechanical strength to facilitate patient insertion and removal, while also yielding the desire hormone release profile, and is an excellent example of the complex product engineering that can be achieved with the melt extrusion platform.

Another recent example of a directly shaped drug product that illustrates the synergy between product, device, and delivery mode is Ozurdex®, an intravitreal

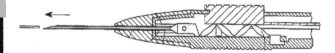

Fig. 11   Ozurdex delivery system schematic.
Source: From Weber et al.[112]

implant that delivers dexamethasone over an extended period to treat macular degeneration.[111] Unlike vaginal implants that can be manually removed by patients, products that are implanted intravitreally cannot be removed by the patient and would potentially require significant invasive surgical procedures to remove. As a result, products such as Ozurdex must utilize biodegradable polymers to eliminate the need for post-delivery removal. To facilitate this, dexamethasone is processed with a biodegradable PLGA polymer to form a cylindrical extrudate with a diameter of 460 μm. The resulting cylinder is then loaded into a 22-guage hypodermic needle capable of delivering the drug product to the back of the eye without the need for post-implantation sutures. A representative schematic of the delivery device is shown in Fig. 11. It is also important to minimize immune response of the drug product, which can be achieved through the implementation of terminal sterilization. Ozurdex is sterilized using gamma irradiation of the finished drug product, which does not negatively impact release characteristics of the product.

CONCLUSIONS

Melt extrusion has been established over the last 30 years as a viable platform for the production of solid dispersions to aid in the delivery of pharmaceutical compounds. Illustrated in this section, the technology can be used to support a range of delivery aspects for a number of new therapies. Being a low cost manufacturing technology, capable of achieving unique delivery attributes and supporting new production paradigms, it will remain a pivotal platform for the development of novel drug products. As the industry continues to evolve in many new areas, including personalized medicine and continuous manufacturing, melt extrusion is uniquely suited to support these paradigms. By virtue of its well-established nature in a number of industries, scalability, and excellent process control, melt extrusion meets the needs of both biotechnology and small molecule discovery pipelines. Combining this with an ever-expanding compositional knowledge for the technology, melt extrusion will make a clear difference in the future of drug delivery.

REFERENCES

1. Ku, M.S. Use of the biopharmaceutical classification system in early drug development. AAPS J. **2008**, *10* (1), 208–212.

2. Lipinski, C.A. Drug-like properties and the causes of poor solubility and poor permeability. J. Pharmacol. Toxicol. Met. **2000**, *44* (1), 235–249.

3. Lipinski, C.A. Lead- and drug-like compounds: the rule-of-five revolution. Drug Discov. Today **2004**, *1* (4), 337–341.

4. Lipinski, C.A.; Lombardo, F.; Dominy, B.W.; Feeney, P.J. Experimental and computational approaches to estimate solubility and permeability in drug discovery and development settings. Adv. Drug Deliv. Rev. **2001**, *46* (1–3), 3–26.

5. Alsenz, J.; Kansy, M. High throughput solubility measurement in drug discovery and development. Adv. Drug Deliv. Rev. **2007**, *59* (7), 546–567.

6. Jain, P.; Yalkowsky, S.H. Prediction of aqueous solubility from SCRATCH. Int. J. Pharm. **2010**, *385* (1–2), 1–5.

7. Di, L.; Fish, P.V.; Mano, T. Bridging solubility between drug discovery and development. Drug Discov. Today **2012**, *17* (9–10), 486–495.

8. Ullrich, M. Historical Development of Co-Rotating Twin Screw Extruders. In *Co-Rotating Twin-Screw Extruders*; Kohlgrüber, K., Ed.; Hanser Garnder Publications: Cincinnati, OH, **2008**, 1–7.

9. DiNunzio, J.C.; Martin, C.; Zhang, F. Melt extrusion: Shaping drug delivery in the 21st century. Pharm. Technol. **2010**, *57* (SI), s30–s37.

10. Crowley, M.M.; Zhang, F.; Repka, M.A.; Thumma, S.; Upadhye, S.B.; Kumar Battu, S.; McGinity, J.W.; Martin, C. Pharmaceutical applications of hot-melt extrusion: Part I. Drug Dev. Ind. Pharm. **2007**, *33*(9), 909–926.

11. Thiele, W. Twin-Screw Extrusion and Screw Design. In *Pharmaceutical Extrusion Technology*; Ghebre-Sellassie, I., Martin, C., Eds.; Informa Healthcare: New York, NY, 2003; 69–98.

12. Erdmenger, R. *Mixing and Kneading Machine*; U.S.P.T. Organization: 1954, Farbenfabriken Bayer: United States of America. 6.

13. Rauwendaal, C. *Polymer Extrusion*; Revised 4th Ed., Hanser: Cincinnati, OH, 2001; 781.

14. Tadmor, Z.; Gogos, C.G. *Principles of Polymer Processing*; 2nd Ed., Wiley-Interscience: Hoboken, NJ, 2006; 961.

15. Serajuddin, A.T.M. Solid dispersion of poorly water-soluble drugs: Early promises, subsequent problems, and recent breakthroughs. J. Pharm. Sci. **1999**, *88* (10), 1058–1066.

16. Breitenbach, J.; Mägerlin, M. Melt-Extruded Solid Dispersions. In *Pharmaceutical Extrusion Technology*; Ghebre-Sellassie, I., Martin, C., Eds.; Informa Healthcare: New York, NY, 2003; 245–260.

17. McGinity, J.W.; Repka, M.A.; Koleng, J.J.; Zhang, F. Hot-Melt Extrusion Technology. In *Encyclopedia of Pharmaceutical Technology*; Swarbrick, J., Boylan, J.C., Eds.; Informa Healthcare USA: Hooboken, NJ, **2007**, 2004–2020.

18. McGinity, J.W.; Zhang, F. Melt-Extruded Controlled-Release Dosage Forms. In *Pharmaceutical Extrusion Technology*; Ghebre-Sellassie, I., Martin, C., Ed.; Informa Healthcare: New York, NY, 2003; 183–208.

19. Leuner, C.; Dressman, J. Improving drug solubility for oral delivery using solid dispersions. Eur. J. Pharm. Sci. **2000**, *50* (1), 47–60.

20. DiNunzio, J.C.; Brough, C.; Hughey, J.R.; Miller, D.A.; Williams, R.O. III; McGinity, J.W. Fusion production of

solid dispersions containing a heat-sensitive active ingredient by hot melt extrusion and Kinetisol dispersing. Eur. J. Pharm. Biopharm. **2012**, *74* (2), 340–351.

21. Djuris, J.; Nikolakakis, I.; Ibric, S.; Djuric, Z.; Kachrimanis, K. Preparation of carbamazepine-Soluplus solid dispersions by hot-melt extrusion, prediction of drug-polymer miscibility by thermodynamic model fitting. Eur. J. Pharm. Biopharm. **2013**, *84* (1), 228–237.

22. Ghebremeskel, A.N.; Vemavarapu, C.; Lodaya, M. Use of surfactants as plasticizers in preparing solid dispersions of poorly soluble API: Selection of polymer–surfactant combinations using solubility parameters and testing the processability. Int. J. Pharm. **2007**, *328* (2), 119–129.

23. Lin, S.-Y.; Yu, H.-L. Thermal stability of methacrylic acid copolymers of eudragits L, S, and L30D and the acrylic acid polymer of carbopol. J. Polym. Sci. Part A **1999**, *37* (13), 2061–2067.

24. Repka, M.A.; Battu, S.K.; Upadhye, S.B.; Thumma, S.; Crowley, M.M.; Zhang, F.; Martin, C.; McGinity, J.W. Pharmaceutical applications of hot-melt extrusion: Part II. Drug Dev. Ind. Pharm. **2007**, *33* (10), 1043–1057.

25. Repka, M.A.; Majumdar, S.; Kumar Battu, S.; Srirangam, R.; Upadhye, S.B. Applications of hot-melt extrusion for drug delivery. Expert Opin. Drug Deliv. **2008**, *5* (12), 1357–1376.

26. Janssens, S.; de Armas, H.N.; Roberts, C.J.; Van den Mooter, G. Characterization of Ternary Solid Dispersions of Itraconazole, PEG 6000, and HPMC 2910 E5. J. Pharm. Sci. **2008**, *97* (6), 2110–2120.

27. Repka, M.A.; McGinity, J.W. Influence of Vitamin E TPGS on the properties of hydrophilic films produced by hot-melt extrusion. Int. J. Pharm. **2000**, *202* (1–2): 63–70.

28. Repka, M.A.; McGinity, J.W. Physical-mechanical, moisture absorption and bioadhesive properties of hydroxypropyl cellulose hot-melt extruded films. Biomaterials **2000**, *21* (14), 1509–1517.

29. Repka, M.A.; McGinity, J.W. Bioadhesive properties of hydroxypropyl cellulose topical films produced by hot-melt extrusion. J. Control. Release **2001**, *70* (3), 341–351.

30. Repka, M.A.; Prodduturi, S.; Stodghill, S.P. Production and characterization of hot-melt extruded films containing clotrimazole. Drug Dev. Ind. Pharm. **2003**, *29* (7), 757–765.

31. Crowley, M.M.; Fredersdorf, A.; Schroeder, B.; Kucera, S.; Prodduturi, S.; Repka, M.A.; McGinity, J.W. The influence of guaifenesin and ketoprofen on the properties of hot-melt extruded polyethylene oxide films. Eur. J. Pharm. Sci. **2004**, *22* (5), 1409–1418.

32. Crowley, M.M.; Schroeder, B.; Fredersdorf, A.; Obara, S.; Talarico, M.; Kucera, S.; McGinity, J.W. Physicochemical properties and mechanism of drug release from ethyl cellulose matrix tablets prepared by direct compression and hot-melt extrusion. Int. J. Pharm. **2004**, *269* (2), 509–522.

33. DiNunzio, J.C.; Zhang, F.; Martin, C.; McGinity, J.W. Melt Extrusion, In *Formulating Poorly Water Soluble Drugs*; Miller, D.A., Watts, A.B., Williams, R.O. III, Eds.; Springer: New York, NY, 2012; 311–362.

34. Bates, S.; Zografi, G.; Engers, D.; Morris, K.; Crowley, K.; Newman, A. Analysis of amorphous and nanocrystalline solids from their X-ray diffraction patterns. Pharm. Res. **2006**, *23* (10), 2333–2349.

35. Newman, A.; Engers, D.; Bates, S.; Ivanisevic, I.; Kelly, R.C.; Zografi, G.Characterization of amorphous api:polymer mixtures using x-ray powder diffraction. J. Pharm. Sci. **2008**, *97* (11), 4840–4856.

36. Shah, B.; Kakumanu, V.K.; Bansal, A.K. Analytical techniques for quantification of amorphous/crystalline phases in pharmaceutical solids. J. Pharm. Sci. **2006**, *95* (8), 1641–1665.

37. Craig, D.Q.M. A review of thermal methods used for the analysis of the crystal form, solution thermodynamics and glass transition behaviour of polyethylene glycols. Thermochim. Acta **1995**, *248* (2), 189–203.

38. Six, K.; Berghmans, H.; Leuner, C.; Dressman, J.; Van Werde, K.; Mullens, J.; Benoist, L.; Thimon, M.; Meublat, L.; Verreck, G.; Peeters, J.; Brewster, M.; Van den Mooter, G. Characterization of solid dispersions of itraconazole and hydroxypropylmethylcellulose prepared by melt extrusion, part II. Pharm. Res. **2003**, *20* (7), 1047–1054.

39. Six, K.; Leuner, C.; Dressman, J.; Verreck, G.; Peeters, J.; Blaton, N.; Augustijns, P.; Kinget, R.; Van den Mooter, G. Thermal properties of hot-stage extrudates of itraconazole and eudragit E100. Phase separation and polymorphism. J. Ther. Anal. Calorimet. **2002**, *68* (2), 591–601.

40. Six, K.; Verreck, G.; Peeters, J.; Brewster, M.; Van Den Mooter, G. Increased physical stability and improved dissolution properties of itraconazole, a class II drug, by solid dispersions that combine fast- and slow-dissolving polymers. J. Pharm. Sci. **2004**, *93* (1), 124–131.

41. van Drooge, D.J.; Hinrichs, W.L.; Visser, M.R.; Frijlink, H.W. Characterization of the molecular distribution of drugs in glassy solid dispersions at the nano-meter scale, using differential scanning calorimetry and gravimetric water vapour sorption techniques. Int. J. Pharm. **2006**, *310* (1–2), 220–229.

42. Weuts, I.; Kempen, D.; Six, K.; Peeters, J.; Verreck, G.; Brewster, M.; Van den Mooter, G.Evaluation of different calorimetric methods to determine the glass transition temperature and molecular mobility below Tg for amorphous drugs. Int. J. Pharm. **2003**, *259* (1–2), 17–25.

43. Gupta, P.; Thilagavathi, R.; Chakraborti,A.K.; Bansal, A.K. Role of molecular interaction in stability of celecoxib-PVP amorphous systems. Mol. Pharm. **2005**, *2* (5), 384–391.

44. Lacoulonche, F.; Chauvet, A.; Masse, J. An investigation of flurbiprofen polymorphism by thermoanalytical and spectroscopic methods and a study of its interactions with poly-(ethylene glycol) 6000 by differential scanning calorimetry and modelling. Int. J. Pharm. **1997**, *153* (2), 167–179.

45. Taylor, L.S.; Zografi, G. Spectroscopic characterization of interactions between PVP and indomethacin in amorphous molecular dispersions. Pharm. Res. **1997**, *14* (12), 1651–1698.

46. Blagden, N.; de Matas, M.; Gavan, P.T.; York, P. Crystal engineering of active pharmaceutical ingredients to improve solubility and dissolution rates. Adv. Drug Deliv. Rev. **2007**, *59* (7), 617–630.

47. Konno, H.; Handa, T.; Alonzo, D.E.; Taylor, L.S. Effect of polymer type on the dissolution profile of amorphous solid dispersions containing felodipine. Eur. J. Pharm. Biopharm. **2008**, *70* (2), 493–499.

48. Matteucci, M.E.; Hotze, M.A.; Johnston, K.P.; Williams, R.O. III.Drug nanoparticles by antisolvent precipitation: Mixing energy versus surfactant stabilization. Langmuir **2006**, *22* (21), 8951–8959.

49. Miller, D.A.; McConville, J.T.; Yang, W.; Williams, R.O. III. McGinity JW. Hot-melt extrusion for enhanced delivery of drug particles. J. Pharm. Sci. **2007**, *96* (2), 361–376.

50. Vertzoni, M.; Dressman, J.; Butler, J.; Hempenstall, J.; Reppas, C. Simulation of fasting gastric conditions and its importance for the *in vivo* dissolution of lipophilic compounds. Eur. J. Pharm. Biopharm. **2005**, *60* (3), 413–417.

51. Vertzoni, M.; Fotaki, N.; Kostewicz, E.; Stippler, E.; Leuner, C.; Nicolaides, E.; Dressman, J.; Reppas, C. Dissolution media simulating the intralumenal composition of the small intestine: Physiological issues and practical aspects. J. Pharm. Pharmacol. **2004**, *56* (4), 453–462.

52. Vogt, M.; Kunath, K.; Dressman, J.B. Dissolution improvement of four poorly water soluble drugs by cogrinding with commonly used excipients. Eur. J. Pharm. Biopharm. **2008**, *68* (2), 330–337.

53. Vogt, M.; Kunath, K.; Dressman, J.B. Dissolution enhancement of fenofibrate by micronization, cogrinding and spray-drying: Comparison with commercial preparations. Eur. J. Pharm. Biopharm. **2008**, *68* (2), 283–288.

54. Guzmán, H.R.; Tawa, M.; Zhang, Z.; Ratanabanangkoon, P.; Shaw, P.; Gardner, C.R.; Chen, H.; Moreau, J.P.; Almarsson, O.; Remenar, J.F. Combined use of crystalline salt forms and precipitation inhibitors to improve oral absorption of celecoxib from solid oral formulations. J. Pharm. Sci. **2007**, *96* (10), 2686–2702.

55. Alonzo, D.E.; Gao, Y.; Zhou, D.; Mo, H.; Zhang, G.G.; Taylor, L.S. Dissolution and precipitation behavior of amorphous solid dispersions. J. Pharm. Sci. **2011**, *100* (8), 3316–3331.

56. Alonzo, D.E.; Zhang, G.G.; Zhou, D.; Gao, Y.; Taylor, L.S. Understanding the behavior of amorphous pharmaceutical systems during dissolution. Pharm.Res. **2010**, *27* (4), 608–618.

57. DiNunzio, J.C.; Miller, D.A.; Yang, W.; McGinity, J.W.; Williams, R.O.,III. Amorphous compositions using concentration enhancing polymers for improved bioavailability. Mol. Pharm. **2008**, *5* (6), 968–980.

58. Miller, D.A. Improved oral absorption of poorly water-soluble drugs by advanced solid dispersion systems. In *Division of Pharmaceutics*; The University of Texas at Austin: Austin, TX, 2007; 312.

59. Miller, D.A.; DiNunzio, J.C.; Yang, W.; McGinity, J.W.; Williams, R.O., III. Targeted intestinal delivery of supersaturated itraconazole for improved oral absorption. Pharm. Res. **2008**, *25* (6), 1450–1459.

60. Baird, J.A.; Van Eerdenbrugh, B.; Taylor, L.S. A classification system to assess the crystallization tendency of organic molecules from undercooled melts. J. Pharm. Sci. **2010**, *99* (9), 3787–3805.

61. Miller, D.A.; DiNunzio, J.C.; Yang, W.; McGinity, J.W.; Williams, R.O. III.Enhanced *in vivo* absorption of itraconazole via stabilization of supersaturation following acidic-to-neutral ph transition. Drug Dev. Ind. Pharm. **2008**, *34* (8), 890–902.

62. Wu, C.; McGinity, J.W. Influence of methylparaben as a solid-state plasticizer on the physicochemical properties of Eudragit® RS PO hot-melt extrudates. Eur. J. Pharm. Biopharm. **2003**, *56* (1), 95–100.

63. Crowley, M.M.; Zhang, F.; Koleng, J.J.; McGinity, J.W. Stability of polyethylene oxide in matrix tablets prepared by hot-melt extrusion. Biomaterials **2002**, *23* (21), 4241–4248.

64. Schenck, L.; Troup, G.M.; Lowinger, M.; Li, L.; McKelvey, C.Achieving a hot melt extrusion design space for the production of solid solutions. In *Chemical Engineering in the Pharmaceutical Industry: R&D to Manufacturing*; am Ende, D.J., Ed.; John Wiley & Sons, Inc.: New York, NY, 2011.

65. Bessemer, B. Shape Extrusion. *Pharmaceutical Extrusion Technology*; Ghebre-Sellassie, I., Martin, C., Ed.; Marcel Dekker, Inc.: New York, 2003; 209–224.

66. Doetsch, W. Material handling and feeder technology. In *Pharmaceutical Extrusion Technology*; Ghebre-Sellassie, I., Martin, C., Eds.; Marcel Dekker, Inc.: New York, 2003; 11–134.

67. Dreiblatt, A. Process design. In *Pharmaceutical Extrusion Technology*; Ghebre-Sellassie, I., Martin, C. Ed.; Marcel Dekker, Inc.: New York, 2003; 149–170.

68. Breitenbach, J. Melt extrusion can bring new benefits to HIV therapy: The example of Kaletra tablets. Am. J. Drug Deliv. **2006**, *4* (2), 61–64.

69. Bauer, J.; Spanton, S.; Henry, R.; Quick, J.; Dziki, W.; Porter, W.; Morris, J. Ritonavir: An extraordinary example of conformational polymorphism. Pharm. Res. **2001**, *18* (6), 859–866.

70. Kessler, T.; Breitenbach, J.; Schmidt, C.;, Degenhardt, M.; Rosenberg, J.; Krull, H.; Berndl, G. Process for Producing a Solid Dispersion of an Active Ingredient, U.S.P.T. Office, 2009, Abbott GmbH & Co., KG: United States of America.

71. Riegelman, S.; Chiou, W.L. Increasing the Absorption Rate of Insoluble Drugs, U.S.P.T. Organization, 1979, The Regents of the University of California: United States of America. 9.

72. Chatterji, A.; Desai, D.; Miller, D.A.; Sandhu, H.K.; Shah, N.H. A Process for Controlled Crystallization of an Active Pharmaceutical Ingredient from Supercooled Liquid State by Hot Melt Extrusion, W.I.P. Organization, 2012: USA.

73. Crew, M.D.; Curatolo W.J.; Friesen, D.T.; Gumkowski, M.J.; Lorenz, D.A.; Nightingale, J.A.S.; Ruggeri, R.B.; Shanker, R.M. Pharmaceutical Compositions of Cholesteryl Ester Transfer Protein Inhibitors, U.S.P.T. Office, 2007, Pfizer, Inc.: United States. 74.

74. Bernard, B.L. Cellulosic films incorporating a pharmaceutically acceptable plasticizer with enhanced wettability, W.I.P. Organization, 2006, Eastman Chemical Company: United States of America.

75. DiNunzio, J.C.; Brough, C.; Miller, D.A.; Williams, R.O.; McGinity, J.W. Applications of KinetiSol® Dispersing for the production of plasticizer free amorphous solid dispersions. Eur. J. Pharm. Sci. **2010**, *40* (3), 179–187.

76. Albano, A.; Desai, D.; Dinunzio, J.; Go, Z.; Iyer, R.M.; Sandhu, H.K.; Shah, N.H. Pharmaceutical Composition with Improved Bioavailability for High Melting Hydrophobic Compound, in World Intellectual Property Organization, W.I.P. Organization, 2013, F. Hoffmann-La Roche: United Stated of America.

77. Young, C.R.; Koleng, J.J.; McGinity, J.W. Production of spherical pellets by a hot-melt extrusion and spheronization process. Int. J. Pharm. **2002**, *242* (1–2), 87–92.

78. Zhang, F.; McGinity, J.W. Properties of sustained-release tablets prepared by hot-melt extrusion. Pharm. Dev. Technol. **1999**, *4* (2), 241–250.

79. Verreck, G.; Six, K.; Van den Mooter, G.; Baert L, Peeters J, Brewster ME. Characterization of solid dispersions of itraconazole and hydroxypropylmethylcellulose prepared by melt extrusion–part I. Int. J. Pharm. **2003**, *251* (1–2), 165–174.

80. Coppens, K.A.; Hall, M.J.; Koblinski, B.D.; Larsen, P.S.; Read, M.D.; Shrestha, M. Controlled Release of Poorly Soluble Drugs Utilizing Hot Melt Extrusion. In Proceedings of the American Association of Pharmaceutical Scientists, 2009, Los Angeles, CA.

81. Alderman, D.A.; Wolford, T.D. Sustained Release Dosage Form Based on Highly Plasticized Cellulose Ether Gels, U.S.P.T. Organization, 1987, The Dow Chemical Company: United States of America. 5.

82. Zhu, Y.; Mehta, K.A.; McGinity, J.W. Influence of plasticizer level on the drug release from sustained release film coated and hot-melt extruded dosage forms. Pharm. Dev. Technol. **2006**, *11* (3), 285–294.

83. Zhu, Y.; Shah, N.H.; Malick, A.W.; Infeld, M.H.; McGinity, J.W. Controlled release of a poorly water-soluble drug from hot-melt extrudates containing acrylic polymers. Drug Dev. Ind. Pharm. **2006**, *32* (5), 569–583.

84. Schilling, S.U.; Lirola, H.L.; Shah, N.H.; Waseem Malick, A.; McGinity, J.W. Influence of plasticizer type and level on the properties of Eudragit® S100 matrix pellets prepared by hot-melt extrusion. J. Microencap. **2010**, *27* (6), 521–532.

85. Schilling, S.U.; Shah, N.H.; Waseem Malick, A.; McGinity, J.W. Properties of melt extruded enteric matrix pellets. Eur. J. Pharm. Biopharm. **2010**, *74* (2), 352–361.

86. Özgüney, I.; Shuwistkul, D.; Bodmeier, R. Development of kollidon SR mini-matrices prepared by hot-melt extrusion. Eur. J. Pharm. Biopharm. **2010**, *73* (1), 140–145.

87. Roth, W.; Setnik, B.; Zietsch, M.; Burst, A.; Breitenbach, J.; Sellers, E.; Brennan, D. Ethanol effects on drug release from Verapamil Meltrex®, an innovative melt extruded formulation. Int. J. Pharm. **2009**, *368* (1–2), 72–75.

88. Debenedetti, P.G.; Stillinger, F.H. Supercooled liquids and the glass transition. Nature **2001**, *410* (259–267), 259.

89. Bartholomaeus, J.H.; Arkenau-Marić, E.; Gali, E. Opioid extended-release tablets with improved tamper-resistant properties. Expert Opin. Drug Deliv. **2012**, *9* (8), 879–891.

90. Terife, G.; Faridi, N.; Wang, P.; Gogos, C.G. Polymeric foams for oral drug delivery—A review. Plast. Eng. **2012**, November/December: 32–39.

91. Listro, T. Innovation in Pharmaceutical Technology, **2012**, 43, 60–62.

92. Brown, C., Mckelvey, C., Faridi, N., Gogos, C., Suwardie, J. Wang, P., Young, M., Zhu, L. Evaluation of foaming of pharmaceutical polymers by CO_2 and N_2 to enable drug products. SPE ANTEC **2011**, 224–1228.

93. Verreck, G.; Decorte, A.; Heymans, K.; Adriaensen, J.; Cleeren, D.; Jacobs, A.; Liu, D.; Tomasko, D.; Arien, A.; Peeters, J.; Rombaut, P.; Van den Mooter, G.; Brewster, M.E. The effect of pressurized carbon dioxide as a temporary plasticizer and foaming agent on the hot stage extrusion process and extrudate properties of solid dispersions of itraconazole with PVP-VA 64. Eur. J. Pharm. Sci. **2005**, *26* (3–4), 349–358.

94. Verreck, G.; Decorte, A.; Heymans, K.; Adriaensen, J.; Liu, D.; Tomasko, D.L.; Arien, A.; Peeters, J.; Rombaut, P.; Van den Mooter, G.; Brewster, M.E. The effect of supercritical CO_2 as a reversible plasticizer and foaming agent on the hot stage extrusion of itraconazole with EC 20 cps. J. Supercrit. Fluids **2007**, *40* (1), 153–162.

95. Verreck, G.; Decortea, A.; Lib, H.; Tomaskob, D.; Ariena, A.; Peetersa, J.; Rombautc, P.; Van den Mooterc, G.; Brewstera, M.E. The effect of pressurized carbon dioxide as a plasticizer and foaming agent on the hot melt extrusion process and extrudate properties of pharmaceutical polymers. J. Supercrit. Fluids **2006**, *38* (3), 383–391.

96. Lyons, J.G.; Hallinan, M.; Kennedy, J.E.; Devine, D.M.; Geever, L.M.; Blackie, P.; Higginbotham, C.L. Preparation of monolithic matrices for oral drug delivery using a supercritical fluid assisted hot melt extrusion process. Int. J. Pharm. **2007**, *329* (1–2), 62–71.

97. Nagy, Z.K.; Sauceau, M.; Nyúl, K.; Rodier, E.; Vajna, B.; Marosi, G.; Fages, J. Use of supercritical CO_2-aided and conventional melt extrusion for enhancing the dissolution rate of an active pharmaceutical ingredient. Polym. Adv. Technol. **2012**, *23* (5), 909–918.

98. Fukuda, M.; Peppas, N.A.; McGinity, J.W. Floating hot-melt extruded tablets for gastroretentive controlled drug release system. J. Control. Release **2006**, *115* (2), 121–129.

99. Streubel, A.; Siepmann, J.; Bodmeier, R. Floating matrix tablets based on low density foam powder: effects of formulation and processing parameters on drug release. Eur. J. Pharm. Sci. **2003**, *18* (1), 37–45.

100. Nakamichi, K.; Yasuura, H.; Fukui, H.; Oka, M.; Izumi, S. Evaluation of a floating dosage form of nicardipine hydrochloride and hydroxypropylmethylcellulose acetate succinate prepared using a twin-screw extruder. Int. J. Pharm. **2001**, *218* (1–2), 103–112.

101. Clarke, A.J. Novel Pharmaceutical Dosage Foams and Method for Producing the Same, U.S.P.a.T. Organization, **2005**: United States of America.

102. Verreck, G.; Brewster, M.E.; Van Assche, I. The use of supercritical fluid technology to broaden the applicability of hot melt extrusion for drug delivery applications. Bull. Tech. Gattef. **2012**, *46* (105), 28–42.

103. Lee, C.H.; Lee, K.J.; Jeong, H.G.; Kim, S.W. Growth of gas bubbles in the foam extrusion process. Adv. Polym. Technol. **2000**, *19* (2), 97–112.

104. Nakamichi, K.; Yasuura, H.; Fukui, H.; Oka, M.; Izumi, S. Evaluation of a floating dosage form of nicardipine hydrochloride and hydroxypropylmethylcellulose acetate succinate prepared using a twin-screw extruder. Int. J. Pharm. **2001**, *218* (1–2), 103–112.

105. Passerini, N.; Craig, D.Q.M. An investigation into the effects of residual water on the glass transition temperature of polylactide microspheres using modulated temperature DSC. J. Control. Release **2001**, *73* (1), 111–115.

106. Wang, Y.; Rodriguez-Perez, M.A.; Reis, R.L.; Mano, J.F. Thermal and thermomechanical behaviour of polycaprolactone and starch/polycaprolactone blends for biomedical applications. Macromol. Mater. Eng. **2005**, *290* (8), 792–801.

107. Li, D.; Guo, G.; Fan, R.; Liang, J.; Deng, X.; Luo, F.; Qian, Z.PLA/F68/Dexamethasone implants prepared by hot-melt extrusion for controlled release of anti-inflammatory drug to implantable medical devices: I. Preparation, characterization and hydrolytic degradation study. Int. J. Pharm. **2013**, *441* (1–2), 365–372.

108. Gosau, M.; Müller, B. Release of gentamicin sulphate from biodegradable PLGA-implants produced by hot melt extrusion. Pharmazie **2010**, *65* (7), 487–492.

109. Ghalanbor, Z.; Körber, M.; Bodmeier, R. Improved lysozyme stability and release properties of poly(lactide-co-glycolide) implants prepared by hot-melt extrusion. Pharm. Res. **2010**, *27* (2), 371–379.

110. De Graaff, W.; Groen, J.S.; Kruft, M.A.B.; Van Laarhoven, J.A.H.; Vromans, H.; Zeeman, R. Drug. Delivery System Based on Polyethylene Vinylacetate Copolymers, U.S.P.a.T. Organization, 2013, N.V. Organon: United States of America. 14.

111. Shiah, J.-G.; Bhagat, R.; Blanda, W.M.; Nivaggioli, T.; Peng, L.; Chou, D.; Weber, D.A. Ocular Implant Made by a Double Extrusion Process, U.S.P.a.T. Organization, 2008, Allergan, Inc.: United States of America. 39.

112. Weber, D., Kane, I., Rehal, M., Lathrop III, R.L., Aptekarev, K., Methods and Apparatus for Delivery of Ocular Implants, U.S.P.a.T. Organization, **2005**, Allergan, Inc.: United States of America. 20.

Melt-Electrospun Fibers

Aleksander Góra
Center for Nanofibers and Nanotechnology, Nanoscience and Nanotechnology Initiative, Faculty of Engineering and Department of Mechanical Engineering, Faculty of Engineering, National University of Singapore, Singapore

Rahul Sahay
Fluid Division, Department of Mechanical Engineering, National University of Singapore, Singapore

Velmurugan Thavasi
Elam Pte Ltd., Singapore

Seeram Ramakrishna
Center for Nanofibers and Nanotechnology, Nanoscience and Nanotechnology Initiative, Faculty of Engineering and Department of Mechanical Engineering, Faculty of Engineering, National University of Singapore, Singapore

Abstract

Melt electrospinning is a technique for the production of micro- and nanofibers, which can be an interesting alternative to conventional solvent electrospinning. In this process, the solvents are replaced by polymer melts. Despite its greater potential, the number of papers published and research groups working in the application of melt electrospinning for novel applications are significantly lower than that of solvent electrospinning. However, it is worth mentioning that attempts were made to improve the design and development of melt electrospinning and fabricate fibrous materials using the setup during last few decades. A lack of interest in melt electrospinning is probably associated with the difficulty of melting the polymers at a precise temperature range. This entry provides an overview on the various design setups employed to achieve melt electrospinning and wide variety of polymers used. This entry has been revised from "Melt-Electrospun Fibers for Advances in Biomedical Engineering, Clean Energy, Filtration, and Separation" *Polymer Reviews*, Vol. 51, Issue 3.

INTRODUCTION

Electrospinning is a technique developed at the turn of the 20th century by Cooley[1] and Morton.[2] They noticed that electrostatically charged fluid or solvent was attracted by the opposite charged collector. Electrospinning[3,4] has been one of the most widely used techniques for the production of nanofibers. This technique of producing nanofibers employs electrostatic forces for stretching the viscoelastic fluid. Rapid progress in research and a high interest in this method in the last two decades led to the development of many modifications, which enabled production of fibers in a wide range of diameters and their arrangements. Simplicity of the device was needed and the relatively easy preparation of the solution is one of the main advantages of this process. As the process is capable of achieving large surface-to-volume ratios with desirable physical and chemical properties, it has been considered for a wide variety of applications, ranging from scaffolds,[5] nanofilters,[6] protective clothing,[7] nanocatalysis[3,7] to energy applications.[8]

MELT ELECTROSPINNING CONCEPT

Electrospinning can be broadly divided into two categories as solvent electrospinning and melt electrospinning. The main difference between these categories is the application of polymer melts in case of melt electrospinning whereas the polymer solution is used in solution electrospinning. The benefits of melt electrospinning over solution electrospinning include the absence of toxic solvent and high throughput. The high throughput makes it a viable technique for biomedical, energy, and environmental applications. The absence of toxic solvent allows for direct deposition of fibers mat on the desired substrate. The design of melt electrospinning is analyzed based on its applicability. Parametric analysis of the melt electrospinning is also performed to get an insight into the abovementioned phenomenon. The main differences between solvent electrospinning and melt electrospinning are summarized in Table 1.

Larrondo & Manley[9–11] were the first to apply electrostatic field to the polypropylene (PP) melt, which led to the production of the microfibers. The benefit of this process is

Table 1 Comparison of solution and melt electrospinning

Parameters	Solution electrospinning	Melt electrospinning
Precursor solution	Polymer dissolved in a solvent	Polymer melt (no solvent in required)
Distance between the needle and collector	Can range from few microns to tens of centimetres	Limited by the solidification of polymer melt
Applied electric field	Low compared to melt electrospinning	High (no solvent, thus reduces availability of free charged particles)
Dominant force	Electrostatic forces for stretching the viscoelastic fluid	Electrostatic forces for stretching the viscoelastic fluid
Throughput	Low (evaporation of solvent)	High (no solvent)
Environment	Ventilation (allows for the evaporation of solvent)	No ventilation required (no solvent)
Viscosity	Low compared to polymer melt (function of polymer concentration and average molecular weight)	High (function of the temperature of the polymer melt)
Temperature of the precursor solution	Ambient temperature	High (close to the melting point of polymer)
Fiber diameter	Usually ranges from nanometers to micron	Usually ranges from 200 nm to 500 μm.

that it can be applied to the polymers that are difficult to dissolve in any given solvent. For instance, polymers such as poly(ethylene terephthalate), polyolefin, and polyamides are only soluble in a group of solvents and even require a high temperature for dissolution. The absence of the solvent also results in high throughput, making it a viable technique for mass production of materials for scaffolds, filter media, etc. In the last decade, a number of publications demonstrated that obtaining fibers with one-micron and even sub-micron diameter is possible.[12–17] Achieved fiber diameters were in the range of 270 ± 100 nm[12] through 800 nm[16,17] to several hundreds of μm.[17] This polymer melt is produced by applying a desired temperature to the polymer reservoir. The need to maintain the reservoir at a certain temperature complicates the system design in comparison to the solution electrospinning. Some researchers added heating chambers[13,16] or heating zones around syringe[14] in electrospinning unit. This approach enabled to control the polymer viscosity in subsequent stages of the process. The application of a polymer melt also restricts the overall electrospinning range of polymers to thermoplastic polymers. Nevertheless, the application of a polymer melt for electrospinning has its own advantages. These include the high throughput and low production cost due to the absence of solvent. Polymer melts also help to avoid toxic solvents and enable the direct deposition of fibers on the desired substrate. Similar to solvent electrospinning, the diameter and morphology of fibers depend on the parameter combinations. Important parameters reported for melt electrospinning are molecular weight, tacticity, and melting point of polymer, electric field strength, distance from nozzle, flow rate, and process temperatures (melting point temperature, electrospinning temperature, cooling/collector temperature, and heating chamber temperature).[12–17] In this method, the characteristics of the

polymer itself affect the process more than in solvent electropsinning.

Nevertheless, the technique also suffers from the drawback such as the production of microfibers. It has been noted that the fiber diameter obtained during the melt electrospinning ranges from sub-micron to micron range.[18–20] These fibers are at least one order of magnitude thicker than the fibers produced by solution electrospinning. This is due to the low surface charge density of the polymer melt in comparison to the polymer solution. The tendency of the polymer to solidify as it flows out of the nozzle also hinders the stretching of the polymer jet, thus resulting in microfibers.

Trajectory of Melt-Electrospun Jet with Respect to Solution Spun Jet

The trajectory of the melt-electrospun jet is appreciably different from that of solution electrospinning as it experiences little to no bending instabilities. It has been noted in the case of solution electrospinning that the surface-charged density is responsible for the occurrence of the instabilities suffered by the electrospun jet.[3,21,22] It was also observed that the solvents contribute a lot to the surface charge density. This implies that absence of the solvent will greatly reduce the surface change density in the case of melt electrospinning. This reduction in the surface charge density will greatly suppress any perturbation appearing on the surface of the jet. Those bending instabilities are also responsible for the reductions of the jet diameter to the order of nanometers in solution electrospinning.[3,4] Suppression of these instabilities in the case of melt electrospinning will appreciably reduce the stretching, resulting in the fiber diameter in the submicron or micron range. Zhmayev, Zhou, and Joo[15] noticed for poly (lactide acid) that the jet whipping motion is more dependent on spinning temperature than applied voltage. Higher voltage causes vibrations of the jet only locally close to collector; however, increasing the spinning temperature results in a stronger whipping motion and thinning of the fiber. Tests on Nylon-6 showed that the thinning region starts from the distance of ~5R_0 (where R_0 is a diameter of nozzle) as an effect of accumulation of internal stress and the alignment of polymer chains. However, the thinning process decreases rapidly below the melting point temperature due to the increasing viscosity of the polymer.[14] Rapid solidification of the jet as it comes out of the nozzle also hinders the stretching of the jet. Possible ways to delay solidification of the polymer and decrease the fiber diameter include additives or jet isolation by applying a hot gas stream around it.[12,13] Nevertheless, the predominately straight line trajectory of the jet allows that the fiber mat can be easily assembled in desired patterns. Such ability to configure in the desired patterns makes it usable in the electronic circuit design and sensor industry. Collectors such as rotating drums or plates, used in solvent electrospinning, can also be introduced to melt electrospinning to achieve various patterns of fibers such as aligned fibers. Brown et al.[22,23] investigated behavior of the poly(ε-caprolactone) (PCL) jet in melt electrospinning. The straight region of the jest was used to produce lines of the polymer on the collector. Axial compression of the jet creates a buckling region close to the collector surface, which results in randomly deposited coils. They can be avoided by introducing the lateral movement of the collector. Researchers investigated minimal distance, polymer flow, and optimum lateral movement to obtain straight fibers and print three-dimensional shapes onto the collector surface. Fibers diameters from 1 up to 50 μm were reported. The additional introduction of the x-y controlled stage allowed the drawing of highly aligned structures like square wave patterns. Some researchers have suggested that melt electrospinning can be treated as bridge technology between solution electrospinning and fused deposition modeling.[22]

Melt Electrospinning Setups

High-voltage direct current (HVDC) power supply, nozzle system, and a collector are the only requirement for solution electrospinning whereas in case of melt electrospinning, a heating element is also included in the setup. This heating element serves the purpose of producing polymer melts employed for electrospinning. Various ways have been adopted to achieve the abovementioned requirements. Kim et al.[24] employed a glass syringe placed inside the stainless steel jacket. The space between the glass syringe and the stainless steel jacket is filled with the liquid medium. Circulation of the medium was employed to control the temperature of the polymer placed inside the glass syringe. On the other hand, Liu et al.[25] employed an electric heating ring around the nozzle system in order to produce the polymer melt and a piston to control polymer flow.

A variety of melt electrospinning setups[16,26] having two or more heating elements has been proposed. These heating elements are so arranged that the temperature of the polymer melt locally increases and decreases from the nozzle melting system to the collector. For instance, Zhou, Green, and Joo[16] devised a four heating element setup to locally obtain changes of the temperature profile from the nozzle exit to the collector surface as shown in Fig. 1A.

It was noted that as the polymer ejects from the heated reservoir, the temperature of the polymer melt starts to reduce exponentially to the room temperature. This implies that the rapid solidification of the jet will put an upper limit on the stretching of the polymer jet. In this system, the polymer is melted above its melting point and in the nozzle zone temperature increases to 20–60°C, causing a local decrease in viscosity. The spinning temperature was maintained at a high level but below the melting point of the polymer. These local changes of the temperature profile prolong the solidification of the jet, thereby achieving improved stretching of the polymer jets. High temperature in the melting chamber could degrade the polymer, but short residence time of the polymer in the needle minimizes this effect.

Li et al.[27] designed a core–shell setup for the production of thermochromic nanofibers. The setup employed consists of concentric syringes with the heating element connected to the outer syringe and HVDC supply connected to the inner syringe. The polymer solution was dispensed from the outer syringe whereas the melt pumped through the inner syringe. This setup was able to achieve the simultaneous solution and melt electrospinning of the outer and inner fluid, respectively, thus resulting in thermochromic nanofibers. The only shortcoming of the abovementioned setup is that the heat is applied to both core and shell materials. This application of heat to both materials makes it applicable for the cases in which the glass transition temperature of the shell polymer and boiling point of its solvent is higher than the melting point of the inner fluid system.

One of the melt electrospinning setups designed by Ogata et al.[28,29] adopted a laser beam to melt rod-like polymer samples as shown in Fig. 1B. The polymer samples were fed in to the laser melting zone at a certain rate with the help of a feeder controlled by a stepping motor. The polymer sample was also connected to an HVDC power supply for the initiation of the electrospinning process. The melt homogeneity was achieved by irradiating the polymer rod with a laser beam from three directions at the same given plane. Laser power significantly influenced the final diameter of fibers and reduced the fiber diameter. However, when power exceeded 20 W, the polymer decomposed and gas-filled grains were produced. The laser increased temperature at the end of the needle without complicated heating system and its area limited activity prevent from polymer decomposition in reasonable power range. This same group produced also poly (lactide acid) (PLA), poly(ethylene-*co*-vinyl alcohol) (EVOH), and PLA/EVOH fibers in their recent work by using this same technique with smallest obtained fiber diameters of 400 nm.[30] Similar work done by Li et al.[31] who employed a similar approach to produce fibers from poly (L-lactide acid) (PLLA) polymers. A CO_2 laser with maximum power of 50 W, wavelength of 10.6 μm, and spot size of 5 mm was used to melt the polymer rod from three directions. The best parameters have been chosen by changing applied voltage, laser current, and collector distance. The best result was shown for applied voltage of 15 kV, laser current of 30 mA, and collector distance of 10 cm. The thinnest obtained fibers had diameters of 2.23 ± 0.35 μm for polymers with melt flow rate of 5.4 g/10 min. The thickest diameter was 7.02 ± 1.05 μm produced under 20 kV of electric potential and laser current of 30 mA on a distance of 14 cm. To observe degradation of the polymer during the laser melting procedure, this same group investigated crystalline structure of PCL before and after process.[32] X-ray diffraction (XRD), differential scanning calorimetry (DSC), and mechanical tests were done to check changes in polymer structure. By using the method described before, researchers produced PCL fibers with diameters ranging from 3 to 12 μm on the distance 14–20 cm, applied voltage 5–25 kV, and laser current of 4–14 mA. Two mechanisms of degradation have been considered: thermal degradation and mechanical scission. Researchers noticed that the formation of polymer rods used in the melt electrospinning equipment changed polymer molecular weight from an initial 80,000 to 74,000 g/mol. The laser current used in the experiment reduced molecular weight further in a linear manner. When 4 mA was applied, the molecular weight decreased to 65,000 g/mol, but with an increase of laser strength to 14 mA, it reduced molecular weight to 20,000 g/mol. The degree of crystallinity stayed at around 60% before and after the process, which suggests that the cooling time of the fibers remain sufficient to recrystalize polymer during process. DSC analysis did not show significant changes in melting enthalpy after the process. Liu et al.[33]

investigated other biodegradable polymer by using unique spinning head construction and process influence to the polymer properties. Umbrella-like spinning head allowed increasing spinning area and increased efficiency of process to 4 g/hr. The average diameter of fibers spun at a temperature of 210°C, voltage of 60 kV, and distance of 11 cm was measured to be 7.65 ± 0.21 μm. Increase of spinning temperature to 245°C resulted in production of short broken fibers. This work showed that spinning at this temperature decreased molecular mass of the polymer by 56.2%. DSC analysis showed that sharp melting peaks at 140°C and 146°C present in pellets form, was smoothened out after spinning process, which suggests not complete crystallization occurred what was confirmed by XRD. In this experiment, scission reaction was reduced by addition of antioxidant 1010 and 168. This allowed to partially terminating reactions of degradation by capturing chain-free radicals generated from thermo-oxidative degradation. Addition of antioxidant 1010 allowed increasing process efficiency to 6 g/hr within these same process parameters. On the other hand, Zhmayev et al.[13] employed a nitrogen in the melt electrospinning chamber to control the trajectory of the electrospun jet as shown in Fig. 1C. It was noted that the air drag induced by the gas helped to reduce the jet diameter. Nevertheless, it was also noted that increase of the temperature due to the air stream resulted in major reduction of the jet diameter. Prolonging of the jet solidification by increase of the air temperature near the jet results in increased stretching. Figure 2 illustrates the formation of PLA fibers with and without the gas-assisted setup.

The highest reduction in fiber diameter was reported for air velocity reaching 300 m/s. This approach resulted in 10% thinning in contrast to heated gas where 20-fold thinning of fiber diameter was observed (Fig. 2B).

A different approach in setup design was presented by Komárek and Martinová.[34] The electrospinning from free surface was used. Poly(ε-caprolactone) and PP were melted on special designed electrodes and a Taylor cone was emitted from surface of melted polymer.

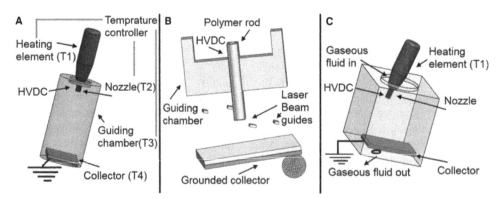

Fig. 1   Design setups of melt electrospinning, (**A**) four heating elements to achieve gradually decreasing temperature profile, (**B**) application of the laser beam for melting polymer rod, and (**C**) external gaseous fluid resulting in the increased stretching of the jet. Expression T1, T2, T3, and T4 represent temperature of a given heating element.

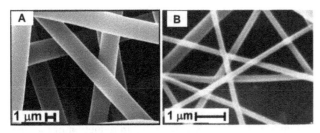

Fig. 2 SEM images of as-melt electrospun PLA fiber mats: (**A**) without gas-assisted system and (**B**) with gas-assisted system. **Source:** From Zhmayev et al.[13] © 2010, with permission from Elsevier.

This "needle-less" electrospinning method simplified the equipment design. "Rod" and "cleft" electrodes served at the same time as melting device and the high potential electrode. Electric field simulations showed that maximum field intensity is concentrated on the edges of the cleft and round edge of the rod electrode. These maximum field areas were the initial places where polymer created multiple Taylor cones.

Special measures are also undertaken in order to design the collector system. One of the setups designed by Dalton et al.[12] employed a stainless steel tweezers as a collector to collect fibers. It was noted that the diameter of the fibers collected between the tweezers arms were smaller than the one's found on the metal rings. This implies that the further stretching took place between the arms of the collector. This may be due to the diffusion of the charged particles from the jet to the metal collectors to minimize the total potential energy of the system in the process stretching the polymer jet. It was also noted that the fibers fuses due to the incomplete solidification of the jet along its trajectory. This implies that the collector should continually transverse in a certain direction to prevent the build-up of molten polymer jet. Use of cylindrical stub also resulted in circular and looped deposition of both examined polymers. Unlike other groups, the heating gun system was introduced to melt high melting point polymers at the end of needle. Difficulty in temperature control was reported in this equipment setting. The glass microscope slide as a nonconducting collector showed multiple-jet formation from a single initial jet. It should be mentioned that from every jet, different fibers were collected; from thin high-quality fibers, up to poor quality thicker structures.

Detta et al.[20] adopted a novel approach by modeling a blend of low-molecular-weight polymer with high molecular weight polymer. This solution in rheological terms acts as large molecular weight polymer dissolved in low molecular weight solvent. They noted that diameter of the fiber obtained is directly related to the flow rate as in case of solution electrospinning. Fibers in micron-range diameters were produced. Nevertheless, it was found to be small enough for the fabrication of scaffolds for tissue engineering. Dalton et al.[12] also employed viscosity-reducing

additive to reduce the fiber diameter. Additives reduce the polymer chain length and therefore the viscosity of the polymer solution, thereby favorably reducing the fiber diameter.

One of the simpler designs was used by Wang et al.[35,36] In this setup, rotating disk is immersed in bath with melted polymer. As in other designs, electric potential is applied to the collector side placed above disk. The whole device was placed in chamber in protective environment of argon. PP was melted in heated bath in temperatures ranged from 280°C to 360°C. Electrospinning in 280°C resulted in discontinuous process and fragmented fibers production due to high viscosity. When temperature reached 360°C degradation of the polymer occurred. The authors noted that in 320°C was the optimal temperature for stable process of electrospinning with no signs of degradation of the material. On the collecting distance of 16 cm and potential of 75 kV good morphological fibers were created. Due to heat transfer differences between various metallic materials, researchers investigated stainless steel and aluminium rods. Use of aluminium with thermal conductivity of 236 W/mK gave better smoother fibers than stainless steel rod (thermal conductivity 31.1 W/mK). Also disk diameter was tested. Larger diameter disks showed bigger electric field intensities at the edge and then smaller ones; however, dissipation of the energy on larger disk is too high to create stable jets. In these conditions fibers of diameters of 1.73 up to 3.63 µm were collected. In order to reduce diameters, dodecyl trimethyl ammonium bromide was added to polymer powder. The best concentration of 3% was shown for reduction of fibers diameters up to 0.40 ± 0.29 µm. In Dasdemir group,[37] melt flow tester was employed to melt polymer and produce fibers that was collected on the aluminium plate connected to high voltage. Polyurethane fibers were produced by solution and melt electrospinning techniques. Efficiency of process was calculated to be 0.6 g/hr for melt electrospinning with final diameters of 4–8 µm and 0.125 g/hr with fiber diameters of 220–280 nm for solution electrospinning.

Electric Field Effect

In solution electrospinning, as the electrospun jet moves along its trajectory, the solvent evaporates, resulting in the solidification of the jet. It means that the time before the polymer jet is solidified will depend upon the volatility of the solvent at the operating temperature. On the other hand, the polymer melts start to solidify as it moves out of the reservoir what puts an upper limit on the distance between needle and collector. This in turn exerts an upper limit on the applied voltage before the breakdown occurs in the air. This phenomenon results in lower stretching compared to solution electrospinning, thus resulting in microfibers in case of melt electrospinning. Larrondo and Manley[9] calculated the critical voltage, at which maximum instability of the jet occurs using Taylor's equation (Eq. 1).

$$V_c^2 = 4\frac{H^2}{L^2}\left(ln\frac{2L}{R} - 1.5\right)(0.117\pi R \gamma) \qquad (1)$$

where V_c is a critical voltage, H is the distance between the electrodes, L is the length of the capillary, R is a radius of capillary and γ is a surface tension. They showed that predictions could be done with sufficient accuracy for PP and polyethylene samples, with observed potential slightly higher than calculated. Zhmayev, Zhou, and Joo[15] developed new model for non-isothermal polymer jets, more suitable for devices with temperature gradient in the chamber. It was based on single filament (1-D) model and coupled momentum continuity, charge conservation equations with the energy and electric field equations. Viscoelastic behavior of the jet at different temperatures was modeled by using Giesekus constitutive model.[15] Later, this same group developed a model for semicrystalline polymers by supplementing previous equations with flow-induced crystallization model by Ziabicki, Cho, and Joo.[14]

The shortcoming of single heating arrangement was solved by Kong et al.[26] upon employing three heating elements along the jet trajectory to control its temperature profile. Similarly, Zhou, Green, and Joo[16] devised four heating element setup to obtain gradually decreasing temperature profile. It was observed that the temperature of the polymer melt decreases exponentially as it pumps out of the reservoir whereas in case of multiheating element arrangements, the temperature profile can be controlled to avoid the rapid solidification of the polymer melt. This increase in the solidification time will favorably affect the needle to collector distance and result in further reduction of the fiber diameter to the submicron range. Also noted that the increase in the applied electric field or increase in the temperature of the polymer melt affects the diameter of melt-electrospun fibers.[16]

The first attempt to model behavior of fiber creation and polymer dropping velocity was done by Liu et al.[38] Dissipative particle dynamics method was used to calculate chain behavior during electrospinning process under various electric field strengths and temperatures. According to authors, increase in electrostatic force accelerates dropping velocity in a nonlinear manner. After crossing some critical value of electrostatic force, dropping velocity of fiber rapidly increases. In elevated temperatures, movement of the particles is faster as well as dropping velocity, causing thickening of collected fibers, which means that increase of temperature can reduce fiber diameter at the same electrostatic force value. The presented model showed changes also in chain length during dropping of the whole jet. The authors noticed that in initial phase of spinning near the melt chamber chains length becomes shorter. Afterward, chain length increases gradually due to rapid elongation of the jet. In the last phase due to small instabilities and already partly arrangement, chains length rapidly increases.

Viscosity and Conductivity of the Precursor Solution

It has been observed that a direct correlation exists between melt temperature and its viscosity for the general thermoplastic materials. It implies that an increase in the applied temperature will appreciably lower the viscosity of the polymer melt. Low viscosity helps in reduction of the final fiber diameter as in the case of solution electrospinning. Nevertheless, lowering of the viscosity can be performed below degradation temperature of the polymer melt. Most optimal approach involves heating polymer melt for a short period of time just before jet solidification and temperature control along with it. So far, only small numbers of polymers and its blends have been processed by this technique. To achieve temperatures gradients, various equipment designs were developed depending on the polymer characteristics. Polymer viscosity in melts was described in reviewed papers by various polymer properties like molecular weight or melt flow index (MFI). Equipment setups with heating methods, achieved fiber diameters, and the polymer properties were summarized in Table 2. The influence of temperature on fiber diameters for two kinds of PP was described by Kong et al. (Fig. 3).[26] As shown before[33,43] additives can reduce polymer viscosity; however, polymer melt conductivity can also influence melt electrospinning process. Nayak et al.[44] investigated influence of ion conducting additives like sodium oleate (SO) and sodium chloride (NaCl). Use of PP with addition of 7% SO and 5% NaCl as material of electrospinnig resulted in fiber diameters of 0.371 ± 0.106 μm and 0.310 ± 0.102 μm, respectively. Increase of conductivity was greater for NaCl due to smaller ionic size of NaCl. When concentration of additives increased from 4% to 12% conductivity at 200°C increased from 10^{-9} to 10^{-6} S/cm. Elongation forces were higher when NaCl were used as an additive due to higher mobility of smaller ions. Higher flow was also noticed for this additive. Spun fibers showed crystallinity of 53% for pure polymer. With additives crystallinity value decreased to 37% with SO and 29% with NaCl. Introduction of additives also marginally decreased hydrophobicity of the polymer. Further work of this group[45] includes combining reduction of melt viscosity and increase of electric conductivity. PP MFI was reduced by addition of PEG and poly(dimethyl siloxane). Addition of SO helped to increase electric conductivity of the melt. Nanofibers with diameter of around 880 nm were obtained by this method, despite slight increase of melt viscosity after addition of SO. In this study also, the shape of the nozzle was investigated. Circular, trilobal, tetralobal, and multilobal dies were used to form final shape of the fibers. Results showed that independently on die shape all produced fibers revealed circular cross-section.

As discussed, melt electrospinning can be compared with stretching of predominately nonconducting liquid under the influence of an applied electrostatic electric field.[46] The suppression of electrostatic instabilities due to the lack of

Melt—Microgels

Table 2 Polymers already used in melt electrospinning

Polymer	Viscosity-related property	Diameter of fibers (μm)	Heating equipment type used	Electrospinning temperature (°C)	References
Isotactic polypropylene	M_w 190,000	~10	Brabender table-top extruder with four heating zones	200	[17]
	M_w 106,000	~6.9		200	[17]
	M_w 12,000	~3.5		200	[17]
	Unknown	20–180	Stainless steel cylindrical chamber with plunger and steel capillary	220–240	[9]
	Zero-shear viscosity (Pa's) 23	8.6 ± 1.0	Heating gun with thermometer above glass syringe	320	[12]
	Zero-shear viscosity (Pa's) 75	35.6 ± 1.7		270	[12]
Isotactic Polypropylene-15 + 1,5% Irgatec	Zero-shear viscosity (Pa's) 33	0.84 ± 0.19	Heating gun with thermometer above glass syringe	270	[12]
Atactic polypropylene	M_w 19,600	~21	Brebender table-top extruder with four heating zones	200	[17]
	M_w 14,000	~13		200	[17]
Polypropylene	M_w 195,100	15–40	Heated syringe, needle, guiding chamber, and collector with independent heating systems	230 syringe, 280–290 needle	[39]
		25–40			
	Melt index 900		Ceramic circular heater for syringe and ring-shaped metal heater for part of spin line	330–410	[26]
	Melt index 1500	~20		330–410	[26]
	Unknown	Unknown	Needle-less electrospinning with "rod" and "cleft" electrodes	180	[34]
Poly(ϵ-caprolactone)	Unknown	Unknown	Needle-less electrospinning with "rod" and "cleft" electrodes	230	[34]
Nylon-6	MFI 3 g/10 min	~1	Two heating zones around MACOR shielding, cooling collector	250–300	[14]
Poly(ethylene terephthalate)	Intristic viscosity 0.512–0.706 dL/g	1.7	Rodlike polymer samples, CO_2 laser melting polymer from three directions, rotating collector	~260	[28]
Polyalirate	Apparent viscosity 3.358 Pa's at 340°C	2.5	Rodlike polymer samples, CO_2 laser melting polymer from three directions, rotating collector	~330	[28]
20% PEG$_{5000}$-block-PCL$_{5000}$ + 80% PCL	M_w~65,960	1.8 ± 0.8	Heated recirculating water tank attached to glass jacket where syringe was seated, syringe pump	90	[20]
10% PEG$_{5000}$-block-PCL$_{5000}$ + 90% PCL	M_w~72,980	2.7 ± 2.0		90	[20]
PEG47-block-PCL95 +30% PCL	Zero-shear viscosity (Pa's) 49	2.00 ± 0.3	Precise temperature control circulatory system	90	[12]
PEG47-block-PCL95 +30% PCL (dual collector)		0.27 ± 0.1		90	[12]

(Continued)

Table 2 (*Continued*) Polymers already used in melt electrospinning

Polymer	Viscosity-related property	Diameter of fibers (µm)	Heating equipment type used	Electrospinning temperature (°C)	References
Polylactide	M_w 186,000 zero-shear viscosity 120 Pa's at 240°C	~0.8	Four zones heating equipment with cooled collector	200 at syringe	[16]
	M_w 186,000	10–25		225	[15]
	M_w 186,000	2.9–57	Gas-assisted melt electrospinning (GAME), an infrared heater at needle	210 at nozzle	[13]
	M_w 147,000	~1	CO_2 laser melting polymer, rod-like samples with N_2 gas blow to melting area, rotating collector	laser power 6-20 W	[29]
	M_w 108,500	1–2		laser power 6–18 W	[29]
Polylactide : EVOH	PLA: MFI = 76.9 at 210°C M_n = 96080 EVOH: MFI = 1.6 at 190°C	0.845 ± 0.5	Rod-like polymer samples, CO_2 laser melting polymer from three directions, rotating collector, heated chamber	laser power 12 W	[40]
Polyethylene	Unknown	10–420	Stainless steel cylindrical chamber with plunger and steel capillary	200–220	[9]
Low-density polyethylene	MFI: 2 g/10 min	5–32	Heating rings around piston with temperature controller	315–355	[19]
PMMA fibers with CBT	PMMA: M_w≈120,000, Bisphenol A, and 1-tetradecanol (CBT) - unknown	Core 0.2–0.4, Shell 0.2–0.5	Co-axial system. PMMA electrospun from solvent in outer syringe and CBT from melt in inner syringe heated by heating tape	45	[27]
TiO_2-PVP sheath loaded with octadecane	TiO_2-PVP – solvent electrospinning octadecane – unknown	~0.15	Co-axial system with glass spinneret placed into needle; electric heater around glass syringe and plastic syringe for sheath material	68	[41]
EVOH	MFI = 4.4 at 190°C	0.8 ± 0.4	CO_2 line like laser; sheets of polymer; multiple Taylore cones	Laser power 45 W	[42]
Poly(hexamethylene dodecanediamide) (Nylon 6/12)	Unknown	0.73 ± 0.2		Laser power 45 W	[42]

free charged particles results in predominately a straight line trajectory of the jet. This stable motion of the jet gives only one or two order of magnitude reduction in the jet diameter compared to the nozzle diameter. The solution for the above-mentioned problem can be addition of the surfactant to reduce viscosity and improve surface charge density.

APPLICATIONS OF MELT ELECTROSPINNING

Although the majority of literature dealt with the research pertaining to solution electrospinning, nevertheless melt electrospinning has its unique advantages. As highlighted earlier, the polymer solution in traditional electrospinning is solvent based. Polymer concentration in electrospinning almost never exceeds 15 wt% and usually lies in the range 4–10 wt%.[47] The performance of the process is further reduced by the need to evaporate the solvent from fiber before reaching the collector. It should be noted that the use of pure solvents is essential to produce good quality fibers and it cannot be recovered after process. In most cases, solvent fumes must be neutralized. As expected, these solvents will increase the overall production cost of

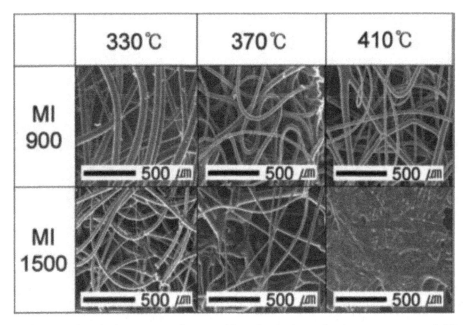

Fig. 3 Representative images of melt-electrospun polymers with various primary heating temperatures of 330–4100°C, at 200°C intervals and MI of 900 and 1500.
Source: From Kong et al.[26] © 2006, with permission from John Wiley and Sons.

the materials. It may also result in the serious environmental problems. Furthermore, nanofibers spun from toxic solvent solution may not be suitable for medical and pharmaceutical applications. This also helps in increasing the throughput and reducing the cost of producing nanofibers. The other advantage includes the direct deposition of the fibers on the desired material as no further processing is required. Melt electrospinning can also be applied in tissue engineering by electrospinning nonwater-soluble polymers and collecting them on live cells.[18]

Biomedical Applications

Properties of native tissue is determined by highly controlled nanoscale assembly of extracellular matrix molecues.[48] Small diameters of fibers produced by electrospinning, small pores between nanofibers, and high surface energy are very useful in the development of scaffolds for tissue engineering and medical devices through better cell absorption on the material. Nanofibrous materials were used to make and examine numerous of devices for medical use like dental restoration, wound dressings[47,50] medical prostheses.[47,49] Tissue engineering principies are shown on Fig. 4. However, an overwhelming majority of these materials were produced by solvent-based electrospinning. Chemicals used to dissolve polymers may leave some residues that can affect biocompatibility of material and induce undesire host response. Melt electrospinning become a viable technology for the production of fibers for biomedical applications as it avoids toxic solvent thereby enabling the fibers to be directly spun on the desired surface or material. For instance, Hunley et al.[52]

spun structured biologically compatible membranes of low molar mass amphiphiles. Poly(D,L-lactide) was synthesized and functionalized with complementary hydrogen bonding groups. It was noted that these thermally reversible hydrogen bonds appreciably increased the viscosity of the precursor, resulting in thick fibers. On the other hand, Detta et al.[20] melt spun a cell invasive scaffolds from PCL and a polymer blend of PEG_{5000}-*block*-PCL_{5000} [poly(ethylene glycol)-*block*-poly(ε-caprolactone)]. An advantage of this method is that melt electrospinning can be performed in the sealed environment as no evaporation of the solvent is required. This makes it a viable fiber producing technique for the applications in tissue engineering.

Kim et al.[24] combined melt and solution electrospinning to fabricate 3D composite scaffolds. This was achieved by employing two syringe systems with a rotating collector. The system consists of syringe pump, syringe, and HVDC supply. The circulation of hot liquid medium was applied to one of those syringe systems to obtain the polymer melt. Composite scaffold was fabricated by simultaneously electrospinning PLGA [poly(D,L-lactide-*co*-glycolide)] solution and PLGA melt in opposite directions facing the grounded rotating drum. The final fiber mat consists of nanofibers produced by solution electrospinning embedded in microfibers produced by melt electrospinning. High degree of cell attachment and in-growth was observed in case of composite scaffolds compared to PLGA microfibers. The other materials employed for tissue scaffold consist of polylactic acid[16] and poly(lactide) and EVOH.[28]

Díaz et al.[53] formulated a novel approach of fabricating a core–shell fiber for drug delivery using solution electrospinning. Although this approach was employed to the

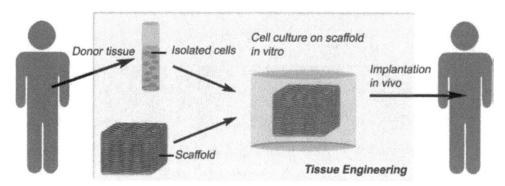

Fig. 4 The principle of tissue engineering and fibrous scaffold usage.
Source: From Stamatialis et al.[51] © 2006, with permission from Elsevier.

polymer solution, it can be easily employed to the polymer melts. The core–shell fiber consists of a hydrophilic polymer solution as its shell, whereas hydrophobic liquid forms its core. The combination employed for this particular case consists of poly(vinyl pyrrolidone) as hydrophilic polymer whereas oil as the hydrophobic counterpart. The morphology of the fiber turns out to be that of bead string structure. In this particular structure, bead represents the hydrophobic liquid trapped inside the polymer solution. This approach can also be applied for the fabrication of composite fibers having both axial and radial anisotropies.

Tian et al.[40] prepared rodlike samples consisting of PLLA and coated with EVOH. PLLA rods were coated with different amounts of EVOH and then placed in the feeder. End of the sample was melted by laser from three different directions and fibers were electrospun from that point. A heating chamber was used to delay solidification of polymer and fibers were collected on moving collector. After removing EVOH from PLLA, fiber morphology was examined. They noticed that the presence of EVOH reduced average fiber diameter from 3 to 1 µm compared to pure PLLA rod. The mechanism of fiber thinning remains unknown so far; however, it was suggested that interfacial boundary of EVOH and PLLA may hold larger electrical charges.

Other kind of biodegradable polymers—polyurethanes (PUs)—were examined as potential materials for melt electrospinning by Karchin et al.[54] Custom-made polyurethane synthetized via standard multistep addition polymerization in *N,N*-dimethylacetamide.[55] Melt electrospinning was done by custom-made equipment with melt chamber ended by 0.5 mm in diameter nozzle and heated with pair of band heaters. Polymer was heated up to 220–240°C and electrospun 13 cm from collector by applying 30 kV potential under the collector surface. No beads were noticed and morphology investigation showed mean fiber diameters of 11.2 ± 2.3 µm. Cytotoxicity of electrospun PUs was tested on primary cell culture taken from anterior cruciate ligament fibroblasts of adolescent Yorkshire pigs. The results showed that after additional purification of synthetized

polymer before spinning and sterilization of ready product, melt-electrospun material did not show any cytotoxicity for the cells. Nevertheless, long-term cytotoxicity as well as toxic influence of the degradation products has not been described. A 4-week culture onto scaffolds showed infiltration of the cell up to thickness of 200–400 µm.

All of the above-described polymers are currently used as biomaterials. The potential of melt-electrospun polymers is dependent on their thermopshysical and rheological properties, and heating method. Especially MFI is a parameter that should be considered during polymer selection. However, it has been proposed that change of some parameters may influence the diameters of the fibers. An increase of melt temperature and spinning temperature as well as the electric field strength cause a decrease in fiber diameter. Likewise, decreasing nozzle diameter and flow rate results in smaller fiber diameter.[16] Table 2 shows some of already electrospun polymers described in this work. Different diameters result from different approaches and electrospinning equipment used. Addition of Irgatec® (BASF, peroxide-free polymer modifier) to decrease viscosity of the melt caused significant reduction in fiber diameters to nanoscale.[12] Gas-assisted melt electrospinning resulted in a 20-fold decrease in the final fiber diameter to 2.9 µm.[13]

More advanced materials have been produced by Brown et al.[56] Tubular scaffolds for various tissue engineering applications like vascular, neural, or urethral were produced by this group by melt electrospinning technique. This group have done first step toward commercialization of electrospinning method in medicine. As mentioned before, lack of toxic solvents may be a big advantege in any medical products. The method described before[22] was updated with tubular collector, in which rotation speed and linear movement could be controlled. Three different architectures of tubular scaffolds were prepared with winding angle of 30°, 45° and 60° between fibers, average diameter of fiber of 60 µm, and overall tube diameter of 6 mm (Fig. 5). Produced tubes were designed to be used in tissue engineering field as long bone model in mices. PCL was intensively investigated as material for bone tissue

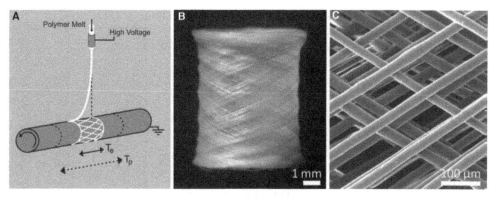

Fig. 5 (**A**) schematic illustration showing fabrication of a porous tube by combining melt electrospinning with direct writing by using a rotating collector on a translating stage. (**B**) porous tube fabricated by combining melt electrospinning of PCL with direct writing. (**C**) SEM image showing smooth uniform fibers interwoven and oriented at 60° to the central axis of the tube.
Source: From Brown et al.[56] © 2012, with permission from Springer Science and Business Media.

engineering due to its mechanical properties and relatively long degradation time. To obtain optimal environment for cell culture, all scaffolds were couvered by immersion with layer of calcium phosphate. To estimate cytotoxicity of the produced tubular scaffolds, primary human ostoblasts, mouse ostobolasts, as well as human mesothelial cells were seeded onto it. Material properties as well as pore size and geometry are the most important factors in cell attachement and proliferation. Diamond size pores are more preferable by the cells due to better collgen matrices quiding. All mentioned cell infiltrated fibrous scaffolds entirely and showed production of collagen natural matrices around, which indicated good cytocompatibility between material and natural tissues. This same group also investigated infiltration of dermal fibroblast on PCL melt-electrospun 3D fibrous scaffodls.[57] Cells taken form the patients were seeded on scaffolds with arrayed structure with fiber diameters 7.5 ± 1.5 µm. The distance between fibers were designed to allow cellular infiltration (46 ± 22 µm). Confolcal microscopy, histology, and computer tomography performed during and after cultivation showed good cell attachement and production of collagen type I, which is the main component of natural extracellular matrix.

Small diameter vascular prostheses were done by Mozalevska et al.[58] by melt electrospinning process from PP and PLA. The smallest average fibers were obtained from PP (Moplen 462R PP) with diameters of 4.80 ± 4.24 µm. Flat surfaces produced from PLA polymer (PLA 406D) had significantly smaller fibers of diameters 3.00 ± 1.18 µm.

The most promising polymers for medical application are PLA, poly(glicolid acid) (PGA), their copolymers, and PCL. These substances are widely used and well described by number of research groups. Increase of its process productivity and limitation of the final cost may be an interesting subject of study, especially for medical prostheses and tissue engineering applications. However, many other bioactive or biostable polymers are being investigated and are

used in some medical devices (Table 3). Melt electrospinning due to high purity of final product can be adopted to improve properties of fibrous materials or reduce its cost. A list of U.S. FDA-approved polymer consist of about 100 devices, which are made partially or completely from polymers. Potential thermoplastic polymers with some physical properties, useful in planning further experiments are listed in Table 3. However, some measurements of viscosity above melting point should be performed to choose the best system for melt electrospinning.

Energy Applications

It has been noted by Sahoo et al.[63] that melt mixing of carbon nanotubes and thermoplastic polymers can result in a nanocomposite having improved electrical and mechanical properties. This nanocomposite melt can be readily electrospun through melt electrospinning producing composite fibers. These fibers can readily be employed in energy application such as solar cells. Due to the high throughput, the melt electrospinning is ideal for the production of large quantity of carbon nanofibers, which can be employed as electrodes in solar cells.

Cho et al.[64] designed superior hydrophobic fiber mat by employing PP melt. It was noted that the fiber mat fabricated from both solution and melt electrospinning of PP exhibited superior hydrophobicity. The only requirement for exhibiting the required property is that the diameter of the fiber obtained should be smaller than 10 µm. This requirement of fiber diameter provides sufficient surface roughness for the fabrication of super hydrophobic surface. It was also observed that the performance of the fiber mat was better than the commercially available nonwoven mats due to their larger diameter in comparison to the electrospun fibers. It was also noted that the PP fibers can be employed as battery separators due to their inherent properties of a large surface area and fine pore sizes. It can be argued that these hydrophobic mats can be applied on the

Table 3 List of some polymers used in biosciences research

Polymer	T_g	T_m	M [g/mol]	ρ [g/cm^3]
cellulose acetate, CA	67	260	–	1.3[a]
polyacrylonitrile, PAN	105	318	53.1	1.184
polyamide, PA (Nylon 6)	40	260	113.2	1.084
polyamide, PA (Nylon 6,6)	50	280	226.3	1.07
poly(ε-caprolactone), PCL	−60	67	114	1.145[a]
polycarbonate, PC	145	225–250	—	1.2–1.26
poly(ethylene oxide), PEO	−67	69	44.1	1.125
polyethylene, PE	−113 to −103	125–135	28.1	0.85
poly(ethylene terephthalate), PET	67–127	245–255	192.2	1.31–1.38
poly(glicolid acid), PGA	45	233	58	1.6
poly(lactid acid), PLA	60–65	173–178	73	1.145[a]
poly(methyl-methacrylate), PMMA	105	160–200	100.1	1.17
polypropylene, PP	−13	187.7	42.1	0.85
polystyrene, PS	100	243.2	104.1	1.05
polysulphone PSU	190–195	297	442.5	<1.24
poly(tetrafluoro ethylene), PTFE	20–22	322–327	100	2.1–2.2
polyurethane, PU	−73 to −23	180–250	466	1.00–1.28
poly(vinyl alcohol), PVA	70–99	248	44.1	1.26
poly(vinyl chloride), PVC	81	273	62.5	1.385

Source: From Ramakrishna et al.,[47] van Krevelen & te Nijenhuis,[59] and Ramakrishna et al.[60]

[a]Data adopted from http://msds.chem.ox.ac.uk/CE/cellulose_acetate.html[61] and www.sigmaaldrich.com.[62]

surface of solar cells to avoid contamination of the surface, thus increasing the longevity of the solar cells. Combination of solvent and melt electrospinning was recently employed by Do, Thuy, and Park.[65] Core shell structure composed of polyethylene glycol core surrounded with polyvinylidene fluoride (PVDF) shell were produced as potential materials for energy storage. PEG is one of the phase change materials. As energy storage capabilities depend on molecular weight of the polymer, molecular weights of 1000, 2000, and 4000 Da have been chosen. Composite material made of 4000 Da PEG and PVDF with PEG content of 42.5% showed the latent heat of 68 J/g. In this technique, syringe with PEG was placed in the oil-heated shield and preheated up to 70°C and pushed through needle with diameter of 0.65 mm. Core material was delivered with help of second pump in form of 20 wt% solution of PVDF in dimethyloformamid to outer nozzle with diameter of 1.20 mm. Shell and core materials were delivered with speed of 1.5 mL/hr and 0.09–0.240 mL/hr respectively. Diameters of the obtained fibers varied depend on flow of the melt core material and rised with feed rate from 408 nm (pure PVDF without core material) up to 911 nm (0.126 μl/min of PEG4000). Differences in melting points of other molecular weight PEG materials suggested a variety of applications for produced materials. PEG1000 because of melting point close to human body temperature (30°C) could be applied as material for smart textiles that could capture heat of human body

and power devices sewn into clothing. Composite materials with PEG2000/PVDF and PEG4000/PVDF with much higher melting points were proposed to be used in thermal protection of electronic devices, heat storage, or solar cooling as well as some heating applications.

Environmental, Filtration, and Separation Applications

Filtration and separation processes are becoming more important in process industries and health care. Due to its highly porous structures, membranes have a lot of potential applications and functions. Depending on the membrane structure, a couple of basic processes can be achieved (Fig. 6). The most popular and practical application is separation where particles of different sizes are separate from each other. Controlled release used, for example, in drug administration is based on releasing specific particles, proteins, or molecules with controlled rate. Membranes processes include immobilization of membranes with other substances to improve among others specific separation or filtration behavior. Contacting membranes are being used as mediators promoting contact between two reactive components. Nanofibrous filter media are being used and investigated by several industries like water (fresh or waste) and air filtration,[66–69] petrochemical,[70] electronic, fuel cells, and clean energy systems[67,71] or biomedical industries.[67,72]

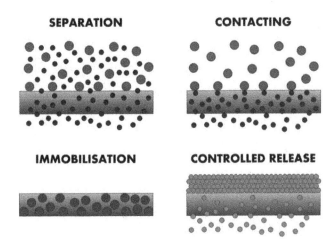

Fig. 6 Some membrane-based mechanisms in filtration and separation.

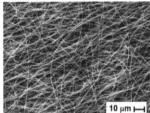

Fig. 7 SEM images of submicron electrospun Nylon-6 fibers. **Source:** From Zhmayev et al.[14] © 2010, with permission from Elsevier.

Demand for novel, efficient, and long usage life filtering and separation media is very high. Nanofibrous materials have some advantages that are not present in typically used materials such as high surface energy, porous structure with controllable small pore size, and relatively high strength. Different mechanisms of filtration and separation can be observed and developed in nanofibrous filtration media field (Fig. 6); this implies less energy consumption during the filtration process. Nanofiltration as a process in water filtration is used as a pretreatment of waste water before the reverse osmosis process, and this significantly reduces power consumption, system maintenance costs, and operates with higher flux.[73] Production of large quantities of solvent electrospun filters for industry or household will cause large increase of produced toxic solvents and its wastes. Melt electrospinning is a good alternative as no solvent is used and no wastes are produced. Till date, some polymers that can be used in filtration process is being produced by melt electrospinning. Nylon-6 fibers widely used in some filtration processes were produced by Zhmayev and Obendorf[14] in submicron diameter range (Fig. 7).

PP widely used in industrial filtration was produced by several groups so far.[9,12,17,26,34,39] The smallest fiber diameter was achieved by Dalton et al.[12] by adding a viscosity reduction component and with relatively simple heating gun method to melt polymer. The average diameter of fibers in the range 0.84 ± 0.19 µm is small enough to obtain nanofiltration membrane. Poly(ethylene terephthalate) melt was electrospun by Ogata et al.[28] with a diameter of about 1.2 µm. These membranes were already developed by solvent electrospinning and may be used as suggested by Veleirinho and Lopes-da-Silva[74] for membranes in apple juice clarification. Other successful electrospun polymers like polyethylene[9,19] or Nylon[14] was being used as filtration media as well.

Lee and Obendorf[39] employed melt electrospinning for the production of protective clothing using PP melts. These protective clothing was intended to be used by the agriculture workers against liquid pesticides. It was observed that the thickness of the fiber mat plays an important role in defining the operating parameters for protective clothing. It was noted that thin mat is adequate enough for the protection against pesticide mixtures of high surface tension whereas thick fiber mat will be required against pesticide mixtures of low surface tension. This may be due to the fact that it is easier for the high surface tension fluid to stick to the surface than flow through the micronized pores. This analysis also took into account that in both the abovementioned cases an adequate level of air permeability is achieved. Li et al.[27] fabricated core–shell nanofibers with transparent poly(methyl-methacrylate) (PMMA) shell and a phase change thermochromic material (PCTM) core. The PTCM employed for the analysis comprises of the crystal violet lactone (CVL), bisphenol A, and 1-tetradecanol. These fibers depicted excellent thermochromic phase change behavior and thus making them applicable as thermoresponsive sensors. As expected, these fibers can be employed for the detection of the foreign materials in a given system, thus improving safety of the concerned surroundings against possible harmful elements.

MELT ELECTROSPINNING: AN INDUSTRIAL PERSPECTIVE

Solvent electrospinning was reported in a number of industrial applications so far. Some devices like NanoSpider™ (Elmarco s.r.o.)[75] or Ultra Web™ (Donaldson Company, Inc.)[76] are already present in the market and widely used in worldwide facilities. Nanofibers for air, water filtration, medical, energy production, and automotive industry are only some possible applications reported. Donaldson Company Inc. focused on nanofiber filter media for various uses like filters for cabin air cleaning, fuel, or oil filtration. The list of companies producing nanofiber filter products numbers more than 50 worldwide using different techniques. Some of these companies with examples of products and main applications are listed in Table 4. Due to the continuous development of nanotechnology and high industry demand for new filtration media, it is impossible to include all industries in this area. Multiple needle systems allowed

Table 4 List of nanofiber filtration related companies

S. No.	Company	Country	Products	Description	Address (accessible on 24.06.2013)
1	AAF International	USA	AstroCel®, MegaCel®, DuraShield, DurKlean, MagnaKlean	air filtration, ceramic filters, HEPA filtration systems	www.aafintl.com
2	Ahlstrom Corporation	Finland	DEFENDER®, SURFCOAT®,	Air, liquid filtration, transportation filtration, composite membranes	www.ahlstrom.com
3	American Air Filter	USA	AAF®, BioCel®, AstroCel®,	industrial dust filtration, waste water, ari filtration, pharmceutical, cleanrooms filters	http://www.aafintl.com
4	Applied Sciences Inc.	USA	Pyrograf I, Pyrograf III	graphitic fibers, water treatment, graphite filters	http://www.apsci.com
5	Argonide Corporation	USA	NanoCeram®,	liquid filters, water filtration, NASA space station filters	http://www.argonide.com/
6	Ballarad	Canada	AvCarb®	gas diffusion layers, fuel cells	www.ballarad.com
7	Blue Nano Inc. (Silver nanowires)	USA	SLV-NW-90, SLV-NP-100,	silver nanowires, membranes for fuel cell catalyst technology	http://www.bluenanoinc.com
8	Canfil Farr	USA	HemiPleat Nano®,	air filtration with lower pressure drop,	http://www.farrapc.com
9	Charge Injection Technologies Inc	USA			www.chargedinjection.com
10	Clarcor Inc	USA	AirGuard®	engine filtration, air, oil, fuel/water separation, air filters,	www.clarcor.com
11	Clarkor Company	USA	ProTura®,	dust collection, industrial filtration, turbine gas cartridges	http://www.clarkfilter.com
12	Clean and Science Co	Korea	PP PET, paper membranes	melt blown, composite polymer fibers, air, oil, fuel filters	www.cands.co.kr
13	Donaldson Company Inc.	USA	Tetratex®, Ultra-Web®, Spider-Web®, Torit®,	air-filtration, process filtration, nanofibrous filters	www.donaldson.com
14	Dupont	USA	FM350, FM100, PT350, Microban™, OGuard™	water filtration, air filtration, medical device filters,	www.dupont.com
15	Elmarco	Czech Republic	NanoSpider™	nanofiber equipment, complete production lines	www.elmarco.cz
16	Engineered Fibers Technology	USA	EFTec™, Spectracarb ™	synthetic and natural based polymer fibers	www.eftfibers.com
17	Esfil Tehno	Republic of Estonia	NEL-3, FPP-D-4,	filtering polymeric materials, non-woven materials, fibrous materials, separation membranes, gas, aerosols filtration	www.esfiltehno.ee
18	Espin Technologies Inc.	USA	exceed®, nWeb®	nanofiber media for air, liquid, molecular filtration, functional Apparel, drug delivery, regenerative medicine	www.espintechnologies.com
19	Espinex Inc	Japan	unknown	unknown	www.espinex.com

(Continued)

Table 4 (*Continued*) List of nanofiber filtration related companies

S. No.	Company	Country	Products	Description	Address (accessible on 24.06.2013)
20	Evanite Fiber (Glass fibers)	USA			www.evanite.com
21	Finetex Technology	S. Korea	Nonwoven Coats™, Membrane Mats™, Finetex Technoweb Nonwoven Coats™	nanofiber membranes, automotive, air, liquid filtration	www.finetextech.com
22	Fredenberg	Germany	Viledon®, micronAir®,	air, liquid filtration, air intake systems for turbomachinery, dust removal filters,	www.freudenberg.com
23	Harmsco Inc.	USA	SureSafe™, Hurricane™, Calypso Blue™	liquid filtration, reverse osmosis membranes, liquid filter bags	www.harmsco.com
24	Helsa-automotive	Germany	Rev-O BRO, Mem-Bio System	air filter systems, diesel particulate filters, high pressure filters, air-oil separation, water and biomedical filtration	www.helsa-automotive.com
25	Hengst GmbH & Co. KG	Germany	ProClean System, high dust retention systems, long-life filters	industrial filtration systems, fuel and air filtration	www.hengst.de
26	Hills Inc.	USA		Island-In-The-Sea fibers, MEMS	www.hillsinc.net
27	Hollingsworth and Vose (glass fiber)	USA	NanoWeb®, NanoWave®, ViaMat®	Air filter media, liquid filter media, performance barrier nonwoven, industrial filtration, fiber composites	www.hollingsworth-vose.com
28	Hollingsworth Co. Ltd.	USA	Technostat®, HiPerm Plus,	HEPA, ULPA filters, HVAC systems, liquid filtration, vacuum filters	www.hollingsworth-vose.com
29	Japan Vilene Company Ltd.	Japan	Ecoalpha®	air filters, nanofibers	www.vilene.co.jp
30	Johns Manville (glass fibers)	Germany	EcoMat®, DuraSpun®, CombiFil®, DeltaAire®	Glass Fiber Nonwoven, Glass and Polyester combination, Meltblown for Filtration	www.jmeurope.com
31	Johns Manville Sales GmbH	Germany	MICRO-AIRE®, Delta-Aire™, Dynaweb™, 35 Tex	air filtration, liquid filtration, mist elimination for Sub-Micron Particles	www.jmeurope.com
32	Kimberly Clark Co	USA	INTERPID®, EVOLUTION®, CYCLEAN®, POWER LOFT®	air and liquid filtration media, dust collectors, HVAC systems	www.kcfiltration.com
33	KX industries	USA	MATRIKX® Carbon Filters, PLEKX® Composite Web	extruded carbon filters, custom filtration equipment	www.kxindustries.com
34	Lydall Filtration /Separation (glass fibers)	USA	Ariso™, SOLUPOR®, LydAir®, Actipure®	air filtration products, microglass filtration media, drug delivery, fuel cells applications	www.lydallfiltration.com
35	Mikropor	Turkey	3S filter	in line filters, air inlet filters, air-oil separators,	www.mikroporamerica.com
36	Nano109	Turkey	FilmWeb™, Nanoparma™, Sound-Spider™	air filter media, liquid filter media, dust filtration, HEPA/ULPA, reverse osmosis, sound absorbing	http://www.nano109.com
37	Nanostatics	USA			www.nanostatics.com
38	Neenah Gessner	Germany	Paper/meltblown, cellulose filter media,	air, oil, fuel filters, vacuum cleaner bags, cabin filters	www.neenah-gessner.de
39	Nicast Ltd.	Israel	Avflo, NovaMesh™	fibrous materials for biomedical applications	www.nicast.com

#	Company	Country	Products	Application	Website
40	Nippon Muki Co., Ltd. (Glass fibers)	Japan	Atomos, Nouvelle, Vanish, Neuec	air filter products, semiconductor/LCD industry, air-conditioning	www.nipponmuki.co.jp
41	Nonwoven Technologies Inc	USA	Advantex®, TruPave®	Industrial fibrous materials, constructions nowoven isolators, electronics	www.nonwoventechnologies.com
42	NSG group, (glass fibers)	Japan	Glass Mat, MICROGLAS® CORD,	high performance separator for flooded-type batteries, nanoporous films, AGM separators	www.nsg.co.jp
43	Physical Sciences Inc	USA	unknown	nanofibrous membranes	www.psicorp.com
44	RoboVent	USA	Endurex N15, W13,E16,P11	welding fumes and smoke filtration	www.robovent.com
45	Silverstar Corporation	Korea	ULPA, HEPA Filters	meltblown micro/nanofibers, core-shell structures, cabin air filters	www.silverstar2000.com
46	Teijin Fibers Ltd.	Japan	BEWELL® SOLOTEX®, ECOSTORM® ELASTY®, NANOFRONT®	ecological materials, comfortable materials	www.teijinfiber.com
47	Thueringer Filter Glass GmbH	Germany	glass fibers		filter.glas@arcor.de
48	Toray	Japan	ultrasuede®	nanofibers, nano-laminated films,	www.toray.com
49	Total Separation Solutions, LLC	USA	NanoSol™	complete line for post-treatment of waste water, water filtration nanofibrous membranes	www.totalsep.com
50	US Global Nanospace	USA	NanoFilter™, NanoFilterCX	Pathogen Air Filter Purification System, Cigarette filters	www.usgn.com
51	WOLTZ GmbH (Glass fiber equipment supplier)	Germany	glass fiber filters	glass fiber products, glass wool, textile glass, microfibers,	www.woltz.de

increasing production efficiency to the economically reasonable level.

Melt electrospinning industrial setups that are presently used were not reported. Polymer flows in syringe are significantly lower for melt electrospinning than solvent electrospinning. Available data shows polymer flow in range 0.003–0.6 mL/hr;[12–16] however, 100% of polymer is collected compared to 2–10% of total volume in solvent electrospinning. Achieving similar flow rates like in solvent electrospinning will increase process efficiency several times and allow for wide industrial applications. Nevertheless, melt electrospinning devices requires equipment that cannot be found in solvent electrospinning like extruders, heating, and cooling systems or precise temperature control systems, which can affect the final price of product.

CONCLUSIONS

As discussed, melt electrospinning provides scores of benefits over solution electrospinning. This includes the absence of the solvent avoiding any toxicity, which may affect the collecting substrate and avoids collection of solvent fumes and recycling them. Also faster and simpler material preparation or the ability to electrospun directly on the desired substrate is an advantage of this technique. Potential high throughput of the melt electrospinning enables the mass production of the microfiber mats. The melt electrospinning process as expected is highly suited for the fabrication of the scaffolds used in bioengineering and filter media employed in environmental or energy applications. This implies a possible reduction in the price of the final product and easier industrial implementation. Due to the absence of the bending instabilities, the trajectory of the melt-electrospun jet is predominantly a straight line, which allows the fabrication of patterned fiber mats. Nanofiber mats can be introduced in the electronic circuit, solar cell and sensors, etc. Designing one universal device with a wide range of controllable temperature and accurate melt transportation through needles may result in the production of nanofibers from the various polymers. Viscosity reduction additives or other medium molecular weight polymers can be helpful in further fiber diameter reduction.

REFERENCES

1. Cooley, J.F. Apparatus for Electrically Dispersing Fluids. U.S. Patent 692631, Feb 04, 1902.
2. Morton, W.J. Method of Dispersing Fluids. U.S. Patent 705691 A, Jul 29, 1902.
3. Yarin, A.L.; Koombhongse, S.; Reneker, D.H. Taylor cone and jetting from liquid droplets in electrospinning of nanofibers. J. Appl. Phys. **2001**, *90* (9), 4836–4846.
4. Yarin, A.L.; Koombhongse, S.; Reneker, D.H. Bending instability in electrospinning of nanofibers. J. Appl, Phys. **2001**, *89* (5), 3018–3026.
5. Chew, S.Y.; Mi, R.; Hoke, A.; Leong, K.W. The effect of the alignment of electrospun fibrous scaffolds on Schwann cell maturation. Biomaterials **2008**, *29* (6), 653–661.
6. Ahn, Y.C.; Park, S.K.; Kim, G.T.; Hwang, Y.J.; Lee, C.G.; Shin, H.S.; Lee, J.K. Development of high efficiency nanofilters made of nanofibers. Curr. Appl. Phys. **2006**, *6* (6), 1030–1035.
7. Huang, Z.M.; Zhang, Y.Z.; Kotaki, M.; Ramakrishna, S. A review on polymer nanofibers by electrospinning and their applications in nanocomposites. Compos. Sci. Technol. **2003**, *63* (15), 2223–2253.
8. Gratzel, M. Dye-sensitized solid-state heterojunction solar cells. Mrs Bull. **2005**, *30* (1), 23–27.
9. Larrondo, L.; Manley, R.S.J. Electrostatic fiber spinning from polymer melts.1. experimental-observations on fiber formation and properties. J. Polym. Sci. B Polym. Phys. **1981**, *19* (6), 909–920.
10. Larrondo, L.; Manley, R.S.J. Electrostatic fiber spinning from polymer melts.2. examination of the flow field in an electrically driven jet. J. Polym. Sci. B Polym. Phys. **1981**, *19* (6), 921–932.
11. Larrondo, L.; Manley, R.S.J. Electrostatic fiber spinning from polymer melts.3. electrostatic deformation of a pendant drop of polymer melt. J. Polym. Sci. B Polym. Phys. **1981**, *19* (6), 933–940.
12. Dalton, P.D.; Grafahrend, D.; Klinkhammer, K.; Klee, D.; Möller, M. Electrospinning of polymer melts: Phenomenological observations. Polymer **2007**, *48* (23), 6823–6833.
13. Zhmayev, E.; Cho, D.; Joo, Y.L. Nanofibers from gas-assisted polymer melt electrospinning. Polymer **2010**, *51* (18), 4140–4144.
14. Zhmayev, E.; Cho, D.; Joo, Y.L. Modeling of melt electrospinning for semi-crystalline polymers. Polymer **2009**, *51* (1), 274–290.
15. Zhmayev, E.; Zhou, H.; Joo, Y.L. Modeling of non-isothermal polymer jets in melt electrospinning. J. Nonnewton. Fluid Mech. **2008**, *153* (2–3), 95–108.
16. Zhou, H.J.; Green, T.B.; Joo, Y.L. The thermal effects on electrospinning of polylactic acid melts. Polymer **2006**, *47* (21), 7497–7505.
17. Lyons, J.; Li, C.; Ko, F. Melt-electrospinning part I: Processing parameters and geometric properties. Polymer **2004**, *45* (22), 7597–7603.
18. Dalton, P.D.; Klinkhammer, K.; Salber, J.; Klee, D.; Möller, M. Direct *in vitro* electrospinning with polymer melts. Biomacromolecules **2006**, *7* (3), 686–690.
19. Deng, R.J.; Liu, Y.; Ding, Y.; Xie, P.; Luo, L.; Yang, W. Melt electrospinning of low-density polyethylene having a low-melt flow index. J. Appl. Polym. Sci. **2009**, *114* (1), 166–175.
20. Detta, N.; Brown, T.D.; Edin, F.K.; Albrecht, K.; Chiellini, F.; Chiellini, E.; Dalton, P.D.; Hutmacher, D.W. Melt electrospinning of polycaprolactone and its blends with poly(ethylene glycol). Polym. Int. **2010**, *59* (11), 1558–1562.
21. Yarin, A.L.; Kataphinan, W.; Reneker, D.H. Branching in electrospinning of nanofibers. J. Appl. Phys. **2005**, *98* (6), 1–12.
22. Brown, T.D.; Dalton, P.D.; Hutmacher, D.W. Direct writing by way of melt electrospinning. Adv. Mater. **2011**, *23* (47), 5651–5657.

23. Brown, T.D.; Dalton, P.D.; Hutmacher, D.W. Melt electrospinning in a direct writing mode. J. Tissue Eng. Regen. Med. **2012**, *6* (SI 1), 383–384.

24. Kim, S.J.; Jang, D.H.; Park, W.H.; Min, B.M. Fabrication and characterization of 3-dimensional PLGA nanofiber/microfiber composite scaffolds. Polymer **2010**, *51* (6), 1320–1327.

25. Liu, Y.; Deng, R.; Hao, M.; Yan, H.; Yang, W. Orthogonal design study on factors effecting on fibers diameter of melt electrospinning. Polym. Eng. Sci. **2010**, *50* (10), 2074–2078.

26. Kong, C.S.; Jo, K.J.; Jo, N.K.; Kim, H.S. Effects of the spin line temperature profile and melt index of poly(propylene) on melt-electrospinning. Polym. Eng. Sci. **2009**, *49* (2), 391–396.

27. Li, F.Y.; Zhao, Y.; Wang, S.; Han, D.; Jiang, L.; Song, Y. Thermochromic core-shell nanofibers fabricated by melt coaxial electrospinning. J. Appl. Polym. Sci. **2009**, *112* (1), 269–274.

28. Ogata, N.; Shimada, N.; Yamaguchi, S.; Nakane, K.; Ogihara, T. Melt-electrospinning of poly(ethylene terephthalate) and polyalirate. J. Appl. Polym. Sci. **2007**, *105* (3), 1127–1132.

29. Ogata, N.; Yamaguchi, S.; Shimada, N.; Lu, G.; Iwata, T.; Nakane, K.; Ogihara, T. Poly(lactide) nanofibers produced by a melt-electrospinning system with a laser melting device. J. Appl. Polym. Sci. **2007**, *104* (3), 1640–1645.

30. Shimada, N.; Ogata, N.; Nakane, K.; Ogihara, T. Spot laser melt electrospinning of a fiber bundle composed of poly(lactide)/poly(ethylene-*co*-vinyl alcohol) pie wedge fibers. J. Appl. Polym. Sci. **2012**, *125* (S2), E384–E389.

31. Li, X.Y.; Liu, H.; Liu, J.; Wang, J.; Li, C. Preparation and experimental parameters analysis of laser melt electrospun poly(L-lactide) fibers via orthogonal design. Polym. Eng. Sci. **2012**, *52* (9), 1964–1967.

32. Li, X.Y.; Liu, H.; Wang, J.; Li, C. Preparation and characterization of poly(epsilon-caprolactone) nonwoven mats via melt electrospinning. Polymer **2012**, *53* (1), 248–253.

33. Liu, Y.; Zhao, F.; Zhang, C.; Zhang, J.; Yang, W. Solvent-free preparation of poly(lactic acid) fibers by melt electrospinning using an umbrella-like spray head and alleviation of the problematic thermal degradation. J. Serb. Chem. Soc. **2012**, *77* (8), 1071–1082.

34. Komárek, M.; Martinová, L.; T. Ltd. Design and Evaluation of Melt-Electrospinning Electrodes, In *Nanocon*, 2nd International Conference, Tanger Ltd: Slezska, Oct 12–14, 2010; p. 72–77.

35. Fang, J.; Zhang, L.; Sutton, D.; Wang, X.; Lin, T. Needleless melt-electrospinning of polypropylene nanofibres. J. Nanomater. **2012**, *2012*, 1–9.

36. Niu, H.; Lin, T.; Wang, X. Needleless electrospinning. I. A comparison of cylinder and disk nozzles. J. Appl. Polym. Sci. **2009**, *114* (6), 3524–3530.

37. Dasdemir, M.; Topalbekiroglu, M.; Demir, A. Electrospinning of thermoplastic polyurethane microfibers and nanofibers from polymer solution and melt. J. Appl. Polym. Sci. **2013**, *127* (3), 1901–1908.

38. Liu, Y.; Wang, X.; Yan, H.; Guan, C.; Yang, W. Dissipative particle dynamics simulation on the fiber dropping process of melt electrospinning. J. Mater. Sci. **2011**, *46* (24), 7877–7882.

39. Lee, S.; Obendorf, S.K. Developing protective textile materials as barriers to liquid penetration using melt-electrospinning. J. Appl. Polym. Sci. **2006**, *102* (4), 3430–3437.

40. Tian, S.; Ogata, N.; Shimada, N.; Nakane, K.; Ogihara, T.; Yu, M. Melt electrospinning from poly(L-lactide) rods coated with poly(ethylene-*co*-vinyl alcohol). J. Appl. Polym. Sci. **2009**, *113* (2), 1282–1288.

41. McCann, J.T.; Marquez, M.; Xia, Y.N. Melt coaxial electrospinning: A versatile method for the encapsulation of solid materials and fabrication of phase change nanofibers. Nano Lett. **2006**, *6* (12), 2868–2872.

42. Shimada, N.; Tsutsumi, H.; Nakane, K.; Ogihara, T.; Ogata, N. Poly(ethylene-*co*-vinyl alcohol) and nylon 6/12 nanofibers produced by melt electrospinning system equipped with a line-like laser beam melting device. J. Appl. Polym. Sci. **2010**, *116* (5), 2998–3004.

43. Zhao, F.W.; Liu, Y.; Ding, Y.M.; Yan, H.; Xie, P.C.; Yang, W.N. Effect of plasticizer and load on melt electrospinning of PLA. In *Progress in Polymer Processing*; Zhang, C., Ed.; Trans Tech Publications: Qingdao, China, 2012; 32–36.

44. Nayak, R.; Kyratzis, I.L.; Truong, Y.B.; Padhye, R.; Arnold, L. Melt-electrospinning of polypropylene with conductive additives. J. Mater. Sci. **2012**, *47* (17), 6387–6396.

45. Nayak, R.; Padhye, R.; Kyratzis, I.L.; Truong, Y.B.; Arnold, L. Effect of viscosity and electrical conductivity on the morphology and fiber diameter in melt electrospinning of polypropylene. Text. Res. J. **2013**, *83* (6), 606–617.

46. Hohman, M.M.; Shin, M.; Rutledge, G.; Brenner, M.P. Electrospinning and electrically forced jets. I. Stability theory. Phys. Fluid **2001**, *13* (8), 2201–2220.

47. Ramakrishna, S.; Murugan, R.; Sampath Kumar, T.S.; Soboyejo, W.O. *Biomaterials: A Nano Approach*; CRC Press: New York, 2010.

48. Polak, J. *Advance in tissue engineering*; Imperial College Press: London, 2008.

49. Tian, M.; Gao, Y.; Liu, Y.; Liao, Y.; Xu, R.; Hedin, N.E.; Fong, H. Bis-GMA/TEGDMA dental composites reinforced with electrospun nylon 6 nanocomposite nanofibers containing highly aligned fibrillar silicate single crystals. Polymer **2007**, *48* (9), 2720–2728.

50. Rho, K.S.; Jeong, L.; Lee, G.; Seo, B.M.; Park, Y.J.; Hong, S.D.; Roh, S.; Cho, J.J.; Park, W.H.; Min, B.M. Electrospinning of collagen nanofibers: Effects on the behavior of normal human keratinocytes and early-stage wound healing. Biomaterials **2006**, *27* (8), 1452–1461.

51. Stamatialis, D.F.; Papenburg, B.J.; Gironés, M.; Saiful, S.; Bettahalli, S.N.; Schmitmeier, S.; Wessling, M. Medical applications of membranes: Drug delivery, artificial organs and tissue engineering. J. Memb. Sci. **2008**, *308* (1–2), 1–34.

52. Hunlely, M.T.; Karikari, A.S.; McKee, M.G.; Mather, B.D.; Layman, J.M.; Fornof, A.R.; Long, T.E. Taking advantage of tailored electrostatics and complementary hydrogen bonding in the design of nanostructures for biomedical applications. Macromol. Symp. **2008**, *270* (1), 1–7.

53. Diaz, J.E.; Barrero, A.; Márquez, M.; Loscertales, I.G. Controlled encapsulation of hydrophobic liquids in hydrophilic polymer nanofibers by co-electrospinning. Adv. Funct. Mater. **2006**, *16* (*16*), 2110–2116.

54. Karchin, A.; Simonovsky, F.I.; Ratner, B.D.; Sanders, J.E. Melt electrospinning of biodegradable polyurethane scaffolds. Acta Biomater. **2011**, *7* (9), 3277–3284.

55. Simonovsky, F.; Zaplatin, A.; Samigullin, F.; Khudyak, E. Method for Producing Polyetherurethanes in Solution. S.U.P. 1746688, Editor, 1989.

56. Brown, T.D.; Slotosch, A.; Thibaudeau, L.; Taubenberger, A.; Loessner, D.; Vaquette, C.; Dalton, P.D.; Hutmacher, D.W. Design and fabrication of tubular scaffolds via direct writing in a melt electrospinning mode. Biointerphases **2012**, *7* (1–4), 1–16.

57. Farrugia, B.L.; Brown, T.D.; Upton, Z.; Hutmacher, D.W.; Dalton, P.D.; Dargaville, T.R. Dermal fibroblast infiltration of poly(epsilon-caprolactone) scaffolds fabricated by melt electrospinning in a direct writing mode. Biofabrication **2013**, *5* (2), article number: 025001 (11p).

58. Mazalevska, O.; Struszczyk, M.H.; Krucinska, I. Design of vascular prostheses by melt electrospinning-structural characterizations. J. Appl. Polym. Sci. **2013**, *129* (2), 779–792.

59. van Krevelen, D.W.; te Nijenhuis, K. *Properties of Polymers*; Elsevier B.V.: the Netherlands, 2009.

60. Ramakrishna, S.; Huang, Z.-M.; Kumar, G.V.; Batchelor, A.W.; Mayer, J. *An Introduction to Biocomposites*; Imperial College Press: London, 2004; 47–51.

61. The University of Oxford Chemistry Department. http://msds.chem.ox.ac.uk/CE/cellulose_acetate.html, 2011 (accessed March 2011).

62. Aldrich, S. Sigma Aldrich Pvt Ltd. www.sigmaaldrich.com, 2011 (accessed March 2011).

63. Sahoo, N.G.; Rana, S.; Cho, J.W.; Li, L.; Chan, S.H. Polymer nanocomposites based on functionalized carbon nanotubes. Prog. Polym. Sci. **2010**, *35* (7), 83–867.

64. Cho, D.; Zhou, H.; Cho, Y.; Audus, D.; Joo, Y.L. Structural properties and superhydrophobicity of electrospun polypropylene fibers from solution and melt. Polymer **2010**, *51* (25), 6005–6012.

65. Do, C.V.; Thuy, T.T.N.; Park, J.S. Fabrication of polyethylene glycol/polyvinylidene fluoride core/shell nanofibers via melt electrospinning and their characteristics. Sol. Energ. Mater. Sol. Cells **2012**, *104*, 131–139.

66. Goncharuk, V.V.; Kavitskaya, A.A.; Skil'skaya, M.D. Nanofiltration in drinking water supply. J. Water Chem. Technol. **2011**, *33* (1), 37–54.

67. Ramakrishna, S.; Jose, R.; Archana, P.S.; Nair, A.S.; Balamurugan, R.; Venugopal, J.; Teo, W.E. Science and engineering of electrospun nanofibers for advances in clean energy, water filtration, and regenerative medicine. J. Mater. Sci. **2010**, *45* (23), 6283–6312.

68. Sutherland, K.; Air filtration: Using filtration to control industrial air pollution. Filtr. Separat. **2008**, *45* (9), 16–19.

69. Sutherland, K. Developments in filtration: What is nanofiltration? Filtr. Separat. **2008**, *45* (8), 32–35.

70. Ravanchi, M.T.; Kaghazchi, T.; Kargari, A. Application of membrane separation processes in petrochemical industry: A review. Desalination **2009**, *235* (1–3), 199–244.

71. Shabani, I.; Hasani-Sadrabadi, M.M.; Haddadi-Asl, V.; Soleimani, M. Nanofiber-based polyelectrolytes as novel membranes for fuel cell applications. J. Memb. Sci. **2011**, *368* (1–2), 233–240.

72. Stamatialis, D.F.; Papenburg, B.J.; Bettahalli, M.S.; Girones, M.; de Boer, J.; van Blitterswijk, C.; Wessling, M. Membranes for bioartificial organs and tissue engineering. Tissue Eng. A **2008**, *14* (5), 755.

73. Van der Bruggen, B.; Manttari, M.; Nystrom, M. Drawbacks of applying nanofiltration and how to avoid them: A review. Separat. Purif. Technol. **2008**, *63* (2), 251–263.

74. Veleirinho, B.; Lopes-da-Silva, J.A. Application of electrospun poly(ethylene terephthalate) nanofiber mat to apple juice clarification. Process Biochem. **2009**, *44* (3), 353–356.

75. Elmarco, Inc. www.elmarco.com, 2011 (accessed March 2011).

76. Donaldson Company, Inc. www.donaldson.com, 2011 (accessed March 2011).

Membranes, Polymeric: Biomedical Devices

Shahram Mehdipour-Ataei
Maryam Oroujzadeh
Iran Polymer and Petrochemical Institute, Tehran, Iran

Abstract

This entry covers the applications of polymeric membranes (natural and synthetic) in biomedical devices. In order to clarify the concepts of biocompatibility and biodegradability, part of the entry is devoted to these two topics. Moreover, some of the most important families of the biodegradable polymers are also introduced. Specific attention has been devoted to dialyzers, membrane oxygenators, and drug delivery devices. In all of the issues, the concept of device or method, the applied polymers, their advantages and drawbacks, and common modifications and strategies are discussed. In the subject of drug delivery, most of the discussion is dedicated to the transdermal drug delivery and strategies used to enhance the permeation of the drug in this technique.

INTRODUCTION

Nowadays, polymeric materials are used in a wide range of biomedical applications including prostheses, artificial organs, tissue engineering, drug delivery, hemodialysis, heart valves, membrane oxygenators, biosensors, and other related themes. For all of the mentioned applications, the biomaterial should possess suitable mechanical and physical properties, biocompatibility, minimum toxicity, appropriate stability in the body, non-carcinogenicity, and sterilizability. Among these characteristics, biocompatibility is the most important one. Polymers used as biomedical devices generally can be divided into two categories, natural and synthetic polymers. Natural polymers include collagen, chitosan, alginate, hyaluranon, dextran, cellulose, etc. and typical synthetic polymers are poly(ethylene glycol) (PEG), polyurethane, polypropylene, poly(sulfone)s, poly(lactic acid) (PLA), poly(vinyl alcohol), poly(lactic-co-glycolic acid) (PLGA), poly(hydroxy ethylmethacrylate) (PHEMA), poly(acrylimide), poly(ether imide), poly(N-vinyl-2-pyrrolidone), etc.[1–3]

Due to the wide range of applications of polymers in medicine and medical devices, this entry is devoted to the application of polymeric membranes in hemodialysis, blood oxygenators, and drug delivery devices. Prior to these discussions the concept of biocompatibility will be explained.

BIOCOMPATIBILITY

For understanding the concept of biocompatibility, it is better to answer this question: What happens when a polymer is placed in contact with the blood or tissue of the body?

When the body recognizes a foreign material, biological reactions including thrombosis, coagulation, and inflammation will occur leading to an unexpected risk for patients.

Before considering the factors that promote body reactions, first let us talk a little about the events that occur after the initial interaction of a polymeric surface and our body.

Blood as a biofluid consists of cellular and fluid components. Approximately 45% of the blood is cellular elements (white and red blood cells, and platelets) and the remaining 55% is plasma. Water is the main component of the plasma (93%) and the remaining 7% is ions, hormones, enzymes, sugars, amino acids, and proteins. Adsorption of the plasma proteins on the surface of the biopolymer is the first event occurring after contact between the blood and the surface.

Albumin, gamma-globulin, and fibrinogen are three important proteins of plasma that possess larger concentrations compared to other plasma proteins. Half of the protein content of plasma is albumin. Adsorption of albumin on the biomaterial surface decreases platelet adhesion and improves biocompatibility of the surface while adsorption of gamma-globulin and fibrinogen increases adhesion of platelets, which is unfavorable.[4,5] So, improvement in biocompatibility of biomedical surfaces is the first goal of researchers. Techniques developed to achieve this target fall into two main categories:[6]

- Controlling and improving the physical and chemical properties of biomedical surfaces.
- Modification of polymer surfaces with bioactive molecules.

Concise Encyclopedia of Biomedical Polymers and Polymeric Biomaterials DOI: 10.1081/E-EBPPC-120049933

Controlling and Improving the Physico-Chemical Properties of the Surface

There are many research studies in this topic. Some of the most important techniques are noted below.

Increasing the surface free energy

It has been proven that increasing the surface free energy of a polymer makes it more biocompatible. To modify surfaces in this way, different techniques such as plasma etching, corona discharge, UV irradiation, and chemical treatment have been used.[7,8] Meanwhile, it has been shown that high water levels on the surface cause less protein adsorption and cell adhesion, and consequently more biocompatibility. In this way, polymer surfaces can be coated with a hydrogel. A hydrogel is a physically or chemically cross-linked network of a polymer, which is partially or completely soluble in water. This behavior of hydrogels is related to water molecules placed around the polymer chains. Adsorption of proteins on a hydrogel surface is possible only when displacement of water molecules happens and since such displacement is not energetically favorable, hydrogels are reluctant to adsorb blood proteins.[5]

Establishing a balance between hydrophobicity and hydrophilicity of the surface

It has been proven that large amounts of blood proteins tend to adsorb on the hydrophobic surfaces. So, increase in hydrophilicity of the surface can improve the biocompatibility of the surface.[9,10] There are three techniques to make more hydrophilic surfaces:

- Chemical modification of the hydrophobic surface using hydrophilic polymers such as PEG also termed as poly(ethylene oxide), poly(vinyl pyrrolidone) (PVP), PHEMA, etc.
- Blending of hydrophobic and hydrophilic polymers.
- Using hydrophilic–hydrophobic diblock copolymers (amphiphilic copolymers). Adsorption of these diblock copolymers on hydrophobic surfaces makes them more biocompatible.

Adjustment of surface charges

Neutral and polyanionic surfaces are less toxic and adsorb less plasma proteins. As most of the plasma proteins have a net negative charge, polyanionic surfaces tend to adsorb fewer proteins compared to polycationic surfaces[5,11] (Fig. 1).

Modification of Polymer Surfaces with Bioactive Molecules

Preparation of biologically active surfaces is another technique for improving biocompatibility of biomaterials.

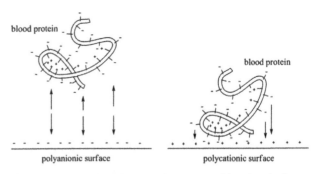

Fig. 1 The behavior of proteins in contact with polyanionic and polycationic surfaces.

In this respect, albumin- and heparin-modified systems have been recognized.

Heparin as a polysaccharide containing sulfonate groups is an important anticoagulant agent. Many research studies have been focused on immobilization of heparin molecules on the biomaterial surfaces. There are three ways for heparin immobilization. The first method is binding of heparin molecules to the surface using ionic or covalent interactions. Nevertheless, ionic binding for some applications like hemodialysis is not appropriate because of the high probability of heparin release, which may increase the level of heparin in blood causing problems for patients. Another way is connection of heparin to a binder and then to the polymeric surface. Finally, the third technique for introducing heparin is chemical cross-linking of heparin within the hydrogel network, which is coated on the biopolymer surface.[12–14] These three techniques are shown in Fig. 2 (A, B, and C).

As it was mentioned earlier, albumin adsorption on the surface makes it more biocompatible. The first way of using this bioactive material is grafting it on the polymeric surface (Fig. 2D), but such systems are susceptible to denaturation of albumin and stability problems may arise. As a result, the next method which is more interesting is designing surfaces to promote albumin adsorption in contact with the blood. It was shown that alkyl chains grafted on the surface tend to adsorb large amounts of albumin. This strategy is based on the hypothesis that albumin transports fatty acids in the blood, and systems containing grafted alkyl chains can promote albumin adsorption (Fig. 2E).[15–18]

Although in polymeric membranes biocompatibility is a necessity, another equally important issue is the ability of polymeric membranes to mimic the function of the real organ. Anyway, in other parts of this entry the applications of polymeric membranes in medicine will be discussed.

HEMODIALYSIS

Kidney is a vital organ in our body. Waste disposal, regulation of acid–base balance, regulation of blood pressure, and blood detoxification are some of the most important

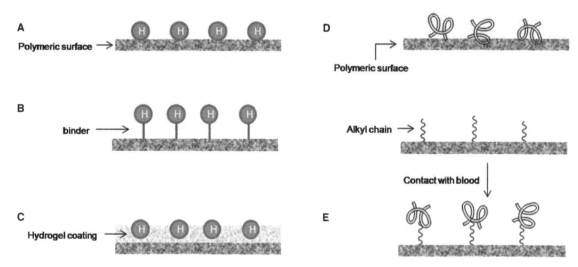

Fig. 2 Modification of polymeric surface with heparin (**A, B, C**) and albumin (**D, E**).

functions of our kidneys. Unfortunately some patients suffer from irreversible kidney failures. The best treatment for these patients is kidney implantation, but finding an appropriate donor is not easy. About 50 years ago and until 1960s when Kole and Berk developed their first successful dialysis device in Netherlands, this condition was fatal. Since that time many patients have survived using these artificial kidneys. Patients with end stage renal disease had to be dialyzed three times a week for 3–5 hr each time.

The central part of a dialyzer is a semipermeable membrane that allows water, urea, creatinine, and glucose to pass through, but blood cells and proteins remain. The first dialysis device consisted of a tubular cellophane membrane wound around a rotating drum immersed in a bath of saline liquid (Fig. 3A). Blood was taken out of the body and passed through the membrane. After waste disposal, the clean blood was returned to the patients' body. One of the most important problems of this early dialyzer was its need for several liters of blood to start the device. In 1950s, Kolf and colleagues developed coil devices, which still had the problem of requiring large amounts of blood to prime, but 10 years later, plate and frame and then hollow fiber dialyzers were introduced in the market, so that today most dialyzer modules are in hollow fiber configuration (Fig. 3B).[19]

After this brief history of dialyzers, let us discuss different types of polymeric membranes used in dialysis. In general, polymeric membranes used in this application are divided into two categories:

- Membranes prepared from cellulosic materials.
- Membranes made of synthetic polymers.

Cellulosic Membranes

Cellulose is a natural polymer made of rings, each bearing three free hydroxyl groups, and is a completely hydrophilic polymer. Unmodified cellulosic membranes (Cuprophan®)

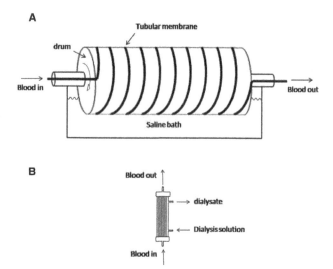

Fig. 3 First (**A**) and last (**B**) generation of dialyzers.

were used for many years. As noted in the previous section, fully hydrophilic polymers cause complement activation. So, manufacturers attempted to substitute some hydroxyl groups of cellulosic membranes with hydrophobic groups.[20] For instance, substitution of approximately 70% of hydroxyl groups with acetate groups produced cellulose diacetate membranes and further acetylation of this product resulted in cellulose triacetate membranes that were more biocompatible. Meanwhile, the pore size of these cellulose di- or tri-acetate membranes was slightly larger than that of the unmodified cellulosic membranes, because the acetate group is more bulky compared to the hydroxyl group causing larger free volumes between polymer chains.[21,22]

The second substitution procedure was the replacement of less than 5% of hydroxyl substituents with bulky groups like tertiary amines (in Hemophan®) and benzyl groups. In this technique, bulky groups reduce the degree of interactions between the membrane and complement

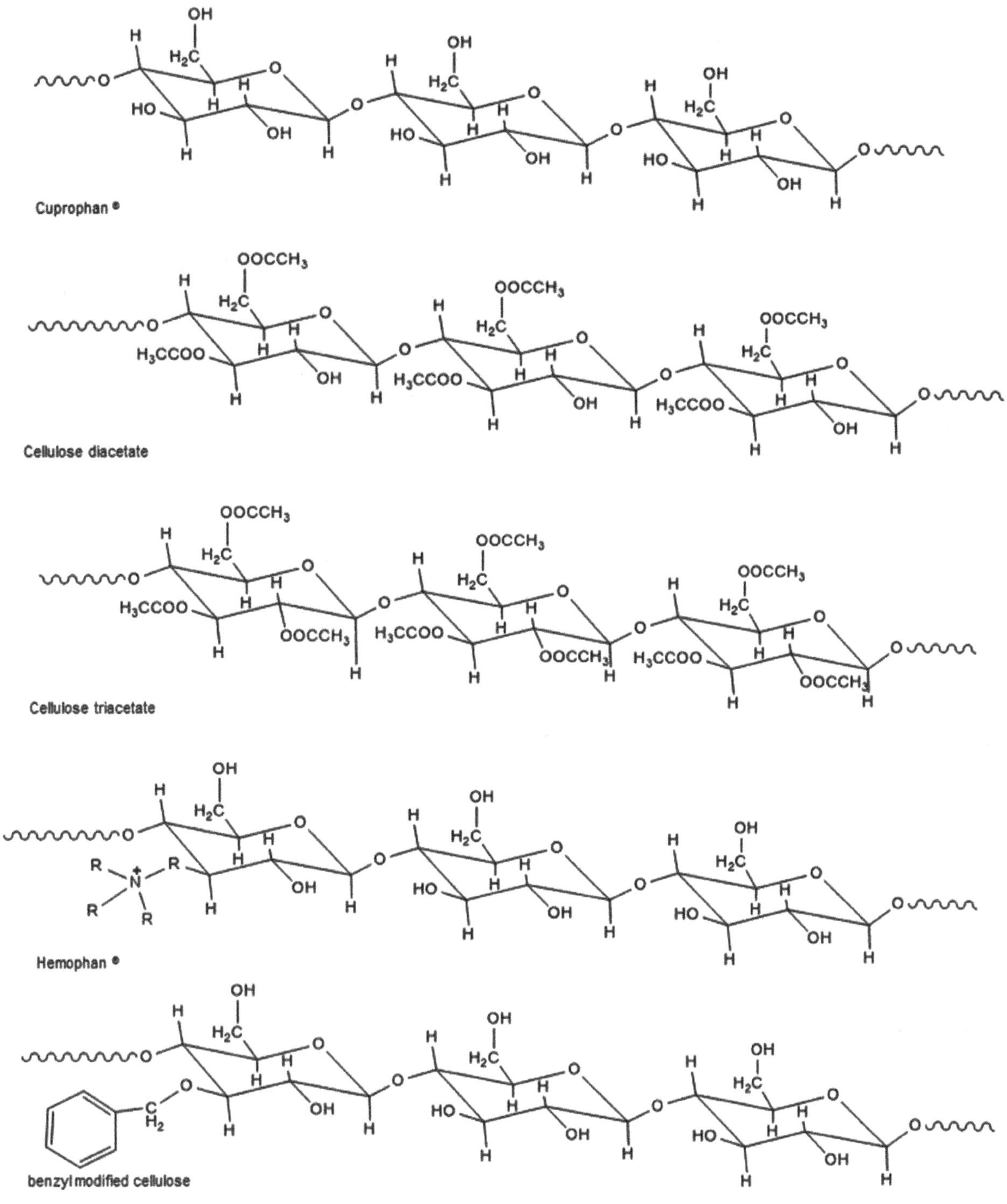

Fig. 4 Structure of unmodified and modified cellulose polymers.

activation products.[23] Structures of unmodified and modified cellulose polymers are depicted in Fig. 4.

Synthetic Membranes

Another major problem with cellulosic membranes is their narrow spectrum of solute removal, which is restricted to small solutes like urea and creatinine. So, metabolites with molecular weights between 1000 Da and 10,000 Da are poorly removed by these membranes. Accumulation of these metabolites causes health risk for patients. One way to solve this problem is modification of cellulosic membranes, for instance, with substitution of hydroxyl groups. Using synthetic polymers is another strategy to overcome this problem. It has been shown that function of these synthetic membranes is closer to the normal kidney.

The first synthetic membrane introduced in the market in 1970s was AN69®, a copolymer of acrylonitrile and sodium methallylsulfonate. After introduction of AN69®, numerous synthetic membranes with different structures were prepared including polysulfone, polyamide, poly(methyl methacrylate), and their derivatives.[24,25] Almost all synthetic membranes had larger pore size compared to cellulosic membranes.

Generally, dialysis membranes, relating to the treatment, are used in three modes:

- Hemodialysis
- Hemofiltration
- Hemodiafiltration

The driving force for waste removal in these three modes is different. In hemodialysis, diffusion controls the metabolite removal and solutes diffuse because of concentration difference, whereas in hemofiltration, convection is the controlling factor and the driving force for waste disposal is pressure difference between both sides of the membrane. Finally, hemodiafiltration is a combination of hemodialysis and hemofiltration, and both diffusion and convection control it.

Dialysis membranes are classified based on water flux or water permeability. Ultrafiltration coefficient is a determining parameter for water flux, and membrane pore size has the largest impact on this parameter. It has been proven that a small increase in pore size has a large effect on permeability. Consequently, dialysis membranes fall into the following classifications:[2,8,26]

- Low-flux membranes that have small pore size (ultrafiltration coefficient of <8 ml/hr/mmHg) and are suitable for small metabolite removal. These membranes are usually used in hemodialysis. Unmodified cellulosic membranes are often placed in this category.
- Mid-flux membranes having medium pore size and ultrafiltration coefficient between 10 and 20 ml/hr/mmHg. Cellulosic di- and tri-acetate membranes fall in this group.
- High-flux membranes that have bigger pores and an ultrafiltration coefficient of more than 20 ml/hr/mmHg. They are often used in hemofiltration. Most of the synthetic membranes are classified in this category.

This is a simplified classification and there are manufacturers who introduced low-flux polysulfones and poly (methyl methacrylate) membranes, high-flux cellulose acetate, or low-flux cellulose triacetate membranes.

It is worthy of note that because of the hydrophobic nature of most synthetic polymers, the prepared membranes are not biocompatible and cause adsorption of blood proteins. One way to avoid blood coagulation during hemodialysis is heparin administration. However, modification of membranes with bound heparin is another

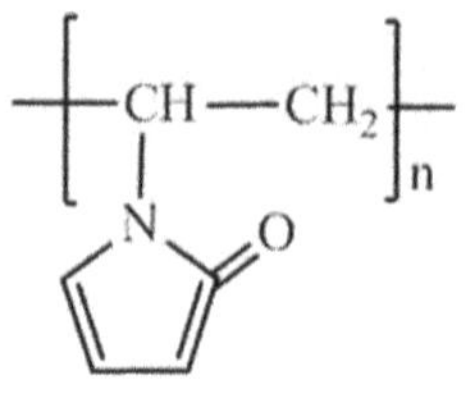

Fig. 5 Poly(vinyl pyrrolidone) (PVP).

technique.[12–14] There are also some research studies in blending of synthetic polymers with hydrophilic polymers like poly(vinyl pyrrolidone)[27] (Fig. 5).

There are also membranes with multilayer structure; for instance, Polyflux S® is composed of three layers of poly(aryl ether sulfone), polyamide, and PVP. In this structure, a thick and sponge-like structure interposed between two thin layers on both the inner and outer sides of the membrane.[28]

BLOOD OXYGENATORS

In our body, the oxygen obtained via breathing the air is carried by hemoglobin in the red blood cells and consumed for oxidizing nutrients. On the other hand, carbon dioxide, water, and heat as the end products of chemical reactions in tissues of our body must be eliminated. Elimination of carbon dioxide is done in our lungs. So, the function of our lung is exchange of oxygen and carbon dioxide. Our lung is a highly efficient gas exchanger with a total surface area of about 80 m^2 and a thickness of 1 μm.[29]

The major application of blood oxygenators is in open heart surgery and in the system of cardiopulmonary bypass in which blood circulation through the heart and lung is prevented.[30–32]

Early blood oxygenators innovated by Gibbon in 1940s led to the first successful open heart surgery in 1953. Gibbon's machine consisted of a small tower with stainless steel screens to contact blood with oxygen flowing in the opposite direction of the blood. Before introduction of membrane oxygenators, these direct oxygenators were used in all devices. The second generation of these direct devices was disk oxygenators that consisted of several rotating disks in a closed cylinder. Some disadvantages of these devices were possible damage to the blood cells and need for large amounts of blood to prime the device. Later, bubble oxygenators were developed. In these devices, oxygenation of blood is accomplished by introduction of oxygen bubbles into the blood. Because of the high surface area of the bubbles, oxygenation was effective but still there was the problem of damage to the blood cells. Since bubbling gas through the blood causes foam formation, which can cause trauma, removing these gas bubbles is very important. As a result, blood must be defoamed using some defoaming compounds and bubble-free blood returned to the body.[8,19,29,30]

In order to avoid direct contact between blood and gas, membrane oxygenators were developed. Membrane oxygenators are classified into two main categories, sheet membranes and hollow fibers.

There are two kinds of sheet membranes, plate and screen, and spiral coil [Fig. 6 (A and B)]. As shown in Fig. 6A, the plate and screen model consists of polypropylene screens placed between microporous polypropylene membranes. Polypropylene screens caused passive mixing of blood and reduced the blood resistance to gas transfer. In the spiral coil layout depicted in Fig. 6B, blood flows between layers of the silicone membrane reinforced with screens. In this type of oxygenators unlike the other ones, the membrane is homogenous that prevents the common problem of plasma leakage from microporous membranes.[33]

First hollow fiber membranes commercialized in 1980s were composed of microporous polypropylene with a nominal pore size of 0.1 μm. There are two forms of hollow fiber membrane oxygenators [Fig. 6 (C and D)]. In the first kind of these membranes, blood flows within the fibers and oxygen around the fibers (Fig. 6C). There are some problems with this type of membranes, which include high pressure drop and high probability of damaged blood cells because of passing of blood inside the capillaries. In other types of these hollow fiber membranes, oxygen flows within the fibers and blood flows outside the hollow fibers (Fig. 6D) in such a way that pressure drop is minimized. Moreover, hollow fibers act as passive mixing elements, which reduce the blood resistance to gas transfer. Today, most of the available membrane oxygenators are of this type.[33,34]

Besides polypropylene and silicone, other polymers like polyethylene, Teflon, and poly(4-methyl pentene) have been used in membrane oxygenators.[35,36]

It should be noted that the driving force in an oxygenator for oxygen is about 15 times more than for carbon dioxide whereas this ratio in our lungs is 13 times, that is our lungs are about 20 times more permeable to carbon dioxide. So, carbon dioxide is the limiting gas in these devices. Required oxygen transfer in a blood oxygenator is about 250 ml/min and for carbon dioxide this value is

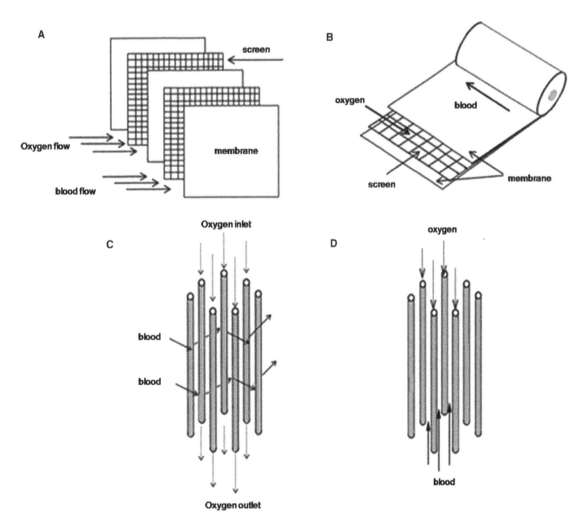

Fig. 6   Sheet membrane (**A, B**) and hollow fiber oxygenators (**C, D**).

around 200 ml/min.[8] So in general, a suitable blood oxygenator should be biocompatible, have adequate oxygen and carbon dioxide exchange properties, low cost, and low priming volume of blood.

DRUG DELIVERY

A multitude of efforts are accomplishing in replacing the conventional and classical medication with controlled drug delivery systems. The objective of such systems is delivering the drug with the controlled rate and concentration to the body or more specifically, to the target tissue or organ of the body. The aims of these systems can be summarized as follows:[37]

- Delivering the drug over a longer period of time.
- Controlled and constant drug concentration during drug therapy and preventing from overdosing and underdosing.
- Reducing the side effects of the drugs and improving the efficacy of drugs.
- Maintaining the drug levels in therapeutic range within the period of drug therapy.

In order to achieve these goals, different drug delivery systems have been developed. These systems can be under control of one of these mechanisms: diffusion of a drug from a polymer matrix or a polymeric membrane, biodegradation of the polymeric matrix, or osmotic effects.

Diffusion-controlled systems can be divided into two categories:

1. Systems in which a polymeric membrane encloses the drug reservoir (Fig. 7A).
2. Systems in which the drug is dispersed in a polymeric matrix (Fig. 7B).

In the first type of these systems, a membrane controls the release of drug. The membrane may be porous or not but covers the drug reservoir. However, the release rate in these systems can be zero- or first-order. In zero-order release systems, the drug in the reservoir exists in saturated form, hence the release rate does not change as long as the solid drug exists. In the first-order release systems, the reservoir contains unsaturated solution of drug so the rate of release reduces with time. On the other hand, in the second category of diffusion-controlled systems there is no membrane and reservoir but the drug is dispersed in a polymeric matrix. In this system, the problem of membrane defects that can affect the release rate does not exist. Drug dispersed in the polymeric matrix can be solid particles or in solution form. The release rate in these systems depends on the geometry of device and also the amount of loaded drug. The mentioned polymeric matrix or membrane in diffusion-controlled systems remains permanently in the body, which is undesirable. Surgical methods should be applied to remove these polymeric parts after finishing the treatment period. To overcome this problem, biodegradable systems were introduced. In these systems, the drug is loaded in a natural or synthetic biodegradable polymer. During drug release, degradation of the polymer also occurs. Two degradation mechanisms control this process comprising *surface erosion* or *bulk degradation*. In systems controlled by surface erosion, degradation starts from the surface of the device and drug releases during this process (Fig. 8A). Decrease in the surface area of these devices with time leads to the reduction in the drug release rate. Two approaches to solve this problem have been developed; the first one is manipulation of the geometry of the device. The second strategy is the use of concentration gradient in the device so that the concentration of drug loading increases from the surface to the core of the device. The second mechanism is using systems controlled by *bulk degradation*. In this system, the initial stage of bulk degradation is permeation of water and hydrolysis of susceptible bonds. In the first steps of degradation, as the number of cleaved bonds is limited, the mechanical properties of the device do not change strongly, but when the chain cession continues, the polymer gradually swells and finally dissolves

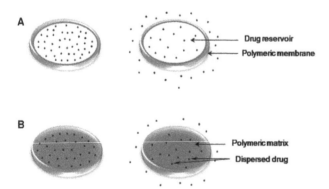

Fig. 7 Diffusion-controlled systems: membrane and reservoir system (**A**), and polymer matrix and dispersed drug system (**B**).

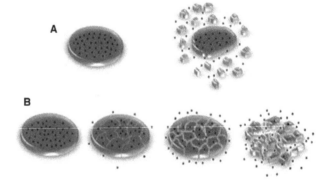

Fig. 8 Schematic of biodegradable drug delivery devices: erosion (**A**) and bulk degradation (**B**).

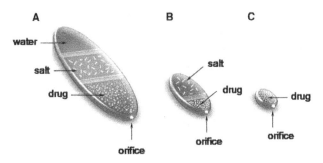

Fig. 9 (A, B, C) Three generations of osmotic pumps.

in water. The process is demonstrated in Fig. 8B schematically. In general, it can be said that *erosion* is a physical mechanism in which dissolution and diffusion control the biodegradation but *bulk degradation* is a chemical process in which decomposition of polymer chains causes degradation of the device.[8,19] Although, both mechanisms take place in every biodegradable device, only one of them is more noticeable and controls the process. Osmotic pumps are another type of drug delivery devices shown in Fig. 9. The first generation of these devices developed by Rose and Nelson consisted of three chambers of water, salt, and drug (Fig. 9A). The water and drug chambers are separated by a semipermeable membrane. Difference in osmotic pressure causes the water to pass from the water chamber to the salt chamber and pushes the drug out of the orifice. The second generation of osmotic pumps was introduced by Higuchi and Leeper. This osmotic pump has no water chamber and water of the surrounding environment activates the device (Fig. 9B). The third generation of these devices that was also the most simplified device was developed by Theeuwes (Fig. 9C). These devices composed of just a drug chamber, and drug itself acts as the osmotic agent.[19,38]

In the following part of the entry biodegradable polymers used in drug delivery devices will be considered.

Biodegradable Polymers

Biodegradable polymers decompose to biocompatible byproducts after degradation in the body; therefore, there is no need to remove the device after releasing the drug. These polymers can be natural or synthetic. In this section some of the applied polymers will be discussed.

Natural biodegradable polymers

The family of polysaccharides like cellulose, starch, chitosan, dextran, and hyaluronic acid is an attractive family of natural polymers for drug delivery due to their good biocompatibility. However, because of some physicochemical limitations of these polymers, alternative synthetic polymers with better properties and diverse range of structures have been introduced.[39,40]

Synthetic polymers

Designing biodegradable polymers with desirable properties and tunable degradation time attracted a great deal of attention for the synthetic polymers. Some of the most important synthetic biodegradable polymers shown in Table 1 are the following.

Poly(ester)s: Aliphatic poly(ester)s are a family of polymers that are extensively employed in biodegradable drug delivery devices because of their biocompatibility. The ester linkage is a labile group in these polymers. PLA, poly(glycolic acid) (PGA) and their copolymers, and also poly(ε-caprolactone) (PCL) are the most common poly (ester)s in this regard. The mechanism of degradation in poly(ester)s is bulk degradation. However, because of more crystalline structures of PLA and also the presence of CH_3 group that causes steric hindrance, its degradation is slower than PGA. It is worthy of note that there is an asymmetric carbon atom in the structure of lactic acid, so lactic acid can exist as D- and L-enantiomers. As naturally occurring lactic acid is L-lactic acid, PLAs prepared from L-lactic acid are more biocompatible. However, copolymers of PLGA show more attractive properties and the ratio of lactic acid to glycolic acid determines the rate of degradation. Copolymers with higher percentage of glycolic acid (>70%) are amorphous and degrade faster. In spite of the advantages of PLGA copolymers, it has been shown that production of acidic substances during degradation of these polymers can affect the drug stability and activity. Another member of the family of biodegradable poly(ester)s is PCL that is a semicrystalline polyester with slow degradation rate. So, it is suitable for long-term drug delivery. Another attractive feature of PCL is its high permeability to many drugs and solubility in a wide range of common organic solvents.[41–44]

Poly(orthoester)s (POE) are a group of hydrophobic biodegradable polymers in which the erosion mechanism controls degradation of the prepared devices. The rate of erosion is very slow because of the hydrophobicity of the prepared devices. Degradation of POE is pH sensitive and can be controlled by using acid or base excipients. There are four classes of these polymers with different structures of I, II, III, and IV that are shown in Table 1. POE I is synthesized through transesterification of an appropriate diol and diethoxytetrahydrofuran. This class of POEs is very sensitive to the acid-catalyzed hydrolysis, and since the product of hydrolysis is an acidic substance (γ-hydrobutyric acid), autocatalytic hydrolysis takes place. To avoid this undesirable reaction, the polymer is stabilized with Na_2CO_3 as a base. POE II is the product of reaction between a diol and a diketene acetal. The initial product of hydrolysis of POE II is a neutral compound that prevents autocatalytic degradation. Another advantage of this class of POEs is the ability to adjust the thermal and mechanical properties by using different diols. In order to achieve appropriate rate of degradation, addition of an acid excipient is necessary in this class of POEs. POE III is synthesized from the reaction between a triol and an orthoester. Due to the high flexibility of the main chain in this class of polymers, the prepared polymer is a gel-like material at room temperature. This property leads to an easier production process of drug delivery devices at room temperature by simple mixing without using any organic solvents. The main disadvantage of POE III is the difficulty in obtaining reproducible results and preparing polymers with the same molecular weights. POE IV was prepared in order to modify POE II to overcome the problem of adding the acidic excipients for appreciable rate of degradation. In this respect, the short segment of lactic acid or glycolic acid is added to the main chain. The rate of degradation is controlled by the length of this segment. Such a segment hydrolyzes to the glycolic acid or lactic acid, which can catalyze degradation of the main chain. This class of poly(orthocenter)s shows the best properties as drug delivery devices and highest potential for commercialization.[45–47]

Poly(anhydride)s: This class of biodegradable polymers with labile anhydride linkages and hydrophobic main chain undergo surface erosion. These biodegradable polymers are of great interest because degradation products are non-mutagenic and non-cytotoxic. Another advantage of poly(anhydride)s is their one-step synthesis that needs no more purification steps. These polymers can be aliphatic, aromatic, or aliphatic–aromatic copolymers (Table 1). The degradation rate of the prepared devices can be controlled by the polymer composition. Aliphatic poly(anhydride)s synthesized from aliphatic diacids are crystalline polymers and are soluble in chlorinated hydrocarbons. These polymers degrade during several weeks. On the other hand, aromatic poly(anhydride)s are crystalline polymers with high melting point (200ºC) and limited solubility in common solvents, consequently they are highly stable. These properties limited their use as drug delivery devices. Aliphatic–aromatic copolymers are between two extremes. Based on the length of the aliphatic segments, the rate of degradation is different in these copolymers. In spite of the advantages of poly(anhydride)s, they have some disadvantages including hydrolytic instability that dictates their storage under moisture free conditions and also spontaneous depolymerization in organic solutions or during storage at room temperature and above.[48–50]

Poly(phosphazene)s: These polymers are composed of nitrogen–phosphorus inorganic main chain. This backbone is very flexible and hydrolytically stable, but they can become unstable by introducing some appropriate side groups in the structure including amino acid esters, glucosyl, glyceryl, glycolate, lactate, imidazole, etc. In all hydrolytically unstable poly(phosphazene)s, the production of hydroxyphosphazene during degradation is responsible for instability of the main chain. As a result, a broad range of biodegradable poly(phosphazene)s can be synthesized depending on their side chains. They undergo both surface erosion and bulk degradation depending on the type and ratio of side chains, and their hydrophobicity and also pH and temperature of the environment. Decomposition of poly(phosphazene)s produces non-toxic molecules like phosphate, ammonia, and the corresponding side groups.[39,51,52]

Transdermal Drug Delivery

Skin is the largest tissue of our body, which acts as a strong barrier that separates the inner organs of our body from the outside. Transdermal drug delivery as a non-invasive technique is an attractive issue in administration of drugs.

Skin consists of two layers, dermis and epidermis. The dermis is the bulk of our skin and epidermis is the top layer of the skin with several layers; the outer layer of epidermis is stratum corneum (SC) that is the main barrier against any foreign material. The rate of penetration of drug through this layer controls the transdermal drug delivery.[53] The simplest form of transdermal drug delivery is a system in which the concentration difference causes the drug to pass through the skin. In all of these patches, a biocompatible polymeric permeable membrane in direct contact to the skin is used. Available transdermal drug delivery systems are limited to several drugs such as clonidine, fentanyl, oxybutynin, nitroglycerin, scopolamine, testosterone, nicotine, estradiol, norelgestromin/ethinyl estradiol, and estradiol/norethindrone acetate. The reason for limited commercial transdermal drug delivery systems is the impermeability of the SC layer. However, to increase the range of these available drugs, several approaches have been used to assist drug passage through the SC layer. These techniques can be divided into two general approaches, the first one is methods used for modification of the SC layer including using chemical penetration enhancers and removing the SC layer by ablation or microneedles. The second way is using methods to force drug to diffuse through the skin such as using electrical current (iontophoresis), ultrasound radiation, or magnetic fields.[54] Regardless of the method used, the following specifications are necessary for every chemical or physical approach:

- It should have no irritating effect on the skin.
- It should be non-toxic.
- The enhancement effect should be completely reversible.
- This enhancement should be unidirectional into the body.
- It should work rapidly.
- It should have no effect on the nature of the drug.

Although no material with all of the above characteristics has been introduced yet, extensive studies in this field are ongoing. In this entry, two approaches of using chemical penetration enhancers and iontophoresis will be explained in more detail.

Table 1 Synthetic biodegradable polymers used in drug delivery

Polymer	Functional group		Structure
Poly(ester)s	$-\overset{O}{\overset{\|}{C}}-O-$	PLA	$+HC-\overset{O}{\overset{\|}{C}}-O+_n$, CH$_3$
		PGA	$+H_2C-\overset{O}{\overset{\|}{C}}-O+_n$
		PLGA	$+H_2C-\overset{O}{\overset{\|}{C}}-O-\overset{H}{\underset{CH_3}{C}}-\overset{O}{\overset{\|}{C}}-O+_n$
		PCL	$+(H_2C)_5\overset{O}{\overset{\|}{C}}-O+_n$
Poly(orthoester)s		I	$+O\diagup{}O-R+_n$
		II	$[\;O\diagup O\diagdown O\diagup O\diagdown O-R\;]_n$
		III	$[\;O\diagup O\diagdown O (CH_2)_4\;]_n$
		IV	$[\;O\diagup O\diagdown O\diagup O\diagdown O-(\overset{H_3C}{\underset{H}{C}}-\overset{O}{\overset{\|}{C}}-O)_{n:\,1\text{-}6}-R'-O\diagup O\diagdown O\diagup O\diagdown O-R\;]_n$
Poly(anhydride)s	$-\overset{O}{\overset{\|}{C}}-O-\overset{O}{\overset{\|}{C}}-$	Aliphatic	$[\;O-\overset{O}{\overset{\|}{C}}-(CH_2)_8-\overset{O}{\overset{\|}{C}}\;]_n$
		Aromatic	aromatic anhydride structure, $_n$
		Aliphatic–aromatic copolymers	$[\;O-\overset{O}{\overset{\|}{C}}-C_6H_4-O-(CH_2)_{n:\,3\text{ or }6}-O-C_6H_4-\overset{O}{\overset{\|}{C}}\;]_n$
Poly (phosphazene)s	$-P=N-$		$+\overset{R}{\underset{R}{P}}=N+_n$

Penetration enhancers

Penetration enhancers are chemical substances that decrease the barrier properties of the SC layer temporarily. A diverse range of chemicals that can achieve this aim are water, sulfoxides [especially dimethyl sulfoxide (DMSO)], azone, pyrrolidones, fatty acids, alcohols, surfactants, urea, essential oils, terpenes, etc.

Although the mechanism of enhancement in these chemicals is complex and unclear yet, it was accepted that most of them react with intracellular lipid structure of the SC layer and open a path for drug penetration. It has been shown that hydration with water may increase the distances between cells in the horny layer and open its compact structure, resulting in enhancement of the drug penetration. Moreover, water can increase the permeability of both hydrophilic and hydrophobic drugs.[55] DMSO is another widely studied substance that can enhance the flux of both hydrophilic and hydrophobic drugs. Some important drawbacks of this enhancer include the need for high concentration for effective enhancement (usually more than 60%) and also bad odor in the breath because of production of dimethyl sulfide in the body. Dimethylacetamide (DMAc) and dimethylformamide (DMF) are two other sulfoxides studied as penetration enhancers. Azone firstly was designed exclusively as a penetration enhancer. The mechanism of the azone enhancement is interaction with lipid domains of the SC layer. Pyrrolidones are another family of enhancers that are more effective for hydrophilic drugs, but the clinical use of these substances is impossible because of some observed adverse effects. Most of the studies on fatty acid enhancers have been focused on the oleic acid. However, research showed that the length of the alkyl chain in fatty acids affects the amount of enhancement. On the other hand, some of the chemical substances used in production of transdermal drug delivery patches can also act as an enhancer. For example, among different alcohols, ethanol which is the common solvent for many drugs received special attention as a penetration enhancer. In addition, surfactants used for dissolution of the lipophilic active agents can act as the penetration enhancers and also terpenes used as a fragrance in transdermal patches can enhance the penetration.[56–59]

Iontophoresis

Iontophoresis is an invasive technique to enhance drug penetration into the body. In this approach, a small electric current (usually <500 μA/cm) is applied to force a charged drug to pass across the skin. An iontophoretic drug delivery patch consists of two electrodes and a power source. The charged drug is placed in an electrode with the same charge and the other electrode is a reference electrode that completes the circuit. Both the electrodes are placed on the skin and connected to the power source. After applying the potential, the charged drug enters into the skin through the permeable polymeric membrane. The presence of an electrolyte is essential for the electrochemical functioning of the device.[60,61] These small species (Na^+ and Cl^-) can compete with drug species and due to their smaller size, they are highly efficient charge carriers, therefore, incorporation of them must be carefully adjusted (Fig. 10). Generally, positively charged drugs are favored because at physiological pH of the body, the skin has negative net charge and penetration of positively charged species is preferable. Fewer side effects, ease of application, and

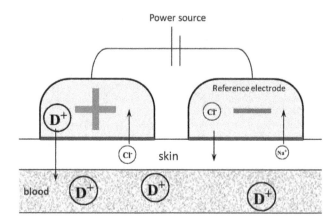

Fig. 10   A schematic representation of an iontophoretic device.

appropriate enhancement of the drug penetration are some of the important characteristics of this technique.

In this technique, the applied charge helps the drug penetration via two mechanisms: first, the electric repulsion that drives charged molecules into the body and second, increase in the permeation of the skin resulted from the flow of the electric current. The density of electric current determines the number of ions transported across the skin, and consequently, the drug flux in iontophoretic devices.[62–64]

Today, iontophoretic devices for lidocaine, epinephrine, pilocarpine, and some other drugs are available in the market. The limited range of drugs is related to some drawbacks of this technique including its need for a bulky and rather heavy power supply and also the possibility of damage to the hair follicles.[55]

CONCLUSIONS

Biomedical polymeric membranes comprise more than half of the industrial polymeric membrane market. They are applied in a large number of medical devices including membrane oxygenators, dialyzers, various drug delivery devices, artificial pancreas, and also artificial liver. Another application of polymeric membranes that is of great interest in recent years is the creation of artificial membranes that mimic cell membranes and insertion of membrane proteins in them. Both natural and synthetic polymers with different structures and also different modifications are used in each of these applications. Despite the extensive efforts of investigators for finding better candidates as substitutes to real organs of the body, most of the polymeric membranes used in these systems have been selected empirically. However, more correlation and collaboration between physicians, polymer engineers, and chemists can lead to the designing of polymers with desired properties and also faster development of this field and consequently reduction of pain and suffering of patients all around the world.

REFERENCES

1. Onuki, Y.; Bhardwaj, U.; Papadimitrakopoulos, F.; Burgess, D.J. A review of the biocompatibility of implantable devices: current challenges to overcome foreign body response. J. Diabetes Sci. Technol. 2008, 2 (6), 1003–1015.

2. Deppisch, R.; Storr, M.; Buck, R.; Gohl, H. Blood material interactions at the surfaces of membranes in medical applications. Sep. Purif. Technol. 1998, 14 (1–3), 241–254.

3. Seifert, B.; Mihanetzis, G.; Groth, T.; Albrecht, W.; Richau, K.; Missirlis, Y.; Paul, D.; Von Sengbusch, G. Polyetherimide: A new membrane-forming polymer for biomedical applications. Artif. Organs 2008, 26 (2), 189–199.

4. Lamba, N.M.K.; Woodhouse, K.A.; Cooper, S.L. Host responses to polyurethanes. In Polyurethanes in Biomedical Applications; CRC Press LLC: Boca Raton, 1997; 147–171.

5. Elbert, D.L.; Hubbell, J.A. Surface treatments of polymers for biocompatibility. Annu. Rev. Mater. Sci. 1996, 26, 365–394.

6. Kawakami, H. Polymeric membrane materials for artificial organs. J. Artif. Organs 2008, 11 (4), 177–181.

7. Wang, Y.X.; Robertson, J.L.; Spillman, W.B.; Claus, R.O. Effects of the chemical structure and the surface properties of polymeric biomaterials on their biocompatibility. Pharm. Res. 2004, 21 (8), 1362–1373.

8. Stamatialis, D.F.; Papenburg, B.J.; Girons, M.; Saiful, S.; Bettahalli, S.N.M.; Schmitmeier, S.; Wessling, M. Medical applications of membranes: drug delivery, artificial organs and tissue engineering. J. Membr. Sci. 2008, 308 (1–2), 1–34.

9. Williams, D.F. On the mechanisms of biocompatibility. Biomaterials 2008, 29 (20), 2941–2953.

10. Klee, D.; Hocker, H. Polymers for biomedical applications: improvement of the interface compatibility. In Biomedical Applications Polymer Blends; Eastmond, G.C., Höcker, H., Klee, D., Eds.; Springer Berlin Heidelberg: New Jersey, 1999; Vol. 149, 1–57.

11. Andrade, J.; Hlady, V. Protein adsorption and materials biocompatibility: a tutorial review and suggested hypotheses. Biopolymers/Non-Exclusion HPLC. Springer Berlin Heidelberg, 1986, 79, 1–63.

12. Chanard, J.; Lavaud, S.; Randoux, C.; Rieu, P. New insights in dialysis membrane biocompatibility: relevance of adsorption properties and heparin binding. Nephrol. Dial. Transplant. 2003, 18 (2), 252–257.

13. Lin, W.C.; Liu, T.Y.; Yang, M.C. Hemocompatibility of polyacrylonitrile dialysis membrane immobilized with chitosan and heparin conjugate. Biomaterials 2004, 25 (10), 1947–1957.

14. Lavaud, S.; Paris, B.; Maheut, H.; Randoux, C.; Renaux, J.L.; Rieu, P.; Chanard, J. Assessment of the heparin-binding AN69 ST hemodialysis membrane: II. Clinical studies without heparin administration. ASAIO J. 2005, 51 (4), 348–351.

15. Keogh, J.R.; Eaton, J.W. Albumin binding surfaces for biomaterials. J. Lab. Clin. Med. 1994, 124 (4), 537–45.

16. Keogh, J.R.; Velander, F.F.; Eaton, J.W. Albumin-binding surfaces for implantable devices. J. Biomed. Mater. Res. 2004, 26 (4), 441–456.

17. Pitt, W.; Cooper, S. Albumin adsorption on alkyl chain derivatized polyurethanes: I. the effect of C-18 alkylation. J. Biomed. Mater. Res. 2004, 22 (5), 359–382.

18. Pitt, W.; Grasel, T.; Cooper, S. Albumin adsorption on alksyi chain derivatized polyurettianes: II. The effect of alkyl chain length. Biomaterials 1988, 9 (1), 36–46.

19. Baker, R.W. Medical application of membranes. In Membrane Technology and Applications, 3rd Ed.; John Wiley & Sons: Chichester, 2004; 465–489.

20. Cheung, A.K. Biocompatibility of hemodialysis membranes. J. Am. Soc. Nephrol. 1990, 1 (2), 150–61.

21. Clark, W.R.; Hamburger, R.J.; Lysaght, M.J. Effect of membrane composition and structure on solute removal and biocompatibility in hemodialysis. Kidney Int. 1999, 56 (6), 2005–2015.

22. Idris, A.; Yet, L.K. The effect of different molecular weight PEG additives on cellulose acetate asymmetric dialysis membrane performance. J. Membr. Sci. 2006, 280 (1–2), 920–927.

23. Hakim, R.M. Clinical implications of hemodialysis membrane biocompatibility. Kidney Int. 1993, 44 (3), 484–494.

24. Rogiers, P.; Zhang, H.; Pauwels, D.; Vincent, J.L. Comparison of polyacrylonitrile (AN69) and polysulphone membrane during hemofiltration in canine endotoxic shock. Crit. Care Med. 2003, 31 (4), 1219–1225.

25. Liu, T.Y.; Lin, W.C.; Huang, L.Y.; Chen, S.Y.; Yang, M.C. Hemocompatibility and anaphylatoxin formation of protein-immobilizing polyacrylonitrile hemodialysis membrane. Biomaterials 2005, 26 (12), 1437–1444.

26. Humes, H.D.; Fissell, W.H.; Tiranathanagul, K. The future of hemodialysis membranes. Kidney Int. 2006, 69 (7), 1115–1119.

27. Hayama, M.; Yamamoto, K.; Kohori, F.; Sakai, K. How polysulfone dialysis membranes containing polyvinylpyrrolidone achieve excellent biocompatibility?. J. Membr. Sci. 2004, 234 (1–2), 41–49.

28. Krause, B.; Storr, M.; Ertl, T.; Buck, R.; Hildwein, H.; Deppisch, R.; Gohl, H. Polymeric membranes for medical applications. Chemie Ingenieur Technik 2003, 75 (11), 1725–1732.

29. Galletti, P.M.; Colton, C.K. Artificial lungs and blood-gas exchange devices. In The Biomedical Engineering Handbook; Bronzino, J.D., Eds.; CRC Press LLC: Boca Raton, 2000, chapter 129.

30. Zwischenberger, B.A.; Clemson, L.A.; Zwischenberger, J.B. Artificial lung: progress and prototypes. Expert Rev. Med. Devices 2006, 3 (4), 485–497.

31. Noora, J.; Lamy, A.; Smith, K.M.; Kent, R.; Batt, D.; Fedoryshyn, J.; Wang, X. The effect of oxygenator membranes on blood: a comparison of two oxygenators in open-heart surgery. Perfusion 2003, 18 (5), 313–320.

32. Oto, T.; Rosenfeldt, F.; Rowland, M.; Pick, A.; Rabinov, M.; Preovolos, A.; Snell, G.; Williams, T.; Esmore, D. Extracorporeal membrane oxygenation after lung transplantation: evolving technique improves outcomes. Ann. Thorac. Surg. 2004, 78 (4), 1230–1235.

33. Gaylor, J.D.S.; Hickey, S.; Bell, G.; Pei, J.M. Membrane oxygenators: influence of design on performance. Perfusion 1994, 9 (3), 173–180.

34. Wickramasinghe, S.R.; Han, B. Designing microporous hollow fibre blood oxygenators. Chem. Eng. Res. Des. 2005, 83 (3), 256–267.

35. Formica, F.; Avalli, L.; Martino, A.; Maggioni, E.; Muratore, M.; Ferro, O.; Pesenti, A.; Paolini, G. Extracorporeal membrane oxygenation with a poly-methylpentene

oxygenator (Quadrox D). The experience of a single Italian centre in adult patients with refractory cardiogenic shock. ASAIO J. **2008**, *54* (1), 89–94.

36. Khoshbin, E.; Roberts, N.; Harvey, C.; Machin, D.; Killer, H.; Peek, G.J.; Sosnowski, A.W.; Firmin, R.K. Poly-methyl pentene oxygenators have improved gas exchange capability and reduced transfusion requirements in adult extracorporeal membrane oxygenation. ASAIO J. **2005**, *51* (3), 281–287.

37. Langer, R. Invited review polymeric delivery systems for controlled drug release. Chem. Eng. Commun. **1980**, *6* (1–3), 1–48.

38. Herrlich, S.; Spieth, S.; Messner, S.; Zengerle, R. Osmotic micropumps for drug delivery. Adv. Drug Deliv. Rev. **2012**, *64* (14), 1617–1627.

39. Park, J.; Ye, M.; Park, K. Biodegradable polymers for microencapsulation of drugs. Molecules **2005**, *10* (1), 146–161.

40. Panyam, J.; Labhasetwar, V. Biodegradable nanoparticles for drug and gene delivery to cells and tissue. Adv. Drug Deliv. Rev. **2012**, *64* (Suppl.), 61–71.

41. Ikada, Y.; Tsuji, H. Biodegradable polyesters for medical and ecological applications. Macromol. Rapid Commun. **2000**, *21* (3), 117–132.

42. Pitt, C.G.; Gratzl, M.M.; Jeffcoat, A.R.; Zweidinger, R.; Schindler, A. Sustained drug delivery systems II: Factors affecting release rates from poly (ε-caprolactone) and related biodegradable polyesters. J. Pharm. Sci. **1979**, *68* (12), 1534–1538.

43. Arshady, R. Preparation of biodegradable microspheres and microcapsules: 2. Polyactides and related polyesters. J. Control. Release **1991**, *17* (1), 1–21.

44. Amass, W.; Amass, A.; Tighe, B. A review of biodegradable polymers: uses, current developments in the synthesis and characterization of biodegradable polyesters, blends of biodegradable polymers and recent advances in biodegradation studies. Polym. Int. **1999**, *47* (2), 89–144.

45. Heller, J.; Barr, J. Poly (ortho esters) from concept to reality. Biomacromolecules **2004**, *5* (5), 1625–1632.

46. Heller, J.; Barr, J.; Ng, S.Y.; Abdellauoi, K.S.; Gurny, R. Poly (ortho esters): synthesis, characterization, properties and uses. Adv. Drug Deliv. Rev. **2002**, *54* (7), 1015–1039.

47. Heller, J. Poly (ortho esters). In *Handbook of Biodegradable Polymers: Isolation, Synthesis, Characterization and Applications*; Wiley-VCH Verlag: Weinheim, 2011; 77–105.

48. Kumar, N.; Langer, R.S.; Domb, A.J. Polyanhydrides: an overview. Adv. Drug Deliv. Rev. **2002**, *54* (7), 889–910.

49. Göpferich, A.; Teßmar, J. Polyanhydride degradation and erosion. Adv. Drug Deliv. Rev. **2002**, *54* (7), 911–931.

50. Leong, K.; Brott, B.; Langer, R. Bioerodible polyanhydrides as drug-carrier matrices. I: characterization, degradation, and release characteristics. J. Biomed. Mater. Res. **1985**, *19* (8), 941–955.

51. Lakshmi, S.; Katti, D.S.; Laurencin, C.T. Biodegradable polyphosphazenes for drug delivery applications. Adv. Drug Deliv. Rev. **2003**, *55* (4), 467–482.

52. Uhrich, K.E.; Cannizzaro, S.M.; Langer, R.S.; Shakesheff, K.M. Polymeric systems for controlled drug release. Chem. Rev. **1999**, *99* (11), 3181–3198.

53. Prow, T.W.; Grice, J.E.; Lin, L.L.; Faye, R.; Butler, M.; Becker, W.; Wurm, E.M.T.; Yoong, C.; Robertson, T.A.; Soyer, H.P. Nanoparticles and microparticles for skin drug delivery. Adv. Drug Deliv. Rev. **2011**, *63* (6), 470–491.

54. Prausnitz, M.R. Microneedles for transdermal drug delivery. Adv. Drug Deliv. Rev. **2004**, *56* (5), 581–587.

55. Barry, B.W. Novel mechanisms and devices to enable successful transdermal drug delivery. Eur. J. Pharm. Sci. **2001**, *14* (2), 101–114.

56. Williams, A.C.; Barry, B.W. Penetration enhancers. Advanced Drug Delivery Reviews **2012**, *64* (Suppl), 128–137.

57. Finnin, B.C.; Morgan, T.M. Transdermal penetration enhancers: applications, limitations, and potential. J. Pharm. Sci. **1999**, *88* (10), 955–958.

58. Karande, P.; Jain, A.; Ergun, K.; Kispersky, V.; Mitragotri, S. Design principles of chemical penetration enhancers for transdermal drug delivery. Proc. Natl. Acad. Sci. U.S.A. **2005**, *102* (13), 4688–4693.

59. Yamane, M.; Williams, A.; Barry, B. Effects of terpenes and oleic acid as skin penetration enhancers towards 5-fluorouracil as assessed with time; permeation, partitioning and differential scanning calorimetry. Int. J. Pharm. **1995**, *116* (2), 237–25.

60. Eljarrat-Binstock, E.; Domb, A.J. Iontophoresis: a noninvasive ocular drug delivery. J. Control. Release **2006**, *110* (3), 479–489.

61. Singh, P.; Maibach, H.I. Iontophoresis in drug delivery: basic principles and applications. Crit. Rev. Ther. Drug Carrier Syst. **1994**, *11* (2–3), 161–213.

62. Kalia, Y.N.; Naik, A.; Garrison, J.; Guy, R.H. Iontophoretic drug delivery. Adv. Drug Deliv. Rev. **2004**, *56* (5), 619–658.

63. Dixit, N.; Bali, V.; Baboota, S.; Ahuja, A.; Ali, J. Iontophoresis-An approach for controlled drug delivery: a review. Curr. Drug Deliv. **2007**, *4* (1), 1–10.

64. Singh, S.; Singh, J. Transdermal drug delivery by passive diffusion and iontophoresis: a review. Med. Res. Rev. **1993**, *13* (5), 569–621.

Metal–Polymer Composite Biomaterials

Takao Hanawa
Institute of Biomaterials and Bioengineering, Tokyo Medical and Dental University, Tokyo, Japan

Abstract
This entry explains metal–polymer composite materials for biomedical use and the corresponding research is reviewed. To understand metallic materials against polymeric materials, an outline of metals is first presented. Then, surface properties of metals are described regarding the reaction of metal surface with polymers that govern the manufacture of metal–polymer composites, followed by the main subject, metal–polymer composites.

INTRODUCTION

The fast technological evolution of polymers has made it possible to apply these materials to medical devices over the last three decades. In particular, because of their excellent biocompatibility and biofunctions, polymers are expected to show excellent properties for use as biomaterials; in fact, many devices made from metals have been replaced by others made from polymers. In spite of this fact, about 80% of implant devices are still made from metals, and this percentage remains unchanged because of their high strength, toughness, and durability. Medical devices consisting of metals cannot be replaced with polymers at present.

The use of metals as raw materials has a long history, and it can be said that present materials science and engineering have been based on research into metals. However, metals are sometimes thought of as "unfavorite materials" for biomedical implants because of memories of the environmental and human damages caused by heavy metals. Since an improvement in the safety of metals for medical use is vital, strenuous efforts have been made to improve corrosion resistance and mechanical durability. In addition, metals are typically artificial materials and have no biofunctions, which makes them fairly unattractive as biomaterials.

On the other hand, polymers are widely used as biomaterials because of their high degree of flexibility, biocompatibility, and technologic properties.[1] Also, it is relatively easy to design biofunctional polymers based on biomimic technique, because biofunctional polymers exist in the human body as parts of biomolecules, cells, tissues, and organs. Therefore, some polymeric materials are widely known as biocompatible and biofunctional materials.

If these advantages of metals and polymers are mixed and disadvantages are eliminated by manufacturing metal–polymer composite materials, then humankind will obtain ideal materials with excellent mechanical properties and biofunctions. Two metal–polymer composite materials are feasible to design: one of them is a combination of bulk polymeric materials and bulk metallic materials, the other is immobilization of polymers to metal surfaces. In this entry, these metal–polymer composite materials for biomedical use are explained and the corresponding researches are reviewed. To understand metallic materials against polymeric materials, an outline of metals is first presented. Then, surface properties of metals are described to know the reaction of metal surface with polymers that govern the manufacture of metal–polymer composites, followed by the main subject, metal–polymer composites.

PROPERTIES OF METALS AGAINST POLYMERS

Materials are categorized to metals, ceramics, and polymers (Fig. 1). Metallic materials are generally multicrystal bodies consisting of metal bonds. For example, metal oxide, metal salt, metal complex, etc. contain metal elements; however, these are compounds consisting of ionic bonds or covalent bonds; these properties are completely different from those of metals consisting of metal bonds. Therefore, in the field of materials engineering, ceramics and metals are clearly distinguished, although they are categorized together in inorganic compounds. Each material has its own advantages and disadvantages, and the applications are determined according to their properties.

Metals have been utilized for dental restoration and bone fixation since 2500 years ago; they have a long history as biomaterials. Advantages of metals as biomaterials are listed as follows. These properties are caused by metal bonds:

1. Great strength
2. Great ductility, easy working
3. Fracture toughness
4. Elasticity and stiffness
5. Electroconductivity

Concise Encyclopedia of Biomedical Polymers and Polymeric Biomaterials DOI: 10.1081/E-EBPPC-120052255

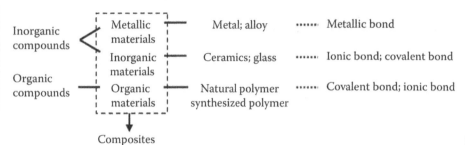

Fig. 1 Category of materials.

Metals and alloys are widely used as biomedical materials and are indispensable in the medical field. Advantages of metals compared with ceramics and polymers are great strength and difficulty to fracture. In particular, toughness, elasticity, rigidity, and electrical conductivity are essential properties for metals used in medical devices. Figure 2 shows stress–strain curves of typical metals, ceramics, and ultra high molecular weight polyethylene (UHMWPE). The strength and elongation of metals against ceramics and polymers are clearly understood.

Conventionally, metals are essential for orthopedic implants: Bone fixators, artificial joints, external fixators, etc. can substitute for the mechanical function of hard tissues in orthopedics. Stents and stent grafts are placed at angusty in blood vessels for dilatation. Therefore, elasticity or plasticity for expansion and rigidity for maintaining dilatation are required for the devices. In dentistry, metals are used for restorations, orthodontic wires, and dental implants.

For mechanical reliability, metals must be used and cannot be replaced with ceramics or polymers. In noble metals and alloys, gold (Au) marker for the imaging of stent, platinum (Pt) for embolization wire, and Au alloys and silver (Ag) alloys for dental restoratives are utilized. In base metals and alloys, austenitic stainless steels, cobalt–chromium (Co-Cr) alloys, titanium (Ti), and Ti alloys, whose corrosion resistance is maintained with surface oxide film as passive film, are utilized for implant materials. On the other hand, wear resistance is required to decrease the generation of wear debris. Co-Cr-molybdenum (Mo) alloys have good wear resistance and are used for sliding parts of artificial joints. Comparison of various properties of metals used for implant is summarized in Table 1. The metals used for such devices are listed in Table 2. Safety to the human body is essential in biomaterials; no toxic material is used for biomaterials. Metals implanted in tissues do not show any toxicity without metal ion dissolution by corrosion and generation of wear debris by wear. Therefore, corrosion resistance is absolutely necessary for metals in biomedical use, resulting in the uses of noble or corrosion-resistant metals and alloys for medicine and dentistry.

A disadvantage of using metals as biomaterials is that they are typically artificial materials and have no biofunction. Therefore, additional properties are required for metals. Requests for metals for biomedical use are summarized in Table 3. To respond to these requests, new design of alloys and many techniques for surface modification of metals are

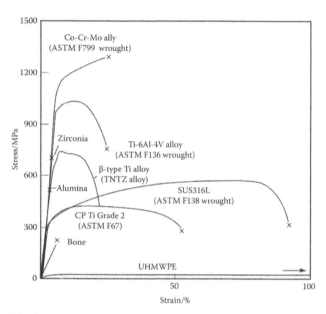

Fig. 2 Stress–strain curves of various materials.

attempted on a research stage and some of them are commercialized. Surface modification is necessary because biofunction cannot be added during manufacturing processes of metals such as melting, casting, forging, and heat treatment. Surface modification is a process that changes a material's surface composition, structure, and morphology, leaving the bulk mechanical properties intact. Reviews on surface modification of Ti have already been published on plasma spraying[2] and electrochemical treatments.[3]

The most important difference between a metal and polymer is cohesive and noncohesive bodies. In other words, the interface of those is a boundary between a rigid and condensed solid and a flexible and mobile molecule. Of course, metals mainly consist of metal bonds; polymers mainly consist of covalent bonds. Therefore, design and fabrication of metal–polymer composites are to decrease the mismatch of their properties at the interface. In addition, the mechanical and chemical properties of metallic materials are not identified, even though the compositions are determined. The phase of metallic materials is easily changed by working and heat treatment, and various factors influence the resultant structure. Factors governing the mechanical and chemical properties of metallic materials are summarized in Fig. 3.

Table 1 Comparison of properties of metals for implant

Materials		Mechanical property	Workability			Corrosion resistance
		Tensile strength	Wear resistance	Plasticity	Machinability	Pitting
Stainless steel	Type316L	Good	Excellent→Poor	Excellent	Excellent	Excellent→Poor
Co-Cr alloy	Cast	Good	Good→Good	Poor	Poor	Good
	Annealed	Excellent	Good→Good	Good	Good	Fair
Ti; Ti alloy	CP Ti	Excellent→Poor	Excellent→Poor	Fair	Fair	Excellent
	Ti-6Al-4V	Excellent	Excellent→Poor	Fair	Excellent→Poor	Excellent

Table 2 Metals used for medical devices

Clinical division	Medical device	Material
Orthopedic surgery	Spina fixation	316L stainless steel; Ti; Ti-6Al-4V; Ti-6Al-7Nb
	Bone fixation (bone plate, screw, wire, bone nail, mini-plate, etc.)	316L stainless steel; Co-Cr-W-Ni; Ti; Ti-6Al-4V; Ti-6Al-7Nb
	Artificial joint; bone head	316L stainless steel; Fe-Cr-Ni-Co; Co-Cr-Mo; Ti-6Al-4V; Ti-6Al-7Nb; Ti-15Mo-5Zr-3Al: Ti-6Al-2Nb-1Ta-0.8Mo
	Spina spacer	316L stainless steel; Ti-6Al-4; Ti-6Al-7Nb
Cardiovascular medicine and surgery	Implant-type artificial heart (housing)	Ti
	Pace maker (case)	Ti; Ti-6Al-4V
	(electric wire)	Ni-Co
	(electrode)	Ti; Pt-Ir
	(terminal)	Ti; 316L stainless steel; Pt
	Artificial valve (frame)	Ti-6Al-4V
	Stent	316L stainless steel; Co-Cr-Fe-Ni; Co-Ni-Cr-Mo; Co-Cr-Ni-W-Fe; Ti-Ni; Ta
	Guide wire	316L stainless steel; Ti-Ni; Co-Cr
	Embolization wire	Pt
	Clip	Ti-6Al-4V; 630 stainless steel; 631 stainless steel; Co-Cr-Ta-Ni; Co-Cr-Ni-Mo-Fe; Ti; Ti-6Al-4V
Otorhinology	Artificial inner ear (electrode)	Pt
	Artificial eardrum	316L stainless steel
Dentistry	Filling	Au foil; Ag-Sn(-Cu) amalgam
	Inlay, crown; bridge; clasp; denture base	Au-Cu-Ag; Au-Cu-Ag-Pt-Pd; Ag-Pd-Cu-Au; Ag-(Sn-In-Zn); Co-Cr-Mo; Co-Cr-Ni; Co-Cr-Ni-Cu; Ti; Ti-6Al-7Nb; 304 stainless steel; 316L stainless steel
	Thermosetting resin facing crown; porcelain-fused-to-metal	Au-Pt-Pd; Ni-Cr
	Solder	Au-Cu-Ag; Au-Pt-Pd; Au-Cu; Ag-Pd-Cu-Zn
	Dental implant	Ti; Ti-6Al-4V; Ti-6Al-7Nb; Au
	Orthodontic wire	316L stainless steel; Co-Cr-Fe-Ni; Ti-Ni; Ti-Mo
	Magnetic attachment	Sm-Co; Nd-Fe-B; Pt-Fe-Nb; 444 stainless steel; 447J1 stainless steel; 316L stainless steel
	Treatment device (bar, scaler, periodontal probe, dental tweezers, raspatory, etc.)	304 stainless steel
General surgery	Needle of syringe	304 stainless steel
	Scalpel	304 stainless steel; 316L stainless steel; Ti
	Catheter	304 stainless steel; 316L stainless steel; Co-Cr; Ti-Ni; Au; Pt-In
	Staple	630 stainless steel

Table 3   Requests of metals for medical devices

Required property	Target medical devices	Effect
Elastic modulus	Bone fixation; spinal fixation	Prevention of bone absorption by stress shielding
Superelasticity shape memory effect	Multipurpose	Improvement of mechanical compatibility
Wear resistance	Artificial joint	Prevention of generation of wear debris; improvement of durability
Biodegradability	Stent; artificial bone; bone fixation	Elimination of materials after healing; unnecessity of retrieval
Bone formation bone bonding	Stem and cup of artificial hip joint; dental implant	Fixation of devices in bone
Prevention of bone formation	Bone screw; bone nail	Prevention of assimilation
Adhesion of soft tissue	Dental implant; trans skin device; external fixation; pacemaker housing	Fixation in soft tissue; prevention of inflectional disease
Inhibition of platelet adhesion	Devices contacting blood	Prevention of thrombus
Inhibition of biofilm formation	All implant devices; treatment tools and apparatus	Prevention of infectious disease
Low magnetic susceptibility	All implant devices; treatment tools and apparatus	No artifact in MRI

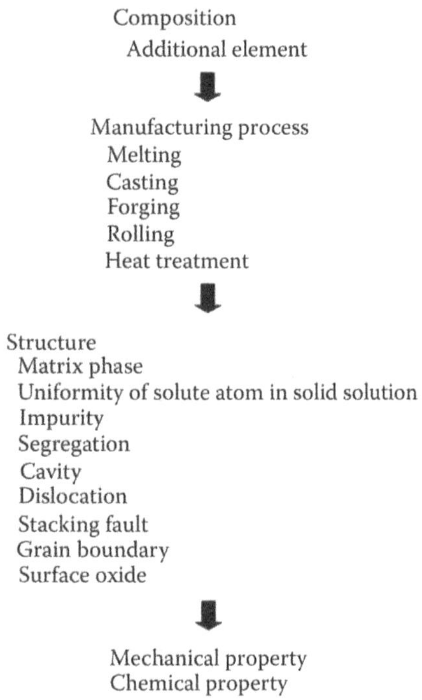

Composition
Additional element

Manufacturing process
Melting
Casting
Forging
Rolling
Heat treatment

Structure
Matrix phase
Uniformity of solute atom in solid solution
Impurity
Segregation
Cavity
Dislocation
Stacking fault
Grain boundary
Surface oxide

Mechanical property
Chemical property

Fig. 3   Factors governing the mechanical and chemical properties of metallic materials through the manufacturing process.

SURFACE OF METALLIC BIOMATERIALS

Importance of Metal Surface

When a metallic material is implanted into a human body, immediate reaction occurs between its surface and the living tissues. In other words, immediate reaction at this initial stage straightaway determines and defines a metallic material's tissue compatibility. Since conventional metallic biomaterials are usually covered with metal oxides, surface oxide films on metallic materials play an important role not only against corrosion but also in tissue compatibility. To fabricate metal–polymer composites, knowledge of the material's surface composition is absolutely necessary. Adhesion of polymers and immobilization of molecules to metals are governed by the surface property of the substrate metals. Immobilization of molecules is a kind of surface modification of metals, because surface modification is a process that improves surface property by changing the composition and structure, while leaving the mechanical properties of the material intact. Surface properties of a metallic material may be controlled with surface modification techniques.

Surface of Metals

Atoms in metallic materials located at the surface are considered partly reactive to the environment because atomic configuration terminates at the surface. The surface represents a property different from the inside of the material. Due to high surface energy, a single molecular layer forms easily on solid surface, where gas molecules are adsorbed at 1 Pa in 10^{-4} s. For example, in the presence of oxygen atoms, oxygen atoms and metal atoms chemically bond together to form an oxide layer. This phenomenon occurs even at the surface of Au—which is the most noble metal. Unlike polymers and ceramics, enrichment of component elements occurs easily at metal surfaces. This means that the surface composition of a metal is different from its inside composition in the order of nanometers. Therefore, the variant surface composition of a metallic material contributes significantly to defining the overall properties of the material.

A metal surface is usually covered with a surface oxide film. The surface oxide layer, on the other hand, is always covered with surface hydroxyl groups that are adsorbed by water, as shown in Fig. 4. In particular, the surface oxide film and surface active hydroxyl groups are important to understand the surface reaction of metals.

Surface Oxide Film

Except in reduction environments, the corrosion process always causes a reaction film to form on metallic materials. Passive film is one such reaction film, and it is particularly significant for corrosion protection. When solubility is extremely low and pores are absent, adhesion of film—which is formed in an aqueous solution—to the substrate will be strong. The film then becomes a corrosion-resistant or passive film. Passive film is about 1–5 nm thick and is transparent. Due to the tremendously fast rate at which it is formed, passive film readily becomes amorphous. For example, film on a Ti metal substrate was generated in 30 ms. This was estimated from the time transient of current of the Ti at 1 V versus a saturated calomel electrode (SCE) after exposing the metal surface, as shown in Fig. 5. Since amorphous films hardly contain grain boundary and structural defects, they are corrosion resistant. However, corrosion resistance decreases with crystallization.

Fortunately, passive films contain water molecules that promote and maintain amorphousness.

Metallic materials such as stainless steels, Co-Cr alloys, commercially pure Ti, and Ti alloys used for biomedical devices are covered by their characteristic passive films. These films self-repair when they are disrupted by some causes. Noble metals and alloys such as dental alloys are also covered with an oxide layer. While the oxide layer protects against corrosion, it is not chemically strong like the passive film.

Titanium

When Ti is polished in deionized water and analyzed using X-ray photoelectron spectroscopy (XPS), the Ti 2p spectrum obtained from the Ti gives four doublets according to valence: Ti^0, Ti^{2+}, Ti^{3+}, and Ti^{4+}. Published data[4] are used to determine the binding energy of each valence. Figure 6 shows an example of the decomposition of Ti 2p spectrum. A distinct Ti^0 peak at metallic state is observed, which accounts for a very thin surface oxide film at less than a few nanometers only. Besides, Ti^{4+} (TiO_2), Ti^{3+} (Ti_2O_3), and Ti^{2+} (TiO) are detected. Though Ti^{2+} oxide exists in the surface oxide film, Ti^{2+} formation is always thermodynamically less favorable than Ti^{3+} formation at the surface. On the other hand, Fig. 7 shows O 1s spectrum obtained from polished Ti, which is decomposed to three peaks originating from oxide (O^{2-}), hydroxide or hydroxyl group (OH^-), and water (H_2O). The surface oxide film on Ti contains these chemical states. Therefore, the surface oxide film on Ti consists mainly of nonstoichio-metric TiO_2 containing the hydroxyl group and water.

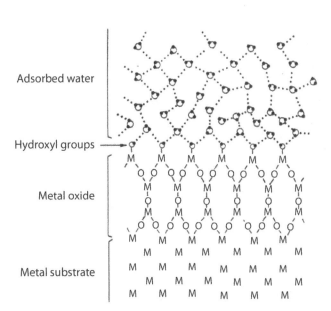

Fig. 4 Schematic model of the structure of surface layer consisting of oxide layer, hydroxyl group layer, and adsorbed water layer on metals and alloys.

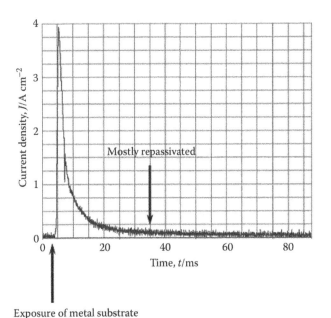

Fig. 5 Time transient of anodic current of titanium in Hanks' solution under the 1 V charge versus SCE. Anodic current is generated with the dissolution and repassivation of titanium.

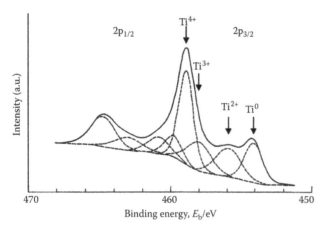

Fig. 6 Decomposition of Ti 2p XPS spectrum obtained from titanium abraded and immersed for 300 s in water into eight peaks (2p3/2 and 2p1/2 electron peaks in four valences). Numbers with arrows are valence numbers.

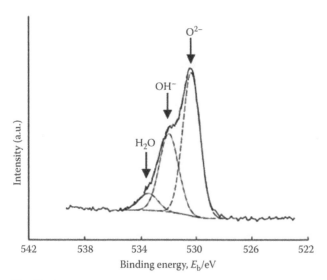

Fig. 7 Typical O 1s spectrum obtained from polished Ti and its deconvolution into O^{2-}, OH^-, and H_2O components.

Since a considerable portion of oxidized titanium stays as Ti^{2+} and Ti^{3+} in the surface film, the oxidation process may proceed to the end just at the uppermost part of the surface film. As shown in Fig. 8, the proportion of Ti^{4+} among titanium cations (Ti^{2+}, Ti^{3+}, and Ti^{4+}) in the film decreases with an increase in photoelectron take-off angle,[5] indicating that more Ti^{4+} exists near the outer layer in the film. Deducing from the take-off angle dependence of $[OH^-]/[O^{2-}]$ in Fig. 8, oxygen atoms in the hydroxyl group are mainly located in the outer part of the surface film. This means that dehydration proceeds inside the surface film and only partly for Ti^{4+} oxide.

Thickness of the film is about 2 nm just after polishing and about 5 nm at 1 week after polishing (Fig. 9). Note too that the thickness of the surface oxide film increases according to the logarithmic rule, which is common in the initial growth of oxide films of metallic materials.

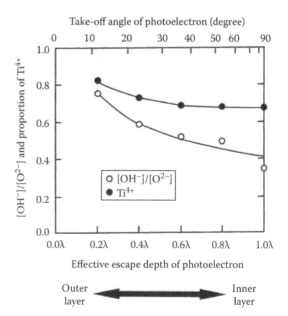

Fig. 8 The ratio of the proportion of the concentration of OH^- to that of O^{2-}, $[OH^-]/[O^{2-}]$, and proportion of cationic fraction of Ti^{4+} among titanium species, in surface oxide film on Ti polished in water plotted against the average effective escape depth of photoelectrons for angle-resolved XPS measurements. Lambda (λ) is the average escape depth of O 1s and Ti $2p_{3/2}$ photoelectrons, and the effective escape depth is the escape depth times sin (take-off-angle). The values at small take-off angle of photoelectron or effective escape depth of photoelectron represent outer region information, and the larger ones represent inner region information.

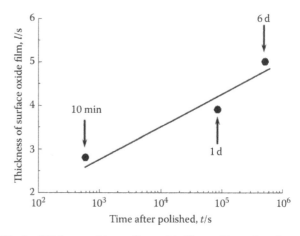

Fig. 9 Thickness of the surface oxide film on Ti as a function of time after polishing.

Titanium alloy

The film on Ti-6 aluminum (Al)-4 vanadium (V) alloy is almost the same as that on Ti containing a small amount of aluminum oxide.[6,7] In other words, the surface oxide film on Ti-6Al-4V is a TiO_2 containing small amounts of Al_2O_3, hydroxyl groups, and bound water. V contained in the alloy is not detected in the oxide film after the alloy is polished.

The Ti-56 nickel (Ni) shape-memory alloy is covered by TiO_2-based oxide, with minimal amounts of Ni in both the oxide and metallic states.[6,7] The film on Ti-56Ni is a TiO_2 containing $Ni(OH)_2$, hydroxyl groups, and bound water. In Ti-zirconium (Zr) alloy, the surface oxide film consists of titanium and zirconium oxides.[8] The relative concentration ratio of Ti to Zr in the film is almost the same as that in the alloy. The thickness of the film increased with increase in Zr content. The chemical state of Zr is more stable than that of Ti in the film.

Stainless steel

Compositions of surface oxide films on stainless steels are well understood in the field of engineering. In an austenitic stainless steel, the surface oxide film consists of iron (Fe) and chromium containing a small amount of molybdenum. However, it does not contain nickel while in the air and in chloride solutions.[9,10]

On the other hand, surface oxide film on 316L steel polished mechanically in deionized water consists of oxide species of Fe, Cr, Ni, Mo, and manganese (Mn), and its thickness is about 3.6 nm.[11] The surface film contains a large amount of OH^-—the oxide which is hydrated or oxyhydroxidized. The surface oxide film is also enriched with Fe, while the alloy substrate just under the film is enriched with Ni, Mo, and Mn.

Cobalt–chromium alloy

Surface oxide film of a Co-Cr-Mo alloy is characterized as containing oxides of cobalt and chromium without molybdenum.[12] On the other hand, the surface oxide film on another Co-Cr-Mo alloy polished mechanically in deionized water consists of oxide species of cobalt, chromium, and molybdenum, and its thickness is about 2.5 nm.[13] This surface film contains a large amount of OH^-—the oxide which is hydrated or oxyhydroxidized. There are also more traces of Cr and Mo distributed at the inner layer of the film. In the surface oxide on Co-Cr-Ni Mo alloy for metallic stent known as MP35N, Cr is enriched and Co and Ni are depleted.[14]

Dental precious alloys

Au-copper (Cu)-Ag alloys and Ag-palladium (Pd)-Cu-Au alloys for dental restoration are covered by copper oxide and silver oxide.[15] An Ag-indium (In) alloy is covered by zinc oxide and indium oxide, and an Ag-Sn-zinc (Zn) alloy is covered by tin oxide and zinc oxide. While these oxides serve as a protection film against corrosion, they are not as chemically strong as the passive film.

Surface Active Hydroxyl Group

Formation

The surface of oxide reacts with moisture in air, and hydroxyl groups are rapidly formed. In the case of Ti, the surface oxide immediately reacts not only with water molecules in aqueous solutions but also with moisture in air and is covered by hydroxyl groups.[16,17] The surface oxide is always formed on conventional metallic biomaterials, and the surface of the surface oxide is active because of the same reason described earlier. Therefore, the oxide surface immediately reacts with water molecules and hydroxyl groups are formed, as shown in Figs. 10A. The surface

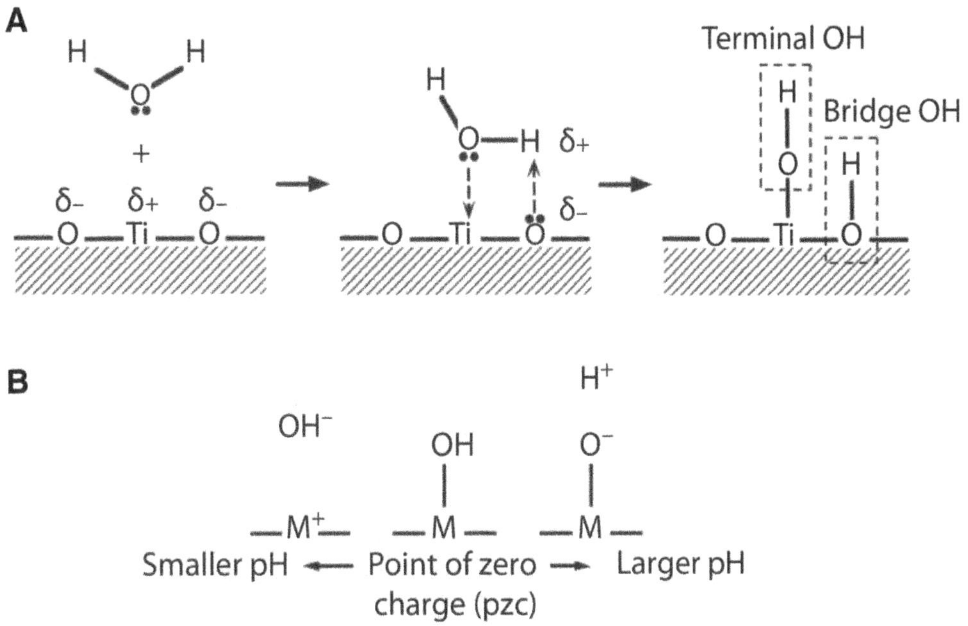

Fig. 10 Formation process of hydroxyl group on titanium oxide (A) and dissociation of the hydroxyl group in aqueous solution and pzc (B).

hydroxyl groups contain both terminal OH and bridge OH in equal amounts. Concentration of hydroxyl groups on the unit area of the surface is determined with various techniques.

Acidic and basic properties

Active surface hydroxyl groups dissociate in aqueous solutions and form electric charges, as shown in Figs. 10B.[16–19] Positive or negative charge due to the dissociation is governed by the pH of the surrounding aqueous solution: Positive and negative charges are balanced and apparent charge is zero at a certain pH. This pH is the point of zero charge (pzc). The pzc is the unique value for an oxide and an indicator that the oxide shows acidic or basic property. For example, in the case of TiO_2, the pzc of rutile is 5.3 and that of anatase is 6.2.[16] In other words, anatase surface is acidic at smaller pH and basic at larger pH than 6.2. Active surface hydroxyl groups and electric charges formed by the dissociation of the groups play important roles for the bonding with polymers and immobilization of molecules. Therefore, the concentration of surface hydroxyl group and pH are important factors for the bonding with polymeric materials and immobilization of molecules.

ADHESION OF POLYMERS TO METALS IN DENTISTRY

Adhesive Reagents

Metal restorations and prostheses such as inlay, crown, and bridge must be retained in a fixed position in a mouth. For this purpose, dental cements are used: glass ionomer cements, zinc phosphate cements, polyacrylate cements, resin-based cements, etc. In particular, resin-based cements generate chemical adhesion between tooth and metal. Cements based on resin composites are now used for cementation of crown and bridges and direct bonding of orthodontic brackets to enamel. Polymer-based filling, restorative, and luting materials are specified in ISO 4049.

Resin cements based on methyl methacrylate (MMA) have been available since 1952 for the use of cementation of inlays, crowns, etc. The resin composite for crown and resin was invented in the early 1970s. Cementation of alloy restorations is performed with self-cured composite cements. Two-paste systems are adopted for self-cured composite cements. One paste mainly consists of a diacrylate oligomer diluted with lower molecular weight dimethacrylate monomers. The other consists of silanated silica or glass. Peroxide amine is used for the initiator accelerator.

Although many efforts have been made to use industrial adhesives such as epoxy resins and cyanoacrylate as dental adhesives, adequate adhesion is not achieved under the extreme conditions of the oral cavity. Much effort has been

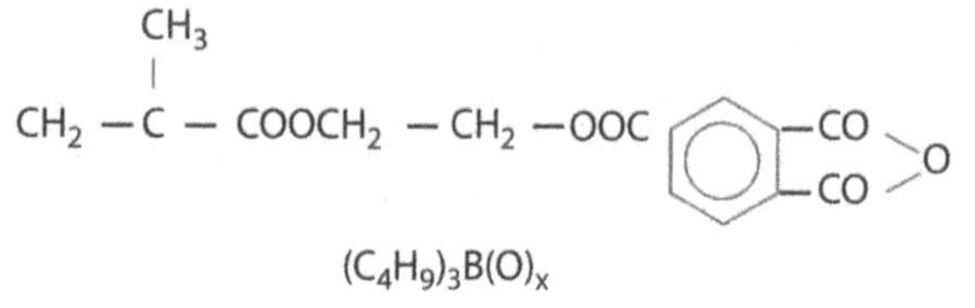

Fig. 11 4-Methacryloxyethyl trimellitate anhydride (4-META) (top) and TBB (bottom).

expended to achieve adhesion between artificial compounds and tooth substances. In 1978, Takeyama et al.[20] synthesized a new dental adhesive that meets the requirements stated here. The dental adhesives contained 4-methacryloxyethyl-trimellitic anhydride (4-META) as an adhesive monomer with MMA. In 1983, Omura et al.[21] synthe-sized an adhesive containing a dimethacrylate monomer bis-glycidyldimethacrylate (Bis-GMA) and phosphate monomers. 4-META is an ethyl anhydride of trimellitic acid and hydroxyethyl methacrylate and has both hydrophilic and hydrophobic groups, as shown in Fig. 11. The 4-META cement is formulated with MMA monomer and acrylic resin filler and is catalyzed by tributylborane (TBB). Another adhesive resin cement is phosphonate cement supplied with a two-paste system, containing Bis-GMA resin and silanated quartz filler. Phosphonate molecule is very sensitive to oxygen, so a gel is provided to coat the margins of a restoration until setting has occurred.

One of the main problems in dental adhesion techniques is that bonds between adhesives and metals are strong in a dry environment and weaken in a wet environment. For example, there are clinical reports of dental adhesion bridges breaking off from teeth after extended use, and this has led to a loss of confidence in dental adhesion techniques. Hence, a major problem awaiting solution in dentistry is the durability of bonding joints exposed to water. Research on dental adhesives is well reviewed elsewhere.[22]

Bonding to Base Metal Alloys

The phosphate end of the phosphonate reacts with calcium of the tooth or with a metal oxide. The double-bonded ends of both 4-META and phosphonate cements react with other double bonds when available. Setting of resin cements results from self- or light-cured polymerization of carbon–carbon double bonds.

There are numerous studies that have been made on the adhesion to dental alloys; pioneering work by Tanaka on Ni-Cr alloys[23] and gold alloys;[24] studies on Ni-Cr alloys, Co-Cr alloys, and precious metal alloys.[25] There are several surface pretreatment methods for Ni-Cr alloys: immersion in concentrated HNO_3 solution after etching with HCl,[23] immersion in solutions containing an oxidizing agent after sandblasting, etching and passivating by an electrochemical method, and spraying of molten metal on the alloy surface. These methods increase bond strength by increasing the mechanical retention and chemical affinity between the adhesive and alloys. However, it is necessary to distinguish the contribution of these two effects, mechanical and chemical, to evaluate how the alloy surface structure improves the resin adhesion.

The adhesion of 4-META resin to Co-Cr and Ni-Cr alloys is examined by tensile test with and without thermally induced stress.[26] The resin bond to the as-polished surface of Co-Cr and Ni-Cr alloys is stronger than that to the oxidized surfaces. The thermal cycles cause clear differences in the surface states that affected the adhesion. The adhesive ability of 4-META resin to Ni-Cr alloy improves remarkably when the alloy surface is treated by concentrated HNO_3 solution. Adhesion to the alloy surfaces treated by concentrated HNO_3 is excellent, comparable with as-polished Co-Cr alloy, and is resistant to and protects against thermal cycling using liquid nitrogen.

The effect of a thick water layer adsorbed on the top of oxide surfaces on the bonding ability of 4-META resin is examined with a Co-Cr alloy and 18–8 stainless steel.[27] The alloys are heated 500°C in air and dehydrated by heating to 700°C at 1×10^{-4} Pa in silica glass tube. Dental adhesive resin containing 4-META is bonded to these alloy surfaces, and after thermal cycle treatment the bonding strengths and failure types for the dehydrated surfaces are compared with surfaces heated at 500°C. In both alloys, the specimens oxidized at 500°C in air show partial interface (alloy/resin) failure at the periphery. However, except for a few cases, dehydrated specimens display cohesive failure and the bonding strength is similar to that of the as polished Co-Cr alloy, showing excellent bonding ability. The bonding ability of the 4-META resin to the dehydrated oxide layer surface is excellent when adhesion procedures are performed in an atmosphere excluding water vapor. The cleaned metal surface obtained by hydrogen gas reaction shows excellent adhesive ability on Cr, Co, and 304 type stainless steel, showing a passive oxide film on the metal substrates.[28] Figures 12 shows the possible adhesion mechanism models of 4-META: Models (a), (b), and (c) are for the as-polished surface and models (d), (e), and (f) are for the oxidized surface.

Durability in Water

Generally, joints of dental adhesive resin bonded to dental alloys weaken in a wet environment, although adhesion is strong in a dry environment. Durability of the adhesion in a wet environment, such as the oral cavity, is predominantly important in the bonding of dental adhesive materials to teeth and dental alloys.

With aluminum/epoxy resin,[29] aluminum/Li-polysulfone,[30] and mild-steel/4-META resin[31] in wet environments, water molecules enter the adhesion interface by diffusion through the adhesive resin rather than by

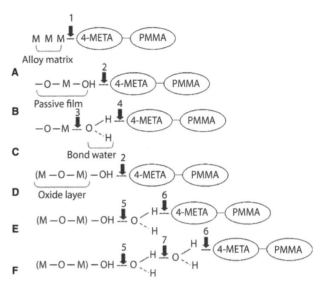

Fig. 12 Suggested adhesion model of PMMA resin containing 4-META on as polished (**A, B,** and **C**) and oxidized (**D, E,** and **F**) surface: M is metal atom—is hydrogen bond. The chemical bonds that are similar in bonding strength are labeled by the same number at the bonding position.

passage along exposed bond lines. Polymethyl methacrylate (PMMA) film bonded by 4-META dental adhesive resin to mild steel is used as a specimen to study the mechanism of water permeation into the adhesive interface. Water enters the interface by diffusion through the resin rather than by passage along the interface. The water content penetrated to the interface is calculated from the solution to Fick's second equation.[31] After 3 days of immersion, no change is observed on the mild steel surface through the clear resin layer, but XPS analysis reveals the same chemical state as in the 2 week immersion specimen.[32] The hydrogen bonds appear to be destroyed with water penetrating through the resin layer. This was followed by corrosion, resulting in a complete destruction of the adhesion interface.

The intensity of thermal stress due to thermal shock, calculated by the three-dimensional finite element method, increases with increasing resin film thickness, indicating that the resin film separates without degradation due to water when the resin film is thick. Total interface failure occurs on specimens with resin thicker than 0.5 mm. The critical thermal shearing stress was calculated as 22.5 MPa. When the adhesion interface degrades, the adhesion is broken by lower shear stresses than the value for 0.25 mm layers, 16 MPa. Immersion time to reach equilibrium water content is 0.5 days for a 0.25 mm resin film.[33]

Bonding to Precious Alloys

4-META adheres to hydroxyl groups on metal surface as explained already, and noble metals and precious alloys have a smaller number of hydroxyl groups on their surfaces. Therefore, the bonding strength of 4-META with noble metals and precious alloys is weaker than base metal alloys. The resins bond independently to dental precious metal alloys because they have low chemical affinity for the precious metals. Several surface modification methods have been developed for improvement of adhesion to dental alloys: high-temperature oxidation, immersion in oxidizing agent, immersion in concentrated nitric acid,[23] anodizing, electropolishing with tin, SiOx coating,[34] and ion coating.[35] The principle of these techniques is to form oxide layer on precious metal alloys. However, these methods have drawbacks, such as complicated procedures, expensive equipment, and degradation of chemical agent.

The other method to obtain bonding of 4-META to precious metal alloy is to design new alloys containing In, Zn, and tin (Sn).[36] The water durability and bonding strength of 4-META resin to binary alloys of Au, Ag, Cu, or Pd containing In, Zn, or Sn are studied. With In in Au-based alloys, the In_2O_3 on the alloy surface plays an important role in the adhesion with 4-META. To obtain excellent adhesion, the element in an oxide with chemical affinity for 4-META must cover at least 50% of the alloy surface. The poor adhesive ability of 4-META resin to pure gold is considered to be caused by chemisorbed H_2O molecules on the surface.[37] The adhesion ability of binary alloy was improved by adding In, Zn, or Sn.

STENTS AND STENT GRAFTS

Stents

Stents and stent grafts are tubes used for dilatation to counteract decreases in vessel or duct diameter and to maintain localized flow in stenotic blood vessels. Stents are also applied to vessels of the bile duct, esophagus, and other passages for dilation. Therefore, elasticity or plasticity for expansion and rigidity for the maintenance of dilatation and resistance to elastic recoil are required. The expandability and plasticity of a balloon-expandable stent and the elasticity of a self-expandable stent, as well as rigidity and resistance to the elastic recoil of blood vessels, are required for stent materials. Conventional metals cover these properties when proper metals are selected.

Stents are expandable tubes of metallic mesh that were developed to address the negative sequelae of balloon angioplasty. At the beginning of the twentieth century, glass tubes that became a prototype of stents were implanted into blood vessels of animals. In the next stage, the origin of percutaneous transluminal angioplasty, the expansion of blood vessels obtained by increasing the diameter of catheter tube, was attempted.[38] This trial failed because of migration and the development of thrombus. In 1985, Palmaz and his colleagues developed the first balloon-expanded stent,[39] and just 1 year later, Gianturco developed a balloon-expanded coil stent.[40] These stents were made of stainless steel. The first self-expandable

stent, Wallstent®, made of a Co-Cr alloy (Elgiloy), was clinically applied in 1986.[41] Another popular self-expandable stent, SMART®, consists of a superelastic Ni-Ti alloy.[42] Since the 1990s, stents have been used in coronary arteries. Typical popular stents for this purpose are the Palmaz-Schatz® stent and the Gianturco-Roubin® stent. New designs and functions of stents have also been developed. Today, for example, the majority of patients undergoing percutaneous transluminal coronary angioplasty receive a stent. Since the mid-1990s, stents have also been applied to the treatment of cerebrovascular disease, and carotid stenting to prevent stroke, recoil, and restenosis has been attempted.[43–45] In addition, the first treatment for abdominal aortic aneurysms with stent grafts started in the 1990s.[46] Stent grafts are generally constructed of stainless steel or a Ni-Ti alloy and coated with compounds such as expanded polytetrafluoroethylene (ePTFE) that is a metal–polymer composite. Typical bare stent and stent graft are shown in Fig. 13.

Metals are the main materials utilized for stents because of their mechanical properties and visibility by X-ray imaging. Metallic stents and stent grafts are now important devices in noninvasive medicine. New techniques for treatments using metallic stents are continuously being developed, and such development will continue into the foreseeable future. While pioneering ideas for treatment are constantly being announced, the development of proper materials for stents is backward from the viewpoint of materials science.

Drug Eluting Stents

Initial complications associated with stenting generally involve subacute stent thrombosis, which occurs in 1–3% of patients within 7–10 days. The major long-term

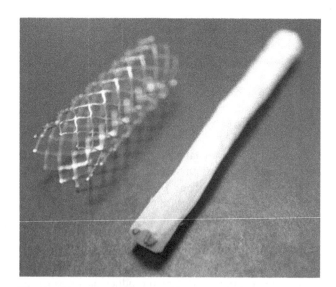

Fig. 13 Examples of bare metallic stent and stent graft.

complication is instent restenosis, which occurs within 6 mo in 50% of patients. Although it is unknown whether the original cause of these problems is the metal from which the stents are made, the use of polymer-coated drug-eluting stents (DESs) is increasing.[47–49]

The most promising results have been attained with polymer-coated DESs[1] and DES is one of the metal–polymer composites. Two of the drugs currently in clinical trials are rapamycin and paclitaxel. These drugs are embedded in a polymer matrix (such as a copolymer of poly-n-butyl methacrylate and polyethylene-vinyl acetate or a gelatin–chondroitin sulfate coacervate film) that is coated onto the stent. The drug is related by diffusion and/or polymer degradation over varying periods of time that can be engineered by the specifics of the polymer-drug system. These coated stents have had excellent initial success, virtually eliminating restenosis over time periods of 2 years and longer and are felt to represent a major breakthrough in the treatment of coronary artery disease. Commercial DESs and their structures are summarized in Table 4.

In DES, chemicals such as immunosuppressants are coated on metal frames. Because direct immobilization of drugs to metals or coating of the polymers and ceramics impregnating the drugs is necessary to control their release rate, attempts have been made to coat the photosetting gelatin impregnating the drug.[50] To improve the biofunction of stents, a stent coated with a phospholipid-like 2-methacryloyloxyethyl phosphorylcholine (MPC) polymer to inhibit the adsorption of proteins and the adhesion of platelets has been developed.[51] For the same purpose, the immobilization of poly(ethylene glycol) (PEG) and compounding with medical polymers are effective.[52]

In DESs, all fractures occurred around areas of increased rigidity due to the overlapping of metals, which may have formed a fulcrum for metal deformation due to vessel movement.[53]

Stent Grafts

The other metal–polymer composite is stent graft. Stents for treatment of peripheral vascular disease are generally constructed of stainless steel or nitinol and may be coated with compounds such as ePTFE. Stent grafts are composed of a metallic frame covered by a fabric tube and combine the features of stents and vascular grafts; they can be deployed endovascularly. Stent grafts are used to treat aortic aneurysms, where the aortic wall has been weakened and threatens to rupture, as well as stenosis of other arterial sites. The graft portion, usually composed of polyester or ePTFE, can sit on either the luminal or abluminal (outside) aspect of the metallic stent and is intended to provide a mechanical barrier to prevent intravascular pressure from being transmitted to the weakened wall of the aneurysm, thus excluding the aneurysm from the flow of blood. These stents and stent grafts are developed in a similar manner to those in the coronary circulation, either as self-expanding units or covering an inflatable balloon. The stent is used for a given application unit or over an inflatable balloon. The stent used for a given application is selected by diameter, length, and geometry of the lesion and location of side braches or branch points.

Current synthetic vascular grafts are typically fabricated from poly(ethylene terephthalate) (Dacron) or ePTFE, with the Dacron graft being used for larger vessel applications and the ePTFE to bypass smaller vessels. These grafts can be made porous to enhance healing, but they are then impregnated with connective tissue proteins to aid clotting, reduce the blood loss through the pores of the graft upon implantation, and stimulate tissue in growth, and with antibiotics to reduce the risk of infection of the grafts that are not impregnated, which need to be preclotted with the patient's own blood for this same purpose.

In light of the complications associated with vascular grafts, current research has focused on improvement of synthetic vascular grafts and on alternatives such as tissue-engineered blood vessels. The purpose of obtaining the luminal surface is as follows: 1) prevent coagulation; 2) prevent platelet adhesion/aggregation; 3) promote fibrinolysis; 4) inhibit smooth muscle cell adhesion/proliferation; and 5) promote endothelial cell adhesion and proliferation. Endotherialization of the entire graft may also help prevent bacterial attachment to the graft and subsequent infection. Engineering an artery muscle and endotherial cells is a promising approach to solving

Table 4 Various drug eluting stents

| Manufacturer | Coating | | | Stent | |
	Brand name	Drug	Material	Name	Material
Johnson & Johnson	Cypher™	Sirolimus	Nonabsorbable polymer	Bx Velocity	316L stainless steel
Boston Scientific	Taxus™	Paclitaxel	Nonabsorbable polymer	Express 2	316L stainless steel
Medtronics	Endeavor™	Zotarolimus	PC coating	Driver	Co-Cr alloy
Abbot/Boston Scientific	Xience™ V	Everolimus	Acryl/fluorine polymer	ML-Vision	Co-Cr alloy
Biosensors/Termo	Nobori™	Biolimus	Absorbable polymer (PLA)	S-Stent	316L stainless steel

the problem of developing an adequate small-diameter vascular graft.[54,55]

Severe pitting and crevice corrosion originating from crevice corrosion are observed in Ni-Ti alloys in stent grafts, as shown in Fig. 14.[56] These problems may occur in Ni-Ti alloys and stainless steel at crevices between an artificial blood vessel and a metallic stent because of the electrochemical properties of the alloys. Corroded sites may then become the initiation site of a fracture. Alloys that are resistant to pitting and corrosion must be developed and used.

IMMOBILIZATION OF PEG

Poly(Ethylene Glycol)

PEG is an oligomer or polymer of ethylene oxide, but historically PEG has tended to refer to oligomers and polymers with a molecular weight below 20,000. PEG has the structure shown in Fig. 15. PEGylation is the act of covalently coupling a PEG structure to another larger molecule, for example, a therapeutic protein (which is then referred to as PEGylated). PEG is soluble in water, methanol, benzene, and dichloromethane and is insoluble in diethyl ether and hexane. It is coupled to hydrophobic molecules to produce nonionic surfactants. This property, combined with the availability of PEGs with a wide range of end functions, contributes to the wide use of PEGs in biomedical research: drug delivery, tissue engineering

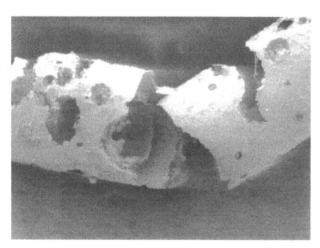

Fig. 14 Severe crevice corrosion observed in Ni-Ti alloy used as stent graft for 6 months.
Source: From Heintz et al.[56]

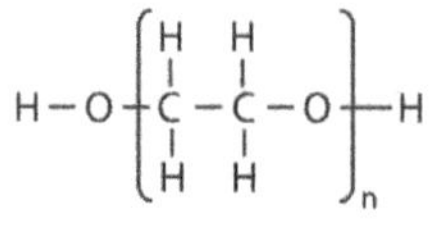

$$H-O\left[\begin{array}{c} H \\ | \\ C \\ | \\ H \end{array} \begin{array}{c} H \\ | \\ C \\ | \\ H \end{array} -O \right]_n H$$

Fig. 15 Chemical structure of PEG.

scaffolds, surface functionalization, and many other applications.[57]

Chemical Immobilization

The immobilization of biofunctional polymers on noble metals such as Au is usually conducted by using the bonding –SH or –SS– group; however, this technique can only be used for noble metals. The adhesion of platelets and adsorption of proteins, peptides, antibodies, and DNA are controlled by modifications of this technique. On the other hand, PEG is a biofunctional molecule on which adsorption of proteins is inhibited. Therefore, immobilization of PEG to metal surface is an important event to biofunctionalize the metal surface. Examples of immobilization of PEG to oxide surface are shown in Figs. 16. A class of copolymers based on poly(L-lysine)-g-poly (ethylene glycol), PLL-g-PEG, has been found to spontaneously adsorb from aqueous solutions onto TiO_2, $Si_{0.4}Ti_{0.6}O_2$, and Nb_2O_5 to develop blood-contacting materials and biosensors.[52,58] In another case, TiO_2 and Au surfaces are functionalized by the attachment of PEG-poly(DL-lactic acid), PEG-PLA, and copolymeric micelles. The micelle layer can enhance the resistance to protein adsorption to the surfaces up to 70%.[59] A surface of stainless steel was first modified by a silane coupling agent, SCA, (3-mercaptopropyl)trimethoxysilane. The silanized stainless steel, SCA-SS, surface was subsequently activated by argon plasma and then subjected to UV-induced graft polymerization of poly(ethylene glycol) methacrylate (PEGMA). The PEGMA graft-polymerized stainless-steel coupon, PEGMA-g-SCA-SS, with a high graft concentration and, thus, a high PEG content was found to be very effective to prevent the absorption of bovine serum albumin and γ-globulin.[60] These processes require several steps but are effective for immobilization; however, no promising technique for the immobilization of PEG to a metal surface has been so far developed. Photoreactive PEG is photoimmobilized on Ti.[61]

Electrodeposition

Both terminals of PEG (MW: 1000) are terminated with –NH_2 (NH_2–PEG–NH_2), but only one terminal is terminated with –NH_2 (NH_2–PEG). The cathodic potential is charged to Ti from the open circuit potential to –0.5 V versus an SCE and maintained at this potential for 300 sec. During charging, the terminated PEGs electrically migrate to and are deposited on the Ti cathode, as shown in Fig. 17. Not only electrodeposition but also immersion leads to the immobilization of PEG onto a Ti surface. However, more terminated amines combine with Ti oxide as an NH-O bond by electrodeposition, while more amines randomly exist as NH_3^+ in the PEG molecule by immersion (Fig. 18).[62,63] A scanning probe microscopic image is shown in Fig. 19. The amounts of the PEG layer immobilized onto the metals are governed by the concentrations of

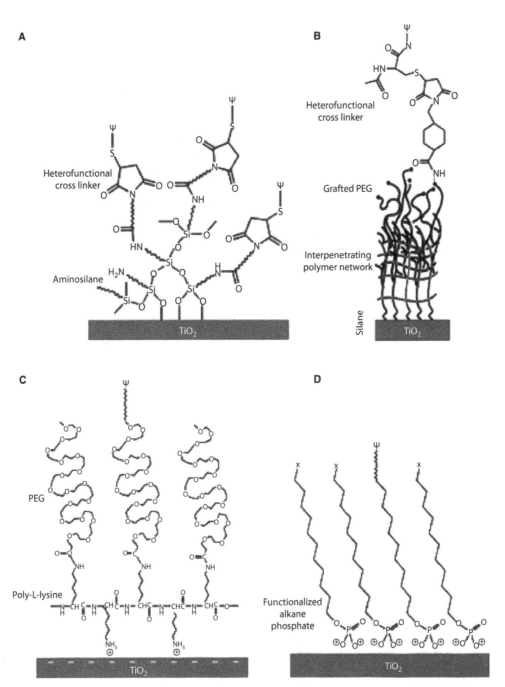

Fig. 16 Various techniques for immobilization of PEG to titanium oxide. (**A**) Silanization using APTES followed by covalent attachment of maleimide; (**B**) silanized surface followed by formation of a non-adhesive P(AAm*co*EG/AA); (**C**) polycationic poly(amino acid) grafted with poly(ethylene grycol) side chain; (**D**) self-assembled monolayers of long-chain alkanephosphate.
Source: From Xiao et al.[129]

the active hydroxyl groups on each surface oxide in the case of electrodeposition, which is governed by the relative permittivity of the surface oxide in the case of immersion.[64] The PEG-immobilized surface inhibits the adsorption of proteins and cells, as well as the adhesion of platelets[65] and bacteria[66] (Figs. 20), indicating that this electrodeposition technique is useful for the biofunctionalization of metal surfaces. It is also useful for all electroconductive materials and materials with complex surface topography.

IMMOBILIZATION OF BIOMOLECULES

Metal–Biomolecule Interface

Organic coatings have been scarcely used on metallic implants. Nevertheless, it is expected that these materials will significantly improve their market share as improved biocompatibility becomes the decisive criterion for patients and surgeons in the future. This approach, which is based

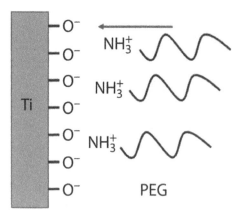

Fig. 17 Attraction of PEG with positively charged terminal to cathodic Ti surface by electrodeposition.

on the latest medical and cellular biological results, employs biopolymers (proteins) that have been immobilized on the surface of metallic implants. The intention is to reduce the region character of the implant for the body. This is accomplished by coating with substances that are normally found on the surface or in the vicinity of the tissue that has to be substituted by the implant. It has been found that these coatings act as local mediators of cell adhesion and, in consequence, as a stimulating factor for the growth and proliferation of the cells normally found around the substituted tissue. The tight attachment at the oxide-coated surface of the metallic implant and the conservation of the biological function of the proteins involved

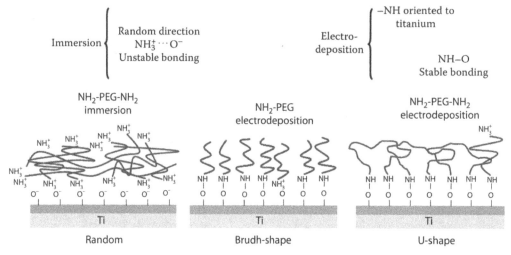

Fig. 18 Schematic model of the deposition manner and chemical bonding state of PEG by immersion and electrodeposition.

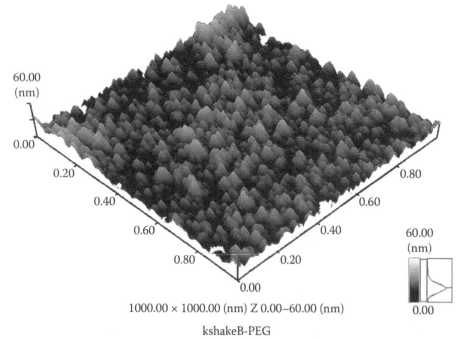

1000.00×1000.00 (nm) Z 0.00–60.00 (nm)

kshakeB-PEG

Fig. 19 Scanning probe microscopic image of electrodeposited PEG to Ti surface.

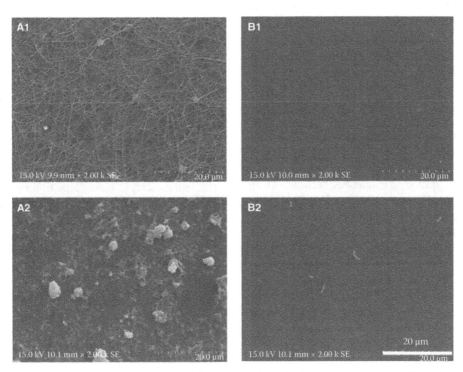

Fig. 20 Platelet adhesion and fibrin network formation (**A1**) and bacterial adhesion (**A2**) are active on Ti, while they are inhibited on PEG-electrodeposited Ti surfaces (**B1** and **B2**).

are prerequisites for obtaining these highly desirable properties.

Since the natural environment around the implant is aqueous while the surface of the implant is either bare or oxidized metal, specific demands are imposed on the coating to mediate successfully between these different structural entities. The purpose of these demands is to obtain the native conformation of all proteins and cells that are in contact with the coating and to avoid all forms of aggregation and other conformational changes that might lead to protein denaturalization or cell death.

One approach is the immobilization of biological molecules (growth factors, adhesive proteins) onto the implant surface in order to induce a specific cellular response and promote osseo-integration. The application of large extracellular matrix proteins, however, can be unpractical due to their low chemical stability, solubility in biological fluids, and high cost. In addition, entire ECM molecules are usually of allogenetic or xenogenetic origin and thus associated with the risk of immune reaction and pathogen transfer. To immobilize biomolecules to metal surfaces, the following techniques are used: modification through silanized titania (thiol-directed immobilization, amino- and carboxyl-directed immobilization), modification through photochemistry, electrochemical techniques, and chemical modifications based on self-assembled monolayers.

The immobilization of biomolecules to a metallic surface can be achieved using self-assembled monolayers as crosslinkers. Self-assembled monolayers provide chemically and

structurally well-defined surfaces that can often be manipulated using standard synthetic methodologies.[67] Thiol on self-assembled monolayers[68,69] and siloxane-anchored self-assembled monolayers[70] has been particularly well studied. A problem related to the application of immobilized biomolecules via silanization techniques is the hydrolysis of siloxane films when exposed to aqueous (physiological) conditions.[71] More recently, alkyl phosphate films that remain robust under physiological conditions[72] have been used to provide an ordered monolayer on tantalum oxide surfaces,[73,74] and alkylphosphonic acids have been used to coat the native oxide surfaces of metals and their alloys inducing iron,[75] steel,[76] and Ti.[77]

Peptides

In a living tissue, the most important role played by the extracellular matrix has been highlighted to favor cell adhesion.[78] Studies have shown that interactions occur between cell membrane receptors and adhesion proteins (or synthetic peptides) derived from the bone matrix, such as type I collagen or fibronectin.[79] These proteins are characterized by an RGD (Arg-Gly-Asp) motif with special transmembrane connections between the actin cytoskeleton and the RGD motif, and the whole system can activate several intracellular signaling pathways modulating cell behavior (e.g., proliferation, apoptosis, shape, mobility, gene expression, and differentiation).[80] Due to the main role of the RGD sequence in cell adhesion, several research groups have developed biofunctionalized surfaces by immobilization of RGD peptides. Grafting RGD peptides has been performed on different biomaterials, such as Ti,[81–83] and has been shown to improve osteoconduction *in vitro*. Methodologies differ by the conformation of RGD (cyclic or linear) and by the technique used for peptide immobilization.[78,79,82–84] Since the graft of an RGD peptide is known to be efficient in bone reconstruction,[85] the challenge is to develop simple and cheap methods to favor cell anchorage on biomaterial surfaces.[83,84]

Self-assembled molecular monolayers bearing RGD moieties have been grafted to numerous surfaces, using either silanes,[86] phosphonates on oxidized surfaces,[83] or thiols on Au[84] but have analyzed some application problems for large-scale production. Phosphonates are known to adsorb on Ti. To be mechanically and physiologically stable, phosphonate layers have to be covalently bound to the material surface by using drastic conditions[77,87] (anhydrous organic solvents, high temperature), which are not compatible with biomolecule stability. Monolayers of RGD phosphonates have been achieved using a complex multistep process, which necessitates tethering a primer onto Ti surface, then a linker, and finally the peptide.[88] To immobilize RGD to the electrodeposited PEG on Ti, PEG with an $-NH_2$ group and a $-COOH$ group ($NH_2-PEG-COOH$) must be employed. One terminal group, $-NH_2$, is required to bind stably with a surface oxide on a metal. On the other hand, the other terminal group, $-COOH$, is useful to bond biofunctional molecules such as RGD, as shown in Fig. 21.[89] This RGD/PEG/Ti surface accelerates calcification by MC3T3-E1 cell.[90] The calcification is the largest on the RGD/PEG/Ti surface (Fig. 22).

Glycine (G)-arginine (R)-glycine (G)-asparaginic acid (D)-serine (S) sequence peptide, GRGDS peptide, is coated with chloride activation technique to enhance adhesion and

migration of osteo-blastic cells.[91] The expression levels of many genes in MC3T3-E1 cells are altered.

Protein and Collagen

Among the relevant molecules involved in biochemical modification of bone-contacting surfaces, growth factor, such as bone morphogenetic protein-2 (BMP-2), is of primary interest. BMP-2 has been known to play an important role in bone healing processes and to enhance therapeutic efficiency. Ectopic bone formation by BMP-2 in animals has been well established following the first reports of BMP-2 by the Urist research group.[92–94] Synthetic receptor binding motif mimicking BMP-2 is covalently linked to Ti surfaces through a chemical conjunction process.[95] A complete and homogeneous peptide overlayers on the Ti surfaces; the content is further measured by gamma counting. Biological evaluations show that the biochemically modified Ti were active in terms of cell attachment behavior. Ti surfaces can enhance the rate of bone healing as compared with untreated Ti surface. Bone morphogenetic protein-4 (BMP-4) is immobilized on a Ti-6Al-4V alloy through lysozyme to improve the hard tissue response.[96] Proteins are silane coupled to the oxidized surfaces of the Co-Cr-Mo alloy, the Ti-6Al-4V alloy, Ti, and the Ni-Ti alloy to improve tissue compatibility.[97]

Biomolecules are also used to accelerate bone formation and soft tissue adhesion on a material. Type I collagen is immobilized by immersion in the collagen solution.[98] Type I collagen production increases with modification by ethane-1,1,2-triphosphonic acid and methylenediphosphonic acid grafted onto Ti.[99] Type I collagen is grafted through glutaraldehyde as a crosslinking agent.[100] For electrodeposition, it is found that an alternating current between -1 V and $+1$ V versus SCE with 1 Hz is effective to immobilize type I collagen to Ti and durability in water is high.[101]

Fibronectin is immobilized directly on Ti using the tresyl chloride activation technique.[102] L-threonine and

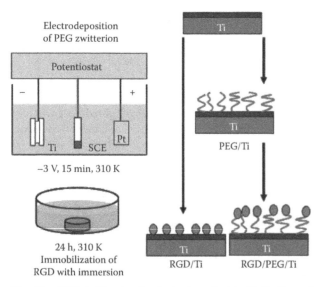

Fig. 21 PEG twitter ion is electrodeposited to Ti firstly and RGD is immobilized on the PEG.

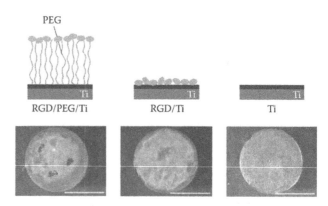

Fig. 22 Calcification (dark regions) by MC3T3-E1 cells is more active on RGD/PEG/Ti specimen than on RGD/Ti and Ti. Scale bar represents 5 mm.

O-phospho-L-threonine are immobilized on acid-etched Ti surface.[103]

Hydrogel and Gelatin

Immobilization or coating of hydrogel to metal surface is currently attempting to add a drug delivery ability to orthopedic implant and stents or fluorescent sensing ability to microchips. Currently, synthetic polymeric hydrogels like poly(hydroxyethylmethacrylate) (pHEMA) and poly(hydroxyethylacrylate) are widely used as compliant materials particularly in the case of contact with blood or other biological fluids.[104] Despite hydrogel's good flexibility in the swollen state, hydrogels usually lack suitable mechanical properties, and this could greatly impair their use as coating materials for surgical procedure. Moreover, in case of inadequate adhesion between the hydrogel coating and the metal surface, a breakage at the coating–steel interface might occur.[105] A spray-coated method has been set up with the aim to control the coating of pHEMA onto the complex surface of a 316L steel stent for percutaneous coronary intervention (PCI).[106] The pHEMA coating evaluation of roughness wettablity together with its morphological and chemical stability after three cycles of expansion-crimping along with preliminary results after 6 months demonstrates the suitability of the coating for surgical implantation of stent.

An alternative promising synthetic route is represented by electrochemical polymerization, which leads to thin film coatings directly on the metal substrates with interesting applications either for corrosion protection or for the development of bioactive films.[107–110] As far as the orthopedic field is concerned, in recent years, many procedures based on surface modification have been suggested to improve the biocompatibility and biofunctionality of Ti-based implant.[111] 2-Hydroxy-ethyl-methacrylate, a macromer PEGDE, and PEGDE copolymerized with acrylic acid were used to obtain hydrogels. A model protein and a model drug were entrapped in the hydrogel and released according to pH change.[112]

OTHER METAL–POLYMER COMPOSITES

Bonding of Polymers with Metal through SCA

The interfacial chemical structure governing the bonding strength, especially at the nanometer level, is one of the most challenging aspects of the development of composite materials. The combination of a Ti alloy with a resin for crown facings has been attempted.[113] In particular, SCAs containing S-H groups and Si-O-CH$_3$ groups are comprehensively used to combine dental alloys with resins.[114] The S-H groups work as a bonding agent with polymers; the Si-O-CH$_3$ works as a bonding agent with metals. The mechanical properties and durability of composite resin

increase with silanized filler.[115–118] However, in most studies about materials using SCAs in the field of dentistry, only the bonding strength is evaluated and discussed, and there are few reports that examine and discuss the chemical structures at the bonding interface and how they influence the bonding strength. Other studies on SCAs to combine polymers with metals have been performed in other fields. An aluminum–vegetable oil composite using a SCA has been developed.[119,120] Rubber-to-metal bonding by a SCA was investigated.[121] In addition, the surface modification of stainless steel by grafting of PEG using a SCA has been reported.[60] However, only the chemical structure is investigated in these studies. In other words, the relationship between the bonding strength and the interfacial chemical structure containing a SCA layer has not been studied.

The unequivocal relationship between the shear bonding strength and the chemical structure at the bonding interface of a Ti-segmentated polyurethane (SPU) composite through a SCA (γ-mercaptopropyl trimethoxysilane [γ-MPS]) is investigated.[122] Schematic models of the fractured region before and after the shear bonding test in the case of a thin γ-MPS layer and a thick γ-MPS layer are shown in Fig. 23. On the other hand, the shear bond strength of the Ti/SPU interface increased with ultraviolet (UV) irradiation according to the increase in the cross-linkage in

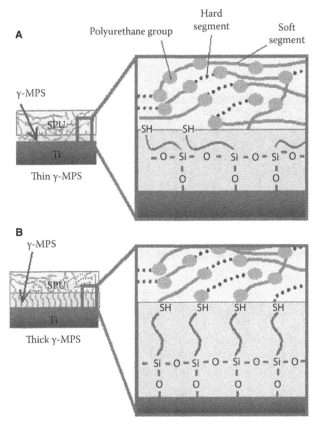

Fig. 23 Schematic model of the fractured region before and after the shear bonding test in the case of a thin γ-MPS layer (**A**) and a thick γ-MPS layer (**B**).

SPU. Platelet adhesion to Ti is inhibited by SPU, as shown in Fig. 24. This technique is used for the creation of a new meta-based material with high strength, high toughness, and biofunction. UV irradiation to a Ti-SPU composite is clearly a factor governing the shear bond strength of the Ti/SPU interface.[123] In addition, active hydroxyl groups on the surface oxide film are clearly factors governing the shear bond strength.[124] After a good bonding between metal and polymer is produced, biofunctionalization techniques developed in the field of polymers could be applied to the composite materials.

Polymers Condensed in Porous Titanium

The Young's modulus of metallic materials is relatively larger than that of cortical bone: about 200 GPa in stainless steels and Co-Cr-Mo alloys, about 100 GPa in Ti and Ti alloys, and 15–20 GPa in cortical bone. When fractured bone is fixed with metallic bone fixator such as bone plate and screws and bone nail, during healing a load to fixation part is mainly received by metallic fixators because of the difference in their Young's modulus. This phenomenon is well known as the so-called "stress shielding" in orthopedics. This large Young's modulus generates other problems. When a metal is used as a metallic spacer in spinal fixation, the spacer is mounted in matrix bone. In the case of a dental implant, occlusal pressure is not absorbed by the implant and directly conducts to jaw bone.

To solve these problems, metals with low Young's modulus are required. Two approaches are feasible: the decrease

in the Young's modulus of a metallic material itself and decrease in the apparent Young's modulus by forming a porous body. In the latter case, the pores are sometimes filled by polymers to control the apparent Young's modulus. UHMWPE[125] and PMMA[126] are attempted to fill the pores in porous Ti. Figure 25 shows porous Ti whose pores are filled by UHMWPE.

Metallization of Polymers

The properties of polymer surfaces with regard to their chemical, electrical, mechanical, and other properties are modified by metallization, and this technique could be applied to biomaterials.[127] In the metallization process, the metal atoms arrive as mobile individuals at a polymer surface. The metal layer formation on polymer will be influenced by the type and strength of metal–polymer interaction and the structure of the polymer. Metallization techniques are categorized by evaporation, sputtering, chemical vapor deposition, and electrochemical deposition.

A number of conditions must be determined to accomplish metal penetration into polymers. The metal deposition rate must be low enough to prevent metal cluster formation at the polymer surface in an initial stage of the metallization process. The metal bonds in such clusters immobilize the atoms, and a strong interaction between the metal atom and polymer chain reduces the diffusion. The metal atoms must not react with the polymer. The polymer substrate temperature must be well above the glass transition of the polymer surface region. The polymer segment

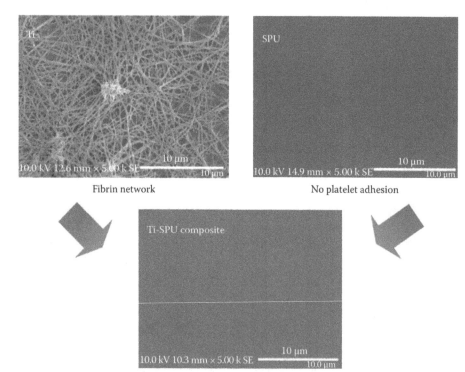

Fibrin network No platelet adhesion

No platelet adhesion

Fig. 24 Platelet adhesion on Ti is inhibited by SPU layer.

Melt—Microgels

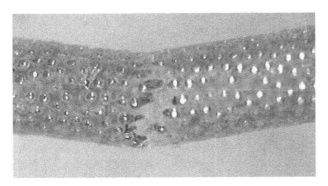

Fig. 25 Porous Ti whose pores are filled by UHMWPE after bending test.

motions assist the metal atom transport deep into the material.

The diffusion front of a metal is not smooth, but it is rough and creates mechanical interlocking, which in turn enhances the mechanical strength between metal overlayer and polymer. The mechanical properties increase in the composite region compared with the pure polymer. Electric conductivity may be affected and the thermal expansion coefficients are matched to some extent by interphase.

CONCLUSIONS

Metallic materials are widely used in medicine for not only orthopedic and dental implants but also cardiovascular devices and other purposes. The metal surface may be biofunctionalized by various techniques such as dry and wet processes, the immobilization of biofunctional molecules, and the creation of metal–polymer composites. These techniques make it possible to apply metals to a scaffold in tissue engineering. Artificial materials such as metal–polymer composites will continue to be used as biomaterials in the future, because of their excellent biocompatibilities and biofunctions. Some metal–polymer composites are reviewed in other textbooks.[127–129]

REFERENCES

1. Ratner, B.D.; Hoffman, A.S.; Schoen, F.J.; Lemons, J.E. *Biomaterials Science: An Introduction to Materials in Medicine*; Elsevier: Amsterdam, the Netherlands, 2004.
2. Yang, Y.; Kim, K.H.; Ong, J. A review on calcium phosphate coatings produced using a sputtering process—An alternative to plasma spraying. Biomaterials **2005**, *26* (3), 327–337.
3. Kim, K.H.; Ramaswany, N. Electrochemical surface modification of titanium in dentistry. Dent. Mater. J. **2009**, *28* (1) 20–36.
4. Asami, K.; Chen, S.C.; Habazaki, H.; Hashimoto, K. The surface characterization of titanium and titanium-nickel alloys in sulfuric acid. Corros. Sci. **1993**, *35* (1–4), 43–49.
5. Hanawa, T.; Asami, K.; Asaoka, K. Repassivation of titanium and surface oxide film regenerated in simulated bioliquid. J. Biomed. Mater. Res. **1998**, *40* (4), 530–538.
6. Hanawa, T.; Ota, M. Calcium phosphate naturally formed on titanium in electrolyte solution. Biomaterials **1991**, *12* (8), 767–774.
7. Hanawa T. Titanium and its oxide film: a substrate for formation of apatite. In *The Bone-Biomaterial Interface*; J.E. Davies, Ed.; University of Toronto Press: Toronto, Ontario, Canada, 1991; pp. 49–61.
8. Hanawa, T.; Okuno, O.; Hamanaka, H. Compositional change in surface of Ti-Zr alloys in artificial bioliquid. J. Jpn. Inst. Metals **1992**, *56* (10), 1168–1173.
9. Bruesch, P.; Muller, K.; Atrens, A.; Neff, H. Corrosion of stainless-steels in chloride solution-an XPS investigation of passive films. Bruesch. Appl. Phys. **1985**, *38* (1), 1–18.
10. Jin, S.; Atrens, A. ESCA-studies of the structure and composition of the passive film formed on stainless steels by various immersion times in 0.1 M NaCl solution. Appl. Phys. A **1987**, *42* (2), 149–165.
11. Hanawa, T.; Hiromoto, S.; Yamamoto, A.; Kuroda, D.; Asami, K. XPS characterization of the surface oxide film of 316L stainless samples that were located in quasi-biological environments. Mater. Trans. **2002**, *43* (12), 3088–3092.
12. Smith, D.C.; Pilliar, R.M.; Metson, J.B.; McIntyre, N.S. Preparative procedures and surface spectroscopic studies. J. Biomed. Mater. Res. **1991**, *25* (9), 1069–1084.
13. Hanawa, T.; Hiromoto, S.; Asami, K. Characterization of the surface oxide film of a Co-Cr-Mo alloy after being located in quasi-biological environments using XPS. Appl. Surf. Sci. **2001**, *183* (1), 68–75.
14. Nagai, A.; Tsutsumi, Y.; Suzuki, Y.; Katayama, K.; Hanawa, T.; Yamashita, K. Characterization of air-formaed surface oxide film on a Co-Ni-Cr-Mo alloy (MP35N) and its change in Hanks' solution. Appl. Surf. Sci. **2012**, *258* (14), 5490–5498.
15. Endo, K.; Araki, Y.; Ohno, H. *In vitro* and *in vivo* corrosion of dental Ag-Pd-Cu alloys. In *Transaction of International Congress on Dental Materials, November 1–4, 1989, Honolulu, Hawaii*; Okabe, T., Takahashi, S., Eds.; The Academy of Dental Materials and The Japanese Society for Dental Materials and Devices, University of Hawaii: Hawaii, 1989; pp. 226–227.
16. Parfitt, G.D. The surface of titanium dioxide. Prog. Surf. Membr. Sci. **1976**, *11*, 181–226.
17. Westall, J.; Hohl, H. A comparison of electrostatic models for the oxide/solution interface. Adv. Colloid Interface Sci. **1980**, *12* (4), 265–294.
18. Healy, T.W.; Fuerstenau, D.W. The oxide-water interface-Interreaction of the zero point of charge and the heat of immersion. J. Colloid Sci. **1965**, *20* (4), 376–386.
19. Boehm, H.P. Acidic and basic properties of hydroxylated metal oxide surfaces. Discuss Faraday Soc. **1971**, *52*, 264–289.
20. Takeyama, M.; Kashibuti, S.; Nakabayashi, N.; Masuhara, E. Studies on dental self-curing resins(17)—Adhesion of PMMA with bovine enamel or dental alloys. J. Jpn. Soc. Dent. Appar. Mater. **1978**, *19* (47), 179–185.
21. Omura, I.; Yamauchi, J.; Nagase, Y.; Uemura, F. Jpn Published Unexamined Patent Application, 58-21607, 1983.

22. Ikemura, K.; Endo, T. A review of our development of dental adhesives – Effects of radical polymerization initiators and adhesive monomers on adhesion. Dent. Mater. J. **2010**, *29* (2), 109–121.

23. Tanaka, T.; Nagata, K.; Takeyama, M.; Atsuta, M.; Nakabayashi, N.; Masuhara, E. 4-META opaque resin—A new resin strongly adhesive to nickel-chromium alloy. **1981**, J. Dent. Res. *60* (9), 1697–1707.

24. Tanaka, T.; Nagata, K.; Takeyama, M.; Nakabayashi, N.; Masuhara, E. Heat treatment of gold alloys to get adhesion with resin. J. Jpn. Soc. Dent. Appar. Mater. **1980**, *21* (54), 95–102.

25. Varga, J.; Matsumura, H.; Tabata, T.; Masuhara, E. Adhesive behavior of the alloys 'ALBABOND E' containing large percentage of Pd after various surface treatments. Dent. Mater. J. **1985**, *4* (2), 181–190.

26. Ohno, H.; Araki, Y.; M. Sagara. The adhesion mechanism of dental adhesive resin to the alloy—Relationship between Co-Cr alloy surface structure analyzed by ESCA and bonding strength of adhesive resin. Dent. Mater. J. **1986**, *5* (1), 46–65.

27. Ohno, H.; Araki, Y.; Sagara, M.; Yamane, Y. The adhesion mechanism of dental adhesive resin to the alloy—Experimental evidence of the deterioration of bonding ability due to adsorbed water on the oxide layer. Dent. Mater. J. **1986**, *5* (2), 211–216.

28. Ohno, H.; Araki, Y.; Endo, K.; Kawashima, K. The adhesion mechanism of dental adhesive resin to the alloy—Adhesive ability of dental adhesive resin to the clean metal surface obtained by hydrogen gas reduction method. Dent. Mater. J. **1989**, *8* (1), 1–8.

29. Brewis, D.M.; Comyn, J.; Tegg, J.L. The durability of some epoxide adhesive-bonded joints on exposure to moist warm air. Int. J. Adhes. Adhes. **1980**, *1* (1), 35–39.

30. Ko, C.U.; Wightman, J.P. Experimental analysis of moisture intrusion into the Al/Li-polysulfone interface. J. Adhes. **1988**, *25* (1) 23–29.

31. Ohno, H.; Endo, K.; Araki, Y.; Asakura, S. Destruction of metal-resin adhesion due to water penetrating through the resin. J. Mater. Sci. **1992**, *27* (19), 5149–5153.

32. Ohno, H.; Endo, K.; Araki, Y.; Asakura, Y. ESCA study on the destruction mechanism of metal-resin adhesion due to water penetrating through the resin. J. Mater. Sci. **1993**, *28* (14), 3764–3768.

33. Ohno, H.; Araki, K.; Endo, K.; Yamane, Y.; Kawashima, I. Evaluation of water durability at adhesion interface by peeling test of resin film. Dent. Mater. J. **1996**, *15* (2), 183–192.

34. Musil, R. Clinical verification of the Silicoater technique, results of three-years' experience. Dent. Lab. **1987**, *35*, 1709–1715.

35. Tanaka, T.; Hirano, M.; Kawahara, H.; Matsumura, H.; M. Atsuta. A new io-coating surface treatment of alloys for dental adhesive resins. J. Dent. Res. **1988**, *67* (11), 1376–1380.

36. Ohno, H.; Araki, K.; Endo, K. A new method for promoting adhesion between precious metal alloys and dental adhesives. J. Dent. Res. **1992**, *71* (6), 1326–1331.

37. Ohno, H.; Endo, K.; Yamane, Y.; Kawashima, I. XPS study on the weakest zone in the adhesion structure between resin containing 4-META and precious metal alloys treated with different surface modification methods. Dent. Mater. J. **2001**, *20* (1), 330–337.

38. Dotter, C.T.; Judkins, M.P. Transluminal treatment of arteriosclerotic obstruction of a new technique and preliminary report of its application. Circulation **1964**, *30* (5), 654–670.

39. Palmaz, J.C.; Sibbitt, R.R.; Tio, F.O.; Reuter, S.R.; Peters, J.E.; Garcia, F. Expandable intra-luminal vascular graft. A feasibility study. Surgery **1986**, *99* (2), 199–205.

40. Roubin, G.S.; Robinson, K.A.; King, S.B.; Gianturco, C.; Black, A.J.; Brown, J.E.; Siegel, R.J.; Douglas, J.S. Early and late results of intracoronary arterial stenting after coronary angioplasty in dog. Circulation **1987**, *76* (4), 891–897.

41. Sigwart, U.; Puel, J.; Mirkovitch, V.; Joffre, F.; Kappenberger, L. Intravascular stents to prevent occlusion and restenosis after trans-luminal angioplasty. N. Eng. J. Med. **1987**, *316* (12), 701–706.

42. Phatouros, C.C.; Higashida, R.T.; Malek, A.M. Endovascular stenting for carotid artery stenosis: Preliminary experience using the shape-memory-alloy-recoverable-technology (SMART) stent. Am. J. Neuroradiol. **2000**, *21* (4), 732–738.

43. Roubin, G.S.; Yadav, S.; Iyer, S.S.; J. Vitek. Carotid stent-supported angioplasty: A neurovascular intervention to prevent stroke. Am. J. Cardiol. **1996**, *78* (3), 8–12.

44. Dietrich, E.B. Aortic endografting: Visions of things to come. J. Endovasc. Surg. **1996**, *3* (4), R21–R23.

45. Wholey, M.H.; Wholey, M.; Bergeron, P.; Diethrich, E.B.; Henry, M.; Laborde, J.C.; Mathias, K.; Myla, S.; Roubin, G.S.; Shawl, F.; Theron, J.G.; Yadav, J.S.; Dorros, G.; Guimaraens, J.; Higashida, R.; Kumar, V.; Leon, M.; Lim, M.; Londero, H.; Mesa, J.; Ramee, S.; Rodriguez, A.; Rosenfield, K.; Teitelbaum, G.; Vozzi, C. Current global status of carotid artery stent placement. Cathet. Cardiovasc. Diagn. **1998**, *44* (1), 1–6.

46. Richter, G.M.; Palmaz, J.C.; Allenberg, J.R.; Kauffmann, G.W. Percutaneous stent grafts for aortic-aneurysms—preliminary experience with a new procedure. Radiology **1994**, *34* (9), 511–518.

47. Fattori, R.; Piva, T. Drug-eluting stents in vascular intervention. Lancet **2003**, *361* (9353), 247–249.

48. Sousa, J.E.; Serruys, P.W.; Costa, M.A. New frontiers in cardiology: Drug-eluting stents – Pt. I. Circulation **2003**, *107* (17), 2274–2279.

49. Sousa, J.E.; Serruys, P.W.; Costa, M.A. New frontiers in cardiology: Drug-eluting stents – Pt. II. Circulation **2003**, *107* (18), 2283–2289.

50. Nakayama, Y.; Kim, J.Y.; Nishi, S.; Ueno, H.; Matsuda, T. Development of high-performance stent: Gelatinous photogel-coated stent that permits drug delivery and gene transfer. J. Biomed. Mater. Res. *57* (4), **2001**, 559–566.

51. Ishihara, K.; Ueda, T.; Nakabayashi, N. Preparation of phospholipids polymers and their properties as hydrogel membrane. Polym. J. **1990**, *22* (5), 355–360.

52. Huang, N.P.; Michel, R.; Voros, J.; Textor, M.; Hofer, R.; Rossi, A.; Elbert, D.L.; Hubbell, J.A.; Spencer, N.D. Poly(L-lysine)-g-poly(ethylene glycol) layers on metal oxide surfaces: Surface-analytical characterization and resistance to serum and fibrinogen adsorption. Langmuir **2001**, *17* (2) 489–498.

53. Sianos, G.; Hofma, S.; Lighthart, J.M. R.; Saia, F.; Hoye, A.; Lemos, P.A.; Serruys, P.W. Stent fracture and restenosis in the drug-eluting stent era. Cathet. Cardiovasc. Interv. **2004**, *61* (1), 111–116.

54. Consgny, P.M. Endotherial cell seeding on prosthetic surfaces. J. Long Term Eff. Med. Implants **2000**, 10 (1–2), 79–95.

55. Seifalian, A.M.; Tiwari, A.; Hamilton, G.; Salacinski, H.J. Improving the clinical patency of prostetic vascular and coronary artery bypass grafts: The role of seeding and tissue engineering. Artif. Organs **2001**, *26* (4), 489–495.

56. Heintz, C.; Riepe, G.; Birken, L.; Kaiser, E.; Chakfe, N.; Morlock, M.; Delling, G.; Imig, H. Corroded nitinol wires in explanted aortic endografts. J. Endovasc. Ther. **2001**, *8* (3) 248–253.

57. Mahato, R.I. *Biomaterials for Delivery and Targeting of Proteins and Nucleic Acids*; CRC Press: Boca Raton, FL, 2005.

58. Kenausis, G.L.; Vörös, J.; Elbert, D.L.; Huang, N.; Hofer, R.; Ruiz-Taylor, L.; Textor, M.; Hubbell, J.A.; Spencer, N.D. Poly(L-lysine)-g-poly(ethylene glycol) layers on metal oxide surfaces: Attachment mechanism and effects of polymer architecture on resistance to protein adsorption. J. Phys. Chem. B **2000**, *104* (14), 3298–3309.

59. Huang, N.P.; Csucs, G.; Emoto, K.; Nagasaki, Y.; Kataoka, K.; Textor, M.; Spencer, N.D. Covalent attachment of novel poly(ethylene glycol)-poly(DL-lactic acid) copolymeric micelles to TiO_2 surfaces. Langmuir **2002**, *18* (1), 252–258.

60. Zhang, F.; Kang, E.T.; Neoh, K.G.; Wang, P.; Tan, K.L. Surface modification of stainless steel by grafting of poly(ethylene glycol) for reduction in protein adsorption. Biomaterials **2001**, *22* (12), 1541–1548.

61. To, Y.; Hasuda, H.; Sakuragi, M.; Tsuzuki, S. Surface modification of plastic, glass and titanium by photoimmobilization of polyethylene glycol for antibiofouling. Acta Biomater. **2007**, *3* (6), 1024–1032.

62. Tanaka, Y.; Doi, H.; Iwasaki, Y.; Hiromoto, S.; Yoneyama, T.; Asami, K.; Imai, H.; T. Hanawa. Electrodeposition of amine-terminated-poly(ethylene glycol) to Ti surface. Mater. Sci. Eng. C **2007**, *27* (2), 206–212.

63. Tanaka, Y.; Doi, H.; Kobayashi, E.; Yoneyama, T.; Hanawa, T. Determination of the immobilization manner of amine-terminated poly(ethylene glycol) electrodeposited on a Ti surface with XPS and GD-OES. Mater. Trans **2007**, *48* (3), 287–292.

64. Tanaka, Y.; Saito, H.; Tsutsumi, Y.; Doi, H.; Imai, H.; Hanawa, T. Active hydroxyl groups on surface oxide film of Ti, 316l stainless steel, and cobalt-chromium-molybdenum alloy and its effect on the immobilization of poly(ethylene glycol). Mater. Trans. **2008**, *49* (4), 805–811.

65. Tanaka, Y.; Matsuo, Y.; Komiya, T.; Tsutsumi, Y.; Doi, H.; Yoneyama, T.; Hanawa, T. Characterization of the spatial immobilization manner of poly(ethylene glycol) to a titanium surface with immersion and electrodeposition and its effects on platelet adhesion. J. Biomed. Mater. Res. **2010**, *92A* (1): 350–358.

66. Tanaka, Y.; Matin, K.; Gyo, M.; Okada, A.; Tsutsumi, Y.; Doi, H.; Nomura, N.; Tagami, J.; Hanawa, T. Effects of electrodeposited poly(ethylene glycol) on biofilm adherence to titanium. J. Biomed. Mater. Res. **2010**, *95A* (4), 1105–1113.

67. Balachander, N.; Sukenik, C.N. Monolayer transformation by nucleophilic substitution: Applications to the creation of new monolayer assemblies. Langmuir **1990**, *6* (11), 1621–1627.

68. Bain, C.D.; Troughton, Y.; Tao, Y.T.; Evall, J.; Whitesides, G.M.; Nuzzo, R.G. Formation of monolayer films by the spontaneous assembly of organic thiols from solution onto gold. J. Am. Chem. Soc. **1989**, 111 (1), 321–335.

69. Dubois, L.H.; Nuzzo, R.G. Synthesis, structure, and properties of model organic surfaces. Ann. Rev. Phys. Chem. **1992**, *43* (1), 437–463.

70. UlMan, A. Formation and structure of self-assembled monolayers. Chem. Rev. **1996**, *96* (4), 1533–1554.

71. Xiao, S.J.; Textor, M.; Spencer, N.D. Covalent attachment of cell-adhesive, (Arg-Gly-Asp)-containing peptides to titanium surfaces. Langmuir **1998**, *14* (19), 5507–5516.

72. Gawalt, E.S.; Avaltroni, M.J.; Danahy, M.P.; Silverman, B.M.; Hanson, E.L.; Midwood, K.S.; Schwarzbauer, J.E.; Schwartz, J. Bonding organics to Ti alloys: Facilitating human osteoblast attachment and spreading on surgical implant materials. Langmuir **2003**, *19* (1), 200–204.

73. Brovelli, D.; Hahner, G.; Ruis, L.; Hofer, R.; Kraus, G.; Waldner, A.; Schlosser, J.; Oroszlan, P.; Ehart, M.; Spencer, N.D. Highly oriented, self-assembled alkanephosphate monolayers on tantalum(V) oxide surfaces. Langmuir **1999**, *15* (13), 4324–4327.

74. Textor, M.; Ruiz, L.; Hofer, R.; Rossi, K.; Feldman, K.; Hahner, G.; Spencer, N.D. Structural chemistry of self-assembled monolayers of octadecylphosphoric acid on tantalum oxide surfaces. Langmuir **2000**, *16* (7), 3257–3271.

75. Fang, J.L.; Wu, N.J.; Wang, Z.W.; Li, Y. XPS, AES and Raman studies of an antitarnish film on tin. Corrosion **1991**, *47* (3), 169–173.

76. Van Alsten, J.G. Self-assembled monolayers on engineering metals: structure, derivatization, and utility. Langmuir **1999**, *15* (22), 7605–7614.

77. Gawalt, E.S.; Avaltroni, M.J.; Koch, N.; Schwartz, J. Self-assembly and bonding of alkane-phosphonic acids on the native oxide surface of titanium. Langmuir **2001**, *17* (19), 5736–5738.

78. Verrier, S.; Pallu, S.; Bareille, R.; Jonczyk, A.; Meyer, J.; Dard, M.; Amedee, J. Function of linear and cyclic RGD-containing peptides in osteoprogenitor cells adhesion process. Biomaterials **2002**, *23* (2), 585–596.

79. Reyes, C.D.; Petrie, T.A.; Burns, K.L.; Schwartz, Z.; Garcia, A.J. Biomolecular surface coating to enhance orthopaedic tissue healing and integration. Biomaterials **2007**, *28* (21), 3228–3235.

80. Hynes, R.O. Integrins: bidirectional, allosteric signaling machines. Cell **2002**, *110* (6), 673–687.

81. Bagno, A.; Piovan, A.; Dettin, M.; Chiarion, A.; Brun, P.; Gambaretto, R.; Fontana, G.; Di Bello, C.; Palu, G.; Castagliuolo, I. Human osteoblast-like cell adhesion on titanium substrates covalently functionalized with synthetic peptides. Bone **2007**, *40* (3), 693–699.

82. Elmengaard, B.; Bechtold, J.E.; Soballe, K. *In vivo* study of the effect of RGD treatment on bone ongrowth on press-fit titanium alloy implants. Biomaterials **2005**, *26* (17), 3521–3526.

Mett—Microgels

83. Rammelt, S.; Illert, T.; Bierbaum, S.; Scharnweber, D.; Zwipp, H.; Schneiders, W. Coating of titanium implants with collagen. RGD peptide and chondroitin sulfate. Biomaterials **2006**, *27* (32), 5561–5571.

84. Auernheimer, J.; Zukowski, D.; Dahmen, C.; Kantlehner, M.; Enderle, A.; Goodman, S.L.; Kessker, H. Titanium implant materials with improved biocompatibility through coating with phosphate-anchored cyclic RGD peptides. ChemBiochem **2005**, *6* (11), 2034–2040.

85. Ferris, D.M.; Moodie, G.D.; Dimond, P.M.; Gioranni, C.W.; Ehrlich, M.G.; Valentini, R.F. RGD-coated titanium implants stimulate increased bone formation *in vivo*. Biomaterials **1999**, *20* (23–24), 2323–2331.

86. Xiao, S.J.; Textor, M.; Spencer, N.D.; Wieland, M.; Keller, B.; Sigrist, H. Immobilization of the cell-adhesive peptide Arg-Gly-Asp-Cys (RGDC) on titanium surfaces by covalent chemical attachment. J. Mater. Sci. Mater. Med. **1997**, *8* (12), 867–872.

87. Silverman, B.M.; Wieghaus, K.A.; Schwartz, J. Comparative properties of siloxane vs phosphonate monolayers on a key titanium alloy. Langmuir **2005**, *21* (1), 225–228.

88. Schwartz, J.; Avaltroni, M.J.; Danahy, M.P.; Silverman, B.M.; Hanson, E.L.; Schwarzbauer, J.E.; Midwood, K.S.; Gawalt, E.S. Cell attachment and spreading on metal implant materials. Mater. Sci. Eng. C **2003**, *23* (3), 395–400.

89. Tanaka, Y.; Saito, H.; Tsutsumi, Y.; Doi, H.; Nomura, N.; Imai, H.; Hanawa, T. Effect of pH on the interaction between zwitterion and titanium oxide. J. Colloid Interface Sci. **2009**, *330* (1), 138–143.

90. Oya, K.; Tanaka, Y.; Saito, H.; Kurashima, K.; Nogi, K.; Tsutsumi, H.; Tsutsumi, Y.; Doi, H.; Nomura, N.; Hanawa, T. Calcification by MC3T3-E1 cells on RGD peptide immobilized on titanium through electrodeposited PEG. Biomaterials **2009**, *30* (7), 1281–1286.

91. Yamanouchi, N.; Pugdee, K.; Chang, W.J.; Lee, S.Y.; Yoshinari, M.; Hayakawa, T.; Abiko, Y. Gene expression monitoring in osteoblasts on titanium coated with fibronectin-derived peptide. Dent. Mater. J. **2008**, *27* (5), 744–750.

92. Urist, M.R. Bone: Formation by autoinduction. Science **1965**, *150* (3698), 893–899.

93. Lee, Y.M.; Nam, S.H.; Seol, Y.J.; Kim, T.I.; Lee, S.J.; Ku, Y.; Rhyu, I.C.; Chung, C.P.; Han, S.B.; Choi, S.M. Enhanced bone augmentation by controlled release of recombinant human bone morphogenetic protein-2 from bioabsorbable membranes. J. Periodontol. **2003**, *74* (6), 865–872.

94. Wikesjo, U.M.; Lim, W.H.; Thomson, R.C.; Cook, A.D.; Wozney, J.M.; Hardwick, W.R. Periodontal repair in dogs: Evaluation of a bioabsorbable space-providing macroporous membrane with recombinant human bone morphogenetic protein-2. J. Periodontol. **2003**, *74* (5), 635–647.

95. Seol, Y.J.; Park, Y.J.; Lee, S.C.; Kim, K.H.; Lee, J.Y.; Kim, T.I.; Lee, Y.M.; Ku, Y.; Rhyu, I.C.; Han, S.B.; Chung, C.P. Enhanced osteogenic promotion around dental implants with synthetic binding motif mimicking bone morphogenetic protein (BMP)-2. J. Biomed. Mater. Res. **2006**, *77A* (3), 599–607.

96. Puleo, D.A.; Kissling, R.A.; Sheu, M.S. A technique to immobilize bioactive proteins, including bone

morphogenetic protein-4 (BMP-4), on titanium alloy. Biomaterials **2002**, *23* (9), 2079–2087.

97. Nanci, A.; Wuest, J.D.; Peru, L.; Brunet, P.; Sharma, V.; Zalzal, S.; McKee, M.D. Chemical modification of titanium surfaces for covalent attachment of biological molecules. J. Biomed. Mater. Res. **1998**, *40* (2) 324–335.

98. Nagai, M.; Hayakawa, T.; Fukatsu, A.; Yamamoto, M.; Fukumoto, M.; Nagahama, F.; Mishima, H.; Yoshinari, M.; Nemoto, K.; Kato, T. *In vitro* study of collagen coating of titanium implants for initial cell attachment. Dent. Mater. J. **2002**, *21* (3), 250–260.

99. Viornery, C.; Guenther, H.L.; Aronsson, B.O.; Péchy, P.; Descouts, P.; Grätzel, M. Osteoblast culture on polished titanium disks modified with phosphonic acids. J. Biomed. Mater. Res. **2002**, *62* (1), 149–155.

100. Chang, W.J.; Qu, K.L.; Lee, S.Y.; Chen, J.Y.; Abiko, Y.; Lin, C.T.; Huang, H.M. Type I collagen grafting on titanium surfaces using low-temperature grow discharge. Dent. Mater. J. **2008**, *27* (3), 340–346.

101. Kamata, H.; Suzuki, S.; Tanaka, Y.; Tsutsumi, Y.; Doi, H.; Nomura N.; Hanawa, T.; Moriyama, K. Effects of pH, potential, and deposition time on the durability of collagen electrodeposited to titanium. Mater. Trans. **2011**, *52* (1), 81–89.

102. Pugdee, K.; Shibata, Y.; Yamamichi, N.; Tsutsumi, H.; Yoshinari, M.; Abiko, Y.; Hayakawa, T. Gene expression of MC3T3-E1 cells on bibronectin-immobilized titanium using tresyl chloride activation technique. Dent. Mater. J. **2007**, *26* (5), 647–655.

103. Abe, Y.; Hiasa, K.; Takeuchi, M.; Yoshida, Y.; Suzuki, K.; Akagawa, Y. New surface modification of titanium implant with phosphor-amino acid. Dent. Mater. J. **2005**, *24* (4), 536–540.

104. Cadotte, A.J.; DeMarse, T.B. Poly-HEMA as a drug delivery device for *in vitro* neutral network on micro-electrode arrays. J. Neural. Eng. **2005**, *2* (4), 114–122.

105. Belkasm J.S.; Munro, C.A.; Shoichet, M.S.; Johnston, M.; Midha, R. Long-term *in vivo* bio-chemical properties and biocompatibility of poly(2-hydroxyethyl methacrylate-*co*-methyl methacrylate) nerve conduits. Biomaterials **2005**, *26* (14), 1741–1749.

106. Indolfi, L.; Causa, F.; Netti, P.A. Coating process and early stage adhesion evaluation of poly(2-hydroxy-ethyl-methacrylate) hydrogel coating of 316L steel surface for stent applications. J. Mater. Sci. Mater. Med. **2009**, *20* (7) 1541–1551.

107. Fenelon, A.M.; Breslin, C.B. The electropolymerization of pryrole at a CuNi electrode: Corrosion protection properties. Corros. Sci. **2003**, *45* (12), 2837–2850.

108. Mengoli, G. Feasibility of polymer film coatings through electroinitiated polymerization in aqueous medium. Adv. Polym. Sci. **1979**, *33*, 1–31.

109. De Giglio, E.; Guascito, M.R.; Sabbatini, L.; Zambonin, G. Electropolymerization of pyr-role on titanium substrates for the future development of new biocompatible surfaces. Biomaterials **2001**, *22* (19), 2609–2616.

110. Rammelt, U.; Nguyen, P.T.; Plieth, W. Corrosion protection by ultrathin films of conducting polymers. Electrochem. Acta **2003**, *48* (9), 1257–1262.

111. De Giglio, E.; Gennaro, I.; Sabbatini, L.; Zambonin, G. Analytical characterization of collagen-and/or hydroxyapatite-modified polypyrrole films electrosynthesized on

Ti-substrates for the development of new bioactive surfaces. J. Biomater. Sci. Polym. Ed. **2001**, *12* (1), 63–76.

112. De Giglio, E.; Cometa, S.; Satriano, C.; Sabbatini, L.; Zambonin, G. Electrosynthesis of hydrogel films on metal substrates for the development of coatings with tunable drug delivery performance. J. Biomed. Mater. Res. **2009**, *88A* (4), 1048–1057.

113. Taira, Y.; Imai, Y. Primer for bonding resin to metal. Dent. Mater. **1995**, *11* (1), 2–6.

114. Smith, N.A.; Antoun, G.G.; Ellis, A.B.; Crone, W.C. Improved adhesion between nickel-titanium shape memory alloy and polymer matrix via silane coupling agents. Compos A Appl. Sci. Manuf. **2004**, *35* (11), 1307–1312.

115. Abboud, M.; Casaubieilh, L.; Morval, F.; Fontanille, M.; Duguet, E. PMMA-based composite materials with reactive ceramic fillers: IV. Radiopacifying particles embedded in PMMA beads for acrylic bone cements. J. Biomed. Mater. Res. **2000**, *53* (6), 728–736.

116. Yoshida, K.; Tanagawa, M.; Atsuta, M. Effects of filler composition and surface treatment on the characteristics of opaque resin composites. J. Biomed. Mater. Res. **2001**, *58* (5), 525–530.

117. Kanie, T.; Arikawa, H.; Fujii, K.; Inoue, K. Physical and mechanical properties of PMMA resins containing γ-meth acryloxypropyltrimethoxysilane. J. Oral Rehabil. **2004**, *31* (2), 161–171.

118. Ferracane, J. L.; Berge, H. X.; Condon, J.R. *In vitro* aging of dental composites in water-effect of degree of conversion, filler volume, and filler matrix coupling. J. Biomed. Mater. Res. **1998**, *42* (3), 465–472.

119. Bexell, U.; Olsson, M.; Jhansson, M.; Samuelsson, J.; Sundell, P.E. A tribological study of a novel pre-treatment with linseed oil bonded to mercaptosilane-treated aluminum. Surf. Coat. Tech. **2003**, *166* (2–3), 141–152.

120. Bexell, U.; Olsson, M.; Sundell, P. E.; Jhansson, M.; Carlsson, P.; Hellsing, M.A. ToF-SIMS study of linseed oil bonded to mercaptosilane-treated aluminum. Appl. Surf. Sci. **2004**, *231–232*, 362–365.

121. Jayaseelan, S.K.; Ooji, W.J.V. Rubber-to-metal bonding by silanes. J. Adhes. Sci. Technol. **2001**, *15* (8), 967–991.

122. Sakamoto, H.; Doi, H.; Kobayashi, E.; Yoneyama, T.; Suzuki, Y.; T. Hanawa. Structure and strength at the bonding interface between a titanium-segmentated polyurethane composite through 3-(trimethoxysilyl) propyl methacrylate for artificial organs. J. Biomed. Mater. Res. **2007**, *82A* (1), 52–61.

123. Sakamoto, H.; Hirohashi, Y.; Saito, H.; Doi, H.; Tsutsumi, Y.; Suzuki, Y.; Noda, K.; Hanawa, T. Effect of active hydroxyl groups on the interfacial bond strength of titanium with segmented polyurethane through γ-mercaptopropyl trimethoxysilane. Dent. Mater. J. **2008**, *27* (1), 81–92.

124. Sakamoto, H.; Hirohashi, Y.; Doi, H.; Tsutsumi, Y.; Suzuki, Y.; Noda, K.; Hanawa, T. Effect of UV irradiation on the shear bond strength of titanium with segmented polyurethane through γ-mercapto propyl trimethoxysilane. Dent. Mater. J. **2008**, *27* (1), 124–132.

125. Nomura, N.; Baba, Y.; Kawamura, A.; Fujinuma, S.; Chiba, A.; Masahashi, N.; Hanada, S. Mechanical properties of porous titanium compacts reinforced by UHMWPE. Mater. Sci. Forum **2007**, *539–543*, 1033–1037.

126. Nakai, M.; Niinomi, M.; Akahori, T.; Yamanoi, H.; Itsuno, S.; Haraguchi, N.; Itoh, Y.; Ogasawara, T.; Onishi, T.; Shindo, T. Effect of silane coupling treatment on mechanical properties of porous pure titanium filled with PMMA for biomedical applications. J. Jpn. Inst. Metals **2008**, *72* (10), 839–845.

127. Possart, W. Adhesion of polymers. In *Metals as Biomaterials*; Hansen, J.A., Breme, H.J.; Eds.; Wiley: New York, 1998; pp. 197–218.

128. Worch, H. Special thin organic film. In *Metals as Biomaterials*, Hansen, J.A., Breme, H.J., Eds.; Wiley: New York, 1998; pp. 177–196.

129. Xiao, S.J.; Kenausis, G.; Textor, M. Biochemical modification of titanium surfaces. In *Titanium in Medicine*; Brunrtte, D.M., Tenvall, P., Textor, M., Thomsen, P., Eds.; Springer: Amsterdam, the Netherlands, 2001; pp. 417–455.

Microcomponents: Polymeric

Can Yang
Xiao-Hong Yin
College of Engineering, Zhejiang Normal University, Zhejiang, China

Abstract

Possessing superior properties such as light weight, low cost, and good biocompatibility, the polymeric materials have become more and more important in fabricating micro/nano components for a broad range of applications. This entry selectively describes some key aspects relating to polymeric microcomponents for biomedical applications. Specifically, starting with some basic conceptions, the biomedical-purposed polymeric materials and microcomponent fabrication technologies including photo defining and mold defining as well as ultra-precision machining techniques are introduced. Furthermore, a case study on a micromixer in terms of its manufacturing process and mixing performance testing is presented. Finally, the main conclusions are given before the future direction of the polymeric microcomponents is pointed out.

INTRODUCTION

Microelectromechanical systems (MEMS) or microsystem technology is a rapidly growing domain with a variety of applications and a strong potential for further development. The volume of MEMS products grew 17% in 2011 reaching US$10.2 billion, and is forecasted to grow to approximate US$21 billion by 2017.[1] With the advancement of the biomedical technology and increased concern on the health issues, the MEMS products have extended their applications to biological and medical fields as they allow fulfilling maximal functionalities with minimal space. The aforementioned new MEMS branch (i.e., for biomedical applications) is known as bio-microelectromechanical systems (BioMEMS), which requires a number of microcomponents suitable for biomedical usage.

In the BioMEMS industry, there are several materials such as metal, glass, quartz, silicon, and polymer, available for manufacturing microcomponents offering various functionalities. Each material has its own characteristics and is suitable for certain applications. In recent years, the polymer materials have received increased attention and been utilized more and more in BioMEMS, being mainly motivated by two factors described as follows:[2] 1) many polymer materials provide superior thermal, mechanical, and electrical properties. In addition, some provide unique chemical, structural, and biological functionalities not available in any other material systems. For instance, a microlancet needle has been successfully accomplished adopting a biodegradable polymer, poly lactic acid (PLA), and its effectiveness in collecting blood sample from human body has been demonstrated;[3] 2) polymer-based microcomponents are of significantly lower cost, because of not only the cheap materials themselves, but also the

corresponding processing without cleanroom confinement. Considering the polymer's promising applications in BioMEMS, this entry intends to selectively review the state of the art of microcomponents made of polymer materials for biomedical applications to highlight both the current status and potential developments in this specific field. In this entry, the basic concepts and principal functionalities of the polymeric microcomponents are firstly introduced. Then the material characteristics and micromanufacturing techniques are presented and discussed. Moreover, a case study on a micromixer for fluid mixing at microscale is given, followed by some main conclusions and some comments on the prospects of this specific field.

POLYMERIC MICROCOMPONENTS

Basic Concepts

A common although interesting question one may ask would be how small is really small, or how do we define small? In fact, as we talk about polymeric microcomponents, we are defining them by comparing with macro components. However, the "micro" and "macro" are relative concepts with no absolute boundary between them, and they are being continuously renewed, thanks to the rapid development of the micro/nano technologies. For example, a human hair is tiny enough in most cases, while it is rather larger than a three-dimensional "HAIR" structure with a height of around 10 μm (see Fig. 1).[4] Microcomponents here don't necessarily mean that all dimensions fall into the range of micron. They also cover those with microfeatures and micron-range precision. Since first proposed by Kukla et al.,[5] the criterion as to what can

Concise Encyclopedia of Biomedical Polymers and Polymeric Biomaterials DOI: 10.1081/E-EBPPC-120050010

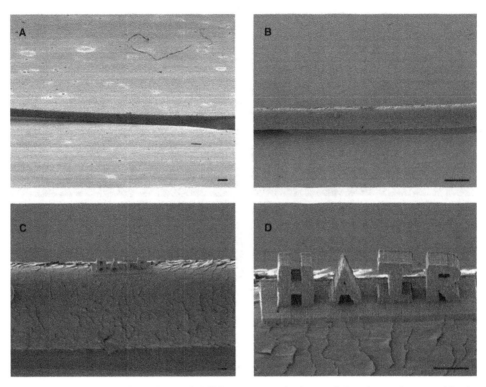

Fig. 1 Size comparison between a three-dimensional "HAIR" structure and a human hair at increasing magnification from (**A**) to (**D**). Scale bars are 100 μm in (**A**) and (**B**) and 10 μm in (**C**) and (**D**).
Source: From Baldacchini et al.[4] © 2004, with permission from AIP Publications.

be considered microcomponents has been continuously renewed, and the most acceptable ones currently include:[6]

1. Microparts: parts have a weight of a few mg or being a fraction of a polymer pellet approximately spherical in shape and 3 mm in diameter.
2. Microfeatured parts: conventional-sized parts with microfeatures having a characteristic dimension typically in several hundred microns.
3. Microprecision parts: parts have any dimensions with tolerances in micrometer range, typically between 2 and 5 μm.

The microcomponents discussed in this entry belong to at least one of the above definitions. As a matter of fact, microproducts weighing only 0.12 mg have been successfully microinjection molded by MTD Micro Molding.[7] In other words, these components are so tiny that a sesame-seed-sized plastic pellet is enough to produce 520 pieces. As examples, Fig. 2 shows some microcomponents available from the industry or academia manufactured using processes such as microinjection molding (μIM) and microextrusion.

Figure 2A: The biodegradable microstent for medical application, which weighs only 1.08 mg vs. the corresponding runner 160 mg (one mold four cavities).[8] Figure 2B: The sensor housing for a hearing aid weighing only 2.2 mg, which means ten pieces of such microcomponents spend only one 3-mm-diameter polymeric pellet. Figure 2C: The microballoon serving in angioplasty, whose sizes

can be as small as 500 μm.[9] Figure 2D: An array of pyramidal dissolving microneedles for transdermal drug delivery, in which each has a base width and height around 300 and 600 μm, respectively.[10]

With the component size decreasing, the size-relevant physical parameters change to different extents, depending on the strength of their dependence on the dimensional factor. Table 1 lists some key physical parameters as well as their dimensional effects.[11] It can be deduced that the surface-to-volume ratio (*S/V*) of the object increases with decreased dimension. As an example, the *Struthio camelus* and wood nymph have an *S/V* value of 10^{-3} and 10^{-1} mm^{-1}, respectively, according to their body size. Larger *S/V* value means more effective contact between the bird and the air, which facilitates flying in the sky. This may explain why the wood nymph is so good at flying, even to a height up to 5 km. In addition, large *S/V* makes the nanostructures excellent candidates for ultra-high-sensitivity detection of chemicals and biological molecules. Toward this direction, researchers at the University of British Columbia have performed some meaningful work aimed at understanding and diagnosing disease using biosensor capable of sensitively detecting RNA.[12] In the microscale dimension, some factors normally neglected in the macroscale should be considered. Surface roughness is one of such key factors, since in cases with microscale this becomes comparable with the microfeature dimensions themselves. In this regard, the corresponding surface roughness for the mold tools used to

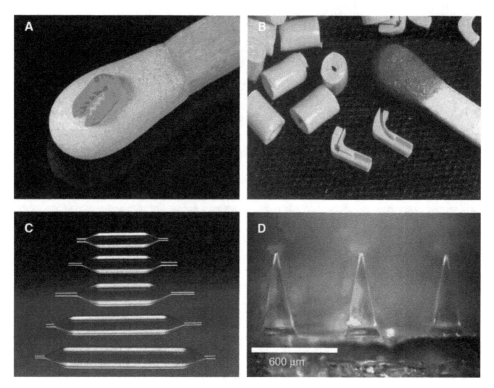

Fig. 2   Representative of available microcomponents: (**A**) biodegradable microstent fabricated using μIM, (**B**) sensor housing for a hearing aid fabricated using μIM with the material polyoxymethylene (POM, acetal), (**C**) microangioplasty balloon, and (**D**) microneedles made of bovine serum albumin.
Source: (A) Courtesy of Wittmann-Battenfeld (Source: Microsystems UK). (B) Courtesy of Wittmann-Battenfeld. (C) Courtesy of Precision Extrusion Inc. (Glens Falls, NY, All right reserved). (D) From Lee et al.[10] © 2008, with permission from Elsevier.

form the microcomponents needs to be as small as 30 nm (reaching a mirror surface).[13] Figure 3 shows an aluminum (6061) mold insert utilized in μIM, which has an arithmetical mean roughness value (Ra) of only 11.82 nm.

Microcomponent Functions in Biomedical Industry

Apart from the size- and weight-based definitions discussed previously, the polymeric microcomponents can also be classified with respect to their applications. In this entry, we focus on the polymeric microcomponents designed for biomedical application, and will not discuss other exciting markets, such as automotive, telecommunication, electronic, micromechanics, and optical industries. Significant researches have been devoted to investigating a number of categories of polymeric microcomponents in biomedical industry, particularly biodegradable ones. These include diagnostic devices, tissue engineering components, implantable devices, microfluidic devices, drug delivery and therapeutic systems, and so forth. Based on the authors' previous work,[14] the specific functions as well as typical examples for each category are summarized as shown in Table 2. The above aspects constitute an entire technical chain being able to provide functionalities such

as health status monitoring, medical diagnosing, analyzing, and even curing. Such functionalities benefit from every great progress in material innovation, process optimization, and microfabrication technique advancement.

The fact that the microdiagnostic devices, in most cases including microfluidic devices, have drawn considerable attention in biomedical application mainly owing to the following reasons: 1) Higher sensitivity and short testing cycle can be realized due to small device dimensions comparable to that of the testing sample; 2) Unlike the conventional lab-based analysis appliances, the credit card-sized analysis system, which is also called Lab-on-Chip, enables point-of-care usage; 3) Adopting biodegradable polymeric material in fabricating such devices exhibits great cost advantage. For instance, a portable blood test device that can diagnose an HIV infection within 15 min is available as a breakthrough in the fight against AIDS. The credit card-sized plastic device (called mChip, as shown in Fig. 4) developed by scientists at the University of Columbia in New York, costs only $1 to make, which has great potential application in developing countries.[15] In another research,[16] a contact lens allowing continuous noninvasive monitoring of tear glucose has been realized (Fig. 5). It was made of biodegradable material polydimethyl siloxane (PDMS) and hydrophilic PMEH [a copolymerization

Table 1 Physical parameters and their dimensional effects as the size decreases

Parameters	Expressions	Dimensional effect	Note
Length	L	L	
Surface area	$S \propto L^2$	L^2	
Volume	$V \propto L^3$	L^3	
Mass	$m \propto L^3$	L^3	
Gravity	mg	L^3	
Inertial force	ma	L^4	
Electrostatic force	$S\varepsilon E^2/2$	L^2	ε: Dielectric constant; E: Electric field intensity
Electromagnetic force	$S\mu H^2/2$	L^4	μ: Magnetoconductivity; E: Magnetic field intensity
Linear elastic coefficient	K	L	
Natural frequency	$(K/m)^{1/2}$	L^{-1}	
Moment of inertia	I	L^5	
Reynolds number	Re	L^2	
Heat transfer	$\lambda \Delta TA/d$	L	λ: Thermal conductivity; ΔT: Temperature difference A: Cross sectional area
Heat convection	$h\Delta TS$	L^2	h: Heat transfer coefficient

Source: From Tian[11] © 2009, with permission from Xidian University Press.

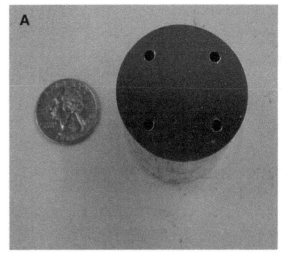

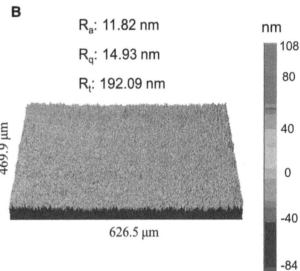

Fig. 3 An aluminum (6061) mold insert for µIM: (**A**) a picture of the precision machined insert, and (**B**) the corresponding surface roughness of the insert.

product of 2-methacryloyloxyethyl phosphorylcholine (MPC) and 2-ethylhexyl methacrylate (EHMA)]. With the film electrodes formed onto the PMDS membrane, the efficacy of the soft lens was experimentally validated using a rabbit.

MATERIALS

Polymeric Materials Suitable for Micro Molding

Polymeric materials have become the key drivers of innovations and industry development, as their applications prevail in a variety of markets such as medical, automotive, aerospace, and telecommunication fields. Though there are about 86,000 polymer materials available from over 880 different resin suppliers in the global market,[17] our understanding of material characteristics are basically obtained from the conventional polymer processing, where

it is known that material properties usually vary from batch to batch as supplied. However, this batch-to-batch or even pellet-to-pellet variation in material properties may cause unacceptable fluctuations in the micro-molding process.[18] Accordingly, to fully understand the polymer characteristics including thermal and rheological behaviors at microscale is still very challenging. In this regard, many efforts have been made dedicating to exploring polymeric material behaviors at microscale. In an attempt, Yang and co-workers[19] carried out a study using a semicrystalline polymer polypropylene (PP) and an amorphous one polymethyl methacrylate (PMMA) in terms of their filling capability into microcavities with a constant width 55 µm and a maximal aspect ratio 2.0. The experimental results show that there is a wide gap between the replicated heights (defined as the height ratio of molded microstructure and the corresponding microcavity) with these two materials (range of over 92.3% for PP vs. 4.4% ~ 99.5%

Table 2 Microcomponent categories for biomedical applications

Categories	Functions	Examples
Diagnostic devices	Test sample collecting, status monitoring, and characterization	Point-of-care microchip for HIV detecting Biodegradable microneedles to collect blood sample for clinic assay Contact lenses capable of monitoring tear glucose level Biosensors enabling on-chip cell characterization
Tissue engineering	Serving as supporting structures providing biofunctionalities	Blood vessel substitutes to reestablish vascular continuity Artificial structures capable of supporting tissue growth Biodegradable polymer coatings enable mitigating tissue reaction
Implantable devices	Components acting as or working with organs	Sensor housing of a hearing aid Deep-brain stimulators Microelectrodes for neural recording and stimulation Man-made organs
Microfluidics	Handling of microfluid for biomedical testing and analysis	Micropump/valve for microflow controlling Micromixer allowing thorough fluid mixing at micro and nano scale Microfilter for component separation
Drug delivery and therapeutic systems	Capable of delivering medication, vaccination and providing curing functionalities	Microvehicle/robot for accurate medication delivery and release Micromedical tubing for drug delivery Cardiac monitoring-purpose microstents Precision balloon catheter for vessel inflation Biodegradable polymeric nanoparticles for carrying various therapeutic agents

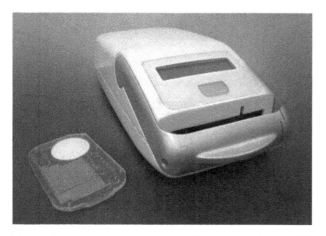

Fig. 4 The portable microdevice (left) capable of testing HIV with an accuracy of almost 100%. It comes with a cheap detector (right) if clarification is needed.
Source: From Webb[45] © 2010, with permission from BioTechniques.

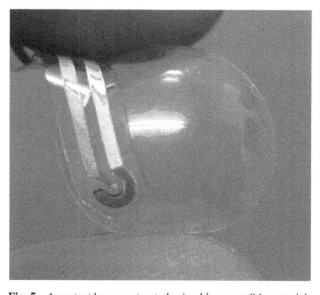

Fig. 5 A contact lens constructed using biocompatible materials (PDMS and hydrophilic PMEH) with flexible electrodes.
Source: From Chu et al.[16] © 2010, with permission from Elsevier.

for PMMA). Though this demonstration does not tell us how one could fill the microcavities fully with every material, this does give some information on how different materials affect microproduct quality with a fixed mold design. In another study of this series work, Yang et al.[20] placed emphasis on investigating the filling behavior of high-density polyethylene (HDPE) melt inside high-aspect-ratio (up to 200) microchannels as well as the formation mechanism of surface defects by means of short shot experiments. It is concluded that the defect morphology strongly depends on the melt temperature and

injection velocity during molding. Figure 6 gives a comparison of surface morphology between two groups of condition settings, from which one observes that increasing melt temperature and injection velocity greatly smooth the surfaces defects (see circled regions) of all investigated microstructures.

Theoretically speaking, most of the 86,000 polymer materials can be adopted for micromolding. However,

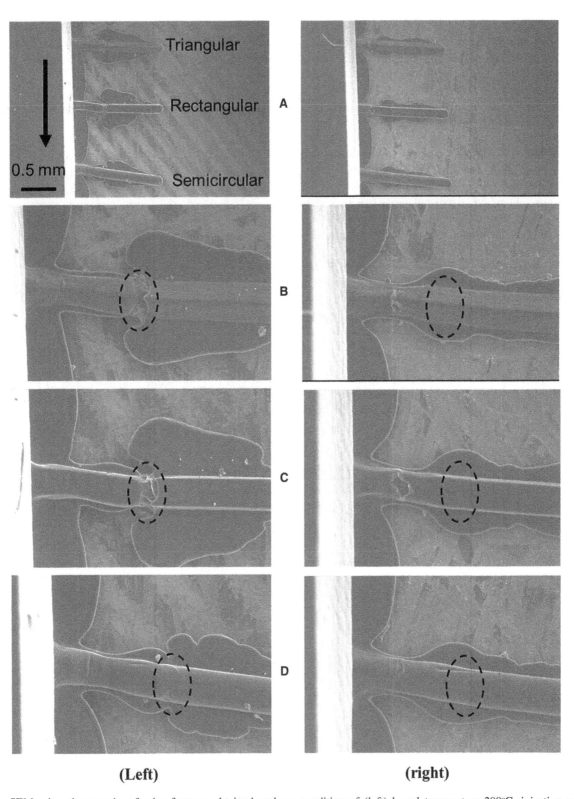

Fig. 6 SEM microphotographs of microfeatures obtained under a condition of (left) barrel temperature 200°C, injection velocity 100 mm/s, and (right) barrel temperature 250°C, injection velocity 250 mm/s for (**A**) overall view, (**B**) triangular, (**C**) rectangular, and (**D**) semicircular shapes. The arrow represents the melt flow direction in the bulk flow; all micrographs use the same scale bar.
Source: From Yang et al.[20] © 2011, with permission from John Wiley and Sons.

thermoplastics with low viscosity and thus good flow properties during filling, high mechanical strength while demolding, as well as superior functionalities in serving are preferred. Selecting a polymer material for a specific case involves many considerations such as the easiness of availability, cost, and physical and chemical properties required in the specific application and so forth. A broad range of polymer materials have been successfully employed in the existing researches, among which the often-used ones are summarized in Table 3. Their availability allows one to select a suitable polymeric material for numerous applications associated with a variety of situations as needed. For example, low-cost polymers PP and PE are good choices for common usages without any special requirements, and on the other hand, for some cases where the products have very thin sections, relatively expensive liquid–crystal polymer (LCP) can serve alternately, thanks to its high ease for mold-detail replication. There are also polymers that have a melting temperature up to 343°C (e.g., PEEK) providing excellent mechanical properties and high resistance to thermal degradation, and others that have excellent abrasion and chemical resistance [e.g., polyester-based thermoplastic polyurethane (TPU)]. Moreover, microcomponents can be molded either soft and elastic with polymers such as POM or hard and brittle with

polymers such as polysulfone (PSU).[21] Besides, characteristics such as optical transparency from PMMA and piezo-electrical effect from polyvinylidene fluoride (PVDF) can be taken full advantage of as needed.

Biomaterials for Polymeric Microcomponents

Biomaterials refer to those interfacing with biological systems in order to evaluate, treat, augment, or replace any tissue, organ, or function of the body. They can be implanted in the body or contact body fluids outside the body, serving as functional components for modifying the cellular response or supporting structures (e.g., scaffolds) for growing cells and tissues. Though a material with good flow ability is preferred for molding polymeric microcomponents, the essential prerequisite to qualify it as a biomaterial is that it should has biocompatibility, which is the ability to perform with an appropriate host response in a specific application.[22,23]

Placing emphasis on their degradation ability with *in vitro* and *in vivo* environments, the polymeric biomaterials discussed in this entry can be broadly classified into two types: 1) biostable polymers; and 2) biodegradable polymers. The latter ones have been receiving increased attention, in that once implanted into the body, there is no

Table 3 The often-used polymeric materials for micromolding

Family abbreviation	Full name	Melting/glass transition temperature[a] (°C)	Main advantages
PP	Polypropylene	130~170	Cheap and easily available
PE	Polyethylene	105~130	Cheap and easily available
POM	Polyoxymethylene	175	High stiffness and low friction
PMMA	Polymethylmethacrylate	85~165	High transparency
PC	Polycarbonate	147	High transparency, balanced features between commodity and engineering plastics
ABS	Acrylonitrile butadiene styrene	105	Good combination of strength, rigidity, and toughness
TPU	Thermoplastic polyurethane	n/a	Excellent abrasion and chemical resistance, good mechanical properties and injectability[b]
COC	Cyclo-olefine copolymer	70~200	High dimensional stability and transparency
PS	Polystyrene	100	Transparency, good thermal stability and flowability
PA	Polyamide	180~280	Good mechanical properties and abrasion-resistance
PEEK	Polyetheretherketone	343	Excellent mechanical properties, and high resistance to thermal degradation
PSU	Polysulfone	185	Good toughness and stability at high temperatures
PVDF	Polyvinylidene fluoride	177	Chemically inert, piezoelectric
PBT	Polybutylene terephthalate	225~235	Mechanically strong, heat and solvent resistant
LCP	Liquid-crystal polymer	>300	Extremely inert, highly fire resistant, and high ease of forming
PFA	Perfluoralkoxy copolymer	305	High dielectric strength

[a]Numbers in this column represent melting temperature and glass transition temperature for semicrystalline and amorphous polymer, respectively.
[b]These advantages are based on polyester-based TPU; chemical resistance is inapplicable to polyether-based TPU.

need for a second surgery for getting them out and no concern of the long-term biocompatibility.[22] Both types of biomaterials stated previously include natural and synthetic materials in terms of the origins of the corresponding polymers. Reports on the applications of natural polymers as biomaterials trace back to thousands of years ago. However, the application of synthetic polymers to medicine is a far recent phenomenon.[22] One of the first attempts was the use of the biostable synthetic PMMA as an artificial corneal substitute. Readers who are interested in polymers as biomaterials are directed to the existing topical reviews.[22,24]

FABRICATION TECHNOLOGIES FOR POLYMERIC MICROCOMPONENTS

Based on the principles with which the polymeric materials are processed, the existing microfabrication technologies can be broadly divided into three main categories: photo defining method, mold defining approach, and ultra-precision micromachining technology.[25] They will be discussed separately, with emphasis on the second category due to its wide usage and versatile variants.

Photo Defining Method

Benefiting from the LIGA technology (a series of steps including lithography, electroplating, and molding) previously used by integrated circuit manufacturers over the past decades, nowadays the photo defining method (a.k.a.,

photolithography) is usually employed to realize micropatterning in manufacturing polymeric microcomponents. Particularly, this method is suitable for forming microstructures from polymer thin films, which are typically used as membranes in biomedical microdevices.[26] With this kind of microfabrication technologies, the desired microstructures are defined and constructed by curing a photosensitive substance. As shown in Fig. 7, the starting point for a specific fabrication sequence is the deposition of a polymer thin film as the photoresist on a substrate (e.g., silicon or glass). Following this step, the generation of a mask (a.k.a., photomask) is needed for the subsequent photographic process. After light exposure with a photomask, the photoresist is developed by washing off the exposed or the unexposed regions, depending on whether the resist material is "positive" or "negative," respectively.[27] For positive resist material, it is exposed to light wherever the material is to be removed because exposure to light changes its chemical structure and hence makes it more soluble in the developer solution (Fig. 7A). In contrast, negative resist material becomes polymerized upon exposure to light, and as a result, more difficult to dissolve. Consequently, the developer solution removes only the unexposed portions of the negative resist, meaning that only the surface wherever it is exposed remains (Fig. 7B).[28] Negative resists were popular in the early history of integrated circuit processing, while now positive resists dominate in very-large-scaled integration fabrication processes owing to their better process controllability for small geometry features. Finally, the desired microstructures are obtained with the same or opposite patterns

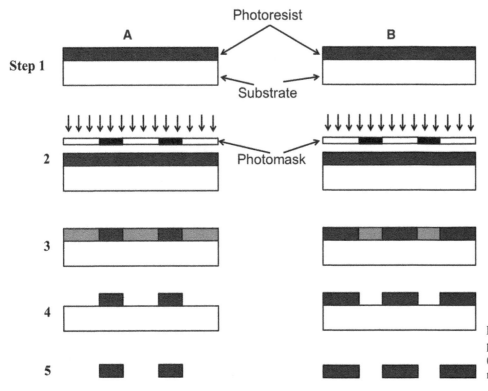

Fig. 7 Schematic drawing of the photolithography steps with a (**A**) positive photoresist and (**B**) negative photoresist.

as in masks associated with the positive and negative resists, respectively. Figure 8 illustrates a microvalve fabricated using lithography technology with a negative photoresist material SU-8.[29]

Unlike the traditional microfabrication, one issue to be addressed in relation to the lithography process as a fabrication method for biomedical microcomponents is that it is necessary to release the patterned thin films from the substrate for subsequent processing or as the final components. This is usually accomplished by using a substrate with low surface energy or by pretreating the surface with a suitable chemical agent that will minimize the adhesion of the resist to the substrate.[26] This can be also realized by dissolving the sacrificial layer in a certain chemical solution (e.g., silicon in KOH solution). It is reported that releasing can be accelerated by dissolving a thin chromium layer precoated between the resist and substrate layer as the sacrificial material.[29]

Mold Defining Approach

This kind of approaches mostly involves melting the polymeric material through mechanical and/or thermal means, confining the molten material with a hard master mold fabricated using standard MEMS micro-fabrication techniques, cooling the formed microcomponents, and finally demolding of them for desired usages. Of the available mold defining approaches, the hot embossing, μIM, micro-extrusion as well as their variants have become the most widely used technologies in industry due to their mass-production capability. From the viewpoint of academia, thermoforming and casting (or soft lithography) also gain popularity in manufacturing microproducts from polymeric materials.

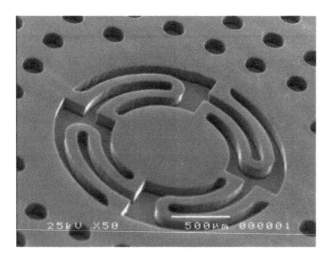

Fig. 8 Microvalve fabricated using lithography technology with a negative photoresist material SU-8.
Source: From Nguyen & Truong[29] © 2003, with permission from Elsevier.

Hot embossing

Developed nearly two decades ago, hot embossing has become a broadly used technology for manufacturing polymeric microcomponents. As shown in Fig. 9, this process starts with heating a polymer sheet (thickness varying from tens of microns to several millimeters)[30] positioned in-between two mold platens of the hot embossing machine. A temperature above the glass transition temperature (T_g) or melting temperature is needed for amorphous and semi-crystalline materials, respectively. Later, the softened polymer sheet is compressed upon the movement of one platen with precision velocity/pressure control. In this situation, deformation of the heated material takes place in such a way that the microstructures on the mold insert (also called master structure) are transferred onto the plastic sheet. In this step, creation of vacuum inside the mold cavity is desired in order to avoid air disturbance. Following microstructure formation, procedure of cooling is carried out prior to the final demolding process in which the molded microcomponent is released from the open mold platens.

This simple but flexible technique enables fulfilling microstructures with high aspect ratio and very tiny sizes down to nanometer range. Particularly, this process exhibits low flow velocity, slow cooling rate, as well as low flow pressure, which gives the advantage of rather low internal stresses inside the molded products and feasibility of forming more delicate structures and high-aspect-ratio microstructures, comparing with other micro-molding techniques such as μIM (to be discussed later).[31,32] Since characteristics such as low birefringence and small warpage associated with low residual stress are important features for optical components, the hot embossed polymeric microcomponents exhibit superiorities in biomedical applications if optical detection technologies are employed.[25] Unfortunately, relatively long cycle time up to 30 min and too much need for manual manipulation become big inconveniences of hot embossing, which limits it as a more prototyping rather than a molding method. Accordingly, this process is mostly used in academic field and low-volume industry application.

Injection process family

In addition to hot embossing, injection-based processes also provide solutions for molding microcomponents. First of all, as a reliable micro-fabrication process, μIM has attracted great attention from both academia and industry. As μIM is the miniaturized version of conventional injection molding (CIM), it shares the same process steps including plasticizing granular material, metering molten material, injecting melt into the mold cavity, packing the injected melt for shrink compensation, cooling the melt, and finally demolding the molded products (Fig. 10).[14] However, this doesn't mean that μIM is the simple scaling down of the CIM in that some negligible factors become more and more important and, therefore, need to be taken into account.

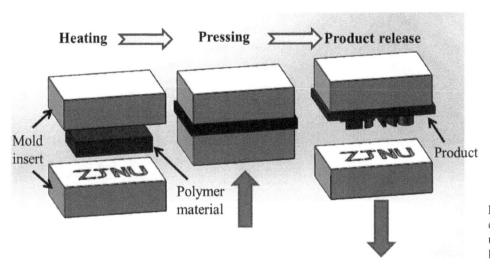

Fig. 9 Schematic representation of the hot embossing process using a mold insert with a "ZJNU" logo.

Comparing with CIM, the main differences encountered in µIM process can be summarized as follows:

1. For the purpose of precision melt metering, most µIM machines adopt separate plasticizing and injection units (usually positioned with an angle as shown in Fig. 10).
2. Due to the extremely high cooling rate caused by tiny cavity size, the mold temperature (T_m) needs to be increased over T_g of the material of interest to avoid premature solidification of the melt ($T_m > T_g + 30~40°C$ as recommended in Ref. [33]).
3. Thorough venting (mostly in the form of vacuuming) is indispensable during µIM in order to prevent the melt from being stopped and even burnt by high-temperature gas trapped inside the microcavities.
4. The surface roughness of the molding tools cannot be ignored any more since its size becomes comparable to those of the microfeatures to be molded. Furthermore, size effect of melt viscosity, wall slip phenomenon, and the surface tension effect also need to be considered.

On the basis of the standard µIM process, several variants have been also established and find their specific application fields. For instance, combining hot embossing and µIM techniques gives birth to microinjection–compression molding, in which the main difference is that mold cavity is only partially closed during the injection process (vs. completely closed in CIM) and that packing force is applied through the movable mold platen instead of the injection gate. This novel packing pattern associated with special packing force distribution leads to more uniform internal stress inside the molded parts and gives more opportunities to control and tailor the microstructures.[34–36] Similarly, with an approach known as microreaction injection molding (µRIM), a mixture of two polymers rather than one is injected into the mold cavity for curing reaction and then forms a mold-defined product. This method extends the molding materials from thermoplastics to thermosets and elastomers. At the early stage of its development, µRIM was found difficult to carry out since a good mixture of two polymers at the microscale for chemical reaction is hard to achieve. Fortunately, the emergence of UV-curing has greatly improved this process, and therefore, making it possible to be a rapid prototyping method.[21] Furthermore, the multicomponent micro-injection molding (MC-µIM) enables getting rid of packaging and assembly steps in certain applications where two or more materials/components are required for integrated functionalities. This process offers decisive advantages for effective production of interesting material combinations such as electrically conductive/insulating or hard/ductile ones.[37]

The success of this process family (injection-based processes) is mainly because of the following: 1) process simplicity and flexibility; 2) short cycle time and low cost; 3) broad range of used materials; 4) easiness for full automation; and 5) accurate dimensional control.

There are also other mold defining techniques available for micro-component fabrication intended for biomedical application. As was mentioned previously, film-like microcomponents can be accomplished using photo defining method. However, this kind of techniques is limited to materials that are photocurable. Alternatively, the microthermoforming offers an easy way for forming film-like products from photo incurable materials. Additionally, another simple process known as casting (a.k.a., soft lithography) instead of injection-based methods enables microfabrication from soft polymers and gels. A good example is the casting of PDMS for microfluidic application due to the material's unique advantages of cost-efficiency, biocompatibility, as well as excellent sealing properties (as shown in Fig. 11).[27]

Microextrusion

The last but not least mold defining method discussed in this entry for forming polymeric microcomponents is the microextrusion. What makes this process different from all

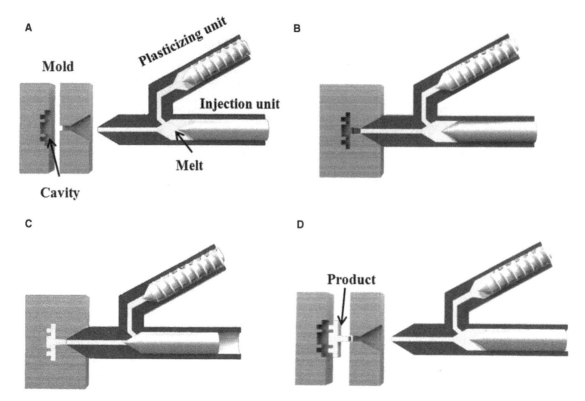

Fig. 10 Process demonstration of μIM with separate plasticizing and injection units: (**A**) plasticizing, (**B**) mold closing, (**C**) injection and packing, and (**D**) demolding and re-plasticizing for next cycle.
Source: From Yang et al.[14] © 2013, with permission from IOP Publishing Ltd.

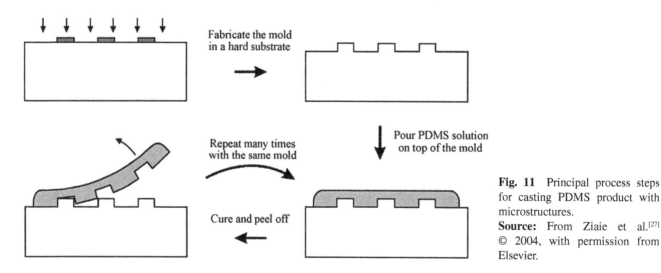

Fig. 11 Principal process steps for casting PDMS product with microstructures.
Source: From Ziaie et al.[27] © 2004, with permission from Elsevier.

aforementioned methods is its capability of producing microproducts having constant cross-sectional shape and very large size in one dimension. The most typical example is the microextrusion of the microtubing, which is shown in Fig. 12. First of all, the barrel of the extruder is heated electrically to a predetermined temperature suitable for the material of interest. Later, through continuous rotation of the screw, the material is transmitted from the hopper in the form of granules to the mold (also called die in this process) side where the uniform melt is formed. This unique process allows production of one-dimensional hollow microcomponents (tubing) with relatively simple set-ups. Comparing with injection-based processes, microextrusion requires larger length-to-diameter ratio of the machine screw for complete melting. This is because of a lack of sequential steps such as injection and packing with μIM. Currently, microtubing with outer diameter of only 100 μm is commercially available. The usage of such products in medical application has the potential to considerably reduce trauma and risk, particularly for

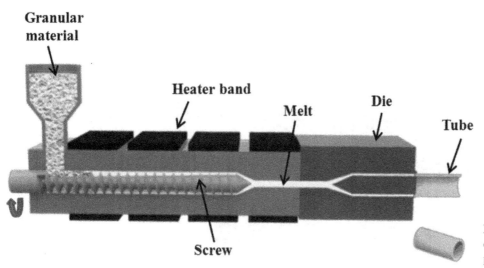

Fig. 12 A three-dimensional demonstration of the micro-extrusion process.

pediatric patients.[38] Originated from the standard micro-extrusion, a process named precision extruding deposition (PED) has also been developed for manufacturing polymeric products with microstructures. In this technique, the extrusion unit is designed in a vertical manner, and the position of the nozzle can be adjusted in three dimensions via prewritten codes. In this way, the filar melt coming from the nozzle of the precision extruder is able to form layer-by-layer polymer structures. Figure 13 illustrates a polycaprolactone (PCL) scaffold for bone tissue engineering having numerous 200 μm-micropores in it fabricated using the PED method.[39]

Precision/Ultra-Precision Machining Technology

Another category of micro-fabrication approaches is the direct computer numerical control (CNC) precision/ultra-precision machining technology with which raw polymeric materials typically in the form of rod or slab is modified to achieve the desired microstructures. This technique follows the routine of conventional CNC machining technologies with improvement in machine control accuracy and microtools making in order to meet the requirements for the precision or ultra-precision fabrication. To date, a microend mill with a diameter as small as 5 μm is commercially available and has been used in several manufacturing applications as well as for advanced academic studies.[40] Figure 14 shows a 25 μm-diameter microend mill demonstrating its minuteness comparable to a human hair tied in a knot. With ultra-precision machining, in polymer blanks materials along a specific path are cut away, leaving corresponding micropatterns on them. In this direction, extensive efforts have been made by researchers all over the world. Yi and co-authors, for example, have carried out a series of studies on understanding the characteristics of the ultra-precision machining process and the corresponding surface quality (e.g., optical scattering characteristics and the surface roughness), possibility of making 3D freeform

geometries from both soft and hard materials, etc. Figure 15 gives a picture of a microprism array for a 3D artificial compound eye. The transparent PMMA component, which has a total of 596 microprisms on the sphere cap, was fabricated utilizing the combination of diamond turning, diamond broaching, and micromilling processes.[41]

It should be noted that apart from the aforementioned three categories of micro-fabricating techniques, other established methods are also available. For example, with plasma etching, micropatterns on a thermoset polymer (e.g., polyimide) can be formed similarly as used in photo-defining methods.[25]

CASE STUDY

To better understand the issues covered in this entry, this section presents a case study on the manufacturing process for an indispensable component, micromixer, in biomedical application. One of the previously discussed mold defining technologies, namely μIM, is selected as the main micro-fabrication method due to its short cycle time and low manufacturing cost. Since the detailed procedures including product design, molding tool fabrication, injection molding, micromixer assembling, and mixing performance testing, are available in a previous work,[42] only the main steps and conclusions are briefly reviewed here.

Though micromixers are broadly divided into two classifications (i.e., active and passive), there are numerous structures available depending on specific application fields (see topical reviews[43,44] for details). The so-called split-and-recombine static micromixers belonging to the passive classification were designed for this case study owing to its high mixing efficiency, which benefits from an exponential increase in fluid interface similar to the chaotic stretching through a series of splitting and recombination steps.[44] The micromixer consisted of two identical components, which allow simplicity of the fabrication

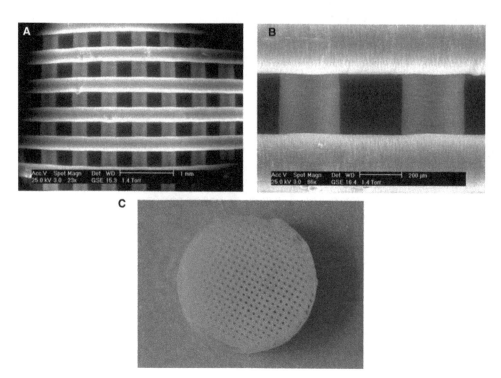

Fig. 13 The PCL scaffold with 200-μm micropores fabricated using PED method: (**A**) and (**B**) represent lower and higher SEM magnifications, respectively; (**C**) a picture showing the overview of the scaffold.
Source: From Shor et al.[39] © 2009, with permission from IOP Publishing Ltd.

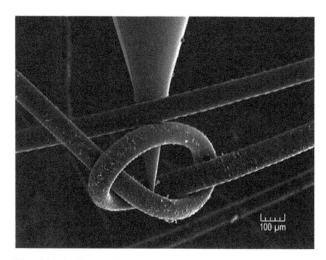

Fig. 14 A 25 μm-diameter micro end mill comparing with a human hair tied in a knot.
Source: Courtesy of Performance Micro Tool.

procedure. One of the designed components having 16 micromixerlets is shown in Fig. 16. This novel structure enables the fluids to split through dividing each channel into two subchannels (one leads to the upper left and the other leads to the lower right direction) and recombine repeatedly, which rapidly mixes two fluids in the form of reducing each fluid width by half in each step (Fig. 17). Therefore, an extremely tiny microfluid with a width of only about 8 nm can be obtained after mixing, passing a total of 16 mixing units.

A combination of ultra-precision CNC machining and μIM was utilized in order to take full advantage of both fabrication processes. The former was adopted to manufacture the aluminum mold insert, and the latter to transfer microstructures from the mold to the product. An optical grade PMMA (Plexiglas1 V825) was employed mainly because of its excellent transparency (see Table 3), which facilitates real-time observation of fluid-mixing process. With optimal μIM conditions (i.e., melt temperature: 280°C; injection velocity: 100 mm/s; packing pressure: 90 MPa; packing time: 2 sec), the mold-defined micromixers with the best replication quality were assembled (Fig. 18) for subsequent fluid mixing testing.

A mixing experiment aiming to reveal the micromixer's performance was then performed with two water solutions as the original fluids to be mixed. Prior to the mixing experiment, both water solutions were colored with different dyes (green and red, respectively) to make them visible under fluorescent light. Figure 19 shows the mixing experiment results from the original fluids to the mixed ones at various mixing steps. Although minor aggregations still exist, the uniform fluid distribution at the outlet of the mixing channel demonstrates a successful design and fabrication of the micromixer, as expected.

CONCLUSIONS AND FUTURE DIRECTION

As a promising branch of the MEMS industry, BioMEMS has been paid great attention to and developed rapidly in

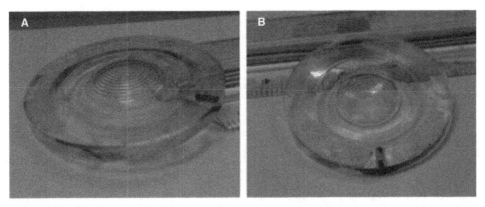

Fig. 15 Ultra-precision machined PMMA components having 596 microprisms: (**A**) front side view and (**B**) back side view. Each microprism has a characteristic dimension of 600 μm.
Source: From Li & Yi[41] © 2010, with permission from Optical Society of America.

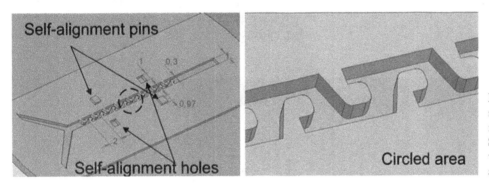

Fig. 16 Structure and dimensions of the micromixer in millimeter.
Source: From Li et al.[42] © 2010, with permission from John Wiley and Sons.

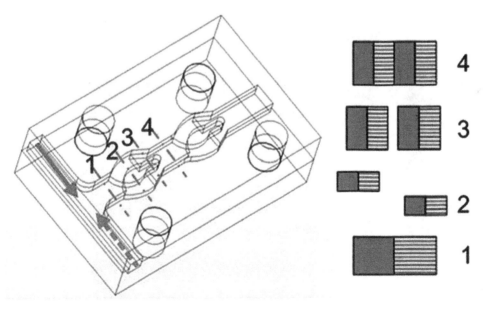

Fig. 17 Structure of the static micromixer (left) and the cross-sectional view of the flows in the micromixer at different stages (right), only two mixerlets are shown.
Source: From Li et al.[42] © 2010, with permission from John Wiley and Sons.

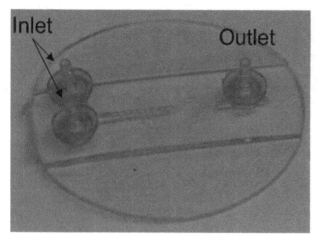

Fig. 18 The assembled PMMA static micromixer with an overall size of 40 mm.
Source: From Li et al.[42] © 2010, with permission from John Wiley and Sons.

recent years due to the advancement of the biomedical technology and increased concern on health issues. With superior properties, especially biological functionalities not being available in other materials, polymers have been playing a more and more important role in manufacturing BioMEMS components. As a category of micro-fabrication methods, photo-defining approach, which has been developed from the integrated circuit manufacturing technique, provides an accurate solution in defining micro/nano patterns. Although having relatively low manufacturing accuracy, the mold-defining methods give more flexibilities and diversities, and receive more attention from both academia and industry. Moreover, precision/ultra-precision direct machining offers a simple way to obtain microstructures from the polymer blanks, mostly with low-volume demand.

On the practical side, there is still a large gap between microcomponents we can make and those the market requires. The existing researches/productions related to polymeric microcomponents are mostly limited to fabrication

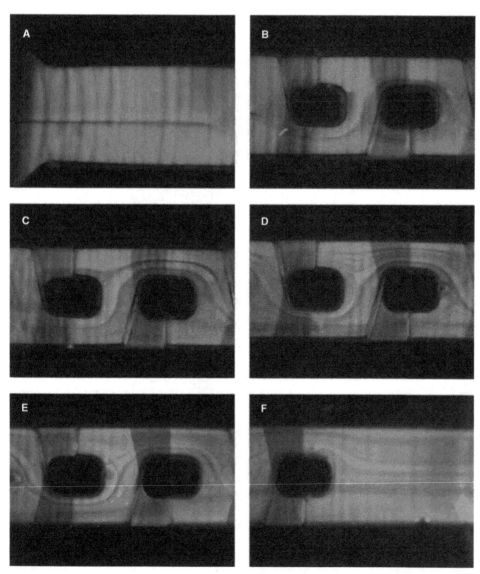

Fig. 19 Mixing experiment results: (**A**) inlet, (**B**) mixerlets 1 and 2, (**C**) mixerlets 9 and 10, (**D**) mixerlets 11 and 12, (**E**) mixerlets 13 and 14, (**F**) mixerlet 16 and outlet.
Source: From Li et al.[42] © 2010, with permission from John Wiley and Sons.

processes themselves, that is, the possibility and feasibility of making certain components by various methods. Recently, some applications demand material properties that are not achievable using pure polymer resins, promoting an increasing interest in considering polymer composites in micro fabrication. For instance, the use of nanoparticle-filled composites is capable of improving the mechanical, electrical, and thermal properties of certain functional microcomponents. In this specific aspect, future research may focus on characterizing the distinct morphology evolution of the composites, and then tailoring the morphology of the microparts according to their desired applications. For example, microfluidic devices with particular functions such as the permeability for selected fluids in medical application may be fabricated using the polymer blending technology.

ACKNOWLEDGMENT

This material is partially based on work supported by Qianjiang Talent Project Foundation (Grant No. QJD1202005), Zhejiang Provincial Natural Science Foundation of China (Grant No. LQ13E050008), and Zhejiang Provincial Department of Education (Grant No. Y201225768).

REFERENCES

1. Eloy, J.-C. Status of the MEMS industry new drivers: the path to new opportunities, 2011. http://euripides.esystems.at/data/ppt-speaker/session3_Jean-Christophe-Eloy.pdf (accessed November 2012).
2. Liu, C. Recent developments in polymer MEMS. Adv. Mater. 2007, 19 (22), 3783–3790.
3. Aoyagi, S.; Izumi, H.; Aoki, T.; Fukuda, M. Development of a micro lancet needle made of biodegradable polymer for low-invasive medical treatment. In Solid-State Sensors, Actuators and Microsystems, 13th International Conference on Seoul, South Korea, Jun 5–9, 2005; IEEE: Piscataway, 2005.
4. Baldacchini, T.; LaFratta, C.N.; Farrer, R.A.; Teich, M.C.; Saleh, B.E.; Naughton, M.J.; Fourkas, J.T. Acrylic-based resin with favorable properties for three-dimensional two-photon polymerization. J. Appl. Phys. 2004, 95 (11), 6072–6076.
5. Kukla, C.; Loibl, H.; Detter, H.; Hannenheim, W. Micro-injection moulding-the aims of a project partnership. Kunststoffe Plast Eur. 1998, 88 (9), 6–7.
6. Attia, U.M.; Marson, S.; Alcock, J.R. Micro-injection moulding of polymer microfluidic devices Microfluid. Nanofluid. 2009, 7 (1), 1–28.
7. MTD micro molding. http://www.mtdmicromolding.com (accessed November 2012).
8. Jin, G. High quality, high efficiency and low consumption, a parade of unique injection technologies, 2011. http://www.adsalecprj.com/Publicity/ePub/lang-eng/article-7594/asid-1/Printing.aspx (accessed November 2012).
9. Custom plastic medical tubing extrusion. http://www.precisionextrusion.com/ (accessed November 2012).
10. Lee, J.W.; Park, J.-H.; Prausnitz, M.R. Dissolving microneedles for transdermal drug delivery. Biomaterials 2008, 29 (13), 2113–2124.
11. Tian, W.-C. Principle, design and analysis of MEMS, 1st Ed.; Zhang, W., Gao, W.-Y., Eds.; Xidian Univ. Press: Xi'an, China, 2009.
12. Bio sensor. http://www.mina.ubc.ca/project_controlled-nanofabrication (accessed November 2012).
13. Yang, C.; Su, L.; Huang, C.; Huang, H.X.; Castro, J.M.; Yi, A.Y. Effect of packing pressure on refractive index variation in injection molding of precision plastic optical lens. Adv. Polym. Tech. 2011, 30 (1), 51–61.
14. Yang, C.; Yin, X.-H.; Cheng, G.-M. Microinjection molding of microsystem components: New aspects in improving performance. J. Micromech. Microeng. 2013, 23 (9), 21, Article ID 093001.
15. The 'credit card' that can tell you if you have HIV within minutes and costs just $1. http://www.dailymail.co.uk/sciencetech/article-2022795 (accessed November 2012).
16. Chu, M.X.; Miyajima, K.; Takahashi, D.; Arakawa, T.; Sano, K.; Sawada, S.; Kudo, H.; Iwasaki, Y.; Akiyoshi, K.; Mochizuki, M.; Mitsubayashi, K. Soft contact lens biosensor for in situ monitoring of tear glucose as non-invasive blood sugar assessment. Talanta 2011, 83 (3), 960–965.
17. Plastic materials. http://www.ides.com/industry/plastics.asp (accessed November 2012).
18. Whiteside, B.R.; Martyn, M.T.; Coates, P.D. Introduction to micromolding. In Precision Injection Molding: Process, Materials, and Applications; Greener, J., Wimberger-Friedl, R., Eds.; Hanser Gardner Publication: Cincinnati, 2006; 239–264.
19. Yang, C.; Li, L.; Huang, H.-X.; Castro, J.M.; Yi, A.Y. Replication characterization of microribs fabricated by combining ultraprecision machining and microinjection molding. Polym. Eng. Sci. 2010, 50 (10), 2021–2030.
20. Yang, C.; Huang, H.-X.; Castro, J.M.; Yi, A.Y. Replication characterization in injection molding of microfeatures with high aspect ratio: Influence of layout and shape factor. Polym. Eng. Sci. 2011, 51 (5), 959–968.
21. Heckele, M.; Schomburg, W.K. Review on micro molding of thermoplastic polymers. J. Micromech. Microeng. 2004, 14 (3), R1–14.
22. Nair, L.S.; Laurencin, C.T. Polymers as biomaterials for tissue engineering and controlled drug delivery. Adv. Biochem. Eng. Biotechnol. 2006, 102, 47–90.
23. Biomaterials based tissue engineering: Biomaterials. http://biomed.brown.edu/Courses/BI108/BI108_2007_Groups/group12/Biomaterials.html (accessed November 2012).
24. Nair L.S.; Laurencin C.T. Biodegradable polymers as biomaterials. Prog. Polym. Sci. 2007, 32 (8–9), 762–798.
25. Becker, H.; Gärtner, C. Polymer microfabrication technologies for microfluidic systems. Anal. Bioanal. Chem. 2008, 390 (1), 89–111.
26. Ochoa, M.; Mousoulis, C.; Ziaie, B. Polymeric microdevices for transdermal and subcutaneous drug delivery. Adv. Drug Deliv. Rev. 2012, 64 (14), 1603–1616.
27. Ziaie, B.; Baldi, A.; Lei, M.; Gu, Y.; Siegel, R.A. Hard and soft micromachining for BioMEMS: Review of techniques and examples of applications in microfluidics and drug delivery. Adv. Drug Deliv. Rev. 2004, 56 (2), 145–172.
28. Photolithography. http://www.ece.gatech.edu/research/labs/vc/theory/photolith.html (accessed November 2012).

29. Nguyen, N.-T.; Truong, T.-Q. A fully polymeric micropump with piezoelectric actuator. Sensor Actuat. B Chem. **2004**, *97* (1), 137–143.

30. Kolew, A.; Münch, D.; Sikora, K.; Worgull, M. Hot embossing of micro and sub-micro structured inserts for polymer replication. Microsyst. Technol. **2011**, *17* (4), 609–618.

31. Heckele, M; Bacher, W.; Müller K.D. Hot embossing—The molding technique for plastic microstructures. Microsyst. Technol. **1998**, *4* (3), 122–124.

32. Giboz, J.; Copponnex, T.; Mélé, P. Microinjection molding of thermoplastic polymers: A review. J. Micromech. Microeng. **2007**, *17* (6), R96–R109.

33. Su, Y.-C.; Shah, J.; Lin, L. Implementation and analysis of polymeric microstructure replication by micro injection molding. J. Micromech. Microeng. **2004**, *14* (3), 415–422.

34. Yang, C.; Huang, H.-X.; Li, K. Investigation of fiber orientation states in injection-compression molded short-fiber-reinforced thermoplastics. Polym. Composite **2010**, *31* (11), 1899–1908.

35. Guan, W.-S.; Huang, H.-X.; Wang, B. Poiseuille/squeeze flow-induced crystallization in microinjection-compression molded isotactic polypropylene. J. Polym. Sci. B Polym. Phys. **2013**, *51* (5), 358–367.

36. Guan, W.-S.; Huang H.-X.; Wu Z. Manipulation and online monitoring of micro-replication quality during injection-compression molding. J. Micromech. Microeng. **2012**, *22* (11), 10, Article ID 115003.

37. Islam A.; Hansen H.N.; Tang P.T.; Jørgensen, M.B.; Ørts, S.F. Two-component microinjection moulding for MID fabrication. Plast. Rubber Composite **2010**, *39* (7) 300–307.

38. Micro-extrusions. http://www.microspecorporation.com/micro_extrusions.php (accessed November 2012).

39. Shor, L.; Guceri, S.; Chang, R.; Gordon, J.; Kang, Q.; Hartsock, L.; An, Y.; Sun, W. Precision extruding deposition (PED) fabrication of polycaprolactone (PCL) scaffolds for bone tissue engineering. Biofabrication **2009**, *1* (1), 10, Article ID 015003.

40. Performance Micro Tool now offers end mills smaller than .001". http://www.pmtnow.com/end_mills/tools/nano.htm (accessed November 2012).

41. Li, L.; Yi, A.Y. Development of a 3D artificial compound eye. Opt. Express **2010**, *18* (17), 18125–18137.

42. Li, L.; Yang, C.; Shi, H.; Liao, W.C.; Huang, H.; Lee, L.J.; Castro, J.M.; Yi, A.Y. Design and fabrication of an affordable polymer micromixer for medical and biomedical applications. Polym. Eng. Sci. **2010**, *50* (8), 1594–1604.

43. Nguyen, N.-T.; Wu, Z. Micromixers—A review. J. Micromech. Microeng. **2005**, *15* (2), R1–R16.

44. Hessel, V.; Lowe, H.; Schonfeld, F. Micromixers—A review on passive and active mixing principles. Chem. Eng. Sci. **2005**, *60* (8–9), 2479–2501.

45. Webb, S.A. Returning to diagnostic basics. Biotechniques *49* (1), 491–493.

Microgels: Smart Polymer and Hybrid

Zahoor H. Farooqi
Institute of Chemistry, University of the Punjab, New Campus, Lahore, Pakistan

Mohammad Siddiq
Department of Chemistry, Quaid-I-Azam University, Islamabad, Pakistan

Abstract
Microgels are polymeric cross-linked colloidal particles that can swell and de-swell in a suitable solvent by changing external stimuli such as temperature, pH, and ionic strength of the medium. Owing to fast response as compared to macrogels, microgels have been attracting much attention of scientists during the past 20 years. This entry describes the fundamental aspects of smart polymer microgels. The responsive behavior of microgels with respect to temperature, pH, and glucose concentration change is elaborated with the help of recent research progress in this field. The concept of biomedical and nanotechnological applications of polymer microgels is provided here.

INTRODUCTION

Recent developments in the area of colloids and polymers have widely diverted the focus of scientists from macro to micro systems. Inorganic and organic materials are largely used to build polymer networks of micro dimensions. Polymer microgels are functional materials of such nature. These microgels possess tunable functionalities, physicochemical properties, large surface areas, polymer composition, size, molecular weight, and narrow molecular weight distribution. The aforementioned characteristics of microgels make them significantly different from bulk materials and other polymeric materials of smaller dimensions because such materials have limited control over architecture, functionalities, and size. Moreover, microgels possess large solvent-holding capacity along with sensitivity to pH, temperature, ionic strength, and light radiations. Often, sensitive microgels are fabricated with nanomaterials such as nanorods or nanoparticles, and multiresponsive fascinating materials are synthesized, which are equally applicable for biological and medical fields. Thus, microgels and hybrid microgels are largely employed for tissue culturing, tissue engineering, drug delivery, medical imaging, biomedical implantation, biosensing, *in vivo* diagnosis, bio-implantation, and biotechnology. Parallel applications of these microgels in *in vivo* and *ex vivo* media are in general trend of scientists nowadays. In this entry, fundamentals and applications of smart polymer microgels are described in detail. The entry covers synthesis, characterization, and applications of hybrid materials based on smart polymers.

POLYMER GELS

"A gel is a three-dimensional cross-linked network, which is swollen by a suitable liquid or by a gas." Both phases are homogeneously mixed and gels can be seen as intermediate between solids and liquids or solids and gases.[1] A polymer gel consists of an elastic cross-linked polymer network with a fluid filling the internal spaces of the network. They are cross-linked colloidal particles with a network structure that is swollen in a suitable solvent.[2] Hydrogels are recently known advanced polymeric materials that can swell in water.[3] On the basis of rheology, hydrogels are defined as cross-linked polymers that exhibit viscoelastic or pure elastic behavior. Hydrogels have very high swelling capacity, because they can absorb water about thousand times more than their dry weight.[4,5] Macrogels and microgels are two different categories of polymer gels based on particle size. Macrogels[6–8] or bulk gels respond very slowly to external parameters.[9] Macrogels[6–8] are bulk gels where the size can be varied from millimeters and above. They show response to the change in external condition in hours due to its dimension in several millimeters. When concentration of monomers is large enough, gels fill the whole volume and are called macrogels. These bulk gels consist of a huge macromolecule formed by the combination of polymer chains via cross-linker. Microgels[10] have colloidal dimensions ranging from 100 nm to 5 μm and respond rapidly to the changes in external parameters owing to their microscopic size. They are intramolecularly cross-linked, high-molecular-weight soluble polymers and were first described by Staudinger and Husemann.[11] Without cross-linking, polymeric chains

Concise Encyclopedia of Biomedical Polymers and Polymeric Biomaterials DOI: 10.1081/E-EBPPC-120050004

in microgels would dissolve in solvent molecularly due to the dominance of polymer–solvent interaction over polymer–polymer interactions; microgels rely on the cross-linking to maintain their dimension stability and integrity.[12] Inside the microgel particles, polymeric chains are cross-linked chemically and/or physically, and the microgels are swollen by water due to their hydrophilicity. These are stable hydrogels.[13]

SMART POLYMER MICROGELS

Polymer microparticles that exhibit quick swelling and shrinking response to slight change in external stimuli are often referred to as "smart" microgels or stimuli-sensitive microgels. They are fascinating materials with great potential for a variety of applications.[14] The swelling ability of a microgel depends on the type and its affinity to the solvent, which in turn is dependent on the monomer (and/or comonomer) composition/concentration, as well as the degree of cross-linking. The nature of the monomer used in the preparation of the microgels will determine the overall properties of the final product. The addition of monomers with different functionalities to the microgel can create particles with a wide range of properties, which, in turn, has a wide range of applications.[15] These applications are based on change in their structure in response to external stimuli such as temperature,[16–20] electric field,[21] biomolecules,[22,23] and pH.[16,24]

TYPES OF RESPONSIVE MICROGELS

Stimulus-responsive polymer microgels and hydrogels can be classified on the basis of stimuli as temperature, pH, ionic strength, light, electric field, and magnetic field-sensitive microgels. They undergo fast, reversible changes in their structure from a hydrophilic to a hydrophobic state.[25] The responsive behavior of microgels is measured in change of hydrodynamic radius of particles as shown in Fig. 1. The stimuli–responsive behavior occurs in aqueous solution, so these microgels have many uses in different fields.

Thermo-Responsive Microgels

Temperature-sensitive polymers exhibit a coil to globule transition in solution where phase separation is induced by increasing temperature. Polymers of such type undergo a thermally induced reversible phase transition; they are soluble in a solvent (water) at low temperatures but become insoluble as the temperature rises above the lowest critical solution temperature (LCST).[26] Poly (*N*-vinylcaprolactam) (PVCL), poly (*N*-isopropyl acrylamide) (PNIPAM), poly(vinyl methyl ether) (PVME), methyl cellulose (MC), etc. are examples of temperature-sensitive polymers.[27] LCSTs of PVCL, PNIPAM, PVME, and MC are 32, 32,

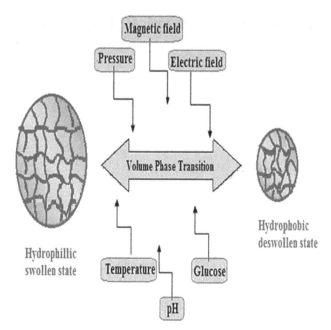

Fig. 1 The responsive behavior of microgels in various stimuli.

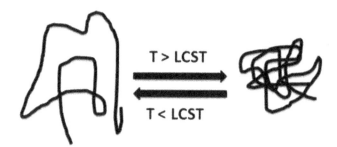

Fig. 2 Coil to globule transition in P(NIPAM).

33.8 and 50°C, respectively. NIPAM-based linear polymers and microgel particles have raised a specific interest for their response to temperature changes.[28]

At low temperatures, P(NIPAM) polymer has a random coil structure and the solvent–polymer interactions are stronger than the polymer–polymer interactions. At higher temperatures, the hydrogen bonds with the water molecules break and there is an entropically favored release of water.[29] As a result, the polymer–polymer interactions become stronger than the polymer–solvent interactions and phase separation occurs. At a certain temperature called as LCST, it changes from a coil to a globule form, as shown in Fig. 2. At a temperature below its LCST ≈32°C, P(NIPAM) is soluble in water and is hydrophilic. The polymer macromolecules are hydrated and fully extended, with extensive intermolecular hydrogen bonding between water molecules and polymer chains. At temperatures above LCST, the hydrogen bonds between water molecules and polymer chains are broken and, instead, intramolecular and intermolecular hydrogen bonds between the C=O and N–H groups are formed. The formation of intramolecular bonds causes the molecule to coil up, exposing its hydrophobic core. This renders the polymers hydrophobic and insoluble in water.[30]

Similarly, microgels also show phase transition at a certain temperature called volume phase transition temperature (VPTT). When temperature is greater than VPTT, microgels shrink and the hydrodynamic radius decreases. At a temperature below the VPTT of P(NIPAM) microgels, water has hydrogen bonding with the amide in the side chains of P(NIPAM) microgels. As microgels are cross-linked moieties, so there is also some structured water around the isopropyl groups, as shown in Fig. 3. Hence the gel is swollen due to strong polymer–solvent interactions. At a temperature above the VPTT of P(NIPAM) microgel, there is an entropically favored release of water from the interior of the microgel and the polymer–polymer interactions become stronger, therefore de-swelling occurs in the gel.[17] P(NIPAM) microgels exist in a swollen state below the VPTT, and when the temperature is greater than the VPTT then the microgels decrease in size due to expulsion of water from the interior of the microgels.[31,32]

The VPTT of the microgels is a function of cross-linking density,[33] nature of solvent,[34] and composition of microgels.[17] For example, the introduction of hydrophilic comonomer (methacrylic acid, acrylic acid, etc.) shifts the VPTT toward higher temperature. NIPAM, acrylamide (AAm), and acrylic acid (AAc) have been polymerized to prepare P(NIPAM-AAm-AAc) microgel, whose VPTT is greater than that of P(NIPAM) microgels due to the hydrophilic nature of AAc and AAm.

pH-Responsive Microgels

These can be defined as cross-linked polymer particles that swell when the pH approaches the pK_a/pK_b of the ionic monomer incorporated within the particles. The pH-responsive microgel is a widely studied class of smart polymer microgels.[35] NIPAM can be copolymerized with a variety of comonomers bearing various functionalities such as acid groups, fluorescent labels, charges, ligands, groups with thermal response or different solubility, and chemically reactive and photoreactive groups by giving the multistimulus sensitivity to microgel particles.[28] For example, the monomer NIPAM has been copolymerized with AAc,[36] methacrylic acid,[37] maleic acid,[38] and vinyl-acetic acid to get pH-responsive microgels. These comonomer molecules are of special interest because they introduce charges to the polymer network. Therefore, electrostatic interactions inside the gel are enhanced and the swelling behavior can be changed in different ways. The presence of ionic groups in polymer microgel network makes it sensitive to the pH and ionic strength and we can tune the VPTT of microgels by varying the composition and pH of the medium. Copolymerization with the more hydrophilic comonomer such as AAc led to an increase in the VPTT, relative to that of the P(NIPAM) microgel. The increase in VPTT depends upon the amount of hydrophilic comonomer used, while incorporation of hydrophobic co-monomer inside the polymer network can decrease the VPTT of microgels.[39] Debord & Lyon investigated the effect of the

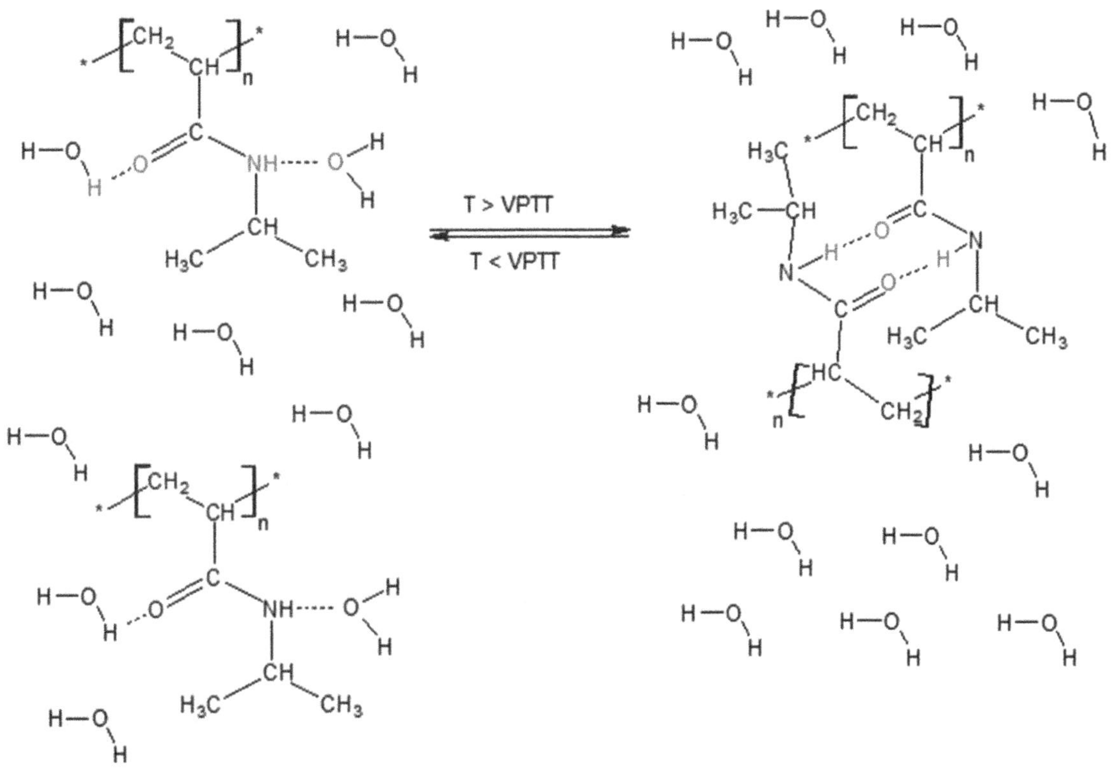

Fig. 3 Structural rearrangement of P(NIPAM) and water molecules.

content of tert-butylacrylamide (TBAm) in P(NIPA-AA-TBAm) microgels at various pH values and found that the increase in TBAm contents decreases the VPTT of the microgels.[40]

The polymer microgels having pH sensitivity can be classified into three: anionic, cationic, and amphiphilic microgels. In anionic or acidic microgels, NIPAM is copolymerized with an acidic moiety, like AAc, etc. while in cationic or basic microgels, NIPAM is copolymerized with basic functionalities like methacrylamide.[41] Amphiphilic microgels may have zwitterionic comonomers or a network having different comonomers; one is acidic and other is basic. Volume phase transition of anionic and cationic microgels occurs in basic and acidic medium, respectively. Amphiphilic microgels show two-stage volume phase transition.[42]

Volume phase transition occurs in all microgels when the pH of the medium approaches pK_a/pK_b values of ionic monomers, as shown in Fig. 4. In anionic microgels, repulsion due to negative charges is the cause of swelling. In acidic medium, microgel carries no charge and exists in a shrunken state (Eq. 1). As the pH of the medium increases, the acidic monomers begin to ionize and negative charges are developed on the network (Eq. 2).

Negative charges repel each other and the microgel swells. The change in hydrodynamic radius occurs till all the acidic monomers ionize. When all monomers become ionized, the microgel attains maximum hydrodynamic radius, and further increase in pH only increases ionic concentration of medium; as a result, osmotic potential outside the microgel decreases and water diffuses out from microgel toward the medium and a small decrease in radius of microgel has been seen.[37] The P(NIPAM-AAc) copolymer microgel consists of carboxylic acid groups in polymer network and it is an example of anionic polymer microgels.

$$[RCOO^{-1}]_{Gel} + [H^{+1}]_{aq} \rightarrow [RCOOH]_{Gel} \qquad (1)$$

$$[OH^{-1}]_{aq} + [RCOOH]_{Gel} \rightarrow [RCOO^{-1}]_{Gel} + H_2O \qquad (2)$$

Cationic microgels have basic functional groups. At low pH, amino groups in microgel network are present in protonated form and there is electrostatic repulsion between these positively charged groups which causes swelling (Eq. 3). With increase in pH, protonation of amines, etc. decreases as hydroxyl ions combine with proton and form water molecules (Eq. 4). Hence, repulsion due to positively charged protonated amines decreases and the microgel shrinks. P(NIPAM-4VP) copolymer system is a cationic polymer microgel.[43,44]

$$[RNH_2]_{Gel} + [H^{+1}]_{aq} \rightarrow [RNH_3^{+1}]_{Gel} \qquad (3)$$

$$[OH^{-1}]_{aq} + [RNH_3^{+1}]_{Gel} \rightarrow [RNH_2]_{Gel} + H_2O \qquad (4)$$

Amphiphilic microgel particles containing acidic and basic groups exhibit two-phase transitions as shown in Fig. 4. First volume phase transition is seen when pH approaches pK_b of basic monomer and second transition in volume is seen when pH approaches pK_a of acidic monomer. The protonation of basic groups at very low pH of medium causes swelling of microgels. The deprotonation of basic groups occurs with the increase of pH and the microgel shrinks. With further increase in pH, the acidic monomers deionize and cause swelling. It means earlier transition in volume is shown due to deprotonation of basic groups and later transition is shown by the ionization of acidic monomers.[45]

Photoresponsive Microgels

Photoresponsive groups are introduced into the microgels system and are made light sensitive. These microgels are of three kinds: microgels modified with photoresponsive groups, supramolecular hydrogel formation of photoresponsive low-molecular-weight gelators, and modification of polymer with supramolecularly interacting groups that can respond to photo-irradiation.[46] These groups interact with light radiations and bring about a change in elasticity and many other properties of microgels. Even cross-linking of these gels is influenced by light radiations.

1. Microgels modified with photoresponsive groups are most convenient to form photoresponsive microgels. Simply a desired group that is responsive to light is attached to the polymer network. For example, Ishihara et al. attached light-sensitive azobenzene groups to poly(2-hydroxyethyl methacrylate).[47] These gels deswelled upon interaction with ultraviolet radiations,

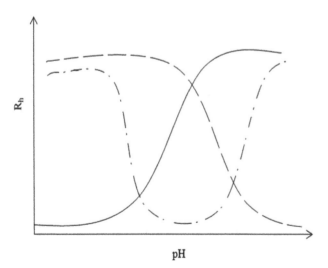

Fig. 4 pH Sensitivity of polymer microgels; anionic microgels (solid line); cationic microgels (dashed line); and amphiphilic microgels (dotted-dashed line).

while the swelling could be induced with visible light.[48] These photoresponsive hydrogels and their swelling and de-swelling behavior have been engineered to fabricate photoactive microcantilevers.[49]

2. Photoresponsive hydrogels formed by low-molecular-weight gelators (LMWGs) have been designed for a diverse range of compounds. The fibril structure of LMWGs is strengthened by hydrogen bonding, leading to high viscosity and elasticity. Their interaction with light is controlled by introduction of photoresponsive moieties. For example, combination of ionic interaction, π–π stacking, hydrogen bonding, and van der Waals interaction helped to maintain the structure of oligopeptide fibrils. Maleic–fumaric acid is photoresponsive, *bis*(phenylalanine) attached to these fibers by these photoresponsive groups. When they interact with ultraviolet radiations, sol-to-gel transition occurs.[50]

3. Polymers cross-linked by a photoresponsive supramolecular assembly are prepared by using photoresponsive cross-linker. Thus, the photoresponsive moiety is within a polymer chain of hydrogel. For example, an azobenzene–cyclodextrin complex has been used to create reversible noncovalent cross-linking. *Trans*-azobenzene can be firmly attached to cyclodextrins, while *cis* configuration is not suitable for binding. Thus, polymer cyclodextrin functionalized by azobenzene has multiple photoresponsive cross-linking points.[51]

Glucose-Responsive Microgels

The chemical-responsive polymer microgels have to be functionalized with a receptor having the ability to recognize the target molecule. The complexation between the receptor and the target molecule induces a physical modification in polymer microgel, which can cause swelling or shrinking.[52,53]

The most extensive efforts in this area have been made for developing insulin release systems in response to high glucose levels.[54] It is known that boronic acid can make a complex with diols through reversible boronate ester formation and shifts the ionization equilibrium toward the charged form (Scheme 1), which causes gel swelling upon increase of glucose concentration; i.e., it exhibits diol

sensitivity.[55] Boronic acid carrying polymers have been used in the recognition, detection, isolation, and transport of diol compounds such as carbohydrates, vitamins, coenzymes, and ribonucleic acids.[56–62]

During the past decade, some glucose-responsive microgel systems were obtained by copolymerization[53,63] and by grafting carboxylate microgels.[54,64,65] Kataoka et al. showed the delivery of insulin amounts according to the presence of glucose from glucose-sensitive microgel bearing phenylboronic acid groups.[62] Lapeyre and coworkers reported on a comprehensive study of volume phase transition and glucose-responsive behavior of core–shell microgels with a thermo-responsive core and glucose-responsive shell made of PNIPAM and poly (*N*-isopropylacrylamide-*co*-acrylamidophenylboronic acid) (PNIPAM-PBA), respectively.[64] Nearly mono-disperse glucose-sensitive P(NIPAM-APBA) microgels have been synthesized and characterized for glucose responsive behavior.[66] Ge et al. prepared P(NIPAM-PBA) microgels by free radical precipitation polymerization in water and investigated the release of glucose and Alizarin Red S from the microgels as a function of temperature using laser light scattering.[67] But these microgels are not stable in a wide range of pH and temperature due to hydrophobic behavior of P(NIPAM) and PBA groups.

SYNTHESIS OF MICROGELS

Microgels prepared by chemical means are stable, and are easier to deal with. The microgels prepared by physical methods are quite unstable and cause several problems. Three different strategies are commonly employed to prepare microgels; from monomers, from polymers, and from macrogels.

From Monomers

The approach at the monomer level is used for synthesis of microgels but monomers should have double bond as a polymerization site. Hence, monomers, cross-linker, initiator, and surfactant are added to synthesize microgels framework. Monomers of all types, such as cationic, anionic, and

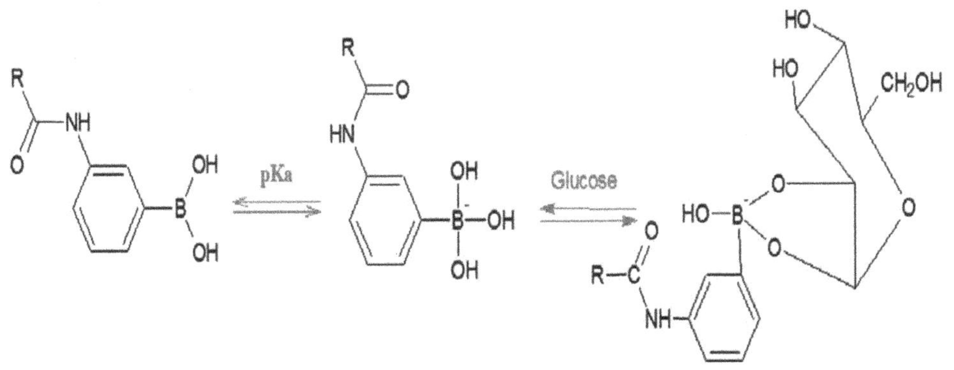

Scheme 1 Representation of the complexation between phenylboronic acid derivative and D-glucose in aqueous medium.

nonionic monomers, can be used. This is a simple method and is adopted by most scientists because monomers of desired functional groups at desired positions are easily imparted in its polymer structure. Two different monomers can also be polymerized by this method. By controlling the amounts of monomers, the size of the microgel can be controlled. Jones et al. have used NIPAM and AAc monomers and prepared P(NIPAM-AAc) core–shell microgel.[17] NIPAM has been copolymerized with methacrylic acid, allylacetic acid, maleic acid, etc. and microgels have been prepared.

From Polymers

In this method, polymer electrolytes having reactive functional groups are used. Long polymeric chain has many functional groups that react with functional groups of other polymers to make a cross-linked polymeric network. In addition to this, the location of the functional group on the chain also directs where bonding occurs and where not. Normally, polymer electrolytes of different charges are used in which electrostatic attractive forces become operative for combining these polymers. Sometimes a polymer chain possesses a single kind of functional group as a side branch and these groups react with each other and form framework. Cao et al.[68] has used this approach. He prepared linear poly [N-isopropylacrylamide-co-3-(trimethoxysilyl)propylmethacrylate] [P(NIPAM-co-TMSPMA)] chains. Then aqueous solution of these linear chains was heated, Si–O–CH$_3$ methoxysilyl groups ionize and form Si-OH silanol groups. Chemical bonding occurs among chains and microgels are synthesized.[68] This method was also used by Kuckling to prepare micro and nanogels.[69]

From Macrogels

Breaking of macrogels is done by some mechanical approach to synthesize microgels. Shearing forces break down the bigger particle size of macrogels into smaller particles of micro level. This is not a reliable method as particle size and shape cannot be easily controlled and a monodispersed colloidal solution cannot be prepared. Breaking of macrogels is a physical method and also needs special instruments. Miao et al. obtained microgels of irregular shapes when they tried grinding of poly(vinyl amine) [p(VAm)] macrogels.[70]

Polymerization Methods

Free radical, atom transfer radical, and reversible addition-fragmentation chain transfer polymerization are three different chemical methods used for preparation of microgels. The most conventional method is free radical polymerization, but atom transfer radical polymerization is a more advanced method than free radical method, and reversible addition-fragmentation chain transfer polymerization is the latest method of present era.

Free radical polymerization

In free radical polymerization, free radicals of monomers add up and form a polymer. Free radical initiator is used, which reacts with monomers and generates monomer free radicals. Then these monomer free radicals react with monomers and a polymer chain begins to grow. Polymer chains assemble together and yield microgel particles. Potassium persulfate and ammonium persulfate[17] are common initiators used for polymerization. Random, alternate, and graft copolymers can be synthesized by this method. In the latter approach, preformed side chains are attached covalently onto the main polymer chain. Many scientists have reported this method and prepared a variety of microgels. For example, P(NIPAM-AAc) core–shell microgels has been prepared by Jones and his coworkers using ammonium persulfate as initiator.[17] P(NIPAM-AAm-AAc) microgels of controlled size were prepared by Farooqi and his coworkers by free radical precipitation polymerization. He used ammonium persulfate as initiator.[71]

Atom transfer radical polymerization

In addition to monomer, solvent, and initiator, a transition metal-based catalyst is also involved in atom transfer radical polymerization. The catalyst helps to form a carbon–carbon covalent bond between monomers. It is controlled polymerization, hence leading to less polydispersity than free radical polymerization. Chain extension is the main merit of this polymerization. It is possible due to halide end-functionality of catalyst, which helps to build block copolymers. Individual particles prepared by this method are uniformly cross-linked. Atom transfer radical polymerization gives such particles whose colloidal stability, swelling behavior, and many other properties are much better than conventional polymerization methods.[72] AAm, AAc, and methacrylic acid have been polymerized with this method. Block and graft copolymers are largely synthesized by this approach. Graft copolymer can be synthesized by this method but preformed side chains are not used here. Side chains are copolymerized onto the main chain during synthesis. Hence, the length of grafted chains can be controlled, or we can say that the molecular weight of side chains can be monitored easily. By this method, prepared microgel product has transition metal complexes that are toxic and interfere in many proceeding reactions. So these complexes have to be removed by dialysis before their application or further processing.

Reversible addition-fragmentation chain transfer polymerization

Initiator, monomer, reversible chain transfer agent, and solvent are used in reversible addition-fragmentation chain transfer polymerization. Initiator forms monomer radicals, which on polymerization form a long polymer chain. This polymer chain reacts with reversible chain transfer agent and adduct is synthesized. This adduct decomposes and generates a radical species that reacts with monomer and a new polymer chain is formed. In this way, chain transfer agent helps to fabricate a large number of chains of polymer in a controlled manner; hence, monodisperse particles are synthesized. This is the advantage of this method in which initiation occurs again and again, and the number of termination steps decreases. Synthesis of reversible chain transfer agent is a tough task. It is synthesized via a long multistep method. Storage of this chain transfer agent is also difficult. Chain transfer agents such as thiocarbonyl compounds are colored, which create problems during biomedical applications. These are the handicaps of this method. But this method is still used widely due to its homogeneous product formation. Graft, alternate, and block copolymers are commonly fabricated by this method. Ethylene oxide and NIPAM-based copolymer were synthesized using macro-chain transfer agent by this method.[73]

HYBRID MICROGELS

Owing to easy synthesis, simple functionalization, tunable dimensions, greater sensitivity, and larger size, microgels are used as the most ideal microreactors for the synthesis of metal nanoparticles.[67] The microgels containing inorganic nanoparticles are termed as hybrid microgels. We can get metal nanoparticles of uniform size using polymer microgels. Gong et al. synthesized monodisperse CdTe nanoparticles within P(NIPAM) microgels.[74] Hydrogen bonding between ligands surrounding the CdTe nanoparticles and P(NIPAM) helps in its formation. Gold-, silver-, and iron-oxide-containing microreactors gather significant attention. The swelling of polymer microgels in the presence of some external stimuli can increase the interparticle distance causing color change.[75] Suzuki and Kawaguchi prepared Au nanoparticles in P(NIPAM) microgels and studied their optical properties.[76] Dong et al. synthesized silver nanoparticles in P(NIPAM-AA) microgels and explained charge transfer from carbonyl groups to silver nanoparticles.[77] Juxiang et al. prepared CuS-P(NIPAM-AA) hybrid microgels to form patterned surface morphologies.[78] Wang et al. also prepared P(NIPAM-AA) microgels having Fe_3O_4 nanoparticles to form superparamagnetic composite microspheres.[79] Li et al. prepared Ph-controllable self-assembly of CdTe in P(NIPAM-AA) microgels.[80]

The stimuli-responsive microgels not only stabilize metal nanoparticles but also control their optical properties by changing its surrounding atmosphere in response to external stimuli such as pH,[81] temperature,[82] and light irradiation.[83] The hybrid systems can be used in different fields such as biotechnology, molecular biology, and drug delivery due to their versatile properties. Phosphor europium-doped yttrium oxide $(Y_2O_3:Eu)$ particles were synthesized by Snowden et al. in P(NIPAM-AA) microgel system for cathode ray tube and field emissive displays applications.[84] These particles were less than 100 nm in diameter. Kim and

Dong used hybrid P(NIPAM-AA) shell and Au nanoparticles core-type microgel in drug delivery.[85–88]

SYNTHESIS OF HYBRID MICROGELS

Hybrid microgels can be synthesized by following four methods: mixing of microgel and nanoparticle suspensions, growing nanoparticles in the presence of preformed microgels, growing microgel in the presence of preformed nanoparticles acting as seeds, and layer-by-layer and core–shell assembly.

Mixing of Microgel and Nanoparticle Suspensions

Both microgel and nanoparticles are separately preformed and mixed. The presence of charged species on the surface of microgel and electrostatic interactions helps to adsorb nanoparticles on the surface of microgels. But nanoparticles should be small enough to penetrate into the microgel network.

Growing Nanoparticles in the Presence of Microgels

Here a relevant metal salt and reducing agents are added into microgel dispersion, metal ions move into the microgel particle where they are reduced and metal atoms are formed. Later these metal atoms aggregate and nanoparticles are synthesized.

Growing Microgel in the Presence of Preformed Nanoparticles

In this method, preformed metal nanoparticles are used as nucleus for synthesis of microgel network and hybrid gels are formed. The nanoparticles act as seeds for the growth of polymer network around these particles. We can obtain core–shell type hybrid microgels having metallic core and polymer shell.

Layer-by-Layer and Core–Shell Assembly

Nanoparticles are covered with a layer of counterions. Microgels have the tendency to replace this layer, thus resulting in the formation of a coat or layer around nanoparticles. Same technique can be used to coat microgels with nanoparticles.

TYPES OF HYBRID MICROGELS

There are three types of hybrid microgels, as shown in Fig. 5:

- Core–shell type hybrid microgels
- Inorganic nanoparticles-filled hybrid microgels
- Inorganic nanoparticles-covered hybrid microgels

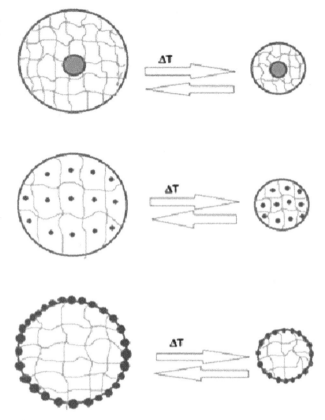

Fig. 5 Schematic diagram showing three types of hybrid microgels.

TUNABLE SURFACE PLASMON RESONANCE IN HYBRID MICROGELS

Surface plasmon resonance (SPR) is a tunable property that is dependent on several factors such as the size of nanoparticles, refractive index of microgel suspension, and interparticle distance.[89,90] Hence, the value of SPR wavelength (λ_{SPR}) can be tuned by changing external stimuli such as temperature and pH.

The hydrodynamic radius of microgels changes due to solvent amount, electrostatic interactions, and ionic strength. On swelling, the volume of solvent (mostly water) increases. If ionic monomer is already incorporated within a network of microgel, then electrostatic repulsion increases, which is also a factor for their swelling. During de-swelling, water molecules move out and electrostatic repulsion vanishes. It means the environment within microgels network changes during swelling and de-swelling. Nanoparticles are present within sieves of polymeric mesh of microgels. Refractive index of environment of nanoparticles decreases in swollen state due to intrusion of water molecules. pH-sensitive swelling of microgels causes change in the ionic strength of medium also. If stimulus of swelling is pH change, then ionic strength also varies within microgels. These changes cause SPR wavelength to change. Generally, SPR wavelength shows red shift on de-swelling. Electrons are

continuously in spin and orbital motion. Owing to their motion, the dipole moment is continuously changes and surface plasmons (waves) are produced.[91] Electrons oscillate with a characteristic frequency, which resonates with characteristic frequency of incident radiation, and absorption occurs. This wavelength of resonance is called λ_{SPR} and nanoparticles show λ_{SPR} in UV-Visible range (200–800 nm). In shrunken state, ionic monomers are protonated and refractive index of the surrounding increases. These factors are forces that shift λ_{SPR} and red shift results. Of the three metals, Ag, Au,[92] and copper (Cu), that display plasmon resonances in the visible spectrum, Ag exhibits the highest efficiency of plasmon excitation. Au has less near-field, so it less scatters light as compared to Ag.[93] Silver possesses most efficient optical excitation plasmon resonance whose nanoparticles efficiently interact with radiations and show efficient optical excitations as compared to organic or inorganic chromophores. Many researchers have reported this tuning of λ_{SPR} in AgNps in response to pH, temperature, etc. Wu et al. have studied λ_{SPR} of Ag and Ag/Au bimetallic nanoparticles incorporated in poly(N-isopropylacrylamide–acrylic acid–acrylamide) polymer microgel. They observed red shift in λ_{SPR} with increase in pH and temperature of medium.[94] Xu et al. has studied effect of refractive index on λ_{SPR} of silver nanoparticles present in corona of P(NIPAM) microgels. In response to high temperature, the λ_{SPR} of Ag nanoparticles, prepared at different dosages of $AgNO_3$, increases.[95]

APPLICATIONS OF MICROGELS

Microgels made from "smart materials" have potential applications in several fields including drug delivery,[96,97] biosensing,[98,99] chemical separations,[100,101] catalysis,[102,103] and optics.[18,19]

Microgels in Biotechnology

Thermoresponsive microgels are used in the field of biotechnology. In one system, P(NIPAM) bearing carboxyl groups at the end was polymerized, and then functionalized on to the microgels. The end carboxylic group was then used to attach an enzyme trypsin, resulting in a particle having two different kinds of P(NIPAM) chains on the surface: one with trypsin and the other without it. The two different chains had different transition temperatures. The free chains collapsed at a lower temperature than the trypsin conjugated chains, hence exposing the enzyme for substrate binding. Thus, by this simple construct, they were able to control the enzyme activity by controlling the temperature.[104]

Kawaguchi, Fujimoto, and Mizuhara[105] studied interaction of P(NIPAM) with biological molecules. He described that sorption and desorption of human γ-globulin is dependent upon temperature. Protein binds on isopropyl groups of P(NIPAM) due to hydrophobic interactions. Hence, maximum adsorption occurs at low pH and high temperature at which maximum isopropyl groups are exposed and free from bonding of water molecules. Desorption of protein can be triggered by increase in pH and decrease of temperature. Both these physical factors cause swelling in microgels and protein release. So regions of body having high pH and low temperature are targeted sites for delivery of proteins.[105] He later performed work related to adsorption of four proteins with polystyrene latex and P(NIPAM). Adsorbing ability of polystyrene was found to be higher than that of P(NIPAM) at 40°C.

Proteins are not only physically attached onto microgels but also covalently bonded to polymer network of microgels. Trypsin and peroxidase were immobilized on P(NIPAM) microgel by Shiroya.[106] When temperature was higher than VPTT of microgel, then the grafted enzyme molecules did not remain in contact with substrate and enzyme activity dropped. In this way, by controlling temperature, catalysis by enzyme can be controlled *in vivo* and vitro media. Poly(ethylene glycol) (PEG) spacers help to obtain temperature-independent enzyme activity. Spacers isolate the enzyme from thermosensitive P(NIPAM) network and thus breakdown by trypsin will be temperature independent.

Microgels in Drug Delivery

One of the key areas of intensive research is the application of microgels in controlled drug delivery. The open network structure of microgels can be used to incorporate small molecules such as drugs in their interiors while their large swelling/de-swelling transitions may be employed as physical and chemical triggers to direct release of the drugs. In addition to pH, ionic strength, or temperature-responsive volume transitions, microgels loaded with a drug can interact with biological components or events such as enzymatic processes that would activate the release of the drug.[107]

Microgels as carrier for drugs are much better than liposomes. Liposomes have been used as carriers for a long time. Efficiency of delivery is hampered by liposomes as in them drugs are only surrounded by a lipid bilayer only. Passive diffusion of drugs is not effective by this method.[108]

The P(NIPAM-AAc) microgels has been used for targeting cells. In this case, the acid groups were converted to amine groups by reacting with ethylene diamine. The amine groups were then used for conjugating galactose to the particles. Hepatocytes have asialoglycoprotein receptors that recognize galactose and hence these galactosylated particles are internalized in these cells. This construct can be used for targeted delivery of drug carrier.[109] Functionalization of microgels allows one to change their volume phase transitions in physiologically relevant conditions. Furthermore, by attaching receptor-specific proteins to the microgel surface, one can achieve selective targeting ability designed to treat specific diseases or specific tumor cells.[110]

Li, Wang, and Wu[111] reported that poly(vinyl alcohol) hydrogel particles can be used for delivery of bovine serum albumin (BSA). This hydrogel was prepared without use of any cross-linker by emulsion polymerization. BSA is a model protein that was incorporated during hydrogel synthesis. At higher temperature, this protein was released out of network. As this is physically cross-linked, this release was attributed to decrease in number of cross-links at higher temperature by Li. Li called this system "open" in state at high temperature. He said that open system has high potential for release of protein.

Photoresponsive Microgels in Cell Culturing

Photoresponsive hydrogels possess photoresponsive groups that interact with radiations, and in turn microgels show response. Properties like elasticity, shape, viscosity, and swelling of photoresponsive microgels change after irradiation. These properties are under a lot of consideration by scientists for biomedical applications. A photoresponsive model is designed by keen choice of photoreactive groups for microgel structure. The hydrated living tissues are similar in structure to microgels, so these microgels can be used as growth medium for cells.[112] Microgels are hollow matrices, whose cross-linking and many other characteristics are stimulated by light. So cells growth can be triggered easily.

The development of cells is recently stimulated by these microgels.[113–115] Hydrogel matrices have RGDS (Arg-Gly-Asp-Ser) peptides that are photocaged by *O*-nitrobenzyl moieties. This assembly is nonadhesive for attachment of cells. On exposure to UV radiations, the cage is removed and exposes the RGDS peptides. Now fibroblasts and many other cells can successfully attach to RGDS peptides and proliferate.[116] In addition to fibroblasts, primary rat dorsal root ganglia cells and DRG neurons are also proliferate by this integrated mechanism.[117,118] Anseth et al.[119,120] presented hydrogel materials that were able to control stem cells migration. This hydrogel shows change in chemical and mechanical patterns through photodegradation.

Reduction-Sensitive Biodegradable Cationic Polymers for Gene Delivery

Various cationic polymers are known as reduction sensitive nowadays. These polymers are stable at physiological conditions; 37°C and 7.34 pH. So they can be used for extracellular and intracellular gene delivery. Their biodegradability helps to convert high-molecular-weight polymer material into monomers within the body, so the body does not act as graveyard of polymers. These polymers have disulfide bridges that break and degradation occurs. In the cytosol of a living cell, different enzymes and compounds such as glutathione are present in high concentrations, which cause thiol–disulfide exchange, and breakdown occurs.[121] The transport of genes is confirmed as polymers break inside the cell. These polymers are quite stable during transportation, which also is a factor for successful transport of genes, RNA, DNA, etc. During degradation, the polymers condense, which leads to breakage of disulfide bridges. As soon as S–S bonds break, the monomers are released and the polymer network of gel degrades. Then the entrapped gene or DNA becomes free and moves out into the cellular matrix. In this way, biodegradability helps in successful and targeted transport of biomaterials. This targeted transport was studied by Yui, who prepared a biocleavable Polyrotaxane with a necklace-like structure. Cationic dimethylaminoethyl-modified a-CDs (DMAE-a-CDs) were grafted onto PEG chains in Polyrotaxane.[122,123] Benzyloxycarbonyl tyrosine was bonded at both termini of PEG chain (DMAE-SS-PRX) via disulfide linkages. PDNA was entrapped within this polyrotaxane. Presence of a counterpolyanion and 10 mM DTT polyanion caused disulfide cleavage. This cleavage caused the release of pDNA from the DMAE-SS-PRX polyplex, then noncovalently linked a-CDs dissociated from PEG chains. Transfection activity of pDNA is related to numbers of a-CD and amino groups of DMAE-SSPRX.[124] The transfection efficiency of biocleavable polyrotaxane was found to be much higher than many other nonbiocleavable polymers.

You et al. reported the preparation of a multiblock copolymer by reversible addition fragmentation chain transfer (RAFT) polymerization.[125] P(NIPAM) and poly(2-(dimethylamino)ethyl methacrylate) (PDMAEMA) were initially prepared, then both were mixed and multiresponsive redox multiblock copolymer was synthesized. This polymer has a hydrophobic core of PNIPAM and cationic shell of PDMAEMA. They reported that reducing agents cause breakdown of polymer. Reducing agent, glutathione, is present within cell in mM concentration. Its concentration is in μM in the extracellular space. Concentration of glutathione affects the degradation of polymer gel. Hence, low concentration of reducing agent does not trigger breaking of S–S bonds, so polymer is stable outside the cell. But as soon as it enters into the cell, the environment of the polymer is changed. High concentration causes breakdown of the polymer microgel into monomers.[125] High-molecular-weight diblock copolymer polystyrene and polyacrylic acid has been prepared and their degradation have been also studied.[126]

In Vivo Diagnosis and Therapy by Implantable Materials

Many implantable devices have been fabricated by scientists. Implantable devices should have anti-biofouling and biosensing properties. Anti-biofouling means they should be anti-inflammatory, antibacterial, and anti-adhesive in nature. Biosensing means these materials should respond according to physiological conditions of body. Artificial or synthetic materials when implanted in body cause biofouling. These materials do not degrade easily and their presence cause swelling and soreness in vessels and in other parts of body. To avoid this, biomaterials are used in implantable materials. Implantable materials are solely built from biomaterials or synthetic materials are coated with compatible biomaterials. Stent was coated with collagen by Chen.[127] The collagen was cross-linked by genipin. This cross-linking avoids dissolution of collagen in various physiological environments. The coating of collagen acts as anti-inflammatory and it also helps in controlled release of drug. The coating of stent with sirolimus also helps in suppressing hyperplasia or thickening of blood vessels. Sirolimus is a natural immunosuppressive agent. It also helps in controlling the release of drug along with collagen. He developed this system for treatment of coronary arterial stenosis. This stent works to open the bore of cardiac vessels that were previously opened by balloon angioplasty or atherectomy. These stents prevented the constriction of vessels. This layer-by-layer assembly of implantable material is present in blood. Due to the velocity of blood, the drug is released from this assembly. Chen[127] studied the loading and unloading of drug at various dosage concentrations. He inferred, by synchronizing the insertion of dose of the drug into body with the velocity of blood, that the release and uptake of drug by stent assembly can be controlled and maintained.[127]

Photoresponsive Hybrid Microgels for Drug Delivery

Hybrid micrgels possess metal nanoparticles, nanorods, or nanotubes within their structure. Their swelling and de-swelling can be induced by thermal energy. Metal nanoparticles absorb energy and transfer their heat to microgel network, consequently de-swelling occurs. Hence, swelling/de-swelling is induced by switching off/on the source of energy. This behavior of hybrid microgels is used in delivery of medicines. Oldenburg et al. studied λ_{SPR} of nanoparticles consisting of a silica core coated with a thin gold shell. The surface plasmon of gold is varied by relation between the diameter of the SiO_2 core and the Au shell thickness. He modulated λ_{SPR} in the spectral range from 700 to 1050 nm.[128] Later, Gorelikov and other scientists used this shifting of λ_{SPR}. Gorelikov, Field, L. and Kumacheva[129] synthesized Au nanorods containing P(NIPAM-AAc) hybrid microgels. Au nanorods were photosensitive, so they tuned the absorption of hybrid microgel by changing the aspect ratio longitudinally of Au nanorods in near-IR range (800–1200 nm). It is in water window region. Microgel polymeric material and water molecules of medium do not absorb in this region but metals do. When hybrid microgel was irradiated by near-IR radiations, Au nanorods converted light energy to heat energy by radiationless relaxation process. Due to this heat energy, microgel network become de-swelled. Hence, by

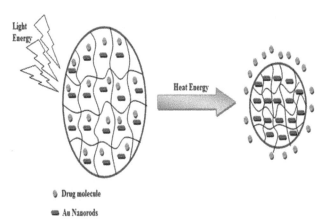

Fig. 6 Schematic diagram of drug release in photosensitive hybrid microgels.

introduction of metal nanorods into microgels, hydrodynamic radius of microgels was controlled.[129] In addition, the swelling or de-swelling behavior induced in microgel network because of the photosensitive metallic rods can be used for drug delivery, as shown in Fig. 6.

Sershen et al.[130] also introduced photothermal swelling/de-swelling in P(NIAPM-AAm) microgels. This hydrogel is thermoresponsive and can release soluble material held within matrix in response to heat. This heating is transferred by source to nanoparticles and then to microgel matrix. He designed a new class of materials, gold–gold sulfide nanoshells, which strongly absorb at 1064 nm. He introduced these nanoparticles into P(NIPAM-AAm) microgels, so that irradiation of light on nanoparticles brought a change in temperature of system. Body tissues do not absorb in water window region. Hence, irradiation of this region helps in delivery of drugs. Enhanced delivery of soluble material in matrix has been studied using 1064-nm wavelength. He studied release of methylene blue and proteins (of varying molecular weights) using this microgel system.

CONCLUSION

This entry highlighted the types, structure, and functioning of pure and hybrid microgels. Different types of responsive microgels are described in this entry. It signifies that microgels are responsive to a number of stimuli such as pH, temperature, light, and glucose concentration. Microgels swell and de-swell as a result of intrusion and expulsion of solvent molecules and the size of microgel particles changes. Free-radical polymerization, reversible addition-fragmentation chain transfer polymerization, and atom transfer radical polymerization are synthetic procedures for microgels that are currently used. Microgels are employed in many biomedical fields for various purposes using their sensitivity to a variety of stimuli. Glucose-responsive microgels have been used for curing diabetes.

Temperature- and pH-responsive microgels have been employed for drug delivery, controlled culturing of cells, biotechnology, and gene transfer. Biodegradable microgels have gained much importance for applications in biomedical field. These microgels are broken down into harmless products, which cause no damage to body and are easily removed from body. Delivery of genes and culturing of cells are widely known applications of these microgels. Properties of microgels have been further refurbished by making hybrid microgels. Combination of metal nanoparticles and smart polymer microgels is used widely for tuning of SPR wavelength of nanoparticles. Hybrid microgels are photoresponsive. Hence, absorption of light by microgels has made its way in biomedical field. Light controls the swelling/de-swelling of microgels in a fascinating way. All these applications are enlisted and explained in detail in entry. This entry also covers many perspectives of pure and hybrid microgels that help to use microgels in many diverse applications in future.

ACKNOWLEDGMENTS

We are thankful to United States Agency for International Development and Higher Education Commission Pakistan for financial support under Pak-US Science and technology cooperative program.

REFERENCES

1. Karga, M.; Hellweg, T. New "smart" poly(NIPAM) microgels and nanoparticle microgel hybrids: Properties and advances in characterization. Curr. Opin. Colloid Interface Sci. **2009**, *14* (6), 438–450.
2. Matsuda, A.; Gong, J.P.; Osada, Y. Effects of water and cross-linkage on the formation of organized structure in the hydrogels. Polym. Gels Networks **1998**, *6* (5), 307–317.
3. Hoffman, A.S. Hydrogels for biomedical applications. Ann. N.Y. Acad. Sci. **2001**, *944* (9), 62–73.
4. Kamath, K.; Park, K. Biodegradable hydrogels in drug delivery. Adv. Drug Deliv. Rev. **1993**, *11* (1–2), 59–84.
5. Hoffman, A.S. Hydrogels for biomedical applications. Adv. Drug Deliv. Rev. 2002, 54, 3–12.
6. Li, Y.; Tanaka, T. Study of the universality class of the gel network system. J. Chem. Phys. **1989**, *90* (10), 5161–5166.
7. Li, Y.; Tanaka, T. Kinetics of swelling and shrinking of gels. J. Chem. Phys. **1990**, *92* (2), 1365–1371.
8. Tanaka, T.; Fillmore, D.J.; Sun, S.T.; Nishio, I.; Swislow, G.; Shah, A. Phase transitions in ionic gels. Phys. Rev. Lett. **1980**, *45* (20), 1636–1639.
9. Pelton, R. Temperature-sensitive aqueous microgels. Adv. Colloid Interface Sci. **2000**, *85* (1), 1–33.
10. Das, M.; Zhang, H.; Kumacheva, E. Microgels: Old materials with new applications. Annu. Rev. Mater. Res. **2006**, *36* (1), 117–142.

11. Staudinger, H.; Husemann, E. Über hochpolymere verbindungen, 116. mitteil.: über das begrenzt quellbare poly-styrol. Eur. J. Inorg. Chem. **1935**, *68* (8), 1618–1634.

12. Saunders, B.R.; Vincent, B. Microgel particles as model colloids: Theory, properties and applications. Adv. Colloid Interface Sci. **1999**, *80* (1), 1–25.

13. David, G.; Simionescu, B.C.; Albertsson, A.C. Rapid deswelling response of poly(N-isopropylacrylamide)/poly(2-alkyl-2-oxazoline)/poly(2-hydroxyethyl methacrylate) hydrogels. Biomacromolecules **2008**, *9* (6), 1678–1683.

14. Seiffert, S.; Weitz, D.A. Controlled fabrication of polymer microgels by polymer-analogous gelation in droplet microfluidics. Soft Matter **2010**, *6* (14), 3184–3190.

15. Pich, A.; Bhattacharya, S.; Lu, Y.; Boyko, V.; Adler, H.P. Temperature-sensitive hybrid microgels with magnetic properties. Langmuir **2004**, *20* (24), 10706–10711.

16. Zhang, J.; Peppas, N.A. Synthesis and characterization of pH- and temperature-sensitive poly(methacrylic acid)/poly(N-isopropylacrylamide) interpenetrating polymeric networks. Macromolecules **2000**, *33* (1), 102–107.

17. Jones, C.D.; Lyon, L.A. Synthesis and characterization of multiresponsive core-shell microgels. Macromolecules **2000**, *33* (22), 8301–8306.

18. Jones, C.D.; Serpe, M.J.; Schroeder, L.; Lyon, L.A. Microlens formation in microgel/gold colloid composite materials via photothermal patterning. J. Am. Chem. Soc. **2003**, *125* (18), 5292–5293.

19. Jones, C.D.; Lyon, L.A. Photothermal patterning of microgel/gold nanoparticle composite colloidal crystals. J. Am. Chem. Soc. **2003**, *125* (2), 460–465.

20. Jones, C.D.; McGrath, J.G.; Lyon, L.A. Characterization of cyanine dye-labeled poly(N-isopropylacrylamide) core/shell microgels using fluorescence resonance energy transfer. J. Phys. Chem. B **2004**, *108* (34), 12652–12657.

21. Tanaka, T.; Nishio, I.; Sun, S.T.; Ueno-Nishio, S. Collapse of gels in an electric field. Science **1982**, *218* (4571), 467–469.

22. Miyata, T.; Asami, N.; Uragami, T. A reversibly antigen-responsive hydrogel. Nature **1999**, *399* (6738), 766–769.

23. Ogawa, K.; Wang, B.; Kokufuta, E. Enzyme-regulated microgel collapse for controlled membrane permeability. Langmuir **2001**, *17* (16), 4704–4707.

24. Ogawa, K.; Nakayama, A.; Kokufuta, E. Preparation and characterization of thermosensitive polyampholyte nanogels. Langmuir **2003**, *19* (8), 3178–3184.

25. Galaev, I.Y.; Mattiasson, B. Use smart polymers for bioseparations. Chem. Tech. **1996**, *17* (12), 335–340.

26. Taylor, L.D.; Cerankowski, L.D. Preparation of films exhibiting a balanced temperature dependence to permeation by aqueous solutions-a study of lower consolute behavior. J. Polym. Sci. Part A **1975**, *13* (11), 2551–2570.

27. Liu, R.; Fraylich, M.; Saunders, B.R. Thermoresponsive copolymers: From fundamental studies to applications. Colloid Polym. Sci. **2009**, *287* (6), 627–643.

28. Dagallier, C.; Dietsch, H.; Schurtenberger, P.; Scheffold, F. Thermoresponsive hybrid microgel particles with intrinsic optical and magnetic anisotropy. Soft Matter **2010**, *6* (10), 2174–2177.

29. Naeem, H.; Farooqi, Z.H.; Shah, L.A.; Siddiq, M. Synthesis and characterization of P(NIPAM-AA-AAm) microgels for tuning of optical properties of silver nanoparticles. J. Polym. Res. **2012**, *19* (9), 9950.

30. Burdukova, E.; Li, H.; Bradshaw, D.J.; Franks, G.V. Poly(N-isopropylacrylamide) (PNIPAM) as a flotation collector: Effect of temperature and molecular weight. Miner. Eng. **2010**, *23* (11), 921–927.

31. Gan, D.; Lyon, L.A. Tunable swelling kinetics in core-shell hydrogel nanoparticles. J. Am. Chem. Soc. **2001**, *123* (31), 7511–7517.

32. Gan, D.; Lyon, L.A. Interfacial nonradiative energy transfer in responsive core-shell hydrogel nanoparticles. J. Am. Chem. Soc. **2001**, *123* (34), 8203–8209.

33. Jones, C.D.; Lyon, L.A. Shell-restricted swelling and core compression in poly(N-isopropylacrylamide) core-shell microgels. Macromolecules **2003**, *36* (6), 1988–1993.

34. Pelton, R.; Richardson, R.; Cosgrove, T.; Ivkov, R. The effects of temperature and methanol concentration on the properties of poly(N-isopropylacrylamide) at the air/solution interface. Langmuir **2001**, *17* (16), 5118–5120.

35. Shen, Y.; Zhang, X.; Lu, J.; Zhang, A.; Chen, K.; Li, X. Effect of chemical composition on properties of pH-responsive poly(acrylamide-co-acrylic acid) microgels prepared by inverse microemulsion polymerization. Colloids Surf. A **2009**, *350* (1), 87–90.

36. Kratz, K.; Hellweg, T.; Eimer, W. Influence of charge density on the swelling of colloidal poly(N-isopropylacrylamide-co-acrylic acid) microgels. Colloids Surf. A **2000**, *170* (2), 137–149.

37. Zhou, S.; Chu, B. Synthesis and volume phase transition of poly(methacrylic acid-co-N-isopropylacrylamide) microgel particles in water. J. Phys. Chem. B **1998**, *102* (8), 1364–1371.

38. Das, M.; Sanson, N.; Fava, D.; Kumacheva, E. Microgels loaded with gold nanorods: Photothermally triggered volume transitions under physiological conditions. Langmuir **2007**, *23* (1), 196–201.

39. Yoshida, R.; Yamaguchi, T.; Ichijo, H. Novel oscillating swelling-deswelling dynamic behaviour for pH-sensitive polymer gels. Mater. Sci. Eng. C **1996**, *4* (2), 107–113.

40. Debord, J.D.; Lyon, L.A. Synthesis and characterization of pH-responsive polymer microgels with tunable volume phase transition temperatures. Langmuir **2003**, *19* (18), 7662–7664.

41. Charalambopoulou, A.; Bokias, G.; Staikos, G. Template copolymerisation of N-isopropylacrylamide with cationic monomer: Influence of the template on the solution properties of the product. Polymer **2002**, *43* (9), 2637–2643.

42. Tan, B.H.; Ravi, P.; Tan, L.N.; Tam, K.C. Synthesis and aqueous solution properties of sterically stabilized pH-responsive polyampholyte microgels. J. Colloid Interface Sci. **2007**, *309* (2), 453–463.

43. Ho, E.; Lowman, A.; Marcolongo, M. Synthesis and characterization of an injectable hydrogel with tunable mechanical properties for soft tissue repair. Biomacromolecules **2006**, *7* (11), 3223–3228.

44. Pinkrah, V.T.; Beezer, A.E.; Chowdhry, B.Z.; Gracia, L.H.; Cornelius, V.J.; Mitchell, J.C.; Castto-Lopez, V.; Snowden, M. Swelling of cationic polyelectrolyte colloidal microgels: Thermodynamic considerations. J. Colloids Surf. A **2005**, *262* (1–3), 76–80.

45. Richter, A.; Paschew, G.; Klatt, S.; Lienig, J.; Arndt, K.F.; Adler, H.J.P. Review on Hydrogel-based pH sensors and microsensors. Sensors **2008**, *8* (1), 561–581.

46. Tomatsu, I.; Peng, K.; Kros, A. Photoresponsive hydrogels for biomedical applications. Adv. Drug Deliv. Rev. **2011**, *63* (14–15), 1257–1266.

47. Ishihara, K.; Hamada, N.; Kato, S.; Shinohara, I. Photoinduced swelling control of amphiphilic azoaromatic polymer membrane. J. Polym. Sci. Part A **1984**, *22* (1), 121–128.

48. Moniruzzaman, M.; Fernando, G.F.; Talbot, J.D.R.J. Synthesis and characterization of an azobenzene- and acrylamide-based photoresponsive copolymer and gel. Polym. Sci., Part A **2004**, *42* (12), 2886–2896.

49. Watanabe, T.; Akiyama, M.; Totani, K.; Kuebler, S.M.; Stellacci, F.; Wenseleers, W.; Braun, K.; Marder, S.R.; Perry, J.W. Photoresponsive hydrogel microstructure fabricated by two-photon initiated polymerization. Adv. Funct. Mater. **2002**, *12* (9), 611.

50. Frkanec, L.; Jokic, M.; Makarevic, J.; Wolsperger, K.; Zinic, M. Bis(PheOH) maleic acid amide-fumaric acid amide photoizomerization induces microsphere-to-gel fiber morphological transition: The photoinduced gelation system. J. Am. Chem. Soc. **2002**, *124* (33), 9716–9717.

51. Takashima, Y.; Nakayama, T.; Miyauchi, M.; Kawaguchi, Y.; Yamaguchi, H.; Harada, A. Complex formation and gelation between copolymers containing pendant azobenzene groups and cyclodextrin polymers. Chem. Lett. **2004**, *33* (7), 890–891.

52. Lapeyre, V.; Gosse, I.; Chevreux, S.; Ravaine, V. Monodispersed glucose-responsive microgels operating at physiological salinity. Biomacromolecules **2006**, *7* (12), 3356–3363.

53. Hoare, T.; Pelton, R. Engineering glucose swelling responses in poly(N-isopropylacrylamide)-based microgels. Macromolecules **2007**, *40* (3), 670–678.

54. Kikuchi, A.; Okano, T. Pulsatile drug release control using hydrogels. Adv. Drug Deliv. Rev. **2002**, *54* (1), 53–57.

55. Farooqi, Z.H.; Wu, W.; Zhou, S.; Siddiq, M. Engineering of phenylboronic acid based glucose-sensitive microgels with 4-vinylpyridine for working at physiological pH and temperature. Macromol. Chem. Phys. **2011**, *212* (14), 1510–1514.

56. Gao, S.; Wang, W.; Wang, B. Building fluorescent sensors for carbohydrates using template-directed polymerizations. Bioorg. Chem. **2001**, *29* (5), 308–320.

57. Appleton, B.; Gibson, T.D. Detection of total sugar concentration using photoinduced electron transfer materials: Development of operationally stable, reusable optical sensors. Sens. Actuators B **2000**, *65* (1–3), 302–304.

58. Westmark, P.R.; Gardiner, S.J.; Smith, B.D. Selective monosaccharide transport through lipid bilayers using boronic acid carriers. J. Am. Chem. Soc. **1996**, *118* (45), 11093–11100.

59. Gabai, R.; Sallacan, N.; Chegel, V.; Bourenko, T.; Katz, E.; Willner, I. Characterization of the swelling of acrylamidophenylboronic acid-acrylamide hydrogels upon interaction with glucose by faradaic impedance spectroscopy, chronopotentiometry, quartz-crystal

microbalance (QCM), and surface plasmon resonance (SPR) experiments. J. Phys. Chem. B **2001**, *105* (34), 8196–8202.

60. Cannizzo, C.; Amigoni-Gerbier, S.; Larpent, C. Boronic acid-functionalized nanoparticles: Synthesis by micro-emulsion polymerization and application as a re-usable optical nanosensor for carbohydrates. Polymer **2005**, *46* (4), 1269–1276.

61. Choi, Y.K.; Jeong, S.Y.; Kim, Y.H.; Jeong, S.Y. A glucose-triggered solubilizable polymer gel matrix for an insulin delivery system. Int. J. Pharm. **1992**, *80* (1–3), 9–16.

62. Kataoka, J.P.; Miyazki, H.; Bunya, M.; Okano, T.; Sakurai, Y. Totally synthetic polymer gels responding to external glucose concentration: Their preparation and application to on-off regulation of insulin release. J. Am. Chem. Soc. **1998**, *120* (48), 12694–12695.

63. Matsumoto, A.; Ikeda, S.; Harada, A.; Kataoka, K. Glucose-responsive polymer bearing a novel phenylborate derivative as a glucose-sensing moiety operating at physiological pH conditions. Biomaromolecules **2003**, *4* (5), 1410–1416.

64. Lapeyre, V.; Ancla, C.; Catargi, B.; Ravaine, V. Glucose-responsive microgels with a core-shell structure. J. Colloid Interface Sci. **2008**, *327* (2), 316–323.

65. Zhang, Y.; Guan, Y.; Zhou, S. Synthesis and volume phase transitions of glucose-sensitive microgels. Biomacromolecules **2006**, *7* (11), 3196–3201.

66. Hoare, T.; Pelton, R. Charge-switching, amphoteric glucose-responsive microgels with physiological swelling activity. Biomacromolecules **2008**, *9* (2), 733–740.

67. Ge, H.; Ding, Y.; Ma, C.; Zhang, G. Temperature-controlled release of diols from N-isopropylacrylamide-co-acrylamidophenylboronic acid microgels. J. Phys. Chem. B **2006**, *110* (41), 20635–20639.

68. Cao, Z.; Du, B.; Chen, T.; Nie, J.; Xu, J.; Fan, Z. Preparation and properties of thermo-sensitive organic/inorganic hybrid microgels. Langmuir **2008**, *24* (22), 12771–12778.

69. Gupta, S.; Kuckling, D.; Kretschmer, K.; Choudhary, V.; Adler, H.J. Synthesis and characterization of stimuli sensitive micro- and nanohydrogels based on photocrosslinkable poly(dimethylaminoethyl methacrylate). J. Polym. Sci., Part A: Polym. Chem. **2007**, *45* (4), 669–679.

70. Miao, C.; Chen, X.; Pelton, R. Adhesion of poly(vinylamine) microgels to wet cellulose. Ind. Eng. Chem. Res. **2007**, *46* (20), 6486–6493.

71. Farooqi, Z.H.; Khan, A.; Siddiq, M. Temperature-induced volume change and glucose sensitivity of poly[(N-isopropylacrylamide)-acrylamide-(phenylboronic acid)] microgels. Polym. Int. **2011**, *60* (10), 1481–1486.

72. Oh, J.K.; Siegwart, D.J.; Lee, H.I.; Sherwood, G.; Peteanu, L.; Hollinger, J.O.; Kataoka, K.; Matyjaszewski, K. Biodegradable nanogels prepared by atom transfer radical polymerization as potential drug delivery carriers: Synthesis, biodegradation, *in vitro* release, and bioconjugation. J. Am. Chem. Soc. **2007**, *129* (18), 5939–5945.

73. Qin, S.; Geng, Y.; Discher, D.E.; Yang, S. Temperature-controlled assembly and release from polymer vesicles of poly(ethylene oxide)-*block*- poly(*N*-isopropylacrylamide). Adv. Mater. **2006**, *18* (21), 2905–2909.

74. Gong, Y.; Gao, M.; Wang, D.; Mohwald, H. Incorporating fluorescent CdTe nanocrystals into a hydrogel via

hydrogen bonding: Toward fluorescent microspheres with temperature-responsive properties. Chem. Mater. **2005**, *17* (10), 2648–2653.

75. Garcia, A.; Marquez, M.; Cai, T.; Rosario, R.; Hu, Z.; Gust, D.; Hayes, M.; Vail, S.A.; Park, C.D. Photo-thermally, and pH-responsive microgels. Langmuir **2007**, *23* (1), 224–229.

76. Suzuki, D.; Kawaguchi, H. Hybrid microgels with reversibly changeable multiple brilliant color. Langmuir **2006**, *22* (8), 3818–3822.

77. Dong, Y.; Ma, Y.; Zhai, T.Y.; Shen, F.; Zeng, Y.; Fu, H.; Yao, J. Silver nanoparticles stabilized by thermoresponsive microgel particles: Synthesis and evidence of an electron donor-acceptor effect. Macromol. Rapid Commun. **2007**, *28* (24), 2339–2345.

78. Juxiang, Y.; Fang, Y.; Chaoliang, B.; Daodao, H.; Ying, Z. CuS-poly (N-isopropylacrylamide-co-acrylic acid) composite microspheres with patterned surface structures: Preparation and characterization. Chin. Sci. Bull. **2004**, *49* (19), 2026–2032.

79. Wang, G.; Zhang, Y.; Fang, Y.; Gu, Z. Flower-like SiO_2-coated polymer/Fe_3O_4 composite microspheres of super-paramagnetic properties: Preparation via a polymeric microgel template method. J. Am. Ceram. Soc. **2007**, *90* (7), 2067–2072.

80. Li, J.; Liu, B.; Li, J.H. Controllable self-assembly of CdTe/poly(N-isopropylacrylamide-acrylic acid) Microgels in response to pH stimuli. Langmuir **2006**, *22* (2), 528–531.

81. Annaka, M.; Tanaka, T. Multiple phases of polymer gels. Nature **1992**, *355*, 430–432.

82. Schlid, H.G. Poly(N-isopropylacrylamide): Experiment, theory and application. Prog. Polym. Sci. **1992**, *17* (12), 163–249.

83. Juodkazis, S.; Mukai, N.; Wakaki, R.; Yamaguchi, A.; Matsuo, S.; Misawa, H. Reversible phase transitions in polymer gels induced by radiation forces. Nature **2000**, *408* (6809), 178–181.

84. Martinez-Rubio, M.I.; Ireland, T.G.; Fern, G.R.; Silver, J.; Snowden, M.J. A new application for microgels: Novel method for the synthesis of spherical particles of the Y_2O_3:Eu phosphor using a copolymer microgel of NIPAM and acrylic acid. Langmuir **2001**, *17* (22), 7145–7149.

85. Kim, J.H.; Lee, T.R. Thermo- and pH-responsive hydrogel-coated gold nanoparticles. Chem. Mater. **2004**, *16* (19), 3647–3651.

86. Kim, J.H.; Lee, T.R. Discrete thermally responsive hydrogel-coated gold nanoparticles for use as drug-delivery vehicles. Drug Dev. Res. **2006**, *67* (1), 61–69.

87. Kim, J.H.; Lee, T.R. Hydrogel-templated growth of large gold nanoparticles: Synthesis of thermally responsive hydrogel-nanoparticle composites. Langmuir **2007**, *23* (12), 6504–6509.

88. Dong, Y.; Ma, Y.; Zhai, T.Y.; Zeng, Y.; Fu, H.B.; Yao, J.N. Incorporation of gold nanoparticles within thermoresponsive microgel particles: Effect of crosslinking density. J. Nanosci. Nanotechnol. **2008**, *8* (12), 6283–6289.

89. Liz-Marzan, L.M.; Giersig, M.; Mulvaney, P. Synthesis of nanosized gold-silica core-shell particles. Langmuir **1996**, *12* (18), 4329–4335.

90. Mazumdar, S.D.; Hartmann, M.; Kampfer, P.; Keusgen, M. Rapid method for detection of Salmonella in milk by surface plasmon resonance (SPR). Biosens. Bioelectron. **2007**, *22* (9–10), 2040–2046.

91. Wiley, B.J.; Im, S.H.; Li, Z.Y.; McLellan, J.; Siekkinen, A.; Xia, Y. Maneuvering the surface plasmon resonance of silver nanostructures through shape-controlled synthesis. J. Phys. Chem. B **2006**, *110* (32), 15666–15675.

92. Basu, S.; Ghosh, S.K.; Kundu, S.; Panigrahi, S.; Praharaj, S.; Pande, S.; Jana, S.; Pal, T. Biomolecule induced nanoparticle aggregation: Effect of particle size on inter-particle coupling. J. Colloid Interface Sci. **2007**, *313* (2), 724–734.

93. Yguerabide, J.; Yguerabide, E.E. Light-scattering sub-microscopic particles as highly fluorescent analogs and their use as tracer labels in clinical and biological applications I. Theory. Anal. Biochem. **1998**, *262* (2), 137–156.

94. Wu, W.; Zhou, T.; Zhou, S. Tunable photoluminescence of Ag nanocrystals in multiple-sensitive hybrid microgels. Chem. Mater. **2009**, *21* (13), 2851–2861.

95. Xu, H.; Xu, J.; Zhu, Z.; Liu, H.; Liu, S. In-situ formation of silver nanoparticles with tunable spatial distribution at the poly(N-isopropylacrylamide) corona of unimolecular micelles. Macromolecules **2006**, *39* (24), 8451–8455.

96. Jeong, B.; Bae, Y.H.; Lee, D.S.; Kim, S.W. Biodegradable block copolymers as injectable drug-delivery systems. Nature **1997**, *388* (6645), 860–862.

97. Brondsted, H.; Kopecek, J. Hydrogels for site-specific drug delivery to the colon: *In vitro* and *in vivo* degradation. Pharm. Res. **1992**, *9* (12), 1540–1545.

98. Brahim, S.; Narinesingh, D.; Guiseppi-Elie, A. Bio-smart hydrogels: Co-joined molecular recognition and signal transduction in biosensor fabrication and drug delivery. Biosens. Bioelectron. **2002**, *17* (11–12), 973–981.

99. Holtz, J.H.; Asher, S.A. Polymerized colloidal crystal hydrogel films as intelligent chemical sensing materials. Nature **1997**, *389* (6653), 829–832.

100. Umeno, D.; Kawasaki, M.; Maeda, M. Water-soluble conjugate of double-stranded DNA and poly (N-isopropylacrylamide) for one-pot affinity precipitation separation of DNA-binding proteins. Bioconjug. Chem. **1998**, *9* (6), 719–724.

101. Kawaguchi, H.; Fujimoto, K. Smart latexes for bioseparation. Bioseparation **1998**, *7* (4–5), 253–258.

102. Bergbreiter, D.E.; Case, B.L.; Liu, Y.S.; Caraway, J.W. Poly(N-isopropylacrylamide) soluble polymer supports in catalysis and synthesis. Macromolecules **1998**, *31* (18), 6053–6062.

103. Bergbreiter, D.E.; Liu, Y.S.; Osburn, P.L. Thermomorphic rhodium(I) and palladium(0) catalysts. J. Am. Chem. Soc. **1998**, *120* (17), 4250–4251.

104. Kawaguchi, H.; Kisara, K.; Takahashi, T.; Achiha, K.; Yasui, M.; Fujimoto, K. Versatility of thermosensitive particles. Macromol. Symp. **2000**, *151* (1), 591–598.

105. Kawaguchi, H.; Fujimoto, K.; Mizuhara, Y. Hydrogel microspheres III. Temperature-dependent adsorption of proteins on poly-N-isopropylacrylamide hydrogel microspheres. Colloid Polym. Sci. **1992**, *270* (1), 53–57.

106. Shiroya, T.; Tamura, N.; Yasui, M.; Fujimoto, K.; Kawaguchi, H. Enzyme immobilization on thermosensitive hydrogel microspheres. Colloids Surf. B **1995**, *4* (5), 267–274.

107. Bromberg, L.; Temchenko, M.; Hatton, T.A. Dually responsive microgels from polyether-modified poly (acrylic acid): Swelling and drug loading. Langmuir **2002**, *18* (12), 4944–4952.

108. Barratt, G. Colloidal drug carriers: Achievements and perspectives. Cell. Mol. Life Sci. **2003**, *60* (1), 21–37.

109. Choi, S.H.; Yoon, J.J.; Park, T.G. Galactosylated poly(*N*-isopropylacrylamide) hydrogel submicrometer particles for specific cellular uptake within hepatocytes. J. Colloid Interface Sci. **2002**, *251* (1), 57–63.

110. Standley, S.M.; Mende, I.; Goh S.L.; Kwon Y.J.; Beaudette, T.T.; Engleman, E.G.; Frechet, J.M.J. Incorporation of CpG oligonucleotide ligand into protein-loaded particle vaccines promotes antigen-specific CD8 T-cell immunity. Bioconjug. Chem. **2007**, *18* (1), 77–83.

111. Li, J.K.; Wang, N.; Wu, X.S. Poly(vinyl alcohol) nanoparticles prepared by freezing-thawing process for protein/peptide drug delivery. J. Control. Release **1998**, *56* (1), 117–126.

112. Fedorovich, N.E.; Alblas, J.; De Wijn, J.R.; Hennink, W.E.; Verbout, A.J.; Dhert, W.J.A. Hydrogels as extracellular matrices for skeletal tissue engineering: State-of-the-art and novel application in organ printing. Tissue Eng. **2007**, *13* (8), 1905–1925.

113. Dankers, P.Y.W.; Meijer, E.W. Supramolecular biomaterials. A modular approach towards tissue engineering. Bull. Chem. Soc. Jpn. **2007**, *80* (11), 2047–2073.

114. Cushing, M.C.; Anseth, K.S. Hydrogel cell cultures. Science **2007**, *316* (1), 1133–1134.

115. Wang, C.M.; Varshney, R.R.; Wang, D.A. Therapeutic cell delivery and fate control in hydrogels and hydrogel hybrids. Adv. Drug Deliv. Rev. **2010**, *62* (7), 699–710.

116. Goubko, C.A.; Majumdar, S.; Basak, A.; Cao, X.D. Hydrogel cell patterning incorporating photocaged RGDS peptides. Biomed. Microdevices **2010**, *12* (3), 555–568.

117. Luo, Y.; Shoichet, M.S. A photolabile hydrogel for guided three-dimensional cell growth and migration. Nat. Mater. **2004**, *3* (4), 249–253.

118. Luo, Y.; Shoichet, M.S. Light-activated immobilization of biomolecules to agarose hydrogels for controlled cellular response. Biomacromolecules **2004**, *5* (6), 2315–2323.

119. Kloxin, A.M.; Kasko, A.M.; Salinas, C.N.; Anseth, K.S. Photodegradable hydrogels for dynamic tuning of physical and chemical properties. Science **2009**, *324* (5923), 59–63.

120. Kloxin, A.M.; Tibbitt, M.W.; Kasko, A.M.; Fairbairn, J.A.; Anseth, K.S. Tunable hydrogels for external manipulation of cellular microenvironments through controlled photodegradation. Adv. Mater. **2010**, *22* (1), 61–66.

121. Meng, F.; Hennink, W.E.; Zhong, Z. Reduction-sensitive polymers and bioconjugates for biomedical applications. Biomaterials **2009**, *30* (12), 2180–2198.

122. Ooya, T.; Choi, H.S.; Yamashita, A.; Yui, N.; Sugaya, Y.; Kano, A. Biocleavable polyrotaxane–plasmid DNA polyplex for enhanced gene delivery. J. Am. Chem. Soc. **2006**, *128* (12), 3852–3853.

123. Yamashita, A.; Yui, N.; Ooya, T.; Kano, A.; Maruyama, A.; Akita, H. Synthesis of a biocleavable polyrotaxane-plasmid DNA (pDNA) polyplex and its use for the rapid non-viral delivery of pDNA to cell nuclei. Nat. Protoc. **2006**, *1* (6), 2861–2869.

124. Yamashita, A.; Kanda, D.; Katoono, R.; Yui, N.; Ooya, T.; Maruyama, A. Supramolecular control of polyplex dissociation and cell transfection: Efficacy of amino groups and threading cyclodextrins in biocleavable polyrotaxanes. J. Control. Release **2008**, *131* (2), 137–144.

125. You, Y.Z.; Zhou, Q.H.; Manickam, D.S.; Wan, L.; Mao, G.Z.; Oupicky, D. Dually responsive multiblock copolymers via RAFT polymerization: Synthesis of temperature- and redox-responsive copolymers of PNIPAM and PDMAEMA. Macromolecules **2007**, *40* (24), 8617–8624.

126. You, Y.Z.; Zhou, Q.H.; Manickam, D.S.; Wan, L.; Mao, G.Z.; Oupicky, D. A versatile approach to reducible vinyl polymers via oxidation of telechelic polymers prepared by reversible addition fragmentation chain transfer polymerization. Biomacromolecules **2007**, *8* (6), 2038–2044.

127. Chen, M.C.; Liang, H.F.; Chiu, Y.L.; Chang, Y.; Wei, H.J.; Sung, H.W. A novel drug-eluting stent spray-coated with multi-layers of collagen and sirolimus. J. Control. Release **2005**, *108* (1), 178–189.

128. Oldenburg, S.L.; Jackson, J.B.; Westcott, S.L.; Halas, N. Infrared extinction properties of gold nanoshells. J. Appl. Phys. Lett. **1999**, *75* (19), 2897–2899.

129. Gorelikov, I.; Field, L.M.; Kumacheva, E. Hybrid microgels photoresponsive in the near-infrared spectral range. J. Am. Chem. Soc. **2004**, *126* (49), 15938–15939.

130. Sershen, S.R.; Westcott, S.L.; Halas, N.J.; West, J.L.J. Temperature-sensitive polymer–nanoshell composites for photothermally modulated drug delivery. Biomed. Mater. Res. **2000**, *51* (3), 293–298.

Molecular Assemblies

Saeed Jafarirad
Research Institute for Fundamental Sciences (RIFS), University of Tabriz, Tabriz, Iran

Abstract

In recent years, the self-assembly of novel copolymer-based nanostructures in solution have been extensively applied in biomedical field. The interest in such nanomaterials lies in the fact that their dimensions, in the nanoscale range (near to 100 nm), and physicochemical factors make it possible to fabricate materials with exclusive properties and technological aspects. In this entry, we describe the most significant biomedical applications of polymeric nanospheres, nanotubes, and nanoparticles, in particular, what can be achieved through molecular self-assembly of copolymers. First, the basic principles of self-assembly of block copolymers in solution will be discussed. An overview of the physicochemical properties of amphiphilic copolymers will be then given, including selected drug–polymer aspects. Finally, we will concentrate on applications such as delivery, therapeutics, and theranostics in biomedical field.

INTRODUCTION

Although the first report on the synthesis of copolymers appeared in 1950s, the regeneration of block copolymers did not take place until 1990s, when controlled polymerization procedures were developed.[1] Extraordinarily, in 10 years, several mechanistically individual radical polymerization methodologies, including atom transfer radical polymerization (ATRP),[2] reversible addition-fragmentation chain transfer (RAFT),[3] polymerization and nitroxide mediated radical polymerization (NMRP),[4] were developed. In contrast to small molecular amphiphiles (up to 10^3 Da), copolymers offer numerous possibilities to change designing factors such as chemical composition, molecular weight, number of blocks, topology, and biodegradability. This infinite combinatorial possibility of block copolymer design has remarkable effects on several disciplines, especially in biomedical applications and the health care industry.

SELF-ASSEMBLY

An extensive range of parameters are regarded as having self-assembly characteristics and the definitions applied differ. In chemistry, self-assembly depicts molecular assemblies resulting by noncovalent forces, for instance, in micelles and Langmuir-Blodgett multilayers. However in biology, self-assembly implies construction of complex structures that occur under environmental inter- and intramolecular interactions without the effect of exterior parameters, for example, protein folding and formation of lipid double layers. In general, the term "self-assembly" states a phenomena in which an organized structure is built spontaneously owing to specific, local interactions among the components. It is important to point out that kinetic and thermodynamics parameters play a significant task in the self-assembly phenomena. From the nanoscale point of view, self-assembly of polymeric structures are frequently used as necessary molecular tools for designing of nanomaterials. It is well known that, in nature, a lot of smart nanomaterials are formed by the organization of different types of biomaterials, such as proteins and DNA, without external factors. Accordingly, self-assembly presents an efficient, well-established, and low-cost method to functional hierarchically organized nanocompounds that are complex if not infeasible to produce by other pathways. The production of molecular assemblies can occur in various methods such as self-assembly of polymers adsorbed on interface (Langmuir-Blodgett technique) or self-assembly of macromolecules in solution into a liquid crystal structure and self-assembly of molecular assemblies such as micelles and vesicles.

The complexity and the range of possible self-assembled morphologies with block copolymers are astounding, particularly with the increasing complexity of the block composition and architectures. Apart from the most explored spherical micelles, other exciting morphologies[5] like elongated micelles, vesicles, nanotubes, nanoparticles, and many other complex structures have been formed by the self-assembly of copolymers.[6–8] The final morphology of the copolymeric aggregate dictates its properties and applications. So far, spherical micelles, elongated micelles, and vesicles form the three distinct types of self-assembled morphologies that have been frequently exploited as delivery, therapeutic, and theranostic agents.

Concise Encyclopedia of Biomedical Polymers and Polymeric Biomaterials DOI: 10.1081/E-EBPPC-120050046

PREPARATION OF POLYMERIC NANOSPHERES

The incompatibility of each blocks of a copolymer in solution causes associative properties in macromolecular chains. In other words, in a AB-type diblock copolymer, when a liquid that is energetically a selective solvent for block A, however a non-solvent for the block B, the copolymer chains undergo self-assembly to lower thermodynamically unfavorable interactions. Thus, polymeric nanosphers (PNSs) can be built by precise arrangement of self-assembly components, such as number of blocks, block sequences, relative sizes of the corresponding blocks, solvent composition, and solution concentration.

Self-Assembly of Amphiphilic Diblock Copolymers

Theoretical aspects of micellization and gelation

According to its definition, amphiphilicity (from Greek, *amphi* meaning both and *philicity* meaning attraction) is attraction for two dissimilar surroundings. These differences between the physicochemical entity of two or more blocks can induce microphase separation of amphiphilic block copolymers in aqueous and nonaqueous solutions.[9] Block copolymers undergo two main phenomena in media: micellization and gelation. Micellization take places when the block copolymer is dissolved in a good solvent for one of the blocks. Under such condition, the macromolecular chains tend to assemble with each other in some structures such as micelles, elongated micelles, vesicles, and other morphologies. The soluble block will be directed toward the context solvent as the "corona" or "shell" of the recently formed micelle, while the insoluble chains will be covered from the context solvent in the "core" of the micelle. Compared to micellization, gelation takes place from the semidilute to the high concentration of copolymer in media. The compacted and ordered arrangement of micellar aggregates is a typical particularity of gelation. On the other hand, in micellization, if the soluble block is larger than the insoluble one, the recently formed micelles consist of a small core and a very large corona which are called star micelles. In contrast, micelles having a large insoluble segment with a short soluble corona are referred to as crew-cut micelles.[10]

Micellar unimolecules and aggregates

The micellization of block copolymers depends on two important factors: the critical micelle temperature (CMT) and the critical micelle concentration (CMC). It means that, for example, if the temperature or concentration of solution is less than CMT or CMC, respectively, self-assembly will not happen. Accordingly, the micelles in the solution remain as unimolecules. In contrast, vice versa, the micelles will form micellar aggregates in which they are in equilibrium with unimolecular micelles (Fig. 1).

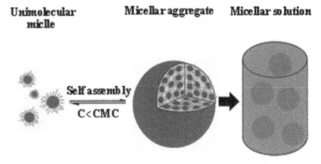

Fig. 1 A schematic representation of micellization of block copolymers in aqueous medium.

Other interdependent parameters to depict a micellar system are the equilibrium constant, overall molar mass of the micelle (M_w), and aggregation number of the micelle (Z). These factors influence the hydrodynamic radius (R_H), the radius of gyration (R_G), the ratio of R_H to R_G (which depends on the micellar shape), the core radius R_C, and the thickness L of the corona.

Experimental determination of micellar characteristics

The use of fluorescent probes is the most used method for the determination of the CMC. UV-absorption spectroscopy has also been reported as a powerful method for the determination of the CMC. Other techniques, such as static light scattering (SLS), dynamic light scattering (DLS), or small-angle x-ray scattering (SAXS), extensively used for small surfactants, can in principle be used for block copolymer micelles. On the other hand, several techniques have been applied for characterization of typical micelle parameters such as shape, size, and aggregation number. SLS and DLS are excellent techniques to calculate M_w and R_H of self-assembled structures, respectively.[11] SAXS has been utilized in obtaining overall and internal sizes of micellar solutions. Finally, small-angle neutron scattering (SANS) was used to estimate micellar shape and the cross-section. Other noncscattering techniques such as transmission electron microscopy (TEM) and atomic force microscopy (AFM) afford precise information about morphological images containing size, shape, or internal structure of the micelles. Further techniques include dilute solution capillary viscometry, membrane osmometry, and ultracentrifugation.[12]

Spherical micelles

Micelles can be divided to three main groups based on types with morphological deference including spherical, elongated, and vesicular micelles (Fig. 2). In addition, some other less common structures were reported such as inverse micelles, bilayers, or cylinders. Among all the morphologies, copolymer-based spherical micelles are simplest and most

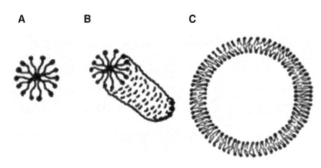

Fig. 2 Three types of micelles: (**A**) spherical, (**B**) elongated, and (**C**) vesicular.

studied ones. With the emergence of micellar systems, plenty of drawbacks in controlled release systems such as low drug incorporation content and nonspecific toxicity of the drugs have been omitted to a great extent. Currently, nonocarriers of micellar drug delivery systems are prepared from amphiphilic diblock copolymers, with polyethylene glycol (PEG) or other hydrophilic segment as shell. In order to reach targeted delivery goals toward tissues or cells, the targeting agent can be connected onto the shell. In recent years, several various stimuli-responsive techniques also have been developed to manage drug delivery systematically.[13]

Elongated micelles

Access to elongated or worm-like micelles, in contrast to the spherical micelles, has been feasible recently[14] (Fig. 2). By comparing the self-assembly behaviors of block copolymers with different weight ratios of blocks, the specific methodology to result in elongated micelles has been discovered.[15] However, several issues have fundamental effects on the morphological outcome.[16] For example, depending upon the block copolymer concentration of the same poly(ethylene oxide)-b-poly(caprolacton) (PEG-b-PCL), micelles with different shapes such as long rod-like, short rod-like, or spherical can be self-assembled.[17] To date, a variety of block copolymers such as coil-coil, rod-coil, and crystalline-coil have been shown to be self-assembled as elongated micelles.[18] It was well known that the flexible worm-like micelles can penetrate into nanopores.[19] This was as a result of the fact that often nonspherical morphologies were favored only in a very narrow compositional space. Furthermore, targeted delivery by using biological signaling agent was feasible with such elongated micelles.

Vesicles

A vesicle or polymersome (polymer-based liposomes) is a bag in nanoscale having double layer in which its external membrane encloses an interior space. Therefore, polymersomes are regarded as nonbiological liposomes.[20] Amphiphilic block copolymers can be self-assembled in various morphologies such as uniform common vesicles,

large polydisperse vesicles, entrapped vesicles, or hollow concentric vesicles (Fig. 2). Due to the higher molecular weight of vesicles than bioactive amphiphilic agents (e.g. lipids, ≤1 kDa), copolymer-based amphiphiles offer the sufficient toughness and multiproposal applications as delivery vehicles. For example, vesicles have the capability to incorporate both the hydrophilic drugs into the aqueous box and the lipophilic boxes within the membrane, offering unique flexibility of being able to incorporate different groups of drugs in the same carriers.

ABC triblock copolymers of polyethylene-b-poly (dimethyl siloxane)-b-poly(methyl oxazoline) have been self-assembled as vesicles with sizes between 60 and 300 nm in water. Thus, access to the vesicles has accelerated numerous applications in pharmaceutical sciences.[21]

Self-Assembly of Amphiphilic Stimuli-Responsive Copolymers

Stimuli-responsive polymers undergo fairly large intramolecular conformation or intermolecular aggregation in response to small external signals. These stimuli could be classified as either physical or chemical stimuli. Physical stimuli, such as temperature, electric, or magnetic fields, and mechanical stress, will influence the level of various energy sources and thereby change molecular interactions. The chemical stimuli, such as pH, ionic factors and chemical agents, will alter the interactions between macromolecular chains. These responses of recently formed systems are useful in biomedical applications such as drug delivery and chemotherapy.[22] Stimuli-responsive micelles are one of the various shapes of stimuli-responsive copolymers that form PNSs. Stimuli-responsive micelles can be designed by two methods. First, amphiphilic blocks in one structure undergo the assembly/disassembly by the alternation of hydrophilicity to lipophilicity balance (HLB), which can be adjusted by stimuli. Secondly, one or both hydrophilic and lipophilic segments would be replaced by stimuli-responsive component, resulting in stimuli-responsive shells or cores in micelle nanostructures. As opposed to systems that undergo stimuli-responsive supramolecular assembly, where the external signals only mediate assembly/disassembly processes, stimuli may drive the transformation or reconstruction of assembled structures in the case of stimuli-triggered morphogenesis. Therefore, these processes can produce entities with new structures that are unattainable through the self-assembly of classic amphiphiles alone. For instance, it has been well known that tumor cells have an acidic pH compared to healthy tissues. Such acidic environment-triggered disassembly of pH-responsive PNSs for modulating drug release selectively.[23,24]

Self-Assembly of Rod–Coil Block Copolymers

The term rod–coil block copolymer refers to a polymer in which one block has a rigid and elongated shape.

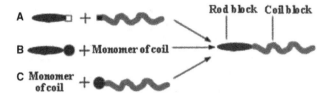

Fig. 3 Synthetic routes for preparing rod-coil block copolymers; (**A**) synthesis based on grafting to method, (**B**) and (**C**) synthesis based on grafting from method.

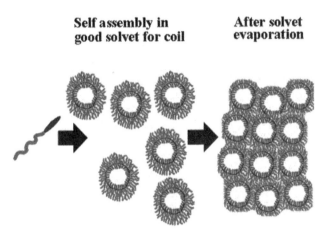

Fig. 4 A schematic representation of micellization of rod-coil block copolymers.

This rigidity can be the result of π-conjugation along the backbone, or of secondary structure, as for helical polypeptide derivatives. In the synthesis route, briefly, the functional groups of rod–coil block copolymers can easily react with the small molecules or oligomers (so-called graft to method) or directly initiate the grafting polymerization of monomers (so-called graft from method) to prepare amphiphilic rod–coil block copolymers[25] (Fig. 3).

In self-assembly of rod–coil block copolymers in rod-selective solvents, nanostructures were self-assembled with the rod blocks on the outside of the vesicle. As a result, when the coil block is smaller than the rod block, strong separation of the coil from the rod occurs and solvent results in segregation of the coil into the compacted cores of spherical or wormlike micelles.

In coil-selective solvents, coil-rich block copolymers self-assembled as micelles with the rods compacted in the core, where the final geometry may be spherical, wormlike, elliptical, or puck-like because of various packing morphologies of the rod segment (Fig. 4).

Self-Assembly of Polypeptide-Based Copolymers

It is no doubt that the polypeptide-based copolymers have profound impact not only in the field of drug delivery, but also in biomedical and tissue engineering. Besides the arginine, other most frequently used amino acid residues are applied in the synthesis of copolymer including lysine, leucine, aspartic acid, glutamic acid, valine, alanine, proline, tryptophan, glycine, and phenylalanine. Polypeptide-based copolymers present intrinsic motifs relative to other classic copolymers. For example, it is possible to design special properties of the polypeptide to produce stable self-assembled nanostructures with tailored functionalities. They have been self-assembled in some shapes as spherical aggregates, wormlike micelles, and vesicles. Depending on hydrophobicity to hydrophilicity balance, the polypeptide chain can be found in the core or shell of the assembly. In both situations, the ability of the peptide chains to stabilize a well-defined conformation (generally α-helix or β-sheet)- and to reversibly shift this secondary structure in response to stimuli allows increased control over shapes and accordingly properties. For instance, the use of charged hydrophilic polypeptide sequences such as poly(L-lysine) (PLLys) and poly(L-glutamic acid) (PLGA) in shell-forming micelles or vesicles let the formation of stimuli-responsive systems with presence of salts relative to traditional polyelectrolytes. Stimuli-responsive micelles and vesicles are of particular interest for drug delivery applications. For example, the vesicles of self-assembled poly(butadiene)–poly(L-lysine) (PB-b-PLLys) copolymer could be modified by changes in pH owing to the rod-coil transition of the PLLYs secondary structure.[26,27]

Self-Assembly of Copolymer Pairs with Specific Intermolecular Interactions

It is possible to produce PNSs by means of self-assembly of copolymers pairs due to difference in solution properties between the covalently connected blocks. Secondary interactions between the macromolecular chains, including charge transfer interaction, ligand binding, H-bonding, and dipolar interaction, can form specific intermolecular configurations as the driving forces. For example, PNSs were produced through self-assembly of PBD-b-PCMA (PS and PCMA stand for polybutadiene and poly(cesium methacrylate) respectively) and PS-b-PMVPI (PS and PMVPI stand for polystyrene and poly(1-methyl-4-vinyl pyridinium iodide) respectively). Thus, by mixing the copolymer pairs in solution, electrostatic attraction between the acid and base groups facilitate the construction of vesicles with a PBD inner membrane and a PS outer membrane. In this method, PNSs 200 nm with a diameter and shell thickness of 200 and 30 nm were produced respectively.[28,29]

Self-Assembly of Hyperbranched Polymers

Hyperbranched polymers (HBPs) are highly branched and have a spherical structure. A HBP could have a lot of pores in its structure as well as functional groups, such as hydroxyl, carboxyl, and amine, which facilitate incorporation of bioactive agents (bioactive agents could be a drug, protecting or targeting agent). Also they could be used in directing the self-assembly of inorganic precursors through supramolecular interactions such as host–guest interaction, hydrogen bonding, or electrostatic adsorption. Thus, HBPs are unique polymers to create host–guest systems and hybrid nanomaterials via complexation or self-assembly.[30,31] Similar to the amphiphilic linear copolymers, amphiphilic HBPs can self-assemble into PNSs.

PREPARATION OF POLYMERIC NANOTUBES: SELF-ASSEMBLY OF COPOLYMERS

The Gibbs free energy of self-assembly of amphiphilic diblock copolymers, on one hand, is contributed by the stretching energies of the core and corona chains. On the other hand, an increase in core/solvent interfacial tension will lead to an increase in micelle size and number, as well as stretching of the core and corona chains. Thus, above a critical interfacial tension, spherical-to-cylindrical micelle transition occurred. In addition, the cylinder-vesicle transition of micelles is affected by temperature of the system and polydispersity of self-assembled amphiphilic copolymers. The self-assembly of polymer–polypeptide conjugates also play an key role in the formation of tubular nanostructures owing to the well-defined secondary structure, such as α-helix or β-sheet of peptides. Polymeric nanotubes (PNTs) can be self-assembled through rod-coil block copolymers by H boding and π–π interaction.[32,33]

PREPARATION OF POLYMERIC NANOPARTICLES: SELF-ASSEMBLY OF COPOLYMERS

As pointed out, the selective interaction of a good solvent for a block of a block copolymer forms core-shell nanospheres in solution. Therefore, self-assembly of such block copolymer gives rise to monodisperse polymeric nanoparticles (PNPs). Unsymmetrical amphiphilic metallo-triblock copolymer of PAAC-b-PMA–Pd–PS is an excellent example (PAAC, PMA, and PS stand for poly(acrylic acid), poly(methyl acrylate) and polystyrene, respectively). In this triblock copolymer the PS block end is connected to PAAC-b-PMA by complex formation of palladium and pyridine. Self-assembly of the copolymer give rise to monodisperse core–shell nanoparticles with a hydrophobic PS core and noncovalently bonded particle shell.

PHYSICOCHEMICAL PARAMETERS EFFECTIVE ON INCORPORATION OF CONTENT

Drug delivery through macromolecular nanocarriers can be achieved by chemical conjugation or physical incorporation. Chemical conjugation is not ideal because some therapeutic agents do not have suitable functional groups for conjugation or because of altering in pharmacologic effect. On the contrary, physical incorporation is suitable for various bioactive molecules. However, the incorporation content of guest molecules depends on some physicochemical parameters.

Drug–Core Compatibility

Enhancement in the degree of compatibility between drug and core forming segment of the carrier increases the incorporation content.[34] Drug–carrier compatibility may be represented by Flory–Huggins interaction parameter (χ_{FH}), which gives an idea of the drug–core interactions. The low values of χ_{FH} imply that the carrier's environment is a nice solvent for the drug. χ_{FH} is calculated through Eqs 1 and 2:

$$\Delta = [(\delta_d - \delta_c)^2_{polarity} + (\delta_d - \delta_c)^2_{dispersion} + (\delta_d - \delta_c)^2_{hydrogen}]^{1/2} \quad (1)$$

$$\chi_{FH} = \Delta 2 V_d / RT \quad (2)$$

where Δ^2 is the solubility difference between the drug (d) and the core of the carrier (c). V_d is the molar volume of the drug, T is the temperature in Kelvin, and R is the gas constant. $(\delta_x)_p$, $(\delta_x)_d$ and $(\delta_x)_h$ are partial solubility factors of Hansen for the drug and the hydrophobic portion of the block copolymer. In general, the best drug incorporation content can only be obtained when a special drug is matched with an special carrier.

Core Length and Core Crystalizability

On one hand, for a constant hydrophilic block length, an increase in hydrophobic core length increases the incorporation content.[35] It is due to the fact an increase in the hydrophobic block length increases the volume of the core, creating a larger hydrophobic reservoir, which enhances the core–solvent partition coefficient of hydrophobic drug molecule. Accordingly, this increase in partition coefficient between the carrier and the solvent results in higher drug incorporation content. Thus, a higher drug incorporation content, in turn, permits the administration of an equivalent dose of the drug with a smaller amount of polymeric carrier, which is vital to reduce side effects as well as cytotoxicity of the polymeric carriers as delivery vehicles. On the other hand, compared to core block length, core crystalizability in crystallizable macromolecules has a negative effect on the accessible volume of the carrier for incorporation. It is due to the fact that the core crystallizability increases with increasing molecular weight.[36]

Drug/Carrier Concentration Ratio

It is documented that for a special drug/carrier system, there exists an optimum drug/carrier concentration ratio, based on the chemical structure of drug and carrier, at which maximum incorporation content is achieved. Above this point, the carrier core is saturated with drug and consequently cannot reach a higher drug incorporation content.[37,38]

Solvent Effect

The selection of the organic solvent, in respect of polarity, used for carrier preparation is effective on drug incorporation content. It is owing to the fact that amphiphilic copolymers have various conformations depending to polarity of solvents. Therefore, the solvent selection is a significant task during determination of the most efficient drug incorporation method.

NANOMEDICINAL APPLICATIONS

In general, ideal characteristics of a nanodevice for application as a delivery vehicle or therapeutic/theranostic agent are that they should have the following characteristics:[39]

- Nontoxic, biodegradable and biocompatible
- Nonimmunogenic
- Stable in blood (no aggregation and dissociation)
- Prolonged blood circulation time

In spite of the different classes of polymer-based nanocarriers reported so far, it is considerable that only micelles, elongated micelles, and vesicles have been extensively developed for nanomedicinal applications.[40] The biomedical advantages of these nanocarriers over other types of nanocarriers could be explained by the following:

(i) Because of the high molecular weights of these carriers compared to the carriers with low molecular weights, the dissociation time upon dilution is usually longer.[41]

(ii) When their corona contains PEG chains, the micelles are able to circulate in the blood for longer periods of time without being cleared from circulation (PEG has a low toxicity, however, in many cases the advantages outweigh its toxic properties).

(iii) The introduction of PEG chains in the architecture of delivery, therapeutic, and theranostic agents enhance the concentration of entrapped bioactive molecule in the blood, which results in a remarkable increase in concentration of bioactive molecule in the targeted tissues.[42]

(iv) They have a long shelf-life because of their stability in other biological fluids.[43]

PNSs as Delivery Vehicles

The permeability of the PNSs membranes is an interesting and important specialty in respect of controlled release.[44] Some factors such as the hydrodynamic radius of molecules, size of nanocarrier, temperature, and pH of the medium control the release behavior from PNSs. The functional groups in amphiphilic copolymers, as another significant factor, enhance the incorporation content of many promising drugs that are limited by their poor bioavailability.

Example of drug incorporation

Cyclosporin A, due to its low water solubility and its special cyclic peptide-based structure, has a scientific peculiarity to release. The core of the polyionic micelles were able to encapsulate the modified protein, which is formed from ionic interactions of PEGylated cationic block copolymer and the modified protein. Eventually, the protein as guest macromolecule, is released by rapid degradation and dissociation of the micellar carrier at acidic endosomal pH, due to charge conversion. Thus, the designing of structural parameters such as the PEG chain length, the type, and number of functional groups in relation to the PEG is important to reach desired incorporation content and particle size. This example obviously reveals that by smart introduction of drug-complementary functionalities in the copolymer backbone, various categories of bioactive molecules could be encapsulated by micellar nanocarriers.

Among the elongated micelles the biodegradable ones are promising candid as carrier in terms of superior therapeutic payload, as a result of their unique degradation mechanism and flexibility. Biodegradable blocks in block copolymer could be an excellent issue. Depending on the molecular weight, the degradation kinetics of the biodegradable blocks could be tunable for a desired period of time. Therefore, the degradation takes place at expected contents through chain-end cleavage to result in the constituent monomer and finally the formation of spherical micelles from the ends of the elongated micelles. It is due to the fact that the elongated micelles have only at very narrow range of amphiphilicity and such degradation-induced enhancement in hydrophilicity supports elongated-to-sphere transition. Unlike the micelles and elongated micelles, polymeric vesicles have the potential of encapsulation and delivery of multiple drugs in respect of polarity. In other words, vesicles have the unique flexibility of being able to incorporate both the hydrophilic and hydrophobic components within the membrane of the polymersome. On the contrary, in the micellar system resulting from the same block copolymers, only the hydrophobic drug could be incorporated and released. On the other hand, the release of high molecular weight polymeric bioactive agents from the hydrophilic internal hollow is challengeable. It arises from the low permeability, due to the thicker (in relative to liposomes) hydrophobic membrane. Nevertheless, the delivery of a large biomacromolecule like peptide to brain, by using vesicles, was successful and improved the learning and memory disorders.[45,46]

Example of nucleic acid incorporation

Micelles and elongated micelles also can be exploited as a nanocarrier to incorporate nucleic acids, which are used in the treatment of genetic diseases, viral infections, or cancers. Nucleic acid must be capable of accessing the distal sites of disease. Delivery of DNA by encapsulation in PNSs can prevent DNA from enzymatic degradation. Thus, the interactions between negatively charged cell membrane and the tunable functionalities of micelle surface let the effective delivery of DNA.

The other member of PNSs, polymeric vesicle, also enables targeting of unregulated receptors and molecules on affected cells. Both *in vitro* and *in vivo* investigations demonstrate that to encapsulate sensitive hydrophilic therapeutics such as siRNA, vesicular membrane could have complete protection from the harsh conditions.[47]

PNSs as Diagnostics Agents

Diagnostic agents are employed to delineate organs and tissues on the images upon parenteral administration. Three main groups of diagnostic methodologies including complex compounds of paramagnetic metals, such as Gd, for magnetic resonance imaging (MRI), complex compounds of metallic radioisotopes, such as ^{111}In or ^{99m}Tc, for scintigraphy scan, and organic iodine for x-ray computed tomography (CT) scan. In clinical imaging, at present time, the most frequent used imaging agents are based on low-molecular-weight contrast medium. However, in several issues, an application of high-molecular-weight media is inevitable. For example, rapid disappearance of low-molecular-weight contrast agents into interstitial space, during the imaging process, is an unwanted issue. Accordingly, by exploiting the high-molecular-weight media of PNSs the contrast media would be restricted in the vasculature or lymphatic and efficiently delineate it. In brief, PNSs as nanocarriers of diagnostic agents could be applied in the abovementioned three main imaging methodologies. Several significant prerequisites such as the protective polymeric layer on the nanocarrier to prevent interaction with (reticuloendothelial system) RES cells as well as the size of resulting PNSs have also high importance.[48]

PNTs as Delivery Vehicles

PNTs are ideal candidates as the advanced delivery vehicle in cancer and infectious viruses. For instance, their distinct features such as easiness of functionalization with bioactive molecules, remarkable drug incorporation content, extended blood circulation, and tailored size in nanoscale make them as new injectable anticancer vaccines. PNTs also were employed in applications such as *in vivo* gene delivery and gene expression, and single-stranded DNA carriers.[49]

PNPs as Delivery Vehicles

Modification of drug-incorporated PNPs with ligands or antibodies increased therapeutic efficiency and reduced side effects compared to the drug alone or the nontargeting drug-carrying PNPs. Such property, which allows for active targeting of the particular cells in the body, is well known as smart delivery. For example, in cancer therapy, the existence of targeting ligands on the surface of the PNPs led to enhanced tumor permeability, accumulation in the cancerous tissue, and finally increased cancer-cell uptake of the drug-loaded PNPs through receptor-mediated endocytosis. Accordingly, the higher intracellular drug concentration enhanced therapeutic performance and minimized harmful side effects. Although PNPs carrying a repetitive unit on their peripheral surface are recognized to trigger an immune response,[50] their immune response can be lower by introducing PEG residues on the surface of the PNPs.

In gene therapy application also, siRNA could be localized in the cytosol of target cells to cause the gene silencing effect. However, free siRNA itself cannot go through the cytoplasm of cells owing to its anionic charge. In order to deliver siRNA to target cells, various delivery nanocarriers based on PNPs have been used effectively.[51]

PNPs as Theranostic Agents

Theranostics based on PNPs is a process that allows the imaging and recording from the reactivity of therapeutic delivery systems.[52] Theranostic agents of PNPs can employed as valuable tools to investigate the process of drug delivery after cellular internalization of nanocarriers, which could afford a deep approach into the superior design of nanocarriers for personalized nanomedicine.

CONCLUSION

In this entry, we have described in a systematic way the methods employed to form nanoscale-sized structures from copolymers. The self-assembly principles that direct the organization of copolymers in solution as well as physicochemical properties of amphiphilic copolymers have been briefly reported from both theoretical and experimental points of view. The biomedical applications of these functional structures are of growing interest for nanomedicine as nonviral gene vectors and theranostic tools. The use of polymeric nanomaterials as delivery vehicles for anticancer drugs has already moved into a clinical trial. Indeed, self-assembled structures may promise a new range of advanced materials, with unique functions through fine-tuning of molecular parameters.

REFERENCES

1. Smid, J. Historical perspectives on living anionic polymerization. J. Polym. Sci. Part A: Polym. Chem. **2002**, *40* (13), 2101–2107.
2. Matyjaszewski, K.; Xia, J. Atom transfer radical polymerization. Chem. Rev. **2001**, *101* (9), 2921–2990.

3. Moad, G.; Rizzardo, E.; Thang, S.H. Toward living radical polymerization. Acc. Chem. Res. **2008**, *41* (9), 1133–1142.

4. Hawker, C.J.; Bosman, A.W.; Harth, E. New polymer synthesis by nitroxide mediated living radical polymerizations. Chem. Rev. **2001**, *101* (12), 3661–3688.

5. Zhang, L.; Eisenberg, A. Multiple morphologies of "crew-cut" aggregates of polystyrene-b-poly(acrylic acid) block copolymers. Science **1995**, *268* (5218), 1728–1731.

6. Qian, J.; Zhang, M.; Manners, I.; Winnik, M.A. Nanofiber micelles from the selfassembly of block copolymers. Trends Biotechnol. **2009**, *28* (2), 84–92.

7. Discher, D.E.; Eisenberg, A. Polymer vesicles. Science **2002**, *297* (5583), 967–973.

8. Li, Z.; Hillmyer, M.A.; Lodge, T.P. Morphologies of multicompartment micelles formed by ABC miktoarm star terpolymers. Langmuir **2006**, *22* (22), 9409–9417.

9. Förster, S.; Antonietti, M. Amphiphilic block copolymers in structure-controlled nanomaterial hybrids. Adv. Mater. **1998**, *10* (3), 195–217.

10. Moffitt, M.; Khougaz, K.; Eisenberg, A. Micellization of ionic block copolymers. Acc. Chem. Res. **1996**, *29* (2), 95–102.

11. Riess, G. Micellization of block copolymers Prog. Polym. Sci. **2003**, *28* (7), 1107–1170.

12. Förster, S.; Plantenberg, T. From self-organizing polymers to nanohybrid and biomaterials. Angew. Chem. Int. Ed. **2002**, *41* (5), 688–714.

13. Matsumura, Y.; Kataoka, K. Preclinical and clinical studies of anticancer agentincorporating polymer micelles. Cancer. Sci. **2009**, *100* (4), 572–579.

14. Won, Y.-Y.; Davis, H.T.; Bates, F.S. Giant wormlike rubber micelles. Science **1999**, *283* (5404), 960–963.

15. Jain, S.; Bates, F.S. On the origins of morphological complexity in block copolymer. Surfactants. Science **2003**, *300* (5618), 460–464.

16. Hu, Y.; Jiang, Z.; Chen, R.; Wiu, W.; Jiang, X. Degradation and degradation-induced re-assembly of PVP-PCL micelles. Biomacromolecules **2010**, *11* (2), 481–488.

17. Fairley, N.; Hoang, B.; Allen, C. Morphological control of poly(ethylene glycol)-blockpoly(ε-caprolactone) copolymer aggregates in aqueous solution. Biomacromolecules **2008**, *9* (9), 2283–2291.

18. Lazzari, M.; López-Quintela, M.A. Micellization phenomenon in semicrystalline block copolymers: reflexive and critical views on the formation of cylindrical micelles. Macromol. Rapid Commun. **2009**, *30* (21), 1785–1791.

19. Kim, Y.; Dalhaimer, P.; Christian, D.A.; Discher, D.E. Polymeric worm micelles as nano-carriers for drug delivery. Nanotechnology **2005**, *16* (7), S1–S8.

20. Soo, P.L.; Eisenberg, A. Preparation of block copolymer vesicles in solution. J. Polym. Sci. Part B: Polym. Phys. **2004**, *42* (6), 923–938.

21. Discher, B.M.; Won, Y.-Y.; Ege, D.S.; Lee, J.C.; Bates, F.S.; Discher, D.E.; Hammer, D.A. Polymersomes: tough vesicles made from diblock copolymers, Science **1999**, *284* (5417), 1143–1146.

22. Petros, R.A.; De Simone, J.M. Stratefies in the design of nanoparticles for therapeutic applications. Nat. Rev. Drug Discov. **2010**, *9* (8), 615–627.

23. Zhang, J.; Li, X.; Li, X. Stimuli-triggered structural engineering of synthetic and biological polymeric assemblies. Prog. Polym. Sci. **2012**, *37* (8), 1130–1176.

24. Kumara, A.; Srivastavaa, A.; Galaev, I.Y.; Mattiasson, B. Smart polymers: Physical forms and bioengineering applications. Prog. Polym. Sci. **2007**, *32* (10), 1205–1237.

25. Lim, Y.B.; Moon, K.S.; Lee, M. Rod-coil block molecules: their aqueous self assembly and biomaterials applications. J. Mater. Chem. **2008**, *18* (25), 2909–2918.

26. Sun, J.; Shi, Q.; Chen, X.; Guo, J.; Jing, X. Self assembly of a hydrophobic polypeptide containing a short hydrophilic middle segment: vesicles to large compound micelles. Macromol. Chem. Phys. **2008**, *209* (11), 1129–1136.

27. Carlsen, A.; Lecommandoux, S. Self assembly of polypeptide-based block copolymer amphiphiles. Curr. Opin. Colloid Interface Sci. **2009**, *14* (5), 329–339.

28. Zhang, L.; Eisenberg, A. Multiple morphologies of "crew-cut" aggregates of polystyrene-b-poly (acrylic acid) block copolymers. J. Am. Chem. Soc. **1996**, *118* (13), 368–381.

29. Schrage, S.; Sigel, R.; Schlaad, H. Formation of amphiphilic polyion complex vesicles from mixtures of oppositely charged block ionomers. Macromolecules **2003**, *36* (5), 1417–1420.

30. Zhou, Y.; Yan, D.Y. Supramolecular self assembly of amphiphilic hyperbranched polymers at all scales and dimensions: progress, characteristics and perspectives. Chem. Commun. **2009**, (10), 1172–1188.

31. Mai, Z.Y.; Yan, D. Real-time hierarchical self assembly of age compound vesicles from amphiphilic hyperbranched multiarm copolymer. Small **2007**, *3* (7), 1170–1173.

32. Han, Y.; Cui, J.; Jiang, W. Effect of polydispersity on the self assembly structure of diblock copolymers under various confined states: a Monte Carlo study. Macromolecules **2008**, *41* (16), 6239–6245.

33. Georgakilas, V.; Pellarini, F.; Prato, M.; Guldi, D.M.; Melle-Franco, M.; Zerbetto, F.; Melle-Franco, M.; Zerbetto, F. Supramolecular self assembled fullerene nanostructures. Proc. Natl. Acad. Sci. USA. **2002**, *99* (8), 5075–5080.

34. Mahmud, A.; Patel, S.; Molavi, O.; Choi, P.; Samuel, J.; Lavasanifar, A. Self-associating poly(ethylene oxide)-b-poly(alpha-cholesteryl carboxylate-epsilon-caprolactone) block copolymer for the solubilization of STAT-3 inhibitor cucurbitacin I. Biomacromolecules **2009**, *10* (3), 417–478.

35. Shuai, X.; Ai, H.; Nasongkla, N.; Kim, S.; Gao, J. Micellar carriers based on block copolymers of poly(-caprolactone) and poly(ethylene glycol) for doxorubicin delivery. J. Control. Release **2004**, *98* (3), 415–426.

36. Shuai, X.; Merdan, T.; Schaper, A.; Xi, F.; Kissel, T. Core-cross-linked polymeric micelles as paclitaxel carriers. Bioconjug. Chem. **2004**, *15* (3), 441–448.

37. Namazi, H.; Jafarirad, S. Controlled release of linear-dendritic hybrids of carbosiloxane dendrimer: The effect of hybrid's amphiphilicity on drug-incorporation; hybrid–drug interactions and hydrolytic behavior of nanocarriers. Int. J. Pharm. **2011**, *407* (1–2), 167–173.

38. Namazi, H.; Jafarirad, S. *In vitro* photo-controlled drug release system based on amphiphilic linear-dendritic diblock copolymers; Self assembly behavior and application as nanocarrier J. Pharm. Pharm. Sci. **2011**, *14* (2), 162–180.

Molecular—Nanogels

Molecular—Nanogels

39. Koo, Y.-E.L.; Reddy, G.R.; Bhojani, M.; Schneider, R.; Philbert, M.A.; Rehemtulla, A.; Ross, B.D.; Kopelman, R. Brain cancer diagnosis and therapy with nanoplatforms, Adv. Drug Deliv. Rev. **2006**, *58* (14), 1556–1577.

40. Garcia-Garcia, E.; Andrieux, K.; Gil, S.; Couvreur, P. Colloidal carriers and blood-brain barrier (BBB) translocation: a way to deliver drugs to the brain. Int. J. Pharm. **2005**, *298* (2), 274–292.

41. Soma, C.E.; Dubernet, C.; Bentolila, D.; Benita, S.; Couvreur, P. Reversion of multidrug resistance by co-encapsulation of doxorubicin and cyclosporin a in polyalkylcyanoacrylate nanoparticles, Biomaterials **2000**, *21* (1), 1–7.

42. Chung, J.E.; Yokoyama, M.; Yamato, M.; Aohagi, T.; Sakurai, Y.; Okano, T. Thermo-responsive drug delivery from polymeric micelles constructed using block copolymers of poly(n-isopropylacrylamide) and poly (butylmethacrylate). J. Control. Release **1999**, *62* (1–2), 115–127.

43. Kabanov, A.V.; Batrakova, E.V.; Melik-Nubarov, N.S.; Fedoseev, N.A.; Dorodnich, T.Y.; Alakhov, V.Y.; Chekhonin, V.P.; Nazarova, I.R.; Kabanov, V.A. A new class of drug carriers: micelles of poly(oxyethylene)–poly(oxypropylene) block copolymers as microcontainers for targeting drugs from blood to brain. J. Control. Release **1992**, *22* (2), 141–158.

44. Leson, A.; Hauschild, S.; Rank, A.; Neub, A.; Schubert, R.; Förster, S.; Mayer, C. Molecular exchange through membranes of PVP-b-PEG vesicles. Small **2007**, *3* (6), 1074–1083.

45. Venkataraman, S.; Hedrick, J.L.; Ong, Z.Y.; Yang, C.; Ee, P.L.; Hammond, P.T.; Yang, Y.Y. The effects of polymeric nanostructure shape on drug delivery. Adv. Drug Delivery Rev. **2011**, *63* (14–15), 1228–1246.

46. Pang, Z.; Lu, W.; Gao, H.; Hu, K.; Chen, J.; Zhang, C.; Gao, X.; Jiang, X.; Zhu, C. Preparation and brain delivery property of biodegradable polymersomes conjugated with OX26. J. Control. Release **2008**, *128* (2), 120–127.

47. Kim, Y.; Tewari, M.; Pajerowski, J.D.; Cai, S.; Sen, S.; Williams, J.H.; Sirsi, S.R.; Lutz, G.J.; Discher, D.E. Polymersome delivery of siRNA and antisense oligonucleotides. J. Control. Release **2009**, *134* (2), 132–140.

48. Fu, G.D.; Li, G.L.; Neoh, K.G.; Kang, E.T. Hollow polymeric nanostructures-Synthesis, morphology and function. Prog. Polym. Sci. **2011**, *36* (1), 127–167.

49. Abidian, M.R.; Kim, D.H.; Martin, D.C. Conducting-polymer nanotubes for controlled drug release. Adv. Mater. **2006**, *18* (4), 405–409.

50. Nah, J.W.; Yu, L.; Han, S.O.; Ahn, C.H.; Kim, S.W.; Artery wall binding peptide-poly(ethylene glycol)-grafted-poly (L-lysine)-based gene delivery to artery wall cells. J. Control. Release **2002**, *78* (1–3), 273–284.

51. Parker, A.L.; Seymour, L.W. Targeting of polyelectrolyte RNA complexes to cell surface integrins as an efficient cytoplasmic transfection mechanism, J. Bioact. Compat. Polym. **2002**, *17* (4), 229–238.

52. Lee, J.H.; Lee, K.; Moon, S.H.; Lee, Y.; Park, T.G.; Cheon, J. All-in-one targetcell-specific magnetic nanoparticles for simultaneous molecular imaging and siRNA delivery. Angew. Chem. Int. Ed. **2009**, *48* (23), 4174–4179.

Mucoadhesive Polymers: Basics, Strategies, and Trends

Andreas Bernkop-Schnürch
Institute of Pharmacy, University of Innsbruck, Innsbruck, Austria

Abstract

Mucoadhesive polymers, by definition, are polymers that are able to adhere to the mucus gel layer that covers various body tissues. These mucoadhesive properties are in many cases advantageous, and they render such polymers interesting tools for pharmaceutical applications. This entry will review research and trends concerning mucoadhesive polymers and should provide a good platform for ongoing development in this field.

INTRODUCTION

In the early 1980s, academic research groups pioneered the concept of mucoadhesion as a new strategy in order to improve the therapeutic effect of various drugs. *Mucoadhesive polymers are able to adhere to the mucus gel layer* which covers various tissues of the body. These mucoadhesive properties are in many cases advantageous, rendering such polymers interesting tools for various pharmaceutical reasons:

1. Mediated by mucoadhesive polymers, the residence time of dosage forms on the mucosa can be prolonged, which allows a sustained drug release at a given target site in order to maximize the therapeutic effect. Robinson and Bologna, for instance, have reported that a polycarbophil gel is capable of remaining on the vaginal tissue for 3–4 days and thus provides an excellent vehicle for the delivery of drugs such as progesterone and nonoxynol-9.[1]
2. Furthermore, drug delivery systems can be localized on a certain surface area for purpose of local therapy or of drug liberation at the "absorption window." The absorption of riboflavin, for instance, which has its "absorption window" in the stomach as well as the small intestine, could be strongly improved in human volunteers by oral administration of mucoadhesive microspheres versus non-adhesive microspheres as illustrated in Fig. 1.[2]
3. In addition, mucoadhesive polymers can guarantee an intimate contact with the absorption membrane providing the basis for a high-concentration gradient as driving force of drug absorption (IIIa), for the exclusion of a presystemic metabolism such as the degradation of orally given (poly)peptide drugs by luminally secreted intestinal enzymes (IIIb),[3,4] and for interactions of the polymer with the epithelium such as a permeation enhancing effect[5,6] or the inhibition of brush border membrane-bound enzymes (IIIc).[7,8]

Because of all these benefits, research work in the field of mucoadhesive polymers has strongly increased within the last two decades, resulting in numerous promising ideas, strategies, systems, and techniques based on a more and more profound basic knowledge. An overview reflecting the status quo as well as future trends concerning mucoadhesive polymers should provide a good platform for ongoing research and development in this field.

MUCUS GEL COMPOSITION

As mucoadhesive macromolecules adhere to the mucus gel layer, it is important to characterize first of all this polymeric network representing the target structure for mucoadhesive polymers. The most important component building up the mucus structure are glycoproteins with a relative molecular mass range of $1–40 \times 10^6$ Da. These so-called mucins possess a linear protein core, typically of high serine and threonine content that is glycosylated by oligosaccharide side chains that contain blood group structures. The protein core of many mucins exhibits furthermore N- and/or C-terminally located cysteine-rich subdomains, which are connected with each other via intra-and/or intermolecular disulfide bonds. This presumptive structure of the mucus is illustrated in Figs. 2. The water content of the mucus gel has been determined to be 83%.[9] Generally, mucins may be classified into two classes, *secretory* (I) and *membrane-bound* (II) forms. Secretory mucins are secreted from mucosal absorptive epithelial cells as well as specialized goblet cells. They constitute the major component of mucus gels of the gastrointestinal, respiratory, ocular, and urogenital surface. The mucus layer based on secretory mucins represents not only a

Concise Encyclopedia of Biomedical Polymers and Polymeric Biomaterials DOI: 10.1081/E-EBPPC-120052253

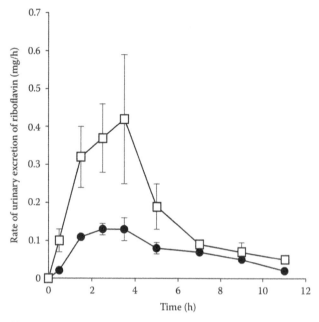

Fig. 1 Improvement in the oral uptake of riboflavin by utilizing a mucoadhesive delivery system (□) in comparison to the same delivery system without mucoadhesive properties (•) in human volunteers.
Source: Adapted from Akiyama et al.[2]

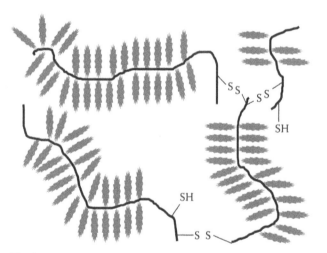

Fig. 2 Schematic presentation of the three-dimensional network of the mucus gel layer; protein core: dark line, glycosidic side-chains: gray shade.

physical barrier but also a protective diffusion barrier.[9,10] Membrane-bound mucins possess a hydrophilic membrane-spanning domain and they are attached to cell surfaces. Up to now, eight different types of human mucins have been discovered and characterized, which are listed in Table 1.

Secreted mucins are continuously released from cells as well as glands undergoing immediately thereafter a polymerization process which is mainly based on an intermolecular disulfide bond formation. This so-formed mucus layer, on the other hand, is continuously eroded by

enzymatic and mechanical challenges on the luminal surface. Although the turnover time of the mucus gel layer seems to be crucial in order to estimate for how long a mucoadhesive delivery system can remain on the mucosa in maximum, there is no accurate information on this time scale available. A clue was given by Lehr, Poelma, and Junginger, who determined the turnover time of the mucus gel layer of chronically isolated intestinal loops in rats to be in the order of approximately 1–4 hr.[21] Both mucus secretion and mucus erosion, however, are influenced by so many factors such as mucus secretagogues, mechanical stimuli, stress, calcium concentration, or the enzymatic activity of luminal proteases leading to a highly variable turnover time, which is therefore quite complex and difficult to evaluate.

PRINCIPLES OF MUCOADHESION

Since the concept of mucoadhesion has been introduced into the scientific literature, considerably many attempts have been undertaken in order to explain this phenomenon. So far, however, no generally accepted theory has been found. A reason for this situation can be seen in the fact that many parameters seem to have an impact on mucoadhesion (Fig. 3), which makes a unique explanation impossible. Although there are many controversies in case of which theories should be favored, at least two basic steps are generally accepted. In step I, the *contact stage*, an intimate contact between the mucoadhesive and the mucus gel layer is formed. In step II, the *consolidation stage*, the adhesive joint is strengthened and consolidated, providing a prolonged adhesion.

Chemical Principles

Formation of non-covalent bonds

Non-covalent chemical bonds include hydrogen bonding (I), which is based on hydrophilic functional groups such as hydroxylic groups, carboxylic groups, amino groups, and sulfate groups, ionic interactions (II) such as the interaction of the cationic polymer chitosan with anionic sialic acid moieties of the mucus, and van der Waals forces (III) based on various dipole–dipole interactions.

Formation of covalent bonds

In contrast to secondary bonds, covalent bonds are much stronger and are not any more influenced by parameters such as ionic strength and pH value. Functional groups that are able to form covalent bonds to the mucus layer are overall thiol groups. The way how such functional groups can form covalent bonds with mucus glycoproteins is illustrated in Fig. 4. Thiolated polymers are able to mimic the natural occurring mechanism how mucus glycoproteins are immobilized in the mucus.

Table 1 Synopsis of human mucins

Mucin	Characteristics	High-level expression	Cysteine-rich subdomains	References
MUC1	Membrane-bound epithelial mucin	Breast, pancreas		[12]
MUC2	Secreted intestinal mucin	Intestine, tracheobronchus	Yes	[13]
MUC3	Secreted intestinal mucin	Intestine, gallbladder, pancreas	Yes	[14]
MUC4	Tracheobronchial secretory mucin	Tracheobronchus, colon, uterine endocervix		[15]
MUC5A/C	Secretory mucin	Tracheobronchus, stomach, ocular, uterine endocervix	Yes	[16]
MUC5B	Secretory mucin	Tracheobronchus, salivary	Yes	[17]
MUC6	Major secretory mucin of the stomach	Stomach, gallbladder	Yes	[18]
MUC7		Salivary		[19]
MUC8		Tracheobronchus, reproductive tract		[20]

Source: From Campbell.[11]

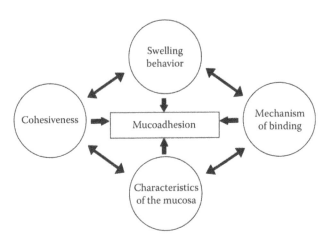

Fig. 3 Schematic presentation of effects influencing mucoadhesion.
Source: From Bernkop-Schnürch & Steininger.[22]

Physical Principles

Interpenetration

One theory in order to explain the phenomenon of mucoadhesion is based on a macromolecular interpenetration effect. The mucoadhesive macromolecules interpenetrate with mucus glycoproteins as illustrated in Figs. 5. The resulting consolidation provides the formation of a strong stable mucoadhesive joint. The theory can be substantiated by the observation that chain flexibility favoring a polymeric interpenetration is a crucial parameter for mucoadhesion. The cross-linking of various polymers or the covalent attachment of large sized ligands[23] leading to a

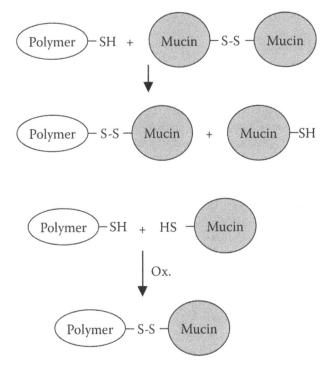

Fig. 4 Thiolated polymers forming covalent bonds with the mucus layer.

reduction in chain flexibility results in a strong decrease in mucoadhesive properties.

Rheological approaches—as discussed in detail later on—demonstrating a synergistic increase in the resistance to elastic deformation by mixing mucus with mucoadhesive polymer, i.e., the consolidation of the adhesive joint and attenuated total reflection Fourier transform infrared

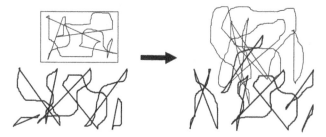

Fig. 5 Interpenetration of a mucoadhesive matrix tablet (light line) and the mucus gel layer (dark line).

(ATR-FTIR) studies showing changes in the spectrum of a poly(acrylic acid) film because of interpenetrating mucin molecules within 6 min, provide further evidence for this theory.[24]

Recently, Imam et al. evaluated the degree of interpenetration of fluorescence labeled polyacrylates (PAA) of increasing molecular mass with various mucus gel layers utilizing confocal laser microscopy and fluorescence quantification techniques.[25] Results of these studies are illustrated in Fig. 6, demonstrating a clear correlation between the molecular mass of the mucoadhesive polymer and the degree of interpenetration.

Mucus dehydration

Dehydration of a mucus gel layer increases its cohesive nature, which was shown by Mortazavi and Smart.[26] Dehydration essentially alters the physicochemical properties of a mucus gel layer, making it locally more cohesive and promoting the retention of a delivery system. The theory can be substantiated by studies with dialysis tubings. Bringing dry mucoadhesive polymers wrapped in dialysis tubings into contact with a mucus layer leads to its dehydration rapidly.[26] Dehydrating a mucus gel increases its cohesive nature and subsequently its adhesive behavior, which could be shown by tensile studies.[26] An objection to the theory that no interpenetration but exclusively mucus dehydration occurs is given by various rheological studies and tensile studies carried out by Caramella, Rossi, and Bonferoni[27] who observed a significant increase in the total work of adhesion (TWA) by mixing the polymer directly with mucin before tensile measurements. It is therefore likely that glycoproteins of the mucus are carried with the flow of water into the mucoadhesive polymer which leads to the already described interpenetration; an explanation which allows the combination of both the interpenetration and mucus dehydration theory.

Entanglements of polymer chains

The mucoadhesive as well as cohesive properties of polymers can also be explained by physical entanglements of polymer chains. The difference in cohesive as well as

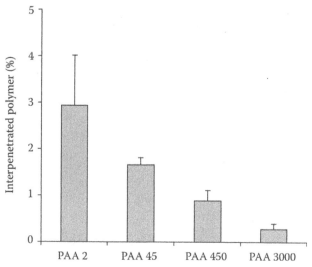

Fig. 6 Size-dependent interpenetration of PAA of indicated molecular mass (2–3000 kDa) in a mucus layer.
Source: Adapted from Imam et al.[25]

mucoadhesive properties of lyophilized and precipitated mucoadhesive polymers gives strong evidence for this theory. Precipitated polymers display high cohesive as well as mucoadhesive properties with a likely high number of polymer chain entanglements, whereas these properties are comparatively lower for the corresponding lyophilized polymers, which do not exhibit such a high extent of entanglements.[28]

METHODS TO EVALUATE MUCOADHESIVE PROPERTIES

In Vitro Methods

The selection of the mucoadhesive material is the first step in developing a mucoadhesive drug delivery system. A screening of the adhesive properties of polymeric materials can be done by various *in vitro* methods such as visual tests, tensile studies, and rheological methods, which are often accompanied by additional spectroscopic techniques. Apart from these well-established methods which are described in detail later, some novel methods such as magnetic[29] and direct force measurement techniques[30] have been introduced into the literature as well.

Visual tests

Rotating Cylinder. In order to evaluate the duration of binding to the mucosa as well as the cohesiveness of muco-adhesive polymers, the rotating cylinder seems to be an appropriate method. In particular, tablets consisting of the test polymer can be brought into contact with freshly excised intestinal mucosa (e.g., porcine), which has been spanned on a stainless steel cylinder (diameter: 4.4 cm; height: 5.1 cm; apparatus 4—cylinder, USP XXII). Thereafter, the cylinder is placed in the dissolution apparatus according to the USP containing an artificial gastric or intestinal fluid at 37°C. The experimental setup is illustrated in Fig. 7. The fully immersed cylinder is agitated with 250 rpm. The time needed for detachment, disintegration, and/or erosion of test tablets can be determined visually.[22]

Rinsed Channel. At this method, freshly excised mucosa is spread out on a lop-sided channel with the mucus gel layer facing upward and placed in a thermostatic chamber, which is kept at 37°C. After applying the test material on the mucosa, the rinse is flushed with an artificial gastric or intestinal fluid at

a constant flow rate and the residence time of the mucoadhesive polymer is determined visually or quantified utilizing insoluble markers such as fluorescein diacetate.[31] The experimental setup is illustrated in Fig. 8.[32,33]

Tensile tests

Tensile Studies with Dry Polymer Compacts. Test polymers are thereby compressed to flat-faced discs. Tensiometer studies with these test discs are carried out on freshly excised mucosa. Test discs are therefore attached to the mucosa. After a certain contact time between test disc and mucosa, the mucosa is pulled at a certain rate (mm s⁻¹) from the disc. The TWA representing the area under the force/distance curve and the maximum detachment force (MDF) are determined. The experimental setup is illustrated in Fig. 9. It represents one of the best established *in vitro* test systems which are used by numerous research groups.[23,34,35]

Tensile Studies with Hydrated Polymers. In order to minimize the influence of an "adhesion by hydration," tensile studies can also be carried out with hydrated polymers as described by Ch'ng et al.[36] Hydrated test polymers are thereby spread in a uniform monolayer over excised mucosa which has been fixed on a flat surface. In an artificial gastric or intestinal fluid at 37°C, the hydrated polymer is brought in contact with the mucus layer of a second mucosa, which is fixed on a flat surface

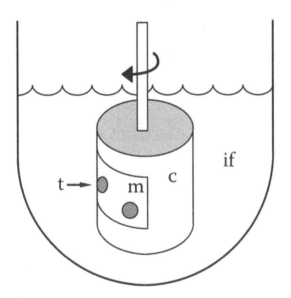

Fig. 7 Schematic presentation of the test system used to evaluate the mucoadhesive properties of tablets based on various polymers. c, cylinder; if, intestinal fluid; m, porcine mucosa; t, tablet.

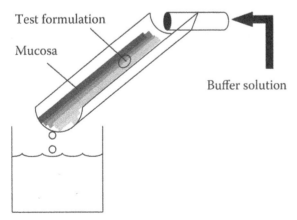

Fig. 8 Experimental setup in order to evaluate the mucoadhesive properties of test formulations on a freshly excised mucosa spread out on a lop-sided channel.

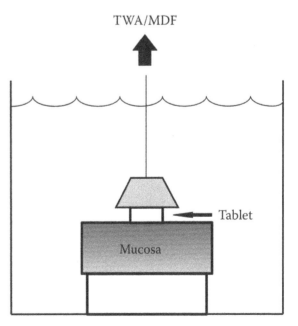

Fig. 9 Experimental setup for tensile studies with dry polymer compacts.

of a weight hanging on a balance. The TWA and MDF are then determined as described earlier.

Tensile Studies with Microspheres. Tensile studies as described earlier are not designed for measuring microscopic interactions such as those that may occur between microparticles and the mucus gel layer. Hence, a method was developed for measuring mucoadhesive properties of microspheres. *In vivo* interactions are thereby mimicked utilizing a miniature tissue chamber, which is heated by a water jacket. Thermoplastic microspheres are mounted to the tips of fine iron wires using a melting technique. The unloaded ends of the wires are then attached to a sample clip and suspended in the microbalance enclosure. The freshly excised section of mucosa is clamped in the buffer-filled chamber at 37°C and the microsphere is brought into contact with the tissue. To fracture the adhesive interactions, the tissue is pulled off the microsphere and certain mucoadhesive parameters are calculated and graphs of force versus position and time are plotted using appropriate software for the microbalance. The method provides valuable information concerning the adhesive properties of microspheres. So far, however, it is limited to microspheres not smaller than 300 μm.[37]

Rheological techniques

During the chain interpenetration of mucoadhesive polymers with mucin macromolecules, physical entanglements, conformational changes, and chemical interactions occur. Thereby, changes in the rheological behavior of the two macromolecular species are produced. An evaluation of the resulting synergistic increase in viscosity, which is supposed to be in many cases directly proportional to results obtained with tensile studies,[38] can be achieved by mixing the mucoadhesive polymer with mucus and measuring viscosity. The rheological behavior can be determined either by a classical rotational viscometry test at a certain shear rate or by dynamic oscillatory measurements, which give useful information about the structure of the polymer–mucin network.[39]

Spectroscopic techniques

Since mucoadhesive properties of polymers have been investigated, these tests were accompanied by additional spectroscopic analyses.[40] Kerr et al.,[41] for instance, using ^{13}C nuclear magnetic resonance spectroscopy have provided evidence of hydrogen bonding between mucus glycoproteins and the carboxylic acid groups of polyacrylic acid. Moreover, Tobyn, Johnson, and Gibson using Fourier transform infrared spectroscopy, reported also interactions between the pig gastric mucus glycoproteins and the test mucoadhesive.[42] Jabbari et al. could confirm the chain

interpenetration theory by investigating a mucin and polyacrylic acid interface via attenuated total reflection Fourier transform infrared spectroscopy (ATR-FTIR).[24] Mortazavi using infrared and ^{13}C-NMR spectroscopy, suggested the formation of hydrogen bonds between the mucoadhesive polyacrylic acid and the terminal sugar residues on the mucus glycoprotein.[43] In another study, the nature of interactions between the mucus gel and polyacrylic acid was investigated by tensile studies (I), dynamic oscillatory rheology (II), and ^{13}C-NMR spectroscopy (III) as well. The addition of hydrogen bond breaking agents resulted thereby in a decrease in mucoadhesive strength, a reduction in viscoelastic properties of polymer/mucin mixtures, and a positional change in the chemical shift of the polyacrylic acid signal.[44]

In Vivo Methods

To date, the mucoadhesion of dosage forms on mucosal membranes has been evaluated *in vivo* by direct observation (I), by gamma scintigraphy (II), and by using insoluble markers (III). Thereby either the time period of mucoadhesion is determined or in case of the GI-tract to which extent the transit time of the mucoadhesive dosage form can be prolonged.

The direct observation offers the advantage that neither radionuclides nor insoluble markers are needed and that a pretreatment of the test formulation is in most cases not necessary. The technique can be used to evaluate mucoadhesion on various tissues in animal studies. Akiyama and Nagahara for instance, administered mucoadhesive microspheres orally to rats. After 2.5 hr, the extent of the adhesion of these microspheres to the gastric mucosa was evaluated visually, demonstrating a high amount of mucoadhesive microspheres being present in the stomach compared to non-adhesive microspheres.[45] For studies in humans, however, the direct observation is limited only to a few mucosal tissues such as the oral cavity. Bouckaert, Lefebvre, and Remon for instance, determined the adhesion of tablets in the region of the upper canine. Test tablets were thereby fixed for 1 min with a slight manual pressure on the lip followed by moistening with the tongue to prevent sticking of the tablet to the lip. The adhesion time was defined as the time after which the mucoadhesive tablet was no longer visible under the lip upon control at 30 min interval.[46] In another study, which was carried out in the same way, volunteers participating in this study were asked to record the time and circumstances of the end of adhesion (erosion or detachment of the tablet).[47]

In contrast, no tissue limitations seem to exist for gamma scintigraphic methods. Using these techniques makes it even possible to evaluate the increase in the GI-transit time of mucoadhesive formulations. Radionuclides used for imaging studies include ^{99}Tcm, ^{111}Inm, ^{113}Inm, and ^{81}Krm. Among them technetium-99m represents the most commonly used radionuclide, as it displays no beta or alpha radiation and a comparatively short half-life of 6.03 hr. Indium-113m, which has an even shorter half-life of

1.7 hr, has a different energy to technetium-99m and can therefore be used in double-labeling studies. For many applications, the longer lived isotope indium-111m (half-life 2.8 days) seems to be more appropriate. Whereas a strongly prolonged GI-transit time of mucoadhesive polymers was demonstrated in various animal studies,[36] the same effect cannot in all studies be shown in human volunteers (effect;[2] no effect[48,49]).

MUCOADHESIVE POLYMERS

A systematic for mucoadhesive polymers can be based on their origin (e.g., natural–synthetic), the type of mucosa on which they are mainly applied (e.g., ocular–buccal), or their chemical structure (e.g., cellulose derivatives–PAA). Apart from these approaches, mucoadhesive polymers can also be classified according to their mechanism of binding as shown within this entry. An overview in the mucoadhesive properties of these different types of mucoadhesive polymers was provided by Grabovac, Guggi, and Bernkop-Schnürch. A rank order of the 10 most mucoadhesive polymers as listed in Table 2 was established by testing more than 50 polymers under standardized conditions.[50]

Non-Covalent Binding Polymers

According to their surface charge which is important for the mechanism of adhesion, they can be divided into anionic, cationic, non-ionic, and ambiphilic polymers.

Anionic polymers

For this group of polymers, mainly –COOH groups are responsible for their adhesion to the mucus gel layer. Carbonic acid moieties are supposed to form hydrogen bonds with hydroxyl groups of the oligosaccharide side chains on mucus proteins. Further anionic groups are sulfate as well as sulfonate moieties, which seem to be more of theoretical than of practical relevance. Important representatives of this group of mucoadhesive polymers are listed in Table 3. Among them one can find the most adhesive non-covalent binding polymers such as PAA and NaCMC.[34,62]

Because of their high charge density, these polymers display a high buffer capacity which might be beneficial for various reasons as discussed in "Mucoadhesion: Buffer System." In contrast to non-ionic polymers, their swelling behavior, which is also crucial for mucoadhesion (Fig. 3), strongly depends on the pH value. The lower the pH value, the lower is the swelling behavior leading to a quite insufficient adhesion in many cases. On the contrary, a too rapid swelling of such polymers at higher pH values can lead to an over-swelling, which causes a strong decrease in the cohesive properties of such polymers. Even if the polymer sticks to the mucus layer, in this case, the drug delivery system will not be any more mucoadhesive, as the adhesive bond fails within the mucoadhesive polymer itself. This effect is illustrated in Fig. 10. A further drawback of anionic mucoadhesive polymers, however, is their incompatibility with multivalent cations such as Ca^{2+}, Mg^{2+}, and Fe^{3+}. In the presence of such cations, these polymers precipitate and/or coagulate,[65] leading to a strong reduction in their adhesive properties.

Cationic polymers

The strong mucoadhesion of cationic polymers can be explained by ionic interactions between these polymers and anionic substructures such as sialic acid moieties of the mucus gel layer. In particular, chitosan, which can be produced in high amounts for a reasonable price, seems to be a promising mucoadhesive excipient. Apart from its mucoadhesive properties, chitosan is also reported to display permeation enhancing properties.[66,67] The most

Table 2 Overview on the mucoadhesive properties of the 10 most mucoadhesive polymers

Polymer	pH	Total work of adhesion (μJ); Means ± SD (n = 5–8)
Chitosan-thiobutylamidine	pH 3 lyophilized	740.0 ± 146.7
Polyacrylic acid (450 kDa)-cysteine	pH 3 lyophilized	412.3 ± 27.3
Chitosan-thiobutylamidine	pH 6.5 precipitated	408.0 ± 67.9
Polycarbophil	pH 7 precipitated	342.0 ± 11.2
Carbopol 980	pH 7 precipitated	311.8 ± 42.0
Sodium carboxymethylcellulose-cysteine	pH 3 precipitated	261.7 ± 32.8
Carbopol 980	pH 3 precipitated	256.9 ± 49.0
Carbopol 974	pH 3 precipitated	219.3 ± 36.6
Polycarbophil-cysteine	pH 3 lyophilized	212.9 ± 14.4
Carbopol 974	pH 7 precipitated	211.2 ± 22.0

Source: From Grabovac et al.[50]

Table 3   Anionic mucoadhesive polymers

Polymer	Chemical structure	Additional information	References
Alginate			[51–53]
Carbomer		Cross-linked with sucrose	[34,53,54]
Chitosan–EDTA		Optionally cross-linked with EDTA	[55,56]
Hyaluronic acid			[57–59]
NaCMC		0.3–1.0 carboxymethyl groups per glucose unit	[34,52,60]
Pectins		R = OH or methyl	[61,62]
Polycarbophil		Cross-linked with divinylglycol	[34,60,63,64]

Fig. 10 Adhesive bond failure in case of insufficient cohesive properties of the mucoadhesive polymer: (**A**) matrix tablets and (**B**) microspheres.

important cationic mucoadhesive polymers are listed in Table 4. Their swelling behavior is strongly pH-dependent as well. In contrast to anionic polymers, however, their swelling behavior is improved at higher proton concentrations. Chitosan, for instance, is hydrated rapidly in the gastric fluid, leading to a strong reduction of its cohesive properties, whereas it does not swell at all at pH values above 6.5 causing a complete loss in its mucoadhesive properties.

Non-ionic polymers

The adhesion of anionic as well as cationic polymers strongly depends on the pH value of the surrounding fluid. On the contrary, non-ionic mucoadhesive polymers are mostly independent from this parameter. Whereas the

Table 4 Cationic mucoadhesive polymers

Polymer	Chemical structure	Additional information	References
Chitosan		Primary amino groups can be acetylated to some extent	[3]
Trimethylated chitosan			[68]
Polylysine			[61]

Table 5 Non-ionic mucoadhesive polymers

Polymer	Chemical structure	Additional information	References
Hydroxypropyl-cellulose		R = H or hydroxypropyl	[35,70]
Hydroxypropyl-methylcellulose		R = H or methoxy or hydroxypropyl	[53,62]
Poly(ethylene oxide)			[35,53]
Poly(vinyl alcohol)			[54,62]
Poly(vinyl pyrrolidone)			[62]

formation of secondary chemical bonds due to ionic interactions can be completely excluded for this group of polymers, some of them such as poly(ethylene oxide) are capable of forming hydrogen bonds. Apart from these interactions, their adhesion to the mucosa seems to be rather based on an interpenetration followed by polymer chain entanglements. These theoretical considerations are in good accordance with mucoadhesion studies, demonstrating almost no adhesion of non-ionic polymers, if they are applied to the mucosa already in the completely hydrated form, whereas they are adhesive if applied in dry form.[69] Hence, non-ionic polymers are in most cases less adhesive than anionic as well as cationic mucoadhesive polymers. Well-known representatives of this group of mucoadhesive polymers are listed in Table 5. In contrast to ionic polymers, non-ionic polymers are not influenced by electrolytes of the surrounding milieu. The addition of 0.9% NaCl, for instance, to carbomer leads to a tremendous decrease in its cohesiveness and subsequently to a strong reduction of its mucoadhesive properties, whereas these electrolytes have no influence on non-ionic mucoadhesive polymers.

Ambiphilic polymers

Ambiphilic mucoadhesive polymers display cationic as well as anionic substructures on their polymer chains. On the one hand, mucoadhesion of the cationic polymers is referred to be caused by electrostatic interactions with negatively charged mucosal surfaces and for anionic polymers, on the other hand, mucoadhesion can be explained by the formation of hydrogen bonds of carboxylic acid groups with the mucus gel layer. The combination of positive as well as negative charges on the same polymer, however, seems to compensate both effects leading to strongly reduced adhesive properties of ambiphilic polymers. Mucoadhesion studies of chitosan–EDTA conjugates with increasing amounts of covalently attached EDTA can clearly show this effect. Whereas the exclusively anionic chitosan–EDTA conjugate (Table 3) exhibiting no remaining cationic moieties and the exclusively cationic polymer chitosan displayed the highest mucoadhesive properties, the mucoadhesion of chitosan–EDTA conjugates showing both cationic moieties of remaining primary amino groups and anionic moieties of already covalently attached EDTA was much lower. In addition, Lueßen et al. could show a strongly increased intestinal buserelin bioavailability in rats using chitosan HCl as mucoadhesive excipient. A mixture of this cationic polymer with the anionic polymer carbomer, however, led to a significantly reduced bioavailability of the therapeutic peptide.[71] Representatives of this type of mucoadhesive polymers are mainly proteins such as gelatin, which is reported as mucoadhesive in various studies.[62,72]

Due to the combination of cationic as well as anionic mucoadhesive polymers leading to ionic interactions, however, the cohesiveness of delivery systems can be strongly improved.[57,61] If the adhesive bond of a delivery system fails rather within the mucoadhesive polymer itself, than between the polymer and the mucus layer as illustrated in Fig. 10, this effect is more important than the mucoadhesive properties of the polymer in order to improve the adhesiveness of the whole dosage form.

Covalent Binding Polymers

Recently, a presumptive new generation of mucoadhesive polymers has been introduced into the pharmaceutical literature.[73] Whereas the attachment of mucoadhesive polymers to the mucus layer has to date been achieved by non-covalent bonds, these novel polymers are capable of forming covalent bonds.[39,74] The bridging structure most commonly encountered in biological systems—the disulfide bond—has thereby been discovered for the covalent adhesion of polymers to the mucus layer of the mucosa. *Thio*lated poly*mers* or so-called *thiomers* are mucoadhesive basis polymers, which display thiol-bearing side chains (Fig. 11). Based on thiol/disulfide exchange reactions[75] as illustrated in Fig. 12 and/or a simple oxidation process, disulfide bonds are formed between such

polymers and cysteine-rich subdomains of mucus glycoproteins (Table 1). Hence, thiomers mimic the natural mechanism of secreted mucus glycoproteins, which are also covalently anchored in the mucus layer by the formation of disulfide bonds. Due to the covalent attachment of thioglycolic acid to chitosan, for instance, the adhesive properties of this polymer could be strongly increased.[76] Results of this study are shown in Fig. 13. Apart from these improved mucoadhesive properties, which could meanwhile also be shown for various other thiolated polymers, thiomers exhibit strongly improved cohesive properties as well. Whereas, for example, tablets consisting of polycarbophil disintegrate within 2 hr, tablets based on the corresponding thiolated polymer remain stable even for days in the disintegration apparatus according to the *Pharmacopoeia Europea*.[77] The result as shown in Fig. 14 can be explained by the continuous oxidation of thiol moieties on

Fig. 11 Thiomer (thiolated polymer); mucoadhesive polymers, which display thiol moieties bearing side chains.

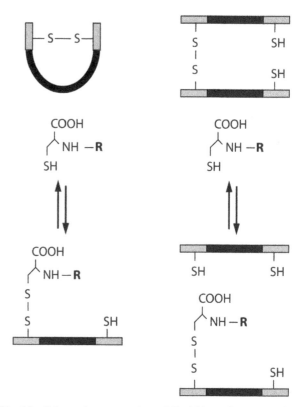

Fig. 12 Schematic presentation of disulfide exchange reactions between a poly(peptide) and a cysteine derivative according to G. H. Snyder.[75] The poly(peptide) stands here for a mucin glycoprotein of the mucus and the cysteine derivative is a polymer–cysteine conjugate (R, mucoadhesive basis polymer).
Source: Adapted from Bernkop-Schnürch et al.[73]

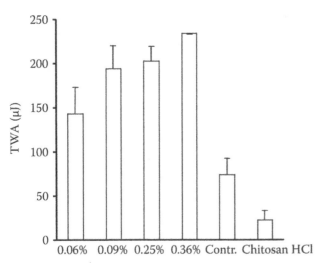

Fig. 13 Comparison of the adhesive properties of chitosan-thiobutylamidine displaying increasing amounts of covalently attached thiol groups (%) and controls of unmodified chitosan. Represented values are means of the TWA determined in tensile studies with dry compacts of indicated test material.
Source: From Roldo et al.[76]

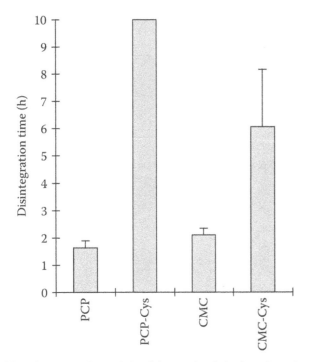

Fig. 14 Comparison of the disintegration behavior of matrix-tablets (30 mg; 5 mm i.d.) containing indicated lyophilized polymers (PCP, polycarbophil; CMC, carboxymethylcellulose; PCP-Cys, polycarbophil–cysteine conjugate; CMC-Cys, carboxymethylcellulose–cysteine conjugate). Studies were carried out with a disintegration test apparatus (Pharm. Eur.) in 50 mM TBS pH 6.8 at 37°C. Indicated values are means ±SD of at least three experiments. Polycarbophil-cysteine tablets did not disintegrate even after 48 hr of incubation.
Source: From Bernkop-Schnürch et al.[77]

thiomers which takes place in aqueous solutions at pH values above 5. Due to thehigh density of negative charges within anionic mucoadhesive polymers, they can also function as ion exchange resins displaying a high buffer capacity (section "Mucoadhesion: Buffer System"). According to this, the formation of disulfide bonds within polymer–cysteine conjugates can be controlled by adjusting the pH value of the system a priori. Adhesion of many quick swelling polymers, as already mentioned, is limited by an insufficient cohesion of the polymer resulting in a break within the polymer network rather than between the polymer and mucus layer (Fig. 10). Although thiolated polymers are rapidly hydrated, they are able to form highly cohesive and viscoelastic gels due to the formation of additional disulfide bonds. The formation of an over-hydrated slippery mucilage can thereby be excluded completely. Using the rotating cylinder ("Visual Tests") in order to evaluate the mucoadhesive properties of various formulations, for instance, revealed a comparatively much longer adhesion of tablets consisting of thiolated polymers.[22] Meanwhile, various anionic as well as cationic thiolated polymers have been synthesized as listed in Table 6. They all display strongly improved mucoadhesive properties compared to the corresponding unmodified polymers. The potential of thiolated PAA could be demonstrated even in clinical trials. As illustrated in Fig. 15, a sustained release of sodium fluorescein over several hours was achieved with thiolated polyacrylate minitablets adhering to the eye of volunteers.[86]

MULTIFUNCTIONAL MUCOADHESIVE POLYMERS

In the 1980s, the adhesive properties of polymers played a central role in the field of mucoadhesion, whereas numerous scientists began in the 1990s to focus their interest also on additional features of mucoadhesive polymers. These properties include enzyme inhibition, permeation enhancement, high buffer capacity, and controlled drug release. For some of these properties, mucoadhesion is substantial, and for others various synergistic effects can be expected due to adhesion.

Mucoadhesion: Enzyme Inhibition

Due to the great progress in biotechnology as well as gentechnology, the industry is capable of producing a large number of potential therapeutic peptides and proteins in commercial quantities. The majority of such drugs is most commonly administered by the parenteral routes that are often complex, difficult, and occasionally dangerous. Besides so-called alternative routes of application such as the nasal or transdermal route, there is no doubt that the peroral route is one of the most favored, as it offers the greatest ease of application. A presystemic metabolism of therapeutic peptides and proteins caused by proteolytic enzymes of the GI-tract, however, leads to a very poor

Table 6 Thiolated mucoadhesive polymers

Polymer	Chemical structure	Additional information	References
Chitosan-cysteine		21 up to 100 µmol thiol-groups per gram polymer	[78]
Chitosan-glutathione			[79]
Chitosan-thiobutylamidine			[80]
Chitosan-thioethylamidine			[81]
Chitosan-thioglycolic acid		11 up to 25 µmol thiol-groups per gram polymer	[82]
NaCMC-cysteine		22 up to 1280 µmol thiol-groups per gram polymer	[22,77]
Polyacrylate-cystamine		1 up to 20 µmol thiol-groups per gram polymer	[83]
Polyacrylate-cysteine		1 up to 142 µmol thiol-groups per gram polymer	[22,39,74,84,85]

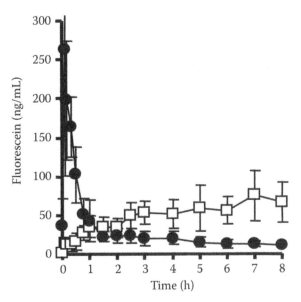

Fig. 15 Prolonged ocular residence time of fluorescein being embedded in thiolated polyacrylate ocular inserts (□) in comparison to the same formulation but containing the unmodified polyacrylate (•).
Source: Adapted from Hornof et al.[89]

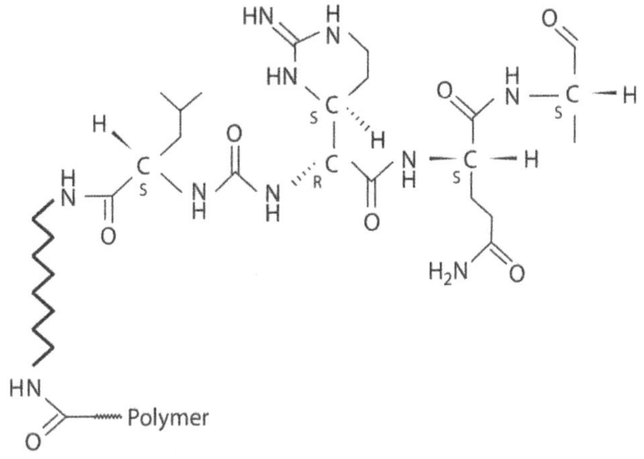

Fig. 16 Example for a mucoadhesive polymer–inhibitor conjugate. The elastase inhibitor elastatinal is thereby covalently attached via a C8-spacer to polymers like polycarbophil or NaCMC.
Source: From Bernkop-Schnürch et al.[89]

bioavailability after oral dosing. Attempts to reduce this barrier include the use of analogs, prodrugs, formulations such as nanoparticles, microparticles, and liposomes that shield therapeutic peptides and proteins from luminal enzymatic attack, and the design of delivery systems targeting the colon where the proteolytic activity is relatively low. Moreover, considerable interest is shown in the use of mucoadhesive polymers, since due to such excipients various *in vivo* studies could demonstrate a significantly improved bioavailability of peptide and protein drugs after oral dosing. As formulations containing mucoadhesive polymers provide an intimate contact with the mucosa, a presystemic degradation of these drugs on the way between the delivery system and the absorbing membrane can be excluded. Takeuchi et al., for instance, demonstrated a significantly stronger reduction in the plasma calcium level after oral administration of calcitonin-loaded liposomes which were coated with a mucoadhesive polymer in comparison to the same formulation without the mucoadhesive coating.[4]

It has been demonstrated by various studies that certain mucoadhesive polymers display also an enzyme inhibitory effect. In particular, poly(acrylic acid) was shown to exhibit a pronounced inhibitory effect toward trypsin.[87,88] Additionally, the immobilization of enzyme inhibitors to mucoadhesive polymers acting only in a very restricted area of the intestine seems to be a promising approach in order to improve their enzyme inhibitory properties. Due to the immobilization of the inhibitor, it remains concentrated on the polymer, which should certainly make a reduced share of this auxiliary agent in the dosage form sufficient. Side effects of the inhibitor such as systemic toxicity, a

disturbed digestion of nutritive proteins and pancreatic hypersecretion caused by a luminal feedback regulation can be avoided. Hence, the covalent attachment of enzyme inhibitors to mucoadhesive polymers such as those shown in Fig. 16 represents the combination of two favorable strategies for the oral administration of poly(peptide) drugs, offering additional advantages compared to a simple combination of both excipients without the covalent linkage. So far, various polymer–inhibitor conjugates have been generated as listed in Table 7. Their efficacy could be verified by *in vivo* studies in diabetic mice showing a significantly reduced glucose level after the oral administration of (PEGylated-)insulin tablets containing a polymer–inhibitor conjugate.[96,97] The results of this study are shown in Fig. 17.[97] In another study, the potential of the combination of mucoadhesive polymers and polymer–enzyme inhibitor conjugates was demonstrated by the improved oral uptake of salmon calcitonin in rats utilizing thiolated chitosan in combination with a chitosan–pepstatin conjugate as illustrated in Fig. 18.[98]

Mucoadhesion: Permeation Enhancement

A number of mucoadhesive polymers have also promising effects on the modulation of the absorption barrier by opening of the intestinal intercellular junctions.[64,99] In contrast to permeation enhancers of low molecular size such as sodium salicylate and medium-chain glycerides,[100,101] these polymers might not be absorbed from the intestine, which should exclude systemic side effects of these auxiliary agents. Permeation studies, for instance, carried out in Ussing chambers on Caco-2 monolayers demonstrated a strong permeation enhancing effect of chitosan and carbomer.[99,102] This permeation enhancing effect on these mucoadhesive polymers can even be significantly

Molecular—Nanogels

Table 7　Comparison of various mucoadhesive polymer–inhibitor conjugates

Polymer–inhibitor conjugate	Inhibited enzymes		Mucoadhesive properties	References
	Based on complexing properties	Based on competitive inhibition		
Carboxymethylcellulose–elastatinal conjugate		Elastase	+	[89]
Carboxymethylcellulose–pepstatin conjugate		Pepsin	n.d.	[90]
Chitosan–antipain conjugate		Trypsin	++	[91,92]
Chitosan–chymostatin conjugate		Chymotrypsin	n.d.	[92]
Chitosan–elastatinal conjugate		Elastase	n.d.	[92]
Chitosan–ACE conjugate		Trypsin, chymotrypsin, elastase	n.d.	[92]
Chitosan–EDTA	Aminopeptidase N, carboxypeptidase A/B		+++	[8,55,56]
Chitosan–EDTA–antipain conjugate	Aminopeptidase N, carboxypeptidase A/B	Trypsin	n.d.	[92]
Chitosan–EDTA–chymostatin conjugate	Aminopeptidase N, carboxypeptidase A/B	Chymotrypsin	n.d.	[92]
Chitosan–EDTA–elastatinal conjugate	Aminopeptidase N, carboxypeptidase A/B	Elastase	n.d.	[92]
Chitosan–EDTA–ACE conjugate	Aminopeptidase N, carboxypeptidase A/B	Trypsin, Chymotrypsin, Elastase	+	[92]
Chitosan–EDTA–BBI conjugate	Aminopeptidase N, carboxypeptidase A	Trypsin, Chymotrypsin, Elastase	+	[93]
Poly(acrylic acid)-Bowman–Birk inhibitor conjugate		Chymotrypsin	++	[94]
Poly(acrylic acid)–chymostatin conjugate		Chymotrypsin	+++	[23]
Poly(acrylic acid)–elastatinal conjugate		Elastase	+++	[89]
Polycarbophil–elastatinal conjugate		Elastase	+++	[89]
Poly(acrylic acid)–bacitracin conjugate	Aminopeptidase N		n.d.	[95]
Polycarbophil–cysteine	Carboxypeptidase A/B		++++	[85]

Mucoadhesive properties are classified in poor (+), good (++), very good (+++), and excellent (++++).
Abbreviations: ACE, antipain/chymostatin/elastatinal; BBI, Bowman–Birk inhibitor.

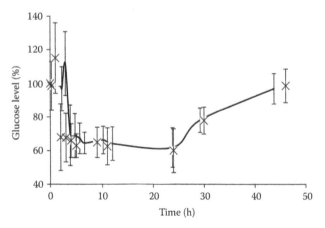

Fig. 17　Decrease in blood glucose level in diabetic mice after oral administration of PEGylated insulin being incorporated in a thiolated polyacrylate.
Source: Adapted from Caliceti et al.[97]

improved due to the immobilization of cysteine on these polymers.[5,74] This improved permeation across the mucosa was accompanied by a decrease in the TEER, indicating a loosening of the tightness of intercellular junctions, i.e., the opening of the paracellular route across the epithelium for otherwise non-absorbable hydrophilic compounds such as peptides. Mechanisms that are responsible for the permeation enhancing effect of mucoadhesive polymers, however, are still unclear.

The permeation enhancing effect of cationic polymers such as chitosan might be based on the positive charges of these polymers which interact with the cell membrane, resulting in a structural reorganization of tight junction-associated proteins.[6]

In case of thiolated polymers, protein tyrosine phosphatase might be involved in the underlying mechanism. This thiol-dependent enzyme mediates the closing process of

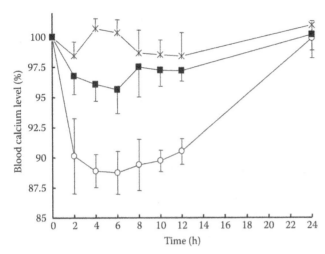

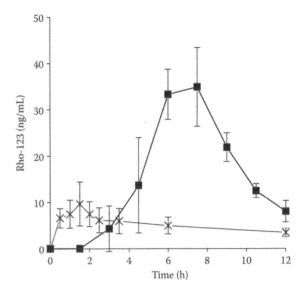

Fig. 18 Comparison in the oral uptake of salmon calcitonin in rats determined via the decrease in blood calcium level after oral administration of the therapeutic peptide with chitosan (×), with chitosan/chitosan–inhibitor conjugate (■) and with thiolated chitosan/chitosan–inhibitor conjugate (○).
Source: Adapted from Guggi et al.[98]

Fig. 19 Efflux pump inhibitory effect of thiolated chitosan (×) in comparison to unmodified chitosan (■) in rats using rhodamine 123 as model Pgp-substrate.
Source: Adapted from Föger et al.[106]

tight junctions by dephosphorylation of tyrosine groups from the extracellular region.[103] The inhibition of PTP by specificinhibitors such as vanadate or pervanadate causes an enhanced opening of the tight junctions. As it is also inhibited by sulfhydryl compounds such as glutathione forming a mixed disulfide with Cys215,[104] it is likely that thiolated polymers might also lead to an inhibition of this enzyme.

Mucoadhesion: Efflux Pump Inhibition

Transmembrane efflux pump proteins are one of the factors influencing and restricting the absorption of non-invasively administered drugs. These proteins can effectively alter the pharmacokinetics of various drugs such as anticancer drugs, immune suppressive drugs, and antibiotics. Efflux pump inhibitors have therefore gained a considerable attention as tool to improve clinical efficacy of drugs that act as substrates for efflux pumps. Among efflux pump inhibitors, polymeric excipients such as polyethyleneglycols, pluronics, and certain polysaccharides have gained considerable attention, as due to their comparatively high molecular mass these compounds are not absorbed from mucosal membranes. Consequently, systemic toxic side effects—as it is the case for low molecular mass efflux pump inhibitors—can be excluded for such mucoadhesive polymers. More recently, Werle et al. revealed efflux pump inhibitory capability of thiolated polymers.[105] The transmucosal transport of the P-gp substrate rhodamine 123, for instance, was strongly improved in the presence of thiolated chitosan. In the following, these *in vitro* results were confirmed by *in vivo* studies in rats. Föger, Schmitz, and Bernkop-Schnürch showed that the oral bioavailability of rhodamine 123 is even 3.0-fold improved when this model P-gp substrate is

embedded in thiolated chitosan minitablets administered orally to rats (Fig. 19).[106] Furthermore, poly(acrylic acid)-cysteine showed to inhibit effectively Mrp2 efflux pump transporter, improving the permeation of sulforhodamine 101 4.67-fold.[107] In another study, tumor growth in rats was significantly reduced when paclitaxel was given orally in combination with a thiolated polymer.[108] In contrast to PEGs showing a P-gp inhibitory effect, which is completely independent from the molecular mass of the applied PEG,[109] the inhibitory effect of thiolated polymers is dependent on their molecular mass. Thiolated PAA of 250 kDa exhibit the highest P-pg inhibitory effect, for instance, whereas thiolated poly(acrylic acid) of higher and lower molecular mass displays significantly lower inhibitory activity. Studies performed *in vivo* in rats confirmed the efficacy of thiolated PAA of 250 kDa as Mrp2 inhibitor.[110] The postulated mechanism of efflux pump inhibition is based on an interaction of thiolated polymers with the channel forming transmembrane domain of efflux pumps. Thiolated polymers seem to enter in the channel and likely form subsequently one or two disulfide bonds with cysteine subunits located within the channel. Due to this covalent interaction, the allosteric change of the transporter being essential to move drugs outside of the cell seems to be blocked.[111]

Mucoadhesion: Buffer System

A further advantage of mucoadhesive polymers can be seen in the high buffer capacity of ionic polymers. As these polymers can act as ion exchange resins, they are able to maintain their pH value inside the polymeric network over a considerable period of time. Matrix tablets based on neutralized carbomer, for instance, can buffer the pH value inside the swollen carrier system even for hours in an artificial gastric fluid of pH 2.[112] This high buffer capacity seems to be highly beneficial for various reasons. For example, the epidermal growth factor (EGF) is recognized as an important agent for acceleration of ulcer healing and has a peculiar biological property to repair tissue damage by an enhanced proliferation and differentiation of epithelial tissues.[113] Itoh and Matsuo demonstrated in a double-blind controlled clinical study the enhanced healing of rat gastric ulcers after oral administration of EGF. This effect could even be drastically increased by using the mucoadhesive polymer hydroxypropyl cellulose as drug carrier matrix.[114] As EGF is strongly degraded by pepsin,[115] the use of mucoadhesive polymers providing an additional protective effect toward pepsinic degradation might therefore be helpful. It can be achieved by a comparatively higher pH value inside the drug carrier matrix based on its high buffer capacity at which penetrating pepsin is already inactive.

Apart from this likely advantage for the poly(peptide) administration, the high buffer capacity of neutralized anionic polymer might also be highly beneficial in treatment of *Helicobacter pylori* infection in peptic ulcer disease, as common antibiotics such as amoxicillin or metronidazole display poor stability in the acidic pH of the stomach. The incorporation of these therapeutic agents in mucoadhesive polymers displaying a high buffer capacity might improve their stability in the acidic milieu.

Mucoadhesion: Controlled Release

If the therapeutic agent is incorporated in the mucoadhesive polymer, the excipient can act both as mucoadhesive and as a matrix system providing a controlled drug release. The release behavior of drugs embedded in mucoadhesive polymers depends thereby mainly on their molecular size and charge. According to the equation determining the diffusion coefficient, in which the radius of a molecule indirectly correlates with the diffusion coefficient, small-sized drugs will be released faster than greater ones. Apart from their size, the release of therapeutic agents from mucoadhesive polymers can be simply controlled by raising or lowering the share of the polymer in the delivery system. Whereas a low amount of the mucoadhesive polymer in the carrier matrix can guarantee a rapid drug release, a sustained drug release can be provided by raising the share of the polymer in the delivery system.[77] In addition, the drug release from mucoadhesive polymers can also be controlled by the extent of cross-linking. The higher the polymer is cross-linked, the lower is the release rate of the drug. Such a cross-linking can be achieved by the formation of covalent bonds, for example, the cross-linking of gelatin with glutaraldehyde[72] or on the basis of ionic interactions. The release rate of insulin from matrix tablets consisting of the anionic mucoadhesive polymer carbomer (C934P), for instance, can be strongly reduced by the addition of the

divalent cationic amino acid lysine.[28] In case of ionic drugs, a sustained release can also be guaranteed by the use of an ionic mucoadhesive polymer displaying the opposite charge of the therapeutic agent. On the basis of an ion exchange resin, for instance, a sustained release of the therapeutic agent α-lipoic acid over a period of more than 12 hr can be provided by the incorporation of this anionic drug in the cationic mucoadhesive polymer chitosan.[116]

CONCLUDING REMARKS

Motivated by the great benefits that can be provided by mucoadhesive polymers such as a prolonged residence time and an intimate contact of the dosage form on the mucosa, considerably intensive research and development has been performed in this field since the concept of mucoadhesion has been pioneered in the early 1980s. Merits of these efforts are the establishment of various useful techniques in order to evaluate the mucoadhesive properties of different polymers as well as the design of novel polymers displaying improved mucoadhesive properties. In addition, the development of multifunctional mucoadhesive polymers exhibiting also features such as enzyme inhibitory properties, permeation enhancing effects, high buffer capacity, and the possibility to control the drug release made them even more promising auxiliary agents. Although there are already numerous formulations based on mucoadhesive polymers on the market, the number of delivery systems making use of these advantages will certainly increase in the near future.

REFERENCES

1. Robinson, J.R.; Bologna, W.J. Vaginal and reproductive system treatments using a bioadhesive polymer. J. Control. Release **1994**, *28* (1), 87–94.
2. Akiyama, Y.; Nagahara, N.; Nara, E.; Kitano, M.; Iwasa, S.; Yamamoto, I.; Azuma, J.; Ogawa, Y. Evaluation of oral mucoadhesive microspheres in man on the basis of the pharmacokinetics of furosemide and riboflavin, compounds with limited gastrointestinal absorption sites. J. Pharm. Pharmacol. **1998**, *50* (2), 159–166.
3. Takeuchi, H.; Yamamoto, H..; Toshiyuki, N.; Tomoaki, H.; Kawashima, Y. Enteral absorption of insulin in rats from mucoadhesive chitosan-coated liposomes. Pharm. Res. **1996**, *13* (6), 896–901.
4. Takeuchi, H.; Matsui, Y.; Yamamoto, H.; Kawashima, Y. Mucoadhesive properties of carbopol or chitosan-coated liposomes and their effectiveness in the oral administration of calcitonin to rats. J. Control. Release **2003**, *86* (2), 235–242.
5. Clausen, A.; Bernkop-Schnürch, A. *In vitro* evaluation of the permeation enhancing effect of thiolated polycarbophil. J. Pharm. Sci. **2000**, *89* (10), 1253–1261.
6. Schipper, N.G.M.; Olsson, S.; Hoogstraate, J.A.; de Boer, A.G.; Varum, K.M.; Artursson, P. Chitosans as

absorption enhancers for poorly absorbable drugs 2: Mechanism of absorption enhancement. Pharm. Res. **1997**, *14* (7), 923–929.

7. Lueßen, H.L.; Bohner, V.; Perard, D.; Langguth, P.; Verhoef, J.C.; de Boer, A.G.; Merkle, H.P.; Junginger, H.E. Mucoadhesive polymers in peroral peptide drug delivery. V. Effect of poly(acrylates) on the enzymatic degradation of peptide drugs by intestinal brush border membrane vesicles. Int. J. Pharm. **1996**, *141* (1), 39–52.

8. Bernkop-Schnürch, A.; Paikl, C.; Valenta, C. Novel bioadhesive chitosan-EDTA conjugate protects leucine enkephalin from degradation by aminopeptidase. N. Pharm. Res. **1997**, *14* (7), 917–922.

9. Matthes, I.; Nimmerfall, F.; Sucker, H. Mucusmodelle zur Untersuchung von intestinalen Absorptionsmechanismen. Pharmazie **1992**, *47* (8), 505–515.

10. Bernkop-Schnürch, A; Fragner, R. Investigations into the diffusion behaviour of polypeptides in native intestinal mucus with regard to their peroral administration. Pharm. Sci. **1996**, *2* (8), 361–363.

11. Campbell, B.J. Biochemical and functional aspects of mucus and mucin-type glycoproteins, bioadhesive drug delivery systems. In *Bioadhesive Drug Delivery Systems: Fundamentals, Novel Approaches, and Developments*; Mathiowitz, E.; Chickering, D.E. III, Lehr, C.-M., Eds.; Marcel Dekker: New York, 1999; pp. 85–130.

12. Gendler, S.J.; Spicer, A.P. Epithelial mucin genes. Ann. Rev. Physiol. **1995**, *57* (1), 607–634.

13. Gum, J.R.; Hicks, J.W.; Toribara, N.W.; Rothe, E.M.; Lagace, R.E.; Kim, Y.S. The human MUC2 intestinal mucin has cysteine-rich subdomains located both upstream and downstream of its central repetitive region. J. Biol. Chem. **1992**, *267* (30), 21375–21383.

14. Gum, J.R.; Hicks, J.W.; Swallow, D.M.; Lagace, R.L.; Byrd, J.C.; Lamport, D.T.A.; Siddiki, B.; Kim, Y.S. Molecular cloning of cDNAs derived from a novel human intestinal mucin gene. Biochem. Biophys. Res. Commun. **1990**, *171* (1), 407–415.

15. Porchet, N.; Nguyen, V.C.; Dufosse, J.; Audie, J.P.; Guyonnet-Duperat, V.; Gross, M.S.; Denis, C.; Degand, P.; Bernheim, A.; Aubert, J.P. Molecular cloning and chromosomal localisation of a novel human tracheo-bronchial mucin cDNA containing tandemly repeated sequences of 48 base pairs. Biochem. Biophys. Res. Commun. **1991**, *175* (2), 414–422.

16. Guyonnet-Duperat, V.; Audie, J.P.; Debailleul, V.; Laine, A.; Buisine, M.P.; Galiegue-Zouitina, S.; Pigny, P.; Degand, P.; Aubert, J.P.; Porchet, N. Characterisation of the human mucin gene MUC5AC: A consensus cysteine rich domain for 11p15 mucin genes? Biochem. J. **1995**, *305*, 211–219.

17. Thornton, D.J.; Howard, M.; Khan, N.; Sheehan, J.K. Identification of two glycoforms of the MUC5B mucin in human respiratory mucus. Evidence for a cysteine-rich sequence repeated within the molecule. J. Biol. Chem. **1997**, *272* (14), 9561–9566.

18. Toribara, N.W.; Roberton, A.M.; Ho, S.B.; Kuo, W.L.; Gum, E.T.; Gum, J.R.; Byrd, J.C.; Siddiki, B.; Kim, Y.S. Human gastric mucin: Identification of unique species by expression cloning. J. Biol. Chem. **1993**, *268* (8), 5879–5885.

19. Bobek, L.A.; Tsai, H.; Biesbrock, A.R.; Levine, M.J. Molecular cloning, sequence, and specificity of expression of the gene encoding the low molecular weight human salivary mucin (MUC7). J. Biol. Chem. **1993**, *268* (27), 20563–20569.

20. Shankar, V.; Pichan, P.; Eddy, R.L.; Tonk, V.; Nowak, N.; Sait, S.N.; Shows, T.B.; Shultz, R.E.; Gotway, G.; Elkins, R.C.; Gilmore, M.S.; Sachdev, G.P. Chromosomal localisation of a human mucin gene (MUC8) and cloning of the cDNA corresponding to the carboxy terminus. Am. J. Respir. Cell Mol. Biol. **1997**, *16* (3), 232–241.

21. Lehr, C.-M.; Poelma, F.G.J.; Junginger, H.E. An estimate of turnover time of intestinal mucus gel layer in the rat in situ loop. Int. J. Pharm. **1991**, *70* (3), 235–240.

22. Bernkop-Schnürch, A.; Steininger, S. Synthesis and characterisation of mucoadhesive thiolated polymers. Int. J. Pharm. **2000**, *194* (2), 239–247.

23. Bernkop-Schnürch, A.; Apprich, I. Synthesis and evaluation of a modified mucoadhesive polymer protecting from α-chymotrypsinic degradation. Int. J. Pharm. **1997**, *146* (2), 247–254.

24. Jabbari, E.; Wisniewski, N.; Peppas, N.A. Evidence of mucoadhesion by chain interpenetration at a poly(acrylic acid)/mucin interface using ATR-FTIR spectroscopy. J. Control. Release **1993**, *26* (2), 99–108.

25. Imam, M.E.; Hornof, M.; Valenta, C.; Reznicek, G.; Bernkop-Schnürch, A. Evidence for the interpenetration of mucoadhesive polymers into the mucous gel layer. STP Pharma. Sci. **2003**, *13* (3), 171–176.

26. Mortazavi, S.A.; Smart, J.D. An investigation into the role of water movement and mucus gel dehydration in mucoadhesion. J. Control. Release **1993**, *25* (3), 197–203.

27. Caramella, C.M.; Rossi, S.; Bonferoni, M.C. A rheological approach to explain the mucoadhesive behavior of polymer hydrogels. In *Bioadhesive Drug Delivery Systems: Fundamentals*; Mathiowitz, E., Chickering, D.E. III, Lehr, C.-M., Eds.; **1999**, pp. 25–65. New York: Marcel Dekker.

28. Bernkop-Schnürch, A.; Humenberger, C.; Valenta, C. Basic studies on bioadhesive delivery systems for peptide and protein drugs. Int. J. Pharm. **1998**, *165* (2), 217–225.

29. Hertzog, B.A.; Mathiowitz, E. Novel magnetic technique to measure bioadhesion. In *Novel Approaches, and Developments*; Mathiowitz, E., Chickering, D.E. III, Lehr, C.-M., Eds.; Marcel Dekker: New York, 1999; pp. 147–175.

30. Schneider, J.; Tirrell, M. Direct measurement of molecular-level forces and adhesion in biological systems *Novel Approaches, and Developments*, Mathiowitz, E.; Chickering, D.E. III, Lehr, C.-M., Eds.; Marcel Dekker: New York, 1999; pp. 223–261.

31. Bernkop-Schnürch, A.; Weithaler, A.; Albrecht, K.; Greimel, A. Thiomers: Preparation and *in vitro* evaluation of a mucoadhesive nanoparticulate drug delivery system. Int. J. Pharm. **2006**, *317* (1), 76–81.

32. Nielsen, L.S.; Schubert, L.; Hansen, J. Bioadhesive drug delivery systems—I. Characterisation of mucoadhesive properties of systems based on glyceryl mono-oleate and glyceryl monolinoleate. Eur. J. Pharm. Sci. **1998**, *6* (3), 231–239.

Molecular—Nanogels

33. Rango Rao, K.V.; Buri, P. A novel in situ method to test polymers and coated microparticles for bioadhesion. Int. J. Pharm. **1989**, *52* (3), 265–270.

34. Tobyn, M.J.; Johnson, J.R.; Dettmar, P.W. Factors affecting *in vitro* gastric mucoadhesion II. Physical properties of polymers. Eur. J. Pharm. Biopharm. **1996**, *42* (1), 56–61.

35. Mortazavi, S.A.; Smart, J.D. An *in vitro* evaluation of mucosa-adhesion using tensile and shear stresses. J. Pharm. Pharmacol. **1993**, *45* (Suppl. 2), 1108–1111.

36. Ch'ng, H.S.; Park, H.; Kelly, P.; Robinson, J.R. Bioadhesive polymers as platforms for oral controlled drug delivery II: Synthesis and evaluation of some swelling, water-insoluble bioadhesive polymers. J. Pharm. Sci. **1985**, *74* (4), 399–405.

37. Chickering, D.E., III; Santos, C.A.; Mathiowitz, E. Adaptation of a microbalance to measure bioadhesive properties of microspheres. In *Novel Approaches, and Developments*; Mathiowitz, E.; Chickering, D.E. III, Lehr, C.-M., Eds.; Marcel Dekker: New York, 1999; pp. 131–147.

38. Hassan, E.E.; Gallo, J.M. A simple rheological method for the *in vitro* assessment of mucinpolymer bioadhesive bond strength. Pharm. Res. **1990**, *7* (5), 491–495.

39. Leitner, V.; Walker, G.F.; Bernkop-Schnürch, A. Thiolated polymers: Evidence for the formation of disulphide bonds with mucus glycoproteins. Eur. J. Pharm. Biopharm. **2003**, *56* (2), 207–214.

40. Kellaway, I.W. *In vitro* test methods for the measurement of mucoadhesion. In *Bioadhesion Possibilities and Future Trends*; Gurny, R.; Junginger, H.E., Eds.; Wissenschaftliche Verlagsgesellschaft: Stuttgart, Germany, 1990; pp. 86–97.

41. Kerr, L.J.; Kellaway, I.W.; Rowlands, C.; Parr, G.D. The influence of poly(acrylic) acids on the rheology of glycoprotein gels. Proc. Int. Symp. Control. Release Bioact. Mater. **1990**, *17*, 122–123.

42. Tobyn, M.J.; Johnson, J.R.; Gibson, S.A. Investigations into the role of hydrogen bonding in the interaction between mucoadhesives and mucin at gastric pH. J. Pharm. Pharmacol. **1992**, *44* (Suppl. 1), 1048–1048.

43. Mortazavi, S.A.; Carpenter, B.G.; Smart, J.D. An investigation into the nature of mucoadhesive interactions. J. Pharm. Pharmacol. **1993**, *45* (Suppl.), 1141.

44. Mortazavi, S.A. *In vitro* assessment of mucus/mucoadhesive interactions. Int. J. Pharm. **1995**, *124* (2), 173–182.

45. Akiyama, Y.; Nagahara, N. Novel formulation approaches to oral mucoadhesive drug delivery systems. In *Bioadhesive Drug Delivery Systems: Fundamentals, Novel Approaches, and Developments*; Mathiowitz, E., Chickering, D.E. III, Lehr, C.-M., Eds.; Marcel Dekker: New York, **1999**, pp. 477–507.

46. Bouckaert, S.; Lefebvre, R.A.; Remon, J.-P. *In vitro/in vivo* correlation of the bioadhesive properties of a buccal bioadhesive miconazole slow-release tablet. Pharm. Res. **1993**, *10* (6), 853–856.

47. Bottenberg, P.; Cleymaet, R.; de Muynck, C.; Remon, J.-P.; Coomans, D.; Michotte, Y.; Slop, D. Development and testing of bioadhesive, fluoride-containing slow-release tablets for oral use. J. Pharm. Pharmacol. **1991**, *43* (7), 457–464.

48. Khosla, L.; Davis, S.S. The effect of polycarbophil on the gastric emptying of pellets. J. Pharm. Pharmacol. **1987**, *39* (1), 47–49.

49. Harris, D.; Fell, J.T.; Sharma, H.L.; Taylor, D.C. GI transit of potential bioadhesive formulations in man: A scintigraphic study. J. Control. Release **1990**, *12* (1), 45–53.

50. Grabovac, V.; Guggi, D.; Bernkop-Schnürch, A. Comparison of the mucoadhesive properties of various polymers. Adv. Drug Deliv. Rev. **2005**, *57* (11), 1713–1723.

51. Witschi, C.; Mrsny, R.J. *In vitro* evaluation of microparticles and polymer gels for use as nasal platforms for protein delivery. Pharm. Res. **1999**, *16* (3), 382–390.

52. Evans, I.V. Mucilaginous substances from macroalgae: An overview. Symp. Soc. Exp. Biol. **1989**, *43*, 455–461.

53. Mortazavi, S.A.; Smart, J.D. An *in vitro* method for assessing the duration of mucoadhesion. J. Control. Release **1994**, *31* (2), 207–217.

54. El Hameed, M.D.; Kellaway, I.W. Preparation and *in vitro* characterization of mucoadhesive polymeric microspheres as intra-nasal delivery systems. Eur. J. Pharm. Biopharm. **1997**, *44* (1), 53–60.

55. Bernkop-Schnürch, A.; Krajicek, M.E. Mucoadhesive polymers as platforms for peroral peptide delivery and absorption: Synthesis and evaluation of different chitosan-EDTA conjugates. J. Control. Release **1998**, *50* (1), 215–223.

56. Bernkop-Schnürch, A.; Freudl, J. Comparative *in vitro* study of different chitosan-complexing agent conjugates. Pharmazie **1999**, *54* (5), 369–371.

57. Takayama, K.; Hirata, M.; Machida, Y.; Masada, T.; Sannan, T.; Nagai, T. Effect of interpolymer complex formation on bioadhesive property and drug release phenomenon of compressed tablets consisting of chitosan and sodium hyaluronate. Chem. Pharm. Bull. **1990**, *38* (7), 1993–1997.

58. Hadler, N.M.; Dourmashikin, R.R.; Nermut, M.V.; Williams, L.D. Ultrastructure of hyaluronic acid matrix. Biochemistry **1982**, *79* (2), 307–309.

59. Sanzgiri, Y.D.; Topp, E.M.; Benedetti, L.; Stella, V.J. Evaluation of mucoadhesive properties of hyaluronic acid benzyl esters. Int. J. Pharm. **1994**, *107* (2), 91–97.

60. Madsen, F.; Eberth, K.; Smart, J.D. A rheological assessment of the nature of interactions between mucoadhesive polymers and a homogenised mucus gel. Biomaterials **1998**, *19* (11), 1083–1092.

61. Liu, P.; Krishnan, T.R. Alginate-pectin-poly-l-lysine particulate as a potential controlled release formulation. J. Pharm. Pharmacol. **1999**, *51* (2), 141–149.

62. Smart, J.D.; Kellaway, I.W.; Worthington, H.E.C. An *in vitro* investigation of mucosa-adhesive materials for use in controlled drug delivery. J. Pharm. Pharmacol. **1984**, *36* (5), 295–299.

63. Longer, M.A.; Ch'ng, H.S.; Robinson, J.R. Bioadhesive polymers as platforms for oral controlled drug delivery III: Oral delivery of chlorothiazide using a bioadhesive polymer. J. Pharm. Sci. **1985**, *74* (4), 406–411.

64. Lehr, C.-M.; Bouwstra, J.A.; Schacht, E.H.; Junginger, H.E. *In vitro* evaluation of mucoadhesive properties of chitosan and some other natural polymers. Int. J. Pharm. **1992**, *78* (1), 43–48.

65. Valenta, C.; Christen, B.; Bernkop-Schnürch, A. Chitosan-EDTA conjugate: A novel polymer for topical used gels. J. Pharm. Pharmacol. **1998**, *50* (5), 445–452.

66. Artursson, P.; Lindmark, T.; Davis, S.S.; Illum, L. Effect of chitosan on the permeability of monolayers of intestinal epithelial cells (Caco-2). Pharm. Res. **1994**, *11* (9), 1358–1361.

67. Lueßen, H.L.; Rentel, C.-O.; Kotzé, A.F.; Lehr, C.-M.; de Boer, A.G.; Verhoef, J.C.; Junginger, H.E. Mucoadhesive polymers in peroral peptide drug delivery. IV. Polycarbophil and chitosan are potent enhancers of peptide transport across intestinal mucosae *in vitro*. J. Control. Release **1997**, *45*, 15–23.

68. Thanou, M.; Florea, B.I.; Langemeÿer, M.W.; Verhoef, J.C.; Junginger, H.E. N-trimethylated chitosan chloride (TMC) improves the intestinal permeation of the peptide drug buserelin *in vitro* (Caco-2 cells) and *in vivo* (rats). Pharm. Res. **2000**, *17* (1), 27–31.

69. Lehr, C.-M. From sticky stuff to sweet receptors—Achievements, limits and novel approaches to bioadhesion. Eur. J. Drug Metabol. Pharmacokinet. **1996**, *21* (2), 139–148.

70. Rillosi, M.; Buckton, G. Modelling mucoadhesion by use of surface energy terms obtained from the Lewis acid—Lewis base approach. II. Studies on anionic, cationic, and unionisable polymers. Pharm. Res. **1995**, *12* (5), 669–675.

71. Lueßen, H.L.; de Leeuw, B.J.; Langemeyer, M.W.; de Boer, A.G.; Verhoef, J.C.; Junginger, H.E. Mucoadhesive polymers in peroral peptide drug delivery. VI. Carbomer and chitosan improve the intestinal absorption of the peptide drug buserelin *in vivo*. Pharm. Res. **1996**, *13* (11), 1668–1672.

72. Matsuda, S.; Iwata, H.; Se, N.; Ikada, Y. Bioadhesion of gelatin films crosslinked with glutaraldehyde. J. Biomed. Mater. Res. **1999**, *45* (1), 20–27.

73. Bernkop-Schnürch, A.; Schwarz, V.; Steininger, S. Polymers with thiol groups: A new generation of mucoadhesive polymers? Pharm. Res. **1999**, *16* (6), 876–881.

74. Bernkop-Schnürch, A. Thiomers: A new generation of mucoadhesive polymers. Adv. Drug Deliv. Rev. **2005**, *57* (11), 1569–1582.

75. Snyder, G.H. Intramolecular disulfide loop formation in a peptide containing two cysteines. Biochemistry **1987**, *26* (3), 688–694.

76. Roldo, M.; Hornof, M.; Caliceti, P.; Bernkop-Schnürch, A. Mucoadhesive thiolated chitosans as platforms for oral controlled drug delivery: Synthesis and *in vitro* evaluation. Eur. J. Pharm. Biopharm. **2004**, *57* (1), 115–121.

77. Bernkop-Schnürch, A.; Scholler, S.; Biebel, R.G. Development of controlled drug release systems based on polymer-cysteine conjugates. J. Control. Release **2000**, *66* (1), 39–48.

78. Bernkop-Schnürch, A.; Brandt, U.-M.; Clausen, A. Synthese und *in vitro* Evaluierung von Chitosan-Cystein Konjugaten. Sci. Pharm. **1999**, *67*, 197–208.

79. Kafedjiiski, K.; Föger, F.; Werle, M.; Bernkop-Schnürch, A. Synthesis and *in vitro* evaluation of a novel chitosan-glutathione conjugate. Pharm. Res. **2005**, *22* (9), 1480–1488.

80. Bernkop-Schnürch, A.; Hornof, M.; Zoidl, T. Thiolated polymers—Thiomers: Modification of chitosan with 2-iminothiolane. Int. J. Pharm. **2003**, *260* (2), 229–237.

81. Kafedjiiski, K.; Hoffer, M.; Werle, M.; Bernkop-Schnürch, A. Improved synthesis and *in vitro* characterization of chitosan-thioethylamidine conjugate. Biomaterials **2006**, *27* (1), 127–135.

82. Kast, C.E.; Bernkop-Schnürch, A. Thiolated polymers—Thiomers: Development and *in vitro* evaluation of chitosan-thioglycolic acid conjugates. Biomaterials **2001**, *22* (17), 2345–2352.

83. Kast, C.E.; Bernkop-Schnürch, A. Polymer-cystamine conjugates: New mucoadhesive excipients for drug delivery? Int. J. Pharm. **2002**, *234* (1–2), 91–99.

84. Leitner, V.; Marschütz, M.; Bernkop-Schnürch, A. Mucoadhesive and cohesive properties of poly(acrylic acid)-cysteine conjugates with regard to their molecular mass. Eur. J. Pharm. Sci. **2003**, *18* (1), 89–96.

85. Bernkop-Schnürch, A.; Thaler, S. Polycarbophil-cysteine conjugates as platforms for peroral poly(peptide) delivery systems. J. Pharm. Sci. **2000**, *89* (7), 901–909.

86. Hornof, M.D.; Weyenberg, W.; Ludwig, A.; Bernkop-Schnürch, A. A mucoadhesive ocular insert: Development and *in vivo* evaluation in humans. J. Control. Release **2003**, *89* (3), 419–428.

87. Lueßen, H.L.; Verhoef, J.C.; Borchard, G.; Lehr, C.-M.; de Boer, A.G.; Junginger, H.E. Mucoadhesive polymers in peroral peptide drug delivery. II. Carbomer and polycarbophil are potent inhibitors of the intestinal proteolytic enzyme trypsin. Pharm. Res. **1995**, *12* (9), 1293–1298.

88. Walker, G.F.; Ledger, R.; Tucker, I.G. Carbomer inhibits tryptic proteolysis of luteinizing hormone-releasing hormone and N-α-benzoyl-l-arginine ethyl ester by binding the enzyme. Pharm. Res. **1999**, *16* (7), 1074–1080.

89. Bernkop-Schnürch, A.; Schwarz, G.; Kratzel, M. Modified mucoadhesive polymers for the peroral administration of mainly elastase degradable therapeutic poly-(peptides). J. Control. Release **1997**, *47* (2), 113–121.

90. Bernkop-Schnürch, A.; Dundalek, K. Novel bioadhesive drug delivery system protecting poly(peptides) from gastric enzymatic degradation. Int. J. Pharm. **1996**, *138* (1), 75–83.

91. Bernkop-Schnürch, A.; Bratengeyer, I.; Valenta, C. Development and *in vitro* evaluation of a drug delivery system protecting from trypsinic degradation. Int. J. Pharm. **1997**, *157* (1), 17–25.

92. Bernkop-Schnürch, A.; Scerbe-Saiko, A. Synthesis and *in vitro* evaluation of chitosan-EDTA-protease-inhibitor conjugates which might be useful in oral delivery of peptides and proteins. Pharm. Res. **1998**, *15* (2), 263–269.

93. Bernkop-Schnürch, A.; Pasta, M. Intestinal peptide and protein delivery: Novel bioadhesive drug carrier matrix shielding from enzymatic attack. **1998**, J. Pharm. Sci. *87* (4), 430–434.

94. Bernkop-Schnürch, A.; Göckel, N.C. Development and analysis of a polymer protecting from luminal enzymatic degradation caused by α-chymotrypsin. Drug Dev. Ind. Pharm. **1997**, *23* (8), 733–740.

95. Bernkop-Schnürch, A.; Marschütz, M.K. Development and *in vitro* evaluation of systems to protect peptide drugs from aminopeptidase. N. Pharm. Res. **1997**, *14* (2), 181–185.

96. Marschütz, M.K.; Caliceti, P.; Bernkop-Schnürch, A. Design and *in vivo* evaluation of an oral delivery system for insulin. Pharm. Res. **2000**, *17* (12), 1468–1474.

Molecular–Nanogels

97. Caliceti, P.; Salmaso, S.; Walker, G.; Bernkop-Schnürch, A. Development and *in vivo* evaluation of an oral insulin-PEG delivery system. Eur. J. Pharm. Sci. **2004**, *22* (4), 315–323.

98. Guggi, D.; Krauland, A.H.; Bernkop-Schnürch, A. Systemic peptide delivery via the stomach: *In vivo* evaluation of an oral dosage form for salmon calcitonin. J. Control. Release **2003**, *92* (1), 125–135.

99. Borchard, G.; Lueßen, H.L.; Verhoef, J.C.; Lehr, C.-M.; de Boer, A.G.; Junginger, H.E. The potential of mucoadhesive polymers in enhancing intestinal peptide drug absorption III: Effects of chitosan-glutamate and carbomer on epithelial tight junctions *in vitro*. J. Control. Release **1996**, *39* (2), 131–138.

100. Lee, V.H.L. Protease inhibitors and permeation enhancers as approaches to modify peptide absorption. J. Control. Release **1990**, *13* (2), 213–223.

101. Aungst, B.J.; Saitoh, H.; Burcham, D.L.; Huang, S.M.; Mousa, S.A.; Hussain, M.A. Enhancement of the intestinal absorption of peptides and non-peptides. J. Control. Release **1996**, *41* (1), 19–31.

102. Illum, L.; Farraj, N.F.; Davis, S.S. Chitosan as a novel nasal delivery system for peptide drugs. Pharm. Res. **1994**, *11* (8), 1186–1189.

103. Rao, R.K.; Baker, R.D.; Baker, S.S.; Gupta, A.; Holycross, M. Oxidant-induced disruption of intestinal epithelial barrier function: Role of tyrosine phosphorylation. Am. J. Physiol. **1997**, *273* (4), G812–G823.

104. Barret, W.C.; DeGnore, J.P.; Konig, S.; Fales, H.M.; Keng, Y.F.; Zhang, Y.; Yim, M.B.; Chock, P.B. Regulation of PTP1B via glutathionylation of the active site cysteine 215. Biochemistry **1999**, *38* (20), 6699–6705.

105. Werle, M.; Hoffer, M. Glutathione and thiolated chitosan inhibit multidrug resistance P-glycoprotein activity in excised small intestine. J. Control. Release **2006**, *111* (1), 41–46.

106. Föger, F.; Schmitz, T.; Bernkop-Schnürch, A. *In vivo* evaluation of polymeric delivery systems for P-glycoprotein substrates. Biomaterials **2006**, *27* (23), 4250–4255.

107. Bernkop-Schnürch, A.; Grabovac, V. Polymeric efflux pump inhibitors in oral drug delivery. Am. J. Drug Deliv. **2006**, *4* (4), 263–272.

108. Föger, F.; Malaivijitnond, S.; Wannaprasert, T.; Huck, C.; Bernkop-Schnürch, A.; Werle, M. Effect of a thiolated polymer on oral paclitaxel absorption and tumor growth in rats. J. Drug Target. **2008**, *16* (2), 149–155.

109. Shen, Q.; Lin, Y.L.; Handa, T.; Doi, M.; Sugie, M.; Wakayama, K.; Okada, N.; Fujita, T.; Yamamoto, A. Modulation of intestinal P-glycoprotein function by polyethylene glycols and their derivatives by *in vitro* transport and in situ absorption studies. Int. J. Pharm. **2006**, *313* (1), 49–56.

110. Greindl, M.; Föger, F.; Hombach, J.; Bernkop-Schnürch, A. *In vivo* evaluation of thiolated poly(acrylic acid) as a drug absorption modulator for Mpr2 efflux pump substrates. Eur. J. Pharm. Biopharm **2009**, *72* (3), 561–566.

111. Gottesman, M.M.; Pastan, I. The multidrug transporter, a double-edged sword. J. Biol. Chem. **1988**, *263* (25), 12163–12166.

112. Bernkop-Schnürch, A.; Gilge, B. Anionic mucoadhesive polymers as auxiliary agents for the peroral administration of poly(peptide) drugs, Influence of the gastric fluid. Drug Dev. Ind. Pharm. **2000**, *26* (2), 107–113.

113. Skov, O.P.; Poulsen, S.S.; Kirkegaard, P.; Nexo, E. Role of submandibular saliva and epidermal growth factor in gastric cytoprotection. Gastroenterology **1984**, *87* (1), 103–108.

114. Itoh, M.; Matsuo, Y. Gastric ulcer treatment with intravenous human epidermal growth factor: A double-blind controlled clinical study. J. Gastroenterol. Hepatol. **1994**, *9* (Suppl. 1), S78–S83.

115. Slomiany, B.L.; Nishikawa, H.; Bilski, J.; Slomiany, A. Colloidal bismuth subcitrate inhibits peptic degradation of gastric mucus and epidermal growth factor *in vitro*. Am. J. Gastroenterol. **1990**, *85* (4), 390–393.

116. Bernkop-Schnürch, A.; Schuhbauer, H.; Pischel, I. α-Liponsäure(-Derivate) enthaltende Retardform, Deutsche Patentschrift **1999**, 1999-09-30.

Mucoadhesive Systems: Drug Delivery

Ryan F. Donnelly
Elizabeth Ryan
Thakur Raghu Raj Singh
A. David Woolfson
School of Pharmacy, Medical Biology Center, Queen's University of Belfast, Belfast, U.K.

Abstract

This entry provides a full overview of mucoadhesive systems for drug delivery purposes, beginning with a detailed description of bioadhesion and mucoadhesion, the bioadhesion of polymers, the nature of mucus, the theories of mucoadhesion, and the factors affecting mucoadhesion.

INTRODUCTION

This entry provides a full overview of mucoadhesive systems for drug delivery purposes, beginning with a detailed description of bioadhesion and mucoadhesion, the bioadhesion of polymers, the nature of mucus, the theories of mucoadhesion, and the factors affecting mucoadhesion. Also described are the techniques used for the determination of mucoadhesion, the most common routes of mucoadhesive administration, mucoadhesive materials, the most examples of formulations studied and alternative uses of mucoadhesive materials. An overview of the recent advances and discoveries in mucoadhesive polymers and mucoadhesive delivery systems is also reviewed.

Bioadhesion is the state in which two materials, at least one biological in nature, are held together for an extended period of time by interfacial forces. Bioadhesion can be classified into three types. Type one is adhesion between two biological phases, that is, wound healing. Type two is the adhesion of a biological phase to an artificial substrate, that is, biofilms formation on prosthetic devices, and type three is the adhesion of an artificial material to a biological substrate, that is, adhesion of synthetic hydrogels to soft tissue. In drug delivery, bioadhesion suggests the attachment of a drug carrier system to a specified biological location such as epithelial tissue or the mucus coat on the surface of a tissue.

Within drug delivery, the term bioadhesion suggests the attachment of a drug carrier system to a specified biological location. The biological surfaces in question are usually epithelial tissue or the mucus coat on the surface of a tissue. In drug delivery, it is important to distinguish between mucoadhesion and bioadhesion, that is, bioadhesion is where the polymer is attached to a biological membranes, and if the substrate is mucus membrane, the term mucoadhesion is used.

A number of definitions exist for mucoadhesion, the most general being, the adhesion between two materials, at least one of which is a mucosal surface. Khutoryanskiy[1] defined mucoadhesion as the attractive interaction at the interface between a pharmaceutical dosage form and a mucosal membrane, whereas Leung and Robinson described mucoadhesion as the interaction between a mucin surface and a synthetic or natural polymer.[2] The development of mucoadhesive dosage forms in drug delivery has prompted research into the prolongation of their residence time on mucosal surfaces. Many mucoadhesive-based topical and local systems allow enhanced bioavailability of the drug being delivered due to the prolonged retention of the mucoadhesive delivery system at the site of delivery. Mucoadhesive delivery systems also allow rapid absorption of drug at the site of action. This is due to the considerable surface area of hydrogel mucosal systems. Further advantages of drug delivery across the mucosa include the bypassing of the first-pass hepatic metabolism and the avoidance of degradation by gastrointestinal enzymes. Possible applications include the delivery of high-molecular-weight sensitive molecules such as peptides and oligonucleotides. Possible disadvantages of mucoadhesive delivery systems include the occurrence of local ulcerous effects due to prolonged contact of a drug possessing ulcerogenic properties and patient acceptability in terms of taste, irritancy, and comfort. A major limitation in the development of oral mucosal delivery is the lack of a good model for *in vitro* screening to identify drugs suitable for such administration.[3] Many mucoadhesive polymers also present significant formulation challenges. As mucoadhesive polymers are generally hydrophilic, they form high viscosity, often pH sensitive, aqueous solutions at low concentrations.[4]

The concept of mucoadhesion in drug delivery was introduced in the field of controlled-release drug delivery systems in the early 1980s. Much research has been focused on the investigations of the interfacial phenomena

Concise Encyclopedia of Biomedical Polymers and Polymeric Biomaterials DOI: 10.1081/E-EBPPC-120050247

of mucoadhesive hydrogels (and other types of mucoadhesive compounds) with the mucus. A mucoadhesive hydrogel must satisfy certain criteria to be used as a drug delivery systems. The mucoadhesive hydrogel should:

1. Be loaded substantially by the active compound;
2. Not interact physicochemically with the active compound or create a hostile artificial environment that would lead to inactivation or degradation of the active compound;
3. Swell in the aqueous-biological environment of the delivery-absorption site;
4. Interact with mucus or its components for adequate adhesion;
5. Allow, when swelled, controlled release of the active compound;
6. Be compatible (biomaterial) with the underlying epithelia by means of complete absence of cytotoxicity, ciliotoxicity, or other type, or irreversible alterations of the cell membrane components;
7. Have the appropriate molecular size and conformation to escape systemic absorption from the administration site; and
8. Be excreted, unaltered, or biologically degraded to inactive, nontoxic oligomers/monomers that will be further subject of physical clearance.

A mucoadhesive delivery system designed for controlled release of active compounds should be localized at specific sites of administration and absorption, and should prolong the residence time of the active compound at the site of action to permit, if possible for once-daily dosing. The mucoadhesive delivery system, designed for the administration of macromolecular therapeutics like peptide or protein drugs, should have permeation-enhancing properties by means of alteration of the permeability properties of the underlying epithelium, and should protect the peptide drug from degradation by inhibiting the proteolytic enzymes usually present at the site of administration or by stabilizing the intrinsic environment of the delivery system by sustaining the suitable pH. Solid dosage forms based on mucoadhesive polymers are used mainly for buccal delivery of drugs, whereas micro- or nanoparticulate formulations are preferred for the delivery of therapeutics in the nasal and intestinal tracts.[5] Much research has focused on the delivery of mucoadhesive dosage forms in the gastrointestinal tract (GIT). This route is the most favorable with regard to patient compliance and ease of application. Solid dosage forms are more suitable for smaller cavities in the human body like the oral cavity, either for systemic absorption of compounds or for local treatment of inflammatory diseases.[5] Various administration routes, such as ocular, nasal, buccal and gingival, gastrointestinal (oral), vaginal, and rectal, make mucoadhesive drug delivery systems attractive and flexible in dosage form development. Reports have suggested that the market for mucoadhesive drug delivery systems is expanding rapidly.[1]

MUCUS

Mucus is the protective gel layer that covers epithelia. Mucus consists of the mucus glycoprotein mucin (0.5–5.0%) dispersed in water (95.0–99.5%). Mucins are the components that are responsible for the gel-like consistency of mucus. Mucins are divided into two categories, membrane bound and secretory forms. Membrane-bound mucins are attached to cell surfaces and may affect immune responses or inflammation, whereas secretory mucins emanate from mucosal absorptive cells and specialized goblet cells. The four characteristics of the mucus layer that are related to mucoadhesion have been described by Leung and Robinson.[2] The first characteristic is that mucus is a network of linear, flexible, and random-coil macromolecules. Second, mucin is negatively charged due to sialic acid and sulfate residues. Third, mucus is a cross-linked network connected by disulfide bonds between mucin molecules, and finally, mucin is heavily hydrated.[6]

THEORIES OF MUCOADHESION

It is widely agreed that the theoretical models of bioadhesion can only explain a limited range of the interactions that constitute a bioadhesive bond.[7] However, four main theories exist. Donnelly et al.[6] recently reviewed the four main theories of mucoadhesion.

Wetting Theory of Mucoadhesion

The wetting theory of mucoadhesion is the oldest established theory of its kind and can be applied most successfully to liquid or low-viscosity bioadhesives. This theory explains adhesion as an embedding process where the adhesive agent penetrates into surface irregularities of the substrate and harden. This process produces many adhesive anchors. The adhesive on the surface of the substrate must overcome the surface tension effects at the interface.

The wetting theory calculates the contact angle and the thermodynamic work of adhesion.

The work done is related to the surface tension of both the adhesive and the substrate, as given by Dupre's equation.[8]

$$\omega_A = \gamma_b + \gamma_t - \gamma_{bt} \tag{1}$$

ω_A is the specific thermodynamic work of adhesion,

γ_b, γ_t, and γ_{bt} represent the respective surface tensions of the bioadhesive polymer, the substrate, and the interfacial tension.

The sum of the surface tensions of the two adherent phases minus the interfacial tensions apparent between both phases is the adhesive work done.[9]

Figure 1 illustrates a drop of liquid bioadhesive spreading over a soft tissue surface.

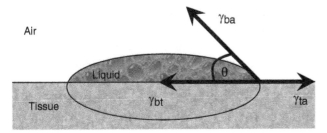

Fig. 1 A liquid bioadhesive spreading over a typical soft tissue surface.
Source: Adapted from Shaikh et al.[6]

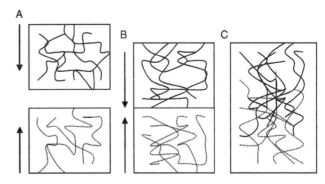

Fig. 2 Schematic representation of the diffusion theory of bioadhesion. Polymer layer and mucus layer before contact (**A**); upon contact (**B**); the interface becomes diffuse after contact for a period of time (**C**).
Source: Adapted from Shaikh et al.[6]

Horizontal resolution of the forces gives the Young equation:

$$\gamma_{ta} = \gamma_{bt} + \gamma_{ba} \cos \theta \tag{2}$$

θ is the angle of contact,
γ_{bt} is the surface tension between the tissue and polymer,
γ_{ba} is the surface tension between polymer and air, and
γ_{ta} is the surface tension between tissue and air.
Equation 3 states that if the angle of contact, θ, is greater than zero, the wetting will be incomplete.
If the vector γ_{ta} greatly exceeds $\gamma_{bt} + \gamma_{ba}$, that is:

$$\gamma_{ta} \geq \gamma_{bt} + \gamma_{ba} \tag{3}$$

then θ will approach zero and wetting will be complete.

For a bioadhesive material to successfully adhere to a biological surface, barrier substances must first be dispelled and then the bioadhesive material must spontaneously spread across the underlying substrate, that is, tissue or mucus.

The spreading coefficient, S_b, can be defined as:

$$S_b = \gamma_{ta} - \gamma_{bt} - \gamma_{ba} > 0 \tag{4}$$

Equation 4 states that bioadhesion is successful if S_b is positive.

Therefore, bioadhesion is favored by large values of γ_{ta} or small values of γ_{bt} and γ_{ba}.[9]

Electrostatic Theory of Mucoadhesion

The electrostatic theory of mucoadhesion theory explains how electrons are transferred across the adhesive interface and adhering surfaces. The transfer of electrons results in the establishment of the electric double layer at the interface and a series of attractive forces responsible for maintaining contact between the two layers.[10]

Diffusion Theory of Mucoadhesion

The diffusion theory of mucoadhesion describes that polymeric chains from the bioadhesive interpenetrate into glycoprotein mucin chains. They then research a sufficient depth within the opposite matrix to allow formation of a

semipermanent bond.[11] Concentration gradients drive the polymer chains of the bioadhesive into the mucus network and the glycoprotein mucin chains into the bioadhesive matrix until an equilibrium penetration depth is achieved (Fig. 2).

The estimated depth needed for good bioadhesive bonds is 0.2–0.5 μm.[12] The mean diffusion depth of the bioadhesive polymer segments, s, may be represented in Eq. 5:

$$s = \sqrt{2tD} \tag{5}$$

D is the diffusion coefficient,
t is the contact time.
Equation 5 was adapted by Duchene[12] to determine the time, t, to bioadhesion of a particular polymer. Equation 6 illustrates this equation:

$$t = l^2/D_b \tag{6}$$

l represents the interpenetrating depth,
D_b is the diffusion coefficient of a bioadhesive through the substrate.

Following intimate contact, the substrate and adhesive chains move along their respective concentration gradients into the opposite phases. The depth of diffusion is dependent on the diffusion coefficient of both phases. It has been reported by Reinhart and Peppas[13] that the diffusion coefficient is dependent on the molecular weight of the polymer strand and that it decreases with cross-linking density.

Adsorption Theory

The adsorption theory of mucoadhesion comes into play after the initial contact between two surfaces, the materials adhere due to the surface forces acting between the chemical structures at the two surfaces.[14] Reorientation of polar molecules occurs at the interface, when such groups are present.[9] When adhesion is particularly strong chemisorption can occur. Chemisorption is the form of adsorption where the forces involved are valence forces like those operating in the formation of chemical compounds. The adsorption theory maintains that adherence to tissue is due to the net result of one or more secondary forces (van der Waals forces, hydrogen bonding, and hydrophobic bonding).[15–17]

Another relating theory, called the fracture theory of adhesion, describes the force required for the separation of the two surfaces following adhesion. The fracture strength is equivalent to the adhesive strength through the following equation:

$$\sigma = (E \times \varepsilon/L)^{1/2} \tag{7}$$

σ is the fracture strength,
ε is the fracture energy,
E is Young's modulus of elasticity, and
L is the critical crack length.[18]

Equation 7 allows the study of bioadhesion by tensile apparatus.

THE FACTORS AFFECTING MUCOADHESION

Mucoadhesion is affected by a number of factors including hydrophilicity, molecular weight, cross-linking, swelling, pH, and the concentration of the active polymer.[11,14,19]

Hydrophilicity

A hydrophilic molecule is one that has the tendency to interact or be dissolved by water and other polar substances. Bioadhesive polymers possess numerous hydrophilic functional groups. These hydrophilic functional groups (i.e., hydroxyl and carboxyl groups) allow hydrogen bonding with the substrate and swelling in aqueous media. These occurrences allow maximal exposure of the potential anchor sites. The stronger the hydrogen bonding, the stronger the adhesion, in general.[20] Polymers that have swollen have the maximum distance between their chains, which leads to increased chain flexibility and increases the efficiency of substrate penetration.[6]

Molecular Weight

Polymer molecular weight is important as it determines many physical properties. The optimum molecular weight for maximum mucoadhesion depends on the type of polymer. Interpenetration of polymer molecules is favored by low-molecular-weight polymers and entanglements are favored at high molecular weights. Generally, bioadhesive forces increases with the molecular weight of the polymer up to 100,000. Above this, there is no further gain.

Cross-Linking and Swelling

Polymers consist of repeating structural subunits (monomers) connected by covalent bonds. In addition to the bonds that hold monomers together in a polymer chain, many polymers form bonds between neighboring chains. These bonds can be formed directly between the neighboring chains, or two chains may bond to a third common molecule. This phenomenon is called cross-linking. Although not as strong or rigid as the bonds within the chain, these cross-links have an important effect on the polymer. Cross-linking density is inversely proportional to the degree of swelling.[21] Low cross-linking density results in higher flexibility and hydration rate. In addition, increasing the surface area of the polymer improves mucoadhesion. A lightly cross-linked polymer is most suitable to obtain a high degree of cross-linking. However, an excess of moisture will result in a degree of cross-linking that is too great, and the resulting slippery mucilage can be easily removed from the substrate.[22] The addition of adhesion promoters such as free polymer chains and polymers grafted onto the performed network can improve the mucoadhesion of the cross-linked polymers.[19]

Spatial Conformation

The spatial conformation of the polymers is also extremely important. An example of the importance of spatial conformation to bioadhesion can be seen when comparing the molecular weight of dextrans (19,500,000) to that of polyethylene glycol (PEG) (200,000). Although the molecular weight of dextran is much higher, both polymers have a similar adhesive strength. This phenomenon is due to the helical conformation of dextran, which may shield many adhesively active groups, unlike PEG that has a linear conformation.

pH

The pH at the bioadhesive to substrate interface can influence the adhesion of polymers that contain ionizable groups. Many polymers used for bioadhesive purposes in drug delivery are polyanions possessing carboxylic acid functionalities. The approximate pK_a for many of these polymers is pH 4–5. The maximum adhesive strength of these polymers is between 4 and 5 and decreases gradually above pH 6. It has been proven that the protonated carboxyl groups, rather than the ionized carboxyl groups, react with the mucin molecules. It has been suggested that the mechanism of this action is the simultaneous formation of many hydrogen bonds.[23]

Concentration of Active Polymer

It has been reported by Ahuja, Khar, and Ali[14] that an optimal concentration of corresponding to optimal mucoadhesion exists in liquid mucoadhesive formulations. In highly concentrated systems, mucoadhesion drops above this optimal concentration. This has been explained by the fact that in concentrated solutions, coiled molecules become solvent poor and the chains available for interpenetration are sparse. For solid dosage forms such as tablets, Duchěne, Touchard, and Peppas[12] showed that the higher the polymer concentration, the stronger the mucoadhesion.

Drug/Excipient Concentration

The ratio of drug to excipient may affect the mucoadhesive effects. The effect of propranolol hydrochloride to Carbopol® hydrogels adhesion was investigated by Blanco-Fuente.[24] Increased adhesion was noted when water was limited in the system. This was due to an increase in elasticity, caused by complex formation between the drug and polymer. In the presence of large amounts of water, the complex precipitated out which led to a decrease in the adhesive character. Increasing toluidine blue O (TBO) concentration in mucoadhesive patches based on poly(methyl vinyl ether/maleic acid) significantly increased mucoadhesion to porcine cheek tissue.[25] It was suggested that this was due to increased internal cohesion within the patches, caused by electrostatic interactions between the cationic drug and anionic polymer.

Other Factors Affecting Mucoadhesion

The initial force of application may affect mucoadhesion.[26] High bioadhesive strength and enhanced interpenetration are induced by higher forces.[27] The initial contact time also influences mucoadhesion. The greater the initial contact time between the bioadhesive and the substrate, the greater the swelling and interpenetration of polymer chains.[28] Physiological variables such as the rate of mucus turnover can affect mucoadhesion.[29] The mucoadhesive surface can vary depending on the site in the body and the presence of local or systemic disease and this in turn will affect the mucoadhesion.[28]

METHODS OF DETERMINING MUCOADHESION

The evaluation of bioadhesive and mucoadhesive properties is fundamental to the development of novel bioadhesive delivery systems. Mucoadhesive formulations are usually either in hydrogel or film form.

Determination of Mucoadhesion in Polymer Gel Systems

The last decade has seen a range of methods employed to study mucoadhesion phenomena and mucoadhesive properties of hydrogel systems. Table 1 lists some of the methods employed and how they are carried out. These methods specifically pertain to hydrogel systems.

Study of the flow and deformation of polymer gel formulations (i.e., their rheological characteristics) may be useful in predicting the mucoadhesive ability of a polymeric formulation. Hassan and Gallo[30] first suggested

Table 1 Methods of studying mucoadhesion

Direct assays	
Tensiometry	Force required to dislodge two surfaces, one coated with mucus and the other solid dosage form consisting of mucoadhesive hydrogel
Flow through	Flow rate dV/dt required to dislodge two surfaces; useful for microparticulate dosage forms
Colloidal gold staining	Measures the "adhesion number"
In vivo techniques	Endoscopy, gamma scintigraphy
Molecular mucin-based assays	
Viscometry and rheology	Intrinsic viscosity [Z] can be related to complex size through MHKS (Mark–Houwink–Kuhn–Sakurada) a, coefficient
Dynamic light scattering	Diffusion coefficient, D, can be related to complex size through MHKS c. coefficient
Turbidity, light scattering	SEC MALLS (Size Exclusion Chromatography–Multi-Angle Laser Light Scattering) particularly useful for determining MW of mucin, turbidity, semiquantitative indicator
Analytical ultracentrifugation	Change in MW (sedimentation equilibrium), sedimentation coefficient ratio of complex to mucin
Surface plasmon resonance imaging methods	Needs mobile and immobile phase, atomic force microscopy (conventional and gold labeled), scanning tunneling microscopy

Source: Adapted from Junginger et al.[5]

this approach by determining the rheological interaction between a polymer gel and a mucin solution. This experiment proved that a polymer gel and mucin solution mixture evoked a larger rheological response than the sum of the values of polymer and mucin. Varying rheological results may be attributed to differences in mucin type and concentration[31,32] and in polymer concentrations.[32,33] It has been recommended that rheological methods should not be used as a standalone method for determining mucoadhesive properties of polymer gels.[34]

There is a much higher proportion of *in vitro* mucoadhesion studies compared to *in vivo* due to the time, cost, and ethical constraints of *in vivo* studies. However, some work into *in vivo* mucoadhesive testing has been completed, specifically in monitoring mucoadhesion on tissue surfaces such as GIT or the buccal cavity. A study by Ch'ng et al. focused on the *in vivo* transit time for bioadhesive beads in the rat GIT.[35] A ^{51}Cr-labeled bioadhesive were inserted at selected time intervals, GIT was removed and cut into sections and radioactivity was measured. Davis completed a noninvasive study to determine the transit of the mucoadhesive agent. The formulation used contained a gamma-emitting radionuclide, and the release characteristics and polymer position was examined by gamma scintigraphy.[36] MRI has also been recently used to detect the time and location of the release of a mucoadhesive formulation using dry Gd-DOTA powder.[37]

Determination of Mucoadhesion in Film Systems

It has have recently pointed out that no standard apparatus is available for testing bioadhesive strength, and therefore, an inevitable lack of uniformity between test methods has arisen. However, three main testing methods have arisen for bioadhesive films.[6] The three main testing methods are (*i*) tensile test, (*ii*) shear strength, and (*iii*) peel strength.

Tensile and shear tests

The most common technique for the determination of force of separation in bioadhesive testing is the application of a force perpendicularly to the tissue/adhesive interface. During this process, a state of tensile stress is set up. During the shear stress, the direction of the forces is reorientated so that it acts along a joint interface. In both tensile and shear modes, an equal pressure is distributed over the contact area.[38] A recent literature review has found that the tensile strength method is the most common technique used for bioadhesion testing. A commercially available texture profile analyzer (TA-XT Plus, stable microsystems, Surrey, UK) operating in bioadhesive test mode can be used to

measure the force required to remove bioadhesive films from excised tissue *in vitro*.[22,39–42] Figure 3 illustrates the texture profile analyzer in bioadhesion mode.

Peel strength

The peel strength is based on the calculation of energy required to detach the patch from the substrate. It is of most use when the bioadhesive system is formulated as a patch.[39] The texture analyzer (as illustrated in Fig. 3) operating in tensile test mode and coupled with a sliding lower platform can also be used to determine peel strength of such formulations.[39] Figure 4 illustrates a typical set-up used to determine the peel strength of bioadhesive films.

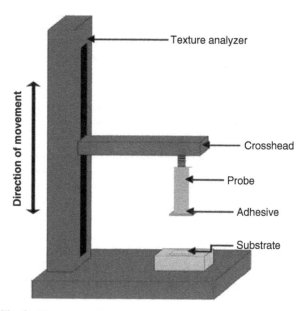

Fig. 3 Texture profile analyzer in bioadhesion test mode.
Source: Adapted from Shaikh et al.[6]

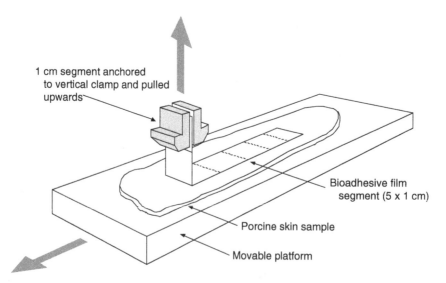

Fig. 4 Simplified representation of the test set-up used to determine peel strength of bioadhesive films.
Source: Adapted from McCarron et al.[39]

ROUTES OF ADMINISTRATION FOR MUCOADHESIVE-BASED DRUG DELIVERY SYSTEMS

The mucus membrane or mucosa is the moist tissue that lines organs and body cavities such as the mouth, gut, rectum, genital area, nose, and eye lid. The anatomical differences of mucus membrane are outlined in Table 2.

Examples of mucoadhesive drug delivery systems, which have been formulated, include powders, compacts, sprays, semisolids, and films. Developments in delivery systems include compacts for drug delivery into the oral cavity,[47] powders and nanoparticles for delivery to nasal mucosa,[48,49] and oral strips for the tongue and buccal cavity.[50] More advanced and alternative mucoadhesive formulations have been suggested including buccal films that

Table 2 Anatomical differences of the mucus membrane

Mucus membrane	Relevant anatomical features
Buccal[42]	Buccal mucosa surface area approximately 30 cm^2
	Comprised of three distinct layers—epithelium, basement membrane, and connective tissues
	Buccal mucosa and sublingual are soft palate nonkeratinized tissue, and gingival are hard palate keratinized tissue
	Thickness of buccal epithelium is in the range of 500–800 µm, 40–50 cells thick
	Mucus secreted by salivary glands, as a component of saliva, forming a 0.1–0.7 mm thick layer
	Turnover time for buccal epithelium 5–6 days
	Permeability barrier property of oral mucosa due to intercellular materials derived from membrane-coating granules
Nasal[43]	Nasal cavity surface area 160 cm^2
	Lined with mucus membrane containing columnar cells, goblet cells, and basal cells
	Columnar cells are covered with cilia, apart from the anterior part of the nasal cavity
	Both keratinized and nonkeratinized epithelial cells present depending upon location within nasal cavity
	Cilia responsible for mucociliary clearance
	Mucus secreted by the submucosal glands and the goblet cells, forming a mucus layer approximately 5–20 µm thick
	Nasal cavity length approximately 60 mm
	Nasal cavity volume approximately 20 mL
	Turnover time for mucus is usually 10–15 min
Ocular[44]	Cornea is composed of five layers—epithelium, Bowman's layer, stroma, Descemet's membrane, and endothelium
	Epithelium consists of 5–6 layers of cells, with the cells of the basal layer being columnar, and the outermost cells flattened polygonal cells
	Tight junctions present between the basal cells of the corneal epithelium
	At the corneal margin, the conjunctiva is structurally continuous with the corneal epithelium
	The conjunctival tissue is permeable to molecules up to 20,000 Da, whereas the cornea is impermeable to molecules greater than 5,000 Da
	The conjunctiva contains around 1.5 million goblet cells, which synthesize secretory mucins and peptides
	A volume of about 2–3 µL of mucus secreted daily
	A turnover of the mucus layer occurs in approximately 15–20 hr
	Exposed part of the eye is covered by a thin fluid layer—precorneal tear film
Vaginal[45]	Tear film thickness is approximately 3–10 µm
	Length of vagina varies from 6 to 10 cm
	The epithelial layer consists of the lamina propria and stratified squamous epithelium
	A cell turnover of about 10–15 layers is estimated to be in the order of 7 days
	Although there are no glands in the vagina mucosa, the surface is usually covered with vaginal fluid
	Major components of vaginal fluid are cervical mucus and vaginal fluid from the well-vascularized mucosa
	The volume, viscosity, and pH of the cervical mucus vary with age and during the menstrual cycle
Rectal[46]	Length approximately 15–20 cm
	Surface area of approximately 300 cm^2
	Epithelium consists of a single layer of cylindrical cells and goblet cells secreting mucus
	Flat surface, without villi, and with three major fold, the rectal valves
	Approximately 3 mL of mucus with a neutral pH spread over the surface

Source: Adapted from Shaikh et al.[6]

Molecular—Nanogels

may provide a greater degree of flexibility and comfort than adhesive tablets. They may also improve the residence time compared to oral gels.[51] The most commonly used polymers for such formulations are cellulose derivatives, poly(acrylic acids) such as Carbopol® and Gantrez® and copolymers such as poly(methyl vinyl ether/maleic anhydride).[51–74]

Oral Mucoadhesive Delivery Systems

The accessibility of mucoadhesive delivery systems to the oral mucosa has resulted in its significant interest as a delivery route. The most commonly used routes are the buccal and sublingual. The nonkeratinized epithelium in the oral cavity, that is, the soft palate, the mouth floor, the ventral side of the tongue, and the buccal mucosa offer a relatively permeable barrier for drug transport.[75] Oral mucosae are comprised of multiple layers of cells, which show various patterns of differentiation. The functions of the different regions in the oral cavity affect the patterns of differentiation.[76] Blood supply to the oral cavity tissue is provided by the external carotid artery. This artery branches to the maxillary lingual and facial artery. The oral mucosa does not contain mucus secreting goblet cells; however, mucins are found in human saliva. Saliva consists of 99% water, and the mucins dissolved within it form a gel of 20–100 μm thickness.[76] Paracellular transport is generally followed by hydrophilic compounds and large or highly polar molecules, whereas lipophilic drugs follow transcellular transport through the lipid bilayer.[77] Advantages of the oral route over others include the bypassing of the hepatic first-pass metabolism, improvement in drug bioavailability, improved patient compliance, excellent accessibility, unidirectional drug flux, and improved barrier permeability that intact skin.[78,79] The application of mucoadhesive delivery systems to the oral cavity allows both local and systemic delivery. Local therapy is use to treat conditions such as aphthous ulceration, gingivitis, periodontal disease, and xerostomia. Common dosage forms include adhesive gels, tablets, films, patches, ointments, mouth washes, and pastes.

The most frequently used dosage form for buccal drug delivery up to now has been adhesive tablets. Tablets can be applied to many regions of the oral cavity, such as cheeks, lips, gums, and the palate. Such tablets allow the drinking, eating, and speaking without any major discomfort. A number of recent studies have examined the suitability of adhesive tablets as an oral mucoadhesive delivery system. Perioli studied the influence of compression force on tablet behavior and drug release rate for mucoadhesive buccal tablets. Tablets were prepared from hydroxyethyl cellulose (HEC) and Carbopol 940 in a 1:1 ratio at varying compression forces. It was found that mucoadhesion performance and drug release were affected by compression force as increasing compression force decreased the *in vitro* and *in vivo* drug release profiles. An increased compression force also resulted in the best *in vivo* mucoadhesive and hydration

time. Tablets prepared at the highest forces caused pain during *in vivo* application and had to be detached by human volunteers.[80] Another recent study carried out by Shermer evaluated the efficacy and tolerability of a mucoadhesive patch compared with a pain relieving oral solution for the treatment of aphthous stomatitis. The mucoadhesive patch was more successful than the oral solution in terms of healing time and pain intensity after 12 and 24 hr. After 1 hr of treatment, the local adverse effects were significantly less frequent with the mucoadhesive patch patients as compared with the oral solution patients. Donnelly reported on a mucoadhesive patch containing TBO as a potential system for use in photodynamic antimicrobial chemotherapy (PACT) of oropharyngeal candidiasis. The patches were prepared from aqueous blends of poly(methyl vinyl ether/ maleic anhydride) and tri(propylene glycol) methyl ether. It was found that short application times of TBO-loaded patches allowed the treatment of recently acquired oropharyngeal candidiasis, if they are caused solely by planktonic cells. For biofilms, it was suggested that longer application times would be needed.[25]

Desai developed oral mucoadhesive patch formulations for the delivery of fenretinide for site-specific chemoprevention of oral cancer. *In vitro* and *in vivo* release studies were investigated. The solubilization of fenretinide in simulated saliva was studied by incorporating nonionic surfactants (Tween® 20 and 80, and Brij® 35 and 98), bile salts (sodium salts of cholic, taurocholic, glycocholic, and deoxycholic acids), phospholipid (lecithin), and novel polymeric solubilizer (Souplus®). Adhesive and drug release layers were prepared using solvent casting. The adhesive layers were prepared from hydroxypropyl methylcellulose (HPMC), and the drug release layers were prepared from Eudragit® RL PO with or without solubilizers with incorporated fenretinide. The oral mucoadhesive patches were formed by attaching drug and adhesive layers onto a backing layer (Tegaderm™ film). The study found that solubilizer-free patch exhibited poor *in vitro-* and *in vivo*-controlled drug release. The incorporation of single or mixed solubilizers in fenretinide/Eudragit® patches significantly improved continuous *in vitro/in vivo* fenretinide release. The paper concluded that Fenretinide/ Eudragit® RL PO patches with 20 wt% Tween® 80+40 wt% sodium deoxycholate solubilizers exhibit excellent release behavior for further preclinical and/or clinical evaluation in oral cancer chemoprevention.[81]

A number of mucoadhesive buccal delivery formulations are currently being commercialized or undergoing clinical trials. Generex Biotechnology has recently developed a novel liquid aerosol formulation called Oralin, which is now in clinical Phase II trials.[82] It allows precise insulin dose delivery through a metered dose inhaler in the form of fine aerosolized droplets directed into the mouth. The oral aerosol formulation is rapidly absorbed through the buccal mucosal epithelium and it provides the plasma insulin levels necessary to control postprandial glucose rise in diabetic patients. Advantages of this novel delivery system include rapid absorption, a user-friendly administration technique, precise dosing control, and bolus delivery. Other products undergoing clinical trials include BioAlliance Pharma's miconazole tablet (Lauriad®), which is in Phase III clinical trials, and Aphtach®, a triamcinolone acetonide buccal tablets from Teijin Ltd., which is now commercially available.[83]

Nasal Mucoadhesive Delivery Systems

The human nasal mucosa is a highly dense vasculature network and is a relatively permeable membrane structure. The area of the nasal mucosa is normally approximately 150 cm^2.[84] The nasal epithelium exhibits high permeability as only two cell layers separate the nasal lumen from the dense blood vessel network in the lamina propria. The main lining of the nasal cavity is the respiratory epithelium. It allows the clearance of mucus by the mucociliary system and is composed of ciliated and non-ciliated columnar cells, goblet cells, and basal cells. The respiratory epithelium is covered by a mucus layer, which can be divided into the periciliary layer and a gel-like upper layer. The periciliary layer is less viscous than the gel-like layer. Mucociliary clearance allows the removal of foreign substances and particles from the nasal cavity, therefore preventing them from reaching the upper airways. This process is facilitated by the cilia which propel the mucous layer toward the nasopharynx.[85] Advantages of the nasal route of delivery include rapid uptake and the avoidance of first-pass hepatic metabolism. Disadvantages include local toxicity and irritation, mucociliary clearance of 5 min, the presence of proteolytic enzymes, and the possible influence of pathological conditions (cold and allergies). The application of bioadhesive delivery systems such as liquids, semisolids, and solids may significantly increase retention time.

The delivery of proteins and peptides through the nasal route can be compromised by the brief residence time at the mucosal surface. Some bioadhesive polymers have been suggested to extend residence time and improve protein uptake across the nasal mucosa. A recent study by McInnes used gamma scintigraphy to quantify nasal residence time of bioadhesive formulations.[86] This study also investigated the absorption of insulin. A conventional nasal spray was compared with three lyophilized nasal insert formulations (1–3% w/w HPMC) in a four-way crossover study with six male volunteers. It was found that the 2% w/w HPMC lyophilized formulation was most successful at achieving an extended residence time. This concentration had the ability to adhere to the nasal mucosa rapidly without overhydration.

Viscosity-enhancing mucosal delivery systems for the induction of an adaptive immune response against a viral antigen were examined by Coucke, Schotsaert, and Libert.[87] Spray-dried formulations of starch (Amioca®) and poly(acrylic acid) (Carbopol® 974P) were used as carriers of the vial antigen. An *in vivo* rabbit model was used to compare these formulations for intranasal delivery of heat-inactivated influenza virus combined with LTR 192G adjuvant. The formulations tested successfully induced a systemic anti-HA antibody response after intranasal vaccination with a whole-virus influenza vaccine.

The development, characterization, and safety aspects of nasal drug products consisting of functionalized mucoadhesive polymers (i.e., polycarbophil, hyaluronan, and amberlite resin) has been examined recently. Work into the use of mucosal vaccines for the induction of a systemic immune response has recently been carried out and it has been found that the addition of a mucoadhesive polymer to the vaccine formulation increases the affinity for the mucus membrane and enhances the stability of the preparation. Examples of such vaccines include influenza vaccines, diphtheria vaccines, and tetanus.[88]

A recent pilot study into the use of a nasal morphine-chitosan formulation for the treatment of pain in cancer patients concluded that the systems were well tolerated and accepted by patients and that it could lead to a rapid onset of pain relief.[89] A recent study by Tzachev compared a commercially available decongestant solution, which was nonmucoadhesive, with a mucoadhesive solution of xylometazoline in human patients with perennial allergic rhinitis. The study found that the mucoadhesive formulation produces a more prolonged clinical effect than the commercially available produced. A study by Luppi investigated the preparation of albumin nanoparticles carrying cyclodextrins for the nasal delivery of the anti-Alzheimer drug tacrine. In this instance, the nasal route selected as peroral administration of this drug is associated with low bioavailability. Bovine serum albumin nanoparticles were prepared using a coacervation method followed by thermal cross-linking. The nanoparticles were then loaded by soaking from solutions of tacrine hydrochloride and lyophilizing. The authors concluded that the study indicated that albumin nanoparticles carrying native and hydrophilic derivatives can be employed for the formulation of mucoadhesive nasal formulations.[90]

Ocular Mucoadhesive Delivery Systems

The ocular route is mainly used for local treatment of eye pathologies. Conventional delivery methods to the eye are generally unsuccessful due to the inherent protective mechanisms of the eye (tear production, tear flow and blinking), the limited area of absorption, and the lipophilic character of the corneal epithelium.[91] The precorneal tear film is the first structure encountered by an ocular dosage form. It consists of three distinct layers. The outer layer is of oily and lipid nature and prevents tear evaporation. The middle layer contains an aqueous salt solution, and the inner layer is a mucus layer, which is secreted by the conjunctiva goblet cells and lacrimal gland. This layer maintains moisture in the corneal and conjunctival epithelia. The ocular membranes comprise the cornea, which is non-vascularized, and the conjunctiva, which is vascularized. The major pathway for ocular drug penetration is considered to be the corneal epithelium. It consists of five or six layers of nonkeratinized squamous cells.[92] Solutions and suspensions are swiftly washed from the cornea, and ointments can alter the tear refractive index and blur the vision. Therefore, prolonging the residence time by mucoadhesion may provide the required conditions of successful ocular delivery.

Recently, Sensoy prepared bioadhesive sulfacetamide sodium microspheres to increase residence time on the ocular surface, thereby attempting to enhance treatment efficacy of ocular keratitis.[93] Polymer mixtures containing pectin, polycarbophil and HPMC at different ratios were used to prepare microparticle formulations through a spray drying method. An *in vivo* model consisting of New Zealand rats with keratitis caused by *Pseudomonas aeruginosa* and *Staphylococcus aureus* was used in the study. An optimized formulation containing sulfacetamide sodium-loaded polycarbophil microspheres at a polymer drug ratio of 2:1 was found to be most suitable for an ocular application. Advances in gene transfer technology have allowed its application for the treatment of several chronic diseases that affect the ocular surface. The efficacy and mechanism of action of a bioadhesive DNA nanocarrier prepared from hyaluronan (HA) and chitosan (CS) were investigated by De la Fuente, Seijo, and Alonso.[94] The formulation was specifically designed for topical ophthalmic gene therapy. An *in vivo* study using rabbits found that the nanoparticles entered the corneal and conjunctival epithelial cells and were assimilated by the cells. The nanoparticles also provided efficient delivery of the associated plasmid DNA inside the cells, reaching significant transfection levels. Work into reducing the movement of ocular films across the eye with the addition of mucoadhesive polymers has been examined by Alza.[88] It has been found that films have a tendency to move across the eye without the addition of mucoadhesive polymers. The improvements associated with the addition of mucoadhesive polymers have been found to minimize ocular irritation and burning sensations.

A recent clinical study was conducted by Baeyens[95] into the use of soluble bioadhesive ophthalmic drug inserts (BODI®) in dogs with external ophthalmic diseases (conjunctivitis, superficial corneal ulcers, or keratoconjunctivitis sicca). These results were compared with classical Tiacil® eye drops from Virbac Laboratories. The BODI® inserts reduced the treatment to a single application, therefore improving patient compliance. A number of mucoadhesive polymers have been incorporated into ophthalmic gels to increase gel efficacy. Examples include NyoGel® (timolol, Novartis) and Pilogel® (pilocarpine hydrochloride, Alcon Laboratories).[96]

Vaginal Mucoadhesive Delivery Systems

The vagina is a fibrovascular tube connecting the uterus to the outer surface of the body. The vaginal epithelium consists of a stratified epithelium and lamina propria. The vagina offers a substantial area for drug absorption because the numerous folds in the epithelium increase in total surface area. A rich vascular network surrounds the vagina, whereas the vaginal epithelium is covered by a film of moisture consisting mainly of cervical mucus and fluid secreted from the vaginal wall. The dosage forms that are usually used for the vaginal route are solutions, gels, suspensions, suppositories, creams, and tablets. They all have a short residence time.[97–99] Bioadhesives may control drug release and extend the residence time of such formulations. They may contain drug or act in conjunction with moisturizing agents.

Recent advances in polymeric technology have increased the potential of vaginal gels. Vaginal gels are semisolid polymeric matrices comprising small amounts of solid, dispersed in relatively large amounts of liquid that have been used in systems for microbiocides, contraceptives, labor inducers, and other substances. An acid-buffering bioadhesive vaginal clotrimazole (antifungal) and metronidazole (antiprotozoal and antibacterial) tablets for the treatment of genitourinary tract infections were developed by Alam et al.[100] Polycarbophil and sodium carboxymethyl cellulose were found to be a good combination for an acid-buffering bioadhesive vaginal tablet from bioadhesion and release studies. It was also found that these bioadhesive polymers held the tablet for over 24 hr inside the vagina. The *in vitro* spreadability of a commercially available conventional tablet (Infa-V®) was compared to the aforementioned formulation, and it was found to be comparable to the commercially available gel. It was also found that the acid-buffering bioadhesive tablet produced better antimicrobial action than marketed intravaginal drug delivery systems (Infa-V®, Candid-V®, and Canesten® 1). Gel formulations consisting of clomiphene citrate (CLM) for the local treatment of human papilloma virus infections were prepared by Cevher et al.[101] Formulations were prepared including the polyacrylic acid (PAA) polymers such as Carbopol® 934P (C934P), Carbopol® 971P (C971P), and Carbopol® 974P (C974P) in various concentrations, and their conjugates containing thiol groups were prepared. The gels which contained C934P-Cys showed the highest adhesiveness and mucoadhesion, and increasing polymer concentration significantly decreased drug release.

A number of clinical trials are underway on microbicide gels. Microbicide gels are intended to improve mucosal permeation rate of microbiocides for the prevention of sexually transmitted diseases. Currently, a 1% tenofovir gel is being investigated in Phase II clinical trials for determining the safety and acceptability of vaginal microbiocides.[102] Clinical trials for contraceptive gels such as Buffergel® are in Phase II and III clinical trials, comparing it to Gynol 2®

marketed product.[102] The Prostin E2® suppository containing dinoprostone found that administration of prostaglandin E2 gel was more effective in inducing labor, when tested in clinical trials.[102] Another Phase III clinical trial was conducted by Janssen Pharmaceutica of a mucoadhesive system based on intraconazole vaginal cream containing cyclodextrins and other ingredients. The clinical tests found that the cream was well tolerated and was an effective delivery system for selected vaginal delivery.[103]

Rectal Mucoadhesive Delivery Systems

The rectum is a part of the colon. It is 10 cm in length and has a surface area of 300 cm^2. The main function of the rectum is the removal of water. The rectum has a relatively small surface area for drug absorption. The absorption of drugs through the rectum is generally achieved by a simple diffusion process through the lipid membrane. Advantages of the rectal route of delivery include the avoidance of first-pass metabolism. The use of bioadhesive delivery systems in the rectum can also decrease the drug migration distance.

A recent study by Kim et al.[104] developed a thermoreversible flurbiprofen liquid suppository base composed of a poloxamer and sodium alginate for the improvement of the rectal bioavailability of flurbiprofen. Cyclodextrin derivatives enhanced the aqueous solubility of flurbiprofen. An *in vivo* study in rats showed that flurbiprofen liquid suppository containing the cyclodextrin derivative HP-beta-CD showed excellent bioavailability. It was concluded that HP-beta-CD could successfully be used as a solubility enhancer in liquid suppositories containing poorly water-soluble drugs.

Cervical and Vulval Delivery Systems

A number of recent studies have been done into the application of mucoadhesive delivery systems to the cervix and vulva. A novel bioadhesive cervical patch containing 5-fluorouracil for the treatment of cervical intraepithelial neoplasia (CIN) was developed by Woolfson. The patch was a drug-loaded bioadhesive prepared from a gel containing 2% w/w Corbopol® 981 plasticized with 1% w/w glycerin. The casting solvent used was ethanol: water 30:70. The film was bonded directly to a backing layer formed from thermally cured poly(vinyl chloride) emulsion. Substantial drug release through human cervical tissue samples was observed over approximately 20 hr.[52,105]

The design, physiochemical characterization, and clinical evaluation of bioadhesive delivery systems for photodynamic therapy (PDT) difficult to manage vulval neoplasis and dysplasis were carried out by Donnelly et al.[106] Aminolevulinic acid (ALA) is commonly delivered to the vulva using creams or solutions and covered by an occlusive dressing. These types of formulations are poor at staying in place, therefore the production of a bio-

adhesive patch was suggested as an alternative delivery system. The patches were shown to release more ALA over 6 hr than the proprietary cream (Porphin®, 20% w/w ALA). Clinically, the patch was extensively used in the successful PDT of vulval intraepithelial neoplasia, lichen sclerosus, squamous hyperplasia, Paget's disease, and vulvodynia.

Gastrointestinal Mucoadhesive Delivery Systems

The oral route is undoubtedly the most favored route of administration. It represents the most convenient route of drug administration, being characterized by high patient compliance. The mucosal epithelium along the GIT varies. In the stomach, the surface epithelium consists of a single layer of columnar cells whose apical membrane is covered by a conspicuous glycocalyx. A thick layer of mucus covers the surface to protect against aggressive luminal content. The small intestine in characterized by an enormous surface area available for the absorption of nutrients and drugs. The intestinal epithelium consists of a single layer of three types of columnar cells: enterocytes, goblet cells, and enteroendocrine cells. The large intestine (colon) has the same cell populations as the small intestine, and its main function is the absorption of water and electrolytes. The role of mucus in the intestine is to facilitate the passage of food along the intestinal tract and to protect the gut from bacterial infections.[107] Problems associated with the oral route include hepatic first-pass metabolism, degradation of drug during absorption, mucus covering GI epithelia, high turnover of mucus covering GI epithelia, and the high turnover of mucus. Recently, the GIT delivery has emerged as a very important route of administration. Bioadhesive retentive system involves the use of bioadhesive polymers, which can adhere to the epithelial surface in the GIT. The use of bioadhesive systems would increase GI transit time and increase bioavailability.

Gastric retention formulations (GRFs) made of naturally occurring carbohydrate polymers and containing riboflavin *in vitro* were studied by Ahmad et al.[100] The swelling and dissolution characteristics of the formulations were examined *in vitro*, and the *in vivo* behavior in fasting dogs for gastric retention was also investigated. The bioavailability of riboflavin from the GRFs was studied in fasted healthy humans and compared to an immediate-release formulation. It was found that on immersion in gastric juice, the GRFs swelled rapidly and released their drug load in zero-order fashion over a period of 24 hr. The GRF stayed in the stomach of fasted dogs for up to 9 hr, then disintegrated and reached the colon in 24 hr. In addition, the bioavailability of riboflavin from a large-size GRF was more than triple of that measured after administration of an immediate-release formulation.

Nanoparticle carriers with bioadhesive properties were evaluated for their adjuvant potential for oral vaccination by Salma et al.[108] Thiamine was used as a specific ligand-nanoparticle conjugate (TNP) to target specific sites within the GIT, namely, enterocytes and Peyer's patches. The affinity of the nanoparticles to the gut mucosa was studied in orally inoculated rats. The thiamine-coated nanoparticles showed promise as a particulate vectors for oral vaccination and immunotherapy.

MUCOADHESIVE POLYMERS

Polymers for bioadhesive drug delivery systems must have certain characteristics.[11–14] The polymer and its degradation products should be nontoxic, nonabsorbable, and nonirritant. The polymer should form strong noncovalent bonds with the mucus or epithelial cell surface and should adhere quickly to moist tissue. It should possess site specificity, allow easy incorporation of the drug, and offer no hindrance to drug release. The polymer must resist decomposition on storage or during the shelf life of the dosage form. Table 3 presents some of the most commonly used polymers for mucoadhesive purposes with their chemical structures.

There are a number of broad categories of polymers that possess the suitable characteristics for adherence to biological surfaces. The first category is polymers that adhere through nonspecific, noncovalent interactions, which are primarily electrostatic in nature. The second is polymers possessing hydrophilic functional groups, which bond with similar groups on biological substrates. The third group is polymers that bind to specific receptor sites on the cell or mucus surface.

Examples of polymers in the third category are lectins and thiolated polymers. Lectins have the ability to bind sugars selectively in a noncovalent manner. They are proteins or glycoprotein complexes of nonimmune origin.[26] Lectins have been used for drug targeting applications due to their ability to attach themselves to carbohydrates on the mucus or epithelial cell surface.[109,110] Thiolated polymers are hydrophilic macromolecules exhibiting free thiol groups on the polymeric backbone. Thiolated polymers have many advantageous mucoadhesive properties including improved tensile strength, rapid swelling, and water uptake behavior. These improved properties are due to the presence of thiol groups in the polymer. They allow the formation of stable covalent bonds with the cysteine subdomains of mucus glycoproteins, which results in increased residence time and improved bioavailability.[19]

ALTERNATIVE USES OF MUCOADHESIVE POLYMERS

Poly(methyl vinyl ether-co-maleic anhydride) (PMVE/MAH) has many applications including its use as a thickening agent, film former, dispersing agent, emulsion stabilizer, and denture adhesive.[111] Films cast from aqueous blends of PMVE/MAH are known to possess

Table 3 Chemical structures of some bioadhesive polymers used in drug delivery

Chemical name (Abbreviation)	Chemical structure
Poly(ethylene glycol) (PEG)	
Poly(vinyl alcohol) (PVA)	
Poly(vinyl pyrrolidone) (PVP)	
Poly(acrylic acid) (PAA or Carbopol®)	
Poly(hydroxyethyl methacrylate) (PHEMA)	
Chitosan	
Hydroxyethyl cellulose (HEC): R = H and CH_2CH_2OH Hydroxypropyl cellulose (HPC): R = H and $CH_2CH(OH)CH_3$ Hydroxypropyl methylcellulose (HPMC): R = H, CH_3 and $CH_2CH(OH)CH_3$ Methylcellulose: R = H and CH_3 Sodium carboxymethyl cellulose (NaCMC): R = H and CH_2COONa	

Source: Adapted from Shaikh et al.[6]

moisture-activated bioadhesive properties.[22] PMVE/MA films containing a novel plasticizer, tripropylene glycol methyl ether (TPME), maintained their adhesive strength and tensile properties. It was also found that films plasticized with commonly used polyhydric alcohols underwent an esterification reaction that led to copolymer cross-linking. This esterification reaction resulted in a poly(ester) network and an accompanying profound alteration in the physical characteristics. Films became brittle over time and lost aqueous solubility and the ability adhere to porcine skin.

PMVE/MAH is a prime example of a commonly used bioadhesive polymer that has many more far reaching applications within drug delivery. Thakur et al.[112] investigated the swelling and network parameters of poly(ethylene glycol)–cross-linked poly (methyl vinyl ether-co-maleic acid) hydrogels. The influence of poly(ethylene glycol) (PEG) plasticizer content and molecular weight on the physico-chemical properties of films cast from aqueous blends of poly(methyl vinyl ether-co-maleic acid) was investigated using thermal analysis, swelling studies, scanning electron microscopy (SEM), and attenuated total reflectance (ATR)-Fourier transform infrared (FTIR) spectroscopy.[113] It was found that FTIR spectroscopy revealed a shift of the C=O peak from 1708 to 1731 cm^{-1}, indicating that an esterification reaction had occurred upon heating, thus producing cross-linked films. Higher molecular weight PEGs (10,000 and 1,000 Da, respectively) have greater chain length and produce hydrogel networks with lower cross-link densities and higher average molecular weight between two consecutive cross-links. The author concluded that hydrogels based on PMVE/MA/PEG 10,000 could be used for rapid delivery of drug, due to their low cross-link density. Moderately cross-linked PMVE/MA/PEG 1000 hydrogels or highly cross-linked PMVE/MA/PEG 200 systems could then be used in controlling the drug delivery rates. Another paper by this group investigated solute permeation across the same type of hydrogels.[114] The swelling kinetics and solute permeation (theophylline, vitamin B12, and fluorescein sodium) of hydrogels composed of poly(methyl vinyl ether-co-maleic acid) (PMVE/MA) and poly(ethylene glycol) (PEG) were examined. The paper concluded that the hydrogels changed in swelling behavior, cross-link density, and

solute permeation with change in PMVE/MA content, therefore suggesting a potential application in controlled drug delivery systems.

Following these extensive characterization studies, PMVE/MA hydrogels were applied to microneedle fabrication. In a study carried out by Donnelly et al.,[115] PMVE/MA was compared to other commonly used pharma polymers for microneedle array fabrication. Microneedle arrays are minimally invasive devices that can be used to by-pass the *stratum corneum* barrier and thus achieve enhanced transdermal drug delivery. The aim of the study was to optimize polymeric composition and to assess the performance of these microneedle devices. Microneedles were prepared using laser-engineered silicone micromolds, and hydrogels of the various polymers were incorporated into these molds at ambient conditions and allowed to dry. Microneedles micromolded from 20% w/w aqueous blends of the Gantrez® AN-139 were found to possess superior physical strength than those produced from commonly used pharma polymers. Theophylline was successfully delivered across neonatal porcine skin from 600 μm microneedles, 83% of the incorporated drug was delivered over 24 hr. Figure 5 illustrates microneedle arrays prepared from a range of polymers, before and after a compression force test.

Optical coherence tomography (OCT) showed that drug-free 600 μm Gantrez® AN-139 microneedles punctured the

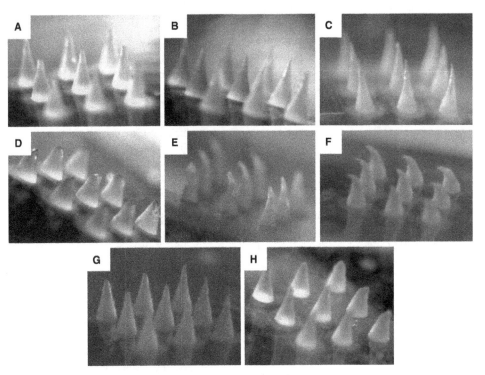

Fig. 5 Light microscope images of polymeric MNs prepared from aqueous blends containing Carbopol® 971-P NF, alginic acid, poly(vinyl) alcohol, and Gantrez® AN-139 before, (**A**, **B**, **C**, and **G**, respectively), and after (**D**, **E**, **F**, and **H**, respectively) the application of a compression force.
Source: Adapted from Donnelly et al.[115]

stratum corneum barrier of human skin *in vivo* and extended approximately 460 μm into the skin. Another study by Donnelly[116] OCT was used to investigate the effect that microneedle geometry and force of application had upon penetration characteristics of soluble PMVE/MA microneedle arrays into neonatal porcine skin *in vitro*. This study concluded that, at an application force of 11.0 N/array, it was found that, in each case, increasing microneedle height from 350 to 600 to 900 μm led to a significant increase in the depth of microneedle penetration achieved. Microneedles dissolved quickly, with an approximate 34% reduction in microneedle height in the first 15 min. Only 17% of the microneedle height remained after a 3-hr period. There was a significantly greater cumulative amount of theophylline delivered after 24 hr from a microneedle array of 900 μm height (292.23 ± 16.77 μg) in comparison to an microneedle array of 350 μm height (242.62 ± 14.81 μg). The authors concluded that OCT could prove to be a key development for polymeric microneedle research, accelerating their commercial exploitation.

Another study by this group assess the feasibility of transdermal macromolecule delivery using novel laser-engineered dissolving microneedles prepared from aqueous blends of 20% w/w PMVE/MA *in vitro* and *in vivo*.[117] This study found that PMVE/MA could successfully load insulin into the polymeric microneedle system. Circular dichroism analysis showed that encapsulation of insulin within the polymeric matrix did not lead to change in protein secondary structure. *In vitro* studies revealed significant enhancement in insulin transport across the neonatal porcine skin. The percutaneous administration of insulin-loaded microneedle arrays to rats resulted in a dose-dependent hypoglycemic effect. These studies show that the properties of some mucoadhesive polymers can be altered to give them other uses in drug delivery.

CONCLUSION

The mucoadhesive dosage form offers prolonged contact at the site of administration, low enzymatic activity, and patient compliance. The formulation of mucoadhesive drug delivery systems depends on the selection of suitable polymer with excellent mucosal adhesive properties and biocompatibility. Researchers are now looking beyond traditional polymers and focusing on next-generation mucoadhesive polymers such as lectins, thiols, and so on. These polymers offer greater attachment and retention of dosage forms. However, these novel mucoadhesive formulations require much more work to deliver clinically for the treatment of both topical and systemic diseases. The use of mucoadhesive polymers for other drug delivery applications, that is, polymeric microneedle-mediated transdermal drug delivery, illustrates the broad scope of such polymers within the drug delivery field.

REFERENCES

1. Khutoryanskiy, V.V. Advances in mucoadhesion and mucoadhesive polymers. Macromol. Biosci. **2011**, *11* (6), 748–764.
2. Leung, S.H.; Robinson, J.R. The contribution of anionic polymer structural features to mucoadhesion. J. Control. Release **1988**, *5* (3), 223–231.
3. Tangri, P.; Khurana, S.; Madhav, N.V.S. Mucoadhesive drug delivery: Mechanism and methods of evaluation. Int. J. Pharm. Bio. Sci. **2011**, *2* (1), 458–467.
4. Smart, J.D. The basics and underlying mechanisms of mucoadhesion. Adv. Drug Deliv. Rev. **2005**, *57* (11), 1556–1568.
5. Junginger, H.E.; Verhoef, J.C.; Thanou, M. Drug delivery: Mucoadhesive hydrogels. In *Encyclopedia of Pharmaceutical Technology*; 3rd Ed. 2006; 1169–82.
6. Shaikh, R.; Raj Singh, T.R.; Garland, M.J.; Woolfson, A.D.; Donnelly, R.F. Mucoadhesive drug delivery systems. J. Pharm. Bioallied Sci. **2011**, *3* (1), 89–100.
7. Longer, M.A.; Robinson, J.R. Fundamental aspects of bioadhesion. Pharm. Int. **1986**, *7*, 114–117.
8. Pritchard, W.H. *Aspects of Adhesion 6*, 3rd Edn.; Alder, D., Ed. London University Press: London, 1970: 11–23.
9. Wake, W.C. Adhesion and the Formulation of Adhesives. London: Applied Science Publishers, 1982.
10. Deraguin, B.V.; Smigla, V.P. *Adhesion: Fundamentals and Practice*; McLaren: London, 1969.
11. Jiménez-castellanos, M.R.; Zia, H.; Rhodes, C.T. Mucoadhesive drug delivery systems. Drug Dev. Ind. Pharm. **1993**, *19* (1–2), 143–194.
12. Duchěne, D.; Touchard, F.; Peppas, N.A. Pharmaceutical and medical aspects of bioadhesive systems for drug administration. Drug Dev. Ind. Pharm. **1988**, *14* (2–3), 283–318.
13. Reinhart, C.T.; Peppas, N.A. Solute diffusion in swollen membranes. Part II. Influence of crosslinking on diffusive properties. J. Membr. Sci. **1984**, *18*, 227–239.
14. Ahuja, A.; Khar, R.K.; Ali, J. Mucoadhesive drug delivery systems. Drug Dev. Ind. Pharm. **1997**, *23* (5), 489–515.
15. Huntsberger, J.R. Mechanisms of adhesion. J. Pain. Technol. **1967**, *39*, 199–211.
16. Kinloch, A.J. The science of adhesion. J. Mater. Sci. **1980**, *15* (9), 2141–2166.
17. Yang, X.; Robinson, J.R. Bioadhesion in mucosal drug delivery. In *Biorelated Polymers and Gels*; Okano, T., Ed. Academic Press: London, 1998.
18. Gu, J.M.; Robinson, J.R.; Leung, S.H. Binding of acrylic polymers to mucin/epithelial surfaces: Structure-property relationships. Crit. Rev. Ther. Drug Carrier Syst. **1988**, *5* (1), 21–67.
19. Peppas, N.A.; Little, M.D.; Huang, Y. Bioadhesive controlled release systems. In *Handbook of Pharmaceutical Controlled Release Technology*; Wise, D.L., Ed. Marcel Dekker: New York, 2000; 255–269.
20. Roy, S.; Pal, K.; Anis, A.; Pramanik, K.; Prabhakar, B. Polymers in mucoadhesive drug delivery system: A brief note. Des. Monomers Polym. **2009**, *12* (6), 483–495.
21. Gudeman, L.F.; Peppas, N.A. Preparation and characterization of pH-sensitive, interpenetrating networks of poly(vinyl alcohol) and poly(acrylic acid). J. Appl. Polym. Sci. **1995**, *55* (6), 919–928.

22. McCarron, P.A.; Woolfson, A.D.; Donnelly, R.F.; Andrews, G.P.; Zawislak, A.; Price, J.H. Influence of plasticizer type and storage conditions on properties of poly(methyl vinyl ether-co-maleic anhydride) bioadhesive films. J. Appl. Polym. Sci. **2004**, *91* (3), 1576–1589.

23. Park, H.; Robinson, J.R. Physico-chemical properties of water insoluble polymers important to mucin/epithelial adhesion. J. Control. Release **1985**, *2*, 47–57.

24. Blanco-Fuente, H. *In vitro* bioadhesion of carbopol hydrogels. Int. J. Pharm. **1996**, *142* (2), 169–174.

25. Donnelly, R.F.; McCarron, P.A.; Tunney, M.M.; David Woolfson, A. Potential of photodynamic therapy in treatment of fungal infections of the mouth. Design and characterisation of a mucoadhesive patch containing toluidine blue O. J. Photochem. Photobiol. B Biol. **2007**, *86* (1), 59–69.

26. Smart, J.D. An *in vitro* assessment of some mucosa-adhesive dosage forms. Int. J. Pharm. **1991**, *73* (1), 69–74.

27. Peppas, N.A.; Buri, P.A. Surface, interfacial and molecular aspects of polymer bioadhesion on soft tissues. J. Control. Release **1985**, *2*, 257–275.

28. Kamath, K.R.; Park, K. Mucosal Adhesive Preparation. In *Encyclopedia of Pharmaceutical Technology*, Swarbrick, J., Boylan, J.C., Eds.; Marcel Dekker: New York, 1992; 133.

29. Lehr, C-M.; Poelma, F.G.J.; Junginger, H.E.; Tukker, J.J. An estimate of turnover time of intestinal mucus gel layer in the rat in situ loop. Int. J. Pharm. **1991**, *70* (3), 235–240.

30. Hassan, E.E.; Gallo, J.M. A Simple rheological method for the *in vitro* assessment of mucin-polymer bioadhesive bond strength. Pharm. Res. **1990**, *7* (5), 491–495.

31. Rossi, S. Influence of mucin type on polymer-mucin rheological interactions. Biomaterials **1995**, *16* (14), 1073–1079.

32. Hägerström, H.; Paulsson, M.; Edsman, K. Evaluation of mucoadhesion for two polyelectrolyte gels in simulated physiological conditions using a rheological method. Eur. J. Pharm. Sci. **2000**, *9* (3), 301–309.

33. Rossi, S.; Ferrari, F.; Bonferoni, M.C.; Caramella, C. Characterization of chitosan hydrochloride–mucin rheological interaction: Influence of polymer concentration and polymer: Mucin weight ratio. Eur. J. Pharm. Sci. **2001**, *12* (4), 479–485.

34. Hägerström, H.; Edsman, K. Limitations of the rheological mucoadhesion method: The effect of the choice of conditions and the rheological synergism parameter. Eur. J. Pharm. Sci. **2003**, *18* (5), 349–357.

35. Ch'ng, H.S.; Park, H.; Kelly, P.; Robinson, J.R. Bioadhesive polymers as platforms for oral controlled drug delivery II: Synthesis and evaluation of some swelling, water-insoluble bioadhesive polymers. J. Pharm. Sci. **1985**, *74* (4), 399–405.

36. Davis, S.S. The design and evaluation of controlled release systems for the gastrointestinal tract. J. Control. Release **1985**, *2*, 27–38.

37. Kremser, C.; Albrecht, K.; Greindl, M.; Wolf, C.; Debbage, P.; Bernkop-Schnürch, A. *In vivo* determination of the time and location of mucoadhesive drug delivery systems disintegration in the gastrointestinal tract. Magn. Reson. Imaging **2008**, *26* (5), 638–643.

38. Park, K.; Park, H. Test methods of bioadhesion. In *Bioadhesive Drug Delivery Systems*; Lenaerts, V., Gurney, R., Eds.; CRC Press: Florida, Boca Raton, 1990.

39. McCarron, P.A.; Donnelly, R.F.; Zawislak, A.; Woolfson, A.D.; Price, J.H.; McClelland, R. Evaluation of a water-soluble bioadhesive patch for photodynamic therapy of vulval lesions. Int. J. Pharm. **2005**, *293* (1–2), 11–23.

40. McCarron, P.A.; Donnelly, R.F.; Zawislak, A.; Woolfson, A.D. Design and evaluation of a water-soluble bioadhesive patch formulation for cutaneous delivery of 5-aminolevulinic acid to superficial neoplastic lesions. Eur. J. Pharm. Sci. **2006**, *27* (2–3), 268–279.

41. Donnelly, R.F.; McCarron, P.A.; Zawislak, A.A.; Woolfson, A.D. Design and physicochemical characterisation of a bioadhesive patch for dose-controlled topical delivery of imiquimod. Int. J. Pharm. **2006**, *307* (2), 318–325.

42. Salamat-Miller, N.; Chittchang, M.; Johnston, T.P. The use of mucoadhesive polymers in buccal drug delivery. Adv. Drug Deliv. Rev. **2005**, *57* (11), 1666–1691.

43. Ugwoke, M.I.; Agu, R.U.; Verbeke, N.; Kinget, R. Nasal mucoadhesive drug delivery: Background, applications, trends and future perspectives. Adv. Drug Deliv. Rev. **2005**, *57* (11), 1640–1665.

44. Ludwig, A. The use of mucoadhesive polymers in ocular drug delivery. Adv. Drug Deliv. Rev. **2005**, *57* (11), 1595–1639.

45. Valenta, C. The use of mucoadhesive polymers in vaginal delivery. Adv. Drug Deliv. Rev. **2005**, *57* (11), 1692–1712.

46. Edsman, K.; Hägerström, H. Pharmaceutical applications of mucoadhesion for the non-oral routes. J. Pharm. Pharmacol. **2005**, *57* (1), 3–22.

47. Ponchel, G.; Touchard, F.; Duchêne, D.; Peppas, N.A. Bioadhesive analysis of controlled-release systems. I. Fracture and interpenetration analysis in poly(acrylic acid)-containing systems. J. Control. Release **1987**, *5* (2), 129–141.

48. Nagai, T.; Konishi, R. Buccal/gingival drug delivery systems. J. Control. Release **1987**, *6* (1), 353–360.

49. Sayin, B.; Somavarapu, S.; Li, X.W.; Thanou, M.; Sesardic, D.; Alpar, H.O.; Şenel, S Mono-N-carboxymethyl chitosan (MCC) and N-trimethyl chitosan (TMC) nanoparticles for non-invasive vaccine delivery. Int. J. Pharm. **2008**, *363* (1), 139–148.

50. Ilango, R. *In vitro* studies on buccal strips of glibenclamide using chitosan. Ind. J. Pharm. Sci. **1997**, *59* (5), 232.

51. Anders, R.; Merkle, H.P. Evaluation of laminated mucoadhesive patches for buccal drug delivery. Int. J. Pharm. **1989**, *49* (3), 231–240.

52. Woolfson, A. Development and characterisation of a moisture-activated bioadhesive drug delivery system for percutaneous local anaesthesia. Int. J. Pharm. **1998**, *169* (1), 83–94.

53. Boyapally, H.; Nukala, R.K.; Bhujbal, P.; Douroumis, D. Controlled release from directly compressible theophylline buccal tablets. Colloids Surf. B Biointerfaces **2010**, *77* (2), 227–233.

54. Petelin, M.; Pavlica, Z.; Bizimoska, S.; Šentjurc, M. *In vivo* study of different ointments for drug delivery into oral mucosa by EPR oximetry. Int. J. Pharm. **2004**, *270* (1), 83–91.

55. Rossi, S.; Marciello, M.; Bonferoni, M.C.; Ferrari, F.; Sandri, G.; Dacarro, C.; Grisoli, P.; Caramella, C. Thermally sensitive gels based on chitosan derivatives for the treatment of oral mucositis. Eur. J. Pharm. Biopharm. **2010**, *74* (2), 248–254.

56. Nafee, N.A.; Ismail, F.A.; Boraie, N.A.; Mortada, L.M. Mucoadhesive buccal patches of miconazole nitrate: *in vitro/in vivo* performance and effect of ageing. Int. J. Pharm. **2003**, *264* (1), 1–14.

Molecular—Nanogels

57. Diaz del Consuelo, I.; Falson, F.; Guy, R.H.; Jacques, Y. *Ex vivo* evaluation of bioadhesive films for buccal delivery of fentanyl. J. Control. Release **2007**, *122* (2), 135–140.

58. Nasal Ointment. Electronic medicines compendium. Available from: http://www.Medicines.Org.Uk/Emc/Medicine/2027/Spc/Jun2010.htm (Last accessed Nov 2011).

59. Jain, A.K.; Khar, R.K.; Ahmed, F.J.; Diwan, P.V. Effective insulin delivery using starch nanoparticles as a potential trans-nasal mucoadhesive carrier. Eur. J. Pharm. Biopharm. **2008**, *69* (2), 426–435.

60. Wu, J.; Wei, W.; Wang, L.-Y.; Su, Z.-G.; Ma, G.-H. A thermosensitive hydrogel based on quaternized chitosan and poly(ethylene glycol) for nasal drug delivery system. Biomaterials **2007**, *28* (13), 2220–2232.

61. Luppi, B.; Bigucci, F.; Abruzzo, A.; Corace, G.; Cerchiara, T.; Zecchi, V. Freeze-dried chitosan/pectin nasal inserts for antipsychotic drug delivery. Eur. J. Pharm. Biopharm. **2010**, *75* (3), 381–387.

62. Hornof, M.; Weyenberg, W.; Ludwig, A. Bernkop-Schnürch, A. Mucoadhesive ocular insert based on thiolated poly(acrylic acid): development and *in vivo* evaluation in humans. J. Control. Release **2003**, *89* (3), 419–428.

63. Grześkowiak, E. Biopharmaceutical availability of sulphadicramide from ocular ointments *in vitro*. Eur J Pharm Sci **1998**, *6* (3), 247–253.

64. Qi, H.; Chen, W.; Huang, C.; Li, L.; Chen, C.; Li, W.; Wu, C. Development of a poloxamer analogs/carbopol-based in situ gelling and mucoadhesive ophthalmic delivery system for puerarin. Int. J. Pharm. **2007**, *337* (1), 178–187.

65. Jain, D.; Carvalho, E.; Banerjee, R. Biodegradable hybrid polymeric membranes for ocular drug delivery. Acta Biomater. **2010**, *6* (4), 1370–1379.

66. Lux, A. A comparative bioavailability study of three conventional eye drops versus a single lyophilisate. B. J. Ophthalmol. **2003**, *87* (4), 436–440.

67. Perioli, L.; Ambrogi, V.; Pagano, C.; Scuota, S.; Rossi, C. FG90 chitosan as a new polymer for metronidazole mucoadhesive tablets for vaginal administration. Int. J. Pharm. **2009**, *377* (1), 120–127.

68. Khanna, N.; Dalby, R.; Tan, M.; Arnold, S.; Stern, J.; Frazer, N. Phase I/II clinical safety studies of teramepro-col vaginal ointment. Gynecol. Oncol. **2007**, *107* (3), 554–562.

69. Kim, Y.-T.; Shin, B.-K.; Garripelli, V.K.; Kim, J.K.; Davaa, E.; Jo, S.; Park, J.S. A thermosensitive vaginal gel formulation with HPgammaCD for the pH-dependent release and solubilization of amphotericin B. Eur. J. Pharm. Sci. **2010**, *41*, 399–406.

70. Yoo, J.W.; Dharmala, K.; Lee, C.H. The physicodynamic properties of mucoadhesive polymeric films developed as female controlled drug delivery system. Int. J. Pharm. **2006**, *309* (1), 139–145.

71. Yahagi, R.; Machida, Y.; Onishi, H. Mucoadhesive suppositories of ramosetron hydrochloride utilizing Carbopol®. Int. J. Pharm. **2000**, *193* (2), 205–212.

72. Available from: http://www.medicines.org.uk/emc/medicine/7162/spc/Anusol+Ointment/#excipients/Jun2010.htm (Last accessed November 2011)

73. Koffi, A.A.; Agnely, F.; Ponchel, G.; Grossiord, J.L. Modulation of the rheological and mucoadhesive properties of thermosensitive poloxamer-based hydrogels intended for

74. the rectal administration of quinine. Eur. J. Pharm. Sci. **2006**, *27* (4), 328–335.

74. de Leede, L.G.J.; de Boer, A.G.; Pörtzgen, E.; Feijen, J.; Breimer, D.D. Rate-controlled rectal drug delivery in man with a hydrogel preparation. J. Control. Release **1986**, *4* (1), 17–24.

75. Leung, S.H.S.; Robinson, J.A. Polyanionic polymers in bioadhesive and mucoadhesive drug delivery. ACS Symp. Ser. **1992**, *480*, 269–284.

76. Squier, C.A.; Wertz, P.W. Structure and function of the oral mucosa and implications for drug delivery. Drugs Pharm. Sci. **1996**, *74* (1), 1–26.

77. Shojaei, A.H.; Berner, B.; Li, X. Transbuccal delivery of acyclovir: I. *In vitro* determination of routes of buccal transport. Pharm. Res. **1998**, *15* (8), 1182–1188.

78. al-Achi, A.; Greenwood, R. Buccal administration of human insulin in streptozocin-diabetic rats. Res. Commun. Chem. Pathol. Pharmacol. **1993**, *82* (3), 297–306.

79. Bouckaert, S.; Remon, J.-P. *In vitro* bioadhesion of a buccal, miconazole slow-release tablet. J. Pharm. Pharmacol. **1993**, *45* (6), 504–507.

80. Perioli, L.; Ambrogi, V.; Giovagnoli, S.; Blasi, P.; Mancini, A.; Ricci, M.; Rossi, C. Influence of compression force on the behavior of mucoadhesive buccal tablets. AAPS PharmSciTech **2008**, *9* (1), 274–281.

81. Desai, K.-G.H.; Mallery, S.R.; Holpuch, A.S. Schwendeman SP. Development and *in vitro-in vivo* evaluation of fenretinide-loaded oral mucoadhesive patches for site-specific chemoprevention of oral cancer. Pharm. Res. **2011**, *28* (10), 2599–2609.

82. 3M, 3M buccal drug delivery system [UKICRS Newsletter, 1998, Page 7].

83. Modi, P.; Mihic, M.; Lewin, A. The evolving role of oral insulin in the treatment of diabetes using a novel RapidMist system. Diabetes. Metab. Res. Rev. **2002**, *18* (Suppl. 1), S38–S42.

84. Nagai, T.; Machida, Y. Bioadhesive dosage forms for nasal administration. In *Bioadhesive Drug Delivery Systems*, Lenaerts, V., Gurney, R., Eds.; CRC Press: Florida, Boca Raton, 1990.

85. Marttin, E.; Schipper, N.G.M.; Verhoef, J.C.; Merkus, F.W.H.M. Nasal mucociliary clearance as a factor in nasal drug delivery. Adv. Drug Deliv. Rev. **1998**, *29* (1), 13–38.

86. McInnes, F.J.; O'Mahony, B.; Lindsay, B.; Band, J.; Wilson, C.G.; Hodges, L.A.; Stevens, H.N. Nasal residence of insulin containing lyophilised nasal insert formulations, using gamma scintigraphy. Eur. J. Pharm. Sci. **2007**, *31* (1), 25–31.

87. Coucke, D.; Schotsaert, M.; Libert, C.; Spray-dried powders of starch and crosslinked poly(acrylic acid) as carriers for nasal delivery of inactivated influenza vaccine. Vaccine **2009**, *27* (8), 1279–1286.

88. Kharenko, E.A.; Larionova, N.I.; Demina, N.B. Mucoadhesive drug delivery systems (Review). Pharm. Chem. J. **2009**, *43* (4), 200–208.

89. Pavis, H.; Wilcock, A.; Edgecombe, J.; Carr, D.; Manderson, C.; Church, A.; Fisher, A. Pilot study of Nasal Morphine-Chitosan for the relief of breakthrough pain in patients with cancer. J. Pain Symp. Manage. **2002**, *24* (6), 598–602.

90. Luppi, B.; Bigucci, F.; Corace, G.; Delucca, A.; Cerchiara, T.; Sorrenti, M.; Zecchi, V. Albumin nanoparticles carrying

cyclodextrins for nasal delivery of the anti-Alzheimer drug tacrine. Eur. J. Pharm. Sci. **2011**, *44* (4), 559–565.

91. Saettone, M.F.; Burgalassi, S.; Chetoni, P. Ocular bioadhesive drug delivery systems. In *Bioadhesive Drug Delivery Systems*, Mathiowitz, E., Chickering, D.E. III, Lehr, C.-M., Eds.; Marcel Dekker, Inc: New York, 1999: 601–640.

92. Joshi, A. Microparticulates for ophthalmic drug delivery. J. Ocul. Pharmacol. **1994**, *10* (1), 29–45.

93. Sensoy, D.; Cevher, E.; Sarici, A.; Yılmaz, M.; Özdamar, A.; Bergişadi, N. Bioadhesive sulfacetamide sodium microspheres: Evaluation of their effectiveness in the treatment of bacterial keratitis caused by *Staphylococcus* aureus and *Pseudomonas aeruginosa* in a rabbit model. Eur. J. Pharm. Biopharm. **2009**, *72* (3), 487–495.

94. de la Fuente, M.; Seijo, B.; Alonso, M.J. Bioadhesive hyaluronan-chitosan nanoparticles can transport genes across the ocular mucosa and transfect ocular tissue. Gene. Ther. **2008**, *15* (9), 668–676.

95. Baeyens, V. Clinical evaluation of bioadhesive ophthalmic drug inserts (BODI®) for the treatment of external ocular infections in dogs. J. Control. Release **2002**, *85* (1), 163–168.

96. Batchelor, H. Formulation strategies in mucosal-adhesive drug delivery. CRS Newslett. **2005**, *22* (1), 4–5.

97. Robinson, J.R.; Bologna, W.J. Vaginal and reproductive system treatments using a bioadhesive polymer. J. Control. Release **1994**, *28* (1), 87–94.

98. O'Hagan, D.T.; Rafferty. D.; Wharton, S.; Illum, L. Intravaginal immunization in sheep using a bioadhesive microsphere antigen delivery system. Vaccine **1993**, *11* (6), 660–664.

99. Bonucci, E.; Ballanti, P.; Ramires, P.A.; Richardson, J.L.; Benedetti, L.M. Prevention of ovariectomy osteopenia in rats after vaginal administration of Hyaff 11 microspheres containing salmon calcitonin. Calcif. Tissue Int. **1995**, *56* (4), 274–279.

100. Alam, M.A.; Ahmad, F.J.; Khan, Z.I.; Khar, R.K.; Ali, M. Development and evaluation of acid-buffering bioadhesive vaginal tablet for mixed vaginal infections. AAPS PharmSciTech **2007**, *8* (4), E109.

101. Cevher, E.; Taha, M.A.M.; Orlu, M.; Araman, A. Evaluation of mechanical and mucoadhesive properties of clomiphene citrate gel formulations containing carbomers and their thiolated derivatives. Drug Deliv. **2008**, *15* (1), 57–67.

102. Available from: http://www.clinicaltrials.gov (Last accessed November 2011)

103. Francois, M.; Snoeckx, E.; Putteman, P.; A mucoadhesive, cyclodextrin-based vaginal cream formulation of itraconazole. AAPS PharmSci **2003**, *5* (1), E5.

104. Kim, J.K.; Kim, M.S.; Park, J.S.; Kim, C.K. Thermoreversible flurbiprofen liquid suppository with HP-beta-CD as a solubility enhancer: improvement of rectal bioavailability. J. Incl. Phenom. Macrocycl. Chem. **2009**, *64*, 265–272.

105. Woolfson, A.D.; McCafferty, D.F.; McCarron, P.A.; Price, J.H. A bioadhesive patch cervical drug delivery system for the administration of 5-fluorouracil to cervical tissue. J. Control. Release **1995**, *35* (1), 49–58.

106. Donnelly, R.F.; McCarron, P.A.; Zawislak, A.; Woolfson, A.D. Photodynamic therapy of vulval neoplasias and dysplasias. In *Design and Evaluation of Bioadhesive Photosensitiser Delivery Systems*, Verlag, M., Ed.; VDM Publication: Saarbrücken, 2009; 78–400.

107. Schumacher, U.; Schumacher, D. Functional histology of epithelia relevant for drug delivery. In *Bioadhesive Drug Delivery Systems*, Mathiowitz, E., Chickering, D.E. III, Lehr, C.-M., Eds.; Marcel Dekker, Inc: New York, 1999; 67–83.

108. Salman, H.H.; Gamazo, C.; Agüeros, M.; Irache, J.M. Bioadhesive capacity and immunoadjuvant properties of thiamine-coated nanoparticles. Vaccine **2007**, *25* (48), 8123–8132.

109. Naisbett, B.; Woodley, J. The potential use of tomato lectin for oral drug delivery. 1. Lectin binding to rat small intestine *in vitro*. Int. J. Pharm. **1994**, *107* (3), 223–230.

110. Nicholls, T. Lectins in ocular drug delivery: An investigation of lectin binding sites on the corneal and conjunctival surfaces. Int. J. Pharm. **1996**, *138* (2), 175–183.

111. Chung, K.H.; Wu, C.S.; Malawer, E.G. Glass transition temperatures of poly(methyl vinyl ether-co-maleic anhydride) (PMVEMA) and poly(methyl vinyl ether-co-maleic acid) (PMVEMAC) and the kinetics of dehydration of PMVEMAC by thermal analysis. J. Appl. Polym. Sci. **1990**, *41* (3–4), 793–803.

112. Raj Singh, T.R.; McCarron, P.A.; Woolfson, A.D.; Donnelly, R.F. Investigation of swelling and network parameters of poly(ethylene glycol)-crosslinked poly(methyl vinyl ether-co-maleic acid) hydrogels. Eur. Polym. J. **2009**, *45* (4), 1239–1249.

113. Singh, T.R.R.; McCarron, P.A.; Woolfson, A.D.; Donnelly, R.F. Physicochemical characterization of poly(ethylene glycol) plasticized poly(methyl vinyl ether-co-maleic acid) films. J. Appl. Polym. Sci. **2009**, *112* (1), 2792–2799.

114. Raj Singh, T.R.; Woolfson, A.D.; Donnelly, R.F. Investigation of solute permeation across hydrogels composed of poly(methyl vinyl ethercomaleic acid) and poly(ethylene glycol). J. Pharm. Pharmacol. **2010**, *62* (7), 829–837.

115. Donnelly, R.F.; Majithiya, R.; Singh, T.R.R.; Morrow, D.I.; Garland, M.J.; Demir, Y.K.; Woolfson, A.D. Design, optimization and characterisation of polymeric microneedle arrays prepared by a novel laser-based micromoulding technique. Pharm. Res. **2011**, *28* (1), 41–57.

116. Donnelly, R.F.; Garland, M.J.; Morrow, D.I.J.; Migalska, K.; Singh, T.R.R.; Majithiya, R.; Woolfson, A.D. Optical coherence tomography is a valuable tool in the study of the effects of microneedle geometry on skin penetration characteristics and in-skin dissolution. J. Control. Release **2010**, *147* (3), 333–341.

117. Migalska, K.; Morrow, D.I.J.; Garland, M.J.; Thakur, R.; Woolfson, A.D.; Donnelly, R.F Laser-engineered dissolving microneedle arrays for transdermal macromolecular drug delivery. Pharm. Res. **2011**, *28* (8), 1919–1930.

Molecular—Nanogels

Muscles, Artificial: Sensing, Transduction, Feedback Control, and Robotic Applications

Mohsen Shahinpoor
Artificial Muscle Research Institute (AMRI), School of Engineering and School of Medicine, University of New Mexico, Albuquerque, New Mexico, U.S.A.

Kwang J. Kim
College of Engineering, University of Nevada–Reno, Reno, Nevada, U.S.A.

Mehran Mojarrad
Pharmaceutical Delivery Systems, Eli Lilly and Co., Indianapolis, Indiana, U.S.A.

Abstract

This entry covers sensing, transduction, feedback control, and robotic actuation capabilities and issues related to these topics. Ionic polymer–metal nanocomposites (IPMNCs) and ionic polymer conductor nanocomposites (IPCNCs) are amazing tools for soft robotic actuation and built-in sensing and transduction in a distributed manner. One can even see the distributed biomimetic, noiseless nanosensing, nanotransduction, and nanoactuation capabilities of IPMNCs and IPCNCs.

INTRODUCTION

This entry covers sensing, transduction, feedback control, and robotic actuation capabilities and issues related to these topics. Ionic polymer-metal nanocomposites (IPMNCs) and ionic polymer conductor nanocomposites (IPCNCs) are amazing tools for soft robotic actuation and built-in sensing and transduction in a distributed manner. One can even see the distributed biomimetic, noiseless nanosensing, nanotransduction, and nanoactuation capabilities of IPMNCs and IPCNCs.

SENSING CAPABILITIES OF IPMNCS

This section presents a brief description and testing results of ionic polymer-metal composites (IPMNCs) as dynamic sensors. As previously noted, a strip of IPMNC can exhibit large dynamic deformation if placed in a time-varying electric field. Conversely, dynamic deformation of such ionic polymers produces dynamic electric fields. The underlying principle of such a mechanoelectric effect in IPMNCs can be explained by the linear irreversible thermodynamics in which ion and solvent transport are the fluxes and electric field and solvent pressure gradient are the forces. Important parameters include the material capacitance, conductance, and stiffness, which are related to material permeability.

The dynamic sensing response of a strip of IPMNC under an impact type of loading is also discussed. A damped electric response is observed that is highly repeatable with a broad bandwidth to megahertz frequencies. Such direct mechanoelectric behaviors are related to the endo-ionic mobility due to imposed stresses. This means that, if one imposes a finite solvent flux without allowing a current flux, the material creates a certain conjugate electric field that can be dynamically monitored. IPMNCs are observed to be highly capacitive at low frequencies and highly resistive under high-frequency excitations. Current efforts are to study the low- and the high-frequency responses and sensitivity of IPMNCs that might conceivably replace piezoresistive and piezoelectric sensors with just one sensor for broad frequency range sensing and transduction capabilities.

Basics of Sensing and Transduction of IPMNCs and IPCNCs

It is so far established that ionic polymers (such as a perfluorinated sulfonic acid polymer, i.e., Nafion™) in a composite form with a conductive metallic medium (here called IPMNCs) can exhibit large dynamic deformation if placed in a time-varying electric field (see Fig. 1). Conversely, dynamic deformation of such ionic polymers produces dynamic electric fields (see Fig. 2). A recently presented model by de Gennes et al.[1] presents a plausible description of the underlying principle of electrothermodynamics in ionic polymers based on internal ion and solvent transport and electrophoresis. It is evident that IPMNCs show great potential as dynamic sensors, soft robotic actuators, and artificial muscles in a broad size range of nano- to micro- to macroscales.

A recent study by de Gennes and coworkers[1] has presented the standard Onsager formulation on the underlining principle of IPMNC actuation/sensing phenomena using linear irreversible thermodynamics: When static conditions are imposed, a simple description of *mechanoelectric effect* is possible based upon two forms of transport: *ion transport* (with a current density, J, normal to the material) and *electrophoretic solvent transport*. (With a flux Q, one can assume that this term is the water flux.) The conjugate forces include the electric field $\vec{E}$ and the pressure gradient $-\nabla p$. The resulting equation has the concise form of

$$J = \sigma \vec{E} - L_{12} \nabla p \qquad (1)$$

$$Q = L_{21} \vec{E} - K \nabla p \qquad (2)$$

where σ and K are the membrane conductance and the Darcy permeability, respectively. A cross-coefficient is usually $L_{12} = L_{21} = L$, estimated to be on the order of 10^{-8} $(ms^{-1})/(volt\text{-}meters^{-1})$.[2,3] The simplicity of the preceding equations provides a compact view of underlining principles of actuation and sensing of IPMNCs.

Fig. 1 Successive photographs of an IPMNC strip that shows very large deformation (up to 4 cm) in the presence of low voltage. The sample is 1 cm wide, 4 cm long, and 0.2 mm thick. The time interval is 1 sec. The actuation voltage is 2 V DC.

Figure 2 shows dynamic sensing response of a strip of an IPMNC (thickness of 0.2 mm) subject to a dynamic impact loading in a cantilever configuration. A damped electric response is observed that is highly repeatable with a high bandwidth of up to 100 Hz. Such direct mechanoelectric behaviors are related to the endo-ionic mobility due to imposed stresses. This implies that, if we impose a finite solvent (= water) flux, $|Q|$—not allowing a current flux, $J = 0$—a certain conjugate electric field $\vec{E}$ is produced that can be dynamically monitored.

From Eqs. 1 and 2, one imposes a finite solvent flux Q while having zero current ($J = 0$) and nonzero bending curvature. This situation certainly creates an intrinsic electric field $\vec{E}$, which has a form of

$$\vec{E} = \frac{L}{\sigma} \nabla p = \frac{12(1-v_p)}{(1-2v_p)} \left\{ \frac{L}{\sigma h^3} \right\} \Gamma \qquad (3)$$

Note that notations v_p, h, and Γ are, respectively, the Poisson ratio, the strip thickness, and an imposed torque at the built-in end produced by a force F applied to the free end multiplied by the free length of the strip l_g.

Electrical Properties

In order to assess the electrical properties of the IPMNC, the standard AC impedance method that can reveal the equivalent electric circuit has been adopted. A typical measured impedance plot, provided in Fig. 3, shows the frequency dependency of impedance of the IPMNC.

Overall, it is interesting to note that the IPMNC is nearly resistive (>50 Ω) in the high-frequency range and fairly capacitive (>100 μF) in the low-frequency range. IPMNCs generally have a surface resistance, R_{ss}, of about a few

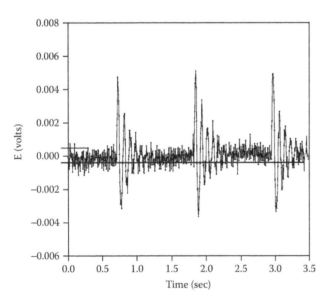

Fig. 2 A typical sensing response of an IPMNC. The IPMNC (5 × 20 × 0.2 mm) in a cantilever mode as depicted in Fig. 1 is connected to an oscilloscope and is manually flipped to vibrate and come to rest by vibrational damping.

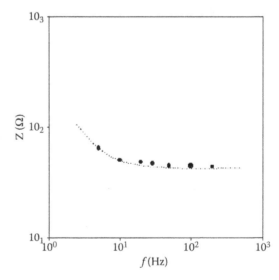

Fig. 3 The measured AC impedance characteristics of an IPMNC sample (dimension = 5-mm width, 20-mm length, and 0.2-mm thickness).

ohms per centimeter, near-boundary resistance, R_s, of a few tens of ohms per centimeter, and cross-resistance, R_P of a few hundreds of ohms per millimeter; typical cross capacitance, C_g, is a few hundreds of microfarads per millimeter.

Based upon these findings, we consider a simplified equivalent electric circuit of the typical IPMNC such as the one shown in Fig. 4.[1] In this approach, each single unit circuit (i) is assumed to be connected in a series of arbitrary surface resistance (R_{ss}) on the surface.

This approach is based upon the experimental observation of the considerable surface electrode resistance (see Fig. 4). We assume that there are four components to each single unit circuit: the surface electrode resistance (R_s), the polymer resistance (R_p), the capacitance related to the ionic polymer and the double layer at the surface–electrode/electrolyte interface (C_d), and an intricate impedance (Z_w) due to a charge transfer resistance near the surface electrode. For the typical IPMNC, the importance of R_{ss} relative to R_s may be interpreted from $\Sigma R_{ss}/R_s \approx L/t \gg 1$, where notations L and t are the length and thickness of the electrode. Note that the problem now becomes a two-dimensional one; the fact that the typical value of t is ~1–10 μm makes the previous assumption.

Thus, a significant overpotential is required to maintain the effective voltage condition along the surface of the typical IPMNC. An effective technique to solve this problem is to overlay a thin layer of a highly conductive metal (such as gold) on top of the platinum surface electrode.[1]

Note that Fig. 4 depicts a digitized rendition of the equivalent circuit for IPMNC, which is a continuous material. Figure 5 depicts the measured surface resistance, R_s, of a typical IPMNC strip as a function of platinum particle penetration depth. Note that SEM was used to estimate the penetration depth of platinum into the membrane.

The four-probe method was used to measure the surface resistance, R_s, of the IPMNCs. Obviously, the deeper the

penetration of metallic particles is, the lower the surface resistance is.

Figure 6 depicts measured chronoamperometry responses of a typical IPMNC sample in which the current response is recorded after a step potential is applied. It should be noted that the important physical phenomena occurring in the vicinity of the electrodes and the associated processes, particularly within a few micron depth, play an important role in the sensing characteristics of IPMNCs.

Mass transfer of cations and their hydrated water molecules involved as they move into and out of the bulk material and through the porous electrodes is another important feature of the sensor. Diffusion can be considered as the sole transport process of electroactive species (cations and hydrated water). The plausible treatment of diffusion is to use the Cottrell equation having a form of

$$i(t) = \frac{nFAD^{1/2}C}{(\pi t)^{1/2}} = Kt^{1/2} \tag{4}$$

where $i(t)$, n, A, D, C, and F are the current at time t and the number of electrons involved in the process, the surface area, diffusivity, concentration, and Faraday constant, respectively. Eq. 4 states that the current is inversely proportional to the square root of time.

Figure 6 shows the overall current versus the square root of time. Also, it states that the product $i(t) \times t$ should be a constant K for a diffusional electrode. Deviation from this constancy could be caused by a number of factors, including slow capacitive charging of the electrode during the step voltage input and coupled chemical reactions (hydrolysis). This figure is constructed under a step potential of –3 V for a typical IPMNC. As can be seen, the characteristics clearly follow the simple Cottrell equation

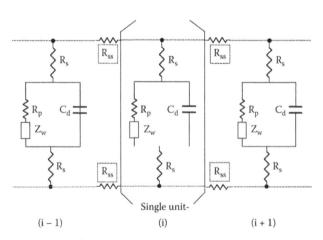

Fig. 4 An equivalent electronic circuit for a typical IPMNC strip obtained by an impedance analyzer.

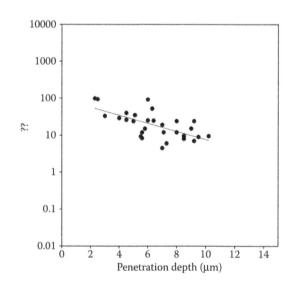

Fig. 5 Measured surface resistance, R_s, of a typical IPMNC strip, as a function of platinum particle penetration depth.

that confirms the fact that the electrochemical process is diffusion controlled.

A more recent equivalent circuit proposed by Paquette and Kim[4] is shown in Fig. 7. The two loops that include R1, C1, R3, and C2 represent the two composited effective electrodes of the IPMNC. R2 represents the effective resistance of the polymer matrix. The value of E is the electric field applied across the material for actuation: $E = V/h$, where V is the voltage applied and h is the membrane thickness.

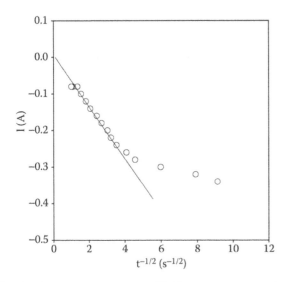

Fig. 6 The current, $i(t)$, versus $t^{-1/2}$ (chronoamperometry data). Note that $A = 6.45$ cm^2 and $E = -3$ V.

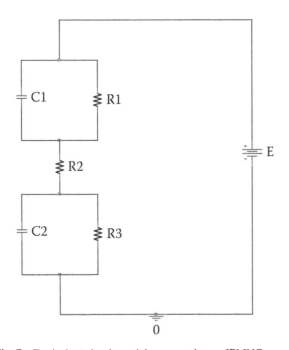

Fig. 7 Equivalent circuit model representing an IPMNC.

The effective capacitance values are C1 = C2 and R1 = R3. Upon examination of the proposed circuit, the current for the equivalent circuit is

$$i(t) = V_A \cdot \left[\frac{1}{(R_2 + 2R_1)} + 2 \cdot \frac{R_1}{(R_2 - 2R_1)} \right.$$

$$\left. \cdot \frac{\exp\left[-(R_2 + 2 \cdot R_1) \cdot \dfrac{t}{R_2 \cdot R_1 \cdot C_1} \right]}{R_2} \right] \tag{5}$$

where V_A is the applied step voltage (at time $t \geq 0$ sec). The resulting voltage across either capacitive loop is

$$V_C(t) = \frac{V_A - V_A \cdot R_2 \cdot \left[\dfrac{1}{(R_2 + 2 \cdot R_1)} + 2 \cdot \dfrac{R_1}{(R_2 + 2 \cdot R_1)} \cdot \dfrac{\exp\left[-(R_2 + 2 \cdot R_1) \cdot \dfrac{t}{R_2 \cdot R_1 \cdot C_1} \right]}{R_2} \right]}{2} \tag{6}$$

Experiment and Discussion

As discussed extensively before and to refresh the reader's memory, the manufacturing of an IPMNC starts with an ionic polymer subjected to a chemical transformation (REDOX), which creates a functionally graded composite of the ionic polymer with a conductive phase. *Ionic polymeric material* selectively passes through ions of a single charge (cations or anions). They are often manufactured from polymers that consist of fixed covalent ionic groups. The currently available ionic polymers are

1. perfluorinated alkenes with short side-chains terminated by ionic groups (typically sulfonate or carboxylate [SO_3^- or COO^-] for cation exchange)
2. styrene/divinylbenzene-based polymers in which the ionic groups have been substituted from the phenyl rings where the nitrogen atoms are fixed to ionic groups

Figure 8 is an SEM micrograph showing the cross section of a typical IPMNC used in this study. In Fig. 9, a preliminary quasistatic DC sensing data is provided in terms of the voltage produced at different displacement. Note that the displacement is shown in terms of the deformed angle relative to standing position in degree. The dimension of the IPMNC sample sensor is $5 \times 25 \times 0.12$ mm. Such direct mechanoelectric effect is convenient in that the produced voltage is large and applicable displacement is large.

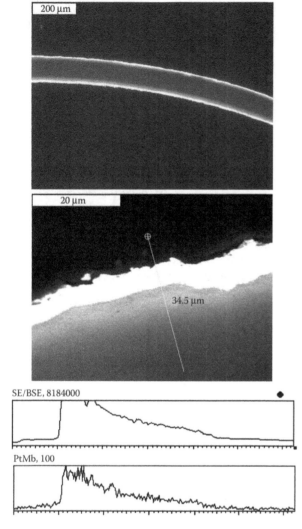

Fig. 8   An SEM micrograph shows the cross-section of an IPMNC sensor. It depicts a cross-section (top) of an IPMNC strip, its close-up (middle), and the x-ray line scan (bottom).

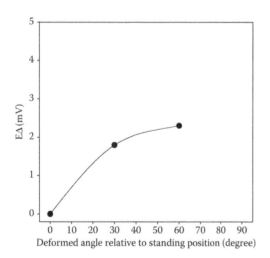

Fig. 9   DC sensing data in terms of produced voltages, ΔE, versus displacement. Note that the displacement is shown in terms of the deformed angle relative to standing position in degree. The dimension of the sample sensor is $5 \times 25 \times 0.12$ mm.

Comparing such unique features of IPMNCs as sensing devices relative to other current state-of-the-art sensing technologies such as piezoresistive or piezoelectric devices, one can find more flexibility in connection with IPMNCs.

The slow current leakage due to redistribution of ions is often observable. Additional investigations of the current leakage are necessary to stabilize the voltage output of the IPMNCs in a sensing mode.

EVALUATION OF IPMNCS FOR USE AS NEAR-DC MECHANICAL SENSORS

Introduction

Henderson and coworkers[5] and Shahinpoor et al.[6] offer information on using IPMNCs as near-DC mechanical sensors. IPMNC active elements enable near-DC acceleration measurement devices with modest power, volume, mass, and complexity requirements, provided their unique properties are accounted for in the design. Advantages over conventional piezoelectric elements are documented for some applications.

Acceleration measurements are necessary for various dynamics experiments and often serve as sensing inputs in structural control systems. In conventional practice, a piezoelectric element, as part of a single degree-of-freedom harmonic system, is often used to sense acceleration. Figures 10A and B are representations of two approaches. Figure 10A depicts the most basic embodiment, a piezoelectric ceramic element, such as lead zirconium titanate (PZT), in line with a mass. When the base is accelerated vertically, the inertia of the mass causes strain in the active element that, when electroded, generates a charge proportional to the strain. Damping of the first-order system may be tuned to provide a nearly flat mechanical response over a desirable range below the system resonance frequency. Since piezoceramic materials are brittle and cannot support significant tension loads, a mechanical or electrical preload must be applied in this embodiment to avoid such a situation.

Figure 10B shows a slightly more common approach for an acceleration sensor. A mass is suspended from three sides

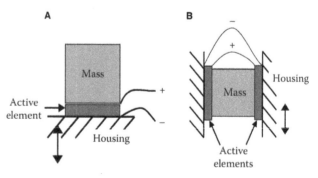

Fig. 10   Accelerometer implementations using PZT in (A) 3–3 mode and (B) 3–1 mode.

by piezoelectric elements that are affixed to the boundary of the accelerometer enclosure. In this case, as the enclosure is vibrated, the active elements undergo a shear strain, which also generates a proportional voltage. The piezoceramic material can support shear loads in either direction, so in this embodiment no vertical preload is necessary.

In either of the conventional embodiments with realistic active element sizes, low-frequency response (<1 Hz) is generally poor without the use of additional electronics. Piezoelectric elements are largely capacitive and, when connected directly to an oscilloscope ($\approx$1 MΩ input impedance), charge tends to bleed off quickly at low frequencies in the effective RC circuit. Voltage followers or charge amplifiers are usually included in the circuit to lower this frequency and eliminate measurement sensitivity to cable noise and environmental parameters. However, low-frequency performance comes at the expense of size, weight, and complexity. Furthermore, useful performance at frequencies below 1 Hz is extremely difficult to achieve. For this reason, piezoelectric accelerometers make a poor choice for near-DC acceleration measurements. Here, we investigate the use of a fairly new material, IPMNC, for use as an active element in near-DC accelerometers.

Background

IPMNCs consisting of a thin Nafion-117 sheet plated with gold on both sides have received much attention for their possible applications to sensing and actuation.[2] The materials work via internal ionic transport phenomena, requiring water internal to the system and providing a coupling between electrical field and bending deformation. Not only do such materials exhibit relatively large deformations with a mild voltage input, but they also have a significant electrical response to bending deformation, even at sub-hertz frequencies. The long response time of IPMNCs—on the order of 4–10 sec—limits their use in applications requiring a short response time. Although this limitation may be mitigated to some extent through the use of feed-back control, it becomes an advantage when considering the material for low-frequency applications. The useful-ness of IPMNC material in a near-DC accelerometer appli-cation, as well as potential problems and design considerations, is examined here.

Experiment Setup

The complex impedance of the IPMNC was measured by the well-known voltage divider method. The impedance was measured over a range from 0.05 to 5000 Hz, using a 15-kΩ resistor in series and a Siglab data acquisition system to record the magnitude and phase of the voltages.

For the dynamic experiments, a Ling 5-lb shaker was mounted on an optical table in the vertical orientation with the stinger pointing down. A machined aluminum block was mounted on the end of the stinger to provide a flat

movable surface. An eddy current probe was mounted such that the head was held beneath the block, near one edge, with a 0.25-mm standoff distance. The probe was cali-brated to provide 1.2-mm/V displacement response near an aluminum surface. For the control case, a piezoelectric patch (PZT-5A with nickel electrodes, $11 \times 29 \times 0.26$ mm, 0.646-g mass, 20-nF capacitance) was used as the active element. In the experiment case, an IPMNC patch (with gold electrodes, 0.237-g mass) was used as the active element and cut to the same planar dimensions as the piezoelectric patch, but with 0.32-mm thickness.

For the control case, the mounting apparatus consisted of an aluminum-base block with a piezoelectric patch (PZT) sandwiched between the base block and the shaker block. The shaker and block were lowered onto the optical table mounting post to generate an initial compressive preload in the patch and to ensure that it remained in compression over the entire range of motion applied to the shaker. Both blocks were covered with Kapton tape where they contacted the piezoelectric element to prevent current bleed-off into the optical table ground. Copper shims were cut and affixed to the top and bottom of the piezoelectric patch with conduc-tive grease to provide external leads. When the shaker was actuated, the piezoelectric element was compressed in the 3–3 direction, generating a voltage between the copper leads. The leads were connected via BNC cable to an input channel of a HP digital signal analyzer (DSA 35665A) with 1-MΩ input impedance. No conditioning circuitry was applied. The DSA source channel was used to actuate the shaker with a 5-V peak-to-peak sine sweep input over a fre-quency range from 0.015 to 30 Hz. The frequency response function from eddy current displacement to piezoelectric element voltage output was measured over this frequency range and converted to a volts/strain spectrum.

The IPMNC mounting apparatus consisted of a Plexiglas cantilevered clamping device with conductive contact points and wire leads. Again, the leads were connected by BNC cable to the DSA input, without any conditioning circuitry. The apparatus was positioned under the shaker block with an initial static displacement such that the tip of the cantile-vered IPMNC patch was deflected by the actuation of the shaker over its entire range of motion. The block was cov-ered with Kapton tape at the contact point to avoid bleed-off of the current into the optical table during the experiment. The response of the IPMNC in cantilevered mode to a sine sweep of the shaker was measured over a frequency range of 0.015 to 50 Hz. The frequency response spectrum was calculated in terms of voltage output versus end deflection.

Experiment Results

Any conditioning circuitry design for an accelerometer will require an accurate understanding of the complex imped-ance of the active element over the frequency range of appli-cation. Hence, this spectrum was measured for the IPMNC. Figures 11 and 12 represent the complex impedance of the sample. This IPMNC patch had been left to dry in the open atmosphere for approximately one month before these data were taken. Note that the element is fairly capacitive at low frequencies, but is almost entirely resistive above 100 Hz. Despite the simple appearance of the curve, this impedance response cannot be modeled simply as a three-element resistor and capacitor circuit, but rather requires a series solution.

Figures 13 and 14 depict the impedance curves for similarly sized IPMNC samples immediately after having been removed from their water-filled storage bags. Note that the magnitudes between wet and dry samples differ by almost three orders of magnitude. Moreover, Shahinpoor and Kim[3] measured an impedance magnitude spectrum (not shown) that lies between these two extremes. This surprising variability underscores the importance of accurately knowing or controlling the moisture state of the active element before IPMNC materials may be used effectively in practice.

Figure 15 is the frequency response magnitude of the piezoelectric patch in 3–3 compression. Note that the break frequency (the point at which the response is 0.707 of the maximum value) occurs at approximately 10 Hz. This corresponds closely to the theoretical value of 8 Hz for a pure RC circuit ($\omega = 1/RC$), based on the measured capacitance of the element (20 nF) and the input impedance of the DSA (1 MΩ).

In conventional practice, the useful range of a piezoelectric element for an accelerometer is considered to include only the region where the response deviates from the norm by no more than 5%. This further limits the usefulness of the PZT for low-frequency accelerometer applications. In a real implementation, a voltage follower or charge amplifier circuit would be used to lower this break frequency. However,

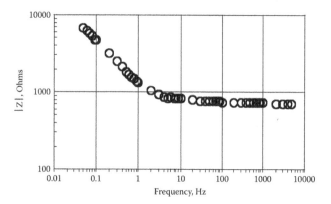

Fig. 11 Dry IPMNC impedance magnitude.

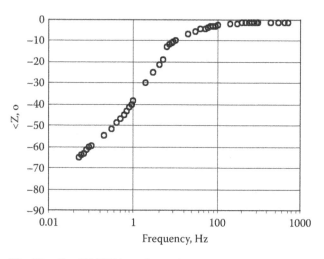

Fig. 12 Dry IPMNC impedance phase.

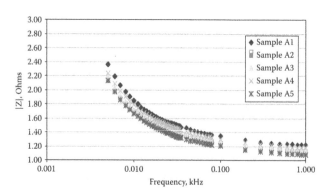

Fig. 13 Wet IPMNC impedance magnitude.

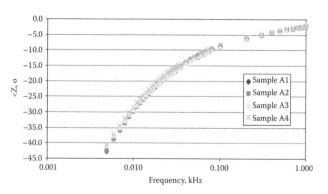

Fig. 14 Wet IPMNC impedance phase.

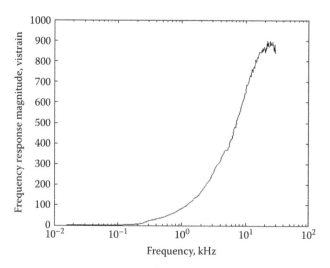

Fig. 15 PZT frequency response magnitude.

this requires increased mass, volume, and complexity, and there are practical limits to how low this frequency may be set. In general, subhertz conditioning with a piezoelectric element is rare and difficult to achieve.

Figure 16 is the frequency response magnitude of the IPMNC patch under cantilever excitation.

The voltage magnitude of the IPMNC output is much lower than that of the piezoelectric element, although a fair comparison cannot be made from these data since they were actuated in different modes. However, note that the break frequency for the IPMNC cantilevered beam occurs at approximately 0.03 Hz. This is significantly better than the 10-Hz break frequency for the comparably sized piezoelectric patch for low-frequency accelerometer applications. Recall that this low break frequency was achieved without any conditioning electronics. However, the IPMNC response begins to roll off above 40 Hz. This high-frequency limit, which depends on the geometry of the patch, must be taken into consideration when designing IPMNC accelerometers for specific applications.

Discussion and Conclusions

Based on the frequency response spectrum of an IPMNC patch under cantilever excitation, it seems likely that this material would be useful in low-frequency accelerometer applications. Unlike piezoelectric elements, the IPMNC specimen tested produced a useful output in the subhertz frequency range without any conditioning electronics. Care must be taken when designing circuits utilizing the material as an active element due to their extreme sensitivity to moisture. In response to this concern, Shahinpoor has developed a polymer-coated IPMNC patch that severely retards moisture evaporation, allowing for consistently performing devices. However, the moisture-retaining performance of such specimens must be quantified in detail before device shelf-life estimations may be made.

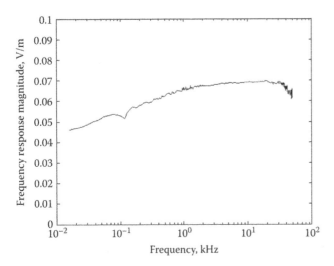

Fig. 16   IPMNC frequency response magnitude.

The nature of the material requires that accelerometers utilizing IPMNC active elements must include a mechanism to induce bending in the material under acceleration. This mechanism may take the form of a cantilever in the manner of the experiments described here, a drumhead type membrane with a mass in the center such as that discussed by Sadeghipour, Salomon, and Neogi[7] or even a low-profile flat spring.

Overall, IPMNC active elements enable near-DC acceleration measurement devices with modest power, volume, mass, and complexity requirements, provided their unique properties are accounted for in the design. The advantages of IPMNC over conventional piezoelectric approaches open up the possibility of thin-film or extremely lightweight accelerometers or, if integrated, devices with rough position-sensing capability.

Advances in Sensing and Transduction

A recent paper[8] presents a review on sensing and transduction properties of ionic polymer conductor nanocomposites. Shahinpoor[9,10] reported that, by themselves and not in hydrogen pressure electrochemical cells, as reported by Sadeghipour and coworkers,[7] IPMNCs can generate electrical power like an electromechanical battery if flexed, bent, or squeezed. He also reported the discovery of a new effect in ionic polymeric gels—namely, the *ionic flexogelectric* effect in which flexing, compression, or loading of IPMNC strips in air created an output voltage like a dynamic sensor or a transducer converting mechanical energy to electrical energy. Keshavarzi and colleagues[11] applied the transduction capability of IPMNCs to the measurement of blood pressure, pulse rate, and rhythm measurement using thin sheets of IPMNCs.

Motivated by the idea of measuring pressure in the human spine, Ferrara et al.[12] applied pressure across the thickness of an IPMNC strip while measuring the output voltage. Typically, flexing of such material in a cantilever form sets them into a damped vibration mode that can generate a similar damped signal in the form of electrical power (voltage or current) as shown in Fig. 3. The experimental results for mechanoelectrical voltage generation of IPMNCs in a flexing mode are shown in Figs. 17A and B. Figure 17A also depicts the current output for a sample of thin sheets of IPMNCs. Figure 16B depicts the power output corresponding to the data presented in Fig. 17A.

The experimental results showed that almost a linear relationship exists between the voltage output and the imposed displacement of the tip of the IMPC sensor (Fig. 17). IPMNC sheets can also generate power under normal pressure. Thin sheets of IPMNC were stacked and subjected to normal pressure and normal impacts and were observed to generate large output voltage. Endo-ionic motion within IPMNC thin sheet batteries produced an induced voltage across the thickness of these sheets when a normal or shear load was applied.

A material testing system (MTS) was used to apply consistent pure compressive loads of 200 and 350 N across the surface of an IPMNC 2- × 2-cm sheet. The output pressure response for the 200-N load (73 psi) was 80 mV in amplitude; for the 350-N load (127 psi), it was 108 mV. This type of power generation may be useful in the heels of boots and shoes or places where there is a lot of foot or car traffic. Figure 18 depicts the output voltage of the thin sheet IPMNC batteries under 200-N normal load. The output voltage is generally about 2 mV/cm length of the IPMNC sheet.

SIMULATION AND CONTROL OF IONOELASTIC BEAM DYNAMIC DEFLECTION MODEL

An effort to model the dynamic motion of an IPMNC elastic beam was undertaken. Development of the static portion of the model was begun by assuming the beam behaves in accordance with the nonlinear equation used to describe large-angle deflection of elastic cantilever beams.

A Simulink simulation was developed to estimate the final deflection of a beam due to a constant moment. The dynamic portion of the model was developed by assuming that each segment of the beam could be represented as a simple second-order system. Future efforts to model beam motion more accurately by modifying the method used to model the forcing moment, by expanding the model to predict the performance of beams of all dimensions, and by validating model performance against actual beam motion were recommended.

Introduction

This section presents a summary of the effort to model the deflection dynamics of iono-elastic beams made of IPMNCs. Strips of these composites can undergo large bending and flapping displacement if an electric field is imposed across their thickness. IPMNC beams show large deformation in the presence of low applied voltage and exhibit low impedance. They have been modeled as capacitive and resistive element actuators that behave like biological muscles and provide an attractive means of actuation as artificial muscles for biomechanic and biomimetic applications. Essentially, the polyelectrolyte membrane inside the composite possesses ionizable groups on its molecular backbone. These groups have the property of disassociating and attaining a net charge in a variety of solvent media. In particular, if the interstitial space of a polyelectrolyte network is filled with liquid containing ions, then the electrophoretic migration of those ions inside the structure due to an imposed electric field can cause the macromolecular network to deform accordingly.

Static Deflection

The ionoelastic beam is depicted as an elastic cantilever beam, shown in Fig. 19. The nonlinear equation for large-angle deflections in elastic cantilever beams is

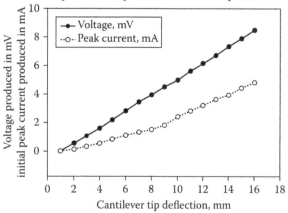

A Cantilever sheet battery (1 cm × 3 cm × 0.3 mm) voltage and peak current produced for various tip deflections

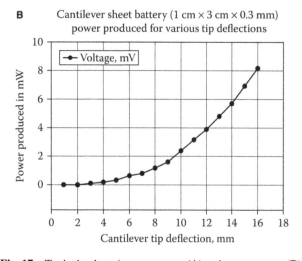

B Cantilever sheet battery (1 cm × 3 cm × 0.3 mm) power produced for various tip deflections

Fig. 17 Typical voltage/current output (**A**) and power output (**B**) of IPMNC samples.

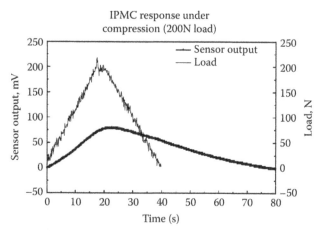

Fig. 18 Outvoltage due to normal impact of 200-N load on a 2-cm × 2-cm × 0.2-mm IPMNC sample.

$$\frac{\frac{\partial^2}{\partial x^2}(v)}{\left(1+\left(\frac{\partial v}{\partial x}\right)^2\right)^{3/2}} = -\frac{M}{EI} \quad (7)$$

Equation 7 can be rearranged algebraically to

$$\frac{\partial^2}{\partial x^2}(v) = -\frac{M}{EI}\left(1+\left(\frac{\partial v}{\partial x}\right)^2\right)^{3/2} \quad (8)$$

Solving Eq. 7 for a constant moment, M, will produce a function, $v(x)$, that is the beam deflection of each point on the beam as a function of the distance from the wall. Determining the solution for Eq. 7 subject to a constant moment, M, can be accomplished in five steps:

1. Change the independent variable (temporarily) from x to t:

$$\frac{\partial^2}{\partial t^2}(v) = -\frac{M}{EI}\left(1+\left(\frac{\partial v}{\partial t}\right)^2\right)^{3/2} \quad (9)$$

This can be rewritten as

$$v'' = -\left(M/EI\right)\left[1+\left(v'\right)^2\right]^{3/2} \quad (10)$$

2. Change the second-order differential Eq. 10 into a set of first-order differential equations:

$$\text{set } x_1 = v \quad (11)$$

$$\text{then } x_1' = v' \quad (12)$$

$$\text{set } x_2 = x_1' = v' \quad (13)$$

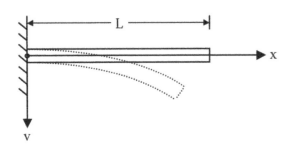

Fig. 19   IPMNC cantilever beam.

then

$$x_2' = x_1'' = v'' = -\left(M/EI\right)\left[1+\left(v'\right)^2\right]^{3/2} \quad (14)$$

or

$$v'' = -\left(M/EI\right)\left[1+\left(x_2\right)^2\right]^{3/2} \quad (15)$$

3. Model the set of first-order differential equations using MATLAB's® Simulink tool, as shown in Fig. 20.
4. Integrate the Simulink model. Simulink defaults to a fourth-order Runge–Kutta integration method, which was used for this effort. The time step used during integration was constrained to be constant, since this would correspond (after the transformation described in step 5) to a fixed-length fraction of the beam.
5. Change the independent variable back to x from t:

$$v(x) = v(t) \quad (16)$$

A hypothetical deflection case, subject to the following boundary conditions, was run through the five-step modeling process and resulted in the beam deflection shown in Fig. 21:

$$M = 0.5v\left(t_0\right) = 0 \quad (17a)$$

$$E = 1v'\left(t_0\right) = 0 \quad (17b)$$

$$I = 1v''\left(t_0\right) = 0 \quad (17c)$$

$$L = 1dx = 0.01 \quad (17d)$$

One of the implications of solving Eq. 7 is the "stretching" of the beam. As depicted in Fig. 22, the beam tip will deflect a distance, v, and the resulting curved beam will be longer by a small distance than it originally was (though no extension force has been placed on the beam).

To rectify this discrepancy, the straight-line displacement of each portion of the beam is summed together until the length of the curved beam equals that of the original beam, as shown in Fig. 23. Once that length is reached, the rest of the beam is not included in the deflection plot. As a result of this beam "reshortening," the deflection of the hypothetical beam shown in Fig. 23 will eventually look as shown in Fig. 24.

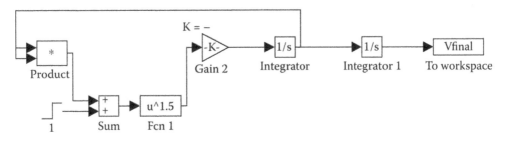

Fig. 20   Simulink model.

For the static case, there is a distinct final deflection solution for each moment value. The plot in Fig. 25 shows the distinct solutions for various moment values.

Dynamic Case

The previous section describes the initial and final positions of the beam, but does not describe the dynamics it undergoes to reach the final position. This section will describe the effort to build the dynamic model.

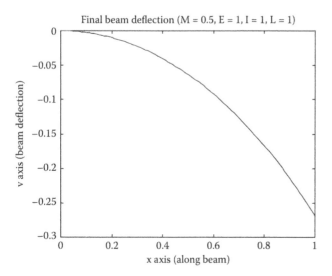

Fig. 21   Hypothetical deflection plot.

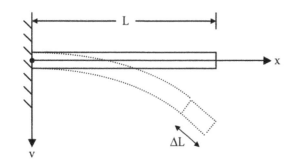

Fig. 22   Beam "lengthening."

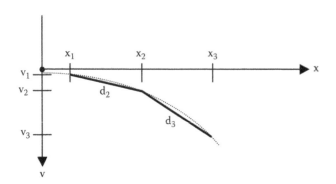

Fig. 23   Segmented beam.

For a step input of 2 V on a 1- × 0.25 in.-strip (0.2-mm thick), the moment will be constant; such a constant moment will produce a distinct final deflection. The step response of the tip to such a deflection command is shown by Shahinpoor.[13] From that step response, it would appear that the beam could be modeled as a simple second-order transfer function.

The step response of a second-order transfer function of the following form will approximate the step response if $\omega = 0.364$ Hz and $\zeta = 0.3$:

$$G(s) = \omega^2 / s^2 + 2\zeta\omega s + \omega^2 \tag{18}$$

The step response of this model is shown in Fig. 26.

The approximate transfer function's step response does correspond to that presented by Shahinpoor & Alvarez.[14] The only significant discrepancy is in the maximum value during the initial portion of the step response. Shahinpoor[15] report the maximum value at 0.170; the approximate solution reports the maximum value at 0.146.

The frequency domain parameters that characterize the model's open-loop transfer function should be assumed to

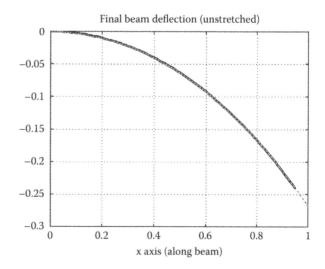

Fig. 24   Unstretched hypothetical deflection plot.

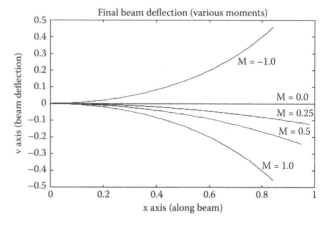

Fig. 25   Deflections due to various moments.

apply to only the 1- × 0.25-in. beam. The natural frequency and damping coefficient of beams with other lengths, widths, and thickness will undoubtedly be very different. The function that relates those frequency domain parameters to beam dimensions was not researched during this modeling effort; an effort to determine this function must be undertaken before model development can be considered complete.

Once the shape of the open-loop transfer function of the tip was determined, a rather significant assumption was made. For the purposes of model development, it was assumed that each point on the beam responded like the tip; that is, the beam was actually a set of second-order transfer functions responding independently to their own final deflection commands. Given that assumption, the process to simulate the deflection of the entire beam dynamically is detailed graphically in Fig. 27.

In this graphic, the modeling input is the moment value (assumed constant for this portion of the effort). The moment

is used in the solution of the nonlinear large-angle deflection equation; the final deflection position is the solution to that equation and is in the form of a single vector (with individual elements for each portion of the curved beam). Each of the final deflection vector elements is used as a step command input into its own second-order open-loop transfer function. The outputs of the independent transfer functions are the individual deflection time responses (one for each portion of the curved beam). Those outputs are collected into a deflection time response matrix; each column of that matrix represents the beam deflection at a certain simulation time.

Figure 28 depicts the beam deflection as a function of time for the hypothetical case initially described in support of Fig. 21. The darker line in the figure represents the final deflection vector command (solution to the nonlinear large-angle deflection equation).

Variable Moments

The previous sections have all constrained the input moment to a constant value. The model developed, however, will work equally well under variable moment conditions. To model the beam deflection properly under variable moment conditions, the moment at each time step is used to create a final deflection command vector at each time step. That variable command vector replaces the constant step command used as an input to the open-loop transfer functions.

In order to develop the variable moment deflection plot shown in Fig. 29, a square wave—switching from +0.5 to −0.5 and back every second for 10 sec—was used as the moment input. The resultant motion is the expected "flapping" motion seen during hardware tests.

The beam model presented in the last sections is only an initial model. In order to upgrade the fidelity of this model, additional work in three areas must be performed as outlined next.

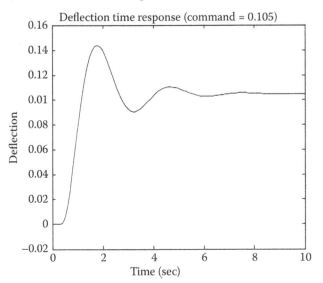

Fig. 26 Beam tip step response.

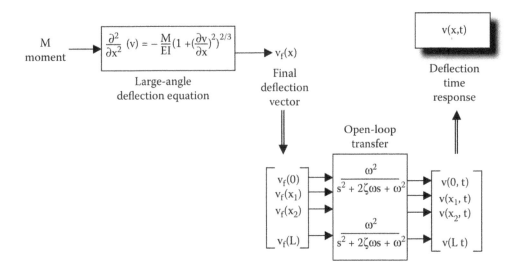

Fig. 27 Beam modeling process.

Moment modification

The simplifying assumption inherent to the implementation of the large-angle deflection Eq. 5 is that the moment, which produces the deflection, is simply the moment at the wall. Though this assumption is reasonable for a first-order model, it is easy to appreciate the errors induced by making such an assumption. Shahinpoor[15–17] has shown that the electric field produces a unique moment upon each segment of the beam.

The moment produced by an electric field can be approximated using a parabola with a maximum value near the wall and a minimum value (zero) at the beam tip. Figure 30 presents a hypothetical moment.

The current model implementation reduces the complex moment model to a single composite moment at a given distance from the wall. The equivalence of those representations is presented in Fig. 31. The moments at the wall for both cases in Fig. 31 are equal, so the deflection of both beams would be identical. Yet, even a cursory analysis of the steady-state deflection of each of those cases would show that the beams would indeed deflect very differently.

The future work that must be undertaken in this area involves dividing the beam into numerous small segments, calculating the deflection of each segment due to the moment applied there, and integrating the individual segment dynamics to produce a composite beam deflection picture.

Extension

The dynamics portion of the first-order model is currently tied to the step response of the 1- × 0.25-in. beam. To make the model usable for a larger swath of the beam population requires that three specific tasks be performed:

1. Data describing the step responses of a representative number of different length and width beams must be collected and archived.

2. The frequency and damping coefficient parameters that would produce similar step responses for each of the beams must be selected.
3. The simple functions that relate the model's parameters of all the various beams (dependent only upon the dimension of each beam) must be derived.

Validation

Finally, and most importantly, the performance of the upgraded first-order model must be validated against actual beam performance. Dynamic responses of a number of beams must be run; conditions must be input into the model to drive equivalent simulation runs. The data collected and

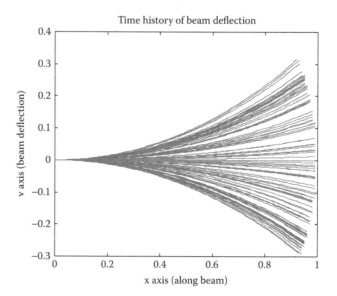

Fig. 29 "Flapping" beam.

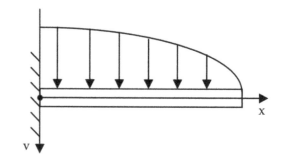

Fig. 30 Moment induced by electric field.

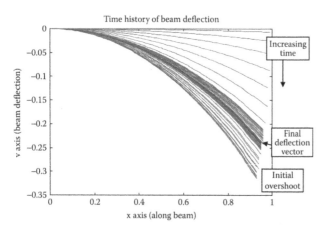

Fig. 28 Time history of beam deflection.

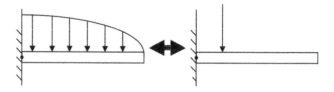

Fig. 31 Hypothetical moment simplification.

archived during the muscle and simulation tests must then be compared toward a goal of modifying model parameters or algorithms until the simulation runs match the beam responses throughout the population of potential beam dimensions and forcing functions.

Summary

An effort to model the dynamic motion of an IPMNC elastic beam was undertaken. Development of the static portion of the model was begun by assuming the beam behaves in accordance with the nonlinear equation used to describe large-angle deflection of elastic cantilever beams.

A Simulink simulation was developed to estimate the final deflection of a beam due to a constant moment. The dynamic portion of the model was developed by assuming that each segment of the beam could be represented as a simple second-order system. Future efforts to model beam motion more accurately by modifying the method used to model the forcing moment, by expanding the model to predict the performance of beams of all dimensions, and by validating model performance against actual beam motion were recommended.

Feedback Control in Bending Response of IPMNC Actuators

One of the disadvantages of ionic polymer materials is that their rather slow time constant limits the actuation bandwidth. A feedback controller has been developed by Mallavarapu and colleagues[18] in order to reduce the open-loop response time of cantilevered actuators from 4 to 10 sec to closed-loop response time from 0.1 to 1.5 sec using linear quadratic regulator (LQR) control.

Figure 32 shows an open-loop response for a 40- × 10- × 0.2-mm IPMNC actuator for an unsupported length of 30 mm. The inset in Fig. 29 shows the resonant modes on closer observation of the open-loop step response. Figure 33 shows the control of a 10- × 20- × 0.2-mm IPMNC actuator polymer.

This section demonstrates the use of feedback control to overcome these resonance modes. An empirical control model was developed after measuring the open loop step response of a 40- × 10- × 0.2-mm IPMNC actuator in a cantilever configuration. A compensator was designed using a linear observer–estimator in state space. The design objectives were to constrain the control voltage to less than 2 V and minimize the settling time by using feedback control. The controller was designed using LQR techniques, which reduced the number of design parameters to one variable. This LQR parameter was varied and simulations were performed, which reduced the settling time for the closed loop. The controller was later used in experimentation to

check simulations. Results obtained were consistent to a high degree. The electromechanical impedance of five sample actuators was measured.

The purpose of the test was to determine the voltage-to-current relationship in the actuator over the frequency range from 5 to 1000 Hz. Multiple actuators were tested to determine the uniformity of the impedance properties over the surface of the sheet sent to us.

Results

Figures 34A and B are pictures of the test setup used by Mallavarapu and colleagues.[18] The setup consisted of a small fixture to hold the IPMNC samples. The fixture was electroded and connected to a BNC jack to facilitate actuator testing. An HP 4192A LF (for low frequency)

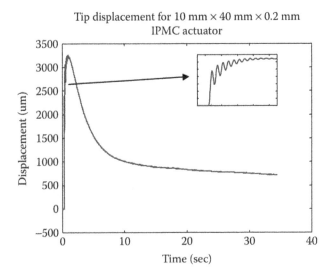

Fig. 32 Tip displacement of IPMNC actuator for 1 V.

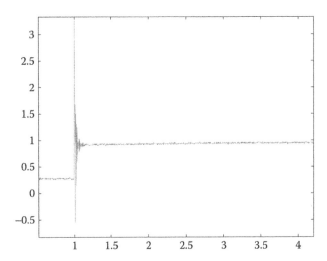

Fig. 33 Experimental closed-loop tip response for 10- × 20-mm IPMNC actuator for a step voltage of 1 V.

impedance analyzer was used to collect the impedance data.

Five samples from the Artificial Muscle Research Institute were cut from the material sent to Mallavarapu and coworkers at Virginia Tech. The samples were labeled A1 through A5 and placed in water-filled baggies to keep them hydrated. Each sample was tested in the impedance analyzer by manually sweeping through a range of frequencies and measuring the magnitude and phase of the impedance at each frequency.

The impedance of the samples was on the order of 2–5 Ω. The impedance of the test fixture was on the order of 1 Ω; therefore, the measured data were corrected to account for the impedance of the fixture. Assuming that the sample was in series with the fixture, the data were corrected by first transforming the measured data into real and imaginary components through the expressions:

$$\mathrm{Re}\left(Z_{meas}\right) = \left|Z_{meas}\right|\cos\left(\angle Z_{meas}\right)$$

$$\mathrm{Im}\left(Z_{meas}\right) = \left|Z_{meas}\right|\sin\left(\angle Z_{meas}\right)$$

(19)

The data were corrected by subtracting the impedance of the fixture from the measured impedance of the actuator, and transforming back into magnitude and phase.

$$\mathrm{Re}\left(Z_{corr}\right) = \mathrm{Re}\left(Z_{meas}\right) - \mathrm{Re}\left(Z_{fixture}\right)$$

$$\mathrm{Im}\left(Z_{corr}\right) = \mathrm{Im}\left(Z_{meas}\right) - \mathrm{Im}\left(Z_{fixture}\right)$$

(20)

Figures 35 and 36 are the magnitude and phase plots for the corrected data. The impedance magnitude varies between 1.1 and 2.4 Ω over the frequency range of 5 Hz to 1 kHz. The reactive component of the impedance is approximately equal to the active component below approximately 10 Hz.

Conclusions

1. The electromechanical impedance of the ionic polymer material was consistent over the five samples tested.
2. The impedance of the samples had a significant reactive component in the range of 5–10 Hz, but become primarily real (resistive) above approximately 100 Hz. The resistance of the samples was on the order of 1.2 Ω in the range of 100–1000 Hz.
3. The samples were highly capacitive at low frequencies. This attribute could complicate the development of the actuator power electronics.

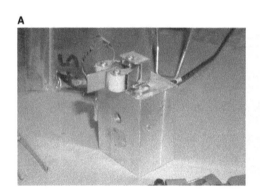

Fig. 34   (A) Fixture for impedance test; (B) Impedance analyzer.

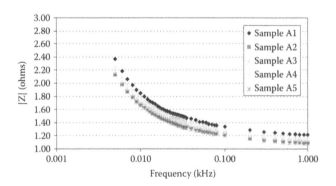
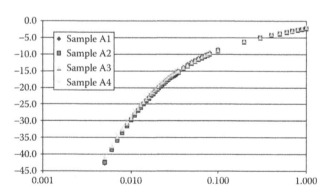

Fig. 35   Measured impedance at an input voltage of 0.5 V.

Fig. 36   Phase of the electromechanical impedance at 0.5 V.

REFERENCES

1. de Gennes, P.G.; Okumura, K.; Shahinpoor, M.; Kim, K.J. Mechanoelectric effects in ionic gels. Europhys. Lett. **2000**, *50* (4), 513–518.

2. Shahinpoor, M.; Kim, K.J. The effect of surface-electrode resistance on the performance of ionic polymer–metal composite (IPMC) artificial muscles. Smart Mater. Struct. **2000**, *9*, 543–551.

3. Shahinpoor, M.; Kim, K.J. Ionic polymer–metal composites—I. Fundamentals. Smart Mater. Struct. Int. J. **2001**, *10*, 819–833.

4. Paquette, J.; Kim, K.J. An electric circuit model for ionic polymer–Metal composites. In Proceedings of the first world congress on biomimetics and artificial muscles. Dec 9–11, 2002. Albuquerque, NM.

5. Henderson, B.; Lane, S.; Shahinpoor, M.; Kim, K.; Leo, D. Evaluation of ionic polymer–metal composites for use as near-DC mechanical sensors. In AIAA space 2001—conference and exposition. AIAA paper 2001-4600, 2001.

6. Shahinpoor, M.; Kim, K.J.; Griffin, S.; Leo, D. Sensing capabilities of ionic polymer–metal composites. In *Proceedings of SPIE 8th Annual International Symposium on Smart Structures and Materials*, Newport Beach, CA, Mar 2001; Publication No. 4329-28.

7. Sadeghipour, K.; Salomon, R.; Neogi, S. Development of a novel electrochemically active membrane and "smart" material based vibration sensor/damper. Smart Mater. Struct. **1992**, *1*, 172–179.

8. Shahinpoor, M. Ionic polymer conductor composite materials as distributed nanosensors, nanoactuators and artificial muscles—A review. In *Proceedings of the Second World Congress on Biomimetics and Artificial Muscle (Biomimetics and Nano-Bio 2004)*, Albuquerque, NM, Dec 5–8, 2004.

9. Shahinpoor, M. A new effect in ionic polymeric gels: The ionic flexogelectric effect. In Proceedings SPIE (1995) North American conference on smart structures materials. Feb 28–Mar 2, San Diego, CA, 2441, paper no. 05.

10. Shahinpoor, M. Intelligent materials and structures revisited. In Proceedings of SPIE conference on intelligent structures and materials. Feb 26–29, San Diego, CA, 238–250.

11. Keshavarzi, A.; Shahinpoor, M.; Kim, K.J.; Lantz, J. Blood pressure, pulse rate, and rhythm measurement using ionic polymer–metal composite sensors. 1999; 369–376, Electroactive Polymers SPIE Publication No. 3669-36.

12. Ferrara, L.; Shahinpoor, M.; Kim, K.J.; Schreyer, B.; Keshavarzi, A.; Benzel, E.; Lantz, J. Use of ionic polymer–metal composites (IPMCs) as a pressure transducer in the human spine. Electroactive Polym. 1999; 394–401, SPIE Publication No. 3669-45.

13. Shahinpoor, M.; Ed. Nonhomogeneous Large Deformation Theory of Ionic Polymeric Gels in Electric and pH Fields. *Proceedings of the First World Congress on Biomimetics and Artificial Muscles*, Albuquerque, NM, Dec 8–11, 2003; ERI Press: Albuquerque, NM.

14. Shahinpoor, M.; Alvarez, R. Simulation and control of ionoelastic beam dynamic deformation model. In *Proceedings of SPIE 9th Annual International Symposium on Smart Structures and Materials*, San Diego, CA, Mar 2002, Publication No. 4695-40.

15. Shahinpoor, M. Ion-exchange membrane-metal composite as biomimetic sensors and actuators. In *Polymer Sensors and Actuators*, Osada, Y., de Rossi, D, Eds.; Springer–Verlag: Heidelberg, 2000; 325–359.

16. Shahinpoor, M. Mechnoelectric effects in ionic polymers. In *Proceedings of 14th U.S. National Congress on Theoretical and Applied Mechanics*, Special presentation in honor of Professor Millard F. Beatty, contemporary issues in mechanics special symposium, Virginia Polytech, Blacksburg, VA, Jun 23–28, 2002.

17. Shahinpoor, M. Electrically controllable deformation in ionic polymer metal nano-composite actuators. In MECE2002-39037, *Proceedings of ASME 2002 International Mechanical Engineering Congress & Exhibition*, New Orleans, November 17–22, 2002.

18. Mallavarapu, K.; Leo, D.J. Feedback control of the bending response of ionic polymer actuators. J. Intelligent Mater. Syst. Struct. **2001**, *12*, 143–155.

Molecular—Nanogels

Nanocomposite Polymers: Functional

Radhakrishnan Sivakumar
Sambandam Anandan
Nanomaterials and Solar Energy Conversion Lab, Department of Chemistry, National Institute of Technology, Tiruchirappalli, India

Abstract

During the past few decades, significant efforts have been made to design polymers with desired properties through new synthetic strategies involving a wide range of chemical reactions towards various applications. In particular, polymers possessing biocompatible and biodegradable features have been attracting enormous attention in multitude of biomedical applications. This is mainly due to structure, morphology, physical, chemical, surface, and biomimetic properties of a polymer, which play a major role in determining its specific applications in the biomedical field. In addition to this, the metal nanoparticles, which have been increasingly recognized as important doping material, are introduced into the polymer composites for enhanced antibacterial activities. In this review, development of various bionanocomposites derived from silver nanoparticles and polymers is discussed with regard to their potential application in the biomedical field. The nature of methodologies adopted to prepare metal nanoparticles-embedded polymers reveals the antibacterial activity of the resultant bionanocomposites. More fundamental investigation about the formation of nanoparticles, their interaction with the polymer matrix, and release rate of nanoparticles in the bacterial environment would be very useful in engineering the polymer nanocomposite systems with excellent antimicrobial activity and in future for the clinical applications in the biomedical field.

INTRODUCTION

The applicability of polymers is rapidly emerging towards biomedical/biotechnological fields such as tissue engineering, implantation of medical devices and artificial organs, prostheses, ophthalmology, dentistry, bone repair, and many other medical fields. In this context, natural polymers such as polysaccharides, polypeptides, collagen, chitosan (Cs), and/or synthetic biodegradable polymers have been employed as the matrix, since such polymers do not produce any harmful products as a result of their biodegradation. On the other hand, producing bionanocomposites based on biomimetic approaches has been a focus of researchers and among them hydroxyapatite, a major constituent of hard tissue, exhibits undesirable mechanical properties if directly employed, whereas such undesirable properties may be avoided if polymer-based matrix composites were utilized. Consequently, researchers investigated polymer nanocomposites derived from hydroxyapatite $(Ca_{10}(PO_4)_6(OH)_2)$ such as hydroxyapatite-based nanocomposites with collagen-derived gelatin and poly-2-hydroxyethylmethacrylate/poly(3-caprolactone) for bone repair and implantation.

In addition to this, attempts were made to prepare polymer composites containing various metal nanoparticles-doped polymers in view of their specific application in the biomedical field due to their shape and size-dependent behavior. This is because small changes in the configuration of the composites concerning the percentage of metal as well as the size and the shape of the nanoparticles can lead to dramatic changes in the biomedical properties of the material. It is well known that silver nanoparticles have long been used in biomedical applications due to their antimicrobial property. The antibacterial activities of silver nanoparticles were dependent on the surface oxidation, optimal particle dispersion, and silver cation Ag^+, which binds strongly to electron donor groups in biological molecules containing sulfur, oxygen, or nitrogen. Hence, silver nanoparticles are considered as a good candidate for surface coating in medical devices. Several preparation procedures are available of which the most preferred approach involves in-situ preparation of nanoparticles in the polymer matrix by thermal reduction of metal salts in the polymer matrix or nanoparticles were first prepared by means of chemical methods and introduced in the polymer/monomer solution to achieve polymer–metal nanocomposites. Further, researchers are keen in preparing uniform nanometer-sized particles containing polymer matrix because of their technological and fundamental scientific importance. Consequently, nanoparticles possess several different morphologies (flakes, spheres, dendritic shapes, etc.) and each one has its specific properties. Hence for

Concise Encyclopedia of Biomedical Polymers and Polymeric Biomaterials DOI: 10.1081/E-EBPPC-120049984

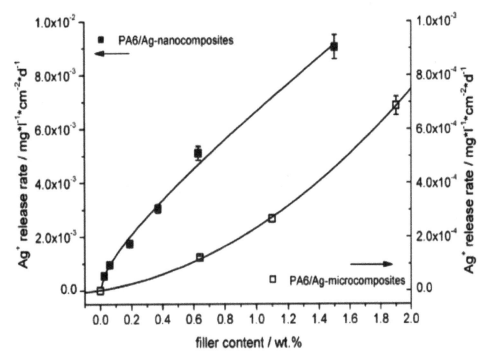

Fig. 1   Rate of silver ion release from PA6/Ag-nanocomposites as well as from PA6/Ag-microcomposites as a function of the filler content.
Source: From Damm et al.[9] © 2008, with permission from Elsevier.

the specific application in biomedical technology, they are generally designed and manufactured with *in vivo* properties tailored to meet the needs of the specific application. The challenge is to control the nanoparticle size, size distribution, morphology, concentration, and release rate to assemble the nanoparticles within the polymer matrix for biomedical applications. It was reported that the incorporation of nanoparticles into the polymer matrix allows slower and controlled release of drug and reduced swelling and improved mechanical integrity of hydrogel-based nanocomposites. Particularly, iron oxide nanoparticles have been investigated for various applications including drug delivery, magnetic resonance imaging contrast enhancement, immunoassay, and cellular therapy. Nonetheless, iron and cobalt nanoparticles encapsulated in polydimethylsiloxane were also noted for treating retinal detachment disorders.

POLYMER/METAL NANOCOMPOSITES IN BIOMEDICAL APPLICATIONS

Though many biocompatible polymers are nowadays in common use in therapeutic and clinical applications, improving their properties for biomedical applications is becoming a challenge for researchers in this field. For applications in the biomedical field, one needs to develop a novel method to attain well defined morphology of the polymer and hence metal nanoparticles are embedded in the polymer matrix. Among various metal nanoparticles-embedded systems, the silver nanoparticles-embedded

polymer matrix has been attracting much attention in the field of biomedical technology.[1–8] In this context, silver nanoparticles have been embedded in polymer films by a variety of chemical and physical methods and then their antibacterial activity was evaluated. Hence in this review, we will discuss biomedical applications of the polymer/silver nanocomposites and their importance in the biomedical field.

Damm, Munstedt, and Rosch[9] investigated the antimicrobial efficacy of polyamide 6 (PA6)/silver nano- and micro-composites against *Escherichia coli* as a function of the varying silver (Ag) content. PA6 filled with 0.06 wt% silver nanocomposites is able to eliminate the bacteria completely within 24 hr. However, the PA6 filled with 1.9 wt% silver microcomposites kills only about 80% of the bacteria during the same time. The silver ion release rate test reveals that the rate of silver ion release is one magnitude higher for nanocomposites compared to microcomposites because of the much larger specific surface area of the nanoparticles. Figure 1 shows the rate of silver ion release for PA6/Ag nanocomposites and PA6/Ag microcomposites as a function of filler content. Consequently, for the nanocomposites the silver ion release rate is proportional to the square root of the filler content, whereas for the microcomposites the silver ion release rate increases parabolically with the filler content. From Fig. 2, it is clear that a silver ion release rate of about 9.5×10^{-4} mg L^{-1} cm^{-2} day^{-1} killed all the bacteria within 24 hr. However, in the case of PA6/Ag microcomposites, the silver ion release rate is less than 9.5×10^{-4} mg L^{-1} cm^{-2} day^{-1} and for that reason they are not able to kill the bacteria completely

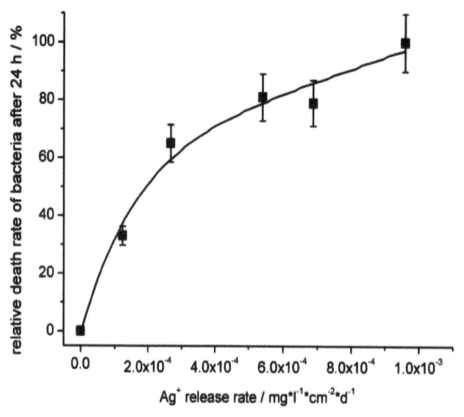

Fig. 2 Relative death rate of *E. coli* within 24 hr as a function of the silver ion release rate. The concentration of eliminated bacteria is related to the initial concentration.
Source: From Damm et al.[9] © 2008, with permission from Elsevier.

within 24 hr. Accordingly, for the PA6/Ag nanocomposites containing 0.06 wt% or more silver a threshold value of 9.5×10^{-4} mg L^{-1} cm^{-2} day^{-1} for the silver ion release rate is achieved or exceeded. Thus, the much better antimicrobial efficacy of the PA6/Ag nanocomposites is explained by their more efficient silver ion release, which may be due to the much larger specific surface area of the silver nanoparticles. Sedaghat and Nasseri[10] reported a novel fabrication procedure, which yielded a stable silver nanoparticles-coated polyamide (nylon 6,6) in a very simple and cost-effective manner with complete control of the silver loading level on the polymer. It was realized that silver-loaded nylon 6,6 nanocomposites show excellent antibacterial activity against *E. coli,* whereas only nylon 6,6 does not show any antibacterial activity.

In continuation, a series of nanocomposites from a polyether-type waterborne polyurethane (PU) incorporated with various amounts (15.1–113 ppm) of silver nanoparticles (approximately 5 nm) were prepared for their potential application in the biomedical field by Hung and Hsu.[11] It was observed that the average domain size of each PU–Ag nanocomposite was smaller than that of the original PU. All the PU nanocomposites with 15.1–113 ppm of Ag showed increased cell proliferation, reduced monocyte activation, decreased bacterial adhesion, and greater free radical scavenging ability than the original PU.

In particular, PU–Ag 30.2 ppm had the best performance among the nanocomposites followed by PU–Ag 50 ppm, which may be due to the generation of microphase-separated morphologies by the addition of a small amount of Ag nanoparticles, and these effects were similar to those in the presence of an appropriate amount (43.5 ppm) of Au nanoparticles.[12] All the PU–Ag nanocomposites show effective reduced bacterial adhesion due to the formation of novel microphase-separated films. In addition, due to the better dispersity of Ag in PU such a reduced effect can be even more pronounced than in the PU–Au system. Hsu, Tseng, and Lin[13] reported nanocomposites derived from waterborne PU with varying concentration of silver for their potential applications in the biomedical field. It was observed that the microstructure of PU was modified by the presence of nanoAg, which further influenced the thermal, chemical, and biological stabilities. As a result, PU containing 30 ppm of nanoAg showed superior physicochemical properties, cellular response, and a bacteriostatic effect. Such biocompatibility was confirmed by rat subcutaneous as well as jugular vein implantation. These provide evidence for PU-nanoAg as a potential cardiovascular biomaterial. In a modified approach, *Mesua ferrea* L. seed oil-based antimicrobial, biocompatible, hyperbranched and linear PU/Ag nanocomposites were reported by Deka et al.[14] These nanocomposite films exhibited excellent antimicrobial property towards *Staphylococcus aureus, E. coli, and Candida albicans*, even though they also exhibited adequate bacterial biodegradation by the right strain of *Pseudomonas aeruginosa* (MTCC 7814 and MTCC 7815). The cytocompatibility test based on the inhibition of RBC hemolysis showed that the materials lack cytotoxicity. However, the RBC hemolysis inhibition assay indicates the cytocompatibility of the materials, but before using these materials in actual field they need to be tested *in vivo*. Kaali et al.[15] realized the fact that the addition of silver ion exchanged zeolite into PU and silicone rubber improved the antimicrobial activity. It was established that the antimicrobial effect was found to increase with the zeolite content, and to be material and strain specific. Since zeolite is a highly hygroscopic material, the water uptake proved to have a significant role as the hydrophobicity of both PU and silicone gets decreased significantly as a function of exposure time.

Alt et al.[16] reported *in vitro* antibacterial activity against multiresistant bacteria and *in vitro* cytotoxicity of poly(methyl methacrylate) (PMMA) bone cement loaded with 5–50 nm metallic silver nanoparticles. Their antibacterial activity against *Staphylococcus epidermidis* EDCC 5245, MRSE EDCC 5130, and MRSA EDCC 5246 was studied by microplate proliferation tests. It was realized that higher silver concentration exhibits higher antibacterial effect in all the cases. In summary, nanosilver bone cement was the only bone cement that completely inhibited the proliferation of bacteria under investigation. Gentamicin-loaded cement could inhibit the proliferation of *S. epidermidis* EDCC 5245 but was not effective against MRSE EDCC 5130 or against MRSA EDCC 5246. However, plain PMMA bone cement does not show any antimicrobial activity against all tested strains. In order to verify the biocompatibility of the silver bone cement, its cytotoxicity was examined in both quantitative and qualitative manner and compared with that of the non-toxic control group and no significant difference was found in cytotoxicity between the silver bone cement and the non-toxic control group. It suggests that this nanosilver was free of *in vitro* cytotoxicity and showed high effectiveness against multiresistant bacteria.

Kong and Jang[17] reported PMMA nanofiber containing silver nanoparticles synthesized by radical-mediated dispersion polymerization as a potential antibacterial agent. Ultraviolet–visible (UV-vis) spectroscopic analysis indicates that when these nanocomposites dispersed in aqueous solution, silver nanoparticles were released continuously from the polymer nanofibers and the rate of release of silver nanoparticles was also calculated to be approximately 0.43 μg/(mL h). They realized the fact that lower glass transition temperature (T_g) of the PMMA film and porous nature sufficiently allow the water to pass through the polymer nanofiber and as a result the silver nanoparticles can diffuse out of the PMMA nanofiber. Antibacterial properties against Gram-negative bacteria

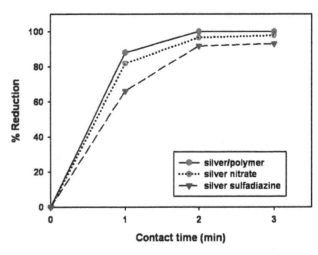

Fig. 3 The plot of percent reduction versus contact time (min) of different silver compounds on *E. coli*. The percent reduction was calculated as $(A - B)/A \times 100$ (where A is the number of surviving *E. coli* colonies in the blank solution and B is the number of surviving *E. coli* colonies in silver compound solution).
Source: From Kong & Jang[17] © 2008, with permission from American Chemical Society.

(*E. coli*) and Gram-positive bacteria (*S. aureus*) were studied for the as-prepared silver/nanofibers. As control, the antibacterial activity of silver nitrate and silver sulfadiazine was also studied. Antibacterial kinetic studies were analyzed by a plot of percent reduction vs. contact time, which provides the effectiveness for releasing the silver nanoparticle in aqueous solution (Fig. 3). The kinetic studies suggest that the nanofibers containing silver nanoparticles have higher antibacterial activities than silver nitrate and silver sulfadiazine. Though silver nitrate and silver sulfadiazine show antibacterial activity by releasing silver ions, salt formation as a result of precipitation deteriorated their antibacterial ability, which was not observed in the case of PMMA nanofibers containing silver nanoparticles. The minimum inhibitory concentration (MIC) of silver/PMMA nanofiber is three-fold lower than that of silver nitrate and nine-fold lower than that of silver sulfadiazine for 24 hr incubation as shown in Table 1. On the other hand, the silver/polymer nanofiber has better sustained activity than silver nitrate and silver sulfadiazine in the 48 hr incubation test. From this result, it can be concluded that the silver/polymer nanofiber has superior antibacterial activity to that of AgNO₃ and silver sulfadiazine, and the enhanced antibacterial performance is preserved. The contact antibacterial property can be measured by the clear zone of inhibition (ZOI) around the pellet after 24 hr incubation. The diameter of the ZOI for the silver/PMMA nanofiber is ca. 45 mm, whereas that of silver sulfadiazine is ca. 16 mm (the size of both pellets is ca. 13 mm). This result indicates that the silver/PMMA nanofiber has a more effective contact biocidal property than silver sulfadiazine. According to these results, it is anticipated that this

Table 1 Minimum inhibitory concentration (MIC) results of various silver compounds

Silver compounds[a]	Concentration of Ag (ng/mL)[b]	E. coli		S. aureus	
		1 day	2 day	1 day	2 day
Ag/polymer	6933	−	−	−	−
	2311	−	−	−	−
	770	−	+	−	+
	257	+	+	+	+
	86	+	+	+	+
AgNO₃	7056	−	−	−	−
	2352	−	+	−	+
	784	+	+	+	+
	261	+	+	+	+
	87	+	+	+	+
SSD[c]	7047	−	+	−	+
	2349	+	+	+	+
	783	+	+	+	+
	261	+	+	+	+
	87	+	+	+	+

[a]+: growth, −: no growth.
[b]The concentration of Ag in the Ag/polymer was measured by an inductively coupled plasma-atomic emission spectrometer.
[c]SSD: silver sulfadiazine.
Source: From Kong & Jang[17] © 2008, with permission from American Chemical Society.

silver nanoparticle/polymer nanofiber can be used in various applications such as clinical wound dressing, bioadhesive, biofilm, and the coating of biomedical materials.

Surface-modified silver nanoparticles were encapsulated into a polystyrene (PS) matrix by in-situ miniemulsion polymerization.[18] Silver nanoparticles were modified with 3-aminopropyltrimethoxysilane that acts as a coupling agent and co-stabilizer in the polymerization reaction. The PS-Ag nanocomposites synthesized via miniemulsion polymerization were made at two different concentrations of the initiator (0.7 g/L and 2.5 g/L in H₂O). Table 2 shows the antimicrobial activity of the PS-Ag nanocomposites studied against E. coli and S. aureus. The inhibitory zone increased from 17% to 21% for E. coli and from 9% to 13% for S. aureus, while the amount of initiator increased from 0.7 g/L to 2.5 g/L. The nanocomposite with higher concentration of initiator showed higher percentage inhibition for both microorganisms and this behavior can be attributed to the size of particles as well as the percentage of silver in the nanocomposite. This means that the decrease in the number of colonies produced an increase in the biocide effect of the microorganisms. Interestingly, the synthesized PS-Ag nanocomposites possess excellent biocidal potential against E. coli and S. aureus. According to these results, these nanocomposites can be used in different applications such as clinical paints and coatings of biomedical materials.

The antibacterial activities of silver nanoparticles embedded into polydimethylsiloxane–chitosan–clay (PDMS/Cs/clay) composites against E. coli (ATCC 25922), P. aeruginosa (ATCC27853), S. aureus (ATCC25923), and C. albicans (ATCC14053) were studied.[19] Silver ion release test, ZOI, and MIC tests suggest that PDMS/Cs/clay-Ag nanocomposites possess excellent antimicrobial activity and controlled drug release characteristics in indwelling urinary catheters. In a modified approach, silver nanoparticles were incorporated into the lamellar space of montmorillonite (MMT)/Cs by following a green physical route and utilizing the UV light irradiation reduction method in the absence of any reducing agent or heat treatment.[20] These Ag nanoparticles were found to be stabilized by the low molecular weight of Cs. It was observed that on increasing UV irradiation time, particle size of Ag nanoparticles was found to be decreased. The as-prepared nanocomposites were characterized by transmission electron microscopy (TEM), X-ray diffraction (XRD), scanning electron microscopy (SEM), and UV spectral analysis. The antibacterial activity of Ag/MMT/Cs bionanocomposites was demonstrated, and showed strong antibacterial activity against Gram-positive and Gram-negative bacteria.

Antibacterial efficacy of silver nanoparticles deposited alternatively layer by layer on the Cs polymer in the form of a thin film over a quartz plate and stainless steel strip has been studied by Ghosh, Ranebennur, and Vasan[21] The bactericidal properties of $(Cs/Ag)_8$, $(Cs/Ag^+)_8$, and control Cs/citrate on the quartz plate were studied in saline solution under biological pH 7.4. In the first 1 hr, the percentage survival on the $(Cs/Ag^+)_8$ hybrid is almost the same as the $(Cs/Ag)_8$ hybrid, and this survival remains even after 4 hr. Whereas in the case of the test sample, there is no significant change in percentage survival (100%) of bacteria throughout the contact time observed. This may be explained by the way the microbes come in contact with the hybrid. As the experiments are bacteriostatic in nature, it was presumed that the contact is mainly due to the diffusion of silver nanoparticles from the hybrid into the solution and physically attacking the microbes. The diffusion of Ag nanoparticles into the saline solution is rather slow compared to Ag^+ ions, though the initial Ag concentration is more or less the same in both the samples. This is understandable as the Cl^- ions present in saline solution have greater affinity for Ag^+ compared to Ag nanoparticles. After the first 1 hr contact, most of the Ag^+ present in $(Cs/Ag^+)_8$ have diffused into the solution and made it ineffective on further contact. The bacterial morphology changes were examined by AFM before treatment (0 hr) [Fig. 4(A)] and after 24 hr treatment [Fig. 4(D)] with the $(Cs/Ag)_8$ hybrid on a quartz plate. It reveals that the physical contact of silver nanoparticles and the bacteria results in complete destruction of the outer membrane of the bacteria. The root mean square (rms) roughness of the bacterial membrane and the height of the cell also decreased

Table 2 The antimicrobial activity of PS-Ag nanocomposites against *E. coli* and *S. aureus*

Sample	[I] (g/L of H_2O)	Inhibition (%)		Colony formation units (CFU) $(mL^{-1}) \times 10^{-7}$		%Ag
		E. coli	*S. aureus*	*E. coli*	*S. aureus*	
1	0.7	17	9	2	1.68	0.17
2	2.5	21	13	1.2	1.54	1.23

Source: From Galindo et al.[18] © 2012.

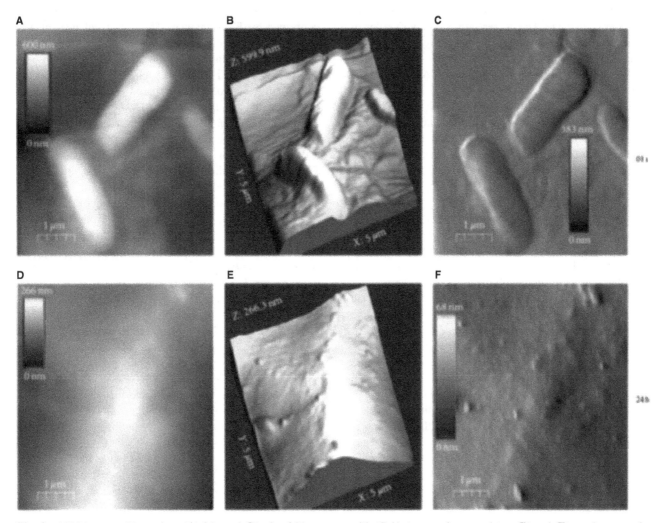

Fig. 4 AFM images of bacteria at (**A**) 0 hr and (**D**) after 24 hr contact with $(Cs/Ag)_8$-coated quartz plates, (**B**) and (**E**) are the respective 3D images, (**C**) and (**F**) are phase images of (**A**) and (**D**), respectively.
Source: From Ghosh et al.[21] © 2011.

with treatment. These phenomena may be attributed to the damage of lipopolysaccharide, which serves as a permeability barrier and is believed to contribute to the structural rigidity of the cell envelope.

The ZOI analysis carried out for the $(Cs/Ag)_8$ hybrid film on stainless steel strips reveals that ZOI of the aged (3 months) film is similar to ZOI of the as-prepared one, which accounts for the mechanical stability and retention of antibacterial activity of the $(Cs/Ag)_8$ hybrid film even on

aging. The cytotoxicity assay results for the fabricated $(Cs/Ag)_8$ hybrid film show that the number of live cells after exposure to the film was found to be almost the same as those exposed to control. Thus the attachment and viability of cells vis-a-vis exposure of quartz plate-coated $(Cs/Ag)_8$ indicate that the coated film has satisfactory biocompatibility. One can argue that the antibacterial property of the hybrid would be better if one increases the number of bilayers further, but this would be at the cost of cytotoxicity of the material. The recyclability of the bioactive films was done and it was found that after the fourth cycle the bacterial survival was nearly 50%. This increase in survival may be due to depletion of Ag nanoparticles in the film after each cycle. This is evident from the intensity decrease and increase in full width at half maxima of the Ag plasmon band. It is interesting to note that the percentage weight loss of Ag NPs in the hybrid (estimated from the UV-vis spectral intensities) after each cycle follows the same trend as the percentage survival of bacteria. And also the broadening of peak indicates the possible agglomeration of Ag NPs after each cycle. The decrease in intensity of color of the film further confirms the depletion of Ag nanoparticles in the film. The diffusion of Ag NPs into the saline solution may be due to the concentration gradient of Ag between the film and the solution.

Copolymeric silver nanocomposite hydrogels were synthesized by using acryloyl phenylalanine, N'-isopropylacrylamide and cross-linked by N,N-methylenebisacrylamide via radical redox polymerization.[22] The as-prepared silver nanoparticles in the hydrogel matrix were characterized by UV-vis spectroscopy, XRD, SEM, TEM, and thermogravimetric (TG) analysis. Antibacterial activity screen studies of the placebo copolymeric hydrogel and its silver nanocomposite hydrogel were carried out by the paper disk method. A quantity of 5 mL of nutrient agar (NA) medium (pH = 6.8) was poured into the sterilized plates and allowed to solidify. The plates were inoculated with spore suspensions of three Gram-positive (*Micrococcus luteus, Bacillus subtilis,* and *S. aureus*) and three Gram-negative (*Proteus vulgaris, Proteus mirabilis,* and *E. coli*) bacteria and by using a sterilized cork borer, paper disks (6 mm diameter) were dug inside the culture plates. The results suggest that the silver nanocomposite copolymeric hydrogel showed more toxic effect than pure copolymeric hydrogel under similar conditions (Table 3). The possible mode of increased toxicity of the silver nanocomposite copolymeric hydrogel may be due to easy release of silver nanoparticles and their interaction with the lipid layer of the cell membrane. From the results, it was revealed that the silver nanocomposite copolymeric

hydrogel showed better activity towards Gram-positive bacteria than Gram-negative. Figure 5 confirms that both antibacterial activity and concentrations are directly proportional, i.e., normally, the biocidal activity increases with increase in concentration.

Hybrid materials based on polyvinylpyrrolidone (PVP) with silver nanoparticles (AgNps) were synthesized applying two different strategies based on thermal or chemical reduction of silver ions to silver nanoparticles using PVP as a stabilizer.[23] The formation of spherical silver nanoparticles with diameter ranging from 9 nm to 16 nm was confirmed by TEM analysis. The UV-vis analysis clearly indicated the formation of silver nanoparticles by the appearance of strong absorption bands at 420 nm in both cases (Fig. 6A and 7A). TEM analysis suggests that the size of silver nanoparticles prepared via thermal reduction is smaller (6.4 ± 1.1 nm) than that via the chemical reduction method (13.8 ± 3.8 nm) (Fig. 6B and 7B). The antibacterial activity of the synthesized AgNPs/PVP against three different groups of bacteria as well as against spores of *B. subtilis* was studied by using the disk diffusion method. It was found that all AgNPs stabilized with PVP exhibit bactericidal and sporocidic activity by the appearance of an inhibition zone in the range 7.5–8.5 mm. Higher antibacterial activity was observed against both Gram-negative *E. coli* and Gram-positive *S. aureus*, as well as against spores of *B. subtilis* and lowest antibacterial activity was detected for *P. aeruginosa*. These results were in good agreement with the previous investigations done by Bryaskova et al.[24] on the antibacterial activity of hybrid material based on PVA/AgNps/TEOS thin films. Also, it was confirmed that even at lowest silver concentration of 0.015 mg/mL strong bactericidal activity was demonstrated by the lack of any bacterial growth. The lack of any sporocidic activity of *B. subtilis* indicated that silver nanoparticles with Ag concentrations ranging from 0.49 mg/mL to 0.015 mg/mL are not effective antibacterial agent for spores of *B. subtilis* (Table 4).

Apparently, thermally prepared silver nanoparticles showed higher antifungicidal activity in comparison to Ag nanoparticles (AgNps) prepared via chemical reduction.

Table 3 Antibacterial activity data of silver nanocomposite copolymeric hydrogels against Gram-positive and Gram-negative bacteriocides

Concentration of Ag nanoparticle (mg/mL)	Gram-positive			Gram-negative		
	M. luteus	*B. subtilis*	*S. aureus*	*P. vulgaris*	*P. mirabilis*	*E. coli*
	Inhibition zone (mm)[a] + paper disk width (mm)[b]					
1.0	11.8 ± 0.3	8.7 ± 0.3	9.1 ± 0.3	6.8 ± 0.3	8.7 ± 0.3	9.3 ± 0.3
3.0	12.7 ± 0.3	9.3 ± 0.3	9.2 ± 0.3	8.2 ± 0.3	9.0 ± 0.3	9.5 ± 0.3
5.0	14.8 ± 0.2	10.4 ± 0.3	9.6 ± 0.3	10.8 ± 0.3	9.3 ± 0.3	10.6 ± 0.3

[a] $\bar{x} \pm \dfrac{ts}{\sqrt{n}}$, $n = 5$ at 95% confidence level.

[b] Paper disks of width of 6 mm.

Source: From Kim et al.[22] © 2011, with permission from Korean Chemical Society.

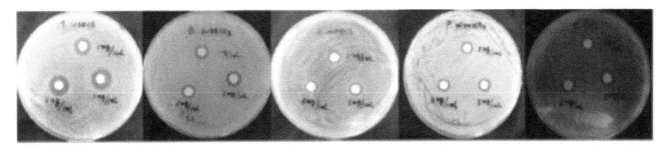

Fig. 5 Antibacterial activity picture of silver nanocomposite copolymeric hydrogels against Gram-positive and Gram-negative bacteriocides.
Source: From Kim et al.[22] © 2011, with permission from Korean Chemical Society.

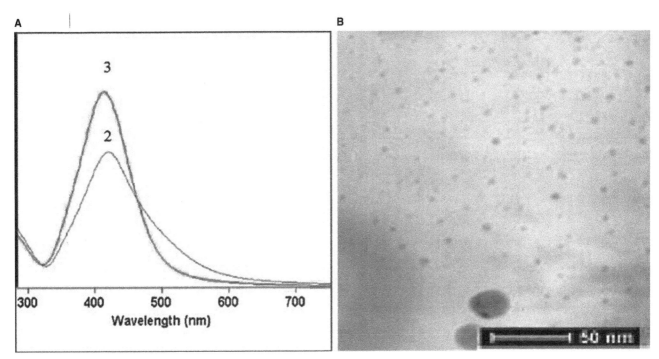

Fig. 6 (**A**) UV spectra of AgNps/PVP obtained via thermal reduction at concentration of silver precursor: (2) 1.96 mg/mL and (3) 3.92 mg/mL. (**B**) TEM of thermally reduced silver nanoparticles stabilized by PVP at a concentration of silver precursor of 3.92 mg/mL.
Source: From Bryaskova et al.[23] © 2011, with permission from Springer Science + Business Media.

The higher antifungicidal activity of thermally prepared Ag nanoparticles is probably due to their smaller size as seen by TEM in comparison to the silver nanoparticles prepared via chemical reduction, thus providing better contact with the environment. Moreover, thermal reduction of silver precursors in the PVP solution allows the complete transformation of Ag^+ into Ag^0 without producing unwanted by-products of the reducing agent, which can affect their antimicrobial properties. AgNps/PVP was tested for the presence of fungicidal activity against different yeasts and mold such as *C. albicans, Candida krusei, Candida tropicalis, Candida glabrata, and Aspergillus brasiliensis* (Table 5). The hybrid materials showed a strong antimicrobial effect against the tested bacterial and fungal strains and therefore have potential applications in biotechnology and biomedical science.

Li et al.[25] developed a facile microwave-assisted method to fabricate cellulose–silver nanocomposites by reducing silver nitrate in ethylene glycol (EG), where EG acts as solvent, reducing agent, and microwave absorber. The products were characterized by XRD, Fourier transform infrared (FT-IR), SEM, TG analysis, and differential scanning calorimetric analysis. The thermal stability of cellulose–silver nanocomposites in nitrogen and air indicates that the heating time and temperature in the preparation had slight effect on the thermal stability of the cellulose–silver nanocomposites. The antibacterial activity was demonstrated with two different concentrations (0.075 g and 0.150 g) of cellulose–silver nanocomposites and microcrystalline cellulose as control. It was found that the inhibition zone is much higher at higher concentration of cellulose–silver nanocomposites for both the microbes

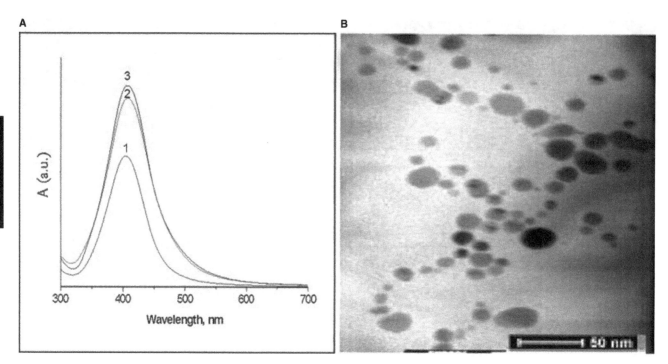

Fig. 7 (A) UV spectra of PVP/AgNps obtained via chemical reduction at concentration of silver precursor: (1) 0.98 mg/mL, (2) 1.96 mg/mL, and (3) 3.92 mg/ mL. (B) TEM of silver nanoparticles obtained via chemical reduction of $AgNO_3$ in the presence of $NaBH_4$ at a concentration of silver precursor of 3.92 mg/mL.
Source: From Bryaskova et al.[23] © 2011, with permission from Springer Science + Business Media.

Table 4 Antibacterial activity of silver nanoparticles stabilized by PVP

Strain	Ag concentration (mg/mL)					
	0.49	0.24	0.12	0.06	0.03	0.015
S. aureus ATCC 25923	–	–	–	–	–	–
E. coli ATCC 25922	–	–	–	–	–	–
P. aeruginosa ATCC 27853	–	–	–	–	–	–
Spore of B. subtilis ATCC 6633	+	+	+	+	+	+

+ (positive) presence of bacterial growth, – (negative) lack of bacterial growth.
Source: From Bryaskova et al.[23] © 2011, with permission from Springer Science + Business Media.

(Fig. 8). Particularly, the antimicrobial activity against *E. coli* is higher than that against *S. aureus*, probably due to the difference in cell walls between Gram-negative and Gram-positive bacteria. However, no inhibition zones were observed for the control microcrystalline cellulose, which indicates that the antibacterial activity is only due to the presence of silver nanoparticles, which were impregnated inside microcrystalline cellulose.

In a modified approach, silver nanoparticles were embedded into polyrhodanine nanofibers by oxidative polymerization of the rhodanine monomer in which silver ions get reduced and in turn induce the polymerization simultaneously.[26] The TEM image indicates that silver nanoparticles of about 3 nm are embedded in the polyrhodanine nanofiber. The XPS spectrum indicates that two peaks are observed at 368 eV and 374 eV with 6.0 eV separation, corresponding to Ag $3d_{5/2}$ and Ag $3d_{3/2}$ binding energy of Ag^0, respectively. Also, UV-vis spectra of the as-prepared nanofibers show the characteristic peak for silver at 400 nm, which evidenced the presence of silver nanoparticle in the polyrhodanine nanofiber. The antibacterial efficacy was evaluated by surface coating 1 mL of 1 wt% silver/polyrhodanine solution in methanol on a glass slide and subsequently spraying the *E. coli* suspension. It was observed that the silver/polyrhodanine nanofiber-coated glass slide inhibited the growth of *E. coli* compared to uncoated glass slide. Further, the kinetics of antimicrobial investigation against Gram-negative *E. coli*, Gram-positive *S. aureus*, and yeast (*C. albicans*) suggests that the silver/ polyrhodanine nanofiber had better antimicrobial efficacy than pure silver and polyrhodanine homopolymer due to the combined antimicrobial activity of silver and polyrhodanine nanofiber. The contact antimicrobial property of silver/polyrhodanine was investigated by a modified Kirby–Bauer technique and compared with standard silver sulfadiazine in order to compare the release and diffusion of silver from silver/polyrhodanine. In this regard, samples with similar concentration of silver were used and the ZOI was observed after 24 hr incubation. It was revealed that

Table 5 Fungicidal activity of AgNps obtained via thermal reduction after 48 hr of incubation

Control strain	AgNps (mf\mL)							MFC (mg/mL)
	1.96	0.98	0.5	0.25	0.12	0.06	0.03	
C. albicans	−	−	−	+	+	+	+	0.5
C. glabrata	−	+ (1 cfu)	+ (3 cfu)	+	+	+	+	0.98 ± 0.5
C. tropicalis	−	−	−	−	−	−	−	<0.03
C. krusei	−	−	−	−	−	−	−	<0.03
C. brasiliensis	−	+	+	+	+	+	+	1.96

Source: From Bryaskova et al.[23] © 2011, with permission from Springer Science + Business Media.

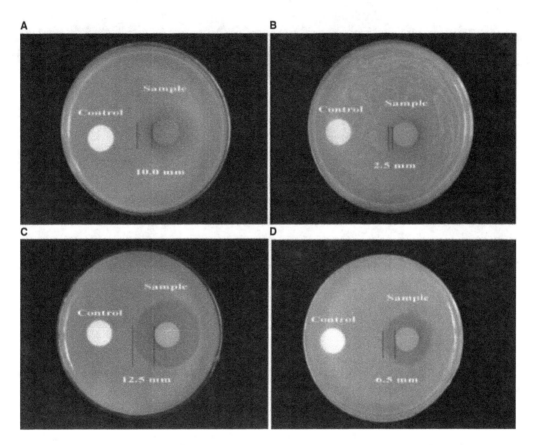

Fig. 8 Antimicrobial activities of the cellulose–Ag nanocomposites: (**A** and **C**) *Escherichia coli* and (**B** and **D**) *Staphylococcus aureus*. **Source:** From Li et al.[25] © 2011, with permission from Elsevier.

the ZOI of silver/polyrhodanine nanofiber is twice as large as that of silver sulfadiazine against *E. coli* and *S. aureus* (Fig. 9). This result indicates that the silver/polyrhodanine nanofiber has the enhanced contact antimicrobial efficacy than silver sulfadiazine.

Shah et al.[27] reported silver-impregnated polyethylene glycol–PU–titania nanocomposite polymer films having excellent antibacterial properties with recycling usage. Initially, different kinds of TiO_2 (commercial titania (CT), as-synthesized TiO_2 (ST), catechol-coated TiO_2 (CCT), carbon-doped TiO_2 (CaT)) were used to prepare the PEG-PU-TiO_2 nanocomposites and silver impregnation was achieved following UV or visible irradiation. Various

characterization techniques (FT-IR, SEM, XRD, and XPS) were applied to confirm the formation and morphology of the as-prepared polymer nanocomposites. SEM-EDAX analysis reveals that the amount of silver impregnated in the films is about 2–4 at%. The antibacterial tests of the Ag-PEGPU-TiO_2 films were carried out by the standard disk diffusion assay on Luria Bertani (LB) agar medium using *E. coli* and *B. subtilis* as model bacteria. In order to avoid the effect of titania under any irradiation, the antibacterial activities were studied in daylight conditions. It was realized that all the silver-doped samples were shown to have excellent antibacterial activities by exhibiting ZOI, whereas the samples without silver did not create a ZOI in

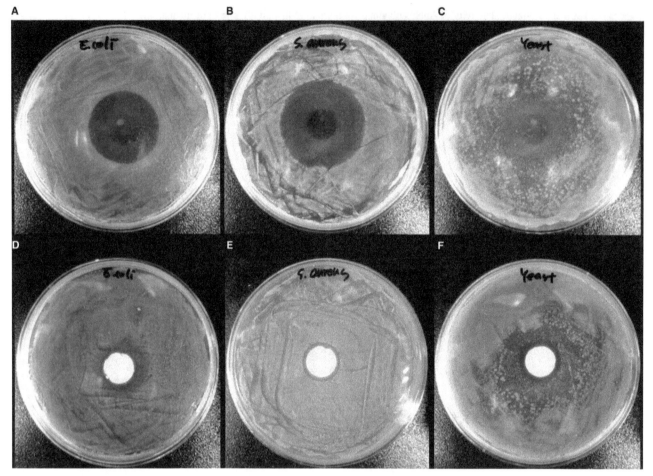

Fig. 9 Photograph images of the Kirby–Bauer plates of silver/polyrhodanine nanofiber (**A, B,** and **C**) and silver sulfadiazine (**D, E,** and **F**) against *E. coli, S. aureus*, and *C. albicans* (yeast).
Source: From Kong & Jang[17] © 2008, with permission from American Chemical Society.

the bacterial growth (Fig. 10). Also, these disks showed reasonable activity even after being used second time (*B. subtilis*-II). A plausible reason for this excellent antibacterial activity exhibited by the films were most likely the presence of Ag nanoparticles, which create redox imbalance causing extensive bacterial death, and the large surface area of the nanoparticles makes this action more effective. Also, it was proposed that the Ag nanoparticles released in a controlled manner by an oxidation mechanism to form silver ions and it determines the extent of zone of activity as seen as ZOI.

Shameli et al.[28] reported a simple way to prepare silver/poly(lactic acid) nanocomposite (Ag/PLA-NC) films and studied their antibacterial activities against Gram-negative bacteria (*E. coli* and *Vibrio parahaemolyticus*) and Gram-positive bacteria (*S. aureus*) by the diffusion method using Muller–Hinton agar. The as-prepared nanocomposites were characterized by FT-IR, XRD, and TEM analysis. TEM analysis indicates that average diameters of the Ag-NPs were between 3.27, 3.83, and 4.77 nm with well-crystallized structures, however, on increasing the

percentage of Ag particle diameters and standard deviation are increased. The silver ion release was investigated in the phosphate buffer solution and it was found that the accumulative amount of the silver ions released could reach up to 60 ppm of the polymeric films at the beginning stage of release. Antibacterial efficacies against Ag/PLA-NC films indicate that AgNPs are responsible for the antibacterial activity and it increases as the concentration of Ag increases. Further studies investigate the bactericidal effect of Ag/PLA-NC films on the types of bacteria for potentially widening such wound dressings or as anti-adhesion membranes.

CONCLUSION

This comprehensive review on silver/polymer-based nanocomposites and their applications in biomedical field reveals that a real-time transport, biocompatibility, and non-toxicity of Ag nanoparticles in the polymer matrix play very important roles in the development of excellent

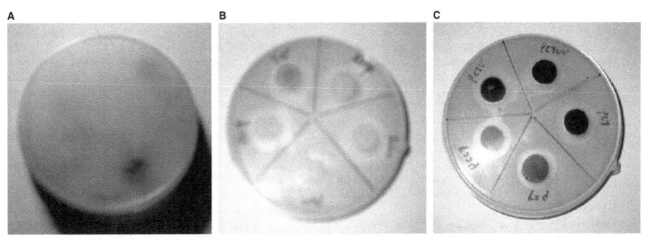

Fig. 10 Visual images of zone of inhibition of (**A**) control, (**B**) with antibiotics, and (**C**) with synthesized Ag-PEG-PU-TiO$_2$ films.
Source: From Shah et al.[27] © 2008, with permission from American Chemical Society.

antimicrobial materials for clinical applications. Though many synthetic methods were investigated, progress in this area will give new green paths in the development of controlled shape and size of Ag NPs in the polymer matrix. Hence the main idea is to prepare novel bionanocomposites, which possess biocompatible, biodegradable, and non-toxic matrix, through non-toxic monomers and precursors with well attached silver nanoparticles exhibiting a low mean diameter and narrow size distribution. Also, it was realized that the antibacterial properties of silver nanoparticles not only depend on their size and shape but also on the nature of the polymer matrix, which influences the potentiality of silver ion release in the bacterial environment.

ACKNOWLEDGMENTS

The authors thank DST, New Delhi (SR/S1/PC-49/2009), for sanctioning major research project and India–Spain Joint Collaborative project (DST/INT/Spain/P-37/11).

REFERENCES

1. Furno, F.; Morley, K.S.; Wong, B.; Sharp, B.L.; Arnold, P.L.; Howdle, S.M.; Bayston, R.; Brown, P.D.; Winship, P.D.; Reid, H.J. Silver nanoparticles and polymeric medical devices: a new approach to prevention of infection. J. Antimicrob. Chemother. **2004**, *54* (6), 1019–1024.
2. Ghosh, S.; Kaushik, R.; Nagalakshmi, K.; Hoti, S.L.; Menezes, G.A.; Harish, B.N.; Vasan H.N. Antimicrobial activity of highly stable silver nanoparticles embedded in agar-agar matrix as a thin film. Carbohyd Research **2010**, *345* (15), 2220–2227.
3. Jain, P.; Pradeep, T. Potential of silver nanoparticle-coated polyurethane foam as an antibacterial water filter. Biotechnol. Bioeng. **2005**, *90* (1), 59–63.
4. Lansdown, A.B.G. Silver I: Its antibacterial properties and mechanism of action. J. Wound. Care **2002**, *11* (4), 125–130.
5. Morones, J.R.; Elechiguerra, J.L.; Camacho, A.; Holt, K.; Kouri, J.B.; Ramírez, J.T.; Yacaman, M.J. The bactericidal effect of silver nanoparticles. Nanotechnology **2005**, *16* (10), 2346–2353.
6. Nguyen, P.M.; Zacharia, N.S.; Verploegen, E.; Hammond, P.T. Extended release antibacterial layer-by-layer films incorporating linear dendritic block co-polymer micelles. Chem. Mater. **2007**, *19* (23), 5524–5530.
7. Shrivastava, S.; Bera, T.; Roy, A.; Singh, G.; Ramachandrarao, P.; Dash, D. Characterization of enhanced antibacterial effects of novel silver nanoparticles. Nanotechnology **2007**, *18* (22), 225–103.
8. Kim, J.S.; Kuk, E.; Yu, K.N.; Kim, J.H.; Park, S.J.; Lee, H.J.; Kim, S.H.; Park, Y.K.; Park, Y.H.; Hwang, C.Y.; Kim, Y.K.; Lee, Y.S.; Jeong, D.H.; Cho, M.H. Antimicrobial effects of silver nanoparticles.Nanomed. Nanotechnol. Bio. Med. **2007**, *3* (1), 95–10.
9. Damm, C.; Munstedt, H.; Rosch, A. The antimicrobial efficacy of polyamide 6/silver-nano- and microcomposites. Mater. Chem. Phys. **2008**, *108* (1), 61–66.
10. Sedaghat, S.; Nasseri, A. Synthesis and stabilization of Ag nanoparticles on a polyamide (nylon 6,6) surface and its antibacterial effects. Int. Nano Lett. **2011**, *1* (1), 22–24.
11. Hung, H.S.; Hsu, S. Biological performances of poly(ether) urethane–silver nanocomposites. Nanotechnology **2007**, *18* (47), 475101.
12. Hsu, S.H.; Tang, C.M.; Tseng, H.J. Biocompatibility of poly(ether)urethane-gold nanocomposites. J. Biomed. Mater. Res. **2006**, *79* (4), 759–770.
13. Hsu, S.; Tseng, H.J.; Lin, Y.C. The biocompatibility and antibacterial properties of waterborne polyurethane-silver nanocomposites. Biomaterials **2010**, *31* (26), 6796–6808.
14. Deka, H.; Karak, N.; Kalita, R.D.; Buragohain, A.K. Bio-based thermostable, biodegradable and biocompatible hyperbranched polyurethane/Ag nanocomposites with antimicrobial activity. Polym. Degrad. Stabil. **2010**, *95* (9), 1509–1517.
15. Kaali, P.; Strömberg, E.; Aune, R.E.; Czél, G.; Momcilovic, D.; Karlsson, S. Antimicrobial properties of Ag+ loaded zeolite polyester polyurethane and silicone rubber and long-term properties after exposure to *in vitro* ageing. Polym. Degrad. Stabil. **2010**, *95* (9), 1456–1465.

16. Alt, V.; Bechert, T.; Steinrucke, P.; Wagener, M.; Seidel, P.; Dingeldein, E.; Domann, E.; Schnettler, R. An *in vitro* assessment of the antibacterial properties and cytotoxicity of nanoparticulate silver bone cement. Biomaterials **2004**, *25* (18), 4383–4391.

17. Kong, H.; Jang, J. Antibacterial properties of novel poly (methyl methacrylate) nanofiber containing silver nanoparticles. Langmuir **2008**, *24* (5), 2051–2056.

18. Galindo, R.B.; Miranda, C.C.; Urbina, B.A.P.; Facio, A.C.; Valdes, S.S.; Padilla, J.M.; Cerda, L.A.G.; Perera, Y.A.; Fernandez, O.S.R. Encapsulation of silver nanoparticles in a polystyrene matrix by miniemulsion polymerization and its antimicrobial activity. Nanotechnology **2012**, *2012*, Article ID 186851.

19. Zhou, N.; Liu, Y.; Li, L.; Meng, N.; Huang, Y.; Zhang, J.; Wei, S.; Shen, J. A new nanocomposite biomedical material of polymer/Clay–Cts–Ag nanocomposites. Curr. Appl. Phys. **2007**, *7S1*, e58–e62.

20. Shameli, K.; Ahmad, M.B.; Yunus, W.M.Z.W.; Rustaiyan, A.; Ibrahim, N.A.; Zargar, M.; Abdollahi, Y. Green synthesis of silver/montmorillonite/chitosan bionanocomposites using the UV irradiation method and evaluation of antibacterial activity. Int. J. Nanomed **2010**, *5*, 875–887.

21. Ghosh, S.; Ranebennur, T.K.; Vasan, H.N. Study of antibacterial efficacy of hybrid chitosan-silver nanoparticles for prevention of specific biofilm and water purification. Int. J. Carbohydr. Chem. **2011**, *2011*, Article ID 693759.

22. Kim, Y.; Ramesh Babu, V.; Thangadurai, D.T.; Krishna Rao, K.S.V.; Cha, H.; Kim, C.; Joo, W.; Lee, Y. Synthesis, characterization, and antibacterial applications of novel copolymeric silver nanocomposite hydrogels. Bull. Korean Chem. Soc. **2011**, *32* (2), 553–558.

23. Bryaskova, R.; Pencheva, D.; Nikolov, S.; Kantardjiev, T. Synthesis and comparative study on the antimicrobial activity of hybrid materials based on silver nanoparticles (AgNps) stabilized by polyvinylpyrrolidone (PVP). J. Chem. Biol. **2011**, *4* (4), 185–191.

24. Bryaskova, R.; Pencheva, D.; Kale, G. M.; Lad, U.; Kantardjiev, T. Synthesis, characterisation and antibacterial activity of PVA/TEOS/Ag-Np hybrid thin films. J. Colloid Interface Sci. **2010**, *349* (1), 77–85.

25. Li, S.M.; Jia, N.; Ma, M.G.; Zhang, Z.; Liu, Q.H.; Sun, R.C. Cellulose–silver nanocomposites: Microwave-assisted synthesis, characterization, their thermal stability, and antimicrobial property. Carbohyd. Polym. **2011**, *86* (2), 441–447.

26. Kong, H.; Jang, J. Synthesis and antimicrobial properties of novel silver/polyrhodanine nanofibers. Biomacromolecules **2008**, *9* (10), 2677–2681.

27. Shah, M.S.A.S.; Nag, M.; Kalagara, T.; Singh, S.; Manorama, S.V. Silver on PEG-PU-TiO$_2$ polymer nanocomposite films: An excellent system for antibacterial. Appl. Chem. Mater. **2008**, *20* (7), 2455–2460.

28. Shameli, K.; Ahmad, M.B.; Yunus, W.M.Z.W.; Ibrahim, N.A.; Rahman, R.A.; Jokar, M.; Darroudi, M. Silver/poly (lactic acid) nanocomposites: preparation, characterization, and antibacterial activity. Int. J. Nanomed. **2010**, *5*, 573–579.

Molecular—Nanogels

Nanogels: Chemical Approaches to Preparation

Sepideh Khoee
Hamed Asadi
Polymer Chemistry Department, School of Science, University of Tehran, Tehran, Iran

Abstract

Nanogels are swollen nanosized networks composed of hydrophilic or amphiphilic polymer chains that have attracted considerable attention as multifunctional polymer-based drug delivery systems. They possess high water content, biocompatibility, and desirable mechanical properties. A tunable size, a large surface area, and an interior network make nanogels a promising platform that has the characteristics of an ideal drug delivery vehicle. Present and future nanogel applications require a high degree of control over properties that are possible by preparation strategies. This entry describes the recent developments of nanogel particles as drug delivery carriers for biological and biomedical applications. Various synthetic strategies for the preparation of nanogels with restriction of chemical approaches are detailed, including the formation of nanogels using preformed polymers and via direct polymerization of monomers using heterogeneous free radical polymerization and heterogeneous controlled/living radical polymerization.

INTRODUCTION

Brief Overview of Nanogels

Polymer-based drug delivery systems (DDS) have attracted significant attention in biomedicine, pharmaceutics, and bio-nanotechnology. In particular, polymer-based DDS with controllable release of therapeutics and cell targeting have the potential to treat numerous diseases, including cancers, with a reduction in the side effects of the drugs. Several types of polymer-based DDS have been explored and nanogels are among the most promising DDS.[1–4] Nanogels are hydrogel networks that are confined to submicron-size. In other words, nanogels are defined as nanosized networks of chemically or physically cross-linked polymers composed of hydrophilic or amphiphilic chains.[5] Like hydrogels, nanogels are three-dimensional biocompatible materials with high water content. For drug delivery applications, key features including high water content/swellability, biocompatibility, and adjustable chemical/mechanical properties are particularly attractive. The large surface area provides space for functionalization and bioconjugation. In addition, they have a tunable size from submicrons to tens of nanometers, and their size[3,6] can be tuned to an optimal diameter for increased blood circulation time *in vivo* after IV administration. A smaller diameter (<200 nm) enables better cellular uptake and reduced NP uptake by mononuclear phagocyte system.[7,8] Finally, the interior network allows for the encapsulation of therapeutics.

These unique properties made nanogels as a topic of increasing interest across multi-disciplinary fields over the past decade as evidenced by the appearance of a number of excellent reviews on the preparation, properties, and applications of nanogels.[1,9–23] For example, the physical entrapment of bioactive molecules (including drugs, proteins, carbohydrates, and nucleic acids in the polymeric network) and their *in vitro* release kinetics have been extensively investigated. In addition, the incorporation of inorganic materials has been reported. Examples include quantum dots[24,25] and magnetic NPs[26,27] for optical and magnetic imaging, and gold nanorods for photodynamic therapy.[28]

PREPARATION OF NANOGELS

Given the immense application potentials of nanogels in various applications, significant efforts have been devoted to the design and synthesis of nanogels. Nanogels are prepared by various methods of copolymerization of hydrophilic or water-soluble monomers in the presence of difunctional or multifunctional cross-linkers.[13] They include photolithographic, micromolding, microfluidic, reverse micellar, membrane emulsification, and homogeneous gelation methods.[13] They are also prepared by heterogeneous free-radical polymerization in various media including precipitation,[29–31] inverse (mini)-emulsion, and inverse microemulsion.[13] Controlled/living radical polymerization (CRP) techniques[32–35] have been explored in various media including water, (inverse) miniemulsion, and dispersion for the preparation of cross-linked particles and gels with well-controlled polymer segments.[36–44] Here we only focus on chemical approaches that can synthesize nanogels in large quantities. With this restriction, we divide existing synthetic strategies into two main

Concise Encyclopedia of Biomedical Polymers and Polymeric Biomaterials DOI: 10.1081/E-EBPPC-120050693

categories. One strategy involves the formation of nanogels using preformed polymers whereas the other entails the formation of nanogels via the direct polymerization of monomers. These two approaches are illustrated in Fig. 1.[45]

Preparation of Nanogels from Polymer Precursors

Amphiphilic copolymers are known to self-assemble in solution to form various nanoscopic structures, thus providing a versatile platform to synthesize nanogels by simply locking the assembly. In this section, several examples of nanogel preparation through this methodology will be given and details discussed based on cross-linking reaction types employed to fix the self-assembled polymer.

Disulfide-based cross-linking

Disulfide bonds are found in natural peptides and proteins and play an important role in keeping the structural stability and rigidity.[46,47] Due to its stability under certain conditions and its reversibility under others, disulfide-thiol chemistry is becoming increasingly popular in conventional polymer syntheses and provides a facile route to the preparation of recyclable cross-linking of micelles.[48,49] Moreover, disulfide can reversibly undergo reduction to thiols depending on the environmental thiol concentration. For example, the thiol concentration varies in our body, depending on the location as well as pathological conditions. It is about 10 μM in the plasma but about 10 mM in the cytosol. It is seven times higher around some tumors.[50,51] In addition, there exist other methods of reducing disulfide bonds to thiols, including dithiothreitol, zinc dust, and UV light.[52] The thiol groups can be subsequently reacted with reactive groups such as activated disulfides, maleimides, iodoacetyl groups, and some thiol-containing biomolecules (e.g., antisense oligonucleotides).[53,54]

These unique properties suggest that the well-defined functional nanogels hold great potential as carriers for controlled drug delivery scaffolds to target specific cells. Drugs can be loaded into the nanogels. In a reducing environment, the nanogels will degrade to individual polymeric chains with a molecular weight below the renal threshold, thus it leads to controllably releasing the encapsulated drugs over a desired period of time.

Thayumanavan and coworkers prepared the nanogels system based on RAFT-synthesized copolymers of oligo(ethylene glycol) methacrylate (OEGMA) and pyridyl disulfide-derived methacrylate (PDS-MA) of different compositions and molecular weights.[55–57] The addition of deficient amounts of dithiothreitol (DTT) reduces a controlled percentage of PDS groups to thiols, which subsequently reacted with an equivalent amount of the remaining PDS groups to generate disulfide cross-links, and as a result nanogels were formed (Fig. 2).[55] The authors

demonstrated nanogels with different sizes can be easily obtained by varying polymer concentration and utilizing the lower critical solution temperature (LCST) behavior of polymers. The authors encapsulated a hydrophobic dye, Nile red, into the nanogels to test the encapsulation capability and the responsive release of hydrophobic

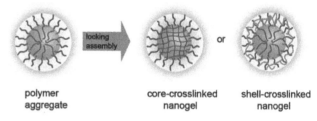

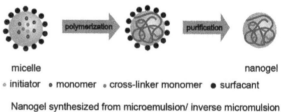

Fig. 1 Methods of nanogel synthesis: the polymer precursor method and the emulsion method.
Source: From Chacko et al.[45] © 2012, with permission from Elsevier.

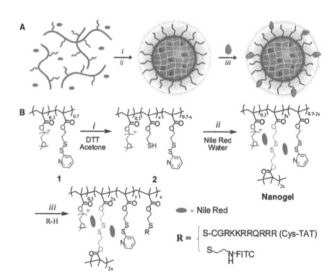

Fig. 2 Design and synthesis of the polymer nanoparticles. (**A**) Schematic representation of the preparation of biodegradable nanogels with surface modification. (**B**) Structure of the polymer and nanogel. (i) Cleavage of specific amount of PDS group by DTT. (ii) Nanogel formation by inter/intrachain cross-linking. (iii) Surface modification of nanogels with thiol-modified Tat peptide or FITC.
Source: From Ryu et al.[55] © 2010, with permission from American Chemical Society.

molecules and also demonstrated the possibility of further functionalizing the nanogels using the remaining PDS groups. For this purpose, the authors functionalized the nanogels with thiol-modified fluorescein isothiocyanate (FITC) and a modified cell-penetrating peptide, Tat-SH, containing a C-terminal cysteine. They also showed that the multifunctional nanoparticles are indeed interesting materials for drug delivery applications.

Control of the stability of the polymer associates as DNA delivery systems has been a long-standing challenge. The associate must maintain its structure during the circulation and efficiently dissociate to release DNA in order to exert biological effects inside the cell. The core-shell-type polyion complex (PIC) micelle with a disulfide cross-linked core was recently proposed, which has the ability to dissociate in response to chemical stimuli given at the site of the drug action. Kataoka, Harada, and Kataoka[58] reported the preparation of PIC composed of poly(ethylene glycol)-*block*-poly (L-lysine) (PEG-*b*-PLL) copolymers modified with thiol groups using *N*-succinimidyl 3-(2-pyridyldithio) propionate (SPDP) to construct PIC micelles with a disulfide cross-linked core for the delivery of plasmid DNA and antisense oligo-DNA (Fig. 3).

In an another study, they reported a PIC micelle siRNA delivery system prepared from the block copolymer poly(ethylene glycol)-block-poly(L-lysine) (PEG-b-PLL) modified with the cross-linking reagent 2-iminothiolane (2-IT, Traut's reagent) (Fig. 4).[59]

The resulting block copolymer, termed PEG-b-PLL(IM), was designed to contain cationic amidine groups for PIC formation with anionic siRNAs and also free sulfhydryls to allow disulfide cross-linking in the micelle core for improved stability. Covalent disulfide cross-links are particularly attractive for micelle core stabilization because they are reversible and more susceptible to cleavage (reduction) at the subcellular site of activity where the levels of natural disulfide-reducing agents are higher than in the

bloodstream. Disulfide cross-linked PIC micelle formation between siRNA and thiol-modified cationic block copolymer is shown in Fig. 5.[60]

Amine-based cross-linking

One of the most commonly used groups in the preparation of nanogels is amine groups due to their reactivity toward carboxylic acids, activated esters, isocyanates, iodides, and others. The Wooley group utilized amine cross-linkers to develop a methodology for the preparation of shell-cross-linked knedel-like structures (SCKs).[61] SCKs are essentially unimolecular polymer micelles, which are prepared by stabilizing the basic structure of the spherical micellar assembly through linking together of the hydrophilic portions of the chains within the micelle shell.

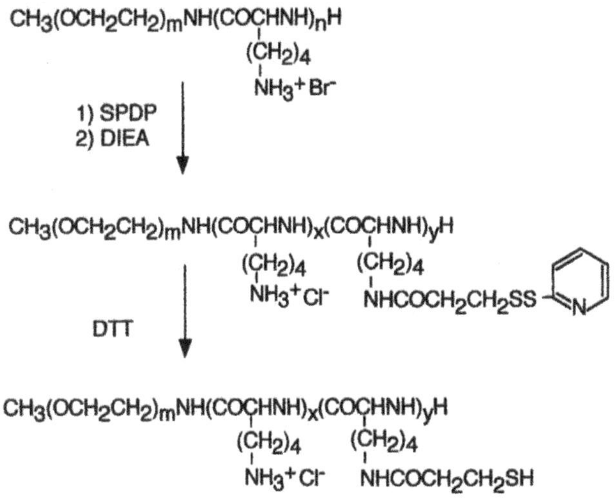

Fig. 3 Introduction of thiol groups into the lysine residues of PEG-PLL.
Source: From Kakizawa et al.[58] © 2001, with permission from American Chemical Society.

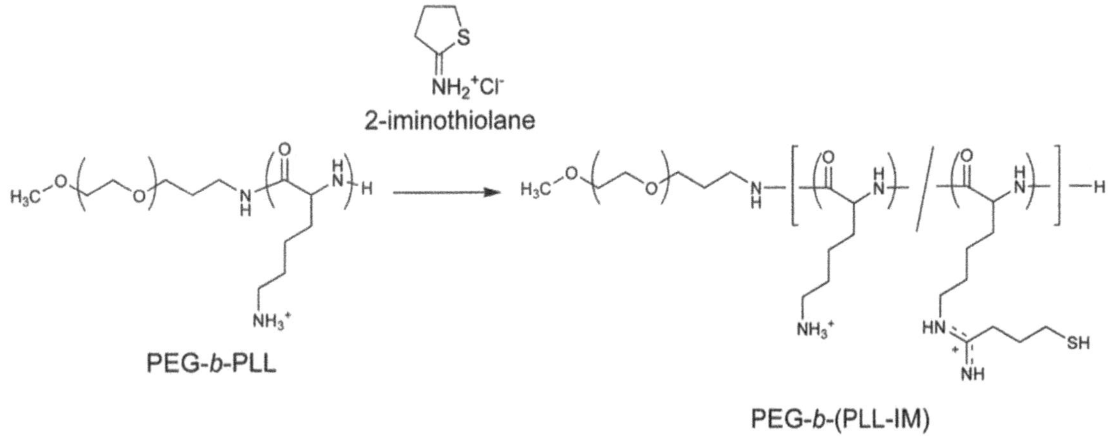

Fig. 4 Preparation of iminothiolane-modified poly(ethylene glycol)-block-poly(L-lysine) [PEG-b-(PLL-IM)].
Source: From Matsumoto et al.[59] © 2009, with permission from American Chemical Society.

They synthesized a variety of amphiphilic block copolymers in which poly(acrylic acid) were employed as the hydrophilic and cross-linkable block. Following the self-assembly of the block copolymers, the amidation of carboxylic acid with diamine cross-linkers resulted in cross-linking of the micellar assemblies and formation of nanogel networks. They also demonstrated that the remaining carboxylic groups on the shell can be converted to other functionalities for orthogonal surface modification.[62] In another study, they utilized diamine cross-linker containing acetal group for the preparation of pH-responsive SCKs. As detailed, after being allowed to assemble supramolecularly into micelles in water, the amphiphilic block copolymers of poly(acrylic acid) (PAA) and polystyrene (PS) were cross-linked through amidation reactions with a unique acetal-containing cross-linker (Fig. 6).[63]

By using reaction between amines and carboxylic acid, Zhang and coworkers reported the preparation of ferrocene-based shell cross-linked (SCL) thermoresponsive hybrid micelles with antitumor efficacy. The SCL micelle consisting of a cross-linked thermoresponsive hybrid shell and a hydrophobic core domain was fabricated via a two-step process: micellization of poly(N-isopropylacrylamide-co-aminoethyl methacrylate)-b-polymethyl methacrylate P(NIPAAm-co-AMA)-b-PMMA in aqueous solution followed by cross-linking of the hydrophilic shell layer via the amidation reaction between the amine groups of AMA units and the carboxylic acid functions of 1,1'-ferrocenedicarboxylic acid.[64]

In addition to carboxylic acids, activated esters also can be used for cross-linking with amines. In this regard, McCormick et al.[65] reported the synthesis of reversible SCL micelles by cross-linking of reactive N, N-acryloxysuccinimide (NAS) units incorporated within an ABC tri-block copolymer, poly(ethylene oxide)-block-[(N, N-dimethyl-acry-lamide)-stat-(N-acryloxysuccinimide)]-block (N-isopropylacrylamide), [PEO-b-P(DMA-stat-NAS)-b-NIPAM], with cystamine, a reversible cross-linking agent. The McCormick group also reported the synthesis of pH-responsive shell cross-linked micelles consisting of α-methoxypoly (ethylene oxide)-b-poly[N-(3-aminopropyl) methacrylamide]-b-poly[2-(diisopropylamino)ethyl methacrylate] (mPEO-PAPMA PDPAEMA) by reaction with an amine-reactive polymeric cross-linking agent, NHS-PNIPAM-NHS (Fig. 7).[66]

Pentafluorophenyl acrylate (PFPA) is another activated ester used for the preparation of nanogels. In this regard, Thayumanavan et al.[67] proposed a facile methodology for the preparation of water-dispersible nanogels based on pentafluorophenyl acrylate and polyethylene glycol methacrylate random copolymer and diamine cross-linkers. They envisaged that the addition of a calculated amount of diamine to a solution of the PPFPA-r-PPEGMA random copolymer will cause inter- and intrachain cross-linking amidation reactions to afford the nanogel. They also demonstrated the possibility of further functionalizing the

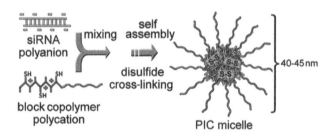

Fig. 5 Preparation of disulfide cross-linked PIC micelles containing siRNA.
Source: From Christie et al.[60] © 2011, with permission from American Chemical Society.

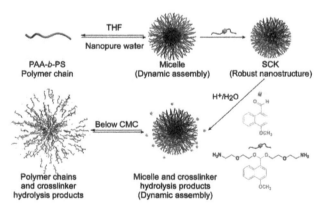

Fig. 6 Illustration of SCKs that contain hydrolytically labile cross-links to allow for pH-triggered hydrolysis and nanostructure disassembly.
Source: From Li et al.[63] © 2008, with permission from American Chemical Society.

NHS-PNIPAM-NHS

Aqueous
25 °C
pH 10.0

Shell Cross-linked Micelles

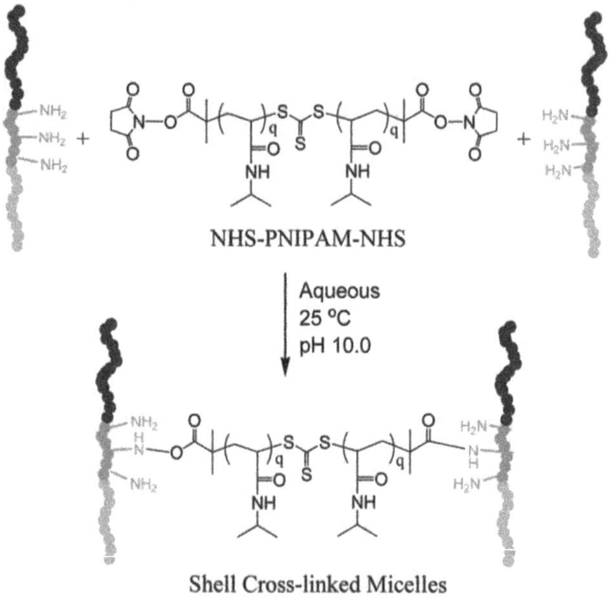

Fig. 7 Cross-linking of the PAPMA shell via polymeric cross-linking agent NHS-PNIPAM-NHS in aqueous solution at 25°C.
Source: From Xu et al.[66] © 2008, with permission from American Chemical Society.

nanogels using the remaining reactive PFP functionalities, as illustrated in Fig. 8.

Furthermore, Davis, Boyer, and coworkers described the synthesis of pH-sensitive core-shell nanoparticles that contain both methanethiosulfonate and activated ester PFPA pendant functionalities within core and poly oligo ethylene glycol (POEG) in shell. They used activated esters in the copolymer to cross-link the nanoparticles with difunctional amino cross-linkers bearing an acid cleavable bond (ketal) to generate pH-sensitive nanogels.[68]

Additionally, reactions of isocyanate with amines provide another cross-linking approach to make nanogels.[69] Cross-linked micelles with pH-responsive features were obtained by the addition of excess amounts of 1,8-diaminooctane to micellar aggregates consisting of poly (poly (ethylene glycol) methyl ether methacrylate)-block-poly (methyl methacrylate-co-poly(3-isopropenyl-a, a-dimethylbenzyl isocyanate)) (PPEGMEMA-b-P(MMA-co-TMI)) (Fig. 9).

Click chemistry–based cross-linking

In the past several years, "click chemistry,"[70–76] as termed by Sharpless et al., has gained a great deal of attention due to its high specificity, quantitative yield, tolerance to a broad variety of functional groups, and applicability under mild reaction conditions. The Wooley and Hawker groups have reported a nanogel fabrication method utilizing click

chemistry.[77,78] Alkynyl shell functionalized block copolymer micelles consist of diblock poly(acrylic acid)$_{80}$-b-poly(styrene)$_{90}$ were utilized as click-readied nanoscaffolds for the formation of nanogels. The following click reaction between click-readied micelles and azido dendrimers resulted in nanogel networks (Fig. 10).[77]

They also found that only the first generation of azido terminated dendrimers successfully cross-linked the hydrophilic shell of the alkynyl functionalized micelles and that is because of the hydrophobic nature of the dendrimer of generations greater than one that is proved to be incompatible with the hydrophilic nature of the micelle corona within the aqueous reaction conditions, and as a result did not behave as effective cross-linkers. They exploited this incompatibility by changing the cross-linking site from the hydrophilic shell to the hydrophobic core of the polymer micelle and synthesized the core click cross-liked micelles of amphiphilic diblock copolymers of poly(acrylic acid)-b-poly(styrene) (PAA-b-PS) that contained alkynyl functionality throughout the hydrophobic PS block.[79] The alkynyl core-functionalized block copolymer micelles and azido-terminated dendrimers were employed to construct nanogels with click reaction (Fig. 11).

Liu and coworkers utilized click chemistry to prepare core cross-linked PIC micelles with thermoresponsive coronas.[80] As detailed, azide-containing monomer was incorporated into two oppositely charged backbones of two graft ionomers, P(MAA-co-AzPMA)-g-PNIPAM and

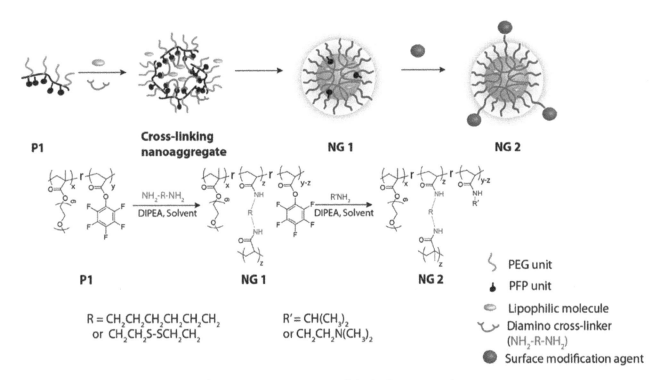

Fig. 8 Schematic representation of design and synthesis of the cross-linked polymer nanogels.
Source: From Zhuang et al.[67] © 2012, with permission from American Chemical Society.

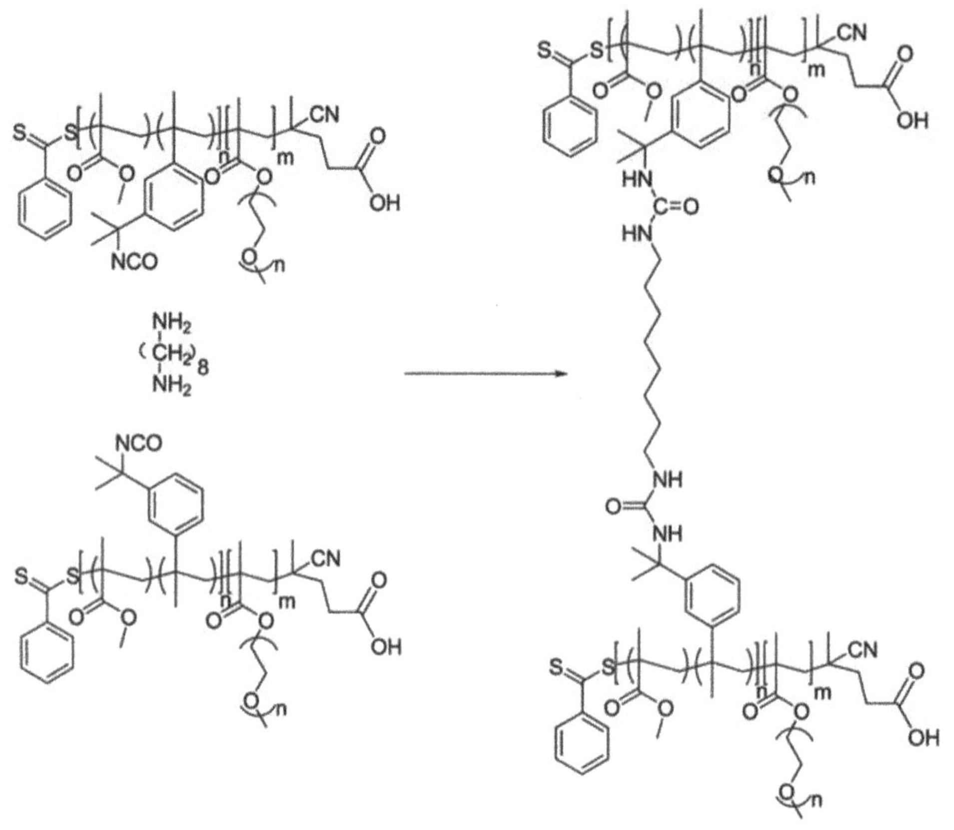

Fig. 9 Synthesis of core cross-linked block copolymer micelles.
Source: From Kim et al.[69] © 2011, with permission from The Royal Society of Chemistry.

P(QDMA-*co*-AzPMA)-*g*-PNIPAM, containing thermo-sensitive PNIPAM graft chains. The self-assembled PIC micelles in aqueous solution were subsequently core-stabilized via "click" reactions upon the addition of a difunctional reagent, propargyl ether (Fig. 12).

Stenzel et al.[81] reported the preparation of click-cross-linked micelle structure as a carrier to deliver cobalt pharmaceuticals. They synthesized block copolymers of poly (propargyl methacrylate)-*block*-poly(poly(ethylene glycol) methyl ether methacrylate) (P(PAMA)-*b*-P(PEGMA)) with pendant alkyne groups (Fig. 13), which self-assembled in aqueous solution into micelles that alkyne groups in the core took on two functions, acting as a ligand for $Co_2(CO)_8$ to generate a derivative of the antitumor agents based on (alkyne) $Co_2(CO)_6$ as well as an anchor point for the cross-linking of micelles via click chemistry.

They also synthesized core-cross-linked micelles by clicking bi-functional Pt(IV) anticancer drugs to isocyanate groups of micelle cores.[82] In a one-pot reaction, the incorporation of anticancer drug and core cross-linking was simultaneously carried out by using the highly effective reaction of isocyanate groups in the core of the polymeric micelles poly(oligo(ethylene glycol) methyl ether methacrylate)-*block*-poly(styrene-*co*-3-isopropenyl-R,R-dimethylbenzyl isocyanate) (POEGMA-*block*-P(STY-*co*-TMI)) with amine groups in the prepared platinum(IV) drug.

Liu et al.[83] reported the preparation of two types of shell-cross-linked (SCL) micelles with inverted structures via click chemistry starting from a well-defined schizophrenic water-soluble triblock copolymer in purely aqueous solution. They present an efficient synthesis of two types of SCL micelles with either pH- or temperature-sensitive cores from a novel poly (2-(2-methoxyethoxy) ethyl methacrylate)-*b*-poly (2-(dimethylamino) ethyl methacrylate)-*b*-poly (2-(diethylamino) ethyl methacrylate) (PMEO$_2$MA-*b*-P(DMA-*co*-QDMA)-*b*-PDEA) triblock copolymer. First, PMEO$_2$MA-*b*-PDMA-*b*-PDEA was prepared and the DMA blocks were partially converted to a quaternized DMA (QDMA) block with click-cross-linkable moieties to form novel schizophrenic water-soluble triblock copolymers. The pH- or temperature-induced micellization and subsequent shell cross-linking of the P(DMA-*co*-QDMA) inner shell with the tetra-(ethylene glycol) diazide via click chemistry resulted in nanogel networks (Fig. 14).

Recently, Liu's group also reported the fabrication of thermoresponsive cross-linked hollow poly(N-isopropylacrylamide) (PNIPAM) nanocapsules with controlled shell thickness via the combination of surface-initiated atom transfer radical polymerization (ATRP) and "click" cross-linking.[84] Cross-linked PNIPAM nanocapsules were fabricated by the "click" cross-linking of PNIPAM shell layer with a tri-functional molecule, 1,1,1-tris (4-(2-propynyloxy)phenyl)ethane. Due to the thermo-responsiveness

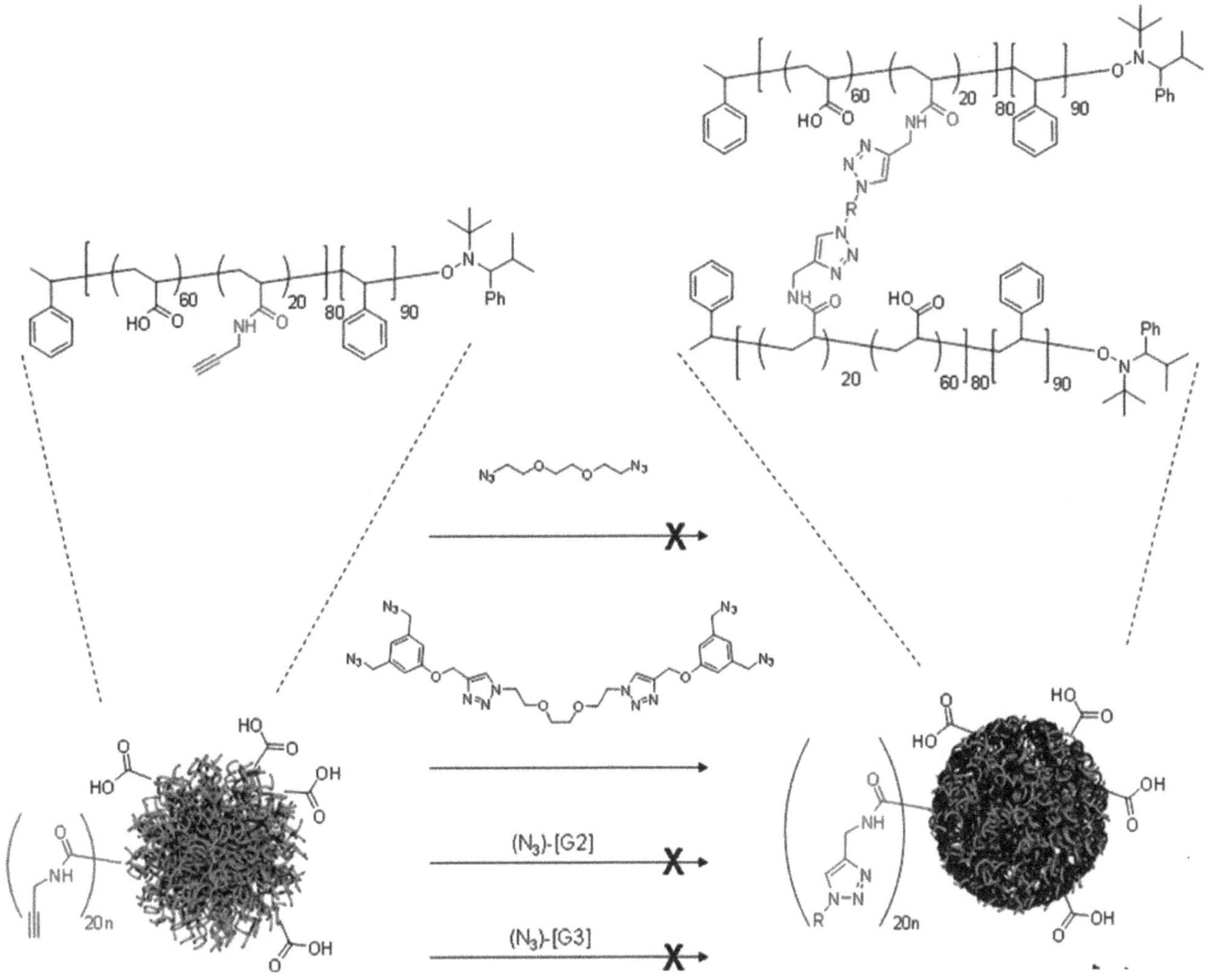

Fig. 10 Synthesis of shell click cross-linked nanoparticles from click-readied micelles and dendrimers, where R represents the dendritic cross-linking unit, having the possibility of multiple cross-linkages and remaining azido functionalities.
Source: From Joralemon et al.[77] © 2005, with permission from American Chemical Society.

of PNIPAM, cross-linked PNIPAM nanocapsules exhibit thermo-induced collapse/swelling transitions that make it possible to classify them as nanogels.

Imine bonds–induced cross-linking

Dynamic covalent bonds have been used to endow polymeric systems with the abilities to adapt their structures or compositions in response to external stimuli.[85] In this regards, Fulton et al. showed that dynamic covalent imine bonds can be used to cross-link linear polymer chains into core cross-linked star (CCS) polymer and nanogel nanoparticles.[86] Imine condensation reactions are a particularly appealing dynamic covalent reaction as it is possible to tune the stability of the imine bond by altering stereo electronic characteristics of the reaction partners, in particular the carbonyl-derived part.[87] They utilized RAFT polymerization to prepare novel aldehyde and amine functional styrenic- and methyl methacrylate-based copolymers and have demonstrated that these copolymers can cross-link through imine bond

formation to prepare core cross-linked star polymers and spherical cross-linked nanogels. As a continuing effort, Fulton and coworkers reported the preparation of nanogels whose disassembly into their component polymer chains is triggered by the simultaneous application of two different stimuli by combination of imine and disulfide bonds.[88] They used acrylamide-based linear copolymers displaying pyridyl disulfide appendages and either aldehyde or amine functional groups for nanogel preparation (Fig. 15).

Additionally, McCormick et al, utilized imine bonds for the preparation of reversible imine SCL micelles.[89] They synthesized a temperature-responsive triblock copolymer, α-methoxypoly(ethylene oxide)-b-poly(N-(3-aminopropyl) methacrylamide)-b-poly(N-isopropylacrylamide) (mPEO-PAPMA-PNIPAM), via aqueous RAFT polymerization. By increasing the solution temperature above the LCST of the PNIPAM block, the polymers self-assembled into micelles. Subsequently, the PAPMA shell was cross-linked with terephthaldicarboxaldehyde to generate SCL micelles with cleavable imine linkages (Fig. 16).

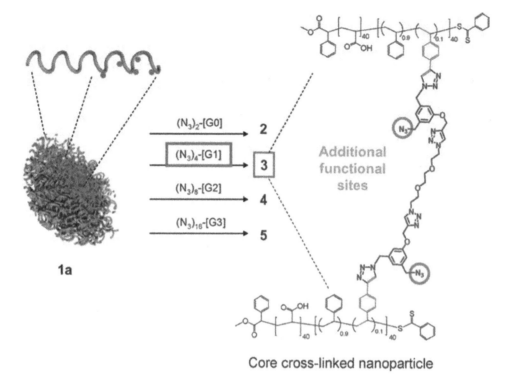

Fig. 11   Synthesis of polymer nanoparticles from alkynyl-functionalized micelles and azido-terminated dendrimers.
Source: From O'Reilly et al.[79] © 2007, with permission from The Royal Society of Chemistry.

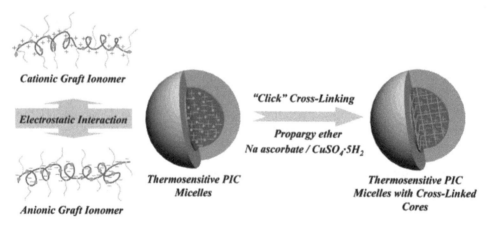

Fig. 12   Schematic illustration of formation of thermosensitive polyion complex (PIC) micelles and their core cross-linking via click chemistry.
Source: From Zhang et al.[80] © 2008, with permission from American Chemical Society.

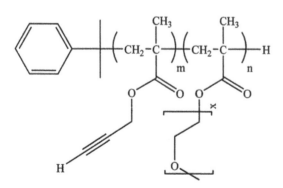

Fig. 13   The structure of P(PAMA)-b-P(PEGMA).

Photo-induced cross-linking

All of the cross-linking methods mentioned earlier need a cross-linking agent and/or catalyst, and the cross-linked micelles need to be purified to remove unreacted cross-linking agents and byproducts. In general, photo-cross-linking is a clean method, compared to the chemical methods, because no cross-linking agents are needed, and no byproducts are formed. As an alternative to the preceding cross-linking techniques, photo-induced cross-linking has been utilized to stabilize polymer assemblies that are functionalized with polymerizable or dimerizable units (Fig. 17).[90]

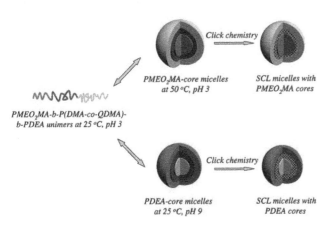

Fig. 14 Schematic illustration of the schizophrenic micellization behavior of PMEO2MA45-b-P(DMA0.65-*co*-QDMA0.35) 47-b-PDEA36 and the preparation of two types of SCL micelles with inverted structures in aqueous solution.
Source: From Jiang et al.[83] © 2009, with permission from American Chemical Society.

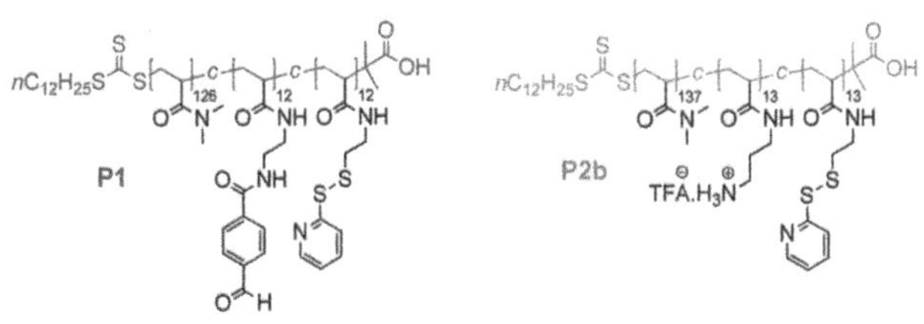

Fig. 15 Polymer chains P1 and P2b are used for nanogel preparation through imine bond.
Source: From Jackson & Fulton[88] © 2012, with permission from American Chemical Society.

For instance, He, Tong, and Zhao[91] reported the preparation of photoresponsive nanogels based on photocontrollable cross-links. In this regards, they used double hydrophilic block copolymers containing coumarin, which is known to dimerize when treated with UV light >310 nm. The design of such photoresponsive nanogels is schematically illustrated in Fig. 18. They combined the use of block copolymer self-assembly and a reversible photo-crosslinking reaction. Basically, with a water-soluble diblock copolymer, one block of which displays a LCST and bears photochromic side groups, micelles can be obtained by heating the solution to T > LCST and cross-linked by the photoreaction of the chromophore upon illumination at λ_1. By cooling the solution to T < LCST, the preparation of nanogel is completed with cross-linked water-soluble nanoparticles. Because of the reversibility of the photoreaction, the cross-linking degree of the nanogel can be reduced in a controlled fashion by illumination at λ_2, which leads to the swelling of nanogel particles.

In another study, the Zhao group as a continuing effort reported the synthesis of both core and shell-cross-linked nanogels with the incorporation of photo-cross-linkable moieties to either the core or shell of nanogels.[92]

Additionally, Yusa et al.[93] prepared a stimuli-responsive nanogel in water by a CCL technique using AB diblock

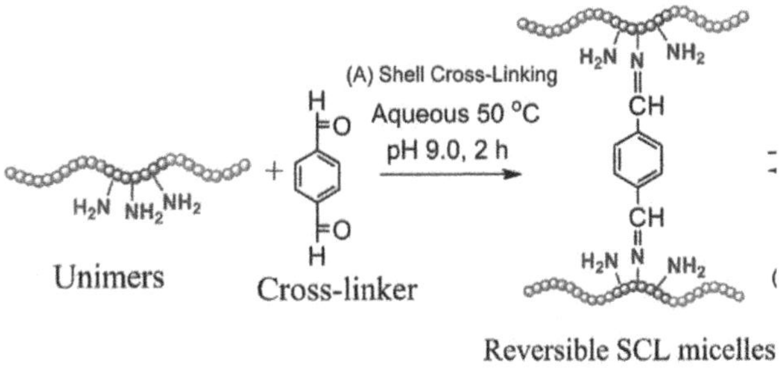

Fig. 16 Cross-linking of mPEO113-PAPMA12-PNIPAM136 triblock copolymer micelles in aqueous solution.
Source: From Xu et al.[89] © 2011, with permission from American Chemical Society.

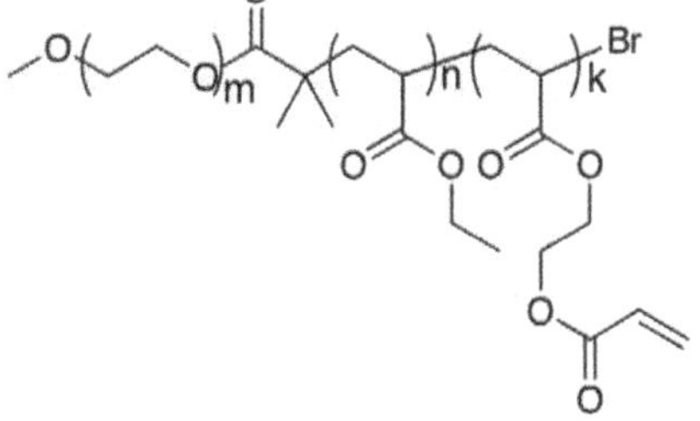

Fig. 17 Polymer precursor with photopolymerizable functionality.

copolymer micelles by photo-cross-linking of the micelle core (Fig. 19). They utilized poly (ethylene glycol)-b-poly (2-(diethylamino) ethyl methacrylate-*co*-2-cinnamoyloxyethyl acrylate) (PEG-b-P (DEAEMA/CEA)), a pH-responsive block copolymer, for nanogel synthesis. While the solubility of DEAEMA depends on the pH, increasing the pH results in the formation of micelles that subsequently cross-link via UV irradiation made nanogels.

The photo-based cross-linking was also used for preparation of thermosensitive nanogels prepared from photocross-linkable copolymers of *N*-isopropylacrylamide (NIPAAm) and 2-dimethylmaleinimido ethylacrylamide (DMIAAm) (Fig. 20).[94]

As mentioned earlier, the principle of this method is based on the phase transition phenomena of temperaturesensitive polymers in combination with photochemistry. At elevated temperatures, the phase-separated structure of aqueous PNIPAAm copolymer solutions was fixed by UV irradiation.

Sugihara et al.[95] introduced an approach for the synthesis of thermoresponsive SCL micelles via living cationic polymerization and UV irradiation. The block copolymers are based on poly(2-ethoxyethyl vinyl ether)-block-poly(2-hydroxyethyl vinyl ether) (PEOVE-b-PHOVE) with methacryloyl groups derived from VEM in the PHOVE segment (shown in Fig. 21). They showed that the resulting SCL micellar core was reversibly hydrated or dehydrated, depending on the solution temperature, because an aqueous PEOVE solution undergoes LCST-type phase separation around 20°C.

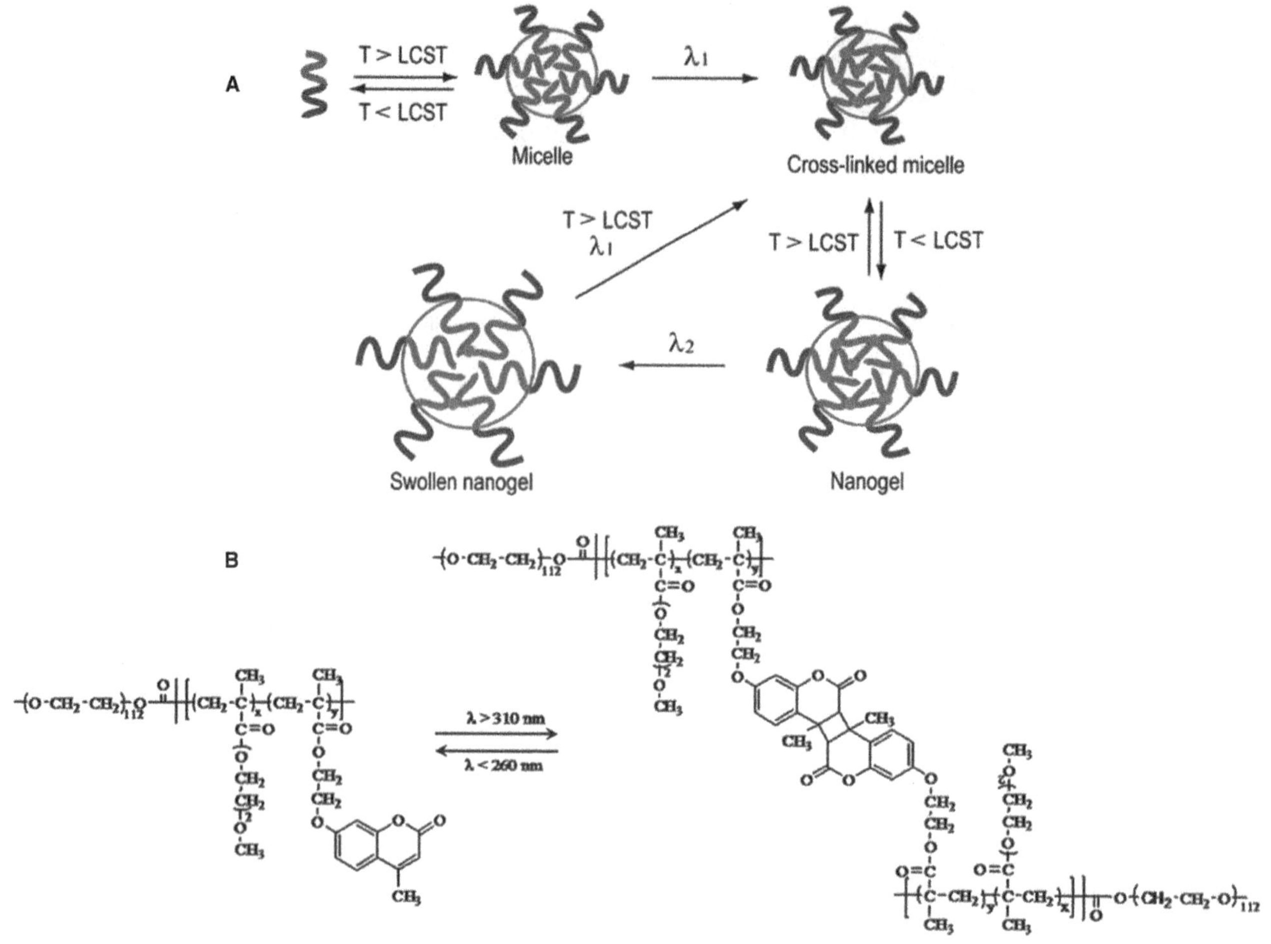

Fig. 18 (**A**) Schematic illustration of the preparation and photocontrolled volume change of nanogel. (**B**) Designed diblock block copolymer bearing coumarin side groups for the reversible photo-cross-linking reaction.
Source: From He et al.[91] © 2009, with permission from American Chemical Society.

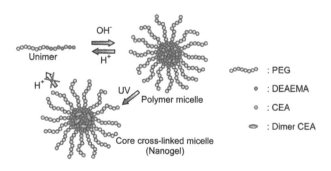

Fig. 19 Schematic illustration of pH-dependent micellization of PEG-b-P(DEAEMA/CEA) and the formation of a nanogel with UV-irradiation.
Source: From Yusa et al.[93] © 2009, with permission from American Chemical Society.

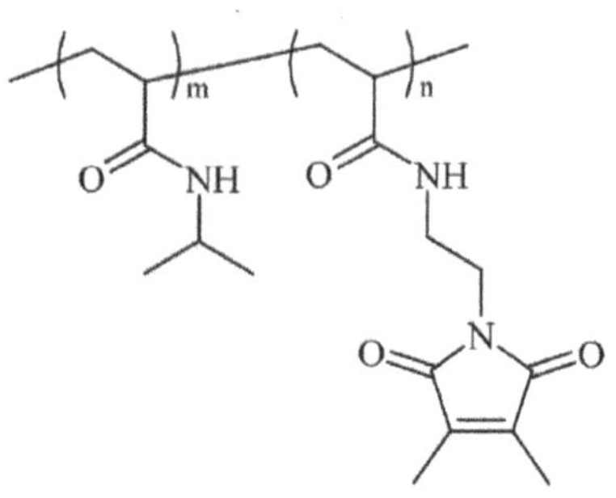

Fig. 20 The structure of NIPAAm-DMIAAm copolymer.

They also demonstrated that it is possible to obtain a response to one stimulus without interfering with the other stimulus.

Physical cross-linking

The physical self-assembly of polymers was used by several research groups to produce various nanogels. This method usually involves a controlled aggregation of hydrophilic polymers capable of hydrophobic or electrostatic

By using of this technique, Kuckling, Vo, and Wohlrab[96] reported the preparation of temperature-responsive colloidal nanogels with a more complex structure. They prepared temperature-responsive colloidal nanogels with a pH-responsive shell by photo-cross-linking of poly(*N*-isopropylacrylamide) (PNIPAAm) graft terpolymers (Fig. 22).

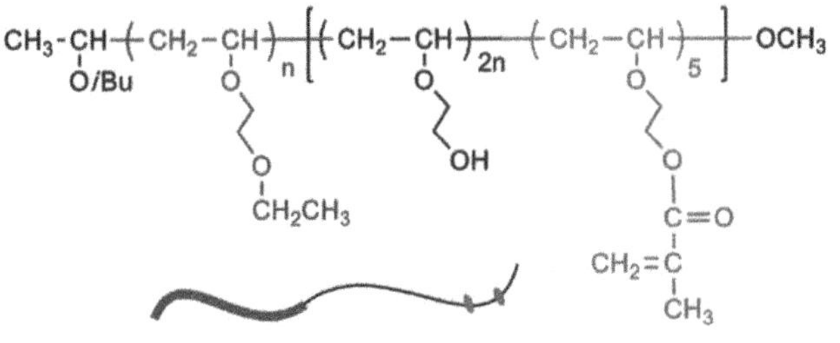

Fig. 21 The structure of PEOVE-b-P(HOVE/VEM) synthesized by living cationic polymerization.

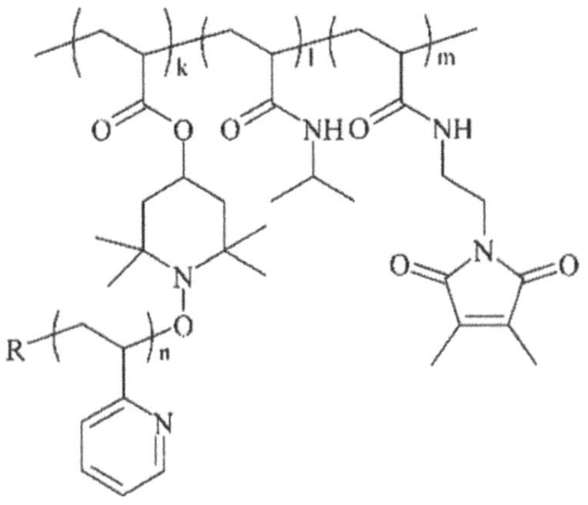

Fig. 22 The structure of PNIPAAm graft terpolymer used for nanogel preparation by photo-induced cross-linking.

interactions and/or hydrogen bonding with each other. The preparation of nanogels is conducted in mild conditions and in aqueous media. A review regarding nanogel preparation from these associating polymers has recently been published.[19]

Preparation of Nanogels via Monomer Polymerization

While cross-linking preformed polymers, physically or chemically, represent an important strategy for the preparation of nanogels, especially for naturally occurring water-soluble polymers, the preparation of nanogels via direct monomer polymerization combines the two processes of polymerization and the formation of nanogels in a one-pot protocol. Thus, in a sense, synthesizing nanogels via monomer polymerization highlights an accelerated strategy, which in some cases involves polymerization-induced self-assembly of the in situ formed polymers. In some other cases, the formation of nanogels via polymerization requires the application of templating methods to realize the desired size and colloidal stability of the nanogels. Given the current intensive pursuit for highly efficient chemical approaches to the preparation of polymers and polymeric nanostructures, such an accelerated strategy of preparing nanogels plays an increasingly important role in modern polymer chemistry and materials.[97]

The predominantly used polymerization technique to fabricate nanogels is free-radical polymerization, because of its ease of manipulation, efficiency, tolerance of functionality, and adaptability to water-based heterogeneous systems. Recently, other polymerization techniques such as ring opening polymerization have been used for the synthesis of polyether-based nanogels.[23] By considering the recent development of polymerization techniques, it is not unreasonable to expect that the scope of polymerization techniques that can be used to fabricate nanogels will be significantly expanded.

Among the mentioned techniques, nanogel synthesis in water via direct polymerization of monomer is obviously advantageous because such a process uses water as the dispersant. However, this process requires the polymer to have a transition in solubility, a property that not every hydrophilic polymer has. A more general process for the synthesis of nanogels of hydrophilic polymers is inverse phase polymerizations, in which the continuous phase is oil and the dispersed phase is water containing hydrophilic monomers. The dispersed aqueous phase is stabilized by surfactant, which also serves as templates for the formation of nanogels.[98] But for some monomers, nanogel synthesis by direct polymerization in an aqueous dispersion system is impossible due to the good water solubility of both the monomer and the corresponding polymer.

Heterogeneous free radical polymerization

Various heterogeneous polymerization reactions of hydrophilic or water-soluble monomers in the presence of either difunctional or multifunctional cross-linkers have been mostly utilized to prepare well-defined synthetic nanogels. They include precipitation, inverse (mini)emulsion, and inverse microemulsion polymerization utilizing an uncontrolled free radical polymerization process.

Precipitation polymerization

Precipitation polymerization involves the formation of a homogeneous mixture at its initial stage and the occurrence of initiation and polymerization in the homogeneous solution. As the formed polymers are not swellable but soluble in the medium, the use of a cross-linker is necessary to cross-link polymer chains for the isolation of particles. As a consequence, the resulting cross-linked particles often have an irregular shape with high polydispersity (PDI).[99–102]

The preparation of microgels and nanogels based on PNIPAM and its derivatives by precipitation polymerization in water has been extensively explored for biomedical applications.[103–109] This is due to the thermosensitive properties of PNIPAM-nanogels that undergo volume change at LCST in water, at around 32°C. Above the LCST, PNIPAM hydrogels and nanogels are hydrophobic and expel water; below LCST, they are hydrophilic and swollen in water. These unique properties facilitate the loading of drugs into nanogels as well as enhance the controllable release of drugs encapsulated in nanogels.[110] However, the use of PNIPAM-based nanogels shows a certain limitation to biomedical applications due to a relatively narrow range of physical and chemical properties of PNIPAM.

Inverse (mini) emulsion polymerization

Inverse miniemulsion polymerization is a water-in-oil heterogeneous polymerization process that forms kinetically stable macro-emulsions at, below, or around the critical micellar concentration (CMC). This process contains aqueous droplets (including water-soluble monomers) stably dispersed, with the aid of oil-soluble surfactants, in a continuous organic medium. Stable inverse miniemulsions are formed under high shear by either a homogenizer or a high-speed mechanical stirrer. Oil-soluble nonionic surfactants with hydrophilic-lipophilic balance value around four are used to implement colloidal stability of the resulting inverse emulsion. Upon addition of radical initiators, polymerization occurs within the aqueous droplets producing colloidal particles (Fig. 23).[111]

Several reports have demonstrated the synthesis of hydrophilic or water-soluble particles of PHEMA,[112] PAA,[113] and PAAm,[112] temperature-sensitive hollow microspheres of PNIPAM,[114] core-shell nanocapsules with hydrophobic shell and hydrophilic interior,[115] and polyaniline nanoparticles.[116] This method has been explored to prepare cross-linked nanogels in the presence of difunctional cross-linkers for effective drug delivery. This is due to a facile confinement of water-soluble drugs in aqueous droplets dispersed in continuous organic solvents. Temperature- and pH-sensitive P(NIPAM-co-AA) minigels[117] and PAAm/PAA interpenetrating polymer network nanogels[118] were prepared in the presence of N,N'-methylenebisacrylamide. Their swelling properties were studied in water by measuring particle diameter upon change in temperature and pH. Stable, cross-linked, amphiphilic nanoparticles based on acrylated triblock copolymer of poly(ethylene glycol)-b-poly(propylene glycol)-b-poly(ethylene glycol) (PEO-b-PPO-b-PEO) were

prepared by inverse emulsion photopolymerization (Fig. 24). The hydrophobic PPO-rich domains enhanced the incorporation of doxorubicin (Dox), an amphiphilic anticancer drug, up to 9.8 wt%. The resulting microgels had a diameter of 50 and 500nm.[119]

Inverse microemulsion polymerization

While inverse (mini)emulsion polymerization forms kinetically stable macro-emulsions at, below, or around the CMC, inverse microemulsion polymerization produces thermodynamically stable microemulsions upon further addition of emulsifier above the critical threshold. This process also involves aqueous droplets, stably dispersed with the aid of a large amount of oil-soluble surfactants in a continuous organic medium; polymerization occurs within the aqueous droplets, producing stable hydrophilic and water-soluble colloidal nanoparticles having a diameter of less than 50–100 nm.[120,121]

This method has been also explored for the preparation of well-defined nanoparticles,[122–127] magnetic polymeric particles,[27,128] and nanogels in the presence of difunctional cross-linkers.[129–134]

Poly(vinylpyrrolidone)-based nanogels incorporated with Dex as a water-soluble macromolecular carbohydrate drug were prepared.[129,130] Cationic nanogels of poly (HEA-co-AETMAC) were prepared in the presence of oligo(ethylene glycol) dimethacrylate (OEGDMA) as a cross-linker. The diameter of the resulting nanogels decreased from 150 to 40 nm as the amount of cross-linker increased. The resulting nanogels had potential for gene delivery, since the presence of quaternary ammonium ion side groups appeared to enhance the incorporation of DNA into nanogels via electrostatic association with the phosphate groups.[131]

Poly (amino acid)-based nanoparticles with different surface PEGylation were prepared. a,b-Poly (N-2-hydroxyethyl)-D,L-aspartamide (PHEAS) and PEG-modified PHEA (PHEAS-PEG) were functionalized with a methacrylate group and then polymerized by UV irradiation in inverse microemulsion. The resulting nanoparticles

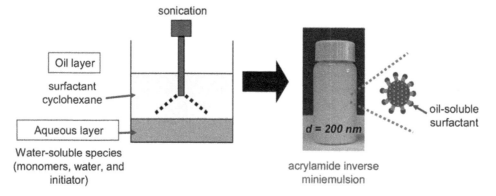

Fig. 23 A schematic representation of inverse miniemulsion or microemulsion polymerization for the preparation of nanometer-sized particles of water-soluble and water-swellable polymers as well as cross-linked particles in the presence of cross-linkers.
Source: From Oh et al.[1] © 2009, with permission from Elsevier.

had a size of around 250 nm in diameter by TEM. The fluorescein-loaded PHEA-based nanoparticles were prepared in the presence of fluorescein sodium salts, and examined for cellular uptake using macrophage cells.[132]

Ultrafine hydrophilic PAAm-based nanogels incorporated with meta-tetra (hydroxyphenyl) chlorine (mTHPC) as a photosensitizer were prepared for photodynamic therapy. In the process, both AAm and N,N-methylene(bis acrylamide) as a cross-linker were directly emulsified into hexane/aerosol OT, without water, which allowed for the preparation of tiny PAAm nanoparticles with a diameter of 2–3 nm.[133] The presence of water in hexane/aerosol OT produced 20 nm diameter PAAm particles with a relatively broad size distribution.[134]

The resulting nanogels with both polymerization methods mentioned earlier are produced in the form of dispersions in organic solvents with oil-soluble surfactants. Biomedical applications require the removal of residual monomers, oil-soluble surfactants, and organic solvents. Nanogels are then redispersed in water before use. For comparison, the aqueous precipitation polymerization of water-soluble monomers in the presence of difunctional cross-linkers is also allowed for the preparation of microgels/nanogels.[13,44,135–138] However, in order to avoid microscopic gelation, the concentration of monomers in water (reaction media) should be kept low. In addition, this method also needs the purification of the resulting polymers by removal of residual monomers and initiators from reaction mixtures.

Heterogeneous controlled/living radical polymerization

CRP provides a versatile route for the preparation of (co) polymers with controlled molecular weight, narrow molecular weight distribution (i.e., Mw/Mn, or PDI < 1.5), designed architectures, and useful end-functionalities.[32–34] Various methods for CRP have been developed; however, the most successful techniques include ATRP,[35,139] stable free radical polymerization,[140] and reversible addition fragmentation chain transfer (RAFT) polymerization.[141,142] CRP techniques have been explored for the synthesis of gels[38,44] and cross-linked nanoparticles of well-controlled polymers in the presence of cross-linkers.

Atom-transfer radical polymerization

ATRP is one of the most successful CRP techniques, enabling the preparation of a wide spectrum of polymers with predetermined molecular weight and relatively narrow molecular weight distribution (Mw/Mn < 1.5).[35,139] ATRP also allows for the preparation of copolymers with different chain architectures, such as block, random, gradient, comb-shaped, brush, and multi-armed star copolymers.[143–145] In addition, ATRP has been utilized to prepare polymer–protein/peptide bioconjugates,[146–148] polymer modified polysaccharides,[149,150] micellar nanoparticulates,[151,152] hydrogels and nanogel,[153] ligands stabilizing metal nanocrystals for cellular imaging,[154,155] surface-initiated brushes,[156,157] and emulsion particles.[158,159]

The mechanism of ATRP is based on a rapid dynamic equilibration between a minute amount of growing radicals and a large majority of dormant species.[160] In a normal ATRP process (Fig. 25), transition metal complexes in a lower oxidation state (Cu(I)/Lm) are added directly to the reaction as an activator, which reacts reversibly with the dormant species (RX) to generate a deactivator (X–Cu(II)/Lm) and an active radical (R). The radical can propagate with the addition of monomers (M) and is rapidly deactivated by

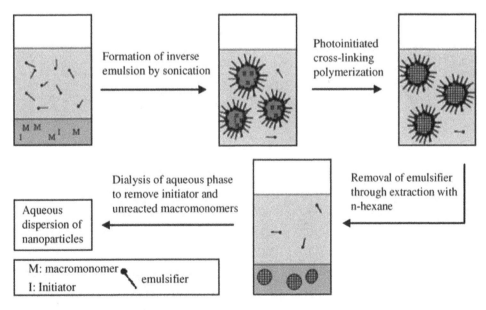

Fig. 24 Illustration of preparation for microgels of PEO-b-PPO-b-PEO via inverse emulsion polymerization.
Source: From Missirlis et al.[119] © 2005, with permission from American Chemical Society.

reacting with X–Cu(II)/ Lm, regenerating Cu(I)/Lm and a halogen-terminated polymeric chain.

Recently, a versatile method for preparing and functionalizing well-defined biodegradable nanogels for targeted drug delivery applications has been developed.[13,155,161,162] The method utilizes ATRP, inverse miniemulsion polymerization,[111] and disulfide–thiol exchange.[163,164] The novel approach allows the preparation of biomaterials with many useful predeterminable site-specific features. As illustrated in Fig. 26, the features include uniform network, high loading efficiency, novel distributed functionality including bromine end groups, and degradation by hydrolysis or through a disulfide–thiol exchange.[1]

Hydrogel NPs of PNIPAAm were prepared by precipitation polymerization via ATRP in water.[165] OEOMA, an analog of PEG has been polymerized by AGET ATRP in homogenous aqueous solution[166] and in heterogeneous conditions.[167] In this context, biodegradable cross-linked nanogels of well-controlled hydrophilic polymers were synthesized using ATRP in inverse miniemulsion in the presence of a disulfide-functionalized dimethacrylate (DMA) cross-linker.[167] The nanogels preserved a high degree of halide end-functionality that enabled further functionalization, including chain extension to form functional block copolymers. The nanogels were nontoxic to cells and degraded in a reducing environment to individual polymeric chains with a relatively narrow molecular weight distribution (Mw/Mn < 1.5), indicating the formation of a uniformly cross-linked network within the NPs. This uniform structure is expected to improve the controlled release of encapsulated species. The measured swelling ratio, degradation behavior, and colloidal stability of nanogels prepared by ATRP were superior to those prepared by conventional free radical inverse miniemulsion polymerization. In another report, these nanogels were loaded with Dox, an anticancer drug.[155] The nanogels released Dox *in vitro* upon exposure to glutathione, which degraded the nanogels.

Since ATRP results in polymers with a high degree of halide end-functionality, facile functionalization with various molecules is possible. Also, the preparing of functional nanogels using ATRP in inverse miniemulsion using functional ATRP initiators and copolymerization with functional monomers offer two approaches toward bioconjugation.[168]

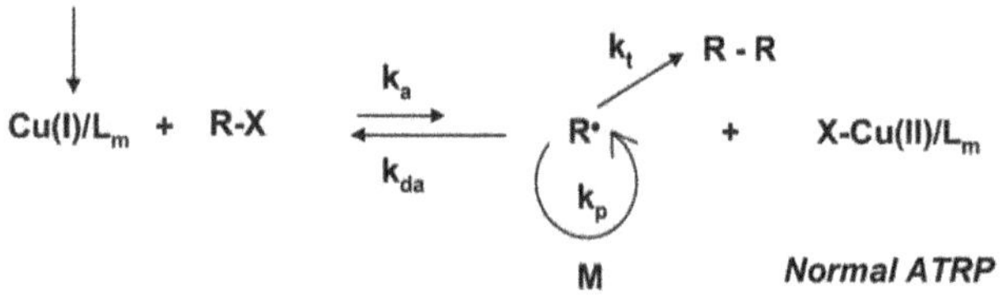

Reversible addition–fragmentation chain transfer

The development of emerging technologies has imposed ever-increasing demand on the synthesis of well-defined polymer architectures with predictable properties and functions. Fortunately living/controlled radical polymerizations are becoming powerful tools to meet these challenges and have been widely used in the design and preparation of architecturally defined and functional macromolecules with predetermined molecular weight, composition, and narrow molecular weight distribution.[176–171] Among the

$$Cu(I)/L_m + R\text{-}X \underset{k_{da}}{\overset{k_a}{\rightleftharpoons}} R^\bullet \overset{k_t}{\rightarrow} R\text{-}R + X\text{-}Cu(II)/L_m$$

Normal ATRP

Fig. 25 Mechanism of normal ATRP.

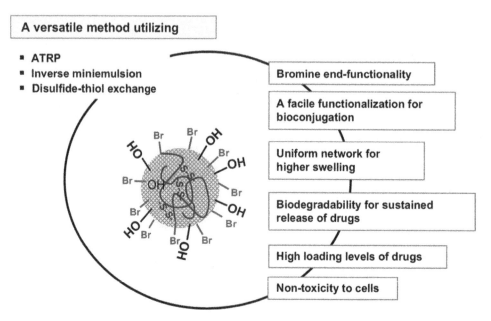

Fig. 26 Illustration of unique features of biodegradable nanogels prepared by atom transfer radical polymerization in inverse miniemulsion of oligo(ethylene glycol) monomethyl methacrylate in the presence of a disulfide-functionalized dimethacrylate.
Source: From Oh et al.[1] © 2009, with permission from Elsevier.

major living/controlled radical polymerizations, reversible addition–fragmentation chain transfer (RAFT)[172–175] polymerization has been advocated by many to be the most versatile polymerization technique.

The major strengths of the RAFT approach include the following:

1. An ability to control the polymerization of a wide range of monomers in varying solvents, including water, using only chain transfer agents (CTAs) and common free radical initiators (without the need for any additional polymerization component such as metal catalysts).[172,176,177]
2. The tolerance to a wide variety of functional groups, allowing the facile synthesis of polymers with pendant, and alpha and omega end-group functionalities (an important feature for biological applications).[178–184]
3. The ability to synthesize a wide variety of architectures such as telechelic, block copolymers, graft copolymers, gradient copolymers, nanogels, stars, and dendritic structures.[185–188]
4. The compatibility of RAFT with a variety of established polymerization methods such as bulk, solution, suspension,[189] emulsion,[190,191] and dispersion[192,193] polymerizations.
5. The ability to perform polymerizations from a wide variety of substrates, allowing the modification of surfaces and the in situ generation of polymer conjugates.[178,194–197]

Nanogels via direct RAFT polymerization

The most widely used approach for nanogel preparation is direct radical polymerization of vinyl monomers in the presence of cross-linkers, because it is easy to carry out, rapid in polymerization rate, amenable to various reaction conditions, compatible with a variety of functionalities, and flexible in initiation methods. As an avenue to nanogels, the RAFT polymerization largely inherits the majority of the advantages of nanogel synthesis by traditional free-radical polymerization. More importantly, it offers additional merits that are unmatchable by the traditional free-radical approach. Firstly, the RAFT approach offers excellent control over molecular weight and molecular weight distribution of the constitutional polymers of the nanogels, which is an important structural parameter that determines the mechanical and responsive properties of the nanogels. Secondly, the RAFT approach allows uniform incorporation of cross-linkers into the nanogels, a feature that is in stark contrast to the nanogels synthesized by the free-radical process. Thirdly, the RAFT approach allows precise localization of functional groups into the nanogels and the direct formation of well-defined nanogel architectures, providing a versatile technique in the design of multi-functional and exquisite nanogel structures for value-added applications.[97]

Nanogel synthesis by RAFT polymerization in water

An interesting approach that combines polymerization and self-assembly of the in situ produced polymers is nanogel synthesis by RAFT polymerization in water in the form of precipitation/dispersion polymerization. More importantly, it has three distinct features that distinguish it from other approaches: 1) In aqueous dispersion/precipitation polymerization systems, water is used as the sole dispersant without the addition of any organic solvents. This is a particularly important feature considering the current intensive pursuit for green polymerization processes; 2) The polymerization can be carried out at high solid content, much higher than many other nanogel synthetic approaches; 3) No surfactant is added to provide colloidal stability of the nanogels. Such a surfactant-free process is not only economic but also environmentally benign.

The first example of nanogel synthesis by direct RAFT polymerization under precipitation/dispersion polymerization condition was reported by An and coworkers in 2007 (Fig. 27).[44] Two types of poly(N,N'-dimethylacrylamide)s (PDMAs) bearing a trithiocarbonate group were first synthesized by RAFT solution polymerization and were subsequently used as both stabilizers and RAFT agents for nanogel synthesis by RAFT precipitation/dispersion polymerization. These two types of PDMAs were either hydrophilic or amphiphilic due to the different R groups of the RAFT agents. When the cross-linker N,N'-methylenebisacrylamide was used, thermosensitive nanogels were produced. When no cross-linker was used, after cooling to room temperature, the nanoparticles dissociated into double hydrophilic block copolymers, which allowed for convenient characterization of the polymers. The use of both types of PDMAs permitted the production of well-defined block copolymers with low values of Mw/Mn, indicating a controlled nature of the RAFT process under precipitation/dispersion conditions.

In addition, the RAFT precipitation/dispersion polymerization process transferred the functional groups of the RAFT agents to those of the produced nanogels, allowing for precise localization of the functionalities into either the core or the shell. By using the carboxylic acid groups at the nanogel surface, bioconjugation of the nanogels with albumin was produced in this study.

Taking advantage of the combination of the RAFT process and click chemistry, heterofunctional nanogels was demonstrated by An and coworkers.[198] By using an azide-functionalized CTA, telechelic PDMAs and PNIPAMs bearing the azide group at their α-end and the trithiocarbonate group at their ω-end were produced by RAFT synthesis. The ω-end trithiocarbonate group was subjected to a one-pot aminolysis and thiol-ene addition, and subsequently, the α-end azide group was reacted with functional alkyne by click chemistry. This approach to obtain hetero-functionalized nanogels indeed demonstrated that multi-functional nanogels with precise location can be produced

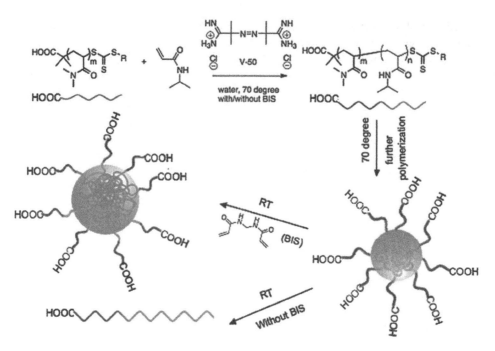

Fig. 27   Macro-CTAs and schematic representation of the RAFT precipitation polymerization process.
Source: From An et al.[44] © 2007, with permission from American Chemical Society.

by RAFT precipitation/dispersion polymerization and that further functionalization is possible to the functional groups installed by the tailored CTAs.

PNIPAM is perhaps the mostly studied thermosensitive polymers and as such the majority of nanogels studied are based on PNIPAM. Recently, there has been much interest in developing PNIPAM alternatives.[190–204] Polymers based on OEGMAs have come to the forefront as a new generation of thermosensitive polymers with improved biocompatibility, sharpness of responsiveness, antifouling properties below LCST, and bio-inertness.[205]

In this regard, Shen and coworkers recently developed a novel type of biocompatible, antifouling, and thermosensitive core-shell nanogels by RAFT aqueous dispersion polymerization.[206] Using both linear poly(ethylene glycol) (PEG) and nonlinear poly(ethylene glycol) methyl ether methacrylate (PEGMA, Mn = 475) polymers bearing either trithiocarbonate or dithioester CTAs, they performed RAFT dispersion polymerizations of di(ethylene glycol) methyl ether methacrylate (MEO$_2$MA) and PEGMA of varied ratios to obtain nanogels of adjustable thermosensitivities. The hydrophilic PEG or PPEGMA polymers became the shell, and the thermosensitive MEO$_2$MA-PEGMA copolymers constituted the core of the nanogels.

Rieger and coworkers investigated dispersion polymerization of N,N'-diethylacrylamide (DEAAm) for the formation of PEGylated thermally responsive block copolymers and nanogels using amphiphilic Macro-CTAs.[135] Two types of Macro-CTAs were studied both of which had a hydrophobic dodecyl tail. The first Macro-CTA was

poly(ethylene oxide) monomethyl ether (Mn = 2000) functionalized with a trithiocarbonate (PEO-TTC). The second one is a block copolymer of PEO-b-PDMA synthesized from PEO-TTC with increased hydrophilic chain length. Both Macro-CTAs exhibited controlled polymerization of DEAAm under dispersion polymerization in water. While the use of PEO-TTC alone produced rather large and heterogeneous gel particles due to the insufficient stabilization of the short PEO chains, the use of PEO-b-PDMA-TTC afforded well-defined nanogels with PEO shell when the PDMA block had a sufficient chain length. Increasing the PDMA length decreased the nanogel size. The influence of cross-linker and monomer concentration on the nanogel formation was also studied in this work, which was further investigated in a recent study by the same research group.[207]

The thermosensitivity of poly($N,N0$-dimethylaminoethyl methacrylate) (PDMAEMA) was also used for nanogel synthesis. By using an amphiphilic Macro-CTA similar to the one used by Rieger and coworkers but with a much higher PEO molecular weight (Mn = 24,000), Yan and Tao reported the synthesis of PDMAEMA nanogels via dispersion polymerization in water.[208] Since PDMAEMA is also a pH-sensitive polymer, the prepared PEGylated cationic nanogels may find application in gene delivery.

The preparation of nanogels by direct RAFT polymerization is still a relatively young area though successful examples have been demonstrated and clear advantages have been established. Surprisingly, although RAFT dispersion polymerization has been well studied in organic media[209–215] and supercritical CO_2,[216–218] their study in water has been limited to a few monomers including those aforementioned and 2-hydroxypropyl methacrylate that was recently studied by Li and Armes.[219] Considering the fact that a variety of thermosensitive and other stimuli-responsive polymers have been extensively studied, significant expansion of nanogel synthesis by direct RAFT polymerization can be envisaged.

Nanogels synthesis by inverse RAFT miniemulsion

Qi and coworkers reported the first study combining RAFT and inverse miniemulsion.[220] Their inverse miniemulsion system comprised cyclohexane as the continuous phase, B246SF as the surfactant, and aqueous solution containing a CTA, acrylamide, and costabilizer MgSO4. They found that using a water-soluble initiator, 4,4′-azobis(4-cyanovaleric acid), afforded better control of the polymerization of acrylamide than using a lipophilic one, 2,2′-azobis(2-methylpropionitrile) (AIBN). RAFT control was realized up to 50% monomer conversion and after that significant deviation from RAFT control was observed. More recently, they have extended their RAFT inverse miniemulsion polymerization approach to other hydrophilic (co)polymers.[221–224]

Nanocapsules with hydrophilic polymer walls can be considered as hollow nanogels, which are useful for encapsulation of hydrophilic materials in the aqueous interior. Using RAFT-confined interfacial inverse miniemulsion polymerization, Lu and coworkers developed a novel approach to the synthesis of nanocapsules (Fig. 28).[225] In their approach, an amphiphilic Macro-CTA was added to a conventional inverse miniemulsion such that both the surfactant and the amphiphilic Macro-CTA together stabilized the formed inverse miniemulsion. The location of the amphiphilic Macro-CTA at the oil–water interface was the initiating site for polymerization. After initiation of polymerization with the use of a water-soluble initiator, the formed oligomeric radicals would migrate to the interface for RAFT-controlled polymerization, and thus an interfacial layer of RAFT polymers was formed, which functioned as the wall of the nanocapsule. It was expected that the functionality and the thickness of the nanocapsules could be controlled through the RAFT polymerization process. In the control experiment where no Macro-CTA was added, polymerization led to the formation of solid particles instead of nanocapsules.

In a recent report, Wang and coworkers nicely extended Lu's approach by synthesizing SCL nanocapsules to improve the stability.[226] In this case, they used an amphiphilic PDMAEMA as the Macro-CTA to interfacially polymerize methylacrylic acid (MAA) in an inverse miniemulsion. The cross-linker they used was bis(acryloyloxyethyl) disulfide, which was dissolved in the continuous oil phase. The disulfide cross-linker could be degraded by reducing agents such as DTT as already demonstrated in numerous examples for the destruction of polymer nanoparticles.

APPLICATIONS OF NANOGELS

Nanogels show a lot of promise as delivery systems in particular due to their encapsulation stability, in addition to water solubility and biocompatibility. These nanocarriers have been utilized in a variety of fields including cancer drug delivery. The following examples from the literature demonstrate the diversity in applications and the versatility of these delivery systems. In 2010, Du et al. designed a pH-responsive charge-conversional nanogel for promoted tumoral-cell uptake and Dox delivery.[227] These nanogels were prepared from poly(2-aminoethyl methacrylate hydrochloride) (PAMA) and were subsequently treated with 2,3-dimethylmaleic anhydride (DMMA) to produce a negatively charged nanogel.

In another application, the delivery of therapeutic molecules to treat inflammatory disorders such as rheumatoid arthritis was investigated. Macrophage cells, of the immune system, are targeted for photodynamic therapy as a treatment for this disorder. Schmitt et al. developed chitosan-based nanogels decorated with hyaluronate to target macrophages, which were then loaded with one of three different photosensitizers.[228] Their studies show that photodynamic therapy using these hyaluronate-chitosan nanogels is effective in the treatment of inflamed articular joints.

Nanogels have also been utilized for the delivery of local anesthetic drugs. Yin et al. designed a biodegradable delivery system where lidocaine was encap-sulated into poly (ε-caprolactone)-poly (ethylene glycol)-poly (ε-caprolactone) (PCL-PEG-PCL or PCEC) nanoparticles.[229] Carriers were coated with hydrophilic thermosensitive Pluronic F-127 hydrogels to form the composite carrier nanogel system. The studies showed that nanogels effectively infiltrate into the wound for prolonging the effects of these anesthetics, which is useful for treatment during post-operative periods.

Physically cross-linked nanogels composed of hexadecyl groups bearing cationic cycloamylose have been developed for the delivery of pDNA.[230] Cycloamylose, a large cationic cyclodextrin ring, was employed for the complexation of pDNA. Phospholipase A2 was co-delivered with pDNA to help in endosome disruption by catalyzing the hydrolysis of membrane phospholipids triggering the release of the pDNA in the cytoplasm. It was found that the levels of protein expression due to the delivery of pDNA were enhanced with the co-delivery of PLA2 using the C16-catCA nanogels/PLA2 system. Hemolysis studies using sheep red blood cells showed that the activity for the native PLA2 was maintained when co-delivered with pDNA into the nanogels.

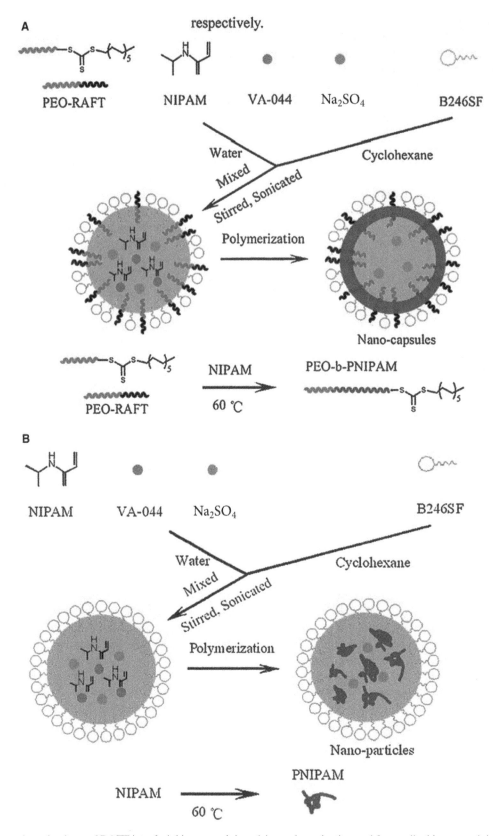

Fig. 28 Suggested mechanisms of RAFT interfacial inverse miniemulsion polymerization and free-radical inverse miniemulsion polymerization with NIPAM as the monomer as illustrated in Part (**A**) and (**B**), respectively.
Source: From Lu et al.[225] © 2010, with permission from American Chemical Society.

Tamura et al. studied the delivery of siRNA using a PIC based on PEGylated polyamine nanogels containing a chemically cross-linked core.[231]

In nucleic acid delivery complexes between amine-based carriers and nucleic acid, drugs are defined in terms of the ratio of amines on the carrier to the number of phosphates on the nucleic acid backbone. This ratio is referred to as the N/P ratio with a higher N/P ratio implying a greater number of amine-based complexing agents per nucleic acid. Gene silencing activity against firefly luciferase in HuH-7 cells showed that these nanogel/siRNA complexes possess high transfection efficiencies at low nitrogen to phosphate (N/P) ratios when compared with the uncross-linked complexes. However, cellular uptake of the nanogels/siRNA complexes was found to be lower than that of oligofectamine/siRNA complexes. This was suggested to be due to steric hindrance between PEG chains and the cell membrane. The chemically cross-linked nanogel was composed of poly[2-(N,N-diethylaminoethyl) methacrylate] (PDEAMA), which served as a siRNA complexing functionality and an aid for endosomal escape, bifunctional ethylene glycol dimethacrylate as the cross-linker, and heterobifunctional PEG macromonomer. Due to their size, most DDSs are taken up into cells by endocytosis. However, once inside the endosomes, the drugs are either ineffective because their sites of action are at the cytosol or are eventually degraded in the lysosomes. Thus, in most cases in order to achieve efficacious drug delivery a vehicle should not only internalize into the cell but also be able to escape the endosome effectively. To achieve endosomal escape, proton sponge functionality was introduced into the nanogel so that during internalization and subsequent acidification of the endosome, these functionalities buffer the endosomal pH eventually causing disruption of the endosome. This is presumably due to the relative increase in osmotic pressure inside the endosome. Thus, it was reasoned that the polyamine core of this nanogel was protonated as a result of the decrease in pH in the endosomal compartment facilitating the escape of the nanocarriers into the cytoplasm through the proton sponge effect. Thus, PEGylated polyamine siRNA nanogels have potential applications in cancer therapy and for the treatment of genetic disorders.

RELEASE MECHANISMS FROM NANOGELS

Biological agents can be incorporated in nanogels by 1) physical entrapment; 2) covalent conjugation; or 3) controlled self-assembly and released from nanogels through 1) simple diffusion; 2) degradation of the nanogel; 3) a shift in the pH value; 4) displacement by counterions present in the environment; or 5) transitions induced by an external energy source.[5]

Examples include diffusion-controlled release of Dox from pluronic-based hydrogels.[232] There is also increased interest in developing nanogels that can release biological agents in response to environmental signals at the disease site. A change in pH value or the presence of a reducing environment can serve as chemical signals that trigger the release. For example, an acrylamide-based nanogel with acetal cross-links is stable at an extracellular pH value of 7.4, but degrades and releases entrapped protein at pH 5.0.[233] Similarly, A poly-[oligo(ethylene oxide)-methyl methacrylate] nanogel with disulfide cross-links degraded in the presence of a glutathione tripeptide commonly found in cells.[155] The degradation of these nanogels was shown to trigger the release of the encapsulated low-molecular-mass solutes rhodamine 6G and Dox. In another study, the dissolution of disulfide cross-linked HA nanogels and the release of siRNA was induced by adding glutathione.[234] Clearly, the release kinetics can be fine-tuned in each case by altering the number of cross-links.

Polyelectrolyte hydrogels that incorporate biological agents through electrostatic bonds can also release biological agents in response to environmental changes. For example, pH-sensitive nanogels based on PAA can release an oppositely charged protein in tumor sites or endosomal compartments upon acidification.[235–238] A different mechanism was proposed for the release of nucleotide drugs from cationic PEG-cl-PEI nanogels.[239] In this case, negatively charged biomacromolecules bound to nanogels can be displaced by negatively charged cellular components. For example, the interaction of cationic nanogels with cellular membranes can trigger the release of anionic 5′-triphosphates of nucleoside analogues.[240]

In summary, the combination of the release approaches can provide a very useful means for the control of the drug-release characteristics of the nanogel carriers. For example, in the nanogels, drug release can be decreased by cross-linking the polymer chains, and it can be adjusted and made responsive to environmental changes by introducing cleavable cross-links. Furthermore, this technology offers the possibility to control the drug-release profiles. In contrast to liposomes and insoluble nanoparticles, the hydrophilic nanogels swell as the drug is released, which should sustain the release of the drug from the inner layers of the nanogels as the amount released increases. This can be used to modify or eliminate batch release or even to achieve zero-order release kinetics of the drug from nanogels delivered to the disease site.[231,240]

CONCLUSION

Recent developments of nanogels as drug delivery carriers for biological and biomedical applications were reviewed in this entry. Among major synthetic strategies for the preparation of nanogels, here we only focus on chemical approaches. With this restriction, we divide existing synthetic strategies into two main categories. One strategy involves the formation of nanogels using preformed

Molecular—Nanogels

polymers whereas the other entails the formation of nanogels via direct polymerization of monomers.

The preparation of nanogels from polymer precursor is based on the simply locking of self-assembled amphiphilic copolymers. This versatile methodology is discussed in detail based on cross-linking reaction types employed to fix the self-assembled polymer in this entry.

Other strategies used for nanogel preparation via the direct polymerization of monomers are heterogeneous free radical polymerization and heterogeneous controlled/living radical polymerization. Heterogeneous polymerization of hydrophilic or water-soluble monomers in the presence of either difunctional or multifunctional cross-linkers is the most general method used for the preparation of well-defined synthetic nanogels. This method as well as free radical precipitation, inverse (mini)emulsion, and inverse microemulsion polymerization have been explored.

CRP techniques including ATRP and RAFT, which are used to prepare stable nanogels of well-controlled polymers, are discussed, and among the ATRP methods, a new method utilizing ATRP, inverse miniemulsion polymerization, and thiol-disulfide exchange reaction allowed for the development and functionalization of stable biodegradable nanogels of well-controlled water-soluble polymers in the presence of a disulfide-functionalized DMA is comprehensively explored. This approach provides the opportunity for the preparation of biomaterials with many useful features, such as the preservation of a high degree of halide end functionality for facile functional block copolymerization and bioconjugation; biodegradation in the presence of a biocompatible glutathione tripeptide; formation of uniform networks; larger and better swelling ratio, degradation behavior, and colloidal stability compared to conventional counterparts; high loading level of anticancer drugs; and enhanced circulation time in the blood.

Nanogel synthesis by direct RAFT polymerization is a promising route to industrial setup because of high monomer conversion and high solids content of such processes. The control of nanogel architectures can be realized through the rational selection of polymerization systems such as dispersion polymerization or inverse miniemulsion. Nanogel synthesis by dispersion polymerization in water is particularly attractive for obvious reasons and more efforts should be directed to expand this approach to a wider range of monomers and to the control of other nanogel architectures and morphologies. Although the use of organic solvents as the continuous phase in RAFT inverse miniemulsion polymerization is an inherent limitation, this method has provided a nice set of tools that can be used by researchers from multidisciplinary fields in the design, synthesis, and application of tailored nanogels for bioapplications.

In the last section, some examples in literature are mentioned that demonstrate diverse nanogels and versatile applications as DDSs. In addition, the drug release mechanisms from nanogels are discussed.

About the future of nanogels, it can be said that one future goal of nanogel research should be the improved the design of nanogels with specific targeting residues to enable highly selective uptake into specific cells, particularly cancer cells. Polymer chemists and biologists can learn from each other to elucidate the specific interactions between biomolecules and cellular integrin receptors, which in turn can be carefully attached to advanced delivery systems. Through collaboration, advanced nanogels with careful control over stability, size, biodegradability, and functionality for bioconjugation can be realized.

ACKNOWLEDGMENTS

The authors would like to thank the University of Tehran for partial support during preparation of this entry.

REFERENCES

1. Oh, J.K.; Bencherif, S.A.; Matyjaszewski, K. Atom transfer radical polymerization in inverse miniemulsion: A versatile route toward preparation and functionalization of microgels/nanogels for targeted drug delivery applications. Polymer 2009, 50 (19), 4407–4423.

2. Zhang, H.; Mardyani, S.; Chan, W.C.W.; Kumacheva, E. Design of biocompatible chitosan microgels for targeted pH-mediated intracellular release of cancer therapeutics. Biomacromolecules 2006, 7 (5), 1568–1572.

3. Jung, T.; Kamm, W.; Breitenbach, A.; Kaiserling, E.; Xiao, J.X.; Kissel, T. Biodegradable nanoparticles for oral delivery of peptides: is there a role for polymers to affect mucosal uptake? Eur. J. Pharm. Biopharm. 2000, 50 (1), 147–160.

4. Raemdonck, K.; Demeester, J.; De Smedt, S. Advanced nanogel engineering for drug delivery. Soft Matter 2009, 5 (4), 707–715.

5. Kabanov, A.V.; Vinogradov, S.V. Nanogels as pharmaceutical carriers: Finite networks of infinite capabilities. Angew. Chem. Int. Ed. 2009, 48 (30), 5418–5429.

6. Rao, J.P.; Geckeler, K.E. Polymer nanoparticles: Preparation techniques and size-control parameters. Prog. Polym. Sci. 2011, 36 (7), 887–913.

7. Seymour, L.W.; Duncan, R.; Strohalm, J.; Kopecek, J. Effect of molecular weight (Mw) of N-(2-hydroxypropyl) methacrylamide copolymers on body distribution and rate of excretion after subcutaneous, intraperitoneal, and intravenous administration to rats. J. Biomed. Mater. Res. Part A 1987, 21 (11), 1341–1358.

8. Davis, M.E.; Chen, Z.; Shin, D.M. Nanoparticle therapeutics: An emerging treatment modality for cancer. Nat. Rev. Drug Discov. 2008, 7 (9), 771–782.

9. Lemieux, P.; Vinogradov, S.V.; Gebhart, C.L.; Guerin, N.; Paradis, G.; Nguyen, H.K.; Ochietti, B.; Suzdaltseva, Y.G.; Bartakova, E.V.; Bronich, T.K.; St-Pierre, Y.; Alakhov, V.Y.; Kabanov, A.V. Block and graft copolymers and nanogel copolymer networks for DNA delivery into cell. J. Drug Target. 2000, 8 (2), 91–105.

10. Vinogradov, S.V.; Bronich, T.K.; Kabanov, A.V. Nanosized cationic hydrogels for drug delivery: Preparation,

properties and interactions with cells. Adv. Drug Deliv. Rev. **2002**, *54* (1), 135–147.

11. Nayak, S.; Lyon, L.A. Soft nanotechnology with soft nanoparticles. Angew. Chem. Int. Ed. **2005**, *44* (47), 7686–7708.

12. Vinogradov, S.V. Polymeric nanogel formulations of nucleoside analogs. Expert Opin. Drug Deliv. **2007**, *4* (1), 5–17.

13. Oh, J.K.; Drumright, R.; Siegwart, D.J.; Matyjaszewski, K. The development of microgels/nanogels for drug delivery applications. Prog. Polym. Sci. **2008**, *33* (4), 448–477.

14. Oh, J.K.; Lee, D.I.; Park, J.M. Biopolymer-based microgels/nanogels for drug delivery applications. Prog. Polym. Sci. **2009**, *34* (12), 1261–1282.

15. Motornov, M.; Roiter, Y.; Tokarev, I.; Minko, S. Stimuli-responsive nanoparticles, nanogels and capsules for integrated multifunctional intelligent systems. Prog. Polym. Sci. **2010**, *35* (1–2), 174–211.

16. Oh, J. K. Engineering of nanometer-sized cross-linked hydrogels for biomedical applications. Can. J. Chem. **2010**, *88* (3), 173–184.

17. Oishi, M.; Nagasaki, Y. Stimuli-responsive smart nanogels for cancer diagnostics and therapy. Nanomedicine **2010**, *5* (3), 451–468.

18. Sanson, N.; Rieger, J. Synthesis of nanogels/microgels by conventional and controlled radical crosslinking copolymerization. Polym. Chem. **2010**, *1* (7), 965–977.

19. Sasaki, Y.; Akiyoshi, K. Nanogel engineering for new nanobiomaterials: From chaperoning engineering to biomedical applications. Chem. Rec. **2010**, *10* (6), 366–376.

20. Tamura, A.; Nagasaki, Y. Smart siRNA delivery systems based on polymeric nanoassemblies and nanoparticles. Nanomedicine **2010**, *5* (7), 1089–1102.

21. Zha, L.; Banik, B.; Alexis, F. Stimulus responsive nanogels for drug delivery. Soft Mater. **2011**, *7* (13), 5908–5916.

22. Hamidi, M.; Azadi, A.; Rafiei, P. Hydrogel nanoparticles in drug delivery. Adv. Drug Deliv. Rev. **2008**, *60* (15), 1638–1649.

23. Sisson, A.L.; Haag, R. Polyglycerol nanogels: Highly functional scaffolds for biomedical applications. Soft Mater. **2010**, *6* (20), 4968–4975.

24. Hasegawa, U.; Nomura S.I.M.; Kaul, S.C.; Hirano, T.; Akiyoshi, K. Nanogel quantum dot hybrid nanoparticles for live cell imaging. Biochem. Biophys. Res. Commun. **2005**, *331* (4), 917–921.

25. Fukui, T.; Kobayashi, H.; Hasegawa, U.; Nagasawa, T.; Akiyoshi, K.; Ishikawa, I. Intracellular delivery of nanogel-quantum dot hybrid nanoparticles into human periodontal ligament cells. Drug Metab. Lett. **2007**, *1* (2), 131–135.

26. Chatterjee, J.; Haik, Y.; Chen C.J. Biodegradable magnetic gel: Synthesis and characterization. Colloid Polym. Sci. **2003**, *281* (9), 892–896.

27. Gupta Ajay, K.; Wells, S. Surface-modified superparamagnetic nanoparticles for drug delivery: Preparation, characterization, and cytotoxicity studies. IEEE Trans. Nanobiosci. **2004**, *3* (1), 66–73.

28. Das, M.; Sanson, N.; Fava, D.; Kumacheva, E. Microgels loaded with gold nanorods: Photothermally triggered volume transitions under physiological conditions. Langmuir **2007**, *23* (1), 196–201.

29. Kim, J.; Nayak, S.; Lyon, L.A. Bioresponsive hydrogel microlenses. J. Am. Chem. Soc. **2005**, *127* (26), 9588–9592.

30. Thornton, P.D.; McConnell, G.; Ulijn, R.V. Enzyme responsive polymer hydrogel beads. Chem. Commun. **2005**, *47* (47), 5913–5915.

31. Ulijn, R.V.; Bibi, N.; Jayawarna, V.; Thornton, P.D.; Todd, S.J.; Mart R.J.; Smith, A.M.; Gough, J.E. Bioresponsive hydrogels. Mater. Today **2007**, *10* (4), 40–48.

32. Matyjaszewski, K., Davis, T.P. *Handbook of Radical Polymerization*; John Wiley & Sons Ltd.: New York, 2002.

33. Davis, K.A.; Matyjaszewski, K. Statistical, gradient, block, and graft copolymers by controlled/living radical polymerizations. Adv. Polym. Sci. **2002**, *159*, 1–169.

34. Coessens, V.; Pintauer, T.; Matyjaszewski, K. Functional polymers by atom transfer radical polymerization. Prog. Polym. Sci. **2001**, *26* (3), 337–377.

35. Braunecker, W.A.; Matyjaszewski, K. Controlled/living radical polymerization: Features, developments, and perspectives. Prog. Polym. Sci. **2007**, *32* (1), 93–146.

36. Ide, N.; Fukuda, T. Nitroxide-controlled free-radical copolymerization of vinyl and divinyl monomers. Evaluation of pendant-vinyl reactivity. Macromolecules **1997**, *30* (15), 4268–4271.

37. Shim, S.E.; Oh, S.; Chang, Y.H.; Jin, M.J.; Choe, S. Solvent effect on TEMPO-mediated living free radical dispersion polymerization of styrene. Polymer **2004**, *45* (14), 4731–4739.

38. Tsarevsky, N.V.; Matyjaszewski, K. Combining atom transfer radical polymerization and disulfide/thiol redox chemistry: A route to well-defined (bio)degradable polymeric materials. Macromolecules **2005**, *38* (8), 3087–3092.

39. Li, W.; Gao, H.; Matyjaszewski, K. Influence of initiation efficiency and polydispersity of primary chains on gelation during atom transfer radical copolymerization of monomer and cross-linker. Macromolecules **2009**, *42* (4), 927–932.

40. Min, K.; Gao, H.; Yoon; J.A; Wu, W.; Kowalewski, T.; Matyjaszewski, K. One-pot synthesis of hairy nanoparticles by emulsion ATRP. Macromolecules **2009**, *42* (5), 1597–1603.

41. Gao, H.; Matyjaszewski, K. Synthesis of functional polymers with controlled architecture by CRP of monomers in the presence of cross-linkers: From stars to gels. Prog. Polym. Sci. **2009**, *34* (4), 317–350.

42. Gao, H.; Miasnikova, A.; Matyjaszewski, K. Effect of cross-linker reactivity on experimental gel points during ATRP of monomer and cross-linker. Macromolecules **2008**, *41* (21), 7843–7849.

43. Huang, J.; Cusick, B.; Pietrasik, J.; Wang, L.; Kowalewski, T.; Lin, Q.; Matyjaszewski, K. Synthesis and in situ atomic force microscopy characterization of temperature-responsive hydrogels based on poly(2-(dimethylamino)ethyl methacrylate) prepared by atom transfer radical polymerization. Langmuir **2007**, *23* (1), 241–249.

44. An, Z.; Shi, Q.; Tang, W.; Tsung, C.K.; Hawker, C.J.; Stucky, G.D. Facile RAFT precipitation polymerization for the microwave-assisted synthesis of well-defined, double hydrophilic block copolymers and nanostructured hydrogels. J. Am. Chem. Soc. **2007**, *129* (46), 14493–14499.

45. Chacko, R.T.; Ventura, J.; Zhuang, J.; Thayumanavan, S. Polymer nanogels: A versatile nanoscopic drug delivery platform. Adv. Drug Deliv. Rev. **2012**, *64* (9), 836–851.

Molecular—Nanogels

46. Sun, K.H.; Sohn, Y.S.; Jeong, B. Thermogelling poly(ethylene oxide-b-propylene oxide-b-ethylene oxide) disulfide multiblock copolymer as a thiol-sensitive degradable polymer. Biomacromolecules **2006**, *7* (10), 2871–2877.

47. Castellani, O.F.; Martinez, E.N.; Anon, M.C. Role of disulfide bonds upon the structural stability of an amaranth globulin. J. Agric. Food Chem. **1999**, *47* (8), 3001–3008.

48. Christensen, L.V.; Chang, C.W.; Kim, W.J.; Kim, S.W.; Zhong, Z.Y.; Lin, C.; Engbersen, J.F.J.; Feijen, J. Reducible poly(amido ethylenimine)s designed for triggered intracellular gene delivery. Bioconjug. Chem. **2006**, *17* (5), 1233–1240.

49. Kakizawa, Y.; Harada, A.; Kataoka, K. Environment-sensitive stabilization of core–shell structured polyion complex micelle by reversible cross-linking of the core through disulfide bond. J. Am. Chem. Soc. **1999**, *121* (48), 11247–11249.

50. Saito, G.; Swanson, J.A.; Lee, K.D. Drug delivery strategy utilizing conjugation via reversible disulfide linkages: Role and site of cellular reducing activities. Adv. Drug. Deliv. Rev. **2003**, *55* (2), 199–205.

51. Lee, Y.; Koo, H.; Jin, G.W.; Mo, H.; Cho, M.Y.; Park, J.-Y.; Choi, J.S; Park, J.S. Poly(ethylene oxide sulfide): New poly(ethylene glycol) derivatives degradable in reductive conditions. Biomacromolecules **2005**, *6* (1), 24–26.

52. Kolano, C.; Helbing, J.; Bucher, G.; Sander, W.; Hamm, P. Intramolecular disulfide bridges as a phototrigger to monitor the dynamics of small cyclic peptides. J. Phys. Chem. B **2007**, *111* (38), 11297–11302.

53. Torchilin, V.P. Recent advances with liposomes as pharmaceutical carriers. Nat. Rev. Drug Discov. **2005**, *4* (2), 145–160.

54. Wang, L.; Kristensen, J.; Ruffner, D.E. Delivery of antisense oligonucleotides using HPMA polymer: synthesis of a thiol polymer and its conjugation to water-soluble molecules. Bioconjug. Chem. **1998**, *9* (6), 749–757.

55. Ryu, J.H.; Jiwpanich, S.; Chacko, R.; Bickerton, S.; Thayumanavan, S. Surface-functionalizable polymer nanogels with facile hydrophobic guest encapsulation capabilities. J. Am. Chem. Soc. **2010**, *132* (24), 8246–8247.

56. Ryu, J.H.; Chacko, R.T.; Jiwpanich, S.; Bickerton, S.; Babu, R.P.; Thayumanavan, S. Self-cross-linked polymer nanogels: A versatile nanoscopic drug delivery platform. J. Am. Chem. Soc. **2010**, *132* (48), 17227–17235.

57. Jiwpanich, S.; Ryu, J.H.; Bickerton, S.; Thayumanavan, S. Noncovalent encapsulation stabilities in supramolecular nanoassemblies. J. Am. Chem. Soc. **2010**, *132* (31), 10683–10685.

58. Kakizawa, Y.; Harada, A.; Kataoka, K. Glutathione-sensitive stabilization of block copolymer micelles composed of antisense DNA and thiolated poly(ethylene glycol)-block-poly(L-lysine): A potential carrier for systemic delivery of antisense DNA. Biomacromolecules **2001**, *2* (2), 491–497.

59. Matsumoto, S.; Christie, R.; Nishiyama, N.; Miyata, K.; Ishii, A.; Oba, M.; Koyama, H.; Yamasaki, Y.; Kataoka, K. Environment-responsive block copolymer micelles with a disulfide cross-linked core for enhanced siRNA delivery. Biomacromolecules **2009**, *10* (1), 119–127.

60. Christie, R.J.; Miyata, K.; Matsumoto, Y.; Nomoto, T.; Menasco, D.; Lai. T. ch.; Pennisi, M.; Osada, K.; Fukushima, S.H.; Nishiyama, N.; Yamasaki, Y.; Kataoka, K. Effect of polymer structure on micelles formed between siRNA and cationic block copolymer comprising thiols and amidines. Biomacromolecules **2011**, *12* (9), 3174–3185.

61. Huang, H.; Remsen, E.E.; Wooley, K.L. Amphiphilic core–shell nanospheres obtained by intramicellar shell cross-linking of polymer micelles with poly(ethylene oxide) linkers. Chem. Commun. **1998**, *13* (13), 1415–1416.

62. Joralemon, M.J.; Smith, N.L.; Holowka, D.; Baird, B.; Wooley, K.L. Antigen decorated shell cross-linked nanoparticles: Synthesis, characterization, and antibody interactions. Bioconjug. Chem. **2005**, *16* (5), 1246–1256.

63. Li, Y.; Du, W.; Sun, G.; Wooley, K.L. PH-responsive shell cross-linked nanoparticles with hydrolytically labile cross-links. Macromolecules **2008**, *41* (18), 6605–6607.

64. Wei, H.; Quan, C.Y.; Chang, C.; Zhang, X.Z.; Zhuo, R.X. Preparation of novel ferrocene-based shell cross-linked thermoresponsive hybrid micelles with antitumor efficacy. J. Phys. Chem. B **2010**, *114* (16), 5309–5314.

65. Li, Y.; Lokitz, B.S.; Armes, S.P.; McCormick, C.L. Synthesis of reversible shell cross-linked micelles for controlled release of bioactive agents. Macromolecules **2006**, *39* (8), 2726–2728.

66. Xu, X.; Smith, A.E.; Kirkland, S.E.; McCormick, C.L. Aqueous RAFT synthesis of pH-responsive triblock copolymer mPEO-PAPMA-PDPAEMA and formation of shell cross-linked micelles. Macromolecules **2008**, *41* (22), 8429–8435.

67. Zhuang, J.; Jiwpanich, S.; Deepak, V.D.; Thayumanavan, S. Facile preparation of nanogels using activated ester containing polymers, ACS Macro Lett. **2012**, *1* (1), 175–179.

68. Duong, H.T.T.; Marquis, C.P.; Whittaker, M.; Davis, T.P.; Boyer, C. Acid degradable and biocompatible polymeric nanoparticles for the potential codelivery of therapeutic agents. Macromolecules **2011**, *44* (20), 8008–8019.

69. Kim, Y.; Pourgholami, M.H.; Morris, D.L.; Stenzel, M.H. Triggering the fast release of drugs from crosslinked micelles in an acidic environment, J. Mater. Chem. **2011**, *21* (34), 12777–12783.

70. Kolb, H.C.; Finn, M.G.; Sharpless, K.B. Click chemistry: Diverse chemical function from a few good reactions. Angew. Chem. Int. Ed. **2001**, *40* (11), 2004–2021.

71. Tsarevsky, N.V.; Sumerlin, B.S.; Matyjaszewski, K. Step-growth "click" coupling of telechelic polymers prepared by atom transfer radical polymerization. Macromolecules **2005**, *38* (9), 3558–3561.

72. Sumerlin, B.S.; Tsarevsky, N.V.; Louche, G.; Lee, R.Y.; Matyjaszewski, K. Highly efficient "click" functionalization of poly(3-azidopropyl methacrylate) prepared by ATRP. Macromolecules **2005**, *38* (18), 7540–7545.

73. Wu, P.; Feldman, A.K.; Nugent, A.K.; Hawker, C.J.; Scheel, A.; Voit, B.; Pyun, J.; Frehet, J.M.J.; Sharpless, K.B.; Fokin, V.V. Efficiency and fidelity in a click-chemistry route to triazole dendrimers by the copper(I)-catalyzed ligation of azides and alkynes. Angew. Chem. Int. Ed. **2004**, *43* (30), 3928–3932.

74. Hawker, C.J.; Wooley, K.L. The convergence of synthetic organic and polymer chemistries. Science **2005**, *309* (5738), 1200–1205.

75. Demko, Z.P.; Sharpless, K.B. A click chemistry approach to tetrazoles by huisgen 1,3-dipolar cycloaddition: Synthesis of 5-sulfonyl tetrazoles from azides and sulfonyl cyanides. Angew. Chem. Int. Ed. 2002, 41 (12), 2110–2113.

76. Demko, Z.P.; Sharpless, K.B. A click chemistry approach to tetrazoles by huisgen 1,3-dipolar cycloaddition: Synthesis of 5-acyltetrazoles from azides and acyl cyanides. Angew. Chem. Int. Ed. 2002, 41 (12), 2113–2116.

77. Joralemon, M.J.; O'Reilly, R.K.; Hawker, C.J.; Wooley, K.L. Shell click-crosslinked (SCC) nanoparticles: a new methodology for synthesis and orthogonal functionalization. J. Am. Chem. Soc. 2005, 127 (48), 16892–16899.

78. O'Reilly, R.K.; Joralemon, M.J.; Wooley, K.L.; Hawker, C.J. Functionalization of micelles and shell cross-linked nanoparticles using click chemistry. Chem. Mater. 2005, 17 (24), 5976–5988.

79. O'Reilly, R.K.; Joralemon M.J.; Hawker C.J.; Wooley K.L. Preparation of orthogonally-functionalized core Click cross-linked nanoparticles. New J. Chem. 2007, 31 (5), 718–724.

80. Zhang, J.; Zhou Y.; Zhu, Z.H.; Ge, Z.H.; Liu, S.H. Polyion complex micelles possessing thermoresponsive coronas and their covalent core stabilization via "click" chemistry. Macromolecules 2008, 41 (4), 1444–1454.

81. Withey, A.B.J.; Chen, G.; Nguyen, T.L.U.; Stenzel, M.H. Macromolecular cobalt carbonyl complexes encapsulated in a click-cross-linked micelle structure as a nanoparticle to deliver cobalt pharmaceuticals. Biomacromolecules 2009, 10 (12), 3215–3226.

82. Duong, H.T.T.; Huynh, V.T.; de Souza, P.; Stenzel, M.H. Core-cross-linked micelles synthesized by clicking bifunctional Pt(IV) anticancer drugs to isocyanates. Biomacromolecules 2010, 11 (9), 2290–2299.

83. Jiang, X.; Zhang, G.; Narain, R.; Liu, S.H. Fabrication of two types of shell-cross-linked micelles with "Inverted" structures in aqueous solution from schizophrenic water-soluble ABC triblock copolymer via click chemistry. Langmuir 2009, 25 (4), 2046–2054.

84. Wu, T.; Ge, Z.H.; Liu, S.H. Fabrication of thermoresponsive cross-Linked poly(N-isopropylacrylamide) nanocapsules and silver nanoparticle-embedded hybrid capsules with controlled shell thickness. Chem. Mater. 2011, 23 (9), 2370–2380.

85. Lehn, J.M. From supramolecular chemistry towards constitutional dynamic chemistry and adaptive chemistry. Chem. Soc. Rev. 2007, 36 (2), 151–160.

86. Jackson, A.W.; Stakes, C.H.; Fulton, D.A. The formation of core cross-linked star polymer and nanogel assemblies facilitated by the formation of dynamic covalent imine bonds. Polym. Chem. 2011, 2 (11), 2500–2511.

87. Corbett, P.T.; Leclaire, J.; Vial, L.; West, K.R.; Wietor, J.-L.; Sanders, J.K.M.; Otto, S. Dynamic combinatorial chemistry. Chem. Rev., 2006, 106 (9), 3652–3711.

88. Jackson, A.W.; Fulton, D.A. Triggering polymeric nanoparticle disassembly through the simultaneous application of two different stimuli. Macromolecules 2012, 45 (6), 2699–2708.

89. Xu, X.; Flores, J.D.; McCormick, C.L. Reversible imine shell cross-linked micelles from aqueous RAFT synthesized thermoresponsive triblock copolymers as potential nanocarriers for "pH-triggered" drug release. Macromolecules 2011, 44 (6), 1327–1334.

90. Pioge, S.; Nesterenko, A.; Brotons, G.; Pascual, S.; Fontaine, L.; Gaillard, C.; Nicol, E. Core cross-linking of dynamic diblock copolymer micelles: Quantitative study of photopolymerization efficiency and micelle structure. Macromolecules 2011, 44 (3) 594–603.

91. He, J.; Tong, X.; Zhao, Y. Photoresponsive nanogels based on photocontrollable cross-links. Macromolecules 2009, 42 (13), 4845–4852.

92. He, J.; Yan, B.; Tremblay, L.; Zhao, Y. Both core- and shell-cross-linked nanogels: Photoinduced size change, intraparticle LCST, and interparticle UCST thermal behaviors. Langmuir 2011, 27 (1), 436–444.

93. Yusa, S.I.; Sugahara, M.; Endo, T.; Morishima, Y. Preparation and characterization of a pH-responsive nanogel based on a photo-cross-linked micelle formed from block copolymers with controlled structure. Langmuir 2009, 25 (9), 5258–5265.

94. Kuckling, D.; Vo, C.D.; Adler, H.-J.P.; Vollkel, A.; Collfen, H. Preparation and characterization of photo-cross-linked thermosensitive PNIPAAm nanogels. Macromolecules 2006, 39 (4), 1585–1591.

95. Sugihara, S.H.; Ito, S.; Irie, S.; Ikeda, I. Synthesis of thermoresponsive shell cross-linked micelles via living cationic polymerization and UV irradiation. Macromolecules 2010, 43 (4), 1753–1760.

96. Kuckling, D.; Vo, C.V.; Wohlrab, S.E. Preparation of nanogels with temperature-responsive core and pH-responsive arms by photo-cross-linking. Langmuir 2002, 18 (11), 4263–4269.

97. An, Z.H.; Qiu, Q.; Liu, G. Synthesis of architecturally well-defined nanogels via RAFT polymerization for potential bioapplications. Chem. Commun. 2011, 47 (46), 12424–12440.

98. Cheng, G.; Mi, L.; Cao, Z.Q.; Xue, H.; Yu, Q.M.; Carr, L.; Jiang, S.Y. Functionalizable and ultrastable zwitterionic nanogels. Langmuir 2010, 26 (10), 6883–6886.

99. Guha, S.; Ray, B.; Mandal, B.M. Anomalous solubility of polyacrylamide prepared by dispersion (precipitation) polymerization in aqueous tert-butyl alcohol. J. Polym. Sci. A 2001, 39 (19), 3434–3442.

100. Liu, T.; Desimone, J.M.; Roberts, G.W. Continuous precipitation polymerization of acrylic acid in supercritical carbon dioxide: The polymerization rate and the polymer molecular weight. J. Polym. Sci. A 2005, 43 (12), 2546–2555.

101. Bai, F.; Yang, X.; Zhao, Y.; Huang, W. Synthesis of core–shell microspheres with active hydroxyl groups by two-stage precipitation polymerization. Polym. Int. 2005, 54 (1), 168–174.

102. Li, W.-H.; Stover, H.D.H. Mono- or narrow disperse poly(methacrylate-co-divinylbenzene) microspheres by precipitation polymerization. J. Polym. Sci. A 1999, 37 (15), 2899–2907.

103. Duracher, D.; Elaissari, A.; Pichot, C. Preparation of poly(N-isopropylmethacrylamide) latexes kinetic studies and characterization. J. Polym. Sci. A 1999, 37 (12), 1823–1837.

104. Hazot, P.; Chapel, J.P.; Pichot, C.; Elaissari, A.; Delair, T. Preparation of poly(N-ethyl methacrylamide) particles via an emulsion/precipitation process: The role of the cross-linker. J. Polym. Sci. A 2002, 40 (11), 1808–1817.

Molecular—Nanogels

105. Huang, G.; Gao, J.; Hu, Z.; St John, J.V.; Ponder, B.C.; Moro, D. Controlled drug release from hydrogel nanoparticle networks. J. Control. Release **2004**, *94* (2–3), 303–311.

106. William, H.; Blackburn, L.; Lyon, A. Size-controlled synthesis of monodisperse core/shell nanogels. Colloid Polym. Sci. **2008**, *286* (5), 563–569.

107. Gan, D.; Lyon, L.A. Tunable swelling kinetics in core–shell hydrogel nanoparticles. J. Am. Chem. Soc. **2001**, *123* (31), 7511–7517.

108. Jones, C.D.; Lyon, L.A. Synthesis and characterization of multiresponsive core-shell microgels. Macromolecules **2000**, *33* (22), 8301–8306.

109. Jones, C.D.; Lyon, L.A. Shell-restricted swelling and core compression in poly(N-isopropylacrylamide) core–shell microgels. Macromolecules **2003**, *36* (6), 1988–1993.

110. Huang, X.; Lowe, T.L. Biodegradable thermoresponsive hydrogels for aqueous encapsulation and controlled release of hydrophilic model drugs. Biomacromolecules **2005**, *6* (4), 2131–2139.

111. Antonietti, M.; Landfester, K. Polyreactions in miniemulsions. Prog. Polym. Sci. **2002**, *27* (4), 689–757.

112. Landfester, K.; Willert, M.; Antonietti, M. Preparation of polymer particles in nonaqueous direct and inverse miniemulsions. Macromolecules **2000**, *33* (7), 2370–2376.

113. Kriwet, B.; Walter, E.; Kissel, T. Synthesis of bioadhesive poly(acrylic acid) nano- and microparticles using an inverse emulsion polymerization method for the entrapment of hydrophilic drug candidates. J. Control. Release **1998**, *56* (1–3), 149–158.

114. Sun, Q.; Deng, Y. In situ synthesis of temperature-sensitive hollow microspheres via interfacial polymerization. J. Am. Chem. Soc. **2005**, *127* (23), 8274–8275.

115. Sankar, C.; Rani, M.; Srivastava, A. K.; Mishra, B. Chitosan based pentazocine microspheres for intranasal systemic delivery: Development and biopharmaceutical evaluation. Pharmazie **2001**, 56 (3), 223–226.

116. Marie, E.; Rothe, R.; Antonietti, M.; Landfester, K. Synthesis of polyaniline particles via inverse and direct miniemulsion. Macromolecules **2003**, *36* (11), 3967–3673.

117. Dowding, P.J.; Vincent, B.; Williams, E. Preparation and swelling properties of poly(NIPAM) "Minigel" particles prepared by inverse suspension polymerization. J. Colloid Interface Sci. **2000**, *221* (2), 268–272.

118. Owens III, D.E.; Jian, Y.; Fang, J.E.; Slaughter, B.V.; Chen, Y.-H.; Peppas, N.A. Thermally responsive swelling properties of polyacrylamide/poly(acrylic acid) interpenetrating polymer network nanoparticles. Macromolecules **2007**, *40* (20), 7306–7310.

119. Missirlis, D.; Tirelli, N.; Hubbell, J.A. Amphiphilic hydrogel nanoparticles. preparation, characterization, and preliminary assessment as new colloidal drug carriers. Langmuir **2005**, *21* (6), 2605–2613.

120. Lovell, P.; El-Aasser, M.S. *Emulsion Polymerization and Emulsion Polymers*; John Wiley & Sons Ltd; West Sussex, England, 1997; p. 723.

121. Nagarajan, R.; Wang, C.-C. Theory of surfactant aggregation in water/ethylene glycol mixed solvents. Langmuir **2000**, *16* (12), 5242–5251.

122. Braun, O.; Selb, J.; Candau, F. Synthesis in microemulsion and characterization of stimuli-responsive polyelectrolytes and polyampholytes based on N-isopropylacrylamide. Polymer **2001**, *42* (21), 8499–8510.

123. Fernandez, V.V.A.; Tepale, N.; Sanchez-Diaz, J.C.; Mendizabal, E.; Puig, J.E.; Soltero, J.F.A. Thermoresponsive nanostructured poly(N-isopropylacrylamide) hydrogels made via inverse microemulsion polymerization. Colloid Polym. Sci. **2006**, *284* (4), 387–395.

124. Juranicova, V.; Kawamoto, S.; Fujimoto, K.; Kawaguchi, H.; Barton, J. Inverse microemulsion polymerization of acrylamide in the presence of N,N-dimethylacrylamide. Angew. Makromol. Chem. **1998**, *258* (1), 27–31.

125. Barton, J. Inverse microemulsion polymerization of oil-soluble monomers in the presence of hydrophilic polyacrylamide nanoparticles. Macromol. Symp. **2002**, *179* (1), 189–208.

126. Renteria, M.; Munoz, M.; Ochoa, J.R.; Cesteros, L.C.; Katime, I. Acrylamide inverse microemulsion polymerization in a paraffinic solvent: Rolling-M-245. J. Polym. Sci. Part A Polym. Chem. **2005**, *43* (12), 2495–2503.

127. Kaneda, I.; Sogabe, A.; Nakajima, H. Water-swellable polyelectrolyte microgels polymerized in an inverse microemulsion using a nonionic surfactant. J. Colloid Interface Sci. **2004**, *275* (2), 450–457.

128. Deng, Y.; Wang, L.; Yang, W.; Fu, S.; Elaissari, A. Preparation of magnetic polymeric particles via inverse microemulsion polymerization process. J. Magn. Magn. Mater. **2003**, *257* (1), 69–78.

129. Gaur, U.; Sahoo, S.K.; De, T.K.; Ghosh, P.C.; Maitra, A.; Ghosh, P.K. Biodistribution of fluoresceinated dextran using novel nanoparticles evading reticuloendothelial system. Int. J. Pharm. **2000**, *202* (1–2), 1–10.

130. Bharali, D.J.; Sahoo, S.K.; Mozumdar, S.; Maitra, A. Cross-linked polyvinylpyrrolidone nanoparticles: A potential carrier for hydrophilic drugs. J. Colloid Interface Sci. **2003**, *258* (2), 415–423.

131. McAllister, K.; Sazani, P.; Adam, M.; Cho, M.J.; Rubinstein, M.; Samulski, R.J.; DeSimone, J.M. Polymeric nanogels produced via inverse microemulsion polymerization as potential gene and antisense delivery agents. J. Am. Chem. Soc. **2002**, *124* (51), 15198–15207.

132. Craparo, E.F.; Cavallaro, G.; Bondi, M.L.; Mandracchia, D.; Giammona, G. PEGylated nanoparticles based on a polyaspartamide. preparation, physico-chemical characterization, and intracellular uptake. Biomacromolecules **2006**, *7* (11), 3083–3092.

133. Gao, D.; Xu, H.; Philbert, M.A.; Kopelman, R. Ultrafine hydrogel nanoparticles: Synthetic approach and therapeutic application in living cells. Angew. Chem. Int. Ed. **2007**, *46* (13), 2224–2227.

134. Gao, D.; Agayan, R.R.; Xu, H.; Philbert, M.A.; Kopelman, R. Nanoparticles for two-photon photodynamic therapy in living cells. Nano Lett. **2006**, *6* (11), 2383–2386.

135. Rieger, J.; Grazon, C.; Charleux, B.; Alaimo, D.; Jerome, C. Pegylated thermally responsive block copolymer micelles and nanogels via in situ RAFT aqueous dispersion polymerization. J. Polym. Sci. Part A Polym. Chem. **2009**, 47 (9), 2373–2390.

136. Das, M.; Mardyani, S.; Chan, W.C.W.; Kumacheva, E. Biofunctionalized pH-responsive microgels for cancer cell targeting: Rational design. Adv. Mater. **2006**, *18* (1), 80–83.

137. Thronton, P.D.; Mart, R.J.; Ulijn, R.V. Enzyme-responsive polymer hydrogel particles for controlled release. Adv. Mater. **2007**, *19* (9), 1252–1256.

138. Kim, J.; Serpe, M.J.; Lyon, L.A. Photoswitchable microlens arrays. Angew. Chem. Int. Ed. **2005**, *44* (9), 1333–1336.

139. Matyjaszewski, K.; Xia, J. Atom transfer radical polymerization. Chem. Rev. **2001**, *101* (9), 2921–2990.

140. Hawker, C.J.; Bosman, A.W.; Harth, E. New polymer synthesis by nitroxide mediated living radical polymerizations. Chem. Rev. **2001**, *101* (12), 3661–3688.

141. Chiefari, J.; Chong, Y.K.; Ercole, F.; Krstina, J.; Jeffery, J.; Le, T.P.T.; Mayadunne, R.T.A.; Meijs, G.F.; Moad, C.L.; Moad, G.; Rizzardo, E.; Thang, S.H. Living free-radical polymerization by reversible addition-fragmentation chain transfer: The RAFT process. Macromolecules **1998**, *31* (16), 5559–5562.

142. Moad, G.; Rizzardo, E.; Thang, S.H. Living radical polymerization by the RAFT process—A third update. Aust. J. Chem. **2012**, *65* (8), 985–1076.

143. Sheiko, S.S.; Sumerlin, B.S.; Matyjaszewski, K. Cylindrical molecular brushes: Synthesis, characterization, and properties. Prog. Polym. Sci. **2008**, *33* (7), 759–785.

144. Yagci, Y.; Tasdelen, M.A. Mechanistic transformations involving living and controlled/living polymerization methods. Prog. Polym. Sci. **2006**, *31* (12), 1133–1170.

145. Hadjichristidis, N.; Iatrou, H.; Pitsikalis, M.; Mays, J. Macromolecular architectures by living and controlled/living polymerization. Prog. Polym. Sci. **2006**, *31* (12), 1068–1132.

146. Broyer, R.M.; Quaker, G.M.; Maynard, H.D. Designed amino acid ATRP initiators for the synthesis of biohybrid materials. J. Am. Chem. Soc. **2008**, *130* (3), 1041–1047.

147. Le Droumaguet, B.; Velonia, K. In situ ATRP-mediated hierarchical formation of giant amphiphile bionanoreactors. Angew. Chem. Int. Ed. **2008**, *47* (33), 6263–6266.

148. Loschonsky, S.; Couet, J.; Biesalski, M. Synthesis of peptide/polymer conjugates by solution ATRP of butylacrylate using an initiator-modified cyclic D-alt-L-peptide. Macromol. Rapid. Commun. **2008**, *29* (4), 309–315.

149. Ostmark, E.; Harrisson, S.; Wooley K.L.; Malmstrom E.E. Comb polymers prepared by ATRP from hydroxypropyl cellulose. Biomacromolecules **2007**, *8* (4), 1138–1148.

150. Ifuku, S.; Kadla, J.F. Preparation of a thermosensitive highly regioselective cellulose/N-isopropylacrylamide copolymer through atom transfer radical polymerization. Biomacromolecules **2008**, *9* (11), 3308–3313.

151. Xu, P.; Van Kirk, E.A.; Murdoch, W.J.; Zhan, Y.; Isaak, D.D.; Radosz, M.; Shen, Y. Anticancer efficacies of cis-plastin-releasing pH-responsive nanoparticles. Biomacromolecules **2006**, *7* (3), 829–835.

152. Joralemon, M.J.; Smith, N.L.; Holowka, D.; Baird, B.; Wooley, K.L. Antigen-decorated shell cross-linked nanoparticles: Synthesis, characterization, and antibody interactions. Bioconjug. Chem. **2005**, *16* (5), 1246–1256.

153. Oh, J.K.; Siegwart, D.J.; Lee, H.I.; Sherwood, G.; Peteanu, L.; Hollinger, J.O.; Kataoka, K.; Matyjaszewski, K. Biodegradable nanogels prepared by atom transfer radical polymerization as potential drug delivery carriers: Synthesis, biodegradation, *in vitro* release, and bioconjugation. J. Am. Chem. Soc. **2007**, *129* (18), 5939–5945.

154. Wang, M.; Dykstra, T.E.; Lou, X.; Salvador, M.R.; Scholes, G.D.; Winnik, M.A. Colloidal CdSe nanocrystals passivated by a dye-labeled multidentate polymer: Quantitative analysis by size-exclusion chromatography. Angew. Chem. Int. Ed. **2006**, *45* (14), 2221–2224.

155. Wang, M.; Felorzabihi, N.; Guerin, G.; Haley, J.C.; Scholes, G.D.; Winnik, M.A. Water-soluble CdSe quantum dots passivated by a multidentate diblock copolymer. Macromolecules **2007**, *40* (17), 6377–6384.

156. Huang, J.; Koepsel, R.R., Murata, H.; Wu, W.; Lee, S.B.; Kowalewski, T.; Russell, A.J.; Matyjaszewski, K. Non-leaching antibacterial glass surfaces via "grafting onto": The effect of the number of quaternary ammonium groups on biocidal activity. Langmuir **2008**, *24* (13), 6785–6795.

157. Huang, J.; Murata, H.; Koepsel, R.R.; Russell, A.J.; Matyjaszewski, K. Antibacterial polypropylene via surface-initiated atom transfer radical polymerization. Biomacromolecules **2007**, *8* (5), 1396–1399.

158. Oh, J.K. Recent advances in controlled/living radical polymerization in emulsion and dispersion. J. Polym. Sci. Part A Polym. Chem. **2008**, *46* (21), 6983–7001.

159. Cunningham, M.F. Controlled/living radical polymerization in aqueous dispersed systems. Prog. Polym. Sci. **2008**, *33* (4), 365–398.

160. Matyjaszewski, K.; Tsarevsky, N.V.; Braunecker W.A.; Dong, H.; Huang, J.; Jakubowski, W.; Kwak, Y.; Nicolay, R.; Tang, W.; Yoon, J.A. Role of Cu⁰ in controlled/"living" radical polymerization. Macromolecules **2007**, *40* (22), 7795–7806.

161. Oh, J.K.; Dong, H.; Zhang, R.; Matyjaszewski, K.; Schlaad, H. Preparation of nanoparticles of double-hydrophilic PEO-PHEMA block copolymers by AGET ATRP in inverse miniemulsion. J. Polym. Sci., Part A Polym. Chem. **2007**, *45* (21), 4764–4772.

162. Oh, J.K.; Siegwart, D.J.; Matyjaszewski, K. Synthesis and biodegradation of nanogels as delivery carriers for carbohydrate drugs. Biomacromolecules **2007**, *8* (11), 3326–3331.

163. Tsarevsky, N.V.; Matyjaszewski, K. Reversible redox cleavage/coupling of polystyrene with disulfide or thiol groups prepared by atom transfer radical polymerization. Macromolecules **2002**, *35* (24), 9009–9014.

164. Tsarevsky, N.V.; Bernaerts, K.V.; Dufour, B.; Du Prez, F.E.; Matyjaszewski, K. Well-defined (co)polymers with 5-vinyltetrazole units via combination of atom transfer radical (co)polymerization of acrylonitrile and "click chemistry"-type postpolymerization modification. Macromolecules **2004**, *37* (25), 9308–9313.

165. Kim, K.H.; Kim, J.; Jo, W.H. Preparation of hydrogel nanoparticles by atom transfer radical polymerization of N-isopropylacrylamide in aqueous media using PEG macro-initiator. Polymer **2005**, *46* (9), 2836–2840.

166. Oh, J.K.; Min, K.; Matyjaszewski, K. Preparation of poly(oligo(ethylene glycol) monomethyl ether methacrylate) by homogeneous aqueous AGET ATRP. Macromolecules **2006**, *39* (9), 3161–3167.

167. Oh, J.K.; Tang, C.B.; Gao, H.F.; Tsarevsky, N.V.; Matyjaszewski, K. Inverse miniemulsion ATRP: A new method for synthesis and functionalization of well-defined water-soluble/cross-linked polymeric particles. J. Am. Chem. Soc. **2006**, *128* (16), 5578–5584.

168. Siegwart, D.J.; Oh, J.K.; Gao, H.; Bencherif, S.A.; Perineau, F.; Bohaty, A.K.; Hollinger, J.O.; Matyjaszewski, K. Biotin-, pyrene-, and GRGDS-functionalized polymers and nanogels via ATRP and end group modification. Macromol. Chem. Phys. **2008**, *209* (21), 2180–2193.

169. Moad, G.; Rizzardo, E.E.; Thang, S.H. Radical addition–fragmentation chemistry in polymer synthesis. Polymer **2008**, *49* (5), 1079–1131.

170. Kakwere, H.; Perrier, S. Design of complex polymeric architectures and nanostructured materials/hybrids by living radical polymerization of hydroxylated monomers. Polym. Chem. **2011**, *2* (2), 270–288.

171. Boyer, C.; Stenzel, M.H.; Davis, T.P. Building nanostructures using RAFT polymerization. J. Polym. Sci. Part A **2011**, *49* (3), 551–595.

172. Moad, G.; Rizzardo, E.E.; Thang, S.H. Living radical polymerization by the RAFT process. Aust. J. Chem. **2005**, *58* (6), 379–410.

173. Moad, G.; Rizzardo, E.E.; Thang, S.H. Toward living radical polymerization. Acc. Chem. Res. **2008**, *41* (9), 1133–1142.

174. Moad, G.; Rizzardo, E.E.; Thang, S.H. Living radical polymerization by the RAFT process—A first update. Aust. J. Chem. **2006**, *59* (10), 669–692.

175. Moad, G.; Rizzardo, E.E.; Thang, S.H. Living radical polymerization by the RAFT process—A second update. Aust. J. Chem. **2009**, *62* (11), 1402–1472.

176. Convertine, A.J.; Lokitz, B.S.; Vasileva, Y.; Myrick, L.J.; Scales, C.W.; Lowe, A.B.; McCormick, C.L. Direct synthesis of thermally responsive DMA/NIPAM diblock and DMA/NIPAM/DMA triblock copolymers via aqueous, room temperature RAFT polymerization. Macromolecules **2006**, *39* (5), 1724–1730.

177. McCormick, C.L.; Lowe, A.B. Aqueous RAFT polymerization: Recent developments in synthesis of functional water-soluble (co)polymers with controlled structures. Acc. Chem. Res. **2004**, *37* (5), 312–325.

178. Boyer, C.; Bulmus, V.; Priyanto, P.; Teoh, W.Y.; Amal, R.; Davis, T.P. The stabilization and bio-functionalization of iron oxide nanoparticles using heterotelechelic polymers. J. Mater. Chem. **2009**, *19* (1), 111–123.

179. Boyer, C.; Liu, J.; Bulmus, V.; Davis, T.P.; Barner-Kowollik, C.; Stenzel, M.H. Direct synthesis of well-defined heterotelechelic polymers for bioconjugations. Macromolecules **2008**, *41* (15), 5641–5650.

180. Liu, J.; Bulmus, V.; Barner-Kowollik, C.; Stenzel, M.H.; Davis, T.P. Direct synthesis of pyridyl disulfide-terminated polymers by RAFT polymerization. Macromol. Rapid Commun. **2007**, *28* (3), 305–314.

181. Postma, A.; Davis, T.P.; Evans, R.A.; Li, G.; Moad, G.; O'Shea, M.S. Synthesis of well-defined polystyrene with primary amine end groups through the use of phthalimido-functional RAFT agents. Macromolecules **2006**, *39* (16), 5293–5306.

182. Scales, C.W.; Convertine, A.J.; McCormick, C.L. Fluorescent labeling of RAFT-generated poly(N-isopropylacrylamide) via a facile maleimide–thiol coupling reaction. Biomacromolecules **2006**, *7* (5), 1389–1392.

183. Wong, L.; Boyer, C.; Jia, Z.; Zareie, H.M.; Davis, T.P.; Bulmus, V. Synthesis of versatile thiol-reactive polymer scaffolds via RAFT polymerization. Biomacromolecules **2008**, *9* (7), 1934–1944.

184. York, A.W.; Scales, C.W.; Huang, F.; McCormick, C.L. Facile synthetic procedure for ω, primary amine functionalization directly in water for subsequent fluorescent labeling and potential bioconjugation of RAFT-synthesized (co)polymers. Biomacromolecules **2007**, *8* (8), 2337–2341.

185. Stenzel, M.H.; Davis, T.P. Star polymer synthesis using trithiocarbonate functional β-cyclodextrin cores (reversible addition–fragmentation chain-transfer polymerization). J. Polym. Sci. Part A **2002**, *40* (24), 4498–4512.

186. Quinn, J.F.; Chaplin, R.P.; Davis, T.P. Facile synthesis of comb, star, and graft polymers via reversible addition–fragmentation chain transfer (RAFT) polymerization. J. Polym. Sci. Part A **2002**, *40* (17), 2956–2966.

187. Liu, J.Q.; Liu, H.Y.; Jia, Z.F.; Bulmus, V.; Davis, T.P. An approach to biodegradable star polymeric architectures using disulfide coupling. Chem. Commun. **2008**, *48* (48), 6582–6584.

188. Jia, Z.F.; Wong, L.J.; Davis, T.P.; Bulmus, V. One-pot conversion of RAFT-generated multifunctional block copolymers of HPMA to doxorubicin conjugated acid- and reductant-sensitive crosslinked micelles. Biomacromolecules **2008**, *9* (11), 3106–3113.

189. Biasutti, J.D.; Davis, T.P.; Lucien, F.P.; Heuts, J.P.A. Reversible addition–fragmentation chain transfer polymerization of methyl methacrylate in suspension. J. Polym. Sci. Part A **2005**, *43* (10), 2001–2012.

190. Simms, R.W.; Davis, T.P.; Cunningham, M.F. Xanthate-mediated living radical polymerization of vinyl acetate in miniemulsion. Macromol. Rapid Commun. **2005**, *26* (8), 592–596.

191. Lansalot, M.; Davis, T.P.; Heuts, J.P.A. RAFT miniemulsion polymerization: Influence of the structure of the RAFT agent. Macromolecules **2002**, *35* (20), 7582–7591.

192. Barner, L.; Li, C.E.; Hao, X.; Stenzel, M.H.; Barner-Kowollik, C.; Davis, T.P. Synthesis of core-shell poly(divinylbenzene) microspheres via reversible addition fragmentation chain transfer graft polymerization of styrene. J. Polym. Sci. Part A **2004**, *42* (20), 5067–5076.

193. Chan, Y.; Bulmus, V.; Zareie, M.H.; Byrne, F.L.; Barner, L.; Kavallaris, M. Acid-cleavable polymeric core–shell particles for delivery of hydrophobic drugs. J. Control. Release **2006**, *115* (2), 197–207.

194. Barsbay, M.; Gueven, O.; Davis, T.P.; Barner-Kowollik, C.; Barner, L. RAFT-mediated polymerization and grafting of sodium 4-styrenesulfonate from cellulose initiated via γ-radiation. Polymer **2009**, *50* (4), 973–982.

195. Barsbay, M.; Gueven, O.; Stenzel, M.H.; Davis, T.P.; Barner-Kowollik, C.; Barner, L. Verification of controlled grafting of styrene from cellulose via radiation-induced RAFT polymerization. Macromolecules **2007**, *40* (20), 7140–7147.

196. Boyer, C.; Bulmus, V.; Liu, J.; Davis, T.P.; Stenzel, M.H.; Barner-Kowollik, C. Well-defined protein–polymer conjugates via in situ RAFT polymerization. J. Am. Chem. Soc. **2007**, *129* (22), 7145–7154.

197. Liu, J.; Bulmus, V.; Herlambang, D.L.; Barner-Kowollik, C.; Stenzel, M.H.; Davis, T.P. In situ formation of protein-polymer conjugates through reversible addition fragmentation chain transfer polymerization. Angew. Chem. Int. Ed. **2007**, *46* (17), 3099–3103.

198. An, Z.S.; Tang, W.; Wu, M.H.; Jiao, Z.; Stucky, G.D. Heterofunctional polymers and core–shell nanoparticles via cascade aminolysis/Michael addition and alkyne–azide click reaction of RAFT polymers. Chem. Commun. 2008, 48 (48), 6501–6503.

199. Lutz, J.F. Polymerization of oligo(ethylene glycol) (meth)acrylates: Toward new generations of smart biocompatible materials. J. Polym. Sci. Part A 2008, 46 (11), 3459–3470.

200. Hoogenboom, R. Poly(2-oxazoline)s: A polymer class with numerous potential applications. Angew. Chem. Int. Ed. 2009, 48 (43), 7978–7994.

201. Schlaad, H.; Diehl, C.; Gress, A.; Meyer, M.; Demirel, A.L.; Nur, Y.; Bertin, A. Poly(2-oxazoline)s as smart bioinspired polymers. Macromol. Rapid Commun. 2010, 31 (6), 511–525.

202. Li, W.; Zhang, A.; Chen, Y.; Feldman, K.; Wu, H.; Schluter, A.D. Low toxic, thermoresponsive dendrimers based on oligoethylene glycols with sharp and fully reversible phase transitions. Chem. Commun. 2008, 45 (45), 5948–5950.

203. Li, W.; Zhang, A.; Feldman, K.; Walde, P.; Schluter, A.D. Thermoresponsive dendronized polymers. Macromolecules 2008, 41 (10), 3659–3667.

204. Fernandez-Trillo, F.; van Hest, J.C.M.; Thies, J.C.; Michon, T.; Weberskirch, R.; Cameron, N.R. Fine-tuning the transition temperature of a stimuli-responsive polymer by a simple blending procedure. Chem. Commun. 2008, 19 (19), 2230–2232.

205. Lutz, J.F.; Akdemir, O.; Hoth, A. Point by point comparison of two thermosensitive polymers exhibiting a similar LCST: is the age of poly(NIPAM) over? J. Am. Chem. Soc. 2006, 128 (40), 13046–13047.

206. Shen, W.Q.; Chang, Y.L.; Liu, G.Y.; Wang, H.F.; Cao, A.N.; An, Z.S. Biocompatible, antifouling, and thermosensitive core–shell nanogels synthesized by RAFT aqueous dispersion polymerization. Macromolecules 2011, 44 (8), 2524–2530.

207. Grazon, C.; Rieger, J.; Sanson, N.; Charleux, B. Study of poly(N,N-diethylacrylamide) nanogel formation by aqueous dispersion polymerization of N,N-diethylacrylamide in the presence of poly(ethylene oxide)-b-poly(N,N-dimethylacrylamide) amphiphilic macromolecular RAFT agents. Soft Matter 2011, 7 (7), 3482–3490.

208. Yan, L.F.; Tao, W. One-step synthesis of pegylated cationic nanogels of poly(N,N′-dimethylaminoethyl methacrylate) in aqueous solution via self-stabilizing micelles using an amphiphilic macroRAFT agent. Polymer 2010, 51 (10), 2161–2167.

209. Bathfield, M.; D'Agosto, F.; Spitz, R.; Charreyre, M.T.; Pichot, C.; Delair, T. Sub-micrometer sized hairy latex particles synthesized by dispersion polymerization using hydrophilic macromolecular RAFT agents. Macromol. Rapid Commun. 2007, 28 (15), 1540–1545.

210. Houillot, L.; Bui, C.; Save, M.; Charleux, B.; Farcet, C.; Moire, C.; Raust, J.A.; Rodriguez, I. Synthesis of well-defined polyacrylate particle dispersions in organic medium using simultaneous RAFT polymerization and self-assembly of block copolymers. A strong influence of the selected thiocarbonylthio chain transfer agent. Macromolecules 2007, 40 (18), 6500–6509.

211. Saikia, P.J.; Lee, J.M.; Lee, K.; Choe, S. Reaction parameters in the RAFT mediated dispersion polymerization of styrene. J. Polym. Sci. Part A 2008, 46 (3), 872–885.

212. Chen, Z.; Wang, X.L.; Su, J.S.; Zhuo, D.; Ran, R. Branched methyl methacrylate copolymer particles prepared by RAFT dispersion polymerization. Polym. Bull. 2010, 64 (4), 327–339.

213. Huang, C.Q.; Pan, C.Y. Direct preparation of vesicles from one-pot RAFT dispersion polymerization. Polymer 2010, 51 (22), 5115–5121.

214. Raust, J.A.; Houillot, L.; Save, M.; Charleux, B.; Moire, C.; Farcet, C.; Pasch, H. Two dimensional chromatographic characterization of block copolymers of 2-ethylhexyl acrylate and methyl acrylate, P2EHA-b-PMA, produced via RAFT-mediated polymerization in organic dispersion. Macromolecules 2010, 43 (21), 8755–8765.

215. Wan, W.M.; Pan, C.Y. One-pot synthesis of polymeric nanomaterials via RAFT dispersion polymerization induced self-assembly and re-organization. Polym. Chem. 2010, 1 (9), 1475–1484.

216. Gregory, A.M.; Thurecht, K.J.; Howdle, S.M. Controlled dispersion polymerization of methyl methacrylate in supercritical carbon dioxide via RAFT. Macromolecules, 2008, 41 (4), 1215–1222.

217. Lee, H.; Terry, E.; Zong, M.; Arrowsmith, N.; Perrier, S.; Thurecht, K.J.; Howdle, S.M. Successful dispersion polymerization in supercritical CO_2 using polyvinylalkylate hydrocarbon surfactants synthesized and anchored via RAFT. J. Am. Chem. Soc. 2008, 130 (37), 12242–12243.

218. Zong, M.M.; Thurecht, K.J.; Howdle, S.M. Dispersion polymerisation in supercritical CO_2 using macro-RAFT agents. Chem. Commun. 2008, 45 (45), 5942–5944.

219. Li, Y.T.; Armes, S.P. RAFT synthesis of sterically stabilized methacrylic nanolatexes and vesicles by aqueous dispersion polymerization. Angew. Chem. Int. Ed. 2010, 49 (24), 4042–4046.

220. Qi, G.G.; Jones, C.W.; Schork, F.J. RAFT inverse miniemulsion polymerization of acrylamide. Macromol. Rapid Commun. 2007, 28 (9), 1010–1016.

221. Qi, G.; Eleazer, B.; Jones, C.W.; Schork, F.J. Mechanistic aspects of sterically stabilized controlled radical inverse miniemulsion polymerization. Macromolecules 2009, 42 (12), 3906–3916.

222. Ouyang, L.; Wang, L.S.; Schork, F.J. RAFT inverse miniemulsion polymerization of acrylic acid and sodium acrylate. Macromol. React. Eng. 2011, 5 (3–4), 163–169.

223. Ouyang, L.; Wang, L.S.; Schork, F.J. Synthesis and nucleation mechanism of inverse emulsion polymerization of acrylamide by RAFT polymerization: A comparative study. Polymer 2011, 52 (1), 63–67.

224. Liu, O.Y.; Wang, L.S.; Schork, F.J. Synthesis of well-defined statistical and diblock copolymers of acrylamide and acrylic acid by inverse miniemulsion raft polymerization. Macromol. Chem. Phys. 2010, 211 (18), 1977–1983.

225. Lu, F.J.; Luo, Y.W.; Li, B.G.; Zhao, Q.; Schork, F.J. Synthesis of thermo-sensitive nanocapsules via inverse miniemulsion polymerization using a PEO–RAFT agent. Macromolecules 2010, 43 (1), 568–571.

226. Wang, Y.; Jiang, G.H.; Zhang, M.; Wang, L.; Wang, R.J.; Sun, X.K. Facile one-pot preparation of novel shell cross-linked nanocapsules: Inverse miniemulsion RAFT

polymerization as an alternative approach. Soft Matter **2011**, *7* (11), 5348–5352.

227. Du, J.Z.; Sun, T.M.; Song, W.J.; Wu, J.; Wang, J. A tumor-acidity-activated chargeconversional nanogel as an intelligent vehicle for promoted tumoral-cell uptake and drug delivery. Angew. Chem. Int. Ed. **2010**, *49* (21), 3621–3626.

228. Schmitt, F.; Lagopoulos, L.; Kauper, P.; Rossi, N.; Busso, N.; Barge, J.; Wagnieres, G.; Laue, C.; Wandrey, C.; Juillerat-Jeanneret, L. Chitosan-based nanogels for selective delivery of photosensitizers tomacrophages and improved retention in and therapy of articular joints. J. Control. Release **2010**, *144* (2), 242–250.

229. Yin, Q.Q.; Wu, L.; Gou, M.L.; Qian, Z.Y.; Zhang, W.S.; Liu, J. Long-lasting infiltration anaesthesia by lidocaine-loaded biodegradable nanoparticles in hydrogel in rats. Acta Anaesthesiol. Scand. **2009**, *53* (9), 1207–1213.

230. Toita, S.; Sawada, S.; Akiyoshi, K. Polysaccharide nanogel gene delivery system with endosome-escaping function: Co-delivery of plasmid DNA and phospholipase A2. J. Control. Release **2011**, *155* (1) 54–59.

231. Tamura, A.; Oishi, M.; Nagasaki, Y. Enhanced cytoplasmic delivery of siRNA using a stabilized polyion complex based on PEGylated nanogels with a cross-linked polyamine structure. Biomacromolecules **2009**, *10* (7), 1818–1827.

232. Missirlis, D.; Kawamura, R.; Tirelli, N.; Hubbell, J.A. Doxorubicin encapsulation and diffusional release from stable, polymeric, hydrogel nanoparticles. Eur. J. Pharm. Sci. **2006**, *29* (2), 120–129.

233. Murthy, N.; Xu, M.; Schuck, S.; Kunisawa, J.; Shastri, N.; Frechet, J.M. A macromolecular delivery vehicle for protein-based vaccines: Acid-degradable protein-loaded microgels. Proc. Natl. Acad. Sci. USA **2003**, *100* (9), 4995–5000.

234. Lee, H.; Mok, H.; Lee, S.; Oh, Y.K.; Park, T.G.; Target-specific intracellular delivery of siRNA using degradable hyaluronic acid nanogels. J. Control. Release **2007**, *119* (2), 245–252.

235. Yu, S.; Hu, J.; Pan, X.; Yao, P.; Jiang, M. Stable and pH-sensitive nanogels prepared by self-assembly of chitosan and ovalbumin. Langmuir **2006**, *22* (6), 2754–2759.

236. 236. Varga, I.; Szalai, I.; Meszaros, R.; Gilanyi, T. Pulsating pH-responsive nanogels. J. Phys. Chem. B **2006**, *110* (41), 20297–20301.

237. Chang, C.; Wang, Z.C.; Quan, C.Y.; Cheng, H.; Cheng, S.X.; Zhang, X.Z.; Zhuo, R.X. Fabrication of a novel pH-sensitive glutaraldehyde cross-linked pectin nanogel for drug delivery. J. Biomater. Sci. Polym. Ed. **2007**, *18* (12), 1591–1599.

238. Oishi, M.; Sumitani, S.; Nagasaki, Y. On-off regulation of 19F magnetic resonance signals based on pH-sensitive PEGylated nanogels for potential tumor-specific smart 19F MRI probes. Bioconjug. Chem. **2007**, *18* (5), 1379–1382.

239. Vinogradov, S.V.; Kohli, E.; Zeman, A.D. Cross-linked polymeric nanogel formulations of 5'-triphosphates of nucleoside analogues: Role of the cellular membrane in drug release. Mol. Pharm. **2005**, *2* (6), 449–461.

240. Ng, E.Y.; Ng, W.K.; Chiam, S.S. Optimization of nanoparticle drug microcarrier on the pharmacokinetics of drug release: A preliminary study. J. Med. Syst. **2008**, *32* (2), 85–92.

Nanomaterials: Conducting Polymers and Sensing

Murat Ates

Department of Chemistry, Faculty of Arts and Sciences, Namik Kemal University, Degirmenalti Campus, Tekirdag, Turkey

Abstract

This entry investigates the hot topics by presenting the latest advances in metal oxides and nanomaterials (gold nanoparticles, silver nanoparticles, etc.) on carbon surfaces such as graphene, carbon nanotubes, conducting polymers with nanomaterials, and sensing behavior of conducting polymers (oxidation of dopamine, gas sensors, etc.). In the literature, many studies on the use of nanomaterials as conducting polymers in biosensor applications have been reported. We have reviewed the related topics especially in nanomaterials together with sensing applications.

INTRODUCTION

The aim of this entry is to provide a brief overview of the various nanomaterials (carbon surfaces, graphene, carbon fiber microelectrodes, carbon nanotubes (CNTs), gold nanoparticles (AuNPs), silver nanoparticles (AgNPs), and metal oxides and nanoparticles), conducting polymers with nanomaterials and sensing behavior of conducting polymers [oxidation of dopamine, gas sensors, ascorbic acid, and uric acid (UA)], investigations that have been taken to construct conducting polymer nanostructures and to apply different conducting polymer nanostructures for using as biosensors. Some excellent papers, reviews with more detailed descriptions of the fundamental aspects of conducting polymers and their nanostructures are available with 446 references.

NANOMATERIALS

Carbon Surfaces

Carbon nanostructures and composite structures[1] are highly interesting for applications such as in sensing[2] and in energy storage.[3,4] New types of nanocarbon materials such as nanotubes,[5] nanoanions,[6] nanodiamond,[7] graphene flakelets,[8] and hydrothermal[9] or pyrolytic[10] carbons are now widely investigated and employed as a substrate for surface modification and optimization.[11] Interaction and nano characterization of thin conjugated polymeric films on carbon surfaces are important to understand the interfacial properties between carbon surface and matrix.[12]

The functional mesoporous polymer/carbon materials with different shapes are the highly promising multifunctional materials for diverse applications.[13] Several research groups have been studied with carbon materials such as graphene,[14,15] CNTs,[16] and fullerene-modified electrodes[17] for its simple, fast, and sensitive voltammetric determination of various compounds of physiological importance nanosheets for biosensing applications. Kushch, Kujunko, and Tarasov reported carbon materials with graphene structures such as single- and multiwalled CNTs and nanofibers after oxidizing by a mixture of sulfuric acid and nitric acids and presumable introducing of carboxyl groups.[18] Carbon materials have unique properties because of their high specific surface area and the possibility of including functional groups. Carbon materials can provide glucose oxidase (GOx) immobilization sites and interfacial affinity between carbon and GOx. Carbon materials, such as CNTs and carbon nanofibers, have been applied as electrodes in the electrochemistry field due to their excellent electrical properties and stability in severe conditions, such as at high temperature and under acidic or basic conditions.[19,20]

Graphene

Graphene-related materials have received significant attention for electrochemistry as new and advantageous materials.[21] Graphene has attracted intense scientific interest because of its exceptional chemical,[22] mechanical,[23] and electrical[24] properties in the latest years. Graphene, a single layer of carbon atoms, is a promising material for use in nanoelectronics, sensors, and optoelectronics due to unique two-dimensional carbon nanostructures.[25–28] Graphene has received increasing attention during recent years due to its unique physicochemical properties including high surface area,[29] excellent electric conductivity,[24] strong mechanical strength, biocompatibility, ease of functionalization, and mass production.[30] Graphene comprises

Concise Encyclopedia of Biomedical Polymers and Polymeric Biomaterials DOI: 10.1081/E-EBPPC-120049912

1035

Nanomaterials—Nanoparticles

a single layer of sp²-hybridized carbon atoms joined by covalent bonds to form a flat hexagonal lattice.[31] It is completely a two-dimensional structure, which is complicated by the difficulty in isolating single layers of graphene in a controlled manner. Since 2004 after the first report of isolation of graphene from graphite,[24] this hexagonal lattice of carbon atoms as a nanosheet with an atomic thickness has increasingly attracted researchers' attention. The low cost of precursors, various methods of fabrication, and novel properties of graphene lead to vast applications in different areas.[32] Irajizad et al. studied hybrid of the graphene oxide (GO) sheets with TiO₂ nanoparticles using photo-reduction process and decorated the hybrid structure with catalytic nanoparticles.[33]

Recently, the potential application of graphene sheets in detecting gas atoms has been numerically investigated.[34] Most biosensors reported to date have used electrochemical reactions involving various types of "graphenes" to detect biomolecules.[35] For example, Lu et al. created graphitic electrode materials for sensitively measuring glucose concentration.[36,37] There are several methods of formation of graphene such as micromechanical cleavage,[25] chemical exfoliation,[38] and epitaxial growth by chemical vapor deposition (CVD).[39] They have been utilized to fabricate high-quality large-area "monolayer" graphene on various substrates. Studies on functionalized graphene and bare graphene have revealed that graphene-based materials as promising candidates as gas sensors.[40–44] Wang et al. studied the photoelectrochemical behavior of graphene–TiO₂ (G-TiO₂) nanohybrids in the visible region and a new photoelectrochemical sensor for sensitive determination of nicotinamide adenine dinucleotide.[45]

Graphene, which is a zero band gap carbonaceous nanomaterial,[46] possesses many unique properties that predispose it for use in photo-conversion devices.[47] Galal group[48] published similar work that graphene supported Pt-M-(M = Ru or Pd) for electrocatalytic methanol oxidation. However, in that study, the peak current of methanol oxidation is about 1.0 V in 0.1 M H₂SO₄.[49]

Electrochemical impedance spectroscopy (EIS) is also an efficient tool for studying the interface properties of surface-modified electrodes. The charge transfer resistance (R_{ct}) can be used to describe the interface properties of the electrode. Therefore, the capability of electron transfer of different electrodes has been investigated by EIS technique. Qazi, Vogt, and Koley[50] studied NO₂ at levels as low as 60 ppb by measuring the changes in surface work function, or in electrical conductivity, or flakes of thin graphite. Most biosensors reported to date have used electrochemical reactions involving various types of graphenes to detect biomolecules.[36,37] For example, Lu et al. created graphitic electrode materials for sensitively measuring glucose concentration.[36,37] Quyang and Guo[51] studied carbon-based nanomaterials and compared them as the contact materials for graphene nanoribbons by atomistic quantum transport simulations. The high mobility of monolayer graphene has stimulated strong interest in high-performance graphene device applications.[25,52,53]

The effect of various p- and n-doped gases on the gas sensing properties of pristine graphene has been investigated in the past few years.[54] In order to improve the gas sensing capability of graphene, it is important to understand the adsorbed gas sensing capability of graphene and the interaction between structurally defective graphene and the adsorbed gas molecules. Hajati et al.[55] reported that by using the focused ion beam technique, defects can be introduced into graphene, and the corresponding evolution of gas sensing and structural properties of defected graphene in an ambient environment.

Graphene, a monolayer of sp² hybridized carbon atoms packed into a dense honeycomb crystal structure,[46] has received increasing attention during recent years due to its unique physicochemical properties including high surface area,[29] excellent electric conductivity,[24] strong mechanical strength, biocompatibility, ease of functionalization, and mass production.[30] Graphene dispersions with concentrations up to 0.5 mg/ml were produced by dispersion in *N,N*-dimethylformamide with the help of polyacrylic acid[56] even at higher concentration (17 mg/ml), and this property was used for the large-scale synthesis of graphene.

Carbon Fiber Microelectrodes

Carbon fibers (CFs) are of particular interest as they are economic and not only improve the conductivity but also enhance the mechanical properties such as strain capacity and fracture toughness.[57–65] Wightman et al.[66] studied two types of CFs, which are named as Thornel P55 and T650

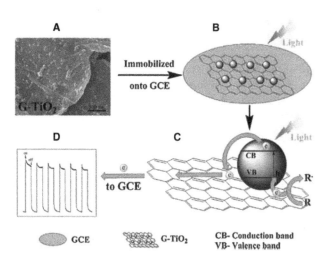

Scheme 1 Schematic illustration of the photo-induced processes of electron–hole generation and charge-transfer processes of G-TiO₂ nanohybrids.
Source: From Wang et al.[45] © 2012, with permission from Elsevier.

models. Thornel P55 fibers are manufactured from a pitch precursor. However, Thornel P650 fibers are made from polyacrylonitrile. The main difference between Thornel P650 and P55 models lead to subtle changes in surface chemistry. Both types have been studied in microelectrodes to investigate neurotransmission events in biological matrixes.[67] CFs, which are constructed from cylindrical electrodes, are relatively easy to fabricate and biosensor applications.[68–70] Wightman et al. have studied the use of carbon fiber cylindrical electrodes with overoxidized polypyrrole.[71] CFs used as carriers to assemble aligned CF-CNTs can provide multifunctional nanomaterials for a wide range of potential applications.[72] One of the most important properties of chemically modified electrodes have been their ability to catalyze the electrode process via significant decreasing of overpotential respect to relatively selective interaction of the electron mediator with the target analyte in a coordination fashion.[73]

Fong et al. studied the effect of relatively uniform and randomly overlaid PAN nanofibers with diameters of ~350 nm, which were introduced onto fiber surfaces through immersion of the PAN nano-felt in NH_2OH aqueous solution.[74] Park, Lee, and Devries studied the electrical resistance for carbon fiber composite, which was measured to evaluate interfacial properties and nondestructive strain–stress sensing depending on different experimental conditions, such as curing temperature, curing agent composition, fiber surface treatment, and fiber-embedded length under cyclic loading.[75]

Carbon Nanotubes

As a promising material in nanoscience and nanotechnology, one-dimensional nanoscale materials such as CNT, are attracting significant interest for its superior mechanical and electrical characteristics.[76]

In this study, the morphology and network structure of MWCNT nanopaper were given by scanning electron microscope (SEM) (ZEISS Ultra-55) at 10 kV. The typical surface views of raw MWCNT arrays about 1 μm and 100 nm, respectively. There is no large particle that came from the aggregates can be found, due to MWCNTs being well dispersed in suspension as shown in Fig. 1A. An individual MWCNT with a diameter of 10–20 nm, and length of 1–15 μm are given in Fig. 1B. As a result, the individual nanotube gather together to form bundles (ropes) through Van der Waals forces. They are mostly arranged in bundles with close-packing stocking, which is formed by molecular interaction and mechanical interlocking among nanotubes.

CNTs are considered by many scientists as used reinforcement material for high-performance structural and multifunctional composites,[77,78] and strong research interests exist in developing CNT-reinforced nanocomposites.[79,80]

CNTs are a novel type of carbon material and can be considered as the result of folding graphite layers into carbon cylinders.[81] Using CNTs is a growing trend in the field of

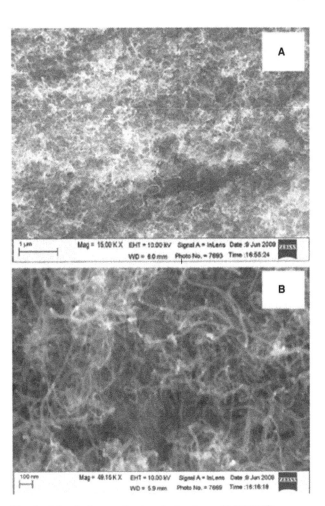

Fig. 1 Morphology and network structure of MWCNT nanopaper. (**A**) At a scale of 1 μm and (**B**) at a scale of 100 nm.
Source: From Lu et al.[76] © 2011, with permission from Elsevier.

physical chemistry.[82] These are multiwalled CNTs and single-walled CNTs.[83] CNTs have attracted a great deal of attention as model systems for nanoscience and for various potential applications, including composite materials,[80,84,85] battery electrode materials,[86] field emitters,[87] nanoelectronics[88] and, obviously, nanoscale sensors.[89,90] CNTs are used for stronger and tougher cement and concrete materials due to their extraordinary mechanical properties and extremely high aspect ratios.[91–93,57] CNTs are one of the most important nanotechnological field and receive widespread attentions in science.[94,95] CNTs/nanofibers have been studied for gas and/or biosensing applications in recent years.[96]

CNTs have high aspect ratios, high mechanical strength, high surface areas, excellent chemical and thermal stability, and rich electronic and optical properties.[97] The latter properties make CNTs important transducer materials in biosensors: high conductivity along their length means they are excellent nanoscale electrode materials.[98–100] Kiani, Ghaffari, and Mehri[101] studied nano-objects, which are attached to the single-walled CNTs (SWCNTs) considered as rigid solids. Some inclusive review studies regarding chemical, physical, and biological sensors based on CNTs could be found in literature.[102–105] Britto, Santhanam, and Ajayan[106] studied one of the most promising electrode materials with CNTs since the first electrode application in the oxidation of dopamine.

In the past several years, there has been a steady increase in activities that incorporate nanotubes into biological systems, including proteins, DNA, and living cells. An electrochemical DNA biosensor with a favorable performance for rapid detection of specific hybridization has been obtained by functionalizing CNTs with a carboxylic acid group.[107,108] CNTs have been brought a wide range of applications for them including nanofluidic conveying,[109,110] drug delivery,[111,112] hydrogen storage,[113,114] and nanosensors.[115,116] Shindo et al. have investigated the electrical resistance-based strain sensing behavior of MWCNT/polycarbonate composites under tension using analytical and experimental approaches.[117] Many researchers have studied the electronic characteristics of CNTs to develop electrochemical analytical methods and sensing devices with improved characteristics.[118] The use of CNT-modified microelectrodes has been suggested for *in vivo* applications. Swamy and Venton studied carbon fiber microelectrodes modified with (SWCNTs/CFMEs) to co-detect dopamine and serotonin in the striatum of anesthetized rats.[119] Multiwalled CNT (MWCNT)-based CFMEs were also used to detect ascorbic acid (AA) in rat brain. The electrocatalyic activity of the electrodes toward AA oxidation ($E_{1/2} = -0.06$ V vs. Ag/AgCl) allowed its determination in the presence of 3,4-dihydroxyphenylacetic acid (DOPAC, $E_{1/2} = 0.19$ V), UA ($E_{1/2} = 0.28$ V), and 5-hydroxytryptamine (5-HT, $E_{1/2} = 0.26$ V).[120] Bai et al. studied vertically aligned CNTs, which were synthesized on SiC microplates by a CVD process.[121] CNTs can serve as not only the structural reinforcement but also the excellent in situ sensor in structural composites.[105] Single-walled carbon nanotubes (SWCNTs) have become one of the most intensively studied nanostructures due to their unique properties.[122] Lay et al. focused on progress toward functionalization of SWCNTs with metal nanoparticles using electrochemical methods and the applications of metal decorated CNTs in energy-related applications.

Single-walled carbon nanotubes (SWCNTs) are the strongest known material.[123–126] SWCNTs are hundreds of times stronger than the highest grade high carbon steel commercially available.[127–131] The ability of CNT to increase electron transfer kinetics,[132] electroactive surface area,[133] and the simplicity of electrochemical[134] and conductometric sensors[135] is well-known. The oriented assemblies of short SWCNTs onto electrodes with different methods have been developed.[136–139] Deng, Matsuda, and Goddard[140] revealed that different linkers, such as –S–, –O–, –N–, –COOH– and –CON–, which anchored CNTs onto Pt(III), presented different resistances.

Schneider et al.[141] reported the Ni-catalyzed CVD synthesis of few-layered graphene and the spatial characterization of the few-layered transparent graphene by micro Raman spectroscopy. Its electrical characterization shows p-semiconductor behavior as well as studies on the gas sensing properties toward low concentrations of CO and H_2. Most studies on CNT-based biosensors are categorized into one of the following: a CNT polymer electrode,[142–146] a CNT–nanoparticle composite electrode,[147–150] a CNT modification on universal electrodes,[151–153] or vertically grown CNTs on silicon wafer as the electrode.[154] There are many difficulties in the fabrication of uniform and reproducible sensing electrode for DNA monitoring that limits mass production of DNA sensing electrode for commercial purposes.

Several articles were published on high-sensitive detection using of CNT-based detectors in microfluidics.[155,156] Graphene is regarded as the electrochemical materials of the future with reported advantageous electrochemical properties for sensing and energy storage.[157,158]

The MWCNT films with controlled thickness are firstly assembled onto ITO electrodes with the help of polyelectrolytes via a layer by layer (LbL) self-assembly technique, which was first reported by Decher and Hong.[159]

Gold Nanoparticles

Gold and platinum are usually considered to be the most appropriate electrode materials for the electrocatalytic oxidation of glucose since their catalytic performances have already been proven.[160–164] However, the series poisoning of Au and Pt electrodes by adsorbed intermediates, low sensitivity, and poor selectivity force researchers to develop more effective electrode materials for the electro-oxidation of glucose.[165–167]

Xie et al.[168] have studied tannic acid–doped poly(pyrrole) (PPy-TA) films on gold electrodes for selective electrochemical detection of dopamine (DA). Electrochemical quartz crystal microbalance (EQCM) studies have shown that the PPy/TA film has redox processes that involve only cation transport when the solution pH is greater than 3–4. The electroanalysis of Tramadol (TRA) was obtained by different electrochemical techniques.[169] A composite-modified glassy carbon electrode (GCE) is performed by electrodeposition of mono-dispersed AuNPs onto carbon nanotubes (AuNPs/CNTs/GCE). TRA was successfully determined in pharmaceutical dosage forms. Zhang et al. have been studied the preparation of the proposed nanohybrid-structured microsensor, which does not require any sophisticated equipment and skills.[170] After that, further investigations on enzymeless glucose sensing optimization and search for other possible applications, such as in fuel cells. Shahrokhian and Rastgar[171] studied the electrochemical reduction of tinidazole (TNZ) on gold nanoparticles/carbon nanotubes (AuNPs/CNTs) modified GCE using the linear sweep voltammetry. The modified electrode showed a wide linear response toward the concentration of TNZ in the range of 0.1–50 μM with a detection limit of 10 nM. Taking into account the advantages of AuNPs and CNTs, AuNPs/CNTs has more potential applications in the generation of the electrochemical sensors, due to the greatly promoted electron transfer rate and electrocatalytic ability.[172–175] The electroanalysis of Tramadol (TRA) was achieved by different electrochemical techniques of mono-dispersed AuNPs onto CNTs (AuNPs/CNTs/GCE).[169]

In comparison to other noble metals, gold nanomaterials are promising as a catalyst for glucose oxidation because of their high electrocatalytic activity, good biocompatibility, and stable chemical and antifouling properties.[176,177] Several approaches including LbL assembly have been applied to construct AuNPs into 2D or 3D structures.[178,179] Reduced graphene oxide (RGO) is an emerging material with attractive electronic, catalytic, and mechanical properties. Thus, it is a good support to load nanoparticles through surface modification for various applications.[180–185]

Gold nanoparticle colorimetric probes have been widely used for colorimetric assays because their extinction coefficients are high relative to common organic dyes.[186,187] Although Au and AgNPs have received extensive attention in plenty of analytical areas such as chemical sensing and biosensing,[188,189] as well as in medicine due to their immense potential for cancer diagnosis and therapy since their size is similar to the biological molecules and structures.[190,191]

Regarding the synthesis, the first scientifically based chemical synthesis of spherical AuNPs was utilized by Faraday in 1857.[192] Citrate is commonly used as reducing agent in AuNPs synthesis.[193] Soon, the syntheses were extended to other materials (e.g., Ag),[194] and shapes such as nanorods,[195] triangular structures,[196] cubes,[197] cages,[198] and nanoshells.[199]

Silver Nanoparticules

Silver shows the highest electrical conductivity among the metals and AgNPs have traditionally been used as catalysts in various reactions.[200,201] Hu et al.[202] studied a novel process for publication of a silver nanoparticle modified electrode using silver ion implantation. In recent years, silver's high electrocatalytic activity in glucose oxidation has made itself a prime electrode material for the development of nonenzymatic electrochemical glucose sensor.[203] AgNPs/ITO presented acceptable reproducibility and stability with a relative standard deviation of 2.3% in eight successive measurements of 5 mM glucose. Zhang et al.[204] studied fabrication of poly(alizarin yellow R) modified electrode via electropolymerization by CV. AgNPs were then electrodeposited on the surface of the poly(alizarin yellow R)-modified electrode. Mohammadrezaei et al.[205] investigated the effects of AgNPs dopants on the ethanol sensing properties of zinc oxide (ZnO) bulk sensors. Zhang studied

the silver coupled-particle system, which has higher plasmon resonance sensitivity as compared to the gold coupled-particle system.[206] The silver-coupled particle may be more suited for sensing applications as compared to the gold-coupled particle. Wang et al.[207] studied a nanostructured biosensor with uniformly deposited AuNPs as the sensing electrode for fast and low serum consuming detection of the IgE in an allergy patient's serum. Therefore, the blood serum detection results show that the nanostructured biosensor is able to detect a patient's allergy level with low sample consumption, short sample preparation time, and quick processing. Blood serum samples with known allergy levels that had been examined by the commercially available ImmunoCAP were used for the verification of the sensor. AgNPs coated on the surface of multiwalled CNTs showed good catalytic properties to trichloroacetic acid and realized selective detection of trichloroacetic acid. Therefore, Sun et al.[208] proposed an innovative way of making AgNP-doped chitosan (SNP-CS) film for electrochemical sensing of trichloroacetic acid. Gryczynski et al.[199] have reported that evaporation included self-assembly of colloidal silver nanoparticles on metallic films (Ag-SACs) form unique, fractal-like surfaces, which display even stronger fluorescence enhancement than SIFs on metal mirrors. In addition, by use of a long lifetime fluorophore Gryczynski et al. are able to provide a quantitative account of the correlation between the local MEF effect and change in fluorescence lifetime. Mohsenifar et al.[209] studied a new effective, pH, and thermally stable GOx-AgNPs bioconjugate was designed using two simple procedures. AgNPs incorporated poly(3,4-ethylenedioxythiophene-sodium dodecyl sulfate) (PEDOT$_{SDS}$) modified electrode was prepared by electrochemical deposition method. The PEDOT$_{SDS}$-AgNPs immobilizes electrochemically active mediator.[210]

AgNPs exhibit surface plasmon absorption in the visible region, which falls around 420 nm.[211] The solution containing AgNPs exhibits bright color whose hue and intensity depend on the setting provided by the reaction parameters.[212] Rahman et al. prepared silver oxide nanoparticles by a simple solution method using reducing agents in alkaline medium.[213] The first cycle of cyclic voltammograms (CV) of bare and different layers of MWCNT-modified indium tin oxide (ITO) electrodes were examined electrochemically in 1.0 mM AgNO$_3$ solution as shown in Fig. 2.[214]

The MWCNT multilayer films are highly conductive for electron transfer in the potential scan range from +0.7 V to −0.7V.[215] As a result, the nonconductive polymers used for MWCNT surface modification (PSS) and film preparation (PDDA) have little effect on the electrochemical behavior of MWCNT. The redox peaks are attributed to the reduction of silver during the forward sweep and subsequent oxidation of deposited silver during the reverse sweep.

A composite sensing system of an optimal microsphere resonator and AgNPs based on surface-enhanced Raman

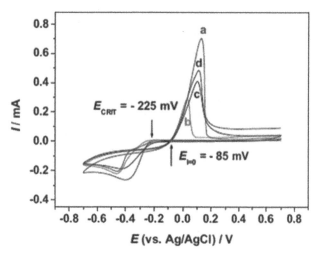

Fig. 2 Cyclic voltammograms of 1.0 M $AgNO_3$ in 0.1 M KNO_3 solution at bare ITO electrode (a) ITO electrodes modified with [PEI/PSS/(PDDA/MWCNT)$_3$], (b) [PEI/PSS/(PDDA/MWCNT)$_6$], (c) [PEI/PSS/(PDDA/MWCNT)$_9$], (d) Scan rate: 100 mV s^{-1}.
Source: From Yu et al.[214] © 2012, with permission from Elsevier.

scattering (SERS) and stimulated Raman scattering (SRS) techniques toward a point of care diagnostic system for AMI using the cTnI biomarker in HEPES-buffered solution is proposed by Saliminasab, Bahrampour, and Zandi et al.[216] The presence of AgNPs on the microsphere surface provides a tremendous enhancement of the resulting Raman signal through an electromagnetic enhancement of both the laser excitation and Stokes-shifted light of the order of 10^{10}.

Wang et al. studied the reduced GO/PAMAM- AgNPs nanocomposite (RGO-PAMAM-Ag), which was synthesized by self-assembly of carboxyl-terminated PAMAM dendrimer (PAMAM-G3.5) on GO as growing template, and in-situ reduction of both $AgNO_3$ and GO under microwave irradiation.[217]

Dawood et al.[218] studied the structural and vibrational characterization of silicon nanowire arrays synthesized by metal-assisted chemical etching (MACE) of Si deposited with metal nanoparticles. Gold and silver metal nanoparticles were synthesized by glancing angle deposition, and MACE was performed in a mixture of H_2O_2 and HF solution.

Shi et al.[219] studied the solvent-induced morphological charge of silver nanoparticles, which were studied with a combination of optical spectroscopy and atomic force microscopy.

Noble metal nanoparticles (NPs) with gold and silver display specific optical characteristics in the visible region (380–750 nm). Using localized surface plasmon resonance (LSPR) optical characteristics, previously, Endo et al.[220] detected the biomolecular interactions such as enzymatic reaction,[221] DNA hybridizations,[222,223] and antigen antibody reactions[224–226] LSPR optical characteristics, which can be observed using polyvinylpyrrolidone (PVP)-coated

AgNPs, are useful for biosensing applications with their high stability and dispersibility in aqueous solutions.[227]

Kanwal et al.[228] studied the conventional synthesis of AgNPs employs a reducing agent and a capping agent. Ag-nanoprobe allows this technique to be greatly robust, highly sensitive, and cost-efficient. In consideration of the fact that the pH stability of synthesized AgNPs matches with most of cellular fluids, these nanoparticles might be used as suitable biolabels for *in vivo* imaging.

Hybrid nanomaterials provide an efficient strategy for promoting the functionality of materials. Many papers have been published, which include graphene nanosheets as diverse metallic nanoparticles such as gold,[229] silver,[230–232] platinum,[233] and palladium.[234] Generally, reductants such as $NaBH_4$ have been applied to reduce GO nanosheets, accompanied by generation of AgNPs.[235] Ag–graphene composite nanosheets were also synthesized using PVP as reductant and stabilizer by heating to 60°C.[236]

The use of a new methodology was used to modify OFTs with AgNPs. The results show that the AgNPs modified optical fiber tip (AgNPs-MOFT) devices fabricated in this way have very good sensitivity and reproducibility.[237] The AgNPs-MOFT device can be used as a single use remote sensing element, since it is easy to fabricate and provide good sample-to-sample reproducibility. It can be envisioned that this kind of device could also be used as a SERS-based biosensor. Compared to AuNPs, AgNPs do not offer long-term chemical stability. Therefore, the applications of AgNPs for SRS sensing are not as plentiful.[238]

Metal Oxides and Nanoparticles

Metal nanoparticles have been used in the fabrication of sensors as nanocomposites or when covalently bonded with CNTs.[239] Yang, Kim, and Hong[240] studied ZnO nanoparticle-coated SWNT network sensors, which were fabricated and their dimethylphosponate (DMMP) gas sensing characteristics. Recently, metal nanoparticles received increasing popularity because of their unique size and shape dependent properties different from those of their bulk counterparts.[241–245] Among many types of metal alloy NPs, gold (Au) and silver (Ag) alloy NPs have been widely researched in the field of optics,[246] biosensing,[247] drug delivery,[248] and catalysis.[249] Metal nanoparticles play a significant role in many applications such as sensors, catalysts, field-emitters, etc. Carbon nanotube supports due to their unique geometry.[122]

Nanomaterials, particularly carbon materials, have a significant role to play in new development in each of the biosensor size domains. Various kinds of zero-, one-, two-, and three-dimensional nanomaterials are helping to meet these challenges. Examples of such materials include semiconductor quantum dots,[250] metallic nanoparticles,[251] metallic or semiconductor nanowires,[252,253] CNTs,[254,255] nanostructured conductive polymers or nanocomposites,[103] mesoporous materials,[256] and various other materials.[257,258]

Electrochemical deposition is an economical and convenient choice for preparing uniform and size controllable nanomaterials,[259] especially preparation of metallic nanomaterials (MNPs).[260,261] Hieu et al. studied a simple and practicable method for the scalable fabrication of various metal oxide nanowires and composites using SWCNTs even those with high melting temperatures, such as TiO_2, SnO_2, ZnO, CuO, and WO_3 by using porous SWCNT film as template.[262] Rossinyal et al.[263] reported the synthesis of cerium and tungsten oxide in two-dimensional hexagonal 6 mm and three-dimensional cubic Ia3d mesostructures.

A good response for the reducing gases was observed with both the In_2O_3 and SnO_2 sensors at temperatures higher than 400°C. This behavior is typical for metal oxides, which generally require the presence of suitable metal promoters to decrease the operating temperature substantially.[264] MNPs can be loaded into coatings for sensing purposes thanks to their special optical properties, making possible the tuning of the coating refractive index, or the apparition of some optical resonant phenomena.[265] Fu, Zhang, and Chen studied CNT–CuNP hybrid, which was prepared by chemical reduction for electrochemical sensing of carbohydrates as shown SEM images in Fig. 3. SEM images show that the surface morphology of the hybrid is much different from those of CNTs and CuNPs. It can be seen that CuNPs are well dispersed and embedded throughout the CNT matrix and interconnected hybrid network is formed.[266]

At present, metal oxide conductometric gas sensors are the most used and studied devices designed for the control of toxic and inflammable gases in technological processes and surrounding atmosphere.[267–271] In the field of sensors based on semiconducting metal oxides (SMOs), SnO_2 is still the most popular candidate for the detection of reducing gases like CO and H_2. A huge effort has been made to improve their performance such as selectivity or sensitivity, most of the times in an empirical manner.[272] Shen et al. reported the task specific ionic liquid for direct electrochemical detection of heavy metal oxides including cadmium oxide, copper oxide, and lead oxide at room temperature.[273] The novel sensor platform opens new pathways for in-situ monitoring of metal oxide particulates for environmental sensing and decontamination applications. MNPs are technologically important materials because of their unique optical, electronic, magnetic, and catalytic properties,[274] which have significant application in display, microelectronics, data storage, drug delivery, imaging, and sensing.[275] Transition metal oxides offer a wide spectrum of properties, which provide the foundation for a broad range of potential applications.[276] The native calf-thymus DNA was successfully immobilized from water solutions onto several kinds of carbon electrodes under controlled dc potentials for study.[277]

Fig. 3 SEM images of CNTs (**A**), CuNPs (**B**), and CNT–CuNP hybrid (**C**) prepared by chemical reduction. Conditions: accelerating voltage, 20 kV, magnification, 20000×.
Source: From Fu et al.[266] © 2012, with permission from Elsevier.

Phanichphat et al. studied the performance toward flammable gases of unloaded and metal (Pt, Sn, Ru, Nb, and W) catalyzed metal oxide (ZnO, WO_3, SnO_2, and TiO_2) nanoparticle thick films fabricated by spin coating is reviewed, discussed, and compared to MOXs prepared by other methods.[278] Several single, binary, and tertiary oxides have been prepared using sol-gel chemistry: SiO_2,[279] TiO_2, ZrO_2, SiO_2,[280] TiO_2, ZrO_2, Al_2O_3,[281] SiO_2, TiO_2, ZrO_2, Al_2O_3, Fe_2O_3, Sb_4O_6, WO_3, and $YZrO_2$.[282]

CONDUCTING POLYMERS WITH NANOMATERIALS

Conducting polymers have been the subject of intensive research since the late 1970s.[283–285] Conducting polymers are a class of functional groups that have alternating single and double carbon–carbon bonds along the polymeric chains.[286] Conducting polymer sensors are based on the variation of their capacity, optical properties; nanostructures with large specific surface area and porous structure are therefore predicted to be excellent sensing materials.[287] To date, a number of conducting polymer-based sensors,[288] such as gas sensors,[289–291] hydrogen sensors,[292] and acetic acid sensors,[293] have been investigated in literature.[294]

In literature, polycarbazole–Au composite was synthesized by the oxidation of carbazole using $HAuCl_4$ oxidant.[295] Poly(3-hexylthiophenes) (P3HT) have been studied for various applications due to their superior optoelectronic properties and processability,[296,297] effect transistors, solar cells, electrochromic devices, and chemical sensors.[298–300] Liu, Reccius, and Craighead[301] and Jenekhe et al.[302,303] studied the field effect transistor based on P3HT electrospun (ES) fibers.

Patil et al. studied the enzyme GOx, which was entrapped into the poly(O-anisidine) (POA) film by a physical adsorption method. POA-GOx films exhibited a fast amperometric response (1–5) s and a linear response in the range of 2–200 mM glucose.[304] The glucose biosensors have received great attention due to their many applications in clinical chemistry, biotechnology, and the food industry.[305,306]

Electropolymerization is well known as a very convenient way to immobilize polymers at the electrode surface, and the thickness, permeation, and charge transport characteristics of the polymeric films can be easily and precisely controlled by modulating the electrochemical parameters for various electrochemical methods.[168]

Nanostructured materials or nanomaterials area new class of materials provide a great potential for improving the performance of biosensing systems and extend their applications in various fields of material sciences and biomedical sciences[307,308] between others.

Polypyrrole was first electrochemically synthesized more than two decades ago by Diaz et al.[309] It has been intensively investigated on different electrode surfaces. In literature, carbon fiber surface modification was performed electrochemically with conducting polymers as thin film coated/CFME and their possible applications[310] such as capacitors,[311] electrochromic devices,[312] and batteries.[313] Polymerization of pyrrole onto CFs has been previously studied;[314–317] however, it has not been considered as a possible material for microsensor applications.[318] Using small diameter (6–7 μm) CFs as a substrate for pyrrole polymerization, flexible and inexpensive nitrate microsensors with characteristics competitive with commercial analogs were prepared.

Functional carbazole monomers were chemically or electrochemically polymerized to yield materials with interesting properties with a number of applications such as electroluminescent applications,[319] photoactive devices,[320] sensors,[321,322] electrochromic displays,[323,324] and rechargeable batteries.[325] Polycarbazole is attributed to good electroactivity, and useful thermal, electrical, and photophysical properties.[326,327] Polycarbazole has been immobilized on different transducer surfaces and its morphology and electrochemical behavior has been studied using different analytical techniques.[328–330]

Polyaniline (PANI) has been one such material of interest. PANI's conducting properties, stability, ease of electron transfer, and ease of synthesis are key features that make it attractive for its use in electrochemical biosensing.[331] Polyaniline (PANI) is known organic semiconductor or conducting polymer. In addition, it is promising host material for various inorganic semiconductors and carbon materials due to the relatively high conductivity, excellent chemical and the electrochemical stability.[332,333] The PANI/Gr composite-modified electrodes were utilized for fabricating the highly sensitive, reliable, and reproducible hydrazine chemical sensor.[334] Bhansali et al.[335] studied polyaniline-protected AuNPs, which were electrophoretically deposited onto a gold electrode, and utilized to fabricate an electrochemical cortisol biosensor.

Polythiophene has received great scientific attention in the last 20 years.[336] Numerous polyalkyl derivatives of thiophene have been synthesized as far by chemical and electrochemical approaches, resulting in conducting polymers with better solubility and higher capacitor behaviors.[337]

SENSORS AND BIOSENSORS

Szunerits et al. described a novel platform for preparing LSPR sensing surfaces.[338] Ferapontova and Gorton[339] studied alkanethiol-modified gold electrodes and was compared with that of some previously studied complex heme enzymes, specifically, cellobiose dehydrogenase (CDH) and sulfite oxidase (SOx). Complex cofactor-containing enzymes and enzyme complexes embedded in biological membranes play an important role in the living cell.

Attempts have been made toward immobilization of biomaterials onto the transducing part of the biosensors.[340,341] Immobilization of GOx on various surfaces especially by its aldehyde group via oxidation of sugar moieties, or carboxylic/amino groups and coenzyme FAD are the effective techniques for developing GOx biosensors on different surfaces. The assay of various substrates such as L-and D-lactate ions,[342] L-carnitine,[343] L-malic acid[344] for different applications such as sports medicine, control of alcoholic and malo-lactic fermentations during wine making, control of cellular culture reactors, etc. Cabrera et al. have studied an electrochemical DNA sensor, which was constructed using single-walled carbon nanotubes (SWCNTs) attached to a self-assembled monolayer of 11-amino-1-undecanethiol on gold surface.[345] Patolsky Weizmann, and Willner and collaborators have studied the first report on the attachment of SWCNTs to a gold surface; this type of integration method has become popular.[346] This type of surface covalently modified with SWCNTs has been used to attach and detect enzymes,[137] UA,[347] metal complexes,[348] semiconductor nanoparticles,[349] and DNA.[350,351]

The glucose biosensors based on the nonconducting poly(O-phenylenediamine),[352,353] poly(ethecridine), overoxidized polypyrrole,[354] poly(phenol),[355] and poly(O-aminophenol) (POAP)[356] films have been reported and showed the good anti-interferent ability. Yao et al.[357] studied GOx, which was immobilized in nonconducting poly(O-aminophenol) film at Prussian blue (PB)–modified platinum (Pt) electrode and a new amperometric glucose biosensor.

Lee et al. studied the glucose sensor electrode, which was prepared from electrospun spherical-type carbon materials.[358] The GOx enzyme was immobilized on a prepared electrode for efficient glucose sensing. Among many methods using glucose oxides, electrochemical sensors hold much promise due to their low cost, simplicity, and high sensitivity. Li et al.[359] studied the electrochemical biosensor to assay purine nucleoside phosphorylase (PNP) activity. Electrochemical sensors are the most widely studied sensing units.[360,361] Several works on amperometric and voltammetric sensors have also been reported.[362,363]

Oxidation of Dopamine

In literature, there are many studies aimed at understanding catecholamine and quinine electrochemistry in biological systems.[364] Dopamine (DA) is a neurotransmitter that is released at the synaptic cleft between two neurons and is found in many discrete regions throughout the central nervous system.[365] *In vivo* electrochemical procedures are ideally suited for the measurement of several physiological parameters, including net DA concentration released per action potential and the rate of DA uptake into neurons afterward. DA → DOQ + 2e⁻ + 2H⁺ where DOQ is the o-quinone form of DA. DA adsorption was also examined with high-speed chronoamperometry.[366] The electrochemical methods have more advantages over other methods since the electrodes can be made conveniently to sense the neurotransmitters in the living organism.[367] Ohsaka et al.[368] studied a gold electrode modified with a monopositive self-assembled monolayer of cyctamine (CYST) for the determination of dopamine (DA) in the presence of ascorbic acid (AA).[369]

Multiwalled carbon nanotube–based CFMEs have been used to detect AA in rat brain. The electrocatalytic activity of the electrodes toward AA oxidation ($E_{1/2}$ = −0.66 V Ag/AgCl) allowed its determination in the presence of 3,4-dihydroxyphenylacetic acid (DOPAC, $E_{1/2}$ = 0.19 V) UA ($E_{1/2}$ = 0.28 V), and 5-hydroxytryptamine (5-HT, $E_{1/2}$ = 0.26 V).[120] Li and Lin[370] have been fabricated through an electrochemical oxidation procedure of a novel covalently modified GCE with poly(vinylalcohol) to electrochemi-

cally detect dopamine (DA), ascorbic acid (AA), UA and their mixture by a differential pulse voltammetry (DPV) method. DA is an important neurotransmitter compound widely distributed in the brain for message transfer in the mammalian central nervous system. Low levels of DA are related to neurological disorders, such as schizophrenia, Parkinson's disease, and HIV infection.[371,372]

In biological environments (pH 7.4), DA exists as cation (pK_a = 8.87) while AA exists as anion (pK_a = 4.10).[373–375] The negatively charged polymer film modified electrodes have been used to suppress the interference of AA with the determination of AA.[376] Rubianes and Rivas utilized an anionic polymer, Nafion, which repels AA and other negatively charged interferences, while attracting positively charged DA. Although Nafion coating introduced the desired specificity, it has some disadvantages, e.g., relatively thick film is required and thus delayed response time of the sensor.[377]

The efficiency of voltammetric dopamine detection is not significantly affected by changes in temperature, pH, or oxygen that may occur under physiological or pathological conditions.[378,379] For instance, pH changes are clearly discernible from dopamine using fast-scan cyclic voltammetry.[380] Recently, DuVall and McCreery[381] reported that modification of carbon electrodes with different quinines, either physisorbed or chemisorbed, significantly reduced adsorption of DA to GC electrodes without altering the heterogeneous kinetics of the electrode reaction or significantly changing the sensitivity of the measurement.

Gas Sensors

SWCNT has been studied as attractive candidates for gas sensor, and responds to both oxidizing and reducing gases under ambient conditions.[382,383] Dimethyl methylphosphonate (DMMP) is a simulant for the nerve agent sarin and various types of sensors. DMMP have been devised to detect the nerve agent including surface acoustic wave (SAW), QCM, and SMO sensors.[384,385] Among them, SMO sensors are a promising technique due to its high sensitivity, reversibility, and low cost.[386,240] The SWCNT-based network sensors were applied to the detection of DMMP at room temperature, but it exhibited a slow response and a long recovery.[387,388] To overcome these shortcomings and improve selectivity, several surface modifications have been explored with catalytic metal nanoparticules and polymers.[389] Hellmich et al.[390] studied the sensitivity of micromachined gas sensor elements with respect to CO, NO, NO_2, CH_4, H_2O, and oxygen partial pressure changes. The gas sensing mechanism of graphene is expressed as the adsorption and desorption of gas molecules, acting as donors and acceptors, on the surface.[54,391] Many researchers have reported the fabrication and testing of graphene based on gas sensors.[392–395] Johnson et al. reported the ammonia sensing behavior of graphitic nanoribbons decorated with platinum (Pt) nanoparticles.[396] Chu et al. used Pt-coated graphene surface to detect H_2 gas in different concentrations. Randeniya et al.[397] and Penza et al.[398] reported the NH_3 gas sensing behavior of CNTs functionalized with AuNPs and Au/Pt nano clusters, respectively. The gas sensors have been studied based on quasi-one-dimensional oxide semiconductor nanomaterials, such as nanofibers,[399] nanorods,[400] and nonospheres.[401]

The gas sensing with functionalized CNTs was mainly directed toward nanostructures with either organic polymers, biomolecules or catalytic metal, metal-oxide and nanoparticles attached to the CNT sidewalls.[89] Respiratory monitoring of carbon dioxide (CO_2) is an important medical diagnosis tool.[402–404] Rajamani et al.[405] fabricated carbon dioxide gas sensors by self-assembing single-walled carbon nanotube films on a SAW delay line operating at 286 MHz. As a result, polymer functionalization was used to enhance the sensitivity of the CNTs to carbon dioxide. According to the results, the sensor tracks carbon dioxide concentration changes between 0% and 10%. The final sensor is sensitive, small and wirelessly interrogable, thus making it potentially useful for respiratory monitoring. Another important application for CO_2 sensors is indoor air quality monitoring where there is a clear need for developing small, inexpensive, and wireless gas sensors. The adsorption electron withdrawing (e.g., NO_2 and O_2) or electron donating (e.g., NH_3) molecules to a CNT can cause charge transfer between the CNT and the molecules.[406] The adsorption of electron-donating molecules causes the number of holes to decrease and the resistance to increase. Kong et al.[382] first reported the fabrication of sensors using single-walled CNTs. Studies have been conducted to create chemical sensors based on CNTs.[407–413] Pandey et al.[414] studied aqueous ammonia sensing of polymer/AgNPs nanocomposite by optical method based on SPR. The response time of 2–3 s and the detection limit of ammonia solution (1 ppm) were found at room temperature. Hellmich et al.[390] investigated the sensitivity of micromachined gas sensor elements with respect to CO, NO, NO_2, CH_4, H_2O, and oxygen partial pressure changes. It is difficult to group such sensors into arrays that offer much higher levels of gas selectivity than individual sensor devices.[415]

Lee et al. studied the sensor sensitivity for NO and CO gases, which was improved about five times based on the effects of chemical activation, carbon black additives, and fluorination treatments.[416] The detection of NO and CO gases has recently become a critical issue due to the some of the most common air pollutants. NO gas was formed during combustion engines such as those in automobiles. CO is especially dangerous since it possesses no odor or color and is therefore undetectable by humans. Vehicle exhaust is a major source of environmental CO emissions, adding to the formation of smog. Therefore, the development of highly selective and stable NO and CO sensors is an important goal that will assist in the study of environmental impacts. Several types of gas sensors have been

reported in the literature.[417–419] Many types of NO and CO sensors have been investigated with different basic techniques, using materials such as metal oxide semiconductors (MOS), solid electrolytes (SE), and CNTs.[420–422]

For the environmental sensing behavior, two models are used to explain the interaction of CO with a metal oxide surface and adsorbed species.[423] The first model assumes that oxygen molecules adsorb on the surface of metal oxides in a form of O_2^-. When CO reacts with O_2^- at high temperature, the reaction ($CO + O_2^- \rightarrow CO_2 + 2e^-$) proceeds on the surface.[424] In the second model, oxygen adsorbs as O_2^-. At low temperature, CO molecules do not have sufficient thermal energy to react with O_2^-, while at a temperature above 300°C. O_2^- is converted to O^- and CO reacts via $CO + O^- \rightarrow CO_2 + e^-$. In both cases, the detection scheme for CO involves the reaction with preadsorbed oxygen, and normal CO sensors based on metal oxides work only at high temperature (>300°C) using its reducing effect.[425,426]

Nanostructured materials present new opportunities for enhancing the properties and performances of gas sensors due to the much higher surface to bulk ratio in nanomaterials compared to coarse micrograined materials.[427] Korotcenkov[428] reviewed the influence of morphology and crystallographic structure on gas-sensing characteristics of metal oxide conductometric-type sensors.

Nanowires of different metal oxides, employed as sensing material of Schottky diodes based on gas sensor, have been produced evidenced the presence of substoichiometric ZnO compound with a deficit of oxygen.[429] Zhou et al.[430] reported CO gas response in the concentration range of 1% at high temperature on La-doped $BaTiO_3$ perovskite as a chemical sensor. Lee and co-workers[431,432] investigated the sensitivity, selectivity, and response time for CO gas in air with various compositions of SnO_2/Pt and at different sintering temperatures. Ryu et al.[426] reported CO gas sensing property on ZnO porous film.

Recently, results were published on the effect of CO with Pt- and Pd-doped SnO_2 sensors in very low oxygen backgrounds (around 20 ppm).[433] The findings indicate that there is a direct reaction between the gas and SnO_2, which does not involve the lattice oxygen ions.[434] The surface of metal oxides is responsible for receptor function of solid-state gas sensors,[435,436] and therefore this part will be devoted to brief overview of some surface properties of metal oxides important in gas sensor operation.

Ascorbic Acid and UA

Ascorbic acid (AA) and UA are the most important interferences for nonenzymatic detection of glucose. In the physiological sample, glucose concentration (3–8 mM) is generally much higher than those of UA (0.1 mM) and AA (0.1 mM).[437,438] Electrocatalytic oxidation of vitamins such as ascorbic acid (vitamin C, AA) has been more interested in medicine, veterinary science, and toxicology, diagnosis

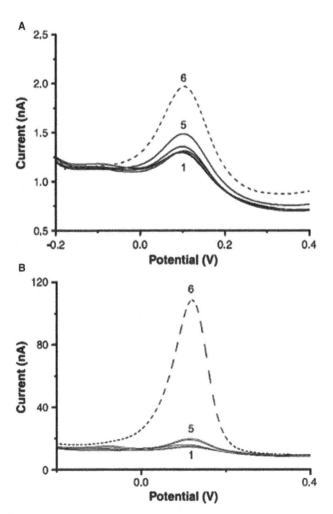

Fig. 4 DPV response for five additions of 100 μM ascorbic acid (1–5), and addition of 44 μM dopamine (6) (**A**) using an uncoated carbon fiber electrode as control, and (**B**) a PCz-modified CFE in LiClO$_4$/AN.
Source: From Ates et al.[442] © 2008, with permission from Springer-Verlag.

of certain metabolic disorders and the determination of nutritional value of foods.[439–441] Single CFMEs were electrocoated polycarbazole and poly(carbazole-*co-p*-tolylsulfonyl pyrrole) films by oxidation of carbazole or mixtures of *p*-tolylsulfonyl pyrrole/carbazole (100:1) dissolved in CH$_3$CN. The modified CFMEs were used as sensors for the detection of dopamine in the presence of ascorbic acid as electrochemically interfering compound.[442] The prepared sensors displayed a sensitivity of 2.5 nA μM^{-1} for PCz-modified CFME as 0.05 nA μM^{-1} for P(Cz-*co-p*-Tsp) modified CFME toward dopamine and reject the interference of ascorbic acid to a concentration of 500 μM.

UA is the primary end product of purine metabolism. Abnormal levels of UA are symptoms of several diseases, such as gout, hyperuricemia, and Lesch-Nyan disease.[443] Electroactive UA can be reversibly oxidized in aqueous solution and the major product is allantoin.[444,445] Kim, Kim, and Jeon[446] studied electrochemical biosensor for the detection

of 5-HT in the presence of DA and AA by various types of graphene obtained from different reductants and several methods.

CONCLUDING REMARKS

The advantages of the detectors include higher sensitivity, satisfactory stability surface renewability, bulk modification, and low expense of fabrication. Scientists should find applications in microchip CE, flowing-injection analysis, and other microfluidic analysis systems. In future applications, the interplay of surface charge and surface hydrophobicity will be important in controlling the pore reactivity in carbon nanoparticle aggregates.[447] Nanomaterials with conducting polymers increase the applicability of the material especially in biosensor evaluations.

REFERENCES

1. Tisch, U.; Haick, H. Nanomaterials for cross-reactive sensor arrays. MRS Bull. **2010**, *35* (10), 797–803.
2. Qureshi, A.; Kang, W.P.; Davidson, J.L.; Gurbuz, Y. Review on carbon derived solid state, micro and nanosensors for electrochemical sensing applications. Diamond Relat. Mater. **2009**, *18* (12), 1401–1420.
3. Su, D.S.; Schlogl, R. Nanostructured carbon and carbon nanocomposites for electrochemical energy storage applications. Chemsuschem **2010**, *3* (2), 136–168.
4. Tao, Y.S.; Endo, M.; Inagaki, M.; Kaneko, K. Recent progress in the synthesis and applications of nanoporous carbon films. J. Mater. Chem. **2011**, *21* (2), 313–323.
5. Varghese, S.H.; Nair, R.; Nair, B.G.; Hanajiri, T.; Maekawa, T.; Yoshida, Y.; Kumar, D.S. Sensors based on carbon nanotubes and their applications: A Review. Curr. NanoSci. **2010**, *6* (4), 331–346.
6. Kovalenko, I.; Bucknall, D.G.; Yushin, G. Detonation nanodiamond and onion-like carbon-embedded polyaniline for supercapacitors. Adv. Mater. **2010**, *20* (22), 3979–3986.
7. Holt, K.B. Diamond at the nanoscale: Application of diamond nanoparticles from cellular biomarkers to quantum computing. Phil. Trans. Royal Soc. A **2007**, *365* (1861), 2845–2861.
8. Pumera, M. Graphene-based nanomaterials and their electrochemistry. Chem. Soc. Rev. **2010**, *39* (11), 4146–4157.
9. Lu, B.P.; Bai, J.; Bo, X.; Zhu, L.D.; Guo, L.P. A simple hydrothermal synthesis of nickel hydroxide-ordered mesoporous carbons nanocomposites and its electrocatalytic application. Electrochim. Acta **2010**, *55* (28), 8724–8730.
10. Vuorema, A.; Sillanpaa, M.; Rasaei, L.; Wasbrough, M.J.; Edler, K.J.; Thielenmans, W.; Dale, S.E.C.; Bending, S.; Wolverson, D.; Marken, F. Ultrathin carbon film electrodes from vacuum-carbonised cellulose nanofibril composite. Electroanalysis **2010**, *22* (6), 619–624.
11. Barriere, F.; Downard, A.J. Covalent modification of graphitic carbon substrates by non-electrochemical methods. J. Solid State Electrochem. **2008**, *12* (10), 1231–1244.
12. Sarac, A.S. *Enyclopedia of Polymer Science and Technology*, part 3, 3rd Ed.; Mark, H.F., Ed.; 2004; Vol. 9.

Nanomaterials—
Nanoparticles

13. Banerjee, S.; Paira, T.K.; Kotal, A.; Mndal, T.K. Surface-confined atom transfer radical polymerization from sacrificial mesoporous silica nanospheres for preparing mesoporous polymer/carbon nanospheres with faitful shape replication: Functional mesoporous materials. Adv. Funct. Mater. **2012**, *22* (22), 4751–4762.

14. Alwarappan, S.; Erdem, A.; Liu, C.; Li, C. Probing the electrochemical properties of graphene nanosheets for biosensing applications. J. Phys. Chem. C **2009**, *113* (20), 8853–8857.

15. Li, F.; Chai, J.; Yang, H.; Han, D.; Niu, L. Synthesis of Pt/ionic liquid/graphene nanocomposite and its simultaneous determination of ascorbic acid and dopamine. Talanta **2010**, *81* (3), 1063–1068.

16. You, J.; Jeong, Y.N.; Ahmed, M.S.; Kim, S.K.; Choi, H.C.; Jeon, S. Reductive determination of hydrogen peroxide with MWCNTs-Pd nanoparticles on a modified glassy carbon electrode. Biosens. Bioelectron. **2011**, *26* (5), 2287–2291.

17. Wei, L.; Lei, Y.; Fu, H.G.; Yao, J. Fullerene hollow microspheres prepared by buble-templates as sensitive and selective electrocatalytic sensor for biomolecules. ACS Appl. Mater. Interfaces **2012**, *4* (3), 1594–1600.

18. Kushch, S.D.; Kujunko, N.S.; Tarasov, B.P. Platinum nanoparticles on carbon nanomaterials with graphene structure as hydrogenation catalysts. Rus. J. Gen. Chem. **2009**, *79* (4), 706–710.

19. Lee, D.; Lee, J.; Kim, J.; Kim, J.; Na, H.B.; Kim, B.; Shin, C.; Kwak, J.H.; Dolnalkova, A.; Grate, J.W.; Hyeon, T.; Kim, H. Simple fabrication of a highly sensitive and fast glucose biosensor use in enzymes immobilized in mesocellular carbon foam. Adv. Mater. **2005**, *17* (23), 2828–2833.

20. Kang, X.; Wang, J.; Wu, H.; Aksay, I.A.; Liu, J.; Lin, Y. Glucose oxidase-graphene-chitosan modified electrode for direct electrochemistry and glucose sensing. Biosens. Bioelectron. **2009**, *25* (4), 901–905.

21. Chua, C.K.; Ambrosi, A.; Pumera, M. Graphene based nanomaterials as electrochemical detectors in Lab-on-a-chip devices. Electrochem. Commun. **2011**, *13* (5), 517–519.

22. Elias, D.C.; Nair, R.R.; Mohiuddin, T.M.G.; Morozov, S.V.; Blake, P.; Halsall, M.P.; Ferrari, A.C.; Baukhvalov, D.W.; Katsnelson, M.I.; Geim, A.K.; Novoselov, K.S. Control of Graphene's properties by reversible hydrogenetaion: Evidence for graphene. Science **2009**, *323* (5914), 610–613.

23. Lee, C.; Wei, X.; Kysar, J.W.; Hone, J. Measurement of the elastic properties and intrinsic strength of monolayer graphene. Science **2008**, *321* (5887), 385–388.

24. Geim, A.K.; Novoselov, K.S. The rise of graphene. Nat. Mater. **2007**, *6* (3), 183–191.

25. Novoselov, K.S.; Geim, A.K.; Morozov, S.V.; Jiang, D.; Zhang, Y.; Dubonos, S.V.; Grigorieva, I.V.; Dubonos, S.V.; Firsov, A.A. Electric field effect in atmocically thin carbon films. Science **2004**, *306* (5696), 666–669.

26. Novoselov, K.S.; Geim, A.K.; Morozov, S.V.; Jiang, D.; Katsnelson, M.I.; Grigorieva, I.V.; Dubonos, S.V.; Firsov, A.A. Two-dimensional gas of massless Dirac fermions in graphene. Nature 2005, *438* (7065), 197–200.

27. Eda, G.; Fanchini, G.; Chhowalla, M. Large-area ultrathin films of reduced graphene oxide as a transparent and flexible electronic materials. Nat. Nanotechnol. **2008**, *3* (5), 270–274.

28. Wang, X.; Zhi, L.; Mullen, K. Transparent, conductive graphene electrodes for dye-sensitized solar cells. Nano Lett. **2008**, *8* (1), 323–327.

29. Li, D.; Muller, M.B.; Gilje, S.; Kaner, R.B.; Wallace, G.G. Processable aqueous dispersions of graphene nanosheets. Nat. Nanotechnol. **2008**, *3* (2), 101–105.

30. Shao, Y.; Wang, J.; Engelhard, M.; Wang, C.; Lin, Y. Facile and controllable electrochemical reduction of graphene oxide and its applications. J. Mater. Chem. **2010**, *20* (4), 743–748.

31. Castro Neto, A.H.; Guinea, F.; Peres, N.M.R.; Novoselov, K.S.; Geim, A.K. The electronic properties of graphene. Rev. Mod. Phys. **2009**, *81* (1), 109–162.

32. Zhu, Y.; Murali, S.; Cai, W.; Li, X.; Suk, J.W.; Potts, J.R.; Ruoff, R.S. Graphene and graphene oxide: Synthesis, properties and applications. Adv. Mater. **2010**, *22* (35), 3906–3924.

33. Esfandiar, A.; Ghasemi, S.; Irajizad, A.; Akhavan, O.; Gholami, M.R. The decoration of TiO2/reduced graphene oxide by Pd and Pt nanoparticles for hydrogen gas sensing. Int. J. Hydrogen Energy **2012**, *37* (20), 15423–15432.

34. Arash, B.; Wang, Q.; Duan, W.H. Detection of gas atoms via vibration of graphenes. Phys. Lett. A **2011**, *375* (24), 2411–2415.

35. Yang, W.; Ratinac, K.R.; Ringer, S.P.; Thordarson, P.; Gooding, J.J.; Braet, F. Carbon nanomaterials in biosensors: Should you use nanotubes or graphene? Angew. Chemie. Int. Ed. **2010**, *49* (12), 2114–2138.

36. Lu, J.; Drzal, L.T.; Worden, R.M.; Lee, I. Simple fabrication of a highly sensitive glucose biosensor using enzymes immobilized in exfoliated graphite nanoplatelets Nafion membrane. Chem. Mater. **2007**, *19* (25), 6240–6246.

37. Lu, J.; Do, I.; Drzal, L.T.; Worden, R.M.; Lee, I. Nano-metal-decorated exfoliated graphite nanoplatelet based glucose biosensors with high sensitivity and fast response. ACS Nano **2008**, *2* (9), 1825–1832.

38. Xu, Y.; Liu, Z.; Zhang, X.; Wang, Y.; Tian, J.; Huang, Y.; Ma, Y.; Zhang, X.; Chen, Y. A graphene hybrid material covalently functionalized with porphyrin: Synthesis and optical limiting property. Adv. Mater. **2009**, *21* (12), 1275–1279.

39. Kim, K.S.; Zhao, Y.; Jang, H.; Lee, S.Y.; Kim, J.M.; Kim, K.S.; Ahn, J.H.; Kim, P.; Choi, J.Y.; Hong, B.H. Large-scale pattern growth of graphene films for stretchable transparent electrodes. Nature **2009**, *457* (7230), 706–710.

40. Gautam, M.; Jayatissa, A.H. Gas sensing properties of graphene synthesized by chemical vapor deposition. Mater. Sci. Eng. C **2011**, *31* (7), 1405–1411.

41. Joshi, R.K.; Gomez, H.; Alvi, F.; Kumar, A. Graphene films and ribbons for sensing of O_2 and 100 ppm of CO and NO_2 in practical conditions. J. Phys. Chem. C **2010**, *114* (14), 6610–6613.

42. Zhang, L.S.; Song, W.D.; Wang, W.; Wu, Z.Y. Characterization of partially reduced graphene oxide as room temperature sensor for H_2, Nanoscale **2011**, *3* (6), 2458–2460.

43. Wu, W.; Liu, Z.; Jauregui, L.A.; Yu, Q.; Pillai, R.; Cao, H.; Bao, J.; Chen, Y.P.; Pei, S.S. Wafer-scale synthesis of graphene by chemical vapor deposition and its application in hydrogen sensing. Sens. Actuators B **2010**, *150* (1), 296–300.

44. Sing, G.; Choudhary, A.; Haranath, D.; Joshi, A.G.; Singh, N.; Pasracha, R. ZnO decorated luminescent graphene as a

potential gas sensor at room temperature. Carbon **2012**, *50* (2), 385–394.

45. Wang, K.; Wu, J.; Liu, Q.; Jin, Y.; Yan, J.; Cai, J. Ultrasensitive photoelectrochemical sensing of nicotinamide adenine dinucleotide based on graphene-TiO_2 nanohybrids under visible irradiation. Anal. Chim. Acta **2012**, *745* (1), 131–136.

46. Allen, M.J.; Tung, V.C.; Kaner, R.B. Honeycomb carbon: A review of graphene. Chem. Rev. **2010**, *110* (1), 132–145.

47. Bell, N.J.; Yun, H.N.; Du, A.J.; Coster, H.; Smith, S.C.; Amal, R. Understanding the enhancement in photoelectrochemical properties of photocatalytically prepared TiO_2-reduced oxide composite. J. Phys. Chem. C. **2011**, *115* (13), 6004–6009.

48. Galal, A.; Ha, A.; Hassan, N.F. Graphene supported Pt-M (M=Ru or Pd) for electrocatalytic methanol oxidation. Int. J. Electrochem. Sci. **2012**, *7* (1), 768–784.

49. Zheng, Z.X.; Du, Y.L.; Wang, Z.; Zhang, F.; Wang, C. Concise route to prepared graphene-CNTs nanocomposite supported Pt nanoparticles and used as new electrode material for electrochemical sensing. J. Mol. Catalysis A **2012**, *363–364*, 481–488.

50. Qazi, M.; Vogt, T.; Koley, G. Trace gas detection using nanostructured graphite layers. Appl. Phys. Lett. **2007**, *91* (23), 233101.

51. Quyang, Y.; Guo, J.; Carbon-based nanomaterials as contacts to graphene nanoribbons. Appl. Phys. Lett. **2010**, *97* (26), 2631115.

52. Zhang, Y.; Tan, Y.; Stormer, H.L.; Kim, P. Experimental observation of the quantum Hall effect and Berry's phase in graphene. Nature **2005**, *438* (7065), 201–204.

53. Berger, C.; Song, Z.; Li, X.; Wu, X.; Brown, N.; Naud, C.; Mayou, D.; Li, T.; Hass, J.; Marchenkov, A.N.; Conrad, E.H.; First, P.N.; de Heer, W.A. Electronic confinement and coherence in patterned epitaxial graphene. Science **2006**, *312* (5777), 1191–1196.

54. Schedin, F.; Geim, A.K.; Morozov, S.V.; Hill, E.W.; Blake, P.; Katsnelson, M.I.; Novoselov, K.S. Detection of individual gas molecules adsorbed on graphene. Nature Mater. **2007**, *6* (9), 652–655.

55. Hajati, Y.; Blom, T.; Jafri, S.H.M.; Haldar, S.; Bhandary, S.; Shoushtari, M.Z.; Eriksson, O.; Sanyal, B.; Leifer, K. Improved gas sensing activity in structurally defected bilayer graphene. Nanotechnology **2012**, *23* (50), 505501–505507.

56. Xu, X.; Qin, J.; Li, Z. Large-scale preparation of graphene sheets and their easy incorporation with other nanomaterials. Polym. Bull. **2012**, *69* (8), 899–910.

57. Azhari, F.; Banthia, N. Cement-based sensors with carbon fibers and carbon nanotubes. Cement Concrete Composites **2012**, *34* (7), 866–873.

58. Wen, S.; Chung, D.D.L. Piezoresistivity-based strain sensing in carbon fiber-reinforced cement. ACT Mater. J. **2007**, *104* (2), 171–179.

59. Wen, S.; Chung, D.D.L. Model of piezoresistivity in carbon fiber. Cem. Conc. Res. **2006**, *36* (10), 1879–1885.

60. Chacko, R.M.; Banthia, N.; Mufti, A.A. Carbon fiber reinforced cement based-sensors. Can. J. Civ. Eng. **2007**, *34* (3), 284–290.

61. Bontea, D.M.; Chung, D.D.L.; Lee, G.C. Damage in Carbon fiber-reinforced concrete monitored by electrical resistance measurement. Cem. Conc. Res. **2000**, *30* (4), 651–659.

62. McCarter, W.J.; Starrs, G.; Chrisp, T.M.; Banfill, P.F.G. Activation energy and conduction in carbon fibre reinforced cement matrices. J. Mater. Sci. **2007**, *42* (6), 2200–2203.

63. Wen, S.; Chung, D.D.L. Self-sensing characteristics of carbon fiber cement. In *Proceedings of ConMat'05 and Mindess Symposium*; The University of British Colombia, 2005.

64. Zhu, S.; Chung, D.D.L. Theory of piezoresistivity for strain sensing in carbon fiber reinforced cement under flexure. J. Mater. Sci. **2007**, *42* (15), 6222–6233.

65. Reza, F.; Batson, G.B.; Yamamuro, J.A.; Lee, J.S. Resistance changes during compression of carbon fiber cement composites. J. Mater. Civ. Eng. **2003**, *15* (5), 476–483.

66. Bath, B.D.; Michael, D.J.; Trafton, B.J.; Joseph, J.D.; Runnels, P.L.; Wightman, R.M. Subsecond adsorption and desorption of dopamine at carbon-fiber microelectrodes. Anal. Chem. **2000**, *72* (24), 95–102.

67. Logman, M.J.; Budygin, E.A.; Gainetdinov, R.R.; Wightman, R.M. Quantitation of vivo measurements with carbon fiber microelectrodes. J. Neurosci. Methods **2000**, *95* (2), 95–102.

68. Gerhardt, G.A.; Oke, A.F.; Nagy, G.; Moghaddam, B.; Adams, R.N. Nafion-coated electrodes with high selectivity for CNS electrochemistry. Brain Res. **1984**, *290* (2), 390–395.

69. Gonon, F.; Buda, M.; Cespuglio, R.; Jouvet, M.; Pujol, J.F. *In vivo* electrochemical detection of catechols in the neostriatum of anesthetized rats-dopamine or dopac. Nature **1980**, *286* (5776), 902–904.

70. Gonon, F.G.; Fombarlet, C.M.; Buda, M.J; Pujol, J.F. Electrochemical treatment of pyrolytic carbon fiber electrodes. Anal. Chem. **1981**, *53* (9), 1386–1389.

71. Pihel, K.; Walker, Q.D.; Wightman, R.M. Overoxidized polypyrrole-coated carbon fiber microelectrodes for dopamine measurements with fast-scan cyclic voltammetry. Anal. Chem. **1996**, *68* (13), 2084–2089.

72. Qu, L.T; Zhao, Y.; Dai, L.M. Carbon microfibers sheathed with aligned carbon nanotubes: Towards multidimensional, multicomponent, and multifunctional nanomaterials. Small **2006**, *2* (8–9), 1052–1059.

73. Raoof, J.B.; Ojani, R.; Beitollahi, H.; Hossienzadeh, R. Electrocatalytic determination of ascorbic acid at the surface of 2.7-Bis (ferrocenyl ethyl)fluoren-9-one modified carbon pasteelectrode. Electroanalysis **2006**, *18* (12), 1193–1201.

74. Zhang, L.; Wang, X.; Zhao, Y.; Zhu, Z.; Fong, H. Electrospun carbon nano-felt surface-attached with Pd nanoparticles for hydrogen sensing application. Mater. Lett. **2012**, *68* (1), 133–136.

75. Park, J.M.; Lee, S.I.; DeVries, K.L. Nondestructive sensing evaluation of surface modified single-carbon fiber reinforced epoxy composites by electrical resistivity measurements. Composites Part B **2006**, *37* (7–8), 612–626.

76. Lu, H.; Liu, Y.; Gou, J.; Leng, J.; Du, S. Surface coating of multi-walled carbon nanotube nanopaper on shape-memory polymer for multifunctionalization. Composites Sci. Technol. **2011**, *71* (11), 1427–1434.

77. Lao, K.T.; Hui, D. The revolutionary creation of new advanced materials-carbon nanotube composites. Composite Part B **2002**, *33* (4), 263–277.

Nanomaterials—
Nanoparticles

78. Lu, H.B.; Liu, Y.J.; Gou, J.; Leng, J.S.; Du, S.Y. Synergistic effect of carbon nanofiber and carbon nanopaper on shape memory polymer composite. Appl. Phys. Lett. **2010**, *96* (8), 084102.

79. Biercuk, M.J.; Llaguno, M.C.; Radosavljevic, M.; Hyun, J.K.; Johnson, A.T.; Fischer, J.E. Carbon nanotube composites for thermal management. Appl. Phys. Lett. **2002**, *80* (15), 2767–2769.

80. Schadler, L.S.; Giannaris, S.C.; Ajayan, P.M. Load transfer in carbon nanotube epoxy composites. Appl. Phys. Lett. **1998**, *73* (26), 3842–3844.

81. Li, G.; Liao, J.M.; Hu, G.Q.; Ma, N.Z.; Wu, P.J. Study of carbon nanotube modified biosensor for monitoring total cholesterol in blood. Biosens. Bioelectron. **2005**, *20* (10), 2140–2144.

82. Baughman, R.H.; Zakhidov, A.A.; de Heer, W.A. Carbon nanotubes-the route toward applications. Science **2002**, *297* (5582), 787–792.

83. Zhao, Q.; Gan, Z.; Zhuang, Q. Electrochemical sensors based on carbon nanotubes. Electroanalysis **2002**, *14* (23), 1609–1613.

84. Qian, D.; Dickey, E.C.; Andrews, R.; Rantell, T. Load transfer and deformation mehanisms in carbon nanotube-polystyrene composites. Appl. Phys. Lett. **2000**, *76* (20), 2868–2870.

85. Wagner, H.D.; Lourie, O.; Feldman, Y.; Tenne, R. Stress-induced fragmentation of multiwall carbon nanotubes in a polymer matrix. Appl. Phys. Lett. **1998**, *72* (2), 188–190.

86. Ajayan, P.M.; Zhou, O.Z. Carbon nanotubes. Top. Appl. Phys. **2001**, *80*, 391–425.

87. Choi, W.B.; Chung, D.S.; Kang, J.H.; Kim, H.Y.; Jin, J.W.; Han, I.T.; Lee, Y.H.; Park, G.S.; Kim, J.M. Fully sealed, high-brightness carbon nanotube field emission display. Appl. Phys. Lett. **1999**, *75* (20), 3129–3131.

88. McEuen, P.L. Single-wall carbon nanotube. Phys. World **2000**, *13* (6), 31–36.

89. Sinha, N.; Ma, J.; Yeow, J.T.W. Carbon nanotube-based sensors. J. Nanosci. Nanotechnol. **2006**, *6* (3), 573–590.

90. Goldoni, A.; Petaccia, L.; Lizzit, S.; Larciprete, R. Sensing gases with carbon nanotubes: A review of the actual situation. J. Phys. **2010**, *22* (1), 013001.

91. Li, G.Y.; Wang, P.M.; Zhao, X. Pressure-sensitive properties and microstructure of carbon nanotube reinforced cement composites. Cement Concrete Composite **2007**, *29* (5), 377–382.

92. Makar, J.M.; Beaudoin, J.J. Carbon Nanotubes and their application in the construction industry. Proceedings of 1st International Symposium on Nanotechnology in Construction: Paisley, Scotland, 2003.

93. Raki, L.; Beaudoin, J.; Alizadeh, R.; Makar, J.; Sato, T. Cement and concrete nanoscience and nanotechnology. Mater. Open Access J. **2010**, *3* (2), 918–942.

94. Schultz-Drost, C.; Sgobba, V.; Gerhards, C.; Leubner, S.; Calderon, R.M.K.; Ruland, A.; Guldi, D.M. Innovative inorganic-organic nanohybrid materials: Coupling quantum dots to carbon nanotubes. Angew. Chem. Int. Ed., **2010**, *49* (36), 6425–6429.

95. Miao, Y.; Yuping, L.; Yong, C.; Ning, Z.; Yu, L. A supramolecular approach of selective zinc ions sensing in living cells. Chin. J. Chem. **2012**, *30* (9), 1948–1952.

96. Huang, J.; Liu, Y.; You, T. Carbon nanofiber based electrochemical biosensors: A review. Anal. Methods **2010**, *2* (3), 202–211.

97. Ajayan, P.M. Nanotubes from carbon. Chem. Rev. **1999**, *99* (7), 1787–1799.

98. Heller, I.; Kong, J.; Heering, H.A.; Williams, K.A.; Lemay, S.G.; Dekker, C. Individual single-walled carbon nanotubes as nanoelectrodes for electrochemistry. Nano Lett. **2005**, *5* (1), 137–142.

99. Krapf, D.; Quinn, B.M.; Wu, M.Y.; Zandbergen, H.W.; Dekker, C.; Lemay, S.G. Experimental observation of nonlinear ionic transport at the nanometer scale. Nano Lett. **2006**, *6* (11), 2531–2535.

100. Gooding, J.J.; Chou, A.; Liu, J.Q.; Losic, D.; Shapter, J.G.; Hibbert, D.B. The effects of the lengths and orientations of single-walled carbon nanotubes on the electrochemistry of nanotube-modified electrodes. Electrochem. Commun. **2007**, *9* (7), 1677–1683.

101. Kiani, K.; Ghaffari, H.; Mehri, B. Application of elastically supported single walled carbon nanotubes for sensing arbitrarily attached nano-objects. Curr. Appl. Phys. **2013**, *13* (1), 107–120.

102. Huang, X.J.; Choi, Y.K. Chemical sensors based on nanostructured materials. Sens. Actuators B **2007**, *122* (2), 659–671.

103. Ahuja, R.T.; Kumar, D. Recent progress in the development of nano-structured conducting polymers/nanocomposites for sensor applications. Sens. Actuators B **2009**, *136* (1), 275–286.

104. Capek, I. Dispersions, novel nanomaterials sensors and nanoconjugates based on carbon nanotubes. Adv. Colloid Interface Sci. **2009**, *150* (2), 63–89.

105. Li, C.; Thostenson, E.T.; Chou, T.W. Sensors and actuators based on carbon nanotubes and their composites: A review. Composite Sci. Technol. **2008**, *68* (6), 1227–1249.

106. Britto, P.J.; Santhanam, K.S.V.; Ajayan, P.M. Carbon nanotube electrode for oxidation of dopamine. Bioelectrochem. Bioenerg. **1996**, *41* (1), 121–125.

107. Wang, J.; Liu, G.D.; Jan, M.R. Ultrasensitive electrical biosensing of proteins and DNA: Carbon nanotube derived amplification of the recognition and transduction events. J. Am. Chem. Soc. **2004**, *126* (10), 3010–3011.

108. He, P.; Dai, L. Aligned carbon nanotube-DNA electrochemical sensors. Chem. Commun. **2004**, *3* (3), 348–349.

109. Skoulidas, A.I.; Ackerman, D.M.; Johnson, J.K.; Sholl, D.S. Rapid transport of gases in carbon nanotubes. Phys. Rev. Lett. **2002**, *89* (18), 185901.

110. Majumder, M.; Chopra, N.; Andrews, R.; Hinds, B.J. Nanoscale hydrodynamics enhanced flow in carbon nanotubes. Nature **2005**, *438* (7064), 44–44.

111. Bianco, A. Carbon nanotubes for the delivery of therapeutic molecules. Expert Opin. Drug Deliv. **2004**, *1* (1), 57–65.

112. Hughes, G.A. Nanostructure-mediated drug delivery. Nanomed: Nanotechnol. Biol. Med. **2005**, *1* (1), 22–30.

113. Zhou, L.G.; Shi, S.Q. Molecular dynamics simulations on tensile mechanical properties of single-walled carbon nanotube with and without hydrogen storage. Comput. Mater. Sci. **2002**, *23* (1–4), 166–174.

114. Hirscher, M.; Becher, M.; Haluska, M.; Quintel, A.; Shakalova, V.; Choi, Y.M.; Dettlaff-Weglikowska, U.; Roth, S.; Stepanek, I.; Bernier, P.; Leonhardt, A.; Fink, J.

Hydrogen storage in carbon nanostructures. J. Alloys Comp. **2002**, *330*, 654–658.

115. Yonzon, C.R.; Stuart, D.A.; Zhang, X.; McFarland, A.D.; Haynes, C.L.; Van Duyne, R.P. Towards advanced chemical and biological nanosensors-An overview. Talanta **2005**, *67* (3), 438–448.

116. Senesac, L.; Thundat, T.G. Nanosensors for trace explosive detection: A review. Mater. Today **2008**, *11* (3), 28–36.

117. Kuronuma, Y.; Takeda, T.; Shindo, Y.; Narita, F.; Wei, Z. Electrical resistance-based strain sensing in carbon nanotube/polymer composites under tension: Analytical modeling and experiments. Composites Sci. Technol. **2012**, *72* (14), 1678–1682.

118. Yanez-Sedeno, P.; Riu, J.; Pingarron, J.M.; Rius, F.X. Electrochemical sensing based on carbon nanotubes. Trends Anal. Chem. **2010**, *29* (9), 939–953.

119. Swamy, B.E.; Venton, B.J. Subsecond detection of physiological adenosine concentrations using fast-scan cyclic voltammetry. Anal. Chem. **2007**, *79* (2), 744–750.

120. Zhang, M.N.; Liu, K.; Xiang, L.; Lin, Y.Q.; Su, L.; Mao, L.Q. Carbon nanotube-modified carbon fiber microelectrodes for *in vivo* voltammetric measurements of ascorbic acid in rat brain. Anal. Chem. **2007**, *79* (17), 6559–6565.

121. Li, W.; Yuan, J.; Dichiara, A.; Lin, Y.; Bai, J. The use of vertically aligned carbon nanotubes grown on SiC for in situ sensing of elastic and plastic deformation in electrically percolative epoxy composites. Carbon **2012**, *50* (11), 5291–4301.

122. Vairavapandian, D.; Vichchulada, P.; Lay, M.D. Preparation and modification of carbon nanotubes: Review of advances and applications in catalysis and sensing. Anal. Chim. Acta **2008**, *626* (2), 119–129.

123. Tan, E.P.S.; Lim, C.T. Mechanical characterization of nanofibers: A review. Composite Sci. Technol. **2006**, *66* (9), 1102–1111.

124. Bhattacharyya, S.; Salvetat, J.P.; Saboungi, M.L. Reinforcement of semicrystalline polymers with collegan-modified single walled carbon nanotubes. Appl. Phys. Lett. **2006**, *88* (23), 23119.

125. Harris, P.J.F. Carbon nanotube composites. Int. Mater. Rev. **2004**, *49* (1), 31–43.

126. Huang, H.J.; Maruyama, R.; Noda, K.; Kajiura, H.; Kadano, K. Preferential destruction of metallic single-walled carbon nanotubes by laser irridation. J. Phys. Chem. B. **2006**, *110* (14) 7316–7320.

127. Ruoff, R.S.; Lorents, D.C. Mechanical and thermal properties of carbon nanotubes. Carbon **1995**, *33* (7), 925–930.

128. Yu, M.F.; Lourie, O.; Dyer, M.J.; Moloni, K.; Kelly, T.F.; Ruoff, R.S. Strength and breaking mechanism of multiwalled carbon nanotubes under tensile load. Science **2000**, *287* (5483), 637–640.

129. Daenen, M.; de Fouw, R.D.; Hamers, B.; Janssen, P.G.A.; Schouteden, K.; Veld M.A.J. *The Wondrous World of Carbon Nanotubes: A Review of Current Carbon Nanotube Technologies*; Eindhoven University of Technology: Eindhoven, the Netherlands, 2003; 1.

130. Yu, M.F.; Files B.S.; Arepalli, S.; Ruoff, R.S. Tensile loading of ropes of single wall carbon nanotubes and their mechanical properties. Phys. Rev. Lett. **2000**, *84* (24), 5552–5555.

131. Krishnan, A.; Dujardin, E.; Ebbesen, T.W.; Yianilas, P.N.; Treacy, M.M. Young's modules of single-walled nanotubes. J. Phys. Rev. B **1998**, *58* (20), 14013–14019.

132. Liu, J.; Chou, A.; Rahmat, W.; Paddon-Row M.N.; Goodling, J.J.; Achieving direct electrical connection to glucose using aligned single walled carbon nanotube arrays. Electroanalysis **2005**, *17* (1), 38–46.

133. Valentini, F.; Amine, A.; Orlanducci, S.; Terranova, M.L.; Palleschi, G. Carbon nanotube purification: Preparation and characterization of carbon nanotube paste electrodes. Anal. Chem. **2003**, *75* (20), 5413–5421.

134. Carrara, S.; Shumyantseva, V.V.; Archakov, A.I.; Samori, B. Screen-printed electrodes based on carbon nanotubes and cytochrome P450scc for highly sensitive choloesterol biosensors. Biosens. Bioelectron. **2008**, *24* (1), 148–150.

135. Bavastrello, V.; Stura, E.; Carrara, S.; Erokhin, V.; Nicolini, C. Poly(2,5-dimethylaniline)-MWCNTs nanocomposite: A new material for conductometric acid vapours sensor. Sens. Actuators B. **2004**, *98* (2–3) 247–253.

136. Wu, B.; Zhang, J.; Wei, Z.; Cai, S.M.; Liu, Z.F. Chemical alignments of oxidatively shortened single-walled carbon nanotubes on silver surface. J. Phys. Chem. B. **2001**, *105* (22), 5075–5078.

137. Gooding, J.J.; Wibowo, R.; Liu, J.Q.; Yang, W.R.; Losic, D.; Orbons, S.; Mearns, F.J.; Shapter, J.G.; Hibbert, D.B. Protein electrochemistry using aligned carbon nanotube arrays. J. Am. Chem. Soc. **2003**, *125* (30), 9006–9007.

138. Kim, B.; Sigmund, W.M. Density control of self-aligned shortened single-wall carbon nanotubes on polyelectrolyte-coated substrates. Colloid Surf. A **2005**, *266* (1), 91–96.

139. Wildgoose, G.G.; Banks, C.E.; Leventis, H.C.; Compon, R.G. Chemically modified carbon nanotubes for use in electroanalysis. Microchim. Acta **2006**, *152* (3–4), 187–214.

140. Deng, W.Q.; Matsuda, Y.; Goddard, W.A. Bifunctional achors connecting carbon nanotubes to metal electrodes for improved nanoelectronics. J. Am. Chem. Soc. **2007**, *129* (32), 9834–9835.

141. Kayhan, E.; Prasad, R.M.; Gurlo, A.; Yilmazoglu, O.; Engstler, J.; Lonescu, E.; Yoon, S.; Weidenkaff, A.; Schneider, J.J. Synthesis, characterization, electronic and gas-sensing properties towards H_2 and CO of transparent, large-area, low-layer graphene. Chem. Eur. J. **2012**, *18* (47), 14996–15003.

142. Bahr, J.L.; Tour, J.M. Covalent chemistry of single-wall carbon nanotubes. J. Mater. Chem. **2002**, *12* (7), 1952–1958.

143. Merkoci, A.; Pumera, M.; Llopis, X.; Perez, B.; del Valle, M.; Alegret, S. New materials for electrochemical sensing VI: carbon nanotubes, TrAC. Trends Anal. Chem. **2005**, *24* (9), 826–838.

144. Yang, T.; Zhang, W.; Du, M.; Jiao, K. A PDDA/poly(2,6-pyridine dicarboxylic acid)-CNTs composite film DNA electrochemical sensor and its application for the detection of specific sequences related to PAT gene and NOS gene. Talanta **2008**, *75* (4), 987–994.

145. Min, K.; Yoo, Y.J. Amperometric detection of dopamine based on tyrosinase-SWCNTs-PPy composite electrode. Talanta **2009**, *80* (2), 1007–1011.

Nanomaterials—
Nanoparticles

146. Zhou, N.; Yang, T.; Jiang, C.; Du, M.; Jiao, K. Higly sensitive electrochemical impedance spectroscopic detection of DNA hybridization based on Au-nano-CNT/PAN (nano) films. Talanta **2009**, *77* (3), 1021–1026.

147. Xing, Y.C.; Li, L.; Chusuei, C.C.; Hull, R.V. Sonochemical oxidation of multiwalled carbon nanotubes. Langmuir **2005**, *21* (9), 4185–4190.

148. Lee, T.M.H.; Li, L.L.; Hsing, I.M. Enhanced electrochemical detection of DNA hybridization based on electrode surface modification. Langmuir **2003**, *19* (10), 4338–4343.

149. Fang, H.T.; Liu, C.G.; Chang, L.; Feng, L.; Min, L.; Cheng, H.M. Purification of single-wall carbon nanotubes by electrochemical oxidation. Chem. Mater. **2004**, *16* (26), 5744–5750.

150. Zhao, W.; Wang, H.; Qin, X.; Wang, X.; Zhao, Z.; Miao, Z.; Chen, L.; Shan, M.; Fang, Y.; Chen, Q. A novel nonenzymatic hydrogen peroxide sensor based on multi-wall carbon nanotube/silver nanoparticle nanohybrids modified gold electrode. Talanta **2009**, *80* (2), 1029–1033.

151. Wang, J.X.; Li, M.X.; Shi, Z.J.; Li, N.Q.; Gu, Z.N. Electrocatalytic oxidation of nanophrine at a glassy carbon electrode modified with single wall carbon nanotubes. Electroanalysis **2002**, *14* (3), 225–230.

152. Yadegari, H.; Jabbari, A.; Heli, H.; Moosavi-Movahedi, A.A.; Kariman, K.; Khodadadi, A. Electrocatalytic oxidation of deferiprone and its determination on a carbon nanotube-modified glassy carbon electrode. Electrochim. Acta **2008**, *53* (6), 2907–2916.

153. Wang, J.X.; Li, M.X.; Shi, Z.J.; Li, N.Q.; Gu, Z.N. Investigation of the electrocatalytic behavior of single-wall carbon nanotube films on an Au electrode. Microchem. J. **2002**, *73* (3), 325–333.

154. Lee, C.S.; Baker, S.E.; Marcus, M.S.; Yang, W.S.; Eriksson, M.A.; Hamers, R.J. Electrically addressable biomolecular functionalization of carbon nanotube and carbon nanofiber electrodes. Nano Lett. **2004**, *4* (9), 1713–1716.

155. Wang, J.; Chen, G.; Chatrathi, M.P.; Musameh, M. Capillary electrophoresis microchip with a carbon nanotube-modified electrochemical detector. Anal. Chem. **2004**, *76* (2), 298–302.

156. Wei, B.G.; Wang, J.; Chen, Z.; Chen, G. Carbon-nanotube-alginate composite modified electrode fabricated by in-situ gelation for capillary electrophoresis. Chem. Eur. J. **2008**, *14* (31), 9779–9785.

157. Pumera, M. Graphene-based nanomaterials and their electrochemistry. Chem. Soc. Rev. **2010**, *39* (11), 4146–4157.

158. Brownson, D.A.C.; Banks, C.E. Graphene electrochemistry: An overview of potential applications. Analyst **2010**, *135* (11), 2768–2778.

159. Decher, G.; Hong, J.D. Buildup of ultrathin multilayer films by a self-assembly process 2. Consecutive adsorption of anionic and cationic bipolar amphiphiles and polyelectrolytes on charged surfaces, Berichte der Bunsen-Gesellschaft Phys. Chem. Chem. Phys. **1991**, *95* (11), 1430–1434.

160. Cherevko, S.; Chung, C.H. Gold nanowire array electrode for non-enzymatic voltammetric and amperometric glucose detection. Sens. Actuators B **2009**, *142* (1), 216–223.

161. Hrapovic, S.; Liu, Y.L.; Male, K.B.; Luong, J.H.T. Electrochemical biosensing platforms using platinum nanoparticles and carbon nanotubes. Anal. Chem. **2004**, *76* (4), 1083–1088.

162. Jena, B.K.; Raj, C.R. Enzyme-free amperometric sensing of glucose by using gold nanoparticles. Chem. Eur. J. **2006**, *12* (10), 2702–2708.

163. Bo, X.J.; Ndamanisha, J.C.; Bai, J.; Guo, L.P. Nonenzymatic amperometric sensor of hydrogen peroxide and glucose based on Pt nanoparticles/ordered mesoporous carbon nanocomposite. Talanta **2010**, *82* (1), 85–91.

164. Zhu, H.; Lu, X.; Li, M.; Shao, Y.; Zhu, Z. Nonenzymatic glucose voltammetric sensor based on gold nanoparticles/carbon nanotubes/ionic liquid nanocomposite. Talanta **2009**, *79* (5), 1446–1453.

165. Zhang, J.; Qiu, C.; Ma, H.; Liu, X. Facile fabrication and unexpected electrocatalytic activity of palladium thin films with hierchical architectures. J. Phys. Chem. C. **2008**, *112* (36), 13970–13975.

166. Tominaga, M.; Nagashima, M.; Nishiyama, K.; Taniguchi, I. Surface poisining during electrocatalytic monosaccharide oxidation reactions at gold electrodes in alkaline medium. Electrochem. Commun, **2007**, *9* (8), 1892–1898.

167. Vassilyev, Y.B.; Khazova, O.A.; Nikolaeva, N.N. Kinetics and mechanism of glucose electrooxidation on different electrode-catalyts 2. Effect of the nature of the electrode and the electro-oxidation mechanism. J. Electroanal. Chem. **1985**, *196* (1), 127–144.

168. Jiang, L.; Xie, Q.; Li, Z.; Li, Y.; Yao, S. A Study on Tannic acid-doped polypyrrole films on gold electrodes for selective electrochemical detection of dopamine. Sensors **2005**, *5* (4), 199–208.

169. Atta, N.F.; Ahmed, R.A.; Amin, H.M.A.; Galal, A. Mono dispersed gold nanoparticles decorated carbon nanotubes as an enhanced sensing platform for nanomolar detection of tramadol. Electroanalysis **2012**, *24* (11), 2135–2146.

170. Sebez, B.; Su, L.; Ogarevc, B.; Tang, Y.; Zhang, X. Aligned carbon nanotube modified carbon fibre coated with gold nanoparticles embedded in a polymer film. Voltammetric microprobe for enzyme less glucose sensing. Electrochem. Commun. **2012**, *25*, 94–97.

171. Shahrokhian, S.; Rastgar, S. Electrochemical deposition of gold nanoparticles on carbon nanotube coated glassy carbon electrode for the improved sensing of tinidazole. Electrochim. Acta **2012**, *78*, 422–429.

172. Guo, Y.; Guo, S.H.; Fang, Y.; Dong, S.H. Gold nanoparticle/carbon nanotube hybrids as an enhanced materials for sensitive amperometric determination of tryptophan. Electrochim. Acta **2010**, *55* (12), 3927–3931.

173. Hung, V.W.S.; Kerman, K. Gold electrodeposition on carbon nanotubes for the enhanced electrochemical detection of homocysteine. Electrochem. Commun. **2011**, *13* (4), 328–330.

174. Alexeyeva, N.; Kozlova, J.; Sammelselg, V.; Ritslaid, P.; Mandar, H.; Tammeveski, K. Electrochemical and surface characterization of gold nanoparticle decorated multi-walled carbon nanotubes, Appl. Surf. Sci. **2010**, *256* (10), 3040–3046.

175. Bui, M.P.N.; Pham, X.H.; Han, K.N.; Li, C.A.; Lee, E.K.; Chang, H.J.; Seong, G.H. Electrochemical sensing of hydroxylamine by gold nanoparticles on single-walled carbon nanotube films. Electrochem. Commun. **2010**, *12* (2), 250–253.

Nanomaterials—
Nanoparticles

176. Zhao, J.; Kong, X.; Shi, W.; Shao, M.; Han, J.; Wei, M.; Evans, D.G.; Duan, X. Self-assembly of layered double hydroxide nanosheets/Au nanoparticles ultrathin films for enzyme-free electrocatalysis of glucose. J. Mater. Chem. **2011**, *21* (36), 13926–13933.

177. Li, C.; Su, Y.; Lv, X.; Xia, H.; Shi, H.; Yang, X.; Zhang, J.; Wang, Y. Controllable anchoring of gold nanoparticles to polypyrrole nanofibers by hydrogen bonding and their application in nonenzymatic glucose sensors. Biosens. Bioelectron. **2012**, *38* (1), 402–406.

178. Shipway, A.; Lahav, M.; Willner, I. Nanostructured gold colloid electrodes. Adv. Mater. **2000**, *12* (13), 993–998.

179. Kim, J.; Lee, S.W.; Hammond, P.T.; Shao-Horn, Y. Electrostatic layer by layer assembled Au nanoparticle/MWNT thin films: Microstructure, optical property, and electrocatalytic activity for methanol oxidation. Chem. Mater. **2009**, *21* (13), 2993–3001.

180. Mao, S.; Lu, G.; Yu.; K.; Bo, Z.; Chen, J. Specific protein detection using thermally reduced graphene oxide sheet decorated with gold nanoparticle-antibody conjugates, Adv. Mater. **2010**, *22* (32), 3521–3526.

181. Choi, Y.; Bae, H.S.; Seo, E.; Jong, S.; Park, K.H.; Kim, B.S. Hybrid gold nanoparticle-reduced graphene oxide nanosheets as active catalysts for highly efficient reduction of nitroarenes. J. Mater. Chem. **2011**, *21* (39), 15431–15436.

182. Huang, X.; Qi, X.; Boey, F.; Zhang, H. Graphene-based composites. Chem. Soc. Rev. **2012**, *41* (2), 666–686.

183. Zhu, C.; Han, L.; Hu, P.; Dong, S. In-situ loading of well-dispersed gold nanoparticles on two-dimensional graphene oxide/SiO$_2$ composite nanosheets and their catalytic properties. Nanoscale **2012**, *4* (5), 1641–1646.

184. Yang, M.H.; Choi, B.G.; Park, H.; Hong, W.H.; Lee, S.Y.; Park, T.J. Development of a glucose biosensor using advanced electrode modified by nanohybrid composing chemically modified graphene and ionic liquid. Electroanalysis **2010**, *22* (11), 1223–1228.

185. Yang, M.H.; Choi, B.G.; Park, H.; Park, T.J.; Hong, W.H.; Lee, S.Y. Directed self-assembly of gold nanoparticles on graphene-ionic liquid hybrid for enhancing electrocatalytic activity. Electroanalysis **2011**, *23* (4), 850–857.

186. Zhang, F.Q.; Zeng, L.Y.; Zhang, Y.X.; Wang, H.Y.; Wu, A.G. A colorimetric assay method for Co+2 based on thioglycolic acid functionalized hexadecyl trimethyl ammonium bromide modified Au nanoparticles (NPs). Nanoscale **2011**, *3* (5), 2150–2154.

187. Kalluri, J.R.; Arbreshi, T.; Khan, S.A.; Neely, A.; Candice, P.; Varisli, B.; Washington, M.; McAfee, S.; Robinson, B.; Banerjee, S.; Singh, A.K.; Senapati, D.; Ray, P.C. A colorimetric assay method for Co+2 based on thioglycolic acid functionalized hexadecyl trimethyl ammonium bromide modified Au nanoparticles (NPs). Angew. Chem. Int. Ed. **2009**, *48* (51), 9668–9671.

188. Lewis, L.N. Chemical catalysis by colloids and clusters. Chem. Rev. **1993**, *93* (8), 2693–2730.

189. Wildgoose, G.G.; Banks, C.E.; Compton, R. Metal nanoparticles and related materials supported on carbon nanotubes: Methods and applications. Small **2006**, *2* (2), 182–193.

190. An-Hui, L.; Salabas, E.L.; Schüth, F, Magnetic nanoparticles: Synthesis, protection, functionalization, and application. Angew Chem. Int. Ed. **2007**, *46* (8), 1222–1244.

191. Hong, H.; Gao, T.; Cai, W.B. Molecular imaging with single-walled carbon nanotubes. Nano Today **2009**, *4* (3), 252–261.

192. Faraday, M. Experimental relations of gold (and other metals) to light. Philos. Trans. R. Soc. Lond. **1857**, *147*, 145–181.

193. Turkevich, J.; Stevenson, P.C. A study of the nucleation and growth processes in the synthesis of colloidal gold. Discuss. Faraday Soc. **1951**, *11* (11), 55–75.

194. Heard, S.M.; Grieser, F.; Barraclough, C.G. The characterization of Ag sols by electron microscopy, optical absorption, and electrophoresis. J. Colloid Interface Sci. **1983**, *93* (2), 545–555.

195. Murphy, C.J.; Sau, T.K.; Gole, A.M.; Orendorff, C.J.; Gao, J.; Gou, L.; Hunyadi, S.E.; Li, T. Anisotropic metal nanoparticles: Synthesis, assembly, and optical applications. J. Phys. Chem. B. **2005**, *109* (29), 13857–13870.

196. Jin, R.; Cao, Y.; Mirkin, C.A.; Kelly, K.L.; Schatz, G.C.; Zheng, J.G. Photoinduced conversion of silver nanospheres to nanoprisms. Science **2001**, *294* (5548), 1901–1903.

197. Sun, Y.; Xia, Y. Shape-controlled synthesis of gold and silver nanoparticles. Science **2002**, *298* (5601), 2176–2179.

198. Chen, J.; Saeki, F.; Wiley, B.J.; Cang, H.; Cobb, M.J.; Li, Z.Y.; Au, L.; Zhang, H.; Kimmey, M.B.; Li, X.; Xia, Y. Gold nanocages: Bioconjugation and their potential use as optical imaging contrast agents. Nano Lett. **2005**, *5* (3), 473–477.

199. Sorensen, T.J.; Laursen, B.W.; Luchowski, R.; Shtoyko, T.; Akopova, I.; Gryczynski, Z.; Gryczynski, I. Enhanced fluorescence emission of Mc-ADOTA by self-assembled silver nanoparticles on a gold film. Chem. Phys. Lett. **2009**, *476* (1), 46–50.

200. Rad, A.S.; Mirabi, A.; Binaian, E.; Tayebi, H. A review on glucose and hydrogen peroxide biosensor based on modified electrode included silver nanoparticles. Int. J. Electrochem. Sci. **2011**, *6* (8), 3671–3683.

201. Tammeveski, L.; Erikson, H.; Sarapuu, A.; Kozlova, J.; Ritslaid, P.; Sammelselg, V.; Tammeveski, K. Electrocatalytic oxygen reduction on silver nanoparticle/multi-walled carbon nanotube modified glassy carbon electrodes in alkaline solution. Electrochem. Commun. **2012**, *20* (1), 15–18.

202. Jia, M.; Wang, T.; Liang, F.; Hu, J. A novel process for the fabrication of a silver-nanoparticle modified electrode and its application in nonenzymetic glucose sensing. Electroanalysis **2012**, *24* (9), 1864–1868.

203. Quan, H.; Park, S.U.; Park, J. Electrochemical oxidation of glucose on silver-nanoparticle-modified composite electrodes. Electrochim. Acta **2010**, *55* (7), 2232–2237.

204. Zhang, K.; Zhang, N.; Zhang, L.; Xu, J.; Wang, H.; Wang, C.; Geng, T. Amperometric sensing of hydrogen peroxide using a glassy carbon electrode modified with silver nanoparticles on poly(alizarin yellow R). Microchim. Acta **2011**, *173* (1–2), 135–141.

205. Mohammadrezaei, A.; Afzalzadeh, R.; Hosseini-Galgoo, S.M. Assessing three different ranges of amounts of silver nanoparticle dopants on the ethanol sensing properties of zinc oxide. Meas. Sci. Technol. **2012**, *23* (3), 035106.

206. Zhang, Y.J. Comparing the interparticle coupling effect on sensitivities of silver and gold nanoparticles. J. Quant. Spectrosc. Radat. Transfer **2012**, *113* (8), 578–581.

Nanomaterials—Nanoparticles

207. Liu, Y.F.; Tsai, J.J.; Chin, Y.T.; Liao, E.C.; Wu, C.C.; Wang, G.J. Detection of allergies using a silver nanoparticle modified nanostructured biosensor. Sensors Actuators B **2012**, *171–172*, 1095–1100.

208. Liu, B.; Deng, Y.; Hu, X.; Gao, Z.; Sun, C. Electrochemical sensing of trichloroacetic acid based on silver nanoparticles doped chitosan hydrogel film prepared with controllable electrodeposition. Electrochim. Acta **2012**, *76* (1), 410–415.

209. Hashemiford, N.; Mohsenifar, A.; Ranjbar, B.; Allameh, A.; Lotfi, A.S.; Etemadikia, B. Fabrication and kinetic studies of a novel silver nanoparticles-glucose oxidase bioconjugate. Anal. Chim. Acta **2010**, *675* (2), 181–184.

210. Balamurugan, A.; Ho, K.C.; Chen, S.M.; Huang, T.Y. Electrochemical sensing of NADH based on meldola blue immobilized silver nanoparticle-conducting polymer electrode. Colloids Surf. A **2010**, *362* (1), 1–7.

211. Doty, R.C.; Tshikhudo, T.R.; Brust, M.; Fernig, D.G. Extremely stable water-soluble Ag nanoparticles. Chem. Mater. **2005**, *17* (18), 4630–4635.

212. Grubbs, R.B. Nanoparticle assembly-solvent-tuned structures. Nat. Mater. **2007**, *6* (8), 553–555.

213. Rahman, M.M.; Khan, S.B.; Jamal, A.; Faisal, M.; Asiri, A.M. Highly sensitive methanol chemical sensor based on undoped silver oxide nanoparticles prepared a solution method. Microchim. Acta **2012**, *178* (1–2), 99–106.

214. Yu, A.; Wang, Q.; Yonk, J.; Mahon, P.J.; Malherbe, F.; Wang, F.; Zhang, H.; Wang, J. Silver nanoparticle-carbon nanotube hybrid films: Preparation and electrochemical sensing. Electrochim. Acta **2012**, *74* (1), 111–116.

215. Zhang, M.G.; Smith, A.; Gorski, W. Carbon nanotube-chitason system for electrochemical sensing based on dehydrogenase enzymes. Anal. Chem. **2004**, *76* (17), 5045–5050.

216. Saliminasab, M.; Bahrampour, A.; Zandi, M.H. Human cardiac troponin I sensor based on silver nanoparticle doped microsphere resonator. J. Opt. **2012**, *14* (12), 122301–122308.

217. Luo, Z.; Yuwen, L.; Han, Y.; Tian, J.; Zhu, X.; Weng, L.; Wang, L. Reduced graphene oxide/PAMAM silver nanoparticles nanocomposite modified electrode for direct electrochemistry of glucose oxidase and glucose sensing. Biosens. Bioelectron. **2012**, *36* (1), 179–185.

218. Dawood, M.K.; Tripath, S.; Dolmanan, S.B.; Ng, T.H.; Tan, H.; Lam, J. Influence of catalytic gold and silver metal nanoparticles on structural, optical and vibrational properties of silicon nanowires synthesized by metal-assisted chemical etching. J. Appl. Phys. **2012**, *112* (7), 073509–073517.

219. Wang, P.; Wang, R.Y.; Jin, J.Y.; Xu, L.; Shi, Q.F. The morphological change of silver nanoparticles in water. Chin. Phys. Lett. **2012**, *29* (1), 017805–017808.

220. Endo, T.; Shibata, A.; Yanagida, Y.; Higo, Y.; Hatsuzawa, T. Localized surface plasmon resonance optical characteristic for hydrogen peroxide using polyvinylpyrrolidone coated silver nanoparticles. Mater. Lett. **2010**, *64* (19), 2105–2108.

221. Endo, T.; Ikeda, R.; Yanagida, R.; Hatsuzawa, T. Stimuli-responsive hydrogel-silver nanoparticles composite for development of localized surface plasmon resonance-based optical biosensor. Anal. Chim. Acta **2008**, *611* (2), 205–211.

222. Endo, T.; Kerman, K.; Nagatani, N.; Takamura, Y.; Tamiya, E. Label-free detection of peptide nucleic acid DNA hybridization using localized surface plasmon resonance-based optical biosensor. Anal. Chem. **2005**, *77* (21), 6976–6984.

223. Endo, T.; Kerman, K.; Nagatani, N.; Tamiya, E. Excitation of localized surface plasmon resonance using a core-shell structured nanoparticle layer substrate and its application for label-free detection of biomolecular interactions. J. Phys. Condens. Matter. **2007**, *19* (21), 215201–215211.

224. Endo, T.; Kerman, K.; Nagatani, N.; Hiep, H.M.; Kim, D.K.; Yonezawa, Y, et al. Multiple label-free detection of antigen-antibody reaction using localized surface plasmon resonance based core-shell structured nanoparticle layer nanochip. Anal. Chem. **2006**, *78* (18), 6465–6475.

225. Ento, T.; Yamamura, S.; Nagatani, N.; Morita, Y.; Takamura, Y.; Tamiya, E. Localized surface plasmon resonance based optical biosensor using surface modified nanoparticle layer for label-free monitoring of antigen-antibody reaction. Sci. Technol. Adv. Mater. **2005**, *6* (5), 491–500.

226. Ento, T.; Yamamura, S.; Kerman, K.; Tamiya, E. Label-free cell-based assay using localized surface plasmon resonance biosensor. Anal. Chim. Acta **2008**, *614* (2), 182–189.

227. Endo, T.; Takizawa, H.; Yanagida, Y.; Hatsuzawa, T. Enhancement of thermal properties of polyvinylpyrrolidone (PVP)-coated silver nanoparticles bu using plasmid DNA and their localized surface plasmon resonance (LSPR) characteristics. Nanobiotechnology **2008**, *4* (1–4), 36–42.

228. Salman, M.; Iqbal, M.; Ashry, S.E.; Kanwal, S. Robust one pot synthesis of colloidal silver nanoparticles by simple redox method and absorbance recovered sensing. Biosens. Bioelectron. **2012**, *36* (1), 236–241.

229. Guardia, L.; Villar-Rodil, S.; Paredes, J.I.; Rozada, R.; Martinez-Alonso, A.; Tascon, J.M.D. UV light exposure of aqueous graphene oxide suspensions to promote their direct reduction, formation of graphene-metal nanoparticle hybrids and dye degradation. Carbon **2012**, *50* (3), 1014–1024.

230. Liu, R.; Zhang, Y.; Zhang, S.Y.; Qiu, W.; Gao, Y. Silver enhancement of gold nanoparticles for biosensing: From qualitative to quantitative. Appl. Spectrosc. Rev. **2014**, *49* (2), 121–138.

231. Chen, J.; Zheng, X.; Wang, H.; Zheng, W. Graphene oxide-Ag nanocomposite: In situ photochemical synthesis and application as a surface-enhanced Raman scattering substrate. Thin Solid Films **2011**, *520* (1), 179–185.

232. Miao, P.; Liu, T.; Li, X.X.; Ning, L.M.; Yin, J.; Han, K. Highly sensitive, label-free colorimetric assay of trypsin using silver nanoparticles. Biosens Bioelectron. **2013**, *49*, 20–24.

233. Yin, Z.Y.; He, Q.Y.; Huang, X.; Zhang, J.; Wu, S.X.; Chen, P. Lu, G.; Chen, P.; Zhang, Q.C.; Yan, Q.Y.; Zhang, H. Real-time DNA detection using Pt-nanoparticle decorated reduced graphene oxide field-effect transistors. Nanoscale **2012**, *4* (1), 293–297.

234. Xiang, G.L.; He, J.; Li, T.Y.; Zhuang, J.; Wang, X. Rapid preparation of noble metal nanocrystals via facile coreduction with graphene oxide and their enhanced catalytic properties. Nanoscale **2011**, *3* (9), 3737–3742.

Nanomaterials–
Nanoparticles

235. Shen, J.F.; Shi, M.; Li, N.; Yan, B.; Ma, H.W.; Hu, Y.Z., Ye, M.X. Facile synthesis and application of Ag-chemically converted graphene nanocomposite. Nano Res. **2010**, *3* (5), 339–349.

236. Zhang, Z.; Xu, F.; Yang, W.; Guo, M.Y.; Wang, X.D.; Zhanga, B.L.; Thang, J.L. A facile one-pot method to high-quality Ag-nanosheets for efficient surface-enhanced Raman scattering. Chem. Commun. **2011**, *47* (22), 6440–6442.

237. Andrade, G.F.S.; Fan, M.; Brolo, A.G. Multi layer silver nanoparticles-modified optical fiber tip for high performance SERS remote sensing. Biosens. Bioelectron. **2010**, *25* (10), 2270–2275.

238. Liang, S.; Pierce, D.T.; Amiot, C.; Zhao, X. Photoactive Nanomaterials for sensing trace analytes in biological samples. Synth. React. Inorg. Metal-Org. Nano-metal Chem. **2005**, *35* (9), 661–668.

239. Takenaka, Y.; Kiyosu, T.; Mori, G.; Choi, J.C.; Sakakura, T.; Yasuda, H. Selective hydrogenation of nitroalkane to N-alkylhydroxylamine over supported palladium catalysts. Catalysis Today **2011**, *164* (1), 580–584.

240. Yang, M.; Kim, H.C.; Hong, S.H. DMMP gas sensing behavior of ZnO-coated single-wall carbon nanotube network sensors. Mater. Lett. **2012**, *89* (1), 312–315.

241. Burda, C.; Chen, X.B.; Narayaran, R.; El-Sayed, M.A. Chemistry and properties of nanocrystals of different shapes. Chem. Rev. **2005**, *105* (4), 1025–1102.

242. Hodak, J.H.; Henglein, A.; Hartland, G.V. Photophysics of nanometer sized metal particles: Electron-phonon coupling and coherent excitation of breathing vibrational modes. J. Phys.Chem. B **2000**, *104* (43), 9954–9965.

243. Link, S.; Burda, C.; Wang, Z.L.; El-Sayed, M.A. Electron Dynamics in gold-silver alloy nanoparticles: The influence of a nonequilibrium electron distribution and the size dependence of the electron-phonon relaxation. J. Chem. Phys. **1999**, *111* (3), 1255–1264.

244. Wilson, O.M.; Scott, R.W.J; Garcia-Martinez, J.C.; Crooks, R.M. Synthesis characterization, and structure selective extraction of 1-2 nm diameter Au-Ag dendrimer-encapsulated bimetallic nanoparticles. J. Am. Chem. Soc. **2005**, *127* (3), 1015–1024.

245. Chung, Y.M.; Rhee, H.K. Dendrimer-templated Ag-Pd bimetallic nanoparticles. J. Colloid Interface Sci. **2004**, *271* (1), 131–135.

246. Mallin, M.P.; Murphy, C.J. Solution-phase synthesis of sub-10 nm Au-Ag alloy nanoparticles. Nano Lett. **2002**, *2*, 1235–1237.

247. Cao, Y.W.; Jin, R.; Mirkin, C.A. DNA-modified core-shell Ag/Au nanoparticles. J. Am. Chem. Soc. **2001**, *123* (32), 7961–7962.

248. Li, D.X.; Li, C.F.; Wang, A.H.; He, Q.A.; Li, J.B. Hierarchical gold/copolymer nanostructures as hydrophobic nanotanks for drug encapsulation. J. Mater. Chem. **2010**, 20 (36), 7782–7787.

249. Wang, C.; Yin, H.; Chan, R.; Peng, S.; Dai, S.; Sun, S. One-pot synthesis of oleylamine coated Au-Ag alloy NPs and their catalysis for CO oxidation. Chem. Mater. **2009**, *21* (3), 433–435.

250. Sapsford, K.E.; Pons, T.; Medintz, I.L.; Mattoussi, H. Biosensing with luminescent semiconductor quantum dots. Sensors **2006**, *6* (8), 925–953.

251. Pingarron, J.M.; Yanez-Sedeno, P.; Gonzalez-Cortes, A. Gold nanopartcle-based electrochemical biosensors. Electrochim. Acta **2008**, *53* (19), 5848–5866.

252. Patolsky, F.; Zheng, G.F.; Lieber, C.M. Nanowire-based biosensors. Anal. Chem. **2006**, *78* (13), 4260–4269.

253. He, B.; Morrow, T.J.; Keating, C.D. Nanowire sensors for multiplexed detection of biomolecules. Curr. Opin. Chem. Biol. **2008**, *12* (5), 522–528.

254. Kauffman, D.R.; Star, A. Electronically monitoring biological interactions with carbon nanotube field-effect transistors. Chem. Soc. Rev. **2008**, *37* (6), 1197–1206.

255. Maehashi, K.; Matsumoto, K. Label-free electrical detection using carbon nanotube based biosensors. Sensors **2009**, *9* (7), 5368–5378.

256. Kilian, K.A.; Boecking, T.; Gooding, J.J. The importance of surface chemistry in mesoporous materials: Lessons from porous silicon biosensors. Chem. Commun. **2009**, *14* (6), 630–640.

257. Qi, H.L.; Peng, Y.; Gao, Q.; Zhang, C.X. Applications of nanomaterials in electrogenerated chemiluminescence biosensors. Sensors **2009**, *9* (1), 674–695.

258. Sarma, A.K.; Vatsyayan, P.; Goswami, P.; Minteer, S.D. Recent advances in material science for developing enzyme electrodes. Biosens. Bioelectron. **2009**, *24* (8), 2313–2322.

259. Fu, C.P.; Zhou, H.H.; Peng, W.C.; Chen, J.H.; Kuang, Y.F. Comparison of electrodeposition of silver in ionic-liquid microemulsions. Electrochem. Commun. **2008**, *10* (5), 806–809.

260. Xu, C.W.; Wang, H.; Shen, P.K.; Jiang, S.P. Higly ordered Pd nanowire arrays as effective electrocatalysts for ethanol oxidation in direct alcohol fuel cells. Adv. Mater. **2007**, *19* (23), 4256–4259.

261. Hanzo, I.; Djenizian, T.; Ortiz, G.F.; Knauth, P. Mechanistic study of Sn electrodeposition on TiO2 nanotube layers: Thermodynamics, kinetics, nucleation, and growth modes. J. Phys. Chem. C **2009**, *113* (48), 20568–20575.

262. Hoa, N.D.; Quang, V.V.; Kim, D.; Hieu, N.V. General and scalable route to synthesize nanowire-structured semiconducting metal oxides for gas-sensor applications. J. Alloys Comp. **2013**, *549* (1), 260–268.

263. Rossinyal, E.; Arbiol, J.; Peiro, F.; Corret, A.; Morante, J.R.; Tian, B.; Bo, T.; Zhao, D. Nanostructured metal oxides synthesized by hard template method for gas sensing applications. Sens. Actuators B **2005**, *109* (1), 57–63.

264. Kohl, D. The role of noble-metals in the chemistry of solid-state gas sensors. Sens. Actuators B **1990**, *1* (1–6), 158–165.

265. Anker, J.N.; Hall, W.P.; Lyandres, O.; Shah, N.C.; Zhao, J.; Van Duyne, R.P. Biosensing with plasmonic nanosensors. Nat. Mater. **2008**, *7* (6), 442–453.

266. Fu, Y.; Zhang, L.; Chen, G. Preparation of a carbon nanotube-copper nanoparticle hybrid by chemical reduction for use in the electrochemical sensing of carbohydrates. Carbon **2012**, *50* (7), 2563–2570.

267. Moseley, P.T.; Tofield, B.C. Eds, *Solid State Gas Sensors*; Adam Hilger: Bristol, UK, 1987.

268. Sberveglieri, G. Ed. *Gas sensors; Principles, Operation and Developments*; Kluwer: Dordrecht, the Netherlands, 1992.

269. Baltes, H.; Gopel, W.; Hesse, J. Eds, *Sensors Update*; Wiley-VCH Verlag: Weinheim, 1996–2003; Vol. 1–13.

Nanomaterials—
Nanoparticles

270. Comini, E.; Faglia, G.; Sberveglieri, G. Eds, *Solid State Gas Sensing*; Springer: Berlin, 2009.

271. Korotcenkov, G. Ed. *Chemical Sensors*; Momentum Press: New York, 2010–2011; vol. 1–6.

272. Barsan, N.; Koziej, D.; Weimer, U. Metal oxide-based gas sensor research: How to?. Sens. Actuators B Chem. **2006**, *121* (1), 18–35.

273. Lu, D.; Shomali, N.; Shen, A. Task specific ionic liquid for direct electrochemistry of metal oxides. Electrochem. Commun. **2010**, *12* (9), 1214–1217.

274. Jeon, S.H.; Xu, P.; Mack, N.H.; Chiang, L.Y.; Brown, L.; Wang, H.L. Understanding and controlled N-methylpyrrolidone as a reducing agent. J. Phys. Chem. C **2010**, *114* (1), 36–40.

275. Ragupath, D.; Gopalan, A.I.; Lee, K.P. Electrocatalytic oxidation and determination of ascorbic acid in the presence of dopamine at multiwalled carbon nanotube-silica network gold nanoparticles based nanohybrid modified electrode. Sens. Actuators B **2010**, *143* (2), 696–703.

276. Cao, J.; Wu, J. Strain effects in low-dimensional transition metal oxides. Mater. Sci. Eng. R. **2011**, *71* (2), 35–52.

277. Lin, X.; Jiang, X.; Lu, L. DNA deposition on carbon electrodes under controlled dc potentials. Biosens. Bioelectron. **2005**, *20* (9), 1709–1717.

278. Samerjai, T.; Tamaekong, N.; Wetchakun, K.; Kruefu, V.; Liewhiran, C.; Siriwong, C.; Wisitsoraat, A.; Phanichphat, S. Flame-spray-made metal-loaded semiconducting metal oxides thick films for flammable gas sensing. Sens. Actuators B **2012**, *171–172*, 43–61.

279. Velev, O.D.; Jede, T.A.; Lobo, R.F.; Lenhoff, A.M. Microstructured porous silica obtained via colloidal crystal templates. Chem. Mater. **1998**, *10* (11), 3597–3602.

280. Imhof, A.; Pine, D.J. Ordered macroporous materials by emulsion templating. Nature **1997**, *389* (6654), 948–951.

281. Holland, B.T.; Blanford, C.F.; Stein, A. Synthesis of macroporous minerals with highly ordered three dimensional arrays of spheroidal voids. Science **1998**, *281* (5376), 538–540.

282. Holland, B.T.; Blanford, C.F.; Do, T.; Stein, A. Synthesis of highly ordered three-dimensional, macroporous structures of amorphous or crystalline inorganic oxides. phosphates, and hybrid composites. Chem. Mater. **1999**, *11* (3), 795–805.

283. Shirakawa, H.; Louis, E.; McDiarmid, A.; Chiang, C.; Heeger, A.J. Synthesis of electrically conducting organic polymers- Halogen derivatives of polyacetylene, (CH)x. J. Chem. Soc. Chem. Commun. **1977**, *16* (1), 578–580.

284. Skotheim, T.A.; Elsenbaumer, R.L.; Reynolds, J.R. Eds. *Handbook of Conducting Polymers*; 2nd edn; Marcel Dekker: New York, 1998.

285. Nalwa, H.S. *Handbook of Organic Conductive Molecules and Polymers*; John Wiley & Sons: Chichester, 1997.

286. Xia, L.; Wei, Z.; Wan, M. Conducting polymer nanostructures and their application in biosensors. J. Colloid Interface Sci. **2010**, *341* (1), 1–11.

287. An, K.H.; Jeong, S.Y.; Hwang, H.R.; Lee, Y.H. Enhanced sensitivity of a gas sensor incorporating single-walled carbon nanotube-polypyrrole nanocomposites. Adv. Mater. **2004**, *16* (12), 1005–1009.

288. Ates, M.; Sarac, A.S.; Turhan, C.M.; Ayaz, N.E. Polycarbazole modified carbon fiber microelectrode: Surface

289. characterization and dopamine sensor. Fibers Polym. **2009**, *10* (1), 46–52.

289. Sharma, S.; Nirkhe, C.; Pethkar, S.; Athawale, A.A. Chloroform vapour sensor based on copper/polyaniline nanocomposite. Sens. Actuators B **2002**, *85* (1–2), 131–136.

290. Anderson, M.R.; Mattes, B.R.; Reiss, H.; Kaner, R.B. Conjugated polymer films for gas separations. Science **1991**, *252* (5011), 1412–1415.

291. Wang, J.; Chan, S.; Carlson, R.R.; Luo, Y.; Ge, G.; Ries, R.S.; Heath, J.R.; Tseng, H.R. Electrochemically fabricated polyaniline nanoframework electrode junctions that function as resistive sensors. Nano Lett. **2004**, *4* (9), 1693–1697.

292. Virji, S.; Kaner, R.B.; Weiller, B.H. Hydrogen sensors based on conductivity changes in polyaniline nanofibers. J. Phys. Chem. B, **2006**, *110* (44), 22266–22270.

293. Ko, S.; Jang, J. Controlled amine functionalization on conducting polypyrrole nanotubes as effective transducers for volatile acetic acid. Biomacromolecules **2007**, *8* (1), 182–187.

294. Ates, M.; Karazehir, T.; Sarac, A.S. Conducting polymers with their applications. Curr. Phys. Chem. **2012**, *2* (1), 224–240.

295. Gupta, B.; Joshi, L.; Prakash, R. Novel synthesis of polycarbazole-gold nanocomposite. Macromol. Chem. Phys. **2011**, *212* (15), 1692–1699.

296. McCullough, R.D. The chemistry of conducting polythiophenes. Adv. Mater. **1998**, *10* (2), 93–116.

297. Perepichka, I.F.; Perepichka, D.F.; Meng, H. Wudl, F. Light emitting polythiophenes. Adv. Mater. **2005**, *17* (19), 2281–2305.

298. Li, G.; Shrotriya, V.; Huang, J.; Yao, Y.; Moriarty, T.; Emery, K.; Yang, Y. High-efficiency solution processable polymer photovoltaic cells by self-organization of polymer blends. Nat. Mater **2005**, *4* (11), 864–868.

299. Liu, J.; Tanaka, T.; Sivula, K.; Alivisatos, A.P.; Frecket, J.M.J. Employing end-functional polythiophene to control the morphology of nanocrystal-polymer composites in hybrid solar cells. J. Am. Chem. Soc. **2004**, *126* (21), 6550–6551.

300. Li, B.; Sauve, G.; Lovu, M.C.; Lambeth, D.N. Volatile organic compound detection using nano-structured copolymers. Nano Lett. **2006**, *6* (8),1598–1602.

301. Liu, H.; Reccius, C.H.; Craighead, H.G. Single electrospun regioregular poly(3-hexylthiophene) nanofiber fieldeffect transistor. Appl. Phys. Lett. **2005**, *87* (25), 253106.

302. Li, D.; Babel, A.; Jenekhe, S.A.; Xia, Y. Nanofibers of conjugated polymers prepared by electrospinning with a two-capillary spinneret. Adv. Mater. **2004**, *16* (22), 2062–2066.

303. Babel, A.; Li, D.; Xia, Y.; Jenekhe, S.A. Electrospun nanofibers of blends of conjugated polymers: Morphology, optical properties, and field-effect transistors. Macromolecules **2005**, *38* (11), 4705–47011.

304. Patil, D.; Gaikwad, A.B.; Patil, P. Poly(o-anisidine) films on mild steel: Electrochemical synthesis and biosensor application. J. Phys. D **2007**, 40 (8), 2555–2562.

305. Wilson, R.; Turner, A.P. Glucose-oxidase-an ideal enzyme. Biosens. Bioelectron. **1992**, *7* (3), 165–185.

306. Turner, A.P.; Karube, I.; Wilson, G. Ed. *Biosensors Fundamentals and Applications*; Oxford Science Publication: UK, 1987.

307. Sanvicens, N.; Pastells, C.; Pascual, N.; Marco, M.P. Nanoparticle-based biosensors for detection of pathogenic bacteria. Trac-Trends Anal. Chem. **2009**, *28* (11), 1243–1252.

308. Agasti, S.S.; Rana, S.; Park, M.H.; Kim, C.K.; You, C.C.; Rotello, V.M. Nanoparticles for detection and diagnosis. Adv. Drug Deliv. Rev. **2010**, *62* (3), 316–328.

309. Kanazawa, K.K.; Diaz, A.F.; Gill, W.D.; Grant, P.M.; Street, G.; Gardini, G.P., Polypyrrole an electrochemically synthesized conducting organic polymer. Synth. Met. **1980**, *1* (3), 329–336.

310. Kaiser, A.B.; Liu, C.J.; Gilberd, P.W.; Chapman, B.; Kemp, N.T.; Wessling, B.; Partridge, A.C.; Smith, W.T.; Shapiro, J.S. Comparison of electronic transport in polyaniline blends, polyaniline and polypyrrole. Synth. Met. **1997**, *84* (1–3), 699–702.

311. Rudge, A.; Davey, J.; Raistrick, I.; Gottsfeld, S.; Feraris, J.P. Conducting polymers as active materials in electrochemical capacitors. J. Power Sources **1994**, *47* (1–2), 89–107.

312. Altgeld, W.; Beck, F. Ed. *GDCh. Monographite*; Elsevier Science Publisher: Frankfurt, 1996; Vol. 5, p. 552.

313. Sarac, A.S.; Sezgin, S.; Ates, M.; Turhan, C.M. Electrochemical impedance spectroscopy and morphological analyses of pyrrole, phenylpyrrole and methoxyphenylpyrrole on carbon fiber microelectrodes. Surf. Coat. Technol. **2008**, *202* (16), 3997–4005.

314. Li, H.; Shi, G.; Ye, W.; Li, C.; Liang, Y. Polypyrrole-carbon fiber composite film prepared by chemical oxidation polymerization of pyrrole. J. Appl. Sci. **1997**, *64* (11), 2149–2154.

315. Lazzaroni, R.; Dujardin, S.; Riga, J.; Verbist, J.J. Electrochemical polymerization of pyrrole on carbon fibers surface. Surf. Interface Anal. **1985**, *7*, 252–253.

316. Chiu, H.T.; Lin, J.S. Electrochemical deposition of polypyrrole on carbon fibers for improved adhesion to the epoxy resin matrix. J. Mater. Sci. **1992**, *27* (2), 319–327.

317. Iroh, J.O.; Wood, G.A. Physical and chemical properties of polypyrrole-carbon fiber interfaces formed by aqueous electrosynthesis. J. Appl. Polym. Sci. **1996**, *62* (10), 1761–1769.

318. Bendikov, T.A.; Kim, J.; Harmon, T.C. Development and environmental application of a nitrate selective microsensor based on doped polypyrrole films. Sens. Actuators B **2005**, *106* (2), 512–517.

319. Trapattur, S.; Belletete, M.; Drolet, N.; Leclerc, M.; Durocher, G. Steady-state and time-resolved studies of 2,7-carbazole based conjugated polymers in solution and as thin films: Determination of their solid state fluorescence quantum efficiencies. Chem. Phys. Lett. **2003**, *370* (5–6), 799–804.

320. Taoudi, H.; Bernede, J.C.; Del Valle, M.A.; Bonnet, A.; Morsli, M. Influence of the electrochemical conditions on the properties of polymerized carbazole. J. Mater. Sci. **2001**, *36* (3), 631–634.

321. Castey, M.C.; Olivero, C.; Fischer, A.; Mousel, S.; Michelson, J.; Ades, A.; Sievo, A. Polycarbazole microcavities: Towards plastic blue lasers. Appl. Surf. Sci. **2002**, *197* (1), 822–825.

322. Abe, S.Y.; Ugalde, L.; Del Valle, M.A.; Tregouet, Y.; Bernede, J.C. Nucleation and growth mechanism of poly-

carbazole deposited by electrochemistry. J. Brazilian Chem. Soc. **2007**, *18* (3), 601–606.

323. Sarac, A.S.; Yavuz, O.; Sezer, E. Electrosynthesis and study of carbazole-acrylamide copolymer electrodes. Polymer **2000**, *41* (3), 839–847.

324. Donat-Bouillud, A.; Mazerolle, L.; Gagnon, P.; Goldenberg, L.; Petty, M.C.; Leclerc, M. Synthesis, characterization, and processing of new electroactive and photoactive polyesters derived from oligothiophenes. Chem. Mater. **1997**, *9* (12), 2815–2821.

325. Saraswathi, R.; Gerard, M.; Mahotra, B.D. Characteristics of aqueous polycarbazole batteries. J. Appl. Polym. Sci. **1999**, *74* (1), 145–150.

326. Nishino, H.; Yu, G.; Heeger, A.J.; Chen, T.A.; Rieke, R.D. Electroluminescence from blend films of poly(3-hexylthiophene) and poly(N-vinylcarbazole). Synth. Met. **1995**, *68* (3), 243–247.

327. McCormick, C.L.; Hoyle, C.E.; Clark, M.D. Water soluble copolymers .35. photophysical and rheological studies of the copolymer of metacylic acid with 2-(1-naphthylacetyl) ethyl acrylate, Macromolecules, **1990**, *23* (12), 3124–3129.

328. Sarac, A.S.; Ates, M.; Parlak, E.A. Electrochemical copolymerization of *N*-Methylpyrrole with carbazole. Int. J. Polym. Mater. **2004**, *53*, 785–798.

329. Sarac, A.S.; Ates, M.; Parlak, E.A. Comparative study of chemical and electrochemical copolymerization of N-Methylpyrrole with N-ethylcarbazole spectroscopic and cyclic voltammetric analysis. Int. J. Polym. Mater. **2005**, *54* (9), 883–897.

330. Serantoni, M.; Sarac, A.S.; Sutton, D. FIB-SIMS investigation of carbazole-based polymer and copolymers electrocoated onto carbon fibers, and on AFM morphological study. Surf. Coat. Technol. **2005**, *194* (1), 36–41.

331. Dhand, C.; Das, M.; Datta, M.; Malhotra, B.D. Recent advances in polyaniline based biosensors. Biosens. Bioelectron. **2011**, *26* (6), 2811–2821.

332. Ameen, S.; Akhtar, M.S.; Husain, M. A review on synthesis processing chemical and conductivity on properties of polyaniline and its nanocomposites. Sci. Adv. Mater. **2010**, *2* (4), 441–462.

333. Ameen, S.; Akhtar, M.S.; Kim, Y.S.; Shin, H.S. Synthesis and electrochemical impedance properties of CdS nanoparticles decorated polyaniline nanorods. Chem. Eng. J. **2012**, *181* (1), 806–812.

334. Ameen, S.; Akhtar, M.S.; Shin, H.S.; Hydrazine chemical sensing by modified electrode based on in situ electrochemically synthesized polyaniline/graphene composite thin films. Sens. Actuators B **2012**, *173* (1), 177–183.

335. Arya, S.K.; Day, A.; Bhansalis, S. Polyaniline protected gold nanoparticles based mediator and label-free electrochemical cortisol biosensor. Biosens. Bioelectron. **2011**, *28* (1), 166–173.

336. Kutsche, C.; Targove, J.; Haaland, P.J. Microlithographic patterning of polythiophene films. J. Appl. Phys. **1993**, *73* (5), 2602–2604.

337. Hotta, S.; Rughoeputh, D.D.V.; Heeger, A.J.; Wudl, F. Spectroscopic studies of soluble poly(3-alkylthienylenes). Macromolecules **1987**, *20* (1), 212–215.

338. Galopin, E.; Touahir, L.; Niedziolka-Jönsson, J.; Buukherroub, R.; Gouget-Laemmel, A.C.; Chazelviel, J.N.; Ozanam, F.; Szunerits, S. Amorphous silicon-carbon alloy for

efficient localized surface plasmon resonance sensing. Biosens. Bioelectron. **2010**, *25* (5), 1199–1203.

339. Ferapontova, E.E.; Gorton, L. Direct electrochemistry heme-multicofactor –containing enzymes on alkanethiol-modified gold electrodes. Bioelectrochemistry **2005**, *66* (1), 55–63.

340. Zhou, X.C.; O'Shea, S.J.; Li, S.F.Y. Amplified microgravimetric gene sensor using Au nanoparticle modified oligonucleotides. Chem. Commun. **2000**, *11* (1), 953–954.

341. Gooding, J.J. Electrochemical DNA hybridization biosensors. Electroanalysis **2002**, *14* (17), 1149–1156.

342. Montagre, M.; Durliat, H.; Comtat, M. Simultaneous use of dehydrogenases and haxacyanoferrate (III) ion in electrochemical biosensors for L-lactate, D-lactate and L-glutamate ions. Anal. Chim. Acta **1993**, *278* (1), 25–33.

343. Comtat, M.; Galy, M.; Goulas, P.; Souppe, J. Amperometric bienzyme electrode for L-carnitine. Anal. Chim. Acta **1988**, *208* (1), 295–300.

344. Gilis, M.; Durliat, H.; Comtat, M. Amperometric biosensors for L-alanine and pyruvate assays in biological fluids. Anal. Chim. Acta **1997**, *355* (2), 235–240.

345. Santiago-Rodriguez, L.; Sanchez-Pomalez, G.; Cabrera, C.R. Electrochemical DNA sensing at single-walled carbon nanotubes chemically assembled on gold surfaces. Electroanalysis **2010**, *22* (23), 2817–2824.

346. Patolsky, F.; Weizmann, Y.; Willner, I. Long-range electrical contacting of redox enzymes by SWCNT connectors. Angew. Chem. Int. Ed. **2004**, *43* (16), 2113–2117.

347. Huang, X.J.; Im, H.S.; Yarimaga, O.; Kim, J.H.; Lee, D.H.; Kim, H.S.; Choi, Y.K.; Direct electrochemistry of uric acid at chemically assembled carboxylated single-walled carbon-nanotubes netlike electrode. J. Phys. Chem. B. **2006**, *110* (43), 21850–21856.

348. Ozoemena, K.I.; Nyokong, T.; Nkosi, D.; Chambrier, I.; Cook, M.J. Insights into the surface and redox properties of single-walled carbon nanotube-cobalt (II) tetraamino phthalocyanine self-assembled on gold electrode. Electrochim. Acta **2007**, 52 (12), 4132–4143.

349. Katz, E.; Sheeney-Haj-Ichia, L.; Basnar, B.; Felner, I.; Willner, I. Magnetoswitchable controlled hydrophilicity/hydrophobicity of electrode surfaces using alkyl-chain functionalized magnetic particles: Application for switchable electrochemistry. Langmuir **2004**, *20* (22), 9714–9719.

350. Rios-Pagan, A.M.; Santiago-Rodriquez, L.; Cabrera, C.R. Study of self-assembled monolayers of 4-aminothiolphenol over gold electrodes. Abstr. Papers Am. Chem. Soc. **2007**, *233*, 260–260.

351. Santiago-Rodriquez, L.; Sanchez-Pomales, G.; Vargas-Barbosa, N.M.; Cabrera, C.R. Bioelectrochemical sensing based on single stranded deoxyribonucleic acid-carbon nanotubes covalently attached on gold electrodes. J. Nanosci. Nanotech. **2009**, *9* (4), 2450–2455.

352. Garjonyte, R.; Malinauskas, A. Amperometric glucose biosensor based on glucose oxidase immobilized in poly(o-phenylenediamine) layer. Sens. Actuators B **1999**, *56* (1–2), 85–92.

353. Malitesta, C.; Palmisano, F.; Torsi, L.; Zambonin, P.G. Glucose fast-response amperometric sensor based on glucose-oxidase immobilized in an electropolymerized poly(ortho-phenylenediaine) film. Anal. Chem. **1990**, *62* (24), 2735–2740.

354. Groom, C.A.; Luong, J.H.T. Improvement of the selectivity of amperometric biosensors by using a permselective electropolymerized film. Anal. Lett. **1993**, *26* (7), 1383–1390.

355. Bartlett, P.N.; Caruana, D.J. Electrochemical immobization of enzymes .5. microelectrodes for the detection og glucose based on glucose-oxidase immobilized in a poly(phenol) film. Analyst **1992**, *117* (8), 1287–1292.

356. Zhang, Z.N.; Liu, H.Y.; Deng, J.Q. A glucose biosensor based on immobilization of glucose-oxidase in electropolymerized o-aminophenol film on platinized glassy carbon electrode. Anal. Chem. **1996**, *68* (9), 1632–1638.

357. Pan, D.; Chen, J.; Nie, L.; Tao, W.; Yao, S. Amperometric glucose biosensor based on immobilization of glucose oxidase in electropolymerized o-aminophenol film at Prussian blue-modified platinum electrode. Electrochim. Acta **2004**, *49*, 795–801.

358. Im, J.S.; Kim, J.G.; Bae, T.S.; Yu, H.R.; Lee, Y.S. Surface modification of electrospun spherical activated carbon for a high-performance biosensor electrode. Sens. Actuators B **2011**, *158* (1), 151–158.

359. Cao, Y.; Wang, J.; Xu, Y.; Li, G. Sensing purine nucleoside phosphorylase activity bu using silver nanoparticles. Biosens. Bioelectron. **2010**, *25* (5), 1032–1036.

360. Riul, A.; dos Santos, D.S.; Wohnrath, K.; Di Tommazo, R.; Carvalho, A.C.P.L.F.; Fonseca, F.J.; Oliveira, O.N.; Taylor, D.M.; Mattoso, L.H.C. Artificial taste sensor: Efficient combination of sensors made from Langmuir-Blodgett films of conducting polymers and a ruthenium complex and self-assembled films of an azobenzene-containing polymer. Langmuir **2002**, *18* (1), 239–245.

361. Yu, R.Q.; Zhang, Z.R.; Shen, G.L. Potentiometric sensors: Aspects of the recent development. Sens. Actuators B **2000**, *65* (1–3), 150–153.

362. Nguyen, T.A.; Kokot, S.; Ongarato, D.M.; Wallace, G.G. The use of chronoamperometry and chemometrics for optimization of conducting polymer sensor arrays. Electroanalysis **1999**, *11* (18), 1327–1332.

363. Winquist, F.; Lundstrom, I.; Wide, P. The combination of an electronic tongue and an electronic nose. Sens. Actuators B **1999**, *58* (1–3), 512–517.

364. Adams, R.N. *In vivo* electrochemical measurements in the CNS. Prog. Neurobiol. **1990**, *35* (4), 297–311.

365. Bath, B.D.; Michael, D.J.; Trafton, J.; Joseph, J.D.; Runnels, P.L.; Wrightman, R.M. Subsecond adsorption and desorption of dopamine at carbon-fiber microelectrodes. Anal. Chem. **2000**, *72* (24), 5994–6002.

366. Forster, R.J. Electron transfer Dynamics and surface coverages of binary anthraquinane monolayers on mercury microelectrodes. Langmuir **1995**, *11* (6), 2247–2255.

367. Adams, R.N. Probing brain chemistry with electroanalytical techniques. Anal. Chem. **1976**, *48* (14), 1126–1138.

368. Raj, C.R.; Ohsaka, T. Electroanalysis of a ascorbate and dopamine at a gold electrode modified with a positively charged self-assembled monolayer. J. Electroanal. Chem. **2001**, *496* (1–2), 44–49.

369. Raj, C.R.; Ohsaka, T. Simultaneous detection of ascorbic acid and dopamine at gold electrode modified with a self-assembled monolayer of cystamine. Electrochemistry **1999**, *67* (12), 1175–1177.

370. Li, Y.; Lin, X. Simultaneous electroanalysis of dopamine, ascorbic acid and uric acid by poly(vinyl alcohol)

covalently modified glassy carbon electrode. Sens. Actuators B **2006**, *115* (1), 134–139.

371. Wightman, R.M.; May, L.J.; Michael, A.C. Detection of dopamine dynamics in the brain. Anal. Chem. **1988**, *60* (13), 769A–779A.

372. Mo, J.W.; Ogarevc, B. Simultaneous measurement of dopamine and ascorbate at their physiological level using voltammetric microprobe based on overoxidized poly(1,2-phenylenedaimine)-coated carbon fiber. Anal. Chem. **2001**, *73* (6), 1196–1202.

373. Hsueh C, Brajter-Tpth A. Electrochemical preparation and analytical applications of ultrathin overoxidized polypyrrole films. Anal. Chem. **1994**, *66* (15), 2458–2464.

374. Zhao H.; Zhang Y.; Yuan Z. Study on the electrochemical behavior of dopamine with poly(sulfosalicylic acid) modified glassy carbon electrode. Anal. Chim. Acta **2001**, *441* (1), 117–122.

375. Malem F, Mandler D. Self-assembled monolayers in electroanalytical chemistry application of omega mercapto carboxylic acid monolayers for the electrochemical detection of dopmaine in the presence of a high-concentration of ascorbic acid. Anal. Chem. **1993**, *65* (1), 37–41.

376. Hulthe P.; Hulthe B.; Johannessen K.; Engel, J. Decreased ascorbate sensitivity with nafion coated carbon fiber electrodes in combination with copper (II) ions for the electrochemical determination of electroactive substances *in vivo*. Anal. Chim. Acta **1987**, *198* (1), 197–206.

377. Rubianes, M.D.; Rivas, G.A. Highly selective dopamine quantification using a glassy carbon electrode modified with a melanin-type polymer. Anal. Chim. Acta **2001**, *440* (2), 99–108.

378. Kawagoe, K.T.; Garris, P.A.; Wightman, R.M. pH dependent processes at nafion®-coated carbon fiber microelectrodes. J. Electroanal. Chem. **1993**, *359* (1–2), 193–207.

379. Gerhardt, G.A.; Hoffman, A.F. Effects of recording media composition on the responses of Nafion-coated carbon fiber microelectrodes measured using high-speed chronoamperometry. J. Neurosci. Medhods **2001**, *109* (1), 13–21.

380. Venton, B.J.; Michael, D.J.; Wightman, R.M. Correlation of local changes in extracellular oxygen and pH that accompany dopaminergic terminal activity in the rat caudate-putamen. J. Neurochem. **2003**, *84* (2), 373–381.

381. DuVall, S.H.; McCreery, R.L. Self-catalysis by catechols and quinones during heterogeneous electron transfer at carbon electrodes. J. Am. Chem. Soc. **2000**, *122* (28), 6759–6764.

382. Kong, J.; Franklin, N.R.; Zhou, C.W; Chapline, M.G.; Peng, S.; Cho, K.J.; Dai, H.J. Nanotube molecular wires as chemical sensors. Science **2000**, *287* (5453), 622–625.

383. Collins, P.G.; Bradley, K.; Ishigami, M.; Zetti, A. Extreme oxygen sensitivity of electronic properties of carbon nanotubes. Science **2000**, *287* (5459), 1801–1804.

384. Thomas, R.C.; Yang, H.C.; DiRubio, C.R.; Ricco, A.J.; Crooks, R.M. Chemically sensitivity surface acoustic wave devices employing a self-assembled composite monolayer film: Molecular specificity and effects due to self-assembled monolayer adsorption time and gold surface morphology. Langmuir **1996**, *12* (9), 2239–2246.

385. Ying, Z.; Jiang, Y.; Du, X.; Xie, G.; Yu, J.; Wang, H. PVDF coated quartz crystal microbalance sensor for DMMP vapor detection. Sens. Actuators B **2007**, *125* (1), 167–172.

386. Lee, S.C.; Choi, H.Y.; Lee, W.S.; Huh, J.S.; Lee, D.D. The development of SnO_2-based recoverable gas sensors for the detection of DMMP. Sens. Actuators B **2009**, *137* (1), 239–245.

387. Snow, E.S.; Perkins, F.K.; Houser, E.J.; Bodescu, S.C.; Reinecke, T.L. Chemical detection with a single-walled carbon nanotube capacitor. Science **2005**, *307* (5717), 1942–1945.

388. Novak, J.P.; Snow, E.S.; Houser, E.J.; Park, D.; Stepnowski, J.L.; McGill, R.A. Nerve agents detection using Networks of single-walled carbon nanotubes. Appl. Phys. Lett. **2003**, *83* (19), 4026–4028.

389. Kim, Y.; Lee, S.; Choi, H.H.; Noh, J.S.; Lee, W. Detection of a nerve agent simulant using single-walled carbon nanotube Networks: Dimethyl-methyl-phosphonate. Nanotechnology **2010**, *21* (49), 495501.

390. Hellmich, W.; Müller, G.; Bosch-v. Braunmühl, C.H.; Doll, T.; Eisele, I. Field effect-induced gas sensitivity changes in metal oxides. Sens. Actuators B **1997**, *43* (1–3), 132–139.

391. Lu, G.; Ocala, L.E.; Chen, J. Gas detection using low-temperature reduced graphene oxide sheets. Appl. Phys. Lett. **2009**, *94* (8), 083111.

392. Goutam, M.; Jayatissa, A.H. Gas sensing properties of graphene synthesized by chemical vapor deposition. Mater. Sci. Eng. C **2011**, *31* (7), 1405–1411.

393. Jeong, H.J.; Lee, D.S.; Choi, H.K.; Lee, D.H.; Kim, J.E.; Lee, J.Y.; Lee, W.J.; Kim, S.O.; Choi, S.Y. Flexible room-temperature [NO. [sub 2. gas sensors based on carbon nanotubes/reduced graphene hybrid films. Appl. Phys. Lett. **2010**, *96* (21), 213105.

394. Ko, G.; Kim, H.Y.; Ahn, J.; Park, Y.M.; Lee, K.Y.; Kim, J. Graphene-based nitrogen dioxide gas sensors. Curr. Appl. Phys. **2010**, *10* (4), 1002–1004.

395. Don, Y.; Lu, Y.; Kybert, N.J.; Luo, Z.; Johnson, A.T.C. Intrinsic response of graphene vapor sensors. Nano Lett. **2009**, *9* (4), 1472–1475.

396. Johnson, J.L.; Behnam, A.; An, Y.; Pearton, S.J.; Ural, A. Experimental Study of graphitic nanoribbon films for ammonia sensing. J. Appl. Phys. **2011**, *109* (12), 124301.

397. Randeniya, L.K.; Martin, P.J.; Bendavid, A.; McDonnell, J. Ammonia sensing characteristics of carbon nanotube yarns decorated with nanocrystalline gold. Carbon **2011**, *49* (15), 5265–5270.

398. Penza, M.; Cassano, G.; Rossi, R.; Alvisi, M.; Rizzo, A.; Signore, M.A.; Dikonimos, T.; Serra, E.; Giorgi, R. Enhancement of sensitivity in gas chemiresistors based on carbon nanotube surface functionalized with noble gas metal (Au, Pt) nanoclusters. Appl. Phys. Lett. **2007**, *90* (17), 173123.

399. Zhao, M.G.; Wang, X.C.; Cheng, J.P.; Zhang, L.W.; Jia, J.F.; Li, X.J. Synthesis and ethanol sensing properties of Al-doped ZnO nanofibers. Curr. Appl. Phys. **2013**, *13* (2), 403–407.

400. Rai, P.; Song, H.M.; Kim, Y.S.; Song, M.K.; Oh, P.R.; Yoon, J.M. Yu, Y.T. Microwave assisted hydrothermal synthesis of single cyrstalline ZnO nanorods for gas sensor application. Mater. Lett. **2012**, *68*, 90–93.

401. Wang, W.C.; Tian, Y.T.; Li, X.J.; Wang, X.C.; He, H.; Xu, Y.R.; He, C. Enhanced ethanol sensing properties of Zn-doped SnO_2 porous hollow microspheres. Appl. Surf. Sci. **2012**, *261*, 890–895.

402. Folke, M.; Cernerud, L.; Ekstrom, M.; Hok, B. Critical review of non-invasive respiratory monitoring in medical care. Med. Biol. Eng. Comput. **2003**, *41* (4), 377–383.

403. CaO, W.Q.; Duan, Y.X. Breath analysis: Potential for clinical diagnosis and exposure assessment. Clin. Chem. **2006**, *52* (5), 800–811.

404. Pavlou, A.K.; Turner, A.P. Sniffling out the truth: Clinical diagnosis using the electronic nose. Clin. Chem. Lab. Med. **2000**, *38* (2), 99–112.

405. Sivaramakrishnan, S.; Rajamani, R.; Smith, C.S.; McGee, K.A.; Mann, K.R.; Yamashita, N. Carbon nanotube-coated surface acoustic wave sensor for carbon dioxide sensing. Sens. Actuators B **2008**, *132* (1), 296–304.

406. Jung, H.Y.; Jung, S.M.; Kim, J.; Suh, J.S.; Chemical sensors for sensing gas adsorbed on the inner surface of carbon nanotube channels. Appl. Phys. Lett. **2007**, *90* (15), 153114.

407. Modi, A.; Koratkar, N.; Lass, E.; Wei, B.; Ajayan, P.M. Miniaturized gas ionization sensors using carbon nanotubes. Nature **2003**, *424* (6945), 171–174.

408. Valentini, L.; Armentano, I.; Kenny, J.M.; Cantalini, C.; Lozzi, L.; Santucci, S. Sensors for sub-ppm NO2 gas detection based on carbon nanotube thin films. Appl. Phys. Lett. **2003**, *82* (6), 961–963.

409. Qi, P.; Vermesh, O.; Grecu, M.; Javey, A.; Wang, Q.; Dai, H.; Peng, S.; Cho, K.J. Toward large arrays of multiplex functionalized carbon nanotube sensors for highly sensitive and selective molecular detection. Nano Lett. **2003**, *3* (3), 347–351.

410. Jang, Y.T.; Moon, S.I.; Ahn, J.H.; Lee, Y.H.; Ju, B.K. A single approach in fabricating chemical sensor using laterally grown multi-walled carbon nanotubes. Sens. Actuators B **2004**, *99* (1), 118–122.

411. Liang, Y.X.; Chen, Y.J.; Wang, T.H. Low resistance gas sensors fabricated from multi-walled carbon nanotubes coated with a tin oxide layer. Appl. Phys. Lett. **2004**, *85* (4), 666–668.

412. Zhang, J.; Boyd, A.; Tselev, A.; Paranjape, M.; Barbara, P. Mechanism of NO_2 detection in carbon nanotube field effect transistor chemical sensors. Appl. Phys. Lett. **2006**, *88* (12), 123112.

413. Cho, W.S.; Moon, S.I.; Paek, K.K.; Lee, Y.H.; Park, J.H.; Ju, B.K. Patterned multiwall carbon nanotube films as materials of NO_2 gas sensors. Sens. Actuators B **2006**, *119* (1), 180–185.

414. Pandey, S.; Goswami, G.K.; Nanda, K.K. Green synthesis of biopolymer-silver nanoparticle nanocomposite: An optical sensor for ammonia detection. Int. J. Biol. Macromol. **2012**, *51* (4), 583–589.

415. Hoefer, U.; Steiner, K.; Wagner, E. Contact and sheet resistances of SnO_2 thin films from transmission-line model measurements. Sens. Actuators B **1995**, *26* (1–3), 59–63.

416. Im, J.S.; Kang, S.C.; Lee, S.H.; Lee, Y.S. Improved gas sensing of electrospun carbon fibers based on pore structure, conductivity and surface modification. Carbon **2010**, *48* (9), 2573–2581.

417. Suri, K.; Annapoorni, S.; Sarkar, A.K.; Tandon, R.P. Gas and humidity sensors based on iron-oxide-polypyrrole nanocomposites. Sens. Actuator B Chem. **2002**, *81* (2–3), 277–282.

418. Liu, C.Y.; Chen, C.F.; Leu, J.P.; Lu, C.C.; Liao, K.H. Fabrication and carbon monoxide sensing characteristics of mesostructured carbon gas sensors. Sens. Actuatot B Chem. **2009**, *143* (1), 12–16.

419. Varghese, O.K.; Kichambre, P.D.; Gong, D.; Ong, K.G.; Dickey, E.C.; Grimes, C.A. Gas sensing characteristics of multi-wall carbon nanotubes. Sens. Actuator B Chem. **2001**, *81* (1), 32–41.

420. Park, Y.H.; Song, H.K.; Lee, C.S. Fabrication and its characteristics of metal-loaded TiO_2/SnO_2 thick film gas sensor for detecting dichloromethane. J. Ind. Eng. Chem. **2008**, *14* (6), 818–823.

421. Suehiro, J.; Zhou, G.; Imakiire, H.; Ding, W.; Hara, M. Controlled fabrication of carbon nanotube NO_2 gas sensor using dielectrophoretic impedance measurement. Sens. Actuator B Chem. **2005**, *198* (1–2), 398–403.

422. Durner, L.; Helbling, T.; Zenger, C.; Jungen, A.; Stampfer, C.; Hierold, C. SWNT growth by CVD on ferritin-based iron catalyst nanoparticles towards CNT sensors. Sens. Actuator B Chem. **2008**, *132* (2), 485–490.

423. Wang, C.Y.; Kinzer, M.; Youn, S.K.; Ramgir, N.; Kunzer, M.; Köhler, K. Oxidation behavior of carbon monoxide at the photostimulated surface at ZnO nanowires. J. Phys. D **2011**, *44* (30), 305302–305309.

424. Miura, N.; Raisen, T.; Lu, G.; Yamazoe, N. Highly selective CO sensor using stabilized zirconia and a couple of oxide electrodes. Sens. Actuators B **1998**, *47* (1–3), 84–91.

425. Chang, J.F.; Kuo, H.H.; Leu, I.C.; Hon, M.H. The effects of thisckness and operation temperature on ZnO: Al thin film CO gas sensor. Sens. Actuators B **2002**, *84* (2–3), 258–264.

426. Ryu, H.W.; Park, B.S.; Akbar, S.A.; Lee, W.S.; Hong, K.J.; Seo, J.Y.; Shin, D.C.; Park, J.S.; Choi, G.P. ZnO sol-gel derived porous film for CO gas sensing. Sensors Actuators B. Chem. **2003**, *96* (3), 717–722.

427. Chen, N.S.; Yang, X.J.; Liu, E.S.; Huang, J.L. Reducing gas sensing properties of ferrite compounds MFe_2O_4 (M= Cu, Zn, Cd, and Mg). Sens. Actuators B Chem. **2000**, *66* (1–3), 178–180.

428. Korotcenkov, G. The role of morphology and crystallographic structure of metal oxides in response of conductometric-type gas sensors. Mater. Sci. Eng. R. **2008**, *61* (1–6), 1–39.

429. Kaciulis, S.; Pandolfi, L.; Comini, E.; Faglia, G.; Ferroni, M.; Sberveglieri, G.; Kandasamy, S.; Shafiei, M.; Wladarski, W. Nanowires of metal oxides for gas sensing applications. Surf. Interface Anal. **2008**, *40* (3–4), 575–578.

430. Zhou, Z.G.; Tang, Z.L.; Zhang, Z.T.; Wlodarski, W. Perovskite oxide of PTCR ceramics as chemical sensors. Sens. Actuators B Chem. **2001**, *77* (1–2), 22–26.

431. Lee, D.D.; Sohn, B.K.; Ma, D.S. Low-power thick film CO gas sensors. Sens. Actuators **1987**, *12* (4), 441–447.

432. Lee, D.D.; Chung, W.Y. Gas-sensing characteristics of SnO2-X thin-film with added Pt fabricated by the dipping method. Sens. Actuators **1989**, *20* (3), 301–305.

433. Hübner, M.; Pavelko, R.; Kemmler, J.; Barsan, N.; Weimar, U. Influence of material properties on Hydrogen sensing for SnO2 nanomaterials. Proc. Chem. **2009**, *1* (1), 1423–1426.

434. Hahn, S.H.; Barsan, N.; Weimar, U.; Ejakov, S.G.; Visser, J.H.; Soltis, R.E. CO sensing with SnO2 thick film

sensors: Role of oxygen and water vapour. Thin Solid Films **2003**, *436* (1), 17–24.

435. Henrich, V.E.; Cox, P.A. *The Surface Science of Metal Oxides*; Cambridge University Press: Cambridge, 1994.

436. Cox, P.A. *Transition Metal oxides: An Introduction to Their Electronic Structure and Properties*; Clarendon press: Oxford, UK, 1992.

437. Shim, J.H.; Cha, A.; Lee, Y. Lee, C. Nonenzymatic amperometric glucose sensor based on nanoporous gold/ruthenium electrode. Electroanalysis **2011**, *23* (9), 2057–2062.

438. Dinga, Y.; Liu, Y.; Zhang, L.; Wang, Y.; Bellagamba, M.; Parisi, J.; Li, C.M.; Lei, Y. Sensitive and selective nonenzymatic glucose detection using functional NiO-Pt hybrid nanofibers. Electrochim. Acta **2011**, *58* (1), 209–214.

439. Amini, M.K.; Shahrokhian, S.; Tangestaninejad, S. PVC-based Mn(III) porphyrin membrane coated graphite electrode for determination of histidine. Anal. Chem. **1999**, *71* (13), 2502–2505.

440. Yu, M.; Dovichi, N.J. Attomole amino-acid determination by capillary zone electrophoresis with thermooptical absorbance detection. Anal. Chem. **1989**, *61* (1), 37–40.

441. Raoof, J.B.; Ojani, R.; Kiani, A. Apple-modified carbon paste electrode: A biosensor for selective determination of dopamine in pharmaceutical formulations. Bull. Electrochem. **2005**, *21* (5), 223–228.

442. Ates, M.; Castillo, J.; Sarac, A.S.; Schuhmann, W. Carbon fiber microelectrodes electrocoated with polycarbazole and poly(carbazole-co-p-tolylsulfonyl pyrrole) films for the detection of dopamine in presence of ascorbic acid. Microchim. Acta **2008**, *160* (1–2), 247–251.

443. Dutt, V.S.E.; Mottola, H.A. Determination of uric acid at the microgram level by a kinetic procedure based on a pseudo-induction period. Anal. Chem. **1974**, *46* (12), 1777–1781.

444. Zen, J.M.; Tang, J.S. Square-wave voltammetric determination of uric acid by catalytic oxidation at a perfluorosulfonate ionomer/ruthenium oxide perchlorate chemically modified electrode. Anal. Chem. **1995**, *67* (1), 1892–1895.

445. Zhang, L.; Lin, X. Covalent modification of glassy carbon electrode with glutamic acid for simultaneous determination of uric acid and ascorbic acid. Analyst **2001**, *126* (3), 367–370.

446. Kim, S.K.; Kim, D.; Jeon, S. Electrochemical determination of serotonin on glassy carbon electrode modified with various graphene nanomaterials. Sens. Actuators B **2012**, *174* (1), 285–291.

447. Watkins, J.D.; Lawrence, K.; Taylor, J.E.; James, T.D.; Bull, S.D.; Marken, F.; Carbon nanoparticle surface electrochemistry: High-density covalent immobilisation and pore-reactivity of 9,10-Anthraquinone. Electroanalysis **2011**, *23* (6), 1320–1324.

Nanomaterials—
Nanoparticles

Nanomaterials: Theranostics Applications

M. Sasidharan
S. Anandhakumar
S. Vivekananthan
Department of Physics and Nanotechnology, Sri Ramaswamy Memorial (SRM) University, Kattankulathur, Chennai, India

Abstract

The term *theranostics*, which is derived from "diagnostics" and "therapy," refers to a combined treatment strategy of a diagnostic agent and a therapy. The integration of diagnostic imaging capability with therapy is critical for addressing the challenges for future biomedical applications. Multifunctional nanomedicine holds considerable promise as the next generation of medicine that enables the early detection of disease, monitoring treatment, and targeted therapy with minimal toxicity. Recently, there has been a tremendous development of a variety of nanotechnology platforms to diagnose and treat diseases. Especially, polymeric nanoplatforms such as polymeric micelles, vesicles, dendrimers, and polymersomes have been intensively investigated for multifunctional theranostic applications. Compared with traditional molecular-based contrast agents or therapeutic drugs, polymeric nanoparticle-based nanomedicine paradigm enables a highly integrated design that incorporates multiple functions, such as cell targeting, ultrasensitive bioimaging/diagnostics, and therapy, in a single system. In particular, the combinations of various nanostructured materials with diverse properties offer synergetic multifunctional nanomedical platforms, which make it possible to accomplish multimodal imaging, and simultaneous diagnosis and therapy. In this entry, we summarize the recent developments on the fabrication strategies of polymeric multifunctional nanoplatforms for simultaneous diagnosis and therapy.

INTRODUCTION

Concept of Theranostics

The term *theranostic* was coined in 2002 by Funkhouser and is defined as a material that combines the modalities of therapy and diagnostic imaging.[1] In contrast to the development and use of separate materials for these two objectives, theranostics combine these features into one "package," which has the potential to overcome undesirable differences in biodistribution and selectivity. The ultimate goal of the theranostic approach is to gain the ability to image and monitor the diseased tissue, delivery kinetics, and drug efficiency. The most promising aspects of utilizing nanoparticles as therapeutics, diagnostics, and theranostics are their potential to localize (or be targeted) in a specific manner to the site of disease, which reduces or eliminates the possible untoward side effects. The nanometric size of these materials precludes them from being readily cleared through the kidneys, thereby extending circulation in the blood pool depending on their surface functionalization characteristics.

Polymeric Nanoparticles

Nanoparticles as colloidal systems can be fabricated from a multitude of materials in a variety of compositions, including quantum dots (QDs), polymeric nanoparticles, gold nanoparticles, paramagnetic nanoparticles, and so on. Among the many exciting nanoplatforms, polymer-based dendrimers, vesicles, and polymeric micelles provide unique nanocarrier systems.[2–5] Highly ordered self-assembled polymer nanoparticles present several distinctive advantages for the development of multifunctional delivery systems, including the relatively small size, the versatile capacity to integrate diagnostic and therapeutic functions. Polymeric micelles, an important class of nanoparticles, are composed of amphiphilic block copolymers that contain distinguished hydrophobic and hydrophilic segments, which form core–shell nanostructures. This unique architecture enables the hydrophobic micelle core to serve as a nanoscopic depot for therapeutic or imaging agents and the shell as biospecific surfaces for targeting applications. A variety of drugs with diverse characteristics, including genes and proteins, were incorporated into the core by engineering the structure of the core-forming segment of the block copolymer to improve interaction with drug molecules. Since these nanoparticles are 100- to 1000-fold smaller than cancer cells, they can easily pass through cell barriers. Current clinical studies of several polymeric nanoparticle (micelles)-based formulations indicates that these carriers showed reduced side effects and high effectiveness to various intractable tumors including triple-negative breast cancers.[6] The detailed account of polymeric nanoparticles for drug delivery applications is

Concise Encyclopedia of Biomedical Polymers and Polymeric Biomaterials DOI: 10.1081/E-EBPPC-120050397

reviewed by several research groups[7–9] and therefore we limit our discussion exclusively to theranostic applications.

Apart from drug delivery applications, polymeric nanoparticles were also exploited as specific molecular probes or contrast agents to detect and characterize early-stage diseases and provide a rapid method for evaluating treatment. When compared to small-molecule imaging probes, polymer-based imaging agents have many advantages such as improved plasma half-life and stability, less toxicity, long blood circulation times, and improved targeting. Hence, the use of combinations of different nano-structured polymeric materials allows the development of novel multifunctional nanomedical platforms for multi-modal imaging. A variety of nanoparticles have already been explored for various imaging modalities such as magnetic resonance imaging (MRI), computed tomography, ultrasound, optical imaging (bioluminescence and fluorescence), single photon emission computed tomography, and positron emission tomography (PET). Diverse imaging modalities or combination of different imaging modalities using various polymeric nanomaterials are beyond the scope of this entry, which are well-documented in literature.[10–13]

Polymeric Nanoparticles for Theranostics

Clever combinations of different kinds of functional polymeric nanomaterials will enable the development of nanomedical platforms for multimodal imaging or simultaneous diagnosis and therapy. In many cases, the polymeric nanoparticles (NPs) are comprised of a hydrophobic core containing the anticancer agent and a hydrophilic surface layer for the stabilization of the NPs in aqueous environment. The polymeric NPs have been applied as effective drug delivery systems capable of enhancing the therapeutic efficacy and reducing the side effects of drugs. So far, numerous polymeric nanoparticles formulations have been developed exclusively for either imaging purpose or therapeutic applications.[7–13] In order to effectively detect and treat cancer or other diseases, we need to develop novel integrated nanoparticle systems that simultaneously image and treat cancer in a single nanoparticle system. In this entry, we present a recent development of the theranostic strategies using polymeric nanoparticles for different types of cancer treatments including gene therapy, chemotherapy, hyperthermia treatment (photothermal ablation), photodynamic therapy, and radiation therapy.

THERANOSTIC APPROACHES

Simultaneous Imaging and Chemotherapy

Superparamagnetic iron oxide–based theranostics systems

Theranostic nanoassemblies should be able to diagnose and deliver targeted therapy and monitor the response to the therapy by addressing the challenges of cancer alterations and heterogeneities. The use of polymer nanoparticles with therapeutic modalities also affords many advantages, such as a long circulation time and targeting capability. Moreover, the uptake of polymeric nanoparticles containing drugs within tumor cells is believed to circumvent the multidrug resistance (MDR) effect. MDR means that when the cancer cells are treated with the free drug, the P-glycoprotein pump in the cell membrane is expressed and hampers the drug action by pumping out the drug molecules from the cytosol to the extracellular area. In this context, multifunctional polymeric nanoparticles comprising iron oxide nanoparticles or fluorescent probes, and anticancer drug have been studied extensively.

Gao and co-workers developed a multifunctional polymeric micelles with cancer-targeting capability for controlled drug delivery and MRI[14] using a mixture of amphiphilic block copolymers of maleimide– and methoxy–terminated poly(ethylene glycol-block-D,L-lactide) (MAL–PEG–PLA). These novel micelles are composed of a chemotherapeutic agent doxorubicin (DOXO), a superparamagnetic iron oxide (SPIO) nanoparticle, and a cRGD ligand that can target $\alpha_v\beta_3$ integrins on tumor endothelial cells and subsequently induce receptor-mediated endocytosis for cell uptake. During the evaporation of the organic solvents, hydrophobic iron oxide nanoparticles and hydrophobic doxorubicin (DOX) were spontaneously incorporated into the hydrophobic PLA core part and hydrophilic PEG was exposed to the outer environment. The amounts of magnetic nanoparticles and DOX in the polymer micelles were 6.7 wt% and 2.7 wt%, respectively. RGD peptide was conjugated to the micelle surface through a covalent thiol–maleimide linkage and concentration of cRGD is therefore controlled by density of the terminal maleimide groups. The $\alpha_v\beta_3$-targeting ability and MRI visibility of cRGD-SPIO-DOXO micelles were investigated over tumor SLK endothelial cells with overexpression of $\alpha_v\beta_3$ integrin. Uptake of SPIO nanoparticles inside SLK cells shortens the spin–spin relaxation time (T_2) by dephasing the spins of neighboring water protons and results in darkening of T_2-weighted images. Rapid and efficient $\alpha_v\beta_3$–mediated endocytosis led to significant darkening of MR images from cRGD-encoded micelles compared to cRGD-free micelles.

Flow cytometry and confocal laser scanning microscopy (CLSM) were used to evaluate the effect of cRGD on micelle targeting and uptake in $\alpha_v\beta_3$-expressing SLK tumor endothelial cells. Figure 1A shows the mean fluorescence intensity of 0% and 16% cRGD micelles incubated with SLK cells for 1 h. At micelle doses corresponding to 2, 5, and 10 μg/mL of DOXO, an approximate 2.5-fold increase in cell uptake was observed over 0% cRGD micelles. The enhancement of cell uptake with 16% cRGO micelles (Fig. 1E) over 0% cRGD micelles (Fig. 1B) was observed and $\alpha_v\beta_3$–mediated endocytosis facilitates the internalization of cRGD-encoded micelles inside SLK cells and

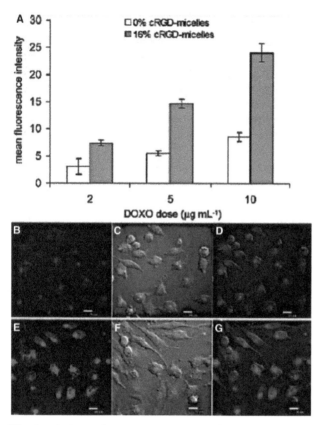

Fig. 1 (A) Mean fluorescence intensity of SLK cells as a function of micelle dose (represented by DOXO concentration between 2 and 10 μg/mL) by flow cytometry. (**B–G**) Confocal laser scanning microscopy of DOXO fluorescence, DIC, and merged images of DOXO fluorescence and DIC in SLK cells treated with 0% (**B–D**) and 16% cRGD (**E–G**) micelles after incubation for 1 hr. DOXO fluorescent images were acquired at λ_{ex} = 485 nm and λ_{em} = 495 nm. The scale bars are 20 μm in all images.
Source: From Nasongkla et al.[14] © 2006, with permission from Wiley-VCH Verlag GmbH & Co. KGaA.

majority of polymeric micelles are localized in the cytoplasmic compartment. Thus, the cancer-targeted polymer micelles exhibited higher toxicity in cancer cells via the release of loaded doxorubicin. In a separate study by Gao et al., polymeric micelles were modified with a lung cancer-targeting peptide (LCP) that binds specifically to $\alpha_v\beta_6$ receptors over expressed on lung cancer cells.[15] The specificity of LCP for $\alpha_v\beta_6$-expressing H2009 cells (lung cancer cells) was evaluated by comparative exposure to $\alpha_v\beta_6$-negative H460 cells and by using micelles containing a scrambled peptide (SP, a peptide with the same amino acid composition as LCP but scrambled sequence) in H2009 cells using [³H] radioactivity measurements, confocal microscopy, and MRI. The multifunctional nanomedicine showed superb T_2 relaxivity for ultrasentive MR detection. The integrated $\alpha_v\beta_6$-targeting, superparamagnetic iron oxide, DOX functions in the nanoplatform open up many exciting opportunities for image-guided, targeted therapy of lung cancer.

Yang et al. also synthesized multifunctional polymer micelles using similar method and further demonstrated their *in vivo* use for simultaneous diagnosis and therapy.[16] Superparamagnetic $MnFe_2O_4$ nanoparticles and DOX were embedded in carboxyl-terminated poly(ethylene glycol)-block-poly(D,L-lactic-*co*-glycolic acid) (PEG-PLGA) micelles and the HER2 antibody was conjugated at the terminal carboxyl groups. The estimated polymeric micelle size was about 70 nm, as determined by dynamic light scattering (DLS). The amounts of magnetic nanoparticles and DOX in the polymer micelles were 41.7 wt% and 3.3 wt%, respectively. These multifunctional magnetic–polymeric nanohybrids (MMPNs) containing anticancer drug allow simultaneous MR imaging and treatment. Nude mice with the tumor model of the NIH3T6.7 cell lines were injected intravenously with the HER2-conjugated multifunctional polymer micelles to confirm the efficacy of T_2 MRI and drug delivery. After injection of the polymer micelles, a signal decrease in T_2 MRI was observed at the tumor site, whereas there was an insignificant change when the control IgG antibody conjugated polymer micelles were administrated. The therapeutic efficacy was also increased remarkably, owing to the targeted release of DOX at the tumor site. The HER2-conjugated multifunctional polymer micelles were target-specifically delivered to the overexpressed HER2/neu receptors of the NIH3T6.7 cells in the mouse model and were taken up by a receptor-mediated endocytosis process. In addition, HER-MMPNs are more cytotoxic than the conventional combination of DOX + HER (11.3 ± 3.2% vs 22.9 ± 1.4%, respectively) on NIH3T6.7 cells. These noticeable results show that HER-MMPNs easily enter into the intercellular cytoplasm by a tumor-target-mediated endocytosis pathway and efficiently release DOX. The cytotoxicity is further enhanced by the additional inhibitory effect of HER on cell growth and proliferation. This enhanced efficacy indicates that effects of HER and DOX mediated by MMPNs are synergistic, thus making HER-MMPNs effective for the treatment of breast cancer.

Recently, thermally cross-linked superparamagnetic iron oxide nanoparticles (TCL-SPIONs) coated with anti-biofouling polymer conjugated with a drug exhibit enhanced permeability and retention (EPR) effect in the absence of any targeting ligands.[17] The excellent passive targeting efficiency of TCL-SPION allowed detection of tumors by MR imaging and at the same time delivery of sufficient amounts of anticancer drugs. Dox@TCL-SPION not only shows exceptional antitumor effects but also the location of tumor, precise drug delivery, and therapeutic response information after drug release.[17]

Theranostic systems with multimodal imaging probes

Most popular nanostructured multimodal imaging probes are combinations of MRI and optical imaging modalities. MR imaging offers high spatial resolution and the capacity to simultaneously obtain physiological and anatomical informations, whereas optical imaging allows for rapid screening. Not only MRI agent has been used as the imaging modality of theranostic nanodevices, but also QD-based theranostic nanoassemblies were recently reported. Hyeon, Park, and their co-workers developed multifunctional polymer nanomedical platform with multiple diagnosing agents.[18,19] This platform is composed of (Fig. 2A): (i) biodegradable poly(D,L-lactic-*co*-glycolic acid) (PLGA) nanoparticles as a matrix for loading hydrophobic drugs; (ii) secondly, superparamagnetic magnetite nanoparticles (MNP) for both magnetically guided delivery and T_2 MRI contrast agent, and semiconductor nanoparticles (QD) for optical imaging; (iii) doxorubicin (DOXO) as a anticancer drug, and (iv) folate (FOL) ligand for cancer targeting. The biodegradable PLGA nanoparticles containing the inorganic nanocrystals and drug molecules were prepared by an oil-in-water emulsion technique using F127 as a surfactant. The density of the iron oxide nanoparticles embedded in the polymer nanoparticles could be easily controlled. Cancer-targeting folate was conjugated onto the PLGA nanoparticles by means of the PEG groups to target KB cancer cells that have overexpressed folate receptors on their cell surface.

The T_2-weighted MR image of the KB cells treated with PLGA(MNP/DOXO)-FOL nanoparticles was much darker than that of the KB cells treated with naked PLGA(MNP/DOXO) and PLGA(MNP/DOXO)-PEG nanoparticles and the control cells (Fig. 2B). This observation indicates that the folate groups on the polymer nanoparticles possessed a specific targeting ability for KB cells that overexpressed the folate receptors. Furthermore, the application of an external magnetic field during incubation produced an even

darker image indicating that PLGA(MNP/DOXO) nanoparticles could serve as cancer-targeted, T_2 contrast agent in MRI and that the combination of folate targeting groups and external magnetic field synergetically enhanced the cancer targeting efficiency. Thus the use of both magnetic targeting and folate targeting enhanced the MRI signal and cytotoxicity toward the cancer cells *in vitro* compared to use of only folate targeting. The comparison of CLSM images of DOXO fluorescence in KB cells treated with PLGA(MNP/DOXO)-FOL showed enhanced cellular uptake compared to naked nanoparticles. Furthermore, comparison of photoluminescence spectrum of PLGA(QD/DOXO) and PLGA(QD/DOXO)-FOL revealed that the latter (active-targeted) exhibited a much stronger emission than the former. These results demonstrate that PLGA(QD/DOXO)-FOL could simultaneously act as an optical imaging agent and drug delivery system.

Sailor and co-workers have reported an interesting theranostic platform of micellar hybrid nanoparticles (MHNs) that contain magnetic nanoparticles (MN), QD, and anticancer drug doxorubicin (DOX) within a single poly(ethylene glycol-phospholipid) micelle.[20] Transmission electron microscopic (TEM) images and dynamic light scattering measurements revealed that the MHNs consist of clusters of both MNs and QDs within a micellar coating with a hydrodynamic size of 60–70 nm. The MN/QD ratio within the individual micelles could be adjusted by changing the mass ratio of MNs to QDs during the synthesis. To enable the specific targeting of tumor cells by the nanoassemblies, the MHNs were conjugated with the targeting ligand F3, a peptide (F3-MHNs) known to target cell-surface nucleolin in endothelial cells in tumor blood vessels and in tumor cells. The ability of MHNs to target

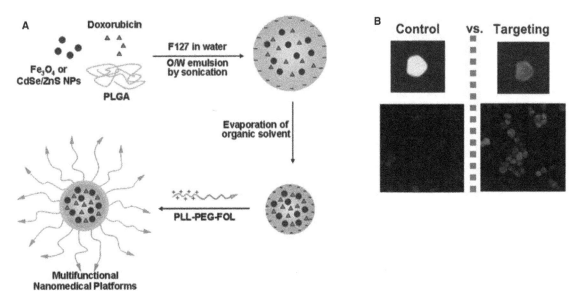

Fig. 2 (**A**) synthetic procedure for the multifunctional polymer nanoparticles. (**B**) T_2-weighted MR images and CLSM images of control versus targeting multifunctional nanoparticles.
Source: From Kim et al.[18] © 2008, with permission from Wiley-VCH Verlag GmbH & Co. KGaA.

and dual-mode image tumor cells was tested on MDA-MB-435 human cancer cells. After incubation with DOX-loaded F3-MHNx (DOX-MHN-F3), the DOX fluorescence signal appeared mainly in the cytoplasm and showed the colocalization of DOX with endosomes. In contrast, when free DOX was added, almost the entire DOX fluorescence signal was observed in the cell nuclei. In addition, F3-MHNs in which DOX was incorporated displayed significantly greater cytotoxicity than that of equivalent quantities of free DOX or MHNs containing DOX without the targeting ligand.

Santra et al., in a separate study, developed a novel synthetic method to prepare biodegradable and biocompatible poly(acrylic acid)-iron oxide nanoparticles (PAA-IONPs). Schematic representation of synthesis of theranostics and multimodal IONPs is shown in Scheme 1.[21] This water-based green chemical approach has five key components: (i) an encapsulated chemotherapeutic agent (Taxol) for cancer therapy, (ii) folic acid for cancer targeting, (iii) click-chemistry-based conjugation of targeting ligands, (iv) an encapsulated NIR dye for fluorescent imaging capabilities, and (v) a superparamagnetic iron oxide core for MRI. Cytotoxicity and target specificity of folate-functionalized IONP were examined in lung carcinoma cells (A549, overexpress folate receptor) and cardiomyocytes (H9c2, do not overexpress folate receptor) via a MTT assay. The results from the MTT assay indicated almost 100% survival, for both A549 and H9c2 cells remained unchanged. These results indicate specificity of PAA-IONPs toward tumor cells that over express folate receptor. Drug/dye release was only observed in acidic pH or in the presence of esterase, a degradative enzyme. Taken together, these results make folate-decorated INOP an important drug carrier, as it can rapidly release Taxol and therefore induce cell death only upon targeted cell internalization.

Organic dyes for optical imaging

Instead of MN, organic dyes can be employed to prepare optical imaging and therapeutic particles. Hsiue and co-workers designed polymer micelles functionalized with

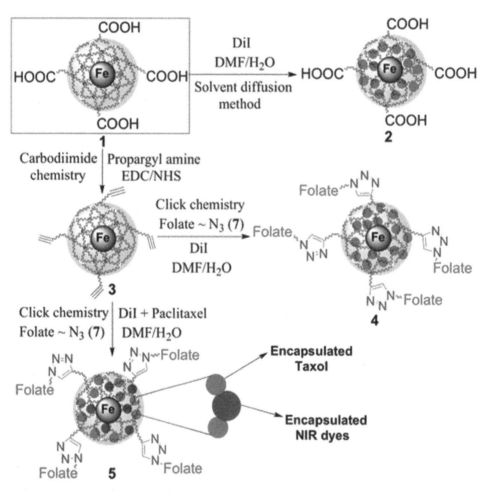

Scheme 1 Schematic representation of the synthesis of theranostics and multimodal IONPs. Click chemistry and carbodiimide chemistry have been used for the synthesis of a library of functional IONPs. NIR dyes and paclitaxel co-encapsulated IONPs were prepared in water using the modified solvent diffusion methods.
Source: From Santra et al.[21] © 2009, with permission from Wiley-VCH Verlag GmbH & Co. KGaA.

flurocein isothiocyanate (FITC), an organic dye, using a grafted polymer and diblock copolymer.[22] A multifunctional micelle was prepared from four components, including FITC-PEG-PLA, Gal-PEG-PLA, Block III (mPEG-PLA), and graft copolymer. The graft copolymer in multifunctional micelles could encapsulate anticancer drugs and control drug release by moderating pH and temperature changes. Block III in micelles helped control the core–shell structure and obtain uniform micellar distribution. The fluorescence dye conjugated diblock copolymer FITC-PEG-PLA micelles provided a direct evidence of where micelles accumulated after cell uptake. On the other hand, the targeting moiety (Gal) conjugated diblock copolymer Gal-PEG-PLA could combine with the asialoglycoprotein of HepG2 cells proceeding to the active tumor targeting. The DOX loading level was about 31 wt% in weight, which was determined by UV/Vis spectrophotometer after dissolving the multifunctional micelles in dimethylsulfoxide. The multifunctional micelle particle size was approximately 160 nm as evaluated from TEM observation. Confocal laser scanning microscopy clearly shows the fluorescence images of multifunctional micelles and released DOX after HeLa cells uptake. The fluorescence images of multifunctional micelles incubated with HeLa cells for 6 hr showed green fluorescence in the cytoplasm, indicating that the multifunctional micelles were located there (Fig. 3A). Additionally, the released DOX, with a red fluorescence, was localized in both the cytoplasm and the nucleus. The exact location and distribution of carrier was observed by the FITC-labeled micelles. To evaluate the functionality of multifunctional micelles in specific tumor targeting, multifunctional micelles were incubated with HepG2 (hepatocellular carcinoma) cells to confirm drug cytotoxicities. The viability (percentage of surviving cells) of HepG2 cells after 24 hr and 48 hr incubation was compared with multifunctional micelles (Fig. 3B). The multifunctional micelles had lower cell viabilities than those without Gal; this is because the multifunctional micelles bound with asialoglycoprotein and then internalized into cancer cells to release Dox by intracellular pH changes. Thus tumor targeting assay and CLSM measurements revealed that multifunctional micelles exhibited a high cytotoxicity by receptor-mediated endocytosis and showed clear fluorescence imaging of their distribution.

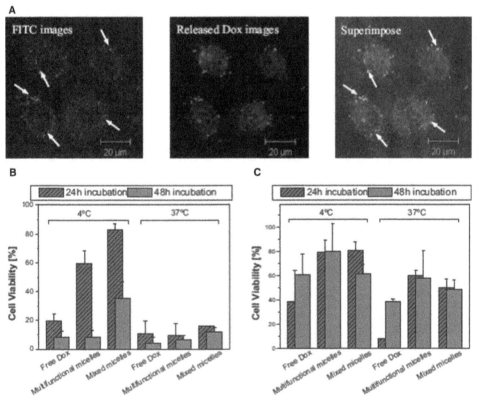

Fig. 3 Evaluation of multifunctional micelle functionalities in drug delivery applications. (A) Confocal images of HeLa cells incubated with multifunctional micelles (50 μg/mL) showing the particulate distribution and localization of released Dox after six hour internalization. (B) Free Dox cytotoxicity, multifunctional micelle cytotoxicity and mixed micelle cytotoxicity after incubation with HepG2 cells for 50 μg/mL. (C) Free Dox cytotoxicity, multifunctional micelle cytotoxicity and mixed micelle cytotoxicity after incubation with HepG2 cells for 24 hr and 48 hr in presence of 150 mM galactose. Dox dosage was 50 μg/mL.
Source: From Huang et al.[22] © 2007, with permission from Wiley-VCH Verlag GmbH & Co. KGaA.

Gadolinium and manganese based contrast agents

Recently, Kataoka and co-workers[23] developed a micelle based theranostic system incorporating gadolinium-diethylenetriaminepentaacetic acid (Gd-DTPA), a widely used T_1-weighted MRI (T_1W) contrast agent and (1,2-diaminocyclohexane)platinum(II) (DACHPt), the parent complex of the potent anticancer drug oxaliplatin, in their core by reversible complexation between DACHPt, Gd-DTPA, and poly(ethylene glycol)-*b*-poly(glutamic acid) [PEG-*b*-P(Glu)]. The Gd-DTPA/DACHPt-loaded micelles show extended circulation of their cargo in the blood stream, whereas free oxaliplatin and free Gd-DTPA were rapidly cleared from plasma. Moreover, the micelles delivered the drugs to solid tumors due to the increased accumulation and retention at the cancer site because of the EPR effect. The enhanced accumulation of Gd-DTPA/DACHPt-loaded micelles indeed improve the antitumor activity of the incorporated Pt drug as DACHPt complexes can exert their cytotoxicity after being released from the Gd-DTPA/DACHPt-loaded micelles. The T_1-weighted MR images after i.v. administration of the Gd-DTPA/DACHPt-loaded micelles clearly showed specific contrast enhancement. Whereas no change in contrast was observed in the tumor region after the administration of free Gd-DTPA, and the

signal intensity was higher in the liver, kidney, or spleen than in tumor as suggested from the tumor-to-organ ratios of the MR intensity.

With the increasing concerns of nephrogenic system fibrosis due to gadolinium-based blood pool agents in patients with renal disease or with recent liver transplant, several approaches have been recently taken to substitute gadolinium with manganese (Mn).[24] For instance, Pan et al. developed a fibrin-targeting "nanobialys"-based contrast agent and drug delivery vehicle containing a porphyrin, which chelates Mn as shown in Fig. 4. "Nanobialys" are toroidal-shaped, nearly monodisperse reverse micellar structures formed from amphiphilic branched polyethyleneimine (PEI) in anhydrous chloroform with a core containing a porphyrin-Mn(III) complex, which is a T_1 contrast agent for MRI. Amphiphilic PEI was modified by reaction of amines with hydrophobic linoleic acid via EDC coupling. The core of the micelles was loaded with hydrophobic (camptothecin) or hydrophilic (doxorubicin) chemotherapeutic drug, and the drug release profile was studied *in vitro* for three days. The MR imaging of fibrin was performed with fibrin-rich clots supported on silk sutures in phosphate-buffered saline with the help of sealed polystyrene test tubes. The MR imaging results for targeted ligand-containing nanobialys (containing a fibrin-specific monoclonal antibody) were

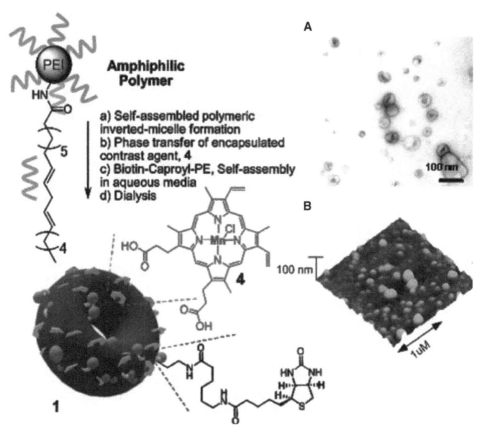

Fig. 4 Preparation of Nanobialys and (**A**) TEM image of nanobialys and (**B**) AFM image of nanobialys.
Source: From Pan et al.[24] © 2008, with permission from the American Chemical Society (ACS).

compared to nanobialys without any targeting ligand or without any metal inside as controls and demonstrated the role of the targeting agent in achieving contrast enhancement. Incorporation of drugs with high efficiency and their retention in dissolution supports the theranostic potential of this platform technology.

Simultaneous Imaging and Photodynamic Therapy

Photodynamic therapy (PDT) is a form of light-activated chemotherapy using light-sensitive drugs, that is, photosensitizers (PSs). The commonly used PSs are usually of porphyrinic origin and, once excited by light of the appropriate wavelength, are capable of producing cytotoxic reactive oxygen species, such as singlet oxygen ($^1O_2$), free radicals or peroxides, through interactions between the excited porphyrinic molecules and O_2. This leads to the irreversible destruction of diseased cells and tissues. One of the main advantages of PDT is that treatment takes place only under the irradiation of light, which can be localized. Kopelman and co-workers developed 30–60 nm polyacrylamide (PA) nanoparticles encapsulating iron oxide nanoparticles and Photofrin, a kind of commercial photosensitizer (PS) for imaging and treatment of brain tumors.[25] It is further functionalized with PEG groups and the RGD targeting peptide on the surface of these nanoparticles. The advantages of encapsulating Photofrin within a polymeric nanoparticles are that it (i) reduces the direct exposure of the drug to the physiological milieu, (ii) reduces cutaneous photosensitivity posttreatment, and (iii) significantly reduces the wait time (from 24 to 1 hr) between intravenous injection of the drug and subsequent laser irradiation. The role of targeting ligand was clearly seen when RGD peptide-bearing nanoparticles showed higher accumulation in $\alpha_v\beta_3$-expressing cancer cells (MDA-435) than $\alpha_v\beta_3$-negative MCF-7 cells.

Ross et al. reported an F3-targeted polymeric nanoparticle formulation consisting of encapsulated imaging agent and PS.[26] These polymeric nanoparticles contain Photofrin and iron oxide crystals within the core matrix. Polyethylene glycol was attached to the surface of the nanoparticle along with the targeting peptide. When these nanoparticles were incubated with MDA-435 cells for a very brief period (five minutes), and irradiated with laser, no Photofrin-mediated cytotoxicity was observed. In fact, most cells were live (calcein-AM positive) with no detectable loss of cell membrane integrity. Furthermore, when the nanoparticles were incubated for 4 hr without irradiation with laser light, no detectable cytotoxicity was observed, suggesting that nanoparticles alone were not cytotoxic to MDA-435 cells. The combination of 4 hr of incubation of F3-tagged nanoparticles and laser light resulted in cytotoxicity wherein 90% of MDA-435 cells were killed as monitored by the presence of propidium iodide–stained red nuclei. However, the combination of

4 hr of incubation and irradiation with laser did not induce cell death when similar nanoparticles lacking the F3 peptide were incubated with MDA-435 cells.

Recently, in a separate study, Kopelman et al. have developed multifunctional biodegradable polyacrylamide nanoplatforms for a "see and treat" strategy, by a combination of PDT and fluorescence imaging for cancer theranostics.[27] The construction of these novel carriers involves adding primary amino groups and biodegradable cross-linkers during the NP polymerization, while incorporating photodynamic and fluorescent imaging agents into the NP matrix, and conjugating PEG and tumor-targeting ligands onto the surface of the NPs. *In vitro* targeting studies on human breast cancer cells indicate that the targeted NPs can be transported efficiently into tumor cells. Incubating the multifunctional nanocarriers into cancer cells enabled strong fluorescence imaging. Irradiation of the photosensitizing drug, incorporated within the NPs, with light of a suitable wavelength, causes significant but selective damage to the impregnated tumor cells. The PDT efficacy is largely determined by intactness of the drug, the drug's efficiency to produce singlet oxygen. The combination photosensitive drug and polymeric nanoparticles did not kill cells when incubated in the dark (Fig. 5). However, both 9 L human glioma and MDA-MB-435 breast cancer cells were effectively killed by exposure to 647 nm light. These results indicate the significance of light and even at a very low concentration (1 μM) result in significant damage to the cells. As expected, modification of the nanoplatform with PEG and tumor targeting cell penetrating peptide enabled guiding the particles to specifically targeted cancer cells and significantly reduced the dark toxicity.

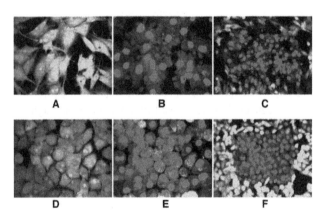

Fig. 5 Images of 9 L (upper row) and MDA-MB-435 (lower row) tumor cells treated with F3-Cys targeted PEGylated CD-linked HPPH-conjugated AFPAA NPs with 1 μM HPPH concentration, illuminated with 647 nm laser light (46 μW) for 1 min, and monitored every minute thereafter. 9 L: (**A**) 0 min (40x magnification); (**B**) 15 min; (**C**) 20x zoom out image. MDA-MB-435: (**D**) 0 min (40x magnification); (**E**) 15 min; (**F**) 20x zoom out image.

Source: From Wang et al.[27] © 2012, with permission from the American Chemical Society (ACS).

McCarthy, Jaffer, and Weissleder have reported dextran based PDT system consisting of iron oxide nanoparticles with the cross-linked dextran coating (CLIO), which was further modified by conjugating a potent PS, 5-(4-carboxyphenyl)-10,15,20-triphenyl-2,3-dihyroxychlorin (TPC), and a NIR fluorescent dye, Alexa Fluor 750, onto the surface amine groups of the dextran coating.[28] The energy transfer was minimized by the 100 nm difference between the longest wavelength absorption for TPC (650 nm) and that for Alexa Fluor 750 (750 nm), thus minimizing the risk of killing the cells while performing fluorescent imaging. The uptake of the nanoparticles within the macrophage cells was observed by fluorescent microscopy. After incubation, the TPC-loaded nanoparticles were irradiated with light at 650 nm, a dose-dependent phototoxicity was observed. These magnetofluorescent nanoparticles conjugated with PSs have the potential to be used for simultaneous MRI, NIR fluorescent imaging, and PDT.

Simultaneous Imaging and Photothermal Therapy

Of late, noninvasive photothermal therapy for the selective treatment of tumor cells gains much importance. This type of therapy utilizes the large absorption cross-section of nanomaterials in the NIR region. Owing to its weak absorption by tissues, NIR radiation is able to penetrate the skin without causing much damage to normal tissues and thus can be used to treat specific cells targeted by the nanomaterials. Several nanomaterials that strongly absorb NIR radiation, including gold nanoshells, gold nanorods, and single-walled carbon nanotubes, were explored for potential therapeutic applications. The radiation absorbed by these nanomaterials is converted efficiently into heat, causing cell destruction.

Park et al. proposed a biodegradable polymer–metal nanoplatforms for combined drug delivery, photothermal therapy, and MRI imaging.[29] In this system, biocompatible polymeric nanomaterials were loaded with hydrophobic anticancer drugs followed by deposition of magnetic and gold metal layers. (Fig. 6). Using this method, they have fabricated poly(lactide-co-glycolic acid) (PLGA)-magnetic (Mn)/gold (Au) half-shell nanoparticles (PLGA-Mn/Au H-S NPs), which contained rhodamine as a model drug. The absorption peak of PLGA-Mn/Au H-S NPs was located in the NIR region. The increase in temperature of PLGA-Mn/Au H-S NP solution was dependent on NIR irradiation time, power density, and nanoparticle concentration. Interestingly, under irradiation, the release of rhodamine from the PLGA nanoparticles was twice as high as the release when NIR irradiation was not used. Due to the presence of the magnetic layer, PLGA-Mn/Au H-S NPs can also be used as MRI contrast agents. These results suggest that PLGA-Mn/Au H-S NPs can be used for photothermally controlled drug delivery and MR imaging.

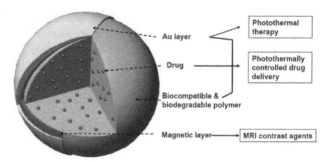

Fig. 6 Schematic diagram of drug-loaded polymer–metal multilayer H-S NPs. The drug is loaded into biocompatible and biodegradable polymer nanoparticles, and magnetic and Au layers are deposited on the polymer nanoparticles.
Source: From Park et al.[29] © 2008, with permission from Wiley-VCH Verlag GmbH & Co. KGaA.

Franchini et al. used bovine serum albumin (BSA)- and poly(lactic-co-glycolic acid) (PLGA)-coated magnetic nanocarriers containing a cobalt-iron oxide (PLGA-BSA-CoFe$_2$O$_4$) core[30] for photothermal therapy. In the presence of a high-frequency magnetic field (168 kHz, 21 kA.m), these particles can heat the surrounding cells up to 48°C. At these conditions, an impressive 82% cell death is induced within 1 hr of treatment. It is proposed that introducing structural anisotropy (such as via cobalt) into the iron oxide MN increases their hyperthermia efficiency and possibly also their magnetic contrast property. *In vivo* studies on the brain and liver of rats indicated that these PLGA-BSA-CoFe$_2$O$_4$ particles show higher image contrast compared to a commercial agent, Endorem.

Recently, Liu et al. developed a theranostic system by combining ultrasound contrast imaging and photothermal therapy.[31] They have developed a multifunctional theranostic agent based on gold-nanoshelled microcapsules (GNS-MCs) by electrostatic adsorption of gold nanoparticles as seeds onto the polymeric microcapsule surfaces. The polymeric microcapsules were generated from PLA and polyvinyl alcohol materials by employing the water-in-oil-in-water (W/O/W) double-emulsion method. The crucial features required for nanoparticles to act as ultrasound contrast agents (UCAs) are that they should exhibit excellent echogenic properties when exposed to ultrasound. The basic contrast-enhancing principle of UCAs is that the backscattering echo intensity is proportional to the change in acoustic impedance between the tissues and the injected agents. Thus, the acoustic enhancement of obtained GNS-MCs was evaluated both *in vitro* and *in vivo*. The experiment was performed by injecting a suspension of GNS-MCs was injected into a latex tube containing a circulating saline. UCAs with and without agent injection in both pulse-inversed harmonic imaging (PIHI) mode and conventional B mode are shown in Fig. 7A, B. In the presence of GNS-MCs, obvious grayscale imaging enhancement in the PIHI contrast mode was seen within

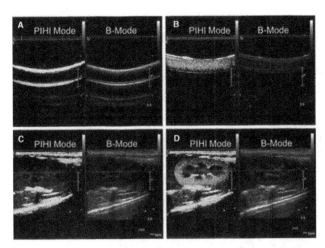

Fig. 7 *In vitro* ultrasound contrast-enhanced images in a latex tube: (**A**) without and (**B**) with GNS-MCs; *in vivo* ultrasonograms in the rabbit right kidney, (**C**) pre- and (**D**) post-administration of GNS-MCs. Both PIHI (MI = 0.42) and conventional B-mode images are shown.
Source: From Ke et al.[31] © 2011, with permission from Wiley-VCH Verlag GmbH & Co. KGaA.

the lumen of the tube, thus suggesting the good contrast-enhancing ability of the GNS-MCs. *In vivo* study using New Zealand white rabbits shows excellent enhancements of rabbit kidney images in a few seconds after bolus intravenous injection of the agent (Fig. 7C, D). These results suggest that GNS-MCs were able to traverse pulmonary capillaries to achieve systemic enhancement and is consistent with the results of *in vitro* ultrasonography. The photothermal cytotoxicity of GNS-MCs with and without laser irradiation on HeLa cells is evaluated using a MTT assy. The HeLa cells treated with GNS-MCs without laser illumination remained more than 80% viable at concentrations ranging from 0.001 to 0.5 mg/mL, thus suggesting that GNS-MCs had little impact on cell survival even at higher concentrations. However, when simultaneously treated with GNS-MCs and laser irradiation, the cell viabilities significantly decreased as the concentration increased. At lower concentrations (0.001–0.025 mg/mL), no obvious photohyperthermia effect was observed and over 80% of cancer cells remained alive. In contrast, at higher concentrations (0.05–0.5 mg/mL), the cell viabilities decreased rapidly with laser irradiation.

Simultaneous Imaging and Gene Therapy

Multifunctional nanoparticles have the potential to deliver DNA or siRNA molecules into cells for therapeutic purposes. In particular, polymer and lipid systems have been used as nonviral gene delivery carriers. Several strategies have been reported to track the gene delivery using nanoparticles of imaging modality like iron oxide nanoparticles combined with therapeutic genes. Small interfering RNA (siRNA) has emerged as an attractive approach for

gene therapy using the well-known cross-linked iron oxide system as a platform. Moore and co-workers reported the development of a dual purpose nanoplatform for simultaneous noninvasive imaging and siRNA delivery to tumors (Fig. 8).[32] They synthesized dual purpose nanoplatforms using iron oxide nanoparticles with CLIO as the core, with NIR fluorescent Cy5.5 dyes and siRNA molecules decorated on the surface of the dextran coating. Furthermore, these nanoparticles were modified with myristoylated polyarginine peptide, a membrane translocation peptide, for intracellular delivery. Upon the intravenous injection of the nanoparticles into nude mice bearing subcutaneous human colorectal carcinoma tumors, the tumor was able to be imaged via T_2 MR and NIR fluorescent imaging (Fig. 8A, B). The nanoparticles might have been accumulated at the tumor site via passive targeting through the permeable tumor vasculature and the nanoparticles were also observed in other organs such as the liver and spleen. Gene silencing was observed only when the multimodal nanoparticles were used (Fig. 8C, D, E). The combination of the favorable biodistribution of these nanoparticles to tumors and their imaging properties represents an exciting possibility for the simultaneous delivery and detection of siRNA-based therapeutic agents *in vivo*.

Bartlett et al. used PET to quantify the *in vivo* biodistribution and siRNA delivery of multifunctional nanoparticles formed with cyclodextrin-containing polycations and siRNA.[33] They conjugated 1,4,7,10-tetraazacyclododec-ane-1,4,7,10-tetraacetic acid (DOTA) to the 5′ end of the siRNA molecules. DOTA allowed the labeling of the nanoparticles with ^{64}Cu for PET imaging. The multifunctional nanoparticles used for PET and siRNA delivery were synthesized through the electrostatic interaction between cyclodextrin-containing polycations and ^{64}Cu-DOTA-siRNA conjugates. The *in vivo* PET imaging of mice bearing luciferase-expressing Neuro2A subcutaneous tumors indicated that both the nontargeted and transferrin-targeted siRNA nanoparticles exhibit a similar biodistribution and tumor localization via the EPR effect. However, the transferrin-targeted siRNA nanoparticles reduce the tumor luciferase activity by almost 50% relative to the nontargeted siRNA nanoparticles one day after injection, indicating that the transferrin-targeted nanoparticles delivered siRNA into the tumor cells owing to their enhanced internalization. Very recently, Tan and co-workers developed a sensitive and selective approach for combined mRNA detection and gene therapy using molecular beacon micelle flares (MBMFs, Scheme 2).[34] MBMFs are easily prepared by self-assembly of diacyllipid-molecular-beacon conjugates (L-MBs), without any biohazardous materials. Just like pyrotechnic flares that produce brilliant light when activated, MBMFs undergo a significant burst of fluorescence enhancement upon target binding. This hybridization event subsequently induces gene silencing, leading to apoptosis of cancer cells. The advantages of MBMFs include easy probe synthesis,

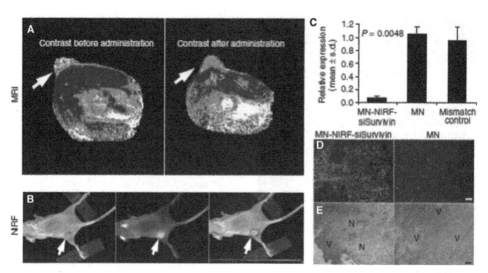

Fig. 8 Application of MN-NIRF-siSurvivin in a therapeutic tumor model. (**A**) *In vivo* MRI of mice bearing subcutaneous LS 174T human colorectal adenocarcinoma (arrows). There was a significant drop in T_2 relaxivity in images acquired after administration of the contrast agent ($P = 0.003$) indicating probe delivery. (**B**) A high-intensity NIRF signal on *in vivo* optical images associated with the tumor following injection of MN-NIRF siSurvivin confirmed the delivery of the probe to this tissue(light-pale spot; left, white light; middle, NIRF; right, dark-spot with white contrast). (**C**) Quantitative RT-PCR analysis of surviving expression in LS 174T tumors after injection with either MN-NIRF-siSurvivin, a mismatch control, or the parental magnetic nanoparticle (MN). Data are representative of three separate experiments. (**D**) Note distinct areas with a high density of apoptotic nuclei in tumors treated with MN-NIRF siSurvivin (left). (**E**) H&E staining of frozen tumor sections revealed considerable eosinophilic areas of tumor necrosis (N) in tumors treated with MN-NIRF-siSurvivin (left). Tumors treated with magnetic nanoparticles were devoid of necrotic tissue (right). Hematoxiphilic regions (V) indicate viable tumor tissues. Scale bar, 50 mM.
Source: From Mearova et al.,[32] with permission from Nature Publishing Group, 2007.

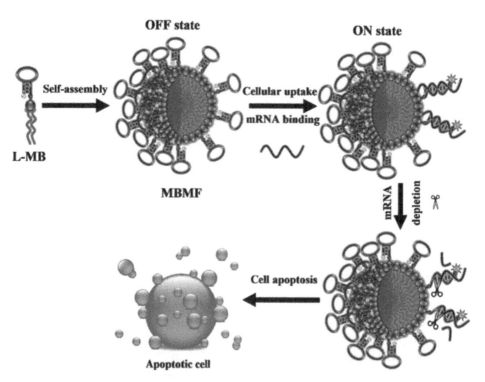

Scheme 2 Schematic illustration of MBMFs for intracellular mRNA detection and gene therapy, Diacyllipid-molecular-beacon conjugates (L-MBs) self-assemble into MBMFs and enter living cells.
Source: From Chen et al.[34] © 2013, with permission from Wiley-VCH Verlag GmbH & Co. KGaA.

efficient cellular uptake, enhanced enzymatic stability, high signal-to-background ratio, excellent target stability, and superior biocompatibility. Instead of incorporating potentially biohazardous materials for efficient nucleic acid probe delivery, simple modification of MBs with a diacyllipid group provide the resulting MBMFs with new properties that MBs do not have, such as self-delivery and enhanced intracellular stability. In addition to their use in the context of mRNA imaging and gene therapy, MBMFs possess a hydrophobic cavity that could be filled with additional hydrophobic materials, such as magnetic contrast agents or anticancer drugs.

CONCLUSIONS

The field of theranostics remains relatively young but is developing at an incredibly fast rate. The design of prodrugs and polymers carrying defined amounts of conjugated diagnostic and therapeutic agents is possible, and several examples have been discussed in this entry. Researchers have developed numerous polymer nanoparticle-based systems that can be used for theranostics applications as in the case of chemothermotherapy, PDT, phototherapy, and gene therapy. More importantly, the requisite concentration for diagnosis and therapy can be controlled during synthesis of these nanoparticles. The ultimate goal in developing these multifunctional nanomaterials is the efficient and specific treatment as well as the diagnosis of diseases. In order to use them for future clinical applications, many issues including the biocompatibility, toxicity, *in vivo* targeting efficacy, and long-term stability of the multifunctional nanoparticles need to be addressed. Especially, toxicological studies of the nanoparticles are necessary before they can be subjected to clinical trials. Despite tremendous advances in cancer therapy, many scientific, technological and clinical challenges remain unanswered, which need a highly interdisciplinary and collaborative approach to overcome. Low molecular weight (<40,000 Da) biodegradable polymers are preferable for theranostic applications, and this is clearly something to keep in mind when designing functional polymers. With advances in cancer biology and explosive developments in materials science and imaging technology, we have a reason to be optimistic that we are at the critical threshold of a major breakthrough in the treatment of cancer. As the capabilities of multifunctional nanoplatforms continue to increase, the integration of cancer biology, diagnostic imaging, and materials science in the future will be essential, not just for theranostic nanomedicine, but for cancer therapy overall.

ACKNOWLEDGMENTS

We kindly thank Prof. D. Narayana Rao, Director (Research), for his valuable support toward manuscript preparation.

REFERENCES

1. Funkhouser, J. Reinventing pharma: The theranostic revolution. Curr. Drug Discov. **2002**, *2*, 17–19.
2. Croy, S.R.; Kwon, G.S. Polymeric micelles for drug delivery. Curr. Pharm. Des. **2006**, *12* (36), 4669–4684.
3. Nishiyama, N.; Kataoka, K. Current state, achievements, and future prospects of polymeric micelles as nanocarriers for drug and gene delivery. Pharmacol. Ther. **2006**, *112* (3), 630–648.
4. Sutton, D.; Nasongkla, N.; Blanco, E.; Gao, J. Functionalized micellar systems for cancer targeted drug delivery. Pharm. Res. **2007**, *24* (6), 1029–1046.
5. Torchilin, V.P. PEG-based micelles as carriers of contrast agents for different imaging modalities. Adv. Drug Deliv. Rev. **2002**, *54* (2), 235–252.
6. Matsumura, Y.; Kataoka, K. Preclinical and clinical studies of anticancer agent-incorporating polymer micelles. Cancer Sci. **2009**, *100* (4), 572–579.
7. Duncan, R. The Dawning era of polymer therapeutics. Nat. Rev. **2003**, *2* (5), 347–360.
8. Cabral, H.; Kataoka, K. Multifunctional nanoassemblies of block copolymers for future cancer therapy. Sci. Technol. Adv. Mater. **2010**, *11* (1), 14109–14117.
9. Oerlemans, C.; Bult, W.; Bos, M.; Storm, G.; Niijsen, J.F.W.; Hennink, W.E. Polymeric micelles in anticancer therapy: Targeting, imaging and triggered release. Pharm. Res. **2010**, *27* (12), 2569–2589.
10. Fernandez, A.F.; Manchanda, R.; McGoron, A.J. Theranostic applications of nanomaterials in cancer: Drug delivery, image-guided therapy, and multifunctional platforms. App. Biochem. Biotechnol. **2011**, *165* (7–8), 1628–1651.
11. Janib, S.M.; Moses, A.S.; MacKay, J.A. Imaging and drug delivery using theranostic nanoparticles. Adv. Drug Deliv. Rev. **2010**, *62* (11), 1052–1063.
12. Kelkar, S.S.; Reineke, T.M. Theranostics: Combining imaging and therapy. Bioconjug. Chem. **2011**, *22* (10), 1879–1903.
13. Cabral, H.; Nishiyama, N.; Kataoka, K. Supramolecular nanodevies: From design validation to theranostic nanomedicine. Acc. Chem. Res. **2011**, *44* (10), 999–1008.
14. Nasongkla, N.; Bey, E.; Ren, J.; Ai, H.; Khemtong, C.; Guthi, J.S.; Chin, S.F.; Sherry, A.D.; Boothman, D.A.; Gao, J. Multifunctional polymeric Micelles as cancer-targeted, MRI-ultrasensitive drug delivery systems. Nano Lett. **2006**, *6* (11), 2427–2430.
15. Guthi, J.S.; Yang, S.G.; Huang, G.; Li, S.; Khemtong, C.; Kessinger, C.W.; Peyton, M.; Minna, J.D.; Brown, K.C.; Gao, J. MRI-visible micellar nanomedicine for targeted drug delivery to lung cancer cells. Mol. Pharm. **2009**, *7* (1), 32–40.
16. Yang, J.; Lee, C.H.; Ko, H.J.; Suh, J.S.; Yoon, H.G.; Lee, K.; Huh, Y.M.; Haam, S. Multifunctional magneto-polymeric nanohybrids for targeted detection and synergistic therapeutic effects on breast cancer. Angew. Chem. Int. Ed. **2007**, *46* (46), 8836–8839.
17. Yu, M.K.; Yeong, Y.Y.; Park, J.; Park, S.; Kim, J.W.; Min, J.J.; Kim, K.; Jon, S. Drug-loaded superparamagentic iron oxide nanoparticles for combined cancer imaging and therapy *in vivo*. Angew. Chem. Int. Ed. **2008**, *47* (29), 5362–5365.

18. Kim, J.; Lee, J.E.; Lee, S.H.; Yu, J.H.; Lee, J.H.; Park, T.G.; Hyeon, T. Designed fabrication of a multifunctional polymer nanomedical platform for simultaneous cancer-targeted imaging and magnetically guided drug delivery. Adv. Mater. **2008**, *20* (3), 478–483.

19. Kim, J.; Piao, Y.; Hyeon, T. Multifunctional nanostructured materials for multimodal imaging, and simultaneous imaging and therapy. Chem. Soc. Rev. **2009**, *38* (2), 372–390.

20. Park, J.H.; von Malizahn, G.; Ruoslahti, E.; Bhatia, S.N.; Sailor, M.J. Micellar hybrid nanoparticles for simultaneous magnetofluorescent imaging and drug delivery. Angew. Chem. Int. Ed. **2008**, *47* (38), 7284–7288.

21. Santra, S.; Kaittainis, C.; Grimm, J.; Perez, J.M. Drug/dye-loaded multifunctional iron oxide nanoparticles for combined targeted cancer therapy and dual optical/magnetic resonance imaging. Small **2009**, *5* (16), 1862–1868.

22. Huang, C.K.; Lo, C.L.; Chen, H.H.; Hsiue, G.H. Multifunctional micelles for cancer cell targeting, distribution, imaging, and anticancer drug delivery. Adv. Funct. Mater. **2007**, *17* (14), 2291–2297.

23. Kaida, S.; Cabral, H.; Kumagai, M.; Kishimura, A.; Terada, Y.; Sekino, M.; Aoki, I.; Nishiyama, N.; Tani, T.; Kataoka, K. Visible drug delivery by supramolecular nanocarriers directing to single-platformed diagnosis and therapy of pancreatic tumor model. Cancer Res. **2010**, *70* (18), 7031–7041.

24. Pan, D.; Caruthers, S.D.; Hu, G.; Senpan, A.; Scottd, M.J.; Gaffney, P.J.; Wickline, S.A.; Lanza, G.M. Ligand-directed nanobialys as theranostic agent for drug delivery and manganese-based magnetic resonance imaging of vascular targets. J. Am. Chem. Soc. **2008**, *130* (29), 9186–9187.

25. Kopelman, R.; Koo, Y-E.L.; Philbert, M.; Moffat, B.A.; Reddy, R.G.; McConville, P.; Hall, D.E.; Chenevert, T.L.; Bhojani, M.S.; Buck, S.M.; Rehemtulla, A.; Ross, B.D. Multifunctional nanoparticle platforms for *in vivo* MRI enhancement and photodynamic therapy of a rat brain cancer. J. Magn. Magn. Mater. **2005**, *293* (1), 404–410.

26. Reddy, G.R.; Bhojani, M.S.; McConville, P.; Moody, J.; Moffat, B.A.; Hall, D.E.; Kim G.; Koo, Y-E. L.; Woolliscroft, M.J.; Sugai, J.V.; Johnson, T.D.; Philbert, M.A.; Kopelman, R.; Rehemtulla, A.; Ross, B.D. Vascular targeted nanoparticles for imaging and treatment of brain tumors. Clin. Cancer Res. **2006**, *12* (22), 6677–6686.

27. Wang, S.; Kim, G.; Lee, Y-E.K.; Hah, H.J.; Ethirajan, M.; Pandey, R.K.; Kopelman, R. Multifunctional biodegradable polyacrylamide nanocarriers for cancer theranostics-A "see and treat" strategy. ACS Nano **2012**, *6* (8), 6843–6851.

28. McCarthy, J.R.; Jaffer, F.A.; Weissleder, R. A macrophage-targeted theranostic nanoparticle for biomedical applications. Small **2006**, *2* (8–9), 983–987.

29. Park, H.; Yang, J.; Seo, S.; Kim, K.; Suh, J.; Kim, Haam, S.; Yoo, K.H. Multifunctional nanoparticles for photothermally controlled drug delivery and magnetic resonance imaging enhancement. Small **2008**, *4* (2), 192–196.

30. Franchini, C.M.; Baldi, G.; Bonacchi, D.; Gentili, D.; Gludetti, G.; Lascialfari, A.; Corti, M.; Marmorato, P.; Ponti, J.; Micotti, E.; Guerrini, U.; Sironi, L.; Gelosa, P.; Ravagli, C.; Ricci, A. Bovine serum albumin-based magnetic nanocarrier for MRI diagnosis and hyperthermic therapy: A potential theranostic approach against cancer. Small **2010**, *6* (3), 366–370.

31. Ke, H.; Wang, J.; Dai, Z.; Jin, Y.; Qu, E.; Xing, Z.; Guo, C.; Yue, X.; Liu, J. Gold-nanoshelled microcapsules: A theranostic agent for ultrasound contrast imaging and photothermal therapy. Angew. Chem. Int. Ed. **2011**, *50* (13), 3017–3021.

32. Mearova, Z.; Pham, W.; Farrar, C.; Petkova, V.; Moore, A. *In vivo* imaging of siRNA delivery and silencing in tumors. Nat. Med. **2007**, *13* (3), 372–377.

33. Bartlett, D.W.; Su, H.; Hildebrandt, I.J.; Weber, W.A.; Davis, M.E. Impact of tumor-specific targeting on the biodistribution and efficacy of siRNA nanoparticles measured by multimodality *in vivo* imaging. Proc. Natl. Acad. Sci. U.S.A. **2007**, *104* (39), 15549–15554.

34. Chen, T.; Wu, C.S.; Jimenez, E.; Zhu, Z.; Dajac, J.G.; You, M.; Han, D.; Zhang, X.; Tan, W. DNA micelle flares for intracellular mRNA imaging and gene therapy. Angew. Chem. Int. Ed. **2013**, *52* (7), 2012–2016.

Nanomaterials—
Nanoparticles

Nanomaterials: Therapeutic Applications

Dominik Witzigmann
Marine Camblin
Jörg Huwyler
Vimalkumar Balasubramanian
Division of Pharmaceutical Technology, University of Basel, Basel, Switzerland

Abstract

In recent years, uses of engineered nanomaterials have increased in day-to-day life, especially in biomedical applications. In this direction, advances in polymer science have significantly contributed to the development of polymeric nanomaterials for drug delivery applications. Particularly intensive efforts have been made to develop new types of nanomaterials through self-assembly techniques. Polymers that can spontaneously self-assemble in aqueous solution into various nanoscale structures such as micelles, nanoparticles, and vesicles have a huge potential to serve as nanocarriers for various therapeutic applications. By controlling the number of physicochemical parameters that can influence the self-assembly process, it is possible to tailor the desired morphology of the nanostructures and to engineer different properties of the nanostructures. This entry reviews the fundamental and physicochemical properties of self-assembled morphologies like micelles, nanoparticles, vesicles (polymersomes), and layer-by-layer capsules that are driven by template-directed assembly. We cover formulation characteristics such as loading efficiency, stability, and release properties of polymeric nanocarrier systems with recent examples. We emphasize the biological properties of the polymeric nanomaterials and their therapeutic applications from the delivery of small drug molecules to proteins and gene delivery.

INTRODUCTION

Drug molecules have often unfavorable pharmacokinetic properties. Encountered problems may include low bioavailability, poor distribution to target tissues, accumulation in non-target tissues and low metabolic stability. Consequently, their therapeutic efficiency is decreased and off target toxicity is more pronounced in conventional approaches. Those issues raise the pressing need of more sophisticated and smart drug delivery systems (DDSs) to improve the performance of classical therapeutics. Nanomaterial-based formulations offer the possibility to modulate pharmacokinetic properties of delivered drugs. Unique nanoscale properties of nanomaterials such as high surface-to-volume ratio due to their nanoscale size (20–200 nm) and enhanced physicochemical properties make them an ideal candidate for nanocarriers. They can, for example, be loaded with a drug of interest. As a result, increase in the apparent solubility of the administered drug and change in the pharmacokinetic properties has the potential to improve the therapeutic efficiency and efficacy.

The main aim of the present review is to discuss the therapeutic use of nanocarriers as DDSs. A wide range of nanomaterials made of organic, inorganic, lipid, polymer, and peptide-based compounds have been proposed as nanocarriers for such therapeutic applications. As compared to clinically established nanocarriers such as liposomal DDSs, polymeric nanomaterials have recently gained attention due to their unique properties. First, physicochemical properties and chemical versatility nature of polymers can be used to combine polymers with drugs (covalent and noncovalent approach) that can dramatically improve the drug characteristics in terms of solubility, stability, and permeability. Second, polymers can be chemically modified to conjugate with targeting vectors such as antibodies, peptides, and aptamers to guide them to specific tissue or cells in the body. Third, environmental responsiveness such as pH and temperature of polymers can be used to control the release profile of the drug in a specified area of interest. It has to be emphasized that polymers can offer the biological properties to be nontoxic, biodegradable, and biocompatible. These are essential requirements for therapeutic applications. In this entry, we discuss recent developments of polymeric nanomaterials forming self-assembled nanostructures such as micelles, nanoparticles, polymersomes, and template-directed assembly-forming layer-by-layer (LbL) capsules for various therapeutic applications (Fig. 1).

POLYMERIC MICELLES

General Characteristics

Polymeric micelles are self-assembled nanostructures consisting of a compact hydrophobic core surrounded by a hydrophilic corona. Micelles are dynamic structures

Concise Encyclopedia of Biomedical Polymers and Polymeric Biomaterials DOI: 10.1081/E-EBPPC-120050055

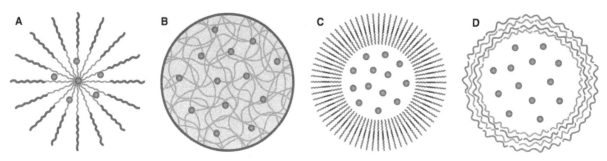

Fig. 1 Schematic illustration of different types of polymeric nanomaterial based nanocarriers (**A**) polymeric micelle, (**B**) polymeric NP, (**C**) polymersome, and (**D**) LbL capsule. Gray dots represent the transported drug.

composed of amphiphilic block copolymers possessing both hydrophilic and hydrophobic blocks. Amphiphilic copolymers begin to self-assemble in solution into micelles at a specific and narrow concentration range, which is known as critical micelle concentration (CMC). One of the main reasons why amphiphilic copolymers self-assemble themselves in aqueous solution is their tendency to isolate the hydrophobic blocks from the aqueous environment and to reach a state of minimum energy. This process is driven by an increase in entropy of water molecules interacting with hydrophobic molecules.[1] Above the CMC, micelle structures are thermodynamically stable and below the CMC, micelles are unstable. The relative size of the hydrophobic block of the copolymer is the most important factor that affects the process of micelle formation.

Amphiphilic block copolymers can form a variety of self-assembled structures in solutions where the solvent preferentially solvates one of the blocks. The most common structures formed by these amphiphilic copolymers are spherical micelles, cylindrical micelles, and vesicles. Depending on the length of the hydrophobic block, spherical micelles can be classified as star-like or crew-cut micelles. When the length of the hydrophilic block is longer than the hydrophobic block, it forms the star-like micelles. Crew-cut micelles are formed when the hydrophilic block is shorter than the hydrophobic block. As the hydrophilic block length decreases, the morphology of the micelles change from spherical to cylindrical (also called worm-like or rod micelles). Recent approaches showed that by controlling the self-assembly of the amphiphilic block copolymers different types of micellar morphologies can be obtained. These types of micellar morphologies include disk-like, toroidal, bi-continuous, cross-linked, and Janus micelles.[2,3] In most cases, polymeric micelles are prepared by two different methods. In the first method, amphiphilic copolymers are directly dissolved in water. During spontaneous self-assembly, drugs can be loaded simply by adding them to the polymers. In the second method, copolymers are dissolved in organic solvent. The solvent is removed by evaporation resulting in a thin film of copolymers. Drugs can then be loaded during the rehydration and self-assembly process of the film exposed to an aqueous solution. These techniques are referred to as direct dissolution and film rehydration method, respectively.[4]

Formulation Characteristics

Classically, polymeric micelles have been utilized to improve the solubilization and loading of hydrophobic drug molecules. When maximum solubilization capacity of the micelles is achieved, drugs cannot be loaded further into the micelles. Thus, solubilization and drug loading efficiency are directly linked parameters.[1] To improve drug solubilization, hydrotropic polymers can be used instead of conventional amphiphilic copolymers.[5,6] Furthermore, solubilization and loading efficiency can also be dependent on the hydrophobicity level of the drugs. For instance, conventional poly(ethylene glycol)–polylactide acid (PEG–PLA) micelles improved the solubilization of paclitaxel (PTX) by a factor of 50, compared to the less hydrophobic nifedipine (factor of 20), due to a higher hydrophobicity level. Similarly, hydrotropic polymer based poly(ethylene glycol)–poly(2-amino-2-methylbutyl)acrylamide (PEG-PDMBA) micelles showed the dramatic solubility enhancement for PTX (6000-fold increased solubility) compared with nifedipine (60-fold increased solubility). Interestingly, poly(ethylene glycol)–poly [2-(4-vinylbenzyloxy)-*N,N*-diethylnicotinamide] (PEG-PDENA) micelles showed high specificity toward the solubilizing properties for PTX (about 9000-fold increased loading) than for the other drugs.[6]

Compatibility between drug molecules and the micelle core is the major determinant of drug solubility and loading efficiency.[5] In one study, the influence of copolymer was investigated on the solubilization of PTX. In this study, PEG-PDENA (hydrotropic polymer) micelles showed the highest loading capacity of PTX (37.4%) compared to the conventional poly(ethylene glycol)–polyphosphoramidate (PEG-PPA) and PEG-PLA micelles that showed 14.7% and 27.6% of PTX loading due to the compatibility between PTX and PDENA.[5] In addition to the single drug loading, micelles offer the possibility to load multiple drugs into single micelles at clinically relevant doses. PTX, etoposide (ETO), docetaxel (DTXL), and 17-allylamino-17-demethyoxygeldanamycin (17-AAG) were solubilized in PEG-PLA micelles within the combinations PTX/17-AAG, ETO/17-AAG, DTXL/17-AAG, and PTX/ETO/17-AAG. Two-drug and three-drug combinations in micelles retained the 94% and 97% of loaded drugs, respectively, compared to the single drug micelles that retained 16% to 32% of the loaded drugs. The presence of 17-AAG in those micelles helped to stabilize the formulation and to offer greater stability of the drugs at the same level of solubilization as single drug micelles.[7]

Stability of micelles is crucial for drug delivery applications. To improve the stability of drug loaded micelles, several strategies have been explored such as enhancing the compatibility between the drug and the block copolymer, cross-linking of the micelle core/corona, and lowering the CMC by altering the polymer.[4] Classical PEG-PLA micelles loaded with PTX showed stability over a period of 4–5 days. To overcome this limitation, hydrotropic polymers have been linked to PEG to develop PEG-PDENA micelles that increase the micellar stability for a period of 25 days due to the compatibility of PDENA and PTX.[5] Polymeric micelles cross-linked with ionic cores by using block ionomer complexes of poly(ethylene oxide)–poly(methacrylic acid) (PEO-PMA) and divalent metal cations showed colloidal stability for a prolonged period of time.[8] Moreover, cross-linking of micelles increased the stability without affecting the drug loading capacity.[4]

Therapeutic Applications

Polymeric micelles are one of the widely used nano-carrier systems due to their unique properties, usually ranging from 20 to 200 nm in size for drug delivery application.[1] High colloidal stability of polymeric micelles is mainly due to the low CMC of the polymers that prevents the dissociation of micelles and offers high stability in biological systems, where it is diluted after intravenous injection. Polymeric micelles have better rigidity and stability than phospholipid micelles due to the covalent link between the polymeric blocks.[4] The hydrophobicity of the micellar core permits the incorporation and stabilization of wide range of small drug molecules, such as doxorubicin (DOX), PTX, amphothericin B, cisplatin, cyclosporine, and hydroxyl camptothecin.[9] Hydrophilic corona of the micelle offers the stealth properties to escape the reticuloendothelial system (RES). These properties lead to a significant increase in the blood residence time of micelles, allow them to permeate the blood vessels, and to accumulate efficiently at the tumor site.[10] The nano-size of polymeric micelles may also facilitate the deep penetration of carrier to tumor tissue and ease of their uptake by tumor cells.[9]

Anticancer drugs such as DOX and PTX are commonly used cancer therapeutics. However, low water solubility, acute toxicity to normal cells, and multidrug resistance are the major issues that can be surmounted by using polymeric micelles. For instance, poloxamer micelles [poly(ethylene oxide–poly(propylene oxide)] loaded with DOX entered clinical trials in Canada in 1999 and completed phase II in esophageal adenocarcinoma.[11] Pharmacokinetic and biodistribution of these micelles in healthy mice showed

a 2.1-fold increase of area under the curve (AUC), 2.1-fold decrease in clearance, and 1.5-fold decrease in volume of distribution compared to the free DOX. *In vivo* (in mice) and at equal dose, poloxamer micelles showed to be more effective than free DOX in solid tumor models. In another approach, use of polymeric micelles made of PEO–poly(D,L-lactic acid) [PDLLA] hamper the unwanted toxic side effects associated with solubilizing agent, Cremophor EL®, which is present in the commercial PTX formulations (Taxol®). Intravenous administration of PTX micelles in nude mice caused 91% decrease in the tumor volume that showed high therapeutic efficiency. Additionally, polymeric micelles offer the possibility of introducing high dose of PTX without any potential side effects.[12] In another study, PEO-poly(4-phenyl-L-butanoate)-L-aspartamide) (PPBA) micelles loaded with PTX (NK-105) showed 87-fold increase in AUC, 86-fold decrease in clearance, and 15-fold decrease in volume of distribution compared with marketed product Taxol® after intravenous injection. Micelles had a long circulation time in blood and were able to evade serum protein binding. Enhanced accumulation in the tumor (25-fold) and stronger antitumor activity in C-26 tumor bearing mice model at a single administration of NK-105 leads to the efficient tumor regression.[13]

In addition to the delivery of small hydrophobic molecules, micelles can serve to deliver large therapeutic proteins. Different PEG-PLA diblock copolymer-based micelles have been investigated for the delivery of recombinant human erythropoietin (rhEPO) in Sprague-Dawley rats.[14] This study showed a twofold increase of AUC with the micelles (from 23 µg/L·hr to 33 µg/L·hr) compared to the native rhEPO (16 µg/L·hr). Plasma concentration of rhEPO was twofold higher with the micelles (i.e., 39.88 ng/mL) than with the native rhEPO. Micelles increased the half-life of rhEPO in blood circulation from 2.0 hr (native rhEPO) to 4.4 hr due to the stealth property of the PEG.[14] Micellar formulation of rhEPO enhanced the pharmacological effect of the drug (i.e., increased level of hemoglobin by 48.72%) demonstrating that micellar delivery of the protein did not affect protein functionality. In an alternative approach, human serum albumin (HSA) was loaded into poly(ethylene glycol)–poly(amino ester) (PEG-PAE) micelles and the delivery was enhanced by a pH-dependent stimulus at reduced pH in ischemic tissue. Accumulation of labeled albumin in ischemic brain areas of rats after IV injection indicated that this type of formulation can be used for targeted protein delivery in experimental models of cerebral ischemia (Fig. 2).[15]

Polycationic-based polymeric micelles have an ability to form complexes with negatively charged plasmid DNA (pDNA). Thus, DNA can be packed into micelles for gene delivery. DNA is thereby protected from enzymatic and hydrolytic degradation. Intravenous injection of pDNA/PEG–poly-L-lysine (PLL) micelles showed circulation of intact pDNA in the blood circulation for 3 hr when compared to naked pDNA that is eliminated from the

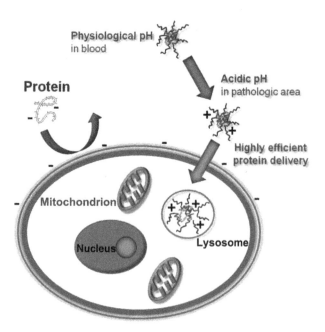

Fig. 2 Schematic representation of pH-tunable micelles loaded with a therapeutic protein to implement a drug targeting strategy. At appropriate pH for noninvasive administration (pH 7 or over), the protein-loaded micelles are stable. If the pH turns acidic (pH < 7), the ionized amino groups of the PAE block forming the micelles are positively charged. This promotes uptake into a diseased tissue, which is characterized by a low pH (e.g., ischemic tissue). At low pH, block copolymers from micelles dissolve and release the transported protein.
Source: From Gao et al.[15] © 2012, with permission from Elsevier.

circulation within 5 min.[16] Micelles are also capable of co-delivering DNA and other drugs. Micelles composed of copolymer poly(N-methyldietheneamine sebacate)-[(cholesteryl oxocarbonylamido ethyl) methyl bis(ethylene) ammonium bromide] sebacate (PMDS-CES) were shown to transport PTX within the hydrophobic core and pDNA within the corona. Combined delivery of PTX and pDNA suppressed the tumor growth in a 4T1 mouse breast cancer model three times more efficiently than the single delivery of drug and pDNA in micelles.[17] Furthermore, a combination of small interfering ribonucleic acid (siRNA) and PTX demonstrated synergistic effects. MDA-MB-231 human breast carcinoma cells were incubated with micelles containing drug and siRNA. After 4 hr, cell viability decreased from 78% to 58%. As the siRNA cytotoxicity was only about 8%, there was a clear synergistic effect associated with the co-delivery of PTX and siRNA.[17]

POLYMERIC NANOPARTICLES

General Characteristics

Polymeric nanoparticles (NPs) are spherical, colloidal particles consisting of macromolecular materials, in which the

active pharmaceutical ingredient (API) is dissolved, entrapped, encapsulated, and/or adsorbed or attached.[18] Although NP size ranges from 10 to 1000 nm, the optimal requirement for drug delivery applications is <200 nm. Three approaches are used to prepare polymeric NPs: (i) dispersion of preformed polymers; (ii) polymerization of monomers for synthetic polymers; and (iii) ionic gelation or coazervation for hydrophilic polymers. Depending on the method of preparation, nanospheres or nanocapsules are designed to tune the release profile (Fig. 3). In general, the process of drug release from NPs is mediated by (i) drug solubility; (ii) diffusion through the NP matrix/polymer wall; (iii) NP matrix erosion; (iv) desorption of adsorbed drug; or (v) a combination of all these factors.[20]

Nanospheres are matrix systems, in which the drug is physically and uniformly dispersed and released by diffusion and erosion.[21] If the erosion of the matrix is faster than the diffusion process, the release mainly depends on the degradation kinetics of the material and results typically in a first order kinetics. Biodegradable polymers have the advantage that they degrade in vivo either enzymatically or nonenzymatically to nontoxic, biocompatible products. One of the disadvantages of nanospheres is the direct contact of the API to the environment at the surface, which can lead to degradation of the compound and burst release. From a drug delivery point of view, nanocapsules can overcome these negative characteristics. Nanocapsules are vesicular systems, in which the drug is encapsulated in a cavity surrounded by a polymer membrane.[22] Depending on the inner core liquid, aqueous or lipophilic, variations of drug solubility are possible. As compared to nanospheres, nanocapsules have the advantage of (i) increased drug protection; (ii) advanced controlled release; and (iii) higher drug loading capacity. Nanocapsules release drugs by a zero order diffusion driven process, which is influenced by the preparation method for these nanocarriers.[23]

Formulation Characteristics

Two different types of polymers have been employed for the development of NPs, namely natural and synthetic materials. The most commonly used natural polymers are gelatin, alginate, albumin, and chitosan, which are considered to be biodegradable and nontoxic.[24] To prepare albumin-bound nanoparticles (NAB), the API is mixed with HSA in a solvent and pressed through a jet to form NPs in the size range of 50–200 nm. The first polymeric NP product, which is used clinically, is the NAB-formulation Abraxane® (ABI-007) containing the mitotic inhibitor PTX. This technology is approved for the treatment of breast cancer and non-small cell lung cancer (NSCLC). Additional clinical trials are ongoing.[25,26]

Another commonly used natural polymer is chitosan, which offers additional advantages like increased paracellular permeability and excellent mucoadhesive properties.[27,28] pDNA-loaded chitosan NPs demonstrated nasal

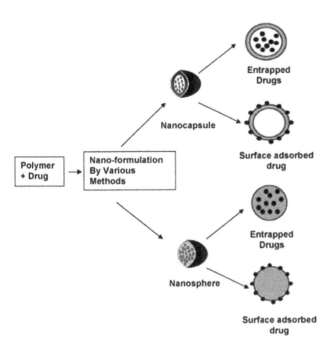

Fig. 3 Schematic representation of methods used to prepare different types of NPs (e.g., nanocapsules and nanospheres). Drug is entrapped within particles or absorbed to the surface. Black dots represent the transported drug.
Source: From Kumari et al.[19] © 2010, with permission from Elsevier.

mucosal immunization against hepatitis B at low pH.[29] However, at physiological pH, the permeation effect by interactions with components of the tight junctions is limited due to the decreased charge (pK_a of amine groups is 6.2). Therefore, quaternary chitosan derivatives such as N-trimethyl chitosan (TMC) have been evaluated to overcome this limitation. This derivative has a positive charge independent of the pH and is soluble over a wide pH range, which affects the mucoadhesive and penetration-enhancing properties.[30] Nasal administration of mucoadhesive and rapid antigen-releasing TMC NPs demonstrated the promise of noninvasive vaccination.[31] Furthermore, chitosan has a high loading capacity for nucleic acids with an encapsulation efficiency of 96.2%, owing to ionic interactions.[29]

The use of synthetic polymers in place of natural polymers can offer a wide range of chemical versatility. This includes improved loading efficiency, functionalization properties, and pharmacokinetic profiles. Synthetic polymers like PLA, polyglycolide acid (PGA), polylactide-co-glycolide acid (PLGA), poly(ε-caprolactone) (PCL), polyalkylcyanoacrylate (PACA), polyanhydrides, or polymethacrylates have been prepared from a great pool of synthetic and readily available monomers. Depending on the material, different release characteristics from hours up to weeks could be achieved.[19] The polyester PLGA/PLA and PCL are the most commonly used synthetic materials since they are biocompatible, biodegradable, and nontoxic. They are approved for clinical use by the Food and Drug Administration (FDA) and European Medicines Agency (EMA).[32,33] In presence of water, the ester linkages of PLGA and PLA are hydrolyzed and the drug is released. The metabolites (monomers), lactide acid and glycolide acid, are not toxic and are removed by endogenous metabolic pathways such as the citric acid or cori cycle.[34] An interesting property of PLGA is its degradation kinetic, which is a function of the copolymer ratio. Higher glycolide content lowers the time required for its degradation.[35] A significant decrease in the release rate and enhanced entrapment efficiency of estradiol in PLGA NPs were observed with increase in lactide content and molecular weight.[36] However, an acidic microenvironment is generated during degradation that may negatively affect the stability of the loaded API.[37] Common PLGA copolymer ratios (lactide/glycolide molar ratio) are 50:50 and 75:25.[38] To overcome the limitation of hydrolytic stability of polylactides, PCL is often preferred for long-lasting DDSs. A drug release of up to 20 days was achieved with vinblastin loaded PCL-NPs.[39]

Therapeutic Applications

During the last decades, polymeric NPs were used to deliver small molecules, larger peptides and proteins, and expression plasmids.[40] For instance, the natural compound curcumin has chemotherapeutic activity in cancer but a poor bioavailability and suboptimal pharmacokinetic characteristics are limiting factors. However, the encapsulation into PLGA-NPs enhanced the therapeutic efficacy in vitro compared to free curcumin.[41] The IC50 of curcumin-NPs was 9.1 µM in MDA-MB-231 cells compared to 16.4 µM for the free curcumin. In another study, coencapsulation of the anticancer drug vincristine and chemosensitizer verapamil into PLGA-NPs showed moderate reversion of multidrug resistance in breast cancer cells in vitro.[42] In an alternative approach, DOX combined with the thermal-optical agent indocyanine green (ICG) allows dual application of chemotherapy and hyperthermia for improved cytotoxicity. The DOX-ICG-PLGA-NPs resulted in enhanced cytotoxicity in Dx5 human uterine cancer cells in vitro.[43]

NPs have to be sterically protected in order to prolong their half-life in the circulation. Using PEG, a hydrophilic protective layer can be created to block and delay the opsonization process.[44] To further increase the therapeutic efficacy, targeted delivery of the NPs to specific cells is performed. A aptamer-funtionalized polymeric NP (BIND-014), which is a PEG-PLGA formulation of DTXL, currently completed a phase I clinical study (Fig. 4).[45] In another study, stealth PEG-PLGA-NPs were used as a delivery vehicle for cisplatin,[46] resulting in an 80 times increased toxicity of the cytotoxic agent. Another aptamer-funtionalized PEG-PLGA-NP containing PTX showed a significant decrease in tumor growth and an increase in survival in a glioma xenograft model compared to the

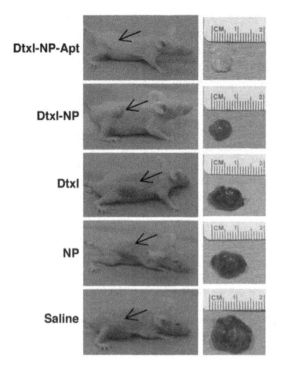

Fig. 4 Comparative prostate tumor regression study in a xenograft mouse model. The left panel shows mice of each treatment group. Black arrows indicate the position of the prostate cancer cell tumor. The right side indicates the tumor size at the endpoint. In the DTXL-NP-Apt group (i.e., mice treated with DTXL loaded PEG-NPs with prostate specific membrane antigen (PSMA) targeting aptamers) the tumor was completely regressed compared with the control treated mice, which received placebo (Saline), empty nanoparticles (NP), DTXL, or nontargeting DTXL-NPs.
Source: From Farokhzad et al.[45] © 2006, with permission from National Academy of Sciences.

unmodified counterparts.[47] Therapeutic proteins and peptides have very limited gastrointestinal bioavailability. They are sensitive to proteolytic enzymes and do not cross passively the biological barriers. NPs can be used to overcome issues associated with peptide and protein delivery. Oral administered insulin-Eudragit-PCL-NPs showed a 52% reduction in blood glucose level in diabetic rats. Their mucoadhesive properties were proposed to enhance gastrointestinal bioavailability.[48] In another study, erythropoietin loaded into oligochitosan NPs was investigated in rats for neuroprotection.[49,50]

Polymeric NPs also have a great potential for the application of nucleic acid gene delivery, because they can overcome the major drawbacks of their viral counterparts like immunogenicity and toxicity. Additionally, they offer the possibility of cheaper large-scale production. The natural polysaccharide chitosan and the noncationic polymer PLGA are the most commonly used nonviral gene vectors for siRNA and pDNA. The inhibition of tumor angiogenesis and growth by PLGA-NP-mediated p53 gene therapy in mice was evaluated recently.[51] In this study, xenografts of p53 mutant tumors were treated with a single intratumoral injection of p53-gene-loaded NPs. Greater levels of apoptosis, antiproliferative activity, decreased angiogenic activity, and finally improved survival indicate that PLGA has potential for sustained gene delivery. The effectiveness of gene silencing with siRNA loaded PLGA-NPs have been evaluated to overcome tumor drug resistance.[52] Silencing of the multidrug resistance protein 1 (MDR1) gene sensitized resistant tumor cells to chemotherapy, as demonstrated by increased PTX accumulation. In a recent study, RGD (Arg-Gly-Asp) peptide-labeled chitosan NPs loaded with siRNA were used for targeted silencing of multiple growth-promoting genes.[53] An inhibition of tumor growth (87% reduction, $P < 0.001$) in an $\alpha v \beta 3$-integrin tumor mouse model was shown with siRNA targeting plexin domain-containing protein 1 (PLXDC-1) that is upregulated in ovarian cancer vasculature. A promising approach for the treatment of NSCLC based on the inhibition of telomerase in cancer cells was carried out with chitosan-coated PLGA-NPs. These NPs with a mean size of 160 nm enhanced the delivery of the antisense oligonucleotide 2'-O-methyl-RNA to human lung cancer cells. Telomerase inhibition (80%) and telomere shortening (from 5.8 kb to 4 kb) in A549 indicated the potential for *in vivo* applications. The cationic NPs showed no cytotoxicity and could be easily modified for active tumor-cell targeting.[54]

POLYMERSOMES

General Characteristics

Polymersomes are artificial vesicles made of synthetic amphiphilic block copolymers and form via a self-assembly process. Generally speaking, polymersomes are hollow spheres, containing an aqueous core that is surrounded by a bilayer membrane. This membrane is composed of a hydrophobic layer sandwiched between internal and external hydrophilic layer forming coronas.[55] Polymersomes were developed as DDS for the transport of either hydrophilic and/or hydrophobic molecules. Compared with liposomes that are lipid analogs, polymersomes have a thick and strong synthetic membrane, conferring a better physicochemical stability as well as lower elasticity and permeability.[2] Polymersomes can be prepared from various types of block copolymers such as nonbiodegradable polymers, degradable polymers, and biocompatible polymers.[56] Depending on the type of polymerization used for synthesis, di-block copolymers (AB), tri-block copolymers (ABA, BAB, or ABC), or even multiblock copolymers can be designed, allowing to tune different properties of the polymersomes.[2] Many factors influence the formation of polymersomes including hydrophilic/hydrophobic block lengths of the copolymer, solvent ratios, concentration of the polymer, and method of preparation. More importantly, volume ratio of hydrophilic to hydrophobic block copolymer fraction influences the formation of polymersomes.

It has been reported that hydrophilic to hydrophobic ratio of less than 1:2 highly favors the formation of polymersomes followed by the ratio of less than 1:3 and ratio more than 1:1 favor the micelle formation.[55] However, the hydrophilic to hydrophobic ratio is not the only determining parameter for polymersome formation.

Although several techniques have been employed to prepare polymersomes, solvent switch and film rehydration techniques are the commonly used methods. In solvent switch amphiphilic block copolymer is dissolved in a suitable organic solvent that is slowly exchanged by an aqueous solution either by adding water to organic polymer solution or vice versa. This technique makes the hydrophobic blocks insoluble, generating copolymer self-assembly into polymersomes as a result of increasing interfacial tension between the hydrophobic blocks and water. In film rehydration, amphiphilic copolymers are dissolved in an organic solvent that is removed by evaporation to form a thin film. An aqueous solution is used to rehydrate the film. Upon mixing, the aqueous phase permeates through defects in the film layers that inflate and finally form vesicles upon separation from the surface.[55,57] Different mechanisms have been reported and proposed for the formation of polymersomes.[58,59] Conventionally, copolymers form a bilayer and close-up to form a vesicular structure. In another mechanism, spherical micelles are formed and then changed into rods, which become flattened and form paddle shape structures. Then these structures transform to circular lamellae that finally close-up to form polymersomes.[60]

Formulation Characteristics

High loading efficiency is a prerequisite for polymersome-based formulation in drug delivery. Loading efficiency can be dependent on preparation method, formation mechanism, and molecular composition of the copolymers. In one study, PEO-poly(benzyl-L-aspartate) (PEO-PBLA) and PEO-poly(2,4,6-trimethoxybenzylidene pentaerythritol carbonate) (PEO-PTMBPEC) based polymersomes were loaded with DOX and showed loading efficiencies of 12% and 8%, respectively.[61] Loading efficiency can be improved using poly(trimethylene carbonate) if the pH of the loading solution is raised to a pH above the pK_a of DOX during the loading process.[62] Using direct dissolution of PEG-polypropylene sulfide (PPS) copolymers, polymersomes achieved higher encapsulation efficiencies for larger protein molecules such as ovalbumin (37%), bovine serum albumin (19%), and bovine gamma-globulin (15%). This method requires a short time for encapsulation, is solvent-free, and simple as compared to methods used for loading of liposomes and other polymersomes.[63] It should be noted that the hollow core of the polymersomes encapsulate the surrounding environment during the self-assembly. Therefore, encapsulation efficiency can also be dependent on the formation mechanism of the polymersomes.

Polymersomes usually exhibit higher stability than polymeric micelles and liposomes due to their thicker membrane. For example, polymersomes made of poly(γ-benzyl L-glutamate)-hyaluronan (PBLG-HYA) loaded with DTXL have shown excellent colloidal stability both at room temperature and at 4°C for one month. More than 90% of DTXL was recovered after storage of DTXL-loaded polymersomes and can be readily lyophilized without alternation in loading content of DTXL for a period of six months storage at 4°C. However, redispersion of lyophilized polymersomes needed a sonication step to accelerate the dispersion time and eliminate the few aggregates still present in solution.[64]

Polymersomes release the loaded drugs through various mechanisms. Furthermore, recently third generation smart systems were designed, where release of drugs from polymersomes can be controlled with external environmental stimuli such as UV light, temperature, and oxidation-reduction.[65] The widely used mechanism for release is degradation of hydrophobic blocks of the polymersomes mediated by simple hydrolysis. This type of degradation is accelerated at acidic pH as found within the endolysosomal compartments of cells.[65] Another possible release mechanism involves the diffusion based permeation of the loaded drug through the polymersome membrane. PEO-PCL polymersomes loaded with DOX showed immediate burst release of 20% of DOX (from 0 hr to 8 hr) by diffusion, followed by a sustained release (up to 14 days) facilitated by pH-driven hydrolysis of the polymersome membrane.[66] The advantages of using stimuli-responsive polymersomes for DDSs have been reviewed extensively elsewhere.[67]

Therapeutic Applications

Polymersomes can readily accommodate a number of drug molecules in the aqueous compartment as well as in their hydrophobic membrane. For instance, to overcome the limitations of DTXL such as low solubility and side effects, this chemotherapeutic drug was incorporated in PBLG-HYA polymersomes. New Zealand rabbits received an IV injection of either polymersome-encapsulated DTXL or free drug. Comparing the plasma profiles, the DTXL-loaded PBLG-HYA polymersomes exhibited a higher maximal concentration (17.97 μg/mL) than the DTXL solution (11.72 μg/mL). The polymersome formulation significantly improved half-life and AUC of DTXL ($t_{1/2}$ = 19.90 hr and AUC = 209.32 μg/mL) compared to the DTXL solution ($t_{1/2}$ = 4.79 hr and AUC = 60.6 μg/mL). Hemolysis tests were performed on human blood and about 30% red blood cell hemolysis (DTXL side-effect) was observed with DTXL solution while less than 5% hemolysis occurred with DTXL-loaded PBLG-HYA polymersomes. Biodistribution studies showed an accumulation of the polymersome formulation at the tumor site in BalB/c mice, and a significantly higher uptake by tumor cells with the polymersomes than with the free drug.[64]

In another series of experiments, biodegradable polymersomes based on PEO-PLA block copolymers were prepared. A hydrophobic drug (PTX) was incorporated into the vesicle membrane. Alternatively, a hydrophilic drug (DOX) was encapsulated within the vesicle core. DOX-PTX loaded into polymersomes showed higher maximum tolerated doses (MDT) of 3 mg/kg for DOX and 7.5 mg/kg for PTX, which is increased by the factor of two compared to free DOX and PTX. *In vitro* studies in MDA-MB231 human breast cancer cells with DOX-PTX polymersomes showed a potent antitumor activity without any other toxic side-effects. Polymersomes loaded with DOX-PTX showed 16-fold increased toxicity in tumor cells as compared to both free drugs, leading to a high tumor apoptosis and ultimately increased therapeutic efficacy (Fig. 5).[68]

Therapeutic applications of proteins are challenging due to poor tissue distribution, rapid elimination by renal clearance, and enzymatic degradation. Polymersomes may be promising candidate for delivering a wide range of proteins without affecting their functionality. Polymersomes formed by film rehydration are able to incorporate transmembrane proteins in the bilayer as well as encapsulate proteins in the aqueous core without loss of their functional conformation.[65,69]

Recently, polymersomes were utilized to encapsulate DNA for delivering into cells. Biomimetic, pH-sensitive polymersomes were prepared using poly[2-(methacryloyloxy) ethyl-phosphorylcholine]-*co*-poly[2-(diisopropylamino) ethyl methcrylate] (PMPC-PDPA) diblock copolymers.[70] Under mild acidic condition, pDNA formed a complex with PMPC-PDPA copolymer by electrostatic interaction and

the pDNA was encapsulated into the polymersomes at pH 7.5.[70] The pDNA-encapsulated polymersomes used the endocytic pathway to enter the cells and the pDNA escaped to the cytosol due to low pH in the endosomal environment. Experiments with DNA-encapsulated polymersomes clearly showed that the pH-dependent release of plasmid and the transfection efficiency was comparable with the commercially available transfection agent Lipofectamine® (Fig. 6).[70] In addition, the polymeric vesicles exhibited less toxicity and no proinflammatory response with less leakage. The biomimetic surface reduces nonspecific interactions with blood plasma proteins and thereby increases the half-life in the circulation of the polymersomes.[70] In another approach, polymersomes forming diblock copolymer poly(oligoethylene glycol methacrylate)–poly(2-(diisopropylamino)ethyl methacrylate) (POEGMA–PDPA) were used to incorporate pDNA into multicompartment capsules via a LbL technique. By this method, very high loading (60% efficiency) was achieved due to cationic amine groups on the PDPA block, which are protonated at low pH.[71] In another experiment, siRNA was encapsulated into PEO-b-PLA and antisense oligonucleotides (AON) were encapsulated into PEG-b-PCL by a co-solvent method. Loaded oligos were released and escaped from the endosomes by hydrolytic degradation of the polymersomes.[72] In gene silencing experiments, siRNA-(PEO-b-PLA) based polymersomes showed a knockdown efficiency of 40% and

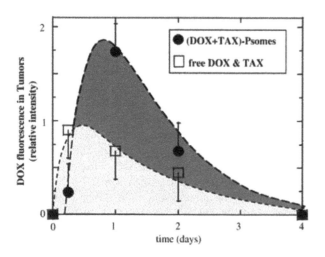

Fig. 5 Tissue accumulation profile of DOX loaded polymersomes in breast cancer tumors. Fluorescence intensity analysis of DOX reveals that accumulation of polymersomes in tumors regions was higher (dark grey) compared to the free drugs (DOX & PTX). Maximal release and tumor tissue accumulation of DOX + PTX loaded polymersomes occur within one day.
Source: From Ahmed et al.[68] © 2006, with permission from Elsevier.

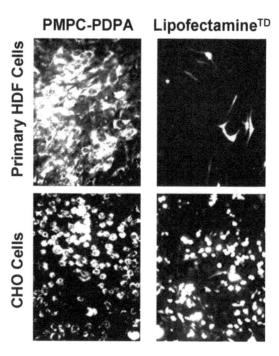

Fig. 6 Delivery of pDNA using pH-dependent polymersomes made of PMPC-PDPA. Polymersomes showed the efficient expression of green fluorescent protein in primary HDF cells and CHO cells and compared with the commercially available LipofectamineTD.
Source: From Lomas et al.[70] © 2007, with permission from WILEY-VCH.

the result was similar to that of Lipofectamine 2000. AON polymeric vesicles were successfully delivered *in vivo* to dystrophin-deficient mice with delivery efficiency above 50% and significantly expressed the dystrophin.[72] It was also reported that PMPC-PDPA polymersomes loaded with enhanced green fluorescent protein (EGFP) siRNA showed significant decrease in the EGFP production in cells. After 48 hr, a 70% EGFP expression silencing was observed without affecting the proliferation of the cells.[73]

LAYER-BY-LAYER CAPSULES

General Characteristics

LbL capsules are polymeric multilayer capsules (PMLC) prepared by alternate adsorption of polymers onto a template via complementary interactions such as electrostatic interactions, hydrogen bonding, and covalent linkage.[74–77] In brief, the colloidal template is immersed into an aqueous solution of polymers to build the first surrounding layer. Subsequently, the excess polymer is removed and a second layer of an interacting polymer is absorbed.[78] These two steps are repeated alternately, resulting in a core–shell particle. By altering the number of layer depositions, the film thickness can be easily regulated. The template is finally dissolved to form a multilayered hollow capsule.[79] Dissolution of organic templates like polystyrene or melamine formaldehyde requires strong acids or organic solvents that can hamper the applicability for therapeutics and particularly biomolecules. Moreover, organic solvents can influence the carrier stability by destroying the capsule shell.[80] To overcome these problems, inorganic templates like silica (0.5–5 μm), calcium carbonate (3–5 μm), manganese carbonate, or metal particles are preferred. Inorganic templates are easily dissolved under mild conditions that do not affect the encapsulated drug.[81,82] The initial template strongly influences size distribution and shape of the PMLC.

The constituents of the capsule shell can tailor the chemical and mechanical properties of the PMLCs. For polyelectrolyte capsules, a large variety of charged compounds like synthetic polymers, polypeptides, polysaccharides, nucleic acids, or NPs are used.[83–85] The negative charge of the polyanions is often achieved by sulfonate or carbonate groups whereas polycations mostly have amino or imino groups.[86] Thus, the selected buffer for the synthesis should maintain a high degree of ionization. Phosphate-buffered saline at pH 7.4 offers the maintenance of the charge as well as the physiological condition for many biotherapeutics. Well-studied polyanion/polycation systems for multilayer assembly are polystyrene sulfonate (PSS)/poly allylamine (PAH), PSS/chitosan, heparin/albumin, heparin/poly(ethylene imine) (PEI), dextran sulfate (DS)/chitosan, DS/poly (L-arginine), DS/PEI, and alginate/PLL. To avoid cytotoxic effects of some polycations, hydrogen bonding

offers an alternative interaction for LbL assembly. PEI and PLL showed mitochondrial-mediated cell death in a range of human cell lines *in vitro*.[87–89] Therefore, poly(*N*-vinylpyrrolidone (PVPON) as hydrogen-bond acceptor and PMA as hydrogen-bond donor have been successfully used. At the end of the production process, it is necessary to cross-link the capsule shell to prevent its destruction at physiological pH and to entrap the drug inside the NP cavity. Carbodiimide chemistry and thiol oxidation are commonly used methods for cross-linking.[90,91]

Formulation Characteristics

Drugs are encapsulated into the carrier mainly by two procedures: (i) postloading or (ii) preloading (Fig. 7). Capsules can be prepared without a compound and the capsule shell can be permeabilized for a postloading process.[93] The capsule wall can be opened by changing pH, ionic strength, solvent polarity, or temperature to allow the diffusion of therapeutics into the capsule interior.[94–97] Returning to the original medium conditions can entrap the drugs inside the capsules. For example, the permeability of biodegradable dextran/chitosan capsules for protein encapsulation can be regulated by variations in pH. At pH values >8, electrostatic repulsion of the polymers leads to an opening of the capsule wall. However, in acidic conditions, the membrane tightens and drugs of interest are entrapped inside the micron-sized hollow capsules.[98] In general, the encapsulation efficiency of the postloading procedure is low. Additionally, the conditions for inducing pores may not be suitable for many biomolecules. An improved variation of this technique is the preloading of the capsule with a sequestering agent, which has a high affinity to the drug. Therefore, the loading process during the diffusion can be increased. Examples include the use of dextran sulfate to increase the encapsulation of DOX.[99]

In contrast to the postloading procedure, porous inorganic templates like mesoporous silica particles offer the possibility of drug preloading due to their large surface area. The polymeric layers adsorb onto the preloaded template.[100,101] Drug crystals as initial template have attracted attention owing to their high encapsulation efficiency.[102–106] For instance, proteins can be directly encapsulated by LbL assembly onto protein crystals.[107] Moreover, other techniques are reported for the drug incorporation. Hydrophobic association is an encapsulation strategy, which was successfully used for natural polyphenols. The anticancer compounds were incorporated into gelatin-based 200 nm NPs surrounded by a 5–20 nm thick LbL shell of polyelectrolytes. The polyphenol loading varied from 20% to 70%.[108] Apart from single drug delivery, polymeric capsules offer the possibility to formulate a DDS, which can release two or more incorporated drugs gradually from a single carrier.[109] Encapsulation of two proteins in separate positions of the capsule showed a time-modulated release. One protein was

Nanomaterials— Nanoparticles

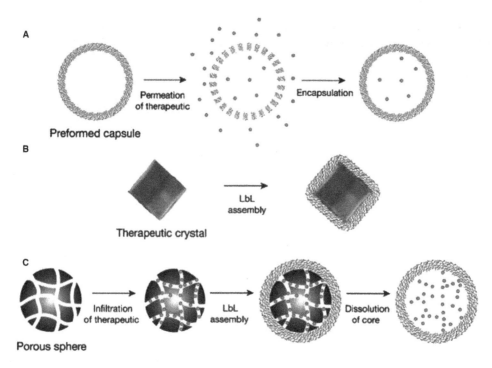

Fig. 7 Different encapsulation methods for loading drugs into LbL capsules: (**A**) postloading procedure of a preformed capsule, (**B**) encapsulation of drug crystals, and (**C**) preloading of porous templates. Gray dots represent loaded drug.
Source: From Johnston et al.[92] © 2006, with permission from Elsevier.

incorporated inside the cavity and the second one between the layers of the shell.[110]

Therapeutic Applications

A versatile spectrum of therapeutics can be incorporated into LbL capsules. Recently, the effect of anticancer drug-loaded capsules has been investigated. Gold NPs coated with a PAH/PSS multilayer were modified with N-(2-hydroxypropyl)methacrylamide (HPMA) and a pro-drug of DOX was linked onto the surface.[111] Specific cleavage of the peptide linker by cathepsin B could release DOX from the stealth core-shell-particle after endocytosis *in vitro*. Another successfully used encapsulation strategy for DOX is emulsion templating.[112] DOX/oleic acid–loaded PMA capsules (500 nm) demonstrated an enhanced toxicity in a human colorectal cancer cell line *in vitro* as compared to the free drug. A further improvement of LbL capsules is the smart system that releases the payload by environmental stimuli.[113] Different formulation tech-niques and polymer combinations can trigger drug release depending on pH or reducing conditions. Redox-responsive capsules demonstrated a novel release method for DOX PMLCs. The disulfide cross-linked shell was stabilized in an oxidizing environment but disrupted under reducing conditions releasing an anticancer drug into the cytoplasm. By this technique, the cytotoxicity of the drug was increased 5000 times compared with the free drug.[114] In a pH-triggered release study LbL-containing quantum dots with

a pH-responsible layer were passively navigated to hypoxic regions of solid tumors *in vivo* and showed a pH-triggered response.[115] PEG was linked to the PLL layer using an iminobiotin-neutravidin interaction, which can be destablized at low pH.[116]

Macromolecules like proteins can also be loaded into PMLCs. Biodegradable multilayered capsules loaded with the cytokine bFGF prolonged the proliferation of L929 fibroblast cells due to sustained release.[98] In general, the LbL technique used for proteins results in micrometer-sized particles. This size is big compared to other nanoparticulate technologies and therefore not suitable for intravenous application.[117] The delivery of nucleic acids like pDNA and siRNA with multilayered capsules is a promising research area. The first successful gene transfection using functional DNA-loaded PMLCs was demonstrated with silica-assisted LbL assembly of dextran and protamine.[118] Recently, single-polyelectrolyte capsules were investigated for the transfection of melanoma cells with pDNA *in vitro*.[119] Cross-linked PLL particles were co-loaded with pDNA of a nuclear transcription factor and alpha-melanocyte-stimulating hormone as a reporter hormone. The gene expression was significantly increased (70%). In another study, RNA interference was studied by incorporating siRNA into PEI layers on gold NPs.[120] The production of EGFP in CHO-K1 cells was knocked down to about 28%. The versatility and modularity of LbL assembly can be used to introduce multiple functionalities. Drug encapsulation, triggered release, active targeting, and imaging could be achieved with one single carrier.[121] Furthermore, polymer capsules containing liposomal subcompartments, so-called capsosomes, can co-encapsulate different drugs and trigger the release by encapsulated enzymatic catalysis.[122,123] Nevertheless, the production of capsules <1 μm without aggregation is quite challenging and would be not acceptable for various applications.[124] Further improvements are necessary to produce capsules less than 200 nm in diameter, which is a common size used in DDSs.

CONCLUSION

Over the last two decades, considerable efforts were paid to the development of self-assembling and template-directed copolymer-based nanostructures. These new and innovative nanomaterials can be used for different types of therapeutic drug delivery applications. Polymeric nanomaterial-based DDSs including micelles, NPs, polymersomes and LbL capsules have been extensively studied. Mechanistic studies of the self-assembly process of different types of copolymers have demonstrated the possibility to modulate the physicochemical properties and morphology of resultant nanostructures. Numerous strategies have been explored to improve drug loading, to control drug release, and to improve the stability of nanocarriers in biological systems. Advanced technologies include the design of polymeric nanomaterials with the possibility to introduce multifunctionalities, responsiveness to environmental stimuli, active targeting, biocompatible, and biodegradable properties. Although numerous studies have been reported using polymeric nanomaterials, the clinical realization of drug targeting strategies remains a challenge.

ACKNOWLEDGMENTS

Financial support of the Swiss Centre of Applied Human Toxicology (SCAHT) and the "Freie Akademische Gesellschaft Basel (FAG)" is acknowledged. We thank Annette Roulier for graphical assistance with the figures.

REFERENCES

1. Letchford, K.; Burt, H. A review of the formation and classification of amphiphilic block copolymer nanoparticulate structures: Micelles, nanospheres, nanocapsules and polymersomes. Eur. J. Pharm. Biopharm. 2007, 65 (3), 259–269.

2. Blanazs, A.; Armes, S.P.; Ryan, A.J. Self-assembled block copolymer aggregates: From micelles to vesicles and their biological applications. Macromol. Rapid Commun. 2009, 30 (4–5), 267–277.

3. Holder, S.J.; Sommerdijk, N.A.J.M. New micellar morphologies from amphiphilic block copolymers: Disks, toroids and bicontinuous micelles. Polym. Chem. 2011, 2 (5), 1018–1028.

4. Kulthe, S.S.; Choudhari, Y.M.; Inamdar, N.N.; Mourya, V. Polymeric micelles: Authoritative aspects for drug delivery. Des. Monomers Polym. 2012, 15 (5), 465–521.

5. Huh, K.M.; Lee, S.C.; Cho, Y.W.; Lee, J.; Jeong, J.H.; Park, K. Hydrotropic polymer micelle system for delivery of paclitaxel. J Control. Release 2005, 101 (1–3), 59–68.

6. Kim, J.Y.; Kim, S.; Pinal, R.; Park, K. Hydrotropic polymer micelles as versatile vehicles for delivery of poorly water-soluble drugs. J. Control. Release 2011, 152 (1), 13–20.

7. Shin, H.-C.; Alani, A.W.; Rao, D.A.; Rockich, N.C.; Kwon, G.S. Multi-drug loaded polymeric micelles for simultaneous delivery of poorly soluble anticancer drugs. J. Control. Release 2009, 140 (3), 294–300.

8. Bronich, T.K.; Keifer, P.A.; Shlyakhtenko, L.S.; Kabanov, A.V. Polymer micelle with cross-linked ionic core. J. Am. Chem. Soc. 2005, 127 (23), 8236–8237.

9. Aliabadi, H.M.; Lavasanifar, A. Polymeric micelles for drug delivery. Expert Opin. Drug Deliv. 2006, 3 (1), 139–162.

10. Kim, S.; Shi, Y.; Kim, J.Y.; Park, K.; Cheng, J.X. Overcoming the barriers in micellar drug delivery: Loading efficiency, in vivo stability, and micelle-cell interaction. Expert Opin collaborators have studied Drug Deliv. 2010, 7 (1), 49–62.

11. Batrakova, E.V.; Kabanov, A.V. Pluronic block copolymers: Evolution of drug delivery concept from inert nanocarriers to biological response modifiers. J. Control. Release 2008, 130 (2), 98–106.

12. Leung, S.Y.L.; Jackson, J.; Miyake, H.; Burt, H.; Gleave, M.E. Polymeric micellar paclitaxel phosphorylates Bcl-2 and induces apoptotic regression of androgen-independent LNCaP prostate tumors. Prostate 2000, 44 (2), 156–163.

13. Hamaguchi, T.; Matsumura, Y.; Suzuki, M.; Shimizu, K.; Goda, R.; Nakamura, I.; Nakatomi, I.; Yokoyama, M.; Kataoka, K.; Kakizoe, T. NK105, a paclitaxel-incorporating micellar nanoparticle formulation, can extend in vivo antitumour activity and reduce the neurotoxicity of paclitaxel. Br. J. Cancer 2005, 92 (7), 1240–1246.

14. Shi, Y.; Huang, W.; Liang, R.; Sun, K.; Zhang, F.; Liu, W.; Li, Y. Improvement of in vivo efficacy of recombinant human erythropoietin by encapsulation in PEG-PLA micelle. Int. J. Nanomed. 2013, 8 (1), 1–11.

15. Gao, G.H.; Park, M.J.; Li, Y.; Im, G.H.; Kim, J.H.; Kim, H.N.; Lee, J.W.; Jeon, P.; Bang, O.Y.; Lee, J.H.; Lee, D.S. The use of pH-sensitive positively charged polymeric micelles for protein delivery. Biomaterials 2012, 33 (35), 9157–9164.

16. Nishiyama, N.; Jang, W.D.; Date, K.; Miyata, K.; Kataoka, K. Photochemical enhancement of transgene expression by polymeric micelles incorporating plasmid DNA and dendrimer-based photosensitizer. J. Drug Target. 2006, 14 (6), 413–424.

17. Wang, L.; Chierico, L.; Little, D.; Patikarnmonthon, N.; Yang, Z.; Azzouz, M.; Madsen, J.; Armes, S.P.; Battaglia, G. Encapsulation of biomacromolecules within polymersomes by electroporation. Angew. Chem. Int. Ed. 2012, 51 (44), 11122–11125.

18. Kreuter, J. Nanoparticles—a historical perspective. Int. J. Pharm. 2007, 331 (1), 1–10.

19. Kumari, A.; Yadav, S.K.; Yadav, S.C. Biodegradable polymeric nanoparticles based drug delivery systems. Colloids Surf. B Biointerfaces **2010**, *75* (1), 1–18.

20. Singh, R.; Lillard, J.W., Jr. Nanoparticle-based targeted drug delivery. Exp. Mol. Pathol. **2009**, *86* (3), 215–223.

21. Soppimath, K.S.; Aminabhavi, T.M.; Kulkarni, A.R.; Rudzinski, W.E. Biodegradable polymeric nanoparticles as drug delivery devices. J. Control. Release **2001**, *70* (1–2), 1–20.

22. Brigger, I.; Dubernet, C.; Couvreur, P. Nanoparticles in cancer therapy and diagnosis. Adv. Drug Deliv. Rev. **2002**, *54* (5), 631–651.

23. Mora-Huertas, C.E.; Fessi, H.; Elaissari, A. Polymer-based nanocapsules for drug delivery. Int. J. Pharm. **2010**, *385* (1–2), 113–142.

24. Kratz, F. Albumin as a drug carrier: Design of prodrugs, drug conjugates and nanoparticles. J. Control. Release **2008**, *132* (3), 171–183.

25. Kratz, F.; Elsadek, B. Clinical impact of serum proteins on drug delivery. J. Control. Release **2012**, *161* (2), 429–445.

26. Uchegbu, I.F.; Siew, A. Nanomedicines and nanodiagnostics come of age. J. Pharm. Sci. **2013**, *102* (2), 305–310.

27. Takeuchi, H.; Yamamoto, H.; Kawashima, Y. Mucoadhesive nanoparticulate systems for peptide drug delivery. Adv. Drug Deliv. Rev. **2001**, *47* (1), 39–54.

28. Thanou, M.; Verhoef, J.C.; Junginger, H.E. Oral drug absorption enhancement by chitosan and its derivatives. Adv. Drug Deliv. Rev. **2001**, *52* (2), 117–126.

29. Khatri, K.; Goyal, A.K.; Gupta, P.N.; Mishra, N.; Vyas, S.P. Plasmid DNA Loaded chitosan nanoparticles for nasal mucosal immunization against hepatitis B. Int. J. Pharm. **2008**, *354* (1–2), 235–241.

30. Amidi, M.; Mastrobattista, E.; Jiskoot, W.; Hennink, W.E. Chitosan-based delivery systems for protein therapeutics and antigens. Adv. Drug Deliv. Rev. **2010**, *62* (1), 59–82.

31. Slutter, B.; Bal, S.; Keijzer, C.; Mallants, R.; Hagenaars, N.; Que, I.; Kaijzel, E.; van Eden, W.; Augustijns, P.; Löwik, C.; Bouwstra, J.; Broere, F.; Jiskoot, W. Nasal vaccination with *N*-trimethyl chitosan and PLGA based nanoparticles: Nanoparticle characteristics determine quality and strength of the antibody response in mice against the encapsulated antigen. Vaccine **2010**, *28* (38), 6282–6291.

32. Chawla, J.S.; Amiji, M.M. Biodegradable poly(ε-caprolactone) nanoparticles for tumor-targeted delivery of tamoxifen. Int. J. Pharm. **2002**, *249* (1–2), 127–138.

33. Irache, J.M.; Esparza, I.; Gamazo, C.; Agüeros, M.; Espuelas, S. Nanomedicine: Novel approaches in human and veterinary therapeutics. Vet. Parasitol. **2011**, *180* (1–2), 47–71.

34. Athanasiou, K.A.; Niederauer, G.G.; Agrawal, C.M. Sterilization, toxicity, biocompatibility and clinical applications of polylactic acid/polyglycolic acid copolymers. Biomaterials **1996**, *17* (2), 93–102.

35. Shive, M.S.; Anderson, J.M. Biodegradation and biocompatibility of PLA and PLGA microspheres. Adv. Drug Deliv. Rev. **1997**, *28* (1), 5–24.

36. Mittal, G.; Sahana, D.K.; Bhardwaj, V.; Ravi Kumar, M.N. Estradiol loaded PLGA nanoparticles for oral administration: Effect of polymer molecular weight and copolymer composition on release behavior *in vitro* and *in vivo*. J. Control. Release **2007**, *119* (1), 77–85.

37. Estey, T.; Kang, J.; Schwendeman, S.P.; Carpenter, J.F. BSA degradation under acidic conditions: A model for protein instability during release from PLGA delivery systems. J. Pharm. Sci. **2006**, *95* (7), 1626–1639.

38. Astete, C.E.; Sabliov, C.M. Synthesis and characterization of PLGA nanoparticles. J. Biomater. Sci. Polym. Ed. **2006**, *17* (3), 247–289.

39. Prabu, P.; Chaudhari, A.A.; Dharmaraj, N.; Khil, M.S.; Park, S.Y.; Kim, H.Y. Preparation, characterization, *in vitro* drug release and cellular uptake of poly(caprolactone) grafted dextran copolymeric nanoparticles loaded with anticancer drug. J. Biomed. Mater. Res. A **2009**, *90A* (4), 1128–1136.

40. Couvreur, P.; Vauthier, C. Nanotechnology: Intelligent design to treat complex disease. Pharm. Res. **2006**, *23* (7), 1417–1450.

41. Yallapu, M.M.; Gupta, B.K.; Jaggi, M.; Chauhan, S.C. Fabrication of curcumin encapsulated PLGA nanoparticles for improved therapeutic effects in metastatic cancer cells. J. Colloid. Interface Sci. **2010**, *351* (1), 19–29.

42. Song, X.R.; Cai, Z.; Zheng, Y.; He, G.; Cui, F.Y.; Gong, D.Q.; Hou, S.X.; Xiong, S.J.; Lei, X.J.; Wei, Y.Q. Reversion of multidrug resistance by co-encapsulation of vincristine and verapamil in PLGA nanoparticles. Eur. J. Pharm. Sci. **2009**, *37* (3–4), 300–305.

43. Tang, Y.; Lei, T.; Manchanda, R.; Nagesetti, A.; Fernandez-Fernandez, A.; Srinivasan, S.; McGoron, A.J. Simultaneous delivery of chemotherapeutic and thermal-optical agents to cancer cells by a polymeric (PLGA) nanocarrier: An *in vitro* study. Pharm. Res. **2010**, *27* (10), 2242–2253.

44. Owens, D.E.; Peppas, N.A. Opsonization, biodistribution, and pharmacokinetics of polymeric nanoparticles. Int. J. Pharm. **2006**, *307* (1), 93–102.

45. Farokhzad, O.C.; Cheng, J.; Teply, B.A.; Sherifi, I.; Jon, S.; Kantoff, P.W.; Richie, J.P.; Langer, R. Targeted nanoparticle-aptamer bioconjugates for cancer chemotherapy *in vivo*. Proc. Natl. Acad. Sci. U.S.A. **2006**, *103* (16), 6315–6320.

46. Dhar, S.; Gu, F.X.; Langer, R.; Farokhzad, O.C.; Lippard, S.J. Targeted delivery of cisplatin to prostate cancer cells by aptamer functionalized Pt(IV) prodrug-PLGA-PEG nanoparticles. Proc. Natl. Acad. Sci. U.S.A. **2008**, *105* (45), 17356–17361.

47. Guo, J.; Gao, X.; Su, L.; Xia, H.; Gu, G.; Pang, Z.; Jiang, X.; Yao, L.; Chen, J.; Chen, H. Aptamer-functionalized PEG-PLGA nanoparticles for enhanced anti-glioma drug delivery. Biomaterials **2011**, *32* (31), 8010–8020.

48. Damge, C.; Socha, M.; Ubrich, N.; Maincent, P. Poly(epsilon-caprolactone)/eudragit nanoparticles for oral delivery of aspart-insulin in the treatment of diabetes. J. Pharm. Sci. **2010**, *99* (2), 879–889.

49. Wang, T.; Hu, Y.; Leach, M.K.; Zhang, L.; Yang, W.; Jiang, L.; Feng, Z.Q.; He, N. Erythropoietin-loaded oligochitosan nanoparticles for treatment of periventricular leukomalacia. Int. J. Pharm. **2012**, *422* (1–2), 462–471.

50. Liu, W.; Shen, Y.; Plane, J.M.; Pleasure, D.E.; Deng, W. Neuroprotective potential of erythropoietin and its derivative carbamylated erythropoietin in periventricular leukomalacia. Exp. Neurol. **2011**, *230* (2), 227–239.

51. Prabha, S.; Sharma, B.; Labhasetwar, V. Inhibition of tumor angiogenesis and growth by nanoparticle-mediated p53 gene therapy in mice. Cancer Gene Ther. **2012**, *19* (8), 530–537.

52. Patil, Y.B.; Swaminathan, S.K.; Sadhukha, T.; Ma, L.; Panyam, J. The use of nanoparticle-mediated targeted gene silencing and drug delivery to overcome tumor drug resistance. Biomaterials **2010**, *31* (2), 358–365.

53. Han, H.D.; Mangala, L.S.; Lee, J.W.; Shahzad, M.M.; Kim, H.S.; Shen, D.; Nam, E.J.; Mora, E.M.; Stone, R.L.; Lu, C.; Lee, S.J.; Roh, J.W.; Nick, A.M.; Lopez-Berestein, G.; Sood, A.K. Targeted gene silencing using RGD-labeled chitosan nanoparticles. Clin. Cancer Res. **2010**, *16* (15), 3910–3922.

54. Beisner, J.; Dong, M.; Taetz, S.; Nafee, N.; Griese, E.U.; Schaefer, U.; Lehr, C.M.; Klotz, U.; Mürdter, T.E. Nanoparticle mediated delivery of 2'-O-methyl-RNA leads to efficient telomerase inhibition and telomere shortening in human lung cancer cells. Lung Cancer **2010**, *68* (3), 346–354.

55. Lee, J.S.; Feijen, J. Polymersomes for drug delivery: Design, formation and characterization. J. Control. Release **2012**, *161* (2), 473–483.

56. Jain, J.P.; Ayen, W.Y.; Kumar, N. Self assembling polymers as polymersomes for drug delivery. Curr. Pharm. Des. **2011**, *17* (1), 65–79.

57. Howse, J.R.; Jones, R.A.; Battaglia, G.; Ducker, R.E.; Leggett, G.J.; Ryan, A.J. Templated formation of giant polymer vesicles with controlled size distributions. Nat. Mater. **2009**, *8* (6), 507–511.

58. Uneyama, T. Density functional simulation of spontaneous formation of vesicle in block copolymer solutions. J. Chem. Phys. **2007**, *126* (11), 114902.

59. He, X.; Schmid, F. Dynamics of spontaneous vesicle formation in dilute solutions of amphiphilic diblock copolymers. Macromolecules **2006**, *39* (7), 2654–2662.

60. Chen, L.; Shen, H.; Eisenberg, A. Kinetics and mechanism of the rod-to-vesicle transition of block copolymer aggregates in dilute solution. J. Phys. Chem. B **1999**, *103* (44), 9488–9497.

61. Kataoka, K.; Matsumoto, T.; Yokoyama, M.; Okano, T.; Sakurai, Y.; Fukushima, S.; Okamoto, K.; Kwon, G.S. Doxorubicin-loaded poly(ethylene glycol)–poly(β-benzyl-l-aspartate) copolymer micelles: Their pharmaceutical characteristics and biological significance. J. Control. Release **2000**, *64* (1–3), 143–153.

62. Sanson, C.; Schatz, C.; Le Meins, J.F.; Soum, A.; Thévenot, J.; Garanger, E.; Lecommandoux, S. A simple method to achieve high doxorubicin loading in biodegradable polymersomes. J. Control. Release **2010**, *147* (3), 428–435.

63. O'Neil, C.P.; Suzuki, T.; Demurtas, D.; Finka, A.; Hubbell, J.A. A novel method for the encapsulation of biomolecules into polymersomes via direct hydration. Langmuir **2009**, *25* (16), 9025–9029.

64. Upadhyay, K.K.; Bhatt, A.N.; Castro, E.; Mishra, A.K.; Chuttani, K.; Dwarakanath, B.S.; Schatz, C.; Le Meins, J.F.; Misra, A.; Lecommandoux, S. *In vitro* and *in vivo* evaluation of docetaxel loaded biodegradable polymersomes. Macromol. Biosci. **2010**, *10* (5), 503–512.

65. Christian, D.A.; Cai, S.; Bowen, D.M.; Kim, Y.; Pajerowski, J.D.; Discher, D.E. Polymersome carriers: From self-assembly to siRNA and protein therapeutics. Eur. J. Pharm. Biopharm. **2009**, *71* (3), 463–474.

66. Ghoroghchian, P.P.; Li, G.; Levine, D.H.; Davis, K.P.; Bates, F.S.; Hammer, D.A.; Therien, M.J. Bioresorbable vesicles formed through spontaneous self-assembly of amphiphilic poly(ethylene oxide)-block-polycaprolactone. Macromolecules **2006**, *39* (5), 1673–1675.

67. Onaca, O.; Enea, R.; Hughes, D.W.; Meier, W. Stimuli-responsive polymersomes as nanocarriers for drug and gene delivery. Macromol. Biosci. **2009**, *9* (2), 129–139.

68. Ahmed, F.; Pakunlu, R.I.; Brannan, A.; Bates, F.; Minko, T.; Discher, D.E. Biodegradable polymersomes loaded with both paclitaxel and doxorubicin permeate and shrink tumors, inducing apoptosis in proportion to accumulated drug. J. Control. Release **2006**, *116* (2), 150–158.

69. Tanner, P.; Baumann, P.; Enea, R.; Onaca, O.; Palivan, C.; Meier, W. Polymeric vesicles: From drug carriers to nanoreactors and artificial organelles. Acc. Chem. Res. **2011**, *44* (10), 1039–1049.

70. Lomas, H.; Canton, I.; MacNeil, S.; Du, J.; Armes, S.P.; Ryan, A.J.; Lewis, A.L.; Battaglia, G. Biomimetic pH sensitive polymersomes for efficient DNA encapsulation and delivery. Adv. Mater. **2007**, *19* (23), 4238–4243.

71. Lomas, H.; Johnston, A.P.; Such, G.K.; Zhu, Z.; Liang, K.; Van Koeverden, M.P.; Alongkornchotikul, S.; Caruso, F. Polymersome-loaded capsules for controlled release of DNA. Small **2011**, *7* (14), 2109–2119.

72. Kim, Y.; Tewari, M.; Pajerowski, J.D.; Cai, S.; Sen, S.; Williams, J.H.; Sirsi, S.R.; Lutz, G.J.; Discher, D.E. Polymersome delivery of siRNA and antisense oligonucleotides. J. Control. Release **2009**, *134* (2), 132–140.

73. Lewis, A.L.; Battaglia, G.; Canton, I.; Stratford, P.W. Amphiphilic Block Copolymers For Nucleic Acid Delivery, http://patent.ipexl.com/U2S/20110151013.html (accessed Mar 27, 2013).

74. Quinn, J.F.; Johnston, A.P.; Such, G.K.; Zelikin, A.N.; Caruso, F. Next generation, sequentially assembled ultrathin films: Beyond electrostatics. Chem. Soc. Rev. **2007**, *36* (5), 707–718.

75. Zhang, Y.J.; Guan, Y.I.N.G.; Yang, S.; Xu, J.; Han, C.C. Fabrication of hollow capsules based on hydrogen bonding. Adv. Mater. **2003**, *15* (10), 832–835.

76. Kozlovskaya, V.; Ok, S.; Sousa, A.; Libera, M.; Sukhishvili, S.A. Hydrogen-bonded polymer capsules formed by layer-by-layer self-assembly. Macromolecules **2003**, *36* (23), 8590–8592.

77. Such, G.K.; Tjipto, E.; Postma, A.; Johnston, A.P.; Caruso, F. Ultrathin, responsive polymer click capsules. Nano Lett. **2007**, *7* (6), 1706–1710.

78. Decher, G.; Hong, J. Buildup of ultrathin multilayer films by a self-assembly process .1. Consecutive adsorption of anionic and cationic bipolar amphiphiles on charged surfaces. Makromol. Chem. Macromol. Symp. **1991**, *46* (1) 321–327.

79. Caruso, F.; Caruso, R.A.; Möhwald, H. Nanoengineering of inorganic and hybrid hollow spheres by colloidal templating. Science **1998**, *282* (5391), 1111–1114.

80. Dejugnat, C.; Sukhorukov, G.B. pH-responsive properties of hollow polyelectrolyte microcapsules templated on various cores. Langmuir **2004**, *20* (17), 7265–7269.

Nanomaterials—
Nanoparticles

81. Volodkin, D.V.; Larionova, N.I.; Sukhorukov, G.B. Protein encapsulation via porous CaCO3 microparticles templating. Biomacromolecules **2004**, *5* (5), 1962–1972.

82. Itoh, Y.; Matsusaki, M.; Kida, T.; Akashi, M. Enzyme-responsive release of encapsulated proteins from biodegradable hollow capsules. Biomacromolecules **2006**, *7* (10), 2715–2718.

83. Peyratout, C.S.; Dahne, L. Tailor-made polyelectrolyte microcapsules: From multilayers to smart containers. Angew. Chem. Int. Ed. **2004**, *43* (29), 3762–3783.

84. De Geest, B.G.; Sanders, N.N., Sukhorukov, G.B.; Demeester, J.; De Smedt, S.C. Release mechanisms for polyelectrolyte capsules. Chem. Soc. Rev. **2007**, *36* (4), 636–649.

85. De Cock, L.J.; De Koker, S.; De Geest, B.G.; Grooten, J.; Vervaet. C.; Remon, J.P.; Sukhorukov. G.B.; Antipina, M.N. Polymeric multilayer capsules in drug delivery. Angew. Chem. Int. Ed. **2010**, *49* (39), 6954–6973.

86. Ai, H.; Jones, S.A.; Lvov, Y.M. Biomedical applications of electrostatic layer-by-layer nano-assembly of polymers, enzymes, and nanoparticles. Cell Biochem. Biophys. **2003**, *39* (1), 23–43.

87. Fischer, D.; Li, Y.; Ahlemeyer, B.; Krieglstein, J.; Kissel, T. *In vitro* cytotoxicity testing of polycations: Influence of polymer structure on cell viability and hemolysis. Biomaterials **2003**, *24* (7), 1121–1131.

88. Moghimi, S.M.; Symonds, P.; Murray, J.C.; Hunter, A.C.; Debska, G.; Szewczyk, A. A two-stage poly(ethylenimine)-mediated cytotoxicity: Implications for gene transfer/therapy. Mol. Ther. **2005**, *11* (6), 990–995.

89. Symonds, P.; Murray, J.C.; Hunter, A.C.; Debska, G.; Szewczyk, A.; Moghimi, S.M. Low and high molecular weight poly(L-lysine)s/poly(L-lysine)-DNA complexes initiate mitochondrial-mediated apoptosis differently. FEBS Lett. **2005**, *579* (27), 6191–6198.

90. Kozlovskaya, V.; Kharlampieva, E.; Mansfield, M.L.; Sukhishvili, S.A. Poly(methacrylic acid) hydrogel films and capsules: Response to pH and ionic strength, and encapsulation of macromolecules. Chem. Mater. **2006**, *18* (2), 328–336.

91. Zelikin, A.N.; Li, Q.; Caruso, F. Disulfide-stabilized poly(methacrylic acid) capsules: Formation, cross-linking, and degradation behavior. Chem. Mat. **2008**, *20* (8), 2655–2661.

92. Johnston, A.P.R.; Cortez, C.; Angelatos, A.S.; Caruso, F. Layer-by-layer engineered capsules and their applications. Curr. Opin. Colloid Interface Sci. **2006**, *11* (4), 203–209.

93. De Koker, S.; Hoogenboom, R.; De Geest, B.G. Polymeric multilayer capsules for drug delivery. Chem. Soc. Rev **2012**, *41* (7), 2867–2884.

94. Shutava, T.; Prouty, M.; Kommireddy, D.; Lvov, Y. pH responsive decomposable layer-by-layer nanofilms and capsules on the basis of tannic acid. Macromolecules **2005**, *38* (7), 2850–2858.

95. Ibarz, G.; Dähne, L.; Donath, E.; Moehwald, H. Smart micro- and nanocontainers for storage, transport, and release. Adv. Mater. **2001**, *13* (17), 1324–1327.

96. Lvov, Y.; Antipov, A.A.; Mamedov, A.; Möhwald, H.; Sukhorukov, G.B. Urease encapsulation in nanoorganized microshells. Nano Lett. **2001**, *1* (3), 125–128.

97. Koehler, K.; Sukhorukov, G.B. Heat treatment of polyelectrolyte multilayer capsules: A versatile method for encapsulation. Adv. Funct. Mater. **2007**, *17* (13), 2053–2061.

98. Itoh, Y.; Matsusaki, M.; Kida, T.; Akashi, M. Locally controlled release of basic fibroblast growth factor from multilayered capsules. Biomacromolecules **2008**, *9* (8), 2202–2206.

99. Khopade, A.J.; Caruso, F. Stepwise self-assembled poly(amidoamine) dendrimer and poly(styrenesulfonate) microcapsules as sustained delivery vehicles. Biomacromolecules **2002**, *3* (6), 1154–1162.

100. Wang, Y.; Yu, A.; Caruso, F. Nanoporous polyelectrolyte spheres prepared by sequentially coating sacrificial mesoporous silica spheres. Angew. Chem. Int. Ed. **2005**, *44* (19), 2888–2892.

101. Yu, A.; Wang, Y.; Barlow, E.; Caruso, F. Mesoporous silica particles as templates for preparing enzyme-loaded biocompatible microcapsules. Adv. Mater. **2005**, *17* (14), 1737–1741.

102. Zheng, Z.; Zhang, X.; Carbo, D.; Clark, C.; Nathan, C.; Lvov, Y. Sonication-assisted synthesis of polyelectrolyte-coated curcumin nanoparticles. Langmuir **2010**, *26* (11), 7679–7681.

103. Chen, Y.; Lin, X.; Park, H.; Greever, R. Study of artemisinin nanocapsules as anticancer drug delivery systems. Nanomed. Nanotechnol. Biol. Med. **2009**, *5* (3), 316–322.

104. Zhang, F.; Wu, Q.; Chen, Z.C.; Zhang, M.; Lin, X.F. Hepatic-targeting microcapsules construction by self-assembly of bioactive galactose-branched polyelectrolyte for controlled drug release system. J. Colloid Interface Sci. **2008**, *317* (2), 477–484.

105. Pargaonkar, N.; Lvov, Y.M.; Li, N.; Steenekamp, J.H.; de Villiers, M.M. Controlled release of dexamethasone from microcapsules produced by polyelectrolyte layer-by-layer nanoassembly. Pharm. Res. **2005**, *22* (5), 826–835.

106. Caruso, F.; Trau, D.; Möhwald, H.; Renneberg, R. Enzyme encapsulation in layer-by-layer engineered polymer multilayer capsules. Langmuir **2000**, *16* (4), 1485–1488.

107. Balabushevitch, N.G.; Sukhorukov, G.B.; Moroz, N.A.; Volodkin, D.V.; Larionova, N.I.; Donath, E.; Mohwald, H. Encapsulation of proteins by layer-by-layer adsorption of polyelectrolytes onto protein aggregates: Factors regulating the protein release. Biotechnol. Bioeng. **2001**, *76* (3), 207–213.

108. Shutava, T.G.; Balkundi, S.S.; Vangala, P.; Steffan, J.J.; Bigelow, R.L.; Cardelli, J.A.; O'Neal, D.P.; Lvov, Y.M. Layer-by-layer-coated gelatin nanoparticles as a vehicle for delivery of natural polyphenols. ACS Nano **2009**, *3* (7), 1877–1885.

109. Matsusaki, M.; Akashi, M. Functional multilayered capsules for targeting and local drug delivery. Expert Opin. Drug Deliv. **2009**, *6* (11), 1207–1217.

110. Itoh, Y.; Matsusaki, M.; Kida, T.; Akashi, M. Time-modulated release of multiple proteins from enzyme-responsive multilayered capsules. Chem. Lett. **2008**, *37* (3), 238–239.

111. Schneider, G.F.; Subr, V.; Ulbrich, K.; Decher, G. Multifunctional cytotoxic stealth nanoparticles. A model approach with potential for cancer therapy. Nano Lett. **2009**, *9* (2), 636–642.

Nanomaterials— Nanoparticles

112. Sivakumar, S.; Bansal, V.; Cortez, C.; Chong, S.F.; Zelikin, A.N.; Caruso, F. Degradable, surfactant-free, monodisperse polymer-encapsulated emulsions as anticancer drug carriers. Adv. Mater. **2009**, *21* (18), 1820–1824.

113. Sukhishvili, S.A. Responsive polymer films and capsules via layer-by-layer assembly. Curr. Opin. Colloid Interface Sci. **2005**, *10* (1–2), 37–44.

114. Yan, Y.; Johnston, A.P.; Dodds, S.J.; Kamphuis, M.M.; Ferguson, C.; Parton, R.G.; Nice, E.C.; Heath, J.K.; Caruso, F. Uptake and intracellular fate of disulfide-bonded polymer hydrogel capsules for doxorubicin delivery to colorectal cancer cells. ACS Nano **2010**, *4* (5), 2928–2936.

115. Poon, Z.; Chang, D.; Zhao, X.; Hammond, P.T. Layer-by-layer nanoparticles with a ph-sheddable layer for *in vivo* targeting of tumor hypoxia. ACS Nano **2011**, *5* (6), 4284–4292.

116. Wohl, B.M.; Engbersen, J.F.J. Responsive layer-by-layer materials for drug delivery. J. Control. Release **2012**, *158* (1), 2–14.

117. Staedler, B.; Price, A.D.; Zelikin, A.N. A critical look at multilayered polymer capsules in biomedicine: Drug carriers, artificial organelles, and cell mimics. Adv. Funct. Mater. **2011**, *21* (1), 14–28.

118. Reibetanz, U.; Claus, C.; Typlt, E.; Hofmann, J.; Donath, E. Defoliation and plasmid delivery with layer-by-layer coated colloids. Macromol. Biosci. **2006**, *6* (2), 153–160.

119. Zhang, X.; Oulad-Abdelghani, M.; Zelkin, A.N.; Wang, Y.; Haîkel, Y.; Mainard, D.; Voegel, J.C.; Caruso, F.; Benkirane-Jessel, N. Poly(L-lysine) nanostructured particles for gene delivery and hormone stimulation. Biomaterials **2010**, *31* (7), 1699–1706.

120. Elbakry, A.; Zaky, A.; Liebl, R.; Rachel, R.; Goepferich, A.; Breunig, M. Layer-by-layer assembled gold nanoparticles for siRNA delivery. Nano Lett. **2009**, *9* (5), 2059–2064.

121. del Mercato, L.L.; Rivera-Gil, P.; Abbasi, A.Z.; Ochs, M.; Ganas, C.; Zins, I.; Sönnichsen, C.; Parak, W.J. LbL multilayer capsules: Recent progress and future outlook for their use in life sciences. Nanoscale **2010**, *2* (4), 458–467.

122. Staedler, B.; Chandrawati, R.; Price, A.D.; Chong, S.F.; Breheney, K.; Postma, A.; Connal, L.A.; Zelikin, A.N.; Caruso, F. A microreactor with thousands of subcompartments: Enzyme-loaded liposomes within polymer capsules. Angew. Chem. Int. Ed. **2009**, *48* (24), 4359–4362.

123. Chandrawati, R.; Odermatt, P.D.; Chong, S.F.; Price, A.D; Städler, B.; Caruso, F. Triggered cargo release by encapsulated enzymatic catalysis in capsosomes. Nano Lett. **2011**, *11* (11), 4958–4963.

124. De Geest, B.G.; Sukhorukov, G.B.; Möhwald, H. The pros and cons of polyelectrolyte capsules in drug delivery. Expert Opin. Drug Deliv. **2009**, *6* (6), 613–624.

Nanomaterials—
Nanoparticles

Nanomedicine: Review and Perspectives

Ali Zarrabi
Department of Biotechnology, Faculty of Advanced Sciences and Technologies, University of Isfahan, Isfahan, Iran

Arezoo Khosravi
Institute for Nanoscience and Nanotechnology, Sharif University of Technology, Tehran, Iran

Ali Hashemi
Chemical and Petroleum Engineering Department, Sharif University of Technology, Tehran, Iran

Abstract

Advances in nanotechnology have had remarkable effects on biomedical sciences, specifically in the fields of drug delivery, diagnostics, monitoring, and control of biological systems. Different nano-systems with unique compositions and biological properties have been investigated for drug and gene delivery applications. Among them, polymeric nanomedicines have been comprehensively utilized in the design of nano-sized delivery vehicles of chemotherapeutics for clinical cancer/non-cancerous therapy. In this entry, we provide a review on the development of polymeric nanomedicine for therapy. This entry describes the progress that has been made in the field of polymeric nanomedicine that brings the science closer to clinical realization of nano-polymeric therapeutics for their application in disease treatment. Early products were developed as anticancer agents, but treatments for a range of non-cancerous diseases and different routes of administration followed. Despite the broad applications of polymeric nanomedicine, the need for progression of an appropriate regulatory framework is at the forefront of the scientific discussion. Several regulatory checkpoints should be taken into account before a polymeric nanomedicine is commercially produced. The need for controlled release and emergence of different biodegradable polymers are highlighted in this entry. Smart polymeric nanomaterials built of stimuli-responsive polymers are discussed with their various applications in the biomedical field. Stimuli-responsive polymeric carriers show a sharp change in properties upon a slight change in environmental conditions such as temperature, light, magnetic field, or pH. This behavior can be employed for the preparation of the so-called "smart" drug delivery systems, which enable the drug delivery systems to release their drug payload at the specific site of disease, which is called targeted drug delivery. Polymeric nano-carriers, with bioactive molecules being either conjugated or encapsulated, have been developed into a variety of different architectures, including polymer–drug conjugates, dendrimers, micelles, vesicles, liposomes, polymersomes, and polysaccharide-based nanoparticles. In addition to the incorporation of drug molecules inside a polymer particle, the polymer itself could act as a drug without any incorporated drug molecules. Polymeric drugs are defined as functional high molecular weight polymers that selectively recognize, sequester, and remove disease-causing species. Here, we summarize the developments of polymeric drugs such as sequestrants of phosphates, iron, bile acids, as well as toxins. Finally, successful applications of theoretical methods to the development of new drug formulations have been discussed as well as the future prospects of polymeric nanomedicine.

INTRODUCTION

Since the 1960s, a large number of researchers have tried to design synthetic drugs based on polymers as well as polymer–drug and polymer–protein conjugates.[1] During the 1970s, the number of articles per year was abated; however, Helmut Ringsdorf's seminal reviews on block copolymer micelles and polymer–drug conjugates[2] emphasized their potential for use as therapeutics and carriers, and the first PEG–protein conjugates prepared by Davis and colleagues' innovative work stirred more interest in the field.[3] Furthermore, in the 1980s, another door was opened by the realization that gene therapy could be

an alternative to drug product development. It was then that "non-viral vectors" for cytosolic delivery of genes, protein, and small interfering ribonucleic acids (siRNAs) came into existence (reviewed in Vasir and Labhasetwar[4]). Clinical development has been hindered by poor transfection efficiency, complexity of the product, and toxicity in spite of more than 20 years of efforts worldwide. In the 1990s, the term "polymer therapeutics" was used to describe "new chemical entities" by the Regulatory Agencies for the first time. Based on the definition offered by European Science Foundation's Forward Look, nano-sized constructs with more than one component should be regarded as first generation "Nanomedicines,"

Concise Encyclopedia of Biomedical Polymers and Polymeric Biomaterials DOI: 10.1081/E-EBPPC-120050031

too. These therapeutics were not welcomed at first, but later products showed that they could meet the stringent needs of industrial development and regulatory authority approval. Polymer–protein conjugates, Zinostatin Stimaler® and Adagen®, were the first to be approved for routine clinical use in 1990. From then on, a large number of polymeric drugs and sequestrants, PEGylated proteins, peptides, and aptamers have been introduced into the market. In spite of the fact that earlier products were anticancer agents (the risk–benefit was deemed acceptable at first), later on, polymer therapeutics were developed to treat infectious diseases (e.g., PEG–interferons) and aging population illnesses (age-related macular degradation and arthritis). In addition, clinical development continues to grow (Fig. 1).

Because of its good safety profile, hydrophilicity, and ability to monofunctionalize for protein conjugation, PEG has been extensively used for conjugation.[5] Figure 2 summarizes some polymer therapeutic products in the market.

The first synthetic polymer anticancer drug conjugate, an N-(2-hydroxypropyl)methacrylamide (HPMA) copolymer conjugate of doxorubicin (FCE 28028), was introduced for clinical trial in 1994. Six HPMA copolymer-based anticancer conjugates followed into "Good Clinical Practice"

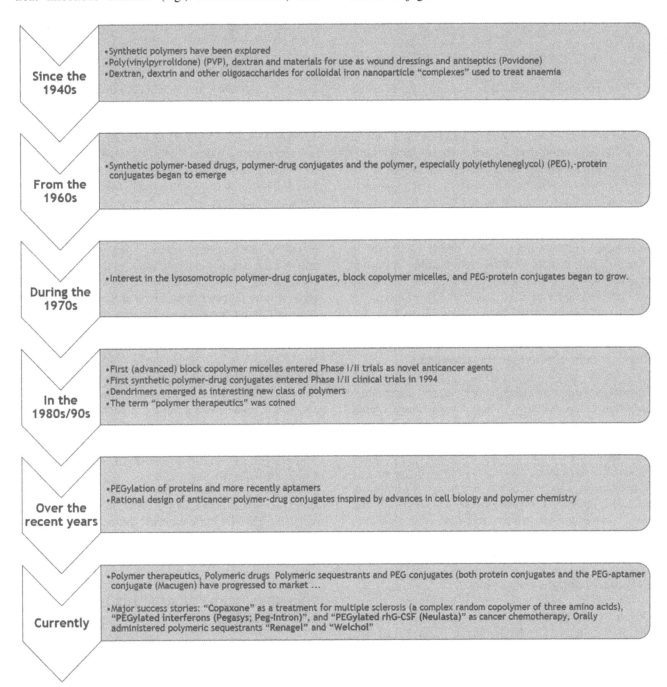

Fig. 1 Evolution of polymeric nanomedicine: a brief review of progress made in the field.

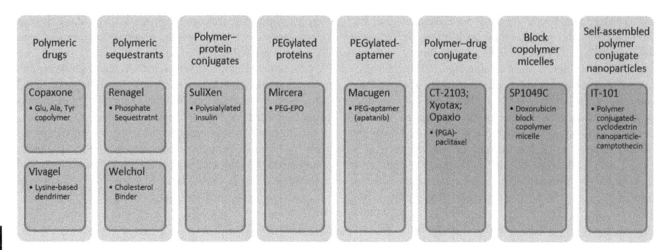

Fig. 2 Polymer therapeutic products in the market.

(GCP) and controlled clinical development.[6] Then Duncan reviewed the historical perspective and the lessons learnt from these studies in *"Development of HPMA copolymer anticancer conjugates: Clinical experience and lessons learnt."* A large number of new polymers are being used in order to prepare the second-generation polymer therapeutics.[7] They are used as micelles to entrap drug, as covalent conjugates, and are even self-assembled into nanoparticles in certain cases. Two such carriers that have entered clinical development are the polymer–cyclodextrin nanoparticle-containing camptothecin (IT-101), which is already in Phase II clinical trials, and the biodegradable polyacetal [poly(1-hydroxymethylene hydroxymethyl formal) named as Fleximer] conjugate of camptothecin (XMT-1001) undergoing Phase I clinical trials.

Polymeric Carriers: Implications for Design and Development of Polymer Therapeutics

Polymer–protein conjugates as well as polymeric drugs and sequestrants, initially developed as anticancer agents, have been introduced into the market as novel polymer-based therapeutics or first generation "nanomedicines" since early 1990s. Treatment of a large number of debilitating and life-threatening diseases has been targeted by way of intravenous, subcutaneous, or oral routes of administration. With regard to controversies over the second-generation polymeric therapeutics regarding the safety of nanomedicines per se, an appropriate regulatory framework seems to be inevitable.

Polymer Toxicity: Checkpoints

Obviously, any polymeric carrier suggested for parenteral administration ought to be non-immunogenic, non-toxic, and preferably biodegradable; it should also show an inherent body distribution allowing appropriate tissue targeting, not only to the desired therapeutic site, but also away from sites of toxicity. An indication of general

cytotoxicity (72 hr), hematocompatibility, and the likely rate of hydrolytic and enzymatic degradation can be checked by prescreening novel polymers *in vitro* using dextran (Mw 70,000 g/mol) as a negative reference control, and polyethylenimine (PEI) or poly-L-lysine (PLL) as a positive one. Moreover, scanning electron microscopy (SEM) has proved to be a proper tool to check more subtle changes of cell aggregation or morphology[8] while being incubated with polymers. Though the results are strongly influenced by polymer molecular weight, cell-type used, period of incubation, conditions, and impurities in the polymer sample, they can help us judge the justifiability of ethics in this regard. Furthermore, it is crucial to have early *in vitro* polymer antigenicity and cellular immunogenicity information since inappropriate bioactivity could make the development of the polymeric carrier impossible. Only polymers with acceptable properties can be optimized in terms of their molecular weight and functionality for conjugate synthesis. It is a necessity that conjugation of drug(s), targeting residues, and/or imaging agents be chemically refined, and a library of conjugates is used in biological testing in order to provide a definition for the structure–activity relationships in terms of conjugate pharmacokinetics, toxicity/immunogenicity, and *in vivo* pharmacological activity.[8]

There are many other regulatory points such as environmental impact, efficacy, and pharmacokinetic–pharmacodynamic (PK–PD) correlations to be taken into account after a polymer passes the safety tests and before it is commercially produced.

POLYMERS USED FOR CONTROLLED DRUG RELEASE

Overview

Because of the inherent diversity of structures, it can be hard to categorize polymers in drug delivery applications.

Nevertheless, the classification is useful since it can high-light the common properties in polymer groups. Polymers are either biodegradable or non-biodegradable. The former have attracted more attention in drug delivery systems compared to the latter that need retrieval after they are used. Within the first group, there is a subclassification based on the mechanism of erosion. In this regard, a differentiation should be made between degradation that is a chemical process referring to bond cleavage and erosion, a physical phenomenon reliant on dissolution and diffusion processes that refers to depletion of the material. Polymer erosion can be identified as surface and bulk, both representing extremes. Both mechanisms will occur for most biodegradable polymers; however, their relative extent varies radically with the chemical structure of the polymer backbone. Surface erosion takes place when the rate of water permeation into the bulk of the polymer is less than that of erosion. Due to the fact that the kinetics of erosion and the rate of drug release are highly reproducible, this is a desirable mechanism of erosion in drug delivery. Moreover, it is possible to change the magnitude of erosion by changing the surface area of the drug delivery device thus protecting water-labile drugs up to the time of drug release. These classes of biodegradable polymers, namely, poly(anhydrides) and poly(ortho esters), have highly labile groups ensuring rapid hydrolysis of polymer chains that encounter water molecules. By designing polymers with hydrophobic monomers, water permeation can be retarded. Alternatively, hydrophobic inert substances can be added to make the polymer bulk stable. In ideal surface erosion, the rate of erosion is directly proportional to the external surface area. If diffusional release is restricted and the overall shape remains constant, surface erosion may result in zero-order drug release. Bulk erosion takes place when molecules of water can permeate into the bulk of the polymer matrix faster than erosion; as a result of which, the polymer molecules may be hydrolyzed and the kinetics of polymer degradation/erosion are more complex than those of surface eroding polymers. Most biodegradable polymers, including the poly(ester) materials, used in controlled drug delivery undergo bulk erosion. The more limited predictability of erosion and the lack of protection of drug molecules are hidden drawbacks of bulk eroding devices; however, they do not prevent their successful employment as drug delivery devices. Moreover, a large number of new applications in controlled release use nano- or microparticle formulations with massive surface areas, which lead to bulk, and surface eroding materials that have similar erosion kinetics. Among biodegradable systems, natural polymers like starch, cellulose, and chitosan have already been studied.[9] The synthesis of such polymers, referred to as biopolymers, is limited to the manipulation of the bulk material to enhance their viability. New opportunities have been opened for synthetic chemists in the field of controlled release by designing biomaterials with specified release, mechanical, and processing properties.

Poly(esters)

The best characterized and the most widely studied biodegradable system is poly(esters) whose synthesis has attracted as much attention as their degradation. The use of poly(lactic acid) (PLA) as a resorbable suture material was first patented in 1967. In poly(ester) materials, the degradation mechanism is classified as bulk degradation with random hydrolytic scission of the polymer backbone. Poly(esters) have been vastly used in drug delivery applications, and thoroughly reviewed.[10,11] The predominant synthetic pathway for producing poly(esters) is from ring-opening polymerization of the corresponding cyclic lactone monomer. Prominent poly(esters) and their starting materials are illustrated in Fig. 3.

Poly(lactic acid), poly(glycolic acid), and their copolymers

With regard to their design and performance, poly(esters) based on poly(lactic acid) (PLA), poly(glycolic acid) (PGA), and their copolymers, poly(lactic acid-co-glycolic acid) (PLGA), are among the best defined biomaterials. Lactic acid has an asymmetric R-carbon, typically described as the D or L form and sometimes as the R and S form, respectively. Optically active PDLA and PLLA have almost the same physiochemical properties while the racemic PLA possesses very different characteristics. These polymers are

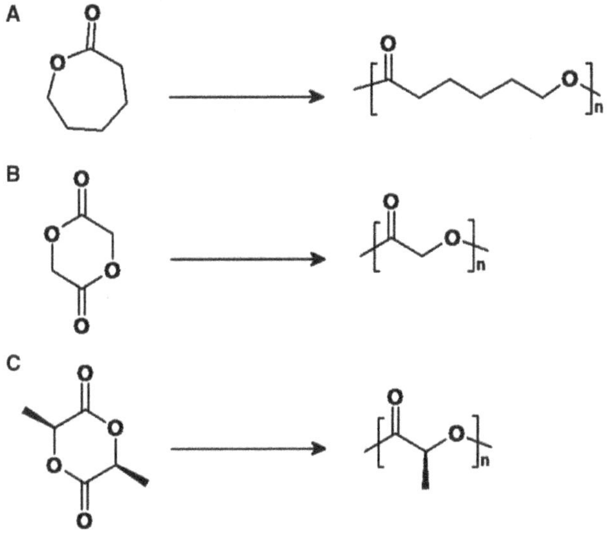

Fig. 3 Ring-opening polymerization of selected cyclic lactones to give the following: (**A**) poly(caprolactone) (PCL), (**B**) poly(glycolic acid) (PGA), and (**C**) poly(L-lactic acid) (PLA).

Nanomaterials—Nanoparticles

made up of monomers, which are natural metabolites of the body; hence, their degradation yields hydroxyl acid making them safe for *in vivo* use. Biocompatibility of degradable polymer systems is due to the biocompatibility of the monomer constituents. Though PLGA is vastly used and represents the gold standard of degradable polymers, increased local acidity because of degradation can result in irritation at the site of the polymer employment. The use of basic salts has been studied as a technique to control the pH in the local environment of PLGA implants.[12,13] From a physical point of view, poly(esters) undergo bulk degradation. Vast degradation surveys have also been reported for PLA, poly(caprolactone) (PCL), and their copolymers both *in vitro* and *in vivo*.[14] Studies on hydrolytic degradation of poly(esters) have concentrated on understanding the impacts of changes in polymer chain composition.

Poly(ethylene glycol) block copolymers

Biocompatible poly(ethylene glycol) (PEG) is also called poly(ethylene oxide) (PEO) at high molecular weights. PEG with a molecular weight of 4000 amu is 98% excreted in humans, and is protein resistant.[15] PEG is so hydrophilic that the hydrogen in water bonds tightly with the polymer chain excluding or inhibiting protein adsorption. A large number of researchers are studying the attachment of PEG chains to therapeutic proteins. Such chains prolong biological events; thus allowing prolonged protein circulation in the body. The inclusion of PEG in copolymer systems adds useful properties inside the body due to its ability to push back proteins in aqueous environments. This inhibits the adsorption of proteins onto polymer surface and subsequently many polymer–cell

interactions. PEG can be made with a range of terminal functionalities helping its incorporation into copolymer systems, and is commonly terminated with chain-end hydroxyl groups providing a ready handle for synthetic modification. Its free hydroxyl groups are ring-opening initiators for lactide in forming diblock or triblock materials (Fig. 4). Further investigations showed that PLA–PEG nanoparticles were inert to proteins of the coagulation system.[16]

Poly(ortho esters)

The reason for designing poly(ortho esters) for drug delivery was to develop biodegradable polymers inhibiting drug release by diffusion and allowing it only after the hydrolysis of polymer chains at the surface of the device.[17] The majority of research on poly(ortho esters) has concentrated on the synthesis of polymers by adding polyols to diketene acetals. Acid-catalyzed hydrolysis of such polymers is controllable by introducing acidic or basic excipients into matrices. This approach has been used in the temporally controlled release of tetracycline over a period of weeks in treating periodontal diseases.[18] Lately, some changes in the diol structure have been attempted in order to avoid the need for acidic excipients. These structures address the problem of acidic excipient diffusion from matrices leading to unpredictable degradation kinetics.

Poly(anhydrides)

The polymer should be hydrophobic and have water-sensitive linkages in order to get a heterogenous eroding

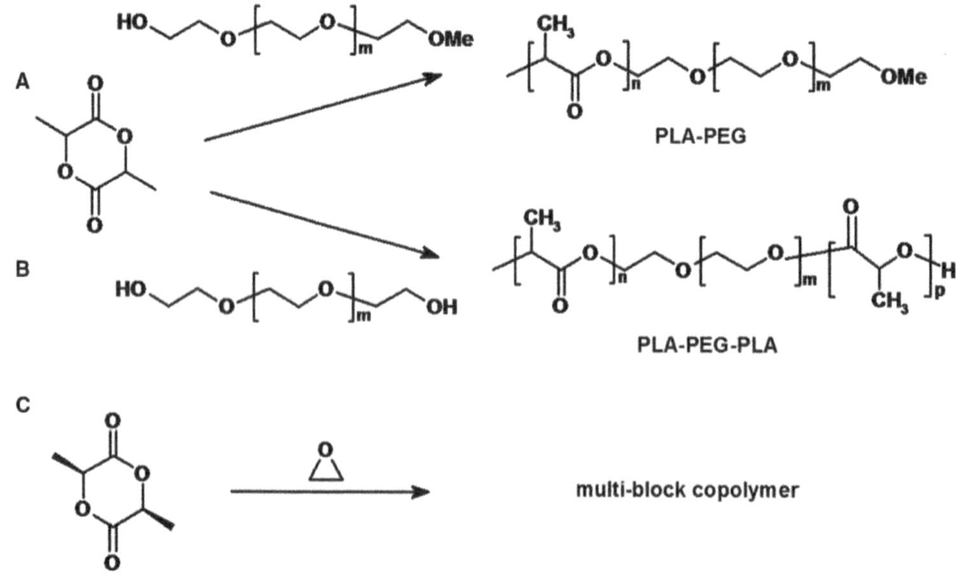

Fig. 4 Synthesis of PLA–PEG copolymers: (**A**) PLA/PEG, (**B**) PLA/PEG/PLA, and (**C**) a multiblock copolymer of L-lactide and ethylene oxide.

device. A polymer system meeting this requirement is the poly(anhydride), which undergoes hydrolytic bond cleavage in order to form water-soluble degradation products dissolvable in an aqueous environment, thus resulting in polymer erosion. Because of their high water liability of the anhydride bonds on the surface and the hydrophobicity which prevents water penetration into the bulk, poly(anhydrides) undergo surface erosion.[19] Most poly(anhydrides) are prepared by melt-condensation polymerization. A prepolymer of a mixed anhydride is formed with acetic anhydride, and the final polymer is obtained by heating the polymer under vacuum in order to remove the acetic anhydride by-product. The most commonly studied poly(anhydrides) are based on sebacic acid (SA), p-(carboxyphenoxy)propane (CPP), and p-(carboxyphenoxy)hexane (CPH) (Fig. 5). Degradation rates of these polymers are controllable by variations in polymer composition. The hydrophobicity of the monomer is directly proportional to the stability of the anhydride bond to hydrolysis. Aliphatic poly(anhydrides) degrade in days while aromatic poly(anhydrides) degrade in several years. Due to their high melting temperatures, poly(anhydrides) are best formed into drug-loaded devices by compression-molding or microencapsulation. A lot of drugs and proteins have been incorporated into poly(anhydride) matrices, and their *in vitro* and *in vivo* release characteristics have been evaluated.[20] Evaluation of the toxicity of poly(anhydrides) reveals that they have excellent *in vivo* biocompatibility.[21,22] *In vitro* and *in vivo* elimination of the polymers depend on monomer solubility.

Poly(amides)

Poly(amino acids) are the most interesting poly(amides) for controlled drug release. Poly(amino acids), used to deliver low molecular weight drugs, are usually tolerated when they are implanted in animals, and are metabolized to non-toxic products. Such results show good biocompatibility, but their antigenic nature causes their widespread use to be uncertain. Another point regarding poly(amino acids) is the intrinsic hydrolytic stability of the amide bond that

relies on enzymes for bond cleavage, which results in poor controlled release *in vivo*.

Poly(amino acids) are generally hydrophilic with degradation rates that depend on hydrophilicity of the amino acids. Amino acids are interesting because of the functionality they can provide in a polymer.

Poly(amino acids) can be modified to increase the release or bioavailability of the drug by attaching the drug molecule to the polymer through carboxylate bonds. The polymer conjugates were designed as insoluble particles for prolonged drug release, and act by penetrating the cells, and then releasing the drug by the action of lysosomal enzymes.[23]

Phosphorus-Containing Polymers

Poly(phosphazenes) and poly(phosphoesters) are two classes of phosphorus-containing polymers that have been of interest for several research works over the past decade because of their applications in medicine and biomedical engineering. Extensive research in this field has led to polymers of different characteristics ranging from viscous gels to amorphous particles with potential applications in drug delivery. In addition, phosphorus pentavalency eases the covalent linking of the drug. Nowadays several applications such as vaccine delivery, delivery of oncolytic and CNS therapeutics, DNA delivery, and tissue engineering are being considered for these classes of polymers.[24,25]

Poly(phosphazenes)

The development of poly(phosphazenes) provides an alternative to poly(ester), poly(ortho ester), and poly(anhydride) systems in terms of biodegradation kinetics and properties developed by structural changes in the side-chain structure instead of the polymer backbone.[26] Poly(phosphazenes) are interesting due to their unique inorganic phosphorus–nitrogen backbone as well as their marked synthetic versatility. They provide covalent and coordinate drug binding sites and decompose into non-toxic products like phosphate, ammonia, amino acids, and ethanol. Biodegradable poly(phosphazenes) have been used in the temporally

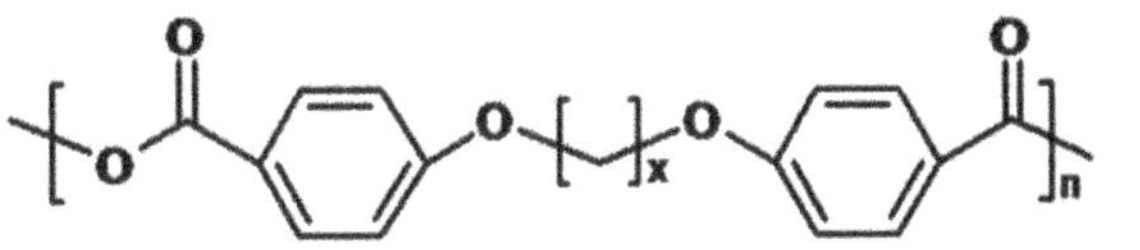

x= 3 (propane) (CPP)
6 (hexane) (CPH)

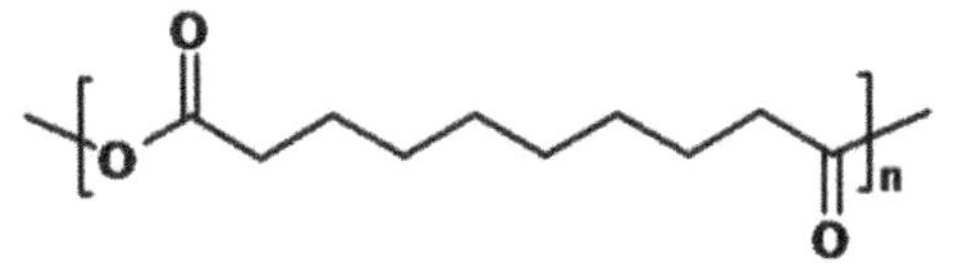

(SA)

Fig. 5   Structures of widely used aromatic poly(anhydrides) based on monomers of p-(carboxyphenoxy)propane (CPP), p-(carboxyphenoxy)hexane (CPH), and aliphatic poly(anhydrides) based on sebacic acid (SA).

Nanomaterials—
Nanoparticles

controlled release of many drug classes because they are insoluble in water before hydrolysis.[27]

Poly(phosphoesters)

Phosphoester groups have been incorporated into poly(urethane) by Leong, Brott, and Langer[19] Poly(urethanes) have already been used as blood-contacting biomaterials due to the physical properties that can be achieved. They were designed to be inert biomaterials, but controlled biodegradation is desirable for some applications. Leong introduced a phosphoester linkage into the poly(urethanes) to provide biodegradable materials that keep the mechanical properties hidden in them. Poly(phosphoester urethanes) are obtained by the reaction of diisocyanates and polyols with phosphites added as chain extenders. Phosphoester bonds are easily cleaved under physiological conditions (Fig. 6). Moreover, the pentavalency of the phosphorus provides a site for future functionalization because the release kinetics of poly(phosphoester urethanes) were affected by the side chains attached by way of the phosphoester of the polymer backbone. The release mechanism was a combination of diffusion, swelling, and degradation.[28,29]

SMART POLYMERS: STIMULI RESPONSIVENESS AND TARGETING

Compared to the conventional dosage drugs, controlled release systems have many advantages in terms of improved efficacy, reduced toxicity, and higher patient compliance and convenience. They often use synthetic polymers as carriers for drugs to treat diseases that would otherwise be impossible to treat. This improvement can increase the therapeutic activity and reduce the number of drug administrations or eliminate the need for specialized drug administration. Polymers that are naturally excreted from the body, whether directly or indirectly, are desirable for many controlled release applications.[30] Non-degradable polymers are acceptable in applications in which the delivery system can be recovered after drug release or for oral applications in which the polymer passes through the gastrointestinal (GI) tract.[31,32] It is important to know that different mechanisms of controlled release require polymers with various physicochemical properties. The main mechanisms are as follows.

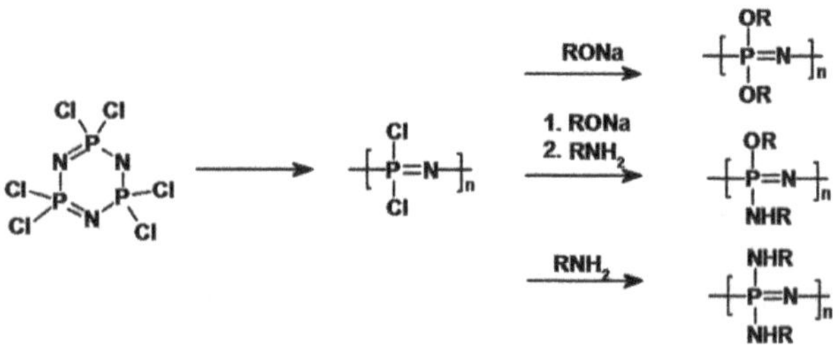

Stimuli-Responsive Polymers

The strategy behind polymer-containing response systems is the physicochemical change caused by stimuli. At the macromolecular level, polymer chains can be altered in different ways, which will cause detectable behavioral changes in self-assembled structures.[33] Many designs that vary the location of responsive moieties or functional groups are possible; the response may be reversible or not depending on the strategy used. Stimuli are categorized into three classes: physical, chemical, or biological (Fig. 7).[34] Physical stimuli usually modify chain dynamics while chemical stimuli modulate molecular interactions, either between polymer and solvent molecules or between polymer chains. Biological stimuli relate to the actual functioning of molecules. Moreover, there are dual stimuli-responsive polymers that respond to more than one stimulus at the same time.[35]

Physically dependent stimuli

Because magnetic fields and ultrasound techniques have been used only for the release of compounds that have been encapsulated in polymer assemblies, only temperature, electric field, and light are covered here.

Temperature-responsive polymers: These polymers have attracted great attention for use in bioengineering and biotechnology since certain diseases manifest temperature changes.[36] Normally, these copolymers are characterized by a critical solution temperature around which the hydrophobic and hydrophilic interactions between the polymeric chains and the aqueous media abruptly change within a small temperature range. This induces the disruption of intra- and intermolecular electrostatic and hydrophobic interactions resulting in chain collapse or expansion. Typically, these polymer solutions have an upper critical solution temperature (UCST) above which one polymer phase exists, and below which a phase separation appears.

Fig. 6 Formation of poly(phosphazenes) and examples of backbone modification.

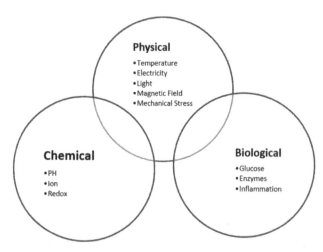

Fig. 7 Classification of stimuli in stimuli-responsive polymers.

Alternatively, polymer solutions that appear as monophasic below a specific temperature and biphasic above it generally have a lower critical solution temperature (LCST). Depending on the mechanism and chemistry of the groups, different temperature-responsive polymers have been reported. For example, poly(N-isopropyl acrylamide) (PNIPAAm) is clearly soluble in water at room temperature. Above the LCST, the solution becomes opaque and finally turns into a gel.[37,38] The LCST can be tuned by copolymerizing hydrophobic monomers or by controlling the polymer molecular weight. More hydrophobic monomers and higher molecular weight increase the LCST. Furthermore, incorporation of hydrophilic monomers forming hydrogen bonds with thermo-sensitive monomers lowers the LCST point. Block copolymers of poly (ethylene glycol)/poly(lactic-co-glycolic acid) (PEG/PLGA, ReGel®)[39] and poly(ethylene oxide)/poly(propylene oxide) (PEO/PPO, Pluronic®) have also thermosensitivity. A solution of each copolymer can be injected into the body, which forms a soft gel at body temperature. Those thermosensitive polymers are very attractive for protein/peptide drug delivery since no organic solvent is used during their formulation.[40] However, they can also solubilize highly hydrophobic drugs like paclitaxel, and have been an excellent formulation of poorly water-soluble drugs.[41,42]

Electro-responsive polymers: Electrical and electrochemical stimuli are broadly used in research and applications due to their advantages of precise control via the magnitude of the current, the duration of an electrical pulse, or the interval between pulses.[43] Electrically responsive polymers such as polythiophene (PT) or sulfonated-polystyrene (PSS) are normally conducting polymers, which can show swelling, shrinking, or bending in response to external fields.[44] Various effects on electrochemical stimulation are: (a) an influx of counterions and solvent molecules leading to an increase in osmotic pressure in the polymer, and consequently a volumetric expansion, (b) control of loading/adsorption of polyelectrolyte onto oppositely charged porous materials, and (c) formation and swelling of redox-active polyelectrolyte multilayers. For instance, in applying an electrochemical stimulus to multilayer polyacrylamide films, the combined effects of H ions migrating to the cathode and the electrostatic attraction between the anode surface and the acrylic acid groups, with negative charges, make the film on the anode side shrink.[45,46]

Photo-responsive polymers: Light, applicable instantaneously and under specific conditions with high accuracy, renders light-responsive polymers highly advantageous to applications.[38] It can be directly used on the polymer surface or delivered to far away locations using optical fibers. The wavelength of the laser is tuned to the near infrared part of the spectrum, which is less harmful and has a deeper penetration into tissues than the visible light. In this case, the light is minimally absorbed by cells/tissues and maximally so by the polymers. Most photo-responsive polymers contain light-sensitive chromophores like azobenzene groups,[47] spiropyran groups,[48] or nitrobenzyl groups.[49,50] Many azobenzene or spiropyran-containing photo-responsive polymers have been reported.

Magneto-responsive polymers: Magnetic nanoparticles could be used alone or in conjugation with other materials such as polymers in drug/cell delivery. The ability to steer and switch polymeric carriers using external stimuli is an attractive research field.[51] State-of-the-art strategies for use of magnetic stimulation toward cellular therapies are summarized in Fig. 8.

Chemically dependent stimuli

Chemically dependent stimuli include pH, ionic strength, and redox activity.

pH-responsive polymers: Since pH changes happen in many specific or pathological compartments, pH is an important environmental parameter for biological applications. In a healthy human, pH of the body tissue is kept around 7.4. However, several mechanisms exist to modulate pH inside the body. At first, the GI tract changes the pH along the tube, and the lysosomes inside the cells have a much lower pH than neutral pH by continuously pumping protons into the vesicles in order to digest foreign molecules. In some cases, solid tumors with malformed capillary blood vessels show low blood pressure, local hypoxia, and accumulation of acidic metabolites, which result in local fluctuation of pH ranging 5.7–7.8. Each mechanism is used as a triggering signal for pH-responsive drug delivery. Ionizable moieties like carboxylic acid, amine, azo, phenylboronic acid, imidazole, pyridine, sulfonamide, and thiol groups can afford pH sensitivity. The main element for pH responsive polymers is attachment of ionizable, weakly acidic or basic moieties to a hydrophobic backbone like polyelectrolytes.[52] Upon ionization, the electrostatic repulsions of the generated charges cause a great extension of coiled chains. The ionization of the pendant acidic or basic groups on polyelectrolytes can be partial due to the electrostatic effect of other adjacent ionized groups.[53] Another pH-responsive polymer shows protonation/deprotonation events by distributing the charge over ionizable groups of molecules like carboxyl or amino groups.[54] pH suddenly induces a phase transition in pH-responsive polymers. Usually, the phase switches within 0.2–0.3 U of pH.[55] pH-responsive polymers include chitosan,[56] albumin, poly(N,N-diakylaminoethylmethacrylates) (PDAAEMA),[57] gelatin,[58] poly(acrylic acid) (PAAC)/chitosan IPN,[59] poly(methacrylic acid-g-ethylene glycol) [P(MAA-g-EG)],[60] poly(ethylene imine) (PEI),[61] and poly(lysine)(PL).[62]

pH-sensitive polymers used for anticancer drug delivery should have a narrow pH range for modulating their physical properties. If not, pH-sensitive drug carriers can induce severe toxicity by drug burst or poor therapeutic efficacy by incomplete drug release at a target site. PEG-b-poly(L-histidine) (PEG-Phis) spontaneously generated the micelle

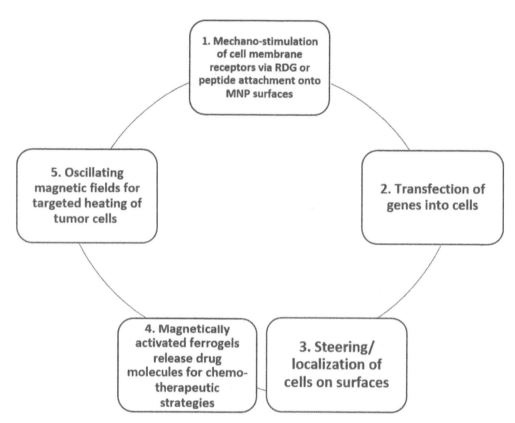

Fig. 8 Strategies for use of magnetic stimulation toward cellular therapies.

structure at high pH over the pK_b of histidine (6.5–7.0) and loaded the anticancer drug doxorubicin inside the micelle core. At pH 6.8, doxorubicin-loaded PEG–PHis micelle could eliminate multidrug-resistant MCF-7 cells, which might work under *in vivo* conditions due to the slightly different pH of the breast tumor tissue compared to the normal tissue.

Ion-responsive polymers: Responsiveness to ionic strength is a property of polymers with ionizable groups. Such polymer systems exhibit unusual rheological behavior because of Coulombic interactions between oppositely charged species, which may render the polymer insoluble in deionized water but soluble in a critical concentration of added electrolytes where the attractive charge/charge interactions are shielded.[63] Therefore, changes in the ionic strength change the length of polymer chains, polymer solubility, and fluorescence quenching kinetics of chromophores bound to electrolytes.[64]

Redox-responsive polymers: Polymers with labile groups have an opportunity to develop redox-responsive biodegradable or bio-erodible systems. Acid-labile moieties in polyanhydrides, poly(lactic/glycolic acid) (PLGA), and poly(b-amino esters) (PbAEs)[65] induce redox responsiveness just like disulfide groups due to their instability in a reducing environment, being cleaved in favor of thiol groups. Polymers with disulfide cross-links degrade upon exposure to cysteine or glutathione, which are reductive amino acid-based molecules.[66] Another redox-responsive

polymer is poly(NiPAAm-*co*-Ru(bpy)), which can generate a chemical wave by the periodic redox change of Ru(bpy)$_3$ into an oxidized state of a lighter color. This redox reaction changes the hydrophobic and hydrophilic properties of polymer chains resulting in swelling of the polymer.

Biologically dependent stimuli

Biologically dependent stimuli involve analytes and biomacromolecules like glucose, glutathione, enzymes, receptors, and overproduced metabolites in inflammation.

Glucose-responsive polymers: Precisely engineered glucose-sensitive polymers have huge potential in the quest to generate self-regulated modes of insulin delivery.[67] For glucose-responsive polymers, glucose oxidase (GOx) is conjugated to a smart, pH-sensitive polymer. GOx oxidizes glucose to gluconic acid, causing a pH change in the environment.[68] Then, the pH-sensitive polymer shows a volume transition in response to the decreased pH. Thus, changes in the polymer conformation are regulated by the body's glucose level, greatly affecting enzyme activity and substrate access.

Enzyme-responsive polymers: In nature, bacteria located in the colon produce special enzymes, reductive or hydrolytic, capable of degrading different kinds of polysaccharides like pectin, chitosan, amylase/amylopectin, cyclodextrin, and dextrin.[69,70] In most enzyme-responsive polymer systems, enzymes destroy the polymer or its assemblies. The biggest advantage of enzyme-responsive polymers is that they do not need an external trigger for decomposition, exhibit high selectivity, and work under mild conditions. For instance, polymer systems based on alginate/chitosan or DEXS/chitosan microcapsules are responsive to chitosanase,[71] and azoaromatic bonds are sensitive to azoreductase. In this respect, they have great potential for *in vivo* biological applications, but the main disadvantage is the establishment of a precise initial response time.

Inflammation-responsive polymers: The inflammatory process is initiated by T- and B-lymphocytes, but amplified and perpetuated by polymorphonuclear (PMN) leukocytes and macrophages. Different chemical mediators in the process can cause tissue damage. For inflammation-responsive systems, the reactive oxygen metabolites released by PMNs and macrophages during the initial phase of inflammation are the stimuli. Such mediators have been successfully used as stimuli for responsive drug delivery. For instance, *in vivo* implantation experiments revealed that hyaluronic acid (HA) cross-linked with glycidyl ether can degrade in response to inflammation.

Dual stimuli: For biomedical applications, if the smart materials respond simultaneously to more than one stimulus, there is a step forward. So, an increase in the efficacy of drug therapies may require polymeric materials responsive to several kinds of stimuli. These will support the diagnosis of patients by immediately checking several physiological changes. The dual-stimuli-responsive approach is ideally suited for theranostics (therapy and diagnostics) because some functionalities can provide on-site feedback and diagnostics, while others could start curing and therapy.[72,73] Various physical, chemical, and biological stimuli must be available for multiple response functions. Hence, multistimuli-responsive polymers, especially dual temperature and pH-responsive systems, have attracted increasing attention for their advantages in biotechnological and biomedical applications. For instance, by using both pH and glutathione-responsive polymeric modules, a dual-stimuli-responsive delivery system was developed to therapeutically deliver medicinal molecules.[74] It was possible to tune the release kinetics by systematically varying the composition of the pH-sensitive hydrophobic moiety, by modifying the glutathione-responsive moiety, or by modifying both of them.

Table 1 summarizes stimuli-responsive polymers grouped by stimulus response, and contains information about the synthesis method and application.

Smart Polymeric Systems: Targeting Strategies

Smart polymers have introduced distinctive systems to enhance the efficacy of drug delivery and therapeutic efficiency through targeting strategies. Passive targeting based on the enhanced permeation and retention effect (EPR) uses a unique physiological property of a disease (physiological targeting), while active targeting, conjugation of targeting moiety to drug carriers, is a biochemical targeting strategy.[75–78] Smart polymeric systems also provide a targeting strategy activated and triggered by a specific environmental signal (triggered targeting). Combined methods can result in a synergistic effect to maximize disease treatment.[79–81] For instance, a pH-sensitive micelle tagged by biotin showed excellent selectivity to target tumor cells only when the environmental pH was less than 7.2. In the micelle, biotin was put inside a hydrophilic PEG shell under normal physiological pH, popping up at low pH. But, the primary goal of a micellar drug carrier might be to prolong the drug circulation time, which would be localized at the tumor site by the EPR effect. Therefore, a combination of these strategies can provide a powerful therapeutic effect under *in vivo* conditions.

Why and when should targeting be done?

Considering the issues that come along with certain diseases, different approaches can be taken in terms of targeting:

- Targeting might not be needed
  - For solid tumors, the enhanced permeability and retention (EPR) effect and the manipulation of molecular weight may be sufficient for the design of effective conjugates.
- Targeting might be beneficial, but more data need to be acquired for
  - The role of addition of ligands is not yet obvious and might be only effective as they increase the molecular weight of the carrier.
- Targeting is beneficial to
  - Treatment of blood cancers, such as non-Hodgkins lymphoma.
  - Drug-free macromolecular therapeutics.

Passive targeting

A polymeric vascular drug delivery system should facilitate controlled delivery and release of the active agent at the target site. The rationale for using polymeric macromolecules as carriers for the delivery of therapeutic agents is based on the EPR effect. In certain conditions, blood vessel walls might become leaky. In tumors, blood vessels have poorly aligned defective endothelial cells with wide fenestration and lacking a smooth muscle layer. This defective vascular architecture is created by rapid vascularization because tumor cells develop so fast demanding a large supply of nutrients and oxygen. Macromolecules or leukocytes can be drained through the leaky blood vessels and be retained, the EPR effect.[82] These anatomical and functional abnormalities allow superior extravasation of

Table 1 Stimuli-responsive polymers grouped by stimulus–response: Notes on synthesis method and applications

Stimuli		Polymer	Application
Physically dependent stimuli	Temperature	Poly(*N*-isopropyl acrylamide)	Doxorubicin release from hydrogel
	Ultrasound	Polyanhydride, polyglycolide, polylactide poly(hydroxyethyl methacrylate-*co*-*N,N′*-dimethylaminoethyl methacrylate)	Ultrasound-enhanced biodegradation Ultrasound-enhanced drug release rate
	Magnetic field	Poly(ethylene-*co*-vinyl acetate)	Prompted BSA release from matrix in magnetic field
	Mechanical stress	Dihydrazide-crosslinked polyglucoronate poly(methyl methacrylate)/poly(vinyl alcohol)/cellulose ether	Pressure-sensitive hydrogel Pressure-sensitive adhesive
	Light	Poly(*N,N*-dimethyl acrylamide-*co*-4-phenyl-azophenyl acrylate) poly(*N,N*-dimethyl acrylamide-*co*-4-phenyl-azophenyl acrylamide)	Photo-sensitive active site-gating of streptavidin Photo-sensitive active site-gating of streptavidin
	Electricity	Poly(ethylenediamine-*co*-1,10-bis (chlorocarbonyl)decane) Polyethyl oxazoline/poly(methacrylate)	Electric-sensitive capsule Electrically erodible matrix for insulin delivery
Chemically dependent stimuli	Oxidation	PEG-*b*-poly(propylene sulfide)-*b*-PEG	Oxidation-sensitive polymer vesicle disintegration
	pH	Poly(acrylic acid)-g-PEG PEG–b–poly(L-histidine)	Oral insulin delivery Doxorubicin release
	Ionic strength	Poly(NIPAAm-*co*-benzo-18-crown[6]-acrylamide)	Ba^{2+} sensitive membrane pore
Biologically dependent stimuli	Enzymes	PEG–peptide linker–doxorubicin Poly(*N*-(2-hydroxypropyl)methacrylamide)–peptide linker–doxorubicin	Doxorubicin release by lysosomal enzyme-mediated peptide degradation
	Biomolecules	PEO-*b*-poly(2-glucosyloxyethyl acrylate) Thiolate PEG-*b*-poly(L-lysine) Poly(RCOOH-*co*-butyl acrylate-*co*-pyridyl disulfide acrylate)	Glucose-sensitive micelle for insulin delivery

macromolecules, nanoparticles, and lipid particles into the tumor tissue. After the macromolecules enter the interstitium, they are retained there by lack of intertumoral lymphatic clearance, accumulating at high concentrations.[83] As opposed to macromolecules, low-molecular weight compounds diffuse quickly and indiscriminately into normal and tumor tissues through the endothelial cell layer of blood capillaries, causing undesirable systemic side effects followed by quick renal clearance. When designing a polymeric drug delivery system, an important parameter is the size of the polymeric carrier, which influences the pharmacokinetic profile and the degree of accumulation at the tumor site. There is a correlation between the plasma half-life, the renal clearance, and the accumulation at the tumor site of the respective macromolecule. The normal renal threshold is in the range of 30–50 kDa; therefore, to achieve an optimal balance between these key elements, polymeric carriers with molecular weights in the range of 20 to 200 kDa are often chosen as the backbone of the drug

delivery system. Factors dictating the biodistribution profile of the macromolecule are charge, conformation, hydrophobicity, and immunogenicity of the polymeric carrier.[84] The EPR effect is affected by (i) anatomical and (ii) permeability-enhancing factors. Besides these factors, active therapeutic agents affecting the blood vessels can influence the EPR effect depending on their activity. Compounds enhancing the EPR effect include pro-inflammatory anticancer agents that generate superoxide radicals and NO and drugs found to up-regulate VEGF-like doxorubicin.[85] Cancer therapy based on the EPR effect is still under investigation. Some studies have shown that drug delivery to a target tumor cannot be achieved merely by the EPR effect. That is why, after accumulation, it is difficult for drugs or drug carriers to interact with tumor cells or to access the deeper place of tumor tissues. In addition, many drug-loading carriers should be administered to achieve a therapeutic dose since just small portions of initially given carriers are expected to be accumulated at the tumor site.

Active targeting & targeting ligands

An active targeting strategy can improve the efficacy of the therapy and diminish side effects associated with drugs. To increase the delivery of a given drug to a specific target site, targeting ligands such as antibodies, carbohydrates, aptamers, and small molecules are conjugated to carriers like liposomes, micelles, and particles.[86]

Since 1975 when monoclonal antibodies (mAb) were developed, many researchers have conjugated mAbs to drugs or to drug carriers. Conjugation of brain-specific antibodies to haloperidol-containing micelles resulted in 5- to 20-fold higher neuroleptic activity than that of non-targeted micelles and free drug. Further investigations proved that mAb-attached PLGA nanoparticles had the ability to recognize and target specific antigens to invasive breast tumors. While normal nanoparticles were taken up by both cells, the immune-nanoparticles entered only breast tumor cells cocultured with human colon adenocarcinoma cells, which demonstrated their ability to target specific cells.[87]

Because carbohydrates like galactose, lactose, and mannose were known as specific ligands of liver cell receptors, they have been used to target drug delivery for treatment of liver diseases. Aptamers are oligonucleotides or peptide molecules that can bind to specific target antigens. They have many favorable characteristics like non-immunogenicity, stability in a wide range of pH 4–9 and temperature, and tumor specificity. Moreover, aptamer synthesis is a chemical process without biological systems, allowing us to easily scale-up for clinical trial. Farokhzad et al.[88] developed aptamer-conjugated biodegradable PEG–PLGA and PEG–poly(lactic acid) (PLA) micelles encapsulating docetaxel to target the prostate-specific membrane antigen (PSMA) expressed on the surface of prostate tumor cells. These micelles showed specific affinity to PSMA-expressed prostate LNCaP epithelial cells with 77-fold binding increment compared to the control group. The aptamer-conjugated PEG–PLAGA micelles encapsulating docetaxel resulted in dramatically increased *in vitro* cytotoxicity over normal PEG–PLGA micelles. Injection of the aptamer-conjugated PEG–PLGA micelles encapsu-lating docetaxel into tumor-bearing mice exhibited low systemic toxicity as determined by mean body weight loss.

Folate has been vastly investigated for targeting various tumor cells overexpressing folate receptors, such as lung, kidney, ovary, breast, brain, uterus, and testis. The use of folate in targeted drug delivery systems is due to the convenient and easy conjugation step and the high affinity for the folate receptor after conjugation. Yoo and Park[89] introduced folate-conjugated PEG–PLGA or PEG–PCL micelles for tumor targeting. Folate-conjugated PEG–PLGA micelles encapsulating doxorubicin effectively bound KB cells *in vitro* and showed an enhanced cytotoxicity when compared to non-targeting micelles.

POLYMER-BASED NANOCARRIERS: FOCUS ON CANCER DRUG DELIVERY

The current challenge of drug delivery is to design vehicles that can carry sufficient amounts of drugs, efficiently cross various physiological barriers to reach disease sites, and cure diseases in a less toxic and sustained manner.[90] As most physiological barriers prohibit the permeation or internalization of particles or drug molecules with large sizes and undesired surface properties, the main input of nanotechnology on nanomedicine is to miniaturize and multifunctionalize drug carriers for improved drug delivery in a time and disease-specific manner.[91] In cancer drug delivery, delivery strategies can be categorized as either lipid based or polymer based. In this entry we will focus on various polymer-based nanocarriers that have been developed for cancer therapy.

Dendrimers

Dendrimers are highly branched macromolecules with small size and three dimensional structures. It has been the research interest of several researchers after the first report in 1970s. They contain a multifunctional core on which several branches are attached. The high density of branches and surface functional groups made dendrimers a potential carrier in drug delivery systems.[92] Two possible mechanisms by which the drug molecules could be delivered by dendrimers are:

- Covalent attachment of drug to the surface functional groups of dendrimers
- Drug encapsulation inside the dendrimers

The former method is easier and straightforward and thus the procedure of interest for several researchers to attach the drug to dendrimers. Dendrimers have also been modified by ligands and other chemical entities to facilitate postsynthetic manipulations such as drug attachments, PEGylation, and the installation of other molecules.[93] It has been shown that the solubility and stability of dendrimers could be enhanced by the attachment of a PEG moiety to the dendrimer molecule.[94]

Micelles

Polymeric micelles could be fabricated by self-assembly of amphiphilic copolymers if their concentrations are above the critical micellar concentration (CMC). Their assembly in aqueous solution forms a condensed inner core serving as a container for hydrophobic compounds. Usually polymeric micelles are more stable than hydrocarbon-based micelles. Therefore, they could be appropriate choices for controlled drug delivery systems. Micelles' physical properties such as size, hydrophilicity, and surface properties deeply influence their biodistribution. In order to have a micellar drug carrier with a prolonged circulation time, the micelle surface should be modified with hydrophilic and biocompatible shells such as PEG. Physiological and biological properties of micelles can be modified by the functionalization of their outer surface.[95,96] Micelles have also been widely used in ocular drug delivery systems.[97]

Vesicles

In addition to micelles, amphiphilic block copolymers could self-assemble into vesicles by forming bilayers similar to phospholipids. These polymeric structures are more stable than liposomes due to strong hydrophobic interactions between polymers than the hydrocarbon segments of liposomes.[98] In addition to polymeric vesicles, peptide vesicles have also been formed through polypeptide self-assembly. Besides drug delivery, vesicles are prone to be used as carriers of magnetic nanoparticles for MRI applications.[99]

Liposomes and Polymersomes

Liposomes and polymersomes are nano-containers fabricated through self-assembly of bilayers of phospholipids and block copolymers, respectively. These nanostructures have a hydrophilic inner cavity suitable for hydrophilic drug delivery and the hydrophobic membrane with potential for incorporating poorly soluble drugs.[100–102] Among all the nano-sized drug delivery systems, liposomes have demonstrated the most clinical applications in cancer treatment. Besides drug delivery, liposomes and polymersomes have been used in MRI applications by delivering contrast agents such as gadolinium (Gd). Moreover, they have also been suggested as promising candidates for ultrasound-triggered drug delivery.[103,104]

Polysaccharide-based Nanoparticles

Polysaccharides are the polymers of monosaccharides. In Nature, polysaccharides have various resources from algal origin (e.g., alginate), plant origin (e.g., pectin, guar gum), microbial origin (e.g., dextran, xanthan gum), and animal origin (chitosan, chondroitin). Most of the natural polysaccharides have hydrophilic groups such as hydroxyl, carboxyl, and amino groups, which could form non-covalent bonds with biological tissues and could prolong the residence time and therefore increase the absorbance of loaded drugs.[105] From the viewpoint of polyelectrolytes, polysaccharides can be divided into polyelectrolytes and non-polyelectrolytes, the former can be further divided into positively charged polysaccharides (chitosan) and negatively charged polysaccharides (alginate, heparin, hyaluronic acid, pectin, etc.).[106,107]

According to structural characteristics, polysaccharide-based nanoparticles are prepared mainly by four mechanisms namely covalent crosslinking, ionic crosslinking, polyelectrolyte complexation (PEC), and self-assembly of hydrophobically modified polysaccharides (Fig. 9).

The covalent approach is based on intermolecular crosslinking of chitosan nanoparticles through the use of biocompatible crosslinkers and biocompatible covalent crosslinking. Ionic cross linking, on the other hand, has the inherent advantages of mild preparation conditions and simple procedures compared with covalent crosslinking. Considering chitosan itself, its ioinically crosslinked derivatives can easily dissolve in neutral aqueous media, avoiding the potential toxicity of acids, and hence protecting the bioactivity of loaded biomacromolecules. Polyelectrolyte polysaccharides can form PEC with oppositely charged polymers by intermolecular electrostatic interaction.[108] Since chitosan is the only natural polycationic polysaccharide that satisfies the qualifications of a polycation in the field, major efforts have been focused on negative polymers to be complexed with chitosan to form PEC nanoparticles. Regarding the self-assembly of hydrophobically modified polysaccharides (Fig. 10), chitosan-based self-aggregates are difficult to be widely applied for drug delivery systems because chitosan aggregates are insoluble in biological solution (pH 7.4) and water-soluble chitosan derivatives have been developed in response.

POLYMERS AS DRUGS: FOCUS ON NON-CANCEROUS DISEASES

In the common sense, polymers are plastic materials that are used to produce everyday materials and known as environmental pollutants. However, by developing biomaterials, polymer applications have extended to medicine. With the development of nanobiotechnology, polymers have been used in drug delivery systems or used as sequestrants for cancerous/non-cancerous disease therapy.[109]

Polymeric Nanomedicine in Cancer Drug Delivery

Cancer nanotechnology, the application of nanotechnology and nanomaterials in cancer treatment, has been of interest for several scientists. Cancer drug delivery systems include lipid-, metal-, and polymer-based systems among which polymer-based systems have gained more attention due to the potential of polymers to be tailored. Polymeric carriers have the ability of improving pharmacokinetics and pharmacodynamics of drugs such as their aqueous solubility. Moreover, they could decrease the severe toxic side effects of cancer drugs through incorporation of the targeting moieties so that the polymeric

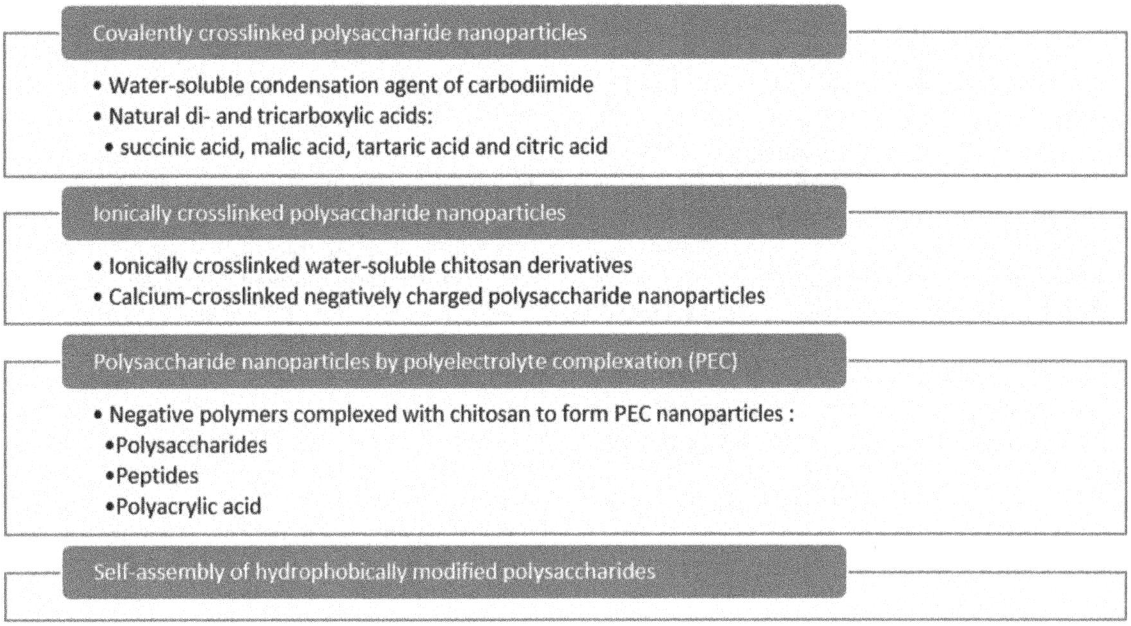

Covalently crosslinked polysaccharide nanoparticles

- Water-soluble condensation agent of carbodiimide
- Natural di- and tricarboxylic acids:
  - succinic acid, malic acid, tartaric acid and citric acid

Ionically crosslinked polysaccharide nanoparticles

- Ionically crosslinked water-soluble chitosan derivatives
- Calcium-crosslinked negatively charged polysaccharide nanoparticles

Polysaccharide nanoparticles by polyelectrolyte complexation (PEC)

- Negative polymers complexed with chitosan to form PEC nanoparticles :
  - Polysaccharides
  - Peptides
  - Polyacrylic acid

Self-assembly of hydrophobically modified polysaccharides

Nanomaterials—Nanoparticles

Fig. 9 Main mechanism for preparation of polysaccharide-based nanoparticles.

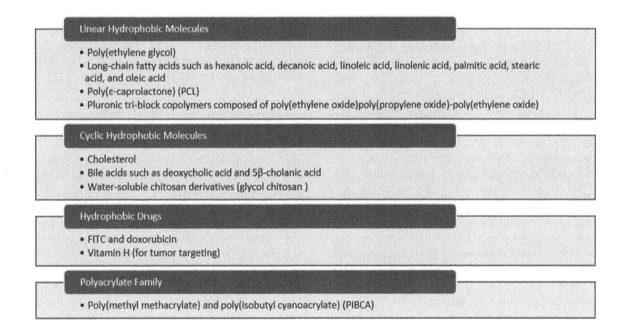

Linear Hydrophobic Molecules

- Poly(ethylene glycol)
- Long-chain fatty acids such as hexanoic acid, decanoic acid, linoleic acid, linolenic acid, palmitic acid, stearic acid, and oleic acid
- Poly(ε-caprolactone) (PCL)
- Pluronic tri-block copolymers composed of poly(ethylene oxide)poly(propylene oxide)-poly(ethylene oxide)

Cyclic Hydrophobic Molecules

- Cholesterol
- Bile acids such as deoxycholic acid and 5β-cholanic acid
- Water-soluble chitosan derivatives (glycol chitosan)

Hydrophobic Drugs

- FITC and doxorubicin
- Vitamin H (for tumor targeting)

Polyacrylate Family

- Poly(methyl methacrylate) and poly(isobutyl cyanoacrylate) (PIBCA)

Fig. 10 Hydrophobic molecules used to modify self-assembling polysaccharides.

nanoparticles could deliver the drug specifically to the cancerous site.

Due to their multifunctional nature, polymeric carriers could also serve as carriers of imaging agents. When integrating therapeutic and diagnostic functions, the delivery system is entitled as "theranostic nanomedicine."

Polymer–Drug Nanomedicine

Polymer drug nanomedicine could be assembled through conjugation of drugs to polymer molecules or physical encapsulation of drugs inside the polymer matrix. In polymer–drug conjugation, the drug molecule could be released through bond cleavage, whereas in the drug encapsulation method, the drug diffuses from the polymer matrix. One of the most interesting polymers used as a carrier of nanomedicine is N-(2-hydroxypropyl)methacrylamide (HPMA) copolymer studied by Kopeček, Duncan, Ghandehari and other scientists.[110] Their research led to several classes of HPMA–drug conjugates that entered the clinical trials.[1] The first polymer–protein conjugate approved for treatment of human cancer was SMANCS, obtained by covalent conjugation of the anticancer drug Neocarzinostatin (NCS) to two styrene maleic anhydride (SMA) polymers, which entered the Japanese market in 1994.

The first report on the controlled release of drug from the polymer matrix was the release of encapsulated soybean trypsin inhibitor from an ethyl–vinyl acetate copolymer matrix developed by Langer and Folkman.[111] These research studies continued till 1996 when the first controlled release cancer treatment system, Gliadel®, was approved by the FDA for brain tumor treatment.

Polymeric Sequestrants as Efficient Nanomedicine

Various potentially harmful substances exist in human GI tract. Overconcentration of these substances could cause cardiovascular, renal, and genetic diseases. A strategy to treat such disease is to remove the causative agents from the GI tract.[99] Polymeric drugs are defined as polymers that are active pharmaceutical ingredients, i.e., they are neither drug carriers nor prodrugs. In general, the high molecular weight and functional characteristics of such polymers let them selectively recognize, sequester, and remove low molecular weight and macromolecular disease-causing species in the intestinal fluid. The high molecular weight nature of these therapeutically relevant polymers makes them systemically non-absorbed, thus providing a number of advantages including long-term safety profiles over traditional small molecule drug products. Furthermore, multiple functional groups in the polymers incorporate polyvalent binding interactions that can result in pharmaceutical properties not found in small molecule drugs.[112]

Polymers as phosphate sequestrants: treatment of chronic renal failure

Increase in serum phosphorus levels could lead to end-stage renal disease (ESRD). As a consequence, these patients may face other diseases such as soft tissue calcification (leading to cardiac calcification and other cardiac-related complications), renal bone disease leading to reduced bone density, and secondary hyperparathyroidism. Naturally, the kidney removes the excess phosphate from the body. Patients with impaired kidney function should undergo phosphate binder therapy using agents such as metal salts which form insoluble phosphates. Nevertheless, long-term exposure to metal salts can have toxic side effects such as neurological disorders. An alternative to these phosphate sequestrants could be cationically charged polymers. Several research studies have been done to develop polycationic compounds such as cationic hydrogels, oligomeric or macrocyclic amines, or other compounds as receptors of phosphate-containing anionic molecules.[113] By tailoring polymer structures and functional group densities, they are confined to the GI tract acting as effective therapeutic agents through absorption of phosphate ions.[114]

Polymers as iron sequestrants: treatment of iron overload disorder

Although iron is a crucial element for many biological systems, it has toxic effects when present in excess concentration in the body, e.g., it could catalyze the molecular oxygen transformation to toxic radicals. An attractive approach to decrease the iron concentration is by using oral iron chelators such as desferrioxamine to remove iron overload from the GI tract. However, these chelators have several deficiencies such as poor oral bioavailability. An alternative moiety that could selectively sequester iron from the GI tract could be polymeric materials. An appropriate polymeric sequestrant for excess iron could have high affinity and binding capacity for iron ions while it is biocompatible and nonabsorbable through the GI tract.[115] Polymeric hydrogels containing catechol moieties and hydroxamic acid have been proposed as iron chelators as well as dendrimers containing surface hydroxypyridinone and catechol moieties.

Polymers as bile acid sequestrants: cholesterol-lowering drugs

Increased incidence of cardiovascular disease in the modern society is in part due to elevated levels of serum cholesterol. One of the most widely used classes of drugs to lower cholesterol concentration is 3-hydroxy-3-methylglutaryl-coenzyme A (HMG-CoA) reductase inhibitors known as the statins.[116] Safety issues associated with statins shed light on research for alternative therapy.

The biotransformation of cholesterol in the liver produces bile acid which is a biological surfactant aiding the digestion of food in the GI tract. Since removal of bile acid from the GI tract could up-regulate its synthesis from cholesterol, bile acid sequestration is introduced as an attractive therapeutic approach to treat cardiovascular diseases. Cationic polymeric gels could bind anionic bile acid in the GI tract. These polymeric drugs are finally removed from the body through fecal excretion thus eliminating the systemic side effects of statins. The structure and physicochemical properties of bile acid reveal that it has an anionic group and a hydrophobic core. Therefore, the polymeric bile acid sequestrants should have an optimal density of cationic groups and a balanced combination of hydrophobicity and hydrophilicity for enhanced capacity and secondary interactions, respectively.

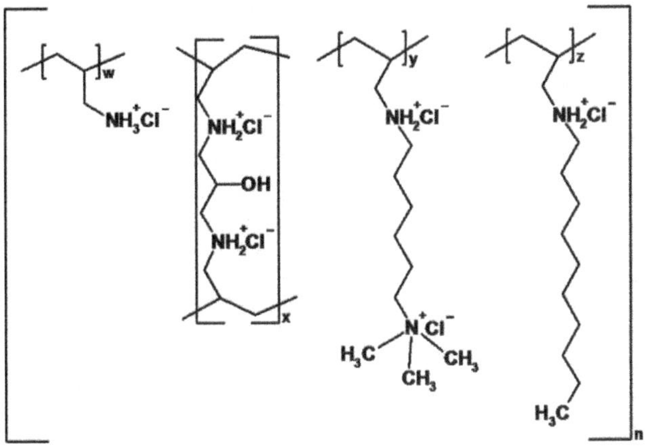

Fig. 11 Structural features of Welchol™, a new generation bile acid sequestrant.

Considering the aforementioned specifications it is proposed to use cationic hydrogels including vinyl polymers, methacrylates, methacrylamide, or polyether as the polymer backbone.[117] A novel marketed polymer approved as cholesterol lowering therapy is WelChol™ which is used for treatment of hypercholesterolemia (Fig. 11).

Polymers as toxin sequestrants: polyvalent interactions

Microbe-derived toxins are the main causatives of most bacterial infections. A traditional approach to remove toxins is by using antibiotics. Although effective in treating the infection, antibiotics have deficiencies such as sterilizing the gut thus increasing the chance of further infections and inducing antibiotic resistance.[118,119]

These protein toxins have multiple sites thus facilitating their conjugation to polymeric sequestrants. These sequestrants have the benefit of covalently conjugating different toxin molecules to different surface ligands.[120]

COMPUTATIONAL APPROACHES TO THE RATIONAL DESIGN OF POLYMERIC CARRIERS

Computational Methods to Predict Physicochemical Properties: Analytical Models vs. Molecular Simulations

Analytical models: identifying drug–material pairs

Analytical models are preliminary tools to reduce the number of evaluations to be conducted which rapidly eliminate material combinations that are unlikely to be miscible. Promising drug–material pairs can be rapidly identified by predicting the strength of drug–material interactions through the estimation of parameters such as solubility, lipophilicity, and chemical compatibility.

Solubility and solubility parameters: Solubility parameter (SP) is a scalar value that gives an indication of the predicted miscibility of two components. Materials with similar SPs are indicated to be miscible. SPs are useful for rapidly ranking materials based on their predicted relative abilities to solubilize a drug. Main approaches for calculation of SPs have been proposed by Hansen and colleagues.[121–123]

Flory–Huggins interaction parameter: Represents the interactions that contribute to the enthalpy of mixing a polymer and a solvent. The miscibility of compounds could be experimentally determined using enthalpy changes associated with creation of a binary mixture. Theoretically, this miscibility is obtained from Monte Carlo and Molecular Dynamics simulations. Using group contribution implementations of Hildebrand and Hansen SPs is another approach to calculate the miscibility.[124,125]

Lipophilicity: Lipophilicity represents the affinity of a molecule for a lipidic environment.[126] Lipophilicity can be determined by measuring the partition coefficient, P, which is the ratio of solute concentrations in binary phases of organic and aqueous solvents. Because the value of P ranges widely, the lipophilicity of compounds is represented as the logarithm of P, $\log P$. The value of $\log P$ is also related to hydrophobicity, a phenomenon that drives the association of non-polar groups or molecules in an aqueous environment. Lipophilicity and hydrophobicity can also be used to predict drug retention in a formulation or drug permeability through a membrane.

Molecular simulations: architectural and conformational contributions to drug–material compatibility

Efficient and stable encapsulation of hydrophobic compounds into nanoparticles is governed not only by solubilities of drugs and materials but also by rigidity, conformation, and MW of the drug and the material. Molecular simulations will give additional information on size, conformation, and the interfacial structure of the drug–material or material–material pairs. Simulations give insights into fundamental interactions governing the drug–polymer assembly, elucidating the physical and chemical features that can be modified to influence drug loading or release. The optimizing materials are as follows:

- *Linear Polymers*: A random-flight model has been proposed for free linear polymers, in which the polymer size depends on both polymer length and solvent suitability. As an extension of the random flight model, De Gennes theory has been proposed for polymers attached to a planar surface. Daoud and Cotton theory extends de Gennes theory to star-shaped polymer models.[127]

- *Micelles Formed from Linear Block Copolymers*: Molecular simulations have been employed to investigate the structure, dynamics, and self-aggregation properties of polymeric micelles, with or without drugs.

Simulation studies on micelles lead to insights into preferred interactions and localization of a drug among a variety of distinct chemical environments within the micellar delivery system. The results can be used to rationally design a new linker between the hydrophobic and hydrophilic blocks.[128] Although implicit water molecules are used in many studies to lower the calculation expenses, water molecules have a significant influence on the conformation and aggregation morphology of many delivery materials.[129] In many cases, therefore, simulations must be conducted in the presence of explicit molecular water so as to obtain results that are comparable to experiments.

- *Dendrimers*: Stable PAMAM dendrimer–PEG conjugates have been designed using CG simulations based on the MARTINI parameterization. In this study, a total of 11 dendrimers were simulated in CG water and counterions. Binary aggregation was studied for G4 dendrimer–PEG conjugates.[130] Although CG methods are unable to provide information on interactions at the atomistic level, such as hydrogen-bonding interactions, they provide excellent insight into large systems such as those investigating the aggregation of complex materials. Molecular dynamic investigations can in turn be of use in the interpretation of drug inclusion and release properties in dendritic architectures.
- *Predicting biological performance*: In general, computational studies can predict biological performance of therapeutics through investigations into cytotoxicity, biodistribution, and interaction with lipid bilayers. It is clear that structural properties of nanoparticles, such as shape and internal structure, can influence cytotoxicity. Importantly, the shape of nanoparticles also influences pharmacokinetics and tumor accumulation. Further, the permeability of lipid bilayers into nanoparticles is theoretically predictable. Overall, atomistic and CG simulations are excellent methods for investigating the nanoparticle structure, which can then be correlated to experimentally determined biological responses. Correlations between molecular or physicochemical properties and biological behavior may also be conducted with group contribution-based methods.

CONCLUSION AND FUTURE PROSPECTS

Polymeric biomaterials have shown good potential as polymer therapeutics over the past few decades. With the advent of "nano" it became more efficient and promising so that higher payloads of drugs could be localized in the targeted tissues/cells where needed. The context of polymeric nanomedicine is now widely being explored by researchers. Over the last decade several new companies have been born working on novel polymeric nanomedicines that are designed for a broad range of diseases. As the molecular basis of disease is interpreted, new polymeric nanomedicine choices have emerged. Successful progression of these polymeric

nanomedicines toward clinical development requires a wide collaboration between preclinical and clinical experts.

A major advantage of polymeric nanomedicines is their multifunctionality. Their surface could be tailored to reach specific targets in the body while minimizing the nonspecific uptake thus decreasing the drug's side effects. These nanostructures could be used for a variety of applications simultaneously including therapeutics and diagnostics. By developing the multifunctional materials and increasing the applications of polymeric nanomedicine the need for an updated regulatory framework would be inevitable. A new regulatory environment should be introduced specifically for nano-sized particles to ensure the safe access of these innovative therapeutics by patients.

Researchers have shown more interest in developing innovative routes for

- Polymer synthesis including reversible addition–fragmentation chain transfer polymerization (RAFT), atom transfer radical polymerization (ATRP) and "one-pot" synthesis methods for hyperbranched polymers
- Linker chemistry development including click chemistry and peptide linkers
- Characterization and monitoring methods including total internal reflection fluorescence (TIRF), fluorescence correlation spectroscopy (FCS), electron spin resonance spectroscopy (ESR), and fluorescence microscopy.

Combination therapy is another exciting development in the field of drug delivery. Combination therapy involves polymer–drug conjugates in addition to low molecular weight chemo/radiotherapy drugs. There are growing numbers of *in vitro* and *in vivo* studies confirming the possibilities of combination therapy in treatment of disease.

Another promising field is molecularly imprinted polymers (MIPs) in which a polymer network is formed through polymerization of functional monomers on a desired template molecule. Then the template is removed and the product is a MIP. There could be a great affinity between polymer pendant groups and the drug template resulting in zero order release of the drug with enhanced circulation time. MIPs have great potential in ocular drug delivery owing to improved bioavailability, longer retention time, and slow drug release.

Subcellular drug delivery is also an encouraging field in which biotherapeutics are conveyed to nucleus, cytosol, or mitochondria. Such drug delivery systems should evade from extracellular and intracellular trafficking barriers. To this end, development of biomimetic polymers capable of emerging with an endosomal membrane could be a scientific proposal. Depending on the ionic nature of polymers, anionic polymers bearing carboxyl groups, or cationic polymers bearing amine groups, diverse mechanisms of endosomal disruption exist.

In summary, the continuing advances in computer performance are allowing atomistic simulations of macromolecules over ever-growing time scales. Concurrently, analytical methods can be applied to larger libraries of drugs and materials. This promises a bright future for the role of theoretical methods in the development of polymeric drug delivery nanomaterials.

REFERENCES

1. Duncan, R. The dawning era of polymer therapeutics. Nat. Rev Drug Discov. **2003**, *2* (5), 347–360.
2. Vicent, M.J.; Ringsdorf, H.; Duncan, R. Polymer therapeutics: Clinical applications and challenges for development. Adv. Drug Deliv. Rev. **2009**, *61* (13), 1117–1120.
3. Abuchowski, A.; McCoy, J.R.; Palczuk, N.C.; van Es, T.; Davis, F.F. Effect of covalent attachment of polyethylene glycol on immunogenicity and circulating life of bovine liver catalase. J. Biol. Chem. **1977**, *252* (11), 3582–3586.
4. Vasir, J.K.; Labhasetwar, V. Polymeric nanoparticles for gene delivery. Expert. Opin. Drug Deliv. **2006**, *3* (3), 325–344.
5. Webster, R.; Didier, E.; Harris, P.; Siegel, N.; Stadler, J.; Tilbury, L.; Smith, D. PEGylated proteins: Evaluation of their safety in the absence of definitive metabolism studies. Drug Metab. Dispos. **2007**, *35* (1), 9–16.
6. Thanou, M.; Duncan, R. Polymer-protein and polymer-drug conjugates in cancer therapy. Curr. OpinInvestig. Drugs. **2003**, *4* (6), 701–709.
7. Duncan, R. Polymer conjugates as anticancer nanomedicines. Nat. Rev. Cancer. **2006**, *6* (9), 688–701.
8. Malik, N.; Wiwattanapatapee, R.; Klopsch, R.; Lorenz, K.; Frey, H.; Weener, J.W.; Meijer, E.W.; Paulus, W.; Duncan, R. Dendrimers: Relationship between structure and biocompatibility *in vitro*, and preliminary studies on the biodistribution of 125I-labelled polyamidoaminedendrimers *in vivo*. J. Control. Release **2000**, *65* (1–2), 133–148.
9. Kaplan, D.L.; Wiley, B.J.; Mayer, J.M.; Arcidiacono, S.; Keith, J.; Lombardi, S.J.; Ball, D.; Allen, A.L. Bioabsorbable poly(ester-amides). In *Biomedical Polymers: Designed-to-Degrade Systems*; Shalaby, S.W., Ed.; Hanser Publications: Munich, 1994; 189–212.
10. Gopferich, A. Mechanisms of polymer degradation and erosion. Biomaterials **1996**, *17* (2), 103–114.
11. Holland, T.A.; Mikos, A.G. Review: Biodegradable polymeric scaffolds. Improvements in bone tissue engineering through controlled drug delivery. Adv. Biochem. Eng. Biotechnol. **2006**, *102*, 161–185.
12. Agrawal, C.M.; Athanasiou, K.A.; Heckman, J.D. Biodegradable PLA-PGA polymers for tissue engineering in orthopaedics. Mater. Sci. Forum, **1997**, *250*, 115–128.
13. Athanasiou, K.A.; Niederauer, G.G.; Agrawal, C.M. Sterilization, toxicity, biocompatibility and clinical applications of polylactic acid/polyglycolic acid copolymers. Biomaterials **1996**, *17* (2), 93–102.
14. Pitt, G.G.; Gratzl, M.M.; Kimmel, G.L.; Surles, J.; Sohindler, A. Aliphatic polyesters II. The degradation of poly (DL-lactide), poly (ε-caprolactone), and their copolymers *in vivo*. Biomaterials **1981**, *2* (4), 215–220.
15. Andrade, J.D.; Hlady, V.; Jeon, S.-I. Poly (ethylene oxide) and protein resistance. In *Hydrophilic Polymers, Performance with Environmental Acceptance*; Glass, J.E. Eds.; Advances in Chemistry Series: Washington, DC, 1996; 248, 51–59.
16. Tirelli, N.; Lutolf, M.P.; Napoli, A.; Hubbell, J.A. Poly (ethylene glycol) block copolymers. J Biotechnol. **2002**, *90* (1), 3–15.
17. Chasin, M.; Langer, R.S. *Biodegradable Polymers as Drug Delivery Systems*; Marcel Dekker, Inc.: New York, 1990; 121–161.
18. Roskos, K.V.; Fritzinger, B.K.; Rao, S.S.; Armitage, G.C.; Heller, J. Development of a drug delivery system for the treatment of periodontal disease based on bioerodible poly (ortho esters). Biomaterials **1995**, *16* (4), 313–317.
19. Leong, K.W.; Brott, B.C.; Langer, R. Bioerodible polyanhydrides as drug-carrier matrices. I: Characterization, degradation, and release characteristics. J. Biomed. Mater. Res. **1985**, *19* (8), 941–955.
20. Mathiowitz, E.; Jacob, J.S.; Jong, Y.S.; Carino, G.P.; Chickering, D.E.; Chaturvedi, P.; Santos, C.A.; Vijayaraghavan, K.; Montgomery, S.; Bassett, M.; Morrell, C. Biologically erodable microspheres as potential oral drug delivery systems. Nature **1997**, *386* (6623), 410–414.
21. Laurencin, C.; Domb, A.; Morris, C.; Brown, V.; Chasin, M.; McConnell, R.; Lange, N.; Langer, R. Poly (anhydride) administration in high doses *in vivo*: Studies of biocompatibility and toxicology. J. Biomed. Mater. Res. **1990**, *24* (11), 1463–1481.
22. Katti, D.S.; Lakshmi, S.; Langer, R.; Laurencin, C.T. Toxicity, biodegradation and elimination of polyanhydrides. Adv. Drug Deliv. Rev. **2002**, *54* (7), 933–961.
23. Kopeček, J. Controlled biodegradability of polymers—A key to drug delivery systems. Biomaterials **1984**, *5* (1), 19–25.
24. Chaubal, M.V.; Gupta, A.S.; Lopina, S.T.; Bruley, D.F. Polyphosphates and other phosphorus-containing polymers for drug delivery applications. Crit. Rev. Ther. Drug Carrier Syst. **2003**, *20* (4), 295–315.
25. Monge, S.; Canniccioni, B.; Graillot, A.; Robin, J.J. Phosphorus-containing polymers: A great opportunity for the biomedical field. Biomacromolecules **2011**, *12* (6), 1973–1982.
26. Uhrich, K.E.; Cannizzaro, S.M.; Langer, R.S.; Shakesheff, K.M. Polymeric systems for controlled drug release. Chem Rev. **1999**, *99* (11), 3181–3198.
27. Teasdale, I.; Brüggemann, O. Polyphosphazenes: Multifunctional, Biodegradable Vehicles for Drug and Gene Delivery. Polymers **2013**, *5* (1), 161–187.
28. Dahiyat, B.; Richards, M.; Leong, K. Controlled release from poly(phosphoester) matrices. J. Controlled Release. **1995**, *33* (1), 13–21.
29. Wang, Y.C.; Yuan, Y.Y.; Du, J.Z.; Yang, X.Z.; Wang, J. Recent progress in polyphosphoesters: From controlled synthesis to biomedical applications. Macromol. Biosci. **2009**, *9* (12), 1154–1164.
30. Kopeček, J. Polymer–drug conjugates: Origins, progress to date and future directions. Adv. Drug Deliv. Rev. **2013**, *65* (1), 49–59.
31. Chung, J.H.; Simmons, A.; Poole-Warren, L.A. Nondegradable polymer nanocomposites for drug delivery. Expert Opin. Drug Deliv. **2011**, *8* (6), 765–778.

32. Fu, Y.; Kao, W.J. Drug release kinetics and transport mechanisms of non-degradable and degradable polymeric delivery systems. Expert Opin. Drug Deliv. **2010**, *7* (4), 429–444.

33. Schmaljohann, D. Thermo-and pH-responsive polymers in drug delivery. Adv. Drug Deliv. Rev. **2006**, *58* (15), 1655–1670.

34. Gil, E.S.; Hudson, S.M. Stimuli-reponsive polymers and their bioconjugates. Prog. Polym. Sci. **2004**, *29* (12) 1173–1222.

35. Cheng, R.; Meng, F.; Deng, C.; Klok, H.A.; Zhong, Z. Dual and multi-stimuli responsive polymeric nanoparticles for programmed site-specific drug delivery. Biomaterials **2013**, *34* (14), 3647–3657.

36. Raucher, D.; Moktan, S. Thermo-Responsive Polymers. In *Drug Delivery in Oncology: From Basic Research to Cancer Therapy*; Kratz, F.,Senter, P., Steinhagen, H. Eds.; Wiley-VCH Verlag GmbH & Co. KGaA: Weinheim, Germany, 2011; 667–700.

37. Shaikh, R.P.; Pillay, V.; Choonara, Y.E.; du Toit, L.C.; Ndesendo, V.M.; Bawa, P.; Cooppan, S. A review of multi-responsive membranous systems for rate-modulated drug delivery. AAPS Pharm. Sci. Tech. **2010**, *11* (1), 441–459.

38. Qiu, Y.; Park, K. Environment-sensitive hydrogels for drug delivery. Adv Drug Deliv Rev. **2001**, *53* (3), 321–339.

39. Roy, D.; Cambre, J.N.; Sumerlin, B.S. Future perspectives and recent advances in stimuli-responsive materials. Prog. Polym. Sci. **2010**, *35* (1), 278–301.

40. Delcea, M.; Möhwald, H.; Skirtach, A.G. Stimuli-responsive LbL capsules and nanoshells for drug delivery. Adv. Drug Deliv. Rev. **2011**, *63* (9), 730–747.

41. Liechty, W.B.; Kryscio, D.R.; Slaughter, B.V.; Peppas, N.A. Polymers for drug delivery systems. Annu. Rev. Chem. Biomol. Eng. **2010**, *1*, 149–173.

42. Ganta, S.; Devalapally, H.; Shahiwala, A.; Amiji, M. A review of stimuli-responsive nanocarriers for drug and gene delivery. J. Control Release. **2008**, *126* (3), 187–204.

43. Mendes, P.M. Stimuli-responsive surfaces for bio-applications. Chem. Soc. Rev. **2008**, *37* (11), 2512–2529.

44. Stuart, M.A.; Huck, W.T.; Genzer, J.; Müller, M.; Ober, C.; Stamm, M.; Sukhorukov, G.B.; Szleifer, I.; Tsukruk, V.V.; Urban, M.; Winnik, F.; Zauscher, S.; Luzinov, I.; Minko, S. Emerging applications of stimuli-responsive polymer materials. Nat Mater. **2010**, *9* (2), 101–113.

45. Kagatani, S.; Shinoda, T.; Konno, Y.; Fukui, M.; Ohmura, T.; Osada, Y. Electroresponsive pulsatile depot delivery of insulin from poly(dimethylaminopropylacrylamide) gel in rats. J. Pharm. Sci. **1997**, *86* (11), 1273–1277.

46. Murdan, S. Electro-responsive drug delivery from hydrogels. J. Control. Release. **2003**, *92* (1–2), 1–17.

47. Ichimura, K.; Oh, SK.; Nakagawa, M. Light-driven motion of liquids on a photoresponsive surface. Science **2000**, *288* (5471), 1624–1626.

48. Wang, S.; Song, Y.; Jiang, L. Photoresponsive surfaces with controllable wettability. J. Photochem. Photobiol. C. Photochem. Rev. **2007**, *8* (1), 18–29.

49. Li, Y.; Jia, X.; Gao, M.; He, H.; Kuang, G.; Wei, Y. Photoresponsivenanocarriers based on PAMAM dendrimers with ao-nitrobenzyl shell. J. Polym. Sci., Part A: Polym. Chem. **2010**, *48* (3), 551–557.

50. Jiang, X.; Lavender, C.A.; Woodcock, J.W.; Zhao, B. Multiple micellization and dissociation transitions of thermo-and light-sensitive poly (ethylene oxide)-b-poly (ethoxytri (ethylene glycol) acrylate-co-o-nitrobenzyl acrylate) in water. Macromolecules, **2008**, *41* (7), 2632–2643.

51. Dai, S.; Ravi, P.; Tam, K.C. pH-Responsive polymers: Synthesis, properties and applications. Soft Matter **2008**, *4* (3), 435–449.

52. Murthy, N.; Campbell, J.; Fausto, N; Hoffman, A.S.; Stayton, P.S. Bioinspired pH-responsive polymers for the intracellular delivery of biomolecular drugs. Bioconjug Chem. **2003**, *14* (2), 412–419.

53. Kwon, Y.M.; Kim, S.W. Thermosensitive biodegradable hydrogels for the delivery of therapeutic agents. In *Polymeric Drug Delivery Systems*; Kwon, G.S., Eds.; Taylor and Francis Group: Boca Raton, FL, 2005; 251–274.

54. Zentner, G.M.; Rathi, R.; Shih, C.; McRea, J.C.; Seo, M.H.; Oh, H.; Rhee, B.G.; Mestecky, J.; Moldoveanu, Z.; Morgan, M.; Weitman, S. "Biodegradable block copolymers for delivery of proteins and water-insoluble drugs. J Control. Release. **2001**, *72* (1–3), 203–215.

55. Kost, J.; Langer, R. Responsive polymeric delivery system. Adv. Drug Deliv. Rev. **2001**, *46* (1–3), 125–148.

56. Abdelaal, M.Y.; Abdel-Razik, E.A.; Abdel-Bary, E.M.; El-Sherbiny, I.M. Chitosan-based interpolymeric pH-responsive hydrogels for *in vitro* drug release. J. Appl. Polym. Sci. **2007**, *103* (5), 2864–2874.

57. Gupta, P.; Vermani, K.; Garg, S. Hydrogels: From controlled release to pH-responsive drug delivery. Drug Discov. Today. **2002**, *7* (10), 569–579.

58. Narayani, R.; Rao, K.P. pH-responsive gelatin microspheres for oral delivery of anticancer drug methotrexate. J. Appl. Polym. Sci. **1995**, *58* (10), 1761–1769.

59. Sokker, H.H.; Abdel Ghaffar, A.M.; Gad, Y.H.; Aly, A.S. Synthesis and characterization of hydrogels based on grafted chitosan for the controlled drug release. Carbohydr. Polym. **2009**, *75* (2), 222–229.

60. Langer, R.; Peppas, N.A. Advances in biomaterials, drug delivery, and bionanotechnology. AIChE J. **2003**, *49* (12), 2990–3006.

61. Sideratou, Z.; Tsiourvas, D.; Paleos, C.M. Quaternized poly (propylene imine) dendrimers as novel pH-sensitive controlled-release systems. Langmuir **2000**, *16* (4), 1766–1769.

62. Burke, S.E.; Barrett, C.J. pH-responsive properties of multilayered poly (L-lysine)/hyaluronic acid surfaces." Biomacromolecules **2003**, *4* (6), 1773–1783.

63. Magnusson, J.P.; Khan, A.; Pasparakis, G. Ion-sensitive "isothermal" responsive polymers prepared in water. J. Am. Chem. Soc. **2008**, *130* (33), 10852–10853.

64. Szczubiałka, K.; Moczek, Ł.; Błaszkiewicz, S.; Nowakowska, M. Photocrosslinkable smart terpolymers responding to pH, temperature, and ionic strength. J. Polym. Sci. A Polym. Chem. **2004**, *42* (15), 3879–3886.

65. Shenoy, D.; Little, S.; Langer, R.; Amiji, M. Poly (ethylene oxide)-modified poly (β-amino ester) nanoparticles as a pH-sensitive system for tumor-targeted delivery of hydrophobic drugs: Part 2. *In vivo* distribution and tumor localization studies. Pharm. Res. **2005**, *22* (12), 2107–2114.

66. Matsumoto, S.; Christie, R.J.; Nishiyama, N.; Miyata, K.; Ishii, A.; Oba, M.; Koyama, H.; Yamasaki, Y.; Kataoka, K. Environment-responsive block copolymer micelles with a disulfide cross-linked core for enhanced siRNA delivery. Biomacromolecules **2008**, *10* (1), 119–127.

67. Park, J.H.; Lee, S.; Kim, J.H.; Park, K.; Kim, K.; Kwon, I.C. Polymeric nanomedicine for cancer therapy. Prog. Polym. Sci. **2008**, *33* (1), 113–137.

68. Hein, C.D.; Liu, X.M.; Wang, D. Click chemistry, a powerful tool for pharmaceutical sciences. Pharm. Res. **2008**, *25* (10), 2216–2230.

69. Ulijn, R.V. Enzyme-responsive materials: A new class of smart biomaterials. J. Mater. Chem. **2006**, *16* (23) 2217–2225.

70. Thornton, P.D.; McConnell, G.; Ulijn, R.V. Enzyme responsive polymer hydrogel beads. Chem. Commun. **2005**, *47*, 5913–5915.

71. Itoh, Y.; Matsusaki, M.; Kida, T.; Akashi, M. Enzyme-responsive release of encapsulated proteins from biodegradable hollow capsules. Biomacromolecules **2005**, *7* (10), 2715–2718.

72. Kelkar, S.S.; Reineke, T.M. Theranostics: Combining imaging and therapy. Bioconjugate chem. **2011**, *22* (10), 1879–1903.

73. Janib, S.M.; Moses, A.S.; MacKay, J.A. Imaging and drug delivery using theranostic nanoparticles. Adv. Drug Deliv. Rev. **2010**, *62* (11), 1052–1063.

74. Miyata, K.; Kakizawa, Y.; Nishiyama, N.; Harada, A.; Yamasaki, Y.; Koyama, H.; Kataoka, K. Block catiomer-polyplexes with regulated densities of charge and disulfide cross-linking directed to enhance gene expression. J. Am. Chem. Soc. **2004**, *126* (8), 2355–2361.

75. Greish, K. Enhanced permeability and retention effect for selective targeting of anticancer nanomedicine: Are we there yet? Drug Discovery Today: Technologies **2012**, *9* (2), e161–e166.

76. Greish, K. Enhanced permeability and retention (EPR) effect for anticancer nanomedicine drug targeting. Methods Mol. Biol. **2010**, *624*, 25–37.

77. Maeda, H.; Wu, J.; Sawa, T.; Matsumura, Y.; Hori, K. Tumor vascular permeability and the EPR effect in macromolecular therapeutics: A review. J. Control. Release. **2000**, *65* (1–2), 271–284.

78. Iyer, A.K.; Khaled, G.; Fang, J.; Maeda, H.l. Exploiting the enhanced permeability and retention effect for tumor targeting. Drug Discov. Today. **2006**, *11* (17–18), 812–818.

79. Langer, R. Drug delivery and targeting. Nature **1998**, *392* (Suppl 6679), 5–10.

80. Torchilin, V.P. Passive and active drug targeting: Drug delivery to tumors as an example. Handb. Exp. Pharmacol. **2010**, (197), 3–53.

81. Greco, F.; Vicent, M.J. Combination therapy: Opportunities and challenges for polymer–drug conjugates as anticancer nanomedicines. Adv. Drug Deliv. Rev. **2009**, *61* (13), 1203–1213.

82. Greish, K. Enhanced permeability and retention of macromolecular drugs in solid tumors: A royal gate for targeted anticancer nanomedicines. J. Drug Targeting **2007**, *15* (7–8), 457–464.

83. Maeda, H.; Sawa, T.; Konno, T. Mechanism of tumor-targeted delivery of macromolecular drugs, including the EPR effect in solid tumor and clinical overview of the prototype polymeric drug SMANCS. J. Control. Release **2001**, *74* (1–3), 47–61.

84. Duncan, R. Designing polymer conjugates as lysosomotropicnanomedicines. Biochem. Soc. Trans. **2007**, *35* (1), 5660.

85. Segal, E.; Satchi-Fainaro, R. Design and development of polymer conjugates as anti-angiogenic agents. Adv. Drug Deliv. Rev. **2009**, *61* (13), 1159–1176.

86. Byrne, J.D.; Betancourt, T.; Brannon-Peppas, L. Active targeting schemes for nanoparticle systems in cancer therapeutics. Adv. Drug Deliv. Rev. **2008**, *60* (15), 1615–1626.

87. Kim, S.; Kim, J.H.; Jeon, O.; Kwon, I.C., Park, K. Engineered polymers for advanced drug delivery. Eur. J. Pharm. Biopharm. **2009**, *71* (3), 420–430.

88. Gu, F.; Zhang, L.; Teply, B.A.; Mann, N.; Wang, A.; Radovic-Moreno, A.F.; Langer, R.; Farokhzad, O.C. Precise engineering of targeted nanoparticles by using self-assembled biointegrated block copolymers. PNAS **2007**, *105* (7), 2586–2591.

89. Yoo, H.S.; Park, T.G. Folate-receptor-targeted delivery of doxorubicin nano-aggregates stabilized by doxorubicin–PEG–folate conjugate, J. Control. Release **2004**, *100* (2), 247–256.

90. Ferrari, M. Cancer nanotechnology: Opportunities and challenges. Nat. Rev. Cancer. **2005**, *5* (3), 161–171.

91. Farokhzad, O.C.; Langer, R. Nanomedicine: Developing smarter therapeutic and diagnostic modalities. Adv. Drug Deliv. Rev. **2006**, *58* (14), 1456–1459.

92. Dufès, C. Dendrimer-based drug delivery systems: From theory to practice. ChemMedChem **2013**, *8* (2), 336–336.

93. Menjoge, A.R.; Kannan, R.M.; Tomalia, D.A. Dendrimer-based drug and imaging conjugates: Design considerations for nanomedical applications. Drug Discov. Today. **2010**, *15* (5–6), 171–185.

94. Astruc, D.; Boisselier, E.; Ornelas, C. Dendrimers designed for functions: From physical, photophysical, and supramolecular properties to applications in sensing, catalysis, molecular electronics, photonics, and nanomedicine. Chem. Rev. **2010**, *110* (4), 1857–959.

95. Kataoka, K.; Harada, A.; Nagasaki, Y. Block copolymer micelles for drug delivery: Design, characterization and biological significance. Adv. Drug Deliv. Rev. **2001**, *47* (1), 113–131.

96. Lavasanifar, A.; Samuel, J.; Kwon, G.S. Poly(ethylene oxide)-block-poly(L-amino acid) micelles for drug delivery. Adv. Drug Deliv. Rev. **2002**, *54* (2), 169–190.

97. Lin, H.R.; Chang, P.C. Novel pluronic-chitosan micelle as an ocular delivery system. J. Biomed. Mater Res. B Appl. Biomater. **2013**, *in press*.

98. Ren, T.; Liu, Q.; Lu, H.; Liu, H.; Zhang, X.; Du, J. Multifunctional polymer vesicles for ultrasensitive magnetic resonance imaging and drug delivery. J. Mater. Chem. **2012**, *22* (24), 12329–12338.

99. Arruebo, M.; Fernández-Pacheco, R.; Ibarra, M.R.; Santamaría, J. Magnetic nanoparticles for drug delivery. Nano Today **2007**, *2* (3), 22–32.

100. Barenholz, Y. Liposome application: Problems and prospects. Curr. Opin. Colloid Interface Sci. **2001**, *6* (1), 66–77.

101. Discher, D.E.; Ahmed, F. Polymersomes. Annu. Rev. Biomed. Eng. **2006**, *8*, 323–341.

102. Discher, B.M.; Won, Y.Y.; Ege, D.S.; Lee, J.C.; Bates, F.S.;Discher, D.E.; Hammer, D.A. Polymersomes: Tough vesicles made from diblock copolymers. Science **1999**, *284* (5417), 1143–1146.

Nanomaterials— Nanoparticles

103. Park, J.W.; Benz, C.C.; Martin, F.J.; Future directions of liposome-and immunoliposome-based cancer therapeutics. Semin. Oncol. **2004**, *31* (6 Suppl. 13), 196–205.

104. Noble, C.O.; Kirpotin, D.B.; Hayes, M.E.; Mamot, C.; Hong, K.; Park, J.W.; Benz, C.C.; Marks, J.D.; Drummond, D.C. Development of ligand-targeted liposomes for cancer therapy. Expert Opin. Ther. Targets. **2004**, *8* (4), 335–353.

105. Matricardi, P.; Di Meo, C.; Coviello, T.; Hennink, W.E.;Alhaique, F. Interpenetrating polymer networks polysaccharide hydrogels for drug delivery and tissue engineering. Adv. Drug Deliv. Rev. **2013**, *in press.*

106. Alvarez-Lorenzo, C.; Blanco-Fernandez, B.; Puga, A.M.; Concheiro, A. Crosslinked ionic polysaccharides for stimuli-sensitive drug delivery. Adv. Drug Deliv. Rev. **2013**, *in press.*

107. Bhattarai, N.; Gunn, J.; Zhang, M. Chitosan-based hydrogels for controlled, localized drug delivery. Adv. Drug Deliv. Rev. **2010**, *62* (1), 8399.

108. Mizrahy, S.; Peer, D. Polysaccharides as building blocks for nanotherapeutics. ChemSoc Rev. **2012**, *41* (7), 2623–2640.

109. Labarre, D.; Ponchel, G.; Vauthier, C. *Biomedical and Pharmaceutical Polymers*; Pharmaceutical Press: London; 2010; 1–17.

110. Duncan, R.; Kopeček, J. Soluble synthetic polymers as potential drug carriers. In *Polymers in Medicine*; Springer: Berlin, Heidelberg, 1984; 51–101.

111. Langer, R.; Folkman, J. Polymers for sustained-release of proteins and other macromolecules. Nature. **1976**, *263*, 797–800.

112. Dhal, P.K.; Holmes-Farley, S.R.; Huval, C.C.; Jozefiak, T.H. Polymers as drugs. Adv. Polym. Sci. **2006**, *192*, 9–58.

113. Dhal, P.K.; Huval, C.C.; Holmes-Farley, S.R. Biologically active polymeric sequestrants: Design, synthesis, and therapeutic applications. Pure Appl. Chem. **2007**, *79* (9), 1521–1530.

114. Dhal, P.K.; Huval, C.C.; Holmes-Farley, S.R. Functional polymers as human therapeutic agents. Ind. Eng. Chem. Res. **2005**, *44* (23), 8593–8604.

115. Dhal, P.K.; Polomoscanik, S.C.; Avila, L.Z.; Holmes-Farley, S.R.; Miller, R.J. Functional polymers as therapeutic agents: Concept to market place. Adv. Drug Deliv. Rev. **2009**, *61* (13), 1121–1130.

116. Ast, M.; Frishman, W.H. Bile acid sequestrants. J. Clin. Pharmacol. **1990**, *30* (2), 99–106.

117. Mendonça, P.V.; Serra, A.C.; Silva, C.L.; Simões, S.; Coelho, J.F. Polymeric bile acid sequestrants—Synthesis using conventional methods and new approaches based on "controlled"/living radical polymerization. Prog. Polym. Sci. **2013**, *38* (3–4), 445–461.

118. Simpson, L.S.; Burdine, L.; Dutta, A.K.; Feranchak, A.P.; Kodadek, T. Selective toxin sequestrants for the treatment of bacterial infections. J. Am. Chem. Soc. **2009**, *131* (16), 5760–5762.

119. Bertrand, N.; Gauthier, M.A.; Bouvet, C.; Moreau, P.; Petitjean, A.; Leroux, J.C.; Leblond, J. New pharmaceutical applications for macromolecular binders. J. Control Release. **2011**, *155* (2), 200–210.

120. Dhal, P.K.; Huval, C.C.; Holmes-Farley, S.L. Polymeric sequestrants as nonabsorbed human therapeutics. In *Drug Discovery and Development*; Chorghade, M.S., Ed.; John Wiley & Sons, Inc.: Hoboken, NJ, USA, 2006; Vol. 1, 383–404.

121. Hansen, C.M. *Hansen solubility parameters: A user's handbook*, 2nd Ed.; CRC Press: Boca Raton, FL, 2007; 27–44.

122. Hancock, B.C.; York, P.; Rowe, R.C. The use of solubility parameters in pharmaceutical dosage form design. Int. J. Pharm. **1997**, *148* (1), 1–21.

123. Marsac, P.J.; Shamblin, S.L.; Taylor, L.S. Theoretical and practical approaches for prediction of drug—Polymer miscibility and solubility. Pharm. Res. **2006**, *23* (10), 2417–2426.

124. Huynh, L.; Neale, C.; Pomes, R.; Allen, C. Computational approaches to the rational design of nanoemulsions, polymeric micelles, and dendrimers for drug delivery. Nanomed. Nanotechnol. **2012**, *8* (1), 20–36.

125. Jennings, A.; Tennant, M.; Discovery strategies in a pharmaceutical setting: The application of computational techniques. Expert Opin. Drug Discov. **2006**, *1* (7), 709–721.

126. Haasnoot, C.A.G.; Kier, L.B.; Muller, K.; Rose, S.V.; Weber, J.; Wibley, K.S.; Wold, S.; Van Lenten, E.J. Glossary of terms used in computational drug design. Pure Appl. Chem. **1997**, *69* (5), 1137–1152.

127. Daoud, M.; Cotton, J.P. Star shaped polymers: A model for the conformation and its concentration dependence. J. Phys. **1982**, *43* (3), 531–538.

128. Costache, A.D.; Sheihet, L.; Zaveri, K.; Knight, D.D.; Kohn, J. Polymer—Drug interactions in tyrosine-derived triblock copolymer nanospheres: A computational modeling approach. Mol. Pharmaceutics **2009**, *6* (5), 1620–1627.

129. Patel, S.K.; Lavasanifar, A.; Choi, P. Molecular dynamics study of the encapsulation capability of a PCL–PEO based block copolymer for hydrophobic drugs with different spatial distributions of hydrogen bond donors and acceptors. Biomaterials **2010**, *31* (7), 1780–1786.

130. Lee, H.; Larson, R.G. Molecular dynamics study of the structure and interparticle interactions of polyethylene glycol-conjugated PAMAM dendrimers. J. Phys. Chem. B **2009**, *113* (40), 13202–13207.

Nanomaterials—
Nanoparticles

Nanoparticles: Biological Applications

Stanislav Rangelov
Laboratory of Polymerization Processes, Scientific Council of the Institute of Polymers, Bulgarian Academy of Sciences, Sofia, Bulgaria

Stergios Pispas
Theoretical and Physical Chemistry Institute of the National Hellenic Research Foundation (TPCI-NHRF), Athens, Greece

Abstract

This entry focuses on present and potential biological applications of polymer-containing nanoparticles and nano assemblies with emphasis given to drug and gene delivery nanocarriers, imaging and diagnostics, and others. The general ideas and principles that allow the use of different polymeric nanoparticles according to the specific application are also briefly discussed. The focus is placed on polymeric nanostructures and nanocolloids where potential applications have been demonstrated already to a significant degree.

This entry focuses on present and potential biological applications of polymer-containing nanoparticles and Nano assemblies with emphasis given to drug and gene delivery nanocarriers, imaging and diagnostics, and others. The general ideas and principles that allow the use of different polymeric nanoparticles according to the specific application are also briefly discussed. The focus is placed on polymeric nanostructures and nanocolloids that their potential applications have been demonstrated already to a significant degree. There are a number of excellent reviews discussing the applications of polymeric nanoparticles, especially of block copolymer micelles and other self-assembled structures, on drug delivery.[1–11]

DELIVERY OF LOW-MOLECULAR-WEIGHT DRUGS

Several pharmaceutical compounds, which show therapeutic activity toward a number of diseases and especially toward cancer tumors, are hydrophobic, having very low solubility in aqueous media. Some of them also show side effects, like nonspecific cytotoxicity, so their administration should be kept within certain concentration windows.[12] The use of specifically designed polymeric nanocarriers for the solubilization, transfer, and delivery/release of hydrophobic drugs is a rapidly developing field in nanomedicine. Among such nanodelivery systems, polymeric micelles formed by amphiphilic block copolymers (AmBC) stand as a highly interesting and promising class of self-assembled nanocarriers.[1,9,10] The interest is primarily due to the great versatility of block copolymer-based micelles through the flexible manipulation of their chemical

structure and properties via synthetic chemistry. In such nanocarrier systems, the hydrophobic core plays the role of the drug-containing compartment, and the micellar shell acts as the interface toward the surrounding biological media providing stabilization, stealth, and targeting properties. In principle, other parts of the micellar carrier, like the micellar corona and the core/corona interface, can be utilized for drug entrapment inside block copolymer micelles (Fig. 1).

There are several aspects of the structure of AmBC micelles that should be considered and optimized in order to obtain an advanced nanocarrier system.[10] The designing principles should also take into account the chemical nature of the drug to be carried. Incorporation of a drug into the nanocarrier can be achieved by physical entrapment of the pharmaceutical, for example, into the hydrophobic core of the micelle, or via chemical conjugation of the drug onto one of the blocks of the copolymer. In the latter case, stability of the polymer-drug bond should be such that it allows release of the drug under certain conditions, usually met at the desired point of delivery (Fig. 2). This approach is strongly connected to the requirement that a specific drug nanocarrier should be prepared for a specific drug, something that cannot always be achieved, since it requires specific chemistry schemes and it is usually laborious, time consuming, and cost inefficient. However, several examples of drug conjugation on the block copolymer chains can be found in the literature.

One of the early studies on micelle-forming block copolymer-drug conjugates for tumor therapy has been reported by Kataoka and coworkers.[13] They investigated the physicochemical properties and biodistribution of poly (ethylene oxide-*b*-aspartate) block copolymer-adriamycin conjugates

Concise Encyclopedia of Biomedical Polymers and Polymeric Biomaterials DOI: 10.1081/E-EBPPC-120052544

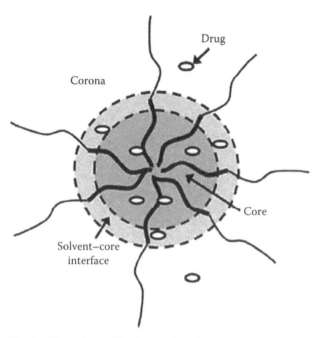

Fig. 1 Illustration of the three regions for drug solubilization in a block copolymer micelle.
Source: From Tyrrell et al.,[8] with permission from Elsevier.

[PEO-*b*-PAsp (ADR)] in murine colon adenocarcinoma 26 (C-26) tumor-bearing mice after intravenous injection. Long circulation times in blood and a parallel reduced uptake in the major organs of the reticuloendothelial system (i.e., liver and spleen) were observed. This was consistent with the *in vivo* persistence of a micellar core/shell structure in which the block copolymer-drug conjugate has been shown to adopt in aqueous media with the PAsp (ADR) blocks forming the core and the PEO chains the micellar corona. Enhanced accumulation of the micellar PEO-*b*-PAsp (ADR) conjugate in tumors evident after 24 hr (ca. 10% dose per g tumor), relative to free ADR (ca. 0.90% dose per g tumor), was demonstrated. Further, peak levels of PEO-PAsp (ADR) conjugate in the heart were lower than for free ADR.

It has been reported by Hennink and coworkers that core cross-linked biodegradable polymeric micelles composed of poly (ethylene glycol)-&-poly[*N*-(2-hydroxy-propyl) methacrylamide-lactate] [PEG-*b*-P(HPMAm-Lac$_n$)] diblock copolymers show prolonged circulation in the bloodstream upon intravenous administration and enhanced tumor accumulation through the enhanced permeation and retention (EPR) effect. In order to fully exploit the EPR effect for drug targeting, the group developed a doxorubicin methacrylamide (DOX-MA) derivative that was covalently incorporated into the micellar core by free-radical polymerization. The structure of the DOX derivative was susceptible to pH-sensitive hydrolysis, enabling controlled release of the drug in acidic conditions (in either the intratumoral environment or the endosomal vesicles).[14]

Between 30% and 40% w/w of the added drug was covalently entrapped, and the micelles with covalently attached DOX had an average diameter of 80 nm. The entire drug payload was released within 24 hr incubation at pH 5 and 37°C, whereas only around 5% release was observed at pH 7.4 (Fig. 3). DOX-conjugated micelles showed higher cytotoxicity in B16F10 and OVCAR-3 cells compared to DOX-MA, likely due to cellular uptake of the micelles via endocytosis and intracellular drug release in the acidic organelles. The micelles showed better antitumor activity than free DOX in mice bearing B16F10 melanoma carcinoma.

Gong and coworkers have recently presented the case of an end-functionalized ABA triblock that can be used for drug conjugation. In particular, an anticancer drug was covalently connected onto the middle block by pH-sensitive hydrazone bonds. Moreover, the polymer is functionalized with a folate group at one end and with an acrylate group at the other. The drug-conjugated triblock could self-assemble into stable vesicles in aqueous solution. The folate groups have been found to be located at the outer vesicle layer, providing active tumor targeting, while acrylate groups were located at the inner layer, able to provide stability through cross-linking. Moreover, magnetic nanoparticles could be encapsulated into the aqueous core of the stable vesicles, allowing ultrasensitive magnetic resonance imaging (MRI) detection. The described system can offer desirable drug release, stability, and excellent tumor-targeting activity in line with MRI ability, suitable for diagnosis purposes,[15] as will be discussed in the following sections.

Well-defined hyperbranched double-hydrophilic block copolymer of poly (ethylene oxide)-hyperbranched-polyglycerol (PEO-hb-PG) was developed as an efficient drug delivery nanocarrier for the hydrophobic anticancer drug DOX.[16] The authors demonstrated that the hydrophilic hyperbranched PEO-hb-PG copolymer formed micellar structures after conjugation with DOX. The drug was chemically linked to the copolymer by pH-sensitive hydrazone bonds, resulting in pH-responsive controlled release of DOX. Structural characterization of the macromolecular drug conjugate with dynamic light scattering (DLS), atomic force microscopy (AFM), and transmission electron microscopy (TEM) showed that the material assembled in spherical core-shell-type micelles with an average diameter of ca. 200 nm. Studies on the pH-responsive release of DOX and *in vitro* cytotoxicity experiments revealed the stimuli-responsive and controlled drug delivery character of the nanosystem. Taking into account the several beneficial features of hyperbranched double-hydrophilic block copolymers, such as enhanced biocompatibility, increased water solubility, and drug-loading efficiency, as well as the improved clearance of the copolymer after drug release, the authors proposed that double-hydrophilic block copolymers are able to provide versatile platforms for the development of efficient drug delivery systems for effective treatment of cancer.

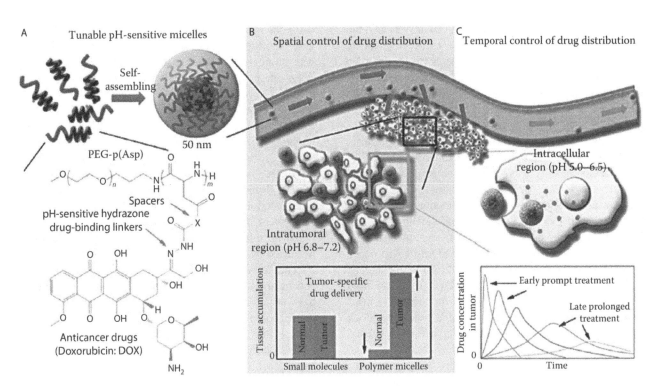

Fig. 2 Desired spatial (**A,B**) and temporal (**C**) control of drug delivery to tumor microenvironment using tunable block polymer micelles with conjugated drug.
Source: From Ponta & Bae,[7] with permission from Springer Science + Business Media.

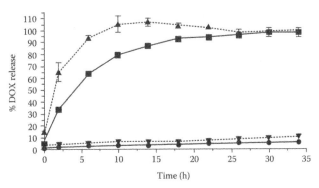

Fig. 3 Release of core-covalently bound DOX from methacrylated mPEG-*b*-p[(50% HPMAm-Lac1)-co-(50<% HPMAm-Lac2)] cross-linked micelles at pH 5 (■) and pH 7.4 (●) and 37°C and of DOX from the DOX-MA monomer containing the acid-sensitive hydra-zone linker at pH 5 (▲) and pH 7.4 (▼) at 37°C.
Source: From Talelli et al.[56] © 2011, with permission from Elsevier.

The approach of physical entrapment allows for greater flexibility in nanodrug formulation, since the same micellar nanocarrier can, in principle, be used for the encapsulation of a number of different hydrophobic drugs. However, also in this case, certain design aspects should be considered and completed successfully.

The first aspect concerns drug loading into the micellar nanocarrier. Drug loading should be as a high as possible,

since in this way the amount of drug to be delivered can be maximized and therapeutic drug levels can be sustained for a longer time upon systemic administration. The physicochemical affinity of the pharmaceutical compound toward the hydrophobic core of the micelles plays a crucial role. The higher is the affinity, the higher should be the loading ability of the micellar nanocarrier. A measure of the aforementioned affinity is the Flory-Huggins interaction parameter (χ_{sp}); the lower the value of the parameter, the higher the affinity between the micellar core and the drug. The interaction parameter is dramatically affected by the characteristics of the drug, such as hydrophobicity, polarity, or degree of ionization in respect to the same parameters of the micellar core.[17] Therefore, a systematic chemical design for preparing the desired core blocks should be followed.

Lavasanifar and coworkers have demonstrated both theoretically and experimentally that the use of block copolymer micelles with different hydrophobic blocks as the cores leads to changes of the micellar drug capacity. The Flory-Huggins interaction parameter was calculated by the group contribution method between a drug, cucurbitacin, and three different block copolymers based on PCL-*b*-PEO copolymers, namely, regular PCL-*b*-PEO copolymer and those with poly (ε-caprolactone) (PCL) blocks functionalized with benzyl and cholesteryl pendant groups. It was found that the pair with the lower χ_{sp}, that is, the cucurbitacin/cholesteryl-PCL-b-PEO pair, was also the system showing the highest drug loading. It is noteworthy that the release profile of the three systems was found to be influenced not only by the interaction parameter but also by the core viscosity.[18,19] Chemically modified PCL-b-PEO copolymers were also investigated for the encapsulation of amphotericin B (AmB) in block copolymer micelles. Pendant benzyl, carboxyl, stearyl, palmitoyl, and cholesteryl groups were attached on the PCL hydrophobic block. The drug-loading capacity was found to increase in the order cholesteryl < stearyl < palmitoyl < carboxyl. In the carboxyl group case, hydrogen bond formation may be also active and leads to the greater increase in drug loading compared to the precursor block copolymer micelles.[20,21] In another work, micelles based on PEO-b-PLAsp block copolymers were utilized as drug encapsulating agents. The poly (l-aspartic acid) blocks were chemically modified with fatty acid groups increasing the hydrophobicity of this block in respect to the benzyl modified precursor block copolymer. AmB was again the drug that was encapsulated in all block copolymers investigated.[22] Results showed that the incorporation of fatty acid groups in the core-forming block increased substantially the encapsulation efficiency of the micellar nanocarriers.

In one of the early and successful studies on encapsulation of hydrophobic anti-cancer drugs into block copolymer micelles, the concept of physicochemical affinity between the micellar core and the drug was utilized.[23] DOX was loaded into micelles formed by poly (ethylene glycol)-b-poly (b-benzyl-l-aspartate) (PEG-b-PBLA) block copolymer following an oil/water (O/W) emulsion methodology. Rather high drug-loading levels were achieved (ca. 15–20 w/w%). This was correlated to the low water solubility of DOX and the possible interactions with the benzyl side groups of PBLA segments through n-n stacking. Drug-loaded micelles had a narrow size distribution and diameters in the range 50–70 nm. It was observed that the release process for DOX from the micelles was comprised of two steps. First, a rapid release of the drug took place followed by a slow and long-lasting release step. Lowering of the media pH from 7.4 to 5.0 resulted in an acceleration of DOX release. The observed pH-sensitive release of DOX from the block copolymer micelles was connected to the protonation of the amine group of the drug, which is expected to result in an increase of its water solubility. Blood circulation of DOX was substantially improved due to its encapsulation into the micellar nanocarrier. DOX loaded in the PEG-b-PBLA micelles showed a considerably higher antitumor activity compared to free DOX against mouse C26 tumor by intravenous injection, indicating the potential of PEG-b-PBLA/DOX nanoensemble to act as a long-circulating carrier system for modulated drug delivery. Related studies have been reported by the Kataoka group even earlier.[24,25]

Along the same lines, camptothecin and paclitaxel (PTX) were efficiently solu-bilized into poly (isoprene-b-ethylene oxide) micelles. This was the first time that this particular type of block copolymer micelles was proposed as nanocarriers for a hydrophobic drug. Polyisoprene block is highly hydrophobic and the presence of C = C bonds allows for n-n interactions with several drugs carrying unsaturated groups. It has a low T_g and also shows chemical structure similarities with natural terpenoids. Solubilization of the drugs resulted in an increase in the size of the micelles, while no change was observed in the surface potential of the micelles. Micellar suspensions of these drugs containing up to 30 μg of the drug per mL of final solution could be prepared. Micelles loaded with drugs were stable for more than 2 weeks.[26]

The affinity between drug and polymeric carrier can be increased by introducing specific interactions between the micellar core and the drug. Specific interactions can be enhanced by, for example, hydrogen bonding. Tan et al. have described the synthesis of a block copolymer with urea pendant groups. The existence of urea groups not only changes the physical properties of the polymer but also allows the increased loading of an anticancer drug, due to hydrogen bonding, without considerable changes on the micellar size. Furthermore, experimental results have indicated that, even though the copolymer is nontoxic, the drug-loaded micelles were able to destroy cancer cells.[17]

Poly (β-lactam-isoprene-b-ethylene oxide) copolymers have been recently utilized for the solubilization of the novel hydrophobic anticancer drug curcumin.[27] It was expected that the hydrogen-bonding interaction between lactam groups of the copolymer and curcumin would enhance curcumin encapsulation. The size of the drug-loaded copolymer micelles was found to be smaller than the size of the empty micelles. Drug encapsulation efficiency and loading capacity increased by an increase in the poly (β-lactam-isoprene) block in the copolymer. Encapsulation efficiency reached 78% and maximum loading capacity was ca. 2.5 mg of drug/mL of solution. Empty and loaded micelles were found substantially stable for a period of nearly 2 months. In vitro release of the drug was slower for the micelles possessing the highest poly (β-lactam-isoprene) content, suggesting the determining role of block copolymer characteristics on the release profile of the drug. The cytotoxic activity of the drug was also improved.

The interaction between block copolymers and pharmaceuticals can be also enhanced through electrostatic interactions, that is, when a charged block copolymer and a complementary charged drug molecule are paired. Typically, the addition of an oppositely charged drug molecule leads to the formation of a water-insoluble polyion complex between the drug and the charged block, while the second block (usually a neutral PEG/PEO chain) ensures the solubility of the complex. Following this concept, Li et al. reported the complexation of the cationic drugs dibucaine, tetracaine, and procaine with anionic poly (methacrylic acid-b-ethylene oxide) (PMAA-b-PEO) copolymers to give complex micelles incorporating the ionic drugs in their cores.[28] The block copolymer-drug pairs self-assembled to form micelle-like nanoaggregates comprised of a core of neutralized polyions (PMAA/drug complex) surrounded by the PEO corona. It was found that the properties of the aggregates strongly depended on the hydrophobicity of the anesthetics and that besides electrostatic interactions, hydrophobic interactions as well play an important role. The aggregates induced by dibucaine were found to have a highly viscous interior, whereas those induced by tetracaine had a fluid interior, due to the lesser hydrophobic effect of the latter drug. It was observed that procaine did not induce aggregate formation due to its low hydrophobicity. All these characteristics are expected to influence significantly drug encapsulation and release in such systems that show good potential as nanocarriers for ionic drugs.

More recently, dibucaine has been successfully incorporated into the inner part of the corona of amphiphilic poly (ethylene oxide-*b*-sodium 2-(acrylamido)-2-methyl-1-propanesulfonate-*b*-styrene) (PEO-*b*-PAMPS-*b*-PS) triblock copolymer micelles through electrostatic interactions by the same group.[29] Spherical micelles with zero surface charge and a stealth PEO corona were formed for up to 100% incorporation of the drug (in respect to the anionic sites of the corona chains), indicating that the amount of incorporated drug can be modulated by the length of the charged block.

The preparation of micelles with a complex core containing a drug molecule was achieved by complex formation between the model cationic drug, diminazene diaceturate (DIM), and a series of novel diblock copolymers, that is, carboxymethyldextran-poly (ethylene glycols) (CMD-PEGs).[30] Micellar properties were found to be dependent on the ionic charge density (or degree of substitution, DS) of the CMD block and the ratio, [+]/[−], of positive charges of the drug to the negative charges of the copolymers. Micelles were formed at [+]/[−] = 2 and incorporated up to 64% by weight DIM. Their sizes were in the nanometer range (R_h values from 36 to 50 nm were observed) depending on the molecular weight and DS of CMD-PEG. The critical aggregation concentration was also found to depend on DS. The micelles were stable within the 4 < pH < 11 range. A more detailed investigation at pH 5.3 proved that the micelles were not destabilized for an ionic strength increase up to almost 0.4 M NaCl in the case of CMD-PEG copolymers of high DS. Micelles of CMD-PEG of low DS (ca. 30%) disintegrated in solutions containing more than 0.1 M NaCl. Sustained *in vitro* DIM release was observed for micelles of CMD-PEG of high DS ([+]/[−] = 2) (Fig. 4).

Besides physicochemical affinity, other parameters can simultaneously influence the loading of a drug into a polymeric micellar nanocarrier. In particular, the solubility of a drug molecule into the hydrophobic compartment of a block copolymer aggregate is also highly influenced by the chosen incorporation method. A systematic encapsulation study of two drugs into aggregates of a series of block

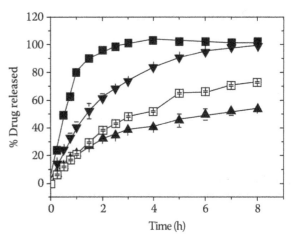

Fig. 4 Release of DIM evaluated by the dialysis bag method from (■) DIM alone in tris-HCl, 25 mM [NaCl] = 150 mM and pH 7.4; (▼) DIM/85-CMD40-PEG140 micelles, [+]/[−] = 2, in 25 mM tris-HCl, [NaCl] = 150 mM and pH 7.4; (▲) DIM/85-CMD40-PEG140 micelles, [+]/[−] = 2, in 25 mM tris-HCl [NaCl] = 0 mM and pH 5.3; and (⊞) DIM/30-CMD68-PEG64 at [+]/[−] = 2, in tris-HCl, 25 mM [NaCl] = 0 mM and pH 5.3.
Source: From Soliman & Winnik[30] © 2008, with permission from Elsevier.

copolymers with variable hydrophobicity has been implemented following two distinct encapsulation procedures, namely, dialysis and the O/W emulsion method. The results indicate the great influence of both the interaction parameter between different drugs and copolymer chains on drug loading, as well as that of the followed encapsulation protocol.[31]

Another study on a similar context was reported recently by Harada and coworkers.[32] They incorporated the anticancer agent, camptothecin, into poly (ethylene glycol)-*b*-poly (aspartic acid-*co*-benzyl aspartate) {PEG-P[Asp (Bzl)]} polymeric micelle carriers by using two different solvents [tetrafluoroethylene (TFE) and chloroform] using a solvent-evaporation drug-incorporation process. They have observed significant differences in the drug-incorporation behavior, in the morphologies of the incorporated drug and the loaded block copolymer micelles, and in the pharmacokinetic behavior of the nanocarriers by using the two different solvents. The block copolymer micelles, which were prepared with TFE as the incorporation solvent, exhibited more stable circulation in the bloodstream than those prepared with chloroform. This contrast indicates a novel technological perspective regarding the drug incorporation into polymeric micelle carriers. Morphological analysis of the inner core, using AFM in conjunction with fluorescence anisotropy measurements, revealed the presence of a directed alignment of the drug molecules and camptothecin crystals in the micelle inner core. These observations were correlated with the observed pharmacokinetic behavior. This was also the first report on the morphologies of the incorporated drug into the cores of block copolymer micelles.

Architecture of the block copolymer chain may also influence drug loading. Zhang et al. have reported on the comparison of drug-loading capacities between hyperbranched copolymers and linear analogues. The experimental data reveal that the hyperbranched architecture is superior in terms of both drug-loading and entrapment efficiency. Moreover, the hyperbranched copolymer had a more sustained drug-release behavior.[33]

Controlled release of the encapsulated drugs is also of great importance in the design of a nanocarrier system. In many cases, drug-loading and drug-release processes are closely interrelated and mutually influenced procedures, in the sense that the way the drug was incorporated into a micellar nanocarrier would influence its release out of the polymeric nanostructure. For example, the use of a common solvent for drug encapsulation may influence the viscosity of the micellar core where the drug is incorporated, and this may also influence drug diffusion out of the core. More importantly, triggered release of encapsulated drugs becomes more important especially in cases that the external stimulus is related to the specific characteristics of the targeted organ/tissue for drug administration within the body.

In a rather informative study involving poly (ethylene oxide)-b-poly (N-hexyl stearate-l-aspartamide) (PEO-b-PHSA) block copolymer micelles incorporating AmB in their cores,[34] it was observed that the increase in the hexyl group substitution of the PHSA block increased the incorporation of the specific drug but decreased the releasing ability of the micellar nanocarriers. In an analogous case study, PTX was found to be released slower from PEO-b-PCL copolymers containing benzyl groups on the PCL block compared to the original nonfunctionalized PEO-b-PCL block copolymer micelles. The authors claimed that the observed behavior should be attributed to the higher rigidity of the benzyl containing blocks resulting in a greater viscosity of the micellar cores.[35]

Triggered release as a response to pH changes is highly useful for the release of anticancer drugs in the acidic microenvironment of tumor cells. Acidic microenvironment of tumor cells results from the increased aerobic and anaerobic glucose metabolism taking place inside such cells. Additionally, cancer tumors are known to have an increased temperature, in comparison to healthy tissues, due also to their increased metabolic rate. In any case, drug levels should be kept within certain limits in order to achieve the desired therapeutic levels. It has been shown that cancer cells are more sensitive to exposure to low drug levels for longer time periods than the case of exposure to high drug levels for shorter periods. Amphiphilic block copolymers of PEG as the first block and a second block composed of hydrophobic methacrylate monomers and methacrylic acid (MAA) were utilized for the solubilization of the poorly water-soluble model drugs, indomethacin, fenofibrate, and progesterone. Drug loadings of <6% and 6–14% w/w were achieved by the dialysis and emulsion methods, respectively. Evaluation of progesterone

release in vitro has demonstrated that the drug release from the block copolymer micelles increased when the pH of the release medium was raised from 1.2 to 7.2.[31]

pH-responsive drug conjugate micelles have been also proposed as a tool for achieving triggered site-specific release of the carried drug. Usually the drug is chemically bonded to the core-forming block via an acid-labile bond. This bond should be stable at physiological pH and should be easily cleaved at the acidic pH of a tumor's extracellular space or in its cell endosomes, allowing for the release of the therapeutic agent. Poly(ethylene oxide)-b-poly (l-lactic acid) copolymers having DOX linked on the PLLA terminus via hydrazone bond were investigated and showed effective drug release at acidic pH. DOX has been also linked to poly (aspartic acid) blocks via the same hydrazone linker, and pH-sensitive drug release was also observed in this case.[13] In an analogous study, PTX was attached to the poly (aspartic acid) block of a PEO-b-PAsp block copolymer formerly functionalized with levulinic acid moieties. The hydrazone-type linker was cleavable at low pH and the drug could be released in a controlled way. However, the preparation of mixed micelles containing levulinic acid and 4-acetyl benzoic acid moieties with attached DOX showed a clearly moderated release profile for the drug compared to the original pure micelles. So, blending of block copolymer-drug conjugates can be another way of fine tuning the release of a drug in such cases.[36]

Gillies and Frechet have utilized an acid-labile cyclic benzylidene acetal linker for the chemical conjugation of drugs on the poly (aspartic acid) block of PEO-b-poly (aspartic acid) block copolymers.[37] The authors found that this type of linkers was rapidly hydrolyzed at pH 5, showing a half-life of ca. 60 min at normal body temperature (37°C). Hydrolysis of the acetal linkers resulted in cleavage of the hydrophobic groups of the copolymer chains that in turn led to the disruption of the copolymer micelles and the subsequent release of the carried hydrophobic drugs. In a similar fashion, but taking advantage of the protonation effect, encapsulated hydrophobic drugs can be released from the nanocarrier by the dissociation of the micelles due to an increase in copolymer solubility. Poly (ethylene glycol)-b-poly (histidine) (PEG-b-PHis) copolymers form stable, narrow-sized micelles at physiological pH because of the hydrophobicity of the PHis block. Therefore, DOX could be encapsulated into the PHis core of the micelles at loading levels close to 20% by weight. When the drug-loaded micelles were present in the lower extracellular pH of the tumor or into the endosomal compartments of tumor cells, they were disintegrated, due to protonation of histidine groups, leading to the release of DOX.[38,40]

In a more recent study, a series of synthetic block copolymers consisting of poly (2-hydroxyethyl methacrylate) (PHEMA) and PHis, showing biocompatibility and membranolytic activity, were investigated as pH-sensitive drug carrier for tumor targeting.[40] DOX was successfully encapsulated in the nanosized micelles and delivered under different pH conditions. DOX release was investigated *in vitro* according to the pH of the surrounding medium. DOX was released from the micelles in a controlled and sustained manner and the release rate of DOX was accelerated at low pH 5.5 compared to pH 7.4. The empty micelles could be effectively internalized by human embryonic kidney 293T cells and HCT116 cells. The viability of both cell types was higher than 80% at a micelle concentration range 1–50 µg/mL. The drug-loaded PHEMA-*b*-PHis micelles could also be effectively internalized by HCT116 human colon carcinoma cells. It was found that they slowly released the encapsulated DOX molecules, showing effective cell proliferation inhibition compared to free DOX, specifically at an acidic environment.

Nanosized drug carriers with stealth properties and long circulation times are preferentially accumulated to tumors through the EPR effect, due to the differences in the microstructure/physiology of the tumor tissues compared to normal ones. In order to effectively drive the nanocarriers toward the specific sites for release of their cargo, targeting capabilities should be introduced onto the nanocarriers. This is more frequently achieved by the introduction of targeting chemical groups/moieties at the free end of the corona-forming block of the diblock copolymer forming the micel-lar nanocarrier. Targeting ligands are helpful in attaching the drug nanovehicles to the target cells or tissues and for enhancing crossing of the cell membrane. Various ligands can be utilized depending on the target and may include small organic molecules as chain-end connected chemical groups, carbohydrates, peptides, and antibodies.

In the category of small organic molecules targeting moieties, folic acid (FA) groups are very commonly used because its receptor is overexpressed in many human cancer cells. Folate-PEO-*b*-PCL micelles (the folate group being attached to the free end of the PEG block) loaded with PTX exhibited distinctly higher cytotoxicity for two types of human adenocarcinoma cells compared to normal human fibroblast cells.[41] A mixed block copolymer micellar system containing folate-PEO-*b*-PLGA and PEO-*b*-PLGA-DOX conjugate chains, having one DOX molecule attached to the poly (d,l-lactic-*co*-glycolic acid) block, was more effective than DOX alone in terms of cytotoxicity toward tumor cells *in vitro*.[42]

Complex amino acid-based core cross-linked star-block copolymers, that is, [poly (l-lysine)$_{arm}$poly (l-cystine)$_{core}$] copolymers, bearing peripheral allyl functionalities, which were further functionalized with a poly (ethylene glycol)-folic acid (PEG-FA) conjugate at the periphery, forming FA-PEG$_{arm}$-[poly (l-lysine)$_{arm}$poly (l-cystine)$_{core}$] stars, were evaluated as drug carriers for targeted drug delivery *in vitro*. Confocal microscopy and flow cytometry revealed that the functionalized stars could be internalized into MDA-MB-231 cells, while cytotoxicity studies indicate that the stars are nontoxic to cells at concentrations of up to 50 µg/mL (Fig. 5). These results show that these amino

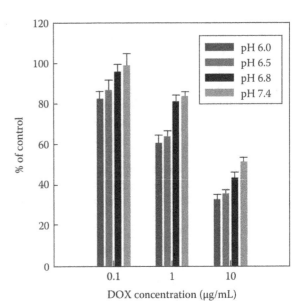

Fig. 5 Dose-dependent antitumor activity of DOX-loaded p (HEMA)$_{25}$-*b*-p(His)$_{15}$ micelles to HCT-116 cells after 48 hr incubation.
Source: From Johnson et al.[40] © with permission from Wiley-VCH Verlag GmbH & Co. KGaA.

acid-based star polymers can be utilized in targeted drug delivery applications including chemotherapy.[43]

Aiming at developing block copolymer micellar nanocarriers with targeting and pH-responsive drug release, Armes and coworkers have presented a series of FA corona-functionalized block copolymer micelles with pH-responsive blocks. They found that two different drugs can be loaded in the respective micelles forming well-defined nanocarriers at physiological pH. However, due to dissociation (or not) of the polymeric micelles, the release profile of the drug was found to be strongly depended on the environmental pH value.[44]

Folate-functionalized polymeric micelles with pH-triggered drug-release properties have been also presented by Yung and coworkers. In particular, the use of mixed micelles composed of two different copolymers has been proposed. One of the copolymers is suitable for the drug loading, while the other copolymer offers pH sensitivity and therefore determines the desired drug-release profile. Notably, the experimental results indicate that by tuning the ratio between the block copolymers utilized for micelle formation, the desirable physicochemical characteristics can be achieved for the nanocarrier system.[45]

The use of carbohydrates as active targeting ligands on the surface of block copolymer micelles has been also reported.[46,47] Glucose, galactose, mannose, and lactose can be utilized for this purpose since the targeting groups can bind to glycol receptors of cells. It has been demonstrated that lectins, ConA- and RCA-1-type receptors, can recognize mannose and P-d-glucose residues.

Polycaprolactone-g-dextran (Gal-PCL-g-Dex-FITC) polymers carrying galactose groups for active targeting and fluorescein isothiocyanate (FITC) groups for fluorescence labeling were self-assembled into stable, spherical micelles in aqueous medium and in serum. The anti-inflammatory drug prednisone acetate was loaded in the polymeric micelles through hydrophobic interactions. The *in vitro* drug-release investigations showed that the galactosylated micelles could be selectively recognized by and subsequently accumulate in HepG2 cells. It was demonstrated that the relative uptake of the micelles by liver was much higher than the other tissues, indicating that the galactosylated micelles have great potential as a liver-targeting drug carrier, as a result of their design and structure[48] (Fig. 6).

Small peptide sequences and proteins have been also utilized for introducing targeting properties to polymeric nanoparticles and nanoassemblies. PEO-c-PCL block copolymer micelles having a cyclic pentapeptide (cyclic-Arg-Gly-Asp-d-Phe-Lys, cRGDfK) attached to the free ends of the PEO blocks, and carrying the hydrophobic drug DOX in their PCL core, were prepared following a multi-step synthetic protocol.[49] Such a peptide, which contains the RGD sequence, can recognize $a_v p3$-integrins that are overexpressed on the surface of tumor cells or the angiogenic endothelial cells of the tumor vasculature. First, maleimide-terminated PEO-c-PCL copolymers were synthesized by a well-established end functionaliza-tion reaction on the -OH terminus of PEO blocks. After preparation of the micelles, the cRGDfK peptide was attached on the micelle surface via reaction of its thiol group with the maleimide groups of the copolymer chains. A ca. 76% yield of the coupling reaction was achieved. Nevertheless, confocal laser scanning microscopy (CLSM) observations indicated that the drug-loaded micelles with the targeting peptide could be accumulated into human Kaposi's

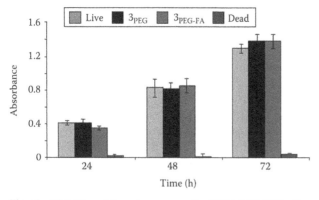

Fig. 6 Viability of breast cancer cells (MDA-MB-231) after incubation with FA-PEG$_{arm}$-[poly(l-lysine)$_{arm}$poly(l-cystine)$_{core}$] stars, 3PEG-FA and 3PEG stars at 5.0 μg/mL for 24, 48, and 72 hr, as assessed by MTS assays. Absorbance is proportional to MTS reduction by metabolically active cells.
Source: From Sulistio et al.[43] © 2011, with permission from American Chemical Society.

sarcoma tumor endothelial SLK cells in a 30 times greater extend compared to non-peptide-modified micelles.

In another work, acetal-terminated-PEO-c-PCL block copolymers were used for the formation of micelles in aqueous media. Then the acetal group on the free PEO terminus was transformed to an aldehyde group by hydrolysis under acidic conditions. The aldehyde group on the micellar surface was utilized for the attachment of the GRGDS targeting peptide by a Schiff-type reaction. Fluorescence spectroscopy/microscopy experiments showed that the uptake of GRGDS carrying micelles by mouse melanoma B16-F10 cells was almost five times greater than that of the PEO-PCL micelles that did not carry the targeting peptide moiety.[50] In an analogous study, an acetal-PEO-*b*-poly (α-carboxyl-ε-caprolactone) (ac-PEO-*b*-PCCL) block copolymer was used for the chemical conjugation of DOX through the α-carboxyl groups of the PCCL block forming an amide group between the polymer chain and the drug. The terminal acetal group was again transformed to an aldehyde, and this end group enabled the covalent attachment of RGD-containing peptides, including the GRGDS peptide, in order to functionalize the micellar surface. The RGD-functionalized micelles with the conjugated DOX presented higher toxicity toward B16-F10 cells compared to micelles without the peptide.[51]

Polymeric micelle-like nanoparticles based on PEGylated poly (trimethylene carbonate) (PEG-c-PTMC) and decorated with the cyclic RGD peptide were prepared for active targeting to integrin-rich cancer cells.[52] The amphiphilic diblock copolymer, α-carboxyl poly (ethylene glycol)-poly (trimethylene carbonate) (HOOC-PEG-c-PTMC), was synthesized by ring-opening polymerization (ROP). The c(RGDyK) ligand was conjugated to the NHS-activated PEG terminus of the copolymer. The c(RGDyK)-functionalized PEG-c-PTMC micellar nanoparticles were loaded with PTX using the emulsion/solvent-evaporation technique, and they were found to have nanometer-scale sizes. Cellular uptake of c(RGDyK)-PEG-c-PTMC/PTX nanoassemblies was found to be higher than that of the non-functionalized micelles containing PTX, due to the integrin-mediated endocytosis effect, taking place in the former case as a result of the presence of the targeting functionalities. *In vitro* cytotoxicity, cell apoptosis, and cell cycle arrest studies were also employed in order to assess the efficiency of the nanodelivery systems and revealed that the c(RGDyK)-PEG-b-PTMC/PTX exhibited significantly stronger *in vitro* anti-angiogenic activity than PEG-b-PTMC/PTX and also Taxol. A pharmacokinetic study in rats demonstrated that the micellar nanoparticles significantly enhanced the bio-availability of PTX than Taxol, by increasing the blood circulation time of the nanocarrier. Enhanced accumulation of c(RGDyK)-PEG-b-PTMC/PTX in tumor tissue *in vivo* xenograft tumor-bearing model was also observed, by real-time fluorescence imaging, highlighting the high specificity and efficiency of the functionalized micellar nanoparticles for tumor active targeting (Fig. 7).

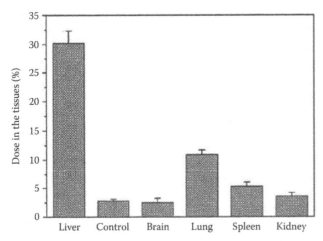

Fig. 7 Overall uptake of Gal-PCL-g-Dex-FITC micelles into different tissues after 2 hr injection. The uptake of PCL-g-Dex-FITC micelles in liver was used as a control.
Source: From Wu et al.[48] © 2009, with permission from Elsevier.

Since it has been realized that transferring receptors are considerably elevated on various types of cancer cells, transferrin itself was proposed as a potential targeting protein to be conjugated to the surface of block copolymer micelles carrying hydrophobic anticancer drugs. Following this concept, transferrin, a serum glycoprotein of 80 kDa molecular weight, was attached to the corona of PEO-b-polyethylenimine (PEI) polyion complex micelles, with a PEO corona and a core composed of the electrostatic complex of PEI and phosphorothioate antisense oligonucleotides (ODNs), via an avidin/biotin structure. Fluorescent-labeled transferrin-conjugated micelles were found to accumulate with a high efficiency into resistant human oral epider-moid carcinoma cells.[53] Some additional examples pertaining to the delivery of relatively higher molecular weight nucleotides using block copolymer micelles and other polymeric nanoparticles containing targeting moieties will be presented in the following sections.

Monoclonal antibodies or their Fab fragments have also shown promise as targeting ligands on the surface of block copolymer micelles. Kabanov et al. have developed immunomicelles based on Pluronic P-85 micelles, to which a murine polyclonal antibody against α_2-glycoprotein was chemically attached on the micel-lar corona, in an effort to deliver the neuroleptic agent haloperidol to the brain.[54] The monoclonal antibody C225 against epidermal growth factor receptors (EGFRs) was attached to the PEG terminus of a poly(ethylene glycol)-b-poly(l-glutamic acid) (PEG-b-PLG) block copolymer. The attachment was accomplished by reaction of the sulfhydryl group introduced to C225 with the vinylsulfone group on the PEG end of the copolymer. The PLG block was chemically conjugated to DOX. The micelles carrying the C225 antibody were selectively bound to human vulvar squamous carcinoma A431 cells that overexpress EGFR, in contrast to

non-modified micelles. Receptor-mediated micelle uptake was found to occur within 5 min where the non-modified DOX micellar conjugates required ca. 24 hr to be internalized into cells. The modified micelles were found to be more effective in inhibiting growth of A431 cells in comparison to free DOX after a 6 hr exposure period.[55]

PEG-b-P(HPMAm-Lac$_n$) core cross-linked thermosensitive biodegradable polymeric micelles suitable for active tumor targeting, by coupling the anti-EGFR (EGFR) EGa1 nanobody to their surface as targeting ligand, were developed by Hennink and coworkers.[56] In the first synthetic step, PEG was functionalized with N-succinimidyl 3-(2-pyridyldithio)-propionate (SPDP) to yield a PDP-PEG-b-P(HPMAm-Lac$_n$) block copolymer. Micelles composed of 80% PEG-b-P(HPMAm-Lac$_n$) and 20% PDP-PEG-b-P(HPMAm-Lac$_n$) were prepared from mixtures of the two copolymers, and then lysozyme (utilized as a model protein) was modified with N-succinimidyl-S-acetylthioacetate, deprotected with hydroxylamine hydrochloride and subsequently coupled to the micellar surface. The micellar conjugates were characterized using SDS-PAGE and gel permeation chromatography showing the success of the synthetic protocol. Following an analogous conjugation step, the EGa1 nanobody was coupled to PEG/PDP-PEG micelles. Conjugation was successful as demonstrated by Western blot and dot blot analysis. Rhodamine-labeled EGa1-micelles showed substantially higher binding, as well as uptake by EGFR-overexpressing cancer cells (A431 and UM-SCC-14C) than rhodamine-labeled micelles without the targeting ligand. No binding of the nanobody-functionalized micelles was observed to EGFR-negative cells (3T3) as well as to 14C cells in the presence of an excess of free nanobody. The latter result demonstrates that the binding of the nanobody micelles was indeed through interaction with the EGF receptor. Although these micelles did not contain any active pharmaceutical compound, the presented studies illustrate the ability of the particular ligand to act as a valuable and specific moiety for tumor targeting of block copolymer micellar nanocarriers.

Nucleic acid ligands (aptamers) are another class of ligands potentially well suited for the selective targeting of drug-encapsulated controlled-release polymer particles in a cell- or tissue-specific manner. Bioconjugates composed of controlled-release polymer nanoparticles and aptamers were synthesized from a poly (lactic acid)-b-poly (ethylene glycol) (PLA-b-PEG) copolymer with a terminal carboxylic acid functional group at the PEG free end (PLA-b-PEG-COOH) and were examined in terms of their efficiency for targeted delivery to prostate cancer cells. Rhodamine-labeled dextran was encapsulated within the micellar nanoparticles as a model drug. The particular nanoparticles showed several desirable characteristics including: 1) substantial negative surface charge (corresponding to a surface zeta-potential of ca. −50 mV), which may minimize nonspecific interaction with the negatively charged nucleic acid aptamers; 2) carboxylic acid groups on the particle surface for modification via covalent conjugation to amine-modified aptamers; and 3) presence of PEG chains on the particle surface, which is expected to increase circulating half-life, while contributing to decreased uptake in nontargeted cells. Nanoparticle-aptamer bioconjugates contained RNA aptamers that bind to the prostate-specific membrane antigen, a well-known prostate cancer tumor marker that is overexpressed on prostate acinar epithelial cells.[57] The authors demonstrated that these bioconjugates could efficiently target and be incorporated into the prostate LNCaP epithelial cells, which express the prostate-specific membrane antigen protein. A 77-fold increase in binding was observed for the aptamer-functionalized micellar nanoparticles compared to reference nanoparticles with no specific binding ligands. In contrast to LNCaP cells, the uptake of the aptamer-functionalized block copolymer nanoparticles was not enhanced in cells that do not express the prostate-specific membrane antigen protein.

It is evident from the former discussion that for the construction of a polymeric nanocarrier with a success potential for drug delivery applications, a number of basic functionalities and responsive properties should be taken into consideration and be physically accommodated into the nanostructure. This holds true in particular for those based on block copolymers for the delivery of small drug molecules. However, some design and construction principles are universal and should be considered also in the case of delivery of large molecules although some fine tuning should always take place. In the following, we continue with the presentation of several elucidating examples of such designed nanocarriers for small drugs that give a better picture of the breadth of the field and some of the advances accomplished so far.

A rather complex block copolymer, poly(N-isopropyl acrylamide-co-N,N-dimethylacrylamide-co-2-aminoethyl methacrylate)-b-poly(10-undecenoic acid) [P(NIPAM-co-DMAAm-co-AEMA)-b-PUA], containing four chemically different segments, each one carrying a specific function, was synthesized and studied in the context of an advanced drug nanocarrier.[58] FA was conjugated to the hydrophilic random copolymer corona-forming block, through the amine group in AEMA units. The copolymer self-assembled into micelles in aqueous media, which exhibited pH-induced responsiveness and temperature sensitivity. The micelles had a rather small size and a well-defined PUA core, while the folate-targeting moieties were concentrated in the corona. The anticancer drug, DOX, was encapsulated into the micelles by a membrane dialysis method. DOX release was pH dependent, being faster at low pH (as is the case of endosomes and lysosomes). Therefore, it was concluded that DOX could be readily released from the micelles into the cell nucleus after being incorporated into cells. The IC50 value of DOX-loaded micelles with folate-targeting groups against folate receptor-expressing 4T1 and KB cells was much lower than that of the DOX-loaded micelles without folate (3.8 vs. 7.6 mg/L for 4T1 cells and 1.2 vs. 3.0 mg/L for KB cells). *In vivo* experiments conducted in a 4T1 mouse breast cancer model demonstrated that DOX-loaded micelles had a longer blood circulation time than free DOX ($t_{1/2}$ was 30 and 140 min, respectively). In addition, the micelles delivered an increased amount of DOX to the tumor when compared to free DOX.

Micelles formed by specially designed amphiphilic 6-arm star-block copolymers with PCL inner blocks and zwitterionic poly (2-methacryloyloxyethyl phosphorylcholine) outer blocks (6sPCL-*b*-PMPC) have been utilized for the encapsulation and delivery of PTX.[59] The star copolymers synthesized by a combination of ROP and atom transfer radical polymerization (ATRP) formed spherical micelles with PCL hydrophobic cores. Micelle size increased by 30–80% after incorporation of PTX. Star-block micelles were labeled with FITC in order to follow their internalization profile into cancer cells. Fluorescence microscopy observations confirmed that the loaded micelles have been efficiently internalized by tumor cells. It was possible to directly visualize the micelles within tumor cells using TEM, and it was concluded that the 6sPCL-*b*-PMPC drug-loaded micelles were more efficiently uptaken by tumor cells compared to diblock PCL-*b*-PEG micelles. The 6sPCL-*b*-PMPC micelles carrying PTX showed much higher cytotoxicity against HeLa cells than PCL-b-PEG micelles loaded with the same drug, a result that was attributed to the higher efficiency of cellular uptake of the particular micelles.

Dendron-like poly (ε-benzyloxycarbonyl-l-lysine)/linear PEO block copolymer micelles encapsulating DOX were tested for their ability to carry and release the drug under different pH environmental conditions.[60] Drug loading in the range 5–10%wt were achieved depending on the composition of the copolymers. The *in vitro* experiments indicated that the drug-loaded micelles showed a triphasic drug-release profile at aqueous pH 7.4 or 5.5 and at 37°C. The authors argued that the initial fast drug-release behavior was probably due to the large surface area of nanoparticles, and the subsequent slower drug release was mainly controlled by the drug diffusion out of the nanoparticles. The nanoparticles sustained a longer drug-release period for about 2 months. The drug-release profile at aqueous pH 5.5 similar to endosome and/or lysosome compartments inside cells was slightly faster, which was due to the relatively increased aqueous solubility of DOX at mildly acidic pH.

Multicomponent micellar nanocarriers based on ABC miktoarm block copolymers (where A = PEG, B = PCL, and C = a short triphenylphosphonium bromide residue) were used for the delivery of coenzyme Q10 (CoQ10) in mitochondria.[61] The micellar nanoparticles displayed sizes of 25–60 nm and a drug-loading capacity reaching 60 wt%. The nanoassemblies were stable in solution for more than 3 months. The authors attributed the extraordinarily high CoQ10 loading capacity of the miktoarm micelles to the good chemical compatibility between CoQ10 and the PCL arm of the copolymer. Confocal microscopy observations of the fluorescently labeled block copolymer analogue, together with the mitochondria-specific vital dye label, showed that the nanocarriers could actually reach the mitochondria within the cells. The high CoQ10 loading efficiency allowed testing of loaded micelles within a broad concentration range and provided evidence for CoQ10 effectiveness in two different experimental paradigms: oxidative stress and inflammation. The experimental results obtained suggested that the miktoarm-based carrier could effectively deliver CoQ10 to mitochondria without loss of drug effectiveness. This particular work shows that controlled and targeted delivery of low molecular weight drugs to intracellular organelles is possible through the use of specially designed block copolymer micellar nanocarriers.

Amphiphilic, shell cross-linked, knedel-like (SCK) polymer nanoparticles were studied as potential nanocarriers of DOX.[62] The main goals were to investigate the effects of the core and shell dimensions on the loading and release of DOX as a function of the solution pH. SCKs were constructed from poly(acrylic acid)-c-polystyrene (PAA-c-PS) amphiphilic diblock copolymers having different relative block lengths and overall degrees of polymerization. A series of different SCK nanoparticle samples with hydrodynamic diameters from 14 to 30 nm were prepared and studied. Block copolymer compositions employed gave SCKs with ratios of shell to core volumes ranging from 0.44 to 2.1. The SCKs were capable to sustain drug loadings in the range 1500–9700 DOX molecules per particle, with larger numbers of DOX molecules encapsulated within the larger core SCKs. The volume occupied by the PAA shell relative to the volume occupied by the polystyrene core was found to correlate inversely with the diffusion-based release of DOX. SCKs having smaller cores and higher acrylic acid corona volume to styrene core volume ratios showed lower final extents of release. Higher final extents of release and faster rates of release were observed for all DOX-loaded SCK particles at pH 5.0 versus pH 7.4, respectively, ca. 60% versus 40% at 60 hr, a promising characteristic for enhanced DOX delivery within tumors and cancer cells (Fig. 8). Quantitative determination of the kinetics of release was made by fitting the release profile data to the Higuchi model. The release rate constants determined ranged from 0.0431 to 0.0540 $hr^{-1/2}$ at pH 7.4 and 0.106 to 0.136 $hr^{-1/2}$ at pH 5.0 for SCKs, compared to the non-cross-linked block copolymer micelle analogues that exhibited rate constants for release of DOX of 0.245 and 0.278 $hr^{-1/2}$ at pH 7.4 and 5.0, respectively.

Nanoparticles or less well-defined micellar aggregates formed by random copolymers have been investigated in parallel for use as delivery nanovehicles for small drug molecules. Ulbrich and coworkers utilized A-(2-hydroxypropyl)methacrylamide (HPMA) random copolymers with covalently bonded cholesteryl groups and their conjugates with DOX.[63] DOX and cholesterol derivatives were bound by pH-sensitive hydrazone bonds to the main polymer chain. The presence of the hydrophobic cholesteryl groups

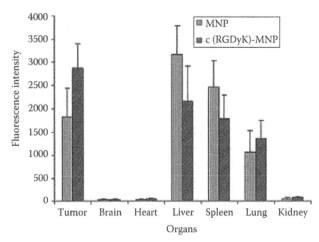

Fig. 8 Fluorescence intensity of Dir-loaded MNP and c (RGDyK)-MNP in various organs.
Source: From Jiang et al.[52] © 2011, with permission from Elsevier.

results in the formation of aggregates in aqueous media, and the covalently bonded drug resides in the hydrophilic corona of the assemblies. DOX could be released from the aggregates at endosomal pH (pH = 5) due to hydrolysis of the covalent bonds. The rate of hydrolysis was strongly dependent on the microenvironment around the polymer-drug bond. Although DOX could be released at first, it was observed that the high molecular weight supramolecular polymer carrier was disintegrated very slowly, forming relatively short polymer fragments that are small enough to be eliminated from the body by glomerular filtration.

A random copolymer composed of thermosensitive A-isopropyl acrylamide, hydrophilic A, A-dimethylacrylamide, and pH-responsive undecenoic acid segments, P(NIPAM-co-DMAAm-co-UA), has been assembled in micelles in aqueous media, at pH 7.4 and at 37°C, and DOX has been encapsulated into the copolymer nanoassemblies.[64–66] NIPAM and DMAAM units were forming the corona of the micelles. Protonation of the UA units at lower pH (e.g., 5 or 6.6) made them more hydrophobic and resulted in a destabilization of the micelles at 37°C and release of DOX. Release of the anticancer drug was found to be slow at pH 7.4, but it was increased at lower pH values due to the deformation of the core/shell structure. Attachment of cholesterol groups on the UA units made the copolymer more hydrophobic, and PTX could be now easily encapsulated into the micelles. PTX release was found to depend on environmental pH and temperature as in the case of DOX-loaded P(NIPAM-co-DMAAm-co-UA) micelles without cholesterol groups. Comparison with thermoresponsive drug-loaded block copolymer micelles showed that pH sensitivity played a major role in the extent of internalization of the micellar nanocarriers. FA groups were also attached for targeting on the particular copolymer. Active targeting increased considerably DOX uptake from 4T1 mouse breast cancer cells compared to drug-loaded copolymer micelles without targeting functionalities. The IC50 value of DOX against KB cells that also overexpress folate receptors was also found to be increased for the case of folate-conjugated copolymer micelles encapsulating the drug.

Besides self-assembled micelles, vesicles formed by block copolymers have been successfully utilized as drug nanocarriers in several cases. Xu et al. have presented the case of a linear triblock copolymer, namely, PEO-PAA-PNIPAM. This copolymer could be easily dissolved molecularly at room temperature, but it was self-assembled into nanosized vesicles upon increasing the solution temperature (temperature higher than 32°C). The formed aggregates could be stabilized by cross-linking at the interface through the PAA block. Next, the encapsulation of biologically interesting molecules took place at relatively high loading efficiency. It was demonstrated that the formed carrier was stable against a number of environmental changes, like temperature, salinity, and dilution. However, dissociation was observed under reductive conditions, leading to the conclusion that such systems can be particularly useful for intracellular delivery of biopharmaceuticals.[66] Another example of complex nanocarrier has been presented by Wang and coworkers. A triblock copolymer with a cross-linkable middle block was used for the release of an entrapped anticancer drug into the cell cytoplasm. The triblock copolymer forms onion-type micelles upon dissolution in aqueous media, while cross-linking can be realized at the shell, through disulfide bonds. The formed aggregate was stable during circulation and it was found to be effective in terms of drug delivery, under the reductive intracellular environment, where cross-linking is damaged.[67]

Hollow core spherical micelles formed by stereocomplexation of AB$_2$ miktoarm copolymers of the types poly (ethylene glycol)-[poly (l-lactide)]$_2$ [PEG-(PLLA)$_2$] and poly (ethylene glycol)-[poly (D-lactide)]$_2$ [PEG-(PDLA)$_2$] and nanofibers formed by stereocomplexation between poly(ethylene glycol)-b-poly(D-lactide) (PEG-b-PDLA) and poly(ethylene glycol)-c-poly(l-lactide) (PEG-c-PLLA) were used for the encapsulation of PTX.[68] Drug loading of 12 and 40%wt were achieved, respectively, showing the effect of nanostructure morphology on the loading capacity of the nanostructure in respect to hydrophobic drugs. Release of PTX was continuous in both cases without an initial rapid increase of the released drug.

DELIVERY OF MACROMOLECULAR DRUGS

Gene Delivery

Gene therapy refers to treatment of diseases by modifying gene expression within specific cells. It has the potential to shift the way numerous human diseases such as sickle cell anemia, HIV, Parkinson's disease, Huntington's disease, and Alzheimer's disease to which a genetic identity has been given are managed and treated. The concept behind is simple—exogenous delivery of therapeutic genes to the body to treat genetic-based diseases. Engineered viruses were the first gene delivery agents. Some of the earliest clinical trials produced promising results. However, although viral therapies have proven to be efficient, they appeared quite risky to the patient[69–72] as severe side effects were made strikingly evident. The documented dangers of using the efficient recombinant viruses as carriers have motivated the exploration for safer, less pathogenic, and immunogenic gene delivery alternatives including lipid-based vectors, chemically modified viruses, inorganic materials, and polymer-based gene delivery systems. As stated elsewhere,[73] in addition to the potential safety benefits, such nonviral systems offer greater structural and chemical versatility for manipulating physicochemical properties, vector stability upon storage and reconstitution, and a larger gene capacity compared to their viral counterparts. In the following, we focus on the polymer-based systems that have been used as nonviral carriers for complexing and intracellular delivery of genetic material (DNA, RNA, other nucleotides).

Biological Barriers to Gene Delivery

The current understanding of the various biological barriers that face efficient gene delivery has been discussed in a recent review.[73] Figure 9 schematically represents the whole process of gene delivery—from DNA packaging to gene expression—as well as all barriers and obstacles that should be overcome. The fundamental criteria for any synthetic gene delivery system require the ability to (I) package therapeutic genes, (II) gain entry into cells, (III) escape the endolysosomal pathway, (IV) effect DNA/vector release, (V) traffic through the cytoplasm and into the nucleus, (VI) enable gene expression, and, last but not least, remain biocompatible (Fig. 9).

The packaging of DNA is the first stage of the gene delivery process. It is typically achieved by three main strategies—electrostatic interactions, encapsulation, and adsorption—aiming at 1) neutralizing the negatively charged phosphate backbone of DNA to prevent charge repulsion against the anionic cell surface; 2) condensing the bulky structure of DNA to appropriate length scales for cellular internalization; and 3) protecting the DNA from both extracellular and intracellular nuclease degradation.[74–76] The polymeric vectors developed to date have exploited the anionic nature of DNA and other nucleotides to drive complexation via electrostatic interactions. As a result, a polyplex is formed, in which the DNA is condensed but remains active and capable of transfecting cells. The size of polyplexes is typically in the range 30–500 nm. Their characteristics can be controlled through the proper selection of the (co)polymer and through the N/P ratio.

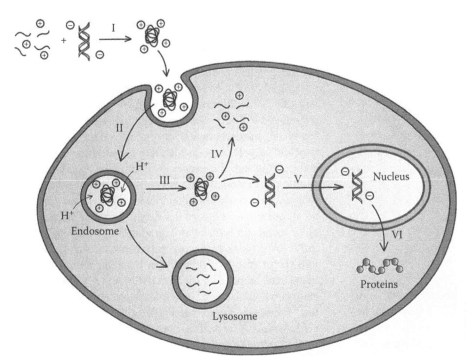

Fig. 9 Barriers to gene delivery.
Source: From Wong et al.[73] © 2007, with permission from Elsevier.

A variety of polyplexes of DNA, RNA, and other ODNs with natural and synthetic (co)polymers is described in Section 3.6. It has been documented that positively charged particles are taken up by cells faster than negatively or neutrally charged ones. In addition, the excess positive charge results in a final positive zeta-potential for the polyplex and the plasmid is protected by the polymer. That is why a surplus of positive charge or polymer added to the polyplex is beneficial as far as gene delivery is concerned.

Another possible way to protect genes from enzymatic degradation is encapsulation within nano- or microspherical structures. The latter are typically prepared from biodegradable polymers such as polyphosphoesters, polyphosphazenes, poly (β-amino ester)s, poly (lactic acid), poly (lactide-co-glycolide), and polyorthoesters, which, owing to the presence of hydrolytically unstable linkages, can degrade to shorter oligomeric and monomeric components. The degradation kinetics as well as the size of the DNA-encapsulated complex can be controlled by varying polymer properties and method of particle formation.[77–80] Drawbacks related to exposure to high shear stress, organic solvent and extreme temperatures, low encapsulation efficiency and DNA bioavailability, and potential DNA degradation have been frequently reported.[81–84] A "hybrid" approach is the chemical conjugation of cationic surfactants or polymers to the surface of biodegradable particles. DNA can be electrostatically bound to the latter, thus enhancing its availability for immediate release.[85–87]

Cellular entry is the first obstacle encountered by the polyplex to be overcome. The strategies developed to surmount this barrier have mimicked the endocytosis—the process by which cells internalize nutrients, bacteria, etc., from the extracellular space. The endocytic uptake is known to occur by at least five pathways—phagocytosis, macropinocytosis, clathrin-mediated endocytosis, caveolae-mediated endocytosis, and clathrin- and caveolae-independent endocytosis.[88] All of them share the common uptake mode of enclosing the internalized polyplexes within transport vesicles derived from the plasma membrane. Differences arise, however, in the subsequent intracellular processing and routing of internalized polyplexes, which consequently affect the extent to which the polyplexes can deliver their genetic cargo efficiently.[73] The cellular entry can ideally be achieved by targeted uptake via receptor-mediated endocytosis and size exclusion-mediated phagocytosis. The former consists of attaching of targeting ligands to the vectors, which induce endocytosis upon binding to their cognate surface receptors. Table 5.1 lists some examples of the most commonly used ligands that have provided selective targeting to various cell types. Preferential internalization of DNA-encapsulated microspheres (1–10 μm size range) by phagocytic cells is the core of the size exclusion-mediated phagocytosis. Such a strategy has been adopted for DNA vaccine applications where antigen-presenting cells of the immune system are the target cells.[87,89]

The nonspecific uptake, that is, in the absence of targeting strategies, is another possible way to gain entry in the cells. It involves ionic interactions with membrane-bound proteoglycans, lipophilic interactions with phospholipid membrane, and cell-penetrating peptide (CPP)-mediated uptake. Proteoglycans are composed of a membrane-associated core protein from which a chain of sulfated or carboxylated glycosaminoglycans extend into the

Table 1 Endogenous ligands, carbohydrates, antibodies, and cpp employed to promote cellular uptake

Endogenous Ligands	Carbohydrates	Antibodies	CPP
Transferrin	Galactose	Anti-cd3	HIV Tat
TGF-α	Mannose	Anti-EGF	Penetratin
Folate	Lactose	Anti-HER2	Transportan
FGF		Anti-pIgR	Polyarginine
EGF		Anti-JL-1	
VEGF			
Asialoorosomucoid			
RGD peptide			

extracellular space.[90] These highly anionic GAG units determine much of the interactions between the cell surface and the extracellular macromolecules and are responsible for the overall negative charge of the plasma membrane;[91] they are believed to play a central role in the endocytic uptake of many non-targeted, positively charged gene delivery systems.[92,93] It is important to consider, however, that differences in proteoglycan distribution between cell types can lead to varying degrees of internalization from one cell type to another and, in addition, the nonspecificity of this mode of interaction can lead to indiscriminate uptake and undesirable gene expression by unanticipated cell populations.[73]

Polyplexes have been able to induce cellular entry through interactions between lipophilic residues bound to the vector and the phospholipid layer of the cell membrane.[94] In particular, increased transfection efficiency of branched PEI modified with long lipophilic chains has been reported elsewhere.[95] The results of these authors indicate that even the position of the substituents (primary or tertiary amines of PEI) can have effects on the extent of this interaction and thus transfection efficiencies. CPPs are shown in Table 1. Typically they are positively charged, amphiphilic in nature, and composed of 5–40 amino acids. They are able to simultaneously bind DNA and serve as cell-penetrating agents. The mechanism of their action is still unclear. The largely accepted hypotheses include; 1) the formation of peptide-lined pores within the membrane; 2) direct penetration through the membrane and into the cytoplasm; 3) transient uptake into a membrane-bound micellar structure that inverts to release the CPP and its genetic cargo inside the cytosol; and 4) the induction of endocytosis.[96]

After internalization, the polyplex is believed to be contained within an endocytic vesicle. It is then either: 1) recycled back to the cell surface; 2) sorted to acidic, degradative vesicles (e.g., lysosomes); or 3) delivered to an intracellular organelle (e.g., Golgi apparatus, endoplasmic reticulum).[97] The intracellu-lar itinerary that the endocytic

vesicles follow depends on the uptake pathway (i.e., phagocytosis, macropinocytosis, clathrin-mediated endocytosis, caveolae-mediated endocytosis, and clathrin- and caveolae-independent endocytosis), which in turn is dependent on a number of factors including cell line, polyplex type, and the conditions under which the polyplex is formulated.[98,99] The clathrin-mediated endocytosis is currently the most characterized pathway. Within this pathway, the polyplex is sequestered in endosomal vesicles and shuttled through the endolysosomal pathway. The escape from endolysosomal pathway is of paramount importance to avoid enzymatic degradation within lysosomal compartment; otherwise, the gene delivery fails at this stage. Strategies to escape lysosomal degradation have exploited the influx of protons by incorporating pH-sensitive moieties into vector designs. These are protonatable amines such as linear and branched PEI; PDMAEMA; polyamidoamine; endosomal viral or synthetic release proteins such as melittin, influenza virus hemagglutinin, diphtheria toxin, GALA, KALA, and JTS-1; and alkylated carboxylic acids such as poly (propylacrylic acid), poly (ethylacrylic acid), and poly (ethylacrylate-co-acrylic acid). The ultimate effect of these vector components is the disruption of the membrane of the endolysosomal vesicles leading to subsequent release of the vesicles' content in the cytosol.

Before the double-layer membrane surrounding the nucleus is reached, the polyplexes (or naked DNA) that have escaped from the endolysosomal pathway must be trafficked through the cytosol. The cytosolic environment is both physically and metabolically hostile to the polyplexes. The mesh-like structure of the cytoskeleton can severely impede the diffusion of both naked DNA greater than 250 bp with an extended linear length of approximately 85 nm[100,101] and polyplexes that are typically on the order of 150–200 nm in size. In addition, nucleolytic enzymes ready to degrade unprotected nucleic acids are interspersed among the microtubules, intermediate filaments, and microfilaments that build the network-like structure of the cytoskeleton. Strategies have been elaborated to shorten the residence time within the cytosol and promote transport toward and into nucleus. They are typically based on nuclear import machinery naturally utilized by cells to transport proteins into nucleus. Cytosolic proteins that are destined for the nucleus contain a distinct amino acid sequence known as nuclear localization signal (NLS), which is recognized by import proteins (importins). The latter direct their subsequent transport and shuttle them through the nuclear pore complexes that perforate the nuclear membrane.[102] The NLSs are short, cationic peptide segments and can be directly used to interact and condense DNA.[103] They can be also attached to a polymer vector that is subsequently complexed with gene material.[104,105] NLSs used to promote cytosolic transport and nuclear import are, for example, human T-cell leukemia virus (HTLV) type 1, M9, HIV type-1 viral protein R (Vpr), and SV40 large T antigen. The limitations of this approach are related to the size and type of DNA used (i.e., linear, plasmid), the method of NLS incorporation (i.e., covalent conjugation to DNA, electrostatic complexation with DNA, or conjugation to a polymer vector), the type of NLS peptide employed, the number of NLSs incorporated, and the type of polymer vector used (e.g., liposomes, PEI).[106,107] Various natural or recombinant proteins, for example, protamine, histone, and high-mobility group proteins, contain NLS sequences and a net positive charge. They fall into two broad categories—histones and nonhistone proteins. The former are involved with DNA binding and condensation as well as regulating transcription and cell cycle progression,[108,109] whereas the latter function similarly and support various DNA-related activities including transcription, replication, and recombination.[109] Carbohydrates such as lactose, mannose, and A-acetylglucosamine can also mediate cytosolic transport and nuclear import: similar to the NLS-based transport, the carbohydrate-mediated transport is an energy- and signal-dependent process that imports the cargo into the nucleus through nuclear pores.[73] In spite of some conflicting reports,[110–112] the relative ease of synthesis, inherent biocompatibility, and reduced immunogenic concerns associated with carbohydrate-based moieties have contributed to its advantageous utility in gene vector design. Polyplexes may also gain nuclear entry by taking advantages of nuclear membrane breakage that occurs during cell mitosis.[73,113,114] This approach, however, is somewhat limited since many applications of nonviral gene delivery target nondividing or slow-dividing cells.[115]

The final step of the journey of the exogenous genes is to express their encoded therapeutic proteins. Genome insertion has been attempted by incorporating DNA transposons within the delivery system. Transposons are naturally occurring DNA sequences capable of enzymatically excising themselves out of the one chromosomal locus and reinserting itself into another locus.[116] An important limitation of transposon-mediated genome integration is the random insertion into the host genome leading to inadvertent gene disruption and unwanted side effects. Therefore, systems providing a certain degree of site specificity have been explored. These are integrase enzymes derived from bacteriophages and hybrid systems comprising a transposable element and a DNA sequence recognition element. Transgene insertion occurs through their ability to recognize and dock the transgenic plasmid to a known genome locus.[73] Furthermore, both physical and chemical stimuli have been explored to provide an additional level of control over transcription.

Polymer-based nonviral gene carriers

There are a number of hybrid co-assembled nanoparticles, the constituents of which are a synthetic or natural (co) polymer and oligo- or polynucleotides. In the following,

these and similar structures are discussed from the perspective of gene therapy, though gene delivery via polyplexes has some issues to overcome such as low transfection, nonspecific short-term delivery, rapid clearance, and toxicity concerns and has to achieve a significant presence in a clinical capacity based on the small percentage of nonviral vectors currently used in clinical trials.[73,117] The examples given as follows demonstrate how research is progressing to meet the demands of nonviral polymer-based gene delivery. When possible, information about the functional effects of a particular structural feature and mechanisms that govern structure-function relationships and, in turn, gene delivery is provided.

Nonviral transfection systems based on polymer vectors have been periodically reviewed by various authors.[73,117–122] Typically, polyplexes between polycations and DNA (RNA or other ODNs) have been evaluated with respect to their effectiveness, toxicity, and cell type dependence. Not only have commercially available polycations such as linear and branched PEIs, polyamidoamine dendrimers, and linear and dendritic polypropyleneimines been examined but also other polymer vectors. These are polymethacrylates, carbohydrate-based polymers (chitosan, dextran, cyclodextrins, polyglycoamidoamine), biodegradable polymers {poly (4-hydroxy-1-proline ester), poly[α-(4-aminobutyl)-1-glycolic acid], poly (amino ester), phosphorus-containing polymers}, and polypeptides (Tat-based peptides, antennapedia homeodomain peptide, MPG peptide, transportan peptide). These systems exhibit activity higher or comparable to "standard" lipid-based transfection reagents. Other studies demonstrate good potential of structurally diverse polymers and polyplex systems as transfection reagents with low cytotoxicity.

Linear (Co)Polymers. Zheng and coworkers have prepared hydrophobically modified 1.8 kDa PEI conjugates with lipoic acid (6,8-dithiooctanoic acid) using carbodiimide chemistry.[123] The DNA-binding ability was impaired by lipoylation and much smaller (84–183 nm) particles were formed. The MTT assays demonstrated that all PEI-lipoic acid conjugates and polyplexes were essentially nontoxic to HeLa and 293T cells up to a tested concentration of 50 μg/mL and an N/P ratio of 80/1, respectively. The *in vitro* gene transfection studies in HeLa and 293T cells showed that lipoylation of 1.8 kDa PEI markedly boosted its transfection activity.

A library of 39 strictly linear PEG-PEI diblock copolymers has been synthesized.[124] The copolymers were composed of PEG moieties with fixed molar mass of 2, 5, or 10 kDa, whereas the PEI molar mass ranged from 1.5 to 10.8 kDa. They were expected to build small and stable polyplexes, as the two blocks were clearly segregated, which implied that the PEG might not sterically counteract the interaction between the nucleic acid and PEI. The results indicated that the PEG domain had a greater influence on the physicochemical properties of the polyplexes than PEI. A PEG content higher than 50% led to small (<150 nm), nearly neutral polyplexes with favorable stability. The transfection efficacy of these polyplexes was significantly reduced compared to the PEI homopolymer but was restored by the application of the corresponding degradable copolymer, which involved a redox triggerable PEG domain.

Quaternized poly[3,5-bis (dimethylaminomethylene)-p-hydroxyl styrene] homopolymer (QNPHOS), having two permanently charged cationic sites per monomer unit as well as its block copolymer with PEO (QNPHOS-b-PEO), were studied as small interfering RNA (siRNA) and plasmid DNA carriers.[125] Comparisons with the standard transfectant PDMAEMA, in terms of nucleic acid-binding strength, gene silencing, and transfection activities of the complexes, were made. It was shown that siRNA complexes based on QNPHOS and QNPHOS-b-PEO dissociate in the presence of a fourfold higher heparin concentration than necessary to destabilize PDMAEMA-based complexes. Under the same conditions, complexes of DNA and QNPHOS or QNPHOS-c-PEO did not show any dissociation, in contrast to PDMAEMA polyplexes. The DNA polyplexes based on QNPHOS or QNPHOS-c-PEO did not show transfection activity, which was mainly ascribed to their high physicochemical stability. On the other hand, siRNA complexes based on QNPHOS and QNPHOS-c-PEO were found to show a low cytotoxicity and an improved siRNA delivery and high gene-silencing activity, even higher than those based on PDMAEMA. The authors concluded that the observed behavior must be due to the excellent binding characteristics of QNPHOS and QNPHOS-c-PEO toward siRNA, which in turn can be correlated with the presence of two permanently charged cationic groups per monomer unit in the QNPHOS chains. Based on the results of this study, it can be concluded that formation of strong siRNA nanosized complexes with polymers containing double charges per monomer is advantageous.

In a recent study, oligomers and polymers of N-ethyl pyrrolidine methacrylamide as well as its copolymers with DMAAm have been synthesized.[126] Cell viability and proliferation after contact with polymer and polyplexes were studied using 3T3 fibroblasts, and the systems showed an excellent biocompatibility at 2 and 4 days. Transfection studies were performed with plasmid Gaussia luciferase (GLuc) kit and were found that the highest transfection efficiency in serum-free was obtained with oligomers from the P/N ratio of 1/6 to 1/10.

Polyphosphonium polymers can be considered as an efficient and nontoxic alternative to polyammonium carriers. A study allowing for direct comparison of DNA binding and gene transfection of phosphonium-containing macromolecules and their respective ammonium analogues

has been performed.[127] Conventional free-radical polymerization of quaternized 4-vinylbenzyl chloride monomers afforded phosphonium- and ammonium-containing homopolymers for gene transfection experiments of HeLa cells. DNA gel shift assays and luciferase expression assays revealed that the phosphonium-containing polymers bound DNA at lower charge ratios and displayed improved luciferase expression relative to the ammonium analogues. The triethyl-based vectors for both cations failed to transfect HeLa cells, whereas tributyl-based vectors successfully transfected HeLa cells similar to SuperFect demonstrating the influence of the alkyl substituent lengths on the efficacy of the gene delivery vehicle. Cellular uptake of Cy5-labeled DNA highlighted successful cellular uptake of triethyl-based polyplexes, showing that intracellular mechanisms presumably prevented luciferase expression. Endocytic inhibition studies demonstrated the caveolae-mediated pathway as the preferred cellular uptake mechanism for the delivery vehicles examined. The authors noted that changing the polymeric cation from ammonium to phosphonium enables an unexplored array of synthetic vectors for enhanced DNA binding and transfection that may transform the field of nonviral gene delivery.[127]

In another paper, a series of polymers based on poly (ethoxytriethylene glycol acrylate) bearing ammonium or phosphonium quaternary groups have been prepared.[128] The triethylphosphonium polymer showed transfection efficiency up to 65% with 100% cell viability, whereas the best result obtained for the ammonium analogue reached only 25% transfection with 85% cell viability. The nature of the alkyl substituents on the phosphonium cations (tert-butyl, hydroxypropyl, phenyl were studied) was shown to have an important influence on the transfection efficiency and toxicity of the polyplexes. The results presented by the authors showed that the use of positively charged phosphonium groups was a worthy choice to achieve a good balance between toxicity and transfection efficiency in gene delivery systems.[128]

Phosphonium-based diblock copolymers for nonviral gene delivery have been synthesized by reversible addition fragmentation transfer (RAFT).[129] They were composed of a stabilizing block of either oligo (ethylene glycol$_9$) methyl ether methacrylate or 2-(methacryloxy)ethyl phosphorylcholine and a phosphonium-containing cationic block of 4-vinylbenzyltributylphosphonium chloride, which induced electrostatic complexation with DNA (Fig. 10). All block copolymers exhibited low cytotoxicity (>80% cell viability) and generated stable polyplexes with hydrodynamic radii between 100 and 200 nm. The cellular uptake by COS-7 and HeLa cells and, consequently, the transfection in these cell lines were reduced, whereas serum transfection in HepaRG cells, which are a predictive cell line for *in vivo* transfection studies, showed successful transfection using all diblock copolymers with luciferase expression.

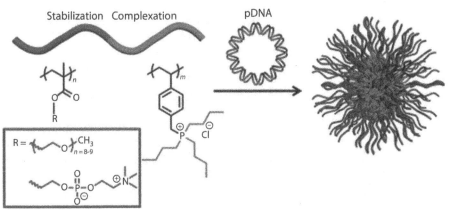

Fig. 10 Phosphonium-containing diblock copolymers for enhanced colloidal stability and efficient nucleic acid delivery. The stabilizing blocks were of molecular weight M_n = 25,000 g/mol. By subsequent chain extension, phosphonium-containing blocks with DPs of 25, 50, or 75 were synthesized.
Source: From Hemp et al.[129] ©2012, with permission from American Chemical Society.

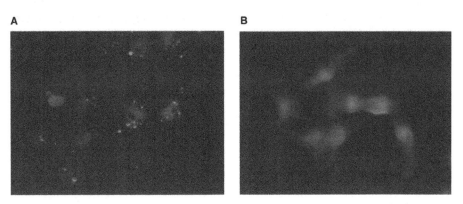

Fig. 11 Polymer-enhanced intracellular delivery of FAM-labeled siRNA. Representative images illustrating (**A**) punctate staining (dots of strong contrast) in the samples treated with lipofectamine/siRNA complexes alone and (**B**) dispersed fluorescence within the cytosol following delivery of poly(DMAEMA)-*b*-poly[(BMA)-co-(DMAEMA)-co-(propylacrylic acid)]/siRNA complexes. Samples were treated for 15 min with 25 nM FAM-siRNA and prepared for microscopic examination following DAPI nuclear staining (areas of low contrast).
Source: From Convertine et al.[130] © 2010, with permission from American Chemical Society.

A micellar delivery system based on "diblock" copolymers of poly(DMAEMA)-*b*-poly[(butyl methacrylate)-co-(DMAEMA)-co-(propylacrylic acid)] (Fig. 10) is described elsewhere in this encyclopedia. As noted, the copolymer was composed of a polycationic block of PDMAEMA, which mediated siRNA binding and a second block of dimethylaminoethyl methacrylate (DMAEMA) and propylacrylic acid in roughly equimolar ratios with butyl methacrylate (BMA), which was pH sensitive and endosome releasing.[130] Delivery systems were formed upon binding siRNA to the corona of the preformed micelles at theoretical + /− charge ratios of 4/1 and greater. Their ability to deliver siRNA through the endosomal pathway into the cytoplasm was investigated by conducting mRNA knockdown experiments against GAPDH. Transfection experiments in HeLa cells were conducted at siRNA concentrations between 12.5 and 100 nM under high serum conditions. Flow cytometry analysis of cell uptake properties indicated enhancement of the knockdown potential of the micellar system: an uptake in 90% of cells and a threefold increase in siRNA per cell compared to a standard lipid transfection agent. The fluorescence micrographs (Fig. 11) showed that labeled siRNA is dispersed throughout the cellular cytoplasm in contrast to the commercial reagent that was primarily punctate.

Amino poly(glycidyl methacrylate)s have been shown to form stable complexes with antisense ODNs that prevented the latter from nuclease degradation. In addition, the polyplexes of linear and starlike poly(glycidyl methacrylate)s of M_n in the range 15–20 kDa modified with methy-lethylamine as well as linear poly(glycidyl methacrylate) modified with 4-amino-1-butanol exhibited higher transfection efficiency *in vitro* compared to the "golden" standard of PEI 25 k.[131]

The potential of a series of triblock copolymers for messenger RNA (mRNA)-based strategies has recently been demonstrated.[132] The materials were composed of a cationic DMAEMA segment to mediate mRNA condensation, a hydrophilic poly(ethylene glycol) methyl ether methacrylate (PEGMA) segment to enhance stability and biocompatibility, and a pH-responsive endosomolytic copolymer of DEAEMA and BMA designed to facilitate cytosolic entry. They were able to condense mRNA into 86–216 nm particles, and the polyplexes formed from polymers with the PEGMA segment in the center of the polymer chain displayed the greatest stability to heparin displacement; they were associated with the highest transfection efficiencies in two immune cell lines, RAW 264.7 macrophages (77%) and DC2.4 dendritic cells (50%). Transfected DC2.4 cells were shown to be capable of subsequently activating antigen-specific T cells.

The efficacy of using a polysorbitol-based osmotically active transporter (PSOAT) system (Fig. 12A for the chemical formula of the copolymer) for siRNA delivery and its specific mechanism for cellular uptake to accelerate targeted gene silencing has been reported.[133] The authors found that PSOAT functioned via a caveolae-mediated uptake mechanism due to its hyperosmotic activity. An example of silencing effect of PSOAT/siLuc and PSOAT/siGFP in A549 cells is demonstrated in Fig. 12A, right panel: luciferase- or green fluorescent protein (GFP)-expressing A549 cells at 80% confluence were transfected with polymer/siRNA polyplexes. Graphical presentation of the caveolae-mediated endocytosis is shown in Fig. 12B. Detailed description of the different stages and events of the caveolae-mediated uptake mechanism is given by the authors:[133] (1–2) PSOAT/siRNA complex recognizes and binds to the sorbitol- transporting channel (STC) on the extracellular membrane and creates a hyperosmotic environment. (3–5) The osmotic pressure-sensitive PSOAT/siRNA complex induces caveolin (Cav)-1 expression and selectively stimulates caveolae-mediated endocytosis. This event subsequently generates COX-2 expression under osmotic stress. At this stage, Cav-1 expression augments and a mature caveolae structure is formed where COX-2 might work on disrupting the joint surface of particle deposition and accelerate caveolae-mediated endocytosis. (6) Selective caveolae endocytosis allows the PSOAT/siRNA complex containing caveolae endosome (caveosome) to avoid lysosomal fusion. (7–8) The endosome containing PSOAT/siRNA complex swells and eventually bursts due to the proton sponge effect of LPEI, which allows the complex to escape into the cytosol. (9) Due to the nature of degradable linkages, PSOAT degradation occurs and siRNAs are released. (10–12) The released siRNA recognizes and breaks down the target mRNA at the posttranscriptional level through an RNAi mechanism.

The transfection and local tissue distribution of plasmid DNA using different cationic polymers have been analyzed after intradermal injection in mice.[134] Cationic polymers with distinct chemical structures: branched PEI, linear poly (2-aminoethyl methacrylate) (PAEM), and diblock copolymer PEG-b-PAEM were selected to study the structure-function relationship of the polymer carriers in the context of in vivo administration (Fig. 13). The authors found that naked DNA dispersed quickly in hours within the skin and showed limited colocalization with APCs. On the other hand, polyplexes formed depots at the injection site and

persisted for days to engage and transfect skin cells. PEGylated polyplexes, in particular, possessed superior stability against aggregation than non-PEGylated polyplexes, and they disseminated well in the skin that promoted interaction with both the antigen-presenting cells and dermal fibroblasts. These findings provide in vivo evidence to support the use of PEGylated polymer carriers for DNA vaccine delivery and suggest possible approaches to further improve polymer design.

In a recent paper,[135] a comprehensive study has been conducted to evaluate the colloidal properties of PAEM/plasmid DNA polyplexes, the uptake and subcellular trafficking of polyplexes in antigen-presenting dendritic cells, and the biological performance of PAEM as a potential DNA vaccine carrier. PAEM of different chain length (45, 75, and 150 repeating units) showed varying strength in condensing plasmid DNA into narrowly dispersed nanoparticles with very low cytotoxicity. Longer polymer chains resulted in higher levels of overall cellular uptake and nuclear uptake of plasmid DNA, but shorter polymer chains favored intracellular and intranuclear release of free plasmid from the polyplexes.

Packaging of plasmid DNA within both rod- and sphere-shaped polyplex micelles has been investigated, with a focus on DNA rigidity and folding.[136] Selective preparation of rod or spherical structures was accomplished by modulating the polylysine segment length of the PEG-polylysine block copolymers. The correlation of these packaging structures to gene expression efficiency was determined both in vitro and in vivo, with improved gene expression resulting from folded plasmid DNA demonstrated in cultured cells as well as in skeletal muscle following intravenous injection. The authors concluded that the enhanced gene expression demonstrated by folded plasmid DNA packaging may be attributed to its nuclease tolerability and transcription-active nature. Results obtained for regularly folded plasmid DNA polyplex micelles were in stark contrast with collapsed plasmid DNA polyplex micelles, probably due to limited transcription efficiency, clearly showing that controlled packaging of plasmid DNA is crucial for achieving effective gene transfer.[136]

Nonlinear (Co)Polymers. Besides molecular weight and composition, the chain architecture of gene delivery vectors plays a critical role in DNA complexation and, hence, gene expression. Generally, the nonlinear (graft, branched, starlike) polymer exhibits superior gene delivery abilities compared to their linear analogues.

Reversibly hydrophobilized 10 kDa PEI based on rapidly acid-degradable acetal-containing hydrophobe has been designed for nontoxic and highly efficient nonviral gene transfer.[137] Water-soluble PEI derivatives with average 5, 9, and 14 units of pH-sensitive 2,4,6-trimethoxybenzylidene-tris (hydroxymethyl)ethane (TMB-THME) hydrophobe per molecule, denoted as PEI-g-(TMB-THME)n, were readily

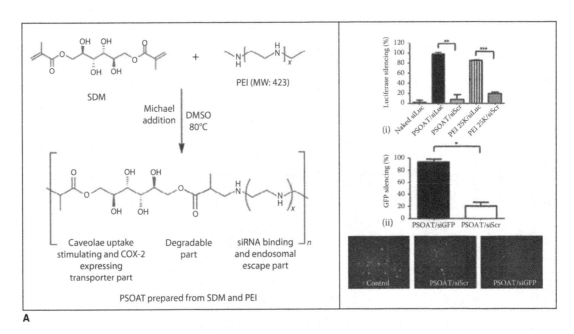

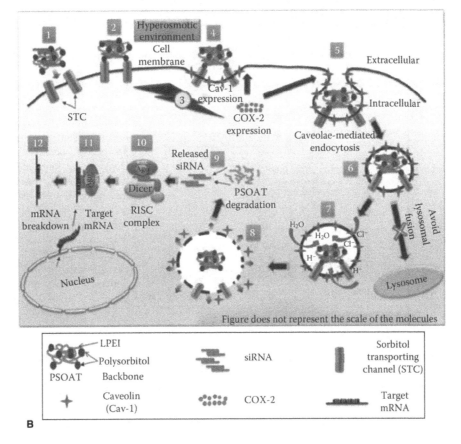

Fig. 12 **(A)** Left panel: proposed reaction scheme for the synthesis of PSOAT from sorbitol dimethacrylate (SDM) and PEI. PSOAT was prepared through a Michael addition reaction in DMSO at 80°C. Right panel: silencing efficacy of PSOAT/siLuc and PSOAT/siGFP in A549 cells. (i) Luciferase silencing by PSOAT/siLuc and PEI 25K/siLuc complexes compared to the respective scrambled siRNA (siScr) group. (ii) GFP silencing efficacy of PSOAT/siGFP complexes compared to the respective siScr group at a siRNA concentration of 100 pmol/L (n = 3, error bar represents SD) (*P < 0.05; **P < 0.01; ***P < 0.001, one-way ANOVA) and fluorescent images of control, PSOAT/siScr- and PSOAT/siGFP-treated A549 cells (magnification: 10x). **(B)** Schematic representation on the function of PSOAT. Graphical presentation shows selective caveolae-mediated endocytosis of the PSOAT/siRNA complex following hyperosmotic pressure that induces caveolin (Cav-1) and cyclooxygenase (COX-2) expression, which accelerates the efficacy of the transporter for RNAi silencing. **Source:** From Islam et al.[133] © 2012, with permission from Elsevier.

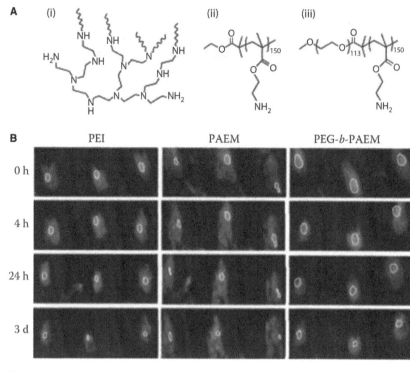

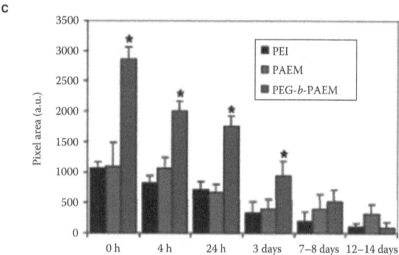

Fig. 13 (**A**) Chemical structures of (i) branched PEI, (ii) linear PAEM, and (iii) diblock copolymer PEG-c-PAEM. (**B**) Tissue distribution of polyplexes in live animals after intradermal injection. Three mice were each injected with polyplexes containing 10 ig of Cy3-labeled DNA in the right hind quadriceps region and were imaged together. Plasmid distribution in three injected mice was shown at indicated time points as marked by a white outline (A), and the area of signal was quantified (B). *PEGylated polyplexes showed statistically larger area of spreading than PEI- and PAEM-based polyplexes (*t*-test, p < 0.001).
Source: From Palumbo et al.[134] © 2012, with permission from Elsevier.

obtained by treating 10 kDa PEI with varying amounts of TMB-THME-nitrophenyl chloroformate. Polyplexes of smaller size (100–170 nm) and higher surface charges (+25 to +43 mV) were obtained. *In vitro* transfection experiments performed at N/P ratios of 10/1 and 20/1 in HeLa, 293T, HepG2, and KB cells using plasmid pGL3 expressing luciferase as the reporter gene showed that reversibly hydrophobilized PEIs had superior transfection activity to 25 kDa PEI control. For example, polyplexes of PEI-g-(TMB-THME)14 showed about 235-fold and 175-fold higher transfection efficiency as compared to 10 kDa PEI in HeLa cells in serum-free and 10% serum media, respectively, which were approximately 7-fold and 16-fold higher than 25 kDa PEI formulation at its optimal N/P ratio under otherwise the

same conditions. Hydrophobic modification of 10 kDa PEI enhances its DNA condensation ability and cellular interactions, while reversal of hydrophobic modification in endosomes facilitates intracellular release of DNA (Fig. 14).

Star copolymers have been developed as siRNA carriers by Matyjaszewski and coworkers.[138] The authors utilized ATRP methodologies and an "arm-first" approach for the synthesis of the copolymers having PEG arms and a degradable cationic core of PEGMA macromonomer, 2-(dimethylamino)ethyl meth-acrylate (DMAEMA), and a disulfide dimethacrylate (acting as the cross-linker). The star polymers had a diameter of ca. 15 nm in aqueous solutions and were found to be degraded under redox conditions by glutathione treatment into individual polymeric chains, due to

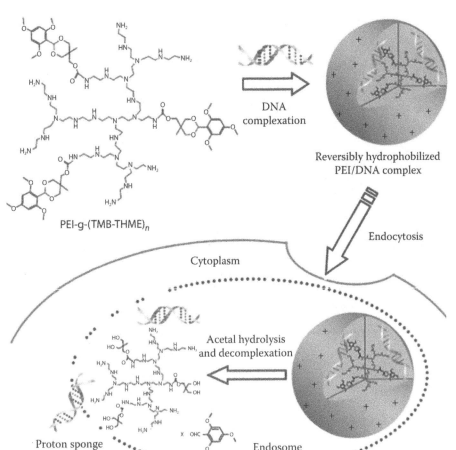

PEI-g-(TMB-THME)$_n$

DNA complexation

Reversibly hydrophobilized PEI/DNA complex

Endocytosis

Cytoplasm

Acetal hydrolysis and decomplexation

Proton sponge effect of PEI

Endosome

Fig. 14 Illustration of reversibly hydrophobilized 10 kDa PEI for efficient intracellular delivery and release of DNA. TMB-THME denotes 2,4,6-trimethoxybenzyli-dene-tris (hydroxymethyl) ethane.
Source: From Liu et al.[137] © 2011, with permission from Elsevier.

cleavage of the disulfide cross-linker. Culture tests with mouse calvarial preosteoblast-like cells, embryonic day 1, subclone 4 (MC3T3-E1.4) indicated that the star copolymers were biocompatible, with more than 80% cell viability after 48 hr of incubation even at high concentration of the copolymer (800 Lig/mL). Surface charge of the nanoassemblies formed by electrostatic interaction with siRNA at varying N/P ratios confirmed the formation of star copolymer/siRNA complexes. Confocal microscopy and flow cytometry measurements showed the cellular uptake of the nanocomplexes in MC3T3-E1.4 cells after 24 hr of incubation.

A copolymer composed of dendritic polyamidoamine, PEG, and poly(l-lysine) has been designed for siRNA delivery.[139] Each moiety played a specific role: the polyamidoamine dendrimer worked as a proton sponge and participated in the endosomal escape and cytoplasmic delivery of siRNA; PEG rendered nuclease stability, whereas poly(l-lysine) provided primary amines to form polyplexes with siRNA and also acted as penetration enhancer. The copolymer provided excellent cellular uptake by A2780 human ovarian cancer cells and ensured homogeneous and uniform distribution of the polyplex in different cellular layers from the top of the cell to the bottom as evidenced by Fig. 15.

Poly(cyclooctene-graft-oligolysine)s, having a comb-like architecture, have been prepared by ring-opening metathesis polymerization of the corresponding oligopeptide-substituted cyclic olefin monomers.[140] The reconfiguration of polylysine into short oligolysine grafts, strung from a hydrophobic polymer backbone, gave transfection reagents greatly superior to poly-lysine, despite having the identical cationic functional groups. Altering the oligolysine graft length modulated DNA-polymer interactions and transfection efficiency, while incorporating the PKKKRKV heptapeptide (the Simian virus SV40 large T-antigen nuclear localization sequence) pendent groups onto the polymer backbone led to even greater transfection efficiency over the oligolysine-grafted structures. The relative strength of the polymer/DNA complex was key to the transfection performance, as judged by serum stability and PicoGreen analysis. Moreover, the polyplexes exhibited low cytotoxicity, contributing to the therapeutic promise of these reagents.

Chemical conjugation of hydrophobic deoxycholate moieties to a polyamido-amine-diethylenetriamine (DET) dendrimer has been shown to improve the stability of the resulting polyplexes with DNA against ionic strength.[141] The transfection efficiency of the latter is higher than that

Top

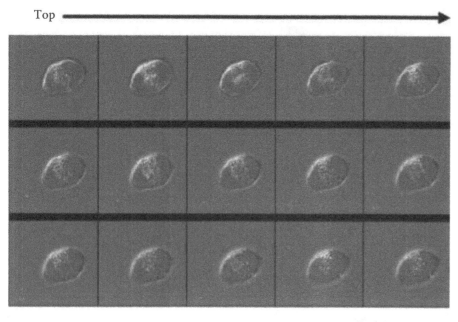

Bottom

Fig. 15 Optical section z series of A2780 human ovarian cancer cells incubated with siRNA complex with polyamidoamine-PEG-poly(l-lysine) copolymer.
Source: From Patil et al.[139] © 2011, with permission from American Chemical Society.

of the polyplex formed from the unmodified dendrimer, but its cytotoxicity remains the same.

Cationic polymeric amphiphiles composed of a poly (l-lactide) middle block and two flanking dendritic poly (l-lysine) moieties of the second generation have been prepared and characterized with respect to CAC, self-aggregation and plasmid DNA-binding affinities, and the effects of poly(l-lactide) block length on particle size, morphology, and ζ-potential.[142] Toxicities of these amphiphiles and their polyplexes were assayed by MTT with HeLa, SMMC-7721 and COS-7 cells, and COS-7 cell luciferase and eGFP gene transfection efficacies with these amphiphiles as the delivery carriers were investigated as well. The copolymers exhibited low toxicities. The polyplex nanoparticles were shown to form via two different aggregation mechanisms depending on the length of the middle poly(l-lactide) block (Fig. 16).

Zhang and coworkers reported on the synthesis of polyaspartamide-based disulfide containing brushed PEI derivatives, P(Asp-Az)$_x$-SS-PEI.[143] They were reduction sensitive and able to condense DNA into small positive nanoparticles. *In vitro* experiments revealed that the reducible P(Asp-Az)$_x$-SS-PEI not only had much lower cytotoxicity but also posed high transfection activity (both in the presence and absence of serum) as compared to the control nondegradable 25 kDa PEI.

The influence of graft densities of PCL-*b*-PEG chains on physicochemical properties, DNA complexation, and transfection efficiency of amphiphilic copolymers prepared by grafting PCL-b-PEG chains onto hyperbranched PEI has been studied.[144] It was found that the transfection efficiencies of these copolymers increased at first toward an optimal graft density (n = 3) and then decreased. The

buffer-capacity test showed almost exactly the same tendency as transfection efficiency. Cytotoxicity depended on the collective effect of PEG molecular weight and graft density of PCL-*b*-PEG chains. The cytotoxicity, zeta-potential, affinity with DNA, stability of the polyplexes, and critical micellization concentration (CMC) values were reduced strongly and regularly with increasing graft density. Increasing the excess of polymer over DNA was shown to result in a decrease of the observed particle size to 100–200 nm. In an earlier paper, essentially the same authors have created a library of PEI-graft-(PCL-block-PEG) copolymers to establish structure-function relationships for siRNA delivery.[145] It was found that longer PEG chains, longer PCL segments, and higher graft density beneficially affected the stability and formation of polyplexes and reduced the zeta-potential of siRNA polyplexes. Significant siRNA-mediated knockdown was observed for hyperbranched PEI25k-(PCL900-mPEG2k)1 at N/P 20 and 30, implying that the PCL hydrophobic segment played a very important role in siRNA transfection.

Zhong and coworkers have grafted PEO with 1.8 kDa branched PEI and prepared copolymers with varying compositions.[146] The copolymers were able to effectively condense DNA into small (80–245 nm) particles with moderate positive (+ 7.2 to + 24.1 mV) surface charges. Reduced cytotoxicity (MTT assay in 293T cells, cell viability >80%) with increasing PEO molecular weight and decreasing PEI graft densities was observed. PEO(13k)-g-10PEI (the numbers correspond to molecular weight and grafting densities, respectively) polyplexes formed at an N/P ratio of 20/1, which were essentially nontoxic (100% cell viability), displayed over three- and fourfold higher transfection efficiencies in 293T cells than 25 kDa PEI standard under

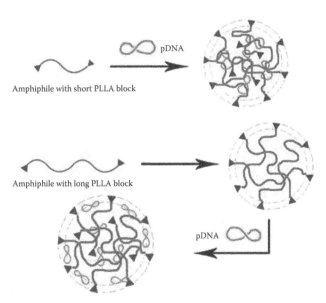

Fig. 16 Schematic presentation of polyplex structures depending on the length of the middle poly(l-lactide) block of dendritic poly(l-lysine)-linear poly(l-lactide)-dendritic poly(l-lysine) symmetric polymeric amphiphiles.
Source: From Zhu et al.[169] © 2012, with permission from American Chemical Society.

serum-free and 10% serum conditions, respectively. CLSM studies using Cy5-labeled DNA confirmed that these PEO-g-PEI copolymers could efficiently deliver DNA into the perinuclei region as well as into nuclei of 293T cells at an N/P ratio of 20/1 following 4 hr transfection under 10% serum conditions. Enhanced gene delivery efficacy and decreased cytotoxicity of polyplexes of branched 25 kDa PEI grafted with PEO-PPO-PEO copolymers with varying molecular weights and PEO contents have been reported.[147] The (PEO-PPO-PEO)-g-PEI copolymers showed lower cytotoxicity in three different cell lines (HeLa, MCF-7, and HepG2) than PEI 25 kDa. pGL3-lus was used as a reporter gene, and the transfection efficiency was *in vitro* measured in HeLa cells.

The role of boronic acid moieties in polyamidoamines has been investigated in the application of these polymers as gene delivery vectors.[148] The polymers contained 30% of phenylboronic acid side chains and exhibited improved polyplex formation abilities with plasmid DNA since smaller and more polydisperse polyplexes were formed as compared to their non-boronated counterparts. The transfection efficiency was approximately similar to that of commercial PEI; however, the polyplexes showed increased cytotoxicity most probably caused by increased membrane-disruptive interactions.

Lin and Engbersen have recently demonstrated that a one-pot Michael addition reaction using an aminoterminated PEG as a comonomer in an equimolar mixture of bisacrylate (N,N'-cystaminebisacrylamide) and primary amine monomers (4-amino-1-butanol) is a versatile

approach to obtain PEGylated polyamidoamines with disul-fide linkages.[149] The copolymers condensed DNA into nanoscaled PEGylated polyplexes (<250 nm) with near neutral (2–5 mV) or slightly positive (9–13 mV) surface charge that remained stable in 150 mM buffer solution over 24 hr. The PEGylated polyplexes showed very low cytotoxicity in MCF-7 and NIH 3T3 cells and induced appreciable transfection efficiencies in the presence of 10% serum, although those were lower than those of the corresponding unPEGylated complexes. The notably lower transfection was likely caused by limited endosomal escape. In another paper, the cellular dynamics of the polyplexes was studied *in vitro* on retinal pigment epithelium cells.[150] It was shown that these net cationic polyplexes required a charge-mediated attachment to the sulfate groups of cell-surface heparan sulfate proteoglycans in order to be efficiently internalized. In addition, the involvement of defined endocytic pathways in the internalization of the polyplexes in ARPE-19 cells by using a combination of endocytic inhibitors, RNAi depletion of endocytic proteins, and live-cell fluorescence colocalization microscopy, was assessed. The authors found that the polyplexes entered RPE cells via both flotillin-dependent endocytosis and PAK1-dependent phagocytosis-like mechanism. The capacity of polyplexes to transfect cells was, however, primarily dependent on a flotillin-1-dependent endocytosis pathway.

Stimuli-Sensitive Systems, Targeting: Clinical therapies have long used control of some environmental parameters at the therapeutic target area. Thermoresponsive nonviral gene carriers based on PNIPAM have been shown to enhance their transfection efficiency upon applying hypothermia, since DNA is assumed to be released as PNIPAM undergoes a globule to coil conformational transition on cooling below its lower critical solution temperature (LCST).[151–154] However, contradictory observations, that is, enhanced transfection obtained by raising the temperature from below to above the LCST, have been reported.[155,156] The enhanced transfection above the LCST has been attributed to increased cation numbers in the endosomal compartment resulting from polyplex aggregation,[156] whereas the former authors attributed this observation to greater cationic presentation on the surface of polyplexes. Recombinant, protein-based polymer analogues of PNIPAM are the elastin-like polypeptides (ELPs). They are a class of genetically engineered, thermoresponsive polymers that consist of VPGXG pentapeptide monomers, where X is a guest residue that can be any amino acid except proline.[157] The ELPs are water soluble, biocompatible, and thermally responsive with an inverse temperature phase transition.[158] Preparation and evaluation of a hybrid recombinant material that possess a thermoresponsive ELP segment and a DET-modified poly-l-aspartic acid segment have been described elsewhere.[159] The reaction scheme is presented in Fig. 17. The polyplexes formed by the copolymer and pGL4 plasmid were of dimensions between 90 and

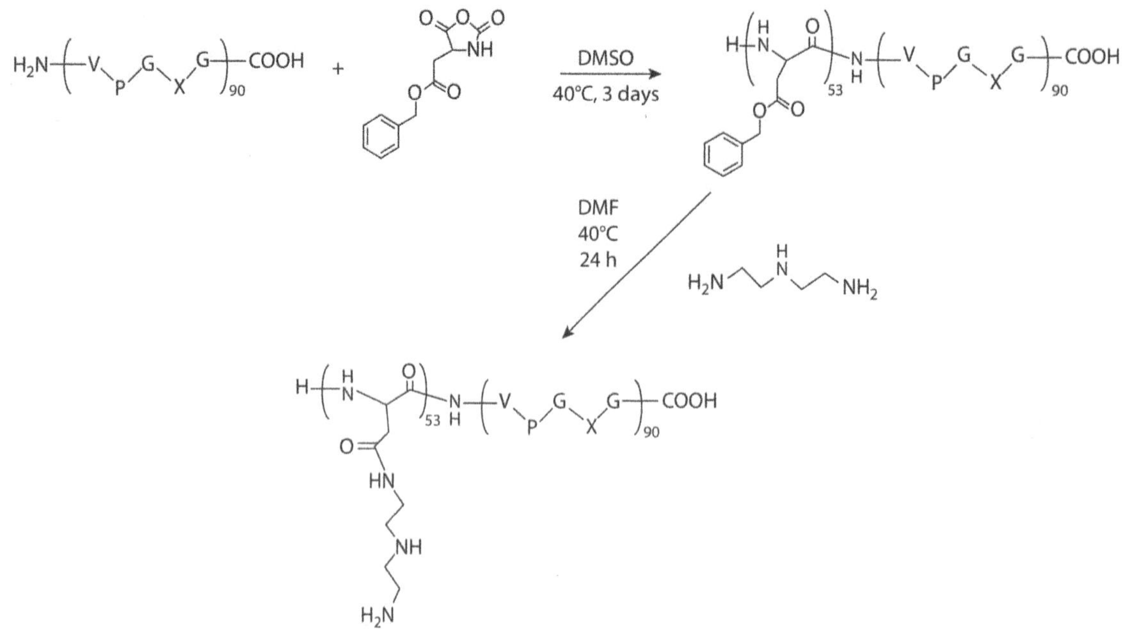

Fig. 17 Reaction scheme for the synthesis of a diblock copolymer possessing a thermoresponsive ELP block and a DET-modified poly-l-aspar-tic acid block.
Source: From Chen et al.[159] © 2012 with permission from Elsevier.

100 nm in diameter and neutral even at N/P > 2. The latter indicated that the ELP block was able to effectively shield the positive charges. The polyplexes also showed appreciable transfection efficiency with low cytotoxicity. The polyplexes practically retained the thermal phase transition behavior conferred by the copolymer; however, they exhibited a two-step transition process: an irreversible primary aggregation below the copolymer's transition temperature, followed by a reversible secondary aggregation at temperatures about the transition temperature. The aggregation is critical for thermal targeting, as particles aggregate at the disease site in response to external heating, which facilitates the particles deposition. Both the copolymer and the polyplexes aggregated at elevated temperatures (~52°C). Although this temperature is not clinically relevant, it may be modulated by changing the ELP block composition or molecular weight. Therefore, the authors concluded the thermal responsiveness of this copolymer can be used for thermal targeting of therapeutic pDNA to hyperthermic disease sites.

Thermosensitive properties and sharp phase transition behavior have been observed for two types of copolymers—random and block-random copolymers of poly (NIPAM-co-hydroxyethyl methacrylate-co-DMAEMA).[160] Cell viability assays indicated low cytotoxicity of block-random copolymers and almost no cytotoxicity of the random copolymers that was ascribed to the presence of low positive charges along copolymer backbones. The random copolymers were used as DNA delivery vectors to deliver GFP gene into HEK293T cells. The transfection efficiency was evaluated by GFP protein expression, which was affected by the charge content and transfection operation temperature. Lowering the temperature to 25°C during the transfection course improves transfection efficiency, which results from the polyplexes dissociation at the temperature below the LCST.[160]

Poly[oligo (ethylene glycol) methacrylate] is another thermosensitive polymer exhibiting LCST properties. Zhang and coworkers have exploited its properties to prepare a family of thermosensitive cationic copolymers that contain branched PEI 25 K as the cationic segment and poly[methoxy (diethylene glycol) methacrylate-co-oligo (ethylene glycol) methacrylate] as the thermosensitive block.[161] All of these copolymers were able to condense DNA and formed poly-plexes with diameters of 150–300 nm and zeta-potentials of 7–32 mV at N/P ratios between 12 and 36. The key factor for shielding the positive surface charge of the polyplexes and protecting them against protein adsorption was the length of thermosensitive block. Lower cytotoxicity and the best transfection performance were achieved with copolymers with, respectively, a higher content and longest chains of the thermosensitive moieties.

Ternary polyplexes comprising plasmid DNA, PEG-based block cationomer PEG-b-{N-[N-(2-aminoethyl)-2-amnoethyl]aspartamide} (PEG-b-P[Asp (DET)], and poly (ethylene oxide)-b-poly (propylene oxide)-b-poly (ethylene oxide)-b-P[Asp (DET)] (P(EPE)-b-P[Asp (DET)]) have been prepared by Lai and coworkers aiming at maintaining adequate transfection efficiency and solving stability issues.[162] The ternary complexes exhibited improved stability against salt-induced aggregation, although the gene delivery ability dropped with increasing amount of

PEG-*b*-P[Asp(DET)]. Reducible ternary complexes, prepared by substituting P(EPE)-*b*-P[Asp(DET)] with a corresponding copolymer possessing a redox potential-sensitive disulfide linkages, P(EPE)-SS-P[Asp(DET)], achieved higher transfection compared to the nonreducible polyplexes while exhibiting comparable stability.

Sensitive to light are polyplexes constructed via host-guest interactions between β-cyclodextrin and azobenzene.[163] The dePEGylation of the complex upon light irradiation facilitated the release of DNA and its entry into the nucleus (Fig. 18), which resulted in efficient transfection.

Magnetic gene vectors for targeting gene delivery have been described elsewhere.[164] They were prepared by surface modification of Fe_3O_4 nanoparticles with carboxymethyl dextran and PEI and used to deliver GFP gene into BHK21 cells. The transfection efficiency and gene expression efficiency of those transfected with a magnet were much higher than that of standard transfection. Sun and coworkers have prepared PEI-decorated magnetic nanoparticles and determined their potency for efficiently complexing and delivering DNA *in vitro* with the help of a magnetic field.[165] PEI was associated with PAA-bound

superparamagnetic iron oxide (PAAIO) through electrostatic interactions (PEI-PAAIO). PEI-PAAIO formed stable polyplexes with pDNA in the presence and absence of 10% fetal bovine serum (FBS) and could be used for magnetofection. The effect of a static magnetic field on the cytotoxicity, cellular uptake, and transfection efficiency of PEI-PAAIO/pDNA was evaluated with and without 10% FBS. Magnetofection efficacy in HEK 293T cells and U87 cells containing 10% FBS was significantly improved in the presence of an external magnetic field.

Other magnetic targeting gene delivery systems have been described by Han and coworkers.[166] The systems prepared by modification of magnetosomes with polyamidoamine dendrimers and Tat peptides were designed to cross the blood-brain barrier and deliver therapeutic genes to brain cancerous tissues. Transfection efficiencies of Tat-magnetosome-polyamidoamine polyplexes with pGL-3 were studied using U251 human glioma cells *in vitro*. The results showed that the incorporation of external magnetic field and Tat peptides could significantly improve transfection efficiency of delivery system. Furthermore, biodistribution *in vivo* demonstrated that Tat-magnetosome-polyamidoamine could

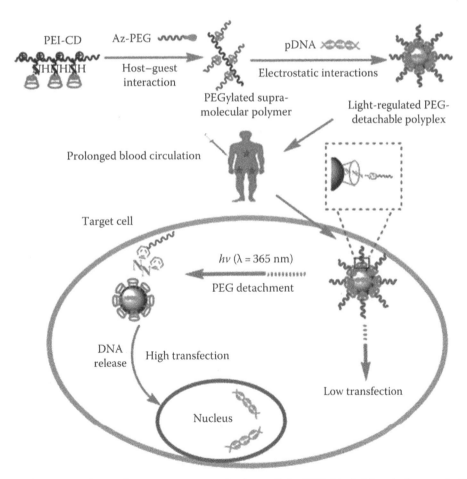

Fig. 18 Light-regulated host-guest interaction as a new strategy for intracellular PEG-detachable polyplexes to facilitate nuclear entry. PEI, CD, Az, and PEG correspond to PEI, p-cyclodextrin, azobenzene, and PEG, respectively.
Source: From Li et al.,[163] with permission from The Royal Society of Chemistry.

efficiently transport across the blood-brain barrier and assemble at brain tissue of rat detected by single-photon emission computed tomography (SPECT).

Branched PEI-hyaluronic acid copolymers have been successfully prepared using carbodiimide chemistry and subsequently functionalized with mannose to improve the transfection efficiency, cell viability, and cell specificity in macrophages.[167] Formation of polyplexes at polymer to DNA weight ratio >2 was observed. The nanohybrids exhibited significantly ($p < 0.05$) lower cytotoxicity than that of unmodified branched PEI. Mannose functionalization of these nanohybrids showed specificity for both murine and human macrophage-like cell lines RAW 264.7 and human acute monocytic leukemia cell line (THP1), respectively, with a significant level ($p < 0.05$) of expression of GLuc and green fluorescent reporter plasmids. Internalization studies indicate that a mannose-mediated endocytic pathway is responsible for this higher transfection rate.

The use of hydroxyethyl starch for controlled shielding/ deshielding of polyplexes has recently been reported.[168] Hydroxyethyl starch of different molar masses was grafted to PEI, and the resulting conjugates were used to generate polyplexes with the luciferase-expressing plasmid DNA pCMVluc. Deshielding was tested *in vitro* by ζ-potential measurements and erythrocyte aggregation assay upon addition of α-amylase to the hydroxyethyl starch-decorated particles. The addition of α-amylase led to gradual increase in the zeta-potential of the nanoparticles over 0.5–1 hr and to a higher aggregation tendency for erythrocytes due to the degradation of the hydroxyethyl starch—coat and exposure of the polyplexes' positive charge. *In vitro* transfection experiments were conducted in two cell lines ± a-amylase in the culture medium. The amylase-treated hydroxyethyl starch-decorated complexes showed up to 2 orders of magnitude higher transfection levels compared to the untreated hydroxyethyl starch-shielded particles, while a-amylase had no effect on the transfection of PEG-coated or uncoated polyplexes.

Reduction-sensitive reversibly shielded DNA polyplexes based on PDMAEMA-SS-PEG-SS-PDMAEMA triblock copolymers have been prepared and shown highly promising for nonviral gene transfection.[169] They were prepared by RAFT polymerization with controlled compositions of 6.6-6-6.6 and 13-6-13 kDa. The resulting polyplexes (diameters < 120 nm, 0 to + 6 mV) showed excellent colloidal stability against 150 mM NaCl. In the presence of 10 mM of dithiothreitol, however, they were rapidly deshielded, unpacked, and DNA was released. Notably, *in vitro* transfection studies showed that reversibly shielded polyplexes afforded up to 28 times higher transfection efficacy as compared to stably shielded control under otherwise the same conditions. Confocal laser scanning microscope studies revealed that reversibly shielded polyplexes efficiently delivered and released pDNA into the perinuclei region as well as nuclei of COS-7 cells.[169]

Folate-conjugated tercopolymers have been shown to deliver gene materials into target tumor cell via folate receptor-mediated endocytosis.[170] The copolymers were based on PEI-g-PCL-b-PEG conjugated with folate. They were able to condense DNA completely at N/P ratio >2 and polyplexes of N/P ratio 10 with sizes of about 120 nm, and positive zeta-potentials were selected for biological evaluations due to their stability. An enhancement of both cellular uptake of PEI-g-PCL-*b*-PEG-Fol/pDNA polyplexes and their transfection efficiency was observed in folate receptor-overexpressing cells in comparison to unmodified PEI-g-PCL-*b*-PEG.

An attractive target for tumor-targeting therapies is the EGFR. A gene delivery system based on PEGylated linear PEI has been developed.[171] Peptide sequences directly derived from the human EGF molecule enhanced transfection efficiency with concomitant EGFR activation. Only the EGFR-binding peptide GE11, which has been identified by phage display technique, showed specific enhancement of transfection on EGFR-overexpressing tumor cells including glioblastoma and hepatoma but without EGFR activation. In a clinically relevant orthotopic prostate cancer model, intratumorally injected GE11 polyplexes were superior in inducing transgene expression when compared with untargeted polyplexes. Induction of tumor-selective iodide uptake and therapeutic efficacy of ^{131}I in a hepa-tocellular carcinoma (HCC) xenograft mouse model, using the same polyplexes, has been reported in another paper.[172]

Other authors have isolated and covalently linked to the distal end of PEG (3500)-PEI(25 kDa) an anti-DF3/Mucin1 (MUC1) nanobody with high specificity for the MUC1 antigen, which is an aberrantly glycosylated glycoprotein overexpressed in tumors of epithelial origin.[173] The resulting macromolecular conjugate successfully condensed plasmids coding a transcriptionally targeted truncated-Bid (tBid) killer gene under the control of the cancer-specific MUC1 promoter. The engineered polyplexes exhibited favorable physicochemical characteristics for transfection and dramatically elevated the level of Bid/tBid expression in both MUC1 overexpressing caspase-3-deficient (MCF7 cells) and caspase-3-positive (T47D and SKBR3) tumor cell lines and, concomitantly, induced considerable cell death.

A newly developed delivery vector has been designed to impart bioreducibility for greater intracellular pDNA release, higher serum stability, and efficient complexing ability by incorporating disulfide linkage, PEG, and low molecular weight PEI, respectively.[174] RVG peptide as a targeting ligand for neuronal cells was used to deliver genes to mouse brain overcoming the blood-brain barrier. The physiochemical properties of the polyplexes formed with plasmid DNA, their cytotoxicity, and the *in vitro* transfection efficiency on Neuro2a cell were studied prior to the successful *in vivo* study. *In vivo* fluorescence assay substantiated the permeation of the plasmid DNA-loaded polymeric vector through the blood-brain barrier.

Nam and coworkers have described the development of a cardiomyocyte-targeting nonviral gene carrier and anti-apoptotic candidate siRNAs to inhibit cardiomyocyte apoptosis.[175] The newly synthesized bioreducible polymers were based on cystamine bisacrylamide-diaminohexane and incorporated a cardiomyocyte-targeting peptide (PCM) and the CPP HIV Tat 49–57 (Fig. 19A(i)). Gel retardation results demonstrated that peptide-modified polymers could effectively condense anionic genes at a weight ratio over 5 (Fig. 19A(ii)). As shown in Fig. 19B, the cells treated with polyplexes of polymer modified by both Tat and PCM peptides, whose plasmid DNA was labeled with YOYO-1 iodide, showed a more intense green signal in the cytoplasm with an additional 2 hr of treatment. These results demonstrate that conjugation with the Tat peptide increases transfection efficiency, while addition of the PCM peptide results in a further increase in transfection efficiency in H9C2 cardiomyocytes. In addition, *in vitro* cytotoxicity by MTT assay demonstrated good cell viability after transfection using the PCM-CD-Tat polyplexes, which verified that PCM conjugation facilitated cardiomyocyte targeting, while Tat peptide conjugation facilitated gene delivery into the cells without significant cytotoxicity.

Sugars, Amino Acids, Glycopolymers, Peptides, Hybrids, Polysaccharides.

Wu and coworkers have focused on the chain flexibility of novel arginine-based polycations with incorporated ethylene glycol units (Fig. 20) in the gene delivery efficiency and the relationships between polyplex size, ζ-potential, and transfection efficiency.[176] The

polymers were synthesized by the solution polycondensation reaction of the p-toluenesulfonic acid salt of l-arginine diester from oligoethylene glycol and di-p-nitrophenyl esters of dicarboxylic acids. The transfection results obtained from luciferase and GFP assays from a wide range of cell lines, primary cells, and stem cells showed that the introduction of flexible segments was beneficial for the improvement of the transfection efficiency, which was comparable or better than that of commercial transfection reagents, but at a much lower cytotoxicity.

In a recent work, Aldawsari and coworkers have demonstrated that the conjugation of amino acids such as arginine, lysine, and leucine to the third generation of polypropyleneimine dendrimers led to an enhanced antiproliferative activity of the polyplexes *in vitro*.[177] *In vivo*, the intravenous administration of amino acid-bearing polypropyleneimine dendrimer polyplexes resulted in a significantly improved tumor gene expression, with the highest gene expression level observed after treatment with lysine-modified polyplex.

Anderson and coworkers have recently demonstrated that very subtle structural changes in polymer chemistry, such as end groups, can yield significant differences in the biological delivery efficiency and transgene expression of polymers used for plasmid DNA delivery.[178] These authors synthesized six analogues of a trehalose-pentaethylenehexamine glycopolymer (Tr4) that contained (1A) adamantane, (1B) carboxy, (1C) alkynyl-oligoethyleneamine, (1D) azido trehalose, (1E) octyl, or (1F) oligoethyleneamine end groups and evaluated the effects of polymer end group chemistry on the ability of these systems to bind, compact, and deliver plasmid DNA to cultured HeLa cells. All polymers studied were able to bind and compact plasmid DNA at similarly low N/P ratios and form polyplexes. The effects

Nanomaterials—
Nanoparticles

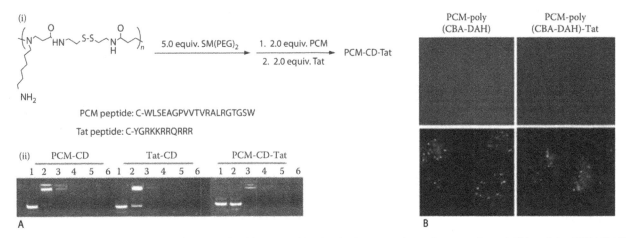

Fig. 19 **(A):** (i) Schematic diagram of a cystamine bisacrylamide-diaminohexane polymer incorporating a PCM and the CPP HIV Tat 49-57. (ii) Agarose gel electrophoresis of polymers complexed with plasmid DNA at the various weight ratios of polymer/plasmid DNA = 0, 1, 5, 10, 20, and 40 (lanes 1, 2, 3, 4, 5, and 6, respectively). **(B)** Intracellular trafficking of YOYO-1 iodide-labeled plasmid DNA polyplex in H9C2 cells incubated with PCM-conjugated or PCM and Tat-conjugated polymer using confocal microscopy after a 2 hr treatment. **Source:** From Nam et al.[175] © 2011, with permission from Elsevier.

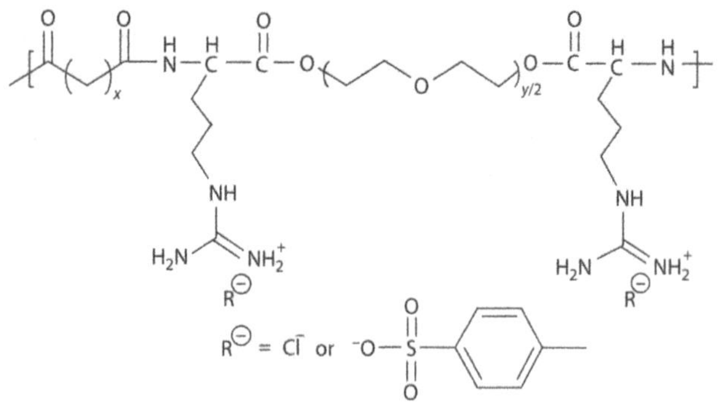

Fig. 20 Chemical formula of L-arginine-based polycations.
Source: From Wu et al.[176] © 2012, with permission from The Royal Society of Chemistry.

of the different end group structures were most evident in the polyplex internalization and transfection assays in the presence of serum as determined by flow cytometry and luciferase gene expression, respectively. The Tr4 polymers end capped with carboxyl groups (1B) (N/P = 7), octyne (1E) (N/P = 7), and oligoethyl-eneamine (1F) (N/P = 7) were taken into cells as polyplex and exhibited the highest levels of fluorescence, resulting from labeled plasmid. Similarly, the polymers end functionalized with carboxyl groups (1E at N/P = 7), octyl groups (1E at N/P = 15), and in particular oligoethyleneamine groups (1F at N/P = 15) yielded dramatically higher reporter gene expression in the presence of serum.[178]

RAFT polymerization has been used to prepare a series of diblock glycopolymers composed of 2-deoxy-2-methacrylamido glucopyranose and the primary amine-containing A-(2-aminoethyl) methacrylamide.[179] The polyplexes with plasmid DNA were found to be stable against aggregation in the presence of salt and serum over the 4 hr time period studied. Delivery experiments were performed *in vitro* to examine the cellular uptake, transfection efficiency, and cytotoxicity of the glycopolymer/plasmid DNA polyplexes in cultured HeLa cells: the diblock copolymer with the shortest A-(2-aminoethyl) methacrylamide block was found to be the most effective. The gene knockdown efficacy (delivery of siRNA to U-87 glioblastoma cells) of the diblock copolymer with the longest *N*-(2-aminoethyl) methacryl-amide block was found to be similar to commercial transfecting agents.

Reilly and coworkers have found that histone H3 peptides (peptides containing transcriptionally activating modifications) conjugated to PEI can effectively bind and protect plasmid DNA.[180] The H3/PEI hybrid polyplexes did not compromise the cell viability. They were found to transfect a substantially larger number of CHO-K1 cells *in vitro* compared to both polyplexes that were formed with only the H3 peptides and those that were formed with only PEI at the same total charge ratio. Transfections with the endolysosomal inhibitors chloroquine and bafilomycin A1 indicated that the H3/PEI hybrid polyplexes exhibited

slower uptake and a reduced dependence on endocytic pathways that trafficked to the lysosome, indicating a potentially enhanced reliance on caveolar uptake for efficient gene transfer.[180]

A panel of HPMA-oligolysine copolymers with varying peptide length and polymer molecular weight has been built using RAFT polymerization.[181] The panel was screened for optimal DNA binding, colloidal stability in salt, high transfection efficiency, and low cytotoxicity. Increasing polyplex stability in phosphate buffered saline (PBS) correlated with increasing polymer molecular weight and decreasing peptide length. Copolymers containing chains of 5 and 10 (K_5 and K_{10}) oligocations transfected cultured cells with significantly higher efficiencies than copolymers of K_{15}.

By exploiting the self-assembly properties and net positive charge of ELP, hollow spheres of sizes of 100, 300, 500, and 1000 nm have been fabricated.[182] The microbial transglutaminase cross-linking provided robustness and stability to the hollow spheres while maintaining surface functional groups for further modifications. The resulting hollow spheres showed a higher loading efficiency of plasmid DNA by using polyplex (~70 µg plasmid DNA/mg of hollow sphere) than that of self-assembled ELP particles and demonstrated controlled release triggered by protease and elastase. Moreover, polyplex-loaded hollow spheres showed better cell viability than polyplex alone and yielded higher luciferase expression by providing protection against endosomal degradation.

Hyperbranched cationic glycopolymers have been prepared by RAFT using 2-aminoethyl methacrylamide (AEMA) and 3-glucoamidopropyl methacryl-amide (GAPMA) or 2-lactobionamidoethyl methacrylamide (LAEMA) as shown in Fig. 21.[183] The hyperbranched polymers were synthesized with varying molecular weights and compositions, and their gene expressions were evaluated in two different cell lines. The copolymers formed complexes with β-galactosidase plasmid at varying molar ratios. Regardless of the molecular weight, the polyplexes produced in water were about 70–200 nm in diameter (Fig. 21). The high polydispersities of ζ-potential also observable in Fig. 21 may be associated with polydisperse nature of the hyperbranched copolymers (M_w/M_n in the range 1.70–11.2!). All copolymers of high molecular weight (50–60 kDa) showed very low gene expression, regardless of composition and polymer/plasmid ratio used. The copolymers of molecular weights 5–30 kDa exhibited high gene expression (Fig. 21), along with high cell viability. Furthermore, the cellular uptake and gene expression were studied in two different cell lines in the presence of lectins. It was found that polyplexes-lectin conjugates showed enhanced cellular uptake *in vitro*; however, their gene expression was cell line- and lectin-type dependent.[183]

The same authors have synthesized a library of cationic glycopolymers of predetermined molar masses (3–30 kDa) and narrow polydispersities using RAFT polymerization.[184]

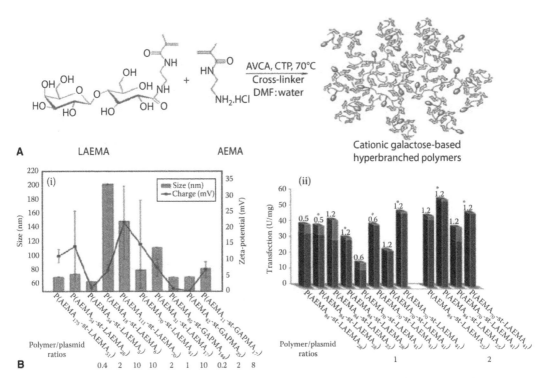

Fig. 21 (**A**) Synthesis of hyperbranched cationic glycopolymers by RAFT. LAEMA, AEMA, AVCA, and CTP correspond to LAEMA, AEMA, 4,4'-azobis (4-cyanovaleric acid), and 4-cyanopentanoic acid dithiobenzoate, respectively. (**B**): (i) DLS and zeta-potential data for the hyperbranched gly-copolymer-DNA polyplexes. All samples are prepared in deionized water at varying polymer/DNA molar ratios. (ii) Gene expression of galactose-based polymers of varying molecular weights, in the presence (*) and absence of serum using Hep G2 cells. Gene expression is evaluated using β-galactosidase assay at DNA doses 0.6 and 1.2 μg and varying polymer/plasmid molar ratios as shown.
Source: From Ahmed & Narain[183] © 2012, with permission from Elsevier.

The polymers contained pendant sugar moieties and differed from each other in their architecture (block vs. random), molecular weights, and monomer ratios (carbohydrate to cationic segment). It was shown that the aforementioned parameters can largely affect the toxicity, DNA condensation ability, and gene delivery efficacy of these polymers. For instance, random copolymers of high degree of polymerization were found to be the ideal vector for gene delivery purposes. They showed lower toxicity and higher gene expression in the presence and absence of serum, as compared to the corresponding diblock copolymers. The effect of serum proteins on random copolymer-based and diblock copolymer-based polyplexes and hence gene delivery efficacy was studied as well: the diblock copolymer-based polyplexes showed lower interactions with serum proteins, lower cellular uptake, and lower gene expression in both Hep G2 and HeLa cells in comparison to the polyplexes based on the random copolymers.

Cyclodextrin-modified PEIs with different cyclodextrin grafting levels have been synthesized.[185] The polycation containing an average of 15 cyclodextrin moieties per PEI chain could protect DNA completely above N/P ratio of 2. The particle sizes of these polyplexes were about 120 nm. Excellent stability in physiological conditions, prob-

ably due to the hydration shell of cyclodextrin, was observed at N/P ratios of 8 and 10. Uptake inhibition experiments indicated that PEI/DNA polyplexes were internalized by HEK293T cells by both clathrin-mediated endocytosis and caveolae-mediated endocytosis. The route of caveolae-mediated endocytosis was significantly promoted after cyclodextrin modification; hence, the cell uptake and transfection efficiency of polyplexes were significantly improved for HEK293T cells. However, the uptake and transfection efficiency in HepG2 cells were similar for the PEI/DNA polyplexes and the polyplexes of cyclodextrin-modified PEI, probably due to the lack of endogenous caveolins. A β-cyclodextrin-PEI-based polymer has been modified with the Tat peptide for plasmid DNA delivery to placenta mesenchymal stem cells.[186] Nanoparticles (~150 nm) were formed with DNA at the optimal N/P ratio of 20/1. The conjugation of the Tat peptide onto PEI β-cyclodextrin was demonstrated to improve the transfection efficiency in cells after 48 and 96 hr of post-transfection incubation. The viability of the cells was shown to be over 80% after 5 hr of treatment and 24 hr of posttreatment incubation.

Water-soluble chitosan-graft-(PEI-β-cyclodextrin) cationic copolymers have been synthesized via reductive amination between oxidized chitosan and low molecular

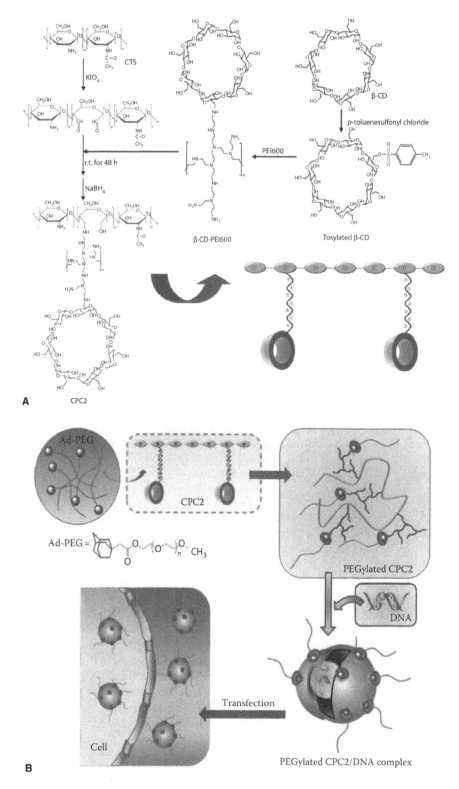

Fig. 22 (**A**) Synthetic procedure for the preparation of chitosan-graft-(PEI-β-cyclodextrin) cationic copolymers. (**B**) Illustration of PEGylation through supramolecular self-assembly between pendant β-cyclodextrin moieties and adamantyl-modified PEG and the formation of supramolecular PEGylated copolymer/DNA complex. CTS, chitosan; β-CD, β-cyclodextrin; Ad-PEG5k, adamantyl-modified PEG; CPC, chitosan-graft-(PEI-β-cyclodextrin) copolymer.
Source: From Ping et al.[187] © 2011, with permission from Elsevier.

weight PEI-modified p-cyclodextrin.[187] The synthetic procedure is presented in Fig. 22A. The polycations exhibited good ability to condense both plasmid DNA and siRNA into compact and spherical nanoparticles. Their gene transfection activity showed improved performance in comparison with native chitosan in HEK293, L929, and COS-7 cell lines. The pendant β-cyclodextrin moieties of the copolymers allowed the supramolecular PEGylation through self-assembly with adamantyl-modified PEG (Fig. 22B), which significantly improved their stability under physiological conditions. The supramolecular PEGylated poly-plexes showed decreased transfection efficiency in all tested cell lines; however, their silencing efficiency in HEK293 and L929 cells was higher (up to 84%) compared to that of commercial agents.

(Coixan polysaccharide)-graft-PEI-folate has been prepared as an effective vector for *in vitro* and *in vivo* tumor-targeted gene delivery.[188] The cytotoxicity of the copolymers was significantly lower than that of PEI 25 kDa and close to that of PEI 1200. The *in vitro* transfection, tested in both FR-positive cells (C6 and HeLa cells) and FR-negative cells (A549 cells), showed a high targeting specificity and good gene transfection efficiency in FR-positive cells.

Arima and coworkers have prepared complexes of siRNA with polyamidoamine dendrimer of the third generation conjugated with α-cyclodextrin.[189] siRNA appeared to be well protected in the complexes against degradation by serum. The complexes showed negligible cytotoxicity and hemolytic activity and were found to deliver fluorescent-labeled siRNA to cytoplasm, not nucleus, after transfection in NIH3T3-luc cells. The authors concluded that the dendritic polyamidoamine conjugated with α-cyclodextrin could be potentially used as a siRNA carrier to provide the RNAi effect on endogenous gene expression with negligible cytotoxicity.

Ternary Complexes: Loosening of the polyplex, weakening of the adsorption of serum proteins, and improving of cellular uptake have been considered as important factors leading to high transfection efficiency of DNA ternary complexes. They have been specifically investigated for such complexes composed of DNA, 25 kDa PEI, and different types of biocompatible polyanions—polysaccharides and polypeptides containing carboxyl groups or sulfonic acid groups such as heparin sodium salt, alginic acid sodium salt, poly (aspartic acid), and PLG.[190] The results obtained indicated that the low pK_a and flexible structure of the polyanions tended to loosen the compact DNA polyplexes. The ternary polyplexes exhibited lower binding affinities and less adsorption to serum proteins compared with the binary ones. In addition, they maintained high levels of cellular uptake and intracellular accumulation in serum-containing medium that correlated with their high transfection efficiency. These results provide a basis for the development of polyanion/DNA/polycation ternary polyplexes gene delivery.

Complexes designed to increase the stability of mRNA, to improve transfection efficiency, and to reduce the cytotoxicity have been prepared.[191] These are composite polyplexes with mRNA consisting of PEI and PEI-PEG copolymers. Stable complexes were formed even at low N/P ratios. Most of them showed small particle dimensions (<200 nm) and positive zeta-potentials of + 20 to + 30 mV. Polyplexes with mRNA Luc and blends of low molecular weight PEI (5 kDa) and PEI (25 kDa)-PEG (20 kDa) block copolymer showed protein expression as high as polyplexes with PEI (25 kDa). Moreover, luciferase expression was significantly higher than that obtained with one of the components alone.

Integration of homo-catiomer, poly {*N'*-[*N*-(2-aminoethyl)-2-aminoethyl]aspar-tamide}, PAsp (DET), into PEGylated polyplex micelles prepared from the corresponding block copolymers with PEG, (PEG)-b-PAsp (DET), and DNA has been attempted aiming at enhancing the cell transfection efficiency.[192] *In vivo* anti-angiogenic tumor suppression evaluations validated the feasibility of the approach. The integration promoted gene transfection to the affected cells via systemic administration. The loaded anti-angiogenic gene remarkably expressed in the tumor site, thereby imparting significant inhibitory effect on the growth of vascular endothelial cells, ultimately leading to potent tumor growth suppression. In a subsequent paper, it has been shown that the formulation of both homo PAsp (DET) and copolymer (PEG)-b-PAsp (DET) in the polyplex can help to improve gene therapy for the respiratory system because both effective PEG shielding of polyplexes and functioning of PAsp (DET) polycations to enhance endosomal escape were achieved.[193]

Copolymers of HPMA and methacrylamido-functionalized oligo-l-lysine peptide monomers with either a nonreducible 6-aminohexanoic acid linker or a reducible 3-[(2-aminoethyl)dithiol]propionic acid linker have been prepared and shown able to form polyplexes with DNA.[194] The copolymer containing the reducible linkers was less efficient at transfection than the nonreducible polymer and was prone to flocculation in saline and serum-containing conditions but was also not cytotoxic at charge ratios tested. Optimal transfection efficiency and toxicity were attained with mixed 1:1 formulation of copolymers.

Ternary complexes formed by mixing (1) chitosan with lipoplex of A-[1-(2,3-dioleyloxy)propyl]-AAA-trimethyl-ammonium chloride (DOTAP) and plasmid DNA and (2) DOTAP with chitosan/plasmid DNA polyplex have been prepared.[195] The DOTAP/chitosan/plasmid DNA complexes were in compacted spheroids and irregular lump of larger aggregates in structure, while the short rodlike and toroid-like and donut shapes were found in chitosan/DOTAP/pDNA complexes by AFM. The transfection efficiency of the lipopolyplexes showed higher GFP gene expression than lipoplex and polyplex controls in Hep-2 and HeLa cells and luciferase gene expression two- to threefold than lipoplex control and 70–120fold than polyplex control in Hep-2 cells. The intracellular trafficking was examined by CLSM. Rapid plasmid DNA delivery to the nucleus enhanced by chitosan was achieved after 4 hr transfection.

A facile approach to constructing ternary gene carriers by adding hyaluronic acid to preformed PEI/DNA polyplexes has been demonstrated.[196] Spherical particles with diameter about 250 nm and shifting of ζ-potential from positive to negative were observed upon the formation of hyaluronic acid-shielded polyplexes. Although the electrostatic complexation was loosened, DNA disassembly was not observed. The stability of PEI/DNA/hyaluronic acid ternary polyplexes in physiological condition was improved, and the cytotoxicity was reduced. Comparing with PEI/DNA polyplexes, the uptake and transfection efficiency of hyaluronic acid-shielded polyplexes was lower for HEK293T cells probably due to the reduced adsorptive endocytosis, whereas it was higher for HepG2 cells due to HA receptor-mediated endocytosis.

By adding mannosylated and histidylated liposomes to mRNA-PEGylated histidylated polylysine polyplex, ternary lipopolyplexes have been formed.[197] The lipopolyplexes enhanced the transfection of dendritic cells in vivo and the anti-B16F10 melanoma vaccination in mice. In a subsequent paper, it has been shown by confocal microscopy study that with DC2.4 cells expressing Rab5-EGFP or Rab7-EGFP, DNA uptake occurred through clathrin-mediated endocytosis. The transfection of DC2.4 cells with mannosylated and histidylated lipopolyplexes containing DNA encoding luciferase gene gave luciferase activity two to three times higher than with non-mannosylated ones. In contrast to the latter, it was inhibited by 90% in the presence of mannose.[198]

Amine-terminated generation 5 polyamidoamine dendrimers have been utilized as templates to synthesize gold nanoparticles (Fig. 23). The dendrimer-entrapped gold nanoparticles were prepared in different Au atom/dendrimer molar ratios (25:1–100:1) and were used as nonviral gene delivery vectors.[199] Gel retardation assay, AFM imaging, and DLS experiments demonstrated that dendrimer-entrapped gold nanoparticles were able to effectively compact pDNA to form polyplexes with a smaller size when compared to the initial poly-amidoamine dendrimers without gold nanoparticles entrapped. The results clearly showed that dendrimer-entrapped gold nanoparticles with an appropriate composition (Au atom/dendrimer molar ratio = 25:1) enabled enhanced gene delivery, with a gene transfection efficiency more than 100 times higher than that of the dendrimers without nanoparticles entrapped (Fig. 23). The authors believed that the entrapment of gold nanoparticles within dendrimer templates helped preserve 3D spherical shape of dendrimers, enabling high compaction of DNA to form smaller particles and consequently resulting in enhanced gene delivery.

A two-step process for preparation of ternary magnetoplexes has recently been described.[200] The sixth generation of polyamidoamine dendrimer modified with superparamagnetic nanoparticles was introduced to PEI/DNA polyplexes. The ternary magnetoplexes exhibited enhanced transfection efficiency in COS-7, 293T, and HeLa cells when a magnetic field was applied. Importantly, time-resolved and dose-resolved transfection indicated that high-level transgene expression was achievable with a relatively short incubation time and low DNA dose when magnetofection was employed. Further evidence from Prussian blue staining, quantification of cellular iron concentration, and cellular uptake of Cy-3-labeled DNA demonstrated that the magnetic field could quickly gather the magnetoplexes to the surface of target cells and consequently enhance the uptake of magnetoplexes by the cells.

Somewhat untypical but worth mentioning are the hybrid polyplexes composed of DNA and polyhedral oligomeric silsesquioxanes (siloxane materials that have a cage-like structure) loaded with hydrophobic drugs such as PTX, which is encapsulated within the hydrophobic core of the silsesquioxane.[201] They exhibited superior transfection efficiency in human breast cancer cells than the non-drug-loaded polyplexes.

Delivery and Encapsulation of Proteins and Enzymes: Enzymatic Nanoreactors

Protein therapy is a promising strategy to fight protein-deficiency diseases by using in vitro produced proteins to intracellularly replace or complement the faulty ones.[202–205] Administration of protein drugs and therapeutic peptides poses some additional issues as compared with conventional pharmaceutical compounds because of the high molecular weight of proteins and peptides and their relatively short half-life in plasma due to opsonization processes. It is now well known that protein delivery can be significantly improved by using designed nanocarriers with targeting abilities. Polymer-based nanoparticles of different structure can be utilized for this purpose since, because of their subcellular size, they can cross the fenestration of the vascular epithelium and penetrate tissues. Furthermore, nanocarriers can be confined at the location of choice by conjugation to molecules that strongly bind to the target cells. Enzymatic conversions can take place in the interior of such nanocarriers, and their membranes or outer shells can be used to confine and tune reaction pathways. Despite the significant progress made in the design and engineering of nanoparticles tailored to the targeted delivery of proteins, these nanocarriers seldom succeed in delivering proteins directly inside the cell cytosol. These nanosystems also have a significant drawback, namely, the potential to be highly immunogenic.[206] Therefore, extensive efforts have been devoted to the development of polymer-based nanocarriers for the delivery of drugs of protein nature. In the following, we give a brief overview of polymer-based nanocarriers of proteins and polypeptides that have been constructed from synthetic and/or natural building

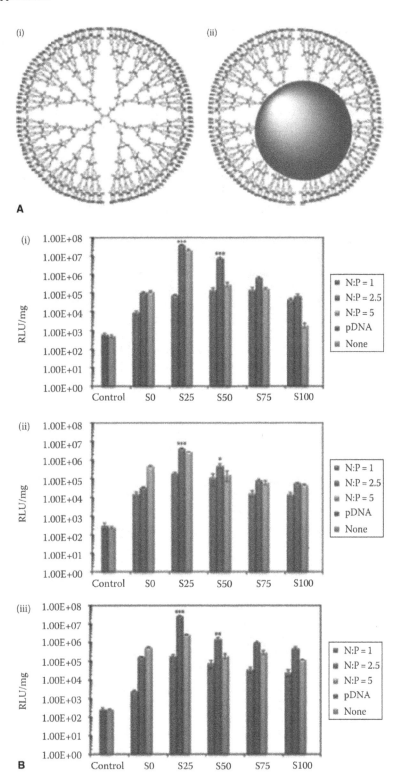

Fig. 23 (**A**) Schematic illustration of the structures of (i) aminoterminated polyamidoamine dendrimers of the fifth generation and (ii) dendrimer-entrapped gold nanoparticles. (**B**) Luciferase gene transfection efficiency of dendrimer-entrapped gold nanoparticles/DNA polyplexes determined in (i) HeLa, (ii) COS-7, and (iii) 293T cells at the N/P ratios of 1:1, 2.5:1, and 5:1, respectively. Transfection was performed at a dose of 1 *ig/well of DNA (mean ± SD, n = 3). Cells without treatment (None) and cells treated with vector-free pDNA (pDNA) were used as controls. Statistical differences between Au DENPs (S25, S50, S75, and S100, respectively) versus G5. NH$_2$ dendrimers (S0) at an N/P ratio of 2.5:1 were compared and indicated with (*) for $p < 0.05$, (**) for $p < 0.01$, and (***) for $p < 0.001$, respectively.
Source: From Shan et al.[199] © 2012, with permission from Elsevier.

segments using the self-assembly and co-assembly approaches. Liposomes, capsules solely based on proteins (viral capsides), microspheres formed by precipitation polymerization or polymerization-induced phase separation, layer-by-layer capsules, as well as unimolecular dendrimer-based containers are not covered or are just occasionally mentioned.

Polymeric protein delivery systems generally rely on the formation of micrometer- or nanometer-size polymer particles containing protein cargo. In these delivery systems, the delivery particles are synthesized by assembling polymer and protein cargo through noncovalent interactions such as electrostatic interactions, hydrogen bonding, and hydrophobic interactions. Proteins can be adsorbed onto polymer particles or dendrimers. Alternatively, when polymers with heterogeneous molecular structure are used, they can co-assemble with protein molecules through appropriate moieties via electrostatic or hydrophobic interactions or other noncovalent interactions (e.g., hydrogen bonding). Other sections of the polymer may serve to provide a stabilizing effect of the complex in solution. Particularly common, although limited to anionic proteins, is the formation of protein-polymer complexes from block copolymers containing a cationic block and a PEG block.

Proteins can be readily complexed to the hydrophobic moieties of amphiphilic copolymer through hydrophobic interactions. For example, cholesterol-bearing pullulan or its derivative was assembled with proteins, resulting in the formation of cationic nanogels with diameters less than 50 nm.[207] In this case, hydrophobic domains formed by the cholesterol moieties on the polymer chains served as complexing sites for the proteins. Enhanced cellular uptake, as well as retained intracellular enzymatic activity, was reported. Effective endosomal release of the proteins was also observed 18 hr after transduction.

Interactions different from electrostatic forces or hydrophilic/hydrophobic interactions can also be utilized to form polymer-protein complexes. These are affinity binding, which has the advantage of minimizing protein denaturation and well-controlled complex structure. However, the approach is limited to proteins that are able to fuse with their binding ligands.[208] A good example of this approach are the PEI-glutathione conjugates that bind with glutathione S-transferase-fused proteins and induce cellular uptake in mammalian cells.[209]

As other delivery systems, the protein delivery by polymers also faces a dilemma: On one side, the weak noncovalent interactions make the complexes vulnerable to dissociation, while strong interactions may alter the protein structure and even cause denaturation, on the other side. The encapsulation of proteins or enzymes in the interior or corona of polymeric micelles or in the lumen of polymersomes represents a possible approach to find the right balance between the magnitude of the interactions and the stability of the loaded carriers.

Lee and coworkers have demonstrated the use of polyion complexes for antibody delivery into cytoplasm. The antibody alone cannot penetrate into the cytoplasm. However, the complexation with a carefully designed block copolymer, consisting of a water-soluble PEO block and a polyelectrolyte block, can facilitate the endosomal delivery. In the described system, the formed polyion complex was stable at pH 7.4, but it could be dissociated at pH 5.5. This behavior led to the targeted release of the antibody into the cytoplasm, where the pH is lower.[210]

Bulmus and coworkers have reported on novel pH-responsive polymeric nanocarriers for the enhanced cytoplasmic delivery of enzyme susceptible drugs, such as proteins and peptides as well as antisense ODNs.[211] A new functionalized monomer, namely, pyridyl disulfide acrylate, was specifically synthesized and incorporated into an amphiphilic copolymer containing also MAA and butyl acrylate segments with the aid of free-radical polymerization techniques. This resulted in a glutathione- and pH-sensitive, membrane-disruptive terpolymer with functional groups, which allowed thiol-containing molecules to be readily conjugated. In this way, the polymeric nanocarriers posed two main functions: 1) pH-dependent, endosomal membrane disruption and escape into the cytoplasm, followed by 2) reaction of disulfide-conjugated drug with glutathione, a normal constituent of the cytoplasm of cells, causing release of the drug from the nanocarrier. Chemical conjugation and/or ionic complexation with oligopeptides (or antisense ODNs) was performed in order to load the corresponding macromolecular drugs on the nanocar-rier. Hemolytic activity at low pH remained high even after the conjugation/complexation with oligopeptides. The terpolymer itself showed no toxicity even at high concentrations, as determined with tests involving mouse 3T3 fibroblasts and human THP-1 macrophage-like cells. Uptake of the radiolabeled polymer and enhanced cytoplasmic delivery of FITC-ODN was also studied in THP-1 macrophage-like cells.

Poly (ethylene glycol) dimethacrylate (PEGDMA) and MAA-based nanoparticles and microparticles have been prepared by emulsion polymerization. Particles of different size were obtained by using different concentration of sodium lauryl sulfate as the emulsifying agent. The particles were further evaluated as carrier systems for the oral delivery of insulin.[212] Insulin loading efficiency of the particles was found to be directly proportional to the particle size and inversely proportional to the acid content of the particles. *In vitro* insulin release studies from insulin-loaded particles were performed by simulating the gastrointestinal tract conditions using HPLC. At pH 2.5, the release of insulin from polymeric particles was observed in the range of 5–8%, while a significantly higher release (20–35%) was observed at pH 7.4 during the first 15 min of *in vitro* release. Larger size particles (of sizes ca. 8.3 µm) showed the highest efficiency to reduce the blood glucose level in diabetic rabbits.

Two types of blend formulations, poly (lactic-*co*-glycolic acid) (PLGA)/linear PEO-PPO-PEO and PLGA/x-shaped PEO-PPO, have been studied as model nano- carriers for insulin in potential oral administration.[213] The results showed that the interactions with the digestive enzymes were considerably reduced in the blend formulations. The net charge of the encapsulated protein showed a clear effect in the final size of the nanoparticles, while the encapsulation efficiency was controlled by the polyoxyethylene derivative. The authors concluded that the carriers formed with encapsulated insulin in PLGA/linear PEO-PPO-PEO particles are capable, at least *in vitro*, to overcome the gastrointestinal barrier.

He and coworkers have recently focused on the size-dependent oral absorption mechanism of polymeric nanoparticles loaded with protein drugs.[214] The authors prepared rhodamine B-labeled carboxylated chitosan-grafted nanoparticles with particle sizes of 300, 600, and 1000 nm and similar ζ-potentials (~−35 mV), which were loaded with FITC-labeled bovine serum albumin. The smallest particles demonstrated elevated intestinal absorption, as mechanistically evidenced by higher mucoadhesion in rat ileum, release amount of the payload into the mucus layer, Caco-2 cell internalization, transport across Caco-2 cell monolayers and rat ileum, and systemic biodistribution after oral gavage. Peyer's patches could play a role in the mucoadhesion of nanoparticles, resulting in their close association with the intestinal absorption of nanoparticles.

For insulin-loaded nanoparticles prepared using trimethyl chitosan chloride modified with a targeting peptide (CSKSSDYQC) has been reported elsewhere.[215] The authors demonstrated the effects of goblet cell-targeting nanoparticles on the oral absorption of insulin *in vitro*, *ex vivo*, and *in vivo*. The article is mainly focused on the targeting, permeation, uptake, and internalization of the particles and showed that the orally administered peptide-modified nanoparticles produced a better hypoglycemic effect with a 1.5-fold higher relative bioavailability compared with unmodified ones.

Chitosan-carrageenan nanoparticles have shown promising properties as carriers of therapeutic macromolecules.[216] By using these natural marine-derived polymers, nanoparticles in the 350–650 nm size range and positive zeta-potentials of 50–60 mV were obtained in a hydrophilic environment, under very mild conditions, avoiding the use of organic solvents or other aggressive technologies for their preparation. They exhibited a noncytotoxic behavior in biological *in vitro* tests performed using L929 fibroblasts, which is critical regarding the biocompatibility of those carriers. Using ovalbumin as model protein, nanoparticles evidenced loading capacity varying from 4% to 17% and demonstrated excellent capacity to provide a controlled release for up to 3 weeks.

Core/corona nanoparticles based on various monomers, prepared by emulsion polymerization methods, have been utilized by Akagi and coworkers as therapeutic peptide carriers.[217] Salmon calcitonin (sCT) was used as the model peptide drug. After the oral administration of mixtures of sCT and nanopar-ticles to rats, it was found that the blood ionized calcium concentration decreased indicating that the use of nanoparticles as sCT carriers enhanced peptide adsorption through the gastrointestinal tract. sCT absorption was larger for nanoparticles with PNIPAM chains on their surfaces.[218] Nanoparticles with PMAA or poly (vinyl amine) chains on the surface showed a smaller effect. A further enhancement of nanoparticle peptide delivery efficiency was observed for nanopar-ticles with polystyrene cores and coronas of mixed PNIPAM and poly (vinyl amine) chains.[219]

The polymersomes have an ideal architecture that can be used as a scaffold for precise positioning of biomacromolecules. By encapsulation of proteins, incorporation of membrane proteins in the polymeric bilayer, and surface conjugation with, for example, cellular signaling molecules, the polymersomes can be applied for various purposes. The first encapsulation experiments have been performed with proteins such as myoglobin, hemoglobin, and albumin via the addition of the solid block copolymer polybutadiene (PBD)-PEG to an aqueous solution of the desired solute, after which the mixture was incubated for a day.[220] The efficiency of this method varied considerably with the type of the solute, ranging from about 5% for albumin to more than 50% for myoglobin.

Pang and coworkers have created biodegradable polymersomes for brain delivery of peptides.[221] They used the thiols present in the monoclonal antibody OX26 to decorate PEG-PCL polymersomes with maleimide functions on their surface. The antibody was shown able to initiate endogenous receptor-mediated transcytosis of the polymersomes across the blood-brain barrier. The polymersomes should degrade with time to release their contents. Biodegradable and biocompat-ible polymersomes as oxygen carriers have been developed as well.[222,223] They were based on PEG-PCL and PEG-PLA in which bovine or human hemoglobin was encapsulated. By varying the hemoglobin concentration or the block copolymer concentration, the aggregate size distribution and encapsulation efficiency were tuned. Due to the higher membrane thickness, the oxygen affinity of the polymersome-encapsulated hemoglobin was lower than that of red blood cells but still consistent with what is required for efficient oxygen delivery in the systematic circulation.

Albumin and tetanus toxoid have been successfully encapsulated by solvent evaporation in PLA nanoparticles of sizes about 100–120 nm.[224] In order to retain the secondary and tertiary structures of the protein drugs, a slow and intermittent sonication at low temperatures was applied. Bovine serum albumin and immuno-γ-globulin (IgG) encapsulated in nanoparticles prepared from PEO-PLGA using solvent diffusion method exhibited encapsulation efficiency of 58.9% and slow *in vitro* release rate.

However, the problem of protein hydration, pH reduction derived from polymer degradation, and presence of the hydrophobic interfaces in PLGA-based delivery devices led to protein inactivation or irreversible aggregation inside PLGA or PLA nanoparticles.[225] These problems have been prevented by incorporation of stabilizers such as PEO and its derivatives. It is noteworthy that the type of the PEO derivative and pH of internal aqueous phase were the most important factors influencing BSA protein encapsulation and release kinetics.[226] Degradation and release characteristics of polyester particles can be improved by the incorporation of polyoxyethylene derivatives with different hydrophilic-lipophilic balance.[225] The extended half-life of bovine serum albumin encapsulated in PEG-PLGA nanoparticles was found to alter the protein biodistribution in rats compared to that of the same protein loaded in PLGA nanoparticles.[227] Similar results have been reported for tetanus toxoid encapsulated in PLA and PLA-PEG nanoparticles of comparable dimensions (137–156 nm) but differing in their hydrophobicity.[228] In addition, the PLA-PEG nanoparticles led to greater penetration of tetanus toxoid into the blood circulation and in lymph nodes than PLA-encapsulated tetanus toxoid.[229] The results suggested that PLA nanoparticles suffered an immediate aggregation upon incubation with lysozyme, whereas the PEG-coated nanoparticles remained totally stable.[230] The antibody levels elicited following in administration of PEG-coated nanoparticles were significantly higher than those corresponding to PLA nanoparticles.[231]

Porous Tat-functionalized polymersomes prepared by co-assembly of polystyrene$_{40}$-block-poly[l-isocyanoalanine (2-thiophen-3-ylethyl)amide]$_{50}$ and polystyrene-block-poly (ethylene glycol)-oxanorbornadiene (Fig. 24) have been described elsewhere.[232] The polymersomes had an average diameter of about 114 nm with no obvious size variation between polymersomes with different protein types and contents. Efficient cellular uptake of polymersomes loaded with GFP by a variety of cell lines was observed. After their uptake into cells, a considerable fraction of the GFP signal colocalized with the acidic vesicles; however, almost equally large population of GFP-containing punctuate structures retained a neutral pH. Furthermore, the same polymersomes loaded with horseradish peroxidase functioned as nanoreactors inside cells. They maintained their activity to a much higher degree than what was reported for free horseradish peroxidase trafficked to lysosomes.[232]

The potential of polymeric vesicles as a light-triggered delivery system has been studied.[233] The polymersomes were prepared by self-assembly of a photocleavable AmBC composed of PAA and poly (methyl caprolactone) as hydrophilic and hydrophobic blocks, respectively, linked by O-nitrobenzyl photocleavable segment (Fig. 25). The vesicles disintegrated upon UV irradiation and rearranged into small micelle-like structures, thus releasing their payload. Graphical summary of the mechanism of polymersome

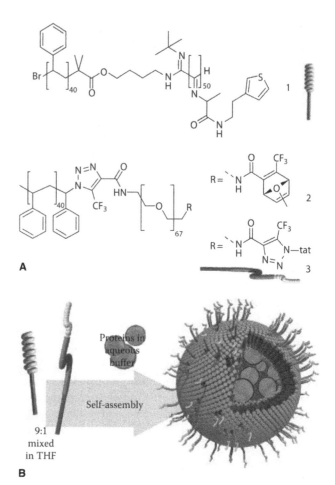

Fig. 24 (A) Structures and representations of polystyrene$_{40}$-block-poly[l-isocyano-alanine (2-thiophen-3-ylethyl)amide]$_{50}$ (1) and polystyrene-block-poly (ethylene glycol)-oxanor-bornadiene (2). (B) Formation of Tat-functionalized polymersomes loaded with proteins by co-assembly of 1 and 3 in 9:1 ratio.
Source: From van Dongen et al.[232] © 2010, with permission from Wiley-VCH Verlag GmbH & Co. KGaA.

degradation is presented in Fig. 25. Enhanced GFP was encapsulated in the polymersomes with encapsulation efficiency of 22% and releases upon reorganization of the polymer chains.

PIC micelles based on PEG-poly (aspartic acid) copolymers with biological polyions or enzymes such as lysozyme and trypsin have been formed.[234,235] By derivatization of the aspartate moiety with citraconic anhydride using ethylene diamine as a linker, a copolymer that was stable at neutral and basic pH but degraded at acidic pH was formed.[236] The derivatized copolymer formed PIC micelles that could contain lysozyme in their cores and degraded rapidly when brought into an acidic environment (Fig. 26).

Intercellular adhesion molecule-1 (ICAM-1)-targeted nanocarriers have been used to deliver acid α-glucosidase into cells to address the specific enzyme deficiency in Pompe's disease.[237] ICAM-1 is a protein involved in

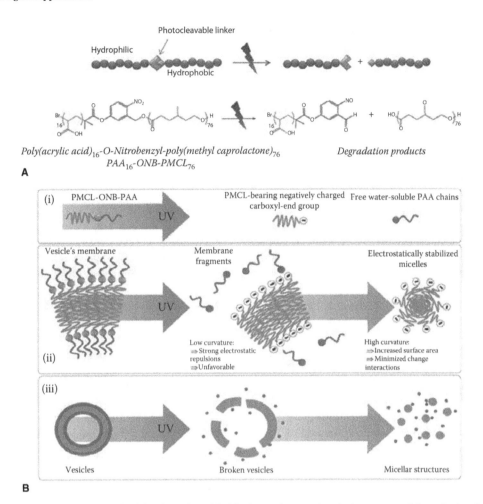

Fig. 25 (**A**) Schematic view of the amphiphilic photocleavable block copolymer, chemical structure of the poly (methyl caprolactone)-O-nitrobenzyl-poly (acrylic acid) diblock copolymer, and its degradation products upon UV irradiation. (**B**) Graphical summary depicting the proposed mechanism of polymersome degradation. (i) Molecular level (the diblock copolymer is rapidly cleaved upon UV irradiation), (ii) supramolecular level (as a result of chain scission, packing of the PMCL chains forming the membrane is progressively destabilized and evolves into a more favorable arrangement), and (iii) aggregate morphology (from vesicles to broken vesicles to stabilized micellar structures).
Source: From Cabane et al.,[233] with permission of The Royal Society of Chemistry.

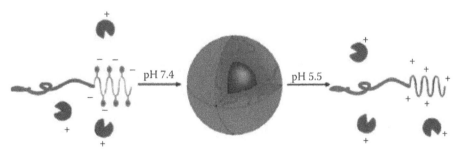

Fig. 26 Schematic representation of the micellar assembly of the PEG-poly[(N'-citraconyl-2-aminoethyl)aspartamide] diblock copolymer with lysozyme and its degradation with time at acidic pH.
Source: From Lee et al.[236] © 2007, with permission from American Chemical Society.

inflammation and overexpressed on most cells under pathological conditions. The deficiency of acid α-glucosidase causes Pompe's disease, which leads to excess glycogen storage throughout the body, mainly in the liver and striated muscle. To improve delivery of acid α-glucosidase, the latter was coupled to polymer nanocarriers (average dimensions ~180 nm), which were coated with an antibody specific to ICAM-1. Fluorescence microscopy showed specific targeting to cells, with efficient internalization and lysosomal transport, enhancing glycogen degradation over nontargeted acid α-glucosidase. Radioisotope tracing in mice demonstrated enhanced acid α-glucosidase accumulation in all organs, including Pompe targets. In an earlier paper, markedly enhanced delivery in brain, kidney, heart, liver, lung, and spleen of mice was reported;[238] TEM showed attachment and internalization into vascular endothelium. Fluorescence microscopy proved targeting, endocytosis, and lysosomal transport of the loaded particles in macro- and microvascular endothelial cells and a marked enhancement of globotriaosylceramide degradation. The lysosomal accumulation of the latter in multiple tissues, due to deficiency of α-galactosidase, causes Fabry disease. Therefore, this ICAM-1-targeting strategy may help improve the efficacy of therapeutic enzymes for Fabry disease and Pompe's disease.

Encapsulation of antioxidant enzymes in PEG-coated liposomes has been shown to increase the antioxidant enzyme bioavailability and enhance protective effects in animal models. Micelles based on PEO-PPO-PEO copolymers showed even more potent protective effect.[239] Such nanocarriers protected encapsulated antioxidant enzymes from proteolysis and improved delivery to the target cells, such as the endothelium lining the vascular lumen. A recent work of Simone and coworkers has been focused on protective encapsulation of the potent antioxidant enzyme, catalase, by filamentous polymer nanocarriers.[230] The authors maintained the same molecular weight ratio of PEG to PLA in a series of PEG-b-PLA diblock copolymers and varied the total copolymer molecular weight from about 10,000 to 100,000 g/mol. All diblock copolymers formed filamentous particles upon processing, which encapsulated active enzyme. The latter proved resistant to protease degradation.

A nanogel formed by self-assembly of a conjugate of heparin and PEO-PPO-PEO block copolymer has been prepared.[240] Heparin possesses advantageous structural features such as negatively charged structure and presence of specific binding domains for proteins. The nanogel solution of the conjugate exhibited high stability, hydrodynamic size below 100 nm, and ability to encapsulate both small molecules and proteins. Circular dichroism and gel electrophoresis demonstrated that the stability of monoclonal antibodies (3D8 scFv) encapsulated into the nanogel had been maintained.

Advanced protein delivery systems for treatment of lysosomal storage diseases have recently been described.[241] Polyelectrolyte complexes between trimethyl chitosan and the lysosomal enzyme α-GAL via self-assembly and ionotropic gelation were prepared. The particles were characterized with an average particle size below 200 nm, polydispersity index <0.2, and a protein loading efficiency of about 65%. The polyelectrolyte nanoparticles were stable and active under physiological conditions and able to release the enzyme at acidic pH, as demonstrated by in situ AFM (Fig. 27). They were further functionalized with Atto 647N and their cellular uptake and fate were tracked using high-resolution fluorescence microscopy. In contrast to their precursor, the polyelectrolyte complexes were efficiently internalized by human endothelial cells and mostly accumulated in lysosomal compartments.

Bovine serum albumin has been encapsulated in levan-based nanocarriers.[242] Microbial levans are biopolymers produced from sucrose-based substrates by a variety of microorganisms. For this study, levan produced by Halomonas sp. was used. Encapsulation efficiency 49.3–71.3%, particle dimensions in the 200–537 range, and slightly positive ζ-potentials were reported.

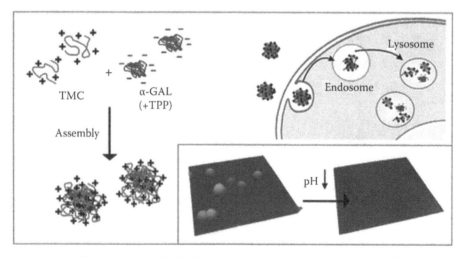

Fig. 27 Schematic representation of the formation of polyelectrolyte complexes between trimethyl chitosan and a-GAL, their cellular uptake and accumulation in lysosomal compartments, and the release of the enzyme at acidic pH.
Source: From Giannotti et al.[241] © 2011, with permission from American Chemical Society.

APPLICATIONS IN IMAGING AND DIAGNOSTICS

Bioimaging techniques have been widely applied in modern biomedical field in order to make disease diagnosis more accurate and rapid.[243–245] Due to their significant potential for disease analysis and diagnosis at the molecular and cellular level, imaging technologies, including techniques like optical imaging, MRI, and ultrasonic imaging, combined with bioimaging and molecular cell biology, have been developed to meet the demands of modern biomedical research. Through these noninvasive methods, the organs' and cell functions *in vivo* can be inspected, and the molecular interactions can be monitored on line. Because the signal between the normal and pathological organs is too low to be detected, fine medical images cannot be readily obtained usually without the use of functional and assisting chemicals and molecular nanodevices designated as the "imaging agent" or "imaging probe." Such agents have been developed for enhancing the image contrast and spatial resolution, which in turn can significantly improve the accuracy of diagnosis. Research on suitable imaging agents therefore became a significant research area in life sciences, medical sciences, and material sciences. With the development of nanotechnology, novel nanoimaging agents have been developed. Nanoparticles, including polymeric, inorganic, and hybrid nanoparticles with sizes lower than 100 nm, have unique properties that can be fine tuned and utilized for bioimaging applications. Generally some basic properties/functionalities should be incorporated into the nanoparticle in order to function properly as a contrast agent: the nanoparticle should possess: 1) an inorganic or organic core/active material that can interact with incoming radiation for imaging enhancement; 2) a hydrophilic shell for improving the stability of the nanoconstruct within the probed medium/site; and 3) functional outer ligands that can offer targeting properties to the nanoparticulate imaging agent. The nature of each component may vary according to the technique implemented for imaging as well as the application environment.

Broadly speaking, the field of biomedical imaging can be divided into several categories based upon the electromagnetic spectrum utilized in each case, that is, magnetic resonance, optical/near infrared (NIR), and ionizing radiation (x-rays and y-rays).[246] Imaging based on ionizing radiation generally refers to the detection of high-frequency emissions from radioactive elements such as the gamma-ray emitters ^{111}In or [99]mTc or the passage of x-rays through the body. The main technologies involved in this category are positron emission tomography (PET), SPECT, and x-ray computed tomography (CT). MRI tends to operate on the other end of the spectrum in the MHz frequency range, relying upon contrast agents such as gadolinium or superparamagnetic iron oxide nanoparticles (SPIONs) to modify the relaxivity of water molecules for providing soft tissue contrast. Long-wavelength and NIR imaging has been developed due to the desire to perform whole animal/body and deep tissue imaging. NIR imaging applications are also growing rapidly due to the availability of targeted biological agents for diagnosis and basic medical research that can be imaged *in vivo*. The wavelength range of 650–1450 nm falls in the region of the spectrum with the lowest absorption in tissue and therefore enables the deepest tissue penetration. The rapid growing interest in multimodal imaging, which involves simultaneous delivery of actives, targeting, and imaging, requires nanoparticles or supramolecular assemblies. Well-defined and smart-designed nanoparticles for diagnostics also have advantages in increasing circulation time and increased imaging brightness relative to single-molecule imaging agents. This has led to rapid advances in nanocarriers for long-wavelength, NIR imaging. Some examples pertaining to polymer-based nanoparticles for bioimaging will be discussed in the following.

Ghoroghchian and coworkers utilized PBD-*b*-PEO block copolymer vesicles in water in order to encapsulate multi[(porphinato)zinc (II)]-based supermolecular fluorophores. This led to the formation of emissive polymersomes with a rich photophysical diversity enabling emission energy modulation over a broad spectral domain.[247] They have shown that by controlling polymer-to-fluorophore noncovalent interactions, the bulk photophysical properties of these soft, supramolecular, optical materials could be finely tuned allowing utilization of the nanoassemblies in NIR bioimaging protocols.

Prud'homme and coworkers reported the preparation of hybrid nanoparticles for NIR imaging composed of rare earth ion-doped phosphors nanocrystals (NaYF$_4$:Yb^{3+}, Er^{3+}) surface modified by PAA homopolymers and poly (ethylene glycol)-*b*-poly (caprolactone) (PEG-*b*-PCL), poly (ethylene glycol)-*b*-poly (lactic-*co*-glycolic acid) (PEG-*b*-PLGA), and poly ((ethylene glycol)-*b*-lactic acid) (PEG-*b*-PLA) AmBC.[248] Initially hexagonal phase nanophosphors were prepared using one-step cothermolysis utilizing oleic acid (OA) and trioctylphosphine (TOP) ligands. Then direct ligand exchange with PAA and using an amphiphilic copolymer encapsulation via flash nanoprecipitation took place.[249] Both surface modification routes produced colloidally stable dispersions in water and in buffers and serum media (Fig. 28). The PEG block confers biocompatibility to the nanostructures and ensures long circulation times in the bloodstream. These polymer-modified upconverting nanophosphors provide promising new nanomaterials also for photodynamic therapy (PDT) applications.

An MRI contrast agent based on polymeric nanoparticles was developed by conjugation of gadolinium (Gd) chelate groups onto the biocompatible poly (l-lactide-c-ethylene glycol) (PLA-c-PEG) block copolymers.[250] The MRI contrast agent was targeted to liver. PLA-c-PEG copolymers were conjugated with diethylenetriaminepentaacetic acid (DTPA). The PLA-c-PEG-DTPA nanoparticles were prepared by the solvent diffusion method, and

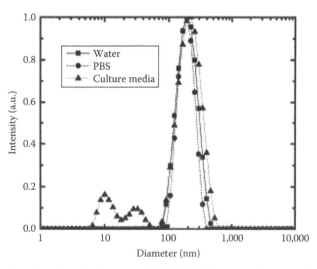

Fig. 28 Size distributions of PEG-*N*-PLA-protected rare earth ion-doped phosphors nanocrystals in water (■), PBS (●), and culture media (▲). The peaks at 10 and 30 nm are proteins in the FBS culture media.
Source: From Budijono et al.[248] © 2010, with permission from American Chemical Society.

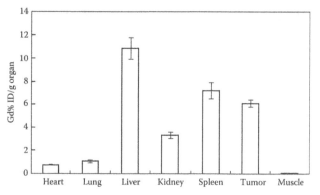

Fig. 29 Biodistribution of PEG-P(Lys-DOTA-Gd) micelle 24 hr after injection at a dose of 0.05 mmol Gd/kg.
Source: From Shiraishi et al.[251] © 2009, with permission from Elsevier.

then Gd was loaded onto the nanoparticles due to complexation with the DTPA moieties on the surface of the PLA-*b*-PEG-DTPA nanoparticles. The mean size of the nanoparticles was ca. 269 nm. The relaxivity of the Gd-labeled polymeric nanoparticles was determined against the conventional contrast agent Magnevist. Compared to Magnevist, the Gd-labeled PLA-PEG nanoparticles showed significant enhancement on imaging signal intensity. The T_1 and T_2 relaxivities per Gd atom of the Gd-labeled nanoparticles were found to be 18.865 and 24.863 mM^{-1} s^{-1} at 3 T, respectively. In addition, the signal intensity *in vivo* was stronger comparing with the Gd-DTPA contrast agent alone, and the T_1 weight time was lasting for 4.5 hr. The distribution of nanoparticles *in vivo* was evaluated in rats, and it was found to be also enhanced compared to Magnevist. The liver-targeting efficiency of the Gd-labeled PLA-*b*-PEG-based nanoparticles in rats was 14.6 comparing with Magnevist injection. Therefore, the Gd-labeled nanoparticles showed the potential as targeting molecular MRI contrast agent for further clinical utilization.

A polymeric micellar system was utilized for creating a novel MRI contrast agent by Shiraishi et al.[251] The block copolymer PEG-*b*-poly (l-lysine) was used as the micelle-forming copolymer, as well as the matrix for the incorporation of MRI-active gadolinium ions through polymer-linked chelating moieties, namely, DOTA. The DOTA moieties were chemically linked to the primary amine groups of the lysine segments in the copolymer. The functionalized PEG-b-poly (l-lysine-DOTA) block copolymer could form micelles in aqueous media in the absence of Gd ions. The polymeric micelle structure was maintained even after partial gadolinium chelation (ca. 40%) to the existing DOTA moieties. The prepared polymeric

micelle MRI contrast agent was injected into a mouse tail vein at a dose of 0.05 mmol Gd/kg and was found to exhibit stable blood circulation. A considerable amount (6.1 ± 0.3% of ID/g of the polymeric micelle MRI contrast agent) was found to accumulate at solid tumors 24 hr after intravenous injection by means of the EPR effect alone (Fig. 29). An MRI analysis revealed that the signal intensity of the tumor was enhanced by twofold via the use of the new hybrid polymeric nanoparticle contrast agent.

A similar approach based on mixed micellar polymeric systems was also reported for the preparation of nanoparticle MRI contrast agents.[252] Polymeric micelles that were formed from cationic polymers (polyallylamine or protamine) and anionic block copolymers (poly (ethylene glycol)-*b*-poly (aspartic acid) derivative) could bind Gd ions, creating hybrid mixed polymeric micelles. These micelles were found to provide high contrasts in MRI experiments by shortening the T_1 longitudinal relaxation time of protons of water. The PEG-*b*-PAsp block copolymer with bound Gd ions showed high relaxivity (T_1-shortening ability) values (from 10 to 11 mol^{-1} s^{-1}), while the mixed polymeric micelles incorporating Gd ions exhibited low relaxivity values (from 2.1 to 3.6 mol^{-1} s^{-1}). The authors concluded that these findings illustrate the feasibility of the novel MRI micellar contrast agent that selectively provides high contrasts at solid tumor sites owing to a dissociation of the micelle structures, while selective delivery to the tumor sites is achieved in the polymeric micelle form.

Iron oxide (IO) magnetic nanoparticles embedded into polymeric nanoparticles and micelles have been also tested as MRI contrast agents. Ujiie et al. reported the preparation of IO nanoparticles in the presence of poly (ethylene glycol)-*b*-poly (4-vinylbenzylphosphonate) (PEG-*b*-PVBP) block copolymer Hao et al.[253] The magnetic nanoparticles were obtained through alkali coprecipitation of iron salts in the presence of the copolymers. In this way, the surface of the particles was coated with a relatively immobilized PEG layer with a high density of PEG chains (referred as PEG-protected iron oxide nanoparticles, PEG-PIONs). The PEG chains were bound to the IO surface via multipoint anchoring of the phosphonate groups present in the PVBP block. The surface density of PEG chains in the PEG-PIONs was varied through variation of the [PEG-*b*-PVBP]/[iron salts] feed-weight ratio in the coprecipitation reaction. PEG-PIONs prepared at an optimal feed-weight ratio in this study showed a hydrodynamic diameter of ca. 50 nm. The PEG-PIONs could be dispersed in PBS that contains 10% serum without any change in their hydrodynamic diameters over a period of 1 week, indicating that PEG-PIONs possess high dispersion stability under *in vivo* physiological conditions as well as excellent anti-biofouling properties. PEG-PIONs showed a long blood circulation time and significant tumor accumulation (more than 15% ID/g of tumor) without the aid of any surface targeting ligand in mouse tumor models (Fig. 30). It was observed that the majority of the PEG-PIONs accumulated in the tumor in 96 hr after administration, whereas those in normal tissues were smoothly eliminated in the 96 hr period. This observation illustrates the enhancement of tumor selectivity in the localization of PEG-PIONs due to the physicochemical characteristics of the nanoparticles. The performance characteristics of such hybrid nanoparticles should promote their use as MRI contrast agents.

On a similar context, a series of pH-responsive polymeric micelles based on a pH-responsive poly (p-amino ester)/(amido amine) block and PEG were developed to act as carriers for the delivery/accumulation of magnetic IO (Fe_3O_4) nanoparticles that respond rapidly to an acidic tumor environment for effective MRI of tumors.[254] At physiological conditions, the Fe_3O_4 nanoparticles could be well encapsulated into the poly(β-amino ester)/(amido amine) core of the polymeric micelles, due to hydrophobic interactions, shielded by the PEG coronal shell. In an acidic tumor environment, the pH-responsive block, carrying ionizable tert-amino groups on its backbone, can become protonated and therefore soluble. In this case, the hydrophobic Fe_3O_4 nanoparticles are released, thus enhancing their accumulation into targeting tumors. The Fe_3O_4-loaded polymeric micelles were tested in a disease rat model of cerebral ischemia that produces acidic tissue, due to its pathological condition, in respect to the ability of this MRI probe as a pH-triggered agent, using a 3.0 T MRI scanner. It was observed that the IO particles were gradually accumulated in the brain ischemic area, indicating that the pH-triggered MRI probe may be effective for targeting the acidic environment and for diagnostic imaging of pathological tissues. The same group utilized a similar system where the red fluorescent dye sulforhodamine 101 (SR101) was conjugated to the poly(β-amino ester)/(amido amine) block allowing for simultaneous MRI and optical imaging of pathological tissues.[255] CLSM observations demonstrated the cellular uptake of SR101-labeled, Fe_3O_4-loaded polymeric micelles by breast cancer cells.

Hydrophobic OA-coated SPIONs (diameter 5–10 nm) were encapsulated into biodegradable thermosensitive polymeric micelles resulting in a system fulfilling the requirements for systemic administration and bioimaging.[256] The micelles were composed of amphiphilic, thermosensitive, and biodegradable block copolymers of poly(ethylene glycol)-*b*-poly[*N*-(2-hydroxypropyl) methacrylamide dilactate] (mPEG-*b*-p(HPMAm-Lac$_2$)). The encapsulation was performed by addition of one volume of SPIONs in tetrahydrofuran to nine volumes of a cold aqueous mPEG-*b*-p(HPMAm-Lac$_2$) solution (at 0°C, which is below the cloud point of the polymer), followed by rapid heating of the resulting mixture to 50°C, in order to induce micelle formation. Hybrid nanoparticles with ca. 200 nm diameter and rather low polydipersity were obtained. TEM analysis demonstrated that clusters of

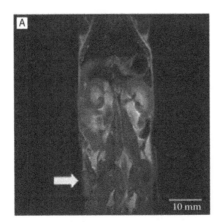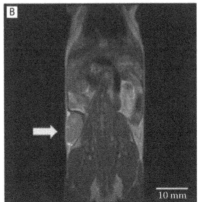

Fig. 30 T_2-weighted images of C-26 tumor-bearing mice at 24 h postinjection of 84 mg Fe kg^{-1} of (**A**) PEG-PION4 and (**B**) control. The arrow denotes tumor site.
Source: From Ujiie et al.[253] © 2011, with permission from Elsevier.

SPIONs were present in the core of the block copolymer micelles. A maximum loading of 40% was reported, while MRI scanning of the samples demonstrated that the SPION-loaded micelles had high r_2 and r_2^* relaxivities. The r_2^* values were determined to be at least twofold higher than the r_2 values, confirming the clustering of the SPIONs in the micellar core. The particles showed excellent stability under physiological conditions for 1 week, even in the presence of FBS. The authors proposed that due to the stability of the hybrid micelles together with their ease of preparation and their nanoscale size, these systems are highly suitable for image-guided drug delivery.

Polymeric nanoparticles carrying fluorine atoms have been also explored in MRI applications since they have large structural design potential when compared to traditional systems such as emulsions and solutions of smaller molecules. There is generally a growing interest in the development of different fluorinated compounds capable of being tracked *in vivo* using ${}^{19}F$-MRI techniques. By the use of commercially available ${}^{19}F$-MRI coils, ${}^{19}F$ imaging is easily achievable in the clinical setting. When the ${}^{19}F$ image is superimposed on the familiar ${}^1H$ density image, the location of a fluorinated compound or a fluorinated particle can be determined in a noninvasive manner. ${}^{19}F$ is an attractive tracking nucleus for MRI due to its high sensitivity and the absence of a confounding ${}^{19}F$ background signal within the body. The requirements for a successful ${}^{19}F$-MRI tracking agent include that it should have high fluorine content; the fluorine nuclei should have appropriate NMR properties, for example, sufficient mobility; and preferably there should be a single peak in the ${}^{19}F$-NMR spectrum. Such compounds have the potential to provide high image intensity without the need for selective excitation sequences.

Core/shell micelles from well-defined diblock copolymers, with acrylic acid hydrophilic blocks and hydrophobic blocks composed of partially fluorinated acrylate and methacrylate monomers synthesized using ATRP, were evaluated as potential ${}^{19}F$-MRI agents.[257] The diblock copolymers spontaneously self-assembled into stable micelles, in aqueous solutions or in mixed organic/aqueous solvents, with diameters from approximately 20 to 45 nm. These micelles had a fluorine-rich core that provides a strong signal for MRI. The observed MRI image intensities were related to the NMR longitudinal and transverse relaxation times and were found to depend on polymer structure and method of micelle formation. Two distinct T_2 relaxation times were observed and measured. By comparison of expected MRI image intensities with those observed experimentally, it was concluded that methacrylate-based polymers show systematically lower signal intensity than acrylate polymers. This was related to the presence of a population of nuclear spins having very short T_2 relaxation times that generally cannot be detected under high-resolution NMR and MRI conditions.

A number of hyperbranched fluoropolymers were synthesized and their micelles in water were studied as potential ${}^{19}F$-MRI agents.[258] A hyperbranched starlike core was first prepared via atom transfer radical self-condensing vinyl (co) polymerization (ATR-SCVCP) of 4-chloromethyl styrene, lauryl acrylate, and 1,1,1-tris (4'-(2''-bromoisobutyryloxy)phenyl)ethane (TBBPE). The polymerization gave a small core that served as a macroinitiator. Trifluoroethyl methacrylate (TFEMA) and tert-butyl acrylate (tBA) in different ratios were then "grafted" from the core to give hyperbranched starlike polymers with M_n of about 120 kDa and polydispersity values of about 1.6–1.8, having statistical copolymer chains as arms. After acidic hydrolysis of the tert-butyl ester groups, amphiphilic, hyperbranched starlike polymers with M_n of about 100 kDa were obtained. These branched copo-lymers were used for micelle formation in aqueous solutions, giving micelles with TEM-measured diameters ranging from 3 to 8 nm and DLS-measured hydrody-namic diameters from 20 to 30 nm. These micelles gave a narrow, single resonance as determined by ${}^{19}F$-NMR spectroscopy, with a half width of approximately 130 Hz (Fig. 31). The T_1 and T_2 relaxation times of the micellar systems were about 500 and 50 ms, respectively, and were not significantly affected by the composition and the overall size of the micelles. ${}^{19}F$-MRI phantom images of these fluorinated micelles were also acquired. These images demonstrated that the particular fluorinated polymeric micelles may be useful as novel ${}^{19}F$-MRI agents for utilization in a variety of biomedical applications.

Fluorinated nanoparticles, constructed by AmBCs based on precursor copolymers of styrene (PS) and 2,3,4,5,6-pentafluorostyrene (PPFS), as well as from block copolymers with tBA segments, that is, (PtBA)-*b*-PS-*co*-PPFS, were evaluated as potential MRI contrast agents.[259] Reversible addition-fragmentation chain transfer polymerization was successfully employed in order to synthesize the aforementioned copolymers with control over molecular weight and composition and with narrow polydispersity. It was found that the copolymerization of styrene and PFS allowed for the preparation of gradient copolymers with opposite levels of monomer consumption, depending on the feed ratio. Conversion to AmBCs, namely, PAA-b-(PS-co-PPFS), was achieved

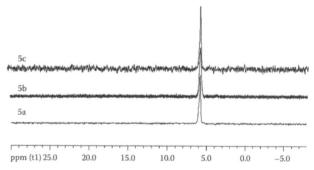

Fig. 31 19F-NMR spectra of hyperbranched fluoropolymer micelles.
Source: From Du et al.[258] © 2008, with permission American Chemical Society.

by removing the protecting groups via hydrolysis, followed by chemical linking of with monomethoxy PEG chains for the enhancement of hydrophilicity and water solubility. Subsequently solution-state assembly and intramicellar cross-linking afforded SCK block copolymer nanoparticles with fluorinated segments. These final fluorinated nanoparticles (having hydrodynamic diameters of ca. 20 nm) were studied as potential MRI contrast agents based on the existing ^{19}F nuclei. It was found that packing of the hydrophobic fluorinated polymers into the core domain restricted the mobility of the chains and prohibited ^{19}F-NMR spectroscopy when the particles were dispersed in water without an organic cosolvent. The authors succeeded in encapsulating perfluoro-15-crown-5-ether (PFCE) into the polymeric micelles with good uptake efficiency, in order to increase fluorine content and mobility of core chains. Nevertheless, they observed that it was necessary to swell the core with a good solvent (dimethyl sulfoxide) to increase the mobility sufficiently enough to observe the ^{19}F-NMR signal of the PFCE blocks. These studies elucidate some aspects on the chemical design and properties of fluorinated polymeric nanoparticles with the potential to be used as ^{19}F-MRI contrast agents.

In another approach, a series of partly fluorinated polyelectrolytes were evaluated for their applicability as corona-forming components in ^{19}F-MRI-active nanoparticles in aqueous solutions.[260] The polyelectrolytes were statistical and block copolymers of TFEMA and DMAEMA. The statistical copolymers were directly dissolved in water, whereas the block copolymers assembled into micellar nanoparticles with poly-TFEMA cores and poly (TFEMA-co-DMAEMA) coronas in aqueous media. The ^{19}F-spin-lattice (T_1) and spin-spin (T_2) relaxation times and ^{19}F image intensities of solutions of the polymers were measured and related to polymer chemical structure, as well as the chain conformation of the coronas of the micellar nanoparticles in aqueous solutions. The ^{19}F-NMR T_2 relaxation times were found to be highly indicative of the ^{19}F imaging performance of the dispersed nanomaterials. It was concluded that maintaining sufficient mobility of the ^{19}F nuclei was important for obtaining images of high intensity. ^{19}F mobility could be increased by preventing their aggregation in water as it was done in the case of statistical copolymers or by incorporating fluorine-containing segments within the corona of the micellar nanoparticles, in the case of block copolymers, where electrostatic repulsion between monomer units was present.

POLYMERIC NANOPARTICLES FOR PDT

PDT is a treatment that utilizes the combined action of photosensitizer molecules and a specific light source for various types of cancer.[261] In comparison to chemotherapy, PDT can be regarded as a noninvasive therapy and in principle induces less harmful effects to the healthy tissues. Photosensitizer molecules utilized in PDT act as strong absorbing agents, in a wide spectrum of wavelengths, and convert oxygen to highly reactive oxygen species (ROS) that in turn induce damage to tumor cells. It has been shown that photodynamic therapies destroy tumors through three main mechanisms: 1) have a direct effect on cancer cells leading to cell death such as apoptosis and necrosis;[262] 2) produced ROS damage the vascular and subsequent deprivation of oxygen and nutrients to the tumor;[263] and 3) have several effects on the immune system causing an immunosup-pressive response to the tumor.[264] In order to increase therapeutic efficiency of the photosensitizer in cancer therapy, a suitable carrier is needed for enhancing the usually poor solubility of such agents and to improve targeting. A variety of nanocarriers have been proposed and developed for PDT, including polymer-photosensitizer conjugates, liposomes, and polymeric micelles.[265–269]

These nanocarriers have been developed mainly because they show tumor-selective accumulation, due to the enhanced, microvascular permeability and impaired lymphatic drainage. However, the aggregation of photosensitizer molecules within the drug carrier is expected to reduce formation of ROS, due to light-quenching effects. Therefore, the release of the photosensitizer is needed for ensuring maximum therapeutic efficiency. The situation then resembles in many ways the design and properties requirements for drug delivery nanocarriers. Some illustrative examples of the application of polymeric nanoparticles in PDT are discussed in the following.

Thermosensitive mPEG-b-p (HPMAm-Lac2) micelles loaded with a hydropho-bic solketal-substituted phthalocyanine [Si (sol)$_2$Pc] photosensitizer were studied by Hennink and coworkers.[266] It was shown that the phthalocyanine molecule could be loaded efficiently in the micelles up to a concentration of ca. 2 mg/ mL. The resulting nanoparticles had a diameter of ca. 75 nm. UV/Vis and fluorescence spectroscopy measurements indicated that at low concentrations (<0.05 llM, 0.45 mg/mL polymer), the photosensitizer molecules were molecularly dissolved within the micellar core, whereas Si (sol)$_2$Pc aggregates were formed at higher photosensitizer concentrations. In vitro studies with B16F10 and 14C cells showed that the photocytotoxicity of Si (sol)2Pc-loaded micelles (at a photosensitizer molar concentration of 0.05 llM) was similar to free Si (sol)$_2$Pc (IC50 values of 3.0 ± 0.2 nM were obtained in solutions containing 10% serum). The cellular uptake of the highly loaded micelles (with a 10 l M Si (sol)$_2$Pc concentration) was low and independent of the serum concentration (Fig. 32). The nanoaggregates of Si (sol)$_2$Pc loaded in the micellar core could be released only during hydrolysis-induced micellar dissociation, which was observed after 5 hr at pH 8.7 and at body temperature. The authors did not present in vivo studies with the particular nanoparticles, but the stability of the highly loaded micellar Si (sol)$_2$Pc formulation in the presence of serum, the controlled release of the photosensitizer molecules upon micellar disintegration,

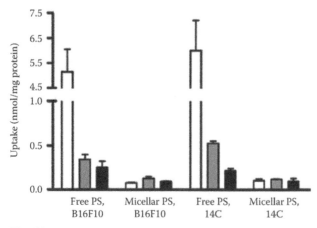

Fig. 32 Cellular uptake of Si (sol)$_2$Pc (nmol/mg protein) by B16F10 and 14C cells after 6 hr of incubation, either administered as free drug or encapsulated in mPEG-b-p (HPMAm-Lac$_2$) micelles (10 *lM Si (sol)$_2$Pc, 0.45 mg/mL polymer). White bars represent the uptake from medium without FBS, gray bars represent the uptake from medium supplemented with 10% FBS for B16F10 cells and 5% FBS for 14C cells, and the black bars show the uptake from medium supplemented with 50% or 25% FBS for B16F10 and 14C cells, respectively.
Source: From Rijcken et al.[266] © 2007, with permission from Elsevier.

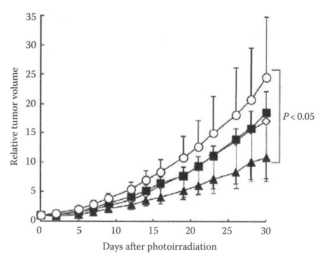

Fig. 33 Growth curves of subcutaneous A549 tumors in control mice (open circle) and mice administered with 0.37 imol/kg DPc (closed square), 0.37 imol/kg DPc/m (closed triangle), and 2.7 imol/kg Photofrin® (PHE) (open diamond) (n = 6). Twenty-four hours after administration of photosensitizing agents, the tumors were photoirradiated using a diode laser (fluence: 100 J/cm^2).
Source: From Nishiyama et al.[269] © 2009, with permission from Elsevier.

and the high photodynamic activity of Si (sol)$_2$Pc make these micelles interesting candidates for PDT protocols.

Polymeric micelles encapsulating dendrimer phthalocyanine (DPc) as the photosensitizer were developed by Nishiyama and coworkers.[269] These micelles were based on poly (ethylene glycol-b-l-lysine) (PEG-b-PLL) block copolymer carrying cationic charges and a second generation aryl ether dendrimer with a Zn (II) phthalocyanine center and 32 carboxylic groups on its periphery. Electrostatic complexation between the two building blocks results in complex poly-ion micelles with the DPc photosensitizer in the core (complexes with PLL blocks of the copolymers) and PEG coronas. The DPc-loaded micelles induced efficient and very rapid light-induced cell death, accompanied by characteristic morphological changes of the tumor cells, such as blebbing of cell membranes, when the cells were photoirradiated using a low-power halogen lamp or a high-power diode laser (Fig. 33). Fluorescent microscopy observations, using organelle-specific dyes, demonstrated that, following internalization by endocytosis, DPc/PEG-c-PLL nanoparticles accumulate in the endo-/lysosomes. However, upon photo-irradiation, DPc/PEG-c-PLL is translocated to the cytoplasm and induces photodamage to the mitochondria, which may account for the enhanced photocytotoxicity of DPc/ PEG-c-PLL nanoparticles. Additional studies demonstrated that DPc/PEG-c-PLL showed significantly higher *in vivo* PDT efficacy than the clinically used Photofrin® (polyhematoporphyrin esters, PHEs) in mice bearing human lung adenocarcinoma A549 cells. Furthermore, unlike the PHE-treated mice, the

DPc/PEG-c-PLL-treated mice showed no sign of skin phototoxicity, under the tested conditions. These results strongly suggest that the DPc/PEG-b-PLL nanosystem is expected to serve as an innovative photosensitizer formulation for improved effectiveness and safety of current PDT protocols.

Mixed polymeric micelles based on non-pH-sensitive and pH-sensitive graft copolymers, namely, [poly (N-vinylcaprolactam)-g-poly (d,l-lactide) (PVCLm-g-PDLLA) and poly (N-vinylcaprolactam-co-N-vinyl imidazole)-g-poly (d,l-lactide)] [P(VCLm-co-NVIM)-g-PDLLA], and PEG-b-PDLLA diblock copolymers, with or without fluorescent groups on the corona, were utilized for the encapsulation of protoporphyrin IX (PPIX) for *in vitro* and *in vivo* PDT-oriented studies by Tsai and coworkers.[270] Photochemical internalization was utilized to study the localization of pH- and non-pH-sensitive micelles uptake in the lysosome. After nontoxic light treatment, confocal microscopy observations indicated that PPIX molecules encapsulated in pH-sensitive micelles were found in the nucleus, while PPIX encapsulated in the non-pH-sensitive micelles was still localized in the lysosomal compartments. Formation of singlet oxygen was observed for both the block and graft copolymer micelles. Differences in the cell viability were ascribed to the damage occurring at the region where the PPIX was located. *In vivo* studies revealed that PPIX-loaded graft and diblock copolymer micelles presented prolonged blood circulation and enhanced tumor-targeting ability, especially in the case of micelles with targeting functionalities. The PPIX released from g-CIM micelles on tumor sites was also observed by *ex vivo* confocal imaging techniques. Mice treated with non-pH-sensitive micelle/PPIX nanoparticles showed a better repression of tumor growth than mice treated with PPIX alone. The latter result was attributed to the larger amount of photosensitizer localized in the tumor region still exhibiting therapeutic effects. Effective PDT-induced inhibition of tumor growth was found in mice treated with pH-sensitive micelle/PPIX.

Shan and coworkers presented the preparation and properties of hybrid nanoparticles comprised of tetraphenylporphyrin inorganic upconverting phosphors encapsulated within poly (lactic acid)-b-poly (ethylene oxide) block copolymers utilizing the flash nanoprecipitation method.[271] The nanoassemblies were found to be stable under physiological conditions and demonstrated strong cancer cell deactivating abilities via the production of ROS under 980 nm illumination.

MULTIFUNCTIONAL POLYMERIC NANOPARTICLES

In recent years, the concept of multifunctional block copolymer micelles or polymeric nanoparticles for simultaneous therapy and diagnostics is gaining increased interest among researchers in the field. The basic idea behind these efforts concerns the incorporation into the polymeric nanoassembly or nanoparticle several functionalities/modalities that can be used for the delivery of therapeutic agents and other functional entities that are useful for imaging and diagnostics. This requires the combination of several design principles already discussed in previous sections and makes the construction of such nanodevices more difficult but at the same time more interesting. Some works exemplifying developments in this new rapidly increasing research area are presented as follows.

A nonviral nanoparticle gene carrier was developed by Liu et al.[272] and its efficiency for siRNA delivery and transfection was validated both *in vitro* and *in vivo*. The nanocarrier was constructed by a core of IO nanoparticles and a shell of alkylated PEI of 2000 kDa molecular weight (alkyl-PEI2k), abbreviated as alkyl-PEI2k-IO. The hybrid nanocarrier was able to bind with siRNA, resulting in well-dispersed nanoparticles with controlled structure and narrow size distribution. Electrophoresis studies show that the alkyl-PEI2k-IOs could retard siRNA completely at N:P ratios (i.e., PEI nitrogen to nucleic acid phosphate) above 10, protect siRNA from enzymatic degradation in serum, and release complexed siRNA efficiently in the presence of polyanionic heparin. The knockdown efficiency of the siRNA-loaded nanocarriers was tested with 4T1 cells stably expressing luciferase (fluc-4T1) and further with a fluc-4T1 xenograft model. Significant downregulation of luciferase was observed, and unlike high molecular weight analogues, the alkyl-PEI2k-coated IOs demonstrated good biocompatibility, highly efficient delivery of siRNA, and

an innocuous toxic profile, making them a potential nanocarrier for gene therapy.

Nasongkla and coworkers described the development of multifunctional polymeric micelles based on poly (ethylene glycol)-b-poly (d,l-lactide) AmBCs.[273] The PEG surface of the micelles was functionalized with cRGD-type peptide with targeting capability toward $\alpha_v\beta_3$-integrins, for controlled drug delivery of DOX to cancer cells. To this nanoconstruct, narrow distribution hydrophobically modified SPIONs were also encapsulated for achieving efficient MRI contrast characteristics. DOX and the SPIONs clusters were loaded successfully inside the PDLLA micelle core utilizing a solvent-evaporation methodology, which allowed for a loading of ca. 7 w/w% for the superparamagnetic iron oxide (SPIO) particles and ca. 3 w/w% for DOX, respectively (Fig. 34). Micellar sizes were in the range 40–50 nm for loaded and empty micelles. The presence of cRGD targeting peptide on the micelle surface resulted in the selective incorporation of the multifunctional micelles and the targeted drug delivery to SLK tumor endothelial cells as evidenced by flow cytometry and CLSM. A 2.5-fold increase of micellar uptake by cells, in comparison to nonfunctionalized micelles, was attributed to the presence of the targeting peptide. *In vitro* MRI and cytotoxicity studies demonstrated the ultrasensitive MRI (down to nanomolar levels) and the $\alpha_v\beta_3$-specific cytotoxic response of these multifunctional hybrid polymeric micelles (Fig. 35). Longer incubation times were found to lead to an overall decrease of cancer cell viability for all of the micelle samples investigated.

Micellar hybrid nanoparticles for simultaneous magnetofluorescent imaging and anticancer drug delivery were reported by Sailor and coworkers.[274] This group utilized a solvent-evaporation method for the encapsulation of spherical OA-coated IO magnetic nanoparticles of ca. 11 nm diameter and elongated TOP-coated quantum dots (QDs) into PEG-lipid micellar nanoassemblies. The SPIO/QD ratio within the individual micelles could be adjusted by changing the mass ratio of magnetic nanoparticles to QDs during the preparation step. TEM imaging demonstrated the successful construction of the hybrid nanoassemblies. Imaging of the multifunctional nanoassemblies was possible even in the NIR region and in subnano-molar concentrations, despite the observed fluorescent quenching and adsorption due to the presence of the magnetic nanoparticles in close proximity to the QDs.

The ability of the novel multifunctional micelles to target and image in dual-mode (via MRI and fluorescence) tumor cells was tested against MDA-MB-435 human cancer cells. Specific targeting to tumor cells was made possible by chemical attachment of the targeting ligand F3 (a peptide known to target cell-surface nucleolin in endothelial cells in tumor blood vessels and in tumor cells and to become internalized in these cells) on the surface of the micelles (via chemical bond formation with the terminus of the PEG corona chains). Simultaneous imaging and drug delivery

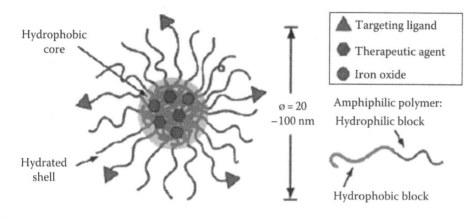

Fig. 34 Multifunctional polymeric nanoparticle platform for targeted drug delivery and MRI.
Source: From Nasongkla et al.[273] © 2006, with permission from American Chemical Society.

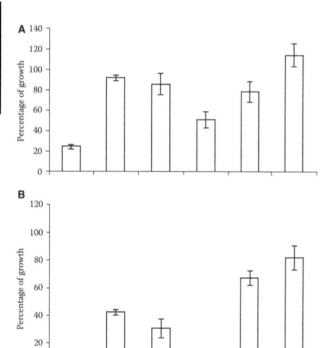

Fig. 35 Inhibition of SLK cell growth in the presence of different formulations of SPIO-loaded micelles with or without 1 μM DOXO concentration after (**A**) 1 hr or (**B**) 4 hr incubation times. The percent inhibition of cell growth was calculated as the ratio of cell number in the treated sample divided by that in the untreated control.
Source: From Nasongkla et al.[273] © 2006, with permission from American Chemical Society.

were demonstrated with the anticancer drug DOX, which was also loaded into the MHNs during preparation of the hybrid nanoassemblies (up to a DOX/micelle ratio of 0.093:1 w/w). The intrinsic fluorescence of DOX enabled the independent imaging of both DOX and QDs contained in the hybrid micelles. The intact hybrid micelles were observed to colocalize in some areas of MDA-MB-435 cells *in vitro* upon incubation for 2 hr. It is interesting to note that

although they were composed of relatively toxic QDs, no significant toxicity of the nanoassemblies was observed. On the other hand, F3-functionalized micelles in which DOX was encapsulated displayed significantly greater cytotoxicity than that of equivalent quantities of free DOX or hybrid micelles containing DOX without the targeting ligand. Biodistribution measurements on mice indicated that the multifunctional micelles were accumulated mainly in the liver, with no significant quantities being observed in other organs.

The preparation of organic-inorganic hybrid micelles by the comicellization of poly (ε-caprolactone)-*b*-poly (glycerol monomethacrylate) (PCL-*b*-PGMA) and poly (ε-caprolactone)-*b*-poly (oligoethylene glycol methacrylate-co-oligoethylene glycol methacrylate folic acid conjugate) [PCL-*b*-P(OEGMA-co-FA)] AmBCs, which physically encapsulate hydrophobic drugs within their micellar cores and with SPIONs embedded within their hydrophilic coronas, has been reported by Liu and coworkers.[275] These nanoparticles possess integrated functions for chemotherapeutic drug delivery and MRI contrast enhancement. The model hydrophobic anticancer drug, PTX, and 4 nm SPIONs were loaded into micellar cores and hydrophilic coronas, respectively, taking advantage of the hydrophobicity of micellar cores and strong affinity between 1,2-diol moieties in PGMA and Fe atoms at the surface of SPIONs. A drug-loading content of ca. 8.5 w/w% was possible. Controlled and sustained release of PTX from the hybrid micelles was achieved, exhibiting a cumulative release of ca. 61% of encapsulated drug over a period of 130 hr. The clustering of magnetic nanoparticles within the coronas led to considerably enhanced T_2 relaxivity, strongly suggesting that the particular hybrid micelles can serve as a T_2-weighted MRI contrast enhancer with improved performance. Preliminary *in vivo* MRI experiments indicated that this type of hybrid multifunctional polymeric micellar nanoparticles can act as a new nanoplatform integrating targeted drug delivery, controlled release, and disease diagnostic functions.

The same group also reported on the fabrication of an analogous mixed block copolymer micellar system composed of two types of amphiphilic diblock copolymers, namely, PCL-*b*-P(OEGMA-FA) and PCL-*b*-P(OEGMA-Gd), consisting of a hydrophobic PCL block and a hydrophilic poly (oligo (ethylene glycol) monomethyl ether methacrylate) (POEGMA) block, to which some segments carried covalently attached FA moieties and DOTA-Gd (Gd) side groups.[276] FA groups and Gd complexes provide synergistic functions for targeted delivery and MRI contrast enhancement. Loading of hydrophobic chemotherapeutic drugs, like PTX, could be achieved within the hydrophobic PCL cores of the mixed micelles. The as-prepared nano-sized mixed micelles were capable of physically encapsulating PTX at a loading content of ca. 5.0 w/w%, exhibiting controlled release of up to ca. 60% of loaded drug and for a time period of almost 130 hr. *In vitro* cell viability assays indicated that drug-free mixed micelles are almost noncytotoxic up to a concentration of 0.2 g/L, whereas PTX-loaded micelles could effectively kill HeLa cells at the same concentration. *In vitro* MRI experiments revealed an increased T_1 relaxivity (almost sevenfold increase) for mixed micelles compared to that of small molecular weight complex counterpart, alkynyl-DOTA-Gd. Additional, *in vivo* MRI experiments in rabbits indicated considerably enhanced signal intensity, prominent positive contrast enhancement, improved accumulation and retention, and extended blood circulation duration for FA-labeled mixed micellar nanoparticles within the rabbit liver, as compared to those for FA-free mixed micelles and alkynyl-DOTA-Gd complexes.

Micelles from fluorine-containing amphiphilic poly (hexafluorobutyl methacrylate-g-oligoethylene glycol methacrylate) (PHFMA-g-PEGMA) graft copolymers have been also utilized for the encapsulation of magnetic nanoparticles and a hydrophobic drug simultaneously.[277] Encapsulated OA-modified magnetite (Fe_3O_4) nanoparticles were found to form clusters in the PHFMA-g-PEGMA micellar cores with a mean diameter of 100 nm, and the hybrid micelles showed high stability in aqueous media due to the highly hydrophobic fluorine segments in the graft copolymers that enhance the stability of the micelles. The hybrid micelles showed good cytocompatibility based on MTT cytotoxicity assay, and they possessed paramagnetic properties with saturation magnetization of ca. 17.14 emu/g. The hydrophobic drug 5-fluorouracil could be loaded into the hybrid micellar nanoparticles, using an emulsion method, with a loading efficiency of ca. 21 wt%. Controlled release of the drug was also achieved. The magnetic micelles had satisfactory transverse relaxivity rates and exhibited high efficacy as a negative MRI agent in T_2-weighted imaging. *In vivo* MRI studies demonstrated that the contrast between liver and spleen was enhanced by the magnetic copolymer micelles.

An interesting hybrid star poly[N-(2-hydroxypropyl) methacrylamide] (PHPMA)-based nanoplatform for simultaneous therapy and diagnosis was presented by Liu and coworkers.[278] pH-disintegrable micellar nanoparticles were fabricated from asymmetrically functionalized β-cyclodextrin (β-CD) core PHPMA star copolymers covalently

conjugated with DOX, FA, and DOTA-Gd moieties for integrated cancer cell-targeted drug delivery and MRI contrast enhancement. The functionalized β-CD cores with 7 azide functionalities and 14 α-bromopropionate moieties, $(N_3)_7$-CD-$(Br)_{14}$, were used for the ATRP polymerization of HPMA through the bromide moieties and the click reaction with alkynyl-(DOTA-Gd) complex. DOX and FA groups were covalently bonded to the hydroxyl groups of HPMA segments. These reactions resulted in the formation of a (DOTA-Gd)$_7$-CD-(PHPMA-FA-DOX)$_{14}$ star copolymer possessing 7 DOTA-Gd complex moieties and 14 PHPMA arms carrying DOX and FA functionalities via acid-labile carbamate linkages and ester bonds, respectively. The covalent fixation of DOX molecules onto the PHPMA star copolymer arms (with ca. 14 wt% drug-loading content) makes the initially hydrophilic polymer amphiphilic, leading to the formation of micellar nanoparticles with sizes of several tens of nanometers in aqueous solution and at pH 7.4. The *in vitro* obtained DOX release profile was highly pH dependent (Fig. 36). For a time period of 42 hr, cumulative releases of 10%, 53%, and 89% conjugated DOX at pH 7.4, 5.0, and 4.0, respectively, were observed. Additionally, the pH-modulated release of conjugated DOX from micellar nanoparticles is accompanied with micelle disintegration due to the loss of the hydrophobic component in the star copolymer molecule. *In vitro* cell viability assays revealed that the (DOTA-Gd)$_7$-CD-(PHPMA$_{15}$)$_{14}$ star copolymer was almost noncytotoxic up to a concentration of 0.5 g/L, whereas DOX-conjugated micellar nanoparticles of (DOTA-Gd)7-CD-(PHPMA-FA-DOX)14 could effectively enter and kill HeLa cells at a concentration higher than 80 mg/L. *In vitro* MRI experiments showed an enhanced T_1 relaxivity for micellar nanoparticles compared to that for the neat alkynyl-DOTA-Gd complex. An *in vivo* MR imaging assay was also made in rats. The

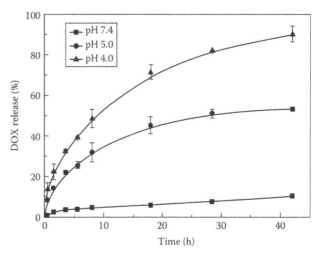

Fig. 36 *In vitro* DOX release profiles at varying pH conditions (37°C, 20 mM buffer solution) from multifunctional micellar nanoparticles of (DOTA-Gd)7-CD-(PHPMA-FADOX) 14-star copolymers.
Source: From Liu et al.[278] © 2012, with permission from Elsevier.

experiments revealed an enhanced accumulation of hybrid copolymer nanoparticles within the liver and kidney of the animals and prominent positive contrast enhancement.

Novel hybrid magnetic/drug nanocarriers were prepared via in situ synthesis of magnetic IO nanocrystallites in the presence of poly (methacrylic acid)-graft-poly (ethylene glycol methacrylate) [p (MAA-g-EGMA)] able for interaction-induced self-assembly into hybrid micelles.[279] The particular hybrid nanoparticles possessed bio-repellent properties due to the presence of EGMA macromonomer segments, pronounced magnetic response, and high loading capacity for the anticancer drug DOX, to an extent not observed before in such hybrid colloids. High magnetic responses were accomplished by engineering the size of the magnetic nanocrystallites (to ca. 13.5 nm) following an aqueous single-ferrous precursor route and through adjustment of the number of the cores in each colloidal assembly. The magnetic response of the hybrid nanocarriers was evaluated by conventional magnetometry and magnetophoretic experiments providing insight on the internal organization of the inorganic particles in respect to the organic/polymeric part and on their response to magnetic manipulation. The structural organization of the graft copolymer, locked on the surface of the IO nanocrystallites, was further probed by small-angle neutron scattering on single-cored colloids. Analysis showed that the MAA segments are selectively populating the area around the magnetic nanocrystallites, while the PEG-grafted chains are arranged as protrusions, pointing toward the aqueous environment, a structure similar to conventional linear block copolymer micelles. The nanocarriers were found to be stable at various pH and at highly salted aqueous media, as evidenced by DLS and electrophoretic light scattering measurements. Their colloidal stability was enhanced, as compared to uncoated nanocrystallites, presumably due to the presence of the external protective PEG canopy. The nanoparticles evaluated in terms of their bio-repellent properties, by assaying their stability using human blood plasma as the surrounding medium. DOX loading was achieved in the protonated form of the drug through physical adsorption onto the nanoparticles' core and reached 22 wt%. Initial experiments showed that DOX could be released in a sustainable manner in a number of aqueous media including blood serum.

A multifunctional nanostructure of nanometer dimensions self-assembled from a hydrophobic poly (DL-lactide-co-glycolide) (PLGA) core and a hydrophilic paramagnetic-folate-coated PEGylated lipid shell (PFPL; PEG) possessing simultaneous MRI and targeted therapeutics properties was also presented.[280] The nanoassembly was found to have a well-defined core/shell structure as revealed by CLSM observations. The paramagnetic diethylenetriaminepentaacetic acid-gadolinium (DTPA-Gd) chelated to the shell layer showed significantly higher spin-lattice relaxivity (r_1) than the clinically used low molecular weight MRI contrast agent Magnevist. The PLGA core served as a nanocontainer

to load and release the hydrophobic drugs. Drug-release studies indicated that the modification of the PLGA core with a polymeric liposome shell can reduce the drug-release rate. Folate functionalities enhanced cellular uptake of the nanocomplex compared to the non-functionalized analogue.

Shuai and coworkers presented the case of a multifunctional block copolymer micellar nanosystem for simultaneous tumor-targeted intracellular drug release and fluorescent imaging.[281] The micelles were formed by a triblock-like copolymer comprised of PEG, poly (N-(N', N'-diisopropylaminoethyl) aspartamide), and a cholic acid hydrophobic relatively large end group (PEG-b-PAsp (DIP)-b-CA). The copolymer formed micelles in aqueous media, at pH 5.0, with a CA core. PAsp (DIP) forms a pH-sensitive interlayer that is expected to undergo hydrophobic/hydrophilic transition in the acidic lysosomes of cells. PTX and QDs were then incorporated into the copolymer nanoassembly. PTX (or fluorescein diacetate) was loaded due to hydrophobic interactions into the CA core, reaching a loading cargo of 2.6 wt%. The negatively charged QDs were incorporated in the positively charged PAsp (DIP) layer through electrostatic interaction (loading content 2.7%). The assembly had a size ca. 50 nm. At pH 7.4, the PAsp (DIP) layer was partially solvated and could act as an "on-off" pH switch controlling the release of the encapsulated drug. PTX release was very small at pH 7.4 (less than 5%), while at pH 5.0, a rapid release of the drug was observed. Folate groups were also incorporated at the PEG terminus for active targeting. In vitro experiments with Bel-7402 cells as well as in rats demonstrated the great potential of these hybrid micellar nanocarriers for anticancer therapy and imaging.

It is important to discuss the case of multifunctional nanocarriers able to deliver two different types of therapeutic agents (Fig. 37). In this context, Yu and coworkers utilized pH-responsive block copolymer micelles from poly (2-(dimethylamino) ethyl methacrylate)-block-poly[2-(diisopropylamino)ethyl methacrylate] (PDMA-b-PDPA) diblock copolymers for the intracellular (endosomal) delivery of AmB and siRNA.[282] AmB was loaded into the hydrophobic PDPA core at pH 7.4, and siRNA was complexed with the positively charged PDMA shell to form the micelleplexes. CLSM experiments showed the cellular uptake of the micellar nanoparticles. The PDMA-b-PDPA/siRNA micelleplexes were found to dissociate in early endosomes, and AmB was released from the nanocarriers. Live-cell imaging studies demonstrated that released AmB significantly increased the ability of siRNA to overcome the endosomal barrier, due to the ability of the drug to increase membrane permeability through the formation of transmembrane pores. Transfection studies also showed that AmB-loaded micelleplexes resulted in significant increase in luciferase knockdown efficiency over the AmB-free control. The enhanced luciferase knockdown efficiency was abolished by bafilomycin A1, a vacuolar ATPase inhibitor that inhibits the acidification of the endocytic organelles. The reported results support the basic

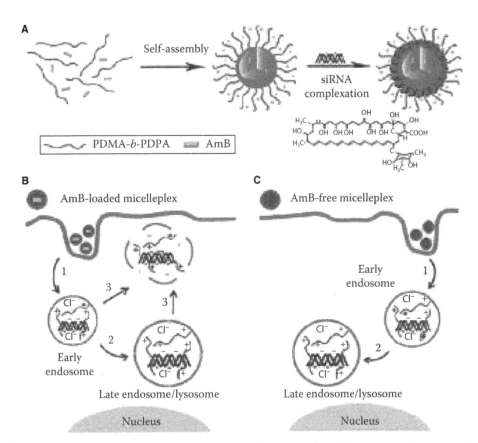

Fig. 37 Preparation and action of AmB-loaded dual pH-responsive micelleplexes for siRNA delivery with enhanced siRNA endosomal escape ability. (**A**) Production of AmB-loaded PDMA-*b*-PDPA micelleplexes. AmB was loaded in the hydrophobic PDPA core, and siRNA was complexed with the PDMA corona shell. (**B**) AmB-facilitated endosome disruption and siRNA cytoplasmic release (1: AmB-loaded micelleplexes dissociated in early endosomes after cell uptake, and AmB molecules are inserted into endosomal membranes; 2: protonated PDMA-*b*-PDPA unimers complexed with siRNA and trafficked from early endosomes into late endosome/lysosomes, causing vesicle swelling; 3: AmB enhanced siRNA release from endosomes into cytoplasm via membrane destabilization). (c) In the case of AmB-free micelleplexes, polymer/siRNA complexes were entrapped in late endosomes or lysosomes without efficient cytoplasmic siRNA release. **Source:** From Yu et al.[282] © 2011, with permission from American Chemical Society.

hypothesis that membrane poration by AmB and increased endosomal swelling and membrane tension by a "proton sponge"-type polymer (like the particular PDMA-*b*-PDPA copolymer) provided a synergistic strategy to disrupt endosomes for improved intracellular delivery of siRNA.

A rather complex nanosystem composed ofOA surface-modified magnetic nanoparticles, the cationic poly[2-(dimethylamino)ethyl methacrylate] end capped with a cholesterol moiety (Chol-PDMAEMA30), DNA, and the anionic poly[poly (ethylene glycol)methyl ether methacrylate]-b-PMA diblock carrying partial mercapto groups on the PMA block (PPEGMA-b-PMAASH), has been reported by.[283] The magnetic nanoparticles were first coated with the hydrophobically modified Chol-PDMAEMA30 polycation and then complexed with DNA chains through electrostatic interaction. The three component complexes were further physically binded with the brush-type PPEGMA-b-PMAASH polyanion. The resulting magnetic particle/DNA/polyion complexes could be further stabilized against disintegration by oxidizing the

mercapto groups of PPEGMA-b-PMAASH to form a cross-linked PMAA shell with bridging disulfide bonds between the PPEGMA-b-PMAASH outer chains. It was shown that the presence of peripheral PPEGMA-b-PMAASH chains on the gene vector can reduce the nonspecific adsorption of blood proteins. The disulfide bonds on the periphery could be cleaved in a cell reducing environment enhancing the release of DNA. The combined results of zeta-potential, DNA-binding capacity, cytotoxicity assay, and *in vitro* transfection tests indicated that the specific gene vector possessed magnetic responsiveness, anti-nonspecific protein adsorption, low cytotoxicity, and good transfection efficiency on HEK 293T and HeLa cells.

A block copolymer-based nanosystem capable of synergistic chemotherapy and PDT toward cancer cells has been presented by Shieh and coworkers.[267] Amphiphilic 4-armed star-shaped 5,10,15,20-tetrakis (4-aminophenyl)-21H,23H-chlorin (TAPC)-core block copolymers based on methoxy poly (ethylene glycol) (mPEG) and PCL were synthesized and studied. The photosensitizer-centered amphiphilic star-block copolymer formed micelles in aqueous media (CSBC micelles) that could encapsulate PTX in their hydrophobic inner PCL cores (PTX-loaded CSBC micelles denoted as PCSBC micelles). The star-block copolymer micelles exhibited efficient singlet oxygen generation, whereas the hydrophobic photosensitizer alone (without the diblock arms conjugation) failed, due to considerable aggregation in aqueous solution. The chlorin-core micelles alone exhibited obvious phototoxicity in MCF-7 breast cancer cells with 7 or 14 J/cm² light irradiation at a chlorin concentration of 125 mg/mL. After PTX loading, the size of the micelles increased from ca. 70–103 nm as a result of the incorporation of the hydrophobic drug. The PTX-loaded micelles were found to have an improved cytotoxicity toward MCF-7 cells after irradiation, most likely through a synergistic chemotherapy/PDT effect (Fig. 38).

The same group also reported a star-block copolymer with a TAPC core and poly[ethylene glycol-*b*-(ε-caprolactone)] arms (CSBC), where the PCL blocks are directly connected to the core, that self-assembled into nanosized micelles, which acted as nanocarriers for the photosensitizing core moiety and the hydrophobic anticancer drug SN-38 (7-ethyl-10-hydroxy-camptothecin).[268] SN-38 was encapsulated into the star-block micelles by using a lyophilization-hydration protocol. Experiments showed a prolonged plasma residence time of SN-38/CSBC micelles as compared to free CPT-11, permitting the increased tumor accumulation and consequently improved antitumor activity. The combined effects of SN-38/CSBC-loaded micelles were evaluated in an HT-29 human colon cancer xenograft model. SN-38/CSBC-mediated simultaneous chemotherapy and PDT action successfully inhibited tumor growth, resulting in up to 60% complete regression of well-established tumors after three consecutive treatments. These treatments also decreased the microvessel density and cell proliferation within the subcutaneous tumors.

A similar system was also reported more recently by[284] They synthesized a star-shaped PCL-*b*-PEO amphiphilic copolymer with a tetrakis-(4-aminophenyl)-terminated porphyrin core. PTX was in situ encapsulated into the star-block copolymer micelles. The fluorescent characteristic of the central porphyrin moiety allowed for monitoring the cellular uptake and biodistribution of the PTX-loaded micelles by fluorescent imaging in cell and live mice. It was concluded that the PTX-loaded micelles could be readily internalized by cancer cells and they showed a slightly higher cytotoxicity than the clinical PTX injection Taxol. *In vivo* real-time fluorescent imaging revealed that the micelles could accumulate at tumor sites via the blood circulation in tumor-bearing mice. *In vivo* antitumor efficacy examinations indicated that the PTX-loaded micelles had significantly superior efficacy in impeding tumor growth than Taxol and low toxicity to the living mice.

Micelles from amphiphilic star-block copolymers, having a hydrophobic hyperbranched core and amphiphilic fluoropolymer arms, were constructed as drug delivery agent assemblies as well as potential nanoparticulate MRI contrast agents.[285] Several polymer structures were prepared from consecutive copolymerizations of 4-chloromethylstyrene with dodecyl acrylate and then 1,1,1-trifluoroethyl methacrylate with tBA, followed by acidolysis in order to produce the hydrophilic acrylic acid residues on the polymer chains. These polymer chains were labeled with cascade blue as a fluorescence reporter, allowing for a simultaneous optical imaging function of the nanoparticles. The series of materials differed primarily in the ratio of 1,1,1-trifluoroethyl methacrylate to acrylic acid units, to give differences in fluorine loading

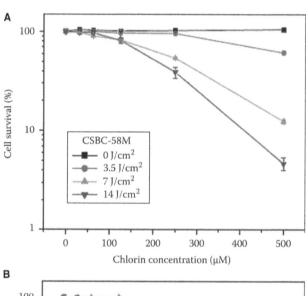

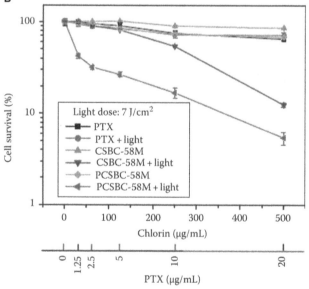

Fig. 38 (**A**) Cytotoxicity of CSBC-58M under different irradiation and (**B**) PCSBC-58M in MCF-7 cells with 7 J/cm² irradiation. **Source:** From Peng et al.[267] © 2008, with permission from Elsevier.

and hydrophobicity/hydrophilicity balance. DOX was used as a model hydrophobic drug to study the loading, release, and cytotoxicity of these polymer micellar constructs on an U87-MG-EGFRvIII-CBR cell line. The micelles (with TEM-measured diameters ranging 5–9 nm and DLS-measured hydrodynamic diameters 20–30 nm) had rather low loading capacities of ca. 4 wt% of DOX. The DOX-loaded micelles exhibited cytotoxicity with cell viabilities of 60–25% at 1.0 LLg/mL effective DOX concentrations, depending upon the polymer composition, as determined by MTT assays. These cell viability values are comparable to that of free DOX, suggesting an effective release of the cargo and delivery to the cell nuclei, which was further confirmed by fluorescence microscopy of the cells. ^{19}F-NMR spectroscopy indicated a partial degradation of the surface-available trifluoroethyl ester linkages of the micelles, which may have caused acceleration of the release of DOX. ^{19}F-NMR spectroscopy was also employed to confirm and to quantify the cell uptake of the drug-loaded micelles. Evidently, these dual-fluorescent and ^{19}F-labeled and chemically functional micelle carriers may be used in a variety of applications, such as cell labeling, imaging, and therapeutic delivery.

REFERENCES

1. Kwon, G.S.; Forrest, M.L. Amphiphilic block copolymer micelles for nanoscale drug delivery. Drug Dev. Res. **2006**, *67* (1), 15–22.

2. Wiradharma, N.; Zhang, Y.; Venkataraman, S.; Hedrick, J.L.; Yang, Y.Y. Self-assembled polymer nanostructures for delivery of anticancer therapeutics. Nano Today **2009**, *4* (4), 302–317.

3. Du, F.-S.; Wang, Y.; Zhang, R.; Li, Z.-C. Intelligent *nucleic* acid delivery systems based on stimuli-responsive polymers. Soft Matter **2010**, *6* (5), 835–848.

4. Horgan, A.M.; Moore, J.D.; Noble, J.E.; Worsley, G.J. Polymer- and colloid-mediated bioassays, sensors and diagnostics. Trends Biotechnol. **2010**, *28* (9), 485–494.

5. Kedar, U.; Phutane, P.; Shidhaye, S.; Kadam, V. Advances in polymeric micelles for drug delivery and tumor targeting. Nanomed. Nanotechnol. Biol. Med. **2010**, *6* (6), 714–729.

6. Mishra, B.; Patel, B.B.; Tiwari, S. Colloidal nanocarriers: A review on formulation technology, types and applications toward targeted drug delivery. Nanomed. Nanotechnol. Biol. Med. **2010**, *6* (1), 9–24.

7. Ponta, A.; Bae, Y. PEG-poly (amino acid) block copolymer micelles for tunable drug release. Pharm. Res. **2010**, *27* (11), 2330–2342.

8. Tyrrell, Z.L.; Shen, Y.; Radosza, M. Fabrication of micellar nanoparticles for drug delivery through the self-assembly of block copolymers. Prog. Polym. Sci. **2010**, *35* (9), 1128–1143.

9. Miyata, K.; Christie, R.J.; Kataoka, K. Polymeric micelles for nano-scale drug delivery. React. Funct. Polym. **2011**, *71* (3), 227–234.

10. Xiong, X.B.; Falamarzian, A.; Garg, S.M.; Lavasanifar, A. Engineering of amphiphilic block copolymers for polymeric micellar drug and gene delivery. J. Control. Release **2011**, *155* (2), 248–261.

11. Wagner, E. Polymers for siRNA delivery: Inspired by viruses to be targeted, dynamic, and precise. Acc. Chem. Res. **2012**, *45* (7), 1005–1013.

12. Langer, R. Drug delivery and targeting. Nature **1998**, *392* (6672), 5–10.

13. Bae, Y.; Fukushima, S.; Harada, A.; Kataoka, K. Design of environment-sensitive supramolecular assemblies for intracellular drug delivery: Polymeric micelles that are responsive to intracellular pH change. Angew. Chem. Int. Ed. **2003**, *42* (38), 4640–4646.

14. Talelli, M.; Iman, M.; Varkouhi, A.K.; Rijcken, C.J.F.; Schiffelers, R.M.; Etrych, T.; Ulbrich, K.; van Nostrum, C.F.; Lammers, T.; Storm, G.; Hennink, W.E. Core-crosslinked polymeric micelles with controlled release of covalently entrapped doxorubicin. Biomaterials **2010**, *31* (30), 7797–7804.

15. Yang, X.; Grailer, J.J.; Rowland, I.J.; Javadi, A.; Hurley, S.A.; Matson, V.Z.; Steeber, D.A.; Gong, S. Multifunctional stable and pH-responsive polymer vesicles formed by heterofunctional triblock copolymer for targeted anticancer drug delivery and ultrasensitive MR imaging. ACS Nano **2010**, *4* (11), 6805–6817.

16. Lee, S.; Saito, K.; Lee, H.-R.; Lee, M.J.; Shibasaki, Y.; Oishi, Y.; Kim, B.-S. Hyperbranched double hydrophilic block copolymer micelles of poly (ethylene oxide) and polyglycerol for pH-responsive drug delivery. Biomacromolecules **2012**, *13* (4), 1190–1196.

17. Tan, J.P.K.; Kim, S.H.; Nederberg, F.; Fukushima, K.; Coady, D.J.; Nelson, A.; Yang, Y.Y.; Hedrick, J.L. Delivery of anticancer drugs using polymeric micelles stabilized by hydrogen-bonding urea groups. Macromol. Rapid Commun. **2010**, *31* (13), 1187–1192.

18. Molavi, O.; Ma, Z.; Mahmud, A.; Alshamsan, A.; Samuel, J.; Lai, R.; Kwon, G.S.; Lavasanifar, A. Polymeric micelles for the solubilization and delivery of STAT3 inhibitor cucurbitacins in solid tumors. Int. J. Pharm. **2008**, *347* (1–2) 118–127.

19. Mahmud, A.; Patel, S.; Molavi, O.; Choi, P.; Samuel, J.; Lavasanifar, A. Self-associating poly (ethylene oxide)- b-poly (α-cholesteryl carboxylate-ε-caprolactone) block copolymer for the solubilization of STAT-3 inhibitor cucurbitacin I. Biomacromolecules **2009**, *10* (3), 471–478.

20. Falamarzian, A.; Lavasanifar, A. Chemical Modification of hydrophobic block in poly (ethylene oxide) poly (caprolactone) based nanocarriers: Effect on the solubilization and hemolytic activity of amphotericin B. Macromol. Biosci. **2010**, *10* (6), 648–656.

21. Falamarzian, A.; Lavasanifar, A. Optimization of the hydrophobic domain in poly (ethylene oxide)-poly (ε-caprolactone) based nano-carriers for the solubilization and delivery of Amphotericin B. Colloids Surf. B Biointerfaces **2010**, *81* (1), 313–320.

22. Lavasanifar, A.; Samuel, J.; Sattari, S.; Kwon, G.S. Block copolymer micelles for the encapsulation and delivery of amphotericin B. Pharm. Res. **2002**, *19* (4),418–422.

23. Kataoka, K.; Matsumoto, T.; Yokoyama, M.; Okano, T.; Sakurai, Y.; Fukushima, S. Doxorubicin-loaded poly (ethylene glycol)–poly (β-benzyl-l-aspartate) copolymer

Nanomaterials— Nanoparticles

micelles: Their pharmaceutical characteristics and biological significance. J. Control. Release **2000**, *64* (1–3), 143–151.

24. Kwon, G.; Suwa, S.; Yokoyama, M.; Okano, T.; Sakura, Y.; Kataoka, K. Enhanced tumor accumulation and prolonged circulation times of micelle-forming poly (ethylene oxide-aspartate) block copolymer-adriamycin conjugates. J. Control. Release **1994**, *29* (1–2), 17–23.

25. Kwon, G.S.; Naito, M.; Yokoyama, M.; Okano, T.; Sakurai, Y.; Kataoka, K. Physical entrapment of adriamycin in AB block copolymer micelles. Pharmacol. Res. **1995**, *12* (2), 192–195.

26. Levchenco, T.S.; Mountrichas, G.; Torchilin, V.P.; Pispas, S. *Proceedings of the International Conference on Nanomedicine*; Chalkidiki, Greece, September 9-11, 2007, pp.160–163.

27. Gardikis, K.; Dimas, K.; Georgopoulos, A.; Kaditi, E.; Pispas, S.; Demetzos, C. Lactam functionalized poly (isoprene-b-ethylene oxide) amphiphilic block copolymer micelles as a new nanocarrier system for curcumin. Curr. Nanosci. **2010**, *6* (3), 277–284.

28. Li, Y.; Ikeda, S.; Nakashima, K.; Nakamura, H. Nanoaggregate formation of poly (ethylene oxide)-b-polymethacrylate copolymer induced by cationic anesthetics binding. Colloid Polym. Sci. **2003**, *281* (6), 562–568.

29. Bastakoti, B.P.; Guragain, S.; Yoneda, A.; Yokoyama, Y.; Yusab, S.; Nakashima, K. Micelle formation of poly (ethylene oxide-b-sodium 2-(acrylamido)-2-methyl-1-propane sulfonate-b-styrene) and its interaction with dodecyl trimethyl ammonium chloride and dibucaine. Polym. Chem. **2010**, *1* (3), 347–353.

30. Soliman, G.M.; Winnik, F.M. Enhancement of hydrophilic drug loading and release characteristics through micellization with new carboxymethyldextran-PEG block copolymers of tunable charge density. Int. J. Pharmacol. **2008**, *356* (1), 248–258.

31. Sant, V.P.; Smith, D.; Leroux, J.C. Novel pH-sensitive supramolecular assemblies for oral delivery of poorly water soluble drugs: preparation and characterization. J. Control. Release **2004**, *97* (2), 301–312.

32. Harada, Y.; Yamamoto, T.; Sakai, M.; Saiki, T.; Kawano, K.; Maitani, Y.; Yokoyama, M. Effects of organic solvents on drug incorporation into polymeric carriers and morphological analyses of drug-incorporated polymeric micelles. Int. J. Pharm. **2011**, *404* (1–2), 271–280.

33. Zhang, X.; Cheng, J.; Wang, Q.; Zhong, Z.; Zhuo, R. Miktoarm copolymers bearing one poly (ethylene glycol) chain and several poly (ε-caprolactone) chains on a hyperbranched polyglycerol core. Macromolecules **2010**, *43* (16), 6671–6677.

34. Lavasanifar, A.; Samuel, J.; Kwon, G.S. The effect of fatty acid substitution on the *in vitro* release of amphotericin B from micelles composed of poly (ethylene oxide)-block-poly (N-hexyl stearate-L-aspartamide). J. Control. Release **2002**, *79* (1–3), 165–172.

35. Shahin, M.; Lavasanifar, A. Novel self-associating poly (ethylene oxide)-poly (ε-caprolactone) based drug conjugates and nano-containers for paclitaxel delivery. Int. J. Pharm. **2010**, *389* (1), 213–222.

36. Alani, A.W.; Bae, Y.; Rao, D.A.; Kwon, G.S. Polymeric micelles for the pH-dependent controlled, continuous

low dose release of paclitaxel. Biomaterials **2010**, *3* (7), 1765–1772.

37. Gillies, E.R.; Frechet, J.M. J. pH-Responsive copolymer assemblies for controlled release of doxorubicin. Bioconjug. Chem. **2005**, *16* (2), 361–366.

38. Lee, E.S.; Shin, H.J.; Na, K.; Bae, Y.H. Poly (L-histidine)-PEG block copolymer micelles and pH-induced destabilization. J. Control. Release **2003**, *90* (3), 363–368.

39. Lee, E.S.; Na, K.; Bae, Y.H. Super pH-sensitive multifunctional polymeric micelle. Nano Lett. **2005**, *5* (2), 325–329.

40. Johnson, R.P.; Jeong, Y.-I. Choi, E.; Chung, C.-W.; Kang, D.H.; Oh, S.-O.; Suh, H.; Kim, I. biocompatible poly (2-hydroxyethyl methacrylate)-b-poly (l-histidine) hybrid materials for ph-sensitive intracellular anticancer drug delivery. Adv. Funct. Mater. **2012**, *22* (5), 1058–1068.

41. Park, E.K.; Kim, S.Y.; Lee, S.B.; Lee, Y.M. Folate-conjugated methoxy poly (ethylene glycol)/poly (ε-caprolactone) amphiphilic block copolymeric micelles for tumor-targeted drug delivery. J. Control. Release **2005**, *109* (1–3), 158–168.

42. Yoo, H.S, Park, T.G. Folate-receptor-targeted delivery of doxorubicin nano-aggregates stabilized by doxorubicin–PEG–folate conjugate J. Control. Release **2004**, *100* (2), 247–256.

43. Sulistio, A.; Lowenthal, J.; Blencowe, A.; Bongiovanni, M.N.; Ong, L.; Gras, S.L.; Zhang, X.; Qiao, G.G. Folic acid conjugated amino acid-based star polymers for active targeting of cancer cells. Biomacromolecules **2011**, *12* (10), 3469–3477.

44. Licciardi, M.; Giammona, G.; Du, J.; Armes, S.P.; Tang, Y.; Lewis, A.L. New folate-functionalized biocompatible block copolymer micelles as potential anti-cancer drug delivery systems. Polymer **2006**, *47* (9), 2946–2955.

45. Zhao, H.; Duong, H.H.P.; Yung, L.Y.L. Folate-conjugated polymer micelles with pH-triggered drug release properties. Macromol. Rapid Commun. **2010**, *31*(13), 1163–1169.

46. Nagasaki, Y.; Yasugi, K.; Yamamoto, Y.; Harada, A.; Kataoka, K. Sugar-installed block copolymer micelles: Their preparation and specific interaction with lectin molecules. Biomacromolecules **2001**, *2* (4), 1067–1070.

47. Jule, E.; Nagasaki, Y.; Kataoka, K. Lactose-installed poly (ethylene glycol)-poly (d,l-lactide) block copolymer micelles exhibit fast-rate binding and high affinity toward a protein bed simulating a cell surface. A surface plasmon resonance study. Bioconjug. Chem. **2003**, *14* (1), 177–186.

48. Wu, D.Q.; Lu, B.; Chang, C.; Chen, C.S.; Wang, T.; Zhang, Y.Y. Galactosylated fluorescent labeled micelles as a liver targeting drug carrier. Biomaterials **2009**, *30* (7), 1363–1369.

49. Nasongkla, N.; Shuai, X.; Ai, H.; Weinberg, B.D.; Pink, J.; Boothman, D.A.; Gao, J. M. cRGD-Functionalized polymer micelles for targeted doxorubicin delivery. Angew. Chem. Int. Ed. **2004**, *43* (46), 6323–6327.

50. Xiong, X.B.; Mahmud, A.; Uludag, H.; Lavasanifar, A. Conjugation of arginine-glycine-aspartic acid peptides to poly (ethylene oxide)-b-poly (ε-caprolactone) micelles for enhanced intracellular drug delivery to metastatic tumor cells. Biomacromolecules **2007**, *8* (3), 874–884.

51. Xiong, X.B.; Mahmud, A.; Uludag, H.; Lavasanifar, A. Multifunctional polymeric micelles for enhanced intracellular delivery of doxorubicin to metastatic cancer cells. Pharm. Res. **2008**, *25* (11), 2555–2566.

52. Jiang, X.; Sha, X.; Xin, H.; Chen, L.; Gao, X.; Wang, X.; Law, K.; Gu, J.; Chen, Y.; Jiang, Y.; Ren, X.; Ren, Q.; Fang, X. Goblet cell-targeting nanoparticles for oral insulin delivery and the influence of mucus on insulin transport. Biomaterials **2011**, *32* (35), 9457–9469.

53. Vinogradov, S.; Batrakova, E.; Li, S.; Kabanov, A. Polyion complex micelles with protein-modified corona for receptor-mediated delivery of oligonucleotides into cells. Bioconjug. Chem. **1999**, *10* (5), 851–860.

54. Kabanov, A.; Zhu, J.; Alakhov, V. Pluronic block copolymers for gene delivery. Adv. Genet. **2005**, *53*, 231–261.

55. Vega, J.; Ke, S.; Fan, Z.; Wallace, S.; Charsangavej, C.; Li, C. Targeting doxorubicin to epidermal growth factor receptors by site-specific conjugation of C225 to poly (L-glutamic acid) through a polyethylene glycol spacer. Pharm. Res. **2003**, *20* (5), 826–832.

56. Talelli, M.; Rijcken, C.J.F.; Oliveira, S.; van der Meel, R.; van Bergen en Henegouwen, P.M.P.; Lammers, T.; van Nostrum, C.F.; Storm, G.; Hennink, W.E.Nanobody— Shell functionalized thermosensitive core-crosslinked polymeric micelles for active drug targeting. J. Control. Release **2011**, *151* (2), 183–192.

57. Farokhzad, O.C.; Jon, S.; Khademhosseini, A.; Tran, T.N.; Lavan, D.A, Langer, R. Nanoparticle-aptamer bioconjugates: A new approach for targeting prostate cancer cells. Cancer Res. **2004**, *64* (21), 7668–7675.

58. Liu, S.Q.; Wiradharma, N.; Gao, S.J.; Tong, Y.W.; Yang, Y.Y. Bio-functional micelles self-assembled from a folate-conjugated block copolymer for targeted intracellular delivery of anticancer drugs. Biomaterials **2007**, *28* (7), 1423–1428.

59. Tu, S.; Chen, Y.-W.; Qiu, Y.-B.; Zhu, K.; Luo, X.-L. Enhancement of cellular uptake and antitumor efficiencies of micelles with phosphorylcholine. Macromol. Biosci. **2011**, *11* (10), 1416–1425.

60. Xu, Y.-C.; Dong, C.-M. Dendron-like poly (ε-benzyloxycarbonyl-L-lysine)/linear PEO block copolymers: Synthesis, physical characterization, self-assembly, and drug-release behavior. J. Polym. Sci. A Polym. Chem. **2012**, *50* (6), 1216–1225.

61. Sharma, A.; Soliman, G.M.; Al-Hajaj, N.; Sharma, R.; Maysinger, D.; Kakkar, A. Design and evaluation of multifunctional nanocarriers for selective delivery of coenzyme Q10 to mitochondria. Biomacromolecules **2012**, *13* (1), 239–252.

62. Lin, L.Y.; Lee, N.S.; Zhu, J.; Nystrom, A.M.; Pochan, D.J.; Dorshow, R.B.; Wooley, K.L. Tuning core vs. shell dimensions to adjust the performance of nanoscopic containers for the loading and release of doxorubicin. J. Control. Release **2011**, *152* (1), 37–48.

63. Chytil, P.; Etrych, T.; Kostka, L.; Ulbrich, K. Hydrolytically degradable polymer micelles for anticancer drug delivery to solid tumors. Macromol. Chem. Phys.**2012**, *213* (8), 858–867.

64. Soppimath, K.S.; Tan, D.C.W.; Yang, Y. Y. pH-Triggered Thermally Responsive Polymer Core–Shell Nanoparticles for Drug Delivery. Adv. Mater. **2005**, *17*(3), 318–323.

65. Soppimath, K.S.; Liu, L.H.; Seow, W.Y.; Liu, S.Q.; Chan, P.; Yang, Y.Y. Multifunctional core/shell nanoparticles self-assembled from pH-induced thermosensitive polymers for targeted intracellular anticancer drug delivery. Adv. Funct. Mater. **2007**, *17* (3), 355–361.

66. Xu, H.; Meng, F.; Zhong, Z. Reversibly crosslinked temperature-responsive nano-sized polymersomes: synthesis and triggered drug release. J. Mater. Chem.**2009**, *19* (24), 4183–4190.

67. Wang, Y.C.; Li, Y.; Sun, T.M.; Xiong, M.H.; Wu, J.; Yang, Y.Y.; Wang, J. Core-shell-corona micelle stabilized by reversible cross-linkage for intracellular drug delivery. Macromol. Rapid Commun. **2010**, *31* (13), 1201–1206.

68. Tan, J.P.K.; Kim, S.H.; Nederberg, F.; Appel, E.A.; Waymouth, R.M.; Zhang, Y.; Hedrick, J.L.; Yang, Y.Y. Hierarchical supermolecular structures for sustained drug release. Small **2009**, *5* (13), 1504–1507.

69. Jooss, K.; Yang, Y.; Fisher, K.J.; Wilson, J.M. Transduction of dendritic cells by DNA viral vectors directs the immune response to transgene products in muscle fibers. J. Virol. **1998**, *72*, 4212–4223.

70. Marshall, E. Clinical Trials: Gene therapy death prompts review of adenovirus vector. Science **1999**, *286* (5448), 2244–2245.

71. Boyce, N. Trial halted after gene shows up in semen. Nature **2001**, *414* (6865), 677–677.

72. Check, E. Gene therapy: A tragic setback. Nature **2002**, *420* (6912), 116–118.

73. Wong, S.Y.; Pelet, J.M.; Putnam, D. Polymer systems for gene delivery—Past, present, and future. Prog. Polym. Sci. **2007**, *32* (8–9), 799–837.

74. Schaffer, D.V.; Lauffenburger, D.A. Optimization of cell surface binding enhances efficiency and specificity of molecular conjugate gene delivery. J. Biol. Chem.**1998**, *273* (43), 28004–28009.

75. Lechardeur, D.; Sohn, K.J.; Haardt, M.; Joshi, P.B.; Monck, M.; Graham, R.W.; Beatty, B.; Squire, J.; O'Brodovich, H.; Lukacs, G.L. Metabolic instability of plasmid DNA in the cytosol: A potential barrier to gene transfer. Gene Ther. **1999**, *6* (4), 482–497.

76. Abdelhady, H.G.; Allen, S.; Davies, M.C.; Roberts, C.J.; Tendler, S.J.; Williams, P.M. Direct real-time molecular scale visualisation of the degradation of condensed DNA complexes exposed to D Nase I. Nucleic Acids Res. **2003**, *31* (14), 4001–4005.

77. Santos, C.A.; Freedman, B.D.; Leach, K.J.; Press, D.L.; Scarpulla, M.; Mathiowitz, E. Poly (fumaric–co-sebacic anhydride): A degradation study as evaluated by FTIR, DSC, GPC and X-ray diffraction. J. Control. Release **1999**, *60* (1), 11–22.

78. Panyam, J.; Labhasetwar, V. Biodegradable nanoparticles for drug and gene delivery to cells and tissue. Adv. Drug Deliv. Rev. **2003**, *55* (3), 329–347.

79. Hedley, M. Gene delivery using poly (lactide-co-glycolide) microspheres. In Polymeric Gene Delivery: Principles and Applications. Amiji, M. M., Ed.; CRC Press: Boca Raton, FL, 2005; pp. 451–466.

80. Jang, J.H.; Shea, L.D. Intramuscular delivery of DNA releasing microspheres: Microsphere properties and transgene expression. J. Control. Release **2006**, *112*(1), 120–128.

Nanomaterials— Nanoparticles

81. Ando, S.; Putnam, D.; Pack, D.W.; Langer, R. PLGA microspheres containing plasmid DNA: Preservation of supercoiled DNA via cryopreparation and carbohydrate stabilization. J. Pharm. Sci. **1999**, *88* (1), 126–130.

82. Walter, E.; Moelling, K.; Pavlovic, J.; Merkle, H.P. Micro-encapsulation of DNA using poly (dl-lactide-co-glycolide): stability issues and release characteristics. J. Control. Release **1999**, *61* (3), 361–374.

83. Wang, D.; Robinson, D.R.; Kwon, G.S.; Samuel, J. Encapsulation of plasmid DNA in biodegradable poly (d,l-lactic-co-glycolic acid) microspheres as a novel approach for immunogene delivery. J. Control. Release **1999**, *57* (1), 9–18.

84. Fu, K.; Pack, D.W.; Klibanov, A.M.; Langer, R. Visual evidence of acidic environment within degrading poly (lactic-co-glycolic acid) (PLGA) microspheres. Pharm. Res. **2000**, *17* (1), 100–106.

85. Kasturi, S.P.; Sachaphibulkij, K.; Roy, K. Covalent conjugation of polyethyleneimine on biodegradable microparticles for delivery of plasmid DNA vaccines. Biomaterials **2005**, *26* (32), 6375–6385.

86. Munier, S.; Messai, I.; Delair, T.; Verrier, B.; Ataman-Onal, Y. Cationic PLA nanoparticles for DNA delivery: Comparison of three surface polycations for DNA binding, protection and transfection properties. Colloids Surf B Biointerfaces **2005**, *43* (3–4), 163–173.

87. O'Hagan, D.T.; Singh, M.; Ulmer, J.B. Microparticle-based technologies for vaccines. Methods **2006**, *40* (1), 10–19.

88. Conner, S.; Schmid, S. Regulated portals of entry into the cell. Nature **2003**, *422* (6927), 37–44.

89. Wang, C.; Ge, Q.; Ting, D.; Nguyen, D.; Shen, H.R.; Chen, J.; Eisen, H.N.; Heller, J.; Langer, R.; Putnam, D. Molecularly engineered poly (ortho ester) microspheres for enhanced delivery of DNA vaccines. Nat. Mater. **2004**, *3* (3), 190–196.

90. Hardingham, T.E.; Fosang, A.J.Proteoglycans: Many forms and many functions. FASEB J. **1992**, *6* (3), 861–870.

91. Yanagishita, M.; Hascall, V.C. Cell surface heparan sulfate proteoglycans. J. Biol. Chem. **1992**, *267*(14), 9451–9454.

92. Mislick, K.A.; Baldeschwieler, J.D. Evidence for the role of proteoglycans in cation-mediated gene transfer. Proc. Natl. Acad. Sci. USA **1996**, *93* (22),12349–12354.

93. Kichler, A.; Mason, A.J.; Bechinger, B. Cationic amphipathic histidine-rich peptides for gene delivery. Biochim. Biophys. Acta **2006**, *1758* (3), 301–307.

94. Takigawa, D.Y.; Tirrell, D.A. Interactions of synthetic polymers with cell membranes and model membrane systems. Part 6. Disruption of phospholipid packing by branched poly (ethylenimine) derivatives. Macromolecules **1985**, *18* (3), 338–342.

95. Thomas, M.; Klibanov, A.M. Enhancing polyethylenimine's delivery of plasmid DNA into mammalian cells. Proc. Natl. Acad. Sci. USA **2002**, *99* (23),14640–14645.

96. El-Andaloussi, S.; Holm, T.; Langel, U. Cell-penetrating peptides: Mechanisms and applications. Curr. Pharm. Des. **2005**, *11* (28), 3597–3611.

97. Khalil, I.; Kogure, K.; Akita, H.; Harashima, H. Uptake pathways and subsequent intracellular trafficking in non-viral gene delivery. Pharmacol. Rev. **2006**, *58* (1), 32–45.

98. Rejman, J.; Bragonzi, A.; Conese, M. Role of clathrin-and caveolae-mediated endocytosis in gene transfer mediated by lipo-and polyplexes. Mol. Ther. **2005**, *12* (3), 468–474.

99. von Gersdorff, K.; Sanders, N.; Vandenbroucke, R.; De Smedt, S.; Wagner, E.; Ogris, M. The internalization route resulting in successful gene expression depends on both cell line and polyethylenimine polyplex type. Mol. Ther. **2006**, *14* (5), 745–753.

100. Lukacs, G.L.; Haggie, P.; Seksek, O.; Lechardeur, D.; Freedman, N.; Verkman, A.S. Size-dependent DNA mobility in cytoplasm and nucleus. J. Biol. Chem. **2000**, *275* (3), 1625–1629.

101. Dauty, E.; Verkman, A.S. Actin Cytoskeleton as the Principal Determinant of Size-dependent DNA Mobility in Cytoplasm a new barrier for non-viral gene delivery. J. Biol. Chem. **2005**, *280* (9), 7823–7828.

102. Pante, N.; Kann, M. Nuclear pore complex is able to transport macromolecules with diameters of 39 nm. Mol. Biol. Cell **2002**, *13* (2), 425–434.

103. Kichler, A.; Pages, J.C.; Leborgne, C.; Druillennec, S.; Lenoir, C.; Coulaud, D.; Delain, E.; Le Cam, E.; Roques, B.P.; Danos, O. Efficient DNA transfection mediated by the C-terminal domain of human immunodeficiency virus type 1 viral protein R. J. Virol. **2000**, *74* (12), 5424–5431.

104. Moffatt, S.; Wiehle, S.; Cristiano, R.J. A multifunctional PEI-based cationic polyplex for enhanced systemic p53-mediated gene therapy. Gene Ther. **2006**, *13*(21), 1512–1523.

105. Talsma, S.S.; Babensee, J.E.; Murthy, N.; Williams, I.R. Development and *in vitro* validation of a targeted delivery vehicle for DNA vaccines. J. Control. Release **2006**, *112* (2), 271–279.

106. Bremner, K.H.; Seymour, L.W.; Logan, A.; Read, M.L. Factors influencing the ability of nuclear localization sequence peptides to enhance nonviral gene delivery. Bioconjug. Chem. **2004**, *15* (1), 152–161.

107. Bergen, J.M.; Pun, S.H. Peptide-enhanced nucleic acid delivery. MRS Bull. **2005**, *30* (9), 663–667.

108. Haberland, A.; Bottger, M. Nuclear proteins as gene-transfer vectors. Biotechnol. Appl. Biochem. **2005**, *42* (Pt 2), 97–106.

109. Kaouass, M.; Beaulieu, R.; Balicki, D.Histonefection: Novel and potent non-viral gene delivery. J. Control. Release **2006**, *113* (3), 245–254.

110. Klink, D.T.; Chao, S.; Glick, M.C.; Scanlin, T.F. Nuclear translocation of lactosylated poly-L-lysine/cDNA complex in cystic fibrosis airway epithelial cells. Mol. Ther. **2001**, *3* (6), 831–841.

111. Monsigny, M.; Rondanino, C.; Duverger, E.; Fajac, I.; Roche, A.C. Glyco-dependent nuclear import of glycoproteins, glycoplexes and glycosylated plasmids. Biochim. Biophys. Acta **2004**, *1673* (1–2), 94–103.

112. Grosse, S.; Thevenot, G.; Monsigny, M.; Fajac, I. Which mechanism for nuclear import of plasmid DNA complexed with polyethylenimine derivatives? J. Gene Med. **2006**, *8* (7), 845–851.

113. Brunner, S.; Sauer, T.; Carotta, S.; Cotton, M.; Saltik, M.; Wagner, E. Cell cycle dependence of gene transfer by lipoplex, polyplex and recombinant adenovirus. Gene Ther. **2000**, *7* (5), 401–407.

114. Pack, D.W.; Hoffman, A.S.; Pun, S.; Stayton, P.S. Design and development of polymers for gene delivery. Nat. Rev. Drug Discov. **2005**, *4* (7), 581–593.

115. Parker, A.L.; Eckley, L.; Singh, S.; Preece, J.A.; Collins, L.; Fabre, J.W. (LYS)16-based reducible polycations provide stable polyplexes with anionic fusogenic peptides and efficient gene delivery to post mitotic cells. Biochim. Biophys. Acta **2007**, *1770* (9), 1331–1337.

116. Yant, S.R.; Meuse, L.; Chiu, W.; Ivics, Z.; Izsvak, Z.; Kay, M.A. Somatic integration and long-term transgene expression in normal and haemophilic mice using a DNA transposon system. Nat. Genet. **2000**, *25* (1), 35–41.

117. O'Rorke, S.; Keeney, M.; Pandit, A. Polyphosphonium Polymers for siRNA Delivery: An Efficient and Nontoxic Alternative to Polyammonium Carriers. Prog. Polym. Sci. **2010**, *35* (4), 441–458.

118. Gebhart, C.L.; Kabanov, A.V. Evaluation of polyplexes as gene transfer agents. J. Control. Release **2001**, *73* (2–3), 401–406.

119. Mintzer, M.A.; Simanek, E.E. Nonviral vectors for gene delivery. Chem. Rev. **2009**, *109* (2), 259–302.

120. Tamboli, V.; Mishra, G.P.; Mitra, A.K. Polymeric vectors for ocular gene delivery. Ther. Deliv. **2011**, *2* (4), 523–536.

121. Kesharwani, P.; Gajbhiye, V.; Jain, N.K. A review of nanocarriers for the delivery of small interfering RNA. Biomaterials **2012**, *33* (29), 7138–7150.

122. Mastrobattista, E.; Hennink, W.E. Polymers for gene delivery: Charged for success. Nat. Mater. **2012**, *11* (1), 10–12.

123. Zheng, M.; Zhong, Y.; Meng, F.; Peng, R.; Zhong, Z. Lipoic acid modified low molecular weight polyethylenimine mediates nontoxic and highly potent *in vitro* gene transfection. Mol. Pharm. **2011**, *8* (6), 2432–2443.

124. Bauhuber, S.; Lieble, R.; Tomasetti, L.; Rachel, R.; Goepferich, A.; Breunig, M. A library of strictly linear poly (ethylene glycol)–poly (ethylene imine) diblock copolymers to perform structure–function relationship of non-viral gene carriers. J. Control. Release **2012**, *162* (2), 446–455.

125. Varkouhi, A.K.; Mountrichas, G.; Schiffelers, R.M.; Lammers, T.; Storm, G.; Pispas, S.; Hennink, W.E. Polyplexes based on cationic polymers with strong nucleic acid binding properties. Eur. J. Pharm. Sci. **2012**, *45* (4), 459–466.

126. Velasco, D.; Collin, E.; San Roman, J.; Pandit, A.; Elvira, C. End functionalized polymeric system derived from pyrrolidine provide high transfection efficiency. Eur. J. Pharm. Biopharm. **2011**, *79* (3), 485–494.

127. Hemp, S.T.; Allen Jr., M.H., Green, M.D.; Long, T.E. Phosphonium-containing polyelectrolytes for nonviral gene delivery. Biomacromolecules **2012**, *13* (1), 231–238.

128. Ornelas-Megiatto, C.; Wich, P.R.; Frechet, J.M.J.J. Polyphosphonium polymers for siRNA delivery: An efficient and nontoxic alternative to polyammonium carriers. Am. Chem. Soc. **2012**, *134* (4), 1902–1905.

129. Hemp, S.T.; Smith, A.E.; Bryson, J.M.; Allen, M.H.; Long, T.E. Phosphonium-containing diblock copolymers for enhanced colloidal stability and efficient nucleic acid delivery. Biomacromolecules **2012**, *13* (8), 2439–2445.

130. Convertine, A.J.; Diab, C.; Prieve, M.; Paschal, A.; Hoffman, A.S.; Johnson, P.H.; Stayton, P. S. pH-responsive polymeric micelle carriers for siRNA drugs. Biomacromolecules **2010**, *11* (11), 2904–2911.

131. Gao, H.; Lu, X.; Ma, Y.; Yang, Y.; Li, J.; Wu, G.; Wang, Y.; Fan, Y.; Ma, J. Amino poly (glycerol methacrylate)s for oligonucleic acid delivery with enhanced transfection efficiency and low cytotoxicity. Soft Matter **2011**, *7* (19), 9239–9247.

132. Cheng, C.; Convertine, A.J.; Stayton, P.S.; Bryers, J.D. Multifunctional triblock copolymers for intracellular messenger RNA delivery. Biomaterials **2012**, *33*(28), 6868–6876.

133. Islam, M.A.; Shin, J.-Y.; Firdous, J.; Park, T.-E.; Choi, Y.-J.; Cho, M.-H.; Yun, C.-H.; Cho, C.-S. The role of osmotic polysorbitol-based transporter in RNAi silencing via caveolae-mediated endocytosis and COX-2 expression. Biomaterials **2012**, *33* (34), 8868–8880.

134. Palumbo, R.N.; Zhong, X.; Panus, D.; Han, W.; Ji, W.; Wang, C. Transgene expression and local tissue distribution of naked and polymer-condensed plasmid DNA after intradermal administration in mice. J. Control. Release **2012**, *159* (2), 232–239.

135. Ji, W.; Panus, D.; Palumbo, R.N.; Tang, R.; Wang, C. Poly (2-aminoethyl methacrylate) with Well-Defined Chain Length for DNA Vaccine Delivery to Dendritic Cells. Biomacromolecules **2011**, *12* (12), 4373–4385.

136. Osada, K.; Shiotani, T.; Tockary, T.A.; Kobayashi, D.; Oshima, H.; Ikeda, S.; Christie, R.J.; Itaka, K.; Kataoka, K.Enhanced gene expression promoted by the quantized folding of pDNA within polyplex micelles. Biomaterials **2012**, *33* (1), 325–332.

137. Liu, Z.; Zheng, M.; Meng, F.; Zhong, Z. Non-viral gene transfection *in vitro* using endosomal pH-sensitive reversibly hydrophobilized polyethylenimine. Biomaterials **2011**, *32* (34), 9109–9119.

138. Cho, H.Y.; Srinivasan, A.; Hong, J.; Hsu, E.; Liu, S.; Arun Shrivats, A.; Kwak, D.K.; Bohaty, A.K.; Paik, H.; Jeffrey, O.Hollinger, J.O.; Matyjaszewski, K. Synthesis of biocompatible PEG-based star polymers with cationic and degradable core for siRNA delivery. Biomacromolecules **2011**, *12* (10), 3478–3486.

139. Patil, M.L.; Zhang, M.; Minko, T. Multifunctional triblock nanocarrier (PAMAM-PEG-PLL) for the efficient intracellular siRNA delivery and gene silencing. ACS Nano **2011**, *5* (3), 1877–1887.

140. Parelkar, S.S.; Chan-Seng, D.; Emrick, T. Reconfiguring polylysine architectures for controlling polyplex binding and non-viral transfection. Biomaterials **2011**,*32* (9), 2432–2444.

141. Jeong, Y.; Jin, G.-W.; Choi, E.; Jung, J.H.; Park, J.-S. Effect of deoxycholate conjugation on stability of pDNA/poly-amidoamine-diethylentriamine (PAM-DET) polyplex against ionic strength. Int. J. Pharm. **2011**, *420* (2), 366–370.

142. Zhu, Y.; Sheng, R.; Luo, T.; Li, H.; Sun, W.; Li, Y.; Cao, A. Amphiphilic cationic [Dendritic poly (L-lysine)]-block-poly (L-lactide)-block-[dendritic poly (L-lysine)]s in aqueous solution: self-aggregation and interaction with DNA as gene delivery carriers. Macromol. Biosci. **2011**, *11* (2), 147–186.

143. Zhang, G.; Liu, J.; Yang, Q.; Zhuo, R.; Jiang, X. Disulfide-containing brushed polyethylenimine derivative synthesized by click chemistry for nonviral gene delivery. Bioconjug. Chem. **2012**, *23* (6), 1290–1299.

144. Zheng, M.; Liu, Y.; Samsonova, O.; Endres, T.Merkel, O.; Kissel, T. Amphiphilic and biodegradable hy-PEI-g-PCL-b-PEG copolymers efficiently mediate transgene

expression depending on their graft density. Int. J. Pharm. **2012**, *427*(1), 80–87.

145. Liu, Y.; Samsonova, O.; Sproat, B.; Merkel, O.; Kissel, T. Biophysical characterization of hyper-branched polyethylenimine-graft-polycaprolactone-block-mono-methoxyl-poly (ethylene glycol) copolymers (hy-PEI-PCL-mPEG) for siRNA delivery. J. Control. Release **2011**, *153* (3), 262–268.

146. Zhong, Z.; Zheng, M.; Zhong, Z.; Zhong, Z.; Zhou, L.; Meng, F.; Peng, R. Poly (ethylene oxide) grafted with short polyethylenimine gives dna polyplexes with superior colloidal stability, low cytotoxicity, and potent *in vitro* gene transfection under serum conditions. Biomacromolecules **2012**, *13* (3), 881–888.

147. Liang, W.; Gong, H.; Yin, D.; Lu, S.; Fu, Q. High-molecular-weight polyethyleneimine conjuncted pluronic for gene transfer agents. Chem. Pharm. Bull. **2011**, *59* (9), 1094–1101.

148. Piest, M.; Engbersen, J.F.J. Role of boronic acid moieties in poly (amido amine) s for gene delivery. J. Control. Release **2011**, *155* (2), 331–340.

149. Lin, C.; Engbersen, J.F.J. PEGylated bioreducible poly (amido amine)s for non-viral gene delivery. Mater. Sci. Eng. C **2011**, *31* (7), 1330–1337.

150. Vercauteren, D.; Piest, M.; van der Aa, L.J.; Al Soraj, M.; Jones, A.T.; Engbersen, J.F.J.; De Smedt, S.C.; Braeckmans, K. Flotillin-dependent endocytosis and a phagocytosis-like mechanism for cellular internalization of disulfide-based poly (amido amine)/DNA polyplexes. Biomaterials **2011**, *32* (11), 3072–3084.

151. Kurisawa, M.; Yokoyama, M.; Okano, T. Gene expression control by temperature with thermo-responsive polymeric gene carriers. J. Control. Release **2000**, *69* (1), 127–137.

152. Sun, S.J.; Liu, W.G.; Cheng, N.; Zhang, B.Q.; Cao, Z.Q.; Yao, K.D.; Liang, D.C.; Zuo, A.J.; Guo, G.; Zhang, J.Y. A thermoresponsive chitosan-NIPAAm/vinyl laurate copolymer vector for gene transfection. Bioconj. Chem. **2005**, *16* (4), 972–980.

153. Lavigne, M.D.; Pennadam, S.S.; Ellis, J.; Alexander, C.; Gorecki, D.C. Enhanced gene expression through temperature profile-induced variations in molecular architecture of thermoresponsive polymer vectors. J. Gene Med. **2007**, *9* (1), 44–54.

154. Turk, M.; Dincer, S.; Piskin, E. Smart and cationic poly (NIPA)/PEI block copolymers as non-viral vectors: *in vitro* and *in vivo* transfection studies. J. Tissue Eng. Regen. Med. **2007**, *1* (5), 377–388.

155. Bisht, H.S.; Manickam, D.S.; You, Y.Z.; Oupicky, D. Temperature-controlled properties of DNA complexes with poly(ethylenimine)-graft-poly(N-isopropylacrylamide) Biomacromolecules **2006**, *7* (4), 1169–1178.

156. Zintchenko, A.; Ogris, M.; Wagner, E. Temperature dependent gene expression induced by PNIPAM-based copolymers, potential of hyperthermia in gene transfer. Bioconj. Chem. **2006**, *17* (3), 766–772.

157. Meyer, D.E.; Chilkoti, A. Purification of recombinant proteins by fusion with thermally-responsive polypeptides. Nat. Biotechnol. **1999**, *17* (11), 1112–1115.

158. Urry, D.W. Physical chemistry of biological free energy transduction as demonstrated by elastic protein-based polymers. J. Phys. Chem. B **1997**, *101* (51), 11007–11028.

159. Chen, T.-H., Bae, Y.; Furgeson, D.Y.; Kwon, G.S. Biodegradable hybrid recombinant block copolymers for non-viral gene transfection. Int. J. Pharm. **2012**, *427* (1), 105–112.

160. Shen, Z.; Shi, B.; Zhang, H.; Bi, J.; Dai, S. Exploring low-positively charged thermosensitive copolymers as gene delivery vectors. Soft Matter **2012**, *8* (5),1385–1394.

161. Zhang, R.; Wang, Y.; Du, F.-S.; Wang, Y.-L.; Tan, Y.-X.; Ji, S.-P.; Li, Z.-C. Thermoresponsive gene carriers based on polyethylenimine-graft-poly[oligo (ethylene glycol) methacrylate]. Macromol. Biosci. **2011**, *11* (10), 1393–1396.

162. Lai, T.C.; Kataoka, K.; Kwon, G.S. Bioreducible polyether-based pDNA ternary polyplexes: Balancing particle stability and transfection efficiency. Colloids Surf. B Biointerfaces **2012**, *99*, 27–37.

163. Li, W.; Wang, Y., Chen, L.; Huang, Z.; Hu, Q.; Ji, J. Light-regulated host–guest interaction as a new strategy for intracellular PEG-detachable polyplexes to facilitate nuclear entry. Chem. Commun. **2012**, *48* (81), 10126–10128.

164. Zheng, S.W.; Liu, G.; Hong, R.Y.; Li, H.Z.; Li, Y.G.; Wei, D.G. Preparation and characterization of magnetic gene vectors for targeting gene delivery. Appl. Surf. Sci. **2012**, *259*, 201–207.

165. Sun, S.-L., Lo, Y.-L., Chen, H.-Y., Wang, L.-F. Hybrid polyethylenimine and polyacrylic acid-bound iron oxide as a magnetoplex for gene delivery. Langmuir **2012**, *28* (7), 3542–3552.

166. Han, L.; Zhang, A.; Wang, H.; Pu, P.; Kang, C.; Chang, J. Construction of novel brain-targeting gene delivery system by natural magnetic nanoparticles. J. Appl. Polym. Sci. **2011**, *121* (6), 3446–3454.

167. Mahor, S.; Dash, B.C.; O'Connor, S.; Pandit, A. Mannosylated polyethyleneimine–hyaluronan nanohybrids for targeted gene delivery to macrophage-like cell lines. Bioconjug. Chem. **2012**, *23* (6), 1138–1148.

168. Noga, M.; Edinger, D.; Rodl, W.; Wagner, E.; Winter, G.; Besheer, A. Controlled shielding and deshielding of gene delivery polyplexes using hydroxyethyl starch (HES) and alpha-amylase. J. Control. Release **2012**, *159* (1), 92–103.

169. Zhu, C.; Zheng, M.; Meng, F.; Mickler, F.M.; Ruthardt, N.; Zhu, X.; Zhong, Z. Reversibly shielded DNA polyplexes based on bioreducible PDMAEMA-SS-PEG-SS-PDMAEMA triblock copolymers mediate markedly enhanced nonviral gene transfection. Biomacromolecules **2012**, *13* (3), 769–778.

170. Liu, L.; Zheng, M.; Renette, T.; Kissel, T. Modular synthesis of folate conjugated ternary copolymers: Polyethylenimine-graft-Polycaprolactone-block-Poly(ethylene glycol)-folate for targeted gene delivery. Bioconjug. Chem. **2012**, *23* (6), 1211–1220.

171. Schafer, A.; Pahnke, A.; Schaffert, D.; Van Weerden, W.M.; De Ridder, C.M.A.; Rodl, W.; Vetter, A.; Spitzweg, C.; Kraaij, R.; Wagner, E.; Ogris, M.Disconnecting the yin and yang relation of epidermal growth factor receptor (EGFR)-mediated delivery: a fully synthetic, EGFR-targeted gene transfer system avoiding receptor activation. Hum. Gene Therapy **2011**, *22* (12), 1463–1473.

172. Klutz, K.; Willhauck, M.J.; Dohmen, C.; Wunderlich, N.; Knoop, K.; Zach, C.; Senekowitsch-Schmidtke, R.; Gildehaus, F.-J.; Ziegler, S.; Furst, S.; Goke, B.; Wagner, E.;

Ogris, M.; Spitzweg, C. Image-guided tumor-selective radioiodine therapy of liver cancer after systemic nonviral delivery of the sodium iodide symporter gene. Hum. Gene Therapy **2011**, *22* (12), 1563–1574.

173. Sabeqzadeh, E.; Rahbarizadeh, F.; Ahmadvand, D.; Rasaee, M.J.; Parhamifar, L.; Moghimi, S.M.Combined MUC1-specific nanobody-tagged PEG-polyethylenimine polyplex targeting and transcriptional targeting of tBid transgene for directed killing of MUC1 overexpressing tumour cells. J. Control. Release **2011**, *156* (1), 85–91.

174. Son, S.; Hwang, D.W.; Singha, K.; Jeong, J.H.; Park, T.G.; Lee, D.S.; Kim, W.J. RVG peptide tethered bioreducible polyethylenimine for gene delivery to brain. J. Control. Release **2011**, *155* (1), 18–25.

175. Nam, H.Y.; Kim, J.; Kim, S.; Yockman, J.W.; Kim, S.W.; Bull, D.A. Cell penetrating peptide conjugated bioreducible polymer for siRNA delivery. Biomaterials **2011**, *32* (22), 5213–5222.

176. Wu, J.; Yamanouchi, D.; Liu, B.; Chu, C.-C. Biodegradable arginine-based poly (ether ester amide)s as a nonviral DNA delivery vector and their structure–function study. J. Mater. Chem. **2012**, *22* (36), 18983–18991.

177. Aldawsari, H.; Edrada-Ebel, R.; Blatchford, D.R.; Tate, R.J.; Tetley, L.; Dufes, C. Enhanced gene expression in tumors after intravenous administration of arginine-, lysine- and leucine-bearing polypropylenimine polyplex. Biomaterials **2011**, *32* (25), 5889–5899.

178. Anderson, K.; Sizovs, A.; Cortez, M.; Waldron, C.; Haddleton, D.M.; Reineke, T.M. Effects of trehalose polycation end-group functionalization on plasmid DNA uptake and transfection. Biomacromolecules **2012**, *13* (8), 2229–2239.

179. Smith, A.E.; Sizovs, A.; Grandinetti, G.; Xue, L.; Reineke, T.M. Diblock glycopolymers promote colloidal stability of polyplexes and effective pDNA and siRNA delivery under physiological salt and serum conditions. Biomacromolecules **2011**, *12* (8), 3015–3022.

180. Reilly, M.J.; Larsen, J.D.; Sullivan, M.O. Polyplexes traffic through caveolae to the Golgi and endoplasmic reticulum en route to the nucleus. Mol. Pharm. **2012**, *9* (5), 1031–1040.

181. Johnson, R.N.; Chu, D.S.H.; Shi, J.; Schellinger, J.G.; Carlson, P.M.; Pun, S.H. HPMA-oligolysine copolymers for gene delivery: Optimization of peptide length and polymer molecular weight. J. Control. Release **2011**, *155* (2), 303–311.

182. Dash, B.C.; Mahor, S.; Carroll, O.; Mathew, A.; Wang, W.; Woodhouse, K.A.; Pandit, A. Tunable elastin-like polypeptide hollow sphere as a high payload and controlled delivery gene depot. J. Control. Release **2011**, *152* (3), 382–392.

183. Ahmed, M.; Narain, R. The effect of molecular weight, compositions and lectin type on the properties of hyperbranched glycopolymers as non-viral gene delivery systems. Biomaterials **2012**, *33* (15), 3990–4001.

184. Ahmed, M.; Narain, R. The effect of polymer architecture, composition, and molecular weight on the properties of glycopolymer-based non-viral gene delivery systems. Biomaterials **2011**, *32* (22), 5279–5290.

185. Li, W.; Chen, L.; Huang, Z.; Wu, X.; Zhang, Y.; Hu, Q.; Wang, Y. The influence of cyclodextrin modification on cellular uptake and transfection efficiency of polyplexes. Org. Biomol. Chem. **2011**, *9* (22), 7799–7806.

186. Lai, W.-F.; Tang, G.-P.; Wang, X.; Li, G.; Yao, H.; Shen, Z.; Lu, G.; Poon, W.S.; Kung, H.-F.; Lin, M.C.M. Cyclodextrin-PEI-tat polymer as a vector for plasmid DNA delivery to placenta mesenchymal stem cells. Bionanoscience **2011**, *1* (3), 89–96.

187. Ping, Y.; Liu, C.; Zhang, Z.; Liu, K.L.; Chen, J.; Li, J. Chitosan-graft-(PEI-β-cyclodextrin) copolymers and their supramolecular PEGylation for DNA and siRNA delivery. Biomaterials **2011**, *32* (32), 8328–8341.

188. Jiang, Q.; Shi, P.; Li, C.; Wang, Q.; Xu, F.; Yang, W.; Tang, G. (Coixan polysaccharide)-graft-polyethylenimine folate for tumor-targeted gene delivery. Macromol. Biosci. **2011**, *11* (3), 435–444.

189. Arima, H.; Tsutsumi, T.; Yoshimatsu, A.; Ikeda, H.; Motoyama, K.; Higashi, T.; Hirayama, F.; Uekama, K. Inhibitory effect of siRNA complexes with polyamidoamine dendrimer/α-cyclodextrin conjugate (generation 3, G3) on endogenous gene expression. Eur. J. Pharm. Sci. **2011**, *44* (3), 375–384.

190. Wang, C.; Luo, X.; Zhao, Y.; Han, L.; Zeng, X.; Feng, M.; Pan, S.; Wu, C. Influence of the polyanion on the physicochemical properties and biological activities of polyanion/DNA/polycation ternary polyplexes. Acta Biomater. **2012**, *8* (8), 3014–3026.

191. Debus, H.; Baumhof, P.; Probst, J.; Kissel, T. Delivery of messenger RNA using poly (ethylene imine)–poly (ethylene glycol)-copolymer blends for polyplex formation: Biophysical characterization and *in vitro* transfection properties. J. Control. Release **2011**, *148* (3), 334–343.

192. Chen, Q.; Osada, K.; Ishii, T.; Oba, M.; Uchida, S.; Tockary, T.A.; Endo, T.; Ge, Z.; Kinoh, H.;Kano, M.R.; Itaka, K.; Kataoka, K. Homo-catiomer integration into PEGylated polyplex micelle from block-catiomer for systemic anti-angiogenic gene therapy for fibrotic pancreatic tumors. Biomaterials **2012**, *33* (18), 4722–4730.

193. Uchida, S.; Itaka, K.; Chen, Q.; Osada, K.; Ishii, T.; Shibata, M.-A.; Harada-Shiba, M.; Kataoka, K. PEGylated polyplex with optimized PEG shielding enhances gene introduction in lungs by minimizing inflammatory responses. Mol. Therapy **2012**, *20* (6), 1196–1203.

194. Shi, J.; Johnson, R.N.; Schellinger, J.G.; Carlson, P.M.; Pun, S.H. Reducible HPMA-co-oligolysine copolymers for nucleic acid delivery. Int. J. Pharm. **2012**, *427* (1), 113–122.

195. Wang, B.; Zhang, S.; Cui, S.; Yang, B.; Zhao, Y.; Chen, H.; Hao, X.; Shen, Q.; Zhou, J. Chitosan enhanced gene delivery of cationic liposome via non-covalent conjugation. Biotechnol. Lett. **2012**, *34* (1), 19–28.

196. Wang, Y.; Xu, Z.; Zhang, R.; Li, W.; Yang, L.; Hu, Q. A facile approach to construct hyaluronic acid shielding polyplexes with improved stability and reduced cytotoxicity. Colloids Surf. B Biointerfaces **2011**, *84* (1), 259–266.

197. Perche, F.; Benvegnu, T.; Berchel, M.; Lebegue, L.; Pichon, C.; Jaffres, P.-A., Midoux, P. Enhancement of dendritic cells transfection *in vivo* and of vaccination against B16F10 melanoma with mannosylated histidylated lipopolyplexes loaded with tumor antigen messenger RNA. Nanomed. Nanotechnol. Biol. Med. **2011**, *7* (4), 445–453.

198. Perche, F.; Gosset, D.; Mevel, M.; Miramon, M.L.; Yaouanc, J.-J.; Pichon, C.; Benvegnu, T.; Jaffres, P.-A.; Midoux, P. Selective gene delivery in dendritic cells with

Nanomaterials—
Nanoparticles

mannosylated and histidylated lipopolyplexes. J. Drug Target. **2011**, *19* (5), 315–325.

199.   Shan, Y.; Luo, T.; Peng, C.; Sheng, R.; Cao, A.; Cao, X.; Shen, M.; Guo, R.; Tomas, H.; Shi, X. Gene delivery using dendrimer-entrapped gold nanoparticles as nonviral vectors. Biomaterials **2012**, *33* (10), 3025–3035.

200.   Liu, W.-M.; Xue, Y.-N.; Peng, N.; He, W.-T.; Zhou, R.-X.; Huang, S.-W. Dendrimer modified magnetic iron oxide nanoparticle/DNA/PEI ternary magnetoplexes: A novel strategy for magnetofection. J. Mater. Chem. **2011**, *21* (35), 13306–13315.

201.   Loh, X.J.; Zhang, Z.-X.; Mya, K.Y.; Wu, Y.-L.; He, C.B.; Li, J. Efficient gene delivery with paclitaxel-loaded DNA-hybrid polyplexes based on cationic polyhedral oligomeric silsesquioxanes. J. Mater. Chem. **2011**, *20* (47), 10634–10642.

202.   De Duve, C.; Wattiaux, R. Functions of lysosomes. Annu. Rev. Physiol. **1966**, *28* (1), 435–492.

203.   Barton, N.W.; Brady, R.O.; Dambrosia, J.M.; Di Bisceglie, A.M.; Doppelt, S.M.; Hill, S.C.; Mankin, H.J.; Murray, G.J.; Parker, R.I.; Argoff, C.E.; Grewal,R.P.; Yu, K. T. Replacement therapy for inherited enzyme deficiency—Macrophage-targeted glucocerebrosidase for Gaucher's disease. N. Engl. J. Med. **1991**, *324* (21), 1464–1470.

204.   Rohrbach, M.; Clarke, J.T.R. Treatment of lysosomal storage disorders. Drugs **2007**, *67* (18), 2697–2716.

205.   Christian, D.A.; Cai, S.; Bowen, D.M.; Kim, Y.; Pajerowski, J.D.; Discher, D.E. Polymersome carriers: From self-assembly to siRNA and protein therapeutics. Eur. J. Pharm. Biopharm. **2009**, *71* (3), 463–474.

206.   Solaro, R. Targeted delivery of proteins by nanosized carriers. J. Polym. Sci. A Polym. Chem. **2008**, *46* (1), 1–11.

207.   Ayame, H.; Morimoto, N.; Akiyoshi, K. Self-assembled cationic nanogels for intracellular protein delivery. Bioconjug. Chem. **2008**, *19* (4), 882–890.

208.   Du, J.; Jin, J.; Yan, M.; Lu, Y. Synthetic nanocarriers for intracellular protein delivery. Curr. Drug Metabol. **2012**, *13* (1), 82–92.

209.   Murata, H.; Futami, J.; Kitazoe, M.; Yonehara, T.; Nakanishi, H.; Kosaka, M.; Tada, H.; Sakaguchi, M.; Yagi, Y.; Seno, M.; Huh, N.-H.; Yamada, H.Intracellular delivery of glutathione S-transferase-fused proteins into mammalian cells by polyethylenimine-glutathione conjugates. J. Biochem. **2008**, *144* (4),447–455.

210.   Lee, Y.; Ishii, T.; Kim, H.J.; Nishiyama, N.; Hayakawa, Y.; Itaka, K.; Kataoka, K. Efficient delivery of bioactive antibodies into the cytoplasm of living cells by charge-conversional polyion complex micelles. Angew. Chem. Int. Ed. **2010**, *49* (14), 2552–2555.

211.   Bulmus, V.; Woodward, M.; Lin, L.; Murthy, N.; Stayton, P.; Hoffman, A. A new pH-responsive and glutathione-reactive, endosomal membrane-disruptive polymeric carrier for intracellular delivery of biomolecular drugs. J. Control. Release **2003**, *93* (2), 105–120.

212.   Tomar, L.K.; Tyagi, C.; Lahiri, S.S.; Singh, H. Poly (PEG-DMA-MAA) copolymeric micro and nanoparticles for oral insulin delivery. Polym. Adv. Technol. **2011**, *22* (12), 1760–1767.

213.   Santander-Ortega, M.J.; Bastos-Gonzalez, D.; Ortega-Vinuesa, J.L.; Alonso, M.J. Insulin-loaded PLGA nanoparticles for oral administration: An *in vitro* physico-chemical characterization. J. Biomed. Nanotechnol. **2009**, *5* (1), 45–53.

214.   He, C.; Yin, L.; Tang, C.; Yin, C. Size-dependent absorption mechanism of polymeric nanoparticles for oral delivery of protein drugs. Biomaterials **2012**, *33* (33), 8569–8578.

215.   Jin, Y.; Song, Y.; Zhu, X.; Zhou, D.; Chen, C.; Zhang, Z.; Huang, Y. Goblet cell-targeting nanoparticles for oral insulin delivery and the influence of mucus on insulin transport. Biomaterials **2012**, *33* (5), 1573–1582.

216.   Grenha, A.; Gomes, M.E.; Rodrigues, M.; Santo, V.E.; Mano, J.F.; Neves, N.M.; Reis, R.L. Development of new chitosan/carrageenan nanoparticles for drug delivery applications. J. Biomed. Mater. Res. Part A **2010**, *92* (4), 1265–1272.

217.   Akagi, T.; Baba, M.; Akashi, M. Preparation of nanoparticles by the self-organization of polymers consisting of hydrophobic and hydrophilic segments: Potential applications. Polymer **2007**, *48* (23), 6729–6747.

218.   Sakuma, S.; Suzuki, N.; Kikuchi, H.; Hiwatari, K.; Arikawa, K.; Kishida, A. Oral peptide delivery using nanoparticles composed of novel graft copolymers having hydrophobic backbone and hydrophilic branches. Int. J. Pharm. **1997**, *149* (1), 93–106.

219.   Sakuma, S.; Suzuki, N.; Sudo, R.; Hiwatari, K.; Kishida, A.; Akashi, M. Optimized chemical structure of nanoparticles as carriers for oral delivery of salmon calcitonin. Int. J. Pharm. **2002**, *239* (1), 185–195.

220.   Lee, J.C.M.; Bermudez, H.; Discher, B.M.; Sheehan, M.A.; Won, Y.Y.; Bates, F.S.; Discher, D.E.Preparation, stability, and *in vitro* performance of vesicles made with diblock copolymers. Biotechnol. Bioeng. **2001**, *73* (2), 135–145.

221.   Pang, Z.Q.; Lu, W.; Gao, H.L.; Hu, K.L.; Chen, J.; Zhang, C.L.; Gao, X.L.; Jiang, X.G.; Zhu, C.Q. Preparation and brain delivery property of biodegradable polymersomes conjugated with OX26. J. Control. Release **2008**, *128* (2), 120–127.

222.   Arifin, D.R.; Palmer, A.F. Polymersome encapsulated hemoglobin: A novel type of oxygen carrier. Biomacromolecules **2005**, *6* (4), 2172–2181.

223.   Rameez, S.; Alosta, H.; Palmer, A.F. Biocompatible and biodegradable polymersome encapsulated hemoglobin, a potential oxygen carrier. Bioconj. Chem. **2008**, *19* (5), 1025–1032.

224.   Soppimath, K.S.; Aminabhavi, T.M.; Kulkarni, A.R.; Rudzinski, W.E. Biodegradable polymeric nanoparticles as drug delivery devices. J. Control. Release **2001**, *70* (1), 1–20.

225.   Santander-Ortega, M.J.; Csaba, N.; Gonzalez, L.; Bastos-Gonzalez, D.; Ortega-Vinuesa, J.L.; Alonso, M. Protein-loaded PLGA–PEO blend nanoparticles: encapsulation, release and degradation characteristics. J. Colloid Polym. Sci. **2010**, *288* (2), 141–150.

226.   Yadav, S.Ch.; Kumari, A.; Yadav, R. Development of peptide and protein nanotherapeutics by nanoencapsulation and nanobioconjugation. Peptides **2011**, *32*(1), 173–187.

227.   Li, Y.; Pei, Y.; Zhang, X.; Gu, Z.; Zhou, Z.; Yuan, W.; Zhou, J.; Zhu, J.; Gao, X. PEGylated PLGA nanoparticles as protein carriers: Synthesis, preparation and biodistribution in rats. J. Control. Release **2001**, *71* (2), 203–211.

228. Simone, E.A.; Dziubla, T.D.; Colon-Gonzalez, F.; Discher, D.E.; Muzykantov, V.R. Effect of polymer amphiphilicity on loading of a therapeutic enzyme into protective filamentous and spherical polymer nanocarriers. Biomacromolecules **2007**, *8* (12), 3914–3921.

229. Vila, A.; Sanchez, A.; Tobio, M.; Calvo, P.; Alonso, M.J. Design of biodegradable particles for protein delivery. J. Control. Rel. **2002**, *78* (1–3), 15–24.

230. Simone, E.A.; Dziuba, T.D.; Discher, D.E.; Muzykantov, V.R. Filamentous polymer nanocarriers of tunable stiffness that encapsulate the therapeutic enzyme catalase. Biomacromolecules **2009**, *10* (6), 1324–1330.

231. Vila, A.; Sanchez, A.; Janes, K.; Behrens, I.; Kissel, T.; Vila Jato, J.L, Alonso, M. Low molecular weight chitosan nanoparticles as new carriers for nasal vaccine delivery in mice. J. Eur. J. Pharm. Biopharm. **2004**, *57* (1), 123–131.

232. van Dongen, S.F.M.; Verdurmen, W.P.R.; Peters, R.J.R.W.; Nolte, R.J.M.; Brock, R.; van Hest, J.C.M. Cellular integration of an enzyme-loaded polymersome nanoreactor. Angew. Chem. Int. Ed. **2010**, *49* (40), 7213–7216.

233. Cabane, E.; Malinova, V.; Menon, S.; Palivan, C.G.; Meier, W. Photoresponsive polymersomes as smart, triggerable nanocarriers. Soft Matter **2011**, *7* (19), 9167–9176.

234. Harada, A.; Kataoka, K. Novel polyion complex micelles entrapping enzyme molecules in the core: Preparation of narrowly-distributed micelles from lysozyme and poly (ethylene glycol)–poly (aspartic acid) block copolymer in aqueous medium. Macromolecules **1998**, *31* (2), 288–294.

235. Yuan, X.F.; Harada, A.; Yamasaki, Y.; Kataoka, K. Stabilization of lysozyme-incorporated polyion complex micelles by the ω-end derivatization of poly (ethylene glycol)–poly (α,β-aspartic acid) block copolymers with hydrophobic groups. Langmuir **2005**, *21* (7), 2668–2674.

236. Lee, Y.; Bae, Y.; Hiki, S.; Ishii, T.; Kataoka, K. A protein nanocarrier from charge-conversion polymer in response to endosomal pH. J. Am. Chem. Soc. **2007**, *129* (17), 5362–5363.

237. Hsu, J.; Northrup, L.; Bhowmick, T.; Muro, S. Enhanced delivery of α-glucosidase for Pompe disease by ICAM-1-targeted nanocarriers: Comparative performance of a strategy for three distinct lysosomal storage disorders. Nanomed. Nanotechnol. Biol. Med. **2012**, *8* (5), 731–739.

238. Hsu, J.; Serrano, D.; Bhowmick, T.; Kumar, K.; Shen, Y.; Kuo, Y.C.; Garnacho, C.; Muro, S. Enhanced endothelial delivery and biochemical effects of α-galactosidase by ICAM-1-targeted nanocarriers for Fabry disease. J. Control. Release **2011**, *149* (3), 323–331.

239. Hood, E.; Simone, E.; Wattamwar, P.; Dziubla, T.; Muzykantov, V. Nanocarriers for vascular delivery of antioxidants. Nanomedicine **2011**, *6* (7), 157–172.

240. Choi, J.H.; Joung, Y.K.; Bae, J.W.; Choi, J.W.; Park, K.D. Self-assembled nanogel of pluronic-conjugated heparin as a versatile drug nanocarrier. Macromol. Res. **2011**, *19* (2), 180–188.

241. Giannotti, M.I.; Esteban, O.; Oliva, M.; Garcia-Parajo, M.F.; Sanz, F. pH-responsive polysaccharide-based polyelectrolyte complexes as nanocarriers for lysosomal delivery of therapeutic proteins. Biomacromolecules **2011**, *12* (7), 2524–2533.

242. Sezer, A.; D., Kazak, H.; Oner, E.T. Levan-based nanocarrier system for peptide and protein drug delivery: Optimization and influence of experimental parameters on the nanoparticle characteristics. Carbohydr. Polym. **2011**, *84* (1), 358–363.

243. Li, Y.Y.; Dong, H.Q.; Wang, K.; Shi, D.L.; Zhang, X.Z.; Zhuo, R.X. Stimulus-responsive polymeric nanoparticles for biomedical applications. Sci. China Chem. **2010**, *53* (3), 447–457.

244. Koo, H.; Huh, M.S.; Sun, I.-C.; Yuk, S.H.; Choi, K.; Kim, K.; Kwon, I.C. *In vivo* targeted delivery of nanoparticles for theranosis. Acc. Chem. Res. **2011**, *44* (10), 1018–1028.

245. Parveen, S.; Misra, R.; Sanjeeb, K.Sahoo, S.K.Nanoparticles: A boon to drug delivery, therapeutics, diagnostics and imagin. Nanomed. Nanotechnol. Biol. Med. **2012**, *8* (2), 147–166

246. Pansare, V.J.; Hejazi, S.; Faenza, W.J.; Prud'homme, R.K. Review of long-wavelength optical and NIR imaging materials: Contrast agents, fluorophores, and multifunctional nano carriers. Chem. Mater. **2012**, *24* (5), 812–827.

247. Ghoroghchian, P.P.; Frail, P.R.; Susumu, K.; Park, T.-H.; Wu, S.P.; Uyeda, H.T.; Hammer, D.A.; Therien, M.J. Broad spectral domain fluorescence wavelength modulation of visible and near-infrared emissive polymersomes. J. Am. Chem. Soc. **2005**, *127* (44), 15388–15390.

248. Budijono, S.J.; Shan, J.; Yao, N.; Miura, Y.; Hoye, T.; Austin, R.H.; Ju, Y.; Prud'homme, R.K. Synthesis of stable block-copolymer-protected NaYF 4: Yb 3 +, Er 3 + Upconverting phosphor nanoparticles Chem. Mater. **2010**, *22* (2), 311–318.

249. Akbulut, M.; Ginart, P.; Gindy, M.E.; Theriault, C.; Chin, K.H.; Soboyejo, W.; Prud'homme, R.K. Generic method of preparing multifunctional fluorescent nanoparticles using flash nanoprecipitation. Adv. Funct. Mater. **2009**, *19* (5), 718–725.

250. Chen, Z.; Yu, D.; Liu, C.; Yang, X.; Zhang, N.; Ma, C.H.; Song, J.B.Lu, ZZZ. J. Gadolinium-conjugated PLA-PEG nanoparticles as liver targeted molecular MRI contrast agent. J. Drug Targeting **2011**, *19* (8), 657–665.

251. Shiraishi, K.; Kawano, K.; Minowa, T.; Maitani, Y.; Yokoyama, M. Preparation and *in vivo* imaging of PEG-poly (L-lysine)-based polymeric micelle MRI contrast agents. J. Control. Release **2009**, *136* (1), 14–20.

252. Nakamura, E.; Makino, K.; Okano, T.; Yamamoto, T.; Yokoyama, M. A polymeric micelle MRI contrast agent with changeable relaxivity. J. Control. Release **2006**, *114* (3), 325–333.

253. Ujiie, K.; Kanayama, N.; Asai, K.; Kishimoto, M.; Ohara, Y.; Akashi, Y.; Yamada, K.; Hashimoto, S.; Oda, T.; Ohkohchi, N.; Yanagihara, H.; Kita, E.;Yamaguchi, M.; Fujii, H.; Nagasaki, Y. Preparation of highly dispersible and tumor-accumulative, iron oxide nanoparticles: Multi-point anchoring of PEG-b-poly (4-vinylbenzylphosphonate) improves performance significantly. Colloids Surf. B **2011**, *88* (2), 771–778.

254. Gao, G.H.; Lee, J.W.; Nguyen, M.K.; Im, G.H.; Yang, J.; Heo, H.; Jeon. P., Park, T.G.; Lee, J.H.; Lee, D. S. pH-responsive polymeric micelle based on PEG-poly (β-amino ester)/(amido amine) as intelligent vehicle for magnetic resonance imaging in detection of cerebral ischemic area. J. Control. Release **2011**, *155* (1), 11–17.

Nanomaterials—
Nanoparticles

255. Gao, G.H.; Heo, H.; Lee, J.H.; Lee, D.S. An acidic pH-triggered polymeric micelle for dual-modality MR and optical imaging. J. Mater. Chem. **2010**, *20* (26), 5454–5461.

256. Talelli, M.; Rijcken, C.J.F.; Lammers, T.; Seevinck, P.R.; Storm, G.; van Nostrum, C.F.; Hennink, W.E. Superparamagnetic iron oxide nanoparticles encapsulated in biodegradable thermosensitive polymeric micelles: Toward a targeted nanomedicine suitable for image-guided drug delivery. Langmuir **2009**, *25*, 2060–2067.

257. Peng, H.; Blakey, I.; Dargaville, B.; Rasoul, F.; Rose, S.; Whittaker, A.K. Synthesis and evaluation of partly fluorinated block copolymers as MRI imaging agents. Biomacromolecules **2009**, *10* (2), 374–381.

258. Du, W.; Nystrom, A.M.; Zhang, L.; Powell, K.T.; Li, Y.; Cheng, C.; Wickline, S.A.; Wooley, K.L. Amphiphilic hyperbranched fluoropolymers as nanoscopic 19F magnetic resonance imaging agent assemblies. Biomacromolecules **2008**, *9* (10), 2826–2833.

259. Nystrom, A.M.; Bartels, J.W.; Du, W.; Wooley, K.L. Perfluorocarbon-loaded shell crosslinked knedel-like nanoparticles: Lessons regarding polymer mobility and self-assembly. J. Polym. Sci. A Polym. Chem. **2009**, *47* (4), 1023–1037.

260. Nurmi, L.; Peng, H.; Seppala, J.; Haddleton, D.M.; Blakey, I.; Whittaker, A.K. Synthesis and evaluation of partly fluorinated polyelectrolytes as components in 19F MRI-detectable nanoparticles. Polym. Chem. **2010**, *1* (7), 1039–1047.

261. Dolmans, D.; Fukumura, D.; Jain, R.K. Photodynamic therapy for cancer. Nat. Rev. Cancer **2003**, *3* (5), 380–387.

262. Pogue, B.W.; Pitts, J.D.; Mycek, M.A.; Sloboda, R.D.; Wilmot, C.M.; Brandsema, J.F. *In vivo* NADH fluorescence monitoring as an assay for cellular damage in photodynamic therapy. Photochem. Photobiol. **2001**, *74* (6), 817–824.

263. Krammer, B. Vascular effects of photodynamic therapy. Anticancer Res. **2001**, *21*, 4271–4277.

264. van Duijnhoven, F.H.; Aalbers, R.; Rovers, J.P.; Terpstra, O.T.; Kuppen, P.J.K. The immunological consequences of photodynamic treatment of cancer, a literature review. Immunobiology **2003**, *207* (2), 105–113.

265. Sortino, S.; Mazzaglia, A.; Monsu Scolaro, L.; Marino Merlo, F.; Valveri, V.; Sciortino, M.T. Nanoparticles of cationic amphiphilic cyclodextrins entangling anionic porphyrins as carrier-sensitizer system in photodynamic cancer therapy. Biomaterials **2006**, *27* (23), 4256–4265.

266. Rijcken, C.J.F.; Hofman, J.-W.; van Zeeland, F.; Hennink, W.E.; van Nostrum, C.F. Photosensitiser-loaded biodegradable polymeric micelles, preparation, characterisation and *in vitro* PDT efficacy. J. Control. Release **2007**, *124* (3), 144–153.

267. Peng, C.-L.; Shieh, M.-J.; Tsai, M.-H.; Chang, C.-C.; Lai, P.-S. Self-assembled star-shaped chlorin-core poly (ε-caprolactone)–poly (ethylene glycol) diblock copolymer micelles for dual chemo-photodynamic therapies. Biomaterials **2008**, *29* (26), 3599–3608.

268. Peng, C.-L.; Lai, P.-S.; Lin, F.-H.; Yueh-Hsiu Wu, S.; Shieh, M.-J. Dual chemotherapy and photodynamic therapy in an HT-29 human colon cancer xenograft model using SN-38-loaded chlorin-core star block copolymer micelles. Biomaterials **2009**, *30* (21), 3614–3625.

269. Nishiyama, N.; Nakagishi, Y.; Morimoto, Y.; Lai, P.-S.; Miyazaki, K.; Urano, K. Enhanced photodynamic cancer treatment by supramolecular nanocarriers charged with dendrimer phthalocyanine. J. Control. Release **2009**, *133* (3), 245–251.

270. Tsai, H.C.; Tsai, C.-H.; Lin, S.-L.; Jhang, C.-R.; Chiang, Y.-C.; Hsiue, G.-H. Stimulated release of photosensitizers from graft and diblock micelles for photodynamic therapy. Biomaterials **2012**, *33* (6), 1827–1837.

271. Shan, J.; Budijono, S.J.; Hu, G.; Yao, N.; Kang, Y.; Ju, Y.; Prud'homme, R.K. Pegylated composite nanoparticles containing upconverting phosphors and meso-tetraphenyl porphine (TPP) for photodynamic therapy. Adv. Funct. Mater. **2011**, *21* (13), 2488–2495.

272. Liu, G.; Xie, J.; Zhang, F.; Wang, Z.; Luo, K.; Zhu, L.; Quan, Q.; Niu, G.; Lee, S.; Ai, H.; Chen, X. N-Alkyl-PEI-functionalized iron oxide nanoclusters for efficient siRNA delivery. Small **2011**, *7* (19), 2742–2749.

273. Nasongkla, N.; Bey, E.; Ren, J.; Ai, H.; Khemtong, C.; Guthi, J.S. Multifunctional polymeric micelles as cancer-targeted, mri-ultrasensitive drug delivery systems. Nano Lett. **2006**, *6* (11), 2427–2432.

274. Park, J.-H.; von Maltzahn, G.; Sangeeta, E.R.; Michael, N.B.; Sailor, J. Micellar hybrid nanoparticles for simultaneous magnetofluorescent imaging and drug delivery. Angew. Chem. Int. Ed. **2008**, *47* (38), 7284–7289.

275. Hu, J.; Qian, Y.; Wang, X.; Liu, T.; Liu, S. Drug-loaded and superparamagnetic iron oxide nanoparticle surface-embedded amphiphilic block copolymer micelles for integrated chemotherapeutic drug delivery and mr imaging. Langmuir **2012**, *28* (4), 2073–2082.

276. Liu, T.; Qian, Y.; Hu, X.; Ge, Z.; Liu, S. Mixed polymeric micelles as multifunctional scaffold for combined magnetic resonance imaging contrast enhancement and targeted chemotherapeutic drug delivery. J. Mater. Chem. **2012**, *22* (11), 5020–5030.

277. Li, X.; Li, H.; Liu, G.; Deng, Z.; Wu, S.; Li, P.; Xu, Z.; Xu, H.; Chu, P.K. Magnetite-loaded fluorine-containing polymeric micelles for magnetic resonance imaging and drug delivery. Biomaterials **2012**, *33* (10), 3013–3024.

278. Liu, T.; Li, X.; Qian, Y.; Hua, X.; Liu, S. Multifunctional pH-Disintegrable micellar nanoparticles of asymmetrically functionalized β-cyclodextrin-Based star copolymer covalently conjugated with doxorubicin and DOTA-Gd moieties. Biomaterials **2012**, *33* (8), 2521–2531.

279. Bakandritsos, A.; Papagiannopoulos, A.; Anagnostou, E.N.; Avgoustakis, K.; Zboril, R.; Pispas, S.; Tucek, J.; Ryukhtin, V.; Bouropoulos, N.; Kolokithas-Ntoukas, A.; Steriotis, T.A.; Keiderling, U.; Winnefeld, F. Merging high doxorubicin loading with pronounced magnetic response and bio-repellent properties in hybrid drug nanocarriers. Small **2012**, *8* (15), 2381–2393.

280. Liao, Z.; Wang, H.; Wang, X.; Zhao, P.; Wang, S.; Su, W.; Chang, J. Multifunctional nanoparticles composed of a poly (dl-lactide-coglycolide) core and a paramagnetic liposome shell for simultaneous magnetic resonance imaging and targeted therapeutics. Adv. Funct. Mater. **2011**, *21* (6), 1179–1186.

281. Wang, W.; Cheng, D.; Gong, F.; Miao, X.; Shuai, X. Design of multifunctional micelle for tumor-targeted

intracellular drug release and fluorescent imaging. Adv. Mater. **2012**, *24* (1), 115–120.

282. Yu, H.; Zou, Y.; Wang, Y.; Huang, X.; Huang, G.; Sumer, B.D.; Boothman, D.A.; Gao, J. Overcoming endosomal barrier by amphotericin B-loaded dual pH-responsive PDMA-b-PDPA micelleplexes for siRNA delivery. ACS Nano **2011**, *5* (11), 9246–9255.

283. Hao, Y.; Zhang, M.; He, J.; Ni, P. Magnetic DNA Vector constructed from PDMAEMA polycation and pegylated brush-type polyanion with cross-linkable shell. Langmuir **2012**, *28* (15), 6448–6460.

284. Zhang, L.; Lin, Y.; Zhang, Y.; Chen, R.; Zhu, Z.; Wu, W.; Jiang, X. Fluorescent micelles based on star amphiphilic copolymer with a porphyrin core for bioimaging and drug delivery. Macromol. Biosci. **2012**, *12* (1), 83–92.

285. Du, W.; Xu, Z.; Nystrom, A.M.; Zhang, K.; Leonard, J.R.; Wooley, K.L. 19 F- and Fluorescently labeled micelles as nanoscopic assemblies for chemotherapeutic delivery. Bioconjug. Chem. **2008**, *19* (12), 2492–2498.

Nanoparticles: Biomaterials for Drug Delivery

Abhijit Gokhale
Thomas Williams
Jason Vaughn
Product Development Services, Patheon Pharmaceuticals Inc., Cincinnati, Ohio, U.S.A.

Abstract

Pharmaceutical drug delivery systems have become very sophisticated to comply with unique drug delivery requirements, *in vivo* imaging needs, and delivery of genes and other large molecules. With increasing complexity of the drug delivery systems, the need for novel biomaterials to achieve successful formulations has become imperative. This article reviews latest nanoparticle-based drug delivery systems such as drug–polymer conjugates, solid lipid nanoparticles, liposomes, lipid–polymer hybrids, and dendrimers. The review of the formulations and the clinical trials shows how biomaterials have evolved from simple basic structures to very complex structures to satisfy the formulation needs.

INTRODUCTION

Biodegradable nanoparticles have many pharmaceutical applications in different areas of human activities like biology and neurology such as targeted drug delivery into the brain for cancer treatment.[1] The developments in bionanoparticles and the understanding of their properties show an optimistic picture in the development of future medicines and related products.

Bionanoparticles are used as functional excipients in a variety of pharmaceutical applications such as drug delivery with enhanced bioavailability, biological labels, and tissue engineering to name a few.[2] Drugs of nanosize are often unstable and encapsulation or conjugation of nanodrugs with biopolymer particles of size less than 200 nm can provide the required stability and better control of drug release. For more sophisticated targeted delivery applications, the required particle size needs to be less than 80 nm. Because of the smaller size, nanoparticles can penetrate through smaller capillaries, sustaining the intracellular drug levels and deliver the drug into the tissues with stable release kinetics.[3]

This article reviews several cutting-edge advanced drug delivery systems. Biomaterials with a wide range of physico-chemical properties such as hydrophilic-lipophilic balance value, solubility parameter, polarity, hydrogen bonding capacity, molecular weight and biological properties such as the ability to inhibit transporters (e.g., p-glycoprotein) and genes (e.g., CYP3A) are used as excipients in these drug delivery systems.

DRUG–POLYMER CONJUGATES

Description and Method of Manufacturing

Polymers offer several advantages in drug delivery systems due to their versatile nature. Polymeric functional groups can be engineered to produce hydrophilic or hydrophobic components to alter the solubility and permeability of the polymers. Therefore, the solubility and permeability of drug molecules attached to the polymer can be increased. Also, the drug delivery systems (drug + polymer) can be formulated to achieve the delivery of two or more drugs from the same formulation, contain readily clearable polymer carriers, control the release of highly toxic drugs, and improve drug targeting to tissues or cells. Drug–polymer conjugates have been used to deliver several biologically active compounds like drug molecules, proteins, peptides, hormones, enzymes, etc.[4] Selecting the proper polymeric excipient for a given drug is very important to achieve the desired therapeutic effect.

Chemical conjugation of drugs to biopolymers can form stable bonds such as ester, amide, and disulfide. Hydrogen bonds (e.g., ester or amide) are stable and can deliver the drug to the targeted site without releasing it during its transport and prior to the cellular localization of the drug.

Polymer properties can be altered by the polymer-specific structural characteristics including molecular weight, coil structure, copolymer composition, variable polyelectrolyte charges, flexibility of polymer chains, and microstructure. The advantages of this type of drug delivery

Concise Encyclopedia of Biomedical Polymers and Polymeric Biomaterials DOI: 10.1081/E-EBPPC-120050070

systems include diversity in composition (molecular structure and molecular weight), polymer water solubility and reactivity for conjugation with drug molecules, as these are typically used in hydrophilic systems with polar or ionic groups.

Ringsdorf[5] first proposed the idea of drug–polymer conjugates in 1975. His model was based on the covalent bond between the drug and a macromolecular backbone through a biodegradable linkage. The hydrophilic functional group makes the entire molecule soluble, the drug molecule is attached to the backbone through covalent bonds and then the third area is comprised of a transport system where a specific functional group helps the entire macromolecule to be transported to the target cells or the site of pharmacological action. The basic structure is shown in Fig. 1.

The drug–polymer conjugates can be classified into two types based on the composition and the structure:

1. Composition: The conjugates can further be divided into three subtypes:[6]
   a. Prodrugs: This type of conjugate reacts with components inside the cells to form active substance(s). The selection of a suitable polymer and a targeting moiety is essential for the effectiveness of the prodrugs since the targeting ability vastly depends on the choice of antibody, ligand internalization and receptor expression.
   b. Combination of substances: This type of conjugate has two or more substances that react under specific intracellular conditions, forming an active drug.
   c. Targeted drug delivery system: This type of conjugate includes three components: a targeting moiety, a carrier, and one or more active component(s).
2. Structure: The drug–polymer conjugates can further be divided into six subtypes based on the structure of the polymer as well as the structure of the drug molecule used:[7]
   a. Linear polymer systems (LP)
   b. Linear polymer–linear spacer systems (LPLS)
   c. Linear polymer–branched spacer systems (LPBS)
   d. Branched polymer systems (BP)
   e. Branched polymer–linear spacer systems (BPLP)
   f. Branched polymer–branched spacer systems (BPBP)

Polymer–drug conjugates generally exhibit prolonged half-life, higher stability, water solubility, lower immunogenicity, and antigenicity. Other important aspects of the synthetic biopolymers used in this type of applications are (i) non-toxicity, (ii) clearable from the body without any accumulation in tissues or organs, (iii) controlled average molecular weight and molecular weight distribution, thereby avoiding any undesirable biological response due to low or high molecular weight polymer–drug conjugates, (iv) high purity of polymers, and (v) possible sterilization of polymers with biodegradable characteristics.

Formulations

The commonly used functional groups for synthesizing these drug–polymer conjugates include carbonate, anhydride, urethane, orthoester, amide, and ester.[8] Since molecules with a low molecular weight (less than ~40,000 Da) are susceptible to excretion, the conjugates are synthesized with a molecular weight higher than 40,000 Da. At the same time high molecular weight conjugates cause problems with permeation and pinocytic capture by target cells other than phagocytes.

The first conjugate synthesized was in the field of cancer research in 1986. Anticancer drugs daunorubicin and doxorubicin were targeted to the tumor using this delivery system.[9] Since then several polymer–protein conjugates have been tested in clinical studies as part of the development phase.[10] The drug–polymer conjugate system is one of the most effective drug delivery systems for targeted drug delivery. Most of the research has been conducted to deliver anticancer or antitumor agents. High molecular weight prodrugs containing cytotoxic components have been developed to decrease peripheral side effects and to obtain a more specific administration of the drugs to the cancerous tissues. Polymer–drug conjugates have been tailored for activation by extra- or intracellular enzymes to release the parent drug in situ.

The polymeric drug delivery system can be designed for passive or active targeting. For cancer treatment, active targeting reduces side effects and enhances efficacy. Because the drug does not distribute throughout the body homogenously but gets delivered directly to the tumor cells, systemic side effects are reduced while enhancing the bioavailability, even for drugs with a relatively short half-life. Drug–polymer conjugates have been very successful in active targeting.[12]

Some of the most widely used polymers for these drug delivery systems are polysaccharides such as dextran and insulin. Cellulose-based polymers and polyarabogalactan

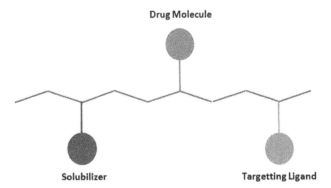

Drug Molecule

Solubilizer

Targetting Ligand

Fig. 1 Ringsdorf model of synthetic polymer drugs.

are also seeing increased use. Other polymers like poly(D,L-lactic-*co*-glycolic acid), polylactic acids, *N*-(2-hydroxypropyl)methacrylamide, vinylic, acrylic polymers, and poly(α-amino acids) have been successfully used for drug conjugations. Poly(ethylene oxide)-β-poly(ε-caprolactone)-based drug conjugates for paclitaxel delivery were successfully demonstrated.[11] Star-shaped drug delivery systems with a poly[(p-iodomethyl) styrene] core and poly(*tert*-butyl acrylate) arms (PScorePtBuAarm) were used to conjugate cisplatin.[13]

Due to higher reactivity at physiological pH, the *N*-hydroxysuccinimide (NHS) ester makes an excellent choice for amine coupling reactions in bioconjugation synthesis. NHS ester compounds react with nucleophiles to release the NHS-leaving group and form an acylated product. Carboxyl groups activated with NHS esters are highly reactive with amine nucleophiles.[6]

Polymers containing hydroxyl groups (e.g., polyethylene glycol, PEG, etc.) can be modified to obtain anhydride compounds. PEG can be acetylated with anhydrides to form an ester terminating with free carboxylate groups. PEG and its succinimidyl succinate and succinimidyl glutarate derivatives can be further used for conjugation with drugs or proteins.[11]

In order to achieve good bioavailability, the conjugate needs to have balanced lipophilicity and hydrophilicity and to be of relatively low molecular weight to achieve the required absorption. The drug polymer–conjugate combines the advantages of both the hydrophilic and hydrophobic segments. Selecting polymeric structures like dendrimers, dendronized polymers, graft polymers, block copolymers, branched polymers, multivalent polymers, stars and hybrid glycol, and peptide derivatives increases the chance of success for producing adequate formulations.[14]

Although these conjugates have shown excellent results in various clinical trials, synthesizing such conjugates has been very difficult due to the complex nature of the drug molecules and the organic synthesis expert needed to produce these types of conjugates. It is always difficult to achieve good permeability with a minimal first-pass effect with macromolecules.[6]

SOLID LIPID NANOPARTICLES AND NANOSTRUCTURED LIPID CARRIERS

Solid Lipid Nanoparticles

Description and method of manufacturing

Nanoparticles made from solid lipids are gaining major attention in the field of oral drug delivery for small drug molecules as well as protein and peptides due to their lipophilic nature. Since most of the drugs coming out of drug discovery are poorly soluble, solid lipid nanoparticles (SLNs) provide an attractive alternative to deliver the drugs

which can be readily absorbed along with SLNs through lipolysis in the small intestine.[15] There are several advantages of using SLNs for drug delivery including, but not limited to enhanced drug stability, improved bioavailability, controlled or targeted release, and aqueous solvent processing to name a few.

Excipients used in SLNs are mostly lipids with low melting points and naturally occurring in food or in the human body. Hence, they have better absorption characteristics. SLNs were initially manufactured in early 1990s using high shear mixing but this resulted in several processing issues including poor reproducibility and wide particle size distributions. To overcome these problems, techniques such as high pressure homogenization or microemulsion were used (with processes operating at elevated temperatures above the melting points of the lipids used in the formulation).[16] Typically much smaller particle sizes (less than 200 nm) are possible with the high pressure homogenization technique.

In the microemulsion method, the lipid component (fatty acids and/or glycerides) is melted, a mixture of water, cosurfactant(s) and a surfactant is heated to the same temperature as the lipid and added under mild stirring to the lipid melt. Once a thermodynamically stable microemulsion is produced, it is dispersed in cold aqueous media under constant mechanical mixing to avoid any agglomeration of the solidified nanoparticles. The use of surfactants and cosurfactants in the cold aqueous media helps avoid agglomerate formation. However, in this approach, the formulation is diluted due to the presence of the cold aqueous media and hence not desirable when high solid contents are required.[17] Figure 2 shows the typical phase diagram to prepare the microemulsions at elevated temperature which in turn transform into nanoparticles upon cooling.

In high pressure homogenization, the formulation containing an emulsified lipid–drug mixture is cooled by the cavitation effect during pressure drop and it is not necessary to add another processing step to cool/solidify the lipid particles. A formulation with a high solid content is possible with this technique.[18]

Formulation

In order to formulate the SLN-based drug delivery system, the choice of lipids, surfactant, and cosurfactant is very important. The drug molecule should dissolve in molten lipid(s) to form the solid solution. The surfactant and cosurfactant should be able to emulsify the molten mixture of drug–lipid solution/dispersion in a nanometer particle size range (ideally less than 200 nm) to obtain a thermodynamically stable suspension. In general the loading capacity of drug in the lipid depends on the solubility of drug in the molten lipid.[17]

Commonly, lipids like glycerides (mono-, di-, or tri-), waxes (e.g., carnauba wax, beeswax, cetyl alcohol, emulsifying wax, cholesterol, cholesterol butyrate, etc.), and

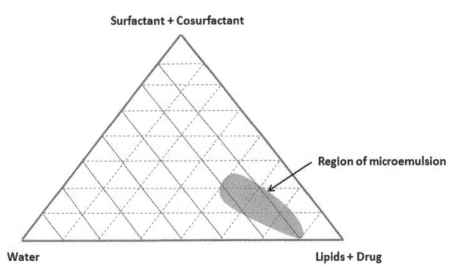

Fig. 2 Phase diagrams to produce microemulsions of SLNs.

vegetable oils are used as carriers to dissolve/disperse the drug molecules. The surfactants (e.g., polysorbates, sodium dodecyl sulfate, castor oil, bile salts, etc.) and cosurfactants (e.g., alcohols, polyethylene glycol, cremophor, poloxamer, etc.) are used to emulsify the molten lipid mixture in aqueous media. The SLNs can then be lyophilized with excipients to obtain the dried powder. The list of excipients and formulations of the marketed products is found in Mehnert and Mader[19] and Pardeike et al.[20]

The oral delivery of proteins and peptides is very difficult due to the poor stability of large molecules in the gastrointestinal environment. Diffusion transport of such molecules through epithelial barriers is slow resulting in poor absorption in the blood stream. Further, these molecules are subject to the hepatogastrointestinal first-pass elimination. Hence, the physico-chemical properties of large molecules prohibit effective delivery through the oral route. SLNs show promising options for oral delivery of these molecules since the molecules can be completely "masked" in the gastrointestinal tract.[21] In the past two decades, several papers have been published detailing the successful formulation of proteins and peptides using SLNs.[22]

The use of SLNs is one of the potential drug delivery methods for targeted brain drug delivery.[23,24] Tightness of the endothelial vascular lining, referred to as the blood–brain barrier (BBB), provides a challenge to development of brain drug delivery. In the BBB, capillary endothelial cells protect the brain from foreign substances in the blood stream while letting required nutrients pass through for proper brain functioning. The BBB strictly limits transport into the brain through both physical (tight junctions) and metabolic (enzymes) barriers. Hence, passing the BBB to deliver the required drug molecules is an extremely challenging task. With the increase in lipophilicity exhibited by SLNs, permeability through the BBB can be increased. However, the permeability of drugs with molecular weight

larger than 400 Da is low irrespective of its lipophilicity. Overall, SLNs have several advantages due to their size (nanometer-sized particles can easily bypass liver and spleen filtration) and lipophilicity (enhanced permeation and controlled release up to several weeks). Attaching ligands to SLNs for targeted drug delivery is relatively easy due to the versatile chemical properties of the lipids. With mixtures of lipids, it is possible to incorporate hydrophilic as well as lipophilic drugs.

Nanostructured Lipid Carrier

The second generation SLNs were developed to overcome the limitations of SLNs and they were termed as the nanostructured lipid carriers (NLCs). The primary difference between a SLN and a NLC is that the NLC also contains lipid in a liquid form. The NLC matrix is a solid at room temperature, but the formulation contains one or more liquid lipids adsorbed onto the solid lipids. Hence, to dissolve or disperse the drug molecule, NLCs offer more versatility in physico-chemical properties than SLNs. Since a larger pool of lipids can be used for formulation, drug loading capacity of NLCs is greater than SLNs and drug absorption and targeting can be greatly enhanced by the greater number of available lipids.

Several papers have been published presenting successful formulations of pharmaceutical drugs using the NLC approach. Itraconazole was formulated for pulmonary delivery using a hot high pressure homogenization technique.[20] The stability and physical properties of drug-loaded NLCs were found to be stable during nebulization. Hence, this drug delivery system was found to be promising for the pulmonary application of this poorly soluble drug.[20]

A comparison of the SLN and NLC drug delivery system was done by formulating clotrimazole using both systems. In both types of drug delivery systems, the formulation was developed by producing a single phase solid solution of drug and lipids. The drug release from NLCs was found to be faster than that of SLNs and also the stability of the NLC formulation was better than SLNs at room temperature.[25]

LIPOSOMES, POLYMEROSOMES, AND NIOSOMES

Liposomes, polymerosomes, and niosomes are part of the same structural family. Their properties, methods of manufacturing, and typical biopolymers used in formulations are discussed in this section.

Liposomes

Description and method of manufacturing

Liposomes are spherical vesicles consisting of a lipid bilayer that encapsulates water at the center. Water soluble drugs can be incorporated into the aqueous center of the liposome and lipophilic drugs can be incorporated into the lipid bilayer and hence liposomes can carry both hydrophilic as well as lipophilic drugs. The typical diameter of multilamellar vesicles (consisting of several concentric bilayers) ranges in size from 500 nm to 5000 nm, while the typical diameter of unilamellar (formed by a single bilayer) vesicles is around 100 nm in size. Figure 3 shows a typical liposome structure of multilamellar and unilamellar vesicles.

The predominant physical and chemical properties of a liposome are based on the net properties of the constituent phospholipids, including permeability, charge density, and steric hindrance. Physico-chemical properties of liposomes such as size, charge, and surface properties can be easily manipulated by changing the composition of excipients in the lipid mixture before liposome preparation and/or by varying preparation methods.

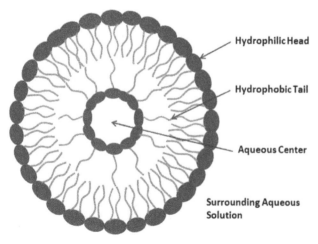

Fig. 3 Liposome structure used for drug delivery systems.

Hydrophilic Head

Hydrophobic Tail

Aqueous Center

Surrounding Aqueous Solution

Bangham published in 1967[26] that phospholipids in aqueous systems can form closed bilayered structures and since then numerous attempts have been made to use liposomes for pharmaceutical drug delivery. There are four commonly used methods to produce drug-loaded liposomes.

> Liposomes are formed in an aqueous solution saturated with a soluble drug.
> Use of organic solvents and solvent exchange mechanisms.
> Use of lipophilic drugs.
> pH gradient methods.

Due to interactions between water molecules and the hydrophobic phosphate groups of the phospholipids, the lipid bilayer closes in on itself. This process of liposome formation is spontaneous because the amphiphilic phospholipids self-associate into bilayers. Initially liposomes were made of phospholipids from the egg yolk but now with advances in materials science, a variety of synthetic materials are being used to produce liposomes.

Since liposomes are typically made from mostly naturally occurring lipids, they are biocompatible making them attractive for drug delivery. Liposome-incorporated pharmaceutical drugs are protected from the inactivating effect of external conditions; therefore drug stability and permeability are enhanced. One of the drawbacks of early liposomal systems in biological models was rapid excretion due to the small sizes. This challenge was overcome by attaching the polyethylene glycol molecule (PEGylation) to the liposome which makes it a so-called "stealth" molecule reducing opsonization (destruction by a phagocyte) and hence reduces clearance by the MPS, increasing the circulation half-life. Opsonization presents such a problem to the development of therapeutically useful liposomes that nearly all research reported in the literature involves PEG-coated or PEGylated liposomes.[27]

Formulations

Thousands of papers have been published presenting liposome-based drug delivery systems to overcome issues associated with pharmaceutical drugs like poor bioavailability, targeting of the drug to specific tissues, food effect, and poor stability in the gastrointestinal tract during oral drug delivery. It is well known that membranes of various kinds of phospholipids undergo phase transitions, such as a gel-to-liquid crystalline transition and a lamellar-to-hexagonal transition. These phase transitions have been used for the design of stimuli-sensitive liposomes.[28] The, temperature sensitization of liposomes has been attempted using thermo-sensitive polymers that exhibit a lower critical solution temperature (CST). Attachment of these polymers to liposomes could give temperature-sensitive functionalities to the liposomes.[28]

Karchemski et al.[29] attached carbon nanotubes (CNTs) covalently to a liposome to produce CNTs–liposomes conjugates (CLC). This novel approach has a unique advantage of being able to load large amounts of drug that can be delivered into cells by the CLC system, thus preventing potential adverse systemic effects of CNTs when administered at high doses. This system provided a versatile and controlled means of enhanced delivery of one or more agents incorporated with the liposomes. With this approach, binding different drugs to the CNTs can also be possible by binding liposomes loaded with different contents on the same CNT.[29]

Ultrasound-controlled drug and gene delivery prepared using echogenic liposomes is gaining popularity. More and more articles are being published in this research field due to the unique advantages provided by these delivery systems. Echogenic liposomes containing entrapped therapeutic agents can be targeted to specific disease sites, allowing high local concentrations and low systemic toxicity. These types of liposomes carry an inert gas in the center of the liposome, which causes a burst due to ultrasonic pulses. Cavitation caused by ultrasound can destroy liposomes and cause local dose dumping locally at the targeted cell.[30]

Date et al.[31] presents a comprehensive review of attempts made to deliver drugs using liposomes and various biomaterials used to prepare complex liposomes in the treatment of parasitic diseases including malaria, leishmaniasis, and trypanosomiasis. Proteins and peptides are increasingly recognized as potential leads for the development of new therapeutics for a variety of human ailments. Proteins and peptides are typically delivered intravenously due to their poor stability in gastrointestinal fluids and poor permeability through gastrointestinal membranes. Liposomes provide an excellent alternative to deliver proteins and peptides orally since they can "hide" them during transport and limit the first-pass effect. Forssen and Willis.[32] reviewed current efforts in using liposomes for targeted drug delivery by site-specific targeting using folate receptors, tailoring the properties of the liposomes to make them cell adhesion molecules, and targeting methods such as antibody targeting.

Liposomes are also used to deliver drugs to the lungs, as reviewed by Kellaway and Farr.[33] Lymphatic drug delivery using liposomes was reviewed by Forrest et al.[34] Drug delivery to macrophages using liposomes and the role of the physico-chemical properties such as surface charge, liposome size, concentration, composition, and inclusion of ligands on liposomes for uptake by macrophages are reviewed extensively by Ahsan et al.[35]

Polymerosomes

Description and method of manufacturing

The basic structure of polymerosomes is similar to liposomes (Fig. 3). Since polymers offer higher stability and structural variability compared to lipids, polymeric vesicles can eliminate the disadvantages associated with liposomes such as short half-life and fewer excipient choices. Polymers offer versatility of form and function due to the possibility of modification of tectons and hydrophilic block length, composition, and chemical structure. Since composition and molecular weight of the polymers can be varied, it allows not only the preparation of polymerosomes with different properties and responsiveness to stimuli but also with different membrane thicknesses and permeabilities. Polymerosomes typically have relatively thick and robust membranes (>50 nm) formed by amphiphilic block copolymers with a relatively high molecular weight. Relatively long blood circulation times of polymerosomes can be accomplished by the introduction of a hydrophilic surface layer [e.g., poly(ethylene glycol) (PEG) blocks].[36]

Polymerosomes are synthesized using various methods including self-assembly and polymer rehydration techniques. In the self-assembly technique, the block copolymers are first dissolved in organic solvents to form a single phase solution. The aqueous phase is then slowly added to precipitate the polymer in the form of a polymerosome. Since the polymer precipitation is uncontrolled, the particle size distribution varies depending on the rate of addition of water and the inherent properties of the polymer to precipitate out of the particular organic solvent. In the case of the polymer rehydration technique, the amphiphilic block copolymer film is hydrated to induce self-assembly. The block copolymers are dissolved in a suitable organic solvent similar to the self-assembly technique. The organic solvent is then evaporated by a technique such as rotovap to produce a thin dry polymeric film. The film is then hydrated by the addition of water.[37]

Formulations

The skin permeability of polymerosomes is excellent, and several articles have been published detailing drug delivery through the skin.[38] For a topical gel formulation, triblock copolymers of PEG and PCL (PCL–PEG–PCL) were synthesized by ring opening polymerization of ε-caprolactone with PEG. The particle size of these vesicles was found to be around 75–140 nm. The self-assembly technique using an acetone–water system was used. A skin permeation study of these polymer vesicles revealed the accumulation of these vesicles in the furrows of the stratum corneum. This research showed that drug delivery through the skin in the form of topical gel is possible.[39]

Rastogi, Anand, and Koul[40] published a study on enhancing the pharmacological efficacy of protein-loaded PCL–PEG-based polymerosomes. Insulin was selected as the model protein and was complexed with sodium deoxycholate, a naturally occurring bile salt. The prepared surfoplexes were efficiently encapsulated in PCL- and PEG-based polymer vesicles. It was found that with this drug delivery system, the initial burst effect was greatly reduced, and the therapeutic effect of insulin was increased by 2 hr.

Nanoparticles—Nerve

Niosomes

Description and method of manufacturing

Niosomes are similar to liposomes (Fig. 3) and can be used to deliver both hydrophilic as well as lipophilic drug molecules. Niosomes are non-ionic surfactant vesicles. Niosome-based drug delivery is a relatively new field and few articles have been published. Similar to liposomes, niosomes can be formulated to produce prolonged systemic circulation of the drug and enhanced penetration into targeted tissues. Niosomes generally consist of a vesicle formed using non-ionic surfactants (e.g., Span-60) stabilized by the addition of cholesterol and a small amount of an anionic surfactant.[41]

Formulations

Kumar et al.[41] presented a study demonstrating the successful preparation of aceclofenac niosomes and their evaluation. Non-ionic surfactants (Span 60 and Span 20) along with cholesterol were used to produce niosomes. Excellent drug entrapment efficiency and controlled *in vitro* release were presented in this research article. Das and Palei[42] published the research on the encapsulation of rofecoxib in sorbitan ester niosomes. Niosomes prepared using Span 60, Span 20, and cholesterol were incorporated into a carbopol gel to produce a topical gel-based drug delivery system. The lower flux value of the niosomal gel as compared to the plain drug gel in pig skin indicated that the lipid bilayer of niosome was rate limiting in drug permeation and hence the controlled release was achieved.[42]

Sathali et al.[43] formulated a topical gel containing clobetasol propionate niosomes to prolong the duration of action and prevent side effects. The clobetasol propionate niosomes were prepared by altering the ratio of various non-ionic surfactants (Span 40, 60, and 80) to cholesterol by the thin film hydration method. The *in vivo* results showed that the niosomal gel had a sustained as well as a prolonged action.

Terzano et al.[44] developed Polysorbate 20 niosomes containing beclomethasone dipropionate for pulmonary delivery to patients with chronic obstructive pulmonary disease. Singh, Jain, and Kumar[45] prepared niosomes containing the anti-inflammatory drug nimesulide and tested its physicochemical properties including stability and *in vitro* release. It was shown that *in vitro* sustained drug release can be achieved with this approach. The research literature for niosomes in ocular drug delivery has been reviewed by I. Kaur et al.[46]

Niosome-based drug delivery for the oral delivery of peptides was investigated by Dufes et al.[47] Due to the blood brain barrier (BBB), vasoactive intestinal peptide (VIP), which is used for the treatment of Alzheimer's disease, has poor bioavailability. Also the VIP has a very short half-life which makes it necessary to increase the dose frequency. Dufes et al. developed glucose-bearing niosomes that encapsulate the VIP for delivery to specific brain areas.

The authors concluded that glucose-bearing vesicles represent a novel tool for delivery of drugs across the BBB.[47]

LIPID–POLYMER HYBRIDS

Description and Method of Manufacturing

In lipid–polymer hybrid drug delivery systems, the drug is incorporated into a polymer in a matrix form or in a solubilized form and then the polymer particle is coated with a lipid film (Fig. 4). Since drugs typically leak through polymeric matrix systems showing an initial burst in the release, such lipid film over a polymeric particle inhibits the drug diffusion outward. Polymeric nanoparticles provide robustness to the drug delivery system while the lipid coating provides the necessary lipophilicity to block the drug leakage.[48] Drug delivery of water soluble drugs to produce controlled or targeted release can be achieved using this system. Thus, in this drug delivery system, the polymeric core functions to encapsulate hydrophilic or lipophilic drug molecules and provides a robust structure, whereas the external lipid coating functions as a biocompatible shield, a template for surface modifications, and a barrier for preventing the fast leakage of water-soluble drugs.[49]

Several methods are used to produce lipid–polymer hybrids. Spray drying and spray freeze drying are two of the most attractive methods since they are cost-effective and readily scalable. In these methods, the drug and the polymer are dissolved in an organic solvent and then spray dried/spray freeze dried to produce matrix particles. The particles are then coated with a lipid film to produce the lipid–polymer hybrids.[50] Spray freeze drying has been found to be more effective than spray drying since it produces porous nano-agglomerates which have superior properties (e.g., aqueous reconstitubility and aerosolization efficiency).[50]

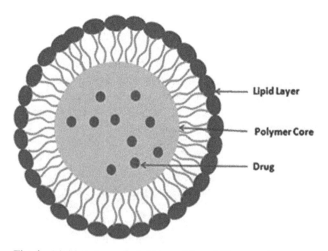

Fig. 4 Lipid–polymer hybrid-based drug delivery system.

Formulations

Lipid–polymer hybrid-based drug delivery systems have been studied largely to deliver water soluble drugs for targeted drug delivery. Cheow and Hadinoto[51] described the preparation of stable lipid–polymer hybrid nanoparticles encapsulating LEV, CIP, and OFX antibiotics. In this study, an extensive study of the effects of factors such as process parameters and excipients on drug loading and stability of the drug delivery system was carried out. The use of solvents and surfactants and the ionic interaction between the drug and the lipid were found to be most influential in preparing successful lipid–polymer hybrids with optimum drug loading.

Farokhzad et al.[52] reported the immunological characterization of lipid–polymer hybrid nanoparticles and proposed a method to control the levels of complement activation induced by these nanoparticles. In this method, the nanoparticle surface was modified by attaching methoxyl, carboxyl, and amine groups. It was found that the surface chemistry significantly affects human plasma and serum protein adsorption patterns.[52]

Wu et al.[53] demonstrates that a lipid–polymer hybrid drug delivery system loaded with doxorubicin is effective for tumor treatment in a well-established animal model. Tumor growth delay and tumor necrosis were observed in tumors treated with the lipid–polymer hybrid formulation of doxorubicin. It was found that these lipid–polymer hybrids carrying anticancer agents were useful for locoregional treatment of breast cancer with an improved therapeutic index.

DENDRIMERS

Description and Method of Manufacturing

Dendrimers are globular in shape and have a three-dimensional highly controlled structure. These nanostructures have a narrow mass and size distribution.[54] Dendrimers are the fourth new class of polymer architectures after traditional linear, crosslinked, and branched types. Dendrimers can be used as well-defined scaffolding or as nanocontainers to conjugate, complex, or encapsulate therapeutic drugs or imaging moieties. Due to their structural and highly diversified physico-chemical properties, dendrimers are gaining popularity in pharmaceutical and biotech industries.

Dendrimers are synthesized using a bottom-up approach where polymeric layers are added stepwise around a central core. Each additional iteration leads to higher generation of dendrimers. Dendrimers can be constructed with a range of molecular weights and have a polyfunctional surface that facilitates the attachment of drugs and pharmacokinetic modifiers such as PEG or targeting moieties. Two distinct methods exist to synthesize dendrimers: The divergent approach developed by Tomalia[55] and Newkome[56] and the convergent approach developed by Hawker and Frechet.[57]

1. Divergent synthesis: In this type of synthesis the dendrimer growth starts from a polyfunctional core and proceeds radially outward with the stepwise addition of successive layers of building blocks. This approach involves assembling of monomeric modules into a radial, branch-upon-branch motif according to certain dendritic rules and principles.[55] Figure 5 shows a typical reaction and growth of the dendrimer.
2. Convergent synthesis: In this type of synthesis several dendrons (small functional groups) are reacted with a multifunctional core to obtain a final dendrimeric structure. Figure 6 shows a typical reaction and growth of the dendrimer. Hence, the growth begins at the surface of the dendrimer, and continues inward by gradually linking surface units together with more monomers.[55]

Since the conception of the dendrimers, a variety of dendrimeric structures have evolved as evidenced by the vast research in this field. Typical structures are liquid crystalline dendrimers, tecto-dendrimers, chiral dendrimers, PAMAM dendrimers, PAMAMOS dendrimers, hybrid dendrimers, peptide dendrimers, and glycodendrimers. Nanjwade et al.[58] reviewed these structures, biomaterials used, their properties, and applications. Various theoretical and computational approaches have been devised to predict the properties of dendrimeric structures so that the effective drug delivery system can be produced. These approaches include solubility parameters, Flory–Huggins theory, analytical predictions of partition coefficients, and molecular simulations. Huynh et al.[59] provided a review of the approaches for the optimization of drug–material pairs using important performance-related parameters including

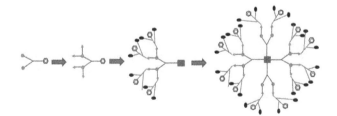

Fig. 5 Divergent synthesis of dendrimers.

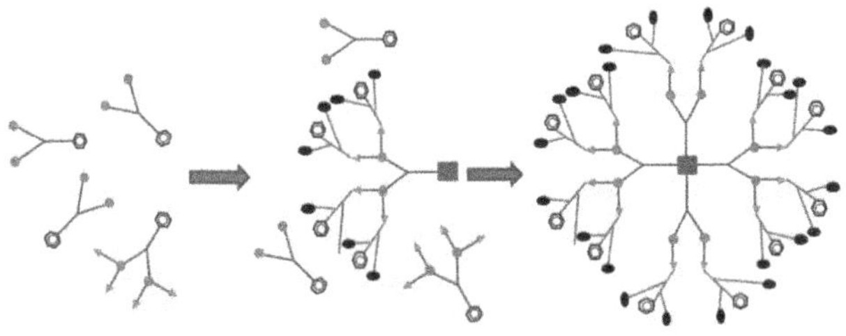

Fig. 6 Convergent synthesis of dendrimers.

the size of the delivery particles, their surface properties, and the compatibility of the materials with the drug to be sequestered. Huynh et al.[59] found that the analytical models are useful for fast prescreening during the development of delivery systems and drug derivatives. Molecular simulations were found to be more reliable.

Formulations

Dendrimer-based drug delivery systems have been successful in both improving bioavailability as well as targeting drugs. The main advantages for the use of dendrimers as drug carriers reside in their multivalent design and their well-defined structure that results from the stepwise, iterative approach used in their preparation. The anionic and neutral dendrimers have been shown to be less toxic than the cationic dendrimers but at the same time they suffer from lower permeability compared to cationic dendrimers. Typically the smaller size of dendrimers (<8 nm) facilitates rapid renal clearance which in turn reduces toxicity. Also, surface coverage of the dendrimers with biocompatible groups such as PEG chains greatly reduced their cytotoxicity.[60]

Drugs are encapsulated in dendrimers primarily using two methods: by covalent dendrimer–drug conjugation or non-covalent encapsulation of drugs.

1. Covalent dendrimer–drug conjugation: In this method, a degradable covalent linkage between the drug and the dendrimer is formed to obtain a stronger relationship between the dendrimer and the drug. The drug release from a conjugate is governed largely by the nature of the linking bond or spacer between the drug and the scaffold and the targeted physiological domain for intended release. Ester and amide bonds might be cleavable by enzymes or under hydrolytic conditions; however, ester cleavage is generally more facile than amide cleavage for releasing drugs.[61]
2. Non-covalent drug encapsulation: In this method, the drug is physically encapsulated in a dendrimeric structure by virtue of a micellar structure. Although it is relatively easy to produce a drug delivery system with this approach, the drug release is often uncontrolled.[61]

Yang and Lopina[62] used Starburst® PAMAM dendrimers for delivery of Penicillin V. Drug delivery scaffolds were designed and built based on the polyethylene glycol polyamidoamine (PEG–PAMAM) star polymer. Penicillin V was coupled with the PAMAM dendrimer through a PEG spacer. The drug delivery scaffold based on the PEG–polyamidoamine star polymer proved to be feasible. The star polymers provided accessible ends which were used to chemically couple the desired small molecules. Ihre et al.[63] designed and synthesized dendritic polyester systems based on the 2,2-bis(hydroxymethyl) propanoic acid monomer unit as a possible versatile drug carrier. These systems were designed to meet the needs of their intended use as drug carriers, including water solubility, non-toxicity, and stability of the polymeric backbone. Several formulations were tried successfully that led to an improved solubility. Attachment of drugs to high molecular weight polymers can significantly improve both tumor targeting and therapeutic efficacy due to the enhanced permeability and the retention effect observed in the tumor tissue.

Kaminskas and Porter[64] reviewed the research work published on lymphatic drug delivery using dendrimers. It was concluded that after subcutaneous administration, larger dendrimers (>8 nm) were found to be drained from the injection site into the peripheral lymphatic capillaries and therefore have potential as lymphatic imaging agents for magnetic resonance and optical fluorescence lymphangiography and as vectors for drug-targeting to lymphatic sites of disease progression. Increasing hydrophilicity and reducing surface charge were found to enhance the drainage from subcutaneous injection sites. Specific access to the lymphatics from the interstitium is dictated by size where larger macromolecules or colloids are precluded from ready access to the vascular capillaries and instead enter the lymphatics via the large inter-endothelial gaps that open under the conditions of lymphatic filling.

Tremendous efforts have been taken to deliver proteins, peptides, genes, and antiretroviral drugs using dendrimers. Several authors[65–67] have reviewed various targeted drug delivery systems using dendrimers like folic acid–dendrimer conjugates, peptide–dendrimer conjugates, monoclonal antibody conjugated to PAMAM, glycosylation of dendrimers, boron neutron capture therapy, photothermal therapy, and gene therapy. Ojewole et al.[68] provided review articles on dendrimer-based antiretroviral drug delivery and Oliveir et al.[69] reviewed articles on dendrimer-based drug delivery for regenerative medicine.

Despite all the research efforts by the research community, it is still a challenge to prepare dendritic polymers that circulate in the blood long enough to accumulate at target sites, but that can also be eliminated from the body at a reasonable rate to avoid long-term buildup. In addition, the tissue localization of dendritic polymers is still difficult to predict in advance and more studies are required to determine the effect of peripheral dendritic groups on these properties.[70]

CONCLUSIONS

Each of the drug delivery system described in this review has unique requirements for biomaterials. For instance, drug–polymer conjugate systems offer unique advantages like ease of synthesis but the success of the formulation depends on the stability of hydrogen bonds between the drug and the polymer. Solid lipid nanoparticles and nanostructured lipid carriers can mask the drug molecules completely and therefore are excellent in enhancing the drug permeability. However, an in-depth knowledge of lipids and surfactants is essential to form a solid solution of drug–lipid and nano-sized particles. Drug delivery systems like liposomes have been shown to be excellent carriers for targeted drug delivery as well as for enhancing bioavailability of drug molecules. PEGylation makes the drug delivery system "stealth" reducing the first-pass effect and increasing the permeability through the gastrointestinal lumen. Hence, PEGylation of drugs and carrier polymers is one of the most successful approaches to enhance bioavailability. Dendrimers provide the highest possible flexibility to the formulation scientist and can handle a variety of drug molecules, genes, imaging markers, etc.

REFERENCES

1. Hans, M.L.; Lowman, A.M. Biodegradable nanoparticles for drug delivery and targeting. Curr. Opin. Solid State Mat. Sci. 2002, 6 (4), 319–327.
2. Salata, O.V. Applications of nanoparticles in biology and medicine. J. Nanobiotechnol. 2003, 2 (3), 1–6.
3. Schiavone, H.; Palakodaty, S.; Clark, A.; York, P. Evaluation of SCF engineered particle-based lactose blends in passive dry powder inhalers. Int. J. Pharm. 2004, 281 (1–2), 55–66.
4. Ringsdorf, H. Synthetic polymeric drugs. In Polymeric Delivery Systems; Gordon and Breach Science Publishers, Inc: New York, 1978; 197–225.
5. Ringsdorf, H. Structure and properties of pharmacologically active polymers. J. Polym. Sci. 1975, 51 (1), 135–153.
6. Pasut, G.; Veronese, F.M. Polymer–drug conjugation, recent achievements and general strategies. Prog. Polym. Sci. 2007, 32 (8–9), 933–961.
7. Minko, T. Soluble polymer conjugates for drug delivery. Drug Discov. Today Technol. 2005, 2 (1), 15–20.
8. Elvira, C.; Gallardo, A.; Cifuentes. A. Covalent polymer-drug conjugates. Molecules 2005, 10 (1), 114–125.
9. Vasey, P.; Kaye, S.; Morrison, R.; Twelves, C.; Wilson, P.; Duncan, R.; Thomson, A.; Murray, L.; Hilditch, T.; Fraier, D.; Frigerio, E.; Cassidy, J. Phase I clinical and pharmacokinetic study of PK1 [N-(2-hydroxypropyl)methacrylamide copolymer doxorubicin]: First member of a new class of chemotherapeutic agents-drug-polymer conjugates. Clin. Cancer Res. 1999, 5 (1), 83–94.
10. Meerum Terwogt, J.M.; ten Bokkel Huinink, W.W.; Schellens, J.H.; Schot, M.; Mandjes, I.A.; Zurlo, M.G.; Rocchetti, M.; Rosing, H.; Koopman, J.; Beijnen, J. Phase I clinical and pharmacokinetic study of PNU166945, a novel water soluble polymer conjugated prodrug of paclitaxel. Anticancer Drugs. 2001, 12 (4), 315–323.
11. Khandare, J.; Minko, T. Polymer–drug conjugates: Progress in polymeric prodrugs. Prog. Polym. Sci. 2006, 31 (4), 359–397.
12. Segal, E.; Satchi-Fainaro, R. Design and development of polymer conjugates as anti-angiogenic agents. Adv. Drug Deliv. Rev. 2009, 61 (13), 1159–1176.
13. Kowalczuk, A.; Stoyanova, E.; Mitova, V.; Shestakova, P.; Momekov, G.; Momekova, D.; Koseva. N. Star-shaped nano-conjugates of cisplatin with high drug payload. Int. J. Pharm. 2011, 404 (1–2), 220–230.
14. Tomalia, D.A.; Baker, H.; Dewald, J.; Hall, M.; Kallos, G.; Martin, S.; Roeck, J.; Ryder, J.; Smith, P. A new class of polymers—starburst-dendritic macromolecules. Polym. J. 1985, 17 (1), 117–132.
15. Souto, E.B.; Muller, R.H. Lipid nanoparticles: Effect on bioavailability and pharmacokinetic changes. Handb. Exp. Pharmacol. 2010, 197, 115–141.
16. Speiser, P. Lipid nanopellets als Tragersystem fur Arzneimittel zur peroralen Anwendung. European Patent EP0167825, 1990.
17. Gasco, M. Method for producing solid lipid microspheres having a narrow size distribution. United States Patent USS188837, 1993.
18. Domb, A. Liposheres for controlled delivery of substances. United States Patent USS188837, 1993.
19. Mehnert, W.; Mader, K. Solid lipid nanoparticles, production, characterization and applications. Adv. Drug Deliv. Rev. 2012, 64 (Suppl.), 83–101.
20. Pardeike, J.; Weber, S.; Haber, T.; Wagner, J.; Zarfl, H.; Plank, H.; Zimmer, A. Development of an Itraconazole-loaded nanostructured lipid carrier (NLC) formulation for pulmonary application. Int. J. Pharm. 2011, 419 (1–2), 329–338.
21. Ugwoke, M.; Agu, R.; Verbeke, N.; Kinget, R. Nasal mucoadhesive drug delivery: Background, applications, trends and future perspectives. Adv. Drug Deliv. Rev. 2005, 57 (11), 1640–1665.
22. Almeida, A.; Souto. E. Solid lipid nanoparticles as a drug delivery system for peptides and proteins. Adv. Drug Deliv. Rev. 2007, 59 (6), 478–490.
23. Blasi, P.; Giovagnoli, S.; Schoubben, A.; Ricci, M.; Rossi. C. Solid lipid nanoparticles for targeted brain drug delivery. Adv. Drug Deliv. Rev. 2007, 59 (6), 454–477.
24. Fundaro, A.; Cavalli, R.; Bargoni, A.; Vighetto, D.; Zara, G.; Gasco, M. Non-stealth and stealth solid lipid nanoparticles (SLN) carrying doxorubicin: Pharmacokinetics and tissue distribution after i.v. administration to rats. Pharmacol. Res. 2000, 42 (4), 337–343.
25. Das, S.; Ng, W.K.; Tan, R.B. Are nanostructured lipid carriers (NLCs) better than solid lipid nanoparticles (SLNs): Development, characterizations and comparative evaluations of clotrimazole-loaded SLNs and NLCs? Eur. J. Pharm. Sci. 2012, 47 (1), 139–151.
26. Bangham, A.; Horne, R. Negative staining of phospholipids and their structural modification by surface-active agents as observed in the electron microscope. J. Mol. Biol. 1964, 8 (5), 660–668.
27. Moghimia, S.; Szebeni, J. Stealth liposomes and long circulating nanoparticles: Critical issues in pharmacokinetics, opsonization and protein-binding properties. Prog. Lipid Res. 2003, 42 (6), 463–478.
28. Yatvin, M.; Weinstein, J.; Dennis, W.; Blumenthal, R. Design of liposomes for enhanced local release of drugs by hyperthermia. Science 1978, 202 (4374), 1290–1293.
29. Karchemski, F.; Zucker, D.; Barenholz, Y.; Regev, O. Carbon nanotubes-liposomes conjugate as a platform for drug delivery into cells. J. Control. Rel. 2012, 160 (2), 339–345
30. Huang, S.L. Liposomes in ultrasonic drug and gene delivery. Adv. Drug Deliv. Rev. 2008, 60 (10), 1167–1176.
31. Date, A.; Joshi, M.; Patravale, V. Parasitic diseases: Liposomes and polymeric nanoparticles versus lipid nanoparticles. Adv. Drug Deliv. Rev. 2007, 59 (6), 505–521.

Nanoparticles—Nerve

32. Forssen, E.; Willis, M. Ligand-targeted liposomes. Adv. Drug Deliv. Rev. **1998**, *29* (3), 249–271.

33. Kellaway, I.; Farr, S. Liposomes as drug delivery systems to the lung. Adv. Drug Deliv. Rev. **1990**, *5* (1–2), 149–161.

34. Cai, S.; Yang, Q.; Taryn, R.; Bagby, M.; Forrest, L. Lymphatic drug delivery using engineered liposomes and solid lipid nanoparticles. Adv. Drug Deliv. Rev. **2011**, *63* (10–11), 901–908.

35. Ahsan, F.; Rivas, I.P.; Khan, M.A.; Torres Suarez, A.I. Targeting to macrophages: role of physicochemical properties of particulate carriers-liposomes and microspheres-on the phagocytosis by macrophages. J. Control. Rel. **2002**, *79* (1–3), 29–40.

36. Romberg, B.; Oussoren, C.; Snel, C.; Hennink, W.; Storm, G. Effect of liposome characteristics and dose on the pharmacokinetics of liposomes coated with poly(amino acid)s. Pharm. Res. **2007**, *24* (12), 2394–2401.

37. Kita-Tokarczyk, K.; Grumelard, J.; Haefele, T.; Meier, W. Block copolymer vesicles – using concepts from polymer chemistry to mimic biomembranes. Polymer **2005**, *46* (11), 3540–3563.

38. Torchilin, V. Micellar nanocarriers: Pharmaceutical perspectives. Pharm. Res. **2007**, *24* (1), 1–16.

39. Rastogi, R.; Ananda, S.; Koul, V. Flexible polymerosomes - An alternative vehicle for topical delivery. Colloids Surf. B Biointerfaces **2009**, *72* (1), 161–166.

40. Rastogi, R.; Anand, S.; Koul, V. Evaluation of pharmacological efficacy of 'insulin–surfoplex' encapsulated polymer vesicles. Int. J. Pharm. **2009**, *373* (1–2), 107–115.

41. Srinivas, S.; Kumar, Y.; Hemanth, A.; Anitha, M. Preparation and evaluation of niosomes containing aceclofenac. Dig. J. Nanomater. Biostructures **2010**, *5* (1), 249–254.

42. Das, M.; Palei, N. Sorbitan ster niosomes for topical delivery of rofecoxib. Indian J. Exp. Biol. **2011**, *49* (6), 438–445.

43. Lingan, M.; Sathali, A.; Kumar, M.; Gokila, A. Formulation and evaluation of topical drug delivery system containing clobetasol propionate niosomes. Sci. Revs. Chem. Commun. **2011**, *1* (1), 7–17.

44. Terzano, C.; Allegra, L.; Alhaique, F.; Marianecci, C.; Carafa, M. Non-phospholipid vesicles for pulmonary glucocorticoid delivery. Eur. J. Pharm. Biopharm. **2005**, *59* (1), 57–62.

45. Singh, C.; Jain, C.; Kumar, B. Formulation, characterization, stability And *in vitro* evaluation of nimesulide niosomes. Pharmacophore **2011**, *2* (3), 168–185.

46. Kaur, I.; Garg, A.; Singl, A.; Aggarwal, D. Vesicular systems in ocular drug delivery: an overview. Int. J. Pharm. **2004**, *269* (1), 1–14.

47. Dufes, C.; Gaillard, F.; Uchegbu, I.F.; Schatzlein, A.G.; Olivier, J.C.; Muller, J.M. Glucose-targeted niosomes deliver vasoactive intestinal peptide (VIP) to the brain. Int. J. Pharm. **2004**, *285* (1–2), 77–85.

48. Alphandary, P.; Andremont, A.; Couvreur, P. Targeted delivery of antibiotics using liposomes and nanoparticles: research and applications. Int. J. Antimicrob. Agents **2000**, *13* (3), 155–168.

49. Cheow, W.; Hadinoto, K. Enhancing encapsulation efficiency of highly watersoluble antibiotic in poly(lactic-*co*-glycolic acid) nanoparticles: modifications of standard nanoparticle preparation methods. Colloids Surf. A Physicochem. Eng. Asp. **2010**, *370* (1–3), 79–86.

50. Wang, Y.; Kho, K.; Cheow, W.S.; Hadinoto, K. A comparison between spray drying and spray freeze drying for dry powder inhaler formulation of drug-loaded lipid–polymer hybrid nanoparticles. Int. J. Pharm. **2012**, *424* (1–2), 98–106.

51. Cheow, W.S.; Hadinoto, K. Factors affecting drug encapsulation and stability of lipid–polymer hybrid nanoparticles. Colloids Surf. B Biointerfaces **2011**, *85* (2), 214–220.

52. Salvador-Morales, C.; Zhang, L.; Langer, R.; Farokhzad, O. Immunocompatibility properties of lipid–polymer hybrid nanoparticles with heterogeneous surface functional groups. Biomaterials **2009**, *30* (12), 2231–2240.

53. Wong, H.; Rauth, A.M.; Bendayan, R.; Wu, X.Y. *In vivo* evaluation of a new polymer-lipid hybrid nanoparticle (PLN) formulation of doxorubicin in a murine solid tumor model. Eur. J. Pharm. Biopharm. **2007**, *65* (3), 300–308.

54. Kolhe, P.; Khandare, J.; Pillai, O.; Kannan, S.; Lieh-Lai, M.; Kannan, R. Preparation, cellular transport, and activity of polyamidoamine-based dendritic nanodevices with a high drug payload. Biomaterials **2006**, *27* (4) 660–669.

55. Tomalia, D.A. A new class of polymers: Starburst-dendritic macromolecules. Polym. J. **1985**, *17* (1), 117–132.

56. Newkome, G. Cascade molecules: A new approach to micelles. J. Org. Chem. **1985**, *50* (11), 2003–2004.

57. Hawker, C. Frechet, J. Preparation of polymers with controlled molecular architecture. A new convergent approach to dendritic macromolecules. J. Am. Chem. Soc. **1990**, *112* (21), 7638–7647.

58. Nanjwade, B.K.; Bechraa, H.M.; Derkara, G.K.; Manvi, F.V.; Nanjwade, V.K. Dendrimers: Emerging polymers for drug-delivery systems. Eur. J. Pharm. Sci. **2009**, *38* (3), 185–196.

59. Huynh, L.; Neale, C.; Pomes, R.; Allen, C. Computational approaches to the rational design of nanoemulsions, polymeric micelles, and dendrimers for drug delivery. Nanomedicine **2012**, *8* (1), 20–36.

60. Kurtoglu, Y.; Mishra, M.; Kannan, S.; Kannan, R. Drug release characteristics of PAMAM dendrimer-drug conjugates with different linkers. Int. J. Pharm. **2010**, *384* (1–2), 189–194.

61. Liu, M.; Frechet, J. Designing dendrimers for drug delivery. Pharm. Sci. Technolo. Today **1999**, *2* (10), 393–401.

62. Yang, H.; Lopina, S. Penicillin V-conjugated PEG-PAMAM star polymers. J. Biomater. Sci. Polym. Ed. **2003**, *14* (10), 1043–1056.

63. Ihre, H.R.; Padilla De Jesus, O.L.; Szoka, F.C. Jr.; Frechet, J.M. Polyester dendritic systems for drug delivery applications: Design, synthesis, and characterization. Bioconjug. Chem. **2002**, *13* (3), 443–452.

64. Kaminskas, L.; Porter, C. Targeting the lymphatics using dendritic polymers (dendrimers). Adv. Drug Deliv. Rev. **2011**, *63* (10–11), 890–900.

65. Wolinsky, J.; Grinstaff, M. Therapeutic and diagnostic applications of dendrimers for cancer treatment. Adv. Drug Deliv. Rev. **2008**, *60* (9), 1037–1055.

66. Wong, A.; DeWit, M.A.; Gillies, E. Amplified release through the stimulus triggered degradation of self-immolative oligomers, dendrimers, and linear polymers. Adv. Drug Deliv. Rev. **2012**, *64* (11), 1031–1045.

67. Dufes, C.; Uchegbu, I.; Schatzlein, A. Dendrimers in gene delivery. Adv. Drug Deliv. Rev. **2005**, *57* (15), 2177–2202.

68. Ojewole, E.; Mackraj, I.; Naidoo, P.; Govender, T. Exploring the use of novel drug delivery systems for antiretroviral drugs. Eur. J. Pharm. Biopharm. **2008**, *70* (3), 697–710.

69. Oliveir, J.M.; Salgado, A.J.; Sous, N.; Mano, J.P.; Reis, R.L. Dendrimers and derivatives as a potential therapeutic tool in regenerative medicine strategies—A review. Prog. Polym. Sci. **2010**, *35* (9), 1163–1194.

70. Gillies, E.; Frechet, J.M. Dendrimers and dendritic polymers in drug delivery. Drug Discov. Today. **2005**, *10* (1) 35–43.

Nanoparticles: Cancer Management Applications

Shyamasree Ghosh
Soham Saha
Abhinav Sur
School of Biological Sciences, National Institute of Science, Education, and Research, Bhubaneswar, India

Abstract

Nanoparticles (NP, 1–100 nm) find immense applications in research. Cancer is a deadly disease known to claim millions of lives globally wherein the main challenge revolves around its early detection, targeted delivery of anticancerous agents and targeted cell killing, overcoming drug resistance, and saving the healthy cells from toxic anticancer agents. Our current study includes the potential promising application of nanotechnology in the arena of cancer management, including NP in cancer research, management, and toxicity with special emphasis on polymer nanoparticles and their application in cancer management. Different types of polymeric nanoparticles together with engineered molecules such as antibodies, DNA, and receptors have been reported as efficient drug delivery devices and detection tools. The potential of nanoparticles in the detection, imaging, and targeting of cancer is advantageous over conventional methods. However, their toxicity and passage into tissues still raise concerns. Overcoming toxicity will enable these molecules to be important and promising tools in effective cancer management.

INTRODUCTION

Cancer is a complex disease claiming millions of lives.[1] An estimate of 12.7 million new cases, 7.9 million deaths in 2007, and projected incidence of 12 million deaths by 2030, by the World Health Organization raises concern over cancer treatment and management. The current cancer therapy includes radiotherapy, chemotherapy, and surgery either in single or in combination with other therapeutic forms. However, conventional treatment suffers from the major limitations of early detection, targeted and adequate drug delivery, multidrug resistance (MDR), relapse, and prevention of metastasis. Nanotechnology is being exploited for its applications to circumvent some of the problems in cancer treatment, such as targeted delivery, increased intracellular concentration of anticancer agent, reduced cytotoxicity of normal cells, and prevention of MDR. Their unique size (10–100 nm), large surface-to-volume ratios, biocompatible nature and properties as self-assembly, stability, specificity, and capability to bypass the P-glycoprotein efflux pump system, thereby preventing MDR makes them potential applicants in cancer management.[2] Coupled with macromolecules, such as anticancer drugs, peptides, proteins, and nucleic acids, they enable cancer detection and targeting over conventional methods. Different types of nanoparticles (NPs) are being largely exploited for application in cancer management, including, gold, silver, polymer, carbon, iron oxide NPs. Of all the different types of NPs employed in cancer research,

polymeric NPs (PNPs) form a major tool for cancer management. Research in the field of polymer chemistry and engineering have enabled the generation of smart biocompatible polymeric molecules that can respond to changes in environmental condition, including pH and temperature, and can sense biomolecules by swelling/deswelling to degradation thus making them promising tools in drug delivery. Engineered PNPs find application in molecular imaging. Although the application of NPs in cancer research is a new and developing domain, in this entry we focus our attention to different types of NPs and cancer research and management with special reference to polymeric NPs.

NPs IN CANCER RESEARCH AND MANAGEMENT

NPs with a general structure of hydrophilic shell providing stability in aqueous environments and a hydrophobic core containing the anticancer agent by physical entrapment or by chemical conjugation together with its function of fast cellular uptake resulted in higher therapeutic efficacy and finds application as nanomedicine (Fig. 1).[3,4] Polymers of natural and synthetic origin with unique properties of biocompatibility, biodegradability, and modification with one or more groups are largely exploited to deliver drugs and macromolecules, such as genes and proteins to cancer cells.[5] The different types of NPs (Fig. 1) include liposomes (25 nm–10 μm) constituting of lipid bilayers with three structural variations, namely, multilamellar vesicles,

Concise Encyclopedia of Biomedical Polymers and Polymeric Biomaterials DOI: 10.1081/E-EBPPC-120050073

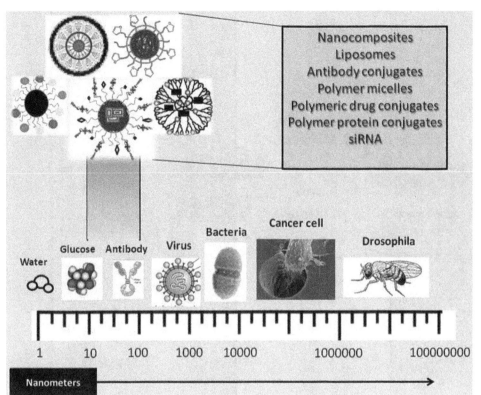

Fig. 1 Nanoparticles in cancer research.

large unilamellar vesicles, and small unilamellar vesicles, which are capable of encapsulating drugs and drug–antibody conjugations. Doxil, a liposomal formulation of doxorubicin (DOX), is reported to be efficient than DOX in targeting acute myeloid leukemia (AML).[6–8]

Quantum Dots (QDs) glow on ultraviolet light stimulation. Latex beads with QDs emit colors on light stimulation and can light up a specific sequence. Different combinations of QDs within a single bead generate probes releasing a distinct spectrum. A spectral bar code thus obtained finds potential application in cancer detection and treatment.[3] The highly branched three-dimensional structures of dendrimers (1–10 nm) can accommodate different chemical structures and functional groups. Commercially available polyamidoamine and polypropylenimine find application in gene and drug delivery.[9] Polyethylene glycol (PEG) dendrimers with increased retention properties find potential application in drug delivery. Polyester-based dendrimer–PEO–DOX conjugate inhibited DOX-insensitive C-26 tumor progression in BALB/c mice.[10] Nanopores are hole-like structures enabling passage of DNA, thus allowing efficient DNA sequencing and monitoring of shape and electrical properties of each base on the strand. With unique properties of each of the four bases of the genetic code, the passage of DNA through a nanopore enables deciphering errors in the code associated with cancer.[8]

Super paramagnetic iron oxide NPs (SPIONs, 2–3 nm) with biocompatible and biodegradable properties, easily coated dextran surfaces enable chemical linkage of functional groups and ligands that can be detected by light and electron microscopy. They can be manipulated magnetically with properties that are size dependent and find immense application in cancer detection.[11] Nanoshells (20–60 nm) comprising layer-by-layer self-assembly of NPs, form a core/shell structure. Gold nanoshells (10–300 nm) with better penetration are exploited in targeting prostate cancer.[3] Biocompatible nanoshells (10–15 nm) with reduced toxicity have been reported to deliver anticancer agents into adjoining healthy cells.[10] Nanoshells comprising PEGylated phospholipids, such as PEG2000 linked with 1,2-distearoyl phosphoethanolamine, (DSPE-PEG2000), and DSPE-PEG2000 coupled with egg phosphatidylcholine have been reported to be promising vehicles for delivery of anticancer drug E7070, N-(3-chloro-7-indolyl)-1,4-benzenedisulfonamide (Indisulam), in breast cancer.[6] Carbon nanotubes comprising hexagonal arrangement of carbon atoms in sp2 hybridizations, appearing as single- or multi-walled can immobilize macromolecules, such as antibodies, DNA, and drugs, and are being tested in drug delivery.[5,12] Phage NPs were designed and conjugated to an anticancer drug DOX.[13] Significant progress in cancer research has been reported by NPs (Fig. 2, Table 1). In this entry, we highlight their application in 1) cancer detection and imaging; and 2) cancer targeting and monitoring.

NPs in cancer detection and imaging

Liposomes, dye-molecule-doped silica NPs, QDs,[8] gold NPs (GNPs),[8] nanoshells,[13] immunotargeted nanoshells, perfluorocarbon nanoparticles (PFCN), and magnetic

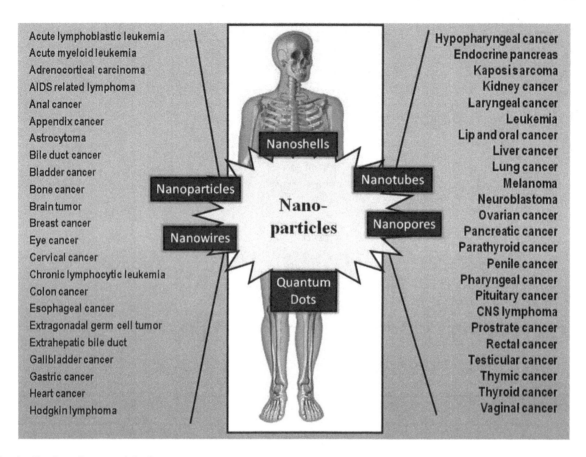

Fig. 2 Application of nanoparticles in cancer.

nanocrystals[13] are being tested for cancer imaging and found effective over conventional methods in different types of cancer (Table 1).[14–143] Hybrid organic or inorganic NPs are applied as novel intravascular probes for cancer diagnostics through molecular imaging, cell or DNA labeling, immunohistochemistry, and tumor vessel imaging.[7,144] Neuroelectronic interfaces and other nanoelectronics-based photoluminiscent sensors are being exploited for cancer detection.[12] Cationic lipid-based SPIONs with reduced cytotoxicity and high loading efficiency have been reported to enhance the imaging potential of colorectal tumor in mice.[144] Aptamer-conjugated magnetic and fluorescent NPs with enhanced signaling capabilities are also reported for the detection of leukemic cells.[145,146] High-potential probes for labeling, such as flourescein isothiocyanate-doped silica NPs are also being exploited in cancer detection due to their high-intensity luminescence with enhanced photostability and low oxidative photodamaging potential. Fluorescent dye (Rubpy) doped inside silica NPs are being synthesized.[147] NPs in conjugation with the specificity of antibody-mediated recognition of tumor antigen markers, is being tested toward ultrasensitive target detection of cancer.[8] Antibody-conjugated paramagnetic liposomes (300–350 nm) with binding specificity for the $\alpha v \beta 3$, that is a receptor for vitronectin, are being applied to visualize tumor angiogenesis in rabbit model of squamous cell carcinoma *in vivo* by magnetic resonance imaging (MRI).[13] PFCNs designed to target

$\alpha v \beta 3$ receptors on endothelial cells is reported to aid in imaging tumor neovascular system by MRI.[144]

QDs with unique properties such as high quantum yield, broad absorption with narrow and symmetric photoluminescence spectra, to size-tune fluorescence emission as a function of core size, high resistance to photo-bleaching, and resistance to photo- and chemical degradation find extensive *in vitro* and *in vivo* applications in cancer cell detection.[8] The selective targeting of peptide-coated ZnS-capped CdSe QDs to mouse tumor vasculature cells *in vivo* enables effective imaging and cancer monitoring.[144] These methods of cancer cell tracking *in vivo* is a substantial improvement over current methods as is independent of animal sacrifices.[148] However, the high toxicity of cadmium may limit the application and needs to be solved before QDs can be used in clinical setting. Fluorescent silica NPs have also been widely studied and used in clinical setting.[147,148]

The polymeric NPs formed from water-soluble hyperbranched polyhydroxyl polymer that allows for dual encapsulation of a hydrophilic protein and an amphiphilic fluorescent dye has been used to ferry endogenous protein Cytochrome *c*, which upon cytoplasmic release, initiates an apoptotic response leading to programmed cell death. Such PNPs serve as potential therapeutic agents by inducing apoptosis upon intracellular delivery. Cytochrome *c* tagged to folic acid could be targeted to folate receptor–positive cancer

Table 1 Cancer types and nanotechnology approaches

| Cancer | Conventional | | Nanotechnology | | References |
	Diagnosis/treatment	Disadvantage	Applications	Advantages	
Acute lympho-blastic leukemia (ALL) and chronic lympho-blastic leukemia (CLL)	Molecular targeting with kinase inhibitors that selectively inhibit molecular aberrations involved in pathogenesis Non-myeloablative stem cell transplantation	Cytotoxicity	Platinum-based chemo-therapy, gold nanoparticles (GNPs)	Low cytotoxicity and higher efficiency in delivery	[14–16]
Acute myeloid leukemia (AML)	New molecularly targeted agents for use in combination with conventional chemotherapy drugs	Cytotoxicity and side effects	Functionalized nanotubes, Realgar nanoparticles (RNP) showed growth inhibitory effect	Versatile new delivery platforms Multiple target recognition Amplify the transport to cancer cells	[17–19]
Adrenocorti-cal carcinoma	Modern day imaging, including ultrasound, computed tomography (CT) and magnetic resonance imaging (MRI) Surgical removal of a localized tumor Medical therapy with mitotane	Cytotoxicity and side effects	NPs	Enhancement of the *in vivo* transcriptional activity through stabilization of the oligomer formation	[19–21]
AIDS-related lymphoma	Highly active antiretroviral therapy low-dose multiagent chemotherapy, with central nervous system prophylaxis and antiretroviral therapy	Opportunistic infections	DOX non-PEGylated liposomal NP formed with biodegradable polyisohexylcyanoacrylate polymer	Avoid the efflux pump thereby overcome MDR.	[22–24]
Anal cancer	Radiotherapy and chemotherapy in combination	Anal ulcers were more frequently observed in the combined-treatment arm	Biodegradable self-assembled NPs targeted drug delivery vehicles for chemotherapeutic agents or therapeutic genes into malignant cells while sparing healthy cells	Easy detection Effective targeted therapeutics at the earliest stages of the disease	[25,26]
Appendix cancer	Surgical debulking and intraperitoneal chemotherapy: IP 5-fluoro-2′-deoxyuridine plus leukovorin	Malignant carcinoids more common in women having younger average age at diagnosis	SERS-active NPs Delivery of paclitaxel by PEG-graft carbon nanotubes	Optical detection Cancer diagnosis	[27–30]
Astrocytoma, childhood cerebellar or cerebral	Radiation therapy following initial surgery and temozolomide treatment	Recur closer to the initial tumor margin Moderate risk of distant brain failure in addition to the high rate of local failure	Treated with pioglitazone (PPAR gamma agonist) with metallic and nonmetallic NP	Pioglitazone evades NP-associated cytotoxicity	[31,32]

(Continued)

Table 1 (*Continued*) Cancer types and nanotechnology approaches

| Cancer | Conventional | | Nanotechnology | | References |
	Diagnosis/treatment	Disadvantage	Applications	Advantages	
Bile duct cancer	Surgery in proximal bile duct carcinoma Liver transplantation	Surgical damage of the biliary system performed as hepatojejunostomy	In combination, emodin and celecoxib delivery by dendrimers.	Caspases-9 and -3 activation	[33–35]
	Multivisceral resections Oral 5-fluorouracil (5-FU) or with oral 5-FU plus streptozotocin (Stz) or oral 5-FU + Methyl-chloroethylcyclohexylnitrosourea	Severe drug toxicity		Apoptosis induction Enhanced suppression of Akt activation	
Bladder cancer	Intravesical chemotherapy after the first tumor recurrence were excluded from an analysis of progression	Lamina propria invasion, atypia elsewhere in the bladder	Therapeutic nanocarriers such as dendrimers and GNPs	Low cytotoxicity	[36,37]
	Gemcitabine plus cisplatin (GC) and methotrexate, vinblastine, DOX, and cisplatin (MVAC)	Tumor multiplicity were associated		Higher bioavailability Increased potency	
Bone cancer, osteosarcoma malignant fibrous histiocytoma	Synergistic effect of cucurbitacin and methotrexate Insulin-like growth factor binding protein 5 suppresses tumor growth and metastasis of human osteosarcoma Podoplanin is a candidate molecule for therapeutic targeting	Cytotoxicity	Curcumin encapsulated in cyclodextrins followed by a second encapsulation in liposomes: 2-hydroxypropyl-γ-cyclodextrin/curcumin–liposome complex, nano-hydroxyapatite particles	Biodegradable polymers Personalized treatment	[38–41]
Brain tumor, cerebral astrocytoma/ malignant glioma	Lenalidomide administered orally	The drug could not cross phase II trials because of its lethality	Nanoemulsion of omega-3 fatty acid-containing oils and enhance brain delivery of therapeutics	Effective in crossing the blood–brain barrier	[42,43]
Breast cancer	Vincristine Quercetin Iodinamine Paclitaxel Surgical interventions	Lower bioavailability Greater side effects and adverse effects Low potency Surgery may exert adverse effects	Liposome formulation of co-encapsulated vincristine and quercetin Lonidamine/paclitaxel-loaded, epidermal growth factor receptor (EGFR)-targeted NPs	Significant antitumor activity at two-thirds of the maximum tolerated dose of vincristine, without significant body weight loss in the animals	[44,45]

Nanoparticles—Nerve

Table 1 (*Continued*)   Cancer types and nanotechnology approaches

Cancer	Conventional		Nanotechnology		References
	Diagnosis/treatment	Disadvantage	Applications	Advantages	
Eye cancer, retino-blastoma	Uveal melanoma detected by binocular indirect ophthal-microscopy using indocya-nine green angiography to visualize the tumor and tumor margins. Treatment by enucleation, local resection, brachytherapy. Chemotherapy by the use of dacarbazine (DTIC) cisplatin Dartmouth regimen (cisplatin + carmustine + DTIC + tamoxifen) chemoimmuno-therapy (α-interferon/interleukin-2)	High failure rates and complica-tions with large tumors or those near the optic nerve	Liposomes, polymeric NPs, dendrimers, nanosuspensions	Effective and high delivery of drug.	[45–48]
Cervical cancer	Development of vaccines targeted to genital human papillomavirus. Radiotherapy to a pelvic and para-aortic field with that of pelvic radiation and concurrent chemotherapy with fluorouracil and cisplatin	Serious side effects. Higher rate of reversible hematologic effects	GNPs conjugated to EGFR antibodies. Gadolinium-containing NPs coated with folate or PEG accumulate in tumors	Effective diagnosis without large scale side effects and easy biopsy	[49–51]
Chronic lympho-cytic leukemia (CLL)	Combinations of purine analogs, fludarabine or 2-chlorodeoxy-adenosine, with alkylating agents are emerging as new treatments. Rituximab (Rituxan; IDEC Pharmaceuticals, San Diego, CA, and Genentech, Inc, San Francisco, CA; directed against CD20) and Campath-1H (directed against CD52 in CLL)	Cytotoxicity	Vascular endothelial growth factor antibody (AbVF) to the GNPs	Biocompatible. High surface area. Easy characterization and surface functionalization	[14,52]
Colon cancer	Drugs targeting downregulation of transcriptional activation mediated by β-catenin and T-cell transcription factor-4	Cytotoxicity and dosage varia-tions in different age groups	Nanostructures with surface-bound ligands for the targeted delivery and ablation of colorectal cancer	Incorporation of iron or iron oxide into such structures would provide advantages for MRI	[53–55]
Esophageal cancer	Surgery or radiation therapy, cisplatin and fluorouracil combined therapy, chemora-diotherapy, poly(L-glutamic acid)-paclitaxel and concurrent radiation	Toxic effects. Acute morbidity. Toxicities of gastritis, esophagitis, neutropenia, and dehydration	P450-activated bio reductive drug and conjugated dendrimer bonded therapeutics	Efficiency. Minimal toxicity	[56–59]

Nanoparticles—Nerve

Table 1 (*Continued*)   Cancer types and nanotechnology approaches

Cancer	Conventional		Nanotechnology		References
	Diagnosis/treatment	Disadvantage	Applications	Advantages	
Extragonadal germ cell tumor	Cisplatin containing combination chemotherapy regimens. Tumors remaining after chemotherapy were surgically removed	Disease recurrence	Cyclophosphamide and DOX Cisplatin-based nano-carriers and dendrimers	Optimal dose and combination of cytotoxic drugs, duration of therapy, and choice of platinum compound	[15,60–64]
	Vinblastine, bleomycin, *cis*-diamminedichloroplatinum Ifosfamide 11 drugs used	Drug toxicity	Paclitaxel, in combination with platinum nanobots		
	TI-CE regimen (paclitaxel [T] plus ifosfamide [I] followed by high-dose carboplatin [C] plus etoposide [E] with stem-cell support)				
Extrahepatic bile duct cancer	Intraductal ultrasonography, radiotherapy	Gastroduodenal complications increased	Calecoxib and emodin: inhibitors of epidermal growth factor receptor 2 protein, over expression and COX-2 up-regulation associated nanosized delivery systems	High bioavailability Reduced toxicity Increased potency	[35,65–67]
Gallbladder cancer	Hepatic wedge resection or segmentectomy trisegmentectomy lobectomy chemotherapy and radiotherapy	Early detection necessary as the tumor in gall bladder may rapidly spread to liver Damage to other vital organs during surgery	COX-2 is a potential new and effective therapy alone or in combination with other therapeutic agents using GNPs	Downregulated the activation of ERBB 2 and EGFR gene, resulting in decreased levels of phosphorylated Akt: Specificity	[68–70]
Gastric (Stomach) cancer	Endoscopic mucosal resection (EMR), laparoscopic gastrectomy, modified gastrectomy A and B (MG A and B), standard gastrectomy, extended gastrectomy, chemotherapy, radiotherapy, multimodality therapy (including neoadjuvant and adjuvant chemotherapy, immunochemotherapy, hyperthermochemotherapy), and terminal care	High incidence of postoperative complications and high operative mortality rates no or marginal survival benefit, with moderate or severe toxicity	Tumor-specific ligands, antibodies, anticancer drugs, and imaging probes with NP formulation liposomes and other lipid-based carriers (such as lipid emulsions and lipid–drug complexes), polymer–drug conjugates, polymer microspheres, micelles, and various ligand-targeted products (such as immunoconjugates) liposomal DOX [Doxil], liposomal DOX [Myocet], and liposomal daunorubicin [DaunoXome] NP formulation of paclitaxel bound to albumin	Greater targeting selectivity and better delivery efficiency Enhanced sensitivity and specificity Selectively bind and target cancer cells Gold nanocages are easy to synthesize and manipulate	[13,71,72]

(Continued)

Table 1 (***Continued***)   Cancer types and nanotechnology approaches

	Conventional		Nanotechnology		
Cancer	**Diagnosis/treatment**	**Disadvantage**	**Applications**	**Advantages**	**References**
Heart cancer	Chemotherapy and radiation therapy Commercially available antisera polychemotherapy	Overall prognosis is dismal Patients develop hypertrophic cardiomyopathy, moderate pressure gradient on the pulmonary artery valve, pericardial effusion and two cardiac tamponades (serous punctuate)	QDs in detection and imaging of cancer nanoconjugated drugs	More efficient and accurate drug delivery and rationally designed, targeted drugs	[73–75]
Hepatocellular (liver) cancer	Hepatitis B virus (HBV) vaccines to prevent the chronic carriage of HBsAg and the development of primary liver cancer when vaccination takes place at birth	High lethality Complex interaction between alcohol and viral infections in the etiology of primary liver cancer	Fe_2O_3 NPs combined with magnetic fluid hyperthermia novel organic dye-doped silica NPs are prepared with a water-in-oil microemulsion technique.	Inhibit the proliferation and induce apoptosis in cell lines. Significant inhibitory effect on the weight and volume of xenograft liver cancer Can identify the target cells selectively and efficiently The fluorescent NP label also exhibits high photostability	[48,76–78]
Hodgkin's lymphoma	Modern combination chemotherapy and radiotherapy gallium-67 scintigraphy radiation techniques and combination chemotherapeutic agents like (DOX, bleomycin, vinblastine, dacarbazine)	Longer follow-up has shown serious long-term adverse effects of the treatment, including heart and lung disease, and secondary malignancies	Anti-CCR4mAb could be a novel treatment modality for patients with CCR4-positive HL Dendrimers and QDs	Lower their risk of treatment failure Avoid unnecessary toxicity Increase the chance of long-term survival enhance antibody-dependent cellular cytotoxicity	[79,80]
Hypopharyngeal cancer	Surgery and postoperative radiotherapy (RT), alternative treatment approaches include induction chemotherapy and radiation therapy	Poor outcome Prevalence of bilateral metastases	Poly(ethylene glycol)-*block*-poly(D,L-lactic acid) (PEG-*b*-PLA) micelles act as a 3-in-1 nanocontainer for three poorly water-soluble drugs—paclitaxel, 17-allylamino-17-demethoxygeldanamycin, and rapamycin (PTX/17-AAG/RAPA)–cancer therapy	Specificity in targeted drug delivery and higher potency	[81,82]

Nanoparticles—Nerve

(*Continued*)

Table 1 (*Continued*)　　Cancer types and nanotechnology approaches

Cancer	Conventional		Nanotechnology		References
	Diagnosis/treatment	**Disadvantage**	**Applications**	**Advantages**	
Islet cell carcinoma (endocrine pancreas)	Subclinical, micrometastatic activity prior to surgery molecularly targeted, radioactive peptides (somatostatin, secretin, bombesin) employed in tumor imaging	Dysregulation of genes at metastatic sites independent of the primary tumor	NP conjugate targeted to bombesin (BN) receptors present on normal acinar cells of the pancreas BN peptide-NP conjugate, BN-CLIO(Cy5.5)	Enhanced the ability to visualize tumor in a model of pancreatic cancer by MRI promising approach to imaging.	[83,84]
Kaposi sarcoma	Pegylated–liposomal DOX	Toxicity	Micro- and nanodrug delivery systems	Cancer cell–specific delivery vehicles	[72,85,86]
	A combination of DOX (Adriamycin, Adria Laboratories, Columbus, Ohio, USA), bleomycin, and vinblastine (ABV)		Pegylated liposomal DOX conjugated nanodrug	Do not interact with targeted tumor-specific proteins	
			liposome-based formulations of several anticancer agents (Stealth liposomal DOX [Doxil], liposomal DOX [Myocet], and liposomal daunorubicin [DaunoXome]		
Kidney cancer (renal cell cancer)	Chemotherapeutic drugs and cumulative dose	Excess bladder cancer risk following treatment	Delivery of antineoplastic and contrast agents, neutron capture therapy, photodynamic therapy, and photothermal therapy, dendrimers	Biologically relevant properties such as lipid bilayer interactions, cytotoxicity, internalization, blood plasma retention time, biodistribution, and tumor uptake	[87,88]
	Cyclophosphamide alone, radiation alone, or both therapies	Increased risk of bladder malignancy			
	Other late sequels of therapy				
Laryngeal cancer	Induction chemotherapy with cisplatin plus fluorouracil followed by RT	Mucosal toxicity of concurrent RT and cisplatin poor response to chemotherapy, chemotherapy-related toxic effects (e.g., neutropenia and nausea or vomiting) and increased rates of severe radiation-related mucosal, pharyngeal, and esophageal effects	Metallic-, dielectric-, magnetic-, liposomal-, and carbon-based structures gold, iron oxide, carbon, dielectric materials, molecular, liposomal) and shapes including solid NPs (e.g., spheres, rods, triangles, cubes), nanoshells (within inner and outer cores), nanocages, nanowire, nanotubes, branched dendrimers, and polymeric and organic lipid NPs	Multifunctional devices capable of sensitive, specific diagnosis and simultaneous therapy / Toxicity of attached drugs is significantly reduced	[89–91]
Leukemia	Chemotherapy (daunorubicine, cytarabine) prognostic factors (not only cytogenetical but also molecular biological characers of the disease), allogenic stem cell transplantation	Cytotoxic	Imatinib mesylate and other tyrosine kinase inhibitors (TKI) bound to gold dendrimers	Most suitable for vaccine development and evaluated their appropriate characteristics	[92–94]

Nanoparticles—Nerve

Table 1 (*Continued*)   Cancer types and nanotechnology approaches

Cancer	Conventional		Nanotechnology		References
	Diagnosis/treatment	Disadvantage	Applications	Advantages	
Lip and oral cavity cancer	Cryosurgical techniques surgery and RT	High risk for surgical treatment because of coagulopathy or severe cardiopulmonary disease Depressive symptomatology Significant deterioration of physical functioning and symptoms	Selective laser photothermal therapy of epithelial carcinoma using anti-EGFR antibody conjugated GNPs	Novel class of selective photothermal agents using a laser at low powers	[95–97]
Lung cancer, small cell	Carboplatin, a cisplatin analog Radiation treatment was identified as either beam radiation or any combination involving implants or radioisotopes Chemotherapy was identified as single agent or multiagent. The use of hormone treatment and biologic response modifiers also was recorded	Nausea or vomiting occurred Leukopenia Thrombocytopenia	Multifunctional NPs gold-plated dendrimers conjugated	Less toxicity and better tolerance than cisplatin	[98,99]
Melanoma	Dose-escalating Phase I/II study of retrovirus-mediated herpes simplex virus type 1 thymidine kinase (HSV-1-TK) suicide gene therapy for metastatic melanoma	Cellular toxicity, poor gene transfer efficiency, and lower potential for treatment efficacy	Super paramagenetic iron oxide nanoparticles (SPION) coated with polyvinyl alcohol (PVA) modified by functional groups (amino-SPION, carboxy-SPION, and thiol-SPION)	The same particles generate heat, which can be used, e.g., for local hyperthermic treatment of tumor cells	[100,101]
Neuroblastoma	Antisurvivin antibody at stage III of the treatment RT	Necrosis and local tissue damage	Lipidic gold porphyrin NPs SPIONs, GD2 mAb (anti-GD2) was conjugated to acidified CNTs	Decrease systemic toxicity	[102,103]
Ovarian cancer	Induction chemotherapy, combination chemotherapy, therapy with platinum and a taxane; these drugs include topotecan, etoposide, pegylated liposomal DOX, epirubicin, gemcitabine, altretamine, oxaliplatin, and vinorelbine	Complications but no postoperative deaths	Liposome-encapsulated formulations of DOX Albumin conjugate paclitaxel Transferrin receptor and folate receptor-targeted drug delivery molecules	Noncytotoxic compounds exquisite modification for binding to cancer cell membranes, the microenvironment, or to cytoplasmic or nuclear receptor sites	[104–109]

(Continued)

Table 1 (*Continued*) Cancer types and nanotechnology approaches

| Cancer | Conventional | | Nanotechnology | | References |
	Diagnosis/treatment	Disadvantage	Applications	Advantages	
Pancreatic cancer	Chemotherapy	Cytotoxicity and side effects	Nanopolymers: micellar aggregates of cross-linked and random copolymers of nisopropylacrylamide, with *N*-vinyl-2-pyrrolidone polymeric and poly(ethyleneglycol) monoacrylate (PEG-A)	Nanocurcumin inhibits the growth of pancreatic cancer cell lines	[110–112]
Parathyroid cancer	Surgical, comprising an initial en bloc resection of the tumor and adjacent structures chemoprotective and hypocalcemic agent WR-2721, S-2-(3-aminopropylamino) ethyl-phosphoro-thioic acid	Familial hyper-parathyroidism, multiple endocrine neoplasia type 1 high recurrence rates	Docetaxel-loaded solid lipid nanoparticles (DSNs)	DSNs induced more minor decreases in body weight gains, slighter hemotoxicity (changes in some clinical hematology and biochemistry parameters), cardiac toxicity, hepatotoxicity, and myelosuppression than Taxotere	[113–116]
Penile cancer	Sentinel lymph node (sN) biopsy combined technique, adjuvant chemotherapy	Neonatal circumcision and potential resulting complications are associated with penile cancer	Nanocolloid around the primary tumor	Reveal disease that might otherwise go undetected by conventional surgical and pathological methods	[117–120]
Pharyngeal cancer	Larynx-preserving treatment, induction chemotherapy, radiation therapy in patients who showed a complete response or surgery in those who did not respond) with conventional treatment (total laryngectomy with partial pharyngectomy, radical neck dissection, and postoperative irradiation)	High rates of recurrence and cytotoxicity due to high energy irradiations A partial response in the neck	Induction chemotherapy (cisplatin) given as a bolus intravenous injection, followed by infusion of FU conjugated with NPs of gold or platinum	Lowered cytotoxicity and reduced re-emergence of the carcinoma Various functional abnormalities (pharyngeal paresis, epiglottic dysfunction, dysfunction of the laryngeal vestibule, cricopharyngeal incoordination, and cervical esophageal webs)	[121–123]
Pituitary adenoma	Surgery and postoperative radiation as initial therapy radiation dose, and preoperative tumor size Permanent hormone replacement therapy Transsphenoidal microsurgery Surgery and RT	The major complication, hypopituitarism requiring hormonal replacement with thyroxine, glucocorticoid, and sex hormone	Development of nanomedicine, a medical specialty based on the use of intelligent nanoinstruments (nanobots) with pharmacological therapies, with dopamine agonists such as bromocriptine, cabergoline, and quinagolide	Self-destruction and nontoxic elimination of nanobots	[32,124,125]

Nanoparticles—Nerve

Table 1 (*Continued*) Cancer types and nanotechnology approaches

| Cancer | Conventional | | Nanotechnology | | References |
	Diagnosis/treatment	Disadvantage	Applications	Advantages	
Primary central nervous system lymphoma	Chemotherapy with high-dose intravenous methotrexate, cyclophosphamide, hydroxy-daunomycin/DOX, Oncovin (vincristine), and prednisone, high-dose cytosine arabinoside, and intra-arterial methotrexate	High rate of neurotoxicity, with older age and long follow-up as the primary risk factors reducing relapse rates and improving disease control while minimizing neurotoxicity	Differentiating tumor, neural lesions, and necrosis from healthy brain tissue; methods of delivery of imaging agents across the blood–brain barrier, and new iron oxide–based NP contrast agents for MRI	Nanotechnology used in imaging, diagnosis, and therapy in the central nervous system	[126–128]
Prostate cancer	Radiation or androgen deprivation after radical prostatectomy Definitive local therapy for prostate cancer	Increased rates of PSA recurrence and secondary treatment were calculated for patients with positive and negative surgical margins	Lymphotropic superparamagnetic NPs high-resolution MRI	Gain access to lymph nodes by means of interstitial–lymphatic fluid transport, could be used in conjunction with high-resolution MRI	[129,130]
Rectal cancer	Performance of lower anterior resections has limited the rate of abdominoperineal excision with permanent colostomy	Three pelvic recurrences have developed but there have been no staple-line recurrences in patients who had "curative" surgery	Nanoscale size of quantum dots, metal colloids, superparamagnetic iron oxide, and carbon-based nanostructures	"Theranostic" NPs that are both diagnostic and therapeutic by design	[131,132]
Retinoblastoma	Carboplatin, vincristine, etoposide) or chemothermo-therapy DTIC and cisplatin	Enucleation remained the treatment of choice in those eyes with total retinal detachment and diffuse vitreous seeding	Drug release nanodevices are made of polymeric (PLLA, PLGA, poly-acrylic acid, or polyamidoamine) or biological materials (oligosaccharides, albumin, or chitosan)	Low toxicity and high potency Low bone marrow toxicity Intratumor injection, intratumor implantation, and targeted delivery Enhance the ocular biocompatibility	[133,134]
Testicular cancer	Three-drug combination consisting of *cis*-diamminedichloroplatinum, vinblastine, and bleomycin Standard chemotherapy consisted of dactinomycin, alone or in combination with chlorambucil and methotrexate	Significant toxicity during the first 12 weeks of this therapeutic regimen	Cisplatin plus vinblastine plus bleomycin chemotherapy Salvage therapy with cisplatin plus etoposide	Reduction in neuromuscular toxicity Patients were continuously disease-free	[135–137]

Nanoparticles—Nerve

(*Continued*)

Table 1 (*Continued*)   Cancer types and nanotechnology approaches

| Cancer | Conventional | | Nanotechnology | | References |
	Diagnosis/treatment	Disadvantage	Applications	Advantages	
Thymoma and Thymic carcinoma	Initial chemotherapy postoperative bronchoscopy Cisplatin, vinblastine, and ifosfamide Chemotherapy including cisplatin, carboplatin, and ifosfamide	Cytokeratin and vimentin antibodies reacted positively in the neoplastic cell	Therapeutic siRNA and a receptor-binding RNA aptamer into individual pRNAs Protein-free nanoscale particles containing the receptor-binding aptamer or other ligands resulted in the binding and co-entry of the trivalent therapeutic particles into cells	Promise for the repeated long-term treatment of chronic diseases Lowered cytotoxicity	[138,139]
Thyroid cancer	Sorafenib multikinase inhibitor of the BRaf, vascular endothelial growth factor receptor-2, and platelet-derived growth factor receptor-B kinase Oncaspar, Neulasta	Anthracycline-related toxicity cardiotoxicity	N-(2-hydroxypropyl) methacrylamide copolymer-DOX	Fivefold reduction in anthracycline-related toxicity, and no cardiotoxicity was observed Polymer anthracycline conjugation can bypass MDR	[140,141]
Vaginal cancer	Operative therapy Radiation therapy Surgical and radiation techniques Topical 5-FU Interstitial therapy	Failure to prevent recurrences and by their tendency to reduce vaginal function	Stilbestrol tamoxifen conjugated nanobots, quantum dots	Facilitate local control without increasing the complication rate	[142,143]

cells, whereas the NP's theranostic properties were conferred via the coencapsulation of Cytochrome *c* and a fluorescent dye indocyanine green dye (ICG), highlighting their theranostic potential facilitating tumor regression monitoring.[149]

IT-101, a cyclodextrin polymer-based NP containing camptothecin, modified through the attachment of a 1,4,7,10-tetraazacyclododecane-1,4,7-tris-acetic acid ligand to bind $^{64}Cu^{2+}$. PET data from ^{64}Cu-labeled IT-101 are used to quantify the *in vivo* biodistribution in mice bearing Neuro2A subcutaneous tumors.[150]

Magnetic nanocrystals modified to conjugate with a cancer-targeting anticarcinoembryonic antigen monoclonal antibody (mAb)[144] have also shown great potential in molecular imaging of tumors by MRI and finds application in cancer detection *in vivo*. Magnetic iron oxide nanocrystals combined with QDs can lead to better clinical imaging by generating intense MRI signals.[7] These probes additionally labeled with an antibody against polysialic acid molecules have shown potential for more specific and efficient detection and targeting of lung tumors.[148] SPIONs capable of entering into lymph nodes enable efficient detection of undetectable lymph node metastases by high-resolution MRI and find application in the detection of nodal metastases in prostate cancer patients with surgical lymph node resection or biopsy by

MRI.[8] Occult lymph node metastases in patients with prostate cancer have been identified by this technique prior to salvage radiation therapy.[8]

NPs in cancer targeting and monitoring

Tumor markers, reported in the tumor tissues and body fluids, including transcription factors, cell surface receptors, and secreted proteins are indicative of tumor state, disease status, and response to therapies and are essential diagnostic and prognostic tools in cancer.[148] GNPs[97] conjugated to antiepidermal growth factor receptor (anti-EGFR) mAbs specifically bind to cancer cells with greater affinity over normal cells thus proving useful in diagnosis and detection of oral epithelial cancer cells[8] and cervical cancer (Table 1). Nontoxic to normal cells, it offers advantage over conventional methods of detection and diagnosis of cancer.

Theragnostic multifunctional NPs with improved biodistribution in tumor tissues act as ultrasensitive personalized biosensors with high specificity hold great promise in simultaneous diagnosis of disease, targeted drug delivery with minimal toxicity, and monitoring of treatment at an early stage, and delivering anticancer agents over an extended period for enhanced therapeutic efficacy.[144,151]

Tumor-homing chitosan-based NPs (CNPs) with unique properties of stability in serum, deformability, and rapid uptake by tumor cells are applied in cancer diagnosis and therapy.[13] These properties are critical in increasing their tumor-targeting specificity and reducing their nonspecific uptake by normal tissues.[4] The use of CNPs containing a near-infrared fluorescent dye and an anticancer drug for monitoring of the cancer cells have been reported.[4] Simultaneous diagnosis by NPs of multiple trace biomarkers or cancer cells and monitoring in the early stages of tumor development hold promise[145,152,153] in cancer detection and targeting. An external Coulomb's force between negatively charged DNA backbones and negatively charged GNP probes stability of DNA double-helix with potential application toward the detection of single nucleotide polymorphisms[151] in cancer.

POLYMERIC NPs IN CANCER RESEARCH AND MANAGEMENT

The biodegradable nature of PNPs, including poly(D,L-lactic acid) (PLA), poly(D,L-lactic-co-glycolic acid) (PLGA), and poly(ε-caprolactone) and their copolymers diblocked or multiblocked with PEG finds application in encapsulated drug delivery. These include polymeric micelles, capsules, colloids, and dendrimers. Genexol-PMTM, a PLGA-b-methoxyPEG NP encapsulating paclitaxel, is in Phase II clinical trial for cancer.[154] Formulations of self-assembled copolymer blocks of two or more polymer chains with different hydrophobicity are being tested due to their ability to spontaneously assemble into a core-shell structure in an aqueous environment thereby minimizing the system's free energy. While the hydrophobic blocks minimize their exposure to aqueous environment, the hydrophilic blocks of shell stabilize the core through direct contact with water and altering the drug release rates, thus making the PNPs promising tools for cancer management. PLGA-b-PEG NPs have been reported to be promising in prostate cancer for controlled drug release. NPs functionalized using the A10 2-fluoropyrimidine ribonucleic acid (RNA) aptamers as a model targeting moiety binds to the prostate-specific membrane antigen (PSMA), prostate tumor marker revealed targeted delivery of polymeric NPs in human prostate cancer cell lines *in vitro* and tumors in xenograft mouse models of prostate cancer *in vivo*.[154]

Sub-200 nm particulates from PLGA and the cationic surfactant dimethyl dioctadecylammonium bromide (DDAB) enabled greater transfection rates for anionic DNA carrying genes *in vitro*. Multiple particle tracking, revealed higher average transport rate of PLGA–DDAB/DNA, stability in mucus, and the ability to transfect cells make PLGA–DDAB/DNA NPs candidates for mucosal DNA vaccines and gene therapy.[155] Encapsulation of poly lactic acid-co-glycolic acid by siRNA prepared by particle replication in nonwetting templates technique is being used as prostate cancer therapeutics *in vitro* to knock down genes relevant to prostate cancer.[156]

The main challenge of treatment of bone metastases revolves around the in situ poor distribution of the drug. Zoledronate (ZOL) with a strong affinity to bone together with docetaxol has been reported to show synergistic effects in bone targeting and efficient management of bone metastasis over conventional methods. ZOL-conjugated PLGA NPs with more cellular uptake by MCF-7 and BO2 cell line have been reported to enhance cell cytotoxicity, cell cycle arrest, and upregulate apoptotic activity by blocking mevalonate pathway and increase the accumulation of apoptotic metabolites, such as ApppI. *In vivo* animal studies revealed reduced liver uptake and a significantly higher retention of ZOL-tagged NPs at the bone site with enhanced tumor retention thus conferring promises to targeting and treatment of bone cancer.[157]

Synergistic effects of DOX and CUR in a single NP formulation have been shown to effectively deliver drug leading to inhibition of MDR, enhancing antiproliferative activity and promoting cytotoxicity and apoptotic effects, thus contributing significantly to clinical management of leukemia.[158] Noscapine, which is a tubulin-binding anticancer drug when loaded with PEG-grafted gelatine NPs provides promise in breast cancer management due to long-circulating injectable property over conventional methods of oral administration and intravenous administration of noscapine.[159]

(−)-Epigallocatechin 3-gallate EGCG-loaded NPs consisting of biocompatible polymers, functionalized with small molecules targeting PSMA, have been reported in effective targeting of PSMA-expressing prostrate cancer cells.[160]

Biocompatible hyperbranched poly(amine-ester) macromers with terminal CC modification are being exploited in making injectable hydrogels as a multidrug delivery system. Combination of more than one drug, including DOX hydrochloride, 5-fluorouracil (5FU), and leukovorin calcium have been used to deliver together. Such a device enabled controlled local release of drugs simultaneously and holds promise for locally delivering single and/or multiple drugs in cancer chemotherapy.[161]

Nanoconstructs of polyethylenimine (bPEI, 25 kDa) and chondroitin sulfate have been reported to impart site-specific property when injected in Ehrlich ascites tumor-bearing mice, thus demonstrating their role as carriers for nanomedicine for efficient management of solid tumor.[162]

Liposomes, nanostructures, and natural and synthetic polymers have been largely exploited in the design and delivery of taxol and taxane anticancer agents.[163] Polybutylcyanoacrylate NPs of diallyl trisulfide (DATS–PBCA–NP) had significant antitumor effects on orthotopic-transplanted HepG2 hepatocellular carcinoma in nude mice without causing weight loss, upregulation of apoptotic index, downregulated expression of proliferation cell nuclear antigen and

Bcl-2 proteins, with good prolonged release effect *in vivo* and hepatic-targeted activity.[164]

Polybutylcyanoacrylate nanoparticles (PBCA-NP) loaded with mitomycin C (MMC-PBCA–NP), have been reported to form a highly effective anticancer drug system as it accumulated more in the liver when injected into mice and revealed antiproliferative effects on the tumor cells.[165]

Improved chemical stability of encapsulated drug, enhanced accumulation in tumors, prolonged drug exposure, and linked to mAb fragments against cancer-associated antigens delivery of drug by pegylated liposome have been reported. Immunoliposomes prepared by combining antibody-mediated tumor recognition with liposomal delivery and when designed for target cell internalization, provide intracellular drug release. An immunoliposome consisting of novel anti-HER2 scFv F5 conjugated to pegylated liposomal DOX (Doxil/Caelxy PLD), have been reported to selectively bind to and internalize in HER2-overexpressing tumor cells. The modular organization of immunoliposome technology enables a combinatorial approach in which a repertoire of mAb fragments can be used in conjunction with a series of liposomal drugs to yield a new generation of improved molecularly targeted agents.[166]

Different polymeric nanoparticle formulations carrying the anticancer compound Curcumin and delivery devices are being tested with promising results as effective agents in cancer management.[167]

Thiolated chitosan with efficient transfection and mucoadhesive properties informulations of NAC-C (*N*-acetyl cysteine-chitosan) and NAP-C (*N*-acetyl penicillamine-chitosan) are being used in anticancer drug delivery targeting EGFR. The drug DOX and antisense oligonucleotide-loaded on such PNPs by gelation process have been reported of being efficient in drug delivery on a breast cancer cell line (T47D).[168]

Echogenic PNPs approved by the US Food and Drug Administration composed either of polymers PLGA or polystyrene have been reported for gene delivery with the help of ultrasound.[169] PBCA NPs coated with polysorbate-80 have been reported to deliver chemotherapeutic drug, temozolomide, into the brain revealing its therapeutic applications.[170] Poly(ethylene glycol)-modified poly(D,L-lactide-*co*-glycolide) nanoparticles (PLGA–PEG–NPs) loading 9-nitrocamptothecin (9-NC) as a potent anticancer drug.[171] Engineered PLGA NPs incorporating 1) two major histocompatibility complex (MHC) class I-restricted peptides; 2) an MHC class II-binding peptide; and 3) a nonclassical MHC class I-binding peptide loaded onto human and murine dendritic cells (DCs) were found to improve peptide presentation and efficient generation of peptide-specific cytotoxic T lymphocyte (CTL) and T-helper cell responses thereby activating the immune system, which highlights its significant role in the development of synthetic vaccines, and the induction of CTL for adoptive immunotherapy.[172]

Biodegradable core-shell structured NPs with a poly(β-amino-ester) core enveloped by a phospholipid bilayer shell are being developed for *in vivo* mRNA delivery, with a view toward delivery of noninvasive mRNA-based vaccines.[173] EGFR-targeted, polymer blend nanocarriers are being developed for the treatment of MDR in cancer using paclitaxel and lonidamine acting as a mitochondrial hexokinase-2 inhibitor.[174]

Novel chemopreventive combination polymers such as chitosan, pectin, and hydroxypropyl methylcellulose were found to improve drug delivery efficiency as compared with aspirin in combination of calcium and folic acid in colon cancer cells.[175]

Biopolymeric poly(L-γ-glutamylglutamine)–docetaxel conjugate made by direct esterification was found to be effective in targeting human non–small cell lung cancer cell line NCI-H460 in mice.[176] Linear, cyclodextrin-containing polymers (CDPs) are being tested for loading of prodrug/drug.[177]

Biocompatible superparamagnetic iron oxide NPs and the anticancer drug, DOX hydrochloride were encapsulated in PLGA–PEG NPs for local treatment of lung cancer. The magnetic properties conferred help to maintain the NPs in the joint with an external magnet, and the drug release kinetics were controlled.[178] Preclinical tests in ovarian tumor-bearing mice to evaluate by cationic biodegradable poly(β-amino ester) polymer as a vector for nanoparticulate delivery of DNA encoding a diphtheria toxin suicide gene (DT-A) when administered intraperitoneally could suppress tumor growth more effectively than by cisplatin and paclitaxel. The DT-A encoding DNA, combined with transcriptional regulation enabled gene expression thereby proving an effective therapy for advanced-stage ovarian cancer.[179]

Cisplatin has been delivered successfully platinum (Pt) (IV)-encapsulated PSMA targeted NPs of PLGA–PEG-functionalized controlled release polymers. By using PLGA–b–PEG NPs with PSMA targeting aptamers (Apt) on the surface as a vehicle for the Pt(IV) compound c,t,c-$[Pt(NH_3)_2(O_2CCH_2CH_2CH_2CH_2CH_3)_2Cl_2]$ (1), a lethal dose of cisplatin could be delivered specifically to prostate cancer cells.[180] Cisplatin NPs with high encapsulation efficiency and reduced toxicity after using cisplatin-cross-linked carboxymethyl cellulose (CMC) core NPs made from poly(lactide-*co*-glycolide)-monomethoxy-poly (ethylene glycol) copolymers (PLGA–mPEG) have been proven effective in drug delivery in the treatment of ovarian cancer.[181]

Stimulation of CD40 and Toll-like receptor (TLR) leads to DC and CD8+ T-cell activation of the immune system. B16F10 melanoma tumors when injected with plasmid DNA encoding a multimeric, soluble form of CD40L (pSP-D-CD40L) either alone or combined with an agonist for TLR1/2 (Pam(3)CSK(4)), TLR2/6 (FSL-1 and MALP2), TLR3 (polyinosinic-polycytidylic acid, poly(I:C)), TLR4 (monophosphoryl lipid A, MPL), TLR7 (imiquimod), or TLR9 (Class B CpG phosphorothioate oligodeoxynu-cleotide, CpG) revealed slowed tumor growth, whereas a combination of these two TLR agonists was more effective than either agent alone. The triple combination of intratu-moral pSP-D-CD40L + CpG + poly(I:C) markedly slowed down tumor growth and prolonged survival, also leading to the reduction in intratumoral CD11c+ dendritic cells and an influx of CD8+ T cells[182] proving it to be an important tool for cancer prevention and treatment.

NP AND TOXICITY ISSUES

Despite promises, NPs are toxic to environment and man.[183,184] Their size enables their entry into the human body, and large surface area to unit mass ratios may lead to proinflammatory effects causing production of reactive oxygen species and free radicals resulting in oxidative stress, inflammation, cytokine production, and consequent damage to proteins, membranes, and DNA.[185,186]

Entering via inhalation, ingestion, and penetration by skin, crossing biological membranes and their large sur-face area enables easy access to tissue and fluids, thus affecting regulatory mechanisms of enzymes and other proteins[183,187,188] leading to diseases such as acne, eczema, shaving wounds, or severe sunburns. Trans-ported through blood, they are taken up by organs and tissues, including the brain, heart, liver, kidneys, spleen, bone marrow, and nervous system[188] causing overall toxicity and cell death.[189] NPs can activate the immune system inducing inflammation, immune responses, and allergy responses. Hence further studies on kinetics and biochemical interactions of NPs, their translocation pathways, accumulation, short- and long-term toxicity, interaction with cells and receptors and signaling path-ways involved, cytotoxicity, and their surface functional-ization for an effective phagocytosis[185] are imperative.

DISCUSSION

Cancer is a very complex disease and its treatment requires constant evaluation of progress of applied therapies on per-sonalized bases. At the same time monitoring levels of tumor markers provides insight into success of the thera-peutic approaches, and requires sensitive and specific sensors.

Theragnostic NPs show great promise as personalized medicine allowing detection as well as monitoring of an individual patient's cancer. Cancer researchers have been actively exploring various tumor-targeting NPs made of lipid-based micelles, natural/synthetic polymeric particles, and inorganic particles for cancer theragnosis.[5,9] But suc-cessful clinical applications in cancer management needs discovery of highly efficient tumor-homing NPs, which can diagnose and deliver targeted therapy and offer a prom-ising new technology in cancer treatment.[4] Polymer nanoparticles have shown promise in the better manage-ment of cancer over conventional methods and some are being tested at the levels of clinical trials (Table 2).

Although nanotechnology has immense application in early detection, diagnosis, and treatment of cancer, it is mandatory to study the side effects of NPs on human health and environment. Very few studies are reported on the migration of NPs and its systemic effects on other organs such as heart, liver, kidney, brain, and skin.[11]

Table 2 Polymeric nanoparticles in clinical trials

Polymeric NPs formulation	Phase of drug treatment	Targets	References
Polyalkylcyanoacrylate (PACA) conjugated with doxorubicin (DOX)	Phase II	Intra-arterial administration to treat patients with advanced hepatocellular carcinoma	[143]
Magnetic iron oxide conjugated Np with 4'-epidoxo-rubicin	Phase I	Monitoring brain tumors	[143]
BIND-014 with docetaxel CALAA-01 (cyclodextrin-polymer hybrid) with transferrin	Phase I	Solid tumors	[190,191]
SEL-068	Phase I	Smoking cessation and relapse prevention vaccine	[192]
Genexol-PM (paclitaxel-loaded polymeric micelle)	Phase II	Breast cancer treatment	[193]
SP1049C (DOX-entrapping hydrophobic core and a hydrophilic polymer)	Phase II	Metastatic cancer of the esophagus and esophageal junction	[194]
MCC-465 (liposome encapsulated DOX)	Phase I	Human stomach cancer cells against GAH-positive xenografts	[195]
SGT53-01 (transferring receptor targeting polymer)	Phase I	Targets the TfR on the surface of cancer cells and results in the expression of p53 gene in the targeted cancer cells	[196]

(Continued)

Table 2 (*Continued*)   Polymeric nanoparticles in clinical trials

Polymeric NPs formulation	Phase of drug treatment	Targets	References
Onco TCS (Transmembrane carrier systems)	Phase I/ Phase II	Treatment of aggressive non-Hodgkin's lymphoma	[197]
NKTR-118 (PEG–naloxol)	Phase I	Treating opioid-induced constipation	[198]
Hepacid (PEG–arginine deaminase)	Phase II	Hepatocellular carcinoma	[198]
Puricase (PEG–uricase)	Phase III	Hyperuricemia	[198]
N-(2-hydroxypropyl)methacrylamide (HPMA) copolymer–DACH platinate	Phase II	Ovarian cancers	[198]
L-leucine, L-glutamate copolymer, and insulin	Phase II	Type-I diabetes	[198]
PEG-anti tumor necrosis factor-α antibody fragment	Phase III	Rheumatoid arthritis and Crohn's disease	[198]
Pluronic block-copolymer DOX	Phase II	Esophageal carcinoma	[198]
FCE28069 (HMPA copolymer–DOX-galactosamine)	Phase II	Hepatocellular carcinoma	[199]
PNU166945 (HPMA copolymer–paclitaxel)	Phase I	Toxicity and pharmacokinetics for treating refractory solid tumors	[200]
Phosphatidylcholine conjugated polymeric NP	Phase IIa/b	Lipid supplementation in the colon led to mucosal healing	[201]

Due to their unique properties of small size, they have been reported to remain suspended in air for longer time and their greater surface area enables them to adsorb pollutants, oxidant gases, organic compounds, and transition metals thereby pointing toward potential source of environmental pollutants. Thus despite promises, challenges exist over applications of NPs, their toxicity, and their removal from the system, and it would need an integrated research to overcome the pitfalls in the application of NPs and increase their efficacy in successful cancer management.

REFERENCES

1. Bode, A.M.; Dong, Z. Cancer prevention research - then and now. Nat. Rev. Cancer **2009**, *9* (7), 508–516.
2. Ehdaie, B. Application of nanotechnology in cancer research: Review of progress in the National Cancer Institute's Alliance for Nanotechnology. Int. J. Biol. Sci. **2007**, *3* (2), 108–110.
3. Wang, L.; Wang, K.M.; Santra, S.; Zhao, X.; Hilliard, L.R.; Smith, J.E.; Wu, Y.; Tan, W. Watching silica nanoparticles glow in the biological world. Anal. Chem. **2006**, *78* (3), 646–654.
4. Kim, K.; Kim, J.H.; Park, H.; Kim, Y.S.; Park, K.; Nam, H.; Lee, S.; Park, J.H.; Park, R.W.; Kim, I.S.; Choi, K.; Kim, S.Y.; Park, K.; Kwon, I.C. Tumor-homing multifunctional nanoparticles for cancer theragnosis: Simultaneous diagnosis, drug delivery, and therapeutic monitoring. J. Control. Release **2010**, *146* (2), 219–227.
5. Cesur, H.; Rubinstein, I.; Pai, I.; Onyüksel, H. Self-associated indisulam in phospholipidbased nanomicelles: Potential nanomedicine for cancer. Nanomedicine **2009**, *5* (2), 178–183.
6. Park, J.W. Liposome-based drug delivery in breast cancer treatment. Breast Cancer. Res. **2002**, *4* (3), 95–39.
7. Huang, H.C.; Chang, P.Y.; Chang, K.; Chen, C.Y.; Lin, C.W.; Chen, J.H.; Mou, C.Y.; Chang, Z.F.; Chang, F.H. Formulation of novel lipid-coated magnetic nanoparticles as the probe for *in vivo* imaging. J. Biomed. Sci. **2009**, *16* (1), 86–95.
8. Jain, K.K. Advances in the field of nanooncology. BMC Med. **2010**, *8* (1), 83–93.
9. Nishiyama, N.; Kataoka, K. Current state, achievements, and future prospects of polymeric micelles as nanocarriers for drug and gene delivery. Pharmacol. Ther. **2006**, *112* (3), 630–648.
10. Martin, F.J. Clinical pharmacology and antitumor efficacy of DOXIL (pegylated liposomal doxorubicin) In *Medical Applications of Liposomes*; Lasic, D.D., Papahadjopoulos, D., Ed.; New York: Elsevier Science BV, 1998, 635–688.
11. Artemov, D.; Mori, N.; Okollie, B.; Bhujwalla, Z.M. MR molecular imaging of the Her-2/neu receptor in breast cancer cells using targeted iron oxide nanoparticles. Magn. Reson. Med. **2003**, *49* (3), 403–408.
12. Wang, X.; Wang, Y.; Chen, Z.G.; Shin, D.M. Advances of cancer therapy by nanotechnology. Cancer. Res. Treat. **2009**, *41* (1), 1–11.
13. Peer, D.; Karp, J.M.; Hong, S.; Farokhzad, O.C.; Margalit, R.; Langer, R. Nanocarriers as an emerging platform for cancer therapy. Nat. Nanotechnol. **2007**, *2* (12), 751–760.
14. Mukherjee, P.; Bhattacharya, R.; Bone, N.; Lee, Y.K.; Patra, C.R.; Wang, S.; Lu, L.; Secreto, C.; Banerjee, P.C.; Yaszemski, M.J.; Kay, N.E.; Mukhopadhyay, D. Potential therapeutic application of gold nanoparticles in B-chronic lymphocytic leukemia (BCLL): Enhancing apoptosis. J. Nanobiotechnol. **2007**, *5* (4), 4–17.
15. Travis, L.B.; Holowaty, E.J.; Bergfeldt, K.; Lynch, C.F.; Kohler, B.A.; Wiklund, T.; Curtis, R.E.; Hall, P.; Andersson, M.; Pukkala, E.; Sturgeon, J.; Stovall, M. Risk of leukemia

after platinum-based chemotherapy for ovarian cancer. N. Engl. J. Med. **1999**, *340* (5), 351–357.

16. Ma, P.; Dong, X.; Swadley, C.L.; Gupte, A.; Leggas, M.; Ledebur, H.C.; Mumper, R.J. Development of idarubicin and doxorubicin solid lipid nanoparticles to overcome Pgp-mediated multiple drug resistance in leukemia. J. Biomed. Nanotechnol. **2009**, *5* (2), 151–161.

17. Reilly, R.M. Carbon nanotubes: Potential benefits and risks of nanotechnology in nuclear medicine. J. Nucl. Med. **2007**, *48* (7), 1039–1042.

18. Ning, N.; Peng, Z.F.; Yuan, L.; Gou, B.D.; Zhang, T.L.; Wang, K. Realgar nano-particles induce apoptosis and necrosis in leukemia cell lines K562 and HL-60. Zhongguo. Zhong. Yao. Za. Zhi. **2005**, *30* (2), 136–140.

19. Ng, L.; Libertino, J.M. Adrenocortical carcinoma: Diagnosis, evaluation and treatment. J. Urol. **2003**, *169* (1), 5–11.

20. Allolio, B.; Fassnacht, M. Clinical review: Adrenocortical carcinoma: Clinical update. J. Clin. Endocrinol. Metab. **2006**, *91* (6), 2027–2037.

21. Kamada, R.; Yoshino, W.; Nomura, T.; Chuman, Y.; Imagawa, T.; Suzuki, T.; Sakaguchi, K. Enhancement of transcriptional activity of mutant p53 tumor suppressor protein through stabilization of tetramer formation by calix[6]arene derivatives. Bioorg. Med. Chem. Lett. **2010**, *20* (15), 4412–4415.

22. Levine, A.M.; Wernz, J. C.; Kaplan, L.; Rodman, N.; Cohen, P.; Metroka, C.; Bennett, J.M.; Rarick, M.U.; Walsh, C.; Kahn, J.; Miles, S.; Ehmann, W.C.; Feinberg, J.; Nathwani, B.; Gill, P.S.; Mitsuyasu, R. Low-dose chemotherapy with central nervous system prophylaxis and zidovudine maintenance in AIDS-related lymphoma. A prospective multiinstitutional trial. JAMA **1991**, *266* (1), 84–88.

23. Lim, S.T.; Karim, R.; Tulpule, A.; Nathwani, B.N.; Levine, A.M. Prognostic factors in HIV-related diffuse large-cell lymphoma: Before versus after highly active antiretroviral therapy. J. Clin. Oncol. **2005**, *23* (33), 8477–8482.

24. Heidel, J.D.; Davis, M.E. Clinical developments in nanotechnology for cancer therapy. Pharm. Res. **2011**, *28* (2), 187–199.

25. Bartelink, H.; Roelofsen, F.; Eschwege, F.; Rougier, P.; Bosset, J.F.; Gonzalez, D.G.; Peiffert, D.; van Glabbeke, M.; Pierart, M. Concomitant radiotherapy and chemotherapy is superior to radiotherapy alone in the treatment of locally advanced anal cancer: Results of a phase III randomized trial of the European Organization for Research and Treatment of Cancer Radiotherapy and Gastrointestinal Cooperative Groups. J. Clin. Oncol. **1997**, *15* (5), 2040–2049.

26. Sinha, R.; Kim, G.J.; Nie, S.; Shin, D.M. Nanotechnology in cancer therapeutics: Bioconjugated nanoparticles for drug delivery. Mol. Cancer. Ther. **2006**, *5* (8), 1909–1917.

27. Culliford, A.T., IV; Brooks, A.D.; Sharma, S.; Saltz, L.B.; Schwartz, G.K.; O'Reilly, E.M.; Ilson, D.H.; Kemeny, N.E.; Kelsen, D.P.; Guillem, J.G.; Wong, W.D.; Cohen, A.M.; Paty, P.B. Surgical debulking and intraperitoneal chemotherapy for established peritoneal metastases from colon and appendix cancer. Ann. Surg. Oncol. **2001**, *8* (10), 787–795.

28. McCusker, M.E.; Coté, T.R.; Clegg, L.X.; Sobin, L.H. Primary malignant neoplasms of the appendix: A population-based study from the surveillance, epidemiology and end-results program, 1973–1998. Cancer **2002**, *94* (12), 3307–3312.

29. Sha, M.Y.; Xu, H.; Penn, S.G.; Cromer, R. SERS nanoparticles: A new optical detection modality for cancer diagnosis. Nanomed. (Lond) **2007**, *2* (5),725–734.

30. Lay, C.L.; Liu, H.Q.; Tan, H.R.; Liu, Y. Delivery of paclitaxel by physically loading onto poly(ethylene glycol) (PEG)-graftcarbon nanotubes for potent cancer therapeutics. Nanotechnology **2010**, *21* (6), 1–10.

31. Dobelbower, M.C.; Burnett Iii, O.L.; Nordal, R.A.; Nabors, L.B.; Markert, J.M.; Hyatt, M.D.; Fiveash, J.B. Patterns of failure for glioblastoma multiforme following concurrent radiation and temozolomide. J. Med. Imaging. Radiat. Oncol. **2011**, *55* (1), 77–81.

32. Randall, R.V.; Laws, E.R., Jr.; Abboud, C.F.; Ebersold, M.J.; Kao, P.C.; Scheithauer, B.W. Transsphenoidal microsurgical treatment of prolactin-producing pituitary adenomas. Results in 100 patients. Mayo Clin. Proc. **1983**, *58* (2), 108–121.

33. Pichlmayr, R.; Weimann, A.; Klempnauer, J.; Oldhafer, K.J.; Maschek, H.; Tusch, G.; Ringe, B. Surgical treatment in proximal bile duct cancer. A single-center experience. Ann. Surg. **1996**, *224* (5), 628–638.

34. Falkson, G.; MacIntyre, J.M.; Moertel, C.G. Eastern Cooperative Oncology Group experience with chemotherapy for inoperable gallbladder and bile duct cancer. Cancer **1984**, *54* (6), 965–969.

35. Lai, G.H.; Zhang, Z.; Sirica, A.E. Celecoxib acts in a cyclooxygenase-2-independent manner and in synergy with emodin to suppress rat cholangiocarcinoma growth *in vitro* through a mechanism involving enhanced Akt inactivation and increased activation of caspases-9 and -3. Mol. Cancer. Ther. **2003**, *2* (3), 265–271.

36. Heney, N.M.; Ahmed, S.; Flanagan, M.J.; Frable, W.; Corder, M.P.; Hafermann, M.D.; Hawkins, I.R. Superficial bladder cancer: Progression and recurrence. J. Urol. **1983**, *130* (6), 1083–1086.

37. Shelley, M.; Cleves, A.; Wilt, T.J.; Mason, M. Gemcitabine for unresectable, locally advanced or metastatic bladder cancer. Cochrane Database Syst. Rev. **2011**, (4), 1–26.

38. Su, Y.; Wagner, E.R.; Luo, Q.; Huang, J.; Chen, L.; He, B.C.; Zuo, G.W.; Shi, Q.; Zhang, B.Q.; Zhu, G.; Bi, Y.; Luo, J.; Luo, X.; Kim, S.H.; Shen, J.; Rastegar, F.; Huang, E.; Gao, Y.; Gao, J.L.; Yang, K.; Wietholt, C.; Li, M.; Qin, J.; Haydon, R.C.; He, T.C.; Luu, H.H. Insulin-like growth factor binding protein 5 suppresses tumor growth and metastasis of human osteosarcoma. Oncogene **2011**, *30* (37), 3907–3917.

39. Kunita, A.; Kashima, T.G.; Ohazama, A.; Grigoriadis, A.E.; Fukayama, M. Podoplanin is regulated by AP-1 and promotes platelet aggregation and cell migration in osteosarcoma. Am. J. Pathol. **2011**, *179* (2), 1041–1049.

40. Dhule, S.S.; Penfornis, P.; Frazier, T.; Walker, R.; Feldman, J.; Tan, G.; He, J.; Alb, A.; John, V.; Pochampally, R. Curcumin-loaded gamma-cyclodextrin liposomal nanoparticles as delivery vehicles for osteosarcoma. Nanomedicine **2012**, *8* (4), 440–451.

Nanoparticles—Nerve

41. Duncan, R. Polymer therapeutics as nanomedicines: New perspectives. Curr. Opin. Biotechnol. **2011**, *22* (4), 492–501.

42. Warren, K.E.; Goldman, S.; Pollack, I.F.; Fangusaro, J.; Schaiquevich, P.; Stewart, C.F.; Wallace, D.; Blaney, S.M.; Packer, R.; Macdonald, T.; Jakacki, R.; Boyett, J.M.; Kun, L.E. Phase I trial of lenalidomide in pediatric patients with recurrent, refractory, or progressive primary CNS tumors: Pediatric Brain Tumor Consortium study PBTC-018. J. Clin. Oncol. **2011**, *29* (3), 324–329.

43. Spyratou, E.; Makropoulou, M.; Mourelatou, E.A.; Demetzos, C. Biophotonic techniques for manipulation and characterization of drug delivery nanosystems in cancer therapy. Cancer Lett. **2012**, *327* (1–2): 111–122.

44. Wong, M.Y.; Chiu, G.N. Liposome formulation of co-encapsulated vincristine and quercetin enhanced antitumor activity in a trastuzumab-insensitive breast tumor xenograft model. Nanomedicine **2011**, *7* (6), 834–840.

45. Milane, L.; Duan, Z.F.; Amiji, M. Pharmacokinetics and biodistribution of lonidamine/paclitaxel loaded, EGFR-targeted nanoparticles in an orthotopic animal model of multi-drug resistant breast cancer. Nanomed. Nanotechnol. **2011**, *7* (4), 435–444.

46. Vijayakrishnan, R.; Shields, C.L.; Ramasubramanian, A.; Emrich, J.; Rosenwasser, R.; Shields, J.A. Irradiation toxic effects during intra-arterial chemotherapy for retinoblastoma: Should we be concerned? Arch. Ophthalmol. **2010**, *128* (11), 1427–1431.

47. Polednak, A.P.; Flannery, J.T. Brain, other central nervous system, and eye cancer. Cancer **1995**, *75* (1 Suppl.) 330–337.

48. Lawrence, T.S.; Robertson, J.M.; Anscher, M.S.; Jirtle, R.L.; Ensminger, W.D.; Fajardo, L.F. Hepatic toxicity resulting from cancer treatment. Int. J. Radiat. Oncol. Biol. Phys. **1995**, *31* (5), 1237–1248.

49. Bosch, F.X.; Manos, M.M.; Muñoz, N.; Sherman, M.; Jansen, A.M.; Peto, J.; Schiffman, M.H.; Moreno, V.; Kurman, R.; Shah, K.V. Prevalence of human papillomavirus in cervical cancer: A worldwide perspective. International biological study on cervical cancer (IBSCC) Study Group. J. Natl. Cancer Inst. **1995**, *87* (11), 796–802.

50. Morris, M.; Eifel, P.J.; Lu, J.; Grigsby, P.W.; Levenback, C.; Stevens, R.E.; Rotman, M.; Gershenson, D.M.; Mutch, D.G. Pelvic radiation with concurrent chemotherapy compared with pelvic and para-aortic radiation for high-risk cervical cancer. N. Engl. J. Med. **1999**, *340* (15), 1137–1143

51. Cuenca, A.G.; Jiang, H.; Hochwald, S.N.; Delano, M.; Cance, W.G.; Grobmyer, S.R. Emerging implications of nanotechnology on cancer diagnostics and therapeutics. Cancer **2006**, *107* (3), 459–466.

52. Keating, M.J. Chronic lymphocytic leukemia. Semin. Oncol. **1999**, *26* (5 Suppl. 14), 107–114.

53. Ricci-Vitiani, L.; Lombardi, D.G.; Pilozzi, E.; Biffoni, M.; Todaro, M.; Peschle, C.; De Maria, R. Identification and expansion of human colon-cancer-initiating cells. Nature **2007**, *445* (7123), 111–115.

54. Morin, P.J.; Sparks, A.B.; Korinek, V.; Barker, N.; Clevers, H.; Vogelstein, B.; Kinzler, K.W. Activation of beta-catenin-Tcf signaling in colon cancer by mutations in beta-catenin or APC. Science **1997**, *275* (5307), 1787–1790.

55. Fortina, P.; Kricka, L.J.; Graves, D.J.; Park, J.; Hyslop, T.; Tam, F.; Halas, N.; Surrey, S.; Waldman, S.A. Applications of nanoparticles to diagnostics and therapeutics in colorectal cancer. Trends Biotechnol. **2007**, *25* (4), 145–52.

56. Eisterer, W.; DE Vries, A.; Kendler, D.; Spechtenhauser, B; Königsrainer, A.; Nehoda, H.; Virgolini, I.; Lukas, P.; Bechter, O.; Wöll, E.; Ofner, D. Triple induction chemotherapy and chemoradiotherapy for locally advanced esophageal cancer. A phase II study. Anticancer Res. **2011**, *31* (12), 4407–12.

57. Harvin, J.A.; Lahat, G.; Correa, A.M.; Lee, J.; Maru, D.; Ajani, J.; Marom, E.M.; Welsh, J.; Bhutani, M.S.; Walsh, G.; Roth, J.; Mehran, R.; Vaporciyan, A.; Rice, D.; Swisher, S.; Hofstetter, W. Neoadjuvant chemoradiotherapy followed by surgery for esophageal adenocarcinoma: Significance of microscopically positive circumferential radial margins. J. Thorac. Cardiovasc. Surg. **2012**, *143* (2), 412–420.

58. Dipetrillo, T.; Milas, L.; Evans, D.; Akerman, P.; Ng, T.; Miner, T.; Cruff, D.; Chauhan, B.; Iannitti, D.; Harrington, D.; Safran, H. Paclitaxel poliglumex (PPX-Xyotax) and concurrent radiation for esophageal and gastric cancer: A phase I study. Am. J. Clin. Oncol. **2006**, *29* (4), 376–379.

59. McFadyen, M.C.; Melvin, W.T.; Murray, G.I. Cytochrome P450 enzymes: Novel options for cancer therapeutics. Mol. Cancer. Ther. **2004**, *3* (3), 363–371.

60. Kumano, M.; Miyake, H.; Hara, I.; Furukawa, J.; Takenaka, A.; Fujisawa, M. First-line high-dose chemotherapy combined with peripheral blood stem cell transplantation for patients with advanced extragonadal germ cell tumors. Int. J. Urol. **2007**, *14* (4), 336–338.

61. Giannis, M.; Aristotelis, B.; Vassiliki, K.; Ioannis, A.; Konstantinos, S.; Nikolaos, A.; Georgios, P.; Georgios, P.; Pantelis, P.; Meletios-Athanasios, D. Cisplatin-based chemotherapy for advanced seminoma: Report of 52 cases treated in two institutions. J. Cancer. Res. Clin. Oncol. **2009**, *135* (11), 1495–1500.

62. Deville, J.L.; Gravis, G.; Salem, N.; Savoie, P.H.; Esterni, B.; Walz, J.; Thomas, P.; Goncalves, A.; Viens, P.; Bladou, F. Resection of residual masses after chemotherapy for advanced non-seminomatous germ cell tumours, a monocentric analysis of preoperative prognosticators. Eur. J. Cancer. Care. (Engl). **2010**, *19* (6), 827–832.

63. Tanaka, T.; Kitamura, H.; Takahashi, A.; Masumori, N.; Tsukamoto, T. Long-term outcome of chemotherapy for advanced testicular and extragonadal germ cell tumors: A single-center 27-year experience. Jpn. J. Clin. Oncol. **2010**, *40* (1), 73–78.

64. Feldman, D.R.; Sheinfeld, J.; Bajorin, D.F.; Fischer, P.; Turkula, S.; Ishill, N.; Patil, S.; Bains, M.; Reich, L.M.; Bosl, G.J.; Motzer, R.J. TI-CE high-dose chemotherapy for patients with previously treated germ cell tumors: Results and prognostic factor analysis. J. Clin. Oncol. **2010**, *28* (10), 1706–1713.

65. Tamada, K.; Ido, K.; Ueno, N.; Kimura, K.; Ichiyama, M.; Tomiyama, T. Preoperative staging of extrahepatic bile

Nanoparticles—Nerve

duct cancer with intraductal ultrasonography. Am. J. Gastroenterol. **1995**, *90* (2), 239–246.

66. Kamada, T.; Saitou, H.; Takamura, A.; Nojima, T.; Okushiba, S.I. The role of radiotherapy in the management of extrahepatic bile duct cancer: An analysis of 145 consecutive patients treated with intraluminal and/or external beam radiotherapy. Int. J. Radiat. Oncol. Biol. Phys. **1996**, *34* (4), 767–74.

67. Tio, T.L.; Cheng, J.; Wijers, O.B.; Sars, P.R.; Tytgat, G.N. Endosonographic TNM staging of extrahepatic bile duct cancer: Comparison with pathological staging. Gastroenterology **1991**, *100* (5 Pt. 1); 1351–1361.

68. Giang, T.H.; Ngoc, T.T.; Hassell, L.A. Carcinoma involving the gallbladder: A retrospective review of 23 cases - pitfalls in diagnosis of gallbladder carcinoma. Diagn. Pathol. **2012**, *7* (1), 10–18.

69. Bartlett, D.L.; Fong, Y.; Fortner, J.G.; Brennan, M.F.; Blumgartm, L.H. Long-term results after resection for gallbladder cancer. Implications for staging and management. Ann. Surg. **1996**, *224* (5), 639–646.

70. Kiguchi, K.; Ruffino, L.; Kawamoto, T.; Franco, E.; Kurakata, S.; Fujiwara, K.; Hanai, M.; Rumi, M.; DiGiovanni, J. Therapeutic effect of CS-706, a specific cyclooxygenase-2 inhibitor, on gallbladder carcinoma in BK5. ErbB-2 mice. Mol. Cancer. Ther. **2007**, *6* (6), 1709–1717.

71. Shimada, Y. JGCA (The Japan Gastric Cancer Association). Gastric cancer treatment guidelines. Jpn. J. Clin. Oncol. **2004**, *34* (1), 58.

72. Wang, X.; Yang, L.; Chen, Z.G.; Shin, D.M. Application of nanotechnology in cancer therapy and imaging. CA Cancer J. Clin. **2008**, *58* (2), 97–110.

73. Kolschmann, S.; Ibrahim, K.; Strasser, R.H. Primary aortic sarcoma with multiple metastatic sites. Eur. Heart. J. **2012**, *33* (16), 1995.

74. Blagova, O.V.; Nedostup, A.V.; Dzemeshkevich, S.L.; Sinitsyn, V.E.; Sedov, V.P.; Gagarina, N.V.; Parshin, V.D.; Chernyĭ, S.S.; Noskova, M.V.; Troitskaia, M.P. Primary lymphoma of the heart: Difficulties in diagnosis and treatment. Ter. Arkh. **2011**, *83* (4), 17–23.

75. Kawasaki, E.S.; Player, A. Nanotechnology, nanomedicine, and the development of new, effective therapies for cancer. Nanomedicine **2005**, *1* (2), 101–109.

76. Bosch, F.X.; Ribes, J.; Díaz, M.; Cléries, R. Primary liver cancer: Worldwide incidence and trends. Gastroenterology **2004**, *127* (5), 5–16.

77. Yan, S.; Zhang, D.; Gu, N.; Zheng, J.; Ding, A.; Wang, Z.; Xing, B.; Ma, M.; Zhang, Y. Therapeutic effect of Fe^2O^3 nanoparticles combined with magnetic fluid hyperthermia on cultured liver cancer cells and xenograft liver cancers. J. Nanosci. Nanotechnol. **2005**, *5* (8), 1185–1192.

78. He, X.; Duan, J.; Wang, K.; Tan, W.; Lin, X.; He, C. A novel fluorescent label based on organic dye-doped silica nanoparticles for HepG liver cancer cell recognition. J. Nanosci. Nanotechnol. **2004**, *4* (6), 585–589.

79. Hutchings, M.; Loft, A.; Hansen, M.; Pedersen, L.M.; Buhl, T.; Jurlander, J.; Buus, S.; Keiding, S.; D'Amore, F.; Boesen, A.M.; Berthelsen, A.K.; Specht, L. FDG-PET after two cycles of chemotherapy predicts treatment failure and progression-free survival in Hodgkin lymphoma. Blood **2006**, *107* (1), 52–59.

80. Ishida, T.; Ishii, T.; Inagaki, A.; Yano, H.; Kusumoto, S.; Ri, M.; Komatsu, H.; Iida, S.; Inagaki, H.; Ueda, R. The CCR4 as a novel-specific molecular target for immunotherapy in Hodgkin lymphoma. Leukemia 2006, 20 (12), 2162–2168.

81. Zelefsky, M.J.; Kraus, D.H.; Pfister, D.G.; Raben, A.; Shah, J.P.; Strong, E.W.; Spiro, R.H.; Bosl, G.J.; Harrison, L.B. Combined chemotherapy and radiotherapy versus surgery and postoperative radiotherapy for advanced hypopharyngeal cancer. Head Neck **1996**, *18* (5), 405–411.

82. Buckley, J.G.; MacLennan, K. Cervical node metastases in laryngeal and hypopharyngeal cancer: A prospective analysis of prevalence and distribution. Head Neck **2000**, *22* (4), 380–385.

83. Lin, Y.; Goedegebuure, P.S.; Tan, M.C.; Gross, J.; Malone, J.P.; Feng, S.; Larson, J.; Phommaly, C.; Trinkaus, K.; Townsend, R.R.; Linehan, D.C. Proteins associated with disease and clinical course in pancreas cancer: A proteomic analysis of plasma in surgical patients. J. Proteome. Res. **2006**, *5* (9), 2169–2176.

84. Montet, X.; Weissleder, R.; Josephson, L. Imaging pancreatic cancer with a peptide-nanoparticle conjugate targeted to normal pancreas. Bioconjug. Chem. **2006**, *17* (4), 905–911.

85. Northfelt, D.W.; Dezube, B.J.; Thommes, J.A.; Miller, B.J.; Fischl, M.A.; Friedman-Kien, A.; Kaplan, L.D.; Du Mond, C.; Mamelok, R.D.; Henry, D.H. Pegylated-liposomal doxorubicin versus doxorubicin, bleomycin, and vincristine in the treatment of AIDS-related Kaposi's sarcoma: Results of a randomized phase III clinical trial. J. Clin. Oncol. **1998**, *16* (7), 2445–2451.

86. Laubenstein, L.J.; Krigel, R.L.; Odajnyk, C.M.; Hymes, K.B.; Friedman-Kien, A.; Wernz, J.C.; Muggia, F.M. Treatment of epidemic Kaposi's sarcoma with etoposide or a combination of doxorubicin, bleomycin, and vinblastine. J. Clin. Oncol., **1984**, *2* (10), 1115–1120.

87. Travis, L.B.; Curtis, R.E.; Glimelius, B.; Holowaty, E.J.; Van Leeuwen, F.E.; Lynch, C.F.; Hagenbeek, A.; Stovall, M.; Banks, P.M.; Adami, J.; Gospodarowicz, M.K.; Wacholder, S.; Inskip, P.D.; Tucker, M.A.; Boice, J.D., Jr. Bladder and kidney cancer following cyclophosphamide therapy for non-Hodgkin's lymphoma. J. Natl. Cancer. Inst., **1995**, *87* (7), 524–530.

88. Wolinsky, J.B.; Grinstaff, M.W. Therapeutic and diagnostic applications of dendrimers for cancer treatment. Adv. Drug. Deliv. Rev. **2008**, *60* (9), 1037–1055.

89. Induction chemotherapy plus radiation compared with surgery plus radiation in patients with advanced laryngeal cancer. The Department of Veterans Affairs Laryngeal Cancer Study Group. N. Engl. J. Med., **1991**, *324* (24), 1685–1690.

90. Forastiere, A.A.; Goepfert, H.; Maor, M.; Pajak, T.F.; Weber, R.; Morrison, W.; Glisson, B.; Trotti, A.; Ridge, J.A.; Chao, C.; Peters, G.; Lee, D.J.; Leaf, A.; Ensley, J.; Cooper, J. Concurrent chemotherapy and radiotherapy for organ preservation in advanced laryngeal cancer. N. Engl. J. Med. **2003**, *349* (22), 2091–2098.

91. Study of peptide T for HIV infection. Am. Fam. Physician **1991**, *43* (6), 2270.

92. Telek, B.; Rejto, L.; Kiss, A.; Batár, P.; Reményi, G.; Szász, R.; Ujj, Z.Á.; Udvardy, M. Current treatment of acute myeloid leukaemia in adults. Orv. Hetil. **2012**, *153* (7), 243–249.

Nanoparticles—Nerve

93. Chen, J.; Zheng, Z.; Shen, J.; Peng, L.; Zhuang, H.; Liu, W.; Zhou, Y. Secondary acute myeloid leukemia occurring after successful treatment of acute promyelocytic leukemia. Int. J. Hematol. **2012**, *95* (3), 327–328.

94. Smahel, M. Antigens in chronic myeloid leukemia: Implications for vaccine development. Cancer. Immunol. Immunother. **2011**, *60* (12), 1655–1668.

95. Gage, A.A. Cryosurgery in the treatment of cancer. Surg. Gynecol. Obstet. **1992**, *174* (1), 73–92.

96. de Graeff, A.; de Leeuw, J.R.; Ros, W.J.; Hordijk, G.J.; Blijham, G.H.; Winnubst, J.A. A prospective study on quality of life of patients with cancer of the oral cavity or oropharynx treated with surgery with or without radiotherapy. Oral. Oncol. **1999**, *35* (1), 27–32.

97. El-Sayed, I.H.; Huang, X.; El-Sayed, M.A. Selective laser photo-thermal therapy of epithelial carcinoma using anti-EGFR antibody conjugated gold nanoparticles. Cancer. Lett. **2006**, *239* (1), 129–135..

98. Smith, I.E.; Harland, S.J.; Robinson, B.A.; Evans, B.D.; Goodhart, L.C.; Calvert, A.H.; Yarnold, J.; Glees, J.P.; Baker, J.; Ford, H.T. Carboplatin: A very active new cisplatin analog in the treatment of small cell lung cancer. Cancer. Treat. Rep. **1985**, *69* (1), 43–46.

99. Fry, W.A.; Phillips, J.L.; Menck, H.R. Ten-year survey of lung cancer treatment and survival in hospitals in the United States: A national cancer data base report. Cancer **1999**, *86* (9), 1867–1876.

100. Klatzmann, D.; Chérin, P.; Bensimon, G.; Boyer, O.; Coutellier, A.; Charlotte, F. Boccaccio, C.; Salzmann, J.L.; Herson, S. A phase I/II dose-escalation study of herpes simplex virus type 1 thymidine kinase "suicide" gene therapy for metastatic melanoma. Study Group on Gene Therapy of Metastatic Melanoma. Hum. Gene. Ther. **1998**, *9* (17), 2585–2594.

101. Weinstein, J.S.; Varallyay, C.G.; Dosa, E.; Gahramanov, S.; Hamilton, B.; Rooney, W.D.; Muldoon, L.L.; Neuwelt, E.A. Superparamagnetic iron oxide nanoparticles: Diagnostic magnetic resonance imaging and potential therapeutic applications in neurooncology and central nervous system inflammatory pathologies, a review. J. Cereb. Blood. Flow. Metab. **2010**, *30* (1), 15–35.

102. Adida, C.; Berrebi, D.; Peuchmaur, M.; Reyes-Mugica, M.; Altieri, D.C. Anti-apoptosis gene, survivin, and prognosis of neuroblastoma. Lancet **1998**, *351* (9106), 882–883.

103. Wang, C.H.; Huang, Y.J.; Chang, C.W.; Hsu, W.M.; Peng, C.A. *In vitro* photothermal destruction of neuroblastoma cells using carbon nanotubes conjugated with GD2 monoclonal antibody. Nanotechnology **2009**, *20* (31), 315101.

104. Griffiths, C.T.; Parker, L.M.; Fuller, A.F., Jr. Role of cytoreductive surgical treatment in the management of advanced ovarian cancer. Cancer Treat. Rep. **1979**, *63* (2), 235–240.

105. Harries, M.; Gore, M. Part II: Chemotherapy for epithelial ovarian cancer-treatment of recurrent disease. Lancet Oncol. **2002**, *3* (9), 537–545.

106. Khandare, J.; Calderón, M.; Dagia, N.M.; Haag, R. Multifunctional dendritic polymers in nanomedicine: Opportunities and challenges. Chem. Soc. Rev. **2012**, *41* (7), 2824–2848.

107. Ferrari, M. Cancer nanotechnology: Opportunities and challenges. Nat. Rev. Cancer **2005**, *5* (3), 161–171.

108. Haley, B.; Frenkel, E. Nanoparticles for drug delivery in cancer treatment. Urol. Oncol. **2008**, *26* (1), 57–64.

109. Moore, M.J.; Goldstein, D.; Hamm, J.; Figer, A.; Hecht, J.R. Gallinger, S.; Au, H.J.; Murawa, P.; Walde, D.; Wolff, R.A.; Campos, D.; Lim, R.; Ding, K.; Clark, G.; Voskoglou-Nomikos, T.; Ptasynski, M.; Parulekar, W. National Cancer Institute of Canada Clinical Trials Group. Erlotinib plus gemcitabine compared with gemcitabine alone in patients with advanced pancreatic cancer: A phase III trial of the National Cancer Institute of Canada Clinical Trials Group. J. Clin. Oncol., **2007**, *25* (15), 1960–1966.

110. Kalser, M.H.; Ellenberg, S.S. Pancreatic cancer. Adjuvant combined radiation and chemotherapy following curative resection. Arch. Surg. **1985**, *120* (8), 899–903.

111. Bisht, S.; Feldmann, G.; Soni, S.; Ravi, R.; Karikar, C.; Maitra, A. Polymeric nanoparticle-encapsulated curcumin ("nanocurcumin"): A novel strategy for human cancer therapy. J. Nanobiotechnol. **2007**, *5* (3), 3–20.

112. Aldinger, K.A.; Hickey, R.C.; Ibanez, M.L.; Samaan, N.A. Parathyroid carcinoma: A clinical study of seven cases of functioning and two cases of nonfunctioning parathyroid cancer. Cancer **1982**, *49* (2), 388–397.

113. Thompson, S.D.; Prichard, A.J. The management of parathyroid carcinoma. Curr. Opin. Otolaryngol. Head Neck Surg. **2004**, *12* (2), 93–97.

114. Glover, D.J.; Shaw, L.; Glick, J.H.; Slatopolsky, E.; Weiler, C.; Attie, M.; Goldfarb, S. Treatment of hypercalcemia in parathyroid cancer with WR-2721, S-2-(3-aminopropylamino)ethyl phosphorothioic acid. Ann. Intern. Med. **1985**, *103* (1), 55–57.

115. Gao, Y.; Yang, R.; Zhang, Z.; Chen, L.; Sun, Z.; Li, Y. Solid lipid nanoparticles reduce systemic toxicity of docetaxel: Performance and mechanism in animal. Nanotoxicology **2011**, *5* (4), 636–649.

116. Burgers, J.K.; Badalament, R.A.; Drago, J.R. Penile cancer. Clinical presentation, diagnosis, and staging. Urol. Clin. North. Am. **1992**, *19* (2), 247–256.

117. Rotolo, J.E.; Lynch, J.H. Penile cancer: Curable with early detection. Hosp. Pract. (Off Ed). **1992**, *26* (6), 131–138.

118. Maden, C.; Sherman, K.J.; Beckmann, A.M.; Hislop, T.G.; Teh, C.Z.; Ashley, R.L.; Daling JR. History of circumcision, medical conditions, and sexual activity and risk of penile cancer. J. Natl. Cancer Inst. **1993**, *85* (1), 19–24.

119. Gipponi, M. Clinical applications of sentinel lymph-node biopsy for the staging and treatment of solid neoplasms. Minerva. Chir. **2005**, *60* (4), 217–233.

120. Lefebvre, J.L.; Chevalier, D.; Luboinski, B.; Kirkpatrick, A.; Collette, L.; Sahmoud, T. Larynx preservation in pyriform sinus cancer: Preliminary results of a European Organization for Research and Treatment of Cancer phase III trial. EORTC Head and Neck Cancer Cooperative Group. J. Natl. Cancer. Inst. **1996**, *88* (13), 890–899.

121. Armstrong, J.; Pfister, D.; Strong, E.; Heimann, R.; Kraus, D.; Polishook, A.; Zelefsky, M.; Bosl, G.; Shah, J.; Spiro, R.; Harrison, L. The management of the clinically positive neck as part of a larynx preservation approach. Int. J. Radiat. Oncol. Biol. Phys. **1993**, *26* (5), 759–765.

122. Ekberg, O.; Nylander, G. Pharyngeal dysfunction after treatment for pharyngeal cancer with surgery and radiotherapy. Gastrointest. Radiol. **1983**, *8* (1), 97–104.

123. Tsang, R.W.; Brierley, J.D.; Panzarella, T.; Gospodarowicz, M.K.; Sutcliffe, S.B.; Simpson, W.J. Radiation therapy for pituitary adenoma: Treatment outcome and prognostic factors. Int. J. Radiat. Oncol. Biol. Phys. **1994**, *30* (3), 557–565.

124. Goya, R.G.; Sarkar, D.K.; Brown, O.A.; Hereñú, C.B. Potential of gene therapy for the treatment of pituitary tumors. Curr. Gene Ther. **2004**, *4* (1), 79–87.

125. Clarke, J.L.; Deangelis, L.M. Primary central nervous system lymphoma. Handb. Clin. Neurol. **2012**, *105*, 517–527.

126. Muldoon, L.L.; Tratnyek, P.G.; Jacobs, P.M.; Doolittle, N.D.; Christoforidis, G.A.; Frank, J.A.; Lindau, M.; Lockman, P.R.; Manninger, S.P.; Qiang, Y.; Spence, A.M.; Stupp, S.I.; Zhang, M.; Neuwelt, E.A. Imaging and nanomedicine for diagnosis and therapy in the central nervous system: Report of the eleventh annual Blood-Brain Barrier Disruption Consortium meeting. AJNR Am J Neuroradiol. **2006**, *27* (3), 715–721.

127. Valerio, M.; Vaucher, L.; Cerantola, Y.; Berthold, D.; Herrera, F.; Jichlinski, P. [Active surveillance in prostate cancer]. Rev. Med. Suisse. **2011**, *7* (320), 2388–2391.

128. Grossfeld, G.D.; Chang, J.J.; Broering, J.M.; Miller, D.P.; Yu, J.; Flanders, S.C.; Henning, J.M.; Stier, D,M.; Carroll, P.R. Impact of positive surgical margins on prostate cancer recurrence and the use of secondary cancer treatment: Data from the CaPSURE database. J. Urol. **2000**, *163* (4), 1171–1177.

129. Harisinghani, M.G.; Barentsz, J.; Hahn, P.F.; Deserno, W.M.; Tabatabaei, S.; van de Kaa, C.H.; de la Rosette, J.; Weissleder, R. Noninvasive detection of clinically occult lymph-node metastases in prostate cancer. N. Engl. J. Med. **2003**, *348* (25), 2491–2499.

130. Heald, R.J.; Ryall, R.D. Recurrence and survival after total mesorectal excision for rectal cancer. Lancet **1986**, *1* (8496), 1479–1482.

131. Hartman, K.B.; Wilson, L.J.; Rosenblum, M.G. Detecting and treating cancer with nanotechnology. Mol. Diagn. Ther. **2008**, *12* (1), 1–14.

132. Brichard, B.; De Bruycker, J.J.; De Potter, P.; Neven, B.; Vermylen, C.; Cornu, G. Combined chemotherapy and local treatment in the management of intraocular retinoblastoma. Med. Pediatr. Oncol. **2002**, *38* (6), 411–415.

133. Nair, A.; Thevenot, P.; Hu, W.; Tang, L. Nanotechnology in the treatment and detection of intraocular cancers. J. Biomed. Nanotechnol. **2008**, *4* (4), 410–418.

134. Einhorn, L.H.; Donohue, J. Cis-diamminedichloroplatinum, vinblastine, and bleomycin combination chemotherapy in disseminated testicular cancer. Ann. Intern. Med. **1977**, *87* (3), 293–298.

135. Einhorn, L.H.; Donohue, J.P. Improved chemotherapy in disseminated testicular cancer. J. Urol. **1977**, *117* (1), 65–69.

136. Einhorn, L.H. Treatment of testicular cancer: A new and improved model. J. Clin. Oncol. **1990**, *8* (11), 1777–1781.

137. Weide, L.G.; Ulbright, T.M.; Loehrer, P.J., Sr.; Williams, S.D. Thymic carcinoma. A distinct clinical entity responsive to chemotherapy. Cancer. **1993**, *71* (4), 1219–1223.

138. Khaled, A.; Guo, S.; Li, F.; Guo, P. Controllable self-assembly of nanoparticles for specific delivery of multiple therapeutic molecules to cancer cells using RNA nanotechnology. Nano Lett. **2005**, *5* (9), 1797–1808.

139. Duncan, R.; Vicent, M.J.; Greco, F.; Nicholson, R.I. Polymer-drug conjugates: Towards a novel approach for the treatment of endrocine-related cancer. Endocr. Relat. Cancer **2005**, *12* (Suppl. 1), S189–S199.

140. Kim, S.; Yazici, Y.D.; Calzada, G.; Wang, Z.Y.; Younes, M.N.; Jasser, S.A.; El-Naggar, A.K.; Myers, J.N. Sorafenib inhibits the angiogenesis and growth of orthotopic anaplastic thyroid carcinoma xenografts in nude mice. Mol. Cancer. Ther. **2007**, *6* (6), 1785–1792.

141. Woodruff, J.D.; Parmley, T.H.; Julian, C.G. Topical 5-fluorouracil in the treatment of vaginal carcinoma-in-situ. Gynecol. Oncol. **1975**, *3* (2), 124–132.

142. Hadji, P.; Kauka, A.; Bauer, T.; Tams, J.; Hasenburg, A.; Kieback, D.G. Effects of exemestane and tamoxifen on hormone levels within the Tamoxifen Exemestane Adjuvant Multicentre (TEAM) Trial: Results of a German substudy. Climacteric **2012**, *15* (5), 460–466.

143. Tosi, G.; Costantino, L.; Ruozi, B.; Forni, F.; Vandelli, M.A. Polymeric nanoparticles for the drug delivery to the central nervous system. Expert Opin. Drug Deliv. **2008**, *5* (2), 155–174.

144. Liu, Y.; Miyoshi, H.; Nakamura, M. Nanomedicine for drug delivery and imaging: A promising avenue for cancer therapy and diagnosis using targeted functional nanoparticles. Int. J. Cancer **2007**, *120* (12); 2527–2537.

145. Chen, X.; Estévez, M.C.; Zhu, Z.; Huang, Y.F.; Chen, Y.; Wang, L.; Tan, W. Using aptamer conjugated fluorescence resonance energy transfer nanoparticles for multiplexed cancer cell monitoring. Anal. Chem. **2009**, *81* (16); 7009–7014.

146. Ludwig, J.A.; Weinstein, J.N. Biomarkers in cancer staging, prognosis and treatment selection. Nat. Rev. Cancer **2005**, *5* (11), 845–856.

147. Ye, Z.Q.; Tan, M.Q.; Wang, G.L.; Yuan, J.L. Preparation, characterization, and time resolved fluorometric application of silica-coated terbium (III) fluorescent nanoparticles. Anal. Chem. **2004**, *76* (3), 513–518.

148. Xia, X.; Xu, Y.; Zhao, X.; Li, Q. Lateral flow immunoassay using europium chelate-loaded silica nanoparticles as labels. Clin. Chem. **2009**, *55* (1), 179–182.

149. Santra, S.; Kaittanis, C.; Perez, J.M. Cytochrome C encapsulating theranostic nanoparticles: A novel bifunctional system for targeted delivery of therapeutic membrane-impermeable proteins to tumors and imaging of cancer therapy. Mol. Pharm. **2010**, *7* (4), 1209–1222.

150. Schluep, T.; Hwang, J.; Hildebrandt, I.J.; Czernin, J.; Choi, C.H.; Alabi, C.A.; Mack, B.C.; Davis, M.E. Pharmacokinetics and tumor dynamics of the nanoparticle IT-101 from PET imaging and tumor histological measurements. Proc. Natl. Acad. Sci. U.S.A. **2009**, *106* (27), 11394–11399.

151. Esener, S.; Ortac, I.; Zlatanovic, S.; Liu, Y.T.; Carson, D. Cancer monitoring with nanoparticles and microfluidics. Proc. of SPIE **2008**, *7035*, 1–8.

152. Herr, J.K.; Smith, J.E.; Medley, C.D.; Shangguan, D.; Tan,W. Aptamer-conjugated nanoparticles for selective collection and detection of cancer. Anal. Chem. **2006**, *78* (9), 2918–2924.

153. Smith, J.E.; Medley, C.D.; Tang, Z.; Shangguan, D.; Lofton, C.; Tan, W. Aptamer conjugated nanoparticles for the collection and detection of multiple cancer cells. Anal. Chem. **2007**, *79* (8), 3075–3082.

154. Chan, J.M.; Valencia, P.M.; Zhang, L.; Langer, R.; Farokhzad, O.C. Polymeric nanoparticles for drug delivery. In *Cancer Nanotechnology, Methods in Molecular Biology 624*; Grobmyer, S.R., Moudgil, B.M., Eds,; Chapter 11; © Springer Science Business Media, LLC 2010

155. Dawson, M.; Krauland, E.; Wirtz, D.; Hanes, J. Transport of polymeric nanoparticle gene carriers in gastric mucus. Biotechnol. Prog. **2004**, *20* (3), 851–857

156. Hasan, W.; Chu, K.; Gullapalli, A.; Dunn, S.S.; Enlow, E.M.; Luft, J.C.; Tian, S.; Napier, M.E.; Pohlhaus, P.D.; Rolland, J.P.; DeSimone, J.M. Delivery of multiple siRNAs using lipid-coated PLGA nanoparticles for treatment of prostate cancer. Nano Lett. **2012**, *12* (1), 287–292.

157. Ramanlal Chaudhari, K.; Kumar, A.; Megraj Khandelwal, V.K.; Ukawala, M.; Manjappa, A.S.; Mishra, A.K.; Monkkonen, J.; Ramachandra Murthy, R.S. Bone metastasis targeting: A novel approach to reach bone using zoledronate anchored PLGA nanoparticle as carrier system loaded with Docetaxel. J. Control. Release **2012**, *158* (3), 470–478.

158. Misra, R.; Sahoo, S.K. Coformulation of doxorubicin and curcumin in poly (D,L-lactide-*co*-glycolide) nanoparticles suppresses the development of multidrug resistance in K562 cells. Mol. Pharm. **2011**, *8* (3), 852–866.

159. Madan, J.; Dhiman, N.; Sardana, S.; Aneja, R.; Chandra, R.; Katyal, A. Long-circulating poly(ethylene glycol)-grafted gelatin nanoparticles customized for intracellular delivery of noscapine: Preparation, *in vitro* characterization, structure elucidation, pharmacokinetics, and cytotoxicity analyses. Anticancer Drugs **2011**, *22* (6), 543–555.

160. Sanna V, Pintus G, Roggio AM, Punzoni S, Posadino AM, Arca A, Marceddu, S.; Bandiera, P.; Uzzau, S.; Sechi, M. Targeted biocompatible nanoparticles for the delivery of (−)-epigallocatechin 3-gallate to prostate cancer cells. J. Med. Chem. **2011**, *54* (5), 1321–1332.

161. Zhang, H.; Zhao, C.; Cao, H.; Wang, G.; Song, L.; Niu, G.; Yang, H.; Ma, J.; Zhu, S. Hyperbranched poly(amine-ester) based hydrogels for controlled multi-drug release in combination chemotherapy. Biomaterials **2010**, *31* (20), 5445–5454.

162. Pathak, A.; Kumar, P.; Chuttani, K.; Jain, S.; Mishra, A.K.; Vyas, S.P.; Gupta, K.C. Gene expression, biodistribution, and pharmacoscintigraphic evaluation of chondroitin sulfate-PEI nanoconstructs mediated tumor gene therapy. ACS Nano. **2009**, *3* (6), 1493–1505.

163. Safavy, A. Recent developments in taxane drug delivery. Curr. Drug Deliv. **2008**, *5* (1), 42–54

164. Zhang, Z.M.; Yang, X.Y.; Deng, S.H.; Xu, W.; Gao, H.Q. Anti-tumor effects of polybutylcyanoacrylate nanoparticles of diallyl trisulfide on orthotopic transplantation tumor model of hepatocellular carcinoma in BALB/c nude mice. Chin. Med. J. (Engl). **2007**, *120* (15), 1336–1342.

165. Xi-Xiao, Y.; Jan-Hai, C.; Shi-Ting, L.; Dan, G.; Xv-Xin, Z. Polybutylcyanoacrylate nanoparticles as a carrier for mitomycin C in rabbits bearing VX2-liver tumor. Regul. Toxicol. Pharmacol. **2006**, *46* (3), 211–217.

166. Park, J.W.; Benz, C.C.; Martin, F.J. Future directions of liposome- and immunoliposome-based cancer therapeutics. Semin. Oncol. **2004**, *31* (6 Suppl 13), 196–205.

167. Ghosh, S.; Pal, S.; Prusty, S.; Girish, K.V.S. Curcumin and cancer, recent developments. J. Res. Biol. **2012**, *2* (3), 251–272

168. Talaei, F.; Azizi, E.; Dinarvand, R.; Atyabi, F. Thiolated chitosan nanoparticles as a delivery system for antisense therapy: Evaluation against EGFR in T47D breast cancer cells. Int J Nanomedicine. **2011**, *6*, 1963–1975.

169. Figueiredo, F.; Esenaliev, R. PLGA nanoparticles for ultrasound-mediated gene delivery to solid tumors. J. Drug Deliv. **2012**, *2012*, 1–20.

170. Tian, X.H.; Lin, X.N.; Wei, F.; Feng, W.; Huang, Z.C.; Wang, P.; Ren, L.; Diao, Y. Enhanced brain targeting of temozolomide in polysorbate-80 coated polybutylcyanoacrylate nanoparticles. Int. J. Nanomed. **2011**, *6*, 445–452.

171. Derakhshandeh, K.; Soheili, M.; Dadashzadeh, S.; Saghiri, R. Preparation and *in vitro* characterization of 9-nitrocamptothecin-loaded long circulating nanoparticles for delivery in cancer patients. Int. J. Nanomed. **2010**, *5*, 463–471.

172. Ma, W.; Smith, T.; Bogin, V.; Zhang, Y.; Ozkan, C.; Ozkan, M.; Hayden, M.; Schroter, S.; Carrier, E.; Messmer, D.; Kumar, V.; Minev, B. Enhanced presentation of MHC class Ia, Ib and class II-restricted peptides encapsulated in biodegradable nanoparticles: A promising strategy for tumor immunotherapy. J. Transl. Med. **2011**, *9* (1), 34.

173. Su, X.; Fricke, J.; Kavanagh, D.G.; Irvine, D.J. *In vitro* and *in vivo* mRNA delivery using lipid-enveloped pH-responsive polymer nanoparticles. Mol. Pharm. **2011**, *8* (3), 774–787.

174. Milane, L.; Duan, Z.; Amiji, M. Therapeutic efficacy and safety of paclitaxel/lonidamine loaded EGFR-targeted nanoparticles for the treatment of multi-drug resistant cancer. PLoS One **2011**, *6* (9), e24075.

175. Kanthamneni, N.; Chaudhary, A.; Wang, J.; Prabhu, S. Nanoparticulate delivery of novel drug combination regimens for the chemoprevention of colon cancer. Int. J. Oncol. **2010**, *37* (1), 177–185.

176. Yang, D.; Van, S.; Shu, Y.; Liu, X.; Ge, Y.; Jiang, X.; Jin, Y.; Yu, L. Synthesis, characterization, and *in vivo* efficacy evaluation of PGG-docetaxel conjugate for potential cancer chemotherapy. Int. J. Nanomed. **2012**, *7*, 581–589.

177. Heidel, J.D.; Schluep, T. Cyclodextrin-containing polymers: Versatile platforms of drug delivery materials. J. Drug Deliv. **2012**, *2012*, 1–17.

178. Akbarzadeh, A.; Mikaeili, H.; Zarghami, N.; Mohammad, R.; Barkhordari, A.; Davaran, S. Preparation and *in vitro* evaluation of doxorubicin-loaded Fe_3O_4 magnetic nanoparticles modified with biocompatible copolymers. Int. J. Nanomed. **2012**, *7*, 511–526.

179. Huang, Y.H.; Zugates, G.T.; Peng, W.; Holtz, D.; Dunton, C.; Green, J.J.; Hossain, N.; Chernick, M.R.; Padera, R.F., Jr.; Langer, R.; Anderson, D.G.; Sawicki, J.A. Nanoparticle-delivered suicide gene therapy effectively reduces ovarian tumor burden in mice. Cancer Res. **2009**, *69* (15), 6184–6191.

180. Dhar, S.; Gu, F.X.; Langer, R.; Farokhzad, O.C.; Lippard, S.J. Targeted delivery of cisplatin to prostate cancer cells by aptamer functionalized Pt(IV) prodrug-PLGA-PEG nanoparticles. Proc. Natl. Acad. Sci. U.S.A. **2008**, *105* (45), 17356–17361.

181. Cheng, L.; Jin, C.; Lv, W.; Ding, Q.; Han, Xu. Developing a highly stable PLGA-mPEG nanoparticle loaded with cisplatin for chemotherapy of ovarian cancer. PLoS One **2011**, *6* (9), 1–9.

182. Stone, G.W.; Barzee, S.; Snarsky, V.; Santucci, C.; Tran, B.; Langer, R.; Zugates, G.T.; Anderson, D.G.; Kornbluth, R.S. Nanoparticle-delivered multimeric soluble CD40L DNA combined with toll-like receptor agonists as a treatment for melanoma. PLoS One **2009**, *4* (10), e7334

183. Marquis, B.J.; Love, S.A.; Braun, K.L.; Haynes, C.L. Analytical methods to assess nanoparticle toxicity. Analyst. **2009**, *134* (3), 425–439.

184. Poland, C.A.; Duffin, R.; Kinloch, I.; Maynard, A.; Wallace, W.A.; Seaton, A.; Stone, V.; Brown, S.; MacNee, W.; Donaldson, K. Carbon nanotubes introduced into the abdominal cavity of mice show asbestos-like pathogenicity in a pilot study, Nat. Nanotechnol. **2008**, *3* (7), 423–428.

185. Nel, A.; Xia, T.; Mädler, L.; Li, N. Toxic potential of materials at the nanolevel. Science **2006**, *311* (5761), 622–627.

186. Buzea, C.; Pacheco, I.I.; Robbie, K. Nanomaterials and nanoparticles: Sources and toxicity. Biointerphases **2007**, *2* (4), MR17–MR71.

187. Ryman-Rasmussen, J.P.; Riviere, J.E.; Monteiro-Riviere, N.A. Penetration of intact skin by quantum dots with diverse physicochemical properties. Toxicol. Sci. **2006**, *91*(1), 159–165.

188. Oberdörster, G.; Maynard, A.; Donaldson, K.; Castranova, V.; Fitzpatrick, J.; Ausman, K.; Carter, J.; Karn, B.; Kreyling, W.; Lai, D.; Olin, S.; Monteiro-Riviere, N.; Warheit, D.; Yang, H.; ILSI Research Foundation/Risk Science Institute Nanomaterial Toxicity Screening Working Group. Principles for characterizing the potential human health effects from exposure to nanomaterials: Elements of a screening strategy. Part. Fibre Toxicol. **2005**, *2* (1), 2–8.

189. Gwinn, M.R.; Vallyathan, V. Nanoparticles: Health effects–pros and cons. Environ. Health Perspect. **2006**, *114* (12), 1818–1825.

190. http://clinicaltrials.gov/ct2/show/NCT01300533

191. Davis, M.E. The first targeted delivery of siRNA in humans via a self-assembling, cyclodextrin polymer-based nanoparticle: from concept to clinic. Mol. Pharm. **2009**, *6* (3), 659–668.

192. http://www.clinicaltrials.gov/ct2/show/NCT01478893

193. Kim, T.Y.; Kim, D.W.; Chung, J.Y.; Shin, S.G.; Kim, S.C.; Heo, D.S.; Kim, N.K.; Bang, Y.J. Phase II study of a cremophor-free, polymeric micelle formulation of paclitaxel for patients with advanced urothelial cancer previously treated with gemcitabine and platinum. Clin. Cancer Res. **2004**, *10*, 3708–3716.

194. Valle, J.W.; Armstrong, A.; Newman, C.; Alakhov, V.; Pietrzynski, G.; Brewer, J.; Campbell, S.; Corrie, P.; Rowinsky, E.K.; Ranson, M. A phase 2 study of SP1049C, doxorubicin in P-glycoprotein-targeting pluronics, in patients with advanced adenocarcinoma of the esophagus and gastroesophageal junction. Invest. New Drugs **2011**, *29* (5), 1029–1037.

195. Hamaguchi, T.; Matsumura, Y.; Nakanishi, Y.; Muro, K.; Yamada, Y.; Shimada, Y.; Shirao, K.; Niki, H.; Hosokawa, S.; Tagawa, T.; Kakizoe, T. Antitumor effect of MCC-465, pegylated liposomal doxorubicin tagged with newly developed monoclonal antibody GAH, in colorectal cancer xenografts Cancer Sci. **2004**, *95* (7), 608–613.

196. http://clinicaltrials.gov/ct2/show/NCT00470613

197. Simoes, S.; Moreira, J.N.; Fonseca, C.; Duzgunes, N.; de Lima, M.C. On the formulation of pH-sensitive liposomes with long circulation times. Adv. Drug Deliv. **2004**, Rev. *56* (7), 947–965.

198. Zhang, L.; Gu, F.X.; Chan, J.M.; Wang, A.Z.; Langer, R.S.; Farokhzad, O.C. Nanoparticles in medicine: therapeutic applications and developments. Clin. Pharmacol. Ther. **2008**, *83* (5), 761–769.

199. Bissett, D.; Cassidy, J.; de Bono, J.S.; Muirhead, F.; Main, M.; Robson, L.; Fraier, D.; Magnè, M.L.; Pellizzoni, C.; Porro, M.G.; Spinelli, R.; Speed, W.; Twelves, C. Phase I and pharmacokinetic (PK) study of MAG-CPT (PNU 166148): A polymeric derivative of camptothecin (CPT). Br. J. Cancer **2004**, *91* (1), 50–55.

200. Hoekstra, R.; Dumez, H.; Eskens, F.A.; van der Gaast, A.; Planting, A.S.; de Heus, G.; Sizer, K.C.; Ravera, C.; Vaidyanathan, S.; Bucana, C.; Fidler, I.J.; van Oosterom, A.T.; Verweij, J. Phase I and pharmacologic study of PKI166, an epidermal growth factor receptor tyrosine kinase inhibitor, in patients with advanced solid malignancies. Clin Cancer Res. **2005**, *11* (19 Pt 1), 6908–6915.

201. Stremmel, W.; Braun, A.; Hanemann, A.; Ehehalt, R.; Autschbach, F.; Karner, M. Delayed release phosphatidylcholine in chronic-active ulcerative colitis: A randomized, double-blinded, dose finding study. J. Clin. Gastroenterol. **2010**, *44* (5), e101–e107.

Nanoparticles—Nerve

Natural Polymers: Tissue Engineering

Kunal Pal
Sai Sateesh Sagiri
Vinay K. Singh
Beauty Behera
Indranil Banerjee
Krishna Pramanik
Department of Biotechnology and Medical Engineering, National Institute of Technology, Rourkela, India

Abstract

In recent times, with technological advancements in medicine, materials science, and biology, the field of tissue engineering has evolved. Tissue engineering deals with either the creation of tissues under *ex vivo* conditions or the regeneration of tissues under *in vivo* conditions. This entry has been designed to discuss the various approaches being taken to develop scaffolds from natural polymers. Attempts have also been made to discuss in brief the applications of the natural polymers in tissue engineering.

INTRODUCTION

The term "polymer" has been derived from the Greek words *polys* (meaning many) and *meros* (part or unit).[1] Hence, polymers may be defined as large molecules, which are made up of small and simple chemical units, usually regarded as monomers.[2] Depending upon the nature of the repetition of monomers, the polymers may be categorized either as linear (like a chain) or branched polymers.[3,4] A large number of polymers have been widely used for developing products of biomedical importance that include medical devices, drug delivery systems, cell and/or enzyme immobilization, and tissue implants.[5–12] This has been possible because of the diverse physical and chemical properties presented by the polymers. The molecular structure, chemical and physical properties of the polymers determine their utilization in a particular medical application. The polymers generally used in medical applications are of high purity and biocompatible.[13] Both natural and synthetic polymers are now being used for such a purpose. Some common examples of polymers used in medical devices have been listed in Table 1.

Polymers can be classified into two distinct types depending on the processing technology, namely, thermoset and thermoplastic.[14,15] Thermosetting polymers have crosslinked structures and do not melt with heat, instead undergo degradation. Once the polymer has been formed, the polymers cannot be worked upon unlike thermoplastic where the polymers can be molded and reshaped, time and again, by the application of heat.[16,17] A thermoplastic has linear or branched polymer chains, which are not cross-linked. When heated the polymer melts and starts to flow. This is due to the sliding of the polymer chains against each other when they are heated.[18,19]

The disadvantages of the polymers include their lower strengths than metals or ceramics, deformation of the structure with time, chemical decomposition during processing, and leaching of toxic by-products under *in vivo* conditions.[20–22]

Tissue engineering is an interdisciplinary field that involves the expertise of material science, life science, and medicine for the development of tissue/organ substitutes. These substitutes help to restore, maintain, or improve tissue or organ function.[23–26] They have three basic components, namely, cells, scaffolds and cell-cell signaling.[27] Typically, the field of tissue engineering involves the seeding of either the patient's or donor's cells on a scaffold. The interaction amongst the cells and scaffolds help in generating new tissues.[28] Currently, hydrogels are commonly being used for the fabrication of scaffolds for tissue engineering applications. Every year, millions of people suffer from tissue loss and organ failure. They are mainly treated by performing surgical reconstruction, transplantation, supplementation of metabolic products of diseased tissues/organ, or by implanting supportive devices for the organ. Due to the shortage of the organ donors, efforts are being made to replace the organ with tissue-engineered tissues and grafts.[29,30] The premise of tissue engineering relies on the successful reconstruction of three dimensional (3D) polymeric solid support matrix that governs the restoration/regeneration of damaged tissues.

The materials used for the development of the solid support system for artificial restoration of tissue function can be categorized into three classes, namely, bioinert, bioresorbable/biodegradable, and bioactive.[31] Bioinert

Concise Encyclopedia of Biomedical Polymers and Polymeric Biomaterials DOI: 10.1081/E-EBPPC-120050408

Table 1 Some commonly used polymers and their applications in healthcare

Polymers		Proposed applications	References
Synthetic polymers			
Nylon		Surgical sutures, tracheal tubes	[14–15]
Silicones		Breast implants, facial devices	[16–19]
Polyester		Scaffolds, wound dressings, drug delivery	[20–24]
Poly(methyl methacrylate)		Bone cements, denture bases, intra-ocular lenses, contact lenses	[25–29]
Poly(vinyl chloride)		Tubing, facial devices	[30–31]
Poly(propylene)		Sutures, heart valves	[32–33]
Polyethylene		Acetabular cup prosthesis, facial bone augmentation, orbital implant	[34–36]
Polyurethane		Breast prosthesis, heart valve cusps, intra-aortic implants	[37–38]
Poly lactide-*co*-glycolide		Drug delivery	[39–40]
Polystyrene		Tissue culture, drug delivery	[41–42]
Natural polymers			
Proteins	**Polysaccharides**		
Collagen		Wound dressings, scaffolds, drug delivery	[43–45]
Gelatin		Wound dressings, scaffolds, drug delivery	[46–48]
Silk protein		Wound dressings, scaffolds, drug delivery	[49–52]
Fibrin and Fibrinogen		Tissue engineering, drug delivery	[53–56]
	Alginate	Wound dressings, scaffolds, drug delivery, sutures	[57–58]
	Chitosan	Wound dressings, scaffolds, drug delivery	[23,59–61]
	Hyaluronic acid	Postoperative adhesion prevention lubricant, drug delivery, scaffold	[62–65]
	Agarose	Wound dressings, scaffolds, drug delivery	[66–68]
	Cellulose	Wound dressings, scaffolds, drug delivery	[69–71]
	Starch	Wound dressings, scaffolds, drug delivery	[72–74]
	Gellan gum	Wound dressings, scaffolds, drug delivery	[75–77]
	Dextran	Wound dressings, scaffolds, drug delivery	[78–80]

materials do not initiate any physiological response up on implantation; rather it complies to mechanical/structural requirements. Most of the bioinert polymers are derived via a synthetic route and are rarely used in tissue engineering. High-density polyethylene and ultra-high density polyethylene are some of the common examples of bioinert materials.[32] Bioresorbable materials are degraded in the physiological system. The degradation products are either excreted directly out of the physiological system or enter into the metabolic pathways and are broken down into simpler molecules before it is being finally excreted out of the physiological system. Bioactive materials initiate a favorable physiological response when implanted *in vivo*. Bioactive materials interact with living cells by providing receptor-mediated "solid factor signals" that modulate cell adhesion, proliferation, and differentiation during neo-tissue genesis. For tissue engineering, bioresorbable and bioactive polymers are often considered as material of choice. It is important to mention that these two properties are not mutually exclusive and many natural (e.g., collagen, gelatin, and chitosan) and synthetic polymers (e.g., poly gly-colic-*co*-lactic acid (PLGA) or RGD (Arg-Gly-Asp) grafted polymers) possess both properties. In tissue engineering,

three different solid support systems, namely, 1) hydrogel matrices, 2) hydrogel films, and 3) scaffolds, are used for the reconstruction or restoration of damaged tissues.

Hydrogels

Hydrogels are a crosslinked polymer network that has the ability to hold water within its porous network structure.[33] They have been extensively used in various biomedical applications such as drug delivery and tissue engineering.[1] Hydrogels are prepared by crosslinking polymer chain by irradiation or chemical or physical methods and can be formed from both monomers and macromers (precursors).[34–36] Hydrogels can be categorized into three main groups, namely, homopolymers (having one monomer), copolymers (contains more than one monomer), and semi-interpenetrating networks (one monomer is polymerized throughout an already crosslinked network).[37–41] The properties of hydrogels could be tailored to have a desired interaction with cells and tissues.[42,43]

Physical hydrogels are formed when a polymer chain reacts with an ionic species or through the polymer–polymer interactions (hydrogen bonding, molecular entanglements) and are generally heterogenous in nature. A classic example of physical gellation is the gellation of alginic acid in the presence of calcium ions.[44,45]

Hydrogels prepared by chemical reaction have covalently linked polymer chains. This method involves the use of chemical crosslinkers and multifunctional reactive agents.[46–48] Irradiation of polymer or polymer blends can also lead to the formation of covalently linked polymer chains.[49] A classic example of chemical gellation is the preparation of poly(methyl methacrylate).[50,51]

The important factors that should be considered before fabricating a hydrogel are swelling properties, mechanical properties, degradability, and finally the biocompatibility of the designed hydrogels.[52,53] Due to the presence of water within the porous network of the crosslinked polymers, the hydrogels have the disadvantage of poor mechanical properties as compared to other polymers.[54] For hydrogels to be used for tissue engineering applications, it is generally desired to match the mechanical properties of the hydrogels to those of the intended tissue. The mechanical properties of the hydrogels could be tailored by changing the crosslinking density and by changing the polymerization conditions.[55,56]

The swelling kinetics of hydrogels is dependent on the mechanical property and cross-linking density of the network.[57] Swelling kinetics and the extent of swelling plays an important role in the viability of encapsulated cells and/ or enzymes and the release of drug entrapped within the crosslinked structure of the polymer.[58,59] Hydrogels may exhibit a swelling behavior dependent on the external environment. Environment-sensitive hydrogels (ESH) show a drastic change in the equilibrium swelling ratios with a change in external conditions. The swelling of ESH could be triggered with the change in temperature, pH, ionic strength, and electric or magnetic stimuli. The most important property of ESH is the reversibility of the swelling ratio to the previous state when the environmental conditions are reversed.[60] This behavior of ESH makes them a suitable material for designing self-regulated, pulsatile drug delivery systems and *in vitro* tissue cultures.[61–63]

Hydrogel films

Films are basically sheet-like structures. Hydrogels may be drawn into films. Hydrogel films have been extensively used as moist wound dressings. These films help promoting the migration and proliferation of the fibroblast cells, responsible for wound healing.[64]

Hydrogel films have also been tried for *in vitro* cell growth.[65]

Scaffolds

Scaffolds may be defined as porous 3D solid support systems suitable for cell growth. The pore diameter generally ranges from 50 to 400 μm depending on the type of cells used for culturing.[66] There are certain criteria that must be fulfilled to create an ideal scaffold. The materials used for scaffold development should not only promote cell adhesion and proliferation, necessary for the preparation of tissue-engineered tissues/organs, but preferably also be resorbable. As the scaffold is reabsorbed within the host environment, it should not release toxic byproducts. During the proliferation of the cells, the scaffold should allow the growth of the cells not only at the surface of the scaffolds but also within the core of the scaffolds. As the cells grow toward the core of the scaffold, the pores in the scaffolds help in the transportation of nutrients to the cells growing in the core of the scaffold by promoting angiogenesis (growth of blood vessels).[67–69] After the development of the tissue-engineered tissues/ organs, the scaffold should promote the integration of the tissue-engineered tissues/organs with the host tissues and morphogenesis.[70,71] This, in turn, minimizes the migration of the implants within the host. This may be ensured by using materials such that the rate of degradation of the scaffold material should allow sufficient time for the integration of the tissue-engineered tissue/organs with that of the host. Another important parameter that plays an important role in the selection of the material for scaffold fabrication is the mechanical property. Conscious attempts have to be made such that the mechanical properties of the materials are close to the mechanical properties of the host tissues so as to eliminate the chances of stress shielding.[72–74]

Drug delivery and its applications for tissue engineering have gained much attention. The scaffolds may be used to deliver biologically active agents to enhance tissue repair and regeneration. The scaffolds should be able to accommodate sufficient amount of the bioactive agents, release the same as per the requirement and should be stable for a sufficient period of time so as to adequately promote tissue repair and regeneration.[75,76]

Generally, polymers are used to regenerate soft tissues, namely, skin and connective tissues, but can also be used for regenerating cartilages and bones. Apart from synthetic materials, scientists are paying attention to natural polymers (e.g., gelatin, collagen, chitin, and alginates) for the development of the scaffolds due to the inherited biocompatibility of the natural polymers. The current review has been dedicated to have an extensive insight on the fabrication of scaffolds using natural polymers for tissue engineering applications. Before going into the details of the natural polymers and their various tissue engineering applications, it becomes customary to have an outlook about the various scaffold preparation techniques.

METHODS OF SCAFFOLD FABRICATION

It is quite obvious that the properties of scaffolds should be different for different applications. For example, the properties for the scaffolds for bone-tissue regeneration should be different from that of the scaffolds for the skin-tissue regeneration and facial-tissue reconstruction. With the advent in the progress in polymer science, there has been a continuous improvement in polymer processing

Table 2 Common techniques used for scaffold fabrication

Conventional technique	Ref	Advanced technique	Ref
Fiber meshes/fiber bonding	[77]	Electrospinning	[78]
Salt leaching	[79]	Rapid prototyping	[80]
Phase separation	[81]	3D printing	[82]
Gas foaming	[83]	Stereolithography	[84]
Freeze drying	[85]	Fused deposition modeling	[86]
Melt molding	[87]	3D plotter	[88]
Solution casting	[89]	Phase change jet printing	[84]
Emulsion freeze drying	[90]	Controlled molecular self-assembly	[91]

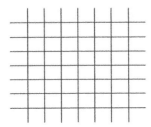

Polymer fiber mesh structure

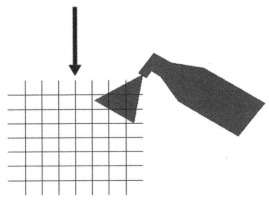

Atomizing the fiber mesh structure with binder polymer solution

Evaporation of the binding polymer solvent

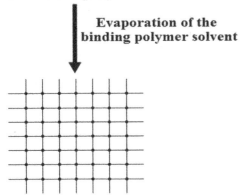

Polymer fibers mechanically bonded at the cross-points

Fig. 1 Scaffold preparation by mechanically binding the polymer fibers.

technologies. This has opened up a new area that deals with the fabrication of the scaffolds with desirable properties. In this section, an attempt has been made to get an insight on the commonly used techniques for the fabrication of the polymeric scaffolds. In Table 2, a list of common techniques used for scaffold fabrication is provided. A few of them have been discussed in brief in this section.

The simplest way to fabricate a scaffold is to use a nonwoven polymer fiber mesh structure. The stability of the polymer fiber matrix is increased by mechanically bonding the polymer fibers at the cross-points (Fig. 1). This is achieved by atomizing a binder polymer solution over the polymer fiber matrix followed by the evaporation of the solvent of the binder polymer solution. This is followed by freeze drying for removing residual solvents.[77] Poly glycolic acid (PGA) scaffolds have been successfully prepared and tested by this method. PLGA or poly L-lactide (PLLA) polymers have been used as binder polymers, which were dissolved in chloroform.[77,78] The preparation of the scaffolds by this method is quite simple and can easily be implemented. Unfortunately, the scaffolds prepared by this method lack sufficient mechanical properties to be used under *in vivo* conditions.[79]

The use of a porogen has been used since long for the development of the scaffolds, both ceramic and polymeric.[80,81] The method involves the preparation of polymer-porogen composite at particular proportions. After the successful preparation of the composite, the porogen is extracted by using a suitable solvent to form a scaffold (Fig. 2). For example, PLGA scaffolds have been reported by this method.[82] The authors dissolved the PLGA polymer in dioxane, an organic solvent, and subsequently dispersed the sodium chloride, used as porogen, in the PLGA solution. The dispersion was properly homogenized so as to obtain a uniform dispersion of sodium chloride in the PLGA solution. This dispersion was used to prepare PLGA-sodium chloride composite by the conventional solvent casting method.[83–92] The composite, so obtained, was then extracted with water thereby resulting in the leaching of the porogen, which resulted in the formation of interconnecting pores in the polymer matrices. The polymer matrices were vacuum dried to obtain polymeric scaffolds.[82] In a similar manner,

Nanoparticles—Nerve

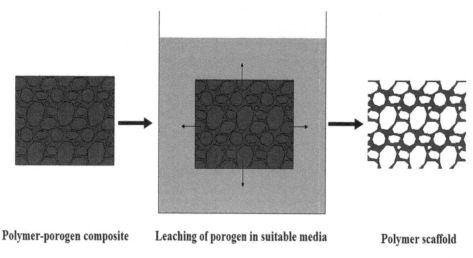

Polymer-porogen composite   **Leaching of porogen in suitable media**   **Polymer scaffold**

Fig. 2 Method of preparation of scaffold using porogen.

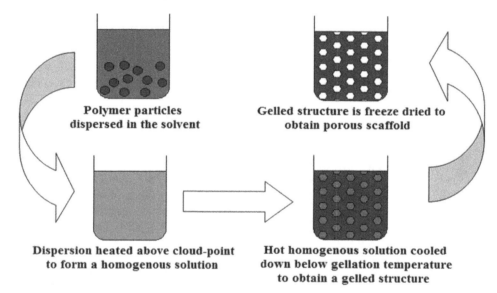

Polymer particles dispersed in the solvent   **Gelled structure is freeze dried to obtain porous scaffold**

Dispersion heated above cloud-point to form a homogenous solution   **Hot homogenous solution cooled down below gelation temperature to obtain a gelled structure**

Fig. 3 Thermally induced phase separation technique for the fabrication of the scaffold.

polyethylene and hydroxyapatite-polyethylene scaffolds have also been successfully prepared.[80,81,93] The only difference being that polyethylene-sodium chloride and hydroxyapatite-polyethylene-sodium chloride composites have been made by hot-isostatic pressure method.

The thermally induced phase separation technique employs a change in the solubility parameter of the polymer with a change in temperature. The polymers are dissolved in a suitable solvent at a temperature above the cloud-point. The hot solutions are then poured in suitable molds and allowed to cool much below the cloud-point. This initiates a change in the solubility parameter of the polymer and subsequent precipitation of the polymer, which in turn forms a gelled structure. The semi-solid gelled structure is then freeze dried to extract the solvent. This results in the formation of the polymeric scaffold (Fig. 3). For example, scaffolds based on PLGA and PLLA polymers may be prepared by using dioxane as the solvent.[79,94]

The foaming of the polymer solution as the polymer is being cured has been studied extensively for the fabrication of the scaffolds. The foaming may be achieved either by adding chemicals that might result in the formation of gas bubbles or by using supercritical CO_2.[79] Quite often, ammonium bicarbonate is used as a gas-forming agent for the fabrication of macroporous scaffolds. Ammonium bicarbonate is mixed with the polymer solution so as to form a polymer-ammonium bicarbonate suspension followed by the gellation of the mixture. The gelled product is dried to obtain polymer-ammonium bicarbonate composite, which is subsequently immersed in boiling water so as to release CO_2. This, in turn, induces foaming within the polymer matrix thereby forming a macroporous scaffold (Fig. 4).[95] Alternatively, foam may also be generated by using sodium bicarbonate. Sodium bicarbonate is added to the polymer solution followed by the addition of acids (e.g., citric acid). This induces the formation of CO_2 as the

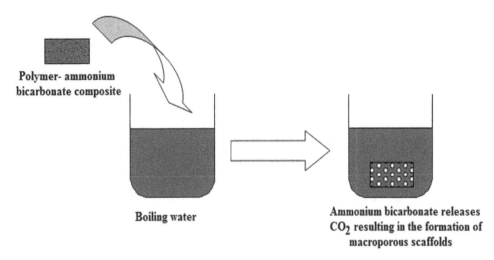

Fig. 4 Schematic diagram of scaffold preparation using ammonium bicarbonate as a gas-producing component.

consolidation of the polymer matrix takes place by a suitable method thereby resulting in the formation of macroporous scaffolds (Fig. 5).[96] Another effective method for the development of scaffolds by foaming technique involves the use of supercritical CO_2. CO_2 at a temperature of 304.1 K and a pressure of 73.8 bar is said to be in supercritical state.[97] At these conditions, CO_2 is said to possess the properties of both liquids and gases.[98] The supercritical CO_2 serves as an excellent solvent in the polymer industry and has been successfully used to develop porous materials.[99] The supercritical CO_2 has the ability to diffuse within the polymer chains thereby resulting in the swelling of the polymer matrix.[100,101] This process of swelling of the polymer in the presence of supercritical CO_2 is regarded as plasticization.[102] Upon the release of the pressure, the CO_2 is instantly removed from the swollen structure, which gives rise to the formation of a porous structure (scaffold, Fig. 6).[103–105] The pore sizes of the scaffolds have been found to be in the range of 50–400 μm.[79]

Freeze drying of the hydrogels has also been widely used for the development of the scaffolds. The method for the preparation of the scaffolds by this method involves the fabrication of 3-dimensional polymeric networked structures, capable of imbibing and holding water within its structure. This is followed by the freezing of hydrogel below −20°C. The ice is sublimed from the polymer network by the process of the freeze-drying method, which results in the formation of the pores within the polymeric matrices (Fig. 7).[106] In a recent study, the freezing condition of the hydrogel was altered so as to obtain a unidirectional freezing of the hydrogel. The hydrogels were kept in plastic tubes, which were subsequently placed within a Styrofoam container such that the bottom surfaces of the plastic tubes were exposed. Thereafter, the Styrofoam containers were placed over a metallic plate that was then immersed in liquid nitrogen. This ensured a unidirectional freezing of the hydrogel, from bottom to top (Fig. 8). The frozen samples were then freeze dried. The authors reported that the pore size of the gelatin scaffolds may be

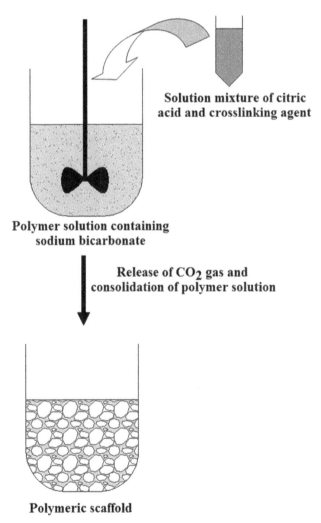

Fig. 5 Schematic representation of scaffold fabrication using sodium bicarbonate as gas producing component.

varied by altering the gelatin (polymer) concentration. The compressive mechanical properties of the scaffolds were superior longitudinally as compared to the transverse direction.[107]

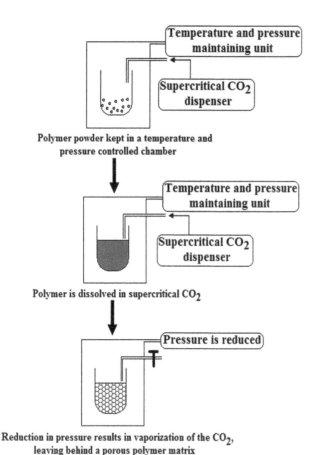

Fig. 6 Schematic diagram of scaffold preparation using super-critical CO_2.

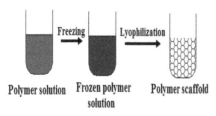

Fig. 7 Schematic representation of fabrication of polymeric scaffold by lyophilization/freeze-drying technique.

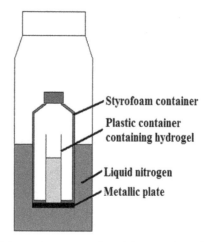

Fig. 8 Schematic representation of the setup for unidirectional freezing of hydrogels.

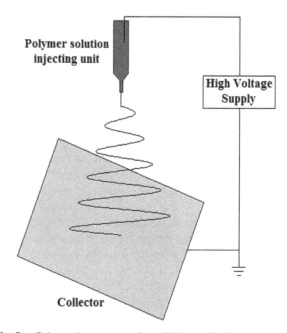

Fig. 9 Schematic representation of electrospinning apparatus.

Similar to the freeze drying, the freeze-extraction method has also been successfully explored for the development of polymeric scaffolds. The freeze-extraction technique involves the freezing of the polymer solution in dioxane or chloroform to form a solid mass. The frozen solvent is then extracted with ethanol aqueous solution. The process takes advantage that the solvents (dioxane and chloroform) are miscible with ethanol. The ethanol treatment results in the exchange of the solvent with the ethanol aqueous solution. The samples are then dried at room-temperature to obtain polymeric scaffold.[108]

In the last decade, the fabrication of the scaffolds by forming polymeric fibers using an electrical field has gained

much importance. The method is popularly known as the electrospinning method (Fig. 9). The method is said to involve two phenomena, namely, electrospraying and solution drying spinning, for the preparation of the fibers.[109] In this method, the polymer solution is passed through a needle (connected to a high voltage source) so as to fall in a collector device (usually connected to the ground). The presence of high voltage converts the spherical droplets of the polymer solution to a cone-like structure known as a Taylor cone. Further increase in voltage helps to overcome the surface active forces acting in the polymer solution thereby resulting in the formation of a jet of polymer solution. As the polymer solution jet is formed, the liquid becomes charged with a subsequent increase in the temperature of the solution. This increase in the temperature of the solution promotes the evaporation of the solvent and subsequent precipitation of the polymer as fibers, which are being collected by the collector device.[110] Various polymers like collagen, PLA, PGA, PLGA, and polycaprolactone have been used to fabricate the scaffold by the said method.[110–112]

Physicochemical Characterization of the Scaffold

The prepared scaffolds have to be characterized before it can be reduced to practice. The molecular weight and polydispersity index of the polymer can be determined by gel permeation chromatography. The chemical information of the polymer could be found from Carbon-Hydrogen-Nitrogen analysis, Nuclear Magnetic Resonance spectroscopy, X-ray diffraction patterns, small angle neutron scattering spectroscopy, Fourier transform infrared spectroscopy, Raman spectroscopy, and solid state UV-spectroscopy.[1,113] Thermal properties can be found from differential scanning calorimetry, thermo-gravimetric analysis, and dynamic mechanical analysis (DMA). DMA can also give an insight on the crosslink density within the polymer network and the mechanical stability of the polymer.[33,114] Mechanical properties can be found using a universal mechanical tester. Mercury porosimetry has been useful in determining the porosity profile within the polymer matrix. Scanning electron microscopy (SEM) has been used to view pore structure and morphology.[115–118] The surface chemistry and composition can be determined using electron spectroscopy for chemical analysis. SEM and atomic forced microscopy can reveal the surface morphology of the scaffold.[119,120] The *in vitro* degradability of the polymeric material could be analyzed by placing the scaffold in simulated physiological solutions and subsequently determining the changes in the properties. The *in vivo* degradation pattern is important to predict the attachment of cells with the host tissues and release of the active agents.[121]

NATURAL POLYMERS IN TISSUE ENGINEERING

Naturally occurring polymers (e.g., cellulose, gellan, chitosan, alginic acid, dextran, etc.), extracted from nature, have been used since long in various biomedical applications.[122–125] These polymers serve a variety of functions, when present in the biological systems. The properties of these materials may be tailored by chemical modification of the polymers.[126–129] Taking inspiration from nature, it is quite feasible to develop products with natural polymers having a variety of physical properties. In the current section, applications of the various natural polymers having potential to serve as a support matrix for tissue engineering will be discussed.

Collagen

The extracellular matrix (ECM), a connective tissue responsible for providing structural support to cells, primarily consists of collagen in all multicellular organisms.[130] The tensile strength of the tissues are highly dependent on the amount of the collagen. Higher the collagen content within the ECM, higher the mechanical properties of the tissues.[131,132] Collagen, rather being a single protein, is composed of group of proteins consisting of three α-peptides. The α-peptides have a left-handed helix structure, which forms a right-handed coiled structure (triple helix). The triple helix structure is stabilized by the intermolecular hydrogen bonding thereby resulting in the formation of a highly organized structure. Due to the highly organized structure, collagen has very good mechanical properties.[133] Apart from this, collagen has been reported to be highly biocompatible with negligible antigenicity.[134,135] The low antigenic properties of the biomedical grade collagen (obtained from animal source) may be attributed to the enzymatic treatment of the collagen. The enzymatic treatment causes elimination of the telopeptides, responsible for immunogenic properties.[136–138] The chances of immunogenic response may be completely eliminated by using collagen and its derivatives obtained from human collagen. The collagen products are prone to biodegradation when implanted. The degradability of the products may be altered either my chemical modification of the collagen or by crosslinking the products. Due to the inherent biocompatibility and ease of chemical modifications of collagen, they have been extensively used in scaffold preparation.[139]

Collagen scaffolds have been used in the treatment of the meniscal cartilage defects under both *in vitro* and *in vivo* studies. The collagen scaffolds were found to be safe after implantation. The results indicated that as time progressed, the meniscal cartilage tissue was able to use the scaffold as the supporting matrix for growth and the scaffold was slowly reabsorbed. These clinical results support that the collagen-based scaffolds may be used for meniscal cartilage repair/replacement.[140] In a separate study, type-II collagen scaffolds were seeded with articular chondrocytes and cultured under *in vitro* conditions for a period of 4 weeks. The cultured products were then implanted into chondroital defects of canines. The study indicated that there was a decrease in the fibrous tissue filling in the defects though the mechanical properties of the repaired tissues were 20 times lower than the healthy existing tissue.[141] Mechanical loading of the tissue-engineered cartilage tissues have indicated that the biosynthetic response by these tissues are similar to that of the tissues in the native environment.[142]

The collagen scaffolds have also been used in correcting bone defects.[143] Collagen scaffolds were used to deliver the platelet-rich plasma to the defect site apart from supporting the cell growth. The authors reported that the platelet-rich plasma does not promote new bone formation if the defect site has low regenerative potential.[144] Collagen scaffolds have been found to support the differentiation of the mesenchymal stem cells into osteogenic cells in the presence of various supplements (e.g., osteopontin and osteocalcin).[145,146] The only limitation of the collagen-based scaffolds for bone-tissue engineering is the short degradation time of the collagen. With advances in polymer modification techniques, this limitation may be taken care of in the near future.[146] Attempts have also been made to use hydroxyapatite-collagen composite as a matrix for bone-tissue engineering.[147] The incorporation of bone morphogenic proteins within the scaffolds has been found to heal bone defects at relatively quicker rates.[148]

Collagen nanofibrous scaffolds, whose fibers are aligned linearly, may be designed by the electrospinning technique.[149,150] The stability of these scaffolds may be improved by glutaraldehyde treatment.[151] The linearly arranged fibers in the scaffolds may be used for the *in vitro* culture of fibroblasts (e.g., conjunctiva fibroblast). It has been found that cell adhesions on these scaffolds are poorer as compared to the scaffolds having random arrangement. On the other hand, it was found that the rate of proliferation of these cells were higher and were attributed to the aligned fibers.[152] Collagen scaffolds having a structure similar to a honeycomb have also been tried for culturing human fibroblast cells. The scaffolds were elastic and hard in nature and had pore sizes in the range of 100–1000 μm. The scaffold supported the growth of the fibroblast cells indicating their probable use in the development of fibroblast cell-based tissue-engineered products. Apart from the fibroblast cells, the scaffolds also supported the growth of the CHO-K1 (hamster ovary cell line), BHK-21 (baby hamster kidney cell line), and bovine endothelial cells.[139] The penetration of the fibroblast cells within the scaffold core has been a great challenge. This has been attributed to the presence of a limited amount of oxygen and nutrients at the core. The cell penetration may be improved by inducing vascularization. The incorporation of vascular endothelial growth factors within the collagen scaffold has been found to promote vascularization, which, in turn, promotes the penetration of the fibroblast cells in the scaffold core.[153]

Collagen-based scaffolds have also been developed in conjunction with other polymers. Chitosan/collagen scaffolds have been prepared by the freeze-drying process. The scaffolds were loaded with plasmid and adenoviral vectors encoding human transforming growth factor-β1 (TGF-β1). Human periodontal ligament cells were cultured on the scaffolds. This resulted in the expression of type-I and type-III collagen. When these scaffolds were implanted, they promoted tissue in-growth within the scaffold thereby resulting in the integration of the scaffold with the host tissue. The authors concluded that the devised scaffolds may be used for periodontal tissue engineering.[154]

Gelatin

Gelatin is obtained by inducing the degradation of the collagen, either by thermal or chemical methods and is often regarded as a partial derivative of collagen.[155] In general, the antigenic properties of gelatin are much lower as compared to its precursor, collagen.[156] Literature indicate that gelatin does not show any antigenicity.[157] Apart from the above, gelatin is relatively cheaper than most of the biopolymers and has been used in designing various biopolymeric products having biomedical applications.[158–160] Gelatin has been found to improve the cell proliferation under *in vivo* conditions, which makes it a suitable candidate for devising scaffolds for tissue engineering.[71,161–163] The only disadvantage, often associated with gelatin-based products, is the rapid degradation of the gelatin-based products when placed in the physiological environment.[65,164–165] But the rate of degradation may be modulated by employing various polymer processing technologies, namely, crosslinking gelatin matrices, composites, and/or polymer blend fabrications.[166–169] In general, a scaffold should mimic the ECM properties so as to promote the process of regeneration of the traumatized tissues.[162] In the present section, various applications of gelatin-based scaffolds will be discussed.

Scaffolds based on gelatin nanofibers have been developed by the electrospinning technique. The average diameter of the nanofibers was 300 nm and could be altered by modulating the various parameters associated with the electrospinning technique (e.g., polymer concentration and voltage applied). The mechanical and degradation properties of the scaffolds were improved by crosslinking with glutaraldehyde. Other crosslinking agents that may be used include genipin, 1-ethyl-3,3-dimethyl aminopropyl carbodiimide, 2-chloro-1-methylpyridinium iodide, and glyceraldehydes.[170–171] Glutaraldehyde crosslinked gelatin scaffolds have been found to promote mineralization (apatite formation) of the gelatin fibers.[172] The rate of mineralization of the scaffolds may be improved by electrodeposition.[173] The glutaraldehyde crosslinked scaffolds have been found to not only promote fibroblast cell growth but also promote the penetration of cells within the scaffolds. The penetration of the cell growth may be improved by employing a template polymer (e.g., polyethylene glycol) during the fabrication of the scaffolds. The template polymer is then extracted using a suitable solvent (e.g., tert-butanol) after the scaffold fabrication using the electrospinning technique.[174] The crosslinking may also be carried out by using glutaraldehyde vapor.[175] The crosslinking of the gelatin structures improves the water-resistant ability of the scaffolds.[175] Hydroxyapatite (HA)-coated gelatin scaffolds have been found to induce osteogenic differentiation in the bone marrow–derived mesenchymal stem cells.[176] Apart from the electrospinning technique, nanostructured scaffolds may be developed by using a thermally induced phase separation technique.[177] Macroporous gelatin scaffolds may be prepared by combining the thermally induced phase separation technique and the porogen leaching method. The macroporous scaffolds showed better dimensional structural stability as compared to the commercially available gelatin foam for tissue culture.[177] The gelatin-based scaffolds have shown promises in the regeneration of the bone tissues under *in vivo* conditions in rabbits.[178] In a recent study, methacrylamide-modified gelatin 3D CAD (computer aided design) scaffolds were prepared by using photon polymerization. Irgacure 2959 was used as the photoinitiator. The process resulted in the polymerization of the methacrylamide groups, which, in turn, resulted in the crosslinking of the methacrylamide-modified gelatin. The developed scaffolds were enzymatically degradable and the rates of degradation of the scaffolds were found to be comparable as against gelatin scaffolds. The scaffolds were found to promote osteogenic differentiation.[179,180]

The incorporation of ceramics (e.g., magnesium calcium phosphate) has been found to improve the mechanical properties, like compressive modulus and resistance to structural change, of the scaffolds. In order to promote the ECM properties of the bone, gelatin-apatite nanofibrous composite scaffolds have been developed by using the thermally induced phase separation technique. The composite scaffolds showed improved mechanical stability as compared to the commercially available gelatin foam. Also, the composite scaffolds promoted osteogenic differentiation indicating its probable use in bone tissue engineering.[162] The activity of the mesenchymal stem cells (obtained from the rat bone marrow) was found to be higher in the scaffolds with higher magnesium calcium phosphate content with a subsequent improvement in the proliferation and osteogenic differentiation of the cells.[181] In a similar study, bone morphogenic protein was incorporated within gelatin/β-tricalcium phosphate so as to release the same in a controlled manner. Human bone marrow–derived stem cells and adipose-derived stem cells were cultured over the scaffolds. The results indicated that both types of stems cells showed improved osteogenic differentiation.[182]

Gelatin-based scaffolds with microtubule orientation have been developed. Unidirectional freezing method was employed for the aligned scaffolds. The porosity of the scaffolds was varied by altering the gelatin concentration. In general, as the concentration of the gelatin was increased there was a subsequent decrease in porosity. The scaffolds promoted cartilage cell growth.[107] Genipin-crosslinked gelatin scaffolds have been developed for cartilage tissue engineering. The effect of temperature on the pore size of the scaffolds was studied. It was found that as the crosslinking temperature was reduced, there was a subsequent decrease in the pore size of the scaffolds and increase in the mechanical properties. The scaffolds were found to undergo reversible deformation as during human locomotion. Based on the results, the authors concluded that the genipin-crosslinked gelatin scaffolds may be used for tissue engineering of articular cartilages.[183]

Myelomeningocele, a congenital defect characterized by the protrusion of a sac in the mid-lower back, has been treated with gelatin scaffold. The authors reported that the amniotic fluid cells were able to proliferate over the scaffold and concluded that gelatin-based injectable scaffolds may be used for the treatment of the myelomeningocele.[184]

Angiogenesis is an important factor in wound healing. An important factor for inducing angiogenesis has been attributed to the controlled delivery of bioactive agents like angiogenic cytokines and cells in the ischemic regions. The human umbilical vein endothelial cells showed directional growth along the fiber of the scaffold prepared by electrospinning technique. The incorporation of the basic fibroblast growth factor into the scaffold was found to increase the capillary formation.[185]

Blending of other polymers with gelatin has resulted in scaffolds with improved mechanical and physical properties. Poly (ε-caprolactone)-gelatin coaxial nanofiber-based scaffolds were developed. The proliferation studies of the human umbilical vein endothelial cells indicated that the cell adherence onto the scaffold nanofibers was much higher as compared to the scaffolds prepared from poly (ε-caprolactone). The results also indicated that the scaffold was able to reduce the apoptosis of the human umbilical vein endothelial cells.[186] A blend of sodium alginate and gelatin solution was freeze dried to obtain a 3D porous structure, which was subsequently incubated in calcium chloride solution to get alginate-gelatin scaffold. The scaffold promoted the growth and differentiation of the human bone marrow–derived mesenchymal stem cells. The results indicated that the scaffold may be used for the tissue engineering of the bone, cartilage, and adipose tissues.[187] Chitosan-gelatin scaffolds were developed for tissue engineering of the bone, cartilage, and other tissues. It has been found that the blending of the chitosan result in the improved properties of scaffolds.[188–190] The rate of proliferation of chondrocytes for cartilage tissue repair may be improved by entrapping plasmid DNA, which expresses the formation of TGF-β1.[191] Incorporation of the silver nanoparticles has resulted in the scaffolds with improved cell proliferation.[192]

Silk Fibroin

Silk contains two proteins, the hydrophobic fibroin and the hydrophilic sericin.[193] Fibroin is the structural protein whereas the hydrophilic sericin helps bind the fibroin fibers. Silk fibroin (SF) is a natural biopolymer obtained from two species of silk worms, namely, *Bombyx mori* and *Antheraea pernyi*.[194] SF promotes cell adherence and proliferation, which may be attributed to the presence of RGD sequences within the SF.[195,196] Scaffolds of SF have been used as carriers for vascular endothelial cells, cardiomyocytes and chondrocytes.[197,198] SF scaffolds not only support the growth and proliferation of the vascular cells but also improve the antithrombogenecity.[199] Freeze-gelled SF scaffolds were found to promote the cell attachment and proliferation of human keratocyte cells in the skin tissue engineering.[193]

Chondrocyte/SF constructs have been used as an implant for *in vivo* repair of cartilage defect.[200,201] Immobilization of the enzyme alkaline phosphatase on SF scaffolds has been found to enhance the bone formation and may be tried in bone tissue engineering.[202] Fast degradable alginate/SF microbeads were used for the encapsulation of human umbilical cord mesenchymal stem cells (hUC-MSCs). The released cells showed excellent osteodifferentiation and cell proliferation properties thus making it suitable for bone tissue regeneration.[203] Microporous SF scaffold embedded with PLGA microparticles for the delivery of growth factors in tissue engineering have been prepared.[204]

SF scaffolds can be used for the reconstruction of choriocapillaries and hence may be tried in ocular tissue engineering.[205] SF scaffolds may promote the survival of Schwann cells. The results suggested that the developed matrices may be used for nerve tissue engineering.[206] SF graft has the ability to repair peripheral nerve defects.[207] The SF scaffolds have also promoted the cell adhesion and growth of keratinocytes thus acting as a potential candidate for the alternative graft in myringoplasty.[208]

The mechanical integrity and rate of degradation of SF scaffolds can be modulated for use in soft tissue reconstructs.[209] The chitosan/SF films have been explored as supporting materials for skin tissue engineering. The scaffolds allowed the proliferation of fibroblast cells.[210]

Fibrin and Fibrinogen

Fibrinogen is a soluble protein (MW ~340 kDa), which is present in blood and plays an important role in blood clotting and platelet aggregation.[211] Fibrinogen-fibronectin constructs exhibit excellent chemotactic and cell adhesion properties, which makes them suitable candidates for the preparation of scaffolds for tissue engineering.[212] The fibrin scaffolds have been used as matrices for stem cells.[213] Fibrinogen scaffolds have been regarded as an excellent candidate for the regeneration of adipocytes and endothelial cells.[214] The fibrin gels are being extensively used in orthopedic surgery.[215,216] The use of human platelet-rich fibrin for the development of scaffolds enhances the proliferation and re-differentiation of chondrocytes thereby making it suitable for cartilage tissue engineering.[217]

Macroporous fibrin constructs allowed the differentiation of hUCMSCs into myogenic lineage to form multinucleated myotubes and was used for muscle tissue engineering.[218] The transplantation of fibrin/bone marrow stromal cells (BMSCs) matrix into the cortical lesions was found to be suitable for nerve tissue regeneration.[219] The fibrin scaffolds can support smooth muscle cells and have been used in urinary tract tissue regeneration.[220] The fibrinogen scaffold allows the migration of cells and deposition of connective tissue.[221] Electrospun fibrinogen

scaffolds have the properties of potential scaffold to be used as wound dressing and soft tissue regeneration.[222]

Fibrin gels can serve as a scaffold for cardiovascular tissue engineering.[223] The fibrin gels may sustain the release of growth factors [e.g., vascular endothelial growth factor (VEGF)] and have been tried successfully for neural tissue regeneration applications.[224] Novel fibrin magnetic hydrogel scaffolds have been prepared. These scaffolds may be visualized (magnetic visualization) after being implanted into the human body.[225] The scaffolds developed from fibrinogen-fibronectin-vitronectin hydrogels have been used for the development of cell-based therapies in patients with end stage pulmonary diseases.[226]

Fibrinogen composite materials have also been used to prepare constructs for soft tissue engineering.[212] The fibrinogen composite scaffolds have been found to enhance the cell proliferation and growth.[227] Composites of fibrinogen and collagen scaffolds when implanted into mice were found to be useful for tissue regeneration and angiogenesis.[228] The poor mechanical properties of the fibrinogen-based scaffolds can be improved by crosslinking either with synthetic polymers (e.g., polydioxanone, polyethylene glycol, and poly glycerol sebacate) or with chemical crosslinkers (e.g., carbodiimide derivative and genipin).[229] The versatile physical properties of PEG-fibrinogen hydrogels have generated a new area in tissue engineering scaffolds. They are easy to produce and provide biofunctionality for cell culture.[230] The hydrogels composed of mono PEGylated albumin and PEGylated fibrinogen have shown potential for controlled drug delivery in tissue engineering. These composite hydrogels have the ability to attach certain drug molecules for controlled delivery in tissue engineering applications.[231] Macroporous sponge-like hydrogel of gelatin-fibrinogen scaffolds prepared by cryogelation technology served as a template for dermal regeneration.[232] PGS/fibrinogen scaffolds showed improved mechanical properties, which makes them suitable for the development of cardiac patches for the treatment of the myocardium infraction through cardiac tissue regeneration.[233] Hybrid fibrin/poly(glycolic acid) sponges have been used for the repair of the damaged skin, cartilage, bone, and adipose tissue.[234] The fibrin composites may be designed to deliver the growth factors (e.g., angiogenic growth factor, vascular endothelial growth factor, and basic fibroblast growth factor).[235] Injectable scaffolds fabricated from a mixture of β-Tricalcium phosphate (β-TCP) particles, fibrinogen, and thrombin have shown improved bone regeneration.[236]

Alginate

Alginic acid or alginate is an anionic linear polysaccharide, composed of β-D-mannuronic acid (M) and α-L-guluronic acid (G) residues. The mechanical properties of the alginate products are dependent on the ratio of M and G units.[237] Due to the anionic nature of the alginates, the alginates inherently do not support mammalian cell adhesion. This is associated with the inhibition of protein absorption due to electrostatic repulsion.[238] Because of this reason, it is necessary to use appropriate ligands to promote cell-scaffold interactions, necessary for tissue engineering applications.[239] This can be achieved by preparing derivatives of alginate, namely, amphiphilic alginate and cell interactive alginate. Amphiphilic alginate derivatives were being synthesized by introducing the hydrophobic group (alkyl chains and polymers) to the alginate backbone, whereas the cell interactive alginate contains peptides in the side chain. One such example is the RGD (Arg-Gly-Asp) immobilized alginate, which helps in cell recognition and adhesion thereby promoting the attachment of the cell with the ECM.[240] The RGD-immobilized alginate scaffold helps in the reorganization of cardiomyocytes during cardiac tissue engineering.[241] The scaffolds may be modulated to contain growth factors (e.g., vascular endothelial growth factor, platelet-derived growth factor-BB, and transforming growth factor-β1) for enhanced cardiomyocyte regeneration.[242]

Alginates have been extensively used in cartilage and bone tissue engineering. The alginate construct serves as a grafting material during regenerative procedures in bone tissue engineering.[243] Alginate microfiber and microparticle-aggregated scaffolds have been found to contain sufficient mechanical strength and elastic modulus for the regeneration of the trabecular bone under *in vitro* conditions.[244] The sustained release of TGFβ1 and bone morphogenetic protein-2 (BMP-2) have been achieved successfully for bone tissue regeneration.[245] Chitosan and HA-blended alginate scaffolds have been found to be more osteoconductive than the pure alginate scaffolds.[246] This type of composite scaffold enables rapid vascularization during cell growth with improved mechanical properties of the developed bone.[247] Composite scaffolds prepared with nano-bioglass enhances biomineralization and is suitable for periodontal tissue regeneration.[243] Alginate scaffolds have shown promising results in the repair of intervertebral discs (IVDs). The alginate scaffolds help improve the mechanical properties of the nucleus pulposus of the IVD.[248] Apart from the bone and cartilage tissue engineering, microparticles of alginates have supported the proliferation of adipose tissue stromal cells.[249] Scaffolds developed with the alginate crosslinked with galactosylated chitosan have shown a spheroidal morphology of the hepatocytes with good integration with the scaffold.[250]

Chitosan

Chitosan is derived from the natural heteropolysaccharide polymer, chitin.[251–253] Chitin is predominantly found in the exoskeleton of crabs and shrimps.[254] The conversion of chitin to chitosan involves partial deacetylation of the heteropolysaccharide either by chemical or enzymatic treatment.[255] Chitosan-based products have poor mechanical properties.[256] The mechanical properties of

the chitosan-based products may be improved either by blending chitosan with other polymers and/or ceramics or by crosslinking the chitosan products.[257–260] In general, chitosan scaffolds do not elicit either inflammatory responses or allow the growth of infecting microbes, and hence the accumulation of toxins associated with the disease-causing microbes.[261] The scaffolds are meant to serve as a cell support for the proliferating cells in study and should also be able to accommodate and subsequently release bioactive agents, necessary for cell proliferation and angiogenesis, in a controlled manner.[261,249] The current section has been devoted to the use of chitosan for developing matrices in various tissue engineering applications.

The properties of the chitosan scaffolds may be varied by changing the freezing conditions during the lyophilization.[262,263] Porcine chondrocytes have been successfully proliferated on the scaffold. The results indicated the synthesis of ECM, rich in proteoglycan and type-II collagen and may be tried for cartilage tissue engineering.[264] Incorporation of the SF into the scaffold structure help improving the cell adhesion.[265] Coating of the chitosan scaffold surfaces with poly(L-lactide-*co*-ε-caprolactone) improves the surface wettability of the scaffolds and supported differentiation of the human bone marrow derived mesenchymal stem cells into chondrocytes. The modification of the chitosan scaffold resulted in the formation of cartilages having better quality.[266] Chitosan-poly(butylene succinate) scaffolds were prepared using compression molding followed by salt leaching method. Bovine articular chondrocytes were seeded onto the chitosan-poly(butylene succinate) scaffolds. The chondrocytes were able to grow throughout the scaffold and generate ECM around itself indicating that the scaffolds may be used in cartilage tissue regeneration.[267]

Chitosan-based scaffolds have shown promising results for culturing hepatocytes. Chitosan-gelatin blend scaffolds have been prepared by the lyophilization technique. The functionality of the scaffolds was improved by using well-organized architecture molds developed by stereolithographic technique. The hepatocytes were found to proliferate readily.[268] The bioactivity of the hepatocytes may be improved by the surface modification of the chitosan with galactose ligands. The surface modification also improves the mechanical stability of the hepatocytes under culture conditions. The nanofibrous scaffold prepared by the electrospinning method showed slow degradation rate in addition to sufficient mechanical properties to be used as substrates for hepatocyte culture.[269]

The use of chitosan scaffolds in bone tissue engineering was limited due to the poor mechanical properties of chitosan scaffolds. Keeping this in mind, alginates have been used for improving the mechanical properties of the chitosan scaffolds. The ionic interaction amongst the chitosan and alginate molecules results in the formation of a complex structure, which in turn helps to improve the mechanical properties of the chitosan scaffolds. Osteoblasts readily

Nanoparticles—Nerve

adhered to the scaffold matrix and the rate of proliferation was quick as compared to the chitosan scaffolds. Development of alginate-chitosan scaffolds by the coacervation technique resulted in the formation of scaffolds with larger pore size.[247] Incorporation of HA within the chitosan scaffolds has been found to improve the mechanical properties of the scaffolds. Apart from this, there was a subsequent increase in the osteoconductivity during the rat calvarial defect study. There was an increase in the osteocalcin production in the chitosan-hydroxyapatite composite scaffolds as compared to the chitosan scaffolds.[270] Similar to the chitosan-hydroxyapatite composite scaffolds, attempts have also been made to develop calcium phosphate and bioactive glass ceramics to improve the growth and proliferation rate of the osteoblasts.[259,271] Though the chitosan-bioactive ceramics may not be used in bone regeneration in load bearing joints, they may be used in maxillofacial bone defects and other moderate load-bearing joint defects.[272] Functionalized multiwalled carbon nanotubes have also been used as reinforcing agents for improving the mechanical properties of the scaffolds. There was an improvement in the cell growth and proliferation as compared to the chitosan scaffolds.[273] In a separate study, chitosan-hydroxyapatite composite scaffolds were grafted with functionalized multiwalled carbon nanotubes. The scaffolds were seeded with MG-63 cells, osteosarcoma cell line. The results showed that the grafted scaffolds promoted the proliferation of the cells indicating that the scaffold may be used for bone tissue engineering.[274] Scaffolds developed by melt blending of chitosan with poly(butylene succinate) was found to induce osteogenic differentiation of human bone marrow mesenchymal stem cells.[275] Chitosan scaffolds coated with fibrillar collagen type-I was analyzed as a matrix for the proliferation and differentiation of the human bone marrow stromal cells as compared to the chitosan scaffolds used as control. The study indicated that the rate of proliferation and differentiation of the stromal cells into osteogenic cells was much higher in the collagen-coated scaffolds.[276]

The repair of soft tissue defects is a great challenge. The scaffolds developed from SF-chitosan blends have been studied as probable matrices for the regeneration and repair of soft tissues. Human adipose tissue-derived stem cells have been used for the regeneration of the fibrovascular, endothelial, and epithelial layers in the soft tissue layer. The findings of the study suggest that the SF-chitosan scaffolds promote wound healing of the soft tissues.[277]

Hyaluronic Acid

Hyaluronic acid (hyaluronan or hyaluronate) (HA) can be isolated from natural sources (e.g., rooster combs) or from bacteria during the biofermentation process.[278] It is a major component of the ECM in connective tissues and is particularly abundant in vitreous and synovial fluids. HA is a linear high-molecular weight non-sulfated anionic polysaccharide consisting of glycosaminoglycan as the

repeating unit. Each repeating unit is composed of a disaccharide: N-acetyl-D-glucosamine and D-glucuronic acid with unbranched units ranging from 500 to several thousand.[279] It is a non-immunogenic and non-adhesive glycosaminoglycan that plays a significant role in several cellular processes including angiogenesis and the regulation of inflammation.[280] The high water-binding capacity of HA facilitates the pure and composite HA scaffolds to be used in dermal implantations and as wound-healing matrices.[281] The porosity of the HA scaffolds can be tailored by preparing biocomposites and/or by changing the scaffold fabrication methods. The stability of the HA scaffolds may be enhanced by chemical crosslinking.[282,283] HA/chitosan/gelatin biocomposite scaffolds have been extensively studied for skin tissue engineering.[284] Composite scaffolds of HA and polypyrrole (an electrically conducting polymer) were investigated in the wound-healing application. These scaffolds were helpful in enhancing vascularization.[285]

Apart from dermal applications, HA-based scaffolds have applications in bone and cartilage tissue engineering. Chitosan-based HA hybrid polymer fibrous scaffold sheets were reported to culture articular chondrocytes for cartilage tissue engineering.[286] The adhesion of chondrocytes was improved when hybrid HA-derived scaffolds were used.[42] HA-based biocomposites have been explored in bone tissue engineering. Pretreatment of the gels with fibronectin has been found to improve cell attachment. This has been associated with an integrin-mediated cell attachment mechanism.[287] The cell adhesion properties may be modulated by coating them with glycosaminoglycan.[288] The coating of synthetic polymer (e.g., PLGA) scaffolds with HA has shown improved cell attachment as compared to the uncoated scaffolds.[289]

Agarose

Agar is a polysaccharide obtained from sea weeds, mainly red algae.[290] Usually agar is a mixture of linear polysaccharides, agarose, and heterogeneous agaropectin. Agarose is made up of repeating units of agarobiose, which is a disaccharide of D-galactose and 3,6-anhydro-L-galactopyranose.[291] The gelling properties of agarose have been exploited in the preparation of scaffolds for tissue engineering.[292] The biocompatibility and biodegradability of agarose also facilitates its use as the material of choice for the fabrication of scaffolds in tissue engineering.[293]

Agarose-based biocomposites have shown potential applications in stem cell cultures. The differentiation and growth of human adipose-derived adult stem cells (hADAS) depends on the cell-biomaterial interactions. The hADAS cells were seeded and cultured on the agarose, alginate, and gelatin scaffolds. The results suggested that the composition of the scaffolds affected the mechanical properties of the constructs. The cell adhesion property of the agarose scaffolds may be tailored by conjugating the agarose with different types of peptides (e.g., AG73, a laminin peptide, was conjugated with agarose scaffolds).[294] The tailored matrices were found to be suitable for the culture of the fibroblast cells, neuronal cells, endothelial cells, and salivary gland cells. Hence, these laminin-conjugated agarose matrices can be regarded as the multifunctional biomaterials for tissue engineering applications. The incorporation of collagen I into the agarose scaffolds showed improved viability of embryonic rat cortical neurons.[295] The agarose hydrogels promote the growth of the neuronal cells and have been used as a guide for the local spinal cord axons after spinal cord injury.[296] Agarose hydrogels have also been designed to induce proteoglycan production by human IVD cells.[297,298]

Agarose hydrogels have also been employed in bone and cartilage tissue engineering.[299] Human hyaline chondrocytes, articular cartilage, bovine chondrocytes, and multipotent mesenchymal stem cells were cultured on the agarose scaffolds.[300–310] Articular cartilages contain depth-dependent inhomogeneity. The agarose scaffolds were fabricated in a similar fashion. These inhomogeneous bilayered agarose scaffolds have found applications in investigating the behavior of the chondrocytes and development of articular cartilage constructs.[308,311] Agarose hydrogels have also been tried in the cell patterning of mammalian osteoblasts.[312] The pattern technique may be employed for the high throughput cytotoxicity and genotoxicity screening applications.[313]

Agarose biocomposites (e.g., agarose-chitosan, agarose-fibrin, agarose-hyaluronic acid, and agarose-gelatin) scaffolds have promised to be good candidates for tissue engineering applications.[314,315] The agarose-fibrin composite scaffolds were used to bioengineer human corneal constructs at the nanoscale. These scaffolds may lead to the regeneration of artificial cornea.[316] Agarose-gelatin and chitosan-agarose-gelatin scaffolds have supported the vascularization[317] and proliferation of fibroblasts and cardiac cells[318] The biodegradation of the scaffolds developed using agarose biocomposites may be altered by chemical crosslinking using chemical crosslinkers (e.g., novel epichlorohydrin)[319]

Agarose hydrogels incorporated with biphasic calcium phosphate have been developed. These hydrogels can help avoid the inherent brittleness often associated with bioceramics. Apart from this, these composites may enhance the compression resistance of the hydrogels.[320] Composite scaffolds of agarose in combination with HA, and TCP have also been developed for bone and cartilage tissue engineering applications.[321,322] The properties of these scaffolds may be tailored to modulate the release properties of the incorporated bioactive agents (e.g., vancomycin).[323]

Cellulose

Cellulose is a fascinating and sustainable biopolymer and is widely available in nature. Cellulose is a linear homopolymer, formed by repeated β(1→4) linked D-glucose units

$(C_6H_{10}O_5)_n$, with n ranging from 500 to 5000.[324] Cellulose and its derivatives have been employed in both hard and soft tissue engineering applications. Non-woven cellulose II has been predominantly explored in tissue engineering applications.[325] The degradability of the cellulose in animals and humans is limited due to the lack of enzymes (hydrolases) to dissociate β(1→4) linkages.[326] Since the 1950s, cellulose sponges/hydrogels are being used as implantable matrices.[327] Nowadays, high-purity cellulose sponges are available in the market for biomedical applications.[328] Cellulose-based products have been used as sealing materials as a substitute for acrylic bone cement for the fixation of the femoral component of hip prostheses.[329,330] When Cellspon® (trade name of a viscose cellulose sponge) was implanted in the rats, moderate foreign body reaction was seen for the first 16 weeks. Thereafter, the foreign body reaction was subsided with improved healing. The implant was completely degraded out of the body within a span of 60 weeks suggesting that the material is slowly biodegradable.[331] Viscose cellulose sponges have shown promises in bone tissue engineering.[332] Complete osseointegration has not been achieved with pure cellulosic scaffolds. A slightly better osseointegration has been reported with phosphorylated cellulose hydrogels.[333,334] Phosphorylated cellulose promotes mineralization during the bone healing process, thereby forming a calcium phosphate layer on the surface that has a closer resemblance with the bone functionality.[335,336] This layer acts as an interface between the hard bone tissue and soft biomaterial.[336] Oxidized regenerated celluloses have been employed to improve the bone tissue compatibility and cellulose degradation. These celluloses have been used in maxillary reconstructive surgery to promote osseous and soft tissue regeneration.[337] The adhesion of cells (e.g., human osteoprogenetor cells) and subsequent proliferation may be improved by incorporating short peptide sequences (e.g., RGD: Arg-Gly-Asp).[338] In addition to bone tissue engineering, cellulose sponges have also been tested for *in vivo* articular chondrocytes regeneration.[339]

Various cellulose derivatives that have been examined for tissue engineering applications include methyl cellulose, ethyl cellulose, aminoethyl cellulose, hydroxyethyl cellulose, and cellulosic polyanion complexes.[340] The blends of cellulose with natural (e.g., SF and starch) and synthetic polymers (e.g., polyvinyl alcohol and polyvinyl pyrrolidone) have been tried for the development of scaffolds for tissue engineering applications.[341–343]

Bacterial Cellulose

Bacteria of *Acetobacter* sp. synthesize pure cellulose and secrete the same into the extracellular spaces. The chemical structure of bacterial cellulose (BC) is similar to that of the plant-derived cellulose, i.e., $(C_6H_{10}O_5)_n$. BC possess high purity, crystallinity, water-holding capacity, ultra-fine network structure, and mechanical properties as compared to the plant-derived cellulose. The mechanical and thermal properties of BC-based products may be improved by using blends of BC and other polymers (e.g., tamarind xyloglucan, chitosan).[344] BC can be easily molded into different shapes.[345,346] BC and it's derivatives have been found to be biocompatible in nature and have been widely used for the development of constructs for biomedical applications.[347] BCs have been successfully tried in cartilage tissue engineering. Scaffolds prepared with native BC showed higher levels of chondrocyte growth as compared to scaffolds prepared from modified BC and calcium alginate without any significant proinflammatory cytokine production.[348] These results led to the further exploration of native BC using human chondrocytes.

The ability to easily mold BC into different shapes has been explored in the fields of microsurgery and reconstructive problems associated with vascular diseases. Tubular BC can be directly prepared during the cultivation of bacteria, which may be used as artificial blood vessels including carotid arteries.[349,350] BC membrane scaffolds can be used as wound dressing.[351] BC/alginate mucosal flaps have demonstrated to support the growth of human keratinocytes and gingival fibroblasts for oral tissue regeneration.[352]

Starch

Starch is a polymer composed of anhydro-glucose units, which occurs naturally in plants.[353] Native starch granules are composed of two polysaccharides, the linear amylose and the highly branched amylopectin, which organize to form semi-crystalline granules.[354,355] Starch is a potentially useful material for the preparation of biodegradable scaffolds because of its natural abundance and low cost. Although starch possesses useful properties (e.g., biocompatibility, biodegradable, water-holding property) required for scaffold preparation, native starch has very limited applications in the preparation of scaffolds. This is due to its poor processability, poor dimensional stability, and mechanical properties of its end products. Therefore, native starch is blended with natural or synthetic polymers to prepare scaffolds.[356,357] Starch originated from the potato, corn, and sweet potato has been extensively used for the preparation of scaffolds for bone tissue engineering applications. Recently, more focus has been given on developing various novel starch-based scaffolds and microparticles for the delivery of drugs like non-steroidal anti-inflammatory agents, corticosteroids, etc., for bone tissue engineering applications.[358,359] More recently, it was identified that the differentiation of stem cells into specific cell types has been affected by the culture environment.[360] Starch-based scaffolds made a significant impact on the differentiation of bone marrow-derived stem cells and amniotic fluid-derived stem cells into osteocytes.[360] Starch/chitosan scaffolds were designed to study the gradual pore formation in natural hybrid scaffolds using biomimetic calcium phosphate coating and lysozyme in a rat subcutaneous implantation model.[361]

Gellan Gum

Gellan gum is chemically deacetylated extracellular polysaccharide.[362] It is a linear anionic extracellular heteropolysaccharide gum produced aerobically by the fermentation of carbohydrates with *Pseudomonas paucimobilis*.[363–365] It consists of repeating tetrasaccharide unit comprising of one α-L-rhamnose, one β-D-glucuronic acid, and two β-D-glucose units. Gellan gum is Food and Drug Administration (FDA) approved food additive and has been extensively used as a stabilizer and thickener in food industries and as a carrier of bioactive agents in pharmaceutical industries.[366,367]

The gel-forming nature of gellan gum has been exploited in tissue engineering in the preparation of scaffolds.[368] Gellan gum usually imparts good swelling properties to the scaffolds.[369] Gellan gum hydrogels were tested for their ability to be used as subcutaneous implantable injectable systems for cartilage tissue-engineering application. Injectable modified gellan gum–based hydrogels were found to promote chondrocytes proliferation, enhance the ECM secretion and maintain normal phenotype.[370,371] 3D bioplotting of the injectable gellan gum scaffolds have shown promise in spinal cord injury regeneration.[372] Subcutaneous implantation of gellan gum hydrogels combined with human articular chondrocytes (hACs) have shown adequate growth and ECM deposition for hACs.[373,374] Quite often, gellan gum has been used in combination with either natural polymers (e.g., collagen) or synthetic polymers (e.g., carboxy methyl chitosan and methacrylate) for the fabrication of composite scaffolds.[375–378] The composite scaffolds have found applications in bone and cartilage tissue engineering.

Dextran

Dextran is a hydrophilic polysaccharide of bacterial origin. It is composed of mainly α-1,6 linked d-glucopyranose residues with a few percent of α-1,2-, α-1,3-, or α-1,4-linked side chains.[379] Dextran hydrogels can support endothelial cells proliferation, which makes it suitable for vascular and cartilage tissue engineering.[380] Dextran-based injectable hydrogels (e.g., dextran tyramin) have been used to repair cartilage defects due to their ability to support the proliferation of the chondrocytes.[381] The encapsulation of chondrocytes or embryonic stem cells within dextran/PEG hydrogels resulted in the development of new cartilaginous tissue. Hence, these hydrogels may be tried in cartilage tissue engineering.[382] Dextran/PLGA nanofibrous scaffolds promote the growth of dermal fibroblasts and have been used in wound healing.[383] Dextran/PLGA scaffolds loaded with VEGF showed a sustained release of the growth factors and have been used for vascular tissue engineering.[384] Cryogels based on gelatin crosslinked oxidized dextran have been used in bone tissue engineering.[385] Dextran hydrogel scaffolds have been found to enhance neovascularization and skin regeneration. These scaffolds were used in the treatment of third-degree burns.[386] Alginate/dextran hybrid hydrogels have been found to promote angiogenesis.[387]

CONCLUSION

Natural polymers have been used in the fabrication of the scaffolds for various tissue-engineering applications including soft tissue, cartilage, skin and bone regeneration. The use of natural polymers offers the advantage of the availability of a wide range of polymer chemistry and cheap availability apart from, in general, biocompatibility and non-irritant nature. The use of natural polymer-ceramic composites for the fabrication of scaffolds not only alters the mechanical properties but also helps in modulating the rate of proliferation and differentiation of cells. Apart from this, the chemical modification of the natural polymers may also help in improving the properties of the biopolymeric scaffolds or constructs. This has opened up a new area in tissue regeneration. The availability of a large number of natural polymers, which have been used scarcely in regenerative medicine, may provide a new prospective in the field of tissue engineering.

ACKNOWLEDGMENT

The authors acknowledge the financial and logistical support provided by the National Institute of Technology, Rourkela, during the completion of the review.

REFERENCES

1. Pal, K.; Banthia, A.; Majumdar, D. Polymeric hydrogels: Characterization and biomedical applications. Des. Monomers Polym. **2009**, *12* (3), 197–220.
2. Roy, S.; Pal, K.; Anis, A.; Pramanik, K.; Prabhakar, B. Polymers in mucoadhesive drug-delivery systems: A brief note. Des. Monomers Polym. **2009**, *12* (6), 483–495.
3. Al Samman, M.; Radke, W.; Khalyavina, A.; Lederer, A. Retention behavior of linear, branched, and hyperbranched polyesters in interaction liquid chromatography. Macromolecules **2010**, *43* (7), 3215–3220.
4. Wang, Z.; Chen, X.; Larson, R.G. Comparing tube models for predicting the linear rheology of branched polymer melts. J. Rheol. **2010**, *54* (2), 223.
5. Pathak, C.P. Implantable medical devices with fluorinated polymer coatings, and methods of coating thereof. 2009, US patents, 13-01-2009.
6. Palasis, M.; Naimark, W.; Richard, R.E. Medical devices for delivery of therapeutic agents. 2010, US20100087783 A1, 8-4-2010.
7. Sastri, V.R. *Plastics in Medical Devices: Properties, Requirements and Applications*; Andrew, W., Ed.; Elsevier: NY, USA, 2009.
8. Gancel, F.; Montastruc, L.; Liu, T.; Zhao, L.; Nikov, I. Lipopeptide overproduction by cell immobilization on iron-enriched light polymer particles. Process Biochem. **2009**, *44* (9), 975–978.
9. Wang, Z.G.; Wan, L.S.; Liu, Z.M.; Huang, X.J.; Xu, Z.K. Enzyme immobilization on electrospun polymer nanofibers: An overview. J. Mol. Catalysis B **2009**, *56* (4), 189–195.
10. Rothwell, S.A.; Killoran, S.J.; O'Neill, R.D. Enzyme immobilization strategies and electropolymerization conditions to control sensitivity and selectivity parameters of a polymer-enzyme composite glucose biosensor. Sensors **2010**, *10* (7), 6439–6462.
11. Cordeiro, A.L.; Pompe, T.; Salchert, K.; Werner, C. Enzyme immobilization on reactive polymer films. In *Methods in Molecular Biology: Bioconjugation Protocols*; Springer: New York, 2011.
12. Suraniti, E.; Studer, V.; Sojic, N.; Mano, N. Fast and easy enzyme immobilization by photoinitiated polymerization for efficient bioelectrochemical devices. Anal. chem. **2011**, *83* (7), 2824–2828.
13. Ellä, V.; Nikkola, L.; Kellomäki, M. Process-induced monomer on a medical-grade polymer and its effect on short-term hydrolytic degradation. J. Appl. Polym. Sci. **2011**, *119* (5), 2996–3003.
14. Mukherji, D.; Abrams, C.F. Anomalous ductility in thermoset/thermoplastic polymer alloys. Phys. Chem. Chem. Phys. **2009**, *11* (12), 2113–2115.
15. López, J.; Rico, M.; Montero, B.; Díez, J.; Ramírez, C. Polymer blends based on an epoxy-amine thermoset and a thermoplastic. J. Ther. Anal. Calorimetry **2009**, *95* (2), 369–376.
16. Foix, D.; Yu, Y.; Serra, A.; Ramis, X.; Salla, J.M. Study on the chemical modification of epoxy/anhydride thermosets using a hydroxyl terminated hyperbranched polymer. Eur. Polym. J. **2009**, *45* (5), 1454–1466.
17. Meng, Y.; Zhang, X.H.; Du, B.Y.; Zhou, B.X.; Zhou, X.; Qi, G.R. Thermosets with core-shell nanodomain by incorporation of core crosslinked star polymer into epoxy resin. Polymer **2011**, *52* (2), 391–399.
18. Taguet, A.; Huneault, M.A.; Favis, B.D. Interface/morphology relationships in polymer blends with thermoplastic starch. Polymer **2009**, *50* (24), 5733–5743.
19. Bonderer, L.J.; Feldman, K.; Gauckler, L.J. Platelet-reinforced polymer matrix composites by combined gel-casting and hot-pressing. Part II: Thermoplastic polyurethane matrix composites. Composites Sci. Technol. **2010**, *70* (13), 1966–1972.
20. Todo, M.; Park, J.E.; Kuraoka, H.; Kim, J.W.; Taki, K.; Ohshima, M. Compressive deformation behavior of porous PLLA/PCL polymer blend. J. Mater. sci. **2009**, *44* (15), 4191–4194.
21. Kumar, A.P.; Depan, D.; Singh Tomer, N.; Singh, R.P. Nanoscale particles for polymer degradation and stabilization—Trends and future perspectives. Prog. Polym. Sci. **2009**, *34* (6), 479–515.
22. Lee, J.A.; Krogman, K.C.; Ma, M.; Hill, R.M.; Hammond, P.T.; Rutledge, G.C Highly reactive multilayer-assembled TiO2 coating on electrospun polymer nanofibers. Adv. Mater. **2009**, *21* (12), 1252–1256.
23. Boehler, R.M.; Graham, J.G.; Shea, L.D. Tissue engineering scaffolds have emerged as a powerful tool within regenerative medicine. These. Tissue Eng. **2011**, *51* (4), 239–240.

Nanoparticles—Nerve

24. Patrick Jr, C.W.; Mikos, A.G.; McIntire, L. *Frontires in Tissue Engineering*; Elsevier: NY, USA, 1998; 3–14.

25. Shieh, S.J.; Vacanti, J.P. Surgical research review State-of-the-art tissue engineering: From tissue engineering to organ building. Surgery **2005**, *137* (1), 1–7.

26. Kaihara, S.; Vacanti, J.P. Tissue engineering: Toward new solutions for transplantation and reconstructive surgery. Arch. Surg. **1999**, *134* (11), 1184.

27. Grayson, W.L.; Martins, T.P.; Vunjak-Novakovic, G. *Biomimetic Approach to Tissue Engineering*; Elsevier: 2009.

28. Shimizu, T.; Sekine, H.; Yamato, M.; Okano, T. Cell sheet-based myocardial tissue engineering: New hope for damaged heart rescue. Curr. Pharm. Des. **2009**, *15* (24), 2807–2814.

29. Badylak, S.F.; Taylor, D.; Uygun, K. Whole-organ tissue engineering: Decellularization and recellularization of three-dimensional matrix scaffolds. Annu. Rev. Biomed. Eng. **2011**, *13*, 27–53.

30. Bishop, G.A.; Sharland, A.F.; Ierino, F.L.; Sandrin, M.S.; Hall, B.M.; Alexander, S.I.; Coates, P.T.; McCaughan, G.W. Operational tolerance in organ transplantation versus tissue engineering: Into the future. Transplantation **2011**, *92* (8), e39.

31. Hench, L.L.; Polak, J.M. Third-generation biomedical materials. Science **2002**, *295* (5557), 1014.

32. Nath, S.; Bodhak, S.; Basu, B. HDPE-Al2O3-HAp composites for biomedical applications: Processing and characterizations. J. Biomed. Mater. Res. B **2009**, *88* (1), 1–11.

33. Thakur, G.; Mitra, A.; Rousseau, D.; Basak, A.; Sarkar, S.; Pal, K. Crosslinking of gelatin-based drug carriers by genipin induces changes in drug kinetic profiles *in vitro*. J. Mater. Sci. Mater. Med. **2011**, *22* (1), 1–9.

34. Mohsen, M.; Aly, E.H.; Gomma, E.; Hegazy, E.A.; Mahmmoud, G.A. The variations in thermal and nano free-volumes properties of some gamma-irradiated hydrogels. J. Mater. Sci. Eng. **2010**, *4* (11), 75–84.

35. Umeda, S.; Nakade, H.; Kakuchi, T. Preparation of super-absorbent hydrogels from poly (aspartic acid) by chemical crosslinking. Polym. Bull. **2011**, *67* (7), 1–8.

36. Bode, F.; da Silva, M.A.; Drake, A.F.; Ross-Murphy, S.B.; Dreiss, C.A. Enzymatically cross-linked tilapia gelatin hydrogels: Physical, chemical and hybrid networks. Biomacromolecules **2011**, *12* (10), 3741–3752.

37. Kadlubowski, S.; Henke, A.; Ulañski, P.; Rosiak, J.M. Hydrogels of polyvinylpyrrolidone (PVP) and poly (acrylic acid)(PAA) synthesized by radiation-induced crosslinking of homopolymers. Radiat. Phys. Chem. **2010**, *79* (3), 261–266.

38. Vogt, A.P.; Sumerlin, B.S. Temperature and redox responsive hydrogels from ABA triblock copolymers prepared by RAFT polymerization. Soft Matter **2009**, *5* (12), 2347–2351.

39. Kopeček, J. Hydrogels: From soft contact lenses and implants to self-assembled nanomaterials. J. Polym. Sci. A **2009**; *47* (22), 5929–5946.

40. Mandal, B.B.; Kapoor, S.; Kundu, S.C. Silk fibroin/polyacrylamide semi-interpenetrating network hydrogels for controlled drug release. Biomaterials **2009**, *30* (14), 2826–2836.

41. Moya-Ortega, M.D.; Alvarez-Lorenzo, C.; Sigurdsson, H.H.; Concheiro, A.; Loftsson, T. γ-Cyclodextrin hydrogels and semi-interpenetrating networks for sustained delivery of dexamethasone. Carbohydr. Polym. **2010**, *80* (3), 900–907.

42. Tan, H.; Chu, C.R.; Payne, K.A.; Marra, K.G. Injectable in situ forming biodegradable chitosan-hyaluronic acid based hydrogels for cartilage tissue engineering. Biomaterials **2009**, *30* (13), 2499–2506.

43. Frisman, I.; Shachaf, Y.; Seliktar, D.; Bianco-Peled, H. Stimulus-responsive hydrogels made from biosynthetic fibrinogen conjugates for tissue engineering: Structural characterization. Langmuir **2011**, *27* (11), 6977–6986.

44. Dhert, W.J.A.; Pescosolido, L.; Vermonden, T.; Malda, J.; Censi, R.; Alhaique, F.; Hennink, W.E.; Matricardi, P. In situ forming IPN hydrogels of calcium alginate and dextran-HEMA for biomedical applications. Acta Biomater. **2011**, *7* (4), 1627–1633.

45. Chueh, B.; Zheng, Y.; Torisawa, Y.S.; Hsiao, A.Y.; Ge, C.; Hsiong, S.; Huebsch, N.; Franceschi, R.; Mooney, D.J.; Takayama, S. Patterning alginate hydrogels using light-directed release of caged calcium in a microfluidic device. Biomed. Microdevices **2010**, *12* (1), 145–151.

46. Zhou, Y.; Sharma, N.; Deshmukh, P.; Lakhman, R.K.; Jain, M.; Kasi, R.M. Hierarchically-structured free-standing hydrogels with liquid crystalline domains and magnetic nanoparticles as dual physical crosslinkers. J. Am. Chem. Soc. **2012**, *134* (3), 1630–1641.

47. Barbucci, R.; Pasqui, D.; Giani, G.; De Cagna, M.; Fini, M.; Giardino, R.; Atrei, A. A novel strategy for engineering hydrogels with ferromagnetic nanoparticles as cross-linkers of the polymer chains. Potential applications as a targeted drug delivery system. Soft Matter **2011**, *7* (12), 5558–5565.

48. Takafuji, M.; Yamada, S.; Ihara, H. Strategy for preparation of hybrid polymer hydrogels using silica nanoparticles as multifunctional crosslinking points. Chem. Commun. **2011**, *47* (3), 1024–1026.

49. Singh, B.; Pal, L. Radiation crosslinking polymerization of sterculia polysaccharide-PVA-PVP for making hydrogel wound dressings. Int. J. Biol. Macromol. **2011**, *48* (3), 501–510.

50. Kumar, M. Hydrogels used as a potential drug delivery system: A review. Int. J. Pharm. Biol. Arch. **2011**, *2* (4), 1068–1076.

51. Malmsten, M. Antimicrobial and antiviral hydrogels. Soft Matter **2011**, *7* (19), 8725–8736.

52. Zhang, Q.Y.; Wang, Z.Y.; Lim, J.; Li, J.; Wen, F.; Teoh, S.H. Tailoring of poly (vinyl alcohol) hydrogels properties by incorporation of crosslinked acrylic acid. IEEE **2011**, 1–3

53. Patenaude, M.; Hoare, T. Injectable, mixed natural-synthetic polymer hydrogels with modular properties. Biomacromolecules **2012**, *13* (2), 369–378.

54. Kai, D.; Prabhakaran, M.P.; Stahl, B.; Eblenkamp, M.; Wintermantel, E.; Ramakrishna, S. Mechanical properties and *in vitro* behavior of nanofiber–hydrogel composites for tissue engineering applications. Nanotechnology **2012**, *23* (9), 095705.

55. Frisman, I.; Seliktar, D.; Bianco-Peled, H. Nanostructuring biosynthetic hydrogels for tissue engineering: A cellular and structural analysis. Acta Biomater. **2011**, *8* (1), 51–60.

56. Limpoco, F.T.; Bailey, R.C. Real-time monitoring of surface-initiated atom transfer radical polymerization using silicon photonic microring resonators: Implications for combinatorial screening of polymer brush growth conditions. J. Am. Chem. Soc. **2011**, *133* (38), 14864–14867.

57. Wei, Q.B.; Luo, Y.-L.; Gao, L.-J.; Wang, Q.; Wang, D.- J. Synthesis, characterization and swelling kinetics of thermoresponsive PAM-g-PVA/PVP semi-IPN hydrogels. Polym. Sci. Ser. A **2011**, *53* (8), 707–714.

58. Renard, D.; Robert, P.; Lavenant, L.; Melcion, D.; Popineau, Y.; Guéguen, J.; Duclairoir, C.; Nakache, E.; Sanchez, C.; Schmitt, C. Biopolymeric colloidal carriers for encapsulation or controlled release applications. Int. J. Pharm. **2002**, *242* (1–2), 163–166.

59. Phelps, E.A.; Enemchukwu, N.O.; Fiore, V.F.; Sy, J.C.; Murthy, N.; Sulchek, T.A.; Barker, T.H.; García, A.J. Maleimide cross-linked bioactive PEG hydrogel exhibits improved reaction kinetics and cross-linking for cell encapsulation and in situ delivery. Adv. Mater. **2012**, *24* (1), 64–70.

60. Jeong, K.J.; Panitch, A. Interplay between covalent and physical interactions within environment sensitive hydrogels. Biomacromolecules **2009**, *10* (5), 1090–1099.

61. Bhattarai, N.; Gunn, J.; Zhang, M. Chitosan-based hydrogels for controlled, localized drug delivery. Adv. Drug Deliv. Rev. **2010**, *62* (1), 83–99.

62. Jagur-Grodzinski, J. Polymeric gels and hydrogels for biomedical and pharmaceutical applications. Polym. Adv. Technol. **2010**, *21* (1), 27–47.

63. Zhu, J. Bioactive modification of poly (ethylene glycol) hydrogels for tissue engineering. Biomaterials **2010**, *31* (17), 4639–4656.

64. Kirker, K.R. Glycosaminoglycan hydrogel films as biointeractive dressings for wound healing. Biomaterials **2002**, *23* (17), 3661–3671.

65. Shu, X.Z.; Lui, Y.; Palumbo, F.; Prestwick, G.D. Disulfide-crosslinked hyaluronan-gelatin hydrogel films: A covalent mimic of the extracellular matrix for *in vitro* cell growth. Biomaterials **2003**, *24* (21), 3825–3834.

66. Sosnowski, S.; Woźniak, P.; Lewandowska-Szumieł, M. Polyester scaffolds with bimodal pore size distribution for tissue engineering. Macromol. Biosci. **2006**; *6* (6), 425–434.

67. Sung, H.J.; Meredith, C.; Johnson, C.; Galis, Z.S. The effect of scaffold degradation rate on three-dimensional cell growth and angiogenesis. Biomaterials **2004**, *25* (26), 5735–5742.

68. Nillesen, S.; Geutjes, P.J.; Wismans, R.; Schalkwijk, J.; Daamen, W.F.; van Kuppevelt, T.H. Increased angiogenesis and blood vessel maturation in acellular collagen-heparin scaffolds containing both FGF2 and VEGF. Biomaterials **2007**, *28* (6), 1123–1131.

69. Kaigler, D.; Wang, Z.; Horger, K.; Mooney, D.J.; Krebsbach, P.H. VEGF scaffolds enhance angiogenesis and bone regeneration in irradiated osseous defects. J. Bone Miner. Res. **2006**, *21* (5), 735–744.

70. Yang, X.B.; Whitaker, M.J.; Sebald, W.; Clarke, N.; Howdle, S.M.; Shakesheff, K.M.; Oreffo, R.O. Human osteoprogenitor bone formation using encapsulated bone morphogenetic protein 2 in porous polymer scaffolds. Tissue Eng. **2004**, *10* (7–8), 1037–1045.

71. Powell, H.; Boyce, S. Fiber density of electrospun gelatin scaffolds regulates morphogenesis of dermal–epidermal skin substitutes. J. Biomed. Mater. Res. A **2008**, *84* (4), 1078–1086.

72. Lee, J.H.; Park, T.G.; Park, H.S.; Lee, D.S.; Lee, Y.K.; Yoon, S.C.; Nam, J.-D. Thermal and mechanical characteristics of poly (L-lactic acid) nanocomposite scaffold. Biomaterials **2003**, *24* (16), 2773–2778.

73. Peter, S.J.; Miller, S.T.; Zhu, G.; Yasko, A.W.; Mikos, A.G.*In vitro* degradation of a poly (propylene fumarate)/β-tricalcium phosphate composite orthopaedic scaffold. Tissue Eng. **1997**, *3* (2), 207–215.

74. Spoerke, E.D.; Murray, N.G.; Li, H.; Brinson, L.C.; Dunand, D.C.; Stupp, S.I. A bioactive titanium foam scaffold for bone repair. Acta Biomater. **2005**, *1* (5), 523–533.

75. Kim, H.W.; Knowles, J.C.; Kim, H.E. Hydroxyapatite/poly (ε-caprolactone) composite coatings on hydroxyapatite porous bone scaffold for drug delivery. Biomaterials **2004**, *25* (7–8), 1279–1287.

76. Kretlow, J.D.; Klouda, L.; Mikos, A.G. Injectable matrices and scaffolds for drug delivery in tissue engineering. Adv. Drug Deliv. Rev. **2007**, *59* (4–5), 263–273.

77. Kang, S.-W.; Son, S.M.; Lee, J.S.; Lee, E.S.; Lee, K.Y.; Park, S.G.; Park, J.H.; Kim, B.S. Regeneration of whole meniscus using meniscal cells and polymer scaffolds in a rabbit total meniscectomy model. J. Biomed. Mater. Res. A **2006**, *77A* (4), 659–671.

78. Kim, B.-S.; Mooney, D.J. Engineering smooth muscle tissue with a predefined structure. J. Biomed. Mater. Res. **1998**, *41* (2), 322–332.

79. Sokolsky-Papkov, M.; Agashi, K.; Olaye, A.; Shakesheff, K.; Domb, A.J. Polymer carriers for drug delivery in tissue engineering. Adv. Drug Deliv. Rev. **2007**, *59* (4–5), 187–206.

80. Pal, K.; Pal, S. Development of porous hydroxyapatite scaffolds. Mater. Manufacturing Process. **2006**, *21* (3), 325–328.

81. Pal, K.; Bag, S.; Pal, S. Development of porous ultra high molecular weight polyethylene scaffolds for the fabrication of orbital implant. J. Porous Mater. **2008**, *15* (1), 53–59.

82. Yue, H.; Zhang, L.; Wang, Y.; Liang, F.; Guan, L.; Li, S.; Yan, F.; Nan, X.; Bai, C.; Lin, F.; Yan, Y.; Pei, X. Proliferation and differentiation into endothelial cells of human bone marrow mesenchymal stem cells (MSCs) on poly DL-lactic-co-glycolic acid (PLGA) films. Chinese Sci. Bull. **2006**, *51* (11), 1328–1333.

83. Pal, K.; Banthia, A.K.; Majumdar, D.K. Preparation and characterization of polyvinyl alcohol-gelatin hydrogel membranes for biomedical applications. AAPS PharmSciTech **2007**, *8* (1), 142–146.

84. Pal, K.; Banthia, A.; Majumdar, D. Polyvinyl alcohol—Gelatin patches of salicylic acid: Preparation, characterization and drug release studies. J. Biomater. Appl. **2006**, *21* (1), 75.

85. Pal, K.; Banthia, A.; Majumdar, D. Biomedical evaluation of polyvinyl alcohol–gelatin esterified hydrogel for wound dressing. J. Mater. Sci. **2007**, *18* (9), 1889–1894.

86. Pal, K.; Banthia, A.; Majumdar, D. Effect of heat treatment of starch on the properties of the starch hydrogels. Mater. Lett. **2008**, *62* (2), 215–218.

Nanoparticles—Nerve

87. Pal, K.; Banthia, A.; Majumdar, D. Starch based hydrogel with potential biomedical application as artificial skin. Afr. J. Biomed. Res. 2009, 9 (1), 23–29.

88. Pal, K.; Banthia, A.K.; Majumdar, D.K. Polyvinyl alcohol–glycine composite membranes: Preparation, characterization, drug release and cytocompatibility studies. Biomed. Mater. 2006, 1 (2), 49.

89. Pal, K.; Banthia, A.K.; Majumdar, D. Development of carboxymethyl cellulose acrylate for various biomedical applications. Biomed. Mater. 2006, 1 (2), 85.

90. Sutar, P.B., et al., Development of pH sensitive polyacrylamide grafted pectin hydrogel for controlled drug delivery system. J. Mater. Sci. 2008, 19 (6), 2247–2253.

91. Pal, K.; Banthia, A.; Majumdar, D. Preparation of transparent starch based hydrogel membrane with potential application as wound dressing. Trends Biomater. Artif. Organs. 2006, 20 (1), 59–67.

92. Pal, K.; Banthia, A.; Majumdar, D. Preparation of novel pH-sensitive hydrogels of carboxymethyl cellulose acrylates: A comparative study. Mater. Manufacturing Process. 2006, 21 (8), 877–882.

93. Roy, S.; Pal, K.; Thakur, G.; Prabhakar, B. Synthesis of novel hydroxypropyl methyl cellulose acrylate—A novel superdisintegrating agent for pharmaceutical applications. Mater. Manufacturing Process. 2010, 25 (12), 1477–1481.

94. Nam, Y.S.; Park, T.G. Biodegradable polymeric microcellular foams by modified thermally induced phase separation method. Biomaterials 1999, 20 (19), 1783–1790.

95. Nazarov, R.; Jin, H.-J.; Kaplan, D.L. Porous 3-D scaffolds from regenerated silk fibroin. Biomacromolecules 2004, 5 (3), 718–726.

96. Kempen, D.H.R., Controlled drug release from a novel injectable biodegradable microsphere/scaffold composite based on poly(propylene fumarate). J. Biomed. Mater. Res. A 2006, 77A (1), 103–111.

97. DeSimone, J.; Maury, E.E.; Menceloglu, Y.Z.; McClain, J.B.; Romack, T.J.; Combes, J.R. Dispersion polymerizations in supercritical carbon dioxide. Science 1994, 265 (5170), 356.

98. DeSimone, J.; Guan, Z.; Elsbernd, C. Synthesis of fluoropolymers in supercritical carbon dioxide. Science 1992, 257 (5072), 945.

99. Kendall, J.L.; Canelas, D.A.; Young, J.L.; DeSimone, J.M. Polymerizations in supercritical carbon dioxide. Chem. Rev. 1999, 99 (2), 543–564.

100. Kelly, C.A.; Naylor, A.; Illum, L.; Shakesheff, K.M.; Howdle, S.M. Supercritical CO2: A clean and low temperature approach to blending PDLLA and PEG. Adv. Funct. Mater. 2012, 22 (8), 1684–1691.

101. Asada, M.; Gin, P.; Endoh, M.K.; Satija, S.K.; Taniquchi, T.; Koga, T. Directed self-assembly of nanoparticles at the polymer surface by highly compressible supercritical carbon dioxide. Soft Matter 2011, 7 (19), 9231–9238.

102. Liu, J.; Thompson, M.R.; Balogh, M.P.; Speer, R.L.; Fasulo, P.D.; Rodgers, W.R. Influence of supercritical CO2 on the interactions between maleated polypropylene and alkyl-ammonium organoclay. J. Appl. Polym. Sci. 2011, 119 (4), 2223–2234.

103. Barry, J.J.; Silva, M.M.; Popov, V.K.; Shakesheff, K.M.; Howdle, S.M. Supercritical carbon dioxide: Putting the fizz into biomaterials. Philos. Trans. A Math. Phys. Eng. Sci. 2006, 364 (1838), 249–261.

104. Tai, H.; Mather, M.L.; Howard, D.; Wang, W.; White, L.J.; Crowe, J.A.; Morgan, S.P.; Chandra, A.; Williams, D.J.; Howdle, S.M.; Shakesheff, K.M. Control of pore size and structure of tissue engineering scaffolds produced by supercritical fluid processing. Eur. Cell. Mater. 2007, 14 (1), 64–77.

105. Wang, X.; Li, W; Kumar, V. A method for solvent-free fabrication of porous polymer using solid-state foaming and ultrasound for tissue engineering applications. Biomaterials 2006, 27 (9), 1924–1929.

106. Kang, H.W.; Tabata, Y.; Ikada, Y. Fabrication of porous gelatin scaffolds for tissue engineering. Biomaterials 1999, 20 (14), 1339–1344.

107. Wu, X.; Liu, Y.; Li, X.; Wen, P.; Zhang, Y.; Long, Y.; Wang, X.; Guo, Y.; Xing, F.; Gao, J. Preparation of aligned porous gelatin scaffolds by unidirectional freeze-drying method. Acta Biomater. 2010, 6 (3), 1167–1177.

108. Ho, M.H.; Kuo, P.Y.; Hsieh, H.J.; Hsien, T.Y.; Hou, L.T.; Lai, J.Y.; Wang, D.M. Preparation of porous scaffolds by using freeze-extraction and freeze-gelation methods. Biomaterials 2004, 25 (1), 129–138.

109. Shenoy, S.L.; Bates, W.D.; Frisch, H.L.; Wnek, G.E. Role of chain entanglements on fiber formation during electrospinning of polymer solutions: Good solvent, non-specific polymer–polymer interaction limit. Polymer 2005, 46 (10), 3372–3384.

110. Huang, Z.M.; Zhang, Y.-Z.; Kotaki, M.; Ramakrishna, S. A review on polymer nanofibers by electrospinning and their applications in nanocomposites. Composites Sci. Technol. 2003, 63 (15), 2223–2253.

111. Boland, E.D.; Pawlowski, K.J.; Barnes, C.P.; Simpson, D.G.; Wnek, G.E.; Bowlin, G.L. Electrospinning of Bioresorbable Polymers for Tissue Engineering Scaffolds, in Polymeric Nanofibers; American Chemical Society: 2006; p. 188–204.

112. Reneker, D.H.; Chun, I. Nanometre diameter fibres of polymer, produced by electrospinning. Nanotechnology 1996, 7 (3), 216.

113. Ray, M.; Pal, K.; Anis, A.; Banthia, A.K. Development and characterization of chitosan-based polymeric hydrogel membranes. Desig. Monomers Polym. 2010, 13 (3), 193–206.

114. Thakur, G.; Pal, K.; Rousseau, D.; Mitra, A.; Basak, A. Genipin-crosslinked gelatin solid emulsion gels as a matrix for controlled delivery. 22nd European conference on Biomaterials, 2009.

115. Han, N.; Rao, S.S.; Johnson, J.; Parikh, K.S.; Bradley, P.A.; Lannutti, J.J.; Winter, J.O. Hydrogel–electrospun fiber mat composite coatings for neural prostheses. Front. Neuroeng. 2011, 4.

116. Omidian, H.; Park, K. Engineered high swelling hydrogels. In Biomedical Applications of Hydrogels Handbook; Springer: 2010; 351–374.

117. Whang, K.; Thomas, C.H.; Healy, K.E.; Nuber, G. A novel method to fabricate bioabsorbable scaffolds. Polymer; 1995, 36 (4), 837–842.

118. Pham, Q.P.; Sharma, U.; Mikos, A.G. Electrospun poly (ε-caprolactone) microfiber and multilayer nanofiber/microfiber scaffolds: Characterization of scaffolds and

measurement of cellular infiltration. Biomacromolecules **2006**, *7* (10), 2796–2805.

119. Zhao, Z.; Liu, Y.; Yan, H. Organizing DNA origami tiles into larger structures using pre-formed scaffold frames. Nano Lett. **2011**, *11* (7), 2997–3002.

120. McLaughlin, C.K.; Three-dimensional organization of block copolymers on 'DNA-Minimal'scaffolds. J. Am. Chem. Soc. **2012**, *134* (9), 4280–4286.

121. Lam, C.X.F.; Hutmacher, D.W.; Schantz, J.T.; Woodruff, M.A.; Teoh, S.H. Evaluation of polycaprolactone scaffold degradation for 6 months *in vitro* and *in vivo*. J. Biomed. Mater. Res. Part A **2009**, *90* (3), 906–919.

122. Sharifi, M.S.; Ebrahimi, D.; Hibbert, D.B.; Hook, J.; Hazell, S.L. Bio-activity of natural polymers from the genus pistacia: A validated model for their antimicrobial action. Global J. Health Sci. **2011**, *4* (1), p149.

123. Espíndola-González, A.; Martínez-Hernández, A.L.; Fernández-Escobar, F.; Castaño, V.M.; Brostow, W.; Datashvili, T.; Velasco-Santos, C. Natural-synthetic hybrid polymers developed via electrospinning: The effect of PET in chitosan/starch system. Int. J. Mol. Sci. **2011**, *12* (3), 1908–1920.

124. Du Plessis, T.; Phillips, G.; Al-Assaf, S. Radiation modification of natural polymers to enhance structure and bio-activity. Mol. Cryst. Liquid Cryst. **2012**, *555* (1), 225–231.

125. Davis, H.E.; Miller, S.L.; Case, E.M.; Leach, J.K. Supplementation of fibrin gels with sodium chloride enhances physical properties and ensuing osteogenic response. Acta Biomater. **2011**, *7* (2), 691–699.

126. Peter, M.G. Chemical modifications of biopolymers by quinones and quinone methides. Angew. Chem. Int. Ed. Engl. **1989**, *28* (5), 555–570.

127. Prestwich, G.D.; Marecak, D.M.; Marecek, J.F.; Vercruysse, K.P.; Ziebell, M.R. Controlled chemical modification of hyaluronic acid: Synthesis, applications, and biodegradation of hydrazide derivatives. J. Control. Release **1998**, *53* (1–3), 93–103.

128. Reddy, B.; Patil, R.; Patil, S. Chemical modification of biopolymers to design cement slurries with temperature-activated viscosification. SPE Dril. Completion **2011**, *27* (1), 94–102.

129. Reddy, B. *Viscosification-on-Demand: Chemical Modification of Biopolymers to Control their Activation by Triggers in Aqueous Solutions*; SPE International Symposium on Oilfield Chemistry: 2011; 17.

130. Vink, J. 295. Anthrax toxin receptor 2 is involved in normal human uterine smooth muscle cell function and modulates extracellular matrix homeostasis. Am. J. Obst. Gynecol. **2012**, *206* (1), S142.

131. Rosenbaum, A.J.; Wicker, J.F.; Dines, J.S.; Bonasser, L.; Razzano, P.; Dines, D.M.; Grande, D.A. Histologic stages of healing correlate with restoration of tensile strength in a model of experimental tendon repair. HSS J. **2010**, *6* (2), 164–170.

132. Calve, S.; Lytle, I.F.; Grosh, K.; Brown, D.L.; Arruda, E.M. Implantation increases tensile strength and collagen content of self-assembled tendon constructs. J. Appl. Physiol. **2010**, *108* (4), 875–881.

133. Bella, J. A new method for describing the helical conformation of collagen: Dependence of the triple helical twist on amino acid sequence. J. Struct. Biol. **2010**, *170* (2), 377–391.

134. Pati, F.; Dhara, S.; Adhikari, B. Fish collagen: A potential material for biomedical application. IEEE. **2010**, 34–38.

135. Shi, C.; Chen, W.; Zhao, Y.; Chen, B.; Xiao, Z.; Wei, Z.; Dai, J. Regeneration of full-thickness abdominal wall defects in rats using collagen scaffolds loaded with collagen-binding basic fibroblast growth factor. Biomaterials **2011**, *32* (3), 753–759.

136. Song, J.W.; Kim, K.H.; Song, J.M.; Chun, B.D.; Kim, Y.D.; Kim, U.K.; Shin, S.H. Clinical study of correlation between C-terminal cross-linking telopeptide of type I collagen and risk assessment, severity of disease, healing after early surgical intervention in patients with bisphophonate-related osteonecrosis of the jaws. J. Korean Assoc. Oral Maxillofac. Surg. **2011**, *37* (1), 1–8.

137. López, B.; González, A.; Díez, J. Circulating biomarkers of collagen metabolism in cardiac diseases. Circulation **2010**, *121* (14), 1645–1654.

138. Fleming, B.C.; Magarian, E.M.; Harrison, S.L.; Paller, D.J.; Murray, M.M. Collagen scaffold supplementation does not improve the functional properties of the repaired anterior cruciate ligament. J. Orthop. Res. **2010**, *28* (6), 703–709.

139. Itoh, H.; Aso, Y.; Furuse, M.; Noishiki, Y.; Miyata, T. A honeycomb collagen carrier for cell culture as a tissue engineering scaffold. Artif. Organs **2001**, 25 (3), 213–217.

140. Kevin, R.S.; Steadman, J.R.; Rodkey, W.G.; Li, S.T. Regeneration of meniscal cartilage with use of a collagen scaffold. Analysis of Preliminary Data*. J. Bone Joint Surg. **1997**, *79* (12), 1770–1777.

141. Lee, C.; Grodzinsky, A.J.; Hsu, H.P.; Spector, M. Effects of a cultured autologous chondrocyte-seeded type II collagen scaffold on the healing of a chondral defect in a canine model. J. Orthop. Res. **2003**, *21* (2), 272–281.

142. Lee, C.; Grodzinsky, A.; Spector, M. Biosynthetic response of passaged chondrocytes in a type II collagen scaffold to mechanical compression. J. Biomed. Mater. Res. A **2003**, *64* (3), 560–569.

143. Wojtowicz, A.M., Coating of biomaterial scaffolds with the collagen-mimetic peptide GFOGER for bone defect repair. Biomaterials **2010**, *31* (9), 2574–2582.

144. Sarkar, M.R.; Augat, P.; Shefelbine, S.J.; Schorlemmer, S.; Huber-Lang, M.; Claes, L.; Kinzl, L.; Ignatius, A. Bone formation in a long bone defect model using a platelet-rich plasma-loaded collagen scaffold. Biomaterials **2006**, *27* (9), 1817–1823.

145. Polini, A.; Pisignano, D.; Parodi, M.; Quarto, R.; Scaglione, S. Osteoinduction of human mesenchymal stem cells by bioactive composite scaffolds without supplemental osteogenic growth factors. PloS One **2011**, *6* (10), e26211.

146. Donzelli, E.; Salvadè, A.; Mimo, P.; Viganò, M.; Morrone, M.; Papagna, R.; Carini, F.; Zaopo, A.; Miloso, M.; Baldoni, M.; Tredici, G. Mesenchymal stem cells cultured on a collagen scaffold: *In vitro* osteogenic differentiation. Arch. Oral Biol. **2007**, *52* (1), 64–73.

147. Curtin, C.M.; Cunniffe, G.M.; Lyons, F.G.; Bessho, K.; Dickson, G.R.; Duffy, G.P.; O'Brien, F.J. Innovative collagen nano-hydroxyapatite scaffolds offer a highly efficient non-viral gene delivery platform for stem cell-mediated bone formation. Adv. Mater. **2012**, *24* (6), 749–754.

148. Hu, Y.; Zhang, C.; Zhang, S.; Xiong, Z.; Xu, J. Development of a porous poly(L-lactic acid)/hydroxyapatite/collagen scaffold as a BMP delivery system and its use in healing canine segmental bone defect. J. Biomed. Mater. Res. A **2003**, *67A* (2), 591–598.

149. Liu, T.; Teng, W.K.; Chan, B.P.; Chew, S.Y. Photochemical crosslinked electrospun collagen nanofibers: Synthesis, characterization and neural stem cell interactions. J. Biomed. Mater. Res. A **2010**, *95* (1), 276–282.

150. McClure, M.J.; Sell, S.A.; Simpson, D.G.; Walpoth, B.H.; Bowlin, G.L. A three-layered electrospun matrix to mimic native arterial architecture using polycaprolactone, elastin, and collagen: A preliminary study. Acta Biomater. **2010**, *6* (7), 2422–2433.

151. Lee, H.; Yeo, M.; Ahn, S.; Kang, D.; Jang, C.H.; Lee, H.; Park, G.; Kim, G.H. Designed hybrid scaffolds consisting of polycaprolactone microstrands and electrospun collagen-nanofibers for bone tissue regeneration. J. Biomed. Mater. Res. B **2011**, *97* (2), 263–270.

152. Zhong, S.; Teo, W.E.; Zhu, X.; Beuerman, R.W.; Ramakrishna, S.; Yung, L.Y. An aligned nanofibrous collagen scaffold by electrospinning and its effects on *in vitro* fibroblast culture. J. Biomed. Mater. Res. A **2006**, *79* (3), 456–463.

153. Shen, Y.H.; Shoichet, M.S.; Radisic, M. Vascular endothelial growth factor immobilized in collagen scaffold promotes penetration and proliferation of endothelial cells. Acta Biomater. **2008**, *4* (3), 477–489.

154. Zhang, Y.; Cheng, X.; Wang, J.; Wang, Y.; Shi, B.; Huang, C.; Yang, X.; Liu, T. Novel chitosan/collagen scaffold containing transforming growth factor-[beta] 1 DNA for periodontal tissue engineering. Biochem. Biophys. Res. Commun. **2006**, *344* (1), 362–369.

155. Lien, S.-M.; Ko, Huang, T.-J. Effect of pore size on ECM secretion and cell growth in gelatin scaffold for articular cartilage tissue engineering. Acta Biomater. **2009**, *5* (2), 670–679.

156. Panzavolta, S.; Fini, M.; Nicoletti, A.; Bracci, B.; Rubini, K.; Giardino, R.; Bigi, A. Porous composite scaffolds based on gelatin and partially hydrolyzed α-tricalcium phosphate. Acta Biomater. **2009**, *5* (2), 636–643.

157. Ratanavaraporn, J.; Damrongsakkul, S.; Sanchavanakit, N.; Banaprasert, T.; Kanokpanont, S. Comparison of gelatin and collagen scaffolds for fibroblast cell culture. J. Metal. Mater. Miner. **2006**, *16* (1), 31–36.

158. Mao, J.S.; Yin, Y.J.; Yao, K.D. The properties of chitosan-gelatin membranes and scaffolds modified with hyaluronic acid by different methods. Biomaterials **2003**, *24* (9), 1621–1629.

159. Olsen, D.; Yang, C.; Bodo, M.; Chang, R.; Leigh, S.; Baez, J.; Polarek, J. Recombinant collagen and gelatin for drug delivery. Adv. Drug Deliv. Rev. **2003**. *55* (12), 1547–1567.

160. Smart, J.; Kellaway, I.; Worthington, H. An *in vitro* investigation of mucosa-adhesive materials for use in controlled drug delivery. J. Pharm. Pharmacol. **1984**, *36* (5), 295–299.

161. Chong, E.; Phan, T.T.; Lim, I.J.; Zhang, Y.Z.; Bay, B.H.; Ramakrishna, S.; Lim, C.T. Evaluation of electrospun PCL/gelatin nanofibrous scaffold for wound healing and layered dermal reconstitution. Acta Biomater. **2007**, *3* (3), 321–330.

162. Liu, X.; Smith, L.A.; Hu, J.; Ma, P.X. Biomimetic nanofibrous gelatin/apatite composite scaffolds for bone tissue engineering. Biomaterials **2009**, *30* (12), 2252–2258.

163. Lee, S.B.; Kim, Y.H.; Chong, M.S.; Hong, S.H.; Lee, Y.M. Study of gelatin-containing artificial skin V: Fabrication of gelatin scaffolds using a salt-leaching method. Biomaterials **2005**, *26* (14), 1961–1968.

164. Angele, P.; Kujat, R.; Nerlich, M.; Yoo, J.; Goldberg, V.; Johnstone, B. Engineering of osteochondral tissue with bone marrow mesenchymal progenitor cells in a derivatized hyaluronan-gelatin composite sponge. Tissue Eng. **1999**, *5* (6), 545–553.

165. Yao, C.H.; Liu, B.S.; Chang, C.J.; Hsu, S.H.; Chen, Y.S. Preparation of networks of gelatin and genipin as degradable biomaterials. Mater. Chem. Phys. **2004**, *83* (2–3), 204208.

166. Ulubayram, K.; Aksu, E.; Gurhan, S.I.D.; Serbetci, K.; Hasirci, N. Cytotoxicity evaluation of gelatin sponges prepared with different cross-linking agents. J. Biomater. Sci. Polym. Ed. **2002**, *13* (11), 1203–1219.

167. Zhao, F.; Yin, Y.; Lu, W.W.; Leong, J.C.; Zhang, W.; Zhang, J.; Yao, K. Preparation and histological evaluation of biomimetic three-dimensional hydroxyapatite/chitosan-gelatin network composite scaffolds. Biomaterials **2002**, *23* (15), 3227–3234.

168. Li, M.; Guo, Y.; Wei, Y.; MacDiarmid, A.G.; Lelkes, P.I. Electrospinning polyaniline-contained gelatin nanofibers for tissue engineering applications. Biomaterials **2006**, *27* (13), 2705–2715.

169. Li, M.; Mondrinos, M.J.; Chen, X.; Gandhi, M.R.; Ko, F.K.; Lelkes, P.I. Co-electrospun poly (lactide-co-glycolide), gelatin, and elastin blends for tissue engineering scaffolds. J. Biomed. Mater. Res. A **2006**, *79* (4), 963–973.

170. Sisson, K.; Zhang, C.; Farach-Carson, M.C.; Chase, D.B.; Rabolt, J.F. Evaluation of cross-linking methods for electrospun gelatin on cell growth and viability. Biomacromolecules **2009**, *10* (7), 1675–1680.

171. Yeh, M.K; Liang, Y.M.; Hu, C.S.; Cheng, K.M.; Hung, Y.W.; Young, J.J.; Hong, P.D. Studies on a novel gelatin sponge: Preparation and characterization of cross-linked gelatin scaffolds using 2-Chloro-1-Methylpyridinium Iodide as a zero-length cross-linker. J. Biomater. Sci. Polym. Ed. **2011**, *23* (7), 973–990.

172. Zhao, J.; Zhao, Y.; Guan, Q.; Tang, G.; Zhao, Y.; Yuan, X.; Yao, K. Crosslinking of electrospun fibrous gelatin scaffolds for apatite mineralization. J. Appl. Polym. Sci. **2011**, *119* (2), 786–793.

173. He, C.; Zhang, F.; Cao, L.; Feng, W.; Qiu, K.; Zhang, Y.; Wang, J. Rapid mineralization of porous gelatin scaffolds by electrodeposition for bone tissue engineering. J. Mater. Chem. **2012**, *22* (5), 2111–2119.

174. Skotak, M.; Ragusa, J.; Gonzalez, D.; Subramanian, A. Improved cellular infiltration into nanofibrous electrospun cross-linked gelatin scaffolds templated with micrometer-sized polyethylene glycol fibers. Biomed. Mater. **2011**, *6* (5), 055012.

175. Wu, S.C.; Chang, W.H.; Dong, G.C.; Chen, K.Y.; Chen, Y.S.; Yao, C.H. Cell adhesion and proliferation enhancement by gelatin nanofiber scaffolds. J. Bioactive Compatible Polym. **2011**, *26* (6), 565–577.

Nanoparticles—Nerve

176. Zandi, M.; Mirzadeh, H.; Mayer, C.; Urch, H.; Eslamine-jad, M.B.; Bagheri, F.; Mivehchi, H. Biocompatibility evaluation of nano-rod hydroxyapatite/gelatin coated with nano-HAp as a novel scaffold using mesenchymal stem cells. J. Biomed. Mater. Res. A 2010, 92 (4), 1244–1255.

177. Liu, X.; Ma, P.X. Phase separation, pore structure, and properties of nanofibrous gelatin scaffolds. Biomaterials 2009, 30 (25), 4094–4103.

178. Pan, Y.; Dong, S.; Hao, Y.; Chu, T.; Li, C.; Zhang, Z.; Zhou, Y. Demineralized bone matrix gelatin as scaffold for tissue engineering. Afr. J. Microbiol. Res. 2010, 4 (9), 865–870.

179. Ovsianikov, A.; Deiwick, A.; Van Vlierberghe, S.; Dubruel, P.; Möller, L.; Dräger, G.; Chichkov, B. Laser fabrication of three-dimensional CAD scaffolds from photosensitive gelatin for applications in tissue engineering. Biomacromolecules 2011, 12 (4), 851–855.

180. Ovsianikov, A.; Deiwick, A.; Van Vlierberghe, S.; Pflaum, M.; Wilhelmi, M.; Dubruel, P.; Chichkov, B. Laser fabrication of 3D gelatin scaffolds for the generation of bioartificial tissues. Materials 2011, 4 (1), 288–299.

181. Hussain, A.; Bessho, K.; Takahashi, K.; Tabata, Y. Magnesium calcium phosphate as a novel component enhances mechanical/physical properties of gelatin scaffold and osteogenic differentiation of bone marrow mesechymal stem cells. Tissue Eng. 2011, 18 (7–8), 768–774.

182. Weinand, C.; Nabili, A.; Khumar, M.; Dunn, J.R.; Ramella-Roman, J.; Jeng, J.C.; Tabata, Y. Factors of osteogenesis influencing various human stem cells on third-generation gelatin/β-tricalcium phosphate scaffold material. Rejuvenation Res. 2011, 14 (2), 185–194.

183. Lien, S.M.; Ko, L.Y.; Huang, T.J. Effect of crosslinking temperature on compression strength of gelatin scaffold for articular cartilage tissue engineering. Mater. Sci. Eng. C 2010, 30 (4), 631–635.

184. Watanabe, M.; Li, H.; Roybal, J.; Santore, M.; Radu, A.; Jo, J.I.; Flake, A. A Tissue engineering approach for prenatal closure of myelomeningocele: Comparison of gelatin sponge and microsphere scaffolds and bioactive protein coatings. Tissue Eng. Part A 2011, 17 (7–8), 1099–1110.

185. Montero, R.B.; Vial, X.; Nguyen, D.T.; Farhand, S.; Reardon, M.; Pham, S.M.; Andreopoulos, F.M. bFGF-containing electrospun gelatin scaffolds with controlled nano-architectural features for directed angiogenesis. Acta Biomater. 2011, 8 (5), 1778–1791.

186. Yan, M.; Xu, K.D.; Zheng, X.X.; Chen, Z.J.; Jiang, H.L. Enhanced cellular function of human vascular endothelial cell on poly (ε-caprolactone)/gelatin coaxial-electrospun scaffold. Appl. Mech. Mater. 2012, 138, 900–906.

187. Petrenko, Y.A.; Ivanov, R.V.; Petrenko, A.Y.; Lozinsky, V.I. Coupling of gelatin to inner surfaces of pore walls in spongy alginate-based scaffolds facilitates the adhesion, growth and differentiation of human bone marrow mesenchymal stromal cells. J. Mater. Sci. 2011, 22 (6), 1–12.

188. Huang, Y.; Onyeri, S.; Siewe, M.; Moshfeghian, A.; Madihally, S.V. In vitro characterization of chitosan-gelatin scaffolds for tissue engineering. Biomaterials 2005, 26 (36), 7616–7627.

189. Tan, H.; Wu, J.; Lao, L.; Gao, C. Gelatin/chitosan/hyaluronan scaffold integrated with PLGA microspheres for cartilage tissue engineering. Acta Biomater. 2009, 5 (1), 328–337.

190. Xia, W.; Liu, W.; Cui, L.; Liu, Y.; Zhong, W.; Liu, D.; Cao, Y. Tissue engineering of cartilage with the use of chitosangelatin complex scaffolds. J. Biomed. Mater. Res. B 2004, 71 (2), 373–380.

191. Guo, T.; Zhao, J.; Chang, J.; Ding, Z.; Hong, H.; Chen, J.; Zhang, J. Porous chitosan-gelatin scaffold containing plasmid DNA encoding transforming growth factor-[beta] 1 for chondrocytes proliferation. Biomaterials 2006, 27 (7), 1095–1103.

192. Rana, V.; Kushwaha, O.S.; Singh, R.; Mishra, S.; Ha, C.S. Tensile properties, cell adhesion, and drug release behavior of chitosan-silver-gelatin nanohybrid films and scaffolds. Macromol. Res. 2010, 18 (9), 845–852.

193. Bhardwaj, N.; Kundu, S.C. Silk fibroin protein and chitosan polyelectrolyte complex porous scaffolds for tissue engineering applications. Carbohydr. Polym. 2011, 85 (2), 325–333.

194. Wu, L.; Li, M.; Zhao, J.; Chen, D.; Zhou, Z. [Preliminary study on polyvinyl alcohol/wild antheraea pernyi silk fibroin as nanofiber scaffolds for tissue engineered tendon]. Zhongguo xiu fu chong jian wai ke za zhi=Zhongguo xiufu chongjian waike zazhi=Chin. J. Reparative Reconstr. Surg. 2011, 25 (2), 181.

195. Fang, Q.; Chen, D.; Yang, Z.; Li, M. In vitro and in vivo research on using Antheraea pernyi silk fibroin as tissue engineering tendon scaffolds. Mater. Sci. Eng. C 2009, 29 (5), 1527–1534.

196. Kearns, V.; MacIntosh, A.C.; Crawford, A.; Hatton, P.V. Silk-based biomaterials for tissue engineering. Top. Tissue Eng. 2008, 4 (1), 1–10.

197. Talukdar, S., Nguyen, Q.T.; Chen, A.C.; Sah, R.L.; Kundu, S.C. Effect of initial cell seeding density on 3D-engineered silk fibroin scaffolds for articular cartilage tissue engineering. Biomaterials 2011, 32 (34), 8927–8937.

198. Unger, R.E.; Peters, K.; Wolf, M.; Motta, A.; Migliaresi, C.; Kirkpatrick, C.J. Endothelialization of a non-woven silk fibroin net for use in tissue engineering: Growth and gene regulation of human endothelial cells. Biomaterials 2004, 25 (21), 5137–5146.

199. Liu, H.; Li, X.; Zhou, G.; Fan, H.; Fan, Y. Electrospun sulfated silk fibroin nanofibrous scaffolds for vascular tissue engineering. Biomaterials 2011, 32 (15), 3784–3793.

200. Wang, Y.; Bella, E.; Lee, C.S.; Migliaresi, C.; Pelcastre, L.; Schwartz, Z.; Motta, A. The synergistic effects of 3-D porous silk fibroin matrix scaffold properties and hydrodynamic environment in cartilage tissue regeneration. Biomaterials 2010, 31 (17), 4672–4681.

201. Yan, L.-P.; Oliveira, J.M.; Oliveira, A.L.; Caridade, S.G.; Mano, J.F.; Reis, R.L. Macro/microporous silk fibroin scaffolds with potential for articular cartilage and meniscus tissue engineering applications. Acta Biomater. 2012, 8 (1), 289–301.

202. Osathanon, T.; Giachelli, C.M.; Somerman, M.J. Immobilization of alkaline phosphatase on microporous nanofibrous fibrin scaffolds for bone tissue engineering. Biomaterials 2009, 30 (27): 4513–4521.

203. Zhou, H.; Xu, H.H.K. The fast release of stem cells from alginate-fibrin microbeads in injectable scaffolds for bone tissue engineering. Biomaterials 2011, 32 (30), 7503–7513.

204. Wenk, E.; Meinel, A.J.; Wildy, S.; Merkle, H.P.; Meinel, L. Microporous silk fibroin scaffolds embedding PLGA

microparticles for controlled growth factor delivery in tissue engineering. Biomaterials **2009**, *30* (13), 2571–2581.

205. Harkin, D.G.; George, K.A.; Madden, P.W.; Schwab, I.R.; Hutmacher, D.W.; Chirila, T.V. Silk fibroin in ocular tissue reconstruction. Biomaterials **2011**, *32* (10), 2445–2458.

206. Yang, Y.; Chen, X.; Ding, F.; Zhang, P.; Liu, J.; Gu, X. Biocompatibility evaluation of silk fibroin with peripheral nerve tissues and cells *in vitro*. Biomaterials **2007**, *28* (9), 1643–1652.

207. Yang, Y.; Ding, F.; Wu, J.; Hu, W.; Liu, W.; Liu, J.; Gu, X. Development and evaluation of silk fibroin-based nerve grafts used for peripheral nerve regeneration. Biomaterials **2007**, *28* (36), 5526–5535.

208. Levin, B.; Redmond, S.L.; Rajkhowa, R.; Eikelboom, R.H.; Atlas, M.D.; Marano, R.J. Utilising silk fibroin membranes as scaffolds for the growth of tympanic membrane keratinocytes, and application to myringoplasty surgery. J. Laryngol. Otol. **2012**, 127 (Suppl. 1), 1–8.

209. Mauney, J.R.; Nguyen, T.; Gillen, K.; Kirker-Head, C.; Gimble, J.M.; Kaplan, D.L. Engineering adipose-like tissue *in vitro* and *in vivo* utilizing human bone marrow and adipose-derived mesenchymal stem cells with silk fibroin 3D scaffolds. Biomaterials **2007**, *28* (35), 5280–5290.

210. Luangbudnark, W.; Viyoch, J.; Laupattarakasem, W.; Surakunprapha, P.; Laupattarakasem, P. Properties and biocompatibility of chitosan and silk fibroin blend films for application in skin tissue engineering. ScientificWorldJournal **2012**, *2012*, doi: 10.1100/2012/697201.

211. Linnes, M.P.; Ratner, B.D.; Giachelli, C.M. A fibrinogen-based precision microporous scaffold for tissue engineering. Biomaterials **2007**, *28* (35), 5298–5306.

212. Bak, H.; Afoke, A.; McLeod, A.J.; Brown, R.; Shamlou, P.A.; Dunnill, P. The impact of rheology of human fibronectin–fibrinogen solutions on fibre extrusion for tissue engineering. Chem. Eng. Sci. **2002**, *57* (6), 913–920.

213. Kolehmainen, K.; Willerth, S.M. Preparation of 3D fibrin scaffolds for stem cell culture applications. J. Vis. Exp. **2012**, (61), e3641, doi:10.3791/3641.

214. Xu, M.; Wang, X.; Yan, Y.; Yao, R.; Ge, Y. An cell-assembly derived physiological 3D model of the metabolic syndrome, based on adipose-derived stromal cells and a gelatin/alginate/fibrinogen matrix. Biomaterials **2010**, *31* (14), 3868–3877.

215. Bensaïd, W.; Triffitt, J.T.; Blanchat, C.; Oudina, K.; Sedel, L.; Petite, H. A biodegradable fibrin scaffold for mesenchymal stem cell transplantation. Biomaterials **2003**, *24* (14), 2497–2502.

216. Larsen, A.; Clausen, K.; Ostler, K.; Everland, H.; Lind, M. 17.3 Cartilage regeneration with chondrocytes in fibrinogen gel scaffold and polylactate porous caffold. n in ivo study in oats. Osteoarthritis and Cartilage **2007**, *15* (Suppl. B), B74.

217. Chien, C.S.; Ho, H.O.; Liang, Y.C.; Ko, P.H.; Sheu, M.T.; Chen, C.H Incorporation of exudates of human platelet-rich fibrin gel in biodegradable fibrin scaffolds for tissue engineering of cartilage. J. Biomed. Mater. Res. B Appl. Biomater. **2012**, *100* (4), 948–955.

218. Liu, J.; Xu, H.H.; Zhou, H.; Weir, M.D.; Chen, Q.; Trotman, C.A. Human umbilical cord stem cell encapsulation in novel macroporous and injectable fibrin for muscle tissue engineering. Acta Biomater. **2012**, *9* (1), 4688–4697.

219. Yasuda, H.; Kuroda, S.; Shichinohe, H.; Kamei, S.; Kawamura, R.; Iwasaki, Y. Effect of biodegradable fibrin scaffold on survival, migration, and differentiation of transplanted bone marrow stromal cells after cortical injury in rats. J. Neurosurg. **2010**, *112* (2), 336–344.

220. Carlisle, C.R.; Coulais, C.; Namboothiry, M.; Carroll, D.L.; Hantgan, R.R.; Guthold, M. The mechanical properties of individual, electrospun fibrinogen fibers. Biomaterials **2009**, *30* (6), 1205–1213.

221. Lokmic, Z.; Thomas, J.L.; Morrison, W.A.; Thompson, E.W.; Mitchell, G.M. An endogenously deposited fibrin scaffold determines construct size in the surgically created arteriovenous loop chamber model of tissue engineering. J. Vasc. Surg. **2008**, *48* (4), 974–985.

222. McManus, M.C.; Boland, E.D.; Koo, H.P.; Barnes, C.P.; Pawlowski, K.J.; Wnek, G.E.; Bowlin, G.L. Mechanical properties of electrospun fibrinogen structures. Acta Biomater. **2006**, *2* (1), 19–28.

223. Ye, Q.; Zünd, G.; Benedikt, P.; Jockenhoevel, S.; Hoerstrup, S.P.; Sakyama, S.; Turina, M. Fibrin gel as a three dimensional matrix in cardiovascular tissue engineering. Eur. J. Cardiothorac. Surg. **2000**, *17* (5), 587–591.

224. Lee, Y.-B.; Polio, S.; Lee, W.; Dai, G.; Menon, L.; Carroll, R.S.; Yoo, S.S. Bio-printing of collagen and VEGF-releasing fibrin gel scaffolds for neural stem cell culture. Exp. Neurol. **2010**, *223* (2), 645–652.

225. Ziv-Polat, O.; Skaat, H.; Shahar, A.; Margel, S. Novel magnetic fibrin hydrogel scaffolds containing thrombin and growth factors conjugated iron oxide nanoparticles for tissue engineering. Int. J. Nanomed. **2012**, *7*, 1259–1274.

226. Ingenito, E.P.; Sen, E.; Tsai, L.W.; Murthy, S.; Hoffman, A. Design and testing of biological scaffolds for delivering reparative cells to target sites in the lung. J. Tissue Eng. Regen. Med. **2010**, *4* (4), 259–272.

227. Zhang, X.; Chen, B.; Fu, W.; Fang, Z.; Liu, Z.; Lu, W.; Chen, T. The research and preparation of a novel nano biodegradable polymer external reinforcement. Appl. Surf. Sci. **2011**, *258* (1), 196–200.

228. Harding, S.I.; Afoke, A.; Brown, R.; MacLeod, A.; Shamlou, P.; Dunnill, P. Engineering and cell attachment properties of human fibronectin-fibrinogen scaffolds for use in tissue engineered blood vessels. Bioprocess Biosyst. Eng. **2002**, *25* (1), 53–59.

229. Sell, S.A.; McClure, M.J.; Garg, K.; Wolfe, P.S.; Bowlin, G.L. Electrospinning of collagen/biopolymers for regenerative medicine and cardiovascular tissue engineering. Adv. Drug Deliv. Rev. **2009**, *61* (12), 1007–1019.

230. Almany, L.; Seliktar, D. Biosynthetic hydrogel scaffolds made from fibrinogen and polyethylene glycol for 3D cell cultures. Biomaterials **2005**, *26* (15), 2467–2477.

231. Oss-Ronen, L.; Seliktar, D. Polymer-conjugated albumin and fibrinogen composite hydrogels as cell scaffolds designed for affinity-based drug delivery. Acta Biomater. **2011**, *7* (1), 163–170.

232. Dainiak, M.B.; Allan, I.U.; Savina, I.N.; Cornelio, L.; James, E.S.; James, S.L.; Galaev, I.Y. Gelatin–fibrinogen cryogel dermal matrices for wound repair: Preparation, optimisation and *in vitro* study. Biomaterials **2010**, *31* (1), 67–76.

233. Ravichandran, R.; Venugopal, J.R.; Sundarrajan, S.; Mukherjee, S.; Sridhar, R.; Ramakrishna, S. Expression of

cardiac proteins in neonatal cardiomyocytes on PGS/fibrinogen core/shell substrate for Cardiac tissue engineering. Int. J. Cardiol. **2013**, 167 (4), 1461–1468.

234. Hokugo, A.; Takamoto, T.; Tabata, Y. Preparation of hybrid scaffold from fibrin and biodegradable polymer fiber. Biomaterials **2006**, 27 (1), 61–67.

235. Losi, P.; Briganti, E.; Magera, A.; Spiller, D.; Ristori, C.; Battolla, B.; Balderi, M.; Kull, S.; Balbarini, A.; Di Stefano, R.; Soldani, G. Tissue response to poly(ether) urethane-polydimethylsiloxane-fibrin composite scaffolds for controlled delivery of pro-angiogenic growth factors. Biomaterials **2010**, 31 (20), 5336–5344.

236. Zhao, H.; Ma, L.; Gao, C.; Wang, J.; Shen, J. Fabrication and properties of injectable β-tricalcium phosphate particles/fibrin gel composite scaffolds for bone tissue engineering. Mater. Sci. Eng. C **2009**, 29 (3), 836–842.

237. Drury, J.L.; Mooney, D.J. Hydrogels for tissue engineering: Scaffold design variables and applications. Biomaterials **2003**, 24 (24), 4337–4351.

238. Khalil, S.E.D. *Deposition and Structural Formation of 3D Alginate Tissue Scaffolds*; Drexel University: 2005.

239. Lee, K.Y.; Mooney, D.J. Alginate: Properties and biomedical applications. Progr. Polym. Sci. **2012**, 37 (1), 106–126.

240. Niu, X., Wang, Y.; Luo, Y.; Xin, J.; Li, Y. Arg-Gly-Asp (RGD) modified biomimetic polymeric materials. J. Mater. Sci. Technol.-Shenyang- **2005**, 21 (4), p. 571.

241. Shachar, M.; Tsur-Gang, O.; Dvir, T.; Leor, J.; Cohen, S. The effect of immobilized RGD peptide in alginate scaffolds on cardiac tissue engineering. Acta Biomater. **2011**, 7 (1), 152–162.

242. Freeman, I.; Cohen, S. The influence of the sequential delivery of angiogenic factors from affinity-binding alginate scaffolds on vascularization. Biomaterials **2009**, 30 (11), 2122–2131.

243. Srinivasan, S.; Jayasree, R.; Chennazhi, K.P.; Nair, S.V.; Jayakumar, R. Biocompatible alginate/nano bioactive glass ceramic composite scaffolds for periodontal tissue regeneration. Carbohydr. Polym. **2012**, 87 (1), 274–283.

244. Valente, J.F.A.; Alves, V.P.; Ferreira, A.; Silva, A.; Correia, I.J. Alginate based scaffolds for bone tissue engineering. Mater. Sci. Eng. C **2012**, 32 (8), 2598–2603.

245. Reyes, R.; Delgado, A.; Sánchez, E.; Fernández, A.; Hernández, A.; Evora, C. Repair of an osteochondral defect by sustained delivery of BMP-2 or TGFbeta1 from a bilayered alginate-PLGA scaffold. J. Tissue Eng. Regen. Med. **2012**, doi: 10.1002/term.1549.

246. Jin, H.-H.; Kim, D.H.; Kim, T.W.; Shin, K.K.; Jung, J.S.; Park, H.C.; Yoon, S.Y. *In vivo* evaluation of porous hydroxyapatite/chitosan-alginate composite scaffolds for bone tissue engineering. Int. J. Biol. Macromol. **2012**, 51 (5), 1079–1085.

247. Li, Z.; Ramay, H.R.; Hauch, K.D.; Xiao, D.; Zhang, M. Chitosan–alginate hybrid scaffolds for bone tissue engineering. Biomaterials **2005**, 26 (18), 3919–3928.

248. Bron, J.L.; Vonk, L.A.; Smit, T.H.; Koenderink, G.H. Engineering alginate for intervertebral disc repair. J. Mech. Behav. Biomed. Mater. **2011**, 4 (7), 1196–1205.

249. Abbah, S.A.; Lu, W.W.; Chan, D.; Cheung, K.M.; Liu, W.G.; Zhao, F.; Li, Z.Y.; Leong, J.C.; Luk, K.D. *In vitro* evaluation of alginate encapsulated adipose-tissue stromal cells for use as injectable bone graft substitute. Biochem. Biophys. Res. Commun. **2006**, 347 (1), 185–191.

250. Chen, F.; Tim, M.; Zhang, D.; Wang, J.; Wang, Q.; Yu, X.; Zhang, X.; Wan, C. Preparation and characterization of oxidized alginate covalently cross-linked galactosylated chitosan scaffold for liver tissue engineering. Mater. Sci. Eng. C **2012**, 32 (2), 310–320.

251. Li, X.; Xia, W. Effects of concentration, degree of deacetylation and molecular weight on emulsifying properties of chitosan. Int. J. Biol. Macromol. **2011**, 48 (5), 768–772.

252. Kumari, R.; Dutta, P. Physicochemical and biological activity study of genipin-crosslinked chitosan scaffolds prepared by using supercritical carbon dioxide for tissue engineering applications. Int. J. Biol. Macromol. **2010**, 46 (2), 261–266.

253. Ko, Y.G.; Shin, S.S.; Choi, U.S.; Park, Y.S.; Woo, J.W. Gelation of chitin and chitosan dispersed suspensions under electric field: Effect of degree of deacetylation. ACS Appl. Mater. Interfaces **2011**, 3 (4), 1289–1298.

254. Shin, S.Y.; Park, H.N.; Kim, K.H.; Lee, M.H.; Choi, Y.S.; Park, Y.J. Chung, C.P. Biological evaluation of chitosan nanofiber membrane for guided bone regeneration. J. Periodontol. **2005**, 76 (10), 1778–1784.

255. Pal, K.; Behera, B.; Roy, S.; Sekhar Ray, S.; Thakur, G. Chitosan based delivery systems on a length scale: Nano to macro. Soft Mater. **2013**, 11 (2), 125–142.

256. Marcasuzaa, P.; Reynaud, S.; Ehrenfeld, F.; Khoukh, A.; Desbrieres, J. Chitosan-graft-Polyaniline-Based Hydrogels: Elaboration and Properties. Biomacromolecules **2010**, 11 (6), 1684–1691.

257. Massouda, D.F.; Visioli, D.; Green, D.A.; Joerger, R.D. Extruded blends of chitosan and ethylene copolymers for antimicrobial packaging. Packaging Technol. Sci. **2011**, 25 (6), 321–327.

258. Shao, H.J.; Lee, Y.T.; Chen, C.S.; Wang, J.H.; Young, T.H. Modulation of gene expression and collagen production of anterior cruciate ligament cells through cell shape changes on polycaprolactone/chitosan blends. Biomaterials **2010**, 31 (17), 4695–4705.

259. Peter, M.; Binulal, N.S.; Nair, S.V.; Selvamurugan, N.; Tamura, H.; Jayakumar, R. Novel biodegradable chitosan-gelatin/nano-bioactive glass ceramic composite scaffolds for alveolar bone tissue engineering. Chem. Eng. J. **2010**, 158 (2), 353–361.

260. Jayakumar, R.; Menon, D.; Manzoor, K.; Nair, S.V.; Tamura, H. Biomedical applications of chitin and chitosan based nanomaterials—A short review. Carbohydr. Polym. **2010**, 82 (2), 227–232.

261. VandeVord, P.J.; Matthew, H.W.; DeSilva, S.P.; Mayton, L.; Wu, B.; Wooley, P.H. Evaluation of the biocompatibility of a chitosan scaffold in mice. J. Biomed. Mater. Res. **2002**, 59 (3), 585–590.

262. Madihally, S.V.; Matthew, H.W.T. Porous chitosan scaffolds for tissue engineering. Biomaterials **1999**, 20 (12), 1133–1142.

263. Ratakonda, S.; Sridhar, U.M.; Rhinehart, R.R.; Madihally, S.V. Assessing viscoelastic properties of chitosan scaffolds and validation with cyclical tests. Acta Biomater. **2011**, 8 (4), 1566–1575.

264. Nettles, D.L.; Elder, S.H.; Gilbert, J.A. Potential use of chitosan as a cell scaffold material for cartilage tissue engineering. Tissue Eng. **2002**, *8* (6), 1009–1016.

265. Bhardwaj, N.; Nguyen, Q.T.; Chen, A.C.; Kaplan, D.L.; Sah, R.L.; Kundu, S.C. Potential of 3-D tissue constructs engineered from bovine chondrocytes/silk fibroin-chitosan for *in vitro* cartilage tissue engineering. Biomaterials **2011**, *32* (25), 5773–5781.

266. Yang, Z.; Wu, Y.; Li, C.; Zhang, T.; Zou, Y.; Hui, J.H.; Lee, E.H. Improved mesenchymal stem cells attachment and *in vitro* cartilage tissue formation on chitosan-modified poly (L-lactide-co-epsilon-caprolactone) scaffold. Tissue Eng. **2011**, *18* (3–4), 242–251.

267. Alves da Silva, M.; Crawford, A.; Mundy, J.M.; Correlo, V.M.; Sol, P.; Bhattacharya, M.; Neves, N.M. Chitosan/polyester-based scaffolds for cartilage tissue engineering: Assessment of extracellular matrix formation. Acta Biomater. **2010**, *6* (3), 1149–1157.

268. Jiankang, H.; Dichen, L.; Yaxiong, L.; Bo, Y.; Hanxiang, Z.; Qin, L.; Yi, L. Preparation of chitosan–gelatin hybrid scaffolds with well-organized microstructures for hepatic tissue engineering. Acta biomater. **2009**, *5* (1), 453–461.

269. Feng, Z.Q.; Chu, X.; Huang, N.P.; Wang, T.; Wang, Y.; Shi, X.; Gu, Z.Z. The effect of nanofibrous galactosylated chitosan scaffolds on the formation of rat primary hepatocyte aggregates and the maintenance of liver function. Biomaterials **2009**, *30* (14), 2753–2763.

270. Chesnutt, B.M.; Yuan, Y.; Buddington, K.; Haggard, W.O.; Bumgardner, J.D. Composite chitosan/nano-hydroxyapatite scaffolds induce osteocalcin production by osteoblasts *in vitro* and support bone formation *in vivo*. Tissue Eng. A **2009**, *15* (9), 2571–2579.

271. Chesnutt, B.M.; Viano, A.M.; Yuan, Y.; Yang, Y.; Guda, T.; Appleford, M.R.; Bumgardner, J.D. Design and characterization of a novel chitosan/nanocrystalline calcium phosphate composite scaffold for bone regeneration. J. Biomed. Mater. Res. A **2009**, *88* (2), 491–502.

272. Moreau, J.L.; Xu, H.H.K. Mesenchymal stem cell proliferation and differentiation on an injectable calcium phosphate-chitosan composite scaffold. Biomaterials **2009**, *30* (14), 2675–2682.

273. Venkatesan, J.; Ryu, B.; Sudha, P.N.; Kim, S.K. Preparation and characterization of chitosan-carbon nanotube scaffolds for bone tissue engineering. Int. J. Biol. Macromol. **2012**, *50* (2), 393–402.

274. Venkatesan, J.; Qian, Z.J.; Ashok Kumar, N.; Kin, S.K. Preparation and characterization of carbon nanotube-grafted-chitosan-Natural hydroxyapatite composite for bone tissue engineering. Carbohydrate Polymers, **2011**, *83* (2), 569–577.

275. Costa-Pinto, A.R.; Correlo, V.M.; Sol, P.C.; Bhattacharya, M.; Charbord, P.; Delorme, B.; Neves, N.M. Osteogenic differentiation of human bone marrow mesenchymal stem cells seeded on melt based chitosan scaffolds for bone tissue engineering applications. Biomacromolecules **2009**, *10* (8), 2067–2073.

276. Heinemann, C.; Heinemann, S.; Lode, A.; Bernhardt, A.; Worch, H.; Hanke, T. *In vitro* evaluation of textile chitosan scaffolds for tissue engineering using human bone marrow stromal cells. Biomacromolecules **2009**, *10* (5), 1305–1310.

277. Altman, A.M.; Yan, Y.; Matthias, N.; Bai, X.; Rios, C.; Mathur, A.B.; Alt, E.U. IFATS collection: Human adipose-derived stem cells seeded on a silk fibroin-chitosan scaffold enhance wound repair in a murine soft tissue injury model. Stem Cells **2009**, *27* (1), 250–258.

278. Kogan, G.; Šoltés, L.; Stern, R.; Gemeiner, P. Hyaluronic acid: A natural biopolymer with a broad range of biomedical and industrial applications. Biotechnol. Lett. **2007**, *29* (1), 17–25.

279. Lanza, R.P.; Vacanti, J. *Principles of Tissue Engineering*; Academic Press via Elsevier: NY, USA, 2007.

280. Leach, J.B.; Schmidt, C.E. Characterization of protein release from photocrosslinkable hyaluronic acid-polyethylene glycol hydrogel tissue engineering scaffolds. Biomaterials **2005**, *26* (2), 125–135.

281. Burgess, C.M. *Cosmetic Dermatology*; Springer Verlag: 2005.

282. Park, S.N.; Lee, H.J.; Lee, K.H.; Suh, H. Biological characterization of EDC-crosslinked collagen-hyaluronic acid matrix in dermal tissue restoration. Biomaterials **2003**, *24* (9), 1631–1641.

283. Park, S.N.; Park, J.C.; Kim, H.O.; Song, M.J.; Suh, H. Characterization of porous collagen/hyaluronic acid scaffold modified by 1-ethyl-3-(3-dimethylaminopropyl) carbodiimide cross-linking. Biomaterials **2002**, *23* (4), 1205–1212.

284. Liu, H.; Yin, Y.; Yao, K. Construction of chitosan—gelatin—hyaluronic acid artificial skin *in vitro*. J. Biomater. Appl. **2007**, *21* (4), 413–430.

285. Collier, J.H.; Camp, J.P.; Hudson, T.W.; Schmidt, C.E. Synthesis and characterization of polypyrrole–hyaluronic acid composite biomaterials for tissue engineering applications. J. Biomed. Mater. Res. **2000**, *50* (4), 574–584.

286. Yamane, S.; Iwasaki, N.; Majima, T.; Funakoshi, T.; Masuko, T.; Harada, K.; Nishimura, S.I. Feasibility of chitosan-based hyaluronic acid hybrid biomaterial for a novel scaffold in cartilage tissue engineering. Biomaterials **2005**, *26* (6), 611–619.

287. Aigner, J.; Tegeler, J.; Hutzler, P.; Campoccia, D.; Pavesio, A.; Hammer, C.; Naumann, A. Cartilage tissue engineering with novel nonwoven structured biomaterial based on hyaluronic acid benzyl ester. J. Biomed. Mater. Res. **1998**, *42* (2), 172–181.

288. Hemmrich, K.; von Heimburg, D.; Rendchen, R.; Di Bartolo, C.; Milella, E.; Pallua, N. Implantation of preadipocyte-loaded hyaluronic acid-based scaffolds into nude mice to evaluate potential for soft tissue engineering. Biomaterials **2005**, *26* (34), 7025–7037.

289. Yoo, H.S.; Lee, E.A.; Yoon, J.J.; Park, T.G. Hyaluronic acid modified biodegradable scaffolds for cartilage tissue engineering. Biomaterials **2005**, *26* (14), 1925–1933.

290. Jang, S.; Lim, G.O.; Song, K.B. Use of nano-clay (Cloisite Na+) improves tensile strength and vapour permeability in agar rich red algae (Gelidium corneum)–gelatin composite films. Int. J. Food Sci. Technol. **2010**, *45* (9), 1883–1888.

291. Wang, T.P.; Chang, L.L.; Chang, S.N.; Wang, E.C.; Hwang, L.C.; Chen, Y.H.; Wang, Y.M. Successful preparation and characterization of biotechnological grade agarose from indigenous gelidium amansii of taiwan. Process Biochem. **2012**, *47* (3), 550–554.

292. Sakai, S.; Hashimoto, I.; Kawakami, K. Agarose–gelatin conjugate for adherent cell-enclosing capsules. Biotechnol. Lett. **2007**, *29* (5), 731–735.

293. Tabata, M.; Shimoda, T.; Sugihara, K.; Ogomi, D.; Serizawa, T.; Akashi, M. Osteoconductive and hemostatic properties of apatite formed on/in agarose gel as a bone-grafting material. J. Biomed. Mater. Res. B **2003**, *67* (2), 680–688.

294. Yamada, Y.; Hozumi, K.; Aso, A.; Hotta, A.; Toma, K.; Katagiri, F.; Nomizu, M. Laminin active peptide/agarose matrices as multifunctional biomaterials for tissue engineering. Biomaterials **2012**, *33* (16), 4118–4125.

295. O'Connor, S.M.; Stenger, D.A.; Shaffer, K.M.; Ma, W. Survival and neurite outgrowth of rat cortical neurons in three-dimensional agarose and collagen gel matrices. Neurosci. Lett. **2001**, *304* (3), 189–193.

296. Gros, T.; Sakamoto, J.S.; Blesch, A.; Havton, L.A.; Tuszynski, M.H. Regeneration of long-tract axons through sites of spinal cord injury using templated agarose scaffolds. Biomaterials **2010**, *31* (26), 6719–6729.

297. Gruber, H.E.; Hoelscher, G.L.; Leslie, K.; Ingram, J.A.; Hanley Jr, E.N. Three-dimensional culture of human disc cells within agarose or a collagen sponge: Assessment of proteoglycan production. Biomaterials **2006**, *27* (3), 371–376.

298. Gruber, H.E.; Leslie, K.; Ingram, J.; Norton, H.J.; Hanley Jr, E.N. Cell-based tissue engineering for the intervertebral disc: *In vitro* studies of human disc cell gene expression and matrix production within selected cell carriers. Spine J. **2004**, *4* (1), 44–55.

299. Gu, W.Y.; Yao, H.; Huang, C.Y.; Cheung, H.S. New insight into deformation-dependent hydraulic permeability of gels and cartilage, and dynamic behavior of agarose gels in confined compression. J. Biomech. **2003**, *36* (4), 593–598.

300. Knight, M.M.; Ghori, S.A.; Lee, D.A.; Bader, D.L. Measurement of the deformation of isolated chondrocytes in agarose subjected to cyclic compression. Med. Eng. Phys. **1998**, *20* (9), 684–688.

301. Mauck, R.L.; Soltz, M.A.; Wang, C.C.; Wong, D.D.; Chao, P.H.G.; Valhmu, W.B.; Ateshian, G.A. Functional tissue engineering of articular cartilage through dynamic loading of chondrocyte-seeded agarose gels. J. Biomech. Eng. **2000**, *122* (3), 252–260.

302. Mauck, R.L.; Yuan, X.; Tuan, R.S. Chondrogenic differentiation and functional maturation of bovine mesenchymal stem cells in long-term agarose culture. Osteoarthritis Cartilage **2006**, *14* (2), 179–189.

303. Lima, E.G.; Tan, A.R.; Tai, T.; Bian, L.; Ateshian, G.A.; Cook, J.L.; Hung, C.T. Physiologic deformational loading does not counteract the catabolic effects of interleukin-1 in long-term culture of chondrocyte-seeded agarose constructs. J. Biomech. **2008**, *41* (15), 3253–3259.

304. Buckley, C.T.; Tan, A.R.; Tai, T.; Bian, L.; Ateshian, G.A.; Cook, J.L.; Hung, C.T. The effect of concentration, thermal history and cell seeding density on the initial mechanical properties of agarose hydrogels. J. Mech. Behav. Biomed. Mater. **2009**, *2* (5), 512–521.

305. Chahine, N.O.; Albro, M.B.; Lima, E.G.; Wei, V.I.; Dubois, C.R.; Hung, C.T.; Ateshian, G.A. Effect of dynamic loading on the transport of solutes into agarose hydrogels. Biophys. J. **2009**, *97* (4), 968–975.

306. Buckley, C.T.; Vinardell, T.; Thorpe, S.D.; Haugh, M.G.; Jones, E.; McGonagle, D.; Kelly, D.J. Functional properties of cartilaginous tissues engineered from infrapatellar fat pad-derived mesenchymal stem cells. J. Biomech. **2010**, *43* (5), 920–926.

307. Kelly, T.-A.N.; Ng, K.W.; Wang, C.C.B.; Ateshian, G.A.; Hung, C.T. Spatial and temporal development of chondrocyte-seeded agarose constructs in free-swelling and dynamically loaded cultures. J. Biomech. **2006**, *39* (8), 1489–1497.

308. Hung, C.T.; Lima, E.G.; Mauck, R.L.; Taki, E.; LeRoux, M.A.; Lu, H.H.; Ateshian, G.A. Anatomically shaped osteochondral constructs for articular cartilage repair. J. Biomech. **2003**, *36* (12), 1853–1864.

309. Weisser, J.; Rahfoth, B.; Timmermann, A.; Aigner, T.; Bräuer, R.; Von der Mark, K. Role of growth factors in rabbit articular cartilage repair by chondrocytes in agarose. Osteoarthritis Cartilage **2001**, *9* (Suppl. 1), S48–S54.

310. Buckley, C.T.; Vinardell, T.; Kelly, D.J. Oxygen tension differentially regulates the functional properties of cartilaginous tissues engineered from infrapatellar fat pad derived MSCs and articular chondrocytes. Osteoarthritis Cartilage **2010**, *18* (10). 1345–1354.

311. Ng, K.W.; Wang, C.C.B.; Mauck, R.L.; Kelly, T.A.N.; Chahine, N.O.; Costa, K.D.; Hung, C.T. A layered agarose approach to fabricate depth-dependent inhomogeneity in chondrocyte-seeded constructs. J. Orthop. Res. **2005**, *23* (1), 134–141.

312. Stevens, M.M.; Mayer, M.; Anderson, D.G.; Weibel, D.B.; Whitesides, G.M.; Langer, R. Direct patterning of mammalian cells onto porous tissue engineering substrates using agarose stamps. Biomaterials **2005**, *26* (36), 7636–7641.

313. Mercey, E.; Obeïd, P.; Glaise, D.; Calvo-Muñoz, M.-L.; Guguen-Guillouzo, C.; Fouqué, B. The application of 3D micropatterning of agarose substrate for cell culture and in situ comet assays. Biomaterials **2010**, *31* (12), 3156–3165.

314. Teng, S.-H.; Wang, P.; Kim, H.-E. Blend fibers of chitosan–agarose by electrospinning. Mater. Lett. **2009**, *63* (28), 2510–2512.

315. Sakai, S.; Hashimoto, I.; Kawakami, K. Synthesis of an agarose-gelatin conjugate for use as a tissue engineering scaffold. J. Biosci. Bioeng. **2007**, *103* (1), 22–26.

316. Ionescu, A.-M.; Alaminos, M.; de la Cruz Cardona, J.; de Dios García-López Durán, J.; González-Andrades, M.; Ghinea, R.; Campos, A.; Hita, E.; del Mar Pérez, M. Investigating a novel nanostructured fibrin–agarose biomaterial for human cornea tissue engineering: Rheological properties. J. Mech. Behav. Biomed. Mater. **2011**, *4* (8), 1963–1973.

317. Bloch, K.; Vanichkin, A.; Damshkaln, L.G.; Lozinsky, V.I.; Vardi, P. Vascularization of wide pore agarose–gelatin cryogel scaffolds implanted subcutaneously in diabetic and non-diabetic mice. Acta Biomater. **2010**, *6* (3), 1200–1205.

318. Bhat, S.; Kumar, A. Cell proliferation on three-dimensional chitosan–agarose–gelatin cryogel scaffolds for tissue engineering applications. J. Biosci. Bioeng. **2012**, *114* (6), 663–670.

319. Zhang, L.-M.; Wu, C.X.; Huang, J.Y.; Peng, X.H.; Chen, P.; Tang, S.Q. Synthesis and characterization of

a degradable composite agarose/HA hydrogel. Carbohydr. Polym. **2012**, *88* (4), 1445–1452.

320. Puértolas, J.A.; Vadillo, J.L.; Sanchez-Salcedo, S.; Nieto, A.; Gomez-Barrena, E.; Vallet-Regí, M. Compression behaviour of biphasic calcium phosphate and biphasic calcium phosphate–agarose scaffolds for bone regeneration. Acta Biomater. **2011**, *7* (2), 841–847.

321. Khanarian, N.T.; Haney, N.M.; Burga, R.A.; Lu, H.H. A functional agarose-hydroxyapatite scaffold for osteochondral interface regeneration. Biomaterials **2012**, *33* (21), 5247–5258.

322. Sánchez-Salcedo, S.; Nieto, A.; Vallet-Regí, M. Hydroxyapatite/β-tricalcium phosphate/agarose macroporous scaffolds for bone tissue engineering. Chem. Eng. J. **2008**, *137* (1), 62–71.

323. Cabañas, M.V.; Peña, J.; Román, J.; Vallet-Regí, M. Tailoring vancomycin release from β-TCP/agarose scaffolds. Eur. J. Pharm. Sci. **2009**, *37* (3–4), 249–256.

324. Müller, F.A.; Müller, L.; Hofmann, I.; Greil, P.; Wenzel, M.M.; Staudenmaier, R. Cellulose-based scaffold materials for cartilage tissue engineering. Biomaterials **2006**, *27* (21), 3955–3963.

325. Klemm, D., et al., Cellulose: Fascinating biopolymer and sustainable raw material. Angew. Chem. Int. Ed. **2005**, *44* (22), 3358–3393.

326. Flieger, M.; Kantorova, M.; Prell, A.; Řezanka, T.; Votruba, J. Biodegradable plastics from renewable sources. Folia Microbiol. **2003**, *48* (1), 27–44.

327. Pajulo, O.; Viljanto, J.; Hurme, T.; Saukko, P.; Lönnberg, B.; Lönnqvist, K. Viscose cellulose sponge as an implantable matrix: Changes in the structure increase the production of granulation tissue. J. Biomed. Mater. Res. **1996**, *32* (3), 439–446.

328. Märtson, M.; Viljanto, J.; Laippala, P.; Saukko, P. Connective tissue formation in subcutaneous cellulose sponge implants in the rat. Eur. Surg. Res. **1998**, *30* (6), 419–425.

329. Poustis, J.; Baquey, C.; Chauveaux, D. Mechanical properties of cellulose in orthopaedic devices and related environments. Clin. Mater. **1994**, *16* (2), 119–124.

330. Barbié, C.; Chauveaux, D.; Barthe, X.; Baquey, C.; Poustis, J. Biological behaviour of cellulosic materials after bone implantation: Preliminary results. Clin. Mater. **1990**, *5* (2–4), 251–258.

331. Märtson, M.; Viljanto, J.; Hurme, T.; Laippala, P.; Saukko, P. Is cellulose sponge degradable or stable as implantation material? An *in vivo* subcutaneous study in the rat. Biomaterials **1999**, *20* (21), 1989–1995.

332. Märtson, M.; Viljanto, J.; Hurme, T.; Saukko, P. Biocompatibility of cellulose sponge with bone. Eur. Surg. Res. **1998**, *30* (6), 426–432.

333. Fricain, J.C.; Granja, P.L.; Barbosa, M.A.; De Jeso, B.; Barthe, N.; Baquey, C. Cellulose phosphates as biomaterials. *In vivo* biocompatibility studies. Biomaterials **2002**, *23* (4), 971–980.

334. Burg, K.J.L.; Porter, S.; Kellam, J.F. Biomaterial developments for bone tissue engineering. Biomaterials **2000**, *21* (23), 2347–2359.

335. Müller, F.A.; Jonášová, L.; Cromme, P.; Zollfrank, C.; Greil, P. Biomimetic apatite formation on chemically modified cellulose templates. Key Eng. Mater. **2004**, *254*, 1111–1114.

336. Granja, P.L.; Barbosa, M.A.; Pouysegu, L.; De Jéso, B.; Rouais, F.; Baquey, C. Cellulose phosphates as biomaterials. Mineralization of chemically modified regenerated cellulose hydrogels. J. Mater. Sci. **2001**, *36* (9), 2163–2172.

337. Skoog, T. The use of periosteum and Surgicel for bone restoration in congenital clefts of the maxilla. A clinical report and experimental investigation. Scand. J. Plastic Reconstructive Surg. **1967**, *1* (2), 113–130.

338. Bartouilh de Taillac, L.; Porté-Durrieu, M.C.; Labrugère, C.; Bareille, R.; Amédée, J.; Baquey, C. Grafting of RGD peptides to cellulose to enhance human osteoprogenitor cells adhesion and proliferation. Composites Sci. Technol. **2004**, *64* (6), 827–837.

339. Pulkkinen, H.; Tiitu, V.; Lammentausta, E.; Hämäläinen, E.R.; Kiviranta, I.; Lammi, M.J. Cellulose sponge as a scaffold for cartilage tissue engineering. Biomed. Mater. Eng. **2006**, *16* (0), S29–S35.

340. Miyamoto, T.; Takahashi, S.I.; Ito, H.; Inagaki, H.; Noishiki, Y. Tissue biocompatibility of cellulose and its derivatives. J. Biomed. Mater. Res. **1989**, *23* (1), 125–133.

341. Shang, S.; Zhu, L.; Fan, J. Physical properties of silk fibroin/cellulose blend films regenerated from the hydrophilic ionic liquid. Carbohydr. Polym. **2011**, *86* (2), 462–468.

342. Salmoria, G.V., et al., Structure and mechanical properties of cellulose based scaffolds fabricated by selective laser sintering. Polym. Test. **2009**, *28* (6), 648–652.

343. He, M.; Chang, C.; Peng, N.; Zhang, L. Structure and properties of hydroxyapatite/cellulose nanocomposite films. Carbohydr. Polym. **2012**, *87* (4), 2512–2518.

344. de Souza, C.F.; Lucyszyn, N.; Woehl, M.A.; Riegel-Vidotti, I.C.; Borsali, R.; Sierakowski, M.R. Property evaluations of dry-cast reconstituted bacterial cellulose/tamarind xyloglucan biocomposites. Carbohydr. Polym. **2013**, *93* (1), 144–153.

345. Kobayashi, S.; Kashiwa, K.; Shimada, J.; Kawasaki, T.; Shoda, S.I. Enzymatic polymerization: The first *in vitro* synthesis of cellulose via nonbiosynthetic path catalyzed by cellulase. Makromole. Chem. Macromol. Symp. **1992**, *54-55* (1), 509–518.

346. Jonas, R.; Farah, l.F. Production and application of microbial cellulose. Polym. Degrad. Stability **1998**, *59* (1–3), 101–106.

347. Helenius, G.; Bäckdahl, H.; Bodin, A.; Nannmark, U.; Gatenholm, P.; Risberg, B. *In vivo* biocompatibility of bacterial cellulose. J. Biomed. Mater. Res. A **2006**, *76A* (2), 431–438.

348. Svensson, A.; Nicklasson, E.; Harrah, T.; Panilaitis, B.; Kaplan, D.L.; Brittberg, M.; Gatenholm, P. Bacterial cellulose as a potential scaffold for tissue engineering of cartilage. Biomaterials **2005**, *26* (4), 419–431.

349. Klemm, D.; Schumann, D.; Udhardt, U.; Marsch, S. Bacterial synthesized cellulose—Artificial blood vessels for microsurgery. Prog. Polym. Sci. **2001**, *26* (9), 1561–1603.

350. Schumann, D.; Wippermann, J.; Klemm, D.O.; Kramer, F.; Koth, D.; Kosmehl, H.; Salehi-Gelani, S. Artificial vascular implants from bacterial cellulose: Preliminary results of small arterial substitutes. Cellulose **2009**, *16* (5), 877–885.

351. Rambo, C.R.; Recouvreux, D.O.S.; Carminatti, C.A.; Pitlovanciv, A.K.; Antônio, R.V.; Porto, L.M. Template assisted synthesis of porous nanofibrous cellulose membranes for tissue engineering. Mater. Sci. Eng. C **2008**, *28* (4), 549–554.

Nanoparticles—Nerve

352. Chiaoprakobkij, N.; Sanchavanakit, N.; Subbalekha, K.; Pavasant, P.; Phisalaphong, M. Characterization and biocompatibility of bacterial cellulose/alginate composite sponges with human keratinocytes and gingival fibroblasts. Carbohydr. Polym. **2011**, *85* (3), 548–553.

353. Wang, S.; Huffman, J. Botanochemicals: Supplements to petrochemicals. Econ. Bot. **1981**, *35* (4), 369–382.

354. Takeda, C.; Takeda, Y.; Hizukuri, S. Physicochemical properties of lily starch. Cereal Chem. **1983**, *60* (3), 212–216.

355. Vliegenthart, J.F.G.; van der Burgt, Y.E.; Bergsma, J.; Bleeker, I.P.; Mijland, P.J.; Kamerling, J.P. Structural studies on methylated starch granules. Starch-Stärke **2000**, *52* (2–3), 40–43.

356. Choi, E.J.; Kim, C.H.; Park, J.K. Synthesis and characterization of starch-g-polycaprolactone copolymer. Macromolecules **1999**, *32* (22), 7402–7408.

357. Stenhouse, P.; Kaplan, D.L.; Mayer, J.M.; Ball, D.; McCassie, J.; Allen, A.L. *Biodegradable Polymers and Packaging*; Ching, C., Ed.; Springer: 1993; 151.

358. Malafaya, P.B.; Stappers, F.; Reis, R.L. Starch-based microspheres produced by emulsion crosslinking with a potential media dependent responsive behavior to be used as drug delivery carriers. J. Mater. Sci. Mater. **2006**, *17* (4), 371–377.

359. Torres, F.G.; Boccaccini, A.R.; Troncoso, O.P. Microwave processing of starch-based porous structures for tissue engineering scaffolds. J. Appl. Polym. Sci. **2007**, *103* (2), 1332–1339.

360. Rodrigues, M.T.; Lee, S.J.; Gomes, M.E.; Reis, R.L.; Atala, A.; Yoo, J.J. Amniotic fluid-derived stem cells as a cell source for bone tissue engineering. Tissue Eng. A **2012**, *18* (23–24), 2518–2527.

361. Martins, A.M.; Kretlow, J.D.; Costa-Pinto, A.R.; Malafaya, P.B.; Fernandes, E.M.; Neves, N.M.; Alves, C.M.; Mikos, A.G.; Kasper, F.K.; Reis, R.L. Gradual pore formation in natural origin scaffolds throughout subcutaneous implantation. J. Biomed. Mater. Res. A **2012**, *100* (3), 599–612.

362. Grasdalen, H.; Smidsrød, O. Gelation of gellan gum. Carbohydr. Polym. **1987**, *7* (5), 371–393.

363. Sutherland, I.W. Structure-function relationships in microbial exopolysaccharides. Biotechnol. Adv. **1994**, *12* (2), 393–448.

364. Crescenzi, V. Microbial polysaccharides of applied interest: Ongoing research activities in Europe. Biotechnol. Prog. **1995**, *11* (3), 251–259.

365. Fialho, A.M.; Martins, L.O.; Donval, M.L.; Leitao, J.H.; Ridout, M.J.; Jay, A.J.; Morris, V.J.; Sa-Correia, I.I. Structures and properties of gellan polymers produced by sphingomonas paucimobilis ATCC 31461 from lactose compared with those produced from glucose and from cheese whey. Appl. Env. Microbiol. **1999**, *65* (6), 2485–2491.

366. Nussinovitch, A. *Hydrocolloid Applications: Gum Technology in the Food and Other Industries*; Springer: U.K, 1997; 354.

367. de Valdez, G.F.; Torino, M.I.; Mozzi, F. *Encyclopedia of Biotechnology in Agriculture and Food*; Heldman, D.R., Hoover, D.G., Wheeler, M.B., Eds.; CRC press: 2010; 784.

368. Smith, A.M.; Shelton, R.M.; Perrie, Y.; Harris, J.J. An initial evaluation of gellan gum as a material for tissue engineering applications. J. Biomater. Appl. **2007**, *22* (3), 241–254.

369. Bertoni, F.; Barbani, N.; Giusti, P.; Ciardelli, G. Transglutaminase reactivity with gelatine: Perspective applications in tissue engineering. Biotechnol. Lett. **2006**, *28* (10), 697–702.

370. Oliveira, J.T.; Santos, T.C.; Martins, L.; Picciochi, R.; Marques, A.P.; Castro, A.G.; Neves, N.M.; Mano, J.F.; Reis, R.L. Gellan gum injectable hydrogels for cartilage tissue engineering applications: *In vitro* studies and preliminary *in vivo* evaluation. Tissue Eng. Part A **2009**, *16* (1), 343–353.

371. Gong, Y.; Wang, C.; Lai, R.C.; Su, K.; Zhang, F.; Wang, D.-A. An improved injectable polysaccharide hydrogel: Modified gellan gum for long-term cartilage regeneration *in vitro*. J. Mater. Chem. **2009**, *19* (14), 1968–1977.

372. Silva, N.A.; Salgado, A.J.; Sousa, R.A.; Oliveira, J.T.; Pedro, A.J.; Leite-Almeida, H.; Cerqueira, R.; Almeida, A.; Mastronardi, F.; Mano, J.F.; Neves, N.M.; Sousa, N.; Reis, R.L. Development and characterization of a Novel Hybrid Tissue Engineering–based scaffold for spinal cord injury repair. Tissue Eng. Part A **2009**, *16* (1), 45–54.

373. Oliveira, J.; Santos, T.C.; Martins, L.; Silva, M.A.; Marques, A.P.; Castro, A.G.; Neves, N.M.; Reis, R.L. Performance of new gellan gum hydrogels combined with human articular chondrocytes for cartilage regeneration when subcutaneously implanted in nude mice. J. Tissue Eng. Regenerative Med. **2009**, *3* (7), 493–500.

374. Oliveira, J.; Santos, T.C.; Martins, L.; Picciochi, R.; Marques, A.P.; Castro, A.G.; Neves, N.M.; Mano, J.F.; Reis, R.L. Gellan gum injectable hydrogels for cartilage tissue engineering applications: *in vitro* studies and preliminary *in vivo* evaluation. Tissue Eng. Part A **2009**, *16* (1), 343–353.

375. Manikoth, R.; Kanungo, I.; Fathima, N.N.; Rao, J.R. Dielectric behaviour and pore size distribution of collagen–guar gum composites: Effect of guar gum. Carbohydr. Polym. **2012**, *88* (2), 628–637.

376. Coutinho, D.F.; Sant, S.V.; Shin, H.; Oliveira, J.T.; Gomes, M.E.; Neves, N.M.; Reis, R.L. Modified Gellan Gum hydrogels with tunable physical and mechanical properties. Biomaterials **2010**, *31* (29), 7494–7502.

377. Shin, H.; Olsen, B.D.; Khademhosseini, A. The mechanical properties and cytotoxicity of cell-laden double-network hydrogels based on photocrosslinkable gelatin and gellan gum biomacromolecules. Biomaterials **2011**, *33* (11), 3143–3152.

378. Silva-Correia, J.; Oliveira, J.M.; Caridade, S.G.; Oliveira, J.T.; Sousa, R.A.; Mano, J.F.; Reis, R.L. Gellan gum-based hydrogels for intervertebral disc tissue-engineering applications. J. Tissue Eng. Regenerative Med. **2011**, *5* (6), e97–e107.

379. Lévesque, S.G.; Shoichet, M.S. Synthesis of cell-adhesive dextran hydrogels and macroporous scaffolds. Biomaterials **2006**, *27* (30), 5277–5285.

380. Liu, Y.; Chan-Park, M.B. Hydrogel based on interpenetrating polymer networks of dextran and gelatin for vascular tissue engineering. Biomaterials **2009**, *30* (2), 196–207.

Nanoparticles—Nerve

381. Moreira Teixeira, L.S.; Leijten, J.C.; Wennink, J.W.; Chatterjea, A.G.; Feijen, J.; van Blitterswijk, C.A.; Karperien, M. The effect of platelet lysate supplementation of a dextran-based hydrogel on cartilage formation. Biomaterials **2012**, *33* (14), 3651–3661.

382. Jukes, J.M., van der Aa, L.J.; Hiemstra, C.; van Veen, T.; Dijkstra, P.J.; Zhong, Z.; de Boer, J. A newly developed chemically crosslinked dextran-poly(ethylene glycol) hydrogel for cartilage tissue engineering. Tissue Eng. Part A **2010**, *16* (2), 565–573.

383. Pan, H.; Jiang, H.; Chen, W. Interaction of dermal fibroblasts with electrospun composite polymer scaffolds prepared from dextran and poly lactide-co-glycolide. Biomaterials **2006**, *27* (17), 3209–3220.

384. Jia, X.; Zhao, C.; Li, P.; Zhang, H.; Huang, Y.; Li, H.; Fan, Y. Sustained release of VEGF by coaxial electrospun dextran/PLGA fibrous membranes in vascular tissue engineering. J. Biomater. Sci. Polym. Ed. **2011**, *22* (13), 1811–1827.

385. Inci, I.; Kirsebom, H.; Galaev, I.Y.; Mattiasson, B.; Piskin, E. Gelatin cryogels crosslinked with oxidized dextran and containing freshly formed hydroxyapatite as potential bone tissue-engineering scaffolds. J. Tissue Eng. Regen. Med. **2012**, *7* (7), 584–588.

386. Sun, G.; Zhang, X.; Shen, Y.I.; Sebastian, R.; Dickinson, L.E.; Fox-Talbot, K.; Gerecht, S. Dextran hydrogel scaffolds enhance angiogenic responses and promote complete skin regeneration during burn wound healing. Proc. Natl. Acad. Sci. USA **2011**, *108* (52), 20976–20981.

387. Sun, G.; Shen, Y.I.; Kusuma, S.; Fox-Talbot, K.; Steenbergen, C.J.; Gerecht, S. Functional neovascularization of biodegradable dextran hydrogels with multiple angiogenic growth factors. Biomaterials **2011**, *32* (1), 95–106.

Nanoparticles—Nerve

Nerve Guides: Multi-Channeled Biodegradable Polymer Composite

Wesley N. Sivak
Trent M. Gause
Department of Plastic Surgery, University of Pittsburgh, Pittsburgh, Pennsylvania, U.S.A.

Kacey G. Marra
Departments of Plastic Surgery and Bioengineering, McGowan Institute for Regenerative Medicine, University of Pittsburgh, Pittsburgh, Pennsylvania, U.S.A.

Abstract

When direct repair of a severed nerve cannot be accomplished without undue tension, the standard clinical treatment remains autologous nerve grafting. Due to the limited availability of donor nerve and limited functional recovery following autografting, neural tissue engineering has focused on the development of artificial nerve guides. A nerve guide is a means of directing axonal regeneration and repair across a gap defect; the most basic objective of a nerve guide is to combine physical, chemical, and biological cues under conditions that will foster functional tissue formation. Whether fashioned from biologic or synthetic materials, the nerve guide facilitates communication between the proximal and distal ends of the nerve gap, blocks external inhibitory factors, retains neurotrophic factors at the site of regeneration, and provides physical constraints on axonal regrowth. Nerve guide technology is constantly evolving and now exists as a combination of many unique elements including scaffolds and biomimetic materials. Innovative fabrication methods have been developed and researchers are now producing biodegradable nerve guides with longitudinally aligned channels, more accurately mimicking the native nerve architecture. Multichanneled nerve guides function based on the premise that enclosing nerve stumps and spanning an intervening gap with a biomimetic scaffold create a protected, proregenerative environment. The choice of which physical, chemical, and biological cues to use is based on the properties of the nerve environment, which is critical in creating the most desirable environment for axon regeneration and subsequent return of function.

INTRODUCTION

Damage to the peripheral nervous system is surprisingly common. Peripheral nerve injury is thought to occur in 2–3% of all major traumas and also regrettably results from complications of surgery.[1] Although complete recovery of nerve function routinely occurs after mild injury, outcomes are often unsatisfactory following complete nerve transection. The emergence of the field of vascularized composite allo-transplantation has also ushered in a new era where entire limbs are being transplanted; recipient nerves are connected to donor nerves at the level of prior amputation. Whether necessitated by trauma, surgical complication, or transplant, the results of peripheral nerve repair are dependent upon the body's ability to precisely coordinate reinnervation of the target tissues.

Once cut, nerves must be able to sprout, identify, and reconnect to the distal segment. When an axon is disconnected from it cell body by injury, its distal segment gradually degenerates through a process called Wallerian degeneration.[2] If regrowth and reinnervation do not occur within 12–18 months, sensation and motion are permanently lost and there is little chance for a functional recovery.[3,4] Surgeons routinely repair severed nerves and can now transplant limbs above the elbow, but poor sensation and motion continue to result due to inadequate nerve regrowth and reinnervation. Maximizing sensation and motion in an injured or transplanted limb appears to lie within optimizing the quantity, quality, speed, and accuracy of nerve regeneration following repair.

With respect to repair of peripheral nerve injury, end-to-end anastomosis of severed nerve stumps is the preferred method.[5] Unfortunately, inelastic recoil occurs within injured nerve stumps. Usually, if the repair is performed within 24 hours following injury, it is much easier to overcome this problem without undue tension.[1] However, the longer the delay, the more likely it is that a graft will be required to span the defect. The nerve stumps will also become embedded in scar tissue and increasing amounts of collagen deposition will occur the longer the delay to definitive repair.

Nerve autografting is the most commonly used surgical technique for repair of peripheral nerve gaps and remains the clinical gold standard.[6] The graft is a nerve generally harvested from a distant site on the body. The success of autologous nerve grafting arises from the presence of Schwann cells and basal lamina endoneural tubes, which provide neurotrophic factors and surface adhesion molecules to regenerating axons.[7] Some studies have reported complete restoration of sensation after nerve autografting, but the recovery rate of motor function is less than 40%.[8] Autografting also depends upon donor nerve availability and compatibility in terms of tissue size and structure. Major disadvantages include limited graft material, loss of function at the donor site, and donor site morbidity (e.g., scarring and painful neuroma formation). In addition, the individual endoneural tubes cannot be exactly approximated in size or number, leading to mismatching of regenerating axons and nonspecific, incomplete reinnervation of the target tissues. Allografts and xenografts have also been studied, but undesired immune responses or the requirement for preprocessing to remove immunogenic material limit their effectiveness.[1]

A nerve conduit comprised of biodegradable materials is a promising strategy for spanning gap defects and promoting long-term nerve recovery—a fact confirmed both experimentally and clinically. A well-engineered biodegradable guide initially provides an adequate scaffold for regeneration while retaining the necessary mechanical stiffness to remain open under the pressure from surrounding tissues without causing damage to the nerve stumps and adjacent structures. Secondly, an ideal guide is semipermeable and allows for controlled exchange between intra- and extra-channel environments; the ability to retain neurotrophic factors within the lumen has been identified as a key factor in increasing nerve growth response. Lastly, the structure should degrade at an appropriate rate as the nerve regenerates. Too quickly, and the walls of the lumen can collapse on the growing nerve or fibrous tissue can infiltrate and disrupt regeneration. In order to be acceptable for clinical use, an artificial nerve guide must ensure satisfactory nerve repair at least to a level equivalent with autografts. Several biodegradable nerve guide implants consisting of collagen, polyglycolic acid (PGA), polylactic acid (PLA), and polylactide–caprolactone (PLA-PCL) have been approved by the Food and Drug Administration (FDA) for human application and are in clinical use, but indications are limited to gaps spanning less than 3 cm. Silk-based guides are a relatively new alternative and have shown promise in experimental models, but are still in development for clinical use.[9]

Many studies have shown that it is possible to improve upon the performance of nerve conduits by modifying the physical configuration of the material backbone to more closely resemble native nerve architecture. In its most basic form, a nerve is simply a series of tubes within a larger tube (Fig. 1). Peripheral nerves have multiple layers of connective tissue surrounding axons, with the endoneurium surrounding individual axons, perineurium binding axons into fascicles, and epineurium binding the fascicles into a nerve. Blood vessels, called vasa nevorum, are contained within the nerve. Innovative fabrication methods have been developed and researchers are now producing biodegradable nerve guides with longitudinally aligned channels, more accurately mimicking the native nerve architecture (Fig. 2). Oriented channels provide the regenerating axons a defined pathway for growing straight through the scaffold and increase the surface area available for cell adhesion. In a scaffold with multichannel

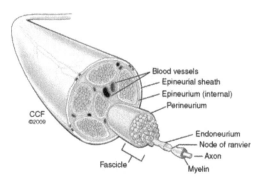

Fig. 1 The normal nerve architecture consisting of multilayered structures.
Source: From Siemionow & Brzezicki[96] © 2009, with permission from Elsevier.

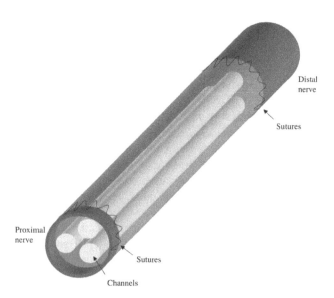

Fig. 2 Schematic diagram of trichanneled nerve guide designed as a conduit for attaching via sutures the proximal and distal ends of a severed nerve.
Source: From Bender et al.[14] © 2004, with permission from Elsevier.

architecture, regenerating axons are able to extend through open longitudinal channels as they would normally extend through endoneurial tubes of grafted autologous peripheral nerves. Furthermore, nerves fascicles may segregate according to destinations or function within the central and peripheral nervous systems. Spinal cord tracts often relay information to similar regions of the brain or may be grouped by action such as motor neurons. Moreover, many peripheral nerves such as the ulnar nerve also contain sensory and motor axons in distinctly separate bundles. In the future, multichanneled conduits with complex computer-designed patterns may group and guide bundles of axons to their respective targets limiting the dispersion of axons and the potential for synkinetic connections. What follows is a review of the scientific and engineering literature detailing the fabrication, evaluation, and use of multichanneled, biodegradable polymer composites as nerve guides.

FABRICATION TECHNIQUES

The advent of artificial nerve conduits began in the 1980s using nonabsorbable materials such as silicone to join disconnected nerve ends. While continuity was achieved in cases of a 10 mm nerve gap, recovery of locomotor function proved to be a more elusive outcome, and autografting remained the preferred treatment. The lack of degradation in the early conduits may have led to compression of a regenerating neurites or blood vessels, which may have sequestered axons from necessary growth factors, explaining the underwhelming results. Research interests then focused on biodegradable substances in later 1980s. The materials comprising nerve guides can broadly be divided into artificial and natural substances. Advances in synthesis procedures yielded artificial polymers such as PLA, PGA, and PLGA all of which are FDA approved and widely used in nerve guide fabrication. Additional synthetics such as PGA and polyhydroxybuturate are currently under investigation. Naturally occurring scaffolds include collagen, chitosan, and agarose for both single and multichannel nerve guides, and alginate and polysialic acid for single channel lumens. It is hypothesized that the biocompatible materials such as collagen, which are components of the extracellular matrix, are better able to guide axons along the channel and provide a source of traction for growth cones. A summary of studies detailing fabrication techniques for multichannel nerve guides can be found in Table 1.

Injection Molding

A variety of methods for constructing symmetrical longitudinally oriented channels within a conduit exist. One of the early techniques present in the literature was low-pressure injection molding. Developed by Hadlock et al., a 10% polymer solution of PLGA 85:15 monomer ratio was added

to 99.9% acetic acid.[10] Stainless steel wires with diameters ranging from 60 to 500 μm were used to create the longitudinal channels. Wire mesh was then used to fix the channels in place. The mesh also affords the ability to configure the rods in different positions and allows for flexibility within the intraluminal construct. Under low pressure, the polymer solution is then injected and the temperature is brought below the freezing point of PLGA and acetic acid. After freezing, the conduit mold is removed, lyophilized in a vacuum, and stored in a dry environment. In the study, Hadlock et al. fabricated conduits with 5, 16, and 57 channels with diameters defined by the stainless steel inserts.

As a slight variation on low-pressure injection molding, Moore et al. detailed injection molding with solvent evaporation, which has the added property of crafting a porous scaffold.[11] Cylindrical Teflon molds with Delrin spacers were used with seven stainless steel wires to create the channels. One gram of PLGA with a ratio of 85:15 lactide:glycolide was added to 2.0 mL of dichloromethane. Upon mixing, the solution was injected to fill each mold. The polymer molds underwent vacuum drying. The evaporation of the solvent left pores in the conduit scaffold. Computer assisted design (CAD) and solid freeform fabrication methods can be used if more intricate channel designs are desired. de Ruiter et al. also used this technique; however, various polymer ratios of 50:50, 75:25, and 85:15 were fabricated with methylene chloride as the solvent.[12]

A study by He et al. in 2008 added the concept of thermally induced phase separation (TIPS) to injection molding as a means to create macro-/microporous architecture within a conduit.[13] In this process, a 75:25 PLGA polymer was used with diethylene dioxide and granular ammonium bicarbonate. Similar to the above procedures, the solution was injected into a mold at −40°C and stored at low temperature until completely frozen. After removal of the wires, the solidified mold was lyophilized at 0.945 mbar for 5 days. In order to create a porous scaffold and remove the ammonium bicarbonate, conduits were submerged in distilled water. The water was changed every 6 hours, and Nessler's reagent was used to ensure that there was no remaining ammonium bicarbonate within the scaffold. The injection molding with TIPS resulted in channel diameters of 0.8 mm and 500 μm in 7- and 16-channel guides, respectively (Fig. 3). Leeching the ammonium bicarbonate from the scaffold left pores of 125–200 μm in diameter.

Mandrel Coating

Aside from injection molding, mandrel coating or mandrel adhesion is another effective means of fabrication. Detailed by Bender et al. in 2004, PVA fibers were used as opposed to stainless steel wires for channel formation.[14] This process allows for selective dissolution of the mandrels and subjects the conduit to less physical trauma than pulling the wires out by force. In the study, mandrels were produced by

Nanoparticles—Nerve

Table 1 Published studies detailing fabrication and characterization of multichanneled nerve conduits

Author	Material(s)	Fabrication method(s)	Characterization test(s)	References
Bender et. al.	PCL (PVA-coated mandrels)	Wire mesh, mandrel adhesion	Morphometry, Young's modulus, tensile strength, degradation rate	[14]
Arcaute et al.	PEG	Stereolithography	Compression testing, suture pullout resistance, swelling	[20–22]
Bozkurt et al.	Porcine collagen	Unidirectional freezing	Cross-linking	[16,17]
Chen et al.	Various polymers	Bilayer fabrication	Morphometry, porosity	[29]
de Ruiter et al.	PLGA (various copolymer ratios)	Injection molding, selective dissolution	Morphometry, porosity, 3-point bending, swelling, degradation rate	[12]
Romero-Ortega and Galvan-Garcia	Agarose	Injection molding	Morphometry	[36]
Hadlock et al.	PLGA 85:15	Injection molding low pressure	Morphometry, porosity	[10]
Moore et al.	PLGA 85:15	Injection molding solvent evaporation	Morphometry, porosity, degradation rate, drug release kinetics	[11]
He et al.	PLGA 72:25	Injection molding with TIPS	Morphomtery, porosity, degradation rate, compression modulus	[13]
Huang et al.	PLGA 50:50, chitosan	Lyophilizing and wire heating	Morphometry, porosity	[15]
Hu et al.	Collagen, chitosan	Unidirectional freezing	Morphometry, porosity	[19]
George et al.	Polypyrrole	Electroplating	Morphomtery	[23]
Jeffries and Wang	PCL	Electrospinning	Morphometry, porosity	[25]
Wang et al.	Chitosan	Injection molding with CAD	Morphometry, porosity, swelling, degradation, compressive strength, suture retention rate	[31]
Yao et al.	Collagen	Mandrel adhesion	Morphometry, porosity, shrink temperature, degradation rates, compressive test, tensile strength, 3-point bending	[33]
Yu and Shoichet	Hydrogel P(HEMA-co-AEMA)	Injection molding selective dissolution	Morphometry, porosity, swelling, compression testing, elastic modulus	[34]
Flynn et al.	Hydrogel P(HEMA)	Injection molding selective dissolution	Morphometry, porosity, swelling, compressive modulus	[35]
Koh et al.	PLLA	Electrospinning	Morphometry, drug release kinetics	[26]
Pabari et al.	Silsesquioxane, polyhexanolactone, and poly(carbonate-urea) urethane	Electrospinning	Morphometry	[28]

removing 28-gage metal wire and soaking in PVA solution. The coated mandrels, still in its coiled configuration from the spool, were then suspended and allowed to straighten and dry. Vibrating the coils helped to remove any droplets of PVA that collected. After mandrel fabrication, Bender et al. described two methods of creating multichannel scaffolds from PCL polymers, denoted as the wire mesh and mandrel adhesion techniques (Fig. 4). In the wire mesh method, mandrels were inserted through a triangular wire mesh and fixed in place with super glue. The process of securing with wire mesh was repeated on the opposite end of the mandrels. The framework was then placed in a solution of PCL and tetrahydrofuran (THF) and allowed to coat the mandrels until the desired cylindrical shape and channel diameter was achieved. After THF evaporation and trimming, the conduits were placed in 37°C water bath to dissolve the PVA

coated mandrels to craft a trichanneled nerve guide. In the mandrel adhesion procedure, PVA-coated mandrels were immersed individually until a thin coating of PCL formed along its length. Mandrels were then placed in PCL/THF solution, and while the THF evaporated, a second mandrel was adhered to the first. The process was repeated and a third mandrel was added. As per the wire mesh method, the three mandrels were resubmerged in the PCL/THF solution until the desired channel diameter was obtained.

Lyophilizing and Wire Heating

Multichanneled nerve conduits were also created by Huang et al. using a lyophilizing and wire heating process.[15] In this process, nickel-chromium (Ni-Cr) wires served as the

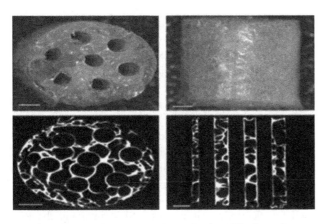

Fig. 3 Photographs (top panels) and microCT slices (bottom panels) of seven-channel PLGA scaffold created by Moore et al. using injection molding. Left: cross-sectional view. Right: Longitudinal view. Scale bar ~500 µm.
Source: From Moore et al.[11] © 2006, with permission from Elsevier.

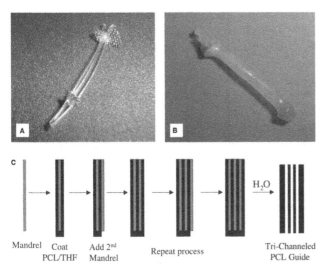

Fig. 4 Wire mesh and Mandrel method to fabricate tri-channeled nerve guides. (**A**) Three PVA fibers, to serve as mandrels, are suspended between two pieces of wire mesh. (**B**) PCL-coated wire mesh scaffold with three PVA mandrels. (**C**) Schematic diagram of the mandrel adhesion method for the fabrication of tri-channeled nerve guides.
Source: From Bender et al.[14] © 2004, with permission from Elsevier.

mandrels, and separate scaffolds were made from chitosan and PLGA. The Ni-Cr wires were inserted into a rubber tube, which was capped with paraffin at both ends. Either the chitosan/acetic acid or PLGA/dixoane solution was injected through a hole at one end of the tube. The mold was solidified in a nitrogen trap. After freezing, the Ni-Cr wires were electrically heated for approximately 1 min and removed from the tubes with pliers. The acetic acid or dixoane was the sublimated before the conduits were stored in a dry environment.

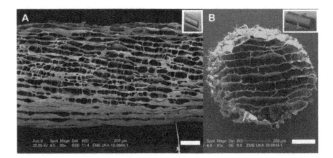

Fig. 5 Perimaix nerve guides prepared using a patented unidirectional freezing process as described by Bozkurt et al. (**A**) SEM of the longitudinal microstructure of the Perimaix nerve guide and (**B**) SEM of the transverse microstructure.
Source: From Bozkurt et al.[16,17] © 2007, with permission from Elsevier.

Unidirectional Freezing

Bozkurt et al. developed a nerve guide (i.e., Perimaix) with longitudinally oriented microchannels from a unidirectional freezing technique.[16,17] Previously described by Schoof et al. in freeze-dried collagen sponges, porcine collagen was harvested with 10–15% elastic to prepare a 1.5% aqueous solution.[18] During the solidification process, ellipsoidal ice crystals formed, which determined the morphology of channels within the scaffold (Fig. 5). The anisotropic properties of this method allows for several small channels ranging from 20 to 100 µm. Either acetic acid or ethanol may be used as an additive to form the ice crystals; the solute concentration determines the diameter of the pores. Vacuum drying may be used to sublimate the solvent revealing the longitudinal pores. A single layer of collagen interconnects the pores, which is defined by the slow growing short axis of the elliptical ice crystals. According to Bozkurt et al. the distance between the collagen layers must be at least large enough to promote cell seeding and migration, which can be tailored using the unidirectional freezing procedure.[16]

Unidirectional freezing has also been utilized by Hu et al. with longitudinal microchannels of chitosan-collagen (CCH), and compared to conduits of the same material with randomly oriented channels.[19] Type I collagen and chitosan was mixed in a 0.05 M acetic acid solution at 4°C to avoid collagen denaturation. The CCH suspension was degassed and injected into a copper mold of 50 mm in length and 2.0 mm in diameter. The suspension and mold was then undirectionally frozen by vertically passing it through a liquid nitrogen container at a velocity of 2.0×10^{-5} m/s. The ice crystals were then freeze-dried and sublimated over a 24-hr period. Next, the scaffolds were cross-linked with a 1% wt solution of genipin for 48 hr. The conduits were then washed, dehydrated by air drying for a week, and then sterilized. In order to craft the CCH conduit with randomly oriented channels, the

Nanoparticles—Nerve

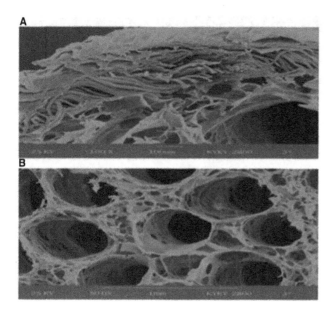

Fig. 6 SEM of transverse sections of nerve guide by Wang et al. using porous chitosan. Showing (**A**) knitted outer wall, and (**B**) highly porous inner matrix and longitudinally oriented guidance channels.
Source: From Wang et al.,[97] with permission from Wiley InterScience.

mold and suspension were directly immersed in the liquid nitrogen chamber. The researchers were able to reproducibly manufacture novel nerve-guiding collagen–CCH scaffolds with inner dimensions resembling the basal lamina microchannels of normal nerves. The scaffold has a number of structural advantages, including longitudinally orientated microchannels and extensive interconnected pores between the parallel microchannels.

Stereolithography

More technological demanding than the injection molding technique, Arcuate et al. detailed the use of sterolithography in crafting multichannel nerve conduits.[20] The process utilizes CAD to build a 3D representation of the conduit design (Fig. 6).[21,22] Poly(ethylene glycol) diacrylate was chosen as the scaffold material, and a circular laser beam of 250 μm diameter was chosen. To assure interlayer bonding had occurred, there were 300 μm of overlapping between two contiguous layers. Since there was some variability between batches of PEG in the amount of cross-linking, the investigators had to empirically derive the correct scan velocity to achieve an inner scaffold channels of 400 μm in diameter. After fabrication, the conduits were placed in PBS for over 2 days and washed to remove any unlinked polymer; it was also noted that the samples swelled as expected. The conduits were then freeze-dried in liquid nitrogen and sterilized with hydrogen peroxide.

Electroconductive Conduits Using Electroplating

As an induced electrical field has been demonstrated to promote axon regeneration, George et al. and Biazar et al. developed a method to create a multichannel scaffold with electroconductive properties.[23,24] Multichannel nerve guides were formed with cylindrical and square Teflon molds with evenly distributed stainless steel wires of 250 and 350 μm. The molds and wires were immersed in a 0.2 M polypyrrole (PPy)/0.2 M sodium dodecyclbeneze sulfonate (NaDBS) aqueous solution, and electroplated using a galavanostat at a current of 0.5 mA for 14 hr; blanket of N_2 was used for electrodeposition. After SEM imaging, the PPy tubes were removed by applying a –10 V difference to the wires and reference platinum mesh for 2 min. The multiarray tubes were then cut, sterilized, and washed in PBS.

Electrospinning

Electrospinning has also become another suitable means to produce nerve guides. As published by Jeffries and Wang 14% PCL pellets were dissolved in 5:1 triflouroethanol:water at room temperature.[25] Fibers were then deposited in an aluminum collector. Sutures were then placed upon the bed of woven fibers already on the collector, a comb held the sutures in a parallel regularly spaced pattern. After applying a second coat of PCL at a fixed volume, the construct was placed under a vacuum overnight. The sutures were then cut to release the electrospun film and rolled in parallel to the sutures. The rolled conduit was then secured with collagen suture, but paraffin is also an alternative to collagen. A thin sheath of radially aligned PCL fibers was applied by rotating the nerve guide to reinforce the entire conduit. Sutures were removed to form the inner channels, soaked in deionized water, frozen on dry ice, and the guide trimmed (Fig. 7). A paper by Koh et al. outlined the fabrication of PLLA, blended laminin–PLLA, PLGA, and blended nerve growth factor (NGF)-PLGA conduits with nanofibers made with electrospinning.[26] Slightly different from electrospinning, Pabari offered a nanofiber conduit of polyhedral oligomeric silsesquioxane, polyhexanolactone, and poly(carbonate-urea) urethane (i.e., UCL-nanobio) made using liquid ultrasonic atomization.[27,28]

Bilayer Multichannel

A bilayer multichannel conduit was patented by Chen et al.[29] The process involves creating an outer porous hollow tube and then placed a multichannel filer in different shapes within the hollow tube. A number of polymers may be used including PCL, PLA, PGA, PLGA, or PCL–PLA copolymer. To create the multichannel matrix, a polymer is dissolved in an organic solvent and poured onto a mold that has an uneven surface resulting in a film-shaped solution. Next, a coagulant is used to form interconnected pores on

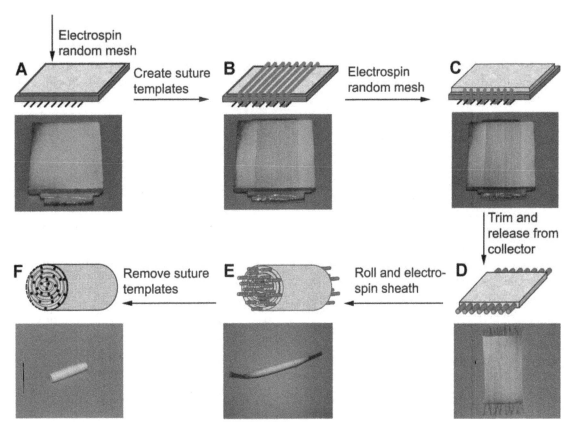

Fig. 7 Fabrication scheme for suture-templated electrospinning nerve guide by Jeffries and Wang. (**A**) Deposition of a thin layer of electrospun PCL fibers. (**B**) Creation of channel template aligning sutures across PCL layer. (**C**) Deposition of a second layer of electrospun PCL fibers. (**D**) After drying in vacuum, cut sutures trim and flat sheet to length. (**E**) Roll up nerve guide and secure with electrospun PCL around outside of guide. (**F**) Sutures are removed.
Source: From Jeffries & Wang[25] © 2012, with permission from John Wiley and Sons.

the uneven mold. Coagulants used in this method are usually combination of water and an organic solvent with the solvent representing 10–50% of the solution. The coagulant is usually comprised of a ketone and alcohol, for instance acetone and methanol, but can also be an amide. The multichannel filler can then be shaped into a single layer, multiple layered, folded, or spiraled form. The process of creating the outer tube is essentially the same as the multichannel filer, except that a solid mandrel is used to obtain a hollow cylinder. The multichannel filler is then placed into hollow tube, resulting in a bilayer and porous conduit.

Microchannels

As an extension of the logic behind multichannel nerve guides, several authors described small grooves etched into the surfaces of lumens as microchannels. The microchannels are believed to help orient the growth cone of axons in a longitudinal direction, thus limiting dispersion in the same manner as the aforementioned multichannel. Using CAD photolithographic techniques on photosensitive polyimide, Mahoney et al. carved small channels into the surface of a

planar film of polyimide; however, the film was never rolled into the cylindrical shape of a conduit.[30]

Novel Materials

Chitosan and CAD

In 2006, Wang et al. offered a multichannel conduit design using CAD and chitosan as the scaffold material.[31] Firstly, a porous hollow tube of chitosan fibers was crafted based upon an industrial knitting and lyophilization technique.[32] The knitting procedures yielded a single lumen chitosan conduit with a variable inner diameter and wall thickness. Two patches with precise perforations were made using CAD and fixed at both ends to two moveable blocks on a pedestal. The distance between the two moveable blocks determined the length of the conduit, allowing for flexibility in design. Commercially available acupuncture needles, with diameters of 140–500 μm, were threaded through corresponding holes in the fixed patches and the hollow chitosan tubes. The specifics of the CAD-crafted patches defined the distribution, number, and diameter of matrix channels in a two-dimensional plane, enabling a variety of scaffolds to be created. The moveable blocks were apposed to the ends of the conduits, and 2% chitosan solution was injected filling the hollow tube. After storing at −20°C for 12 hr, the conduit was lyophilized, deionized, rinses, and dried prior to removing the needle mandrels.

Cross-linked collagen based conduit

A multichannel design by Yao et al. utilized collagen as the scaffold material to construct conduits with four and seven channels of 530 and 410 μm in diameter, respectively (Fig. 8).[33] Negative molds were constructed with four or seven holes for wire fixation. Stainless steel wires were inserted in parallel through holes in the circular negative molds at both ends. The distance between two ends was measured at 1 cm, and two wires were passed into the molds. The apparatus was then placed in a 12 mg/mL type I collagen solution derived from bovine Achille's tendon and the coated steel wires were then allowed to air dry. Two additional wires were inserted into the adjacent channels for a total of four channels after submersion in the collagen solution. In order to fabricate a seven-channeled scaffold, this process was repeated twice with two wires and one final mandrel. Both the four and seven channel conduits were cross-linked overnight EDC and NHS in 2-morpholinoethansulfonic acid solution at a pH of 5.5. Distilled water and NaH_2PO_4 was used to wash, and the collagen was freeze dried. Molds and stainless steel wires were then removed, resulting in a multichanneled nerve guide.

Hydrogel-based conduit

To promote guided neural cell regeneration using hydrogels, Yu and Shoichet proposed a method of polymerizing 2-hydroxylethyl methacrylate (HEMA) with 2-aminoethyl-methacrylate (AEMA).[34] The HEMA-*co*-AEMA polymer was achieved by using a redox initiator accelerator system. Ammonium persulfate served as the initiator, and tetramethylene diamine as the accelerator. Ethylene dimethacrylate (EDMA) was used as the cross-linker. To fabricate a multichanneled nerve guide, PCL fibers were submerged in a P(HEMA-*co*-AEMA) gel. Acetone with sonication was then used to selectively dissolve the PCL fibers, creating longitudinal channels according to a protocol previously published by Flynn, Dalton, and Shoichet[35]

Another use of hydrogels was described in a 2003 patent by Romero-Ortega and Galvan-Garcia which offered the advantage of a transparent nerve conduit, enabling the visualization of Schwann cells and neurons in three dimensions.[36] Pure agarose (1–1.5%) was dissolved in PBS and injected into a mold with the desired numbers of wire of fibers channels. Fibers were pulled from the gel after polymerization had occurred at either room temperature or 4°C for 15 min. It should be noted that the aforementioned technique by Yu et al. also resulted in transparent scaffold material.[34]

PROPERTIES AND CHARACTERIZATION

While a plethora of methods exist for fabricating nerve conduits, the mechanical properties and structural arrangement

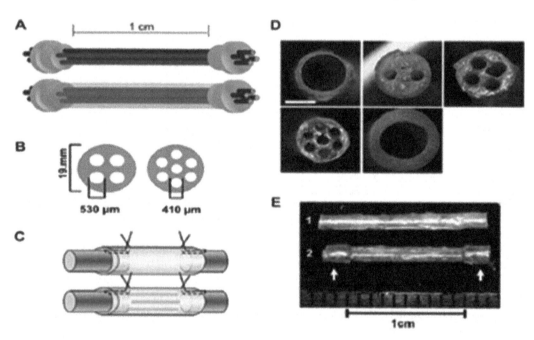

Fig. 8 (**A**) Fabrication of collagen based conduit by Yao et al. using stainless steel wires. (**B**) Channel morphometry. (**C**) Demonstrating *in vivo* placement and suturing. (**D**) Transverse imaging of collagen conduits, scale = 1 mm. (**E**) Images of (1) one-channel and (2) four-channel conduits. Arrows point to lips at each end for nerve guide insertion and suturing.
Source: From Yao et al.[33,98] © 2010, with permission from Elsevier.

of each design largely determine their clinical utility. Biodegradable materials have supplanted nonbiodegradable conduits as the latter may lead to compression of the regenerating neuritis.[37,38] A synthetic material is usually preferred over natural as synthetic conduits have a lower probability of eliciting an undesirable foreign body reaction.[39–41] On the other hand, natural materials are able to better mimic the native extracellular matrix, and in doing so may promote cellular adhesion and growth. Although the balance between natural and synthetic material has yet to be clearly defined, in designing an ideal conduit, one must take into account the following physical properties: permeability or porosity, flexibility, propensity for the material to swell, and biodegradation rates.[42] As many authors do not report every mechanical property of conduits, comparison of two different fabrication methods can be challenging. The exact properties that optimize neural regeneration with regards to multichannel scaffolds have yet to be discerned; however, generally desirable principles of multichannel conduit fabrication can be gleaned from the current state of research. A summary of studies detailing characterization techniques for multichannel nerve guides can be found in Table 1.

Permeability

Several studies have demonstrated that permeable conduits lead to improved regeneration rates as opposed to impermeable conduits.[43–45] The importance of choosing a scaffold with the appropriate porosity is twofold. First, the interconnected pores within the conduit allow for the exchange of gas, fluid, and nutrients between the inner lumens and the outer environment.[46] The diffusion of these molecules is vital to the survival of neurite prior to capillary infiltration and ensures that there is ample influx of

required growth and factors and outflow of fluid.[47] Second, conduit permeability influences the development of a fibrin matrix early in nerve regeneration, which eventually leads to capillary in-growth.[48] Therefore, porosity allows proper nutrient diffusion and the establishment of a blood supply while serving as a barrier to scar-forming fibroblasts. Many of the aforementioned fabrication techniques yield a conduit with interconnected pores, namely injection molding solvent evaporation,[12] uni-directional freezing,[19] and fiber spinning.[49] Other methods of creating pores involves mesh rolling,[50,51] the addition of salt[52] or sugars[53] that are then leached, and cutting holes into the lumen wall.[54] With salt or sugar leaching, both the pore diameter and porosity percentage may be controlled.[55] The benefits of porosity have been demonstrated in vitro, where increased axonal growth was correlated with higher degrees of porosity and nonporous conduits had decreased growth.[56] Kokai found that a porosity of at least 80% is generally desirable for the diffusion of glucose, oxygen, and protein, while porosity less than 50% limits the flux of nutrients across the single lumen conduits.[57] In addition to porosity, the permeability of a conduit design is also influenced directly by the hydrophilic properties of the scaffold.[58]

Biomechanics

The ideal nerve guide must have sufficient strength to resist permanent deformation, kinking, tearing, and yet maintain sutures while regeneration occurs (Fig. 9). The nerve guide also must have sufficient flexibility to accommodate some degree of motion, especially across a joint. However, a higher degree of flexibility may result in mechanical weakness and hinder axonal regeneration. Thus, a balance must be struck between the two properties. The ideal synthetic conduit must mirror the mechanical properties of a native

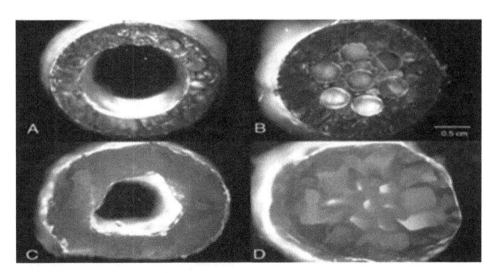

Fig. 9 Demonstration of *in vitro* swelling and degradation of 75:25 PLGA single channel and multichannel scaffolds by de Ruiter et al. (**A** and **B**) tubes at day 0 and dry. (**C** and **D**) after 12 weeks in PBS.
Source: From de Ruiter et al.,[12] with permission from Wiley InterScience.

nerve, and many researchers have calculated the Young's modulus of peripheral nerves. A value of 0.05 MPa in the longitudinal direction has been generally accepted as the appropriate modulus for the innate rat tibial nerve.[59] The load of human ulnar and median nerves was found to be 66–155 and 73–220 N, respectively.[60] Bender et al. also investigated the tensile strength and modulus of a PCL conduit, which demonstrated a decreasing tensile strength and modulus with increasing porosity.[14] Using cross-linked collagen-based conduits, Yao and co-workers compared the mechanical properties of a commercial conduit, a single-channeled, four-, and seven-channel membrane.[33] The transverse compressive force, tensile strength, and three-point bending of the various conduits were determined. With respect to each test parameter, the multichannel conduit outperformed the single lumen scaffolds, demonstrating that multichannel conduits show greater mechanical stiffness and strength. The increased mechanical properties of mutlichannel nerve guides may support an increased porosity thereby promoting diffusion and the infusion of growth factors within the conduit. In addition to the material comprising the conduit, other factors such as wall thickness, porosity, lumen diameters, and embedded cytokines can influence the physical integrity of the nerve guide.

Biocompatibility and Biodegradation

In order to support axonal regeneration, biocompatibility is essential. Williams defines biocompatibility as the ability to aid in neural regeneration through the promotion of appropriate cellular behaviors and *in vivo* signaling systems, without incurring a local or systemic inflammatory response and any long-term undesirable effects on neural tissue.[61] The principle of biocompatibility can be further stratified into blood compatibility, histocompatibility, and mechanical compatibility. Blood compatibility refers to the lack of hemolysis, coagulation, thrombus formation, or any destruction of blood constituents upon contact with the scaffold material. Histocompatibility requires that the scaffold material have no cytotoxicity, teratogenicity, or gene mutation on the tissue surrounding the conduit. Furthermore, native tissues should not elicit an immune rejection or have unanticipated corrosive effects upon the conduit. Lastly, mechanical compatibility was discussed in the preceding section as the endeavor to match the mechanical properties of artificial conduits with those of the native neural tissue.

As the underwhelming results of conduit fabrication with nondegrading materials such as silicone and polyethylene demonstrated, nerve guides must also possess a degree of *in vivo* biodegradability as the nerve regenerates. A lack of biodegradation also requires that an artificial conduit be a two-stage procedure, one operation for both implantation and explantation, as conduits can become toxic or constrictive over time as the nerve regenerates.[62]

As such, researchers have focused on the use of the biodegradable polymers listed in the fabrication section. Ideally, the degradation kinetics of a guide should match the rate of regeneration. Too rapid degradation may release toxic by-products and fail to support the axons, and too slow can lead to compression and a foreign body response. Degradation rate may also contribute to the swelling of scaffolds since by-products can be osmotically active.[12,63] As the conduit swells, the inner diameter of channels decreases and ultimately has the detrimental effect of occluding axonal growth. Therefore, it is of interest to construct a conduit with tunable degradation rates and minimal swelling *in vivo*. Manipulating both nerve tube dimensions and the biomaterial composition can lead to the optimization of both degradation rates and conduit swelling.[12,63]

DEVICE PERFORMANCE AND EVALUATION

In the last 30 years, the concept of the nerve conduit has evolved from a tool to investigate nerve regeneration to a device that is now being used clinically rather than autologous nerve graft repair in select instances. Clinical use of nerve conduits is limited to the repair of defects <3 cm in small-caliber peripheral sensory nerves.[64–66] However, the potential for extending clinical application to the repair of larger defects and larger mixed and/or motor nerves has rendered the development of an ideal nerve conduit appealing for entrepreneurs in the medical device industry.[67] Currently, several biomaterial-based nerve conduits are being marketed; the most successful to date include Neurotube (Synovis), Neurolac (Ascension), and NeuraGen (Integra). Yet, the basic design of these tubes remains a single hollow lumen in which the severed nerve stumps are inserted. The results of all these conduits remain modest at best and do not always appear to achieve a level of performance equivalent to autograft repair.[68] Many strategies have developed in an attempt to improve upon the performance of the nerve conduit in hopes of replacing the autologous nerve graft, and multichanneled devices have emerged as a popular area of research and development among researchers in the field of neural tissue engineering. Decellularized allografts also have great potential for enhancing clinical outcomes following severe peripheral nerve injury, with over 5000 Avance allografts (Axogen, Inc.) implanted to date worldwide in patients.[69]

Improved performance characteristics of the multichanneled conduit are related to the unique geometries achieved through the advanced fabrication techniques developed since the original introduction of the single lumen nerve conduit. Hadlock and co-workers were among the first to produce multichanneled nerve conduits using a novel foam-processing technique and further seeded them with Schwann cells; numerous studies have followed in an effort to improve upon their initial report.[10,70] Direct comparison of published studies detailing the fabrication and use of multichanneled nerve conduits proves difficult, as material design and microchannel architectures differ significantly from study to study. Furthermore, many researchers augment nerve conduits with biologically active or cellular components with the intent of improving the regenerative response as described above, which skews the results, thus rendering it challenging to determine the sole effects of the microchannel design.

Although it can be difficult to attribute observed responses solely to the conduit materials and their biomimetic architecture, key principles and design criteria can be derived from examination of published efforts. Prior studies have suggested the efficacy of regeneration through autografts and allografts may be due to the presence of aligned conduits in conjunction with retained neurotrophins.[71] Decellularization of nerve tissue, another popular strategy for nerve repair, also results in hollow, aligned conduits that support nerve regeneration.[72,73] It is also known that axons will track along fibers formed by extrusion or electrospinning in a fashion dependent on the diameter of the fibers.[74,75] Importantly, neural scaffolds consisting only of electrospun fibers are known to affect regenerative processes *in vitro* and have led to axonal bridging of large neural defects.[75,76] Studies in other areas of tissue engineering have repeatedly shown that well-defined, structured topographical cues such as grooves, ridges, pores, and nodes influence cell–substrate interaction by promoting cell adhesion, migration, proliferation, and can even lead to differentiation into new functional tissue.[77,78]

An important mechanism that appears to differentiate the multichannel nerve guide from its single lumen counterparts relates to understanding the mechanism of propagation of the "growth cone" of regenerating neurites through the conduit. Scott et al. nicely demonstrated how multichannel fibrin guides augmented the curvature and penetration observed in axonal growth cones at various timepoints in an *in vitro* embryonic chicken dorsal root ganglion (DRG) model of nerve regeneration.[79] DRG cells migrated into and through the multichanneled constructs, revealing a convex shaped growth cone compared to a concave shape for a conduit (Fig. 10). The authors of the study attribute this observation to the differing surface areas presented by the multi-channel conduits to the migrating DRG cells. In a single lumen channel, cellular migration can only proceed along the walls of the conduit and the entire central portion must wait and then catch up to the front of the concave regenerating cone. Conversely in the multichanneled conduit, the leading edge of the migratory growth cone appears to originate from the center to the conduit, resulting in a convex shape. Recent studies have also demonstrated the relative importance of fillers for the single lumen nerve guides, highlighting their ability to internally splint the conduit and boost cellular migration and speed of nerve regeneration.[80–84] Scott et al. further speculate that one advantage to multichanneled devices with small diameter conduits of <100 µm may be the ability to exclude infiltration of inflammatory cells (e.g., neutrophils, monocytes, and

105 µm Template

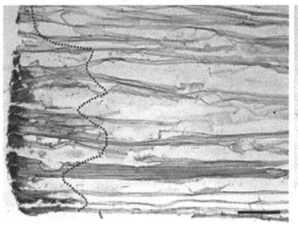

Random Porous Control

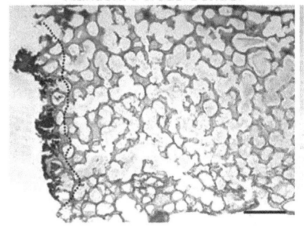

Fig. 10 Axon infiltration of 105 µm fibrin scaffold and porous control by embryonic chicken DRG cells after 1-day culture *in vitro*. Dotted lines have been added to highlight the progression of axon infiltration. Note the convex appearance in the channeled scaffold. Scale bar = 500 µm.
Source: From Scott et al.[79] © 2011, with permission from Elsevier.

macrophages), allowing for unimpeded infiltration and migration of neurites and axon precursors.[79]

In vitro Analysis of Multichanneled Conduits

Cellular assays play an important and pivotal role in the early evaluation of any nerve conduit-based technologies. Nerve conduits must first be proven nontoxic in cell culture conditions. Conduits must be able to support cell growth and proliferation by allowing for adequate surface attachment. Finally, conduits must be able to support cellular attachment migration by providing adequate surfaces and surface area, respectively. Cell-based assays can provide much of this information and provide proof of concept to many technologies prior to embarking upon animal studies. Numerous methods for testing and evaluation have been developed by many researchers; again, lack of standardized methods renders it difficult to compare results from existing studies.[85] A summary of studies detailing *in vitro* techniques for characterizing multichannel nerve guides can be found in Table 2.

Published and well-accepted studies examine nerve conduits or representative samples of the conduit materials in cell culture prior to initiating animal experimentation. In a study examining multichanneled biodegradable polymer composites, Bender et al. utilized rat cortical neurons and rat Schwann cells (SC) to examine neuron adhesion, neurite extension, neuron viability, and Schwann cell adhesion to PCL/collagen-based CultiSpher scaffolds.[14] To assess cortical neuron attachment to PCL as compared to PCL/CultiSpher composite materials, 40,6-diamidino-2-phenylindole dihydrochloride (DAPI) was used to stain nuclei, and fluorescein isothiocyanate (FITC)-phalloidin was used to counterstain the actin cytoskeleton of the cells. Neurite outgrowth was assessed by FITC-phalloidin staining of the actin cytoskeleton using fluorescent microscopy with digital image capture. To assess cell viability, the live/dead staining was used to evaluate the presence of live and dead cells on the composite surface. For adhesion studies, rat Schwann cells were plated on laminin-coated PCL, PCL/CultiSpher, and the control tissue culture polystyrene plates. Utilizing these methods the researchers were able to demonstrate adequate biocompatibility of materials and sufficient performance to warrant further investigation and support preclinical animal studies.

Similarly, Bozkurt et al. isolated Schwann cells from the sciatic nerve of terminally anesthetized GFP-positive rats, expanded the cells in culture, and seeded a Perimax nerve guide constructs.[16] After periods of incubation ranging from 16 to 24 hr allowed for adequate attachment, confocal laser scanning microscopy was used to evaluate the presence of GFP-positive Schwann cells without the need for tissue manipulation, such as tissue sectioning and immunohistochemistry (Fig. 11). Jeffries and Wang used a similar protocol to employing cultured SC to evaluate cellular infiltration into microchanneled guides with restricted migration barriers.[25] Hadlock and co-workers, the first to produce a multichanneled nerve conduits using a novel polyurethane foam-processing technique, chose to demonstrate cellular compatibility by seeding the conduits with Schwann cells isolated from neonatal rats and subsequently detected cellular viability with the MTT assay.[10] Also, Hsu and Ni utilized the glioma cell line BCRC-60046 (C6 cells) for demonstrating cellular alignment on grooved PLA substrates prior to implantation into rat sciatic nerve defects.[86]

Several *in vitro* studies involving multichannel nerve guides have also tried to mimic the structure and alignment of structures that occur in the central nervous system. Yu and Shoichet fabricated nerve guidance channels with longitudinally oriented tracks, and the adhesion and outgrowth of DRG neurons *in vitro* were shown to be increased on these scaffolds, especially in the presence of laminin (LN)-derived peptides.[34] In another *in vitro* guidance channel study, DRG explants were cultured on polypropylene filaments within a polyvinyl chloride hollow fiber membrane, and the effect of filament diameter on neurite outgrowth and SC migration was measured.[87] Fibers ranged from supracellular (100–500 μm) to cellular (30 μm) to subcellular (5 μm) in diameter, and an inverse relationship between filament diameter and both SC migration and neurite outgrowth was found. In addition, when fibers were uncoated, SCs migrated ahead of neurite

Table 2 Published studies detailing *in vitro* testing of multichanneled nerve conduits

Author	Material	Cell type	Animal	Measurement	References
Bender et al.	PCL, PCL/Cultispher	SC, cortical neuron	Rat	Absorbance, Immunohistochemistry	[14]
Bozkurt et al.	Porcine collagen	Dorsal root ganglion	Rat	Immunohistochemistry	[16,17]
Hadlock et al.	PLGA	SC	Rat	MTT assay	[10,70]
Moore et al.	PLGA	SC	Rat	Immunohistochemistry	[11]
He et al.	PLGA	Mesenchymal stem cells, SC	Rat	Immunohistochemistry	[13]
Scott et al.	Fibrin gel	DRG	Chick	Immunohistochemistry	[79]
Yu et al.	Hydrogel P(HEMA-*co*-AEMA)	DRG	Chick	Immunohistochemistry	[34]
Jeffries and Wang	PCL	SC	Rat	Immunohistochemistry; MTT assay	[25]
Wang et al.	Chitosan	Neuroblastoma	Mouse	Immunohistochemistry	[31]
Yao et al.	Collagen	DRG	Rat	Immunohistochemistry	[33]

Abbreviations: SC, Schwann cell; DRG, Dorsal root ganglion; MTT, 3-(4,5-dimethylthiazol-2-yl)-2,5-diphenyltetrazolium bromide.

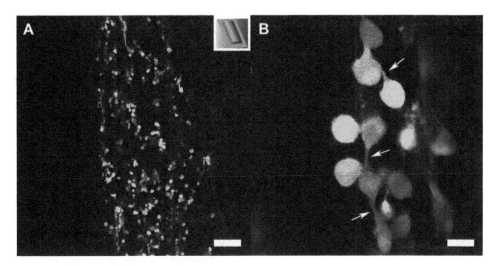

Fig. 11 Confocal laser scanning microscopy of GFP+ Schwann cells seeded into the Perimaix nerve guide. (**A**) Low magnification longitudinal image demonstrating the dense and near homogeneous seeding of SC into the nerve guide. (**B**) Higher magnification of (**A**) showing rounded SC bodies and processes extension (see white arrows) that follow the orientation of nerve guide's microstructure. Scale bars: **A** = 200 μm, **B** = 20 μm.
Source: From Bozkurt et al.[17] © 2007, with permission from Elsevier.

outgrowth, indicating that SCs may be establishing a pathway for neurite growth, but when fibers were coated with LN or fibronectin (FN), the neurites actually grew past the SCs, indicating that with permissive ECM support, neurite growth may be independent of SC migration. These results agree with previous work, in which poly-L-lactic acid (PLLA) microfilaments with a relatively large diameter of 375 μm guided neurite outgrowth and SC migration, and neurite outgrowth was significantly increased in the presence of LN, past the SC leading edge.[88] Wen and Tresco have noted that when DRGs were cultured on aligned fibers, there was likely an optimal size range for the filament diameter, in which filaments that were very large or very small relative to the size of the growth cone did not present discernible differences in different growth directions.[87]

Multichanneled nerve conduits are biomimetic devices and depending upon the fabrication technique and resulting channel size may sometimes require specialized techniques in order to gain meaningful results. For example, Scott et al. added 1% 5-fluoro-20-deoxyuridine (FDU) and uridine to isolated sciatic nerve cultures in order to suppress proliferation of non-neuronal cells, especially targeting Schwann cells.[79] Inhibiting Schwann cell proliferation allowed for assessment of axonal extension solely in response to the multichannel conduits diameters rather than in response to Schwann cells and other cell-based trophic cues. Specialized assays such as this allow researchers to begin to directly probe the impact of multichannel structure and material architecture on nerve regeneration in order to boost observed cellular responses prior to undertaking animal testing.

In vivo Analysis of Multichanneled Conduits

Biocompatibility and chemical safety profiles are studied prior to clinical application using both *in vitro* and *in vivo* studies. As previously discussed, *in vitro* testing is used principally as a first phase test for acute toxicity and cytocompatibility while minimizing use of animals. *In vitro* testing offers information regarding cytotoxicity, genotoxicity, cell proliferation, and differentiation, and can be more easily standardized and quantifiable than *in vivo* testing.[12,89] However, *in vitro* assays cannot establish tissue reaction to materials because systemic factors such as foreign body or immune system response, vascularization, oxygen and nutrient supply, and waste elimination are absent. For these reasons, animal models are essential for evaluating biocompatibility, tissue response, and mechanical function of any nerve conduit prior to clinical application. *In vivo* studies in diverse animal species permit the evaluation of material over long periods of time and under clinically relevant biological conditions. Although animal models might closely mimic the mechanical and physiological human clinical conditions, one must always consider that these models represent solely an approximation of the human response to pathologic factors.[39–41] Researchers must realize each animal model has distinct benefits and drawbacks when used in an experimental study of peripheral nerve injury. A summary of studies detailing *in vitro* techniques for characterizing multichannel nerve guides can be found in Table 3.

The rat sciatic model has been the most commonly used animal model for study of synthetic nerve scaffolds, including multichannel conduit designs (Fig. 12). Rats are economical, simple to handle and care for, very resistant to

Table 3 Published studies detailing *in vivo* testing of multichanneled nerve conduits

Author	Material	Animal	Nerve	Gap size	Methods	Follow-up	References
de Ruiter et al.	75:25 PLGA	Rat	Sciatic	8 mm	RT, CMAP, NM, MM	8 weeks = RT 12 weeks = CMAP, NM, MM	[94]
Hadlock et al.	85:15 PLGA	Rat	Sciatic	7 mm	Histology	6 weeks	[10]
Moore et al.	85:15 PLGA	Rat	Spinal cord	2 mm	Histology and immunohistochemistry	4 weeks	[11]
He et al.	72:25 PLGA	Rat	Spinal cord	2 mm	Histology and immunohistochemistry	8 weeks	[13]
Yao et al.	Collagen	Rat	Sciatic	5 mm	NM, CMAP, ankle motion, RT	6 weeks = NM 16 weeks = CMAP, ankle motion, RT	[33]
Koh et al.	PLLA	Rat	Sciatic	10 mm	Pain sensation, NCV, immunohistochemistry	Every 2 weeks up to week 12	[26]
Hsu and Ni	PLA	Rat	Sciatic	10 mm	Histology	4 and 6 weeks	[86]
Hu et al.	Collagen-Chitosan	Rat	Sciatic	15 mm	Histology and immunohistochemistry	4 and 12 weeks	[19]
George et al.	Ppy	Rat	Sciatic	10 mm	Histology and SEM	4 and 8 weeks	[23]

Abbreviations: RT, Retrograde tracing; CMAP, Compound muscle action potential; NM, Nerve morphometry; MM, Muscle morphometry; SEM, Scanning electron microscopy; NCV, Nerve conduction velocity.

surgical infections, and can be investigated in large groups. Furthermore, the model can provide information regarding electrophysiology, functional recovery, muscle and nerve morphology, and other routine assessments of nerve regeneration.[40,90,91] A major disadvantage is that relatively short gaps from 1 to 50 mm must be used, with the majority of studies using a gap of 10 mm or less. This gap is small compared with target human nerve lesions. Researchers must also be aware that nerve axotomy (i.e., simple cut without a resulting gap) in the rat will undergo complete recovery and this does not occur in human nerve injuries. The difference is further complicated by the fact that the rate of peripheral axonal regeneration is slower in humans than in rodents.[92,93]

Published and well-received studies animal studies on multichanneled nerve conduits focus almost exclusively on sciatic nerve injuries in the rat. Hadlock et al. evaluated nerve regeneration across 7 mm sciatic nerve gaps in Fisher rats.[10] At 6 weeks following repair, the percentage of neural tissue seen on cross-sectional area through conduits was similar to that of autograft. However, the mean axon diameter in the conduits was significantly larger than the autografts (3.7 vs. 2.3 μm, respectively). This led the authors to suggest that larger diameter myelinated fibers may give preference to faster nerve regeneration, which may lead to better functional outcomes through better recovery of motor and sensory control.

In 2010, a study by Yao et al. investigated the influence of channel number (either 1, 2, 4, or 7) on the axonal

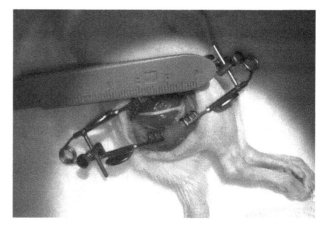

Fig. 12 Surgical exposure of the sciatic nerve in the Lewis rat. Exposure was obtained using a lateral approach, splitting the biceps femoris muscle which directly overlies the sciatic nerve.

regeneration in multichannel collagen conduits by implanting them in 1-cm rat sciatic nerve defects.[33] After 4 months, quantitative results of regeneration were evaluated with nerve morphometry and the accuracy of regeneration was assessed using retrograde tracing in the tibial and peroneal nerves. Researchers concluded that the fabricated two-channel and four-channel collagen conduits are superior with respect to axonal regeneration and limiting axonal dispersion compared to single lumen guides. A similar study by de Ruiter et al. examined the accuracy of motor axon regeneration across autografts and single-lumen or multichannel PLGA nerve tubes.[94] They employed simultaneous tracing of the tibial and peroneal nerves with fast blue and diamidino yellow 8 weeks after the repair of a 1-cm nerve gap in the rat sciatic nerve to determine the percentage of double-projecting motoneurons (Fig. 13). More motoneurons had double projections to both the tibial and peroneal nerve branches after single-lumen nerve tube repair (21.4%) than after autograft repair (5.9%); following multichannel nerve tube repair, this percentage was slightly

reduced (16.9%). Both of the aforementioned studies reveal that retrograde tracing techniques provide new insights into the process of regeneration across multichanneled nerve tubes, which appear to be able to limit axonal dispersion.

A 2010 study by Koh et al. introduced a novel nanofibrous-filled tube used to bridge a sciatic nerve gap in rats, with good functional and histological results.[95] The external conduit was comprised of PLLA fibers (250–1000 nm in diameter) with the outermost layers consisting of randomly oriented nanofibers to minimize conduit collapse while the

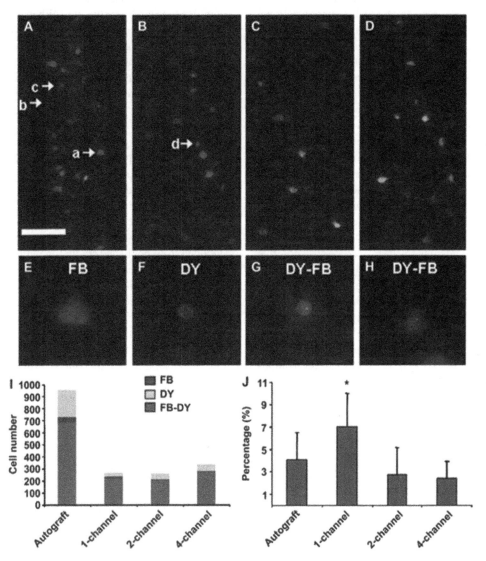

Fig. 13   Dispersion of regenerating axons across autografts and collagen based conduits by Yao et al. (A–D) Showing fluorescent microscopic images at 10× magnification and a DAPi filter. Longitudinal sections taken at anterior horn after (A) autograft, (B) one-channel, (C) two-channel, and (D) four-channel. (E-H) Fluorescent images at 20× magnification and with DAPi filter of (E) single labeled FB (cytoplasm and nucleus, arrow a), (F) single-labeled DY (nucleus and cytoplasm, arrow b), (G, H) Double-labeled DY-FB profile (nucleus and cytoplasm, G, arrow c; H, arrow d). (I) Results of simultaneous tracing showing the mean number of FB, DY, and DY-FB labeled profiles after autograft, one-, two-, and four-channel conduit repair at 16 weeks. (J) Results of simultaneous tracing demonstrating the percentage of double projections after autograft, one-, two-, and four-channel conduit repair at 16 weeks.
Source: Reprinted from Yao et al.[98] © 2010, with permission from Elsevier.

innermost consisting of longitudinally aligned nanofibers to guide axonal regrowth. Intraluminal guidance channels comprised of PLGA fibers (200–600 nm in diameter) were engineered to result in a faster degradation rate than PLLA fibers. NGF were added in an effort to enhance regeneration. Male Wistar rats were subjected to resection of 10 mm segments of sciatic nerve and subsequently repaired by conduit or autograft. Sensory recovery was evaluated using a hotplate at 56°C and motor recovery was assessed by weighing the gastrocnemius muscles. At 12 weeks, nerve conduction studies (NCS) were performed as well as histological analysis. The results of the study showed that animals that implanted with tubes containing intraluminal nanofibrous channels had faster and better functional recovery than those rats implanted with hollow tubes and autografts. However, the overall Schwann cell number and axonal concentrations were higher in the autograft group. Hsu and Ni demonstrated greater degrees of remyelination at 4 and 16 weeks in asymmetric microgrooved and/or microporous PLA peripheral nerve conduits.[86] Moreover, regenerated nerves in the conduits with surface microgrooves demonstrated the highest degree of myelination at 4 weeks and the greatest number of blood vessels at 6 weeks. The authors also performed walking track analysis that demonstrated that the PLA conduits with surface microgrooves displayed the highest degree of functional recovery. In addition to peripheral nerve repair, there have been studies examining the use of multichanneled conduit devices for spinal cord injury in the rat. Moore et al. describe multiple-channeled, biodegradable PLGA copolymer scaffolds and simultaneously investigate the effects on axon regeneration of scaffold architecture, transplanted cells, and locally delivered molecular agents.[11] The authors used female Sprague-Dawley rats and resected a 1.0 mm of spinal cord at level of T9 postlaminectomy. Conduits were treated with Matrigel alone or SC suspended in Matrigel; animals were followed for 1 month with histology used as the main endpoint of the study. Micro-channeled devices showed a larger centralized core of regenerating axons and capillaries surrounded by fibrous tissue relative to control devices. He et al. also reported the examination of Sprague-Dawley rats undergoing spinal cord injury by 2 mm transection at T9-T10 level following postlaminectomy.[13] The authors repaired the defect with 16 channel PLGA copolymer conduits fashioned by TIPS techniques interposing the injured spinal cord segments. The implantation of the PLGA conduit in the spinal cord showed that it had good biocompatibility, and no obvious inflammatory response was detected.

CONCLUSION

When a nerve is transected, axonal sprouts are no longer restrained by the basal lamina during regeneration, and growth along the original pathway becomes difficult. The lack of sufficient endoneurial guidance can ultimately hinder the ability of an axon to reinnervate its original target, and may be partially responsible for unsatisfactory functional recovery. Numerous studies have suggested multichannel nerve conduits can better support the proper segregation of fascicles and remove the dispersion of axonal sprouts compared to single lumen guides. Perhaps the end goal of multichannel nerve guides is to mirror the *in vivo* situation of neural regeneration, where many smaller channels form the bands of Büngner. The optimization of lumen diameter has yet to be defined in the literature, and the crafting of smaller channels requires more complex fabrication techniques. When comparing the multichannel conduits to single lumen guides, it is evident that the multichannel guide offers increased surface area for cellular attachment and the elution of embedded growth factors. Theoretically, the added mechanical stability of multichannel scaffolds can better support the establishment of the fibrin matrix across a longer gap than hollow guides. However, the ideal conduit design does not rely solely on increased surface area as the properties of permeability, flexibility, swelling, and degradation rates influence neural regeneration as well. While the eventual clinical utility of multichannel guides is promising, the superiority of increased channels compared single lumens still requires conclusive evidence.

It is currently challenging to compare the results obtained from published studies on multichanneled devices because different nerve gaps and evaluation techniques are used. Functional results in most studies are not included and physical properties were not determined for all conduits in great detail. The research on peripheral nerve regeneration has achieved considerable advances in a short time, but complete functional recovery over large gaps has remained elusive. From the start, the aims of artificial conduit design were simple to define: maintain the remaining axons after injury, establish a microenvironment suitable for neurite extension, and ensure the accuracy of axons as they approach their distal targets. Yet promoting neuronal outgrowth, in either the PNS or CNS, is an exceedingly intricate process, one that we do not completely understand and autografting remains as the treatment of choice for bridging nerve gaps. In light of these findings, the construction of an engineered conduit remains worthwhile. Autologous nerve grafting entails a more invasive surgical procedure and results in donor site morbidity as direct result of the nerve harvest. If a particular conduit design can yield similar functional outcomes as an autologous graft, the conduit can be utilized without the cost of neural sacrifice at one location for another. Current studies suggest there remains enormous potential for improvement of multichanneled nerve conduits for the repair of peripheral nerve gaps and demonstrates progress toward the goal of autograft equivalency.

REFERENCES

1. Siemionow, M.; Bozkurt, M.; Zor, F. Regeneration and repair of peripheral nerves with different biomaterials: Review. Microsurgery **2010**, *30* (7), 574–588.

2. Lundborg, G. A 25-year perspective of peripheral nerve surgery: Evolving neuroscientific concepts and clinical significance. J. Hand Surg. Am. **2000**, *25* (3), 391–414.

3. Ngeow, W.C. Scar less: A review of methods of scar reduction at sites of peripheral nerve repair. Oral Surg. Oral Med. Oral Pathol. Oral Radiol. Endod. **2010**, *109* (3), 357–366.

4. Dow, D.E.; Cederna, P.S.; Hassett, C.A.; Kostrominova, T.Y.; Faulkner, J.A.; Dennis, R.G. Number of contractions to maintain mass and force of a denervated rat muscle. Muscle Nerve **2004**, *30* (1), 77–86.

5. Battal, M.N.; Hata, Y. A review on the history of end-to-side neurorrhaphy. Plast. Reconstr. Surg. **1997**, *99* (7), 2110–2111.

6. Ichihara, S.; Inada, Y.; Nakamura, T. Artificial nerve tubes and their application for repair of peripheral nerve injury: An update of current concepts. Injury **2008**, *39* (Suppl 4), 29–39.

7. Myckatyn, T.; MacKinnon, S. A review of research endeavors to optimize peripheral nerve reconstruction. Neurol. Res. **2004**, *26* (2), 124–138.

8. Nichols, C.M.; Brenner, M.J.; Fox, I.K.; Tung, T.H.; Hunter, D.A.; Rickman, S.R.; Mackinnon, S.E. Effects of motor versus sensory nerve grafts on peripheral nerve regeneration. Exp. Neurol. **2004**, *190* (2), 347–355.

9. Ghaznavi, A.M.; Kokai, L.E.; Lovett, M.L.; Kaplan, D.L.; Marra, K.G. Silk fibroin conduits: A cellular and functional assessment of peripheral nerve repair. Ann. Plast. Surg. **2011**, *66* (3), 273–279.

10. Hadlock, T.; Sundback, C.; Hunter, D.; Cheney, M.; Vacanti, J.P. A polymer foam conduit seeded with Schwann cells promotes guided peripheral nerve regeneration. Tissue Eng. **2000**, *6* (2), 119–127.

11. Moore, M.J.; Friedman, J.A.; Lewellyn, E.B.; Mantila, S.M.; Krych, A.J.; Ameenuddin, S.; Knight, A.M.; Lu, L.; Currier, B.L.; Spinner, R.J.; Marsh, R.W.; Windebank, A.J.; Yaszemski, M.J. Multiple-channel scaffolds to promote spinal cord axon regeneration. Biomaterials **2006**, *27* (3), 419–429.

12. de Ruiter, G.C.; Onyeneho, I.A.; Liang, E.T.; Moore, M.J.; Knight, A.M.; Malessy, M.J.; Spinner, R.J.; Lu, L.; Currier, B.L.; Yaszemski, M.J.; Windebank, A.J. Methods for *in vitro* characterization of multichannel nerve tubes. J. Biomed. Mater. Res. A **2008**, *84* (3), 643–651.

13. He, L.; Zhang, Y.; Zeng, C.; Ngiam, M.; Liao, S.; Quan, D.; Zeng, Y.; Lu, J.; Ramakrishna, S. Manufacture of PLGA multiple-channel conduits with precise hierarchical pore architectures and *in vitro*/vivo evaluation for spinal cord injury. Tissue Eng. Part C Methods **2009**, *15* (2), 243–255.

14. Bender, M.D.; Bennett, J.M.; Waddell, R.L.; Doctor, J.S.; Marra, K.G. Multi-channeled biodegradable polymer/Culti-Spher composite nerve guides. Biomaterials **2004**, *25* (7–8), 1269–1278.

15. Huang, Y.C.; Huang, Y.Y.; Huang, C.C.; Liu, H.C. Manufacture of porous polymer nerve conduits through a lyophilizing and wire-heating process. J. Biomed. Mater. Res. B Appl. Biomater. **2005**, *74* (1), 659–664.

16. Bozkurt, A.; Lassner, F.; O'Dey, D.; Deumens, R.; Bocker, A.; Schwendt, T.; Janzen, C.; Suschek, C.V.; Tolba, R.; Kobayashi, E.; Sellhaus, B.; Tholl, S.; Eummelen, L.; Schugner, F.; Damink, L.O.; Weis, J.; Brook, G.A.; Pallua, N. The role of microstructured and interconnected pore channels in a collagen-based nerve guide on axonal regeneration in peripheral nerves. Biomaterials **2012**, *33* (5), 1363–1375.

17. Bozkurt, A.; Brook, G.A.; Moellers, S.; Lassner, F.; Sellhaus, B.; Weis, J.; Woeltje, M.; Tank, J.; Beckmann, C.; Fuchs, P. *In vitro* assessment of axonal growth using dorsal root ganglia explants in a novel three-dimensional collagen matrix. Tissue Eng. **2007**, *13* (12), 2971–2979.

18. Schoof, H.; Apel, J.; Heschel, I.; Rau, G. Control of pore structure and size in freeze-dried collagen sponges. J. Biomed. Mater. Res. **2001**, *58* (4), 352–357.

19. Hu, X.; Huang, J.; Ye, Z.; Xia, L.; Li, M.; Lv, B.; Shen, X.; Luo, Z. A novel scaffold with longitudinally oriented microchannels promotes peripheral nerve regeneration. Tissue Eng. A **2009**, *15* (11), 3297–3308.

20. Arcaute, K.; Mann, B.K.; Wicker, R.B. Fabrication of off-the-shelf multilumen poly(ethylene glycol) nerve guidance conduits using stereolithography. Tissue Eng. Part C Methods **2010**, *17* (1), 27–38.

21. Arcaute, K.; Mann, B.K.; Wicker, R.B. Stereolithography of three-dimensional bioactive poly(ethylene glycol) constructs with encapsulated cells. Ann. Biomed. Eng. **2006**, *34* (9), 1429–1441.

22. Arcaute, K.; Mann, B.K.; Wicker, R.B. Practical use of hydrogels in stereolithography for tissue engineering applications. In *Stereolithography*; Springer: New York, 2011; 299–331.

23. George, P.M.; Saigal, R.; Lawlor, M.W.; Moore, M.J.; LaVan, D.A.; Marini, R.P.; Selig, M.; Makhni, M.; Burdick, J.A.; Langer, R.; Kohane, D.S. Three-dimensional conductive constructs for nerve regeneration. J. Biomed. Mater. Res. A **2009**, *91* (2), 519–527.

24. Biazar, E.; Khorasani, M.T.; Montazeri, N.; Pourshamsian, K.; Daliri, M.; Rezaei, M.; Jabarvand, M.; Khoshzaban, A.; Heidari, S.; Jafarpour, M.; Roviemiab, Z. Types of neural guides and using nanotechnology for peripheral nerve reconstruction. Int. J. Nanomedicine. **2010**, *5*, 839–852.

25. Jeffries, E.M.; Wang, Y. Biomimetic micropatterned multichannel nerve guides by templated electrospinning. Biotechnol. Bioeng. **2012**, *109* (6), 1571–1582.

26. Koh, H.; Yong, T.; Chan, C.; Ramakrishna, S. Enhancement of neurite outgrowth using nano-structured scaffolds coupled with laminin. Biomaterials **2008**, *29* (26), 3574–3582.

27. Pabari, A.; Yang, S.Y.; Seifalian, A.M.; Mosahebi, A. Modern surgical management of peripheral nerve gap. J. Plast. Reconstr. Aesthet. Surg. **2010**, *63* (12), 1941–1948.

28. Pabari, A.; Sedaghati, T.; Yang, S.; Mosahebi, A.; Seifalian, A. Development of a new multichannel biodegradable conduit for peripheral nerve regeneration using nanocomposite polymer. Proceedings of the International Symposium on Peripheral Nerve Repair and Regeneration. Turin, Italy, December 4–5, 2009.

29. Chen, J.-H.; Yang, J.-D.; Shen, H.-H.; Hsieh, Y.-L.; Tsai, B.-H.; Yang, C.-L.; Liu, M.-J. Multi-channel Bioresorbable Nerve Regeneration Conduit and Process for Preparing the Same. US 20030176876, September 18, 2003.

Nanoparticles—Nerve

30. Mahoney, M.J.; Chen, R.R.; Tan, J.; Saltzman, W.M. The influence of microchannels on neurite growth and architecture. Biomaterials **2005**, *26* (7), 771–778.

31. Wang, S.; Lu, L.; Kempen, D.H.; de Ruiter, G.; Nesbitt, J.J.; Gruetzmacher, J.A.; Knight, A.M.; Currier, B.; Hefferan, T.; Windebank, A.J. A Novel Injectable Polymeric Biomaterial Poly(propylene fumarate-co-caprolactone) with Controllable Properties for Bone and Nerve Regenerations. In *Proceedings of AIChE annual meeting*; AIChE headquarters: New York, 2006.

32. Xie, F.; Li, Q.F.; Gu, B.; Liu, K.; Shen, G.X. *In vitro* and *in vivo* evaluation of a biodegradable chitosan–PLA composite peripheral nerve guide conduit material. Microsurgery **2008**, *28* (6), 471–479.

33. Yao, L.; Billiar, K.L.; Windebank, A.J.; Pandit, A. Multichanneled collagen conduits for peripheral nerve regeneration: Design, fabrication, and characterization. Tissue Eng. Part C Methods **2010**, *16* (6), 1585–1596.

34. Yu, T.T.; Shoichet, M.S. Guided cell adhesion and outgrowth in peptide-modified channels for neural tissue engineering. Biomaterials **2005**, *26* (13), 1507–1514.

35. Flynn, L.; Dalton, P.D.; Shoichet, M.S. Fiber templating of poly(2-hydroxyethyl methacrylate) for neural tissue engineering. Biomaterials **2003**, *24* (23), 4265–4272.

36. Romero-Ortega, M.; Galvan-Garcia, P. A Biomimetic Synthetic Nerve Implant. US 20070100358, May 3, 2006.

37. Braga-Silva, J. The use of silicone tubing in the late repair of the median and ulnar nerves in the forearm. J. Hand Surg. **1999**, *24* (6), 703–706.

38. Merle, M.; Lee Dellon, A.; Campbell, J.N.; Chang, P.S. Complications from silicon-polymer intubulation of nerves. Microsurgery **1989**, *10* (2), 130–133.

39. Schimandle, J.H.; Boden, S.D. Spine update. The use of animal models to study spinal fusion. Spine, **1994**, *19* (17), 1998–2006.

40. Talac, R.; Friedman, J.; Moore, M.; Lu, L.; Jabbari, E.; Windebank, A.; Currier, B.; Yaszemski, M. Animal models of spinal cord injury for evaluation of tissue engineering treatment strategies. Biomaterials **2004**, *25* (9), 1505–1510.

41. Hazzard, D.G.; Bronson, R.T.; McClearn, G.E.; Strong, R. Selection of an appropriate animal model to study aging processes with special emphasis on the use of rat strains. J. Gerontol. **1992**, *47* (3), B63–B64.

42. Gu, X.; Ding, F.; Yang, Y.; Liu, J. Construction of tissue engineered nerve grafts and their application in peripheral nerve regeneration. Prog. Neurobiol. **2011**, *93* (2), 204–230.

43. Chang, C.-J.; Hsu, S.-H. The effect of high outflow permeability in asymmetric poly (DL-lactic acid-*co*-glycolic acid) conduits for peripheral nerve regeneration. Biomaterials **2006**, *27* (7), 1035–1042.

44. Yang, Y.; De Laporte, L.; Rives, C.B.; Jang, J.-H.; Lin, W.-C.; Shull, K.R.; Shea, L.D. Neurotrophin releasing single and multiple lumen nerve conduits. J. Control. Release **2005**, *104* (3), 433–446.

45. Wang, H.; Lineaweaver, W.C. Nerve conduits for nerve reconstruction. Oper. Tech. Plast. Reconstr. Surg. **2002**, *9* (2), 59–66.

46. She, Z.; Zhang, B.; Jin, C.; Feng, Q.; Xu, Y. Preparation and *in vitro* degradation of porous three-dimensional silk fibroin/chitosan scaffold. Polym. Degrad. Stabil. **2008**, *93* (7), 1316–1322.

47. Maquet, V.; Martin, D.; Malgrange, B.; Franzen, R.; Schoenen, J.; Moonen, G.; Jérôme, R. Peripheral nerve regeneration using bioresorbable macroporous polylactide scaffolds. J. Biomed. Mater. Res. **2000**, *52* (4), 639–651.

48. Zhao, Q.; Dahlin, L.B.; Kanje, M.; Lundborg, G. Repair of the transected rat sciatic nerve: Matrix formation within implanted silicone tubes. Restor. Neurol. Neurosci. **1993**, *5* (3), 197–204.

49. Aebischer, P.; Guénard, V.; Valentini, R.F. The morphology of regenerating peripheral nerves is modulated by the surface microgeometry of polymeric guidance channels. Brain Res. **1990**, *531* (1), 211–218.

50. Dellon, A.L.; Mackinnon, S.E. An alternative to the classical nerve graft for the management of the short nerve gap. Plast. Reconstr. Surg. **1988**, *82* (5), 849–856.

51. Molander, H.; Olsson, Y.; Engkvist, O.; Bowald, S.; Eriksson, I. Regeneration of peripheral nerve through a polyglactin tube. Muscle Nerve **1982**, *5* (1), 54–57.

52. Widmer, M.S.; Gupta, P.K.; Lu, L.; Meszlenyi, R.K.; Evans, G.R.; Brandt, K.; Savel, T.; Gurlek, A.; Patrick, C.W., Jr.; Mikos, A.G. Manufacture of porous biodegradable polymer conduits by an extrusion process for guided tissue regeneration. Biomaterials **1998**, *19* (21), 1945–1955.

53. Rodriguez, F.J.; Gomez, N.; Perego, G.; Navarro, X. Highly permeable polylactide-caprolactone nerve guides enhance peripheral nerve regeneration through long gaps. Biomaterials **1999**, *20* (16), 1489–1500.

54. Jenq, C.B.; Coggeshall, R.E. Nerve regeneration through holey silicone tubes. Brain Res. **1985**, *361* (1–2), 233–241.

55. Rutkowski, G.E.; Heath, C.A. Development of a bioartificial nerve graft. I. Design based on a reaction-diffusion model. Biotechnol. Prog. **2002**, *18* (2), 362–372.

56. Rutkowski, G.E.; Heath, C.A. Development of a bioartificial nerve graft. II. Nerve regeneration *in vitro*. Biotechnol. Prog. **2002**, *18* (2), 373–379.

57. Kokai, L.E. Controlled delivery systems for neuronal tissue engineering, University of Pittsburgh, 2010.

58. Busscher, H.; Van Pelt, A.; De Jong, H.; Arends, J. Effect of spreading pressure on surface free energy determinations by means of contact angle measurements. J. Colloid Interface Sci. **1983**, *95* (1), 23–27.

59. Kwan, M.K.; Wall, E.J.; Massie, J.; Garfin, S.R. Strain, stress and stretch of peripheral nerve. Rabbit experiments *in vitro* and *in vivo*. Acta Orthop. Scand. **1992**, *63* (3), 267–272.

60. Sunderland, S.S. *Nerves and Nerve Injuries*, 2nd Ed.; Churchill Livingstone, London, 1978.

61. Williams, D.F. *The Williams Dictionary of Biomaterials*; Liverpool University Press: Liverpool, England, UK, 1999.

62. Deng, M.; Chen, G.; Burkley, D.; Zhou, J.; Jamiolkowski, D.; Xu, Y.; Vetrecin, R. A study on *in vitro* degradation behavior of a poly(glycolide-*co*-L-lactide) monofilament. Acta Biomater. **2008**, *4* (5), 1382–1391.

63. Den Dunnen, W.; Van der Lei, B.; Robinson, P.; Holwerda, A.; Pennings, A.; Schakenraad, J. Biological performance of a degradable poly(lactic acid-ε-caprolactone) nerve guide: Influence of tube dimensions. J. Biomed. Mater. Res. **1995**, *29* (6), 757–766.

64. Bertleff, M.J.; Meek, M.F.; Nicolai, J.P. A prospective clinical evaluation of biodegradable neurolac nerve guides for sensory nerve repair in the hand. J. Hand Surg. Am. **2005**, *30* (3), 513–518.

65. Schlosshauer, B.; Dreesmann, L.; Schaller, H.E.; Sinis, N. Synthetic nerve guide implants in humans: A comprehensive survey. Neurosurgery 2006, 59 (4), 740–748.

66. Weber, R.A.; Breidenbach, W.C.; Brown, R.E.; Jabaley, M.E.; Mass, D.P. A randomized prospective study of polyglycolic acid conduits for digital nerve reconstruction in humans. Plast. Reconstr. Surg. 2000, 106 (5), 1036–1045.

67. Rosson, G.D.; Williams, E.H.; Dellon, A.L. Motor nerve regeneration across a conduit. Microsurgery 2009, 29 (2), 107–114.

68. de Ruiter, G.C.; Malessy, M.J.; Yaszemski, M.J.; Windebank, A.J.; Spinner, R.J. Designing ideal conduits for peripheral nerve repair. Neurosurg. Focus 2009, 26 (2), E5.

69. Szynkaruk, M.; Kemp, S.W.; Wood, M.D.; Gordon, T.; Borschel, G.H. Experimental and clinical evidence for use of decellularized nerve allografts in peripheral nerve gap reconstruction. Tissue Eng. B Rev. 2013, 19 (1), 83–96.

70. Sundback, C.; Hadlock, T.; Cheney, M.; Vacanti, J. Manufacture of porous polymer nerve conduits by a novel low-pressure injection molding process. Biomaterials 2003, 24 (5), 819–830.

71. Nguyen, Q.T.; Sanes, J.R.; Lichtman, J.W. Pre-existing pathways promote precise projection patterns. Nat. Neurosci. 2002, 5 (9), 861–867.

72. Hudson, T.W.; Zawko, S.; Deister, C.; Lundy, S.; Hu, C.Y.; Lee, K.; Schmidt, C.E. Optimized acellular nerve graft is immunologically tolerated and supports regeneration. Tissue Eng. 2004, 10 (11–12), 1641–1651.

73. Hudson, T.W.; Liu, S.Y.; Schmidt, C.E. Engineering an improved acellular nerve graft via optimized chemical processing. Tissue Eng. 2004, 10 (9–10), 1346–1358.

74. Wen, X.; Tresco, P.A. Fabrication and characterization of permeable degradable poly(DL-lactide-co-glycolide) (PLGA) hollow fiber phase inversion membranes for use as nerve tract guidance channels. Biomaterials 2006, 27 (20), 3800–3809.

75. Wang, H.B.; Mullins, M.E.; Cregg, J.M.; McCarthy, C.W.; Gilbert, R.J. Varying the diameter of aligned electrospun fibers alters neurite outgrowth and Schwann cell migration. Acta Biomater. 2010, 6 (8), 2970–2978.

76. Hoffman-Kim, D.; Mitchel, J.A.; Bellamkonda, R.V. Topography, cell response, and nerve regeneration. Annu. Rev. Biomed. Eng. 2010, 12, 203–231.

77. Subramanian, A.; Krishnan, U.M.; Sethuraman, S. Development of biomaterial scaffold for nerve tissue engineering: Biomaterial mediated neural regeneration. J. Biomed. Sci. 2009, 16 (1), 108.

78. Yim, E.K.; Reano, R.M.; Pang, S.W.; Yee, A.F.; Chen, C.S.; Leong, K.W. Nanopattern-induced changes in morphology and motility of smooth muscle cells. Biomaterials 2005, 26 (26), 5405–5413.

79. Scott, J.B.; Afshari, M.; Kotek, R.; Saul, J.M. The promotion of axon extension in vitro using polymer-templated fibrin scaffolds. Biomaterials 2011, 32 (21), 4830–4839.

80. Lin, Y.-C.; Ramadan, M.; Van Dyke, M.; Kokai, L.E.; Philips, B.J.; Rubin, J.P.; Marra, K.G. Keratin gel filler for peripheral nerve repair in a rodent sciatic nerve injury model. Plast. Reconstr. Surg. 2012, 129 (1), 67–78.

81. Chen, M.B.; Zhang, F.; Lineaweaver, W.C. Luminal fillers in nerve conduits for peripheral nerve repair. Ann. Plast. Surg. 2006, 57 (4), 462–471.

82. Hill, P.S.; Apel, P.J.; Barnwell, J.; Smith, T.; Koman, L.A.; Atala, A.; Van Dyke, M. Repair of peripheral nerve defects in rabbits using keratin hydrogel scaffolds. Tissue Eng. A 2011, 17 (11–12), 1499–1505.

83. Jiang, X.; Lim, S.H.; Mao, H.-Q.; Chew, S.Y. Current applications and future perspectives of artificial nerve conduits. Exp. Neurol. 2010, 223 (1), 86–101.

84. Pace, L.A.; Plate, J.F.; Smith, T.L.; Van Dyke, M.E. The effect of human hair keratin hydrogel on early cellular response to sciatic nerve injury in a rat model. Biomaterials 2013, 34 (24), 5907–5914.

85. Angius, D.; Wang, H.; Spinner, R.J.; Gutierrez-Cotto, Y.; Yaszemski, M.J.; Windebank, A.J. A systematic review of animal models used to study nerve regeneration intissue-engineered scaffolds. Biomaterials 2012, 33 (32), 8034–8039.

86. Hsu, S.-H.; Ni, H.-C. Fabrication of the microgrooved/microporous polylactide substrates as peripheral nerve conduits and in vivo evaluation. Tissue Eng. A 2008, 15 (6), 1381–1390.

87. Wen, X.; Tresco, P.A. Effect of filament diameter and extracellular matrix molecule precoating on neurite outgrowth and Schwann cell behavior on multifilament entubulation bridging device in vitro. J. Biomed. Mater. Res. A 2006, 76 (3), 626–637.

88. Rangappa, N.; Romero, A.; Nelson, K.D.; Eberhart, R.C.; Smith, G.M. Laminin-coated poly (L-lactide) filaments induce robust neurite growth while providing directional orientation. J. Biomed. Mater. Res. 2000, 51 (4), 625–634.

89. Hanks, C.T.; Wataha, J.C.; Sun, Z. In vitro models of biocompatibility: A review. Dent. Mater. 1996, 12 (3), 186–193.

90. IJkema-Paassen, J.; Jansen, K.; Gramsbergen, A.; Meek, M. Transection of peripheral nerves, bridging strategies and effect evaluation. Biomaterials 2004, 25 (9), 1583–1592.

91. Vleggeert-Lankamp, C.L. The role of evaluation methods in the assessment of peripheral nerve regeneration through synthetic conduits: A systematic review. J. Neurosurg. 2007, 107 (6), 1168–1189.

92. Buchthal, F.; Kühl, V. Nerve conduction, tactile sensibility, and the electromyogram after suture or compression of peripheral nerve: A longitudinal study in man. J. Neurol. Neurosurg. Psychiatry 1979, 42 (5), 436–451.

93. Dolenc, V.; Janko, M. Nerve regeneration following primary repair. Acta Neurochir. 1976, 34 (1–4), 223–234.

94. de Ruiter, G.C.; Spinner, R.J.; Malessy, M.J.; Moore, M.J.; Sorenson, E.J.; Currier, B.L.; Yaszemski, M.J.; Windebank, A.J. Accuracy of motor axon regeneration across autograft, single-lumen, and multichannel poly(lactic-co-glycolic acid) nerve tubes. Neurosurgery 2008, 63 (1), 144–155.

95. Koh, H.; Yong, T.; Teo, W.; Chan, C.; Puhaindran, M.; Tan, T.; Lim, A.; Lim, B.; Ramakrishna, S. In vivo study of novel nanofibrous intra-luminal guidance channels to promote nerve regeneration. J. Neural Eng. 2010, 7 (4), 1–14.

96. Siemionow, M.; Brzezicki, G. Current techniques and concepts in peripheral nerve repair. Int. Rev. Neurobiol. **2009**, *87*, 141–172.

97. Wang, A.; Ao, Q.; Cao, W.; Yu, M.; He, Q.; Kong, L.; Zhang, L.; Gong, Y.; Zhang, X. Porous chitosan tubular scaffolds with knitted outer wall and controllable inner structure for nerve tissue engineering. J. Biomed. Mater. Res. A **2006**, *79* (1), 36–46, doi:10.1002/jbm.a.30683.

98. Yao, L.; de Ruiter, G.C.; Wang, H.; Knight, A.M.; Spinner, R.J.; Yaszemski, M.J.; Windebank, A.J; Pandit, A. Controlling dispersion of axonal regeneration using a multichannel collagen nerve conduit. Biomaterials **2010**, *31* (22), 5789–5797.

Neural Tissue Engineering: Polymers for

Ashok Kumar
Tanushree Vishnoi
Department of Biological Sciences and Bioengineering, Indian Institute of Technology, Kanpur, India

Abstract
Injury to the nervous system leads to a cascade of events that eventually inhibits regeneration. The repair mechanism involves a lot of challenge and many attempts have been made to regenerate the damaged nerves by various conventional surgical methods like the use of the direct end-to-end repair method as well as the employment of natural nerve grafts. However, longer gaps generated in severed nerve or lesion formations, leading to significant tissue damage, demanded better therapeutic approaches. Recent advances have paved the way for the use of biodegradable three-dimensional polymeric scaffolds that provide support and guidance to the seeded or endogenous cells at the injury site. These polymeric scaffolds apart from being biocompatible and porous also mimicked the extracellular microenvironment of neural tissue, resulting in enhanced recovery. These polymers can also be used as delivery vehicles for bioactive molecules, cells, etc. This contribution describes the target areas for neuroengineering strategies, the limitations in their inbuilt recovery mechanism, and the different formats of the polymeric scaffolds used for improved neural regeneration.

INTRODUCTION

"Activity of the nervous system improves the capacity for activity, just as exercising a muscle makes it stronger," said Dr. Ralph Gerard, quite rightly so. Irreparable damage to the nervous system leads to its functional loss in both the central nervous system (CNS) and the peripheral nervous system (PNS) affecting all the physiological activities of life. The inhibitory microenvironment created at the site of injury (due to trauma, accidents, stroke, etc.) negatively influences the neural regeneration at that site henceforth. It has been estimated that only 50% of patients attain useful nerve function with the present nerve repair techniques. Moreover 265,000 persons in United States alone suffer from spinal cord injury as per 2010 reports, as a result of which they are unable to lead a normal life and have either been bed ridden/restricted days, leading to expenses of $95,000 for treatment and $14,135/year for recurring costs. Therefore there has been an enormous uproar in this field over the past few years and shift from the use of autografts, the gold standard method, and the use of biomaterials for improved neural regeneration has been observed.

Defects of the nervous system within the critical size are capable of regenerating and growing on its own. However, the loss of tropic and cellular support in the case of peripheral nerve regeneration along with the presence of inhibitory microenvironment in the CNS poses a problem for the defects exceeding the critical size. Moreover, transplantation of cells (stem cells, hippocampal, oligodendrocytes, etc.) at the site of injury or disease has shown limited success mainly because of the migration and loss of the cells from the implanted site. Direct exposure of the cells to this inhibitory microenvironment leads to their death. In order to avoid this scaffolds have been used as a three-dimensional (3-D) supporting matrix for cells, which also allows the immigration of the endogenous cells toward the matrix for better healing. Furthermore, the restricted microenvironment allows better cell growth and proliferation. Various methods like electrospinning, solvent extraction method, sol–gel method, freeze-drying, cryogelation, etc., have been used for the synthesis of these scaffolds. Thus, these bioengineered constructs both in the presence or absence of different type of cells have been used for neural regeneration with varying degree of achievement. Therefore, in order to achieve optimal success, a combination of these scaffolds with various growth factors, growth promoting and electrical cues have been integrated. Therefore, parameters that have been outlined in order to synthesize an optimum scaffold for neural regeneration are: biocompatibility with the host tissue, an optimum balance between degradation and regeneration (biodegradable), porous nature to allow diffusion of gases and supporting cells, micro- and nano-structures to guide the axons, ability to deliver bioactive molecules in a controlled manner, oriented matrix to support cell migration and axonal growth, electrical activities for cell–cell signaling, and functionality. In this entry, contribution the main aim is to highlight the various approaches and the advances made using polymeric biomaterials in the field of neural regeneration.

Concise Encyclopedia of Biomedical Polymers and Polymeric Biomaterials DOI: 10.1081/E-EBPPC-120050547

Neural—Rapid Prototyping

NERVOUS SYSTEM

Nervous system consists of CNS and PNS wherein the CNS comprises brain and spinal cord and the PNS sensory neurons, which receive information from both the external and internal environment and transfer it to the CNS where it is processed and further relocated to the effecter cells, like muscles and glands (Fig. 1).[1] Neurons and glial cells are the two primary cells of the nervous system. Neurons, the structural unit of the nervous system, are involved in the transmission of the nerve signals whereas the glial cells referred to as the supporting cells are in contact with the neurons and assist them in their function.[2]

Bundled sensory and motor axons of the PNS covered by the insulating myelin sheath and layers of connective tissue are known as the fascicle. Group of fascicle enclosed by the epineurium along with the supporting connective tissue forms the nerve trunk. Schwann cell is the supporting cell of the PNS and is involved in myelin sheath synthesis.[3]

The brain and spinal cord of the CNS is made up of the gray and the white matter in which the cell bodies form the gray matter and the axon bundles with the myelin sheath comprises the white matter. In brain the gray matter is at the surface and the white matter inside whereas in the spinal cord it is reversed. Oligodendrocytes and astrocytes are the supporting cells of the CNS and the CNS–PNS transition zone is clearly defined by the separation of these glia cells.[4]

TARGET REGIONS FOR NEUROENGINEERING STRATEGIES

Peripheral Nerve Injury

Unlike glia cells that have some capability to divide, neurons cannot undergo mitosis and cell division.[5] They can only regenerate a cut portion or shoot a new process under specific conditions. Peripheral nerve injury has been classified by Campbell into five types based on the site of injury (neuropathy).[6]

Type I: It is mild form of injury wherein the nerve continuity is not lost. It is most likely due to nerve compression or ischemia and is completely reversible (neuropraxia).

Type II: Complete transaction of the axon along with its myelin sheath except its stromal layers (axonomtesis).

Type III/IV/V: Complete transaction of the nerve along with the surrounding connective tissue layers: endoneurium, perineurium, and epineurium leading to type III, IV, and V injuries.

Traumatic Brain and Spinal Injury

Traumatic brain injury (TBI) and spinal injury (TSI) can be associated with physical injuries or posttraumatic stress affecting motor, cognitive functions.[7] It leads to cascade of molecular events and eventually the neuronal cell death depending on the anatomical location, mode, and severity of the injury.[8] The symptoms, depending on the intensity, range from headaches to paralysis and can be classified into mild, moderate, or severe form.[9] Moreover, spinal cord injury has been ranked recently from A to E on American Spinal Injury Association scale, wherein complete spinal cord injury results in lost sensation as well as voluntary mobility below the injury site and is ranked as "A" and in incomplete spinal cord injury there is still some motor and sensory activity (B to E).[10]

Neurodegenerative Diseases

Neurodegenerative diseases are characterized by progressive neural system dysfunction. Parkinson's disease (PD), Alzheimer's disease (AD), etc. are few examples of it. Protein aggregates and inclusion body formation are the characteristic features of these neurodegenerative diseases.

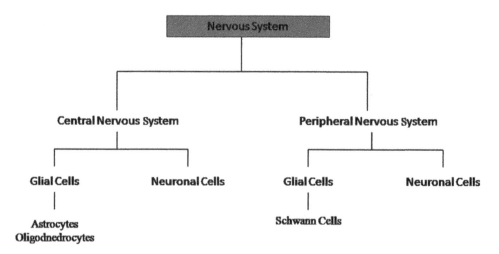

Fig. 1 Nervous system classification based on its cells.

Neural—Rapid Prototyping

These aggregates referred to as amyloid are group of misfolded proteins in β-sheet conformation, which are hereditary and sporadic in nature.[11–13]

Limitations of the Peripheral Nerve Injury

After nerve injury in the PNS, the distal segment undergoes Wallerian degeneration and the cytoskeleton begins to breakdown. Monocytes, macrophages, and proliferating Schwann cells clear the debris from the site and allow the movement of the axon from proximal to distal site if the gap is small.[14,15] Cytokines and neurotrophins secreted by the above cells along with the bands of bunger (formed by the alignment of Schwann cells) guide the axon toward its target at the distal end in order to restore the neural function. Axonal sprouts emanates from the proximal site toward the distal site for functional recovery.[16,17] The rate of regeneration occurs at about 2–5 mm per day, in humans, as a result of which it can take months for regeneration and healing.[18]

Limitations of TBI/TSI

In case of TBI/TSI if the blood brain barrier (BBB)[19,20] and blood spinal cord barrier (BSCB)[21–24] is injured, it leads to infiltration of cells like fibroblast, macrophages, etc. forming a glial scar, which inhibits axonal migration. It is also accompanied by release of cytokines, chemokines, and complement proteins resulting in edema and dysfunctioning of the tissue.

In CNS, the spinal cord injury is manifested by two main events: the physical injury and the secondary damage. The primary injury is followed by cascade of events like the death of the cells at the injury site, shear stress to the axons causing blood flow disruption, enhanced inflammatory response leading to migration of fibroblasts and reactive astrocytes to the epicenter of the injury, resulting in fibrous and glial scar formation. Further, myelin degradation results in release of inhibitory degradation products in the proximity of the scar.[25–27] Molecular inhibitors associated with the scar formation and damaged myelin inhibits the neural regeneration. These reactive fibroblasts and astrocytes further upregulate the expression of semaphorins and chondroitin sulfate proteoglycans (CSPG), which restrict the axonal regeneration.[28]

Although surgical techniques have undoubtedly been successful for the treatment of the above neural defects, they have certain methodological and therapeutic constrains. Direct end-to-end repair is performed when there is no gap generated between the proximal and distal stump and the two ends are sutured against each other. This is referred to as end-to-end neurorrhaphy and suffer from limitations like scar formation (at the suture site), longitudinal tension, etc.[29] For larger defects in nerve, autografts is the gold standard method. However, it suffers from disadvantages like donor site morbidity, significant differences in the structure of the defective and implanted nerve, and

limited availability.[30] Decellularized muscle/vein grafts have also been used in order to combat the shortage of the donor nerves.[31] Allogenic grafts are also an alternative but they suffer from immune rejection and disease transmission from the host. Moreover, the patient has to undergo immunosuppressive therapy for a period of two years.[32] Furthermore implantation of biomedical devices like electrode for neural growth and regeneration has been performed but these do not have a compatible device–tissue interface as well as cannot perform the biological functions of the tissue.[33] Therefore, the limitations of the existing techniques strained the researchers to look into new avenues and thus the field of "tissue engineering" was formally introduced in 1980s. It is an "interdisciplinary field that applies the principles of engineering and life sciences toward the development of biological substitutes that restore, maintain, or improve tissue function or a whole organ."[34] It works[35] in parallel with the medical science to produce end products, which can have clinical applications and thus improve the treatment and lifestyle of the people. Earlier metals like stainless steel and titanium were used for replacement of the organ especially in the orthopedic field.[36] Later, those implants whose mechanical properties could be manipulated to match the site and size of injury apart from being biocompatibile and biodegradable gained importance. There was an increasing shift from replacement to regeneration of the diseased/injured tissue/organ.[37] As a result, third generation of biomaterials were introduced in the market, among which polymers formed a major section. Polymers are extensively used in tissue engineering because of their biocompatibility, high porosity, surface area-to-volume ratio, chemical structure versatility, etc.[38] Moreover, scaffolds synthesized from it can be tailored as per the application. Scaffolds are 3-D networks that provide support framework to the cells. The synthesis of first biologically active scaffold with low immunogenicity and thromboresistant properties was introduced in 1974. Various other scaffolds have been derived from biological sources like human and animal tissues and have shown improvement in regeneration.[39] Apart from this, polymers, because of the above-mentioned properties, have shown significant results when used for bone,[40,41] cartilage,[42,43] skin,[44] tissue engineering. Poly (methyl methacrylate) (PMMA) was first used in synthesizing lenses because of its inert and transparent properties and later paved the way for the incorporation of polymers in the field of tissue engineering and regenerative medicine.[45]

IDEAL PROPERTIES OF BIOMATERIAL FOR NEURAL REGENERATION

Biomaterials are 3-D scaffolds that are used to mimic the native microenvironment both physically as well as chemically. These scaffolds should be optimized to resemble the

Neural—Rapid Prototyping

micro- and nanostructures of the extracellular microenvironment (ECM) to enable cell adherence. Moreover, the scaffold should be porous in nature to allow the mass transfer of nutrients and waste matter and would also assist in proper gaseous exchange and thus revascularization at the injury site. Scaffolds apart from serving as structural support should also provide the cells with certain chemical cues in order to provide them guidance. It is also necessary that the mechanical properties of the synthesized scaffolds can be manipulated in order to suit the injury site.[46] The scaffold should neither be too hard or too soft as it can damage the surrounding tissues like blood vessels, nerve stumps, etc. or it will not be able to withstand the mechanical pressure of the surroundings, respectively. Controlled biodegradation of the scaffold is also essential because if the scaffold undergoes degradation too slowly it can lead to nerve compression and if their rate of degradation is too fast it will not be able to support regeneration of the nerve cells properly. Bioengineered scaffold tuned to match the properties of the microenvironment can be implanted without cells at the site of injury to enable the migration of the exogenous cells toward implant. As a result, the native cells proliferate and secrete the growth factors and signaling molecules at the site and thus allow regeneration of the injured tissue.[47] However, whenever an injury occurs, that site is infiltrated by fibroblasts, resulting in fibrosis at the site like glial scar formation in case of neural cells, which inhibits the regeneration of the surrounding tissue and therefore there is a need to implant the scaffold with cells and/or growth factors so that the chances of synaptic connections between the existing axons is greater even in the inhibitory microenvironment.[48] The ideal properties of a nerve graft are described (Fig. 2).

POLYMERS AS SCAFFOLDS FOR NEURAL TISSUE ENGINEERING

Protein is a polymer composed of amino acids. Purified ECM proteins like laminin, fibronectin, collagen etc. have been used for the synthesis of 3-D scaffolds for CNS and PNS as these matrices have certain receptors for cell adhesion, which further allows growth and signaling events.[49] However, the scaffolds synthesized by these proteins are expensive due to which polymers are also being widely used. Polymers are mainly classified into degradable/nondegradable, natural/synthetic based on its structure, chemical nature, and biological characteristics.[50]

Natural Polymers

Natural polymers are derived from biological sources like gelatin (fish), chitosan (crab shells), etc. These polymers, which are biodegradable in nature, were the first to be used for clinical applications. Owing to their bioactive property, these polymers enable better cell adherence and interactions. A few examples of such polymers are mentioned below.[51]

Hyaluronic Acid

It is negatively charged, highly hydrated glycosaminoglycan (GAG), which forms an important component of ECM, especially the CNS. It is reported that hyaluronic acid leads to wound healing without scar formation and thus promote nerve regeneration.[52]

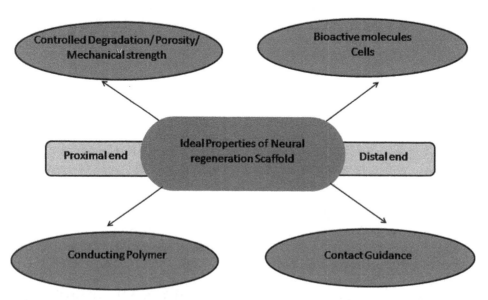

Fig. 2 Ideal properties of scaffold for neural regeneration.

Chitosan

It is a positively charged polymer produced by deacetylation of chitin and allows the adhesion of the cells due to the electrostatic interaction between the polymer and the negatively charged cell membrane of the cells.[53]

Gelatin

Gelatin is the denatured form of collagen. Apart from having RGD motifs as receptors for cell adhesion, high water absorbing capacity enables gelatin to mimic the microenvironment of the soft tissues like neural cells.[35]

Agarose

It is a linear polymer of repeating monomeric units of agarobiose, which physically cross-links to form a gel above 40°C. It does not cause any immunological responses like inflammation after implantation. Reports have shown a relationship between the agarose stiffness and neurite extension.[54]

Alginate

It is an acidic polysaccharide with β-D-mannuronic acid and α-L-gluronic acid as its basic units. It is inert in nature.[55]

Synthetic Polymers

Synthetic polymers also form an important area in tissue engineering field. These polymers have known structural composition and therefore the scaffolds synthesized with these polymers can have controlled properties like degradation, mechanical strength, etc. Moreover, these polymers can be modified with known peptides for better cell adherence and growth.[51]

Poly (2-Hydroxyethyl Methacrylate) (pHEMA)

It consists of hydrophilic copolymers capable of absorbing large amounts of water. It is reported to have its physical properties similar to soft tissues apart from being permeable to gases and entrapment of neural cells.[56]

Polyethylene Glycol (PEG)

Unlike pHEMA, PEG is biodegradable with low toxicity. It resists protein adsorption and thus cell adhesion, which minimizes the immune response elicited by the host.[57]

Polylactic-co-Glycolic Acid (PLGA)

PLGA is the most extensively used biodegradable polymer in tissue engineering. Its most important feature of controlled biodegradation, which can be obtained simply by varying the composition of lactic acid and glycolic acid present in it.[58]

SCAFFOLD SYNTHESIS FOR VARIOUS NEURAL REGENERATION APPLICATIONS

Since 1980s, a lot of research has been done to synthesize 3-D polymeric scaffolds from natural as well as synthetic polymers. The fabrication technique primarily depends on the properties and application of the scaffold.

Hydrogels

Hydrogels are highly cross-linked gels formed by hydrophilic polymers. They have the capacity to absorb large amounts of water enabling proper exchange of nutrients, waste, ions and metabolites to maintain balance with their surroundings, which makes them highly suitable for cell and tissue engineering application.[59] These hydrogels can be synthesized by physical as well as chemical methods. The main advantage of this process is that the properties of these hydrogels can be tailored as per the application because their degradation, mechanical strength, and porosity, can be altered by varying the polymer concentration and cross-linking density.[60] For example, neuronal cells require the compressive moduli of their substrate to be low (0.1–1.0 kPa) as compared to astrocytes and oligodendrocytes, which grow and adhere better on stiffer surfaces (0.5–10 kPa) (Fig. 3).[61,62] The characteristic visco-elastic nature of these hydrogels allows them to retain their 3-D shape. Depending on the technique of synthesis or the molecular weight of the polymer used, porosity can be incorporated in them. The porous nature of the hydrogels further allows cell proliferation, axon outgrowth, and migration, which are required especially in CNS and SCI repair.[63] A combination of the abovementioned natural or synthetic degradable/nondegradable polymers such as pHEMA–gelatin,[64] chitosan-gelatin,[65] etc. have been widely synthesized by various techniques like cryogelation,[66,67] salt leaching,[68] solvent casting,[69] etc. Furthermore to achieve better results, these hydrogels have been coated or grafted with certain peptides to enhance cell–material interaction.[70] For example, these hydrogels, when incorporated with fibronectin-derived sequences, like arginine-glycine-aspartate (RGD), have shown to enhance neural stem cell adhesion and neurite outgrowth.[71] Bioactive scaffolds were prepared by linking dextran or gelatin scaffolds, synthesized by cryogelation technology, with the native ECM protein: laminin. Umbilical cord blood stem cells, when seeded on the cryogel scaffolds, differentiated into neural cells in the presence of the differentiating media. When the bioactive scaffold was implanted in the rat brain, it attracted the neuroblast cells of the host, showing improved neural regeneration properties (Fig. 4).[72]

Neural—Rapid Prototyping

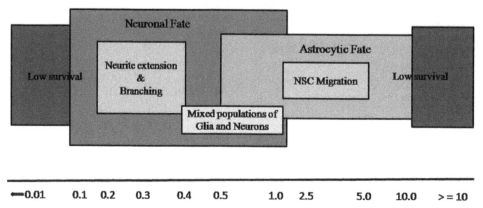

Fig. 3 Neural stem cells survival and fate as a function of substrate stiffness.
Source: From Auranda et al.[62] © 2012, with permission from Elsevier.

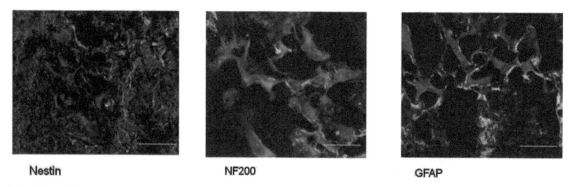

Fig. 4 Bioactive gelatin-linked laminin cryogel scaffold attracted the infiltration of neural cells after implantation in rat brain.
Source: From Jurga et al.[72] © 2011, with permission from Elsevier.

In order to guide the axons across the nerve gap generated in peripheral nerve injury or transaction in the spinal cord, microchannels have been incorporated in these hydrogels, which like, the bands of bunger of PNS, guide the axon through the gap. The microchannels seeded with Schwann cells secrete neurotrophins, which further allows the proximal axons to reach to their distal counterparts.[73] Agarose have been synthesized by freeze-drying method to produce long interconnected porous network, which allows the migration of the axon across the inhibitory microenvironment[74] Similarly, many polymers have been used for the treatment of TBI. It has been reported that pHEMA and poly [N-2-(hydroxypropyl) methacrylamide] (pHPMA) have been implanted between the hippocampus and spetum in lesion cavity. Both the scaffolds showed axonal and astrocyte growth. However, pHEMA showed less glial scar formation as compared to the pHPMA hydrogel.[75] Matrigel is one of the commercially available hydrogel from the ECM of Engelbreth-Holm-Swarm mouse sarcoma. It consists of biologically derived ECM proteins like fibronectin, laminin, and proteoglycans. Tonge et al. has shown to increase neurite outgrowth of neurons and astrocytes, when seeded on it, along with the expression of neuron-specific markers. However, it has hindered differentiation

of neural progenitor cells derived from humans, which might be because of its animal origin.[76]

In-Situ Gelation

In-situ gelation can be achieved by ionic cross-linking, free radical polymerization, phase transition by change in temperature, photo-oxidation, or redox initiation.[77] These smart polymers have gained importance during the last few years and are also referred to as stimuli-responsive/environmental-sensitive polymers. These undergo changes in their structure from extended to coiled conformation and hydrophilic to hydrophobic nature by changes in the environment, which can be either internal or external and as soon these stimuli are removed, the properties are reversed.[78] These hydrogels and their importance is best demonstrated in diabetes experiments wherein the hydrogel releases insulin in response to high glucose, which in return changes the pH of the surrounding tissue and inhibits the further release.[62] Such properties enable implantation of the scaffold with minimum invasive surgery and thus minimal tissue damage, for example, damage to the dura of the brain and spinal cord when the scaffold is implanted at this site. In TBI, the primary injury is followed by a secondary injury, leading to a cascade of events like cellular death and nearby tissue damage. Therefore, to prevent further damage, the polymer is injected to the site of injury, where it undergoes gelation and forms 3-D-like structure in the created lesion, providing guidance and protection to the cells. In such processes, unpolymerized monomer precursor solution or hydrogel solution undergoes polymerization and gelation when injected into the body due to difference in the body temperature or pH, thus filling an irregular lesion created.[79] Methyl cellulose, derived from cellulose by replacement of hydroxyl groups with methoxide group, is an example of thermoresponsive polymer that undergoes phase transition/gelation at or above 60°C. However, for tissue engineering applications, their gelation point can be varied by altering its composition and then used for in-situ gelation. It has properties similar to chitosan.[80] Agarose forms hydrogel by undergoing intramolecular hydrogen bonding when thermally induced. Agarose solution with microtubules and brain-derived nerve growth factor (BDNF) were injected in the spinal cord injury model and liquid nitrogen was used to convert it into gel form and it was shown to have enhanced the regeneration process of nerves.[81]

Electrospun Nanofibers

Nanofibers closely resemble the native microenvironment of the neural tissue wherein it follows the hierarchical fibrillar morphology resembling the laminin, collagen, etc. of ECM proteins.[82] These nanofibers less than 1 μm in diameter can be synthesized by various techniques like electrospinning,[83,84] phase separation,[85] template synthesis,[86] etc. Although these techniques result in synthesis of nanofibers, electospinning is still the most availed technology for their synthesis because of the ease of fabrication. The nanofibers are widely used in neural tissue engineering because of their porosity, controlled mechanical strength and degradation, high surface area to volume ratio. All these properties can be varied by altering the polymer viscosity, voltage, surface tension, distance between the electrodes, etc. Nanofibers because of their size, which is one or two magnitude, less than the conventional size of the fiber holds a significance in the field of neural tissue engineering because of their high surface area and dimensions similar to the native environment.

Both natural and synthetic polymers have been used for the synthesis of the abovementioned nanofibers. Photochemically cross-linked electrospun collagen fibers using acetic acid have shown better stability and helical structure as compared to the other solvents. Moreover, neural stem cells seeded on these fibers showed initiation of neurite outgrowth and cell confluency by seven days of seeding.[87] Synthetic polymers like poly(lactic-co-glycolic acid) (PLGA), poly(ε-caprolactone) (PCL), polyethersulfonate (PES), polyurethane (PU), among others, have also been used for synthesis of nanofibers among which PLGA, PCL, and PLA are extensively used. PU fibers were used as scaffolds for human embryonic stem cell (hESCs) differentiation and it was observed that these nanofibers directed the stem cells to differentiate into dopaminergic-positive neurons.[88] PCL aligned and random fibers demonstrated faster axonal growth and migration on the aligned fibers as compared to the random PCL fibers. The random fibers because of their tortuous path reduced the axonal migration.[89] A combination of natural and synthetic polymers like PCL/collagen nanofibers have been used for the synthesis of electrospun nanofibers and showed glial migration, neuritogenesis, and also cell proliferation.[90] Authors have shown that poly-L-lactic acid (PLLA) nanofibers increased neurite outgrowth of motor neurons along with the formation of its polarity, which further helped in the regeneration and guidance of the native as well as implanted neurons during CNS injury (Fig. 5).[91]

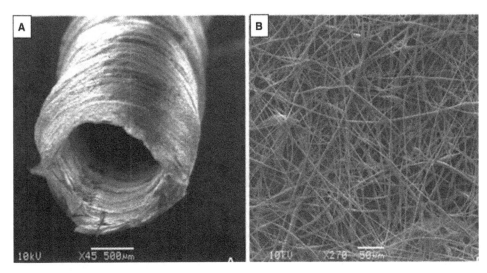

Fig. 5 Electrospun PLGA/PCL nerve conduit for neural regeneration. The nonwoven micro and nano fibers range from approximately 280 nm to 8 μm in diameter with pore size ranging from 700 nm to 20 μm.
Source: From Biazar, et al.[91] © 2010, Dove Medical Press Ltd.

Neural—Rapid Prototyping

CONDUCTING POLYMERS

It dates backs to the era of Galvani when the existence of electric current in animals was proposed.[92] Later, by various experiments, it was shown that these endogenous electric currents were involved in myriad physiological events such as cell division, nervous stimulation, neural development, cell polarity, among others. Defect in these electric fields resulted in abnormal development, loss of proper internal structure formation, etc.[93] Thus, cells that can produce and respond to electrical signals are referred to as excitable cells like neuron, cardiac, pancreatic cells (insulin producing), etc. Various theories have been speculated regarding the effects of electrical stimulation on neurons like calcium channel opening, resulting in signaling followed by a cascade of events,[94] promotes axonal regeneration and neurite orientation,[95] increases cell proliferation of Schwann cells, and thus the secretion of neurotrophic factors and facilitates regeneration even after a delay in treatment.[96]

Piezoelectric polymeric materials have been used as a substratum for growing neural cells, which can be further electrically stimulated by producing certain transient surface charges due to the mechanical deformation of the material.[97] For example, polyvinylidine fluoride (PVDF) membrane showed an increased neural markers expression like nestin and tubulin. However, these PZ materials have limited control over stimulation.[98] Conducting polymers, such as polypyrrole (PPy), polyaniline (PAn), polythiophene, and *trans*-polyacetyleneine, are another group of organic polymers widely approved and used in tissue engineering applications. The prime feature of these polymers responsible for their conductance is the presence of conjugate backbone. Conducting polymers can be synthesized either by electrochemical or chemical method.[99] Various bioactive molecules such as protein and neuroactive molecules have also been added as the dopant in order to increase their biocompatibility with neural cells.[100] It has been shown that electrical stimulation alters the adsorption of protein on the conducting polypyrrole film, leading to enhanced protein adsorption and thus cell adhesion and neurite length.[101] Apart from this, other cells like fast-growing fibroblast cells when cultured on polypyrrole/polylactide conductors showed enhanced IL-6 secretion when they were electrically stimulated compared to their control.[102] Conducting scaffolds synthesized by cryogelation technique have shown increased neural (neuro-2a) and C2C12 proliferation when assayed by MTT method (Fig. 6).[103] Moreover, it has been shown the alignment of astrocytes in the presence of an electric field (500 mV) after 24 hr of exposure (Fig. 7).[104]

PPy/PDLLA/PCL conducting composites have been implanted in rat sciatic nerve after creating an 8 mm lesion. These implants have shown improved axonal regeneration in response to electrical cues. The histological analysis at the end of 2 months showed newly formed Schwann cells and axons covered by myelin sheath.[105] PC12 cells have shown increased neurite length when seeded on conducting scaffolds polypyrrole/poly(2-methoxy-5-aniline sulfonic acid).[106]

The major inhibition associated with using these conducting polymers is their degradability and brittle nature. Therefore, these polymers are used in combination with other degradable polymers like gelatin and chitosan. Recent reports have shown synthesis of degradable conducting polymers, which can have an added advantage in tissue engineering field. Rivers et al. have used degradable ester linkages to connect oligomers of pyrrole and thiophene for creating conducting polymers.[107] It has also been reported that if the size of the degrading conducting polymer is between 200 and 300 nm, then it can be removed from the body.[108]

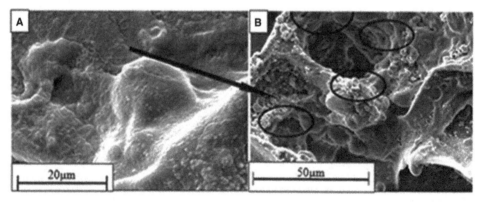

Fig. 6 SEM showing the (**A**) magnified image of neuro 2a cells with extended neurites in the presence of an electric field; (Scale bar 20 μm : Mag. 10000×). (**B**) homogenous distribution of cells on the synthesized conducting cryogel scaffold in the presence of an electric field of 100 mV (Scale bar 50 μm: Mag. 4000×).
Source: From Vishnoi & Kumar.[103] © 2012, with permission from Springer.

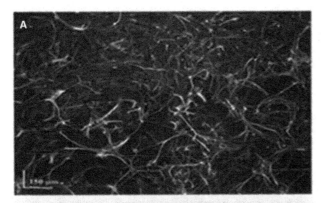

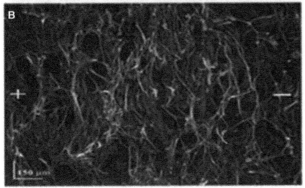

Fig. 7 (**A**) Random alignment of astrocytes in the absence of electric field. (**B**) Aligned astrocytes on exposure to the applied electric field (500 mV).
Source: From Alexander et al.[104] © 2006, with permission from Cambridge University Press.

STRATEGIES FOR NERVE REGENERATION

Cell Therapy

In order to attain an optimum recovery from the injury/diseased site, it is essential to supply the damaged site with potential therapeutic cells either seeded on the 3-D scaffolds or directly at the site. In order to explore this various cell types like glial cells (astrocytes, oligodendrocytes, and Schwann cells), macrophages, stem cells, and olfactory enseathing cells have been used. These cells are being used extensively at the implantation site for recovery.[51,109]

Glial Cells

Glial cells are the supporting cells, which, when transplanted at the injury site, secrete cell adhesion molecules and ECM and neurotrophic factors, which support cell proliferation and provide physical and chemical cues.[110,111] Schwann cells, when transplanted in dorsal column lesion, resulted in remyelination of the axons in the CNS, thus suggesting its potential role in the treatment of multiple sclerosis.[112]

Stem Cells

Stem cells are characterized by self renewal properties as well as their differentiation into varied lineages. Isolation of neural cells from CNS/PNS in sufficient quantity to enable regeneration is a difficult task.[113] Moreover, regeneration at the injured site to re-create cell–cell interaction or synaptic contacts is further difficult to achieve and therefore stem cell differentiation seems to be an alternative wherein stem cells injected in the gap generated due to spinal cord transaction differentiated into neuronal and glial cells and also migrated further across the 8 mm lesion site.[114]

Neural Stem Cells

NSC have gained a lot of attention in the field of neural regeneration as these cells are multipotent, i.e., they have the capacity to differentiate into any of the three types of neural cells: 1) oligodendrocytes; 2) astrocytes; and 3) neurons. These cells can be isolated either from embryonic or adult CNS. NSC–polyglycolic acid (NSC–PGA) complex, when used for treating hypoxic/ischemic rats, did reduce inflammation and scarring along with increasing the reparative process in neurons.[115] Human NSC, when implanted in spinal cord after dorsal laminectomy, showed their differentiation into axons and synapses. These differentiated cells further successfully created contact with native motor neurons (Fig. 8).[116]

Olfactory Ensheathing Cells

OECs can be derived centrally from the olfactory bulb and peripherally through the olfactory epithelium. However, the invasive isolation surgery from olfactory bulb limits their use to the peripheral OECs.[117] These cells resemble Schwann cells phenotypically and have been shown to allow the axonal regeneration extensively by Schwann cells through glial scar.[118]

Macrophages

Debris clearence at the nerve injury site is performed by the macrophages. However, conflicting results have been reported by various research groups in their role of neural regeneration. In a few reports, there is secretion of neurotrophic factors from these cells apart from clearance whereas other reports have shown macrophages to inhibit the axonal regeneration across the gap of the severed nerve.[119,120]

DELIVERY OF BIOACTIVE MOLECULES

The poor regeneration capacity of CNS compared to the PNS has been attributed to the absence of the

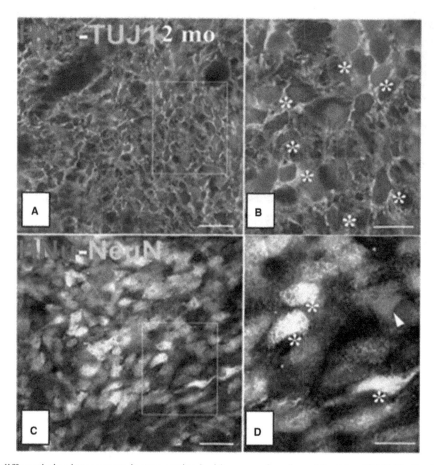

Fig. 8 Human NSC differentiation into neurons immunostained with neuronal markers (Neu N and TUJ 1) after transplantation in transected spinal cord.
Source: From Yan et al.[116] © 2007, Public Library of Science.

neurotrophic factors, which help in providing cues for growth, development, and regeneration of the cells. These polymeric delivery vehicles can either be encapsulated with DNA (due to charge effect) or bioactive molecules like growth factors or modified with peptides for cell adherence or growth factor attachment (Fig. 9).[121] As a result, extensive studies have been done to deliver these growth factors and other bioactive molecules at the site of injury. Polymer-based hydrogels, apart from being used for cell grafting, can also be used as delivery tools. The two prime methods employed for the delivery of these bioactive molecules are embedding/adsorbing these molecules on the scaffolds or incorporating them in microspheres.[122] In the former method, polymeric scaffolds have been embedded with the growth factors during their synthesis such that these can be released in a controlled manner during the life of a graft. Thus compounds with short half-lives like growth factors have 30 min shelf life *in vivo* and can be supplied through microspheres. These microspheres have been synthesized both from natural or synthetic polymers and the compound is encapsulated, entrapped, adsorbed, or covalently attached.[123] The size of the molecule incorporated

significantly affects its release kinetics. Glial cell–derived nerve growth factor (GDNF), a molecule with dimension of 30 Å × 36 Å × 80 Å, can be released from mesh sizes of 80 Å or large whereas larger molecules can only be released once the microspheres have degraded, resulting in different release kinetics.[62] Recent studies have shown that different formulations of poly(lactic-*co*-glycolic) acid (PLGA) microparticles having different rates of degradation were loaded with GDNF and BDNF. This difference lead to different release profiles wherein GDNF microparticles degraded fast and released GDNF early in substantia nigra to provide support to the neuronal cells seeded on the same hydrogel as compared to the BDNF-containing microparticles, which degraded later and increased the neurite outgrowth.[124] These polymers further allows targeted delivery at the site over longer duration as these are stable when they come in contact with biological fluids. PCPP-SA, a biocompatible polyanhydride, was successfully used for treating intra cranial glioma in rats.[125] It has also been reported that polymer-encapsulated NGF-secreting cells can be used to treat neurodegenerative diseases like AD in a controlled manner upto one year.[126]

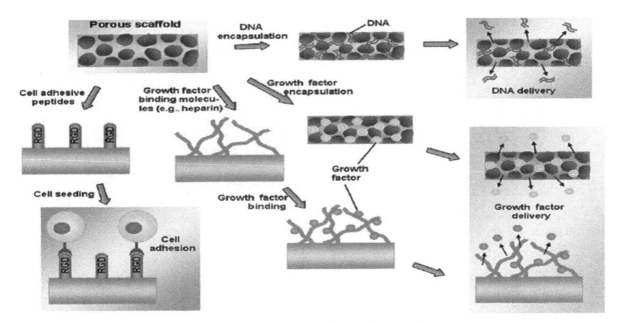

Fig. 9 Different modes of delivery of DNA, growth factors, and cells for neural regeneration.
Source: From Pillay & Choonara[121] © 2012, license InTech.

NERVE GUIDANCE CHANNELS

In the case of a severed nerve, a gap is generated separating the proximal end with the distal end, thus leading to functional loss. When these gaps are below the critical size defect, the presence of glial cells and other intrinsic factors enable the neural regeneration. However, gaps larger than this size needs to undergo external manipulation in order to regain the loss function. Mainly three types of surgical interventions are available to deal with the created small nerve gaps.[127] Epineual repair, the epineural covering of the proximal and distal end, is sutured against each other so that the continuity of the nerve is maintained without causing tension at the suture site.

Perineural repair involves the suturing of the fascicle of the proximal and distal end of the severed nerve. It is also referred to as fascicular/funicular repair whereas group fascicular repairs the grouped fascicles. It is done either when the treatment is delayed or the nerve is crushed and requires trimming before suturing. However, with increasing length of the gap generated, there is an increasing tension at the injury site, which is responsible for compression of the nerve, leading to poor vascular supply. Therefore, there was a need for implantation of nerve conduits between the nerve ends.[128] These tubular nerve guidance channel (NGC) allows the proximal and distal end of the nerves to be encased and accumulate the secreted factors so as to create an environment optimal for neural axonal outgrowth. The presence of nerve conduits also prevents the turning back, branching of the axons and assist them to reach to their distal end (Fig. 10).[129]

It has been reported that pore size from 100 to 220 μm is optimum for nerve tissue engineering. It would also provide passage for the diffusion of factors along as well as across the NGC.[130] Initially nondegradable tubes, e.g., silicon tubes, have been used for implantation. It was observed that the number of regenerated cables in 5, 8, and 10 mm gap created in rats due to diabetic rats were 100%, 87.5% and 70%, respectively. However, the disadvantage of these channels was the requirement of second surgery as well as nerve compression over a period of time.[131] Therefore, the focus shifted to the use of biodegradable NGCs. PLGA- and PCL-eletrospun micro and nano nerve guidance conduit have shown functional recovery in the case of transected sciatic nerve.[86] Asymmetric conduits of PLGA have been synthesized by immersion precipitation phase method, resulting in high outflow rate compared to the inner flow rate, leading to waste drainage and less toxic microenvironment in the tubular NGC.[132] Various bioengineered nerve grafts seeded either with cells like Schwann cells, stem cells, etc. or nerve growth factors and adhesion molecules have also shown enhanced neural recovery. Modification of NGC in terms of providing physical cues has been done by introducing multiple channels of approximately 500 μm in the tubular nerve guidance channel, by the method of injection molding. Moreover, green fluorescent proteins (GFP) transfected cells were used for studying the axonal outgrowth and migration from proximal to distal end in the case of spinal cord regeneration (Fig. 11).[133]

Few commercially available nerve guides are[134]

- *Neuragen*: It is made up of Type I collagen nerve conduits, which is used as an alternate to short nerve grafts for nerve repair.
- *Axogen*: Three types of commercially available axogen grafts are described below.

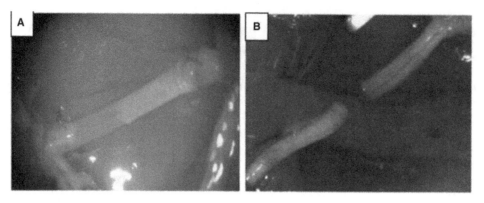

Fig. 10 (**A**) Sciatic nerve repair using a PLGA/PCL nerve guidance channel and (**B**) sham control.
Source: From Panseri et al.[129] © 2008, licensee BioMed Central Ltd.

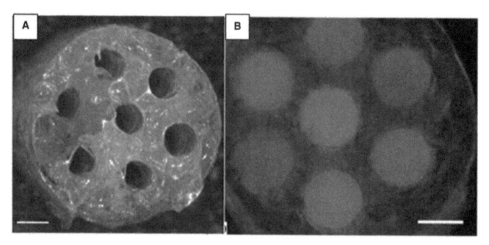

Fig. 11 (**A**) Nerve conduit with multiple channels; (**B**) Nerve conduit seeded with transfected cells for studying the axonal outgrowth and their migration across the nerve conduit with multiple channels.
Source: From Moore et al.[133] © 2006, with permission from Elsevier.

1. *Avanace*: It is able to bridge a gap of 70 mm in size. It consists of bundled tubes wrapped by an outer covering and is made up of decellularized and processed ECM of peripheral nerve.

2. *Axo guard nerve protector*: It wraps the nerve (around 40 mm) and thus protects the soft tissues from the hostile environment during healing process.

3. *Axo guard nerve connector*: It is a hollow tube made up of multilaminar ECM and is used to bridge gaps up to 5 mm.

FACTORS AFFECTING NEURAL REGENERATION

It has been shown that the regeneration of the nerve is indirectly proportional to the patient's age wherein a young patient has a higher regeneration capacity as compared to his or her older counterparts. However, contradicting results have been shown by Sunderland who denied the correlation of patients' recovery ability with its age. The time of repair is also shown to play an important role; reports have shown that repair of the nerve within 6 weeks shows better recovery as compared to the delayed treatments. Early treatment minimizes the scar formation and inhibitions related to it during neural recovery. The level at which the injury occurs is also an important factor and plays a significant role during recovery. The more proximal the injury lesser is the chance of its recovery.

CONCLUSION

In conclusion, neural cells because of their inherited inability to divide cannot regenerate once damaged and therefore there is a need to provide an external support to assist in their regeneration. Polymeric scaffolds, because of their unique properties, have become an important candidate among the other available biomaterials and thus in the field of tissue engineering. Implantation of three-dimensional polymeric scaffolds as well as delivery of cells, growth

Neural—Rapid Prototyping

factors, and bioactive molecules (in polymeric delivery vehicles) to the site of injury has shown improved recovery both in terms of growth and proliferation as well as functionality. Although there has been a continuous thrust to synthesize new scaffolds for improvement in nerve regeneration, there is still a void in the field of neural tissue engineering. Therefore, there is an urge to bridge the gap between the research in the lab and its actual implication in the medical field in order to save many lives.

ACKNOWLEDGMENT

Department of Biotechnology, Ministry of Science and Technology, Govt of India, is acknowledged for the financial support for our projects. T.V. would like to acknowledge CSIR for her Senior Research Fellowship during her PhD work. Authors would like to acknowledge all researchers whose works have been referred here. A. K. would like to acknowledge TATA Innovation Fellowship award from DBT.

REFERENCES

1. Huang, C.; Huang, Y. Tissue engineering for nerve repair. Biomed. Eng. Appl. Basis Comm. **2006**, *18* (3), 100–110.
2. Taylor, J.S.H.; Bampton, E.T.W. Factors secreted by Schwann cells stimulate the regeneration of neonatal retinal ganglion cells. J. Anat. **2004**, *204* (1), 25–31.
3. Topp, K.S.; Boyd, B.S. Structure and biomechanics of peripheral nerves: Nerve responses to physical stresses and implications for physical therapist practice. Phys. Ther. **2006**, *86* (1), 92–109.
4. Fraher, J.P. The transitional zone and CNS regeneration. J. Anat. **2000**, *196*, 137–158.
5. Currais, A.; Hortobagyi, T.; Soriano, S. The neuronal cell cycle as a mechanism of pathogenesis in Alzheimer's disease. Aging **2009**, *1* (4), 363–371.
6. Campbell, W.W. Evaluation and management of peripheral nerve injury. Clin. Neurophysiol. **2008**, *119* (9), 1951–1965.
7. Shoichet, M.S.; Tate, C.C.; Baumann, M.D; LaPlaca, M.C. Strategies for regeneration and repair in the injured central nervous system. In *Indwelling Neural Implants: Strategies for Contending with the In Vivo Environment*; Reichert, W.M., Ed.; CRC Press: Boca Raton, FL, 2008; 221–254.
8. Stoica, B.A.; Faden, A.I. Cell death mechanisms and modulation in traumatic brain injury. Neurotherapeutics **2010**, *7* (1), 3–12.
9. Thatcher, R.W.; North, DM.; Curtin, R.T.; Walker, R.A.; Biver, C.J.; Gomez, J.F.; Salazar, A.M. An EEG severity index of traumatic brain injury. J. Neuropsychiatry Clin. Neurosci. **2001**, *13* (1), 77–87.
10. Wong, M. Functional electrical stimulation (FES): An alternative rehabilitative option for individuals with spinal cord injuries. The Meducator **2006**, *1* (8), 1–2.
11. Naeem, A.; Fazili, N.A. Defective protein folding and aggregation as the basis of neurodegenerative diseases: The darker aspect of proteins. Cell Biochem. Biophys. **2011**, *61* (2), 237–250.
12. Nakamura, T.; Lipton, S.A. Cell death: Protein misfolding and neurodegenerative diseases. Apoptosis **2009**, *14* (4), 455–468.
13. Muchowski, P.J. Protein misfolding, amyloid formation, and neurodegeneration: A critical role for molecular chaperons? Neuron **2002**, *35* (1), 9–12.
14. Gaudet, A.D.; Popovich, P.G.; Ramer, M.S. Wallerian degeneration: Gaining perspective on inflammatory events after peripheral nerve injury. J. Neuroinflammation **2011**, *8* (1), 110.
15. Hoke, A. Mechanisms of disease: What factors limit the success of peripheral nerve regeneration in humans? Nat. Clin. Pract. Neurol. **2006**, *2* (8), 448–454.
16. Wagner, R.; Heckman, H.M.; Myers, R.R. Wallerian degeneration and hyperalgesia after peripheral nerve injury are glutathione-dependent. Pain **1998**, *77* (2), 173–179.
17. Burnett, M. Pathophysiology of peripheral nerve injury: A brief review. Neurosurg. Focus **2004**, *16* (5), 1–7.
18. Schmidt, C.E.; Leach, J.B. Neural tissue engineering: Strategies for repair and regeneration. Annu. Rev. Biomed. Eng. **2003**, *5* (1), 293–347.
19. Chodobski, A.; Zink, B.J.; Szmydynger-Chodobska, J. Blood-brain barrier pathophysiology in traumatic brain injury. Transl. Stroke Res. **2011**, *2* (4), 492–516.
20. Tomkins, O.; Feintuch, A.; Benifla, A.; Cohen, M.; Friedman, A.; Shelef, I. Blood-brain barrier breakdown following traumatic brain injury: A possible role in post-traumatic epilepsy. Cardiovasc. Psychiatry Neurol. **2011**, 11p. doi:10.1155/2011/765923.
21. Werner, C.; Engelhard, K. Pathophysiology of traumatic brain injury. Br. J. Anaesth. **2007**, *99* (1), 4–9.
22. Sharma, H.S. Pathophysiology of blood-spinal cord barrier in traumatic injury and repair. Curr. Pharm. Des. **2005**, *11* (11), 1353–1389.
23. Bartanusz, V.; Jezova, D.; Alajajian, B.; Digicaylioglu, M. The blood-spinal cord barrier: Morphology and clinical implications. Ann. Neurol. **2011**, *70* (2), 194–206.
24. Echeverry, S.; Shi, X.Q.; Rivest, S.; Zhang, J. Peripheral nerve injury alters blood-spinal cord barrier functional and molecular integrity through a selective inflammatory pathway. J. Neurosci. **2011**, *31* (30), 10819–10828.
25. Xanthos, D.N.; Püngel, I; Wunderbaldinger, G; Sandkühler, J. Effects of peripheral inflammation on the blood-spinal cord barrier. Mol. Pain **2012**, *8*, 44. doi:10.1186/1744-8069-8-44.
26. Mautes, A.E.; Weinzierl, M.R.; Donovan, F.; Noble, L.J. Vascular events after spinal cord injury: Contribution to secondary pathogenesis. Phys. Ther. **2000**, *80* (7), 673–687.
27. Oyinbo, C.A. Secondary injury mechanisms in traumatic spinal cord injury: A nugget of this multiply cascade. Acta Neurobiol. Exp. **2011**, *71* (2), 281–299.
28. Jones, L.L.; Sajed, D.; Tuszynski, M.H. Axonal regeneration through regions of chondroitin sulfate proteoglycan deposition after spinal cord injury: A balance of permissiveness and inhibition. J. Neurosci. **2003**, *23* (28), 9276–9288.
29. Liao, W.C.; Chen J.R.; Wang Y.J.; Tseng, G.F. The efficacy of end-to-end and end-to-side nerve repair (neurorrhaphy) in the rat brachial plexus. J. Anat. **2009**, *215* (5), 506–521.
30. Belkas, J.S.; Munro, C.A.; Shoichet, M.S.; Midha, R. Peripheral nerve regeneration through a synthetic hydrogel nerve tube. Restor. Neurol. Neurosci. **2005**, *23* (1), 19–29.

Neural—Rapid Prototyping

31. Neto, H.S. Axonal regeneration through muscle autografts submitted to local anaesthetic pretreatment. Br. J. Plast. Surg. **1998**, *51* (7), 555–560.

32. Chimutengwende-Gordon, M.; Khan, W. Recent advances and developments in neural repair and regeneration for hand surgery. Open Orthop. J. **2012**, *6* (1), 103–107.

33. Ravichandran, R.; Sundarrajan, S.; Venugopal, J.R.; Mukherjee, S.; Ramakrishna, S. Applications of conducting polymers and their issues in biomedical engineering. J. R. Soc. Interface **2010**, *7* (Suppl 5), S559–S579.

34. Chapekar, M.S. Tissue engineering: Challenges and opportunities. J. Biomed. Mater. Res. **2000**, *53* (6), 617–620.

35. Kathuria, N.; Tripathi, A.; Kar, K.K.; Kumar, A. Synthesis and characterization of elastic and macroporous chitosan-gelatin cryogels for tissue engineering. Acta Biomater. **2009**, *5* (1), 406–418.

36. Hermawan, H.R.; Ramdan, D.; Djuansjah,J.R.P. Metals for biomedical applications. In *Biomedical Engineering: From Theory to Applications*; Fazel-Rezai, R., Ed.; InTech: Croatia, 2011; 411–430.

37. Kirkpatrick, C. The paradigm shift from replacement to regeneration. Eur. Med. Dev. Tech. **2010**, *1* (6), 1–1.

38. Kohane, D.S.; Langer, R. Polymeric biomaterials in tissue engineering. Pediatr. Res. **2008**, *63* (5), 487–491.

39. Dhandayuthapani, B.; Yoshida, Y.; Maekawa, T.; Kumar, D.S. Polymeric scaffolds in tissue engineering application: A review. Int. J. Polym. Sci. **2011**, 19. doi: 10.1155/2011/290602.

40. Liu, X.H.; Ma, P.X. Polymeric scaffolds for bone tissue engineering. Ann. Biomed. Eng. **2004**, *32* (3), 477–486.

41. Mishra, R.; Kumar, A. Inorganic/organic biocomposite cryogels for regeneration of bony tissues. J. Biomater. Sci. Polym. Ed. **2011**, *22* (16), 2107–2126.

42. Bhat, S.; Tripathi, A.; Kumar, A. Supermacroprous chitosan-agarose-gelatin cryogels: *In vitro* characterization and *in vivo* assessment for cartilage tissue engineering. J. R. Soc. Interface **2011**, *8* (57), 540–554.

43. Zhao, W.; Jin, X.; Cong, Y.; Liu, Y.; Fu, J. Degradable natural polymer hydrogels for articular cartilage tissue engineering. J. Chem. Technol. Biotechnol. **2013**, *88* (3), 327–339.

44. Mohamed, A.; Xing, M. Nanomaterials and nanotechnology for skin tissue engineering. Int. J. Burns Trauma **2012**, *2* (1), 29–41.

45. Sankar, V.; Kumar, T.S.; Rao, K.P. Preparation, characterization and fabrication of Intraocular lens from photoinitiated polymerized poly (methyl methacrylate). Trends Biomater. Artif. Organs **2004**, *17* (2), 24–30.

46. Subramanian, A.; Krishnan, U.M.; Sethuraman, S. Development of biomaterial scaffold for nerve tissue engineering: Biomaterial mediated neural regeneration. J. Biomed. Sci. **2009**, *16* (1), 1–11.

47. Mallapragada, S.K.; Recknor, J.B. Polymeric biomaterials for nerve regeneration. Adv. Chem. Eng. **2004**, *29*, 47–74.

48. Huang, Y.C.; Huang, Y.Y. Biomaterials and strategies for nerve regeneration. Artif. Organs **2006**, *30* (7), 514–522.

49. Khatun, F.; King, V.; Alovskaya, A.; Brown, R.A.; Priestley, J.V. Fibronectin based biomaterials as a treatment for spinal cord injury. Tissue Eng. **2007**, *13* (7), 1697–1697.

50. Nair, L.S.; Laurencin, C.T. Polymers as biomaterials for tissue engineering and controlled drug delivery. In *Tissue Engineering I. Advances in Biochemical Engineering/ Biotechnology;* Lee, K., Kaplan, D., Eds.; Springer Verlag Review Series: Berlin, 2006; Vol. 102, 47–90.

51. He, J.; Wang, X.M.; Spector, M.; Cui, F.Z. Scaffolds for central nervous system tissue engineering. Front. Mater. Sci. **2012**, *6* (1), 1–25.

52. Wang, X.; He, J.; Wang, Y.; Cui, F.Z. Hyaluronic acid-based scaffold for central neural tissue engineering. Interface Focus **2012**, *2* (3), 278–291.

53. Cho, Y.; Shi, R.; Borgens, R. Chitosan nanoparticle-based neuronal membrane sealing and neuroprotection following acrolein-induced cell injury. J. Biol. Eng. **2010**, *4* (2), 1–11.

54. Cao, Z.; Gilbert, R.J.; He, W. Simple agarose-chitosan gel composite system for enhanced neuronal growth in three dimensions. Biomacromolecules **2009**, *10* (10), 2954–2959.

55. Matyash, M.; Despang, F.; Mandal, R.; Fiore, D.; Gelinsky, M.; Ikonomidou, C. Novel soft alginate hydrogel strongly supports neurite growth and protects neurons against oxidative stress. Tissue Eng. Part A **2012**, *18* (1–2), 55–66.

56. Flynn, L.; Dalton, P.D.; Shoichet, M.S. Fiber ternplating of poly(2-hydroxyethyl methacrylate) for neural tissue engineering. Biomaterials **2003**, *24* (23), 4265–4272.

57. Scott, R.; Marquardt, L.; Willits, R.K. Characterization of poly(ethylene glycol) gels with added collagen for neural tissue engineering. J. Biomed. Mater. Res. A **2010**, *93* (3), 817–823.

58. Makadia, H.K.; Siegel, S.J. Poly lactic-co-glycolic acid (PLGA) as biodegradable controlled drug delivery carrier. Polymers **2011**, *3* (3), 1377–1397.

59. Slaughter, B.V.; , Khurshid, S.S.; Fisher, O.Z.; Khademhosseini, A.; Peppas, N.A. Hydrogels in regenerative medicine. Adv. Mater. **2009**, *21* (32–33), 3307–3329.

60. Gupta, S.; Goswami, S.; Sinha, A. A combined effect of freeze--thaw cycles and polymer concentration on the structure and mechanical properties of transparent PVA gels. Biomed. Mater. **2012**, *7* (1), doi:10.1088/1748-6041/7/1/015006.

61. Pettikiriarachchi, J.T.S.; Parish, C.L.; Shoichet, M.S.; Forsythe, J.S.; Nisbet, D.R. Biomaterials for brain tissue engineering. Aust. J. Chem. **2010**, *63* (8), 1143–1154.

62. Auranda, E.R.; Lampe, K.J.; Bjugstada, K.B. Defining and designing polymers and hydrogels for neural tissue engineering. Neurosci. Res. **2012**, *72* (3), 199–213.

63. Woerly, S. Porous hydrogels for neural tissue engineering. Mater. Sci. Forum. **1997**, *250*, 53–68.

64. Singh, D.; Nayak, V.; Kumar, A. Proliferation of myoblast skeletal cells on three-dimensional supermacroporous cryogels. Int. J. Biol. Sci. **2010**, *6* (4), 371–381.

65. Huang, Y.; Onyeri, S.; Siewe, M.; Moshfeghian, A.; Madihally, S.V. *In vitro* characterization of chitosan-gelatin scaffolds for tissue engineering. Biomaterials **2005**, *26* (36), 7616–7627.

66. Bolgen, N.; Bölgen, N.; Plieva, F.; Galaev, I.Y.; Mattiasson, B.; Pişkin, E. Cryogelation for preparation of novel biodegradable tissue-engineering scaffolds. J. Biomater. Sci. Polym. Ed. **2007**, *18* (9), 1165–1179.

67. Priya, S.G.; Jungvid, H.; Kumar, A. Skin tissue engineering for tissue repair and regeneration. Tissue Eng. Part B Rev. **2008**, *14* (1), 105–118.

68. Reignier, J.; Huneault, M.A. Preparation of interconnected poly(epsilon-caprolactone) porous scaffolds by a combination of polymer and salt particulate leaching. Polymer **2006**, *47* (13), 4703–4717.

69. Suh, S.W.; Shin, J.Y.; Kim, J.; Kim, J.; Beak, C.H.; Kim, D.I.; Kim, H.; Jeon, S.S.; Choo, I.W. Effect of different particles on cell proliferation in polymer scaffolds using a solvent-casting and particulate leaching technique. ASAIO J. **2002**, *48* (5), 460–464.

70. Wojtowicz, A.M.; Shekaran, A.; Oest, M.E.; Dupont, K.M.; Templeman, K.L.; Hutmacher, D.W.; Guldberg, R.E.; García, A.J. Coating of biomaterial scaffolds with the collagen-mimetic peptide GFOGER for bone defect repair. Biomaterials **2010**, *31* (9), 2574–2582.

71. Jeon, W.B.; Park, B.H.; Choi, S.K.; Lee, K.M.; Park, J.K. Functional enhancement of neuronal cell behaviors and differentiation by elastin-mimetic recombinant protein presenting Arg-Gly-Asp peptides. BMC Biotechnol. **2012**, *12* (61), 1–9.

72. Jurga, M.; Dainiak, M.B.; Sarnowska, A.; Jablonska, A.; Tripathi, A.; Plieva, F.M.; Savina, I.N.; Strojek, L.; Jungvid, H.; Kumar, A.; Lukomska, B.; Domanska-Janik, K.; Forraz, N.; McGuckin, C.P. The performance of laminin-containing cryogel scaffolds in neural tissue regeneration. Biomaterials **2011**, *32* (13), 3423–3434.

73. Zhang, Y.G.; Sheng, S.Q.; Qi, F.Y.; Hu, X.Y.; Zhao, W.; Wang, Y.Q.; Lan, L.F.; Huang, J.H.; Luo, Z.J. Schwann cell-seeded scaffold with longitudinally oriented microchannels for reconstruction of sciatic nerve in rats. J. Mater. Sci. Mater. Med. **2013**, *24* (7), 1767–1780.

74. Stokolsa, S.; Tuszynski, M.H. The fabrication and characterization of linearly oriented nerve guidance scaffolds for spinal cord injury. Biomaterials **2004**, *25* (27), 5839–5846.

75. Lesný, P.; De Croos, J.; Prádný, M.; Vacík, J.; Michálek, J.; Woerly, S.; Syková, E. Polymer hydrogels usable for nervous tissue repair. J. Chem. Neuroanat. **2002**, *23* (4), 243–247.

76. Tonge, D.A.; Golding, J.P.; Edbladh, M.; Kroon, M.; Ekström, P.E.; Edström, A. Effects of extracellular matrix components on axonal outgrowth from peripheral nerves of adult animals *in vitro*. Exp. Neurol. **1997**, *146* (1), 81–90.

77. Guvendiren, M.; Lu, H.D.; Burdick, J.A. Shear-thinning hydrogels for biomedical applications. Soft Matter **2012**, *8* (2), 260–272.

78. Kumar, A.; Srivastava, A.; Galaev, I.; Mattiasson, B. Smart polymers: Physical forms and bioengineering applications. Prog. Polym. Sci. **2007**, *32* (10), 1205–1237.

79. Pakulska, M.M.; Ballios, B.G.; Shoichet, M.S. Injectable hydrogels for central nervous system therapy. Biomed. Mater. **2012**, *7* (2), 024101.

80. Kobayashi, K.; Huang, C.I.; Lodge, T.P. Thermoreversible gelation of aqueous methylcellulose solutions. Macromolecules **1999**, *32* (21), 7070–7077.

81. Gao, M.Y.; Lu, P.; Bednark, B.; Lynam, D.; Conner, J.M.; Sakamoto, J.; Tuszynski, M.H. Templated agarose scaffolds for the support of motor axon regeneration into sites of complete spinal cord transection. Biomaterials **2013**, *34* (5), 1529–1536.

82. Neal, R.A.; McClugage, S.G.; Link, M.C.; Sefcik, L.S.; Ogle, R.C.; Botchwey, E.A. Laminin nanofiber meshes that mimic morphological properties and bioactivity of basement membranes. Tissue Eng. Part C Methods **2009**, *15* (1), 11–21.

83. Lee, Y.S.; Arinzeh, T.L. Electrospun nanofibrous materials for neural tissue engineering. Polymers **2011**, *3* (1), 413–426.

84. Xie, J.; MacEwan, M.R.; Schwartz, A.G.; Xia, Y. Electrospun nanofibers for neural tissue engineering. Nanoscale **2010**, *2* (1), 35–44.

85. Dahlin, R.L.; Kasper, F.K.; Mikos, A.G. Polymeric nanofibers in tissue engineering. Tissue Eng. Part B Rev. **2011**, *17* (5), 349–364.

86. Long, Y.Z.; Li, M.M.; Gu, C.; Wan, M.; Duvail, J.L.; Liu, Z.; Fan, Z. Recent advances in synthesis, physical properties and applications of conducting polymer nanotubes and nanofibers. Prog. Polym. Sci. **2011**, *36* (10), 1415–1442.

87. Liu, T.; Teng, W.K.; Chan, B.P.; Chew, S.Y. Photochemical crosslinked electrospun collagen nanofibers: Synthesis, characterization and neural stem cell interactions. J. Biomed. Mater. Res. A **2010**, *95* (1), 276–282.

88. Carlberg, B.; Axell, M.Z.; Nannmark, U.; Liu, J.; Kuhn, H.G. Electrospun polyurethane scaffolds for proliferation and neuronal differentiation of human embryonic stem cells. Biomed. Mater. **2009**, *4* (4), doi:10.1088/1748-6041/4/4/045004.

89. Jahani, H.; Kaviani, S.; Hassanpour-Ezatti, M.; Soleimani, M.; Kaviani, Z.; Zonoubi, Z. The effect of aligned and random electrospun fibrous scaffolds on rat mesenchymal stem cell proliferation. Cell J. **2012**, *14* (1), 31–38.

90. Schnell, E.; Klinkhammer, K.; Balzer, S.; Brook, G.; Klee, D.; Dalton, P.; Mey, J. Guidance of glial cell. migration and axonal growth on electrospun nanofibers of poly-epsilon-caprolactone and a collagen/poly-epsilon-caprolactone blend. Biomaterials **2007**, *28* (19), 3012–3025.

91. Biazar, E.; Khorasani, M.T.; Montazeri, N.; Pourshamsian, K.; Daliri, M.; Rezaei, M.; Jabarvand, M.; Khoshzaban, A.; Heidari, S.; Jafarpour, M.; Roviemiab, Z. Types of neural guides and using nanotechnology for peripheral nerve reconstruction. Int. J. Nanomedicine **2010**, *5*, 839–852.

92. Piccolino, M. Animal electricity and the birth of electrophysiology: The legacy of Luigi Galvani. Brain Res. Bull. **1998**, *46* (5), 381–407.

93. McCaig, C.D.; Rajnicek, A.M.; Song, B.; Zhao, M. Controlling cell behavior electrically: Current views and future potential. Physiol. Rev. **2005**, *85* (3), 943–978.

94. Huang, J.; Ye, Z.; Hu, X.; Lu, L.; Luo, Z. Electrical stimulation induces calcium-dependent release of NGF from cultured Schwann cells. Glia **2010**, *58* (5), 622–631.

95. Huang, J.; Lu, L.; Zhang, J.; Hu, X.; Zhang, Y.; Liang, W.; Wu, S.; Luo, Z. Electrical stimulation to conductive scaffold promotes axonal regeneration and remyelination in a rat model of large nerve defect. Plos One **2012**, *7* (6), e39526.

96. Kim, I.S.; Song, Y.M.; Cho, T.H.; Pan, H.; Lee, T.H.; Kim, S.J.; Hwang, S.J. Biphasic electrical targeting plays a significant role in Schwann cell activation. Tissue Eng. Part A **2011**, *17* (9–10), 1327–1340.

97. Lee, Y.S.; Collins, G.; Arinzeh, T.L. Neurite extension of primary neurons on electrospun piezoelectric scaffolds. Acta Biomater. **2011**, *7* (11), 3877–3886.

Neural—Rapid Prototyping

98. Weber, N.; Lee, Y.S.; Shanmugasundaram, S.; Jaffe, M.; Arinzeh, T.L. Characterization and *in vitro* cytocompatibility of piezoelectric electrospun scaffolds. Acta Biomater. **2010**, *6* (9), 3550–3556.

99. Bendrea, A.D.; Cianga, L.; Cianga, I. Review paper: Progress in the field of conducting polymers for tissue engineering applications. J. Biomater. Appl. **2011**, *26* (1), 3–84.

100. Guimard, N.K.; Gomez, N.; Schmidt; C.E. Conducting polymers in biomedical engineering. Prog. Polym. Sci. **2007**, *32* (8–9), 876–921.

101. Kotwal, A.; Schmidt, C.E. Electrical stimulation alters protein adsorption and nerve cell interactions with electrically conducting biomaterials. Biomaterials **2001**, *22* (10), 1055–1064.

102. Shi, G.X.; Zhang, Z.; Rouabhia, M. The regulation of cell functions electrically using biodegradable polypyrrole-polylactide conductors. Biomaterials **2008**, *29* (28), 3792–3798.

103. Vishnoi, T.; Kumar, A. Conducting cryogel scaffold as a potential biomaterial for cell stimulation and proliferation. J. Mater. Sci. Mater. Med. **2013**, *24* (2), 447–459.

104. Alexander, J.K.; Fuss, B.; Colello, R.J. Electric field-induced astrocyte alignment directs neurite outgrowth. Neuron Glia Biol. **2006**, *2* (2), 93–103.

105. Zhang, Z.; Rouabhia, M.; Wang, Z.; Roberge, C.; Shi, G.; Roche, P.; Li, J.; Dao, L.H. Electrically conductive biodegradable polymer composite for nerve regeneration: Electricity-stimulated neurite outgrowth and axon regeneration. Artif. Organs **2007**, *31* (1), 13–22.

106. Liu, X.; Gilmore, K.J.; Moulton, S.E.; Wallace, GG. Electrical stimulation promotes nerve cell differentiation on polypyrrole/poly (2-methoxy-5 aniline sulfonic acid) composites. J. Neural Eng. **2009**, *6* (6), 065002.

107. Rivers, T.J.; Hudson, T.W.; Schmidt, C.E. Synthesis of a novel, biodegradable electrically conducting polymer for biomedical applications. Adv. Funct. Mater. **2002**, *12* (1) 33–37.

108. Wan, Y.; Wen, D.J. Preparation and characterization of porous conducting poly(DL-lactide) composite membranes. J. Memb. Sci. **2005**, *246* (2), 193–201.

109. Yu, K.W.; Kocsis, J.D. Schwann cell engraftment into injured peripheral nerve prevents changes in action potential properties. J. Neurophysiol. **2005**, *94* (2), 1519–1527.

110. Fawcett, J.W.; Keynes, R.J. Peripheral-nerve regeneration. Annu. Rev. Neurosci. **1990**, *13*, 43–60.

111. Love, S.; Plaha, P.; Patel, N.K.; Hotten, G.R.; Brooks, D.J.; Gill, S.S. Glial cell line-derived neurotrophic factor induces neuronal sprouting in human brain. Nat. Med. **2005**, *11* (7), 703–704.

112. Kohama, I.; Lankford, K.L.; Preiningerova, J.; White, F.A.; Vollmer, T.L.; Kocsis, J.D. Transplantation of cryopreserved adult human Schwann cells enhances axonal conduction in demyelinated spinal cord. J. Neurosci. **2001**, *21* (3), 944–950.

113. Reubinoff, B.E.; Itsykson, P.; Turetsky, T.; Pera, M.F.; Reinhartz, E.; Itzik, A.; Ben-Hur, T. Neural progenitors from human embryonic stem cells. Nat. Biotechnol. **2001**, *19* (12), 1134–1140.

114. Chen, J.R.; Cheng, G.Y.; Sheu, C.C.; Tseng, G.F.; Wang, T.J.; Huang, Y.S. Transplanted bone marrow stromal cells migrate, differentiate and improve motor function in rats with experimentally induced cerebral stroke. J. Anat. **2008**, *213* (3), 249–258.

115. Park, K.I.; Teng, Y.D.; Snyder, E.Y. The injured brain interacts reciprocally with neural stem cells supported by scaffolds to reconstitute lost tissue. Nat. Biotechnol. **2002**, *20* (11), 1111–1117.

116. Yan, J.; Xu, L.; Welsh, A.M.; Hatfield, G.; Hazel, T.; Johe, K.; Koliatsos, V.E. Extensive neuronal differentiation of human neural stem cell grafts in adult rat spinal cord. PLoS Med. **2007**, *4* (2), 318–332.

117. Higginson, J.R.; Barnett, S.C. The culture of olfactory ensheathing cells (OECs)-a distinct glial cell type. Exp. Neurol. **2011**, *229* (1), 2–9.

118. Barnett, S.C.; Riddell, J.S. Olfactory ensheathing cells (OECs) and the treatment of CNS injury: Advantages and possible caveats. J. Anat. **2004**, *204* (1), 57–67.

119. Miyauchi, A.; Kanje, M.; Danielsen, N.; Dahlin, L.B. Role of macrophages in the stimulation and regeneration of sensory nerves by transposed granulation tissue and temporal aspects of the response. Scand. J. Plast. Reconstr. Surg. Hand Surg. **1997**, *31* (1), 17–23.

120. Popovich, P.G.; Whitacre, C.C.; Stokes, B.T. Inflammation, autoimmunity and spinal cord injury. Shock **1999**, *12* (4), 2–2.

121. Pillay, V.K., P. Choonara, Y. , Processing and templating of bioactive-loaded polymeric neural architectures: Challenges and innovative strategies. In *Recent Advances in Novel Drug Carrier Systems*; Sezer, A.D., Ed.; InTech: Croatia, 2012; 355–392.

122. Edlund, U.; Albertsson, A.C. Degradable polymer microspheres for controlled drug delivery. In *Degradable Aliphatic Polyesters*, 157 ed.; Albertsson, A.C., Ed.; Springer: Berlin, UK, 2002; 67–112.

123. Chung, H.J.; Park, T.G. Surface engineered and drug releasing pre-fabricated scaffolds for tissue engineering. Adv. Drug Deliv. Rev. **2007**, *59* (4–5), 249–262.

124. Lampe, K.J.; Kern, D.S.; Mahoney, M.J.; Bjugstad, K.B. The administration of BDNF and GDNF to the brain via PLGA microparticles patterned within a degradable PEG-based hydrogel: Protein distribution and the glial response. J. Biomed. Mater. Res. A **2011**, *96* (3), 595–607.

125. Walter, K.A.; Cahan, M.A.; Gur, A.; Tyler, B.; Hilton, J.; Colvin, O.M.; Burger, P.C.; Domb, A.; Brem, H. Interstitial taxol delivered from a biodegradable polymer implant against experimental malignant glioma. Cancer Res. **1994**, *54* (8), 2207–2212.

126. Winn, S.R.; Lindner, M.D.; Lee, A.; Haggett, G.; Francis, J.M.; Emerich, D.F. Polymer-encapsulated genetically modified cells continue to secrete human nerve growth factor for over one year in rat ventricles: Behavioral and anatomical consequences. Exp. Neurol. **1996**, *140* (2), 126–138.

127. Mafi, P.; Hindocha, S.; Dhital, M.; Saleh, M. Advances of peripheral nerve repair techniques to improve hand function: A systematic review of literature. Open Orthop. J. **2012**, *6*, 60–68.

128. Wang, S.F.; Cai, L. Polymers for fabricating nerve conduits. Int. J.Polym. Sci. **2010**, doi:10.1155/2010/138686.

129. Panseri, S.; Cunha, C.; Lowery, J.; Del Carro, U.; Taraballi, F.; Amadio, S.; Vescovi, A.; Gelain, F. Electrospun micro- and nanofiber tubes for functional nervous regeneration in

sciatic nerve transections. BMC Biotechnol. **2008**, *8* (39), 1–12.

130. Li, H.; Wijekoon, A.; Leipzig, N.D. 3D differentiation of neural stem cells in macroporous photopolymerizable hydrogel scaffolds. PLoS One **2012**, *7* (11), e48824.

131. Tantuwaya, V.S.; Bailey, S.B.; Schmidt, R.E.; Villadiego, A.; Tong, J.X.; Rich, K.M. Peripheral nerve regeneration through silicone chambers in streptozocin-induced diabetic rats. Brain Res. **1997**, *759* (1), 58–66.

132. Chang, C.J.; Hsu, S.H.; Yen, H.J.; Chang, H.; Hsu, S.K. Effects of unidirectional permeability in asymmetric poly(DL-lactic acid-co-glycolic acid) conduits on peripheral nerve regeneration: An *in vitro* and *in vivo* study.

J. Biomed. Mater. Res. B Appl. Biomater. **2007**, *83* (1), 206–215.

133. Moore, M.J.; Friedman, J.A.; Lewellyn, E.B.; Mantila, S.M.; Krych, A.J.; Ameenuddin, S.; Knight, A.M.; Lu, L.; Currier, B.L.; Spinner, R.J.; Marsh, R.W.; Windebank, A.J.; Yaszemski, M.J. Multiple-channel scaffolds to promote spinal cord axon regeneration. Biomaterials **2006**, *27* (3), 419–429.

134. Whitlock, E.L.; Tuffaha, S.H.; Luciano, J.P.; Yan, Y.; Hunter, D.A.;Magill, C.K.; Moore, A.M.; Tong, A.Y.; Mackinnon, S.E.; Borschel, G.H. Processed allografts and type I collagen conduits for repair of peripheral nerve gaps. Muscle Nerve **2009**, *39* (6), 787–799.

Neural—Rapid Prototyping

Non-Viral Delivery Vehicles

Ernst Wagner

Department of Pharmacy, Ludwig Maximilians University, Munich, Germany

Abstract

Polyplexes are based on the condensation of negatively charged DNA by electrostatic attraction with polycationic condensing compounds. The resulting compact particles protect the nucleic acid and also improve the uptake into the cells. Numerous polycations have been used for formulating DNA into complexes. Ideally, the cationic polymer will carry out multiple tasks which include compacting DNA into particles that can migrate to the target tissue, shielding the particles against degradation and undesired interactions, and enhancing cell binding and intracellular delivery into cytoplasm and the nucleus. In practical terms, the polymer is unable to carry out all these tasks. Additional functional domains have to be integrated into the formulation. Advantageously, polymers can be chemically linked to molecules such as cell-targeting ligands, including proteins (antibodies, growth factors) and small molecules (carbohydrates, peptides, vitamins). Various polymer–ligand gene delivery systems have been demonstrated to facilitate receptor-cell-mediated delivery into cultured cells. Targeted delivery to the lung, the liver, or tumors has been achieved in experimental animals, either by localized or systemic application.

TARGETED POLYPLEXES FOR GENE THERAPY

Polycations for DNA Complex Formation

DNA-binding polycations include synthetic polymers such as polylysine, polyethylenimine, cationic dendrimers, carbohydrate-based polymers such as modified chitosan or dextran, and natural DNA-binding proteins such as histones or protamines. The characteristics of these polymers and their use in transfections have been reported extensively.[1–3] Of the "first-generation" cationic polymeric carriers evaluated, polyethylenimine,[4] also termed PEI, has the highest transfection efficiency. This can be explained by its intrinsic ability to facilitate endosomal release. The polymer acts as a "proton sponge," containing protonable amines which after endocytosis slow down endosomal acidification, triggering enhanced endosomal chloride accumulation followed by osmotic swelling and breaking up of endosomes.[5] Biocompatible and biodegradable polymers with further enhanced efficiency would be advantageous. Approaches currently under evaluation include the use of low-molecular weight polymers oligomerized into larger polymeric structures by biodegradable, connecting disulfide linkages[6] or esters.[7,8]

In Vivo Applications of Polyplexes

For targeted delivery into a distant organ, the following factors have to be taken into account (Fig. 1) ideally: 1) DNA polyplexes are stable and inert in blood; 2) they must be able to reach their target tissue and therefore to cross different biologicals barriers, including vascular endothelium, extracellular matrix, and others; 3) once reaching the target cell they should internalize; 4) they should disassemble at the right moment, but still protect the DNA against intracellular degradation; 5) they should release the DNA into the nuclear compartment; and 6) they must elicit as low an inflammatory or immune response as possible. Although no polyplex or other nonviral gene transfer system exists which would fulfill all these requirements, targeted delivery to the lung, the liver, or tumors has already been achieved in experimental animals (Table 1).

Several strategies have been evaluated for gene transfer to the lung. Systemic application of PEI polyplexes resulted in very high gene transfer to the lung.[9] In this application, the linear polymer form of PEI mediates a much higher transfection activity than branched PEI of similar molecular weight.[10] However, a positive charge ratio (cation of polymer to DNA phosphate anion ratio) is required, with a narrow window between efficiency and severe toxicity, as PEI/DNA activates the lung endothelium and forms small aggregates.

As an alternative approach, Ferkol and colleagues generated polylysine polyplexes for targeting the polymeric immunoglobulin receptor by using an antibody Fab fragment as ligand conjugated with polylysine. Systemic delivery of these polyplexes in rats resulted in reporter gene expression in cells of the airway epithelium and submucosal glands.[11] Also, different lung-targeted ligand–polylysine polyplexes were evaluated, using a synthetic peptide ligand for the serpin–enzyme complex receptor. These polyplexes upon nasal administration were able to

Concise Encyclopedia of Biomedical Polymers and Polymeric Biomaterials DOI: 10.1081/E-EBPPC-120029832

Neural—Rapid Prototyping

transiently correct the chloride transport defect in the nasal epithelium of CF mice.[12]

Targeted gene transfer to the liver was reported by Wu and Wu.[13] DNA/asialoorosomucoid–polylysine complexes were administered for targeting to the hepatocyte-specific asialoglycoprotein receptor, and intravenous (i.v.) injection in rats resulted in marker gene expression in rat livers. This work was the very first successful *in vivo* application of a targeted polyplex system for gene transfer. Using the same type of formulation, Nagase analbuminemic rats were injected with a human albumin expression plasmid, followed by partial hepatectomy.[14] Circulating human albumin increased in concentration to a maximum by 2 weeks postinjection and remained stable for further 2 weeks.

A synthetic hepatocyte-directed, multifunctional polyplex system was applied by the group of Nishikawa et al.,[15] consisting of DNA complexed with polyornithine which was modified with galactose (to serve as asialogly-coprotein receptor ligand) and a fusogenic peptide derived from the influenza virus hemagglutinin subunit HA2 (to serve as endosomal release domain). Upon i.v. injection in mice, a large amount of a marker gene product was detected in the liver, with the hepatocytes contributing more than 95% of the total activity in all tissues. In another approach, a conjugate of low-molecular weight PEI with Pluronic 123 (a block copolymer of polyethylene oxide and poly-propylene oxide) was synthesized by the group of Nguyen et al.[16] In combination with unmodified Pluronic 123 and DNA, the conjugate forms small and stable complexes which after i.v. injection into mice exhibit highest gene expression in liver.

Targeting tumors might present a unique opportunity to reach and attack multiple-spread metastases. Direct intratumoral delivery of PEI polyplexes has been investigated. Expression levels were low, and a special form of administration, the local infusion of PEI polyplexes into the tumor mass by a micropump,[17] had to be applied to obtain satisfactory results. As an alternative for lung tumors, Gautam and colleagues successfully delivered PEI polyplexes to lung metastases of melanoma as aerosol through the airways.[18]

A series of targeted DNA polyplex formulations with the potential of systemically targeting tumors have been established. A charge-neutral surface of the DNA particles is essential to minimize nonspecific interactions with blood

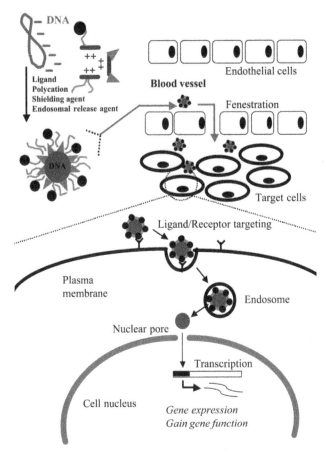

Fig. 1 Targeting opportunities for DNA polyplexes. Polyplex formation with polymer compacting DNA into particles, shielding the polyplex against degradation and undesired interactions, enabling migration to the target tissue, enhancing cell binding and intracellular delivery into cytoplasm and the nucleus.

Table 1 *In vivo* delivery of polyplexes

Polyplex system	Delivery mode	Target organ	Results	References
PEI	Intravenous	Lung	High lung gene expression, toxicity	[9]
Anti pIgR–polylysine	Intravenous	Lung	Highest expression in lung; also in liver	[11]
SECR ligand–polylysine	Intranasal	Nasal epithelium	Functional CFTR gene transfer	[12]
ASOM–polylysine	Intravenous	Liver	First demonstration of targeted gene transfer	[13]
Galactose–polyornithine–HA2 peptide	Intravenous	Liver	Hepatocyte-targeted gene expression	[15]
Pluronic–PEI	Intravenous	Liver	Highest gene expression in liver	[16]
PEI	Infusion into tumor	Tumor	Micropump required	[17]
TfR-targeted, shielded PEI	Intravenous	Tumor	Up to hundredfold higher expression in tumor	[19,21]
EGF–PEI, PEG shielded	Intravenous	Hepatoma	Up to hundredfold higher expression in tumor	[23]

Abbreviations: ASOM, asialoorosomucoid; EGF, epidermal growth factor; PEG, polyethylene glycol; PEI, polyethylenimine; SECR, serpine–enzyme complex receptor; TfR, transferrin receptor.

components, allowing greater i.v. circulation time for the vector to reach its target, and also reducing vector toxicity.[4] The hydrophilic shielding agents investigated include the serum protein transferrin and hydrophilic polymers such as hydroxypropyl methacrylate or polyethylene glycol (PEG). For example, transferrin-shielded polyplexes[19] or PEG-shielded polyplexes[20,21] demonstrated potential for systemic *in vivo* targeting of tumors. Intravenous injection resulted in gene transfer into distant subcutaneous neuroblastoma tumors of syngeneic mice[19–21] with luciferase marker gene expression levels in tumor tissues approximately hundredfold higher than in other organ tissues. Specificity was confirmed by luciferase imaging in living mice.[22] In analogous manner, EGF–PEG-coated polyplexes were successfully applied for systemic targeting of human hepatocellular carcinoma xenografts in SCID mice.[23] Similar observations of *in vivo* hepatoma targeting were made with polylysine polyplexes linked with EGF-derived peptides and an endosomally active peptide.[24]

Therapeutic Approaches Using Polyplexes

Table 2 lists examples of therapeutic concepts evaluated in animal models. Using the first targeted *in vivo* gene delivery system developed,[13] Wu and colleagues demonstrated hepatocyte-specific gene transfer of the LDL receptor in a rabbit animal model for familial hypercholesterolemia. This resulted in a temporary amelioration of the disease phenotype.[25] In analogous manner, the albumin serum levels of Nagase analbuminemic rats were transiently raised by human albumin gene delivery.[14]

For cancer, Gautam and colleagues[18] successfully delivered PEI polyplexes expressing the p53 gene as aerosol through the airways to established melanoma lung metastases, which, especially in combination with 9-nitrocamptothecin chemotherapy, resulted in tumor growth inhibition.

Polyplexes with appropriate surface shielding can target the tumor tissue and exploit the unique neovasculature for extravasation from the blood stream (compare Fig. 1) and delivery of therapeutic genes into the tumor tissue. Repeated systemic application of transferrin-and/or PEG-shielded polyplexes encoding tumor necrosis factor alpha (TNF-alpha) into tumor-bearing mice induced expression in tumor cells close to the feeding blood vessels, which triggers TNF-mediated destruction of the tumor vasculature, tumor necrosis, and inhibition of tumor growth as demonstrated in several murine tumor models.[21] The gene expression of TNF-alpha was localized within the tumor; no significant systemic TNF-related toxicities were observed. Systemic administration of combined p21(WAF-1) and granulocyte macrophage-colony stimulating factor (GM-CSF) genes formulated into EGF receptor-targeted polyplexes inhibited the growth of subcutaneous hepatoma cells and increased the survival rate of tumor-bearing mice.[24] It remains to be demonstrated whether encouraging findings such as those obtained in mice can be transferred to studies in larger animals and humans.

CONCLUSION

Targeted delivery to the lung, the liver, or tumors has been achieved in experimental animals, either by localized or systemic application. Therapeutic effects have been demonstrated, although efficiencies currently appear to be still too low to justify clinical use. The limitations of first-generation polymeric carriers (modest activity and significant toxicity) have to be overcome by developments of new biodegradable polycationic polymers, incorporation of targeting and intracellular transport functions, and polyplex formulations that avoid unspecific adverse interactions with the host. A key future step will be the further development of polyplexes into "artificial viruses," i.e., polyplexes possessing viruslike entry functions which are presented by smart polymers and conjugates. These "smart" polymers in contrast to conventional polymers have to respond in a more dynamic and controlled manner to alterations in their biological microenvironment such as pH or redox environment, and have to undergo programmed structural changes, to more accurately switch on the individual delivery functions only when required in the individual steps (Fig. 1) of the delivery process.

Table 2 Examples for therapeutic strategies using polyplexes

Polyplex system	Delivery mode	Gene (target organ)	Therapeutic result	References
ASOM–polylysine	Intravenous (rat)	Human albumin (liver)	Human albumin in serum of Nagase rat	[14]
ASOM–polylysine	Intravenous (rabbit)	LDL receptor (liver)	Reduced hypercholesterolemia	[25]
PEI	Aerosol (mice)	p53 (lung metastases)	Growth inhibition of established lung metastases	[18]
TfR-targeted, shielded PEI	Intravenous (mice)	TNF-alpha (tumors)	Tumor necrosis, inhibition of tumor growth	[21]
EGFR-targeted polylysine	Intravenous (mice)	p21WAF-1, GM-CSF (hepatoma)	Inhibition of hepatoma growth	[24]

Abbreviations: ASOM, asialoorosomucoid; EGFR, epidermal growth factor receptor; GM-CSF, granulocyte macrophage-colony stimulating factor; LDL, low density lipoprotein; PEI, polyethylenimine; TfR, transferrin receptor; TNF, tumor necrosis factor.

Neural—Rapid Prototyping

REFERENCES

1. Wagner, E. Strategies to improve DNA polyplexes for *in vivo* gene transfer—Will "artificial viruses" be the answer? Pharm. Res. **2004**, *21* (1), 8–14.

2. De Smedt, S.C.; Demeester, J.; Hennink, W.E. Cationic polymer based gene delivery systems. Pharm. Res. **2000**, *17* (2), 113–126.

3. Han, S.; Mahato, R.I.; Sung, Y.K.; Kim, S.W. Development of biomaterials for gene therapy. Mol. Ther. **2000**, *2* (4), 302–317.

4. Kichler, A. Gene transfer with modified polyethylenimines. J. Gene Med. **2004**, *6* (S1), S3–S10.

5. Sonawane, N.D.; Szoka, F.C., Jr.; Verkman, A.S. Chloride accumulation and swelling in endosomes enhances DNA transfer by polyamine–DNA polyplexes. J. Biol. Chem. **2003**, *278* (45), 44826–44831.

6. Kwok, K.Y.; Park, Y.; Yang, Y.; McKenzie, D.L.; Liu, Y.; Rice, K.G. *In vivo* gene transfer using sulfhydryl cross-linked PEG-peptide/glycopeptide DNA co-condensates. J. Pharm. Sci. **2003**, *92* (6), 1174–1185.

7. Lim, Y.B.; Kim, S.M.; Suh, H.; Park, J.S. Biodegradable, endosome disruptive, and cationic network-type polymeras a highly efficient and nontoxic gene delivery carrier. Bioconjug. Chem. **2002**, *13* (5), 952–957.

8. Anderson, D.G.; Lynn, D.M.; Langer, R. Semi-automated synthesis and screening of a large library of degradablecationic polymers for gene delivery. Angew. Chem. Int. Ed. Engl. **2003**, *42* (27), 3153–3158.

9. Zou, S.M.; Erbacher, P.; Remy, J.S.; Behr, J.P. Systemic linear polyethylenimine (L-PEI)-mediated gene delivery inthe mouse. J. Gene Med. **2000**, *2* (2), 128–134.

10. Wightman, L.; Kircheis, R.; Rossler, V.; Carotta, S.; Ruzicka, R.; Kursa, M.; Wagner, E. Different behavior of branched and linear polyethylenimine for gene delivery *in vitro* and *in vivo*. J. Gene Med. **2001**, *3* (4), 362–372.

11. Ferkol, T.; Perales, J.C.; Eckman, E.; Kaetzel, C.S.; Hanson, R.W.; Davis, P.B. Gene transfer into the airway epithelium of animals by targeting the polymeric immunoglobulin receptor. J. Clin. Invest. **1995**, *95* (2), 493–502.

12. Ziady, A.G.; Kelley, T.J.; Milliken, E.; Ferkol, T.; Davis, P.B. Functional evidence of CFTR gene transfer in nasal epithelium of cystic fibrosis mice *in vivo* following luminal application of DNA complexes targeted to theserpin–enzyme complex receptor. Mol. Ther. **2002**, *5* (4), 413–419.

13. Wu, G.Y.; Wu, C.H. Receptor-mediated gene delivery and expression *in vivo*. J. Biol. Chem. **1988**, *263* (29), 14621–14624.

14. Wu, G.Y.; Wilson, J.M.; Shalaby, F.; Grossman, M.; Shafritz, D.A.; Wu, C.H. Receptor-mediated gene delivery *in vivo*. Partial correction of genetic analbuminemia in Nagase rats. J. Biol. Chem. **1991**, *266* (22), 14338–14342.

15. Nishikawa, M.; Yamauchi, M.; Morimoto, K.; Ishida, E.; Takakura, Y.; Hashida, M. Hepatocyte-targeted *in vivo* gene expression by intravenous injection of plasmid DNA complexed with synthetic multi-functional gene delivery system. Gene Ther. **2000**, *7* (7), 548–555.

16. Nguyen, H.K.; Lemieux, P.; Vinogradov, S.V.; Gebhart, C.L.; Guérin, N.; Paradis, G.; Bronich, T.K.; Alakhov, V.Y.; Kabanov, A.V. Evaluation of polyether–polyethyleneimine graft copolymers as gene transfer agents. Gene Ther. **2000**, *7* (2), 126–138.

17. Coll, J.L.; Chollet, P.; Brambilla, E.; Desplanques, D.; Behr, J.P.; Favrot, M. *In vivo* delivery to tumors of DNAcomplexed with linear polyethylenimine. Hum. Gene Ther. **1999**, *10* (10), 1659–1666.

18. Gautam, A.; Waldrep, J.C.; Densmore, C.L.; Koshkina, N.; Melton, S.; Roberts, L.; Gilbert, B.; Knight, V. Growth inhibition of established B16-F10 lung metastases by sequential aerosol delivery of p53 gene and 9-nitrocamptothecin. Gene Ther. **2002**, *9* (5), 353–357.

19. Kircheis, R.; Wightman, L.; Schreiber, A.; Robitza, B.; Rossler, V.; Kursa, M.; Wagner, E. Polyethylenimine/ DNA complexes shielded by transferrin target gene expression to tumors after systemic application. Gene Ther. **2001**, *8* (1), 28–40.

20. Ogris, M.; Brunner, S.; Schuller, S.; Kircheis, R.; Wagner, E. PEGylated DNA/transferrin–PEI complexes: Reduced interaction with blood components, extended circulation in blood and potential for systemic gene delivery. Gene Ther. **1999**, *6* (4), 595–605.

21. Kursa, M.; Walker, G.F.; Roessler, V.; Ogris, M.; Roedl, W.; Kircheis, R.; Wagner, E. Novel shielded transferrin–polyethylene glycol–polyethylenimine/DNA complexes for systemic tumor-targeted gene transfer. Bioconjug. Chem. **2003**, *14* (1), 222–231.

22. Hildebrandt, I.J.; Iyer, M.; Wagner, E.; Gambhir, S.S. Optical imaging of transferrin targeted PEI/DNA complexes in living subjects. Gene Ther. **2003**, *10* (9), 758–764.

23. Wolschek, M.F.; Thallinger, C.; Kursa, M.; Rossler, V.; Allen, M.; Lichtenberger, C.; Kircheis, R.; Lucas, T.; Willheim, M.; Reinisch, W.; Gangl, A.; Wagner, E.; Jansen, B. Specific systemic nonviral gene delivery to human hepatocellular carcinoma xenografts in SCID mice. Hepatology **2002**, *36* (5), 1106–1114.

24. Liu, X.; Tian, P.K.; Ju, D.W.; Zhang, M.H.; Yao, M.; Cao, X.T.; Gu, J.R. Systemic genetic transfer of p21WAF-1 and GM-CSF utilizing of a novel oligopeptide-based EGF receptor targeting polyplex. Cancer Gene Ther. **2003**, *10* (7), 529–539.

25. Wilson, J.M.; Grossman, M.; Wu, C.H.; Chowdhury, N.R.; Wu, G.Y.; Chowdhury, J.R. Hepatocyte-directed gene transfer *in vivo* leads to transient improvement of hypercholesterolemia in low density lipoprotein receptor-deficient rabbits. J. Biol. Chem. **1992**, *267* (2), 963–967.

Neural—Rapid Prototyping

Orthopedic Applications: Bioceramic and Biopolymer Nanocomposite Materials

Clark E. Barrett
Ruth E. Cameron
Serena M. Best
Department of Materials Science and Metallurgy, University of Cambridge, Cambridge, U. K.

Abstract

Polymer–ceramic nanocomposites (PMNCs) are defined as polymers containing a ceramic phase that possesses at least one dimension less than 100 nm. Due to the multiplicity of materials, processing conditions, and geometries available to the experimenter, PMNCs present a wealth of different device design directions and possible applications.

INTRODUCTION

Bone is an organ capable of self-repair following an injury. However, the loss of significant bone volume due to infection or trauma may result in a permanent defect at an injury site. Current surgical techniques employed to repair large defects include bone grafting and metallic implants. Bone grafting is limited by the quantity of bone available from a possible donor site, making the procedure unsuitable for large defects. Metallic implants are of particular use in load-bearing environments due to their high strength and toughness. However, these devices are far from optimal solutions due to stress shielding[1] and associated bone atrophy, which is often a cause of device loosening and ultimate failure.[2] Long-term metallic implant concerns involve potential cytotoxicity arising from heavy metal ion liberation and harmful corrosion products.[3]

Much research has been undertaken with the aim of optimizing osseointegration on metal surfaces that possess nanometer-scale topographic features. Often, these techniques incorporate hydroxyapatite (HA) and various substituted HA types.[4] Nanomanipulation is not limited to the bone—surface nanopatterning is employed in the cartilage, vascular tissue, bladder, and other tissue engineering applications.

A basic literature search reveals a plethora of investigations into polymer-ceramic nanocomposite (PMNC) development for clinical use which is logical considering the number of degrees of freedom in these systems. PMNCs are defined as polymers containing a ceramic phase that possesses at least one dimension less than 100 nm. Due to the manifold variables that determine PMNC behavior, this entry considers only a limited number of materials used in PMNC designs. Emphasis is given to the effects of nanoceramic addition on material properties: degradation kinetics, mechanical and biological effects (Sections "Nanocomposite Degradation Behavior and Acidity Regulation," "Mechanical Behavior," and "Bioactivity"), and the specific changes manifest by nanoparticles as opposed to conventional microparticle inclusions. Furthermore, modeling attempts to account for modifications to material parameters effected by nanoparticle properties are reported (Section "Theoretical Approach").

In order to account for nanoparticle effects in these materials, it is first necessary to describe other material properties and modifications wrought by nanoscale phenomena. To this end, some of the most common polymers and ceramics used in osseous tissue regeneration are elucidated in Sections "Polymers" and "Ceramics" before proceeding to nanoparticle effects on material properties.

MATERIALS

Polymers

The use of polymers in prostheses is long established and widespread. The choice of polymer that may be used in a biological implant is limited by the requirements that the polymer be nontoxic, biocompatible, and biodegradable and the material's degradation products should have no negative effects on the surrounding tissues and organs of a host organism.[5] Most available polymeric devices currently on the market do not degrade *in vivo*, e.g., ultra high molecular weight polyethylene, which is used in total hip replacements.[6] The majority of polymers investigated for osseous tissue regeneration or defect filling pertain to the polyester family to which poly(α-hydroxy acids) [incorporating polylactide (PLA), polyglycolide, and their copolymers] belong.[7] A biopolymer's mechanical properties,

Concise Encyclopedia of Biomedical Polymers and Polymeric Biomaterials DOI: 10.1081/E-EBPPC-120052548

Neural—Rapid Prototyping

biological effects, and degradation kinetics are influenced by its hydrophobicity, molecular weight distribution, glass transition temperature, crystallinity, geometry, temperature, and polymer processing technique. Accordingly, a wide range of possibilities exist to manipulate a polymer's properties to suit different requirements. In addition, radiation is often used as a sterilization method prior to clinical use; therefore, the effects of radiation damage have been investigated to ascertain changes to molecular weight, glass transition temperatures, and other properties with demonstrable effects.[8,9]

Poly(α-hydroxy acids) have a long history of the U.S. Food and Drug Administration (FDA)-approved use in medical devices and a large variation in properties between different poly(lactic-co-glycolic acid) (PLGA) forms, making this class of polymer a preferred candidate for a resorbable tissue engineering scaffold polymeric phase.[10,11] PLGA and other polyesters degrade by hydrolysis (or partly via enzymatic action *in vivo*), producing lactic acid, which is metabolized by the body via the tricarboxylic acid cycle to be excreted as CO_2 and H_2O,[12] and glycolic acid, which is excreted in urine.

Polyester degradation in an aqueous environment is classified as bulk heterogeneous degradation in which de-esterification is catalyzed by $H^+_{(aq)}$ ions generated by acid dissociation of degradation products. The acids also diffuse through the material; hence, degradation is a reaction-diffusion problem, depending on material geometry, oligomer diffusion coefficients, and associated boundary conditions, among other factors. Due to autocatalysis, acid accumulates at the center of the material, whereas acids produced closer to the material boundary may diffuse into the surrounding medium, resulting in position and time-dependent degradation rates with de-esterification proceeding faster at the material center compared to the boundary. Accumulation of acidic degradation products from PLGA has been shown to induce chronic inflammatory responses *in vivo* and a propensity to elicit fibrous tissue encapsulation at the polymer-hose interface;[13] fibrosis at the implant-bone interface can cause device failure at early degradation times[14] For more information regarding PLGA degradation, the reader is directed to the work of Therin et al.[15] and Li, Garreau, and Vert[10]

In this entry, the D,L-lactide/glycolide molar ratio in PLGA studies is given in parentheses. A plethora of other biodegradable polymers have been extensively investigated, and a brief description of some common types is given in Table 1.

The motivations for ceramic nanoparticle addition to polymers include possible improvements to degradation kinetics control offered by polymer-ceramic micro-composites, biological enhancement, and possible nanoparticle stiffening improvements to PMNC mechanical properties affected by nanoparticle addition. PMNCs are viewed as an evolution of traditional micrometer-sized polymer-ceramic composites. Basic properties of the most commonly used ceramics are described in Sections "Ceramics" and "Calcium Phosphate Solubility and pH Buffering Reactions."

Ceramics

The concept of the incorporation of HA or another bioceramic to a polymer was introduced by Bonfield et al.[16] based on the observation that the bone consists of an organic matrix reinforced with a mineral phase. Nanocrystalline HA is similar in composition to the main inorganic constituent of the bone. Consequently, many scaffold designs are based on or have incorporated HA as the primary ceramic phase. Nevertheless, the physical properties and mechanical reliability of pure HA ceramics are poor compared to the bone and, for this reason, is principally used in the form of powders, implant coatings, low-loaded porous implants, and bone cement.[4,17] HA exhibits low biodegradability, with some studies reporting incomplete reabsorption of sintered HA after 9 months *in vivo*.[18] Numerous investigations have attested to enhanced osteo-conductivity and osteoblast metabolism on nanoscale HA.[19,20] Much research has focused on modifying the chemical properties of HA to enhance its osteoconductivity; the interested reader is directed to the summary in the work of Suchanek et al.[21]

For certain applications, high ceramic degradation rates are required; hence, interest in the tricalcium phosphate (TCP) $Ca_3(PO_4)_2$ class of bioceramic has increased. TCP has four known polymorphs: α, β, γ, and super-alpha (a). The a and y phases are only observed at high temperatures and pressures.[22] The p phase is the most stable form at standard temperature and pressure; the a phase is thermodynamically stable between 1120°C and 470°C in the absence of impurities[23] and occurs at room temperature by quenching or rapid cooling from its stable state. Both α- and β-TCPs are prepared by the thermal decomposition of calcium-deficient HA, and often, both TCP phases occur with HA after processing.[24]

Table 1 Material properties of various biopolymers

Polymer	Crystallinity	Tensile Strength (MPa)	Young's Modulus (GPa)	Approximate Degradation Time
Polyglycolide	Semicrystalline	~70	~6	6–12 months
Polycaprolactone	Semicrystalline	20–25	0.4	~3 years
Poly-L-lactide	Semicrystalline	60–70	3	~3 years
Collagen	Variable	50–100	1	—

Calcium phosphate solubility and pH buffering reactions

The acidity solubility regimes and apatite reactions with carboxylic acids are fundamental phenomena in composite behavior and affect local pH variant autocatalysis (see Sections "Polymers" and "Theoretical Approach"). There are only three stable calcium phosphates at standard temperature and pressure: 1) monocalcium phosphate monohydrate $[Ca(H_2PO_4)_2,\ pH < 2.51]$; 2) brushite $(CaHPO_4 \cdot 2H_2O,\ 1.5 < pH < 4.2)$; and 3) HA $(pH > 4.2)$.[25] The pH dependence on solubility is an important factor of any model of these materials' temporal and spatial degradation. The solubility of α and β-TCP at pH 7.4 is $K = 10^{-25.5}$ and $K = 10^{-29.5}$, respectively.[26] HA possesses a much lower solubility product at pH 7.4 $(K = 10^{58.6})$.[27] Dissolved TCP, given a sufficient concentration of calcium and phosphate ions, will reprecipitate as HA, which is the only stable calcium phosphate at standard temperature and pressure, forming an HA bounding layer on TCP crystal surfaces; cf. Fig. 1).[28,29]

A primary reason for the ceramic addition to PMNCs is to buffer acidity and to modify degradation kinetics in the degrading nanocomposite (see Fig. 2). The end scission reaction of PLGA autocatalyzed by H+j may be written as

$$\overset{H^+}{PLGA + H_2O \Leftrightarrow R - COOH + - OH.} \tag{1}$$

The carboxylic acid has a high degree of acid dissociation, and thus, degradation releases H+q, which autocatalyzes further hydrolysis, i.e.,

$$R - COOH \Leftrightarrow R - COO + OH_{aq}^+. \tag{2}$$

In an aqueous medium, the calcium phosphate dissociates to form

$$Ca_3(PO_4)_2 \Leftrightarrow Ca_{aq}^{2+} + (PO_4^{3-})aq. \tag{3}$$

The phosphate ions subsequently react with the hydrons via the buffering reactions, forming monohydrogen phosphate

$$PO_4^{3-} + H^+ \Leftrightarrow HPO_4^{2-} \tag{4}$$

dihydrogen phosphate

$$PO_4^{3-} + H^+ \Leftrightarrow H_2PO_4^- \tag{5}$$

and phosphoric acid

$$H_2PO_4^- + H^+ \Leftrightarrow H_2PO_4. \tag{6}$$

Equations 4 through 6 elucidate the mechanisms that remove catalytic H+q from the solution. TCP and HA dissolution is affected by acidity, physical disintegration into small particles due to preferential dissolution at grain boundaries, and biological factors.[31] Equations 2 through 6 are important in nano-composite material modeling incorporating a ceramic phase (Section "Theoretical Approach").

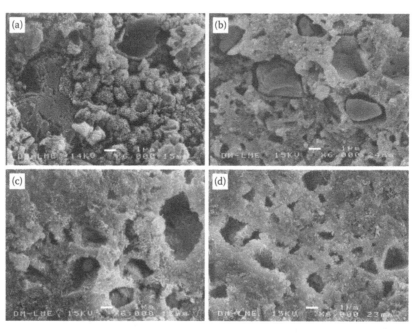

Fig. 1 Evolution of the structure of α-TCP crystals observed using scanning electron microscopy (SEM) after (A) 2 hr, (B) 8 hr, (C) 64 hr, (D) and 360 hr in aqueous medium.
Source: From Ginebra et al.,[29] with permission from Elsevier.

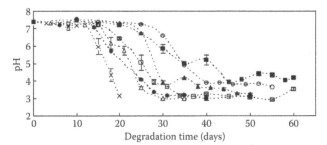

Fig. 2 Variation of pH of solution containing α-TCP-PLGA composite materials as a function of degradation time. x, Pure PLGA; A, 5% TCP; •, 10% TCP; □, 15% TCP; ▲, 20% TCP; O, 30% TCP; ■, 40% TCP.
Source: From Ehrenfried et al.,[30] with permission from Elsevier.

Other Constituents

Further considerations to PMNC design include improved coupling between the mineral and organic phases. Numerous investigations have highlighted the critical role played by polymer-ceramic bonding interactions in reducing nanoparticle agglomeration[32] or improving mechanical properties (either by reductions in stress concentrations in the polymer matrix or indirectly via T_g manipulation), especially when considering device degradation and performance in an aqueous environment. The interaction strength between the two phases has important ramifications for material behavior, specifically mechanical behavior and glass transition temperature modifications, which are further addressed in Sections "Mechanical Behavior" and "Glass Transition Behavior." Often, only physical adsorption is achieved between the polymeric and ceramic phases. To address this, the ceramic surface may be treated to become more chemically compatible with the polymer matrix.[33] A common reactive treatment is the incorporation of silanes to the surface of the ceramic phase, which results in increased end-group availability fomenting stronger polymer-ceramic bonding.[32] Thus, the interaction between the inorganic surface and the matrix polymer can be tuned by selecting appropriate end groups. These considerations, albeit important with regard to nanocomposite behavior, are not considered in detail in this entry. For a treatise on the silane treatment of fillers, the reader is directed to the work of Plueddemann.[34]

Other PMNC topics that will not be discussed include the addition of growth factors such as bone morphogenic proteins[35] to induce new bone formation and a plethora of other factors, including scaffold geometry, that play important roles in degradation kinetics.[36]

MATERIAL PROPERTIES

Nanocomposite Degradation Behavior and Acidity Regulation

TCP-PLGA composite acidity buffering capacity generally improves with TCP content.[6,30,37] degraded α-TCP-PLGA

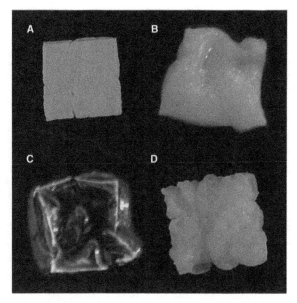

Fig. 3 Morphological state of α-TCP-PLGA(50:50) nanocomposites at different degradation times in phosphate-buffered saline (PBS). (A) and (B) α-TCP (30 wt%) and nanocomposite (20 wt%) after 43 days. (C) Pure PLGA(50:50) at 21 days. (D) Microcomposite (30 wt%) at 36 days.
Source: From Yang et al.,[37] with permission from Wiley.

(50:50) composites with varying weight ceramic loadings and particle sizes in simulated body fluid (SBF). Differences in degradation kinetics are shown in Figs. 3 and 4. figure 3 clearly demonstrates the reduction in composite degradation rates due to reaction rate modification resulting from the ceramic inhibition of the autocatalytic effect. Moreover, Fig. 3 shows ceramic crystal-size effects, with 30 wt% nanocomposites exhibiting a delay of approximately 7 days in pH change, onset of polymer mass loss, and time of maximum water absorption compared to the equivalent microcomposite. A comparison of surface morphologies for both composite types at different times is shown in Fig. 4.

The overall pH change in buffering medium during *in vitro* degradation is independent of ceramic particle size, although the rate of pH change is observed to be particle size dependent, with nanoparticles exhibiting a slower pH decrease during degradation. This is important, as slower changes allow acids to be removed by the body, preventing accumulation and cytotoxicity. However, numerous studies have established that, above particular ceramic weight loadings (approximately 30%), little change in degradation kinetics or acid release profile is observed. This convergence has often been attributed to nanoparticle agglomeration, which is now challenged by new material modeling results, as summarized in the section "Theoretical Approach."

Tang et al.[38] investigated the water absorption of poly(3-hydroxybutyrate-co-3-hydroxyvalerate)/HA nanocomposites using a standard Fickian diffusion description.[39,40] The nano-HA used in the study was modified with a silane coupling agent. Water absorption in the materials is greatly

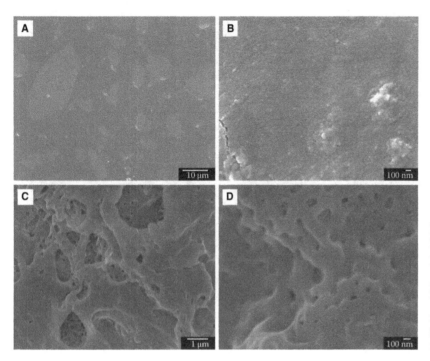

Fig. 4 SEM images of 30 wt% α-TCP-PLGA(50:50) microcomposites and nano-composites at various degradation times *in vitro* in SBF. (A) Undegraded microcomposite. (B) Microcomposite degraded for 36 days. (C) Undegraded nanocomposite. (D) Nanocomposite degraded for 30 days.
Source: From Yang et al.,[37] with permission from Wiley.

affected by HA leaching. Furthermore, water diffusion coefficients have an inverse relationship with nano-HA filler content, although diffusion coefficients remained within the same order of magnitude for all composite types.

Mechanical Behavior

There is much evidence regarding the enhancement of biopolymer mechanical strength by the incorporation of nanoparticulate bioceramics. McManus et al.[41] reported significantly greater bending moduli in PLA containing 40 and 50 wt% nanophase alumina, titania, and HA than composites with coarser grained ceramics. Liu and Webster[42] characterized the compressive and tensile properties of composites composed of dispersed and agglomerated particles of 30 wt% HA and titania contained in PLGA(50:50) matrices. Titania nanocomposites exhibited superior elastic and bulk moduli and greater tensile strength at yield than pure PLGA and agglomerated composites, whereas dispersed nano-HA demonstrated inferior moduli but greater ductility than the agglomerated composite. Nano titania and HA composite differences were attributed to different polymer-ceramic bonding strengths.

PMNC nanoparticle effects may be expressed directly as a matrix stiffener or indirectly by changing the polymer phase's thermodynamic state by modifying the material's glass transition temperature, T_g. Wilberforce et al.[43,44] reported large T_g reductions with increasing nano α-TCP loadings in PLGA(50:50), which were attributed to the large interfacial area and poor polymer-ceramic bonding. Results from expanding the study to gauge changes

wrought by differing polymer-processing methods are presented in Fig. 5, which demonstrates T_g reduction due to polymer-ceramic interface effects (variations between polymer-processing routes probably result from polymer thermal degradation during solvent evaporation). Also shown in Fig. 5 is the dramatic effect of water diffusion and material plasticization, which is detrimental to the composite's mechanical strength—an important factor influencing composite behavior, which is further assessed in Section "Glass Transition Behavior."

The transition to a rubbery state in polymers results in increased polymer chain mobility and significant changes to material properties. Elastic polymers are much more viscoelastic and present reduced storage moduli for $T > T_g$ than in the glassy state. Consequently, systematic studies and empirical descriptions of MPNC T_g variation with the ceramic interface area, polymer-ceramic bonding, water content and other factors affecting T_g are required.

Neglecting material property changes at temperatures close to T_g, numerous studies have demonstrated that, in general, a polymer's bulk and elastic moduli increase with the incorporation of nanoparticles with the increase being proportional to the nanoparticle loading and increasing for smaller particles. The strengthening mechanism is not understood completely.[45] Mechanical deformation theory indicates that the high-volume fraction of interfacial regions compared to bulk materials leads to increased deformation by grain-boundary sliding and short-range diffusion-healing events as the grain size is reduced, thus increasing ductility for nanocrystalline ceramics.[46] This issue is complicated close to T_g and is exemplified in Fig. 6, which shows the storage modulus of α-TCP-PLGA(50:50)

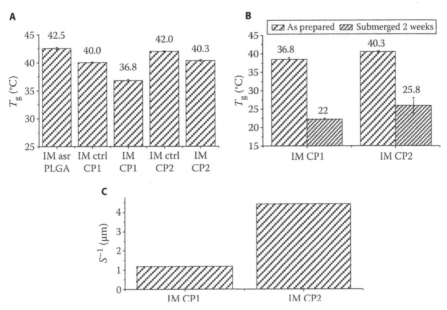

Fig. 5 **(A)** 7g measured using differential scanning calorimetry of injection-molded samples (IM) processed via solvent evaporation (CP1) or twin screw extrusion (CP2). IM asr PLGA denotes injection-molded PLGA as received from the supplier. IM ctrl CP1 and IM ctrl CP2 represent injection-molded PLGA prepared and compacted via solvent evaporation and twin screw extrusion, respectively, which are used as controls. **(B)** 7g of samples after 2 weeks of degradation in PBS solution. **(C)** Reciprocal surface area of α-TCP particles per unit volume of composite.
Source: From Wilberforce et al.,[43] with permission from Springer.

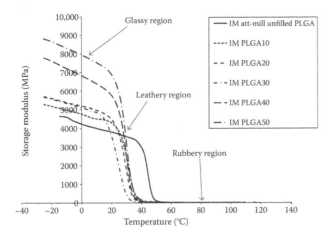

Fig. 6 Storage modulus as a function of temperature measured by dynamical mechanical testing performed at 1 Hz for dry nano α-TCP-PLGA(50:50), with different ceramic weight loadings illustrating the critical role of T_g on mechanical properties of PLGA PMNCs.
Source: From Wilberforce et al.,[43] with permission from Springer.

nanocomposites prepared by injection molding as a function of temperature.[44] The pure polymer exhibits the largest T_g, which is approximately 50°C, and the lowest T_g occurs in the 30 wt% nanocompos-ite. The higher ceramic loaded PMNCs demonstrate higher transition temperatures due to particle agglomeration in these samples,

which indicates that the phenomenon is attributable to nanoparticle effects. Far from T_g, the storage modulus increases with filler content. However, due to nanoparticle-induced changes in T_g and subsequent deterioration of mechanical strength, at body temperature, any enhancement of PLGA(50:50) properties caused by nanoparticle addition is lost, and the pure polymer possesses superior properties. Consequently, nanoparticle addition to some polymers such as PLGA and resulting T_g reduction might be detrimental to the desired materials' mechanical properties compared to pure polymer or conventional microcomposites. This factor is considered further in Section "Glass Transition Behavior." Wilberforce et al.[47] showed that the addition of α-TCP nanoparticles to poly-l-lactide (PLLA), followed by quenching or annealing, reduced the T_g to a minimum of 60°C, making rubber transition effects unimportant at body temperature (pure PLLA exhibits a glass transition at approximately 70°C).

Glass Transition Behavior

Reductions in T_g as a function of film thickness have been well documented for polymer nanofilms.[48] Bulk polymers may be described as having two regions with two different glass transition temperatures for the surface and bulk, T_g^{surf} and T_g^{bulk}, respectively. The surface layer extends to approximately 100 nm from the surface and is hypothesized to represent a liquid-like surface layer with increased polymer mobility compared to a bulk polymer substratum.[49]

Neural—Rapid Prototyping

For a freestanding pure polymer thin film of total thickness h, the contribution of the surface and bulk glass transition temperatures to the total material glass transition temperature as a function of film thickness is

$$T_g(h) = T_g^{bulk} + \frac{2\varepsilon}{h}\left(T_g^{surf} - T_g^{bulk}\right) \qquad (7)$$

where ε represents the surface region thickness. This description assumes a freestanding polymer film that produces the largest change in T_g. Glass transition temperature deviations are observed in supported films whereby the change in T_g depends on the strength of the interaction between the polymer and the substrate. Polymer-substrate interfaces exhibiting attractive interface properties (wetting) and those with weak interactions (nonwetting) manifest T_g increases and decreases, respectively.[50] Interfacial polymer-ceramic properties play a critical role in composite mechanical properties.[51] Moreover, improvements in polymer-ceramic wettability aid nanoparticle dispersion, as demonstrated by Guo et al.,[52] in which improved nanoparticle dispersion and consequent tensile strength increases were observed for Fe_2O_3 nanoparticle-resin composite treated using the bifunctional coupling agent methacryloxypropyltrimethoxysilane compared to uncoupled nanocomposites.

Bansal et al.[53] demonstrated an equivalence principle between thermo-mechanical thin film properties and polymer-nanoparticle surface interactions in PMNCs. The authors calculated three-dimensional average particle separations of silica nanoparticles dispersed in polystyrene (PS) from transmission electron microscopy images and showed a good correlation between T_g as a function of particle separation for the composites and T_g as a function of polymer film thickness for PS-silica nanocomposites (cf. Fig. 7).

PMNCs exhibit a wide variation in glass transition temperature, which depends on factors such as polymer and nanoparticle composition, solvent extraction, and nanoparticle sizes and loadings (Section "Mechanical Behavior"). Generally, increasing the polymer's molecular order and greater intermolecular energy raises T_g.[54] As a general rule, T_g increases for larger polymer cohesive energies and greater molecular order, i.e.,

$$T_g = \frac{\Delta H}{\Delta S} \qquad (8)$$

where ΔH and ΔS represent the enthalpy and entropy changes wrought by the transition to a rubbery state, respectively. Increasing polymer cross-linking results in greater cohesion energy, which raises ΔH. More importantly, in view of water-induced plasticization (cf. Fig. 5), composite water absorption increases the system disorder, which can result in a dramatic T_g decrease.

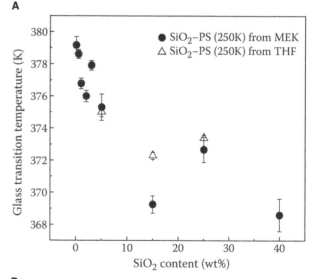

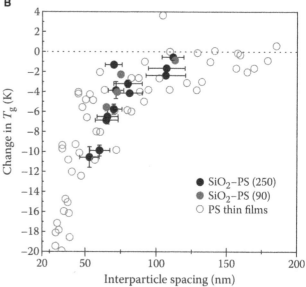

Fig. 7 (a) T_g of SiO_2-PS nanocomposites as a function of silica weight content for composites prepared using the solvents methyl ethyl ketone (MEK) and tetrahydrofuran (THF). Large deviations observed at high weight loadings are caused by particle agglomeration. (b) Comparison between T_g reduction measured for thin PS films as a function of film thickness and T_g reduction of PS-silica nanocomposites as a function of mean interparticle spacing. Numbers in parentheses denote matrix molecular weight.
Source: From Bansal et al.[53] Courtesy of Nature Publishing Group.

The structural relaxation (aging) of polymers is a closely related phenomenon to T_g surface modification, which is influenced by interface effects with relaxation rate changes caused by surface interactions extending more than 100 nm into the film interior.[55] The authors believe these phenomena to be of great importance in designing PMNCs for clinical use, not only in the assessment of device behavior in the laboratory but also in aging-induced changes in properties due to storage time before device implantation in patients. Moreover, these properties must be addressed for all PMNC types and also evaluated as a function of degradation time. Furthermore, nanoparticle-induced changes in composite mechanical properties and aging can be modified by the use of binding agents, which warrants further development.

Bioactivity

Yang et al.[56] evaluated the *in vitro* bioactivity of nanostructured α-TCP-PLGA(75:25) composites by quantifying the rapid formation of bonelike apatite layers on the material's surface while immersed in SBF (the a phase was chosen due to its greater solubility than β-TCP). Enhanced apatite nucleation and dense lamellar-like apatite formation were observed after 7 and 14 days of immersion, respectively. Acellular indicators of improved bioactivity were corroborated by *in vitro* cell culture studies using human osteoblast-like (HOB) cells, elucidating greater cell numbers on the nanocomposite as the culture proceeded and the observation of a confluent lamellar-like apatite layer beneath a proliferating HOB layer at day 16. Enhanced apatite nucleation was attributed to the nanostructured surface providing large numbers of nucleation sites for apatite precipitation or fast dissolution of nanoparticles near the surface. Both of these mechanisms most

likely work in conjunction to create rougher surface topographies. In addition, in a later study, Yang, Best, and Cameron[37] reported the liberation of residual micron-scale ceramic particles (which are known to cause adverse cell response) from the microcomposite into the buffer medium, which did not occur in the nanocomposites.

Biocompatibility enhancement is not confined to calcium phosphate addition; other nanoceramics such as titania and alumina have demonstrated the inducement of improved cell function.[57,58] Nanophase titania-PLGA(50:50) composites investigated by Liu, Slamovich, and Webster[59] demonstrated that the greatest *in vitro* osteoblast adhesion and optimal indicators of bioactivity such as collagen, alkaline phosphatase activity, intracellular and extracellular protein synthesis, and calcium-containing mineral deposition occurred for composites possessing surface roughness values, which most closely matched that of the natural bone (cf. Fig. 8). In addition, the material surface roughness as opposed to ceramic surface coverage was shown to dominate the cell response. The surface topographic effect is not limited to bioceramics. Webster and Ejiofor[60] analyzed osteoblast adhesion on several nanophase metals, which exhibited increased osteoblast adhesion to nanopatterned surfaces relative to conventional surfaces presenting micrometer-sized topographic features. Interestingly, the study observed preferential osteoblast adhesion at grain boundaries.

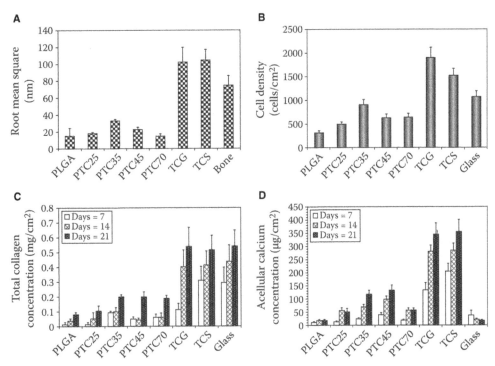

Fig. 8 (A) Surface roughness (root mean square) of PLGA, nanoparticulate titania-PLGA (ultrasonicated at 25%, 35%, 45%, and 70% maximum power), green titania (TCG), sintered titania (TCS), and porcine bone. (B) Cell adhesion to various materials [PLGA, nanoparticulate titania-PLGA (ultrasonicated at 25%, 35%, 45%, and 70% maximum power), TCG, TCS, and glass] after incubation for 4 hr. (C) and (D) Total collagen content and acellular calcium concentration for various surfaces analyzed.
Source: From Liu et al.,[59] with permission from Wiley.

The compendium of studies performed to assay cell responses to nanometer-sized topographic surface modification clearly demonstrates that nanometer topographies alter cell behavior significantly. In the study of Palin Liu, and Webster[61] the authors sought to assess the preponderance of nanoscale topography over extraneous factors such as surface chemistry and surface energy. To this end, nanometer-scale and conventional topographic structures were formed on PLGA by molding and seeded with osteoblasts. The bioactivities of the different surfaces were assessed by quantifying osteoblast proliferation as a function of time, with nanopatterned PLGA demonstrating more than double the initial number of adhered osteoblasts than the conventional surface after 1 day and 5 days in culture.

Enhancing osteoclast activity is also important for nanocomposite clinical performance; rapid bone deposition mediated by osteoblasts and remodeling via osteo-clasts is a vital component of wound healing and the incorporation of the orthopedic device. Webster et al.[62] showed that indicators of osteoclast activity, the number of resorption pits, and tartrate-resistant acid phosphatase increased on nanophase alumina and HA compared to conventional-sized particles. Increased wettability or hydrophilicity is associated with increased protein absorption and, consequently, increased cell adhesion and enhanced function. The study showed increased numbers of absorption pits on nanometer surfaces compared to micro surfaces.

Further biological considerations of PMNC performance include fibroblast activity, because excessive fibrosis at the implant-host interface is a known factor in device failure.[63] Prolonged increased fibroblast numbers at implant interfaces is associated with callous formation and associated soft-tissue formation rather than bone juxtaposition. Thus, a reduction in fibroblast activity (or limited increase in fibroblast activity relative to osteoblasts) might be beneficial to device-mediated wound healing. Promisingly, some studies have shown that the ratio of osteoblast to fibroblast adhesion increased from 1:1 on microparticle alumina surfaces to 3:1 on nanoparticle surfaces.[63]

Qualitative arguments for the improvement in biological action of nanostructured materials assumed that the improvement was due to mimicking the naturally occurring topographies of natural biological tissues; HA crystals in bone are approximately 50 nm long and 5 nm in diameter. More quantitatively, the enhanced bioactivity observed with nanocomposites results from protein interactions with nanopatterned surfaces,[64] whereby cumulative protein absorption has been shown in many studies to be higher on nanograined materials.[57,65] Several proteins that are known to augment osteoblast function have been shown to be deposited much faster and in greater quantity on nanocomposites compared to microcomposites. Dimensionality effects are also observed in cartilage and tendon tissue engineering constructs. Research indicates that chondrocyte and mesenchymal stem cell activities are strongly dependent on scaffold dimensionality.[31]

Nanoparticulate addition to composites and surface coatings is not the only application of nanotechnology to medical devices. Other techniques have been used to create nano-sized surface features via chemical etching, electron beam lithography, and polymer demixing,[66,67] and nanopatterned surfaces produced by electrospraying, which has demonstrated the ability to direct osteoblast growth.[68]

THEORETICAL APPROACH

There has been significant development of mathematical models designed to predict the rate of drug release from pure biopolymers using a variety of modeling techniques, many of which elucidate the observed phenomena to a high degree of accuracy for materials with simple geometries such as planes, cylinders, and spheres. However, modeling descriptions of PMNC behavior and degradation kinetics are underdeveloped due to the complexity of degradation processes. Many polymer degradation models employing a plethora of techniques, ranging from Monte Carlo statistical analyses of PLGA microspheres to direct analytical solutions of Fick's equations, have been described in the literature. Finite element modeling is now being used as an indispensable modeling tool, and models have been developed to predict degradation mechanics and diffusion mechanisms in pure polymer matrices. Pan et al.[69] have advanced model complexity to include the effects of ceramic incorporation. For simplicity, the degradation model of TCP-PLGA composites assumes abundant water absorption and a spherical representative unit volume of a TCP particle surrounded by a PLGA matrix. PLGA degradation is primarily a chain end scission. Wang et al.,[70] Han and Pan,[71] and Han et al.[72] showed that the rate of chain scission dR is equal to

$$\frac{R}{dt} = \frac{dC_{end}}{dt} = k_1 C_e + k_2 C_e C_{H^+} \tag{9}$$

where C_e, C_{H^+}, k_1, and k_2 represent the ester concentration, hydrogen ion concentration, random polymer scission rate (uncatalyzed reaction), and autocatalysis rate constants, respectively. The concentrations of the ions at each step are calculated by dissociation rate constants. Moreover, the model must account for the differing solubilities of the calcium phosphates. Equilibrium in Eq. 3 is not reached due to removal of phosphate ions via processes in Eqs. 4 through 6. Accordingly, there is an undersaturation leading to TCP dissolution. The rate of TCP dissociation may be described in terms of an ion flux J, which depends on the surface area of the particle and the undersaturation, which is typically represented in terms of a power law, i.e.,

$$J = A_m \sigma^n$$

where σ is the undersaturation, and A and n are material constants.[26] Using reaction equilibria constants, mass conservation equations, and the reaction equations, Pan et al.[68] defined a relative rate of calcium phosphate dissolution to autocatalytic hydrolysis S_{cp} as

$$S_{cp} = \frac{A_d \left(\dfrac{A_{cp}}{V_{polymer}} \right)}{k_2 (C_{H^+})_{pH=7.4} \times (C_{ca^{2+}})_{eq}} \qquad (10)$$

where A_d and A_{cp} represent the TCP dissolution rate and the TCP crystal surface area, respectively. Accordingly, due to the surface area term, not only is the autocatalytic reaction rate dependent on the quantity of TCP, but the size of the particles also causes the different degradation kinetics between composites of the same TCP loadings but different morphologies.

Using this approach, Pan et al.[69] predicted that there exists a saturation limit for the effectiveness of α-TCP as defined in terms of S_{cp}. Above a certain limit, the dissolution of calcium phosphate has no effect on the rate of polymer degradation, as any hydrogen ions produced by degradation react with phosphate ions and are removed from the reaction-diffusion process. In the case of S_{cp} saturation, the controlling parameter on the reaction rate is the noncatalyzed reaction rate constant k_1. For completeness, the degradation kinetics of polymers that degrade by uncatalyzed hydrolysis only are not affected by TCP quantity or form.

Graphical representations of applying this formalism to empirical α-TCP-PLGA(50:50) degradation data are shown in Figs. 9 and 10. Figure 9 shows data from the degradation study of Yang, Best, and Cameron,[37] where the onset of pH decrease was hypothesized to herald the simultaneous start of oligomer loss from degrading composites. These mass loss onset times demonstrated close agreement with simulated polymer molecular mass loss shown in Fig. 10 using the formalism described above.

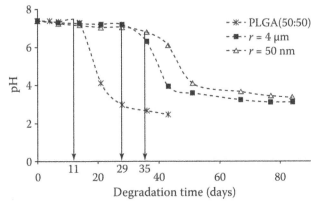

Fig. 9 pH variation as a function of degradation time of buffer medium during degradation of pure PLGA(50:50) and PLGA(50:50) incorporating 30 wt% α-TCP inclusions of 4 μm and 50 nm. Data from Yang et al.,[37] fitted with model results from Pan et al.[69]

Source: From Pan et al.,[69] with permission from Elsevier.

The data suggest that, due to TCP dissolution saturation (cf. Fig. 11), nano-sized α-TCP crystals have a small but noticeable effect on degradation rates compared to microparticles with the same weight loading. However, for lower weight loadings, the difference between microcomposites and nanocomposites is more pronounced.

Note that these models are still in their infancy, and further phenomena such as water transport, swelling, and polymer phase changes are required to complete a total material description. However, these early results are promising and will aid future material development.

CONCLUSION

The theoretical modeling of α-TCP-PLGA(50:50) degradation kinetics raises important questions with regard to

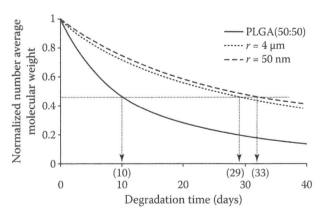

Fig. 10 Predicted half degradation times for pure PLGA(50:50) and α-TCP-PLGA(50:50) composites (30 wt% α-TCP), assuming spherical particles of 4 μm and 50 nm in diameter.
Source: From Pan et al.,[69] with permission from Elsevier.

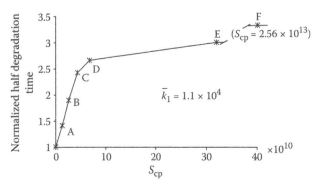

Fig. 11 Predicted normalized degradation times of PLGA(50:50) using noncatalyzed reaction rate constant of 1.1×10^4. Points A through D denote α-TCP-PLGA(50:50) composites containing 30-μm-diameter spherical particles at loadings of 10, 20, 30, and 40 wt%, respectively. Points E and F represent 30 wt% loadings of α-TCP of 4 μm and 50-nm diameter, respectively. N.B. Point F lies off the main axis
Source: From Pan et al.,[69] with permission from Elsevier.

the cost-benefit returns of using ceramic nanoparticles to improve composite degradation times. Moreover, the decrease in T_g observed due to poor bonding at the interface between TCP and PLGA complicates the assumption that nanoparticle inclusions have a positive effect on mechanical properties. On the other hand, a plethora of evidence indicates that biocompatibility is vastly enhanced using nanoceramic fillers. The specific analysis performed in Section "Theoretical Approach" and subsequent discussion is only applicable to TCP-PLGA composites; however, the technique is applicable to other PMNC types and should be extended.

Due to the multiplicity of materials, processing conditions, and geometries available to the experimenter, PMNCs present a wealth of different device design directions and possible applications. The authors believe that, to further the goal of employment in clinical applications, many other issues must be addressed. These include the systematic study of coupling agents to improve the interfacial bonding between ceramic and polymer phases. Moreover, there is a dearth of information regarding T_g and the mechanical property evolution of nanocomposites and micro-composites caused by polymer hydration and possible debonding between the organic and inorganic phases. Much more data are required regarding material properties at physiological temperatures. More practically, for materials intended for clinical use, the choice of sterilization technique such as y irradiation prior to implantation has demonstrable effects on polymer degradation rates.[9,73] Systematic investigations of sterilization effects on mechanical and biological PMNC properties are required. Furthermore, a fundamental prerequisite for commercial PMNC validity are the effects of material aging and changes in performance between fabrication and implantation.

On the biological level, important issues regard the effects of free nanoparticles released into the host if the polymer degrades before total ceramic dissolution. Moreover, although good evidence exists regarding the *in vitro* indications of superior nanocomposite bioactivity, there is a distinct lack of histological data comparing microparticle and nanoparticle polymer composites.

REFERENCES

1. Park, J.B.; Lakes, R.S. *Biomaterials: An Introduction*; Plenum Press: New York, 1992.

2. Wolff, J. *Das Gesetz der Transformation der Knochen*; Herschwald: Berlin, 1987.

3. Williams, D.F. *Fundamental Aspects of Biocompatibility*; CRC Press: Boca Raton, FL, 1981.

4. Best, S.M.; Porter, A.E.; Thian, E.S.; Huang, J. Bioceramics: Past, present and for the future. J. Eur. Ceram. Soc. **2008**, *28* (7), 1319–1327.

5. Yang, F.; Cui, W.; Xiong, Z.; Liu, L.; Bei, J.; Wang, S. Poly(L,L-lactide-co-glycolide)/tricalcium phosphate composite scaffold and its various changes during degradation *in vitro*. Polym. Degrad. Stab. **2006**, *91* (21), 3065–3073.

6. Wang, M. Developing bioactive composite materials for tissue replacement. Biomaterials **2004**, *24* (13), 2133–2151.

7. Balasundaram, G.; Webster, T.J. Overview of nanopolymers for orthopedic applications. Macromol. Biosci. **2007**, *7* (5), 635–642.

8. Yoshioka, S.; Aso, Y.; Kojima, S. Drug release from poly(DL-lactide) microspheres controlled by y-irradiation. J. Control. Release **1995**, 37, 263.

9. Tan, H.Y.; Widjaja, E.; Boey, F.; Loo, S.C. Spectroscopy techniques for analyzing the hydrolysis of PLGA and PLLA. J. Biomed. Mater. Res. Part B Appl. Biomater. **2009**, *91* (1), 433–440.

10. Li, S.M.; Garreau, H.; Vert, M. Structure-property relationships in the case of the degradation of massive poly(a-hydroxy acids) in aqueous media—Part 1: Poly(DL-lactic acid). J. Mater. Sci. Mater. Med. **1990**, *1* (4), 198–206.

11. Lu, L.; Petera, S.J.; Lyman, M.D.; Lai, L.H.; Leite, S.M.; Tamada, J.A.; Mikos, A.G. *In vitro* and *in vivo* degradation of porous poly(lactic-co-glycolic acid) foams. Biomaterials **2000**, *21* (18), 1837–1845.

12. Holland, S.J.; Tighe, B.J.; Gould, P.L. Polymers for biodegradable medical devices—Part 1: The potential of polyesters as controlled macromolecular release systems. J. Control. Release **1986**, *4* (3), 155–180.

13. Lickorish, D.; Guan, L.; Davies, J.E. A three-phase, fully resorbable, polyester/calcium phosphate scaffold for bone tissue engineering: Evolution of scaffold design. Biomaterials **2007**, *28* (8), 1495–1502.

14. Athanasiou, K.; Niederauer, G.G.; Agrawal, C.M. Sterilization, toxicity, biocompatibility and clinical applications of polylactic acid/polyglycolic acid copolymers. Biomaterials **1996**, *17* (2), 93–102.

15. Therin, M.; Christel, P.; Li, S.; Garreau, H.; Vert, M. *In vivo* degradation of massive poly(a-hydroxy acids): Validation of *in vitro* findings. Biomaterials **1992**, *13* (9), 594–600.

16. Bonfield, W.; Grynpas, M.; Tully, A.E.; Bowman, J.; Abram, J. Hydroxyapatite-reinforced polyethylene—A mechanically compatible implant material for bone replacement. Biomaterials **1981**, *2* (3), 185–186.

17. Lewandrowski, K.U.; Gresser, J.D.; Wise, D.L.; White, R.L.; Trantolo, D.J. Osteoconductivity of an injectable and bioresorbable poly(propylene glycol-co-fumaric acid) bone cement. Biomaterials **2000**, *21* (3), 293–298.

18. Klein, C.P.; Driessens, A.A.; de Groot, K.; van den Hooff, A. Biodegradation behavior of various calcium phosphate materials in bone tissue. J. Biomed. Mater. Res. **1983**, *17* (5), 769–784.

19. Huang, J.; Best, S.; Bonfield, W.; Brooks, R.A.; Rushton, N.; Jayasinghe, S.N.; Edirisinghe, M.J. *In vitro* assessment of the biological response to nano-sized hydroxyapatite. J. Mater. Sci. Mater. Med. **2004**, *15* (4), 441–445.

20. Pezzatini, S.; Solito, R.; Morbidelli, L.; Lamponi, S.; Boanini, E.; Bigi, A.; Ziche, M. The effect of hydroxyapatite nano-crystals on microvascular endothelial cell viability and functions. J. Biomed. Mater. Res. Part A **2006**, *76* (3), 656–663.

21. Suchanek, W.; Yashima, M.; Kakihana, M.; Yoshimura, M. Hydroxyapatite ceramics with selected sintering additives. Biomaterials **1997**, *18* (13), 923–933.

22. Nurse, R.W.; Welsh, J.H.; Gutt, W.J. High-temperature phase equilibria in the system dical-cium silicate-tricalcium phosphate. J. Chem. Soc. **1959**, *220*, 1077–1083.

23. Dorozhkin, S.V. Calcium orthophosphates in nature, biology and medicine. Materials **2009**, *2* (2), 399–498.

24. Gibson, I.R.; Best, S.M.; Bonfield, W. Phase transformations of tricalcium phosphates using high-temperature X-ray diffraction. Bioceramics **1996**, *9*, 173–176.

25. Driessens, F.C.M. *Mineral Aspects of Dentistry*; Karger: Basel, 1982.

26. Bohner, M.; Lemaitre, J.; Ring, T.A. Kinetics of dissolution of beta-tricalcium phosphate. J. Colloid Interface Sci. **1997**, *190*, 37–.

27. Fernandez, E.; Gil, F.J.; Ginebra, M.P.; Driessens, F.C.; Planell, J.A.; Best, S.M. Calcium phosphate bone cements for clinical applications—Part I: Solution chemistry. J. Mater. Sci. Mater. Med. **1999**, *10* (3), 169–176.

28. Klein, C.P.; Blieck-Hogemrst, J.M.A.; Wolke, J.G.C.; de Groot, K. Studies of the solubility of different calcium phosphate ceramic particles *in vitro*. Biomaterials **1990**, *11* (7), 509–512.

29. Ginebra, M.P.; Driessens, F.C.M.; Planell, J.A. Effect of the particle size on the microand nanostructural features of a calcium phosphate cement: A kinetic analysis. Biomaterials **2003**, *25* (17), 3453–3452.

30. Ehrenfried, L.M.; Farrar, D.; Morsley, D.; Cameron, R.E. Mechanical behavior of interpenetrating co-continuous beta-TCP-PDLLA composites. Bioceramics **2008**, *20*, 361.

31. Lu, J.; Descamps, M.; Dejou, J.; Koubi, G.; Hardouin, P.; Lemaitre, J.; Proust, J.P. The biodegradation mechanism of calcium phosphate biomaterials in bone. J. Biomed. Mater. Res. **2002**, *63* (4), 408–412.

32. Jancar, J.; Kucera, J. Yield behavior of PP/CaCO3 and PP/Mg(OH)2 composites—Part II: Enhanced interfacial adhesion. Polym. Eng. Sci. **1990**, *30* (12), 714–720.

33. Hong, Z.K.; Zhang, P.B.; He, C.L.; Qiu, X.Y. Nanocomposite of poly(L-lactide) and surface grafted hydroxyapatite: Mechanical properties and biocompatibility. Biomaterials **2005**, *26* (32), 6296–6304.

34. Plueddemann, E.P. *Silane Coupling Agents*, 3rd Ed.; Plenum Press: New York, 1991.

35. Wozney, J.M. Overview of bone morphogenetic proteins. Spine **2002**, *27* (16S), S2–S8.

36. Braunecker, J.; Baba, M.; Milroy, G.E.; Cameron, R.E. The effects of molecular weight and porosity on the degradation and drug release from polyglycolide. Int. J. Pharm. **2004**, *282* (1), 19–34.

37. Yang, Z.; Best, S.M.; Cameron, R.E. The influence of a-tricalcium phosphate nanopar-ticles and microparticles on the degradation of poly(D,L-lactide-coglycolide). Adv. Mater. **2009**, *21* (38-39), 3900–3904.

38. Tang, C.Y.; Chen, D.Z.; Yue, T.M.; Chan, K.C. Water absorption and solubility of PHBHV/HA nanocomposites. Comp. Sci. Technol. **2008**, *68* (7), 1927–1934.

39. Becker, O.; Varley, R.J.; Simon, G.P. Thermal stability and water uptake of high-performance epoxy-layered silicate nanocomposites. Eur. Polym. J. **2004**, *40* (1), 187–195.

40. Chuang, T.H.; Yang, T.C.K; Chen, T.Y.; Chang, A.H. Effects of glass microfillers on the water-transport behavior of poly-urethane composites. Int. J. Polym. Mater. **2004**, *53* (7), 553–564.

41. McManus, A.J.; Doremus, R.H.; Siegel, R.W.; Bizios, R. Evaluation of cytocompatibility and bending modulus of nanoceramic/polymer composites. J. Biomed. Mater. Res. **2005**, *72A* (1), 98–106.

42. Liu, H.; Webster, T.J. Mechanical properties of dispersed ceramic nanoparticles in polymer composites for orthopedic applications. Int. J. Nanomed. **2010**, *5*, 299.

43. Wilberforce, S.I.J.; Finlayson, C.E.; Best, S.M.; Cameron, R.E. The influence of the compounding process and testing conditions on the compressive mechanical properties of poly(D,L-lactide-co-glycolide)/-tricalcium phosphate nanocomposites. J. Mech. Behav. Biomed. Mater. **2001**, *4* (7), 1081–1089.

44. Wilberforce, S.I.J.; Best, S.M.; Cameron, R.E. A dynamic mechanical thermal analysis study of the viscoelastic properties and glass transition temperature behavior of bioresorbable polymer matrix nanocomposites. J. Mater. Sci. Mater. Med. **2010**, *21* (12), 3085–3093.

45. Hu, H.; Onyebueke, L.; Abatan, A. Characterizing and modeling mechanical properties of nanocomposites—Review and evaluation. J. Min. Mater. Characterization Eng. **2010**, *9*, 275.

46. Narayan, R.J.; Kumta, P.N.; Sfeir, C.; Lee, D.H. Nanostructured ceramics in medical devices: Applications and prospects. J. Min. Metals Mater. Soc. **2004**, *56* (10), 38–43.

47. Wilberforce, S.I.J; Finlayson, C.E.; Best, S.M.; Cameron, R.E. A comparative study of the thermal and dynamic mechanical behavior of quenched and annealed bioresorbable poly-L-lactide/a-tricalcium phosphate nanocomposites. Acta Biomater. **2011**, *7* (5), 2176–2184.

48. Ash, B.J.; Schadler, L.S.; Siegel, R.W. Glass transition behavior of alumina/polymethyl methacrylate nanocomposites. Mater. Lett. **2002**, *55* (1), 83–87.

49. Keddie, J.L.; Jones, R.A.L; Cory, R.A. Size-dependent depression of the glass transition temperature in polymer films. Europhys. Lett. **1994**, *27* (1), 59.

50. Ellison, C.J.; Mundra, M.K.; Torkelson, J.M. Impacts of polystyrene molecular weight and modification to the repeat unit structure on the glass transition—Nanoconfinement effect and the cooperativity length scale. Macromolecules **2005**, *38* (5), 1767–1778.

51. Charvet, J.L.; Cordes, J.A.; Alexander, H. Mechanical and fracture behavior of a fiber-reinforced bioabsorbable material for orthopedic applications. J. Mater. Sci. Mater. Med. **2000**, *11* (2), 101–109.

52. Guo, Z.; Lei, K.; Li, Y.; Ng, H.W. Fabrication and characterization of iron oxide nanopar-ticles reinforced vinyl-ester resin nanocomposites. Composite Sci. Technol. **2008**, *68* (6), 1513–1520.

53. Bansal, A.; Yang, H.; Li, C.; Cho, K. Quantitative equivalence between polymer nanocom-posites and thin polymer films. Nature **2005**, *4* (9), 693–698.

54. Donth, E. *The Glass Transition: Relaxation Dynamics in Liquids and Disordered Materials*; Springer: Heidelberg, 2001.

55. Priestley, R.D.; Ellison, C.J.; Broadbelt, L.J.; Torkelson, J.M. Structural relaxation of polymer glasses at surfaces, interfaces, and in between. Science **2005**, *308* (5733), 456–459.

56. Yang, Z.; Thian, E.S.; Brooks, R.A.; Rushton, N. Apatite formation on alpha-tricalcium phosphate/poly(D,L-lactide-

co-glycolide) nanocomposite. Bioceramics **2008**, *20* (1), 459.

57. Webster, T.J.; Siegel, R.W.; Bizios, R. Osteoblasts adhesion on nanophase ceramics. Biomaterials **1999**, *20*, 1221.

58. Webster, T.J.; Ergun, C.; Doremus, R.H.; Siegel, R.W.; Bizios, R. Enhanced functions of osteoblasts on nanophase ceramics. Biomaterials **2000a**, *21* (17), 1803–1810.

59. Liu, H.; Slamovich, E.B.; Webster, T.J. Increased osteoblast functions among nanophase titania/poly(lactide-co-glycolide) composites of the highest nanometer surface roughness. J. Biomed. Mater. Res. Part A **2006**, *78* (4), 798–807.

60. Webster, J.T.; Ejiofor, J.U. Increased osteoblast adhesion on nanophase metals: Ti, Ti6Al4V, and CoCrMo. Biomaterials **2004**, *25* (19), 4731–4739.

61. Palin, E.; Liu, H.; Webster, T.J. Mimicking the nanofeatures of bone increases bone-forming cell adhesion and proliferation. Nanotechnology **2005**, *16* (9), 1828.

62. Webster, T.J.; Ergun, C.D.; Siegel, R.W.; Bizios, R. Enhanced functions of osteoclast-like cells on nanophase ceramics. Biomaterials **2001**, *22* (11), 1327–1333.

63. Vance, R.J.; Miller, D.C.; Thapa, A.K.; Haberstroh, M.; Webster, T.J. Decreased fibroblast cell density on chemically degraded poly(lactic-co-glycolic) acid, polyurethane, and polycaprolactone. Biomaterials **2004**, *25* (11), 2095–2103.

64. Christenson, E.M.; Anseth, K.S.; van, J.J.J.; den Beucken, P.; Chan, C.K. Nanobiomaterial applications in orthopedics. J. Orthop. Res. **2006**, *25* (1), 11–22.

65. Webster, T.J.; Ergun, C.; Doremus, R.H.; Siegel, R.W.; Bizios, R. Specific proteins mediate enhanced osteoblast adhesion on nanophase ceramics. J. Biomed. Mater. Res. **2000b**, *51* (3), 475–483.

66. Schift, H.; Heyderman, L.J.; Padeste, C.; Gobrecht, J. Chemical nanopatterning using hot embossing lithography. Microelec. Eng. **2002**, *61*, 423–428.

67. Thapa, A.; Miller, D.C.; Webster, T.J.; Haberstroh, K.M. Nanostructured polymers enhance bladder smooth muscle cell function. Biomaterials **2003**, *24* (17), 2915–2926.

68. Thian, E.S.; Ahmad, Z.; Huang, J.; Edirisinghe, M.J. The role of electrosprayed apatite nanocrystals in guiding osteoblast behavior. Biomaterials **2008**, *29* (12), 1833–1843.

69. Pan, J.; Han, X.; Niu, W.; Cameron, R.E. A model for biodegradation of composite materials made of polyesters and tricalcium phosphates. Biomaterials **2011**, *32* (9), 2248–2255.

70. Wang, Y.; Pan, J.; Han, X.; Sinka, C.; Ding L. A phenomenological model for the degradation of biodegradable polymers. Biomaterials **2008**, *29* (23), 339–401.

71. Han, X.; Pan, J. A model for simultaneous crystallisation and biodegradation of biodegradable polymers. Biomaterials **2009**, *30* (3), 423–430.

72. Han, X.; Pan, J.; Buchanan, F.; Weir, N.; Farrar, D. Analysis of degradation data of poly(L-lactide-co-D,L-lactide) and poly(L-lactide) obtained at elevated and physiological temperatures using mathematical models. Acta Biomater. **2010**, *6* (10), 3882–3889.

73. Faisant, N.; Siepmann, J.; Oury, P.; Laffineur, V.; Bruna, E.; Haffner, J.; Benoit, J. The effect of gamma-irradiation on drug release from bioerodible microparticles: A quantitative treatment. Int. J. Pharm. **2002**, *242* (1–2), 281–284.

Peptide–Polymer Conjugates: Synthetic Design Strategies

Sabrina Dehn
Robert Chapman
Key Center for Polymers and Colloids, School of Chemistry, University of Sydney, Sydney, New South Wales, Australia

Katrina A. Jolliffe
School of Chemistry, University of Sydney, Sydney, New South Wales, Australia

Sébastien Perrier
Key Center for Polymers and Colloids, School of Chemistry, University of Sydney, Sydney, New South Wales, Australia

Abstract

We review the advantages and drawbacks of the various synthetic strategies for conjugating peptides to synthetic polymers obtained from reversible-deactivation radical polymerization (also known as living radical polymerization, controlled radical polymerization, or controlled/living radical polymerization). By using a selection of examples, we summarize the concept behind divergent and convergent syntheses, which lead to the production of linear and grafted peptide/polymer conjugates. In the second section of this entry, we present an overview of a near-universal convergent approach for the production of peptide/polymer conjugates, by combining reversible addition–fragmentation chain transfer polymerization and copper(I)–catalyzed Huisgen 1,3-dipolar cycloaddition reactions between an azide and an alkyne. This entry has been revised from "Synthetic Strategies for the Design of Peptide/Polymer Conjugates" *Polymer Reviews*, Vol. 51, Issue 2.

INTRODUCTION

Over the past two decades, an emerging research field has focused on combining biological and synthetic polymers to create highly advanced materials with applications as diverse as pharmaceuticals,[1] enzyme recovery,[2] modulation of peptide binding properties,[3] and for the fabrication of nanostructured materials.[4,5] The range of biopolymers that have been targeted for conjugation with synthetic polymers is vast, and includes cofactors, oligo-nucleotides, saccharides, lipids, and proteins/peptides.[6] The latter biomolecules are by far the most studied, and from the wide range of synthetic polymer conjugates available, PEG has received the most attention, with applications in the improvement of a broad range of pharmaceuticals.[1] Recent advances in synthetic tools have enabled an expansion of the number of synthetic polymer conjugates that have been successfully coupled to proteins and peptides, including for instance poly(hydroxyethyl acrylate) (PHEA),[7] poly(butyl acrylate) (PBA),[8–11] poly(hydroxyethyl methacrylate),[12–15] poly(N-isopropyl acrylamide) (PNIPAM),[15–24] poly[poly(ethylene glycol) methyl ether acrylate] [poly(PEGA)],[23–27] poly[N-(2-hydroxypropyl) methacrylamide] (PHPMA),[23,24,28–31] poly[poly(ethylene glycol) methyl ether methacrylate][(poly(PEGMA)],[32–37] poly(N-dimethylaminoethyl methacrylate) (PDMAEMA),[37,38] poly(N-vinylpyrrolidone) (PVP),[39] poly(methyl methacrylate) (PMMA),[23,24,40] polystyrene (PS),[24,41,42] p-nitrophenyl methacrylate (pPNPMA), and poly(diethoxypropyl methacrylate) (PDEPMA),[43,44] and even block copolymers.[12,45]

Proteins and peptides belong to one of the most diverse families of biopolymers, and exhibit a variety of functionalities, accurately defined structures and a wide range in molecular weight. Their structure and functionalities are key to their biological and structural properties, but their use in therapeutic applications is limited, since they can cause immune response and have low solubility and short circulation half-times.[1,2] Synthetic polymers have received renewed interest over the past decades, with the development of new synthetic tools enabling the production of well-controlled polymeric structures.[46] However, despite the impressive progress in methodologies for their preparation, synthetic polymers are still far from matching the structural perfection of proteins and peptides. In recent years, chemists and material scientists have realized that the conjugation of natural and synthetic polymers may overcome the limitations of each constituent, and generate improved materials that synergistically combine the properties of each individual component. The design of nanostructured materials has also greatly benefited from the conjugation of synthetic polymers to peptide segments. Peptide segments introduced in bioconjugates permit direction of the self-assembly of synthetic polymers into complex organized nanostructures, which are impossible to obtain from synthetic polymers alone. The synthetic elements

Concise Encyclopedia of Biomedical Polymers and Polymeric Biomaterials DOI: 10.1081/E-EBPPC-120052047

provide these structures with a range of well-tuned functionalities that would not be available to the natural components. For instance, β-sheet type interactions between peptide segments direct the assembly of peptides/polymer conjugates into fibers, for the production of hydrogels with applications in tissue engineering and as biosensors.[47] Synthetic polymers can also influence the self-assembly properties of the peptide/protein, thus resulting in the modification of their activity. For instance, recent work has focused on the use of synthetic polymers to prevent β-sheet formation of peptides, an area directly relevant to neurologic diseases, such as Alzheimer's and Creutzfeld Jacob.[5,48]

In this entry, we focus on the synthesis of bioconjugates based on peptides. Peptides are particularly difficult to work with, since they carry a vast range of functionalities that can interfere with chemical syntheses. The topic of peptide/polymer conjugates has been extensively reviewed over the past few years, and the number of excellent reviews available can be overwhelming for newcomers to the field.[49–56] Here, we present a critical short review of the state-of-the-art techniques to produce peptide/polymer conjugates. In the first section, we provide an overview of the various approaches reported to date, using a selection of relevant references as examples. In this section, it becomes rapidly clear that the techniques of reversible-deactivation radical polymerization (RDRP, also known as living radical polymerization, controlled radical polymerization, LRP, or controlled/living radical polymerization, CLRP), and in particular atom transfer radical polymerization (ATRP),[57] nitroxide-mediated polymerization (NMP),[58] and reversible addition fragmentation chain transfer (RAFT),[59–62] are powerful tools for the production of bioconjugates. In the second section, we provide a tutorial-type overview of a near-universal approach that is applicable to any peptide, and allows for conjugation to almost any polymer of well-controlled structure.

SYNTHETIC STRATEGIES

The synthesis of peptide–polymer conjugates typically follows so-called divergent or convergent routes, depending on whether one component is built from the other, or both components are built in parallel, then attached to each other (Fig. 1).[51,54,63]

Divergent Synthesis

The sequential synthesis of one block after the other is often referred to as divergent synthesis. This approach includes the polymerization of a peptide-functionalized monomer, the polymerization of a monomer from a peptide initiator, and the synthesis of a peptide by sequential addition of amino acids from a polymeric support.

Divergent Synthesis of Bioconjugates

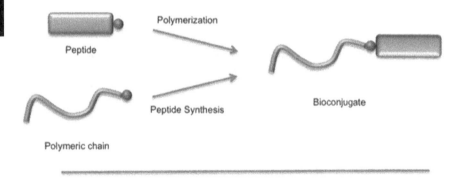

Convergent Synthesis of Bioconjugates

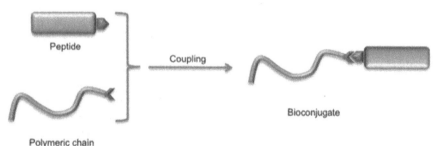

Fig. 1 Schematic representation of divergent and convergent synthesis of peptide–polymer bioconjugates.

Grafting through

When polymerizing a peptide–functionalized monomer, the term *grafted through* is used. The grafting through route yields comb-shaped structures, for which the peptide segments are pendant groups attached to the synthetic polymer backbone. Typically, a peptide will be modified by the introduction of polymerizable group at its chain-end, either *post* peptide synthesis, or during the last step of the solid phase peptide synthesis (SPPS). Typical procedures include the reaction of (meth)acryloyl chloride or (meth)acrylic acid with the *N*-terminus of the peptide, to produce the corresponding peptide-functionalized (meth) acrylamide derivative, or reaction of hydroxyl ethyl (meth)acrylate with the C-terminus of the peptide, to produce the corresponding peptide-functionalized (meth) acrylate derivative, as shown in Fig. 2.[64–66]

Peptide synthesis from polymers

The growth of a peptide from a polymer often requires either a soluble or a solid supported polymer as initiator. A common route uses Tentagel resin where a PEG chain is attached to a solid support via an acid labile linker.[5] PEG, PS, poly(vinyl alcohol), poly(ethyl imine), and poly(acrylic acid) have also been reported as good soluble supports for peptide synthesis.[67] The different solubility of the support material and all other reagents involved in the peptide synthesis facilitates the isolation of the desired conjugate. However, the solubility of the peptide/polymer conjugate is influenced by long peptide segments, thus making the isolation process more difficult.

Polymer synthesis from peptides

The use of peptide and proteins as initiators for polymerization has always been very challenging, as the polymerization conditions (temperature, solvent, etc.) may affect the structure of the biomolecule. The rise of reversible-deactivation radical polymerization techniques, and their versatility in terms of reaction conditions and neutrality to a wide range of functionalities, has made growing a polymeric chain from a peptidic group a much more straightforward approach. Furthermore, in addition to versatile

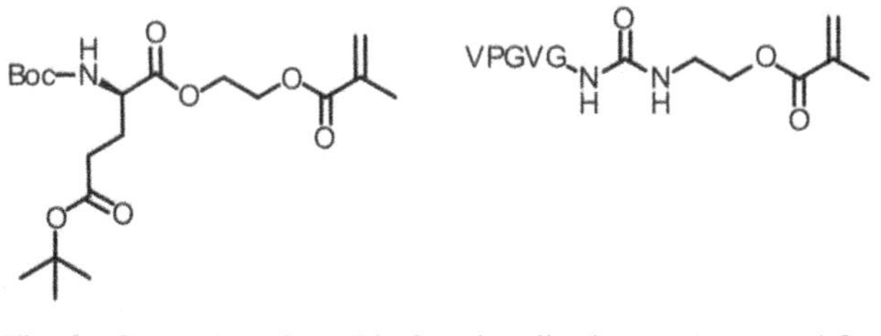

Fig. 2 Examples of peptide functionalised monomers used for the grafting through technique.
Source: From Ayres et al.[65] and Fernandez-Trillo et al.[66]

reaction conditions, reversible-deactivation radical polymerization techniques provide reasonable control over the chain length and molecular weight distribution of the synthetic element of the conjugates. In this strategy, the peptide serves as a RDRP mediator (e.g., initiator or chain transfer agent) and therefore a specific moiety needs to be introduced into the biomolecule. ATRP, NMP, and RAFT polymerization are excellent examples of techniques that enable the generation of peptide/polymer conjugates via this approach.[8–10,40,68–77] For instance, an elegant strategy was presented by Wooley and co-workers, who performed NMP[68,69] and ATRP[9] from solid supported peptide initiators obtained by SPPS, followed by cleavage of the resulting bioconjugates, as shown in Fig. 3. Another interesting example is the use of multiple initiators from cyclic peptides to grow polymers via ATRP, and generate star-shaped polymers.[74,75] The specific case of the polymerization of α-amino-acid-*N*-carboxyanhydrides (NCAs), which proceeds by a ring-opening mechanism initiated by nucleophiles or bases such as primary amines or alkoxides,[78] followed by initiation of RDRP, is an excellent example of the combination of two complementary "living"[79] polymerization systems to afford well-controlled block copolymers.[51,54,80]

Limitations of the divergent synthesis approach

Despite progress in polymerization techniques, the divergent approach suffers from a number of drawbacks.

Reaction Conditions. Despite the versatility of the reaction conditions of RDRP, both polymerization conditions and the peptide functionalities can be an issue. For instance, carboxylic acid groups can deactivate ATRP catalysts[81] and amines can reduce the thiocarbonyl thio group of RAFT agents.[59–61] For all techniques of RDRP, thiol groups on cysteine residues are likely to hinder polymerization, by acting as chain transfer agent, thus leading to unpredictable side products.

Bioconjugate Analysis and Characterization. The direct growth of a synthetic polymer from a peptidic substrate leads to bioconjugates whose unique solution properties make their isolation and purification challenging. It can be very difficult, for example, to accurately determine the polydispersity and molecular weight of a polymer grafted from a peptide. Where the biomolecule is a single graft from a small molecule, traditional techniques of analyzing polymers (such as size exclusion chromatography size exclusion chromatography (SEC) and [1]Nuclear magnetic resonance H-(NMR)) are still relatively useful, but where the biomolecule is large, these techniques are usually inadequate. The problem is further complicated when multiple polymers are grown from a single center, as in the case of cyclic-peptide polymer conjugates.[7,8,11,74–77] To analyze the conjugates, one has to either assume the

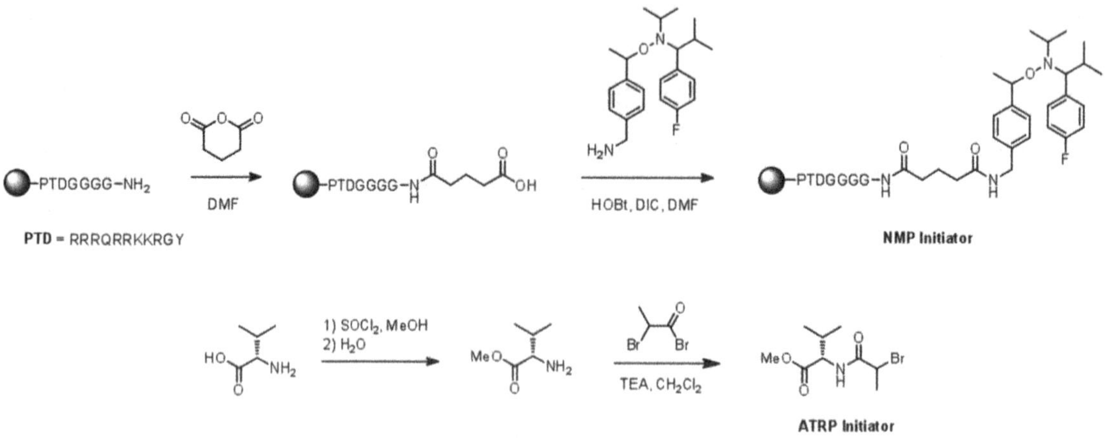

Fig. 3 Peptides anchored to NMP and ATRP initiators.
Source: From Venkataraman & Wooley[9] and Becker et al.,[69] with permission from the American Chemical Society and the Royal Society of Chemistry.

livingness of the polymerization, or cleave the grafted polymer from the conjugate for analysis post synthesis, or introduce a sacrificial initiator into the system and assume that the polymeric chains it generates are representative of the polymers grown from the biomolecule. However, all such approaches are difficult and imprecise. The unavoidable termination reactions and irreversible chain transfer events occurring at the propagating chain end also lead to side products that are almost impossible to remove from the bioconjugates. A potential route to avoid this specific problem is to use RAFT polymerization, and design peptidic RAFT agents in which the peptide/protein is attached to the Z group. In this case, side reactions occurring at the propagating chain end of the synthetic polymer form pure synthetic polymers, while the peptide remains attached to the living polymeric chains.[82]

Process. Further problems include the small quantities in which the peptidic initiator (in the case of e.g., ATRP and NMP) or chain transfer agent (in the case of e.g., RAFT) is available (typically on a mg scale). Such low quantities require very small-scale processes, in which accurate weighting of reagents (and therefore accurate molecular weights predictions) are difficult, and in which even traces of oxygen can inhibit polymerization.

Convergent Synthesis

The numerous disadvantages of using the divergent approach have led many researchers to investigate the synthesis of bioconjugates via a convergent route. In such an approach, the polymer and biomolecule are synthesized and characterized independently, and then coupled together to form the desired product. The convergent approach offers two key advantages over the divergent approach. First, it allows for the separate synthesis and characterization of the peptide and polymer components; the preparation of

components individually can ease their synthesis significantly, and the knowledge of the structure of each element makes the characterization of the conjugate much easier. In addition, for peptides offering multiple points of attachment, graft densities can be precisely controlled by monitoring the efficiency of the coupling reaction. Secondly, the convergent approach allows for the easy preparation of a library of bioconjugates, following a combinatorial approach. Once peptides and synthetic polymers have been created, they can be coupled in any combinations without changing reaction conditions. Access to similar libraries following a divergent approach would require much more effort, as conditions for polymerization would need to be specifically tuned to each reaction.

As in the divergent approach, the convergent synthesis allows for the production of conjugates in which the peptide segment is attached to either the backbone of the polymeric chain, or to one of its chain ends.

Pendant Peptide Conjugates

Pendant peptide conjugates are easily accessible via reaction with reactive monomers. For instance, short sequence (Gly-OMe and Gly-Gly-b-naphthylamide hydrobromide)[31] and long sequence (HTSTYWWLDGAPK)[30] peptides were grafted to statistical copolymers of Hexamethylphosphoramide (HMPA) and N-methacryloyloxysuccinimide obtained via ATRP (for Gly-OMe and Gly-Gly-b-naphthylamide hydrobromide) and RAFT (for HTSTYWWLDGAPK), by exploiting the reactivity of the succinimide moiety toward the peptide N-terminus. An alternative synthetic route saw the polymerization of p-nitrophenyl methacrylate and diethoxypropyl methacrylate monomers via RAFT. The resulting poly(p-nitrophenyl methacrylate) polymer was coupled with Gly-OMe via the activated ester, while the acetal side chains of poly(diethoxypropyl methacrylate) were turned into aldehydes and then conjugated to an aminooxy-RGD peptide.[43,44] It is also possible to couple the peptide via a two-step process, activating an otherwise unreactive polymer prior to the conjugation reaction. An elegant example is the reaction of PPEGMA brushes with p-nitrophenyl chloroformate, forming an activated ester through which a variety of peptide sequences were coupled.[83]

Linear Peptide Conjugates

The polymer end group functionality required for conjugation can either be introduced via the polymerization initiator (or chain transfer agent) on the α-terminus of the polymeric chain, or via the living chain end, the ω-terminus of the polymer. Initial work has used traditional biochemical ligation techniques to functionalize proteins by exploiting amino acid functionalities. Typically, amines (present in lysine and at the N-terminus of peptides) and thiols or disulfide bridges (from cysteine, which is less abundant in proteins, and therefore offers better selectivity) are targeted. Examples of coupling polymers to protein significantly outnumber examples of peptide/polymer conjugations.[54]

Noteworthy examples of coupling peptides to the α-terminus of polymeric chains include the use of an aldehyde–functionalized ATRP initiator[34] or RAFT agent[84] that can be reacted with the N-terminal amino acid of peptides after polymerization and carboxylic acid-functionalized ATRP-made polymers that can be reacted with the amine moiety of lysine residues[8] or the N-terminal amino group on the peptide fragment (Fig. 4).[85]

The presence of an active end group at the ω-terminus of polymeric chains has also been exploited for efficient coupling. The halide chain end of ATRP-made polymers is easily converted to a range of functionalities (e.g., azide groups for click reactions, see later), as is the thiocarbonyl

thio group of RAFT-made polymers. For instance, the thio-carbonyl thio end-group of RAFT–made polymers can be reduced to a thiol by aminolysis, then converted into a pyridyl disulfide moiety, which in turn can be reacted with a cysteine thiol group, e.g., that in the Asn-Gly-Arg (NGR) model peptide GNGRGC.[23,24]

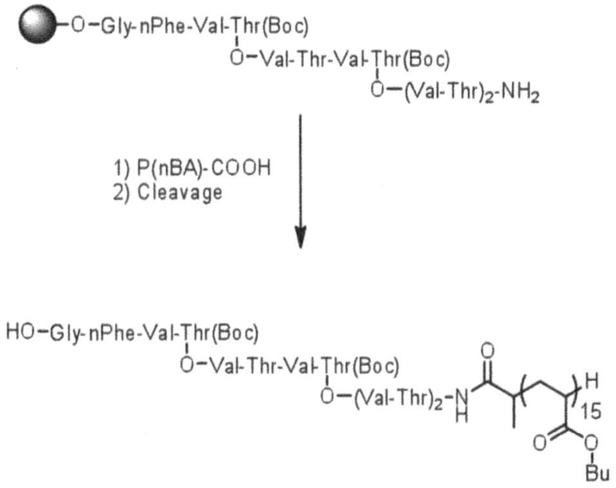

Requirements for Efficient Coupling

The key limitation of the convergent approach lies in the conjugation reaction. Such reaction should be high yielding to avoid polluting the final product with starting materials and it must proceed under relatively mild conditions to prevent damages to the peptide, and if at all needed, catalysts must be used in small quantities and must be easily removable. In addition, the functional groups required for the conjugation reactions must be orthogonal to functionalities found in both the peptide segment and the synthetic polymer, as well as being orthogonal to the synthetic procedure employed to build each component. The traditional biochemical ligation techniques discussed earlier, which exploit native peptide functionalities, cannot fulfill this latter requirement and so additional synthetic steps are usually required to introduce more reactive groups.

In 2001, Sharpless and co-workers proposed a concept in chemistry that remarkably fits these prerequisites. The so-called click chemistry is tailored to generate substances quickly and reliably by joining small units together, and its main attributes include very high yields, mild reaction conditions, and stereospecificity.[86] Reactions meeting these requirements include, but are not limited to, the copper(I)-catalyzed Huisgen 1,3-dipolar cycloaddition reaction between an azide and an alkyne copper-catalyzed azide-alkyne cycloaddition (CuAAC),[47,52,53,87–91] cycloadditions using strained or activated alkynes,[92,93] thiol-ene, thiol-yne and thiol-isocyanate click couplings,[94–96] RAFT hetero Diels-Alder cycloadditions,[92] inverse electron demand diels alder reactions,[97–100] nitrile oxide couplings,[101] Staudinger ligation,[46,50] reductive alkylation,[50] as well as oxime formation, oxidative coupling, and the Suzuki/Heck coupling reactions.[46,50] To date, only CuAAC has been used for the synthesis of peptide/polymer conjugates (see later).

A NEAR-UNIVERSAL STRATEGY FOR THE SYNTHESIS OF PEPTIDE/POLYMER CONJUGATES

When reviewing the literature, it becomes clear that there exist many different routes for the synthesis of bioconjugates, all with their advantages and drawbacks. In almost all cases, both the synthetic and natural components dictate the synthetic strategy. It is, however, possible to isolate synthetic routes for which drawbacks are only limited to a handful of specific systems. In the following section, we propose a strategy based on convergent synthesis that can

Fig. 4 Convergent coupling of PBA to the amine N-terminal group of a peptide.
Source: From Hentschel & Hentschel,[85] with permission from the American Chemical Society.

be applied to almost any peptide and introduce almost any functionality in the synthetic polymer element (Fig. 5).

In this approach, the peptide can either be directly synthesized (via e.g., SPPS, NCA polymerization, etc.), or purchased and used as is. The direct synthesis of the peptide does, however, offer the advantage of including in the overall synthesis the reaction step that introduces the functional groups required for conjugation.

Among the many polymerization techniques available, we find the RAFT process to be one of the most suitable techniques for the generation of the synthetic polymer element. RAFT polymerization is a simple method to create well-controlled and highly functional polymers, which are able to react with suitable functional groups of the peptide. It can be applied to a wide range of polymers and various solvents without the need of any metal catalysts. Architectures such as block co-polymers, graft copolymers, nanogels, stars, and dendritic structures are easily achievable.[59–62] One of the key advantages of RAFT polymerization is its high tolerance to a broad range of functional groups. This allows the incorporation of diverse functionalities on the backbone, or at the α- and β-position (via the R or Z group of the RAFT agent), which can be subsequently used for direct conjugation with biomolecules.[102]

Significant advances have been made in recent years for the use of RAFT in aqueous systems, of importance in numerous biological applications. Ring opening polymerization (ROP) has long provided the ability to polymerize a number of water-soluble monomers, including aliphatic polyesters, polyurethanes, polyanhydrides, polyamides, and polyethers.[103] While ROP is a useful strategy, RAFT polymerization is tolerant to a more versatile range of monomers, more functionality, and requires less stringent reaction conditions than ROP.[104] RAFT has therefore become an important technique for the synthesis of water-soluble stimuli-responsive polymers, which are highly important for the development of controlled drug delivery systems.[105]

As seen in the previous section, the conjugation reaction is key to the bioconjugate synthesis. Among the orthogonal chemical reactions available, the CuAAC seems to be the most versatile. Azide and alkyne groups are orthogonal to functional groups found in both peptides/amino acids and polymers/monomers, they can be easily incorporated into both peptide and polymer moieties, and the conjugation reaction itself is simple to undertake and provide high yields of a single product. There are an enormous number of variables involved in the optimization of CuAAc, which have been extensively reviewed in the literature (Fig. 3).[89,90] A wide variety of different catalyst and ligand systems have been investigated. These include the use of an oxygen sensitive Cu(I) halide, or the more stable $Cu(I)(PPh_3)_3Br$

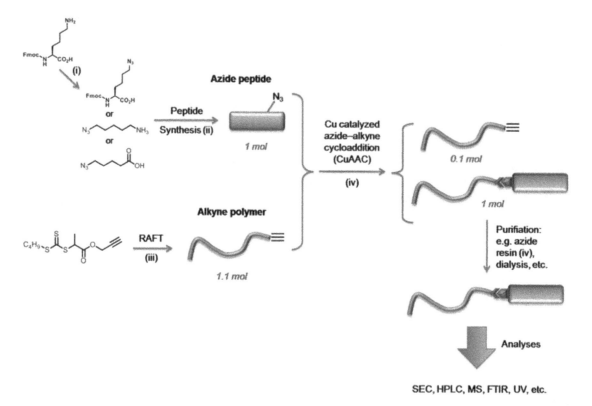

Fig. 5 Scheme summarizing the synthesis of peptide/polymer conjugates using SPPS, RAFT polymerization and CuAAC. Typical reaction conditions: (i) Fmoc-L-Lys-OH, Im-SO$_2$-N$_3$, CuSO$_4$, NaHCO$_3$; (ii) e.g., SPPS: 1) coupling: amino acids, resin, HBTU, Hünig's base, DMF/; 2) Fmoc deprotection: 20% piperidine, DMF/; and 3) cleavage: Trifluoroacetic-acid (TFA), TIPS, thioanisole, H$_2$O; (iii) monomer, AIBN; (iv) microwave irradiation, DMSO, PMDETA, Cu(II)SO$_4$, sodium ascorbate.

complex. Such systems require the use of a stabilizing ligand such as PMDETA, under inert atmospheres. The use of copper(II) species such as $Cu(II)SO_4 \cdot 5H_2O$ or $Cu(OAc)_2$ reduced in situ to Cu(I) with sodium ascorbate has also been widely investigated, as have a number of Cu(0) systems.[106] While other metals, such as Ru, Au(I), Ni, and Pt catalysts, have been studied,[89,107] their use is far less common. The use of microwave reactors has been found to dramatically improve yields, particularly in difficult systems.[108–111] Despite the growing number of reports on the use of CuAAC, only a few researchers have used it to conjugate synthetic polymers obtained from RDRP to peptides (Fig. 6).[7,26,38,42,112]

Efficient removal of the catalyst in CuAAC is clearly a key requirement, especially if the bioconjugate is to be used in biomedical applications. However, as will be discussed in the following section, recent progresses in CuAAC now allow for minimal amount of copper catalyst, and almost no residual copper polluting the conjugate after reaction.

Preparation of Polymer-Reactive Peptides by Introducing an Azide Functionality

The azide group is truly orthogonal to peptide functional groups, as it does not exist among, nor cross-react with, functional groups present in amino acids.[113,114] On the other hand, we have observed that azides, when, for example, attached to a RAFT agent, can undergo 1,3-cycloadditions with monomeric vinyl bonds, in a similar fashion as they do during reaction with acetylene.[115] This side-reaction affects more significantly acrylates and acrylamides, since it depends on the electron withdrawing properties of the vinyl bond substituent. Steric hindrance of the double bond, for instance in the case of methacrylates, limits the reaction. Monomers such as styrene, with delocalization of the double bond to the benzene ring, are less reactive. Since azide groups cannot be protected against such side reactions, they are not ideal candidates to be introduced in a free radical polymerization system. Fortunately, introduction of azide groups onto a peptide is relatively straightforward. The N-terminus chain end of a peptide or the amine group of the lysine residue provide ideal handles for the modification of existing peptides in this fashion.[116] Modification of an amine to an azide can be achieved using triflyl azide, prepared in situ from triflic anhydride and sodium azide,[117] in near-quantitative conversions. The azide functionality can be introduced on the amino acid before reaction, or directly on the formed peptide. The better approach to introduce a pendant azide moiety is to use an azide-modified amino acid (typically lysine), which is incorporated into the peptide sequence at the desired position during SPPS. Because the azide-modified amino acid can be prepared prior to peptide synthesis, quantitative yields to form the azide are not critical, and it is possible to use lower yielding but far less dangerous methods for this modification. For instance, Goddard-Borger and Stick recently reported a simple procedure to synthesize a shelf-stable diazo transfer group.[118] The side chain approach also allows the positioning of the azide functionality at any position and as many times in the peptide sequence as is desired, and is useful for grafting polymers to the side of a peptide chain. The transformation of the N-terminal of a peptide sequence into an azide is more difficult. An alternative path is to couple a synthetically prepared azide-containing molecule to the peptide, for

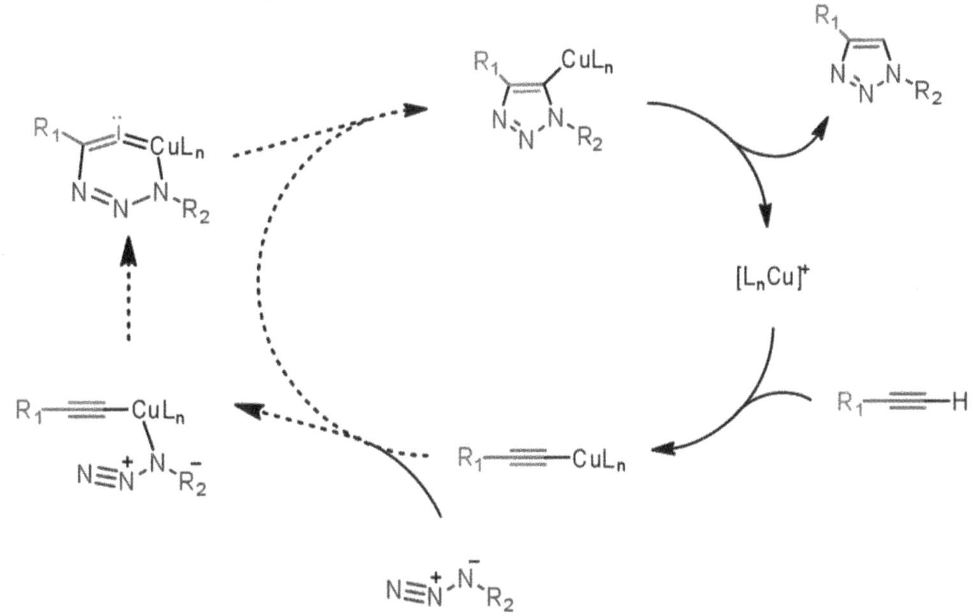

Fig. 6 Proposed catalytic cycle for the Cu(I)-catalyzed ligation, as published by Sharpless and co-workers.
Source: From Rostovtsev et al.,[87] with permission from Wiley-VCH Verlag GmbH.

instance during SPPS, either as the first step (for C-terminus peptides)[119,120] or as the last step (for N-terminus peptides).[117] The end group approach is useful for creating peptides that are reactive at one position and are useful for attaching a polymer to one end, yielding block copolymers. It also permits control of the length and type of linker between the peptide and the polymer.

Azides as end groups

The introduction of an azide end group is efficiently done via an azido alkyl linker, added during the final step of the SPPS [16] For instance, when synthesizing the eicosapeptide GVTSAPDTRPAPGSTAPPAH (a model for the immunogenic peptide sequence from the variable number tandem repeat (VNTR) of the cancer associated glycoprotein MUC1) [121] via Fmoc protected SPPS on Wang resin using PyBOP and N-Methylmorpholine (NMM), an azidoglycine linker was coupled onto the N-terminal amino acid, prior to cleavage from the resin with Trifluoroacetic-acid (TFA).[16] The same strategy was applied during the synthesis of the undecapeptide QQRFQWQFFQQ, which demonstrates a high tendency for β-sheet formation[122,123] and thus is a useful prototype to study amyloid formation. An Fmoc-protected SPPS strategy was used with hydroxybenzotriazole (HOBT), N,N'-diisopropylcarbodiimide (DIPCDI), and Rink resin to assemble the peptide, and in the last step an azido pentanoic acid was added. The subsequent cleavage from the resin by

TFA, thioanisole, triisopropylsilyl (TIPS), and water provided the azide-linked undecapeptide.

Azides as pendant groups

Azide pendant groups are readily introduced in peptide segments by transformation of lysine.[7] The direct transformation of lysine to the corresponding azide prior to SPPS is simpler than the above approach and enables accurate control of the position of the azide group in the final amino acid sequence. For instance, azido D-alt-L cyclic octapeptides with four azide functionalities were produced following this route via standard Fmoc SPPS using 2-chlorotrityl resin, 2-(1H-Benzotriazole-1-yl)-1,1,3,3-tetramethyluronium hexafluorophosphate (HBTU), and N,N-Diisopropylethylamine (DIPEA).[7] The azido Fmoc–L-lysine was synthesized following Stick's protocol using a diazo transfer reagent[118] and coupled alternately with Fmoc-D-Leucine, to form the linear octapeptides (Fig. 7). After cleavage from the resin, the cyclization step was undertaken using similar conditions, at high dilution.

Preparation of Peptide-Reactive Polymers by Introducing an Alkyne Functionality

The alkyne group is not bio-orthogonal as it can react with the thiol of a cysteine group; this can be avoided by

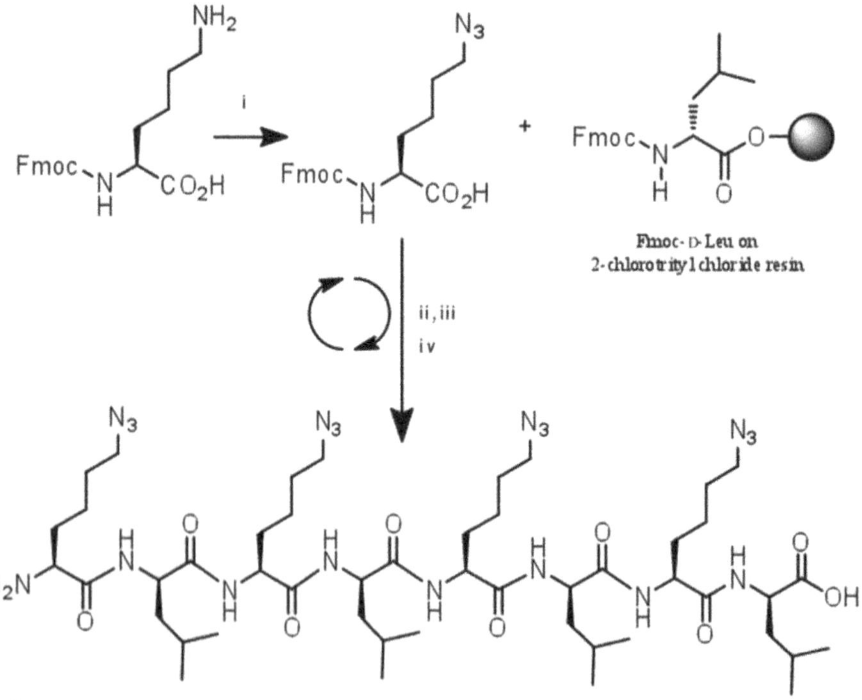

Fig. 7 Synthesis of azide–functionalized amino acid from lysine and subsequent solid phase peptide synthesis; (i) Fmoc-L-Lys-OH, Im-SO$_2$-N$_3$, CuSO$_4$, NaHCO$_3$; (ii) Fmoc-L-Lys(N$_3$)-OH/Fmoc-D-Leu-OH, HBTU, Hünig's base, DMF; (iii) 20% piperidine, DMF; (iv) Trifluoroacetic-acid (TFA), TIPS, thioanisole, H$_2$O.
Source: From Chapman et al.,[7] with permission from CSIRO.

protection of the cysteine, which is often undertaken to prevent the peptide coupling to itself through disulfide bridging. However, a simple synthetic route, and the philosophy of click chemistry, call for minimal use of protection groups. Protection of the alkyne group is however also required if introducing an acetylene group into the polymer to prevent its potential reaction with radicals during a radical polymerization process.[124] Typical protective groups include trimethylsilyl moieties, which can be readily removed via hydrolysis, after the polymerization process. A similar protective approach can also be taken when monomers containing acetylene units are polymerized.[125] However, in the case of acetylene-functionalized RAFT agents, protection can be avoided, but temperatures must remain low, and reaction times must be kept short.[7,126]

The versatility of reversible-deactivation radical polymerization techniques offers a large number of relatively straightforward methods for the introduction of alkyne groups into a polymer chain, as reviewed recently by Gautier and Klok.[50] One can introduce such functionality as a pendant group by copolymerizing an appropriate monomer,[31,43,127–135] or through the use of end groups either in α- or β-position on the polymeric chain.[50,60,136–138] For instance, α-functional polymers are accessible by modification of an ATRP initiator, or the R group of a RAFT agent prior to polymerization, while β-functional polymers are readily accessible through the modification of the RAFT agent Z group,[21,139] or by exchange of the RAFT or ATRP leaving groups for a functional terminator.[136] The halogen terminus of a polymer prepared via ATRP also opens up the possibility of introducing functionality via *post*-polymerization reactions.[50]

A simple approach to design a synthetic polymer conjugate via RAFT polymerization is to employ an alkyne-modified RAFT agent, which can be used to produce well-controlled polymers from a wide range of monomers (e.g., Hydroxyethyl Acrylate (HEA)[7,16], NIPAM,[16] etc.) with various molecular weights, structural and physical properties. The alkyne group is introduced at the ω-end of the polymeric chain by modification of a carboxylic acid–functionalized trithiocarbonate RAFT agent with propargyl alcohol.[42] A TMS protecting group was introduced on the alkyne functionality to prevent any side reactions during polymerization. This approach led to low molecular weight polymers with low polydispersity (e.g., M_n = 2000–7000 g/mol; PDI = 1.1–1.2) (Fig. 8).[16]

Protection of the alkyne group prior to polymerization is, however, not required for short polymeric chains and short reaction times. Indeed, poly(HEA) of molecular weight ranging 1600 to –4000 g/mol with polydispersity index (PDI) below 1.2 were obtained from polymerization of HEA mediated by a similar RAFT agent, for which the alkyne functionality was left unprotected (Fig. 9).[7,16]

Coupling Reaction

The use of CuAAC in presence of peptides often requires strong ligands[42] and/or higher than catalytic amounts of copper, due to strong binding of the Cu catalyst to the amide bonds and other functionalities present in the peptide. Typical conditions in such reactions use CuBr/PMDETA as catalytic system in ratio 1/1 to 2/1 with azide–alkyne functionalities, in Dimethyl sulfoxide (DMSO), over 60–80 hr at temperatures ranging from room temperature to 40°C. For instance, PHEA was efficiently coupled to azido-undecapeptide QQRFQWQFFQQ in 72 hr at 40°C.[140] The completion of the reaction and formation of the desired conjugates was confirmed by[1] H NMR, Fourier transform infrared spectroscopy (FT-IR) (following the disappearance of the characteristic azide peak), SEC and analytical High-performance liquid chromatography (HPLC). Similar conditions were used to efficiently conjugate the much larger eicosapeptide peptide GVTSAPDTRPAPGSTAPPAH to a PNIPAM of ca. 2000 g/mol at room temperature for 60 hr.[16] However, this system is limited by the time required for reaction, and, as always in reactions involving the coupling of macromolecules, molecular weights of reagents, and associated steric hindrance. The effect of steric hindrance on reaction yields was assessed when attempting to attach four PHEA (1500 or 3000 g/mol) to a cyclic octapeptide.[7] The click reaction was complicated by the insolubility of the cyclic peptide in DMSO, which became soluble as polymeric chains were attached. Again, the formation of the desired short and long chain conjugates were confirmed by IR, NMR, and SEC. While conjugation of the shorter PHEA chains proceeded to near complete conversion (up to 90%), conjugation of longer polymeric chains (ca. 3000 g/mol) was not quantitative, with conversion only reaching 50%, thus demonstrating the effect of steric hindrance of the long PHEA. This dramatic decreased efficiency of the reaction was characterized by SEC analysis, and the presence of a dominant azide peak in the IR spectrum of the conjugate, indicating the presence of unreacted azide functionalities.

Using Cu(II)SO$_4$ and sodium ascorbate as reducing agent to generate Cu(I) in situ improves the reaction time when compared to systems based on Cu(I)Br catalysts. An additional advantage of this catalytic system is that it is not critical for the reaction to be performed in oxygen-free conditions. In addition, yields and reaction times can be further improved by using microwave irradiations, especially when clicking multiple polymeric chains to a small centre such as a cyclodextrin or a dendrimer.[108–111] Similar conditions were tested on the reaction involving cyclic peptides described above, leading to a reduction of reaction times from 60 to 80 hr for conventional heating to 15 min, while giving similar yields.

Purification and Analysis

The convergent approach also allows for much easier purification and characterization of the final conjugate.

Neural—Rapid Prototyping

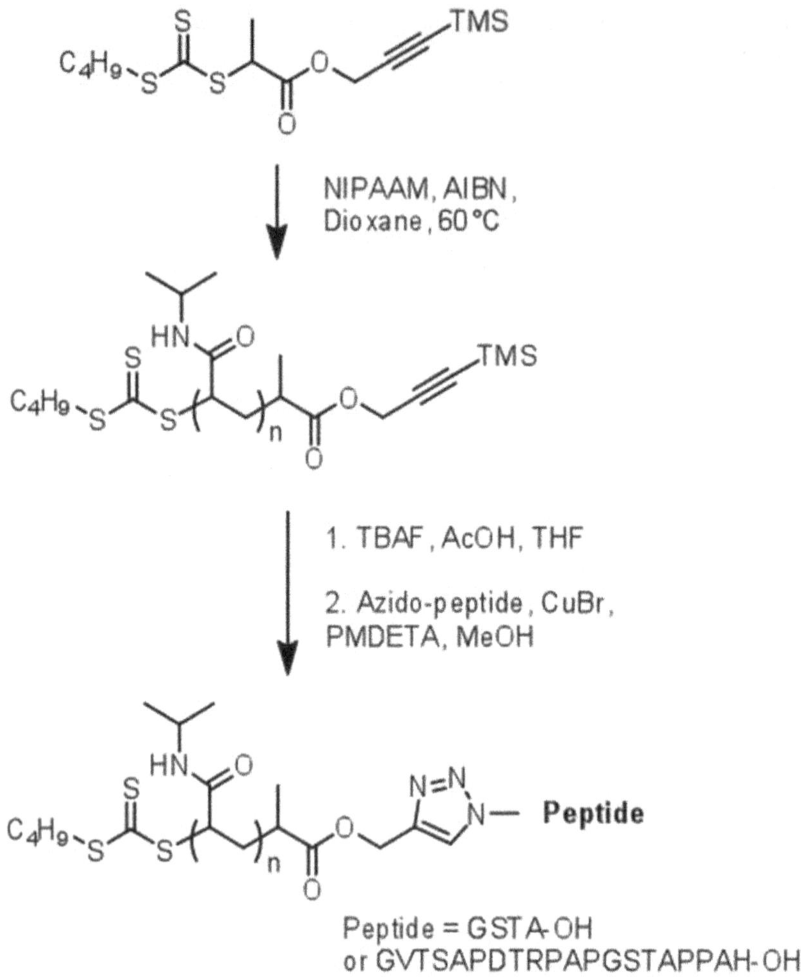

Fig. 8 Synthesis of PNIPAM from TMS–protected alkyne-functionalized RAFT agent, and subsequent CuAAC with azide-functionalized peptides.
Source: From Kakwere et al.,[16] with permission from the Royal Society of Chemistry.

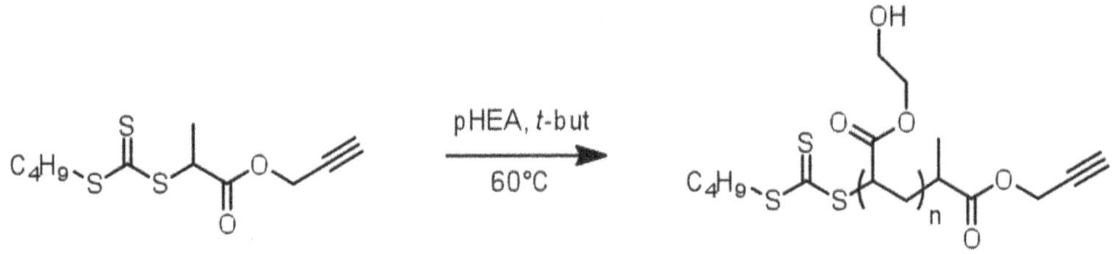

Fig. 9 Example of synthesis of pHEA from unprotected alkyne-functionaliazed RAFT agent.
Source: From Chapman et al.,[7] with permission from CSIRO.

Polymer and peptide are independently synthesized following typical procedures. Polymerization techniques, especially reversible-deactivation radical polymerization, lead to relatively well-defined polymeric chains that can be easily isolated and analyzed via the traditional techniques available to polymer chemists. NCA polymerization typically yields highly pure polypeptides, and SPPS of azido-peptides permits the removal of unreacted amino acids after each addition step, and the yield of each step is monitored via ultraviolet (UV) measurements. Following typical SPPS, the final product is isolated via preparative HPLC after cleaving the peptide off the resin.

The modular nature of the conjugation reaction avoids the formation of side products, thus yielding the final conjugate, and potentially small amounts of the initial reagents as contaminants. Typically, a slight excess of polymer (10 mol%) is used for conjugation to ensure full consumption of the peptide, which is usually available in lower amounts. Purification of excess of alkyne polymer is easily achieved by addition of an azide–functionalized Merrifield[42] or polystyrene[141] resin, which react with the polymer and is easily filtered out after reaction, leaving the pure conjugate in solution. Other purification techniques, which are more system dependant, include dialysis and precipitation into water.[7,16] Complete removal of the copper catalyst can be difficult, especially when ligated to the peptide or the triazole ring, but washing the product with aqueous EDTA,[42] or passing it over alumina is usually quite effective.[142]

The absence of side products after conjugation makes the analysis of the conjugate much easier. Proton and ^{13}C NMR, HPLC-MS or electospray ionization mass spectrometer (ESI-MS), SEC, FT-IR, UV, matrix-assisted laser desorption/ionization-time of flight-mass spectrometry (MALDI-TOF-MS), reverse phase HPLC, Atmospheric pressure chemical ionization (APCI), Gas chromatography–mass spectrometry (GC-MS), and high-resolution mass spectrometry (MS) are all efficient techniques that provide information on the final product.[7,16] Limitations in these techniques do of course exist. For instance, MS is only suitable to certain molecular weights (upper limits for electrospray and lower limits for MALDI), SEC, as for any block copolymer-type system, is limited by the solubility of the conjugate and its precursors in the eluent, FT-IR might be not sensitive enough for certain functional groups. A modular approach to conjugate synthesis also permits a direct comparison of the analytical data from the precursors to those of the conjugate, thus simplifying data analysis. Indeed, in cases when the conjugates are difficult to characterize, successful reaction can be inferred from the absence of precursors in the sample.

CONCLUSION

The growing number of applications of peptide/polymer conjugates, from medicine to material science, calls for efficient and versatile synthetic techniques. Recent advances in polymer chemistry, in particular reversible-deactivation radical polymerization, and in organic synthesis have contributed to provide a range of solutions to the complex problems of combining multifunctional elements together, in true orthogonal reactions. Among the various solutions available, combining RAFT polymerization to CuAAC is an effective, relatively simple, and versatile strategy to generate conjugates that combine almost any amino acid and synthetic polymer functionalities.

REFERENCES

1. Harris, J.M.; Chess, R.B. Effect of pegylation on pharmaceuticals. Nat. Rev. Drug Discov. **2003**, *2* (3), 214–221.
2. Klok, H.-A. Biological-synthetic hybrid block copolymers: Combining the best from two worlds. J. Polym. Sci. A **2005**, *43* (1), 1–17.
3. Hoffman, A.S.; Stayton, P.S. Bioconjugates of smart polymers and proteins: Synthesis and applications. Macromol. Symp. **2004**, *207* (1), 139–151.
4. Waterhouse, S.H.; Gerrard, J.A. Amyloid fibrils in bionanotechnology. Aust. J. Chem. **2004**, *57* (6), 519–523.
5. Roesler, A.; Klok, H.-A.; Hamley, I.W.; Castelletto, V.; Mykhaylyk, O.O. Nanoscale structure of poly(ethylene glycol) hybrid block copolymers containing amphiphilic β-strand peptide sequences. Biomacromolecules **2003**, *4* (4), 859–863.
6. Harris, J.M.; Martin, N.E.; Modi, M. Pegylation. A novel process for modifying pharmacokinetics. Clin. Pharm. **2001**, *40* (7), 539–551.
7. Chapman, R.; Jolliffe, K.A.; Perrier, S. Synthesis of self-assembling cyclic peptide-polymer conjugates using click-chemistry. Aust. J. Chem. **2010**, *63* (8), 1169–1172.
8. Ten Cate, M.G.J.; Severin, N.; Börner, H.G. Self-assembling peptide-polymer conjugates comprising D-alt-L)-cyclopeptides as aggregator domains. Macromolecules **2006**, *39* (23), 7831–7838.
9. Venkataraman, S.; Wooley, K.L. ATRP from an amino acid-based initiator: A facile approach for α-functionalized polymers. Macromolecules **2006**, *39* (26), 9661–9664.
10. Rettig, H.; Krause, E.; Börner, H.G. Atom transfer radical polymerization with polypeptide initiators: A general approach to block copolymers of sequence-defined polypeptides and synthetic polymers. Macromol. Rapid Commun. **2004**, *25* (13), 1251–1256.
11. Loschonsky, S.; Couet, J.; Biesalski, M. Synthesis of peptide/polymer conjugates by solution ATRP of butylacrylate using an initiator-modified cyclic D-alt-L-peptide. Macromol Rapid Commun. **2008**, *29* (4), 309–315.
12. He, X.; Zhong, L.; Wang, K.; Lin, S.; Luo, S. Synthesis of water-soluble ABC triblock copolymers containing polypeptide segments. React. Funct. Polym. **2009**, *69* (9), 666–672.
13. Bontempo, D.; Heredia, K.L.; Fish, B.A.; Maynard, H.D. Cystein-reactive polymers synthesized by atom transfer radical polymerization for conjugation to proteins. J. Am. Chem. Soc. **2004**, *126* (47), 15372–15373.
14. Bontempo, D.; Maynard, H.D. Streptavidin as a macroinitiator for polymerization: In situ protein-polymer conjugate formation. J. Am. Chem. Soc. **2005**, *127* (18), 6508–6509.
15. Heredia, K.L.; Tolstyka, Z.P.; Maynard, H.D. Aminooxy end-functionalized polymers synthesized by atrp for chemoselective conjugation to proteins. Macromolecules **2007**, *40* (14), 4772–4779.
16. Kakwere, H.; Chun, C.K.Y.; Jolliffe, K.A.; Payne, R.J.; Perrier, S. Polymer-peptide chimeras for the multivalent display of immunogenic peptides. Chem. Commun. **2010**, *46* (13), 2188–2190.
17. Vazquez-Dorbatt, V.; Tolstyka, Z.P.; Maynard, H.D. Synthesis of aminooxy end-functionalized pnipaam by raft polymerization for protein and polysaccharide conjugation. Macromolecules **2009**, *42* (20), 7650–7656.
18. Boyer, C.; Liu, J.; Wong, L.; Tippett, M.; Bulmus, V.; Davis, T.P. Stability and utility of pyridyl disulfide functionality in RAFT and conventional radical polymerizations. J. Polym. Sci. A **2008**, *46* (21), 7207–7224.

Neural—Rapid Prototyping

19. Li, M.; De, P.; Gondi, S.R.; Sumerlin, B.S. Responsive polymer-protein bioconjugates prepared by RAFT polymerization and copper-catalyzed azide-alkyne click chemistry. Macromol. Rapid Commun. **2008**, *29* (12–13), 1172–1176.

20. Grover, N.; Alconcel, S.N.S.; Matsumoto, N.M.; Maynard, H.D. Synthesis of thiol terminated acrylate polymers with divinyl sulfone to generate well-defined semi-telechelic michael acceptor polymers. Macromolecules **2009**, *42* (20), 7657–7663.

21. Boyer, L.; Bulmus, V.; Liu, J.; Davis, T.P.; Stenzel, M.H.; Barner-Kowollik, C. Well defined polymer-protein conjugates via in-situ RAFT polymerization. J. Am. Chem. Soc. **2007**, *129* (22), 7145–7154.

22. De, P.; Li, M.; Gondi, S.R.; Sumerlin, B.S. Temperature-regulated activity of responsive polymer-protein conjugates prepared by grafting-from via RAFT polymerization. J. Am. Chem. Soc. **2008**, *130* (34), 11288–11289.

23. Boyer, C.; Liu, J.; Bulmus, V.; Davis, T.P.; Barner-Kowollik, C.; Stenzel, M.H. Direct synthesis of well-defined heterotelechelic polymers for bioconjugations. Macromolecules **2008**, *41* (15), 5641–5650.

24. Boyer, C.; Liu, J.Q.; Bulmus, J.Q.; Davis, T.P. RAFT polymer end-group modification and chain coupling/conjugation via disulfide bonds. Aust. J. Chem. **2009**, *62* (8), 830–847.

25. Bays, E.; Tao, L.; Chang, C.W.; Maynard, H.D. Synthesis of semitelechelic maleimide poly(PEGA) for protein conjugation By RAFT polymerization. Biomacromolecules **2009**, *10* (7), 1777–1781.

26. Lutz, J.-F.; Börner, H.G.; Weichenhan, K. Combining ATRP and "Click" chemistry: A promising platform toward functional biocompatible polymers and polymer bioconjugates. Macromolecules **2006**, *39* (19), 6376–6383.

27. Liu, J.; Bulmus, V.; Herlambang, D.L.; Barner-Kowollik, C.; Stenzel, M.H.; Davis, T.P. In situ formation of protein–polymer conjugates through reversible addition fragmentation chain transfer polymerization. Angew. Chem. Int. Ed. **2007**, *46* (17), 3099–3103.

28. Tao, L.; Liu, J.Q.; Xu, J.T.; Davis, T.P. Synthesis and bioactivity of poly(HPMA)–lysozyme conjugates: The use of novel thiazolidine-2-thione coupling chemistry. Org. Biomol. Chem. **2009**, *7* (17), 3481–3485.

29. Tao, L.; Liu, J.; Davis, T.P. Branched polymer–protein conjugates made from mid-chain-functional P(HPMA). Biomacromolecules **2009**, *10* (10), 2847–2851.

30. Yanjarappa, M.J.; Gujraty, K.V.; Joshi, A.; Saraph, A.; Kane, R.S. Synthesis of copolymers containing an active ester of methacrylic acid by RAFT: Controlled molecular weight scaffolds for biofunctionalization. Biomacromolecules **2006**, *7* (5), 1665–1670.

31. Godwin, A.; Hartenstein, M.; Mueller, A.H.E.; Brocchini, S. Narrow molecular weight distribution precursors for polymer–drug conjugates. Angew. Chem. Int. Ed. **2001**, *40* (43), 594–597.

32. Lecolley, F.; Tao, L.; Mantovani, G.; Durkin, I.; Lautru, S.; Haddleton, D.M. A new approach to bioconjugates for proteins and peptides (pegylation) utilising living radical polymerisation. Chem Commun. **2004**, (18), 2026–2027.

33. Tao, L.; Mantovani, G.; Lecolley, F.; Haddleton, D.M. α-Aldehyde terminally functional methacrylic polymers from living radical polymerization: Application in protein conjugation "pegylation". J. Am. Chem. Soc. **2004**, *126* (41), 13220–13221.

34. Ryan, S.M.; Wang, X.X.; Mantovani, G.; Sayers, C.T.; Haddleton, D.M.; Brayden, D.J. Conjugation of salmon calcitonin to a combed-shaped end functionalized poly(poly(ethylene glycol) methyl ether methacrylate) yields a bioactive stable conjugate. J Control. Release **2009**, *135* (1), 51–59.

35. Tao, L.; Liu, J.; Hu, J.; Davis, T.P. Bio-reversible polyPEGylation. Chem Commun. (43), **2009**, 6560–6562.

36. Lele, B.S.; Murata, H.; Matyjaszewski, K.; Russell, A.J. Synthesis of uniform protein–Polymer conjugates. Biomacromolecules **2005**, *6* (6), 3380–3387.

37. Nicolas, J.; Miguel, V.S.; Mantovani, G.; Haddleton, D.M. Fluorescently tagged polymer bioconjugates from protein derived macroinitiators. Chem. Commun. **2006**, (45), 4697–4699.

38. Agut, W.; Taton, D.; Lecommandoux, S. A versatile synthetic approach to polypeptide based rod-coil block copolymers by click chemistry. Macromolecules **2007**, *40* (16), 5653–5661.

39. McDowall, L.; Chen, G.J.; Stenzel, M.H. Synthesis of seven-arm poly(vinyl pyrrolidone) star polymers with lysozyme core prepared by MADIX/RAFT polymerization. Macromol. Rapid Commun. **2008**, *29* (20), 1666–1671.

40. Ayres, L.; Hans, P.; Adams, J.; Löwik, D.W.P.M.; Hest, J.C.M.v. Peptide-polymer vesicles prepared by atom trandsfer radical polymerization. J. Polym. Sci. A **2005**, *43* (24), 6355–6366.

41. Droumaguet, B.L.; Velonia, K. In Situ ATRP-mediated hierarchical formation of giant amphiphile bionanoreactors. Angew. Chem. Int. Ed. **2008**, *47* (63), 6263–6266.

42. Dirks, A.J.T.; van Berkel, S.S.; Hatzakis, N.S.; Opsteen, J.A.; van Delft, F.L.; Cornelissen, J.; Rowan, A.E.; Hest, J.C.M.v.; Rutjes, F.P.J.T.; Nolte, R.J.M. Preparation of biohybrid amphiphiles via the copper catalysed Huisgen [3 + 2] dipolar cycloaddition reaction. Chem. Commun. **2005**, (33), 4172–4174.

43. Hwang, J.; Li, R.C.; Maynard, H.D. Well-defined polymers with activated ester and protected aldehyde side chains for bio-functionalization. J. Control. Release **2007**, *122* (3), 279–286.

44. Li, R.C.; Hwang, J.; Maynard, H.D. Reactive block copolymer scaffolds. Chem. Commun. **2007**, (35), 3631–3633.

45. Wang, K.; Liang, L.; Lin, S.; He, X. Synthesis of well-defined ABC triblock copolymers with polypeptide segments by ATRP and click reactions. Eur. Polym. J. **2008**, *44* (10), 3370–3376.

46. Canalle, L.A.; Löwik, D.W.P.M.; van Hest, J.C.M. Polypeptide-polymer bioconjugates. Chem. Soc. Rev. **2010**, *41* (39), 329–353.

47. Hamley, I.W. Peptide fibrillization. Angew. Chem. Int. Ed. **2007**, *46* (43), 8128–8147.

48. Krysman, M.J.; Castelletto, V.; Kelaraksi, A.; Hamley, I.W.; Hule, R.A.; Pochan, D.J. Self-assembly and hydrogelation of an amyloid peptide fragment. Biochemistry **2008**, *47* (16), 4597–4605.

49. Hawker, C.J.; Wooley, K.L. The convergence of synthetic organic and polymer chemistries. Science. **2005**, *309* (5738), 1200–1205.

50. Gauthier, M.A.; Klok, H.-A. Peptide/protein–polymer conjugates: Synthetic strategies and design concepts. Chem Commun. **2008**, (23), 2591–2611.

51. Klok, H.-A. Peptide/protein-synthetic polymer conjugates: Quo vadis. Macromolecules **2010**, *42* (21), 7990–8000.

52. Lutz, J.-F.; Börner, H.G. Modern trends in polymer bioconjugates design. Prog. Polym. Sci. **2008**, *33* (1), 1–39.

53. Hartmann, L.; Börner, H.G. Precision polymers: Monodisperse, monomer-sequence-defined segments to target future demands of polymers in medicine. Adv. Mater. **2009**, *21* (32–33), 3425–3431.

54. Le Droumaguet, B.; Nicolas, J. Recent advances in the design of bioconjugates from controlled/living radical polymerization. Polym. Chem. **2010**, *1* (5), 563–598.

55. Loewik, D.W.P.M.; Ayres, L.; Smeenk, J.M.; van Hest, J.C.M. Synthesis of bio-inspired hybrid polymers using peptide synthesis and protein engineering. Adv. Polym. Sci. **2006**, *202*, 19–52.

56. Börner, H.G.; Schlaad, H. Bioinspired functional block copolymers. Soft Matter **2007**, *3* (4), 394–408.

57. Matyjaszewski, K.; Xia, J. Atom transfer radical polymerization. Chem. Rev. **2001**, *101* (9), 2921–2990.

58. Hawker, C.J.; Bosman, A.W.; Harth, E. New polymer synthesis by nitroxide mediated living radical polymerizations. Chem. Rev. **2001**, *101* (12), 3661–3688.

59. Moad, G.; Rizzardo, E.; Thang, S.H. Living radical polymerization by the RAFT process–A first update. Aust. J. Chem. **2006**, *59* (10), 669–692.

60. Moad, G.; Chong, Y.K.; Postma, A.; Rizzardo, E.; Thang, S.H. Advances in RAFT polymerization: The synthesis of polymers with defined end-groups. Polymer **2005**, *46* (19), 8458–8468.

61. Perrier, S.; Takolpuckdee, P. Macromolecular design via reversible addition-fragmentation chain transfer (RAFT)/ xanthates (MADIX) polymerization. J. Polym. Sci. A **2005**, *43* (22), 5347–5393.

62. Barner-Kowollik, C.; Perrier, S. The future of reversible addition fragmentation chain transfer polymerization. J. Polym. Sci. A **2008**, *46* (17), 5715–5723.

63. Börner, H.G. Strategies exploiting functions and self-assembly properties of bioconjugates for polymer and material sciences. Prog. Polym. Sci. **2009**, *34* (9), 811–851.

64. van Hest, J.C.M.; Tirrell, D.A. Protein-based materials, toward a new level of structural control. Chem. Commun. **2001**, (19) 1897–1904.

65. Ayres, L.; Vos, M.R.J.; Adams, P.J.H.M.; Shklyarevskiy, I.O.; Hest, J.C.M.v. Elastin-based side-chain polymers synthesized by ATRP. Macromolecules **2003**, *36* (16), 5967–5973.

66. Fernandez-Trillo, F.; Dureault, A.; Bayley, J.P.M.; Hest, J.C.M.v.; Thies, J.C.; Michon, T.; Weberskirch, R.; Cameron, N.R. Elastin-based side-chain polymers: Improved synthesis via RAFT and stimulus responsive behavior. Macromolecules **2007**, *40* (17), 6094–6099.

67. Gravert, D.J.; Janda, K.D. Organic synthesis on soluble polymer supports: Liquid-phase methodologies. Chem. Rev. **1997**, *97* (2), 489–509.

68. Becker, M.L.; Liu, J.; Wooley, K.L. Functionalized micellar assemblies prepared via block copolymers synthesized by living radical polymerization upon peptide-loaded resins. Biomacromolecules **2005**, *6* (1), 220–228.

69. Becker, M.L.; Liu, B.; Wooley, K.L. Peptide-polymer bioconjugates: Hybrid block copolymers generated via living radical polymerizations from resin-supported peptides. Chem Commun. **2003**, (2), 180–181.

70. Mei, Y.; Beers, K.L.; Byrd, H.C.M.; VanderHart, D.L.; Washburn, N.R. Solid-phase ATRP synthesis of peptide–polymer hybrids. J. Am. Chem. Soc. **2004**, *126* (11), 3472–3476.

71. Broyer, R.M.; Quaker, G.M.; Maynard, H.D. Designed amino acid ATRP initiators for the synthesis of biohybrid materials. J. Am. Chem. Soc. **2008**, *130* (3), 1041–1047.

72. Ten Cate, M.G.J.; Rettig, H.; Bernhardt, K.; Börner, H.G. Sequence-defined polypeptide–polymer conjugates utilizing reversible addition fragmentation transfer radical polymerization. Macromolecules **2005**, *38* (26), 10643–10649.

73. Hentschel, J.; Bleek, K.; Ernst, O.; Lutz, J.-F.; Börner, H.G. Easy access to bioactive peptide-polymer conjugates via RAFT. Macromolecules **2008**, *41* (4), 1073–1075.

74. Couet, J.; Jeyaprakash, J.D.; Samuel, S.; Kopyshev, A.; Santer, S.; Biesalski, M. Peptide-polymer hybrid nanotubes. Angew. Chem. Int. Ed. **2005**, *44* (21), 3297–3301.

75. Couet, J.; Biesalski, M. Surface-Initiated ATRP of *N*-Isopropylacrylamide from initiator-modified self-assembled peptide nanotubes. Macromolecules **2006**, *39* (21), 7258–7268.

76. Couet, J.; Biesalski, M. Polymer-wrapped peptide nanotubes: Peptide-grafted polymer mass impacts length and diameter. Small **2008**, *4* (7), 1008–1016.

77. Hentschel, J.; Ten Cate, M.G.J.; Börner, H.G. Peptide-guided organization of peptide-polymer conjugates: Expanding the approach from oligo- to polymers. Macromolecules **2007**, *40* (26), 9224–9232.

78. Deming, T.J. Facile synthesis of block copolypeptides of defined architecture. Nature **1997**, *390* (6658), 386–389.

79. Since termination reactions cannot be avoided in LRP, i. i. n. a. l. p. t., but it is often defined as a 'pseudo-living' polymerization.

80. Klok, H.-A.; Lecommandoux, S. Solid-state structure, organization and properties of peptide-synthetic hybrid block copolymers. Adv. Polym. Sci. **2006**, *202*, 75–111.

81. Matyjaszewski, K.; Braunecker, W.A. Controlled/living radical polymerization: Features, developments, and perspectives. Prog. Polym. Sci. **2007**, *32* (1), 93–146.

82. Perrier, S.; Zhao, Y.L. Synthesis of well-defined conjugated copolymers by RAFT polymerization using cysteine and glutathione-based chain transfer agents. Chem. Commun. **2007**, (41), 4294–4296.

83. Tugulu, S.; Silacci, P.; Stergiopulos, N.; Klok, H.-A. RGD-Functionalized polymer brushes as substrates for the integrin specific adhesion of human umbilical

vein endothelial cells. Biomaterials **2007**, *28* (16), 2536–2546.

84. Duong, H.T.T.; Nguyen, T.L.U.; Stenzel, M.H. Micelles with surface conjugates RGD peptide and crosslinked-polyurea core via RAFT polymerization. Polym. Chem. **2010**, *1* (2), 171–182.

85. Hentschel, J.; Hentschel, H.G. Peptide-directed microstructure formation of polymers in organic media. J. Am. Chem. Soc. **2006**, *128* (43), 14142–14149.

86. Kolb, H.C.; Finn, M.G.; Sharpless, K.B. Click chemistry: Diverse chemical function from a few good reactions. Angew. Chem. Int. Ed. **2001**, *40* (11), 2004–2019.

87. Rostovtsev, V.V.; Green, L.G.; Fokin, V.V.; Sharpless, K.B. A stepwise huisgen cycloaddition process: Copper(I)-catalyzed regioselective "ligation" of azides and terminal alkynes. Angew. Chem. Int. Ed. **2002**, *41* (14), 2596–2599.

88. Lutz, J.F. 1,3-Dipolar cycloadditions of azides and alkynes: A universal ligation tool in polymer and materials science. Angew. Chem. Int. Ed. **2007**, *46* (7), 1018–1025.

89. Binder, W.H.; Sachsenhofer, R. "Click"—Chemistry in polymer and material science: The update. Macromol. Rapid Commun. **2008**, *29* (12–13), 951–951.

90. Binder, W.H.; Sachsenhofer, R. "Click" Chemistry in polymer and materials science. Macromol. Rapid Commun. **2007**, *28* (1), 15–54.

91. Meldal, M. Polymer "clicking" by CuAAC reactions. Macromol. Rapid Commun. **2008**, *29* (12–13), 1016–1051.

92. Inglis, A.J.; Barner-Kowollik, C. Ultra rapid approaches to mild macromolecular conjugation. Macromol. Rapid Commun. **2010**, *31* (14), 1247–1266.

93. Becer, C.R.; Hoogenboom, R.; Schubert, U.S. Click chemistry beyond metal-catalyzed cycloaddition. Angew. Chem. Int. Ed. **2009**, *48* (27), 4900–4908.

94. Hoyle, C. E.; Bowman, C.N. Thiol-ene click chemistry. Angew. Chem. Int. Ed. **2010**, *49* (9), 1540–1573.

95. Hoyle, C.E.; Lee, T.Y.; Roper, T. Thiol–enes: Chemistry of the past with promise for the future. J. Polym. Sci. A **2004**, *42* (21), 5301–5338.

96. Hoyle, C.E.; Lowe, A.B.; Bowman, C.N. Thiol-click chemistry: A multifaceted toolbox for small molecule and polymer synthesis. Chem. Soc. Rev. **2010**, *39* (4), 1355–1387.

97. Lahue, B.R.; Lo, S.M.; Wan, Z.K.; Woo, G.H.C.; Snyder, J.K. Intramolecular inverse-electron-demand Diels-Alder reactions of imidazoles with 1,2,4-triazines: A new route to 1,2,3,4-tetrahydro-1,5-naphthyridines and related heterocycles. J. Org. Chem. **2004**, *69* (21), 7171–7182.

98. Blackman, M.L.; Royzen, M.; Fox, J.M. Tetrazine ligation: Fast bioconjugation based on inverse-electron-demand diels-alder reactivity. J. Am. Chem. Soc. **2008**, *130* (41), 13518–13519.

99. Royzen, M.; Yap, G.P.A.; Fox, J.M. A photochemical synthesis of functionalized trans-cyclooctenes driven by metal complexation. J. Am. Chem. Soc. **2008**, *130* (12), 3760–3761.

100. Jewett, J.C.; Bertozzi, C.R. Cu-free click cycloaddition reactions in chemical biology. Chem. Soc. Rev. **2010**, *39* (4), 1272–1279.

101. Singh, I.; Zarafshani, Z.; Lutz, J.F.; Heaney, F. Metal-Free "click" chemistry: Efficient polymer modification via 1,3-dipolar cycloaddition of nitrile oxides and alkynes. Macromolecules **2009**, *42* (15), 5411–5413.

102. Boyer, C.; Bulmus, V.; Davis, T.P.; Ladmiral, V.; Liu, J.; Perrier, S. Bioapplications of RAFT polymerization. Chem. Rev. **2009**, *109* (11), 5402–5436.

103. Jerome, C.; Lecomte, P. Recent advances ind the synthesis of aliphatic polyesters by ring-opening polymerization. Adv. Drug Deliv. Rev. **2008**, *60* (9), 1056–1076.

104. Lowe, A.B.; McCormick, C.L. Reversible addition-fragmentation chain transfer (RAFT) radical polymerization and the synthesis of water-soluble (co)polymers under homogeneous conditions in organic and aqueous media. Prog. Polym. Sci. **2007**, *32* (3), 283–351.

105. Smith, A.E.; Xu, X.; McCormick, C.L. Stimuli-responsive amphiphilic (co)polymers via RAFT polymerization. Prog. Polym. Sci. **2010**, *35* (1), 45–93.

106. Urbani, C.N.; Bell, C.A.; Whittaker, M.R.; Monteiro, M.J. Convergent synthesis of second generation AB-type miktoarm dendrimers using "click" chemistry catalyzed by copper wire. Macromolecules **2008**, *41* (4), 1057–1060.

107. Golas, P.L.; Tsarevsky, N.V.; Sumerlin, B.S.; Matyjaszewski, K. Catalyst performance in "Click" coupling reactions of polymers prepared by ATRP: Ligand and metal effects. Macromolecules **2006**, *39* (19), 6451–6457.

108. Appukkuttan, P.; Mehta, V.P.; Van der Eycken, E.V. Microwave-assisted cycloaddition reactions. Chem. Soc. Rev. **2010**, *39* (5), 1467–1477.

109. Kaval, N.; Ermolat'ev, D.; Appukkuttan, P.; Dehaen, W.; Kappe, C.O.; Van der Eycken, E. The application of "click chemistry" for the decoration of 2(1H)-pyrazinone scaffold: Generation of templates. J. Comb. Chem. **2005**, *7* (3), 490–502.

110. Rijkers, D.T.S.; van Esse, G.W.; Merkx, R.; Brouwer, A.J.; Jacobs, H.J.F.; Pieters, R.J.; Liskamp, R. M. Efficient microwave-assisted synthesis of multivalent dendrimeric peptides using cycloaddition reaction (click) chemistry. Chem. Commun. **2005**, (36), 4581–4583.

111. Hoogenboom, R.; Moore, B.C.; Schubert, U.S. Synthesis of star-shaped poly(epsilon-caprolactone) via "click" chemistry and "supramolecular click" chemistry. Chem. Commun. **2006**, (38), 4010–4012.

112. Lutz, J.F.; Börner, H.G.; Weichenhan, K. "Click" bioconjugation of a well-defined synthetic polymer and a protein transduction domain. Aust. J. Chem. **2007**, *60* (6), 410–413.

113. Baskin, J.; Bertozzi, C. Bioorthogonal click chemistry: Covalent labeling in living systems. QSAR Comb. Sci. **2007**, *26* (11–12), 1211–1219.

114. Griffin, R.J. The medicinal chemistry of the azido group. Prog. Med. Chem. **1994**, *31*, 121–232.

115. Ladmiral, V.; Legge, T.M.; Zhao, Y.; Perrier, S. "Click" chemistry and radical polymerization: Potential loss of orthogonality. Macromolecules **2008**, *41* (18), 6728–6732.

116. Thordarson, P.; LeDroumaguet, B.; Velonia, K. Well-defined protein–polymer conjugates—Synthesis and potential applications . Appl. Microbiol. Biotechnol. **2006**, *73* (2), 243–254.

117. Lundquist, J.T.I.V.; Pelletier, J.C. Improved solid-phase peptide synthesis method utilizing α-azide-protected amino acids. Org. Lett. **2001**, *3* (5), 781–783.

118. Goddard-Borger, E.D.; Stick, R.V. An efficient, inexpensive, and shelf-stable diazotransfer reagent: Imidazole-1-sulfonyl azide hydrochloride. Org. Lett. **2007**, *9* (19), 3797–3800.

119. ten Brink, H.T.; Meijer, J.T.; Geel, R.V.; Damen, M.; Loewik, D.W.P.M.; van Hest, J.C.M. Solid-phase synthesis of C-terminally modified peptides. J. Pept. Sci. **2006**, *12* (11), 686–692.

120. O'Donnell, M.J.; Drew, M.D.; Pottorf, R.S.; Scott, W.L. UPS on weinreb resin: A facile solid-phase route to aldehyde and ketone derivatives of "unnatural" amino acids and peptides. J. Comb. Chem. **2000**, *2* (2), 172–181.

121. Hanisch, F.-G.; Mueller, S. MUC1: The polymorphic appearance of a human mucin. Glycobiology **2000**, *10* (5), 439–449.

122. Aggeli, A.; Bell, M.; Boden, N.; Keen, J.N.; Knowles, P.F.; McLeish, T.C.B.; Pitkeathly, M.; Radford. S.E. Responsive gels formed by the spontaneous self-assebly of peptides into polymeric β-sheet tapes. Nature **1997**, *386* (6622), 259–262.

123. Aggeli, A.; Bell, M.; Carrick, L.M.; Fishwick, C.W.G.; Harding, R.; Mawer, P.J.; Radford, S.E.; Strong, A.E.; Boden, N. pH as a trigger of peptide β-sheet self-assembly and reversible switching between nematic and isotropic phases. J. Am. Chem. Soc. **2003**, *125* (32), 9619–9628

124. Benaglia, M.; Chen, M.; Chong, Y.K.; Moad, G.; Rizzardo, E.; Thang, S.H. Polystyrene-*block*-poly(vinyl acetate) through teh Use of Switchable RAFT Agent. Macromolecules **2009**, *42* (24), 9384–9386.

125. Quemener, D.; Hellaye, M.L.; Bissett, C.; Davis, T.P.; Barner-Kowollik, C.; Stenzel, M.H. Graft block copolymers of propargyl methacrylate and vinyl acetate via a combination of RAFT/MADIX and click chemistry: Reaction analysis. J. Polym. Sci. A Polym. Chem. **2008**, *46* (1), 155–173.

126. Ranjan, R.; Brittain, W.J. Combination of living radical polymerization and click chemistry for surface modification. Macromolecules **2007**, *40* (17), 6217–6223.

127. Dispinar, T.; Sanyal, R.; Sanyal, A. A diels-alder/retro diels-alder strategy to synthesize polymers bearing maleimide side chains. J. Polym. Sci. A **2007**, *45* (20), 4545–4551.

128. Hua, D.; Bai, W.; Xiao, J.; Bai, R.; Lu, W.; Pan, C. A strategy for synthesis of azide polymers via controlled/living free radical copolymerization of allyl azide under ^{60}Co γ-ray irradiation. Chem. Mater. **2005**, *17* (18), 4574–4576.

129. Sumerlin, B.S.; Tsarevsky, N.V.; Louche, G.; Lee, R.Y.; Matyjaszewski, K. Highly efficient "Click" functionalization of poly(3-azidopropyl methacrylate) prepared by ATRP. Macromolecules **2005**, *38* (18), 7540–7545.

130. Sanda, F.; Abe, T.; Endo, T. Syntheses and radical polymerizations of optically active (meth)acrylamides having amino acid moieties. J. Polym. Sci. A **1997**, *35* (13), 2619–2629.

131. Trnka, T.M.; Grubbs, R.H. The development of L_2X_2Ru-CHR olefin metathesis catalysts: An organometallic success story. Acc. Chem. Res. **2001**, *34* (1), 18–29.

132. Parrish, B.; Breitenkamp, R.B.; Emrik, T. PEG- and peptide-grafted aliphatic polyesters by click chemistry. J. Am. Chem. Soc. **2005**, *127* (20), 7404–7410.

133. Tanaka, M.; Sudo, A.; Sanda, F.; Endo, T. Samarium enolate on crosslinked polystyrene beads. II. An anionic initiator for the well-defined synthesis of poly(allyl methacrylate) on a solid support. J. Polym. Sci. A **2003**, *41* (6), 853–860.

134. Binder, W.H.; Gruber, H. Block copolymers derived from photoreactive 2-oxazoline, 1.Synthesis and micellization behavior. Macromol. Chem. Phys. **2000**, *201* (9), 949–957.

135. Luxenhofer, R.; Jordan, R. Click chemistry with poly(2-oxazoline)s. Macromolecules **2006**, *39* (10), 3509–3516.

136. Coessens, V.; Pintauer, T.; Matyjaszewski, K. Functional polymers by atom transfer radical polymerization. Prog. Polym. Sci. **2001**, *26* (3), 337–377.

137. Heredina, K.L.; Maynard, H.D. Synthesis of protein–polymer conjugates. Org. Biomol. Chem. **2007**, *5* (1), 45–53.

138. O'Reilly, R.K.; Joralemon, M.J.; Hawker, C.J.; Wooley, K.L. Facile syntheses of surface-functionalized micelles and shell cross-linked nanoparticles. J. Polym. Sci. A **2006**, *44* (17), 5203–5217.

139. Boyer, L.; Bulmus, V.; Herlambang, D.L.; Barner-Kowollik, C.; Stenzel, M.H.; Davis, T.P. In situ formation of protein-polymer conjugates through reversible addition fragmentation chain transfer polymerization. Angew. Chem. Int. Ed. **2007**, *46* (17), 3099–3103.

140. Kakwere, H.; Payne, R.J.; Jolliffe, K.A.; Perrier, S. Self-assembling macromolecular chimerase: Preventing fibrillization of a β-sheet forming peptide by polymer conjugation. Soft Matter **2011**, *7* (8), 3754–3757.

141. Urbani, C.N.; Bell, C.A.; Lonsdale, D.E.; Whittaker, M.R.; Monteiro, M.J. Reactive alkyne and azide solid supports to increase purity of novel polymeric stars and dendrimers via the "click" reaction. Macromolecules **2007**, *40* (19), 7056–7059.

142. Durmaz, H.; Dag, A.; Altintas, O.; Erdogan, T.; Hizal, G.; Tunca, U. One-pot synthesis of ABC type triblock copolymers via in situ click [3 + 2] and diels-alder [4 + 2] reactions. Macromolecules **2007**, *40* (2), 191–198.

Neural—Rapid Prototyping

Pharmaceutical Polymers

Narendra Pal Singh Chauhan
Department of Polymer Science, University College of Science, Mohanlal Sukhadia University, Udaipur, India

Arpit Kumar Pathak
Kumudini Bhanat
Department of Chemistry, University College of Science, Mohanlal Sukhadia University, Udaipur, India

Rakshit Ameta
Department of Chemistry, Pacific College of Basic and Applied Sciences, Pacific Academy of Higher Education and Research Society (PAHER) University,, Udaipur, India

Manish Kumar Rawal
Department of Chemistry, Vidya Bhawan Rural Institute, Udaipur, India

Pinki B. Punjabi
Department of Chemistry, University College of Science, Mohanlal Sukhadia University, Udaipur, India

Abstract

Synthetic and natural-based polymers have found their way into the pharmaceutical and biomedical industries, and their applications are growing at a fast pace. Major applications of polymers in the current pharmaceutical field are for controlled drug release, which will be discussed in detail in the following sections. In the biomedical area, polymers are generally used as implants and are expected to perform long-term service. In general, the desirable polymer properties in pharmaceutical applications are film forming (coating), thickening (rheology modifier), gelling (controlled release), adhesion (binding), pH-dependent solubility (controlled release), solubility in organic solvents (taste masking), and barrier properties (protection and packaging). From the solubility point of view, pharmaceutical polymers can be classified as water-soluble and water-insoluble (oil soluble or organic soluble). In this entry, many water-soluble biomedical and pharmaceutical polymers including polyethylene glycol (PEG), polyvinyl alcohol, polyethylene oxide, polyvinyl pyrrolidone, polyacrylate or polymethacrylate esters, cellulose-based polymers, hydrocolloids, and many plastics and rubbers will be tabulated and their anionic and cationic functionalities will be summarized and discussed. This entry will also focus on PEGylation, defined as the covalent attachment of PEG chains to bioactive substances.

INTRODUCTION

Synthetic and natural-based polymers have found their way into the pharmaceutical and biomedical industries and their applications are growing at a fast pace. The use of polymers for biomedical and pharmaceutical applications has gained an enormous impact during the past decades. Polymers can be applied in drug delivery systems, as scaffolds for tissue engineering and repair, and as novel biomaterials.[1] Polymers are becoming increasingly important in the field of drug delivery. The pharmaceutical applications of polymers range from their use as binders in tablets to viscosity and flow controlling agents in liquids, suspensions, and emulsions. Polymers can be used as film coatings to disguise the unpleasant taste of a drug, to enhance drug stability, and to modify drug release characteristics.

Examples of pharmaceutical polymers:

1. Vinyl polymers
2. Cellulose ethers
3. Polyesters
4. Silicones
5. Polysaccharides and related polymers
6. Miscellaneous polymers

APPLICATION OF BIOMEDICAL PHARMACEUTICAL POLYMERS

In a traditional pharmaceutics area, such as tablet manufacturing, polymers are used as tablet binders to bind the excipients of the tablet. Advanced pharmaceutical dosage forms utilize polymers for drug protection, taste masking, controlled release of a given drug, targeted delivery,

Concise Encyclopedia of Biomedical Polymers and Polymeric Biomaterials DOI: 10.1081/E-EBPPC-120050558

increase drug bioavailability, and so on and so forth. To one side from solid dosage forms, polymers have found application in liquid dosage forms as rheology modifiers.

They are used to control the viscosity of an aqueous solution or to stabilize suspensions or even for the granulation step in preparation of solid dosage forms. Major applications of polymers in the current pharmaceutical field are for controlled drug release, in the biomedical area. Polymers are generally used as implants and are expected to perform long-term service. This requires that the polymers have unique properties that are not offered by polymers intended for general applications. Table 1 provides a list of polymers with their applications in pharmaceutical and biomedical industries. In general, the desirable polymer properties in pharmaceutical applications are film forming (coating), thickening (rheology modifier), gelling (controlled release), adhesion (binding), pH-dependent solubility (controlled release), solubility in organic solvents (taste masking), and barrier properties (protection and packaging). From the solubility standpoint, pharmaceutical polymers can be

Table 1 Polymers in pharmaceutical and biomedical applications

Name of the Polymer	Applications
Water-Soluble Synthetic Polymer	
Poly(ethylene oxide)	Coagulant, flocculent, swelling agent
Poly(vinyl pyrrolidone)	Plasma replacement, tablet granulation
Poly(vinyl alcohol)	Water-soluble packaging, tablet binder, tablet coating
Poly(ethylene glycol)	Plasticizer, base for suppositories
Poly(isopropyl acrylamide) and poly(cyclopropyl methacrylamide)	Thermogelling acrylamide derivatives, its balance of hydrogen bonding, and hydrophobic association changes with temperature
Water-Insoluble Biodegradable Polymers	
(Lactide-co-glycolide) polymers	For protein delivery
Starch-Based Polymer	
Sodium starch glycolate	Superdisintegrant for tablets and capsules in oral delivery
Starch	Glidant, a diluent in tablets and capsules, a disintegrant in tablets and capsules, a tablet binder
Plastics and Rubbers	
Polycyanoacrylate	Biodegradable tissue adhesives in surgery, a drug carrier in nano- and microparticles
Polychloroprene	Septum for injection, plungers for syringes, and valve components
Polyisobutylene	Pressure-sensitive adhesives for transdermal delivery
Silicones	Pacifier, therapeutic devices, implants, medical grade adhesive for transdermal delivery
Polystyrene	Petri dishes and containers for cell culture
Poly(methyl methacrylate)	Hard contact lenses
Poly(hydroxyethyl methacrylate)	Soft contact lenses
Poly(vinyl chloride)	Blood bag, hoses, and tubing
Hydrocolloids	
Carrageenan	Modified release, viscosifier
Chitosan	Cosmetics and controlled drug delivery applications, mucoadhesive dosage forms, rapid release dosage forms
Pectinic acid	Drug delivery
Alginic acid	Oral and topical pharmaceutical products; thickening and suspending agent in a variety of pastes, creams, and gels, as well as a stabilizing agent for oil-in-water emulsions; binder and disintegrant
Cellulose-Based Polymers	
Hydroxypropyl methyl cellulose	Binder for tablet matrix and tablet coating, gelatin alternative as capsule material
Hydroxyethyl and hydroxypropyl cellulose	Soluble in water and in alcohol, tablet coating

Neural—Rapid Prototyping

classified as water-soluble and water-insoluble (oil soluble or organic soluble). The cellulose ethers with methyl and hydroxypropyl substitutions are water-soluble, whereas ethyl cellulose and a group of cellulose esters such as cellulose acetate butyrate or phthalate are organic soluble. Hydrocolloid gums are also used when solubility in water is desirable. The synthetic water-soluble polymers have also found extensive applications in pharmaceutical industries, among them polyethylene glycol (PEG), vinyl alcohol polymers, polyethylene oxide (PEO), polyvinyl pyrrolidone, and polyacrylate or polymethacrylate esters containing anionic and cationic functionalities are well established (Table 1).

CLASSIFICATION OF BIOMEDICAL PHARMACEUTICAL POLYMERS

Polymers have played important roles in the preparation of biomedical pharmaceutical products. Their applications range widely from material packaging to fabrication of the most sophisticated drug delivery devices. Of the many materials used in the pharmaceutical formulations, polymers play the most important roles. The use of polymers ranges from the manufacturing of various drug packaging materials to the development of dosage forms.

Polymers in pharmaceutical applications are classified into three general categories according to their common uses:

1. Polymers in conventional dosage forms
2. Polymers in controlled release dosage forms
3. Polymers for drug packaging

From the solubility point of view, pharmaceutical polymers can be classified into two categories:

1. Water-soluble polymers
2. Water-insoluble polymers (oil soluble or organic soluble)

Polymers in Conventional Dosage Forms

Despite the well-known advantages of controlled release dosage forms, conventional dosage forms are still more widely used, probably because they cost less to manufacture. More than three-quarters of all drug formulations are made for oral administration. Oral dosage forms such as tablets, capsules, and liquids are still most popular. Since tablet is one of the most widely used dosage forms and its preparation requires the incorporation of polymers, we will focus on polymers used in the tableting process.

Polymers Used in Controlled Release Dosage Forms

One of the most important applications of polymers in modern pharmaceutics is the development of new, advanced drug delivery systems, commonly known as controlled release drug delivery systems. Controlled release formulations attempt to alter drug absorption and subsequent drug concentration in blood by modifying the drug release rate from the device. Polymer use in controlled release dosage is reduced fluctuations in the plasma drug concentration, less side effects, and increased patient compliance. Controlled release products consist of the active agent and the polymer matrix or membranes that regulate its release. Advances in controlled release technology in recent years have been possible as a result of advances in polymer science that allow the fabrication of polymers with tailor-made specifications, charge density, such as molecular size, specific functional groups, hydrophobicity, biocompatibility, and degradability. Controlled release dosage forms refresh old drugs by reducing pharmaceutical shortcomings and improving biopharmaceutical properties of the drugs. Polymers used in controlled release dosage forms are an alternative to the development of new drugs, which is extremely costly. The controlled release dosage forms are also important in the delivery of newly developed protein drugs. Currently, most protein drugs are administered by injection. Although protein drugs have delicate bioactivity, its success in treating chronic illness largely depends on the development of new delivery systems for the routine administration other than injection.

The mechanisms of controlled drug delivery can be classified into the following five mechanisms: 1) diffusion; 2) dissolution; 3) osmosis; 4) ion-exchange; and 5) polymeric prodrug. In all cases, polymers function as a principal component that controls the transport of drug molecules and the way this process is utilized in the device determines the primary mechanism for each drug delivery system.

Polymers for Drug Packaging

Many polymers are used as packaging materials for pharmaceutical products. The properties of the plastic packaging materials, such as gas permeability, flexibility, and transparency, are responsible for specific applications. Flexible packages are made by the use of thin and flexible polymer films. When they are wrapped around a product, they can easily adapt their shape to conform to the shape of the contents. The thin, flexible films are usually produced from cellulose derivatives, Ppoly(vinyl chloride) (PVC), polyethylene, polypropylene, polyamide (nylon), polystyrene, polyesters, polycarbonate, poly(vinylidene chloride), and polyurethanes. These polymeric materials are generally heat sealable and are also capable of being laminated to other materials.[2] A tight package can be prepared by wrapping an article with these polymer films followed by a brief heat treatment. Rigid packages such as bottles, boxes, trays, cups, vials, and various closures are made from materials of sufficient strength and inflexibility. Widely used polymers are high-density polyethylene, polypropylene, polybutene, poly(vinyl chloride), acrylic copolymers, polycarbonate, nylon, and polyethylene terephthalate (PET).

Neural—Rapid Prototyping

Biodegradable PET is preferred due to environmental concerns, but it is expensive.

The closing or cap of the container is typically made of polypropylene or polyethylene. The cap is usually lined with moisture-resistant materials such as low-density polyethylene, polyvinyl chloride, or poly(vinylidene dichloride).[3]

Polyisoprene, ethylene propylene/dicylopentadiene copolymer, styrene/butadiene copolymer, polybutadiene, silicone elastomers, and natural rubber are used as polymeric ingredients of the rubber stopper.[4] One of the most important necessities of any packaging material is that it should not release any component into the drug product. The preparation of containers free from any leachable components are especially important for the containers of ophthalmics, parenteral products, and any liquid products. It was shown that di(2-ethylhexyl) phthalate was released from the PVC bags and that caused uncertainty of the taxol solution.[5] United States Pharmacopeia/NF offers the protocol of chemical, spectral, and water vapor access tests and tolerances for plastic containers.[6] Among those, a chemical test is designed to give a quantitative assessment of the extractable materials in both organic solvents and water. When drugs are stored in the polymeric containers or in contact with polymer surfaces, the drug loss by adsorption to the polymer surface should be considered. It was reported that significant portions (ranging from 23% to 55%) of drugs, such as diazepam, isosorbide dinitrate, nitroglycerin, and warfarin sodium, were lost during the 24 hr storage in PVC bags.[7] Caution should be exercised in the use of polymeric containers and other devices for the delivery of protein drugs, since proteins readily adsorb and sometimes denature on the hydrophobic polymer surface. It was shown that more than 90% of tissue necrosis factor was lost when delivered using a burette mixing set simply due to the adsorption onto the surface of the delivery device.[8] It has been projected that insulin molecules aggregate in the bulk solution as a result of adsorption to the container surface and subsequent denaturation.[9] Clearly, the loss of protein drugs by adsorption to the polymeric surface can be significant.[10,11]

Taste Masking

Taste is an important factor in the development of dosage form. Taste masking becomes a requirement for bitter drugs to improve patient agreement especially in the pediatric and old population. Taste-masking technologies are increasingly paying attention on aggressively bitter tasting drugs like the macrolide antibiotics, non-steroidal anti-inflammatory drugs, and penicillin. Taste-masking technologies offer a great scope for invention and patents (Fig. 1). Undesirable taste is one of the significant formulation problems encountered with most drugs. The methods most commonly involved for achieving taste masking include various chemical and physical methods that prevent the drug substance from interaction with taste buds. The simplest method for taste masking is to use of flavor

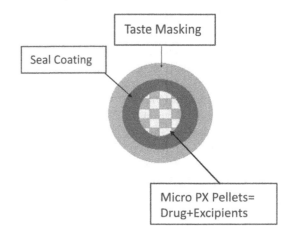

Fig. 1 Taste masking.

enhancers. Where these methods fail, more complex methodologies are adopted. A variety of techniques have been recognized for taste masking, which include polymer coating, inclusion complex formation with cyclodextrin, use of ion exchange resins, solubility-limiting methods, liposome, multiple emulsions, use of anesthetic agents, etc.

Methods usually in use for achieving effective taste masking include various physical and chemical methods that prevent the drug substance from interaction with the taste buds:

1. Use of flavor enhancers.
2. Applying polymer coatings: Polymer coating is one of the most important methods of taste masking of bitter drugs. It avoids direct contact of the bitter drug with the taste buds. The selection of polymer for taste masking is based on the physiochemical property and compatibility with the drug.[12,13] A water-soluble polymer such as a cellulose acetate, cellulose butyrate, hydroxyethyl cellulose is used in taste masking of bitter drug.
3. Complexation with ion exchange resins.
4. Inclusion complex formation with cyclodextrins.

Rheology Modifiers

Rheology modifiers commonly referred to as thickeners or viscosifiers are ever present in most of the products. The use of these additives cut across several process industries including food, pharmaceuticals, cosmetics and personal care, adhesives, textile, ceramics, paper, detergents, paints, inks, and coatings, among others. Rheology can be defined as "the science or study of how things flow." Thickeners come from both natural and synthetic sources. Naturally occurring polymers comprise polysaccharide or amino acid building blocks and are generally water-soluble. Common examples are starch, cellulose, alginate, egg yolk, agar, arrowroot, carrageenan, collagen, gelatin, guar gum, pectin, and xanthan gum.

Barrier Properties

Barrier polymers are used for many packaging and protective applications. As barriers, it is a separate system, such as an article of food or an electronic component, from an environment. Barrier polymers' boundary movement of substances is called permeants. The movement can be through the polymer or, in some cases, merely into the polymer. After crossing the barrier polymer, the permeant moves to the polymer surface, desorbs, and moves away. Permeant movement is a physical process that has both a thermodynamic and a kinetic component. For polymers without special surface treatments, the thermodynamic contribution is in the solution step. The permeant partitions between the environment and the polymer are generally according to thermodynamic rules of solution. The polymers that are good barriers to permanent gases, especially oxygen, have important commercial significance. Vinylidene chloride copolymers are available as resins for extrusion, latices for coating, and resins for solvent coating.

Hydrolyzed ethylene–vinyl acetate copolymers, commonly known as ethylene–vinyl alcohol (EVOH) copolymers, are usually used as extrusion resins, although some may be used in solvent-coating applications. Copolymers of acrylonitrile are used in extrusion and molding applications. Commercially important comonomers for barrier applications include styrene and methyl acrylate. Polyamide polymers can provide a good-to-moderate barrier-to-permeation by permanent gases. Two often-used polymers have adequate properties for some applications. Poly(ethylene terephthalate) is used to make films and bottles. PVC is a moderate barrier-to-permanent gas. Plasticized PVC is used as a household wrapping film. In regard to water–vapor transmission rate (WVTR) values, those polymers that are good oxygen barriers are often poor water–vapor barriers and vice versa. Polymer molecules without dipole–dipole interactions, such as polyolefins, dissolve very little water and have low WVTR and permeability values. The permeation of flavor, aroma, and solvent molecules in polymers follow the same physics as the permeation of small molecules, but with two significant differences. For these larger molecules, the diffusion coefficients are much lower and the solubility coefficients are much higher. Furthermore, the large solubility coefficient can lead to enough sorption of the large molecule so that plasticization occurs in the polymer, which can increase the diffusion coefficient. Generally, vinylidene chloride copolymers and glassy polymers such as polyamides and EVOH are good barriers to flavor and aroma permeation, whereas polyolefins are poor barriers. Several physical factors can affect the barrier properties of a polymer. These include temperature, humidity, orientation, and cross-linking. Typically, the permeability increases 5–10% for every increase of 1°C. When a polymer equilibrates with a humid environment, it absorbs water. This can plasticize the polymer and increase the permeability. The effect of orientation on the permeability of polymers is difficult to assess; diffusion in some polymers is unaffected by orientation, and in others increases or decreases are observed. Cross-linking has been shown in few cases to decrease the diffusion coefficient. Reasonable prediction can be made of the permeabilities of low molecular weight gases such as oxygen, nitrogen, and carbon dioxide in many polymers. The diffusion coefficients are not complicated by the shape of the permeant, and the solubility coefficients of each of these molecules do not vary from polymer to polymer.

Gelling

Gels are semisolid vehicles for drug. The gel has been used as a medium for the delivery of drug for both local treatment and systematic effect. Gel remains one of the most popular and important pharmaceutical dosage forms. There are many types of gelling agents used in pharmaceutical applications. There are some common gelling agents are acacia, alginic acid, bentonite, Carbopols (now known as carbomers), carboxymethylcellulose, ethylcellulose (EC), gelatin, hydroxyethylcellulose, hydroxypropyl cellulose, magnesium aluminum silicate, methylcellulose (MC), poloxamers, polyvinyl alcohol (PVA), sodium alginate, and xanthan gum. Topical gel formulations of diclofenac sodium were prepared by using Carbopol 934, Carbopol 940, sodium carboxymethylcellulose (NaCMC), and polymer as a gel-forming material that is biocompatible and biodegradable. The transdermal and topical delivery of drugs afford advantages over conventional oral administration. Hydroxypropyl methylcellulose (HPMC) is one of the most commonly used hydrophilic biodegradable polymers for developing controlled release formulations, because it works as a pH-independent gelling agent. HPMC is a widely accepted pharmaceutical excipient polymer, because HPMC is available in a wide range of molecular weights and the successful control of gel viscosity is easily possible.[14] The development of in situ gel systems has received significant attention over the past few years. In situ gel forming drug delivery systems are in principle capable of releasing drugs in a sustained manner maintaining relatively constant plasma profiles. Both natural and synthetic polymers can be used for the production of in situ gels. In situ gelling systems are liquid at room temperature but undergo gelation when in contact with body fluids or change in pH. The formation of gel depends on factors like temperature modulation, pH change, presence of ions, and ultra violet irradiation from which the drug gets released in a sustained and controlled manner. In situ gelling systems are liquid at room temperature but undergo gelation when in contact with body fluids or change in pH. Carboxymethyl mungbean starch (CMMS) is also investigated as a gelling agent in the topical pharmaceutical.[15]

POLYETHYLENE GLYCOL

PEG is a polyether compound with many applications from industrial manufacturing to medicine. PEG is also known as PEO or polyoxyethylene (POE), depending on its molecular weight. PEGs contain potential toxic impurities, such as ethylene oxide and 1,4-dioxane. PEGs are nephrotoxic if applied to damaged skin.[16]

They are used commercially in numerous applications, including as surfactants, in foods, in cosmetics, in pharmaceutics, in biomedicine, as dispersing agents, as solvents, in suppository bases, as tablet excipients, and as laxatives. Some specific groups are lauromacrogols, nonoxynols, octoxynols, poloxamers, macrogol used as a laxative, is a form of PEG. PEG is the basis of a number of laxatives (e.g., macrogol-containing products, such as Movicol and PEG 3350, or SoftLax, MiraLAX, or GlycoLax). Whole bowel irrigation with PEG and added electrolytes is used for bowel preparation before surgery or colonoscopy. PEG is used as an incipient in many pharmaceutical products. Lower-molecular-weight variants are used as solvents in oral liquids and soft capsules, whereas solid variants are used as ointment bases, tablet binders, film coatings, and lubricants.[17] PEG is also used in lubricating eye drops.

- PEG, when labeled with a near-infrared fluorophore, has been used in preclinical work as a vascular agent, lymphatic agent, and general tumor-imaging agent by exploiting the enhanced permeability and retention effect of tumors.[18]
- High-molecular-weight PEG (e.g., PEG 8000) has been shown to be a dietary preventive agent against colorectal cancer in animal models.[19]
- The chemoprevention database shows PEG is the most effective known agent for the suppression of chemical carcinogenesis in rats. Cancer prevention applications in humans, however, have not yet been tested in clinical trials.[20]
- The injection of PEG 2000 into the bloodstream of guinea pigs after spinal cord injury leads to rapid recovery through molecular repair of nerve membranes.[21] The effectiveness of this treatment to prevent paraplegia in humans after an accident is not known yet.
- Research is being done in the use of PEG to mask antigens on red blood cells. Various research institutes have reported that using PEG can mask antigens without damaging the function and shape of the cell.
- Research is also being done on the use of PEG in the field of gene therapy.
- PEG is being used in the repair of motor neurons damaged in crush or laceration incidents *in vivo* and *in vitro*. When coupled with melatonin, 75% of damaged sciatic nerves were rendered viable.[22]
- PEG is commonly used as a precipitant for plasmid DNA isolation and protein crystallization. X-ray diffraction of protein crystals can reveal the atomic structure of the proteins.

- In microbiology, PEG precipitation is used to concentrate viruses. PEG is also used to induce complete fusion (mixing of both inner and outer leaflets) in liposomes reconstituted *in vitro*.
- Gene therapy vectors (such as viruses) can be PEG-coated to shield them from inactivation by the immune system and to de-target them from organs where they may build up and have a toxic effect.[23] The size of the PEG polymer has been shown to be important, with larger polymers achieving the best immune protection.
- PEG is a component of stable nucleic acid lipid particles (SNALPs) used to package siRNA for use *in vivo*.[24,25]
- In blood banking, PEG is used as a potentiator to enhance the detection of antigens and antibodies.[26]
- When working with phenol in a laboratory situation, PEG 300 can be used on phenol skin burns to deactivate any residual phenol.

PEGylation

PEGylation has become a well-accepted method for the delivery of biopharmaceuticals especially peptide and protein. PEGylation is the process of covalent attachment of PEG polymer chains to another molecule, normally a drug or therapeutic protein. PEGylation is routinely achieved by incubation of a reactive derivative of PEG with the target molecule. Those in the biomedical, biotechnical, and pharmaceutical communities have become quite familiar. In its most common form poly(ethylene glycol), with the improved pharmacological and biological PEG, is a linear or branched polyether terminated properties that are associated with the covalent of poly(ethylene glycol) or PEG to therapeutically useful polypeptides.

PEGylation, the covalent attachment of PEG to molecules of interest, has become a well-established prodrug delivery system.[27,28] This polymer is nontoxic, non-immunogenic, non-antigenic, and highly soluble in water and FDA approved. The PEG-drug conjugates have several advantages: a prolonged residence in body, a decreased degradation by metabolic enzymes, and a reduction or elimination of protein immunogenicity. PEGylation now plays an important role in drug delivery, enhancing the potentials of peptides and proteins as therapeutic agents.

The covalent attachment of PEG to a drug or therapeutic protein can "mask" the agent from the host's immune system (reduced immunogenicity and antigenicity), and increase the hydrodynamic size (size in solution) of the agent, which prolongs its circulatory time by reducing renal clearance. PEGylation can also provide water solubility to hydrophobic drugs and proteins. PEGylation is a process of attaching the strands of the polymer PEG to molecules most typically peptides, proteins, and antibody fragments that can help to meet the challenges of improving the safety and efficiency of many therapeutics.[29] It produces alterations in the physiochemical properties including changes in conformation, electrostatic binding, hydrophobicity, etc. These physical and chemical changes increase the systemic retention of the therapeutic agent. Also, it can influence the binding affinity of the therapeutic moiety to the cell receptors and can alter the absorption and distribution patterns. Typically, most of the PEG-based prodrugs have been developed for the delivery of anticancer agents such as paclitaxel, methotrexate, and cisplatin. High-molecular-weight prodrugs containing cytotoxic components have been developed to decrease peripheral side effects and to obtain a more specific administration of the drugs to the cancerous tissues.[30]

PEGylation, by increasing the molecular weight of a molecule, can impart several significant pharmacological advantages over the unmodified form, such as:

- Improved drug solubility
- Reduced dosage frequency, without diminished efficacy with potentially reduced toxicity
- Extended circulating life
- Increased drug stability
- Enhanced protection from proteolytic degradation

Starch is one of the superior known natural and biodegradable biopolymers. Starch and its derivatives received much concentration in the food, plastic, and pharmaceutical industries because of its gelling, film forming, and biodegradable properties.[31] Different modifications or derivatization techniques have been tried for developing new polymers from natural biopolymers. One approach is acetylation of starch, which converts the hydrophilic starch to hydrophobic starch acetate. Starch acetate has been broadly used for drug delivery applications. Acetylation of starch has received much attention for varied drug delivery applications, such as in the preparation of coatings for the sustained release of drugs. Hydrophobically modified polymers have attracted much attention in drug delivery applications due to their fine balance of hydrophilic–hydrophobic nature and biodegradability.[32,33] Starch was allowed to react with acetic anhydride (1:4 ratio) with sodium hydroxide as a catalyst as per the reported method.[34] Tight junction visualization studies demonstrated that PEGylated starch acetate nanoparticles are capable of opening tight junctions. The control Caco-2 cells stained with rhodamine phalloidin, to visualize actin, showed uniform staining pattern as shown in Fig. 2(A). However, cells treated with PEGylated starch acetate nanoparticles displayed a disrupted staining pattern [Fig. 2(B)]. The untreated cells were observed as smooth lines at the cell–cell junction [Fig. 2(C)]; whereas, for PEGylated starch acetate nanoparticle-treated cells the staining intensity was weaker at cell–cell contact sites [Fig. 2(D)].

The use of PEGylated starch acetate nanoparticles for controlled drug deliveries utilize its self-aggregation property. Micellar nanoparticles are formed on the self-association of PEGylated starch acetate in the presence of aqueous insulin solution. The hydrophobic core of starch acetate protects insulin from the gastric environment. It has been demonstrated that the insulin released from these nanoparticles was stable with no conformational changes. PEG-modified starch acetate is a good approach for the controlled delivery of insulin or other proteins for various therapeutic applications.[35]

POLYVINYL ALCOHOL AND POLYVINYL PYRROLIDONE

A polyvinyl pyrrolidone was the first synthetic polymer used as a blood substitute. Its solutions were mainly used to treat the shock after the burns and in the case when the blood transfusion was not indicated.[36–38]

Likewise, the solutions of PVA have found applications as blood substitutes. However, they were withdrawn from the list of the blood substitutes as result of their undesirable side effects. The blood substitutes with the therapeutic action have also been elaborated as a result of incorporation of some therapeutic agents (e.g., penicillin, pelentanic acid, p-aminosalicylic chloride) into PVA.[36,38] PVA has a hydroxyl group in its structure. It is synthesized by the polymerization of vinyl acetate to polyvinyl acetate (PVAc), which is then hydrolyzed to get PVA. PVA hydrogels have been used for various biomedical and pharmaceutical applications.[39] PVA hydrogels have certain advantages, which make them ideal candidates for biomaterials. Advantages of PVA hydrogels are that they are nontoxic, non-carcinogenic, and bioadhesive in nature. PVA also shows a high degree of swelling in water (or biological fluids) and a rubbery and elastic nature and therefore closely simulates natural tissue and can be readily accepted into the body. PVA gels have been used for contact lenses, the lining for artificial hearts, and drug-delivery applications. PVA is mainly used in topical pharmaceutical and ophthalmic formulations.[40,41] It is used as a stabilizer in emulsions. PVA is used as a viscosity increasing agent for viscous formulations such as ophthalmic products. It is used as a lubricant for contact lens solutions, in sustained release oral formulations, and transdermal patches.[42]

HYDROCOLLOID

Hydrocolloids refer to a range of polysaccharides and proteins that are nowadays widely used to perform a number of functions in the pharmaceutical and biomedical sector. The specific application of plant-derived hydrocolloids in pharmaceutical formulations include their use in the manufacture of solid monolithic matrix systems, implants, films, beads, microparticles, nanoparticles, inhalable and injectable systems, as well as viscous liquid formulations.[43–47] The hydrocolloids are available in a wide range and variety, which can be modified depending upon research requirement. The selection of appropriate hydrocolloid is dictated by the functional characteristics

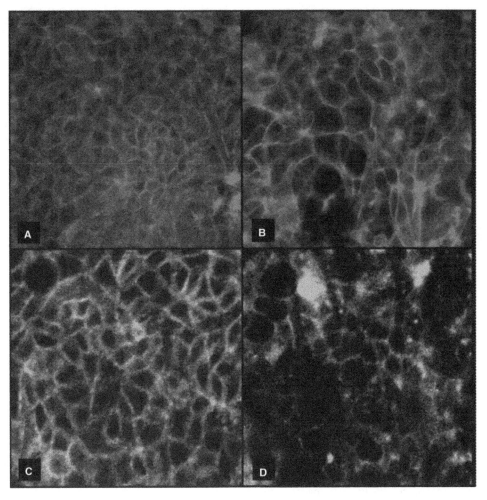

Fig. 2 Confocal images (20×) of Caco 2 cell-actin. (**A**) Caco cells without any treatment (control). (**B**) Caco cells exposed to 5 mg of SA-PEG1900 nanoparticles for 1 hr—tight-junction protein ZO-1. (**C**) Caco cells without any treatment (control). (**D**) Caco cells exposed to 5 mg of SA-PEG1900 nanoparticles for 1 hr.
Source: From Minimol et al.[35] © 2013, with permission from Elsevier.

required; moreover it is inevitably influenced by price. A lot of work has been carried out on various hydrocolloids in the field of scientific research. The biomedical and pharmaceutical applications of hydrocolloid-based polymer hydrogels are of greater importance because of properties such as biocompatibility, non-toxicity, equilibrium swelling, and renewability. Hydrocolloids are often utilized in the design of novel drug delivery systems such as those that target the delivery of the drug to a specific region in the gastrointestinal tract or in response to external stimuli to release the drug. This can be done via different mechanisms including the coating of tablets with polymers having pH-dependent solubilities or incorporating non-digestible polymers that are degraded by bacterial enzymes in the colon. Non-starch, linear polysaccharides are resistant to the digestive action of the gastrointestinal enzymes and retain their integrity in the upper gastrointestinal tract. Matrices manufactured from these hydrocolloids therefore remain intact in the stomach and the small intestine, but once they reach the colon they are

degraded by the bacterial polysaccharidases. This property makes these hydrocolloids exceptionally suitable for the formulation of colon-targeted drug delivery systems.[48]

PLASTICS AND RUBBERS

Plastics are synthetic materials, which mean that they are artificial or manufactured. Synthesis means that "something is put together," and synthetic materials are made of building blocks that are put together in factories. Plastics and natural materials such as rubber or cellulose are composed of very large molecules called polymers. Plastics are polymers, but polymers do not have to be plastics. The way plastics are made is actually a way of imitating nature, which has created a huge number of polymers. Cellulose, the basic component of plant cell walls, is a polymer and so are all the proteins produced in the body and the proteins one eats. Another famous example of a polymer is DNA—the long molecule in the nuclei of ones cells that

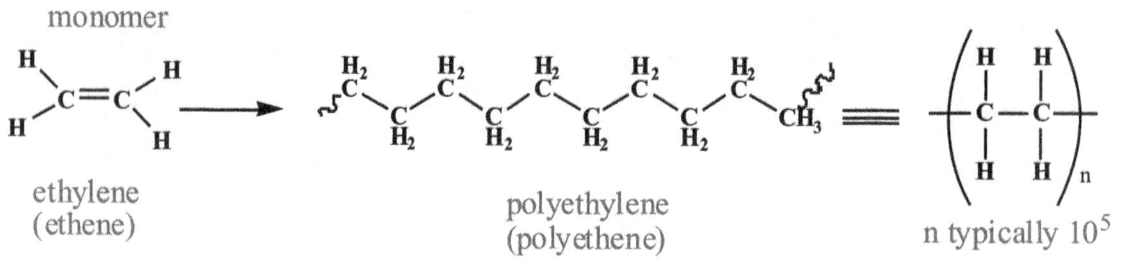

monomer

ethylene
(ethene)

polyethylene
(polyethene)

n typically 10^5

Fig. 3 Polyethylene.

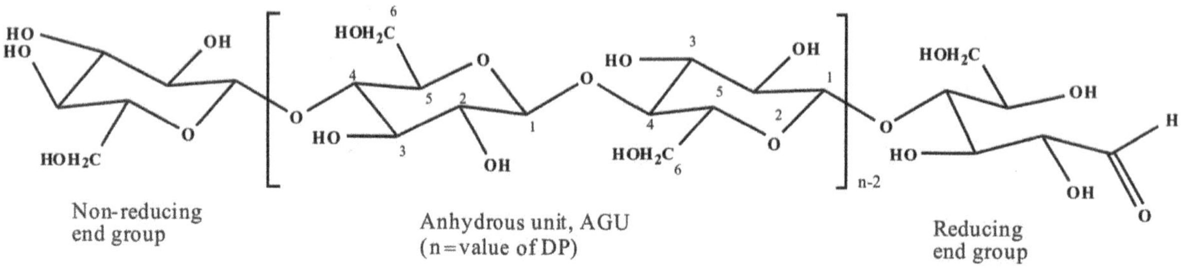

Non-reducing
end group

Anhydrous unit, AGU
(n = value of DP)

Reducing
end group

Fig. 4 Structure of cellulose.

carries all the genetic information about oneself. Wool, cotton, silk, wood, and leather are examples of natural polymers that have been known and used since ancient times. This group includes biopolymers such as proteins and carbohydrates that are constituents of all living organisms. Synthetic polymers, which include the large group known as plastics, came into prominence in the early 20th century. An artificial polymer that is known to everyone in the form of flexible, transparent plastic bags that is polyethylene. It is also the simplest polymer, consisting of random-length (but generally very long) chains made up of two-carbon units (Fig. 3).

The largest single use of rubber is the production of vehicle tires. Tires are highly engineered products that use different kinds of rubber in different parts.

Rubber is used in the pharmaceutical industry to make closures, cap liners, and bulbs for dropper assemblies. The rubber stopper is used mainly for multiple dosage vials and disposable syringes. The rubber polymers most commonly used are natural, neoprene, and butyl rubber. Butyl rubbers and nitrile rubbers are some of the synthetic rubbers used for the manufacturing of closures.

CELLULOSE-BASED POLYMER

Polysaccharides, natural polymers, fabricated into hydrophilic matrices stay accepted biomaterials for controlled-release dosage forms. Cellulose is the most abundant naturally occurring biopolymer.[49,50] A variety of natural fibers such as cotton and higher plants have cellulose as a main constituent.[51,52] It consists of long chains of anhydro-D-glucopyranose units (AGU) with each cellulose molecule having three hydroxyl groups per AGU, with the exception of the terminal ends shown in Fig. 4.

Cellulose is insoluble in water and the most common other solvent. In spite of its poor solubility, cellulose is used in many applications including composites, netting, upholstery, coatings, packing, paper, etc. Cellulosics (cellulose derivatives) are in general strong, reproducible, recyclable, and biocompatible,[53] being used in various biomedical applications such as blood purification membranes and the like.

Pharmaceutical Uses of Cellulose and Cellulose Derivatives

Methylcellulose

MC resembles cotton in appearance and is neutral, odorless, tasteless, and inert. When MC contacts with water, it swells and reproduces a clear to opalescent, viscous, colloidal solution and it is insoluble in most of the common organic solvents. Solutions of MC are stable over a wide range of pH (2 to 12) with no change in viscosity. They can be used as bulk laxatives, so it can be used to treat constipation, and in nose drops, ophthalmic preparations, burn preparations ointments, and like preparations.

Ethylcellulose

EC is the non-ionic, pH-insensitive cellulose ether and is insoluble in water but soluble in many polar organic solvents. It can be used on its own and in combination with water-soluble polymers to prepare sustained release film coatings that are frequently used for the coating of microparticles, pellets, and tablets. Many researchers like Friedman and Golomb[54] and Soskolne et al.[55] have confirmed the ability of EC to sustain the release of drugs.

Sodium carboxymethyl cellulose

It is a low-cost commercial soluble and polyanionic polysaccharide derivative of cellulose that has been used in medicine, as an emulsifying agent in pharmaceuticals, and also used in cosmetics.[56] Carboxymethyl cellulose (CMC) is one of the important cellulose derivatives in industries, which is widely used as an anti-caking agent, emulsifier, stabilizer, dispersing agent, thickener, and gelling agent. Sodium carboxymethyl cellulose is used in both therapeutic and excipient. Therapeutic uses include bulk-forming laxatives in which it may be the primary ingredient. Excipient uses include those of suspending, tablet binding, or viscosity increasing.

Hydroxypropylcellulose

Hydroxypropylcellulose (HPC)-SL and low-viscosity HPC polymers are versatile pharmaceutical excipients. It is widely used in pharmaceutical formulations like in oral products, as a tablet binder, in film-coating, and as a controlled-release matrix. All of the cellulosic polymers described in the previous section have been used as binders for solid dosage forms in wet granulation and dry- or direct-compression tableting processes.[57–59]

REVERSE MICELLES FOR PHARMACEUTICAL APPLICATIONS

Polymeric micelles are core–shell structures formed through the self-assembly of amphiphilic polymers in a solvent that is considered hostile toward either moiety. In water, these micelles are characterized by a hydrophobic core shielded from the external medium by a hydrophilic shell. This particular type of micelle has been extensively studied with a particular attention to their ability to improve the aqueous solubility of hydrophobic therapeutic agents.[60,61] Polyamidoamine-based reverse micelles (RMs) were also proposed for the controlled delivery of the anticancer drug 5-fluorouracil. However, moderate success was obtained, since *in vitro* an aqueous dispersion of the micellar formulation induced less than 5% drug release over 7 hr.[62] Vasopressin (9 amino acids, 1084 Da), also known as the antidiuretic hormone, is naturally secreted in response to changes in blood pressure or volume.[63]

It is highly soluble in water but has incomplete affinity for apolar solvents such as ethyl oleate, as indicated by a low K_{app} (b0.05) (Fig. 5). However, in the presence of RMs, the resemblance of vasopressin for the oil phase was significantly enhanced.[64] Thus, S_A-YC_Z RMs seems better suited for the encapsulation of peptidic molecules with a MW not exceeding a few 1000 Da. All alkylated star-shaped polymers are referred to as S_A-YC_Z, where A, Y, and Z are the degree of branching, alkyl chain length and DA respectively. According to Fig. 5 at an initial loading of

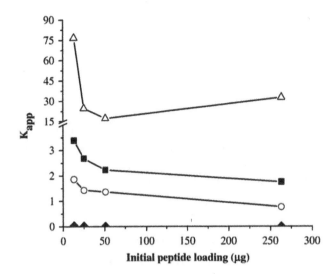

Fig. 5 Partition of vasopressin between ethyl oleate and water in the presence of RMs composed of S4-12C60 (circles), S4-14C60 (squares), and S4-18C60 (open triangles). Blank ethyl oleate was used as control (closed triangles). Kapp values were calculated from the ratio of mean peptide concentrations in oil and water.
Source: From Jones et al.[64] © 2008, with permission from Elsevier.

12.5 μg, the experimental K_{app} was maximal and reached 2, 4, and 77 for the C12, C14, and C18 derivatives, respectively. The ability of RMs to solubilize vasopressin in oil decreased with increasing peptide loading until saturation of the available space was achieved, as indicated in Fig. 5.

FILM FORMING (COATING)

Film coating is one of the most commonly used methods for the preparation of sustained and controlled release dosage forms leading to the evaluation of various materials with film-forming property.[65] The drug delivery applications of polymer or biomaterial films are well established for providing protective coatings and controlled drug release from dosage forms.[66] The main parameters used to characterize the film-forming materials for coating purposes are mechanical properties,[67,68] permeability properties,[69] and WVTRs.[70] Coated formulations for drug release behavior have been investigated for different conditions by various investigators.[71–73]

Aqueous film coating applications are either solutions or dispersions, depending on the water solubility of the film former polymers. Film formation from the polymer solution occurs through a series of phases. When the polymer solution is applied to the surface of the tablet, cohesion forces form a bond between the coating polymer molecules. Film formation from dispersion occurs when polymeric particles coalesce to form a continuous film (Fig. 6), making it a more complex mechanism compared to film formation from solution.[74]

POLYDIMETHYLAMINOETHYLMETHACRYLATE

A poly[2-(dimethylamino)ethylmethacrylate] (PD), biocompatible, water-soluble polymer finds extensive applications as membranes,[75] efficient non-viral gene-delivery vectors,[76] antimicrobial agents,[77] and biosensors.[78]

N-hydroxypropyltrimethylammonium polydimethylaminoethylmethacrylate (PDG) prepared by the quaternization reaction of PD was conducted using N-hydroxypropyltrimethylammonium chloride as the quaternizing agent. Swelling of polymers or drug-polymer matrices in the dissolution test media have been reported according to significant effects on ordinary dosage forms and controlled drug release systems. Atomic force microscopy (AFM) image shown in Fig. 7(A) confirms the morphology and size of PDG. The swelling profile of PDG at pH 1.2 and 6.8 is depicted in Fig. 7(B). PDG particles exhibited a high percentage of swelling at acidic pH whereas the degree of swelling was lower at intestinal pH. P-value as determined by Student's t-test was found to be statistically significant.[79]

Figure 8 shows the *in vitro* release profile of Eudragit-coated insulin-loaded PDG at gastric and intestinal pH. PDG exhibited encapsulation efficiency of 85.5 ± 3.24%. Due to the swelling nature of the particles at pH 1.2, the particles were coated with Eudragit L 100-55. About 1.5% of insulin was released at gastric conditions whereas about 30% of insulin was released at intestinal pH in the first hour. The results are statistically significant as the P-value was found to be less than 0.05. ELISA and Circular dichroism studies proved that the insulin released from PDG was capable of retaining its biological activity.

PDG having high positive charge displayed a calcium binding property, which helped in the loosening of epithelial tight junctions by actin filament dislocation and binding to ZO-1 proteins. The PDG polymer is mucoadhesive thereby making it a suitable carrier for the mucosal delivery of proteins.

Recently, a number of terpolymers derived from 4-acetylpyridineoxime, 4-chloroacetophenone oxime, 8-hydroxyquinoline, formaldehyde, benzoic acids, furfural, and substituted acetophenones have been reported with excellent antimicrobial properties.[80–84]

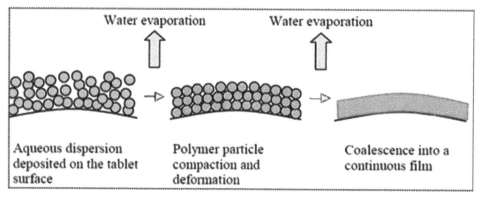

Fig. 6 Film formation from aqueous polymeric dispersion.

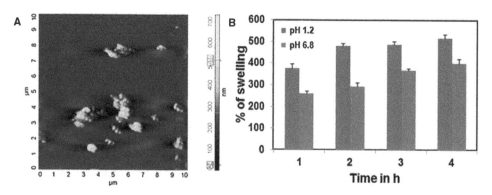

Fig. 7 (**A**) AFM of N-hydroxypropyltrimethylammonium polydimethylaminoethyl methacrylate and (**B**) swelling profile of N-hydroxypropyltrimethylammonium polydimethyl aminoethylmethacrylate at pH 1.2 and 6.8. Indicated values are the mean of three measurements (±SD) (n = 3).
Source: From Sonia & Sharma[79] © 2013, with permission from Elsevier.

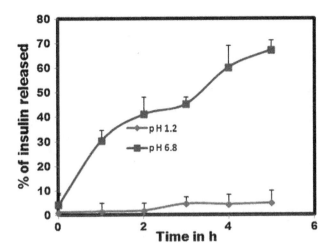

Fig. 8 *In vitro* release profile of N-hydroxypropyltrimethylam-monium polydimethyl aminoethylmethacrylate at pH 1.2 and 6.8. Each data point is the mean of three measurements (±SD) (n = 3). **Source:** From Sonia & Sharma[79] © 2013, with permission from Elsevier.

CONCLUSION

Polymeric substances are in contact with drugs not only as ingredients in final dosage forms but also as processing aids or packaging materials. In conventional dosage forms, the majority of polymers used as excipients are natural polymers and many are included in the GRAS list as a result of a long history of pharmaceutical marketing. In pharmaceutical packaging, polyethylene, polypropylene, and PVC have been used most widely. There is a trend, however, to replace them with more environment-friendly, biodegradable polymers. Of the numerous roles played by polymers in the production of pharmaceutical products, emphasis often has been placed on the use of polymers in the fabrication of controlled release drug delivery systems.

Progress in the area of controlled drug delivery has been possible only as a result of the incorporation of polymer science into pharmaceutics. The development of sophisticated pharmaceutical products requires multidisciplinary efforts. Polymers with special or multiple properties need to be developed to achieve self-regulated drug delivery, long-term delivery of protein drugs, and drug targeting to specific organs in the body. These cannot be achieved by elaborate device design alone. These require the development of smart polymeric systems, which recognize and respond to physiological and pathological processes in the body. Polymers will remain the indispensable component in the development of new pharmaceutical products.

ACKNOWLEDGMENT

One of the authors, Dr. Narendra P. S. Chauhan, is thankful to Elsevier for copyright permission. Authors are also thankful to Prof. Suresh C. Ameta, Department of Chemistry, Pacific College of Basic & Applied Sciences, PAHER University, Udaipur (Raj.) for suggestions.

REFERENCES

1. Deming, T.J. Polypeptide materials: New synthetic methods and applications. Adv. Mater. **1997**, *9* (4), 299–311.
2. Griffin, R.C.J.; Sacharow, S. *Drug and Cosmetic Packaging*; Noyes Data Corporation: Park Ridge, NJ, 1975, 136–148.
3. Liebe, D.C. *Modern Pharmaceutics*, Drugs and the pharmaceutical sciences Ed.; Banker, G.S., Rhodes, C.T., Eds.; Marcel Dekker Inc.: New York, NY, 1990; Vol. 40, 695–740.
4. *Extractables from Elastomeric Closures: Analytical Procedures for Functional Group Characterization/Identification*; Parenteral Drug Association, Inc.: Philadelphia, PA, 1980; Technical Methods Bull. No. 1.
5. Waugh, W.N.; Trissel, L.A.; Stella, V.J. Stability, compatibility, and plasticizer extractionoftaxol (NSC-125973) injection diluted in infusion solutions and stored in various containers. Am. J. Hosp. Pharm. **1991**, *48* (7), 1520–1524.
6. United States Pharmacopeial Convention. USP XXII NF XVII, General tests and assays, <661> Containers. In *United States Pharmacopeia/National Formulary*; United States Pharmacopeial Convention, Inc.: Rockville, MD, 1990; 1570–1576.
7. Martens, H.J.; De Goede, P.N.; Van Loenen, A.C. Sorption of various drugs in polyvinyl chloride, glass, and polyethylene-lined infusion containers. Am. J. Hosp. Pharm. **1990**, *47* (2), 369–373.
8. Geigert, J.J. Overview of the stability and handling of recombinant protein drugs. J. Parenter. Sci. Technol. **1989**, *43* (5), 220–224.
9. Arakawa, T.; Kita, Y.; Carpenter, J.F. Protein-solvent interactions in pharmaceutical formulations. Pharm. Res. **1991**, *8* (3), 285–291.
10. Tarr, B.D.; Campbell, R.K.; Workman, T.M. Stability and sterility of biosynthetic human insulin stored in plastic insulin syringes for 28 days. Am. J. Hosp. Pharm. **1991**, *48* (12), 2631–2634.
11. D'Arcy, P.F. Drug interaction with plastic. Drug. Intell. Clin. Pharm. **1983**, *17* (10), 726–731.
12. Friend, D.R.; Steve, N.G.; Weber, T.P.; Geoffery, J. US Patent, 6, 139.865, 2000.
13. Dell, M.; Reier, S.M.; Stamato, G.E.; Memmo, L.M.D. US Patent, 6099.865, 2000.
14. Siepmann, J.; Peppas, N.A. Modeling of drug release from delivery systems based on hydroxypropyl methylcellulose (HPMC). Adv. Drug Deliv. Rev. **2001**, *48* (2–3), 139–157.
15. Kittipongpatana, O.S.; Burapadaja, S.; Kittipongpatana, N. Carboxymethyl mungbean starch as a new pharmaceutical gelling agent for topical preparation. Drug Dev. Ind. Pharm. **2009**, *35* (1), 34–42.
16. Andersen, F.A. Special report: Reproductive and developmental toxicity of ethylene glycol and its ethers. Int. J. Toxicol. **1999**, *18* (Suppl. 2), 53–67.
17. Smolinske, S.C. *Handbook of Food, Drug, and Cosmetic Excipients*; CRC Press: Boca Raton, 1992; p. 287, ISBN 0-8493-3585-X.
18. Kovar, J.; Wang, Y.; Simpson, M.A.; Olive, D.M. Imaging lymphatics with a variety of near-infrared-labeled optical agents; In *World Molecular Imaging*, 2009.

Neural—Rapid Prototyping

19. Corpet, D.E.; Parnaud, G.; Delverdier, M.; Peiffer, G.; Tache, S. Consistent and fast inhibition of colon carcinogenesis by polyethylene glycol in mice and rats given various carcinogens. Cancer Res. **2000**, *60* (12), 3160–3164.

20. Chemoprevention Database, http://www7.inra.fr/internet/Projets/reseau-nacre/sci-memb/corpet/indexan.html (accessed November 2012).

21. Borgens, R.B.; Bohnert, D. Rapid recovery from spinal cord injury after subcutaneously administered polyethylene glycol. J. Neurosci. **2001**, *66* (6), 1179–1186.

22. Stavisky, R.C.; Britt, J.M.; Zuzek, A.; Truong, E.; Bittner, G.D. Melatonin enhances the *in vitro* and *in vivo* repair of severed rat sciatic axons. Neurosci. Lett. **2005**, *376* (2), 98–101.

23. Kreppel, F.; Kochanek, S. Modification of adenovirus gene transfer vectors with synthetic polymers: A scientific review and technical guide. Mol. Ther. **2007**, *16* (1), 16–29.

24. Rossi, J.J. RNAi therapeutics: SNALPing siRNAs *in vivo*. Gene Ther. **2006**, *13* (7), 583–584.

25. Geisbert, T.W.; Lee, A.C.; Robbins, M.; Geisbert, J.B.; Honko, A.N.; Sood, V.; Johnson, J.C; de Jong, S. Postexposure protection of non-human primates against a lethal Ebola virus challenge with RNA interference: A proof-of-concept study. Lancet **2010**, *375* (9729), 1896–1905.

26. Harmening, D.M. *Modern Blood Banking & Transfusion Practices*, 5th Ed.; F.A. Davis Company: Philadelphia, PA, 2005.

27. Filpula, D.; Zhao, H. Releasable PEGylation of proteins with customized linkers. Adv. Drug Deliv. Rev. **2008**, *60* (1), 29–49.

28. Hinds, K.D. Protein conjugation, cross-linking, and PEGylation. In *Biomaterials for Delivery and Targeting of Proteins and Nucleic Acids*; Mahato, R.I., Ed.; CRC Press: Boca Raton, FL, USA, 2005, 119–185.

29. Veronese, F.M.; Harris, J.M. Introduction and overview of peptide and protein pegylation. Adv. Drug Deliv. Rev. **2002**, *54* (4), 453–456.

30. Kopecek, J. Synthesis of tailor-made soluble polymeric drug carriers. In *Recent Advances in Drug Delivery Systems*; Kluwer Academic/Plenum: New York, NY, USA, 1984, 41–62.

31. Ogura, T. Modified starch and utilization. In *Encyclopedia on Starch-science*; Hidetsugu, F., Komaki, T., Hizukuri, S., Kainuma, K., Eds.; Asakura-shoten: Tokyo, 2004; 393–427.

32. Daoud-Mahammed, S.; Ringard, L.C.; Razzouq, N.; Rosilio, V.; Gillet, B.; Couvreur, P. Spontaneous association of hydrophobized dextran and poly-beta-cyclodextrin into nanoassemblies. Formation and interaction with a hydrophobic drug. J. Colloid Interface Sci. **2007**, *307* (1), 83–93.

33. Wintgens, V.; Amiel, C. Surface plasmon resonance study of the interaction of a beta-cyclodextrin polymer and hydrophobically modified poly(N-isopropylacrylamide). Langmuir **2005**, *21* (24), 11455–11461.

34. Xu, Y.; Miladinov, V.; Hanna, M.A. Synthesis and characterization of starch acetates with high substitution. Cereal Chem. **2004**, *81* (6), 735–740.

35. Minimol, P.F.; Paul, W.; Sharma, C.P. PEGylated starch acetate nanoparticles and its potential use for oral insulin delivery. Carbohydr. Polym. **2013**, *95* (1), 1–8.

36. Janicki, M.P.; Davidson, P.W.; Henderson, C.M. Health characteristics and health services utilization in older adults with intellectual disability living in community residences. J. Intellect Disabil. Res. **2002**, *46*, 287–298.

37. Florjańczyk, Z.; Penczek, S. *Chemia polimerow, Wydawnictwo Politechniki Warszawskiej*; Warsaw, Poland, 1998, ISBN 83-86569-35-2.

38. Zejc, A.; Gorczyca, M. *Chemia Lekow, PZWL*; Warsaw, Poland, 2002, ISBN 83-200-2709-8.

39. Reneker, D.H.; Yarin, A.L.; Fong, H.; Koombhongse, S. Bending instability of electrically charged liquid jets of polymer solutions in electrospinning. J. Appl. Phys. **2000**, *87* (9), 4531–4547.

40. Krishna, N.; Brow, F. Polyvinyl alcohol as an ophthalmic vehicle. Effect on regeneration of corneal epithelium. Am. J. Opthalmol. **1964**, *57*, 99–106.

41. Paton, T.F.; Robinson, J.R. Ocular evaluation of polyvinyl alcohol vehicle in rabbits. J. Pharm. Sci. **1975**, *64* (8), 1312–1316.

42. Wan, L.S.C.; Lim, L.Y. Drug release from heat treated polyvinyl alcohol films. Drug Dev. Ind. Pharm. **1992**, *18* (17), 1895–1906.

43. Pandey, R.; Khuller, G.K. Polymer based drug delivery systems for mycobacterial infections. Curr. Drug Deliv. **2004**, *1* (3), 195–201.

44. Chamarthy, S.P.; Pinal, R. Plasticizer concentration and the performance of a diffusion-controlled polymeric drug delivery system. Colloids Surf. A Physiochem. Eng. Aspects **2008**, *331* (1–2), 25.

45. Sande, A.M.; Teijeiro, D.; Lopez, R.C.; Alonso, M.J. Glucomannan, a promising polysaccharide for biopharmaceutical purpose. Eur. J. Pharm. Biopharm. **2009**, *72* (2), 453–462.

46. Shu, X.Z.; Zhu, K.J. Chitosan/gelatin microspheres prepared by modified emulsification and ionotropic gelation. J. Microencapsul. **2001**, *18* (2), 237–245.

47. Koa, J.A.; Park, H.J.; Hwang, S.J.; Park, J.B.; Lee, J.S. Preparation and characterization of chitosan microparticles intended for controlled drug delivery. Int. J. Pharm. **2002**, *249* (1), 165–174.

48. Prashar, D.; Mittal, H.; Kumar, R. Hydrocolloids based colon specific drug delivery. Asian J. Biochem. Pharm. Res. **2011**, *1* (3), 2231–2560.

49. Hinterstoisser, B.; Salmen, L. Application of dynamic 2D FTIR to cellulose. Vib. Spectrosc. **2000**, *22* (1–2), 111–118.

50. Bochek, A.M. Effect of hydrogen bonding on cellulose solubility in aqueous and nonaqueous solvents. Russ. J. Appl. Chem. **2003**, *76* (11), 1711–1719.

51. Myasoedova, V.V. *Physical Chemistry of Non-Aqueous Solutions of Cellulose and Its Derivatives*; John Wiley and Sons: Chirchester, 2000.

52. Gross, R.A.; Scholz, C. *Biopolymers from Polysaccharides and Agroproteins*. Am. Chem. Soc.: Washington, 2000.

53. Conner, A.H. Size exclusion chromatography of cellulose and cellulose derivatives. In *Handbook of Size Exclusion Chromatography*; Wu, C.-S., Ed.; Marcel Dekker: New York, 1995; 331–352.

54. Friedman, M.; Golomb, G. New sustained release dosage form of chlorhexidine for dental use. J. Periodont. Res. **1982**, *17* (3), 323–328.

55. Soskolne, W.A.; Golomb, G.; Friedman, M.; Sela, M.N. New sustained release dosage form of chlorhexidine for dental use. J. Periodontal. Res. **1983**, *18* (3), 330–336.

Neural—Rapid Prototyping

56. Arion, H. Carboxymethyl cellulose hydrogel-filled breast implants. Our experience in 15 years (in French). Annales de Chirurgie Plastique et Esthétique, **2001**, *46* (1), 55–59.

57. Liu, J.; Lin, S.; Li, L.; Liu, E. Releases of theophylline from polymer blend hydrogels. Int. J. Pharm. **2005**, *298* (1), 117–125.

58. Rodriguez, C.F.; Bruneau, N.; Barra, J.; Alfonso, D.; Doelker, E. Hydrophilic cellulose derivatives as drug delivery carriers: Influence of substitution type on the properties of compressed matrix tablets. In *Handbook of Pharmaceutical Controlled Release Technology*; Wise, D.L., Ed.; Marcel Dekker: New York, 2000; 1–30.

59. Klemm, D.; Eublein, B.; Fink, H.P.; Bonn, A. Cellulose: Fascinating biopolymer and sustainable raw material. Angew. Chem. Int. Ed. **2005**, *44* (22), 3358–3393.

60. Liggins, R.T.; Burt, H.M. Polyether-polyester diblock copolymers for the preparation of paclitaxel loaded polymeric micelle formulations. Adv. Drug Deliv. Rev. **2002**, *54* (2), 191–202.

61. Gaucher, G.; Dufresne, M.H.; Sant, V.; Maysinger, D.; Leroux, J.C. Block copolymer micelles: Preparation, characterization and application in drug delivery. J. Control. Release **2005**, *109* (1–3), 169–188.

62. Tripathi, P.; Khopade, A.J.; Nagaich, S.; Shrivastava, S.; Jain, S.; Jain, N. Dendrimer grafts for delivery of 5-fluorouracil. Pharmazie **2002**, *57* (4), 261–264.

63. Treschan, T.A.; Peters, J. The vasopressin system: Physiology and clinical strategies. Anesthesiology **2006**, *105* (3), 599–612.

64. Jones, M.; Gao, H.; Leroux, J. Reverse polymeric micelles for pharmaceutical applications. J. Control. Release **2008**, *132* (3), 208–215.

65. Phuapradit, W.; Shah, N.H.; Williams, W.L.; Infeld, M.H. *In vitro* characterization of polymeric membrane used for controlled release application. Drug. Dev. Ind. Pharm. **1995**, *21* (8), 955–963.

66. Knaig, J.L.; Goodman, H. Evaluative procedure for film forming materials used in pharmaceutical applications. J. Pharm. Sci. **1962**, *51* (1), 77–83.

67. Nagarsenkar, M.S.; Hegde, D.D. Optimization of the mechanical properties and water vapour transmission properties of free films of hydroxy propyl methylcellulose. Drug. Dev. Ind. Pharm. **1999**, *25* (1), 95–98.

68. Rowe, R.C.; Kotaras, A.D.; White, E.F.T. An evaluation of the plasticizing efficiency of the alkyl phthalates in ethyl cellulose films using the torsional braid pendulum. Int. J. Pharm. **1984**, *27*, 57–62.

69. Sun, Y.; Ghannam, M.; Toijo, K.; Chein, Y.W.; Lee, C.L.; Ulman, K.L.; Larson, K.R. Effect of polymer composition on steroid permeation: Membrane permeation kinetics of androgen and progestins. J. Control. Release **1987**, *5* (1), 69–78.

70. Shogren, R. Water vapour permeability of biodegradable polymers. J. Enviorn. Polym. Degr. **1997**, *5* (2), 91–95.

71. Benita, S.; Dor, P.H.; Aronhime, M.; Marom, G. Permeability and mechanical properties of a new polymer: Cellulose hydrogen phthalate. Int. J. Pharm. **1986**, *33* (1), 71–80.

72. Nath, B.; Kanta Nath, L.; Mazumder, B.; Kumar, P.; Sharma, N.; Sahu, B.P. Preparation and characterization of salbutamol sulphate loaded ethyl cellulose microspheres using water-in-oil-oil emulsion technique. Iran. J. Pharm. Res. **2010**, *9* (2), 97–105.

73. Rosilio, V.; Treupel, R.; Costa, M.L.; Beszkin, A. Physico-chemical characterization of ethyl cellulose drug loaded coast films. J. Control. Release **1988**, *7* (2), 171–180.

74. Obara, S.; Mc Ginity, J.W. Influence of processing variables on the properties of free films prepared from aqueous polymeric dispersions by a spray technique. Int. J. Pharm. **1995**, *126* (1–2), 1–10.

75. Du, R.; Zhao, J. Properties of poly(N,N-dimethylaminoethyl methacrylate)/polysulfone positively charged composite nanofiltration membrane. J. Membr. Sci. **2004**, *239* (2), 183–188.

76. Cheng, H.; Deng, H.; Zhou, L.; Su, Y.; Yu, S.; Zhu, X.; Zhou, Y.; Yan, D. Improvement of gene delivery of PDMAEMA by incorporating a hyperbranched polymer core. J. Control. Release **2011**, *152* (Suppl. 1), 187–188.

77. Hervas, P.J.P.; Lopez, E.; Lopez, B. Amperometric glucose biosensor based on biocompatible poly(dimethylaminoethyl) methacrylate microparticles. Talanta **2010**, *81* (4–5), 1197–1202.

78. Rawlinson, L.A.; O'Gara, J.P.; Jones, D.S.; Brayden, D.J. Resistance of *Staphylococcus aureus* to the cationic antimicrobial agent poly(2-(dimethylamino ethyl)methacrylate) (pDMAEMA) is influenced by cell-surface charge and hydrophobicity. J. Med. Microbiol. **2011**, *60* (Pt. 7), 968–976.

79. Sonia, T.A.; Sharma, C.P. N-hydroxypropyltrimethylammoniumpolydimethylaminoethylmethacrylate sub-microparticles for oral delivery of insulin—An *in vitro* evaluation. Colloids Surf. B Biointerfaces **2013**, *107*, 205–212.

80. Chauhan, N.P.S. Structural and thermal characterization of macro-branched functional terpolymer containing 8-hydroxyquinoline moieties with enhancing biocidal properties. J. Ind. Eng. Chem. **2013**, *19* (3), 1014–1023.

81. Chauhan, N.P.S.; Ameta, R.; Ameta, S.C. Synthesis, characterization, and thermal degradation of pchloroacetophenone oxime based polymers having biological activities. J. Appl. Polym. Sci. **2011**, *122* (1), 573–585.

82. Chauhan, N.P.S. Preparation and thermal investigation of renewable resource based terpolymer bearing furan rings as pendant groups. J. Macromol. Sci. A **2012**, *49* (8), 655–665.

83. Chauhan, N.P.S.; Ameta, S.C. Preparation and thermal studies of self-crosslinked terpolymer derived from 4-acetylpyridine oxime, formaldehyde and acetophenone. Polym. Degrad. Stabil. **2011**, *96* (8), 1420–1429.

84. Chauhan, N.P.S. Isoconversional curing and degradation kinetics study of self-assembled thermo-responsive resin system bearing oxime and iminium groups. J. Macromol. Sci. A **2012**, *49* (9), 706–719.

Neural—Rapid Prototyping

Polyurethanes: Medical Applications

Michael Szycher
Sterling Biomedical, Incorporated, Lynnfield, Massachusetts, U.S.A.

Abstract

Many critical components of medical devices and diagnostic products are fabricated of polyurethane elastomers, since these polymers offer an unsurpassed combination of biocompatibility, performance, and ease of manufacture. This entry will review the use of polyurethane in medical devices and in other biomedical applications.

BACKGROUND

Medical applications of polyurethane elastomers contribute significantly to the quality and effectiveness of the nation's health care system. These products range from nasogastric catheters to the insulation on the leads of electronic pacemakers implanted in many cardiac patients.

The Food and Drug Administration (FDA) estimates there are now 3600 different kinds of medical devices and another 2800 diagnostic products. Many critical components of these devices and diagnostic products are fabricated of polyurethane elastomers, since these polymers offer an unsurpassed combination of biocompatibility, performance, and ease of manufacture.

Sales of medical devices and diagnostic products in 2010 exceeded $55 billion. Yet, many chemists or chemical engineers are unaware of the technical revolution in medicine brought about by the utilization of biomedical polymers. This entry presents the story of one family of biomedical polymers, the polyurethane elastomers, and how these elastomers are helping physicians diagnose disease and save lives of critically ill patients.

HIGH-PERFORMANCE POLYURETHANES

Among the highest-performing biomedical-grade elastomers are the polyurethanes, block copolymers containing blocks of high-molecular-weight polyols linked together by a urethane group. These have the versatility of being rigid, semirigid, or flexible. These materials, in general, have excellent biocompatibility, outstanding hydrolytic stability, superior abrasion resistance, excellent physical strength, and high flexure endurance. They have been described as resistant to gamma radiation, oils, acids, and bases.

Thermoplastic polyurethane elastomers consist of essentially linear primary polymer chains. The structure of these primary chains comprises a preponderance of relatively long, flexible, soft-chain segments that have been joined end to end by rigid, hard-chain segments through covalent chemical bonds. For medical applications, the soft segments are diisocyanate-coupled, low-melting polyether chains. Hard segments are formed by the reaction of diisocyanate with the small glycol chain extender component.

PERITONEAL DIALYSIS APPLICATIONS

In end-stage kidney disease (ESKD), the kidneys cease to perform their function of cleansing and purifying the patient's blood. Systems for treatment of ESKD provide a substitute for kidney function by removing water and toxic waste materials from the blood. The alternative to these substitute waste removal systems is kidney transplantation. However, this cannot be considered as a viable medical treatment for the millions affected by kidney disease because of limited availability of donor kidneys and numerous medical complications inherent in this method of treatment.

Approximately 300,000 patients throughout the world are presently receiving treatment for ESKD. According to published data, approximately 150,000 patients receive treatment in the United States, and approximately 150,000 receive treatment in Japan and Western Europe, with the remaining patients in various other countries.

Financial assistance is sustained primarily by government reimbursement programs. Almost all dialysis patients require financial assistance provided by government programs to pay for dialysis services and materials. The cost of treatment in the United States is approximately $55,000 per year if the patient is treated at a kidney center, and approximately $35,000 per year if treatment is performed at home. Due to the high cost of treatment, the Department of Health and Human Services has promulgated regulations to limit the level of reimbursement for dialysis in the United States. These regulations generally provide for equal rates of payment to hospitals and kidney centers,

Concise Encyclopedia of Biomedical Polymers and Polymeric Biomaterials DOI: 10.1081/E-EBPPC-120052226

Neural—Rapid Prototyping

regardless of whether treatment is provided in a clinic or at the patient's home. Due to the lower overhead and personnel costs associated with dialysis carried out in the home, these regulations may have the effect of encouraging home dialysis.

Hemodialysis is currently the predominant treatment method for ESKD. Hemodialysis requires the continuous external circulation of blood through a dialyzer (artificial kidney) to remove water and toxins during the treatment and is performed in a hospital, clinic, dialysis center, or in the patient's home. Treatments are usually performed three times a week, with each treatment lasting from 4 to 6 hr.

Alternative methods of treatment to hemodialysis include hemofiltration and peritoneal dialysis that could be performed at home. Hemofiltration cleanses the blood by using a filter to separate fluids containing waste products from the blood. These fluids are then replaced with other sterile solutions. Peritoneal dialysis that could be done at home consists of the periodic infusions of sterile fluids into the patient's abdomen through a surgically implanted catheter and the use of the peritoneum (a membrane that surrounds the abdomen and its contents) as a path to remove toxic waste from the blood. Tecoflex® tubing has received FDA approval for use as peritoneal dialysis catheters.

Enteral Feeding

Enteral feeding means, literally, *to provide nutrients through the gut*. Total enteral feeding is gaining acceptance in several clinical states such as postoperative care, anorexia, burn patients, trauma, head and neck cancer, and long-term convalescent care, where swallowing may be painful or impossible to the patient. Until recently, nutritional support to impaired patients could be done only intravenously; however, the intravenous (IV) route is highly prone to infection. Many patients develop an allergic reaction (anaphylactic shock) to the large-scale delivery of nutrients directly to the bloodstream, resulting in fever, convulsions, and other undesirable conditions.

Now, with the advent of enteral feeding, nutritional modular supplements allow physicians the flexibility of customizing a nutritional care plan specifically suited for each patient's need. Enteral feeding tubes are constructed of elastomeric, biomedical-grade elastomers that are nonirritating, flexible, and comfortable. The tubes are designed with a weighed bolus and a stylet, to allow easy transnasal insertion into the esophagus, and into the stomach or small intestine. Once in place, the stylet is removed, and feeding is initiated at desired intervals.

In the manufacture of enteral feeding tubes, a special grade of polyurethane is compounded with 20% by weight of barium sulfate. The inclusion of radiopaque barium sulfate allows physicians to observe the entry and final position of the catheter tube under x-ray confirmation during passage through the gastroesophageal and pyloric sphincters.

PACEMAKER LEADS

Cardiac pacing systems are utilized for patients with some impairment of the heart's natural electrical system, which limits the heart's ability to pump blood throughout the body at a rate suitable to fulfill the body's needs. Under normal circumstances, the rhythmic contractions of the heart are stimulated by small electrical signals emitted by the sinus node, the heart's natural pacemaker. These signals are conducted downward along nerve fibers to the four chambers of the heart, first to the two upper chambers or atria, which serve to prime the two lower chambers or ventricles. These, in turn, perform the principal pumping function of the heart.

When the heart's natural pacing mechanism is impaired, a pacemaker may be used to remedy the problem by electrically stimulating the heart to restore its regular rhythmic muscular contractions. Pacemakers are generally implanted and connected to the heart by means of insulated wire leads and electrodes called lead/electrodes. Since the right and left sides of the heart contract in unison, it is necessary to pace only one side, and only the right side is commonly paced. Most modern pacemakers are hermetically sealed, usually in titanium metal cases containing a long-life lithium-based battery and appropriate electronic circuitry to generate pulses and control their characteristics. Pacemakers currently on the market range from approximately 35 to 100 g (1.25–3.5 oz) in weight, and from approximately 15 to 50 cm^3 in volume. The useful lives of these pacemakers typically range from 8 to 15 years.

The cardiac pacing industry began in the 1960s, with the development of relatively simple, single-chamber pacing devices providing a continuous stream of electrical pulses at a fixed rate, generally to the right ventricle of the heart. By the early 1970s, the industry had almost entirely converted to more advanced single-chamber *demand* pacemakers capable of sensing or monitoring the heart's natural rhythm, and providing stimulation only when the natural rhythm is inadequate.

In the mid-1970s, programmable pacemakers came into increasing use, offering the ability to modify, without surgery, several operating parameters of the pacemaker, including pulse rate and pulse energy, simply by activating a programming unit communications head held against the skin over the implanted pacemaker. The ability to adjust certain operating parameters enabled the physician to regulate the pacemaker to best meet the patient's needs, to extend the expected life of the pacemaker under certain conditions, and to eliminate the need for surgical replacements, in some cases, where the physiological condition of the patient's heart had changed. Later in the 1970s, such external control was extended in multiprogrammable pacemakers, enabling the physician to program additional operating parameters.

Another important development during the late 1970s was the creation of pacing systems that enabled bidirectional

Neural—Rapid Prototyping

telemetry between the implanted unit and an external programmer. This improvement enabled the physician, for the first time, to interrogate the implanted unit and to receive back information confirming that the pacemaker had received and responded correctly to the programming signals, and to ascertain the operating characteristics of the implanted pacemaker. Such telemetry systems also permitted transmission of important physiologic and operating data on how the heart and pacemaker function together.

The next advance in pacing systems was the development of dual-chamber devices, the first practical versions of which were released commercially in the early 1980s. Unlike single-chamber systems, which stimulate only the right ventricle of the heart through a single lead/electrode, dual-chamber systems are capable of synchronized stimulation of both the right atrium and the right ventricle through two separate lead/electrodes. This synchronized stimulation can increase blood flow with less strain on the heart. As a result, physicians are recommending the use of dual-chamber systems for increasing numbers of patients.

Recently, more advanced dual-chamber systems have been developed. These devices not only provide synchronized stimulation of both chambers but are also capable of sensing and tracking in both chambers as well. When dual-chamber pacemaker senses that there is a regular natural atrial rhythm, it will ensure that the ventricles track this rhythm in synchrony. When a regular atrial rhythm is not present, an atrial pulse is delivered by the system followed by a synchronized ventricular pulse. Because dual-chamber systems more closely simulate the heart's actual physiological rhythm, thereby more effectively meeting the individual physiological needs of each patient, these pacemakers are expected to increase from 20% in 1984 to >50% of all implanted pacemakers.

Based upon estimates from industry analysts, the world-wide pacemaker utilization reached 400,000 units in 2000, representing a $2.5 billion market, of which approximately 60% was implanted in the United States. Until recently, silicone elastomers were the polymers of choice for single-chamber pacing; however, with the introduction of dual-chamber pacing, a stronger and more slippery polymer became necessary.

Compared to standard silicone rubbers, polyurethane elastomers offer several advantages, including higher tensile strengths, significantly higher tear resistance, and excellent abrasion resistance. In the manufacture of cardiac pacing leads, these advantages have resulted in the introduction of polyurethane leads with significantly reduced wall thicknesses. These thinner wall leads have resulted in easier surgical insertion, less traumatic introduction of multiple leads when inserted into single veins (for dual-chamber pacing), and greater elasticity for the implanted lead.

For dual-chamber pacemaker lead insulation, a thermoplastic urethane polymer appears to offer the best combination of properties. Polyurethanes display high tensile strength (<5000 psi), are highly elastic, are exceptionally blood compatible, and soften significantly 1 hr after insertion into the human body. This property facilitates the insertion of a semirigid pacemaker lead, which then becomes compliant and less prone to damage the delicate lining of the heart, blood vessels, and other sensitive surfaces of the human body.

NEUROLOGICAL LEADS

In the late 1980s, a team of researchers from Grenoble led by neurosurgeon Benabid et al.,[1] introduced the technique of chronic stimulation of the brain to treat uncontrolled movement disorders, such as Parkinson's disease. Parkinson's disease is a disabling chronic neurodegenerative condition, clinically characterized by akinesia, tremor, rigidity, and postural instability, caused mainly by dopaminergic neuron degeneration of the substantia nigra.

This procedure is now known as deep brain stimulation, and involves the bilateral implantation of polyurethane-insulated electrodes that emit short high-frequency electrical signals to modulate functional neuronal circuits. The tip of each electrode contains at least four poles, permitting a wide range of stimulation from outside the brain. Each electrode is connected via a lead to the impulse generator, which is usually implanted under the collarbone, as seen in Fig. 1.

As explained in U.S. Pat. 7319904, nerve tissue stimulation is used to treat numerous neurological disorders, including, but not limited to, cerebral palsy, multiple sclerosis, amyotrophic lateral sclerosis, dystonia, and torticollis. It is further known that nerve tissue

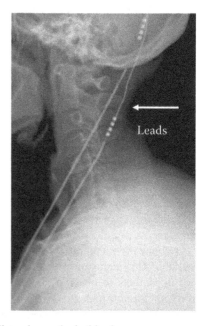

Fig. 1 Bilateral neurological leads.

stimulation is useful to treat intractable malignant and nonmalignant pain. Stimulation of nerve tissue of the spinal cord, for example, is often accomplished through implanted medical leads in the epidural space of the spinal cavity. The implanted lead defines a lead body which includes neural stimulation electrodes that conduct electrical stimulation signals from a stimulation source, such as implantable pulse generators, to targeted nerve fibers in the epidural space. These medical leads may be percutaneous lead bodies which have a cylindrical shape with cylindrical electrodes spaced along the body of the lead.

Neurological lead insulation is critically dependent on aromatic-based polyurethane elastomers. The aromatic poly(ether)urethanes are still dominant in this field, although there is a push to switch to biodurable aromatic and aliphatic poly(carbonate)urethanes.

Schrom MG in Neurostimulating Lead

U.S. Pat. 7047082 explains that conventional neural stimulation therapies rely on electrode catheters for stimulating various regions of the spinal cord that correspond to each physiologic region of the body. Spinal cord stimulation, however, has limited effectiveness for certain pain conditions primarily due to limited accessibility to targeted nerves. In many cases where spinal cord stimulation is inadequate, spinal or peripheral nerves must be specifically stimulated to provide pain relief. However, with the existing technology, nerve-specific stimulation can only be accomplished with a surgical implant, which results in scarring and significant patient discomfort. Therefore, a need exists for a lead that provides greater specificity and increased accessibility to perform a broader array of nerve stimulation, while using less invasive methods to improve treatment outcome.

A variety of medical electrode catheters are available today for the diagnosis and treatment of various disorders of the cardiovascular and neurological systems. These electrode catheters can be used to sense electrical activity within the body and to deliver different forms of energy to stimulate, ablate, cauterize, or pace. The core electrode technology common to all of these catheter designs is the application of one or more metallic bands on a catheter body. Examples of medical catheters using metallic banded electrodes include permanent and temporary cardiac pacing leads, electrophysiologic (EP) catheters, electrocautery probes, and spinal stimulation catheters. The use of preformed metallic band electrodes manufactured from noble metals, such as gold or platinum, and various other conductive alloys has found widespread application despite their functional design and performance limitations. Metallic band electrodes possess several distinct performance problems. When placed on flexible catheter materials, they add significant stiffness that greatly interferes with the steerability of such

catheters. As such, prior art catheters having band electrodes are often restricted to applications where steerability and selective placement are not required. In addition, when direct current (d.c.) or radio frequency (RF) energy is applied to metallic band electrodes, a thermal field is generated which can interfere with energy delivery, increased power consumption, and, in blood environments, create potentially life-threatening blood clots. Finally, the manufacture of catheters utilizing metallic band electrodes is quite labor intensive, resulting in high manufacturing costs.

Placement of leads for both external and implantable RF stimulating devices is quite simple for spinal cord stimulation. Here, a Tuohy needle is inserted into the spinal epidural space and the leads are placed adjacent to the targeted nerves addressing a specific painful region of the body. Relatively high power must be applied when directly stimulating the spinal nerves compared to that required when peripheral nerve stimulation is involved. While this is not a problem when the spinal leads are used with an external stimulator for which battery replacement is relatively easy. It is a major limitation of totally implantable systems in that high power consumption necessarily shortens the time between surgeries for battery replacement. Procedurally, nerve stimulation therapy becomes more challenging when spinal or peripheral nerves of the body are targeted. Due to the fact that many regions of the body cannot be effectively stimulated via the spinal cord, the only alternative in many cases is to surgically implant electrodes. Therefore, a significant need exists for therapeutic access to spinal and peripheral nerves without surgical intervention.

The neurostimulating leads of the Schrom invention eliminate many of the problems encountered with conventional, band electrode leads. The method employed in fabricating leads of the present invention afford the ability to fabricate highly flexible electrodes on extremely small-diameter catheter lead bodies while, if desired, still providing a central lumen permitting such catheters to be advanced over a guidewire or with a stylet until the electrodes disposed thereon are positioned adjacent target tissue.

The present invention provides many improvements and advantages over the related art. The present invention provides an improved method for fabricating electrical stimulating leads of reduced diameter and carrying a plurality of longitudinally spaced electrodes at the distal end thereof, each of the electrodes being individually connected to a connector at the proximal end thereof by conductors that are embedded within the wall of the lead body and insulated from one another. The present invention provides a method of fabricating such a lead while still maintaining a high degree of steerability thereof. The present invention provides an improved neurostimulating lead having a plurality of longitudinally spaced, multilayer, thin-film electrodes proximate to its distal end, where the electrodes are connected by spiral wound wires embedded in the wall of the lead body and where the lead body can, if desired, retain a central lumen through which a guidewire or stylet may pass. The present invention provides a construction of microlead catheters in very small diameters that maximizes the inner lumen space for over-the-wire delivery, stylet insertion, infusion of fluids, multielectrode lead wires, and steering systems. The resulting leads provide enhanced sensitivity to low-level signals, providing improved output clarity and lower energy requirements when delivering stimulating currents to selected nerve tissue. Other improvements and advantages over the related art will be evident to those skilled in the art.

The foregoing improvements and advantages of the present invention are realized by devising a neurostimulating lead having an elongated, thin, flexible body member of a predetermined length and with annular wall defining an internal lumen that either extends from the proximal end to the distal end of that body member or over a sheet segment at the distal end of the body member. A plurality of spiral wound conductors are embedded within the wall of the body member and are electrically insulated from one another. They extend from the distal end to the proximal end of the body member.

A plurality of multilayer thin-film metal electrodes are deposited on the outer surface of the annular wall of the body member at discrete longitudinally spaced locations in a zone proximate the distal end of the body member. To establish an electrical connection between the thin-film electrodes and the buried spiral wound conductors, a plurality of tunnels are formed radially through the body member from the outer surface of the annular wall reaching the buried conductors. Laser etching is a preferred way of forming such tunnels. An electroplating operation or the application of a conductive epoxy is employed to create conductive links that extend through the tunnels from the buried conductors to the wall surface on which the thin-film electrodes are later deposited. The lead further includes at least one connector at its proximal end. The connector includes a plurality of contacts that are electrically joined to the plurality of conductors. The connector is adapted to connect the lead to either an implanted or an external neurostimulator.

Utilizing the manufacturing method described, it has been possible to produce neurostimulating leads having an outer diameter of only 0.026 in (about two French) and with an internal lumen diameter of about 0.012 in, allowing the catheter to be passed over a 0.010 in guidewire or stylet. The thin-film electrodes are typically less than about 250 microns in thickness and, as such, do not detract from the flexibility of the resulting catheter and its ability to be readily steered through the epidural space and out through a selected intervertebral foramen beyond the dorsal or ventral root fibers and into the sheath surrounding the target peripheral nerves to be stimulated by using a guidewire or stylet delivery. This is shown in Fig. 2.

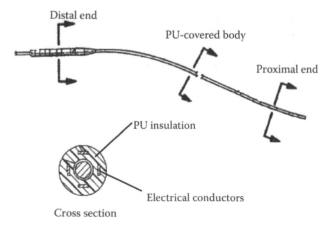

Fig. 2 Neurological lead insulated with polyurethane.

INFUSION PUMPS

New drug delivery systems are under intense development activity for all routes of drug administration. Statistically, it is likely that buccal/intestinal absorption of drugs will remain the predominant form of delivery. Although other routes of administration have included occular, nasal, vaginal, and rectal mucosa as well, in the past years, there has been a growing interest in novel methods for controlled drug delivery to ambulatory (nonhospital) patients.

Controlled drug delivery is a relatively new concept in pharmaceutical product development, offering safer and more effective means of administering drugs. It involves releasing a drug into the bloodstream or delivering it to a target site in the body at predetermined rates over an extended period of time. The mechanisms for controlled drug delivery are varied, but in all cases the goal is to provide more effective drug therapy and patient compliance while reducing or eliminating many of the side effects associated with conventional drug therapy. One method of eliminating side effects, increasing patient compliance, and treating chronic disease is to deliver constant IV therapy by means of infusion pumps.

Infusion pumps consist of a piston, roller, or peristaltic pump that introduces fluids into a patient in a controlled and predictable manner. In a typical application, the pump is primed with a drug solution and connected to a vein via an IV line involving a needle and catheter. Once the IV connection has been accomplished, the pump is turned on, and drug delivery is initiated.

Constant IV therapy on an ambulatory basis has been achieved by the development of infusion pumps that deliver minute amounts of liquid at extremely constant rates over long periods of time. The pump's major use would be to deliver small amounts of highly potent water-soluble medications. Other types of compounds, for example, antibiotics, might seem to be useful applications. However, the amount of antibiotic used would require extremely high concentrations that might not be achievable because of solubility limits. It also would involve the risk of causing considerable pain and inflammation at the site of entry into the vein.

Another major application could be the treatment of insulin-dependent diabetes. Hormones of this type are very soluble, reasonably but not entirely stable, and potent in very low concentrations, so a large reservoir of material could be placed internally and infused without interruption for weeks. Techniques for refilling these infusion pumps without their complete removal also exist. Currently, it is believed that close regulation of diabetes will prevent many of the long-term complications of the disease. One way to achieve close regulation would be the capability of constantly infusing insulin. The danger would be that unexpected periods of little or no food intake could lead to relative insulin excess and shock. The ultimate technology will be the constant monitoring of the blood sugar levels and the capacity to adjust the infusion rate to the blood sugar level.

Insulin-dependent (Type 1) diabetics, at over 1.4 million in the United States alone, constitute the single largest patient population for which evidence has been accumulating rapidly of improved efficacy of drug administration via an ambulatory, external, or wearable syringe pump, as they are variously called. Other large U.S. markets in which interest is quickly rising in formulating pharmaceuticals to enhance clinical effectiveness of chronic infusion include cancer chemotherapy (850,000 new cases per year), chronic or intractable pain syndromes (1.8 million), and chronic antibiotic therapy (0.5 million). The rapidly escalating interest in this field is, likewise, reflected by the variety of pumps either commercially available or slated for market entry—25 models from 17 different manufacturers.

Product development in infusion pumps has advanced and promises to continue to advance at an astounding rate. From purely experimental devices (heavy, complicated to operate, fragile) major strides have been made toward redesign of pumps that are less obtrusive, more reliable, easier to operate, and less sensitive to the wear and tear of normal living.

Table 5 lists some of the drugs, chemicals, hormones, and related diseases for which frequent or chronic administration has been suggested to improve therapeutic effectiveness. The drug list in Table 5 is not exhaustive, but

it is clear from this partial list that the potential impact of chronic infusion pumps in the practice of internal medicine (particularly diabetology, oncology, endocrinology, and neurology) over the next 5–8 years may well become profound.

Several forces are driving this newly emerged technology-linked-to-therapy market. In the broadest, simplest sense, when a drug or chemical is best administered slowly and chronically by reason of limited bioavailability or metabolic half-life, and control over variable delivery rates in the range of microliters to tenths of milliliters per hour is required, ambulatory drug infusion pumps are rapidly gaining recognition as a treatment or drug administration modality of choice. Drug infusion devices are providing a new level of precision over delivery rates, which will allow therapeutic pharmacologists and medicinal chemists to pay attention, as never before, to the implications of pharmacokinetics and pharmacodynamics in new drug development.

The alternative of lying in a hospital bed with an IV line or another route of administration apparatus at upward of $500 per day (assuming only medication and room costs) is not a happy state of emotional or financial affairs for most people. Therein lies a major market force—cost containment. The home health care industry experienced early, explosive growth on the basis of total parenteral and enteral nutrition services alone. Recent projections put this market at >$300 million per year by 1995 plus another $50 million for insulin delivery systems and $50 million for other drug delivery systems, including chemotherapy and antibiotics.

For the 1.4 million insulin-dependent diabetics, continuous insulin infusion means an end to daily cycles of hyper- and hypoglycemia and, with normalized blood sugar levels, a far greater chance of avoiding the devastating microand macrovascular side effects of intermittent insulin replacement therapy. Similarly, on any given day, approximately 300,000 people undergo cancer chemotherapy in this country alone. Add antibiotic and chronic pain therapy, plus others as suggested in Table 1 and, given clinical evidence of improved efficacy with chronic, low-level drug therapy, one cannot escape seeing an impressive picture of large and growing markets representing a major change in medical and pharmaceutical technology. Already, the ethical pharmaceutical industry has responded by reformulating trade-marked drugs for wearable pump use, especially those about to expire from patent life, thereby extending the drug's proprietary life.

It is clear that the pump manufacturers and drug houses have just begun to find ways to jointly respond and shape a market force of major dimensions. Approximately 20 companies manufacture, or are clinically testing, wearable/implantable infusion pumps. While the 1982 market for these innovative pumps did not exceed $10 million, their versatility and clinical usefulness are expected to gain rapid acceptance.

Two principal types of IV catheter placement systems currently are being marketed: over-the-needle systems

Table 1 Properties of carbothane hemodialysis catheters after 9 weeks exposure to lock solution

Lock solution composition	Force at break (N)	Elongation at break (increase from baseline)
Aqueous heparin	192	38
4% Aqueous TSC	230	42
4% TSC/30% ethanol	113	22

Source: Modified after Bell Jayaraman, & Vercaigne.[3]

and through-the-needle systems. In an over-the-needle system, a catheter is placed over the needle by molding or shrinking an external plastic sheath over it. These over-the-needle catheters are made of a large fitting that remains on the skin after needle withdrawal, and a stiff catheter of polyethylene or Teflon piercing the skin. In some cases, the IV fluids are infused by means of a very stiff stainless-steel needle penetrating the skin. Either way, discomfort, insulin aggregation, frequent clogging (particularly with pulsatile pumps), and (worse for the patient) daily catheter changes are the rule. Even though the patient is wearing a pump, frequent needle changing is necessary for patient hygiene and to prevent clogging. Therefore, much of the inconvenience and pain of intermittent insulin therapy remains.

The newest advance in IV/infusion pump catheter systems features a completely removable needle, with the smallest flexible polyurethane catheter available. For instance, a Neonatal Central line is composed of a tiny, soft radiopaque Tecoflex catheter with an outer diameter of only 0.015 in, inside of a breakaway needle. Once the needle has been removed, the pliable catheter remains in the fragile vein, reducing trauma, and ensuring long-term patency. In addition to neonatal care, these systems are also utilized for pediatrics, geriatrics, critical care, long-term control of diabetes, and the treatment of cancer.

The economic cost of diabetes in terms of medical care and losses due to disability or premature death is colossal—$10 billion annually. But just as important is the cost of treating the disease's major complications, which can lead to blindness, kidney failure, heart disease, and peripheral vascular disease.

Insulin infusion devices will be able to control blood glucose levels much more closely than traditional once-a-day injections. There is great hope that this better control will eliminate the long-term problems caused by diabetes. If this happens, cost savings to the health care system could be in the billions of dollars.

Perhaps the most significant contribution that these drug delivery systems are making today is in the treatment of cancer, specifically liver cancer. Side effects are the limiting factors in cancer chemotherapy. In conventional chemotherapy, large doses of medication are injected into the circulatory system, carrying the toxic chemicals not only to the cancer site but to the rest of the body. Since the same medications that are lethal to the tumor are toxic to normal body cells, undesirable side effects are inevitable.

With a medication delivery system, however, the drug is directed through a catheter into a diseased organ's own blood supply. This reduces the dose rate needed to kill tumor cells by confining the toxic effects to the target organ, and diluting the amount delivered to healthy tissue. Thus, side effects can be minimized and sometimes eliminated. Patients are already benefiting from reduced hospital stays and costs and from being able to spend more time at home.

CARDIOVASCULAR CATHETERS

First, the good news. After half a century of steady increases in heart disease fatalities in the United States, the rate now appears to be on the downswing. The death rate due to all types of heart disease slipped from 369 per 100,000 in 1960 to less than 325 per 100,000 in the late 1990s and it continues to decline in 2010.

The bad news is that heart disease is still by far the nation's leading killer, claiming almost a million lives in 2009—nearly twice as many as cancer and accidents combined. About 1.5 million Americans will suffer heart attacks this year, and 500,000 of them will die as a result. Of those, approximately half will not survive long enough to reach medical care.

Cardiologists admit they are not sure why the death rates are falling. Just as heart disease is a multifactorial process, so is its decline. Most of them agree, however, that biomaterials technology is aiding the process. Biomaterials technology is aiding in the form of better diagnostic techniques, utilizing plastic catheters that permit the visualization of coronary arteries under radiography (angiograph), plastic catheters used to dissolve clots (thrombolysis), and the newest technique, transluminal angioplasty. We will discuss each of these important applications for polyurethanes in sequence.

Angiography Catheters

During the early part of this century, when infectious disease was the leading killer, patching up the patient was the most logical approach. The introduction of antibiotics, in the 1940s and 1950s, virtually eliminated infections as major killers in the United States. Since then, heart and blood vessel diseases have soared into the number one position, claiming nearly one million lives in 1983. The important difference is that the development of heart disease (like that of cancer, the second-place killer) is almost always accompanied by presymptomatic clues and warning signs.

As recently as the early 1960s, the most common early warning of heart disease was the chest pain called angina pectoris—the signal that a portion of the heart muscle (myocardium) is receiving inadequate supplies of blood. All too often, even that sign was absent. In any event, the cardiologist could do little more than speculate about the origins of the pain and provide the patient with pain-relieving drugs.

The development of coronary angiography (or arteriography) changed that. By snaking a thin, flexible catheter through a patient's artery to the aorta and injecting a small amount of radiopaque dye, the coronary arteries can now be seen in brilliant detail on a conventional fluoroscope. In most cases, details of coronary blockages can also be visualized. (Blockages are composed of waxy cholesterol, tiny blood clots, or calcified combinations.) These blockages are responsible for the patient's angina symptoms.

More than 400,000 angiograms are performed in the United States yearly, at an average cost of $1500 each. The angiographic catheters utilized in these techniques must be exceptionally blood compatible so as to not to create life-threatening blood clots during the procedure. Tecoflex catheters, because of their inherent blood compatibility, are particularly useful in this application.

Thrombolysis Catheters

Heart attacks come in different sizes. There are big ones and small ones, depending on how much heart muscle is lost once the coronary artery supplying it with blood becomes blocked. In recent years, intense interest has focused on efforts to limit heart muscle destruction once a heart attack has begun, thus reducing the likelihood of long-term disability in survivors. The type of intervention that is now receiving the greatest attention is called *thrombolysis*, which means *dissolving clot*. Thrombolytic therapy is far from standard practice in any hospital at this time, but enough is being learned to warrant a review of what is involved and how the therapy works and when it works.

The central event in a heart attack is blockage of a coronary artery. As a result, blood flow to a portion of the heart muscle is stopped, and death of that tissue results. About 85% of the time, the block is due to the formation of a blood clot within a coronary artery (which is usually already narrowed by cholesterol accumulation in its walls). Once the blockage occurs, the blood-deprived heart tissue dies gradually. (Indeed, muscle fibers at the outskirts of the area that has been deprived may get a small amount of blood from neighboring vessels and be able to live for a number of hours.) If the blocked artery can be opened before the injured heart muscle reaches its point of no return, a substantial amount of heart muscle can be salvaged. A substance called streptokinase can dissolve the clot if it is directed into the clogged vessel. This can be achieved by the technique of coronary artery catheterization which, in this instance, is put to use at the very onset of a heart attack. A thin catheter is threaded (through an arm or leg blood vessel) to the coronary arteries. The site of blockage is determined by squirting a small amount of contrast material through the catheter and taking x-rays. Then, the same catheter provides the route through which streptokinase is fed to the clot, as shown in Fig. 1. Polyurethane catheters can be utilized advantageously in this technique, since they can be made stiff initially, thereby allowing proper placement within the small coronary arteries. Subsequently, the substance softens significantly once it has reached body temperature and absorbed approximately 0.75% by weight of water. Since these catheters are left in place for many hours or days until all the clot has been dissolved, this softening reduces trauma to the delicate inner lining of the coronary arteries.

Transluminal Angioplasty Catheters

The normal adult human heart is a muscular, four chambered organ about the size of a medium grapefruit and weighing less than a pound. It is located almost exactly in the center of the chest (although its lower end, or apex, is tilted slightly toward the left). For all its complexity, it has just one purpose—*to provide the rest of the body with oxygen-carrying blood, at the same time carrying away the waste products for disposal.* The task is carried out by two main pumping chambers, the right and left ventricles, separated by a muscular septum and a series of one-way valves.

Oxygen-depleted blood from throughout the body enters the heart via two large veins, the superior and inferior vena cavae. Both veins empty first into a small chamber called the *right atrium* and then through the *tricuspid valve* into the *right ventricle*. With each contraction of the heart muscle, blood is forced from the right ventricle through the pulmonary valve to the lungs via the pulmonary arteries. In the lungs, the blood exchanges its carbon dioxide for fresh oxygen and returns to the heart through the pulmonary veins.

As these veins empty into the heart's *left atrium*, the oxygenated blood flows through the *mitral valve* into the powerful *left ventricle*. Contractions force the blood through a fourth and final valve (the aortic) into the aorta, the body's main artery. The aorta separates into various arteries to the upper and lower extremities; here, the blood's hemoglobin gives up oxygen to the cells, takes a fresh load of carbon dioxide, and begins its journey back to the right atrium.

The heart's constant activity, consisting of approximately 70 contractions per minute, every minute of our lives, exacts a high price—gram for gram, its need for steady oxygen supply exceeds that of any other body organ. The demand is met by a series of coronary arteries, branching from the aorta and encircling the heart like a crown (in Latin, *corona*, hence *coronary*). These arteries carry freshly oxygenated blood directly to the heart muscle, or myocardium.

In more than one million hearts per year in the United States, these arteries begin to narrow to an internal diameter of only a millimeter or two. The arteries become choked with fatty deposits, and once the flow is finally dampened below a critical point, the symptoms of angina pectoris develop. Until recently, pain-reducing drugs (such as nitroglycerine) or coronary bypass surgery were the only available treatments.

Now, there is an alternative to bypass surgery. In the mid-1970s, Swiss cardiologist Andreas Gruntzig developed an ingenious method of unclogging arteries using a small balloon. In angioplasty, now performed on about 12,000 patients a year, a narrow tube (catheter) is threaded into the diseased artery until it reaches the clogged area. At that point, a tiny balloon at the tip of the catheter is repeatedly inflated so that it flattens the deposits against the arterial wall and widens the channel, as shown in Fig. 2.

Coronary angioplasty is now being used with increased success in patients with single vessel disease and provides a low-cost, lower-risk alternative to coronary bypass surgery. With angioplasty, symptoms are relieved immediately, patients are out of the hospital within 2 days, and the majority are back to work within 1 week.

In coronary angioplasty catheters, the most crucial part of the device is the balloon. The balloon must be repeatedly inflated, without any appreciable change in dimensions, since an overexpansion would rupture the coronary artery wall. In this case, polyurethane is particularly advantageous, since fabric-reinforced balloons can be easily fabricated of the substance while still retaining the blood-compatible properties of this elastomeric polyurethane.

Alcohol-Resistant Polyurethane Catheters

Every polymer and solvent is characterized by a specific solubility parameter and hydrogen bonding index. Polymers and solvents having approximately the same values are mutually miscible (soluble). Those with very different values are mutually immiscible (insoluble). Those polymers and solvents with intermediate solubility values result in swelling behavior.

The Hansen and Beerbower[2] solubility parameter of methanol is 14.5, of ethanol is 12.7, and of isopropyl alcohol is 11.5. The solubility parameter of the polyether soft segment is 10.5. Consequently, Tecoflex exhibits poor resistance when exposed to these alcohols, since they behave as swelling agents. If exposed to alcoholic vapors, the resistance is even poorer.

In contrast, the solubility parameter of the polycarbonate soft segment is greater than 16, therefore CarboThane® is more resistant to alcohols. However, this does not mean that CarboThane is totally insensitive to alcohols; for example, there is a noticeable reduction in mechanical properties after 9 weeks of exposure[3] to ethanol-containing locking solution, (containing trisodium citrate (TSC) or heparin) as shown in Table 1.

As Table 1 indicates, there is a substantial reduction in the force required to break the catheter when ethanol is part of the lock solution, but little reduction when aqueous TSC is used. Thus, these authors concluded that "from a mechanical perspective, our study supports the use of 4% TSC as a locking solution inside Carbothane hemodialysis catheters without adversely affecting the catheter integrity."

Effect of MW on Kink Resistance

Murphy et al. in U.S. Patent Application 2008/0051759 also investigated the effect of alcohol on the kink resistance of weight-average molecular weight of different Carbo-Thane lots fabricated into PICC (peripherally inserted

central catheters). A "catheter" is a medical device that includes a flexible shaft, which contains one (including annular shafts, i.e., tubes) or more lumens, and which may be inserted into the body for introduction of fluids, for removal of fluids, or both. This is illustrated in Fig. 3.

Central venous access may be achieved by direct puncture of the central venous circulation system, for example, via the internal jugular vein, subclavian vein, or femoral vein. Catheters of this type, known as "central catheters" or "central venous catheters," are relatively short, and can generally remain in place for only a short time (e.g., generally less than 7 days).

Other catheters have also been developed which can be inserted into peripheral veins (e.g., the antecubital, basilica, or cephalic vein) and advanced to access the central venous system, with the tip commonly positioned in the superior vena cava or right atrium, thus allowing for rapid dilution of infused fluids. These devices avoid difficulties associated with the direct puncture of the central venous circulation system, and they allow for longterm (e.g., 180 days or more) and repeated access to a patient's vascular system, thereby avoiding multiple injections and minimizing trauma and pain to the patient.

Specific examples of catheters of this type include the so-called PICCs, midline catheters, and peripheral catheters. A typical PICC, midline, or peripheral catheter contains a thin, flexible shaft, which contains one or more lumens and which terminates at the proximal end with a suitable fitting, such as a hub or other fitting. The primary difference between these three devices is the length of the tubing, with the peripheral catheter being the shortest and the PICC being the longest. The rationale for different lengths is driven by the type and duration of the therapy a patient is to receive.

Hemodialysis catheters are another important class of central venous access catheters. Hemodialysis catheters are commonly multilumen catheters in which one lumen is used to carry blood from the body to a dialysis machine, and another lumen returns blood to the body. Central venous access may be attained by puncture of various major blood vessels, including the internal jugular vein, subclavian vein, or femoral vein.

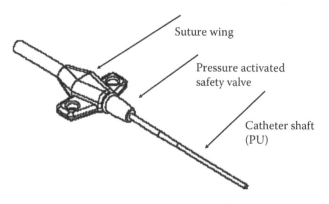

Fig. 3 Venous access catheter U.S. 2008/0051759.

Neural—Rapid Prototyping

Central venous access may also be provided via venous access ports. These specialized catheters typically have the three following components: 1) a catheter; 2) a reservoir, typically formed of a metal or polymer, which holds a small amount of liquid and which is connected to the catheter; and 3) a septum, which covers the reservoir and allows access to the reservoir upon insertion of a needle. The reservoir and covering septum are surgically placed under the skin of the chest or arm, and the catheter extends into a central vein.

Catheter shafts for central venous catheters such as those described above, among others, are typically made from polymers. Suitable polymers are those that can be formed into tubing that is flexible enough to be routed through the vasculature without causing trauma to the patient. When formed into tubing, the polymer chosen should also provide strength sufficient to ensure that the lumen does not collapse in the vasculature, and should resist repeated flexure. Polyurethane-based polymers are commonly employed to meet these criteria.

Polycarbonate polyurethanes are strong, allowing catheters to be formed with thinner walls, regardless of whether the catheter shaft is a single-lumen shaft or a multilumen shaft. Subsequently, catheters made from these materials may be formed with smaller ODs as compared, for example, to other catheter materials such as silicone, or they may be formed having the same OD, but with a larger ID, and therefore having a greater flow rate. Wall thickness for polycarbonate polyurethane catheter shafts will vary with application and may range, for example, from 0.003–0.005″ to 0.01–0.015″, among other thicknesses.

Moreover, while being sufficiently flexible to avoid trauma to the patient, these materials are stiff as compared, for example, to silicone, which helps with the insertion of the catheter.

These materials are also thermoplastics, meaning that a variety of thermoplastic processing techniques, such as extrusion, molding, and so forth, may be employed to form medical devices and medical device components, including catheter shafts, from the same.

Kink resistance (i.e., resistance to failure as a result of repeated kinking) is also an important property of catheter shaft materials. In general, the kink resistance of polycarbonate polyurethanes lessens as the walls of the catheter shaft become thinner and thinner. Moreover, kink resistance also lessens significantly upon exposure to certain materials such as alcohol (i.e., ethanol, isopropyl alcohol, etc.) and alcohol-containing materials such as ChloraPrep® (a product commonly used for skin preparation, which contains 2% chlorhexidine gluconate and 70% isopropyl alcohol), among other materials. Hence, kink resistance is particularly important for central venous access devices such as PICCs, which have thin catheter walls and which are subjected to repeated exposure to alcohol at the entrance site of the body.

The inventors have found that by employing an extruded catheter shaft, which comprises a polycarbonate polyurethane having a weight-average molecular weight in excess of 90,000 g/mol, superior kink resistance is obtained, even upon exposure to substantial amounts of alcohol. Weight-average molecular weight may be within the range, for example, of 90,000 to 95,000 to 100,000 to 105,000 to 110,000 or more.

To be visible under x-ray (e.g., by x-ray fluoroscopy), the catheter shaft may be rendered more absorptive of x-rays than the surrounding tissue. For example, the catheter shaft may contain from 30 to 50 wt% barium, more preferably about 40 wt% barium.

The shaft includes a body section Bo, having length L_{Bo}. The body section Bo also has an outer diameter OD_{Bo} and an inner diameter (lumen diameter) ID_{Bo} at its proximal end. The shaft further includes a tip section Ti. At its distal end, the tip section Ti has an outer diameter OD_{Ti} and an inner diameter (lumen diameter) ID_{Ti}. The shaft further has a tapered section Ta, having length L_{Ta}. L_{Ti} is the combined length of the tip section Ti and the tapered section Ta. The overall length is L_{Ov}. Typical dimensions for some of these values are provided in Table 2.

Table 2 Typical dimensions for extruded catheters

	Catheter type		
	4–3 French	**6–4 French**	**6–5 French**
L_{Bo} (cm)	7.5 min	7.5 min	7.5 min
OD_{Bo} (in)	0.053 ± 0.003	0.079 ± 0.004	0.079 ± 0.004
ID_{Bo} (in)	0.028 ± 0.002	0.047 ± 0.003	0.047 ± 0.003
L_{Ti} (cm)	65 min	65 min	65 min
OD_{Ti} (in)	0.040 ± 0.003	0.053 ± 0.003	0.066 ± 0.003
ID_{Ti} (in)	0.025 ± 0.002	0.034 ± 0.002	0.044 ± 0.0025
L_{Ta} (cm)	5 max	5 max	5 max
Wall thickness (in)	0.003 min	0.003 min	0.003 min

Neural—Rapid Prototyping

Examples

Catheter shafts (extrusions) are formed using a range of CarboThane resins having molecular weights of 76,000, 87,000, 102,000, and 110,000 g/mol. Extrusions of ODs ranging from 4 to 7 French were cut to approximately 2.75 in in length for testing. Where the extrusion is tested in conjunction with an assembly, the extrusion is cut 4.5 cm from the suture wing and the extension tube is cut 1 cm from the suture wing.

Samples are exposed to vapor in a sealed glass jar by placing them on aluminum blocks above a solution of 70% alcohol or ChloraPrep for a designated time. Temperature is controlled by immersing the jar halfway in a water bath at 37 ± 2°C.

Axial cycling tester (built in-house) is used for kink resistance testing. The distance between the grippers is set in the up and down positions such that the sample kinks when in the down position and undergoes 25% elongation.

Where an extrusion is tested in conjunction with an assembly, 10 kinks are formed at the suture wing prior to testing. Also, a 3-min kink at body temperature is performed prior to testing.

The speed controller is adjusted as necessary to ensure that it is operating at 200 ± 5 cycles per minute. The machine is stopped every 1000 cycles and the sample is examined at both kink locations for cracks or holes or for any other sign of failure.

Figure 4 illustrates the effect of alcohol vapor upon cycles to failure for extrusion and extrusion to suture wing interface. As can be seen, a small amount of vapor exposure (1 hr) actually increases the resistance of the sample to failure. However, this benefit is quickly lost. As also seen from this figure, the 102,000 molecular weight sample outperformed the 76,000 molecular weight sample in all cases. These data are also displayed in a bar graph format. Also shown are cycles to failure data where the samples are flexed at body temperature.

Table 3 shows maximum, minimum, and average cycles to failure, both before and after sterilization two times in ethylene oxide using standard protocol for the following: twenty-one 102,000 molecular weight extrusions, twenty-two 110,000 molecular weight extrusions, and twenty-two 87,000 molecular weight extrusions. This table again illustrates the critical advantages of increasing molecular weight vis-a-vis kink resistance.

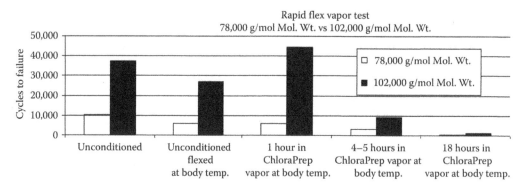

Fig. 4 Alcohol vapor reduces polyurethane flexure endurance. Lower molecular weight polymers suffor disproportionately (ChloraPrep is 20% chlorhexidine gluconte 70% IPA).

Table 3 Average cycles to failure before and after sterlization

Sample ID	Average cycles to failure	Maximum cycles to failure	Minimum cycles to failure
Presterile 87 kMW	6271	8001	4670
Post 2× -sterile 87 kMW	4801	6004	3002
Presterile 102 kMW	15,275	23,002	6000
Post 2× -sterile 102 kMW	17,620	19,000	7000
Presterile 110 kMW	15,184	24,004	8003
110 kMW		26,005	7000

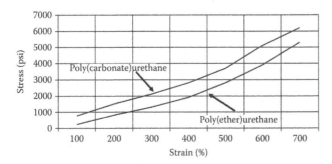

Fig. 5 Calculation of Young's modulus. Ether linkages being more mobile than carbonate linkages, produce polymers with lower stress/strain, thus resulting in more "supple" and "elastic" products.

Elastic Modulus

While the polyurethane industry traditionally refers to the different grades by their surface hardness (durometer), the "hand feel" of the polyurethane tubing is determined by the Young's modulus. Figure 5 is a typical stress–strain curve for a polyurethane elastomer and it shows how the Young's modulus is calculated.

Ether linkages are highly mobile, possessing three degrees of rotational freedom. Thus, a poly(ether)urethane is very "supple" and "elastic." In contrast, carbonate linkages are stiffer with only two degrees of freedom. Thus, poly(carbonate) tubing, durometer for durometer, always feels stiffer than a corresponding poly(ether)urethane, since the Young's modulus is higher for poly(carbonates) than for poly(ethers).

ANTIMICROBIAL POLYURETHANES

Catheters play critical roles in the administration of chemotherapy, antibiotics, blood, blood products, and total parenteral nutrition essential for the successful treatment of many chronic afflictions. Recent advances in catheter technology have enabled their explosive growth in nephrology (hemodialysis catheters) and inpatient interventions (peripherally inserted catheters).

Unfortunately, the hydrophobic catheter surface has a very high potential of allowing microbial colonization that lead to serious and often life-threatening complications, particularly nosocomial infections. To counter this risk, catheter insertion sites are maintained scrupulously clean, which, while reducing the probability of infection cannot completely eliminate infections that significantly increase patient morbidity and mortality. Current treatments to prevent catheter-related infections rely primarily on the use of antimicrobial loaded coatings. However, the trade-offs between thin and thick coatings severely limit the performance of these powerful coatings, since thin coatings only have limited antibacterial life, while thick coatings may last longer but are far more susceptible to cracking, peeling, and flaking-off.

Many approaches have been studied to reduce the incidence of bacterial infections associated with the use of indwelling catheters and transdermal implanted devices, but none have met with more success for limited periods of time. Such infections include nosocomial infections, which are those resulting from treatment in a hospital or a health care service unit, but secondary to the patient's original condition. Infections are considered nosocomial if they first appear 48 hr or more after hospital admission or within 30 days after discharge. By way of background, the term "nosocomial" derives from the Greek word nosokomeion (nosos = disease, komeo = to take care of).

Nosocomial infections are even more alarming in the twenty-first century as antibiotic resistance spreads. The reasons why nosocomial infections are so common include

- Catheter insertions bypass the body's natural protective barriers.
- Hospitals house large numbers of people who are sick and whose immune systems are often in a weakened state.
- Increased use of outpatient treatment means that people who are in the hospital are sicker on average.
- Routine use of antimicrobial agents in hospitals creates selection pressure for the emergence of resistant strains.

In the United States, it has been estimated that as many as one hospital patient in 10 acquires a nosocomial infection, or 2 million patients per year. Estimates of the annual systemic cost resulting from such infections range from $4.5 billion to $11 billion. Nosocomial infections contributed to 88,000 deaths in the United States in 1995.

The risk of contracting an infection in a clinical setting is increasing every year, and the types and virulent nature of such infections continue to rise. Due in part to the loosening of import restrictions into the United States under the North American Free Trade Agreement (NAFTA), as well as widespread international air travel, ecotourism to exotic third-world forests and islands, and massive migration of third-world peoples to Europe and America, hosts of exotic diseases that were once isolated to small areas of the planet are now finding their way into U.S. and European hospitals. Eradicated for almost a century, malaria is once again returning to the United States, and the exotic and deadly Ebola virus has broken out in a lab in Maryland. Shigella (which causes dysentery) was practically unheard of in America before 1990, but it is now being spread from contaminated fruits and vegetables imported into the United States under the auspice of the NAFTA treaty, and is now routinely seen at clinics in California.

Of potentially greater concern is that many common strains of microorganisms have become increasingly resistant to a wide range of antibiotics (due to incomplete kills and simple natural selection). Many strains must be

Neural—Rapid Prototyping

treated with one or two "last-resort" antibiotics and new compounds must be continually developed to combat these evolving strains. By way of example, some common (and dangerous) germs such as *Staphylococcus aureus* (found especially in hospitals) are now known to be resistant to all but one antibiotic—vancomycin—and soon are expected to be vancomycin-resistant as well. According to the Centers for Disease Control, in 1992, 13,300 hospital patients died (in the United States) of bacterial infections that resisted the antibiotics administered to fight them. Generally, antibacterial agents inhibit or kill bacterial cells by attacking one of the bacterium's structures or processes. Common targets are the bacterium's outer shell (called the "cell wall") and the bacterium's intracellular processes that normally help the bacterium grow and reproduce. However, since a particular antibiotic typically attacks one or a limited number of cellular targets, any bacteria with a resistance to that antibiotic's killing mechanism could potentially survive and repopulate the bacterial colony. Over time, these bacteria could make resistance or immunity to this antibiotic widespread.

Silver, platinum, and gold, which are elements of the noble metal group, have long been known to have medicinal properties. For example, platinum is the primary active ingredient in cisplatin, a prominent cancer drug. Similarly, gold is the active agent in some treatments for rheumatoid arthritis. More particularly, unlike its heavy metal counterparts, silver (atomic symbol Ag) with atomic element number 47 and an atomic weight of 108, is surprisingly nontoxic to humans and animals, and has a long history of successful medical and public health use dating back 6000 years. Also, unlike antibiotics, silver has been shown to simultaneously attack several targets in the bacterial cell and therefore, it is less likely that bacteria would become resistant to all of these killing mechanisms and create a new silver-resistant strain of bacteria. This may be the reason that bacterial resistance to silver has not been widely observed despite its centuries-long use. This can be particularly important in hospitals, nursing homes, and other health care institutions where patients are at risk of developing infections.

By way of further background, from 1900 to the beginning of the modern antibiotic era circa 1940 with the introduction of sulfa drugs—silver and its ionic and colloidal compounds (e.g., silver nitrate) was one of the mainstays of medical practice in Europe and America. Various forms of silver were used to treat literally hundreds of ailments: lung infections such as pneumonia, tuberculosis, and pleurisy; sexual diseases such as gonorrhea and syphilis; skin conditions such as cuts, wounds, leg ulcers, pustular eczema, impetigo, and boils; acute meningitis and epidemic cerebrospinal meningitis; infectious diseases such as Mediterranean fever, erysipelas, cystitis, typhus, typhoid fever, and tonsillitis; eye disorders such as dacryocystitis, corneal ulcers, conjunctivitis, and blepharitis; and various forms of septicemia, including puerperal

fever, peritonitis, and postabortion septicemia. An even larger list of the published medical uses for silver in Europe and America exists between 1900 and 1940.

Sparsely soluble silver salts are composed of large microcrystals, usually several microns in diameter or greater. These microcrystals dissolve extremely slowly, thereby limiting the rate and amount of silver ion released over time. By converting the salt's microscaled structure into an atomically nanoscaled structure, it tremendously increases the surface area, thus enhancing silver ion release and efficacy characteristics and thereby making it a more potent antimicrobial agent.

The provision of silver in a releasable form for use in an antimicrobial application is discussed, for example, in U.S. Pat. 6,821,936, "Textiles Having a Wash-Durable Silver-Ion Based Antimicrobial Topical Treatment," by David E. Green et al., the teachings of which are expressly incorporated herein by reference. This patent provided a coating of an antimicrobial silver-salt-based treatment to fabric threads to resist the build-up of bacteria on a fabric. While this approach may be effective for fabrics, it and other silver-based solutions have certain drawbacks when applied to catheters and other invasive devices. First, when a coating is applied, for example, to the lumen or exterior of a catheter, it changes the diameter of that catheter. The insertion of a guidewire, syringe tip, or other close-conforming structure will tend to abrade the antimicrobial coating, again exposing the underlying, unprotected surface of the catheter/device. In addition, these coatings are designed to either last long term, with very few silver atoms/ions released to the environment by the use of sparingly soluble silver compounds, or release all of their exposed soluble silver salt very quickly. This is because they are not adapted to exist within an implanted environment, where there is a constant source of new bacterial infiltration via the open wound channel. Also, because the coatings are relatively thin, they exhaust the available supply of silver salt (which is exposed at the coating surface) in a relatively short time.

It is, thus, desirable to provide a structural polymer, for which the implanted and exposed portion of the device is constructed, that contains the antimicrobial silver compound as an integrated part of its composition. However, the creation of a structural material that contains an embedded supply of silver is not trivial. The embedded silver may either release too slowly or not at all, if the material is not sufficiently hydrophilic. A hydrophilic material allows the needed ion exchange via interaction of the material with adjacent bodily fluids/water. If the infiltration of water is absent, deeper embedded silver will never have the chance to be released, and only the material's surface silver is released. However, if the material is too hydrophilic, it may not exhibit the necessary structural strength to act as an invasive or implanted device or the material may undesirably swell as it absorbs bodily fluids causing the device to fall outside needed size tolerances. It is also not trivial to provide a material with the proper degree of hydrophilicity, while maintaining desired structural characteristics, and releasing silver at a desired rate. This goal typically calls for a polymer blend, and most polymers do not blend well—if at all. Adding a silver-containing compound only complicates the blending. Moreover, while some silver compounds may be blendable, industry often desires that the resulting device be clear or translucent and typically uncolored. Many silver-containing compounds exhibit dark and/or undesired colors upon heating or exposure to light.

The challenge of developing a structural polymer is further complicated by a desire to quickly treat any existing infections or the large initial introduction of bacteria when inserting a device with a large, short-time-released dose of silver, and then providing a lower, continuing dose to ward off any reinfection. Accordingly, it is desirable to provide a silver-compound-containing material that satisfies all of these often-competing goals.

The structural material for use in forming catheters and other devices that come into contact with exposed wounds and internal tissues can be constructed from a combination of four basic components. By way of example, these components can be mixed together in a solution of dimethyl acetamide (or another nonpolar acceptable solvent), which causes the compounds to dissolve and intermix. Notable, it has been determined that a combination of relatively hydrophobic, but structurally durable and ductile poly (carbonate) urethane polymer, also simply termed polyurethane, can be mixed with relatively hydrophilic polyethylene vinyl acetate (PEVA) polymer to form a compatible polymer blend with the requisite degree of hydrophilicity. The PEVA is approximately 0.5–1.0% to 20.0% by weight of the resulting mixture. This combination, when mixed, displays a relatively clear appearance, denoting a complete blending of polymer compounds.

The combination of polyurethane and PEVA is then provided with approximately 0.1–1% to 5.0–10% by weight by weight of a relatively soluble silver compound that releases ionic silver to the wound site over a short period of time based upon the ability for bodily water to ingress into the PEVA. An acceptable quick-release silver compound is silver trifluoroacetate. Alternatively, silver nitrate or another soluble silver salt can be substituted at a suitable concentration within the material, such as silver lactate or silver benzoate.

The compound also contains approximately 0.1–1% to 5.0–15% by weight of a sparingly soluble or nearly insoluble silver salt that releases from the material at a substantially slower, but relatively constant rate. One material that may be employed for this function is silver stearate. Other sparingly soluble silver salt materials can also be employed, such as silver iodate, silver zeolite, silver zirconium phosphate, or silver soluble glass.

In continuous melt production, these materials can be mixed within the gravimetric feeders of a twin-screw extrusion device with an extrusion die having: 1) the ability to produce pellets; 2) any desired catheter profile; or 3) the profile of another device. In an illustrative embodiment, the material from which medical devices are constructed consists of CarboThane PC-3485A-B20 (the structural thermoplastic polyurethane component of the overall material mixture), a product commercially available from Lubrizol Advanced Materials, Inc. of Wilmington, MA, and Elvax 470 (the hydrophilic PEVA component of the material mixture), a product commercially available from the DuPont Company of Wilmington, DE. The illustrative twin-screw compounding extruder intimately mixes the two polymers together, and contemporaneously incorporates the appropriate silver salts all in one operation of the exemplary extruder.

To construct more complex devices other than extruded tubing, appropriate mold cavities, which receive the mixed compound components, can also be employed during the compounding operation.

In an experimental procedure, using a commercially available, medical-grade polyurethane solution (available, e.g., from ChronoFlex® AR, CardioTech International, Inc., Woburn, MA), silver salt(s) is/are manually incorporated into the solution. The silver-loaded ChronoFlex solution is spread onto glass, and dried in a circulating oven at 80°C for 1.5 hr. The loaded polyurethane film was allowed to equilibrate at room temperature (RT) for at least 2 days, and subsequently demolded. The film was manually cut into circles by means of a cork borer, and placed into serially labeled Petri dishes. The microbiology technician was blinded regarding which silver salts were incorporated into the film. Two types of silver salts were tested as described generally above: 1) soluble salts for quick release; and 2) an insoluble salt for longevity. The two candidate soluble salts were silver trifluoroacetate and silver nitrate, and the relatively insoluble salt tested was silver stearate.

A standard *in vitro* agar zone of clearing test (better known as zone of inhibition test), after 24 hr of exposure, was conducted. If the test samples show a "zone of inhibition" (ZOI), the specimen thus displays the capacity to kill the microorganisms tested. The ZOI is measured in mm; the greater the ZOI distance, the more powerful is the microbiocidal effect. Note that, prior to application, the test films were sterilized using short-wave ultraviolet at a distance of 2 in. for 10 sec.

For both subjects, soluble silver salts exhibited a large ZOI clearing against both *S. aureus* and *Pseudomonas aeruginosa*. The insoluble salt showed no microbiocidal activity at 24 hr, which was expected, as its active period should be significantly longer in time, and less aggressive toward the test strains. Table 4 records the observed ZOI measurement for the control polyurethane material (no active compound) and for each active compound within the polyurethane material. Using a 10-mm sample disk, the microbial growth inhibition was measured from the edge of each sample disk, producing antimicrobial results in accordance with Table 4.

Table 4 Microbial killing measured by zone of inhibition

Sample identification	*S. aureus* ZOI (mm)	*P. aeruginosa* ZOI (mm)	MRSA ZOI (mm)
Control polyurethane CarboThane PC-3485A-B20	0	0	0
Polyurethane + silver trifluoroacetate + silver stearate	6	7	6
Polyurethane + EVA + silver trifluoroacetate	12	10	11
Polyurethane + EVA + silver stearate	6	4	6
Polyurethane + EVA + silver nitrate	13	9	11
Polyurethane + EVA + silver iodate	4	3	4
Polyurethane + EVA + silver lactate	2	3	2
Polyurethane + EVA + silver benzoate	2	8	2
Polyurethane + EVA + silver trifluoroacetate + silver iodate	14	12	13
Polyurethane + silver acetate	2	3	2
Polyurethane + silver stearate	0	0	0
Polyurethane + silver trifluoroacetate	8	2	7
Polyurethane + silver nitrate	10	6	10
Polyurethane + silver iodate	2	1	2
Polyurethane + silver lactate	0	2	0
Polyurethane + silver benzoate	0	5	0

It should be noted that of all ionic silver compounds tested, only silver iodate discolored slightly, a significant consideration in the manufacture of light-colored medical devices. Most of the other silver salts were photosensitive. As reported in Table 4, silver trifluoroacetate (a more soluble salt) combined with silver iodate (a less soluble salt) provides a highly effective material that is also substantially free of undesired discoloration.

In the above-described initial tests, the illustrative compounds exhibit rapid antimicrobial activity, killing many organisms within 30 min of application, which is faster than many other commercially available forms of antimicrobial silver. These organisms include Gram-positive (*S. aureus*) and Gram-negative bacteria (*P. aeruginosa*), and also include some antibiotic-resistant strains. It is recognized that the inclusion of PEVA in the proportions defined generally above would serve to enhance the delivery of silver compounds beyond the film surface, allowing the killing action of the compound to be extended for longer periods of time and in greater delivery concentrations for both the soluble and insoluble

salts. Qualitative observation of batches of polyurethane-PEVA materials as described in Table 4 have shown good strength and ductility, without flaking, brittleness, or cracking, making them suitable for the construction of catheters and other invasive and/or implantable medical devices.

While the antimicrobial polymer material of the illustrative embodiments can be employed to construct all or part of a variety of implanted and implantable devices, a highly beneficial aspect is provided to various indwelling catheters where the prevention of bacterial attachment, microbial colonization, and growth of biofilms on any surfaces thereof are critical. Hence, the catheter device can be representative of the general indwelling shaft portion of a variety of catheters, including, but not limited to, 1) chronic dialysis catheter; 2) central venous catheter; 3) PICC; 4) urinary catheter; and 5) gastrostomy catheter. In any illustrative catheter contemplated herein, additional lumens, steering, guiding, and other useful structures can be provided to the depicted shaft without departing from the teachings of this invention.

Of course the illustrative PEVA used herein can be substituted with another material (or plurality of materials) that is miscible with the base (normally hydrophobic) material and provides a hydrophilic conduit-producing structural material. Likewise, while a single layer of uniform material is used in various embodiments to construct a device, conventional techniques can be used to coextrude (or otherwise form) a device wall having a plurality of material layers, each having different material compositions. For example, material exhibiting a slower ionic compound release can be provided near the exterior surface, while materials exhibiting a faster ionic release can be located deeper within the wall or vice versa. Likewise, higher concentrations of ionic compound can be placed in a deeper layer for longer release time. Accordingly, Fig. 6 depicts such a coextruded embodiment of a catheter device. In this embodiment, the catheter shaft, which can be any acceptable type of indwelling device, defines a pair of concentric, inner and outer wall structures, respectively. Each wall structure has an associated wall thickness TO and TI that collectively (unitarily or integrally) define the overall wall thickness DCC between the outer surface and the inner luminal surface (which defines lumen). In this exemplary embodiment, the material of the outer wall structure is provided with generally higher concentrations of one or both ionic silver salts, as depicted by the thickened dashed arrows (rapidly soluble salts) and wavy arrows (slower dissolving salts). This provides higher concentrations of released salts on and from the outer surface, which is in contact with penetrated tissue. The material of the inner wall structure can contain lower concentrations of one or both salts to provide less salt on or from the inner luminal surface, which is generally in communication with sterile fluids. Likewise, the composition of the polymers in each layer can also vary. In general, a lower concentration nearer through the inner surface will produce less migration of both more soluble and less soluble salts as depicted by the thin dashed arrows and thin wavy arrows, respectively. Some compound may cross the margin between wall structures. In alternate embodiments, this factor can be controlled coextruding a thin third impermeable layer (e.g., pure polyurethane) at the margin, or otherwise providing a substantially impermeable barrier between different layers.

It should be clear that a variety of implantable devices, as well as catheters, can be provided with multiple layers of, or differing portions (e.g., catheter tips) having differing compositions of salts and/or polymers. In various embodiments, some layers or portions can omit either the more soluble or the less soluble salt, where the omitted compound's effects are not desired. Also, while the construction of layers or portions with differing antimicrobial characteristics using coextrusion is described, any acceptable manufacturing and/or assembly technique can be employed in alternate embodiments. For example, different portions/layers can be adhered, fastened, comolded, welded together, interlocked, force-fitted, or otherwise joined in alternate embodiments. Note, as used herein, the term "composition" in connection with a portion or layer of the device shall refer to a predetermined mixture at least some of polyurethane, PEVA, soluble ionic silver salt, and sparsely ionic silver salt. Some or all of these components (and other additional components as desired) can be provided to the material of each portion or layer.

Polycarbonate-Based Polyurethane

In 1992, a second-generation biomedical-grade thermoplastic polyurethane elastomer was introduced by Szycher and collaborators in U.S. Pat. 5254662, "Biostable Polyurethane Products." This polyurethane is considered to be a second-generation elastomer since it is a carbinate-based polymer. This type of linear, segmented urethane is a rubbery reaction product of an aliphatic isocyanate, a high-molecular-weight polyol, and a low-molecular-weight chain extender. Polycarbonate-based polyurethanes represent a substantial advance over older, conventional ether-based polyurethanes since the polycarbonate-based polyurethanes are far more durable in the highly corrosive environment of the human body. In this context, *biomedical grade* means *a systemic, pharmacologically inert substance designed for implantation within living systems.*

Wound Dressings

Burns and other related wounds, such as donor sites and the like, present a serious problem in that they tend to produce large amounts of exudate that can cause conventional

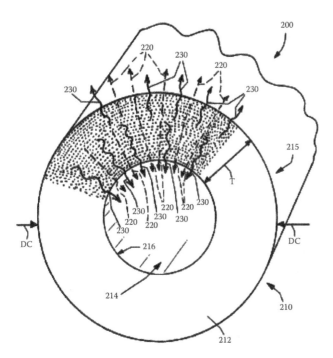

Fig. 6 Co-extruded antimicrobial catheter.

dressings to become saturated, or to stick to the wound, or even to become infected. One method of covering such wounds has been to cover the burns with a material into which new epithelial or fibroblast growth can penetrate. Dressings of this kind are disclosed in U.S. Pats. 3526224, 3648692, and 3949742.

However, such dressings can be extremely painful to remove and often require surgical excision. A fundamentally different approach, requiring a fundamentally different type of dressing, is to employ materials that are designed to reduce the propensity to adhere to the wound. Dressings of this kind are disclosed in U.S. Pat. 3543750. One more recent attempt at nonadherent dressings is U.S. Pat. 3709221, which discloses a dressing having an outer microporous, liquid-repellent fibrous layer, an inner macroporous fibrous layer, and an absorbent intermediate layer, which was also envisaged as normally being fibrous. To reduce the tendency of this material to adhere to the wound, the inner layer had to be treated with an agent to render it nonwetted by body liquid. It is now realized that it would be desirable to provide a dressing in which the wound-facing layer does not require special treatment.

Lang[4] disclosed an invention related to an absorptive wound dressing suitable for use on burns or other wounds, a dressing that has a reduced tendency to adhere to the wound and can act as a bacterial barrier. The present invention also relates to the manufacture and use of such dressings.

Lang discovered that, by avoiding fibrous materials, it is possible to produce a dressing with reduced tendency to adhere to wounds without the need for special treatments. An attempt at producing an absorbent dressing is described in U.S. Pat. 3888748, which describes a dressing fabricated from at least four sheet materials. The wound-facing part of the dressing apparently consists of a grid or scrim coated with polyethylene in such a manner that the polyethylene surrounds the filaments of the grid and collects any loose thread or particle that may be present in the core material. It is now realized that it is desirable to avoid the use of wound-facing layers that can allow such penetration of the central layer to the wound surface. It has also been realized that it would be desirable to provide a material that was highly conformable to the wound so that it is possible to minimize the quantity of exudate between the wound surface and the dressing. U.S. Pats. 3709221 and 3888248 disclose materials that are bonded along their edges that may reflect a desire to improve conformability. The dressing of the present invention allows for bonding over the whole of the operative area while retaining flexibility.

Accordingly, the Lang invention provides a low-adherence wound dressing that comprises a wound-facing layer, an intermediate absorbent layer, and an outer layer. The wound dressing is distinguished in that the wound-facing layer consists of a conformable elastomeric apertured film, the intermediate absorbent layer is made up of a conformable hydrophilic foam, and the outer layer is a continuous moisture vapor transmitting conformable film.

Thermoplastic Medical-Grade Polyurethane

Extensive investigations have been undertaken over many years to find materials that will be biologically and chemically stable toward body fluids. This area of research has become increasingly important with the development of various objects and articles that can be in contact with blood, such as artificial organs, vascular grafts, probes, cannulas, catheters, and the like.

Synthetic plastics have come to the forefront as preferred materials for such articles. However, these materials have the major drawback of being thrombogenic. Thrombogenicity has conventionally been counteracted by the use of anticoagulants such as heparin. Exemplary of procedures for attachment of heparin to otherwise thrombogenic polymeric surfaces are the disclosures in U.S. Pat. 4613517 to Williams et al. and U.S. Pat. 4521564 to Solomon et al.

In general, the most blood-compatible plastics known are the fluorinated polyolefins, such as polytetrafluoroethylene and the silicone polymers. However, while being basically hemocompatible, silicone polymers do not have the desired mechanical strength for most blood-contacting applications. One approach to improving the mechanical properties of silicone polymers has been the addition of appropriate fillers and curing agents. Such additives, although providing strength, are usually themselves thrombogenic so that the improved physical strength is offset by reduced blood compatibility.

Another approach has been to combine the blood compatibility of the silicone with the excellent mechanical properties of polyurethane. U.S. Pat. 3562352 discloses a copolymer consisting of about 90% polyurethane and 10% polydimethylsiloxane. This material, under the trade name Cardiothane® (Kontron Cardiovascular, Inc., Everett, MA), has been widely used in blood-contacting applications but has the major drawback that it is not thermo-plastic and cannot be melt processed.

Thermoplastic polyoxyalkylene polyurethanes having up to 15% of a soft segment formed from a polysiloxane devoid of oxygen atoms bonded to both silicon and carbon are disclosed by Zdrahala and Strand in U.S. Pat. 4647643. Polyurethanes prepared from 1,3-bis(4-hydroxybutyl) tetramethyl disiloxane are reported by Yilgor et al.[5] and are suggested to have possible utility in the biomedical field.

Silicone coatings have been achieved by plasma polymerization of silicon-containing monomers onto various polymeric base materials. Preparation and hemocompatibility studies of such materials are described by Chawla.[6]

Ward, in U.K. Pat. GB 2140437B, disperses up to 5% of a silicone-containing additive in a polymeric base material by mixing the components as a melt or in a solvent. Biomedical devices are prepared therefrom by conventional techniques such as injection molding and by homogeneous extrusion.

While significant advances have been made toward blood-compatible surfaces for fabrication of medical devices, further improvements are needed. In particular, materials having surfaces are needed that are essentially nonthrombogenic for use in devices that will be in contact with blood for prolonged periods. It is toward the fulfillment of this need that this invention is directed.

The Hu, Solomon, and Wells[7] invention discloses an article having a hemocompatible surface. In the present disclosure, the term *hemocompatible* describes a surface that does not induce thrombosis or changes in blood cells, enzymes, or electrolytes; does not damage adjacent tissue; and does not cause adverse immune responses or toxic reactions. Preferred articles are medical devices such as catheters.

The Hu, Solomon, and Wells[7] invention includes a thermoplastic polymeric base material having thereon a layer of copolymer having urethane and silicon-containing segments. In preferred articles, the base material is a polyurethane, and the silicon-containing segment is a siloxane. Particularly preferred articles have siloxane segments in the copolymer which are hydroxyalkyl terminated and which thereby have outstanding stability when in hydrolytic environments.

Fluorinated Polyurethane

Polyetherurethane compositions develop microdomains conventionally termed *hard-segment domains* and *soft-segment domains*. They are $(AB)_n$-type block copolymers, A being the hard segment and B the soft segment, and they are occasionally termed *segmented* polyurethanes. The hard-segment domains form by localization of the portions of the copolymer molecules that include the isocyanate and extender components, whereas the soft-segment domains form from the polyether glycol portions of the copolymer chains. The hard segments are generally more crystalline and hydrophilic than the soft segments, and these characteristics prevail, in general, for the respective domains. One disadvantage of polyurethane resins of the softness desired for many medical devices, for example, resins having Shore A hardness less than about 100, is surface blocking (tack) after extrusion or molding into desired shapes. To avoid this problem, many remedies have been developed in the art, including the use of external mold release agents and the use of various antiblockers or detackifiers in admixture with the polymer. Most antiblocking agents/detackifiers are low-molecular-weight materials that have a tendency to migrate or leach out of the polymer. This represents a problem when the polyurethanes are to be used as biomaterials (tubing, prostheses, implants, and so on). The presence of such low molecular-weight extractables can affect the biocompatibility of the polyurethanes and lead to surface degradation such as assuring or stress cracking.

Fluorine-containing polyurethanes are known. Kato et al.[8] disclose polyurethanes synthesized from fluorinated

isocyanates. Yoon and Ratner[9] disclose polyurethanes synthesized from fluorinated chain extenders. Field et al., in U.S. Pat. 4157358, disclose a randomly fluorinated epoxyurethane resin.

Although some progress has been made toward providing a thermoplastic polyurethane that: 1) is nonblocking without additives; 2) provides a desirable balance of stiffness in air and softness in liquid; and 3) is suitable for blood contact, further improvement is needed. This invention is directed toward fulfillment of this need.

Zdrahala and Strand[10] disclosed nonblocking, hemocompatible, thermoplastic, fluorinated polyetherurethanes, and a method for their preparation from fluorinated polyether glycols, isocyanates, chain extenders, and a nonfluorinated polyol. The method includes two steps in which the fluorinated glycol is reacted initially with the diisocyanate to give a prepolymer having terminal isocyanate groups, and the prepolymer is then reacted with the extender and nonfluorinated polyol. Medical devices are fabricated from the fluorinated polyetherurethane.

POLYURETHANE FEEDING TUBE

In U.S. Patent Application 20030009152, "Polyurethane Feeding Tube and Associated Adaptors," O'Hara and coinventors discussed the development of specialized gastrostomy and jejunostomy tubes.

Gastrostomy and jejunostomy tubes are used to deliver nutritional products to the gastrointestinal tract of a patient having difficulty ingesting food. Gastrostomy tubes deliver the nutritional products percutaneously from an external source, through the patient's abdominal wall, and directly to the patient's stomach, while jejunostomy tubes deliver the nutritional products percutaneously into the patient's jejunum or small bowel. In addition to primary placement percutaneously through a patient's abdominal wall, gastrostomy and jejunostomy tubes can be placed into the patient's gastrointestinal tract through a mature stoma formed in the abdominal wall, or through the nasal passage using nasogastric or nasojejunal tube, respectively. Gastrostomy, jejunostomy, nasogastric, and nasojejunal tubes are referred to collectively herein as "feeding tubes," unless otherwise noted.

The first step for the primary percutaneous placement of a feeding tube in a patient typically involves the passing of an endoscope down the patient's esophagus in order to view the esophagus and determine whether there are any obstructions or lesions in the esophagus that will inhibit or preclude passage of the feeding tube through the esophagus. The endoscope is also used to examine the interior of the stomach and/or the small bowel. Next, the doctor visually selects the site through which the feeding tube will be introduced into the stomach and transilluminates the selected site by directing light outwardly from the endoscope such that the light shines through the patient's abdominal wall, thereby allowing the doctor to identify the entry site from a point outside of the patient's body. The doctor then inserts a catheter or introducer through the patient's abdominal wall and into the stomach at the selected entry site. A first end of a placement wire is then passed through the introducer and into the stomach. The first end of the wire is grasped using a grasping tool associated with the endoscope, and the endoscope and the placement wire are drawn outwardly from the patient's stomach and esophagus through the patient's mouth. Upon completing this step of the procedure, a second end of the wire remains external to the patient's abdominal wall while the first end of the wire extends outwardly from the patient's mouth.

Placement wires can have a variety of forms. In one commercially available embodiment, the placement wire is a doubled wire coated with a biocompatible plastic material. However, other forms of placement wires are well known. These placement wires typically are provided in a sterile package for use by a medical professional. For example, the placement wire can be coiled and placed in a sealed pouch. The wire is removed from the pouch immediately prior to placement in a patient. This packaging methodology presents certain disadvantages in that the wire is prone to entanglement during insertion into the patient. Thus, the wire must be carefully manipulated in order to ensure that it is fed properly through the introducer and into the patient. Such manipulation may result in touch contamination of the wire as it is manipulated. Further, in order to ensure that the wire is properly fed into the patient's stomach, it is sometimes necessary to have one person manipulate the wire while a second person feeds the wire into the patient. This need for additional medical personnel increases the cost of placing the feeding tube in the patient.

In a different commercially available embodiment, the wire is a "silk"-type pull thread that is loosely coiled in a provided holder. The thread extends through a hole in the holder and can be pulled outwardly from the holder through the hole. As the endoscope is withdrawn through the patient's esophagus, an assistant must carefully pull the thread out of the holder and allow it to feed through the catheter. This embodiment also presents certain disadvantages due to the fact that an assistant is required in order to manipulate and feed the thread into the catheter. In addition, it is typically necessary to create a knot at the end of the thread before attaching it to a feeding tube. In some cases, the creation of this knot can be difficult due to the physical characteristics of the silk thread after it has been drawn through the patient's stomach, esophagus, and mouth.

In another commercially available embodiment, the placement wire is retained in a coil of rigid tubing. The wire can be difficult to manipulate and therefore may require the presence of an assistant to withdraw the wire from the coiled tubing. However, this embodiment does tend to reduce tangling of the placement wire during placement of the wire in the patient. Yet another commercially available embodiment includes a placement wire provided

in a circular dispenser. Although this embodiment tends to minimize tangling of the placement wire, the wire still can be difficult to dispense from the circular dispenser and therefore requires the presence of an assistant.

It is preferable to provide a placement wire in such a way that: 1) the possibility of entanglement of the wire is minimized; 2) the possibility of touch contamination of the wire is minimized; and 3) withdrawal of the wire from its packaging does not require additional personnel.

With the wire properly inserted, initial or primary placement of a feeding tube percutaneously into the gastrointestinal tract of the patient can be performed. In one technique for primary feeding tube placement, the first end of the placement wire is attached to a first end of a feeding tube. Attachment of the feeding tube to the first end of the placement wire is facilitated by a loop on the first end of the placement wire and by a complementary loop on the first end of the feeding tube. By pulling on the second end of the wire positioned external to the patient's abdominal wall, the feeding tube is pulled through the patient's mouth and esophagus, and into the stomach. Further pulling of the second end of the wire causes the first end of the feeding tube to exit percutaneously from the stomach through a tract in the abdominal wall formed by the introducer. The feeding tube is pulled outwardly through the tract until a retaining member mounted on the second end of the feeding tube engages the interior of the stomach. This technique is referred to as a "pull" technique.

In an alternative technique for primary feeding tube placement, a channel defined through the feeding tube is positioned over the wire such that the feeding tube can be pushed along the length of the wire. As the feeding tube is pushed over the wire, it passes through the patient's mouth, esophagus, and stomach until the first end of the feeding tube exits through the incision in the abdominal wall. The feeding tube is then drawn outwardly through the abdominal tract until a retaining member on the internal or second end of the feeding tube engages the interior of the stomach. The wire is then withdrawn from the patient through the feeding tube channel. This technique is referred to as a "push" technique.

Yet another method may be used for percutaneous placement of the feeding tube, particularly when passage through the esophagus is precluded. This method, which is commonly known as the "poke" technique, is most often used for placement of a feeding tube having a balloon-type internal retaining member. Once a safe site into the stomach is identified, the poke technique requires that the stomach wall be retracted and secured against the abdominal wall of the patient using a known anchoring device, such as T-Fasteners®. One or more dilators are then used in sequence to form an adequate tract of sufficient size through which the feeding tube and the retaining member, in a deflated state, can be inserted. The retaining member is then inflated in its proper position to secure the placement of the feeding tube.

After a stoma tract has fully formed through the abdominal wall of the patient using any of the techniques summarized above, the initial feeding tube can be removed if secondary placement of a different feeding tube is desired or necessary. In this manner, a new feeding tube can be inserted directly into the mature stoma tract that is formed through the patient's abdominal wall and retained in a conventional manner.

A variety of feeding tube configurations made of different materials are well known, and are each suitable for its intended purpose. There remains a continued need, however, for feeding tube assemblies of enhanced operational characteristics and cost-effective construction.

BISMUTH OXYCHLORIDE AS RADIOPACIFIER

In European Patent Application EP2248542, "Polycarbonate Polyurethane Venous Access Devices Having Enhanced Strength," Davis and Lareau relate vascular catheters that comprise a catheter shaft having one or more lumens. The catheter shaft comprises a polycarbonate polyurethane and bismuth oxychloride.

The bismuth oxychloride is provided in the catheter shafts of the invention in an amount that typically ranges from approximately 5 wt% to 25 wt%. The polycarbonate polyurethane is provided in an amount that typically ranges from approximately 5 wt% to 75 wt%.

As used herein, a "catheter" is a medical device that includes a flexible shaft, which contains one or more lumens (including annular shafts, i.e., tubes), and which may be inserted into a subject (e.g., a vertebrate subject, for instance, a mammalian subject such a human, dog, cat, and horse) for introduction of material (e.g., fluids, nutrients, medications, and blood products), for removal of material (e.g., body fluids), or both.

A catheter may further include various accessory components, for example, molded components, overmolded subassemblies, connecting fittings such as hubs, extension tubes, and so forth. Various catheter tips designs are known, including stepped tips, tapered tips, overmolded tips, and split tips (for multilumen catheters), among others.

Central venous access may be achieved, for instance, by direct puncture of the central venous circulation system, for example, via the internal jugular vein, subclavian vein, or femoral vein. Catheters of this type, known as "central catheters" or "central venous catheters," are relatively short, and may remain in place for months or even years.

Other central venous access catheters have also been developed which are peripheral venous catheters. These catheters can be inserted into peripheral veins (e.g., the antecubital, basilica, or cephalic vein) and advanced to access the central venous system, with the tip commonly positioned in the superior vena cava or right atrium, thus allowing for rapid dilution of infused fluids. These devices avoid difficulties associated with the direct puncture of the central venous circulation system, and are generally for short-term use (e.g., a few days to a few months) to provide repeated access to a patient's vascular system, thereby avoiding multiple injections and minimizing trauma and pain to the patient.

Specific examples of catheters of this type include the so-called PICCs, midline catheters, and peripheral catheters. A typical PICC, midline, or peripheral catheter contains a thin, flexible shaft, which contains one or more lumens and which terminates at the proximal end with a suitable fitting, such as a hub or other fitting. A primary difference between these three devices is the length of the tubing, with the peripheral catheter being the shortest and the PICC being the longest. The rationale for different lengths is driven by the type and duration of the therapy that a patient is to receive. Other differences may include a diameter, a lumen configuration, a catheter configuration, and so on.

Hemodialysis catheters are another important class of central venous access catheters. Hemodialysis catheters are commonly multilumen catheters in which one lumen is used to carry blood from the body to a dialysis machine, and another lumen returns blood to the body. Central venous access may be attained by puncture of various major blood vessels, including the internal jugular vein, subclavian vein, or femoral vein.

Central venous access may also be provided via venous access ports. These specialized catheters typically have three components: 1) a catheter; 2) a reservoir which holds a small amount of liquid and which is connected to the catheter; and 3) a septum, which covers the reservoir and allows access to the reservoir upon insertion of a needle. The reservoir and covering septum may be surgically placed under the skin of the chest or arm, and the catheter extends into a central vein.

Catheter shafts for catheters that provide access to the central venous circulation, including those described above, among others, are typically made from polymers. Suitable polymers are those that can be formed into a shaft, having one or more lumens, which is flexible enough to be routed through the vasculature without causing trauma to the patient. Polymeric materials that balance softness and flexibility may also be desirable. When formed into a shaft, the polymer chosen should also provide strength sufficient to ensure that the lumen does not collapse in the vasculature, and should resist repeated flexure. Recently, there has been a trend to use these devices for power injection of contrast media for use in computed tomography, requiring sufficient burst strength.

In the present invention, these properties are provided, in part, by forming catheter shafts using polyurethanes. Polyurethanes are a family of polymers that are typically synthesized from polyfunctional isocyanates (e.g., diisocyanates, including both aliphatic and aromatic diisocyanates) and polyols, also referred to as macroglycols (e.g., macrodiols). Commonly employed macroglycols include polyester diols, polyether diols, and polycarbonate diols. Typically, aliphatic or aromatic diols are also employed as chain extenders, for example, to enhance the physical properties of the material.

Polyurethanes are commonly classified based on the type of macroglycol employed, with those containing polyester glycols being referred to as polyester polyurethanes, those containing polyether glycols being referred to as polyether polyurethanes, and those containing polycarbonate glycols being referred to as polycarbonate polyurethanes. Polyurethanes are also commonly designated aromatic or aliphatic on the basis of the chemical nature of the diisocyanate component in their formulation.

In the present invention, polycarbonate polyurethanes are preferred polyurethanes, more preferably, aliphatic polycarbonate polyurethanes, although aromatic polycarbonate polyurethanes may be employed.

Macroglycols for use in forming polycarbonate polyurethanes may be selected from suitable members of the following, among others: polycarbonate diols, for example, homopolyalkylene carbonate diols and copolyalkylene carbonate diols such as those containing one or more linear or branched alkylene carbonate monomers, for instance, selected from one or more of the following: methyl carbonate, ethyl carbonate, propyl carbonate (e.g., n-propyl and isopropyl carbonate), butyl carbonate, pentyl carbonate, hexyl carbonate, heptyl carbonate, octyl carbonate, nonyl carbonate, decyl carbonate, undecyl carbonate, dodecyl carbonate, and so forth. Poly(1,6-hexyl-1,2-ethyl carbonate) diol is a common polycarbonate diol for use in forming polycarbonate polyurethanes.

Aromatic diisocyanates for use in forming polycarbonate polyurethanes may be selected from suitable members of the following, among others: 4,4'-methylenediphenyl diisocyanate (MDI), 2,4- and/or 2,6-toluene diisocyanate (TDI), 1,5-naphthalene diisocyanate (NDI), para-phenylene diisocyanate, 3,3'-tolidene-4,4'-diisocyanate, and 3,3'-dimethyl-diphenylmethane-4 and 4,4'-diphenyl diisocyanate.

Aliphatic diisocyanates for use in forming polycarbonate polyurethanes may be selected from suitable members of the following, among others: dicyclohexylmethane-4,4'-diisocyanate (hydrogenated MDI), 1,6-hexamethylene diisocyanate (HDI), 3-isocyanatomethyl-3,5,5-trimethyl-cyclohexyl isocyanate (isophorone diisocyanate or IPDI), cyclohexyl diisocyanate, and 2,2,4-trimethyl-1,6-hexamethylene diisocyanate.

Chain extenders for use in forming polycarbonate polyurethanes may be selected from suitable members of the following, among others: diol chain extenders such as alpha,omega-alkane diols, including ethylene glycol (1,2-ethanediol), 1,3-propanediol, 1,4-butanediol, 1,5-pentanediol, and 1,6-hexanediol.

Commercially available aliphatic polycarbonate polyurethanes include CarboThane (Lubrizol Advanced Materials, Inc., Cleveland, OH) and Chronoflex AL (CardioTech International, Inc., Woburn, MA). Commercially available aromatic polycarbonate polyurethanes include Bionate® (The Polymer Technology Group, Inc., Berkeley, CA) and Chronoflex AR (CardioTech International, Inc.).

Polycarbonate polyurethanes are typically thermoplastics, meaning that a variety of thermoplastic processing techniques, such as extrusion, molding, and so forth, may be employed to form medical devices and medical device components, including catheter shafts, from the same.

As noted above, when formed into a catheter shaft, the polymer chosen should also provide strength sufficient to ensure that the lumen does not collapse in the vasculature. Moreover, where used for power injection of contrast media, the selected polymer should ensure that the catheter shaft does not burst. The polymer should also resist repeated flexure.

Polycarbonate polyurethanes go a long way toward meeting these goals. They are flexible and strong, allowing catheters to be formed with thin walls, regardless of whether the catheter shaft is a single-lumen shaft or a multilumen shaft. Subsequently, catheters made from these materials may be formed with smaller ODs as compared, for example, to other catheter materials such as silicone, or they may be formed having the same OD, but with a larger ID, and therefore provide a greater flow rate.

Generally, polycarbonate polyurethanes are known to soften when exposed to elevated temperatures and aqueous environments for extended periods of time (e.g., 24 hr or more).

Bismuth compounds, such as bismuth subcarbonate, bismuth trioxide, and bismuth oxychloride, among others like barium sulfate, can render a catheter shaft more absorptive of x-rays than the surrounding tissue, allowing the catheter shaft to be viewed using radiographic imaging techniques (e.g., by x-ray fluoroscopy). The amount of bismuth compound required will depend upon the thickness of the catheter wall, among other factors.

Barium sulfate and bismuth oxychloride are both used to add radiopacity to a catheter, which may be formed of a variety of thermoplastic materials. Historically, bismuth salts have not been used in polyurethane due to the potential for polymer degradation such as, for example, in the presence of bismuth subcarbonate. However, the present inventors unexpectedly found that by adding bismuth oxychloride to polycarbonate polyurethane, the softening in mechanical properties that is exhibited upon exposure to aqueous fluids at body temperature for extended time periods is markedly reduced, relative to the same polycarbonate polyurethane containing barium sulfate. This enhancement in mechanical properties (modulus and tensile strength) allows the device to resist forces exerted by routine use, including CT power injection of contrast media.

With tensile properties better retained, the need to move to unfavorably less flexible grades or thicker walls to offset the property reduction in aqueous fluids at body temperature is eliminated. Specific examples of commercially available bismuth oxychloride powders include Biron® (Merck) and PearlGlo UVR (Engelhard). Typical powder sizes may range from less than 2 to 20 microns in width.

Wall thickness for polycarbonate polyurethane central venous access catheter shafts in accordance with the invention will vary with application and may range, for example, from 0.002 to 0.100 in, among other thicknesses. Other portions of the catheter may also be enhanced such as, for example, a molded suture wing, which may have a thickness of up to 0.250 in or more.

In addition to providing enhanced mechanical properties, bismuth oxychloride also reduces surface tack and friction along the length of the catheter shaft when compared to barium sulfate. This may, for example, enhance guidewire trackability in venous access devices and enhance loading of port catheters onto port body stems. Bismuth oxychloride also produces a pearlescent surface finish which may be a desired end use characteristic.

Without wishing to be bound by theory, it is believed that various properties observed upon the addition of bismuth oxychloride, including strength, pearlescent finish, and lubricity characteristics, are likely due to the plate-like nature of the particles and due to the thermoplastic processing techniques that are employed to form the catheter shafts, techniques that create shear forces that tend to orient the particles.

In an alternate embodiment, catheter shafts may be formed using a blend of polyurethane with polycarbonate such as Texin 4210 (Bayer Corporation, Pittsburgh, PA). As it is a polyurethane and polycarbonate blend, the addition of bismuth oxychloride will result in similarly enhanced mechanical properties.

Examples

Samples (tubular extrusions) were formed from the following: 1) 104 wt% CarboThane PC-3585A (available from Lubrizol Advanced Materials, Inc., Cleveland, OH); 2) 60 wt% CarboThane PC-3585A and 40 wt% barium sulfate; 3) 70 wt% CarboThane PC-3585A and 30 wt% bismuth oxychloride (PearlGlo UVR available from Engelhard Corp.); 4) 100 wt% CarboThane PC-3595A; 5) 60 wt% CarboThane PC-3595A and 40 wt% barium sulfate; and 6) 70 wt% CarboThane PC-3595A and 30 wt% bismuth oxychloride. In particular, tubular extrusions of dimension 5F tip/6F body were formed via conventional thermoplastic bump extrusion. The bismuth oxychloride was incorporated into the polyurethane prior to tube extrusion via conventional thermoplastic compounding techniques.

Stress at 100% elongation and Young's modulus were measured for the samples using an Instron Tensile Tester Model 5565. Testing was performed for as-formed samples at room temperature. Testing was also performed after removal from a conditioning bath at body temperature

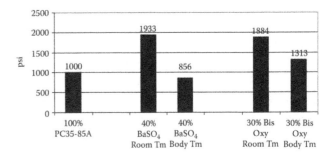

Fig. 7 Effect of radiopacifier and temperature on polyurethane stiffness.

(37°C) for a minimum of 24 hr, immediately after removal from the water.

Test results are presented in Fig. 7. The addition of bismuth oxychloride resulted in an unexpected increase in stress at 100% elongation and an unexpected increase in Young's modulus relative to the addition of traditional barium sulfate. This was observed for all grades of CarboThane and was observed at both room temperature and postconditioning in water at body temperature. It should be noted that the room temperature stress at 100% elongation and Young's modulus did not increase with the addition of bismuth oxychloride as compared to traditional barium sulfate.

In particular, with bismuth oxychloride in CarboThane PC-3585A, there is only 17.8% reduction in stress at 100% elongation upon conditioning versus 40% reduction with barium sulfate. With bismuth oxychloride in CarboThane PC-3585A, there is only 18.2% reduction in Young's modulus upon conditioning versus 50.8% reduction with barium sulfate. With bismuth oxychloride in CarboThane PC-3535A, there is only 29.5% reduction in stress at 100% elongation upon conditioning versus 55% reduction with barium sulfate. With bismuth oxychloride in CarboThane PC-3595A, there is only 46.8% reduction in Young's modulus upon conditioning versus 73.7% reduction with barium sulfate (Table 5).

SUMMARY

The purpose and advantages of the present invention will be set forth in and is apparent from the description that follows, as well as learned by practice of the invention. Additional advantages of the invention will be realized and attained by the methods and systems particularly pointed out in the written description and claims hereof, as well as from the appended drawings.

To achieve these and other advantages and in accordance with the purpose of the invention, as embodied and broadly described, the invention includes a method for placing a feeding tube placement wire in a patient. The method includes the step of providing a placement wire dispenser. The dispenser includes a rotatable placement

Table 5 Therapeutic applications improved by infusion pump administration

Disease	Drug
Acute cardiac failure	Nitroglycerine
Asthma	Theophylline
Cancer	Antineoplastic agents
Cardiac arrythmia	Lidocaine
Chronic infection (i.e., osteomyelitis)	Antimicrobial/antineoplastic agents
Chronic intractable pain	Morphine, opiates, endorphins
Depression	Lithium
Diabetes insipidus	Vasopressin
Diabetes mellitus	Insulin
Glaucoma	Pilocarpine
Growth disorders, infertility, etc.	Hormone replacement (GH, LH/RH)
Hemophilia	Factor VIII
Hypertension	Beta blocker, clonidine
Menopause, female hypogonadism	Estradiol
Parkinson's disease	Dopamine
Respiratory distress	Scopolamine
Rheumatoid arthritis	NSAIDs
Seizure disorders	Dilantine, phenobarbital, etc.
Thalassemia	Deferoxamine
Thrombosis	Heparin
Ulcerative colitis	Azulfidine, steroids

wire receptacle that defines a placement wire outlet. The dispenser further includes a tube extending outwardly from the placement wire receptacle. The tube defines a placement wire inlet that is in communication with the placement wire outlet of the receptacle. A placement wire is wound about the receptacle, and is in mechanical engagement with the receptacle, such that the rotation of the placement wire receptacle causes a first end of the placement wire to be advanced through the placement wire outlet, through the placement wire inlet of the tube, and through the tube. A first end of the tube is constructed for insertion through a patient's abdominal wall and into a patient's stomach. The method further includes the step of placing the first end of the tube through a patient's abdominal wall and into a patient's stomach. Rotational movement is imparted to the receptacle so as to advance the placement wire through the placement wire outlet, through the placement wire inlet, through the tube, and into a patient's stomach.

The present invention also includes a placement wire dispenser. The dispenser includes a rotatable placement wire receptacle that defines a placement wire outlet. The

dispenser further includes a tube extending outwardly from the placement wire receptacle. The tube defines a placement wire inlet that is in communication with the placement wire outlet of the receptacle. A placement wire is wound about the receptacle, and is in mechanical engagement with the receptacle, such that rotation of the placement wire receptacle causes a first end of the placement wire to be advanced through the placement wire outlet, through the placement wire inlet of the tube, and through the tube. A first end of the tube is constructed for insertion through a patient's abdominal wall and into a patient's stomach.

The present invention is further directed to a feeding tube placement kit. The kit includes a placement wire dispenser. The dispenser includes a rotatable placement wire receptacle that defines a placement wire outlet. The dispenser further includes a tube extending outwardly from the placement wire receptacle. The tube defines a placement wire inlet that is in communication with the placement wire outlet of the receptacle. A placement wire is wound about the receptacle, and is in mechanical engagement with the receptacle, such that rotation of the placement wire receptacle causes a first end of the placement wire to be advanced through the placement wire outlet, through the placement wire inlet of the tube, and through the tube. A first end of the tube is constructed for insertion through a patient's abdominal wall and into a patient's stomach. The kit further includes a feeding tube having a first end portion and a second end portion. The feeding tube defines a feeding lumen therethrough. A retaining member is disposed on the second end portion of the feeding tube.

Additionally, the present invention is directed to a feeding tube assembly, including a feeding tube made of a polyurethane, preferably CarboThane. The feeding tube has a first end portion, a second end portion, and at least one lumen defined therein. The feeding tube assembly further includes a feeding tube adaptor having an inlet conduit and an outlet conduit, preferably made of a substantially rigid material, wherein the outlet conduit has an outlet end to be inserted into a lumen at the first end portion of the feeding tube for fluid communication between there. The exterior surface of the outlet conduit defines a retention member to engage an interior surface at the first end portion of the feeding tube when the outlet end of the outlet conduit is inserted within the lumen of the feeding tube. The retention member includes a first section having an exterior peripheral dimension increasing with increasing distance from the outlet end of the outlet conduit, and a second section having an exterior peripheral dimension decreasing with increasing distance from the outlet end of the outlet conduit.

In accordance with another aspect, the present invention is directed to a feeding tube assembly, including a feeding tube, a feeding tube adaptor having a first inlet conduit, and a removable cap made of polyurethane to close selectively an inlet port of the first inlet conduit. The cap has an engagement surface to engage a corresponding surface of the first inlet conduit, wherein the engagement surface of the cap has a surface configuration different than that of the corresponding surface of the first inlet conduit. For example, the inlet conduit can include a port defined by an interior surface and the cap can include a plug to be inserted into the port, wherein the engaging surface of the cap is an exterior surface of the plug and the corresponding surface of the inlet conduit is an interior surface of the inlet conduit. Alternatively, the cap can include a peripheral flange, such that the engaging surface of the cap is an interior surface of the peripheral flange that engages a corresponding surface of the exterior surface of the inlet conduit. The difference in surface configurations can be established by providing one or both of the surfaces with a series of protuberances, a series of indentations, a different cross-sectional shape, or a combination thereof.

REFERENCES

1. Benabid, A.L.; Pollack, P.; Louveau, A.; Henry, S.; deRougemont, J. Combined (thalamotomy and stimulation) stereotactic surgery of the VIM thalamic nucleus for bilateral Parkinson disease. Appl. Neurophysiol. **1987**, *50* (1–6), 344–346.

2. Hansen, C.M.; Beerbower, A. In *Kirk-Othmer Encyclopedia of Chemical Technology*, 2nd Ed.; Standen, A., Ed.; 1971; 889–910.

3. Bell, A.L.; Jayaraman, R.; Vercaigne, L.M. Effect of ethanol/trisodium citrate lock on the mechanical properties of carbothane hemodialysis catheters. Clin. Nephrol. **2006**, *65* (5), 342–348.

4. Lang, S.M. Wound Dressing with Conformable Elastomeric Wound Contact Layer. U.S. Patent 5445604, Aug 29, 1995.

5. Yilgor, I. et al. Polymer Preprint (American Chemical Society) **1982**, *20*, 286.

6. Chawla, A.S. Evaluation of plasma polymerized hexamethylcyclotrisiloxane biomaterials towards adhesion of canine platelets and leucocytes. Biomaterials **1981**, *2* (2), 83–88.

7. Hu, C.B.; Solomon, D.; Wells, S.C. Method of Making an Article Having a Hemocompatible Surface. U.S. Patent 5059269, Oct 22, 1991, assigned to Becton, Dickinson and Company, Franklin Lakes, NJ.

8. Kato, M. et al. Progress in Artificial Organs 1983, 858.

9. Yoon, S.C.; Ratner, B.D. Surface structure of segmented poly(ether urethanes) and poly(ether urethane ureas) with various perfluoro chain extenders. An x-ray photoelectron spectroscopic investigation. Macromolecules **1986**, *19* (4), 1068.

10. Zdrahala, R.J.; Strand, M.A. Fluorinated Poly-etherurethanes and Medical Devices therefrom. U.S. Patent 4841007, Jun 20, 1989.

Rapid Prototyping

Shaochen Chen
Carlos A. Aguilar
Yi Lu
Department of Mechanical Engineering, University of Texas at Austin, Austin, Texas, U.S.A.

Abstract
The term rapid prototyping (RP) refers to a broad category of techniques that can automatically construct physical prototypes from computer-aided design models. This entry breaks down RP fabrication techniques into categories based on their state of material, and then discusses the benefits of this method and its applications.

INTRODUCTION

The term rapid prototyping (RP) refers to a broad category of techniques that can automatically construct physical prototypes from computer-aided design (CAD) models. RP provides the unique opportunity to quickly create functional prototypes of highly complex designs in an additive fashion. This layered manufacturing is also known as solid freeform fabrication (SFF), desktop manufacturing, and computer-aided manufacturing (CAM). There are several methods of RP, but universal to all of them is the basic approach they use, which can be described in three phases:[1,2]

1. A geometric model is constructed on a CAD/CAM system. Because the manufacturing technique relies on a layer-by-layer additive process, the model must be represented as closed surfaces, which define an enclosed volume.
2. Next, the CAD model is converted into a stereolithography (SL) file format, which approximates the surfaces of the model by polygons. The more complex the geometry of the model, the larger the number of polygons needed to characterize the surface. Curved surfaces usually require more polygons to fit and in turn produce larger files.
3. A computer program reads the generated SL file and "slices" the model into a finite set of layered cross sections. Each layer is created individually from liquid, powder, or solid material and stacked onto the previous layer with each layer joined to its neighboring surfaces. The created three-dimensional (3-D) model resembles the geometry of the CAD structure.

There are a variety of methods that involve SFF, but several approaches have found frequent use in the biomedical arena. In this entry, the RP fabrication techniques are categorized by virtue of their state of material. The methods can be classified into three categories: 1) liquid-based systems; 2) powder-based systems; and 3) solid-based systems (see Fig. 1). The liquid-based systems cure liquid material into a solid state through interaction with a laser beam. The most popular liquid-based technique used for biomedical applications is stereolithography (SL). The powder-based systems rely on a laser beam to melt and fuse powder grains or selectively bind powder particles using liquid binders. Frequently used powder-based techniques in biomedical applications are selective laser sintering (SLS) and 3-D printing. The solid-based systems extrude melted material in filament form and build 3-D shapes by layering filament in predesigned patterns. A popular solid-based method, fused deposition modeling (FDM), has gained interest recently for several biomedical engineering uses. Each of the aforementioned techniques will be discussed in further detail in the subsequent sections.

There are several advantages of RP systems, and the most prevalent is the ability to rapidly produce functional prototypes in small quantities.[3] This benefit has greatly improved the efficacy of surgical planning and education by providing a convenient platform for rapidly creating anatomical models.[4] An excellent example of this is demonstrated in a recent study, where a patient was involved in a car accident and suffered from a serious bone fracture of the upper and lateral orbital rim in the skull.[5] As the first reconstructive surgery had proven unsuccessful, it was necessary to perform subsequent surgery to transplant an artificial bone into the hole. This would require the surgeons to carve an artificial bone during the operation and try and fit it into the hole. Given that the operation would be very time-consuming because of the fitting and refitting of the artificial implant, RP was applied to create a prototype of the patient's skull. The model was then used to prepare an artificial bone that would fit the hole properly. The use of

Concise Encyclopedia of Biomedical Polymers and Polymeric Biomaterials DOI: 10.1081/E-EBPPC-120017967

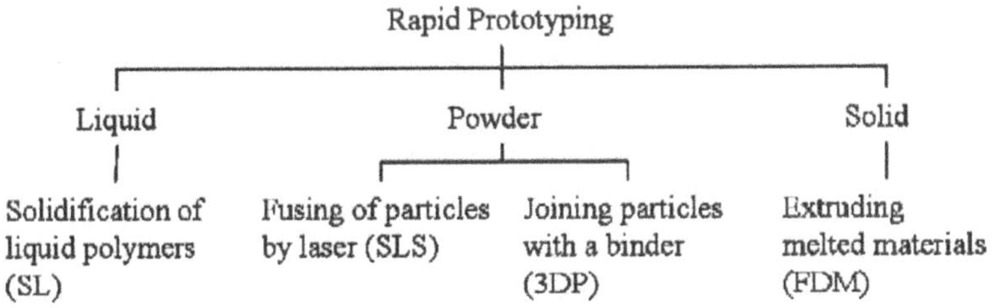

Fig. 1 Rapid prototyping systems.

RP model not only greatly reduced the surgery time and risk of infection but also improved its accuracy. In another example of successful surgical planning using RP, a patient was diagnosed with a cancerous bone tumor in his temple area. Initially, the surgery plan was to access the tumor through the frontal, highly functional area of the brain. As a consequence of the surgery, the patient would lose vision in one eye and several motor functions would be impaired.[6] Using RP technology, a replica of the patient's skull was manufactured from a series of 2-D computed tomography (CT) scans of the patient's skull. With the model, the surgeon was able to recognize that the tumor could be treated by entering through the patient's jawbone and that an alternative approach could be used for treatment. Using the alternative method, the risk of harming the eye and motor functions could be avoided. Other case studies relating to bone tumors have been reported with similar successful results.[7,8]

The uses of RP for biomedical applications are extended beyond surgical planning and education. The single-operative RP processes also offer an excellent tool to aid in the development of new medical devices such as targeted drug delivery devices. An excellent example of this is illustrated in a recent study, where RP was used to manufacture morphologically detailed human tracheobronchial airway models in various disease states based on high-resolution anatomical imaging data.[9] These anatomical detailed casts of respiratory tract structures that corresponded to diseased states from exposure to toxic or pathogenic aerosols were then used to explore new methods of targeted drug delivery via inhaled aerosols. These case studies are exceptional examples of the potential impact of RP for biomedical applications. RP processes also provide avenues to develop custom implants and prostheses, new scaffolds for the guided development of highly structured tissues, organs, and bones, next-generation *in vivo* controlled-release vehicles, and multilayer microfluidic networks. In the following sections, we will further explain several RP techniques and several examples of how RP can play an effective role in the biomedical industry.

STEREOLITHOGRAPHY

Stereolithography is perhaps the most widely used RP technique. It was commercialized by 3D Systems of Valencia, California, U.S.A.

Stereolithography is a liquid-based method that allows 3-D microfabrication in a room-temperature environment.[1,10,11] This method relies on a photosensitive monomer resin or a photocrosslinkable polymer resin, which solidifies when exposed to ultraviolet (UV) or visible light. The photosensitive monomers cure via photopolymerization, while the photocrosslinkable polymers become solids or hydrogels by forming cross-linking networks.

A typical SL system consists of a container filled with liquid photopolymer resin, a computer-controlled stage immersed in the resin, and a laser scanner, as shown in Fig. 2. The process begins with a 3-D solid model designed in CAD. The model is numerically sliced into a series of 2-D layers with an equal thickness in an STL file format.[12] The code generated from each sliced 2-D file is then executed to scan a UV laser beam in the *X–Y* plane on the surface of the resin. The interaction with the laser beam causes the liquid to solidify in areas where the laser strikes. After the layer is completely traced and cured by the laser beam, the stage is lowered into the vat a distance equal to the thickness of a layer. However, as the resin is generally quite viscous, tools were developed to increase the speed of the process. Early SL systems drew a knife-edge over the surface to smoothen it. More recently, pump-driven recoating systems have been utilized. The scanning and recoating steps are repeated until the object is completely fabricated.

The resolution of the conventional SL is limited by the spot size of the laser beam to approximately 25 μm. To fabricate smaller and more complicated geometries, such as tissue engineering scaffolds, minor modifications are made. The laser beam must be focused into a 1–2 μm spot by an objective lens. As it is relatively difficult to scan the objective lens, a 3-D stage is employed to move in the *X–Y–Z* directions instead. A line width of roughly smaller than 5 μm has been achieved using this approach.[13]

This layer-by-layer liquid micromanufacturing system enables the fabrication of complex internal features, such as intricate passageways for microfluidic devices and curved surfaces, to be accurately produced.[14] Furthermore, the approach can easily incorporate different proteins and microparticles containing polymer solutions for each layer (or even for partial layers). This allows SL the unique ability to rapidly create a precise spatial distribution of biochemical microenvironments within a single scaffold or system.

There are several factors that influence the quality of the SL generated products, and the photocurable polymer resin

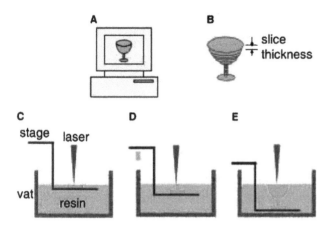

Fig. 2 Principle of stereolithography. (**A**) CAD model design. (**B**) STL file generation. (**C**) Laser scanning to photopolymerize the first layer of the 3-D structure. (**D**) The stage was lowered to produce the second layer of the 3-D structure. (**E**) Layer-by-layer selective photopolymerization to produce the entire 3-D structure.

is the most sensitive to processing. Additionally, as the products are used for biomedical scaffold purposes, the demands of the resin are as follows: 1) biocompatibility and, in some cases, biodegradability; 2) low viscosity; 3) stability under visible light; 4) fast cross-linking kinetics; 5) compatibility with 3-D polymerization; and 6) reasonably good accuracy and mechanical strength in the cured state.

Several systems that meet the above demands have been investigated recently and include, poly(ethylene glycol) (PEG)-modified systems cross-linked with a photoinitiator, typically acetophenone (UV light) or eosin Y, and triethanolamine (visible blue light).[15] Another system, comprising poly(ethylene glycol) dimethacrylate (PEGDMA) hydrogel polymer, has been used in a variety of *in vitro* and *in vivo* applications and has been proven to have no adverse effect on cells for a sufficient period of time. The polymer PEG, which is hydrophilic, highly wettable, and cell-adhesion resistant, can be modified with proteins [i.e., arginine-glycine-aspartic acid (RGD)] to enhance cell adhesion. Ten percent (w/v) resin of PEG has a viscosity close to that of water. Increasing the concentration leads to higher viscosity. The resin is stable under the normal experimental conditions, though sunlight should be minimized because it carries UV light. The curing speed depends on the UV sensitivity of the resin. High sensitivity indicates that a fast processing speed could be achieved. The resin can be cured only at the laser focal point and the uncured resin is reasonably stable under scattered or reflected UV light. Typical lateral resolution ranges from tens to hundreds of micrometers for PEGDMA systems. Though the resulting scaffold is relatively soft and fragile in aqueous environment, it has good tensile strength when water is extracted.

The successful fabrication of scaffolds using PEGDMA (MW 1000) along with cyto-compatible photoinitiator,

1-[4-(2-hydroxyethoxy)-phenyl]-2-hydroxy-2-methyl-1-propane-1-one, Darocure 2959, initiated by a frequency tripled Nd:YAG laser (355 nm in wavelength) was demonstrated recently.[16] Poly(ethylene glycol) acrylates, covalently modified with the cell adhesive peptide RGD or the extracellular matrix (ECM) component heparin sulfate, was incorporated within the scaffolds to facilitate cell attachment and to allow spatial sequestration of heparin-binding growth factors. Fluorescently labeled polymer microparticles and basic fibroblast growth factor (FGF-2) were chosen to illustrate the capability of SL to spatio-temporally pattern scaffolds. The results demonstrated a precise, predesigned distribution of single or multiple factors within the single 3-D structure with specific internal architectures. Functionalization of these scaffolds with RGD and heparan sulfate allowed efficient cell attachment and spatial localization of growth factors. Such patterned scaffolds might provide effective systems to study cell behavior in complex microenvironments and could eventually lead to engineering of complex, hybrid tissue structures through predesigned, multilineage differentiation of a single stem cell population.

The unique advantages of SL as a suitable technique for scaffolding lie in the following bases. First of all, the inherent photochemical process is attractive for the production of tissue engineering scaffolds because it has a controlled reaction initiation and termination, a relative short reaction time, and spatial control. Second, SL can be performed at mild conditions, such as room temperature and under physiologic pH, which allow incorporation of biologically active molecules and in situ cell seeding. Third, photopolymers can be modified with various peptides and copolymers, for instance, to promote or inhibit cell adhesion.

While SL does offer advantages for scaffolding, several issues must be addressed for subsequent use. First, as in situ cell seeding or *in vivo* scaffolding requires nontoxic precursors, initiators, and residues, a low concentration of initiators is essential. The inclusion of initiators frequently results in prolonged fabrication times and reduced control over resolution and composition of the scaffold. In addition, prolonged exposure to UV lasers may cause degradation in some bioactive molecules and reduce the capacity of the scaffolds to promote cell proliferation.

Selective Laser Sintering

Selective laser sintering is a popular RP powder-based technique. It was developed and patented at the University of Texas at Austin and commercialized by the DTM Corporation (now 3D Systems) in 1987.

The SLS process is similar to SL in that it creates 3-D objects layer by layer. However, SLS utilizes materials in powder form instead of liquid and uses heat generated from a CO_2 laser. The method begins by depositing a thin layer of material in powder form into a container. The layer of powder is then selectively irradiated by a CO_2 laser to the point of melting, fusing the powder particles to form a solid mass, as shown in Fig. 3 below. The intensity of the laser beam is modulated to melt the powder only in areas defined by the part's geometry, while the rest of the material remains in its original powder form. When the cross-sectional layer is solidified, sequential layers of powder are deposited via a roller mechanism on top of the previously scanned layer. The steps are repeated with each layer fusing to the layer below it until the 3-D model is completely built. As the material used in the SLS process is in powder form, the powder not melted or fused during the processing serves as a customized, built-in support structure. Therefore, there is no need to create support structures within the design phase or during processing and no support structure to remove when the part is complete. There is a broad range of materials that can be processed using SLS and no post-processing or postcuring is required.

Much like SL, SLS provides an economical and efficient method to construct scaffolds for guided development of human tissues and organs. As SLS is also a layer-by-layer additive process, the design and fabrication of anatomically shaped scaffolds with varying internal and external architectures is possible.[17] Additionally, as SLS uses materials in powder form, the precise control over pore interconnectivity, permeability, and stiffness of the scaffold can be achieved. The control of porosity arises from space between the individual granules of powder. The manipulation of these properties may optimize cell infiltration and adhesion, as well as mass transport of nutrients and metabolic waste throughout the scaffold. The SLS technique was recently exhibited to engineer bone tissue scaffolds using the common biodegradable polymer, poly(ϵ-caprolactone) (PCL).[18] Scaffolds were manufactured in a variety of designs and the biological properties of each were evaluated by seeding each with bone morphogenetic protein-7 (BMP-7) transduced human fibroblasts. The generated tissues were then studied using microcomputed tomography (μCT) scans and histology. The results showed that PCL scaffolds fabricated via SLS possessed

similar mechanical properties to trabecular bone and enhanced tissue in-growth *in vivo*.

It has been demonstrated that a derivative of the SLS approach could be used to fabricate bioactive scaffolds using a layer of carbon, a strong absorber of laser radiation, deposited on a biodegradable polymer, poly(D,L-lactic) acid (DLPLA).[19] In this experiment, near-infrared ($\lambda = 970$ nm) laser radiation, which DLPLA polymer particles do not absorb, was used to initiate the sintering process between a small quantity of carbon microparticles that were homogenously distributed on the surface of the DLPLA particles. This limited the melting process to the surfaces of each particle and possibly opens up the SLS process to a range of polymers that could not previously be processed by conventional SLS. Furthermore, as the laser melts only the surface of the particles, delicate bioactive species such as enzymes or growth factors trapped within each particle can retain their activity throughout the processing.

The SLS technique has also been realized to fabricate controlled-release drug-delivery devices.[20] A device composed of nylon was fabricated by selectively controlling the position and composition of the material. The drug model used for infiltrating the sample matrix was Fluka methylene blue dye. The results showed a gradient release profile and indicated that the release mechanism can be tailored to fit the specific application by controlling the device wall composition, anisotropy, and microstructure.

The SLS process also provides the ability to design and optimize subject-specific prototypes of orthotic devices that serve to compensate for a variety of injuries, disorders, and diseases. This was exhibited recently, where a patient suffered from a large frontal cranium defect after complications from a previous meningioma tumor surgery.[2] The surgery left the patient with a missing cranial section, which caused the patient's head to look slightly deformed. A corrective surgery would have required a titanium-mesh plate to be hand-formed by the surgeon during the operation using trial and error. Using a CT scan of the affected area, a 3-D CAD generated model could be made, and fabricated using SLS. The SLS mold could then be used to mechanically press the titanium-mesh plate to the anatomical profile of the missing cranium section. The SLS technique improved the fit of the implant and significantly reduced the operation time.

Some disadvantages of the SLS technique are surface roughness, which can influence the cellular behavior in a scaffold, and trapped powder, which can affect the scaffold stiffness. Another disadvantage of the SLS process is the smallest attainable feature size, which is primarily governed by powder particle size, focused laser beam diameter, and heat transfer in the powder bed.

3-D Printing

Three-dimensional printing (3-DP) was invented and patented at the Massachusetts Institute of Technology.

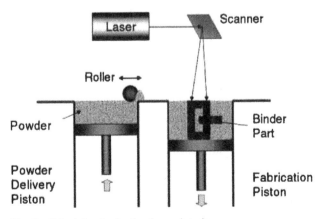

Fig. 3 Principle of selective laser sintering.

It was subsequently licensed and further developed by Z Corporation (Massachusettes, U.S.A.).

3-DP has demonstrated the capability of fabricating microstructures and controlling local composition with a high resolution in the interior of the component.[21] Similar to SL and SLS, a computer model (CAD) of the desired part is generated and a slicing algorithm draws detailed information for every layer. Just as in SLS, the process begins with a thin distribution of powder spread over the surface of a powder bed. Using a technology similar to ink-jet printing, a binder material selectively joins particles where the object is to be formed. A piston that supports the powder bed and the part-in-progress lowers so that the next powder layer can be spread and selectively joined. This layer-by-layer process is repeated until the part is completed (Fig. 4). The local microstructure within the component can be controlled either by changing the binder, which is printed, or by changing the printing parameters during the component construction. Although the system was initially designed for metallic or ceramic particles, powders of various polymeric biomaterials can be used directly. To print tissue engineering scaffolds, liquid solvent binder can be printed onto a powder bed of porogens (dissolvable particles that creates pores in a polymer matrix) and polymer particles. The solvent will dissolve the polymer and evaporate, and the polymer will reprecipitate to form solid structures. The final porosity is achieved after particulate leaching and solvent removal.

A combination of natural and synthetic biomaterials can be selectively bound using organic or water-based binders. Synthetic polymers include poly(lactic-co-glycolic acid) (PLGA), PCL, PLA, polyethylene oxide (PEO), with common organic binders such as chloroform, dichloromethane. Cornstarch, dextran, and gelatin have also been used in conjunction with water-based binders. Fine biomaterial powders are usually made by cryogenic milling with liquid nitrogen. The powders are then vacuum dried and sieved to obtain a uniform size.

Devices consisting of PCL and PEO were fabricated to demonstrate control of drug delivery profiles by controlling the position, composition, and microstructure.[21] The top and bottom layers of the device were constructed by binding PCL powder into thin solid layers. A cellular-type pattern was printed with PEO, which has a faster degradation rate, in the intermediate layers. Dyes, which represent drugs, were selectively placed within the cells manually.

The fabrication sequence is described as follows: first, a bead of PCL powder (45–75 μm) is spread in a thin layer on the piston. Droplets of 20% acid-modified PCL LPS-60HP (PCL-LPS)/chloroform binder solution are deposited on the PCL powder layer through a 45 μm orifice plate to form a solid PCL sheet. Poly(ε-caprolactone) particles bind to each other according to the pattern of the printhead motion as the binder contacts the PCL layer. Lines of PCL are formed by moving the printhead in a linear motion. A dense PCL sheet is formed by repeatedly printing lines directly adjacent to each other. The piston is then lowered and ready for the next layer of powder. The second layer begins by spreading PEO powders (75–150 μm) on the top of the solid PCL sheet. A square grid pattern of 5 cells by 5 cells can be made by printing six parallel lines at 3 mm interline spacing, and then repeating another set of six lines at 90° to the first set by rotating the piston. Twenty-five cells, which serve as drug reservoirs, are formed by 20 layers of PEO grid pattern. Different dyes are selectively dropped into the unbound PEO powder before a top layer of dense PCL sheet is printed to seal the cells. Different release mechanisms can be achieved by controlling device wall composition, anisotropy, and microstructure with this technique.

A unique, heterogenous, osteochondral scaffold, which could be an effective tool for treating articular defects, was developed using the 3-DP process. The scaffold consisted of two distinctive regions that differed in material composition, porosity, macroarchitecture, and mechanical properties. The cartilage region was 90% porous and composed of D,L-PLGA/L-PLA, with macroscopic staggered channels to facilitate homogenous cell seeding. The cloverleaf-shaped bone portion was 55% porous and consisted of a L-PLGA/calcium phosphate tribasic composite, designed to maximize bone ingrowth while maintaining critical mechanical properties. The transition region between these two sections contained a gradient of materials and porosity to prevent delamination. It was shown that chondrocytes preferentially attached to the cartilage portion of the device and cartilage formed during a 6-week *in vitro* culture period. The tensile strength of the bone region was similar

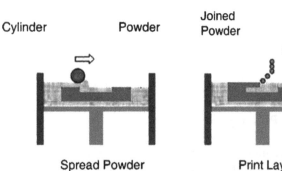
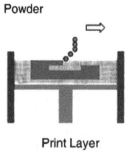
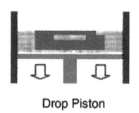

Cylinder   Powder   Joined Powder

Spread Powder   Print Layer   Drop Piston

Fig. 4 Principle of 3-D printing.

in magnitude to fresh cancellous human bone, suggesting that these scaffolds have desirable mechanical properties for *in vivo* applications, including full joint replacement.

The fabrication of complex-shaped scaffolds has also been demonstrated using an indirect approach of the 3-DP process, where molds are printed and the final materials are cast into the mold cavity.[22] To demonstrate the ability of this approach to build complex-shaped scaffolds from common biodegradable polymers with large pore size directly from medical imaging data, a zygomatic arch defect was fabricated. To evaluate the resolution that is possible with the indirect approach, scaffolds with small villi architecture were constructed. As the indirect approach involves the use of molding and mold release materials, the study addressed the removal of these additional materials by evaluating cell proliferation in the demolded scaffolds. To demonstrate the benefits of printing small features, cell attachment and proliferation in various regions of the scaffolds were also studied and showed cell growth in culture. This technique may be a useful adjunct for the fabrication of complex scaffolds for tissue engineering.

Postprocessing is necessary to remove the residual solvents and porogens, enhance the strength of the structures, or absorb functional molecules. Chlorinated solvents are undesirable because they may bring potential health risks. Those solvents are difficult to extract completely. An extraction method based on liquid CO_2 has been developed.[23] Using water as a solvent eliminates the extraction procedure.

The as-solidified structures are vacuum dried at an elevated temperature to enhance the structural rigidity. Chemical modification may be performed, for instance, by infiltrating polymer structures with functional molecules to adjust the hydrophilicity.

In general, 3-DP is a fast and accurate process. It has precise control over composition, anisotropy, and microstructure. A wide range of materials can be used, provided they can be processed into powder form. It is possible to incorporate biomolecules, which may be carried by the binders, in the structures. However, the surface of the structures is usually powdery and the trapped powder is hard to completely remove.

Fused Deposition Molding

FDM is a widely used RP process that was developed by Scott Crump and commercialized by Stratasys (Minnesota, U.S.A.).

In this method (see Fig. 5) a spool of material is heated into a semiliquid state and dispensed into an extrusion head. The semiliquid material is extruded through the head and deposited into the desired shape. The platform, onto which the material is deposited, is maintained at a lower temperature so that the material cures immediately on contact after exiting the heated chamber. After the completion of each layer, the platform lowers, and the extrusion head deposits the next layer on top of the previous one until the

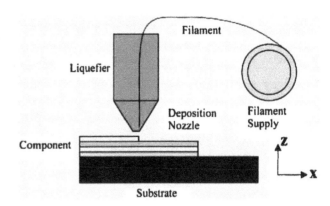

Fig. 5 Principle of fused deposition molding.

3-D model is completed. In newer FDM machines, two materials are fed through the extrusion head. One is used to produce the geometric model and the other is used to construct support structures. The second material forms a bond with the primary model material and is washed away when the 3-D model is processed. The advantages of using FDM include the speed and safety of the machine.

FDM has been demonstrated as a viable method for manufacturing functional scaffolds for tissue engineering. Tissue-engineering scaffolds with fully interconnected pore networks based on PCL were fabricated using FDM.[24] The scaffolds were then seeded with human fibroblasts from the anterior cruciate ligament of a 34-year-old male. The 3-D structures showed excellent cell proliferation, differentiation, and extracellular matrix production. An adapted FDM method was recently applied in a similar fashion to create scaffolds. This method used a microsyringe to expel a dissolved polymer under constant pressure to form a desired pattern. Using this microextrusion approach, resolution down to 20 μm can be achieved.[25] A modified FDM method produced PLGA scaffolds with a lateral resolution of up to 5 μm. The layer-by-layer fabrication method allows the design of a pore morphology that varies across the structure.[26] Researchers have also used FDM to manufacture composite scaffolds of PCL, PCL-Tricalcium phosphate (TCP) and PCL–hydroxyapatite (HA). The scaffolds were seeded with human mesenchymal progenitor cells and proliferation toward and onto the surfaces were detected.

Another version of FDM uses a microsyringe to construct 3-D heterogenous multilayer tissue-like structures inside microchannels.[27] The process involves a three-step approach: 1) immobilization of a cell–matrix assembly; 2) cell–matrix contraction; and 3) pressure-driven microfluidic delivery using a syringe. Using this approach, patterning of biological microstructures can be used to form "neotissue" from different types of cells and biopolymer components in one single system. Thus, the approach provides a novel solution to fabricate hybrid biopolymers and hierarchical tissue structures for tissue engineering and basic cell biology.

Neural—Rapid Prototyping

Table 1 Comparison of different RP technologies in biomedical applications

RP system	Resolution (μm)	Material	Strength	Weakness	Reference
Stereolithography	100	PEG-acrylate	Biofriendly liquid environment; control over reaction; high resolution; can incorporate biomolecules and cells in situ	Relatively slow processing; UV exposure may cause degradation	[10–17]
Selective laser sintering	400	PCL, PEEK–HA	Fast processing; control over porosity; no postprocessing	High temperature; material must be in powder form; rough surface finish; trapped powder	[17–21]
3-D printing	100	PLGA, PEO, starch-based polymer	Fast processing; nonorganic binder is possible; no supporting structure needed; control over porosity	Material must be in powder form; rough surface finish; trapped powder	[17,22]
Fused deposition molding	250	PCL, PCL–TCP	Can be operated in office-type environment; good mechanical strength	High temperature; simple geometry can be obtained	[23–26]

Abbreviation: PEEK, polyetheretherketone.

FDM provides the unique ability to use almost any material but is limited by its narrow processing window, shrinkage, and high operating temperature, which could prevent the inclusion of biomolecules into the material. The approach also requires the input material in a particular size and geometry. (For a comparison of different RP technologies in biomedical applications see Table 1.)

CONCLUSIONS

In addition to the aforementioned techniques, other plausible approaches for developing medical devices are available. One such method was recently developed as a means to create biodegradable microfluidic devices, using a rational scaffold design.[28] The novel approach used a thermal fusion bonding technique to rapidly fabricate 3-D monolithic microdevices with various internal geometries. The researchers stacked micropatterned thin films of the common biodegradable polymer PLGA, to create multilayer, microfluidic networks with extruded inlets and outlets without leaks and occlusions. The method also presents a viable platform for generating drug delivery vehicles with microscale cavities, capsules and microchannels for controlled-release locations, species and kinetics in complex biomolecular environments.

Another excellent study developed a technique derived from conventional integrated circuit technologies to photopattern hydrogels.[29] The method began with a polytetrafluoroethylene (Teflon) base filled with polymer solution loaded with cells. Ultraviolet light was then shone through a patterned template atop the thin film, curing the exposed polymer that was seeded with cells. Using this layer-by-layer method, the construction of complex networks of cells and materials can be built by using various templates and depositing different layers atop one another.

Ultimately, RP platforms serve as an invaluable tool to achieve exciting new functionality in the fields of microfluidics, cell-support scaffold development, and controlled drug-release devices. The techniques are rapid, versatile, and low-cost approaches to precision manufacturing, thus making them ideal for large-scale engineering of complex, high-performance medical devices.

ACKNOWLEDGMENT

The authors appreciated financial support from the U.S. National Science Foundation (DMI 0093364).

REFERENCES

1. Chua, C.K.; Leong, K.F. *Rapid Prototyping: Principles and Applications in Manufacturing*; John Wiley: Singapore, 1997.
2. Chua, C.K.; Leong, K.F.; Lim, C.S. *Rapid Prototyping: Principles and Applications*; World Scientific: Singapore, 2003.
3. Beaman, J.J.; Barlow, J.W.; Bourell, D.L.; Crawford, R.H.; Marcus, H.L.; McAlea, K.P. *Solid Freeform Fabrication: A New Direction in Manufacturing*; Kluwer: Dordrecht, 1997.
4. Liew, C.L.; Leong, K.F.; Chua, C.K.; Du, Z. Dual material rapid prototyping techniques for the development of biomedical devices. Part 1: Space creation. Int. J. Adv. Manuf. Technol. **2001**, *18* (10), 717–723.

5. Adachi, J.; Hara, T.; Kusu, N.; Chiyokura, H. Surgical simulation using rapid prototyping. Proceedings of the 4th International Conference on Rapid Prototyping, Dayton, Ohio, June 14–17, 1993; 135–142.

6. Mahoney, D.P. Rapid prototyping in medicine. Computer Graphics World **1995**, *18* (2), 42–48.

7. Jacobs, A.; Hammer, B.; Niegel, G.; Lambrecht, T.; Schiel, H.; Hunziker, M.; Steinbrich, W. First experience in the use of stereolithography in medicine. Proceedings of the 4th International Conference on Rapid Prototyping, Dayton, Ohio, June 14–17, 1993; 121–134.

8. Swaelens, B.; Kruth, J.P. Medical applications in rapid prototyping. Proceedings of the 4th International Conference on Rapid Prototyping, Dayton, Ohio, June 14–17, 1993; 107–120.

9. Clinkenbeard, R.E.; Johnson, D.L.; Parthasarathy, R.; Altan, M.C.; Tan, K.H.; Park, S.M.; Crawford, R.H. Replication of human tracheobronchial hollow airway models using a selective laser sintering rapid prototyping technique. J. Am. Ind. Hygiene Assoc. **2002**, *63* (2), 141–150.

10. Cho, D.W.; Lee, I.H. An investigation on photo-polymer solidification considering laser irradiation energy in micro-stereolithography. Microsyst. Technol. **2004**, *10* (8–9), 592–698.

11. Cooke, M.N.; Fisher, J.P.; Dean, D.; Rimnac, C.; Mikos, A.G. Use of stereolithography to manufacture critical-sized 3D biodegradable scaffolds for bone ingrowth. J. Biomed. Mater. Res. B Appl. Biomater. **2002**, *64* (B), 65–69.

12. Zhang, X.; Jiang, X.N.; Sun, C. Micro-stereolithography of polymeric and ceramic microstructures. Sens. Actuators **1999**, *77* (2), 149–156.

13. Cabral, J.T.; Hudson, S.D.; Harrison, C.; Douglas, J.F. Frontal photopolymerization for microfluidic applications. Langmuir **2004**, *20* (23), 10020–10029.

14. Fisher, J.P.; Dean, D.; Engel, P.S.; Mikos, A.G. Photoinitiated polymerization of biomaterials. Annu. Rev. Mater. Res. **2001**, *31* (1), 171–181.

15. Mapili, G.; Lu, Y.; Chen, S.C.; Roy, K. Laser-layered microfabrication of spatially patterned, functionalized tissue engineering scaffolds. J. Biomed. Mater. Res. *in press*.

16. Bryant, S.J.; Nuttelman, C.R.; Anseth, K.S. Cytocompatibility of UV and visible light photoinitiating systems on cultured NIH/3T3 fibroblasts *in vitro*. J. Biomater. Sci. Polym. Ed. **2000**, *11* (5), 439–457.

17. Yeong, W.Y.; Chua, C.K.; Leong, K.F.; Chandrasekaran, M. Rapid prototyping in tissue engineering: Challenges and potential. Trends Biotechnol. **2004**, *22* (12), 643–652.

18. Williams, J.M.; Adewunmi, A.; Schek, R.M.; Flanagan, C.L.; Krebsbach, P.H.; Feinberg, S.E.; Hollister, S.J.; Das, S. Bone tissue engineering using polycaprolacone scaffolds fabricated via selective laser sintering. Biomaterials **2005**, *26* (23), 4817–4827.

19. Antonov, E.H.; Bagratashvili, V.N.; Whitaker, M.J.; Barry, J.J.A.; Shakesheff, K.M.; Konovalov, A.N.; Popov, V.K.; Howdle, S.M. Three-dimensional bioactive and biodegradable scaffolds fabricated by surface-selective laser sintering. Adv. Mater. **2005**, *17* (3), 327–330.

20. Cheah, C.M.; Leong, K.F.; Chua, C.K.; Low, K.H.; Quek, H.S. Characterization of microfeatures in selective laser sintered drug delivery devices. Proc. Inst. Mech. Eng. H: J. Eng. Med. **2002**, *216* (H6), 369–383.

21. Wu, B.M.; Borland, S.W.; Giordano, R.A.; Cima, L.G.; Sachs, E.M.; Cima, M.J. Solid free-form fabrication of drug delivery devices. J. Control. Release **1996**, *40* (1–2), 77–87.

22. Lee, M.; Dunn, J.C.D.; Wu, B.M. Scaffold fabrication by indirect three-dimensional printing. Biomaterials **2005**, *26* (20), 4281–4289.

23. Sherwood, J.K.; Riley, S.L.; Palazzolo, R.; Brown, S.C.; Monkhouse, D.C.; Coates, M.; Griffith, L.G.; Landeen, L.K.; Ratcliffe, A. A three-dimensional osteochrondral composite scaffold for articular cartilage repair. Biomaterials **2002**, *23*, 4739–4751.

24. Hutmacher, D.W.; Schantz, T.; Zein, I.; Ng, KW.; Teoh, S.H.; Tan, C.K. Mechanical properties and cell cultural response of polycaprolactone scaffolds designed and fabricated via fused deposition modeling. J. Biomed. Mater. Res. **2001**, *55* (2), 203–216.

25. Vozzi, G.; Flaim, C.; Ahluwalia, A.; Bhatia, S. Fabrication of PLGA scaffolds using soft lithography and microsyringe deposition. Biomaterials **2003**, *24* (14), 2533–2540.

26. Zein, I.; Hutmacher, D.W.; Tan, K.C.; Teoh, S.H. Fused deposition modeling of novel scaffold architectures for tissue engineering applications. Biomaterials **2002**, *23* (4), 1169–1185.

27. Tan, W.; Desai, T.A. Layer-by-layer microfluidics for biomemitic three-dimensional structures. Biomaterials **2004**, *25* (7–8), 1355–1364.

28. King, K.R.; Wang, C.C.J.; Kaazempur-Mofrad, M.R.; Vacanti, J.P.; Borenstein, J.P. Biodegradable micro-fluidics. Adv. Mater. **2004**, *16* (22), 2007–2012.

29. Valerie, A.L.; Sangeeta, N.B. Three-dimesional photo-patterning of hydrogels containing living cells. Biomed. Microdev. **2002**, *4* (4), 257–266.

BIBLIOGRAPHY

1. Curtis, A.; Wilkinson, C. Nanotechniques and approaches in biotechnology. Trends Biotechnol. **2001**, *19* (3), 97–101.

2. LaVan, D.A.; Mcguire, T.; Langer, R. Small-scale systems for *in vivo* drug delivery. Nat. Biotechnol. **2003**, *21* (10), 1184–1191.

3. Lu, Y.; Chen, S.C. Micro and nano-fabrication of biodegradable polymers for drug delivery. Adv. Drug Deliv. Rev. **2004**, *56* (11), 1621–1633.

4. Yeong, W.Y.; Chua, C.K.; Leong, K.F.; Chandrasekaran, M. Rapid prototyping in tissue engineering: Challenges and potential. Trends Biotechnol. **2004**, *22* (12), 643–652.

Neural—Rapid Prototyping

Rapid Prototyping: Tissue Engineering

Giovanni Vozzi
Arti Ahluwalia
E. Piaggio Interdepartmental Research Center, Department of Chemical Engineering, University of Pisa, Pisa, Italy

Abstract

This entry provides an overview of rapid prototyping methods for biomaterial fabrication with the aim of illustrating the basic principles behind the scaffold-based tissue engineering. It describes aspects which must be taken into consideration when realizing a scaffold, such as the resolution time of manufacture ratio, materials employed, and scaffold geometrical design.

INTRODUCTION

Tissue engineering can be defined as the development of biological substitutes to restore, maintain, or improve tissue functions and is based on the application of principles and methods of engineering and life sciences toward a fundamental understanding of structure-function relationships in normal and pathological mammalian tissues. This is an emerging interdisciplinary area of research and technology that has the potential of revolutionizing our methods of health care treatment and dramatically improving the quality of life for millions of people throughout the world. Several approaches to tissue engineering have been established of which the most common approach is based on scaffold-guided tissue formation *in vitro*. A scaffold is a biodegradable and biocompatible three-dimensional (3-D) porous structure, which can support cell adhesion and growth. Typically cells are seeded onto biodegradable polymeric scaffolds, and the constructs in some way reform the intrinsic tissue structures.[1] The scaffold approach to tissue engineering first emerged in the early 1990s, and since then much of the work has focused on scaffolds in the form of sponges or foams, which possess a random microstructure. More recently as a result of a hypothesis that tissues must have a prearranged spatial architecture in order to function correctly, much attention has shifted to designer scaffolds. This hypothesis is based on the fact that a large number of organs possess a well-defined internal structure, which is essential to their function. Thus, in order to realize 3-D constructs for tissue repair and reconstruction, it is necessary to guide cell growth on structures with an established topology, which replicates that of natural tissue. As a result, several different microfabrication techniques for biomaterial scaffolds have been developed. Most of these techniques arise from preexisting manufacturing methodologies such as photolithography, silicon micromachining, and the realization of pseudomechanical microcomponents. We can distinguish two groups of methodologies: one based on the realization of two-dimensional (2-D) structures and the other on the fabrication of 3-D scaffolds. In this entry, we focus exclusively on 3-D methods borrowed from the well-established field of computer-aided design/computer-aided manufacturing (CAD/CAM) and rapid prototyping (RP).

MICROFABRICATION OF THREE-DIMENSIONAL STRUCTURES: RP

RP, which is synonymous with solid free-from fabrication (SFF), refers to the fabrication of 2-D or 3-D structures using a preprogrammed computer graphics file containing layer-by-layer maps of the structures. Figure 1 schematizes this structure in terms of the liver, one of the most complex organs in the body. These maps are reconstructed through computer-aided fabrication (usually an *x-*, *y-*, *z*-positioning system), much as a 3-D printer would do. Each layer is about 0.01–1 mm thick (this depends very much on the manufacturing process as well as the resolution required), and a 3-D object is built-up through the assembly of successive 2-D layers, or what we call a pseudo-3-D (PS3D), sometimes also known as 2 1/2 D. Nowadays RP is the chosen method of production for manufacturing complex objects, surpassing traditional techniques such as milling and turning. However, it should be noted that rapid is a relative term, and it usually takes several minutes to several hours to produce an object.

The RP process can be subdivided into three main units:

- Generation and conversion of the CAD model
- Realization of the prototype
- Postprocessing

In the first phase, a 3-D object is decomposed into a stratified structure using appropriate software, such as AutoCAD. The layers are then codified into step-by-step instructions for the control system, which drives the

Concise Encyclopedia of Biomedical Polymers and Polymeric Biomaterials DOI: 10.1081/E-EBPPC-120052535

x-, *y*-, *z*-positioner. Currently, RP for tissue engineering scaffolds is associated with the use of CAD design files originating from computerized tomography (CT) data. Their most appropriate application is in the bone tissue engineering, where high-resolution micro-CT can reveal structural features of the order of 5 μm. In the case of soft tissues, architecture is not only harder to define, but also more difficult to image with high resolution, because information on structure is reconstructed largely from histological specimens. Conversion of histological images into a layer-by-layer and step-by-step scaffold is hindered by the lack of contrast and nonspecificity of stains (such as eosin and hematoxylin). It is very difficult to identify cell contours and separate them from the contents of the extracellular matrix (ECM). Soft-tissue scaffolds are therefore generally constructed with a repetitive grid consisting of squares, triangles, or hexagons. In fact, one of the most challenging aspects of scaffold design for nonbony biological tissues is the extraction of structural features, and the conversion of these into a repetitive algorithm describing an appropriate locus of points or lines in a given plane.

Once the architecture has been chosen, the three-axis positioners proceed to the actual fabrication of the structure, following the location maps provided by the controller. This process can be quite complicated, since binders, powders, and fillers are usually involved.

The final phase can be described as a cleaning and polishing phase, where the waxes or fillers are melted or dissolved away. These three steps usually vary for each method and will be illustrated individually for the description of the tissue engineering RP application.

As illustrated in Fig. 2, several types of RP methods are available for tissue engineering, and all but one of them have been adapted from existing manufacturing technologies. Almost all RP methods are based on PS3D except those constructs that are poured into 3-D sacrificial molds, which are then destroyed during postprocessing. In this case, the sacrificial mold is inevitably manufactured through PS3D methods.

We will use the classification suggested by Yeong et al.,[2] which subdivides RP methods into PS3D solution or fluid-based systems and printing head and powder-based fabrication, in addition a third class is added, which is based on the use of sacrificial molds. Pressure-assisted microsyringe (PAM), ink-jet organ printing, and fused deposition modeling (FDM) are some of the processes that use solutions. RP methods such as 3-D printing (3-DP) and laser sintering fall into the second class of fabrication, which relies on the use of a printing head emitting a binder such as light, heat, or a solvent.

MATERIALS USED FOR TISSUE ENGINEERING SCAFFOLDS

One of the most critical aspects of tissue engineering is the choice of biomaterial. Typically biomaterials fall into two main categories: biological polymers and synthetic polymers. Biological polymers are rarely used in RP because of their delicate nature; they denature easily, are difficult to sterilize, and often do not possess adequate mechanical properties to allow the creation of freestanding porous 3-D structures. However, some reports using collagen containing RP scaffolds have been published. In particular,

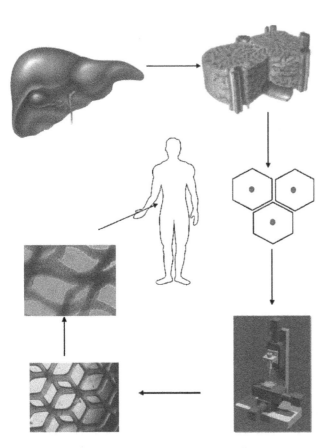

Fig. 1 Rapid prototyping scheme: CAD/CAM SFF.

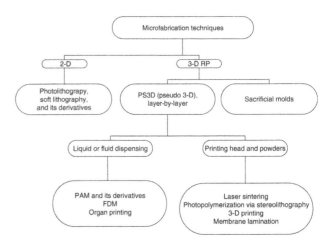

Fig. 2 Classification of RP methods.

Sachlos et al.[3] have used reconstituted collagen composites, which were poured in the rapid-prototyped sacrificial molds. Biological hydrogels such as gelatin and alginate have also been employed, despite the fact that their floppiness makes them difficult to manage and assemble, and in general the resolutions obtained are low.[4,5] Synthetic polymers are, however, the preferred material for most systems. The most commonly used synthetic polymers for the realization of 3-D scaffolds for tissue engineering are the polyesters—polylactide (PLA), lactide/glycolide co polymers (PLGA), and polycaprolactone (PCL). Often synthetic polymers are incapable of supporting an adequate degree of cell adhesion and must be surface treated or modified with appropriate ligands. These synthetic polymers, also perform poorly in mechanical terms, being too compliant for bone tissue but too rigid for soft tissue. It is therefore becoming increasingly common to use composites or blends, and hydroxyapatite is very often used to render synthetic polymers more rigid for bone engineering applications.[6] The positive trend toward the employment of hybrid materials with different chemical and mechanical properties should result in major improvements in this field.

These materials are either used in powder or solid form, as in the case of laser sintering, 3-DP, or FDM or as solutions as with the PAM or photopolymerization. Obviously different materials are suited to different methods, and some methods are limited to a very narrow range of polymers, particularly in the case of photopolymerization and ink-jet organ printing. Table 1 lists some of the different materials employed in RP for tissue engineering.

RESOLUTION AND RESOLUTION/TIME OF MANUFACTURE RATIO AND GEOMETRY

Table 1 clearly shows that RP methods have a wide range of resolutions ranging from 5 to 1000 μm. Prior to discussing the issue of cell response to spatial patterns or geometry and topography, it is necessary to define these terms in the context of this entry. We define geometry as the 2-D or 3-D spatial organization of cells, referring to spatial dimensions over five times greater than that of the cell. Topography, on the other hand, can be defined as surface features and relief with dimensions of the order of cell size. Thus topography is local, whereas geometry is global. Geometry

Table 1 A comparison of well-known microfabrication methods, their resolution, and limits

Technique	Materials used	RTM ratio (cm²/min)	Resolution (μm)	Cells used	Limits
Membrane lamination	Bioerodable polymers (PLA, PLGA, etc.), bioceramics	Low (<1)	1000	Osteoblasts	Structures not really porous, low resolution
3-D printing	Bioerodable polymers (PLA, PLGA, etc.), hydroxyapatite	Medium (about 1)	300	Various types	Presence of polymeric grains and of excess solvent
Laser sintering	Calcium phosphates, polymers (PLA, PLGA, etc.)	Medium to high	<400	Osteoblasts	Presence of polymeric grains and of excess solvent
Photopolymerization	Photopolymeric resins	0.5 (medium)	250	Osteoblasts	Use of photo-sensitive polymers and initiators, which may be toxic
Fused deposition modeling	Bioerodable polymers (PLA, PLGA, etc.)	7 (very high)	200	Various types	Limited to nonthermolabile materials. Layered structure is very evident
Sacrificial molds— multiphase ink-jet printing	Bioerodable polymers (PLA, PLGA, etc.), collagen	0.1 (low)	300	Various types	Complex to realize, build materials limited, low fidelity
Pressure-assisted microsyringe	Bioerodable polymers (PLA, PLGA, etc.), gels (alginate)	1 (medium)	5–10	Neurons, endothelial cells, fibroblasts, hepatocytes, muscle	Highly water soluble materials cannot be used. Extrusion head is very small
Organ printing	Cells and thermoreversible gels	Medium (about 1)	100	Hepatocytes	Limited range of gels available
Bioplotter	Gels and polymers	Medium	50	Various types	Gels have low fidelity

is easier to control and design using RP whereas topography is harder to define and is a dynamic feature in tissue engineering because cells will remodel a substrate.

It has been amply demonstrated that cells react to the topography of their substrate on which they are seeded.[7] Furthermore it is important to note that characteristic topographical dimensions may vary significantly from one cell type to another. In addition, the adhesion and motility of cells may be enhanced by their contact with a surface with topological features. It has also been shown that the dimensions and aspect ratios of structures can affect the reaction of cells to a marked extent, but despite several studies, we are still far from a full understanding of the events involved.[8] However, much effort is being directed toward the study of the phenomena that guide cell reaction to the geometry of the substrate on which they are seeded, and also the reason why some cell functions are either enhanced or suppressed according to the spatial patterns or geometry.[9] The scaffold-based tissue engineering is founded on the principle that the microstructural environment of a cell undoubtedly conditions its behavior. However, there are no hard-and-fast rules as to which type of geometry or topography is most suitable for this process, and as mentioned in Section "Microfabrication of Three-Dimensional Structures: RP," there are no algorithms or rules to define soft-tissue microarchitecture.

At present there is no RP method, which enables scaffolds with a precision of less than a few microns to be fabricated, and it is likely that smaller features would necessitate elimination of the term rapid, since time is sacrificed at the expense of resolution. We can define the resolution/time of manufacture ratio (RTM ratio) as the maximum resolution (expressed as the inverse of minimum feature length, d) divided by the time (t) required to realize a unit volume (V) of scaffold. The RTM ratio is then

$$\text{RTM} = \frac{1/d}{t/V}$$

and has been quantified where possible in Table 1. The higher this number is, the more efficient is the RP method, and the better it lends itself to a high throughput scaffold production. Here we have only used the manufacture time without considering solvent extraction, sterilization, cell seeding, and proliferation times since these times will vary greatly with the selected application.

RP methods can also be characterized in terms of their fidelity, that is, the match between structural features of the scaffolds and the actual CAD design. Obviously no technique has 100% fidelity; in particular, methods that involve melting, swelling, or the use of solvents will show deviations from the input design where usually the scaffold features are larger and less defined than the specified features.

In our experience, the optimum resolution for RP is the order of cell dimensions, or a few tens of microns.[10]

In most tissues the functional element is only a few cell diameters,[11] so any architectural dimensions greater than this diameter would probably comprise spatial control of cell organization, which is the underlying philosophy behind scaffold-based tissue engineering.

FLUID-BASED RP MICROFABRICATION

Fluid-based RP methods use a solution or melt of polymer, which is extruded through a syringe mounted on an arm or on the z-axis of a 3-D micropositioner. The material of the upper layer is bonded to that of the lower layer in order to obtain a 3-D scaffold. Often an intermediate supporting layer is required to avoid collapse of the structure during fabrication, which is then sacrificed in the postprocessing phase.

This group includes all the PAM-like methods, FDM and its variants, fiber spinning, and different types of ink-jet-based organ printing. The resolution is generally a function of the viscosity of the polymeric solution extruded as well as the diameter of the deposition head.

Pressure-Assisted Microsyringe System

The pressure-assisted microsyringe (PAM) technique, developed at the Interdepartmental Research Center "E. Piaggio" at the University of Pisa, is based on the use of a microsyringe that allows the deposition of a wide range of polymers, as well as hydrogels.[12] The system consists of a stainless-steel syringe with a 10–20 µm glass capillary needle, as shown in Fig. 3. A solution of the polymer, in a volatile solvent, is placed inside the syringe and expelled from the tip by the application of filtered, compressed air. The syringe is mounted on the z-axis of a three-axis micropositioning system, which was designed and built in-house. A supporting substrate, usually glass, is placed on the two horizontal motors and is moved relative to the syringe. The control software is developed in C++ (a programming language) with a user-friendly graphical interface and allows a wide range of patterns with a well-defined geometry to be designed and deposited. Within horizontal plane, the resolution of PAM-fabricated parts is 5–600 µm. This resolution depends on the pressure applied to the syringe, the viscosity of the solution, the motor speed, and the dimensions of the syringe tip. After the first layer has been deposited, subsequent layers are deposited by moving the syringe up along the z-axis by an amount corresponding to the height of each layer. To avoid collapse of overhanging structures during the fabrication process, a water-soluble polymeric spacer of about 10–20 µm is deposited between layers. Different layers can be built from different polymers in different patterns, adding to the flexibility of the technique. Once the microfabrication steps are completed, the scaffolds are rinsed with water to remove the polymeric spacer for creating interspaces

Neural—Rapid Prototyping

where the cells can penetrate and adhere to initiate the proliferation process. Typically it takes about 30 s to realize a 2-D layer of 1 cm × 1 cm. In the case of 3-D scaffolds, the postprocessing phase requires about 6–24 hr of rinsing. Examples of the scaffolds realized with PAM are shown in Fig. 4.

The PAM method has the highest resolution among the RP methods and has been adapted by several research groups.[5] Its main drawbacks are the low vertical dimensions (due to the high resolution) and as a result a fairly low RTM ratio, which implies that thick constructs take over an hour to fabricate. Furthermore, the incorporation of the small particles such as hydroxyapatite and nanotubes requires a capillary needle with a larger exit diameter to avoid clogging of the syringe, so precluding high resolutions.

Fused Deposition Modeling

FDM is an RP process that extrudes polymeric materials in molten form through the use of a heated deposition head developed at the National University of Singapore (NUS). The system is composed of a microcontrolled mechanical arm, which moves in the horizontal plane on which the extrusion head is positioned. A filament of polymer is drawn into the head through the use of rollers, where it is heated to near fusion. The semimolten extrudate is driven through a needle having a diameter of 1.27 mm onto a substrate mounted on a platform, and solid

objects are built string-by-string as the platform moves down (Fig. 5). The FDM-fabricated structures have a lateral resolution of a few hundred microns.[13] Several improvements to the FDM system have been reported by the NUS group, in particular, the extrusion head has been modified to enable incorporation of particles or granules rather than filaments and hence renamed as precision extrusion deposition and precision extrusion manufacturing (PED and PEM).[2]

The main limitation of this technique is its resolution since many cell types are smaller than 50 μm. Small scaffolds are difficult to produce because previous layers may melt during heating, whereas larger structures tend to split or peel along layers due to insufficient bonding. Furthermore, because the extrusion is forced by a heating system, this method cannot be used with thermolabile materials. On the other hand, it has the highest RTM ratio of all RP methods because the postprocessing phase is practically nonexistent as there are no intervening layers or binders and solvents to remove as with most other techniques.

Organ Printing

Mironov et al.[14] developed an ink-jet printer that can print gels, single cells, and cell aggregates. This method, which is known as organ printing, relies on the use of a thermosensitive gel to generate sequential layers for cell printing. This nontoxic biodegradable gel is a liquid below 20°C and solidifies above 32°C. Living cells are sprayed onto the solidified thin layers of the thermoreversible gel, which also serves as printing paper. The printers used are

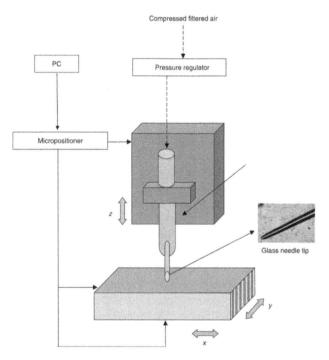

Fig. 3 Block diagram of the pressure-assisted microfabrication (PAM) system. The inset illustrates the capillary needle with an inner diameter of 20 μm.

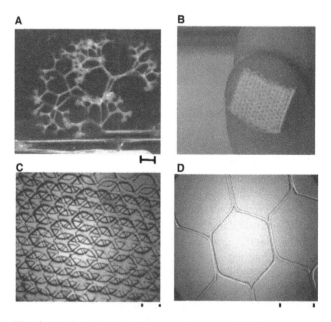

Fig. 4 Typical (**A**) fractal (scale bar 1 mm), (**B**) 3-D hexagonal, (**C**) 3-D hexagonal detail (scale bar 500 μm), and (**D**) high-resolution 2-D hexagonal (scale bar 500 μm) scaffolds of PLGA obtained by PAM.

old commercial printers adapted by washing out the ink cartridges and refilling them with cell suspensions. The software that controls the viscosity, electrical resistances, and temperature of the printing fluids has been reprogrammed and the feed systems altered. As defined by Mironov et al.,[14] organ printing includes the many different printer designs and components of the deposition process, for example, jet-based cell printers, cell dispensors or bioplotters, the different types of 3-D hydrogels, and varying cell types. The concept is schematized in Fig. 6.

With this method the resolution is about 100 μm, and in fact several cell aggregates are formed within the drops. Due to the use of an ink-jet head, cells may be damaged during spraying due to the high shear stress by the piezoelectric actuator, and obviously the choice of materials is also restricted to a narrow range of biocompatible thermoreversible hydrogels.

PRINTING HEAD AND POWDER-BASED MICROFABRICATION

In this section, we include all the techniques where the polymeric powders or liquids are placed on the *x–y* plane of the system, and the working head (laser, ink-jet head, etc.) realizes microstructures in which the resolution is a function of the grain dimensions or of the printing head beam or drop size.

Membrane Lamination

Membrane lamination consists of the realization of membranes having a thickness between 500 and 2000 μm. They are then cut by a laser beam or scissors in the form of predetermined, 2-D shapes. Once the basic elements are ready, the structure is assembled layer-by-layer using micro-controlled and microactuated system controlled by CAD/CAM software. Lamination is obtained by wetting an absorbent material such as paper, cloth, or a sponge with an organic solvent. Light pressure is applied to the exposed surface of each membrane for a sufficient amount of time to wet each surface. The resulting bilayer-laminated structure is gently compressed to ensure sufficient adhesion between the wet surfaces of the first and the second membranes. This procedure is repeated with the bilayer or resulting multilayer-laminated structure until the desired 3-D shape is obtained. The final laminated structure is dried to ensure complete solvent evaporation.[15] Since the resolution of this method is insufficient for most tissue engineering applications, it has been replaced by more precise techniques.

Three-Dimensional Printing

Originally developed for rapid-prototyping ceramics and metal objects such as molds for investment casting, three-dimensional printing (3-DP) has since been applied to the fabrication of drug delivery devices and tissue scaffolds from biomedical polymers. It allows simultaneous fabrication and surface modification of 3-D devices. A schematic description of the 3-DP process is shown in Fig. 7.

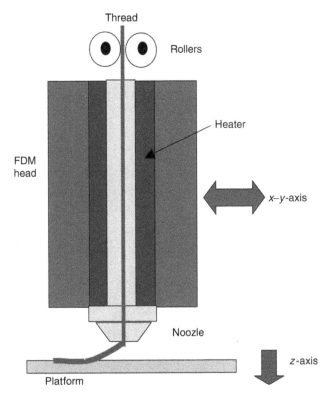

Fig. 5 Schematic representation of fused deposition modeling.

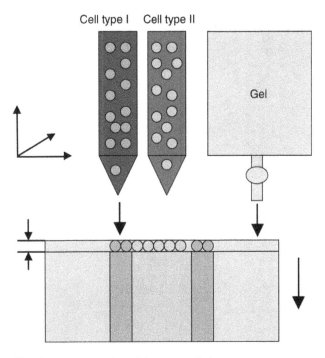

Fig. 6 A representation of the organ printing concept.

A thin layer of powder (0.05–0.20 mm) is first spread on the top of a piston. An ink-jet printhead then prints a liquid binder onto the layer wherever the particles are to be bonded together to form the solid part of the object. Colloidal silica is commonly used as a binder in ceramic systems, while organic solvents such as chloroform, which swell and partially dissolve individual powder particles, are suitable for polymer systems.[16]

Each printed droplet is 50–80 μm in diameter, and a CAD/CAM program controls its position. After the first layer is printed, the piston is dropped, and the sequential steps of powder spreading and printing are repeated. Scaffolds possessing complex internal features can be created because during the building process, the powder bed allows the formation of channels and overhangs as supports for objects. Furthermore, multiple printheads containing different solutions can be readily employed to modify local surface chemistries and compositions. At the end of the process, the scaffold is lifted from the powder bed, and the loose powder is removed. The resulting product is a solid, continuous 3-D structure with well-defined internal microarchitectures. The resolution of features is approximately 300 μm. Interconnected pores can be generated in the walls of polymer devices by inclusion of a porogen in the powder.[17] Therics Inc. uses a 3-DP system to produce bone fillers composed of PLGA and tricalcium phosphate (TCP), the first example of an RP commercial product specifically for tissue engineering.

With 3-DP it is often difficult to remove excess polymer grains, and the residual amount of solvent is only reduced by 10–15% in weight. As mentioned, this system allows the realization of polymeric structures with a lateral resolution of approximately 0.3 mm, which is suboptimal if precise control of cell positioning is required. Its RTM ratio is about 1, thus placing it at the high end of manufacturing efficiency.

Laser Sintering

This microfabrication system uses a laser beam in order to realize polymeric microstructures. A polymer powder layer is placed on the substrate, and an infrared laser beam locally raises a given spot to the temperature of glass transition, thereby fusing the polymeric grains. Once a layer has been appropriately patterned, the substrate is moved down, and a new polymer powder layer is applied. Excess polymer powder is left in loco during the fabrication process because it prevents collapse of overhangs, but then it can be rather difficult to remove during the postprocessing phase.

CAD/CAM software is employed for designing the structures layer-by-layer and for controlling the laser beam and the motion of the micropositioner where the laser beam and substrate for deposition are mounted, as schematized in Fig. 8.[18]

This technique is typically used to realize bone implants with calcium phosphate powders or ceramics but has also been successfully employed for polymers and gels such as PCL and polysaccharides.[19] One of the main advantages of laser sintering is the ability to incorporate particles such as hydroxyapatite into the scaffold with minimum effort where the particles are simply added to the polymer, which fuses around them.[6]

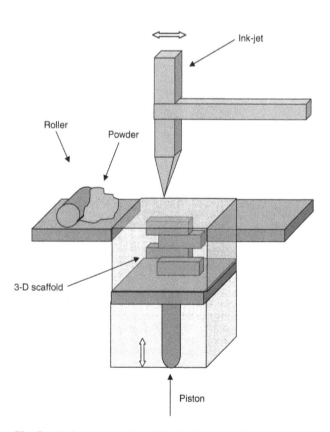

Fig. 7 Basic scheme of the 3-D printing method.

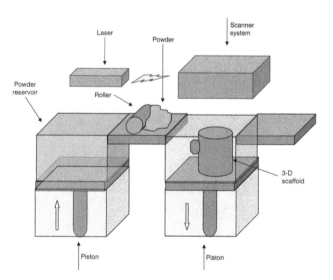

Fig. 8 The laser-sintering process.

In laser sintering, the resolution of the system depends on laser spot size, thermal conductivity and absorption of the polymer, and the grain size. A commercial device, Sinter Station 2500 (DTM, USA), has a laser beam with a 400 µm spot. Heat conduction inevitably makes the precision of the system rather larger than the spot size. At present, new systems with better resolutions are being developed. Much effort is also being dedicated to resolve the problem of the presence of excess powder by using ultrasonic vibrations, compressed air, and particular solvents.

Photopolymerization

Photopolymerization is a method based on the polymerization of photopolymeric resins in a site-specific manner. The photopolymeric resins are mixtures of simple monomers with low molecular weight that form a long chain polymer when they are activated by light. An initiator or a catalyst is often required for polymerization to occur (these are toxic to cells) and the light source must have a high frequency (usually ultraviolet).

Site-specific polymerization takes place either by scanning a laser beam over the surface of the liquid prepolymer—a process known as stereolithography—or by irradiating a mask placed above the surface. In both methods, irradiated zones become solid whereas the other areas remain liquid. Once the layer has polymerized, a fresh layer of prepolymer is applied. In theory, the laser spot could also be focused at different heights, but the depth of action is limited by the absorption of the liquid. The second technique requires the fabrication of a different mask for each layer and cannot strictly be defined as RP. Using the data in Dhariwala, Hunt, and Boland [20] an RTM ratio of about 0.5 can be estimated, which places this method among the more efficient techniques.

The main limitation of RP by photopolymerization is the use of acrylic or epoxy photopolymeric materials, which are often not biocompatible. Currently attempts are being made to extend this methodology to bioerodable polymers and biological polymers. For example, Dhariwala Hunt, and Boland [20] have used a commercial stereolithography machine to polymerize poly(ethylene oxide) and poly(ethylene glycol) dimethacrylate hydrogel using a photoinitiator. In their setup, a 250 µm ultraviolet light spot was rastered over a layer of prepolymer containing living cells, as schematized in Fig. 9. The results demonstrated that the cells were very sensitive to the initiator and remained viable only at very low initiator concentrations.

OTHER RP METHODS

Two further techniques that are worth mentioning are the sacrificial mold method and fiber spinning. The former uses rapid-prototyped molds to cast melts, solutions, or even powders, whereas the latter is not strictly RP at present

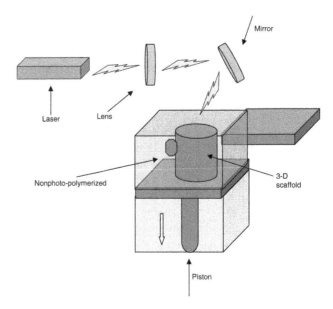

Fig. 9 Scheme of photopolymerization using laser stereolithography.

but can be used to build nanoscale features on microfabricated scaffolds.

Sacrificial Molds

A limited number of techniques using sacrificial molds have been reported, and this is likely due to the complexity of the fabrication process. Firstly, a plastic mold of the microstructure to be realized is constructed through one of the RP methods described above, usually ink-jet printing or 3-DP. Secondly, a deposition head then fills up the empty spaces present in the microstructure with a selected biopolymer thereby forming the scaffold. The polymer mold that simply constructs the support for the 3-D structure is then removed or sacrificed by bathing it in an appropriate solvent. In theory, using more heads, complex and multipolymeric structures can be rapidly realized.

Sachlos et al.[3] reported a two-phase ink-jet-based method to prepare molds of Protobuild (phase 1) and Protosupport (phase 2). The phase 2 material was dissolved away leaving a 3-D structure with interconnected channels. Subsequently collagen dispersion was poured into the channels and solidified through freeze-drying. Finally the phase 1 build material was dissolved away, leaving a collagen containing 3-D scaffold with features of about 200 µm. A very similar method was recently reported in which polypropylene fumarate was injected into a mold fabricated using two-phase ink-jet printing. Phase 1 and phase 2 materials were wax (removed by heating and solvent) and polystyrene (dissolved with acetone), respectively. The final PPF scaffold has a resolution of a few hundred microns and is mainly delimited by the precision of the ink-jet printer.[21] The process is outlined in Fig. 10.

The use of sacrificial molds adds to the complexity of the design, fabrication, and postprocessing of scaffolds but eliminates the pseudo-3-D layered effect, which gives rise to the characteristic steps that are observed in all other RP scaffolds. The final structure is a homogeneous monolith with channels, pores, and overhangs. However, this feature is obtained at the expense of extremely long manufacture times; the RTM ratio is the lowest of all the RP methods described in this entry. Besides its complexity, a major disadvantage of this method is the inability to use different materials in different areas and the limitations on choice of biomaterials, which have to be insoluble in the solvent used to degrade the mold. Furthermore, owing to the multiple steps involved, the fidelity of the scaffold is quite low. On the other hand, it is one of the few RP methods that can be used to produce scaffolds composed of proteins such as collagen.

Electrospinning

Although not strictly an RP method at present, electro or fiberspinning has the potential to become an RP method by appropriate eletrostatic and mechanical control of the deposition plane.

In electrospinning, a syringe is loaded with a polymer solution, which is forced out by the application of a high voltage to the syringe tip. This action causes a high-speed jet of polymer to exit from the tip and land on an appropriate collector, usually a flat plate or a rotating cylinder, as illustrated in Fig. 11. Polymer fibers with diameters ranging from a few nanometers to several microns can thus be produced. A wide range of materials can be spun using this method, including the classic biodegradable polyesters, gels such as gelatin and polyvinyl alcohol, and proteins (the range of possible polymers is much wider than any of the RP methods reported in Table 1). Furthermore, the method is also highly tunable, in the sense that fiber diameter and quality is dependent on several parameters, for example, viscosity, dipole moment, conductivity, and applied potential. Several designs and configurations of needle tips have been also proposed, such as multiple tips and coaxial tips, which allow the strands to be deposited as hollow fibers. Interested readers are encouraged to consult the excellent review by Pham, Sharma, and Mikos.[22]

At present the collector system where fibers are laid is very simple, but we envisage more sophisticated and fast moving collection systems, which could enable site-specific fiber deposition and orientation and render the electrospinning method RP. We predict that its manufacturing efficiency in terms of RTM ratio is likely to be astonishingly high, possibly a degree of magnitude over that of FDM because of the high speed with which the polymer jet is ejected from the needle tip and the high surface to volume ratio produced by very thin fibers.

INTEGRATION OF RP METHODS

None of the techniques so far developed can meet the requirements of resolution, RTM ratio, or flexibility of materials for the realization of a multifunctional scaffold for tissue engineering. An emerging trend now is to use the combination of technologies to integrate features at different scales. In an earlier work, we integrated the PAM technique with site-specific surface modification.[23] Ideally this approach should also combine one or more RP methods. At present, electrospinning, although not strictly RP, is most commonly integrated with other processing methods to produce microscale networks with higher order networks. An example of this method is the integration of fiber bonding with electrospinning. The resulting 3-D scaffolds comprise a macro and a nanofibrillar network, which is able to support cell viability and differentiation into bone cells.[24] The electrospinning technique has also been combined with a wet-spun macrofibrillar network to create structure-mimicking vessels.[25]

Zhang et al.[26] have recently proposed a technique to fabricate scaffold or cell constructs for tissue engineering

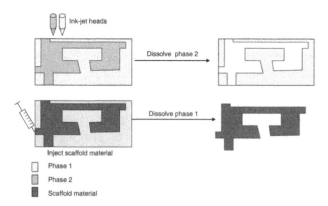

Fig. 10 The process of obtaining an RP scaffold through the use of a sacrificial mold.

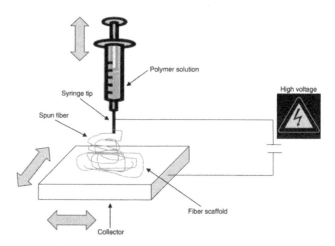

Fig. 11 Schematic of the electrospinning technique.

by the assembly of microscopic building blocks, realized with different methods such as bioplotter and FDM techniques in order to mimic the original topology of tissues.

We are certain that this new trend will bring about improvements in design and production capacity of RP scaffolds.

COMMERCIAL RP SYSTEMS FOR TISSUE ENGINEERING SCAFFOLDS

RP systems appear to be fairly expensive and bulky, but this is not always the case particularly as far as fabrication of tissue engineering products is concerned. The maximum size of scaffold desired is usually no larger than a few centimetres squared, and the fabrication costs are actually only a small fraction of the total cost of a cell-based 3-D structure, not to mention that of future preclinical and clinical trails. Some of the RP systems and products described in this entry are available commercially, but at present these devices and the scaffolds are still very much a niche product used more for research than for medical applications.

Therics Inc. is a company that produces a line of beta-TCP (β-TCP) products as bone fillers using a patented technology, Theriform (www.therics.com). Theriform technology is based on the 3-DP method, which was the first RP technology to be applied for tissue engineering. The company currently has eight products, two of which are FDA approved with positive clinical results.

PAM-fabricated scaffolds are marketed by Biodigit (www.biodigit.it) for use in cardiac tissue engineering and are mainly employed for research purposes. They are available in a variety of materials from biodegradable polyesters to polyurethanes as well as alginates.

Sciperio Inc. (www.sciperio.com) claims to be developing a new *in vivo* direct-deposition machine for the fabrication of *in vivo* tissue using advanced minimally invasive techniques, vision and diagnostic imaging systems, biological sensing systems, ultrashort-pulse-laser tissue ablation, and the fabrication and assembly of 3-D tissue constructs. The system described is essentially a dream tissue engineering factory, which has been envisaged by several researchers. According to Sciperio, the tool will be able to deposit a wide variety of cells (e.g., stem cells, endothelial cells, chondrocytes, T-cells, dendritic cells), growth factors, nutrients, ECM proteins, and biocompatible structural materials with exquisite precision.

Envisiontec (www.envisiontec.de) has developed the bioplotter technology, invented at the Freiburg Materials Research Centre in 1999. The system is based on a 3-D dispenser and allows processing a magnitude of materials including various biochemical systems and even living cells. CAD data handling and machine/process control are done through a system-specific pseudo-3-D CAD/CAM software. The bioplotter principle is based on dispensing a plotting material into a plotting medium to cause solidification of the material and to compensate gravity force through buoyancy. In the presence of a temperature-controlled plotting medium, the solidification of the material during plotting into the medium can be modulated by precipitation reactions, phase transitions, or chemical reactions. The bioplotter is the only commercial RP system available at fairly reasonable cost.

Neatco and Dimarix (www.dimatix.com) have created a cartridge-style printhead that allows users to fill their own fluids and print immediately with different materials. For the moment the systems are still used for liquid printing DNA arrays or sensors, but they can potentially be employed for organ printing. Each single-use cartridge has 16 nozzles linearly spaced at 254 μm with a typical drop size of 10 pL and can be replaced to facilitate printing of a series of fluids.

There is no doubt that in future the number of scaffolds available on the market will increase steadily. It should, however, be kept in mind that the road ahead to the real medical applications of RP scaffolds is still long and arduous. An emerging critical factor that should be considered in all design and processing phases is the application of good manufacturing practice (GMP) principles and standards to RP methods and products for biomedical use. Scaffolds should be considered as biomedical devices for advanced therapeutics, and RP technologies are therefore a key part of the manufacturing and validation processes.

DISCUSSION: LIMITATIONS AND CRITIQUES

In this entry, we have given a general description of the state of the art of RP methods for the realization of scaffolds for tissue engineering applications. There is no doubt that the microstructural environment of a cell conditions its behavior, and scaffold-based tissue engineering is founded on this very principle. However, as yet we do not have any rules or guidelines on just what this micro-structure should be and in general very little effort has been made toward defining structural design parameters. Until we establish architectural canons for biological tissue, the main advantage of RP technologies over other material processing methods is their ability to generate reproducible and repetitive structures with controlled porosity, which do not vary from batch to batch and at present adhering to an industrial rather than biological requirement.

As summarized in Table 1, it is clear that a large variety of RP techniques exist, and the choice of one method with respect to another depends on manufacturing efficiency and the desired resolution as well as on the polymer chosen for fabrication; there is no optimal or unique RP method. Each RP technique must be matched with a particular organ considering cell density, cell size, organizational and nutritional requirements, as well as mechanical matching between scaffold and tissue.

It can also be argued that RP fits the tissue engineering paradigm better than the classical subtractive fabrication methods such as milling or turning. In RP the object or prototype is created through an additive process not dissimilar to the process of biological development in which cells assemble and organize through a process of accumulation and accretion rather than degradation and attrition. However, the differences between tissue development and scaffold-based tissue engineering, some of which are listed in Table 2, should be kept in mind. Cell assembly and organization is an immensely complicated process and depends on the correct orchestration of biochemical signals with spatial and other physical stimuli. Simply seeding cells on a porous scaffold with the same shape as the end organ and hoping for a functional tissue to form has already been shown to be insufficient. Indeed, despite the enormous economic resources injected into tissue engineering, very few products are actually viable from a commercial[27] or even medical point of view.[28]

In this sense, scaffold fabrication techniques through RP are not stand-alone tissue engineering tools but, taking into account the intricate complexity of living systems, must be integrated with other micro and macroscale techniques. For instance, that scaffolds must be accompanied by preor postsurface modification usually thorough the immobilization of adhesion proteins, which is already well established. Other supporting elements such as soluble growth factors, incubators, or bioreactors are also mandatory. We have already mentioned that the nature of the biomaterial as well as its microstructure and topology are additional critical factors. It should not be forgotten that even in mechanical terms, the microfabrication or biomaterial processing method must be matched to each particular tissue to be engineered. Optimized solutions could be found by combining different RP methods and by following biomimetic design principles through the use of biomaterials, which resemble biological materials as much as possible. This implies the development of hybrid materials and structures with a composite nature and high water

content and the integration of two or more RP techniques, for example, with different resolutions and different types of polymers.

CONCLUSION

An overview of RP methods for biomaterial fabrication has been provided with the aim of illustrating the basic principles behind the scaffold-based tissue engineering. Rather than providing exhaustive technical and comparative details, which are available in several excellent reviews that the interested reader is encouraged to refer to Refs.,[2,29,30] we have described aspects which must be taken into consideration when realizing a scaffold such as the resolution time of manufacture ratio, materials employed, and scaffold geometrical design. Furthermore, it is also important for the RP engineer involved in tissue-engineering construct manufacture to be acutely aware of the differences between tissue development and the use of a scaffold to guide cells to appropriate locations on a scaffold. High throughput production of tissue-engineered constructs is still a long way away and not only requires great multidisciplinary effort but also a thorough understanding of the biology, chemistry, and engineering of cells and biomaterials and above all their interaction in complex 3-D environments.

REFERENCES

1. Langer, R.; Vacanti, J.P. Tissue engineering. Science **1993**, *260* (5110), 920–926.
2. Yeong, W.Y.; Chua, C.; Leong, K.F.; Chandrasekaran, M. Rapid prototyping in tissue engineering: Challenges and potential. Trends Biotechnol. **2004**, *22* (12), 643–652.
3. Sachlos, E.; Reis, N.; Ainsley, C.; Derby, B.; Czernuszka, J.T. Novel collagen scaffolds with predefined internal morphology made by solid freeform fabrication. Biomaterials **2003**, *24* (8), 1487–1497.
4. Landers, R.; Hubner, U.; Schmelzeisen, R.; Mulhaupt, R. Rapid prototyping of scaffolds derived from thermoreversible hydrogels and tailored for applications in tissue engineering. Biomaterials **2002**, *23* (23), 4437–4447.
5. Yan, Y.; Wang, X.; Pan, Y.; Liu, H.; Cheng, J.; Xiong, Z.; Lin, F.; Wu, R.; Zhang, R.; Lu, Q. Fabrication of viable tissue-engineered constructs with 3D cell-assembly technique. Biomaterials **2005**, *26* (29), 5864–5871.
6. Tan, K.H.; Chua, C.K.; Leong, K.F.; Cheah, C.M.; Cheang, P.; Abu Bakar, M.S.; Cha, S.W. Scaffold development using selective laser sintering of polyetheretherketone-hydroxyapatite biocomposite blends. Biomaterials **2003**, *24* (18), 3115–3123.
7. Curtis, A.G.; Wilkinson, C.D.W. Reactions of cells to topography. J. Biomater. Sci. Polymer Edn. **1998**, *9* (12), 1313–1329.
8. Chen, C.S.; Mrksich, M.; Huang, S.; Whitesides, G.M.; Ingber, D.E. Geometric control of cell life and death. Science **1997**, *276* (5317), 1425–1428.

Table 2 Differences between biological tissue development and tissue engineering using scaffolds

Tissue development	RP scaffold approach
Not scaffold dependent	A porous biodegradable scaffold is necessary
Cells create their matrix and extend in three dimensions from a "point source"	Takes place through pseudo-3-D fabrication: through layer-by-layer assembly of 2-D
Environment and structure modulated in real time	Preprogrammed architecture
Complex multicomponent materials	Usually one material
Soft and wet	Often hard and dry

9. Ingber, D.E. Cellular mechanotransduction: Putting all the pieces together again. FASEB J. **2006**, *20* (7), 811–827.

10. Francis, K.; Palssojn, B.O. Effective inter-cellular communication distances are determined by the relative cytokine secretion and diffusion time constants. Proc. Natl. Acad. Sci. U.S.A. **1997**, *94* (23), 12258–12262.

11. Mattioli-Belmonte, M.; Vozzi, G.; Kyriakidou, K.; Pulieri, E.; Lucarini, G.; Vinci, B.; Pugnaloni, A.; Biagini, G.; Ahluwalia, A. Rapid-prototyped and salt-leached PLGA scaffolds condition cell morpho-functional behavior. J. Biomed. Mater. Res. A. **2008**, *85* (2), 466–476.

12. Vozzi, G.; Previti, A.; De Rossi, D.; Ahluwalia, A. Microsyringe based deposition of 2 and 3-D polymer scaffolds with a well defined geometry for application to tissue engineering. Tissue Eng. **2002**, *8* (6), 1089–1098.

13. Zein, I.; Hutmacher, D.W.; Tan, K.C.; Teoh, S.H. Fused deposition modeling of novel scaffold architectures for tissue engineering applications. Biomaterials **2002**, *23* (4), 1169–1185.

14. Mironov, V.; Boland, T.; Trusk, T.; Forgacs, G.; Markwald, R.R. Organ printing: Computer-aided jet-based 3D tissue engineering. The layer-by-layer assembly of biological tissues and organs is the future of tissue engineering. Trends Biotechnol. **2003**, *21* (4), 157–161.

15. Mikos, A.G.; Sarakinos, G.; Leite, S.M.; Vacanti, J.P.; Langer, R. Laminated three-dimensional biodegradable foams for use in tissue engineering. Biomaterials **1993**, *14* (5), 323–330.

16. Giordano, R.A.; Wu, B.M.; Borland, S.W.; Cima, L.G.; Sachs, E.M.; Cima, M.J. Mechanical properties of dense poly-lactic acid structures fabricated by three-dimensional printing. J. Biomater. Sci. Polym. Ed. **1996**, *8* (1), 63–75.

17. Zeltinger, J.; Sherwood, J.K.; Graham, D.A.; Mueller, R.; Griffith, L.G. Effect of pore size and void fraction on cellular adhesion, proliferation, and matrix deposition. Tissue Eng. **2001**, *7* (5), 557–572.

18. Berry, E.; Brown, J.M.; Connell, M.; Craven, C.M.; Efford, N.D.; Radjenovic, A.; Smith, M.A. Preliminary experience with medical applications of rapid prototyping by selective laser sintering. Med. Eng. Phys. **1997**, *19* (1), 90–96.

19. Ciardelli, G.; Chiono, V.; Vozzi, G.; Pracella, M.; Ahluwalia, A.; Barbani, N.; Cristallini, C.; Giusti, P. Blends of poly-(epsilon-caprolactone) and polysaccharides in tissue engineering applications. Biomacromolecules **2005**, *6* (4), 1961–1976.

20. Dhariwala, B.; Hunt, E.; Boland, T. Rapid prototyping of tissue-engineering constructs, using photopolymerizable hydrogels and stereolithography. Tissue Eng. **2004**, *10* (9–10), 1316–1322.

21. Lee, K.W.; Wang, S.; Lu, L.; Jabbari, E.; Currier, B.L.; Yaszemski, M.J. Fabrication and characterization of poly(propylene fumarate) scaffolds with controlled pore structures using 3-dimensional printing and injection molding. Tissue Eng. **2006**, *12* (10), 2801–2811.

22. Pham, Q.P.; Sharma, U.; Mikos, A.G. Electrospinning of polymeric nanofibers for tissue engineering applications: A review. Tissue Eng. **2006**, *12* (5), 1197–1211.

23. Bianchi, F.; Vassalle, C.; Simonetti, M.; Vozzi, G.; Ahluwalia, A.; Domenici, C. Endothelial cell function on 2D and 3D micro-fabricated polymer scaffolds: Applications in cardiovascular tissue engineering. J. Biomater. J. Biomater. Sci. Polym. Ed. **2006**, *17* (1–2), 37–51.

24. Tuzlakoglu, K.; Bolgen, N.; Salgado, A.J.; Gomes, M.E.; Piskin, E.; Reis, R.L. Nano- and micro-fiber combined scaffolds: A new architecture for bone tissue engineering. J. Mater. Sci. Mater. Med. **2005**, *16* (12), 1099–1104.

25. Kim, T.G.; Park, T.G. Biomimicking extracellular matrix: Cell adhesive RGD peptide modified electrospun poly(D,L-lactic-co-glycolic acid) nanofiber mesh. Tissue Eng. **2006**, *12* (2), 221–233.

26. Zhang, H.; Hutmacher, D.W.; Chollet, F.; Poo, A.N.; Burdet, E. Microrobotics and MEMS-based fabrication techniques for scaffold-based tissue engineering. Macromol. Biosci. **2005**, *5* (6), 477–489.

27. Lysaght, M.J.; Hazlehurst, A.L. Tissue engineering: The end of the beginning. Tissue Eng. **2004**, *10* (1–2), 309–320.

28. Hunziker, E.B. Commentary. Osteoarthr. Cartilage **2002**, *10* (6), 432–463.

29. Yang, S.; Leong, K.; Zhoahui, D.; Chua, C. The design of scaffolds for use in tissue engineering part II. Rapid prototyping techniques. Tissue Eng. **2002**, *8* (1), 1–11.

30. Leong, K.F.; Cheah, C.M.; Chua, C.K. Solid freeform fabrication of three-dimensional scaffolds for engineering replacement tissues and organs. Biomaterials **2003**, *24* (13), 2363–2378.

Neural—Rapid Prototyping

Scaffolds, Polymer: Microfluidic-Based Polymer Design

Mohana Marimuthu
Sanghyo Kim
Department of Bionanotechnology, Gachon University, Gyeonggi-do, South Korea

Abstract

The challenge for constructing the functional biomimetic artificial organs *in vitro* paves the way for the inventions of several microfabrication technologies such as microfluidics. The most important limitations in developing artificial tissues are allowing hierarchical multi-cell interactions three-dimensionally under perfusion culture for nutrient supply and waste removal mimicking the physiological environment. By combining biology with bio-MEMS techniques, it is highly possible to overcome the above-mentioned limitations to provide a suitable environment for cells to grow and interact *in vitro* and mimic living systems.

INTRODUCTION

Langer and Vacanti defined and summarized the term, tissue engineering, as "an interdisciplinary field that applies the principles of engineering and life sciences toward the development of biological substitutes for the repair or regeneration of tissue or organ functions" in National Science Foundation-sponsored meetings in 1988 and 1993.[1] Until now, the field of tissue engineering has been playing intrinsic and potential roles in replacing or restoring physiological functions of diseased and/or damaged organs. Products of tissue engineering can be of two types: one is a completely functional *ex vivo* developed organ to be implanted and the other is a combination of scaffold with cells that becomes functional tissue/organ after implantation.[2] In either of the types of tissue-engineered products, cells are the potential component for the treatment of diseased tissues where it plays significant roles such as proliferation, differentiation, cellular signaling, development of extracellular matrix, and most importantly, protein production.[3] There have been several tissues such as skin, blood vessels, bone, cartilage, kidney, and heart,[4] developed based on tissue engineering techniques and utilized for tissue regeneration that proves the success of tissue engineers.

The cells that are incorporated into scaffold can become functional tissue only when the scaffold can facilitate cellular development through several factors such as mechanical support, sufficient room for nutrient supply and excretion of cellular waste, gaseous transport, and protein/growth factor supplement. Researchers in earlier days used several porous scaffold fabrication methods such as solvent casting and particulate leaching where the pores are expected to facilitate nutrient supply for the cells in them.[5,6] However, due to lack of controlled internal polymer architecture, these methods were replaced with the rapid prototyping technique by which the three-dimensional (3D) scaffold with spatially controlled porous architecture can be developed using a computer-aided design (CAD) program.[7] Methods based on heat for 3D scaffold fabrication include lamination technique,[8] soft lithography,[9] fused deposition modeling,[10] selective laser sintering,[11] and 3D plotting technology,[12] where the polymeric sheets or particles are fused or deposited to form 3D scaffold with the application of heat and pressure. Similarly, light has been used to initiate the polymerization and solidification of polymeric solutions into a scaffold termed as the photopolymerization technique, which includes stereolithography, photopatterning, and photolithography.[13,14] Table 1 describes the advantages, disadvantages, and applications of various scaffold fabrication techniques.[15] The limitation of acellular scaffold fabrication techniques on multicellular engraftment leads to the development of 3D assembly of cell sheets.[16] Although both the acellular and cellular scaffolds can provide mechanical support and heterogeneous cell engraftment, the main challenge to regenerate an organ is to provide a vascular architecture that plays a vital role in nutrient supply and metabolite transport.

The microfluidics technique has the ability to develop vascular-like flow patterns as small as capillaries of physiological vasculature with controlled flow rate mimicking *in vivo* conditions.[17] The microfluidics technique can also be used as a cell culture platform for controlling cellular microenvironment for multicellular interactions. It is possible using microfluidics to pattern biomolecules or cells three-dimensionally to generate completely functional tissues. The combined use of microfluidic and molding techniques was employed for the patterning of a cell-collagen matrix.[18] Microchannels under perfusion have been extensively studied as a microbioreactor system for the analysis of various cellular biology such as understanding axonal out-growth[19] or hepatocyte development.[20] The study of multicellular interaction has also been carried out using a

Concise Encyclopedia of Biomedical Polymers and Polymeric Biomaterials DOI: 10.1081/E-EBPPC-120052552

Scaffolds—Smart

Table 1 Advantages, disadvantages, and applications of various scaffold fabrication techniques

Fabrication technique	Advantages	Disadvantages	Applications
Thermally induced phase separation	Can control the porosity and pore morphology	Achievable sizes range from only 10–2000 µm in diameter	Proteins and drug delivery and higher drug encapsulation efficiency
Solvent casting and particulate leaching	Simple operation, control of the pore size, and porosity by selecting the particle size and the amount of salt particles	Distribution of salt particles is often not uniform within the polymer solution, and the degree of direct contact between the salt particles is not well controlled, interconnectivity of pores in a final scaffold cannot be well controlled, limited membrane thickness, lack of mechanical strength, problems with residual solvent, residual porogens	Cardiac and vascular tissue engineering applications
Solid freeform fabrication techniques	Customized design, computer-controlled fabrication, anisotrophic scaffold microstructures, processing conditions	Lack of mechanical strength, limited to small pore sizes	Production of scale replicas of human bones and body organs to advanced customized drug delivery devices
Phase separation	Allows incorporation of bioactive agents, highly porous structures	Lack of control over microarchitecture, problems with residual solvent, limited range of pore sizes	Drug release and protein delivery applications
Electrospinning	Easy process, high porosity, high SAV ratio	Limited range of polymers, lack of mechanical strength, problems with residual solvent, lack of control over microarchitecture	Bone, skin, nerve, and cardiac tissue engineering

Source: From Ravichandran et al.[15] © 2012, with permission from Wiley-VCH Verlag GmbH and Co. KGaA.

microfluidic platform that includes micropatterning of hepatocytes and fibroblasts,[21] and multicellular culture for mimicking blood vessels.[22] This entry focuses on various types of polymeric scaffolds that are utilized for tissue regeneration and the role of microfluidics for designing these polymeric scaffolds, the role of a monoculture or multiculture, and the role of a microbioreactor.

POLYMERIC SCAFFOLDS FOR TISSUE ENGINEERING

Selection of polymeric materials suitable for specific tissue regeneration is an important challenge in the fabrication of scaffolds. These polymeric materials should be safe, biocompatible, mechanically stable, nonimmunogenic, and biodegradable, and their degradation products should remain nontoxic until elimination.[23] Polymers utilized for scaffold fabrication can be characterized depending on their source such as natural, synthetic, or hybrid of both with or without inorganic materials. To obtain a completely functional specific phenotype, polymeric scaffolds that are utilized for

tissue engineering should possess the characteristics to incorporate biomolecules such as proteins, growth factors, and cells. Biomolecule or cell-embedded scaffolds could regulate signaling mechanisms for various biological pathways and initiate cellular adhesion, migration, differentiation, and proliferation for generation of biomimetic functional tissue. Based on the type of biomolecules or cells to be incorporated, the scaffold materials can be differentiated. Among the scaffolds made of natural polymers, protein-based polymers such as collagen, gelatin, fibrin, and elastin, and polysaccharide-based polymers such as chitosan, alginate, hyaluronic acid, chondroitin sulfate, dextran, and agarose were widely studied for biomolecule or cell delivery. Collagen-based bone morphogenetic protein (BMP) carriers called INFUSE® Bone Graft (Medtronic, Minneapolis, MI) sponges have been developed and are commercially available in the market for treatment of spinal fusion.[24] Similarly, Surgifoam® (Ethicon, Inc., Somerville, NJ) is a commercially available porous gelatin disk that carries transforming growth factor beta-1 (TGF-pl) for cartilage tissue engineering.[25] Small interference RNA (siRNA) delivery through cationized gelatin was researched for silencing the TGF-h receptor gene for renal tissue engineering.[26] Chitosan- and alginate-based scaffolds were studied for delivery of TGF-pl and BMP, respectively, for bone and cartilage tissue regeneration.[27,28] Hyaluronic acid microsphere and chon-droitin sulfate scaffold were used for the delivery of DNA and vascular endothelial growth factor, respectively, for vascular tissue engineering.[29,30]

For 2 decades, the poor mechanical strength and poor versatility in scaffold fabrication of natural polymers led to the high-order shift of research interest over synthetic polymers where several physical properties can be altered such as hydrophilicity to hydrophobicity or even to amphiphilicity. Among the synthetic polymers, poly(ahydroxy acid)s such as poly(L-lactic acid) (PLLA), poly(glycolic acid), poly(ε-caprolactone), and/or poly(lactic-co-glycolic) acid (PLGA) are widely studied for tissue engineering and drug delivery applications. Indeed, two different PLGA-based scaffolds have been commercially available in the market in the name of Dermagraft and TransCyte (Advanced Bio-Healing, Inc., La Jolla, CA) for skin tissue engineering.[31] Recently, PLGA polymer has also been utilized for plasmid-encoding basic fibroblast growth factor delivery through a porous scaffold.[32] With the help of a modern electrospinning technique, PLLA polymer was electrospun into 0 scaffold incorporating platelet-derived growth factor-BB for complex tissue regneneration.[33] Several other synthetic polymers utilized for fabrication of scaffold include polypyrrole,[34] polyethylene glycol (PEG),[35] poly(carbonate-urea) urethane,[36] poly(propylene fumarate),[37] and polyphosphazenes.[38] Polypyrrole and poly(3,4-ethylenedioxythiophene) are unique polymers possessing conductive and semiconducting properties, which make them extremely appropriate for electrically active tissue regeneration, especially neurite outgrowth that was found to be enhanced further upon incorporation of biomolecules such as neurotrophin 3 and brain-derived neurotrophic factor.[39,40] Among all synthetic polymers, a widely studied polymer for gene delivery system is PEG or PEG-derived copolymers. For instance, a PEG graft polymerized with polyethylene imine was successfully used for DNA delivery in order to obtain extended gene expression in the spinal cord.[41] Elaborate discussions have taken place on a wide range of gene and growth factors incorporated into polymeric scaffolds for tissue regeneration.[23,42–45]

Stimuli-responsive polymers are currently in great demand for scaffold fabrication as their response behavior is similar to the physiological system to some extent. Responses given by these polymers are governed by such change in physical and/or chemical properties as conformational change, solubility change, and the change of hydrophilicity to hydrophobicity and vice versa.[15] The property change of the polymers could be caused by either the formation or breakdown of weak/secondary linkages such as hydrogen bond, hydrophobic interaction, electrostatic force, or acid-base reactions of polymers, or the difference in osmotic pressure.[46] Stimuli-responsive polymers

have been characterized based on the types of stimuli which include internal stimuli such as pH or temperature and external stimuli such as magnetic or electric field, light, or ultrasound.[15] Every organ and tissue in the human body possesses a unique pH that increases the interest of study on pH-sensitive polymer-based scaffold for tissue regeneration. The polymers containing functional groups of weak acids such as carboxylic acids, phosphoric acid, and weak bases such as amines tend to either accept or donate protons, thereby changing the polymer's ionization state depending on environmental pH. Therefore, the polymers with a pKa value of 3–10 are considered as pH-sensitive polymers.

Similarly, polymers that are sensitive to temperature respond by changing their hydration degree under aqueous environment. For instance, a temperature-sensitive polymer swells up and de-swells below and above its transition temperature, respectively, in order to equilibrate the hydration degree. The above transition process is generally reversible, which allows the use of these temperature-sensitive polymers as an on-off system depending on the presence/absence of stimuli.[13] Critically, transition temperature between 20°C and 40°C is optimum for temperature-sensitive polymers, although several factors including solvent type, salt concentration, and their molecular weight govern their transition process.[13] Transition between room temperature and body temperature and vice versa is the widely used phenomenon of making biodegradable and injectable scaffolds for biomedical applications. For instance, a pH-sensitive polymer, sulfamethazine oligomer has been bound to the extreme ends of a thermosensitive block copolymer, poly(ε-caprolactone-co-lactide)-poly(ethylene glycol)-poly(ε-caprolactone-co-lactide) (PCLA-PEG-PCLA), which formed pH- and thermo-sensitive hydrogel for bone tissue engineering.[47] This pH- or thermosensitive polymer remains in the solution under pH 8 and 37°C, whereas upon injection it becomes a hydrogel scaffold at physiological pH 7.4 and 37°C. Delivery of human mesenchymal stem cells and recombinant human BMP-2 was studied on this pH- or thermosensitive injectable scaffold for autologous bone regeneration.[47] Similar to pH and temperature, photosensitive polymers that undergo structural transformations through the change of functional groups of polymer backbone can be induced by a light of a specific wavelength.[48] Polymers that are prone to change their temporary shape to a permanent shape upon any of the stimuli, such as temperature, pH, light, electric/magnetic field, or enzyme, are termed as shape memory polymers. Hydroxylapatite, together with poly(D,L-lactide), is such an example of shape memory polymer, which has been examined by Zheng et al.[49] for hard tissue engineering. Extended discussion on stimuli-responsive polymers and their biomedical applications was found elsewhere.[15,50] Several other stimuli-responsive polymers that were studied for applications along with tissue regeneration, including clinical

Scaffolds—Smart

diagnostics, drug delivery, and sensing, are summarized in Table 2.

In some cases, physical and/or chemical modifications without altering inherent bulk properties of polymer or polymer scaffold would behave biomimetically for enhancing the cellular attachment, proliferation, and differentiation into a desired tissue. Polymer or polymer scaffold surface modification can be carried out using several techniques which include plasma treatment, ion sputtering, oxidation/reduction reactions, and corona discharge.[15] It is possible to change the functional group of the polymer surface to carboxylic or amine groups depending on the type of plasma chosen such as oxygen,

ammonia, or air.[51] Appropriate plasma selection modifies the polymer surface chemically, through which various desired physical properties can be obtained such as hydrophilicity, refractive index, hardness, surface energy, biocompatibility, and even topographical changes.[52] Plasma-modified polymer surfaces can be further treated with various extracellular matrix components such as collagen, elastin, gelatin, and fibronectin for obtaining successful cellular functions.[53] For instance, plasma-treated nanofibrous scaffold of a complex PLLA-co-poly(ε-caprolactone)/gelatin was found to be highly biocompatible for attachment and proliferation of human foreskin fibroblasts for skin tissue engineering.[54

Table 2 Various stimuli-responsive polymers and their properties and applications

Functional polymer	Examples	Properties	Applications
Conducting polymer	PEDOT, PPy, PANI	Possesses high electronic and ionic conductivities	Nerve and cardiac tissue engineering
Glucose-responsive polymer	PBA acid groups	Responses to the by-products that result from the enzymatic oxidation of glucose	Glusoce sensing and insulin delivery
pH-responsive polymer	Poly(acrylamide), poly(acrylic acid), poly(methacrylic acid), PDE-AEMA, PDMAEMA	Exhibits a change in the ionization state upon variation of the pH leading to conformational change	Insulin delivery and drug delivery
Enzyme-responsive polymer	Genetically engineered variant of spider dragline silk	Undergoes macroscopic property changes when triggered by selective enzymatic reactions	Drug delivery
Temperature-responsive polymer	PNIPAM	Undergoes a conformational change upon the change in temperature	Cardiac tissue engineering and drug release
Antigen-responsive polymer	NSA, poly(acrylamide)	Antigen-antibody interactions	Biosensors and drug release
Redox/thiol-responsive polymer	Polymers containing disulfide-functional dimethacrylate	Responds to the interconversion of thiols and disulfides	Bioengineering and drug delivery
Shape memory polymer	Poly(D,L-lactide)	Can rapidly change its shape from temporary to permanent under appropriate stimulus	Biomedical applications
Electro-responsive polymer	Polyacrylamide gels, chondroitin sulfate, hyaluronic acid	Responds to either the presence or the absence of magnetic fields	Soft biomimetic actuators, sensors, cancer therapy agents, artificial muscles
Ultrasound-responsive polymer	Polyglycolides, polylactides, and poly-[bis(p-carboxyphenoxy)] alkane anhydrides with sebacic acid and ethylene vinyl acetate copolymers	Responds to cavitation that results from high and low-pressure waves generated by ultrasound energy	Medical treatment and diagnostics
Photoresponsive polymer	Azobenzene and spiropyran-containing polymer	Macromolecules that change their properties when irradiated with light of appropriate wavelength	Biological applications
Photoluminescent polymer	Chromophores attached to polyesters, polyamides, polyimides, polyurethanes	Polymers with emissive properties	Biosensors and medical diagnostics

NSA, N-succinimidylacrylate; PANI, poly(aniline); PBA, phenylboronic acid; PDEAEMA, poly(diethyl- aminoethyl methacrylate; PDMAEMA, poly(dimethylaminoethyl methacrylate); PEDOT, poly(3,4-ethylenedioxythiophene); PNIPAM, poly(N-isopropylacrylamide); PPy, polypyrrole.
Source: From Ravichandran et al.[15] © 2012, with permission from Wiley-VCH Verlag GmbH and Co. KGaA.

Scaffolds—Smart

MICROFLUIDICS AND POLYMERIC SCAFFOLD FABRICATIONS

Successful tissue regeneration highly depends on the execution of balance between the physiological and synthetic scaffold functions. Two-dimensional and 3D polymeric scaffolds are of great interest in the field of tissue engineering that can be achieved by integrated assembly of tubes, fibers, or micro- or nanoparticles of polymeric materials. Fibers or tubes as 3D scaffolds were conventionally fabricated using extrusion, casting, and layering techniques.[55,56] However, microfluidics has been studied as a cost-effective method for the continuous extrusion of micro-fibers and microtubes, overcoming most of the limitations of conventional methods. "On the fly" photopolymerization under microfluidic flow condition was the first microfluidic-based fiber fabrication technique introduced by Lee's group in which a mixture of photopolymerizable polymers was composed of 4-hydroxybutyl acrylate, acrylic acid, ethyleneglycol dimethacrylate, and 2,2'-dimethoxy-2-phenyl-aceto-nephenone.[57] Lee's group also examined other microfluidic fiber fabrication mechanisms such as diffusion-controlled ionic cross-linking and phase inversion utilizing polymers such as sodium alginate and PLGA.[58,59] Recently, Mohana et al. developed microfluidic device for porous fiber formation utilizing the mechanisms of immersion precipitation and solvent evaporation of an ABA triblock copolymer poly(*p*-dioxanone-*co*-caprolactone)-*block*-poly(ethylene oxide)-*block*-poly(*p*-dioxa-none-*co*-caprolactone) having both hydrophilic and hydrophobic segments for the delivery of biomolecules with lipophobic and lipophilic properties.[60] To some extent, polymeric particles also have an important role in tissue engineering applications. For instance, several research groups examined the scaffolds made of functional aggregation of polymeric microspheres.[61,62] The basic principle involved in developing such a 3D particle-based porous scaffold is packing of microparticles randomly and either physically or chemically inducing agglomeration.[63] Particle synthesis using microfluidic technology can be characterized as droplet-based method, multiphase flow method, photolithography-based method, and supraparticle synthesis using assembly of colloids described elaborately elsewhere.[64] For instance, Chung and coworkers developed a "railed microfluidics" method by which self-assembly of polymeric microstructures inside fluidic channels was demonstrated for tissue engineering applications.[65] In physiological condition, the exchange of nutrients and oxygen takes place throughout the 3D spaces with the help of vascular networks containing highly branched fractal-like morphologies. Recently, Ugaz and coworkers developed bio-inspired 3D microfluidic vascular networks using a microfabrication process in which electron beam irradiation was used to implant electric charge inside the polymer, poly(methyl methacrylate), and followed by controlled discharge of accumulated energy to produce a treelike

vascular microchannel of ~10 µm in diameter.[66] Figure 1 shows the construction mechanism of biomimetic 3D micro-vascular networks. This new method acts as a new tool to embed 3D microchannel networks within the scaffold materials for supporting cell culture with larger volumes directing successful tissue regeneration.

In spite of using microfluidics as a platform for polymeric scaffold fabrication, some polymers with an embedded microfluidic structure itself act as scaffolds.

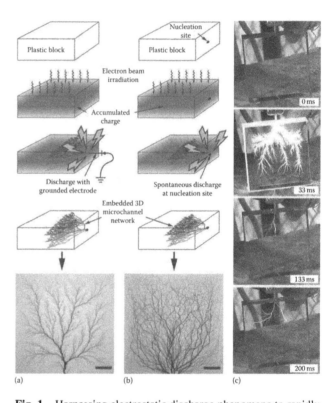

Fig. 1 Harnessing electrostatic discharge phenomena to rapidly construct branched 3D microvascular networks. (**a**) In the grounded contact method, electron beam irradiation is used to implant a high level of internal electric charge inside a dielectric substrate. A grounded electrode is then brought into contact with the substrate surface, initiating sudden energy release that locally vaporizes the surrounding material leaving behind a treelike branched microchannel network. (**b**) In the spontaneous discharge method, a defect (e.g., small hole) is first created on the substrate surface prior to irradiation. When the internal electric charge exceeds a critical level upon exposure to the electron beam, the defect acts as a nucleation site for spontaneous energy release. The grounded contact method yields a more "treelike" morphology, while microchannels produced by spontaneous discharge permeate the substrate more uniformly. (**c**) Image sequence from a video recording of discharge by the grounded contact method shows that energy release is nearly instantaneous with subsequent weaker discharges persisting over longer timescales. All photographs depict microchannel networks in 1 inch × 3 inch × 3 inch polished acrylic blocks. Note that all networks extend in 3D throughout the volume of the blocks. Scale bars = 1 cm.
Source: From Huang et al.[66] © 2009, with permission from Wiley-VCH Verlag GmbH and Co. KGaA.

Such polymers should possess several unique properties such as mechanical stability, biocompatibility, microstructural replicability, ability to bind into an air-tight fluidic network, diffusibility or permeability for small and large molecules, stability under physiological conditions (such as pH, temperature, and osmolarity) and be relatively inexpensive.[67] For instance, a PLGA-based highly branched multilayer microfluidic network was demonstrated as an analog to physiological microvascular network for tissue engineering applications.[68] Likewise, the calcium alginate is another example of a hydrogel with an embedded microfluidic network studied for controlled soluble species concentration and incorporation of multiple cell types within a 3D gel scaffold.[69] It is essential to develop a cell-seeded microfluidic scaffold for obtaining accurate biomimetic environment to direct the specific tissue regeneration. Biopolymer obtained from the *Bombyx mori* silkworm has also been applied for the fabrication of microfluidic scaffold.[70] A biodegradable elastomer, poly(glycerol sebacate), another polymer with excellent chemical and mechanical properties, was fabricated into an endothelialized microfluidic scaffold, an analog to microvasculature.[71] Recently, a new biomaterial, poly(1,3-diamino-2-hydroxypropane-*co*-polyol sebacate) was found to be a biodegradable microfluidic device which possesses a significantly longer degradation profile than other polymer (PLGA or silk fibroin)-based microfluidic scaffolds.[72] A detailed summary on biomaterials-based microfluidic systems was offered by Bettinger and Borenstein.[73]

MICROFLUIDIC CELL CULTURE SYSTEMS FOR TISSUE REGENERATION

A microfluidic system cell culture platform can deliver several types of substances such as growth factors, enzymes, fluorescent dyes, or drug materials with specific targets to single or multiple cells without disturbing them. Multiple streams of laminar flow in parallel with each other combined into a single stream without mixing of each fluid, except the diffusion in the interface. With the help of specifically designed geometries and structures, a microfluidic perfusion culture can mimic *in vivo* physiological conditions by producing fluidic shear stress[74] or interstitial fluid flow.[75] One important feature of a microfluidic cell culture system is surface area-to-volume (SAV) ratio. When this SAV ratio increases, diffusion of gas in the microfluidic system occurs efficiently. Indeed, efficient gas exchangeable design is significant for cells to maintain their metabolic activities.[76] For instance, microfluidic cell culture devices are well known for their high SAV ratio, and therefore enhance diffusion of gases such as oxygen for cell culture without the need of external oxygenation structure.[77] Likewise, maintaining specific temperature for cell culture plays an important role in obtaining unique cell physiology.[78,79] Due to the characteristics of heat transfer,

cell culture under precisely controlled gaseous and temperature environments is significantly attainable with a microfluidic cell culture platform.[80] These characteristics of a microfluidic system make it unique as a cell culture platform for the studies of cell biology, cell-cell interaction, drug-cell interaction, and tissue regeneration. One of the promising roles of a microfluidic cell culture platform is to develop 3D tissue models for the exploration of drug sensitivity, drug toxicity analysis, drug therapeutic property studies, and cellular response related to concentration of drugs. A 3D liver tissue model for drug study has been developed by the co-culture of hepatocytes and endothelial cells in a photosensitive polymer-based microfluidic cell culture platform.[81] Similarly, a 3D HepaTox Chip has been developed by the Institute of Bioengineering and Nanotechnology, Singapore. It is a microfluidic-based multiplexed hepatocyte culture chip that generates linear concentration gradient for the study of drug toxicity.[82] Micropillars were designed at the middle of each culture platform to culture hepatocyte cells, where microchannels on either side allow medium perfusion. Such a design provided a 3D microenvironment for hepatocytes to behave phenotypically as a functional liver for the analysis of drug toxicity.[82] Several other microfluidic co-culture designs for *in vitro* models of liver tissue regeneration were studied elaborately by biomedical engineers.[83,84] In addition to vascular and liver tissue development, neuronal tissue regeneration is highly challengeable for tissue engineers to develop a competent multicell culture platform for facilitating cell-cell interactions, which is significantly vital for metabolic coupling of neuronal cells with astrocytes.[85–87] Varieties of worthwhile approaches based on a microfluidic neural co-culture platform for axon-glial interactions were developed.[88–90] *In vitro* neural tissue regeneration highly involves myelination that helps in the conduction of nerve impulses. *In vitro* myelination can be obtained when axons and oligodendrocytes or Schwann cells are co-cultured to initiate signaling between them.[91,92] Recently, two compartmentalized co-cultures of neurons and oligodendrocytes were studied based on a microfluidic system.[93] Microfluidic channels of two compartments were designed circularly as shown in Fig. 2 to guide the axonal outgrowth and promote the myelination process. The microchannels of this device physically isolate the axons from cell bodies and dendrites (Fig. 3), which is believed to facilitate the long-term co-culture of cells and enhanced axon-glia signaling.[93] A detailed review on microfluidic-based cell culture platforms for tissue engineering applications can be found elsewhere.[94]

MICROFLUIDIC BIOREACTORS FOR TISSUE ENGINEERING APPLICATIONS

Obtaining higher cell density while a cell culture is *in vitro* is an important factor in tissue engineering applications.

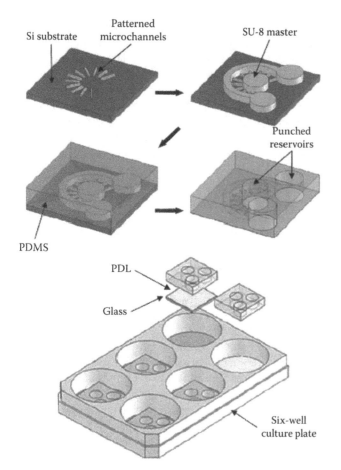

Fig. 2 Fabrication and assembly steps for the microfluidic co-culture device. Two SU-8TM layers with different thicknesses were patterned on the top of a silicon substrate to form the axon-guiding microchannel array and the two cell culture compartments (soma and axon/glia). PDMS devices were replicated from the SU-8TM master using soft lithography process and 7-mm diameter reservoirs were punched out followed by sterilization in 70% ethanol for 30 min and bonding onto poly(d-lysine) (PDL) or Matrigel™-coated substrates (BD Biosciences, San Jose, CA). Each device fits into one-well of a conventional six-well polystyrene culture plate.
Source: From Park et al.,[93] with permission from Springer Science+Business Media.

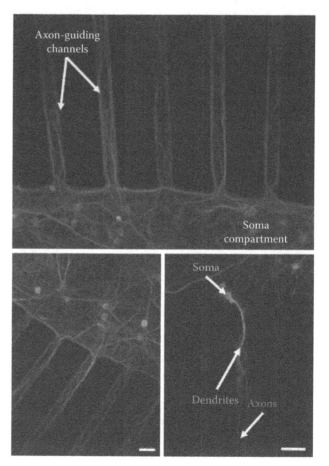

Fig. 3 Immunocytochemistry images of neurons at 2 weeks in culture demonstrate that axons grew from the soma compartment into the axon/glia compartment through the arrays of axon-guiding microchannels, but dendrites and neuronal soma could not reach into the axon/glia compartment due to the length of the microchannels (200–800 μm long) and the height of the microchannels (2.5 μm). Axons were immunostained for neurofilament (NF) and dendrite for microtubule-associated protein 2 (MAP2). Scale bars = 20 μm.
Source: From Park et al.,[93] with permission from Springer Science+Business Media.

To achieve this aspiration, continuous perfusion of nutrition and oxygen to the cells and continuous removal of metabolic waste from cultured cells should be ensured. Such a continuous perfusion culture is attainable using specifically designed microfluidic bioreactor systems. Along with continuous perfusion, some tissues such as blood vessels, cardiac muscles, and cartilage are in need of compression, shear stresses, and pulsative flow for improving mechanical properties; those characteristics can also be afforded by microfluidic bioreactor designs.[2,95] A multilayered polydimethylsiloxane (PDMS)-based microfluidic perfusion bioreactor system with an oxygen chamber for the mammalian cell culture has been successfully demonstrated by high cell densities.[96] A bioreactor design with

continuous perfusion has been examined to obtain a morphogenetically biomimetic 3D liver tissue model.[97] Powers and coworkers[97] revealed that such a continuous perfusion bioreactor system permits hepatocellular aggregate generation much similar to physiological structure, hepatic acinus. In addition, a bioreactor perfusion platform for a neuron-astrocyte co-culture has been inspected for 3D cell-cell and cell-matrix interactions to obtain high-density multicell population for brain tissue regeneration applications.[98] A microbioreactor has been designed significantly for long-term culture of human foreskin fibroblast cells under constant perfusion of culture medium.[99] In addition, a bioreactor perfusion system has also been used for the seeding of human dermal fibroblast cells into porous electrospun poly(ε-caprolactone) nanofibrous mats for the study of skin tissue engineering.[100] Recently, a unique flow perfusion bioreactor embedded with PLLA-collagen scaffold as shown in Fig. 4 was studied for the repair of defects in a complex abdominal wall.[101] It was demonstrated that the culture of dermal fibroblasts under perfusion conditions enhances cell proliferation and distribution within the PLLA-collagen scaffolds. A construct with high cellular density was observed within 7 days of dermal cell culture under perfusion and found increased cellularity by 28th day of culture period. *In vitro* perfusion ensured that a highly cellularized and matured construct was implanted, and its significant survival rate of cells with expression of extracellular protein collagen types I and III was examined. Furthermore, an electrospun poly(ε-caprolactone) microfiber scaffold was also studied after coating with cartilaginous extracellular matrix for cartilage tissue engineering

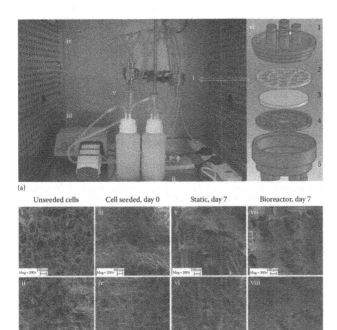

Fig. 4 (**a**) Perfusion bioreactor system consisting of (i) a flow chamber that has a scaffold holder and four media flow channel; (ii) a media container; (iii) a multichannel peristaltic pump (Watson-Marlow 323 series; Watson-Marlow Ltd., Falmouth, UK). (iv) The whole system is housed in a humidified CO2 incubator, allowing gaseous exchange to occur through the silicone rubber tubing, connected to (v) a 0.2-mm sterile filter. The media was pumped through the chamber and continuously perfused around the cell-loaded scaffold from underneath at a flow rate of 6 mL/min for up to 4 weeks. (vi) Schematic representation of 3D model of the bioreactor, indicating (1) screwing top part, (2) top guiding holder, (3) scaffold, (4) bottom guiding holder, and (5) screwing bottom part. (**b**) scanning electron microscopy (SEM) of cell-loaded scaffolds: scaffold without cell seeding (i and ii), cell-seeded for 2 hr only (iii and iv), cell-loaded scaffolds in static condition for 7 days (v and vi), and cell-loaded scaffolds in perfusion bioreactor for 7 days (vii and viii).
Source: From Pu et al.[101] © 2010, with permission from Elsevier.

under a flow perfusion bioreactor system.[102] These studies prove the efficiency of a microfluidic perfusion bioreactor system for the reconstruction of functional large, complex, and/or hard tissues.

CONCLUSION

The challenge for constructing the functional biomimetic artificial organs *in vitro* paves the way for the inventions of several microfabrication technologies such as microfluidics. The most important limitations in developing artificial tissues are allowing hierarchical multicell interactions three-dimensionally under perfusion culture for nutrient supply and waste removal mimicking the physiological environment. By combining biology with bio-MEMS techniques, it is highly possible to overcome the abovementioned limitations to provide a suitable environment for cells to grow and interact *in vitro* and mimic living systems. One bio-MEMS technique is microfluidics, which has high potential for tissue engineering applications. Microfluidics has been utilized to understand fundamental biological functions such as factors influencing cellular proliferation, differentiation or apoptosis, migration, and gene expressions. These understandings help design a suitable culture platform providing physiological microenvironments such as cell-cell interactions, cell-extracellular matrix interactions, soluble signaling phenomena, and mechanical forces for tissue regeneration. Indeed, several researchers are investigating a variety of biodegradable polymeric functional scaffolds together with microfluidics for biomimetic tissue development. Microfluidic-based *in vitro*-developed tissue models were employed not only for regeneration applications but also in drug research such as toxicity analysis, therapeutic activity analysis, and dose analysis. However, utilization of a microfluidic cell culture platform for tissue regeneration applications or drug research is still in the developmental stage, although it is one of the highly emerging fields of interest. In spite of its huge application areas, some technical issues such as complex operation procedure to carry out the experiments or reading out the data in a simple understandable format with an equal or higher level of accuracy than conventional methods should be addressed. To overcome these hurdles, microfluidic chip designers should have a healthy understanding with biologists or users, and therefore, microfluidic scaffolding design and culture platforms would make several steps forward to be commercialized in the near future.

REFERENCES

1. Langer, R.; Vacanti, J. Tissue engineering. Science **1993**, *260* (5110), 920–926.
2. Andersson, H.; Berg, A.V. D. Microfabrication and microfluidics for tissue engineering: State of the art and future opportunities. Lab Chip **2004**, *4* (2), 98–103.

3. Ochoa, E.; Vancanti, J. An overview of the pathology and approaches to tissue engineering. Ann. N.Y. Acad. Sci. **2002**, *979* (1), 10–26.

4. Marler, J.; Upton, J.; Langer, R.; Vacanti, J. Transplantation of cells in matrices for tissue generation. Adv. Drug Deliv. Rev. **1998**, *33* (1–2), 165–182.

5. Liao, C.; Chen, C.; Chen, J.; Chiang, S.; Lin, Y.; Chang, K. Fabrication of porous biodegradable polymer scaffolds using a solvent merging/particulate leaching method. J. Biomed. Mater. Res. **2002**, *59* (4), 676–681.

6. Oh, S.H.; Kang, S.G.; Kim, E.S.; Cho, S.H.; Lee, J.H. Fabrication and characterization of hydrophilic poly(lactic-co-glycolic acid)/poly(vinyl alcohol) blend cell scaffolds by melt-molding particulate-leaching method. Biomaterials **2003**, *24* (22), 4011–4021.

7. Yang, S.; Leong, K.F.; Du, Z.; Chua, C.K. The design of scaffolds for use in tissue engineering: Part II. Rapid prototyping techniques. Tissue Eng. **2002**, *8* (1), 1–11.

8. Borenstein, J.T.; Terai, H.; King, K.R.; Weinberg, E.J.; Kaazempur-Mofrad, M.R.; Vacanti, J.P. Microfabrication technology for vascularized tissue engineering. Biomed. Microdevices **2002**, *4* (3), 167–175.

9. Vozzi, G.; Flaim, C.; Ahluwalia, A.; Bhatia, S. Fabrication of PLGA scaffolds using soft lithography and microsyringe deposition. Biomaterials **2003**, *24* (14), 2533–2540.

10. Zein, I.; Hutmacher, D.W.; Tan, K.C.; Teoh, S.H. Fused deposition modeling of novel scaffold architectures for tissue engineering applications. Biomaterials **2002**, *23* (4), 1169–1185.

11. Lee, G.; Barlow, J.; Fox, W.; Aufdermorte, T. Biocompatibility of SLS-formed calcium phosphate implants. *Proceedings of Solid Freeform Fabrication Symposium*, Austin, TX, August 12–14, 1996.

12. Landers, R.; Hubner, U.; Schmelzeisen, R.; Mulhaupt, R. Rapid prototyping of scaffolds derived from thermoreversible hydrogels and tailored for applications in tissue engineering. Biomaterials **2002**, *23* (23), 4437–4447.

13. Cooke, M.N.; Fisher, J.P.; Dean, D.; Rimnac, C.; Mikos, A.G. Use of stereolithography to manufacture critical-sized 3D biodegradable scaffolds for bone in growth. J. Biomed. Mater. Res. B **2003**, *64B* (2), 65–69.

14. Yu, T.; Chiellini, F.; Schmaljohann, D.; Solaro, R.; Ober, C.K. Microfabrication of hydrogels as polymer scaffolds for tissue engineering applications. Polym. Prepr. **2000**, *41* (2), 1699–1700.

15. Ravichandran, R.; Sundarrajan, S.; Venugopal, J.R.; Mukherjee, S.; Ramakrishna, S. Advances in polymeric systems for tissue engineering and biomedical applications. Macromol. Biosci. **2012**, *12* (3), 286–311.

16. Shimizu, T.; Yamato, M.; Kikuchi, A.; Okano, T. Cell sheet engineering for myocardial tissue reconstruction. Biomaterials **2003**, *24* (13), 2309–2316.

17. El-Ali, J.; Sorger, P.K.; Jensen, K.F. Cells on chips. Nature **2006**, *442* (7101), 403–411.

18. Tan, W.; Desai, T.A. Microfluidic patterning of cells in extracellular matrix biopolymers: Effects of channel size, cell type, and matrix composition on pattern integrity. Tissue Eng. **2003**, *9* (2), 255–267.

19. Taylor, A.M.; Blurton-Jones, M.; Rhee, S.W.; Cribbs, D.H.; Cotman, C.W.; Jeon, N.L. A microfluidic culture platform for CNS axonal injury, regeneration and transport. Nat. Methods **2005**, *2* (8), 599–605.

20. Leclerc, E.; Sakai, Y.; Fujii, T. Perfusion culture of fetal human hepatocytes in microfluidic environments. Biochem. Eng. J. **2004**, *20* (2–3), 143–148.

21. Kane, B.J.; Zinner, M.J.; Yarmush, M.L.; Toner, M. Liver-specific functional studies in a microfluidic array of primary mammalian hepatocytes. Anal. Chem. **2006**, *78* (13), 4291–4298.

22. Tan, W.; Desai, T.A. Microscale multilayer coculture for biomimetic blood vessels. J. Biomed. Mater. Res. A **2004**, *72* (2), 146–160.

23. Marimuthu, M.; Kim, S. Survey of the state of the art in biomaterials, cells, genes and proteins integrated into micro- and nanoscaffolds for tissue regeneration. Curr. Nanosci. **2009**, *5* (2), 189–203.

24. Chevallay, B.; Herbage, D. Collagen-based biomaterials as 3D scaffolds for cell cultures: Application for tissue engineering and gene therapy. Med. Biol. Eng. Comput. **2000**, *38* (2), 211–218.

25. Awad, H.A.; Quinn, W.M.; Leddy, H.A.; Gimble, J.M.; Guilak, F. Chondrogenic differentiation of adipose-derived adult stem cells in agarose, alginate, and gelatin scaffolds. Biomaterials **2004**, *25* (16), 3211–3222.

26. Barquinero, J.; Eixarch, H.; Perez-Melgosa, M. Retroviral vectors: New applications for an old tool. Gene Ther. **2004**, *11* (Suppl. 1), S3–S9.

27. Zhang, Y.; Cheng, X.; Wang, J.; Wang, Y.; Shi, B.; Huang, C.; Yang, X.; Liu, T. Novel chitosan/collagen scaffold containing transforming growth factor-beta1 DNA for periodontal tissue engineering. Biochem. Biophys. Res. Commun. **2006**, *344* (1), 362–369.

28. Grunder, T.; Gaissmaier, C.; Fritz, J.; Stoop, R.; Hortschansky, P.; Mollenhauer, J. Bone morphogenetic protein-2 enhances the expression of type II collagen and aggrecan in chondrocytes embedded in alginate beads. Osteoarthr. Cartil. **2004**, *12* (7), 559–567.

29. Yun, Y.H.; Goetz, D.J.; Yellen, P.; Chen, W.L. Hyaluronan microspheres for sustained gene delivery and site-specific targeting. Biomaterials **2004**, *25* (1), 147–157.

30. Liu, Y.; Yang, H.; Otaka, K.; Takatsuki, H.; Sakanishi, A. Effects of vascular endothelial growth factor (VEGF) and chondroitin sulfate A on human monocytic THP-1 cell migration. Colloids Surf. B **2005**, *43* (3–4), 216–220.

31. Jiang, W.G.; Harding, K.G. Enhancement of wound tissue expansion and angiogenesis by matrix-embedded fibroblast (Dermagraft), a role of hepatocyte growth factor scatter factor. Int. J. Mol. Med. **1998**, *2* (2), 203–210.

32. Rives, C.B.; Rieux, A.D.; Zelivyanskaya, M.; Stock, S.R.; Lowe, W.L.; Shea, L.D. Layered PLG scaffolds for *in vivo* plasmid delivery. Biomaterials **2009**, *30* (3),, 394–401.

33. Wei, G.; Jin, Q.; Giannobile, W.V.; Ma, P.X. Nano-fibrous scaffold for controlled delivery of recombinant human PDGF-BB. J. Control. Release **2006**, *112* (1), 103–110.

34. Stauffera, W.R.; Cuia, X.T. Polypyrrole doped with 2 peptide sequences from laminin. Biomaterials **2006**, *27* (11), 2405–2413.

35. Park, H.; Temenoff, J.S.; Tabata, Y.; Capland, A.I.; Mikos, A.G. Injectable biodegradable hydrogel composites for rabbit marrow mesenchymal stem cell and growth factor

delivery for cartilage tissue engineering. Biomaterials **2007**, *28* (21), 3217–3227.

36. Rashid, S.T.; Salacinski, H.J.; Button, M.J.C., Fuller, B.; Hamilton, G.; Seifalian, A.M. Cellular engineering of conduits for coronary and lower limb bypass surgery: Role of cell attachment peptides and pre-conditioning in optimizing smooth muscle cells (SMC) adherence to compliant poly(carbonate-urea)urethane (MyoLinke) scaffolds. Eur. J. Vasc. Endovasc. Surg. **2004**, *27* (6), 608–616.

37. Fisher, J.P.; Vehof, J.W.M., Dean, D.; van der Waerden, J.; Holland, T.A.; Mikos, A.G.; Jansen, J.A. Soft and hard tissue response to photocrosslinked poly(propylene fumarate) scaffolds in a rabbit model. J. Biomed. Mater. Res. **2002**, *59* (3), 547–556.

38. Conconi, M.T.; Lora, S.; Menti, A.M.; Carampin, P.; Parnigotto, P.P. *In vitro* evaluation of poly[bis(ethyl alanato)phosphazene] as a scaffold for bone tissue engineering. Tissue Eng. **2006**, *12* (4), 811–819.

39. Abidian, M.R.; Corey, J.M.; Kipke, D.R.; Martin, D.C. Conducting-polymer nanotubes improve electrical properties, mechanical adhesion, neural attachment, and neurite outgrowth of neural electrodes. Small **2010**, *6* (3), 421–429.

40. Thompson, B.C.; Richardson, R.T.; Moulton, S.E.; Evans, A.J.; Stephen, O.; Clark, G.M.; Wallace, G.C. Conducting polymers, dual neurotrophins and pulsed electrical stimulation—Dramatic effects on neurite outgrowth. J. Control. Release **2010**, *141* (2), 161–167.

41. Shi, L.; Tang, G.P.; Gao, S.J.; Ma, Y.X.; Liu, B.H.; Li, Y.; Zeng, J.M.; Ng, Y.K.; Leong, K.,W., Wang, S. Repeated intrathecal administration of plasmid DNA complexed with polyethylene glycol-grafted polyethylenimine led to prolonged trans-gene expression in the spinal cord. Gene Ther. **2003**, *10* (14), 1179–1188.

42. Liu, X.; Holzwarth, J.M.; Ma, P.X. Functionalized synthetic biodegradable polymer scaffolds for tissue engineering. Macromol. Biosci. **2012**, *12* (7), 911–919.

43. Bettinger, C.J. Biodegradable elastomers for tissue engineering and cell-biomaterial interactions. Macromol. Biosci. **2011**, *11* (4), 467–482.

44. Jia, X.; Kiick, K.L. Hybrid multicomponent hydrogels for tissue engineering. Macromol. Biosci. **2009**, *9* (2), 140–156.

45. Ravichandran, R.; Venugopal, J.R.; Sundarrajan, S.; Mukherjee, S.; Ramakrishna, S. Applications of conducting polymers and their issues in biomedical engineering. J. R. Soc. Interface **2010**, *7* (Suppl_5), 559–579.

46. Schmaljohann, B. Thermo and pH-responsive polymers in drug delivery. Adv. Drug. Deliv. Rev. **2006**, *58* (15), 1655–1670.

47. Kim, H. K, Shim, W.S.; Kim, S.E.; Lee, K.H.; Kang, E.; Kim, J.H.; Kim, K.; Kwon, I.C.; Lee, D.S. Injectable in situ-forming pH/thermo-sensitive hydrogel for bone tissue engineering. Tissue Eng. A **2009**, *15* (4), 923–933.

48. Dai, S.; Ravi, P.; Tam, K.C. Thermo- and photo-responsive polymeric systems. Soft Matter **2009**, *5* (13), 2513–2533.

49. Zheng, X.; Zhou, S.; Li, X.; Weng, J. Shape memory properties of poly(D,L-lactide)/hydroxyapatite composites. Biomaterials **2006**, *27* (24), 4288–4295.

50. Stuart, M.A.C., Huck, W.T.S., Genzer, J.; Mttller, M.; Ober, C.; Stamm, M.; Sukhorukov, G.B.; Szleifer, I.; Tsukruk, V.V.; Urban, M.; Winnik, F.; Zauscher, S.; Luzinov, I.; Minko, S. Emerging applications of stimuli-responsive polymer materials. Nat. Mater. **2010**, *9* (2), 101–113.

51. Park, H.; Lee, K.Y.; Lee, S.J.; Park, K.E.; Park, W.H. Plasma-treated poly(lactic-co-glycolic acid) nanofibers for tissue engineering. Macromol. Res. **2007**, *15* (3), 238–243.

52. Wan, Y.; Qu, X.; Lu, J.; Zhu, C.; Wan, L.; Yang, J. Characterization of surface property of poly(lactide-co-glycolide) after oxygen plasma treatment. Biomaterials **2004**, *25* (19), 4777–4783.

53. Shen, H.; Hu, X.; Bei, J.; Wang, S. The immobilization of basic fibroblast growth factor on plasma-treated poly(lactide-co-glycolide). Biomaterials **2008**, *29* (15), 2388–2399.

54. Chandrasekaran, A.R.; Venugopal, J.; Sundarrajan, S.; Ramakrishna, S. Fabrication of a nanofibrous scaffold with improved bioactivity for culture of human dermal fibroblasts for skin regeneration. Biomed. Mater. **2011**, *6* (1), 015001.

55. Dalton, P.D.; Flynn, L.; Shoichet, S.C. Manufacture of poly(2-hydroxyethyl methacrylate-co-methyl methacrylate) hydrogel tubes for use as nerve guidance channels. Biomaterials **2002**, *22* (18), 3843–3851.

56. Chou, S.Y.; Krauss, P.R.; Renstrom, P.J. Imprint lithography with 25-nanometer resolution. Science **1996**, *272* (5258), 85–87.

57. Jeong, W.; Kim, J.; Kim, S.; Lee, S.; Mensingc, G.; Beebe, D.J. Hydrodynamic microfabrication via "on the fly" photopolymerization of microscale fibers and tubes. Lab Chip **2004**, *4* (6), 576–580.

58. Shin, S.; Park, J.; Lee, J.; Park, H.; Park, Y.; Lee, K.; Whang, C.; Lee, S. "On the fly" continuous generation of alginate fibers using a microfluidic device. Langmuir **2007**, *23* (17), 9104–9108.

59. Hwang, C.M.; Khademhosseini, A.; Park, Y.; Sun, K.; Lee, S. Microfluidic chip-based fabrication of PLGA microfiber scaffolds for tissue engineering. Langmuir **2008**, *24* (13), 6845–6851.

60. Marimuthu, M.; Kim, S.; An, J. Amphiphilic triblock copolymer and a microfluidic device for porous microfiber fabrication. Soft Matter **2010**, *6* (10), 2200–2207.

61. Borden, M.; El-Amin, S.F.; Attawia, M.; Laurencin, C.T. Structural and human cellular assessment of a novel microsphere-based tissue engineered scaffold for bone repair. Biomaterials **2003**, *24* (4), 597–609.

62. Borden, M.; Attawia M., Laurencin, C.T. The sintered microsphere matrix for bone tissue engineering: *In vitro* osteoconductivity studies. J. Biomed. Mater. Res. **2002**, *61* (3), 421–429.

63. Malafaya, P.B.; Pedro, A.J.; Peterbauer, A.; Gabriel, C.; Redl, H.; Reis, R.L. Chitosan particles agglomerated scaffolds for cartilage and osteochondral tissue engineering approaches with adipose tissue derived stem cells. J. Mater. Sci. Mater. Med. **2005**, *16* (12), 1077–1085.

64. Dendukuri, D.; Doyle, P.S. The synthesis and assembly of polymeric microparticles using microfluidics. Adv. Mater. **2009**, *21* (41),4071–4086.

65. Chung, S.E.; Park, W.; Shin, S.; Lee, S.A.; Kwon, S. Guided and fluidic self-assembly of microstructures using

Scaffolds—Smart

railed microfluidic channels. Nat. Mater. **2008**, *7* (7), 581–587.

66. Huang, J.; Kim, J.; Agrawal, N.; Sudarsan, A.P.; Maxim, J.E.; Jayaraman, A.; Ugaz, V.M. Rapid fabrication of bio-inspired 3D microfluidic vascular networks. Adv. Mater. **2009**, *21* (35), 3567–3571.

67. Cabodi, M.; Choi, N.W.; Gleghorn, J.P.; Lee, C.S.D., Bonassar, L.J.; Stroock, A.D. A microfluidic biomaterial. J. Am. Chem. Soc. **2005**, *127* (40), 13788–13789.

68. King, K.R.; Wang, C.C.J., Kaazempur-Mofrad, M.R.; Vacanti, J.P.; Borenstein, J.T. Biodegradable microfluidics. Adv. Mater. **2004**, *16* (22), 2007–2012.

69. Choi, N.W.; Cabodi, M.; Held, B.; Gleghorn, J.P.; Bonassar, L.J.; Stroock, A.D. Microfluidic scaffolds for tissue engineering. Nat. Mater. **2007**, *6* (11), 908–915.

70. Bettinger, C.J.; Cyr, K.M.; Matsumoto, A.; Langer, R.; Borenstein, J.T.; Kaplan, D.L. Silk fibroin microfluidic devices. Adv. Mater. **2007**, *19* (19), 2847–2850.

71. Fidkowski, C.; Kaazempur-Mofrad, M.R.; Borenstein, J.T.; Vacanti, J.P.; Langer, R.; Wang, Y.D. Endothelialized microvasculature based on a biodegradable elastomer. Tissue Eng. **2005**, *11* (1–2), 302–309.

72. Wang, J.; Bettinger, C.J.; Langer, R.S.; Borenstein, J.T. Biodegradable microfluidic scaffolds for tissue engineering from amino alcohol-based poly(ester amide) elastomers. Organogenesis **2010**, *6* (4), 212–216.

73. Bettinger, C.J.; Borenstein, J.T. Biomaterials-based microfluidics for engineered tissue constructs. Soft Matter **2010**, *6* (20), 4999–5015.

74. Vickerman, V.; Blundo, J.; Chung, S.; Kamm, R.D. Design, fabrication and implementation of a novel multi-parameter control microfluidic platform for three-dimensional cell culture and real-time imaging. Lab Chip **2008**, *8* (9), 1468–1477.

75. Ng, C.P.; Pun, S.H. A perfusable 3D cell-matrix tissue culture chamber for in situ evaluation of nanoparticle vehicle penetration and transport. Biotechnol. Bioeng. **2008**, *99* (6), 1490–1501.

76. Wu, M.; Huang, S.; Lee, G. Microfluidic cell culture systems for drug research. Lab Chip **2010**, *10* (8), 939–956.

77. Prokop, A.; Prokop, Z.; Schaffer, D.; Kozlov, E.; Wikswo, J.; Cliffel, D.; Baudenbacher, F. NanoLiterBioReactor: Long-term mammalian cell culture at nanofabricated scale. Biomed. Microdevices **2004**, *6* (4), 325–339.

78. Chong, S.L.; Mou, D.G.; Ali, A.M.; Lim, S.H.; Tey, B.T. Cell growth, cell-cycle progress, and antibody production in hybridoma cells cultivated under mild hypothermic conditions. Hybridoma **2008**, *27* (2), 107–111.

79. Brandam, C.; Castro-Martinez, C.; Delia, M.L.; Ramon-Portugal, F.; Strehaiano, P. Effect of temperature on Brettanomyces bruxellensis: Metabolic and kinetic aspects. Can. J. Microbiol. **2008**, *54* (1), 11–18.

80. Huang, C.W.; Lee, G.B. A microfluidic system for automatic cell culture. J. Micromech. Microeng. **2007**, *17* (7), 1266–1274.

81. Leclerc, E.; Miyata, F.; Furukawa, K.S.; Ushida, T.; Sakai, Y.; Fujii, T. Effect on liver cells of stepwise microstructures fabricated in a photosensitive biodegradable polymer by soft lithography. Mater. Sci. Eng. C **2004**, *24* (3), 349–354.

82. Toh, Y.C.; Lim, T.C.; Tai, D.; Xiao, G.; van Noort, D.; Yu, H. A microfluidic 3D hepatocyte chip for drug toxicity testing. Lab Chip **2009**, *9* (14), 2026–2035.

83. Toh, Y.C.; Zhang, C.; Zhang, J.; Khong, Y.M.; Chang, S.; Samper, V.D.; van Noort, D.; Hutmacher, D.W.; Yu, H. A novel 3D mammalian cell perfusion-culture system in microfluidic channels. Lab Chip **2007**, *7* (3), 302–309.

84. Leclerca, E.; Baudoina, R.; Corlub, A.; Griscomc, L.; Duvala, J.L.; Legallais, C.C. Selective control of liver and kidney cells migration during organotypic cocultures inside fibronectin-coated rectangular silicone microchannels. Biomaterials **2007**, *28* (10), 1820–1829.

85. Tsacopoulos, M.; Magistretti, P.J. Metabolic coupling between glia and neurons. J. Neurosci. **1996**, *16* (3), 877–885.

86. Aschner, M. Neuron-astrocyte interactions: Implications for cellular energetics and antioxidant levels. Neurotoxicology **2000**, *21* (6), 1101–1107.

87. Tsacopoulos, M. Metabolic signaling between neurons and glial cells: A short review. J. Physiol. **2002**, *96* (3), 283–288.

88. Ng, B.K.; Chen, L.; Mandemakers, W.; Cosgaya, J.M.; Chan, J.R. Anterograde transport and secretion of brain-derived neurotrophic factor along sensory axons promote Schwann cell myelination. J. Neurosci. **2007**, *27* (28), 7597–7603.

89. Millet, L.; Stewart, M.; Sweedler, J.; Nuzzo, R.; Gillette, M. Microfluidic devices for culturing primary mammalian neurons at low densities. Lab Chip **2007**, *7* (8), 987–994.

90. Morin, F.; Nishimura, N.; Griscom, L.; LePioufle, B.; Fujita, H.; Takamura, Y.; Tamiya, E. Constraining the connectivity of neuronal networks cultured on micro-electrode arrays with microfluidic techniques: A step towards neuron based functional chips. Biosens. Bioelectron. **2006**, *21* (7), 1093–1100.

91. Baumann, N.; Pham-Dinh, D. Biology of oligodendrocyte and myelin in the mammalian central nervous system. Physiol. Rev. **2001**, *81* (2), 871–927.

92. Sherman, D.L.; Brophy, P.J. Mechanisms of axon ensheathment and myelin growth. Nat. Rev. Neurosci. **2005**, *6* (9), 683–690.

93. Park, J.; Koito, H.; Li, J.; Han, A. Microfluidic compartmentalized co-culture platform for CNS axon myelination research. Biomed. Microdevices **2009**, *11* (6), 1145–1153.

94. Marimuthu, M.; Kim, S. Microfluidic cell coculture methods for understanding cell biology, analyzing bio/pharmaceuticals, and developing tissue constructs. Anal. Biochem. **2011**, *13* (2), 81–89.

95. Grodzinsky, A.; Levenston, M.; Jin, E.; Frank, H. Cartilage tissue remodeling in response to mechanical forces. Annu. Rev. Biomed. Eng. **2000**, *2* (1), 691–713.

96. Leclerc, E.; Sakai, Y.; Fujii, T. A multi-layer PDMS microfluidic device for tissue engineering applications. *Proceedings of MEMS*, Kyoto, Japan, January 19–23, 2003.

97. Powers, M.; Domansky, K.; Kaazempur-Mofrad, M.; Kalezi, A.; Capitano, A.; Upadhyaya, A.; Kurzawski, P.; et al. A microfabricated array bioreactor for perfused 3D liver culture. Biotech. Bioeng. **2002**, *78* (3), 257–269.

98. Cullen, D.K.; Vukasinovic, J.; Glezer, A.; LaPlaca, M.C. Microfluidic engineered high cell density three-dimensional neural cultures. J. Neural Eng. **2007**, *4* (2), 159–172.

99. Korin, N.; Bransky, A.; Dinnar, U.; Levenberg, S. A parametric study of human fibroblasts culture in a microchannel bioreactor. Lab Chip **2007**, *7* (5), 611–617.

100. Lowery, J.L.; Datta, N.; Rutledge, G.C. Effect of fiber diameter, pore size and seeding method on growth of human dermal fibroblasts in electrospun poly(3-caprolactone) fibrous mats. Biomaterials **2010**, *31* (3), 491–504.

101. Pu, F.; Rhodes, N.P.; Bayon, Y.; Chen, R.; Brans, G.; Benne, R.; Hunt, J.A. The use of flow perfusion culture and subcutaneous implantation with fibroblast-seeded PLLA-collagen 3D scaffolds for abdominal wall repair. Biomaterials **2010**, *31* (15), 4330–4340.

102. Liao, J.; Guo, X.; Grande-Allen, K.J.; Kasper, F.K.; Mikos, A.G. Bioactive polymer/extracellular matrix scaffolds fabricated with a flow perfusion bioreactor for cartilage tissue engineering. Biomaterials **2010**, *31* (34), 8911–8920.

Scaffolds—Smart

Scaffolds, Porous Polymer: Tissue Engineering

Guoping Chen

Tissue Regeneration Materials Unit, International Center for Materials Nanoarchitectonics, National Institute for Materials Science, Tsukuba, Japan

Abstract

Tissue engineering has been used to regenerate functional tissues and organs for the treatment of various diseases and defects. Porous scaffolds play an important role as a temporary support for the implanted cells to support their assembling and to control their functions to guide new tissue and organ formations. The properties of scaffolds such as pore structures and mechanical properties should be carefully designed and well controlled to meet the requirements. Many methods have been developed to prepare porous scaffolds of synthetic polymers, naturally derived polymers, and their hybrids for tissue engineering. Polymeric porous scaffolds have been used for tissue engineering of various tissues and organs. This entry summarizes the recent developments of polymeric porous scaffolds by focusing on the preparation methods using pre-prepared ice particulates and the hybridization of synthetic polymers and naturally derived polymers.

INTRODUCTION

Tissue engineering has been developed as an attractive approach to treat the diseases and defects that are uncurable or difficult to be treated by conventional drug administration, artificial prosthesis, and organ transplantation.[1] Although scaffold-free tissue engineering has been used to regenerate thin or small tissues,[2] generally scaffolds are required for tissue engineering, especially for regeneration of tissues and organs with large scale and complicated structures.[3–5] The scaffolds serve as temporary supports to allow cell adhesion, promote cell proliferation and differentiation, assemble the cells and extracellular matrices, and guide the formation of functional tissues and organs.[6–8] Various porous scaffolds have been developed from synthetic polymers and naturally derived polymers to mimic the *in vivo* extracellular microenvironment for the implanted cells to control their functions.[9–11] Many methods have been developed to control the pore structures and properties of porous scaffolds.[2–19] This entry summarizes the recent developments of polymeric porous scaffolds and their preparation methods.

BIODEGRADABLE POLYMERS FOR TISSUE ENGINEERING

Biodegradable polymers been widely used for tissue engineering because of easy control over biodegradability and processability. There are two types of biodegradable polymers: synthetic and naturally derived polymers.[20] Biodegradable synthetic polymers include polyesters, polyanhydride, polyorthoester, polycaprolactone, polycarbonate, and polyfumarate. Polyesters such as poly(glycolic acid) (PGA), poly(lactic acid) (PLA), and their copolymer of poly(lactic-*co*-glycolic acid) (PLGA) are the principal synthetic polymers used for tissue engineering. They have gained the approval of the US Food and Drug Administration for certain human clinical use, such as surgical sutures and some implantable devices.[21]

The naturally derived biodegradable polymers include proteins of natural extracellular matrices such as collagen and glycosaminoglycan (GAG), alginic acid, chitosan, and polypeptides. Collagen is the most abundant and ubiquitous protein in vertebrates. It accounts for 20–30% of total body proteins and its functions range from serving crucial biomechanical purposes in skin, bone, cartilage, tendon, and ligament to controlling cellular gene expression in development. More than 20 genetically distinct collagens have been identified. The versatile properties of collagen have made it one of the most useful biomaterials for tissue engineering. It can provide biological cues to support cell adhesion, proliferation, and differentiation. GAG is usually attached to proteins as a part of a proteoglycan. GAG can be incorporated in a collagen matrix to improve bioactivity and biomechanical properties to better imitate the biological environment because of its role in retaining water, which helps maintain structural integrity, as well as its importance in interacting with proteins in extracellular matrix (ECM).[22] Alginic acid, a polysaccharide from seaweed, is a family of natural copolymers of β-D-mannuronic acid (M) and α-L-guluronic acid (G). Chitosan is a natural polysaccharide, whose structural characteristics are similar to GAGs. Polypeptides with some amino acid sequences

Concise Encyclopedia of Biomedical Polymers and Polymeric Biomaterials DOI: 10.1081/E-EBPPC-120050755

can be favorable to cell adhesion and function and thus they may have potential for cell attachment and transplantation. These naturally derived biodegradable polymers have been used for scaffolds of a variety of tissues such as bone, liver, neural tissue, vascular tissue, cartilage, and skin.[10–16]

POROUS SCAFFOLDS PREPARED BY POROGEN LEACHING

Particle leaching is one of the most useful methods to prepare porous scaffolds.[12] The method includes dispersion of a porogen agent in a liquid, particulate, or powder-based base material and selective removal of the porogen agent after solidification or compacting of the dispersion mixture. The liquid may be solidified by solvent evaporation, cross-linking, or other reactions, and the powder may be compacted using pressure and temperature. After the shaping process, porogen particulates are dispersed in the solid base matrix and are dissolved by immersion in a solvent specific to the porogen particulates to obtain the porous scaffolds. Advantages of this method include its simplicity, versatility, and ease of control of the pore size, porosity, and geometry. The pore geometry and pore size can be controlled by the shape and size of the porogen particulates. The porosity can be controlled by the ratio of the porogen particulates to the base materials. Porogen agents such as the particulates of sodium chloride, tartrate, citrate, sugar, gelatin, and paraffin can be used. To increase the interconnectivity of the pores, the porogen particulates can be melted to make them interconnected before leaching or creating nano-structure porous structures in the walls of the pores.[22,23] Porous scaffolds of biodegradable synthetic polymers such as PLA, PLGA, and PCL are frequently prepared by this method. The porogen leaching method has also be combined with other methods such as electrospinning and elusion freeze drying to increase the porosity and large pores in the scaffolds to facilitate cell seeding and cell distribution.

POROUS SCAFFOLDS PREPARED WITH ICE PARTICULATES AS A POROGEN MATERIAL

Ice particulates are a good porogen material for the preparation of porous scaffold as ice particulates can be easily and completely removed by freeze-drying.[24] Pre-prepared ice particulates can be formed by spraying or injecting cold deionized water into liquid nitrogen through a capillary. The ice particulates are almost spherical. Their diameters can be controlled by the spraying speed. The preparation procedure of polymer porous scaffolds using pre-prepared ice particulates is shown in Fig. 1. The pre-prepared ice particulates were homogenously mixed with polymer solution in a mold. The mixing temperature should be set at a temperature where the ice particulates do not melt and polymer solution does not freeze. The mixture is then frozen and freeze-dried to remove the solvent and ice particulates for pore structure formation. If the polymers are water-soluble, the porous scaffolds should be cross-linked after freeze-drying.

When the ice particulates are used for the preparation of synthetic polymer scaffolds, mixing of ice particulates with polymer solution is easy because synthetic polymers are usually dissolved in a non-water solvent whose freezing temperature is much lower than melting temperature of ice particulates. The synthetic polymer solution should be pre-cooled at a temperature lower than 0°C to protect ice particulates melting during the mixture. PLGA and PLLA sponges have been prepared by this method.[24] The prepared PLGA and PLLA sponges have a porous structure with evenly distributed and interconnected pores. The pore shapes are almost the same as those of the ice particulates. The degree of interconnection within the sponges increases as the weight fraction of the ice particulates increase. The polymer concentration has some effect on the pore wall structure. Lower polymer concentration results in more porous pore wall structures. The porosity and surface area/weight ratio increase with the increase of the weight ratio of the ice particulates. The scaffold pore structure can be

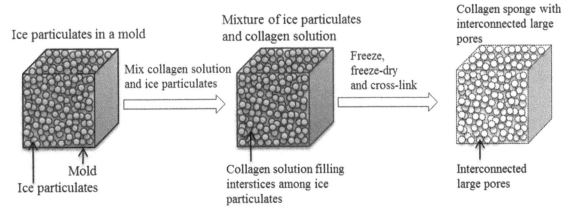

Fig. 1 Preparation procedure of collagen sponge with interconnected large pores by using pre-prepared ice particulates as a porogen material.

manipulated by varying the shape, weight ratio, size of the ice particulates, and the polymer concentration.

To apply the method to prepare porous scaffolds of naturally derived polymers, an appropriate solvent should be chosen to facilitate the mixing of ice particulates in the polymer solution. Unlike synthetic polymers, naturally derived polymers are easily dissolved in water to prepare their aqueous solutions. However, aqueous solutions will freeze around 0°C, which makes mixing with ice particulates difficult. Usually, a mixture solvent of water and ethanol is used to prepare the aqueous solution of naturally derived polymers to lower the freezing temperature of the solution. As one example, the preparation of collagen sponges with ice particulates is described in detail.[25,26]

At first, the aqueous collagen solution is prepared by dissolving freeze-dried collagen in a mixture of ethanol and acetic acid (10:90 v/v, pH 3.0) at 4°C to obtain a designated concentration of collagen aqueous solution. The acetic acid/ethanol mixture was used to decrease the freezing temperature of the collagen solution below −4°C. Subsequently, pre-prepared ice particulates are added to the collagen aqueous solution and well mixed by mixing and pushing with a steel spoon in a −4°C chamber. The collagen aqueous solution is very viscous. The mixing process should be carefully carried out to ensure complete and homogeneous mixing while avoiding the formation of air bubbles in the mixture. Pre-prepared ice particulates can be sieved by sieves with different sizes of mesh pores to obtain ice particulates having a designated range of size. The temperature of collagen aqueous solution is adjusted to −4°C by putting it in a −4°C chamber before mixing (temperature balancing). Finally, the collagen solution/ice particulates mixture is frozen by placing the whole set in a −80°C freezer and the frozen structure is freeze-dried by a freeze-dryer. The freeze-dried collagen sponges are cross-linked and washed.

The collagen sponges have a pore structure of interconnected large pores that are surrounded by some small pores. The large pores are spherical and of the same size as the ice particulates used as porogen material. The small pores have a random morphology and different sizes. The large pores are negative replicas of the pre-prepared ice particulates, while small pores are negative replicas of newly formed ice crystals during the freezing process. When an aqueous solution of synthetic polymers is used, pre-prepared ice particulates not only work as porogens to control the pore size and porosity but also work as nuclei to initiate the formation of new ice crystals in the surrounding aqueous solution, therefore increasing the interconnectivity of the scaffolds. Therefore, the large pores can be controlled by the pre-prepared ice particulates while the surrounding small pores can be controlled by the temperature when the mixture of collagen aqueous solution and pre-prepared ice particulates are frozen.

The effect of ice particulates ratio of 25%, 50%, and 75% has been compared. The density of the large spherical pores is low when 25% ice particulates are used and it increases with increasing percentages of ice particulates. The ice-collagen scaffolds prepared with 50% ice particulates show the most homogenous pore structure. When 75% ice particulates are used, some collapsed large pores are observed due to an insufficient amount of collagen matrix and incomplete mixing. A higher ratio of pre-prepared ice particulates to the collagen aqueous solution results in more compact large pores, and vice versa. The ratio of ice particulates should not be too high or too low. When the ratio of ice particulates becomes too high, homogeneous mixing of ice particulates with collagen solution will become difficult. Nonhomogeneous mixing may result in nonhomogeneous distribution of collagen aqueous solution in the interstices among ice particulates. There will be some collapsed regions in the final porous scaffolds. A too small ratio of ice particulates cannot induce the formation of sufficient large pores to guarantee even distribution of seeded cells.

The concentration of collagen aqueous solution can also affect the pore structures. The effect of the collagen concentration on the pore structure has been investigated by fixing the ice particulate ratio at 50% (w/v) and changing the collagen concentration from 1% to 3% (w/v). Collapsed large pores are observed in ice-collagen scaffolds prepared with 1% and 3% aqueous collagen solutions. The collapsed large pores in collagen scaffolds prepared with the 1% aqueous collagen solution occur because the low concentration results in a less dense collagen matrix surrounding the large pores. When using a 3% aqueous collagen, imperfect structures are formed, which may be due to incomplete mixing because 3% collagen solution is too viscous. Collagen sponges prepared with the 2% collagen solution and an ice particulate/collagen solution ratio of 50% show the most homogeneous pore structure.

Collagen sponges prepared with a different ratio of ice particulates and concentration of collagen aqueous solution have different mechanical properties. Young's modulus of collagen sponges prepared with 2% collagen aqueous solution increases in the following order: 75% < 25% < 50% (ratio of ice particulates, Table 1). Collagen sponges prepared with 50% ice particulates have the highest Young's modulus. The differences in the mechanical properties can be explained by the different pore structures. The spherical

Table 1 Compression mechanical property of collagen porous scaffolds prepared with different ratios of ice particulates

Ratio of ice particulates (%)	Young's modulus (kPa)
0	0.9 ± 0.1
25	12.8 ± 0.9[a]
50	14.1 ± 1.3[a]
75	0.8 ± 0.1

[a]Significant difference compared with collagen scaffold prepared with 0% ice particulates ($p < 0.001$).

shape and well compacting are thought to contribute to high mechanical property. The high mechanical strength of the collagen sponges prepared with 50% ice particulates should be due to the most appropriate packing of the large spherical pores and appropriate filling of the collagen matrix between the large spherical pores. Collagen concentration can also affect mechanical property. Young's modulus increases as the collagen concentration increases, which can be explained by the presence of a dense collagen matrix surrounding the large pores when the collagen concentration increases.

Collagen sponges have been used for three-dimensional culturing of bovine articular chondrocytes for cartilage tissue engineering. The collagen sponges are cut into cylinders (Ø8 mm × H4 mm) and cells are seeded on the cutting surface. The cells can penetrate into the inner regions and show even distribution. The collagen sponges prepared with 2% collagen and 50% ice particulates show the most homogeneous cell distribution due to its homogeneous pore structure. *In vitro* culture and *in vivo* implantation show the well-controlled pore structure, and improved mechanical properties of the collagen sponges facilitate the regeneration of cartilage tissue.

PREPARATION OF OPEN PORE STRUCTURE BY EMBOSSING ICE PARTICULATES

Open and interconnected porous structures are important for scaffolds to ensure smooth cell seeding and spatially uniform cell distribution. Many of the scaffold preparation methods have the problem of closed surface pore structures that inhibit cell penetration into the inner body of the scaffolds and result in uneven cell distribution. The preceding described method by using ice particulates as a porogen material can improve the inner bulk pore structures but not the surface pore structures. Although collagen sponges prepared by this method have well interconnected large pores in the bulk, some of the surface pores are closed. The sponges should be cut to expose the inner bulk pores when being used for cell culture. To solve the openness of surface pores, a method by using embossing ice particulates as a template has been developed.[27]

The method is similar to the previous ice particulates method. However, the ice particulates are not incorporated in polymer solution but kept on the surface of the polymer solution. The preparation procedure is as follows. At first, hemispheric water droplets are prepared by spraying pure water onto a hydrophobic surface or application of moisture onto a hydrophobic surface using an ultrasonic humidifier. The size of the water droplets is controlled by the number of spraying times or the moisture application time. Embossing ice particulates are formed after freezing the water droplets (Fig. 2). Subsequently, an aqueous solution of polymer is eluted onto the embossing ice particulates and kept at a designated freezing temperature to gradually

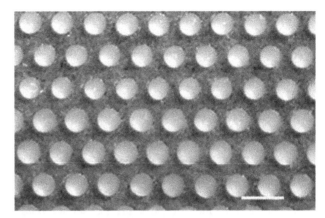

Fig. 2 Photomicrograph of embossing ice particulates on a template. Scale bars = 1000 µm.

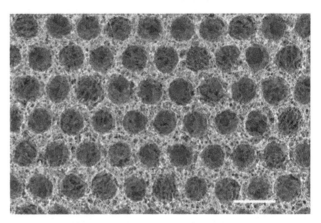

Fig. 3 SEM photomicrograph of funnel-like collagen sponge prepared with embossing ice particulates. Scale bars = 1000 µm.

freeze. Finally, the ice particulate template and ice crystals are removed by freeze-drying. After cross-linking, blocking of any non-reacted cross-linking agent, washing, and a second freeze-drying, porous scaffolds are prepared. The porous scaffolds prepared by this method have funnel-like porous structures. The funnel-like scaffolds have large open surface pores on their top surface and highly interconnected small pores beneath the top surfaces (Fig. 3).

The formation of the unique porous structure may be due to the formation of new ice crystals initiated by the embossing ice particulates. The embossing ice particulates can serve as nuclei to initiate ice crystallization at the freezing interface of the liquid phase polymer solution. This leads to the growth of natural dendritic ice crystals from the surface of the ice particulates into the bulk aqueous solution. The ice particulates together with the newly formed dendritic ice crystal network establish the unique funnel-like porous structure of the scaffolds. The inner bulk pores are well connected with the large surface pores. The formation of the ice crystals is affected by the temperature. Low temperature results in quick formation of dense, small ice crystals and high temperature results in slow formation of sparse, big ice crystals.

Scaffolds—Smart

Funnel-like sponges of collagen, chitosan, hyaluronic acid, and glycosaminoglycan-collagen have been prepared by this method.[27–30] The funnel-like scaffolds show unique hierarchical porous structures of an open surface porous layer and an interconnected bulk porous layer under the surface porous layer (Fig. 2). The morphology, size, and density of the large, top surface pores are only dependent on the embossing ice particulates. The freezing temperature shows no effect on the large, top surface pores. The large, top surface pores are, in effect, replicas of the embossing ice particulates. The freezing temperature does not affect the shape of the ice particulates and shows no effect on the formation of the large, top surface pores. Therefore, the properties of the large, top surface pores can be controlled by adjusting the shape, dimension, and density of the ice particulates.

The funnel-like scaffolds facilitate cell seeding and show high cell seeding efficiency when used for cell culture. Cells are observed both on the top surface and within the inner pores of the funnel-like scaffolds. The cells are distributed throughout the funnel-like scaffolds. The unique porous structures facilitate homogenous cell seeding, improve cell distribution, and promote tissue formation.

HYBRID SCAFFOLDS OF SYNTHETIC BIODEGRADABLE POLYMERS AND NATURALLY DERIVED POLYMERS

Biodegradable synthetic polymer such as PGA, PLA, and PLGA and naturally derived polymers such as collagen have their respective advantages and drawbacks when used to prepare porous scaffolds for tissue engineering. Generally, the synthetic biodegradable polymers are easily formed into the desired shapes with relatively good mechanical strength. Their periods of degradation can also be manipulated by controlling the crystallinity, molecular weight, and copolymer ratio. However, the scaffolds derived from synthetic polymers lack cell-recognition signals and their hydrophobic property hinders smooth cell seeding. In contrast, naturally derived polymers such as collagen have the potential advantages of specific cell interactions and hydrophilicity, but scaffolds constructed entirely of collagen have poor mechanical strength. Therefore, these two kinds of biodegradable polymers have been hybridized to combine the advantageous properties of both constitutes and overcome the drawbacks. A few types of hybrid scaffolds of synthetic polymers and collagen have been developed.[31–33]

Hybrid sponges and meshes of synthetic polymers and collagen have been developed by introducing collagen microsponges into the pores of synthetic polymer sponges or the interstices of synthetic polymer meshes.[31,34] PLGA-collagen hybrid sponge is one of the examples. The PLGA sponge is immersed in a collagen solution under a vacuum so that the sponge pores of the PLGA sponge are filled with collagen solution. The PLGA sponge containing the collagen solution is then frozen and freeze-dried to allow the formation of collagen microsponges in the sponge pores. The collagen microsponges are further cross-linked by treatment with glutaraldehyde vapor. After blocking and washing, PLGA-collagen hybrid sponge is formed. A PLGA-collagen hybrid mesh is another example of hybrid scaffolds. By the similar method, collagen microsponges are formed in the interstices of a PLGA knitted mesh to construct the PLGA-collage hybrid mesh.[35] Both the PLGA-collagen hybrid sponge and hybrid mesh show high mechanical strength and good biocompatibility. The synthetic polymer sponge or mesh serving as a mechanical skeleton allows for easy formation into the desired shapes and provides the hybrid scaffolds with appropriate mechanical strength, while the collagen microsponges formed in openings provide the hybrid scaffolds with a microporous structure, hydrophilicity, and good cell interaction. The hybrid sponge and mesh have been used for three-dimensional cell culture and their application for tissue engineering has been studied.[36–39]

The hybrid mesh can be used to regenerate both thin and thick cartilage tissue.[40] When bovine chondrocytes are cultured in the PLGA-collagen hybrid mesh, the chondrocytes adhere to the hybrid mesh, proliferate and produce cartilaginous matrix, filling the void spaces in the hybrid mesh. The web-like collagen microsponges that are formed in the interstices of the knitted mesh not only prevent the seeded cells going through the hybrid scaffold, but also increase the specific surface area to provide sufficient surfaces for a spatially even chondrocyte distribution. After being cultured in vitro for 1 day, the cell/scaffold construct is used in single form to regenerate thin cartilage implants, or in laminated form to yield thick cartilage implants. The thickness of the implant can be adjusted by changing the number of laminated scaffold sheets. The single sheet and 5-sheet implants are cultured in vitro for another week and implanted subcutaneously in the dorsum of athymic nude mice. Cartilage-like tissue is regenerated after implantation. The thickness of the regenerated cartilage implants of single sheet and 5-sheet implants are 200 μm and 1 mm, respectively. Hematoxylin and eosin histological staining of these implants after 4, 8, and 12 weeks implantation indicates a uniform spatial cell distribution throughout the implants. The implant sheets integrate with each other for the laminated and rolled implants. The chondrocytes showed a natural round morphology in all the implants. Bright safranin-O-positive stain indicates that GAGs were abundant and homogeneously distributed throughout the implants. Toluidine blue staining demonstrates the typical metachromasia of articular cartilage. Immunohistological staining with an antibody to type II collagen shows a homogeneous extracellular staining for type II collagen. The similarity of the results for the 5-sheet implants to those of the single sheet implants suggests that an increase in the implant thickness from 200 μm to 1 mm does not compromise cell viability, cell uniformity, or cellular function. The mechanical properties of the 5-sheet implant after 12 weeks and bovine native articular cartilage are close to those of native bovine articular cartilage.

Recently, semi-type and sandwich type PLGA-collagen hybrid scaffolds have been reported.[41] The semi-type and sandwich-type PLGA-collagen hybrid scaffolds are prepared by forming collagen sponge on one side of a PLGA knitted mesh and collagen sponge formed on both sides of a PLGA knitted mesh, respectively. Bovine chondrocytes are cultured in the semi-type and sandwich-type PLGA collagen hybrid scaffolds and chondrogenesis is examined after subcutaneous implantation in the nude mice. Both hybrid scaffolds show a spatially even cell distribution, natural chondrocyte morphology, abundant cartilaginous ECM deposition, and excellent biodegradation in vivo. The histological structure and mechanical properties of the engineered cartilage using the semi-type and sandwich-type hybrid scaffolds match the native bovine articular cartilage.

Hybridization of biodegradable synthetic polymers with collagen by introducing collagen sponge or microsponges in the pores or interstices of a mechanical skeleton of synthetic polymers can combine the advantages of the two kinds of polymers. However, the porosity of these hybrid scaffolds is less than that of synthetic polymer or natural polymer scaffolds alone. Because the high porosity of porous scaffolds is strongly desired for functional tissue engineering as it provides sufficient space for cell proliferation and tissue formation as well as decreasing the ratio of biomaterials, the hybridization method has been further improved to increase the porosity of hybrid scaffolds. In this case, the mechanical skeleton is not a simple sponge or mesh. A cylinder or cup-shaped porous skeleton of biodegradable synthetic polymers is used. Collagen sponge is formed in the central space of the skeleton and collagen microsponges are formed in the pores or interstices of the porous wall of the cylinder or cup skeleton.[42,43]

A cylinder-type PLLA-collagen hybrid sponge is prepared by enclosing collagen sponge in a cup-shaped PLLA sponge.[42] The PLLA sponge cup is prepared using a method of porogen leaching using a Teflon mold. The hybrid sponge is composed of two parts: the central collagen sponge and the surrounding PLLA-collagen sponge cup. Collagen sponge is formed in the center of the PLLA sponge cup and collagen microsponges are formed in the pores of the wall and bottom of the PLLA sponge cup. The central collagen sponge is connected with the collagen microsponges in the pores of the wall and bottom of the PLLA sponge cup by collagen fibers that pass through the interstices of the PLLA sponge. The porosity of the cylinder-type PLLA-collagen hybrid sponge is significantly higher than that of the PLLA

sponge cup. A combination of the highly porous central collagen sponge and the surrounding PLLA-collagen sponge cup results in the high porosity of the hybrid sponge. The PLLA sponge cup serves as a mechanical skeleton and reinforces the hybrid sponge to keep its shape during cell culture and transplantation. The cylinder-type PLGA-collagen hybrid sponge has another effect that the PLLA-collagen sponge cup protects against cell leakage from the scaffold during cell seeding. The high cell seeding efficiency helps the scaffolds to hold more cells and decreases cell loss. The high porosity provides more space for cell accommodation and tissue formation, decreasing the ratio of materials.

By the same approach, a cell leakproof porous PLGA-collagen hybrid scaffold has been developed by forming collagen sponge in a PLGA mesh cup.[43] All the surfaces of the collagen sponge except the top surface were covered with a PLGA mesh. The PLGA mesh cup has a bi-layered structure: a PLGA knitted mesh having big interstices and a PLGA woven mesh having small interstices. The PLGA knitted mesh is used to prevent the central collagen sponge from shrinking while the PLGA woven mesh protects against cell leakage. The bi-layered skeleton structure also reinforces the hybrid scaffold. The cell leakproof hybrid scaffold has high mechanical strength. The cell leakproof PLGA-collagen scaffold is used for the culture of human bone marrow-derived mesenchymal stem cells. High cell seeding efficiency is achieved. The central collagen sponge provides enough space for cell loading and supports cell adhesion, while the bi-layered PLGA mesh cup protects against cell leakage and provides high mechanical strength for the collagen sponge to maintain its shape during cell culture. The MSCs in the hybrid scaffolds show round cell morphology, expressed type II collagen and cartilaginous proteoglycan after 4 weeks of culture in chondrogenic induction medium. The MSCs differentiate and form cartilage-like tissue when being cultured in the cell leakproof PLGA-collagen hybrid scaffold.

CONCLUSIONS

Many methods have been developed to control the pore structures and other properties of polymeric porous scaffolds to meet the requirements for functional tissue engineering. Open surface structure, interconnected large pores, and appropriate mechanical properties are critical parameters of the scaffolds. By using ice particulates as a porogen material or as an embossing template, both the surface and inner bulk pore structures can be precisely controlled. The hybridization of synthetic and naturally derived polymers creates another unique pore structure to improve the mechanical property, increase porosity and cell seeding efficiency. The polymeric porous scaffolds prepared by these methods promote tissue regeneration and will be useful for tissue engineering.

Scaffolds—Smart

ACKNOWLEDGMENTS

This work was supported by the World Premier International Research Center Initiative on Materials Nanoarchitectonics from the Ministry of Education, Culture, Sports, Science and Technology, Japan.

REFERENCES

1. Griffith, L.G.; Naughton, G. Tissue engineering—Current challenges and expanding opportunities. Science **2002**, *295* (5557), 1009–1014.

2. Hoashi, T.; Matsumiya, G.; Miyagawa, S.; Ichikawa, H.; Ueno, T.; Ono, M.; Saito, A.; Shimizu, T.; Okano, T.; Kawaguchi, N.; Matsuura, N.; Sawa, Y. Skeletal myoblast sheet transplantation improves the diastolic function of a pressure-overloaded right heart. J. Thorac. Cardiovasc. Surg. **2009**, *138* (2), 460–467.

3. Oberpenning, F.; Meng, J.; Yoo, J.J.; Atala, A. De novo reconstitution of a functional mammalian urinary bladder by tissue engineering. Nat. Biotechnol. **1999**, *17* (2), 149–155.

4. Engelmayr, G.C., Jr.; Cheng, M.; Bettinger, C.J.; Borenstein, J.T.; Langer, R.; Freed, L.E. Accordion-like honeycombs for tissue engineering of cardiac anisotropy. Nat. Mater. **2008**, *7* (12), 1003–1010.

5. Lu, H.; Kawazoe, N.; Kitajima, T.; Myoken, Y.; Tomita, M.; Umezawa, A.; Chen, G.; Ito, Y. Spatial immobilization of bone morphogenetic protein-4 in a collagen-PLGA hybrid scaffold for enhanced osteoinductivity. Biomaterials **2012**, *33* (26), 6140–6146.

6. Lutolf, M.P.; Hubbell, J.A. Synthetic biomaterials as instructive extracellular microenvironments for morphogenesis in tissue engineering. Nat. Biotechnol. **2005**, *23* (1), 47–55.

7. Hollister, S.J. Porous scaffold design for tissue engineering. Nat. Mater. **2005**, *4* (7), 518–524.

8. Higuchi, A.; Ling, Q.D.; Chang, Y.; Hsu, S.T.; Umezawa, A. Physical cues of biomaterials guide stem cell differentiation fate. Chem. Rev. **2013**, *113* (5), 3297–328.

9. Chen, G.; Ushida, T.; Tateishi, T. Scaffold design for tissue engineering. Macromol. Biosci. **2002**, *2* (2), 67–77.

10. Kim, Y.B.; Kim, G.H. Collagen/alginate scaffolds comprising core (PCL)–shell (collagen/alginate) struts for hard tissue regeneration: fabrication, characterisation, and cellular activities. J. Mater. Chem. B **2013**, *1* (25), 3185–3194.

11. Yan, S.; Zhang, K.; Liu, Z.; Zhang, X.; Gan, L.; Cao, B.; Chen, X.; Cui, L.; Yin, J. Fabrication of poly(L-glutamic acid)/chitosan polyelectrolyte complex porous scaffolds for tissue engineering. J. Mater. Chem. B **2013**, *1* (11), 1541–1551.

12. Mikos, A.G.; Thorsen, A.J.; Czerwonka, L.A.; Bao, Y.; Langer, R.; Winslow, D.N.; Vacanti, J.P. Preparation and characterization of poly(L-lactic acid) foams. Polymer **1994**, *35* (5), 1068–1077.

13. Zhang, H.; Hussain, I.; Brust, M.; Butler, M.F.; Rannard, S.P.; Cooper, A.I. Aligned two- and three-dimensional structures by directional freezing of polymers and nanoparticles. Nat. Mater. **2005**, *4* (10), 787–793.

14. Faraj, K.A.; Kuppevelt, T.H.V.; Daamen, W.F. Construction of collagen scaffolds that mimic the three-dimensional architecture of specific tissues. Tissue Eng. **2007**, *13* (10), 2387–2394.

15. Yang, F.; Murugan, R.; Ramakrishna, S.; Wang, X.; Ma, Y.X.; Wang, S. Fabrication of nano-structured porous PLLA scaffold intended for nerve tissue engineering. Biomaterials **2004**, *25* (10), 1891–1900.

16. Harris, L.D.; Kim, B.S.; Mooney, D.J. Open pore biodegradable matrices formed with gas foaming. J. Biomed. Mater. Res. **1998**, *42* (3), 396–402.

17. Meng, W.; Kim, S.Y.; Yuan, J.; Kim, J.C.; Kwon, O.H.; Kawazoe, N.; Chen, G.; Ito, Y.; Kang, I.K. Electrospun PHBV/collagen composite nanofibrous scaffolds for tissue engineering. J. Biomater. Sci. Polym. Ed. **2007**, *18* (1), 81–94.

18. Mikos, A.G.; Bao, Y.; Cima, L.G.; Ingber, D.E.; Vacanti, J.P.; Langer, R. Preparation of poly(glycolic acid) bonded fiber structures for cell attachment and transplantation. J. Biomed. Mater. Res. **1993**, *27* (2), 183–189.

19. Sun, W.; Starly, B.; Darling, A.; Gomez, C. Computer-aided tissue engineering: Application to biomimetic modeling and design of tissue scaffolds. Biotech. Appl. Biochem. **2004**, *39* (1), 49–58.

20. Kim, B.S.; Mooney, D.J. Development of biocompatible synthetic extracellular matrices for tissue engineering. Trends Biotechnol. **1998**, *16* (5), 224–230.

21. Nair, L.S.; Laurencin, C.T. Biodegradable polymers as biomaterials. Prog. Polym. Sci. **2007**, *32* (8), 762–798.

22. Chen, V.J.; Ma, P.X. Nano-fibrous poly(L-lactic acid) scaffolds with interconnected spherical macropores. Biomaterials **2004**, *25* (11), 2065–2073.

23. Ma, P.X.; Choi, J.W. Biodegradable polymer scaffolds with well-defined interconnected spherical pore network. Tissue Eng. **2001**, *7* (1), 23–33.

24. Chen, G.; Ushida, T.; Tateishi, T. Preparation of poly(L-lactic acid) and poly(DL-lactic-co-glycolic acid) foams by use of ice microparticulates. Biomaterials **2001**, *22* (18), 2563–2567.

25. Zhang, Q.; Lu, H.; Kawazoe, N.; Chen, G. Preparation of collagen porous scaffolds with a gradient pore size structure using ice particulates. Mater. Lett. **2013**, *107*, 280–283.

26. Zhang, Q.; Lu, H.; Kawazoe, N.; Chen, G. Preparation of collagen scaffolds with controlled pore structures and improved mechanical property for cartilage tissue engineering. J. Bioact. Compat. Polym. **2013**, *28* (5), 426–438.

27. Ko, Y.G.; Kawazoe, N.; Tateishi, T.; Chen, G. Preparation of novel collagen sponges using an ice particulate template. J. Bioact. Compat. Polym. **2010**, *25* (4), 360–373.

28. Ko, Y.G.; Kawazoe, N.; Tateishi, T.; Chen, G. Preparation of chitosan scaffolds with a hierarchical porous structure. J. Biomed. Mater. Res. B Appl. Biomater. **2010**, *93* (2), 341–350.

29. Ko, Y.G.; Oh, H.H.; Kawazoe, N.; Tateishi, T.; Chen, G. Preparation of open porous hyaluronic acid scaffolds for tissue engineering using ice particulate template method. J. Biomater. Sci. Polym. Ed. **2011**, *22* (1–3), 123–138.

30. Ko, Y.G.; Grice, S.; Kawazoe, N.; Tateishi, T.; Chen, G. Preparation of collagen-glycosaminoglycan sponges with open surface porous structures using ice particulate template method. Macromol. Biosci. **2010**, *10* (8), 860–871.

31. Chen, G.; Ushida, T.; Tateishi, T. Hybrid biomaterials for tissue engineering: a preparative method for PLA or PLGA-collagen hybrid sponges. Adv. Mater. **2000**, *12* (6), 455–457.

Scaffolds—Smart

32. Hokugo, A.; Takamoto, T.; Tabata, Y. Preparation of hybrid scaffold from fibrin and biodegradable polymer fiber. Biomaterials 2006, 27 (1), 61–67.

33. Dunn, M.G.; Bellincampi, L.D.; Tria, A.J., Jr.; Zawadsky, J.P. Preliminary development of a collagen-PLA composite for ACL reconstruction. J. Appl. Polym. Sci. 1997, 63 (11), 1423–1428.

34. Chen, G.; Ushida, T.; Tateishi, T. A biodegradable hybrid sponge nested with collagen microsponges. J. Biomed. Mater. Res. 2000, 51 (2), 273–279.

35. Chen, G.; Ushida, T.; Tateishi, T. A hybrid network of synthetic polymer mesh and collagen sponge. Chem. Commun. 2000, 16, 1505–1506.

36. Chen, G.; Sato, T.; Ushida, T.; Ochiai, N.; Tateishi, T. Tissue engineering of cartilage using a hybrid scaffold of synthetic polymer and collagen. Tissue Eng. 2004, 10 (3–4), 323–330.

37. Chen, G.; Sato, T.; Ohgushi, H.; Ushida, T.; Tateishi, T.; Tanaka, J. Culturing of skin fibroblasts in a thin PLGA-collagen hybrid mesh. Biomaterials 2005, 26 (15), 2559–2566.

38. Ochi, K.; Chen, G.; Ushida, T.; Gojo, S.; Segawa, K.; Tai, H.; Ueno, K.; Ohkawa, H.; Mori, T.; Yamaguchi, A.; Toyama, Y.; Hata, J.; Umezawa, A. Use of isolated mature osteoblasts in abundance acts as desired-shaped bone regeneration in combination with a modified poly-DL-lactic-co-glycolic acid (PLGA)-collagen sponge. J. Cell. Physiol. 2003, 194 (1), 45–53.

39. Iwai, S.; Sawa, Y.; Ichikawa, H.; Taketani, S.; Uchimura, E.; Chen, G.; Hara, M.; Miyake, J.; Matsuda, H. Biodegradable polymer with collagen microsponge serves as a new bioengineered cardiovascular prosthesis. J. Thorac. Cardiovasc. Surg. 2004, 128 (3), 472–479.

40. Chen, G.; Sato, T.; Ushida, T.; Hirochika, R.; Shirasaki, Y.; Ochiai, N.; Tateishi, T. The use of a novel PLGA fiber/collagen composite web as a scaffold for engineering of articular cartilage tissue with adjustable thickness. J. Biomed. Mater. Res. A 2003, 67 (4), 1170–1180.

41. Dai, W.; Kawazoe, N.; Lin, X.; Dong, J.; Chen, G. The influence of structural design of PLGA/collagen hybrid scaffolds in cartilage tissue engineering. Biomaterials 2010, 31 (8), 2141–2152.

42. He, X.; Lu, H.; Kawazoe, N.; Tateishi, T.; Chen, G. A novel cylinder-type poly(L-lactic acid)-collagen hybrid sponge for cartilage tissue engineering. Tissue Eng. 2010, 16 (3), 329–338.

43. Kawazoe, N.; Inoue, C.; Tateishi, T.; Chen, G. A cell leak-proof PLGA-collagen hybrid scaffold for cartilage tissue engineering. Biotechnol. Prog. 2010, 26 (3), 819–826.

Semiconducting Polymer Dot Bioconjugates

Garima Ameta

Sonochemistry Laboratory, Department of Chemistry, University College of Science, Mohanlal Sukhadia University, Udaipur, India

Abstract

Nanoparticle-based diagnostic and therapeutic agents have attracted considerable interest because of their potential for applications in clinical oncology and other biomedical research. Versatile nanostructures for *in vivo* applications, such as lipid and polymeric nanocapsules for drug delivery, iron oxide nanoparticles for magnetic resonance imaging, gold nanoparticles for X-ray computed tomography, and quantum dots (Qdots) for fluorescence imaging have been reported. However, the intrinsic toxicity of Qdots is of critical concern, which may impede their final clinical application. Therefore, the design of bright probes with biologically benign materials is highly desirable for many *in vivo* clinical purposes. Semiconducting polymer dot (Pdot) bioconjugates are a new class of ultrabright fluorescent probes because of their exceptional brightness and their nontoxic feature. Researchers have developed various methods to improve the versatility and functions of Pdots for biomedical studies, such as tuning the emission color, exploring new preparation methods, engineering the particle surface, doping functional sensing molecules, encapsulating magnetic materials, and mapping the sentinel lymph node as a first *in vivo* study. Researchers have developed a general method to form semiconducting Pdot bioconjugates, and have demonstrated their applications in specific cellular targeting and bioorthogonal labeling. In this entry, we summarize the preparation and applications of various semiconducting Pdot bioconjugates.

INTRODUCTION

Highly fluorescent nanoparticles have attracted much attention due to a variety of fluorescence-based applications such as biosensing, imaging, and high-throughput assays.[1–2] Conjugated polymers are known to possess high absorption coefficients and high fluorescence efficiency, which have led to a wide range of applications in optoelectronic thin film devices. Ultrasensitive sensing schemes based on super-quenching of water-soluble conjugated polyelectrolyte fluorescence in aqueous solution have also been demonstrated. [3–4] Encapsulated, bioconjugated nanoparticles consisting of conjugated polymer molecules have been developed. These "polymer dot" (Pdot) nanoparticles consist of one or more conjugated polymer molecules encapsulated in a silica shell. Owing to the very high effective chromophore density, these nanoparticles exhibit excellent properties for fluorescence labeling and sensing applications, including small size (~10 nm, typical), high brightness, highly tunable excitation and emission wavelengths, and excellent photostability.[5] Preliminary results also indicate low cytotoxicity. The nanoparticles are also promising for advanced applications such as multiphoton excited fluorescence imaging and fluorescence quenching–based sensing schemes.

BIOCONJUGATION OF POLYMERIC NANOPARTICLES

To apply fluorescent nanoparticles to biosensing and biomedical imaging applications, it is crucial to develop strategies toward their biofunctionalization. These include the proper linkage of biomolecules to nanoparticles (bioconjugation) and the design of appropriate biocompatible coatings.[6] Bioconjugation can be described as any procedure that links a nanoparticle to a biomolecule under mild conditions.[7] As described above, the synthesis of nanoparticles often does not render them capable of attachment to biomolecules, because their surface-chemical properties are not appropriate. Therefore, nanoparticles frequently must undergo surface transformations to create the chemistry needed for coupling to biomolecules under mild (physiological) conditions. There are a few key requirements for successful bioconjugation reactions.[8] Crucially, the conjugation process must avoid compromising the activity of biomolecules. In addition, the bioconjugation ideally should not hinder the signal of the nanoparticle.

Another requirement is the ability to control the number of linkage sites on the nanoparticle surface where

Concise Encyclopedia of Biomedical Polymers and Polymeric Biomaterials DOI: 10.1081/E-EBPPC-120053903

Scaffolds—Smart

biomolecules can bind. This requirement can be quite challenging. In addition, the biomolecule–nanoparticle coupling should be stable and, for crystalline particles, the surface should be covered to avoid free valence states. Finally, the thickness of any nanoparticle shell should remain as small as possible relative to the nanoparticle size.

From a bioconjugation perspective, polymeric nanoparticles are attractive because they can be prepared with various different reactive groups on their surface. The obvious benefit is that a comparably broad spectrum of conjugation chemistry approaches can be used, including covalent and non-covalent ones shown in Fig. 1. These different bioconjugation strategies are reviewed below.

Non-Covalent Approaches

One of the most widely applied approaches in bioconjugating polymeric nanoparticles is simple physisorption of the biomolecules on the particle surface in Fig. 1A.[9–13] Because biomolecule activity is compromised by uncontrolled adsorption, spacer molecules have been introduced (Fig. 1B). For example, nanoparticles first coated with Protein A (isolated from the cell wall of *Staphylococcus aureus*), which attaches specifically to the Fc fragment of the IgG antibody, have proper Fab orientation for antibody binding.[13] Unfortunately, such simple adsorption approaches are only of limited use when applied in serum. This is because serum proteins compete with antibodies for adsorption sites.[9–11] Furthermore, *in vivo* experiments have revealed that such simply designed nanoparticles tend to accumulate in the liver and spleen. Using a similar design strategy, polystyrene nanoparticles have been functionalized

with humanized mAB HuEP5C7.g2.[14] These delivery systems target cells expressing E and P-select in.

Covalent Approaches

Obviously, the activity of ligands bound to nanoparticles strongly depends on the stability of the ligand–nanoparticle linkage and the proper orientation of the ligand. To address these needs, ligands have been bound covalently to polymeric nanoparticles using several different conjugation approaches in Fig. 1E.[15–19] For example, an amide linkage strategy was pursued to bind lactose to the polyvinylamine-grafted nanoparticles.[15] Such lactose-functionalized carrier systems are useful because they specifically bind to lectin RCA120. The same amide-based linkage chemistry was used to bind proteins to dye encoded latex particles.[16] In another approach, transferrin was conjugated to polyethylene glycol (PEG) coated polycyanoacrylate nanocrystals, with the transferrin bound to PEG via periodate oxidation.[17] A hetero bifunctional linker was used to bind antibodies to bovine serum albumin (BSA)-containing nanoparticles.[18,19] In this case, the cross-linker (glutaraldehyde) was bound to BSA to provide free aldehyde groups, which could then be linked to the primary amines of the antibodies. Another interesting strategy is based on formation of ligand-carrying nanoparticles via polymerization of monomers with bound ligands. Such a strategy has been applied to create nanoparticles with biotin groups, which then allowed for specific linkage to avidin.[20,21] The grafting of carbohydrates to polylysine was also reported.[22] This carrier can bind to carbohydrate-binding proteins on the surface of target cells.

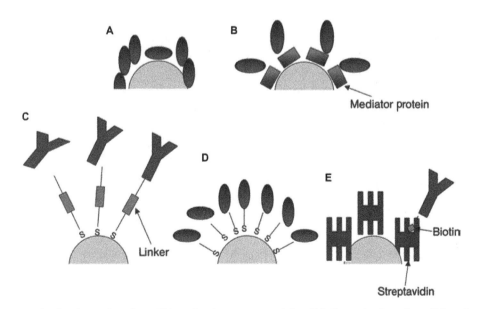

Fig. 1 Common strategies for the conjugation of biomolecules to nanoparticles, (**A**) direct physisorption of biomolecules, (**B**) assisted physisorption using pre-bound molecules, (**C**) chemical linkage of biomolecules to cross-linkers either physisorbed or chemisorbed on the nanoparticle surface, (**D**) direct chemical coupling of biomolecules to nanoparticles, and (**E**) the targeted binding of biotinylated biomolecules to streptavidin-coated nanoparticles via biotin–streptavidin coupling.

Highly fluorescent semiconducting Pdots with functional groups allow for covalent conjugation to biomolecules. The strategy is based on entrapping heterogeneous polymer chains into a Pdot particle, driven by hydrophobic interactions during nanoparticle formation. We have shown that a small amount of amphiphilic polymer bearing functional groups can be co-condensed with the majority of semiconducting polymers to modify and functionalize the nanoparticle surface. Subsequent covalent conjugation to biomolecules such as streptavidin and antibodies was performed using the standard carbodiimide coupling chemistry. These Pdot bioconjugates can effectively and specifically label cell surface receptors and subcellular structures in both live and fixed cells, without any detectable nonspecific binding. We performed single-particle imaging, cellular imaging, and flow cytometry to experimentally evaluate the Pdot performance and demonstrate their high-cellular labeling brightness compared to those of Alexa-IgG and Qdot probes.

Streptavidin and IgGs are widely used in bioconjugation for immunofluorescent labeling of cellular targets. We created Pdot–IgG and Pdot–streptavidin probes and investigated their ability to label a specific cellular target EpCAM, an epithelial cell surface marker in use for the detection of circulating tumor cells.[23] Figure 2A shows that the Pdot–IgG probes successfully labeled EpCAM receptors on the surface of live MCF-7 human breast cancer cells after the cells were incubated with a monoclonal primary anti-EpCAM antibody. When the cells were incubated with just the Pdot–IgG conjugate, in the absence of

the primary antibody, cell labeling was not detected in Fig. 2A (bottom), indicating that the Pdot–IgG conjugates are highly specific for the target.

Pdot–streptavidin conjugates as an alternative probe to detect EpCAM. The Pdot–streptavidin probes, together with the primary anti-EpCAM antibody and biotinylated goat anti-mouse IgG secondary antibody, also effectively labeled EpCAM on the surface of live MCF-7 cells as shown in Fig. 2B. When the cells were incubated with primary antibody and Pdot–streptavidin in the absence of biotin anti-mouse IgG, no fluorescence was observed on the cell surface as seen in Fig. 2B, (bottom), thus again demonstrating the highly specific binding of Pdot–strepavidin. The lack of signal also indicated the absence of nonspecific binding in this biotin–streptavidin labeling system.

Semiconducting Pdots coupled with biomolecules have emerged as a new type of nanoprobe that has been used for bioanalysis areas, for example, the distribution of biomolecules *in vivo*, conformation dynamics, target analyte, and biomolecule interaction.[24] Because conventional organic dyes, such as fluorescein, rhodamine, alexa fluor, and cyanine derivatives possess limited brightness, poor photostability, and poor thermal stability in aqueous solution, various strategies for designing brighter fluorescent nanoprobes have been advanced. Despite improvements, the intrinsic limitations of conventional dyes, requiring high excitation light intensities in the UV-visible light region for long-term imaging, can continue to pose great difficulties.[25] There is also a great deal of interest in the development of quantum dots (Qdots), which have been facilitating

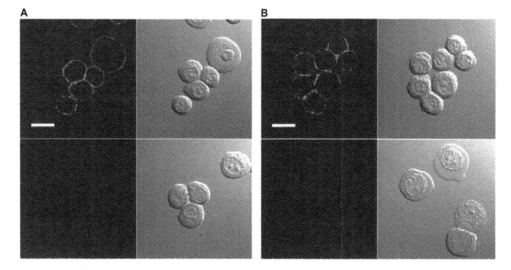

Fig. 2 Fluorescence imaging of cell surface marker (EpCAM) in human breast cancer cells labeled with Pdot bioconjugates. (**A**) Imaging of live MCF-7 cells incubated sequentially with anti-EpCAM primary antibody and Pdot–IgG conjugates. The bottom panels show control samples in which the cells were incubated with Pdot–IgG alone (no primary antibody). The Nomarski (DIC) images are shown to the right of the confocal fluorescence images. Scale bar represents 20 μm. (**B**) Imaging of live MCF-7 cells incubated sequentially with anti-EpCAM primary antibody, biotinylated goat antimouse IgG secondary antibody, and Pdot–streptavidin conjugates. The bottom panels show control samples where the cells were incubated with anti-EpCAM antibody and Pdot–strepavidin (no secondary antibody). The Nomarski (DIC) images are shown to the right of the confocal fluorescence images. Scale bar represents 20 μm.
Source: From Wu, Schneider, et al.[23] © 2010, American Chemical Society. Reprinted with permission.

the development of DNA array technology, immunofluorescence assays, and animal biology during 1995–2015 due to their high quantum yield, narrow and tunable emission spectrum, and good photostability.[26–28] The preparation of monodispersed, luminescent, high brightness, nontoxic, and bioconjugated Qdots has not been successful in high ionic buffer or in biological matrices. Compared with the above-mentioned fluorescence labels using organic dyes and Qdots, green fluoresent protein (GFP) is unique among light-emitting proteins because it does not require the presence of cofactors or substrates for generation of its green light.[29] However, the GFP signal cannot be amplified in a controlled manner, thus preventing detection of low expression levels. Increasing worldwide demand requires simple, stable, and sensitive labels for

biomolecules and evaluation and monitoring of cellular assays. Pdots as fluorescent nanoprobes offer several advantages, such as good biocompatibility for living organisms or biological cells,[30] easy functionalization with antibodies or other biomolecules, and excellent photostability for *in vivo* imaging.[24] Researchers designed a series of fluorescent semiconducting polymers, such as multicolored narrow emissive Pdots based on different boron dipyrromethene units,[31] semiconducting polymer poly (9-vinylcarbazole) doped with europium complexes,[32] highly emissive Pdot bioconjugates,[33] highly luminescent and fluorinated semiconducting Pdots,[34] and hybrid semiconducting Pdots–Qdots,[35] for cellular imaging and analysis in a variety of *in vitro* and *in vivo* applications.

Pdots combined with vancomycin or polymyxin B are specific and sensitive enough for the trace image and detection of bacteria Fig. 3.[36] Because the experimental setup requires only a bioprobe, the developed approach is simple and cost-efficient, which indicates that it can be employed in the on-site detection of gram positive bacteria and gram negative bacteria in hospitals and even in homes for food safety monitoring. Multifunctional Pdots dispersions also exhibited selective antibacterial activity toward gram positive bacteria and gram negative bacteria. Transmission electron microscopy images and dynamic light scattering measurements of Van-Pdots or polymyxin B sulfate are shown in Fig. 4. The bioconjugated Pdots developed possess the ability to selectively recognize, image, and photo-

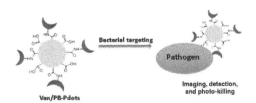

Fig. 3 Schematic of covalently functionalized semiconducting polymer and antibiotic bioconjugation for specific bacterial targeting.
Source: From Wan, Zheng, et al.[36] © 2014, Royal Society of Chemistry. Reprinted with permission.

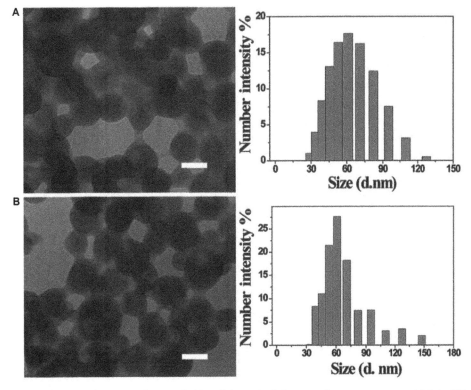

Fig. 4 (**A**) Transmission electron microscopy images (left) and dynamic light scattering measurements (right) of Van-Pdots, (**B**) polymyxin B sulfate. Scale bar is 50 nm.
Source: From Wan, Zheng, et al.[36] © 2014, Royal Society of Chemistry. Reprinted with permission.

Scaffolds—Smart

kill gram positive and gram negative pathogens simultaneously.

Polyhedral oligomeric silsesquioxanes-poly-9,9-bis[2-(2-(2-methoxyethoxy) ethoxy) ethyl]fluorenyl-divinylene (POSS-PFV)-loaded NPs with high PL quantum yield, excellent stability, low cytotoxicity, and tunable surface protein density have been synthesized as shown in Fig. 5 for HER2-positive cancer cell detection. The use of a mixed encapsulation matrix composed of PLGA- b -PEG-NH$_2$ and PLGA-OCH$_3$ at different molar ratio allows control of the –NH$_2$ groups on NP surface. Subsequent conjugation with trastuzumab yields target protein functionalized NPs that can significantly increase their uptake by SKBR-3 breast cancer cells. The specific binding affinity between trastuzumab functionalized POSS-PFV-loaded NPs and SKBR-3 breast cancer cells over MCF-7 breast cancer cells indicates that the POSS-PFV-loaded NPs have great potentials in specific HER2-positive cancer cell detection.[37] By designing different specific protein functionalized conjugated polymer NPs with tunable emission wavelength, simultaneous detection of multiple biological targets under single laser excitation is achievable.

Numerous supramolecular recognition motifs have been applied to the rapidly expanding field of polymer self-assembly. Hydrogen bonding has featured prominently,[38,39] as have metal–ligand coordination,[40,41] p–p stacking,[42,43] and host–guest inclusion complexes.[44,45] Surprisingly, porphyrin-based supramolecular complexes have been largely overlooked for directing polymer self-assembly despite their rich physical properties[46,47] and well-studied supramolecular chemistry.[48,49]

A modular strategy for synthesizing triazole-linked porphyrin–polymer conjugates, which assemble into supramolecular assemblies via porphyrinatozinc–triazole coordination, has been developed.[50] By altering the polymer microenvironment around the porphyrin core we are able to systematically tune the association constant of the porphyrinatozinc–triazole coordination complex, or switch it completely to the disassembled state by introducing a competitive ligand. Porphyrinatozinc–triazole coordination

was shown to alter the properties of the bulk material, manifesting in a 6K increase in the glass transition temperature of (PS20)2–Zn. We envisage that the multitopic design of the porphyrin–polymer conjugates (PPCs) may stabilize the tertiary structure of larger supramolecular assemblies due to potential preorganizational effects. Furthermore, the inter-chromophore distances between the porphyrin subunits may be sufficiently small to enable light-mediated energy-transfer processes along the porphyrinatozinc–triazole backbone. Consequently, PPCs containing ruthenium and cobalt metalloporphyrin cores are being investigated for constructing larger, more stable PPC arrays with potentially interesting photophysical properties.

Novel, multifunctional Pdots can encapsulate drugs and act as an imaging tag. The Pdots were produced from an amphiphilic polymer based on polyethyleneimine (PEI) and hydrophobic polylactide (PLA) in Fig. 6. Drugs can be easily encapsulated into the Pdots through a modified emulsion/evaporation method. Moreover, the Pdots possess an unexpectedly bright, multicolor fluorescence in comparison to the weak fluorescence of PEI alone.[51] Most importantly, due to the unique proton sponge effect of PEI, the drug-loaded Pdots effectively overcame a major hurdle for drug-delivery vehicles[52] by escaping from endosomes/lysosomes in the cytoplasm. In addition, the drug-loaded Pdots have higher antineoplastic activity compared with free drug. Hence, Pdots derived from amphiphilic PEI could be ideal theranostic tools for imaging-guided drug delivery.

STORAGE OF POLYMER DOT BIOCONJUGATES

Poor colloidal stability can be a significant drawback, and it has plagued other types of nanoparticles. Lyophilization, a freeze-drying/dehydration technique,[53] can be used to prepare Pdot bioconjugates for long-term storage or shipping, provided the right conditions are used. Lyophilization is an important practical advance for making Pdots practical to use in biomedical research in Fig. 7.

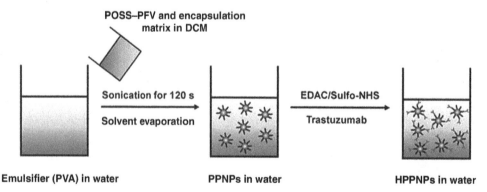

Fig. 5 Schematic illustration of CP NP synthesis through a modified solvent extraction/evaporation single emulsion method and subsequent conjugation to yield trastuzumab functionalized CP NPs.
Source: From Li, Liu, et al.[37] © 2011, John Wiley and Sons. Reprinted with permission.

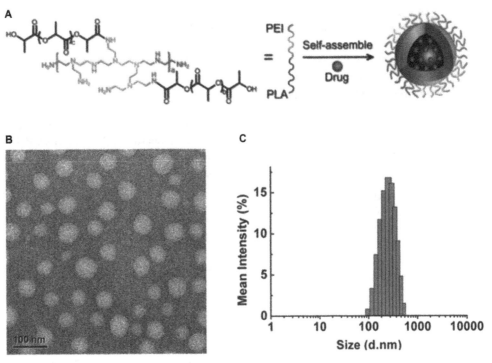

Fig. 6 (**A**) Schematic illustration of the construction of the multifunctional PDs. (**B, C**) TEM and DLS characterization of blank PDs. Scale bar, 100 nm.
Source: From Sun, Cao, et al.[51] © 2013, Nature Publishing Group. Reprinted with permission.

Fig. 7 Cell labeling from lyophilized Pdot.
Source: From Adapted from Sun, Ye, et al.[54] © 2013. American Chemical Society. Reprinted with permission.

Lyophilization is a technique to store Pdot bioconjugates so that they successfully retain their optical properties, colloidal stability, and cell-targeting capability during storage. Hydrodynamic diameter of streptavidin-conjugated Poly[(9,9-dioctylfluorenyl-2,7-diyl)-co-(1,4-benzo-(2,10,3)-thiadiazole)] (PFBT) Pdots (Strep–PFBT) is 32 nm Fig. 8A, which was measured immediately after they were prepared. Then the same Pdot–streptavidin conjugates were lyophilized without adding any reagents and then the Pdots were rehydrated with water. Even after vigorous sonication, the hydrodynamic diameter of the rehydrated Pdots increased to 220 nm (Fig. 8B), indicating that the lyophilization process had caused severe aggregation of the Pdots. Aggregated Pdot bioconjugates are not suitable for biological studies. When the same Pdot-streptavidin conjugates were lyophilized in the presence of 1% (w/v) sucrose, the hydrodynamic diameter of the rehydrated Pdots was 50 nm

Fig. 8C, which was much smaller than the 220 nm obtained without sucrose, but it was still significantly larger than the original size of 32 nm. Sonication did not help further shift the size distribution of rehydrated Pdots to that before lyophilization. Therefore, the Pdot–streptavidin bioconjugates still were partially aggregated, albeit much less severely than without sucrose. When Pdot bioconjugates were lyophilized in the presence of 10% (w/v) sucrose, the rehydrated Pdot bioconjugates did not show any signs of aggregation and exhibited the same hydrodynamic diameters as before lyophilization i.e., 32 nm in Fig. 8A and 8D. The brightness of the lyophilized Pdots was at least as good as before lyophilization, but in some cases, the quantum yield of lyophilized Pdots curiously showed an improvement. Finally, using flow cytometry, we demonstrated that lyophilized Pdot bioconjugates retained their biological targeting properties and were able to effectively label cells; in fact, cells labeled with lyophilized Pdot bioconjugates composed of PFBT, which were stored for six months at −80°C, were ~22% brighter than those labeled with identical but unlyophilized Pdot bioconjugates. These results indicate that lyophilization may be a preferred approach for storing and shipping Pdot bioconjugates,[54] which is an important practical consideration to ensure Pdots are widely adopted in biomedical research.

SPECIFIC CELLULAR IMAGING

A simple strategy that allows the formation of multicolor conjugated polymer nanoparticles bioconjugates with specific biorecognition was demonstrated in Fig. 9, multicolor conjugated polymer nanoparticles formed from such a method would directly have functional groups available for subsequent bioconjugation, thus avoiding the respective

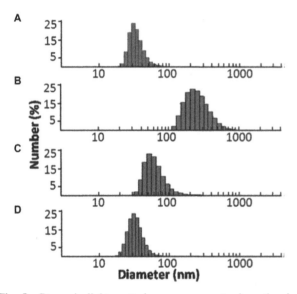

Fig. 8 Dynamic light scattering measurements show the size distributions of streptavidin-conjugated PFBT Pdots. (**A**) The size distribution of Pdots before lyophilization. (**B, C,** and **D**) The size distributions for rehydrated Pdots after lyophilization with (**B**) 0%, (**C**) 1%, and (**D**) 10% sucrose.
Source: From Sun, Ye, et al.[54] © 2013. American Chemical Society. Reprinted with permission.

and sophisticated surface modification of the nanoparticles and the dissociation of functional groups. The multicolor and surface-functionalizable PPVseg-COOH was also successfully applied to prepare specific nanoparticle bioconjugates, thus indicating its general applicability and feasibility for biological imaging and bioanalytical applications. Compared with bare conjugated polymer nanoparticles, the streptavidin-modified PPVseg-COOH-conjugated polymer nanoparticles (PPVseg-SA CPNs) exhibit dramatically improved photostability in buffer solutions and physiologically relevant environments. The multicolor PPVseg-SA CPNs probes have also been applied for targeted cellular imaging, which demonstrates that the ease of such functionalization can impart multicolor and desired properties such as specific recognition or biocompatibility simultaneously.[55]

Multicolor bioconjugated PPVseg-COOH CPNs are useful in biological analysis and bioimaging, for example, PPVseg-SA CPNs can be used for targeted imaging of MCF-7 breast cancer cells. The detection of live MCF-7 breast cancer cells was demonstrated by sequentially staining EpCAM receptors with biotinylated primary anti-EpCAM antibody and PPVseg-SA CPNs, and the corresponding fluorescence images are shown in Fig. 10. It is obvious that PPVseg-SA CPNs are more concentrated along the cell periphery than in the cytoplasm of MCF-7 breast cancer cells, which indicates the effective labeling of MCF-7 breast cancer cells by PPVseg-SA CPNs. In the control experiments without biotinylated primary antibody, no fluorescence on the cell surface was detected as shown in Fig. 10C, confirming the specific interaction between PPVseg-SA CPNs and EpCAM receptors. Furthermore, MCF-7 breast cancer cells treated with biotinylated primary antibody and free PPVseg-COOH CPNs in the absence of streptavidin were also used as a control for specific cellular labeling studies of MCF-7 cancer cells. The confocal laser scanning microscopy (CLSM) image also reveals that no fluorescence was observed on the cell surface, confirming that the bioconjugation of streptavidin to PPVseg-COOH CPNs was covalent and the conjugation of streptavidin onto the cell surface was specific through biotin–streptavidin interaction rather than the nonspecific interaction between nanoparticles and cell surfaces.

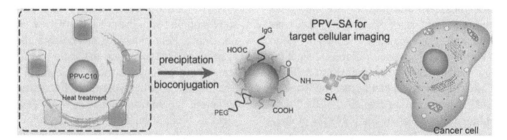

Fig. 9 Schematic illustration of the preparation of multicolor PPVseg CPNs bioconjugates and PPV-SA for target cellular imaging.
Source: From Bao, Ma, et al.[55] © 2014. American Chemical Society. Reprinted with permission.

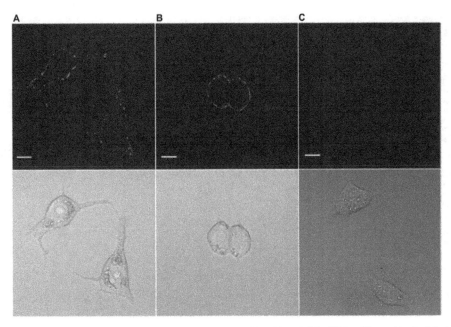

Fig. 10 Confocal images of live MCF-7 cells. (**A**) positively labeled with PPVYel-SA CPNs, (**B**) positively labeled with PPVCya-SA CPNs, and (**C**) negatively labeled with PPVCya-SA CPNs in the absence of biotinylated primary antibody. Scale bar: 10 μm.
Source: From Bao, Ma, et al.[55] © 2014. American Chemical Society. Reprinted with permission.

TARGETING OF BIOCONJUGATED POLYMER NANOPARTICLE

The properties of PEG lipid conjugated polymer nanoparticles such as their high extinction coefficient, bright fluorescence, photostability, and functionalization indicate significant potential for targeted single particle imaging and tracking in living cells. Kandel et al. demonstrated targeted localization of functionalized nanoparticles to individual CD16/32 receptors on the surface of mouse macrophage J774A.1cells.[56] In these experiments, a commercially available biotin-linked rat anti-CD16/32 antibody was bound to CD16/32 on the cell surface, and then labeled with biotin-functionalized PEG lipid PFBT nanoparticles, using streptavidin as a linker in a sandwich format. The result was Ab-conjugated nanoparticles that specifically labeled antibody-tagged receptors on the cell surface. Fig. 11 shows the differential interference contrast and fluorescence images of labeled cells. Localized nanoparticle fluorescence is observed on the periphery of the cell, as is typical for membrane localization. Control experiments performed either without streptavidin or using carboxy modified nanoparticles instead of biotinylated nanoparticles showed no fluorescence, indicating that observed binding of biotinylated PEG lipid PFBT nanoparticles does not represent nonspecific adsorption to the cell membrane. Together, these data demonstrate that PEG lipid CPNs can target specific tagged proteins on the cell surface. To our knowledge, these results represent the first report of targeted delivery of conjugated polymer nanoparticles to individual sites on the cell surface.

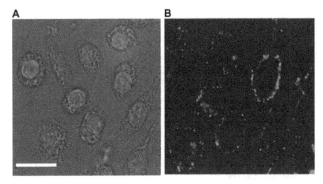

Fig. 11 Biotinylated PEG lipid-PFBT nanoparticles targeted to cell surface receptors. (**A**) Differential interference image (DIC) of fixed J774A.1 cells, (**B**) fluorescence image of fixed J774A.1 cells labeled with biotinylated PFBT nanoparticles. Scale bar is 25 mm. J774 A1 macrophage cells which express CD16/32 (Fc receptor) were paraformaldehyde fixed and incubated with biotinylated anti-CD16/32 antibody. After washing the cell with RB, the cells were next incubated with streptavidin, washed, and labeled with biotinylated PEG lipid–PFBT nanoparticles. Images were obtained with 495 nm excitation, using a 510 nm long pass emission filter.
Source: From Kandel, Fernando, et al.[56] © 2011, Royal Society of Chemistry. Reprinted with permission.

NEAR-INFRA-RED EMITTING POLYMER DOT BIOCONJUGATES FOR POTENTIAL BIOIMAGING

Near-infrared (NIR) emitting Pdots for potential bioimaging MCF-7 cells labeled with 3-NIR–Pdots–SA conjugate were studied using confocal fluorescence microscopy.[57] As shown in Fig. 12A, the 3-NIR–Pdot–SA probes

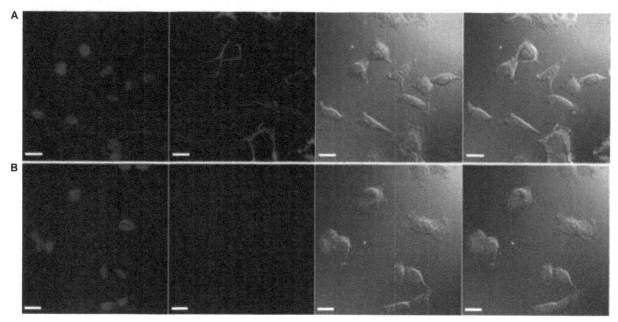

Fig. 12 Confocal fluorescence images of MCF-7 cells labeled with 3-NIR Pdot–SA probes. (**A**) Positive labeling using 3-NIR–Pdot–SA probe. (**B**) Negative labeling carried out under the same condition as (**A**) but in the absence of biotinylated antibody on the surface of the MCF-7 cells. Images from left to right: blue fluorescence from the nuclear stain Hoechst 34580; deep red fluorescence images from 3-NIR–Pdot–SA probes; Nomarski (DIC) images; combined fluorescence images. Scale bars: 20 μm.
Source: Zhang, Yu, et al.[57] © 2013, Royal Society of Chemistry. Reprinted with permission.

effectively labeled EpCAM receptors on the MCF-7 cell surface. In the control experiment, where the biotinylated antibody on the surface of the MCF-7 cells was absent, no fluorescence on the cell surface was detected (Fig. 12B), which further confirmed that the covalent bioconjugation of streptavidin to Pdots was successful and the cellular targeting was specific through biotin–streptavidin interaction.

CONCLUSION

There are several challenges associated with Pdots as *in vivo* probes. The fluorescence brightness of Pdots in the red and NIR is generally limited by their low quantum yields and it is unclear whether Pdot-based probes can be specifically delivered to diseased tissues *in vivo*. Therefore, highly fluorescent Pdot probes that consist of optimally tailored semiconducting polymer bioconjugates can be used for *in vivo* tumor targeting. The development and application of Pdot bioconjugates continue to be an active and important area in fluorescence imaging and biomedical studies. A reliable functionalization and bioconjugation method is highly desirable for preparing small, bright Pdot probes which specifically recognize and label cellular targets. Research in the Pdot field continues to inspire the chemistry community with new discoveries and challenges. It involves cross-disciplinary collaboration between chemists, biologists, and clinicians. The exploration of new Pdot species with improved performance, well-controlled surface properties, and multimodal imaging/sensing functionalities is to be the main focus of the field. When fully optimized, Pdot technology is expected to have a broad and lasting impact on biomedical imaging, diagnostics, and therapeutics.

ACKNOWLEDGMENT

The author is thankful to John Wiley and Sons, American Chemical Society, Royal Society of Chemistry, and Nature Publishing Group for copyright permission.

REFERENCES

1. Bruchez, M.; Moronne, M.; Gin, P.; Weiss, S.; Alivisatos, A.P. Semiconductor nanocrystals as fluorescent biological labels. Science **1998**, *281* (5385), 2013–2016.
2. Chan, W.C.W.; Nie, S.M. Quantum dot bioconjugates for ultrasensitive nonisotopic detection. Science **1998**, *281* (5385), 2016–2018.
3. Jones, R.M.; Lu, L.D.; Helgeson, R.; Bergstedt, T.S.; McBranch, D.W.; Whitten, D.G. Building highly sensitive dye assemblies for biosensing from molecular building blocks. Proc. Natl. Acad. Sci. USA **2001**, *98* (26), 14769–14772.
4. Fan, C.H.; Wang, S.; Hong, J.W.; Bazan, G.C.; Plaxco, K.W.; Heeger, A.J. Beyond superquenching: Hyper-efficient energy transfer from conjugated polymers to gold nanoparticles. Proc. Nat. Acad. Sci. USA **2003**, *100* (11), 6297–6301.
5. McNeill, J.D.; Wu, C.; Szymanski, C.J.; Zheng, Y. Bioconjugate polymer dot nanoparticles. Polym. Prepr. **2007**, *48*, 23–24.

6. Murcia, M.J.; Naumann, C.A. Biofunctionalization of fluorescent nanoparticles: Nanotechnologies for the life sciences. In *Biofunctionalization of Nanomaterials*; Kumar, C.S.S.R., Eds.; WILEY-VCH Verlag GmbH & Co. KGaA: Weinheim, 2005; 1, 1–40.

7. Veronese, F.M.; Morpurgo, M. Bioconjugation in pharmaceutical chemistry. Farmaco **1999**, 54 (8), 497–516.

8. Bagwe, R.P.; Zhao, X.; Tan, W. Bioconjugated luminescent nanoparticles for biological applications. J. Dispers. Sci. Technol. **2003**, *24* (3–4), 453–464.

9. Illum, L.; Jones,P.D.E.; Kreuter, J.; Baldwin, R.W.;Davis, S.S. Adsorption of monoclonal antibodies to polyhexylcyanoacrylate nanoparticles and subsequent immunospecific binding to tumour cells *in vitro*. Int. J. Pharm. **1983**, *17* (1), 65–76.

10. Illum, L.; Jones, P.D.; Baldwin, R.W.; Davis, S.S. Tissue distribution of poly(hexyl 2-cyanoacrylate) nanoparticles coated with monoclonal antibodies in mice bearing human tumor xenografts. J. Pharmacol. Exp. Ther. **1984**, *230* (3), 733–736.

11. Kubiak, C.; Manil, L.; Couvreur, P. Sorptive properties of antibodies onto cyanoacrylic nanoparticles. Int. J. Pharm. **1988**, *41* (3), 181–187.

12. Manil, L.; Roblot-Treupel, L.; Couvreur, P. Isobutyl cyanoacrylate nanoparticles as a solid phase for an efficient immunoradiometric assay. Biomaterials **1986**, *7* (3), 212–216.

13. Couvreur, P.; Aubry, J.Monoclonal antibodies for the targeting of drugs: Application to nanoparticles. In *Topics in Pharmaceutical Sciences*; Breimer, D.D.; Speises, P., Eds.; Elsevier Science Publishers B.V.: Amsterdam, 1983; 305–316.

14. Blackwell, J.E.; Dagia, N.M.; Dickerson, J.B.; Berg, E.L.; Goetz, D.J. Ligand coated nanosphere adhesion to E- and P-selectin under static and flow conditions. Ann. Biomed. Eng. **2001**, *29* (6), 523–533.

15. Serizawa, T.; Uchida, T.; Akashi, M. Synthesis of polystyrene nanospheres having lactose-conjugated hydrophilic polymers on their surfaces and carbohydrate recognition by proteins. Biomater. Sci. Polym. Ed. **1999**, *10* (3), 391–401.

16. Taylor, J.R.; Fang, M.M.; Nie, S. Probing specific sequences on single DNA molecules with bioconjugated fluorescent nanoparticles. Anal. Chem. **2000**, *72* (9), 1979–1986.

17. Li, Y.; Ogris, M.; Wagner, E.; Pelisek, J.; Ruffer, M. Nanoparticles bearing polyethylenglycol-coupled transferrin as gene carriers: preparation and *in vitro* evaluation. Int. J. Pharm. **2003**, *259* (1–2), 93–101.

18. Akasaka, Y.; Ueda, H.; Takayama, K.; Machida, Y.; Nagai, T. Preparation and evaluation of bovine serum albumin nanospheres coated with monoclonal antibodies. Drug Des. Disc. **1988**, *3* (1), 85–97.

19. Rolland, A.; Bourel, D.; Genetet, B.; Le Verge, R. Monoclonal antibodies covalently coupled to polymethacrylic nanoparticles: *in vitro* specific targeting to human T lymphocytes. Int. J. Pharm. **1987**, *39* (1–2), 173–180.

20. Gautier, S.N.; Grudzielski, N.; Goffinet, G.; de Hassonville, S.H.; Delattre, L.; Jerome, R. Preparation of poly (D,L-lactide) nanoparticles assisted by amphiphilic poly (methyl methacrylate-co-methacrylic acid) copolymers. J. Biomater. Sci. Polym. **2001**, *12* (4), 429–450.

21. Gref, R.; Couvreur, P.; Barratt, G.; Mysiakine, E. Surfaceengineered nanoparticles for multiple ligand coupling. Biomaterials **2003**, *24* (24), 4529–4537.

22. Maruyama, A.; Ishihara, T.; Kim, J.S.; Kim, S.W.; Akaike, T. Nanoparticle DNA carrier with poly(L-lysine) grafted polysaccharide copolymer and poly(D,L-lactic acid). Bioconjug. Chem. **1997**, *8* (5), 735–742.

23. Wu, C.; Schneider, T.; Zeigler, M.; Yu, J.; Schiro, P.G.; Burnham, D.R.; McNeill, J.D.; Chiu, D.T. Bioconjugation of ultrabright semiconducting polymer dots for specific cellular targeting. J. Am. Chem. Soc. **2010**, *132* (43), 15410–15417.

24. Wu, C.; Chiu, D.T. Highly fluorescent semiconducting polymer dots for biology and medicine. Angew. Chem. Int. Ed. **2013**, *52* (11), 3086–3109.

25. Resch-Genger, U.; Grabolle, M.; Cavaliere-Jaricot, S.; Nitschke, R.; Nann, T. Quantum dots versus organic dyes as fluorescent labels. Nat. Methods **2008**, *5* (9), 763–775.

26. Medintz, I.L.; Uyeda, H.T.; Goldman, E.R.; Mattoussi, H. Quantum dot bioconjugates for imaging, labelling and sensing. Nat. Mater. **2005**, *4* (6), 435–446.

27. Michalet, X.; Pinaud, F.F.; Bentolila, L.A.; Tsay, J.M.; Doose, S.; Li, J.J.; Sundaresan, G.; Wu, A.M.; Gambhir, S.S.; Weiss, S. Quantum dots for live cells, *in vivo* imaging, and diagnostics. Science **2005**, *307* (5709), 538–544.

28. Abdelhamid, H.N.; Wu, H.F. Probing the interactions of chitosan capped CdS quantum dots with pathogenic bacteria and their biosensing application. J. Mater. Chem. B **2013**, 1 (44), 6094–6106.

29. Chalfie, M.; Tu, Y.; Euskirchen, G.; Ward, W.; Prasher, D. Green fluorescent protein as a marker for gene expression. Science **1994**, 263 (5148), 802–805.

30. Fernando, L.P.; Kandel, P.K.; Yu, J.; McNeill, J.; Ackroyd, P.C.; Christensen, K.A. Mechanism of cellular uptake of highly fluorescent conjugated polymer nanoparticles. Biomacromolecules **2010**, *11* (10), 2675–2682.

31. Rong, Y.; Wu, C.; Yu, J.; Zhang, X.; Ye, F.; Zeigler, M.; Gallina, M.E.; Wu, I.C.; Zhang, Y.; Chan, Y.H.; Sun, W.; Uvdal, K.; Chiu, D.T. Multicolor fluorescent semiconducting polymer dots with narrow emissions and high brightness. ACS Nano **2013**, *7* (1), 376–384.

32. Sun, W.; Yu, J.; Deng, R.; Rong, Y.; Fujimoto, B.; Wu, C.; Zhang, H.; Chiu, D.T. Semiconducting polymer dots doped with Europium complexes showing ultra narrow emission and long luminescence lifetime for time-gated cellular imaging. Angew. Chem. Int. Ed. **2013**, *52* (43), 11294–11297.

33. Wu, C.; Hansen, S.J.; Hou, Q.; Yu, J.; Zeigler, M.; Jin, Y.; Burnham, D.R.; McNeill, J.D.; Olson, J.M.; Chiu, D.T. Design of highly emissive polymer dot bioconjugates for *in vivo* tumor targeting. Angew. Chem. Int. Ed. **2011**, *50* (15), 3430–3434.

34. Zhang, Y.; Yu, J.B.; Gallina, M.E.; Sun, W.; Rong, Y.; Chiu, D.T. Highly luminescent, fluorinated semiconducting polymer dots for cellular imaging and analysis. Chem. Commun. **2013**, *49* (74), 8256–8258.

35. Chan, Y.H.; Ye, F.; Gallina, M.E.; Zhang, X.; Jin, Y.; Wu, I.C.; Chiu, D.T. Hybrid semiconducting polymer dotquantum dot with narrow-band emission, near-infrared fluorescence, and high brightness. J. Am. Chem. Soc. **2012**, *134* (17), 7309–7312.

36. Wan, Y.; Zheng, L.; Sun, Y.; Zhang, D. Multifunctional semiconducting polymer dots for imaging, detection, and photo-killing of bacteria. J. Mater. Chem. B **2014**, *2* (30), 4818–4825.

37. Li, K.; Liu, Y.; Pu, K.Y.; Feng, S.S.; Zhan, R.; Liu, B. Polyhedral oligomeric silsesquioxanes-containing conjugated polymer loaded PLGA nanoparticles with trastuzumab (herceptin) functionalization for HER2-positive cancer cell detection. Adv. Funct. Mater. **2011**, *21* (2), 287–294.

38. Cordier, P.; Tournilhac, F.; Soulie-Ziakovic, C.; Leibler, L. Self-healing and thermoreversible rubber from supramolecular assembly. Nature **2008**, *451* (7181), 977–980.

39. Altintas, O.; Lejeune, E.; Gerstel, P.; Barner-Kowollik, C. Bioinspired dual self-folding of single polymer chains via reversible hydrogen bonding. Polym. Chem. **2012**, *3* (3), 640–651.

40. Nishiyama, N.; Yokoyama, M.; Aoyagi, T.; Okano, T.; Sakurai, Y.; Kataoka, K. Preparation and characterization of self-assembled polymer–metal complex micelle from cis-Dichlorodiammineplatinum(II) and poly(ethylene glycol)–poly(α,β-aspartic acid) block copolymer in an aqueous medium. Langmuir **1999**, *15* (2), 377–383.

41. Grimm, F.; Ulm, N.; Grçhn, F.; Dring, J.; Hirsch, A. Selfassembly of supramolecular architectures and polymers by orthogonal metal complexation and hydrogen-bonding motifs. Chem. Eur. J. **2011**, *17* (34), 9478–9488.

42. Burattini, S.; Colquhoun, H.M.; Fox, J.D.; Friedmann, D.; Greenland, B.W.; Harris, P.J.F.; Hayes, W.; Mackay, M.E.; Rowan, S.J. A self-repairing, supramolecular polymer system: Healability as a consequence of donor–acceptor π–π stacking interactions. Chem. Commun. **2009**, (44), 6717–6719.

43. Greenland, B.W.; Bird, M.B.; Burattini, S.; Cramer, R.; OReilly, R.K.; Patterson, J.P.; Hayes, W.; Cardin, C.J.; Colquhoun, H.M. Mutual binding of polymer end-groups by complementary π–π-stacking: a molecular "Roman Handshake." Chem. Commun. **2013**, *49* (5), 454–456.

44. Stadermann, J.; Komber, H.; Erber,M.; Dbritz, F.; Ritter, H.; Voit, B. Diblock copolymer formation via self-assembly of cyclodextrin and adamantyl end-functionalized polymers. Macromolecules **2011**, *44* (9), 3250–3259.

45. Schmidt, B.V.K.J.; Rudolph, T.; Hetzer,M.; Ritter, H.; Schacher, F.H.; Barner-Kowollik, C. Supramolecular three-armed star polymers via cyclodextrin host–guest self-assembly. Polym. Chem. **2012**, *3* (11), 3139–3145.

46. Suslick, K.S. Shape selective oxidation by metalloporphyrins. In *The Porphyrin Handbook, Biochemistry and Binding: Activation of Small Molecules*; Kadish, K.M., Smith, K.M., Guilard, R., Eds.; Academic Press: San Diego, 2000; Vol. 4; 41–60.

47. Reimers, J.R.; Hush, N.S.; Crossley, M.J. Inter-porphyrin Coupling: How strong should it be for molecular electronics applications? J. Porphyrins Phthalocyanines **2002**, *6* (12), 795–805.

48. Sanders, J.K.M.; Bampos, N.; Clyde-Watson, Z.; Darling, S.L.; Hawley, J.C.; Kim, H.J.; Mak, C.C.; Webb, S.J. Axial coordination chemistry of metalloporphyrins. In *The Porphyrin Handbook*; Kadish, K.M., Smith, K.M., Guilard, R., Eds.; Academic Press: San Diego, 2000; 3, 1–48.

49. Sessler, J.L.; Karnas, E.; Sedenberg, E. Porphyrins and expanded porphyrins as receptors. In *Supramolecular Chemistry: From Molecules to Nanomaterials*; Philip, A., Jonathan, W., Eds.; Wiley: Chichester, 2012; 3.

50. Roberts, D.A.; Schmidt, T.W.; Crossley, M.J.; Perrier, S. Tunable self-assembly of triazole-linked porphyrin–polymer conjugates. Chem. Eur. J. **2013**, *19* (38), 12759–12770.

51. Sun, Y.; Cao, W.; Li, S.; Jin, S.; Hu, K.; Hu, L.; Huang, Y.; Gao, X.; Wu, Y.; Liang, X.J. Ultrabright and multicolorful fluorescence of amphiphilic polyethyleneimine polymer dots for efficiently combined imaging and therapy. Sci. Rep. **2013**, 3, 3036.

52. Mishra, D.; Kang, H.C.; Bae, Y.H. Reconstitutable charged polymeric (PLGA)2-b-PEI micelles for gene therapeutics delivery. Biomaterials **2011**, *32* (15), 3845–3854.

53. Abdelwahed, W.; Degobert, G.; Stainmesse, S.; Fessi, H. Freeze-drying of nanoparticles: Formulation, process and storage considerations. Adv. Drug Delivery Rev. **2006**, *58* (15), 1688–1713.

54. Sun, W.; Ye, F.; Gallina, M.E.; Yu, J.; Wu, C.; Chiu, D.T. Lyophilization of semiconducting polymer dot bioconjugates. Anal. Chem. **2013**, *85* (9), 4316–4320.

55. Bao, B.; Ma, M.; Chen, J.; Yuwen, L.; Weng, L.; Fan, Q.; Huang, W.; Wang, L. Facile preparation of multicolor polymer nanoparticle bioconjugates with specific biorecognition. ACS Appl. Mater. Interfaces **2014**, *6* (14), 11129–11135.

56. Kandel, P.K.; Fernando, L.P.; Ackroyd, P.C.; Christensen, K.A. Incorporating functionalized polyethylene glycol lipids into reprecipitated conjugated polymer nanoparticles for bioconjugation and targeted labeling of cells. Nanoscale **2011**, *3* (3), 1037–1045.

57. Zhang, X.; Yu, J.; Rong, Y.; Ye, F.; Chiu, D.T.; Uvdal, K. High-intensity near-IR fluorescence in semiconducting polymer dots achieved by cascade FRET strategy. Chem. Sci. **2013**, *4* (5), 2143–2151.

Sensing and Diagnosis

Mizuo Maeda
Graduate School of Engineering, Kyushu University, Fukuoka, Japan

Abstract

In this entry, two types of supramolecular assembling phenomena will be discussed in relation to gene mutation assay, which is one of the most important areas of biomedical sensing and diagnosis. First, the entry will review DNA-carrying colloidal nanoparticles prepared from DNA-PNIPAAm graft copolymer through supramolecular assembly. Then, it will discuss DNA-carrying nanoparticles assembled in the presence of complementary DNA to yield visibly detectable aggregates due to cross-linking or colloidal stability change.

INTRODUCTION

Developments in molecular biology make it possible to correlate small mutations of certain genes with heritable disorders and cancers. These findings promoted the study of gene mutation assays, especially in the fields of biomedical sensing and diagnosis. Various methods have been proposed to detect changes in DNA base sequence.[1] One method relies on a sequence-specific enzymatic reaction and the other on an oligonucleotide probe. Enzyme-based methods may be more convenient and reliable in some cases, whereas the methods based on nucleic acid hybridization may be more general and widely applicable. Most of the latter methods take advantage of single-stranded DNA (ssDNA), which is immobilized on a solid surface such as polymer membrane, metal electrode, and latex particle.

A well known example of diagnostic methods that utilize supramolecular systems is the latex agglutination immunoassay.[2] An antigenor antibody-modified latex particle is cross-linked by the complementary antibody or antigen and the resulting particle assembly is detected by the increase in turbidity or light scattering intensity of the dispersion. The DNA diagnosis using the supramolecular assembling phenomenon of colloidal particles was reported by Mirkin et al.[3–7] ssDNA-modified gold nanoparticles were used as a detection probe; the particles were cross-linked by hybridization with the complementary DNA to form particle assemblies. This aggregation behavior is easily detected by the color change of the dispersion due to a red shift of the plasmon band.

In addition to Mirkin's study, ssDNA-carrying colloidal particles have been recognized as useful tools in the field of molecular biology. Most of the particles are latex-based and are used for the sequence-specific separation of DNA[8] and RNA[9] and the detection of DNA[10] and RNA.[11] DNA-carrying colloidal particles reported thus far are commonly prepared by a two-step procedure: plain colloidal particle preparation followed by DNA immobilization to the particle.

The preparation of colloidal particles through the supramolecular assembly of amphiphilic copolymers has received much attention.[12–15] These particles are generally prepared by solvent exchange from nonselective solvent to selective solvent (water). As an alternative method, a lower critical solution temperature (LCST)-showing polymer, poly(N-isopropylacrylamide) (PNIPAAm), was used for the preparation of colloidal particles by simple heating above its LCST (~32°C).[16–18]

We developed a one-step method to preserve DNA-carrying colloidal particles through the self-organization of DNA-PNIPAAm graft copolymer. The copolymer is composed of a PNIPAAm main chain and a DNA $[d(T)_{12}]$ graft chain, and forms a colloidal particle with a PNIPAAm core surrounded by hydrophilic DNA above the LCST. This particle aggregates in the presence of complementary DNA according to the cross-linking mechanism, and is observed as a turbidity increase of the dispersion.

We found another aggregation behavior of this DNA-containing colloidal particle. The mechanism is based on the stability change of the particle induced by the hybridization of the surface DNA with the complementary DNA. The DNA detection with selectivity in the sequence and also in the chain length (the number of nucleic bases) was achieved by the turbidimetric particle aggregation assay according to the supramolecular assembly mechanism.

In this entry, two types of supramolecular assembling phenomena will be discussed in relation to gene mutation assay, which is one of the most important areas of biomedical sensing and diagnosis:

1. DNA-carrying colloidal nanoparticles prepared from DNA-PNIPAAm graft copolymer through supramolecular assembly;

Concise Encyclopedia of Biomedical Polymers and Polymeric Biomaterials DOI: 10.1081/E-EBPPC-120052247

1393

Scaffolds—Smart

2. DNA-carrying nanoparticles assembled in the presence of complementary DNA to yield visibly detectable aggregates due to cross-linking or colloidal stability change.

DNA-CARRYING NANOPARTICLES

DNA graft copolymer (1) was prepared by copolymerization between N-isopropylacrylamide (NIPAAm) and DNA macromonomer (Fig. 1). DNA macromonomer was synthesized by coupling reaction between methacryloyloxysuccinimide and amino-linked DNA having an aminohexyl linker at its 5′-terminus, as reported previously.[19,20] DNA macromonomer (0, 0.15, 0.30, and 0.60 μmol) was mixed with N-isopropylacrylamide (NIPAAm) (140 μmol) in buffer solution [10 mM Tris-HCl (pH 7.4)]. After flushing the solution with nitrogen, aqueous solutions of ammonium persulfate and N,N,N′,N′-tetramethylethylenediamine were added. Polymerization was carried out at 25°C for 1 hr. The unreacted NIPAAm monomer was removed by dialysis (Spectra/Por6, MWCO = 1000). The unreacted DNA macromonomer was separated from 1 on a 1.5 × 10 cm Sephadex G-100 column. The composition of 1 was determined by dry weight and UV absorption at 260 nm due to DNA macromonomer unit using the molar extinction coefficient of 97,800 $M^{-1}cm^{-1}$.[21]

In order to obtain molecular data for graft copolymers, static light scattering (SLS) and dynamic light scattering (DLS) measurements were conducted with a DLS-7000 instrument (Otsuka Electronics) according to the literature.[22] The light source was an Ar ion laser (488 nm, 75 mW). The polymer solutions were filtered through 0.22-μm filters (Millipore) prior to measurements. For the determination of the Mw of each polymer, SLS measurement was carried out over the angular range from 30° to 140° and the concentration range from 0.5 to 2.3 mg/mL at 25°C in 10 mM Tris-HCl buffer (pH 7.4). The refractive index increment (dn/dC) of copolymer was calculated from the additive equation, dn/dC = $w_{PNIPAAm}$(dn/dC)$_{PNIPAAm}$ + w_{DNA}(dn/dC)$_{DNA}$, where

w represents the weight fraction. The dn/dC of DNA macromonomer was measured by DRM-1021 (Otsuka Electronics) at 488 nm and determined to be 0.187 mL/g (25°C) and 0.192 mL/g (40°C). The literature values were used for the dn/dC of PNIPAAm.[21]

The properties of DNA graft copolymers are summarized in Table 1. The conversions of both monomers were about 70%. The molar ratios of DNA macromonomer units in the copolymers agreed with the feed ratios. The Mw values of the polymers were determined from Zimm plots and were almost constant (~2 × 10^5) irrespective of the DNA macromonomer fraction.

The polymer concentration for LCST measurement was 0.1 mg/mL in 10 mM Tris-HCl (pH 7.4)/5 mM MgCl$_2$. The solution turbidity was monitored at 500 nm by raising the solution temperature (0.5°C/min). LCST was defined as the temperature at which the turbidity began to increase. Only one transition was observed for all the polymers. As a result, the LCSTs of polymers increased with increasing DNA macromonomer fractions. Since this tendency is in line with the behavior of copolymers of NIPAAm and hydrophilic comonomers such as acrylamide and acrylic acid,[23,24] the DNA macromonomer was considered to act as a hydrophilic part in the copolymer.

To form the colloidal particle from 1, the temperature of the solution was raised to 40°C, which was higher than the phase transition temperature of the copolymer. The copolymer solution was 0.1 mg/mL in 10 mM Tris-HCl (pH 7.4)/5 mM MgCl$_2$. The colloidal particles were examined by DLS analysis conducted at a detection angle of 90°. Figure 2 shows the hydrodynamic radius (Rh) distributions of colloidal particles. The radii of colloidal particles increased as DNA macromonomer fractions in copolymers decreased. Table 2 summarizes the results of DLS analyses of colloidal particles. The association numbers (copolymers per particle) increased with decreasing fractions of DNA macromonomer in copolymers. The radius of each particle remained constant at least for several hours when heated to 40°C. When the particle dispersions were cooled to room temperature to dissociate the particles into the individual copolymers and the solutions were heated to

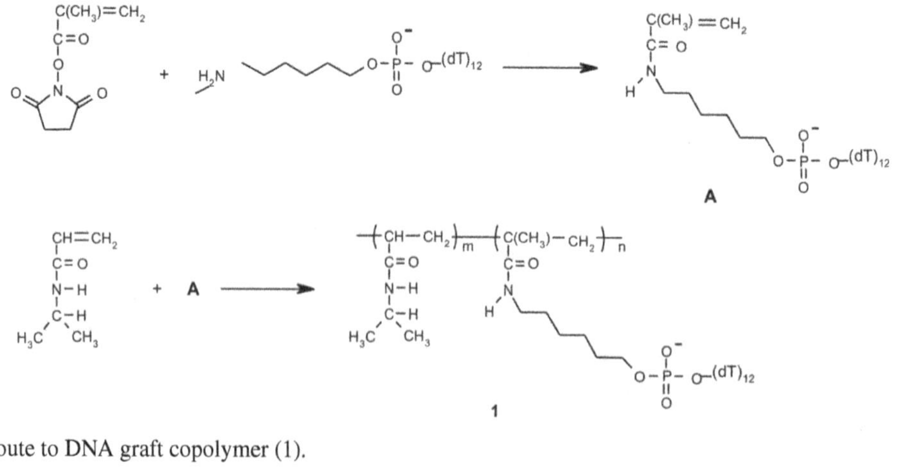

Fig. 1 Synthetic route to DNA graft copolymer (1).

40°C again, particles of the same size were formed (relative difference was ±2%).

The ratios of radius of gyration (R_g) and (R_h) of colloidal particles (R_g/R_h) were calculated to range from 0.72 to 0.81, which is close to the theoretical value for a hard sphere (0.776).[25] The density of core of the colloidal particle (ρ) was calculated from the equation, $\rho = Mw/(4/3r_{core}{}^3N_A)$. Assuming that $r_{core} \sim R_h$, the calculated ρ values roughly agreed with that reported for the collapsed PNIPAAm homopolymer (0.36 g/cm³)[26] as shown in Table 2. However, if the thickness of the DNA layer is assumed to be about 7.8 nm which is estimated from a polymer brush conformation of DNA on the core surface, the calculated values are about two times larger than that of collapsed PNIPAAm homopolymer. These results may indicate that the DNA layer was not detected by the light scattering method due to the low density of this layer or that the DNA existed as a flat conformation on the core surface.

The value of surface area per DNA (S_{DNA}) calculated from the assumption of $r_{core} \sim R_h$, was roughly constant for all the particles. This is very similar to the case of colloidal particles formed from poly(NIPAAm-co-acrylic acid)[21] and should be explained as follows. In the particle formation process, the association of copolymers stops when the surface area per DNA reaches a certain value and this value is constant irrespective of the copolymer composition. This is the reason why the copolymer with a smaller DNA fraction formed a larger particle.

DNA-DEPENDENT ASSEMBLY OF NANOPARTICLES

The complementary DNA recognition by the particle surface DNA [d(T)$_{12}$] was examined. The copolymer solution was 0.1 mg/mL in 10 mM Tris-HCl (pH 7.4)/5 mM MgCl₂. After the formation of colloidal particle from D-0.17 at 40°C, the cross-linking DNA [equimolar mixture of 3′-d(A)$_{12}$(TG)$_6$-5′ and 5′-d(AC)$_6$(A)$_{12}$-3′] having complementary d(A)$_{12}$ regions at both ends or noncross-linking DNA [equimolar mixture of 3′-d(A)$_{12}$(TG)$_6$-5′ and 5′-d(AC)$_6$-3′] having a d(A)$_{12}$ region at only one end was added to the particle dispersion and the turbidity change of the dispersion was monitored. As shown in Fig. 2, the turbidity did not change by the addition of non-crosslinking DNA and increased steeply with cross-linking DNA.

Table 1 Molecular parameters of polymers

Sample	Mw	mol% DNA	DNA graft number	LCST[a] (°C)
D-0	2.1×10^5	—	—	33.4
D-0.085	2.2×10^5	0.085	1.6	34.5
D-0.17	2.3×10^5	0.17	3.2	34.9
D-0.36	2.6×10^5	0.36	7.3	35.8

[a]Determined by turbidity measurements as described in the text.

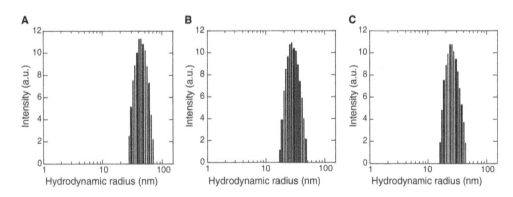

Fig. 2 z-Weighted hydrodynamic radius distributions of colloidal particles formed from copolymers (1) at 40°C in 10 mM Tris-HCl (pH 7.4)/5 mM MgCl₂. (**A**) D-0.085, (**B**) D-0.17, (**C**) D-0.36. Copolymer concentration was 0.1 mg/mL.

Table 2 Results of light scattering analysis of colloidal particles formed from copolymers at 40°C

Sample	Mw of particle	R_g (nm)	$R_h{}^a$ (nm)	Rg/Rh	Association number	S_{DNA} (nm²/DNA)	ρ(g/cm³)
D-0.085	7.6×10^7	35	43	0.81	350	40	0.38
D-0.17	2.0×10^7	21	29	0.72	87	36	0.33
D-0.36	8.6×10^6	18	25	0.72	33	31	0.22

[a]R_h was calculated from Stokes-Einstein equation.[22]

Moreover, when the temperature of the turbid dispersion with cross-linking DNA was raised, the turbidity began to decrease at 44°C and reached the same value before DNA addition. This phenomenon should be explained by assembly and dissociation between the particles according to the hybridization of cross-linking DNA [$d(A)_{12}$ regions] with the surface DNA [$d(T)_{12}$]. In fact, the temperature range of this particle dissociation process was included in that of the melting of free DNA duplex formed from $5'$-$d(AC)_6(A)_{12}$-$3'$and $d(T)_{12}$ (26–47°C).

The effect of DNA addition on the particle size was investigated for D-0.17 by light scattering measurements. On the addition of noncross-linking DNA to the colloidal particles at 40°C, R_h did not change for more than 3 hr. Although the hybridization of noncross-linking DNA with the surface DNA would lead to the increase in overall hydrophilicity of the particle, the particle size remained constant. This indicates a relatively slow rate of copolymer exchange at 40°C. When the particle assembly formed by the addition of cross-linking DNA at 40°C was heated to 50°C, the dispersed colloidal particle was found to have R_h of 31 nm, which was very close to that of the original colloidal particle.

Thus we found that the DNA-carrying colloidal particles are easily prepared from PNIPAAm–graft-DNA by heating through supramolecular assembly. The particle surface DNA recognizes the complementary cross-linking DNA so that the particle assembly formed shows easily

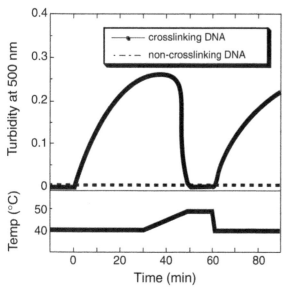

Fig. 3 Turbidity change of colloidal dispersion induced by DNA addition at 40°C and the effect of temperature on dispersion turbidity. The dispersion (700 μL) contains 0.1 mg/mL of D-0.17 [1.0 nmol of $d(T)_{12}$ strand]. At 0 min, 11 μL of cross-linking DNA (44 μM equimolar) mixture of $3'$-$d(A)_{12}(TG)_6$-$5'$ and $5'$-$d[AC)_6(A)_{12}$-$3'$] or non-cross-linking DNA [equimolar mixture of $3'$-$d(A)_{12}(TG)_6$-$5'$ and $5'$-$d(AC)_6$-$3'$] is added. The dispersion temperature is changed between 40 and 50°C periodically.

detectable turbidity changes (see Fig. 3). These particles may be useful for turbidimetric DNA detection.

COLLOIDAL PROPERTIES OF DNA-CARRYING NANOPARTICLES

We studied in detail the stability of the DNA-carrying colloidal particles by changing the concentrations of coexisting salts ($MgCl_2$ and NaCl). Generally, colloidal particles in dispersion are known to aggregate at a certain salt concentration called the critical coagulation concentration. We prepared a graft copolymer having DNA branches of $5'$-GCCACCAGC-$3'$ in order to investigate more precisely DNA sequence recognition using a series of target DNAs (**I** through **V**) which are listed in Table 3.

Figure 4 shows the optical density of particle dispersion at 500 nm after the mixing with NaCl or $MgCl_2$ solution. The increase of optical density resulting from the aggregation of colloidal particle was observed at 20 mM of $MgCl_2$ but not for NaCl in the concentration range of 10–1400 mM. The stability of the colloidal particles was also examined in the presence of complementary DNA (**I**) or its one point-mismatching DNA (**II**). Surprisingly, the optical density of the particle dispersion in the presence of **I** increased steeply at 10 mM for $MgCl_2$ and at 400 mM for NaCl. No change was observed in the presence of **II**; the behavior with the one-point mutant was very similar to the DNA absence. These results suggest that the instability of the colloidal dispersion in the presence of **I** should be caused by the selective hybridization of the particle surface DNA with its complementary DNA (**I**).

To confirm this conjecture, we investigated the temperature effect on the particle aggregate formed in the presence of **I** at an NaCl concentration of 500 mM. As the temperature increased, the absorbance of the dispersion decreased steeply in the narrow temperature range and then reached almost zero, as shown in Fig. 5 (bottom). The R_h of colloidal particles at 55°C was 22 nm, which agreed with that of the particles at that temperature without the addition of DNA, indicating that the aggregate

Table 3 Graft copolymer

Code	Sequence ($3'$–$5'$)	Description
I	CGGTGGTCG	Complementary
II	CGGTAGTCG	Point mutant
III	TTCGGTGGTCG	Different length (two bases longer)
IV	TTTTTCGGTGGTCG	Different length (five bases longer)
V	CGGTGGTCGTT	Different length (two bases longer)

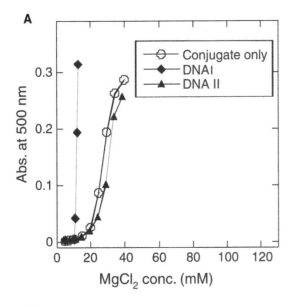

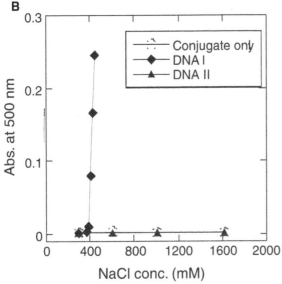

Fig. 4 Absorbance of the colloidal particle dispersion as a function of concentration of (**A**) MgCl$_2$ and (**B**) NaCl in the absence and presence of complementary DNA (**I**) or one point mismatching DNA (**II**) at 40°C. The colloidal particle concentration was 0.1 mg/mL, containing 2.8 μM of DNA unit. The target DNA concentration was 2.8 μM.

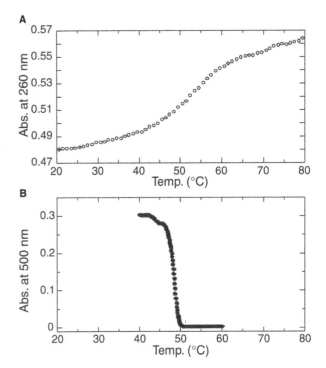

Fig. 5 Temperature dependence of absorbance of the particle aggregate formed in the presence of **I** (**B**) and melting curve of free duplex DNA (**A**) in 500 m*M* NaCl/10 m*M* Tris-HCl buffer (pH 7.4). The concentrations of the colloidal particle and **I** were the same as those in Figure 4. The free duplex DNA concentration was 2.8 μM. Heating rate was 0.5°C/min.

potential of the particle surface should be smaller when the surface DNA was in duplex state, so that the particle became less stable.

It should be noted that the dissociation of aggregate took place at a very narrow temperature range in comparison with the melting of free duplex DNA (Fig. 5). A similar phenomenon was reported for the aggregate from DNA modified-gold nanoparticles cross-linked by the complementary DNA.[3–7] This property is important for the discriminative detection of DNA duplexes based on small differences in melting temperatures between probe DNA and the target DNAs.

We presumed that the discrimination of the target DNA in the chain length would also be possible according to the aggregation mechanism: the hybridization of particle surface DNA with the longer DNAs (**III** and **V**) should result in the protruding of a single-stranded region at the end of the duplex DNA, so that the colloidal particles would become more stable than when the particle hybridized with the same length DNA (**I**) to give a perfect duplex. Figure 6 shows the stability of the dispersion against NaCl. The particle was clearly more stable in the presence of **III** and **V** than in the presence of **I** as we expected. In the presence of **V**, having an additional T$_2$ at its 5′-terminus, the particle became more stable than in the presence of **III** or **IV**, having an additional T$_2$ or T$_5$ at its 3′-terminus, respectively. Since the hybridization of **V** with the particle shell DNA

dissociated into the individual particle. Since the temperature range of this aggregate dissociation process was included in the temperature range of the dissociation of free duplex DNA (38–63°C), the aggregate dissociation should be attributed to the dissociation of DNA duplex on the particle. From these results, it can be concluded that the stability decrease of colloidal dispersion was induced by the duplex formation of the surface DNA. Generally, the binding of Mg^{2+} and Na$^+$ to the duplex of DNA or RNA is stronger than binding to ssDNA or RNA, because of the higher anionic charge density of the duplex.[27] Thus, the absolute value of the electric

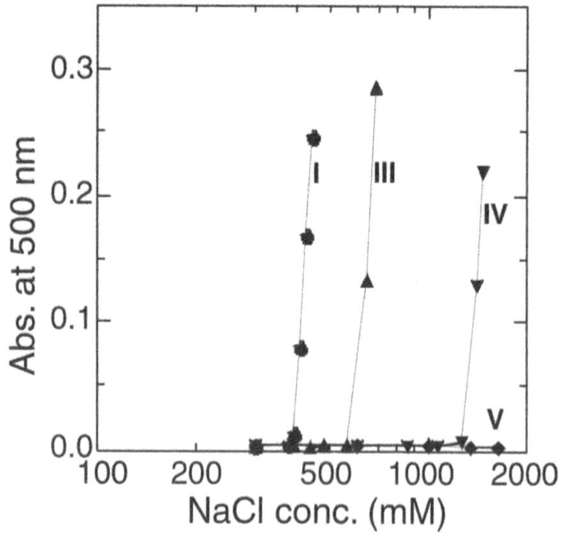

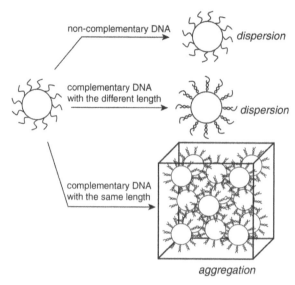

Fig. 6 Absorbance of the colloidal particle dispersion as a function of NaCl concentration in the presence of complementary DNA (**I**) or complementary but longer DNAs (**III** through **V**) at 40°C. The concentrations of colloidal particle and the target DNA were the same as those in Fig. 4.

Fig. 7 DNA sequence and chain length-selective aggregations of the DNA-carrying colloidal particles.

should result in the exposure of the protruding single-stranded region at the outer surface of colloidal particles, the property of the single-stranded region, namely the lower affinity to Na^+, remarkably appeared in this case. The comparison between **III** and **IV** indicates that the particle is more stable in the presence of the longer DNA (**IV**). These results suggest that particle aggregation according to the proposed mechanism would be applicable to chain length-selective DNA detection. It was reported that a few base deletions in a certain region of the genome is closely related to a cancer,[28] and this method may be applicable as a diagnostic tool.

Finally, we demonstrated the turbidimetric detection of DNA with sequence and chain length selectivity. After the addition of DNA (**I** through **III** and **V**), the time dependent absorbance change of the particle dispersion containing 500 mM NaCl was monitored. The absorbance did not change by the addition of either one pointmismatching DNA (**II**) or longer DNAs (**III** and **V**), while the absorbance increased steeply by the addition of complementary DNA **I** and was saturated within 5 min.

CONCLUSION

In this entry, a typical example of supramolecular systems which are useful for biological sensing and diagnosis was explained. We proposed a novel aggregation mechanism of DNA-containing colloidal particles, which is based on the stability decrease of colloidal particles accompanied by the duplex formation of the shell DNA with the complementary DNA. Using the particle aggregation assay according to the above mechanism, sequence and chain

length-selective DNA detection was attained (Fig. 7). However, further research must be carried out before it can be applied to biomedical gene diagnosis since the present results are limited to short oligonucleotides.

REFERENCES

1. Lyamichev, V.I.; Kaiser, M.W.; Lyamicheva, N.E.; Vologodskii, A.V.; Hall, J.G.; Ma, W.P.; Allawi, H.T.; Neri, B.P. Experimental and theoretical analysis of the invasive signal amplification reaction. Biochemistry **2000**, *39* (31), 9523–9532.
2. Gella, F.J.; Serra, J.; Gener, J. Latex agglutination procedures in immunodiagnosis. Pure Appl. Chem. **1991**, *63* (8), 1131–1134.
3. Mirkin, C.A.; Letsinger, R.L.; Mucic, R.C.; Storhoff, J.J. A DNA-based method for rationally assembling nanoparticles into macroscopic materials. Nature **1996**, *382* (6592), 607–609.
4. Elghanian, R.; Storhoff, J.J.; Mucic, R.C.; Letsinger, R.L.; Mirkin, C.A. Selective colorimetric detection of polynucleotides based on the distance-dependent optical properties of gold nanoparticles. Science **1997**, *277* (5329), 1078–1081.
5. Storhoff, J.J.; Elghanian, R.; Mucic, R.C.; Mirkin, C.A.; Letsinger, R.L. One-pot colorimetric differentiation of polynucleotides with single base imperfections using gold nanoparticle probes. J. Am. Chem. Soc. **1998**, *120* (9), 1959–1964.
6. Reynolds, R.A., III, Mirkin, C.A.; Letsinger, R.L. Homogeneous, nanoparticle-based quantitative colorimetric detection of oligonucleotides. J. Am. Chem. Soc. **2000**, *122* (15), 3795–3796.
7. Storhoff, J.J.; Lazarides, A.A.; Mucic, R.C.; Mirkin, C.A.; Letsinger, R.L.; Schatz, G.C. What controls the optical properties of DNA-linked gold nanoparticle assemblies? J. Am. Chem. Soc. **2000**, *122* (19), 4640–4650.

8. Kuribayashi-Ohta, K.; Tamatsukuri, S.; Hikata, M.; Miyamoto, C.; Furuichi, Y. Application of oligo(dT)30-latex for rapid purification of poly(A)+ mRNA and for hybrid subtraction with the in situ reverse transcribed cDNA. Biochim. Biophys. Acta **1993**, *1156* (2), 204–212.

9. Imai, T.; Sumi, Y.; Hatakeyama, M.; Fujimoto, K.; Kawaguchi, H.; Hayashida, N.; Shiozaki, K.; Terada, K.; Yajima, H.; Handa, H. Selective isolation of DNA or RNA using single-stranded DNA affinity latex particles. J. Colloid Interface Sci. **1996**, *177* (1), 245–249.

10. Hatakeyama, M.; Iwato, S.; Hanashita, H.; Nakamura, K.; Fujimoto, K.; Kawaguchi, H. DNA-carrying latex particles for DNA diagnosis 3. Detection of point mutant DNA using an indirect labeled probe. Colloids Surf. A Physicohem. Eng. Asp. **1999**, *153* (1–3), 445–451.

11. Ihara, T.; Kurohara, K.; Jyo, A. Aggregation of DNA-modified nanospheres depending on added polynucleotides. Chem. Lett. **1999**, (10) 1041–1042.

12. Gao, Z.; Varshney, S.K.; Wong, S.; Eisenberg, A. Block copolymer "crew-cut" micelles in water. Macromolecules **1994**, *27* (26), 7923–7927.

13. Prochazka, K.; Martin, T.J.; Munk, P.; Webber, S.E. Polyelectrolyte poly(tert-butyl acrylate)-block-poly(2-vinylpyridine) micelles in aqueous media. Macromolecules **1996**, *29* (20), 6518–6525.

14. Li, M.; Jiang, M.; Zhu, L.; Wu, C. Novel surfactant-free stable colloidal nanoparticles made of randomly carboxylated polystyrene ionomers. Macromolecules **1997**, *30* (7), 2201–2203.

15. Kataoka, K.; Ishihara, A.; Harada, A.; Miyazaki, H. Effect of the secondary structure of poly(l-lysine) segments on the micellization in aqueous milieu of poly(ethylene glycol)–poly(l-lysine) block copolymer partially substituted with a hydrocinnamoyl group at the N ε–position. Macromolecules **1998**, *31* (18), 6071–6076.

16. Topp, M.D.C.; Dijkstra, P.J.; Talsma, H.; Feijen, J. Thermosensitive micelle-forming block copolymers of poly(ethylene glycol) and poly(N-isopropylacrylamide). Macromolecules **1997**, *30* (26), 8518–8520.

17. Qiu, X.; Wu, C. Study of the core–shell nanoparticle formed through the "coil-to-globule" transition of poly(N-isopropylacrylamide) grafted with poly(ethylene oxide). Macromolecules **1997**, *30* (25), 7921–7926.

18. Liang, D.; Zhou, S.; Song, L.; Zaitsev, V.S.; Chu, B. Copolymers of poly(N-isopropylacrylamide) densely grafted with poly(ethylene oxide) as high-performance separation matrix of DNA. Macromolecules **1999**, *32* (19), 6326–6332.

19. Mori, T.; Umeno, D.; Maeda, M. Sequence-specific affinity precipitation of oligonucleotide using poly(N-isopropylacrylamide)-oligonucleotide conjugate. Biotechnol. Bioeng. **2001**, *72* (3), 261–268.

20. Cantor, C.R.; Warshaw, M.M.; Shapiro, H. Oligonucleotide interactions. III. Circular dichroism studies of the conformation of deoxyoligonucleolides. Biopolymers **1970**, *9* (9), 1059–1077.

21. Qiu, X.; Kwan, C.M.S.; Wu, C. Laser light scattering study of the formation and structure of poly(N-isopropylacrylamide-co-acrylic acid) nanoparticles. Macromolecules **1997**, *30* (20), 6090–6094.

22. Harada, A.; Kataoka, K. Novel polyion complex micelles entrapping enzyme molecules in the core: Preparation of narrowly-distributed micelles from lysozyme and poly (ethylene glycol)-poly (aspartic acid) block copolymer in aqueous medium. Macromolecules **1998**, *31* (2), 288–294.

23. Taylor, L.D.; Cerankowski, L.D. Preparation of films exhibiting a balanced temperature dependence to permeation by aqueous solutions—A study of lower consolute behavior. J. Polym. Sci. Polym. Chem. Ed. **1975**, *13* (11), 2551–2570.

24. Kyong Yoo, M.; Kiel Sung, Y.; Moo Lee, Y.; Su Cho, C. Effect of polymer complex formation on the cloud-point of poly(N-isopropylacrylamide) (PNIPAAm) in the poly(NIPAAm-co-acrylic acid): Polyelectrolyte complex between poly(acrylic acid) and poly(L-lysine). Polymer **1998**, *39* (16), 3703–3708.

25. Douglas, J.F.; Roovers, J.; Freed, K.F. Characterization of branching architecture through "universal" ratios of polymer solution properties. Macromolecules **1990**, *23* (18), 4168–4180.

26. Meewes, M.; Ricka, J.; De Silva, M.; Nyffenegger, R.; Binkert, T. Coil-globule transition of poly (N-isopropylacrylamide): A study of surfactant effects by light scattering. Macromolecules **1991**, *24* (21), 5811–5816.

27. Kankia, B.I.; Marky, L.A. DNA, RNA, and DNA/RNA oligomer duplexes: A comparative study of their stability, heat, hydration, and Mg 2+ binding properties. J. Phys. Chem. B **1999**, *103* (41), 8759–8767.

28. Akiyama, Y.; Iwanaga, R.; Saitoh, K.; Shiba, K.; Ushio, K.; Ikeda, E.; Iwama, T.; Nomizu, T.; Yuasa, Y. Transforming growth factor beta type II receptor gene mutations in adenomas from hereditary nonpolyposis colorectal cancer. Gastroenterology **1997**, *112* (1), 33–39.

Scaffolds—Smart

Shape Memory Polymers: Applications

Meymanant Sadat Mohsenzadeh
Mohammad Mazinani
Seyed Mojtaba Zebarjad
Department of Materials Engineering, Faculty of Engineering, Ferdowsi University of Mashhad, Mashhad, Iran

Abstract

A lot of research on biomedical applications of shape memory polymers (SMPs) has been conducted. Due to novel properties, SMPs can be an excellent choice for a number of medical applications. In this entry, the properties of SMPs that have led to its wide acceptance in biomedical applications are discussed. Then, a detailed review of biomedical applications of SMPs in various fields such as surgical, vascular, and urogenital applications, as well as orthopedics, orthodontics, application in the brain, and controlled drug release systems, is provided. The novel applications that have been proposed in recent years are also described.

INTRODUCTION

As shape memory polymers (SMPs) show a variety of novel properties, the field of SMP materials for biomedical application has grown into a very active area of research during the past decades. Several surgical and implantable devices are being developed based on SMPs, such as self-expandable stents, surgical instruments, smart sutures, thrombectomy devices, or controlled drug release systems. A specific combination of material properties and functions is required for each application. When a SMP-based device is implanted in the body for long periods, strict biocompatibility of the SMP is required. Due to SM effect (SME), SMPs have caught much attention for using in minimally invasive surgery. An SMP implant can be inserted in the form of initially small object through a small hole and then unfold to eventually bulky shape in the body.

This entry aims to review biomedical applications of SMPs, including a number of novel applications that have been proposed in recent years. First, the properties of SMPs for biomedical applications are discussed. Next, examples of biomedical applications of SMPs in various fields including cardiovascular system, general surgery, urogenital application, orthopedics, orthodontics, application in the brain, and controlled drug release systems are summarized.

PROPERTIES OF SMPs FOR BIOMEDICAL APPLICATIONS

The properties of SMPs that are important for biomedical applications are discussed in this section.

Biocompatibility is the most important property, especially for an implant designed to remain for a long time in the body. Biocompatibility means that the material does not produce toxic or carcinogenic effects in local tissues. In addition, the degradation products do not interfere with tissue healing.[1]

A number of SMPs perform good biocompatibility. Among thermoplastic SMPs, the polyurethane SMP has better biocompatibility.[2]

Due to their polymeric nature, the SMPs can be biodegradable. The biodegradable SMP, is mainly based on polyglycolide, poly(L-lactide) (PLLA) and poly(ε-caprolactone) (PCL).[3–8] A PCL-based biodegradable SMP has been demonstrated its potential in medical applications.[9] Biodegradability is a valuable additional functionality of an SMP, as a second surgery for explantation could be avoided by using a biodegradable implant. In addition, the need for an additional operation for suture removal could be eliminated by using biodegradable smart suture. Apart from this, biodegradability is a necessary property for SMP scaffolds used in tissue engineering. Scaffolds must degrade into nontoxic products with a controlled degradation rate that matches the regeneration rate of the native tissue.

SME is the unique property of the SMPs that is attractive for biomedical applications particularly minimally invasive surgery. SME is schematically illustrated in Fig. 1. After fabrication of the SMP into the original shape, it is heated above a transition temperature (T_{trans}), and deformed by applying an external force. Then, the SMP is cooled below T_{trans}, and the force is removed. Thus, a temporary predeformed shape is obtained. When heated above T_{trans}, the SMP recovers its original shape.[10] The SM phenomenon in SMP is originated from a dual segment system: cross-links to determine the permanent shape and switching segments with T_{trans} to fix the temporary shape. An extensive review

Concise Encyclopedia of Biomedical Polymers and Polymeric Biomaterials DOI: 10.1081/E-EBPPC-120049941

focused on the SMPs, the available SM functions, and SME mechanisms has been published.[10]

Utilizing the SME, a medical device made of SMP, can easily deploy in the body. If the T_{trans} of the SMP is below body temperature, shape recovery can easily be induced by the heat of the body. The deployment function is used in neuronal electrode, self-expandable stent, and etc.

A pre-stress can be created after deployment, which can be used for the fixation functions in medical applications such as orthodontic wires or medical cast.

Due to the recovery force created during the shape recovery, the SMPs can be used as an actuator in biomedical applications such as a nitinol-core/SMP-shell thermally activated endovascular thrombectomy device which is discussed later. The SMPs also receive much attention in tissue engineering, as the SMP with a temporary shape may be delivered to an area of tissue defect through a small incision, and subsequently revert to an original shape that conforms to the defect.[11]

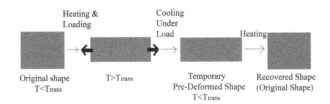

Fig. 1 Schematic illustration of shape fixing to obtain a temporary pre-deformed shape, and shape recovery in an SMP.

BIOMEDICAL APPLICATIONS

Vascular Applications

Among biomedical applications, cardiovascular applications have been the major targets for the evaluation of SMPs. Stents might be the most prominent example.[12] Previously, stainless steel was the dominant material for stent. Today, most of stents are NiTi shape memory alloys (SMAs). Also various self-expandable polymer stents (both biodegradable or biostable) gain attention.[12,13] Polymer stents can be programmed to a small temporary shape by furling and rolling, compressed into a catheter, then pushed out of the catheter and expand at body temperature. The shape recovery of a SMP stent is shown in Fig. 2. It is shown that the recovery time and thermomechanical properties of the stents can be easily controlled by glass transition temperature (T_g) and cross-link density.[14] Generally, polymer stents can bear a large reversible strain compared with NiTi SMAs.

Although biodegradable polymer stents, based on, poly(vinyl alcohol) (PVA), poly lactide (PLA) and poly(ethylene glycol) (PEG), etc., have been invented,[15] the ability of instant removal of stents is a great advantage. Retractable stents can be realized by using the moisture-responsive SMP[16] or the triple-shape SMP.[17] By using triple-shape memory effect, stents could be designed that are delivered into the required location in their first temporary shape, expand upon heating to their second temporary shape and, upon further heating, shrink for easy removal.[17]

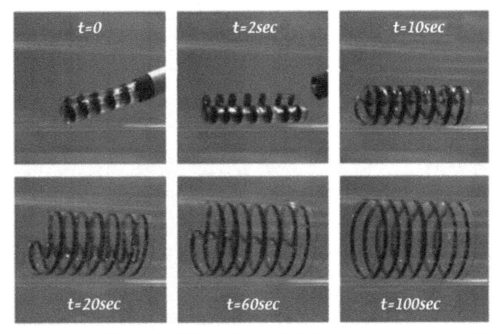

Fig. 2 The shape recovery of a SMP stent into a 37°C water bath, the black rings were drawn on the surface of the stent for better visualization.

Source: From Yakacki et al.[14] © 2007, with permission from Elsevier.

A retractable stent utilizing the moisture-responsive feature in polyurethane SMP has been reported.[16] A tubular stent was fabricated from a prestretched thin film of polyurethane (0.5 thick) and then mechanically folded into a star-like shape. This folded tube could be implanted into the body and expanded at a target site in a standard mechanical manner. After absorbing water (more or less similar to that inside a human body), the stent recovered its original shape. Hence, it shrank and was ready for being taken out.

Most of the stents can recover their original shapes when they are placed at the body temperature. Besides, induction the shape recovery due to indirect heating by light absorption has been evaluated in temperature-sensitive polyurethane SMP stents.[18]

Aneurysm is abnormal dilatation or bulging of a segment of blood vessel. Its highest prevalence is in certain regions of brain arteries and the aorta. Commonly used treatment options include the deposition of a metallic coil (e.g., Pt) in intracranial aneurysms for a controlled thrombus formation, and subsequent closure of the aneurysm[19,20] or the implantation of stents in aortic aneurysm.[21] In addition, SMP coils with T_g at approximately body temperature for treating intracranial aneurysms have been reported.[22] SMP coil with a temporary straightened shape can be implanted into an aneurysm cavity, and subsequently recover its coil shape in order to efficiently reduce the flow velocity and lead to thrombus formation for the treatment of the aneurysm (Fig. 3). The use of polymers instead of Pt coils can be advantageous because of achieving an improved cell adhesion.[22]

The embolic treatment of aneurysms using SMP foams have been proposed.[23] An SMP foam device can be delivered in a compressed form into the human body, and then expand in an aneurysm upon laser triggering. The deployment time, maximum heating temperature, absorption of laser light by the blood and tissue and heat transfer from the device to the blood and tissue require further engineering to avoid thermal damage to arterial tissue. Nevertheless, the prototype SMP foam devices still present an innovative treatment option.[24]

For the treatment of cerebral ischemia in stroke patients, it is necessary to remove the intravascular blood clots. For this purpose, a nitinol wire is guided through a microcatheter into occluded vessel and passed beyond the thrombus. Then, the wire deploys at the end of the catheter to a helical corkscrew shape, and capture the thrombus.[25] Similar to this thrombectomy device, SMP devices for the mechanical removal of blood clots have been designed.[26,27] Figure 4 illustrates removal of a clot in a blood vessel using the laser-activated SMP thrombectomy device.[27] The SMP microactuator, which is coupled to an optical fiber, is in the temporary straight rod form. Indirect heating by laser induces the shape recovery and deploys it into the corkscrew shape. Then, the microactuator is retracted and captures the thrombus. A thermally activated endovascular thrombectomy device consisting of a permanently corkscrew-shaped nitinol wire and SMP coating has been developed in recent years.[28] Below the T_g, the SMP is stiff, consequently the device is fixed in a straight shape. Due to Joule heating, the SMP reaches its T_g and its stiffness reduced, allowing the nitinol wire to transform to a preprogrammed corkscrew form. After turning off the current, the polymer cools below T_g, and returns to its rigid state, ensuring enhanced stiffness to the corkscrew shape for clot removal.

Surgical Applications

Suture is a commonly used device in surgery. However, in minimally invasive surgery, it is difficult to tighten the knots one by one because of limited space available for maneuver. Self-tightening SMP suture provides a good solution to overcome this problem.[29–31] In order to obtain a temporary shape, threads from SMP elongate under controlled stress. This suture can be applied loosely to the wound. When the shape recovery is triggered by human body heat, the suture will shrink and then tighten the knot. These sutures can provide a defined pressured on the wound that is advantageous for optimal healing. Besides, by using the biodegradable SMP materials the need for an additional operation for

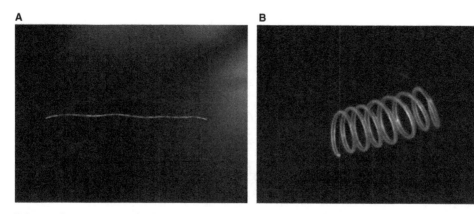

Fig. 3 SMP coil for treating aneurysm: (**A**) Temporary straightened shape prepared for catheter delivery, (**B**) helical coil configuration (after the shape recovery).
Source: From Hampikian et al.[22] © 2006, with permission from Elsevier.

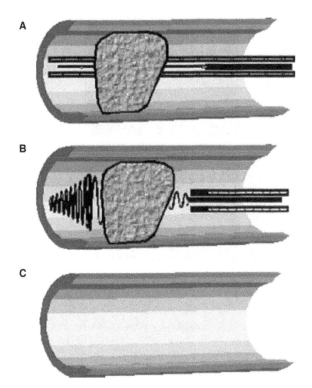

Fig. 4 Removal of a clot in a blood vessel using the laser-activated SMP thrombectomy device.
Source: From Maitland et al.[27] © 2002, with permission from Wiley-Liss, Inc.

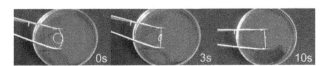

Fig. 5 The shape recovery process of the sample when it was placed in water of 70°C.
Source: From Zhang et al.[32] © 2009, with permission from Elsevier.

suture removal is eliminated. Figure 5 shows automatically knotting of SMP sample when it was placed in water of 70°C.[32] Instead of heating, the thermomoisture responsive polyurethane SMP suture is able to self-tighten a knot upon absorbing moisture in water.[16]

Recently, the use of SMP materials for surgery at the cellular level has been proposed. However, there are many technical challenges for achieving this end.[33] Many polymer micromachines with the size of cells or even smaller have been developed which can be triggered for operation by a laser beam outside the cell.[34,35] By using SMP materials, the device could be delivered into a living cell through a small hole without any significant damage, recover its original shape, and perform a designed operation (Fig. 6). It is proven that the thermomoisture responsive polyurethane SMP could be a good choice for cellular surgery.[33] Figure 7 shows the shape recovery in ultrathin polyurethane SMP film upon heating which confirms the use of SMP based tiny machines for cellular surgery.

SMP materials can also be used for the fixation of tissues or medical devices during surgery.[36] In this application, needles with temporary straight shapes are pierced through the implanted device, and the respective tissue. When the needle is heated to body temperature, the shape recovery is induced and both ends of the needle recover to a helical shape. Consequently, the implant is fixed in the desired place.

Urogenital Applications

In patients with abdominal tumors, it is necessary to keep the ureter open. Figure 8 illustrates a self-anchorable SMP ureteral stent that can be useful for this purpose. As can be seen, the stent can be anchored in the ureter in order to keep it open.[37]

It has been suggested that shape memory concept can also be used in order to capture concretions like renal calculi. During lithotripsy treatment, stones are destroyed

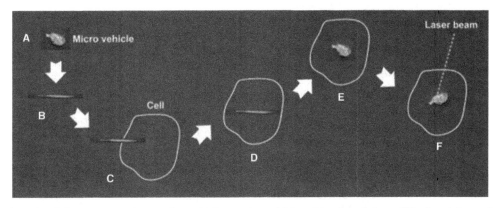

Fig. 6 Illustration of cellular surgery using a microvehicle, a microvehicle with original shape (**A**) after deformation (**B**) could be inserted to a cell (**C, D**), recover its original shape (**E**), and then perform a pre-designed operation (**F**).
Source: From Sun & Huang[33] © 2010, with permission from Elsevier.

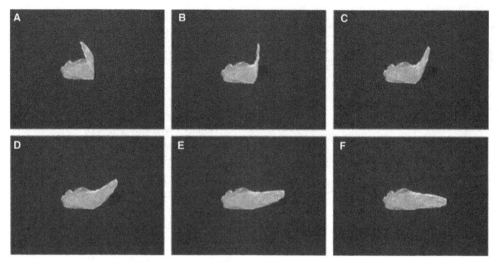

Fig. 7 The shape recovery in ultrathin polyurethane SMP film upon heating.
Source: From Sun & Huang[33] © 2010, with permission from Elsevier.

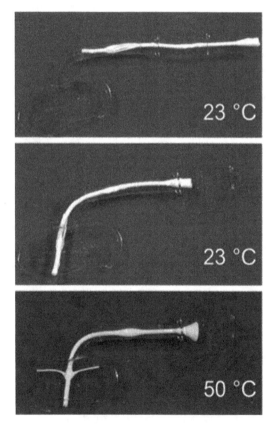

Fig. 8 A self anchorable SMP ureteral stent.
Source: From Neffe et al.[37] © 2009, with permission from WILEY-VCH Verlag GmbH and Co. KGaA, Weinheim.

into small fragments that are supposed to be passed out of the body with the urine. However, in some cases, larger fragments may painfully block urine flow or even migrate back into the kidney. A light-activated SMP device with permanent shape of basket or corkscrew can be used in order to capture such fragments in the cavities of kidney.[38]

Orthopedics

SMPs can be used as a medical cast to heal broken bones, tendon tears, and other injuries to a subject's limb. Compared with traditional casting materials, the use of SM materials can be more advantageous because of their light weight, ease of processing, X-ray transparency, and low cost. Methods have been invented for applying a SMP as a medical cast.[39] A SMP material with a larger diameter is fitted to the injured position. Upon heating to its T_{trans}, the shape recovery is induced and SMP attempts to recover its original shape. Consequently, it conforms to a secondary temporary shape with a diameter smaller than the diameter of the primary temporary shape and larger than that of the original shape, whereas the SMP in the secondary temporary shape conforms to the limb.

It has been suggested that SMPs can be used to fix ruptured tendons or ligaments to bones.[40] When common metal screws are used, the screw threads can cause damage of ligaments during fixation. On the other hand, by using an SMP rod, such damage may be avoided. In this application, a thin-stretched rod expands upon shape recovery to a more compact-shape rod and fixes the loose end of the ligament to the bone by a defined force that is depend on the material composition (Fig. 9).[40]

Orthodontics

Due to their rigidity, flexibility, fatigue resistance, and durability, metallic wires are conventionally used in orthodontics. However, their metallic color poses aesthetic problems. Alternatively, it has been suggested that a thermoplastic SMP containing PCL building blocks can be employed as arch wires.[41] This SMP arch wires can maintain a constant shape recovery force to teeth over a long period of times. In addition to many advantages, such as

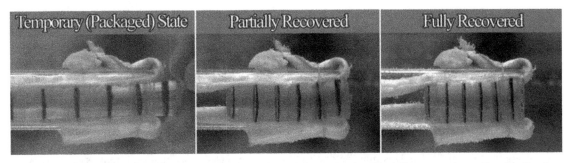

Fig. 9 The shape recovery of an SMP device that can be used for soft tissue fixation, black lines were drawn for visualization.
Source: From Yakacki et al.[40] © 2008, with permission from WILEY-VCH Verlag GmbH and Co. KGaA, Weinheim.

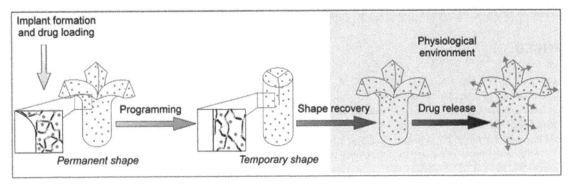

Fig. 10 Schematic illustration of programming, shape recovery and drug release of SMP devices.
Source: From Wischke et al.[46] © 2009, with permission from Elsevier.

low density, high shape recovery, easy processing, and transparency, the SMP orthodontic wires have aesthetically satisfactory appearance.

Applications in the Brain

A self-deploying neuronal electrode made of an amorphous SMP has been reported by Sharp et al.[42] Existing metallic and ceramic probes that are used for chronic recording of brain activity and functional stimulation cannot meet the requirement for long term biocompatibility with brain tissue. On the contrary, SMP probes enhance the long-term biocompatibility. Besides, they provide a suitable slow shape recovery rate that is necessary to avoid insertion-induced tissue damage.[42]

Application of SMP as a Controlled Drug Release System

SMPs can be used as efficient drug delivery systems.[37,43,44] Functionalized with drugs, SMPs are able to deliver incorporated drugs in a controlled manner, preferentially at predefined rates. Figure 10 is a schematic illustration of SME and drug release of drug loaded SMP device. Controlled drug release capability of SMP materials has recently been

studied.[12,45,46] Wache et al.[12] developed a polymer vascular stent with an SMP as the drug delivery system. The use of the SMP stent as a drug delivery system can lead to significant reduction of restenosis and thrombosis. A combination of the SM function with controlled drug delivery is expected to open new application fields of SMPs. However, a detailed analysis of the interplay between drug loading, shape memory and biodegradability has to be conducted in order to prepare a device for a specific application. The relationship between drug loading and degradation as well as drug release before and after actuation of the SME has been reviewed in detail.[47]

In addition, targeted drug release on specified cells has attracted much attention in the current medical research.[48]

CONCLUSIONS

SMPs have attracted extensive interest for biomedical applications which is reflected in the rapidly increasing number of publications on SMPs during recent years. Due to SME, SMPs can be used in the deployment, fixations, and even actuation functions in biomedical applications. Besides, SMPs could be an appropriate choice for cellular surgery, as well as targeted drug release on the specified cells. Besides, biocompatibility of a number of SMPs has

Scaffolds—Smart

been demonstrated which is a necessary property, specially for implants. SMPs can be biodegradable. Biodegradability is a desired feature as the need of a second surgery for explantation can be avoided. Several medical devices have been developed based on SMPs for various applications, including general surgery, as well as intravascular, urogenital, and other application.

Although biomedical applications have been the widely researched area of SMPs in the past decade, there is a limited clinical application of SMPs. This may be due to the challenges that exist for application of a material in the human body such as interaction of the material with biological metabolism and complicated chemical environment. However, in the future, more attention may be paid to the translation of the SMPs into clinical applications.

REFERENCES

1. Athanasiou, K.A.; Niederauer, G.G.; Agrawal, C.M. Sterilization, toxicity, biocompatibility and clinical applications of polylactic acid/polyglycolic acid copolymers. Biomaterials 1996, 17 (2), 93–102.

2. Kim, B.K.; Lee, S.Y.; Lee, J.S.; Baek, S.H.; Choi, Y.J.; Lee, J.O. Polyurethane ionomers having shape memory effects. Polymer 1998, 39 (13), 2803–2808.

3. Min, C.; Cui, W.; Bei, J.; Wang, S. Biodegradable shape-memory polymer-polylactide-co-poly(glycolide-co-caprolactone) multiblock copolymer. Polym. Adv. Technol. 2005, 16 (8), 608–615.

4. Liu, L.; Sheardown, H. Glucose permeable poly(dimethyl siloxane) poly(N-isopropyl acrylamide) interpenetrating networks as ophthalmic biomaterials. Biomaterials 2005, 26 (3), 233–244.

5. Nagata, M.; Kitazima, I. Photocurable biodegradable poly(vepsiln-caprolactone)/poly(ethylene glycol) multiblock copolymers showing shape-memory properties. Colloid Polym. Sci. 2006, 284 (4), 380–386.

6. Kim, Y.B.; Chung, C.W.; Kim, H.W.; Rhee, Y. Shape memory effect of bacterial poly[(3-hydroxybutyrate)-co-(3-hydroxyvalerate)]. Macromol. Rapid. Commun. 2005, 26 (13), 1070–1074.

7. Wang, W.; Ping, P.; Chen, X.; Jing, X. Polylactide-based polyurethane and its shape-memory behavior. Eur. Polym. J. 2006, 42 (6), 1240–1249.

8. Bertmer, M.; Alina, B.; Hofgess, I.B.; Kelch, S.; Lendlein A. Solid-state NMR characterization of biodegradable shape-memory polymer networks. Macromol. Symp. 2005, 230 (1),110–115.

9. Lendlein, A.; Kelch, S.; Shape-memory polymers. Angew. Chem. Int. 2002, 41, 2034–57.

10. Xie, T. Recent advances in polymer shape memory. Polymer 2011, 52 (22), 4985–5000.

11. Karp, J.M.; Langer, R. Development and therapeutic applications of advanced biomaterials. Curr. Opin. Biotechnol. 2007, 18 (5), 454–459.

12. Wache, H.M.; Tartakowska, D.J.; Hentrich, A.; Wagner, M.H. Development of a polymer stent with shape memory effect as a drug delivery system. J. Mater. Sci. Mater. Med. 2003, 14 (2), 109–112.

13. Metcalfe, A.; Desfaits, A.C.; Salazkin, I.; Yahia, L.; Sokolowski, W.M.; Raymond, J. Cold hibernated elastic memory foams for endovascular interventions. Biomaterials 2003, 24 (3), 491–497.

14. Yakacki, C.M.; Shandas, R.; Lanning, C.; Rech, B.; Eckstein, A.; Gall, K. Unconstrained recovery characterization of shape-memory polymer networks for cardiovascular applications. Biomaterials 2007, 28 (14), 2255–2263.

15. O'Brien, B.; Carroll, W. The evolution of cardiovascular stent materials and surfaces in response to clinical drivers: A review. Acta. Biomater. 2009, 5 (4), 945–958.

16. Huang, W.M.; Yang, B.; Liu, N.; Phee, S.J. Water-responsive programmable shape memory polymer devices. In Proceedings of SPIE, international conference on smart materials and nanotechnology in engineering, Harbin, China, July 1, 2007; 6423, 64231s-1-7.

17. Bellin, I.; Kelch, S.; Langer, R.; Lendlein, A. Polymeric triple-shape materials. Proc. Natl. Acad. Sci. U.S.A. 2006, 103 (48), 18043–18047.

18. Baer, G.M.; Small, W.; Wilson, T. Fabrication and in vitro deployment of a laser-activated shape memory polymer vascular stent. Biomed. Eng. Online. 2007, 6, 43.

19. Koebbe, C.J.; Veznedaroglu, E.; Jabbour, P.; Rosenwasser, R.H. Endovascular management of intracranial aneurysms: Current experience and future advances. Neurosurgery 2006, 59 (5 Suppl. 3), S93–S102.

20. Ferns, S.P.; Sprengers, M.E.; van Rooij, W.J.; Coiling of intracranial aneurysms: A systematic review on initial occlusion and reopening and retreatment rates. Stroke 2009, 40 (8), e523–e529.

21. Duarte, M.P.; Maldjian, C.T.; Laskowski, I. Comparison of endovascular versus open repair of abdominal aortic aneurysms: A review. Cardiol. Rev. 2009, 17 (3), 112–114.

22. Hampikian, J.M.; Heaton, B.C.; Tong, F.C.; Zhang, Z.Q.; Wong, C.P. Mechanical and radiographic properties of a shape memory polymer composite for intracranial aneurysm coils. Mater. Sci. Eng. C 2006, 26 (8), 1373–1379.

23. Small, W.; Buckley, P.R.; Wilson, T.S.; Benett, W.J.; Hartman, J.; Saloner, D.; Maitland, D.J. Shape memory polymer stent with expandable foam: A new concept for endovascular embolization of fusiform aneurysms. IEEE Trans. Biomed. Eng. 2007, 54 (6), 1157–1160.

24. Chen, M.C.; Tsai, H.W.; Chang, Y.; Lai, W.Y.; Mi, F.L.; Liu, C.T.; Wong, H.S.; Sung, H.W. Rapidly self-expandable polymeric stents with a shape memory property. Biomacromolecules 2007, 8 (9), 2774–80.

25. Gobin, Y.P.; Starkman, S.; Duckwiler, G.R. MERCI 1: A phase 1 study of mechanical embolus removal in cerebral ischemia. Stroke 2004, 35 (12), 2848–2854.

26. Metzger, M.F.; Wilson, T.S.; Schumann, D.; Matthews, D.L.; Maitland, D.J. Mechanical properties of mechanical actuator for treating ischemic stroke. Biomed. Microdevices 2002, 4 (2), 89–96.

27. Maitland, D.J.; Metzger, M.F.; Schumann, D.; Lee, A.; Wilson, T.S. Photothermal properties of shape memory polymer micro-actuators for treating stroke. Lasers Surg. Med. 2002, 30 (1), 1–11.

28. Small, W.; Wilson, T.S.; Buckley, P.R.; Benett, W.J.; Loge, J.A.; Hart-man, J.; Maitland, D.J. Prototype fabrication and preliminary in vitro testing of a shape memory endovascular

thrombectomy device. IEEE Trans. Biomed. Eng. **2007**, *54* (9), 1657–1666.

29. Lendlein, A.; Langer, R. Biodegradable, elastic shape-memory polymers for potential biomedical applications. Science **2002**, *296* (5573), 1673–1676.

30. Ashley, S. Shape-shifters - shape-memory polymers find use in medicine and clothing. Sci. Am. **2001**, *284* (5), 20–22.

31. Lendlein, A.; Langer, R. Biodegradable shape memory polymeric sutures. U.S. Patent 2004/0015187 A1, January 22, 2004.

32. Zhang, H.; Wang, H.; Zhong, W.; Du, Q. A novel type of shape memory polymer blend and shape memory mechanism. Polymer. **2009**, *50* (6), 1596–1601.

33. Sun, L.; Huang, W.M. Thermo/moisture responsive shape-memory polymer for possible surgery/operation inside living cells in future. Mater. Des. **2010**, *31* (5), 2684–2689.

34. Maruo, S.; Ikuta, K.; Korogi, H. Submicron manipulation tools driven by light in a liquid. Appl. Phys. Lett. **2003**, *82* (1), 133–135.

35. Maruo, S.; Ikuta, K.; Korogi, H. Force-controllable, optically driven micromachines fabricated by single-step two-photon microstereolithography. J. Microelectromech. Syst. **2003**, *12* (5), 533–539.

36. Bettuchi, M.; Heinrich, R. Novel surgical fastener. U.S. Patent 2009/0118747 A1, May 13, 2009.

37. Neffe, A.T.; Hanh, B.D.; Steuer, S.; Lendlein, A. Polymer networks combining controlled drug release, biodegradation, and shape memory capability. Adv. Mater. **2009**, *21* (32–33), 3394–3398.

38. Teague, J.A. Light responsive medical retrieval devices. International Publication, WO 2007/145800 A2, December 21, 2007.

39. Rousseau, I.A.; Berger, E.J.; Owens, J.N.; Kia, H.G. Shape memory polymer medical cast. U.S. Patent 0249682 A1, September 30, 2010.

40. Yakacki, C.M.; Shandas, R.; Safranski, D.; Ortega, A.M.; Sassaman, K.; Gall, K. Strong, tailored, biocompatible shape-memory polymer networks. Adv. Funct. Mater. **2008**, *18* (16), 2428–2435.

41. Jung, Y.C.; Cho, J.W. Application of shape memory polyurethane in orthodontic. J. Mater. Sci. Mater. Med. **2010**, *21* (10), 2881–2886.

42. Sharp, A.A.; Panchawagh, H.V.; Ortega, A.; Artale, R.; Richardson-Burns, S.; Finch, D.S.; Gall, K.; Mahajan, R.L.; Restrepo, D. Toward a self-deploying shape memory polymer neuronal electrode. J. Neural Eng. **2006**, *3* (4), L23–L30.

43. Chen, M.C.; Chang, Y.; Liu, C.T.; Lai, W.Y.; Peng, S.F.; Hung, Y.W. The characteristics and *in vivo* suppression of neointimal formation with sirolimus-eluting polymeric stents. Biomaterials **2009**, *30* (1), 79–88.

44. Wischke, C.; Lendlein, A. Shape-memory polymers as drug carriers – a multifunctional system. Pharm. Res. **2010**, *27* (4), 527–529.

45. Sokolowski, W.; Metcalfe, A.; Hayashi, S.; Yahia, L.; Raymond, J. Medical Applications of Shape Memory Polymers. Biomed. Mater. **2007**, *2* (1), S23–S27.

46. Wischke, C.; Neffe, A.T.; Steuer, S.; Lendlein, A. Evaluation of a degradable shape-memory polymer network as matrix for controlled drug release. J. Control. Release **2009**, *138* (3), 243–250

47. Wischke, C.; Neffe, A.T.; Lendlein, A. Controlled drug release from biodegradable shape-memory polymers. Adv. Polym. Sci. **2010**, *226*, pp. 177–205, doi: 10.1007/12_2009_29.

48. Rivkin, I.; Cohen, K.; Koffler, J.; Melikhov, D.; Peer, D.; Margalit, R. Paclitaxel-clusters coated with hyaluronan as selective tumor-targeted nanovectors. Biomaterials **2010**, *31* (27), 7106–7114.

Scaffolds—Smart

Skin Tissue Engineering

Ilaria Armentano
Luigi Torre
Josè Maria Kenny
Materials Science and Technology Center, University of Perugia, Terni, Italy

Abstract

The present entry is an overview of the application of tissue engineering in skin regeneration. A complete analysis of natural and synthetic polymeric materials is undertaken, in terms of material properties and processing technology, including strategies for fabrication of scaffolds with highly porous and interconnected structures. The selection of the cell type is determined by considering both differentiated and adult stem cells. The results of the *in vitro* cell culture analysis of the cell–scaffold interaction for skin tissue engineering are described here.

INTRODUCTION

The skin is the largest organ in the human body and it represents the critical boundary between the external world and our viscera, working as a barrier to the environment. The skin plays a key role in protecting the body against pathogens and is important in providing insulation, temperature regulation, and hydratation.[1] The skin consists of three main layers: the epidermis, dermis and hypodermis. The epidermis is a multilayered epithelium and its characteristics constitute the main barrier properties of the skin. It is relatively thin (10–100 μm) and is mainly comprised of keratinocytes, which help it to function as a barrier. Below the epidermis is the dermis that confers the right mechanical properties on the skin and is the thickest of the three layers, whereas the hypodermis is composed of adipose tissue and works as the insulator between the skin and other skeletal structures as bone and muscle. The hypodermis is a subcutaneous layer of connective tissue, fat, and large blood vessels, which helps, primarily, in cushioning and thermoregulation.[2] In view of its function and structure, complete skin regeneration is a challenge. The ideal goal would be to regenerate tissues in such a way that both the structural and functional properties of the wounded tissue are restored to the levels before injury.

The socioeconomic burden imposed by limited availability of replacement skin is enormous. Tens of millions annual cases in the United States require clinical intervention for major skin loss. To meet this demand, in the last few decades, the tissue engineering (TE) approach has grown rapidly and is emerging as a significant potential alternative or complementary solution.[3] TE is widely used for skin regeneration; in particular, the use of specific scaffold materials has been applied commercially for wound dressings to improve the rate of healing in chronic wounds.[4] The scheme of the skin TE approach is described in the scheme shown in Fig. 1. Skin regeneration is an important area of research in the field of TE, especially in cases involving the loss of massive areas of skin, where current treatments are not capable of producing permanent satisfactory replacements.

Research in this field over the past 30 years has yielded a number of commercial bioengineered skin equivalents (BSEs).[5] However, none of these products meet the criteria for fully-functional skin.[6] A key obstacle in the development of engineered skin is the control of cellular behavior.[7] The skin is a complex organ that was first studied in TE, and now a lot of commercial products have been developed based on the natural or synthetic porous scaffold materials and the different types of stem cells. The skin was also the first tissue to be successfully engineered in the laboratory, with the development of biodegradable matrix materials that can emulate the dermis. Research studies now focus on the need for permanent coverage in case of extensive skin injury in patients with an insufficient source of autologous skin for grafting.[8]

The aging population (longer lifespan) and the concurrent increase in comorbidities, such as diabetes, obesity, and peripheral vascular diseases, have contributed to the growing incidence of chronic wounds in developed countries. Treatment of such wounds imposes an increasing burden on the healthcare budget of these countries. In the United Kingdom alone, the allocated annual budget for chronic wound management is estimated to top £3 billion in 2005–2006.[9] Advanced wound dressings are capable of supporting cell growth and they may safely be implanted and left in situ in the skin after successful tissue regeneration. To support cell growth and successful integration with

Concise Encyclopedia of Biomedical Polymers and Polymeric Biomaterials DOI: 10.1081/E-EBPPC-120053753

Scaffolds—Smart

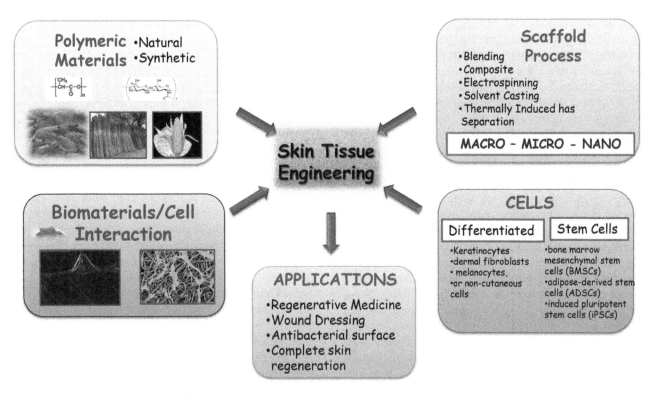

Fig. 1 Skin tissue engineering paradigms.

the natural skin tissue, the wound dressings need to provide mechanical support, be porous, and provide chemotactic signals to the surrounding cells.

Patients with serious skin injuries, such as burns or chronic wounds, require a prompt closure of full-thickness wounds, which is critical for survival.[9] Limited donor sites, donor site morbidity, and reduced surgical procedures provide a major impetus for the development of skin substitutes. Culture of fibroblasts and keratinocytes have been used for skin tissue engineering, to treat skin defects developing a challenging method for wound healing.[10,11] In this approach, a temporary scaffold serves as support for culture of fibroblasts and keratinocytes to construct three-dimensional (3D) skin architecture.[12] The ideal aim of skin regeneration is to find a means to replace or regenerate this complex organ to its normal appearance with complete functionality. This process can be done *in vivo* or *in vitro* and may require cells, natural or synthetic cell-supporting scaffold materials, bioactive molecules, genetic manipulation, or a combination of all of these. Despite the many advances in epidermal biology, regenerative medicine, and tissue engineering, the ideal goal of restoring a functional, cosmetically pleasing skin substitute has remained elusive.

The ideal dermal matrix should be able to provide the right biological and physical environment to ensure homogenous cell and extracellular matrix (ECM) distribution, as well as the right size and morphology of neotissue required.

The multidisciplinary approach in TE, which includes fundamental principles from materials engineering and molecular biology, permits development of biological substitutes for failing tissues and organs.[13] The following text provides a detailed examination of the main components of the TE approach that is applied to skin regeneration: cells and materials.

CELLS

The cell is the first component to be considered in the TE approach, and selection of the cell type is an important factor in skin regeneration. We can use skin derived cells, such as keratinocytes, dermal fibroblasts, embrionic stem cells (ESCs), melanocytes, or non-cutaneous cells, as adult and embryonic stem cells. Autologous differentiated cells are commonly studied; however, a new wave of research involves the use of stem cells (SCs), that are recognized as ideal candidates for tissue regeneration. They have unique features and might represent an effective way to meet the challenges of skin restoration. These features include such characteristics as their potential to be an unlimited source of donor material for grafting along with their ability to switch into a variety of cell phenotypes *in vitro*.[14] Advances in the research on adult stem cells and embryonic stem cells offer hope for therapeutic deficiencies in severe burn treatment using existing skin tissue-engineered products. The therapeutic power of stem cells resides in their clonogenicity and potency,[15] and these can be delivered in conjunction with skin composites or by various other methods, including direct application.[16] SCs are undifferentiated cells with the ability to divide in culture and give rise to different forms of specialized cells. According to their source, SCs are divided into adult and embryonic SCs, the former being multipotent and the latter mostly pluripotent; some cells are totipotent, in the earliest stages of the embryo.[17] Although there is still a large ethical debate related to the use of embryonic stem cells, it is believed that stem cells may be useful for the repair of diseased or damaged tissues or may be used to grow new organs.[18,19] Among adult stem cells, bone marrow mesenchymal stem cells (BMSCs) and adipose-derived stem cells (ADSCs) are perhaps the most promising candidates for skin regeneration.[20] The potential of such mesenchymal SCs to differentiate into dermal cells has been described.[21,22] Human adipose-derived stem cells isolated from liposuction or lipectomy specimens have proven their capacity to differentiate into other lineages and cell types *in vitro*. These cells are a subpopulation of the nonadipocyte cell fraction in lipoaspirates and are also called the stromal vascular fraction (SVF). Unlike adipocytes, the SVF-cells sediment in aqueous medium, and a subset of these cells attach and grow on tissue culture plastic. These cells have surface antigens similar to those of BMSCs and constitute the ADSC subpopulation proper.[23] ADSCs possess many of the traits common to BMSCs including plasticity and a high proliferative potential. ADSCs can be procured easily from the donor, which makes the goal of clinical application more feasible.

More recently, in 2006 inducible pluripotent stem cells (iPSCs) were first discovered by Yamanaka and colleagues, and in 2012 he received the Nobel Prize in Physiology or Medicine (jointly with John B. Gurdon) for the finding that mouse fibroblasts could acquire properties similar to those of embryonic SCs after "reprogramming".[24] iPSCs technology, therefore, allows for patient- and disease-specific stem cells to be used for the development of therapeutics, including more advanced products for skin grafting and treatment of cutaneous wounds.[16] Since this milestone, iPSCs have become a promising new source of stem cells for skin TE without any moral and ethical controversies. Recent publications show that procedures differentiating mouse iPSCs into epidermal keratinocytes are similar to keratinocyte differentiation of embryonic stem cells (ESCs).[25] By using the iPSCs-derived keratinocyte lineage cells, Bilousova et al.,[25] have successfully regenerated epidermis, hair follicles, and sebaceous glands in the skin of the athymic nude mouse. Although this exciting development has been achieved, utility of iPSCs for medical applications is still pending because of the possibility that the transgene technology may cause carcinogenesis and tumor formation.[26] As novel technologies relating to iPSCs are rapidly being developed,[27] the therapeutic applicability of iPSCs in skin TE and regenerative medicine will eventually be a reality.[28] However, the recent suspension of the world's first clinical trial involving iPSCs to treat

age-related macular degeneration continues to raise questions about the safety of this new technology. iPSCs often acquire mutations with epigenetic and chromosomal changes in culture.[29] Hence, human ESCs and mesenchymal SCs remain the more promising options for clinical use to treat severe burns, at least in the near term.

SCAFFOLD

In skin TE, a 3D porous scaffold is necessary to support cell adhesion and proliferation and to guide cells moving into the repair area in the wound healing process. Structurally, the ideal scaffold should have an open and interconnected porous architecture, which helps cells to penetrate the inner part of the scaffold easily and to facilitate homogenous cell distribution and tissue formation and adequate permeability to allow the transfer of nutrients necessary for the cells and wastes produced by the cells.[28,30] Pore size, porosity, and the swelling ability of matrices are predetermining factors of scaffold functions in skin repair and regeneration. The swelling ability of the porous scaffolds depends mainly on the 3D network structure of matrices and their hydrophilicity. Moreover, the scaffolds should be mechanically strong to protect deformation during the formation of new skin.[31]

Scaffolds have to mimic the ECM with an adequate surface chemistry that will enable the attachment and spreading of cells. Functional biomaterial research has been directed toward the development of improved scaffolds for TE.[32] The choice of the biomaterial for scaffold design is of great importance for proper cell adhesion, spreading, migration, and differentiation process. Polymers are the most suitable scaffold materials because of their flexibility and controllable functional properties. The design and selection of a biomaterial is a critical step in the development of scaffolds for TE. Generally, the ideal biomaterial should be nontoxic, biocompatible, promoting favorable cellular interactions and tissue development, while possessing adequate mechanical and physical properties. The material should have appropriate degradation rate and mechanical properties, so that the artificial material is eliminated in time, and the different stresses that may develop during new tissue formation can be handled. Various types of scaffolds have been developed for research and clinical applications of skin TE.[31] Different processing technologies were developed to fabricate scaffolds with the properties mentioned here, including solvent casting/particulate leaching, thermally induced phase separation, electrospinning, rapid prototyping, batch foaming, and microcellular injection molding.[32] It is important, that the technique used to fabricate the scaffolds should not affect the biocompatibility of the material used. To date, scaffolds have been produced and tested for their suitability for TE in different forms such as films, foams, and fibers with widely differing chemistries, and a variety of studies ranging from in situ to clinical have been carried out. Foam and film types proved most popular initially, but in recent times there has been an increase in the use of fibrous scaffolds.[31–33] Some recent studies have shown that cells attach, grow, and organize well on nanofibrous structures even though the fiber diameter is smaller than that of the cells.[33] Furthermore, micro and nanofeatured scaffolds with controlled pore size, geometry, dimension, and spatial orientation are being intensely investigated. Since fibrous scaffolds with diameters of fibers on a nanometer scale have been found to be satisfactory for TE, extensive research is being pursued to develop processes for the fabrication of these fibrous structures.[32]

Polymeric Materials

Polymeric materials are widely applied in TE applications. A number of biodegradable polymers have been exhaustively explored as scaffolds for TE applications. The design of such a polymeric material with controllable properties and biocompatibility along with reliably large scale production, is still a challenging task.

The commonly used materials include synthetic polymers like polycaprolactone,[34] poly (lactic-co-glycolic acid),[35] poly(vinyl alcohol),[36,37] and polyurethane,[38] and natural polymers such as alginate,[39] gelatin,[40] collagen,[6] and chitosan.[41] The main polymers used, which are divided into natural and synthetic materials, are described in Fig. 2, and the main properties, processing technology, and applications are summarized in the Table 1.

In both clinical and preclinical models, collagen and gelatin are the most commonly used scaffolding materials because of their advantageous properties, including low antigenicity and promotion of cell adhesion and proliferation.[42] However, the problem that compromises the effect of collagen- and gelatin-based scaffolds is low mechanical strength. To maintain the scaffold structure and shape after seeding the cells and during implantation, the scaffolds should be mechanically strong, because the dermis needs to provide physical support and flexibility to the skin. Biodegradable synthetic polymer could provide favorable mechanical properties. Owing to the respective advantages of naturally derived collagen or gelatin and biodegradable synthetic polymers, hybridization of these two types of biodegradable polymers has been reported to combine their advantages. Hybrid scaffolds composed of collagen and synthetic polymer retain the advantages of both biodegradable synthetic polymers and collagen with high mechanical strength and good cell-interactions.[42]

Generally, an ideal skin TE material should have high liquid absorbing capacity, proper gas permeation, biocompatibility, and antibacterial properties to protect the skin from infections, dehydration, and subsequent tissue damage. The following text provides a detailed description of the natural and synthetic polymer used in skin TE.

Natural polymeric materials

Natural polymers occur in nature and can be extracted. Living organisms are able to synthesize a vast variety of polymers, which can be divided into major classes

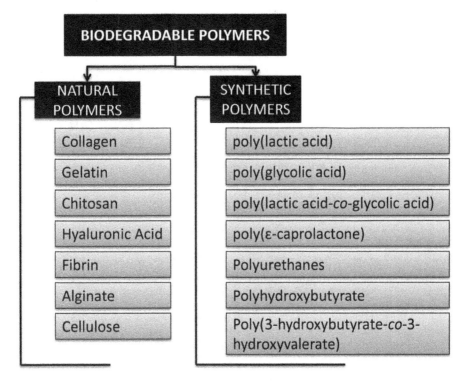

Fig. 2 Scheme of biodegradable polymers used as scaffold for skin tissue engineering: natural and synthetic polymers.

Table 1 Biodegradable polymer properties

Name	Properties	Processing technology	Applications	Commercial products	Refs.
Synthetic Polymers					
Poly(lactic acid) (PLA)	• Semicrystalline polymer • 1 year for degradation • T_m: 150–180°C • Highly Processable • Many Commercial Vendors Available T_g=60–65°C	• Solvent Casting-Particulate Leaching • Thermally Induced phase Separation Process • Electrospinning • Gas Foaming • 3D Printing	• Tissue Engineering • Drug delivery	Fixsorb®	68–70
Poly(glycolic acid) (PGA)	T_m >200°C, T_g ~ 35–40°C Vendors Available	• Fiber mesh technique • Fiber Bonding • Freeze drying	Degradable Suture	Dexon® Dermagraft	32, 33, 35
Poly (lactic-co-glycolic acid) (PLGA)	• Amorphous polymer • Degradation Rate depending on the copolymer ratio (1–6 months)	• Solvent Casting-Particulate Leaching • Thermally Induced phase Separation Process (TIPS)	• Multifilament Absorbable sutures • Skin graft	• Vicryl Mesh • Vicryl, -Vicryl Rapid & CRYL • Polysorb® • Purasorb®	32, 35
Polycaprolactone (PCL)	T_m ~60°C Highly Processable • Many Commercial • Vendors Available • T_g –54 °C • Low degradation rate	• Gas Foaming • Electrospinning • 3D Printing	Tissue Engineering		34, 65, 72
Poly(vinyl alcohol) (PVA)	• Water Soluble Polymer • hydrophilic, • semicrystalline polymer	• Cryogelation	• Woud Dressing • Drug delivery	Aqua-gel' wound dressings	36, 37
Polyurethane (PU)	Thermoplastic and Thermosetting Mechanically strong Handle Physical Stresses Well	• Electrospinning • Electrospraying	• Skin Repair • Prostheses • Tissue Engineering	NovoSorb™ (PolyNovo®)	31, 38
Natural Polymers					
Alginate	• obtained from brown algae • non toxic • non immunogenic • high degree of swelling • pH-sensitive.	• mild procedures • chemical crosslinking	• cell encapsulation • injectable cell delivery vehicle • alginate scaffolds for hepatic TE • alginate hydrogels for cartilage TE	• AlgiSite® • Nu-Derm® • Curasorb®	39, 64, 65
Gelatin	• hydrolysis of collagen • Important rheological properties • low antigenicity	• Chemical crosslinking • freeze-driyng • hydrogel injectable matrix	• Pharmaceutical and Medical Applications • Carrier of biactive component	• Gelfoam® • Gelfilm® • Surgifoam® • CultiSpher-G®	40, 46
Collagen	Main Protein constituent of the skin	• mould-leaching process • Lyophilization • Electrospinning	• dermal matrix • Bioengineered skin equivalents	Apligraf® TransCyte Hyalograft 3D Laserskin CellerateRX OrCel™ PriMatrix	6, 42, 44

(Continued)

Scaffolds—Smart

Table 1 (*Continued*)   Biodegradable polymer properties

Name	Properties	Processing technology	Applications	Commercial products	Refs.
Chitosan	• Antifungal and Bactericidal properties • permeability to oxygen • Brittle behavior • Film forming	• supercritical fluid technology • freezedrying or lyophilizing a chitosan gel solution	• wound healing- • dermal regeneration. • treatment of wounds and burns • Healing, -bacteriostatic • Anti aging	• Beschitin® • GeniaBeads® • HemCon®	41, 48, 54–56
Cellulose	• most abundant inorganic polymer • can be converted into different • poor degradation *in vivo*	• lyophilization • ice-templating process • salt leaching method	• scaffolds for cardiac TE Scaffold for cartilage TE	MyDerm™	51
Fibrin	• Component of ECM	• Gelation Process • Fibrin microbeads	• carrier of active Biomolecules • Tissue Engineering • Cell Carrier in Tissue engineering	• Tisseel VH® • CryoSeal® • Vivostat®	61, 62
Hyaluronic Acid (HA)	• linear polysaccharide • soluble in water • form hydrogels by covalent and photocrosslinking with hydrazide derivatives -• enhanced viscoelastic properties	• photocrosslinked hyaluronic acid hydrogels	• drug delivery, r • dermal, nasal, pulmonary, parenteral, liposome-modified, implantable delivery devices and for gene delivery • as a viscoelastic gel for surgery and wound healing	• HYAFF • Bionect® • Jossalind® • Healon® • Hyalgan® • Hyalubrix®	60

according to their chemical structure: i) polysaccharides; ii) proteins; and iii) polyesters. Naturally derived polymers are of special interest because, as natural components of living structures, they have biological and chemical similarities to natural tissues.[42–44]

Collagen.   Collagen is regarded as an ideal scaffold or matrix for TE as it is the major protein component of the ECM. It is the most abundant protein in the body. Twenty-seven types of collagens have been identified to date, but collagen type I is the most abundant and the most investigated for biomedical applications.[42] Since collagen is the main protein constituent of skin, a significant portion of the commercially available wound management products use collagen to foster skin regeneration.[6] Recently, a broad range of TE products based on animal-sourced collagen scaffolds have been developed and commercialized. Several commercial products are available currently; for example, bilayered collagen gels seeded with human fibroblasts in the lower layer and human keratinocytes in the upper layer have been used as the "dermal" matrix of an artificial skin product, commercialized by Organogenesis in U.S.A. under the name of Apligraf,® and was the first bioengineered skin to receive FDA approval in 1998.[44]

Collagen scaffold can be developed in gel, sponge, film, alone, or in combination with other polymers or inorganic structures, by using different processing technologies, such as mould-leaching process, lyophilization, and electrospinning.

Combinations of collagen with other materials have also been prepared. For example, collagen microsponges can be easily impregnated into previously prepared synthetic polymeric scaffolds; this increases their mechanical performance.[45]

Gelatin.   Gelatin is a natural polymer that is derived from the hydrolysis of collagen and is commonly used for pharmaceutical and medical applications because of its biodegradability and biocompatibility in physiological environments.[40,46] As a result, two different types of gelatin can be produced depending on the method in which collagen is treated, prior to the extraction process.[47] In skin TE, gelatin is used to develop a scaffold, often in combination with other natural or synthetic polymers. Gelatin quality for a particular application depends on the basic physico-chemical properties, such as chemical composition parameters, solubility, transparency, color, odor, and taste; however the main characteristics that best define the overall commercial quality and applications of gelatin are gel strength and thermal stability (gelling and melting temperatures) and, in general, its rheological properties[48,49].

Because of its easy processability and gelationous properties, gelatin has been manufactured in a range of shapes including sponges and injectable hydrogels, but definitively the most used carriers are gelatin microspheres, which are incorporated normally in a second scaffold such as a hydrogel.

The biodegradable hydrogels are prepared by chemical crosslinking of acidic or basic gelatin and are enzymatically degraded in the body with time. The degradation is controlled by changing the extent of crosslinking, which, in turn, produces hydrogels with different water contents. Under specific conditions, such as temperature, solvent or pH, gelatin macromolecules present sufficient flexibility to realize a variety of conformations. For safety and, mainly, to reduce the possibility of polyion complexation, gelatin is been used in drug delivery for TE applications.

The most abundant sources of gelatin are pig skin (46%), bovine hide (29.4%), and pork and cattle bones (23.1%). Gelatin has been reported to be one of the first materials used as a carrier of bioactive components.[50] Enriching gelatin films with natural antioxidants and/or antimicrobial substances will extend the functional properties of these biodegradable films and provide an active packaging biomaterial.

There are several commercially available gelatin based carriers for drug delivery that are being used in TE applications.[51] The most commonly used one is Gelfoam®, commercialized now by Pfizer in the United States (former Pharmacia and Upjohn), which is an absorbable gelatin sponge also available in powder form by milling the gelatin sponges. Gelfoam® is a sterile and workable surgical sponge prepared from specially treated and purified gelatin solution and is used as a hemostatic device.

Chitosan. Chitosan is a linear, semi-crystalline polysaccharide that can be easily derived from the partial deacetylation of a natural polymer: chitin. After cellulose, chitin is the second most abundant biopolymer and is commonly found in invertebrates—as crustacean shells or insect cuticles—but also in some mushroom envelopes, green algae cell walls, and yeast.[52] Chitosan has been popular in TE application as a tissue culture scaffold and wound dressing.[53] Chitosan is a unique biopolymer

that exhibits outstanding properties, besides biocompatibility and biodegradability and nontoxic antibacterial capabilities.[54,55] Chitosan can be used in 3D-scaffolds, as gels and sponges, and in 2D-scaffolds, as films and fibers, with special focus on wound healing application.

The solubility of chitosan is usually tested in acetic acid by dissolving it in 1% or 0.1 M acetic acid. Rinaudo et al., demonstrated that the amount of acid needed depends on the quantity of chitosan to be dissolved.[56] The concentration of protons needed is at least equal to the concentration of $-NH_2$ units involved. In fact, the solubility is a very difficult parameter to control and affects the final material properties. Chitosan is used to prepare hydrogels, films, fibers, or sponges. Chitosan is much easier to process than chitin, but the stability of chitosan materials is generally lower, owing to the fact that they are more hydrophilic in character and possess higher pH sensitivity. To control both their mechanical and chemical properties, various techniques are used.

Chitosan-based scaffolds and sponges have been investigated for wound care in different situations, since it can protect the wound from infection and dehydration.[57] The antibacterial properties of chitosan are an important way to reduce skin infections. Recently, Anisha also prepared chitosan based sponges for wound care.[58] These sponges had a typical porous structure and exhibited enhanced liquid absorbing capacity and cell interaction. However, one major limitation for chitosan materials is its brittle behavior. Its blends with other synthetic materials are believed to be an effective way to develop a suitable TE material.

Suh and Mattew[41] reported that chitosan and its fragments on contact with immune cells may stimulate local cell proliferation and, ultimately, integration of the implanted material with the host tissue. Actually, chitosan possesses the properties favorable for promoting rapid dermal regeneration and accelerates wound healing; it is suitable for applications extending from simple wound coverings to sophisticated artificial skin matrices. Okamoto et al.[59] reported that chitosan influenced all stages of wound repair in experimental animal models. In the inflammatory phase, chitosan has unique hemostatic properties that are independent of the normal clotting cascades. Wound dressing is one of the most promising medical applications of chitin and chitosan. The adhesive nature of chitin and chitosan, together with their antifungal and bactericidal character and their permeability to oxygen, is a very important property associated with the treatment of wounds and burns.[54]

Some commercially available formats of chitosan include the geniaBeads® CN, which are hydrogel beads made from chitosan, commercialized by Genialab in Germany. As a result of its properties in wound healing, a chitosan bandage called HemCon® bandage is a commercially available product from HemCon Medical Technologies Inc. in the United States.

Scaffolds—Smart

Hyaluronic acid

Hyaluronic acid, also known as hyaluronan or hyaluronate, is an important glycosaminoglycans (GAGs) component of connective tissue, synovial fluid (the fluid that lubricates joints), and the vitreous humour of the eye.[60] Hyaluronic acid is one of the main polysaccharide components of dermal ECM and promotes migration and proliferation of skin fibroblasts and keratinocytes. The most common sources of the hyaluronic acid are rooster combs extraction, and recombinant production using *Streptococcus* bacterium.[51]

In solution, hyaluronate occupies a volume approximately 1000 times that in its dry state. Hyaluronate solutions exhibit clear viscoelastic properties that make them excellent biological absorbers and lubricants. In addition, hyaluronic acid can be easily and controllably produced on a large scale through microbial fermentation, enabling the scale-up of derived products and avoiding the risk of animal-derived pathogens. Hyaluronan is soluble in water and can form hydrogels by photocrosslinking with covalent hydrazide derivatives, by esterification and annealing; it is enzymatically degraded by hyaluronidase, which exists in cells and serum. Esterification of hyaluronic acid with benzyl alcohol is used to obtain HYAFF: an ester of hyaluronic acid used in Hyalomatrix PA production. Hyalograft 3D is mainly used in treatment of feet ulcer in combination with Laserskin autologous epidermal bioconstructs.

Fibrin. Fibrin plays an important role in skin regeneration, in hemostasis, and spontaneous tissue repair (it naturally forms during blood coagulation). It is a biopolymer resulting from enzymatic reactions of the coagulation cascade. Activation of this cascade leads to the conversion of soluble fibrinogen by thrombin into a network of insoluble fibrin fibers.[61] The fibrin network stabilizes the platelet plug by binding platelets to the fibrin. The platelets secrete the growth factor, namely platelet-derived growth factor, which stimulates fibroblasts to proliferate, migrate into the wound site, and produce collagen I and III, GAGs, and proteoglycans. Moreover, the fibrin network binds other cell types, such as smooth muscle cells and endothelial cells, through integrin adhesion receptors. In addition, fibrin binds cell adhesion-mediating ECM proteins, fibronectin, and vitronectin, which support the adhesion of fibroblasts.[62]

Fibrin-based biomaterials are well-established biological sealants for skin TE. However, raw blood is used to isolate the fibrin and this has limited its use as a novel skin substitute. Commercial fibrin in kit form can successfully heal a wound but is relatively expensive for the majority of patients. Fibrin gel is continuously in demand in plastics and reconstructive surgery as it minimizes subcutaneous seroma formation and decreases wound morbidity. The combination of fibrin with BMSCs can improve the condition of scalded skin, providing strong self-repair capability and promising an acceptable cosmetic appearance with hair follicle formation. Long-lasting fibrin biomaterials ensure stable and functional angiogenesis by highly tunable and sustained delivery of growth factors.

Fibrin has been a useful cell delivery matrix for cartilage tissue engineering, especially in combination with other biodegradable substances, such as alginate or hyaluronic acid. It has also been used in the regeneration of skin, with considerable success and even in the loading and posterior release of growth factors.[63] Fibrin can be developed as gel, beads, non-woven mesh, and so on by using different processes. Fibrin gel is able to function as both 2D and 3D cell culture scaffolds. Furthermore, the use of fibrin gel as a cell carrying microbeads has been widely investigated in recent decades. The purpose of using fibrin gel as a carrier to deliver cells into a 3D scaffold is to protect cells from the forces applied during the preparation and delivery processes.

Alginate. Alginate is one of the most studied and applied polysaccharide polymers in the TE and drug delivery field. They are abundant in nature and are found as structural components of marine brown algae and as capsular polysaccharides in some soil bacteria. Alginate can be obtained from brown algae. It is composed of two monomers, (1,4)-linked β-D-mannuronate (M) and α-L-guluronate (G). Alginates extracted from different sources have different M and G contents with the length of each block influencing the final properties of the polymer.[64] For example, increasing the length of the G block and molecular weight, the mechanical properties of the alginate will be improved.[65] Also, a high M content makes it immunogenic and more potent in inducing cytokine production as compared to G content. Alginate scaffolds are characterized by a high degree of swelling and shrinking during cationic crosslinking and are pH-sensitive.

Alginate is one of the best-known materials with good scaffold-forming property that can be useful to treat the loss or failure of organs. Alginate is considered to be biocompatible, nontoxic, nonimmunogenic, and biodegradable. In addition to its biocompatibility, source abundance, and low price, it has been widely used in the food industry as a thickener, an emulsifying agent, and TE material. Alginate can be easily modified into any form such as hydrogels, microspheres, microcapsules, sponges, foams, and fibers. This property can increase the applications of alginate in various fields such as TE and drug delivery.[64,65]

As for the commercially available products, alginate has been widely marketed as wound dressings. Some examples are Nu-Derm® commercialized by Johnson & Johnson in the united States, Curasorb® by Kendall or AlgiSite® by Smith & Nephew in the United States, and geniaBeads® CA, which are hydrogel beads made from calcium alginate, marketed by Genialab in Germany.

Scaffolds—Smart

Cellulose. Cellulose is often referred as the most abundant organic polymer in the world. In nature, it is the primary structural component of plant cell walls. However, cellulosic materials exhibit poor degradation *in vivo*. A cellulose sponge has been evaluated in rat femurs for its permissibility for bone formation, and it was found to need more time to regenerate than the control. However, with growth factor, this material can display a desirable enhancement of bone formation. Several works have investigated the use of cellulose for cartilage, bone, and cardiac applications.[51]

Cellulose can be converted into different derivatives, such as carboxymethylcellulose, cellulose nitrate, cellulose acetate, and cellulose xanthate that can be easily moulded or drawn into fibres. Cellulosic based materials can be used as scaffolds for cardiac and cartilage TE applications.

Synthetic polymeric materials

Artificial polymers typically exhibit superior mechanical properties, compared to naturally derived polymeric biomaterials, and can also exhibit variable plasticity.[66,67]

The most often utilized synthetic biodegradable polymers for three dimensional scaffolds in TE are aliphatic poly(α-hydroxy esters), including poly(lactic acid) (PLA), poly(glycolic acid) (PGA), poly(lactic acid-*co*-glycolic acid) (PLGA), and poly(ε-caprolactone) (PCL) and their copolymers.

PGA. Polyglycolic acid (PGA) is a biodegradable, thermoplastic polymer and is considered the simplest linear, aliphatic polyester. It can be prepared starting from glycolic acid by means of polycondensation or ring-opening polymerization. Owing to its hydrolytic instability, however, its use was initially limited.[32] With a melting point greater than 200°C, a glass transition temperature of 35–40°C, and very high tensile strength (12.5 GPa), it was one of the very first degradable polymers ever investigated for biomedical use. Currently polyglycolide and its copolymers [poly(lactic-co-glycolic acid) with lactic acid, poly(glycolide-co-caprolactone) with ε-caprolactone, and poly (glycolide-co-trimethylene carbonate) with trimethylene carbonate] are widely used as materials for the synthesis of absorbable sutures.[35] Polyglycolide is characterized by hydrolytic instability owing to the presence of the ester linkage in its backbone. The degradation process is erosive and appears to take place in two steps during which the polymer is converted back to its monomer glycolic acid: first water diffuses into the amorphous (noncrystalline) regions of the polymer matrix, cleaving the ester bonds; the second step starts after the amorphous regions have been eroded, leaving the crystalline portion of the polymer susceptible to hydrolytic attack. Upon collapse of the crystalline regions, the polymer chain dissolves.[32,33] PGA scaffolds with controlled micro and nanoporosity can be developed by the freeze-drying process, fiber mesh technique, and fiber bonding techniques. PGA found favor as the degradable suture DEXON®, which has been actively used since 1970. From 1984 to 1996, PGA was marketed as an internal bone pin under the name Biofix®, but since 1996 Biofix has been converted to a poly(L-lactide) base for better long-term stability.[32,33,35]

PLA. Poly(lactic acid) (PLA) is an aliphatic polyester approved for human clinical use by the FDA. It is of special interest and has been the main material for implantable meshes for guided tissue regeneration. PLA degrades to form lactic acid which is normally present in the body.

PLA is present in three isomeric forms d(-), l(+), and racimic (d,l), and the polymers are usually abbreviated to indicate the chirality. Poly(l)LA and poly(d)LA are semicrystalline solids, with similar rates of hydrolytic degradation. PLA is more hydrophobic than PGA and is more resistant to hydrolytic attack than PGA. For most applications the (l) isomer of lactic acid (LA) is chosen because it is preferentially metabolized in the body.[68,69] PLA has been widely used in the tissue engineering scaffold field in order to promote tissue regeneration for a large variety of tissue, alone or in combination with other polymers.[70] The rate of degradation depends on the molecular weight of the polymer, its crystallinity, and the porosity of the matrix. The main technique used for PLA-based scaffold development are solvent casting particulate leaching, thermally induced phase separation process, electrospinning, gas foaming and 3D printing technology. PLA is usually mixed with specific nano- and microparticles to add new functionality and improve the properties of the polymer. Furthermore the surface of the scaffolds can be modified by introducing specific functional groups that encourage cell interaction.

PLGA. Random copolymerization of PLA (both L- and D,L-lactide forms) and PGA, known as poly(lactide-*co*-glycolide) (PLGA), is the most investigated degradable polymer for biomedical applications and it has been used in sutures, drug delivery devices, and TE scaffolds.

PLGA forms amorphous polymers, which are very hydrolytically unstable compared to the more stable homopolymers.[32,35] This is evident in the degradation times of 50:50 PLGA, 75:25 PLGA, and 85:15 PLGA being 1–2 months, 4–5 months and 5–6 months, respectively. Different commercial products based on PLGA, are in the market, such as Vicryl Mesh, Vicryl, Vicryl Rapid and CRYL.

Polycaprolactone. Polycaprolactone (PCL) is a semicrystalline, synthetic polymer, extensively used in biomedical applications,[34] that belongs to the group of degradable polymers because of its susceptibility to hydrolytic cleavage of the ester bond. This property, along with good compatibility and easy processing (low melting point at 60°C), high tensile strength and non-toxic nature, makes PCL an interesting substrate for TE.[71] PCL has been investigated as a potential matrix for skin TE.[72] However, like other synthetic polymers, PCL also lacks surface wettability and functional surface groups that improve cell attachment, essential in TE. Furthermore, a blend of natural and synthetic polymers are often developed to combine different properties.

PCL is often blended or copolymerized with other polymers like PLLA, PDLLA, PLGA and polyethers to expedite overall polymer erosion. PCL processability allows for the formation of scaffolds composed of adhered microspheres electrospun fibers, or through porous networks created by porogen leaching. PCL and PCL composites have been used as TE scaffolds for regeneration of bone, ligament, cartilage, skin, nerve, and vascular tissues.

Polyurethane. Polyurethanes (PUs) are polymers composed of organic units joined by carbamate (urethane) links. While most polyurethanes are thermosetting polymers that do not melt when heated, thermoplastic polyurethanes are also available. They have been used for a range of biomedical applications varying from cardiovascular repair, cartilage implant, ligament regeneration, and bone replacement to controlled drug/gene delivery and also as a tissue scaffold.[38]

Polyurethanes are one of the most versatile families of polymers.[31] They can be prepared from a wide variety of materials exhibiting extremely different properties and therefore, have a high variety of biomedical applications. Polyurethanes have been testing in several biomedical fields including pacemaker lead insulation, breast implants, heart valves, vascular prostheses, and bioadhesives.

PU's morphology presents two very different structural phases: alternated hard and soft segments. While the hard segments are responsible for the high mechanical resistance, the soft segments provide the PU an elastomeric behavior. Therefore, their singular molecular structure provides them good properties like elasticity, abrasion resistance, durability, chemical stability and easy processability, which captured the attention of researchers interested in developing biomedical devices.

Nowadays, PUs are widely employed materials in the biomedical field to prepare catheters, blood oxygenators, scaffolds for tissue engineering, cardiac valves, membranes for wound dressings, drug delivery systems.

PUs can be used as topical skin adhesives to replace sutures, staples, and adhesive strips for wound closure, with the advantages of rapid application, unnecessary administration of anesthetics, and no trauma.

Das et al. demonstrated for the first time that vegetable oil-based hyperbranched polyurethane with a combination of good mechanical, thermal, chemical, and biological properties could serve as a prospective scaffold in the niche of TE preferentially as an artificial skin or as a cardiovascular implant.[38]

Thermoplastic polyurethane, as a class of PUs, is a linear segmented block copolymer composed of soft and hard segments. It possesses various properties, such as biodegradability, high elongation, moderate tensile strength, and Young's modulus, and excellent abrasion and tear resistance.[73]

PU has been widely employed in medical applications due to its biocompatibility and flexibility, and has been used in medical devices, they offer high elongation, moderate tensile strength and Young's modulus, and excellent abrasion and tear resistance.[73]

Polyhydroxyalkanoates. Polyhydroxyalkanoates are biodegradable polyesters that can be produced by both bacterial and synthetic routes. The most common polymer is poly(3-hydroxybutyrate) (PHB), a semi-crystalline polymer that undergoes surface erosion due to the hydrophobicity of the backbone and its crystallinity. PHB has a glass transition temperature around 5°C and a melting temperature from 160–180°C. Hydrolytic degradation of PHB results in the formation of D-(−)-3-hydroxybutyric acid, a normal blood constituent. The biocompatibility, processability and degradability of PHB make it an excellent candidate for use in long-term TE applications. Unfortunately, the stability of PHB makes it a poor candidate for controlled delivery applications.[74]

Blend

In view of the complexity involved in wound healing, it is difficult for a single material to provide a comprehensive treatment. The blending approach applies physical integration of natural and synthetic polymers to provide a favorable substrate for growth of skin cells, in order to develop biomaterials with new properties. This approach exploits the cell binding properties of a natural polymer, such as collagen, and the structural properties of a synthetic polymer, such as PCL. The resulting developed material can show high flexibility, which potentially allows effective draping over the wound site and close proximity with the wound bed to improve the prospects for graft take. Tear resistance is also conferred by the PCL component, which facilitates handling and manipulation during surgery. Effective inhibition of wound contraction by contractile fibroblasts has long been considered essential for optimum dermal regeneration in full thickness skin wounds.[75] The structural stability conferred on the material by the PCL phase is expected to be advantageous in this respect. In general, blending PCL with other synthetic/natural polymers is a simple approach to bring desired mechanical and biological properties. In fact, CS and PCL blend was prepared in a modified solvent of formic acid and acetone and used for electrospinning. The mats were evaluated for biocompatibility and its suitability for skin TE

Scaffolds—Smart

and the performance of the blend mats were compared with PCL mats. The results showed that CS-PCL fibrous mats showed enhanced cell adhesion cell spreading and proliferation. Improved cell material interaction of Keratinocyte cell line (HaCaT) on CS-PCL suggests that blending CS with PCL makes it a better substrate for skin TE. The outcome of the research is that CS-PCL is a better scaffold for attachment and proliferation of keratinocytes and is a potential material for skin tissue engineering.[72] Other natural origin materials are also proposed for skin substitution in combination with synthetic polymer. Fibrin-based scaffolds with PGA for guided skin regeneration have been tested *in vitro* and *in vivo*.[76] In addition, fibrin scaffolds proved to be good matrices for the spreading and proliferation of human mesenchymal SCs, enhancing their suitability for wound healing and skin substitution.[77]

Chitosan modified with hyaluronic acid[78] or collagen[79] has also been reported to be suitable for wound healing and skin substitution. Berthod et al. [79] developed a collagen–chitosan sponge on which human fibroblasts, cultured for one month, produced differentiated connective tissue. An *in vivo* study demonstrated that a TE construct of collagen–chitosan and human skin fibroblasts and keratinocytes, transplanted in the back of nude mice, enhanced nerve regeneration, proving it to be a suitable construct for skin substitution also.[80]

Composites/Nanocomposite

To enhance the mechanical properties and cellular adhesion and proliferation, the incorporation of nanoparticles (e.g., apatite component, carbon nanostructures, and metal nanoparticles) has been extensively investigated.[32] Nanocomposites have emerged in the last two decades as an efficient strategy to upgrade the structural and functional properties of synthetic polymers.

They often show an excellent balance between strength and toughness and usually improved characteristics compared to their individual components.

Dai et al. reported the preparation and characterisation of collagen-PCL composites for manufacture of tissue-engineered skin substitutes and models.[75]

Of particular interest are nanocomposites involving biopolymeric matrices and bioactive nanosized fillers.

The fillers with nanosized features can bring about an intense change in the physical properties of the polymer matrix, allowing for the engineering of improved biomaterials that individual materials cannot attain. The nanoparticles have a large surface area when compared to the conventional microsized fillers, which form a tight interface with the polymeric matrices, offering improved mechanical properties, while maintaining the favorable osteoconductivity and biocompatibility of the fillers, thus influencing protein adsorption, cells adhesion, and proliferation and differentiation for new tissue formation.[32]

Biodegradability, high mechanical strength, and osteointegration and formation of ligamentous tissue are properties required for such materials.[32] Composite materials possessing multifunctional properties have been used to mimic ECM for skin TE, with promising results[81]

Processing conditions

Several processing techniques are available to develop the ideal 3D scaffold with micro and nanostructures for skin TE.[32]

Nanostructured materials are advantageously applied for constructing a tissue replacement, because they are more attractive for cell adhesion, proliferation, and differentiation than flat or microstructured surfaces. The nanostructure of the material mimics the nanoarchitecture of natural ECM, for example, its nanofibrous component, and it enhances the cell–material interaction.[69,81,82] Nanostructured materials can promote skin regeneration by improving the adhesion and proliferation of skin cells and the neovascularization of a tissue-engineered skin implant. They enable cells to be supplied with oxygen and nutrition and prevent fluid accumulation at the wound site.[83] From the techniques used to construct biomaterials to be cultivated with cells, electrospinning is the most widely studied and it has also been demonstrated as giving the most promising results to develop scaffolds for TE applications.

Electrospinning

Electrospinning (ES) is a simple and versatile technique that produces scaffolds formed by nano- and microfibers, which offer a favorable microenvironment for cellular development by mimicking the native ECM. This technique provides nonwovens resembling in their fibrillar structures those of the ECM, and offering large surface areas, ease of functionalization for various purposes, and controllable mechanical properties.

The electrospinning process works on the electrostatic principle. The machine is basically composed of a syringe with a nozzle, a counter electrode (normally a metal plate), a source of electrical field, and a pump. The solution to be electrospun is applied to the system via the nozzle of the syringe and is pulsed by the pump. It is then subjected to a difference in electrical voltage present between the nozzle and the counter electrode. This electrical voltage generated by the source causes a cone-shaped deformation of the drop of polymer solution. The solvent in the solution evaporates and at the end of the process, solid continuous filaments are yielded. It is important to note that gravitational forces do not interfere in the process because the acceleration of the fiber formation is up to 600 m/s^2, which is close to two orders of magnitude greater than the acceleration of gravitational forces. Because of this, it is possible to form fibers from top-down, bottom-up, or other types of arrangements.[84] The parameters affecting electrospinning include electrical charges on the emerging jet, charge density and removal, as well as the effects of external perturbations. The solvent and the method of fiber collection also affect the construction of the final nanofibrous architecture.

The important advantages of the ES technique include the production of fibers ranging from few nanometers to micrometers in diameter, high porosity and large surface areas, which can mimic the ECM structure in terms of the chemistry and dimensions. Electrospun fibers are promising TE scaffolds that offer the cells an environment that mimics the native ECM. Fibers with different characteristics can be produced by the electrospinning technique according to the needs of the tissue to be repaired.[85]

The recent developments toward large-scale productions combined with the simplicity of the process render this technique very attractive. Progress concerning the use of electrospinning for TE applications has advanced impressively.[86–88]

BIOMEDICAL APPLICATIONS

The development of cell biology, molecular biology, and material science, has increased the development of biomimic tissue-engineered skins.[89] To improve the safety, durability, elasticity, biocompatibility, and clinical efficacy of tissue-engineered skin, several powerful seed cells have already found their application in wound repair, and a variety of bioactive scaffolds have been discovered to influence cell fate in epidermogenesis. These increased interests provide insights into advanced construction strategies for complex skin mimics. Based on these exciting developments, a complete full-thickness tissue-engineered skin is likely to be generated. Skin is the outer covering of the body and it is the largest organ of the body. Wound or injury to skin can be caused by burn, accidental trauma, or chronic ulcerations (venous stasis, diabetic wounds, and pressure ulcers).[1] Wound healing is a complex and dynamic process comprising four distinct phases, namely, hemostasis, inflammation, proliferation, and remodeling, requiring cellular and biochemical interaction of various cells such as keratinocytes, fibroblasts, and endothelial cells of which the fibroblast cells play a vital role in initiating angiogenesis, epithelialization, and collagen formation. The ability of the skin to regenerate itself following minor epidermal injury is remarkable; however, during bacterial infection or when the injury is severe, such as loss of epidermis and dermis in full-thickness skin wounds, the damaged skin cannot initiate the healing process leading to life threatening complications.[1–6]

Walsh et al., explored nanotopography as a novel strategy to enhance transdermal drug delivery. They demonstrated for the first time that nanotopography applied to microneedles significantly enhances transdermal delivery of Etanercept, a 150 kD therapeutic, in both rats and rabbits, underlining that this effect is mediated by remodeling of the tight junction proteins initiated via integrin binding to the nanotopography, followed by phosphorylation of the myosin light chain (MLC) and activation of the actomyosin complex, which in turn increase paracellular permeability.[90]

Lu et al. showed that hybrid scaffolds, constructed by forming funnel-like collagen or gelatin sponges on one side of a PLLA woven mesh, can improve cell adhesion and proliferation, compared to the control collagen sponges. The PLLA–collagen scaffolds showed more promotive effect than did the PLLA–gelatin scaffolds. The hybrid scaffolds improved the wound healing of full-thickness skin defects in athymic nude mice. The funnel-like structure facilitated the cell penetration and the PLLA mesh improved the mechanical strength of the scaffolds. The funnel-like PLLA–collagen or PLLA–gelatin hybrid scaffold will be a useful tool for skin TE.[91,92]

Subcutaneous implantation showed that the PLLA–collagen and PLLA–gelatin scaffolds promoted the regeneration of dermal tissue and epidermis and reduced contraction during the formation of new tissue. These results indicate that funnel-like hybrid scaffolds can be used for skin tissue regeneration.[92]

A blend of natural and synthetic polymers, as also a blend of two different natural polymers can used to modulate the properties for TE. As the dermis is comprised of a meshwork of ECM fibers, electrospinning has been employed as a tool for generating nanopatterned scaffolds for dermal regeneration. Venugopal et al. used electrospinning to create mesh fiber scaffolds of PCL, collagen, and PCL–collagen composite. The scaffolds could be fabricated on a large scale (11×7 cm^2 scaffolds) with pore diameters ranging from 170 to 250 nm and with high (> 90%) porosity, allowing for nutrient diffusion and cell migration. Although the human dermal fibroblasts were shown to attach to all these scaffolds, the cells exhibited a different morphology on the PCL scaffolds from that of the PCL–collagen scaffolds. The PCL–collagen blended scaffolds also were mechanically more stable than the collagen scaffolds. Moreover, PCL–collagen and collagen both demonstrated higher cell proliferation, indicating increased biocompatibility[93,94]. In parallel, individual PCL fibers were coated with collagen to improve proliferation.[95] A similar experiment using dextran/PLGA electrospun fiber scaffolds and mouse dermal fibroblasts also demonstrated cell attachment without the need for ECM precoating or treatment, with enhanced cell viability (up to 10 days) and penetration into the fiber scaffolds.[35] Bacakova et al. showed that the fibrin nanocoating on PLA nanofibrous membranes strongly influenced the behavior of human dermal fibroblasts. Fibrin promoted adhesion and spreading of cells and increased the expression of β_1-integrins, cell proliferation, and collagen synthesis. The beneficial effect of fibrin on the adhesion and proliferation of cells is probably because of its ability to bind to cells easily through the integrin adhesion receptors to attract the adhesive molecules from the cell culture

Scaffolds—Smart

medium. It is probably also due to its mitogenic effects on the cells. Fibrin increases the mRNA expression of collagen I, the total amount of synthesized collagen, and its deposition as ECM on the membrane surface.[96]

The combination of nanofibrous membranes with a fibrin nanocoating and ascorbic acid seems to be particularly advantageous for skin TE.

Recently Liu et al. proved the ability of the bFGF-loaded alginate microspheres (Ms) incorporated into carboxymethyl chitosan (CMCS)–poly(vinyl alcohol) (PVA) hydrogel to accelerate wound healing in a full-thickness burn model, suggesting its potential for use in dermal tissue regeneration.[37]

CONCLUSION

The primary goal of skin TE is to enable the rapid formation of a construct that will support and enable the complete regeneration of functional skin with all the skin appendages, and the various layers (epidermis, dermit, fatty subcutis).

Improvement of skin substitutes will result from inclusion of additional cell types and from modifications of culture media, biopolymer substrates, and the physical environment (i.e., humidity, mechanical tension, electrical properties) that promote greater fidelity to native skin.

However, it is reasonable to believe that with the development of skin biology, material science, and engineering technology, the skin TE will finally have an equivalent structural and functional therapeutic outcome as autogeneic skin transplantation does.

REFERENCES

1. Clark, R.A.; Ghosh, K.; Tonnesen, M.G. Tissue engineering for cutaneous wounds J. Invest. Dermatol. **2007**, *127* (5), 1018–1029.
2. Orlando, G.; Wood, K.J.; De Coppi, P.; Baptista, P.; Binder, K.W.; Bitar, K.N.; Breuer, C.; Burnett, L.; Christ, G.; Farney, A.; Figliuzzi, M.; Holmes, J.H., 4; Koch, K.; Macchiarini, P.; Mirmalek Sani, S.H.; Opara, E.; Remuzzi, A.; Rogers, J.; Saul, J.M.; Seliktar, D.; Shapira-Schweitzer, K.; Smith, T.; Solomon, D.; Van Dyke, M.; Yoo, J.J.; Zhang, Y.; Atala, A.; Stratta, R.J.; Soker, S. Regenerative medicine as applied to general surgery. Ann. Surg. **2012**, *255* (5), 867–880.
3. Tissue engineering. Despite technical and regulatory challenges, the prospects for tissue engineering are good. Nat. Biotechnol. **2000**, *18*, IT56–IT58.
4. Kamel, R.A.; Ong, J.F.; Eriksson, E.; Junker, J.P.E.; Caterson, E.J. Tissue engineering of skin. J. Am. Coll. Surg. **2013**, *217* (3), 533–555.
5. Shevchenko, R.V.; James, S.L.; James, S.E. A review of tissue-engineered skin bioconstructs available for skin reconstruction J. R. Soc. Interface **2010**, *7* (43), 229–258.
6. Supp, D.M.; Boyce, S.T. Engineered skin substitutes: practices and potentials Clin. Dermatol. **2005**, *23* (4), 403–412.
7. Miller, K.J.; Brown, D.A.; Ibrahim, M.M.; Ramchal, T.D.; Levinson, H. MicroRNAs in skin tissue engineering. Adv. Drug. Deliv. Rev. **2005**, *88*, 16–36.
8. Schulz, J.T.; Tompkins, R.G.; Burke, J.F. Artificial Skin. Annu. Rev. Med. **2000**, *51*, 231–244.
9. Carter, M.J. Economic evaluations of guideline-based or strategic interventions for the prevention or treatment of chronic wounds. Appl. Health Econ. Health Policy **2014**, *12* (5), 373–389.
10. Boyce, S.T. Design principles for composition and performance of cultured skin substitutes. Burns **2001**, *27* (5), 523–533.
11. Ng, K.W.; Khor, H.L.; Hutmacher, D.W. *In vitro* characterization of natural and synthetic dermal matrices cultured with human dermal fibroblasts. Biomaterials **2004**, *25* (14), 2807–2818.
12. Trasciatti, S.; Podestà, A.; Bonaretti, S.; Mazzoncini, V.; Rosini, S. *In vitro* effects of different formulations of bovine collagen on cultured human skin. Biomaterials **1998**, *19* (10), 897–903.
13. Langer, R.; Vacanti. J.P. Tissue engineering. Science **1993**, *260* (5110), 920–926.
14. Hollander, A.; Macchiarini, P.; Gordijn, B.; Birchall, M. The first stem-cell based tissue-engineered organ replacement: implications for regenerative medicine and society. Regen. Med. **2009**, *4* (2), 147–148.
15. Butler, K.L.; Goverman, J.; Ma, H.; Fischman, A.; Yu, Y.M.; Bilodeau, M.; Rad, A.M.; Bonab, A.A.; Tompkins, R.G.; Fagan, S.P. Stem cells and burns: review and therapeutic implications. J. Burn Care Res. **2010**, *31* (6), 874–881.
16. Sun, B.K. Siprashvili, Z.; Khavari, P.A. Advances in skin grafting and treatment of cutaneous wounds. Science. **2014**, *346* (6212):941–945.
17. Martino, S.; D'Angelo, F.; Armentano, I.; Kenny, J.M.; Orlacchio, A. Stem cell-biomaterial interactions for regenerative medicine. Biotechnol. Adv. **2012**, *30* (1), 338–351.
18. Richards, M.; Tan, S.P.; Tan, J.H.; Chan, W.K.; Bongso, A. The transcriptome profile of human embryonic stem cells as defined by SAGE. Stem Cells **2004**, *22* (1), 51–64.
19. Orlacchio, A.; Bernardi, G.; Martino, S. Stem cells and neurological diseases. Discov. Med. **2010**, *9* (49), 546–553.
20. Hodgkinson, T.; Bayat, A. Dermal substitute-assisted healing: Enhancing stem cell therapy with novel biomaterial design. Arch Dermatol Res **2011**, *303* (5), 301–315.
21. Wu, Y.; Chen, L.; Scott, P.G.; Tredget, E.E. Mesenchymal stem cells enhance wound healing through differentiation and angiogenesis. Stem Cells **2007**, *25* (10):2648–2659.
22. Li, H.; Fu, X.; Ouyang, Y.; Cai, C.; Wang, J.; Sun, T. Adult bone-marrow-derived mesenchymal stem cells contribute to wound healing of skin appendages. Cell Tissue Res **2006**, *326* (3), 725–736.
23. Zuk, P.A.; Zhu, M.; Mizuno, H.; Huang, J.; Futrell, J.W.; Katz, A.J.; Benhaim, P.; Lorenz, H.P.; Hedrick, M.H. Multilineage cells from human adipose tissue: implications for cell-based therapies. Tissue Eng **2001**, *7* (2), 211–228.
24. Takahashi, K.; Yamanaka, S. Induction of pluripotent stem cells from mouse embryonic and adult fibroblast cultures by defined factors. Cell **2006**, *126* (4), 663–676.
25. Bilousova, G.; Chen, J.; Roop, D.R. Differentiation of mouse induced pluripotent stem cells into a multipotent

keratinocyte lineage. J. Invest. Dermatol. **2011,** *131* (4):857–864.

26. Okita, K.; Nakagawa, M.; Hyenjong, H.; Ichisaka, T.; Yamanaka, S. Generation of mouse induced pluripotent stem cells without viral vectors. Science **2008,** *322* (5903), 949–953.

27. Uitto, J. Regenerative medicine for skin diseases: iPS cells to the rescue. J. Invest. Dermatol. **2011,** *131* (4), 812–814.

28. Bi, H.; Jin, Y. Current progress of skin tissue engineering: Seed cells, bioscaffolds, and construction strategies. Burns & Trauma **2013,** *1* (2), 63–72.

29. Garber, K. RIKEN suspends first clinical trial involving induced pluripotent stem cells. Nat. Biotechnol. **2015,** *33* (9), 890–891.

30. Poursamar, S.A.; Hatami, J.; Lehner, A.N.; da Silva, C.L.; Ferreira, F.C.; Antunes, A.P.M. Potential application of gelatin scaffolds prepared through *in situ* gas foaming in skin tissue engineering. Int. J. Polym. Mater. Po **2016,** *65* (6), 315–322.

31. Bello, Y.M.; Falabella, A.F.; Eaglstein, W.H. Tissue-engineered skin. Current status in wound healing. Am. J. Clin. Dermatol. **2001,** *2* (5), 305–313.

32. Armentano, I.; Dottori, M.; Fortunati, E.; Mattioli, S.; Kenny, J.M. Biodegradable polymer matrix nanocomposites for tissue engineering: A review. Polym. Degrad. Stab. **2010,** *95* (11), 2126–2146.33.

33. Kim, K.; Luu, Y.K.; Chang, C.; Fang, D.; Hsiao BS, Chu B, Hadjiargyrou M. Incorporation and controlled release of a hydrophilic antibiotic using poly (lactideco-glycolide)-based electrospun nanofibrous scaffolds. J. Control. Release. **2004,** *98* (1), 47–56.

34. Prabhakaran, M.P.; Venugopal, J.R.; Ramakrishna, S. Mesenchymal stem cell differentiation to neuronal cells on electrospun nanofibrous substrates for nerve tissue engineering. Biomaterials **2009,** *30* (28), 4996–5003.

35. Pan, H.; Jiang, A.; Chen, W. Interaction of dermal fibroblasts with electrospun composite polymer scaffolds prepared from dextran and poly lactide-co-glycolide. Biomaterials **2006,** *27* (17), 3209–3220.

36. Vashisth, P.; Nikhil, K.; Roy, P.; Pruthi, P.A.; Singh, R.P.; Pruthi, V. A novel gellan–PVA nanofibrous scaffold for skin tissue regeneration: Fabrication and characterization. Carbohydr. Polym. **2016,** *136*, 851–859.

37. Liu, Q.; Huang, Y.; Lan, Y.; Zuo, Q.; Li, C.; Zhang, Y.; Guo, R.; Xue, W. Acceleration of skin regeneration in full-thickness burns by incorporation of bFGF-loaded alginate microspheres into a CMCS–PVA hydrogel. J. Tissue Eng. Regen. Med. **2015,** doi: 10.1002/term.2057.

38. Das, B.; Chattopadhyay, P.; Mandal, M.; Voit, B.; Karak, N. Bio-based biodegradable and biocompatible hyperbranched polyurethane: A scaffold for tissue engineering. Macromol. Biosci. **2013,** *13* (1), 126–139.

39. Choi, Y.S.; Hong, S.A.; Lee, Y.M.; Song, K.W.; Park, M.H.; Nam, Y.S. Study on gelatin-containing artificial skin: I. Preparation and characteristics of novel gelatin-alginate sponge. Biomaterials **1999,** *20* (5), 409–417.

40. Young, S.; Wong, M.; Tabata, Y.; Mikos, A.G. Gelatin as a delivery vehicle for the controlled release of bioactive molecules. J. Control. Release **2005,** *109* (1–3), 256–274.

41. Suh, J.K.; Matthew, H.W. Application of chitosan-based polysaccharide biomaterials in cartilage tissue engineering: a review. Biomaterials **2000,** *21* (24), 2589–2598.

42. Trasciatti, S.; Podesta, A.; Bonaretti, S.; Mazzoncini, V.; Rosini, S. *In vitro* effects of different formulations of bovine collagen on cultured human skin. Biomaterials **1998,** *19* (10), 897–903.

43. Krajewska, B. Membrane-based processes performed with use of chitin/chitosan materials. Sep. Purif. Technol. **2005,** *41* (3), 305–312.

44. Chunlin, Y.; Hillas, P.J.; Buez, J.A.; Nokelainen, M.; Balan, J.; Tang, J.; Spiro, R.; Polarek, J.W. The application of recombinant human collagen in tissue engineering. BioDrugs **2004,** *18* (2), 103–119.

45. Chen, G.P.; Ushida, T.; Tateishi, T. Hybrid biomaterials for tissue engineering: a preparative method for PLA or PLGA-collagen hybrid sponges. Adv. Mater. 2000, 12 (6), 455–467.

46. Kim, I.Y.; Seo, S.J.; Moon, H.S.; Yoo, M.K.; Park, I.Y.; Kim, B.C.; Cho, C.S. Chitosan and its derivatives for tissue engineering applications. Biotechnol. Adv. **2008,** *26* (1), 1–21.

47. Djagny, K.B.; Wang, Z.; Xu. S. Gelatin: a valuable protein for food and pharmaceutical industries: review. Crit. Rev. Food Sci. Nutr. **2001,** *41* (6), 481–492.

48. Stainsby, G. Gelatin gels. In *Advances in Meat Research, Collagen as a Food*; Pearson, A.M., Dutson, T.R., Bailey, A.J., Eds.; Van Nostrand Reinhold Company Inc.; New York, 1987; Vol. 4, pp. 209–222.

49. Yang, H.; Wang, Y. Effects of concentration on nanostructural images and physical properties of gelatin from channel catfish skins. Food Hydrocoll. **2009,** *23* (3), 577–584.

50. Gennadios, A.; McHugh, T.H.; Weller, C.L.; Krochta, J.M. Edible coatings and films based on proteins. In *Edible Coatings and Films to Improve Food Quality*; Krochta, J.M., Baldwin, E.A., Nísperos-Carriedo, M., Eds.; Technomic Publishing Co.: Lancaster, 1994; pp. 210–278.

51. Malafaya, P.B.; Silva, G.A.; Reis, R.L. Natural—origin polymers as carriers and scaffolds for biomolecules and cell delivery in tissue engineering applications. Adv. Drug Deliv. Rev. **2007,** 59, 207–233.

52. IAranaz, I.; Harris, R.; Heras, A. Chitosan amphiphilic derivatives. Chemistry and applications. Curr. Org. Chem. **2010,** *14* (3), 308–330.

53. Croisier, F.; Jérôme, C. Chitosan-based biomaterials for tissue engineering. Eur. Polym. J. **2013,** *49* (4), 780–792.

54. Jayakumar, R.; Prabaharan, M.; Nair, S.V.; Tokura, S.; Tamura, H.; Selvamurugan, N. Novel carboxymethyl derivatives of chitin and chitosan materials and their biomedical applications. Prog. Mater. Sci. **2010,** *55* (7), 675–709.

55. Muzzarelli, R.A.A. Chitin and chitosans for the repair of wounded skin, nerve, cartilage and bone. Carbohydr. Polym. **2009,** *76* (2),167–82.

56. Rinaudo, M.; Pavlov, G.; Desbrières, J. Influence of acetic acid concentration on the solubilization of chitosan. Polymer **1999,** *40* (25):7029–7032.

57. Denkbaş, E.B.; Oztürk, E.; Ozdem, N.; Keçec, K.; Agalar, C. Norfloxacin-loaded chitosan sponges as wound dressing material. J. Biomater. Appl. **2004,** *18* (4), 291–303.

58. Anisha, B.S.; Sankar, D.; Mohandas, A.; Chennazhi, K.P.; Nair, S.V.; Jayakumar, R. Chitosan–hyaluronan/nano chondroitin sulfate ternary composite sponges for medical use. Carbohydr. Polym. **2013,** *92* (2), 1470–1476.

59. Okamoto, Y.; Shibazaki, K.; Minami, S.; Matsuhashi, A.; Tanioka, S.; Shigemasa, Y. Evaluation of chitin and chitosan

Scaffolds—Smart

in open wound healing in dogs. J. Vet. Med. Sci. **1995**, *57* (5), 851–854.

60. Drury, J.L; Mooney, D.J. Hydrogels for tissue engineering: scaffold design variables and applications. *Biomaterials* **2003**, *24* (24), 4337–4351.

61. Mutsaers, S.E.; Bishop, J.E.; McGrouther, G.; Laurent, G.J. Mechanisms of tissue repair: from wound healing to fibrosis. Int. J. Biochem. Cell Biol. **1997**, *29* (1), 5–17.

62. Laurens, N.; Koolwijk, P.; de Maat, M.P. Fibrin structure and wound healing. J. Thromb. Haemost. **2006**, *4* (5), 932–939.

63. Mano, J.F.; Silva, G.A.; Azevedo, H.S.; Malafaya, P.B.; Sousa, R.A.; Silva, S.S.; Boesel, L.F.; Oliveira, J.M.; Santos, T.C.; Marques, A.P.; Neves, N.M.; Reis, R.L. Natural origin biodegradable systems in tissue engineering and regenerative medicine: present status and some moving trends. J. R. Soc. Interface. **2007**, *4* (17), 999–1030.

64. Tønnesen, H.; Karlsen, J. Alginate in drug delivery systems. Drug Dev. Ind. Pharm. **2002**, *28* (6), 621–630.

65. George, M.; Abraham, T. Polyionic hydrocolloids for the intestinal delivery of protein drugs: Alginate and chitosan-a review J. Control. Rel. **2006**, *114* (1), 1–14.

66. Keane, T.J.; Badylak, S.F. Biomaterials for tissue engineering applications. Semin. Pediatr. Surg. **2014**, *23* (3), 112–118.

67. Gunatillake, P.A.; Adhikari, R. Biodegradable synthetic polymers for tissue engineering. Eur. Cell. Mater. **2003**, *5*, 1–16.

68. Armentano, I.; Bitinis, N.; Fortunati, E.; Mattioli, S.; Rescignano, N.; Verdejo, R.; Lopez-Manchado, M.A.; Kenny, J.M. Multifunctional nanostructured PLA materials for packaging and tissue engineering. Prog. Polym. Sci. **2013**, *38* (10–11), 1720–1747.

69. Fabbri, M.; Soccio, M.; Costa, M.; Lotti, N.; Gazzano, M.; Siracusa, V.; Gamberini, R.; Rimini, B.; Munari, A.; García-Fernández, L.; Vázquez-Lasae, B.; San Román, J. New fully bio-based PLLA triblock copoly(ester urethane)s as potential candidates for soft tissue engineering. Polym. Degrad. Stab. **2016**, *132*, 169–180.

70. Mohiti-Asli, M.; Pourdeyhimi, B.; Loboa E.G. Skin Tissue engineering for the infected wound site: Biodegradable PLA nanofibers and a novel approach for silver ion release evaluated in a 3D coculture system of keratinocytes and *Staphylococcus aureus*. Tissue Eng. Part C Methods **2014**, *20* (10), 790–797.

71. Prasad, T.; Shabeena, E.A.; Vinod, D.; Kumary, T.V.; Kumar, P.R.A. Characterization and *in vitro* evaluation of electrospun chitosan/polycaprolactone blend fibrous mat for skin tissue engineering. J. Mater. Sci. Mater. Med. **2015**, *26* (1), 5352.

72. Ng, K.W.; Hutmacher, D.W.; Schantz, J.T.; Ng, C.S.; Too, H.P.; Lim, T.C.; Phan, T.T.; Teoh, S.H. Evaluation of ultra-thin poly(epsilon-caprolactone) films for tissue-engineered skin. Tissue Eng. **2001**, *7* (4), 441–455.

73. Ma, Z.W.; Hong, Y.; Nelson, D.M.; Pichamuthu, J.E.; Leeson, C.E.; Wagner, W.R. Biodegradable polyurethane ureas with variable polyester or polycarbonate soft segments: Effects of crystallinity, molecular weight, and composition on mechanical properties. Biomacromolecules **2011**, *12* (9), 3265–3274.

74. Ahmed, T.; Marçal, H.; Lawless, M.; Wanandy, N.S.; Chiu, A.; Foster, L.J. Polyhydroxybutyrate and its copoly-mer with polyhydroxyvalerate as biomaterials: influence on progression of stem cell cycle. Biomacromolecules. **2010**, *11* (10), 2707–2715.

75. Dai, N.T.; Williamson, M.R.; Khammo, N.; Adams, E.F.; Coombes, A.G.A. Composite cell support membranes based on collagen and polycaprolactone for tissue engineering of skin. Biomaterials **2004**, *25* (18), 4263–4271.

76. Hokugo, A.; Takamoto, T.; Tabata, Y. Preparation of hybrid scaffold from fibrin and biodegradable polymer fiber. Biomaterials **2006**, *27* (1), 61–67.

77. Bensaïd, W.; Triffitt, J.T.; Blanchat, C.; Oudina, K.; Sedel, L.; Petite, H. A biodegradable fibrin scaffold for mesenchymal stem cell transplantation. Biomaterials **2003**, *24* (14), 2497–2502.

78. Mao, J.S.; Liu, H.F.; Yin, Y.J.; Yao, K.D. The properties of chitosan–gelatin membranes and scaffolds modified with hyaluronic acid by different methods. Biomaterials **2003**, *24* (9), 1621–1629.

79. Berthod, F.; Sahuc, F.; Hayek, D.; Damour, O.; Collombel, C. Deposition of collagen fibril bundles by long-term culture of fibroblasts in a collagen sponge. J. Biomed. Mater. Res. **1996**, *32* (1), 87–93.

80. Gingras, M.; Paradis, I.; Berthod, F. Nerve regeneration in a collagen–chitosan tissue-engineered skin transplanted on nude mice. Biomaterials **2003**, *24* (9), 1653–1661.

81. Liu, H.F.; Yao, F.L.; Zhou, Y.; Yao, K.; Mei, D.; Cui, L.; Cao, Y. Porous poly(DL-lactic acid) modified chitosan–gelatin scaffolds for tissue engineering. J. Biomater. Appl. **2005**, *19* (4), 303–322.

82. Armentano, I.; Montanucci, P.; Morena, F.; Bicchi, I.; Basta, G.; Fortunati, E.; Mattioli, S.; Calafiore, R.; Martino, S.; Kenny, J.M. Nano-engineered PLLA based biomaterial drives stem cell responses. J Tissue Eng. Regen. Med. **2014**, *8*, 501–502.

83. D'Angelo, F.; Armentano, I.; Cacciotti, I.; Tiribuzi, R.; Quattrocelli, M.; Del Gaudio, C.; Fortunati, E.; Saino, E.; Caraffa, A.; Cerulli, G.G.; Visai, L.; Kenny, J.M.; Sampaolesi, M.; Bianco, A.; Martino, S.; Orlacchio, A. Tuning multi/pluri-potent stem cell fate by electrospun poly(L-lactic acid)-calcium-deficient hydroxyapatite nanocomposite mats. *Biomacromolecules* **2012**, *13* (5), 1350–1360.

84. Sridhar, S.; Venugopal, J.R.; Ramakrishna, S. Improved regeneration potential of fibroblasts using ascorbic acid-blended nanofibrous scaffolds. J. Biomed. Mater. Res. A. **2015**, *103* (11), 3431–3440.

85. Wendorff, J.H.; Agarwal, S.; Greiner, A. *Electrospinning—Materials, Processing, and Applications*; Wiley-VCH Verlag GmbH&Co. kGaA; Weinheim; Germany, 2012.

86. Agarwal, S.; Wendorff, J.H.; Greiner, A. Progress in the field of electrospinning for tissue engineering applications. Adv. Mater. **2009**, *21* (32–33), 3343–3351.

87. Khorshidi, S.; Solouk, A.; Mirzadeh, H.; Mazinani, S.; Lagaron, J.M.; Sharifi, S.; Ramakrishna, S. A review of key challenges of electrospun scaffolds for tissue-engineering applications. J Tissue Eng Regen Med. **2016**, *10* (9), 715–738.

88. Teo, W.E.; Inai, R.; Ramakrishna, S. Technological advances in electrospinning of nanofibers. Sci. Technol. Adv. Mater. **2011**, *12* (1), 013002.

89. Kim, H.N.; Jiao, A.; Hwang, N.S.; Kim, M.S.; Kang, D.H.; Kim, D.H.; Suh, K.Y. Nanotopography-guided tissue

engineering and regenerative medicine. Adv. Drug Deliv. Rev. **2013**, *65* (4), 536–558.

90. Poursamar, S.A.; Hatami, J.; Lehner, A.N.; da Silva, C.L.; Ferreira, F.C.; Antunes, A.P.M. Potential application of gelatin scaffolds prepared through in situ gas foaming in skin tissue engineering. Int. J. Polym. Mater. Po **2016**, *65* (6), 315–322.

91. Walsh, L.; Ryu, J.; Bock, S.; Koval, M.; Mauro, T.; Ross, R.; Desai, T. Nanotopography facilitates *in vivo* transdermal delivery of high molecular weight therapeutics through an integrin-dependent mechanism. Nano Lett. **2015**, *15* (4), 2434–2441.

92. Lu, H.; Oh, H.H.; Kawazoe, N.; Yamagishi, K.; Chen, G. PLLA–collagen and PLLA–gelatin hybrid scaffolds with funnel-like porous structure for skin tissue engineering. Sci. Technol. Adv. Mater. **2012**, *13* (6), 064210.

93. Venugopal, R.; Zhang, Y.; Ramakrishna, S. *In vitro* culture of human dermal fibroblasts on electrospun polycaprolactone collagen nanofibrous membrane. Artif. Organs **2006**, *30* (6), 440–446.

94. Venugopal, J.; Ramakrishna, S. Biocompatible nanofiber matrices for the engineering of a dermal substitute for skin regeneration. Tissue Eng. **2005**, *11* (5–6), 847–854.

95. Zhang, Y.Z.; Venugopal, J.; Huang, Z.M.; Lim, C.T.; Ramakrishna, S. Characterization of the surface biocompatibility of the electrospun PCL–collagen nanofibers using fibroblasts. Biomacromolecules **2005**, *6* (5), 2583–2589.

96. Bacakova, M.; JMusilkova, J.; Riedel, T.; Stranska, D.; Brynda, E.; Zaloudkova, M.; Bacakova, L. The potential applications of fibrin-coated electrospun polylactide nanofibers in skin tissue engineering. Int. J. Nanomedicine **2016**, *11*, 771–789.

Smart Polymers: Imprinting

Carmen Alvarez-Lorenzo
Angel Concheiro
Department of Pharmacy and Pharmaceutical Technology, University of Santiago de Compostela (USC), Santiago de Compostela, Spain

Jeffrey Chuang
Boston College, Chestnut Hill, Massachusetts, U.S.A.

Alexander Yu. Grosberg
Department of Physics, University of Minnesota, Minneapolis, Minnesota, U.S.A.

Abstract

The use of materials as so-called smart polymers is motivated by the simple observation that biological molecules perform incredibly complex functions. While the goal of engineering polymers as effectively as nature remains somewhat in the realm of science fiction, in recent years many researchers have sought to find or design synthetic polymeric materials capable of mimicking one or another "smart" property of biopolymers. This entry reviews some basic approaches to molecular imprinting and then moves to discuss in detail the Tanaka equation and its impact on imprinting.

INTRODUCTION

The use of materials as so-called smart polymers is motivated by the simple observation that biological molecules perform incredibly complex functions. While the goal of engineering polymers as effectively as nature remains somewhat in the realm of science fiction, in recent years many researchers have sought to find or design synthetic polymeric materials capable of mimicking one or another "smart" property of biopolymers. Polymer gels are promising systems for such smart functions because of their volume phase transition, predicted theoretically by Dusek and Patterson[1] in 1968 and experimentally demonstrated by Tanaka[2] in 1978. Gel collapse can be driven by any one of the four basic types of intermolecular interactions operational in water solutions and in molecular biological systems,[3] namely, by hydrogen bonds and by van der Waals, hydrophobic, and Coulomb interactions between ionized (dissociated) groups. According to Flory's theory,[4] the degree of swelling of a hydrogel is the result of a competition between the entropy due to polymer conformations, which causes rubber elasticity, and the energy associated with internal attractions and repulsions between the monomers in the gel and the solvent. A change in the environmental conditions, such as temperature, pH, or composition, modifies the balance between the free energy of the internal interactions and the elasticity component, inducing a volume phase transition.[5] The variety of external stimuli that can trigger the phase transition as well as the present possibilities of modulating the rate and the intensity of the response to the stimulus guarantee smart gels a wealth of applications.

The interest in stimuli-sensitive hydrogels, especially in the biomedical field, could be remarkably increased if they could mimic the recognition capacity of certain biomacromolecules (e.g., receptors, enzymes, antibodies). The unique details of the protein's native state, such as its shape and charge distribution, enable it to recognize and interact with specific molecules. Proteins find their desired conformation out of a nearly infinite number of possibilities. In contrast, as known from recent theoretical developments, a polymer with a randomly made sequence will not fold in just one way (see, for example, the review by Pande, Grosberg, and Tanaka[6] and references therein Tanaka and Annaka[7]). Therefore, the ability of a polymer (or polymer hydrogel) to always fold back into the same conformation after being stretched and unfolded, i.e., to thermodynamically memorize a conformation, should be related to properly selected or designed nonrandom sequences. To obtain, under proper conditions, synthetic systems with sequences able to adopt conformations with useful functions, the molecular imprinting technology can be applied.[8–10] The hydrogels can recognize a substance if they are synthesized in the presence of such a substance (which acts as a template) in a conformation that corresponds to the global minimum energy. The "memorization" of this conformation, after the swelling of the network and the washing of the template, will only be possible if the network is always

Concise Encyclopedia of Biomedical Polymers and Polymeric Biomaterials DOI: 10.1081/E-EBPPC-120052186

able to fold into the conformation, upon synthesis, that can carry out its designated function (Fig. 1).[11] This revolutionary idea is the basis of new approaches to the design of imprinted hydrogels and has been developed at different levels, as explained below.

Approaches to Molecular Imprinting

The concept of molecular imprinting was first applied to organic polymers in the 1970s, when covalent imprinting in vinyl polymers was first reported.[10,12] The noncovalent imprinting was introduced a decade later.[13,14] Both approaches are aimed at creating tailor-made cavities shaped with a high specificity and affinity for a target molecule inside or at the surface of highly cross-linked polymer networks. To carry out the process, the template is added to the monomers and cross-linker solution before polymerization, which allows some of the monomers (called "functional" monomers) to be arranged in a configuration complementary to the template. The functional monomers are arranged in position through either covalent bonds or non-covalent interactions, such as hydrogen bond or ionic, hydrophobic, or charge-transfer interactions (Fig. 2). In the first case, the template is covalently

bound to the monomers prior to polymerization; after synthesis of the network, the bonds are reversibly broken for removal of the template molecules and formation of the imprinted cavities. In the noncovalent or *self-assembly* approach, the template molecules and functional monomers are arranged prior to polymerization to form stable and soluble complexes of appropriate stoichiometry by noncovalent or metal coordination interactions. In this case, multiple-point interactions between the template molecule and various functional monomers are required to form strong complexes in which both species are bound as strongly as in the case of a covalent bond. The noncovalent imprinting protocol allows more versatile combinations of templates and monomers, and provides faster bond association and dissociation kinetics than the covalent imprinting approach.[15] By any of these approaches, copolymerization of functional monomer–template complexes with high proportions of cross-linking agents, and subsequent removal of the template, provides recognition cavities complementary in shape and functionality. These vacant cavities are then available for rebinding of the template or structurally related analogues. Excellent reviews about the protocols to create rigid imprinted networks can be found in the literature.[16,17] Nowadays, molecular imprinting is a well-developed tool in the analytical field, mainly for separating and quantifying a wide range of substances contained in complex matrices.[18–20] Additionally, there has been a progressive increase in the number of papers and patents devoted to the application of molecularly imprinted polymers (MIPs) in the design of new drug-delivery systems and of devices useful in closely related fields, such as diagnostic sensors or chemical traps to remove undesirable substances from the body.[21,22] To obtain functional imprinted cavities, several factors must be taken into account. It is essential to ensure that the template does not bear any polymerizable group that could attach itself irreversibly to the polymer network, does not interfere in the polymerization process, and remains stable at moderately elevated temperatures or upon exposure to UV irradiation.

Other key issues are related to the nature and proportion of the monomers and to the synthesis conditions (solvent, temperature, etc.). The cavities should have a structure stable enough to maintain their conformation in the absence of the template and, at the same time, be sufficiently flexible to facilitate the attainment of a fast equilibrium between the release and re-uptake of the template in the cavity. The conformation and the stability of the imprinted cavities are related to the mechanical properties of the network and depend to a great extent on the cross-linker proportion. Most imprinted systems require 50–90% of cross-linker to prevent the polymer network from changing the conformation adopted during synthesis.[23] Consequently, the chances to modulate the affinity for the template are very limited, and it is not foreseeable that the network will have regulatory or switching capabilities. The lack of response

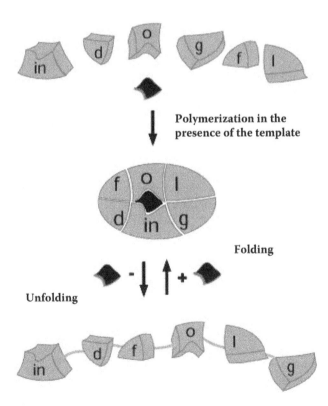

Fig. 1 Intuitive view of the recognition process of a template by an imprinted smart hydrogel based on the educative model drawn by Grosberg and Khokhlov.
Source: From Grosberg et al.[11]

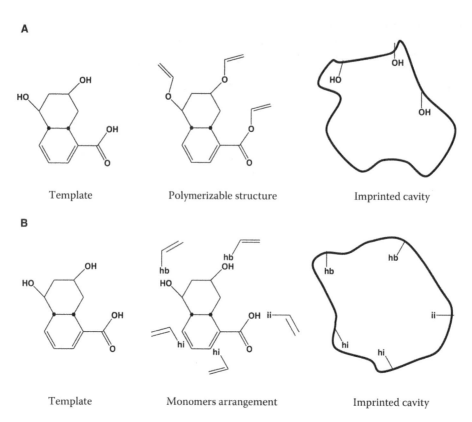

Fig. 2 Schematic view of the imprinting process: (**A**) covalent approach, in which the template is covalently bound to polymerizable binding groups that are reversibly broken after polymerization; and (**B**) noncovalent approach, in which the template interacts with functional monomers through noncovalent interactions (e.g., ionic interaction, ii; hydrophobic interaction, hi; or hydrogen bond, hb) before and during polymerization.

to changes in the physicochemical properties of the medium within the biological range or to the presence of a specific substance notably limits their utility in the biomedical field. A high cross-linker proportion also considerably increases the stiffness of the network, making it difficult to adapt its shape to a specific device or to living tissues.

Imprinting in Gels

A further step in the imprinting technology is the development of stimuli-sensitive imprinted hydrogels. The low-cross-linked proportion required to achieve adequate viscoelastic properties can compromise the stability of the imprinted cavities in the hydrogel structure, resulting in some sacrifice of both affinity and selectivity. Strong efforts are being made to adapt the molecular imprinting technology to materials that are more flexible and thus more biocompatible, such as hydrogels for use in the biomedical field.[24–30] To produce low cross-linked hydrogel networks capable of undergoing stimuli-sensitive phase transitions, the synthesis is carried out in the presence of template molecules, each able to establish multiple contact points with functional monomers.[31–35] Multiple contacts are the key for strong adsorption of the template molecules because of the larger energy decrease upon adsorption as well as the

higher sensitivity due to the greater information provided for recognition. As in the classical noncovalent approach, the monomers and the template molecules are allowed to move freely and settle themselves into a configuration of thermodynamic equilibrium. The monomers are then polymerized in this equilibrium conformation at the collapsed state. As the hydrogel is made from the equilibrium system by freezing the chemical bonds forming the sequence of monomers, we might expect such a hydrogel to be able to return to its original conformation, at least to some degree of accuracy, upon swelling–collapse cycles in which the polymerized sequence remains unchanged (Fig. 3). If the memory of the monomer assembly at the template adsorption sites is maintained, truly imprinted hydrogels will result. The combination of stimuli sensitivity and imprinting can have considerable practical advantages: the imprinting provides a high loading capacity of specific molecules, while the ability to respond to external stimuli modulates the affinity of the network for the target molecules, providing regulatory or switching capability of the loading/release processes.

This entry focuses on the sorption/release properties of imprinted smart gels compared with those of random (non-imprinted) heteropolymer gels. We first discuss the theoretical ideas behind the adsorption properties of random heteropolymer gels and review the experiments that have been done on adsorption of target molecules by random gels, detailing the dependence of the affinity on structural and environmental factors.[36–39] We then show the early successes of the imprinting method and highlight the advantages of the imprinted gels compared with random gels.

THE TANAKA EQUATION

Tanaka and coworkers[31–35] were pioneers in proposing the creation of stimuli-sensitive gels with the ability to

recognize and capture target molecules using polymer networks consisting of at least two species of monomers, each having a different role. One forms a complex with the template (i.e., the functional or absorbing monomers capable of interacting ionically with a target molecule), and the other allows the polymers to swell and shrink reversibly in response to environmental changes (i.e., a smart component such as N-isopropylacrylamide, NIPA) (Fig. 4). The gel is synthesized in the collapsed state and, after polymerization, is washed in a swelling medium. The imprinted cavities develop affinity for the template molecules when the functional monomers come into proximity, but when they are separated, the affinity diminishes. The proximity is controlled by the reversible phase transition that consequently controls the adsorption/release of the template (Fig. 3). A systematic study of the effects of the functional monomer concentration and the cross-linker proportion of the hydrogels (for both imprinted and nonimprinted gels) and of the ionic strength of the medium on the affinity of the hydrogels for different templates led to the development of an equation, called the Tanaka equation in memory of the late Professor Toyoichi Tanaka (1946–2000).[39]

Theoretical Considerations

The Tanaka equation[39] relates the gel affinity for the target molecules with the aforementioned variables as follows:

$$\text{Affinity} \approx \frac{[\text{Ad}]p}{p[\text{Re}]^p} \exp(-p\beta\varepsilon) \exp\left(-(p-1)c\frac{[\text{X1}]}{[\text{Ad}]^{2/3}}\right) \quad (1)$$

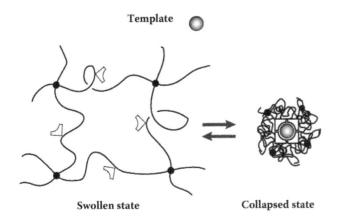

Fig. 3 The volume phase transition of the hydrogel—induced by external stimuli such as a change in pH, temperature, or electrical field—modifies the relative distance of the functional groups inside the imprinted cavities. This alters their affinity for the template. The affinity is recalled when the stimulus reverses and the gel returns to its original conformation.

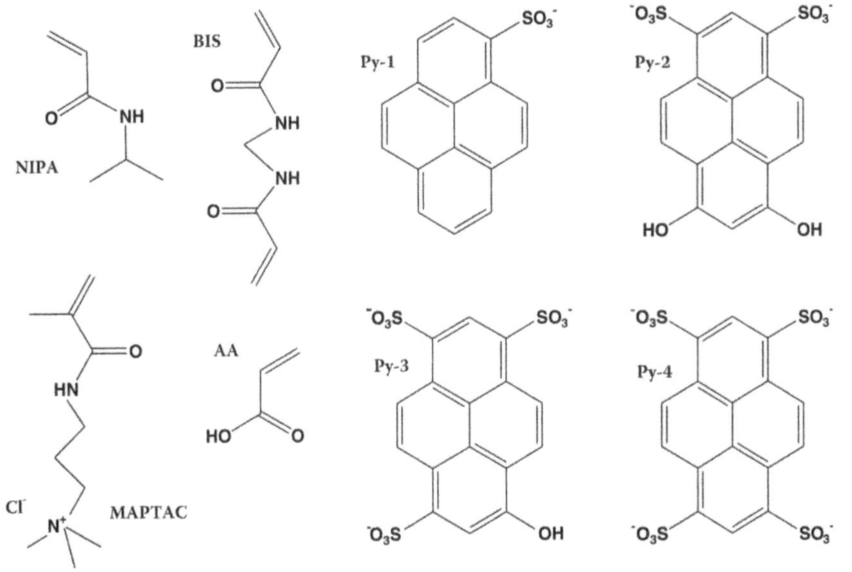

Fig. 4 Chemical structures of some monomers used to create imprinted smart gels: N-isopropylacrylamide (NIPA, temperature-sensitive), N,N'-methylene-bisacrylamide (BIS, cross-linker), methacrylamidopropyl trimethylammonium chloride (MAPTAC, cationic adsorber), and acrylic acid (AA, anionic adsorber). Structures of some ionically charged derivatives of pyranine used as targets are also shown.

where [Ad] represents the concentration of functional monomers in the gel; [Re] is the concentration of replacement molecules, i.e., ions that are bound to the target molecule when it is not bound to the functional monomers (in cases where they have ionic or protonized groups); [Xl] is the concentration of cross-linker; p is the number of bonds that each template can establish with the functional monomers; β is the Boltzmann factor ($1/k_B T$); ε is the difference between the binding energy of an adsorbing monomer to the target molecule and that of a replacement molecule to the target molecule; and c is a constant that can be estimated from the persistence length and concentration of the main component of the gel chains (e.g., NIPA). The main assumption in Eq. 1 is that the adsorption of target molecules is dominated by one value of p at each state of the gel. The value of p changes from 1 in the swollen state to p_{max} in the collapsed state, where p_{max} is the number of functional monomers that simultaneously interact with the target molecule.

The affinity of a network for a given target molecule is usually determined through the analysis of sorption isotherms. The suitability and limitations of different binding models, such as those developed by Langmuir, Freun-dlich, or Scatchard, have been discussed in detail elsewhere.[40–42] The Langmuir isotherm is derived from the point of view of the binding sites, where each binding site can be filled or unfilled:

$$[T_{ads}] = S\frac{K[T_{sol}]}{K[T_{sol}]+1} \tag{2}$$

where $[T_{ads}]$ is the concentration of target molecules adsorbed into the gel, $[T_{sol}]$ is the concentration of target molecules in solution, and S is the concentration of binding sites. K is the binding constant, with units of inverse concentration, and indicates the affinity per binding site. The overall affinity of the binding sites in the gel for the target molecule is defined to be the product SK (also denoted as Q), which is dimensionless.

The overall affinity SK can be determined from the partition function that sums over the different possible states of the target molecule: 0 adsorbers bound, 1 adsorber bound,..., p_{max} adsorbers bound. The partition function will be of the form:

$$Z = Z_0 + Z_1 + Z_2 + \cdots + Z_{P\max} \tag{3}$$

where Z_p indicates the term of the partition function in which a target molecule is bound by p adsorbing monomers, and Z_0 corresponds to the case of the target molecule being completely unbound. In the Langmuir equation, the term $K[T_{sol}]$ is proportional to the fraction of filled binding sites, and $SK[T_{sol}]$ is proportional to the concentration of bound target molecules. Compared with Eq. 3, we see that

$$SK[T_{sol}] \propto Z_0 + Z_1 + Z_2 + \cdots + Z_{P\max} \tag{4}$$

The partition function component Z_0 must be proportional to the number of target molecules in solution (i.e., $[T_{sol}]$). Therefore

$$SK \propto (Z_1 + Z_2 + \cdots + Z_{P\max})/Z_0 \tag{5}$$

To calculate each of the terms Z_p, a "fixed-point model" was developed. We can consider the case of a cross-linked gel made of a major component (i.e., a smart one, such as NIPA) and some functional monomers, prepared in the absence of templates (i.e., nonimprinted). The polymerized monomers can move in the gel to some extent, but are also constrained by the connectivity of the chains. When the cross-linking density is high, there are many constraints on the motion of the monomers. Conversely, at low cross-linker densities, monomers can diffuse more freely. The length scale of monomer localization is determined by the concentration of cross-linker in the gel.

If the gel is immersed in the solution of target molecules and these can diffuse into and out of the gel to form an adsorption binding site of p adsorbing monomers, the monomers have to move in space to properly group together. However, their motions are severely restricted because almost every adsorbing monomer in the gel belongs to a subchain, which means it is connected by the polymer to two cross-links (there might also be a few adsorbing groups on the dangling ends, connected to only one cross-link). Apart from real cross-links, the freedom of subchains is also restricted by the topological constraints, such as entanglements between polymers.

From a qualitative and very simplified approach, one particular adsorbing monomer in the gel can only access some relatively well-defined volume. It is reasonable to assume that in the center of this volume the adsorbing group is free to move, but approaching the periphery of its spherical cage the group feels increasing entropic restrictions. This can be modeled by saying that for each adsorbing monomer in the gel there is a point fixed in space, which is the center of the cage, and then there is a free-energy-potential well (of entropic origin) around this center. Every adsorbing monomer is attached to its corresponding self-consistent center by an effective polymer chain, like a dog on a leash (Fig. 5). The length of the effective chain must be of the same order as the distance between cross-link points. Thus, it will be a decreasing function of the cross-link density. In a simplified way, we can imagine that each one of the adsorbing monomers is at the end of one of these effective chains. At the other end of the chain is one of these points fixed in space, the positions of which are distributed randomly in the gel. Each chain is assumed to be made of n links, where n is inversely proportional to the cross-linking density of the gel, based on the concept that additional cross-links increase the frustration in the gel. The parameter n should be proportional to the ratio of main-component monomers to cross-linker monomers.

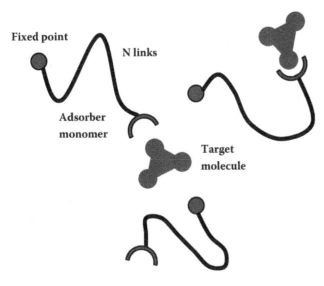

Fixed point

N links

Adsorber monomer

Target molecule

Fig. 5 Schematic drawing of the fixed-point model that places each adsorber monomer at the end of a finite chain of length n, the value of which is inversely proportional to the cross-linker density. In this example, the target molecule can interact with 1–3 adsorber monomers ($p_{max} = 3$).

An advantage of the fixed-point model is that it allows one to determine the entropic properties of the network using the well-known statistics of polymer chains. Adsorption of target molecules in the gel will deform the chain network, and the accompanying entropy loss can be analyzed via the entropy of Gaussian chains. This entropic effect will be affected not only by the cross-linker density, but also by the density of adsorber monomers, which are implicit in the definition of a fixed point. Note that this density can be adjusted via the gel-swelling phase transition. In the swollen state there will be a low density of fixed points, and in the collapsed state there will be a high density.

The dependences shown in Eq. 1 can be qualitatively explained using this model as follows (Figs. 5 and 6).

1. Power-law dependence of the affinity on [Ad]. For a target molecule to be adsorbed, the p adsorbing monomers must be clustered together to simultaneously bind it. The probability of such a cluster existing at a given point in a random (nonimprinted) gel is a product of the probabilities for each of the adsorbing monomers. Therefore, the dependence goes as $[Ad]^p$. Each of these clusters requires p adsorbing monomers; hence SK is proportional to $1/p$.

2. Power-law dependence of the affinity on [Re]. The replacement molecules (typically salt ions) act as competitors to the adsorbing monomers. In solution, a target molecule can either be adsorbed into the gel or bound by p replacement molecules. Binding to the replacement molecules prevents adsorption by the gel. Similar to a mass action law, the p replacement molecules must cluster around the target. This creates a power-law dependence similar to that for target molecule adsorption by adsorbing monomers, but with an opposite sign exponent ($[Re]^{-p}$). Note that it was assumed that each replacement molecule binds

to one site on the target molecules, as the adsorbing monomers do. If the adsorber monomers and the replacement molecules have different valences, these exponents should be modified.

3. The term $\exp(-\beta p\varepsilon)$ represents the enthalpy contribution to the sorption, i.e., the attraction energy of a target molecule to p adsorbing monomers. This is a Boltzmann probability based on a binding energy ε per adsorbing monomer.

4. The term $\exp[-c(p - 1)[Xl]/[Ad]^{2/3}]$ summarizes the entropy restric-tions to the sorption, which are mainly related to the cross-linking density, as mentioned above. The adsorber units in the gel can move rather freely within a certain volume determined by the cross-linking density. Below a certain length scale associated with the cross-linking density, the gel behaves like a liquid, allowing the adsorber groups to diffuse almost freely. Beyond that length scale, however, the gel behaves as an elastic solid body. The adsorber units cannot diffuse further than that length scale. As shown earlier in Fig. 5, we assume that each adsorber is at one end of a fictitious Gaussian chain with a length half the average polymer length between the nearest cross-links, which can be estimated as follows:

$$l = nb = ([NIPA]/2[Xl])a \qquad (6)$$

where n is the number of monomer segments of persistent length b contained in the chain. In the case of a bifunctional cross-linker, i.e., with two polymerizable groups, there are $([NIPA]/2[Xl])$ monomers between the cross-link point and an adsorbing monomer group. Then, $n = ([NIPA]/2m[Xl])$ and $b = ma$, where m is the number of monomers involved in the persistent length, and a is the length of each monomer. At a concentration [Ad] of adsorbing monomers, the average spatial distance between adsorbing monomers is $R = [Ad]^{-1/3}$. For a molar concentration C_{ad}, this corresponds to $R = 1 \text{ cm}/(C_{ad}N_A)^{1/3}$, where N_A is the Avogadro number. This fictitious Gaussian chain represents the restricted ability of the adsorber groups to diffuse within a certain volume in the gel. We expect that the probability for two adsorber monomers to meet should be proportional to the Boltzmann factor of the entropy loss associated with the formation of one pair of adsorbers

$$P = P_0 \exp(-R^2/nb^2) = P_0 \exp(-c[Xl]/[Ad]^{2/3}) \qquad (7)$$

where the quantity c is determined by the persistence length, the number of monomers in a persistence length, and the concentration of the main component of the chains through the relation

$$c = 2m/([NIPA]b^2) \qquad (8)$$

Since the adsorption of a divalent target by two adsorbers brings together each end from two fictitious Gaussian polymers, the affinity should be proportional to this

Scaffolds—Smart

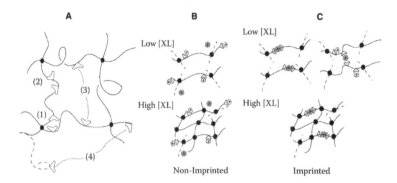

Fig. 6 Binding site formation and gel microstructure. (**A**) Candidate adsorber pairs can be located near a cross-link (1), on a single chain (2), and on distant chains without (3) and with (4) intervening entanglements. (**B**) With increasing cross-linking density [XL], adsorbers are frustrated in non-imprinted gels. (**C**) In imprinted gels, flexibility at low [XL] leads to competition for targets, with possible topological consequences (mispairings) that diminish at high [XL].
Source: From Stancil et al.,[35] with permission from the American Chemical Society.

probability. If more than two contact points are expected, the equation can be generalized as

$$SK \propto \exp[-c(p-1)[X1]/[Ad]^{2/3}] \qquad (9)$$

where p is the number of contact points. If the target molecule is adsorbed only by a single contact ($p = 1$), the affinity should be independent of the cross-linker concentration. In contrast, if the target binding site requires several adsorber monomers ($p > 1$), the cross-links may frustrate the formation of the binding site, and the frustration will increase with p. This will be particularly significant at low concentrations of [Ad] (since only a small fraction of the adsorber monomers will be close enough together to participate in multiple binding sites) and high cross-linking concentration. It should also be taken into account that the adsorption process itself can increase the cross-linking density, since the complexes will act as tie junctions of polymeric chains. This could modify the value of n as the sorption is going on. Nevertheless, usually this last contribution to the cross-linking is low enough to be negligible.

It can be shown by consideration of the relative weights of the terms Z_p that for a given set of experimental parameters, the affinity SK will almost always be dominated by a single value of p, either $p = 1$ or $p = p_{max}$. This can be understood by considering whether the attraction of the target molecule to adsorber monomers (due to energetic and concentration effects) is stronger than the repulsion due to the entropy loss required to deform the gel. If attraction is sufficiently favored, then the target will be bound by as many adsorber monomers as possible and, therefore, $p = p_{max}$. However, if the entropy loss to deform the gel is stronger, then the adsorption will only be possible by single adsorber monomers, i.e., $p = 1$, which would not require deformation of the network (no chain entropy penalty).

The basic concept of gels as smart materials is that they will have high affinity for the target in the collapsed state, but low affinity in the swollen state. By controlling the phase transition of the gel, one will be able to create a switchlike behavior in the affinity. The Tanaka equation allows one to predict the composition of gels that will drastically change affinity during the gel phase transition. If the gel is to have a low affinity in the swollen state, the adsorber monomers should have only a weak attraction to the target molecules, i.e., any adsorption should be single-handed ($p = 1$). To have a high gel affinity, adsorption in the collapsed phase should involve as many adsorber monomers as possible ($p = p_{max}$). The p value transition should occur where the entropic and energetic contributions to the affinity are equal, i.e., the crossover should occur when

$$\ln([Ad]/[Re]) \approx ([Ad]^{2/3} nb^2)^{-1} + \beta\varepsilon \qquad (10)$$

A more detailed description of the Tanaka theory has been published by Ito et al.[39,40]

The experiments discussed below use gels in which p changes across the phase transition. However, the design of such gels still requires a significant amount of research, since it is difficult to know *a priori* the exact value for the binding energy ε.

Since the Tanaka equation is based on the idea that there are many sets of adsorber monomers that work together locally to form binding sites for the target molecules and that the probability for chain stretching drops off exponentially with distance, it can also explain the greater affinity of imprinted gels compared with the nonimprinted ones. The synthesis in the presence of the target molecules leads to the distribution of the adsorbing monomers in groups of the p members required for the binding of each target molecule. In the imprinted gels, the p members are closely fixed during polymerization in the collapsed state due to the template. The template is removed from the gel in the swollen state because of the deformation of the binding sites. Once again, in the collapsed state, the binding sites can be reconstructed, and since each binding site possesses the needed p adsorbing monomers close together, the entropic restrictions to the sorption should be minimized (Fig. 6). Representative examples of the enhancement in affinity observed for imprinted smart gels are given in the following subsections.

Experimental Assessment of the Tanaka Equation

Among the studies carried out to assess the Tanaka equation in nonimprinted heteropolymer gels,[36–38,43] those carried out with pyrene derivatives as target molecules are particularly representative.[36–39] The gels were prepared by free-radical polymerization using $6M$ NIPA as the stimuli-sensitive component, 0–120 mM methacrylamidopropyl ammonium chloride (MAPTAC) as functional monomer, and 5–200 mM N,N'-methylene bis(acrylamide) (BIS) as cross-linker. The monomers were dissolved in dimethyl-sulfoxide and polymerized inside micropipettes (i.d. 0.5 mm). Once washed with water, all gels were collapsed at 60°C and showed the same degree of swelling as during polymerization. As adsorbates or target molecules, several different types of pyrene sulfonate derivatives were used: 1-pyrene sulfonic acid sodium salt (Py-1·Na), 6,8-dihy-droxy-pyrene-1,3-disulfonic acid disodium salt (Py-2·2Na), 8-methoxy pyrene-1,3,6-trisulfonic acid trisodium salt (Py-3·3Na), and 1,3,6,8-pyrene tetrasulfonic acid tetra-sodium salt (Py-4·4Na), portrayed in Fig. 4. These chemicals present 1 (Py-1), 2 (Py-2), 3 (Py-3), or 4 (Py-4) anionic charges, which can interact electrostati-cally with a cationic charged site such as on MAPTAC. Pieces of cylindrical gel (5–20 mg dry weight) were placed in 2- or 4-ml target aqueous solution, the concentration of which ranged from 2–0.5 mM. The solutions also contained NaCl of a prescribed concentration (27–200 mM) to provide chloride ions to replace the target molecules. The samples were kept swollen (20°C) or shrunken (60°C) for 48 hr, and the

adsorption isotherms were analyzed in terms of the Langmuir equation (Eq. 2) to calculate the affinity SK.

Figure 7 shows the dependence of affinity for Py-3 and Py-4 on the MAPTAC concentration. Above a certain MAPTAC concentration (20 mM) and in the collapsed state, both the log–log plots show a straight line, with slope 3 for Py-3 and with slope 4 for Py-4. These power-law relationships are due to three and four adsorption points, respectively. Adsorption sites are formed when three (or four) equivalent adsorbing molecules (MAPTAC) gather to capture one Py-3 (or Py-4) molecule. The obtained power laws are in agreement with the Tanaka equation. At MAPTAC concentrations below 10 mM, the major component of the gel (NIPA) contributes more to the adsorption of pyranine (due to a hydrophobic interaction) than do the MAPTAC groups, and the power-law exponent becomes zero. In the swollen state, the log–log slope becomes 1, indicating that MAPTAC adsorbs the target molecule with a single contact. Single-point adsorption is favored because the MAPTAC monomers are well separated one from another, and it becomes entropically unfavorable for the multipoint adsorption complex to assemble. The slope returns to 3 or 4 upon shrinking, indicating recovery of the multipoint binding sites. The reversible adsorption ability is controlled by the volume phase transition.

The effect of external salt concentration on the binding of target molecules to the gel is shown in Fig. 8. As mentioned above, when these low-cross-linked gels are in the collapsed state, a number of MAPTAC groups equal to the number of charges of the target (p_{max}) can gather to form a binding site. Since the attraction is electrostatic, coexistent ions in the solution can make the target molecule adsorption difficult. Ions with the same charge sign as the target molecules should compete with the target molecules for

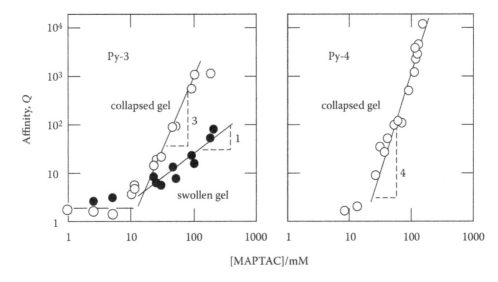

Fig. 7 Dependence of the affinity for Py-3 and Py-4 on the concentration of adsorber monomer. In this experiment, an amount (100 mM) of salt was added to the system to sweep out the Donnan potential.
Source: From Oya et al.,[36] with permission from the American Association for the Advancement of Science.

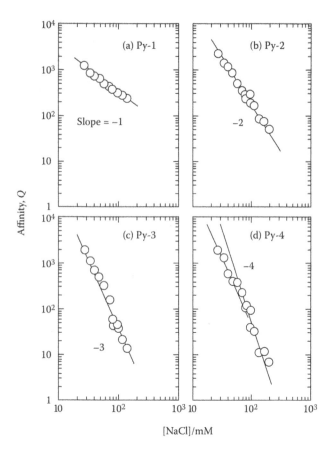

Fig. 8 Dependence of the affinity for pyranines on the replacement ion concentration. The solid line displays a slope with the value shown in each plot. In this experiment, the concentrations of the adsorber (MAPTAC) and the cross-linker (BIS) were fixed at 40 and 10 mM, respectively.
Source: From Watanabe et al.,[37] with permission from the American Institute of Physics.

binding with the adsorber monomers. In the cases Py-1, -2, and -3, the log–log plot showed slopes of −1, −2, and −3, respectively, i.e., −p_{max}, where p_{max} is the number of charged groups on the target molecule (affinity $\propto$ [Re]$^{-p}$). Py-4 followed a similar behavior, though there was a small discrepancy below 40 mM of salt concentration, probably because of the increasing effect of the Donnan potential at lower salt concentrations. In the swollen state ($p = 1$), the log–log plots showed slopes of −1. Therefore, the Tanaka equation fits the results quite well in a wide range of salt concentrations.

As explained above, the multipoint adsorption should lead to an entropy loss in the polymer chains that are bound together by the target molecules. The effect of this entropy loss, which is a function of the concentration of cross-linker and adsorber, is to reduce the affinity of the gel for the target. Several experimental studies have shown that the affinity of a heteropolymer gel involved in multiple contact adsorption decreases with its cross-linking degree. Hsein and Rorrer[44] showed an exponential decrease in calcium adsorption by chitosan as the extent of cross-linking increases. Eichenbaum et al.[45] found that, for alkali earth metal binding in methacrylic acid-co-acrylic acid microgels, the cross-links prevent the carboxylic groups from achieving the same proximity as in a linear polymer, which affects the binding properties of the metals. In the case of a thermosensitive gel, the influence of the degree of cross-linking has also been shown.[24,31]

The volume phase transition of a NIPA gel induced by a stimulus is responsible for separating the adsorber monomers (e.g., MAPTAC) in the swollen state, which decreases their probability of coming close to each other to adsorb a multicharged target. Consequently, in the swollen state, the affinity for the target increases slightly with the degree of cross-linking, since the degree of swelling is reduced. In contrast, in the collapsed state an exponential decrease of the affinity with the cross-linking, which was especially significant for the cases of contact numbers above 2, was observed (Fig. 9). The affinity values predicted by the Tanaka equation agree well with those experimentally obtained. This unfavorability for the affinity has been understood to be a "frustration" to the mobility of the adsorbing sites.[31,46,47] The frustration can also be viewed in terms of the "flexibility" of the polymer chains, which is critical for allowing the adsorber groups to come into proximity for a multipoint adsorption. A detailed analysis of the data shown in Fig. 9 revealed that a plot of the exponential decay rate vs. the number of contact points gives a linear relationship in which the slope is 0.32. This slope associates with the parameter c in Eq. 1, and can be used to calculate the persistence length $b = ma$ for the polymer chains, applying Eq. 8. For $c = 0.32$, the value of b equals 2.9 nm. The persistent length is theoretically predicted to be around 2 nm, e.g., $\approx$10 monomers ($m = 10$) of 2 Å ($a = 2 \times 10^{-10}$ m).[48] Thus, the obtained result is somewhat reasonable. Therefore, the theory predicts and explains well the exponential decay with concentration of the cross-linker. The cross-links and polymer connections create frustrations so that the adsorber groups (MAPTAC) cannot lower the energy of the polymer by forming pairs, triplets, or groups of p members for capturing target molecules. As will be shown below, such frustrations can be overcome using molecular imprinting.

TEMPERATURE-SENSITIVE IMPRINTED HYDROGELS

Like proteins, a heteropolymer gel can exist in four thermodynamic phases:

1. Swollen and fluctuating
2. Shrunken and fluctuating
3. Shrunken and frozen in a degenerate conformation
4. Shrunken and frozen in the global minimum energy conformation

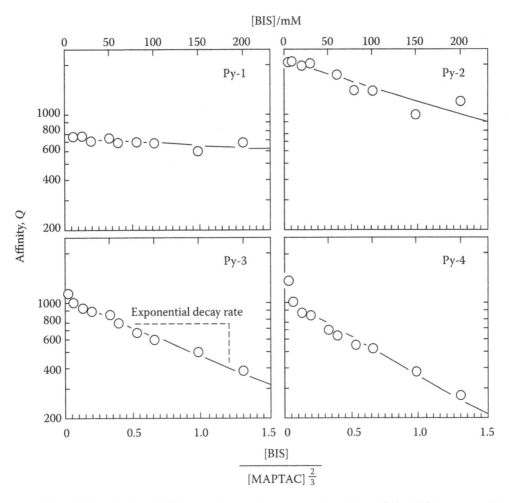

Fig. 9 Dependence of the affinity of gels at 60°C for pyranines on the concentration of cross-linker BIS (upper *x*-scale) or on the ratio of cross-linker concentration to that of adsorber monomer to the two-thirds power (lower *x*-scale).
Source: From Ito et al.,[39] with permission from Elsevier.

The order parameter that describes the phase transition between the first and the second phases is the polymer density or, equivalently, the swelling ratio of the gel. The third and fourth phases are distinguished by another order parameter: the overlap between the frozen conformation and the minimum energy conformation. In the third phase, the frozen conformation is random.[7] In the fourth phase, the frozen conformation is equivalent to that of the global energy minimum. Proteins in this fourth phase take on a specific conformation, which may be capable of performing catalysis, molecular recognition, or many other activities. Tanaka and colleagues strove to recreate such a fourth phase in gels by designing a low-energy conformation and then testing whether the gel could be made to reversibly collapse into this "memorized" conformation. According to developments in the statistical mechanics of polymers, to achieve the memory of conformation by flexible polymer chains, several requisites must be satisfied.[49]

1. The polymer must be a heteropolymer, i.e., there should be more than one monomer species, so that

some conformations are energetically more favorable than others.

2. There must be frustrations that hinder a typical polymer sequence from being able to freeze to its lowest energy conformation (as considered in absence of the frustration). Such frustrations may be due to the interplay of chain connectivity and excluded volume, or they may be created by cross-links. For example, a cross-linked polymer chain will not freeze into the same conformation as the non-cross-linked polymer chain, at least for most polymer sequences.

3. The sequence of monomers must be selected so as to minimize these frustrations,[50] i.e., a particular polymer sequence should be designed such that the frustrating constraints do not hinder the polymer from reaching its lowest energy conformation.

These three conditions allow the polymer to have a global free-energy minimum at one designed conformation.

The nonimprinted gels described in previous sections satisfy the first two conditions, and can be engineered to satisfy the third. The adsorber and the main-component monomers provide heterogeneity in interaction energies, since adsorber interactions are favorably mediated by the target molecules.[51] Frustrations to the achievement of the global energy minimum exist due to the cross-links in the gel as well as chain connectivity. To meet the third condition, the minimization of the frustration, the molecular imprinting technique can be particularly useful. In this section, we discuss the early experimental successes of the imprinting method and the application of the Tanaka equation to imprinted gels.

Two approaches were investigated to achieve elements of conformational memory in gels:

1. Cross-linking of a polymer dispersion or preformed gel in the presence of the target molecule (two-step imprinting or post-imprinted)
2. Simultaneous polymerization and cross-linking of the monomer solution (one-step imprinting)

The first approach was studied in heteropolymer gels consisting of NIPA as the major component sensitive to stimuli and MAPTAC as the charged monomer able to capture pyranine target molecules.[24] The polymer networks were randomly copolymerized, in the absence of pyranines, with a small quantity of permanent cross-links and thiol groups (–SH). The gels were then further cross-linked by connecting thiol group pairs into disulfide bonds (–S–S–). The process was carried out directly (nonimprinted gels) or after the gels were immersed in a solution of pyranines and had all charged groups forming complexes with the target molecules (postimprinting). These post-cross-links were still in very low concentrations,

in the range 0.1–3 mol%, and therefore the gels could freely swell and shrink to undergo the volume phase transition. The postimprinted gels showed higher affinity for the target than those that were randomly post-cross-linked (Fig. 10). However, this post-cross-linking approach has a fundamental drawback. Before the post-cross-linking, the sequence of the components has already been determined and randomly quenched. The minimization of the frustration is allowed only in the freedom of finding best partners among –SH groups. For this reason, the imprinting using a post-cross-linking technique can give only a partial success.

Ideally, the entire sequence of all monomers should be chosen so that the system will be in its global energy minimum. The complete minimization of the frustration can be achieved by polymerizing monomers while they self-organize in space at a low-energy spatial arrangement.[31,32,35] This second approach controls the sequence formation, allowing the monomers to equilibrate in the presence of a target molecule. These monomers are then polymerized with a cross-linker. It is hoped that, upon removal of the template species, binding sites with the spatial features and binding preferences for the template are formed in the polymer matrix. The choice of functional monomers and the achievement of an adequate spatial arrangement of functional groups are two of the main factors responsible for specificity and reversibility of molecular recognition.

The first experiments carried out by Alvarez-Lorenzo et al.[31,32] used gels prepared by polymerization of NIPA, small amounts of methacrylic acid (MAA), and BIS in the absence (nonimprinted) or the presence of divalent cations. MAA was used as the functional monomer able to form complexes in the ratio 2:1 with divalent ions. The effect of

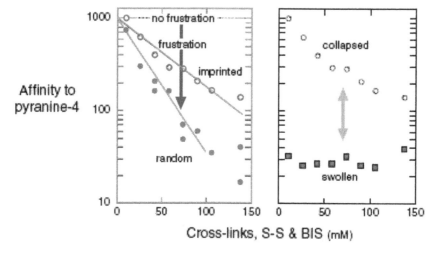

Fig. 10   The affinity for Py-4 is shown as a function of S–S bonds plus BIS (10 mM) concentration within the gels in the collapsed state at 60°C. For random gels, the affinity decays exponentially with S–S concentration (the solid line is a guide to the eye). The loss of ability to form complexes indicates the frustration due to cross-links. For the postimprinted gels, the affinity increases 3–5 times at higher S–S concentrations. This indicates the partial resolution of frustration. When the gels are allowed to swell at 25°C, the affinity becomes very low and independent of S–S concentration, indicating the destruction of the adsorption sites upon gel swelling (right panel). When the imprinted gels are allowed to collapse again, the pyranine adsorption resumes its original high values, indicating the restoration of the original complexes with the same MAP-TAC monomers.
Source: From Enoki et al.,[24] with permission from the American Institute of Physics.

temperature on the adsorption capacity of the imprinted copolymers prepared with different template ions and in different organic solvents was compared with that of the nonimprinted ones. Successful imprinting was obtained using calcium or lead ions as template. After removing the template and swelling in water at room temperature, the affinity for divalent ions notably decreased. When the gels were shrunken by an increase in temperature, the affinity was recovered (Fig. 11). The measurements of the affinity suggested that multipoint adsorption occurs for both imprinted and nonimprinted gels in the collapsed state (saturation $S \approx [Ad]/2$), but that in the imprinted gel, the multipoint adsorption is due to memorized binding sites. After recollapsing, two-point adsorption is recovered in the random gels with a power law of $SK \propto [Ad]^2$, while the

imprinted gels showed a stronger affinity with a power law of $SK \propto [Ad]$.[1] The difference in the power dependence of the affinity is due to a different [Ad] dependence for K. The affinity per adsorption site is actually independent of [Ad] for imprinted gels. This is because the adsorbers are ordered such that they already form sites with their unique partners. Hence, the adsorbers are ordered in groups of two to give the highest possible K at all values of [Ad]. Furthermore, the affinity of the imprinted gels does not change at all with an increase of the cross-linking density, while the affinity decreases for random gels (Fig. 12). The nondependence of the affinity on [Xl] proves the minimization of the frustration with regard to the cross-links and chain connectivity. Thus, the requirements 1, 2, and 3 necessary for obtaining conformational memory have been achieved.

In these hydrogels the relative proportion of functional monomers, compared with the other monomers, is quite low (1 mol%), and it should be difficult for the gel to take on a strong-binding conformation other than the one imprinted during synthesis. For random gels, it was difficult to pair randomly distributed MAA, and their affinity for divalent ions decreased exponentially as a function of cross-linker concentration. In contrast, in the imprinted gels, the local concentration of MAA in the binding site is very high, i.e., the members of each pair are closely fixed by the template during polymerization. If the gel did not memorize such pairs after washing the template out, swelling, and reshrinking, an MAA would have to look for a new partner nearby, and such a probability would be the same as that in a nonimprinted gel. Therefore, it can be concluded that the greater adsorption capacity of the imprinted gels comes from the successfully memorized MAA pairs.[32,35,52]

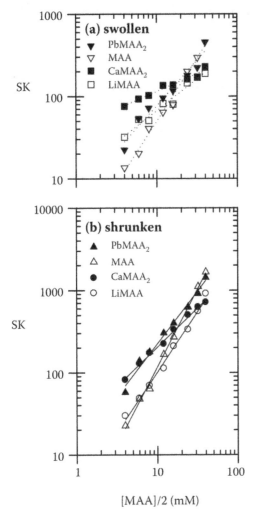

Fig. 11 The overall affinity SK of the nonimprinted and imprinted gels for calcium ions is plotted as a function of MAA monomer concentration for (**A**) the swollen state and (**B**) the shrunken state. The values of the slope for nonimprinted gels duplicate the slope of the imprinted ones prepared with Ca^{2+} as template or with the alternative template Pb^{2+}.
Source: From Alvarez-Lorenzo et al.,[32] with permission from the American Institute of Physics.

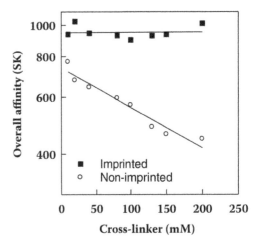

Fig. 12 Influence of the cross-linker (BIS) concentration on the overall affinity for calcium ions of the imprinted (full symbols) and nonimprinted (open symbols) NIPA (6 M) gels in the shrunken state in water. The concentration of functional monomers (MAA) was fixed at 32 mM.
Source: From Alvarez-Lorenzo et al.,[31] with permission from the American Chemical Society.

Güney, Yilmaz, and Pekcan[53,54] followed a similar procedure to prepare temperature-sensitive gels that specifically recognize and sorb heavy metal ions from aqueous media in an effort to develop chemosensors. Kanazawa et al.[55] used another functional monomer N-(4-vinyl)benzyl ethylenediamine) (Vb-EDA) that contains two nitrogen groups that can specifically form coordination bonds with one copper ion; each copper ion requires two functional monomers to complete its bonding capacity. This occurs when the gel is formed at the shrunken state. At the swollen state, the bonds are broken and the affinity for the ions disappears. These gels showed a high specificity for Cu^{2+} compared with Ni^{2+}, Zn^{2+}, or Mn^{2+}. The different coordination structure—square planar for Cu^{2+} and Ni^{2+}, tetrahedral for Zn^{2+}, and octahedral for Mn^{2+}—together with the differences in ion radii explains this specificity, which was not observed in the nonimprinted gels. The amphiphilic character of the functional monomer Vb-EDA made it possible to develop an emulsion polymerization procedure to prepare imprinted microgels. The main advantage of these microgels is that they can adsorb and release the copper ions faster because of their quick volume phase transitions.[56]

Detailed calorimetric studies of the thermal volume transition of polyNIPA (PNIPA) hydrogels and the influence of ligand binding on the relative stability of subchain conformations have been carried out by Grinberg and coworkers.[57-59] The dependence of the critical temperature of PNIPA hydrogels on the proportion of ionic co-monomers is an obstacle to obtaining devices with a high loading capability while still maintaining the PNIPA temperature-sensitive range.[2] This can be overcome by synthesizing interpenetrated polymer networks (IPN) of PNIPA with ionizable hydrophilic polymers. Although only a few papers have been devoted to this topic, the results obtained in those studies have shown the great potential of IPNs.[60,61] A two-step approach to imprint interpenetrated gels with metal ions

was identified by Yamashita et al.[62] It basically consists of 1) polymerization of AA monomers to have a loosely cross-linked (1 mol%) polyAA network; 2) immersion of polyAA in copper solution to enable the ions to act as junction points between different chains; and 3) transfer of polyAA-copper ion complexes to a NIPA solution containing cross-linker (9.1 or 16.7 mol%) and synthesis of the NIPA network in the collapsed state (Fig. 13). The nonimprinted IPNs (i.e., prepared in the absence of copper ions) showed a similar affinity for Cu^{2+} and for Zn^{2+}. In contrast, the imprinted IPNs in the collapsed state could discriminate between the square planar structure of Cu^{2+} and the tetrahedral structure of Zn^{2+}.

NIPA-based imprinted hydrogels have also been prepared using organic molecules as templates. Watanabe et al.[63] observed that NIPA (16 mmol)–acrylic acid (4 mmol) cross-linked (1 mmol) polymers synthesized in dioxane and in the presence of norephedrine (2 mmol) or adrenaline (2 mmol) showed, after template removal, an increase in the swelling ratio in the collapsed state as the target molecule concentration in water increases (Fig. 14). Since the molar concentration of the adsorbing monomers was much higher than the cross-linking density, the cross-links should

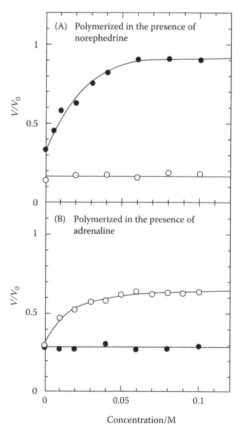

Fig. 14 Equilibrium swelling ratios at 50°C as a function of concentration of either norephedrine (solid symbols) or adrenaline (open symbols) for imprinted gels prepared in the presence of norephedrine (**A**) or adrenaline (**B**).
Source: From Watanabe et al.,[63] with permission from the American Chemical Society.

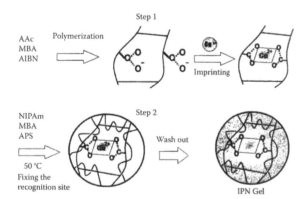

Fig. 13 Two-step procedure to obtain an interpenetrated system comprising a Cu^{2+} imprinted poly(acrylic acid) hydrogel and a poly(*NIPA*) temperature-sensitive hydrogel.
Source: From Yamashita et al.,[62] with permission from the Society of Polymer Science of Japan.

not have created frustration. Thus the imprinting effect may only be due to the phase separation caused by the template during polymerization, which is macroscopically evidenced as a change in the conformation when the network collapses. Liu et al.[64,65] obtained temperature-sensitive imprinted gels for 4-aminopyridine and L-pyroglutamic, which showed the same ability to sorb and release the drug after several shrinking–swelling cycles. These gels had significantly higher saturation and affinity constants than the nonimprinted ones and were also highly selective. These results indicate that temperature-sensitive imprinted gels have a potential application in drug delivery.

Natural polymers, such as chitosan, have also been evaluated as a basis for temperature-sensitive hydrogels instead of synthetic monomers.[66] Chitosan is an aminopolysaccharide (obtained from chitin) that can be chemically cross-linked through the Schiff base reaction between its amine groups and the aldehyde ends of some molecules, such as glutaraldehyde.[67] A recent study has shown that if the reaction is carried out in the presence of target molecules, such as dibenzothiophenes (DBT), imprinted networks with a remarkably greater adsorption capability than nonimprinted ones can be obtained. This effect was particularly important when the gel was collapsed in the same solvent (acetonitrile) and at the same temperature (50°C) as were used during the cross-linking.[66] Additionally, the DBT-imprinted gels showed a high selectivity for the target molecules compared with other structurally related compounds (Fig. 15). These gels have been proposed as traps for organosulfur pollutants.

In general, stimuli-sensitive imprinted gels are very weakly cross-linked (less than 2 mol%) and, therefore, the success of the imprinting strongly depends on the stability of the complexes of template/functional monomers during polymerization and after the swelling of the gels. If the molar ratio in the complex is not appropriate or if the complex dissociates to some extent during polymerization, the functional monomers will be far apart from both the template and each other, and the imprinting will be thwarted. However if the interaction is too strong, it may be difficult to remove the templates completely, which leads to a reduction of the number of free binding sites and template bleeding during the assays.

Efforts have shown the possibilities of using adsorbing monomers directly bonded to each other prior to polymerization, which avoids the use of the template polymerization technique.[33,34] Each adsorbing monomer can be broken after polymerization to obtain pairs of ionic groups with the same charge. Since the members of each pair are close together, they can capture target molecules through multipoint ionic interactions (Fig. 16). The adsorption process was found to be independent of the cross-linking density, and the entropic frustrations were completely resolved. Divalent ions or molecules with two ionic groups in their structures can be loaded in a greater amount and with higher affinity by these "imprinted" hydrogels.[34] Furthermore, the hydrogels prepared with PNIPA and imprinters were gifted with a new ability not observed in common stimuli-sensitive gels: they can re-adsorb, at the

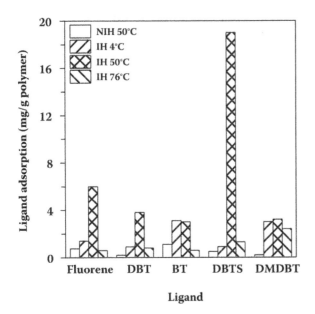

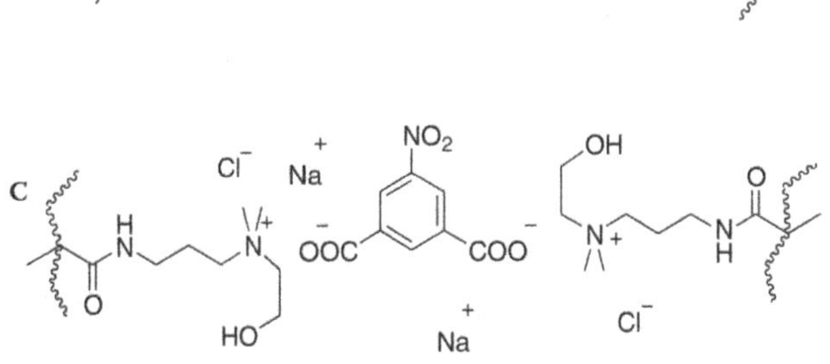

Fig. 15 Effect of temperature and ligand on adsorption by a nonimprinted (NIH) and DBTS-imprinted hydrogel (IH): benzothiophene (BT), dibenzothiophene (DBT), 4,6-dimethyl-DBT (DMDBT), and dibenzothiophene sulfone (DBTS). Cross-linking agent ratio 2:1 mol, 50°C. Uptake conditions: [ligand]/CH_3CN = 4 mM; 16 hr; 300 rpm.
Source: From Aburto & Le Borgne,[66] with permission from the American Chemical Society.

Fig. 16 Structure of Imprinter-Q monomer (**A**) and of the binding sites of the gel made with Imprinter-Q after breakage of the 1,2-glycol bond (**B**), and schematic representation of the capture of a target molecule (**C**).
Source: From Moritani & Alvarez-Lorenzo,[34] with permission from the American Chemical Society.

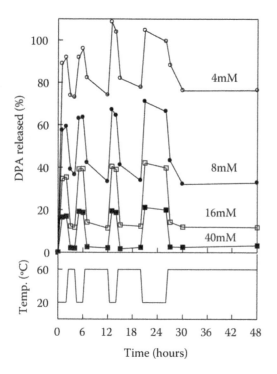

Fig. 17 Influence of temperature on the release and readsorption of disodium 5-nitroisophthalate (DPA) in water by imprinted NIPA (6 *M*) gels prepared with different concentrations of Imprinter-Q. Cross-linker concentration was 40 m*M*. Degrees of swelling at 20°C and 60°C were 6.0–6.5 and 0.9, respectively.
Source: From Moritani & Alvarez-Lorenzo,[34] with permission from the American Chemical Society.

shrunken state, a significantly high amount of the templates previously released at the swollen state. Common (nonimprinted) temperature-sensitive PNIPA hydrogels have a pulsate release behavior that allows a substance entrapped in the polymer network to diffuse out of the hydrogel in the swollen state, but then stops the release when the temperature increases and the network collapses.[68,69] In contrast, a change from the swollen to the shrunken state of the imprinted PNIPA hydrogels not only stops the release, but also promotes a re-adsorption process. This process occurs quickly and in a way that can be reproduced after several temperature cycles (Fig. 17). At 37°C, the high affinity for the template provokes a stop in the release when an equilibrium concentration between the surrounding solution and the hydrogel is reached.

This type of hydrogel has a great potential for development of drug-delivery devices capable of maintaining stationary drug levels in their environment. The gel would stop the release while the released drug has not yet been absorbed or distributed but remains near the hydrogel.

PH-SENSITIVE IMPRINTED GELS

The term "pH-sensitive imprinted gel" can comprise two different behaviors:

1. Imprinted gels with adsorber monomers that show an affinity for the target molecule that can be tuned by changes in pH, but that do not undergo phase transitions
2. Imprinted gels that contain protonizable groups capable of causing pH-induced phase transitions and adsorber groups responsible for the interactions with the target molecule

In the first group we could insert almost all temperature-sensitive gels described in the previous section, as well as many other examples of non-smart imprinted networks,[70] since the recognition relies upon strong ionic interactions between the functional monomers and the template. If a change in pH modifies the degree of protonization of their chemical groups (although no change in degree of swelling takes place), a strong change in binding energy and, therefore, in affinity will occur.[71] In this section, we will mainly focus on gels that undergo pH-sensitive phase transitions, these changes in volume being mainly responsible for the control of the sorption process. Nevertheless, in some cases, the pH can alter both the swelling and the binding energy.

Kanekiyo et al.[72,73] have developed pH-sensitive imprinted gel particles using hydrophobic interactions to sorb the target molecules. The method consists of using a polymerizable derivative of amylose capable of wrapping around a hydrophobic template, such as bisphenol-A. The helical inclusion complex formed is then copolymerized with a cross-linker and a monomer having ionizable groups (e.g., acrylic acid, AA) (Fig. 18). Similarly, other imprinted polymers were prepared with acrylamide instead of acrylic acid. The rebinding ability of MIPs prepared with AA showed a strong dependence on pH: the greater the pH, the lower the binding of bisphenol-A. These results indicate that the binding cavity created through the imprinting process is disrupted by a conformational change in the amylose chain arising from the electrostatic repulsion between the anionic groups. A decrease in pH restores the cavities and the binding affinity.

The group of Peppas has developed an imprinting procedure using star-shaped polyethylene glycols (PEGs) copolymerized with MAA for the recognition of sugars in aqueous solution.[74] Star polymers, also called hyperbranched polymers, have a large number of arms emanating from a central core and, therefore, they can contain a large number of functional groups in a small volume. The star-shaped PEGs provide the multiple hydrogen bonds required to interact with sufficient strength in water with sugars or proteins (Fig. 19). Two different star-PEGs were evaluated. Grade 423 with 75 arms (MW = 6,970) and Grade 432 with 31 arms (MW = 20,000). Imprinted networks were obtained by carrying out the polymerization in the presence of glucose. The imprinted 31-arm star-PEG gel showed a 213% increase in glucose sorption over the nonimprinted polymer (253 vs. 118 mg/g), while for the 75-arm star-PEG gels no improvement in sorption ability by the imprinting was observed (198 vs. 199 mg/g). This

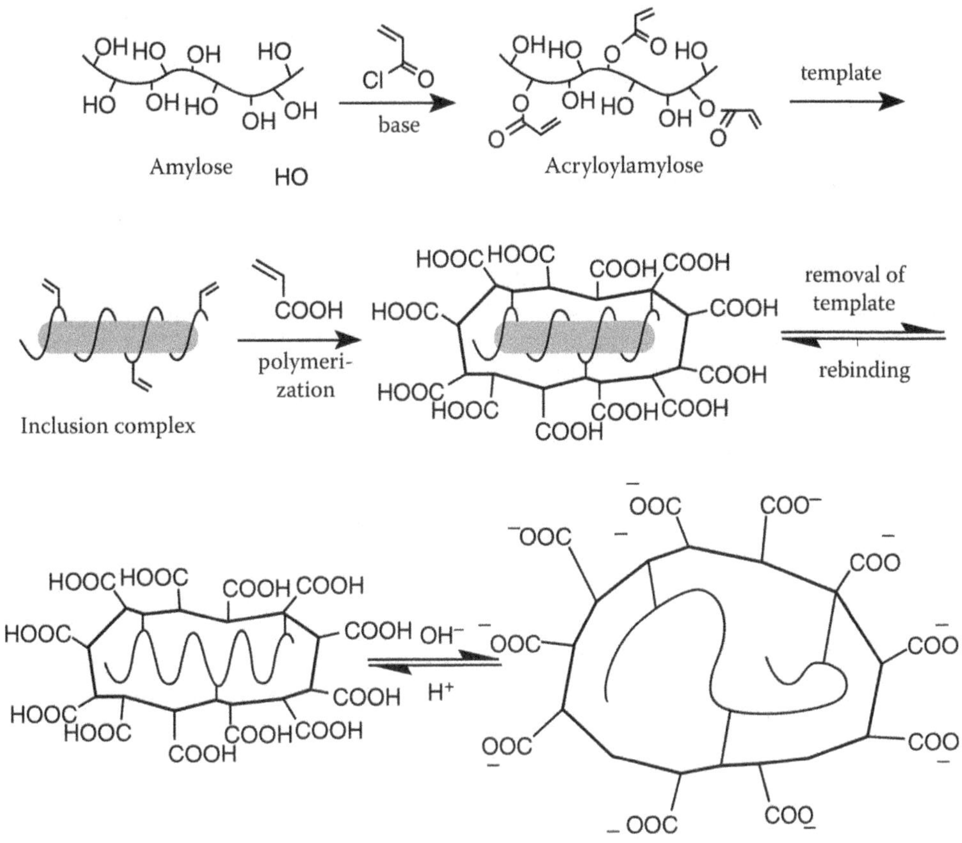

Fig. 18 Synthesis of amylose-based imprinted polymer and its pH-responsive structural change.
Source: From Kanekiyo et al.,[73] with permission from Wiley-VCH Verlag GMbH and Co. KG.

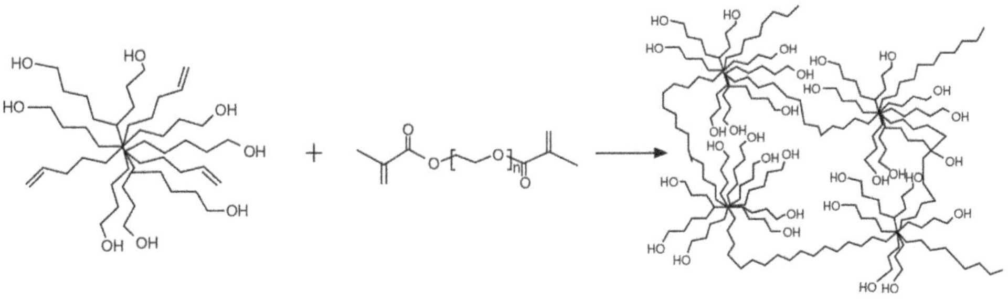

Fig. 19 Scheme of the network formation of PEG star polymers.

last finding could be attributed to the ratio of adsorber groups to the cross-linking density being too high for the cross-linker to create frustrations and therefore resulting in no observable difference between imprinted and nonimprinted systems. The incorporation of MAA enabled the system to become pH-responsive. At pH below the pK_a of MAA (4.5), the gel collapses and the template is strongly held in the network cavities. As the pH rises, the MAA dissociates and the gel swells. This also decreases the hydrogen bonding between the PEGs and the target molecules, allowing them to diffuse out of the gel.

The synthesis of MIPs selective to natural macromolecules, such as peptides and proteins, is not common. It is

difficult because bulky protein cannot easily move in and out through the mesh of a polymer network. The attempts to overcome these limitations have been focused on synthesizing macroporous MIPs[75,76] or creating imprinted cavities at the surface of the network.[77,78] Stimuli-sensitive networks can be particularly useful for overcoming steric impediments, as recently shown by Demirel et al.[79] using pH- and temperature-sensitive gels imprinted for Bovine serum albumin (BSA). The ionic poly(N-tert-butylacrylamide-co-acrylamide/maleic acid) hydrogels synthesized in the presence of BSA showed a remarkably greater affinity for the protein compared with the nonimprinted ones, the adsorption being dependent on both pH and temperature. The hydrogels were synthesized at 22.8°C, at the swollen state. At this temperature, the adsorption is maximal. In contrast, when the gel collapses it is difficult for the protein to diffuse into the gel, the imprinted cavities are distorted, and the nature of the interactions can also be altered. At low temperature, the interactions between BSA and the hydrogel are based on hydrogen bonds. As the temperature rises, hydrogen bonds become weaker, while hydrophobic interactions get stronger. These results clearly highlight the relevance of the memorization of the conformation achieved during polymerization, which provides the gel with the ability to recognize a given template.

CONCLUSIONS

The synthesis of stimuli-sensitive hydrogels in the presence of target molecules enables one to design the sequence of adsorbing or functional monomers in the polymer network. Through such design, cavities with size and chemical properties complementary to the target can be imprinted in the hydrogel. The appropriate polymer sequence is fixed during polymerization, which allows the hydrogel to memorize the desired conformation. After undergoing cycles of swelling and collapse, the gel maintains the ability to recognize and host the target molecules. Adsorption occurs through multiple contacts in the collapsed state, but through single contacts in the swollen state. This multipoint adsorption makes the collapsed-state affinity significantly larger than that in the swollen state. The abrupt change in affinity during the gel volume phase transition allows one to turn gel adsorption on and off. The Tanaka equation can theoretically explain and predict the effects on a gel's affinity for the target molecules of the adsorbing monomer and cross-linker proportions in the network and of the salt or replacement molecules' concentration in the sorption medium. The Tanaka equation is also a useful tool for better understanding the advantages of imprinted stimuli-sensitive hydrogels, compared with non-imprinted ones, and for optimizing their properties for application in different fields.

ACKNOWLEDGMENTS

This work was financed by the Ministerio de Ciencia y Tecnología, FEDER (RYC2001-8; SAF2005-01930), and Xunta de Galicia (PGIDIT03PXIC20303PN), Spain.

REFERENCES

1. Dusek, K.; Patterson, D. Transition in swollen polymer networks induced by intramolecular condensation. J. Polym. Sci. A2 **1968**, *6* (7), 1209–1216.
2. Tanaka, T. Collapse of gels and the critical endpoint. Phys. Rev. Lett. **1978**, *40* (12), 820–823.
3. Ilmain, F.; Tanaka, T.; Kokufuta, E. Volume transition in a gel driven by hydrogen bonding. Nature **1991**, *349* (6308), 400–401.
4. Flory, P.J. *Principles of Polymer Chemistry*; Cornell, New York, 1953.
5. Shibayama, M.; Tanaka, T. Volume phase transition and related phenomena of polymer gels. In *Advances in Polymer Science, Responsive Gels: Volume Transitions I*; Dusek, K., Ed.; Springer: Berlin, 1993; Vol. 109, pp. 1–62.
6. Pande, V.S.; Grosberg, A.Y.; Tanaka, T. Statistical mechanics of simple models of protein folding and design. Biophys. J. **1997**, *73* (6), 3192–3210.
7. Tanaka, T.; Annaka, M. Multiple phases of gels and biological implications. J. Intel. Mater. Syst. Struct. **1993**, *4*, 548–552.
8. Pande, V.S.; Grosberg, A.Y.; Tanaka, T. Folding thermodynamics and kinetics of imprinted renaturable heteropolymers. J. Chem. Phys. **1994**, *101* (9), 8246–8257.
9. Pande, V.S.; Grosberg, A.Y..; Tanaka, T. Thermodynamic procedure to synthesize heteropolymers that can renature to recognize a given target molecule. Proc. Natl. Acad. Sci. **1994**, *91* (26), 12976–12979.
10. Wulff, G. Molecular imprinting in cross-linked materials with the aid of molecular templates: A way towards artificial antibodies. Angew. Chem. Int. Ed. Engl. **1995**, *34* (17), 1812–1832.
11. Grosberg, A.Y.; Khokhlov, A.R. *Giant Molecules*; Academic Press: San Diego, 1997.
12. Wulff, G.; Biffis, A. Molecularly imprinting with covalent or stoichiometric non-covalent interactions. In *Molecularly Imprinted Polymers*; Sellergren, B., Ed.; Elsevier, Amsterdam, 2001; pp. 71–111.
13. Arshady, R.; Mosbach, K. Synthesis of substrate-selective polymers by host-guest polymerization. Makromol. Chem. **1981**, *182* (2), 687–692.
14. Sellergren, B. The non-covalent approach to molecular imprinting. In *Molecularly Imprinted Polymers;* Sellergren, B., Ed.; Elsevier: Amsterdam, 2001; pp. 113–184.
15. Ansell, R.J. Molecularly imprinted polymers in pseudoimmunoassay. J. Chromatogr. B **2004**, *804* (1), 151–165.
16. Yan, M.; Ramström, O. Molecular imprinting: An introduction. In *Molecularly Imprinted Materials,* Yan, M., Ramström, O., Eds.; Marcel Dekker: New York, 2005; pp. 1–12.
17. Mayes, A.G.; Whitcombe, M.J. Synthetic strategies for the generation of molecularly imprinted organic polymers. Adv. Drug Deliv. Rev. **2005**, *57* (12), 1742–1778.
18. Andersson, L.I. Molecular imprinting for drug bioanalysis: A review on the application of imprinted polymers to solid-phase extraction and binding assay. J. Chromatogr. B **2000**, *739* (1), 163–173.
19. Kandimalla, V.B.; Ju, H.X. Molecular imprinting: A dynamic technique for diverse applications in analytical chemistry. Anal. Bioanal. Chem. **2004**, *380* (4), 587–605.
20. Xu, X.; Zhu, L.; Chen, L. Separation and screening of compounds of biological origin using molecularly imprinted polymers. J. Chromatogr. B **2004**, *804* (1), 61–69.
21. Piletsky, S.A.; Turner, A.P.F. Electrochemical sensors based on molecularly imprinted polymers. Electroanalysis **2002**, *14* (5), 317–323.

Scaffolds—Smart

22. Hillberg, A.L.; Brain, K.R.; Allender, C.J. Molecular imprinted polymer sensors: Implications for therapeutics. Adv. Drug Del. Rev. **2005**, *57* (12), 1875–1889.

23. Sibrian-Vazquez, M.; Spivak, D.A. Improving the strategy and performance of molecularly imprinted polymers using cross-linking functional monomers. J. Org. Chem. **2003**, *68* (25), 9604–9611.

24. Enoki, T.; Tanaka, K.; Watanabe, T.; Oya, T.; Sakiyama, T.;Takeoka, Y.; Ito, K.; Wang, G.; Annaka, M.; Hara, K.; Du, R.; Chuang, J.; Wasserman, K.; Grosberg, A.Y.; Masamune, S.; Tanaka, T. Frustrations in polymer conformation in gels and their minimization through molecular imprinting. Phys. Rev. Lett. **2000**, *85* (23), 5000–5003.

25. Byrne, M.E.; Park, K.; Peppas, N.A. Molecular imprinting within hydrogels. Adv. Drug Deliv. Rev. **2002**, *54* (1), 149–161.

26. Miyata, T.; Uragami, T.; Nakamae, K. Biomolecule-sensitive hydrogels. Adv. Drug Deliv. Rev. **2002**, *54* (1), 79–98.

27. Alvarez-Lorenzo, C.; Concheiro, A. Molecularly imprinted polymers for drug delivery. J. Chromatogr. B **2004**, *804* (1), 231–245.

28. Cunliffe, D.; Kirby, A.; Alexander, C. Molecularly imprinted drug delivery systems. Adv. Drug Deliv. Rev. **2005**, *57* (12), 1836–1853.

29. Alvarez-Lorenzo, C.; Concheiro, A. Molecularly imprinted materials as advanced excipients for drug delivery systems. In *Biotechnology Annual Review*; El-Gewely, M.R., Ed.; Elsevier: Amsterdam 2006; Vol. 12, pp. 225.

30. Alvarez-Lorenzo, C.; Concheiro, A. Molecularly imprinted gels and nano- and microparticles: Manufacture and applications. In *Smart Assemblies and Particulates*; Arshady, R., Kono, K., Eds.; Kentus Books: London, 2006; Vol. 7, pp. 275–336.

31. Alvarez-Lorenzo, C.; Guney, O.; Oya, T.; Sakai, Y.; Kobayashi, M.; Enoki, T.; Takeoka, Y.; Ishibashi,T.; Kuroda, K.; Tanaka, K.; Wang, G.; Grosberg, A.Y.; Masamun, S.; Tanaka, T. Polymer gels that memorize elements of molecular conformation. Macromolecules **2000**, *33* (23), 8693–8697.

32. Alvarez-Lorenzo, C.; Guney, O.; Oya, T.; Sakai, Y.; Kobayashi, M.; Enoki, T.; Takeoka, Y.; Ishibashi,T.; Kuroda, K.; Tanaka, K.; Wang, G.; Grosberg, A.Y.; Masamun, S.; Tanaka, T. Reversible adsorption of calcium ions by imprinted temperature sensitive gels. J. Chem. Phys. **2001**, *114* (6), 2812–2816.

33. D'Oleo, R.; Alvarez-Lorenzo, C.; Sun, G. A new approach to design imprinted polymer gels without using a template. Macromolecules **2001**, *34* (14), 4965–4971.

34. Moritani, T.; Alvarez-Lorenzo, C. Conformational imprinting effect on stimuli-sensitive gels made with an imprinter monomer. Macromolecules **2001**, *34* (22), 7796–7803.

35. Stancil, K.A.; Feld, M.S.; Kardar, M. Correlation and cross-linking effects in imprinting sites for divalent adsorption in gels. J. Phys. Chem. B **2005**, *109* (14), 6636–6639.

36. Oya, T.; Enoki, T.; Grosberg, A.Y.; Masamune, S.; Sakiyama, T.; Takeoka, Y.; Tanaka, K.; Wang, G.; Yilmaz, Y.; Feld, M.S.; Dasari, R.; Tanaka, T. Reversible molecular adsorption based on multiple point interaction by shrinkable gels. Science **1999**, *286* (5444), 1543–1545.

37. Watanabe, T.; Ito, K.; Alvarez-Lorenzo, C.; Grosberg, A.Y.; Tanaka, T. Salt effects on multiple-point adsorption of target molecules by heteropolymer gel. J. Chem. Phys. **2001**, *115* (3), 1596–1600.

38. Ito, K.; Chuang, J.; Alvarez-Lorenzo, C.; Watanabe, T.; Ando, N.; Grosberg, A.Y. Multiple contact adsorption of target molecules. Macromol. Symp. **2004**, *207* (1), 1–16.

39. Ito, K.; Chuang, J.; Alvarez-Lorenzo, C.; Watanabe, T.; Ando, N.; Grosberg, A.Y. Multiple point adsorption in a heteropolymer gel and the Tanaka approach to imprinting: Experiment and theory. Prog. Polym. Sci. **2003**, *28* (10), 1489–1515.

40. Rampey, A.M.; Umpleby, R.J.; Rushton, G.T.; Iseman, J.C.; Shah, R.N.; Shimizu, K.D. Characterization of the imprinting effect and the influence of imprinting conditions on affinity, capacity, and heterogeneity in molecularly imprinted polymers using the Freundlich isotherm-affinity distribution analysis. Anal. Chem. **2004**, *76* (4), 1123–1133.

41. Pap, T.;Horvai, G. Binding assays with molecularly imprinted polymers: Why do they work? J. Chromatogr. B **2004**, *804* (1), 167–173.

42. Umpleby, R.J., II.; Baxter, S.C.; Rampey, A.M.; Rushton, G.T.; Chen, Y.; Shimizu, K.D. Characterization of the heterogeneous binding site affinity distributions in molecularly imprinted polymers. J. Chromatogr. B **2004**, *804* (1), 141–149.

43. Alvarez-Lorenzo, C.; Concheiro, A. Reversible adsorption by a pH- and temperature-sensitive acrylic hydrogel. J. Control. Release **2002**, *80* (1), 247–257.

44. Hsein, T.Y.; Rorrer, G.L. Heterogeneous cross-linking of chitosan gel beads: Kinetics, modeling, and influence on cadmium ion adsorption capacity. Ind. Eng. Chem. Res. **1997**, *36* (9), 3631–3638.

45. Eichenbaum, G.M.; Kiser, P.F.; Shah, D.; Meuer, W.P.; Needham, D.; Simon, S.A. Alkali earth metal binding properties of ionic microgels. Macromolecules **2000**, *33* (11), 4087–4093.

46. Takeoka, Y.; Berker, A.N.; Du, R.; Enoki, T.; Grosberg, A.; Kardar, M.; Oya, T.; Tanaka, K.; Wang, G.; Yu, X.; Tanaka, T. First order phase transition and evidence for frustration in polyampholytic gels. Phys. Rev. Lett. **1999**, *82* (24), 4863–4865.

47. Alvarez-Lorenzo, C.; Hiratani, H.; Tanaka, K.; Stancil, K.; Grosberg, A.Y.; Tanaka, T. Simultaneous multiple-point adsorption of aluminum ions and charged molecules in a polyampholyte thermosensitive gel: Controlling frustrations in a heteropolymer gel. Langmuir **2001**, *17* (12), 3616–3622.

48. Grosberg, A.Y.; Khokhlov, A.R. *Statistical Physics of Macromolecules*; AIP: New York, 1994.

49. Pande, V.S.; Grosberg, A.Y.; Tanaka, T. Heteropolymer freezing and design: Towards physical models of protein folding. Rev. Mod. Phys. **2000**, *72* (1), 259–314.

50. Bryngelson, J.D.; Wolynes, P.G. Spin glasses and the statistical mechanics of protein folding. Proc. Natl. Acad. Sci. USA **1987**, *84* (21), 7524–7528.

51. Tanaka, T.; Enoki, T.; Grosberg, A.Y.; Masamune, S.; Oya, T.; Takaoka, Y.; Tanaka, K.; Wang, C.; Wang, G. Reversible molecular adsorption as a tool to observe freezing and to perform design of heteropolymer gels. Ber. Bunsenges. Phys. Chem. **1998**, *102* (11), 1529–1533.

52. Hiratani, H.; Alvarez-Lorenzo, C.; Chuang, J.; Guney, O.; Grosberg, A.Y.; Tanaka, T. Effect of reversible cross-linker, N,N′-bis(acryloyl)cystamine, on calcium ion adsorption by imprinted gels. Langmuir **2001**, *17* (14), 4431–4436.

Scaffolds—Smart

53. Güney, O.; Yilmaz, Y.; Pekcan, O. Metal ion templated che-
     mosensor for metal ions based on fluorescence quenching.
     Sens. Actuat. B **2002**, *85* (1), 86–89.

54. Güney, O. Multiple-point adsorption of terbium ions by lead
     ion templated thermosensitive gel: Elucidating recognition
     of conformation in gel by terbium probe. J. Mol. Recogn.
     2003, *16* (2), 67–71.

55. Kanazawa, R.; Yoshida, T.; Gotoh, T.; Sakohara, S. Prepara-
     tion of molecular imprinted thermosensitive gel adsorbents
     and adsorption/desorption properties of heavy metal ions by
     temperature swing. J. Chem. Eng. Jpn **2004**, *37* (1), 59–66.

56. Kanazawa, R.; Mori, K.; Tokuyama, H.; Sakohara, S. Prepa-
     ration of ther mosensitive microgel adsorbent for quick
     adsorption of heavy metal ions by a temperature change. J.
     Chem. Eng. Jpn **2004**, *37* (6), 804–807.

57. Grinberg, N.V.; Dubovik, A.S.; Grinberg, V.Y.; Kuznetsov,
     D.V.; Makhaeva, E.E.; Grosberg, A.Y.; Tanaka, T. Studies of
     the thermal volume transition of poly(*N*-isopropylacryl-
     amide) hydrogels by high-sensitivity differential scanning
     microcalorimetry, 1: Dynamic effects. Macromolecules
     1999, *32* (5), 1471–1475.

58. Grinberg, V.Y.; Dubovik, A.S.; Kuznetsov, D.V.; Grinberg,
     N.V.; Grosberg, A.Y.; Tanaka, T. Studies of the thermal vol-
     ume transition of poly(*N*-isopropylacrylamide) hydrogels
     by high-sensitivity microcalorimetry, 2: Thermodynamic
     functions. Macromolecules **2000**, *33* (23), 8685–8692.

59. Burova, T.; Grinberg, N.V.; Dubovik, A.S.; Tanaka, K.;
     Grinberg, V.Y.; Grosberg, A.Y. Effects of ligand binding on
     relative stability of subchain conformations of weakly
     charged *N*-isopropylacrylamide gels in swollen and
     shrunken states. Macromolecules **2003**, *36* (24), 9115–9121.

60. Alvarez-Lorenzo, C.; Concheiro, A.; Dubovik, A.S.; Grinberg,
     N.V.; Burova, T.V.; Grinberg, V.Y. Temperature-sensitive
     chitosan-poly(*N*-isopropyl-acrylamide) interpenetrated net-
     works with enhanced loading capacity and controlled release
     properties. J. Control. Release **2005**, *102* (3), 629–641.

61. Yamashita, K.; Nishimura, T.; Nango, M. Preparation of
     IPN-type stimuli responsive heavy-metal-ion adsorbent gel,
     Polym. Adv. Technol. **2003**, *14* (3–5), 189–194.

62. Yamashita, K.; Nishimura, T.; Ohashi, K.; Ohkouchi, H.;
     Nango, M. Two-step imprinting procedure of inter-penetrat-
     ing polymer network-type stimuli-responsive hydrogel
     adsorbents. Polym. J. **2003**, *35* (7), 545–550.

63. Watanabe, M.; Akahoshi, T.; Tabata, Y.; Nakayama, D.
     Molecular specific swelling change of hydrogels in accor-
     dance with the concentration of guest molecules. J. Am.
     Chem. Soc. **1998**, *120*, 5577–5578.

64. Liu, X.Y.; Ding, X.B.; Guan, Y.; Peng, Y.X.; Long, X.P.;
     Wang, X.C.; Chang, K.; Zhang, Y. Fabrication of tempera-
     ture-sensitive imprinted polymer hydrogel. Macromol.
     Biosci. **2004**, *4* (4), 412–415.

65. Liu, X.Y.; Guan, Y.; Ding, X.B.; Peng, Y.X.; Long, X.P.;
     Wang, X.C.; Chang, K. Design of temperature sensitive

66. imprinted polymer hydrogels based n multiple-point hydro-
     gen bonding. Macromol. Biosci. **2004**, *4* (7), 680–684.

66. Aburto, J.; Le Borgne, S. Selective adsorption of diben-
     zothiophene sulfone by an imprinted and stimuli-
     sensitive chitosan hydrogel. Macromolecules **2004**,
     37 (8), 2938–2943.

67. Berger, J.; Reist, M.; Mayer, J.M.; Felt, O.; Peppas, N.A.;
     Gurny, R. Structure and interactions in covalently and
     ionically crosslinked chitosan hydrogels for biomedical
     applications. Eur. J. Pharm. Biopharm. **2004**, *57* (1), 19–34.

68. Qiu, Y.; Park, K. Environment-sensitive hydrogels for drug
     delivery. Adv. Drug Deliv. Rev. **2001**, *53* (3), 321–339.

69. Kikuchi, A.; Okano, T. Pulsatile drug release control using
     hydrogels. Adv. Drug Deliv. Rev. **2002**, *54* (1), 53–77.

70. Puoci, F.; Iemma, F.; Muzzalupo, R.; Spizzirri, U.G.; Trom-
     bino, S.; Cassano, R.; Picci, N. Spherical molecularly
     imprinted polymers (SMIPs) via a novel precipitation
     polymerization in the controlled delivery of salazine.
     Macromol. Biosci. **2004**, *4* (1), 22–26.

71. Chen, Y.B.; Kele, M.; Quiñones, I.; Sellergren, B.;
     Guiochon, G. Influence of the pH on the behavior of an
     imprinted polymeric stationary phase: Supporting evidence
     for a binding site model. J. Chromatogr. A **2001**, *927* (1–2),
     1–17.

72. Kanekiyo, Y.; Naganawa, R.; Tao, H. Molecular imprinting
     of bisphenol-A and alkylphenols using amylose as a host
     matrix. Chem. Commun. **2002**, *2002* (22), 2698–2699.

73. Kanekiyo, Y.; Naganawa, R.; Tao, H. pH-responsive molec-
     ularly imprinted polymers. Angew. Chem. **2003**, *42* (26),
     3014–3016.

74. Oral, E.; Peppas, N.A. Responsive and recognitive hydro-
     gels using star polymers. J. Biomed. Mater. Res. **2004**, *68A*
     (3), 439–447.

75. Guo, T.Y.; Xia, Y.Q.; Wang, J.; Song, M.D.; Zhang, B.H.
     Chitosan beads as molecularly imprinted polymer matrix
     for selective separation of proteins. Biomaterials **2005**, *26*
     (28), 5737–5745.

76. Hawkins, D.M.; Stevenson, D.; Reddy, S.M. Investigation
     of protein imprinting in hydrogel-based molecularly
     imprinted polymers (HydroMIPs). Anal. Chim. Acta **2005**,
     542 (1), 61–65.

77. Shnek, D.R.; Pack, D.W.; Sasaki, D.Y.; Arnold, F.H.
     Specific protein attachment to artificial membranes via
     coordination to lipid-bound copper(II). Langmuir **1994**,
     10 (7), 2382–2388.

78. Rachkov, A.; Minoura, N. Towards molecularly imprinted
     polymers selective to peptides and proteins: The epitope
     approach. Biochim. Biophys. Acta **2001**, *1544* (1–2),
     255–266.

79. Demirel, G.; Ozçetin, G.; Turan, E.; Caykara, T. pH/temper-
     ature-sensitive imprinted ionic poly (*N*-tert-butylacrylamide-
     co-acrylamide/maleic acid) hydrogels for bovine serum
     albumin. Macromol. Biosci. **2005**, *5* (10), 1032–1037.

Smart Polymers: Medicine and Biotechnology Applications

Allan S. Hoffman
Center for Bioengineering, University of Washington, Seattle, Washington, U.S.A.

Abstract
The term "intelligent polymers" refers to soluble, surface-coated, or crosslinked polymer systems that exhibit relatively large and sharp physical or chemical changes in response to small physical or chemical stimuli.

One can define "intelligent" polymers as those polymers that respond with large property changes to small physical or chemical stimuli. These polymers may be in various forms, such as in solution, on surfaces, or as solids. One may also combine "intelligent" aqueous polymer systems with biomolecules, to yield a large family of polymers that respond "intelligently" to physical, chemical or biological stimuli. This entry overviews such interesting and versatile polymer systems.

The term "intelligent polymers" refers to soluble, surface-coated, or crosslinked polymer systems that exhibit relatively large and sharp physical or chemical changes in response to small physical or chemical stimuli. Although the well-known glass and melting transitions of solid polymers can fit within this definition, interest in "intelligent" polymer systems extends to aqueous polymer solutions, interfaces, and hydrogels. "Intelligent" polymers are also sometimes called "smart," "stimuli-responsive," or "environmentally sensitive" polymers. Figure 1 shows schematically such aqueous polymer systems in solution, on surfaces, or as hydrogels.

Many different stimuli have been applied, and they are listed in Table 1. The many potential responses to these stimuli are listed in Table 2. Two typical examples of sharp responses are shown in Fig. 2 for soluble polymers or hydrogels.

Many different properties of the polymer system may change when such sharp responses to stimuli occur. For example, when a soluble polymer is stimulated to precipitate, it will be selectively removed from solution, which will become cloudy. When such polymers are grafted or coated onto a solid support, then one may reversibly change the water absorption into the coated polymer, thus changing the wettability of the surface. When a hydrogel is stimulated to collapse, it will squeeze out its pore water, turn opaque, become stiffer, and shrink in size. One may take advantage of one or more of these "signals" or phase changes for different end-uses. A significant amount of research on these interesting systems has been carried out, producing many publications and patents.[1–15]

There are a number of possible molecular mechanisms that can cause such sharp, sometimes discontinuous transitions in polymer systems, and they are listed in Table 3. Water is involved in most of these mechanisms. There are numerous publications describing and discussing such mechanisms in both natural and synthetic polymers.[16–21]

TEMPERATURE-SENSITIVE POLYMERS AND COPOLYMERS

Temperature-sensitive, "smart" polymers have been extensively studied.[1–3,6–15,22] There are many polymers which exhibit a cloud point (CP) or lower critical solution temperature (LCST) in aqueous solutions. Some examples of thermally sensitive polymers are listed in Table 4. One property which is common to these water soluble-insoluble polymers is that they each have a balance of hydrophilic and hydrophobic groups. The main mechanism of a thermally induced phase separation is the release of hydrophobically bound water. This is the mechanism of precipitation as well as of physical adsorption of a soluble LCST polymer onto a solid polymer substrate.[23–25] If one increases or decreases the relative hydrophilic content of the temperature-sensitive polymer, this will usually cause an increase or decrease, respectively, in the LCST, and it will have a similar effect on its tendency to physically adsorb onto a particular solid polymer substrate.[25–27]

We have studied intelligent copolymer systems with more than one stimulus-response component. These copolymers can exhibit very interesting properties, with many new and novel applications. If one combines temperature-sensitivity with pH-sensitivity in the same "smart" polymer, then the LCST of the co-polymer may be especially sensitive to the pH, due to the strong hydrophilic character of the ionized state of the pH-sensitive component. In a random vinyl copolymer containing both temperature-sensitive

Scaffolds—Smart

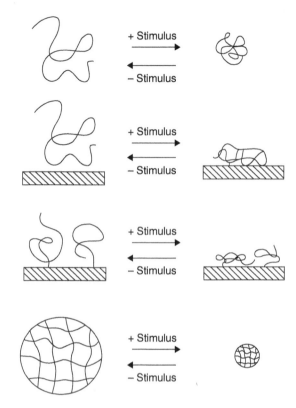

Fig. 1 Schematic examples of "intelligent" polymer systems in solution, on surfaces and as hydrogels.

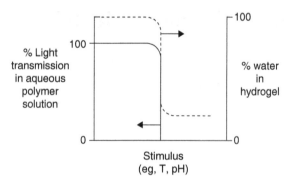

Fig. 2 Typical responses to a stimulus for aqueous-based "intelligent" polymers in solution and as a hydrogel. The transition shown for the solution is called the CP or the LCST.

Table 1 Environmental stimuli

Physical

Temperature

E.m. radiation (UV, visible)

Electrical fields

Mechanical stress, strain

Solvents

Chemical

pH

Salts (specific ions)

Chemical agents

Biochemical agents

Table 2 Intelligent polymer responses can cause changes in:

Phase

Shape

Optics

Mechanics

Electrical fields

Surface energies

Reaction rates

Permeation rates

Recognition

Table 3 Some molecular mechanisms of sharp transitions in natural and synthetic polymer systems

Ionization or neutralization

Ion exchange

Ion-ion repulsion or attraction

Release or formation of hydrophobically bound water

Helix-coil transition

Onset or inhibition of chain mobility

Crystallization or melting

Isomerization between hydrophilic and hydrophobic forms

Counter-ion movement in an electric field

Electron transfer redox reactions

Table 4 Some polymers and surfactants that show thermally-induced, reversible phase-separation in aqueous solutions:

Polymers with amide groups

Poly(*N*-substituted acrylamides)

Poly(*N*-acryloyl pyrrolidine)

Poly(*N*-acryloyl piperidine)

Poly(acryl-L-amino acid amides)

Poly(vinyl lactams)

Polymers with ether groups

PEO-PPO-PEO triblock surfactants

Alkyl-PEO block surfactants

Random (EO/PO) copolymers

Poly(vinyl methyl ether)

Polymers with alcohol groups

Hydroxypropyl acrylate

Hydroxypropyl methylcellulose

Hydroxypropyl cellulose

Methylcellulose

Poly/(vinyl alcohol) derivatives

and pH-sensitive monomers, only a small mole fraction of a pH-sensitive monomer may be sufficient to completely eliminate the LCST of the major, temperature-sensitive component when the pH is raised above the pK of the pH-sensitive component. [28,29]

In order to retain both sensitivities, we have synthesized novel "hybrid intelligent" graft and block copolymer structures where the backbone polymer and grafted or block polymer chains each independently exhibit and retain a different stimulus-response sensitivity, in contrast to the random copolymers.[30] Figures 3 and 4 illustrate this schematically for graft vs. random copolymers prepared from monomers whose homopolymers have either temperature- or pH-sensitivities. Table 5 presents some typical data for homopolymers, random and graft co-polymers in solution.[29,30]

If the pH-sensitive polymer component is poly(acrylic acid) (PAAc), then this type of graft copolymer may be most suitable for topical or oral drug delivery. This is due to the fact that high molecular weight PAAc is a well-known bioadhesive polymer, which "sticks" to the hydrated mucosal cells coating the eye, nose, mouth, lungs, G-I tract, vagina, and anus. In order to prolong the residence time of a drug delivery vehicle in contact with such mucosal surfaces, PAAc is often incorporated into a delivery formulation. Further, if it is also desirable to slow the rate of drug release from the bioadhesive formulation, a more hydrophobic component, such as the temperature-sensitive polymer, may be added. Random copolymers of Acrylic Acid (AAc) and N-isopropyl acrylamide (NIPAAm) won't work, because of the loss of the temperature sensitivity when you increase the content of the AAc component in

order to get bioadhesive properties. Physical mixtures of these two types of polymers are also possible, but they tend to physically separate and release drug too rapidly. Thus, the graft copolymer structure is most effective, since it combines both behaviors in one single molecule and doesn't permit physical separation. We have applied such graft co-polymers for opthalmic drug delivery.[31] Hydrogels based on these graft copolymers may also be synthesized;

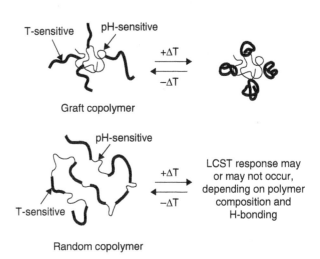

Fig. 4 Temperature-induced responses of random vs. graft copolymers below the pK of the pH-sensitive component.
Source: Data from Nabeshima et al.[29] and Chen & Hoffman.[30]

Table 5 Comparison of LCSTs of random vs. graft co-polymers of NIPAAm and AAc

Type of copolymer	wt% NIPAAm	wt% AAc	LCST (°C) pH 4.0	LCST (°C) pH 7.4
Random				
R-1	100	0	31	33
R-2	93	7	32	64
R-3	88	12	35	>95
R-4	79	21	38	>95
R-5	67	33	61	>95
R-6	57	43	>95	>95
R-7	46	54	>95	>95
Graft[a]				
AN-1	45	55	22	32
AN-2	28	72	22	32
G-20	19	81	16	32
G-25	24	76	16	32
G-30	29	71	16	32
G-50	49	51	16	32

[a]The "AN" samples were prepared by copolymerization of the macromonomer of NIPAAm with AAc; and the "G" samples were prepared by coupling oligoNIPAAm onto PAAc through the reaction of the amino-terminal group of oligoNIPAAm with the carboxyl group of PAAc.
Source: Data from Nabeshima et al.[29] and Chen & Hoffman.[30]

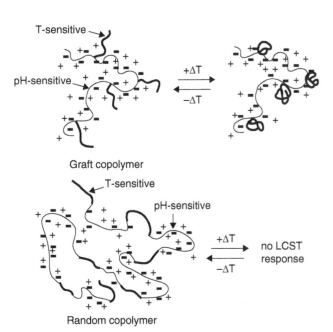

Fig. 3 Temperature-induced responses of random vs. graft copolymers above the pK of the pH-sensitive component.
Source: Data from Nabeshima et al.[29] and Chen & Hoffman.[30]

they exhibit similar properties to the soluble graft polymers. Applications of such "hybrid intelligent" gels are discussed below. We have also synthesized block copolymers of pH and temperature-sensitive components, and the different components behave similarly to the graft copolymers, and retain their individual temperature and pH transitions over a wide range of compositions.[31]

COMBINING BIOMOLECULES AND INTELLIGENT POLYMERS

A large number of biologically active molecules ("biomolecules") may be combined with intelligent polymer systems. Table 6 lists many of these biomolecules. A soluble "intelligent polymer-biomolecule conjugate" is shown schematically in Fig. 5. The biomolecules may be conjugated to pendant groups along a polymer backbone, or to one or both terminal ends of the polymer. In either case, the smart polymer may be: 1) a soluble polymer; 2) a polymer grafted to a solid support; 3) a physically adsorbed polymer on a solid support; or 4) a polymer chain segment within a hydrogel, as shown in Fig. 1.

One of the important aspects of the conjugation of a biomolecule to a polymer molecule is the possibility of conjugating many biomolecules to the same polymer molecule, thereby providing the opportunity for significant amplification of the biological activity. This is illustrated in Fig. 6. The biomolecule may also be physically entrapped within a hydrogel, either permanently, as in the case of a large protein, or temporarily, as in the case of a small drug molecule. There are many diverse biomedical and biotechnological applications of environmentally sensitive "smart" polymeric biomaterials, whether or not they contain immobilized biomolecules. Table 7 lists examples of such applications.

STIMULI-RESPONSIVE POLYMERS IN AQUEOUS SOLUTIONS

Soluble, environmentally sensitive polymers in aqueous solutions can be precipitated at a specific environmental conditions. Such systems can be useful as temperature or pH indicators, or as "on-off" light transmission switches.

Three applications of "intelligent" polymer-biomolecule conjugates being studied are illustrated schematically in Fig. 7. In the first example, a biomolecule is conjugated to the polymer, and then it is selectively phase-separated from the solution by a small change in environmental conditions. In this way, an enzyme in a bioprocess may be easily phase-separated and recycled, also permitting easy recovery of the product at the same time.[32]

In the second example in Fig. 7, a recognition biomolecule or receptor ligand such as a cell receptor peptide or an antibody is conjugated to an intelligent polymer and used in a precipitation-induced affinity separation process. When mixed with a complex solution, the conjugate will

Table 6 Examples of biomolecules which may be immobilized on or within "intelligent" polymeric biomaterials

Proteins/Peptides

Enzymes

Antibodies

Antigens

Cell-adhesion molecules

"Blocking" proteins

Drugs

Anti-thrombogenic agents

Anticancer agents

Antibiotics

Contraceptives

Drug antagonists

Peptide, protein drugs

Saccharides

Sugars

Oligosaccharides

Polysaccharides

Ligands

Hormone receptors

Cell surface receptors (peptides, saccharides)

Avidin, biotin

Lipids

Fatty acids

Phospholipids

Glycolipids

Nucleic acids, nucleotides

Single or double-stranded DNA, RNA(e.g., anti-sense oligonucleotides)

Other

Conjugates or mixtures of the above

Labels

Chromophores

Therapeutic isotopes

"Stealth" molecules

selectively complex its binding partner and then it can be readily and cleanly separated by providing a stimulus (e.g., a small change in temperature), which causes the polymer-ligand/receptor conjugate/complex to precipitate. We have applied such a thermally induced affinity precipitation process to recover IgG from solution using a poly (NIPAAm)-protein A conjugate, and are currently using it to recover CD-44 cell receptors from membrane lysates.[33]

In the third example in Fig. 7, the affinity precipitation principle is extended to an immunoassay, where the affinity precipitate is the conjugate/complex of (polymer-first antibody/antigen/second, labeled antibody).[34,35] Such

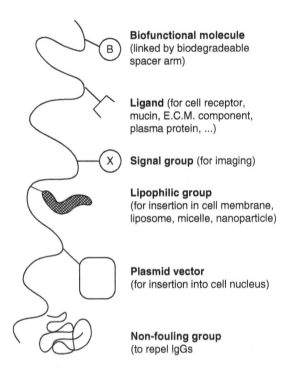

Biofunctional molecule
(linked by biodegradeable spacer arm)

Ligand (for cell receptor, mucin, E.C.M. component, plasma protein, ...)

Signal group (for imaging)

Lipophilic group
(for insertion in cell membrane, liposome, micelle, nanoparticle)

Plasmid vector
(for insertion into cell nucleus)

Non-fouling group
(to repel IgGs)

Fig. 5 Schematic illustration of a variety of natural or synthetic biomolecules which have been conjugated to an "intelligent" polymer backbone. In some uses only one biomolecule may be conjugated, while in other uses more than one biomolecule is needed. Sometimes the "intelligent" polymer molecule may be conjugated via a reactive terminal group directly to the biomolecule.

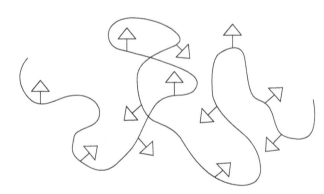

Fig. 6 Illustration of the possibility of amplification of the action of a particular biomolecule (e.g., a drug or a ligand or an enzyme), by conjugating many biomolecules to the same "intelligent" polymer molecule. The polymer may be in solution, on a surface, or in a hydrogel.

"intelligent" polymer-based affinity separations are potentially more efficient than traditional methods, such as affinity chromatography or the ELISA immunoassay, both of which involve solid surfaces and the accompanying problems of antibody or ligand desorption and non-specific adsorption.

When soluble smart polymers are mixed with liposome or cell suspensions, they may be phase-separated by a stimulus, and may interact with liposomal or cell membranes by

Table 7 Some applications of "intelligent" polymers and their combinations with biomolecules

I. *Polymers in solution*
  A. Optical indicators (sensors, switches)
  B. Precipitation separations
  C. Affinity precipitations (separations, sensors and diagnostics)
  D. Phase-transfer catalysis
  E. Binding to and stimulating cells (cell separations, expression, endo-or exocytosis, lysis, etc.)

II. *Polymers on surfaces (physically adsorbed-desorbed polymers, graft co-polymers, gels on surfaces)*
  A. Wettability changes
  B. Cell and protein attachment/detachment (implants, therapeutic devices)
  C. Bioactive surfaces (immobilized enzymes)
  D. Affinity separations
  E. Permeation switches in microporous membranes
  F. Optical indicators (sensors, switches)

III. *Homogeneous or heterogeneous hydrogels*
  A. Separations (size or affinity)
  B. Drug delivery (pulsed, cyclic, controlled release)
  C. Immobilized enzymes, cells (bioprocesses, implants, therapeutic devices)
  D. Permeation switches (molecular pores)
  E. Robotics

hydrophobic interactions. This can cause dramatic changes, such as lysis.[36] We and others have conjugated cell or liposomal membrane components, (e.g., phospholipids) and cell surface receptor ligands (e.g., the Arg-Gly-Asp (RGD)-peptide) to a temperature-responsive polymer.[37–41] After the conjugate interacts with the membranes and is caused to phase- separate, a gel may be formed, and cells may be reversibly cultured on surfaces.[40,41]

Another kind of "intelligent" polymer molecule in solution is based on the random incorporation of the key chemical groups of a known recognition sequence from a natural biomolecule along a water-soluble polymer backbone, and in the same ratio as in the natural recognition molecule.[42] If the polymer is a stimuli-responsive polymer, one might then be able to stimulate this polymer to change its conformation, enhancing the possibility of recognition.

STIMULI-RESPONSIVE POLYMERS ON SURFACES

Stimuli-responsive polymers can also be chemically grafted or physically adsorbed onto solid polymer supports, and then one can rapidly change surface film thickness, wettability, or surface charge in response to small changes in stimuli such as solution temperature, pH, or specific ionic

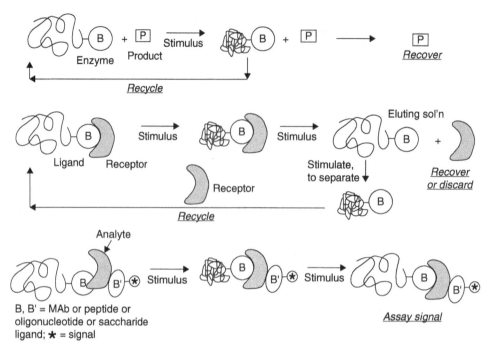

Fig. 7 Three examples of "intelligent" polymer-biomolecule conjugates are shown schematically. The applications of these three examples include bioprocesses and downstream separations, affinity separations, sensors, diagnostics, and environmental water processing. **Source:** From Chen & Hoffman[32,33] and Monji & Hoffman.[34]

concentrations.[25,43] These responses can be much faster than for solids as hydrogels since the surface coatings can be very thin. Permeation "switches" can be prepared by depositing "intelligent" polymers onto the surfaces of pores in a porous membrane, and stimulating their swelling (to block the pore flow) or collapse (to open the pore to flow).[44–47]

If proteins or cells are exposed to "intelligent" polymer surfaces that are in the swollen or collapsed state, they usually will preferentially adsorb on the more hydrophobic surface compositions (Fig. 8).[48,49] One potential use of such a system is to cycle the temperature and thereby reduce fouling by adsorbed proteins or cells on surfaces. Another interesting application is to reversibly culture cells on a chemically grafted LCST polymer surface; after the cells have been cultured for a while, they can be detached without the use of trypsin, simply by reversing the stimulus, and converting the surface to the more hydrophilic condition.[49,50] It is possible that specific cells may be separated on LCST copolymer surfaces by adjusting the polymer composition to the desired cell surface character. If the smart polymer is only physically adsorbed to the surface, unusual effects have been observed when cells are reversibly detached from such surfaces and remove the polymer with them.[51]

We have conjugated a peptide cell receptor ligand, RGD, to temperature-sensitive polymer compositions having LCSTs close to ambient temperature. The compositions have been "matched" to specific polymer support surfaces, to provide reversible adsorption-desorption behavior for thermally reversible cell culture based on temperature-stimulated, polymer adsorption-desorption (Fig. 9).[41]

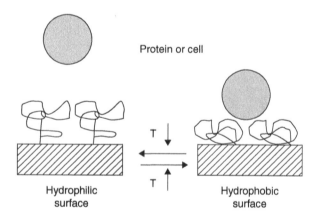

Fig. 8 Illustration of the use of a stimulus to convert a surface coated with an "intelligent" polymer from hydrophilic to hydrophobic, and thus from a protein- or cell-repelling surface to one which is more attractive to the protein or the cell. **Source:** From Kawaguchi et al.[48] and Okano et al.[49]

In related studies, we have conjugated a monoclonal antibody to a similar LCST polymer for use in a novel membrane-based immunoassay. When the assay solution is drawn through the microporous membrane after incubation, the temperature-sensitive polymer component of the polymer-antibody conjugate preferentially adsorbs on a protein-resistant cellulose acetate membrane surface at ambient temperature, both because its LCST is close to ambient temperature and also because its composition is designed to preferentially interact with the cellulose acetate (Fig. 10).[24]

STIMULI-RESPONSIVE HYDROGELS

The most extensive work on "intelligent" polymers has been carried out on stimuli-responsive hydrogels, particularly those based on pH-sensitive monomers and thermally sensitive monomers.[1–3,6–14,22,28,52–57] In a number of cases, enzymes entrapped within the hydrogels create a local pH change when their substrate is converted to product, and this can cause changes in swelling and pore sizes within the gel.[4,58] Stimuli derived from changes in solvent mixtures, specific ions or solutes, pH, temperature, electric fields, or electromagnetic radiation have all been used to effect collapse or swelling of stimuli-responsive hydrogels. This action has led to applications such as desalting and/or dewatering of protein solutions, microrobotics and artificial muscles, and "on-off" immobilized enzyme reactors.[2,3,13,59,60]

Delivery of drugs has been one of the most extensively studied application areas for stimuli-responsive

hydrogels.[4–6,11–14,28,31,40,55,57] Application of cyclic temperature or electric field stimuli have resulted in cyclic delivery of physically incorporated drug molecules.[11,14,61] Figure 11 illustrates many of these applications. Figure 12 shows the new grafted hydrogel structures, which can have unusual combinations of sensitivities to different environmental stimuli.[30] Figure 13 illustrates how such "hybrid intelligent" hydrogels may be utilized to provide controlled, bioadhesive drug delivery.

When enzymes are immobilized within smart hydrogels, then cyclic changes in environmental stimuli can lead to "on-off" activity of the enzyme due to the cyclic collapse and reswelling of the hydrogel pores. This action can also be used to enhance mass transport of substrate into and product out of immobilized enzyme hydrogels (Fig. 11). We have demonstrated significant enhancement of bioreactor productivity in such systems.[59,60]

If the enzyme is immobilized within a pH-sensitive gel, and the enzyme-substrate reaction produces a local microenvironmental pH change, then the actual "stimulus" for the resultant swelling or deswelling of the gel is the substrate concentration in the external solution. This type of gel can be used either as a biosensor or as a permeation switch, for example, to permit release of a drug such as insulin through a swelling gel in response to an increase of a systemic metabolite such as glucose (Fig. 11).[4,58,62]

The pore sizes in an "intelligent" polymer hydrogel can be controlled by the environmental conditions and the hydrogel composition.[63] Thus, such gels can be used to separate molecules on the basis of size.[2,3,63] (See "Separations" in Fig. 11.) If a specific recognition ligand (e.g., an antibody), is immobilized within a stimuli-responsive gel, then its specific binding partner (e.g., antigen) can be selectively removed from solution if it is not too large to enter the gel. This process can be used to selectively recover a desired compound or to selectively remove an undesirable compound such as a toxin.[6,64] The binding alone could shrink the gel if there are multiple binding sites on the immobilized ligand and/or its binding partner.[64]

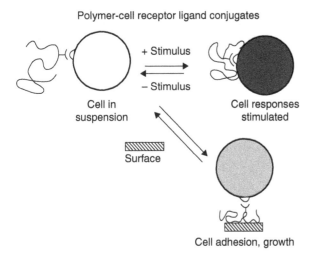

Polymer-cell receptor ligand conjugates

Fig. 9 Schematic illustration of a cell receptor ligand-"intelligent" polymer conjugate binding to the cell membrane receptor, and stimulating various cell responses, including reversible cell culture on a surface, when stimulated to precipitate or redissolve.
Source: From Miura & Hoffman.[41]

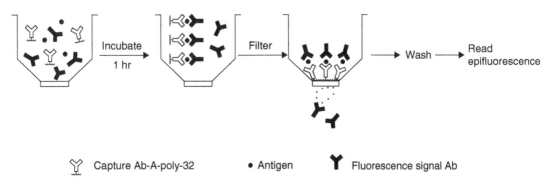

Capture Ab-A-poly-32   • Antigen   Fluorescence signal Ab

Fig. 10 Novel membrane immunoassay based on selective adsorption of the polymer component of a polymer-antibody conjugate onto the surface of the membrane. The membrane itself is resistant to protein adsorption.
Source: From Monji et al.[24]

Scaffolds—Smart

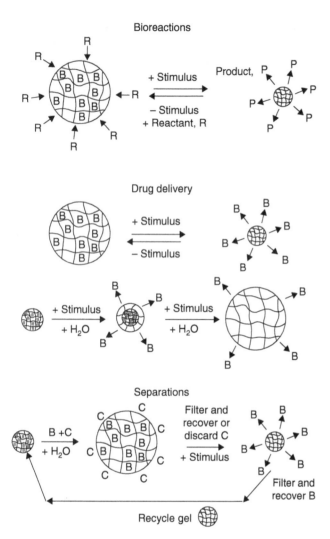

Fig. 11 Schematic illustration of various applications of "intelligent" hydrogels in medicine and biotechnology.

KINETIC CONSIDERATIONS

The kinetics of the various "intelligent" responses described in this entry will be very sensitive to the speed of the stimulus and the dimensions of the system being stimulated. The slowest systems will be the hydrogels, with thicker dimensions than the coatings, which are in turn "thicker" than polymers in solution, which has the highest surface/volume ratio of all systems. A study has shown that if a temperature sensitive hydrogel is constructed with many grafted chains of the same composition as the network polymer, the temperature-stimulated collapse rate is significantly enhanced.[65] The speed of the stimulus will vary according to the resistances it encounters in reaching the intelligent polymer molecules or segments. Clearly, electromagnetic radiation is the fastest stimulus, at least in systems transparent to the radiation, while those stimuli involving diffusion of molecules will be among the slowest.

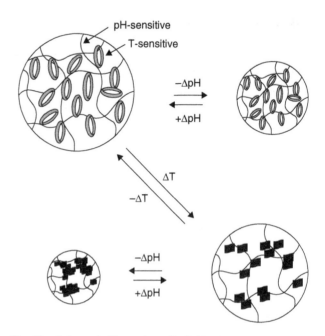

Fig. 12 Schematic illustration of individual responses of a dual-sensitivity "hybrid intelligent" hydrogel to individual temperature or pH stimuli.
Source: From Chen & Hoffman.[30]

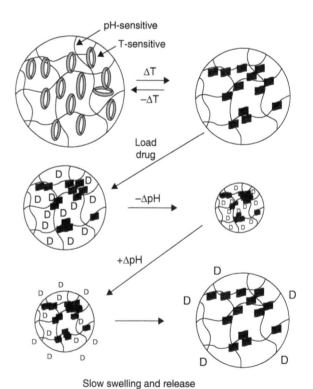

Slow swelling and release

Fig. 13 Illustration of loading and release of a drug from a dual-sensitivity, "hybrid intelligent" hydrogel. The release is stimulated by a rise in pH, (e.g., from gastric to enteric conditions) ionizing the matrix of the hydrogel, while control of release rate is provided by the temperature-sensitive grafted chains which form hydrophobic domains above the LCST that resist rapid swelling of the ionized hydrogel.

CONCLUSIONS

There are many diverse possibilities for the design of compositions, molecular structures and physical properties of "intelligent" polymers and hydrogels. Such diversity provides great opportunities for many diverse and novel applications in medicine and biotechnology.

ACKNOWLEDGMENTS

I would like to acknowledge the "stimulation" I received from the many "intelligent" and creative contributions of my students and colleagues while working in this exciting field. I would also like to note that the reference list is relatively brief and therefore is only illustrative and not comprehensive, and I apologize for all omissions. This entry was previously published in *Macromolecules,* Vol. 98, pp. 645–664, 1995 and appears with permission.

REFERENCES

1. Tanaka, T. Gels. Sci. Am. **1981**, *244* (1), 124, doi:10.1038/scientificamerican0181-124.
2. Cussler, E.L.; Stokar, M.R.; Varberg, J.E. Gels as size selective extraction solvents. AIChE J. **1984**, *30* (4), 578–582.
3. Cussler, E.L. Method of size-selective extraction from solutions. U.S. Patent 4555344, Nov 26, 1985.
4. Ishihara, K.; Kobayashi, M.; Ishimaru, N.; Shinohara, I. Glucose induced permeation control of insulin through a complex membrane consisting of immobilized glucose oxidase and a poly(amine). Polym. J. **1984**, *16* (8), 625–631
5. Heller, J. Med. Dev. Diag. Ind. **1985**, *7*, 34.
6. Hoffman, A.S. Applications of thermally reversible polymers and hydrogels in therapeutics and diagnostics. J. Control. Release **1987**, *6* (1), 297–305.
7. Peppas, N.A., Korsmeyer, R.W., Ed. *Hydrogels in Medicine and Pharmacology*; CRC: Boca Raton, FL, 1987.
8. Bae, Y.H.; Okano, T.; Hsu, R.; Kim, S.W. Thermo-sensitive polymers as on-off switches for drug release. Makromol. Chem. Rapid Commun. **1987**, *8* (10), 481–485.
9. Hoffman, A.S.; Monji, N. Methods for selectively reacting ligands immobilized within a temperature-sensitive polymer gel. U.S. Patent 4912032, Mar 27, 1990.
10. Kost, J., Ed. *Pulsed and Self-Regulated Drug Delivery*; CRC: Boca Raton, FL, 1990.
11. Okano, T.; Bae, Y.H.; Jacobs, H.; Kim, S.W. Thermally on-off switching polymers for drug permeation and release. J. Control. Release **1990**, *11* (1–3), 255–265.
12. Hoffman, A.S. Mater. Res. Soc. Bull. **1991**, *16* (9), 42.
13. DeRossi, D.E.; DeRossi, D., Eds. *Polymer Gels*; Plenum: New York, 1991.
14. Kwon, I.C.; Bae, Y.H.; Kim, S.W. Electrically credible polymer gel for controlled release of drugs. Nature **1991**, *354* (6351), 291–293.
15. Osada, Y.; Okuzaki, H.; Hori, H. A polymer gel with electrically driven motility. Nature **1992**, *355* (6357), 242–244.
16. Dusek, K.; Patterson, D.J. Polym. Sci. A-2 **1968**, *6*, 1209.
17. Verdugo, P. Biophys. J. **1986**, *49*, 231.
18. Tanaka, T.; Sun, S.T.; Hirokawa, Y.; Katayama, S.; Kucera, J.; Hirose, Y.; Amiya, T. Mechanical instability of gels at the phase transition. Nature **1987**, *325* (6107), 796–798.
19. Tanaka, T. In *Structure and Dynamics*; Nicolin, C., Ed.; Nijhoff: Amsterdam, 1987; p. 237.
20. de Gennes, P.G. Exponents for the excluded volume problem as derived by the Wilson method. Phys. Lett A **1972**, *38* (5), 339–340.
21. Urry, D.W.; Harris, R.D.; Prasad, K.U. Chemical potential driven contraction and relaxation by ionic strength modulation of an inverse temperature transition. J. Am. Chem. Soc. **1988**, *110* (10), 3303–3305.
22. Schild, H.G. Poly(N-isopropylacrylamide): Experiment, theory and application. Prog. Polym. Sci. **1992**, *17* (2), 163–249.
23. Heskins, H.; Guillet, J.E. Solution properties of poly (N-isopropylacrylamide). J. Makromol. Sci. Chem. A **1968**, *2* (8), 1441.
24. Monji, N.; Cole, C.A.; Tam, M.; Goldstein, L.; Nowinski, R.C. Application of a thermally-reversible polymer-antibody conjugate in a novel membrane-based immunoassay. Biochem. Biophys. Res. Commun. **1990**, *172* (2), 652–660
25. Miura, M.; Cole, C.A.; Monji, N.; Hoffman, A.S. Temperature-dependent absorption/desorption behavior of lower critical solution temperature (LCST) polymers on various substrates. J. Biomat. Sci. Polym. Ed. **1994**, *5* (6), 555–568
26. Taylor, L.D.; Cerankowski, L.D. Preparation of films exhibiting a balanced temperature dependence to permeation by aqueous solutions—A study of lower consolute behavior. J. Polym. Sci. Polym. Chem. **1975**, *13* (11), 2551–2570.
27. Priest, J.H.; et al. In *Reversible Polymeric Gels and Related Systems*; ACS Symposium Series 350; Russo, P., Ed.; ACS: Washington, DC, 1987; p. 255.
28. Dong, L.C.; Hoffman, A.S. A novel approach for preparation of pH-sensitive hydrogels for enteric drug delivery. J. Control. Release **1991**, *15* (2), 141–152.
29. Nabeshima, Y.; et al. Unpublished results; University of Washington: Seattle, WA, 1992.
30. Chen, G. H.; Hoffman, A.S. Graft copolymers that exhibit temperature-induced phase transitions over a wide range of pH. Nature **1995**, *373* (6509), 49–52.
31. Chen, G.H.; Hoffman, A.S. Temperature-induced phase transition behaviors of random vs. graft copolymers of N-isopropylacrylamide and acrylic acid. Macromol. Rapid Comm. **1995**, *16* (3), 175–182.
32. Chen, G.H.; Hoffman, A.S. Preparation and properties of thermoreversible, phase-separating enzyme-oligo(N-isopropylacrylamide) conjugates. Bioconjug. Chem. **1993**, *4* (6), 509–514.
33. Chen, J.P.; Hoffman, A.S. Polymer-protein conjugates. II. Affinity precipitation separation of human immunogammaglobulin by a poly(N-isopropylacrylamide)-protein A conjugate. Biomaterials **1990**, *11* (9), 631–634.
34. Monji, N.; Hoffman, A.S. A novel immunoassay system and bioseparation process based on thermal phase separating polymers. Appl. Biochem. Biotech. **1987**, *14* (2), 107.
35. Monji, N.; Hoffman, A.S.; Priest, J.H.; Houghton, R.L. Thermally induced phase separation immunoassay. U.S. Patent 4 780 409, Oct 25, 1988.
36. Park, T. G.; Hoffman, A. S. Unpublished results; University of Washington: Seattle, WA, 1990.
37. Nightingale, J.A.S.; et al. Proc. Soc. Biomater. **1987**, *56*.

Scaffolds—Smart

L

38. Ringsdorf, H.; Venzmer, J.; Winnik, F.M. Fluorescence studies of hydrophobically modified poly(N-isopropylacryl-amides). Macromolecules **1991**, *24* (7), 1678–1686.

39. Schild, H.G.; Tirrell, D.A. Microheterogeneous solutions of amphiphilic copolymers of N-isopropylacrylamide. An investigation via fluorescence methods. Langmuir **1991**, *7* (7), 1319–1324.

40. Wu, X.S.; Hoffman, A.S.; Yager, P. Synthesis of and insulin release from erodible poly(N-isopropylacrylamide)-phospholipid composites. J. Intell. Mtls. Syst. Struct. **1993**, *4* (2), 202.

41. Miura, M.; Hoffman, A. S. Unpublished results; University of Washington: Seattle, WA, 1992.

42. Tardieu, M.; Gamby, C.; Avramoglou, T.; Jozefonvicz, J.; Barritault, D. Derivatized dextrans mimic heparin as stabilizers, potentiators, and protectors of acidic or basic FGF. J. Cell Physiol. **1992**, *150* (1), 194–203.

43. Uenoyama, S.; Hoffman, A.H. Radiat. Phys. Chem. **1988**, *32*, 665.

44. Tirrell, D.A. Macromolecular switches for bilayer membranes. J. Control. Release **1987**, *6* (1), 15–21.

45. Iwata, H.; Oodate, M.; Uyama, Y.; Amemiya, H.; Ikada, Y. Preparation of temperature-sensitive membranes by graft polymerization onto a porous membrane. J. Membr. Sci. **1991**, *55* (1–2), 119–130.

46. Osada, Y.O.; Honda, K.; Ohta, M. Control of water permeability by mechanochemical contraction of poly (methacrylic acid)-grafted membranes. J. Membr. Sci. **1986**, *27* (3), 327–338.

47. Yoshida, M.; et al. Rad. Eff. Defects Solids **1993**, *126*, 409, doi:10.1080/10420159308219752.

48. Kawaguchi, H.; Fujimoto, K.; Mizuhara, Y. Hydrogel microspheres III. Temperature-dependent adsorption of proteins on poly-N-isopropylacrylamide hydrogel microspheres. Colloid Polym. Sci. **1992**, *270* (1), 53–57.

49. Okano, T.; Yamada, N.; Sakai, H.; Sakurai, Y. A novel recovery system for cultured cells using plasma-treated polystyrene dishes grafted with poly(N-isopropylacrylamide). J. Biomed. Mater. Res. **1993**, *27* (10), 1243–1251.

50. Yamada, N.; Okano, T.; Sakai, H.; Karikusa, F.; Sawasaki, Y.; Sakura, Y. Thermo-responsive polymeric surfaces; control of attachment and detachment of cultured cells. Makromol. Chem. Rapid Commun. **1990**, *11* (11), 571–576.

51. Takezawa, T.; Mori, Y.; Yoshizato, K. Cell culture on a thermo-responsive polymer surface. Biotechnology **1990**, *8* (9), 854–856.

52. Kuhn, W.; Hargitay, B.; Katchalsky, A.; Eisenberg, H. Reversible dilation and contraction by changing the state of ionization of high-polymer acid network. Nature **1950**, *165* (4196), 514–516.

53. Ilavsky, M. Phase transition in swollen gels. 2. Effect of charge concentration on the collapse and mechanical behavior of polyacrylamide networks. Macromolecules **1982**, *15* (3), 782–788.

54. Yu, H.; Grainger, D. (Thesis Ph.D. Yu, H. Beaverton, Oregon: Oregon Graduate Institute, 1994).

55. Siegel, R.A.; Falamarzian, M.; Firestone, B.A.; Moxley, B.C. pH-Controlled release from hydrophobic/polyelectrolyte copolymer hydrogels. J. Control. Release **1988**, *8* (2), 179–182.

56. Nakamae, K.; Miyata, T.; Hoffman, A.S. Swelling behavior of hydrogels containing phosphate groups Makromol. Chem. **1992**, *193* (4), 983–990.

57. Dong, L.C.; Qi, Y.; Hoffman, A.S. Controlled release of amylase from a thermal and pH-sensitive, macroporous hydrogel. J. Control. Release **1992**, *19* (1–3), 171–182.

58. Albin, G.; Horbett, T.A.; Ratner, B.D. Glucose sensitive membranes for controlled delivery of insulin: Insulin transport studies. J. Control. Release **1985**, *2*, 153–164.

59. Park, T.G.; Hoffman, A.S. Effect of temperature cycling on the activity and productivity of immobilized β-galactosidase in a thermally reversible hydrogel bead reactor. Appl. Biochem. Biotech. **1988**, *19* (1), 1–9.

60. Park, T.G.; Hoffman, A.S. Immobilization of arthrobacter simplex in a thermally reversible hydrogel: Effect of temperature cycling on steroid conversion. Biotech. Bioeng. **1990**, *35* (2), 152–159.

61. Bae, Y.H.; Kim, S.W. Pulsatile drug release by electric stimulus, polymeric drug and drug administration. ACS Sympos. Ser. **1994**, *545*, 98–110.

62. Horbett, T.A.; Ratner, B.D.; Kost, J.; Singh, M. A Bioresponsive memberane for insulin delivery. In *Recent Advances in Drug Delivery Systems*; Anderson, J.M., Kim, S.W., Ed.; Plenum: New York, 1984; p 209.

63. Park, T.G.; Hoffman, A.S. Estimation of temperature-dependent pore size in poly(N-isopropylacrylamide) hydrogel beads. Biotechnol. Progr. **1994**, *10* (1), 82–86.

64. Kokufata, E.; Yong-Qing, Z.; Toyoichi, T. Saccharide-sensitive phase transition of a lectin-loaded gel. Nature **1991**, *351* (6324), 302–304.

65. Yoshida, R.; Katsumi, U.; Yuzo, K.; Kiyotaka, S.; Akihiko, K.; Yasuhisa, S.; Teruo, O. Comb-type grafted hydrogels with rapid deswelling response to temperature changes. Nature **1995**, *374* (6519), 240–242.

Stem Cell: Hematopoietic Stem Cell Culture, Materials for

Akon Higuchi
Department of Chemical and Materials Engineering, National Central University, Jhongli, Taiwan, and Department of Botany and Microbiology, King Saud University, Riyadh, Saudi Arabia

Siou-Ting Yang
Pei-Tsz Li
Ta-Chun Kao
Department of Chemical and Materials Engineering, National Central University, Jhongli, Taiwan

Yu Chang
Yung Hung Chen
Department of Obstetrics and Gynecology, Kaohsiung Medical University Hospital, Kaohsiung Medical University, Kaohsiung, Taiwan

S. Suresh Kumar
Department of Medical Microbiology and Parasitology, Putra University, Slangor, Malaysia

Abstract

Hematopoietic stem and progenitor cells (HSPCs) are multipotent cells that have the specific capacity to self-renew and differentiate into all mature blood cells. Umbilical cord blood is a promising alternative source of HSPCs for allogeneic and autologous hematopoietic stem cell transplantation for the treatment of a variety of hematological disorders and can be used as a supportive therapy for malignant diseases. However, the low number of HSPCs obtainable from a single donor of umbilical cord blood limits direct transplantation of umbilical cord blood to the treatment of pediatric patients because of the small volume of blood collected. Therefore, *ex vivo* HSPC expansion is necessary to produce a sufficient number of cells that can engraft and sustain long-term hematopoiesis. The native bone marrow microenvironment, a complex network of stromal cells, and the extracellular matrix, serve as a stem cell niche that regulates HSPC functions such as self-renewal, proliferation, homing, and fate choice. The importance of the topography of the tissue culture materials and of signaling molecules or specific functional groups immobilized on these materials has been taken into consideration in the development of HSPC culture systems. We review the culture materials available for *ex vivo* HSPC expansion and discuss the effects of their surface chemistry and topography on such expansion. "This entry has been revised from "Polymeric Materials for *Ex vivo* Expansion of Hematopoietic Progenitor and Stem Cells" in *Polymer Reviews* Vol. 49, Issue 3.

INTRODUCTION

Hematopoietic stem/progenitor cells (HSPCs) are multipotent cells that have the specific capacity to self-renew and differentiate into all mature blood cells.[1] Besides bone marrow (BM), umbilical cord blood (UCB) is also a promising alternative source of HSPCs for allogeneic and autologous hematopoietic stem cell transplantation[2] for the treatment of a variety of hematological disorders and as a supportive therapy for malignant diseases.[3] However, the low number of HSPCs obtainable from a single donor of UCB limits direct transplantation of UCB to the treatment of pediatric patients because of the small volume of blood collected.

UCB transplantation has been limited thus far to children with an average weight of 20 kg. The major disadvantage of such transplantations is the low cell dose, resulting in slower time to engraftment and higher rates of engraftment failure than with BM transplantation. Numerous efforts have been made to expand HSPCs *ex vivo* to improve engraftment time and reduce the graft failure rate, particularly in developing this therapy for adult patients.[4-7]

The success of HSPC transplantation is dependent on both the dose of HSPCs and the pluripotency of the HSPCs transplanted.[3,8-10] The goal of *ex vivo* HSPC expansion is to produce a number of cells sufficient for engrafting and to sustain long-term hematopoiesis.[11,12] Therefore, it is necessary to develop appropriate strategies and methods to expand HSPCs *ex vivo*, so that UCB can serve as a source of transplantable HSPCs for adult patients suffering from a variety of hematological disorders.

HSPCs are generally regarded as suspension cells in *ex vivo* expansion. Therefore, a variety of protocols have

Concise Encyclopedia of Biomedical Polymers and Polymeric Biomaterials DOI: 10.1081/E-EBPPC-120052046

Stem Cell—Ultrasound

been developed for HSPC suspension culture in the presence of various combinations of early-acting cytokines such as Flt-3 ligand (FL), stem cell factor (SCF), interleukin-3 (IL-3), interleukin-6 (IL-6), erythropoietin (EPO), thrombopoietin (TPO), and granulocyte-macrophage colony stimulating factor (GM-CSF) under serum-free conditions.[8,9] *Ex vivo* HSPC culture is mainly performed in tissue culture flasks or cell culture bags that do not provide any microarchitecture for cellular interaction with the container materials.[3]

The importance of the topography of the tissue culture materials and of signaling molecules or specific functional groups immobilized on these materials has been not generally taken into consideration in such culture systems. However, suspension culture conditions differ drastically from conditions provided by the natural BM microenvironment. It is generally accepted that the native BM microenvironment, a complex network of stromal cells and extracellular matrix (ECM) serves as a stem cell niche that regulates HSPC functions such as self-renewal, proliferation, homing, and fate choice (Fig. 1).[1,3,13,14]

HSPCs in BM retain the ability to maintain self-renewal and commitment to differentiation, whereas they tend to lose their self-renewal ability after extensive proliferation in *ex vivo* culture.[11] The mechanism that regulates the fate decision of HSPC still remains unclear, but it has been proposed that not only autonomous signals, but signals originating from the HSPC niche play an important role therein.[11,15,16] The HSPC niche is composed of a variety of differentiated cell types (osteoblasts and stromal cells such as endothelial cells, fibroblasts, etc.), with several kinds of ECM molecules and many different soluble and membrane/ECM-bound cytokines or hematopoietic growth factors distributed in this three-dimensional (3D) environment (Fig. 1).[11,15,16]

A growing body of evidence has suggested that the surface chemistry and topography of culture materials

influence the rate of HSPC proliferation and CD34+ cell expansion.[1,3,5,6,11] Several research groups have been trying to develop cell culture conditions mimicking those of the HSPC niche (Fig. 1). Typical methods of evaluating HSPCs after *ex vivo* expansion are summarized in Table 1. We review the materials used for HSPC expansion *ex vivo*, and discuss the effects thereon of the surface chemistry and topography of the culture materials.

HSPC CO-CULTURE SYSTEMS

Stromal cells in HSPC co-culture systems are generally considered to provide signal transduction through two major mechanisms to promote the expansion of HSPCs. First, stromal cells secrete various soluble factors, such as growth factors and cytokines, that stimulate the proliferation and differentiation of HSPCs, and provide an appropriate niche environment for the proliferation and maintenance of HSPCs.[4] Secondly, direct contact with stromal elements may be required for optimal preservation of primitive progenitor capacity during *ex vivo* culture.[17,18] Adhesive interactions between HSPCs and stromal ligands, in addition to determining progenitor localization within the BM stromal microenvironment, may regulate hematopoiesis by directly transducing growth regulatory signals.[17,19,20] The co-culture of UCB cells with murine stromal cells, such as M2-10B4,[21] MS-5,[22] OP9,[23,24] or HESS-5[25,26] cells, is the conventional method for the expansion of HSPCs.[4]

The general consensus among researchers in this field is that direct contact between HSPCs and the cellular microenvironment in the BM niche plays an important role in the maintenance of "stemness." In the *ex vivo* expansion of HSPCs, mesenchymal stem cell (MSC) feeder layers act as a surrogate for this interaction. Specific adhesion molecules are thought to be responsible for this cell–cell contact. Wagner et al. studied the adhesive interactions of HSPCs in several fractions by using a method based on gravimetric force upon inversion of cell culture dishes to define cell–cell contact between HSPC and MSC.[27] Adherent cells were separated from nonadherent cells by the gravimetric force, and analyzed for long-term hematopoietic culture initiating cell (LTC-IC) frequency and gene expression. HSPC subsets with higher self-renewing capacity showed significantly higher adherence to human MSC (CD34+ vs. CD34−, CD34+/CD38− vs. CD34+/CD38+, slow dividing fraction vs. fast dividing fraction). The LTC-IC frequency was significantly higher in the adherent fraction of the cells than in the nonadherent fraction. The adherent subsets of CD34+ cells strongly expressed genes coding for adhesion proteins and ECM molecules [fibronectin (FN) 1, cadherin 11, vascular cell adhesion molecule (CAM)-1, connexin 43, integrin β-like 1, and TGFBI]. It was demonstrated that primitive subsets of HSPCs had a higher affinity for human MSCs. The specific junction proteins involved in the stabilization of cell–cell contacts may

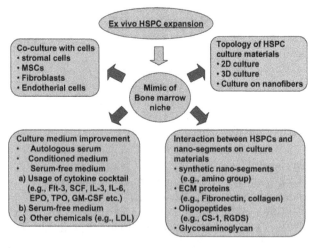

Fig. 1 Strategies for *ex vivo* expansion of HSPCs by mimicking BM niche.

Table 1 Evaluation method of HSPCs after *ex vivo* expansion

1st Stage Cell no. & surface marker	2nd Stage Colony-forming assay	3rd Stage Engraftment assay analysis
No. of mononuclear cells	CFU-GM (colony-forming unit granulocyte macrophage)	Engraftment in NOD/SCID mouse
No. of CD34+ cells	CFU-GEMM (colony-forming unit multilineage colonies)	
Surface marker analysis	BFU-E/CFU-E(burst-forming unit erthroid colony-forming unit erthrocyte)	
VLA4+	LTC-IC	
VLA5+		
CD3+ (T lymphocyte)		
CD13+ (myelomonocytic cells)		
CD14+ (monocyte)		
CD15+ (granulocyte)		
CD19+ (B lymphocyte)		
CD34+ (HSPC)		
CD38-(unactivated cells)		
CD41+ (platelet)		
CD45+ (leukocyte)		
CD133+ (HSPC)		
GlyA+ (red blood cells)		

play a significant role in maintaining the stemness of HSPCs, as indicated by the significantly higher expression of adhesion proteins and ECM molecules in more primitive HSPCs.[27]

However, several difficulties have been encountered in the use of xenogeneic stromal cells to expand HSPCs in a clinical setting. Public health concerns exist regarding zoonosis and other unknown infectious diseases caused by contaminated xenogeneic feeder cells, and regarding the pathogens they may carry, such as bovine spongiform encephalopathy and porcine endogenous retrovirus.[4,28] Furthermore, rejection of or unexpected reactions against transplanted HSPCs could be caused by xenogeneic cell debris or proteins and polysaccharides derived from stromal cells.[4,29] Thus, the development of a co-culture system free from contamination by feeder cells or their debris is highly desirable.[4]

Conditioned media (CM) derived from feeder cells have been evaluated as a possible means of ameliorating the risks pointed out earlier.[4,18,30,31] Although sufficient amounts of essential bioactive substances exist in CM for the initial proliferation of HSPCs, the cultured cells gradually consume these nutrients, and thus the efficiency of CM decreases with time. For the efficient proliferation of HSPCs in culture media supplemented with CM, either the media need to be frequently changed, or bioactive substances need to be supplied continuously.

Fujimoto et al. examined the possibility of using microencapsulated feeder cells to provide a continuous supply of bioactive substances. Spherical aggregates of feeder cells (e.g., microencapsulated murine stromal cells (HESS-5) or immortalized human MSCs) were enclosed in agarose gel microcapsules and co-cultured with HSPCs. The gel prohibited contact between feeder cells and HSPCs,[4] while remaining permeable to various chemicals, such as amino acids, glucose, and growth factors.[4,32,33,34]

Bioactive substances for HSPC proliferation present in conditioned media at the onset of culture are likely consumed by HSPCs over time, while co-culture of HSPCs with microencapsulated feeder cells would lead to a continuous supply of bioactive substances. The number of HSPC progeny cells increased efficiently with the use of microencapsulated feeder cells. The total number of nucleated cells and that of CD34+ cells increased 194- and 7.4-fold, respectively, in the presence of microencapsulated stromal cells in the conditioned medium. Colony-forming cells and cobblestone area-forming cells were well maintained. The transfusion of the progeny obtained from HSPCs cultured with microencapsulated stromal cells in conditioned medium might shorten the time to engraftment by bridging the pancytopenic period and support functional hematopoietic repopulation by their effective expansion of total cells and maintenance of primitive HSPC progenitor cells.[4]

EFFECT OF CULTURE MATERIALS ON *EX VIVO* EXPANSION OF HSPCS

Stromal cell-supported culture has been shown to preserve the ability of human HSPCs to maintain hematopoiesis.

However, it is difficult to obtain a sufficient quantity of autologous stromal cells for clinical applications. When allogeneic stromal cells or MSC-derived cell lines are used in clinical applications instead of autologous stromal cells, there is the definite possibility of cells being contaminated, generating an immune response upon transplantation along with HSPCs in patients.[11,35,36] Therefore, it would be preferable to develop alternate methods of HSPC expansion that are equally as effective as co-culturing with stromal cells. First, we will examine the possibility of using conventional polymers as culture materials for HSPCs or other types of stem cells.

Conventional Synthetic and Natural Polymeric Materials

Synthetic and natural biomaterials play an important role in the treatment of diseases and in modern health care and medicine.[37,38] Conventional polymeric materials used for HSPC culture cannot reasonably be expected to provide a niche similar to the stromal microenvironment in BM. However, it is well known that parameters such as surface topography, chemistry (physicochemical properties), including surface wettability (surface energy) and surface charge, strongly influence the interaction between HSPCs and culture materials.[37] There exist no general principles to predict the cellular behavior of HSPCs on a given biomaterial.[37,38] Therefore, the adhesion, morphology, vitality, proliferation, cytotoxicity, and apoptosis of HSPCs have to be analyzed separately for each type of surface under assessment.

Laluppa et al. have investigated 15 types of commercially available polymeric materials, four types of metal, and glass substrates for their ability to support expansion of HSPCs in a clinical setting. HSPCs were collected by apheresis from cancer patients following stem cell mobilization regimens consisting of treatment with G-CSF. CD34+ cells and mononuclear cells were evaluated for expansion of total cells and colony-forming unit–granulocyte monocytes (CFU-GM; progenitors committed to the granulocyte and/or monocyte lineage). Human hematopoietic cultures were found to be extremely sensitive to the substrate material in serum-free medium.[5]

The efficiency of *ex vivo* expansion of HSPCs on conventional polymeric materials as well as on metal and glass is summarized in Table 2, where total cell numbers and CFU-GM expansion ability are shown relative to polystyrene materials. Comments on the demerits of certain materials used for CFU-GM expansion in this study, such as protein adsorption and/or leaching of toxins, are summarized from Laluppa et al.[5] Of the materials tested, tissue culture polystyrene, cellulose acetate, perfluoroalkoxy Teflon, and titanium supported expansion at or near the levels of polystyrene (Table 2).

CFU-GM expansion was more sensitive to substrate materials than was total cell expansion. The detrimental effects of some materials on HSPC culture were ascribed to protein adsorption and/or leaching of toxins. Factors such as cleaning, sterilization, and reuse also significantly affected the performance of some materials as HSPC culture substrates.[5]

An HSPC cell culture was also used to examine the biocompatibility of gas-permeable cell culture bags and blood storage bags, and several types of tubing commonly used with biomedical equipment. While many of the culture bag materials gave satisfactory results, all of the tubing

Table 2 Efficiency of *ex vivo* expansion of HSPCs on conventional polymer materials, metal and glass compared to that on polystyrene

Materials	Total cells (relative to polystyrene)	CFU-GM (relative to polystyrene)	Cause of adverse effects
Glass	Poor	Poor	Protein adsorption
Stainless 316	Poor	Poor	Protein adsorption/leaching
Titanium	Good/excellent	Good/excellent	
PTFE	Poor	Poor	Protein adsorption/leaching
PFA	Excellent	Excellent	
Cellulose acetate	Excellent	Good	
HDPE	Good	Poor	
Polyethylene	Poor	Poor	Unknown
Polypropylene	Poor	Poor	Unknown
PET	Good/poor	Poor	Protein adsorption/leaching
TCPS	Excellent	Excellent	

Abbreviations: PTFE, polytetrafluoroethylene; PFA, Teflon perfluoralkoxy; HDPE, High density polyethylene; PET, Polyethylene terephthalate; TCPS, Tissue culture polystyrene dishes.
Source: Summarized from Laluppa et al.[5]

materials severely inhibited total cell and CFU-GM expansion. Many materials approved for blood contact or considered biocompatible were nevertheless unsuitable for use in HSPC culture employing serum-free medium.[5]

Neussa et al. reported systematic and combinatorial approaches to identify biomaterials suitable for *ex vivo* culture of embryonic and adult stem cells (e.g., HSPCs, MSCs and dental pulp stem cells). They assessed 140 combinations of seven different stem cell types and 19 different polymers (Table 3), performing systematic screening assays to analyze parameters such as morphology, vitality, cytotoxicity (from lactate-dehydrogenase secretion), apoptosis (from aspase 3/7 activity), and proliferation, using a grid-based platform.[37]

The materials evaluated were as follows:

1. Two fluorine-containing nondegradable polymers, (1) poly(vinylidene fluoride) (PVDF) and (2) polytetrafluoroethylene (PTFE)
2. Seven degradable synthetic polymers, the Resomer types (3) L209S, (4) R203S, (5) RG503, and (6) LT706, the polyesteramides (7) type BAK1095 and (8) type C (PEA-C), and (9) poly(ε-caprolactone) (PCL)
3. Four biopolymers, (10) alginate, (11) collagen, (12) fibrin, and (13) hyaluronic acid
4. Two synthetic biodegradable copolymers, (14) poly(ethylene oxide)-*b*-poly(ε- caprolactone), and (15) poly(ethylene oxide)-*b*-polydimeric D,L-lactic acid; and
5. Three nondegradable synthetic polymers, (16) poly(dimethylsiloxane)(PDMS), (17) an aromatic polyether-based thermoplastic polyurethane, Texin 950, and (18) poly(ethylene terephthalate) (PET).[37]

Certain polymers preferentially allowed attachment by specific stem cell types, e.g., only mouse MSCs adhered to PEA-C, and only dental pulp stem cells and HSPCs adhered to PDMS. A high number of viable cells with a low cytotoxic and apoptotic signal was found for human endothelial progenitor cells on PTFE, and for dental pulp stem cells on

poly(D,L-lactic acid) (Resomer R203S), whereas HSPCs grown on these materials yielded a low number of viable cells with a high cytotoxic and apoptotic signal.[37]

On a given polymer, differences were observed in proliferation rates between stem cells from different tissues and between the same types of stem cells from different species (Table 3). Polyesteramide BAK 1095 supported the proliferation of human dental pulp stem cells, but not of human MSCs, i.e., a cell type-specific influence was observed, although the two stem cells have very similar biological characteristics. Further, PTFE allowed mouse MSCs to proliferate, but not human MSCs, indicating a species-specific influence.[37]

In any basic assessment of biomaterials used for stem cell culture, strict attention must be paid to the analysis of parameters such as material topography, cell adhesion, morphology, viability, proliferation, cytotoxicity, and apoptosis. The following combinations of stem cells and polymers for tissue-engineering applications were recommended, where apoptosis and necrosis were inhibited, while cell adhesion and proliferation were supported: human dental pulp stem cells on poly(D,L-lactic acid) (R203S); human endothelial progenitor cells on PVDF, PTFE, Texin, poly(L-lactic acid) (L209S), Resomer LR705, and collagen; human preadipocytes on Texin; mouse MSCs on PVDF and Texin; and mouse ES cells on fibrin (Table 3). No materials among the conventional synthetic and natural polymers investigated met these criteria for human MSCs or HSPCs.[37]

Materials Manufactured with Nanotechnology and Having Nanosegments

Recent developments in nanotechnology have yielded nanofibers for use as scaffolds in tissue engineering[39–43] and as cell culture materials.[44–49] Electrospinning provides a simple and versatile method to generate nanofibers from various materials including polymers, composites, and

Table 3 Combinations of stem cells and recommended culture materials

Stem cells	Culture materials								
	Collagen	Fibrin	Texin	PDMS	PTFE	PVDF	PEA-C	PLLA	PDLLA
Preadipocytes			O						
Human endotherial progenitor cells	O		O		O	O		O	
Dental pulp stem cells									O
HSPCs									
hMSCs									
mMSCs			O			O			
mESs		O							

Abbreviations: PDMS, poly(dimethyl siloxane); PTFE, poly(tetrafluor ethylene); PVDF, poly(vinylidene fluoride); PEA-C, polyesteramide type C; PLLA, poly(L-lactic acid); PDLLA, poly(D,L-lactic acid).
Source: Summarized Neussa et al.[37]

ceramics,[50,51] and Chua et al.[3] used it to prepare poly-ethersulfone (PES) nanofibers. PES nanofibers as well as PES films were surface-modified by derivatization with carboxylate, hydroxyl, and amine groups (Fig. 2). HSPC expansion was analyzed on these surface-modified nanofibers and films using frozen CD34+ cells as HSPCs obtained from UCB.[3]

Ex vivo expansion of human UCB CD34+ cells was analyzed on unmodified, carboxylated, hydroxylated, and aminated nanofibers and films. Results from 10-day expansion cultures showed that aminated nanofiber mesh and film were most efficient in supporting the expansion of the CD34+CD45+ cells (195- and 178-fold, respectively), compared to tissue culture polystyrene (50-fold). Aminated nanofiber meshes resulted in a slightly higher degree of cell adhesion and percentage of HSPCs relative to aminated films. SEM imaging revealed discrete colonies of cells proliferating and interacting with the aminated nanofibers. These results suggested that HSPC culture materials having amine functional groups were more effective than those carrying carboxylate or hydroxyl groups.[3]

Chua et al.[3] also prepared nanofibers with amine groups conjugated to the fiber surface through different spacers (ethylene, butylene and hexamethylene groups, respectively), and have investigated the effect of spacer length on adhesion and expansion of HSPCs from UCB. Aminated nanofibers with ethylene and butylene spacers showed high expansion efficiencies (773- and 805-fold expansion of total cells; and 200- and 235-fold expansion of CD34+CD45+ cells, respectively). HSPC proliferation on aminated nanofibers with the hexamethylene spacer was significantly lower (210-fold expansion of total cells and 86-fold expansion of CD34+CD45+ cells), but resulted in the highest CD34+CD45+ cell fraction (41.1%). Colony-forming unit granulocyte-erythrocyte-monocyte-megakaryocyte and LTC-IC maintenance showed the same tendencies when HSPCs were expanded on all three aminated nanofibers. However, the engraftment potential in nonbese diabetic/severe combined immunodeficiency

(NOD/SCID) mice of HSPCs expanded on the aminoethyl- and aminobutyl-conjugated nanofibers was found to be significantly higher than for those expanded on aminohexyl-conjugated nanofibers.[1]

The enhanced HSPC adhesion to these fibers suggested that aminated nanofibers were superior substrates for *ex vivo* HSPC expansion. The spacer used to conjugate the amine groups to the nanofiber surface clearly affected the expansion outcome.

These studies highlight the potential of a biomaterials approach to regulate and influence the proliferation and differentiation of HSPCs *ex vivo*.[3] Table 4 summarizes the properties reported in the literature of culture materials having several synthetic nanosegments, and used for *ex vivo* expansion of HSPCs.

Polymeric Materials Modified with Immobilized Proteins and Oligopeptides

HSPCs *in vivo* reside in specialized niches in BM, where they participate in multiple interactions with stromal cells and ECM molecules.[52] This BM stem cell niche provides cardinal signals that regulate HSPC self-renewal, differentiation, migration, and homing.[6,53–67]

ECM components are crucial elements in the HSPC niche. They mediate the adhesive interactions between HSPCs and CAMs, which are of critical importance in the regulation of hematopoiesis by interacting HSPCs in the BM niche, in close contact with osteoblasts, stromal cells, thus exposing them to the signaling molecules they secrete.[11,58,59]

Adhesive interactions between ECM molecules and various integrin receptors on HSPCs are an important part of signaling control. FN, one of the most important ECM molecules, is directly involved in the adhesion and proliferation of HSPCs.[6,60,61] The data from investigations into HSPC expansion on FN-based materials as well as on materials having adhesion motif peptides are summarized in Table 5.

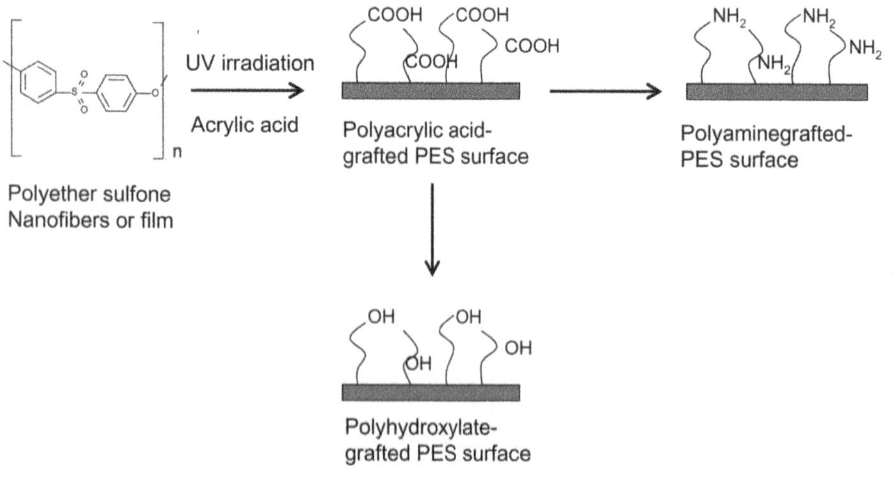

Fig. 2 Chemical modification of PES film.
Source: Adapted from Chua et al. (Fig. 1).[3]

Table 4 Polymeric materials immobilized synthetic nanosegments in *ex vivo* expansion of HSPCs from literatures

S. no.	Nanosegment	Base materials	Immobilization method	Stem cell source	References
1.	PAAc	PES Nanofiber	Covalent bonding	UCB	[1]
2.	EtDA	PES Nanofiber	Covalent bonding	UCB	[1]
3.	BuDA	PES Nanofiber	Covalent bonding	UCB	[1]
4.	HeDA	PES Nanofiber	Covalent bonding	UCB	[1]
5.	PAAc	PES Nanofiber	Covalent bonding	UCB	[3]
6.	EtDA	PES Nanofiber	Covalent bonding	UCB	[3]
7.	EtDA	PES film	Covalent bonding	UCB	[3]
8.	EtOH	PES Film	Covalent bonding	UCB	[3]
9.	AMA	TCPS	Covalent bonding	UCB	[81]

Abbreviations: PAAc, poly(acrylic acid); EtDA, 1,2-ethanediamine; BuDA, 1,4-butanediamine; HeDA, 1,6-hexanediamine; EtOH, hydroxyethylamine; PES, polyethersulfone; UCB, umbilical cord blood; AMA, poly(2-aminoethyl methacrylate).

Table 5 Polymeric materials immobilized nanosegments (protein and oligopeptide) in *ex vivo* expansion of HSPCs from literatures

S. no.	Nanosegment	Base materials	Immobilization method	Stem cell source	References
1.	ELIDVPST (CS-1)	PET	Covalent bonding	UCB	[6]
2.	ELIDVPST (CS-1)	TCPS	Covalent bonding	UCB	[81]
3.	ELIEVPST (CS-1i, dummy)	PET	Covalent bonding	UCB	[6]
4.	GRGDSPC	PET	Covalent bonding	UCB	[6]
5.	GRGESPC (dummy for RGDS)	PET	Covalent bonding	UCB	[6]
6.	FN	PET	Covalent bonding	UCB	[11]
7.	FN	TCPS	Covalent bonding	UCB	[81]
8.	COL	PET	Covalent bonding	UCB	[11]
9.	FN & LN	TCPS	Coating	Murine BM	[67]
10.	FN fragment containing CS1	TCPS	Coating	BM	[17]
11.	FN fragment containing RGDS	TCPS	Coating	BM	[17]
12.	FN fragment containing integrin-binding domain	TCPS	Coating	BM	[17]
13.	FN	PEMA	Covalent bonding	PB	[7]
14.	HEP	PEMA	Covalent binding	PB	[7]
15.	HepS	PEMA	Covalent binding	PB	[7]
16.	Hya	PEMA	Covalent binding	PB	[7]
17.	T-coll	PEMA	Covalent binding	PB	[7]
18.	Col/Hep	PEMA	Covalent binding	PB	[7]
19.	Col/Hya	PEMA	Covalent binding	PB	[7]

Abbreviations: FN, fibronectin; Col, collagen; LN, laminin; Hep, heparin; HepS, heparan sulphate; Hya, hyaluronic acid; T-coll, tropocollagen I; PET, polyethylene terephtharate; TCPS, tissue culture polystyrene flask; PEMA, poly(ethylene-*alt*-maleic anhydride); UCB, umbilical cord blood; BM, bone marrow; PB, peripheral blood.

Franke et al. have cultured HSPCs on poly(ethylene-*alt*-maleic anhydride) films on which several kinds of ECM (FN, heparin, heparan sulphate, hyaluronic acid, tropocollagen I, and co-fibrils of collagen I with heparin or hyaluronic acid) were immobilized, to analyze the role of ECM components in the niche microenvironment. The attachment of human HSPCs from G-CSF mobilized peripheral blood were evaluated *in vitro*.[7] The extent of the adhesion areas of individual cells and the fraction of adherent cells were analyzed by reflection interference contrast microscopy. Tight cell–matrix binding occurred with surface-immobilized FN, heparin, heparan sulfate, and the collagen I based co-fibrils, whereas little HSPC adhesion took place with tropocollagen I or hyaluronic acid. HSPCs adhered most strongly to FN with contact areas of about 7 mm^2. HSPCs also interacted with surfaces modified using heparin, heparan sulfate, and the co-fibrils, resulting in small circular contact zones of 3 mm^2, indicating less efficient adhesion than with FN. Antibody blocking experiments to evaluate the specificity of cell–matrix interactions suggested that the FN bound to an integrin ($\alpha_5\beta_1$)-specific receptor on HSPCs, while HSPC adhesion on heparin was mediated by selectins (CD62L).[7]

The integrin-mediated adhesion of HSPCs on ECM proteins such as FN was found to generate a larger adhesion area compared to the selectin-mediated adhesion on heparin. Force spectroscopy studies showed that the $\alpha_5\beta_1$-integrin binding to FN was stronger than the binding of L-selectin to carbohydrates.[62,63] This might have been because the function of integrins is their intracellular linkage to the actin cytoskeleton and because of their role in inside-out and outside-in signaling of adhesion. Therefore, the interaction of HSPCs with different matrix components, i.e., proteins and glycosaminoglycans, can be regarded as an important switch in stem cell fate control in the BM niche.[64] Cell shape, which is regulated by extracellular linkages to matrix constraints, is directly related to cell function down to the level of protein expression and cell cycle control.[7,65,66]

The work of Franke et al. focused solely on the adhesion area of HSPCs on ECM-derivatized surfaces; the primitiveness and expansion ability of HSPCs on these surfaces was not investigated. Sagar et al. have investigated the expansion of HSPCs from murine BM on FN- and laminin-coated plates in the presence of cytokines to evaluate whether ECM components helped maintain stem cell properties.[67] They observed significant phenotypic and functional improvement of expanded cells [e.g., 800-fold expansion of colony-forming units-granulocyte erythrocyte monocyte megakaryocyte (CFU-GEMM) after 10-day culture] cultured on these coated plates.[67] There was no apparent activation of the cell cycle, although CD29 and very late antigen-4 (VLA-4) expression was increased on both types of coated plates relative to fresh HSPCs in BM. A fraction of the expanded cells became verapamil sensitive, which suggested upregulation of multidrug resistance genes in cells on the coated plates, as found in primitive HSPCs. The HSPC compartment was found to be amplified in a competitive repopulation assay during culture on the coated plates. These studies demonstrated that *ex vivo* culture of murine HSPCs in the presence of FN and laminin resulted in the expansion of primitive stem cells and improvement in marrow engraftibility.[67] It was found that the direct interactions with FN enhanced preservation of primitive hematopoietic progenitors during culture and promoted self-renewing divisions of proliferating primitive hematopoietic cells.

Bhatia, Williams, and Munthe have investigated the role of direct contact of HSPCs with FN protein fragments for the preservation and self-renewal of primitive progenitor

potential. BM CD34$^+$ cells were cultured on several integrin-binding FN fragments (e.g., FN-40, a COOH-terminal FN fragment containing the heparin binding and CS1 domains ($\alpha_4\beta_1$ integrin binding FN fragment); FN-120, a fragment containing the cell binding sequence RGDS (FN-120, $\alpha_5\beta_1$ binding); and FN-CH296, a recombinant FN fragment containing $\alpha_4\beta_1$ and $\alpha_5\beta_1$ integrin-binding domains) in growth factor-containing medium.[17] Primitive progenitors of HSPCs were assessed using week 6 LTC-IC and week-10 extended LTC-IC (ELTC-IC) assays. Increased LTC-IC and ELTC-IC preservation was observed after culture in contact with FN-coated plates. Both $\alpha_4\beta_1$ and $\alpha_5\beta_1$-integrin binding FN fragments (i.e., FN-40, FN-120 and FN-CH296) coated on the plates enhanced LTC-IC preservation. Analysis of single CD34$^+$CD38$^-$ cells cultured on FN fragments-coated plates revealed significantly reduced cell division, but enhanced retention of LTC-IC capacity in divided cells. FN binding of HSPCs cultured on the plates increased LTC-IC frequency in undivided cells.[17]

Direct interactions of HSPCs with FN through β_1 integrin enhanced preservation of primitive hematopoietic progenitors during serum-free culture and promoted self-renewing divisions of proliferating primitive hematopoietic cells, as was found for stromal co-culture of HSPCs.[67,68]

Two of the most important adhesion domains in FN are the connecting segment-1 (CS-1) and the RGD motif, both of which are recognized by surface receptors on early hematopoietic progenitors. The CS-1 domain (EILDVPST) binds to the VLA-4 integrin receptor (very late antigen-4, $\alpha_4\beta_1$, CD49d/CD29) while the RGD sequence binds to the VLA-5 integrin receptor ($\alpha_5\beta_1$, CD49e/CD29) found on these early progenitor HSPCs.[69–71] The FN-VLA-4 interaction plays an important role in hematopoiesis,[72–75] and both the CS-1 and RGD segments are structurally important for the growth-supporting effects of FN.[76] It has also been reported that the VLA-4 binding sequence of FN stimulates the proliferation of human cord blood CD34$^+$ cells.[6,60]

Jiang et al.[6] have prepared PET films carrying immobilized peptides containing the CS-1 binding motif (EILDVPST) and the RGD motif (GRGDSPC) (Fig. 3). The *ex vivo* expansion of human UCB CD34$^+$ cells (HSPCs) on these films was carried out over 10 days in serum-free medium. The greatest cell expansion was observed on the CS-1 peptide-modified films, where total nucleated cells, total colony forming units, and LTC-IC were expanded 590-, 767-, and 3-fold, respectively, compared to unexpanded CD34$^+$ cells. All films carrying immobilized peptides on the surface, including control peptides, more efficiently supported the expansion of CD34$^+$ cells, CFU-GEMMs and LTC-ICs than the tissue culture polystyrene surface. However, only cells cultured on the CS-1-immobilized film could generate positive engraftment after 10-days of *ex vivo* expansion from 600 CD34$^+$ cells. The PET film immobilized with the RGD peptide was less efficient than the corresponding

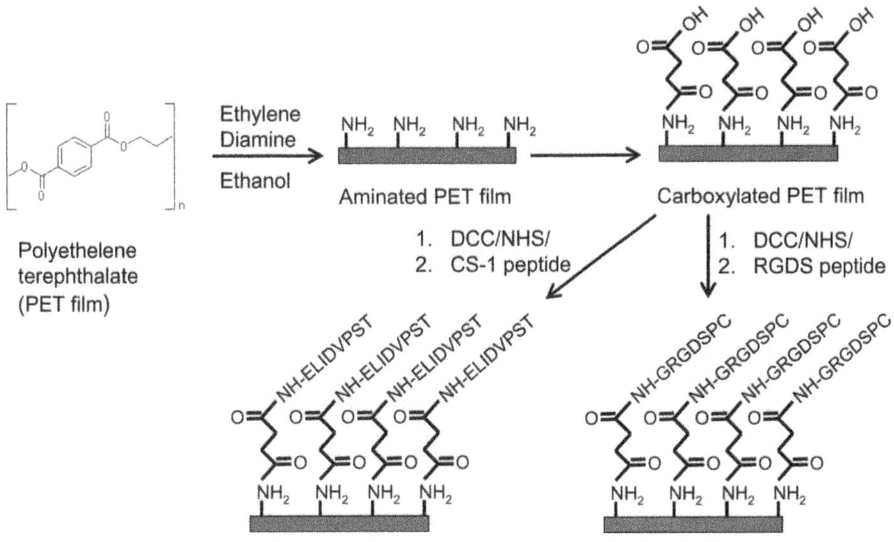

Fig. 3 Surface modification of PET film immobilized with peptides.
Source: Adapted from Jiang et al. (Fig. 1).[6]

CS-1 peptide film. These results suggested that covalently immobilized adhesion peptides could significantly influence the proliferation characteristics of HSPCs cultured from UCB.[6]

MATERIALS FOR 3D CULTURE

An efficient and practical *ex vivo* expansion methodology for human HSPCs is critical for realizing the potential of HSPC transplantation in treating a variety of hematologic disorders and as a supportive therapy for malignant diseases.[11]

Several studies have suggested that 3D culture of HSPCs provides the proper microenvironment to regulate HSPC proliferation and differentiation.[77–80] Bagley et al. have shown that a 3D scaffold of porous tantalum supported the maintenance of primitive CD34+CD38− cells for up to 6 weeks, and yielded a 6.7-fold increase in colony forming cells (CFCs) without supplementing cytokines in serum-free medium.[78] Li et al. have cultured human cord blood cells in a 3D scaffold of nonwoven PET mesh, and have found significantly higher numbers of CD34+ cells and CFCs after 7–9 weeks of culture in the 3D scaffold compared to a 2D substrate.[79] Feng et al. have synthesized 3D PET scaffolds carrying surface-immobilized FN or collagen, thus mimicking the characteristics of the HSPC niche.[11]

Covalent conjugation of FN on a 3D PET mesh resulted in higher expansion efficiency of HSPCs than on an unmodified 3D PET mesh or on PET films conjugated with FN. After 10 days of culture in serum-free medium, CD34+ cells cultured in a 3D scaffold conjugated with FN yielded the highest expansion of CD34+ cells (100-fold) and LTC-IC (47-fold). The expanded human HSPCs successfully reconstituted hematopoiesis in NOD/SCID mice. This study demonstrated the synergistic effect between the three-dimensionality of the scaffold and surface-conjugated FN, and the potential of FN-conjugated 3D scaffolds for *ex vivo* expansion of HSPCs.[11]

CONCLUSIONS AND FUTURE PERSPECTIVES

The key goal in *ex vivo* expansion of HSPCs is to mimic the microenvironment of the BM niche using cell culture materials.

Recent studies have shown that *ex vivo* expansion of HSPCs is possible without co-culture with MSCs.[1,3,6,7,17] The success of *ex vivo* expansion of HSPCs by co-culture with MSCs in growth factor-containing medium has been attributed to the direct contact of HSPCs with FN through the β_1 integrin family such as $\alpha_4\beta_1$ and $\alpha_5\beta_1$ integrins. The direct contact of HSPCs with FN or FN fragments is expected to be sufficient for the expansion and maintenance of the primitiveness of HSPCs *ex vivo*. It was also found possible to expand HSPCs on cell culture dishes having amine groups attached through spacer segments of sufficient length.[1]

It is still unclear whether the direct signaling to HSPCs through $\alpha_4\beta_1$ and $\alpha_5\beta_1$ integrins is absolutely necessary for *ex vivo* expansion or whether human HSPCs cultured on amine-derivatized surfaces could also be used for engraftment.[81–83] It is necessary to compare the expansion efficiency and engraftment efficiency of HSPCs cultured on FN-immobilized surfaces and aminated surfaces to resolve this question.

3D culture materials carrying immobilized FN performed well in the *ex vivo* expansion of HSPCs.[11] However, more detailed studies on 3D culture of HSPCs are necessary to understand why 3D culture seems to be more effective than 2D culture, besides the effect of the larger surface area per unit volume.

In conclusion, advanced polymeric materials are playing an important role in the development of modern regenerative medicine, and stem cell culture technology.[84,85]

ACKNOWLEDGMENTS

We acknowledge the International High Cited Research Group (IHCRG #14-104), Deanship of Scientific Research, King Saud University, Riyadh, Kingdom of Saudi Arabia. Akon Higuchi thanks King Saud University, Riyadh, Kingdom of Saudi Arabia, for the Visiting Professorship.

REFERENCES

1. Chua, K.-N.; Chaib, C.; Lee, P.-C.; Ramakrishna, S.; Leong, K.W.; Mao, H.-Q. Functional nanofiber scaffolds with different spacers modulate adhesion and expansion of cryopreserved umbilical cord blood hematopoietic stem/progenitor cells. Exp. Hematol. 2007, 35, (5), 771–781.

2. Cohen, Y; Nagler, A. Umbilical cord blood transplantation—How, when and for whom?. Blood Rev. 2004, 18 (3), 167–79.

3. Chua, K.N.; Chai, C.; Lee, P.C.; Tang, Y.-N.; Ramakrishna, S.; Leong, K.W.; Mao, H.-Q. Surface-aminated electrospun nanofibers enhance adhesion and expansion of human umbilical cord blood hematopoietic stem/progenitor cells. Biomaterials 2006, 27 (36), 6043–6051.

4. Fujimoto, N.; Fujita, S.; Tsuji, T. Toguchida, J.; Ida, K.; Suginami, H.; Iwata H. Microencapsulated feeder cells as a source of soluble factors for expansion of CD34(+) hematopoietic stem cells. Biomaterials 2007, 28 (32), 4795–4805.

5. Laluppa, J.A.; McAdams, T.A.; Papoutsakis, E.T. Miller, W.M. Culture materials affect ex vivo expansion of hematopoietic progenitor cells. J. Biomed. Mater. Res. 1997, 36 (3), 347–359.

6. Jiang, X.S.; Chai, C.; Zhang, Y.; Zhuo, R.-X.; Mao, H.-Q.; Leong, K.W. Surface-immobilization of adhesion peptides on substrate for ex vivo expansion of cryopreserved umbilical cord blood CD34(+) cells. Biomaterials 2006, 27 (13), 2723–2732.

7. Franke, K.; Pompe, T.; Bornhauser, M.; Werner, C. Engineered matrix coatings to modulate the adhesion of CD133+ human hematopoietic progenitor cells. Biomaterials 2007, 28 (5), 836–843.

8. Eridani, S.; Mazza, U.; Massaro, P.; La Targia, M.L.; Maiolo, A.T.; Mosca, A. Cytokine effect on ex vivo expansion of haemopoietic stem cells from different human sources. Biotherapy 1998, 11 (4), 291–296.

9. Stewart, D.A.; Guo, D.; Luider, J.; Auer, I.; Klassen, J.; Ching, E. Morris, D.; Chaudhry, A.; Brown, C.; Russell, J.A. Factors predicting engraftment of autologous blood stem cells: CD34+ subsets inferior to the total CD34+ cell dose. Bone Marrow Transplant. 1999, 23 (12), 1237–1243.

10. Lee, S.H.; Lee, M.H.; Lee, J.H.; Min, Y, H.; Lee, K.H.; Cheong, J.W. Lee, J.; Park, K.W.; Kang, J.H.; Kim, K.; Kim, W.S.; Jung, C.W.; Choi, S.-J.; Lee, J.-H.; Park, K. Infused CD34+ cell dose predicts long-term survival in acute myelogenous leukemia patients who received allogeneic bone marrow transplantation from matched sibling donors in first complete remission. Biol. Blood Marrow Transplant. 2005, 11 (2), 122–128.

11. Feng, Q.; Chai, C.; Jiang, X.S. Leong, K.W. Mao, H.-Q. Expansion of engrafting human hematopoietic stem/progenitor cells in three-dimensional scaffolds with surface-immobilized fibronectin. J. Biomed. Mater. Res. 2006, 78A (4), 781–791.

12. Heike, T.; Nakahata, T. Ex vivo expansion of hematopoietic stem cells by cytokines. Biochim. Biophys. Acta 2002, 1592 (3), 313–321.

13. Bonnet, D. Biology of human bone marrow stem cells. Clin. Exp. Med. 2003, 3 (3), 140–149.

14. Takagi, M. Cell processing engineering for ex-vivo expansion of hematopoietic cells. J. Biosci. Bioeng. 2005, 99 (3), 189–196.

15. Fuchs, E.; Tumbar, T.; Guasch, G. Socializing with the neighbors: Stem cells and their niche. Cell 2004, 116 (6), 769–778.

16. Lemischka, I.R.; Moore, K.A. Stem cells: Interactive niches. Nature 2003, 425 (6960), 778–779.

17. Bhatia, R.; Williams, A.D.; Munthe, H.A. Contact with fibronectin enhances preservation of normal but not chronic myelogenous leukemia primitive hematopoietic progenitors. Exp. Hematol. 2002, 30 (4), 324–332.

18. Breems, D.A.; Blokland, E.A.W.; Siebel, K.E.; Mayen, A.E.M.; Engels, L.J.A.; Ploemacher, R.E. Stroma-contact prevents loss of hematopoietic stem cell quality during ex vivo expansion of CD34+ mobilized peripheral blood stem cells. Blood 1998, 91 (1), 111–117.

19. Verfaillie, C.; Hurley, R.; Bhatia, R.; McCarthy, J.B. Role of bone marrow matrix in normal and abnormal hematopoiesis. Crit. Rev. Oncol. Hematol. 1994, 16 (3), 201–224.

20. Hurley, R.W.; McCarthy, J.B.; Verfaillie, C.M. Direct adhesion to bone marrow stroma via fibronectin receptors inhibits hematopoietic progenitor proliferation. J. Clin. Invest 1995, 96 (1), 511–519

21. Sutherland, H.J.; Eaves, C.J.; Lansdorp, P.M.; Thacker, J.D.; Hogge, D.E. Differential regulation of primitive human hematopoietic cells in long-term cultures maintained on genetically engineered murine stromal cells. Blood 1991, 78 (3), 666–672.

22. Croisille, L.; Auffray, I.; Katz, A.; Izac, B.; Vainchenker, W.; Coulombel, L. Hydrocortisone differentially affects the ability of murine stromal cells and human marrow-derived adherent cells to promote the differentiation of CD34++/CD38- long-term culture-initiating cells. Blood 1994, 84 (12), 4116–4124.

23. Nakano, T.; Kodama, H.; Honjo, T. Generation of lymphohematopoietic cells from embryonic stem cells in culture. Science 1994, 265 (5175), 1098–1101.

24. Nakano, T.; Kodama, H.; Honjo, T. In vitro development of primitive and definitive erythrocytes from different precursors. Science 1996, 272 (5262), 722–724.

25. Tsuji, T.; Ogasawara, H.; Aoki, Y.; Tsurumaki, Y.; Kodama, H. Characterization of murine stromal cell clones established from bone marrow and spleen. Leukemia 1996, 10 (5), 803–812.

26. Kawada, H.; Ando, K.; Tsuji, T.; Shimakura, Y.; Nakamura, Y.; Chargui, J.; Hagihara, M.; Itagaki, H.; Shimizu, T.; Inokuchi, S.; Kato, S.; Hotta, T. Rapid ex vivo expansion of human umbilical cord hematopoietic progenitors using a novel culture system. Exp. Hematol. 1999, 27 (5), 904–915.

27. Wagner, W.; Weina, F.; Roderburga, C.; Saffricha, R.; Fabera, A.; Krausea, U.; Schuberta, M.; Benesc, V.; Ecksteina, V.; Mauld, H.; Hoa, A.D. Adhesion of hematopoietic progenitor cells to human mesenchymal stem cells

as a model for cell-cell interaction. Exp. Hematol. **2007**, *35* (2), 314–325.

28. Patience, C.; Takeuchi, Y.; Weiss, R.A. Infection of human cells by an endogenous retrovirus of pigs. Nat. Med. **1997**, *3* (3), 282–286.

29. Martin, M.J.; Muotri, A.; Gage, F.; Varki, A. Human embryonic stem cells express an immunogenic nonhuman sialic acid. Nat. Med. **2005**, *11* (2), 228–232.

30. Kögler, G.; Radke, T.F.; Lefort, A.; Sensken, S.; Fischer, J.; Sorg, R.V.; Wernet, P. Cytokine production and hematopoiesis supporting activity of cord blood-derived unrestricted somatic stem cells. Exp. Hematol. **2005**, *33* (5), 573–583.

31. Bilko, N.M.; Votyakova, I.A.; Vasylovska, S.V.; Bilko, D.I. Characterization of the interactions between stromal and haematopoietic progenitor cells in expansion cell culture models. Cell Biol. Int. **2005**, *29* (1), 83–86.

32. Nilsson, K.; Scheirer, W.; Merten, O.W.; Ostberg, L.; Liehl, E.; Katinger, H.W.D.; Mosbach, K. Entrapment of animal cells for production of monoclonal antibodies and other biomolecules. Nature **1983**, *302* (5909), 629–630.

33. Chaikof, E.L. Engineering and material considerations in islet cell transplantation. Annu. Rev. Biomed. Eng. **1999**, *1* (1), 103–127.

34. Avgoustiniatos, E.S.; Colton, C.K. Effect of external oxygen mass transfer resistances on viability of immunoisolated tissue. Ann. N Y Acad. Sci. **1997**, *831* (1), 145–167.

35. Loeuillet, C.; Bernard, G.; Remy-Martin, J.; Saas, P.; Herve, P.; Douay, L.; Chalmers, D. Distinct hematopoietic support by two human stromal cell lines. Exp. Hematol. **2001**, *29* (6), 736–745.

36. Gluckman, E. *Ex vivo* expansion of cord blood cells. Exp. Hematol. **2004**, *32* (5), 410–412.

37. Neussa, S.; Apel, C.; Buttler, P.; Denecke, B.; Dhanasingh, A.; Ding, X.; Grafahrend, D.; Groger, A.; Hemmrich, K.; Herr, A.; Jahnen-Dechent, W.; Mastitskaya, S.; Perez-Bouza, A.; Rosewick, S.; Salber, J.; Woltje, M.; Zenke, M. Assessment of stem cell/biomaterial combinations for stem cell-based tissue engineering. Biomaterials **2008**, *29* (3), 302–313

38. Langer, R.; Tirrell, D.A. Designing materials for biology and medicine. Nature **2004**, *428* (6982), 487–492.

39. Janjanin, S.; Li, W.J.; Morgan, M.T.; Shanti, R.M.; Tuan, R.S. Mold-shaped, nanofiber scaffold-based cartilage engineering using human mesenchymal stem cells and bioreactor. J. Surg. Res. **2008**, *149* (1), 47–56.

40. Venugopal, J.; Low, S.; Choon, A.T.; Ramakrishna, S. Interaction of cells and nanofiber scaffolds in tissue engineering. J. Biomed. Mater. Res. **2008**, *84B* (1), 34–48.

41. Dang, J.M.; Leong, K.W. Myogenic induction of aligned mesenchymal stem cell sheets by culture on thermally responsive electrospun nanofibers. Adv. Mater. **2007**, *19* (19), 2775–2779.

42. Li, W.J.; Jiang, Y.J.; Tuan, R.S. Cell-nanofiber-based cartilage tissue engineering using improved cell seeding, growth factor, and bioreactor technologies. Tissue Eng. A **2008**, *14* (5), 639–648.

43. Han, I.; Shim, K.J.; Kim, J.Y.; Sung, Y.K.; Kim, M.; IKang, I.-K.; Kim, J.C. Effect of poly(3-hydroxybutyrate-*co*-3-hydroxyvalerate) nanofiber matrices cocultured with hair follicular epithelial and dermal cells for biological wound dressing. Artif. Org. **2007**, *31* (11), 801–808.

44. Jeong, S.I.; Jun, I.D.; Choi, M.J.; Nho, Y.C.; Lee, Y.M.; Shin, H. Development of electroactive and elastic nanofibers that contain polyaniline and poly(L-lactide-*co*-epsilon-caprolactone) for the control of cell adhesion. Macromol. Biosci. **2008**, *8* (7), 627–637.

45. Ma, K.; Chan, C.K.; Liao, S.; Hwang, W.Y.K.; Feng, Q.; Ramakrishna, S. Electrospun nanofiber scaffolds for rapid and rich capture of bone marrow-derived hematopoietic stem cells. Biomaterials **2008**, *29* (13), 2096–2103.

46. Vournakis, J.N.; Eldridge, J.; Demcheva, M.; Muise-Helmericks, R.C. Poly-N-acetyl glucosamine nanofibers regulate endothelial cell movement and angiogenesis: Dependency on integrin activation of Ets1. J. Vasc. Res. **2008**, *45* (3), 222–232.

47. van Aalst, J.A.; Reed, C.R.; Han, L.; Andrady, T.; Hromadka, M.; Bernacki, S.; Kolappa, K.; Collins, J.B.; Loboa, E.G. Cellular incorporation into electrospun nanofibers - Retained viability, proliferation, and function in fibroblasts. Ann. Plast. Surg. **2008**, *60* (5), 577–583.

48. Li, W.S.; Guo, Y.; Wang, H.; Shi, D.; Liang, C.; Ye, Z.; Qing, F.; Gong, J. Electrospun nanofibers immobilized with collagen for neural stem cells culture. J. Mater. Sci. Mater. Med. **2008**, *19* (2), 847–854.

49. Tian, F.; Hosseinkhani, H.; Hosseinkhani, M.; Hosseinkhani, M.; Khademhosseini, A.; Yokoyama, Y.; Estrada, G.G.; Kobayashi, H. Quantitative analysis of cell adhesion on aligned micro- and nanofibers. J. Biomed. Mater. Res. **2008**, *84A* (2), 291–299.

50. Jin, Y.; Yang, D.; Zhou, Y.; Ma, G.; Nie, J. Photocrosslinked electrospun chitosan-based biocompatible nanofibers. J. Appl. Polym. Sci. **2008**, *109* (5), 3337–3343.

51. Li, D.; Xia, Y. Electrospinning of Nanofibers: Reinventing the Wheel? Adv. Mater. **2004**, *16* (14), 1151–1170.

52. Rafii, S.; Mohle, R.; Shapiro, F.; Frey, B.M.; Moore, M.A. Regulation of hematopoiesis by microvascular endothelium. Leuk Lymphoma **1997**, *27* (5–6), 375–386.

53. Dexter, T.M.; Allen, T.D.; Lajtha, L.G. Conditions controlling the proliferation of haemopoietic stem cells *in vitro*. J. Cell Physiol. **1977**, *91* (3), 335–344.

54. Tavassoli, M.; Friedenstein, A. Hemopoietic stromal microenvironment. Am. J. Hematol. **1983**, *15* (2), 195–203.

55. Torok-Storb, B. Cellular interactions. Blood **1988**, *72* (2), 373–385.

56. Gordon, M.Y. Extracellular matrix of the marrow microenvironment. Br. J. Haematol. **1988**, *70* (1), 1–4.

57. Boudreau, N.J.; Jones, P.L. Extracellular matrix and integrin signalling: The shape of things to come. Biochem. J. **1999**, *339* (Pt. 3), 481–488.

58. Verfaillie, C.M.; Gupta, P.; Prosper, F.; Hurley, R.; Lundell, B.; Bhatia, R. The hematopoietic microenvironment: Stromal extracellular matrix components as growth regulators for human hematopoietic progenitors. Hematology **1999**, *4* (4), 321–333.

59. Potocnik, A.J.; Brakebusch, C.; Fassler, R. Fetal and adult hematopoietic stem cells require β1 integrin function for colonizing fetal liver, spleen, and bone marrow. Immunity **2000**, *12* (6), 653–663.

60. Schofield, K.P.; Humphries, M.J.; de Wynter, E.; Testa, N.; Gallagher, J.T. The effect of alpha 4beta 1-integrin binding sequences of fibronectin on growth of cells from human hematopoietic progenitors. Blood **1998**, *91* (9), 3230–3238.

61. Verfaillie, C.M.; McCarthy, J.B.; McGlave, P.B. Differentiation of primitive human multipotent hematopoietic progenitors into single lineage clonogenic progenitors is accompanied by alterations in their interaction with fibronectin. J. Exp. Med. **1991**, *174* (3), 693–703.

62. Li, F.; Redick, S.D.; Erickson, H.P.; Moy, V.T. Force measurements of the $\alpha_5\beta_1$ integrin–fibronectin interaction. Biophys. J. **2003**, *84* (2), 1252–1262.

63. Evans, E. Looking inside molecular bonds at biological interfaces with dynamic force spectroscopy. Biophys. Chem. **1999**, *82* (2), 83–97.

64. Ingberm, D.E. Mechanical signaling and the cellular response to extracellular matrix in angiogenesis and cardiovascular physiology. Circ. Res. **2002**, *91* (10), 877–887.

65. Huang, S.; Ingber, D.E. The structural and mechanical complexity of cell-growth control. Nat. Cell Biol. **1999**, *1* (5), E131–E138.

66. McBeath, R.; Pirone, D.M.; Nelson, C.M.; Bhadriraju, K.; Chen, C.S. Cell shape, cytoskeletal tension, and RhoA regulate stem cell lineage commitment. Dev. Cell **2004**, *6* (4), 483–495.

67. Sagar, B.M.M.; Rentala, S.; Gopal, P.N.V.; Sharma, S.; Mukhopadhyay, A. Fibronectin and laminin enhance engraftibility of cultured hematopoietic stem cells. Biochem. Biophys. Res. Comm. **2006**, *350* (4), 1000–1005.

68. Orschell-Travcoff, C.M.; Hiatt, K.; Dagher, R.N.; Rice, S.; Yoder, M.C.; Srour, F.F. Homing and engraftment potential of Sca-1⁺Lin⁻ cells fractionated on the basis of adhesion molecule expression and position in cell cycle. Blood **2000**, *96* (4), 1380–1387.

69. Rosemblatt, M.; Vuillet-Gaugler, M.H.; Leroy, C.; Coulombel, L. Coexpression of two fibronectin receptors, VLA-4 and VLA-5, by immature human erythroblastic precursor cells. J. Clin. Invest. **1991**, *87* (1), 6–11.

70. Teixido, J.; Hemler, M.E.; Greenberger, J.S.; Anklesaria, P. Role of beta 1 and beta 2 integrins in the adhesion of human CD34hi stem cells to bone marrow stroma. J. Clin. Invest. **1992**, *90* (2), 358–367.

71. Kerst, J.M.; Sanders, J.B.; Slaper-Cortenbach, I.C.; Doorakkers, M.C.; Hooibrink, B.; van Oers, R.H.; von dem Borne, A.E.; van der Schoot, C.E. Alpha 4 beta 1 and alpha 5 beta 1 are differentially expressed during myelopoiesis and mediate the adherence of human CD34+ cells to fibronectin in an activationdependent way. Blood **1993**, *81* (2), 344–351.

72. Ryan, D.H.; Nuccie, B.L.; Abboud, C.N.; Winslow, J.M. Vascular cell adhesion molecule-1 and the integrin VLA-4 mediate adhesion of human B cell precursors to cultured bone marrow adherent cells. J. Clin. Invest. **1991**, *88* (3), 995–1004.

73. Williams, D.A.; Rios, M.; Stephens, C.; Patel, V.P. Fibronectin and VLA-4 in haematopoietic stem cellmicroenvironment interactions. Nature **1991**, *352* (6334), 438–441.

74. Yanai, N.; Sekine, C.; Yagita, H.; Obinata, M. Roles for integrin very late activation antigen-4 in stroma-dependent erythropoiesis. Blood **1994**, *83* (10), 2844–2850.

75. Hamamura, K.; Matsuda, H.; Takeuchi, Y.; Habu, S.; Yagita, H.; Okumura, K. A critical role of VLA-4 in erythropoiesis *in vivo*. Blood **1996**, *87* (6), 2513–2507.

76. Yokota, T.; Oritani, K.; Mitsui, H.; Aoyama, K.; Ishikawa, J.; Sugahara, H.; Matsumura, I.; Tsai, S.; Tomiyama, Y.; Kanakura, Y.; Matsuzawa, Y. Growth-supporting activities of fibronectin on hematopoietic stem/progenitor cells *in vitro* and *in vivo*: structural requirement for fibronectin activities of CS1 and cell-binding domains. Blood **1998**, *91* (9), 3263–3272.

77. Banu, N.; Rosenzweig, M.; Kim, H.; Bagley, J.; Pykett, M. Cytokineaugmented culture of haematopoietic progenitor cells in a novel three-dimensional cell growth matrix. Cytokine **2001**, *13* (6), 349–358.

78. Bagley, J.; Rosenzweig, M.; Marks, D.F.; Pykett, M.J. Extended culture of multipotent hematopoietic progenitors without cytokine augmentation in a novel three-dimensional device. Exp. Hematol. **1999**, *27* (3), 496–504.

79. Li, Y.; Ma, T, Kniss, D.A.; Yang, S.T.; Lasky, L.C. Human cord cell hematopoiesis in three-dimensional nonwoven fibrous matrices: *In vitro* simulation of the marrow microenvironment. J. Hematother. Stem Cell Res. **2001**, *10* (3), 355–368.

80. Kim, H.S.; Lim, J.B.; Min, Y.H.; Lee, S.T.; Lyu, C.J.; Kim, E.S.; Kim, H.O. *Ex vivo* expansion of human umbilical cord blood CD34⁺ cells in a collagen bead-containing 3-dimensional culture system. Int. J. Hematol. **2003**, *78* (2), 126–132.

81. Chen, L.Y.; Chang, Y.; Shiao, J.S.; Ling, Q.D.; Chen, Y.H.; Chen, D.C.; Hsu, S.T.; Lee, H.H.; Higuchi, A. Effect of the surface density of nanosegments immobilized on culture dishes on *ex vivo* expansion of hematopoietic stem and progenitor cells from umbilical cord blood. Acta Biomater. **2012**, *8* (5), 1749–1758.

82. Higuchi, A.; Yang, S.T.; Li, P.T.; Tamai, M.; Tagawa, Y.; Chang, Y.; Chang, Y.; Ling, Q.D.; Hsu, S.T. Direct *ex vivo* expansion of hematopoietic stem cells from umbilical cord blood on membranes. J. Memb. Sci. **2010**, *351* (1–2), 104–111.

83. Higuchi, A.; Chen, L.Y.; Shiao, J.S.; Ling, Q.D.; Ko, Y.A.; Chang, Y.; Chang, Y.; Bing, J.T.; Hsu, S.T. Separation and cultivation of hematopoietic stem cells from umbilical cord blood by permeation through membranes with nanosegments. Curr. Nanosci. **2011**, *7* (6), 908–914.

84. Higuchi, A.; Ling, Q.D.; Ko, Y.A.; Chang, Y.; Umezawa, A. Biomaterials for the feeder-free culture of human embryonic stem cells and induced pluripotent stem cells. Chem. Rev. **2011**, *111* (5), 3021–3035.

85. Higuchi, A.; Ling, Q.D.; Hsu, S.T.; Umezawa, A. Biomimetic cell culture proteins as extracellular matrices for stem cell differentiation. Chem. Rev. **2012**, *112* (8), 4507–4540.

Stents: Endovascular

G. Lawrence Thatcher
TESco Associates Incorporated, Tyngsborough, Massachusetts, U.S.A.

Abstract
This entry presents an introduction to the development of polymeric endovascular stents, with the indications-for-use in peripheral and coronary revascularization. Because the ultimate goal is to have a device that is both clinically and commercially viable, the process of product realization is woven into this discussion with examples drawn from clinical and preclinical experience. The objective of this entry is to present the story of polymeric endovascular stents, in a translational approach, putting the body of work into context of the development path, and to draw attention to some of the multitude of interactions that can affect device efficacy and viability.

INTRODUCTION

This entry presents an introduction to the development of polymeric endovascular stents, with the indications-for-use in peripheral and coronary revascularization. Tsuji and Tamai and coworkers stated, "The success of biodegradable stents depends on not only the biocompatibility of the stent materials but also the ability of the *manufactured* stent itself."[1] Because the ultimate goal is to have a device that is both clinically and commercially viable, the process of product realization is woven into this discussion with examples drawn from clinical and preclinical experience. The objective is to present the story of polymeric endovascular stents, in a translational approach, putting the body of work into context of the development path, and to draw attention to some of the multitude of interactions that can affect device efficacy and viability. What are commonly referred to as "stent grafts," for the exclusion of aneurysms of the abdominal aorta, will only be mentioned in regard to the potential for a polymeric primary stent platform, leaving fuller discussion of the polymers used for their outer and inner lining constructs to entries discussing vascular grafts and tissue engineering of blood vessels. Polymer stent platforms may also be considered for the controlled release of active pharmaceutical ingredients (APIs), in addition to their drug eluting coatings, however, details of polymer systems for drug release are addressed in other entries. Since about 2006, concern has been raised regarding the increased risk of late and very late stent thrombosis occurring with drug eluting stents (commonly referred to as DES) constructed of a permanent metal platform with various drug release coatings. Therefore this entry focuses primarily on polymeric fully bioresorbable scaffolding.

CLINICAL PRELUDE

The tools and paradigms for the treatment of atherosclerotic vascular disease have undergone both revolutionary and evolutionary changes that have fueled the desire to develop fully bioresorbable stents.[2,3] Some developments, particularly in the arena of imaging, such as multislice spiral computed tomography (MSCT), quantitative coronary angiography (QCA), intravascular ultrasound (IVUS), and optical coherence tomography (OCT) (Fig. 1), have not only improved diagnoses and interventional treatment but also dramatically improved our means to evaluate new stent designs and stent materials.

Labinaz and coworkers[4] summarized Dr. Russell Ross's response-to-injury model to explain the development of atherosclerosis. Ross suggested that when the endothelium was injured, a complex cascade of events and healing responses formed a plaque comprised of various cells, fats or cholesterol, and fibrous tissue. These lesions could progress, possibly calcify, and eventually form a blockage (stenosis) in the artery, leading to reduced blood flow. Reduced blood flow, as well as a plaque rupture forming a clot, can cause a heart attack. This cascade of events has been the focus of significant research, and the response-to-injury model has been refined not only to help us understand responses to interventional therapies but also to target the development of new designs or materials.

In 1977, a decade after the introduction of coronary artery bypass grafting, balloon angioplasty was introduced. With the use of a percutaneous balloon catheter, blockages in arteries could be treated, improving blood supply. Onuma and Serruys characterized this treatment milestone as the first revolution in the field of revascularization.[3]

Initial catheters were not easy to steer or navigate to the stenosis. Subsequent numerous technical innovations with

Concise Encyclopedia of Biomedical Polymers and Polymeric Biomaterials DOI: 10.1081/E-EBPPC-120052172

OCT of OrbusNeich 3.5 × 18 mm Stent Presenting Great Apposition

Post Expansion   1–2 Hr Follow-up

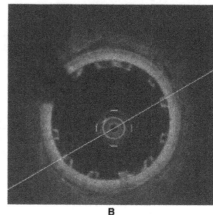

A   B

Fig. 1 OCT image of the OrbusNeich stent in a porcine left internal mammary artery: **(A)** immediately after deployment with great apposition and no thrombus and **(B)** 1 hr postdeployment with great apposition and no thrombus. The stent struts present as highlighted black boxes against the artery wall.
Source: Courtesy of OrbusNeich, Fort Lauderdale, FL.

guide catheters, guide wires, and angioplasty balloons led to increased use of this procedure. Success over previous treatments was profound but suffered some important limitations. Injury to the artery intimal wall would occur when extensive dilation was needed to remodel plaque, particularly with calcified and eccentric lesions. Sometimes, a tear in the blood vessel lining or a dissection of the plaque could occur, creating an obstructive flap with abrupt thrombus formation, resulting in acute blood vessel blockage. Some cases respond to an immediate repeat procedure, while others may require emergency bypass surgery. Successful balloon angioplasty restored vessel patency and allowed vascular remodeling and late luminal enlargement. Frequently, however, restenosis would occur caused by elastic recoil, neointimal hyperplastic healing response to the vessel trauma, and what is termed constrictive remodeling. A variety of adjunctive therapies was examined as well as alternate treatments such as several atherectomy devices and ablative laser catheters to remove atherosclerotic material from inside the artery.

The treatment of atherosclerotic vascular disease evolved, was driven, to address the less-than-perfect outcome as well as treatment induced complications of balloon angioplasty. The concept of placing a metal scaffold to support the artery from recoil and to seal plaque or intimal flaps, thus preventing acute stenosis, was introduced by Charles Dotter in 1967. However, it was not until 1986 that the self-expanding Wallstent™ (Schneider Inc.) was first successfully used in human coronary arteries. This was followed by the balloon expandable Gianturco-Roubin (Cook Inc.) and the Palmaz-Schatz (Johnson & Johnson) stainless steel bare metal stents (BMSs). Various stent designs progressively emerged to reduce the intensity of vessel wall injury and subsequent influence on neointimal

proliferation, as well as improved crossing profiles, improved stent placement, reduced side branch occlusion, and so on.

Providing a valuable solution to some of the complications of "simple" balloon angioplasty also resulted in some new complications. Neointimal hyperplasia was more significant than with balloon angioplasty alone, occurring now inside the stent, frequently requiring repeat treatment. Two advantageous healing responses, late luminal enlargement and vascular remodeling with vasomotion, were now also physically inhibited by a permanent metal scaffold.

Stents coated with a drug-eluting polymer were developed to mitigate the profound neointimal hyperplasia experienced with BMSs.[5,6] The first of these drug-eluting stents (DESs) became commercially available in about 2002. The polymer coatings are compounded with an antiproliferative or anti-inflammatory active pharmaceutical ingredient that elutes over time, inhibiting tissue ingrowth. The first-generation DES employed nonbiodegradable polymers in the coatings. As early as 2004, Virmani and colleagues[7] warned that there might be a risk of late and very late stent thrombosis in patients after implantation of DESs. Her examination of autopsy specimens revealed struts that remained uncovered with new endothelium, as well as some persistent inflammatory reaction.[8] It was suggested that this might have occurred because the polymer caused the inflammatory response or possible persistent drug release. Additional concerns have also been expressed related to discontinuities, or cracks, in the surface coatings, presenting poststent expansion. This has led to the development of a second generation DES based on various bioresorbable polymer coatings. Having a clear clinical need has now renewed the drive toward the development of fully bioresorbable stent platforms and prompted excitement that they will not only mitigate these new complications but also allow for restoration of the endothelial structure and function and vessel remodeling. This reinforces what Stack and Clark called for as the "ideal stent design"[2] in 1994 and has since become called "vascular restoration therapy" and the fourth revolution in interventional cardiology by Wykrzykowska, Onuma, and Serruys.[9] In addition, a fully bioresorbable stent platform would allow for the use of newer diagnostic imaging, such as magnetic resonance imaging (MRI) and MSCT, which are becoming the noninvasive imaging of choice, and easier reintervention, avoiding the full metal jacket syndrome.

POLYMERIC STENT DEVELOPMENT BACKGROUND

In 1988, just 2 years after the first successful use of a metallic self-expanding stent in humans, Stack and Clark at Duke University were making the case for the bioabsorbable stent. They reported on their poly-L-lactide (PLLA), braided-strand construct, self-expanding stent. The results after up to 18 months *in vivo* were judged biocompatible with a low thrombotic response, minimal neointimal response, and low inflammatory reaction. They stated that it was a "specialized form of poly-L-lactide"[2] that proved most suitable. This gave an early indication of how important not only polymer selection is to the outcome but also polymer processing and morphology development to achieve the requirements for a stent.

This foundational work was followed by the Cleveland/ Mayo Clinic healing response study (a joint effort between the Cleveland Clinic, the Mayo Clinic, and Thoraxcenter, Erasmus University), reported in 1992. This study examined the healing response to four bioresorbable polymer films (other than PLLA) solvent cast on a portion of a balloon-expandable tantalum wire coil stent. Table 1 shows the biodegradable test polymers in the Cleveland/

Mayo Clinic healing response study. The outcome from this study was less than encouraging for the bioresorbable polymers tested, which presented significant inflammatory responses. This prompted some confusion as some had anticipated that bioabsorbable polymers would be noninflammatory based on both the Duke study results and then successes with fracture fixation, wound closure, and drug delivery. Further testing repeating the same protocol but with several biodurable polymers [polyether urethane urea, silicone, and polyethylene terephthalate (PET)] attempted to differentiate between an inflammatory response to the polymer degradation products or the presence of a polymer implant itself. The results showed similar inflammatory responses even though these three biodurable polymers had previously demonstrated good graft or implant experience.

Equally conflicting results with biodurable polymeric stents made from PET were also reported in porcine trials. Murphy and coworkers developed and reported on a self-expanding PET stent.[10] It was reported that at 4–6 weeks after implantation, all stent deployment sites were occluded due to neointimal proliferation "with chronic foreign body inflammatory response surrounding the stent filaments and marked neointimal proliferative response in the center of the vessel." van der Giessen and colleagues[11] reported the opposite response, with their PET stent showing patent vessels after 4 weeks with minimal neointimal hyperplasia.

Dennis Jamilokowski shared a favorite axiom of his with participants at the Medical Device and Manufacturing conference on bioresorbable polymers in February 2011: "The devil is in the details." This is evident when confronted with conflicting study outcomes, and we are looking for root causes for better understanding. It is easy to overlook what seem to be "just details" when focusing on the inflammatory response associated with a polymer. We have come to understand, for example, that some implants in these studies were sterilized and others not.

Table 1 Cleveland clinic/mayo clinic biodegradable test polymers

	Polymer name	Degradation products	Degradation rate
PGLA	Poly(D,L-lactide/glycolide): copolymer 85% lactide/15% glycolide	D-Lactic acid L-Lactic acid Hydroxyacetic (glycolide) acid	100% in 60 to 90 days (rat subcutaneous model)
POE	Polyorthoester	Cyclohexanedimethanol 1,6-Hexanediol Pentaerythritol Propionic acid	60% in 46 weeks (in saline, pH 7.4 at 37°C)
PHBV	Poly(hydroxybutyrate/ hydroxyvalerate) copolymer: 22% valerate	Hydroxybutyric acid Hydroxyvaleric acid	0% to 20% in 182 days (rat subcutaneous model)
PCL	Polycaprolactone	Hydroxycaproic acid	52% in 1,491 days (rabbit subcutaneous model)

Source: Adapted from Zidar et al.[2]

Tamai also showed us that even changing stent construct design had an influence on the inflammatory response.[3,12] The presence of residual solvents from solution casting could foster differing cellular responses and stimulus for intimal proliferation. Even though there was an effort to rule this out in the Mayo study, we have since come to understand better how difficult it is to remove trace solvent as it has been shown that a solvent-cast polyactide acid (PLA) sample had trace chloroform even after several years under high vacuum. Later work by Vipule Dave with solvent-cast stent blanks resorted to supercritical CO_2 extraction to remove residual solvent sufficiently.[13]

Differences in the polymers due to their synthesis can also influence outcome. Residual monomers or solvents from the polymerization alone can induce inflammatory responses. Differences in the chain transfer agents used will change a polymer's end group and could precipitate different reactivity and cellular response. Separate work by Lincoff and coworkers[14] and Tamai and colleagues[12] suggested that high molecular weight PLLA is more quiescent in the artery than lower molecular weight polymer. Various nuances in the methods of manufacture of a stent aside from the polymer itself will also influence the morphology and subsequent properties as well as degradation kinetics.

Such apparently conflicting data led some research teams to examine more closely the healing response to proposed bioresorbable polymers before committing to further costly stent and stent delivery system development. Accordingly, Yoklavich and coworkers[15] reported in 1996 a favorable healing response at 52 weeks follow up to a novel bioabsorbable stent (Cordis/TESco) (Fig. 2), injection molded from a hybrid blend of PLLA and a trimethylene carbonate copolymer (TMC), deployed in the iliac artery in both canine and porcine models. This approach used a delivery balloon briefly heated to 70°C, allowing expansion of the stent at above its glass transition temperature T_g, then cooling back to 37°C. Because of questions regarding possible heat damage to the intimal wall, the alternate artery branch was used as a control with only the heated balloon. The concept of the novel hybrid polymer was to provide a mechanism that would be friendly to balloon expansion versus a braided, self-expanding construct used previously by Stack and Clark. The hybrid polymer system was developed also to have a quiescent breakdown even to the extent that the TMC moiety prevailed longer than the lactide, so that on loss of physical integrity the breakup would be soft or compliant within the endothelium.

Reports of early either mixed or disappointing results may have hampered general enthusiasm for bioresorbable stents, just as early disappointments with inflammatory response from the first polyglycolic first proximation pins hampered development of other bioresorbable orthopedic devices. The first to market pins suffered even though polyglycolide had been well received as a suture. The new pins made from the same suture material presented a larger bolus than previously experienced, prompting an inflammatory response to the amount of dye as well as the larger amount of polymer degradation product. It was found that even the cutting of the pins during surgery left ends of the pins that caused localized inflammation. These kinds of responses prompt us to look beyond a small sample of material, breaking out of the research mode and into the reality of manufacturing. In particular, they suggest the need to examine how the ends of braid or other filament-type constructs are cut and sealed and if they present a risk of physical inflammatory response.

Stack continued work with Zidar and colleagues to develop a PLLA stent[2] and in 1999 reported placement for 18 months in canine femoral arteries without inflammatory reaction. At about the same time, Tamai and coworkers[12] reported good biocompatibility, also using high molecular weight PLLA in a self-expanding stent. These studies, along with the Yoklavich (Cordis/TESco) stent, successfully used high molecular weight PLLA or a PLLA hybrid polymer to establish quiescent degradation.

Lincoff and colleagues,[14] in their work on coated stents, also suggested better compatibility with higher molecular weight PLLA. Both of the Stack and Tamai stents used high molecular weight PLLA filaments draw oriented and crystallized to impart the requisite strength and self- or semi-self-expansion. The Tamai stent, however, introduced a significant change in the stent design and delivery, changing from a braided construct to zigzag form elements. The delivery became semi-self-expanding, requiring balloon

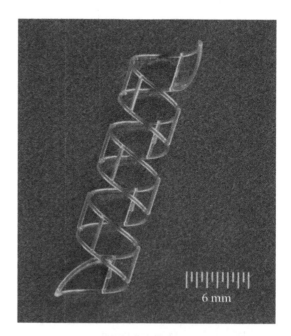

Fig. 2 Cordis-TESco Hybrid PLA/TMC intravascular prototype stent.
Source: Courtesy of TESco Associates Inc., Tyngsborough, MA.

expansion with contrast dye heated to 50°C for 13 sec. The stent would otherwise take considerably longer to self-expand at 37°C. This design change demonstrated less intimal damage by stent implantation (with a concomitant reduction in neointimal proliferation and inflammatory response) and less thrombus formation even though employing a similar polymer. Missing from this snapshot picture are the substantive details in polymer synthesis and processing actually to make the oriented filaments and stent constructs.

These early works seemed focused more on polymer safety, that is, low inflammatory response, than on design efficacy. With the challenges to develop an efficacious bioresorbable stent platform and delivery system seeming formidable, and especially prior to a clear and immediate clinical need, attention appeared directed more toward what appeared to be a less-complicated path of coating current metal platforms developing DESs to resolve BMS problems.

THE DESIGN ENVELOPE

When developing polymeric endovascular scaffolding, one needs to reflect on the complex interactions that influence successful development of a safe, efficacious, and cost-beneficial stent system. "This 'process' of product realization involves understanding the *entire* product life cycle" (italics mine).[16, p. 95] This process is just as applicable to endovascular stents as it was proposed for orthopedic implants. It involves developing an understanding not only of the device's clinical task but also all of its life cycle,

starting with polymer synthesis and ending, as in the case of a fully bioresorbable polymer stent, with its degradation and final elimination of the polymer degradation products from the body. This life-cycle understanding becomes layered with risk assessment of the ability to reach each objective, including that of commercial viability within various cost or time constraints. Oberhauser and coworkers[17] presented a review of the clinical portion of the life cycle divided into three phases: revascularization, restoration, and resorption. To these phases, we need to add an understanding of the manufacturing as well as the intervention procedure, or delivery, of the device. So, as we continue the discussion of polymeric endovascular stents, it is as much about the process as it is the polymer itself. Dave[13] suggested an interaction of polymer materials, stent design, and drug delivery with novel processes, and Oberhauser[18] reinforced the emphasis on polymer processing to deliver clinical utility, not just simple polymer selection. Cottone[19] proposed the formation of a design envelope to guide the development of bioresorbable stent platforms; an adaption of this is shown in Fig. 3.

When we explore the four main groups in this design envelope, we can break some of these into subsets more dependent on when or where the requirement presents within the life cycle. Cottone suggested the main requirement groups in an overview detailed in Table 2.

It is important to establish how entwined these criteria sets are; when one part of an interdisciplinary development team begins to address one issue, it can dramatically affect other requirements or change the constraint window.

One may think, for example, that the process of polymer selection starts with establishing the biocompatibility of a

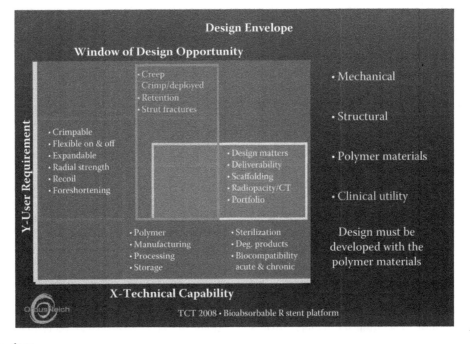

Fig. 3 Design envelope.
Source: Courtesy of Cottone.[46]

Table 2 Requirement groups

Mechanical	Structural	Polymer materials	Clinical utility
Crimpable	Creep	Polymer	Design matters
Expandable	Crimp deployed	Manufacturing	Deliverability
Radial strength	Retention	Processing	Scaffolding
Recoil	Strut fractures	Storage	Radiopacity
Foreshortening		Sterilization	
Flexible on and off (delivery catheter)		Degradation products	Enables a full product portfolio
		Biocompatibility (both acute and chronic)	

Source: Adopted from cottone.[19]

polymer. However, what really needs to be considered is biocompatibility of the stent system. As mentioned, there is a complex cascade of events that represent the healing response to the implant, starting with insult to the intimal wall and inner elastic lamina of the vessel when the stent is delivered and deployed at the lesion. Thus acute biocompatibility is dependent not only on the immediate polymer chemistry but also on many other factors, for example, how the polymer was made; residual contaminants from synthesis; effects of processing on the polymer, on both degradation and contaminants; design of the stent and how it is manufactured and deployed. These interactions become challenges under design control requirements and planning for validation. The team needs to estimate when in the design review process to impose a design lock. This typically occurs when starting long-term preclinical trials in which the data is to be used to support requests for an Investigational Device Exemption (IDE) and the start of human clinical trials. Changes implemented to address manufacturing and performance issues may necessitate additional preclinical trials and the restarting of clinical trials. It follows, then, that early in the process there should be a discussion of the regulatory hurdles or risks as they pertain to the proposed theater of sales and the different and constantly changing regulatory environments. This especially weighs heavy depending on the regulatory theaters considered and the need to demonstrate both safety and efficacy for approval to market. For example, at this date, obtaining a European CE mark requires demonstrating device safety (meeting the essential safety requirement of the EU directive), whereas obtaining U.S. Food and Drug Administration (FDA) approval to market requires establishing both safety and efficacy. In addition, a new device is expected to demonstrate improved efficacy or equivalent efficacy, but at a lower cost, over the current standard of treatment. I learned the cost for a human clinical trial in the United States for a coronary stent is estimated to be between $20 and $30 million for an initial primary indication for use. In addition, a full bioresorbable stent platform may add significant length to the trial follow-up, versus a bioresorbable coated metal DES, to

monitor full stent degradation and absorption as well as potential late responses. Berglund and coworkers[20] referred to the "technical challenges *to develop and commercialize* a successful bioabsorbable stent," (italics mine) (p. F72) underscoring that this is all about commercial product development, not just esoteric research. Further, he drew attention to the decision process regarding the target indication for use, selecting between peripheral and coronary revascularization: "Since SFA (superficial femoral artery) therapies are typically life-enhancing rather than life-sustaining, the regulatory paths and risks associated with developing an unproven technology are greatly reduced compared to coronary indications" (p. F73).

Once the target indication for use is established, we typically hear first the requirements for clinical utility of a finished, sterile, delivered (implanted) stent. These usually include that stents must be a certain strength, must degrade quiescently (i.e., not induce an inflammatory response), and must do their job for a certain minimum period of time. Also, one has to include objectives surrounding deliverability of the stent, such as crossing profile, branch vessel occlusion, tracking ability, ability to overexpand to ensure proper apposition without strut failure, and so on. These statements of clinical utility are only the beginning. Sooner rather than later, one also needs to embrace the concept that we have to be able to make the stent in a commercial and cost-effective manner.

As we explore approaches from the perspective of a design envelope, let us start with the methodology of stent delivery. The stent must traverse an artery and then the lesion and, when appropriately positioned, somehow be deployed, expanding radially from the crimped state. The stent must become larger than the initial vessel diameter, remodeling the plaque burden, to reduce the blockage, thus opening the flow path (revascularization). This can be achieved by either self-expanding, balloon delivery, or a combination of partial self-expansion with final balloon delivery. With balloon expansion or assisted expansion, it is the angioplasty balloon that is doing the work of vessel and lesion dilation. This stent delivery decision directs subsequent stent architecture, polymer choice, and manufacturing.

CURRENT APPROACHES

The publishing of the findings of late stent thrombosis in DES-implanted patients presented a new clear clinical need, igniting enthusiasm to drive investment in a new technology platform to replace DESs. A bioresorbable stent platform seeks to mitigate the issues surrounding DESs as well as enable vascular restoration therapy.[9]

With the emphasis now on fully bioresorbable stents for revascularization, it would be easy to forget the role of a permanent stent for the exclusion of aneurysms and their possible polymeric platforms. The clinical jobs are quite different. Revascularization can be viewed as the opening of a closed (stenosed) or restricted vessel to restore blood flow, whereas exclusion therapy intervenes in an artery that has a bulging, weakened wall threatening to rupture. These types of stent have an inner and outer covering over the support structure, as shown in Fig. 4. Currently, these are self-expanding structures and when in place seal both proximal and distal to the threatened area and provide for a new blood conduit. With the pressure now relieved from the damaged wall section, this portion of the artery wall will eventually reduce and conform to the stent's outer profile. Most of these support structures are currently made from shape memory nitinol wire construct covered with either PET mesh or fluoropolymer membrane derived from vascular graft technology. It remains to be shown that biodurable shape memory polymers[21] have a clinical preference. It has yet to be demonstrated if the damaged artery wall from a reduced aneurysm will remodel itself to fully restored healthy function of both the endothelium and the inner elastic lamina. If this were shown, there would be merit to explore a bioresorbable shape memory polymer[22,23] for such a self-expanding stent structure to provide temporary scaffolding and then disappear.

Various examples of clinical, preclinical, and *in vitro* work with coronary and peripheral revascularization stents are listed in Table 3. Some of these are discussed along with the concept of a design envelope and issues of product realization. At this time, the Igaki-Tamai™ stent has 10 years of follow-up in humans and has obtained the CE mark for peripheral indications. The Abbott/BVS stent, with over 2 years of human clinical data, has also received the CE mark, but for coronary indications (allowing for continued clinical trials). Reva Medical has reported on its initial 25-patient clinical trial experiences and further design optimization. Last, in July 2009, Bioabsorbable Therapeutics completed a 12-month follow-up of its 11-patient first-in-man trial.

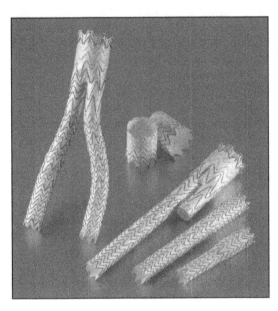

Fig. 4 Covered stent graft for AAA repair.
Source: Reprinted public domain photo from "Covered Stent Graft," Wikipedia, accessed July 20, 2011.

Table 3 Examples of work with coronary and peripheral revascularization stents

Current clinical, preclinical, and *in vitro* work

Igaki-Tamai	• PLLA filament-based construct
Abbott/BVS	• Laser-cut, extruded and oriented PLLA tube
Reva Medical	• Laser-cut, molded tyrosine polycarbonate sheet
Bioabsorbable Therapeutics Inc. (BTI)	• Polyanhydride ester based on salicylic acid
OrbusNeich Medical	• Laser-cut, extruded-tube PLLA/PDLA/CL copolymer blend
Arterial Remodeling Technologies	• Molded amorphous PLDLA
Cordis (J&J)	• Laser-cut, solvent-cast tube, PLGA85:15/PCL-PLG (Monocryl) blend
Tepha	• PLLA/PHB peripheral stent

Additional or alternate material concepts

• Mnemoscience (Aporo Biomedical)	• Bioresorbable, self-expanding, shape memory polymer
• Alternate abbott polymers	• Phase-separated star block copolymers[47,48]
• Bezwada biomedical	• NSAID-functionalized polymer backbone (e.g., naproxen or salicylic acid)

Abbreviation: PLDLA, poly L-*co*-D,L-lactide; PLGA, poly L-lactide-co-glycotide.

Self-Expanding Stents

A fully self-expanding stent is manufactured at its deployed size and then collapsed around a guide catheter. A sheath restrains the stent while it is being delivered and is withdrawn once the stent is positioned for deployment. Either elastic recoil or the response to an applied stimulus like heat to a shape memory polymer is the mechanism for expansion. Polymer creep or stress relaxation while constrained in the protective delivery sheath can diminish the ability to self-expand. A semi-self-expanding stent is manufactured like a fully self-expanding stent, either at its deployed size or at a partial deployed size. When the protective sheath is withdrawn, the stent partially opens and then is fully deployed with the assistance of a balloon catheter. Self-expanding and semi-self-expanding designs, requiring sheathed delivery, are sometimes thought to be more difficult to track, especially in the coronary arteries, potentially somewhat limiting their clinical utility to predominantly peripheral indications.

The Igaki-Tamai stent is considered semi-self-expanding because of the extended length of time it takes to self-expand at physiologic conditions. Therefore to obtain full expansion within a reasonable time frame (e.g., less than 30 sec), the stent is allowed to self-expand partially and is further assisted into place using a balloon catheter employing heated contrast dye as the expansion fluid. The contrast dye is heated to above the T_g of the polymer (in initial clinical trials, heated to 80°C) and the balloon catheter inflated to 6–14 atmospheres, holding for about 30 sec,[24] then cooling back to 37°C. In later trials, this was reduced to 13sec at 50°C to minimize heat injury to the vessel, and the cycle was repeated until there was equilibrium and good apposition to the vascular wall.[1]

Many self-expanding stent architectures are braided constructs similar to the metallic Wallstent and the original Stack and Clark PLLA stent. The Igaki-Tamai design presents a marked departure with a zigzag architecture that reduces the filament overlaps and resulting irritation and intimal injury on expansion. The reduced filament overlap also probably reduces the stent profile and thrombogenicity. Both the Igaki-Tamai and the original Stack-Clark stents are based on higher molecular weight (e.g., >2 IV) PLLA spun into monofilament and drawn to impart molecular orientation, paracrystallinity, and crystallinity to enhance the base polymer's mechanical properties. This especially reduced the polymer elongation and increased tensile strength and modulus, enabling self-expansion, good hoop strength, and low creep. Using higher molecular weight PLLA resulted in noticeably quiescent degradation and resorption.

In 2007, the Igaki-Tamai stent was the first bioresorbable stent to receive a CE mark for peripheral indications. In 2009, a new version called the Remedy™ was introduced with a slightly different architecture. The Igaki-Tamai stent was constructed of zigzag elements connected by fabricated bars, whereas the new Remedy appears fabricated with three filaments forming the zigzag elements with a few overlaps in lieu of hard fabricated connections.

Regardless of the construct architecture, both stents are manufactured from similar high-strength, higher molecular weight PLLA filaments. Manufacturing and processing methods present significant challenges over conventional lower molecular weight polymers. One may consider gel spinning over melt spinning to avoid thermal degradation. However, this presents its own set of challenges to remove the solvent. As with solvent casting, there are questions regarding potential differences in the morphology from gel-spun samples and those prepared from a melt process[25] and any resulting microstructure from extracting the solvent. In addition, different postspinning processing, such as secondary orientation and heat treatment (sometimes incorrectly referred to as "annealing") may have to be considered.[26] These process steps to induce orientation and crystallinity are critical to filament constructs to impart sufficient elastic recoil for deployment and then once deployed to be able to resist both acute stent recoil and longer-term creep.[27]

The orientation imparted during processing is not stable and can relax over time. Further, this relaxation can be accelerated by sterilization temperatures. Accordingly, special low-temperature ethylene oxide (EtO) sterilization cycles are often utilized, as are freezing of parts that are to be exposed to various radiation sources (e.g., E-beam sterilization). This is also a problem for shelf life and shipping of a commercial product, in particular for self-expanding stents that are stored constrained in their delivery sheath. The orientation can be somewhat stabilized by imparting crystallization that acts as a physical cross-link. The polymer specifications may also be influenced by the choice of filament manufacturing process. For example, PLLA selected for melt spinning may employ a different polymerization initiator, or chain transfer agent, to impart better thermal stability for processing not necessary for a solvent process. This in turn could also influence the hydrophobicity of the polymer, its degradation kinetics, and so on. The postpolymerization extraction of unreacted residual monomer and the solvent used to carry the catalyst is also crucial for melt processing stability. Improving processing stability preserves the molecular weight for initial mechanical property integrity and reduces monomer and oligomers resulting from thermal degradation that would accelerate the stent degradation kinetics.[16]

In addition to the polylactides with clinical experience discussed previously, a new class of bioresorbable shape memory polymers has been proposed for self-expanding stents by Mnemoscience, a spin-out from work originally performed at the Massachusetts Institute of Technology.[22,23]

Balloon-Delivered Stents

A balloon-delivered stent presents a whole new field of manufacturing and performance issues. To begin, the stent is manufactured at some size larger than the deflated balloon catheter; it is then crimped onto the balloon. Various design techniques are employed to keep the stent in its proper location on the balloon while it is being delivered down a guide catheter and finally traversing an artery. Typically, a stent is cut from a tube, or a tubular stent is constructed from cut sheet.

There have been a few attempts at injection molding either blanks for further cutting or even a completed stent, as in Cordis/TESco stent (Fig. 2). Injection molding a fully formed stent presents significant issues with the melt viscosity of the higher molecular weight polymer needed for mechanical properties and biocompatibility. Yoklavich and colleagues[15] reported that the injection molded struts were about 0.013 inch thick, almost twice the thickness currently typically considered viable for clinical utility. Not only is there the difficulty of getting the polymer to flow into thin-wall sections without degrading it, but also there is the problem of multiple flow front knit lines leading to strut failure.

In 2009, Lafont and coworkers[28] reported on Arterial Remodeling Technologies (ART)' program to develop a balloon delivered "molded" stent platform based on a poly-L-DL-lactide. They reported on a 6-month follow-up on iliac rabbit arteries, using a 6F balloon-delivered stent manufactured with a proprietary molding and memory-shaping process. Midwest Plastics reported an attempt at injection molding a tubular blank for laser cutting. Even molding a blank without molding a finished strut pattern is challenging, with problems such as core pin shift or bending resulting in a nonuniform wall unacceptable for laser cutting.

Many balloon expanded stent designs, versus self expanding stents, seem to be at greater risk for strut crazing and fracture when stresses are high during both crimping and expansion deployment. Herein is the challenge in manufacturing: to balance the morphology and microstructure developed during processing for radial strength properties while preserving sufficient elongation and a fracture mechanism that averts strut failure. "Stent fracture will lead to premature radial strength loss, vessel recoil, tissue irritation, and inflammation independent of the material biocompatibility."[20, p. F75] Further, clinical utility is greatly diminished if a stent cannot be stretched (i.e., overdilated) to match the dimension of the target vessel or comply with either calcified or eccentric lesions without stent or strut fracture. Even with the advancement of QCA to measure the lesion for appropriate stent size selection, calibration errors can lead to underestimation of the target lesion diameter and result in the need to overdilate the stent. It should be noted that underexpansion of a stent could also cause one not to develop the optimal mechanical properties in some stent design/material combinations, resulting in excessive stent recoil and poor apposition.

Stents fabricated from sheet

The Reva Medical stent has attempted to mitigate some of these issues by using a different approach to design and construction. It is manufactured by cutting sheet made from an amorphous tyrosine (desaminotyrosyltyrosine ethyl ester) polycarbonate. A unique ratcheting design causes the stent struts to slide and then lock from recoil as it is expanded, minimizing the high-strain points that exist in most strut designs cut from a tube blank. Its mechanical properties and degradation kinetics have been shown to be similar to PLLA.[29] Iodine molecules have even been incorporated into the backbone of the polymer to impart radiopacity. The first-generation REVA stent was evaluated in the 25-patient RESORB clinical trial. The trial data demonstrated the need for "further polymer and design optimization"[30, p. F56] for improved performance under real load conditions. The first REVA ratchet lock design has been replaced with the more flexible helical slide-and-lock ReZolve stent,[3] manufactured from a tyrosine polycarbonate slightly modified to improve cyclic or fatigue loading.[30]

The embracing of tyrosine polycarbonate by Reva Medical for its stent represents one of the challenges, or risks, in polymer selection as well as the potential opportunity. Typically, one looks for well-established, validated, commercial scale polymerizations to avoid surprises going into clinical trials and scale-up. The family of tyrosine polycarbonates has seen limited commercial applications in devices, the REVA stent being the most notable, but has had more applications in drug delivery coatings. However, the value of the opportunity to capture a unique polymer to differentiate the product line and the ability to modify the polymer backbone or pendant groups to adapt to the evolving understanding of performance requirements are also significant. I understand that the scale-up of the polymerization had its challenges. Having a unique polymer platform may have significant advantages but needs to be weighed against the risk of manufacturing scale-up taking longer and being more costly than expected, as well as the possibility of a restricted or limited material source.

Stents fabricated from tubes

Fabricating a stent by laser cutting a tube follows the dominant manufacturing and design concepts of the most popular BMS and metal DES platforms. Both stainless steel and cobalt-chrome have well-understood structures and properties and are available in tubes drawn to precise uniform wall sections. Laser cutting of metals has progressed such that beam widths are narrower for smaller kerfs, and pulse rates are ever faster, allowing for significantly less heat affected zone and cleaner cuts. Even with these improvements, laser cut metal stents still typically require electropolishing and other treatments to clean the surface.

Laser cutting of polymer stents is different from the cutting of metals. Metal ablation is mainly by vaporization of the metal due to heat. Polymer ablation also involves chain scission and vaporization, not just melting. To reduce thermal damage and improve cutting efficiency, the output wave-length of eximer lasers can be selected or tuned to match the ultraviolet (UV) absorption spectra of the polymer. Coupling a tuned laser with very high pulse rates has significantly improved cut edge quality, which in metals was frequently dealt with by secondary operations that are not available with polymers. Polymer struts are typically thicker than metal to compensate for lower mechanical properties; thus it takes several more passes with the laser to complete the stent cut pattern.

These new lasers have shorter focal lengths, leading to them to be less forgiving. This has been problematic for manufacturing teams as metal tubing for stents has much tighter wall section and diameter tolerances than can be readily achieved with most polymer tube manufacturing. So if the polymer tube wall section varies too much, it can fall out of the focal range of the laser. Traditional methods for polymer tubing manufacturing, as well as fixture and inspection methods, needed to be reinvented, leading development teams to embark on customized processes to achieve desired quality and properties. As with any polymer cutting, it has been suggested that the molecular weight is reduced at the cut edge.[31] It has yet to be demonstrated if this significantly affects either initial stent mechanical properties or degradation kinetics.

When stents are cut from a tube, one needs to consider the state of expansion of the cut form prior to crimping or loading onto a delivery balloon catheter. This decision occurs early in the development cycle and can easily entrap the development team in one direction. Your extremes are to have a cut pattern in the crimped state or at an overdilated state; however, reality would probably be somewhere in between. For example, a stent may be laser cut from a tube about 2.0-mm diameter then crimped onto a much smaller 4F or 5F delivery balloon catheter to have a final crossing profile of 6F but a target expansion out to maybe 3.5-mm diameter. An alternate approach might be to laser cut the stent from a tube that is the same diameter as the final target expansion; in the first example, this is about 3.5-mm diameter.

The amounts of strain that the stent struts undergo both in crimping and expansion must be taken into account with materials and strut design. Just dealing with the ability to "fold" the stent struts down into a crimped position can be a challenge, as can keeping them in place. It is reasonable to plan to stretch the stent into final apposition to the artery wall. This can preclude cutting in the final deployed diameter as it may not be able to fold back down onto the balloon from this stretched configuration. This all influences the morphology, that is, the degree of crystallinity or orientation that may be required for stent strength as well as the toughness of the polymer system to withstand subsequent crimping and expansion without strut failure or extensive recoil and creep.

The introduction of strain-induced orientation and crystallization during crimping and expansion may be crucial to performance. When manufacturing the stent tube blank for cutting, the degree of orientation or crystallinity at each stage of manufacture and stent delivery must be well understood. The ability of a stent to be overdilated to ensure good apposition, or to comply with a difficult lesion, is critical to clinical success and utility. Strut cracking can occur if the development and manufacturing teams do not plan for sufficient elongation or compliance in the stent material or design. Manufacturing high-tolerance tubing from high molecular weight bioresorbable polymers with just the right orientation and morphology to meet this objective is no small task.

A tube blank for laser cutting can be manufactured by melt extrusion, injection molding, gel extrusion, dip coating, or spray coating. The last three introduce solvents into the process that at some point need to be extracted. This can be a challenging step. As indicated previously, it has been demonstrated that traces of chloroform remained in cast samples even after several years under a vacuum process. The effect at the cellular level is not clear. Earlier studies such as the early Cleveland Clinic trials that attempted to expose any adverse inflammatory reaction from residual solvent have been questioned. Vipul Dave and colleagues[13] reported solvent dip-coating tubes for stent blanks from PLG (85:15) copolymer blended with polycaprolactone (PCL)-PLG copolymer (Monocryl). They reported using supercritical CO_2 extraction to mitigate the solvent from tube casting.

As mentioned previously, there are questions regarding potential differences in the morphology of solvent-cast polymer versus that obtained from melt processing as well as the phase domain stability of solvent-cast materials.[25] It is not clear if this type of processing is beneficial. For example, some copolymers that do not crystallize from the melt have been shown to develop significant crystallinity cast from solution. In addition, layering or laminating effects from dip or spray coating a tube may need to be better understood. The impact strength of PLLA is reported to become higher by the addition of rubbery biodegradable polymers such as PCL.[32] Similarly, the addition of the soft and elastomeric PCL-PGA copolymer in this blend with stiff PLG increases 85:15 increases the blend's ductility to allow the stent structure to deploy and remain open following balloon catheter removal. The PLGA copolymer was selected with a 2.2–2.4 IV for mechanical performance and absorption time. One of the benefits of this solvent process is the ability to easily add both a radiopacier such as barium sulfate and the anti-proliferative sirolimus (rapamycin) into the bulk of the stent structure without resorting to intense thermal compounding that might otherwise compromise the polymer or active agent.

Bioabsorbable Therapeutics has been working with another unique polymer in an attempt to provide therapeutic moieties from polymer breakdown. Their polymer is a polyanhydride ester based on salicylic acid and adipic acid anhydride and is intended to provide anti-inflammatory properties as the polymer degrades. It is also radiopaque. Even though the stents used in the first-in-man trials included a coating that contained the anti-proliferative sirolimus, insufficient neointimal suppression was reported.[3,33] The first-generation design had a large 8F crossing profile and a 65% occlusion ratio; it has been replaced with a second-generation 6F design with thinner struts. Bezwada Biomedical has also proposed additional unique polymers with nonsteroidal anti-inflammatory drug- (NSAID) functionalized polymer backbones based on, for example, naproxen[34] or salicylic acid.[35] The therapeutic utility and the processing/mechanical requirements for stents of these proposed materials have yet to be determined. The mass of these polymers required to release a therapeutic bolus from the backbone of the polymer will have to be carefully considered and could greatly influence the stent design parameters. The release kinetics of the therapeutic moiety may also be governed by the degradation mechanism of the polymer backbone in combination with a number of other factors that influence drug delivery kinetics (e.g., surface area, molecular free volume, etc.), complicating designing for therapeutic efficacy. In addition, the mechanical performance life cycle will need to be determined in conjunction with changes to the polymer bulk related to the release of the therapeutic moiety.

Melt extrusion of a tube profile appears to be the dominant manufacturing method of blanks for laser cutting. As mentioned, there has been some attempt to injection mold tubing blanks for laser cutting; however, this approach has not met with reported success, and the conventional approaches to micromolding, for example, have not yet demonstrated that they can be adapted to the high molecular weight polymers required for this application. Extruding tubing with these bioresorbable polymers typically excludes conventional thin-walled tubing techniques such as under-water vacuum sizing. We have found that the critical wall section control required for both stent fabrication and uniform stent deployment is considerably tighter than typically practiced, even for what is conventionally considered high-tolerance tubing, leading to customized and usually proprietary extrusion processes.

Oberhauser[18] presented compelling reinforcement to the previous statements here about the pivotal importance of polymer processing for successful outcome. The Abbott/ BVS Absorb™ stent program confronted the less-than-optimal mechanical properties, reported in more generalized resources, by concentrating on the property improvements from polymer orientation and crystalline morphology as demonstrated in applications such as fibers, angioplasty balloons, biaxially oriented film, and stretch blow-molded containers.[26,36] This was accomplished by extruding a tube thicker than the final strut and then stretch blow molding to develop its orientation and crystallinity.[18] The Absorb stent utilizes high molecular weight PLLA as the primary stent structure. Selection of an appropriate polymer is a significant factor toward determining the success or failure of the stent design. This is not just simply specifying a certain molecular weight PLLA. As the polymer will undergo melt extrusion and then further thermal/mechanical processes, the molecular weight not only needs to be called out but also other factors that influence processability.

As mentioned previously, polymerization catalyst, initiators, condition parameters, extraction, and even grinding can have an impact on the thermal processability of the polymer. Yuan and coworkers[37] reported up to about 20% loss just in grinding PLLA small enough to feed in a small extruder. Molecular weight losses from chain scission make the polymer even more unstable in a subsequent molding or extrusion process. There are myriad details surrounding the polymer preparation and extrusion process itself that influence the quality and properties of the stent blank. Further, Oberhauser detailed the attention given to the polymer crystal nucleation and propagation as well as to both crystalline orientation and oriented amorphous (paracrystalline) domains to impart significant increases to the strength and modulus over the bulk unoriented polymer properties. However, there is a risk of applying significant orientation and strain-induced crystallization to a polymer that already has inherently low elongation.[20] Additional strain imparted during crimping or expansion can result in strut failure, particularly where strut cut patterns change directions across oriented polymer chains. Glauser and colleagues[38] discussed crack counts for deployed stents, which suggested further process or materials development may be required. The initial Absorb trial precipitated a stent redesign to provide, for example, more uniform support to the artery wall and higher radial strength without changing the polymer or strut thickness.[39]

Cottone[19] has reported on OrbusNeich Medical Incorporated's unique hybrid polymer[40,41] balloon-delivered stent platform (Fig. 5). It is a platform that combines three technologies: a hybrid material, a novel stent design, and

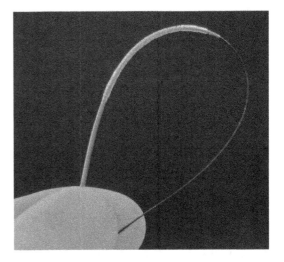

Fig. 5 OrbusNeich R stent crimped on a 6F delivery catheter.
Source: Courtesy of OrbusNeich, Fort Lauderdale, FL.

partitioned coatings for both drug delivery and endothelial progenitor cell capture. The hybrid material and stent design are the focus of the present discussion as they represent the stent "platform," a particular approach that "provides a marriage between the polymer material backbone formulations and the mechanical and structural aspects of the stent's design."[42, p. F66] This stent is fabricated by laser cutting a tube extruded in a proprietary process. The hybrid material is a blend of three lactide polymers used together to address different critical performance criteria during the various stages of the material/stent life cycle, each with known pharmacokinetics: PLLA, poly-D-lactide (PDLA), and a lactide e-caprolactone or trimethylene-carbonate copolymer. The unique blend is suggested to enable polymer mechanics and crystalline orientation at various stages of manufacture and deployment and to present a balance of strength and compliance. It is a polymorphic and polyphasic system in which stereo complexation between PLLA and PDLA is said to enhance the tensile properties of blends compared to those of the nonblended PLLA or PDLA.[43] The increased properties of these blends may also be attributed to dense chain packing in the amorphous regions due to a strong interaction between the L- and D-unit sequences. Multiple crystal forms are suggested through various stages, whether postextrusion, postprocessing, or strain induced during stent crimping and delivery. It is probably that the microstructure formed by stereocomplexation gelation during initial processing increases the number of spherulites per unit mass to enhance the tensile properties of these blends[43] and is followed by further stereocomplex growth and epitaxial homoe nantiomer crystallization. Designed in molecular free volume probably enables enough polymer chain mobility to provide flexibility and elongation while enhancing molecular orientation and strain-induced crystallization during crimp mounting

on a delivery balloon and when stretched at physiologic conditions during deployment. The OrbusNeich program employs a stretched ringlet design to improve strength and recoil resistance, shown in Figs. 6 and 7, which, in combination with the cross-moiety crystallization in the polymer blend, increases resistance to creep.[40–42] The balance of polymer microstructures and stent design appears to present both radial strength and stent strut crack resistance critical to clinical utility.

Van der Giessen and coworkers[44] stated that polyhydroxybutyrate (PHB) and polyhydroxybutyrate valerate (PHBV) are not well-suited materials for immediate and extended blood contact. However, Grabow and coworkers[45] demonstrated promising results with a blend of PLLA with high-elongation poly-4-hydroxybutyrate (P4HB) in porcine iliac arteries. This approach appears built on their work to improve performance of DES coatings by blending poly-D-L-lactide (PDLLA) with P4HB, especially addressing the coating surface quality and mechanical integrity poststent expansion. Their prior work with a PLLA stent prototype prompted "improvements to the expansion behavior of the stent and its in vivo biocompatibility" by blending the PLLA with P4HB.[45, p. 747] The differences in biocompatibility of polyhydroxyalkanoates reported by Grabow and Van der Giessen's teams may well stem from dramatic improvements in the genetic-engineered fermentation process for monomer production as earlier PHB polymers were noted for cell debris that promoted extensive inflammatory responses.

CONCLUSIONS

The vision of leaders in interventional cardiology is now clear. The mandate is to develop a fully bioresorbable stent

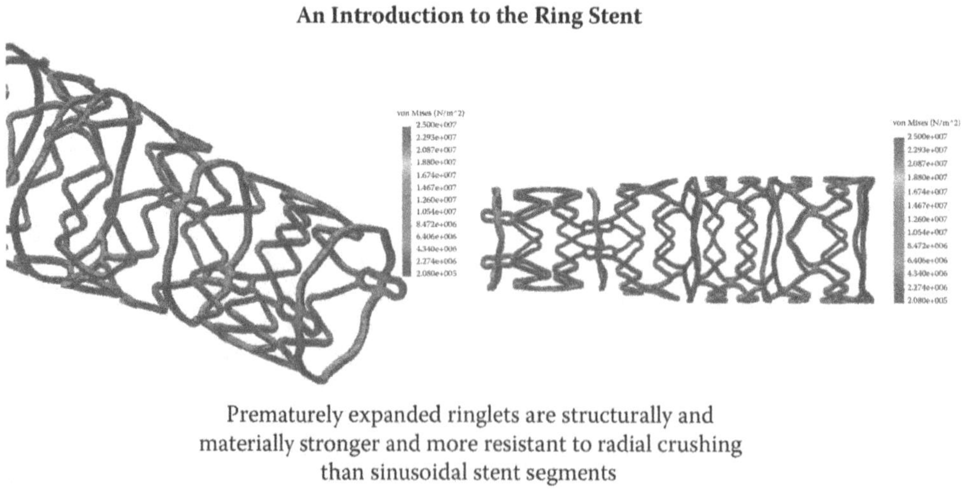

An Introduction to the Ring Stent

Prematurely expanded ringlets are structurally and
materially stronger and more resistant to radial crushing
than sinusoidal stent segments

Fig. 6 Finite element analysis (FEA) introduction to the OrbusNeich ring stent. Prematurely expanded ringlets are structurally and materially stronger and more resistant to radial crushing than sinusoidal stent segments.
Source: Courtesy of OrbusNeich, Fort Lauderdale, FL.

OrbusNeich
Single Ringlet Stent with Tantalum Marker

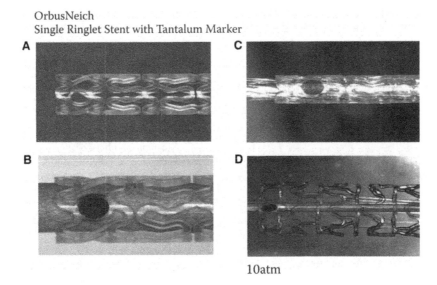

10atm

Fig. 7 OrbusNeich stent: (**A**) as cut, (**B**) as cut with marker dot, (**C**) crimped on balloon, (**D**) deployed at 10 atmospheres.
Source: Courtesy of OrbusNeich, Fort Lauderdale, FL.

platform that performs mechanically and clinically equivalent to DESs—and then disappears, resulting in vascular restoration therapy. The rationale for a bioresorbable stent is to

- Eliminate the chronic presence of a foreign body and allow for "complete healing" or vascular restoration therapy
  - Limiting the need for anticoagulation/antiplatelet therapy with no late adverse events
  - Providing opportunity for reintervention therapy
- Provide for drug elution
  - With greater drug loading possible
  - With greater flexibility in elution characteristics
- Provide improved compatibility with imaging technologies
- Enable broader development of percutaneous local therapeutics for the treatment or prevention of vascular diseases

Undoubtedly, there will be additional polymers, designs, and development programs for fully bioresorbable endovascular stent platforms. Ultimately, fully bioresorbable stents may replace permanent stents, and even though the feasibility and safety of some systems has been established, there remain important issues and improvements before they become both clinically and commercially viable to enjoy widespread clinical use.[46] (Details of the preclinical and clinical trials, introduced in the above text, may be found in the listed references).

REFERENCES

1. Onuma, Y.; Garg, S.; Okamura, T.; Ligthart, J.; Van Geuns, R.J.; De Feyter, P.J. Ten-year follow-up of the IGAKI-TAMAI stent: A posthumous tribute to the scientific work of Dr. Hideo Tamai. EuroIntervention Supplement **2009**, 5 (Suppl. F), F109–F111.
2. Zidar, J.P.; Lincoff, A.M.; Stack, R.S. Biodegradable stents. In *Textbook of Interventional Cardiology*; 2nd Ed.; Topol, E.J., Ed.; Saunders: New York, 1994; 787–802.
3. Onuma, Y.; Serruys, P.W. Bioresorbable scaffold: the advent of a new era in percutaneous coronary and peripheral revascularization. Circulation **2011**, *123* (7), 779–797.
4. Labinaz, M.; Carter, J.; Dossenbach, M.; Sketch, M. New device therapy for coronary artery disease (revisited). New Developments in Medicine and Drug Therapy Sep/Dec, 1996.
5. Costa, M.A.; Sabate, M.; van der Giessen, W.J.; Kay, I.P.; Cervinka, P.; Ligthart, J.M.; Serrano, P.; Coen, V.L.M.A.; Levendag, P.C.; Serruys, P.W. Late coronary occlusion after intracoronary brachytherapy. Circulation **1999**, *100* (8), 789–792.
6. Sousa, J.E.; Costa, M.A.; Abizaid, A.; Abizaid, A.S.; Feres, F.; Pinto, I.M.; Seixas, A.C.; Staico, R.; Mattos, L.A.; Sousa, A.G.; Falotico, R.; Jaeger, J.; Popma, J.J.; Serruys, P.W. Lack of neointimal proliferation after implantation of sirolimus-coated stents in human coronary arteries: A quantitative coronary angiography and three-dimensional intravascular ultra-sound study. Circulation **2001**, *103* (2), 192–195.
7. Virmani, R.; Guagliumi, G.; Farb, A.; Musumeci, G.; Grieco, N.; Motta, T.; Mihalcsik, L.; Tespili, M.; Valsecchi, O.; Kolodgie, F.D. Localized hypersensitivity and late coronary thrombosis secondary to a sirolimus-eluting stent: should we be cautious? Circulation **2004**, *109* (6), 701–705.
8. Joner, M.; Finn, A.V.; Farb, A.; Mont, E.K.; Kolodgie, F.D.; Ladich, E.; Kutys, R.; Skorija, K.; Gold, H.K.; Virmani, R. Pathology of drug-eluting stents in humans: Delayed healing and late thrombotic risk. J. Am. Coll. Cardiol. **2006**, *48* (1), 193–202.
9. Wykrzykowska, J.; Onuma, Y.; Serruys, P.W. Vascular restoration therapy: The fourth revolution in interventional cardiology and the ultimate "Rosy" prophecy. EuroIntervent. Suppl. **2009**, *5* (Suppl. F), F7–F8.

Stem Cell—Ultrasound

10. Murphy, J.G.; Schwartz, R.S.; Edwards, W.D.; Camrud, A.R.; Vlietstra, R.E.; Holmes, D.R. Jr. Percutaneous polymeric stents in porcine coronary arteries: initial experience with polyethylene terephthalate stents. Circulation **1992**, *86*(5), 1596–1604.

11. van der Giessen, W.J.; Slager, C.J.; Gussenhoven, E.J.; van Beusekom, H.M.; Huijts, R.A.; Schuurbiers, J.C.; Wilson, R.A.; Serruys, P.W.; Verdouw, P.D. Mechanical features and *in vivo* imaging of a polymer stent. Int. J. Card. Imag. **1993**, *9* (3), 219–226.

12. Tamai, H.; Igaki, K.; Tsuji, T.; Kyo, E.; Kosuga, K.; Matsui, S.; Komori, H.; Motohara, S.; Uehata, H.; Takeuchi, E. A biodegradable poly-l-lactic acid coronary stent in porcine coronary artery. J. Intervent. Cardiol. **1999**, *12* (6), 443–450.

13. Dave, V.; Overaker, D.; Donovan, R.; Falotico, R. Polymer, process and design elements of a balloon expandable bioabsorbable drug eluting stent. Presentation at the annual meeting of the Society for Biomaterials, San Antonio, TX, April 22–25, 2009.

14. Lincoff, A.M.; Furst, J.G.; Ellis, S.G.; Tuch, R.J.; Topol, E.J. Sustained local delivery of dexa-methasone by a novel intravascular eluting stent to prevent restenosis in the porcine coronary injury model. J. Am. Coll. Cardiol. **1997**, *29* (4), 808–816.

15. Yoklavich, M.F.; Thatcher, G.L.; Sasken, H.F. Vessel healing response to bioabsorbable implant. Proceedings from the Fifth World Biomaterials Congress, Toronto, May 29–June 2, 1996.

16. Thatcher, G.L. Product realization: the processing of bioabsorbable polymers. In *Degradable Polymers for Skeletal Implants*; Wuisman, P.I.J.M., Smit, T.H., Eds.; Nova Science: New York, 2009; 93–122.

17. Oberhauser, J.P.; Hossainy, S.; Rapoza, R.J. Design principles and performance of bioresorbable polymeric vascular scaffolds. EuroIntervent. Suppl. **2009**, 5 (Suppl. F), F15–F22.

18. Oberhauser, J. Engineering bioresorbable polymers into vascular scaffolds: An application in interventional cardiology. Presentation at the MD&M West Conference, Anaheim, CA, February 7–10, 2011.

19. Cottone, R.J. Fully absorbable vascular scaffold with combination CD34 ab cell capture and abluminal sirolimus eluting coating. Presentation at the Transcatheter Cardiovascular Therapeutics, Washington, DC, September 21–25, 2010.

20. Berglund, J.; Guo, Y.; Wilcox, J.N. Challenges related to development of bioabsorbable vascular stents. EuroIntervent. Suppl. **2009**, *5* (Suppl. F), F72–F79.

21. Toensmeier, P.A. Shape memory polymers reshape product design. Plast. Eng. **2005**, 10–11.

22. Venkatraman, S.S.; Tan, L.P.; Joso, J.F.D.; Boey, Y.C.F.; Wang, X. Biodegradable stents with elastic memory. Biomaterials **2006**, *27* (8), 1573–1578.

23. Lendlein, A.; Simon, P.; Kratz, K.; Schnitter, B. Stents for use in the non-vascular field, which comprise an SMP material. U.S. Patent Application Number 2007, 0129784, filed June 9, 2004, issued June 7, 2007.

24. Tsuji, T.; Tamai, H.; Igaki, K.; Kyo, E.; Kosuga, K.; Hata, T.; Okada, M.; Uehata, H. Biodegradable polymeric stents. Curr. Cardiol. Rep. **2001**, *3* (1), 10–17.

25. Manson, J.A.; Sperling, L.H. Diblock and triblock copolymers: Effect of solvent casting on morphology. In *Polymer Blends and Composites*; Plenum Press: New York, 1976, 141–142.

26. Ghosh, S.; Vasanthan, N. Structure development of poly (L-lactic acid) fibers processed at various spinning conditions. J. Appl. Polym. Sci. **2006**, *101* (2), 1210–1216.

27. Zilberman, M.; Nelson, K.D.; Eberhart, R.C. Mechanical properties and *in vitro* degradation of bioresorbable fibers and expandable fiber-based stents. J. Biomed. Mater. Res. A Part B **2005**, *74B* (2), 792–799.

28. Lafont, A.; Durand, E. A.R.T. Concept of a bioresorbable stent without drug elution. EuroIntervent. Suppl. **2009**, *5* (Suppl. F), F83–F87.

29. Tovar, N.; Bourke, S.; Jaffe, M.; Murthy, N.S.; Kohn, J.; Gatt, C.; Dunn, M.G. A comparison of degradable synthetic polymer fibers for anterior cruciate ligament reconstruction. J. Biomed. Mater. Res. Part A **2010**, *93* (2), 738–747.

30. Pollman, M.J. Engineering a bioresorbable stent: REVA programme update. EuroIntervent. Suppl. **2009**, *5* (Suppl. F), F54–F57.

31. Crugnola, A.M.; Radin, E.L.; Rose, R.M.; Paul, I.L.; Simon, S.R.; Berry, M.B. Ultrahigh molecular weight polyethylene as used in articular prostheses (a molecular weight distribution study). J. Appl. Polym. Sci. **1976**, *20* (3), 809–812.

32. Grijpma, D.W.; van Hofslot, R.D.A.; Supèr, H.; Nijenhuis, A.J.; Pennings, A.J. Rubber toughing of poly(lactide) by blending and block copolymerization. Polym. Eng. Sci. **1994**, *34* (22), 1674–1684.

33. Jabara, R.; Pendyala, L.; Geva, S.; Chen, J.; Chronos, N.; Robinson, K. Novel fully bioabsorbable salicylate-based sirolimuseluting stent. EuroIntervent. Suppl. **2009**, *5* (Suppl. F), F58–F64.

34. Bezwada, R.S. Absorbable poly naproxen. Presentation at the annual meeting of the Society for Biomaterials, Chicago, April 18–21, 2007.

35. Bezwada, R.S. Absorbable polymers from functionalized salicylic acid. Presentation at the annual meeting of the Society for Biomaterials, Chicago, April 18–21, 2007.

36. Yu, L.; Liu, H.; Xie, F.; Chen, L.; Li, X. Effect of annealing and orientation on microstructures and mechanical properties of polylactic acid. Polym. Eng. Sci. **2008**, *48* (4), 634–641.

37. Yuan, X.; Mak, A.F.T.; Kwok, K.W.; Yung, B.K.O.; Yao, K. Characterization of poly(L-lactic acid) fibers produced by melt spinning. J. Appl. Polym. Sci. **2001**, *81* (1), 251–260.

38. Glauser, T.; Gueriguian, V.J.; Steichen, B.; Oberhauser, J.; Gada, M.; Kleiner, L. Controlling crystalline morphology of a bioabsorbable stent. International Patent Application Number WO 2011/031872 A2, filed September 9, 2010, publication March 17, 2011.

39. Onuma, Y.; Piazza, N.; Ormiston, J.A.; Serruys, P.W. Everolimus-eluting bioabsorbable stent—Abbot Vascular programme. EuroIntervent. Suppl. **2009**, *5* (Suppl. F), F98–F102.

40. Thatcher, G.L.; Cottone, R.J. Bioabsorbable polymeric composition for a medical device. U.S. Patent Number US7846361 B2, July 20, 2007.

41. Thatcher, G.L.; Cottone, R.J. Bioabsorbable polymeric composition for a medical device. U.S. Patent Number US7897224 B2, July 22, 2009.

42. Cottone, R.J.; Thatcher, G.L.; Parker, S.P.; Hanks, L.; Kujawa, D.A.; Rowland, S.M.; Costa, M.; Schwartz, R.S.; Onuma, Y. OrbusNeich fully absorbable coronary stent platform incorporating dual partitioned coatings. EuroIntervent. Suppl. **2009**, *5* (Suppl. F), F65–F71.

43. Tsuji, H.; Ikada, Y. Stereocomplex formation between enantiomeric poly(lactic acid)s. 11. Mechanical properties and morphology of solution-cast films. Polymer **1999**, *40* (24), 6699–6708.

44. van der Giessen, W.J.; Lincoff, A.M.; Schwartz, R.S. Marked inflammatory sequelae to implantation of biodegradable and nonbiodegradable polymers in porcine coronary arteries. Circulation **1996**, *94* (1), 1690–1697.

45. Grabow, N.; Martin, D.P.; Schmitz, K.; Sternberg, K. Absorbable polymer stent technologies for vascular regeneration. J. Chem. Technol. Biotechnol. **2009**, *85* (6), 744–751.

46. Cottone, R. Bioabsorbable R stent design concepts. Presentation at the Transcatheter Cardiovascular Therapeutics Conference, Washington, DC, October, 2008.

47. Wang, Y. Implantable medical devices fabricated from block copolymers. U.S. Patent Application Number 11/864729, September 28, 2007.

48. Wang, Y.; Gale, D.C.; Gueriguian, V.J. Implantable medical devices fabricated from polymers with radiopaque groups. U.S. Patent Application Number 11/799354, April 30, 2007.

BIBLIOGRAPHY

1. Arterial Remodeling Technologies. Bioresorbable stent. http://www.art-stent. com/index.php (accessed June 2011).

2. Brugaletta, S.; Garcia-Garcia, H.M.; Diletti, R.; Gomez-Lara, J.; Garg, S.; Onuma, Y.; Shin, E.S.; van Geuns, R.J.; de Bruyne, B.; Dudek, D.; Thuesen, L.; Chevalier, B.; McClean, D.; Windecker, S.; Whitbourn, R.; Dorange, C.; Veldhof, S.; Rapoza, R.; Sudhir, K.; Bruining, N.; Ormiston, J.A.; Serruys, P.W. Comparison between the first and second generation bioresorbable vascular scaffolds: a six month virtual histology study. EuroIntervention **2011**, *6* (9), 1110–1116.

3. Buchbinder, M. Biodegradable stents: Future or fancy. Presentation at China Interventional Therapeutics Conference, Beijing, March 31–April 3, 2010.

4. Colombo, A.; Karvouni, E. Biodegradable stents: Fulfilling the mission and stepping away. Circulation **2000**, *102* (4), 371–373.

5. DiMario, C.; Borgia, F. Assimilating the current clinical data of fully bioabsorbable stents. EuroIntervent. Suppl. **2009**, 5 (Suppl. F), F103–F108.

6. Douglas, F.L.; Acharya, S.; Davis, B.L. Value-driven engineering for U.S. global competitiveness: A call for a national platform to advance value-driven engineering. http://www.abiakron.org/Data/Sites/1/pdf/abiawhitepaper6-14-11.pdf. (accessed June 2011)

7. Farooq, V.; Onuma, Y.; Radu, M.; Optical coherence tomography (OCT) of overlapping bioresorbable scaffolds: from bench-work to clinical application. EuroIntervention **2011**, 1–13. http://www.pcronline.com/eurointervention/ahead_of_print/32_04/index.php?ind=1.

8. Fourné, F. Synthetic Fibers: Machines and Equipment, Manufacture, Properties: Handbook for Plant Engineers, Machine Design, and Operation. Munich, Germany: Hanser, 1999.

9. Ge, J. Limus-eluting stents with poly-L-lactic acid coating. Asia Pacif. Cardiol. **2007**, *1* (1), 42–43.

10. Guo, Q.; Lu, Z.; Zhang, Y.; Li, S.; Yang, J. *In vivo* study on the histocompatibility and degradation behavior of biodegradable poly(trimethylene carbonate-co-D,L-lactide). Acta Biochim. Biophys. Sin. **2011**, *43* (6), 433–440.

11. Hietala, E.; Salminen, U.; Stahls, A.; Välimaa, T.; Maasilta, P.; Törmälä, P.; Nieminen, M.S.; Harjula, A.L.F. Biodegradation of the copolymeric polylactide stent. J. Vascu. Res. **2001**, *38* (4), 361–369.

12. Jamiolkowski, D.D. Satisfying product requirements with absorbable polyesters. Presentation at the MD&M West Conference, Anaheim, CA, February 7–10, 2011.

13. Lendlein, A.; Langer, R. Biodegradable, elastic shape-memory polymers for potential biomedical applications. Science **2002**, *296* (5573), 1673–1676.

14. Ormiston, J.A.; Serruys, P.W.S. Bioabsorbable coronary stents. Circ. Cardiovasc. Intervent. **2009**, *2* (3), 255–260.

15. Salemi, T. Can stents pull off a disappearing act? Start-Up **2007**, *12* (1).

16. Behrend, D.; Grabow, N.; Martin, D.P.; Schmitz, K.P.; Sternberg, K.; Williams, S.F. Polymeric, degradable drug-eluting stents and coatings. U.S. Patent No. 7618448, February 6, 2007.

17. Scholz, C. The molecular structure of degradable polymers. In *Degradable Polymers for Skeletal Implants*; Wuisman, P.I.J.M., Smit, T.H., Eds.; Nova Science: New York, 2009; 3–20.

18. Serracino-Inglott, F. Endovascular aneurysm repair—technical aspects. http:// www.stent-graft.com/id2.html (accessed 2008).

19. Shalaby, S.W.; Burg, K.J.L. *Absorbable and Biodegradable Polymers*; CRC Press: Boca Raton, FL, 2004.

20. Soares, J.S.; Moore, J.E., Jr.; Rajagopal, K.R. Theoretical modeling of cyclically loaded, biodegradable cylinders. In *Modeling of Biological Materials*; Mollica, F., Preziosi, L., Rajagopal, K.R., Eds.; Birkhäuser Boston: Boston, 2007

21. Soares, J.S.; Moore, J.E., Jr.; Rajagopal, K.R. Constitutive framework for biodegradable polymers with applications to biodegradable stents. ASAIO J. **2008**, *54* (3), 295–301.

22. Soares, J.S.; Moore, J.E. Jr.; Rajagopal, K.R. Mechanics of deformation-induced degradation of poly(L-lactic acid) endovascular stents. Proceedings of the ASME Summer Bioengineering Conference, Lake Tahoe, CA, June 17–21, 2009.

23. Su, S.; Chao, R.Y.N.; Landau, C.L.; Nelson, K.D.; Timmons, R.B.; Meidell, R.S.; Eberhart, R.C. Expandable bioresorbable endovascular stent. I. Fabrication and properties. Ann. Biomed. Eng. **2003**, *31* (6), 667–677.

24. Tsuji, H. Poly(lactide)s and their copolymers: physical properties and hydrolytic degradation. In *Degradable Polymers for Skeletal Implants*; Wuisman, P.I.J.M., Smit, T.H., Eds.; Nova Science: New York, 2009; 41–70.

25. Vert, M. Bioabsorbable polymers in medicine: An overview. EuroIntervent. Suppl. **2009**, *5* (Suppl. F), F9–F14.

26. Waksman, R. Biodegradable stents: They do their job and disappear. J. Intervent. Radiol. **2006**, *18* (2), 70–74.

Stem Cell—Ultrasound

Stimuli-Responsive Materials

Quazi T. H. Shubhra
Doctoral School of Molecular and Nanotechnologies, Faculty of Information Technology, University of Pannonia, Veszprém, Hungary

A. K. M. Moshiul Alam
Institute of Radiation and Polymer Technology, Bangladesh Atomic Energy Commission, Dhaka, Bangladesh

Abstract
Polymers that respond to environmental conditions are of increasing interest because of their potential applications in biomedical fields. The polymers that undergo physical or chemical changes due to change(s) in the environment are stimulus polymers. This entry outlines the structural phenomena and some properties of stimulus/stimuli-responsive polymers (SRPs). Various temperature, pH, electro and light-sensitive SRPs with their health care applications in fields like drug delivery, tissue engineering, artificial muscles, and wound dressing are also discussed. A special emphasis is given to hydrogels as stimuli-responsive polymeric materials (SRPMs) and their preparation by using a radiation method. Hydrogels are one of the most used radiation-induced SRPMs and can be used widely in various fields of biomedical science. Due to the huge probability of successful application of SRPs in various health care fields, their demand is increasing day by day and it has many opportunities and challenges in biomedical science.

INTRODUCTION

SRPs are defined as polymers that undergo relatively large and abrupt physical or chemical changes in response to small external change(s) in the environmental conditions. Stimuli-responsive polymers (SRPs) can also be called stimuli-sensitive,[1] intelligent,[2] smart,[3] or environmentally sensitive polymers.[4] When SRPs are used for health care, their properties change with body temperature and pH of body fluid.

The preparation of stimuli-responsive materials can be varied by tuning external parameters such as temperature,[5,6] pH,[7] ionic strength,[8] electric field, share rate,[9] light, etc.

Radiation-Induced SRPs

Radiation-induced SRPs are prepared by radiation techniques using both ionizing radiation (gamma (γ), electron beam, etc.) and nonionizing radiation (like UV). It has been demonstrated that processing of natural polymers through radiation is not only simple and effective but also commercially attractive. During a short span, many of these research and development activities have successfully progressed, demonstrating their commercial utility in the areas of environment, health care, and agriculture in some countries.[10] Radiation processed SRPs can provide a variety of applications in biomedical fields. The interest in these polymers has exponentially increased due to their promising potentiality.

Some polymers are responsive to one type of stimuli such as pH, but unresponsive to other types of stimuli, such as

temperature. For example, pH-responsive poly(hydroxyethyl methacrylate-acrylic acid) (p(HEMA-AA)) based hydrogels show negligible volume change during temperature variations. But some systems have been developed to combine two or more stimuli-responsive mechanisms into one polymeric system. For instance, temperature-sensitive polymers may be designed in a way so that they will also respond to pH changes.[11] Kurisawa et al. has reported that two or more signals could be simultaneously applied in order to induce response in so called dual-responsive polymeric systems.[12]

Classification of Stimuli

Stimuli are generally of two types: chemical stimuli and physical stimuli.

Chemical stimuli

Chemical stimuli change the interactions between polymer chains and solvents or between polymer chains only at the molecular level. pH, ionic factors, chemical agents, etc. are examples of some chemical stimuli.

Physical stimuli

Physical stimuli affect the level of various energy sources and alter molecular interactions at critical onset points in various polymers. Temperature, magnetic or electric fields and mechanical stress, are examples of some physical stimuli.

Concise Encyclopedia of Biomedical Polymers and Polymeric Biomaterials DOI: 10.1081/E-EBPPC-120049965

Stem Cell—Ultrasound

Classification of SRPs

SRPs can be both natural and synthetic. Lots of natural polymers are stimuli responsive such as chitosan, gelatin, alginate, and gum acacia. Figure 1 shows structures of some stimuli-responsive natural polymers. There are huge stimuli-responsive synthetic polymers. Poly(N-isopropylacrylamide) (PNIPAAm) and poly(2-carboxyisopropylacrylamide) are common stimuli-responsive synthetic polymeric materials.[13] Figure 2 shows structure of PNIPAAm, which is a stimuli-responsive synthetic polymer.

PREPARATION OF STIMULI-RESPONSIVE POLYMERIC MATERIALS

Traditional Method

Stimuli-responsive polymeric materials (SRPMs) can be prepared by chain or step polymerization reactions.[14] Chain polymerization reactions occur through the initiation

of monomer, propagation of initiated monomer, and termination of propagated polymer. Chain polymerization reactions can be used for the preparation of hydrogel, which is a widely used SRPM. On the other hand, step polymerization reactions occur through the substitution reactions between monomer functionality. The characteristic polymerizations of hydrogel network begin from the reactive center initiated by the polymerization source and terminated upon the loss of the reactivity of the produced radicals. The reactive centers at the initiation stage could be a free radical or ion and thus promotes free radical or ionic polymerization.[15] The polymerization techniques described are possible to carry out by using various curing processes such as thermal, redox, precipitation methods, etc.

Radiation Method

Radiation sources are commonly utilized by several researchers to synthesize SRPMs.[16,17] Ionizing radiation is a high-energy radiation process and involves electronic

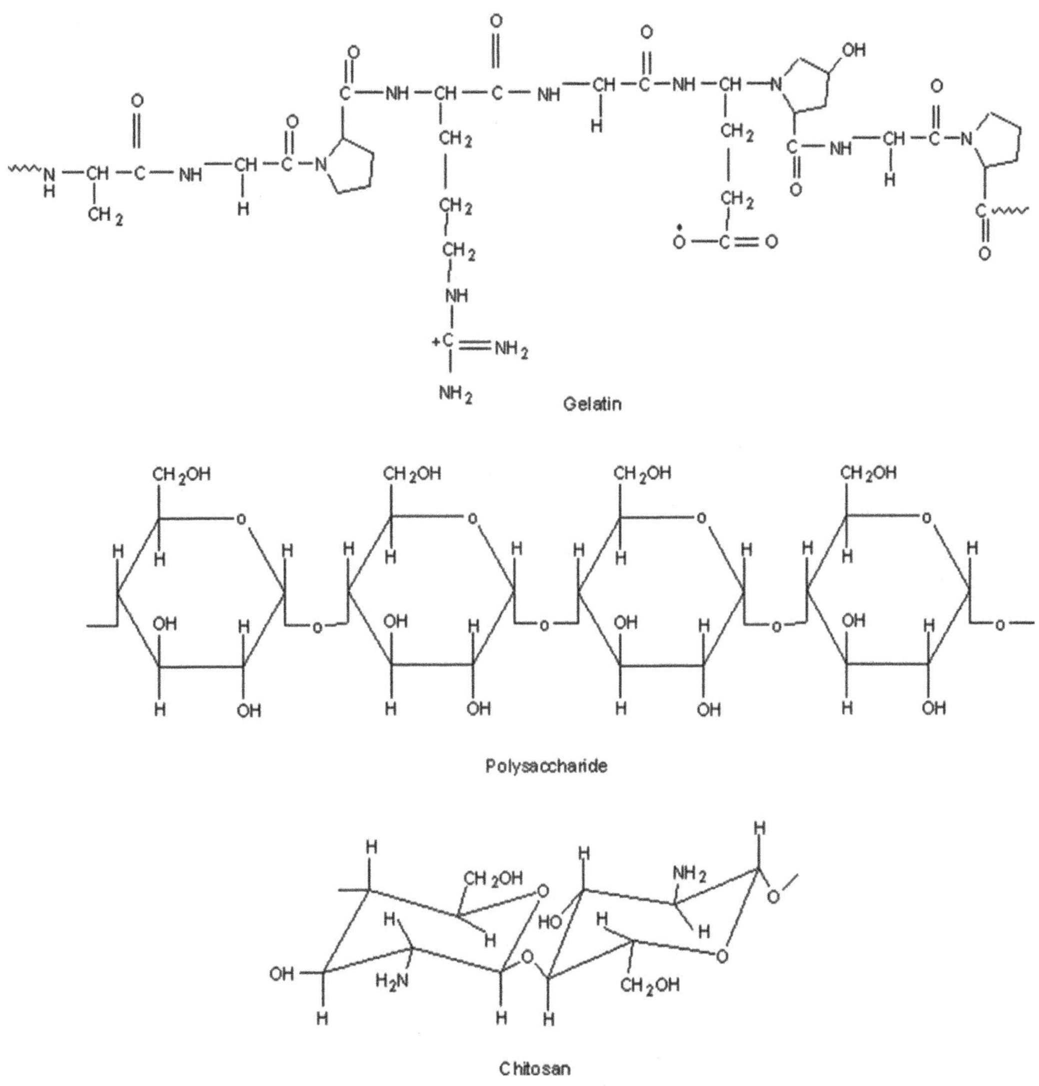

Fig. 1   Structure of some stimuli-responsive natural polymers.

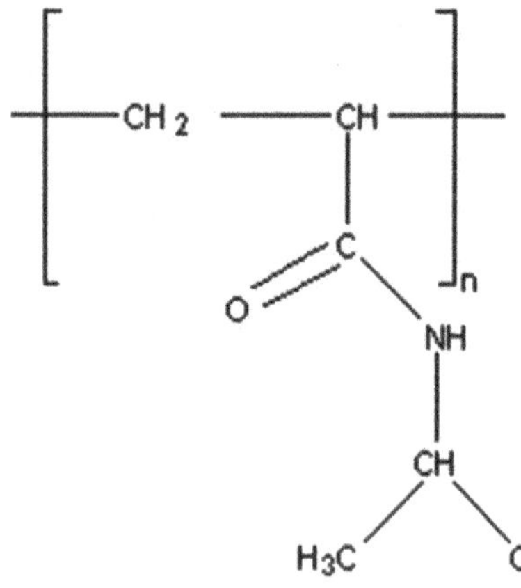

Fig. 2 Structure of stimuli-responsive synthetic polymer (PNIPAAm).

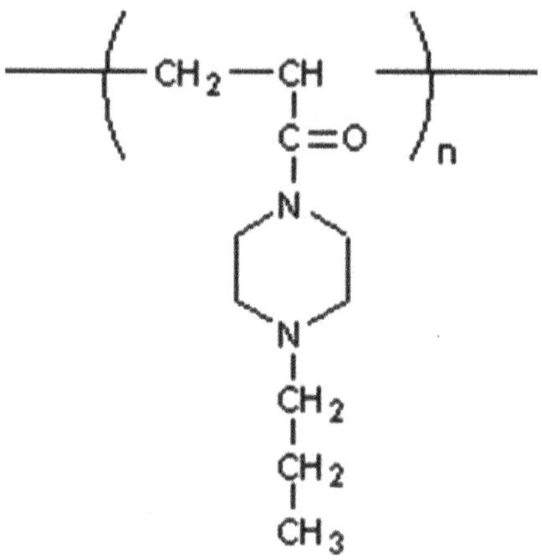

Fig. 3 Structure of dual sensitive poly(N-acryloyl-N-propyl-piperazine) polymer.

radiation of moving particles that carry energy to ionize molecules either in air or water and therefore more penetrative.[18] Electron beam (EB) radiation is a high energy and efficient process and does not require initiators in the reactive mixture. Free radicals are produced in gamma radiation and produced free radicals take part in initiation, propagation and termination reactions. A great advantage of gamma radiation is the removal of any residual initiators that are present after other conventional polymerization processes such as thermal or UV curing. Residual initiator may act as an undesirable contaminant[19] and needs to be removed.

To produce a useable SRPM, the necessary requirement is the formation of cross-links between different polymer chains, which results in a three-dimensional network structure.[20] This requirement is achieved for aqueous solution of polymeric materials poly(vinyl alcohol) (PVA) and poly(N-vinyl pyrrolidone) (PVP) by gamma radiation or EB. During irradiation of aqueous solution of polymeric materials, the major portion of the energy is absorbed by water for the formation of free radicals and molecular products[21] as shown in the following reaction:

$$H_2O \rightarrow H^*, OH^*, e^-(aq.), H_2O_2, H_2, H^+$$

$^*OH$ and $^*H$ radicals are the species responsible for generating cross-linking. The radicals ($^*OH$, $^*H$) abstract hydrogen from a polymer chain and produce a carbon centered radicals in polymer chains, which further decay by forming intermolecular cross-linking.[21]

Many types of SRPMs, especially hydrogels, are prepared from stimuli-responsive polymers by radiation techniques. For example, PVA/polysaccharide biomaterials for wound dressing can be prepared by using radiation technique. PVP-g-TA (poly(N-Vinyl 2-pyrrolidone-g-tartaric acid)) hydrogels are prepared by γ-irradiaiton of ternary mixture of N-vinyl-2–pyrrolidone/tartaric acid/water. A change in ionic strength of swelling solution from 0.01 to 0.2 reduces swelling of this hydrogel.[22] Gelatin

hydrogels are produced by cross-linking using γ-irradiation in the dose range over 16–20 kGy in the gelatin solutions. Carboxymethylated chitosan (CM-chitosan) hydrogels show pH-stimuli response. CM–chitosan hydrogels are synthesized by γ-irradiation. The radiation energy of gamma-ray is absorbed mainly by water in dilute CM–chitosan aqueous solutions and the direct effect of radiation on CM-chitosan can be neglected.[23]

DUAL-SENSITIVE SRPs

Dual-sensitive SRPs are sensitive to more than one type of stimulus. For example, it is possible to obtain polymer structure sensitive to both temperature and pH by the simple combination of ionizable and hydrophobic (inverse thermo sensitive) functional groups. It can mainly be achieved by the copolymerization of monomers bearing these functional groups,[24] combining thermosensitive polymers with polyelectrolytes semi-interpenetrating network, interpenetrating network,[24,25] or by the development of new monomers that respond simultaneously to both stimuli.[26]

Gan et al. studied a new pH and temperature sensitive polymer based on poly(acryloyl-N-propylpiperazine) (PAcrNPP) (Fig. 3) that exhibited a lower critical solution temperature (LCST) in water at 37°C and soluble in water.[11]

TEMPERATURE-SENSITIVE SRPs

The common characteristic of temperature-sensitive polymers is the presence of hydrophobic groups in their structure. The structures of some of temperature sensitive polymers are shown in Fig. 4. Temperature-sensitive hydrogels have been studied most extensively and their unique applications have been reviewed in depth before.[13,27] Lots of hydrogels are known to undergo physical changes in response to changes in temperature such as

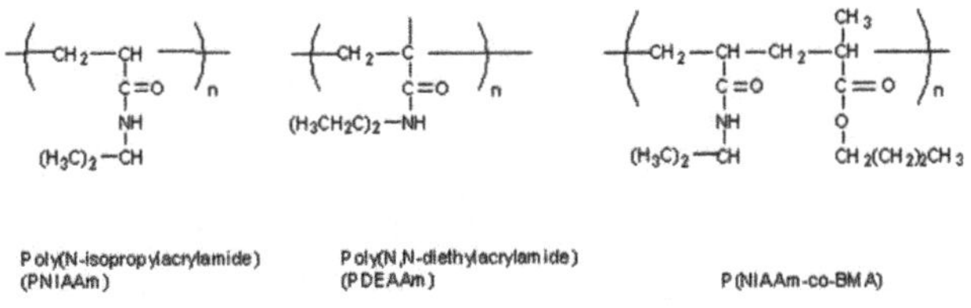

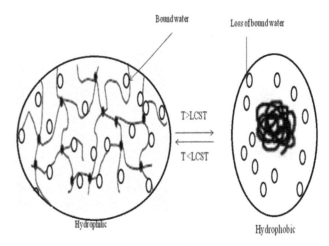

Fig. 4 Structures of some temperature-sensitive polymers.

poly(vinylmethyl ether)s cross-linked by γ-irradiation[28] poly(N-isopropylacrylamide) (PIPAAm) and its derivatives,[29] poly(ethylene oxide)-b-poly(propylene oxide)-b-poly(ethylene oxide) triblock copolymers,[30] poly(alkylvinyl ether)s and their block copolymers, hydroxymethyl cellulose, gelatin, and other more exotic materials. All of these materials share a unique hydration chemical structure in common in aqueous milieu, which is metastable and can be altered radically by increasing the thermal energy in the system. Changing temperature therefore often produces a dramatic and pseudo first-order phase change resulting from the dehydration and rehydration of the materials' chemistry and results in a collapse and expanding behavior in water. This phase transition is reversible and has some characteristic hysteresis upon reversal of the temperature changes. Many materials exhibit this property.[31]

One of the most important and unique properties of temperature responsive polymers is the presence of a critical solution temperature. Most polymers increase their water-solubility with the increase in temperature. But polymers with LCST decrease their water-solubility as the temperature increases. The temperature which induces the polymer to collapse is referred to LCST (also known as "cloud point").[5,32] LCST polymers produce hydrogels which shrink as the temperature increases above the LCST. This type of swelling behavior is known as the inverse (or negative) temperature-dependence. The hydrogen bonding between hydrophilic segments of the polymer chain and water molecules dominates at lower temperature leading to enhance dissolution in water. With the increase in temperature, however, hydrogen bonding becomes weaker while hydrophobic interactions among hydrophobic segments become stronger. The net result is shrinkage of the hydrogels due to inter-polymer chain association through hydrophobic interactions. Some temperature responsive synthetic polymers showing LCST are poly(N-vinylisobutylamide), poly(vinyl methyl ether),[33] poly(N-vinylcaprolactam),[34] poly(dimethylaminoethyl methacrylate), etc. The LCSTs of homopolymers as poly(N-isopropylacrylamide), poly(N,N-diethylacrylamide), poly(N-cyclopropylacrylamide), and poly(N-ethylacrylamide) in distilled water have been reported to be 32°C, 33°C, 58°C, and 74°C, respectively.[35]

On raising the temperature of aqueous solution of SRPMs above LCST, phase separation takes place. The phase transition of temperature-responsive SRPMs are shown schematically in Fig. 5. An aqueous phase containing practically no

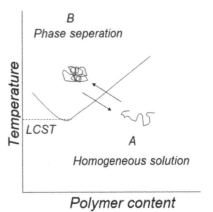

Fig. 5 Schematic of SRPM responsive to temperature.

poly(N-isopropylacrylamide) poly(N-vinylcaprolactum)
LCST 32°C   LCST 35°C

Fig. 6 Temperature response for thermosensitive polymers. **(A)** Soluble phase (below LCST), **(B)** insoluble phase (above LCST).

polymer and a polymer-enriched phase are formed on raising the temperature of the aqueous solution of SRPMs above LCST. Both phases can be easily separated by decanting, centrifugation, or filtration. The temperature of phase transition depends on polymer concentration and molecular weight as shown in Fig. 6. This phase separation is completely reversible and the SRPMs dissolve in water upon cooling.

Temperature-responsive poly(N-isopropylacrylamide) (PNIPAm) undergoes a sharp coil–globule transition in water at 32°C. It changes from a hydrophilic state below this temperature to a hydrophobic state above it.[13] LCST of PNIPAm lies close to body temperature. But LCST of PNIPAm can be increased above and below 37°C by addition of co-monomer units, renders PNIPAm-based materials particularly suitable for biomedical applications.[36] This unique temperature response of PIPAAm in an aqueous solution has been extensively investigated for the use of stimuli-responsive materials in biomedical applications such as in drug delivery systems, bioseparations, bioconjugates, and noninvasive cell manipulations.[37–39]

pH-SENSITIVE SRPs

Polymers containing ionizable functional groups that respond to change in pH are called pH-sensitive polymers. The pH-sensitive polymers contain pendant acidic groups (e.g., carboxylic and sulfonic acids) or basic (e.g., ammonium salts) groups. Such groups either accept or release protons in response to changes in environmental pH. pH-sensitive SRPs are popular for their wide therapeutic applications.[40] Polyacrylic acid, poly(ethylene imine), poly(L-lysine), polymethacrylic acid (PMAA), and poly(N,N-dimethyl aminoethyl methacrylamide) are typical examples of pH-sensitive polymers. Figure 7 shows structure of some pH-sensitive polymers. Poly(acrylic acid) (PAA) becomes ionized at a high pH, whereas poly(N,N′-diethylaminoethyl methacrylate) (PDEAEM) becomes ionized at a low pH.

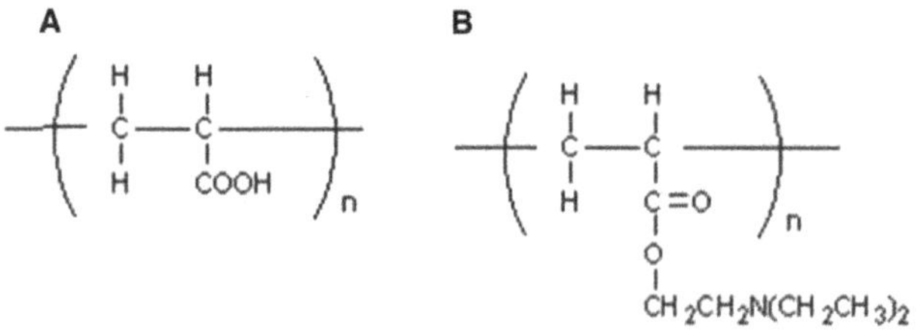

As shown in Fig. 8, cationic polyelectrolytes, such as PDE-AEM, are more soluble at a low pH whereas polyanions, such as PAA, swell at a high pH.[7]

In case of pH-responsive polymers, the driving force behind the transitions is usually a neutralization of charged groups in the polymer by a pH shift. With the change in the environmental pH the degree of ionization in a polymer bearing weakly ionizable groups is dramatically altered at a specific pH, which is called pK_a. This rapid change in net charge of pendant groups causes an alternation of the hydrodynamic volume of the polymer chains. As shown in Fig. 9, hydrogel volume (hydration) increases abruptly at a pH region below its pK_a for basic polymers and above its pK_a for acidic polymer. In a typical pH-sensitive polymer, protonation/deprotonation events occur and impart electrical charge over the molecule (generally on carboxyl or amino groups). The net charge can be decreased by changing pH to neutralize the charges on the polymer, which reduces the hydrophilicity (increases the hydrophobicity) of the polymer and the repulsion between polymer segments.

Precipitation of SRPs is very sharp for pH-induced precipitation and typically requires a change in pH of not more than 0.5 units. Increase in the hydrophobicity of the pH-sensitive polymers through copolymerization with more hydrophobic monomers results in transitions at higher pH.

An important site for the application of pH-sensitive polymers in the body of human being is the gastrointestinal tract (GI-tract). The gastric pH is ~2 whereas intestinal pH is ~7.4 or 7.8. pH-sensitive polymers can be used for various health care purposes like gene delivery and gene therapy research.[41–43]

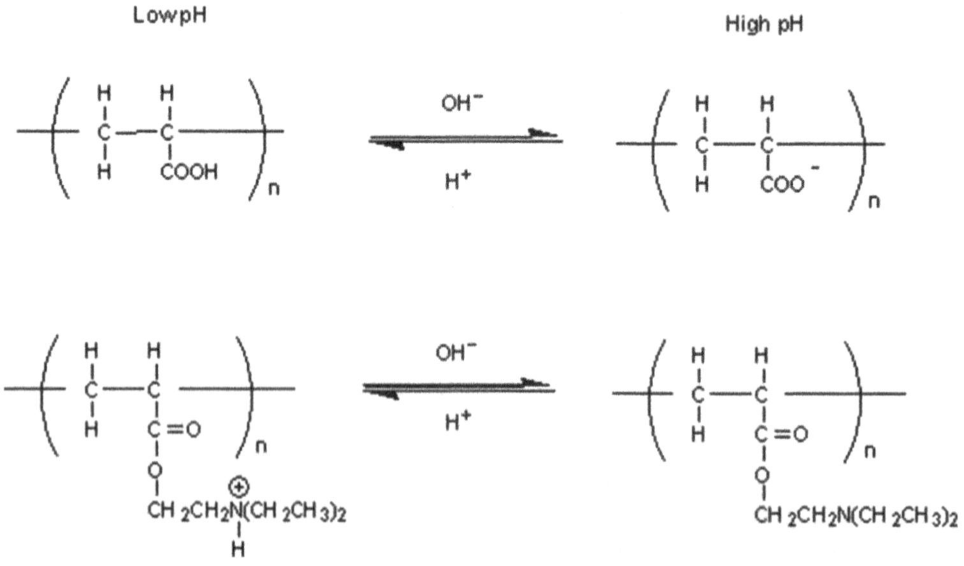

Fig. 7 Structure of pH-sensitive polymers. (**A**) Poly(acrylicacid), (**B**) Poly(N,N′-diethylaminoethyl methacrylate).

Fig. 8 pH-dependent ionization of polyelectrolytes. Poly(acrylicacid) (top) and poly(N,N′-diethylaminoethyl methacrylate) (bottom).

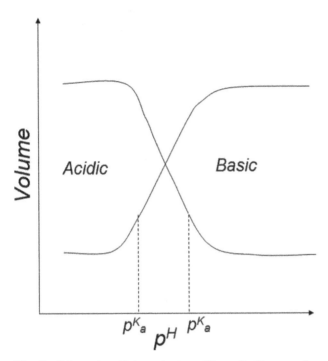

Fig. 9 Schematic pH-dependent swelling of pH-responsive acidic and basic polymers in water.

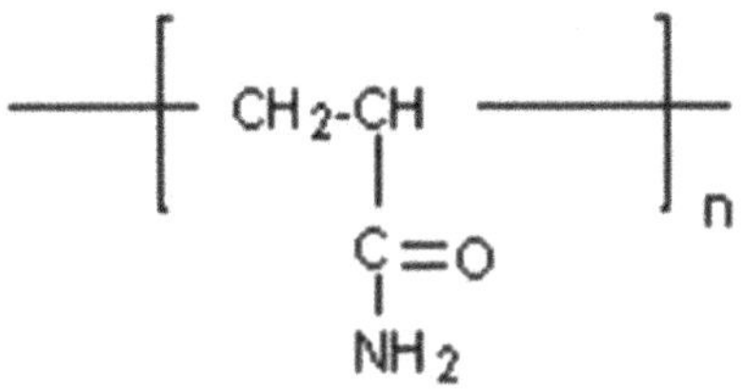

Fig. 10 Structure of electrosensitive stimuli-responsive poly(acrylamide) polymer.

ELECTROSENSITIVE SRPs

Some polymers with ionic groups are sensitive to electric fields due to their ionic charges. Phase transition behavior of these polymers is altered by electric field. This property of such polymers has been applied for bio-related applications such as drug delivery systems, artificial muscle, or biomimetic actuators.[44] It is reported that hydration of cross-linked hyaluronic acid hydrogels show a decrease in hydration as the gel is exposed to a DC voltage of 5 or 10 V/cm, and increases again when the current is switched off.[45] Figure 10 represents structure of electrosensitive poly(acrylamide) (PAM) polymer.

Recent developments reveal that polythiophene-based conductive polymer gels undergo swelling–deswelling transition in response to applied potential over –0.8 to 0.5 square wave potential. When confined in a well, the gel developed a pressure of 10 kPa, which can be used in small-scale actuators or valves in microsystems.[46] Investigation showed that polypyrrole, which is a conducting polymer, can be used for nerve cell regeneration. The nerve cell differentiation was enhanced by passing an electric current through the oxidized form of the polymer in the rats. Nerve cells of rats showed a response to electric current by producing neutral that aided the growth of structural support cells needed for nerve regeneration.[47]

Electro-sensitive hydrogels undergo shrinking or swelling in the presence of an applied electric field. Partially hydrolyzed PAM hydrogels undergo volume collapse by an infinitesimal change in electric potential across the gel

when in contact with both the anode and cathode electrodes. When the potential is applied, hydrated H^+ ions migrate toward the cathode, resulting in loss of water at the anode side. At the same time, electrostatic attraction of negatively charged acrylic acid groups toward the anode surface creates unaxial stress along the gel axis, mostly at the cathode side. Shrinkage of the hydrogel at the anode side occurs due to these two simultaneous events.[48]

When a hydrogel made of sodium acrylic acid–acrylamide copolymer is placed in an aqueous solution (acetone–water mixture) under electric field and without touching the electrodes, hydrogel deformation occurs. This type of hydrogel deformation depends on the concentration of the electrolytes. Application of an electric field in the absence of electrolytes or in the presence of very low concentration of electrolytes causes the hydrogel to shrink. This is due to the migration of Na^+ to the cathode electrode, resulting in changes in the carboxyl groups of the polymer chains from COO^-Na^+ to $–COOH$. In the presence of high concentration of electrolytes in solution, however, more Na^+ enters in the hydrogel than migrates from the hydrogel to the cathode. As a result hydrogel does not shrink.[49]

LIGHT-SENSITIVE SRPs

Light-responsive polymers can change their structures upon UV or visible light irradiation. It originates from their chromophores, i.e., residues with double bonds, triple bonds, other conjugated systems,[50] etc. The mechanism of light sensitivity including *cis–trans* isomerization, ionization, and ring opening that increase polymer hydrophilicity, leading to irradiation-stimulated phase transition (increased hydration). Typical examples of polymers carrying light-sensitive residues are poly(N,N-dimethylacrylamide-4-phenylazophenyl acrylate), partially esterified poly(NIPAM-hydroxyethylacrylamide) and poly(NIPAM-triphenylmethane leuconitrile).[50]

Light-sensitive hydrogels have potential applications in developing optical switches, display units, and ophthalmic drug delivery devices. The light stimulus can be imposed instantly and delivered in specific amounts with high accuracy, which gives light-sensitive hydrogels special advantages over others. For example, the sensitivity of pH-sensitive hydrogels can be limited by hydrogen ion diffusion while the sensitivity of temperature-sensitive hydrogels is rate

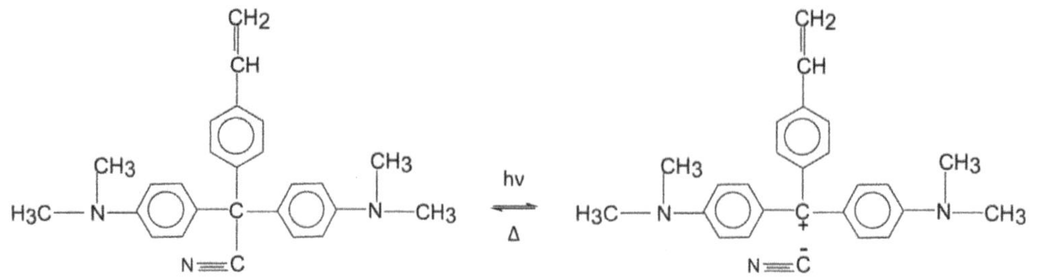

Fig. 11 Structure of leuco derivative molecule bis(4-(dimethylamino)phenyl)(4-vinylphenyl)methylleucocyanide.

limited by thermal diffusion. The capacity for instantaneous delivery of the sol–gel stimulus makes the development of light-sensitive hydrogels important for various applications in both engineering and biochemical fields.

Light-sensitive hydrogels can be separated into UV-sensitive and visible light-sensitive hydrogels.

The UV-sensitive hydrogels are synthesized through the introduction of a leuco derivative molecule like bis(4-dimethylamino)phenylmethyl leucocyanide into the polymer network.[51] As shown in Fig. 11, the leuco derivative molecule can be ionized upon UV irradiation. At a fixed temperature, the hydrogels discontinuously swell in response to UV irradiation but shrink when the UV light is removed.

Visible light-sensitive hydrogels are prepared by introducing a light-sensitive chromophore (e.g., trisodium salt of copper chlorophyllin) to a polymer (e.g., poly(N-isopropylacrylamide) hydrogels).[52] When light (e.g., 488 nm) is applied to the hydrogel, the chromophore absorbs light and then dissipates the absorbed light locally as heat by radiationless transitions, increasing the "local" temperature of the hydrogel. The temperature increase alters the swelling behavior of thermosensitive poly(N-isopropylacrylamide) hydrogels. The temperature increase is proportional to the light intensity and the chromophore concentration. By the incorporation of additional functional group, such as an ionizable group of PAA, the light-sensitive hydrogels become responsive to pH changes also.[8] This type of hydrogel can be activated (i.e., induced to shrink) by visible light and can be deactivated (i.e., induced to swell) by increasing pH. Photopolymerizable hydrogels are suitable for many tissue engineering applications, including growth of vascular tissues, cartilage, bone, other tissues, etc.

WIDELY USED SRPM: HYDROGEL

Hydrogels are three-dimensional swollen networked structures. Hydrogels are able to swell and retain the volume of the adsorbed aqueous medium in their own three-dimensional swollen network. The structure of the hydrogel is shown in Fig. 12. For preparation of hydrogels, radiation techniques are widely used. Using gamma radiation, lots of hydrogels are prepared. For example, poly(vinyl alcohol) hydrogels

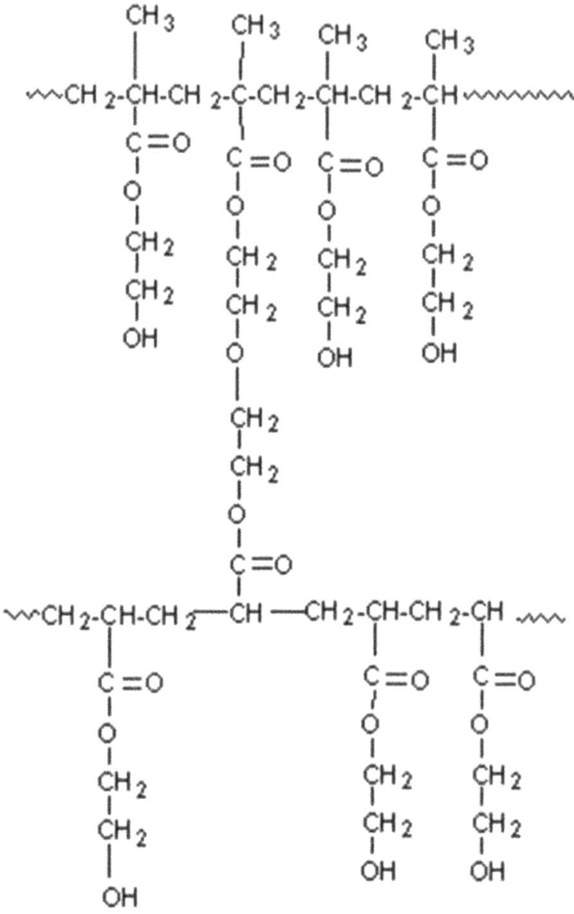

Fig. 12 Structure of a hydrogel.

containing citric or succinic acid,[53] alginate/poly(N-isopropylacrylamide) hydrogels are prepared using γ-irradiation.

Synthetic polymeric hydrogels are generally three-dimensional swollen networks of hydrophilic copolymers or homopolymers, covalently or ionically cross-linked.[18,20,54] The original network of polymeric hydrogel was developed by Wichterle and Lim in Czechoslovakia in 1954[54] which was a copolymer of 2-hydroxyethyl methacrylate (HEMA) and ethyl dimethacrylate and was used as contact lenses.

Since hydrogels are composed of hydrophilic homopolymer or copolymer networks, they can swell in the presence of water or physiological fluids. Chemical cross-links (covalent bonds) or physical junctions (e.g., secondary forces, crystallite formation, chain entanglements) provide both unique swelling behavior and three-dimensional structure of hydrogels.[18,55]

Based on the nature of the side groups, hydrogels can be neutral or ionic. Depending on the physical structure of the networks, they can be semicrystalline, amorphous, supermolecular structures, hydrogen bonded structures, or hydrocolloidal aggregates.[16,18,56–59] Hydrogels have been used extensively in the field of controlled drug delivery.[60] The use of smart hydrogels for drug delivery application has increased significantly over the past few years.[57] For example, PAA–chitosan hydrogels are prepared using γ-irradiation and used for *in vitro* drug release.[61]

HEALTH CARE APPLICATIONS OF SRPMs

Stimuli-responsive polymeric systems are very useful in bio-related applications such as drug delivery and drug targeting systems,[32,62–64] biotechnology,[65–68] chromatography,[39,69] immobilized enzyme systems,[70,71] sensing, fabrication of photonic crystals,[72] nanoparticle templates, separation and purification technologies, controlled cell patterning, DNA separation, and sequencing. Some SRPMs are being implemented in real-life medical practice successfully and some are still being tested in labs. PAM hydrogels are one type of SRPM, which is extensively used for soft tissue augmentation. Approximately 30,000 patients underwent injection of PAM hydrogel for soft tissue augmentation during the last 10 years.[73]

Drug Delivery

The goal of drug delivery is to maintain the concentration of drugs in the body (plasma) within therapeutic limits for long periods of time. The controlled release of drugs from stimuli-responsive polymeric matrices has, however, been very successful. Controlled drug delivery applications include sustained delivery (time) and target delivery systems (insertion at the diseased site). The delivery of protein-based drugs (e.g., growth factor, insulin) and conventional drugs (e.g., steroids, antibiotics) might be possible by using SRPMs through different administration routes for example, cutaneous,[74] and subcutaneous[75] delivery, buccal delivery,[76] delivery to the periodontal pocket,[77] colon,[78] vagina,[79] etc.

Depending on their mode of release, controlled release systems can generally be divided into three sections: chemical erosion, diffusion controlled, and solvent activated.[78–80] In a chemical erosion device, the drug is dispersed in a bioerodible stimuli-responsive polymeric system or linked covalently to a stimuli-responsive polymer backbone via a hydrolyzable linkage. As the polymer or hydrolyzable link degrades, the drug is released. In a diffusion-controlled

device, the drug is surrounded by an inert barrier and diffuses from a reservoir. Another way is the dispersion of drug through a stimuli-responsive polymer and diffusion of the drug from the polymer matrix. In a solvent-activated device, the drug is dispersed within polymeric matrix and the device is swelled with a suitable solvent (generally water). As the device swells, the drug is released.

Hydrogels have been applied widely as intelligent carriers in controlled drug delivery systems.[80] Swelling properties of hydrogel can be controlled, which can be used as a method to trigger drug release.[81] Various delivery and release mechanisms of hydrogels are shown in the Fig. 13. Temperature-responsive hydrogels have been used widely to create systems for drug delivery that exhibit a pulsatile release in response to temperature changes.[1,82] pH-responsive hydrogels have been applied in numerous controlled-release applications also. For example, pH-responsive hydrogels composed of ionic polyethylene glycol (PEG) containing networks have been applied for the oral delivery of proteins such as insulin[83] and calcitonin.[84]

An important class of polymers for drug delivery is glucose-responsive hydrogels. These hydrogels are based on polymers incorporating glucose oxidase within their network.[85] Much of the reported literature centers on insulin release for feedback regulated treatment of diabetes, wherein pH-responsive systems as well as temperature-responsive systems have been evaluated.[5,86] One example of an insulin delivery system is a hydrogel comprising an insulin-containing reservoir within a poly(methacrylic

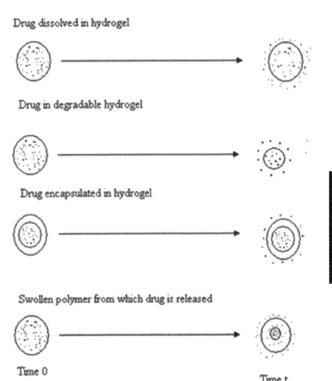

Drug dissolved in hydrogel

Drug in degradable hydrogel

Drug encapsulated in hydrogel

Swollen polymer from which drug is released

Time 0   Time t

Fig. 13 Various delivery and release mechanisms of hydrogels.

acid-*graft*-poly[ethylene glycol]) (P(MAA-g-EG)) copolymer and glucose oxidase is immobilized in that hydrogel.[87] The surface of the polymer contained a series of molecular "entrances," which opened and released insulin, dependent on glucose concentration. Increase of glucose through the polymer layer to the entrapped glucose oxidase results a pH drop as glucose was oxidized to gluconic acid and the released protons caused the pendent PMAA chains of the hydrogel to contract, which results in the opening of the gates to allow insulin transport. An additional feature of this system was the cross-linked polyethylene glycol graft component. In the expanded state of the gel, it was able to adhere to specific regions in the upper intestine. In this way, delivery of insulin could be targeted to the preferred locations in the human body.

Hydrogels made of PAA or PMAA can be used to develop formulations to release drugs in a neutral pH environment.[89] Hydrogels made of polyanions (e.g., PAA) cross-linked with azo-aromatic cross-linkers are developed for colon-specific drug delivery. Such hydrogels swells in the stomach minimally and thus the drug release is also minimal. As the hydrogel passes down through the intestinal tract due to the increase in pH, the extent of swelling increases, leading to ionization of the carboxylic groups. But, only in the colon, the azoaromatic cross-links of the hydrogels can be degraded by azoreductase produced by the microbial flora of the colon,[89] as shown in Fig. 14.

Tissue Engineering

The goal of tissue engineering is to replace, repair, or regenerate tissue(s) or organ(s) functions and create artificial organs and tissue for transplantation. The aim of tissue engineering is to create living, three-dimensional tissue/organs using cells obtained from readily available sources, such as cells obtained directly from the patient. The ultimate goal of tissue engineering is to grow cells specific to a particular organ and then direct the cell growth to form the actual organ. This can be accomplished theoretically through the attachment of specific cells to a scaffolding matrix that directs cell attachment, differentiation, and growth. In order to do this amazing feat, a physical scaffold is required to allow the organization of cells to form the specific organ (to simulate cell growth, differentiation, and migration). These scaffolds need to interact with the cells through specific bioreactions and these bioreactions control cell adhesion and growth factor responses. The scaffolds are ideally biodegradable. Some excellent work in this area has been reported.

An interesting scaffolding material is cell-laden hydrogel. Its high water content, mechanical properties, and biocompatibility resembles natural tissues and makes this hydrogel particularly attractive for tissue engineering applications. Cells are added to a hydrogel before the gelling process distributes the cells homogeneously throughout the resulting scaffold. Cells can also be encapsulated into hydrogels. Combinations of natural and artificial stimuli-responsive polymers can be used to achieve proper scaffold degradation behavior after implantation. Osteoblasts, fibroblasts, vascular smooth muscle cells, and chondrocytes are successfully immobilized and attached to these hydrogel scaffolds for use in tissue engineering purposes. Combination of microfluidic channel technology and photopatterning of hydrogels result in scaffolds that can facilitate increased growth factor delivery and shape sculpting, which is only limited by its molded housing.[90]

Scaffolds made from hydrogels have been used as for tissue engineering[91] and as immunoisolation barriers for microencapsulation technology.[92] In microencapsulation, protection of allogeneic or xenogeneic cells from the host's immune system is done by separating them from the immune

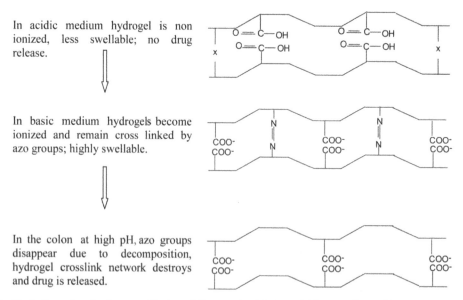

Fig. 14 Schematic illustration of oral colon-specific drug delivery using biodegradable and pH-sensitive hydrogels.

components using a semipermeable membrane. In tissue engineering scaffolds, hydrogels can be used to deliver signals to the cells to act as support structures for cell growth and function and provide space filling.[91,93] Hydrogel scaffolds need to have some desired characteristics like physical parameters and biological properties. Physical parameters include mechanical strength and degradability, while biological properties include biocompatibility and the ability to provide a biologically relevant microenvironment.

PEG is the most commonly used hydrogel for tissue engineering. PEG–based hydrogels can be prepared via γ-radiation or UV photopolymerization.[94] PEG gels are inherently cell repellent. However, with chemical modification it is possible to incorporate various peptides or other signaling molecules into the gels to reduce their cell repulsion behavior.[95]

Artificial Muscles

Artificial muscles are materials that can convert chemical energy directly into mechanical work.[96] Progress toward artificial muscles[97] will lead to great benefits, particularly in the medical area, such as prosthetics that use polymeric hydrogels, which are electroactive to replace damaged human muscles.

Smart hydrogels of electroactive polymers have the potentiality to perform as artificial muscles that could match the force and energy density of biological muscles. This is due to the fact that such hydrogels show a remarkable volumetric change in response to external stimuli and can consequently convert chemical energy into mechanical work. The idea of preparing artificial muscles with smart electroactive polymeric hydrogels is attractive due to the ease of controlling and manipulating of the electric field. The first generation of artificial muscle was developed by Katchalsky and co-workers in the 1950s using polymeric gels. Since then, many interesting works have been done. A cross-linked gel of PVA chains entangled with PAA chains shows rapid electric-field-associated bending deformation: a gel rod of diameter 1 mm bends semicircularly within single second upon the application of an electric field. A mechanical hand composed of four gel fingers could pick up a fragile quail egg (9 g) from a sodium carbonate solution. This hand can hold it without breaking it by the control of an electrical signal.[44] This device performs in solution. Recently, it was demonstrated that an artificial muscle can work normally in the air. In response to an applied electric field, the gel can expand and contract along one axis, rather than simply bending.

Wound Dressing

Hydrogels have also been used as stimuli-responsive wound dressing materials, because most hydrogels are flexible, soft, conform to the wound, biocompatible, and permeable to water vapor and metabolites. As wound dressings, they absorb the exudates, do not stick to the wound, allow the access of oxygen to the wound site, and accelerate healing. Wound dressing hydrogels have been used successfully on burns (first and second degree), external wounds, sunburns, etc. It adheres only to healthy skin and not to the wound skin. Hence dressing is painless. It also cools and relieves the pain immediately. It maintains moist environment and it is nonallergic, nontoxic, eco-friendly, and bio-degradable. Being transparent, the progress of wound healing can be observed without opening the dressing. If desired, hydrogel can transport iodine or water-soluble drugs across its interface to the wound in a sustained manner. It absorbs moderate amount of wound exudates and necrotic tissues adhere gently to the inside of the gel, which comes out when the dressing is removed. This property not only keeps the wound extremely clean but also prevents scar formation. Hydrogel cools the skin at contact point almost instantaneously by 8–10°C and maintains this differential for a very long time (more than 12 hrs). Wound dressing hydrogel contains no extraneous chemicals like antimicrobials, humectants, preservatives, plasticizers and so on, which could retard wound healing. Hydrophilic monomers and polymers used to prepare the hydrogel bandages are based on PVP, poly(ethylene oxide), and PVA. A safe and reliable surgical wound dressing is chitosan–gelatin sponge wound dressing (CGSWD). It provides excellent antibacterial properties with a quicker wound healing speed than vaseline sterile gauze.[98]

PVA/PVP/chitosan hydrogels are prepared by a low-temperature treatment and subsequent γ-ray irradiation. The prepared PVA/PVP/Chi-1 hydrogels showed comprehensive properties suitable for wound dressings.[99] Cellulose supported thermosensitive hydrogel is potentially useful in wound-dressing materials.[100] Elastin-like polypeptide hydrogels are formed by irradiation and have been successfully used for wound treatment.[101] CM-Tec, Inc. (established in 2002, www.cmtec-inc.com) produces a temperature-sensitive hydrogels and delivers them to clients for wound dressing applications. Figure 15 shows trial of wound dressing in the Radiation and Polymer Chemistry Laboratory, Bangladesh Atomic Energy Commission, Bangladesh.

CONCLUSION

Advances in the SRPs were summarized with a view toward some of their molecular structures, properties (related to temperature, pH, electric field and light sensitivity) as well as their biomedical applications. In this entry, only a small portion of the literature is featured from the vast and growing research field. Over the last few years, an impressive number of polymer derivatives have been developed as SRPMs. The applications of radiation technique for chemical modification of natural polymers or

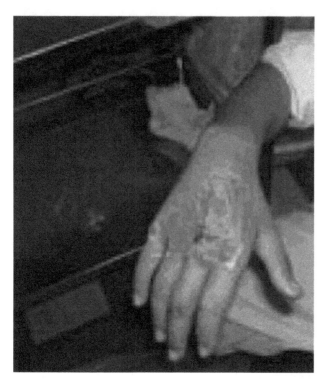

Fig. 15 Trial of wound dressing in the Radiation and Polymer Chemistry Laboratory, Bangladesh Atomic Energy Commission.

grafting natural polymers with other synthetic polymers provide potential importance in the various purposes. So radiation-induced cross-linked/grafted polymers have given interesting and challenging path to the researchers. Recent progress in radiation-induced polymeric synthesis has led to intriguing new stimuli-responsive polymer systems with efficient control of molecular weight and molecular weight distribution. The entry represents only some of the approaches where stimuli-responsive behaviors of irradiated polymers have attracted our attention as versatile bio-related intelligent systems such as drug delivery, tissue engineering, artificial muscle, wound dressing, etc. With continuing advances in material science and biomedical science, we expect continuous and even greater progress of SRPMs in biomaterial researches and their much more clinical applications in the future.

REFERENCES

1. Jeong, B.; Gutowska, A. Lessons from nature: Stimuli-responsive polymers and their biomedical applications. Trends Biotechnol. **2002**, *54* (1), 305–311.
2. Kikuchi, A.; Okano, T. Intelligent thermo responsive polymeric stationary phases for aqueous chromatography of biological compounds. Prog. Polym. Sci. **2002**, *27* (6), 1165–1193.
3. Hoffman, A.S. Hydrogels for biomedical applications. Adv. Drug Deliv. Rev. **2002**, *54* (1), 3–12.
4. Qiu, Y.; Park, K. Environment-sensitive hydrogels for drug delivery. Adv. Drug Deliv. Rev. **2001**, *53* (3), 321–339.
5. Tanaka, T. Phase transitions in gels and single polymer. Polymer **1979**, *20* (11), 1404–1412.
6. Bae, Y.H.; Okano, T.; Kim S.W. Insulin permeation through thermo sensitive hydrogels. J. Control. Release **1989**, *9* (3), 271–279.
7. Kopecek, J.; Vacik, J.; Lim, D. Temperature-responsive polymers for liquid-phase separation. J. Polym. Sci. A Polym. Chem. **1971**, *9*, 2801.
8. Suzuki, A.; Ishii, T.; Maruyama, Y. Optical switching in polymer gels. J. Appl. Phys. **1996**, *80* (1), 131–136.
9. Glass, J.E. *Advances in Chemistry Series*; American Chemical Society: Washington DC, 1989.
10. Clough, R.L. Review of the IRAP 2000 Conference. Nucl. Instrum. Methods Phys. Res. Sect. **2001**, *B* (185), 1–3.
11. Gan, L.H.; Gan, Y. Y.; Deen, G.R. Poly (N-acryloyl-N'-propylpiperazine): A newstimuli-responsive polymer. Macromolecules **2000**, *33* (21), 7893–7897.
12. Kurisawa, M.; Yui, N. Dual-stimuli-responsive drug release from interpenetrating polymer network-structured hydrogels of gelatin and dextran. J. Control. Release **1998**, *54* (2), 191–200.
13. Schild, H.G. Poly(N-isopropylacrylamide): Experiment, theory and application. Prog. Polym. Sci. **1992**, *17* (2), 163–249.
14. Allcock, H.R.; Lampe, F.W. *Contemporary Polymer Chemistry*; Prentice-Hall, Inc.: New Jersey, 1990, 420–442.
15. Nicholson, J.W. *The Chemistry of Polymers*; The Royal Society of Chemistry: Cambridge, UK, 1997, 27–44.
16. Rosiak, J.M.; Ulanski, P. Synthesis of hydrogels by irradiation of polymers in aqueous solution. Radiat. Phys. Chem. **1999**, *55* (2), 139–151.
17. Chowdhury, M.N.K.; Alam, A.K.M.M.; Dafader, N.C.; Haque, M.E.; Akhtar, F.; Ahmed, M.U.; Rashid, H.; Begum, R. Radiation processed hydrogel of poly(vinyl) alcohol with biodegradable polysaccharides. Bio-Med. Mater. Eng. **2006**, *16* (3), 223–228.
18. Peppas, N.A.; Mikos, A.G. *Hydrogels in Medicine and Pharmacy Fundamentals*; CRC Press, Inc.: Florida, 1986, 1–25.
19. Hamilton, C.J.; Tighe, B.J. *Comprehensive Polymer Science: The Synthesis, Characterisation, Reactions & Applications of Polymers*; Pergamon Press: Oxford, 1989.
20. Mack, E.J.; Okano, T.; Kim, S.W. *Hydrogels in Medicine and Pharmacy: Polymers*; CRC Press, Inc.: Florida, 1987, 85–93.
21. Kost, J.; Langer, R. *Hydrogels in Medicine and Pharmacy: Properties and Applications*; CRC Press, Inc.: Florida, 1987, 95–108.
22. Ozyurek, C.; Caykara, T.; Kantoglu, O.; Guven, O. Radiation synthesis of poly(N-vinyl-2-pyrrolidone-g-tartaric acid) hydrogels and their swelling behaviors. Polym. Adv. Technol. **2002**, *13* (2), 87–93.
23. Ling, H.; Maolin, Z.; Jing, P.; Jiuqiang, L.; Genshuan, W. Radiation-induced degradation of carboxymethylated chitosan in aqueous solution. Carbohydr. Polym. **2007**, *67* (3), 305–312.
24. Bulmus, V.; Ding, Z.; Long, C.J.; Stayton, P.S.; Hoffman, A.S. Site-specific polymer-streptavidin bioconjugate for

pH-controlled binding and triggered release of biotin. Bioconjug. Chem. **2000**, *11* (1), 78–83.

25. Verestiuc, L.; Ivanov, C.; Barbu, E.; Tsibouklis, J. Dual-stimuli-responsive hydrogels based on poly(Nisopropylacrylamide)/chitosan semi-interpenetrating networks. Int. J. Pharm. **2004**, *269* (1), 185–194.

26. Gonzalez, N.; Elvira, C.; San Roman, J. Novel dual-stimuli-responsive polymers derived from ethyl pyrrolidine. Macromolecules **2005**, *38* (22), 9298–9303.

27. Hoffman, A.S. Applications of thermally reversible polymers and hydrogels in therapeutics and diagnostics. J. Control. Release **1987**, *6* (1), 297–305.

28. Hirasa, O.; Morishita, Y.; Onomura, R.; Ichijo, H.; Yamauchi, A. Preparation and mechanical properties of thermo-responsive fibrous hydrogels made from poly(vinyl methyl ether)s. Kobunshi Ronbunshu **1989**, *46* (11), 661–665.

29. Dong, L.C.; Hoffman, A.S. Synthesis and application of thermally reversible heterogels for drug delivery. J. Control. Release **1990**, *13* (1), 21–31.

30. Fults, K. A.; Johnston, T.P. Sustained-release of urease from a poloxamer gel matrix. J. Parenter. Sci. Technol. **1990**, *44* (2), 58–65.

31. Sato-Matsuo, E.; Tanaka, T. Kinetics of discontinuous volume phase transition of gels. J. Chem. Phys. **1988**, *89*, 1695–1703.

32. Sershen, S.; West, J. Implantable, polymeric systems for modulated drug delivery. Adv. Drug Deliv. Rev. **2002**, *54* (9), 1225–1235.

33. Maeda, Y. IR spectroscopic study on the hydration and the phase transition of poly(vinyl methyl ether) in water. Langmuir **2001**, *17* (5), 1737–1742.

34. Inoue, T. Temperature sensitivity of a hydrogel network containing different LCST oligomers grafted to the hydrogel backbone. Polym. Gels Networks **1998**, *5* (6), 561–575.

35. Idziak, I. Thermosensitivity of aqueous solutions of poly(N,N-diethylacrylamide). J. Macromol. **1999**, *32* (4), 1260.

36. Chen, G.; Hoffman, A.S. Preparation and properties of thermo reversible, phase-separating enzyme-oligo(N-isopropylacrylamide) conjugates. Bioconjug. Chem. **1993**, *4* (6), 509–514.

37. Okano, T.; Bae, Y.H.; Jacobs, H.; Kim, S.W. Thermally on-off switching polymers for drug permeation and release. J. Control. Release **1990**, *11* (1–3), 255–265.

38. Kanazawa, H.; Sunamoto, T.; Matsushima, Y.; Kikuchi, A.; Okano, T. Temperature-responsive chromatographic separation of amino acid phenylthiohydantoins using aqueous media as the mobile phase. Anal. Chem. **2000**, *72* (24), 5961–5966.

39. Kobayashi, J.; Kikuchi, A.; Sakai, K.; Okano, T. Aqueous chromatography utilizing hydrophobicity-modified anionic temperature-responsive hydrogel for stationary phase. J. Chromatogr. A **2002**, *958* (1–2), 109–119.

40. Kyriakides, T.R.; Cheung, C.Y.; Murthy, N.; Bornstein, P.; Stayton, P.S.; Hoffman, A.S. pH-Sensitive polymers that enhance intracellular drug delivery *in vivo*. J. Control. Release **2002**, *78* (1–3), 295–303.

41. Godbey, W.T. Poly(ethylenimine) and its role in gene delivery. J. Control. Release **1999**, *60* (2–3), 149–160.

42. Godbey, W.T.; Mikos, A.G. Recent progress in gene delivery using non viral transfer complexes. J. Control. Release **2001**, *72* (1–3), 115–125.

43. Tang, M.X. *In vitro* gene delivery by degraded polyamidoamine dendrimers. Bioconjug. Chem. **1996**, *7* (6), 703–714.

44. Shiga, T. Deformation and viscoelastic behavior of polymer gels in electric fields. Adv. Polym. Sci. **1997**, *134*, 131–163.

45. Tomer, R.; Dimitrijevic, D.; Florence, A.T. Electrically controlled-release of macromolecules from cross-linked hyaluronic-acid hydrogel. J. Control. Release **1995**, *33* (3), 405–413.

46. Irvin, D.J.; Goods, S.H.; Whinnery, L.L. Direct measurement of extension and force in a conductive polymer gel actuator. Chem. Mater. **2001**, *13* (4), 1143–1145.

47. Schmidt, C.E.; Shastri, V.R.; Vacanti, J.P.; Langer, R. Stimulation of neurite outgrowth using an electrically conducting polymer. Proc. Natl. Acad. Sci. U.S.A. **1997**, *94* (17), 8948–8953.

48. Tanaka, T.; Nishio, I.; Sun, S.T.; Ueno-Nishio, S. Collapse of gels in an electric field. Science **1982**, *218* (4571), 467–469.

49. Shiga, T.; Hirose, Y.; Okada, A.; Kurauchi, T. Electric field-associated deformation of polyelectrolyte gel near a phase transition point. J. Appl. Polym. Sci. **1992**, *46* (4), 635–640.

50. Wu, X.Y.; Zhang, Q.; Arshady, R. *Stimuli Sensitive Hydrogels*; Citus Books: London, UK, 2003.

51. Mamada, A.; Tanaka, T.; Kungwachakun, D.; Irie, M. Photo induced phase transition of gels. Macromolecules **1990**, *23* (5), 1517–1519.

52. Suzuki, A.; Tanaka, T. Phase transition in polymer gels induced by visible light. Nature **1990**, *346*, 345–347.

53. Aji, Z. Preparation of poly (vinyl alcohol) hydrogels containing citric or succinic acid using gamma radiation. Radiat. Phys. Chem. **2005**, *74* (1), 36–41.

54. Wichterle, O.; Lim, D. Hydrophilic gels for biological use. Nature **1960**, *185*, 117–118.

55. Peppas, N.A. Hydrogels. In *Biomaterials Science. An Introduction to Materials in Medicine*; Academic Press: San Diego, California, 1996, 60–64.

56. Stauffer, S.R.; Peppas, N.A. Poly(vinyl alcohol) hydrogels prepared by freezing-thawing cyclic processing. Polymer **1992**, *33* (18), 3932.

57. Bekturov, E.A.; Bimendina, L.A. Interpolymer complexes. Adv. Polym. Sci. **1981**, *43*, 100.

58. Tsuchida, E.; Abe, K. Interactions between macromolecules in solution and intermacromolecular complexes. Adv. Polym. Sci. **1982**, *45*, 1–119.

59. Bell, C.L.; Peppas, N.A. Biomedical membranes from hydrogels and inter polymer complexes. Adv. Polym. Sci. **1995**, *122*, 125–175.

60. Chen, S.C.; Wu, Y.C.; Mi, F.L.; Lin, Y.H.; Yu, L.C.; Sung, H.W. A novel pH-sensitive hydrogel composed of N, O-carboxymethyl chitosan and alginate cross-linked by genipin for protein drug delivery. J. Control. Release **2004**, *96* (2), 285–300.

61. Shim, J.W.; Nho, Y.C. Preparation of poly(acrylic acid)–Chitosan hydrogels by gamma irradiation and *in vitro* drug release. J. Appl. Polym. Sci. **2003**, *90* (13), 3270–3277.

Stem Cell—Ultrasound

62. Rolland, A. *Pharmaceutical Particulate Carriers: Therapeutic Applications*; Dekker: New York, 1993, 367–421.

63. Yokoyama, M. Gene delivery using temperature-responsive polymeric carriers. Drug Discov. Today **2002**, *7* (7), 426–432.

64. Chilkoti, A.; Dreher, M.R.; Meyer, D.E.; Raucher, D. Targeted drug delivery by thermally responsive polymers. Adv. Drug Deliv. Rev. **2002**, *54* (5), 613–630.

65. Yoshida, R.; Toshikazu, T.; Hisao, I. Self-oscillating gel. J. Am. Chem. Soc. **1996**, *118* (21), 5134–5135.

66. Tabata, O.; Hirasawa, H.; Aoki, S.; Yoshida, R.; Kokufuta, E. Ciliary motion actuator using self-oscillating gel. Sens. Actuators A Phys. **2002**, *95* (2–3), 234–238.

67. Giannos, S.A.; Dinh, S.M.; Berner, B. Polymeric substitution in a pH oscillator. Macromol. Rapid Commun. **1995**, *16* (7), 527–531.

68. Sharma, S.; Kaur, P.; Jain, A.; Rajeswari, M.R.; Gupta, M.N. A smart bioconjugate of chymotrypsin. Biomacromolecules **2003**, *4* (2), 330–336.

69. Anastase-Ravion, S.; Ding, Z.; Pelle, A.; Hoffman, A.S.; Letourneur, D. New antibody purification procedure using a thermally responsive poly(N sopropylacrylamide)-dextran derivative conjugate. J. Chromatogr. **2001**, *B761* (2), 247–254.

70. Galaev, I.Y.; Mattiasson, B. *Smart Polymers for Bioseparation and Bioprocessing*; Taylor & Francis: London and New York, 2002, 257.

71. Galaev, I.Y.; Gupta, M.N.; Mattiasson, B. Use smart polymers for bio-separation. Chem. Tech. **1996**, *12*, 19–25.

72. Xu, S.; Zhang, J.; Paquet, C.; Lin, Y.; Kumacheva, E. From hybrid microgels to photonic crystals. Adv. Funct. Mater. **2003**, *13* (6), 468.

73. Tsung-Hua, Y. Recent applications of polyacrylamide as biomaterials. Recent Pat. Mater. Sci. **2008**, *1* (1), 29–40.

74. Singh, R.; Vyas, S.P. Topical liposomal system for localized and controlled drug delivery. J. Dermatol. Sci. **1996**, *13* (2), 107–111.

75. Barichello, J.M.; Morishita, M.; Takayama, K.; Nagai, T. Absorption of insulin from Pluronic F-127 gels following subcutaneous administration in rats. Int. J. Pharm. **1999**, *184* (2), 189–198.

76. Bremecker, K.D.; Strempel, H.; Klein, G. Novel concept for a mucosal adhesive ointment. J. Pharm. Sci. **1984**, *73* (4), 548–552.

77. Needleman, I.G.; Martin, G.P.; Smales, F.C. Characterization of bioadhesives for periodontal and oral mucosal drug delivery. J. Clin. Periodontol. **1998**, *25* (1), 74–82.

78. Brondsted, H.; Kopecek, J. Hydrogels for site-specific drug delivery to the colon-invitro and invivo degradation. Pharm. Res. **1992**, *9* (12), 1540–1545.

79. Pavelic, Z.; Skalko-Basnet, N.; Schubert, R. Liposomal gels for vaginal drug delivery. Int. J. Pharm. **2001**, *219* (1), 139–149.

80. Peppas, N.A.; Bures, P.; Leobandung, W.; Ichikawa, H. Hydrogels in pharmaceutical formulations. Eur. J. Pharm. Biopharm. **2000**, *50* (1), 27–46.

81. Peppas, N.A.; Langer, R. Origins and development of biomedical engineering within chemical engineering. AIChE J. **2004**, *50* (3), 536–546.

82. Rongbing, Y.; Alexander, V.G.; Fawaz A.; William M.C.; Yury R. An implantable thermoresponsive drug delivery system based on Peltier device. Int. J. Pharm. **2013**, *447* (1–2), 109–114.

83. Morishita, M.; Lowman, A.M.; Takayama, K.; Nagai, T.; Peppas, N.A. Elucidation of the mechanism of incorporation of insulin in controlled release systems based on complexation polymers. J. Control. Release **2002**, *81* (1), 25–32.

84. Torres-Lugo, M.; Garcia, M.; Record, R.; Peppas, N.A. Physicochemical behavior and cytotoxic effects of p(methacrylic acid–g-ethylene glycol) nanospheres for oral delivery of proteins. J. Control. Release **2002b**, *80* (1), 197–205.

85. Podual, K.; Doyle, F.J.; Peppas, N.A. Preparation and dynamic response of cationic copolymer hydrogels containing glucose oxidase. Polymer **2000c**, *41* (1), 3975–3983.

86. Makino, K.; Mack, E.J.; Okano, T.; Sung W.K. A microcapsule self-regulating delivery system for insulin. J. Control. Release **1990**, *12* (3), 235–239.

87. Peppas, N.A.; Keys, K.B.; TorresLugo, M.; Lowman, A.M. Poly(ethylene glycol)-containing hydrogels in drug delivery. J. Control. Release **1999**, *62* (1), 81–87.

88. Lim, F.; Sun, A.M. Microencapsulated islets as bioartificial pancreas. Science **1980**, *210* (4472), 908–910.

89. Ghandehari, H.; Kopeckova, P.; Kopecek, J. *In vitro* degradation of pH-sensitive hydrogels containing aromatic azo bonds. Biomaterials **1997**, *18* (12), 861–872.

90. Sershen, S.; Mensing, G.; Ng, M.; Halas, N.; Beebe, D.; West, J. Independent optical control of microfluidic valves formed from optomechanically-responsive nanocomposite hyderogels. Adv. Mater. **2005**, *17* (11), 1366–1368.

91. Lee, K.Y.; Mooney, D.J. Hydrogels for tissue engineering. Chem. Rev. **2001**, *101* (7), 1869–1879.

92. Lahooti, S.; Sefton, M.V. Agarose enhances the validity of intraperitoneally implanted microcapsulated L929 fibroblasts. Cell Transplant. **2000**, *9* (6), 785–796.

93. Anseth, K.S.; Burdick, J.A. New directions in photopolymerizable biomaterials. MRS Bull. **2002**, *27* (02), 130–136.

94. Williams, C.G; Kim, T.K.; Taboas, A.; Malik, A.; Manson, P. *In vitro* chondrogenesis of bone marrow-derived mesenchymal stem cells in a photopolymerizing hydrogel. Tissue Eng. **2003**, *9* (4), 679–689.

95. Temenoff, J.S.; Mikos, A.G. Injectable biodegradable materials for orthopaedic tissue engineering. Biomaterials **2000**, *21* (23), 2405–2412.

96. Schreyer, H.B. Electrical activation of artificial muscles containing polyacrylonitrile gel fibers. Biomacromolecules **2000**, *1* (4), 642–647.

97. Onoda, M. Artificial muscle using conducting polymers. Elect. Eng. Jpn. **2004**, *149* (4), 7–13.

98. Chun-Mei, D.; Lan-Zhen, H.; Ming, Z.; Dan, Y.; Yi, L. Biological properties of the chitosan-gelatin sponge wound dressing. Carbohydr. Polym. **2007**, *69* (3), 583–589.

99. Yu, H.; Xu, X.; Chen, X.; Hao, J.; Jing, X. Medicated wound dressings based on poly(vinyl alcohol)/poly(N-vinyl pyrrolidone)/chitosan hydrogels. J. Appl. Polym. Sci. **2005**, *101* (4), 2453–2463.

100. Liu, B.; Hu, J. The application of temperature-sensitive hydrogels to textiles: A review of Chinese and Japanese investigations. Fibres Text. East. Eur. **2005**, *13* (6), 45–49.

101. Chow, D.; Nunalee, M.L.; Lim, D.W.; Simnick, A.J.; Chilkoti, A. Peptide-based biopolymers in biomedicine and biotechnology. Mater. Sci. Eng. R **2008**, *62* (4), 125–155.

Stem Cell—Ultrasound

Stimuli-Responsive Materials: Thermo- and pH-Responsive Polymers for Drug Delivery

Jasbir Singh
Department of Pharmacy, University of Health Sciences, Rohtak, India
Harmeet Kaur
Prabhu Dayal Memorial (PDM) College of Pharmacy, Bahadurgarh, India

Abstract

The field of drug delivery has focused on the use of polymers for controlling release and targeting of therapeutics. For years, research endeavors have concentrated mainly on the development of new degradable polymers for delivering proteins/peptides, DNA, and higher generation chemical drugs in an efficient and safe manner. Recent interest in biomaterials that show sharp changes in properties to small or modest changes in the environment have opened new methods to trigger the release of drugs and localize the therapeutic within a particular site. These novel polymers, usually termed "smart" or "intelligent," are stimuli-responsive materials and could potentially elicit a therapeutically effective dose without adverse side effects. Smart polymers responding to different stimuli, such as pH, light, temperature, salt concentration, ultrasound, magnetism, or biomolecules, have been investigated. However, the possible environmental conditions, most applicable or limited to the biomedical application for drug delivery are pH and temperature, which makes these stimuli important among others. This entry describes "smart" drug delivery systems based on polymers or polymer combinations sensitive to either temperature or pH or both.

INTRODUCTION

Polymer science has made revolutionary changes in drug delivery systems and the development of polymers that change their structures and properties in response to environmental stimuli such as pH, temperature, electric fields, and light has attracted a great deal of attention.[1–3] Such polymers have been called "smart polymers," "intelligent polymers," "stimulus-sensitive polymers," or "responsive polymers." Various delivery systems based on the smart polymers have been proposed because of their unique potential in the modulation of drug release and targeting functionality.[4–6] These stimulus-sensitive polymers can show physicochemical changes to physiological variations (whether naturally or pathologically) like temperature and pH in the body. These changes can be observed as a change in conformation, solubility, swelling/collapsing, hydration state, micellization, and hydrophilic/hydrophobic balance of the polymer. The changes may also include a combination of several responses at the same time.

The increases in temperature are associated with several disease states like cancer, inflammation.[4,7] Thermoresponsive polymers have been employed to release their payload within environments above the physiological temperature. For example, poly(*N*-isopropylacrylamide) (PNIPAAm), a well-studied thermoresponsive polymer, undergoes a reversible phase transition in aqueous solution from hydrophilic to hydrophobic at its critical solution temperature

(CST) of approximately 32°C.[5] A variety of polymers, copolymers, and graft polymers, in an aqueous solution, exhibit dramatic changes upon a temperature change below or above the body temperature (Fig. 1). For instance, low molecular weight triblock copolymers of poly(lactic-*co*-glycolic acid) (PLGA) and poly(ethylene glycol) (PEG) have been designed with emphasis on hydrophilic–hydrophobic balance to meet various temperatures. Jeong and co-workers first demonstrated a temperature-induced sol–gel transition of a PEG-PLGA-PEG aqueous solution upon subcutaneous injection in rats.[6,8]

Similar to temperature changes, the pH change in body can be used to direct the delivery system to specific targets. The pH variation in human body is seen in gastrointestinal tract (GIT), intracellular organelles like endosomes/lysosomes, tumors as well as inflamed or wound tissues, etc. In GIT, pH varies from acidic in stomach to basic in intestine (Table 1). The pH increases progressively toward small intestine and reaches a peak level in the distal ileum. After the ileo-cecal junction, the pH falls to slightly acidic due to the presence of short-chain fatty acids and then again rises to neutrality in the colon.[9]

The extracellular pH is often affected by tumor progress. For example, the extracellular pH of tumors is more acidic (~6.5) than in the blood (~7.4).[10] Many studies have exploited the acidic tumor conditions to achieve tumor-targeted gene delivery.[11,12] The production of lactic acid under anaerobic conditions and ATP hydrolysis in an

Concise Encyclopedia of Biomedical Polymers and Polymeric Biomaterials DOI: 10.1081/E-EBPPC-120050042

Stem Cell—Ultrasound

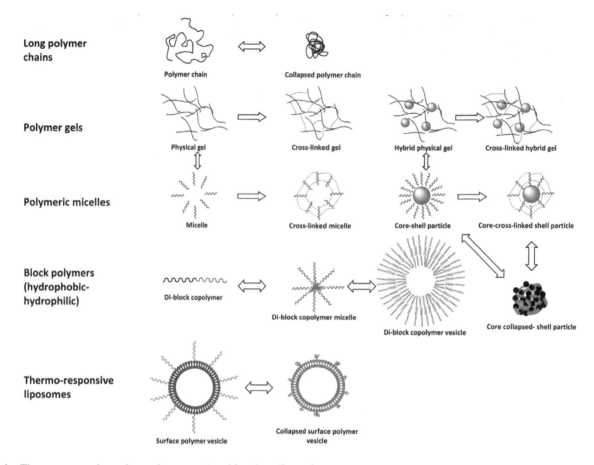

Fig. 1 Thermo-responsive polymers' common transitions/transformations.

Table 1 Physiological pH and transit time within the GIT

GIT Portion	pH	Transit time	Enzymes
Oral cavity	6.0–7.0	10–14 sec	Polysaccharidases
Stomach	1.5	0.2–2.0[a] hr	Proteasaes Lipases
Duodenum	6.4	0.5–0.75 hr	Polysaccharidases Oligosaccharidases Proteases Peptidases Lipases
Jejunum	7.0	1.5–5.5 hr	Oligosaccharidases
Ileum	7.4		Oligosaccharidases
Caecum	5.9	13–68[b] hr	Glycosidases
Rectum	6.5		Azoreductase Polysaccharidases etc.

Abbreviation: GIT, Gastrointestinal tract.
[a]Postprandial time may vary upto 12 hr and dependent upon size of ingested material.
[b]Highly variable.
Source: Data taken from Washington et al.[186]

energy-deficient environment contribute to creation of acidic microenvironment in many tumor tissues.[13] Tumor-targeted delivery can also be achieved by thermoresponsive polymers in response to hyperthermic conditions in tumors.[14,15]

For example, polyethylene imine (PEI) and *N*-isopropyl-acrylamide (NIPAAm) derived cationic thermoresponsive copolymers, with a tunable CST between 37°C and 42°C, effectively condensed DNA and formed small, charged neutral polyplexes. The thermoresponsive polyplexes showed a twofold increase in transfection efficiency for a temperature change from 37°C to 42°C.[14] Upon intravenous (i.v.) injection, the thermoresponsive polyplexes were significantly accumulated in the hyperthermically treated neuroblastoma tumor in mice, resulting in significantly increased DNA transfection *in vivo*.[15] However, an underlying challenge in cancer-targeted gene delivery using stimuli-responsive polymers is to molecularly tune them to be significantly responsive to relatively insubstantial pH and/or temperature changes at varying disease stages.

Thermo- and pH-Responsive Polymers

Temperature/pH-induced phase transition

The human body accomplishes specific settings of environmental conditions in different parts, hence, it is more important to study those polymers that show behavioural changes to these stimuli. For temperature-sensitive polymers, as the solvent in a polymer solution becomes poorer, through a temperature change, a phase transition eventually takes place. Similar transition affects solubility behaviour of pH-sensitive polymers and becomes a significant tool for controlling/targeting drug release. The polymers exhibit a large nonlinear change in properties as a result of a small change in environmental condition. Of course, this is often reversible. The mechanism of transition may involve polymer–polymer and polymer–water interactions and is seen as volume change in thermoresponsive polymers and as polymer assembling and disassembling in pH-responsive polymers. Ilmain and co-workers have classified volume phase transitions according to the nature of the intermolecular forces.[16] These intermolecular forces are Van der Waals interaction, hydrophobic interactions, hydrogen bonding, repulsive, and attractive ionic interactions. The Van der Waals interaction causes a phase transition in hydrophilic polymers in mixed solvents, such as an acrylamide gel in an acetone–water mixture whereas hydrophobic interactions in hydrophobic polymers, such as NIPAAm gels, causes a phase transition with temperature. In NIPAAm, this transition is seen from a swollen state at low temperature to a collapsed state at high temperature. At low temperature, the hydrogen bonding between the polymer and the water molecules is responsible for polymer swelling and its hydrophilicity. When the temperature is raised, the hydrophobic interaction due to polymer backbones dominates, leading to polymer contraction and volume change like precipitation or separation. It can be said that hydrogen bonding responsible for solubility has negative temperature dependence. The backbones of the polymer are hydrophobic in nature and tend to reduce their surface area exposed to the highly polar water molecules by forming aggregates. At normal temperatures, hydrogen bonds between the side groups and water molecules prevent aggregation of the backbone because the hydrogen bond interactions with the water molecules are stronger than the backbone interactions. When the hydrogen bonds are broken by increasing thermal agitation, aggregation takes place, resulting in shrinkage of the thermoresponsive hydrogel with increasing temperature.[17,18] So, thermal transitions in polymers are seen as expansion or collapse of the polymer network structure, and considered responsible for drug release or inducing injury to biological membranes for escape of therapeutic agents.

The transitions in pH-responsive polymers are mainly observed due to polyelectrolyte nature of pH-responsive polymers, which generate ionic interactions upon dissolving in water or other ionizing solvent. The repulsion between charges on the polymer chain causes the chain expansion when it is ionized in a suitable solvent. However, if the solvent prevents ionization of the polyelectrolyte, the dissolved chain remains in a compact, folded state. If the polyelectrolyte chains are hydrophobic when unionized in a poor solvent, they collapse into globules and precipitate from solution. The interplay between hydrophobic surface energy and electrostatic repulsion between charges dictates the transition behaviour of the pH-responsive polymers.[19] Since the degree of ionization of a weak polyelectrolyte is controlled by pH,

and the ionic composition of the aqueous medium, "smart" polymers dramatically change conformation in response to minute changes in the pH of the aqueous environment. Even some mixed polymers, like poly(2-(methacryloyloxy)ethyl-phosphorylcholine)-*co*-poly(2-(diisopropylamino)ethyl methacrylate) (PMPC-*co*-PDPA) diblock copolymers formed stable DNA/PMPC-PDPA polyplexes at a physiological pH but rapidly disassembled at mildly acidic pH (pH 5–6).[20]

LCST and UCST behavior

Thermoresponsive polymers exhibit a phase transition in solution at a temperature known as the CST (Fig. 2). This is that particular temperature where solubility of a polymer changes remarkably due to abrupt changes in interactions between polymer and water. The CST may be lower critical solution temperature (LCST) or upper critical solution temperature (UCST). These systems are often classified into negatively sensitive and positively sensitive polymers. There are also systems, which exhibit both LCST and UCST behavior, but that usually not occur within the setting of the intended biomedical applications. Polymers, which precipitate upon heating, are said to have LCST while which become soluble upon heating, have an UCST. Temperature sensitivity originates from the balance between hydrophobic and hydrophilic segments, molecular weight, presence of cosolvent, or other additives like salts and surfactants. From interaction point of view, phase transition at CST is considered to occur by decrease in the hydration state, which causes volume phase transition, where intra- and intermolecular hydrogen bonding of the polymer molecules are favored compared to a solubilization by water, causing coil-to-globule (C-G) transition. For instance, PNIPAAm having LCST 32°C is clearly soluble below 32°C and above this temperature, the solution becomes opaque and finally turns into gel.[21,22] The LCST can be tuned by copolymerizing hydrophilic/hydrophobic monomers. More hydrophobic monomers and higher molecular weight increase the LCST. In addition, incorporation of

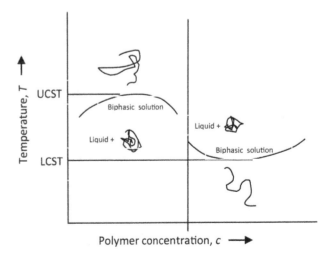

Fig. 2 Phase transitions of thermo-responsive polymers with temperature.

hydrophilic monomers that form hydrogen bonds with thermosensitive monomers increases the LCST point. For example, when NIPAAm is copolymerized with hydrophilic monomers such as acrylamide (AAm), the LCST increases up to about 45°C when 18% of AAm is incorporated in the polymer, whereas LCST decreases to about 10°C when 40% of hydrophobic N-tert-butyl acrylamide (N-tBAAm) is added to the polymer.[23] However, some researchers have proposed PNIPAAm phase transition almost independent of the concentration or molecular weight.[21,22]

The different additives can also affect solubility characteristics of polymers. Surfactants are of particular interest and can alter polymer behavior either by interacting with polymer molecules or by solubilization of polymeric molecules or small aggregates. The effect may be so remarkable that transition temperature can be shifted to a large extent or it can even disappear. Some block copolymers, having LCST, like Pluronic® F127, show thermally reversible gelation at certain concentrations. The gelation behavior is considered dependent upon the weight ratio of hydrophilic/hydrophobic segments. Pluronics® with relatively high poly(ethylene oxide) (PEO) weight ratios are soluble in aqueous solutions due to formation of polymeric micelles and do not show thermo-sensitive behavior.[24] Similarly, the biodegradable graft polyphosphazenes exhibit reversible sol–gel properties. The factors affecting gelation behavior includes composition of substituents, the chain length of hydrophilic α-amino-ω-methoxy-PEG (AMPEG), and hydrophobic amino acid esters, and the concentration of the polymers solutions measured as critical gel concentration (CGC). Increasing hydrophilic AMPEG chain length gave higher gelation temperatures.[25]

PNIPAAm, a well-studied thermoresponsive polymer, undergoes a reversible phase transition in aqueous solution from hydrophilic to hydrophobic at its LCST of approximately 32°C. Chemical modifications of PNIPAAm have been exercised in literature and found helpful in controlling the LCST.[26] For example, Liu and co-workers synthesized poly(N-isopropylacrylamide-co-N,N-dimethylacrylamide)-b-poly(D,L-lactide-co-glycolide) (P(NIPAAm-co-DMAAm)-b-PLGA) micelles for controlled paclitaxel delivery.[27] Paclitaxel release was accelerated when the physiological temperature was raised above the LCST. The paclitaxel-loaded micelles were more effective in killing human breast carcinoma cells at 39.5°C than 37°C. Similarly, De and colleagues developed folate-conjugated, thermoresponsive block copolymer micelles.[28] The drug release studies from folate-conjugated PNIPAAm-DMAAm micelles demonstrated a temperature-responsive drug release. In another study, Needham and co-workers developed temperature-sensitive liposomes containing doxorubicin.[29] They formulated liposomes of a series like 1-palmitoyl-2-hydroxy-sn-glycero-3-phosphocholine (MPPC), 1,2-dipalmitoyl-sn-glycero-3-phosphocholine (DPPC), hydrogenated soy-sn-glycero-3-phosphocholine (HSPC), and

1,2-distearoyl-sn-glycero-3-phosphoethanolamine-N-poly ethylene glycol 2000 (DSPE-PEG-2000), and optimized them to rapidly release the drug under mild hyperthermic temperatures (39–40°C). It was proposed that modifying thermoresponsiveness or CSTs could change drug biodistribution and therapeutic efficacy.

Thermo-responsive polymers

Thermo- or temperature-responsive polymers exhibit a volume phase transition at a certain temperature, which causes a sudden change in the solvation state. The approach has been extended to nucleic acid delivery. For example, PEI and NIPAAm derived cationic thermo-responsive copolymers with a tunable LCST between 37°C and 42°C effectively condensed DNA and formed small, charge-neutral polyplexes.[30] The thermo-responsive polyplexes were effectively uptaken by cells, aggregated, and resulted in a twofold increase in transfection efficiency for a temperature change of 37–42°C. The thermo-responsive polyplexes were significantly accumulated in the hyperthermically treated neuroblastoma tumor in mice, subsequently resulting in significantly increased DNA transfection in vivo.[31] The important classes of polymers that undergo sol–gel transition with temperature are acrylic polymers, block polymers, graft polymers and star copolymers. Typical LCST polymers are based on NIPAAm, N,N-diethylacrylamide (DEAAm), methylvinylether (MVE), and N-vinylcaprolactam (VCL), hydroxypropyl cellulose L, hydroxypropyl cellulose SL, and hydroxypropyl cellulose SSL (Fig. 3).[27,32–38] A typical UCST system is based on a combination of acrylamide (AAm) and acrylic acid (AAc).[39] The special block copolymers that show LCST are: PEG/PPG copolymers, PEO/PPO copolymers (Poloxamers or Pluronics®), PEG/polyester copolymers, and polypeptides/polypeptide-based copolymers. The CSTs of some polymers and copolymers are given in Table 2.

Among amphiphilics, PEO-PPO diblock copolymers, PEG-g-PLGA, PLGA-g-PEG graft polymers, PEO-PPO-PEO, PEO-PLA-PEO, PEO-PLGA-PEO, PLGA-PEO-PLGA, PEG-PLA-PEG, PEG-PLGA-PEG, and PEG-PCL-PEG triblock copolymers are variety of thermogelling materials having LCST.[30,31,40] The PEO-PPO-PEO, having LCST, like Pluronic F127(containing 70% PEO, an hydrophobic segment) is a typical example of thermally reversible gelation material. Its 20% aqueous solution gels spontaneously upon

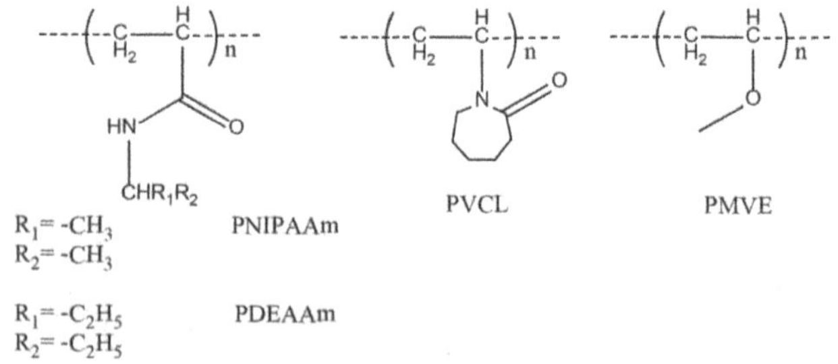

Fig. 3 Structures of common thermo-responsive polymers.

Table 2 CSTs of thermo-responsive polymers

Polymer	Transition temperature	References
Poly(N-isopropylacrylamide) (PNIPAAm)	LCST 32°C	[76]
Poly(N-isopropylacrylamide-co-acrylamide) (P(NIPAAm-co-AAm))	LCST 45°C	[23]
N-isopropylacrylamide-co-N-tert-butyl acrylamide (NIPAAm-co-tBAAm)	LCST 22°C	[23]
Poly(N-ethylacrylamide) (PEAAm)	LCST 73–86°C	[179]
Poly(N,N′-diethylacrylamide) (PDEAAm)	LCST 33°C	[77,179]
Poly(dimethylaminoethylmethacrylate) (PDMAEMA)	LCST ≈50°C	[78]
Poly(methyl vinyl ether) (PMVE)	LCST ≈37°C	[180]
Poly(N-vinylcaprolactam) (PVCL)	LCST ≈32°C	[46]
Poly(N-(DL)-(1-hydroxymethyl)propylmethacrylamide)) (P(DL)-HMPMA))	LCST ≈37°C	[79,179]
Poly(acrylamide-co-butylmethacrylate) (P(AAm-co-BMA))	UCST	[81]
Poly(N-isopropylacrylamide-co-butylmethacrylate-co-acrylic acid) (P(NIPAAm-co-BMA-co-AAc))	–	[85]
Poly(acrylic acid-co-acrylamide) (P(AAc-co-AAm))	UCST ≈25°C	[180]
Poly(N-isopropylacrylamide-co-N,N-dimethylacrylamide)-b-poly(D,L-lactide) (P(NIPAAm-co-DMAAm)-b-PLA)	LCST ≈40°C	[101]
Poly(N-isopropylacrylamide-co-N,N-dimethylacrylamide)-b-poly(ε-caprolactone) (P(NIPAm-co-DMAAm)-b-PCL)	LCST ≈40°C	[102]
Poly(N-isopropylacrylamide-co-acrylamide)-b-poly(D,L-lactide) (P(NIPAAm-co-AAm)-b-PLA)	LCST ≈41°C	[103]
Poly(NIPAAm-co-HMAAm)-b-PMMA)	LCST ≈42.8°C	[105]
Hydroxypropyl cellulose L, SL, SSL	UCST 40–45°C	[38]
Poly(organophosphazenes)	LCST 37–75°C	[25]
PEG-PLA-PEG	LCST 20–60°C	[25]
PEG-PCL-PEG	LCST ≈53°C	[25]
Poloxamer 407	UCST ≈20°C	[41]
PEO-PPO-PEO (Poloxamers or trade names Synperonics/Pluronics/Kolliphor)	LCST 20–85°C	[180]
PEG-PLGA-PEG	LCST ≈37°C	[25,83,181]
PLGA-PEG-PLGA	–	[84]
PLGA-PEO-PLGA	LCST	[182]
1-Palmitoyl-2-hydroxy-sn-glycero-3-phosphocholine (MPPC) 1,2-Dipalmitoyl-sn-glycero-3-phosphocholine (DPPC) Hydrogenated soy-sn-glycero-3-phosphocholine (HSPC) 1,2-Distearoyl-sn-glycero-3-phosphoethanolamine-N-polyethylene glycol 2000 (DSPE-PEG-2000)	LCST 39–40°C	[29]
Pluronic F127-Dn and Pluronic F127-Ln (PL-Dn and PL-Ln)	–	[183]

Abbreviation: CST, Critical solution temperature.

a temperature increase up to 20°C and exhibits no volume change during gelation. The dense packing of polymeric micelles accommodate the entire volume of the polymers solution and bulk water. Upon lowering temperature, the gels show thermally reversible conversion to solution state. This unique characteristic has been utilized in delivery of various labile drugs like proteins and peptides through parenteral route.[24] The poloxamer 407 has been used for incorporation of liposomes without any effect on thermosensitivity of poloxamer gel.[41] As most of poloxamers solutions experience phase transition from sol to gel and to sol with temperature when polymer concentrations were above critical value.

However, because the gel dissociated rapidly in body fluids, poloxamers can be used only for a short time period. The solution to this problem was novel triblock copolymers PEG-PLA-PEG and PEG-PLGA-PEG. As for other block copolymers, both the sol–gel transition temperature and CGS could be controlled by changing molecular composition of hydrophilic PEG and hydrophobic PLA or PLGA components. The PEG-PLGA-PEG, especially made solution at room temperature and became gel at body temperature, retained its integrity for more than one month.[25]

Another development in thermoresponsive polymers was their combination with other stimuli-responsive

Stem Cell—Ultrasound

polymers, like NIPAAm with one of the pH-responsive monomer, which yielded double-responsive copolymer.[42] Similarly, thermoresponsive nonviral vectors have been developed for the disruption of the endosomal/lysosomal compartment. For example, hydrogel nanoparticles prepared by cross-linking poloxamers with PEI exhibited a thermally reversible swelling/shrinking property (i.e., dramatic swelling upon decrease of temperature from 37°C to 20°C).[43] It was demonstrated that the endocytosed poloxamer/PEI nanoparticles, which were encapsulating siRNA–PEG conjugates, disrupted the endosomal compartments and enhanced their cytoplasmic release after a cold-shock treatment at 15°C, resulting in significantly improved gene silencing *in vitro*.[44]

Hinrichs and co-workers studied 2-(dimethylamino)ethyl methacrylate)-*co*-*N*-isopropyl acrylamide (DMAEMA-*co*-NIPAAm) as carrier for DNA delivery. The copolymer–plasmid complexes showed enhanced transfection efficiency and reduced cytotoxicity. Maximum transfection efficiencies were found when the plasmid was maximally covered with copolymer, while the amount of free copolymer was minimal.[45]

PNIPAAm has been most widely used thermoresponsive polymer in drug targeting for solid tumors. Poly(vinylcaprolactum) (PVCL) is another similar type of polymer but putative toxicity of the PVCL has not been established. However, biocompatibility of these polymers has been observed to be improved by grafting hydrophobic backbones with hydrophilic chains (e.g. PEO). Vihola and co-workers showed that PVCL and PVCL grafted with PEO were well tolerated up to a wide range of concentrations as compared to PNIPAAm and PVCL.[46]

pH-responsive polymers

Generally, pH-responsive polymers are weak acids or bases with pK_a values between 3 and 10. Carboxylic, sulfonate, and primary or tertiary amino groups exhibit a change in ionization state as a function of pH. Transitions in solubility, conformation, and swelling arise due to changes in ionization, where specific polymer groups switch between a neutral and charged state (e.g., poly(*N*,*N*-dimethylaminoethyl methacrylate (DMAEMA))[26,29,47] or a hydrophilic and hydrophobic state (e.g., PNIPAAm).[28,48,49] Various pH-responsive polymers of pharmaceutical interest have been given in Table 3.

The pH-responsive polymers can be divided into three main categories: polyelectrolytes, cellulose esters, polyvinyl derivatives. In polyelectrolytes, the functional groups responsible for activity are either acidic (carboxylic and sulfonic groups) or basic amino groups. Poly(acrylic acid) (PAAc), poly(methacrylic acid) (PMMA), poly (ethyl-acrylic acid) (Eudragit®), and poly(sulfonic acids) (PSA) are among the common polyanions having pK_a around 4–6 (Fig. 4). In addition to these polymers, poly(ethylene imine) (PEI) and poly(L-lysine) (PLL) have also been explored for use in drug delivery.[50] The poly[2-(dimethylamino)ethyl

methacrylate] (PDMAEMA), poly(*N*,*N*'-diethyl aminoethyl methacrylate) (PDEAEMA), poly(vinylamine) (PVAm), and poly(vinylpyridine) (PVP) are examples of pH-responsive polycations and have pK_a around 8.0. Another biodegradable pH-sensitive cationic polymer, poly(β-amino ester) (PbAE), which is insoluble at physiological pH 7.4 and becomes soluble with pH reduction below 6.5 by ionization, has been designed for pH-based drug delivery.[51] The polycations, such as PAAc, PMMA, and PSA, become ionized and soluble at a higher pH due to their pK_a around 4–6. The cross-linking of these polymers make them swellable at ionizing pH values. Thus PAAc, which is more soluble at a high pH, become swollen if cross-linked at this pH. On the contrary, PDEAEMA shows opposite behavior and swells at a low pH.[31] In general, the pH-sensitive polymers manifest their sensitivity to changes in pH, as soluble–insoluble phase transition, i.e., swelling–shrinking changes, or as conformational changes. These properties are dependent on the degree of the ionization of the ionizable groups in the polymers, polymer composition, ionic strength, and the hydrophobicity of the polymer backbone, decided by bulkiness of alkyl group. Hoffman and co-workers studied this particular pH-dependent activity of acrylic acid derivatives with increasing alkyl group. The pH-dependent activity of poly(acrylic acid) (PAA), poly(methyl acrylic acid) (PMAA), poly(ethyl acrylic acid) (PEAA), poly(propyl acrylic acid) (PPAA), and poly(butyl acrylic acid) (PBAA) showed that the systematic increase in the length of the hydrophobic alkyl group by methylene unit results in an increase in the polymer's pK_a value and consequently affects the pH at which this polymer switches from a hydrophilic conformation to a hydrophobic. This was evident by the increase in the polymer's pK_a from 6.3 for PEAA to 6.7 for PPAA to 7.4 for PBAA, which limited the utility of PBAA as a carrier for drug delivery because of its nonspecific membrane-destabilizing activity at physiologic pH.[52,53]

Another group of polyanionic polymers, cellulose esters (Aquateric®), and poly(vinyl esters) (Coateric®), for example, hydroxypropylmethylcellulose phthalate (HPMCP), poly(vinyl acetate phthalate) (PVAP) have ionizable phthalic acid group, which dissolves fast at a lower pH than polymers with acrylic or methacrylic acid groups. These are preferred coating materials for small particles and have been used for coating core materials by coacervation method using electrolytes for salting out. In a study, coaservate phases occurred at a pH > 5.2, corresponding to approximately 95% ionization of the phthalyl carboxyl groups for the coacervate formation.[54] Aquateric® is an spray-dried cellulose acetate phthalate (CAP) powder composed of mixture of CAP, Pluronic® F-68, Myvacet 9–45, and Polysorbate 60. This mixed polymer composition have been used with a subcoating material amylopectin. The combined coating resisted behavior of riboflavin sodium phosphate efficiently.[55] Poly(vinyl acetate phthalate) (PVAP) (Coateric®, threshold pH 5.0) is used as an ingredient in coating systems for oral solid dosage forms.[56] The another pH-sensitive polymers group includes anionic poly(sulfonic

Table 3 pH-responsive polymers

Polymer	Transition pH	Transition or drug release mechanism	References
Poly(dimethylaminoethyl methacrylate (PDMAEMA)	7.5	Protonation of amino groups at low pH values	[26,29,47]
Poly(diethylaminoethyl methacrylate (PDEAEMA)	7.2–7.5	Protonation of amino groups at low pH values	[31]
Poly(acrylic acid) (PAAc)	4.5	Ionization of carboxylic acid groups at high pH values along with conformational changes	[50]
Poly(ethylacrylic acid) (PEAAc)	5.9–6.3*		[52,53,57]
Poly(propylacrylic acid) (PPAAc)	~6.7		[52,53]
Poly(butylacrylic acid) (PBAAc)	~7.4		[52,53]
Poly(methacrylic acid) (PMAA)	5.5		[52,53]
Poly(ethyleneimine) (PEI)	—[a]		[50]
Poly(β-amino ester) (PbAE)	~6.5	Neutral-to-ionized state at low pH values	[51, 146]
Hydroxypropylmethylcellulosephthalate (HPMCP)	4.5–4.8	Ionization of carboxylic acid group and ester bond hydrolysis	[54]
Poly(vinyl acetate phthalate) (PVAP)	~5.0		[56]
Poly(1,4-phenyleneacetone dimethylene ketal) (PPADK)		Acid hydrolysis of ketal linkages leading to dissolution of polymer	[65]
Dioleoylphosphatidylethanolamine (DOPE)	5.0–6.0	Conformational changes at acidic pH values causes fusion of DOPE liposomes with endosomal membrane and rupturing	[69]
Poly(L-lactide)/poly(ethylene glycol)-polysulfonamide (PLLA/PEG-PSD) micelles	7.0	Deionization of sulphonamide monomers on the surface and aggregation at lower pH values	[146]
Poly(N-isopropylacrylamide-co-methacrylic acid) (PNIPAAm-MAA)	5.0–6.0	Mixture of thermo/pH-sensitive polymers with transitions resulting from polymer-polymer and polymer-water interactions	[185]
Phosphatidylcholine (PC)/diethylammonium propane (DAP) liposomes	6.7	Change in net charge of liposomes with pH affects drug release	[162]
PEG-b-PDMAEMA-b-PDEAEMA	7.1–7.3	Micelle-to-sol at low pH values	[146]
Poly(L-cystinebisamide-g-sulfadiazine)-b-PEG (PCBS-b-PEG)	6.4–7.3*	Micelle-to-sol at low pH values	[184]
Poly[2-(diisopropylamino) ethyl methacrylate-b-2-methacryloyloxyethyl phosphorylcholine-b-2-(diisopropylamino) ethyl methacrylate] (DPA-MPC-DPA) triblock copolymer	8.0	Triblock copolymer shows transitions between micelles-gel-sol with increase in pH	[24]

[a]Presence of primary, secondary and tertiary amines in same moiety. Primary amines of PEI lead to complex formation with DNA material, while other unprotonated amino groups create osmotic unbalance and the "Proton sponge" effect.
*pH transition was found dependent upon molecular wt.

acid) polymers and their derivatives. The pK_a of this group mainly lies between 2 and 3 thus remain ionized over a wide range of pH values. These are often used as cross-linking agents for polymers like styrene- and vinyl-based polymers. The anionic polymers are often characterized by their ability to switch from a hydrophilic, stealth-like conformation at physiologic pH to a hydrophobic, membrane destabilizing one in response to acidic pH gradients. Thomas and co-workers showed that anionic PEAA (20 kDa; pK_a 5.9) causes disruption of lipid membranes at acidic pH values.[57]

Cationic polymer with amino groups, e.g., DMAEMA, has a tertiary amine functional group with a pK_a of 7.5 and could interact with phospholipid membranes for easy endosomal escape.[58,59] In another study, DMAEMA/2-hydroxyethyl methacrylate (HEMA) combination has shown high volume swelling ratio at low pH values near tumor pH,

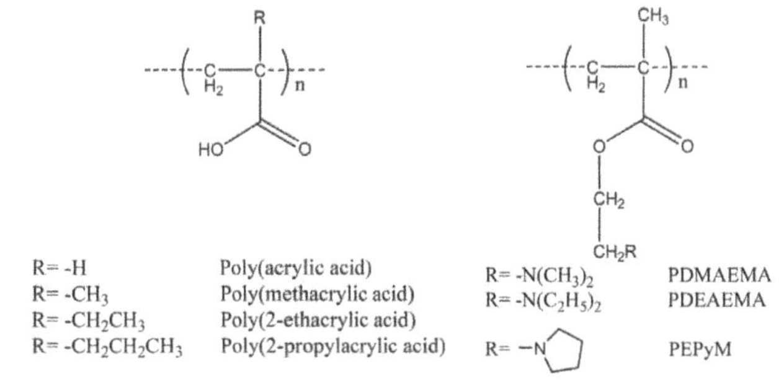

R= -H	Poly(acrylic acid)
R= -CH₃	Poly(methacrylic acid)
R= -CH₂CH₃	Poly(2-ethacrylic acid)
R= -CH₂CH₂CH₃	Poly(2-propylacrylic acid)

R= -N(CH₃)₂	PDMAEMA
R= -N(C₂H₅)₂	PDEAEMA
R= —N⟨ ⟩	PEPyM

Fig. 4 Structures of common pH-responsive polymers.

resulting drug release in an extracellular environment.[3] Other cationic polymers, like poly(ethyleneimine) (PEI), form an ionic complex with negatively charged DNA molecules through electrostatic interactions forming polyplexes.

These particles shield nucleic acids in their cores from serum proteins, nucleases, and other denaturing substances in the systemic circulation. These particles typically carry a net positive charge, which allows them to interact with the negatively charged cell surfaces, trigger endocytosis, and gain access into the cell. After cellular uptake, the complex is located in endosomes derived by fusion between lysosomes and endocytotic vesicles.[60,61]

Many primary, secondary, and tertiary amino groups in the cationic polymer produce dramatically elevated osmotic pressure and unbalance, which finally results in dissociation of therapeutic genes and endosomal rupture. This special effect is also referred to as "proton sponge effect." According to this theory, the lower transfection efficiency of poly(L-lysine) (PLL), a popularly used cationic peptide nonviral vector, than that of PEI can be explained by PLL's limited buffering capacity and noneffectiveness in disrupting endosomal membrane.[37] Modifications like grafting poly(L-histidine) (PLH) segments (pK_a = ~6.0) to PLL greatly enhanced the PLL activity, due to the supplementary buffering capacity of imidazole groups and endosomal escape via the proton sponge mechanism,[39] simultaneously reducing the cytotoxicity of PLL.[30] The PLL and PLH have been used with PDE-AEMA for improving its endosomal membrane-disrupting properties.[59,62]

The use of nonionic pH-sensitive polymers, di-ortho ester, has remained limited because these have been reported to undergo rapid biodegradation, sensitive to small pH decreases from physiological pH to weak acidic conditions (pH 5–7).[63] Recently, two amphiphilic surfactants, N,N'-bis(oleoylcysteinylhistidyl)lysine ethylenediamine monoamide (EKHCO) and N,N'-bis[(oleoylcysteinyl)lysyl] histidylhistidine ethylenediamine monoamide (EHHKCO) were synthesized and optimized for the delivery of plasmid DNA and siRNA. Both showed pH- and concentration-dependent membrane disrupting properties in free as well as complexed form with nucleic acids.[64]

Acid-labile linkages were found efficient in delivering therapeutic agents on targets.[27,32] For example, acid-degradable and targetable PIC micelles were synthesized by assembling PLL with lactosylated-PEG conjugated siRNA via acid-labile β-thiopropionate linkage.[27] In another study, hydrazone linkage-based block copolymer such as block copolymer of ethyl acrylic acid (EAA) methacrylate, and block copolymer of hexyl methacrylate (HMA)-trimethyl aminoethyl methacrylate (TMAEMA) grafted with N-acryloxy succinimide or β-benzyl-L-aspartate N-carboxy-anhydride have shown a robust membrane-destabilizing activity in response to mildly acidic pH in the endosomes.[32] Heffernan and Murthy developed an acid-sensitive biodegradable drug delivery vehicle using poly(1,4-phenyleneacetone dimethylene ketal) (PPADK), which contains ketal linkages allowing for acid-catalyzed hydrolysis of the polymer into low-molecular-weight hydrophilic compounds and accelerating release of drug molecules under acidic

conditions.[65] PEI conjugated with acid-cleavable, primary amine-bearing ketals to the cationic polymeric backbone, facilitated the endosomal escape of both plasmid DNA and siRNA.[33–35] The ketalized PEI induced the endosomal disruption by increased osmotic pressure upon acid hydrolysis of amino ketal branches in addition to dramatic polyplex swelling and the hypothetic "proton sponge effect." The hydroxyl termini of the acid-hydrolyzed branches also lost attractive interactions with nucleic acids that were subsequently released in the cytoplasm. Furthermore, the acid-hydrolyzed ketalized PEI with a lowered cationic density showed greatly diminished cytotoxicity due to significantly reduced interactions with endogenous genes.[66] In vitro studies demonstrated significantly enhanced nucleic acid delivery with reduced cytotoxicity by ketalized PEI, which was demonstrated to be molecularly tunable by ketalization ratio, molecular weight, and topology of PEI. The use of PEI along with poly(methacryloyl sulfadimethoxine) (PSD)-b-PEG diblock copolymer have shown effectiveness in targeting the acidic extracellular matrix of tumors. At tumor pH, the diblock copolymer PSD-b-PEG detached and direct interaction of PEI with cells was made possible. At physiological pH 7.4, the shielding of PEI by PSD-b-PEG was found responsible for reduced transfection efficiency.[12]

The viral-based fusogenic peptides, known for pH-dependent membrane disruptive activity, were considered efficient means for endosomal escape.[67] However, pathogenic and immunogenic problems still remain associated with use of such viral-based approaches. The pH-responsive polymers, which mimic fusogenic peptides, have given a new nonviral approach for successful intracellular delivery. Poly(2-ethyl acrylic acid) (PEAAc) and poly(2-propyl acrylic acid) (PPAAc) was reported to enhance the disruption of endosomal membrane by undergoing pH-responsive hydrophobic changes.[68] The pH-sensitive liposomes prepared with dioleoylphosphatidyl ethanolamine (DOPE) has also been known to destabilize the endosomal membrane and efficiently release the encapsulated macromolecules into the cytoplasm.[69,70] A stable lipid bilayer of DOPE was formed at a physiological pH of 7.0, while a hexagonal-II structure of DOPE formed at acidic pH of 5–6 that induced the fusion of liposome with the endosomal membrane, facilitating endosomal escape of DNA.[69] Phosphatidylethanolamine (PE)-containing liposomes were also employed to deliver antisense oligonucleotides (ODNs) into the cytoplasm by destabilization under a mildly acidic endosomal condition.[70]

The multifunctional pH-responsive polymers have been used to enhance cytosolic delivery of therapeutics. For example, pH-responsive liposomes were obtained from the PNIPAAm copolymers by incorporating pH-sensitive moieties like MAAc and N-glycidylacrylamide.[71] The destabilization of liposome at 37°C under weakly acidic condition was observed. Another strategy to form multifunctional pH-responsive polymers is to develop interpolymer networks (IPNs). Interpolymeric hydrogels containing N-vinyl pyrrolidone (NVP), polyethylene glycol diacrylate (PAC), and chitosan (as pH-responsive polymer) were prepared by a free radical polymerization technique using azobisisobutyronitrile (AIBN) as initiator and N,N'-methylenebisacrylamide (BIS) as cross-linker.[72] In another report, a semi-IPN was obtained by cross-linking chitosan and poly(vinyl pyrrolidone) blend with glutaraldehyde. The freeze-dried hydrogels showed superior pH-dependent swelling properties over nonporous air-dried hydrogels.[73]

Thermoresponsive Biodegradable Hydrogels

Thermoresponsive hydrogels are one of the most commonly studied environment-responsive drug delivery polymer systems. Drug delivery systems responsive to temperature utilize various polymer properties, including the thermally reversible transition of polymer molecules, swelling, change of networks, glass transition, and crystalline melting for drug release at target site.[74] As defined, the temperature-responsive hydrogels show transition at CST, which may be lower as well as upper one (Fig. 2). Based on CST, these temperature-sensitive hydrogels can be classified into negatively thermosensitive, positively thermosensitive, and thermally reversible gels.[75]

The negative temperature-sensitive hydrogels have a LCST, below which the polymer swells in the solution but do not dissolve, while above it the polymer contracts and separates.[18] A well-known negative thermosensitive hydrogel polymer is PNIPAAm.[76] Some other polymers belonging to the PNIPAAm family are poly(N,N'-diethylacrylamide) (PDEAAm), poly(dimethylaminoethylmethacrylate) (PDMAEMA) and poly(N-dl)-1-hydroxymethyl) propylmethacrylamide (P(dl)-HMPMA)).[77–79] LCST systems are relevant for controlled release of drugs and proteins, in particular.[80] The CST of various polymers have been given in Table 2.

A positive temperature-sensitive hydrogel has an UCST. The hydrogel contracts upon cooling below the UCST. Polymer networks of poly(acrylic acid) (PAA) and poly(acrylamide) (PAAm), poly(N-isopropylacrylamide) (PNIPAAm), poly(acrylamide-co-butylmethacrylate) (PAAm-co-BMA), poloxamer 407 exhibit positive temperature dependence of swelling.[81]

The temperature-responsive hydrogels have been used in drug delivery. For example, the thermoresponsive hydrogels made of PNIPAAm/polytetramethylene glycol (PTMG) interpenetrating polymer networks showed pulsatile patterns of indomethacin release in response to temperature changes.[82] In another study, release of hydrophilic and hydrophobic drugs ketoprofen and spironolactone was studied from biodegradable hydrogels of poly(ethylene glycol)-b-poly(DL-lactic acid-co-glycolic acid)-b-poly(ethylene glycol) (PEG-PLGA-PEG) triblock copolymers. Ketoprofen, a hydrophilic drug, was released over 2 weeks with a first-order release profile. Spironolactone, a hydrophobic drug, was released over 2 months with an sigmoidal shaped release profile.[83] Similarly, triblock copolymers of reverse order, PLGA-PEG-PLGA, have been also reported for protein's release like insulin.[84]

For PNIPAAm-based hydrogels, the increase of acrylic acid concentration from 0–10 mol% in poly(N-isopropylacrylamide-co-butylmethacrylate-co-acrylic acid) (P(NIPAAm-co-BMA-co-AAc)) pH/thermo-responsive beads improved the loading efficiency as well as stability of calcitonin. The stability of calcitonin was also preserved.[85] Park and co-workers also studied the insulin release from such pH/thermo-responsive hydrogels made of NIPAAm and DMAPMAAm.[86] The hydrogel showed pH sensitivity around pH 7.4 and insulin release switched from Fickian to zero-order for a temperature change from 32°C to 42°C. Andersson and co-workers determined the diffusional characteristics of PNIPAAm hydrogel for insulin with changing temperature. The gel itself was found to be a critical factor in controlling insulin release which means a small change in temperature changed gel volume considerably. The effective diffusion coefficient for insulin increased from 4.4×10^{-10} to 5.9×10^{-10} m²/s with change of temperature from 10°C to 30°C.[87]

Poloxamer or Pluronics® show phase transitions from sol to gel near body temperature, and are used as injectable implants. To make hydrogels biodegradable, the hydrophobic PPO segments of PEO-PPO-PEO copolymer could be replaced by a biodegradable PLA. However, there are still disadvantages associated to poloxamer gels as drug delivery systems such as limited stability, poor mechanical properties, and short residence times due to its rapid dissolution when placed within biological environments.[80,88,89]

Thermoresponsive Polymeric Micelles

Several amphiphilic block copolymers have been synthesized with a particular interest in drug delivery. The use of temperature-responsive polymers capable of forming micellar structures or combination of thermoresponsive polymers as a part of amphiphilic copolymers have been used to achieve controlled release of drugs from micelles in response to environmental changes. The amphiphilic polymers have also been widely studied mostly using aliphatic polyesters (poly(ε-caprolactone)) (PCL), poly(D,L-lactide) (PLA), poly(glycolide) (PGA)), and their copolymers like poly(lactide-co-glycolide) (PLGA) and PEG. Several other thermoresponsive polymeric micelles that have been extensively studied include block copolymers of PNIPAAm with hydrophilic and hydrophobic polymers,[90] triblock copolymers like poly(ethylene oxide)–poly(propylene oxide)–poly(ethylene oxide) triblocks (PEO–PPO–PEO) (Poloxamer or Pluronic®),[91] poly(ethylene glycol)-poly(ε-caprolactone) (PECL), chitosan glycerolphosphate, and ethyl(hydroxyethyl) cellulose (EHEC) formulated with ionic surfactants[92] and poly(ethylene glycol)–poly(lactic acid)–poly(ethylene glycol) triblocks (PEG-PLA-PEG).[93] The thermosensitivity of these block copolymer micelles could be modulated by the alteration of the concentration of the aqueous solution, the composition of the copolymers, and so forth. Importantly, these block copolymer micelles consisting of PEG and biodegradable polyesters, such as PLA, PLGA, and PCL, were found to be thermoresponsive and biodegradable, making them potential candidate materials for use in drug delivery.[94]

PNIPAAm dissolved in water precipitates in the form of large particles above its LCST. Therefore block copolymers of PNIPAAm and PEG (hydrophilic block) could form micelles above the LCST.[95–97] Conversely, the PNIPAAm block combined with a hydrophobic block could form the micelles at temperatures below the LCST. Some examples of such micelles are diblock copolymers of PNIPAAm with PLA, PMMA, and PBMA as hydrophobic blocks.[98–100] So, NIPAAm copolymerized with hydrophilic monomers yielded polymers with an LCST above 37°C. For example, poly(N-isopropylacrylamide-co-N,N-dimethylacrylamide)-b-poly(D,L-lactide) (P(NIPAAm-co-DMAAm)-b-PLA) and poly(N-isopropylacrylamide-co-N,N-dimethylacrylamide)-b-poly(ε-caprolactone) (P(NIPAAm-co-DMAAm)-b-PCL)) block copolymers were developed with an LCST of 40°C.[101,102] Similarly, P(NIPAAm-co-acrylamide)-b-poly(D,L-lactide) (P(NIPAAm-co-AAm)-b-PLA) was synthesized with an LCST of 41°C and used for the design of thermosensitive polymeric micelles for the triggered release of docetaxel by hyperthermia, resulting in enhanced *in vitro* and *in vivo* effects.[103,104] Copolymerization of NIPAAm with N-hydroxymethylacrylamide (HMAAm) and subsequent coupling to poly(methyl methacrylate) resulted in poly(NIPAAm-co-HMAAm)-b-PMMA) with an LCST of 42.8°C that formed micelles for anticancer drug methotrexate and showed enhanced release when heated to 43°C.[105]

The major drawback of PNIPAAm block copolymers for the development of polymeric micelles for pharmaceutical applications is that even though they are biocompatible, they are not degradable. Moreover, local hypo- or hyperthermia is necessary for micellar destabilization and subsequent release of the encapsulated drug, which is not always feasible in clinical settings. Therefore, copolymers of NIPAAm with biodegradable monomers have been synthesized. In these systems, side chain hydrolysis leads to a phase transition of the thermoresponsive block from the hydrophbic to the hydrophilic, resulting in micelle destabilization and release of the payload.[106] Shah and co-workers copolymerized NIPAAm with N-acryloxy succinimide and observed an increase of the LCST upon conversion of acryloxy succinimide to acrylic acid.[107] Later, NIPAAm was copolymerized with 2-hydroxyethyl methacrylate-monolactate (HEMA-Lac₁) to yield polymers with an LCST that increased upon hydrolysis of the lactate side groups, leading to systems that show a hydrophobic-to-hydrophilic conversion in time.[108,109] Neradovic and co-workers showed that copolymers of NIPAAm and N-(2-hydroxypropyl)-methacrylamide lactate (HPMAAm-Lac$_n$) had an LCST

(~15°C for NIPAAm-co-HPMAAm-Lac$_3$) below body temperature that, upon degradation of the lactate side groups, yielded NIPAAm-HPMAAm, having increased LCST to 45°C, owing to the hydrophilic nature of HPMAAm and made copolymer again soluble.[110] The same authors also showed that block copolymers of PEG-b-P((HPMAAm-Lac$_n$)-co-NIPAAm)) formed micelles with a critical micelle temperature (CMT) below body temperature, which upon side chain hydrolysis increased to above 37°C, resulting in the hydrophilization of the copolymers and concomitant destabilization of the micelles. Therefore, these micelles were expected to stay intact upon administration and destabilize over time due to side chain hydrolysis. In addition, it was further shown that the destabilization times could be predicted by the degradation kinetics of the lactate side groups and also tailored by the number of oligolactate grafts attached on the block copolymers.[111] Later, it was found that, NIPAAm-free homopolymers of HPMAAm-Lac$_n$ also exhibit thermosensitive behavior, with an LCST that can be tuned from 13°C to 65°C by the number of the lactate side chains.[112] In a study diblock copolymers based on PEG-b-P(HPMAAm-Lac$_n$) exhibited biodegradability and thermosensitivity and polymeric micelles based on these polymers were formed above the CMT. After hydrolysis of lactate groups at physiological pH and temperature, the polymer gets converted to hydrophilic PEG-b-PHPMAAm, leading to destruction of the miceller structure. It was also shown that when the structurally related P(HEMAAm-Lac$_n$) was used as a thermosensitive block, the destabilization of the micelles occurred after approximately 8 hr incubation at physiological conditions (pH 7.4, 37°C), matching the circulation time of nanoparticles *in vivo* (8–10 hr),[113] making them potentially suitable for tumor-targeting purpose. Soga and co-workers succeeded in encapsulating the hydrophobic anticancer drug paclitaxel in the core of PEG-P(HPMAAm-Lac$_n$) micelles that increased the drug's aqueous solubility by a factor of 5000.[114] It was also shown that PEG-b-P(HPMAAm-Lac$_n$) micelles loaded with the anticancer drug paclitaxel showed comparable cytotoxicity to Taxol® (the market formulation of paclitaxel). In another study, vitamin K was loaded in PEG-b-P(HPMAAm-Lac$_n$) micelles with an efficiency of more than 95% and it was shown that vitamin K plasma levels increased significantly upon gastric administration of the micelles.[115]

Recently, a thermoresponsive terpolymer poly-(PEG:CPP:SA) composed of PEG, 1,3-bis (carboxyphenoxy) propane (CPP), and sebacic acid (SA) was reported for release of doxorubicin.[116] The polymer had cloud point of 33°C and formed micelles (size ~100 nm) in an aqueous solution at 37°C. At different temperatures, a sustained release was observed with initial rapid burst release.

Another very important class of micelle forming block copolymers exhibiting thermo-sensitive behavior is that of Poloxamers or Pluronic® triblock copolymers that consist of poly(ethylene glycol)-b-poly(propylene oxide)-b-poly(ethylene glycol) (PEG-PPO-PEG).[117,118] In aqueous solutions, above their CMT, these copolymers undergo self-assembly, resulting in the formation of micellar structures with a PPO core and a PEG corona.[119] In a study during Phase I clinical trials, poloxamer micelles loaded with the anticancer drug doxorubicin showed promising results even in advanced stage solid tumors.[120]

One of the most important challenges that polymeric micelles face is either dissolution of micellar structure to unimers below their CMC or interactions of the hydrophobic blocks with plasma proteins, causing disruption of the equilibrium between unimers and micelles.[121,122] To avoid premature destabilization, a variety of strategies can be applied, like physical or chemical cross-linking of their core or shells (Fig. 1).[122,123] For example, Wei and co-workers developed shell cross-linked (SCL) thermoresponsive micelles based on poly(N-isopropylacrylamide-co-3-(trimethoxysilyl)propylmethacrylate)-b-polymethyl methacrylate (P(NIPAAm-co-MPMA)-b-PMMA) amphiphilic block copolymers, via a silica-based cross-linking strategy.[124] The SCL micelles formed demonstrated increased stability, better encapsulation efficiency, and release of the model drug prednisone acetate over a longer period compared with noncross-linked PNIPAAm-b-PMMA micelles. Importantly, shell cross-linked micelles retained their thermosensitivity at various temperatures. The same group also developed SCL micelles based on poly(N-isopropylacrylamide-co-aminoethyl methacrylate)-b-polymethyl methacrylate (P(NIPAAm-co-AMA)-b-PMMA).[125] While the P(NIPAAm-co-AMA)-b-PMMA copolymer alone showed good cytocompatibility when incubated with HeLa cells (cell viability above 70%), the ferrocene SCL micelles showed increased cytotoxicity, which indicated that the conjugated ferrocene moiety still exhibited tumor cell killing activity. Shell cross-linked and stable thermosensitive micelles based on poloxamers have also been developed, using gold nanoparticles or by chemical modification and cross-linking of the outer shell.[126,127]

Another cross-linking strategy applied to stabilize polymeric micelles is the functionalization of the hydrophobic blocks with methacrylates and linking of these groups after micelle formation via radical polymerization (thermal-, chemical- or photo-polymerization) or Michael addition.[128,129] In a study, functionalized PEG-b-P(HPMAm-Lac$_n$) block copolymers were developed, where the hydroxyl groups of the lactic acid side chains were partially methacrylated. The polymerization of the methacrylated groups upon micelle formation formed small core cross-linked micelles, having a much higher stability compared with non-cross-linked micelles and retained their biodegradability.[122,130]

pH-Based Oral Protein/Peptide and Drug Delivery

The pH-responsive polymer as enteric coatings have been well known for controlling release in particular parts of GIT although the effectiveness of this approach remained a point of discussion. The majority of pH-responsive polymers in oral delivery are mainly intended for use in defined parts of the GIT, prevention of irritation, and drug stability. However, proteins and peptides through oral route are for systemic delivery in more physiological manner, which have to survive the acidity of the stomach, withstand the hydrolytic activity of the metabolizing enzymes in the small intestine, permeate across the intestinal epithelium, and reach the systemic circulation to become therapeutically effective.[131] The cationic and anionic polymers have been used as carriers to enhance oral absorption of peptide and protein drugs across the GIT.[132,133] These pH-sensitive carriers shield the encapsulated drug from the metabolizing enzymes and harsh environment of the GIT while providing a mechanism to tune the site and rate of drug release in response to changes in the environment pH.

Several anionic polymers such as methyl acrylic acid (Eudragit® L-100), methyl methacrylate (S-100), and hydroxypropyl methylcellulose phthalate (HP-55) have been commercially used as enteric coatings for oral delivery of peptide and protein drugs, where they release the encapsulated drug molecules into the small intestine in response to the alkaline pH.[134,135] In another study, the release of pindolol was slowed down by enteric coatings of Eudragit S-100 and HP-55 at pH below 4.5 conditions. The drug release was not affected at higher neutral pH values of ileum, hence, allowed the site-specific delivery of pindolol while escaping the drastic environments of upper GIT.[133] Microparticles of poly(methacrylic acid-g-ethylene glycol) (P(MAA-g-EG)) loaded with insulin have exhibited pH-responsive characteristics. The interpolymer complexes were formed in acidic media and dissociated in neutral/basic environments, resulting in retarded insulin release in acidic media and rapid release under neutral/basic conditions.[136] Copolymer networks of P(MAA-g-EG), showing reversible pH-dependent swelling behavior, have also been developed.[137] Gels containing equimolar amounts of MAA/EG exhibited less swelling at low pH and rapid release under alkaline conditions.

Eaimtrakam and coworkers developed a four-layer carrier system incorporating water-insoluble ethylcellulose and an enteric coating mixture of Eudragit® copolymers. The S-100 polymer coat with the pK_a of 6.8 swelled and adhered to the ileum wall for up to 3 h after administration and shielded the encapsulated drug molecules from the hydrolytic enzymes of duodenum.[138] Weak polyacid-based hydrogels have also been utilized for intestinal delivery of polypeptides. The anionic groups were uncharged in the acidic environment of the stomach (pH ≈1.2), leading to collapse of hydrogels. The hydrogels remain collapsed in acidic environment, providing protection to protein therapeutics, and become ionically charged and swollen at neutral pH in the intestine, resulting in targeted release in the intestinal tract. In a concerned study, oral delivery of the salmon calcitonin was reported with a pH-responsive hydrogel made of PMAAc grafted with PEG (P(MAAc-g-EG)).[139]

Stem Cell—Ultrasound

The PEG chains provided enhanced mucoadhesion along with protein protection and the PMAAc backbone, responsible for pH-dependent behavior, enhanced clacitonin penetration. The bioavailability of calcitonin was reported increased with the use of grafted hydrogel. The hydrogel nanospheres of P(MAAc-g-EG) to further improve GIT absorption of polypeptide drugs have also been reported.[140]

The pH-responsive hydrogel microspheres composed of chemically modified polyacrylamide-grafted guar gum were prepared by converting –CONH groups of polyacrylamide to –COOH groups.[141] The –COOH groups provided pH-responsive functionality to hydrogels. Enzymatically degradable cross-links, introduced into weak polyacid-based hydrogels, have been reported for colon-specific drug delivery. In an approach, acrylic acid hydrogels were consisted of a hydrophobic moiety, and 4,4′-di(methacryloylamino) azobenzene for enzymatically degradable cross-links. In acidic conditions, hydrogels remain collapsed and in neutral intestinal pH these become swollen but remain nondegraded. In the colon region, the azoreductases activity degraded swollen gels and released drug.[142]

Hydrogels based on weak polybases have been used to target acidic stomach environment. These hydrogels are uncharged at neutral and higher pH values, hence, insoluble and become charged and swollen at acidic conditions. Due to swollen state in acidic pH, these hydrogels release drug at acidic pH in the stomach. Chitosan, a polybase, has appeared as polymer of particular interest to several workers in oral drug delivery systems especially for negatively charged drug therapeutics such as DNA and proteins due to charged amino groups.[143] The pH-responsive chitosan–poly(vinyl pyridine) (PVP) hydrogel was reported for controlled release of antibiotics.[144] The freeze-dried hydrogels showed more effective release of antibiotics than conventional air-dried hydrogels in acidic conditions. Similarly, poly[N-vinyl-2-pyrrolidone-polyethylene glycol diacrylate]-chitosan pH-responsive hydrogels were investigated for stomach drug delivery.[145] In another study, cross-linked chitosan was studied for release of riboflavin in simulated gastric and intestinal pH environments. A larger swelling ratio and quicker riboflavin release were observed, when the chitosan hydrogel films were formed in lower citrate concentration solutions.

pH Based Parenteral Drug Delivery

The pH-responsive polymers for parenteral drug delivery are mainly used for targeting cancerous tissues. The extracellular pH in cancerous tissue is 6.5–7.2, thus slightly lower than the normal pH of 7.4. For systemic anticancer drug delivery, pH-responsive polymers should have a narrow pH range for modulating their physical property. Otherwise, pH-sensitive drug carriers can induce either severe toxicity by drug burst or poor therapeutic efficacy.

Bae and coworkers used the weak acid sulfonamide as trigger for extracellular delivery of doxorubicin.[146] Upon deprotonation of weak acid sulfonamide in acidic environment of tumor, the micelles containing drug collapsed, releasing doxorubicin in tumor tissue, which was subsequently taken up by the cells. This accumulation of doxorubicin in tissue cells made this a specific tumor delivery. In another study, poly(L-histidine)-b-PEG and PLLA-b-PEG in combination were studied for tumor-specific delivery of adriamycin.[147] Adriamycin was rapidly released at extracellular pH of 6.6 due to ionizing and soluble behavior of polymers.

A ternary nonviral gene delivery system consisting of DNA, PEI, and poly(methacryloyl sulfadimethoxine) (PSD)-b-PEG that effectively targets the acidic extracellular matrix of tumors demonstrated high transfection and cytotoxicity at pH 6.6, attributed to PSD-b-PEG detachment and direct interaction of PEI with cells. At pH 7.4, the nanoparticles showed low transfection due to shielding of PEI by PSD-b-PEG.[12] The effective extracellular delivery of therapeutics is possible by several other strategies also, such as using streptavidin conjugated to a tumor-specific antibody. However, for the intracellular delivery, the therapeutic agents have to be transported from extracellular space to cytoplasm and then to nucleus with protection from unwanted trafficking. While their transportation, at the cell surface, the therapeutic agents are internalized by endocytosis. The therapeutic agents should escape from endosomes quickly before the fusion between endosomes and lysosomes containing high enzymatic activity.

The pH-responsive polymers, that mimic fusogenic peptides activity, for example, PPAAc, disrupted red blood cells 15 times more efficiently at physiological pH than PEAAc at pH 6.1.[68] An in vivo murine excisional wound healing model confirmed that pH-sensitive, membrane-disruptive polymers enhanced the release of plasmid DNA for expression of therapeutics, from the acidic endosomal compartment to the cytoplasm.[148] Pseudo-peptides, poly(L-lysine iso-phthalamide) and poly(L-lysine dodecan-amide) containing carboxylic acid groups, have been utilized as effective DNA carriers. These polymers changed their conformations into hypercoiled structures at endosomal pH, leading to high transfection efficiency.[149]

The polycationic or polybases have also shown increased endosomal escape due to their increased interaction with negatively charged membrane phospholipids.[59] Monodisperse, pH-sensitive DMAEMA/HEMA nanocarriers encapsulating paclitaxel exhibited pH-dependent release kinetics. The particles having high content of DMAEMA showed low cross-linking density and a high volume swelling ratio at low pH.[150] A similar series of particles were used for gene delivery, where triggered release of plasmid DNA within the low pH endosome was optimized.[151] Poly(L-lysine) (PLL), a cationic polymer, has been known to form complexes (polyplexes) with DNA and stabilizing these therapeutics. The combination of PLL as side chains with pH-responsive poly(*N*,*N*-diethyl aminoethyl methacrylate) (PDEAEMA) backbone resulted a comb-type polybase DNA carrier capable of disrupting membranes.[62] However, PDEAEMA shows transition at physiological pH and for endosomal membrane disruption, the transition pH value of PDEAEMA has to be adjusted to the endosomal environment. For this particular purpose the PDEAEMA was incorporated with a hydrophobic group poly(L-histidine) (PLH) into comb-type copolymer with PLL.[59] The PLH-g-PLL-based polyplex particles with DNA showed enhanced transfection efficiency, compared with PLL-based ones.

Noninonic pH-sensitive liposome has also been reported with 10:90 ratio of poly(ethylene glyco)-diortho ester distearoyl glycerol conjugate and fusogenic lipid.[63] The fast degradation kinetics at pH 5–6 was proposed to make the conjugate suitable for applications for triggered drug release systems to endosomes and tumors.

Anionic polymers have been used to enhance the delivery of DNA materials past endosomal membrane into the cytoplasm. Cheung and coworkers evaluated the ability of anionic polymer, poly(propyl acrylic acid) (PPAAc), to improve the transfection efficiency of plasmid DNA (pDNA) from 1,2-dioleoyl-3-trimethylammonium-propane (DOTAP)/DNA complexes and stability against serum proteins. The results showed 20-fold increase in DNA transfection from the DOTAP/DNA complexes against complexes prepared without PPAAc and increased stability against serum proteins.[152,153] Subsequent *in vivo* studies also showed that DOTAP/PPAAc/antisense oligonucleotide (ODN) complexes targeted against the thrombospondin-2 (TSP2) gene have successfully downregulated the desired gene and improved would healing.

The modified polymers or polymers with acid-labile linkers have been used as another approach to induce pH-triggered release. The drug linked or conjugated remains soluble in normal physiologic conditions and dissociates in acidic environments. Most often used linkers are *cis*-aconityl acid, hydrazone, and Schiff's base derivatives.[32,154] For example, adriamycin has been delivered to intracellular environment in cancer cells by attaching to IgM by *cis*-aconityl acid, and showed efficient cell accumulation and therapeutic response.[154] Acid-cleavable hydrazone linker based diblock copolymers has also been utilized to achieve the facilitated release of nucleic acids from polyplexes.[32] These comb-like diblock copolymers have a robust membrane-destabilizing activity in response to a mildly acidic pH in the endosome. The first pH-sensitive block was the copolymer of ethyl acrylic acid (EAAc) and hydrophobic methacrylate, and the second block was the copolymer of hexyl methacrylate (HMA) and trimethyl aminoethyl methacrylate (TMAEMA) grafted with *N*-acryloxy succinimide or β-benzyl-L-aspartate *N*-carboxy-anhydride via acid-cleavable hydrazone linkages. These comb-like copolymers exhibited a high concentration-dependent hemolytic activity in acidic solutions and degraded into

smaller fragments via acid-hydrolysis of hydrazone linkages, resulting in minimized toxicity and facilitated elimination by renal excretion *in vivo*. These polymers formed siRNA complexing polyplexes that were stable even in the presence of serum and nucleases, and efficiently silenced GAPDH expression in MCF-7 breast cancer cells *in vitro*. In another study acid-degradable and targetable PIC micelles were synthesized by assembling PLL with lactosylated-PEG conjugated siRNA via acid-labile β-thiopropionate linkage.[27] The lactosylated-PEG–siRNA/PLL polyplexes were efficiently internalized into hepatoma cells *in vitro* in a receptor-mediated manner and exhibited significant gene silencing.

A number of experimental evidence suggest that polyplex disassembly in a cell is a relatively inefficient process.[155] Therefore, facilitated disassembly of polyplexes with pH-sensitive poly(2-(methacryloyloxy)ethyl-phosphorylcholine)-*co*-poly(2-(diisopropylamino)ethyl methacrylate) (PMPC-*co*-PDPA) diblock copolymers were studied. The diblock copolymers were able to form stable polyplexes at a physiological pH but rapidly disassembled at mildly acidic pHs (pH 5–6) by forming dissolved copolymer chains (i.e., unimers).[103] The phase transition was triggered by protonation of tertiary amino groups on the PDPA in mildly acidic conditions, and the DNA/PMPC-*co*-PDPA polyplexes showed high transfection efficiency with less cytotoxicity compared to commercially available Lipofectamine.

Another interesting approach for enhancing ODNs delivery was incorporation of pyridyldisulfide acrylate (PDSA) monomers in copolymers of EAA or PAA and butylmethacrylate (BMA) along with conjugation to PLL. These copolymer–PLL conjugates condensed ODNs more efficiently while retaining stability and pH-responsive membrane destabilizing activity of parent polymers.[156] In another study, Murthy and co-workers prepared a hydrophobic random copolymer incorporating DMAEMA, butyl acrylate (BA), BMA, and styrene benzaldehyde (SBA), masked by hydrophilic PEG chains, and anchored to the polymer backbone via acid-sensitive acetal linkages.[157] The PEG chains conjugated to cationic PLL chains helped in better condensation of nucleic acids via electrostatic interactions and formation of stable complexes. In the endosomes, the acetal linkage was proposed to be hydrolyzed, unmasking the hydrophobic polymer backbone, causing rupture of the endosomal membrane and releases of the condensed nucleic acid material into the cytoplasm. These encrypted polymers proved ninefold more efficient in endosomal disruption compared to PPAA homopolymers and were effective in delivery of model ODNs into macrophage-like cells *in vitro*.[158]

Complexation of pH-responsive polymers with liposomes is another approach to enhance DNA tranfection efficiency. Ponnappa and coworkers reported that ODNs encapsulated in pH-sensitive liposomes were delivered efficiently to liver cells and produced a 65–70% reduction

Stem Cell—Ultrasound

in plasma levels of TNF-alpha.[159] A pH-responsive polymer with long hydrophobic pendant groups, anchored onto liposomes, improved the ability of liposome complexes to disrupt the lipid bilayer.[160] In a similar strategy, the pH responsive polymer (a copolymer of NIPAAm, N-glycidyl-acrylamide, and N-octadecylacrylamide) was combined with a large unilamellar noisome, bilayer consisting of polyoxyethylene-3-stearyl ether (POE-SE) and cholesterol.[161] In another study, pH-sensitive liposomes composed of phosphatidylcholine (PC) and dimethylammonium propane (DAP) with a pKa of 6.7, allowed the liposome's net charge to become cationic upon decreasing the pH, allowing subsequent intracellular delivery of siRNA within the endosome.[162] In further modification, PEG was linked to the surface of these liposomes by electrostatic attractions of polycationic liposomes at normal physiologic pH. The PEG increased circulation time and in microenvironment of pH 5.5 the PEG was released and siRNA was freed. Lee and co-workers designed super pH-sensitive multifunctional polymeric micelles with mobile cell interacting ligands (i.e., biotin) for tumor targeting.[163] The micelles comprised two block copolymers, polyHis-b-PEG and pLLA-b-PEG-b-polyHis-biotin, where polyHis functioned as a pH-responsive mobile actuator. Above pH 7.0, hydrophobic polyHis segment held biotin down while hydrophilic PEG formed the outer shell. At tumoral pH values (6.5–7.0), ionized polyHis was liberated from the hydrophobic core, leading to extension of biotin beyond the PEG shell, thus enabling the biotin to interact with tumor cells. Furthermore, the micelles dissociated at endosomal pH values (below pH 6.5) and disrupted endosomal membrane to release doxorubicin into cytosol. PEG-b-poly(L-histidine) (PEG-PHis), for instance, spontaneously generated micelle structure at high pH over the pK_a of histidine (6.5–7.0) and loaded anticancer drug, doxorubicin inside the micelle core.[164,165] At pH 6.8, doxorubicin-loaded PEG-PHis micelle could effectively eliminate multidrug-resistant MCF-7 cells, which might be expected to work in vivo condition due to the slightly different pH of breast tumor tissue compared to the normal tissue.

Poly(β-amino ester) (PbAE), an amine containing biodegradable polyester, has been used to release payloads in the acidic microenvironment of tumors as well as within the endosomes of tumor cells. This polymer becomes rapidly soluble at pH < 6.5 and releases drug in an acidic environment. In vitro and in vivo studies have demonstrated that pH-sensitive PbAE-based nanoparticles deliver more payload compared to non pH-sensitive nanoparticles (PCL-based nanoparticles).[90,91,166] Similarly, microspheres composed of poly(β-amino ester) showed a rapid release of encapsulated material within the range of endosomal pH.[167] At physiological pH, the PbAE interacted electrostatically with plasmid DNA and formed polymer–DNA complexes of nanoscale, which biodegraded into nontoxic product.[168] In another study, polymeric micelles made with mPEG-PbAE block copolymer have been developed for tumor-specific release of doxorubicin.[66] The block copolymer formed stable micelles at pH 7.4; however, demicellization occurred due to PbAE dissolution at tumoral acidic pH, leading to instant release of doxorubicin.

Shen et al. have reported pH-responsive multifunctional nanoparticles composed of three layers, a PCL core, a poly[2-(N,N-diethylamino)ethyl methacrylate (PDEA) middle layer, and a PEG outer layer.[169] At pH 7.4, the insoluble PDEA collapsed on the PCL core, preventing premature drug leakage. At tumoral pH, PDEA become cationic and soluble, thus protruding over the PEG layer to enhance interactions with cells. PDEA was further protonated at endosomal pH, facilitating endosomal disruption by proton sponge effects.

In some studies, poly(N-isopropylacrylamide-co-acrylamide)-b-poly(DL-lactide)) (p(NIPAAm-co-AAm)-b-PDLLA) polymeric micelles were loaded with the antitumor drug docetaxel with an encapsulation efficiency of 27% and increased cytotoxicity was observed against tumor cells after hyperthermia.[103,104] The same formulation, when tested in vivo upon intravenous administration, showed reduced toxicity of docetaxel and significantly higher antitumor activity after hyperthermia. Liu and co-workers developed folate functionalized poly(N-isopropylacrylamide-co-N,N-dimethylacrylamide-co-2-aminoethyl methacrylate)-b-poly(10-undecenoic acid) (P(NIPAAm-co-DMAAm-co-AMA)-b-PUA) thermosensitive micelles loaded with doxorubicin. Due to the pH-dependent LCST of the polymer (being above body temperature at physiological pH 7.4 and below body temperature in acidic environment), these micelles showed a pH-dependent destabilization and triggered drug release at the low pH of endo/lysosomal vesicles.[170]

The cationic polymers that interact electrostatically with the nucleic acid phosphate backbone to form stable complexes have been reported for plasmids DNA delivery.[171,172] These cationic complexes in turn bind to anionic proteoglycans present on cell surfaces, enter cells within membrane-coated vesicles. Escape from this pathway relies on the incorporation of endosomolytic functions within the complexes. Creusat and co-workers showed that rendering PEI insoluble, without modifying its cationic charge density, by conjugation with hydrophobic α-amino acids, could lead to effective siRNA delivery within cells. The tyrosine–PEI conjugate disassembly at pH 6.0, a pH value that is found in PEI-buffered endosomes, was considered responsible for intracellular delivery of siRNA.[173]

The polymers showing response to both pH as well as temperature have also been developed by a combination of individual polymers. For example, pH-sensitive and temperature-sensitive fluorescein isothiocyanate-labeled beads of insulin (insulin bovine) were prepared with poly(N-isopropylacrylamide-co-butylmethacrylate-co-acrylic acid) terpolymers. The release of insulin was observed at 37°C for pH 2.0 and 7.4. Minimal insulin release was observed at pH 2.0, while release was found to be dependent upon molecular weight in pH 7.4.[174]

Poloxamer copolymer micelles represent a very important class of thermosensitive micelles used for drug delivery. SP1049C, a product of doxorubicin formulated with two poloxamer block copolymers, Pluronic L61 and Pluronic F127, showed better *in vivo* anti-tumor activity than doxorubicin in both drug-resistant and drug-sensitive tumor models, as well as improved pharmacokinetics and tumor accumulation.[175,176] The higher activity of SP1049C compared with doxorubicin was shown to be due to increased cellular uptake, inhibition of the drug efflux, as well as changes in the intracellular drug trafficking.[175] A clinical trial of this formulation, demonstrated slower clearance of SP1049C than free doxorubicin in patients.[101] In the same study, evidence of antitumor activity was observed in some patients. The same product also very recently proceeded in a Phase II study in patients with advanced adenocarcinoma and the formulation showed notable single agent activity and an acceptable safety profile.[177]

In another study, it was shown that core cross-linking of the micelles substantially increased the PEG-*b*-p(HPMAAm-Lac$_n$) micelle stability in the circulation. The core cross-linked micelles showed increased circulation times (more than 50% of the injected dose remained in the circulation 6 hr postinjection) and tumor (B16F10 melanoma bearing mice) accumulation compared with non-cross-linked micelles.[122] In a study, doxorubicin derivative was covalently linked to the micellar core through hydrazone spacer. An increased cytotoxicity compared with free doxorubicin was observed in two different cell lines, as well as an increased antitumor activity in B16 melanoma bearing mice.[130] Recently, same system was also actively targeted, via conjugation of an anti-EGF receptor nanobody on the micellar surface, and increased cellular binding and uptake was observed by EGF receptor overexpressing cells, as compared with micelles without nanobody.[178] This opens the route to actively targeted thermosensitive polymeric micelles, which were expected to result in even more enhanced tumor accumulation and thus increased efficacy, owing to the targeted moiety coupled on their surface.

CONCLUSION

Temperature- or pH-responsive polymers, instead of acting passively as pure drug carriers, could allow making of drug delivery with superior biodistribution. This would directly reduce toxicity and side effects while improving therapeutic outcomes due to the ability of these responsive systems to deliver higher doses of drug to the site of interest. The stimuli responsiveness reviewed in this entry has provided the utility and complexity of polymers that can sense, process, and respond to stimuli in modulating the release of a drug. Stimuli-responsive drug delivery systems like, liposomes, micelles, hydrogels etc., all have shown delivery of an effective dose of drug at a specific time and place.

Despite many advances, still translation of these polymer systems for clinical use is remaining to be defined. There is a need to develop polymers with greater sensitivity to a more diverse range of stimuli. In terms of therapeutic applications, there is need to develop polymers that are highly sensitive and tuneable since there are only minor differences in temperature and pH between diseased and normal conditions. But, as continuous development of new responsive polymers and combinations takes place, there are expectations that more elaborate and versatile drug carriers will become available in the future.

REFERENCES

1. Chen, G.; Hoffman, A.S. Graft copolymers that exhibit temperature-induced phase transitions over a wide range of pH. Nature **1995**, *373* (6509), 49–52.
2. Brahim, S.; Narinesingh, D.; Guiseppi-Elie, A. Release characteristics of novel pH-sensitive p(HEMA-DMAEMA) hydrogels containing 3-(trimethoxy-silyl) propyl methacrylate. Biomacromolecules **2003**, *4* (5), 1224–1231.
3. Yu, Y.; Nakano, M.; Ikeda, T. Photomechanics: Directed bending of a polymer film by light. Nature **2003**, *425* (6954), 145.
4. Sutton, D.; Nasongkla, N.; Blanco, E.; Gao, J.M. Functionalized micellar systems for cancer targeted drug delivery. Pharm. Res. **2007**, *24* (6), 1029–1046.
5. Wang, M.Z.; Fang, Y.; Hu, D.D. Preparation and properties of chitosan-poly(Nisopropylacrylamide) full-IPN hydrogels. React. Funct. Polym. **2001**, *48* (16), 215–221.
6. Jeong, B.; Bae, Y.H.; Lee, D.S.; Kim, S.W. Biodegradable block copolymers as injectable drug-delivery systems. Nature **1997**, *388* (6645), 860–862.
7. Needham, D.; Anyarambhatla, G.; Kong, G.; Dewhirst, M.W. A new temperature sensitive liposome for use with mild hyperthermia: Characterization and testing in a human tumor xenograft model. Cancer Res. **2000**, *60* (5), 1197–1201.
8. Jeong, B.; Bae, Y.H.; Kim, S.W. In situ gelation of PEG-PLGA-PEG triblock copolymer aqueous solutions and degradation thereof. J. Biomed. Mater. Res. **2000**, *50* (2), 171–177.
9. Sasaki, Y.; Hada, R.; Nakajima, H.; Fukuda, S.; Munakata, A. Improved localizing method of radiopill in measurement of entire gastrointestinal pH profiles: Colonic luminal pH in normal subjects and patients with Crohn's disease. Am. J. Gastroenterol. **1997**, *92* (1), 114–118.
10. Engin, K.; Leeper, D.B.; Cater, J.R.; Thistlethwaite, A.J.; Tupchong, L.; McFarlane, J.D. Extracellular pH distribution in human tumours. Int. J. Hyperthermia **1995**, *11* (2), 211–216.
11. Mok, H.; Veiseh, O.; Fang, C.; Kievit, F.M.; Wang, F.Y.; Park, J.O.; Zhang, M. pH-sensitive siRNA nanovector for

targeted gene silencing and cytotoxic effect in cancer cells. Mol. Pharm. **2010**, *7* (6), 1930–1939.

12. Sethuraman, V.A.; Na, K.; Bae, Y.H. pH-responsive sulfonamide/PEI system for tumor specific gene delivery: An *in vitro* study. Biomacromolecules **2006**, *7* (1), 64–70.

13. Tannock, I.F.; Rotin, D. Acid pH in tumors and its potential for therapeutic exploitation. Cancer Res. **1989**, *49* (16), 4373–4384.

14. Zintchenko, A.; Ogris, M.; Wagner, E. Temperature dependent gene expression induced by PNIPAM-based copolymers: Potential of hyperthermia in gene transfer. Bioconjug. Chem. **2006**, *17* (3), 766–772.

15. Schwerdt, A.; Zintchenko, A.; Concia, M.; Roesen, N.; Fisher, K.; Lindner, L.H.; Issels, R.; Wagner, E.; Ogris, M. Hyperthermia-induced targeting of thermosensitive gene carriers to tumors. Hum. Gene Ther. **2008**, *19* (11), 1283–1292.

16. Ilmain, F.; Tanaka, T.; Kokufuta, E. Transition in a gel driven by hydrogen bonding. Nature **1991**, *349* (6308), 400–401.

17. Tanaka, T.; Fillmore, D.; Sun, S.; Nishio, I.; Swislow, G.; Shah, A. Phase transitions in ionic gels. Phys. Rev. Lett. **1980**, *45* (20), 1636–1639.

18. Heiko, J.; Vander, L.; Sebastiaan, H.; Wonter, O.; Piet, B. Stimulus-sensitive hydrogels and their application in chemical (micro) analysis. Royal Soc. Chem. **2003**, *128* (4), 325–331.

19. Bromberg, I. Intelligent polyelectrolytes and gels in oral drug delivery. Curr. Pharm. Biotechnol. **2003**, *4* (5), 39–49.

20. Lomas, H.; Canton, I.; MacNeil, S.; Du, J.; Armes, S.P.; Ryan, A.J.; Lewis, A.L.; Battaglia, G. Biomimetic pH sensitive polymersomes for efficient DNA encapsulation and delivery. Adv. Mat. **2007**, *19* (23), 4238–4243.

21. Kaneko, Y.; Nakamura, S.; Sakai, K.; Kikuchi, A.; Aoyagi, T.; Sakurai, Y.; Okano, T. Synthesis and swellingdeswelling kinetics of poly(N-isopropylacrylamide) hydrogels grafted with LCST modulated polymers. J. Biomater. Sci. Polym. Ed. **1999**, *10* (11), 1079–1091.

22. Nakayama, M.; Okano, T.; Miyazaki, T.; Kohori, F.; Sakai, K.; Yokoyama, M. Molecular design of biodegradable polymeric micelles for temperature-responsive drug release. J. Control. Release **2006**, *115* (1), 46–56.

23. Hoffman, A.S.; Stayton, P.S.; Bulmus, V. Really small bioconjugates of smart polymers and receptor proteins. J. Biomed. Mater. Res. **2000**, *52* (4), 577–586.

24. Kikuchi, A.; Okano, T. Hydrogels: Stimuli-sensitive hydrogels. In *Polymeric drug delivery systems*; Kwon, G. S., Ed.; Taylor & Francis Group: Boca Raton, 2005, 275–322.

25. Qiu, L.Y.; Bae, Y.H. Polymer architecture and drug delivery. Pharmaceutical Res. **2006**, *23* (1), 1–30.

26. Guice, K.B.; Loo, Y.L. Azeotropic atom transfer radical polymerization of hydroxyethyl methacrylate and (dimethylamino)ethyl methacrylate statistical copolymers and block copolymers with polystyrene. Macromolecules **2006**, *39* (7), 2474–2480.

27. Oishi, M.; Nagasaki, Y.; Itaka, K.; Nishiyama, N.; Kataoka, K. Lactosylated poly(ethylene glycol)-siRNA conjugate through acid-labile β-thiopropionate linkage to construct pH-sensitive polyion complex micelles achieving enhanced gene silencing in hepatoma cells. J. Am. Chem. Soc. **2005**, *127* (6), 1624–1625.

28. Liu, S.Q.; Tong, Y.W.; Yang, Y.Y. Thermally sensitive micelles self-assembled from poly(N-isopropylacrylamide-co-N,N-dimethylacrylamide)-*b*-poly(D,L lactide-co-glycolide) for controlled delivers of paclitaxel. Mol. Biosyst. **2005**, *1* (2), 158–165.

29. Traitel, T.; Cohen, Y.; Kost, J. Characterization of glucose-sensitive insulin release systems in simulated *in vivo* conditions. Biomaterials **2000**, *21* (16), 1679–1687.

30. Putnam, D.; Gentry, C.A.; Pack, D.W.; Langer, R. Polymer-based gene delivery with low cytotoxicity by a unique balance of side chain termini. Proc. Natl. Acad. Sci. U.S.A. **2001**, *98* (3), 1200–1205.

31. Qiu, Y.; Park, K. Environment-sensitive hydrogels for drug delivery. Adv. Drug Deliv. Rev. **2001**, *53* (3), 321–339.

32. Lin, Y.L.; Jiang, G.; Birrell, L.K.; El-Sayed, M.E.H. Degradable, pH-sensitive, membrane-destabilizing, comb-like polymers for intracellular delivery of nucleic acids. Biomaterials **2010**, *31* (27), 7150–7166.

33. Shim, M.S.; Kwon, Y.J. Controlled delivery of plasmid DNA and siRNA to intracellular targets using ketalized polyethylenimine. Biomacromolecules **2008**, *9* (2), 444–455.

34. Shim, M.S.; Kwon, Y.J. Controlled cytoplasmic and nuclear localization of plasmid DNA and siRNA by differentially tailored polyethylenimine. J. Control. Release **2009**, *133* (3), 206–213.

35. Shim, M.S.; Kwon, Y.J. Acid-responsive linear polyethylenimine for efficient, specific, and biocompatible siRNA delivery. Bioconjug. Chem. **2009**, *20* (3), 488–499.

36. Benns, J.M.; Choi, J.S.; Mahato, R.I.; Park, J.S.; Kim, S.W. pH-sensitive cationic polymer gene delivery vehicle: N-Ac-poly(L-histidine)-graft-poly(L-lysine) comb shaped polymer. Bioconjug. Chem. **2000**, *11* (5), 637–645.

37. Midoux, P.; Monsigny, M. Efficient gene transfer by histidylated polylysine/pDNA complexes. Bioconjug. Chem. **1999**, *10* (3), 406–411.

38. Ichikawa, H.; Fukumori, Y. Negatively thermosensitive release of drug from microcapsules with hydroxypropyl cellulose membranes prepared by the Wurster process. Chem. Pharm. Bull. **1999**, *47* (8), 1102–1107.

39. Benns, J.M.; Choi, J.S.; Mahato, R.I.; Park, J.S.; Kim, S.W. pH-sensitive cationic polymer gene delivery vehicle: N-Ac-poly(L-histidine)-graft-poly(L-lysine) comb shaped polymer. Bioconjug. Chem. **2000**, *11* (5), 637–645.

40. Chourasia, M.K.; Jain, S.K. Pharmaceutical approaches to colon targeted drug delivery systems. J. Pharm. Pharm. Sci. **2003**, *6* (1), 33–66.

41. Bochot, A.; Fattal, E.; Grossiord, J.L.; Puisieux, F.; Couvreur, P. Characterization of a new ocular delivery system based on a dispersion of liposomes in a thermosensitive gel. Int. J. Pharm. **1998**, *162* (1–2), 119–127.

42. Dong, L.C.; Hoffman, A.S. A novel approach for preparation of pH-sensitive hydrogels for enteric drug delivery. J. Control. Release **1991**, *15* (2), 141–152.

43. Choi, S.H.; Lee, S.H.; Park, T.G. Temperature-sensitive pluronic/poly(ethylenimine) nanocapsules for thermally triggered disruption of intracellular endosomal compartment. Biomacromolecules **2006**, *7* (6), 1864–1870.

44. Lee, S.H.; Choi, S.H.; Kim, S.H.; Park, T.G. Thermally sensitive cationic polymer nanocapsules for specific cytosolic delivery and efficient gene silencing of

siRNA: Swelling induced physical disruption of endosome by cold shock. J. Control. Release **2008**, *125* (1), 25–32.

45. Hinrichs, W.L.; Schuurmans-Nieuwenbroek, N.; Wetering, V.D.; Hennink, W.E. Thermosensitive polymers as carriers for DNA delivery. J. Control. Release **1999**, *60* (2–3), 249–259.

46. Vihola, H.; LaukkaneN, A.; Valtola, L.; Tenhu, H.; Hirvonen, J. Cytotoxicity of thermosensitive polymers poly(N-isopropylacrylamide), poly(N-vinylcaprolactam) and amphiphilically modified poly(N-vinylcaprolactam). Biomaterials **2005**, *26* (16), 3055–3064.

47. Herber, S.; Eijkel, J.; Olthuis, W.; Bergveld, P.; van den Berg, A. Study of chemically induced pressure generation of hydrogels under isochoric conditions using a microfabricated device. J. Chem. Phys. **2004**, *121* (6), 2746–2751.

48. Budhlall, B.M.; Marquez, M.; Velev, O.D. Microwave, photo- and thermally responsive PNIPAm-gold nanoparticle microgels. Langmuir **2008**, *24* (20), 11959–11966.

49. Shi, J.; Alves, N.M.; Mano, J.F. Chitosan coated alginate beads containing poly(N-isopropylacrylamide) for dual-stimuli-responsive drug release. J. Biomed. Mater. Res. B **2008**, *84* (2), 595–603.

50. Jeong, B.; Gutowska, A. Lesson from nature; stimuli responsive polymers and their biomedical applications. Trends Biotechnol. **2002**, *20* (7), 305–311.

51. Lynn, D.M.; Amiji, M.M.; Langer, R. pH-responsive polymer microspheres: Rapid release of encapsulated material within the range of intracellular pH. Angew. Chem. Int. Ed. Engl. **2001**, *40* (9), 1707–1710.

52. Hoffman, A.S.; Stayton, P.S.; Press, O.; Murthy, N.; Lackey C.A.; Cheung C.; Black F.; Campbell J.; Fausto N.; Kyriakides T.R.; Bornstein P. Bioinspired polymers that control intracellular drug delivery, Biotechnol. Bioprocess Eng. **2001**, *6* (4), 205-212.

53. Murthy, N.; Robichaud, J.R.; Tirrell, D.A.; Stayton, P.S.; Hoffman A.S. The design and synthesis of polymers for eukaryotic membrane disruption, J. Control. Release. **1999**, *61* (1–2), 137–143.

54. Weiß, G.; Knoch, A.; Laicher, A.; Stanislaus, F.; Daniels, R. Simple coacervation of hydroxypropyl methylcellulose phthalate (HPMCP) I. Temperature and pH dependency of coacervate formation. Int. J. Pharm. **1995**, *124* (1), 87–96.

55. Guo, H.X.; Heinamaki, J.; Yliruusi, J. Amylopectin as a subcoating material improves the acidic resistance of enteric-coated pellets containing a freely soluble drug. Int. J. Pharm. **2002**, *235* (1–2), 79–86.

56. Schoneker, D.R.; DeMerlis, C.C.; Borzelleca, J.F. Evaluation of the toxicity of polyvinylacetate phthalate in experimental animals. Food Chem. Toxicol. **2003**, *41* (3), 405–413.

57. Thomas, J.L.; Tirrell, D.A. Polyelectrolyte-sensitized phosholipid vesicles. Acc. Chem. Res. **1992**, *25* (8), 336–342.

58. van de Wetering, P.; Moret, E.E.; Schuurmans-Nieuwenbroek, N.M.E.; van Steenbergen, M.J.; Hennink, W.E. Structure-activity relationships of watersoluble cationic methacrylate/methacrylamide polymers for nonviral gene delivery. Bioconjug. Chem. **1999**, *10* (4), 589–597.

59. Benns, J.M.; Choi, J.; Mahato, R.I.; Park, J.; Kim, S.W. pH-sensitive cationic polymer gene delivery vehicle: N-Ac-poly(L-histidine)-graft-poly(L-lysine) comb shaped polymer. Bioconjug. Chem. **2000**, *11* (5), 637–645.

60. Klemm, A.R.; Young, D.; Lloyd, J.B. Effects of polyethyleneimine on endocytosis and lysosome stability. Biochem. Pharmacol. **1998**, *56* (1), 41–46.

61. Bieber, T.; Meissner, W.; Kostin, S.; Niemann, A.; Elsasser, H.P. Intracellular route and transcriptional competence of polyethylenimine-DNA complexes. J. Control. Release **2002**, *82* (2–3), 441–454.

62. Asayama, S.; Maruyama, A.; Cho, C.; Akaike, T. Design of comb-type polyamine copolymers for a novel pH-sensitive DNA carrier. Bioconjug. Chem. **1997**, *8* (6), 833–838.

63. Guo, X.; Szoka, Jr F.C. Steric stabilization of fusogenic liposomes by a low-pH sensitive PEG-diortho ester-lipid conjugate. Bioconjug. Chem. **2001**, *12* (2), 291–300.

64. Xu, R.; Wang, X.; Lu, Z. New amphiphilic carriers forming pH-sensitive nanoparticles for nucleic acid delivery. Langmuir **2010**, *26* (17), 13874–13882.

65. Heffernan, M.J.; Murthy, N. Polyketal nanoparticles: A new pH-sensitive biodegradable drug delivery vehicle. Bioconjug. Chem. **2005**, *16* (6),1340–1342.

66. Ko, J.; Park, K.; Kim, Y.S.; Kim, M.S.; Han, J.K.; Kim, K.; Park, R.W.; Kim, I.S.; Song, H.K.; Lee, D.S.; Kwon, I.C. Tumoral acidic extracellular pH targeting of pH-responsive mPEG-poly(β-amino ester) block copolymer micelles for cancer therapy. J. Control. Release **2007**, *123* (2), 109–115.

67. Chen, G.; Ito, Y.; Imanishi, Y. Micropattern immobilization of a pH-sensitive polymer. Macromolecules **1997**, *30* (22), 7001–7003.

68. Khan, M.Z.; Prebeg, Z.; Kurjakovic, N. A pH-dependent colon targeted oral drug delivery system using methacrylic acid copolymers. I. Manipulation of drug release using Eudragit L100–55 and Eudragit S100 combinations. J. Control. Release **1999**, *58* (2), 215–222.

69. Wang, C.Y.; Huang, L. pH-sensitive immunoliposomes mediate target-cell-specific delivery and controlled expression of a foreign gene in mouse. Proc. Natl. Acad. Sci. U.S.A. **1987**, *84* (22), 7851–7855.

70. Zhang, Y.P.; Sekirov, L.; Saravolac, E.G.; Wheeler, J.J.; Tardi, P.; Clow, K.; Leng, E.; Sun, R.; Cullis, P.R.; Scherrer, P. Stabilized plasmid-lipid particles for regional gene therapy: Formulation and transfection properties. Gene Ther. **1999**, *6* (8), 1438–1447.

71. Francis, M.F.; Dhara, G.; Winnik, F.M.; Leroux, J. *In vitro* evaluation of pH-sensitive polymer/niosome complexes. Biomacromolecules **2001**, *2* (3), 741–749.

72. Shantha, K.L.; Harding, D.R.K. Preparation and *in vitro* evaluation of poly[N-vinyl-2-pyrrolidone-polyethylene glycol diacrylate]-chitosan interpolymeric pH-responsive hydrogels for oral drug delivery. Int J. Pharm. **2000**, *207* (1–2), 65–70.

73. Risbud, M.V.; Hardikar, A.A.; Bhat, S.V.; Bhonde, R.R. pH sensitive freeze-dried chitosan–polyvinyl pyrrolidone hydrogels as controlled release system for antibiotic delivery. J. Control. Release **2000**, *68* (1), 23–30.

74. Anal, K.A. Stimuli-induced pulsatile or triggered release delivery systems for bioactive compounds. Recent Patents Endocrine Metab. Immune Drug Discov. **2007**, *1* (1), 83–90.

75. Peppas, N.A.; Bures, P.; Leobandung, W.; Chikawa, H. Hydrogels in pharmaceutical formulations. Eur. J. Pharm. Biopharm. **2000**, *50* (1), 27–46.

76. Sopinath, K.S.; Aminabhavi, T.M.; Dave, A.M.; Kumbar, S.G.; Rudzinski, W.E. Stimulus-responsive 'smart' hydrogel as

novel drug delivery systems. Drug Del. Ind. Pharm. **2002**, *28* (8), 957–974.

77. Qiu, Y.; Park, K. Environment sensitive hydrogels for drug delivery. Adv. Drug Deliv. Rev. **2001**, *53* (3), 321–339.

78. Cho, H.S.; Jhon, M.S.; Yuk, S.H.; Lee, H.B. Temperature-induced phase transition of poly(N,N-dimethylaminoethyl methacrylateco-acrylamide). J. Polym. Sci. B: Polym. Phys. **1997**, *35* (4), 595–598.

79. Aoki, T.; Muramatsu, M.; Torii, T.; Sanui, K.; Ogatapp, N. Thermosensitive phase transition of an optically active polymer in aqueous medium. Macromolecules **2001**, *34* (15), 3118–3129.

80. Bromberg, I.F.; Ron, E.S. Temperature responsive gels and thermogelling polymer matrices for protein and peptide delivery. Adv. Drug Deliv. Rev. **1998**, *31* (3),197–221.

81. Ichikawa, H.; Fukumori, Y. Novel positively thermosensitive controlled-release microcapsule with membrane of nano-sized poly(N-isopropylacrylamide) gel dispersed in ethyl-cellulose matrix. J. Control. Release. **2000**, *63* (3), 107–119.

82. Bae, Y.H.; Okano, T.; Kim, S.W. On-off thermocontrol of solute transport. Part 2. Solute release from thermo-sensitive hydrogels. Pharm. Res. **1991**, *8* (5), 624–628.

83. Jeong, B.; Bae, Y.H.; Kim, S.W. Drug release from biodegradable injectable thermosensitive hydrogel of PEG-PGLA-PEG triblock copolymers. J. Control. Release. **2000**, *63* (1–2), 155–163.

84. Rim, Y.J.; Choi, S.; Koh, U.; Lee, M.; Ko, K.S.; Kim, S.W. Controlled release of insulin from injectable biodegradable triblock copolymer. Pharm. Res. **2001**, *18* (4), 548–550.

85. Serres, A.; Baudys, M.; Kim, S.W. Temperature and pH-sensitive polymers for human calcitonin delivery. Pharm. Res. **1996**, *13* (2), 196–201.

86. Park, T.G. Temperature modulated protein release from pH/temperature-sensitive hydrogels. Biomaterials **1999**, *20* (6), 517–521.

87. Andersson, M.; Axelsson, A.; Zacchi, G. Diffusion of glucose and insulin in a swelling N-isopropylacrylamide gel. Int. J. Pharm. **1997**, *157* (2), 199–208.

88. Cohn, D.; Sosnik, A.; Levy, A. Improved reverse thermo-responsive polymeric systems. Biomaterials **2003**, *24* (21), 3707–3714.

89. Nanjawade, B.K.; Manvi, F.V.; Manjappa, A.S. In situ-forming hydrogels for sustained ophthalmic drug delivery. J. Control. Release. **2007**, *122* (2), 119–134.

90. Hoffman, A.S. Applications of thermally reversible polymers and hydrogels in therapeutics and diagnostics. J. Control. Release **1987**, *6* (1), 297–305.

91. Cohn, D.; Lando, G.; Sosnik, A.; Garty, S.; Levi, A. PEO-PPO-PEO based poly(ether ester urethane)s as degradable thermo-responsive multiblock copolymers. Biomaterials **2006**, *27* (9), 1718–1727.

92. Zana, R.; Binanalimb, W.; Kamenka, N. Ethyl(hydroxyethyl) cellulose cationic surfactant interactions-electrical conductivity, self-diffusion, and time-resolved fluorescence, quenching investigations. J. Phys. Chem. **2007**, *96* (13), 5461–5465.

93. Jeong, B.; Bae, Y.H.; Lee, D.S.; Kim, S.W. Biodegradable block copolymers as injectable drug-delivery systems. Nature **1997**, *388* (6645), 860–862.

94. Kataoka, K.; Harada, A.; Nagasaki, Y. Block copolymer micelles for drug delivery: Design, characterization and biological significance. Adv. Drug Deliv. Rev. **2004**, *47* (1), 113–131.

95. Topp, M.D.C.; Dijkstra, P.J.; Talsma, H.; Feijen, J. Thermosensitive micelle-forming block copolymers of poly(ethylene glycol) and poly(N-isopropylacrylamide). Macromolecules **1997**, *30* (26), 8518–8520.

96. Zhang, W.; Shi, L.; Wu, K.; An, Y. Thermoresponsive micellization of poly(ethylene glycol)-b-poly(N-isopropylacrylamide) in water. Macromolecules **2005**, *38* (13), 5743–5747.

97. Qin, S.; Geng, Y.; Discher, D.E.; Yang, S. Temperature-controlled assembly and release from polymer vesicles of poly(ethylene oxide)-block-poly(N-isopropylacrylamide). Adv. Mater. **2006**, *18* (21), 2905–2909.

98. Kohori, F.; Sakai, K.; Aoyagi, T.; Yokoyama, M.; Sakurai, Y.; Okano, T. Preparation and characterization of thermally responsive block copolymer micelles comprising poly(N-isopropylacrylamide-b-DL-lactide). J. Control. Release. **1998**, *55* (1), 87–98.

99. Wei, H.; Zhang, X.Z.; Zhou, Y.; Cheng, S.X.; Zhuo, R.X. Self-assembled thermoresponsive micelles of poly(N-isopropylacrylamide-b -methyl methacrylate). Biomaterials **2006**, *27* (9), 2028–2034.

100. Chung, J.E.; Yokoyama, M.; Yamato, M.; Aoyagi, T.; Sakurai, Y.; Okano T. Thermo-responsive drug delivery from polymeric micelles constructed using block copolymers of poly(N -isopropylacrylamide) and poly(butylmethacrylate). J. Control. Release. **1999**, *62* (1–2), 115–127.

101. Economidis N.V.; Pena D.A.; Smirniotis P.G.; Kohori F.; Sakai K.; Aoyagi T.; Yokoyama M.; Yamato M.; Sakurai Y.; Okano T. Control of adriamycin cytotoxic activity using thermally responsive polymeric micelles composed of poly(N-isopropylacrylamide-co-N,N-dimethylacrylamide)-b-poly(D,L-lactide). Colloid Surface B: Biointerfaces **1999**, *16* (1–4), 195–205.

102. Nakayama, M.; Okano, T.; Miyazaki, T.; Kohori, F.; Sakai, K.; Yokoyama, M. Molecular design of biodegradable polymeric micelles for temperature-responsive drug release. J. Control. Release **2006**, *115* (1), 46–56.

103. Yang, M.; Ding, Y.; Zhang, L.; Qian, X.; Jiang, X.; Liu, B. Novel thermosensitive polymeric micelles for docetaxel delivery. J. Biomed. Mater. Res. A **2007**, *81* (4), 847–858.

104. Liu, B.; Yang, M.; Li, R.; Ding, Y.; Qian, X.; Yu, L.; Jiang, X. The antitumor effect of novel docetaxel-loaded thermosensitive micelles. Eur. J. Pharm. Biopharm. **2008**, *69* (2), 527–534.

105. Lomas, H.; Canton, I.; MacNeil, S.; Du, J.; Armes, S.P.; Ryan, A.J.; Lewis, A.L.; Battaglia, G. Biomimetic pH sensitive polymersomes for efficient DNA encapsulation and delivery. Adv. Mat. **2007**, *19* (23), 4238–4243.

106. Rijcken, C.J.F.; Soga, O.; Hennink, W.E.; van Nostrum, C.F. Triggered destabilisation of polymeric micelles and vesicles by changing polymers polarity: An attractive tool for drug delivery. J. Control. Release **2007**, *120* (3), 131–148.

107. Shah, S.S.; Wertheim, J.; Wang, C.T.; Pitt, C.G. Polymer-drug conjugates: Manipulating drug delivery kinetics using model LCST systems. J. Control. Release **1997**, *45* (1), 95–101.

108. Neradovic, D.; Hinrichs, W.L.J.; Kettenes-Van, D.; Bosch, J.J.; Hennink, W.E. Poly(N-isopropylacrylamide)

with hydrolyzable lactic acid ester side groups: A new type of thermosensitive polymer. Macromol. Rapid Comm. **1999**, *20* (11), 577–581.

109. Lee, B.H.; Vernon, B. Copolymers of N-isopropylacrylamide HEMA-lactate and acrylic acid with time-dependent lower critical solution temperature as a bioresorbable carrier. Polym. Int. **2005**, *54* (2), 418–422.

110. Neradovic, D.; van Nostrum, C.F.; Hennink, W.E. Thermoresponsive polymeric micelles with controlled instability based on hydrolytically sensitive N-isopropylacrylamide copolymers. Macromolecules **2001**, *34* (22), 7589–7591.

111. Neradovic, D.; van Steenbergen, M.J.; Vansteelant, L.; Meijer, Y.J.; van Nostrum, C.F.; Hennink, W.E. Degradation mechanism and kinetics of thermosensitive polyacrylamides containing lactic acid side chains. Macromolecules **2003**, *36* (20), 7491–7498.

112. Soga, O.; van Nostrum, C.F.; Hennink, W.E. Poly(N-(2-hydroxypropyl) methacrylamide mono/di lactate): A new class of biodegradable polymers with tuneable thermosensitivity. Biomacromolecules **2004**, *5* (3), 818–821.

113. Moghimi, S.M.; Hunter, A.C.; Murray, J.C. Long-circulating and target-specific nanoparticles: Theory to practice. Pharmacol. Rev. **2001**, *53* (2), 28–318.

114. Soga, O., van Nostrum, C.F.; Fens, M.; Rijcken, C.J.F.; Schiffelers, R.M.; Storm, G.; Hennink, W.E. Thermosensitive and biodegradable polymeric micelles for paclitaxel delivery. J. Control. Release **2005**, *103* (2), 341–353.

115. Van Hasselt, P.M.; Janssens, G.E.P.J.; Slot, T.K.; van der Ham, M.; Minderhoud, T.C.; Talelli, M.; Akkermans, L.M.; Rijcken, C.J.; van Nostrum, C.F. The influence of bile acids on the oral bioavailability of vitamin K encapsulated in polymeric micelles. J. Control. Release **2009**, *133* (2), 161–168.

116. Zhao, A.; Zhou, S.; Zhou, Q.; Chen, T. Thermosensitive micelles from PEG-based ether-anhydride triblock copolymers. Pharm. Res. **2010**, *27* (8), 1627–1643.

117. Kabanov, A.V.; Batrakova, E.V.; Alakhov, V.Y. Pluronic® block copolymers as novel polymer therapeutics for drug and gene delivery. J. Control. Release **2002**, *82* (2–3), 189–212.

118. Oh, K.T.; Bronich, T.K.; Kabanov, A.V. Micellar formulations for drug delivery based on mixtures of hydrophobic and hydrophilic Pluronic® block copolymers. J. Control. Release **2004**, *94* (2–3), 411–422.

119. Song, M.J.; Lee, D.S.; Ahn, J.H.; Kim, D.J.; Kim, S.C. Thermosensitive sol-gel transition behaviors of poly(ethylene oxide)/aliphatic polyester/poly(ethylene oxide) aqueous solutions. J. Polym. Sci. A1 **2004**, *42* (3), 772–784.

120. Danson, S.; Ferry, D.; Alakhov, V.; Margison, J.; Kerr, D.; Jowle, D.; Brampton, M.; Halbert, G.; Ranson, M. Phase I dose escalation and pharmacokinetic study of Pluronic polymer-bound doxorubicin (SP1049C) in patients with advanced cancer. Br. J. Cancer **2004**, *90* (11), 2085–2091.

121. Liu, J.; Zeng, F.; Allen, C. Influence of serum protein on polycarbonate-based copolymer micelles as a delivery system for a hydrophobic anti-cancer agent. J. Control. Release **2005**, *103* (2), 481–497.

122. Rijcken, C.J.; Snel, C.J.; Schiffelers, R.M.; van Nostrum, C.F.; Hennink, W.E. Hydrolysable core-crosslinked thermosensitive polymeric micelles: Synthesis, characterisation and *in vivo* studies. Biomaterials **2007**, *28* (36), 5581–5593.

123. Zeng, Y.; Pitt, W.G. Poly(ethylene oxide)-b-poly(N-isopropylacrylamide) nanoparticles with cross-linked cores as drug carriers. J. Biomat. Sci. Polym. E. **2005**, *16* (3), 371–380.

124. Wei, H.; Chang, C.; Cheng, C.; Chen, W.Q.; Cheng, S.X.; Zhang, X.Z.; Zhuo, R.X. Synthesis and applications of shell cross-linked thermoresponsive hybrid micelles based on poly(N-isopropylacrylamide-co-3-(trimethoxysilyl)propyl methacrylate)-b-poly(methyl methacrylate). Langmuir **2008**, *24* (9), 4564–4570.

125. Wei, H.; Quan, C.Y.; Chang, C.; Zhang, X.Z.; Zhuo, R.X. Preparation of novel ferrocene-based shell cross-linked thermoresponsive hybrid micelles with antitumor efficacy. J. Phys. Chem. B **2010**, *114* (16), 5309–5314.

126. Bae, K.H.; Choi, S.H.; Park, S.Y.; Lee, Y.; Park, T.G. Thermosensitive pluronic micelles stabilized by shell cross-linking with gold nanoparticles. Langmuir **2006**, *22* (14), 6380–6384.

127. Yang, T.F.; Chen ,C.N.; Chen, M.C.; Lai, C.H.; Liang, H.F.; Sung, H.W. Shell-crosslinked Pluronic L121 micelles as a drug delivery vehicle. Biomaterials **2007**, *28* (4), 725–734.

128. Kim, J.H.; Emoto, K.; Iijima, M.; Nagasaki, Y.; Aoyagi, T.; Okano, T.; Sakurai, Y.; Kataoka, K. Core-stabilized polymeric micelle as potential drug carrier: Increased solubilization of taxol. Polym. Advan. Technol. **1999**, *10* (11), 647–654.

129. Iijima, M.; Nagasaki, Y.; Okada, T.; Kato, M.; Kataoka, K. Core-polymerized reactive micelles from heterotelechelic amphiphilic block copolymers. Macromolecules **1999**, *32* (4), 1140–1146.

130. Talelli, M.; Iman, M.; Varkouhi, A.K.; Rijcken, C.J.; Schiffelers, R.M.; Etrych, T.; Ulbrich, K.; van Nostrum, C.F.; Lammers, T.; Storm, G.; Hennink, W.E. Core-crosslinked polymeric micelles with controlled release of covalently entrapped doxorubicin. Biomaterials **2010**, *31* (30), 7797–7804.

131. Kompella, U.B.; Lee, V.H.L. Pharmacokinetics of peptide and protein drugs. In *Peptide and protein drug delivery*; Lee, V., Eds.; Marcel Dekker: New York, 1991; 391–484.

132. Hu, Z.; Kimura, G.; Ito, Y.; Mawatari, S.; Shimokawa, T.; Yoshikawa, H.; Yoshikawa, Y.; Takada, K. Technology to obtain sustained release characteristics of drugs after delivered to the colon. J. Drug Target. **1998**, *6* (6), 439–448.

133. Venkatesh, H.; Sanghavi, N.M. Controlled drug delivery of pH-dependent soluble drugpindolol. Drug Dev. Ind. Pharm. **1994**, *20* (1), 111–118.

134. Langer, R.; Folkman, J. Polymers for the sustained release of proteins and other macromolecules. Nature **1976**, *263* (5580), 797–800.

135. Sefton, M.V.; Nishimura, E. Insulin permeability of hydrophilic polyacrylate membranes. J. Pharm. Sci. **1980**, *69* (2), 208–209.

136. Morishita, M.; Lowman, A.M.; Takayama, K.; Nagai, P.; Peppas, N.A. Elucidation of the mechanism of incorporation of insulin in controlled release system based on complexation. J. Control. Release **2002**, *81* (1–2), 25–32.

137. Peppas, N.A. Devices based on intelligent biopolymers for oral protein delivery. Int. J. Pharm. **2004**, *277* (1–2), 11–17.

138. Eaimtrakarn, S.; Itoh, Y.; Kishimoto, J.; Yoshikawa, Y.; Shibata, N.; Takada, K. Retention and transit of intestinal mucoadhesive films in rat small intestine. Int. J. Pharm. **2001**, *224* (1–2), 61–67.

Stem Cell—Ultrasound

139. Torres-Lugo, M.; Peppas, N.A. Molecular design and *in vitro* studies of novel pH-sensitive hydrogels for the oral delivery of calcitonin. Macromolecules **1999**, *32* (20), 6646–6651.

140. Torres-Lugo, M.; Garcia, M.; Record, R.; Peppas, N.A. pHsensitive hydrogels as gastrointestinal tract absorption enhancers: Transport mechanisms of salmon calcitonin and other model molecules using the Caco-2 cell model. Biotechnol. Prog. **2002**, *18* (3), 612–616.

141. Soppimath, K.S.; Kulkarni, A.R.; Aminabhavi, T.M. Chemically modified polyacrylamide-g-guar gum-based crosslinked anionic microgels as pH-sensitive drug delivery systems: Preparation and characterization. J. Control. Release **2001**, *75* (3), 331–345.

142. Wang, D.; Dusek, K.; Kopečková, P.; Duskova-Smrckova, M.; Kopeček, J. Novel aromatic azo-containing pH-sensitive hydrogels: Synthesis and characterization. Macromolecules **2002**, *35* (20), 7791–7803.

143. Shu, X.Z.; Zhu, K.J.; Song, W. Novel pH-sensitive citrate crosslinked chitosan film for drug controlled release. Int. J. Pharm. **2001**, *212* (1), 19–28.

144. Risbud, M.V.; Hardikar, A.A.; Bhat, S.V.; Bhonde, R.R. pH-sensitive freeze-dried chitosan–polyvinyl pyrrolidone hydrogels as controlled release system for antibiotic delivery. J. Control. Release **2000**, *68* (1), 23–30.

145. Shantha, K.L.; Harding, D.R.K. Preparation and *in vitro* evaluation of poly[N-vinyl-2-pyrrolidone-polyethyleneglycol diacrylate]-chitosan interpolymeric pH-responsive hydrogels for oral drug delivery. Int. J. Pharm. **2000**, *207* (1–2), 65–70.

146. Na, K.; Bae, Y.H. pH-sensitive polymers for drug delivery. In *Polymeric drug delivery systems*; Kwon, G.S., Eds.; Taylor & Francis, Boca Raton, 2005; 129–194.

147. Lee, E.S.; Na, K.; Bae, Y.H. Doxorubicin loaded pH-sensitive polymeric micelles for reversal of resistant MCF-7 tumor. J. Control. Release. **2003**, *91*, 103–113.

148. Murthy, N.; Robichaud, J.R.; Tirrell, D.A.; Stayton, P.S.; Hoffman, A.S. The design and synthesis of polymers for eukaryotic membrane disruption. J. Control. Release **1999**, *61* (1–2), 137–143.

149. Eccleston, M.E.; Kuiper, M.; Gilchrist, F.M.; Slater, N.K.H. pH responsive pseudo-peptides for cell membrane disruption. J. Control. Release **2000**, *69* (2), 297–307.

150. You, J.O.; Auguste, D.T. Nanocarrier cross-linking density and pH sensitivity regulate intracellular gene transfer. Nano. Lett. **2009**, *9* (12), 4467–4473.

151. You, J.O.; Auguste, D.T. The effect of swelling and cationic character on gene transfection by pH-sensitive nanocarriers. Biomaterials **2010**, *31* (26), 6859–6866.

152. Cheung, C.Y.; Murthy, N.; Stayton, P.S.; Hoffman, A.S. A pH-sensitive polymer that enhances cationic lipid-mediated gene transfer. Bioconjug. Chem. **2001**, *12* (6), 906–910.

153. Kyriakides, T.R.; Cheung, C.Y.; Murthy, N.; Bornstein, P.; Stayton, P.S.; Hoffman, A.S. pH-sensitive polymers that enhance intracellular drug delivery *in vivo*. J. Control. Release **2001**, *78* (1–3), 295–303.

154. Butun, V.; Billingham, N.C.; Armes, S.P. Synthesis of shell crosslinked micelles with tunable hydrophilic/hydrophobic cores. J. Am. Chem. Soc. **1998**, *120* (46), 12135–12136.

155. Schaffer, D.V.; Fidelman, N.A.; Dan, N.; Lauffenburger, D.A. Vector unpacking as a potential barrier for receptor mediated polyplex gene delivery. Biotechnol. Bioeng. **2000**, *67* (5), 598–606.

156. El-Sayed, M.E.H.; Hoffman, A.S.; Stayton, P.S. Rational design of composition and activity correlations for pH-sensitive and glutathione-reactive polymer therapeutics. J. Control. Release **2005**, *101* (1–3), 47–58.

157. Murthy, N.; Campbell, J.; Fausto, N.; Hoffman, A.S.; Stayton, P.S. Design and synthesis of pH-responsive polymeric carriers that target uptake and enhance the intracellular delivery of oligonucleotides. J. Control. Release **2003**, *89* (3), 365–374.

158. Taft, R.W.; Kreevoy, M.M. The evaluation of inductive and resonance effects on reactivity. I. Hydrolysis rates of acetals and non-conjugated aldehydes and ketones. J. Am. Chem. Soc. **1955**, *77* (21), 5590–5595.

159. Ponnappa, B.C.; Dey, I.; Tu, G.C.; Zhou, F.; Aini, M.; Cao, Q. N.; Israel. Y. *In vivo* delivery of antisense oligonucleotides in pH-sensitive liposomes inhibits lipopolysaccharide-induced production of tumor necrosis factor-alpha in rats. J. Pharmacol. Exp. Ther. **2001**, *297* (3), 1129–1136.

160. Leroux, J.; Roux, E.; Garrec, D.L.; Hong, K.; Drummond, D.C. N-isopropylacrylamide copolymers for the preparation of pH-sensitive liposomes and polymeric micelles. J. Control. Release **2001**, *72* (1–3), 71–84.

161. Francis, M.F.; Dhara, G.; Winnik, F.M.; Leroux, J. *In vitro* evaluation of pH-sensitive polymer/niosome complexes. Biomacromolecules **2001**, *2* (3), 741–749.

162. Yeh, P.; Kopečková, P.; Kopeček, J. Biodegradable and pH sensitive hydrogels: Synthesis by crosslinking of N,N-dimethylacrylamide copolymer precursors. J. Polym. Sci. Part A: Polym. Chem. **1994**, *32* (9), 1627–1637.

163. Lee, S.B.; Song, S.; Jin, J.; Sohn, Y.S. Structural and thermosensitive properties of cyclotriphosphazenes with poly(ethylene glycol) and amino acid esters as side groups. Macromolecules **2001**, *34* (21), 7565–7569.

164. Li, H.; Yu, G.; Price, C.; Booth, C.; Hecht, E.; Hoffmann, H. Concentrated aqueous micellar solutions of diblock copoly(oxyethylene/oxybutylene) $E_{41}B_8$: A study of phase behavior. Macromolecules **1997**, *30* (5), 1347–1354.

165. Yu, G.E.; Yang, Y.W.; Yang, Z.; Attwood, D.; Booth, C.; Nace, V. M. Association of diblock and triblock copolymers of ethylene oxide and butylene oxide in aqueous solution. Langmuir **1996**, *12* (14), 3404–3412.

166. Batrakova, E.V.; Kabanov, A.V. Pluronic block copolymers: Evolution of drug delivery concept from inert nanocarriers to biological response modifiers. J. Control. Release **2008**, *130* (2), 98–106.

167. Lynn, D.M.; Amiji, M.M.; Langer, R. pH-responsive biodegradable polymer microspheres: Rapid release of encapsulated material within the range of intracellular pH. Angew. Chem. Int. Ed. Engl. **2001**, *40*, 1707–1710.

168. Lynn, D.M.; Langer, R. Degradable poly(β-amino esters): Synthesis, characterization, and self-assembly with plasmid DNA. J. Am. Chem. Soc. **2000**, *122* (44), 10761–10768.

169. Shen, Y.Q.; Zhan, Y.H.; Tang, J.B.; Xu, P.S.; Johnson, P.A.; Radosz, M.; Van Kirk, E.A.; Murdoch, W.J. Multifunctioning pH-responsive nanoparticles from hierarchical self-assembly of polymer brush for cancer drug delivery. AIChE J. **2008**, *54* (11), 2979–2989.

170. Liu, S.Q.; Wiradharma, N.; Gao, S.J.; Tong, Y.W.; Yang, Y.Y. Bio-functional micelles self-assembled from a folate-conjugated block copolymer for targeted intracellular delivery of anticancer drugs. Biomaterials **2007**, *28* (7), 1423–1433.

171. Lleres, D.; Weibel, J.M.; Heissler, D.; Zuber, G.; Duportail, G.; Mely, Y. Dependence of the cellular internalization and transfection efficiency on the structure and physicochemical properties of cationic detergent/DNA/liposomes. J. Gene Med. **2004**, *6* (4), 415–428.

172. Itaka, K.; Kataoka, K. Recent development of nonviral gene delivery systems with virus-like structures and mechanisms. Eur. J. Pharm. Biopharm. **2009**, *71* (3), 475–483.

173. Creusat, G.; Rinaldi, A.; Weiss, E.; Elbaghdadi, R.; Remy, J.; Mulherkar, R.; Zuber, G. Proton sponge trick for pH-sensitive disassembly of polyethylenimine-based siRNA delivery systems. Bioconjug. Chem. **2010**, *21* (5), 994–1002.

174. Ramkissoon-Ganorkar, C.; Liu, F.; Baudys, M.; Kim, S.W. Modulating insulin-release profile from pH/thermosensitive polymeric beads through polymer molecular weight. J. Control. Release **1999**, *59* (2), 287–98.

175. Alakhov, V.; Klinski, E.; Li, S.; Pietrzynski, G.; Venne, A.; Batrakova, E.; Bronitch, T.; Kabanov, A.V. Block copolymer-based formulation of doxorubicin. From cell screen to clinical trials. Colloid. Surface. B **1999**, *16* (1–4), 113–134.

176. Batrakova, E.V.; Dorodnych, T.Y.; Klinskii, E.Y.; Kliushnenkova, E.N.; Shemchukova, O.B.; Goncharova, O.N. Anthracycline antibiotics non-covalently incorporated into the block copolymer micelles: *in vivo* evaluation of anti-cancer activity. Br. J. Cancer **1996**, *74* (10), 1545–1552.

177. Valle, J.W.; Armstrong, A.; Newman, C.; Alakhov, V.; Pietrzynski, G.; Brewer, J.; Campbell, S.; Corrie, P.; Rowinsky, E.K.; Ranson, M. A phase 2 study of SP1049C, doxorubicin in P-glycoprotein-targeting Pluronics, in patients with advanced adenocarcinoma of the esophagus and gastroesophageal junction. Invest. New Drug. **2011**, *29* (5), 1029–1037.

178. Talelli, M.; Rijcken, C.J.F.; Oliveira, S.; van Meel, R.; van Bergen en Henegouwen, P.M.; Lammers, T.; van Nostrum, C.F.; Storm, G.; Hennink, W.E. Nanobody-shell functionalized thermosensitive core-crosslinked polymeric micelles for active drug targeting. J. Control. Release **2011**, *151* (2), 183–192.

179. Liu, F.; Urban, W. Recent advances and challenges in designing stimuli-responsive polymers. Prog. Polym. Sci. **2010**, *35* (1–2), 3–23.

180. Schmaljohann, D. Thermo- and pH-responsive polymers in drug delivery. Adv. Drug Deliv. Rev. **2006**, *58* (15), 1655–1670.

181. Tyagi, P.; Li, Z.; Chancellor, M.; De Groat, W.C.; Yoshimura, N.; Huang, L. Sustained intravesical drug delivery using thermosensitive hydrogel. Pharm. Re. **2004**, *21* (5), 832–837.

182. Crommelin, D.J.A.; Strom, G.; Jiskot, W.; Stenekes, R.; Mastrobattista, E.; Hennink, W.E. Nanotechnological approaches for the delivery of macromolecules. J. Control. Release **2003**, *87* (1–3), 81

183. Chung, H.J.; Lee, Y.; Park, T.G. Thermo-sensitive and biodegradable hydrogels based on stereocomplexed pluronic multi-block copolymers for controlled protein delivery. J. Control. Release **2008**, *127* (1), 22-30.

184. Sethuraman, V.A.; Lee, M.C.; Bae, Y.H. A biodegradable pH-sensitive micelle system for targeting acidic solid tumors. Pharm. Res. **2008**, *25* (3), 657–666.

185. Chen, J.; Chu, M.; Koulajian, K.; Wu, X.Y.; Giacca, A.; Sun, Y. A monolithic polymeric microdevice for pH-responsive drug delivery. Biomed. Microdevices **2009**, *11* (6), 1251–1257.

186. Washington, N.; Washington, C.; Wilson, C.G. *Physiological pharmaceutics: Barriers to drug absorption*. 2nd Ed, Taylor & Francis: New York, 2001.

Stem Cell—Ultrasound

Sutures

Chih-Chang Chu
Department of Textiles and Apparel and Biomedical Engineering Program, Cornell University, Ithaca, New York, U.S.A.

Abstract

In this entry, the focus is on suture-based wound closure biomaterials because they are the most frequently used and studied. All aspects of suture-based wound closure biomaterials, their classification, and their chemical, physical, mechanical, biological, and biodegradative properties are concisely covered.

INTRODUCTION

Every wound requires the use of biomaterials to close for subsequent successful healing. The complexities involved in wound healing—such as the various degrees of wound strength during the process of healing, the involvement of different type of tissues in a wound, the presence of biomaterials, or the variety of surgical wounds each with its own healing problems—call for different types of wound closure materials. The choice of these biomaterials is based largely on the type of wound and surgeons' preferences.

OVERVIEW

Wound closure biomaterials are generally divided into three major categories: sutures, staples/ligating clips, and tissue adhesives. The sutures have the longest use history, have received the most attention, and are the most widely used in wound closure. Ligating clips and staples facilitate anastomosis with minimal trauma, necrosis, or interruption of tissue function and their use has steadily increased in specific clinical conditions, particularly with the availability of synthetic absorbable ligating clips and staples. Tissue adhesives are the least frequently used for wound closure at the present time, even though they received considerable attention in the 1960s; however, some new tissue adhesives have since received increasing attention. There are several reviews about wound closure biomaterials.[1–9] In this entry, the focus is on suture-based wound closure biomaterials because they are the most frequently used and studied. All aspects of suture-based wound closure biomaterials, their classification, and their chemical, physical, mechanical, biological, and biodegradative properties will be concisely covered.

A suture is a strand of material, either natural or synthetic, used to ligate blood vessels and to fasten tissue together. Textile materials were the earliest and most frequently used materials for surgical wound closure. Linen was used as a suture material as early as 4000 years ago.

Since then, numerous materials have been used as ligatures and sutures: iron wire, gold, silver, dried gut, horse hair, strips of hide, bark fibers, silk, linen, and tendon. Among them, catgut and silk dominated the suture market until 1930. The introduction of steel wire and synthetic nonabsorbable fibers such as nylon, polyester, and polypropylene during and after World War II greatly expanded the chemical composition of suture materials. During the early 1970s, the introduction of two synthetic absorbable suture materials, Dexon® (Davis & Geck, Danby, CT) and Vicryl® (Ethicon, Inc., Somerville, NJ), opened a new milestone for suture materials. Owing to their precisely controlled manufacturing processes and uniform and reproducible properties, these synthetic absorbable sutures have received a great deal of attention from both surgeons and researchers. Since then, several new synthetic absorbable suture materials such as PDSII® (Ethicon), Maxon® (US Surgical/Davis & Geck), Monocryl® (Ethicon), and Biosyn® (US Surgical, Norwalk, CT) have become commercially available. The most important advantage of synthetic absorbable sutures is their reproducible and predictable degradability inside a biological environment. This property enables the sutures to minimize chronic undesirable tissue reactions after the sutures have lost most of their mechanical properties. The latest introduction of sutures is the one having multifunction, and a typical example is Triclosan-coated Vicryl, Vicryl Plus®, which has antimicrobial capability to reduce the chance of wound infection.[10,11] Today, surgeons can choose between a large number of suture materials with various chemical, physical, mechanical, and biological properties.

NEEDLES

The surgical needle, to which a suture is attached, has the primary function of introducing the suture through the tissues to be brought into apposition. Ideally, the needle has no role in wound healing but inappropriate needle selection can prolong the operating time and/or damage tissue

Concise Encyclopedia of Biomedical Polymers and Polymeric Biomaterials DOI: 10.1081/E-EBPPC-120005531

Stem Cell—Ultrasound

Table 1 Needle selection criteria

Minimal tissue trauma

High sharpness (acuity)

Corrosion resistance

High strength

Stable shape

Proper balance

Abrasion resistance

Smooth profile

integrity, leading to such complications as tissue necrosis, wound dehiscence, bleeding, leakage of anastomoses, and poor tissue apposition. Clearly, needle selection is an important factor in suturing and the progress of wound healing. For maximum effectiveness, the surgical needle must be able to carry the suture material through tissue with minimal trauma and, to achieve this goal, needles are required to satisfy several criteria shown in Table 1.

Needles are commonly fabricated from stainless steel, a material that has high strength, is readily available, presents few manufacturing problems, can be polished to a smooth finish, and is relatively inexpensive. Because the gripping action of needle holders and the repeated grip-release action during surgical procedures can damage the needle surface, wear and abrasion resistance is a necessary prerequisite for surgical needles.

A wide variety of surgical needles exists but all types have basically three components: the point, the body, and the eye, swage, or attachment end. Differences between needles arise from variations and modifications within these basic parameters.

The most important biomechanical performance parameters for surgical needles are acuity or sharpness, bending resistance, and ductility. Needle acuity is the most important biomechanical parameter because the needle is required to penetrate tough or fibrous tissue without undue trauma, creating a pathway through which the body of the needle and the attached suture will then pass with minimum drag. Further, the relative ease of tissue penetration with a sharp needle facilitates surgical dexterity, at the same time reducing the risk of needle bending or deformation during use because less force has to be applied to a sharp needle.

COMMERCIAL SUTURES, THEIR SIZES, AND THEIR PHYSICAL CONFIGURATIONS

Suture materials are generally classified into two broad categories: absorbable and nonabsorbable. Absorbable suture materials generally lose their entire or most of their tensile strength within two to three months; those that retain most of their initial strength longer than two to three months are considered nonabsorbable. The absorbable suture materials are catgut (collagen sutures derived from sheep intestinal submucosa), reconstituted collagen, polyglycolide (Dexon®, Dexon II®, Dexon S®), poly(glycolide-lactide) random copolymer (Vicryl®), antimicrobial-coated Vicryl® (Vicryl Plus®), poly-p-dioxanone (PDS®, PDSII®), poly(glycolide-trimethylene carbonate) block copolymer (Maxon®), poly(glycolide-ε-caprolactone) (Monocryl®), and Gycolide-dioxanone-trimethylene carbonate block copolymer (Biosyn®). The nonabsorbable sutures are divided into the natural fibers (i.e., silk, cotton, linen) and man-made fibers (i.e., polyethylene, polypropylene, polyamide, polyester, polytetrafluoroethylene (Gore-Tex®), and stainless steel). Table 2 summarizes all of the commercial suture materials that are available mainly in the United States, Europe and the Pacific area, their generic and trade names, their physical configurations, and their manufacturers. Figure 1 shows the scanning electron images of some of the most popular commercial sutures.

Suture materials are also classified according to their size. Currently, two standards are used to describe the size of suture materials: USP (United States Pharmacopoeia) and EP (European Pharmacopoeia).[9] Table 3 summarizes both EP and USP standards. The USP standard is more commonly used. In the USP standard, the size is represented by a series of combinations of two Arabic numbers: a zero and any number other than zero, such as 2–0 (or 2/0). The larger the first number, the smaller the suture material diameter. Sizes greater than 0 are denoted by 1, 2, 3, etc. This standard size also varies with the type of suture material.

In the EP standard, the code number ranges from 0.1 to 10. The corresponding minimum diameter (mm) can be easily calculated by taking the code number and dividing by 10. The EP standard does not separate natural from synthetic absorbable sutures as the USP standard does.

In terms of the physical configuration of suture threads, they can be classified into monofilament, multifilament, twisted, and braided. Suture materials made of nylon, polyester, and stainless steel are available in both multifilament and monofilament forms. Catgut, reconstituted collagen, and cotton are available in twisted multifilament form; Dexon®, Vicryl®, Monosyn®, Polysorb®, PolySyn FA®, Safil®, BioSorb®, silk, and polyester-based and polyamide-based suture materials are available in the braided multifilament configuration. PDS®, Maxon®, Monocryl®, Biosyn®, Caprosyn®, MonoPlus®, polypropylene, and Gore-Tex® (polytetrafluoroethylene) suture materials exist in monofilament form only. Stainless steel metallic suture materials can be obtained in either monofilament or twisted multifilament configurations. Another unique physical configuration of suture material is available in polyamide (nylon 6) and has the trade name Supramid®; it has a twisted core covered by a jacket of the same material.

Suture materials are frequently coated to enhance their handling properties, particularly to reduce tissue drag when

Table 2 List of commercial sutures, their trade names, and their manufacturers

Generic name	Trade name	Physical configuration	Surface treatment	Manufacturer
Natural absorbable sutures				
Catgut			Plain and chromic	Surgical Specialties Corporation
Catgut	Plain and chromic	Dynek Sutures		
Catgut	Multifilament	Monofilament finish, plain, and chromic	SURU	
Collagen—bovine	Surgical Gut	Plain, chromic, and mild chromic	USS/DG	
Collagen—bovine, ovine	Surgical Gut	Plain and chromic	Ethicon	
Synthetic absorbable sutures				
Glycolic acid and trimethylene carbonate—copolymer	MAXON	Monofilament	Clear or dyed green	USS/DG
Glycolic acid and trimethylene carbonate—copolymer	MAXON CV	Monofilament	Clear or dyed green	USS/DG
Glycolic acid— homopolymer	DEXON II	Braided	Dyed green or bicolored or undyed; coated with polycaprolate	USS/DG
Glycolic acid— homopolymer	DEXON S	Braided	Dyed green or undyed	USS/DG
Glyconate	Monosyn	Mid-term braided	Dyed violet or undyed	B. Braun Melsungen AG
Lactomer	POLYSORB	Braided	Dyed violet or undyed	United States Surgical
Poliglecaprone 25	MONOCRYL	Monofilament	Dyed or undyed	Ethicon
Polydioxanone	PDS II	Monofilament	Dyed or undyed	Ethicon
Polyester—glycolide (60%), dioxanone (14%), trimethylene (26%)	BIOSYN	Monofilament	Dyed violet or undyed	USS/DG
Polyglactin 910	Coated VICRYL	Braided	Dyed or undyed	Ethicon
Polyglactin 910	VICRYL RAPIDE	Braided	Undyed	Ethicon
Polyglactin 910	Coated VICRYL Plus Antibacterial	Braided	Dyed violet or undyed; coated with glycolide and lactide (polyglactin 370) and calcium stearate	Ethicon
Polyglycolic acid	PolySyn FA	Braided	Undyed	Surgical Specialties Corporation
Polyglycolic acid (PGA)	SURUCRYL	Braided	Coated with polycaprolactone calcium stearate	SURU
Poly(l-lactide/glycolide) copolymer	Panacryl	Braided	95% lactide/ 5% glycolide and coated with 90% caprolactone/ 10% glycolide copolymer	Ethicon (discontinued in late 2003)
Polyglycolic acid—low molecular weight	Safil—Quick	Mid-term braided	Coated	B. Braun Melsungen AG
Polyglycolic acid—pure	Safil—green	Mid-term braided	Coated; dyed green	B. Braun Melsungen AG

Stem Cell—Ultrasound

(*Continued*)

Table 2 (*Continued*) List of commercial sutures, their trade names, and their manufacturers

Generic name	Trade name	Physical configuration	Surface treatment	Manufacturer
Polyglycolic acid—pure	Safil—violet	Short-term braided	Coated; dyed violet or undyed	B. Braun Melsungen AG
Polyglytone 6211 synthetic polyester	CAPROSYN	Monofilament	Undyed	United States Surgical
Poly-p-dioxanone	MonoPlus	Long-term monofilament	Dyed violet	B. Braun Melsungen AG
	BioSorb	Braided	Dyed green	Alcon Laboratories
	BioSorb Coated	Braided	Dyed green; coated with polycaprolate	Alcon Laboratories
Nonabsorbable sutures				
316L stainless steel	Surgical Stainless Steel	Monofilament and multifilament		Ethicon
Corrosion-resistant steel	Steelex	Twisted or monofilament		B. Braun Melsungen AG
Nonabsorbable sutures				
Fibroin-natural	SOFSILK	Braided	Dyed black with logwood extract; coated with special wax	USS/DG
Linen	Linatrix	Twisted	Natural white	B. Braun Melsungen AG
Nylon	NYLENE	Monofilament	Dyed blue	Dynek Sutures
Nylon		Monofilament	Dyed black	Surgical Specialties Corporation
Nylon 6 and 6.6	ETHILON		Dyed green or black and undyed clear	Ethicon
Nylon 6 and 6.6	MONOSOF	Monofilament	Dyed black with Logwood extract or undyed; silicone coated	USS/DG
Nylon 6 and 6.6	DERMALON	Monofilament	Dyed blue; silicone coated	USS/DG
Nylon 6 and 6.6	SURGILON	Braided	Dyed blue; silicone coated	USS/DG
Nylon		Monofilament	Dyed black	Alcon Laboratories
Polethylene terephthalate	TI . RON	Braided	Dyed blue or undyed; silicone coated or uncoated	USS/DG
Poly(vinylidene fluoride) and polyvinylidene fluoride-co-hexafluoro-propylene)	PRONOVA		Dyed blue	Ethicon
Polyamide (nylon)	SURULON	Monofilament	Dyed blue/black	SURU
Polyamide 6	Dafilon	Monofilament	Dyed blue	B. Braun Melsungen AG
Polyamide 6.6	Dafilon	Monofilament	Dyed black	B. Braun Melsungen AG
Polyamide 6/6.6	Supramid	Pseudomonofilament-core polyamide 6.6	Cover polyamide 6	B. Braun Melsungen AG
Polybutester	NOVAFIL	Monofilament	Dyed and undyed	USS/DG
Polybutester	VASCUFIL	Monofilament	Dyed blue; coated	USS/DG
Polybutylene terephthalate polyester	Miralene	Monofilament	Dyed blue	B. Braun Melsungen AG
Polyester	SURGIDAC	Braided	Dyed green; coated	USS/DG
Polyester	SURUPOL	Braided	Dyed green/white, silicone coated	SURU
Polyester		Braided or monofilament	Dyed white or green	Alcon Laboratories
Polyester	Polyviolene	Braided	Dyed green or white	Surgical Specialties Corporation

(*Continued*)

Stem Cell—Ultrasound

Table 2 (*Continued*)   List of commercial sutures, their trade names, and their manufacturers

Generic name	Trade name	Physical configuration	Surface treatment	Manufacturer
Polyether	DYLOC	Monofilament	Dyed blue	Dynek Sutures
Polyethylene terephthalate polyester	Dagrofil HRT	Braided	Uncoated; dyed green	B. Braun Melsungen AG
Polyethylene terephthalate polyester	Synthofil	Braided	Coated; dyed green or undyed	B. Braun Melsungen AG
Polyethylene terephthalate polyester	PremiCron	Braided	Silicone coated; dyed green or white	B. Braun Melsungen AG
Polyethylene terephthalate polyester	ETHIBOND	Braided	Dyed green; coated with polybutylate	Ethicon
Polypropylene	Premilene	Monofilament	Dyed blue w/ copper phathalocyanine	B. Braun Melsungen AG
Polypropylene	PROLENE	Dyed blue	Ethicon	
Polypropylene	SURGIPRO	Monofilament	Dyed blue or undyed	United States Surgical
Polypropylene	SURGIPRO II	Monofilament	Dyed blue or undyed	United States Surgical
Polypropylene	SURULENE	Monofilament	Dyed blue	SURU
Polypropylene		Monofilament	Dyed blue	Alcon Laboratories
Polypropylene		Monofilament	Dyed blue	Surgical Specialties Corporations
Polyvinylidene fluoride	RADENE	Monofilament	Dyed blue	Dynek Sutures
Polyvinylidene fluoride	VILENE	Monofilament	Dyed blue	Dynek Sutures
Siliconized polyester	POLYFLEX	Braided	Dyed black	Dynek Sutures
Siliconized polyester	DYFLEX	Braided	Dyed green	Dynek Sutures
Siliconized polyester	TEFLEX	Braided	White	Dynek Sutures
Silk	SURUSIL	Braided	Dyed black, treated with wax	SURU
Silk		Twisted or braided	Dyed black or white	Alcon Laboratories
Silk		Braided	Dyed black	Surgical Specialties Corporation
Silk-fibroin	PERMA-HAND	Natural waxes and gums removed	Dyed black and undyed; special wax	Ethicon
Silk-fibroin	PERMA-HAND-virgin	Sericin gum not removed	Ethicon	
Silk-natural	Silram	Braided	Coated; dyed black	B. Braun Melsungen AG
Silk-natural	Virgin Silk	Twisted	Dyed methylene blue	B. Braun Melsungen AG
Stainless steel	FLEXON	Multifilament	FEB polymer coating	USS/DG
Stainless steel	Steel	Monofilament	USS/DG	
Treated silk	DYSILK	Braided	Dyed black	Dynek Sutures
USP8/0 to 10/0 in various materials	MICROFLEX	Dynek Sutures		

passing through the needle tract and to ease the sliding of knots down during knotting (i.e., knot tie-down). Although nonabsorbable bee's wax, paraffin wax, silicone, and polytetrafluoroethylene (Teflon) are the traditional coating materials, new coating materials have been reported, particularly those that are absorbable. This is because the coating materials used for absorbable sutures must be absorbable; traditional nonabsorbable coating materials such as wax are not appropriate for absorbable sutures.[12–14] Furthermore, absorbable coating

materials have better tissue biocompatibility because of the lack of chronic tissue reaction. There are basically two types of absorbable coating materials: water-soluble and water-insoluble. Water-insoluble coating materials have chemical constituents similar to those in the suture and they are broken down by hydrolysis. They remain on the suture surface longer than water-soluble coatings. A typical example is polyglactin 370 used for Vicryl suture. Dexon II sutures have a polycaprolate coating which is water-insoluble. Water-soluble coating

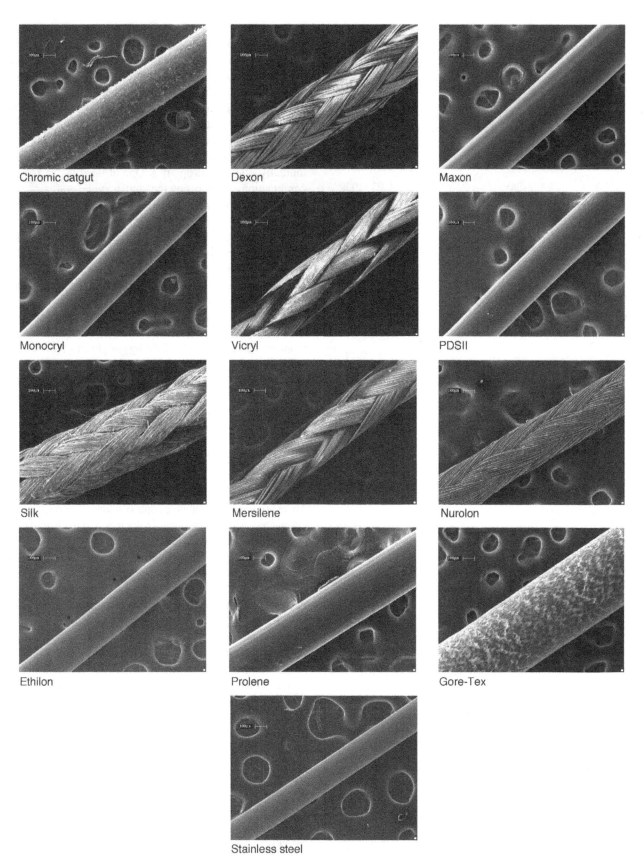

Fig. 1 Scanning electron images of some commercial sutures.

Table 3 USP suture size classification

USP size codes		EP size codes	Suture diameter (mm)	
Nonsynthetic absorbable materials	Nonabsorbable and synthetic absorbable materials	Absorbable and nonabsorbable materials	Min.	Max.
	11/0	0.1	0.01–0.019	
	10/0	0.1	0.02–0.029	
	9.0	0.3	0.03–0.039	
	8/0	0.4	0.04–0.049	
8/0	7/0	0.5	0.05–0.069	
7/0	6/0	0.7	0.07–0.099	
6/0	5/0	1	0.10–0.14	
5/0	4/0	1.5	0.15–0.19	
4/0	3/0	2	0.20–0.24	
3/0	2/0	2.5	0.25–0.29	
2/0	0	3	0.30–0.39	
0	1	4	0.40–0.49	
1	2	5	0.50–0.59	
2	3	6	0.60–0.69	
3	4	7	0.70–0.79	
4	5	8	0.80–0.89	
5	6	9	0.90–0.99	
6	7	10	1.00–1.09	

materials dissolve promptly to reveal the underneath uncoated suture after wound closure. A typical example is poloxamer 188 found on Dexon Plus. There is, however, one technical concern about using water-soluble coating material in actual wound closure. Suture materials are frequently soaked in saline after their removal from packages before use. Some or the bulk of water-soluble coating materials might be removed by this routine soaking practice. Thus, it is important to minimize the time of soaking when dealing with water-soluble coated suture materials. Multifilament sutures are more commonly coated than monofilament sutures. For example, multifilament Vicryl and Dexon Plus or II have coating materials applied, whereas monofilament PDS and Maxon sutures have no coatings.

Although coating of suture materials facilitates easy passage through tissue, it frequently results in poor knot security. For example, Dexon Plus and coated Vicryl require four or five square throws to form secure square knots, whereas the uncoated Dexon and Vicryl sutures form secure knots with only two throws (1 = 1).[15,16] Water-soluble coating materials such as poloxamer 188, found on Dexon Plus, do not suffer from the adverse effects of water-insoluble coating materials on knot security.

There are several other patented procedures and materials reported to improve knot tie-down performance (the ease of sliding a knot down the suture into place during knotting) or/and knot security (the ability of a knot to hold after knotting).[17–20] In general, a coating designed to improve knot tie-down can reduce knot security. It is difficult to achieve both ease of knot tie-down and enhanced knot security of sutures. There are very few reported treatments that would achieve these two contradictory and mutually exclusive properties. One of them is the use of a combination of both coating and textured yarns.[17] The coating materials used in that study included sucrose fatty acid ester, bees wax, paraffin wax, polytetrafluoroethylene, silicone, poly(oxyethylene-co-oxypropylene), polyglactin-370, and gelatin. Other reported absorbable but water-insoluble coating materials that can improve knot tie-down and knot security are high-molecular-weight poly-ε-caprolactone, a copolymer of at least 90% by weight of caprolactone and 10% at most of other biodegradable monomers such as glycolide, lactide, and their derivatives,[18,19] or a random copolymer of 25–75% by weight of glycolide and the rest of trimethylene carbonate.[20] The observed improved knot tie-down and knot security was attributed to deep penetration and even distribution of the coating materials into the interstices of suture filaments.

ESSENTIAL PROPERTIES OF SUTURE MATERIALS

There are four essential properties of suture materials: physical and mechanical, handling, biological, and biodegradative. Table 4 summarizes the characteristics of each of the four essential properties. It is important to recognize that these characteristics are interrelated. For example, the capillarity of a suture material, a physical/mechanical property, is closely related to the ability of the suture to transport bacteria which is a biological property. The modulus of elasticity, a physical/mechanical property, is frequently used to relate to pliability of sutures, a handling property. A brief description of each of those essential properties will be discussed subsequently; they are listed in Tables 5–9. Readers should be aware of the fact that the data in the tables will vary, depending on specific clinical and/or physical environments that suture materials are subjected to and the constant refining of manufacturing processes by suture manufacturers.

Physical and Mechanical Properties

Physical and mechanical properties are probably the most important ones in terms of suture function, i.e., to close wounds and carry physiologic load during healing. They include those related to strength, stiffness, viscoelasticity, coefficient of friction, compliance, size, form (monofilament or multifilament), fluid absorption, transport, etc. Strength includes knotted and unknotted (straight pull)

Table 4 Four major categories of the characteristics of suture materials

Physical/mechanical	Handling	Biocompatibility	Biodegradation
USP vs. EP size (diameter)	Pliability	Inflammatory reaction	Tensile breaking strength and mass loss profiles
Mono vs. multifilament	Memory	Propensity toward wound infection, calculi formation, thrombi formation, carcinogenicity, allergy	Biocompatibility of degradation products
Tensile breaking strength and elongation	Knot tie-down		
Modulus of elasticity	Knot slippage		
Stiffness	Tissue drag		
Stress relaxation and creep			
Capillarity			
Swelling			
Coefficient of friction			

Table 5 Mechanical properties of absorbable sutures[a]

Class (chemical name)	Commercial name	Break strength straight pull (MPa)	Break strength knot pull (MPa)	Elongation to break (%)	Young's modulus (GPa) (psi)[b]
Catgut		310–380	110–210	15–35	2.4 (358,000)
Regenerated collagen					
Poly(p-dioxanone)	PDS®, PDSII®	450–560	240–340	30–38	1.2–1.7 (211,000)
Poly(glycolide-co-trimethylene carbonate)	Maxon®	540–610	280–480	26–38	3.0–3.4 (380,000)
		570–910	300–400	18–25	7–14
Poly(glycolide-co-lactide) or Polyglactin 910	Vicryl®				
Polyglycolide-co-ε-caprol-actone or Polyglecaprone 25	Monocryl®	654–882		67–96	(113,000)
	Dexon S®	760–920	310–590	18–25	7–14
Poly(glycolic acid) or Polyglycolide	Dexon Plus®				

[a]Mechanical properties presented are typical for sizes 0 through 3-0 but may differ for finer or larger sizes.
Source: Partially taken from Casey & Lewis.[14]
[b]Data in ( ) are in psi units.
Source: Adapted from Bezwada et al.[21]

tensile strength, modulus of elasticity (relating to stiffness), elongation at break, and toughness.

The tensile strength character is the most frequently reported and studied physical/mechanical property of suture materials. Because tensile strength is expressed in terms of the cross-sectional area of the material, it is normalized based on the dimension of the material and hence can be used to compare sutures having different chemical structure or/and sizes. Tensile breaking force, however, does not take into account suture size (i.e., diameter). Thus, a larger size suture will have a higher tensile breaking force than the same suture of a smaller size, even though the two sutures may have the same tensile strength. Therefore, a meaningful comparison of tensile breaking force of several sutures should be made using the same suture size (diameter) and form. In addition, knotted tensile strength or breaking force is frequently lower than for the unknotted suture. Strength values are obtained in either dry or wet conditions. The ASTM testing conditions that stipulate the cross-head speed, gage length, temperature (21 ± 1°C), and humidity (65 ± 2%) of testing room should be followed, if possible, for obtaining and reporting strength data.

Bending stiffness is a complex mechanical phenomenon and it closely relates to the handling characteristic of

Table 6 Mechanical properties of nonabsorbable sutures[a]

Class (chemical name)	Commercial name	Break strength straight pull (MPa)	Break strength knot pull (MPa)	Elongation to break (%)	Young's modulus (GPa)
Cotton & linen	Cotton	280–390	160–320	3–6	5.6–10.9
Silk	Silk	370–570	240–290	9–31	8.4–12.9
	Surgical Silk				
	Dermal®				
	Virgin Silk				
Polypropylene	Surgilene®	410–460	280–320	24–62	2.2–6.9
	Prolene®				
Nylon 66 & Nylon 6	Surgilon®	460–710	300–330	17–65	1.8–4.5
	Dermalon®				
	Nurolon®				
	Ethilon®				
	Supramid®				
Poly[(tetramethylene ether)terephthalate-co-tetramethylene terephthalate]	Novafil®	480–550	290–370	29–38	1.9–2.1
Poly(butylene terephthalate)	Miralene®	490–550	280–400	19–22	3.6–3.7
Poly(ethylene) terephthalate	Dacron® Ethiflex®	510–1,060	300–390	8–42	1.2–6.5
	Ti.Cron® Polydek®				
	Ethibond® Tevdek®				
	Mersilene® Mirafil®				
Stainless steel	Flexon®	540–780	420–710	29–65	200
	Stainless Steel				
	Surgical S.S.				

[a]Mechanical properties presented are typical for sizes 0 through 3–0 but may differ for finer or larger sizes.
Source: Adapted from Casey & Lewis.[14]

Table 7 Relative tissue reactivity to sutures

Tissue reactivity level	Nonabsorbable	Absorbable
Most →		Catgut
	Silk, cotton	
	Polyester coated	Dexon and Vicryl
	Polyester uncoated	Maxon, PDS, Monocryl
	Nylon	
Least →	Polypropylene	

Source: Adapted from Bennet.[1]

suture materials, particularly knot security. There are very few reported data describing bending stiffness of sutures. Most reported stiffness of sutures was derived from the modulus of elasticity obtained from a tensile strength test.

Because a knot involves bending of suture strands, stiffness based on modulus of elasticity may not adequately represent the performance of knot strength, security, and tie-down. There are two reported studies of bending stiffness of sutures.[23,24]

Chu et al.'s study[23] of bending stiffness was based on the force required to bend a suture to a predetermined angle. The measured bending force was converted to flexural stiffness in pounds/in[2] according to an ASTM formula. Braided sutures were generally more flexible than monofilament sutures of equivalent size, irrespective of their chemical constituents. Coated sutures had a significantly higher bending stiffness than the corresponding uncoated ones. This increase in bending stiffness is attributable to the loss of mobility of constituent fibers under bending force. An increase in suture size significantly increased their stiffness, and the magnitude of increase depended on the chemical constituent of the suture. The large porous volume inherent in the Gore-Tex monofilament suture was

Table 8 General comparison of absorbable sutures

Trade name	Configuration	Tensile strength	Tissue reactivity	Handling	Knot security	Memory	Absorption	Degradation mode	Comments
Collagen (plain)	Twisted	Poor (0% at 2–3 wks)	Moderate	Fair	Poor	Low	Unpredictable (12 weeks)	Proteolytic	Less impure than surgical gut
Collagen (chromic)	Twisted	Poor (0% at 2–3 wks)	Moderate	Fair	Poor	Low	Unpredictable (12 weeks)	Proteolytic	Less impure than surgical gut
Surgical gut (plain)	Twisted	Poor (0% at 2–3 wks)	High	Fair	Poor	Low	Unpredictable (12 weeks)	Proteolytic	May be ordered as "fast-absorbing gut" (Ethicon) for percutaneous sutures
Surgical gut (chromic)	Twisted	Poor (0% at 2–3 wks)	Moderately high	Fair	Fair	Low	Unpredictable (14–80 days)	Proteolytic	Darker, more visible (Davis & Geck); mild or extra chromatization (Davis & Geck)
Coated vicryl	Braided	Good (50% at 2–3 wks)	Low	Better	Fair	Low	Predictable (80 days)	Hydrolytic	Clear, violet, coated
Dexon "S"	Braided	Good (50% at 2–3 wks)	Low	Fair	Good	Low	Predictable (90 days)	Hydrolytic	Uncoated
Dexon plus	Braided	Good (50% at 2–3 wks)	Low	Better	Fair	Low	Predictable (90 days)	Hydrolytic	Clear, green, coated
PDS	Monofilament	Better (50% at 5–6 wks)	Low	Fair	Poor	High	Predictable (180 days)	Hydrolytic	Violet, clear
Maxon	Monofilament	Better (50% at 4–5 wks)	Low	Fair	Fair	High	Predictable (210 days)	Hydrolytic	Green, clear
Biosyn	Monofilament	Good (5% at 2–3 wks)	Low	Good		Low	Predictable	Hydrolytic	Second lowest in Modulus of elasticity
Monocryl	Monofilament	Good (50% at 1–2 wks)	Low	Good	Fair	Low	Predictable (119 days)	Hydrolytic	Gold

Source: Taken partially from Bennett[1] and Roby et al.[22]

Table 9 General comparison of nonabsorbable sutures

Generic or trade name	Configuration	Tensile strength	Tissue reactivity	Handling	Knot security	Memory	Comments
Cotton	Twisted	Good	High	Good	Good	Poor	Obsolete
Silk	Braid	Good	High	Good	Good	Poor	Predisposes to infection; does not tear tissue; D & G suture is silicone-treated; Ethicon is waxed
Ethilon	Monofilament	High	Low	Poor	Poor	High	Cuts tissue; nylon 6.6; black, clear, or green
Dermalon	Monofilament	High	Low	Poor	Poor	High	Nylon 6.6
Surgamid	Monofilament or braid	High	Low	Poor	Poor	High	Nylon 6.6
Nurolon	Braid	High	Moderate	Good	Fair	Fair	May predispose to infection; black or white; waxed; nylon 6.6
Surgilon	Braid	High	Moderate	Fair	Fair	Fair	Nylon 6.6
Prolene	Monofilament	Fair	Low	Poor	Poor	High	Very low coefficient of friction; cuts tissue; blue or clear
Surgilene	Monofilament	Fair	Low	Poor	Poor	High	–
Dermalene	Monofilament	Good	Low	Poor	Poor	High	–
Novafil	Monofilament	High	Low	Fair	Poor	Low	Blue or clear
Mersilene	Braid	High	Moderate	Good	Good	Fair	Green or white
Dacron	Braid	High	Moderate	Good	Good	Fair	–
Polyviolene	Braid	High	Moderate	Good	Good	Fair	Green or white
Ethibond	Braid	High	Moderate	Good	Good	Fair	Green or white
Tri-Cron	Braid	High	Moderate	Poor	Poor	Fair	–
Polydek	Braid	High	Moderate	Good	Good	Fair	–
Tevdek	Braid	High	Moderate	Poor	Poor	Fair	–
Fore-Tex	Monofilament	Fair–Poor	Low	Very good	Fair	Poor	No packing memory and knot construction has virtually no effect on breaking strength
Stainless steel	Monofilament, twist, or braid	High	Low	Poor	Good	Poor	May kink

Source: Taken partially from Bennett.[1]

the reason for its lowest flexural stiffness. The second bending stiffness study of a few sutures[24] was based on attaching a constant weight to each of the two ends of a bending suture and measuring the distance between these two ends after one minute of loading. Bending stiffness data from these two studies agree with each other that braided sutures are generally more flexible than monofilamentous sutures, and that the Gore-Tex suture has the lowest bending stiffness.

Suture compliance (also referred to as elasticity) is a mechanical property that closely relates to the ease of a suture to elongate under a tensile force. It is believed that

the level of suture compliance should contribute to the compliance of tissues at the anastomotic site. Suture compliance is particularly important in surgery where there is a tubular or vascular anastomosis. Compliance mismatch between a vascular graft and host tissue has long been suggested as one of several factors contributing to graft failure.[25] Compliance mismatch at the anastomotic site constitutes a major component of overall compliance mismatch associated with vascular grafts. Because sutures are the only foreign materials in the anastomotic site, it is expected that a wide range of suture compliance might result in different levels of anastomotic compliance. There is only one reported study that examined the effect of suture compliance on the compliance of arterial anastomotic tissues closed with two sutures vastly different in compliance: 6/0 Novafil and Prolene.[26] Novafil is an elastomeric suture made from polybutester and is characterized by a high elongation at low tensile force, low modulus of elasticity, and high hystersis, whereas the Prolene suture has relatively higher modulus of elasticity, low elongation at low tensile force, and low hystersis. In a clinical condition of minimal tubular compliance and diameter mismatch such as artery-artery anastomoses, a far more compliant anastomosis was achieved with Novafil (5.9 ± 2.0%) than with Prolene (3.3 ± 0.6%) suture. Thus, arterial anastomoses closed with a more compliant suture such as Novafil produced on average over 75% more arterial anastomotic compliance than those closed with the less compliant Prolene suture.

Handling Properties

Handling properties describe those that relate to the feel of suture materials by surgeons during wound closure. It is the only category of suture property that is the most difficult to evaluate objectively. Handling properties include pliability (or stiffness), ease of knot tie-down, knot security, packaging memory, surface friction, viscoelasticity, tissue drag, etc. They are directly and indirectly related to physical/mechanical properties of a suture. For example, the term pliability of a suture is a subjective description of how easily a person can bend it and hence relates to the surgeon's feel of a suture during knot tying. It is directly related to the bending modulus of a suture and indirectly to the coefficient of friction. Packaging memory, another handling property that indirectly relates to pliability, is the ability to retain the kink form of sutures after unpacking them. The ability to retain such kink form after unpacking would make surgeons' handling of sutures more difficult during wound closure, particularly tying a knot. This is because sutures with high memory such as nylon, polypropylene, PDS, and Maxon tend to loosen their knots as they try to return to their kink form from packaging. Thus, packaging memory should be as low as possible. In general, monofilament sutures have more packaging memory than braided ones. The three exceptions are the newly available Monocryl, Biosyn, and Gore-Tex monofilament sutures which were reported to have exceptionally low packaging memory. The easiest means to evaluate packaging memory of sutures is to hang them in air and measure the time required to straighten out the kink form from packaging.

Knot tie-down and security describe how easily a surgeon can slide a knot down to the wound edge and how well the knot will stay in position without untying or slippage. This handling property relates to surface and mechanical properties of sutures. The relatively smooth surface of a monofilament or coated braided suture will have a better knot tie-down than a suture with a rough surface such as an uncoated braided suture, if everything else is equal.

The coefficient of friction of sutures also relates to knot tie-down and security. A linear relationship between knot security and coefficient of friction was reported by Herman.[27] A high coefficient of friction makes knot tie-down difficult but leads to a more secure knot. This is because a high friction suture provides additional frictional force to hold the knot together. This high friction suture surface also makes the passage of suture strands difficult during knot tie-down. It thus appears that knot tie-down and knot security are two contradictory requirements. There is no reported standard test for evaluating knot tie-down capacity. However, Tomita et al. reported a method to objectively quantify the knot tie-down capacity of 2/0 silk, polyester sutures, Gore-Tex, and an experimental ultra-high-molecular-weight polyethylene suture (Nesplon®).[24]

Biological Properties

Biocompatibility of suture materials describes how sutures, which are foreign materials to the body, can affect surrounding tissues and how the surrounding tissues can affect the properties of sutures. Thus, biocompatibility is a two-way relationship. The extent of tissue reactions to sutures depends largely on the chemical nature of the sutures and their degradation products if they are absorbable. Sutures from natural sources such as catgut and silk usually provoke more tissue reactions than synthetic ones because of the availability of enzymes to react with natural biopolymers. Besides the more important chemical factors and physical form, the amount and stiffness of suture materials have been reported to elicit different levels of tissue reactions. For example, a stiff suture will result in stiff projecting ends in a knot where cut. These stiff ends can irritate surrounding tissues through mechanical means, a problem associated with some monofilament sutures but generally not found in braided multifilament sutures.

Because the quantity of a buried suture relates to the extent of tissue reaction, it is a well-known practice in surgery to use as little suture material as possible, such as a smaller knot or a smaller size, to close wounds. The use of a smaller size of suture for wound closure, without detriment to the provision of adequate support to wounds and without cutting through wound tissue, is due to the square relationship between diameter and volume which suggests that a slight increase in suture size or diameter would increase its volume considerably.

There are two basic means to study biocompatibility of suture materials: cellular response and enzyme histochemistry. The former is the most frequently used and provides information about the type and density of inflammatory cells at a suture site. In the cellular response approach, sutures without tension are implanted in the gluteal muscle in small animals such as the rat. This implantation site has given a very consistent reproducible cellular response for valid comparisons, even though it is not a common site for suture in surgery. However, Walton questioned the use of

Stem Cell—Ultrasound

this common test procedure, particularly in orthopedic surgery,[28] due to the observed inflamed nature of the postoperative synovial tissue and the mechanically stressed nature of the suture. Histological stains with a variety of dyes such as the most frequently used H & E are the standard methods of evaluation of cellular activity at the suture sites. Figure 2 is a typical example of histological photomicrographs of PDS and Maxon sutures at 35 days postimplantation in a variety of tissues.[29] In addition to a qualitative description of cellular activities, tissue response can also be graded by the most frequently used and accepted Sewell, Wiland, and Craver method or its modification.[30]

The enzyme histochemical approach is a more objective, quantitative, consistent, and reproducible method than cellular response. Enzyme histochemistry is based on the fact that any cellular response to a foreign material is always associated with the presence of a variety of enzymes; however, this approach is more tedious and requires more sophisticated facilities and better experience. The data obtained provide additional insight into the functions of

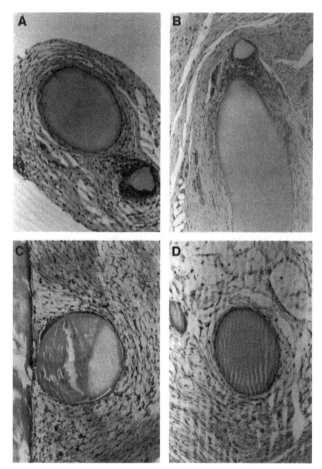

Fig. 2 Light histologic photomicrographs of tissue adjacent to PDS and Maxon sutures at 3 and 5 days postimplantation in a variety of tissues of New Zealand White rabbit (× 130). (**A**) PDS in peritoneum; (**B**) PDS in fascia; (**C**) Maxon in peritoneum; (**D**) Maxon in fascia.
Source: Adapted from Metz et al.[29]

those cells appearing during various stages of wound healing. The enzymatic activity of a suture implant site is quantified by microscopic photometry of a cryostat section of the tissue. For example, van Winkle et al. reported that the high level of cellular response to silk suture observed from histological study is confirmed in their enzyme histochemical study.[31–33] Enzyme histochemistry is also useful for studying the biodegradation mechanism of absorbable sutures.

The normal tissue reaction to sutures can be grouped into three stages, according to the time for the appearance of a variety of inflammatory cells.[1,31,33,34] They are: initial infiltration of polymorphonuclear leukocytes, lymphocytes and monocytes during the first three to four days (i.e., acute response); appearance of macrophages and fibroblasts from day 4 to day 7; and beginning maturation of fibrous connective tissue formation with chronic inflammation after the seventh to the tenth day. During the first seven days postimplantation, there is virtually no difference in normal tissue reaction between synthetic absorbable and nonabsorbable sutures. However, a slightly higher inflammatory reaction to synthetic absorbable sutures can persist for an extended period until they are completely absorbed and metabolized, whereas synthetic nonabsorbable sutures, in general, are characterized by a minimal chronic inflammatory reaction with a thin fibrous connective tissue capsule surrounding the sutures usually by 28 days postimplantation.

In addition to the normal tissue reactions to sutures, there are several adverse tissue reactions that are suture- and site-specific. Some examples include urinary stone or calculi formation, granuloma formation, thrombogenicity, propensity toward wound infection, and recurrence of tumor after radical surgery and allergy.

Monofilament sutures are considered to be a better choice than multifilament ones in closing contaminated wounds. This is because not only do multifilament sutures elicit more tissue reactions which may lessen tissue ability to deal with wound infections but also multifilament sutures have a capillary effect that could transport microorganisms from one region of the wound to another. The reason that multifilament sutures generally elicit more tissue reactions than their monofilament counterparts is because inflammatory cells are able to penetrate into the interstitial space within a multifilament suture and invade each filament. Such an invasion by inflammatory cells, well evident in histological pictures, cannot occur in monofilament sutures. Thus, the available surface area of a suture to tissue should bear a close relationship to the level of tissue reaction that a suture could elicit.

Biodegradation and Absorption Properties

Biodegradation and absorption properties are the most important issue of absorbable sutures, but are far less relevant for most nonabsorbable sutures, particularly those of synthetic nature. This biodegradation property is also responsible for the fact that absorbable sutures do not elicit permanent chronic inflammatory reactions found with

Table 10 Absorption delay of commercial synthetic absorbable sutures

Suture materials	Time to complete loss of tensile strength (days)	Time to complete mass absorption (days)	Absorption delay (days)	Useful lifetime (%)[a]
Dexon®	28	50–140	22–112	20–56
Vicryl®	28	90	62	31
PDS®	63	180–240	117–170	26–35
Maxon®	56	210	155	27
Monocryl®	21	90–119	69–98	18–23

[a]The ratio of [the time to complete loss of tensile strength] to [the time to complete mass absorption]. The higher percentage denotes the better absorbable suture.

nonabsorbable sutures. The most important characteristics in biodegradation and absorption of sutures are the strength and mass loss profiles and biocompatibility of degradation products. Although there is a wide range of strength and mass loss profiles among the available absorbable sutures, they have one common characteristic: Strength loss always occurs much earlier than mass loss, as shown in Table 10. This suggests that absorbable sutures retain a large portion of their mass in tissue while they have already lost all their mechanical properties required to provide support for tissues during wound healing. Thus, an ideal absorbable suture should have matched mass loss and strength loss profiles, and none of the commercial absorbable sutures can achieve this ideal biodegradation property.

The observed wide range of strength and mass loss profiles among the available absorbable sutures is attributable not only to the chemical differences among the absorbable sutures but also to a variety of intrinsic and extrinsic factors, such as pH, electrolytes, stress applied, temperature, γ irradiation, superoxide, microorganisms, and tissue type, to name a few. Among these intrinsic and extrinsic factors, the role of superoxide on the biodegradation of absorbable sutures appears to be one of the most interesting factors because of the unusually fast loss of mechanical integrity and unique surface morphology.[35] For example, at a 0.005 M superoxide ion concentration and room temperature, the five synthetic absorbable sutures retained 20–70% of their original tensile breaking force at the end of 24 hrs as shown in Fig. 3. The bulk of the loss of tensile breaking force of these sutures occurred during the initial two-hour period. The order of tensile breaking force of these five absorbable suture materials at this relatively high superoxide ion concentration was the same as at the lower superoxide ion concentration: Monocryl® > Maxon® > Vicryl® > Dexon® > PDS II®. It is important to know that there would be no change in tensile breaking force of these absorbable sutures in regular buffered saline medium at 25°C for as long as days.[9]

Upon biodegradation, absorbable sutures have shown quite interesting surface morphology and some examples are shown in Fig. 4. For example, multifilament Dexon sutures that were subjected to γ-irradiation treatment and hydrolytic degradation in buffer solution showed very regular

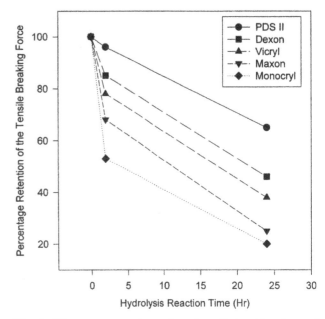

Fig. 3 The percentage of retention of tensile breaking force of five 2/0 synthetic absorbable sutures upon 0.005 M superoxide ion-induced hydrolytic degradation at 25°C. ● - PDS; ■ - Dexon; ◆ - Monocryl; ▲- Vicryl; ▼ - Maxon.
Source: Adapted from Lee & Chu.[35]

circumferential surface cracks along the longitudinal fiber axis and had the appearance of corn-like structure (Figs. 4A and B). Upon γ-irradiation treatment and hydrolytic degradation in buffer solution, monofilament Maxon sutures, however, showed both circumferential and longitudinal surface cracks (Fig. 4C) and subsequent peeling off of these surface cracks (Fig. 4D). The appearance of moon-crater-shaped impressions of various sizes (about 10–100 μm diameter) on a Monocryl® suture at a superoxide ion concentration of > 0.005 M (Fig. 4E) is unique because such circular impressions were never observed in the hydrolytic degradation of all existing absorbable sutures in a conventional buffered saline medium or *in vivo*. The formation of moon crater-shaped impressions on Monocryl and Maxon sutures deviates from the conventional understanding of the anisotropic character of fibers. In the

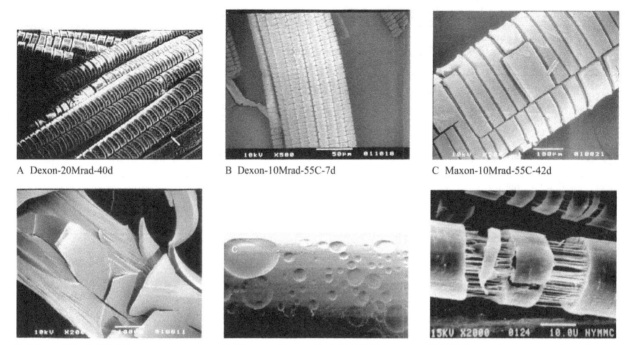

A Dexon-20Mrad-40d

B Dexon-10Mrad-55C-7d

C Maxon-10Mrad-55C-42d

D Maxon-2Mrad-55C-42d

E Monocryl-Superoxide 0.0025M-24 hr

F PGA fibers as the component of vascular graft

Fig. 4 Scanning electron images of degradation of some commercial absorbable sutures. (**A**) 2/0 Dexon after 20 Mrad γ irradiation at room temperature and 40 days *in vitro* in buffer at 37°C; (**B**) 2/0 Dexon after 10 Mrad γ irradiation at 55°C and 7 days *in vitro* in buffer at 37°C; (**C**) 2/0 Maxon after 10 Mrad γ irradiation at 55°C and 42 days *in vitro* in buffer at 37°C; (**D**) 2/0 Maxon after 2 Mrad γ irradiation at 55°C and 42 days *in vitro* in buffer at 37°C; (**E**) 2/0 Monocryl upon 0.005 M superoxide ion-induced hydrolytic degradation at 25°C; (**F**) Polyglycolide fibers as the component of woven vascular grafts upon *in vitro* hydrolytic degradation in buffer of pH 7.4 at 37°C.

reported morphological studies of all existing absorbable sutures in conventional buffer media,[9] the most common surface morphological characteristic upon hydrolytic degradation of suture fibers is the formation of circumferential or/and longitudinal surface cracks that are consistent with the anisotropic character of fibers. It is not fully understood at this stage how superoxide ion-induced degradation could lead to such unusual surface morphology on Monocryl and Maxon sutures.

The biocompatibility of degradation products is usually not a problem because all existing absorbable sutures are made from well-known biocompatible glycolide, lactide, and their derivatives. However, biocompatibility of degradation products also depends on the rate of their accumulation in the surrounding tissues. This implies that the ability of the surrounding tissues to actively remove and metabolize degradation products is essential. Such a metabolism depends on the extent of blood circulation in the tissue. A well-vascularized tissue can remove degradation products as fast as they are released from an absorbable suture and subsequently metabolized, which can minimize tissue reactions to degradation products.

Due to their ability to release degradation products, absorbable sutures have been studied as a vehicle to deliver a variety of biochemicals such as growth factors to facilitate wound healing or antibiotics to combat wound infection. This new approach will increase the value of

absorbable sutures and extend their function beyond the traditional role of wound closure. A typical example is Vicryl Plus which has an antimicrobial agent coating.[10,11]

Biodegradation properties are usually examined *in vitro* or/and *in vivo*. In the *in vitro* environment, the most commonly used medium is phosphate-buffered physiological saline, pH 7.44 at 37°C. However, other buffers such as Tris or body fluids such as urine, bile, and synovial fluids have been used. In the *in vivo* environment, unstressed absorbable sutures are normally implanted in rat gluteal muscle for predetermined periods of implantation. The sutures retrieved at various periods of immersion or implantation are then subjected to evaluation of their mechanical and physical properties to assess their changes with time. The degree of absorption *in vivo* is evaluated by the change in suture cross-sectional area, and the level of tissue reaction is assessed by a histological method and/or enzyme histochemistry.

SUMMARY

Suture materials are the most frequently used biomaterials because every wound must be closed. Suture materials are made from either natural or synthetic fibers and there is a wide range of properties of suture materials available for surgeons to choose from. Among suture properties, four properties are essential: physical and mechanical, handling, biological, and biodegradative. These properties are interrelated. The trend of research development of suture materials is toward value-added products, such as the incorporation of antimicrobial agents into sutures to ward off wound infection.

REFERENCES

1. Bennett, R.G. Selection of wound closure materials. J. Am. Acad. Dermatol. **1988**, *18* (4), 619–637.

2. Guttman, B.; Guttmann, H. Sutures: Properties, uses, and clinical investigation. In *Polymeric Biomaterials*; Dumitriu, S., Ed.; Marcel Dekker: New York, 1994.

3. Stone, I.K. Suture materials. Clin. Obstet. Gynecol. **1988**, *31* (3), 712–717.

4. Chu, C.C. Suture materials. In *Encyclopedia of Materials Science and Engineering*; Beaver, M.B., Ed.; Pergamon Press: New York, 1986; Vol. 6, 4826–4832.

5. Chu, C.C. The degradation and biocompatibility of suture materials. In *CRC Critical Reviews in Biocompatibility*; Williams, D.F., Ed.; CRC Press: Boca Raton, FL, 1985; Vol. 1, 261–322.

6. Chu, C.C. Survey of clinically important wound closure biomaterials. In *Biocompatible Polymers, Metals, and Composites*; Szycher, M., Ed.; Society for Plastics, Engineers, Technomic: Westport, CT, 1983; 477–523.

7. Chu, C.C. Biodegradable suture materials: Intrinsic and extrinsic factors affecting biodegradation phenomena. In *Encyclopedic Handbook of Biomaterials and Bioengineering*; Wise, D.L., Altobelli, D.E., Schwartz, E.R., Yszemski, M., Gresser, J.D., Trantolo, D.J., Eds.; Marcel Dekker: New York, 1995; Vol. 1, 543–688.

8. Chu, C.C. Degradation and biocompatibility of synthetic absorbable suturematerial: General biodegradation phenomena and some factors affecting biodegradation. In *Biomedical Applications of Synthetic Biodegradable Polymers*; Hollinger, J., Ed.; CRC Press: Boca Raton, FL, 1995; 103–128.

9. Chu, C.C.; von Fraunhofer, J.A.; Greisler, H.P. *Wound Closure Biomaterials and Devices*. CRC Press: Boca Raton, FL, 1997.

10. Storch, M.; Perry, L.C.; Davison, J.M.; Ward, J.J. A 28-day study of the effect of coated VIcryl Plus antibacterial suture (coated Polyglactin 910 suture with triclosan) on wound healing in Guinea pig linear incisional skin wounds. Surg. Infect. **2002**, *3* (Suppl. 1), S89–S98.

11. Rothenburger, S.; Spangler, D.; Bhende, S.; Burkley, D. *In vitro* antimicrobial evaluation of coated Vicryl Plus antibacterial suture (coated Polyglactin 910 with riclosan) using zone of inhibition assays. Surg. Infec. **2002**, *3* (Suppl. 1), S79–S87.

12. Conn, J., Jr.; Beal, J.M. Coated Vicryl synthetic absorbable sutures. Surg. Gynecol. Obstet. **1980**, *150* (6), 843–844.

13. Mattei, F.V. Absorbable Coating Composition for Sutures. US Patent 4,201,216, 1980.

14. Casey, D.J.; Lewis, O.G. Absorbable and nonabsorbable sutures. In *Handbook of Biomaterials Evaluation: Scientific, Technical, and Clinical Testing of Implant Materials*; von Recum, A.F., Ed.; Macmillan Publishing: New York, 1986; 86–94.

15. Rodeheaver, G.T.; Thacker, J.G.; Owen, J. Knotting and handling characteristics of coated synthetic absorbable sutures. J. Surg. Res. **1983**, *35* (6), 525–530.

16. Rodeheaver, G.T.; Thacker, J.G.; Delich, R.F. Mechanical performance of polyglycolic acid and polyglactin 910 synthetic absorbable sutures. Surg. Gynecol. Obstet. **1981**, *153* (6), 835–841.

17. Kawai, T.; Matsuda, T.; Yoshimoto, M. Coated Sutures Exhibiting Improved Knot Security. US Patent 4,983,180, 1991.

18. Messier, K.A.; Rhum, J.D. Caprolactone Polymers for Suture Coating. US Patent 4,624,256, 1986.

19. Bezwada, R.S.; Hunter, A.W.; Shalaby, S.W. Copolymers of ε-caprolactone, glycolide and glycolic acid for suture coatings. US Patent 4,994,074, 1991.

20. Wang, D.W.; Casey, D.J.; Lehmann, L.T. Surgical Suture Coating. US Patent 4,705,820, 1987.

21. Bezwada, R.S. *Monocryl, a new ultra-pliable absorbable monofilament suture derived from ε-caprolactone and glycolide*, 1994.

22. Roby, M.S.; Bennett, S.L.; Liu, C.K. Absorbable Block Copolymers and Surgical Articles Fabricated Therefrom. US Patent 5,403,347, April 4 1995.

23. Chu, C.C.; Kizil, Z. Quantitative evaluation of stiffness of commercial suture materials. Surg. Gynecol. Obstet. **1989**, *168* (3), 233–238.

24. Tomita, N.; Tamai, S.; Morihara, T.; Ikeuchi, K.; Ikada, Y. Handling characteristics of braided suture materials for tight tying. J. Appl. Biomater. **1993**, *4* (1), 61–65.

25. Abbott, W.M.; Megerman, J.; Hasson, J.E.; L'Italien, G.; Warnock, D. Effect of compliance mismatch upon vascular graft patency. J. Vasc. Surg. **1987**, *5* (2), 376–382.

26. Megerman, J.; Hamilton, G.; Schmitz-Rixen, T.; Abbott, W.M. Compliance of vascular anastomoses with polybutester and polypropylene sutures. J. Vasc. Surg. **1993**, *18* (5), 827–834.

27. Herman, J.B. Tensile strength and knot security of surgical suture materials. Am. Surg. **1971**, *37* (4), 209–217.

28. Walton, M. Strength retention of chromic gut and monofilament synthetic absorbable suture materials in joint tissues. Clin. Orthop. Relat. Res. **1989**, *242*, 303–310.

29. Metz, S.A.; Chegini, N.; Masterson, B.J. *In vivo* tissue reactivity and degradation of suture materials: A comparison of Maxon and PDS. J. Gynecol. Surg. **1989**, *5* (1), 37–46.

30. Sewell, W.R.; Wiland, J.; Craver, B.N. A new method of comparing sutures of bovine catgut with sutures of bovine catgut in three species. Surg. Gynecol. Obstet. **1955**, *100* (4), 483–494.

31. Van Winkle, W.; Salthouse, T.N. *Biological Response to Sutures and Principles of Suture Selection*. Ethicon: Somerville, NJ, 1976; 1–20.

32. Salthouse, T.N.; Matlaga, B.F. Significance of cellular enzyme activity at nonabsorbable suture implant sites: Silk, polyester and polypropylene. J. Surg. Res. **1975**, *19* (2), 127–132.

33. Salthouse, T.N. Biocompatibility of sutures. In *Biocompatibility in Clinical Practice*; Williams, D.F., Ed.; CRC Press: Boca Raton, FL, 1982; Vol. 1, 12–32.

34. Madsen, E.T. An experimental and clinical evaluation of surgical suture materials, I and II. Surg. Gynecol. Obstet. **1953**, *97* (1), 73–80.

35. Lee, K.-H.; Chu, C.C. The effect of superoxide ions in the degradation of five synthetic absorbable suture materials. J. Biomed. Mater. Res. **2000**, *49* (1), 25–35.

Theranostics: Biodegradable Polymer Particles for

Naveed Ahmed
Department of Pharmacy, Quaid-i-Azam University, Islamabad, Pakistan

Nasir M. Ahmad
Polymer and Surface Engineering Lab, Department of Materials Engineering, School of Chemical and Materials Engineering (SCME), National University of Sciences and Technology (NUST), Islamabad, Pakistan

Asad Ullah Khan
Department of Chemical Engineering, Commission on Science and Technology for Sustainable Development in the South (COMSATS) Institute of Information Technology, Lahore, Pakistan

Haseeb Shaikh
Polymer and Surface Engineering Lab, Department of Materials Engineering, School of Chemical and Materials Engineering (SCME), National University of Sciences and Technology (NUST), Islamabad, Pakistan

Asghari Maqsood
Department of Physics, Center for Emerging Sciences, Engineering and Technology (CESET), Islamabad, Pakistan

Abdul Hamid Elaissari
University of Lyon, Lyon, France

Abstract

Biodegradable polymer particles for theranostics applications with both simultaneous diagnostic and therapeutic actions are vital for present and future health care and well-being. Their multifunctionalities along with tunable sizes and shapes, degree of biodegradability and biocompatibility, responsiveness to stimuli, encapsulation of a wide range of active substances and surface functionalization provide perhaps the most effective approach for in-time diagnosis and treatment of diseases. These particles have the potential to reduce the time lapse between diagnosis and therapy. The richness of availability of monomers and their polymerization can be carried out to prepare a diverse range of biodegradable theranostic polymer particles for theranostic applications including imaging and drug delivery. These theranostic particles have opened up new horizons in biomedical research and consequently will play an increasing role in the future to find viable solutions for diseases through diagnostic and therapeutic functions.

INTRODUCTION TO THERANOSTICS

Current research in various fields brought the combination of different disciplines for the development of new subfields. This merging of different research fields and the emergence of newer fields brought a revolution in medicine for solving many health-related problems. One of the examples of such a new field is biomedical nanotechnology, where nanotechnology has been applied in the biomedical field for searching for remedies for different ailments. Investigations in the field of biomedical nanotechnology brought useful impacts in the treatment of different diseases. Such types of developments result in creation of products that are able to target multiple benefits in one approach specifically in the field of biomedical nanotechnology. The preparation of nanoparticles containing various active ingredients is one of the elementary steps in the development of such multitargeting domaines. The process of incorporating active ingredients into the nanoparticles is termed as nanoencapsulation and its choice can be made on the basis of the properties of the active to be encapsulated as well as the final applications of the nanoparticles. This nanoencapsulation of the actives provided the idea of encapsulating more than one active in the nanoparticles to have the multiple advantages from a single moiety.

The multiple advantages from a single moiety are applied in the diagnosis and treatment of fatal diseases to shorten the time lapse between consecutive steps. Treatment of diseases like cancer is one of the factors that trigger research in this multidisciplinary approach. A newer term for diagnosis as well as treatment at the same time has been coined as theranostics.[1–6] In this newer term, the first part, i.e., "thera," has been taken from therapy and "nostics" has been taken from diagnostics, meaning agents that can be used for therapy as well as for diagnosis. Agents acting as theranostics are termed as theranostic agents. Considering that nanoparticles are used as theranostics agents, various applications can be envisioned as shown in Fig. 1. Currently, the "theranostics" term includes two distinct functions on a single platform based on the combination of

Concise Encyclopedia of Biomedical Polymers and Polymeric Biomaterials DOI: 10.1081/E-EBPPC-120049905

Stem Cell—Ultrasound

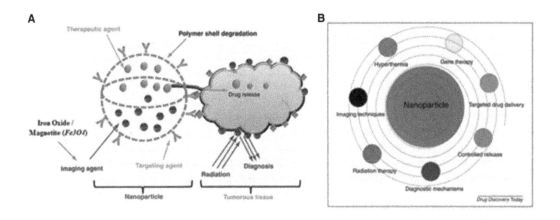

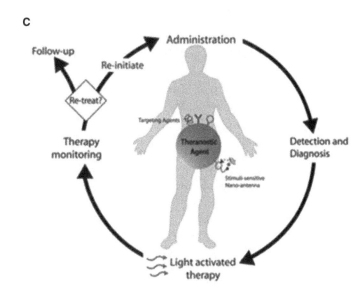

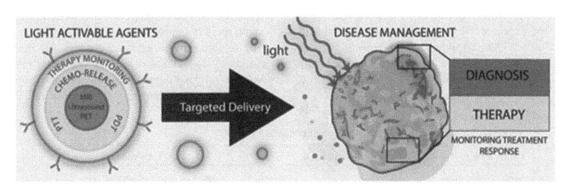

Fig. 1 (**A**) Schematic illustration of theranostics platform for diagnosis and therapy with nanoparticles. (**B**) Applications of nanoparticles as theranostics agents illustrated on different axes. (**C**) The combination of therapeutic and diagnostic mechanisms results in unique application known as theranostics.
Source: (A and B) From Ahmed et al.[1] © 2012, with permission from Elsevier. (C) From Rai et al.[6] © 2010, with permission from Elsevier.

therapeutic and diagnostic agents. This means theranostic nanoparticles (TNPs) are capable of simultaneously monitoring and treating diseases by therapeutic intervention after performing diagnostics. This single platform enables us to find a solution to personalize health care through various

approaches such as ultrasensitive molecular imaging, responsive drug release, and passive and active targeting.[1–6]

The main task in the creation of theranostic agents is to obtain complete understanding of both the therapeutic and the diagnostic mechanisms. This involves understanding the diagnostic therapeutic mechanisms at molecular levels and toxicology approaches of the materials and methods used for the preparation of nanoparticles. In addition, there are others factors involved in this whole process like biodegradability, biocompatibility, and stimuli-responsiveness of the materials.[7–14] Furthermore, conditions for the preparation of nanoparticles through a specific technique and modifications according to the route of administration, material toxicity, pharmacokinetic and pharmacodynamics profiles should also be considered when evaluating benefits. Computed tomography (CT), positron emission tomography (PET), ultrasound (US), X-rays, near infrared fluorescence (NIRF), and magnetic resonance imaging (MRI) are among the various diagnostic techniques for diseased tissues. Any of these diagnostic techniques can be combined with therapeutic methods to prepare the theranostic agents with necessary modifications. Various classes of theranostics agents have been investigated.[4,5] The nanoparticles can be prepared via a modified emulsion evaporation technique by incorporating a model active ingredient for therapy and iron oxide as the contrast agent for the purpose of diagnosis with the help of MRI. Other metals or agents used for diagnostic aspects in theranostics may be gadolinium, silica particles, carbon nanotubes, quantum dots (QDs), etc., which may be used as secondary agents in theranostics. Another interesting field involves the preparation of dual purpose particles for therapy and cosmetology introducing the relatively new field of cosmeto-therapy and stimuli-responsive materials.[11] For cosmeto-therapy, a modified nanoprecipitation technique can be used to encapsulate the anti-inflammatory drugs with argan oil that is normally used in the North African region for cosmetic purposes as well as for therapeutic aspects. Similarly, stimuli-responsive materials that respond to external stimuli of pH, temperature, and magnetic field to perform either diagnostic or therapeutic functions are also expected to lead to solutions for future health care issues.[11–14] Thus, considering the significance of the materials and techniques for theranostics, research in this field has accelerated significantly as summarized in Table 1.

PREPARATION OF BIOMEDCIAL-BIODEGRADEABLE THERNANOSTIC PARTICLES

Several methods have been developed by employing biodegradable polymers to prepare drug-containing microspheres. A few of the important methods are described in this section.

Microspheres Prepared by Polymerization of Monomers

Generally microspheres for drug delivery applications are prepared from linear polymers.[9,15] In certain cases, the preparation through polymerization of monomers with opposite solubility and dispersed in a liquid is also relevant.[16] For example, spherical dispersed droplets of both organic-soluble monomers and aqueous-soluble monomers can be respectively made in aqueous media using oil in water (O/W) or water in oil (W/O) systems. Different polymerization techniques including emulsion, dispersion, and suspension techniques have been developed to synthesize microspheres. The focus has also been to employ supercritical CO_2 for synthesis due to its potential benefits that alleviate toxic solvent effects.[17] The choice of a specific polymerization method is critical to control the properties of microspheres including their sizes as indicated in Table 2. A concise discussion about various methods for synthesizing biodegradable microspheres for potential theranostics applications follows.

Various formulation processes have been used in order to prepare capsules and spheres such as nanoprecipitation, solvent diffusion, polyplex, and double emulsion solvent evaporation. The main principle of particle formation is to transfer a given biodegradable polymer from good solvent conditions to a bad solvent, which leads to polymer precipitation and consequently to colloidal dispersion formation. The solubility of polymers in a given solvent can be discussed by considering the Flory parameter integration.[18] Solvent–polymer interactions can be discussed on the basis of the Flory χ-parameter derived from the well-known

Table 1 Biomedical applications of different theranostics agents

Contrast agent	Drug used	Applications
Manganese oxide	siRNA	MRI plus RNA delivery
Gold	DOX (doxorubicin)	Diagnosis, tumor targeting
Iron oxide	siRNA, Docetaxel (DOX)	Targeting, MRI, and therapy
Silica	Pyropheophorbide (HPPH), DOX	Drug carrier, X-ray/CT imaging, photodynamic therapy
Carbon nanotubes	DNA plasmid, DOX	Diagnosis, DNA, and drug delivery
QDs	DOX, methotrexate (MTX)	Imaging, therapy, and sensing

Abbreviations: siRNA, short interfering ribonucleic acid; CNTs, carbon nanotubes; QDs, quantum dots; DOX, doxorubicin; HPPH, 2-[1-hexyloxyethyl]-2-devinyl pyropheophorbide-alpha; MTX, methotrexate; PTX, paclitaxel; MRI, magnetic resonance imaging; CT, computed topography.
Source: From Ahmed et al.[1] © 2012, with permission from Elsevier.

Table 2 Microspheres sizes obtained from various polymerization methods

Polymerization methods	Size range (mm)
Emulsion	0.1–1.0
Microemulsion	0.01–0.08
Miniemulsion	0.1–0.3
Dispersion	0.5–10
Precipitation	0.05–1.0
Suspension	50–500

thermodynamic of free energy change ($\Delta G = \Delta H - T\Delta S$) per solvent molecule in polymer solution expressed mathematically as

$$\chi = \frac{\Delta G}{\kappa_B T} = \frac{\Delta H - T\Delta S}{\kappa_B T} = \frac{1}{2} - A\left(1 - \frac{\theta}{T}\right)$$

in which k_B and T are Boltzmann constant and temperature, respectively, and A and θ parameters are defined as follows:

$$A = \frac{2\Delta S + \kappa_B}{2\kappa_B}$$

This parameter A is entropy-dependent:

$$\theta = \frac{2\Delta H}{2\Delta S + \kappa_B}$$

This theta parameter is enthalpy- and entropy- dependent. In addition, when, the Flory parameter $\chi < \frac{1}{2}$, the polymer can be considered to be in a good solvent, leading to an expanded polymer chain. But when this Flory parameter is higher than $\chi > 1/2$, the polymer can be considered to be in a poor solvent and consequently the polymer chains collapse under confined coils.

Microspheres Prepared from Linear Preformed Polymers

This method has advantages due to the commercial availability of a wide range of linear polymers.[15] This technique has other advantages, since there are several important biodegradable polymers including polylactide (PLA) and polyglycolide (PGA) that are generally synthesized by an ionic polymerization method instead of emulsion polymerization. Furthermore, one can also employ various natural polymers like chitosan, cellulose, and chitin to synthesize biodegradable microspheres for theranostics.

Preparation by the Solvent Evaporation Method

Several methods have been developed using the evaporation of a solvent to prepare biodegradable microspheres for theranostic applications. One of the methods involves the formation of dispersed oil droplets with polymers and drug followed by the evaporation of an organic solvent.[15] For example, a double emulsion method involves first the formation of W/O emulsion by dissolving drug molecules in the aqueous phase followed by its the dispersion in organic phase of low boiling point solvent and degradable polymer. Using a stabilizer in the aqueous phase, this emulsion is stabilized to form the final O/W emulsion. Upon evaporation of the organic solvent, the polymer hardens and traps the encapsulated drug inside the microspheres. In an interesting work, biodegradable magnetic/polycaprolactone core–shell microparticles were prepared using a modified double emulsion process for *in vivo* diagnostics and therapy-related theranostic applications.[10] Both aqueous and organic ferrofluids having different solid contents of iron oxide nanoparticles were prepared by the co-precipitation method presented in Figs. 2 and 3. A new modified double emulsion evaporation method has been employed to encapsulate a model hydrophilic drug (stilbene) as a contrast agent. This work was mainly divided into two parts: the first part is dedicated to the preparation of aqueous iron oxide and well-dispersed magnetic nanoparticles in the organic phase; the second part is focused on the encapsulation of iron oxide nanoparticle dispersion (aqueous and organic) via water-in-oil in a water modified double emulsion–evaporation process. This study is the first systematic work dedicated to the encapsulation of aqueous and organic magnetic nanoparticles for *in vivo* biomedical diagnosis and therapy. The magnetization power against magnetic field for the nanoparticles prepared by aqueous and organic ferrofluids was also studied as shown in Fig. 2 C and also model drug encapsulations as presented in Fig. 3. All the characterization results showed that the process is very useful for encapsulating iron oxide nanoparticles in both aqueous and organic media along with an active ingredient for theranostics purposes with a final size of about 300 nm.

BIOPOLYMERS WITH RESPONSIVE ACTIVITIES

Stimuli-responsive biopolymers possess smart functionalities by first interacting with the surrounding environment in a passive way and then responding to it in an active way.[5] Because many living systems perform active responses after first interacting with the surrounding media, such smart polymer systems are capable of mimicking the responsiveness of living systems. The responsive polymer can contain functionalities in response to pH, light, heat, solvent, salt, and many others.[19–21] A few examples of various chemical structures and their photo-induced transitions are shown in Fig. 4.[5] A minor change in the environmental stimuli is capable of inducing significant physiochemical transitions in the structure, polarity, and chemical composition. Such transitions in the physical and chemical properties of synthetic and biopolymers have been applied in various

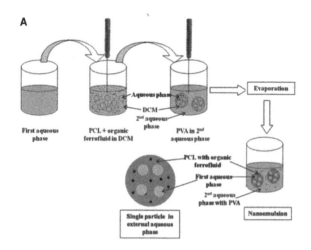

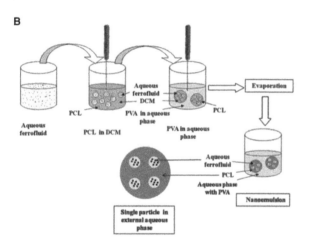

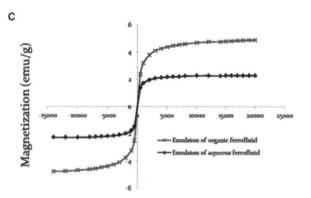

Fig. 2 (**A**) Emulsification process based on the modified double emulsion evaporation method using aqueous ferrofluids. (**B**) Emulsification process based on the modified double emulsion evaporation method using organic ferrofluids. (**C**) Magnetization power against magnetic field for the nanoparticles prepared by aqueous and organic ferrofluids.

Source: From Ahmed et al.[10] © 2012, with permission from RSC Publications.

areas of diagnostics as biosensor and therapeutics for drug delivery applications. Keeping in view the significance of such intelligent systems, there are continuous efforts to develop as well as design new strategies to modify biopolymers for further progress of theranostics. Typical examples of such an intelligent design can be exploited for pH-controlled drug delivery to specific organs by the inclusion of pH-responsive functional groups in the microspheres that can respond to biological surroundings.

TYPES OF BIODEGRADEABLE TNPs

There are extensive efforts underway to develop biodegradable polymers microspheres for theranostics to combine intrinsic diagnostics and therapeutic properties.[2,7] Figure 5 represents the structure of various functionalized nanoparticle for theranostics applications.[2] In many cases, biodegradable polymers of polycaprolactone (PCL), polylactic acid, or their copolymers have been employed. One of the primary focuses of biodegradable microspheres is to induce a controlled release of therapeutic agents. Furthermore, TNPs can be prepared from other biomolecules such as peptides, proteins, and lipids. Furthermore, these polymers or biopolymers can be combined with metal or their oxides to develop hybrid nanoparticles with interesting characteristics. For example, TNPs have therapeutic effects along with the incorporation of gold, enabling to acquire optical imaging (OI) and superparamagnetic iron oxide nanoparticles (SPIONs) to provide MRI. In addition, TNP structures may include functionalization with a targeting moiety, diagnostics, and therapeutic agents as shown in Fig. 5, and also include conjugation with specific drug molecules, core–shell structures, dendrimers, micelles, vesicles, microbubbles, and carbon nanotubes. These TNPs are designed for the co-delivery of drugs as well as imaging agents as summarized in Table 3. Brief discussions about the salient features of few of the important TNP are given below.

Vesicles

Vesicles are extensively investigated for their therapeutic and diagnostic activities due to their available core volume and surface functionality.[2,22] For example, vesicles based on liposomes and polymers are capable of encapsulating both hydrophilic and hydrophobic cargo via their covalent or noncovalent encapsulation. The structures and functions of liposome vesicles provide excellent capability to simultaneously deliver physiochemical agents for successful diagnostic and therapeutic applications. The focus has been on polymeric vesicles or polymersomes because of their clinical potential as co-carriers of drugs and contrast agents. Therefore, both liposome and polymer vesicles can be designed to incorporate a wide range of diagnostic and therapeutic agents for theranostics applications.

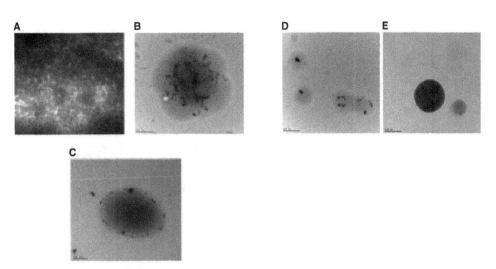

Fig. 3 (**A**) Active molecule/hydrophilic drug nanoparticles prepared via modified double emulsification process. (**B**) Encapsulation of iron oxide particles in aqueous phase; iron oxide particles are well dispersed inside the polymer. (**C**) Encapsulation of iron oxide particles in aqueous phase where iron oxide particles are on surface of polymer. (**D**) Encapsulation of iron oxide particles in organic medium with 500 nm magnification. (**E**) Encapsulation of iron oxide particles in organic medium with different 200 nm magnifications.
Source: From Ahmed et al.[10] © 2012, with permission from RSC Publications.

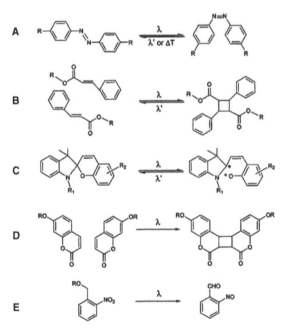

Fig. 4 Photo-induced reversible transitions for various chromophores. (**A**) *trans-cis* (right) isomerization of azobenzene molecules. (**B**) Dimerization of cinnamoyl groups. (**C**) Isomerization of spirobenzopyran derivatives. (**D**) Cross-linking of the coumarin based polymers. (**E**) Dissociation of derivatives of 2-nitrobenzyl.
Source: From Xie et al.[5] © 2010, with permission from Elsevier.

Micelles

Micellar nanoparticles with multiple functionalities have been used as versatile theranostic agents.[2] These can be made from different amphiphilic polymers with uniform size structures and higher solubility of hydrophobic molecules. Both the therapeutic and diagnostic agents are required to bind to the polymer prior to the formation of micellar structures using an anchor molecule for conjugation and entrapping in the hydrophobic core. Polymeric micelles structures are known as successful theranostic systems in clinical investigations because of their facile formation, biodegradability, encapsulation of hydrophobic drugs, and contrast agents. One of the main limitations though is the effect of critical micelle concentration (CMC) on the stability of micelles. The CMC can also influence the degree of exchange of constituents of a media with other biomembranes.

Core–Shell

Core–shell nanoparticles perhaps provide the most versatile and robust therapeutic and diagnostic agents for developing hybrid biodegradable theranostic systems.[2,23] A wide range of materials can be employed either as cores or shells such as polymers, metals, metal oxides, QDs, and carbon nanotubes (CNTs). One study demonstrated the formation of core–shell nanoparticles with co-encapsulation of paclitaxel and a dye in the polymer shell along with a core of iron oxide particles.[24] Dye and iron oxide provide multimodal imaging for the core–shell particles as well as dose-dependent internalization. Feraheme™ nanoparticles are FDA approved and useful for vascular imaging and treatment of anemia. The core of superparamagnetic iron oxide (SPIO) particles can also be labeled with specific cells to monitor *in vivo* behavior of these cells.[25] Although there are many advantages of core–shell structures, they do have certain inherent shortcomings because of issues related to clearance rate, toxicity, weak

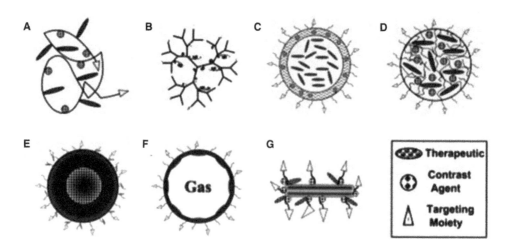

Fig. 5 Structure of functionalized nanoparticle for theranostics applications: (**A**) drug conjugate; (**B**) dendrimer; (**C**) vesicle; (**D**) micelle; (**E**) core–shell nanoparticle; (**F**) microbubble; and (**G**) carbon nanotube.
Source: From Janib et al.[2] © 2012, with permission from Elsevier.

Table 3 Functionalized TNPs and their characteristics. Formulation are based on *N*-(2-hydroxypropyl) methacrylamide (HMPA), poly(amidoamine) (PAMAM)-polyethylene glycol (PEG), polylactic acid (PLA), perfluoropentane (PFP), poly(L-Lactic acid) (PLLA), poly(g-benzyl,t-glutamate)-block-hyaluronan (PBG-Hy), superparamgentic iron oxide (SPIO), QD and Technetium-99m isotopes (Tc-99)

Type of TNP	Material	Size (approx.)	Therapeutic	Diagnostics (contrast agent)	Targeting
Drug conjugate	HPMA	74 kDa	n.a.	DY-615	Active (RGD)
Dendrimer	PAMAM-PEG	100 nm	5-fluorouracil	5-fluorouracil	Active (folic acid)
Vesicle	PBG-Hy	260 nm	Docetaxel	Tc-99	Passive
Micelle	PEG-PLA	45 nm	Dox	SPIO	Active (RGD)
Core–shell	PAA/SPIO	90 nm	Paclitaxel	SPIO	Active (folic acid)
Microbubblle	PEG-PLLA/PFP	125 nm	Dox	PFP	Passive
CNT	Carbon	110 × 10 nm	Cisplatin	QD	Active (epidermal)

Source: From Janib et al.[2] © 2012, with permission from Elsevier.

biodistribution, and less control over biodegradation. There are continuous efforts to overcome these issues to develop next-generation biodegradable core–shell particles.[2]

Microbubbles

Microbubbles are spherical cavities of micron size filled with gas molecules and of varying chemical compositions. Microbubbles are capable of expanding or contracting by the application of certain resonance frequencies. Encapsulated contrast agents such as gases of perfluorocarbons are very common to employ along with the polymer shell, which is biodegradable. If the desirable polymer shells can also be functionalized to constitute an association with therapeutics and diagnostic agents, there are several interesting applications of microbubbles and nanobubbles that have been explored such as drug carriers and imaging contrast agents for angiogenesis, clotting, cancers, and inflammation.[26]

BIOMEDICAL–BIODEGRADABLE POLYMERS AND FACTORS THAT INFLUENCE BIODEGRADATION

Biomaterials are capable of interfacing with biological systems in order to assist body functions, organs, or tissues via diagnosis, treatment, and replacement.[8–10] Thus biomaterials are vital to maintain and improve health. There are many classifications of biomaterials including biopolymers. Furthermore, there are also several types of biopolymers and their usage depends on the applications. For example, certain applications require biodegradation, which involves the process to break backbone bonds in a biomacromolecule to generate shorter and nonharmful degradation products. The degradation process takes place due to the action of living organisms of bacteria and enzymes, or by chemical processes of hydrolysis and oxidation.[27] The biodegradation nature of biopolymers is generally affected by a combination of factors such as

Table 4 Important characteristics and factors that influence the process of hydrolytic biodegradation of polyesters

Characteristics	Factors
Chemical composition	Hydrophilic, hydrophobic, amphiphilic functional group
Aqueous interactions	Permeability, porosity and solubility
Mechanism	Catalytic, non-catalytic and enzymatic
Morphology	Percentage of amorphousness or crystallinity
Thermal characteristics	Glass transition temperature and melting temperature
Average size	Average molecular weight and its distribution.
Physicochemcial characteristics	Ion exchange, ionic strength, pH, temperature, magnetic, etc.
Device factors	Dimension and site of implantation

when wood biodegrades due to oxidation and hydrolysis. There is a great demand to tailor the characteristics of biopolymers to predetermine their service life and control the biodegradation process under a wide range of biological environmental conditions. Biodegradable biopolymers have started to emerge as among the most important components of health care in numerous areas such as medical packaging, disposable, implants, controlled drug delivery, and theranostics.[9,28] For example, controlled drug delivery systems rely on the biodegradation process based on hydrolysis or enzyme to deliver the drug at a controlled rate for a predetermined time. Degradation rates can be controlled through clever tuning of biopolymer chemistry in terms of molecular weight, architecture, morphology, functionalities, and environment. By incorporating pH- or temperature-responsive functional groups, the degradation process can be made responsive to such changes.[29] Furthermore, upon degradation, polymer properties change significantly including decreases in molecular weight and mechanical strength, and possibility of forming monomer molecules. An important process is hydrolytic degradation due to the reaction of water with labile bonds in biopolymers such as polyesters. Thus, by controlling the extent of water absorbance in the biopolymers with labile bonds, the rate of hydrolytic degradation can be determined. Therefore, if biopolymers contain hydrophilic functional groups, then they would degrade faster because of the higher water take-up as compared to the relative hydrophobic polymers.[30] Various factors that influence the biodegradation are summarized in Table 4.

IMPORTANT TYPES OF BIOMEDICAL-BIODEGRADABLE POLYMERS

Polymers when interacting with biological environment can be simply classified into either biostable or biodegradable.[8,9] There are applications that require biostability while others require biodegradation. For example, prostheses demands biostable polymers that maintain their characteristics for

A

R: reactive groups, including protected carboxyl, amino, chloro, ketal, hydroxyl, bromo, carbon-carbon double bonds, etc

B

R = Bn or t-Bu

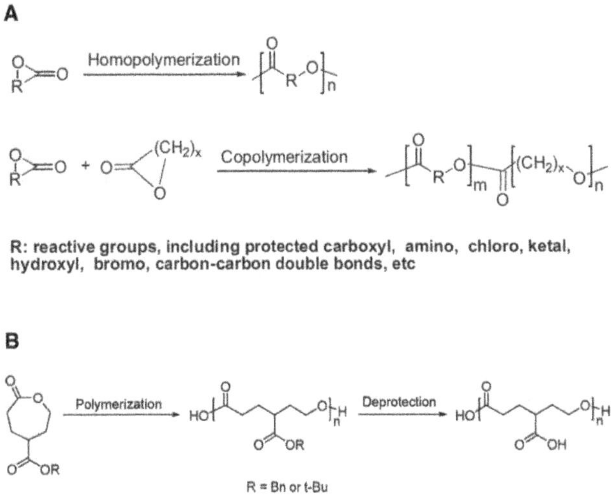

Fig. 6 (**A**) Synthesis of aliphatic polyesters containing reactive group R. (**B**) Synthesis of PCL containing side-chain carboxylic acid groups.
Source: From Tian et al.[8] © 2012, with permission from Elsevier.

several decades. Polyethylene is biostable and used in hip replacements because it is physiologically inert inside the body and also is able to keep the mechanical properties for a long duration. Contrary to long-term usage of biostable polymers, however, biodegradable polymers are used for a shorter time and intended for temporary aids like sutures, drug delivery, and scaffolds for supporting tissue. Biodegradable polymers slowly degrade into degradation products that are later excreted from the body.

The most widely used biodegradable–biocompatible polymers for biomedical applications are poly(ε-caprolactone), poly(lactic acid) (PLA), poly(glycolic acid), and their copolymers.[2,8,9] These and other polymers with certain functional groups of the types of carboxylic, hydroxyl, halogen, amine, and unsaturated carbon–carbon bonds can be synthesized by polymerization of monomers as presented in Fig. 6.[8] As shown in Fig. 6B, PCL can be synthesized by different ionic or radical ring-opening polymerizations of the six-member lactone monomer ε-caprolactone (ε-CL). Enzymes have also been used as catalysts to carry out polymerizations such as polyester synthesis from condensation and ring-opening reactions. Lipase enzymes are capable of polymerizing different lactones including ε-caprolactone γ-caprolactone, β-butyrolactone, and δ-valerolactone. Lactone-based polymers have the potential for controlled drug delivery because of their good biocompatibility. In addition, PCL is also known for blending with other polymers to develop microspheres for various therapeutics applications.

As far as biodegradation of PCL is concerned, its main chain bonds are hydrolyzed in acidic or basic conditions or through enzymatic action as shown in Fig. 7A.[9,31] The degradation process can also take place *in vivo*. The degradation products produce intermediates that are metabolized. The PCL hydrolysis takes place at a slower rate that makes it suitable for long-term drug delivery. A representative example is provided by one-year contraceptive drug delivery systems provided by Capronor®. In order to tune the rate of degradation, PCL can be blended or copolymerized with other monomers. On the other hand, aliphatic polyester degrades at a relatively faster rate via acidic or basic chemical hydrolytic degradation of the backbone ester bonds and also by enzymatic action. A schematic representation of PLA hydrolysis is shown in Fig. 7B.

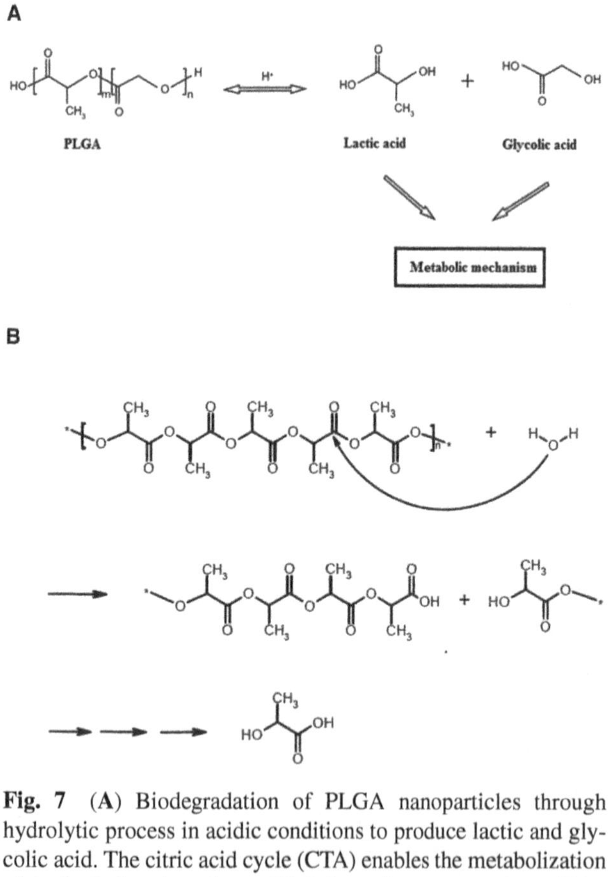

BIOCOMPATIBILITY OF POLYMERS

Applications of polymers in the biomedical field require some important characteristics and especially for their application in theranostics that may also require *in vitro* as well as *in vivo* usage. Biodegradability and biocompatibility with the human the body and various ongoing mechanisms inside the body are key requirements as all the administered polymeric materials are in direct contact with various tissues and other biological compartments. In a similar way, it is important to understand the biodegradability and biocompatibility analysis of the polymeric materials for the design and development of biocompatible vectors. The polymers to be applied and their metabolic products should be free of toxic effects. To bring these above-mentioned properties to the polymers, some modifications must be taken into consideration.[32] PCL degrades slowly and is less biocompatible with soft tissues due to its crystalline and hydrophobic nature. Hence, researchers proposed modifications in PCL and through attachment of poly(ethylene glycol) (PEG) forming PCL–PEG copolymers for improvement in the mechanical, biodegradability, and hydrophilic properties. Other polymers of synthetic nature like polymethyl methacrylate (PMMA), polyglycolic acid (PGA), and polylactic-glycolic acid (PLGA) are widely applied in the biomedical the field due to their various advantages.

Formulations can be evaluated *in vitro* first and then *in vivo* through some general experiments.[33,34] For the investigation of polymer biocompatibility *in vitro*, the tests on specific tissue or cells are done and then the rate of cell death gives the toxicity as well as the biocompatibility values. However, tests like methylthiazol tetrazolium assay (MTT), cell membrane integrity tests, and percentage of DNA synthesis and cell proliferation are commonly applied tests for *in vitro* analysis. For *in vivo* the biocompatibility analysis can be varied depending on the mode and site of administration of the nanoformulations. A general analysis is through hematoxylin-eosin stained paraffin-embedded sections that provide morphology at the tissue level and information about inflammation. To monitor specific biological responses, specific stains and immunohistochemical methods are available, and various microscopic techniques can also be used in studying tissue level changes.

The characteristics of the polymers like molecular weight, structure, surface properties, hydrophilic or hydrophobic nature play an important role in the interaction of polymers and cells or tissues. The effect of surface properties of nanomaterials on biocompatibility and biosafety of presently used materials still must be addressed to evaluate

A

B

Fig. 7 (**A**) Biodegradation of PLGA nanoparticles through hydrolytic process in acidic conditions to produce lactic and glycolic acid. The citric acid cycle (CTA) enables the metabolization of the hydrolysed products. (**B**) Schematic representation of biodegradation of PLA through hydrolysis process.
Source: From Edlund & Albertsson[9] © 2002, with permission from Springer Science and Business Media.

both short- and long-term toxicity effects. Modifications at the tissue and bulk levels are being used for the improvement of compatibility profiles of the materials with surrounding tissues and among different materials. In spite of various benefits obtained through various modifications in the polymer structures or the preparation of copolymers, the important parameters to be noted or addressed are the detailed evaluations of the toxicity of the polymers. For example, polyester exhibits cytotoxic effects that may be induced due to particle size in nanometers, causing internalization by cells (macrophages) and degradation inside cells. Toxicity can also appear due to the accumulation of nonbiodegradable metabolic derivatives in various tissues. Directions for future research in the field of polymeric nanoparticles for biomedical applications require focus studies about degradation rate as well as biosafety and toxicology effects of the polymers and their degradation products. These characteristics can be controlled through modification at the structural level for providing safe polymers to be used *in vitro* and *in vivo*. This objective can be met by devising clever synthetic design for biopolymers with biocompatibility and also new testing methods to evaluate their performances.[7–10]

ENCAPSULATION OF THERAPEUTICS AND DIAGNOSTIC AGENTS

In the preparation of polymeric particles for theranostics applications, an important step is the selection of a diagnostic technique according to the final applications including site and mode of administration. Next is the selection of an encapsulation method for the co-encapsulation of therapeutic and diagnostic agents. For example, diagnosis in cancer involves studying the cancerous cells or tissues through various imaging techniques by employing contrast agents.[1–5] Such techniques may include CT, single-photon emission computed tomography (SPECT), NIRF, PET, US, X-rays, and MRI. The substances responsible for the diagnosis of diseased tissues or cells in the body are the contrast agent used in these above-mentioned techniques. The commonly used contrast agents for diagnosis include iron oxide, gadolinium, and manganese oxide, while others like carbon nanotubes and QDs are used as carriers. Among all these, iron oxide particles are widely explored due to properties like superparamagnetism, biocompatibility, and economic feasibility.[1–11] Various approaches for preparing theranostics containing iron oxide have been applied such as porous hollow iron oxide nanoparticles loaded with cisplatin as an anticancer agent, and particles containing iron oxide nanoparticles and Doxorubicin.[35,36] Another anticancer drug, methotrexate, has been coupled with iron oxide nanoparticles for therapy and diagnostic purposes.[37] Carbon nanotubes are also used in theranostics for diagnostic as well as delivery purposes. Different active ingredients such as drugs and contrast agents are loaded on these carbon nanotubes. For example, gemcitabine can be loaded on carbon nanotubes for cancer therapy where these carbon nanotubes are primarily magnetically functionalized.[38] Another strategy may involve coupling Paclitaxel, an anticancer drug, on carbon nanotubes for therapy of cancerous cells.[39] Similar to carbon nanotubes, silica particles are used for various purposes in theranostics and more specifically as a template for loading active ingredients. Furthermore, silicon nanoparticles can also be used as coating agents on nanoparticles for theranostics applications.[13]

All the above-mentioned agents that are used as templates or as contrast agents in theranostics applications are incorporated into nanoparticles or co-encapsulated with active ingredients through various encapsulation processes.[12–14,40] These methods generally involve either mixing two phases, aqueous and organic, in different ways using preformed polymers or making use of a polymerization technique to synthesize polymers. Interfacial polymerization, condensation polymerization, microemulsion polymerization, and miniemulsion polymerization are a few important methods that make use of monomers forming in situ polymers. On the other hand, nanoprecipitation, multiple emulsion method, emulsion evaporation method, layer by layer method, and supercritical fluid technology have been frequently applied making use of preformed polymers for encapsulating of numerous active ingredients. Various modifications are required to bring in a selected method according to the element or compound to be used for the diagnostic purposes. The main purpose of these modifications is incorporating two active ingredients, i.e., a diagnostic agent and a therapeutic agent, without changing their principal characteristics. There are few facile methods that provide an option for encapsulating two different moieties. As already mentioned, for this purpose, a modified double emulsion (W/O/W) method has been reported to prepare theranostic agents.[1,10] This procedure involves evaporating the solvent and encapsulating a model hydrophilic active ingredient (stilbene derivative) and iron oxide nanoparticles as contrast agent. A relatively new method is also exploring the preparation of multipurpose particles, Janus particles, through polymerization techniques via loading of two different moieties.[41]

THERANOSTICS APPLICATIONS

A combination of diagnostics and therapy results in thenostics to provide "an integrated biomedical system" to diagnose and deliver targeted therapy. Furthermore, in order to fully utilize the advanced requirement in theranostics and then subsequently explore its application, there it is necessary to develop biomaterials of multiple functionalities. In this direction, biodegradable multifunctional polymeric particles offer great potential to develop versatile multifunctional theranostics systems for targeted applications. For example, simultaneous diagnosis and therapy of fatal diseases like cancer can greatly benefit from multimodal tailored approaches offered by theranostics. For this purpose, biodegradable polymer particles contain functions to yield a combination of imaging of tumors for diagnostic purposes as well as therapeutics effects involving chemotherapy and hyperthermia. In view of the above, applications of theranostics in terms of therapeutic involving drug delivery and diagnostics based on imaging are further elaborated below to provide useful insight into these important biomaterials.

Drug Delivery Applications

Drug delivery fully encompasses the concept of theranostics. It involves targeting specific areas that require treatment and diagnosis as well as releasing therapeutic agents in those specific areas.[1–6] A special focus has been on the degradation mechanism caused by externally controlled stimulation such as magnetic, irradiative, ultrasonic, or thermal stimulation as well as environmental or biological stimulation. Anomalous biological phenomena occurring within the body may cause changes in different chemical and physical parameters within the affected area, which can be used for targeting the drug delivery.

There are many mechanisms responsible for the controlled release of a therapeutic agent such as polymer erosion, dispersed drug particles diffusion in the matrix, and therapeutic substance dissolution rate in the surrounding medium.[9] Furthermore, administration routes to deliver therapeutic agents can be critical because the local tissue environment may be influenced by pH or enzyme activity and differs throughout the body. Thus, the performance of a drug delivery device can depend on the site of administration. Earlier systems were based on oral delivery of drug substances. Later on, numerous other procedures were developed including transferring to the rectum or under the tongue. Another system known as parenteral administration deals with delivery other than to the digestive system. One of the key parameters that ultimately determines the theranostics applications of materials is the biodegradability itself. Therefore, in view of its significance, it is vital to know whether the degradation is caused by externally controlled stimulation or biological stimulation.

For example, signature biochemical stimuli have been used to control the distribution of TNPs within the body.[42] An important facet of drug delivery studies is cancer targeting. One study has worked on fabrication of core–shell type nanoparticles with a pH-sensitive functional group layer at the surface that degrades in low pH regions, such as that created due to tumor hypoxia. This exposes the inner surfaces of the theranostics nanoparticle, which can interact with the surface of cells and allow drug injection into these cells.[43] By applying the cellular

phenomenon, a polymeric microbubble can also be used to allow injection of magnetically responsive nanoparticles of iron oxide into the cancer cells.[44] The trigger for this phenomenon is the use of external ultrasound (US), which can cause openings in the cell membrane. The magnetically responsive ferrite particles can consequently be used to trigger cancer cell apoptosis through magnetically stimulated hyperthermia.

In order to exploit the above-mentioned drug delivery systems, numerous naturally active and degradable synthetic polymers have been explored as summarized in Table 5.

Each of these polymers or their copolymers has significant importance to develop next-generation drug delivery systems. Poly(ε-caprolactone) (PCL) is an important biopolymer that exploits hydrolysis for biodegradation. It blends with polymers like polyhydroxybutyrate (PHB), polylactic acid (PLA), and polyglycolic acid (PLGA) and has been studied for preparation of drug delivery microspheres.[9] The drug release rate in such blends has been observed to be controlled by the diffusion rate of the hydrolyzing agents into the microspheres.[45] Thus information from Table 5 would serve to help in tuning the diffusion and hydrolysis processes.[9] For example, blends of PCL/PGLA are useful to synthesize microspheres to treat central nervous system disorders through nerve growth factors. For long-term drug delivery, PCL-based systems are more suited due to slow degradation relative to that of fast degradation polymers based on PLA and PLGA. PCL microspheres containing an antibacterial agent, Nitrofurantoin, can be employed to treat urinary tract infections by controlling release that has been found to be proportional to the square-root of time.[46] PCL microspheres can also be

Table 5 Few common natural and synthetic biodegradable polymers for drug delivery systems

Natural	Synthetic
Polysaccharides	Aliphatic polyesters
Chitin	Poly(glycolide), PGA
Chitosan	Poly(lactide), PLA
Proteins	Poly(glycolide-co-lactide), PLGA
	Polyanhydrides
	Aliphatic polycarbonates
	Poly(orthoesters)
	Poly(ε-caprolactone), PCL
	Poly(3-hydroxybutyrate), PHB
	Poly(3-hydroxybutyrate-co-3-hydroxyvalerate), P(HB-co-HV)
	Poly(amino acids)
	Poly(ethylene oxide)
	Polyphosphazenes

Source: Data from Edlund & Albertsson.[9]

developed to encapsulate and release hydrophobic Nifedipine (anti-calcium) and hydrophilic propranolol (a β-blocker) drugs.[9] PCL is also explored for its function as a barrier to protect the untimely release of labile drugs in the gastrointestinal tract.[47] Microspheres of PCL and copolymers are used for releasing the various types of hormones for prolonging *in vitro* and *in vivo* release.[48]

Micellar TNPs with functionalized contrast agents for image-guided diagnosis and light-responsive drug release mechanisms have also been explored.[49] These TNPs incorporate a light-responsive moiety in their core, which switches from hydrophobic to hydrophilic upon ultraviolet irradiation. The induced hydrophilic nature leads to the hydrolysis of micelles, causing the release of the encapsulated drug, the rate of which can be controlled by controlling the external stimulation. The mechanism for this drug delivery technique is shown in Fig. 8 in the case of Doxorubicin.

Imaging Applications

Various imaging techniques have been employed to observe and diagnose anomalous bodily functions. Important techniques are OI, US, computed tomography (CT), positron emission tomography (PET), single photon CT, and MRI. Figure 9 schematically represents the role of various imaging techniques.

All these imaging techniques are important. Among these techniques, MRI is the most advantageous modality, owing to its high resolution spatial and chronological results, harmlessness for the body, and high blood and tissue sensitivity.[50–53] MRI produces images based on magnetic resonance

of substances within the region tested. To create a strong contrast between features within the body, biopolymer-based particles are injected into the body as probes for MRI. Biopolymer-based MRI probes are available in the market and widely used in two different modalities: paramagnetic gadolinium-based and superparamagnetic ferrite-based. The former type has been researched for stable as well as biodegradable probes. Galodinium-DTPA 1-cystine bisamide copolymers (GCAC) have been tested upon rat models to diagnose cancerous cell growth and observe minimal tissue retention, confirming biodegradability of the probes.[50]

Vesicular ferrite-based probes have also been designed and studied for cell tracking. These novel hybrid vesicular self-assemblies can be based on poly(trimethylene carbonate)-*b*-poly(L-glutamic acid) (PTMC-*b*-PGA) block copolymer vesicles and hydrophobically coated γ-Fe$_2$O$_3$ nanoparticles of magnetic polymersomes.[53] These copolymers are useful theranostic agents because of their capability to load doxorubicin and –Fe$_2$O$_3$ nanoparticles for chemotherapy and MRI, respectively.[3] A nanoprecipitation procedure can be adapted to simultaneously load doxorubicin drug and iron oxide nanoparticles. Figure 10 represents the combined imaging and therapy capabilities for these theranostic systems.[3] The efficiency of vesicles containing ultrasmall superparamagnetic iron oxide (USPIO)–loaded particles can be determined by the measurements of T2 (transverse) and T1 (longitudinal) relaxation times of the proton spins relaxations. Steeper relaxivity slopes have been noted for vesicles with higher Fe loading. Furthermore, T2-weighted MR images for the USPIO-loaded vesicles revealed remarkable negative contrast enhancement

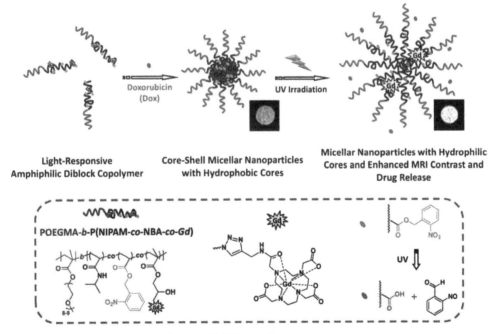

Fig. 8 Core–shell light-responsive micelles based on amphiphilic copolymers.
Source: From Li et al.[49] © 2012, with permission from ACS Publications.

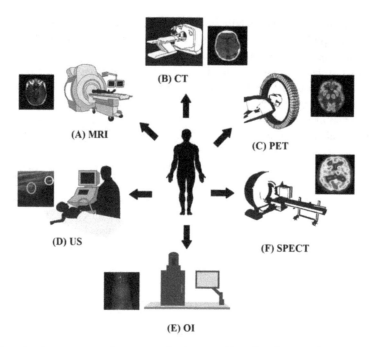

Fig. 9   Molecular sensitive imaging instruments and representative images. (**A**) MRI; (**B**) CT; (**C**) PET; (**D**) US; (**E**) OI; (**F**) SPECT.
Source: From Janib et al.[2] © 2012, with permission from Elsevier.

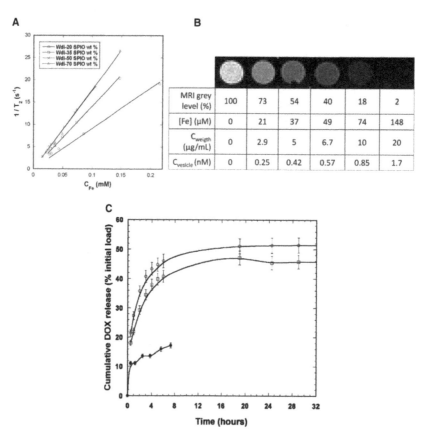

Fig. 10   (**A**) Transverse relaxation rates ($1/T2$, s^{-1}) as a function of iron concentration (mM) for PTMC24-b-PGA19 vesicles loaded with USPIO wt%. of 20, 35, 50, and 70. Relaxivity (r2) values are determined from the slopes and found to increase with USPIO wt%. (**B**) T2-weighted MRI images obtained from T2 measurements (vesicles with Rh = 52 nm PDI = 0.8; USPIO loading = 70 wt%) at various dilution factors. The table provided the molar concentrations of iron ions, the total weight concentrations (Cweight), and the equivalent molar concentrations of vesicles (Cvesicle); (**C**) Kinetic profiles of *in vitro* release of USPIO/DOX for loading of 20 wt% DOX in vesicles with Rh = 61 nm) (O) and Rh = 124 nm) (□) at 37°C and for vesicles Rh = 61 nm) at lower temperature of 23°C (●).
Source: From Sanson[3] © 2011, with permission from ACS Publications.

appeared even at lower vesicle concentrations. Such magnetic polymersomes can be controlled via a magnetic field gradient. Another fascinating potential of such magnetic systems is the controlled delivery of drugs like encapsulated doxorubicin molecules due to radio frequency or magnetic field controlled hyperthermia effects. Thus such multifunctional nanocarriers provide interesting platforms for theranostic systems to enhance tumor regression.

CONCLUSIONS AND FUTURE PERSPECTIVES

Biodegradable polymer particles for theranostic applications with capabilities to perform simultaneous diagnostic and therapeutic functions are critical for present and future health care. Because of their multifunctionality, size, and biodegradability, these particles are very effective for the facile and in-time diagnosis of diseases as well as their treatment. In many cases including the case of cancer, the problem of late diagnosis of the disease is a major concern as this causes it to become incurable. It is expected that these theranostics polymer particles would be of great help in reducing the time-lapse between diagnosis and therapy. Applications of these particles in theranostics are numerous due to their tunable biocompatibility, biodegradability, responsiveness, encapsulation, and surface functionalization. Furthermore, numerous biodegradable polymers are avaliable, and, if required, co-polymers also can be prepared. The use of polymers with control over their biodegradation and responsiveness to stimuli of temperature, pH, light, salt, magnetism, and biologicals has lead to the development of novel types of particles for their applications *in vitro* or *in vivo*. There are already considerable research efforts going on to explore the use of biodegradable theranostic polymer particles in imaging and drug delivery.

In the future, biodegradable polymer particles for theranostic applications will be applied for oncology and other ailments. The potential advantages like noninvasive quantifications and personalized pharmacotherapies of these materials make them an interesting opportunity for the better management of patients and disease. These particles have started to enter in clinical trials and this trend will increase more in the future. For this purpose, the role of nanotechnology will continue to be the most important and open new horizons in the biomedical domain although the toxicity and complexity issues must be addressed earlier.

REFERENCES

1. Ahmed, N.; Fessi, H.; Elaissari, A. Theranostic applications of nanoparticles in cancer. Drug Discov. Today **2012**, *17* (17–18), 928–934.

2. Janib, S.M.; Moses, A.S.; MacKay, J.A. Imaging and drug delivery using theranostic nanoparticles. Adv. Drug Deliv. Rev. **2010**, *62* (11), 1052–1063.

3. Sanson, C. Doxorubicin loaded magnetic polymersomes: Theranostic nanocarriers for MR imaging and magneto–chemotherapy. ACS Nano **2011**, *5* (2), 1122–1140.

4. Kelkar, S.S.; Reineke, T.M. Theranostics: Combining imaging and therapy. Bioconjug. Chem. **2011**, *22* (10), 1879–1903.

5. Xie, J.; Lee, S.; Chen, X. Nanoparticle–based theranostic agents. Adv. Drug Deliv. Rev. **2010**, *62* (11), 1064–1079.

6. Rai, P.; Mallidi, S.; Zheng, X.; Rahmanzadeh, R.; Mir, Y.; Elrington, S.; Khurshid, A.; Hasan, T. Development and applications of photo-triggered theranostic agents. Adv. Drug Deliv. Rev. **2010**, *62* (11), 1094–1124.

7. Anderson, J.M. Biodegradation and biocompatibility of PLA and PLGA microspheres. Adv. Drug Deliv. Rev. **1997**, *28* (1), 5–24.

8. Tian, H.; Tang, Z.; Zhuang, X.; Chen, X.; Jing, X. Biodegradable synthetic polymers: Preparation, functionalization and biomedical application. Progr. Polym. Sci. **2012**, *37* (2), 237–280.

9. Edlund, U.; Albertsson, A.-C. Degradable polymer microspheres for controlled drug delivery. Adv. Polym. Sci. **2002**, *157* (1), 67–1112.

10. Ahmed, N.; Michelin-Jamois, M.; Fessi H.; Elaissari, A. Modified double emulsion process as a new route to prepare submicron biodegradable magnetic/polycaprolactone particles for *in vivo* theranostics. Soft Matter **2012**, *8* (8), 2554–2564.

11. Rosset, V.; Ahmed, N.; Zaanoun, I.; Stella, B.; Fessi, H.; Elaissari. A. Elaboration of Argan oil nanocapsules containing Naproxen Sodium for cosmetic and transdermal local application. J. Colloid Sci. Biotechnol. **2012**, *1* (2), 218–224.

12. Arshad, A.; Ahsan, A.; Aziz, A.; Ali, O.; Ahmad, N.M.; Elaissari, A. Smart magnetically engineering colloids and biothin films for diagnostics applications. J. Colloids Biotechnol. **2013**, *2* (1), 19–26.

13. Bitar, A.; Ahmad, N.M.; Fessi, H.; Elaissari, A. Silica-based nanoparticles for biomedical applications. Drug Discov. Today **2012**, *17* (19), 1147–1154.

14. Rahman, M.M.; Elaissari, A. Multi-stimuli responsive magnetic core–shell particles: Synthesis, characterization and specific RNA recognition. J. Colloids Biotechnol. **2012**, *1* (1), 3–15.

15. Freiberg, S.; Zhu, X.X. Polymer microspheres for controlled drug release. International Journal of Pharmaceutics **2004**, *282*, 1–18.

16. Piirma, I. Colloids. In *Encyclopedia of Polymer Science and Engineering*, 2nd Ed.; Mark, H.F., Bikales, N.M., Overberger, C.G., Menges, G., Kroschwitz, J.I. Eds.; John Wiley and Sons: New York, 1985; pp. 125–130.

17. Benedetti, L.; Bertucco, A.; Pallado, P. Production of microparticles of a biocompatible polymer using supercritical carbon dioxide. Biotechnol. Bioeng. **1997**, *53* (2), 232–237.

18. Elaissari, A. Thermally sensitive latex particles: Preparation, characterization and application in the biomedical field. In *Handbook of Surface and Colloid Chemistry*, 3rd Ed.; Birdi, K.S.; CRC Press, Taylor & Francis Group: 2008; pp. 539–566.

19. Bawa, P.; Pillay, V.; Choonara, Y.E.; du Toit, L.C. Stimuli-responsive polymers and their applications in drug delivery. Biomed. Mater. **2009**, *4* (2), 022001.

Stem Cell—Ultrasound

20. Galaev, I.Y.; Mattiasson, B. Smart polymers and what they could do in biotechnology and medicine. Trends Biotechnol. **1999**, *17* (8), 335–340.

21. Hoffman, A.S. Stimuli-responsive polymers: Biomedical applications and challenges for clinical translation. Adv. Drug Deliv. Rev. **2013**, *65* (1), 10–16.

22. Ren, T.; Liu, Q.; Lu, H.; Liu, H.; Zhang, X.; Du, J. Multifunctional polymer vesicles for ultrasensitive magnetic resonance imaging and drug delivery. J. Mater. Chem. **2012**, *22* (24), 12329–12338.

23. Torchilin, V.P. Micellar nanocarriers: Pharmaceutical perspectives. Pharm. Res. **2007**, *24* (1), 1–16.

24. Santra, S.; Kaittanis, C.; Grimm, J.; Perez, J.M. Drug/dye-loaded, multifunctional iron oxide nanoparticles for combined targeted cancer therapy and dual optical/magnetic resonance imaging. Small **2009**, *5* (16), 1862–1868.

25. Lu, M.; Cohen, M.H.; Rieves, D.; Pazdur, R. FDA report: Ferumoxytol for intravenous iron therapy in adult patients with chronic kidney disease. Am. J. Hematol. **2010**, *85* (5), 315–319.

26. Kiessling, F.; Fokong, S.; Koczera, P.; Lederle, W.; Lammers, T. Ultrasound microbubbles for molecular diagnosis, therapy, and theranostics. J. Nucl. Med. **2012**, *53* (3), 345–348.

27. Lenz, R.W. Biodegradable polymers. Adv. Polym. Sci. **1993**, *107*, 1–40.

28. Amass, W.; Amass, A.; Tighe, B. A review of biodegradable polymers: Uses, current developments in the synthesis and characterization of biodegradable polyesters, blends of biodegradable polymers and recent advances in biodegradation studies. Polym. Int. **1998**, *47* (2), 89–144.

29. Matlaga, B.F.; Yasenchak, L.P.; Salthouse T.N. Tissue response to implanted polymers: The significance of sample shape. J. Biomed. Mater. Res. **1976**, *10* (3), 391–397.

30. Göpferich, A. Mechanisms of polymer degradation and erosion. Biomaterials **1996**, *17*, 103–114

31. Kricheldorf, H.R.; Kreiser-Saunders, I. Polylactides—Synthesis, characterization and medical applications. Macromol. Symp. **1996**, *103* (1), 85–102.

32. Jean-François, L.; Börner, H.G.; Weichenhan, K. Combining ATRP and "Click" chemistry: A promising platform toward functional biocompatible polymers and polymer bioconjugates. Macromolecules **2006**, *39* (19), 6376–6383.

33. Hill, A.; Geißler, S.; Meyring, M.; Hecht, S.; Weigandt, M.; Mäder, K. *In vitro-in vivo* evaluation of nanosuspension release from subcutaneously implantable osmotic pumps. Int. J. Pharm. **2013**, *451* (1–2), 57–66.

34. Park, J.S.; Shim, J.Y.; Park, J.S.; Lee, M.J.; Kang, J.M.; Lee, S.H.; Kwon, M.C.; Choi, Y.W.; Jeong, S.H. Formulation variation and *in vitro–in vivo* correlation for a rapidly swellable three-layered tablet of tamsulosin HCl. Chem. Pharm. Bull. **2011**, *59* (5), 529–535.

35. Cheng, K.; Peng, S.; Xu, C.; Sun, S. Porous hollow Fe_3O_4 nanoparticles for targeted delivery and controlled release of cisplatin. J. Am. Chem. Soc. **2009**, *131* (30), 10637–10644.

36. Kievit, F.M.; Wang, F.Y.; Fang, C.; Mok, H.; Wang, K.; Silber, J.R.; Ellenbogen, R.G.; Zhang, M. Doxorubicin loaded iron oxide nanoparticles overcome multidrug resistance in cancer *in vitro*. J. Control. Release **2011**, *152* (1), 76–83.

37. Kohler, N.; Sun, C.; Wang, J.; Zhang, M. Methotrexate-modified superparamagnetic nanoparticles and their intracellular uptake into human cancer cells. Langmuir **2005**, *21* (19), 8858–8864.

38. Yang, F.; Jin, C.; Yang, D.; Jiang, Y.; Li, J.; Di, Y.; Hu, J.; Wang, C.; Ni, Q.; Fu, D. Magnetic functionalised carbon nanotubes as drug vehicles for cancer lymph node metastasis treatment. Eur. J. Cancer **2011**, *47* (12), 1873–1882.

39. Sobhani, Z.; Dinarvand, R.; Atyabi, F.; Ghahremani, M.; Adeli, M. Increased paclitaxel cytotoxicity against cancer cell lines using a novel functionalized carbon nanotube. Int. J. Nanomed. **2011**, *6* (1), 705–719.

40. Medeiros, S.F.; Santos, A.M.; Fessi, H.; Elaissari, A. Thermally-sensitive and magnetic poly(N-Vinylcaprolactam)-based nanogels by inverse miniemulsion polymerization. J. Colloids Biotechnol. **2012**, *1*, 99–112.

41. Berger, S.; Synytska, A.; Ionov, L.; Eichhorn, K.-J.; Stamm, M. Stimuli-responsive bicomponent polymer Janus particles by "grafting from"/"grafting to" approaches. Macromolecules **2008**, *41* (24), 9669–9676.

42. Poon, Z.; Lee, J.B.; Morton, S.W.; Hammond, P.T. Layer–by–layer nanoparticles with a pH-sheddable layer for *in vivo* targeting of tumor hypoxia. Nano Lett. **2011**, *5* (6), 2096–2103.

43. Poon, Z.; Chang, D.; Zhao, X.; Hammond, P.T. Layer–by–layer nanoparticles with a pH-sheddable layer for *in vivo* targeting of tumor hypoxia. ACS Nano **2011**, *5* (6), 4284–4292.

44. Cai, X.; Yang, F.; Gu, N. Applications of magnetic microbubbles for theranostics. Theranostics **2012**, *2* (1), 103–112.

45. Pitt, C.G. Poly-ε-caprolactone and its copolymers. In *Biodegradable Polymers as Drug Delivery Systems*; Chasin, M., Langer, R., Eds.; Marcel Dekker: New York, 1990; 71.

46. Dubernet, C.; Benoit, J.P.; Couarraze, G.; Duchêne, D. Microencapsulation of nitrofurantoin in poly(∈-caprolactone): Tableting and *in vitro* release studies. Int. J. Pharm. **1987**, *35* (1), 145–156.

47. Allémann, E.; Leroux, J.-C.; Gurny, R. Polymeric nano- and microparticles for the oral delivery of peptides and peptidomimetics. Adv. Drug Deliv. Rev. **1998**, *34* (2), 171–189.

48. Allen, C.; Han, J.; Yu, Y.; Maysinger, D.; Eisenberg, A. Poly-caprolactone-b-poly(ethylene oxide) copolymer micelles as a delivery vehicle for dihydrotestosterone. J. Control. Release **2000**, *63* (3), 275–286.

49. Li, Y.; Qian, Y.; Liu, T.; Zhang, G.; Liu, S. Light-triggered concomitant enhancement of magnetic resonance imaging contrast performance and drug release rate of functionalized amphiphilic diblock copolymer micelles. Biomacromolecules **2012**, *13* (11), 3877–3886.

50. Kaneshiro, T.L.; Ke, T.; Jeong, E.K.; Parker, D.L.; Lu, Z.R. Gd-DTPA l-cystine bisamide copolymers as novel biodegradable macromolecular contrast agents for MR blood pool imaging. Pharm. Res. **2006**, *23* (6), 1285–1294.

51. Muhammed, A.H.; Saqib, M.; Ahmad, N.M.; Elaissari, A. Magnetically engineering smart thin films: Towards ultrasensitive molecular imaging for lab-on-chip. J. Biomed. Nanotechnol. **2013**, *9* (3), 467–474.

52. Nasongkla, N.; Bey, E.; Ren, J.M.; Ai, H.; Khemtong, C.; Guthi, J.S.; Chin, S.F.; Sherry, A.D.; Boothman, D.A.; Gao, J.M. Multifunctional polymeric micelles as cancer-targeted, MRI-ultrasensitive drug delivery systems. Nano Lett. **2006**, *6* (11), 2427–2430.

53. Toyota, T.; Ohguri, N.; Maruyama, K.; Fujinami M.; Saga, T.; Aoki, I. Giant vesicles containing superparamagnetic iron oxide as biodegradable cell-tracking MRI probes. Anal. Chem. **2012**, *84* (9), 3952–3957.

Stem Cell—Ultrasound

Ultrasound Contrast Agents: Sonochemical Preparation

Garima Ameta
Kiran Meghwal
Arpita Pandey
Department of Chemistry, University College of Science, Mohanlal Sukhadia

Narendra Pal Singh Chauhan
Department of Chemistry, B.N.P.G. College, Udaipur, India

Pinki B. Punjabi
Department of Chemistry, University College of Science, Mohanlal Sukhadia University, Udaipur, India

Abstract

Medical ultrasound is a non-invasive, portable, extremely safe, inexpensive, highly valuable diagnostic tool especially when compared with X-ray and magnetic resonance imaging (MRI). But ultrasound has some limitation in distinguishing between diseased and healthy tissue. This limitation has led to the development of contrast agents. The primary goal of these contrast agents is to enhance the diagnostic image, so they can be used for targeted drug delivery. Polymeric microcapsules have many advantages, such as they are stable, provide a good surface to adsorb or carry the drug, and have good shelf life stability. Protein microspheres (PMs) have a wide range of biomedical applications, including their use as echo contrast agents for sonography, MRI, contrast enhancement, and oxygen or drug delivery. In this entry, we will discuss about ultrasound-assisted synthesis and coating of PM and other polymeric ultrasound contrast agents (UCAs). Proteins are dissolved in a liquid, which is then irradiated with intense ultrasound to induce acoustic cavitation. When the bubbles are heated during the rapid collapse, water vapor in the bubble is dissociated into OH radicals. These radicals cause the protein molecules to cross-link and a solid, spherical, protein shell is formed where the bubble once existed. The shells can be manufactured by filling with liquid or gas—liquid-filled spheres can be used for targeted or time released drug delivery and air-filled spheres are used as echo contrast agents in medical ultrasound.

INTRODUCTION

Ultrasound is an oscillating sound pressure wave with a frequency greater than the upper limit of the human hearing range. Ultrasound devices operate with frequencies from 20 KHz up to several gigahertz. Ultrasound is used in many different fields. Ultrasonic devices are used to detect objects and to measure distances. Ultrasonic imaging (sonography) is used in both veterinary medicine and human medicine. Industrially, ultrasound is used for cleaning and mixing and to accelerate chemical processes. Organisms such as bats and porpoises use ultrasound for locating prey and obstacles.[1] Most applications of ultrasound make use of the physical effects of the sound as it is reflected, absorbed, or transmitted through a particular medium. However, ultrasound can also generate chemical effects under certain conditions and this has led to the development of the field of sonochemistry. Contrast agents are utilized in virtually every imaging modality to enhance diagnostic capabilities.

CONTRAST-ENHANCED ULTRASOUND (CEUS)

CEUS is the application of ultrasound contrast medium to traditional medical sonography. Ultrasound contrast agents (UCAs) rely on different ways in which sound waves are reflected from interfaces between substances. This may be the surface of a small air bubble or a more complex structure. Commercially available contrast media are gas-filled microbubbles that are administered intravenously to the systemic circulation as shown in Fig. 1. Microbubbles have a high degree of echogenicity, which is the ability of an object to reflect the ultrasound waves. The echogenicity difference between the gas in the microbubbles and the soft tissue surroundings of the body is immense. Thus, ultrasonic imaging using microbubble contrast agents enhances the ultrasound backscatter, or reflection of the ultrasound waves, to produce a unique sonogram with increased contrast due to the high echogenicity difference. CEUS can be used to image blood perfusion in organs, measure blood flow rate in the heart and other organs, and has other

Stem Cell—Ultrasound

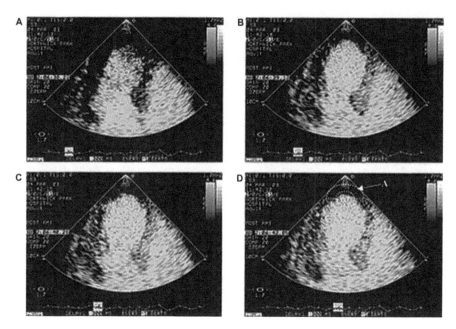

Fig. 1 Contrast-enhanced ultrasound.

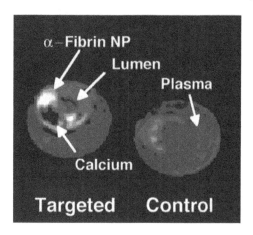

Fig. 2 Color-enhanced magnetic resonance imaging (MRI) of fibrin-targeted and control carotid endarterectomy specimens revealing contrast enhancement (white) of a small fibrin deposit on a symptomatic ruptured plaque. Calcium deposit (black). 3-D fat-suppressed T_1-weighted fast gradient echo.

applications as well. There are two forms of CEUS, namely, untargeted (used in the clinic today) and targeted (under preclinical development), as shown in Fig. 2. The two methods slightly differ from each other.

Untargeted CEUS

Untargeted microbubbles are injected intravenously into the systemic circulation in a small bolus. The microbubbles will remain in the systemic circulation for a certain period of time. During that time, ultrasound waves are directed on the area of interest. When microbubbles in the blood flow passes the imaging window, the microbubbles' compressible gas cores oscillate in response to the high-frequency sonic energy field.

The microbubbles reflect a unique echo that stands in stark contrast to the surrounding tissue due to the orders of magnitude mismatch between microbubble and tissue echogenicity. The ultrasound system converts the strong echogenicity into a contrast-enhanced image of the area of interest. In this way, the bloodstream's echo is enhanced, thus allowing the clinician to distinguish blood from surrounding tissues.

Targeted CEUS

Targeted CEUS works in a similar fashion, with a few alterations. Microbubbles targeted with ligands that bind certain molecular markers that are expressed by the area of imaging interest are still injected systemically in a small bolus. Microbubbles theoretically travel through the circulatory system, eventually finding their respective targets and binding specifically. Ultrasound waves can then be directed on the area of interest. If a sufficient number of microbubbles have bound in the area, their compressible gas cores oscillate in response to the high-frequency sonic energy field. The targeted microbubbles also reflect a unique echo that stands in stark contrast to the surrounding tissue due to the orders of magnitude mismatch between microbubble and tissue echogenicity. The ultrasound system converts the strong echogenicity into a contrast-enhanced image of the area of interest, revealing the location of the bound microbubbles.[2] Detection of bound microbubbles may then show that the area of interest is expressing that particular molecular marker, which can be an indicative of a certain disease state, or identify particular cells in the area of interest.

Targeting ligands that bind to receptors characteristic of intravascular diseases can be conjugated to microbubbles, enabling the microbubble complex to accumulate selectively in areas of interest, such as diseased or abnormal

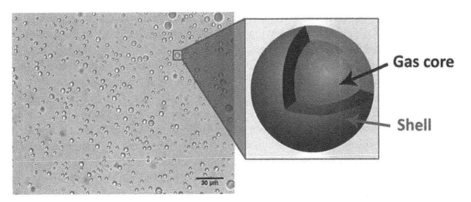

Fig. 3 Microscopic picture of microbubbles from bovine serum albumin (BSA).

tissues. This form of molecular imaging, known as targeted CEUS, will only generate a strong ultrasound signal if targeted microbubbles bind in the area of interest. Targeted CEUS can potentially have many applications in both medical diagnostics and medical therapeutics. However, the targeted technique has not yet been approved for clinical use. There are a variety of microbubbles contrast agents. Microbubbles differ in their shell makeup, gas core makeup, and whether or not they are targeted.

Microbubble shell

Selection of shell material determines how easily the microbubble is taken up by the immune system. A more hydrophilic material tends to be taken up more easily, which reduces the microbubble residence time in the circulation. This reduces the time available for contrast imaging. The shell material also affects microbubble mechanical elasticity. The more elastic the material, the more acoustic energy it can withstand before bursting.[3] Microbubble shells are composed of albumin, galactose, lipid, or polymers[4] as shown in Fig. 3.

Microbubble gas core

The gas core is the most important part of the ultrasound contrast microbubble because it determines the echogenicity. When gas bubbles are caught in an ultrasonic frequency field, they compress, oscillate, and reflect a characteristic echo, which generates the strong and unique sonogram in CEUS. Gas cores can be composed of air, or heavy gases like perfluorocarbon (PFC), or nitrogen as shown in Fig. 4. Heavy gases are less water soluble so they are less likely to leak out from the microbubble leading to microbubble dissolution. As a result, microbubbles with heavy gas cores last longer in circulation. Regardless of the shell or gas core composition, the size of microbubble is fairly uniform. They lie within a range of 1–4 μm in diameter that makes them smaller than red blood cells, which allows them to flow easily through the circulation as well as the microcirculation.

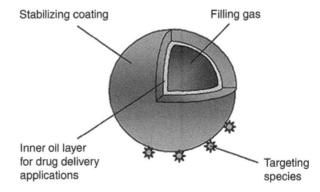

Fig. 4 Schematic representation of a multilayered microbubble.

Another interesting field of research concerning the use of ultrasound in polymeric material synthesis is the preparation of biopolymers, most notably, the synthesis of proteinaceous microspheres. These air- and oil-filled microspheres have a range of biomedical applications such as the use in targeted drug delivery, as echo contrast agent for sonography, as MRI, and as microencapsulation of pharmaceuticals, neutraceuticals, and flavors. The first report on air-filled proteinaceous microspheres synthesized sonochemically was reported by Feinstein and coworkers with their human serum albumin (HSA) air-filled microspheres used as contrast agents in echo-sonography.

SONOCHEMICAL SYNTHESIS OF BIOMATERIALS

Irradiation of liquids with high-intensity ultrasound generates high-energy chemistry through acoustic cavitation—the formation, growth, and collapse of bubbles within the liquid.[5] This collapse generates extreme temperatures and pressures within the bubble sufficient to break all chemical bonds. The primary site of reaction is the gas phase of the collapsing bubble, thus macromolecules such as proteins are effected by sonochemistry in secondary reactions involving the primary sonolysis products of the volatiles present in the solution. In aqueous media, the predominant sonochemistry is that of water itself, which produces hydrogen atoms (H), hydroxyl radicals ($^{\cdot}$OH, H_2, H_2O_2), and in the presence of O_2, the hydroperoxyl radical ($HO_2^{\cdot}$, i.e., protonated superoxide). A new chemical application of ultrasound is discovered, i.e., the sonochemical synthesis of protein microsphere (PM) with liquid cores.[6] These microspheres are small enough to pass unimpeded through the circulatory system. Air-filled microspheres made from HSA have been commercialized by Molecular Biosystems, Inc., under the trade name Albunexm.[7] Applications of these microsphere to *in vivo* MRI, drug delivery, and O_2 transport have been rapidly developed.[8] The sonochemical synthesis of PM yields diameters of ≈ 2.5 μm at 20 kHz. The microsphere shells can be made from a variety of proteins, including serum albumin, hemoglobin (Hb), horseradish peroxidase, lipase, and pepsin. The mechanism of microsphere formation involves both emulsification and chemical cross-linking of cysteine residues between protein molecules from superoxide ($HO_2^{\cdot}$), which is produced from the sonolysis of water in the presence of O_2 during acoustic cavitation.

While the chemical effects of ultrasound on aqueous solutions have been studied for many years, the development of aqueous sonochemistry for biomaterials synthesis is very recent (2005–2014), particularly in the area of microencapsulation. Using high-intensity ultrasound and simple protein solutions, a remarkably easy method to make both air-filled microbubbles and nonaqueous liquid-filled microcapsules has been developed. Fig. 5 shows an electron micrograph of sonochemically prepared microspheres.

These microspheres are stable for months and, being slightly smaller than erythrocytes, can be intravenously injected to pass unimpeded through the circulatory system. The mechanism responsible for microsphere formation is a

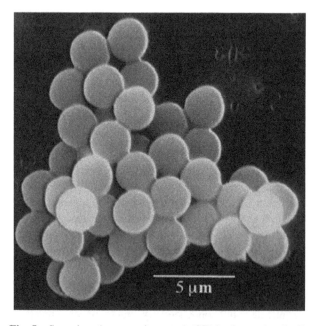

Fig. 5 Scanning electron micrograph (SEM) of sonochemically synthesized Hb microspheres.

combination of *two* acoustic phenomena, namely, emulsification and cavitation. Ultrasonic emulsification creates the microscopic dispersion of the protein solution necessary to form the proteinaceous microspheres. Emulsification alone, however, is insufficient to produce long-lived microspheres. The long life of these microspheres comes from a sonochemical cross-linking of the protein shell. Chemical reactions requiring O_2 are critical in forming the microspheres. Chemical trapping experiments are used to establish that the roteinaceous microspheres are held together by disulfide bonds between protein cysteine residues and that superoxide is the cross-linking agent. The cross-linked shell of the microspheres is only about 10 protein molecules in thickness. These PMs have a wide range of biomedical applications, including their use as echo contrast agents for sonography,[9] MRI,[10] and oxygen or drug delivery,[11] among others.

THERAPEUTIC USE OF UCAS

Ultrasound is best known as a diagnostic imaging tool in clinical medicine. However, the use of ultrasound for therapy actually predated the development of ultrasound for diagnosis. In the 1930s, ultrasound was first introduced in Europe as a modality to generate heat for deep tissue warming (thermal bioeffect).[12] The other important ultrasound nonthermal bioeffect is termed as "cavitation." This is the biophysical interaction between the ultrasonic field in a liquid and a gaseous inclusion (or bubble). There are two types of cavitation: (i) gas body cavitation [or stable cavitation (SC)] and (ii) inertial cavitation (IC or transient cavitation).[13]

In gas body cavitation, the bubble undergoes periodic and regular changes in volume in response to the applied acoustic pressure. This results in mechanical vibration within the tissue, creating eddies of current flow around the oscillating bubble. In contrast, in the case of IC, although the bubble also undergoes periodic changes, these changes are qualitatively different in volume—rapidly increasing in size, becoming unstable, and then imploding violently. This can then result in acoustic microstreaming, hydrodynamic flow around bubbles, and bubble–cell collisions as shown in Fig. 6.

ULTRASOUND AND UCAS

In 1990, Holland and Apfel[14] showed that UCAs can significantly lower the acoustic cavitation production threshold. This work suggested that UCAs can be used to induce cavitation for therapeutic applications with much lower ultrasound energy. The study by Tachibana and Tachibana reported the use of UCAs for therapeutic application.[15] In this study, they demonstrated that albumin microbubbles (Albunex) accelerated urokinase-mediated thrombolysis, shortening lysis time to one-fifth. It is hypothesized that the contrast agent adheres to the clot with resultant shearing effect during bubble destruction. This mechanically erodes

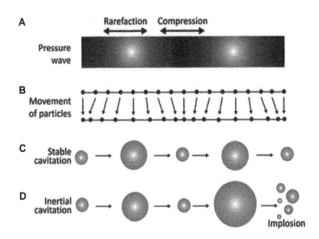

Fig. 6 Schematic representation of possible drug cargo spaces in microbubbles. (A) Drugs can be dissolved in an oil layer between perfluorocarbon core and microbubble shell (B) or incorporated in the microbubble shell (C) Shows stable (gas body) cavitation, where regular, periodic changes in volume result in eddies of current flow around the oscillating bubble. (D) Shows inertial (transient) cavitation, where the bubbles may increase rapidly in size, resulting in acoustic microstreaming, hydrodynamic flow around bubbles, and bubble–cell collisions..

the clot allowing more fibrin to be exposed to the lytic agent. It has been demonstrated that microbubbles can be intentionally ruptured by ultrasound [referred to as ultrasound-targeted microbubble destruction (UTMD)],[16,17] which may then be exploited to enhance drug and gene delivery into cells as shown in Fig. 7. UTMD not only leads to directed release of the therapeutic at the desired site but also produces microjets or microstreaming during bubble collapse that could promote diffusion of the therapeutic into the cell.[18] The resultant shear stress on the cellular surface then results in increased membrane permeability by formation of transient nonlethal perforations on the cell membrane (sonoporation)[19] or an increase in membrane fluidity. Ingress of therapeutics may also result from endocytosis of the microbubble, fusion of the microbubble membrane with the cell membrane, or a combination of these mechanisms. [20] Targeting to position the microbubble close to the diseased tissue is critical, as particle destruction and drug/gene release occurring in the center of the vessel wall, for instance, will only result in the therapeutic being released to the bloodstream with ensuing reduced efficacy. In this regard, the use of UCAs affords another advantage in which delivery to the diseased site can be confirmed by highlighting the pathological tissues before the application of ultrasound energy for triggered drug release.

UCAS FOR DELIVERY OF CARDIOVASCULAR THERAPEUTICS

There are three methods to use UCAs for therapeutic delivery. The two more common applications of this technique are disussed subsequently.

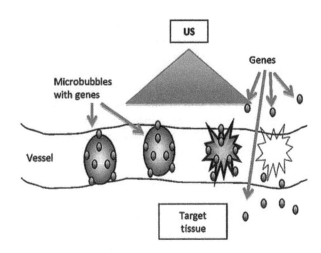

Fig. 7 The mechanism of ultrasound-targeted microbubble destruction for gene delivery.

Coadministration of Microbubbles and Therapeutics

Gene therapy

In 1997, Bao et al.[21] described the use of ultrasound and albumin-coated microbubbles to enhance transfection of luciferase reporter plasmid into cultured hamster cells. In 2000, Shohet et al.[22] demonstrated that ultrasound can be used to direct transgene expression in the rat myocardium using albumin microbubbles. In 2003, Erikson et al.[23] reported the delivery of antisense oligonucleotides and PFC-exposed sonicated dextrose albumin to rat hearts and demonstrated blunted ischemia/reperfusion-induced tumor necrosis factor-α expression. In 2007, Suzuki et al.[24] reported the development of a novel polyethyleneglycol-modified liposome encapsulating nanobubbles of perfluoropropane, which can be used as a vector for gene delivery into mouse femoral arteries. Interestingly, gene uptake using their "bubble liposome" was limited to the area exposed to ultrasound, with implications for ultrasound-directed site-specific gene delivery. These studies, and many others, laid the foundation for *in vivo* work in larger animal models. In 2002, Vannan et al. used intravenous cationic lipid microbubbles containing perfluorobutane gas to deliver plasmid chloramphenicol acetyltransferase, a marker gene, to the myocardium in closed-chest mongrel dogs.[25] Most *in vivo* experiments employed artificial delivery methods, such as isolation of arterial segments or cessation of blood flow,[26] which are not adaptable to clinical studies or chronic disease conditions; as of now, only few studies have been carried out in large animal models.

Drug therapy

The addition of UCAs to augment ultrasound-mediated thrombolysis is a natural progression of this technique. The synergistic effect of ultrasound and microbubbles on sonothrombolysis has been demonstrated in canine models of arteriovenous dialysis graft thrombosis.[27] Another application of ultrasound and UCAs is in the delivery of proteins that induce growth of endothelial cells, such as vascular endothelial growth factor. Ultrasound and UCAs are also being investigated to induce vascular endothelial damage in an innovative strategy for non-invasive, highly targeted therapeutic thrombosis of small vessels.[28]

Stem cell therapy

An exciting novel application of UCAs is for the enhancement of stem cell delivery. Therapeutic approaches to stem cell therapy include stimulating stem cells by restoring the capacity of the bone marrow to produce precursor cells or the administration of exogenous precursor cells to treat the specific disease state. The intramuscular transplantation of bone marrow-derived endothelial progenitor cells has already been investigated in clinic trials to induce angiogenesis in the ischemic limb.[29] Other clinical applications include intracoronary delivery for acute myocardial infarction[30] and intramyocardial injection for ischemic cardiomyopathy.[31]

Advantages

Therapeutic-targeted UCAs for drug and gene delivery afford several advantages, including; (i) low toxicity—as lower concentrations of the drug are given systematically, concentrating the drug only to where it is needed (this improved therapeutic index may allow expansion of the use of drugs with severe systemic effects such as cytotoxic agents); (ii) lower immunogenicity compared with viral gene delivery; (iii) low invasiveness with potential for repeated applications; (iv) organ specificity; (v) broad availability; (vi) broad applicability to organs or tissue amenable to sonication; and (vii) portability and being relatively cheap. Ultrasonography also avoids hazardous ionizing radiation, making repeated "treatments" clinically acceptable. Unlike other imaging modalities, ultrasound can be optimized to allow triggered, controlled, and targeted interventions.

Sonochemically Prepared Fragrance-Encapsulated PMs

Durable fragrances are one of the main aspirations in the cosmetics, food, textile, detergents, and pharmaceutical industries. Fragrance compounds and essential oils are volatile substances that react with other components and are very sensitive to the effects of light, oxygen, high temperature, humidity, and other factors. Encapsulation is an effective technique for protecting active agents against the environment, rapid evaporation, contamination, and also

for releasing the volatile substance from the enclosed capsules, as required.[32] The most common form of controlled release in many industries is encapsulation, where active molecules are surrounded by a layer of material (shell) that prevents their release and the penetration of environmental factors until desired. Thus, encapsulation improves the performance of a fragrance in regard to its tenacity or endurance.[33] Depending on the nature of the shell material, encapsulation can control the agent's release across the capsule to the bulk medium.[34] Suslick and coworkers have carried out pioneering work on the ultrasonic synthesis of PMs.[35,36] The PMs are prepared by sonicating an aqueous solution of a protein with an over-layered organic liquid. The formation of microspheres under the influence of ultrasound is related to the oscillation of acoustic waves and the subsequent disruption of the solvent by ultrasound. The size of the microspheres is mainly determined by the energy input, which is a function of acoustic variables, such as the acoustic power employed in the sonication process, the sonication time, and the volume of the material being subjected to sonication.[37] Suslick proposed that for sulfur-containing proteins [e.g., BSA,[38] HSA, and Hb],[39] the mechanism of the formation of stable oil- and gas-filled microspheres involves the cross-linking of cysteine residues through a disulfide bond formation. Interprotein cross-linking is caused by the superoxide radical generated during the acoustic cavitation process leading to the formation of the PMs. It has been claimed that the one-step sonochemical encapsulation process is extremely effective in producing a core–shell structure with high encapsulation efficiency.[10] PMs prepared by sonication usually have a broad size distribution due to the disruption of acoustic sound. The prepared microspheres are nano- to micro-sized and biocompatible.[40] Generally, they have diameters between 100 nm and 50 Lm with 2.5 Lm as the most abundant size. The sono-prepared PMs are stable at room temperature for several months, which is ascribed to the protection of the stable cross-linking protein shell.[41] BSA is the most commonly used solution for the preparation of PMs due to its availability in a pure form, its biodegradability, non-toxicity, low cost, and nonimmunogenicity.[42] Feczkó and coworkers prepared ethylcellulose (EC) capsules and chitosan (CS)-coated EC capsules using an oil-in-water emulsion solvent evaporation method and compared the release of vanillin from both the capsules. They air dried the capsules at elevated temperatures and sampled them at certain time intervals, after which the samples were dissolved in dichloromethane and the vanillin content was determined by ultraviolet–visible spectrophotometry. The release of vanillin from the capsules was investigated for 3 weeks at 50°C. After 6 days, 80 mg of the CS-coated capsules released 20% of the total vanillin content, while the uncoated capsules lost the same vanillin percentage after 1 day. Their experiment showed that EC could efficiently sustain the delivery of vanillin and an additional CS layer can elongate the release of the fragrance. Gumi and coworkers used the phase-inversion precipitation technique and prepared several polysulfone capsules containing different vanillin concentrations. In order to study the release properties, they added water to 1 g of capsules and stirred at 700 rpm for several days. They periodically sampled the bulk solution that was hermetically stored until analysis and then checked the release of vanillin from the capsules using high-performance liquid chromatography. Their release tendencies presented a rapid increase within the first 10 hours, followed by an equilibrating period of a slow increase until reaching a plateau. Sansukcharearnpon et al.[43] encapsulated six fragrances (camphor, citronellal, eucalyptol, limonene, menthol, and 4-tert-butylcyclohexyl acetate) with a polymer blend of EC, hydroxypropyl methylcellulose, and poly(vinyl alcohol) using the solvent-displacement method (water displacement of ethanol). The release profiles of the centrifuged, air-dried particles were acquired by quantifying the amount of fragrance remaining in the samples that had been left uncovered for specified times. The release profile of the encapsulated fragrances indicated different characteristics among the six fragrances. Limonene showed the fastest release, while eucalyptol and menthol showed the slowest release. The release rate of the encapsulated fragrance was independent of the fragrance's vapor pressure.

The fragrance-encapsulated protein spheres were prepared and their release profile was investigated by using a very simple, in real-time, automated, low-cost, and direct method. Amyl acetate (AA) encapsulated in BSA durable spheres based on the ultrasonic synthesis method was prepared, and the release profile of AA by continuous weight measurements in an open vial at ambient temperatures was examined and comparison was done to the evaporation rates of the free components.

CHARACTERIZATION OF PROTEIN CONTAINERS

The morphology and approximate sizes of the BSA–AA containers were characterized using an environmental SEM, a cryo-SEM, and light microscopy. All three devices revealed the BSA–AA containers as smooth surface-spherical particles, and no surface defects were observed.

During the sonication process, two major types of BSA–AA container size were produced, which are reflected as a clear peak around 1300 nm and a rise in the particle's size toward smaller sized spheres. However, the exact location of the second peak could not be located because of the detection limit of light microscopy (~300 nm). It is likely that the difference in the size of the containers is caused by the uneven distribution of acoustic energy in the ultrasonic vessel. The region with intense energy is restricted close to the sound-emitting surface of the ultrasonic probe. Changing the BSA concentration from 0.006% to 0.02% (w/v) in the aqueous phase, maintaining the 3:2 volume ratio of an aqueous organic layer, does not influence the particle size

and the size distribution, but only affects the concentration of the particles, i.e., increasing the BSA concentration in the aqueous phase increases the amount of BSA–AA microcapsules. This observation is consistent with the work of Makino et al.,[44] who found that as the concentration of BSA increases from 0.005% to 0.02% (w/v), the microencapsulation yield increases. Further experiments were carried out in order to confirm the encapsulation of the AA inside the BSA spheres. The fluorescent dye, Nile red, was dissolved in the AA phase before sonication. Nile red is an uncharged heterocyclic molecule and thus is soluble in organic solvents. However, its solubility in water is negligible, less than 1 µg/ml. This property makes Nile red suitable for probing the encapsulation process.[45,46] Confocal laser scanning microscopic (CLSM) images of f–BSA–AA spheres show that the fluorescence of the red color is distributed homogeneously across the sphere's cross section, indicating that the sphere is loaded with AA containing Nile red. We found that BSA–AA containers are stable for more than 6 months, and they can be stored in a sealed bottle either in the refrigerator (4°C) or at room temperature (~20°C). The spheres were found to be unstable when they were self dried by placing them in an open glass vial at ambient temperature.

RELEASE PROFILE MEASUREMENTS

A very simple and direct method was used for measuring the release profile of the BSA–AA containers. A measured volume of the BSA–AA containers was placed inside an open vial and weighed. It was predicted that as the AA is released from the BSA–AA containers and evaporates, the sample's weight is reduced accordingly. Using computer software, the sample's weight was constantly followed for 60 hours. These weight measurements are the mirror image of the release profile of the BSA–AA containers. The release profile of AA from BSA–AA containers at two different temperatures, $15 \pm 2°C$ and $25 \pm 2°C$, was measured. At the end of the sonication process, the product consists of three layers, namely, water, AA, and BSA containers. Thus, identical measurements were carried out for each of these layers and the results compared with those of the BSA–AA containers.

LIQUID-ENCAPSULATED LYSOZYME MICROSPHERES

Hen egg white lysozyme, an enzyme present in many animal and vegetable tissues, contains four intramolecular disulfide bonds that can be partly broken using reducing agents, such as dithiothreitol. This enhances the probability of intermolecular disulfide bond formation leading to cross-linking between lysozyme molecules. Based on this strategy, lysozyme air-filled microspheres have been successfully synthesized using chemically denatured

Stem Cell—Ultrasound

lysozyme under high-intensity ultrasound.[47] Upon sonication, new disulfide linkages are formed enabling interprotein cross-linkage to occur leading to the formation of lysozyme microspheres with a thick (0.1–0.8 lm) protein shell that are stable for several months. The lysozyme microbubbles represent multifunctional ultrasound-responsive systems endowed with antimicrobial and biodegradability features. When tailored with targeting moieties, the lysozyme-shelled microspheres are promising candidates for controlled drug delivery applications.

MORPHOLOGY OF ENCAPSULATED MICROSPHERES

The oil-filled lysozyme microspheres could be stored for several weeks as aqueous dispersions. The morphology of these microspheres varied with the type of encapsulation material as shown in Fig. 8 and listed in Table 1. It can be

seen from Table 1 that the mean size and size distribution of lysozyme microspheres are significantly different for different encapsulated materials. Among the four solvents used, tetradecane- and perfluorohexane-filled microspheres showed smaller mean sizes and narrow size distributions. The average size of the lysozyme microspheres filled with perfluorohexane and tetradecane was about 2.5 μm and 3 μm, respectively. The size of dodecane-filled lysozyme microspheres is around 3–5.5 μm, while the size has a broad range of 4–13 μm for sunflower oil. This might be due to other additives present in the cooking oil, which could affect the stability of the microemulsion prior to cross-linking lysozyme. The perfluorohexane-filled microspheres were found to be the most stable system when dried (Fig. 8e). Other solvent-filled lysozyme microspheres were broken during air drying or under vacuum in the SEM chamber and appeared as broken shells in the SEM images shown in Fig. 8a. It can be speculated that lysozyme adsorption at the liquid/water interface (when an emulsion is

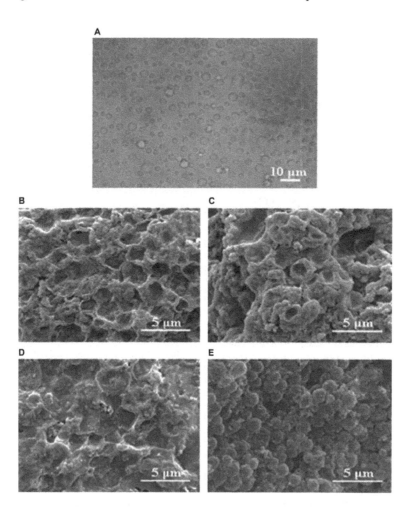

Fig. 8 (**A**) Optical microscopic image and SEM images of (**B**) tetradecane, (**C**) dodecane, (**D**) sunflower oil, and (**E**) perfluorohexane-filled lysozyme microspheres.
Source: Reprinted with permission from Zhou, Leong et al.[45] © 2010 Elsevier.

Stem Cell—Ultrasound

Table 1 Characteristics of sonochemically synthesized liquid-filled lysozyme microspheres (oil concentration in sonication solution equals 4 w/v%).

Liquid	Molecular weight (g/mol)	Boiling point (°C)	Surface tension at 25°C (mN/m)	Density (g/ml)	Average size (µm)	Size distribution
Sunflower oil	—	—	—	0.917	4–13	Nonuniform
Tetradecane	198.40	253.5	26.5	0.762	3	Uniform
Dodecane	170.34	216.3	25.3	0.749	3–5.5	Nonuniform
Perfluorohexane	338.04	56.6	11.9	1.71	2.5	Uniform

Source: Reprinted with permission from Zhou, Leong et al.[45] ©2010 Elsevier.

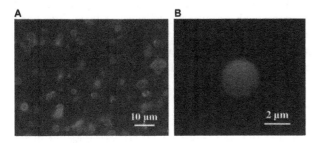

Fig. 9 (A) Fluorescence and (B) CLSM of Nile red/tetradecane-filled lysozyme microspheres.
Source: Reprinted with permission from Zhou, Leong et al.[45] © 2010 Elsevier.

produced by sonication) is effective in perfluorohexane system due to the lower surface tension of this liquid as shown in Table 1, which leads to the formation of a relatively stronger/thicker shell wall when the microspheres are produced.

PROBING THE LIQUIDS ENCAPSULATED INSIDE THE LYSOZYME MICROSPHERES

Further experiments were carried out in order to confirm the encapsulation of the liquid inside the microspheres, where a fluorescent dye, Nile red, was used. Nile red is an uncharged heterocyclic molecule and thus is soluble in organic solvents, but its solubility in water is less than 1 µg/ml. This property makes Nile red suitable for probing the encapsulation process. The fluorescence microscopic image in Fig. 9A shows the uniform size distribution of Nile red/tetradecane-filled microspheres. The CLSM image of Nile red/tetradecane-filled lysozyme microsphere in Fig. 9B shows that the fluorescence color observed is distributed homogeneously across the microsphere's cross section, indicating that the microsphere is filled with tetradecane containing Nile red.

ULTRASOUND-INDUCED RELEASE OF LIQUID

In order to evaluate the possibility of using the ultrasonic encapsulation technique for "delivery" applications (e.g., flavor, drug, etc.), the Nile red/solvent-encapsulated microbubbles were ultrasonically broken and the release of the

liquid into the aqueous medium was monitored using the solvent-dependent fluorescence properties of the dye.[48] The emission spectrum of Nile red exhibits one band in methanol and ethanol,[49,50] whereas two bands in hexane and dodecane.[51] The ratio of the emission intensity of the first to the second band (I_1/I_2) is found to be dependent on the solvent environment. The I_1/I_2 ratio for tetradecane/water at 50:50 is about 1.5, which is similar to that observed in pure tetradecane. However, this ratio increases to about 2.8 when tetradecane/water ratio is changed to 25:75. This change could be due to the direct contact of solvent containing Nile red molecules with water or due to the additional contribution of emission form Nile red dissolved in water. In summary, if Nile red/tetradecane is isolated from water, then the I_1/I_2 ratio can be expected to be closer to 1.5 and if Nile red/tetradecane is in contact with water, then the I_1/I_2 ratio can be more than two. In fact, when Nile red/tetradecane is encapsulated within the protein shell, the solvent is not in direct contact with water and hence I_1/I_2 ratio of about 1.5 can be expected if the encapsulated microspheres are dispersed in 50:50 tetradecane/water mixture. When the microspheres were broken by sonicating at 355 kHz at 30 W for 10 min and the liquid was released into water, the ratio changes to about 2.5, which is similar to that observed when Nile red was dissolved in tetradecane and mixed with 75% water in the absence of the microspheres. Thus, the observation that I_1/I_2 ratios are similar for Nile red in tetradecane/water mixture after the microspheres were broken illustrates that the encapsulated materials can be released into specific sites by ultrasonically breaking them.[45]

ULTRASOUND-MEDIATED DRUG DELIVERY IN CANCER TREATMENT

Cancer is the second leading cause of mortality in the world. Ultrasound is an emerging modality for drug delivery in chemotherapy. The present discussion includes the designs and characteristics of three classes of drug/gene vehicles, microbubble (including nanoemulsion), liposomes, and micelles. In comparison to conventional free drug, the targeted drug release and delivery through vessel wall and interstitial space to cancerous cells can be activated and enhanced under certain sonication conditions.

Stem Cell—Ultrasound

ACOUSTICALLY ACTIVE DRUG VEHICLES

Microbubbles

UCAs, biocompatible microbubbles less than 10 μm in size in order to exit the heart through the pulmonary capillaries, are popular in perfusion monitoring to measure vascular density and microvascular flow rate.[52] A typical dose of UCAs for an echocardiographic evaluation is 109–1010 microbubbles in a 1–2 mL bolus intravenous injection.[53] Microbubbles have a gas core [i.e., perfluoropentane (PFP), sulfur hexafluoride, and nitrogen] and a highly cohesive and insoluble shell (i.e., protein, phospholipid, and polymer), permitting prolonged circulation and preventing nonspecific removal from the circulation by the reticuloendothelial system. The gas core not only has a significant acoustic impedance mismatch for strong echogenicity but also is compressible for bubble cavitation. For SC, a bubble oscillates nonlinearly around its equilibrium radius. However, in IC, a bubble is reduced to a minute fraction of its original size and the gas within dissipates into the surrounding liquid via a violent mechanism. UCAs, regardless of being coadministered or encapsulated with pharmaceutical agents, can be intentionally ruptured by ultrasonic waves at a moderately high acoustic pressure at the target sites.[54]

DRUG-LOADED MICROBUBBLES

There are various ways of entrapping drugs within a microbubble.[55] Drugs may be incorporated into the membrane or in the shell of microbubbles. Strong deposition of the charged drugs in or onto the microbubbles' shell is realized by electrostatic interactions. However, Kupfer cells, leukocytes, and macrophages have the tendency to capture charged microbubbles, which could substantially decrease their half-life. In addition, drugs can be embedded into microbubbles. The advantages of drug-loaded microbubbles are that the loaded drugs can be released at the occurrence of IC into the tissue due to the induced shock wave, microstreaming, and microjet. Meanwhile, the process could be tracked with sonography as the drug carriers are essentially UCAs (Fig. 10).

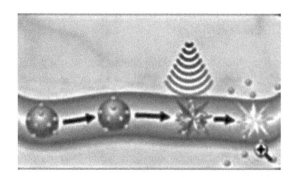

Fig. 10 Ultrasound activated microbubble to fight cancer.

Targeted Microbubbles

Although acoustic radiation force could promote the attachment of microbubbles to endothelial cells in the wave propagation path, it is more attractive to selectively adhere microbubbles to cellular epitopes and receptors of cancer or solid tumors by one or several specific ligands, such as antibodies, carbohydrates, and/or peptides.[56] Monoclonal antibodies have a very high specificity and selectivity to a large range of epitopes. In contrast, peptides are low cost and less immunogenic because of much smaller size (5–15 amino acids). There are two ways of coupling ligands to the microbubble shell, namely, covalent (being attached to the head of phospholipids directly or via an extended polymer spacer arm) and noncovalent by avidin–biotin bridging because of the wide availability and excellent affinity.

Nanoemulsions

Although the use of microbubbles is very attractive in drug delivery, especially for its uniqueness of combined diagnosis and targeted therapy with a high effectiveness to cost ratio, it lacks an essential prerequisite for effective extravasation into tumor and subsequent drug targeting with a sufficient lifetime in the circulation, because the pore size of most tumors is usually smaller (380–780 nm).[57] An alternative solution to the aforementioned problems is developing drug-loaded nanoparticles that could accumulate in tumor and then expand to microbubbles in situ under sonication, such as copolymer-stablilized echogenic PFP nanoemulsions, PFC nanoemulsions, etc. (Fig. 11). The nanoemulsions are produced from drug-loaded poly (ethylene oxide)-co-poly(L-lactide) or poly(ethylene oxide-co-polycaprolactone) micelles.

Liposomes

Liposomes with a typical diameter of 65–120 nm are non-toxic, biodegradable, and nonimmunogenic drug delivery vehicles for both hydrophilic and lipophilic drugs, such as doxorubicin (DOX) and vincristine.[58] Local heating causes the liposome to change from a well-ordered gel to a less-structured liquid crystalline state by inducing thermotropic phase transitions on phospholipids. The *in vivo* stability and accumulation at the tumor site can be enhanced to 50- to 100-fold compared to the free drug,[59] and small polymeric carriers can respond to both mechanical (i.e., fracture) and thermal (i.e., temperature elevation) activations. Poly(ethylene glycol) (PEG) liposomes containing DOX have been used in the treatment of Kaposi's sarcoma, refractory ovarian cancer, and breast cancer.[60] Furthermore, liposomes are also efficient as nonviral gene carriers.[61] Drug release from microbubbles is mainly due to the ultrasound-induced IC, whereas that from a liposome is the consequence of the heat produced by the absorption of acoustic energy. Meanwhile, the gas nuclei will expand and dilate the monolayer boundary or the adjacent bilayer in the strong acoustic field. If the bubble expansion-induced stress is beyond the elastic threshold of the bounding membrane, the liposome will rupture and then the incorporated contents will be released (Fig. 12).

Micelles

A micelle is an aggregate of surfactant molecules dispersed in a liquid colloid with the hydrophilic "head" regions in contact with surrounding aqueous solution and the hydrophobic single-tail regions in the micelle center

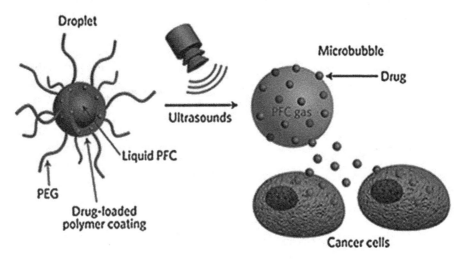

Fig. 11　Drug delivery from echogenic PFC-containing nanoemulsions.

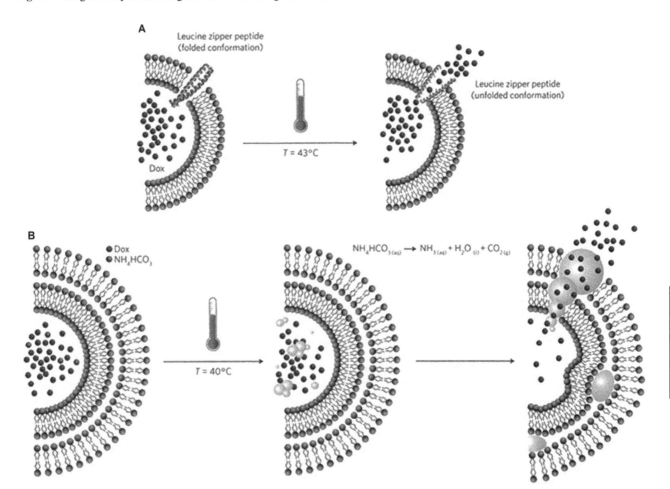

Fig. 12　Temperature-based actuation mechanisms for liposomal drug delivery. (**A**) shows the liposomal drug release, and in (**B**) the liposome ruptures and then the incorporated contents are released.

Stem Cell—Ultrasound

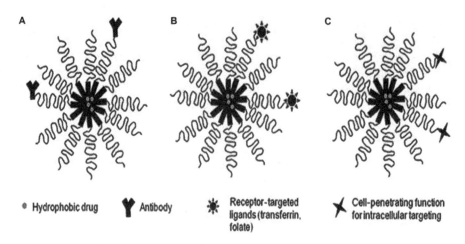

Fig. 13 Drug-loaded polymeric micelles with various targeting functions. (**A**) Antibody-targeted micelles, (**B**) ligand-targeted micelles, and (**C**) micelles with cell-penetrating function.

(Fig. 13). They are small enough to avoid renal excretion but permit extravasation at the tumor site via the enhanced penetration and retention effect.[62] The thermodynamic and kinetic stability of micelle is determined by molecular interactions and the length of hydrophobic blocks and the micelle core type, respectively. The self-assembly of amphiphilic block copolymers is activated thermodynamically and is reversible when the concentration reaches the critical micelle concentration (CMC). The CMC is primarily controlled by the length of a hydrophobic block (the greater the length, the lower the CMC) but less sensitive to the length of a hydrophilic block. However, an excessive concentration of the copolymer will initiate micelle aggregation and precipitation.[63] Antibodies have been the most popular targeting ligands for polymeric micelles because of the diversity of their targets and the specificity of their interaction (Fig. 13). Active targeting especially benefits intracellular delivery of macromolecules like DNA, siRNA, and proteins.

DRUG/GENE DELIVERY APPROACH

The mechanisms for the ultrasound-mediated drug/gene delivery include both thermal and mechanical effects.[20] First, acoustic radiation force can help agents penetrate through the vessel wall to tissue. Second, ultrasound may affect the morphology and properties of the cell membrane due to bubble cavitation for drug permeation and absorption (sonoporation). Third, ultrasound-induced hyperthermia has a significant biological effect on cell activities and drug uptake. Last, ultrasound can alter the performance of a drug, for example, activating light-sensitive materials of hematoporphyrins to kill cancer cells and inhibit restenosis.[64] The efficiency of drug delivery is mainly determined by the acoustic intensity. However, with a higher power, the propensity of cell lysis is increased. Furthermore, cell membrane permeability and drug cytotoxicities can be enhanced at elevated temperatures.[12]

CONCLUSION

The advantages of ultrasound imaging include bedside availability and the relative ease of performing repeated examinations. Imaging is real time and free of harmful radiation. There are no documented side effects and discomfort is minimal. Targeted CEUS is being developed for a variety of medical applications. Untargeted CEUS is applied in echocardiography and radiology. The therapeutic use of UCAs is an emerging technique with high potential for enhanced directed gene and drug delivery. UCAs can be designed as safe vehicles for encapsulating or cotransporting drugs or genes. Furthermore, the contrast agent can be targeted to cell-specific receptors for site-specific delivery. PMs have been prepared by sonicating a mixture of pure fragrant oil (AA) with an aqueous protein (BSA) solution. The prepared protein spheres are nano- to micrometer sized with an encapsulation efficiency of ~97% for the AA present on the surface and inside the BSA capsule. Two disappearance rates were found for AA from the spheres. The higher rate is the sum of a few evaporation rates, also including the water molecules, while the slower rate is due to the pure evaporation of AA resulting from the collapse of the spheres. Liquid-encapsulated lysozyme microspheres were obtained by the sonochemical method. The size and the stability of these microbubbles were found to be dependent on the nature of the encapsulated materials. Using an oil-soluble fluorescent solute, the potential use of this technique for encapsulating specific materials within the microspheres and the controlled release of these materials from the microspheres were discussed. This method can be conveniently adopted in many potential applications in medicine and food industry, since the liquid could be replaced by a great of variety of water-insoluble drugs or food ingredients. The most common cancer treatment, chemotherapy, is often limited by its cytotoxic effects on normal tissues, particularly under high doses. Therefore, it is highly desirable to reduce the dosage or frequency of administration by enhancing the effectiveness of drugs to the specific target. The application of ultrasound in delivery of several therapeutic classes (i.e., chemotherapeutic, thrombolytic, and DNA-based drugs) acting via bubble cavitation (SC and IC), radiation forces, and/or heat has gained impetus. The advantages of this novel technology include non-invasiveness, low cost, easy operation, good focusing and penetration inside the body, and no radiation.

ACKNOWLEDGMENTS

The authors are thankful to Elsevier for copyright permission. Authors are also thankful Professor Suresh C. Ameta, Department of Chemistry, Pacific College of Basic & Applied Sciences, PAHER University, Udaipur (Rajasthan) for his suggestions.

REFERENCES

1. Robert, N. *Squire's Fundamentals of Radiology*; Harvard University Press: Cambridge, 1997; 34–35.
2. Klibanov, A.L. Targeted delivery of gas-filled microspheres, contrast agents for ultrasound imaging. Adv. Drug Deliver. Rev. **1999**, *37*, 139–157.
3. McCulloch, M.; Gresser, C.; Moos, S.; Odabashian, J.; Jasper, S.; Bednarz, J.; Burgess, P.; Carney, D.; Moore, V.; Sisk, E.; Waggoner, A.; Witt, S.; Adams, D. Ultrasound contrast physics: A series on contrast echocardiography. J. Am. Soc. Echocardiog. **2000**, *13* (10), 959–967.
4. Lindner, J.R. Microbubbles in medical imaging: Current applications and future directions. Nat. Rev. Drug Discov. **2004**, *3* (6), 527–532.
5. Suslick, K.S.; Kolbeck, K.J.; Kufner, G.S.; Szewczyk, G.W. Sonochemical preparation of protein microspheres. J. Acoust. Soc. Am. **1998**, *103*, 2923.
6. Grinstaff, M.W.; Suslick, K.S. Air-filled proteinaceous microbubbles: Synthesis of an echo contrast agent. P. Natl. Acad. Sci. U.S.A. **1991**, *88* (17), 7708–7710.
7. Keller, M.W.; Feinstein, S.B.; Briller, R.A.; Powsner, S.M. Automated production and analysis of echo contrast agents. J. Ultras. Med. **1986**, *5* (9), 493–498.
8. Grinstaff, M.W.; Soon-Shiong, P.; Wong, M.; Sandford, P.A.; Suslick, K.S.; Desai, N.P. Methods for the preparation of immunostimulating agents for *in vivo* delivery. U.S. Patent 5665383 A, September 9, 1997.
9. Keller, M.W.; Feinstein, S.B. In Echocardiography in Coronary Artery Disease; Kerber, R.E., Ed.; Future: New York, 1988; 443–465.
10. Webb, A.G.; Wong, M.; Kolbeck, K.J.; Magin, R.; Suslick, K.S. Sonochemically produced fluorocarbon microspheres: A new class of magnetic resonance imaging agent. J. Magn. Reson. Imaging **1996**, *6* (4), 675–683.
11. Wong, M.; Suslick, K.S. Sonochemically produced hemoglobin microbubbles, hollow and solid spheres and microspheres. In *Materials Research Society Symposium Proceedings*; Wilcox, D.L., Berg, M., Bernat, T., Kellerman, D., Corchran, J.K., Eds.; Materials Research Society: Pittsburgh, 1995; Vol. 372, 89–94.
12. Ng, K.Y.; Liu, Y. Therapeutic ultrasound: Its application in drug delivery. Med. Res. Rev. **2002**, *22* (2), 204–223.
13. Miller, M.W.; Miller, D.L.; Brayman, A.A. A review of the *in vitro* bioeffects of inertial ultrasonic cavitation from a mechanistic perspective. Ultrasound Med. Biol. **1996**, *22* (9), 1131–1154.
14. Holland, C.K.; Apfel, R.E. Thresholds for transient cavitation produced by pulsed ultrasound in a controlled nuclei environment. J. Acoust. Soc. Am. **1990**, *88* (5), 2059–2069.
15. Tachibana, K.; Tachibana, S. Albumin microbubble echocontrast material as an enhancer for ultrasound accelerated thrombolysis. Circulation **1995**, *92* (5), 1148–1150.
16. Stride, E.; Saffari, N. On the destruction of microbubble ultrasound contrast agents. Ultrasound Med. Biol. **2003**, *29* (4), 563–573.
17. Marin, A.; Sun, H.; Husseini, G.A.; Pitt, W.G.; Christensen, D.A.; Rapoport, N.Y. Drug delivery in pluronic micelles: Effect of high-frequency ultrasound on drug release from micelles and intracellular uptake. J. Control. Release **2002**, *84* (1–2), 39–47.
18. Liu, Y.; Yang, H.; Sakanishi, A. Ultrasound: Mechanical gene transfer into plant cells by sonoporation. Biotechnol. Adv. **2006**, *24* (1), 1–16.
19. Tachibana, K.; Uchida, T.; Ogawa, K.; Yamashita, N.; Tamura, K. Induction of cell-membrane porosity by ultrasound. Lancet **1999**, *353* (9162), 1409.
20. Dijkmans, P.A.; Juffermans, L.J.; Musters, R.J.; vanWamel, A.; ten Cate, F.J.; van Gilst, W.; Visser, C.A.; de Jong, N.; Kamp, O. Microbubbles and ultrasound: From diagnosis to therapy. Eur. J. Echocardiogr. **2004**, *5* (4), 245–256.
21. Bao, S.; Thrall, B.D.; Miller, D.L. Transfection of a reporter plasmid into cultured cells by sonoporation *in vitro*. Ultrasound Med. Biol. **1997**, *23* (6), 953–959.
22. Shohet, R.V.; Chen, S.; Zhou, Y.T.; Wang, Z.; Meidell, R.S.; Unger, R.H.; Grayburn, P.A. Echocardiographic destruction of albumin microbubbles directs gene delivery to the myocardium. Circulation **2000**, *101* (22), 2554–2556.
23. Erikson, J.M.; Freeman, G.L.; Chandrasekar, B. Ultrasoundtriggered antisense oligonucleotide attenuates ischemia/reperfusion-induced myocardial tumor necrosis factoralpha. J. Mol. Cell Cardiol. **2003**, *35* (1), 119–130.
24. Suzuki, R.; Takizawa, T.; Negishi, Y.; Hagisawa, K.; Tanaka, K.; Sawamura, K.; Utoguchi, N.; Nishioka, T.; Maruyama, K. Gene delivery by combination of novel liposomal bubbles with perfluoropropane and ultrasound. J. Control. Release **2007**, *117* (1), 130–136.
25. Vannan, M.; McCreery, T.; Li, P.; Han, Z.; Unger, E.; Kuersten, B.; Nabel, E.; Rajagopalan, S. Ultrasoundmediated transfection of canine myocardium by intravenous administration of cationic microbubble-linked plasmid DNA. J. Am. Soc. Echocardiog. **2002**, *15* (3), 214–218.
26. Hashiya, N.; Aoki, M.; Tachibana, K.; Taniyama, Y.; Yamasaki, K.; Hiraoka, K.; Makino, H.; Yasufumi, K.; Ogihara, T.; Morishita, R. Local delivery of E2F decoy oligodeoxynucleotides using ultrasound with microbubble agent (Optison) inhibits intimal hyperplasia after balloon injury in rat carotid artery model. Biochem. Bioph. Res. Co. **2004**, *317* (2), 508–514.
27. Xie, F.; Tsutsui, J.M.; Lof, J.; Unger, E.C.; Johanning, J.; Culp, W.C.; Matsunaga, T.; Porter, T.R. Effectiveness of lipid

Stem Cell—Ultrasound

microbubbles and ultrasound in declotting thrombosis. Ultrasound Med. Biol. **2005**, *31* (7), 979–985.

28. Hwang, J.H.; Brayman, A.A.; Reidy, M.A.; Matula, T.J.; Kimmey, M.B.; Crum, L.A. Vascular effects induced by combines 1-MHz ultrasound and microbubble contrast agent treatments *in vivo*. Ultrasound Med. Biol. **2005**, *31* (4), 553–564.

29. Tateishi-Yuyama, E.; Matsubara, H.; Murohara, T.; Ikeda, U.; Shintani, S.; Masaki, H.; Amano, K.; Kishimoto, Y.; Yoshimoto, K.; Akashi, H.; Shimada, K.; Iwasaka, T.; Imaizumi, T; Therapeutic Angiogenesis using Cell Transplantation (TACT) Study Investigators. Therapeutic angiogenesis for patients with limb ischemia by autologous transplantation of bone-marrow cells: A pilot study and a randomised controlled trial. Lancet **2002**, *360* (9331), 427–435.

30. Wollert, K.C.; Meyer, G.P.; Lotz, J.; Ringes-Lichtenberg, S.; Lippolt, P.; Breidenbach, C.; Fichtner, S.; Korte, T.; Hornig, B.; Messinger, D.; Arseniev, L.; Hertenstein, B.; Ganser, A.; Drexler, H. Intracoronary autologous bone-marrow cell transfer after myocardial infarction: The BOOST randomized controlled clinical trial. Lancet **2004**, *364* (9429), 141–148.

31. Fuchs, S.; Satler, L.F.; Kornowski, R.; Okubagzi, P.; Weisz, G.; Baffour, R.; Waksman, R.; Weissmans, N.J.; Cerqueira, M.; Leon, M.B.; Epstein, S.E. Catheter-based autologous bone marrow myocardial injection in no-option patients with advanced coronary artery disease: A feasibility study. J. Am. Coll. Cardiol. **2003**, *41* (10), 1721–1724.

32. Feczkó, T.; Kokol, V.; Voncina, B. Preparation and characterization of ethylcellulose-based microcapsules for sustaining release of a model fragrante. Macromol. Res. **2010**, *18* (7), 636–640.

33. Bhargava, N.U.; Magar, V.P.; Momin, S.A. Controlledrelease mechanisms of fragrances. Cosmet. Toiletries **2010**, *125*, 42–49.

34. Gumi, T.; Gascon, S.; Torras, C.; Garcia-Valls, R. Vanillin release from macrocapsules. Desalination **2009**, *245* (1–3), 769–775.

35. Grinstaff, M.W.; Suslick, K.S. Air-filled proteinaceous microbubbles: Synthesis of an echo-contrast agent. P. Natl. Acad. Sci. U.S.A. **1991**, *88* (17), 7708–7710.

36. Suslick, K.S.; Grinstaff, M.W.; Kolbeck, K.J.; Wong, M. Characterization of sonochemically prepared proteinaceous microspheres. Ultrason. Sonochem. **1994**, *1* (1), S65–S68.

37. Han, Y.; Radziuk, D.; Shchukin, D.; Moehwald, H. Stability and size dependence of protein microspheres prepared by ultrasonication. J. Mater. Chem. **2008**, *18* (42), 5162–5166.

38. Suslick, K.S.; Grinstaff, M.W. Protein microencapsulation of nonaqueous liquids. J. Am. Chem. Soc. **1990**, *112* (21), 7807–7809.

39. Wong, M.; Suslick, K.S. Sonochemically produced hemoglobin microbubbles. Mater. Res. Soc. Symp. P. **1994**, *372*, 89.

40. Han, Y.; Shchukin, D.; Möhwald, H. Drug release of sonochemical protein containers. Chem. Lett. **2010**, *39* (5), 502–503.

41. Han, Y.; Shchukin, D.; Yang, J.; Simon, C.R.; Fuchs, H.; Möhwald, H. Biocompatible protein nanocontainers for controlled drugs release. ACS. Nano. **2010**, *4* (5), 2838–2844.

42. Kratz, F.; Fichtner, I.; Beyer, U.; Schumacher, P.; Roth, T.; Fiebig, H.H.; Unger, C. Antitumor activity of acid labile transferrin and albumin doxorubicin conjugates in *in vitro* and *in vivo* human tumour xenograft models. Eur. J. Cancer **1997**, *33* (8), S175.

43. Sansukcharearnpon, A.; Wanichwecharungruang, S.; Leepipatpaiboon, N.; Kerdcharoen, T.; Arayachukeat, S. High loading fragrance encapsulation based on a polymerblend: Preparation and release behavior. Int. J. Pharm. **2010**, 391 (1–2), 267–273.

44. Makino, K.; Mizorogi, T.; Ando, S.; Tsukamoto, T.; Ohshima, H. Sonochemically prepared bovine serum albumin microcapsules: Factors affecting the size distributing and the microencapsulation yield. Colloid. Surface. B. **2001**, 22, 251–255.

45. Zhou, M.; Leong, T.S.; Melino, S.; Cavalieri, F.; Kentish, S.; Ashokkumar, M. Sonochemical synthesis of liquidencapsulated lysozyme microspheres. Ultrason. Sonochem. **2010**, *17* (2), 333–337.

46. Greenspan, P.; Fowler, S.D. Spectrofluorometric studies of the lipid probe, nile red. J. Lipid Res. **1985**, *26* (7), 781–789.

47. Cavalieri, F.; Ashokkumar, M.; Grieser, F.; Caruso, F. Ultrasonic synthesis of stable, functional lysosome microbubbles. J. Phys. Chem. **2008**, *24* (18), 10078–10083.

48. Sackett, D.L.; Wolff, J. Nile red as a polarity-sensitive fluorescent probe of hydrophobic protein surfaces. Anal Biochem. **1987**, *167* (2), 228–234.

49. Dutta, A.K.; Kamada, K.; Ohta, K. The effects of cyclodextrin on guest fluorescence. J. Photoch. Photobio. **1996**, *93*, 57–64.

50. Hawe, A.; Sutter, M.; Jiskoot, W. Extrinsic fluorescent dyes as tools for protein characterization. Pharm. Res. **2008**, *25* (7), 1487–1499.

51. Yablon, D.G.; Schilowitz, A.M. Solvatochromism of Nile Red in nonpolar solvents. Appl. Spectrosc. **2004**, *58* (7), 843–847.

52. Cosgrove, D. Ultrasound contrast agents: An overview. Eur. J. Radiol. **2006**, *60* (3), 324–330.

53. Dolan, M.S.; Gala, S.S.; Dodla, S.; Abdelmoneim, S.S.; Xie, F.; Cloutier, D.; Bierig, M.; Mulvagh, S.L.; Porter, T.R.; Labovitz, A.J. Safety and efficacy of commercially available ultrasound contrast agents for rest and stress echocardiography: A multicenter experience. J. Am. Coll. Cardiol. **2009**, *53* (1), 32–38.

54. Ferrara, K.; Pollard, R.; Borden, M. Ultrasound microbubble contrast agents: Fundamentals and applications to gene and drug delivery. Annu. Rev. Biomed. Eng. **2007**, *9*, 415–447.

55. Unger, E.C.; Hersh, E.; Vannan, M.; Matsunaga, T.O.; McCreery, T. Local drug and gene delivery through microbubbles. Prog. Cardiovasc. Dis. **2001**, *44* (1), 45–54.

56. Hernot, S.; Klibanov, A.L. Microbubbles in ultrasoundtriggered drug and gene delivery. Adv. Drug Deliver. Rev. **2008**, *60* (10), 1153–1166.

57. Campbell, R.B. Tumor physiology and delivery of nanopharmaceuticals. Anti-Cancer Agents Me. **2006**, *6* (6), 503–512.

58. Schroeder, A.; Kost, J.; Barenholz, Y. Ultrasound, liposomes, and drug delivery: Principles for using ultrasound to control the release of drugs from liposomes. Chem. Phys. Lipids **2009**, *162* (1–2), 1–16.

59. Allen, T.M. Liposomes. Opportunities in drug delivery. Drugs **1997**, *54* (4), 8–14.

60. Ranson, M.R.; Carmichael, J.; O'Byrne, K.; Stewart, S.; Smith, D.; Howell, A. Treatment of advanced breast cancer with sterically stabilized liposomal doxorubicin: Results of a

multicenter phase II trial. J. Clin. Oncol. **1997**, *15* (10), 3185–3191.

61. Lentacker, I.; De Smedt, S.C.; Sanders, N.N. Drug loaded microbubble design for ultrasound triggered delivery. Soft Matter **2009**, *5* (11), 2161–2170.

62. Rapoport, N. Combined cancer therapy by micellarencapsulated drug and ultrasound. Int. J. Pharm. **2004**, *277* (1–2), 155–162.

63. Rapoport, N. Physical stimuli-responsive polymeric micelles for anti-cancer drug delivery: Progress in polymer science. Prog. Polym. Sci. **2007**, *32* (8–9), 962–990.

64. Umemura, S.; Kawabata, K.; Sasaki, K.; Yumita, N.; Umemura, K.; Nishigaki, R. Recent advances in sonodynamic approach to cancer therapy. Ultrason. Sonochem. **1996**, *3* (3), S187–S191.

Stem Cell—Ultrasound

Vascular Grafts: Biocompatibility Requirements

Shawn J. Peniston
Georgios T. Hilas
Poly-Med, Incorporated, Anderson, South Carolina, U.S.A.

Abstract

The search for a patent vascular graft has attracted much attention from researchers and physicians due to the significant number of mortalities in the Western world each year from vascular disease. Vascular bypass grafting for coronary artery and peripheral vascular disease (atherosclerosis) account for the majority of bypass procedures; in addition, the repair of thoracic and abdominal aneurysms is also a prominent use of vascular grafts. This entry begins with a background section detailing the biology of the artery and the basic biology of vascular tissue. It then moves to examine vascular graft requirements, clinically relevant vascular grafts and the insufficiencies and limitations of vascular grafts.

INTRODUCTION

Background and Clinical Relevance

The search for a patent vascular graft has attracted much attention from researchers and physicians due to the significant number of mortalities in the Western world each year from vascular disease. Vascular bypass grafting for coronary artery and peripheral vascular disease (atherosclerosis) account for the majority of bypass procedures; in addition, the repair of thoracic and abdominal aneurysms is also a prominent use of vascular grafts. Several procedures are now available for physicians to treat vascular disease. The development of minimally invasive endovascular procedures such as percutaneous angioplasty, atherectomy, stenting, and laser angioplasty can open occluded vessels. However, these procedures require access using the catheter technique and are not effective if access is physically impossible or extensive occlusion has developed. Consequently, an open bypass surgical procedure using an autologous or prosthetic vascular graft is the only clinical alternative. As a result, 600,000 coronary and peripheral bypass procedures[1] are performed each year in the United States, most commonly with the saphenous vein or internal mammary artery. However, 30% of patients[2] have inadequate or insufficient tissue. These patients require the use of a substitute vascular conduit.

Prosthetic vascular grafts comprised of polyethylene terephthalate (PET) and polytetrafluoroethylene (PTFE) are the primary clinical options. These grafts are biocompatible in the sense that they are stable in the biologic environment and nontoxic, but they are not blood compatible and do not possess the viscoelastic or biomechanical properties of the native artery. The success of using prosthetic vascular grafts has been acceptable for large (12- to 38-mm) arteries, marginal for medium (6- to 12-mm) arteries, and not viable for small arteries (<6 mm). In the case of coronary arteries, the replacement size is less than 6 mm. In part, the lack of a suitable synthetic option accounts for the high fatality rates from coronary artery disease, highlighting the clinical necessity of a vascular graft with improved patency. The development of a patent vascular graft for use in all bypass procedures has enormous commercial and humanitarian benefits. To this end, research has focused on improving vascular grafts by developing more compliant and blood-compatible biomaterials. In addition, the potential of regenerating vascular tissue has received significant attention in the last decade due the advancement of tissue engineering.

BIOLOGY OF THE ARTERY

To better grasp the challenges of duplicating the function of the artery, it is important to understand the basic biology of vascular tissue. The vascular wall functions as a dynamic, closed-loop system rather than an inactive conduit for blood. It is composed primarily of endothelial cells (ECs), smooth muscle cells (SMCs), fibroblasts, collagen, elastin, and glycosaminoglycans (GAGs). The vascular cells and the unique three-dimensional matrix of structural proteins have a high degree of interaction to function and remodel.

The vascular wall is commonly segmented into three layers based primarily on function: the intima, media, and adventitia (Fig. 1). The intima is the blood contact surface and consists of a single layer of polygonal ECs. The vascular endothelium is responsible for several functions: hemostasis, regulation of vascular tone, synthesis of cytokines and growth factors, angiogenesis, selective transmission of

Concise Encyclopedia of Biomedical Polymers and Polymeric Biomaterials DOI: 10.1081/E-EBPPC-120052171

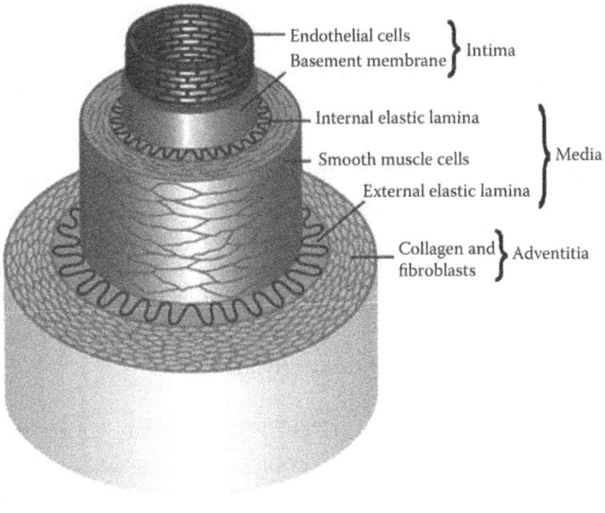

Fig. 1 The anatomy of an artery depicted in layers.
Source: From Sarkar et al.[3]

inflammatory cells (diapedesis), and provision of a barrier function between blood and all other tissues. Under homeostatic conditions, ECs provide continuous chemical and physical signals to circulating blood components to prevent thrombosis and coagulation. Thrombosis formation occurs when platelets encounter a foreign material (collagen or synthetic polymer), resulting in the adhesion and release of adenosine diphosphate (ADP) and thromboxane. The release of ADP signals other platelets, and thromboxane promotes additional platelet aggregation. Furthermore, degranulated platelets initiate the formation of thrombin, which then converts blood-soluble fibrinogen into insoluble fibrin fibers resulting in coagulation. As such, ECs play a unique and critical role in maintaining the balance between being able to react quickly to a site of injury yet prevent potentially life-threatening blood-clotting episodes.

The endothelium is anchored to the media by a tie layer of connective tissue called the basement membrane, which is made up of type IV collagen, laminin, and fibronectin proteins. The internal- and external-elastic lamina are pronounced layers of elastic connective tissue that give arteries their unique balance of compliance, elastic recoil, and strength. The media is comprised of radial aligned sheets of elastin, cross-linked elastic connective tissue fibers, similar to the internal and external lamina, but less pronounced and separated by layers of SMCs. The media is largely responsible for the biomechanics of the artery, generating the contractile forces and participating in the elastic recoil within the arterial wall. The adventitia is constructed of axially aligned type I collagen fibers that provide radial resistance to arterial blood pressure and prevent excessive dilation. Overall, the artery wall is a complex composite structure with layers that each have specific functions yet interact to transport blood to tissue efficiently.

VASCULAR GRAFT FUNCTIONAL REQUIREMENTS

The development of a patent vascular graft has proven to be exceptionally difficult. In part, the specialized nature of the cells and the complex structure of the extracellular matrix (ECM) within the arterial wall are responsible for this challenge. Consequently, vascular grafts must replicate a complex myriad of chemical signals and possess anisotropic, nonlinear biomechanical properties to function with a high degree of biocompatibility. Table 1 provides a summary of loosely categorized requirements for a vascular graft, with an understanding that there is significant interaction and overlap between categories. The degree of biocompatibility increases as the vascular graft design incorporates more of the requirements. In addition, for clinical acceptance the design of vascular grafts must consider the regulatory pathway, meet user needs within the surgical suite, and be commercially cost effective.

CLINICALLY RELEVANT VASCULAR GRAFTS

Materials

Polyethylene terephthalate

Polyethylene terephthalate (PET) is a synthetic polymer, commonly known under the trade name Dacron, that has been a viable vascular graft material for medium and large arterial replacements. PET is a heterochain linear aromatic polymer ($-C_{10}H_8O_4-$) with the characteristic feature of two ester linkages on either side of an aromatic ring in the backbone of the polymer chain. Consequently, the polymer is slightly polar, more hydrophilic, and more hygroscopic than homochain hydrocarbon polymers. The ester groups create chemically liable sites that are susceptible to hydrolysis. However, local hydrophobic moieties and high crystallinity slow this reaction considerably. The strong dipole-dipole-type van der Waals-London forces between adjacent carbonyl groups contribute to its high-strength characteristics.

PET grafts are produced from multifilament yarn, which is woven or knitted into a fabric tube. Grafts produced using a woven structure show minimal fluid permeability and are mechanically stiff. In contrast, knitted grafts are more flexible and have increased porosity, leading to a higher degree of tissue ingrowth.[4] However, these properties are lost shortly after implantation due to tissue integration.[1] In addition, PET grafts are often crimped longitudinally or have external coils added to provide additional compression and kink resistance.[5] Although PET has been used as a vascular graft material since the 1970s, it does induce protein absorption, thrombosis, and leukocyte activation and migration. After 18 months in the *in vivo* environment, a pseudointema containing an inner

Table 1 Vascular graft biocompatibility requirements

Physical and structural

Assortment of calibers and lengths to accommodate the patient population

Adequate porosity for tissue ingrowth and angiogenesis without significant blood loss

Confluent and intact intimal surface

Stratified vascular wall structure

Material

Low inflammatory response

Nonimmunogenic

Noncarcinogenic

Nonthrombogenic

Chemical stability in the biologic environment (permanent scaffolds)

Predictable degradation of absorbable scaffold materials

No leachable low molecular weight fractions (monomer, oligomer, impurities)

Resistance to infection for aseptic implantation

Biomechanical

Kink resistant

Elastic recoil

Mechanical strength to resist luminal burst pressure and axial stress

Viscoelasticity to accommodate efficient pulsatile blood flow

Compliance (radial and axial)

Tear resistance to secure sutures

Creep resistance

Fatigue resistance

Mechanical stability of the intimal surface to resist wall shear stress

Biochemical and cellular signaling

Ability to be repopulated and remodeled by the host cells

Active regulation of vascular tone

Development and maintenance of cell phenotype (biomechanical influence as well)

Continuous production of chemical factors to maintain thromboresistance

Selective permeability to inflammatory cells (diapedesis)

layer of fibrin, an intermediate layer of foreign body giant cells, and an outer connective tissue layer develop to form a relatively thick, compact layer over the graft surface, with the greatest thickness at the proximal and distal ends and less in the midgraft region.[5,6] When PET is used for an aortal or thoracic arterial conduit, the relatively large diameter and high blood flow are forgiving to occlusions. In addition to blood compatibility issues, knitted PET grafts have shown a tendency to dilate with time.[5] This may be attributed to the susceptibility of PET to hydrolysis,

possibly accelerated by enzymatic and oxidative attack by macrophages or foreign body giant cells, which are present after the first week of implantation and remain for the life of the implant.[7] However, due to their construction, PET grafts are more compliant than other commercially available materials. PET grafts used for aortic bypass have 5-year patency rates of 93%, and peripheral grafts have a 3-year patency rate of less than 45%.[6] For these reasons, currently PET grafts are the preferred material for the replacement of large arteries.

Expanded polytetrafluoroethylene

Polytetrafluoroethylene (PTFE) is a linear homochain polymer constructed of a carbon backbone saturated with fluorine atoms ($-CF_2-$). The characteristic feature of PTFE is its inert nature due to the extreme stability of the carbon-fluorine bond. Although highly crystalline, the intermolecular attraction between PTFE molecules is very small, which results in bulk properties that are inferior to other engineered polymers. The extreme perfection of the crystalline lattice of PTFE results in a melting point that is very close to its degradation temperature. As a result, PTFE does not possess a meaningful melt processing window. Accordingly, PTFE is manufactured by stretching and sintering a solid meltextruded tube. This process gives rise to the "expanded" precursor designation (ePTFE). The result is a microporous nodefibril structure in which solid nodes connect through fine fibrils with an average internodal distance of 30 mm for a standard graft.[6] The formation of micropores does increase graft compliance and create a porous substrate for the attachment of a pseudointema. However, standard ePTFE grafts are still 2 or 3 times stiffer than the natural artery. In an attempt to improve compliance, thin-walled ePTFE grafts have been developed that have an elastic stiffness 1.6 times that of the coronary artery.[8]

Small-diameter vascular grafts in low-flow conditions, such as below the knee, are prone to short-term occlusion from acute thrombosis. Although the blood compatibility of ePTFE is considered poor, its electronegative luminal surface lessens the onset of platelet adhesion and produces a thinner pseudointema. As a consequence, ePTFE has gained popularity in peripheral bypass applications for which occlusion results from thrombosis.[9] However, in a multicenter, prospective, randomized trial investigating the use of a PET or ePTFE graft for femoropopliteal bypass surgery, there was no difference in patency rate after 3 years.[10] It was observed that PET performed slightly better than ePTFE in above-knee applications and worse in below knee applications, but with no statistical difference. Roll and coworkers conducted a systemic evaluation and meta analysis of randomized controlled trials comparing PET and ePTFE for peripheral bypass procedures and showed no evidence of an advantage of one synthetic material over the other.[11] The preference of ePTFE grafts over

PET for femoropopliteal bypass surgery is relatively unjustified since no significant difference in patency rates exists between the two materials.[12] Overall, like PET grafts, the compliance mismatch and the lack of a functional endothelium account for ePTFE grafts having poor patency.

Polyurethanes

Polyurethanes are copolymers composed of alternating hard and soft segments, with the hard segments comprised of urethane linkages and the soft segments typically composed of ether or ester moieties. By varying the ratios of hard and soft segments, polyurethanes can be tailored to be elastic and compliant. They have been researched for years as a desirable biomaterial for vascular grafts because of their excellent compliance, fatigue resistance in flexure, tensile strength, high ultimate elongation, and relatively good blood compatibility. Unfortunately, the traditional polymer containing ester or ether soft segment linkages has *in vivo* stability issues with hydrolysis and oxidation, respectively. Consequently, the development of more hydrolytically and oxidatively stable variants that employ a polycarbonate polyol as the polymer soft segment has been investigated.

The first human trials have been initiated in Europe for a poly(carbonate) urethane based prosthesis (Cardiopass, CardioTech International). In past studies on the predicate device to Cardiopass called MyoLink, the *in vivo* biostability of the poly(carbonate) urethane was challenged by exposing the polymer to *in vitro* solutions of hydrolytic, oxidative, peroxidative, and biological media. The findings showed excellent hydrolytic stability, especially when compared to conventional poly(ether) urethane controls.[13,14] In addition, a long-term biocompatibility study was completed *in vivo* on canine models that resulted in acceptable mechanical biostability at 3 years.[15] The compliant nature of the MyoLink graft is due to a micro-porous, single-layer structure comprised of an inner and outer skin with a spongy, elastic middle section. When a bolus of blood enters the graft, the inner diameter expands in an elastic manner, while the outside diameter remains relatively unchanged. This design provides radial compliance and conserves the energy of the pulsatile flow of blood through the prosthesis, which maintains a relatively constant wall shear stress (WSS). In addition, the compliant wall produces minimal bleeding from suture holes because the elastic nature of the poly(carbonate) urethane collapses together around the suture after the needle is pulled through the wall. Figure 2 compares the compliance of an artery, vein, PET, and ePTFE to MyoLink.

Other poly(carbonate) urethane-based vascular grafts have been developed that showed promising patency rates in rabbits.[17,18] This system possessed a microporous luminal lining of absorbable collagen and hyaluron that could incorporate heparin and sirolimus for time controlled release. Heparin functions as an anticoagulant, while sirolimus reduces the proliferation of SMCs in a similar

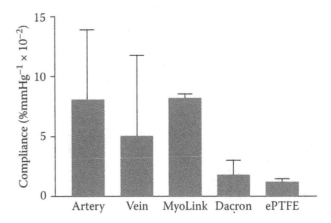

Fig. 2 A comparison of the compliance values for several vascular graft materials and natural tissue.
Source: From Tiwari et al.[16]

manner to its use in drug-eluting coronary stents. However, of primary concern is the development of an endothelium on the graft intimal surface. The use of sirolimus in drug-eluting coronary stents proved relatively unsafe as the drug prevented endothelialization of the stent surfaces, leading to thrombus formation and an increased risk for myocardial infarction.[19] In addition, a potential issue with a drug-eluting graft is that the drug will be exhausted prior to the required period of efficacy.

Poly(dimethyl siloxane) (PDMS) is another attractive soft segment for vascular grafts. PDMS-based polyurethanes show excellent biostability due to their hydrolytically and oxidatively stable siloxane groups.[20,21] In addition, siloxanes produce chemically inert biomaterials, and thrombogenicity testing has revealed good blood compatibility.[22] A preliminary *in vivo* study in a sheep model was conducted to determine the blood and tissue compatibility of a novel compliant small-diameter vascular graft constructed from a poly(ether)urethane-PDMS semi-interpenetrating polymeric network (Fig. 3).[23] The graft had a highly porous inner layer and low-porosity external layer constructed using a spray hase inversion technique. At 6 months, 50% of the ePTFE control grafts were occluded, while all of the PDMS-based grafts were still patent, with neointima formation and no signs of calcification. At 24 months, all PDMS-based grafts were patent, and the luminal surfaces were covered with a neointima. These results indicate that a PDMS-based polyurethane as a vascular graft material may be feasible for use in small-diameter graft applications.

LIMITATIONS AND INSUFFICIENCIES OF CURRENT VASCULAR GRAFTS

Mechanical Compatibility

The importance of matching the compliance of vascular grafts to the native artery was first established in a canine model using autologous tissue.[24] Compliance matching or

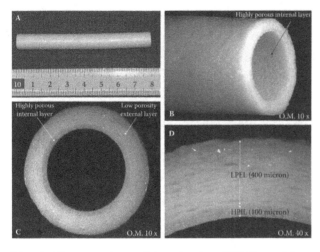

Fig. 3 Representative poly(ether)urethane-PDMS semi-interpenetrating polymeric network used in a sheep model: (**A**) longitudinal view of graft, (**B**) tilted view of graft, (**C**) cross-sectional view of graft showing the sponge-like appearance of the graft wall (original magnification ×10), (**D**) higher magnification of the graft wall (original magnification ×40).

mismatching using carotid artery (1:1) and femoral artery (2:5) autografts resulted in 85% or 37% patency at 90 days postimplantation, respectively. Compliance mismatch is a significant challenge to long-term patency because it causes intimal hyperplasia (IH), which leads to graft occlusion. IH is a condition of tissue ingrowth into the luminal space that causes occlusion of the lumen due to the proliferation of SMCs from the media to the intima. Compliance mismatch between synthetic grafts and natural arteries causes hemodynamic disturbances at the anastomosis that can lead to turbulent flow and changes in WSS.[25] Evidence suggested that both high and low shear stress can stimulate IH.[26] This response is complex, and the mechanisms and interactions are not completely understood; however, IH formation within the *in vivo* period of 2–24 months has been reported to be the primary cause of bypass graft failure.[3] Furthermore, it has been demonstrated that a linear correlation between graft compliance and percentage patency exists (Fig. 4).[27] IH can also be induced from size mismatch.[28] Divergent geometry can reduce WSS and may induce flow separation and turbulence, whereas convergent geometry will increase WSS and may cause EC injury and platelet activation. However, diameter mismatch has been reported to be less of a factor than compliance mismatch in the formation of IH.[29]

ePTFE and PET vascular grafts are incompliant, isotropic, biomechanically inactive, homogeneous conduits compared to the native artery, which is an elastic, anisotropic, dynamic, three-dimensional composite structure. As such, vascular grafts lack the ability to provide mechanical transduction to developing neotissue that controls cell phenotype and ECM production. The viscoelasticity of the arterial wall is an important factor in preserving hemodynamic energy within pulsatile blood flow and maintaining WSS.

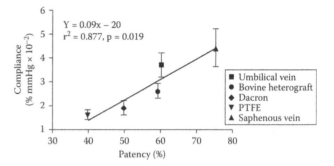

Fig. 4 Data reported for the compliance of various biological and prosthetic grafts versus percentage patency.
Source: From Salacinski et al.[27]

In part, the success of autologous vein grafts results from their ability to respond dynamically to arterial pressure pulse waves in a manner similar to the native artery.[31] Figure 5 shows the modulation of compliance as a function of changes in mean pressure for different vascular conduits. ECs perceive changes in WSS and via physical and chemical pathways transmit signals to the vessel wall that change vessel diameter, tone, SMC proliferation, lumen thrombogenicity, and ECM organization.[32] Synthetic vascular grafts are relatively incompliant and do not reproduce these dynamic characteristics.[33] An artery can increase its compliance in response to a decrease in blood pressure (shock or trauma) to preserve pulsatile energy.[27]

Porosity and Surface Characteristics

Porosity, characterized by pore size and percentage pore area, is a major determining factor for tissue ingrowth at the suture lines and to facilitate the development of a luminal lining.[34] However, the porosity must be balanced between the competing requirements of tissue integration and minimal blood leakage through the graft wall in situ. Preclotting with either the patient's blood or using crosslinked coatings of proteins, such as albumin or gelatin, is used to prevent blood leakage through the graft at the time of implantation. In general, woven PET vascular grafts have smaller pores, while knitted grafts produce looping fibers (velour technique) that result in larger pores and facilitate greater tissue ingrowth.[35] For ePTFE grafts, animal models have indicated that internodal distance is a determining factor for tissue integration.[36] The ideal range of pore sizes has been suggested to be between 10 and 45 mm to facilitate significant fibrovascular tissue infiltration.[37] However, long-term human explants of ePTFE grafts have revealed that the luminal surface is not well covered with tissue. Rather, thin and irregular layers of fibrin are generally found interspersed with areas of exposed ePTFE.[38] In addition, infiltration occurred mainly at the proximal and distal ends of the explanted vascular grafts and increased with the duration of implantation. The surface charge of the luminal surface can be a determining

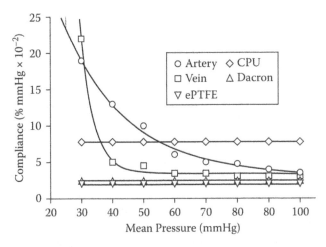

Fig. 5 Modulation of compliance as a function of mean pressure for different vascular conduits.
Source: From Sarkar et al.[30]

factor for the amount and type of cell adhesion and retention under flow conditions. As a result, more electronegative surfaces, such as that of ePTFE, tend to incur lower levels of thrombosis[39] and cell attachment. In addition, the hydrophobicity of ePTFE and PET grafts is a factor in cell proliferation and healing characteristics, with surfaces that are more hydrophilic producing greater proliferation.[40] As such, graft microporosity and surface characteristics influence graft integration, cell proliferation, and the formation of a neointimal layer.

Blood Compatibility

The long-term success of implants is heavily influenced by cell-biomaterial interactions, even more so for vascular grafts as they are in constant contact with platelets, various proteins, complement, and leukocytes. The lack of blood compatibility of PET and ePTFE vascular grafts leads to poor patency when used in medium-diameter applications such as in peripheral vascular surgery, for which 5-year primary patency rates for ePTFE grafts have been reported to be as low as 39%.[41] Therefore in an attempt to create a more blood-compatible surface and increase patency rates, a large amount of research has been focused on the development of thromboresistant coatings and surfaces. Research in this field has been concentrated in two areas: 1) coating the surface of vascular grafts with blood-compatible molecules or polymers; and 2) coating the surface with biomaterials that release antithrombogenic molecules. Each approach attempts to simulate a subset of the properties provided by a functional endothelium. In addition, a confluent, functional EC layer is considered the most ideal "coating" for making synthetic vascular grafts blood compatible. However, protein coatings currently used to seal vascular grafts have not shown evidence of endothelium formation.[42]

The coating of vascular grafts with native proteins to reduce thrombogenicity has been examined since the 1980s with varied success. One such protein, albumin, has been extensively studied after it was discovered that it had the ability to reduce thrombosis. Although *in vitro* studies and *in vivo* animal models have shown promising results, in humans the albumin coating is absorbed into the blood in about 2 months.[6] Coated PET grafts have shown patency rates similar to standard uncoated grafts.[43,44] Consequently, clinical outcomes for albumin have been largely disappointing; therefore, other protein systems have been investigated. The structural protein elastin and other elastic tissue components, namely fibrilin and fibulin, have also been explored as a thrombosis resistant coating. These proteins are interesting as coatings in vascular applications because they: 1) show a reduction in platelet adhesion; 2) have the ability to inhibit the proliferation of SMCs (IH); and 3) facilitate the attachment of ECs.[45,46] This approach provides a pathway for establishing an endothelium on the surface of a synthetic vascular graft.

Ideally an intact endothelium will prevent thrombosis through a number of physical and chemical mechanisms, resist protein and procoagulant deposition, and secrete prostacyclin and nitric oxide (NO), which inhibit platelet adherence and SMC proliferation.[47] For the natural function of the endothelium and SMCs, the regulation of the cell pheno type must be controlled. *In vivo* EC growth, differentiation, barrier function, migration, and survival are regulated in a complex manner by the surrounding ECM, cell-cell contacts, growth factors, and mechanical cues.[48] In particular, hemodynamic forces on ECs stimulate mechanosensor mechanisms and produce junctions, cytoskeleton, and integrin reorganization that alter ECM organization, EC morphology, and EC gene expression. In addition, hemodynamic forces affect the expression and secretion of proteins that control vascular tone, fibrinolysis, surface adhesion, and coagulation promoters and inhibitors. Although research involving the coating of proteins that have the ability to promote endothelialization is promising, sufficient testing is yet to be conducted to show its efficacy in a clinical situation.

Another approach used to create thromboresistant surfaces is the immobilization of anticoagulants on the surface of a synthetic vascular conduit. One such anticoagulant, heparin, has been widely used and shown to be effective *in vitro*, but clinical trials have yielded mixed results.[49] In a randomized multicenter study, patency rates were higher for heparin-bonded PET grafts compared to standard ePTFE at 3 years, but at 5 years the difference was not statistically significant.[50]

In spite of these results, research with heparin as a vascular graft coating is ongoing, with special emphasis placed on the immobilization of heparin to increase efficacy. One such technique, known as end-point immobilization, involves: 1) depolymerization of the reducing end of the linear heparin chain to yield an aldehyde group, which is then; 2) conjugated to a primary amine on the vascular graft surface. *In vitro* studies of graft materials using this method of heparin attachment have shown effective inhibition of initial contact activation enzymes by an antithrombin-mediated mechanism.[51,52] This led to the development of a heparin-coated PTFE vascular graft in which heparin is bound via end-point immobilization (Propaten Graft, W. L. Gore & Associates). Currently, only 1-year patency rates have been reported, but the data indicate that heparin-bound grafts are more patent than standard ePTFE grafts in femoropopliteal bypass surgery.[53] Although initial results appear promising, 5-year patency rates are needed to determine the long-term efficacy of this prosthesis.

NO is a bioactive molecule released from the endothelium that prevents platelet activation. Interestingly, this molecule can be released by polymers that contain diazeniumdiolate groups ($[N(O)NO]^-$), which on decomposi-tion in physiological fluids spontaneously generate NO.[54] A number of approaches for preparing NO-releasing polymers have been studied and include: 1) the physical blending of the $[N(O)NO]^-$ complex into a bulk polymer; 2) the incorporation of a group (nucleophile) containing $[N(O)NO]^-$ as a side chain into the polymer; and 3) the incorporation of $[N(O)NO]^-$ into the polymer backbone (Fig. 6).

A polymer coating composed of a cross-linked poly(ethylenimine) where $[N(O)NO]^-$ was directly attached to the polymer chain was studied in a baboon animal model to examine thrombus formation and show feasibility of such a coating in vascular applications.[55] In this particular study, the polymer containing $[N(O)NO]^-$ was coated onto a ePTFE graft, which was then exposed to flowing blood via an artery-to-vein shunt for 1 hr; results showed substantially less thrombus formation in the graft coated with $[N(O)NO]^-$.

Although promising, one of the major limitations of NO-generating coatings is the eventual exhaustion of NO donors. The maximum reported NO release period is 2 months.[56] Due to this limitation, research has focused on the development of sustainable release mechanisms based on the catalytic generation of NO from endogenous nitrite via polymeric films doped with lipophilic Cu(II) complexes.[57] Using this process, NO has the ability to be generated at therapeutic levels for extended periods far beyond the previous 2-month maximum.

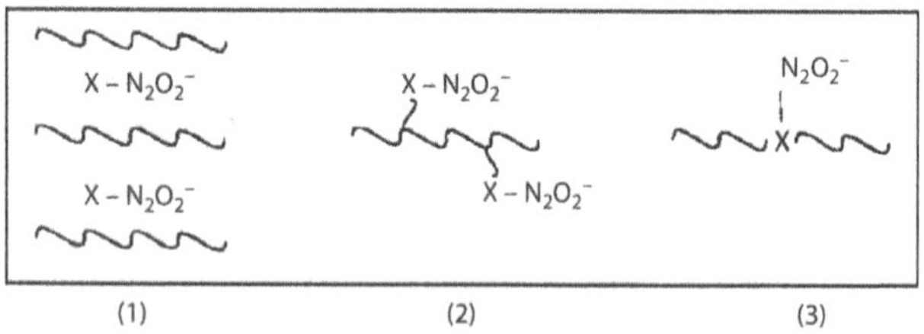

Immune and Inflammatory Response

Along the vascular wall, blood contacts an endothelium, which interacts with a high concentration of leukocytes and complement. For synthetic vascular conduits, the blood-biomaterial interface is under constant surveillance by both the inflammatory and immune systems. As such, the introduction of a foreign surface to blood triggers a series of events, including protein adsorption, complement and leukocyte adhesion, and the acute activation of platelets and coagulation. This intense immune and inflammatory response to synthetic vascular grafts contributes to their poor blood biocompatibility and subsequent complications.

It is well established that polymers of biologic origin activate a strong immune response mediated by the classical pathway of the complement system. As a consequence of the collagen, albumin, gelatin, or other biologic coating used to seal the intimal surface, the immune response is activated by the perceived antigen. Several studies have recognized the humoral and cell-mediated immune response to coated PET vascular grafts.[58–60] On the other hand, the "nonself" surface of polymeric biomaterials often results in spontaneous low-rate complement activation via the alternative pathway. To date, blood compatibility issues have focused on minimizing coagulation and platelet adhesion or activation through systemic drug therapy. Considering that blood clot formation is a special case of inflammation, the direct, local control of thrombosis and coagulation may be possible. However, this requires a more complete understanding of the role of leukocytes in blood-biomaterial interaction.

This subject has been reviewed by Gobert and Sefton with emphasis on complement activation as a prelude to thrombosis and the role of leukocytes in coagulation.[61] To demonstrate, Hong and coworkers showed that without leukocytes the formation of thrombin was negligible when blood was exposed to a biomaterial.[62] Following exposure to polyvinyl chloride, levels of thrombin were negligible in plasma and platelet rich plasma, but significant in whole blood (i.e., presence of leukocytes). Therefore leukocyte activation and cytokine production were a prerequisite for coagulation.

Monocytes and macrophages play a pivotal role in synthetic vascular graft complications. Unfortunately, ePTFE and PET elicit a chronic inflammatory response that starts with early leukocyte adhesion and can ultimately assist in late graft failure.[47] For example, macrophages have been implicated in the chemical signal that stimulates SMC proliferation resulting in IH.[63] In addition, it has also

Fig. 6 Generalized polymer types for the addition of $[N(O)NO]^-$: X = nucleophile residue, $N_2O_2^-$ = NO releasing unit, ~ = polymer backbone. (1) the physical blending of the $[N(O)NO]^-$ complex into a bulk polymer, (2) the incorporation of a group (nucleophile) containing $[N(O)NO]^-$ as a side chain into the polymer, (3) the incorporation of $[N(O)NO]^-$ into the polymer backbone.
Source: From Smith et al.[55]

been suggested that the potent cytokines secreted by chronically active macrophages are inhibitors to angiogenesis, which significantly reduces neovascularization.[64] Constructed from permanent biomaterials, current vascular grafts produce chronic inflammatory responses from the seemingly unlimited local supply of leukocytes within blood. In a subcutaneous rat study investigating the macrophage response to PET and ePTFE, it was shown that PET produced a more intense initial inflammatory response than ePTFE. This may be the result of the more inert surface of ePTFE, which reduces early leukocyte adhesion. However, ePTFE produced a delayed proliferating cell response, with 4% at 3 weeks and 21% at 5 weeks, compared to PET, with 23% at 3 weeks and 8% at 5 weeks. [65] These results underlined the importance of minimizing the long-term inflammatory response, which impedes healing and the population of native vascular cells within vascular grafts. In addition, a better understanding of the role of leukocytes in vascular graft complications may hold the key to improving patency.

TISSUE-ENGINEERED VASCULAR GRAFTS

Scaffolds for the Next Generation

Due to the poor patency of current vascular grafts, motivation to create the ideal prosthesis has stimulated and challenged current tissue-engineering practices. Tissue engineering a patent vascular graft remains an enormous task as the list of prerequisites is long and requires a multidisciplinary approach with input from bioengineers, clinicians, and researchers. The three basic elements that are required to accomplish this task are a structural scaffold, vascular cells, and a nurturing environment.[66] The implementation of these elements is wide and varied due to the many choices for scaffold materials, cell types and sources, and the choice of an *in vitro* or *in vivo* nurturing environment. The vast number of choices provides flexibility but adds to the complexity of finding the ideal solution that meets all, or most, of the prerequisites. To make matters even more difficult, the device design must have clinical acceptance, and the approach must not require excessive preparation time or effort.

Synthetic Absorbable Scaffolds

Synthetic polymers can be engineered to have a range of different mechanical and chemical properties, manufactured easily and consistently, and processed using a variety of methods into various forms (e.g., films, nanofibers, yarns). As a result, polymers, and their copolymers, represent a class of materials that provide extensive design flexibility and are cost effective. Hydrolytically degrading and bioabsorbable polymers for use as tissue-engineered vascular graft scaffolds possess the ability to modulate

temporal degradation and strength retention, support cell ECM production and proliferation, and absorb to be replaced by functional remodeling vascular tissue. In the long term, this approach leaves the implant site absent of any foreign material. However, early attempts to produce fully absorbable vascular grafts from polyglycolide (PG) did not perform well due to aneurysm formation.[6] This can be attributed to the relatively fast loss of mechanical integrity of PG, which preceded the development of sufficient tissue strength. To overcome this issue, partially absorbable scaffolds have been developed that combine a fast-degrading absorbable polymer supported by a nondegradable polymer.[67–69] However, research has focused mainly on the use of compliant, fully absorbable scaffolds based mostly on linear, aliphatic copolyesters with the major molecular component being polylactide (PL) or polycaprolactone (PCL). These polymeric systems provide significantly longer strength retention.

Crapo and Wang investigated two fully absorbable scaffolds with the hypothesis that elastomeric scaffolds under dynamic conditions would develop strong and compliant arterial constructs. To this end, scaffolds composed of either rigid poly(lactide-co-glycolide) (PLGA) or elastomeric poly(glycerol sebacate) (PGS) were seeded with baboon SMCs and *in vitro* cultured under dynamic mechanical stimulation for 10 days.[70] Their results found that initial scaffold compliance was a determining factor in tissue development and postconditioning mechanical properties. Furthermore, the elastic PGS graft produced significant amounts of elastin within days, while the stiff PLGA graft produced no appreciable elastin.

Other attempts to fabricate elastic scaffolds that can withstand mechanically dynamic conditions have been developed using poly(glycolide-co-caprolactone)[71] and poly(lactide-co-caprolactone) (PLCL).[72] In the former, scaffolds were produced by solvent casting and NaCl leaching to create a compliant, porous tube. Mechanical testing showed that radial strains up to 120% provided 98% recovery, and failure occurred at 250% elongation, indicating excellent compliance and viscoelasticity. Cyclic loading at an applied strain of 10% and a frequency of 1 Hz for 2 days produced less than 5% permanent deformation. To develop a mechanically active small-diameter graft, scaffolds of PLCL were prepared using the electrospinning technique to produce fabrics of different wall thickness (50–340 mm).[72] Thinner-walled vessels were shown to pulsate synchronously, approaching the native artery, to dynamic pulsatile flow conditions.

The use of electrospinning is another novel solution to construct a fully absorbable vascular prosthesis that is complaint. This technology offers the benefit of being able to incorporate synthetic and biologic polymers simultaneously and control composition, structure, and mechanical properties.[73,74] Scaffold preparation using collagen type I, elastin, and poly(D,L lactide-co-glycolide) (PDLGA) was electrospun onto a tubular mandrel to a thickness of 1 mm and a length of 12 cm.[73] *In vitro* mechanical testing produced burst pressures of 1,425 mmHg, nearly 12 times systolic pressure, with 40% strain at failure. Tissue compatibility results showed a confluent monolayer of ECs on the inner surface after 4 days and SMCs on the outer surface after 3 days. This approach has advantages in the clinical setting as different geometries, compositions, and tailored mechanical properties can be produced reliably.

Another proposed construction utilizing the mechanical stability of PLGA was a hybrid scaffold using porous freeze-dried marine collagen as a cell interface with a fibrous PLGA electrospun structural scaffold.[75] Following EC and SMC seeding, an *in vitro* pulsatile perfusion system was utilized to provide radial distension. Results indicated that expressions of smooth muscle (SM) α-actin, SM myosin heavy chain, EC von Willebrand factor, and NO were observed. As such, normal cell phenotype activity was preserved.

The ideal tissue-engineered scaffold may incorporate both synthetic and naturally derived tissues to take advantage of the mechanical stability and cell attachment and signaling characteristics, respectively. An improved understanding of cell phenotype regulation and cell responses to chemical and mechanical stimulation will be a key factor in the development of a patent tissue-engineered vascular graft.

PCL is a slow-degrading polyester that is not as widely used as its other absorbable counterparts (PL, PG); however, it offers several advantages for use in vascular graft tissue engineering.[76] For example, PCL provides much longer strength retention to facilitate tissue maturation for elderly patients or those with comorbidities. This greater strength retention may be required for such a high-risk device, for which failure may result in mortality. In addition, due to fewer ester groups, the degradation products are released over a longer period of time, which reduces the likelihood of local pH changes to an acidic environment, as is known to occur with PL and PG.

The use of a PL/PCL-based scaffold seeded with autologous bone marrow-derived mononuclear cells (BMMCs) has shown excellent patency in a pediatric clinical study of patients born with a single-ventricle defect, a life-threatening condition.[77–80] Twenty-five grafts were implanted, and there was no graft-related mortality, evidence of aneurysm formation, graft rupture, infection, or ectopic calcification. Within the first 7 years, the only graft-related complication was stenosis, affecting approximately 16% of the grafts.[78] To better understand the underlying mechanisms, canine,[81] lamb,[82] and mice[83–86] studies were initiated by researchers using a miniaturized model of that used in the clinical trial. Results indicated that BMMCs had an indirect rather than direct impact on the regeneration of vascular tissue. Rather than BMMCs incorporating into the graft, they instead stimulated monocyte activity, which triggered an inflammatory response, leading to the

recruitment of ECs and SMCs from adjacent normal vessels.[87] In a short time, the seeded cells were gone, and the host cells were incorporated. This investigation has shown that the innate repair mechanisms can be used to create new tissues with seeded cells as only precursors to initiate the response.

Overall, tissue-engineered vascular grafts based on absorbable synthetic copolymers have shown feasibility to produce a viable temporary scaffold that will accommodate vascular cells and maintain structural integrity during cell proliferation and ECM development under hemodynamic conditions; however, the engineered scaffold that will meet all prerequisites is still unclear. In addition, the logistics of acquiring cells, the long culturing periods, and the limited or variable cell proliferation capacity are all potential clinical limitations.[1]

Polymers of Natural Origin

The use of biologic materials derived from proteins and polysaccharides has many advantages. For example, biopolymers have the potential to preserve or recreate structural characteristics and mechanical function, the ability to be naturally remodeled, and the ability to provide chemical signals and intrinsic sites for cell attachment and growth. Polymers of natural origin exhibit the ability to augment the healing and remodeling process and are not subject to some of the insufficiencies of synthetic absorbable polymers, which include the release of acidic by-products and marked inflammatory responses.[88] However, like PG, biopolymers have a strength retention profile that is inadequate for many load-bearing applications. Although several methods of incorporating natural materials into scaffolds have been investigated, one primary approach is using decellularized ECM biomaterials.

Decellularized tissues attempt to mimic arterial physiology by decellularizing a donor tissue, typically using a series of physical, chemical, and enzymatic approaches to isolate the ECM and remove all of the highly immunogenic cellular matter. Chemical methods vary greatly and include alkaline/acidic treatments, nonionic and ionic detergents, zwitterionic detergents, hypotonic and hypertonic treatments, tri(n-butyl) phosphate (TBP), and various chelating agents to achieve decellularization.[89] Once decellularized, the ECM graft can then be repopulated with cardiovascular cells using either an *in vitro* seeding or *in vivo* infiltration method to approximate a biomimetic structure. In addition, *in vitro* seeding approaches often involve the use of some type of vessel bioreactor, which functions to mimic the physiological vessel environment, including the cyclic strain and shear stresses seen in the body, with the end goal of having a fully functional vessel at the time of surgery.[90] ECM scaffolds can be derived from human and several animal sources; they are composed of various structural and functional proteins, ranging from collagen, to elastin, to a number of growth factors, and to proteoglycans, making them ideal for use as graft scaffold materials.[91,92] The primary sources of decellularized scaffolds that have been studied include human umbilical arteries and veins, ureters (porcine, bovine, or canine), and carotid arteries (porcine and canine).

Gui and colleagues showed successful decellularization of a human umbilical artery with maintenance of graft mechanical strength using a CHAPS (3-[(3-cholamidopropyl)-dimethylammonio]-1-propane sulfonate)/ sodium dodecyl sulfate buffer treatment followed by incubation in EC growth media.[93] *In vivo* testing using an abdominal aorta rat model revealed retained function for up to 8 weeks. In another study, decellularized canine ureters showed only 20% patency at 1 week when implanted in a canine carotid artery without cell seeding, while the EC-seeded versions showed 100% patency at 24 weeks.[94] Cell seeding prior to implantation appears to be a key factor in determining the success of a decellularized, protein-based vascular graft material.

Although the use of decellularized tissues as scaffolds seems attractive, there are a number of hurdles that have yet to be overcome. One of the largest of these is stabilization of the decellularized ECM. On implantation, unmodified natural polymers will degrade due to chemical and enzymatic degradation, seriously decreasing the life of the prosthesis.[95] In addition, graft calcification and immunological recognition can occur, possibly due to the presence of residual cellular components. An examination of a decellularized bovine ureter graft indicated that graft failure with aneurysmal dilation and thrombus formation may be due to an acute and chronic inflammatory response caused by residual cellular components not removed during the decellularization process.[96]

For these reasons, and the need for sterilization, naturally derived biomaterials typically require chemical pretreatment with cross-linking reagents such as glutaraldehyde. However, this leaves the scaffold susceptible to cytotoxic effects on repopulated cells. Increased cross-linking produces a structure more resistant to degradation, but at the cost of mechanical compliance. Other cross-linking agents, such as polyepoxy compounds, showed reduced levels of calcification, decreased antithrombogenicity, and increased flexibility, resulting in improved patency compared to vascular grafts treated with other chemicals.[95]

Another noted challenge associated with decellularized vessels is incomplete repopulation, generally caused by the inability of seeded cells to penetrate the densely packed collagen-elastin network. In an attempt to solve this issue, Kurane and colleagues preferentially removed the collagen of decellularized porcine carotid artery grafts to increase their porosity in addition to loading the grafts with basic fibroblast growth factor.[97] Cell infiltration was analyzed from the histological evaluation of subdermal implanted tubular constructs. Overall, improved cellular infiltration was noted.

Another investigated avenue for attaining a tissue-engineering scaffold is using biological proteins or polysaccharides as raw materials for constructing vascular grafts. This concept of reconstituting a vascular graft from natural materials was first introduced by Weinburger and Bell over 20 years ago with the formation of a tubular type I collagen gel that was seeded or loaded with various cell types that are contained in natural vascular tissues.[98] Although the general idea remains the same, techniques and material types have since evolved to employ the use of various materials, such as elastin, fibrin, collagen, silk, and hyaluronan. While many of these constructs have excellent cell compatibility, they generally lack the mechanical properties (strength, compliance, and elasticity) required for vascular applications. To address this issue, Wise and colleagues developed a composite graft material using elastin and PCL through the use of an electrospinning technique.[99] The resulting grafts had mechanical properties similar to human internal mammary arteries, with excellent EC attachment/proliferation and blood compatibility.

Fibroin has also shown promise as a vascular graft material. Fibroin is the core protein component of silk that remains after the removal of sericin, the gum-like protein surrounding silk fibers.[100] Enomoto and coworkers constructed small-caliber (1.5-mm diameter) vascular grafts by weaving silk fibroin thread and implanted them in an abdominal aorta rat model.[101] They showed a significant increase in patency rates (85.1% vs. 30%) at 1 year compared to ePTFE controls. In addition, ECs and SMCs successfully migrated into the scaffolds, with a significant increase in SMC migration between 1 and 3 months.[101]

Hyaluronan, or estrified hyaluronic acid, is another versatile tissue-engineering tool due to its many desirable characteristics. From a standpoint of physical properties, it is a GAG with variable molecular weight, a negative charge that attracts water, and controllable cross-linking properties, making load-bearing hydrogels possible.[102] The *in vivo* biological advantages include tissue organization, wound healing, angiogenesis, cell adhesion, regulation of inflammation, and nonimmunogenicity, and when degraded by hyaluronidases during neotissue formation, the resulting products induce production of ECM.[103,104] In addition, hydrogels can be produced by photopolymerization of glycidyl methacrylate and hyaluronic acid to comprise bioactive, degradable, hydrated, and pliable materials that can be modified with peptide moieties for protein therapy.[103] In vascular graft research, hyaluronan-based materials have been shown to promote endothelial and SMC attachment while displaying hemocompatibility properties.[88,104,105] Although these results seem promising, like protein-based biomaterials, hyaluronan vascular grafts may have inadequate *in vivo* mechanical stability with time.

Vascular—Zwitteronic

CONCLUSIONS

Autologous vascular tissue has been used successfully for decades to replace diseased vessels. However, after significant effort by bioengineers and clinicians, synthetic alternatives are not equivalent to autologous vessels. This endeavor has provided an appreciation for blood-biomaterial interactions, the importance of cell integration, and the biomechanical properties required to achieve a highly patent vascular prosthesis. Future advances in the patency of vascular grafts will likely occur as stepwise improvements, rather than a single novel discovery that alleviates all of the complications. In the short term, the next advances will likely come from the use of poly(carbonate) urethane grafts, which have improved blood compatibility and more closely match the compliance of an artery. The development of new materials or coatings that support a functional endothelium will be required to produce the first small-diameter graft for use in coronary bypass procedures and to improve the patency of medium-diameter grafts. While the field of tissue engineering continues to search for a solution to total vascular tissue regeneration, more emphasis on reliably producing an endothelium would have the greatest short-term impact on patient care.

At the current rate of advancements in tissue engineering, it is likely that by the year 2035, vascular grafts as we know them today will be obsolete and replaced by cell-based scaffolds. Although the realization of tissue-engineered devices has been slow, the investment in cell-based technologies will eventually result in a highly patent vascular graft. Although the optimal implementation of tissue-engineered conduits remains uncertain, and significant challenges remain ahead, it is likely that polymeric scaffolds will have a role in supporting the application of a cell-based vascular graft technology.

The translation of a tissue-engineered vascular graft from research to clinical use will be challenging. All cellular, mechanical, and physical characteristics of a tissue-engineered vascular graft must be controlled as they relate to the design, manufacturing processes, and equipment used in the construction of the device. Research is beginning to uncover some of the significant variables for achieving success. However, the translation to clinical use requires that the critical design variables are well understood and controlled, with either validation or verification to ensure efficacy for all high-risk design features. The Food and Drug Administration requires that all design and manufacturing activities are risk based. In other words, as the risk of a design feature failing causes injury or mortality, the level of control and understanding must increase to mitigate the risk. It is easy to anticipate that tissue-engineered vascular grafts would have a significant number of design risks related to controlling cell phenotype, ECM production and quality, and endothelialization. In addition, the process of producing a tissue-engineered

vascular graft must be predictable and reproducible to a high degree of statistical significance. Patient age and comorbidities will mean that cell-seeded grafts will inherently have a high degree of variability. As such, the validation of such a process will be difficult, if not impossible. Therefore verification of each graft produced will be required using an elaborate series of tests that ensure safety and efficacy of the specific graft intended for implantation. With design verification, several issues can arise. How is the failure of a graft to meet predetermined specifications handled? Do you continue with conditioning the implant in a bioreactor or start over? What about the patient? Can the patient wait, or is intervention critical? These are just a few examples of the many challenges ahead as cell-based vascular graft technologies translate from research to the clinical setting.

REFERENCES

1. Ravi, S.; Qu, Z.; Chaikof, E.L. Polymeric materials for tissue engineering of arterial substitutes. Vascular **2009**, *17* (Suppl 1), S45–S54.
2. Veith, F.J.; Moss, C.M.; Sprayregen, S.; Montefusco, C. Preoperative saphenous venography in arterial reconstructive surgery of the lower extremity. Surgery **1979**, *85* (3), 253–256.
3. Sarkar, S.; Salacinski, H.J.; Hamilton, G.; Seifalian, A.M. The mechanical properties of infrainguinal vascular bypass grafts: Their role in influencing patency. Eur. J. Vasc. Endovasc. Surg. **2006**, *31* (6), 627–636.
4. Zilla, P.; Bezuidenhout, D.; Human, P. Prosthetic vascular grafts: Wrong models, wrong questions and no healing. Biomaterials **2007**, *28* (34), 5009–5027.
5. Kannan, R.Y.; Salacinski, H.J.; Butler, P.E.; Hamilton, G.; Seifalian, A.M. Current status of prosthetic bypass grafts: A review. J. Biomed. Mater. Res. B Appl. Biomater. **2005**, *74* (1), 570–581.
6. Xue, L.; Greisler, H.P. Biomaterials in the development and future of vascular grafts. J. Vasc. Surg. **2003**, *37* (2), 472–480.
7. Smith, R.; Oliver, C.; Williams, D. The enzymatic degradation of polymers *in vitro*. J. Biomed. Mater. Res. **1987**, *21* (8), 991–1003.
8. Jorgensen, C.S.; Paaske, W.P. Physical and mechanical properties of ePTFE stretch vascular grafts determined by time-resolved scanning acoustic microscopy. Eur. J. Vasc. Endovasc. Surg. **1998**, *15* (5), 416–422.
9. Venkataramen, S.; Boey, F.; Luciana, L. Implanted cardiovascular polymers: Natural, synthetic, and bio-inspired. Prog. Polym. Sci. **2008**, *33* (9), 853–874.
10. Post, S.; Kraus, T.; Muller-Reinartz, U.; Weiss, C.; Kortmann, H.; Quentmeier, A.; Winkler, M.; Husfeldt, K.J.; Allenberg, J.R. Dacron vs. polytetrafluoroethylene grafts for femoropopliteal bypass: A prospective randomised multicentre trial. Eur. J. Vasc. Endovasc. Surg. **2001**, *22* (3), 226–231.
11. Roll, S.; Muller-Nordhorn, J.; Keil, T.; Scholz, H.; Eidt, D.; Greiner, W.; Willich, S.N. Dacron vs. PTFE as bypass

materials in peripheral vascular surgery—Systematic review and meta-analysis. BMC Surg. **2008**, *8* (1), 22.

12. Lau, H.; Cheng, S.W. Is the preferential use of ePTFE grafts in femorofemoral bypass justified? Ann. Vasc. Surg. **2001**, *15* (3), 383–387.

13. Salacinski, H.J.; Odlyha, M.; Hamilton, G.; Seifalian, A.M. Thermo-mechanical analysis of a compliant poly(carbonate-urea)urethane after exposure to hydrolytic, oxidative, peroxidative and biological solutions. Biomaterials **2002**, *23* (10), 2231–2340.

14. Salacinski, H.J.; Tai, N.R.; Carson, R.J.; Edwards, A.; Hamilton, G.; Seifalian, A.M. *In vitro* stability of a novel compliant poly(carbonate-urea)urethane to oxidative and hydrolytic stress. J. Biomed. Mater. Res. **2002**, *59* (2), 207–218.

15. Seifalian, A.; Salacinski, H.; Tiwari, A.; Edwards, A.; Bowald, S.; Hamilton, G. *In vivo* biostability of poly(carbonate-urea)urethane graft. Biomaterials **2002**, *24* (14), 2549–2557.

16. Tiwari, A.; Salacinski, H.; Seifalian, A.M.; Hamilton G. New prostheses for use in bypass grafts with special emphasis on polyurethanes. Cardiovasc. Surg. **2002**, *10* (3), 191–197.

17. Ishii, Y.; Kronengold, R.T.; Virmani, R.; Rivera, E.A.; Goldman, S.M.; Prechtel, E.J.; Schuessler, R.B.; Damiano, R.J. Novel bioengineered small caliber vascular graft with excellent one-month patency. Ann. Thorac. Surg. **2007**, *83* (2), 517–525.

18. Ishii, Y.; Sakamoto, S.; Kronengold, R.T.; Virmani, R.; Rivera, E.A.; Goldman, S.M.; Prechtel, E.J.; Hill, J.G; Damiano, R.J. A novel bioengineered small-caliber vascular graft incorporating heparin and sirolimus: Excellent 6-month patency. J. Thorac. Cardiovasc. Surg. **2008**, *135* (6), 1237–1245; discussion 1245–1246.

19. Luscher, T.F.; Steffel, J.; Eberli, F.R.; Joner, M.; Nakazawa, G.; Tanner, F.C.; Virmani, R. Drug-eluting stent and coronary throm-bosis: Biological mechanisms and clinical implications. Circulation **2007**, *115* (8), 1051–1058.

20. Martin, D.J.; Warren, L.A.; Gunatillake, P.A.; McCarthy, S.J.; Meijs, G.F.; Schindhelm, K. Polydimethylsiloxane/poly-ether-mixed macrodiol-based polyurethane elastomers: Biostability. Biomaterials **2000**, *21* (10), 1021–1029.

21. Simmons, A.; Padsalgikar, A.D.; Ferris, L.M.; Poole-Warren L.A. Biostability and biological performance of a PDMS-based polyurethane for controlled drug release. Biomaterials **2008**, *29* (20), 2987–2995.

22. Tepe, G.; Schmehl, J.; Wendel, H.P.; Schaffner, S.; Heller, S.; Gianotti, M.; Claussen, C.D.; Duda, S.H. Reduced thrombogenicity of nitinol stents—*in vitro* evaluation of different surface modifications and coatings. Biomaterials **2006**, *27* (4), 643–650.

23. Soldani, G.; Losi, P.; Bernabei, M.; Burchielli, S.; Chiappino, D.; Kull, S.; Briganti, E.; Spiller, D. Long term performance of small-diame-ter vascular grafts made of a poly(ether) urethane-polydimethylsiloxane semi-interpenetrating polymeric network. Biomaterials **2009**, *31* (9), 2592–2605.

24. Abbott, W.M.; Megerman, J.; Hasson, J.E.; L'Italien, G.; Warnock, D.F. Effect of compliance mismatch on vascular graft patency. J. Vasc. Surg. **1987**, *5* (2), 376–382.

25. Sarkar, S.; Sales, K.M.; Hamilton, G.; Seifalian, A.M. Addressing thrombogenicity in vascular graft construction. J. Biomed. Mater. Res. B Appl. Biomater. **2007**, *82* (1), 100–108.

26. Fei, D.Y.; Thomas, J.D.; Rittgers, S.E. The effect of angle and flow rate upon hemo-dynamics in distal vascular graft anastomoses: A numerical model study. J. Biomech. Eng. **1994**, *116* (3), 331–336.

27. Salacinski, H.J.; Goldner, S.; Giudiceandrea, A.; Hamilton, G.; Seifalian, A.M.; Edwards, A.; Carson, R.J. The mechanical behavior of vascular grafts: A review. J. Biomater. Appl. **2001**. *15* (3), 241–278.

28. Weston, M.W.; Rhee, K.; Tarbell, J.M. Compliance and diameter mismatch affect the wall shear rate distribution near an end-to-end anastomosis. J. Biomech. **1996**, *29* (2), 187–198.

29. Perktold, K.; Leuprecht, A.; Prosi, M.; Berk, T.; Czerny, M.; Trubel, W.; Schima, H. Fluid dynamics, wall mechanics, and oxygen transfer in peripheral bypass anastomoses. Ann. Biomed. Eng. **2002**, *30* (4), 447–460.

30. Sarkar, S.; Schmitz-Rixen, T.; Hamilton, G.; Seifilian, A. Achieving the ideal prop-erties for vascular bypass graft using tissue engineering approach: A review. Med. Biol. Eng. Comput. **2007**, *45* (4), 327–336.

31. Baird, R.N.; Abbott, W.M. Pulsatile blood-flow in arterial grafts. Lancet **1976**, *2* (7992), 948–950.

32. Isenberg, B.C.; Williams, C.; Tranquillo, R.T. Small-diameter artificial arteries engineered *in vitro*. Circ. Res. **2006**, *98* (1), 25–35.

33. Tai, N.R.; Salacinski, H.J.; Edwards, A.; Hamilton, G.; Seifalian, A.M. Compliance properties of conduits used in vascular reconstruction. Br. J. Surg. **2000**, *87* (11), 1516–1524.

34. Matsuda, T. Recent progress of vascular graft engineering in Japan. Artif. Organs **2004**, *28* (1), 64–71.

35. Bowald, S.; Busch, C.; Eriksson, I. Arterial regeneration following polyglactin 910 suture mesh grafting. Surgery **1979**, *86* (5), 722–729.

36. Nagae, T.; Tsuchida, H.; Peng, S.K.; Furukawa, K.; Wilson, S.E. Composite porosity of expanded polytetrafluoroethylene vascular prosthesis. Cardiovasc. Surg. **1995**, *3* (5), 479–484.

37. He, H.; Matsuda, T. Arterial replacement with compliant hierarchic hybrid vascular graft: Biomechanical adaptation and failure. Tissue Eng. **2002**, *8* (2), 213–224.

38. Guidon, R.; Chafke, N.; Maurel, S.; How, T.; Batt, M.; Marois, M.; Gosselin, C. Expanded polytetrafluoroethylene arterial prosthesis in humans: A histopathological study of 298 surgically excised grafts. Biomaterials **1993**, *14* (9), 678–693.

39. Akers, D.L.; Du, Y.H.; Kempczinski, R.F. The effect of carbon coating and porosity on early patency of expanded polytetrafluoroethylene grafts: An experimental study. J. Vasc. Surg. **1993**, *18* (1), 10–15.

40. Wildevuur, C.R.; van der Lei, B.; Schakenraad, J.M. Basic aspects of the regeneration of small-calibre neoarteries in biodegradable vascular grafts in rats. Biomaterials **1987**, *8* (6), 418–422.

41. Klinkert, P.; Post, P.N.; Breslau, P.J.; van Bockel, J.H. Saphenous vein versus PTFE for above-knee femoropopliteal

Vascular—Zwitterionic

bypass. A review of the literature. Eur. J. Vasc. Endovasc. Surg. **2004**, *27* (4), 357–362.

42. Francois, S.; Chakfe, N.; Durand, B.; Laroche, G. A poly(L-lactic acid) nanofibre mesh scaffold for endothelial cells on vascular prostheses. Acta Biomater. **2009**, *5* (7), 2418–2428.

43. Kudo, F.A.; Nishibe, T.; Miyazaki, K.; Flores, J.; Yasuda, K. Albumin-coated knitted Dacron aortic prostheses. Study of postoperative inflammatory reactions. Int. Angiol. **2002**, *21* (3), 214–217.

44. al-Khaffaf, H.; Charlesworth, D. Albumin-coated vascular prostheses: A five-year follow-up. J. Vasc. Surg. **1996**, *23* (4), 686–690.

45. Ito, S.; Ishimaru, S.; Wilson, S.E. Application of coacervated alpha-elastin to arterial prostheses for inhibition of anastomotic intimal hyperplasia. ASAIO J. **1998**, *44* (5), M501–M505.

46. Williamson, M.R.; Shuttleworth, A.; Canfield, A.E.; Black, R.A.; Kielty, C.M. The role of endothelial cell attachment to elastic fibre molecules in the enhancement of monolayer formation and retention, and the inhibition of smooth muscle cell recruitment. Biomaterials **2007**, *28* (35), 5307–5318.

47. Parikh, S.A.; Edelman, E.R. Endothelial cell delivery for cardiovascular therapy. Adv. Drug Deliv. Rev. **2000**, *42* (1), 139–161.

48. McGuigan, A.P.; Sefton, M.V. The influence of biomaterials on endothelial cell thrombogenicity. Biomaterials **2007**, *28* (16), 2547–2571.

49. Heyligers, J.M.; Verhagen, H.J.; Rotmans, J.I.; Weeterings, C.; de Groot, P.G.; Moll, F.L.; Lisman, T. Heparin immobilization reduces thrombogenicity of small-caliber expanded polytetrafluoroethylene grafts. J. Vasc. Surg. **2006**, *43* (3), 587–591.

50. Devine, C.; McCollum, C. Heparin-bonded Dacron or polytetrafluoroethylene for femoropopliteal bypass: Five-year results of a prospective randomized multi-center clinical trial. J. Vasc. Surg. **2004**, *40* (5), 924–931.

51. Elgue, G.; Blomback, M.; Olsson, P.; Riesenfeld, J. On the mechanism of coagulation inhibition on surfaces with end point immobilized heparin. Thromb. Haemost. **1993**, *70* (2), 289–293.

52. Sanchez, J.; Elgue, G.; Riesenfeld, J.; Olsson, P. Studies of adsorption, activation, and inhibition of factor XII on immobilized heparin. Thromb. Res. **1998**, *89* (1), 41–50.

53. Lindholt, J.S.; Gottschalksen, B.; Johannesen, N.; Dueholm, D.; Ravn, H.; Christensen, E.D.; Viddal, B.; Fasting, H. The Scandinavian Propaten™ Trial—1-year patency of PTFE vascular prostheses with heparin-bonded luminal surfaces compared to ordinary pure PTFE vascular prostheses—A randomised clinical controlled multi-centre trial. Eur. J. Vasc. Endovasc. Surg. **2011**, *41* (5), 668–673.

54. Vanin, A.F.; Vedernikov Iu, I.; Galagan, M.E.; Kubrina, L.N.; Kuzmanis, I.; Kalvin'sh, I.; Mordvintsev, P.I. [Angeli salt as a producer of nitrogen oxide in animal tissues]. Biokhimiia **1990**, *55* (8), 1408–1413.

55. Smith, D.J.; Chakravarthy, D.; Pulfer, S.; Simmons, M.L.; Hrabie, J.A.; Citro, M.L.; Saavedra, J.E.; Keefer, L.K. Nitric oxide-releasing polymers containing the [N(O)NO]⁻ group. J. Med. Chem. **1996**, *39* (5), 1148–1156.

56. Ho-wook, J.; Taite, J.; West, J. Nitric oxide releasing polyurethanes. Biomacromolecules **2005**, *6*, 838–844.

57. Oh, B.K.; Meyerhoff, M.E. Catalytic generation of nitric oxide from nitrite at the interface of polymeric films doped with lipophilic CuII-complex: A potential route to the preparation of thromboresistant coatings. Biomaterials **2004**, 25 (2), 283–293.

58. Schlosser, M.; Wilhelm, L.; Urban, G.; Ziegler, B.; Ziegler, M.; Zippel, R. Immunogenicity of polymeric implants: Long-term antibody response against polyester (Dacron) following the implantation of vascular prostheses into LEW.1A rats. J. Biomed. Mater. Res. **2002**, *61* (3), 450–457.

59. Schlosser, M.; Zippel, R.; Hoene, A.; Urban, G.; Ueberrueck, T.; Marusch, F.; Koch, A.; Meyer, L.; Wilhelm, L. Antibody response to collagen after functional implantation of different polyester vascular prostheses in pigs. J. Biomed. Mater. Res. A **2005**, *72* (3), 317–325.

60. Wilhelm, L.; Zippel, R.; von Woedtke, T.; Kenk, H.; Hoene, A.; Patrzyk, M.; Schlosser, M. Immune response against polyester implants is influenced by the coating substances. J. Biomed. Mater. Res. A **2007**, *83* (1), 104–113.

61. Gorbet, M.B.; Sefton, M.V. Biomaterial-associated thrombosis: Roles of coagulation factors, complement, platelets and leukocytes. Biomaterials **2004**, *25* (26), 5681–5703.

62. Hong, J.; Nilsson Ekdahl, K.; Reynolds, H.; Larsson, R.; Nilsson, B. A new *in vitro* model to study interaction between whole blood and biomaterials. Studies of platelet and coagulation activation and the effect of aspirin. Biomaterials **1999**, *20* (7), 603–611.

63. Simon, D.I.; Xu, H.; Ortlepp, S.; Rogers, C.; Rao, N.K. 7E3 monoclonal antibody directed against the platelet glycoprotein IIb/IIIa cross-reacts with the leukocyte integrin Mac-1 and blocks adhesion to fibrinogen and ICAM-1. Arterioscler. Thromb. Vasc. Biol. **1997**, *17* (3), 528–535.

64. Salzmann, D.L.; Kleinert, L.B.; Berman, S.S.; Williams, S.K. Inflammation and neovascularization associated with clinically used vascular prosthetic materials. Cardiovasc. Pathol. **1999**, *8* (2), 63–71.

65. Hagerty, R.D.; Salzmann, D.L.; Kleinert, L.B.; Williams, S.K. Cellular proliferation and macrophage populations associated with implanted expanded polytetra-fluoroethylene and polyethyleneterephthalate. J. Biomed. Mater. Res. **2000**, *49* (4), 489–497.

66. Kakisis, J.D.; Liapis, C.D.; Breuer, C.; Sumpio, B.E. Artificial blood vessel: The Holy Grail of peripheral vascular surgery. J. Vasc. Surg. **2005**, *41* (2), 349–354.

67. Yu, T.J.; Chu, C.C. Bicomponent vascular grafts consisting of synthetic absorbable fibers. I. *In vitro* study. J. Biomed. Mater. Res. **1993**, *27* (10), 1329–1339.

68. Yu, T.J.; Ho, D.M.; Chu, C.C. Bicomponent vascular grafts consisting of synthetic absorbable fibers: Part II: *In vivo* healing response. J. Invest. Surg. **1994**, *7* (3), 195–211.

69. Izhar, U.; Schwalb, H.; Borman, J.B.; et al. Novel synthetic selectively degradable vascular prostheses: A preliminary implantation study. J. Surg. Res. **2001**, *95* (2), 152–160.

70. Crapo, P.M.; Wang, Y. Physiologic compliance in engineered small-diameter arterial constructs based on an elastomeric substrate. Biomaterials **2010**, *31* (7), 1626–1635.

71. Lee, S.H.; Kim, B.S.; Kim, S.H.; Choi, S.W.; Jeong, S.I.; Kwon, I.K.; Kang, S.W.; Nikolovski, J.; Mooney, D.J.; Han, Y.K.; Kim, Y.H. Elastic biodegradable poly(glycolide-co-cap-rolactone) scaffold for tissue engineering. J. Biomed. Mater. Res. A **2003**, *66* (1), 29–37.

72. Inoguchi, H.; Kwon, I.K.; Inoue, E.; Takamizawa, K.; Maehara, Y.; Matsuda, T. Mechanical responses of a compliant electrospun poly(L-lactide-co-epsilon-caprolactone) small-diameter vascular graft. Biomaterials **2006**, *27* (8), 1470–1478.

73. Stitzel, J.; Liu, J.; Lee, S.J.; Komura, M.; Berry, J.; Soker, S.; Lim, G.; Van Dyke, M.; Czerw, R.; Yoo, J.J.; Atala, A. Controlled fabrication of a biological vascular substitute. Biomaterials **2006**, *27* (7), 1088–1094.

74. Tillman, B.W.; Yazdani, S.K.; Lee, SJ.; Geary, R.L.; Atala, A.; Yoo, J.J. The *in vivo* stability of electrospun polycaprolactone-collagen scaffolds in vascular reconstruction. Biomaterials **2009**, *30* (4), 583–588.

75. Jeong, S.I.; Kim, S.Y.; Cho, S.K.; Chong, M.S.; Kim, K.S.; Kim, H.; Lee, S.B.; Lee, Y.M. Tissue-engineered vascular grafts composed of marine collagen and PLGA fibers using pulsatile perfusion bioreactors. Biomaterials **2007**, *28* (6), 1115–1122.

76. Woodruff, M.A.; Hutmacher, D.W. The return of a forgotten polymer: Polycaprolactone in the 21st century. Prog. Polym. Sci. **2010**, *35* (10), 1217–1256.

77. Hibino, N.; McGillicuddy, E.; Matsumura, G.; Ichihara, Y.; Naito, Y.; Breuer, C.; Shinoka, T. Late-term results of tissue-engineered vascular grafts in humans. J. Thorac. Cardiovasc. Surg. **2011**, *139* (2), 431–436, 436e1–436e2.

78. Naito, Y.; Shinoka, T.; Duncan, D.; Hibino, N.; Solomon, D.; Cleary, M.; Rathore, A.; Fein, C.; Church, S.; Breuer, C. Vascular tissue engineering: Towards the next generation vascular grafts. Adv. Drug. Deliv. Rev. **2011**, *63* (4), 312–323.

79. Matsumura, G.; Hibino, N.; Ikada, Y.; Kurosawa, H.; Shin'oka, T. Successful application of tissue engineered vascular autografts: Clinical experience. Biomaterials **2003**, *24* (13), 2303–2308.

80. Shin'oka, T.; Matsumura, G.; Hibino, N.; Naito, Y., Watanabe, M., Konuma, T.; Sakamoto, T.; Nagatsu, M.; Kurosawa, H. Midterm clinical result of tissue-engineered vascular autografts seeded with autologous bone marrow cells. J. Thorac. Cardiovasc. Surg. **2005**, *129* (6), 1330–1338.

81. Watanabe, M.; Shin'oka, T.; Tohyama, S.; Hibino, N.; Konuma, T.; Matsumura, G.; Kosaka, Y.; Morita, S.I. Tissue-engineered vascular auto-graft: Inferior vena cava replacement in a dog model. Tissue Eng. **2001**, *7* (4), 429–439.

82. Brennan, M.P.; Dardik, A.; Hibino, N.; Roh, J.D.; Nelson, G.N.; Papademitris, X.; Shinoka, T.; Breuer, C.K. Tissue-engineered vascular grafts demonstrate evidence of growth and development when implanted in a juvenile animal model. Ann. Surg. **2008**, *248* (3), 370–377.

83. Goyal, A.; Wang, Y.; Su, H.; Dobrucki, L.W.; Brennan, M.; Fong, P.; Dardik, A.; Tellides, G.; Sinusas, A.; Pober, J.S.; Saltzman, W.M.; Breuer, C.K. Development of a model system for preliminary evaluation of tissue-engineered vascular conduits. J. Pediatr. Surg. **2006**, *41* (4), 787–791.

84. Lopez-Soler, R.I.; Brennan, M.P.; Goyal, A.; Wang, Y.; Fong, P.; Tellides, G.; Sinusas, A.; Dardik, A.; Breuer, C. Development of a mouse model for evaluation of small diameter vascular grafts. J. Surg. Res. **2007**, *139* (1), 1–6.

85. Roh, J.D.; Nelson, G.N.; Brennan, M.P.; Mirensky, T.L.; Yi, T.; Hazlett, T.F.; Tellides, G.; Sinusas, A.J.; Pober, J.S.; Saltzman, W.M.; Kyriakides, T.R.; Breuer, C.K. Small-diameter biodegradable scaffolds for functional vascular tissue engineering in the mouse model. Biomaterials **2008**, *29* (10), 1454–1463.

86. Roh, J.D.; Sawh-Martinez, R.; Brennan, M.P.; Jay, S.M.; Devine, L.; Rao, D.A.; Yi, T.; Mirensky, T.L.; Nalbandian, A.; Udelsman, B.; Hibino, N.; Shinoka, T.; Saltzman, W.M.; Snyder, E.; Kyriakides, T.R.; Pober, J.S.; Breuer, C.K. Tissue-engineered vascular grafts transform into mature blood vessels via an inflammation mediated process of vascular remodeling. Proc. Natl. Acad. Sci. USA **2011**, *107* (10), 4669–4674.

87. Hibino, N.; Villalona, G.; Pietris, N.; Duncan, D.R.; Schoffner, A.; Roh, J.D.; Yi, T.; Dobrucki, L.W.; Mejias, D.; Sawh-Martinez, R.; Harrington, J.K.; Sinusas, A.; Krause, D.S.; Kyriakides, T.; Saltzman, W.M.; Pober, J.S.; Shin'oka, T.; Breuer, C.K. Tissue-engineered vascular grafts form neovessels that arise from regeneration of the adjacent blood vessel. FASEB J. **2011**, *25* (8), 2731–2739.

88. Turner, N.J.; Kielty, C.M.; Walker, M.G.; Canfield, A.E. A novel hyaluronan-based biomaterial (Hyaff-11) as a scaffold for endothelial cells in tissue engineered vascular grafts. Biomaterials **2004**, *25* (28), 5955–5964.

89. Gilbert, T.W.; Sellaro, T.L.; Badylak, S.F. Decellularization of tissues and organs. Biomaterials **2006**, *27* (1), 3675–3683.

90. Sorrentino, S.; Haller, H. Tissue engineering of blood vessels: How to make a graft. Tissue Eng. **2011**, *2*, 263–278.

91. Piterina, A.V.; Cloonan, A.J.; Meaney, C.L.; et al. ECM-based materials in cardiovascular applications: Inherent healing potential and augmentation of native regenerative processes. Int. J. Mol. Sci. **2009**, *10* (10), 4375–4417.

92. Badylak, S.F.; Freytes, D.O.; Gilbert, T.W. Extracellular matrix as a biological scaffold material: Structure and function. Acta. Biomater. **2009**, *5* (1), 1–13.

93. Gui, L.; Muto, A.; Chan, S.A.; Breuer, C.K.; Niklason, L.E. Development of decellularized human umbilical arteries as small-diameter vascular grafts. Tissue Eng. Part A **2009**, *15* (9), 2665–2676.

94. Narita, Y.; Kagami, H.; Matsunuma, H.; Murase, Y.; Ueda, M.; Ueda, Y. Decellularized ureter for tissue-engineered small-caliber vascular graft. J. Artif. Organs **2008**, *11* (2), 91–99.

95. Schmidt, C.E.; Baier, J.M. Acellular vascular tissues: Natural biomaterials for tissue repair and tissue engineering. Biomaterials **2000**, *21* (22), 2215–2231.

96. Spark, J.I.; Yeluri, S.; Derham, C.; Wong, Y.T.; Leitch, D. Incomplete cellular depopulation may explain the high failure rate of bovine ureteric grafts. Br. J. Surg. **2008**, *95* (5), 582–585.

Vascular—Zwitterionic

97. Kurane, A.; Simionescu, D.T.; Vyavahare, N.R. *In vivo* cellular repopulation of tubular elastin scaffolds mediated by basic fibroblast growth factor. Biomaterials **2007**, *28* (18), 2830–2838.

98. Weinburger, C.B.; Bell, E. A blood vessel model constructed from collagen and cultured vascular cells. Science **1986**, *231* (4736), 397–400.

99. Wise, S.G.; Byrom, M.J.; Waterhouse, A.; Bannon, P.G.; Ng, M.K.; Weiss, A.S. A multilayered synthetic human elastin/polycaprolactone hybrid vascular graft with tailored mechanical properties. Acta. Biomater. **2011**, *7* (1), 295–303.

100. Altman, G.H.; Diaz, F.; Jakuba, C.; Calabro, T.; Horan, R.L.; Chen, J.; Lu, H.; Richmond, J.; Kaplan, D.L. Silk-based biomaterials. Biomaterials **2003**, *24* (3), 401–416.

101. Enomoto, S.; Sumi, M.; Kajimoto, K.; Nakazawa, Y.; Takahashi, R.; Takabayashi, C.; Asakura, T.; Sata, M. Long-term patency of small-diameter vascular graft made from fibroin, a silk-based biodegradable material. J. Vasc. Surg. **2010**, *51* (1), 155–164.

102. Leach, J.B.; Schmidt, C.E. Characterization of protein release from photocrosslinkable hyaluronic acid-polyethylene glycol hydrogel tissue engineering scaffolds. Biomaterials **2005**, *26* (2), 125–135.

103. Hoenig, M.R.; Campbell, G.R.; Rolfe, B.E.; Campbell, J.H. Tissue-engineered blood vessels: Alternative to autologous grafts? Arterioscler. Thromb. Vasc. Biol. **2005**, *25* (6), 1128–1134.

104. Remuzzi, A.; Mantero, S.; Colombo, M.; Morigi, M.; Binda, E.; Camozzi, D.; Imberti, B. Vascular smooth muscle cells on hyaluronic acid: Culture and mechanical characterization of an engineered vascular construct. Tissue Eng. **2004**, *10* (5–6), 699–710.

105. Amarnath, L.P.; Srinivas, A.; Ramamurthi, A. *In vitro* hemocompatibility testing of UV-modified hyaluronan hydrogels. Biomaterials **2006**, *27* (8), 1416–1424.

Vascular—Zwitterionic

Vascular Grafts: Polymeric Materials

Alexandre F. Leitão
Center of Biological Engineering (CBE), University of Minho, Braga, Portugal

Ivone Silva
Department of Angiology and Vascular Surgery, Hospital of Porto, Porto, Portugal

Miguel Faria
Pharmacology Laboratory, Multidisciplinary Biomedical Research Unit, Abel Salazar Institute of Biomedical Science, University of Porto (ICBAS-UP), Oporto, Portugal

Miguel Gama
Center of Biological Engineering (CBE), University of Minho, Braga, Portugal

Abstract

Cardiovascular disease is one of the major leading causes of death in today's world. Among the most common is atherosclerosis, a thickening of the arterial wall due to the buildup of plaque. When the disease causes complications, due to occlusion of the vessel or arterial wall lesions that ultimately lead to thrombosis, usually a bypass is required to redirect blood flow around the occluded vessel. The preferred vessels for the bypass procedure are autografts, veins, and arteries culled from the patient. When these vessels are not available, due to the natural progression of the disease or prior procedures, synthetic alternatives must be used. These alternatives are created from polymeric materials that can substitute the autografts completely and remain in the patients for years and decades or serve as a temporary substitute that will allow tissue engineering. In this entry we intend to show some of the polymers that are currently being studied, and two in particular that have been in use for over 50 years, that can one day serve the purpose of substituting the autografts completely and provide a better future for patients worldwide.

INTRODUCTION

Cardiovascular disease (CVD) is a broad term encompassing several different diseases among which atherosclerosis and hypertension are the most common. The causes, prevention, and/or treatment of all forms of CVD remain active fields of biomedical research, with hundreds of scientific studies being published on a weekly basis. The ongoing research in the various fields that are related to this topic all have one single objective: improving the quality of life of patients worldwide via prevention of the disease by finding better courses of treatment.

The leading cause of death associated with vascular disease is atherosclerosis, a syndrome that affects arterial blood vessels. Atherosclerosis is defined as a chronic inflammatory response in the walls of arteries, closely associated with high cholesterol levels in the blood, and can remain asymptomatic for decades. Eventually the blood flow through the affected arteries can be cut off completely and blood flow must be reestablished around the occluded vessel. In these extreme cases, the most common procedures performed are bypass surgeries.

Since the 1950s, when autologous vessels are not available to perform the bypass procedure, synthetic alternatives have to be used. The most common synthetic, alternatives are known simply as Dacron and expanded polytetrafluoroethylene (ePTFE). They have been used countless times in life-saving procedures. While they perform well, as long as certain conditions are met, they are not perfect and do not provide a definitive solution.

Much research has focused on improving the current grafts and/or discovering new better grafts and material combinations. Research has focused on polymeric materials, either nondegradable polymers that remain *in vivo* for undetermined periods of time or biodegradable polymers that serve as tissue-engineering scaffolds, ultimately allowing for native tissue to form a new blood vessel. Our objective in this entry is to outline the need for the prosthetics, the standardized methodologies to assess their viability, an unbiased review of the state of research being performed on several polymers with the objective of developing new synthetic vascular grafts, and finally what we consider to be the future perspectives of the research performed in this ongoing field.

CVD: THE NEED FOR VASCULAR GRAFTS

The World Health Organization estimates that 17.3 million people died in 2008 of CVD, of which 80% occurred in

Vascular—Zwitterionic

Concise Encyclopedia of Biomedical Polymers and Polymeric Biomaterials DOI: 10.1081/E-EBPPC-120050697

low- and middle-income countries. The American Heart Association estimates that, from 1999 to 2009, the relative rate of death attributable to CVD declined by 32.7%. Yet, it is still responsible for one of every three deaths in the United States (one every 40 seconds).[1] On the basis of CVD is atherosclerosis, a complex disease involving the heart and blood vessels and is responsible for several disorders such as stroke, myocardial infarction, and peripheral arterial disease (PAD).

Atherosclerosis was described for the first time in 1904 by Marchand to define arterial stiffening in association with fatty degeneration. Atherosclerosis is a systemic disease affecting large and medium-sized arteries. Accumulation of lipids in both smooth muscle cells (SMCs) and macrophages leads to inflammatory response and arterial thickening, which plays a major role in atherogenesis.[2]

Atherosclerotic plaques remain asymptomatic, below critical stenosis (>70%), due to compensatory vessel enlargement and to the development of collateral circulation in response to chronic hypoxia. Chronic injury leads to endothelium dysfunction, increased oxidative stress, and reduction in bioactivity or synthesis of endothelium-derived nitric oxide (NO), which results in a reduced vascular tone and ischemia.[3–5] The atherosclerotic plaque may become unstable, ulcerate, and rupture with subsequent distal embolization or vessel occlusion due to exposure of the highly thrombogenic subintimal surface. The exposure of the subintimal surface promotes platelet adherence and activation—atherothrombosis,[6] – leading to the acute clinical presentations or worsening of a chronic controlled disease, followed by a high morbidity and mortality. Despite significant improvement in medical and surgical management of CVD, atherosclerosis still remains a serious life-threatening disease.[2]

Cardiovascular risk factors play a major role in atherosclerosis pathogenesis. They may be hereditary or acquired, related to behavioral or environmental factors. The INTERHEART study, spearheaded by Dr. Yusuf and colleagues in over 52 countries, found that 90% of myocardial infarction risk factors were abnormal lipids, smoking, hypertension, diabetes, abdominal obesity, psychosocial factors, irregular diet, and absence of a planned exercise program.[7] We now have overwhelming evidence that risk factor control is highly cost-effective and significantly reduces CVD morbidity and mortality.

Clinical manifestations of CVD manifest mainly as cerebral vascular disease, coronary heart disease, and PAD, and can cause severe disability with great limitations in patient habits and decreased quality of life or even death. PAD is a chronic disease with prevalence in several epidemiologic studies in the range of 3–10%, increasing to 15–20% in people over 70 years of age.[5] Initially asymptomatic, it can become progressively symptomatic presenting with pain in the calves due to exercise— intermittent claudication – and, when severe, evolves to critical ischemia (CI) with a high risk of limb loss. Patients

with CI have high amputation rates: 25% will be amputated at time of diagnosis and another 30% over the following year.[8] In association with PAD, these patients have other comorbidities as a consequence of diffuse atherosclerosis involvement with the heart, cerebral-vascular, and kidney disease. In PAD's treatment we have to recognize the disease, quantify the extent of local and systemic disease, identify and control risk factors, and establish a comprehensive treatment program.[8]

In vascular surgery, surgical bypass is fundamental in the treatment of arterial and some venous diseases (Fig. 1A). Vascular grafts are used to replace, bypass, or maintain function of damaged, occluded, or diseased blood vessels of small, medium, and large diameter. The chosen conduit and its success depend on several factors such as availability, size, ease of handling and technical facility, thrombogenicity, resistance to infection and dilation, durability, long-term patency, and price.

Depending on the target vessel different options of autografts can be considered. In coronary surgery the best conduits are internal mammary artery, radial artery, and long saphenous veins (LSV). Visceral arteries are also best substituted by LSV or by the iliac vessels, hypogastric, and more uncommonly by the external iliac artery.[9] In PAD surgery, the most common autograft in use is the long saphenous vein (using either the reversed or in situ techniques) and diameters of 3 mm are required for good long-term results.[10] Short saphenous vein or arm veins can be an alternative when no LSV is available. Autologous vein grafts remain the conduit of choice for infrainguinal revascularization, and long-term results are quite good when a good run-off is present.[10] Infrainguinal vein graft failure still occurs in 20–50% of cases, and remains a major problem for vascular surgeons and patients. Recovery of the thrombosed vein graft has a very low patency. Furthermore, replacement of a failed bypass graft with a new one requires that the surgeon finds an alternative graft for limb salvage. Thus, the results of repeat bypass surgery, after failed previous bypass, are inferior to those of primary bypass surgery. It is therefore imperative to maintain the patency of infrainguinal bypass grafts.[11] Autogenous arterial or venous conduits are not always available, due to venous insufficiency or inadequate LSV diameter/length, previous harvest of vein for cardiac surgery, or previous varicose vein surgery. These factors raise the need for alternative grafts. At present, these include autologous or nonautologous biological grafts (obtained using different methods for harvesting and preservation), tissue engineering (using either synthetic- or biological-based scaffolds for cell seeding), endovascular methodologies,[12] gene therapy,[13] and nondegradable synthetic grafts.

For large vessel replacement, graft size match, resistance to infection, and durability are of great importance. Aorto-iliac reconstructions have blood flow properties that allow long patency due to minimal tissue ingrowth and endothelial repopulation.[9] Infra-renal aortic graft

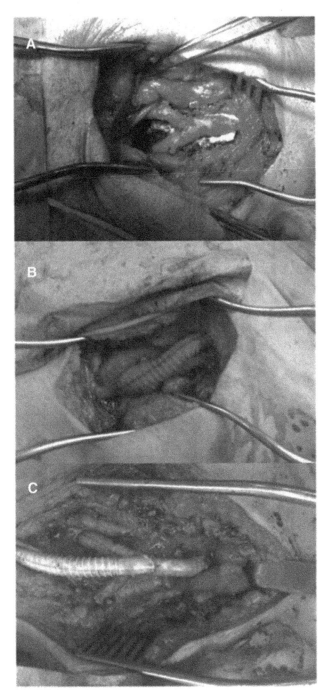

Fig. 1 Photographs of the surgical application of the three most common bypass grafts: (**A**) vein graft; (**B**) Dacron graft; (**C**) ePTFE graft.

diameters range from 16 to 22 mm and, Poly(ethylene tere-phthalate) (PET) or ePTFE prosthesis—single tube or bifurcated—are the most commonly used (Fig. 1B and C). The size of the graft should be as close as possible to that of the native artery that is replaced. Ten-year primary patency is 89.2% for aorto-iliac bypass and 78% for aorto-bifemoral bypass.[14] The replacement of an aortic infected graft can be made with prosthetic grafts, arterial homo-grafts or deep veins of the lower limb. The latter present

a high grade of morbidity and increased operative time. Alternatively, arterial homografts are a solution with good early results but a high grade of late degeneration. A study comparing cryo-preserved arterial homograft with silver-coated Dacron grafts for the treatment of infected aortic grafts demonstrated comparable effectiveness with respect to early mortality and midterm survival. Graft-inherent complications, aneurismal homograft degeneration, and reinfection of the silver-coated grafts have also been observed.[15]

In peripheral lower limb bypass, there are two types of arteries to be replaced: femoral and popliteal arteries, with a medium size (6–8 mm), and distal popliteal, tibial, pero-neal, and foot arteries with small size arteries (<6 mm).[9] With the expanding use and indications for endovascular surgery, the number of vessels needing replacement is decreasing, with a lower rate of preoperative morbidity. Treatment of the common femoral artery has limitations due to increased rate of infection in groin surgeries. Bypasses at and from common femoral artery should, when possible, be performed with an autologous graft.

In small size arteries (<6 mm), intimal hyperplasia at the proximal or distal anastomosis site is of paramount impor-tance. In these grafts, thromboresistance of the inner lining, shear stress, and behavior of the conducts are important aspects in the choice of the graft. In these cases, autologous grafts are again the best choice.

Femoro-popliteal bypass may be made with a synthetic graft when the distal anastomosis is above the knee and the patient has no ulcer that can increase the risk of graft infec-tion. In the case of below-knee bypass, LSV is the preferred choice. A meta-analysis of femoral-popliteal bypass grafts for lower extremity arterial insufficiency refers that primary graft patency was 57.4% for above-knee polytetrafluoroeth-ylene (PTFE), 77.2% for above-knee vein, and 64.8% for below-knee vein at 5 years; there was a significant difference between above-knee grafts at 3, 4, and 5 years.[14] Another study reported a similar patency, to the aforementioned study, for infragenicular bypasses with ePTFE for limb salvage in above-knee (27%) and below-knee (25%) over 5 years.[16]

When LSV is unavailable, alternative techniques have been applied to increase patency, such as the use of a cryo-preserved allografts, composed synthetic-LSV graft, vein cuffs, and ePTFE cuffs at the distal anastomosis in order to reduce the risk of intimal hyperplasia.[10] These techni-cal maneuvers and the use of numerous pharmacological adjuvants, included at the time of production or implanta-tion, are of paramount importance to increase artificial vascular grafts patency. In carotid surgery, Dacron patches are used routinely in most centers for carotid endarterec-tomy and carotid bypasses are used in the treatment of aneurismal disease of the carotid artery. Autogenous LSV, again, is the first choice for carotid bypass, although high rate of restenosis has been described.[17] Alternative PTFE prosthesis can be used due to the high flow and short length, with long-term graft patency of 95–97%.[18]

Arteriovenous grafts, used to connect an artery to a vein, are commonly used in hemodialysis patients. In these patients, it is need for a high flow graft that is also durable, resistant to infection, and to continuous trauma (due to multiple puncture, two to three times a week) to guarantee an efficient hemodialysis for the patient. Autogenous veins are the conduct of choice, with an emerging of new techniques to recover the arm veins, and only in extreme cases, prosthetic grafts are considered.

VASCULAR GRAFT REQUIREMENTS

As described above, by far, the preferred grafts are autogenous, i.e., the donor and the patient are the same individual and this vastly decreases the risk of rejection. However, the progression of the disease in these patients many times limits the availability of viable vessels. When this occurs, two alternatives are available: biological (which can be allografts or xenografts) or synthetic grafts (which consist of polymers that can be either synthetic or biological). The biological alternative to the autogenous vessels are allografts and xenografts. Theoretically, biografts meet all the conditions for the best vascular graft. They promise "off-the-shelf" availability, a wide variety of sizes, excellent handling characteristics, and patency rates similar to those of autogenous vessels.[9] Disappointingly, clinical applications of these grafts have not met the high expectancy awaited. This is in large part due to the lack of long-term patency associated with tissue degeneration.

Therefore, a large amount of study has gone into assessing and discovering a synthetic alternative graft vessel. A viable synthetic vascular graft should combine fundamental characteristics that are inherent to autografts: biocompatibility, namely hemocompatibility, and mechanical compliance. Hemocompatibility is linked with the blood–material interaction, the mechanisms of which are still not completely understood. However, it is well established that the material should provide a minimal risk of hemolysis, clotting, and thrombosis, along with a minimal risk of infection or triggering of the immune response. Biocompatibility, particularly in the case of a tissue engineering approach, ultimately refers to the ability of the biomaterial to stimulate cell adhesion, invasion, and proliferation. The grafts should allow for the migration of cells from the surrounding tissues in order to, at least, produce an endothelium on the luminal surface of the graft. In tissue-engineered approaches, where cell-seeding might be adopted, this will allows for gradual cell migration, proliferation, and production of an extracellular matrix while at the same time the material is slowly degraded. This approach allows the synthetic graft to be completely substituted by newly formed native tissue, thus reducing the risk of long-term deterioration and improve the overall long-term patency.

Finally, and of no less importance, is the matching of mechanical properties. The grafts should provide mechanical characteristics similar to those of the native vessels,

especially in terms of compliance, in order to allow for a consistent and uninterrupted blood flow through the graft. This will ultimately avoid problems such as intimal hyperplasia, which is closely associated with mismatch of mechanical properties.[19,20]

Methods for Analysis and Testing

Over the years, several assays and tests have been suggested in order to determine the viability of synthetic grafts before advancing to any clinical assays. The main guidelines for hemocompatibility testing are laid out in ISO 10993–4. Essentially, the ISO lays out that it is necessary to perform assays for immunology (complement system), hematology, platelets, thrombosis, and coagulation.

Complement activation

Complement activation is the most relevant test for determining the immune response to a blood-contacting material. It can be performed *in vitro* by exposing isolated complement proteins to the material surface and determining the decrease in total CH50 levels, a classical approach, alternatively the determination of C3- and C5-convertase cleavage products formed over time. The amount of cleaved products can be easily determined by semiquantitative Western blot and compared with baseline levels in order to determine activation of the complement system. Currently, there are no clearly determined acceptable levels and data is generally presented as comparisons with other materials. The American Society for Testing and Materials (ASTM) F1984-99 and ASTM F2065-00 standards should be considered when assessing this particular parameter.

Hematology

Hematology is generally assessed by determination of the hemolytic index and can be determined by the method suggested by the Standard Practice for Assessment of Hemolytic Properties of Materials from the ASTM F756-00, 2000. Additionally, leukocyte activation via microscopic analysis or flow cytometry, for determination of L-selectin and CD11b expression, are also indicators of hemolytic activity of a material.

Pro-coagulant activity

Several tests can be explored in order to thoroughly evaluate the potential activation of the coagulation cascade by surface–material interaction, i.e., coagulation triggered by the intrinsic (or contact) activation pathway. The contact activation pathway is a result of direct activation of Factor XII, via conformational change induced by exposure to the material surface. Direct quantification of how much activated Factor XII is produced can be achieved by use of a chromogenic, or fluorogenic, substrate that is

broken down by the activated form of Factor XII.[21,22] Alternatively, indirect assessment of the activation of this pathway can be made by determination of the partial thromboplastin time or plasma recalcification profiles.[23,24]

Thrombogenicity

Platelet–material interactions can lead to both platelet activation and/or adhesion, depending on their state of activation and the presence of ligands.[25,26] Ultimately, the involvement of activated platelets along with the coagulation cascade results in the formation of a thrombus (Fig. 2). The clearest *in vitro* measure of the thrombogenicity is the determination of whole blood clotting time through spectrophotometric analysis. Scanning electron microscopy, and to a lesser extent light microscopy, are also useful tools in determining platelet adhesion and activation, though the cues here are visual (formation of pseudopodia and platelet aggregates) rather than biochemical. Activation and subsequent adhesion of platelets can also be determined by fluorescence and confocal microscopy. Most assays suggested for determination of thrombogenicity must be tested in dynamic (preferably *in vivo*) conditions. Angiography, intravascular ultrasound, Doppler ultrasound, computer-assisted tomography, and magnetic resonance imaging are useful tools for the determination of the *in vivo* performance in regard to percentage of occlusion and flow reduction. Also it is important to analyze, after *in vivo* placement, the gravimetric analysis of the thrombus mass and histological inspection of the prosthesis and surrounding tissues.

Platelet activation

It is paramount that platelet activation be minimized in blood-contacting materials. The extent of free platelet activation can be easily determined by cytometric analysis. Platelets are exposed to the material surface and the presence of certain activation markers (like CD62P or CD63) determined with the use of fluorescent antibodies. Also, the formation of platelet aggregates due to material activation should be determined by means of an aggregometer. These tests can be performed both *in vitro* and *in vivo*. Other *in vivo* tests that are important for assessment of platelet activation are determination of platelet count, template bleeding time (as a measure of platelet function), or alternatively platelet function analysis with collagen filters. Additionally, gamma labeling and platelet life span assays are good measures of platelet viability.

Mechanical properties

The mechanical properties of the graft must assure adequate mechanical compliance. As discussed above, the mechanical compliance is paramount in order to allow an uninterrupted pulsatile flow through the vessel matching the mechanical characteristics of the surrounding tissue. This parameter can be quantified by setting up a pulsatile flow into a graft and measuring the volume changes caused by the increase in pressure. The best possible graft will match the compliance of the native vessel it is intended to substitute. Assuring mechanical strength and compliance will minimize the risk of mechanical failure of the graft.

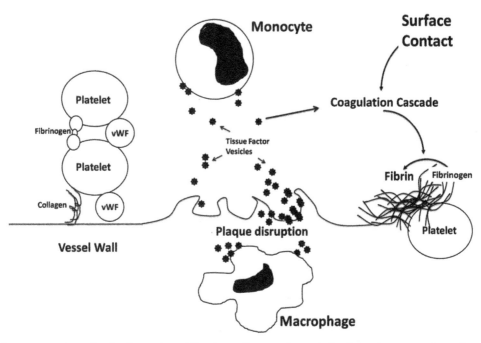

Fig. 2 Schematic representation of activation and amplification pathways of coagulation by activator complexes (generating thrombin for platelet activation and fibrin formation) and thrombus formation on disrupted atherosclerotic plaques.

In vivo tests

The final assays before potentially moving on to a clinical trial are *in vivo* tests. The *in vivo* tests can produce useful information to build on any of the aforementioned assays and also provide a realistic view of the behavior of the grafts in a complex and dynamic system that simply cannot be reproduced *in vitro*. These assays generally are performed simultaneously with some of the above tests in order to provide a back and forth feedback. Care should be taken when choosing an appropriate animal model. A rat or mouse model is an appropriate one in order to determine the feasibility of the graft over a significant portion of the animal's life, since their life cycle is short (circa 24 months), rapidly providing insight into feasibility. However, the mechanics and body response of a small animal model are significantly different from those of a larger one and therefore other models such as dog (though today it is rarely used due to ethical and social issues), sheep and pig should also be considered. These larger animals provide many more surgical implantation options and a longer life cycle that can produce results that are more in line with what can be expected in humans. Another important parameter is precisely the placement of the prosthetic graft. The placement in a high-flow vessel, such as the rat aorta or a sheep's carotid artery, will provide different results from a low-flow vessel, such as the femoral artery of a dog or pig. Nevertheless, translation of *in vivo* results to humans is not always straightforward.

NONDEGRADABLE POLYMERS

The standard polymers for synthetic vascular grafts are PET (Dacron) and ePTFE grafts, synthetic nondegradable polymers that offer a conduit for life and limb saving surgeries when autografts are not available. However, as present in the market today, these materials have inherent problems that need to be overcome. One of the broad possibilities that are currently being researched is the development of novel graft that can be readily implanted. The solution may very well be the development of permanent grafts that can outperform other options in regards to long-term patency (Fig. 3). Toward that end, the following are the latest focus of research in this field.

Poly(Ethylene Terephthalate)

PET is a thermoplastic polymer resin of the polyester family. It can exist as both an amorphous and a semicrystalline polymer depending on processing and thermal history. Since its introduction in 1939 by DuPont, later patented in 1950 as Dacron®, it has found numerous applications namely as a synthetic fiber (over 60% of its applications). Thermoforming applications make up the bulk of remaining uses found for this polymer with a small percentage

being used as an engineering resin often in combination with glass fiber.

It is a relatively inert, hydrophobic, and biostable polymer that presents a high tensile strength (40–180 MPa) and Young's modulus (3–14 GPa), values varying widely based on preparation and dimensions.[27] PET has been widely studied for biomedical applications in cardiovascular surgery, as ligaments, joint replacements, implantable sutures, and surgical meshes.[28–32] Under the aforementioned name Dacron, it has been used as an aortic vascular graft as far back as 1957 when DeBakey performed a substitution of thoracic and abdominal aorta and of proximal peripheral vessels. To this day, PET is still one of the most commonly used synthetic grafts in large caliber (>8 mm) bypass surgery performed above the waist[33] since it is a cheaper alternative to ePTFE. Patency rates for PET have been continuously improved and have been shown to remain patent for up to 8 years after implantation.[34] A prior study actually showed that PET patency could go up to 25 years, though the follow-up rate was under 20% after 10 years.[35]

However, when PET vascular grafts are implanted they trigger the activation of the intrinsic pathway of blood coagulation resulting in surface-induced thrombosis.[36] This is most prevalent when the diameter of PET vascular grafts is smaller than 6 mm. There is a considerable reaction between PET and the native artery, as well as with blood, causing a rapid buildup of fibrin at the blood–graft interface.[37] This is due to both the hydrophobic nature of PET, which has been shown to induce protein adhesion,[38] and also the low-flow of small caliber vessels. This ultimately results in graft failure and poor clinical results. Hence, the patency of small-diameter PET vascular grafts needs to be improved. The PET vascular grafts are produced by two different methods. Woven, in an over-and-under pattern in the lengthwise and circumferential directions, or the more commonly used knit grafts, by looping the fibers in an interlocking chain, into cylindrical and longitudinally crimped grafts.[39] The crimping of the graft increases flexibility, elasticity, and kink resistance. However, after implantation, these properties are lost as a consequence of tissue ingrowth on average after 7 years.[40] The grafts incorporate a velour finish, which increases the surface area. This elevates the number of anchorage points for both fibrin and cells in order to promote tissue integration. The weaving/knitting method of producing the grafts allows them to be shaped and branched so as to fit any needed requirements and mechanical properties can be tailored so as to allow for better compliance with the native arterial wall at the site of anastomosis.

The production method of the grafts has remained essentially the same over the years. The knit PET grafts have been impregnated with gelatin,[41] albumin,[42] or collagen[43] in order to improve their hemocompatibility. Another approach to ameliorate hemocompatibility is surface modification. The chief methods of surface modification of

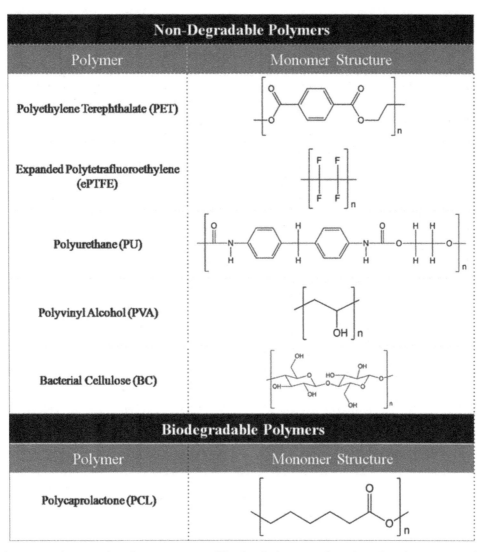

Non-Degradable Polymers	
Polymer	Monomer Structure
Polyethylene Terephthalate (PET)	
Expanded Polytetrafluoroethylene (ePTFE)	
Polyurethane (PU)	
Polyvinyl Alcohol (PVA)	
Bacterial Cellulose (BC)	
Biodegradable Polymers	
Polymer	Monomer Structure
Polycaprolactone (PCL)	

Fig. 3 Chemical structure of nonprotein polymer monomers. The chemical structure for polyurethane is a representation of one possible polyurethane.

blood-contacting biomaterials involve coupling bioagents or increasing hydrophilicity. Heparin has been bonded onto PET grafts showing positive results, namely increasing plasma recalcification times[44] and overall in-patient patency.[45] Silver particles have also been bound to the PET grafts to reduce infection and are currently used in cases where an infected graft has to be substitute.[15,46,47] Amination may also lead to improved patency without any loss of mechanical properties.[48,49] However, the chemically inert nature of PET impedes significant bonding of functional groups and so surface modification, such as plasma treatments, have been explored in order to augment the number of e.g., heparin–PET bonds.

Expanded Polytetrafluoroethylene

PTFE was discovered in 1938, patented in 1941, and its commercial name, Teflon, trademarked in 1945 by Roy Plunkett of the DuPont Co. ePTFE is a processed version of PTFE. PTFE is heated and rapidly stretched, the end result is that a microporous structure with 70% air is obtained. The process was discovered in 1969 and patented, under the name Gore-Tex, in 1976. PTFE and ePTFE have found numerous applications such as high-performance fabrics, medical implants, filter media, wire and cable insulation, gaskets, lubricants, and sealants.

The PTFE molecule is chemically inert and biostable, due to the nature of the carbon–fluorine bond, and has been found to be less prone to biological deterioration than PET.[50] It is also very hydrophobic, due to mitigated London dispersion forces associated with the high electronegativity of fluorine. The electronegativity of the graft surface minimizes its reaction with blood components.[51] Mechanically, ePTFE is stiffer than both the native arterial walls and PET. This can be a cause of graft failure due to pseudo-intimal hyperplasia. The Young's modulus for these grafts is around 3–6 MPa.[52,53] Comparatively in native arteries and veins (in a canine model), the Young's modulus is around 600 (circumferentially) and 900 kPa (longitudinally), while PET registers 12 MPa circumferentially and 0.7 MPa longitudinally. The ePTFE graft for vascular applications, produced by extrusion, is a nontextile porous tube composed of irregular-shaped solid membranes ("nodes"). Between these solid nodes are fibrous regions and so the overall porosity of the graft is defined by its internodal distance (IND) (available options vary between 30 and 90 μm). The first ePTFE grafts were first introduced experimentally in 1972 and clinically in 1975 in a portal vein replacement.[54,55] PET and ePTFE are the most widely used synthetic vascular replacements with little difference in performance between them.[56] Despite the aforementioned mechanical compliance issues, ePTFE grafts are commonly used for peripheral vessel bypass, particularly for femoral artery bypass. However, over 30 years since their initial appearance, research has achieved little improvement in regards to patency rates of ePTFE grafts.[57] Over a 5-year time period, approximately 50% of implanted grafts have been shown to be blocked[58] and over a 10-year time period a maximum of 28% primary patency (in above-knee femoro-popliteal bypass grafts).[59] The problem with patency is exacerbated in small caliber grafts. The main cause of graft failure is strongly associated with thrombosis, due to the lack of endothelial cell coverage, which leads to protein adhesion that ultimately promotes thrombosis.[60–62]

Animal studies have demonstrated rapid endothelization after relatively short periods of time. However, this has not transitioned into humans. After 10 years of implantation in human patients, the same grafts that provided a complete endothelization in animals are almost completely deprived of endothelium.[63] Clinically, when endothelization does occur, it is generally limited to 1–2 cm from the anastomosis site.[64] This is partially due to low porosity of the standard ePTFE grafts (IND of 30 μm). The low porosity does not allow for adequate transanastomic or trans-mural endothelization.[33] Studies have shown that a higher IND (60 or 90 μm) promoted a more rapid endothelization in animal models.[65,66] However, none of these studies have been demonstrated and proven clinically.In order to resolve the patency issues, a two-stage endothelial cell seeding protocol has shown to greatly enhance clinical patency rates to 90%, up to 52 weeks after implantation, on 3 mm diameter coronary grafts.[67] Additionally, surface modification by carbon coating/impregnation[68] heparin coating (with the commercial name Propaten)[69,70] and growth factor coating, with fibrin glue,[71] have been attempted, though the latter seems to have been abandoned. Both carbon-coating and impregnation have been tested clinically and demonstrated no significant improvement in patency after 36 months.[72,73] Heparin coating has been tested as a means to improve nonthrombogenicity and has demonstrated that a luminal coating with immobilized heparin improves primary patency and reduces intimal hyperplasia development over standard ePTFE grafts up to one year after implantation.[74,75]

Polyurethane

Polyurethanes (PUs) belong to a class of compounds called reaction polymers. They arise from the reaction of an isocyanate with a polyol, via carbamate links. PUs are thermosetting polymers that form a foam that does not melt when heated. They can also be manufactured into a harder thermoplastic form that is used in medical applications.

The thermoplastic PU used in medical settings present a high tensile strength, elasticity, and ease of manipulation.[76] These mechanical and physical properties are extremely important in providing good mechanical compliance that ultimately prevents problems such as intimal hyperplasia. Along with its physical/mechanical properties, PU also presents good hemocompatibility, biocompatibility, and microbial resistance.[77,78] Hemocompatibility of PU is, at least, on par with ePTFE.[79] Yet another aspect that has encouraged further study of PU is porosity. Fibrilar PU grafts can be produced by weaving, knitting, electrospinning, or winding[76,78] while foam PU grafts can be cast and porosity controlled by gas expansion and laser perforation.[76] Results have shown good in vivo biocompatibility and accelerated endothelization in small caliber grafts (4 mm) associated with the improved porosity. Incorporation of carbon nanotubes have also been explored in order to improve endothelization, in vitro, showing promising results.[80] Rapid endothelization (12 weeks after implantation) has been shown critical in preventing hyperplasia and improving patency.[81] Another approach, recently studied, involves a PU graft capable of controlled release of rapamycin, which inhibits the migration and proliferation of SMCs, thus avoiding intimal hyperplasia.[82] However, pore interconnectivity, a mandatory requirement in order to provide better tissue ingrowth, remains an issue with no clear solution. Blends of PU with collagen/elastin and also polycaprolactone have shown improvements in both mechanical compliance and cell invasion.[83,84]

First-generation PU grafts suffered hydrolytic biodegradation, which ultimately led to the abortion of clinical trials.[85] Second generation PUs have shown a pressing disadvantage, the material's tendency to oxidize and degrade in vivo, creating problems after implantation, which could be overcome by chemically coating the surface with an antioxidant aids in reducing oxidation.[86] Given the good mechanical compliance (a step up as compared to ePTFE), hemocompatibility, and potential for controlled porosity, and despite in vivo degradation, several PU grafts are or have been studied. Mitrathane, Pulse-Tec, Vectra, and Vascugraft are all brand names for PU grafts. Of these only Vectra is FDA approved; however, the only application is for vascular access during hemodialysis, which is a short-term application.[87] Continued study is necessary in order to produce small-caliber PU graft for clinical applications.

Polyvinyl Alcohol

Polyvinyl alcohol (PVA) is a nontoxic water-soluble synthetic polymer that forms a hydrogel once polymerized. It has excellent film-forming, emulsifying, and adhesive properties. PVA is not prepared by polymerization of the vinyl alcohol monomer but rather partial or complete hydrolysis of poly(vinyl acetate). The removal of the acetate group can be controlled and alters the physical characteristics of the resulting PVA. This polymer presents high tensile strength and its flexibility is dependent on humidity. The water acts as a plasticizer, reducing tensile strength, but increasing elongation and tear strength.

PVA-based materials can be prepared by casting and then cross-linked by irradiation, physical (by thermal cycling) or chemical treatment.[88,89] The resulting hydrogels can be fine-tuned to present varying mechanical and physical properties, depending on number of hydrogen bonds. The hydrogel is biocompatible and nonirritating to soft tissues and, therefore, it has been of great interest in pharmaceutical and biomedical applications.[90–92] The FDA approved the use of PVA microspheres as an embolization agent and further applications have been found as contact lenses, wound dressings, and coating agent for pharmaceuticals and catheters.[93,94]

As a potential vascular graft, PVA has been studied both on its own[95] and composited with other materials.[96–98] Several combinations of PVA with other synthetic and natural polymers, such as gelatin, chitosan, starch, and bacterial cellulose (BC), have been proposed and studied. In general, these studies have all presented positive results in regards to hemocompatibility, biocompatibility, and mechanical compliance. The intent behind the formulation of these composites is generally to enhance endothelial cell adhesion and proliferation. Gelatin and chitosan are probably the most well studied of these combinations and have demonstrated good hemocompatibility and endothelization.[99–102] The PVA/BC composite was conceived as a means to enhance the mechanical properties of the PVA hydrogel, while still maintaining elasticity and malleability.[96–98] One of the rare studies demonstrating the *in vivo* performance of PVA based tubular structures was produced by Chaouat and colleagues.[95] In this study, a PVA sheet was either stitched or glued into a tubular shape and showed good patency after 1 week in a rat model. Extensive *in vivo* studies of vascular grafts of this kind are still lacking and so further work is needed to demonstrate its potential.

Bacterial Cellulose

Cellulose is the most abundant organic compound on earth, a linear polysaccharide with β (1→4)–linked D-glucose monomers. Chemically, bacterial and plant cellulose are identical; however, BC is obtained pure (free of lignin, pectin or hemicellulose). It is produced mainly by the *Gluconacetobacter* genus as a fibrous network hydrogel at air–medium interfaces.[103,104] The bacteria that produce cellulose are obligatory aerobes and produce the hydrogel as a pellicle,[104] which can be exploited to grow into three-dimensional structures given that a three-dimensional oxygen permeable support is provided.[105–107]

The BC hydrogel has several beneficial properties for use in the biomedical field, namely high purity, high water-holding capacity, morphology, tensile strength, malleability, hemocompatibility, and biocompatibility.[103,108,109] These characteristics have led to BC being long studied for biomedical applications such as wound dressings,[110–112] artificial skin,[113] tissue engineering scaffold,[23,104,109,114] synthetic dura mater,[115] and as a vascular graft.[107,108,116–118]

Several studies addressed the use of BC as a vascular graft and have shown its potential. Dieter Klemm and his group have looked at the *in vivo* behavior of small diameter BC tubes. These studies, carried out over 4 weeks and later over 3 months, in a rat model, showed that a small diameter BC tube can remain patent *in vivo*.[108,118] These studies also showed that in the course of the 4 weeks the BC grafts were completely covered with connective tissue and, after 3 months, the formation of a neointima. Another study later demonstrated that, under the registered name BASYC, BC tubes remained patent, in mice, for up to a year. However, *in vitro* studies have shown that, while minimal, BC does slightly trigger coagulation. Indeed, studies by both Andrade et al. and Fink et al.[23,22,119] showed that BC minimally activates coagulation. However, another study from the same group demonstrate that, compared to ePTFE, BC has superior hemocompatibility, as assessed by the analysis of the activation of platelets and of the intrinsic pathway of hemostasis.[24] Attempting to improve the BC hemocompatibility, these authors used engineered cell-adhesion peptides (such as RGD), linked to a carbohydrate binding module, allowing the number and proliferation of endothelial cells on BC to be further improved.[104]

Surface modification via plasma treatment and addition of aminoalkyl groups has been shown to alter the surface properties of BC. Cell adhesion and proliferation was improved with plasma treatment and addition of aminoalkyl groups demonstrated antimicrobial properties similar to chitosan, these approaches may ultimately prove useful for improving vascular grafts.[120,121]

A drawback to BC is that it is not biodegradable; furthermore, its porosity is insufficient for cell ingrowth to take place. This limits the potential in regards to producing a graft that could, eventually, be substituted integrally by newly formed tissue. Despite this, the potential of BC for *in vitro* and *in vivo* tissue regeneration continues to be explored and shows great promise. [107,109,116,117]

DEGRADABLE POLYMERS AND TISSUE-ENGINEERED VESSELS

A large portion of research into the development of novel vascular grafts has been focused on tissue-engineering approaches by prior cell-seeding or by providing a scaffold

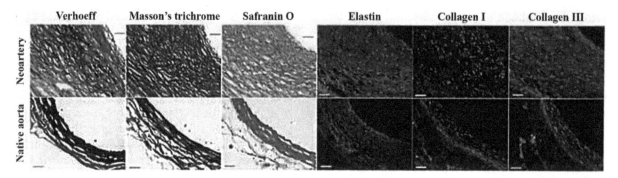

Fig. 4 ECM organization and quantification at 90 days. Verhoeff's, Masson's trichrome, and safranin O staining show elastin, collagen, and glycosaminoglycansin. Immunofluorescent staining shows distribution of elastin, collagen I, and collagen III. Top: day 90 explant, bottom: native aorta.
Source: Adapted from Wu et al.[148] © 2012, with permission from Macmillan Publishers Ltd.

that can be colonized *in vivo* by native cells through migration and adhesion. A permanent vascular graft must offer definitive solutions especially in regards to long-term patency and structural integrity that, at the moment, have yet to be achieved. Therefore, a possible solution has focused on providing a biodegradable scaffold, which temporarily serves as a conduit for blood flow while allowing cells to grow and breakdown the grafts, simultaneously producing native tissue and ultimately a new vessel (Fig. 4). The following degradable polymers are the focus of research in attempts to provide a tissue-engineered approach to the vascular graft development field.

Polycaprolactone

Polycaprolactone (PCL) is a hydrophobic synthetic biodegradable polyester that has received a great deal of attention for use as an implantable biomaterial. This is due to its good mechanical properties, good biocompatibility and also, in large part, due to its slow biodegradability. PCL sheets have good mechanical properties but are somewhat limited in terms of compliance, though it is at least comparable to ePTFE and PET. It is however generally accepted that, as it is degraded and substituted by native tissue, compliance should improve over time due to elastogenesis.[19]

Under physiological condition, such as in the human body, PCL suffer hydrolysis of its ester linkages. Generally, biodegradable graft approaches are viewed with some reserve. If *in vivo* degradation occurs too rapidly, as compared to overall cell growth, mechanical failure may occur ultimately leading to subsequent aneurysms and ruptures, which in turn can lead to internal bleeding and death. Degradation of PCL occurs at slow enough a pace, taking over 24 months to be completely absorbed, such that mechanical strength is guaranteed during cell invasion.[19]

PCL grafts can be easily generated by electrospinning with varying pore sizes and prosthesis thicknesses. This has led to several studies focusing on determining the feasibility of an electrospun PCL vascular graft over the last 6 years.[19,122–126] Several studies, performed in rat, both showed rapid and stable endothelization, which were responsible for good cell invasion and patency over the course of the studies, as compared to ePTFE.[19,126–128] However, the poor mechanical compliance and loose cell attachment showed some development of intimal hyperplasia in all the studies.[19] The long-term studies also revealed a propensity for calcification of the prosthesis over time (usually in the last months of the assays) although they did not affect the end-term patency of the graft.

In order to improve the mechanical compliance, issues that PCL has exhibited *in vivo*, several studies have focused on either preseeding the grafts with vascular endothelium cells[129] or creating nanocomposite grafts, with collagen, elastin, and polylactide (Fig. 5), all showing improved *in vivo* patency.[130–132] Similar approaches involve surface functionalization or the creation of a polymeric system (inclusion of bioactive molecules in the polymer fibers during the electrospinning process), which have also shown promising results, improving endothelial cell adhesion and hemocompatibility.[133,134] Hybrid approaches of both incorporating other polymers and bioactive molecules have shown positive results.[135,136]

Collagen

Collagen is the most abundant group of proteins found in animals, the most common among which is Collagen I. Once irreversibly hydrolyzed, collagen originates gelatin. It is the main component of the body, making up 25–35% of the protein content. As elongated fibrils, collagen can be found in many fibrous tissues of the body, such as tendons, ligaments, and skin, and is also abundant in cartilage, bone, gut, intervertebral discs, and blood vessels.[137] As one of the major body components, it is biocompatible, biodegradable, and has the advantage of being easily synthesized. This has led to it being abundantly researched for application in fields such as skin, cartilage, bone. and vascular engineering and, additionally, as a drug delivery

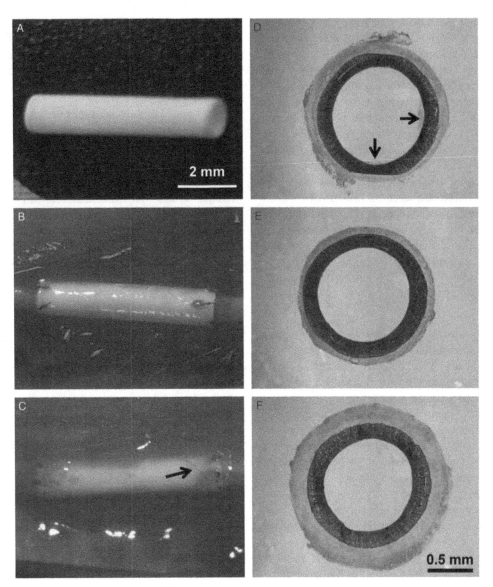

Fig. 5 Macroscale characterization of a tissue engeneering approach using a polycaprolactone/poly(L-lactic acid) grafts. (**A**) A graft before implantation. (**B**) A graft right after implantation. (**C**) A heparin-SDF-1α-treated graft at 4 weeks after implantation. Arrow indicates microvessels in the wall of the graft. (**D–F**) H&E staining of an untreated graft (**D**), a heparin-treated graft (**E**) and a heparin-SDF-1α-treated graft (**F**) at 4 weeks after implantation. Arrows in D indicate thrombus formation.
Source: Adapted from Yu et al.[206] © 2012, with permission from Elsevier.

system.[138–140] The most common method of producing vascular collagen grafts is by electrospinning; however, the chemical nature of collagen makes it sensible to deterioration during the process. The alternate option is to use a cross-linking agent to create the tubular prosthesis. Glutaraldehyde is used for this purpose, and while increasing strength it often carries the risk of increased toxicity and calcification. Also, without the aided mechanical support provided by seeded cells or other stabilizing polymers, collagen tubes are mechanically weak.[141] On top of all this, it is difficult to precisely control the mechanical behavior of the grafts due to a lack of means to do so. Other than the mechanical compliance issues, and most importantly, collagen is also well documented as thrombogenic

and a procoagulant.[142–145] Indeed, while a constituent of the extracellular matrix, collagen is normally not exposed to circulating blood. Once exposed, it triggers the activation of the Hageman factor and promote clotting and healing of an injured site.[145]

Studies have, therefore, mainly focused on combinations of collagen and other polymers,[124,141,146,147] bioactive molecules,[144] or cell seeding.[148] Cells easily adhere to collagen matrices and so many studies have focused on cell seeding (as mentioned above) with the dual intention of enhancing mechanical compliance and especially hemocompatibility. These approaches have provided a scaffold with adequate characteristics for small diameter vessels. A cell seeded collagen substrate with immobilized heparin demonstrated enhanced hemocompatibility and endothelial cell adhesion.[144] Some studies have researched the possibility of pure collagen prosthesis. A 2 mm (outer diameter) collagen vessel, seeded with both endothelial and SMCs, remained patent, *in vivo*, after 12 weeks.[148] Ultimately, collagen I and its blends outperform many synthetic polymers by providing an extracellular matrix that is beneficial to cell differentiation and proliferation, with adequate mechanical behavior, given certain conditions are met.[149] The greatest application of collagen seems to be associated with enhancing other polymer grafts in a "supporting" role. Collagen shows therefore great potential; a construct consisting solely of collagen with no harmful cross-linking agent and adequate mechanical compliance and hemocompatibility would be a serious competitor in vascular tissue engineering.

Elastin

Elastin is a hydrophobic connective tissue protein polymer consisting of posttranslational modified and cross-linked tropoelastin monomers. As its name suggests, it is elastic and allows tissues to retain shape memory, dictating tissue mechanics at low strains and preventing tissue creep.[150] At higher mechanical strains, collagen then assumes the role of providing mechanical resistance, due to the low maximum tensile strength of elastin. Elastin is the dominant extracellular matrix protein in arterial walls, organized as concentric rings around the medial layer of arteries and making up roughly 50% of their dry weight.[151] Elastin exhibits low thrombogenicity and several vascular devices, coated with elastin-based polypeptides, demonstrated reduced platelet adhesion and activation, and overall improvement *in vivo* patency.[152,153] Additionally, elastin is an autocrine regulator of SMC activity that has an essential role in preventing fibrocellular pathology and regulating proliferation.[154] The importance of the elastic properties and SMC regulatory activity has led to many studies aimed at inducing elastin biosynthesis in tissue-engineered vascular grafts. Several studies, using static collagen scaffolds seeded with SMC, did not show adequate elastin production.[155–157] However, further studies have shown that elastogenesis can be significantly improved by changing the scaffold substrate,[150,158] providing a mechanical stimuli[159–162] and also by the use of growth factors.[150] These cell seeding approaches have shown some success in promoting elastin deposition, the end result being the intended improvement in mechanical behavior, biocompatibility, and hemocompatibility. Besides cell seeding, several methods of producing vascular grafts that incorporate elastin have been studied.[163,164] Collagen,[165] polyurethane,[166] polydioxanone,[167] PCL,[168] gelatin, and recently polyethylene glycol,[169] and also combinations of these polymers,[124] have all been electrospun into hybrid tubular grafts, with elastin, to produce viable vascular grafts. These approaches all have

similar objectives, to limit the degradation rate and provide high mechanical resistance while allowing for the beneficial mechanical and chemotactic performance of elastin. As such, they have all shown potential in regards to mechanical, bio- and hemocompatibility. Attempts at incorporating elastin in tissue-engineered products and designing biocompatible synthetic elastic polymers are, however, still insufficient. A major drawback of a viable elastin graft is cost related. Tropoelastin is hydrophilic and easily susceptible to proteolysis, it is hard to isolate and readily reacts with other monomers. The main source of elastin—extraction from animal tissues—provides small amounts of the polymer at relatively high cost due to its insolubility in water, which prevents the use of conventional wet-chemistry techniques.[170] Further studies are needed to understand elastin biosynthesis in order to produce adequate elastin-incorporating and/or elastin-based grafts at cost-efficient methods.

Fibrin

Fibrin is the end-product of the coagulation cascade, produced by the conversion of fibrinogen to fibrin monomers, which polymerizes into an insoluble fibrin hydrogel network.[171] Fibrin seems an ideal scaffold for tissue regeneration since, *in vivo*, it acts as a "physiological scaffold" and plays a role in hemostasis, inflammation, wound healing, and angiogenesis.[172–174] It has also been shown to promote cell migration into tissue-engineered constructs, and the cells then gradually replace the fibrin scaffold with native extracellular matrix.[175,176] Fibrinogen can be isolated from the same patient that would receive the graft, thus lowering the risks of body reaction and infection.[171,177] Conventionally, fibrinogen is isolated from platelet poor plasma after freezing and thawing. However, this method of isolation has disadvantages: the quantity yield is relatively low (20–25%) and the production process is time-consuming (2 days).[178] The low yields would require large amounts of donor blood, which is not a viable option. Alternative precipitation methods have been studied and, among them ethanol precipitation was found to be the most efficient with ~80% of fibrinogen precipitated.[179] Fibrin has been used as a coating in vascular prosthetics in order to enhance the hemo- and biocompatibility of synthetic grafts.[180] Fibrin-based vascular grafts have been studied toward applications in coronary and peripheral artery bypass, arterio-venous access grafts, or for congenital pulmonary artery reconstruction in children.[178] However, the fibrin gel lacks suitable mechanical properties. The grafts generally contain a supporting, bioresorbable, macroporous mesh. The meshes provide the required mechanical support while prior cell seeding remodels the fibrin matrix. Polylactide[181] and polyglactin[182] have been used as support structures. The tubular prosthesis is then generally seeded with different combinations of EC, SMC, and myofibroblasts and incubated before implantation[181] and have been shown to remain patent for up to 6 months.[183] The major drawback with these approaches is the extremely long production time for the graft (ca. 22 days), which prevents its application as an immediate alternative to the synthetic alternatives ePTFE and PET.

Silk

Silk is a natural protein polymer composed of two proteins: fibroin, the insoluble protein that is the main structural center of the silk fiber; and sericin, a water soluble protein that forms a gum coating the fibers and allowing them to stick to each other. Most of the silk produced and used worldwide is obtained from the pupating larvae of the mulberry silkworm *Bombyx mori*. The cocoons are then soaked in boiling water, to soften the sericin, and the fibers unwound to produce a continuous silk thread. Fibroin and sericin are separated in a degumming process. The resulting silk fibroin forms protein fibers with high tensile strength and high toughness, which have long found a home in medicine as a suture. From aqueous or organic solutions of silk fibroin, it's also possible to form films, sponges, powder, gels, and the regenerated fibers. Silk fibroin has many of the requirements of a good biomaterial: it has good mechanical properties, biocompatibility, hemocompatibility, low immunogenicity, and is biodegradable.[184–186] This has led to it being extensively studied for biomedical applications.[185,187–194] For vascular grafts, in particular, the biocompatibility of silk is extremely important. Cell adhesion and growth on silk scaffolds have been shown to be very good[195–198] and, with the aid of biotechnology developments, improvements can be made further enhancing cell adhesion.[199] The inclusion of cell-adhesive sequences into the fibroin amino-acid sequence has been shown to enhance cell adhesion.[194,198,200] This cytocompatibility of silk fibroin has even led it to be studied as a coating for other polymeric materials in order to enhance cell adhesion.[201]

As vascular grafts, silk fibroin can be woven,[202] knitted,[203] braided,[204] or electrospun (using polyethylene oxideas an additive)[198] in order to produce tubular grafts with an open porous structure, this can easily be controlled through control of the flow rate.[205] Knitting and braiding require coating with an aqueous solution of silk fibroin, and usually another polymer such as poly(propylene glycol) dimethyl ether, in order to seal the graft and control permeability, mechanical behavior, and pore size.[203] Some studies have already shown that vascular grafts of silk fibroin can be successful in small diameter applications.[198,203,204] A 1.5 mm (inner diameter) by 10 mm (length) graft was sutured, end-to-end, to a rat abdominal aorta and was shown to remain patent for 12 weeks in 85% of the cases.[204] The patency of the silk graft was largely due to the almost complete cell coverage of the luminal surface, which was achieved after 9 weeks.[204] A similar study was also performed on the carotid artery of a canine model using 3 mm diameter tubes and showing of up to one year in one of the implanted grafts.[203]

NEXT STEPS FOR POLYMERIC GRAFTS

The future of synthetic vascular graft development seems very promising. The field is every growing and the lack of a clear optimal polymer has led, and continues to lead, to many potential options. Either by further development and research into the already existing grafts (PET and ePTFE), the further development of new nondegradable grafts (PU, PVA, and BC) or researching and developing temporary grafts that ultimately allow for a tissue-engineered solution, either by prior cell-seeding or allowing for native cell migration, attachment and ingrowth, ultimately producing a new vessel out of native tissue (collagen, fibrin, elastin, silk, and PCL).

In further years, research seems like it will focus on the end development and fine-tuning of either a nanocomposite or surface-modified solution, with bioactive molecules, through *in vivo* testing. The nanocomposites approach seems to be the most common trend in the development of a tissue-engineered vessel. This method offers the possibility of combining the favorable properties of one polymer with the favorable properties of another, usually providing positive results with little or none of the drawbacks.

Those that have been mentioned here are the basis of the research that is currently ongoing and each has a beneficial aspect to contribute to the field. Alternatively, the surface modification of some of the abovementioned polymers may provide a viable alternative for a ready and off-the-shelf solution. The efforts of all the researchers in these fields will almost certainly provide a new graft that can outperform the current standards and vastly improve the quality of life for millions.

ACKNOWLEDGMENTS

Funding made possible by the FCT through the project BCGrafts (PTDC/EBB-EBI/112170/2009) and the FCT PhD grant (SFRH/BD/66094/2009).

REFERENCES

1. Go, A.S.; Mozaffarian, D.; Roger, V.L.; Benjamin, E.J.; Berry, J.D.; Borden, W.B.; Bravata, D.M.; Dai, S.; Ford, E.S.; Fox, C.S.; Franco, S.; Fullerton, H.J.; Gillespie, C.; Hailpern, S.M.; Heit, J.A.; Howard, V.J.; Huffman, M.D.; Kissela, B.M.; Kittner, S.J.; Lackland, D.T.; Lichtman, J.H.; Lisabeth, L.D.; Magid, D.; Marcus, G.M.; Marelli, A.; Matchar, D.B.; McGuire, D.K.; Mohler, E.R.; Moy, C.S.; Mussolino, M.E.; Nichol, G.; Paynter, N.P.; Schreiner, P.J.; Sorlie, P.D.; Stein, J.; Turan, T.N.; Virani, S.S.; Wong, N.D.; Woo, D.; Turner, M.B.; American Heart Association Statistics Committee and Stroke Statistics Subcommittee. Heart disease and stroke statistics—2013 update: A report from the American Heart Association. Circulation **2012**, *127* (1), e6–e245.

Vascular—Zwitterionic

2. Defraigne, J.-O. Development of Atherosclerosis for the Vascular Surgeon. In *Vascular Surgery*; Liapis, C., Balzer, K., Benedetti-Valentini, F., Fernandes e Fernandes, J., Eds.; Springer: Berlin, Heidelberg, 2007; 23–34.

3. Rajagopalan, S.; Kurz, S.; Munzel, T.; Tarpey, M.; Freeman, B.A.; Griendling, K.K.; Harrison, D.G. Angiotensin II-mediated hypertension in the rat increases vascular superoxide production via membrane NADH/ NADPH oxidase activation. Contribution to alterations of vasomotor tone. J. Clin. Invest. **1996**, *97* (8), 1916–1923.

4. Ohara, Y.; Peterson, T.E.; Harrison, D.G. Hypercholesterolemia increases endothelial superoxide anion production. J. Clin. Invest. **1993**, *91* (6), 2546–2551.

5. Tesfamariam, B.; Cohen, R.A. Free radicals mediate endothelial cell dysfunction caused by elevated glucose. Am. J. Physiol. **1992**, *263* (2 Pt. 2), H321–H326.

6. Fuster, V.; Badimon, J.J.; Chesebro, J.H. Atherothrombosis: Mechanisms and clinical therapeutic approaches. Vasc. Med. **1998**, *3* (3), 231–239.

7. Yusuf, S.; Hawken, S.; Ounpuu, S.; Dans, T.; Avezum, A.; Lanas, F.; McQueen, M.; Budaj, A.; Pais, P.; Varigos, J.; Lisheng, L.; INTERHEART study Investigators. Effect of potentially modifiable risk factors associated with myocardial infarction in 52 countries (the INTERHEART study): Case-control study. Lancet **2004**, *364* (9438), 937–952.

8. Norgren, L.; Hiatt, W.R.; Dormandy, J.A.; Nehler, M.R.; Harris, K.A.; Fowkes, F.G. Inter-Society Consensus for the Management of Peripheral Arterial Disease (TASC II). J. Vasc. Surg. **2007**, *45* (Suppl. S), S5–67.

9. Rutherford, R.B. *Vascular Surgery, 2 Volume Set*; W.B. Saunders Company: Philadelphia, PA, 2000.

10. Panneton, J.M.; Hollier, L.H.; Hofer, J.M. Multicenter randomized prospective trial comparing a pre-cuffed polytetrafluoroethylene graft to a vein cuffed polytetrafluoroethylene graft for infragenicular arterial bypass. Ann. Vasc. Surg. **2004**, *18* (2), 199–206.

11. Nguyen, L.L.; Conte, M.S.; Menard, M.T.; Gravereaux, E.C.; Chew, D.K.; Donaldson, M.C.; Whittemore, A.D.; Belkin, M. Infrainguinal vein bypass graft revision: Factors affecting long-term outcome. J. Vasc. Surg. **2004**, *40* (5), 916–923.

12. White, C.J.; Gray, W.A. Endovascular therapies for peripheral arterial disease: An evidence-based review. Circulation **2007**, *116* (19), 2203–2215.

13. Mughal, N.A.; Russell, D.A.; Ponnambalam, S.; Homer-Vanniasinkam, S. Gene therapy in the treatment of peripheral arterial disease. Br. J. Surg. **2012**, *99* (1), 6–15.

14. Bradbury, A.W.; Adam, D.J.; Bell, J.; Forbes, J.F.; Fowkes, F.G.; Gillespie, I.; Ruckley, C.V.; Raab, G.M. Bypass versus Angioplasty in Severe Ischaemia of the Leg (BASIL) trial: Analysis of amputation free and overall survival by treatment received. J. Vasc. Surg. **2010**, *51* (Suppl. 5), 18S–31S.

15. Bisdas, T.; Wilhelmi, M.; Haverich, A.; Teebken, O.E. Cryopreserved arterial homografts vs silver-coated Dacron grafts for abdominal aortic infections with intraoperative evidence of microorganisms. J. Vasc. Surg. **2011**, *53* (5), 1274–1281 e4.

16. Moawad, J.; Gagne, P. Adjuncts to improve patency of infrainguinal prosthetic bypass grafts. Vasc. Endovasc. Surg. **2003**, *37* (6), 381–386.

17. Fabiani, J.N.; Julia, P.; Chemla, E.; Birnbaum, P.L.; Chardigny, C.; D'Attellis, N.; Renaudin, J.M. Is the incidence of recurrent carotid artery stenosis influenced by the choice of the surgical technique? Carotid endarterectomy versus saphenous vein bypass. J. Vasc. Surg. **1994**, *20* (5), 821–825.

18. Veldenz, H.C.; Kinser, R.; Yates, G.N. Carotid graft replacement: A durable option. J. Vasc. Surg. **2005**, *42* (2), 220–226.

19. de Valence, S.; Tille, J.C.; Mugnai, D.; Mrowczynski, W.; Gurny, R.; Moller, M.; Walpoth, B.H. Long term performance of polycaprolactone vascular grafts in a rat abdominal aorta replacement model. Biomaterials **2012**, *33* (1), 38–47.

20. Engler, A.J.; Sen, S.; Sweeney, H.L.; Discher, D.E. Matrix elasticity directs stem cell lineage specification. Cell **2006**, *126* (4), 677–689.

21. Sperling, C.; Fischer, M.; Maitz, M.F.; Werner, C. Blood coagulation on biomaterials requires the combination of distinct activation processes. Biomaterials **2009**, *30* (27), 4447–4456.

22. Fink, H.; Faxalv, L.; Molnar, G.F.; Drotz, K.; Risberg, B.; Lindahl, T.L.; Sellborn, A. Real-time measurements of coagulation on bacterial cellulose and conventional vascular graft materials. Acta Biomater. **2010**, *6* (3), 1125–1130.

23. Andrade, F.K.; Silva, J.P.; Carvalho, M.; Castanheira, E.M.; Soares, R.; Gama, M. Studies on the hemocompatibility of bacterial cellulose. J. Biomed. Mater. Res. A **2011**, *98* (4), 554–566.

24. Leitão, A.F.; Gupta, S.; Silva, J.P.; Reviakine, I.; Gama, M. Hemocompatibility study of a bacterial cellulose/polyvinyl alcohol nanocomposite. Colloids Surf. B **2013**, *111* (0), 493–502.

25. Godo, M.N.; Sefton, M.V. Characterization of transient platelet contacts on a polyvinyl alcohol hydrogel by video microscopy. Biomaterials **1999**, *20* (12), 1117–1126.

26. Sheppard, J.I.; McClung, W.G.; Feuerstein, I.A. Adherent platelet morphology on adsorbed fibrinogen: Effects of protein incubation time and albumin addition. J. Biomed. Mater. Res. **1994**, *28* (10), 1175–1186.

27. Northolt, M.G.; den Decker, P.; Picken, S.J.; Baltussen, J.J.M.; Schlatmann, R. The Tensile Strength of Polymer Fibres. *Polymeric and Inorganic Fibers*; Springer: Berlin, Heidelberg, 2005; 1–108.

28. Rajendran, S.; Anand, S.C. Developments in medical textiles. JOTP **2002**, *32* (4), 1–42.

29. Ventura, A.; Terzaghi, C.; Legnani, C.; Borgo, E.; Albisetti, W. Synthetic grafts for anterior cruciate ligament rupture: 19–year outcome study. Knee **2010**, *17* (2), 108–113.

30. Bodugoz–Senturk, H.; Macias, C.E.; Kung, J.H.; Muratoglu, O.K. Poly(vinyl alcohol)-acrylamide hydrogels as load-bearing cartilage substitute. Biomaterials **2009**, *30* (4), 589–596.

31. Buchensk, J.; Slomkowski, S.; Tazbir, J.W.; Sobolewska, E. Poly(ethylene terephthalate) yarn with antibacterial properties. J. Biomater. Sci. Polym. Ed. **2001**, *12* (1), 55–62.

32. Zieren, J.; Neuss, H.; Paul, M.; Müller, J. Introduction of polyethylene terephthalate mesh (KoSa hochfest®) for abdominal hernia repair: An animal experimental study. Bio-Med. Mater. Eng. **2004**, *14* (2), 127–132.

33. Menu, P.; Stoltz, J.F.; Kerdjoudj, H. Progress in vascular graft substitute. Clin. Hemorheol. Microcirc. **2013**, *53* (1), 117–129.

34. Prager, M.R.; Hoblaj, T.; Nanobashvili, J.; Sporn, E.; Polterauer, P.; Wagner, O.; Böhmig, H.-J.; Teufelsbauer, H.; Ploner, M.; Huk, I. Collagen- versus gelatine-coated Dacron versus stretch PTFE bifurcation grafts for aortoiliac occlusive disease: Long-term results of a prospective, randomized multicenter trial. Surgery **2003**, *134* (1), 80–85.

35. Nevelsteen, A.; Wouters, L.; Suy, R. Long-term patency of the aortofemoral Dacron graft. A graft limb related study over a 25-years period. J. Cardiovasc. Surg. **1991**, *32* (2), 174–180.

36. Hanson, S.R.; Kotze, H.F.; Savage, B.; Harker, L.A. Platelet interactions with Dacron vascular grafts. A model of acute thrombosis in baboons. Arterioscler. Thromb. Vasc. Biol. **1985**, *5* (6), 595–603.

37. Salacinski, H.J.; Goldner, S.; Giudiceandrea, A.; Hamilton, G.; Seifalian, A.M.; Edwards, A.; Carson, R.J. The mechanical behavior of vascular grafts: A review. J. Biomater. Appl. **2001**, *15* (3), 241–278.

38. Holmberg, M.; Hou, X. Competitive protein adsorption of albumin and immunoglobulin G from human serum onto polymer surfaces. Langmuir **2010**, *26* (2), 938–942.

39. Hake, U.; Gabbert, H.; Iversen, S.; Jakob, H.; Schmiedt, W.; Oelert, H. Evaluation of the healing of precoated vascular dacron prostheses. Langenbecks Archiv für Chirurgie **1991**, *376* (6), 323–329.

40. Van Damme, H.; Deprez, M.; Creemers, E.; Limet, R. Intrinsic structural failure of polyester (dacron) vascular grafts: A general review. Acta Chir. Belg. **2005**, *105* (3), 249–255.

41. Vogt, P.R.; Brunner-LaRocca, H.P.; Lachat, M.; Ruef, C.; Turina, M.I. Technical details with the use of cryopreserved arterial allografts for aortic infection: Influence on early and midterm mortality. J. Vasc. Surg. **2002**, *35* (1), 80–86.

42. Cziperle, D.J.; Joyce, K.A.; Tattersall, C.W.; Henderson, S.C.; Cabusao, E.B.; Garfield, J.D.; Kim, D.U.; Duhamel, R.C.; Greisler, H.P. Albumin impregnated vascular grafts: Albumin resorption and tissue reactions. J. Cardiovasc. Surg. **1992**, *33* (4), 407–414.

43. Scott, S.M.; Gaddy, L.R.; Sahmel, R.; Hoffman, H. A collagen coated vascular prosthesis. J. Cardiovasc. Surg. **1987**, *28* (5), 498–504.

44. Koromila, G.; Michanetzis, G.P.; Missirlis, Y.F.; Antimisiaris, S.G. Heparin incorporating liposomes as a delivery system of heparin from PET-covered metallic stents: Effect on haemocompatibility. Biomaterials **2006**, *27* (12), 2525–2533.

45. Lev, E.I.; Assali, A.R.; Teplisky, I.; Rechavia, E.; Hasdai, D.; Sela, O.; Shor, N.; Battler, A.; Kornowski, R. Comparison of outcomes up to six months of Heparin-Coated with noncoated stents after percutaneous coronary intervention for acute myocardial infarction. Am. J. Cardiol. **2004**, *93* (6), 741–743.

46. Kowalik, Z.; Kucharski, A.; Hobot, J. Use of dacron vascular prosthesis impregnated with salts of silver, in treatment of extraanatomical axilla-femoral by-pass's infection. Infections in vascular surgery. Polim. Med. **2002**, *32* (1–2), 80–84.

47. Jeanmonod, P.; Laschke, M.W.; Gola, N.; von Heesen, M.; Glanemann, M.; Dold, S.; Menger, M.D.; Moussavian, M.R. Silver acetate coating promotes early vascularization of Dacron vascular grafts without inducing host tissue inflammation. J. Vasc. Surg. **2013**, *58* (6), 1637–1643.

48. Noel, S.; Liberelle, B.; Yogi, A.; Moreno, M.J.; Bureau, M.N.; Robitaille, L.; De Crescenzo, G. A non-damaging chemical amination protocol for poly (ethylene terephthalate)–application to the design of functionalized compliant vascular grafts. J. Mater. Chem. B **2013**, *1* (2), 230–238.

49. Li, P.; Cai, X.; Wang, D.; Chen, S.; Yuan, J.; Li, L.; Shen, J. Hemocompatibility and anti-biofouling property improvement of poly(ethylene terephthalate) via self-polymerization of dopamine and covalent graft of zwitterionic cysteine. Colloids Surf. B **2013**, *110* (0), 327–332.

50. Guidoin, R.; Chakfe, N.; Maurel, S.; How, T.; Batt, M.; Marois, M.; Gosselin, C. Expanded polytetrafluoroethylene arterial prostheses in humans: Histopathological study of 298 surgically excised grafts. Biomaterials **1993**, *14* (9), 678–693.

51. Chlupac, J.; Filova, E.; Bacakova, L. Blood vessel replacement: 50 years of development and tissue engineering paradigms in vascular surgery. Physiol. Res. **2009**, *58* (Suppl. 2) S119–S139.

52. Grigioni, M.; Daniele, C.; D'avenio, G.; Barbaro, V. *Biomechanics and Hemodynamics of Grafting*; WIT Press: Billerica, MA, 2003; 41–82.

53. Lee, J.M.; Wilson, G.J. Anisotropic tensile viscoelastic properties of vascular graft materials tested at low strain rates. Biomaterials **1986**, *7* (6), 423–431.

54. Soyer, T.; Lempinen, M.; Cooper, P.; Norton, L.; Eiseman, B. A new venous prosthesis. Surgery **1972**, *72* (6), 864–872.

55. Norton, L.; Eiseman, B. Replacement of portal vein during pancreatectomy for carcinoma. Surgery **1975**, *77* (2), 280–284.

56. Roll, S.; Muller-Nordhorn, J.; Keil, T.; Scholz, H.; Eidt, D.; Greiner, W.; Willich, S.N. Dacron vs. PTFE as bypass materials in peripheral vascular surgery—Systematic review and meta-analysis. BMC Surg **2008**, *8*, 22.

57. Greisler, H.P. *New Biologic And Synthetic Vascular Prostheses*; R G Landes Co: 1991; 86.

58. Veith, F.J.; Gupta, S.K.; Ascer, E.; White-Flores, S.; Samson, R.H.; Scher, L.A.; Towne, J.B.; Bernhard, V.M.; Bonier, P.; Flinn, W.R.; Astelford, P.; Yao, J.S.T.; Bergan, J.J. Six-year prospective multicenter randomized comparison of autologous saphenous vein and expanded polytetrafluoroethylene grafts in infrainguinal arterial reconstructions. J. Vasc. Surg. **1986**, *3* (1), 104–114.

59. van Det, R.J.; Vriens, B.H.; van der Palen, J.; Geelkerken, R.H. Dacron or ePTFE for femoropopliteal above-knee bypass grafting: Short- and long-term results of a multicentre randomised trial. Eur. J. Vasc. Endovasc. Surg. **2009**, *37* (4), 457–463.

60. Pevec, W.C.; Darling, R.C.; L'Italien, G.J.; Abbott, W.M. Femoropopliteal reconstruction with knitted, nonvelour Dacron versus expanded polytetrafluoroethylene. J. Vasc. Surg. **1992**, *16* (1), 60–65.

61. Jensen, L.P.; Lepantalo, M.; Fossdal, J.E.; Roder, O.C.; Jensen, B.S.; Madsen, M.S.; Grenager, O.; Fasting, H.; Myhre, H.O.; Baekgaard, N.; Nielsen, O.M.; Helgstrand, U.; Schroeder, T.V. Dacron or PTFE for above-knee

femoropopliteal bypass. a multicenter randomised study. Eur. J. Vasc. Endovasc. Surg. **2007**, *34* (1), 44–49.

62. Lu, S.; Sun, X.; Zhang, P.; Yang, L.; Gong, F.; Wang, C. Local hemodynamic disturbance accelerates early thrombosis of small-caliber expanded polytetrafluoroethylene grafts. Perfusion **2013**, *28* (5), 440–448.

63. Zilla, P.; Brink, J.; Human, P.; Bezuidenhout, D. Prosthetic heart valves: Catering for the few. Biomaterials **2008**, *29* (4), 385–406.

64. Berger, K.; Sauvage, L.R.; Rao, A.M.; Wood, S.J. Healing of arterial prostheses in man: Its incompleteness. Ann. Surg. **1972**, *175* (1), 118–127.

65. Clowes, A.W.; Kirkman, T.R.; Reidy, M.A. Mechanisms of arterial graft healing. Rapid transmural capillary ingrowth provides a source of intimal endothelium and smooth muscle in porous PTFE prostheses. Am. J. Pathol. **1986**, *123* (2), 220–230.

66. Cameron, B.L.; Tsuchida, H.; Connall, T.P.; Nagae, T.; Furukawa, K.; Wilson, S.E. High porosity PTFE improves endothelialization of arterial grafts without increasing early thrombogenicity. J. Cardiovasc. Surg. **1993**, *34* (4), 281–285.

67. Laube, H.R.; Duwe, J.; Rutsch, W.; Konertz, W. Clinical experience with autologous endothelial cell-seeded polytetrafluoroethylene coronary artery bypass grafts. J. Thorac. Cardiovasc. Surg. **2000**, *120* (1), 134–141.

68. Venkatraman, S.; Boey, F.; Lao, L.L. Implanted cardiovascular polymers: Natural, synthetic and bio-inspired. Prog. Polym. Sci. **2008**, *33* (9), 853–874.

69. Heyligers, J.M.; Lisman, T.; Verhagen, H.J.; Weeterings, C.; de Groot, P.G.; Moll, F.L. A heparin-bonded vascular graft generates no systemic effect on markers of hemostasis activation or detectable heparin-induced thrombocytopenia-associated antibodies in humans. J. Vasc. Surg. **2008**, *47* (2), 324–329, discussion 329.

70. Hoshi, R.A.; Van Lith, R.; Jen, M.C.; Allen, J.B.; Lapidos, K.A.; Ameer, G. The blood and vascular cell compatibility of heparin-modified ePTFE vascular grafts. Biomaterials **2013**, *34* (1), 30–41.

71. Zarge, J.I.; Gosselin, C.; Huang, P.; Vorp, D.A.; Severyn, D.A.; Greisler, H.P. Platelet deposition on ePTFE grafts coated with fibrin glue with or without FGF-1 and heparin. J. Surg. Res. **1997**, *67* (1), 4–8.

72. Kapfer, X.; Meichelboeck, W.; Groegler, F.M. Comparison of carbon-impregnated and standard ePTFE prostheses in extra-anatomical anterior tibial artery bypass: A prospective randomized multicenter study. Eur. J. Vasc. Endovasc. Surg. **2006**, *32* (2), 155–168.

73. Akers, D.L.; Du, Y.H.; Kempczinski, R.F. The effect of carbon coating and porosity on early patency of expanded polytetrafluoroethylene grafts: An experimental study. J. Vasc. Surg. **1993**, *18* (1), 10–15.

74. Pedersen, G.; Laxdal, E.; Ellensen, V.; Jonung, T.; Mattsson, E. Improved patency and reduced intimal hyperplasia in PTFE grafts with luminal immobilized heparin compared with standard PTFE grafts at six months in a sheep model. J. Cardiovasc. Surg. **2010**, *51* (3), 443–8.

75. Saxon, R.R.; Chervu, A.; Jones, P.A.; Bajwa, T.K.; Gable, D.R.; Soukas, P.A.; Begg, R.J.; Adams, J.G.; Ansel, G.M.; Schneider, D.B.; Eichler, C.M. and Rush, M.J. Heparin-bonded, Expanded polytetrafluoroethylene-lined stent graft in the treatment of femoropopliteal artery disease: 1-year results of the VIPER (Viabahn Endoprosthesis with Heparin Bioactive Surface in the Treatment of Superficial Femoral Artery Obstructive Disease) Trial. J. Vasc. Intervent. Radiol. **2013**, *24* (2), 165–173.

76. Zilla, P.; Bezuidenhout, D.; Human, P. Prosthetic vascular grafts: Wrong models, wrong questions and no healing. Biomaterials **2007**, *28* (34), 5009–5027.

77. Peckham, S.M.; Turitto, V.T.; Glantz, J.; Puryear, H.; Slack, S.M. Hemocompatibility studies of surface-treated polyurethane–based chronic indwelling catheters. J. Biomater. Sci. Polym. Ed. **1997**, *8* (11), 847–858.

78. He, W.; Hu, Z.; Xu, A.; Liu, R.; Yin, H.; Wang, J.; Wang, S. The preparation and performance of a new polyurethane vascular prosthesis. Cell Biochem. Biophys. **2013**, *66* (3), 855–866.

79. Maya, I.D.; Weatherspoon, J.; Young, C.J.; Barker, J.; Allon, M. Increased risk of infection associated with polyurethane dialysis grafts. Semin. Dial **2007**, *20* (6), 616–620.

80. Meng, J.; Cheng, X.; Kong, H.; Yang, M.; Xu, H. Preparation and biocompatibility evaluation of polyurethane filled with multiwalled carbon nanotubes. J. Nanosci. Nanotechnol. **2013**, *13* (2), 1467–1471.

81. Hu, Z.-J.; Li, Z.-l.; Hu, L.-Y.; He, W.; Liu, R.-M.; Qin, Y.-S.; Wang, S.-M. The *in vivo* performance of small-caliber nanofibrous polyurethane vascular grafts. BMC Cardiovasc. Disord. **2012**, *12* (1), 1–11.

82. Han, J.; Farah, S.; Domb, A.; Lelkes, P. Electrospun rapamycin-eluting polyurethane fibers for vascular grafts. Pharm. Res. **2013**, *30* (7), 1735–1748.

83. Wong, C.S.; Liu, X.; Xu, Z.; Lin, T.; Wang, X. Elastin and collagen enhances electrospun aligned polyurethane as scaffolds for vascular graft. J. Mater. Sci. **2013**, *24* (8), 1865–1874.

84. Nguyen, T.-H.; Padalhin, A.R.; Seo, H.S.; Lee, B.-T. A hybrid electrospun PU/PCL scaffold satisfied the requirements of blood vessel prosthesis in terms of mechanical properties, pore size, and biocompatibility. J. Biomater. Sci. Polym. Ed. **2013**, *24* (14), 1692–1706.

85. Zhang, Z.; Marois, Y.; Guidoin, R.G.; Bull, P.; Marois, M.; How, T.; Laroche, G.; King, M.W. Vascugraft polyurethane arterial prosthesis as femoropopliteal and femoroperoneal bypasses in humans: Pathological, structural and chemical analyses of four excised grafts. Biomaterials **1997**, *18* (2), 113–124.

86. Stachelek, S.J.; Alferiev, I.; Fulmer, J.; Ischiropoulos, H.; Levy, R.J. Biological stability of polyurethane modified with covalent attachment of di-tert-butyl-phenol. J. Biomed. Mater. Res. A **2007**, *82* (4), 1004–1011.

87. Glickman, M.H.; Stokes, G.K.; Ross, J.R.; Schuman, E.D.; Sternbergh Iii, W.C.; Lindberg, J.S.; Money, S.M.; Lorber, M.I. Multicenter evaluation of a polyurethaneurea vascular access graft as compared with the expanded polytetrafluoroethylene vascular access graft in hemodialysis applications. J. Vasc. Surg. **2001**, *34* (3), 465–473.

88. Chu, K.C.; Rutt, B.K. Polyvinyl alcohol cryogel: An ideal phantom material for MR studies of arterial flow and elasticity. Magn. Reson. Med. **1997**, *37* (2), 314–319.

89. Bourke, S.L.; Al-Khalili, M.; Briggs, T.; Michniak, B.B.; Kohn, J.; Poole-Warren, L.A. A photo-crosslinked

poly(vinyl alcohol) hydrogel growth factor release vehicle for wound healing applications. AAPS PharmSci **2003**, *5* (4), 101–111.

90. Takeuchi, H.; Kojima, H.; Yamamoto, H.; Kawashima, Y. Polymer coating of liposomes with a modified polyvinyl alcohol and their systemic circulation and RES uptake in rats. J. Control. Release **2000**, *68* (2), 195–205.

91. Laurent, A.; Wassef, M.; Saint Maurice, J.P.; Namur, J.; Pelage, J.P.; Seron, A.; Chapot, R.; Merland, J.J. Arterial distribution of calibrated tris-acryl gelatin and polyvinyl alcohol microspheres in a sheep kidney model. Invest. Radiol. **2006**, *41* (1), 8–14.

92. Pal, K.; Banthia, A.K.; Majumdar, D.K. Preparation and characterization of polyvinyl alcohol-gelatin hydrogel membranes for biomedical applications. AAPS PharmSci-Tech. **2007**, *8* (1), 21.

93. Yang, S.-H.; Lee, Y.-S.J.; Lin, F.-H.; Yang, J.-M.; Chen, K.-s. Chitosan/poly(vinyl alcohol) blending hydrogel coating improves the surface characteristics of segmented polyurethane urethral catheters. J. Biomed. Mater. Res. Part B **2007**, *83B* (2), 304–313.

94. Walker, J.; Young, G.; Hunt, C.; Henderson, T. Multicentre evaluation of two daily disposable contact lenses. Cont. Lens Anterior Eye **2007**, *30* (2), 125–133.

95. Chaouat, M.; Le Visage, C.; Baille, W.E.; Escoubet, B.; Chaubet, F.; Mateescu, M.A.; Letourneur, D. A Novel Cross-linked Poly(vinyl alcohol) (PVA) for Vascular Grafts. Adv. Funct. Mater. **2008**, *18* (19), 2855–2861.

96. Millon, L.E.; Wan, W.K. The polyvinyl alcohol-bacterial cellulose system as a new nanocomposite for biomedical applications. J. Biomed. Mater. Res. B Appl. Biomater. **2006**, *79* (2), 245–253.

97. Millon, L.E.; Guhados, G.; Wan, W. Anisotropic polyvinyl alcohol-Bacterial cellulose nanocomposite for biomedical applications. J. Biomed. Mater. Res. B Appl. Biomater. **2008**, *86* (2), 444–452.

98. Mohammadi, H. Nanocomposite biomaterial mimicking aortic heart valve leaflet mechanical behaviour. Proc. Inst. Mech. Eng. H **2011**, *225* (7), 718–722.

99. Chuang, W.Y.; Young, T.H.; Yao, C.H.; Chiu, W.Y. Properties of the poly(vinyl alcohol)/chitosan blend and its effect on the culture of fibroblast in vitro. Biomaterials **1999**, *20* (16), 1479–1487.

100. Koyano, T.; Minoura, N.; Nagura, M.; Kobayashi, K.-I. Attachment and growth of cultured fibroblast cells on PVA/chitosan-blended hydrogels. J. Biomed. Mater. Res. **1998**, *39* (3), 486–490.

101. Mathews, D.T.; Birney, Y.A.; Cahill, P.A.; McGuinness, G.B. Vascular cell viability on polyvinyl alcohol hydrogels modified with water-soluble and -insoluble chitosan. J. Biomed. Mater. Res. Part B **2008**, *84B* (2), 531–540.

102. Tudorachi, N.; Cascaval, C.N.; Rusu, M.; Pruteanu, M. Testing of polyvinyl alcohol and starch mixtures as biodegradable polymeric materials. Polym. Test. **2000**, *19* (7), 785–799.

103. Leitão, A.; Silva, J.; Dourado, F.; Gama, M. Production and characterization of a new bacterial cellulose/Poly(Vinyl Alcohol) nanocomposite. Materials **2013**, *6* (5), 1956–1966.

104. Andrade, F.; Pertile, R.; Douradoa, F. *Bacterial Cellulose: Properties, Production and Applications. Cellulose: Structure and Properties, Derivatives and Industrial Uses*; Nova Science Publishers, Inc., 2010; 427–458.

105. Jia, S.; Tang, W.; Yang, H.; Jia, Y.; Zhu, H. Preparation and Characterization of Bacterial Cellulose Tube. Bioinformatics and Biomedical Engineering, ICBBE 2009. 3rd International Conference on 2009, 1–4.

106. Kaźmierczak, D.; Kazimierczak, J. Biosynthesis of modified bacterial cellulose in a tubular form. Fibres Textiles East. Eur. **2010**, *18* (5), 82.

107. Bäckdahl, H.; Risberg, B.; Gatenholm, P. Observations on bacterial cellulose tube formation for application as vascular graft. Mater. Sci. Eng. C **2011**, *31* (1), 14–21.

108. Klemm, D.; Schumann, D.; Udhardt, U.; Marsch, S. Bacterial synthesized cellulose—Artificial blood vessels for microsurgery. Progr. Polym. Sci. **2001**, 26 (9), 1561–1603.

109. Svensson, A.; Nicklasson, E.; Harrah, T.; Panilaitis, B.; Kaplan, D.L.; Brittberg, M.; Gatenholm, P. Bacterial cellulose as a potential scaffold for tissue engineering of cartilage. Biomaterials **2005**, *26* (4), 419–431.

110. Alvarez, O.; Patel, M.; Booker, J.; Markowitz, L. Original research effectiveness of a biocellulose wound dressing for the treatment of chronic venous leg ulcers: Results of a single center randomized study involving 24 patients. Wounds **2004**, *16* (7), 224–233.

111. Czaja, W.; Krystynowicz, A.; Bielecki, S.; Brown Jr, R.M. Microbial cellulose—the natural power to heal wounds. Biomaterials **2006**, *27* (2), 145–151.

112. Solway, D.R.; Consalter, M.; Levinson, D.J. Microbial cellulose wound dressing in the treatment of skin tears in the frail elderly. Wounds **2010**, *22* (1), 17–19.

113. Fontana, J.D.; de Souza, A.M.; Fontana, C.K.; Torriani, I.L.; Moreschi, J.C.; Gallotti, B.J.; de Souza, S.J.; Narcisco, G.P.; Bichara, J.A.; Farah, L.F. Acetobacter cellulose pellicle as a temporary skin substitute. Appl. Biochem. Biotechnol. **1990**, *24–25* (1), 253–64.

114. Bäckdahl, H.; Helenius, G.; Bodin, A.; Nannmark, U.; Johansson, B.R.; Risberg, B.; Gatenholm, P. Mechanical properties of bacterial cellulose and interactions with smooth muscle cells. Biomaterials **2006**, *27* (9), 2141–2149.

115. Mello, L.R.; Feltrin, L.T.; Fontes Neto, P.T.; Ferraz, F.A.P. Duraplasty with biosynthetic cellulose: An experimental study. J. Neurosurg. **1997**, *86* (1), 143–150.

116. Wippermann, J.; Schumann, D.; Klemm, D.; Kosmehl, H.; Salehi-Gelani, S.; Wahlers, T. Preliminary results of small arterial substitute performed with a new cylindrical biomaterial composed of bacterial cellulose. Eur. J. Vasc. Endovasc. Surg. **2009**, *37* (5), 592–596.

117. Andrade, F.K.; Costa, R.; Domingues, L.; Soares, R.; Gama, M. Improving bacterial cellulose for blood vessel replacement: Functionalization with a chimeric protein containing a cellulose-binding module and an adhesion peptide. Acta Biomater. **2010**, *6* (10), 4034–4041.

118. Schumann, D.A.; Wippermann, J.; Klemm, D.O.; Kramer, F.; Koth, D.; Kosmehl, H.; Wahlers, T.; Salehi-Gelani, S. Artificial vascular implants from bacterial cellulose: Preliminary results of small arterial substitutes. Cellulose **2009**, *16* (5), 877–885.

119. Fink, H.; Hong, J.; Drotz, K.; Risberg, B.; Sanchez, J.; Sellborn, A. An *in vitro* study of blood compatibility of vascular grafts made of bacterial cellulose in comparison

with conventionally-used graft materials. J. Biomed. Mater. Res. A **2011**, *97* (1), 52–58.

120. Flynn, C.N.; Byrne, C.P.; Meenan, B.J. Surface modification of cellulose via atmospheric pressure plasma processing in air and ammonia–nitrogen gas. Surf. Coat. Technol. **2013**, *233*, 108–118.

121. Fernandes, S.C.M.; Sadocco, P.; Alonso-Varona, A.; Palomares, T.; Eceiza, A.; Silvestre, A.J.D.; Mondragon, I.; Freire, C.S.R. Bioinspired antimicrobial and biocompatible bacterial cellulose membranes obtained by surface functionalization with aminoalkyl groups. ACS Appl. Mater. Interfaces **2013**, *5* (8), 3290–3297.

122. Lee, S.J.; Liu, J.; Oh, S.H.; Soker, S.; Atala, A.; Yoo, J.J. Development of a composite vascular scaffolding system that withstands physiological vascular conditions. Biomaterials **2008**, *29* (19), 2891–2898.

123. Ju, Y.M.; Choi, J.S.; Atala, A.; Yoo, J.J.; Lee, S.J. Bilayered scaffold for engineering cellularized blood vessels. Biomaterials **2010**, *31* (15), 4313–4321.

124. McClure, M.J.; Sell, S.A.; Simpson, D.G.; Walpoth, B.H.; Bowlin, G.L. A three-layered electrospun matrix to mimic native arterial architecture using polycaprolactone, elastin, and collagen: A preliminary study. Acta Biomater. **2010**, *6* (7), 2422–2433.

125. Ye, L.; Wu, X.; Mu, Q.; Chen, B.; Duan, Y.; Geng, X.; Gu, Y.; Zhang, A.; Zhang, J.; Feng, Z.G. Heparin-Conjugated PCL scaffolds fabricated by electrospinning and loaded with fibroblast growth factor 2. J. Biomater. Sci. Polym. Ed. **2010**, *22* (1–3),389–406.

126. Pektok, E.; Nottelet, B.; Tille, J.C.; Gurny, R.; Kalangos, A.; Moeller, M.; Walpoth, B.H. Degradation and healing characteristics of small-diameter poly(epsilon-caprolactone) vascular grafts in the rat systemic arterial circulation. Circulation **2008**, *118* (24), 2563–2570.

127. Mugnai, D.; Tille, J.C.; Mrowczynski, W.; de Valence, S.; Montet, X.; Moller, M.; Walpoth, B.H. Experimental non-inferiority trial of synthetic small-caliber biodegradable versus stable vascular grafts. J. Thorac. Cardiovasc. Surg. **2013**, *146* (2), 400–407.

128. Walpoth, B.; Mugnai, D.; de Valence, S.; Mrowczynski, W.; Tille, J.; Montet, X.; Gurny, R.; Moeller, M.; Kalangos, A. Non-inferiority of synthetic small-calibre biodegradable vs. stable ePTFE vascular prosthesis after long-term implantation in the rat aorta. The Thoracic and Cardiovascular Surgeon **2013**, *61* (Suppl. 1), OP56.

129. Singh, S.; Wu, B.M.; Dunn, J.C. Accelerating vascularization in polycaprolactone scaffolds by endothelial progenitor cells. Tissue Eng. Part A **2011**, *17* (13–14), 1819–1830.

130. Tillman, B.W.; Yazdani, S.K.; Lee, S.J.; Geary, R.L.; Atala, A.; Yoo, J.J. The *in vivo* stability of electrospun polycaprolactone-collagen scaffolds in vascular reconstruction. Biomaterials **2009**, *30* (4), 583–588.

131. Mun, C.H.; Kim, S.-H.; Jung, Y.; Kim, S.-H.; Kim, A.-K.; Kim, D.-I.; Kim, S.H. Elastic, double-layered poly(L-lactide-co-ε-caprolactone) scaffold for long-term vascular reconstruction. J. Bioactive Compatible Polym. **2013**, *28* (3), 233–246

132. Diban, N.; Haimi, S.; Bolhuis-Versteeg, L.; Teixeira, S.; Miettinen, S.; Poot, A.; Grijpma, D.; Stamatialis, D. Hollow fibers of poly(lactide-co-glycolide) and poly(ε-caprolactone) blends for vascular tissue engineering applications. Acta Biomater. **2013**, *9* (5), 6450–6458.

133. Yuan, S.; Xiong, G.; Roguin, A.; Teoh, S.H.; Choong, C. Amelioration of blood compatibility and endothelialization of polycaprolactone substrates by surface-initiated atom transfegar radical polymerization. In *Advances in Biomaterials Science and Biomedical Applications*; Pignatello, R., Ed.; InTechOpen: Croatia, 2013.

134. Del Gaudio, C.; Ercolani, E.; Galloni, P.; Santilli, F.; Baiguera, S.; Polizzi, L.; Bianco, A. Aspirin-loaded electrospun poly(ε-caprolactone) tubular scaffolds: Potential small-diameter vascular grafts for thrombosis prevention. J. Mater. Sci. **2013**, *24* (2), 523–532.

135. Wang, S.; Mo, X.; Jiang, B.; Gao, C.; Wang, H.; Zhuang, Y.; Qiu, L. Fabrication of small-diameter vascular scaffolds by heparin-bonded P (LLA-CL) composite nanofibers to improve graft patency. Int. J. Nanomed. **2013**, *8*, 2131–2139.

136. Lu, G.; Cui, S.J.; Geng, X.; Ye, L.; Chen, B.; Feng, Z.G.; Zhang, J.; Li, Z.Z. Design and preparation of polyurethane-collagen/heparin-conjugated polycaprolactone double-layer bionic small-diameter vascular graft and its preliminary animal tests. Chin. Med. J. **2013**, *126* (7), 1310–1316.

137. Lodish, H.; Berk, A.; Zipursky, S.L.; Matsudaira, P.; Baltimore, D.; Darnell, J. *Collagen: The Fibrous Proteins of the Matrix*. 2000.

138. Olsen, D.; Yang, C.; Bodo, M.; Chang, R.; Leigh, S.; Baez, J.; Carmichael, D.; Perala, M.; Hamalainen, E.R.; Jarvinen, M. and Polarek, J. Recombinant collagen and gelatin for drug delivery. Adv. Drug Deliv. Rev. **2003**, *55* (12), 1547–1567.

139. Smith, M.; McFetridge, P.; Bodamyali, T.; Chaudhuri, J.B.; Howell, J.A.; Stevens, C.R.; Horrocks, M. Porcine-derived collagen as a scaffold for tissue engineering. Food Bioproducts Process. **2000**, *78* (1), 19–24.

140. Kose, G.T.; Korkusuz, F.; Ozkul, A.; Soysal, Y.; Ozdemir, T.; Yildiz, C.; Hasirci, V. Tissue engineered cartilage on collagen and PHBV matrices. Biomaterials **2005**, *26* (25), 5187–5197.

141. Sell, S.A.; McClure, M.J.; Garg, K.; Wolfe, P.S.; Bowlin, G.L. Electrospinning of collagen/biopolymers for regenerative medicine and cardiovascular tissue engineering. Adv. Drug Deliv. Rev. **2009**, *61* (12), 1007–1019.

142. Amarnath, L.P.; Srinivas, A.; Ramamurthi, A. *In vitro* hemocompatibility testing of UV-modified hyaluronan hydrogels. Biomaterials **2006**, *27* (8), 1416–1424.

143. Miyata, T.; Taira, T.; Noishiki, Y. Collagen engineering for biomaterial use. Clin. Mater. **1992**, *9* (3–4), 139–148.

144. Wissink, M.J.; Beernink, R.; Pieper, J.S.; Poot, A.A.; Engbers, G.H.; Beugeling, T.; van Aken, W.G.; Feijen, J. Immobilization of heparin to EDC/NHS-crosslinked collagen. Characterization and *in vitro* evaluation. Biomaterials **2001**, *22* (2), 151–163.

145. Wilner, G.D.; Nossel, H.L.; LeRoy, E.C. Activation of Hageman factor by collagen. J. Clin. Invest. **1968**, *47* (12), 2608–2615.

146. McClure, M.J.; Simpson, D.G.; Bowlin, G.L. Tri-layered vascular grafts composed of polycaprolactone, elastin, collagen, and silk: Optimization of graft properties. J. Mech. Behav. Biomed. Mater. **2012**, *10* (1), 48–61.

147. Kumar, V.A.; Caves, J.M.; Haller, C.A.; Dai, E.; Li, L.; Grainger, S.; Chaikof, E.L. A cellular vascular grafts

generated from collagen and elastin analogues. Acta Biomater. **2013**, *9* (9), 8067–8074.

148. Wu, H.; Fan, J.; Chu, C.C.; Wu, J. Electrospinning of small diameter 3-D nanofibrous tubular scaffolds with controllable nanofiber orientations for vascular grafts. J. Mater. Sci. Mater. Med. **2010**, *21* (12), 3207–3215.

149. Berglund, J.D.; Mohseni, M.M.; Nerem, R.M.; Sambanis, A. A biological hybrid model for collagen-based tissue engineered vascular constructs. Biomaterials **2003**, *24* (7), 1241–1254.

150. Long, J.L.; Tranquillo, R.T. Elastic fiber production in cardiovascular tissue-equivalents. Matrix Biol. **2003**, *22* (4), 339–350.

151. Karnik, S.K.; Brooke, B.S.; Bayes-Genis, A.; Sorensen, L.; Wythe, J.D.; Schwartz, R.S.; Keating, M.T.; Li, D.Y. A critical role for elastin signaling in vascular morphogenesis and disease. Development **2003**, *130* (2), 411–423.

152. Jordan, S.W.; Haller, C.A.; Sallach, R.E.; Apkarian, R.P.; Hanson, S.R.; Chaikof, E.L. The effect of a recombinant elastin-mimetic coating of an ePTFE prosthesis on acute thrombogenicity in a baboon arteriovenous shunt. Biomaterials **2007**, *28* (6), 1191–1197.

153. Woodhouse, K.A.; Klement, P.; Chen, V.; Gorbet, M.B.; Keeley, F.W.; Stahl, R.; Fromstein, J.D.; Bellingham, C.M. Investigation of recombinant human elastin polypeptides as non-thrombogenic coatings. Biomaterials **2004**, *25* (19), 4543–4553.

154. Li, D.Y.; Brooke, B.; Davis, E.C.; Mecham, R.P.; Sorensen, L.K.; Boak, B.B.; Eichwald, E.; Keating, M.T. Elastin is an essential determinant of arterial morphogenesis. Nature **1998**, *393* (6682), 276–280.

155. Weinberg, C.B.; Bell, E. A blood vessel model constructed from collagen and cultured vascular cells. Science **1986**, *231* (4736), 397–400.

156. L'Heureux, N.; Germain, L.; Labbe, R.; Auger, F.A. *In vitro* construction of a human blood vessel from cultured vascular cells: A morphologic study. J. Vasc. Surg. **1993**, *17* (3), 499–509.

157. Hirai, J.; Matsuda, T. Venous reconstruction using hybrid vascular tissue composed of vascular cells and collagen: Tissue regeneration process. Cell Transpl. **1996**, *5* (1), 93–105.

158. Ramamurthi, A.; Vesely, I. Evaluation of the matrix-synthesis potential of crosslinked hyaluronan gels for tissue engineering of aortic heart valves. Biomaterials **2005**, *26* (9), 999–1010.

159. Ziegler, T.; Alexander, R.W.; Nerem, R.M. An endothelial cell-smooth muscle cell co-culture model for use in the investigation of flow effects on vascular biology. Ann. Biomed. Eng. **1995**, *23* (3), 216–225.

160. Kolpakov, V.; Rekhter, M.D.; Gordon, D.; Wang, W.H.; Kulik, T.J. Effect of mechanical forces on growth and matrix protein synthesis in the *in vitro* pulmonary artery. Analysis of the role of individual cell types. Circ. Res. **1995**, *77* (4), 823–831.

161. Kim, B.S.; Mooney, D.J. Scaffolds for engineering smooth muscle under cyclic mechanical strain conditions. J. Biomech. Eng. **2000**, *122* (3), 210–215.

162. Kim, B.S.; Nikolovski, J.; Bonadio, J.; Mooney, D.J. Cyclic mechanical strain regulates the development of engineered smooth muscle tissue. Nat. Biotechnol. **1999**, *17* (10), 979–983.

163. Berglund, J.D.; Nerem, R.M.; Sambanis, A. Incorporation of intact elastin scaffolds in tissue-engineered collagen-based vascular grafts. Tissue Eng. **2004**, *10* (9–10), 1526–1535.

164. Sell, S.; McClure, M.J.; Barnes, C.P.; Knapp, D.C.; Walpoth, B.H.; Simpson, D.G.; Bowlin, G.L. Electrospun polydioxanone–elastin blends: Potential for bioresorbable vascular grafts. Biomed. Mater. **2006**, *1* (2), 72.

165. Caves, J.M.; Kumar, V.A.; Martinez, A.W.; Kim, J.; Ripberger, C.M.; Haller, C.A.; Chaikof, E.L. The use of microfiber composites of elastin-like protein matrix reinforced with synthetic collagen in the design of vascular grafts. Biomaterials **2010**, *31* (27), 7175–7182.

166. Sarkar, S.; Schmitz-Rixen, T.; Hamilton, G.; Seifalian, A.M. Achieving the ideal properties for vascular bypass grafts using a tissue engineered approach: A review. Med. Biol. Eng. Comput. **2007**, *45* (4), 327–336.

167. Smith, M.J.; McClure, M.J.; Sell, S.A.; Barnes, C.P.; Walpoth, B.H.; Simpson, D.G.; Bowlin, G.L. Suture-reinforced electrospun polydioxanone–elastin small-diameter tubes for use in vascular tissue engineering: A feasibility study. Acta Biomater. **2008**, *4* (1), 58–66.

168. Wise, S.G.; Byrom, M.J.; Waterhouse, A.; Bannon, P.G.; Ng, M.K.; Weiss, A.S. A multilayered synthetic human elastin/polycaprolactone hybrid vascular graft with tailored mechanical properties. Acta Biomater **2011**, *7* (1), 295–303.

169. Chuang, T.-H.; Stabler, C.; Simionescu, A.; Simionescu, D.T. Polyphenol-stabilized tubular elastin scaffolds for tissue engineered vascular grafts. Tissue Eng. Part A **2009**, *15* (10), 2837–2851.

170. Mecham, R.P. Methods in elastic tissue biology: Elastin isolation and purification. Methods **2008**, *45* (1), 32–41.

171. Weisel, J.W. Fibrinogen and fibrin. Adv. Protein Chem. **2005**, *70* (1), 247–299.

172. Laurens, N.; Koolwijk, P.; de Maat, M.P. Fibrin structure and wound healing. J. Thromb. Haemost. **2006**, *4* (5), 932–939.

173. Lominadze, D.; Dean, W.L. Involvement of fibrinogen specific binding in erythrocyte aggregation. FEBS Lett. **2002**, *517* (1–3), 41–44.

174. Grassl, E.D.; Oegema, T.R.; Tranquillo, R.T. A fibrin-based arterial media equivalent. J. Biomed. Mater. Res A **2003**, *66* (3), 550–561.

175. Ahmed, T.A.; Dare, E.V.; Hincke, M. Fibrin: A versatile scaffold for tissue engineering applications. Tissue Eng. Part B Rev. **2008**, *14* (2), 199–215.

176. Koroleva, A.; Gittard, S.; Schlie, S.; Deiwick, A.; Jockenhoevel, S.; Chichkov, B. Fabrication of fibrin scaffolds with controlled microscale architecture by a two-photon polymerization-micromolding technique. Biofabrication **2012**, *4* (1), 015001.

177. Kaijzel, E.L.; Koolwijk, P.; van Erck, M.G.; van Hinsbergh, V.W.; de Maat, M.P. Molecular weight fibrinogen variants determine angiogenesis rate in a fibrin matrix *in vitro* and *in vivo*. J. Thromb. Haemost. **2006**, *4* (9), 1975–1981.

178. Jockenhoevel, S.; Zund, G.; Hoerstrup, S.P.; Chalabi, K.; Sachweh, J.S.; Demircan, L.; Messmer, B.J.; Turina, M. Fibrin gel—Advantages of a new scaffold in cardiovascular tissue engineering. Eur. J. Cardiothorac. Surg. **2001**, *19* (4), 424–430.

179. Dietrich, M.; Heselhaus, J.; Wozniak, J.; Weinandy, S.; Mela, P.; Tschoeke, B.; Schmitz-Rode, T.; Jockenhoevel, S. Fibrin-based tissue engineering: Comparison of different methods of autologous fibrinogen isolation. Tissue Eng. Part C Methods **2013**, *19* (3), 216–226.

180. Hasegawa, T.; Okada, K.; Takano, Y.; Hiraishi, Y.; Okita, Y. Autologous fibrin-coated small-caliber vascular prostheses improve antithrombogenicity by reducing immunologic response. J. Thorac. Cardiovasc. Surg. **2007**, *133* (5), 1268–1276, 1276 e1.

181. Tschoeke, B.; Flanagan, T.C.; Koch, S.; Harwoko, M.S.; Deichmann, T.; Ella, V.; Sachweh, J.S.; Kellomaki, M.; Gries, T.; Schmitz-Rode, T.; Jockenhoevel, S. Tissue-engineered small-caliber vascular graft based on a novel biodegradable composite fibrin-polylactide scaffold. Tissue Eng. Part A **2009**, *15* (8), 1909–1918.

182. Aper, T.; Teebken, O.E.; Steinhoff, G.; Haverich, A. Use of a fibrin preparation in the engineering of a vascular graft model. Eur. J. Vasc. Endovasc. Surg. **2004**, *28* (3), 296–302.

183. Koch, S.; Tschoeke, B.; Deichmann, T.; Ella, V.; Gronloh, N.; Gries, T.; Tolba, R.; Kellomäki, M.; Schmitz-Rode, T.; Jockenhoevel, S. Fibrin-based tissue engineered vascular graft in carotid artery position-the first *in vivo* experiences. Thorac. Cardiovasc. Surg. **2010**, *58* (Suppl. 1), MP25.

184. Santin, M.; Motta, A.; Freddi, G.; Cannas, M. *In vitro* evaluation of the inflammatory potential of the silk fibroin. J. Biomed. Mater. Res. **1999**, *46* (3), 382–389.

185. Demura, M.; Asakura, T. Porous membrane of Bombyx mori silk fibroin: Structure characterization, physical properties and application to glucose oxidase immobilization. J. Membr. Sci. **1991**, *59* (1), 39–52.

186. Asakura, T.; Kitaguchi, M.; Demura, M.; Sakai, H.; Komatsu, K. Immobilization of glucose oxidase on nonwoven fabrics with bombyx mori silk fibroin gel. J. Appl. Polym. Sci. **1992**, *46* (1), 49–53.

187. Demura, M.; Asakura, T. Immobilization of glucose oxidase with Bombyx mori silk fibroin by only stretching treatment and its application to glucose sensor. Biotechnol. Bioeng. **1989**, *33* (5), 598–603.

188. Mhuka, V.; Dube, S.; Nindi, M.; Torto, N. Fabrication and structural characterization of electrospun nanofibres from Gonometa Postica and Gonometa Rufobrunnae regenerated silk fibroin. Macromol. Res. **2013**, *21* (9), 995–1003.

189. Tamada, Y. New process to form a silk fibroin porous 3-D structure. Biomacromolecules **2005**, *6* (6), 3100–3106.

190. Meinel, L.; Hofmann, S.; Karageorgiou, V.; Zichner, L.; Langer, R.; Kaplan, D.; Vunjak-Novakovic, G. Engineering cartilage-like tissue using human mesenchymal stem cells and silk protein scaffolds. Biotechnol. Bioeng. **2004**, *88* (3), 379–391.

191. Altman, G.H.; Diaz, F.; Jakuba, C.; Calabro, T.; Horan, R.L.; Chen, J.; Lu, H.; Richmond, J.; Kaplan, D.L. Silk-based biomaterials. Biomaterials **2003**, *24* (3), 401–416.

192. Zhang, X.; Wang, X.; Keshav, V.; Johanas, J.T.; Leisk, G.G.; Kaplan, D.L. Dynamic culture conditions to generate silk-based tissue-engineered vascular grafts. Biomaterials **2009**, *30* (19), 3213–3223.

193. Makaya, K.; Terada, S.; Ohgo, K.; Asakura, T. Comparative study of silk fibroin porous scaffolds derived from salt/water and sucrose/hexafluoroisopropanol in cartilage formation. J. Biosci. Bioeng. **2009**, *108* (1), 68–75.

194. Enomoto, S.; Sumi, M.; Kajimoto, K.; Nakazawa, Y.; Takahashi, R.; Takabayashi, C.; Asakura, T.; Sata, M. Long-term patency of small-diameter vascular graft made from fibroin, a silk-based biodegradable material. J. Vasc. Surg. **2010**, *51* (1), 155–164.

195. Rockwood, D.N.; Gil, E.S.; Park, S.H.; Kluge, J.A.; Grayson, W.; Bhumiratana, S.; Rajkhowa, R.; Wang, X.; Kim, S.J.; Vunjak-Novakovic, G. and Kaplan, D.L. Ingrowth of human mesenchymal stem cells into porous silk particle reinforced silk composite scaffolds: An *in vitro* study. Acta Biomater. **2011**, *7* (1), 144–151.

196. Mauney, J.R.; Sjostorm, S.; Blumberg, J.; Horan, R.; O'Leary, J.P.; Vunjak-Novakovic, G.; Volloch, V.; Kaplan, D.L. Mechanical stimulation promotes osteogenic differentiation of human bone marrow stromal cells on 3-D partially demineralized bone scaffolds *in vitro*. Calcif. Tissue Int. **2004**, *74* (5), 458–468.

197. Li, C.; Vepari, C.; Jin, H.-J.; Kim, H.J.; Kaplan, D.L. Electrospun silk-BMP-2 scaffolds for bone tissue engineering. Biomaterials **2006**, *27* (16), 3115–3124.

198. Lovett, M.; Eng, G.; Kluge, J.; Cannizzaro, C.; Vunjak-Novakovic, G.; Kaplan, D.L. Tubular silk scaffolds for small diameter vascular grafts. Organogenesis **2010**, *6* (4), 217–224.

199. Marelli, B.; Achilli, M.; Alessandrino, A.; Freddi, G.; Tanzi, M.C.; Farè, S.; Mantovani, D. Collagen-reinforced electrospun silk fibroin tubular construct as small calibre vascular graft. Macromol. Biosci. **2012**, *12* (11), 1566–1574.

200. Maskarinec, S.A.; Tirrell, D.A. Protein engineering approaches to biomaterials design. Curr. Opin. Biotechnol. **2005**, *16* (4), 422–426.

201. Liu, H.; Li, X.; Niu, X.; Zhou, G.; Li, P.; Fan, Y. Improved hemocompatibility and endothelialization of vascular grafts by covalent immobilization of sulfated silk fibroin on poly (lactic-co-glycolic acid) scaffolds. Biomacromolecules **2011**, *12* (8), 2914–2924.

202. Yang, X.; Wang, L.; Guan, G.; King, M.W.; Li, Y.; Peng, L.; Guan, Y.; Hu, X. Preparation and evaluation of bicomponent and homogeneous polyester silk small diameter arterial prostheses. J. Biomater. Appl. **2014**, *28* (5), 676–687.

203. Aytemiz, D.; Sakiyama, W.; Suzuki, Y.; Nakaizumi, N.; Tanaka, R.; Ogawa, Y.; Takagi, Y.; Nakazawa, Y.; Asakura, T. Small-diameter silk vascular grafts (3 mm Diameter) with a double-raschel knitted silk tube coated with silk fibroin sponge. Adv. Healthcare Mater. **2013**, *2* (2), 361–368.

204. Nakazawa, Y.; Sato, M.; Takahashi, R.; Aytemiz, D.; Takabayashi, C.; Tamura, T.; Enomoto, S.; Sata, M.; Asakura, T. Development of small-diameter vascular grafts based on silk fibroin fibers from bombyx mori for

vascular regeneration. J. Biomater. Sci. Polym. Ed. **2011**, *22* (1–3), 195–206.

205. Soffer, L.; Wang, X.; Zhang, X.; Kluge, J.; Dorfmann, L.; Kaplan, D.L.; Leisk, G. Silk-based electrospun tubular scaffolds for tissue-engineered vascular grafts. J. Biomater. Sci. Polym. Ed. **2008**, *19* (5), 653–664.

206. Yu, J.; Wang, A.; Tang, Z.; Henry, J.; Li-Ping Lee, B.; Zhu, Y.; Yuan, F.; Huang, F.; Li, S. The effect of stromal cell-derived factor-1alpha/heparin coating of biodegradable vascular grafts on the recruitment of both endothelial and smooth muscle progenitor cells for accelerated regeneration. Biomaterials **2012**, *33* (32), 8062–8074.

Vascular—Zwitterionic

Vascular Tissue Engineering: Polymeric Biomaterials

George Fercana
Biocompatibility and Tissue Regeneration Laboratories, Department of Bioengineering, Clemson University, Clemson, and Laboratory of Regenerative Medicine, Patewood/CU Bioengineering Translational Research Center, Greenville Hospital System, Greenville, South Carolina, U.S.A.

Dan Simionescu
Department of Bioengineering, Clemson University, Clemson, and Patewood/CU Bioengineering Translational Research Center, Greenville Hospital System, Greenville, South Carolina, U.S.A., and Department of Anatomy, University of Medicine and Pharmacy, Tirgu Mures, Romania

Abstract
Tissue engineering using appropriate tubular scaffolds seeded with autologous living stem cells provides living tissue substitutes capable of integration, remodeling, and long-term patency. Mechanical conditioning of cell-seeded grafts before implantation is required for cell differentiation and adaptation to dynamic conditions. This entry first examines the necessary properties for the ideal vascular graft before moving to a more in depth discussion of research that is fueling implant improvement, such as small-diameter prosthetic grafts.

CURRENT OPTIONS AND CLINICAL NEED FOR REPLACEMENTS

The total need for vascular grafts has been estimated to be more than 1.4 million in the United States alone.[1] This need can be divided into three categories, in order of decreasing diameter (Table 1). The large- and medium-caliber synthetic grafts (>6-mm diameter) are used in the thoracic and abdominal cavities with good long-term outcomes. Common applications include replacement of the aorta, carotid artery, arch vessels, and iliac and femoral arteries. However, synthetic grafts of smaller diameter do not share the same outcomes and fail relatively early due to occlusion. Since tissue engineering holds great potential for development of viable grafts, this entry delves into research and development of small-diameter vascular grafts using natural polymers.

Almost 200,000 small-caliber grafts (<6 mm) are used every year for vascular access, to relieve peripheral lower limb ischemia, or for coronary artery bypass graft; autologous veins or arteries are the "gold standard" for replacement of small-caliber arteries, but in 30–40% of patients these are not available due to prior harvesting or preexisting conditions. In these last cases, synthetic grafts are used, but they provide poor outcomes as 50% of these will occlude within 5 years,[2] potentially leading to amputation.

In the search for small-caliber vascular substitutes, a range of materials and methodologies has been evaluated,[3] but most exhibited poor midterm performance. These included: 1) synthetic grafts such as expanded polytetrafluoroethylene (ePTFE), polyethylene-terephthalate (PET), polyurethanes, polyglycolic (PG), and polylactide polyglycolide (PL);[4] and 2) bioprosthetic grafts (homologous, heterologous), each with different conformations that are better suited for particular applications (Tables 2 and 3). Most synthetic polymers exhibit good biostability, compliance, and hemocompatibility but are prone to infections and dilation and variable healing patterns and endothelialization potential. Taken together, as multiple bypasses are sometimes required and availability of autologous arteries or veins is limited, there is clearly a significant unmet clinical need for small-caliber vascular grafts.

Tissue engineering using appropriate tubular scaffolds seeded with autologous living stem cells holds promise to solve this need by providing living tissue substitutes capable of integration, remodeling, and long-term patency. Mechanical conditioning of cell-seeded grafts before implantation is required for cell differentiation and adaptation to dynamic conditions. Notably, a tissue-engineered graft scaffold is required to be capable of immediately reestablishing blood flow; thus mechanical properties, including burst pressure and compliance, need to be engineered *in vitro* before implantation.

NECESSARY PROPERTIES FOR THE IDEAL VASCULAR GRAFT

In the realm of biomedical polymeric engineering, a large breadth of materials has been investigated for vascular replacement applications to achieve necessary properties intrinsic to the ideal vascular replacement.[5] The properly

Concise Encyclopedia of Biomedical Polymers and Polymeric Biomaterials DOI: 10.1081/E-EBPPC-120052170

Vascular—Zwitterionic

Table 1 Overview of sizes and preferential choices for vascular grafts

Vascular substitute choice	Vascular regions				
	Large-caliber arteries (≥ 8 mm)	Medium-caliber arteries (6–8 mm)	Small-caliber arteries (≤ 6 mm)	Venous reconstructions	Hemodialysis arteriovenous access
	Aorta, arch vessels, iliac and common femoral arteries	Carotid, subclavian, common femoral, visceral, and above-the-knee arteries	Coronary, below-the-knee, tibial, and peroneal arteries	Superior and inferior vena cava, iliofemoral veins, portal vein, visceral veins	Upper > lower extremity
First choice	Prosthesis (Dacron, ePTFE)	Prosthesis or autograft (equal)	Arterial or venous autograft	Saphenous spiral vein graft, deep venous autograft	Native material
Second choice	Allograft, deep venous autograft	Prosthesis or autograft	Composite graft, vein interposition, prosthesis (ePTFE, Dacron), allograft, biosynthetic	Allografts, ePTFE, Dacron, biografts	ePTTE, PU, xenografts, biografts. TEBV (clinical trial)

Source: Chlupac et al.[12]

functioning vascular graft needs to accommodate physiological properties such as adequate burst pressure, the capacity to recoil, and radial compliance.

Burst pressure is an intrinsic property of tubular materials and is tested by cannulating both ends of the graft and connecting a piezoelectric digital pressure transducer at one end and a peristaltic pump (or syringe pump) at the opposite end. The system is filled with saline, and pressure is slowly increased. Burst pressure is recorded when a sharp drop in internal graft pressure occurs. At times, it is useful to document the pattern of tissue failure by photography, as this may provide clues to the presence of weaker components in the graft structure. Native arteries exhibit a burst pressure of 1500–2200 mmHg, which in theory is redundant by a safety factor of about 100. Currently, it is not known what the most tolerable safety factor for an engineered graft should be, but scientists aim at reaching at least 300–500 mmHg.

Physiological compliance is defined as the increase in graft diameter as pressure increases from 80 to 120 mmHg. Compliance measurement is performed as follows: Grafts are cannulated with one end connected to a piezoelectric digital pressure transducer and the other adapted to a saline reservoir. The reservoir is raised until the transducer reads 80 mmHg hydrostatic pressure, and digital images are taken for external diameter measurement. Then, the pressure is increased to 120 mmHg, and the diameter is measured again (Fig. 1). Typically, a native artery exhibits 10–15% compliance. This is an important parameter to engineer because of known effects of compliance mismatch on graft performance.

In addition to the aforementioned qualities, vascular grafts will require adequate performance regarding implant biocompatibility, such as resistance to calcification, thrombosis, and infection,[6] which are typically modulated using alterations to the material's surface or bulk chemistry via

known anticoagulants, such as variations of heparin,[7,8] or treating the material to resist calcification.[9–11]

Once satisfactory material properties have been established for the proposed vascular graft, the material must also prove to be noncytotoxic and porous and allow retention of shear resistant endothelial cells (ECs) in the intima, vascular smooth muscle cells (VSMCs) in the media, and fibroblasts (FBs) in the adventitia, the three tunics native to blood vessels (Fig. 2). Pending infiltration of these cell types, the material must permit degradation and matrix remodeling by the host's own cells to produce a patient-tailored, autologous replacement according to the paradigm of regenerative medicine.[12]

SMALL-DIAMETER PROSTHETIC GRAFTS: RESEARCH FUELING IMPLANT IMPROVEMENT

The available amount of polymeric options for any surgical procedure is quite large, but over the years ePTFE has remained the material of choice for small-diameter vascular replacement due to its decreased inflammatory and thrombogenic activity in comparison to other implant graft materials. Research on vascular replacements began with Voorhees and colleagues on the introduction of woven Vinyon N as a vascular replacement material.[14] This research sparked studies of other polymeric materials; however, the work of Creech and colleagues established PET (Dacron) and ePTFE as the most resistant to degradation when compared to the other tested materials, including nylon, Orlon, and Ivalon.[15] Since this discovery, much work has been done on improving ePTFE for vascular replacement. Although ePTFE retains properties that decrease thrombogenicity and inflammation to improve overall implant biocompatibility, the material will still fall prey to the body's innate defense mechanisms

Table 2 Brief overview of characteristics and history of biological vascular grafts, where IND denotes internodal distance

| | Synthetic vascular grafts | | | | | |
| | PET (Dacron, Terylen) | | ePTFE (Teflon, Gore-Tex) | | Polyurethane | |
	Woven	Knitted	Low porosity (<30 μm IND)	High porosity (>45 μm IND)	Fibrillar	Foamy
Advantages	Better stability, lower permeability and less bleeding	Greater porosity, tissue ingrowth and radial distensibility	Biostability, no dilation over time	Biostability, better cell ingrowth	Compliance, good hemo- and biocompatibility, less thrombogenicity	
Disadvantages	Reduced compliance and tissue incorporation, low porosity, fraying at edges, infection risk	Dilation over time, infection risk	Stitch bleeding, limited incorporation, infection risk, perigraft seroma formation	Late neointimal desquamation in 90 μm IND, infection risk	Biodegradation in first generation, infection risk, carcinogenic?	
Healing	Inner fibrinous capsule, outer collagenous capsule, scarce endothelial islands	Fibrin luminal coverage, very sporadic endothelium, trans-anastomotic endothelialization in animals	Luminal fibrin and platelet carpet, connective tissue capsule with foreign body giant cells, no transmural tissue ingrowth	Macrophages and polymorphonuclear invasion, capillary sprouting, fibroblast migration, certain angiogenesis, thicker neointima, endothelialization in animals	Thin inner fibrin layer, outside foreign body cells, limited ingrowth	Better ingrowth with bigger pores

Source: Chlupac et al.[12]

of platelet adhesion, protein adsorption, and other processes intrinsic to the coagulation cascade, starting at the blood-biomaterial interface. Problem areas for all polymeric vascular grafts are the interfaces between the: 1) material and blood; 2) tissue and blood; and 3) sites of anastomosis, keeping in mind the "intimal" and "adventitial" regions of the material will both be exposed to the body's defense mechanisms. Improving the biocompatibility of ePTFE and other polymeric vascular replacements includes research into promoting endothelialization of the intimal surface of the implant material and decreasing VSMC proliferation while retaining quiescent phenotype in the tissue immediately adjacent to the implant graft.[16] Several groups have chosen a different route of improvement and dedicated their efforts to surface modification with known anticoagulants commonly utilized in the medical field in hopes of decreasing long-term thrombogenicity of synthetic implant materials.[7,8]

THE ALLURE OF NATURAL BIOPOLYMERS

When deciding on the optimum material for a vascular graft, one should keep in mind the aforementioned "ideal" properties for such a scaffold. The two main categories of available scaffolds are biological and synthetic (Table 4).

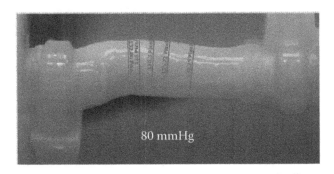

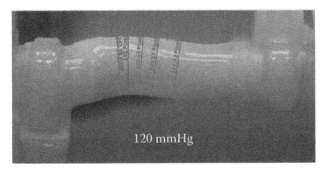

Fig. 1 Setup for measurement of vascular graft compliance. The graft is mounted using barbed adapters and subjected to physiological pressures; digital pictures are taken for diameter measurements.

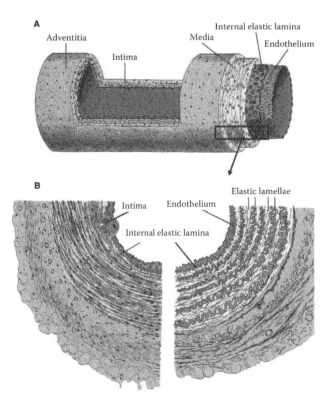

Fig. 2 (**A**) Representation of the three tunics native to a blood vessel. Intima, ECs; media, VSMCs; adventitia, FBs. (**B**) Structural differences between muscular arteries (left) and elastic arteries (right); note differences in elastin content dependent on artery type.
Source: From Junqueira & Cameiro.[13]

Table 3 Brief overview of properties of biological and synthetic scaffolds

	Biological scaffolds	**Synthetic scaffolds**
Advantages	Naturally occurring, nontoxic	Precise control over material properties
	Favorable for cell binding	Easily available and cheap
	Generally biocompatible	Easy to process
		Little or no batch-to-batch variation
Disadvantages	May degrade rapidly	Toxic residual monomers or catalysts and degradation by-products may illicit inflammation
	Weak mechanical property unless cross-linked	Poor cellular interaction
	Inconsistency between different batches	
	Chance of disease transmission	

Table 4 Brief overview of characteristics and history of biological vascular grafts

| | Biological vascular grafts | | | |
| | Autografts | | Allografts (homografts) | Xenografts (heterografts) |
	Arterial	Venous	Arterial	Venous
Advantages	Closest approximation, less diameter mismatch, internal mammary artery anatomically nearby, excellent function	Durable and versatile, good results, infection resistance, relative availability	Off-the-shelf availability, better resistance to infection, transplant-recipient patients	
Disadvantages	Availability, vasospasm (radial artery), donor site morbidity	Availability, harvest injury, vein graft disease	Antigenicity, graft deterioration, early occlusions, chronic rejection, intake of drugs, infection risk	
Healing	Intimal thickening, myointimal hyperplasia (radial artery)	Endothelial desquamation, vein dilation, wall thickening, arterialization, reendothelialization	Endothelial denudation, immune response, fibrotization	

Source: Chlupac et al.[12]

Biological scaffolds are naturally occurring biocompatible molecules to which cells readily bind and onto which cells proliferate; however, biological scaffolds generally have insufficient mechanical properties and have a tendency to degenerate rapidly after implantation. These properties can be better controlled using chemical or physical methods of cross-linking or stabilization, which reduce biodegradability and enhance tensile strength. Synthetic scaffolds share the major advantages that material properties can be tailored to the target tissue or organ, are easy to process, and are consistent between batches. Some polymers exhibit poor cellular interactions and may also leach residual monomers, catalysts, and degradation by-products that may be toxic or elicit inflammation.

Choosing the appropriate source of scaffolds depends greatly on desired properties of the target tissue. If the implant needs to function mechanically immediately after implantation (e.g., heart valve, artery, knee replacement), a strong scaffold is needed that allows few changes in mechanical properties with time as this will impair functionality. For skin and wound dressing regeneration, it is beneficial that the scaffold degrades completely and allows rapid regeneration; thus a rapidly degradable biological scaffold may be optimal. If the targeted tissue is metabolically active (liver, pancreas), one may choose a strong, porous biocompatible scaffold that degrades slowly.

What often leads researchers to use biopolymers such as collagen- or elastin-based proteins are the unique properties of said biopolymers, which often already provide ideal characteristics due to their natural origins. For example, elastin provides elasticity to tissues to achieve necessary compliance and cyclic fatigue strength and has been shown to function as a completely nonthrombogenic surface.[17,18]

Collagen-based scaffolds, conversely, provide rigidity that will not only impart mechanical strength and the potential for rapid degradation and remodeling, but also contribute to thrombus formation due to the material's known high thrombogenicity.

One example of successful biological polymers currently in use would be bioprosthetic heart valves, which essentially are highly cross-linked, nondegradable animal tissue scaffolds. These biomaterials perform very well mechanically and do not require anticoagulant or immunosuppressive therapies but fall prey to degeneration and calcification.[10,19]

With the combination of necessary mechanical properties and coveted *in vivo* biocompatibility, biopolymers are often the "natural" choice for researchers. Interestingly, researchers observe that neither purified collagen- nor elastin-based vascular replacements retain sufficient mechanical properties when used separately for functional vascular replacements in arterial positions.[20] This has prompted investigators to attempt to upregulate elastin biosynthesis to supplement three-dimensional scaffolds in hopes of improving scaffold mechanical and biocompatibility properties, as well as capitalizing on elastin's recently discovered bioactivity and nonthrombogenicity.[18,20,21]

CURRENT BIOPOLYMER OPTIONS: COLLAGEN

The extracellular matrix (ECM) protein collagen exists in the body in numerous forms, including fine fibers, intricate fibrils, delicate bundles, or strong flat sheets. Collagen has a rapid turnover and is deposited rapidly into *in vivo* implants in the form of either a fibrous capsule surrounding the implant, indicating a host response that can lead to implant rejection, or a neocollagen formation within the matrix of the implant. Such neocollagen formation often indicates a positive implant remodeling. As the most abundant protein in the body,[22] collagen is readily available and has been experimented with thoroughly in the literature for the purpose of engineering vascular grafts; an example of such is discussed next.

One methodology for using this ECM protein includes electros-pinning (Fig. 3), a process in which the fibers are spun with varying voltages, distances between the plate and needle, and solvents to dissolve the fiber for the process. Readers are directed to a thorough review of electrospinning technology and the potential applications of the methodology by Prabhakaran and coworkers.[23] With experimentation into hydrogel-based substrates generally producing inadequate mechanical properties for the vascular constructs as discussed by Patel and colleagues,[20] researchers began to investigate mechanical conditioning on hydrogels in hopes of elucidating keys to increased mechanical strength for hydrogel-based vascular constructs. It was later suggested by L'Heureux and coworkers[24] that mechanical forces should be applied to collagen gels to induce alignment of collagen fibrils and seeded smooth muscle cells in circumferential fashion, as seen in natural vessels.[25,26] When this methodology was first applied in attempts to produce a hollow, cylindrical artery using a mandrel-based approach by Hirai and Matsuda,[27] mechanical strength was not sufficient for implantation into arterial positions. Follow-up experimentation utilized glycation mechanisms in addition to lysyl oxidase-based cross-linking of fibers; however, again, mechanical strength of the collagen-based gels proved to be insufficient.[28,29] These findings suggest that vascular grafts based on collagen require supplementation with smooth muscle cells that maintain matrix homeostasis, elastin to improve resilience, and finally an endothelial cell layer to prevent thrombogenicity.[20]

CURRENT BIOPOLYMER OPTIONS: ELASTIN

Elastin is a very hydrophobic, "rubbery," highly cross-linked insoluble protein endowed with one of the slowest turnover rates in the human body. Due to this property, once elastin is degraded or significantly altered (as in aortic aneurysms, Marfan's syndrome, medial arterial elastocalcinosis), elastin is not replenished by resident cells. Elastin provides properties tantamount to a functional tissue-engineered vessel and is widely known for the resilience it imparts to biological tissues.[17] For a tissue engineered vascular graft to be remodeled by the host with acceptable mechanical properties, said construct should induce elastin biosynthesis during remodeling *in vivo* or perhaps somehow graft tropoelastin to existing matrix fibers to endow the implant with increased elasticity and resilience. However, currently there have been no large successes in elastin biosynthesis or tropoelastin grafting for tissue engineered constructs.

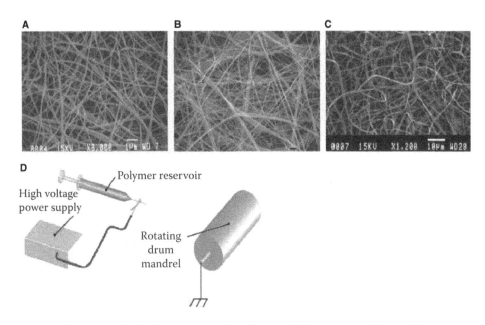

Fig. 3 Scanning electron micrographs of (**A**) electrospun type I collagen (×8000, scale bar 1.0 mm); (**B**) electrospun type III collagen (×4300, scale bar 1.0 mm); and (**C**) electrospun collagen type IV collagen (×1200, scale bar 10 mm). (**D**) Schematic representation of simple electrospinning setup using a grounded rotating drum as a collection mandrel.
Source: (**A**) From Boland et al.[30] (**B**) and (**C**) From Barnes et al.[31] © 2007. (**D**) From Sell et al.[32]

Vascular—Zwitterionic

An example of previous efforts conducted in the field to incite elastin biosynthesis utilized combinations of fibrin and collagen. Other growth factors and conditions involving smooth muscle cells known to produce the coveted elastin were utilized in these experiments, but no elastin was produced by the efforts of said groups.[24,27,33,34] Research into differences between fibrin- and collagen-based hydrogel scaffolds conducted by Long and Tranquillo did, however, elucidate that when murine neonatal smooth muscle cells are seeded into both collagen and fibrin gels, the fibrin gels produce significantly more elastin than those made of collagen, indicating that perhaps fibrin is a much more favorable substrate for elastin biosynthesis.[34] Investigating the effects of another hydrogel substrate on elastin biosynthesis, work done by Ramamurthi's group compared performance of neonatal rat smooth muscle cells in both hyaluronan gels and monolayer conditions on tissue culture polystyrene. The results showed that elastin was indeed produced and in the fenestrated form native to the internal elastic lamina and concentric lamellae inside the media of blood vessels.[35] The research from these groups suggested that elastin biosynthesis is upregulated when used in specific 3D substrates, but the key ingredients necessary to repeatedly and dependably produce elastin are not fully understood.

TISSUE DECELLULARIZATION: THE ALTERNATIVE TO POLYMERIC FABRICATION

The act of gathering specific biologically inspired polymers and integrating them methodically and precisely into a mechanically stable structure via hydrogels, mandrels, or other commonly utilized substrates is a technique favored by many research groups. This particular methodology can easily be included in the category of a "bottom-up" fabrication approach in which a biologically viable scaffold is essentially "created from scratch." This process retains advantages due to the seemingly unlimited variability afforded by such a technique, in which the researcher specifically selects the desired components and implements them into a scaffold.

However, investigators in the field are also adopting a more "top-down" approach for scaffold fabrication. For this process, donor tissue of either xenogenic or human origin is meticulously treated to remove the undesired antigens, soluble proteins, and cellular remnants. This step, referred to as "decellularization," is essential because of the possibility of immune rejection on implantation. A second prerequisite is to ensure that the ECM composition and 3D integrity are fully retained after decellularization. This in turn will determine mechanical properties of the scaffolds. This process is also similarly advantageous in that researchers can fine-tune their approach to retain specific ECM compositions and porosities while further augmenting their process by choosing a different donor tissue as

deemed necessary. An example of such would be the choice of either a muscular or elastic artery for decellularization and scaffold treatment: Each artery type will provide different mechanical properties and porosity, which is discussed more thoroughly in this entry.

With the advent of decellularization for production of tissue-engineered scaffolds came numerous questions regarding the extent of decellularization needed to retain necessary mechanical properties.[36] However, the question of the extent of decellularization with respect to left-over nucleic acids and the consequences of such a presence in a scaffold has been explored thoroughly since then.[37] With this newfound understanding, researchers often are combining several known decellularization methods, such as detergents, acids/bases, alcohols, and biologic agents such as nucleases and proteases. Crapo and coworkers clarified that the choice of the optimum decellularization medium is often determined by four components: 1) cell density; 2) total cell content; 3) lipid content, and 4) thickness.[37] Readers are directed to this excellent review for further details regarding clinical options and mechanisms utilized to deliver the necessary treatment agents to the tissue as well as a general overview of possible sterilization options for biological scaffolds, and their consequences of use, before implantation and use.

Following successful decellularization of a scaffold, the researcher must then characterize the scaffold with respect to mechanical properties, biological properties, and host cell ingrowth potential as well as porosity of the scaffold as described previously (see the section "Necessary Properties for the Ideal Vascular Graft").[6] Pending *in vitro* properties, the scaffold is then ready for the initial *in vivo* biocompatibility test for scaffolds, subcutaneous implantation in a small animal model, to assess the potential for inflammation and calcification on functional implantation. A functional implantation utilizes a large-animal model and necessitates the implantation of the fully prepared scaffold in the position it was originally intended; evaluation for mechanical and biocompatible properties follows. For example, a tissue-engineered aortic valve would be implanted in the position of the aortic valve within a large animal. Pending successful large-animal studies, the potential for clinical trials arises in which the scaffold is implanted in human patients; efficacy is then again evaluated. The Food and Drug Administration (FDA) investigation begins, with positive results holding promise for finally bringing the device to market.

APPLICATION OF DECELLULARIZATION IN VASCULAR TISSUE ENGINEERING

Before the Holy Grail of getting a tissue-engineered scaffold to market, extensive research is necessary (Fig. 4). Research from our group focuses on decellularization mechanisms of scaffold preparation and is applied to

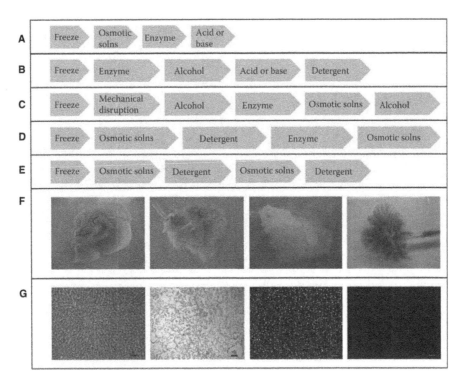

Fig. 4 Example decellularization protocols for (**A**) thin laminates such as pericardium; (**B**) thicker laminates such as dermis; (**C**) fatty, amorphous tissues such as adipose; (**D**) composite tissues or whole simple organs such as trachea; and (**E**) whole vital organs such as liver. Arrow lengths represent relative exposure times for each processing step. Rinse steps for agent removal and sterilization methods are not shown to simplify comparison. (**F**) Representative images of the gross appearance of intact rat liver subjected to decellularization (left to right): before, during, and after decellularization; decellularized liver perfused with blue dye. (**G**) Representative photomicrographs showing no nuclear staining after whole-organ decellularization (left to right): native rat liver hematoxylin and eosin (H&E); decellularized liver ECM H&E; native rat liver DAPI (4',6-diamidino-2-phenylindole); liver ECM DAPI. Scale bars are 50 mm.
Source: From Crapo et al.[37]

vascular grafts, heart valves, myocardium, and the nucleus pulposus found in intervertebral discs,[11,38–42] all of which make use of natural polymers. For vascular scaffolds in particular, this route of scaffold production has the advantage of retaining mature elastin, a highly cross-linked protein no longer turned over past neonatal development,[20] inside the scaffold; the benefits of such have been described previously. However, Isenburg and colleagues clearly demonstrated that simply retaining elastin inside a scaffold is not sufficient for producing a biologically viable scaffold[43] because of its natural tendency to hastily degrade *in vivo*[38] as a result of matrix-degrading proteases.

To circumvent this issue, our approach is to stabilize scaffolds with compounds that bind to matrix components and decrease their degradation *in vivo* but not completely annihilate them. Our group utilizes naturally derived phenolic tannins to achieve this partial and reversible fixation, thereby remaining in the realm of "natural" components for vascular tissue engineering.

Penta-galloyl-glucose (PGG) is a polyphenol[44] naturally derived from tannic acid (Fig. 5) that shows particularly high affinity for proline-rich proteins[45] such as collagen and elastin,[46] to which it binds strongly via hydrophobic and hydrogen bonds. Our group has reported

the use of PGG to stabilize cardiovascular collagen scaffolds[44] in addition to vascular elastin scaffolds.[38] Furthermore, we have showed that treatment with PGG diminishes vascular calcification.[38] PGG is a strong antioxidant,[47,48] inhibits many proteases, reduces inflammation and antigenicity,[49] and is not cytotoxic[50,51] and thus can be used safely in tissue-engineering applications. In addition to the properties that fixation with PGG grants a tissue, the created bonds degrade over time to bestow robust mechanical strength initially on implantation and gradually allow for implant remodeling *in vivo*,[44] a desired biological property of all tissue-engineered constructs.

THE FUTURE OF BIOENGINEERING: CLINICALLY TRANSLATIONAL RESEARCH

Most researchers in the field are familiar with the briefly discussed and highly difficult process of bringing a medical device or implant to clinical trials. Such successful individuals have accomplished the task of producing clinically viable implants. However, few researchers are able to consolidate their science into a small device easily usable by clinicians, nurses, and other medical faculty in a

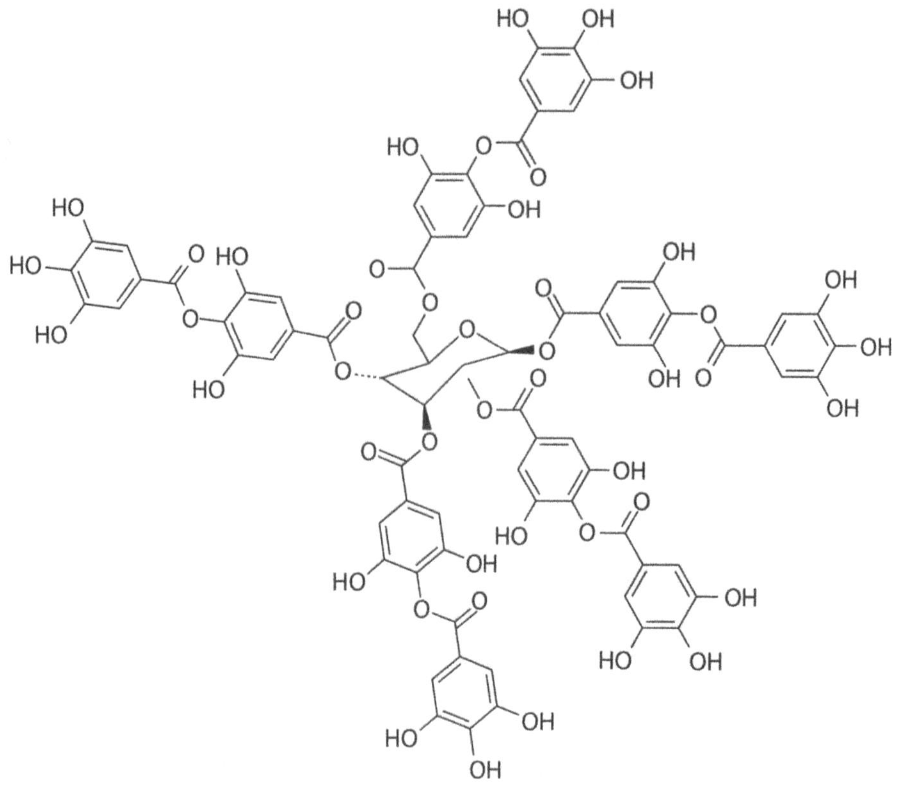

Fig. 5 Structure of tannic acid. A central glucose molecule is esterified with gallic acid at all five hydroxyl moieties, resulting in a penta-galloyl glucose core. This may be further derivatized by gallic acid residues, yielding a deca-galloyl-glucose, as necessary.
Source: From Isenburg et al.[43]

hospital setting without any supervision by the inventor of said technology. With the majority of "translational" literature surrounding the often-discussed topic of stem cell treatments,[52] the potential for improvement in this area is recognized. Furthermore, fully automated machines remove the issue of batch-to-batch reproducibility that is innate to anything that has to be done by hand with human supervision. By incorporating a researcher's science into a theoretical device capable of creating patient-tailored implants at the push of a button, one has created a translational device, a device that successfully bridges the gap between lab bench and patient bedside. Although a lofty goal for all researchers, this is arguably the future of regenerative medicine, one in which implanted devices will be unique to each patient and not require meticulous sizing in operating rooms, as is currently done with the majority of cardiovascular implantation procedures. Time will tell if such dreams are valid and whether we have the resources to bring science to the patient, our prime target as bioengineers.

REFERENCES

1. Niklason, L.E.; Langer, R.S. Advances in tissue engineering of blood vessels and other tissues. Transpl. Immunol. **1997**, *5* (4), 303–306.

2. Veith, F.J.; Gupta, S.K.; Ascer, E.; White-Flores, S.; Samson, R.H.; Scher, L.A.; Towne, J.B.; Bernhard, V.M.; Bonier, P.; Flinn, W.R.; Astelford, P.; Yao, J.S.T.; Bergan, J.J. Six-year prospective multicenter randomized comparison of autologous saphenous vein and expanded polytetrafluoroethylene grafts in infrainguinal arterial reconstructions. J. Vasc. Surg. **1986**, *3* (1), 104–114.

3. Bezuidenhout, D.; Zilla, P. Vascular grafts. In *Encyclopaedia of Biomaterials and Biomedical Engineering*; Bowlin, G.L., Wnek, G.E., Eds.; Informa Healthcare: New York, 2004.

4. Zdrahala, R.J. Small caliber vascular grafts. Part II: Polyurethanes revisited. J. Biomater. Appl. **1996**, *11* (1), 37–61.

5. Barrett, D.G.; Yousaf, M.N. Thermosets synthesized by thermal polyesterification for tissue engineering applications. Soft Matter **2010**, *6* (20), 5026–5036.

6. Sarkar, S.; Schmitz-Rixen, T.; Hamilton, G.; Seifalian, A.M. Achieving the ideal properties for vascular bypass grafts using a tissue engineered approach: A review. Med. Biol. Eng. Comput. **2007**, *45* (4), 327–336.

7. Murugesan, S.; Xie, J.; Linhardt, R.J. Immobilization of heparin: Approaches and applications. Curr. Topics Med. Chem. **2008**, *8* (2), 80–100.

8. Rabenstein, D.L. Heparin and heparan sulfate: Structure and function. Nat. Prod. Rep. **2002**, *19* (3), 312–331.

9. Simpson, C.L.; Lindley, S.; Eisenberg, C.; Basalyga, D.M.; Starcher, B.C.; Simionescu, D.T.; Vyavahare, N.R. Toward cell therapy for vascular calcification: Osteoclast-mediated

demineralization of calcified elastin. Cardiovasc. Pathol. **2007**, *16* (1), 29–37.

10. Simionescu, D.T. Prevention of calcification in bioprosthetic heart valves: Challenges and perspectives. Expert Opin. Biol. Ther. **2004**, *4* (12), 1971–1985.

11. Isenburg, J.; Simionescu, D.T.; Vyavahare, N.R. Tannic acid treatment enhances biostability and reduces calcification of glutaraldehyde fixed aortic wall. Biomaterials **2005**, *26* (11), 1237–1245.

12. Chlupac, J.; Filova, E.; Bacakova, L. Blood vessel replacement: 50 years of development and tissue engineering paradigms in vascular surgery. Physiol. Res. **2009**, *58* (Suppl. 2), S119–S139.

13. Junqueira, L.C.; Cameiro, J.; Eds. *Basic Histology: Text and Atlas*, 11th Ed.; McGraw Hill: New York, 2005; 206–207.

14. Voorhees, A.B., Jr.; Jaretzki, A., III; Blakemore, A.H. The use of tubes constructed from vinyon "N" cloth in bridging arterial defects. Ann. Surg. **1952**, *135* (3), 332–336.

15. Creech, O., Jr.; Deterling, R.A., Jr.; Edwards, S.; Julian, O.C.; Linton, R.R.; Shumacker, H., Jr. Vascular prostheses: Report of the Committee for the Study of Vascular Prostheses of the Society for Vascular Surgery. Surgery **1957**, *41* (62–85), 62–80.

16. Nishibe, T.; Kondo, Y.; Muto, A.; Dardik, A. Optimal prosthetic graft design for small diameter vascular grafts. Vascular **2007**, *15* (6), 356–360.

17. Wise, S.G.; Mithieux, S.M.; Weiss, A.S. Engineered tropoelastin and elastin-based biomaterials. Adv. Protein Chem. Struct. Biol. **2009**, *78*, 1–24.

18. Waterhouse, A.; Wise, S.G.; Ng, M.K.; Weiss, A.S. Elastin as a nonthrombogenic biomaterial. Tissue Eng. B **2011**, *17* (2), 93–99.

19. Bracher, M.; Simionescu, D.; Simionescu, A.; Davies, N.; Human, P.; Zilla, P. Matrix metalloproteinases and tissue valve degeneration. J. Long Term Eff. Med. Implants **2001**, *11* (3/4), 221–230.

20. Patel, A.; Fine, B.; Sandig, M.; Mequanint, K. Elastin biosynthesis: The missing link in tissue-engineered blood vessels. Cardiovasc. Res. **2006**, *71* (1), 40–49.

21. Daamen, W.F.; Veerkamp, J.H.; Van Hest, J.C.M.; Van Kuppevelt, T.H. Elastin as a biomaterial for tissue engineering. Biomaterials **2007**, *28* (30), 4378–4398.

22. Isenburg, J.C.; Karamchandani, N.V.; Simionescu, D.T.; Vyavahare, N.R. Structural requirements for stabilization of vascular elastin by polyphenolic tannins. Biomaterials **2006**, *27* (19), 3645–3651.

23. Prabhakaran, M.P.; Ghasemi-Mobarakeh, L.; Ramakrishna, S. Electrospun composite nanofibers for tissue regeneration. J. Nanosci. Nanotechnol. **2011**, *11* (4), 3039–3057.

24. L'Heureux, N.; Germain, L.; Labbé, R.; Auger, F.A. *In vitro* construction of a human blood vessel from cultured vascular cells: A morphologic study. J. Vasc. Surg. **1993**, *17* (3), 499–509.

25. Barocas, V.H.; Girton, T.S.; Tranquillo, R.T. Engineered alignment in media equivalents: Magnetic prealignment and mandrel compaction. J. Biomech. Eng. **1998**, *120* (5), 660–666.

26. Grassl, E.D.; Oegema, T.R.; Tranquillo, R.T. A fibrin-based arterial media equivalent. J. Biomed. Mater. Res. A **2003**, *66* (3), 550–561.

27. Hirai, J.; Matsuda, T. Venous reconstruction using hybrid vascular tissue composed of vascular cells and collagen: Tissue regeneration process. Cell Transpl. **1996**, *5* (1), 93–105.

28. Girton, T.S.; Oegema, T.R.; Grassl, E.D.; Isenberg, B.C.; Tranquillo, R.T. Mechanisms of stiffening and strengthening in media-equivalents fabricated using glycation. J. Biomech. Eng. **2000**, *122* (3), 216–223.

29. Elbjeirami, W.M.; Yonter, E.O.; Starcher, B.C.; West, J.L. Enhancing mechanical properties of tissue-engineered constructs via lysyl oxidase crosslinking activity. J. Biomed. Mater. Res. A **2003**, *66* (3), 513–521.

30. Boland, E.D.; Matthews, J.A.; Pawlowski, K.J.; Simpson, D.G.; Wnek, G.E.; Bowlin, G.L. Electrospinning collagen and elastin: Preliminary vascular tissue engineering. Front. Biosci. J. Virtual Library **2004**, *9*, 1422–1432.

31. Barnes, C.P.; Sell, S.A.; Boland, E.D.; Simpson, D.G.; Bowlin, G.L. Nanofiber technology: Designing the next generation of tissue engineering scaffolds. Adv. Drug Deliv. Rev. 2007, *59* (14), 1413–1433.

32. Sell, S.A.; McClure, M.J.; Garg, K.; Wolfe, P.S.; Bowlin, G.L. Electrospinning of collagen/biopolymers for regenerative medicine and cardiovascular tissue engineering. Adv. Drug Deliv. Rev. **2009**, *61* (12), 1007–1019.

33. Weinberg, C.B.; Bell, E. A blood vessel model constructed from collagen and cultured vascular cells. Science **1986**, *231* (4736), 397–400.

34. Long, J.L.; Tranquillo, R.T. Elastic fiber production in cardiovascular tissue-equivalents. Matrix Biol. J. Int. Soc. Matrix Biol. **2003**, *22* (4), 339–350.

35. Ramamurthi, A.; Vesely, I. Evaluation of the matrix-synthesis potential of crosslinked hyaluronan gels for tissue engineering of aortic heart valves. Biomaterials **2005**, *26* (9), 999–1010.

36. Gilbert, T.W.; Sellaro, T.L.; Badylak, S.F. Decellularization of tissues and organs. Biomaterials **2006**, *27* (19), 3675–3683.

37. Crapo, P.M.; Gilbert, T.W.; Badylak, S.F. An overview of tissue and whole organ decellularization processes. Biomaterials **2011**, *32* (12), 3233–3243.

38. Chuang, T.H.; Stabler, C.; Simionescu, A.; Simionescu, D.T. Polyphenol-stabilized tubular elastin scaffolds for tissue engineered vascular grafts. Tissue Eng. A **2009**, *15* (10), 2837–2851.

39. Sierad, L.N.; Simionescu, A.; Albers, C.; Chen, J.; Maivelett, J.; Tedder, M.E.; Liao, J.; Simionescu, D.T. Design and testing of a pulsatile conditioning system for dynamic endothelialization of polyphenol-stabilized tissue engineered heart valves. Cardiovasc. Eng. Technol. **2010**, *1* (2), 138–153.

40. Tedder, M.E.; Simionescu, A.; Chen, J.; Liao, J.; Simionescu, D.T. Assembly and testing of stem cell-seeded layered collagen constructs for heart valve tissue engineering. Tissue Eng. A **2011**, *17* (1–2), 25–36.

41. Mercuri, J.J.; Gill, S.S.; Simionescu, D.T. Novel tissue-derived biomimetic scaffold for regenerating the human nucleus pulposus. J. Biomed. Mater. Res. A **2011**, *96* (2), 422–435.

42. Simionescu, A.; Tedder, M.E.; Chuang, T.H.; Simionescu, D.T. Lectin and antibody-based histochemical techniques for cardiovascular tissue engineering. J. Histotechnol. **2011**, *34* (1), 20–29.

43. Isenburg, J.C.; Simionescu, D.T.; Vyavahare, N.R. Elastin stabilization in cardiovascular implants: Improved resistance to enzymatic degradation by treatment with tannic acid. Biomaterials **2004**, *25* (16), 3293–3302.

44. Tedder, M.E.; Liao, J.; Weed, B.; Stabler, C.; Zhang, H.; Simionescu, A.; Simionescu, D.T. Stabilized collagen scaffolds for heart valve tissue engineering. Tissue Eng. A **2009**, *15* (6), 1257–1268.

45. Charlton, A.J.; Baxter, N.J.; Lilley, T.H.; Haslam, E.; McDonald, C.J.; Williamson, M.P. Tannin interactions with a full-length human salivary proline-rich protein display a stronger affinity than with single proline-rich repeats. FEBS Lett. **1996**, *382* (3), 289–292.

46. Luck, G.; Liao, H.; Murray, N.J.; Grimmer, H.R.; Warminski, E.E.; Williamson, M.P.; Lilley, T.H.; Haslam, E. Polyphenols, astringency and proline-rich proteins. Phytochemistry **1994**, *37* (2), 357–371.

47. Piao, X.; Piao, X.L.; Kim, H.Y.; Cho, E.J. Antioxidative activity of geranium (Pelargonium inquinans Ait) and its active component, 1,2,3,4,6-penta-O-galloyl-beta-D-glucose. Phytother. Res. **2008**, *22* (4), 534–538.

48. Choi, B.M.; Kim, H.J.; Oh, G.S.; Pae, H.O.; Oh, H.; Jeong, S.; Kwon, T.O.; Kim, Y.M.; Chung, H.T. 1,2,3,4,6-Penta-O-galloyl-beta-D-glucose protects rat neuronal cells (Neuro 2A) from hydrogen peroxide-mediated cell death via the induction of heme oxygenase-1. Neurosci. Lett. **2002**, *328* (2), 185–189.

49. Haslam, E. Plant polyphenols: Vegetable tannins revisited. In *Chemistry and Pharmacology of Natural Products*; Phillipson, J., Ed.; Cambridge University Press: Cambridge, UK, 1989; 167–195.

50. Isenburg, J.C.; Simionescu, D.T.; Starcher, B.C.; Vyavahare, N.R. Elastin stabilization for treatment of abdominal aortic aneurysms. Circulation **2007**, *115* (13), 1729–1737.

51. Isenberg, B.C.; Williams, C.; Tranquillo, R.T. Small-diameter artificial arter-ies engineered *in vitro*. Circ. Res. **2006**, *98* (1), 25–35.

52. Banerjee, C. Stem cells therapies in basic science and translational medicine: Current status and treatment monitoring strategies. Curr. Pharm. Biotechnol. **2011**, *12* (4), 469–487.

53. Hirai, J.; Matsuda, T. Self-organized, tubular hybrid vascular tissue composed of vascular cells and collagen for low-pressure-loaded venous system. Cell Transpl. **1995**, *4* (6), 597–608.

Wound Care: Natural BioPolymer Applications

Soheila S. Kordestani
ChitoTech Inc., Tehran, Iran

Abstract

Various neutral, basic, acidic, and sulfated polysaccharides have been the focus of interest with respect to biomedical and wound care applications. They are produced in different forms in order to cover all chronic and acute wounds. These bioactive wound dressings act as a chemo-attractant and the process of healing can start straight away by providing a moist environment. A relatively comprehensive review of the function and requirements of wound management aids, their physical forms, and the structural features of the polysaccharides and a brief overview of selected commercially available products is the aim of this entry.

INTRODUCTION

Wound healing is a multifactorial, complicated, physiological process. Cellular and biochemical components as well as enzymatic pathways play pivotal roles during the repair and recovery of a wound tissue. Some natural polymers have excellent structural and physicochemical properties, making them suitable agents for different applications in medical care of which wound management aids are one of the most important and encouraging modality. A variety of neutral (e.g., cellulose), basic (e.g., chitin and chitosan), acidic [e.g., alginic acid and hyaluronic acid (HA)], and sulfated polysaccharides (e.g., heparin, chondroitin, dermatan and keratan sulfates) have been the focus of interest with respect to biomedical/ wound care applications. Furthermore, various researchers have studied more unusual complex heteropolysaccharides, isolated from plant and microbial sources. Their studies have shown that these biopolymers possess potentially useful biological and/or physicochemical characteristics with respect to wound care applications. The present entry aims to conduct a relatively comprehensive review of the function and requirements of wound management aids, their physical forms, and the structural features of the polysaccharides that are usually employed for their preparation and synthesis. Furthermore, a brief overview of selected commercially available products, specifically hydrogels, their applications in wound care and dressings, the pioneering research and manufacturing companies are presented for each compound.

CELLULOSE

Introduction

Cellulose has various distinctive structural properties, making it an outstanding compound for different medical and industrial applications. This substance is the main component of the plant cell wall.[1]

Cellulose use started with the exploratory investigation at Johnson & Johnson[2] and has been applied for covering wounds.[3]

Partial oxidation of the primary hydroxyl groups on the anhydroglucose rings produced oxidized regenerated cellulose (ORC), which in turn can be used to synthesize monocarboxyl cellulose as a natural and topical biomaterial. Those ORC materials containing 16–24% carboxylic acid content act as an important class of biocompatible and bio-absorbable polymers, which are available in a sterilized knitted fabric or powder form and can be used to stop bleeding.

ORC polymers were developed and presented for the first time in the late 1930s (Fig. 1).[4]

Structure

The structure of cellulose is shown in Fig. 1.

Applications in Wound Care

Traditionally, skin tissue repair materials are absorbent and permeable agents. For example, gauze, a traditional dressing material, can adhere to desiccated wound surfaces and induce trauma on removal of the dressing. Various medical and cosmetic applications of bacterial cellulose (BC) synthesized from surface cultures have drawn plenty of research attention in this field. Various potentials of BC originate from the unique properties of this compound, such as high mechanical strength of its never-dried BC membrane. Furthermore, high liquid absorbency, biocompatibility,

Concise Encyclopedia of Biomedical Polymers and Polymeric Biomaterials DOI: 10.1081/E-EBPPC-120049925

Vascular—Zwitterionic

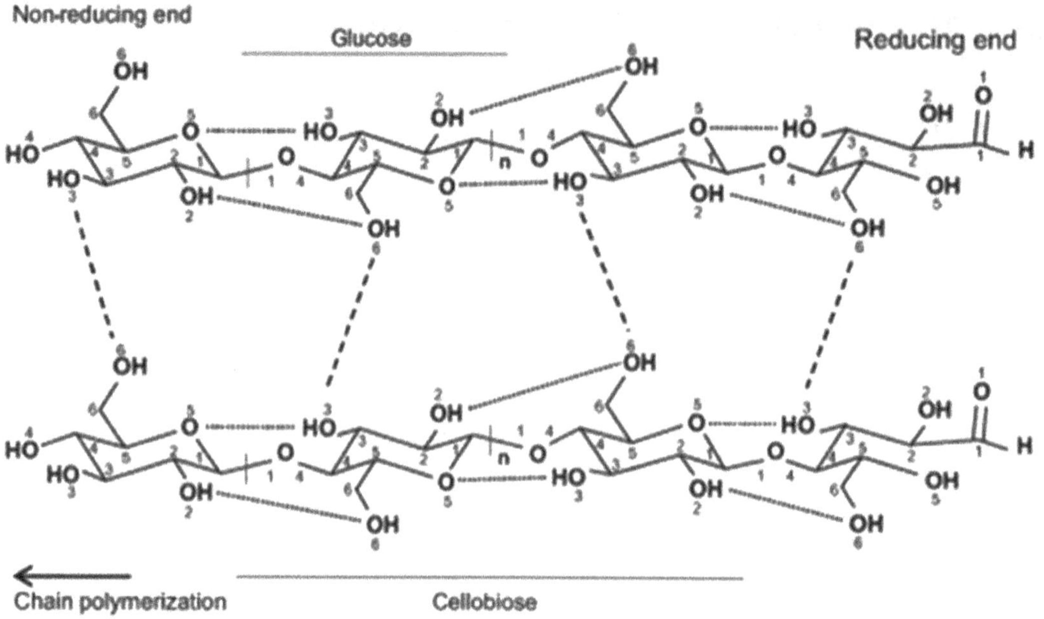

Fig. 1 The structure and the inter- and intra-chain hydrogen bonding pattern in cellulose I, dashed lines: inter-chain hydrogen bonding, Dotted lines: intra-chain hydrogen bonding.
Source: From Festucci-Buselli[1] © 2007, with permission from Brazilian Society of Plant Physiology.

and hygienic nature of this compound makes it an ideal option for specific demands of skin tissue repair.[3]

Manufacturing Companies

Johnson & Johnson Company is one of the pioneering groups in developing industrial-scale oxidation processes using nitrogen dioxide to manufacture an ORC-absorbable hemostat—Surgicel.

Cellulose Solutions Company produces Dermafill and Xylinum Cellulose—Cellulose Membrane Dressing.

CHITIN AND CHITOSAN

Introduction

Chitin is a copolymer of *N*-acetyl-glucosamine and *N*-glucosamine units randomly or block distributed throughout the biopolymer chain depending on the processing method used to derive the biopolymer. The biopolymer is termed chitin or chitosan when the number of *N*-acetyl-glucosamine or *N*-glucosamine units is higher than 50%, respectively. Chitosan has been more frequent studied because of its ready solubility in dilute acids, making it more accessible for use and chemical reactions.[5] However, it is normally insoluble in aqueous solutions above pH 7 because of its rigid crystalline structure and the deacetylation limiting its wide application.[6]

The main biochemical activities of the chitin- and chitosan-based materials are polymorphonuclear cell and fibroblast activation, cytokine production, giant cell migration, and stimulation of type IV collagen synthesis.[7]

Chitosan carries two types of reactive groups that can be grafted: first, the free amino groups on deacetylated units and, second, the hydroxyl groups on the C3 and C6 carbons on acetylated or deacetylated units. Grafting of chitosan facilitates the functional derivatives formation through covalent binding of a molecule, the graft, onto the chitosan backbone.[8]

The chitosan-based nanoparticles possess several advantages over the use of chitosan microspheres and microcapsules for drug-delivery process.[9]

Chitin and chitosan as sources of nutrients have antimicrobial activity with high resistance against environment conditions. Various studies have demonstrated the antimicrobial activity of these two compounds. Chitosan can efficiently control the growth of algae, and to inhibit *in vivo* and *in vitro* plant viral multiplication. Chitosan has been used as an antimicrobial compound through external application (exogenous) to the host, to the substrate or media, and to a physical surface containing microbial population. The chitosanase and chitinase are induced in plants as resistance mechanisms against pathogens, especially fungi. Some mechanisms for antifungal activity of exogenous chitosans are as follows:

- Induction of phenylpropanoid and octadecanoid pathways
- Induction of chitosanase
- Induction of chitosanase with many polypeptides
- Effect of chitosans on the plant enzymes in reaction to plant resistance against fungal pathogens

- Induction of phenolic compounds and/or phytoalexins
- Induction of morphological and/or physiological changes

In addition, some mechanisms suggested for understanding this antibacterial action are as follows:

- Reaction with bacterial teichoic acids, polyelectrolyte complexes
- Chelation of metals present in metalloenzymes
- Alteration of the bacterial adhesion
- Inhibition of the enzymes that link glucans to chitin
- Prevention of nutrients permeation.[10]

Structure

The structure of chitin and chitosan is shown in Fig. 2.

Application in Wound Care

Wound dressing is one of the most promising medical applications for chitin and chitosan. The adhesive nature of chitin and chitosan, as well as their antifungal and bactericidal traits, and finally their permeability to oxygen are the most important characteristics in improving the wound and burn treatment efficiency of the compound. Various derivatives of these two compounds have been proposed for wound treatment in the form of hydrogels, fibers, membranes, scaffolds, and sponges. The present entry aims to conduct a relatively comprehensive review on the wound dressing applications of biomaterials based on chitin, chitosan, and their derivatives in various forms.[12] In the field of veterinary medicine, chitosan has been proven to enhance the functions of polymorphonuclear leukocytes (PMNs) (phagocytosis, and production of osteopontin and leukotriene B4), macrophages

(phagocytosis, and production of interleukin-1, transforming growth factor b1, and platelet-derived growth factor), and fibroblasts (interleukin-8 synthesis). Therefore, chitosan promotes granulation and organization, making it a beneficial agent for treating open wounds; certain PMN functions are enhanced, such as phagocytosis and the production of chemical mediators.[13]

Manufacturing Companies

The ChitoTech Company was developed from a research facility for the study of natural biopolymers and their applications in medicine. ChitoHeal film and ChitoHeal Gel are two important chitosan-based dressings.

HemCon is now a world leader in advanced chitosan research and development, and continues to expand its application.

Medovent is the first company to overcome the technological barriers for processing chitin- and chitosan-based biopolymers; and aim to become a leading specialist for chitin- and chitosan-processing technologies to manufacture medical devices of complex designs with highest quality, reliability, and performance.

Medoderm is a world innovator in developing chitosan-based medical products.

ALGINATE

Introduction

Alginate is a collective term for a family of polysaccharides synthesized from brown algae and bacteria. Alginic acid was first discovered, extracted, and patented by Stanford.

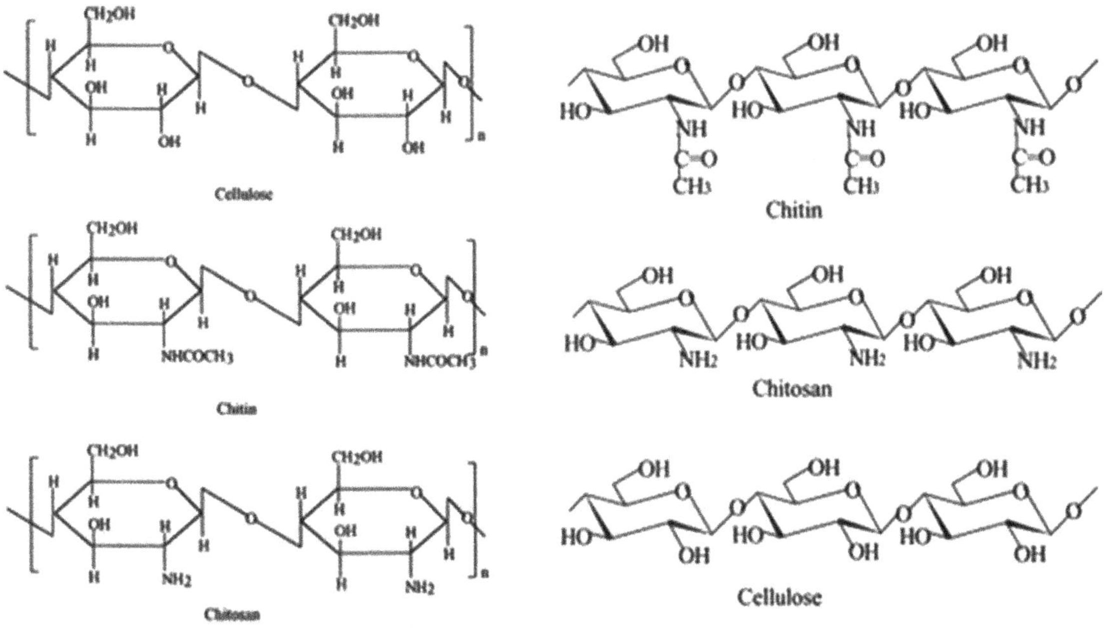

Fig. 2 Structures of cellulose, chitin and chitosan.
Source: From Ravi Kumar[11] © 2000, with permission from Elsevier.

This polysaccharide was recognized as a structural component of marine brown algae, constituting up to 40% of the dry matter and occurs mainly in the intercellular mucilage and algal cell wall as an insoluble mixture of calcium, magnesium, potassium, and sodium salts. The presence of alginate improves the mechanical strength and flexibility of the seaweed as well as acts as water reservoir preventing dehydration when a portion of seaweed is exposed to air. Therefore, alginate can be assumed to have the same morphophysiological properties in brown algae as those of cellulose and pectin in terrestrial plants. Several bacteria such as *Azotobacter vinelandii* and various species of pseudomonas produce an exocellular polymeric material that resembles alginate.

The stability of an alginate molecule is strongly dependent on conditions such as temperature, pH, and presence of contaminants. The glycosidic linkages between the sugar monomers of the polysaccharide are susceptible to cleavage in both acidic and alkaline media.[14]

Structure

Three fractions are traditionally isolated: two of these contain almost exclusively α-L-guluronic acid (G) and β-D-mannuronic acid (M) residues, respectively; the third one is composed of both uronic acids in almost equal proportion (Fig. 3).[14]

Applications in Wound Care

Alginate dressings are absorbent, nonadherent, biodegradable, nonwoven fibers derived from brown seaweed. They are composed of calcium salts of alginic acid and mannuronic and guluronic acids. Alginate dressings in contact with sodium-rich solutions such as wound drainage impose the calcium ions to undergo an exchange for the sodium ions, forming a soluble sodium alginate gel. This gel maintains a moist wound bed and supports a therapeutic healing environment. Alginates can absorb fluid 20 times their own weights that can vary based on the particular product. They are extremely beneficial in managing large draining cavity wounds, pressure cavity ulcers, vascular ulcers, surgical incisions, wound dehiscence, tunnels, sinus tracts, skin graft donor sites, exposed tendons, and infected wounds. Furthermore, their hemostatic and absorptive properties make them efficient treatments for bleeding wounds. Alginates are contraindicated for dry wounds, eschar-covered wounds, surgical implantation, or on third-degree burns. Alginates are available in various sizes and forms, including in sheet, pad, and rope. However, newer versions of calcium alginate dressings contain controlled release of ionic silver. They are usually changed daily or as indicated by the amount of drainage. Early wound care interventions may warrant more frequent dressing changes due to high volume of drainage. The frequency of dressing changes decreases as fluid management is reached.[15–18]

Manufacturing Companies

Coloplast Company produces Comfeel plus Ulcer Dressing, which is consisted of a semipermeable polyurethane film coated with a flexible, cross-linked adhesive mass containing sodium carboxymethylcellulose and calcium alginate as the principal absorbent and gel-forming agents.

Smith & Nephew Company produces Algisite M, which is a calcium–alginate dressing, forming a soft, integral gel in contact with wound exudate.

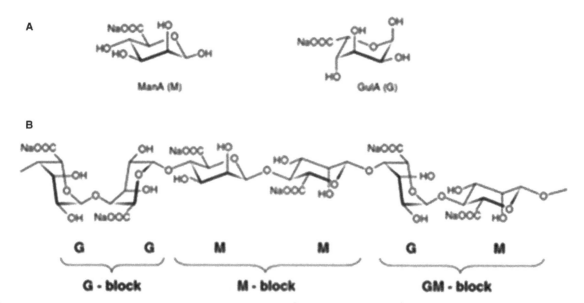

Fig. 3   Alginate chemical structure. (**A**) The 4C1 conformation of β-D-mannuronic acid (M) sodium salt and the 1C4 conformation of α-L-guluronic acid (G) sodium salt. (**B**) The block composition of alginate with G-blocks, M-blocks, and MG-blocks.
Source: From Rehm[14] © 2009, with permission from Springer Science + Business Media.

STARCH

Introduction

Starch is a common constituent of higher plants and the major form of carbohydrates store. Starch in chloroplasts is in transitory state that accumulates during the light period to be utilized during the dark.[19]

Starch industries need high-amylose-content starch in great volume that is increasing by its unique functional properties.[20]

Structure

The structure of starch is shown in Fig. 4.

Applications in Wound Care

Starch is one of the most common and cost-effective polysaccharides. It usually includes about 30% amylose, a linear a-(1, 4) glucan, and 70% amylopectin, dendritically branched version. Chemically modified starches, enjoying outstanding properties such as lower cost and biodegradability, are finding various applications in industry. Various research teams have comprehensively studied chemical modifications of starch through graft copolymerization of vinyl monomers onto starch.[22,23] Polyvinyl alcohol (PVA)/starch blend hydrogels can be prepared by chemical cross-linking technique. Membranes synthesized through cross-linking of corn starch and PVA with glutaraldehyde have sufficient strength. The resulted hydrogel membrane can serve as artificial skin. Using these membranes, concurrently various nutrients or healing factors and medications can be delivered directly onto the site of action.[24]

Manufacturing Companies

PolyMem Dressings protect the wound and facilitate the body's natural healing process.

Aspen Medical Company is one of the pioneering companies with more than 25 years' experience in the medical device. Aquaform is a clear, viscous, sterile gel containing a modified starch polymer, glycerol, preservatives, and water.

Smith & Nephew Company produces Cadesorb, which is a white, starch-based sterile ointment that reduces local wound pH to around 5, thus modulating protease activities.

COLLAGEN

Introduction

Collagen type I is the most abundant proteins available in mammals. Its application in a range of tissues from tendons and ligaments, to skin, cornea, bone, and dentin improves mechanical stability, strength, and toughness. These tissues have quite different mechanical demands, some need to be elastic or to store mechanical energy, while the others need to be stiff and tough. This shows the versatility of collagen

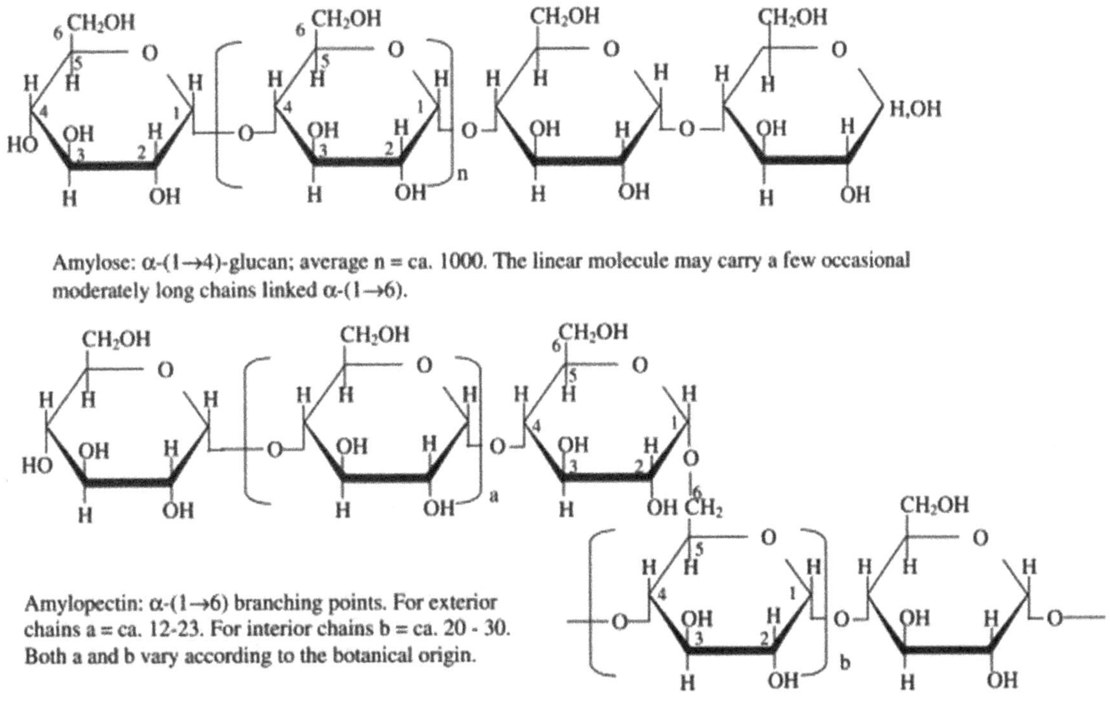

Amylose: α-(1→4)-glucan; average n = ca. 1000. The linear molecule may carry a few occasional moderately long chains linked α-(1→6).

Amylopectin: α-(1→6) branching points. For exterior chains a = ca. 12-23. For interior chains b = ca. 20 - 30. Both a and b vary according to the botanical origin.

Fig. 4 Structure of amylose and amylopectin.
Source: From Kiatkamjornwong[23] © 2000, with permission from Elsevier.

as a building material. While in some cases, including bone and dentin, the stiffness is increased by the inclusion of mineral. In addition, the mechanical properties are adapted by a modification of the hierarchical structure rather than by a different chemical composition. The collagen fibril with 50 to a few hundred nanometer thickness is the basic building block of collagen-rich tissue. These fibrils are then assembled to a variety of more complex structures with very different mechanical properties. As in all collagens, each fibrillar collagen molecule consists of three polypeptide chains, called α chain. Molecules can be homotrimeric, consisting of three identical α chains, as in collagens II and III, or heterotypic, consisting of up three genetically distinct α chains. Individual α chain is identical by the following nomenclature: α n (N), where N is the Roman numeral indicating collagen type and n is the number of α chain.[25]

Structure

The structure of collagen is shown in Fig. 5.

Applications in Wound Care

Collagen is a major protein of the body that is necessary for wound healing and repairing process. Collagen dressings, derived from bovine hide (cowhide), are either 100% collagen or may be combined with alginates or other products. They are a highly absorptive, hydrophilic, and moist wound dressing. Collagen dressings can be used on granulating or necrotic wounds as well as on partial- or full-thickness wounds. A collagen dressing agent should be changed a minimum of every 7 days. If infection in wound is present, daily dressing change is recommended. Collagen dressings require a secondary dressing for securement.[27]

Manufacturing Companies

Johnson & Johnson Company produces Fibracol Plus. Advanced wound care dressing with 90% collagen composition.

Coloplast Woundres Collagen Hydrogel is the main component of skin and connective tissue playing a pivotal

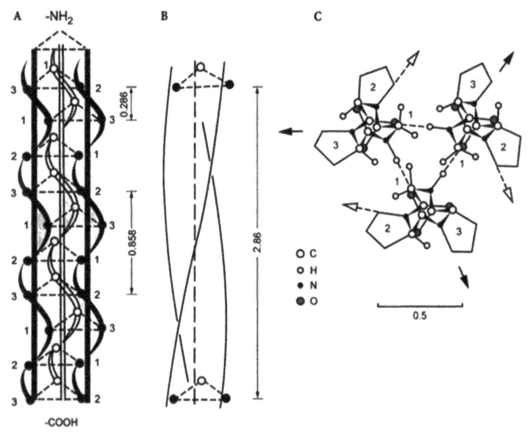

Fig. 5 Model of the collagen triple helix. The structure is shown for (Gly-Pro-Pro)$_n$ in which glycine is designated by 1, proline in X-position by 2 and proline in Y-position by 3: (**A, B**) side views. Three left-handed polyproline-II-type helices are arranged in parallel. For clarity, the right-handed supercoil of the triple helix is not shown in (**A**) but indicated in (**B**). Dashed lines indicate positions of C-atoms (and not hydrogen bonds as in (**C**)). All indicated values for axial repeats correspond to the supercoiled situation; (**C**) top view in the direction of the helix axis. The three claims are connected by hydrogen bonds between the backbone NH of glycin and the backbone CO of proline in Y-position (dashed lines). Arrows indicate the directions in which other side chains than proline rings emerge from the helix. Approximate residue- to residue distances, repeats of the polyproline-II- and triple helix and a scale bar are indicated in nm.
Source: From Fakirov[28] © 2007, with permission from Hanser Publications.

Vascular—Zwitterionic

role in all phases of wound healing. It promotes autolytic debridement through rehydrating and softening dry wounds and necrotic tissue.

SILK

Introduction

Silk is a fibrous protein biopolymer with remarkable mechanical properties. Silk fibers belong to the group of secretion-type animal fibers.[28]

Historically, silkworm silk has been used commercially as biomedical sutures in repairing wound injuries. Furthermore, several studies have been conducted on the feasibility and application of this biomaterial in tissue engineering because of its slow degradation, excellent mechanical properties, and biocompatibility. Advances in technology have made it possible to fabricate silk-based materials with various geometries, including films, sponges, mats, and fibers, from purified silk fibroin solution.[29]

Its composition is a mix of an amorphous polymer, which makes it elastic, and chains of two of the simplest proteins, which make it tough. Out of 20 amino acids, only glycine and alanine serve as a primary constituent of silk. Fibroin consists of about 40% glycine and 25% alanine as the major amino acids. The remaining components are mostly glutamine, serine, leucine, valine, proline, tyrosine, and arginine. The high elasticity of spider silk is because of glycine-rich regions where several ordered multiple amino acids are continuously repeated. A 180° turn (α-turn) occurs after each sequence, resulting in α-spiral or α-helix structure. β-sheets act as a cross-link between the protein molecules where the regular structure of these sheets gives high tensile strength to spider silk.[30]

Structure

The structure of silk is shown in Fig. 6.

Applications in Wound Care

Different studies have shown the wound-healing-facilitating properties of silk and its different derivatives. For instance, studies on animal mouse wound model have demonstrated high efficiency of silk protein–biomaterial wound dressings with epidermal growth factor (EGF) and silver sulfadiazine as a novel wound-healing agent[32] or silk fibroin/alginate-blended sponge for wound healing.[33]

Manufacturing Companies

Zhejiang Huikang Medicinal Articles and Wuxi Wemade Healthcare Products are the two main companies located in China that produce Medical Silk Adhesive Bandage. Furthermore, Jinhua Jingdi Medical Product Company in China produces Silk Medical Tape bandage.

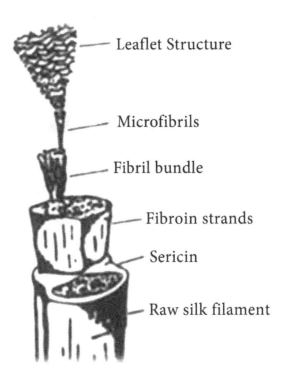

Fig. 6 The structure of raw silk fiber.
Source: This figure was published in Karmakar[31] © 1999, with permission from Elsevier.

HYALURONIC ACID

Introduction

Hyaluronan (known as HA or hyaluronate) is an anionic, nonsulfated glycosaminoglycan distributed widely throughout connective, epithelial, and neural tissues. HA is a big molecule, with molecular weight often reaching millions.[34] HA is an important component of articular cartilage, which coats around each cell (chondrocyte).[35]

As well as a major component of skin that plays an important role in tissue-repairing process.[36]

Structure

HA is a polymer of disaccharides, composed of D-glucuronic acid and D-N-acetylglucosamine, linked via alternating β-1,4 and β-1,3 glycosidic bonds (Fig. 7).[37]

Applications in Wound Care

HA is a naturally occurring polymer within the skin. It has been extensively studied since its discovery in 1934. It has been used extensively in a wide range of medical fields as diverse as orthopedics and cosmetic surgery. However, it is in tissue engineering that it has been primarily advanced for treatment. The breakdown products of this large macromolecule have a range of properties that lend it specifically to this setting as well as to the field of wound healing. It is

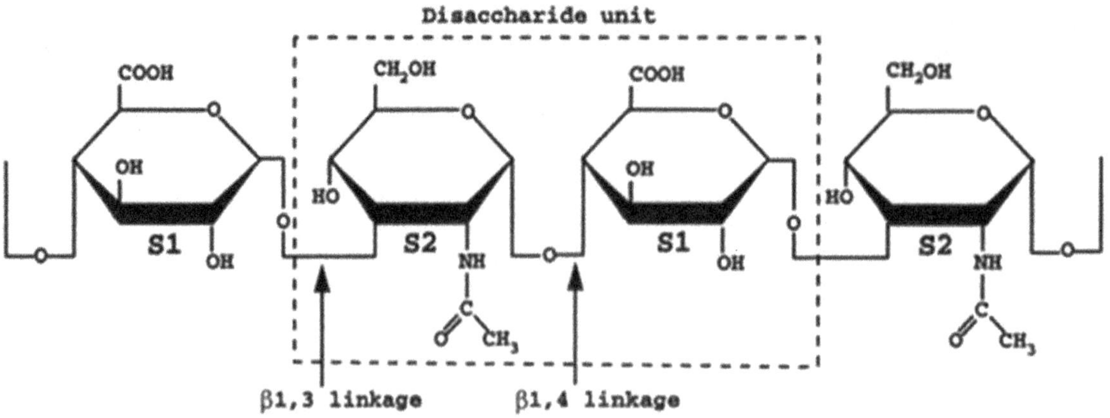

Fig. 7 The polymer is built from alternating units of glucuronic acid (S1) and *N*-acetyl glucosamine (S2). Stereo diagram of ribbon representation of the SpnHL protein with two bound disaccharide units (HA1 and HA2) of HA. The direction of HA1 is such that glucuronic acid residue (UA1) is the non-reducing end, which interacts with the Arg243, Arg300, and Arg355 and the *N*-acetyl glucosamine residue (NAc1) is the reducing end, which interacts with the key catalytic residue Tyr408.
Source: From Ponnuraj[38] © 2000, with permission from Elsevier.

non-antigenic and may be manufactured in a number of forms, ranging from gels to sheets of solid material through to lightly woven meshes. Epidermal engraftment may be the best candidate among available biotechnologies so that has shown great promise in both animal and clinical studies of tissue engineering. Ongoing work centers on the ability of the molecule to enhance angiogenesis and the conversion of chronic wounds into acute wounds.[39]

Manufacturing Companies

Misonix Inc. produces Hyalofill®-F, which is an absorbent, soft, and conformable dressing composed of HYAFF® (HA ester) that provides a moist HA-enriched wound environment.

NovaMatrix™, a business unit of FMC BioPolymer (Philadelphia, PA, USA), produces and supplies well-characterized and documented ultrapure biocompatible and bioabsorbable biopolymers.

KERATIN

Introduction

The term "keratin" originally is referred to the broad category of insoluble proteins that associate as intermediate filaments (IFs) and form the bulk of cytoplasmic epithelia and epidermal appendageal structures. Researchers have classified mammalian keratins into two distinct groups based on their structure, function, and regulation: "hard" and "soft" keratins.[40]

Chains of amino acid groups are the primary structure of keratin proteins that vary in the number and sequence for different keratins. Its sequence influences the properties and functions of the keratin filament. All proteins that form IFs have a tripartite secondary structure.[41]

Structure

The structure of keratin is shown in Fig. 8.

Applications in Wound Care

The keratin dressings are called "gel," "matrix," and "foam." The gel can be used for dry exudate, the matrix for light-to-heavy exudates, and the foam for moderate-to-heavy exudate.[43]

Manufacturing Companies

Keraplast Technologies Company located in the US producing keragelT, keragel, keramatrix, kerasorb and keragelT deressings for wounds caused by Epidermolysis bullosa.

CONCLUSION

Wound healing is a multifactorial, complicated, physiological process that is prone to abnormalities. Various type of natural biopolymers including neutral (cellulose), basic (chitin and chitosan), acidic (alginic acid and HA), and sulfated polysaccharides (heparin, chondroitin, dermatan, and keratan sulfates) have been the focus of interest with respect to biomedical and wound care applications. The present entry presented a relatively comprehensive review on the most available and frequent use of biopolymers in the wound care applications, including cellulose, chitin and chitosan, alginate, starch, collagen, silk, HA, and finally keratin. Different aspects of these natural polymers were included in this entry, including a brief description of each compound, its composition, natural forms, structure, its applications in wound care, and famous companies

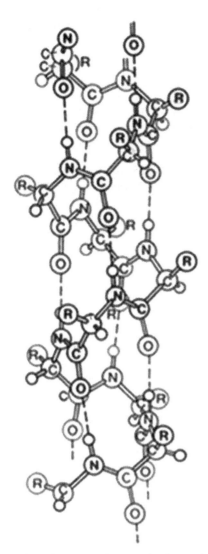

Fig. 8 α-Helical Structure, the organized protein structure, which forms approximately 30% of any - keratin.
Source: From Feughelman,[42] *Mechanical Properties and Structure of Alpha-Keratin Fibres: Wool, Human Hair and Related Fibres* by Max Feughelman (UNSWPress, Sydney, Australia, 1997).

pioneered in manufacturing of that biopolymer. Findings of the present review show the ever-increasing research interest on the development and applications of the natural biopolymers in different medical applications, especially in wound care applications.

REFERENCES

1. Festucci-Buselli, R.A.; Otoni, W.C.; Joshi, C.P. Structure, organization, and functions of cellulose synthase complexes in higher plants. Braz. J. Plant Physiol. **2007**, *19* (1), 1–13.

2. Belgacem, M.N.; Gandini, A., Eds. *Monomers, Polymers and Composites from Renewable Resources*; Elsevier Science: Netherlands, 2011, p. 379.

3. Czaja, W.; Krystynowiczl, A.; Bielecki, S.; Brown, R.M. Microbial cellulose—The natural power to heal wounds. Biomaterials **2006**, *27* (2), 145–151.

4. Wu, Y.; He, J.; Cheng, W.; Gu, H.; Guo, Z.; Gao, S.; Huang, Y. Oxidized regenerated cellulose-based hemostat with microscopically gradient structure. Carbohydr. Polym. **2012**, *88* (3), 1023–1032.

5. Khor, E.; Lim, L.Y. Implantable applications of chitin and chitosan. Biomaterials **2003**, *24* (13), 2339–2349.

6. Ma, G.; Yang, D.; Zhou, Y.; Xiao, M.; Kennedy, J.F.; Nie, J. Preparation and characterization of water-soluble *N*-alkylated chitosan. Carbohydr. Polym. **2008**, *74* (1), 121–126.

7. Muzzarelli, R.A.; Morganti, P.; Morganti, G.; Palombo, P.; Palombo, M.; Biagini, G.; Mattioli, Belmonte, M.; Giantomassi, F.; Orlandi, F.; Muzzarelli, C. Chitin nanofibrils/chitosan glycolate composites as wound medicaments. Carbohydr. Polym. **2007**, *70* (3), 274–284.

8. Jayakumar, R.; Prabaharan, M.; Reis, R.; Mano, J. Graft copolymerized chitosan—Present status and applications. Carbohydr. Polym. **2005**, *62* (2), 142–158.

9. Banerjee, T.; Mitra, S.; Kumar Singh, A.; Kumar Sharma, R.; Maitra, A. Preparation, characterization and biodistribution of ultrafine chitosan nanoparticles. Int. J. Pharm. **2002**, *243* (1), 93–105.

10. Jollès, P.; Muzzarelli, R.A.A. *Chitin and Chitinases*; Walter de Gruyter GmbH; Genthiner Straße 13 D-10785: Berlin, Germany, 1999; p. 316–317.

11. Ravi Kumar, M.N. A review of chitin and chitosan applications. React. Funct. Polym. **2000**, *46* (1), 1–27.

12. Jayakumar, R.; Prabaharan, M.; Sudheesh Kumar, P.; Nair, S.; Tamura, H. Biomaterials based on chitin and chitosan in wound dressing applications. Biotechnol. Adv. **2011**, *29* (3), 322–337.

13. Muzzarelli, R.A. Chitins and chitosans for the repair of wounded skin, nerve, cartilage and bone. Carbohydr. Polym. **2009**, *76* (2), 167–182.

14. Rehm, B. *Alginates: Biology and Applications*; Springer-Verlag: USA, 2009, p. 2–5.

15. Baranoski, S.; Elizabeth, A.; Ayello, P.D. *Wound Care Essentials: Practice Principles*; Lippincott Williams and Wilkins: Philadelphia, 2008, p. 148.

16. Shai, A.; Maibach, H.I. Wound Healing and Ulcers of the Skin: Diagnosis and Therapy—The Practical Approach; Springer: USA, 2005, p. 111–112.

17. William, L.; Wilkins. *Wound Care Made Incredibly Easy!*, 2nd Ed.; Lippincott Williams and Wilkins: Philadelphia, 2007, p. 54.

18. Thomas, S.; Visakh, P.M.; Mathew, A.P. *Advances in Natural Polymers: Composites and Nanocomposites*; Springer: USA, 2012, p. 205.

19. BeMiller, J.N.; Whistler, R.L. *Starch: Chemistry and Technology*; Elsevier Science: Netherlands, 2009, p. 1–9.

20. Schwall, G.P.; Safford, R.; Westcott, R.J.; Jeffcoat, R.; Tayal, A.; Shi, Y.C.; Gidley, M.J.; Jobling, S.A. Production of very-high-amylose potato starch by inhibition of SBE A and B. Nat. Biotechnol. **2000**, *18* (5), 551–554.

21. Tester, R.F.; Karkalas, J.; Qi, X. Starch—Composition, fine structure and architecture. J. Cereal Sci. **2004**, *39* (2), 151–165.

22. Athawale, V.D.; Lele, V. Graft copolymerization onto starch. II. Grafting of acrylic acid and preparation of it's hydrogels. Carbohydr. Polym. **1998**, *35* (1), 21–27.

Vascular—Zwitterionic

23. Kiatkamjornwong, S.; Chomsaksakul, W.; Sonsuk, M. Radiation modification of water absorption of cassava starch by acrylic acid/acrylamide. Radiat. Phys. Chem. **2000**, *59* (4), 413–427.

24. Pal, K.; Banthia, A.; Majumdar, D. Starch based hydrogel with potential biomedical application as artificial skin. ABNF J. **2009**, *9* (1), 23–29.

25. Fratzl, P. *Collagen: Structure and Mechanics*; Springer Science+Business Media, LLC: USA, 2008, p. 1–5.

26. Brinckmann, J.; Notbohm, H.; Müller, P.K. *Collagen: Primer in Structure, Processing, and Assembly*; Springer: USA, 2005, p. 9.

27. Baranoski, S.; Elizabeth, A.; Ayello, P.D. *Wound Care Essentials: Practice Principles*; Lippincott Williams and Wilkins: Philadelphia, 2008, p. 149.

28. Fakirov, S.; Bhattacharya, D. *Handbook of Engineering Biopolymers: Homopolymers, Blends and Composites*; Carl Hanser GmbH: Munich, Berlin, Germany, 2007, p. 485.

29. Humenik, M.; Smith, A.M.; Scheibel, T. Recombinant spider silks—Biopolymers with potential for future applications. Polymers **2011**, *3* (1), 640–661.

30. Gole, R.S.; Kumar, P. Spider's Silk: Investigation of spinning process, web material and its properties. Available from http://www.iitk.ac.in/bsbe/web%20on%20asmi/spider.pdf.

31. Karmakar, S.R. *Chemical Technology in the Pre-Treatment Processes of Textiles*; Elsevier Science: Netherlands, 1999, p. 15.

32. Padol, A.R.; Jayakumar, K.; Shridhar, N.; Swamy, H.N.; Swamy, M.N.; Mohan, K. Safety evaluation of silk protein film (A novel wound healing agent) in terms of acute dermal toxicity, acute dermal irritation and skin sensitization. Toxicol. Int. **2011**, *18* (1), 17–21.

33. Roh, D.H.; Kang, S.Y.; Kim, J.Y.; Kwon, Y.B.; Young Kweon, H.; Lee, K.G.; Park, Y.H.; Baek, R.M.; Heo, C.Y.; Choe, J. Wound healing effect of silk fibroin/alginate-blended sponge in full thickness skin defect of rat. J. Mater. Sci. Mater. Med. **2006**, *17* (6), 547–552.

34. Fraser, J.; Laurent, T.; Laurent, U. Hyaluronan: Its nature, distribution, functions and turnover. J. Intern. Med. **2003**, *242* (1), 27–33.

35. Holmes, M.; Bayliss, M.; Muir, H. Hyaluronic acid in human articular cartilage. Age-related changes in content and size. Biochem. J. **1988**, *250* (2), 435–441.

36. Averbeck, M.; Gebhardt, C.A.; Voigt, S.; Beilharz, S.; Anderegg, U.; Termeer, C.C.; Sleeman, J.P.; Simon, J.C. Differential regulation of hyaluronan metabolism in the epidermal and dermal compartments of human skin by UVB irradiation. J. Invest. Dermatol. **2006**, *127* (3), 687–697.

37. Saari, H.; Konttinen, Y.T.; Friman, C.; Sorsa, T. Differential effects of reactive oxygen species on native synovial fluid and purified human umbilical cord hyaluronate. Inflammation **1993**, *17* (4), 403–415.

38. Ponnuraj, K.; Jedrzejas, M.J. Mechanism of hyaluronan binding and degradation: Structure of *Streptococcus pneumoniae* hyaluronate lyase in complex with hyaluronic acid disaccharide at 1.7 Å resolution. J. Mol. Biol. **2000**, *299* (4), 885–895.

39. Price, R.D.; Myers, S.; Leigh, I.M.; Navsaria, H.A. The role of hyaluronic acid in wound healing: Assessment of clinical evidence. Am. J. Clin. Dermatol. **2005**, *6* (6), 393–402.

40. Rouse, J.G.; Van Dyke, M.E. A review of keratin-based biomaterials for biomedical applications. Materials **2010**, *3* (2), 999–1014.

41. Bragulla, H.H.; Homberger, D.G. Structure and functions of keratin proteins in simple, stratified, keratinized and cornified epithelia. J. Anat. **2009**, *214* (4), 516–559.

42. Feughelman, M. *Mechanical Properties and Structure of Alpha-Keratin Fibers: Wool, Human Hair and Related Fibres*; University of New South Wales: South Wales, 1997, p. 13.

43. Kelly, R.J.; Sigurjonsson, G.F.; Marsh, C.; Smith, R.A.; Ali, M.A. Porous Keratin Constructs, Wound Healing Assemblies and Methods Using the Same. WO2008148109, December 4, 2008.

BIBLIOGRAPHY

1. Albu, M.; Ferdes, M.; Kaya, D.; Ghica, M.; Titorencu, I.; Popa, L.; Albu, L. Collagen wound dressings with anti-inflammatory activity. Mol. Cryst. Liq. Cryst. **2012**, *555* (1), 271–279.

2. Alvarez, O.; Phillips, T.; Menzoian, J.; Patel, M.; Andriessen, A. An RCT to compare a bio-cellulose wound dressing with a non-adherent dressing in VLUs. J. Wound Care **2012**, *21* (9), 448–453.

3. Arockianathan, P.M.; Sekar, S.; Sankar, S.; Kumaran, B.; Sastry, T. Evaluation of biocomposite films containing alginate and sago starch impregnated with silver nano particles. Carbohydr. Polym. **2012**, *90* (1), 717–724.

4. Atluri, P.; Giorgos, C.; Karakousis, M.D.; Paige, M.; Porrett, M.D. *The Surgical Review: An Integrated Basic And Clinical Science Study Guide*; Lippincott Williams and Wilkins: Philadelphia, 2005.

5. Bailey, B.J.; Johnson, J.T.; Newlands. S.D. *Head and Neck Surgery: Otolaryngology*; Lippincott Williams and Wilkins: Philadelphia, 2006.

6. Baranoski, S.; Elizabeth, A.; Ayello, P.D. *Wound Care Essentials: Practice Principles*; Lippincott Williams and Wilkins: Philadelphia, 2008.

7. Barbucci, R. *Integrated Biomaterials Science*; Springer: USA, 2002.

8. Belgacem, M.N.; Gandini, A. *Monomers, Polymers and Composites from Renewable Resources*; Elsevier Science: Netherlands, 2011.

9. Bhatia, S.K. *Engineering Biomaterials for Regenerative Medicine*; Springer: New York, USA, 2012.

10. Bielecki, S.; Kalinowska, H.; Krystynowicz, A.; Kubiak, K.; Kołodziejczyk, M.; De Groeve, M. Wound dressings and cosmetic materials from bacterial nanocellulose. In *Bacterial Nanocellulose: A Sophisticated Multifunctional Material*, **2012**, *9*, 157–174.

11. Binder, T.B. *Modulation of Adult Wound Healing Using Hyaluronic Acid*; Boston University: Boston, MA, 1993.

12. Birgand, G.; Radu, C.; Alkhoder, S.; Al Attar, N.; Raffoul, R.; Dilly, M.P., Nataf, P.; Lucet, J.C. Does a gentamicin-impregnated collagen sponge reduce sternal wound infections in high-risk cardiac surgery patients? Interact. Cardiovasc. Thorac. Surg. **2013**, *16* (2), 134–141.

13. Brown, J. The role of hyaluronic acid in wound healing's proliferative phase. J. Wound Care **2004**, *13* (2), 48–51.

14. Brown, P. *Quick Reference to Wound Care: Palliative, Home, and Clinical Practices*; Jones and Bartlett Learning: Burlington, MA, 2012.

15. Carella, S.; Maruccia, M.; Fino, P.; Onesti, M.G. An atypical case of Henoch-Shönlein Purpura in a young patient: Treatment of the skin lesions with hyaluronic acid-based dressings. *In Vivo* **2013**, *27* (1), 147–151.

16. Cirilo, G.; Iemma, F. *Antioxidant Polymers: Synthesis, Properties, and Applications*; Wiley: New York, USA, 2012.

17. Craver, C.; Carraher, C. *Applied Polymer Science: 21st Century*; Elsevier Science: Netherlands, 2000.

18. Dealey, C. *The Care of Wounds: A Guide for Nurses*; Wiley: New York, USA, 2012.

19. Dereure, O.; Czubek, M.; Combemale, P. Efficacy and safety of hyaluronic acid in treatment of leg ulcers: A double-blind RCT. J. Wound Care **2012**, *21* (3), 131–139.

20. Dolan, B.; Holt, L. *Accident and Emergency: Theory Into Practice*; Baillière Tindall: Netherlands, 2008.

21. Domb, A.J.; Kumar, N. *Biodegradable Polymers in Clinical Use and Clinical Development*; Wiley: New York, USA, 2011.

22. Donaghue, V.M.; Chrzan, J.S.; Rosenblum, B.I.; Giurini, J.M.; Habershaw, G.M.; Veves, A. Evaluation of a collagen-alginate wound dressing in the management of diabetic foot ulcers. Adv. Wound Care **1998**, *11* (3), 114–119.

23. Dumville, J.; O'Meara, S.; Deshpande, S.; Speak, K. Alginate dressings for healing foot ulcers in people with diabetes mellitus. Cochrane Database Syst. Rev. **2012**, *6*, CD009110.

24. Ehrlich, H. *Biological Materials of Marine Origin*; Springer: USA, 2010, p. 10–25.

25. Fan, L.; Wang, H.; Zhang, K.; Cai, Z.; He, C.; Sheng, X.; Xiumie, M. Vitamin C-reinforcing silk fibroin nanofibrous matrices for skin care application. RSC Adv. **2012**, *2* (10), 4110–4119.

26. Flanagan, M. *Wound Healing and Skin Integrity: Principles and Practice*; Wiley: New York, USA, 2013.

27. Foster, A.V.M. *Podiatric Assessment And Management of the Diabetic Foot*; Churchill Livingstone Elsevier: Netherlands, 2006.

28. Foster, L.; Moore, P. Acute surgical wound care 3: Fitting the dressing to the wound. Br. J. Nurs. **1999**, *8* (4), 200–210.

29. Foster, L.; Moore, P. The application of a cellulose-based fibre dressing in surgical wounds. J. Wound Care **1997**, *6* (10), 469–473.

30. Gama, M.; Gatenholm, P.; Klemm, D. Bacterial Cellulose: A Sophisticated Multifunctional Material; Taylor & Francis Group: UK, 2012.

31. Garg, H.G.; Cowman, M.K.; Hales, C.A. *Carbohydrate Chemistry, Biology and Medical Applications*; Elsevier Science: Netherlands, 2011.

32. Gil, E.S.; Panilaitis, B.; Bellas, E.; Kaplan, D.L. Functionalized Silk Biomaterials for Wound Healing. Adv. Healthc. Mater. **2013**, *2* (1), 206–217.

33. Gottrup, F.; Cullen, B.M.; Karlsmark, T.; Bischoff-Mikkelsen, M.; Nisbet, L.; Gibson, M.C. Randomized controlled trial on collagen/oxidized regenerated cellulose/silver treatment. Wound Repair Regen. **2013**, *21* (2), 216–225.

34. Habibi, Y.; Lucia, L.A. *Polysaccharide Building Blocks: A Sustainable Approach to the Development of Renewable Biomaterials*; Wiley: New York, USA, 2012.

35. Hafner, J. *Management of Leg Ulcers*; Hafner, J.; Ramelet, A.-A.; Schmeller, W.; Brunner, U. Eds., S Karger Ag, 1999; Vol. 27, p. X + 294.

36. Hampton, S.; Collins, F. *Tissue Viability*; Wiley: New York, USA, 2006.

37. Hess, C.T. *Clinical Guide to Skin and Wound Care*; Lippincott Williams & Wilkins: Philadelphia, 2012, p. 607.

38. Hess, C.T. *Skin and Wound Care*; Wolters Kluwer Health/Lippincott Williams and Wilkins: Philadelphia, 2008.

39. Hess, C.T. *Wound Care*; Lippincott Williams and Wilkins: Philadelphia, 2005.

40. Holmes, C.; Wrobel, J.S.; MacEachern, M.P.; Boles, B.R. Collagen-based wound dressings for the treatment of diabetes-related foot ulcers: A systematic review. Diabetes Metab. Syndr. Obes. **2013**, *6*, 17–29.

41. Jang, S.I.; Mok, J.Y.; Jeon, I.H.; Park, K.H.; Nguyen, T.T.T.; Park, J.S.; Hwang, H.M.; Song, M.S.; Lee, D.; Chai, K.Y. Effect of electrospun non-woven mats of dibutyryl chitin/poly(lactic acid) blends on wound healing in hairless mice. Molecules **2012**, *17* (3), 2992–3007.

42. Jayakumar, R.; Menon, D.; Manzoor, K.; Nair, S.; Tamura, H. Biomedical applications of chitin and chitosan based nanomaterials—A short review. Carbohydr. Polymers **2010**, *82* (2), 227–232.

43. Kalin, M.; Kuru, S.; Kismet, K.; Barlas, A.M.; Akgun, Y.A.; Astarci, H.M, Ustun, H.; Ertas, E. The effectiveness of porcine dermal collagen (Permacol®) on wound healing in the rat model. Indian J. Surg. **2013**, 1–5.

44. Keogh, S.J.; Nelson, E.A.; Webster, J.; Jolly, J.; Ullman, A.J.; Chaboyer, W.P. Hydrocolloid dressings for treating pressure ulcers. Cochrane Libr. **2013**, (2), DOI: 10.1002/14651858.CD010364

45. Kim, S.K. Marine Cosmeceuticals: Trends and Prospects; CRC Press: USA, 2011.

46. Kneedler, J.A.; Dodge, G.H. *Perioperative Patient Care: The Nursing Perspective*; Jones and Bartlett Pub, 1994.

47. Kondo, S.; Kuroyanagi, Y. Development of a wound dressing composed of hyaluronic acid and collagen sponge with epidermal growth factor. J. Biomater. Science Polym. Ed. **2012**, *23* (5), 629–643.

48. Koyama, T.I.; Yamada, S. Chitosan application in dentistry. *Mar Drugs*. **2013**, *11* (4), 1300–1303.

49. Krieg, T.; Bickers, D.R.; Miyachi, Y. *Therapy of Skin Diseases: A Worldwide Perspective on Therapeutic Approaches and Their Molecular Basis*; Springer: USA, 2010.

50. Longaker, M.T.; Chiu, E.S.; Adzick, N.S.; Stern, M.; Harrison, M.R.; Stern, R. Studies in fetal wound healing. V. A prolonged presence of hyaluronic acid characterizes fetal wound fluid. Ann. Surg. **1991**, *213* (4), 292–296.

51. LPN Expert Guides. *Wound care*; Wolters Kluwer Health/Lippincott Williams and Wilkins: Philadelphia, 2007.

52. Luan, J.; Wu, J.; Zheng, Y.; Song, W.; Wang, G.; Guo, J.; Ding, X. Impregnation of silver sulfadiazine into bacterial cellulose for antimicrobial and biocompatible wound dressing. Biomed. Mater. **2012**, *7* (6), 065006.

53. Maklebust, J.A.; Sieggreen, M. *Pressure Ulcers: Guidelines for Prevention and Management*; Lippincott Williams & Wilkins, 2001, p. 322.

Vascular—Zwitterionic

54. Maneerung, T.; Tokura, S.; Rujiravanit, R. Impregnation of silver nanoparticles into bacterial cellulose for antimicrobial wound dressing. Carbohydr. Polym. **2008**, *72* (1), 43–51.

55. Marks, R.; Plewig, G. *The Environmental Threat To The Skin*; CRC Press: USA, 1991, p. 432.

56. Mary, E.; Klingensmith, M.D. *Surgery WUDo. The Washington Manual of Surgery*; Wolters Kluwer Health/ Lippincott Williams and Wilkins: Philadelphia, 2008.

57. Miloro, M.; Ghali, G.E.; Peterson, L.J.; Larsen, P.E.; Waite, P.D. *Peterson's Principles of Oral and Maxillofacial Surgery*; Decker, 2004.

58. Moran, S.L.; William, P.; Cooney, I. *Soft Tissue Surgery*; Lippincott WilliamsandWilki: Philadelphia, 2008.

59. Mosti, G. Wound care in venous ulcers. Phlebology **2013**, *28* (Suppl. 1), 79–85.

60. Mulder, M. *Basic Principles of Wound Care*; Pearson Education South Africa, 2002, p. 443.

61. Natarajan, S.; Williamson, D.; Stiltz, A.J.; Harding, K. Advances in wound care and healing technology. Am. J. Clin. Dermatol. **2000**, *1* (5), 269–275.

62. Navard, P. Polysaccharide Research: The European Polysaccharide Network of Excellence Views; Springer: Vienna, USA, 2012.

63. Nyanhongo, G.S.; Steiner, W.; Gübitz, G.; Andersen, S. *Biofunctionalization of Polymers and Their Applications*; Springer: USA, 2011.

64. Onesti, M.; Fioramonti, P.; Carella, S.; Fino, P.; Sorvillo, V.; Scuderi, N. A new association between hyaluronic acid and collagenase in wound repair: An open study. Eur. Rev. Med. Pharmacol. Sci. **2013**, *17* (2), 210–216.

65. Ovington, L.G. Wound care products: How to choose. Adv. Skin Wound care **2001**, *14* (5), 259–266.

66. Pachence, J.M. Collagen-based devices for soft tissue repair. J. Biomed. Mater. Res. **1998**, *33* (1), 35–40.

67. Park, S.N.; Kim, J.K.; Suh, H. Evaluation of antibiotic-loaded collagen-hyaluronic acid matrix as a skin substitute. Biomaterials **2004**, *25* (17), 3689–3698.

68. Parsons, D.; Bowler, P.G.; Myles, V.; Jones, S. Silver antimicrobial dressings in wound management: A comparison of antibacterial, physical, and chemical characteristics. Wounds **2005**, *17* (8), 222–232.

69. Pechter, P.M.; Gil, J.; Valdes, J.; Tomic-Canic, M.; Pastar, I.; Stojadinovic, O.; Kirsner, R.S.; Davis, S.C. Keratin dressings speed epithelialization of deep partial-thickness wounds. Wound Repair Regen. **2012**, *20* (2), 236–242.

70. Pereira, R.; Mendes, A.; Bártolo, P. Alginate/aloe vera hydrogel films for biomedical applications. Procedia CIRP **2013**, *5*, 210–215, http://dx.doi.org/10.1016/j.procir.2013.01.042

71. Peter, M.G. Applications and environmental aspects of chitin and chitosan. J. Macromol. Sci. A Pure Appl. Chem. **1995**, *32* (4), 629–640.

72. Pillai, C.; Paul, W.; Sharma, C.P. Chitin and chitosan polymers: Chemistry, solubility and fiber formation. Prog. Polym. Sci. **2009**, *34* (7), 641–678.

73. Pope, E.; Lara-Corrales, I.; Mellerio, J.; Martinez, A.; Schultz, G.; Burrell, R.; Goodman, L.; Coutts, P.; Wagner, J.; Allen, U.; Sibbald, G. A consensus approach to wound care in epidermolysis bullosa. J. Am. Acad. Dermatol. **2012**, *67* (5), 904–917.

74. Portal, O.; Clark, W.A.; Levinson, D.J. Microbial cellulose wound dressing in the treatment of non-healing lower extremity ulcers. Wounds **2009**, *21* (1), 1–3.

75. Price, R.D.; Myers, S.; Leigh, I.M.; Navsaria, H.A. The role of hyaluronic acid in wound healing: Assessment of clinical evidence. Am. J. Clin. Dermatol. **2005**, *6* (6), 393–402.

76. MacDonald, M.G.; Ramasethu, J.; Rais-Bahrami, K. Management of extravasation injuries. In *Atlas of Procedures in Neonatology*, 5th Eds.; Lippincott Williams and Wilkins: 2012; p. 480

77. Rao, NM. Medical Biochemistry; New Age International: New Delhi, 2007.

78. Rehm, B. *Microbial Production of Biopolymers and Polymer Precursors: Applications and Perspectives*; Caister Academic Press: England, 2009.

79. Rodrigues, A.D. *Scarless Wound Healing*; Taylor & Francis: UK, 2000.

80. Ruszczak, Z. Effect of collagen matrices on dermal wound healing. Adv. Drug Deliv. Rev. **2003**, *55* (12), 1595–1611.

81. Santos, T.C.; Höring, B.; Reise, K.; Marques, A.P.; Silva, S.S.; Oliveira, J.M.; Mano, J.F.; Castro, A.G.; Reis, R.L.; van Griensven, M. *In vivo* performance of chitosan/soy-based membranes as wound-dressing devices for acute skin wounds. Tissue Eng. A **2013**, *19* (7–8), 860–869.

82. Sayag, J.; Meaume, S.; Bohbot, S. Healing properties of calcium alginate dressings. J. Wound Care **1996**, *5* (8), 357–362.

83. Scott, G. *Degradable Polymers*; Springer: USA, 2002.

84. Shai, A.; Maibach, H.I. Wound Healing and Ulcers of the Skin: Diagnosis and Therapy—The Practical Approach; Springer: USA, 2005.

85. Sheikh, E.S.; Sheikh, E.S.; Fetterolf, D.E. Use of dehydrated human amniotic membrane allografts to promote healing in patients with refractory non healing wounds. Int. Wound J. **2013**, doi: 10.1111/iwj.12035

86. Snow, S.N.; George, R.; Mikhail, M.D. *Mohs Micrographic Surgery*; University of Wisconsin Press: USA, 2004.

87. Sussman, C.; Bates-Jensen, B.M. *Wound Care: A Collaborative Practice Manual*; Wolters Kluwer Health / Lippincott Williams and Wilkins: Philadelphia, 2007.

88. Sweeney, I.R.; Miraftab, M.; Collyer, G. A critical review of modern and emerging absorbent dressings used to treat exuding wounds. Int. Wound J. **2012**, *9* (6), 601–612.

89. Symposium, C.F. *The Biology of Hyaluronan*; Wiley: New York, USA, 2008.

90. Takayama, Y. *Lactoferrin and Its Role in Wound Healing*; Springer: USA, 2012.

91. Teh, B.M.; Shen, Y.; Friedland, P.L.; Atlas, M.D.; Marano, R.J. A review on the use of hyaluronic acid in tympanic membrane wound healing. Expert Opin. Biol. Ther. **2012**, *12* (1), 23–36.

92. Thomas, S. Alginate dressings in surgery and wound management: Part 2. J. Wound Care **2000**, *9* (3), 115–119.

93. Thomas, S. Alginate dressings in surgery and wound management—Part 1. J. Wound Care **2000**, *9* (2), 56–60.

94. Thomas, S. *Surgical Dressings and Wound Management*; Upfront Publishing Limited: South Wales, 2010.

95. Torrance, C. *Pressure Sores: Aetiology, Treatment and Prevention*; Croom Helm: Scottish, 1983.

96. Trott, A. *Wounds and Lacerations: Emergency Care and Closure*; Mosby: Netherlands, 1997.

97. Trott, A.T. *Wounds and Lacerations: Emergency Care and Closure (Expert Consult—Online and Print)*; Elsevier Health Sciences: Netherlands, 2012.

98. Twersky, J.; Montgomery, T.; Sloane, R.; Weiner, M.; Doyle, S.; Mathur, K.; Francis, M.; Schmader, K. A randomized, controlled study to assess the effect of silk-like textiles and high-absorbency adult incontinence briefs on pressure ulcer prevention. Ostomy Wound Manage. **2012**, *58* (12), 18–24.

99. Ul-Islam, M.; Khan, T.; Khattak, W.A.; Park, J.K. Bacterial cellulose-MMTs nanoreinforced composite films: Novel wound dressing material with antibacterial properties. Cellulose **2013**, *20* (2), 589–596.

100. Vasconcelos, A.; Cavaco-Paulo, A. Wound dressings for a proteolytic-rich environment. Appl. Microbiol. Biotechnol. **2011**, *90* (2), 445–460.

101. J.G. Webster, Prevention of Pressure Sores: Engineering and Clinical Aspects, CRC Press, **2011**.

102. Voigt, J.; Driver, V.R. Hyaluronic acid derivatives and their healing effect on burns, epithelial surgical wounds, and chronic wounds: A systematic review and meta-analysis of randomized controlled trials. Wound Repair Regen. **2012**, *20* (3), 317–331.

103. Weindl, G.; Schaller, M.; Schäfer-Korting, M.; Korting, H. Hyaluronic acid in the treatment and prevention of skin diseases: Molecular biological, pharmaceutical and clinical aspects. Skin pharmacol. Physiol. **2004**, *17* (5), 207–213.

104. Williams, L. Wilkins. *Best of Incredibly Easy!*; Lippincott Williams and Wilkins: Philadelphia, 2006.

105. Williams, L. Wilkins. *Lippincott's Nursing Procedures*; Wolters Kluwer Health/Lippincott Williams and Wilkins: Philadelphia, 2009.

106. Williams, L. Wilkins. *Skillmasters: Wound care*; Lippincott Williams and Wilkins: Philadelphia, 2006.

107. Williams, L. Wilkins. *Wound Care Made Incredibly Easy!*, 2nd Ed.; Lippincott Williams and Wilkins: Philadelphia, 2007.

108. Woo, K.Y.; Coutts, P.M.; Sibbald, R.G. A randomized controlled trial to evaluate an antimicrobial dressing with silver alginate powder for the management of chronic wounds exhibiting signs of critical colonization. Adv. Skin Wound care **2012**, *25* (11), 503–508.

109. Woodings, C. *Regenerated Cellulose Fibres*; Woodhead Pub: UK, 2001.

110. Yeung, S.C.J.; Carmen, P.; Escalante, M.D.; Robert, F.; Gagel, M.D. *Medical Care of Cancer Patients*; People's Medical Publishing House, 2009.

111. Yuvarani, I.; Kumar, S.S.; Venkatesan, J.; Kim, S.K.; Sudha, P. Preparation and characterization of curcumin coated chitosan-alginate blend for wound dressing application. J. Biomater. Tissue Eng. **2012**, *2* (1), 54–60.

112. Zhao, L.; Zhang, W.X.; Chen, D.L.; Qiu, H.; Li, M. Function model construction–on the question of defining optimal Ag-concentration in silver-containing biological wound dressing. Compos. Interfaces **2013**, *20* (2), 1–15.

Vascular—Zwitterionic

Wound Care: Skin Tissue Regeneration

Soonmo Choi
Department of Nano, Medical and Polymer Materials, Yeungnam University, Gyeongsan, South Korea

Deepti Singh
Sung Soo Han
Department of Nano, Medical, and Polymer Materials, Polymer Gel Cluster Research Center, Yeungnam University, Daedong, South Korea

Abstract

The ultimate goal of tissue engineered skin substitutes used for wound healing is to enhance the healing process. Skin is the largest organ that serves as an outer barrier at the interface between the body and its surrounding environment. Accordingly, it provides a protective barrier against microbial invasion and protects the body against mechanical, chemical, and thermal injuries. The most common reason for loss of skin is burns in which a considerable area of skin can be injured without the possibility of tissue regeneration. Great techniques have been developed to create substitutes that imitate human skin. Engineered cell-free as well as cell-containing skin substitutes provide a possible off-the-shelve solution to the problem of donor graft shortage. Tissue engineering (TE) is a concept whereby cells are taken from a patient, the number of cells is expanded *in vitro*, and then seeded into a scaffold. The seeded cells proliferate in scaffold and over a time from few days to months, new tissue is formed. Scaffolds are three-dimensional matrixes and act as a template for the regeneration of tissue. The ideal scaffolds must possess proper microstructures that enable the adherence, proliferation, and differentiation of cells. Moreover, it should have mechanical strength and biodegradability. The tissue-engineered skin substitutes serve as protection from fluid loss and infection. In this entry, we deal with an overview of critical issues related to skin, wound healing, advancement in TE, fabrication techniques, and methods employed in aiding in skin regeneration.

INTRODUCTION

Advanced therapies combating chronic and acute skin wounds are likely to be conveyed about using our limited knowledge of regenerative medicine coupled appropriately with tissue-engineered skin substitutes. Tissue engineering (TE) aims to regenerate new biological material for restoring or replacing diseased or damaged tissues and organs. For achieving this, not only is a source of cells necessary, but a vital step is the fabrication of artificial extracellular matrix (ECM) that supports cell growth and proliferation. Engineering skin substitutes is pivotal as human skin measures around 1/10th of the body mass and any damage effecting this has drastic consequences. Trauma or burn can harm the skin, and since this is a major organ, attempting for restore the skin by potential advance therapy has been a forefront for biomedical researchers. The skin is a dynamic structure and every layer is composed of a specific ECM component along with cells and interplay between these two results in the high-powered structure.

To date, there is no artificial system that can totally mimic the uninjured skin either structurally, functionally, anatomically, physiologically, or aesthetically. Skin surrogate should have some indispensable characteristics including easy handling, transportation, and appropriate configuration to heal the injured site, along with acting as a vital barrier function with suitable water flux, readily adherent, and have appropriate mechanical and physical properties while undergoing controlled degradation, which needs to be in tune with skin regeneration, beside being non-antigenic, non-toxic, and evoking minimal inflammatory reactivity. Moreover, it should also integrate into the host tissue without pain with minimal scarring along with assisting in angiogenesis and should be cost effective. In order to mimic a system as complicated yet as simple as skin, the basic structure and functions have to be explored and understanding of this makes the process of engineering a substitute little less complex.

Function of Skin

Skin is a fundamental organ and any massive loss of skin can endanger individual life. Injury of skin can result from many reasons such as fire, a mechanical accident, cancer, sunburn, and some infection and loss of skin can also occur in a chronic manner or due to skin ulcers. In other words, skin is directly exposed to potentially harmful microbial, thermal, mechanical, and chemical influences. It serves as

Concise Encyclopedia of Biomedical Polymers and Polymeric Biomaterials DOI: 10.1081/E-EBPPC-120050587

Vascular—Zwitterionic

a protective wall-like barrier between the human body and its surrounding environment. Furthermore, it provides a primary defense against infection. It allows passing fluid, electrolytes, and various substances while providing protection against microorganisms, toxic agents, and external stress (Table 1). When this outer barrier collapses, the microorganism can easily invade the body and can cause critical injury.[1–4]

Structure of Skin

Skin is an integumentary system of the human body, which includes the skin, nail, sweat glands, and hair. It constitutes two layers, epidermis also known as the outermost layer and dermis the innermost layer (Fig. 1). The dermal-epidermal junction, commonly referred to as the basement membrane zone, separates the two layers. Under the dermis lies a layer of connective tissue called subcutaneous tissue or hypodermis.[5–7]

Epidermis

The epidermis is a thin outermost avascular layer constantly nourished by diffusion from the underlying dermis layer, constituted with 95% of keratinocytes along with langerhans, melanocytes, Merkel cells, and inflammatory cells. This layer can regenerate itself in 6 weeks from most normal injuries and day-to-day wear and tear. In addition, the epidermis is divided into four sub-layers.[7,8]

Table 1 Details of different layers of skin and its specialized functions

Skin layer	Function
Epidermis	Protective barrier
	Synthesizes vitamin D and cytokines
	Contact with dermis
	Provides pigmentation (contains melanocytes)
	Recognizes allergen (contains langerhans cells)
	Differentiates into hair, nails, sweat glands, sebaceous glands
Dermis	Supports structure
	Provides mechanical strength
	Supplies nutrition
	Resists shearing forces
	Supplies inflammatory response
Hypodermis	Attaches to underlying structure
	Provides thermal insulation
	Stores energy
	Controls body shape
	Serves as mechanical "shock absorber"

Stratum Corneum. The stratum corneum (or horned layer) is the outermost layer of the epidermis, which is made up of layers of hexagonal-shaped, non-viable cells known as corneocytes (final stage of keratinocytes differentiation) that lack nuclei and organelles. The corneocyte is surrounded by a protein envelope and is full of water-retaining keratin proteins. In addition, the layers of lipid bilayers are stacked in the extracellular space of these corneocytes. This structure serves the water-retaining barrier of the skin and the corneocyte layer is able to absorb water thrice higher than its own weight. Moreover, this layer acts as a barrier to protect tissue from infection, dehydration, chemicals, and mechanical stress.[8–11]

Stratum Granulosum. The stratum granulosum (or granular layer) is a thin layer in the epidermis under the stratum corneum. Keratinocytes which loses its nuclei and appears granulated at this stage released into the extracellular space through process called exocytosis to form lipid barriers which lie parallel to the surface of the cells.[7–10]

Stratum Spinosum. The stratum spinosum is a layer of the epidermis that is found between the stratum granulosum and the stratum basale. It is also referred to as the spinous layer. When basal cells reproduce and mature, they start to move toward the outer layer of skin, forming the stratum spinosum. Desmosomes, as intercellular bridges, which look like "prickles" at a microscopic level, connect these cells. This layer is consists of polyhedral keratinocytes, known as cytokeratin. Langerhans cells, immunologically active, are located in the middle of this layer and are responsible for antigen response of the skin.[9–12]

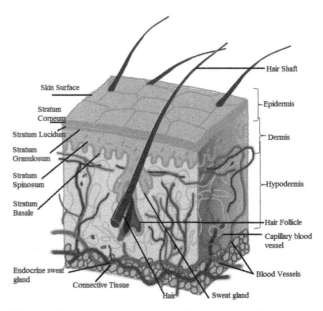

Fig. 1 Diagrammatic sketch of skin showing epidermis and dermis.

Stratum Basale/Germinativum. This is the deepest layer of the epidermis, which is the outer cover of skin. It is consists of basal keratinocyte cells, which can be considered as the stem cells (SC) of the epidermis. When the keratinocyte cells divide and differentiate, they migrate from stratum basale to the surface of skin. Melanocytes are present in this layer and are connected to several keratinocytes in this and other strata layers through dendrites. Merkel cells are also found in large numbers in the strata in touch-sensitive areas such as lips and fingertips.[8–12]

Basement Membrane. The basement membrane is a thin sheet that underlies the epithelium that lines the cavities and surfaces between the epidermis and dermis layers. This is a complex structure composed of two layers, the basal lamina and the reticular lamina (or lamina reticularis). Its structure is highly irregular, with dermal papillae from the papillary dermis protecting rectangularly to the skin surface. It is through diffusion at this junction that the epidermis obtains nutrients and disposes the waste.[11–13]

Dermis

Dermis is a layer between the epidermis and subcutaneous tissues that is responsible for the mechanical properties of skin. It consists of fibers of connective tissue that run in all directions and cushions the body from stress and strain. These fibers are composed of elastin, collagen, and extrafibrillar matrix (called ground substance).[14] The collagen fibers serve strength, whereas the elastin fibers provide recoil strength to the skin. The collagen and elastin fibers are both embedded in an interfibrillar matrix of proteoglycans. The dermis is divided into two layers, a thinner superficial layer adjoining to the epidermis referred to as the papillary region and a thicker known as the reticular dermis, which is made up of bundles of thicker collagen layers. It is tightly connected to the epidermis through a basement membrane. It also contains hair follicles, sweat glands, sebaceous glands, apocrine glands, and lymphatic and blood vessels.[15,16]

Biomechanics of Skin

Biomechanics of skin are of great importance to skin disease, wound healing, structural integrity, and aging. The studies on skin biomechanics have begun in the nineteenth century, which focused on skin mechanical anisotropy.[17] Afterward, other studies were concentrated on the biomechanics of skin in the field of aging, plastic surgery, sun exposure, skin cancer, and burns.[18–21] Biomechanics is described as mechanics applied to biology and it is the feedback of bodies to forces or displacements. Therefore, it is very important to conduct research the mechanics of skin. The skin consist of a composite materials such as a collagen-rich fibrous network embedded in a substance matrix. The proteoglycan-rich matrix serves skin viscous

nature at low loads. The fibrous components of collagen and elastin provide elasticity and structural stiffness to the skin.[22] Collagen fibers provide strength to the dermis and can range from 500 to 10,000 psi (3.4 ~ 68.9 MPa).

Wounded Skin

Contraction of wounded skin

The skin can be injured by diverse mechanical, chemical, and other factors. It provides a vital role as the outermost barrier to protect the inner organ in the body. Therefore, wounded skin should be replaced quickly. During the grafted or wounded skin healing process, the edges of the wound contract to make the wounded area smaller, called as wound contraction. A severe wound heals by forming scar tissue that shrinks and tightens as it forms. This shrinking results in restricted movement of joints. In this case, the wound beds are not in close proximity and thus in order to bring them closer, fibroblasts move to at the edge of the wound. These fibroblasts make wound closure.[23,24] Wound contraction is the main reason for the wound bed to be lessened to 5–10% of size within 6 weeks of the lesion. A scar is fibrous tissue that replaces the injured skin. Scar tissue contains collagen fibers that are much thicker than the fibers in normal skin. This causes scar tissue to be much tougher than normal skin tissue. Also, it has been shown that there is a relationship between the orientation of axis of a fibroblast along which the external force develops and the orientation of collagen fibers synthesized by these cells. During collagen synthesis, fibers are extruded outside the cells with long axes oriented parallel to the long axis of the synthesizing cell. Moreover, in a wound bed sustaining contraction, the axes of contractile fibroblasts were observed to be oriented along the plane of the wound surface.[25]

Type of wound healing

Wound healing is the process by which the cells in the body close the wound along with regeneration of the injured tissue. This is a natural process that can be optimized therapeutically. Wounds heal by primary or secondary intention, the primary wound healing known as "healing by first intention" occurs as the result of surgical closure and is only possible if the edges of the wound are smooth. It is also crucial that the edges are close together and the wound is free from bacteria. Also, secondary wound healing known as "healing by second intention" always takes place where there is a loss and injury of tissue over a large area. It is described as a process in which a wound is left open and allowed to close by epithelialization and contraction.[26–28]

Impact of wound

The depth of wound is defined according to the layers of the skin injured. Partial-thickness wounds involve only the epidermis or the dermis, or both, whereas full-thickness wounds involve not only epidermis but also the entire dermis. The process of healing is entirely different between partial and full-thickness wound. Partial-thickness wounds sustain minimum contraction and re-epithelialize from wound edge and epidermal stem cell around sweat glands and hair follicles. On the other hand, full-thickness wounds rely upon epithelialization from the wound edges owing to the wreckage of the epidermal stem cell populations about a full-thickness wound. Furthermore, it is more likely to sustain contraction during the healing process.[24,28]

Wound healing process

By the definition of the Wound Healing Society, wound healing is "a complex and dynamic process that results in restoration of anatomic continuity and function." The wound healing process is a cascade of events, beginning with injury to tissue. Appropriate wound management is dependent on an understanding of the normal repair process, the factors affecting this process, and the interventions that can impact either positively or negatively on the outcome. Healing progresses in a series of many overlapping phases. The wound healing process can be categorized in 4 phases, which are vascular response phase (hemostatic phase), inflammatory phase, proliferative phase, and maturation phase (Fig. 2).[29]

Hemostatic (Immediately Upon Injury). Hemostasis is described as a physiological reaction occurring after tissue injury and the process for preventing blood loss also referred to as coagulation. Coagulation is the formation of a blood clot that prevents blood loss from damaged tissue and vessels. Hemostasis normally involves three processes: 1) contraction of the damaged blood vessel; 2) formation of a temporary plug; and 3) coagulation of blood at the injured site. The initial two are for stopping blood loss, whereas coagulation of blood provides permanence.[30,31] When tissue is injured, platelets start to congregate at the site of injury and then come into contact with collagen in the walls of the damaged blood vessels leading to further clumping. Platelets do not attach normally to the smooth endothelial lining of blood vessels and its normal function is the formation of a mechanical plug during hemostasis as a response to any vascular injury. The platelets serve three functions such as platelet adherence known to stick to the injured blood vessel, attachment to other platelets to spread the platelet aggregation, and supporting for the processes of the coagulation cascade. The structure of platelets is complex and bounded by a complex plasma membrane. This membrane contains glycoproteins (GP) including GPIa, GPIb, and GPIIb/IIIa that act as adhesion molecules. The platelets contain microfilaments, a dense tubular system and two types of granules including α-granules

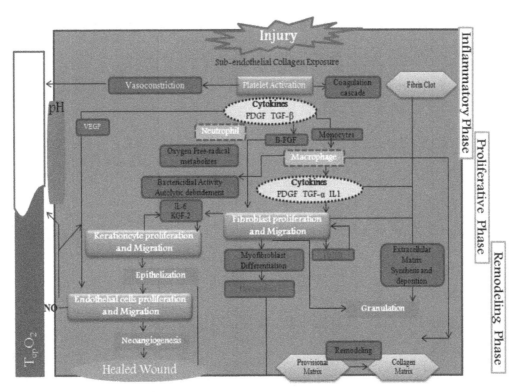

Fig. 2 Schematic representation of wound healing cascade and different signaling pathway involved in each phase that helps the skin to recover from an injured state.

and dense-granules. α-granules contain fibrinogen, a von Willebrand factor (vWF), a heparin antagonist (PF₄), and a platelet growth factor. Dense-granules include adenosine nucleotides [including adenosine-di-phosphate (ADP)] and 5-hydroxytryptamine (5-HT). The platelets combine with collagen through GPIa and the binding is allowed by vWF. The coagulation activates platelets to release chemicals, such as platelet-derived growth factor (PDGF), which initiates the clotting cascade. This process allows the contraction of blood vessel thereby reducing blood flow. The platelet plug serves a nidus around the blood clots, forming a certain hemostatic plug. The basic response in the formation of blood clot is the conversion of the soluble plasma protein fibrinogen into insoluble fibrin under the activation of thrombin (Fig. 3). Thrombin is divided into two polypeptide chains from the fibrinogen molecule for forming a fibrin monomer that synthesizes to form a mesh of strands. The strand is linked together by hydrogen bonding, however the activation of factor XIII by thrombin brings about the formation of covalent bridges between the fibrin chains.[31–33]

Inflammation (Days 4 through 6). The inflammatory phase of wound healing lasts for about 3–5 days in the case of acute wounds, whereas it can be prolonged in chronic wounds. The inflammatory process that starts immediately as soon as the injury is sustained is characterized by the infiltration of leukocytes into the local area. The release of PDGF and proteases aids in attracting chemical agents and involves cells to the injured site. This inflammation phase of wound healing is conceded by erythema, edema, and increased body temperature, which is a result of the enhanced blood flow to the injured site and the blood vessel permeability, and this phase is often associated with pain.[34–36] The increase in extracellular fluid causes swelling as the fluid passes through the wall of vessel.

The extra flow of blood by vasodilatation gives rise to neutrophils that cleanse the wound area and devitalize tissue at the site. In addition, monocytes are attracted to the site by PDGF and convert into macrophages that continue to do the cleaning activity, which is crucial step in the healing process. Macrophage acts as a supervisor in wound healing cascade. The neutrophils not only phagocyte debris and microorganism but also serve the first line of defense against infection, while macrophages that are able to phagocyte bacteria provide a second line of defense. Furthermore, they secrete varied growth factors such as epidermal growth factor (EGF), transforming growth factor (TGF), and interleukin-1, which direct the next phase.[36–39]

Proliferative Phase (Days 4 Through 21). The proliferative phase initiates approximately 4 days after injury and usually continues to 21 days in case of acute wounds according to the size of the injury. The proliferative phase of healing involves the appearance of red granulation

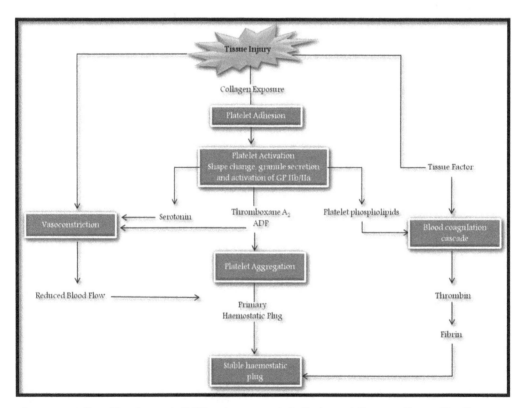

Fig. 3 Schematic representation of involvement of different components in the hemostatic stage of wound healing.

tissue that comprises macrophages, fibroblasts, immature collagen, and newly formed extracellular matrix. Fibroblasts are the obvious cellular mediator during the proliferative phase, aiding endothelial angiogenesis and participating in the contraction of wound and matrix production. Epithelization due to keratinocyte migration from the wound edges finally covers the lost tissue with a new cellular at the end of the proliferative phase. A vital stage in the wound healing is the formation of new blood vessels within the granulation tissue.[40–42] The angiogenesis process is called neo-vascularization and occurs with the proliferation of fibroblasts as endothelial cells migrate to the injured site. Because the activation of fibroblasts and endothelial cells requires nutrients and oxygen, angiogenesis is essential for other stages in wound healing such as the migration of fibroblast and epithelial cells. Hypoxia at the injured site stimulates vasculogenesis, angiogenesis, the proliferation of fibroblast, and collagen deposition. Macrophages respond to hypoxia by the development of inducible nitric oxide synthase (iNOS), tumor necrosis factor-α, and interleukin-6 (IL-6). Neighboring endothelial cells move toward the direction of the wound using pseudopodia followed by the secretion of collagenase and later migrates to the perivascular space on the second day of wound healing (Fig. 4).[42–44]

The endothelial SC from some part of blood vessels without injury allow pseudopodia to develop and push new vessels generation. Endothelial cells start to proliferate and develop a tubular structure in response to exposure to fibronectin, heparin, and growth factors. To migrate, an endothelial cell needs plasminogen activator and collagenase that degrade part of the extracellular matrix and clot. Growth factors, cytokines, and chemokines are sequestered in the extracellular matrix, and as vessel growth proceeds, the secretion of protease remodels the extracellular matrix.[45–47] Furthermore, the extracellular matrix presents a scaffold on which the new blood vessels produce, and the presence of fibronectin, collagen, vitronectin, tenascin, and laminin in the matrix milieu present binding sites for activated integrins on the endothelial cells surface. The interaction between matrix and integrin serves both traction and tension for the migrating endothelial cells, but also produce intracellular signaling cascades that add to the migration and proliferation of blood vessel cells along with the deposition of new matrix components to support the vessel in the developing granulation tissue. When the growth factors that produce other cells and macrophages is no longer in hypoxia and lactic acid-filled environments, they stop producing the angiogenic factor. Followed by migration and proliferation of endothelial cells begin to reduce as the tissue is sufficiently perfused. Blood vessels that are no longer needed eventually die by apoptosis. Fibroblasts are attracted to the injured site by PDGF, a potent chemoattractant and mitogen for fibroblasts, secreted by platelets and macrophages. They start to divide and develop collagen that creates elasticity and strength in the wound. The migration of fibroblasts is mediated by binding of surface integrins to cell adhesion molecules in the extracellular matrix. In the initial stages of healing, they bind fibronectin in the matrix, and as collagen and proteoglycans are

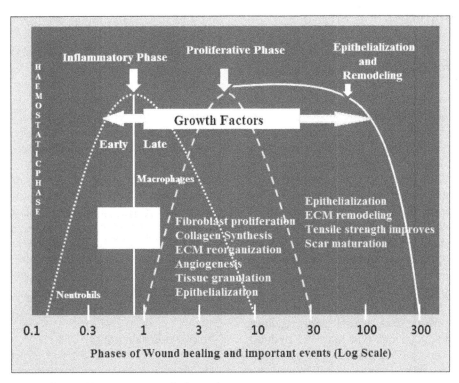

Fig. 4 Phases of wound healing and important events in log scale.

secreted by fibroblasts, they bind and migrate on substrates as well. The components of extracellular matrix have an effect on the proliferation and migration of fibroblast by regulating integrin expression and activation of signaling through these surface contacts.[46,48–52]

Remodeling (Days 8 through 1 Year). The remodeling phase of wound healing may take up to 18–24 months to complete. This phase is sometimes known as the maturation phase of wound healing and the injured site is strengthened during this time and the scar may change color considerably. The contraction of wound continues as fibroblasts become myofibroblasts through the interactions with extracellular matrix proteins and growth factors during the remodeling phase.[53–55] Then, myofibroblasts interact with collagen, vitronectin, and proteins for the contraction of wound. Upon the remodeling phase proceeding, fibronectin and hyaluronic acid are substituted by a bundle of collagen to serve as strength to the tissue. The bundles of collagen in the wound with irregular structure develop to form a more organized structure.[53,56,57] The scar has 3% of its ultimate strength at 1 week after wound formation, and 30% at 3 weeks. Afterward, the scar is 80% of the strength in comparison to uninjured skin at 3 months. There is a connection between the strength increment and the composition and alignment of collagen in the matrix. The small diameter type III collagen built in granulation tissue composed of 30% collagen in the new matrix. This is steadily substituted by thicker type I collagen during maturation to the normal levels of 10% type III and 90% type I.[58–61]

TREATMENT STRATEGIES FOR WOUND HEALING

Conventional Strategies: Grafts

The skin graft is a patch of skin that mimics or protects skin for a brief period of time and grafting is a surgical procedure involving skin transplantation and this transplanted tissue is known as skin graft. The patch is removed by a surgical operation from one area of the body and transplanted or attached to other area. Basically, a skin graft involves removing a patch of skin from an unwounded area of the body, called the donor site, and applied as a cover for a wounded site. A split-thickness skin graft (STSG) is skin graft that is composed of the top layers of skin (the epidermis and some part of the dermis), whereas a full thickness skin graft (FTSG) includes the epidermis, all dermis, and associated dermal appendages.[62–66]

The autologous skin graft is classified based on its thickness as full or partial thickness. All grafts require a vascularized wound with no graft type limit.[67] Full thickness skin graft has some advantages such as a better cosmetic result with less contraction in comparison to thin grafts; however, it is desired for better healing and more

vascularized bed. Also, the size and site of a full thickness skin graft is limited. The donor sites of full thickness skin graft include post- and pre-auricular, supra-clavicular, antecubital fossa, inguinal crease, and volar wrist crease skin. The extensive full thickness grafts can be achieved by tissue expansion of the planned donor site. The full thickness skin graft donor site may need to be closed with a split thickness graft from another site again as one step procedure is not sufficient to close the wound.[68–71]

Autologous STSG are the most practiced form of tissue transplantation in plastic surgery. STSG depends upon the level of harvest through the dermis. There are several advantages that include a large area of available donor sites and better engraftment rates; however, this is more likely to lead to wound contraction, hypertrophic scarring, pigment irregularities, and susceptibility to trauma. Epithelial cells within hair follicles and sweat glands generally regenerate a split thickness skin graft from donor site in 7–21 days. Regardless of its ability to heal, the donor site is usually left with a scar and significant discoloration. Hence, harvest sites should be hidden by clothing, such as the thigh, trunk, and buttocks, as far as possible.[69,70]

Almost all types of skin grafts are applied as sheet type, called unmeshed or meshed type, at a ratio ranging from 1:1 to 4:1. The meshed type allows the loss of serum and blood from wounds, thereby minimizing the risk of hematomas. Also, meshed grafts can be stretched to cover a larger surface area. Sheet type is generally used on the face, neck, and hands whenever possible. In exposed areas, the cosmetic results obtainable with sheet grafts make it more desirable. However, since sheet types have no lacunule, they should be closely monitored and any fluid collections must be squeezed by rolling the graft with a cotton tipped applicator.[71,72]

Autologous skin graft today is clinically the best way especially for full-thickness burn cases. Before transplantation, excision is a significant stage in the treatment of burn, the heat-denatured proteins of the skin is required to be removed to prevent complications including infection, formation of scar, uncontrolled inflammatory response, and contamination. Any infection can be destructive to a burned patient, because the wound can bring about a temporary suppression of cell mediated and humoral immunity. Autologous split skin graft is taken from the donor site with dermatome that detaches the epidermis and a superficial part of the dermis, since the dermis is capable of re-growing and the epidermis donor site is easily healed. The host epidermal cells in the dermis of the donor site will re-grow an epidermis.[70,72] There are different types of grafts such as autologous, allogenic, isogeneic, xenogenic, and prosthetic.

Autologous skin grafts

The up-to-date ideal strategy of wound healing is to employ an autologous skin graft that is transplanted from one part to another of the same individual, namely, a patient's own tissue. This is also known as an autologous graft, meaning the donor tissue is the same with the recipient of the tissue. It can usually be employed for a surgical reconstruction procedure. Grafting is a surgical procedure in which a tissue is transplanted, or adhered to an injured, loss, or defective site of the body and is considered to be the safest and fastest technique. However, harvest of autograft skin might create a secondary surgical site or scar from which the patient should recover. Also, the additional recovery time extends a patient's hospital stay. Besides, the second injured site might be uncomfortable for years after surgery.[70,74,75]

Allogeneic skin grafts

In contrast with autografts, allogeneic grafts, also called allografts, are harvested from a cadaver. Harrison reported the first successful kidney transplantation between identical twins in 1955 and the group reported another successful cadaveric renal transplantation in 1963. The following progress in immunologic typing of tissues and immunosuppression allowed solid type organ allograft transplant to accomplish long-term graft function. One of the possibilities is the use of allografts (cadaveric skin), for a temporary prevention of body fluid loss or contamination of the wound. The allografts are available in nonprofit skin banks; however, skin banks do not have enough skin tissue available to meet demand. It involves ethical as well as safety issues, since the screening for viral diseases and standardized sterilization techniques cannot completely eliminate the possibility of infective agent transmission.[77–79] In comparison to autografts, major disadvantages for allografts are cost, limited supply, and infectious disease transmission. Allografts usually suffer from immunogenic rejection and the site of wound needs to be covered again with an autograft. In the absence of immunosuppression, the rejection process brings about graft destruction or sloughing in 10–14 days. Hence, there is great need for an alternative that can provide a more permanent solution.[75]

Xenogenic skin grafts

Xenografts are defined as the transplantation of tissues or organs from one species to another and have been employed as temporary wound cover. The primary donors are pigs in the United States due to their similar histological structure compared to humans. Though the skin of animals becomes extremely adherent and indeed occasionally incorporated into the healed wound, xenografts are not true transplants.[73,76] The capillary does not grow and the connection between each vessel does not occur after porcine xenograft. To date, there are multiple reports in the literature attesting to the benefits of porcine skin graft in the treatment of larger wounds, including an increase in healing rates for partial thickness burns and granulating wounds, a reduction in pain once placed over burns, and a

decline in heat, fluid, protein, and electrolyte loss. Moreover, porcine skin serves as a physical protective layer for an epithelializing wound and it reduces bacterial overgrowth. It is used to protect the dermis until epithelialization takes place. It is effective as a temporary biological dressing in the treatment of toxic epidermal necrolysis. On the other hand, the major disadvantages affecting the use of porcine skin are bacterial infection, cost, as well as cysticercosis, influenza.[80]

New Advances

Burn injuries are devastating trauma that come along with systemic consequences. Despite increase in the survival rate of patients, burn injuries to date remain a great challenge for the medics. The science of wound healing is progressing fast with results for advanced therapeutic products such as skin substitute, SC, and gene therapy along with growth factors. Various groups have focused on providing treatment for different wound healing stages.

1. **Product for the inflammatory phase** (first stage of wound healing) targets decreasing the elevated production and activity of several proteases that include metalloprotease, neutrophils elastases, and serine proteases. The over production of these proteases that is generally regulated in the body can be detrimental to the wound healing process.[81,82]
2. **Product for proliferative phase** (second stage of wound healing) targets the proliferation of fibroblast, which is a key player in the healing process. Hence, the growth factors or chemo-attractants such as TGF-β or PDGF are used to draw these cells toward the site of injury or wounds. These cells are known to initiate the matrix proteins fibronectin followed by hyaluronan and proceed to the collagen and proteoglycan production. All of these help in reconstructing new extracellular matrices. Similarly the growth factors such PDGF are keratinocytes attractants and promotes the formation of granulation tissues.[83]
3. **Product for epithelialization and remodeling phase** (final stage of wound healing) targets the use of attractants for keratinocytes migration, proliferation, differentiation, and adhesion. Epidermal growth factors or EGF are known stimulants of cells as they bind to the receptor epidermal growth factor receptor (EGFR), and the tropical application of this growth factor effectively induces epithelialization of partial thickness and superficial granulating wounds. Another growth factor like TGF-β2 targets in reducing the deposition of collagen during the remodeling phase, which significantly decreases scar formation.[84]

Even though most of these approaches seem to be feasible and ideal, using growth factors along with recombinant deoxyribonucleic acid (DNA) technology can be extremely expensive, which in turn restricts their widespread usage. Among all these, stem cell therapy is emerging as a promising new approach in most of the medical specialty.[84–86]

Regenerative medicine

Regenerative medicine is the "process of regenerating or replacing human cells, tissues or organs to establish or restoring its normal function." This process enables the researcher to grow tissues or organs in the *in vitro* condition and implant as replacement when the body cannot heal by itself.[87–89] Regenerative medicine refers to an assembly of biomedical advance to clinical therapies that involve the use of SC. Some examples include cell therapies (injection of progenitor cells or SC), immune modulation therapy (biologically active molecules are administered as single or as secretions by infused cells for regeneration), and TE (transplantation of laboratory cultured organs and tissues). The age of cell-based therapy for treating wounds was first born when lab methodology advanced to the point where keratinocytes could be easily cultured and expanded in lab conditions. In 1988, the first ever cultured skin product known as "Epicel" was made by using autologous skin biopsy for culturing an epidermal graft. However, this technique was extremely expensive and also the resulting epidermis was very fragile and tricky to use.[90] The next generation product using a regenerative medicine approach is "Dermagraft," which is composed of neonatal dermal fibroblasts that are cultured on a biodegradable synthetic polymeric mesh. These fibroblasts produce an ECM that integrates into the wound and aids in the dermis healing.[91] Another example is "Apligraf" made from neonatal foreskin that contains both fibroblasts and keratinocytes.[92] The collagen mixed with fibroblast forms a scaffold and the keratinocytes are seeded onto it. Another group cultured cells in bioreactor which much improves upon the two-dimensional (2D) petri dish and using a new spraying technique for delivering cells at site of injury.[93] This combination of the bioreactor to culture cells in 3D format, an exclusive cell spraying system, and a "wound cap" offer potentially new therapies for burn victims. A form of regenerative medicine system that has made it to clinical practice is the usage of heparan sulfate (HS) analogues on wound healing especially the chronic wounds. HS analogues replace the degraded natural HS at the site of injury or wound. It is known that HS assists the damaged tissue in self regenerating by repositioning cytokines and growth factors back into the injured extracellular matrix. However, one most important point for regenerative medicine is not only to accelerate re-epithelialization after skin trauma and injury but also aid in the reconstruction of the fully functional skin with hair follicles, dermal capillaries, and sweat glands. Realistically, these goals could be achieved by stem cell–based therapy or SC transplantation. An approach for stem cell–based therapy includes local recruitment of endogenous SC, which are mostly altered or modified in

lab conditions and these cells can be combined with gene therapy or TE. The use of SC in very precise, efficient, high–quality, and low-morbidity therapy for coverage of burn or injured skin results in the regeneration of skin appendages and also minimal hypertrophic scarring risk. The skin trauma or burn represents the cellular stress at various levels, and hence to achieve successful regeneration of the affected area cell therapy seems to be the most feasible approach.[90,91,94]

Role of SC

Ideally, SC for regenerative medicine could be and should be freely available and accessible by the minimal invasive technique; however, the ground reality is contrasting (Fig. 5). Many research groups across the globe are testing methods for creating specialized cell types derived from pluripotent stem cells (PSCs). Much success has been achieved in the transplantation of undifferentiated hematopoietic stem cells (HSCs) through bone marrow transplant; however, the use of specialized cells created from PSCs is still in the research phase. The tumorigenicity and various ethical considerations have impeded the success story is for clinical application. Given these complications, regenerative medicine research is largely focusing on using induced pluripotent stem cells (iPSCs) and adipose stem cells (ASCs).[94]

1. Induced pluripotent stem cells: They are derived from blood cells or skin (typically an adult somatic cell) that are reprogrammed back into an embryonic-like state (pluripotent) by inducing a "forced" expression of specific genes and that allows the development of an infinite source of any type of human cell that are needed for therapeutic purposes. For example, iPSCs can be prodded into differentiating to beta islet cells for diabetes treatment or neurons to treat neurological disorders, and mature blood cells to create new blood that is cancer cells free for a leukemia patient.[94–96] iPSCs are similar to embryonic stem cells in certain genetic level and chromatin methylation pattern, protein expression, and even in chimera formation. iPSCs were first produced from mouse cells in 2006 and from human cells in 2007 in a series of experiments performed by Shinya Yamanaka's team from Kyoto University, Japan, and simultaneously by James Thomson's team at University of Wisconsin-Madison. The importance of this work won both the lead scientists 2012 Nobel Prize in Physiology or Medicine "for the discovery that mature cells can be reprogrammed to become pluripotent" (Fig. 6). The ability of creating PSCs from the patients' own skin cells holds great promise for what can be a very personalized medicine. The genetic modulation performed by the transduction of four highly expressed reprogramming transcription factors in embryonic stem cells are "cMYC, SOX2, OCT4,

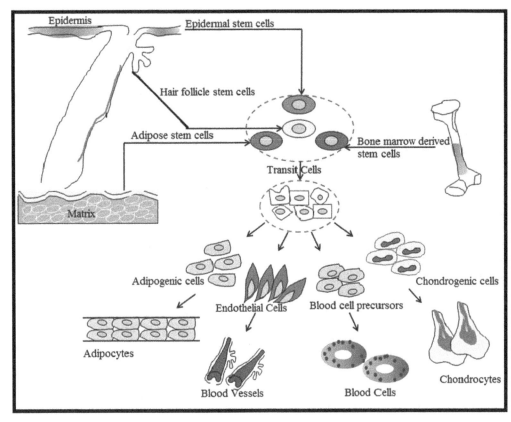

Fig. 5   Role of stem cells in regenerative medicine and potentiality of these cells in various human applications.

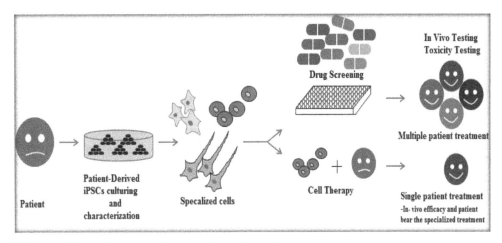

Fig. 6   Induced PSCs and their role in regenerative medicine and drug discovery.

and KLF4" and these reprogramming result in the proliferative and differentiation capability of iPSCs that are similar to ESCs. Transplantation therapies involving human embryonic stem cells began in 2010 (FDA approved) and are performed along with immunosuppressant drugs; however, these drugs have their own side effect that enable the body to fight any infection due to lower immunity. iPSCs hold the potential for generating customized cells and tissues for human transplantation.[96–99]

2. Adipose stem cells: These are derived from the adipose tissue, which is known to be a niche of stromal stem cells (SSCs) and researchers found these cells to have a higher percentage of healing cells compared to bone marrow stem cells. The SSCs found in adipose tissue have lymphocyte common antigen (CD45), CD105, and CD31 along with surface phenotypes from fibroblast-like colonies proliferate, e.g., colony forming unit factor (CFUF). These cells can differentiate into multiple cell linage, which includes

chondrogenic, neurogenic, and oestrogenic. The cytokines secreted by ASCs are seen to promote both fibroblast migration during wound healing cascade along with up-regulating neo-vascularization in animal models. These multipotent cells display significant potential for several applications in skin repair.[100–102] ASCs enhances human dermal fibroblast proliferation via direct cell-cell contact, thereby inducing paracrine activation, which is known to play a role in significantly accelerating re-epithelialization of cutaneous wounds. Adult stem cells have the ability to alter the micro-environment of the tissue by secreting soluble factors that contribute extensively to tissue repair in comparison to multipotent stem cells. Along with this, the adult stem cells can also modulate the immune system and inflammatory responses. This advantage added with the ability of ASC to spontaneously differentiate along the epidermal epithelial, fibroblastic, and vascular endothelial lineages considerably aids in the wound healing process.[97–99,102]

TE APPROACHES

TE targets to replace living tissue construct for repairing damaged or lost tissue or organ by combining the principles of engineering and biology and using the know-how for fabricating a tissue analogue that can help in regaining the function of the lost organs. To achieve the goal, it is indispensable to use artificial ECM in which cells grow, proliferate, and differentiate similar to the condition where cells cognize in their natural tissue.[103–105] Artificial ECMs or scaffolds have been exponentially used to form new tissue from cell. It should be designed to guide the desired cell type and offer proper mechanical strength until new growing tissue is structurally stable. Therefore, it is crucial for fabricating artificial ti syooirt or modulating the cell behavior including proliferation, differentiation, and migration. Also, the artificial ECM has multiple functions such as detection of cell signaling, release of growth factor, and structural support. To promote cellular functions, the ideal scaffolds used for TE must possess the following characteristics:[103,105–108]

Scaffolds: Definition and Requirements of Scaffolds

1. Porous structure
   i. Interconnecting pores for cell migration and adhesion in scaffold
   ii. Exchange of oxygen and nutrients and waste products can be carried away
2. Surface chemistry
   i. For cell attachment and proliferation
   ii. The ability to deliver biomolecular signals and to promote ECM secretion
3. Biocompatibility
   i. No immune response during implantation

4. Physical structure
   i. Desirable shape to guide neo-tissue formation
5. Biodegradability
   i. Rate of degradation should be matched with period of tissue formation
   ii. Tunable with *in vitro/in vivo/ex vivo* tissue regeneration
   iii. Proper biodegradability without toxic by-product
6. Mechanical stability
   i. Should match the strength of normal tissue or organ of interest
7. Shape: 3D design should mimic natural tissue
8. Reproducibility: Easy in fabrication, handling, sterilization with high reproducibility is important

Fabrication Techniques

A variety of techniques have been developed for processing biodegradable polymers into 3D porous scaffolds of high porosity and surface area. The conventional methods embrace fiber melt, fiber bonding, melt molding, solvent casting/particulate leaching, gas foaming/particulate leaching, phase separation, and high-pressure processing.[109] The major characteristic and parameters of present techniques used for achieving high porosity and surface area. Present challenges in TE are not only to fabricate but also to manufacture reproducible biocompatible 3D scaffolds, enabling to function for a desirable period of time under load-bearing conditions.

Solvent casting

The solvent casting for the fabrication of 3D porous scaffold includes dissolving the polymer in an organic solvent and subsequently enabling solvent evaporation. The major benefit of the solvent casting technique is the ability to easily fabricate without the need for any equipment. The rate of scaffold degradation has no effect on the tissue regeneration. However, there are disadvantages to this technology, such as: 1) the limitation in shapes (typically sheets and tubes can be formed); 2) the possibility of retention of toxic organic solvent; and 3) the denaturalization of the proteins and bioactive molecules into the polymer by using toxic solvent. The organic solvents may reduce the activity of biomolecules.[109,112] To overcome the disadvantages of the solvent casting technique, researchers have to combine different techniques to fabricate a nontoxic, 3D porous scaffold.

Particulate leaching

This methodology, in combination with particle leaching, applies for thin membranes or 3D samples with thin wall sections (Fig. 7). The conventional method for the fabrication of macroporous scaffold involves casting a polymer/salt/organic solvent solution, followed by solvent

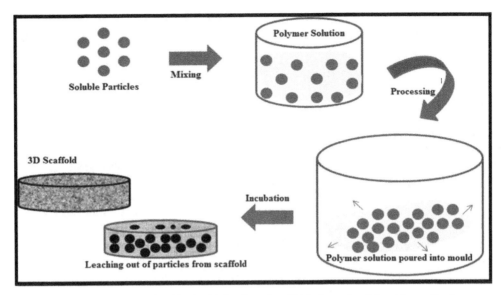

Fig. 7 Schematic representation of steps involved in the fabrication of scaffolds through salt-leaching methods.

vaporization and dissolving the salt particulates in an aqueous solution. Mikos et al., with the described method, fabricated porous sheets and laminated them into 3D scaffolds using chloroform that was used on the attachment interface for the lamination process.[109,110] However, this fabrication technology is time consuming, only fabricates thin membranes, and it is not possible to completely remove soluble traces from the polymer scaffold. Along with limited number of inter-connected pores in sheet, the disadvantage of this technology is the large use of toxic organic solvents that requires a long time for the desired solvent evaporation. Several attempts have led to the fabrication of thin scaffolds with porosity as high as 93%.[110]

Gas foaming

The gas foaming method is widely used to produce porous a scaffold that has a foam structure. The technique was utilized creating gas bubble and nucleation dispersion in a solution polymer to produce a porous structure.[111] The principle of forming foam is to either by using supercritical or carbon dioxide at high pressure, followed by quickly removing carbon dioxide (CO_2) and escaping gas results in pore formation.[112–114] Gas bubbles are created by the chemical reaction of a blowing agent and released from a saturated gas/polymer mixture. A supercritical solution is effective for the creation of a gas-saturated polymer phase, due to tunable density, zero surface tension, and great mass transfer property by various temperature and pressure.[115] Carbon dioxide is extensively employed as a supercritical solution because of nontoxicity and combustibility. Also, it is useful for creating porosity in amorphous or semi-crystalline polymers such as poly(lactic acid) (PLA), poly(lactic-co-glycolic acid) (PLGA), poly(ε-caprolactone) (PCL), and polystyrene.[116–123] Gas foaming by supercritical

CO_2 generally includes two strategies for producing a 3D porous structure: 1) formation of gas-saturated polymer phase; and 2) creation of pore nucleation, growing, and conglutination.[115,123–126] The exposure of high pressure CO_2 to polymer causes plasticization and increases the solubility of CO_2 in the polymer matrix. With the decrease of CO_2 solubility in the polymer, super saturation occurs and the nuclei of gas molecules are formed.[127,128] Various parameters such as depressurization, temperature, and pressure have a crucial effect on the pore characteristic of pore size and interconnectivity.[117] Gas foaming is a rapid solvent-free process and takes place at moderate temperature.[103] This technology can be used for in situ impregnation of active compounds, such as DNA and release from matrix, due to the diffusion and degradation of polymer scaffold.[129,130]

Thermally induced phase separation

Numerous approaches have been developed to fabricate a 3D porous scaffold. To date, thermally induced phase separation (TIPS) has been employed for producing a porous membrane or microcellular foam.[131–133] This process utilizes the variation of thermal energy to induce the separation of homogeneous polymer–solvent or polymer–solvent–nonsolvent solution by liquid–liquid phase separation.[134–136] With this technique, a fabricated single homogeneous polymer solution at critical temperature is converted into two phase-separated areas consisting of a polymer-rich phase and a polymer-lean phase through removal of thermal energy.[137–139] It exhibits the change of polymer foam morphology depending on the ultimate thermodynamic circumstances of the polymer solution to be thermally quenched. The interconnectivity and distribution of pore size in the resultant foam are determined by the balance of variable parameters such as polymer concentration, quenching depth, composition of solvent/non-solvent, and addition agents.[134]

Freeze-drying

The freeze-drying technique has been developed for producing polymer foam of low density.[134] The solution is frozen and subsequently lyophilized, after dissolving the polymer in a solvent such as acetic acid or benzene to make a solution of the desired concentration. This technique is used for several polymers including PLGA, PLGA/polypropylene fumarate converted to porous foam. The foam has a capillary structure depending on the used polymer or solvent. The emulsion freeze-drying technique has also been developed for the fabrication of porous scaffold.[140] In this technique, emulsion is immersed into mold and maintained at liquid nitrogen. After freezing, porous PLGA foam with a porosity of 91–95% is created by variable parameters such as molecular weight, polymer weight fraction, and water volume fraction. It has a pore size of general diameter from 13 to 35 μm and maximum size of over 200 μm. Compared to solvent casting/particulate leaching techniques, this method allows thick foam with high surface area though the pore sizes are smaller.

Electrospinning

The electrospinning technique is essential for manufacturing polymeric nanofiber by applying electrostatic force due to its simplicity and efficiency. A polymer solution is drawn from a nozzle under mechanical pressure with an electric field of high voltage.[140] As the electric charge overcomes the surface tension of the polymer solution drop, the polymer solution is sown into a nanofibrous structure (Fig. 8). It has been widely employed in many applications such as TE, sensors, and protective clothing for over 70 years and was introduced to the TE application several years ago.[141] Li et al. (2002) fabricated a novel PLGA nanofiber mesh with interconnected pores from 500 to 800 nm.[144] 3D nanofibrous porous scaffolds fabricated by

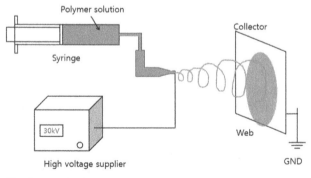

Fig. 8 Detailed scheme of steps involved in the fabrication of scaffolds using the electrospinning technique.

electrospinning have been employed in TE because of good cell adhesion and proliferation.[142–144]

Although many different polymers have been used for the fabrication of 3D scaffold by electrospinning, this technique is challenged due to the unavailability of a suitable solvent for electrospinning.[145–147]

Cells and Scaffolds Interaction

The function of cells and tissues is dependent on spatial interaction with adjacent cells and with ECM. The natural ECM is physically and chemically a cross-linked network of proteins and glycosaminoglycans (GAGs). The ECM carries out innumerable functions including the organizing cells and the processing of environmental signals to manage site-specific cellular recognition. These are underlying reasons of cell adhesion, aiding to designate 3D cellular organization and to perform directly in cell signaling, controlling cell recruitment, growth, differentiation, immune recognition, and modulation of inflammation. Subsequently, summarizing the interactions between ECM and cells is significant for *in vivo* tissue substitution. Many natural and synthetic materials have been employed to develop 3D scaffolds to function as an artificial extracellular matrix. An ideal scaffold should assist in two most important events: 1) cell adhesion; and 2) immobilized cytokines.[147–150]

Cell adhesion

Natural materials including collagen, gelatin, or hyaluronan are favorable for employing as TE matrix owing to their supply for specific cellular interactions. For example, collagen, the major component of mammalian connective tissue, serves as a physical frame to the regenerating tissue and is associated with numerous developmental and physiological functions, such as cell adherence, differentiation, and chemotaxis.[152–154] The interaction between cells and scaffolds occurs by means of a class of receptors called integrins, which bring about morphological modification. Integrins are the cell surface adhesion receptors that mediate the adhesion between cell and scaffolds. Integrins are heterodimeric receptors created by selective pairing between 18α and 8β subunits. The morphological modification is involved in cell spreading, the formation of focal contacts that extend toward the ECM surface, clustering of integrin receptors at the sites of focal contact, and assembly of accessory proteins to the cytoplasmic face of clustered integrins to form connections with the cytoskeleton. Integrins initiate cell-signaling mechanisms, such as the activation of tyrosine kinase, the response of growth factor, and the modification of gene expression (Fig. 9).[155,156]

Synthetic polymer materials have been utilized to serve better versatility than natural materials for developing scaffolds with controlled properties. They could be fabricated

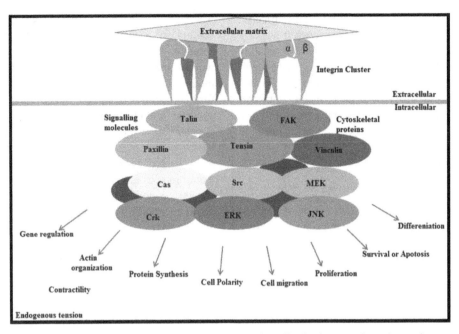

Fig. 9 Interaction between cell, scaffold, and signaling cascade resulting in cell adhesion to polymeric matrix.

reproducibly and processed into exogenous ECMs in which the porous structure, mechanical properties, surface chemical characteristic, and the time of degradation could be controlled. These scaffolds possess an interconnected open-pore structure that permit cells to be attached, grown, and proliferated into the scaffold *in vitro* or to permeate into the scaffold *in vivo*. Even if the synthetic materials are not fundamentally recognized by cell surface receptors, the interaction between cells and scaffolds is sustained via proteins that adsorb to the surface of synthetic scaffolds.[151,154–156] The adsorption is a consequence of an entropic gain as water is released from the surface of hydrophobic materials. The intrinsic complexity in the interaction between the protein and surface entangles the ability to accurately control the concentration, conformation, and the bioactivity of adsorbed proteins. Hence, the fabrication techniques are focused on specifically controlling the surface of the polymers, such as the development of materials that self-assemble on surfaces. In the case of hydrogels, which exhibit little protein adsorption, they could be modified with peptides or proteins to interact with specific receptors and direct cellular responses including proliferation and differentiation. Also, hydrogels exhibit that a cellular can be functionalized using chemical groups along with the gel's backbone. Consequently, oligopeptides that mimic natural adhesion proteins of the extracellular matrix can be chemically incorporated into the surface of the polymer to allow interactions between the cell's receptors and the scaffolds. The Arginine-Glycine-Aspartic acid (RGD) sequence that is found in ECM represents a minimal sequence that is needed to interact with integrin receptors for the purpose of leading acellular adherence to the substrate and this phenomena can be used to adhere the cell to no interactive surface.[157–159]

The density of covalently immobilized peptides or absorbable protein manages cell adherence, morphology, following cell process including cell migration and extracellular deposition. The concentration of adsorbed protein or peptide affects the migration of cells by increasing cell adhesion. At low density of adsorbed protein, an increase in protein adsorption brings about increases in the migration speed of cells, whereas, at high density of adsorbed protein, additional increase in protein adsorption reduces the migration speed of cells. In addition, the density of peptide and protein also affects the deposition of extracellular matrix. The pattern or morphology of scaffold surface can direct to the arrangement or orientation of cells, namely, serves spatial guidance to tissue formation. The surface pattern of scaffolds with RGD molecules spatially controlled the adherence and spreading of cells on line stripes.[160]

Immobilized cytokines

The extracellular matrix locally modulates the concentration and biological activity of the growth factors.[161] The covalent immobilization of cytokines and other growth on surfaces of scaffold sustain the bioactivity of the protein while stimulating cellular responses. The growth factor β2 (TGF-β2) is immobilized to fibrillar collagen to increase the stability and bioactivity. One week later, the response of the immobilized TGF-β2 is analogous to the response of unbound TGF. Nevertheless, immobilized TGF-β2 sustained the activity for 4 weeks compared to 1 week for the activity of unbound TGF.[162] Also equally, the immobilization of insulin on the surface of polymethyl methacrylate proved the increased rate of growth relative to controls with free insulin.[163] The immobilization of insulin and fibronectin was found to act in order to accelerate growth and to enhance adherence. Moreover, hepatocytes seeded on the surface of the scaffold with epidermal growth factor bonded by way of the primary amine, which triggered the responses similar to that of the epidermal growth factor including DNA synthesis, cell motility, and changes in morphology.[164]

IN VIVO TISSUE REGENERATION

Progress in Skin TE

Skin substitution has been a great challenge for not only engineers but also surgeons. The skin grafts have been established in surgery only since the nineteenth century, although skin graft has been employed in medicine about 2500 years ago. Techniques of skin harvesting and the best dressing for the grafts were the most important issues. The history of skin grafts triggered with Baronio in 1804[165] who recorded epidermis grafts in sheep, which was used by surgeons for granulating wound covers in humans. Surgeon Jacques Louis Reverdin who worked at hospital Necker in Paris introduced skin autograft, which was used on the wounds of the arm, and developed new techniques in 1871.[166] The transplantation of the patient's own skin from a healthy donor site as skin graft has become the surgical "gold standard" to cover injured skin. However, the wound surface in massive injury necessitated the demand of variable temporary or permanent skin substitutes, because of insufficient skin resources of the patient's own body. Thus, the clinical use of allograft for wounded skin was introduced by Girdner.[167] He is a pioneer who first reported the use of allogeneic skin as a covering of a burn wound.[168] However, it started to be used in clinical practice the first time and was popularized after Thiersch described the histological anatomy of skin engraftment. He published a report of using partial-thickness grafts on patients in 1874.[169] However, the grafts were so discontented for large areas that the use of the graft was limited to small ulcer applications.

In the three developments including the xenograft of Baronio, the autograft of Reverdin, and the allograft of Girdner, skin grafting led to radical advances in the treatment of injured skin. First, it was obvious that the dermal layer is the most significant layer of a newly produced skin graft. Second, it was established that after removing, the donor-site regenerated from the deep islet of the gland and follicle epithelium. Therefore, the skin grafts had better be cut as thick as possible. The technique of "epithelial cell seeding" was published by Von Mangoldt in 1895 to treat chronic wound and wound with excellent clinical results.[170–172] He harvested epithelial cells by exfoliating superficial epithelium from a patient's arm with a blade. This was then used to cover wounds. He showed a decrease in the donor site wound by harvesting when compared to the report of

Reverdin. The technique was modified in the nineteenth century by Pels Leusden who mixed the epithelial cell-serum with blood suspension and injected it into the chronic wound. However, it was less employed because of the formation of epithelial cell cysts near the wounds.[173] In 1952, further trying to implant epidermal cell suspensions in solution without matrix such as fibrin were published, however they did not show coherent clinical results.[174] Horch et al. reported that the successful transplantation of non-cultured keratinocytes was gained by trypsinization from biopsies and suspended in a fibrin matrix to heal the chronic wound, whereas the control group without fibrin could not result in re-epithelialization.[175]

The combination of cultured autologous keratinocytes with various templates has been studied by various researchers. Yannas et al. reported that trypsinized whole-skin-cell suspensions centrifuged into a collagen-glycosaminoglycan matrix and grafted onto guinea pigs could facilitate the regeneration of the healthy epidermis.[176–180] In experiments from our group, cultured fibroblasts seeded directly into fabricated scaffolds exhibited biocompatibility well in the laboratory.[181,182]

The matrices that have been studied for reconstructive and injured skin for many years comprise a double-layer, otherwise called as bilayer, with the "dermal part" represented by a macroporous cross-linked collagen and GAG and "epidermal part" with that of synthetic polysiloxane polymer (silicone). As the collagen part integrates into the wound, the silicone part starts to separate. The role of the silicone elastomer layer is to control bacterial ingress and liquid loss and provide additional physical support. The collagen and GAG layer were fabricated to make sure rapid wound adherence.[177] The most general approach is to create 3D biodegradable scaffolds in the shape of the tissue.[183] There are two major approaches, *in vitro* and *in vivo*, which have been used to develop engineered tissue. The *in vitro* method is the most common in which efforts are made to fabricate tissue biomimic organs in bioreactors for the transplantation and substitution of the injured tissue. On the other hand, the *in vivo* system uses acellular matrix that includes clues conducive for neo-tissue regeneration due to host cell–matrix interaction. Synthetic or natural matrices are widely employed to rapid wound repair and improve the quality of healing in chronic or acute wounds.[184–186] It has encouraging prospects for clinicians to recover injured skin.[187] In 2003, Ma et al. in Zhejiang University fabricated collagen/chitosan porous scaffolds with improved biostability for skin TE using freeze-drying.[188] Collagen-polycaprolactone (PCL) composites were employed the mimicking of tissue-engineered skin substitutes. The composite films are found to be favorable substrates for the growth of cells and find usage in skin repair.[189] Adekogbe and Ghanem in Canada developed dimethyl 3-3, dithio bis' propionimidate (DTBR)-cross-linked chitosan scaffolds that can be effective, biocompatible, and biodegradable for skin TE.[190]

Success of Skin TE

There are various commercially available products but with restricted success.

1. ***Wound coverage by amniotic membrane***: Human amniotic membranes (AMs) acquired from the placenta soon after delivery have been used for over a decade to cover burn wounds. An AM is readily available in large quantity in most of the hospitals and can be processed as low-cost material. This membrane is similar in characteristics to normal skin and can be an ideal substitute. Properties of the AM, such as outstanding adherence to the wound, extremely low immunogenicity, microbial control, decrease of pain, and translucency, make the inspection of wound easier along with stimulating a normal healing process making them one of the most successful grafts for skin regeneration.[191]

2. ***Acellular human dermis substitute***: *This is* commercially sold by AlloDerm; LifeCell Corporation is basically healthy human dermis with all the cellular components removed.[192]

3. ***Tissue-engineered substitute***: Promogran (Johnson & Johnson Wound management), a commercially available wound dressing material, is a collagen spongy matrix that contains oxidized regenerated cellulose (ORC). It is readily sold in both the US and EU markets.[193]

4. ***Bilayer system***: Burke et al. were the original designers of artificial skin that is presently commercialized by the name of "Integra Dermal Regeneration Template."[194] They fabricated a bilayer artificial skin of which a dermal portion was composed from porous lattice of cross-linked collagen and GAG fibers and the epidermal layer was fabricated using poly-siloxane synthetic polymer (silicone). Another example of bilayer skin substitute used frequently for severe burns is "Biobrane" commercialized by Bertek Pharmaceuticals Inc. It is a biosynthetic wound dressing made of silicon film with partially imbedded nylon fabric into the film. This system presents complex 3D trifilament collagen threads that are cross-linked chemically; when these are placed on wounds, the blood/sera clot into the nylon matrix, leading to a firm adherence of the film to the wound until epithelialization occurs. Recently approved in the United States, a novel bilayer skin membrane called OrCel, developed by Ortec International Inc., includes living allogenic human cells. OrCel is a bilayered system fabricated using bovine collagen (type 1) and dermal fibroblasts along with human allogeneic epidermal keratinocytes cultured separately on each layer. The dermal fibroblasts are cultured on the porous spongy side of the collagen scaffold, while epidermal keratinocytes are cultured on the nonporous coated side of this collagen matrix. Similar to OrCel, another system called "Apligraf" is fabricated and is the only living bilayered cell-based product that has been FDA approved for application in both diabetic foot and venous leg ulcers healing.[195–198]

5. ***Multilayered system***: A new multilayer system named Acticoat or Acticoat-7 marketed by Smith and Nephew is an nanocrystalline silver impregnated antimicrobial barrier dressing consisting of two non-woven rayon/polyester inner cores sandwiched between three outer layers of silver-coated low adhering polyethylene nets, and this system is held in place with ultrasound welds. Nanocrystalline silver-coated high density polyethylene (HDPE) layers act as barriers against microbial infection and the inner core aids in the maintenance of a moist environment that is needed for optimal wound healing.[195,199]

6. ***Bioengineered skin***: TransCyte (Smith & Nephew) is the first FDA-approved human-based, temporary, bioengineered skin substitute for treating excised full and partial-thickness burns made up of silicon membrane and human keratinocytes cells. Providing biocompatible, functional, and ideal replacement for the injured skin is the ultimate aim of skin tissue regeneration, for this field holds the key to have improved quality of life. Despite the physical trauma a person suffers from, it is possible to have a scar-free life in the near future.[200–201]

Future Prospects and Conclusion

The ultimate goal for regenerative medicine and TE of the skin is to create a construct that offers not only complete regeneration but also functional skin that includes all its appendages such as sweat and sensory glands along with hair follicles that can enable patients to have a scar-free life. An artificial skin should also be functionally similar acting as a protective barrier against environment, microbes, radiation, and thermo-regulated one which along with aesthetic and mechanical functional is quite a challenge to achieve. The next generation of 3D skin equivalents produced using normal and PS-iPSC-derived keratinocytes by Itoh et al. showed this construct as multilayered and cornified layers on the surface similar to the epidermis of the normal skin.[202] The parakeratosis and the separation along the basement layer was non-specific the findings showed in all that one 3D skin equivalents can be developed was irrespective of cells source. The production of laminin 5 at the dermal-epidermal junction along with K1 on the suprabasal layers all suggested that iPSCs can be explored for fabricating fully functional 3D skin equivalents generated from the patient's own keratinocytes. This study clearly showed the potential of autologous iPSCs in regenerative medicine especially in specific skin trauma and injuries repair. Another group, Bader et al., showed the regeneration of skin with hair follicle and conical structures in patients suffering from deep second-degree burns when they combined the erythropoietin (EPO) (which is known for its protective roles in a wide range of tissues, organs, and cells during and after trauma) expression along with beta common receptor. Complete skin

Vascular—Zwitterionic

regeneration was achieved with local nano-sized recombinant erythropoietin (rhEPO) injection at the site of burn injury and this finding has serious implications with regard to the potentiality of rhEPO application. However, it is the combination of one or more components that could possibly overcome the limitation posed by each individual field. Combining the autologous stem cells with biomaterial science along with cell biology will hold the key to achieving the possible next-generation 3D skin equivalent.

ACKNOWLEDGMENT

The authors would like to thank all the researchers and groups whose work has been cited here. This work is funded by 2012 Yeungnam University Research Grant.

REFERENCES

1. Ahn, S.H.; Yoon, H.; Kim, G.H.; Kim, Y.Y.; Lee, S.H.; Chun, W. Designed three-dimensional collagen scaffolds for skin tissue regeneration. Tissue Eng.: Part C **2010**, *16* (5), 813–821.
2. Haberzeth, S.B.; Biedermann, T.; Reichmann, E. Tissue engineering of skin. Burns **2010**, *36* (4), 450–460.
3. Weller, C.; Sussman, G. Wound dressings update. J. Pharm. Pract. Res. **2006**, *36* (4), 318–324.
4. Suzuki, S.; Matsuda, K.; Nishimura, Y.; Maruguchi, Y.; Maruguchi, T.; Ikada, Y.; Morita, S.I.; Morota, K. Review of acellular and cellular artificial skins. Tissue Eng. **1996**, *2* (4), 267–276.
5. Tobin, D.J. Biochemistry of human skin. Chem. Soc. Rev. **2006**, *35* (1), 52–67.
6. Koch, S.; Kohl, K.; Klein, E. Skin homing of langerhans cell precursors: Adhesion, chemotaxis, and migration. J. Allergy Clin. Immun. **2006**, *117* (1), 163–168.
7. Blumenberg, M.; Tomic-Canic, M. Human epidermal keratinocyte: Kertinization processes. EXS. **1997**, *78*, 1–29.
8. Freed, I.M.; Tomic-Canic, M.; Komine, M. Keratins and the keratinocyte activation cycle. J. Invest. Dermatol. **2001**, *116* (5), 633–640.
9. Kanitakis, J. Anatomy, histology and immunohistochemistry of normal human skin. Eur. J. Dermatol. **2002**, *12* (4), 390–399.
10. Eckes, B.; Krieg, T. Regulation of connective tissue homeostrasis in the skin by mechanical forces. Clin. Exp. Rheumatol. **2004**, *22* (3), S73–S76.
11. Yosipovitch, G.; Hu, J. The importance of skin pH. Skin Aging **2003**, *11* (3), 88–93.
12. Rippke, F.; Schreiner, V.; Doering, T. Stratum corneum pH in atopic dermatitis: Impact on skin barrier function and colonization with Staphylococcus aureus. Am. J. Clin. Dermatol. **2004**, *5* (4), 217–223.
13. Yilmaz, E.; Borchert, H.H. Effect of lipid-containing, positively charged nanoemulsions on skin hydration, elasticity and erythema-An *in vivo* study. Int. J. Pharm. **2006**, *307* (2), 232–238.
14. Kurabayashi, H.; Tamura, K.; Machida, H.I. Inhibiting bacteria and skin pH in hemiplegis: Effects of washing hands with acidic mineral water. Am. J. Phys. Med. Rehab. **2002**, *81* (1), 40–46.
15. Fisher, G.J. The pathophysiology of photoaging of the skin. Cutis **2005**, *75* (2 Suppl.), 5–8.
16. Wilhelmi, B.J.; Blackwell, S.J.; Phillips, L.G. Langer's lines: To use or not to use. Plast. Reconst. Surg. **1999**, *104* (1), 208–214.
17. Vogel, H.G. Directional variations of mechanical parameter in rat skin depending on maturation and age. J. Invest. Dermatol. **1981**, *76* (6), 493–497.
18. Escoffier, C.; Rigal, J.; Rochefort, A.; Vasselet, R.; Leveque, J.L.; Agache, P.G. Age-related mechanical properties of human skin: An *in vivo* study. J. Invest. Dermatol. **1989**, *93* (3), 353–357.
19. Jemec, G.B.; Jemec, B.; Jemec, B.I.; Serup, J. The effect of superficial hydration on the mechanical properties of human skin *in vivo*: Implications for plastic surgery. Plast. Reconst. Surg. **1990**, *85* (1), 100–103.
20. Lu, W.W.; Ip, W.Y.; Holmes, A.D.; Chow, S.P.; Jing, W.M. Biomechanical properties of thin skin flap after basic fibroblast growth factor (bFGF) administration. Brit. J. Plast. Surg. **2000**, *53* (3), 225–229.
21. Nishimori, Y.; Edwards, C.; Pearse, A.; Matsumoto, K.; Kawai, M.; Marks, R. Degenerative alterations of dermal collagen fiber bundles in photodamaged human skin and uv-irradiated hairless mouse skin: Possible effect on decreasing skin mechanical properties and appearance of wrinkles. J. Invest. Dermatol. **2001**, *117* (6), 1458–1463.
22. Marks, F.; Payne, P.A. *Bioengineering and the Skin*; MTP Press: Hingham, MA, 1981; 93–96.
23. Orgill, D.P. The effects of an artificial skin on scarring and contraction in open wounds. Ph D thesis, Harvard University-MIT division of Health sciences and technology. Program in Medical engineering and Medical Physics. 1983; 214.
24. Kumar, V.; Fausto, N.; Robbins, A.A. *Cotran Pathologic Basis of Disease*; 7th Ed.; Elsevier saunders: Philadelphia, 2005; Vol. 15, 1525.
25. Zitelli, J. Wound healing for the clinician. Adv. Dermatol. **1987**, *2*, 243–267.
26. Krawczyk, W.S. A pattern of epidermal cell migration during wound healing. J. Cell Biol. **1971**, *49* (2), 247–263.
27. Farrell, D.H.; Mondhiry, H.A. Human fibroblast adhesion to fibrinogen. Biochemistry **1997**, *36* (5), 1123–1128.
28. Folkman, J. Fundamental concepts of the angiogenic process. Curr. Mol. Med. **2003**, *3* (7), 643–651.
29. Lodish, H.F.; Berk, A.; Zipursky, S.L. Intergrating cells into tissues. In *Molecular Cell Biology*; WH Freeman: New York, 1999; 968–1002.
30. Schulze, H.; Shivdasani, R.A. Mechanisms of thrompopoiesis. J. Thromb. Haemost. **2005**, *3* (8), 1717–1724.
31. Burkitt, H.G.; Young, B.; Heath, J.W. *Blood in: Wheater's Functional Histology: A Text and Colour Atlas*; Churchill Livingstone: Hong Kong, 1993; 42–60.
32. Ruggeri, Z.M.; Mendolicchio, G.L. Adhesion mechanisms in platelet function. Circ. Res. **2007**, *100* (12), 1673–1685.
33. Ruggeri, Z.M. Platelets in atherothrombosis. Nat. Med. **2002**, *8* (11), 1227–1234.
34. Cook-Mills, J.M.; Deem, T.L. Active participation of endothelial cells in inflammation. J. Leukocyte Biol. **2005**, *77* (4), 487–495.

Vascular—Zwitterionic

35. Luscinskas, F.W.; Cybulsky, M.I.; Kiely, J.M. The role of endothelial cell lateral junctions during leukocyte trafficking. Immunol. Rev. **2002**, *186* (1), 57–67.

36. Gopalan, P.K.; Burns, A.R.; Simon, S.I. Preferential sites for stationary adhesion of neutrophils to cytokine-stimulated HUVEC under flow conditions. J. Leukocyte Biol. **2000**, *68* (1), 7–57.

37. Mamdouh, Z.; Chen, X.; Pierini, L.M. Targeted recycling of PECAM from endothelial surface-connected compartments during diapedesis. Nature **2003**, *421* (6924), 748–753.

38. Muller, W.A. Migration of leukocytes across endothelial junctions: Some concepts and controversies. Microcirculation **2001**, *8* (3), 181–193.

39. Shaw, S.K.; Bamba, P.S.; Perkins, B.N. Real-time imaging of vascular endothelial-cadherin during leukocyte transmigration across endothelium. J. Immunol. **2001**, *167* (4), 2323–2330.

40. Tepper, O.M.; Capla, J.M.; Galiano, R.D. Adult vasculogenesis occurs through the in situ recruitment, proliferation, and tubulization of circulating bone marrow derived cells. Blood **2005**, *105* (3), 1068–1077.

41. Reyes, M.; Dudek, A.; Jahagirdar, B. Origin of endothelial progenitors in human postnatal bone marrow. J. Clin. Invest. **2002**, *109* (3), 337–346.

42. Takahashi, T.; Kalka, C.; Masuda, H. Ischemia- and cytokine-induced mobilization of bone marrow-derived endothelial progenitor cells for neovascularization. Nat. Med. **1999**, *5* (4), 434–438.

43. Asahara, T.; Murohara, T.; Sullivan, A. VEGF contributes to postnatal neovascularization by mobilizing bone marrow-derived endothelial cells. EMBO J. **1999**, *18* (14), 3964–3972.

44. Bauer, S.M.; Bauer, R.J.; Velazquez, O.C. Angiogenesis, vasculogenesis, and induction of healing in chronic wounds. Vasc. Endovasc. Surg. **2005**, *39* (4), 293–306.

45. Bauer, S.M.; Bauer, R.J.; Liu, Z.J. Vascular endothelial growth factor-C promotes vasculogenesis, antiogenesis, and collagen constriction in three-dimensional collagen gels. J. Vasc. Surg. **2005**, *41* (4), 699–707.

46. Hanahan, D. Signaling vascular morphogenesis and maintenance. Science **1997**, *277* (5322), 48–50.

47. Carmeliet, P. Mechanism of angiogenesis and arteriogenesis. Nat. Med. **2000**, *6* (4), 389–395.

48. Velazquez, O.C.; Snyder, R.; Liu, Z.J. Fibroblast-dependent differentiation of human microvascular endothelial cells into capillary-like 3-dimensional networks. FASEB J. **2002**, *16* (10), 1316–1318.

49. Liu, Z.J.; Snyder, R.; Souma, A. VEGF-A and alphaV-beta3 integrin synergistically rescue angiogenesis via N-Ras and P13-K signaling in human microvascular endothelial cells. FASEB J. **2003**, *17* (13), 1931–1933.

50. Ceradini, D.J.; Kulkarni, A.R.; Callaghan, M.J. Progenitor cell trafficking is regulated by hypoxic gradients through HIF-1 induction of SDF-1. Nat. Med. **2004**, *10* (8), 858–864.

51. Aicher, A.; Heeschen, C.; Mildner-Rihm, C. Essential role of endothelial nitric oxide synthase for mobilization of stem and progenitor cells. Nat. Med. **2003**, *9* (11), 1370–1376.

52. Arnold, F.; West, D.; Kumar, S. Wound healing: The effect of macrophage and tumor derived angiogenesis factors on skin graft vascularization. Bri. J. Exp. Pathol. **1987**, *68* (4), 569–574.

53. Juliano, R. Cooperation between soluble factor and integrin-mediated cell anchorage in the control of cell growth and differentiation. Bioessays **1996**, *18* (11), 911–917.

54. Shyy, J.Y.; Chien, S. Role of integrins in cellular responses to mechanical stress and adhesion. Curr. Opin. Cell Biol. **1997**, *9* (5), 707–713.

55. Rosenfeldt, H.; Lee, D.J.; Grinnell, F. Increased c-fos mRNA expression by human fibroblasts stressed collagen matrices. Mol. Cell Biol. **1998**, *18* (5), 2659–2667.

56. Werner, S.; Grose, R. Regulation of wound healing by growth factors and cytokines. Physiol. Rev. **2003**, *83* (3), 835–870.

57. Werb, Z.; Tremble, P.; Damsky, C.H. Regulation of extracellular matrix degradation by cell-extracellular matrix interactions. Cell Differ. Dev. **1990**, *32* (3), 299–306.

58. Circolo, A; Welgus, H.G.; Pierce, G.F. Differential regulation of the expression of proteinases/antiproteinases in fibroblasts: Effects of interleukin-1 and platelet-derived growth factor. J. Biol. Chem. **1991**, *266* (19), 12283–12288.

59. Dipietro, L.A.; Nissen, N.N.; Gamelli, R.L. Thrombospondin 1 synthesis and function in wound repair. Am. J. Pathol. **1996**, *148* (6), 1851–1860.

60. Raugi, G.J.; Olerud, J.E.; Gown, A.M. Thrombospondin in early human wound tissue. J. Invest. Dermatol. **1987**, *89* (6), 551–552.

61. Reed, M.J.; Puolakkainen, P.; Lane, T.F. Differential expression of SPARC and thrombospondin 1 in wound repair: Immunolocalization and in situ hybridization. J. Histochem. Cytochem. **1993**, *41* (10), 1467–1477.

62. Rubis, B.A.; Danikas, D.; Neumeister, M.; Williams, W.G.; Suchy, H.; Milner, S.M. The use of split-thickness dermal grafts to resurface full thickness skin. Burns **2002**, *28* (8), 752–759.

63. Vallet, V.; Cruz, C.; Josse, D.; Bazire, A.; Lallement, G.; Boudry, I. *In vitro* percutaneous penetration of organophosphorus compounds using full-thickness and split-thickness pig and human skin. Toxicol. *In Vitro* **2007**, *21* (6), 1182–1190.

64. Hynes, P.J.; Earley, M.J.; Lawlor, D. Split-thickness skin grafts and negative-pressure dressings in the treatment of axillary hidradenitis suppurativa. Br. J. Plast. Surg. **2002**, *55* (6), 507–509.

65. Sidebottm, A.J.; Moore, S.M.; Magennis, P.; Devine, J.C.; Brown, J.S.; Vaughan, E.D. Repair of the radial free flap donor site with full or partial thickness skin grafts: A prospective randomized controlled trial. Int. J. Oral Max. Surg. **2000**, *29* (3), 194–197.

66. Suliman, M.T. A simple method to facilitate full-thickness skin graft harvest. Burns **2009**, *35* (1), 87–88.

67. Wood, F.M.; Kolybaba, M.L.; Allen, P. The use of cultured epithelial autograft in the treatment of major burn wounds: Eleven years of clinical experience. Burns **2006**, *32* (5), 538–544.

68. Goldstein, S.; Clarke, D.R.; Walsh, S.P.; Black, K.S.; O'Brien, M.F. Transpecies heart valve transplant: Advanced studies of a bioengineered xeno-autograft. Ann. Thorac. Surg. **2000**, *70* (6), 1962–1969.

69. Malak, S.; Anderson, I.A. Orthogonal cutting of cancellous bone with application to the harvesting of bone autograft. Med. Eng. Phys. **2008**, *30* (6), 717–724.

Vascular—Zwitterionic

70. Anderl, H.; Menardi, G.; Hager, J. Closure of gastroschisis by mesh skin grafts in problem cases. J. Pediatr. Surg. **1986**, *21* (10), 870–872.

71. Kreis, R.W.; Mackie, D.P.; Hermans, R.P.; Vloemans, A.R. Expansion techniques for skin grafts: Comparison between mesh and Meek island (sandwich-) grafts. Burns **1994**, *20* (1), 39–42.

72. Peeters, R.; Hubens, A. The mesh skin graft-true expansion rate. Burns **1988**, *14* (3), 239–240.

73. Turner, C.; Fauza, D.O. Fetal tissue engineering. Clin. Perinatol. **2009**, *36* (2), 473–488.

74. Bhat, S.; Kumar, A. Cell proliferation on three-dimensional chitosan-agarose-gelatin cryogel scaffolds for tissue engineering applications. J. Biosci. Bioeng. **2012**, *114* (6), 663–670.

75. Boyce, S.T. Fabrication, quality assurance, and assessment of cultured skin substitutes for treatment of skin wounds. Biochem. Eng. J. **2004**, *20* (2–3), 107–112.

76. Atiyeh, B.S.; Hayek, S.N.; Gunn, S.W. New technologies for burn wound closure and healing-review of the literature. Burns **2005**, *31* (8), 944–956.

77. Snyder, R.J. Treatment of nonhealing ulcers with allografts. Clin. Dermatol. **2005**, *23* (4), 388–395.

78. Conklin, B.S.; Richter, E.R.; Kreutziger, K.L.; Zhong, D.S.; Chen, C. Development and evaluation of a novel decellularized vascular xenograft. Med. Eng. Phys. **2002**, *24* (3), 173–183.

79. Zarbock, A.; Polanowska-Grabowska, R.K.; Ley, K. Platelet-neutrophil-interactions: Liking hemostasis and inflammation. Blood Rev. **2007**, *21* (2), 99–111.

80. Engelhardt, E.; Toksoy, A.; Goebeler, M. Chemokines IL-8, GRO-alpha, MCP-1, IP-10, and Mig are sequentially and differentially expressed during phase-specific infiltration of leukocyte subsets in human wound healing. Am. J. Pathol. **1998**, *153* (6), 1849–1860.

81. Brooks, P.C.; Stromblad, S.; Sanders, L.C. Localization of matrix metalloproteinase MMP-2 to the surface of invasive cells by interaction with integrin alpha v beta 3. Cell **1996**, *85* (5), 683–693.

82. Esser, S.; Wolburg, K.; Wolburg, H. Vascular endothelial growth factor induces endothelial fenestrations *in vitro*. J. Cell Biol. **1998**, *140* (4), 947–959.

83. Watt, F.M.; Hogan, B.L. Out of eden: Stem cells and their niches. Science **2000**, *287* (5457), 1427–1430.

84. Spradling, A.; Drummond-Barbarosa, D.; Kai, T. Stem cells find their niche. Nature **2001**, *414* (6859), 98–104.

85. Sheyn, D.; Mizrahi, O.; Benjamin, S.; Gazit, Z.; Pelled, G.; Gazit, D. Genetically modified cells in regenerative medicine and tissue engineering. Adv. Drug Deliv. Rev. **2010**, *62* (7–8), 683–698.

86. Sundelacruz, S.; Kaplan, D.L. Stem cell- and scaffold-based tissue engineering approaches to osteochondral regenerative medicine. Semin. Cell Dev. Biol. **2009**, *20* (6), 646–655.

87. Kim, H.N.; Jiao, A.; Hwang, N.S.; Kim, M.S.; Kang, D.H.; Suh, K.Y. Nanotopography-guided tissue engineering and regenerative medicine. Adv. Drug Deliv. Rev. **2013**, *65* (4), 536–558.

88. Priya, S.G.; Jungvid, H.; Kumar, A. Skin tissue engineering for tissue repair and regeneration. Tissue Eng. B **2008**, *14* (1), 105–118.

89. Horch, R.E.; Kopp, J.; Kneser, U.; Beier, J.; Bach, A.D. Tissue engineering of cultured skin substitutes. Tissue Eng. Rev. **2005**, *9* (3), 592–608.

90. Unal, S. Treatment options for skin and soft tissue infections: 'oldies but goldies'. Int. J. Antimicrob. Ag. **2009**, *34* (S1), S20–S23.

91. Kim, S.S.; Gawk, S.J.; Choi, C.Y.; Kim, B.S. Skin regeneration using keratinocytes and dermal fibroblasts cultured on biodegradable microspherical polymer scaffolds. J. Biomed. Mater. Res. Part B **2005**, *75* (2), 369–377.

92. Chen, M.; Przyborowski, M.; Berthiaume, F. Stem cells for skin tissue engineering and wound. Crit. Rev. Biomed. Eng. **2009**, *37* (4–5), 399–421.

93. Salahat, M.A.; Hadid, L.A. Autologous adipose stem cells use for skin regeneration and treatment in humans. J. Biol. Agri. Health Care **2013**, *3* (1), 135–139.

94. Atiyeh, B.S.; Ioannovich, J.; Al-Amm, C.A.; EI-Musa, K.A. Management of acute and chronic open wounds: The importance of moist environment in optimal wound healing. Curr. Pharm. Biotechnol. **2002**, *3* (3), 179–195.

95. Wu, Y.; Chen, I.; Scott, P.G.; Tredget, E.E. Mesenchymal stem cells enhance wound healing through differentiation and angiogenesis. Stem Cells. **2007**, *25* (10), 2648–2659.

96. Rhett, J.M.; Ghatnekar, G.S.; Palatinus, J.A.; O'Quinn, M.; Yost, M.J.; Gourdie, R.G. Novel therapies for scar reduction and regenerative healing of skin wounds. Trends Biotechnol. **2008**, *26* (4), 173–180.

97. Weissman, I.L. Stem cells: Units of development, units of regeneration, and units in evolution. Cell **2000**, *100* (1), 157–168.

98. Fu, S.; Liesveld, J. Mobilization of hematopoietic stem cells. Blood Rev. **2000**, *14* (4), 205–218.

99. Kucia, M.; Ratajczak, J.; Reca, R.; Janowska-Wieczorek, A.; Ratajczak, M. Z. Tissue-specific muscle, neural and liver stem/progenitor cells reside in the bone marrow, respond to an SDF-1 gradient and are mobilized into peripheral blood during stress and tissue injury. Blood Cells Mol. Dis. **2004**, *32* (1), 52–57.

100. Badiavas, E.V.; Abedi, M.; Butmare, J.; Falanga, V.; Quesenberry, P. Participation of bone marrow derived cells in cutaneous wound healing. J. Cell Physiol. **2003**, *196* (2), 245–250.

101. Karam, J.P.; Muscari, C.; Montero-Menei, C.N. Combining adult stem cells and polymeric devices for tissue engineering in infracted myocardium. Biomaterials **2012**, *33* (23), 5683–5695.

102. Sanchez-Adams, J.; Athanasiou, K.A. Dermis isolated adult stem cells for cartilage tissue engineering. Biomaterials **2012**, *33* (1), 109–119.

103. Elena, R.; Frederick, J.S. Cardiovascular tissue engineering. Cardiovasc. Pathol. **2002**, *11* (6), 305–317.

104. Catherine, P.; Barnes, S.A.; Sell, E.D.; Boland, D.G.; Simpson, G.B. Nanofiber technology: Designing the next generation of tissue engineering scaffolds. Adv. Drug Deliv. Rev. **2007**, *59* (14), 1413–1433.

105. Joerg, K.; Tessmar, A.M.; Gopferich. Matrices and scaffolds for protein delivery in tissue engineering. Adv. Drug Deliv. Rev. **2007**, *59* (4–5), 274–291.

106. Robert, R.A.; Phillips, J.B. Cell responses to biomimetic protein scaffolds used in tissue repair and engineering. Int. Rev. Cytol. **2007**, *262*, 75–150.

107. Lee, S.H.; Shin, H. Matrices and scaffolds for delivery of bioactive molecules in bone and cartilage tissue engineering. Adv. Drug Deliv. Rev. **2007**, *59* (4–5), 339–359.

108. Tsang, V.L.; Bhatia, S.N. Three-dimensional tissue fabrication. Adv. Drug Deliv. Rev. **2004**, *56* (11), 1635–1647.

109. Hutmacher, D.W. Scaffold in tissue engineering bone and cartilage. Biomaterials **2000**, *21* (24), 2529–2543.

110. Mikos, A.G.; Sarakinos, G.; Leite, S.M.; Vacanti, J.P.; Langer, R. Laminated three dimensional biodegradable foams for use in tissue engineering. Biomaterials **1993**, *14* (5), 323–330.

111. Annabi, N.; Nichol, J.W.; Zhong, X.; Ji, C.; Koshy, S.; Khademhosseini, A.; Dehqhani, F. Controlling the porosity and microarchitecture of hydrogels for tissue engineering. Tissue Eng. B Rev. **2010**, *16* (4), 371–383.

112. Watson, M.S.; Whitaker, M.J.; Howdle, S.M.; Shakesheff, K.M. Incorporation of proteins into polymer materials by a novel supercritical fluid processing method. Adv. Mater. **2002**, *14* (24), 1802–1804.

113. Howdle, S.M.; Watson, M.S.; Whitaker, M.J.; Popov, V.K.; Davies, M.C.; Mandel, F.S.; Wang, J.D.; Shakesheff, K.M. Supercritical fluid mixing: Preparation of thermally sensitive polymer composites containing bioactive materials. Chem. Commun. **2001**, 109–110.

114. Ginty, P.J.; Howard, D.; Rose, F.; Whitaker, M.J. Mammalian cell survival and processing in supercritical CO₂, Proc. Natl. Acad. Sci. USA. **2002**, *103* (19), 7426–7431.

115. Tomasko, D.L.; Guo, Z. Supercritical fluids. In *Kirk-Othmer Encyclopedia of Chemical Technology*; John Wiley and Sons: New York, 2006.

116. Barry, J.; Silva, M.; Popov, V.K.; Shakesheff, K.M.; Howdle, S.M. Supercritical carbon dioxide: Putting the fizz into biomaterials. Philos. Trans. A. Math. Phys. Eng. Sci. **2006**, *364* (1838), 249–261.

117. Tai, H.; Mather, M.L.; Howard, D.; Wang, W.; White, L.J.; Crowe, J.A. Control of pore size and structure of tissue engineering scaffolds produced by supercritical fluid processing. Eur. Cells Mater. **2007**, *14*, 64–77.

118. Xu, Z.; Jiang, X.; Liu, T.; Hu, G.; Zhao, L.; Zhu, Z. Foaming of polypropylene with supercritical carbon dioxide. J. Supercrit. Fluid 2007, *41* (2), 299–310.

119. Nalawade, S.P.; Picchioni, F.; Marsman, J.H.; Grijpma, D.W.; Feijen, J.; Janssen, L.P.B.M. Intermolecular interactions between carbon dioxide and the carbonyl groups of polylactides and poly(e-caprolactone). J. Control. Release **2006**, *116* (2), e38–e40.

120. Kazarian, S.G. Polymer processing with supercritical fluids. Polym. Sci. Ser. C **2000**, *42* (1), 78–101.

121. Nalawade, S.P.; Picchioni, F.; Janssen, L.P.B.M. Supercritical carbon dioxide as a green solvent for processing polymer melts: Processing aspects and applications. Prog. Polym. Sci. **2006**, *31* (1), 19–43.

122. Annabi, N.; Fathi, A.; Mithieux, S.M.; Weiss, A.S.; Dehghani, F. Fabrication of porous PCL/elastin composite scaffolds for tissue engineering applications. J. Supercrit. Fluid **2011**, *59*, 157–167.

123. Harris, L.D.; Kim, B.S.; Mooney, D.J. Open pore biodegradable matrixes formed with gas foaming. J. Biomed. Mater. Res. **1998**, *42* (3), 396–402.

124. Hile, D.D.; Amirpour, M.L.; Akgerman, A.; Pishko, M.V. Active growth factor delivery from poly(D,L-lactide-co-glycolide) foams prepared in supercritical CO₂. J. Control. Release **2000**, *66* (2–3), 177–185.

125. Lopez-Periago, A.M.; Vega, A.; Subra, P.; Argemi, A.; Saurina, J.; Garcia-Gonzalez, C.A. Supercritical CO₂ processing of polymers for the production of materials with applications in tissue engineering and drug delivery. J. Mater. Sci. **2008**, *43* (6), 1939–1947.

126. Gualandi, C.; White, L.J.; Chen, L.; Gross, R.A.; Shakesheff, K.M.; Howdle, S.M. Scaffold for tissue engineering fabricated by non-isothermal supercritical carbon dioxide foaming of a highly crystalline polyester. Acta Biomater. **2010**, *6* (1), 130–136.

127. Cooper, A.I. Polymer synthesis and processing using supercritical carbon dioxide. J. Mater. Chem. **2000**, *10* (2), 207–234.

128. Goel, S.K.; Beckman, E.J. Generation of microcellular polymeric foams using supercritical carbon dioxide I. Effect of pressure and temperature on nucleation. Polym. Eng. Sci. **1994**, *34* (14), 1137–1147.

129. Kanczler, J.M.; Ginty, P.J.; White, L.; Clarke, N.; Howdle, S.M.; Shakesheff, K.M. The effect of the delivery of vascular endothelial growth factor and bone morphogenic protein-2 to osteoprogenitor cell populations on bone formation. Biomaterials **2010**, *31* (6), 1242–1250.

130. Heyde, M.; Partridge, K.; Howdle, S.M.; Oreffo, R.O.C.; Garnett, M.; Shakesheff, K.M. Development of a slow non-viral DNA release system from PDLLA scaffolds fabriacted using a supercritical CO2 technique. Biotechnol. Bioeng. **2007**, *98* (3), 679–693.

131. Aubert, J.H.; Clough, R.L. Low density, microcellular polystyrene foams. Polymer **1985**, *26* (13), 2047–2054.

132. Lloyd, D.R.; Kinzer, K.E.; Tseng, H.S. Microporous membrane formation via thermally induced phase separation. I. Solid-liquid phase separation. J. Membr. Sci. **1990**, *52* (3), 239–261.

133. Lloyd D.R. Microporous membrane formation via thermally induced phase separation. II. Liquid-liquid phase separation. J. Membr. Sci. **1991**, *64* (1–2), 1–11.

134. Nam, Y.S.; Park, T.G. Biodegradable polymeric microcellular foams by modified thermally induced phase separation method. Biomaterials **1999**, *20* (19), 1783–1790.

135. Heijkants, R.G.; van Calck, R.V.; De Groot, J.H.; Pennings, A.J.; Schouten, A.J. Design synthesis and properties of a degradable polyurethane scaffold for meniscus regeneration. J. Mater. Sci. **2004**, *15* (4), 423–427.

136. Lee, S.H.; Kim, B.S.; Kim, S.H.; Kang, S.W.; Kim, Y.H.Thermally produced biodegradable scaffolds for cartilage tissue engineering. Macromol. Biosci. **2004**, *4* (8), 802–810.

137. van de Witte, P.; Dijkstra, P.J.; van den Berg, J.W.A.; Feijen, J. Phase separation processes in polymer solutions in relation to membrane formation. J. Membr. Sci. **1996**, *117* (1–2), 1–31.

138. Mulder, M. *Preparation of Synthetic Membranes*; Kluwer Academic Publishers: Dordrecht, Netherlands, 1996; 71–156.

139. Bansil, R.; Liao, G. Kinetics of spinodal decomposition in homopolymer solutions andgels. Trends Polym. Sci **1997**, *5* (5), 146–154.

140. Chung, H.J.; Park, T.G. Surface engineered and drug releasing pre-fabricated scaffolds for tissue engineering. Adv. Drug Deliv. Rev. **2007**, *59* (4–5), 249–262.

141. Zhong, S.P.; Teo, W.E.; Zhu, X.; Beuerman, R.; Ramakrishna, S.; Lanry-Yang, L.Y. Development of a novel collagen-GAG nanofibrous scaffold via electrospinning. Mater. Sci. Eng. C **2007**, *27* (2), 262–266.

142. Zander, N.F. Hierarchically structured electrospun fibers. Polymer **2013**, *5* (1), 19–44.

143. Huang, Z.M.; Zhang, Y.Z.; Kotaki, M.; Ramakrishna, S. A review on polymer nanofibers by electrospinning and their applications in nanocomposites. Composites Sci. Technol. **2003**, *63* (15), 2223–2253.

144. Li, W.J.; Laurencin, C.T.; Caterson, E.J.; Tuan, R.S.; Ko, F.K. Electrospun nanofibrous structure: A novel scaffold for tissue engineering. J. Biomed. Mater. Res. **2002**, *60* (4), 613–621.

145. Reneker, F.H.; Yarin, A.L.; Fong, H.; Koombhongse, S. Bending instability of electrically charged liquid jets of polymer solutions in electrospinning. J. Appl. Phys. **2000**, *87* (9), 4531–4547.

146. Zong, X.; Fang, D.; Kim, K. S.; Kim, J.; Cruz, S.; Bengamin, S.; Hsiao.; Chu, B. Structure and relationships in bioabsorbable nanofiber membranes by electrospinning. Polymer **2002**, *43* (16), 4403–4412.

147. Rutledge, G.C.; Li, S.F.Y.; Warner, S.B.; Kalayci, P.P. Electrostatic Spinning and Properties of Ultrafine Fibers. National Textile Center Annual Report, 2001.

148. Campbell, A.M.; Briggs, R.C.; Bird, R.E.; Hnilica, L.S. Cell specific antiserum to chromosome scaffold proteins. Nucl. Acids Res. **1979**, *6* (1), 205–218.

149. Mi, S.; Connon, C.J. The formation of a tissue-engineered corean using plastically compressed collagen scaffolds and limbal stem cells. Methods Mol. Biol. **2013**, *1014*, 143–155.

150. Holan, V.; Javorkova, E.; Trosan, P. The growth and delivery of mesenchymal and limbal stem cells using copolymer polyamide 6/12 nanofiber scaffolds. Methods Mol. Biol. **2013**, *1014*, 187–199.

151. Kawano, T.; Nakamichi, Y.; Fujinami, S.; Nakajima, K.; Yabu, H.; Shimomura, M. Mechanical regulation of cellular adhesion onto honeycomb-patterned porous scaffolds by altering the elasticity of material surfaces. Biomacromolecules **2013**, *14* (4), 1208–1213.

152. Ravichandran, R.; Venugopal, J.R.; Sundarrajan, S.; Mukherjee, S.; Ramakrishna, S. Cardiogenic differentiation of mesenchymal stem cells on elastomeric poly(glycerol sebacate)/collagen core/shell fibers. World J. Cardiol. **2013**, *5* (3), 28–41.

153. Mirmalek-Sani, S.H.; Orlando, G.; McQuilling, J.P.; Pareta, R.; Mack, D.L.; Salvatori, M.; Farney, A.C.; Stratta, R.J.; Atala, A.; Opara, E C.; Soker, S. Porcine pancreas extracellular matrix as a platform for endocrine pancreas bioengineering. Biomaterials **2013**, *34* (22), 5488–5495.

154. Da-Silva, A.C.; Rodrigues, R.; Rosa, L.F.; de-Carvalho, J.; Tome, B.; Ferreira, G.N. Acoustic detection of cell adhesion on a quartz crystal microbalance. Biotechnol. Appl. Biochem. **2012**, *59* (6), 411–419.

155. Carr, H.S.; Zuo, Y.; Oh, W.; Frost, J.A. Regulation of FAK activation, breast cancer cell motility and amoeboid invasion by the RhoA GEF Net1. Mol. Cell Biol. **2013**, *33* (14), 2778–2786.

156. Luo, Y.W.; Zhao, Z.; Li, F.N.; Zhang, J. The effect pathways analysis in the abdominal aortic aneurysms. Rev. Med. Pharmacol. Sci. **2013**, *17* (9), 1245–1251.

157. Kaihara, S.; Sumimoto, A.; Fujimoto, K. Photoreactive nanotool for cell aggregation and immobilization. J. Biomater. Sci. Polym. Ed. **2013**, *24* (6), 714–725.

158. Chen, A.K.; Delri, F.W.; Peterson, A.W.; Chung, K.H.; Bhadiragu, K.; Plant, A.L. Cell spreading and proliferation in response to the composition and mechanics of engineered fibrillar extracellular matrices. Biotechnol. Bioeng. **2013**, *110* (10), 2731–2741.

159. Yang, D.; Lu, X.; Hong, Y.; Xi, T.; Zhang, D. The molecular mechanism of mediation of adsorbed serum proteins to endothelial cells adhesion and growth on biomaterials. Biomaterials **2013**, *34* (23), 5747–5758.

160. Pater, N.; Padera, R.; Sanders, G.H.; Gannizzaro, S.M.; Davies, M.C.; Langer, R.; Roberts, C.J.; Tendler, S.J.; Williams, P.M.; Shakesheff, K.M. Spatially controlled cell engineering on biodegradable polymer surfaces. FASEB J. **1998**, *12* (14), 1447–1454.

161. Ramtani, S. Mechanical modeling of cell/ECM and cell/cell interactions during the contraction of a fibroblast populated collagen microsphere: Theory and model simulation. J. Biomech. **2004**, *37* (11), 1709–1718.

162. Benta, H.; Schroeder, J.A.; Estridge, T.D. Improved local delivery of TGF-beta2 by binding to injectable fibrillar collagen via difunctional polyethylebe glycol. J. Biomed. Maer. Res. **1998**, *39* (4), 539–548.

163. Ito, Y.; Inoue, M.; Liu, S.Q.; Imanishi, Y. Cell growth on immobilized cell growth factov. 6. Enhancement of fibroblast cell growth by immobilized insulin and/or fibronectin. J. Biomed. Mater. Res. **1993**, *27* (7), 901–907.

164. Kuhl, P.R.; Griffith-Cima, L.G. Tethered epidermal growth factor as a paradigm for growth factor-induced stimulation from the solid phase. Nat. Med. **1996**, *2* (9), 1022–1027.

165. Baronio, F.; Battisti, L.; Radetti, G. Central hypothyroidism following chemotherapy for acute lymphoblastic leukemia. J. Pediatr. Endocrinol. Metab. **2011**, *24* (11–12), 903–906.

166. Ball, S.G.; Shuttleworth, C.A.; Kielty, C.M. Mesenchymal stem cells and neovascularization: Role of platelet-derived growth factor receptors. J. Cell Mol. Med. **2007**, *11* (5), 1012–1030.

167. Reverdin, J.L. De la greffe epidermique. Paris. **1872**.

168. Thiersch, J.C. On skin grafting. Verhandl 2nd Deutsch Ges Chir **1886**, *15*, 17–20.

169. Macneil, S. Biomaterials for tissue engineering of skin. Mater. Today **2008**, *11* (5), 26-35.

170. Horch, R.E.; Jeschke, M.G.; Spilker, G.; Herndon, D.N.; Kopp, J. Treatment of second-degree facial burns with allografts-preliminary results. Burns **2005**, *31* (5), 597–602.

171. Lee, K.H. Tissue-engineered human living skin substitutes: Development and clinical application. Yonsei Med. J. **2000**, *41* (6), 774–779.

172. Mangoldt, F. Die uberhautung von wundflachen und wundhohlen durch epithelausaat, eine neue Methode der Transplantation. Deut. Med. Wschr. **1895**, *21*, 798–799.

173. Pels-Leusden, F. Die anwendung des spalthautlappens in der chirurgie. Deut. Med. Wschr. **1905**, *31* (2), 99–102.

174. Billinghan, R.; Reynolds, J. Transplantation studies on sheet of pure epidermal epithelium and of epidermal cell suspensions. J. Plast. Surg. **1952**, *5* (1), 25–36.

175. Horch, R.E.; Bannasch, H.; Kopp, J.; Andree, C.; Stark, G.B. Single-cell suspensions of cultured human keratinocytes in fibrin-glue reconstitute the epidermis. Cell. Tranplant. **1998**, *7* (3), 309–317.

176. Yannas, I.V.; Burke, J.F.; Orgill, D.P.; Skrabut, E. M. Wound tissue can utilize a polymeric template to synthesize a functional extension of skin. Science **1982**, *215* (4529), 174–176.

177. Yannas, I.V.; Lee, E.; Orgill, D.P.; Skrabut, E.M.; Murphy, G.F. Synthesis and characterization of a model extracellular matrix that induces partial regeneration of adult mammalian skin. Proc. Natl. Acad. Sci. USA **1989**, *86* (3), 933–937.

178. Antonios, G.M.; Georgios, S.; Susan, M.L.; Vacanti, J.P.; Langer, R. Laminated three-dimensional biodegradable foams for use in tissue engineering. Biomaterials **1993**, *14* (5), 323–330.

179. Puelacher, W.C.; Kim, S.W.; Vacanti, J.P.; Schloo, B.; Mooney, D.; Vacanti, C.A. Tissue-engineered growth of cartilage: The effect of varying the concentration of chondrocytes seeded onto synthetic polymer matrices. Int. J. Oral. Max. Surg. **1994**, *23* (1), 49–53.

180. Murphy, G.F.; Orgill, D.P.; Yannas, I.V. Partial dermal regeneration is induced by biodegradable collagen-glycosaminoglycan grafts. Lab Invest. **1990**, *62* (3), 305–313.

181. Choi, S.M.; Singh, D.; Kumar, A.; Oh, T.H.; Cho, Y.W.; Han, S.S. Porous three-dimensional PVA/Gealtin sponge for skin tissue engineering. Int. J. Polym. Mater. **2013**, *62* (7), 384–389.

182. Choi, S.M.; Singh, D.; Cho, Y.W.; Oh, T.H.; Han, S.S. Three-dimensional porous HPMA-co-DMAEM hydrogels for biomedical application. Colloid Polym. Sci. **2013**, *291* (5), 1121–1133.

183. Stock, U.A.; Vacanti, J.P. Tissue engineering: Current state and prospects. Annu. Rev. Med. **2001**, *52* (1), 443–451.

184. Robson, M.C.; Krizek, T.J.; Hegger, J.P. Biology of surgical infection. Curr. Probl. Surg. **1973**, *10* (3), 1–62.

185. Gokoo, C. A primer on wound bed preparation. J. Am. Coll. Certified Wound Specialists **2009**, *1* (1), 35–39.

186. Purna, S.K.; Babu, M. Collagen based dressings-a review. Burns **2000**, *26* (1), 54–62.

187. Liu, F.; Wu, J. Artificial skin cultured *in vitro*. J. Biomed. Eng. **2000**, *17* (2), 223–225.

188. Ma, L.; Gao, C.; Mao, Z.; Zhou, J.; Shen, J.; Hu, X.; Han, C. Collagen/chitosan porous scaffolds with improved biostability for skin tissue engineering. Biomaterials **2003**, *24* (26), 4833–4841.

189. Dai, N.T.; Williamson, M.R.; Khammo, N.; Adams, E.F.; Coombes, A.G. Composite cell support membranes based on collagen and polycaprolactone for tissue engineering of skin. Biomaterials **2004**, *25* (18), 4263–4271.

190. Adekogbe, I.; Ghanem, A. Fabrication and characterization of DTBP-crosslinked chitosan scaffolds for skin tissue engineering. Biomaterials **2005**, *26* (35), 7241–7250.

191. Mohammadi, A.A.; Seyed-Jafari, S.M.; Kiasat, M.; Tavakkolian, A.R.; Imani, M.T.; Ayaz, M.; Tolide-ie, H.R. Effect of fresh human amniotic membrane dressing on graft take in patients with chronic burn wounds compared with conventional methods. Burns **2013**, *39* (2), 349–353.

192. Patel, N.P.; Cermino, A.L. Keloid treatment: Is there a role for acellular human dermis(Alloderm)?. J. Plast. Reconstr. Aes. **2010**, *63* (8), 1344–1348.

193. Adly, O.A.; Moghazy, A.M.; Abbas, A.H.; Ellabban, A.M.; Ali, O.S.; Mohamed, B.A. Assessment of amniotic and polyurethane membrane dressings in the treatment of burns. Burns **2010**, *36* (5), 703–710.

194. Burke, J.F.; Yannas, I.V.; Quinby, W.C.; Bondoc, C.C.; Jung, W.K. Successful use of a physiologically acceptable artificial skin in the treatment of extensive burn injury. Ann. Surg. **1981**, *194* (4), 413–428.

195. Haberal, M.; Oner, Z.; Bayraktar, U.; Bilgin, N. The use of silver nitrate-incorporated amniotic membrane as a temporary. Burns **1987**, *13* (2), 159–163.

196. Vermette, M.; Trottier, V.; Menard, V.; Saint-Pierre, L.; Roy, A.; Fradette, J. Production of a new tissue-engineered adipose substitute from human adipose-derived stromal cells. Biomaterials **2007**, *28* (18), 2850–2860.

197. Sandoval-Sanchez, J.H.; Zuniga, R.; Anda, S.L.; Lopez-Dellamary, F. A new bilayer chitosan scaffolding as a dural substitute: Experimental evaluation. World Neurosurg. **2012**, *77* (3–4), 577–582.

198. Fu, Y.; Yuan, R.; Chai, Y.; Liu, Y.; Tang, D.; Zhang, Y. Electrochemical impedance behavior of DNA biosensor based on colloidal Ag and bilayer two-dimensional sol-gel as matrices. J. Biochem. Biophys. Methods **2005**, *62* (2), 163–174.

199. Kinikoglu, B.; Rodriguez-Cabello, J.C.; Damour, O.; Hasirci, V. A smart bilayer scaffold of elastin-like recombinamer and collagen for soft tissue engineering. J. Mater. Sci. Mater. Med. **2011**, *22* (6), 1541–1554.

200. Franco, R.A.; Min, Y.K.; Yang, H.M.; Lee, B.T. Fabrication and biocompatibility of novel bilayer scaffold for skin tissue engineering applications. J. Biomater. Appl. **2012**, *27* (5), 605–615.

201. Pham, C.; Greenwood, J.; Cleland, H.; Woodruff, P.; Maddern, G. Bioengineered skin substitutes for the management of burns: A sysmatic review. Burns **2007**, *33* (8), 946–957.

202. Itoh, M.; Umegaki-Arao, N.; Guo, Z.; Liu, L.; Higgins, C.A.; Christiano, A.M. Generation of 3D skin equizalents fully reconstituted from human induced pluripotent stem cells (iPSCs). Plos One **2013**, *8* (10), e77673.

Wound Healing: Hemoderivatives and Biopolymers

Silvia Rossi
Franca Ferrari
Giuseppina Sandri
Maria Cristina Bonferoni
Department of Drug Sciences, University of Pavia, Pavia, Italy

Claudia Del Fante
Cesare Perotti
Immunohaematology and Transfusion Service and Cell Therapy Unit, San Matteo University Hospital, Pavia, Italy

Carla Caramella
Department of Drug Sciences, University of Pavia, Pavia, Italy

Abstract

The present entry deals with the employment of some hydrophilic polymers known for their bioactive properties as enabling excipients of novel therapeutic systems intended for wound healing. After a preliminary description of wound classification and the healing process, the mechanisms underlying tissue repairing properties of chitosan, hyaluronic acid, and chondroitin sulfate have been reviewed. Beside these biopolymers, particular attention has been focused on the use of hemoderivatives, namely platelet derivatives [platelet-rich plasma, platelet gel, and platelet lysate (PL)] in the treatment of epithelial (cutaneous, corneal, and mucosal) lesions. Since the authors have gained experience on the association of the above-mentioned biopolymers with PL, an overview of the newly developed platforms for the delivery of such platelet derivative in skin and corneal lesions is reported.

INTRODUCTION

In the past decades, there was a progressive change in the therapeutic approaches employed for the wound treatment. Traditional treatments provided the use of saline and disinfectant solutions to clean the wound bed, followed by the application of dressings with the unique task of isolating the lesion from the infected environment. Such "inert" dressings have been substituted by modern ones capable of promoting healing. The bioactive properties of modern dressings are due to the use of the so-called biopolymers that are capable of interacting with tissue components, taking part in the healing process.

In the present entry, attention will be focused on three polysaccharides: chitosan, hyaluronic acid, and chondroitin sulfate. In particular, after a brief description of different types of wounds and of the healing process, the roles of such polymers in the cascade of events leading to wound closure will be discussed.

In the past decades, many papers have been published on the use of biopolymers in association with growth factors (GFs) that are generally recognized as the "leading actors" of wound healing. Some of them are available on the market in purified form and have been loaded in biopolymers-based formulations with a consequent improvement in the efficacy of the treatment. Recently, it has been pointed out that tissue repair cannot be effectively mediated by a single agent or GF, as multiple signals are required to complete the regeneration process. Given that, since platelets constitute a potential source of multiple GFs and proteins involved in tissue regeneration, the therapeutic employment of platelet-rich preparations has been suggested.

The second part of this entry will provide an overview of platelet-rich preparations employed in wound healing. In particular, the following hemoderivatives will be considered: 1) platelet-rich plasma (PRP); 2) platelet gel (PG); and 3) platelet lysate (PL). Finally, the entry will bring the reader's attention to the use of such hemoderivatives in association with the above-mentioned biopolymers. In particular, the authors' experience in the preparation and characterization of PL-biopolymers–based formulations will be described.

WOUNDS

A wound is the result of a breach or a tear in the normal anatomic structure and function of the skin. In literature, a distinction between acute and chronic wounds is reported: the former are lesions that heal within 8–12 weeks, the latter do not heal properly and persist beyond 12 weeks and

Concise Encyclopedia of Biomedical Polymers and Polymeric Biomaterials DOI: 10.1081/E-EBPPC-120050387

often recur.[1,2] While acute wounds are caused by mechanical injures, exposure to thermal sources, irradiation, electrical shock, and/or corrosive chemicals, chronic wounds are generally linked to specific diseases such as diabetes, tumors, immunodeficiency, metabolic, and connective tissue disorders.[2] These pathologies interrupt the orderly sequence of events occurring during the healing process, whose phases will be summarized in the subsequent paragraph. Chronic wounds comprise: 1) decubitus (bedsores or pressure) ulcers caused principally by unrelieved pressure; 2) vascular ulcers due to deficits in either quality of arterial flow (as a result of restriction or occlusion of an artery) or derangement of the venous system (as a result of deep vein thrombosis, obesity, congestive heart failure, severe trauma to limbs); and 3) neurophatic ulcers (as a result of insensate foot).[3]

The term "wounds" can be also used in a broad sense to indicate lesions against an integument covering a body cavity or surface; in this perspective, mucositis of the oral cavity as well as lesions of the corneal epithelium are included.

Oral mucositis is a pathological condition induced by chemiotherapy and/or radiotherapy that are widely employed in the treatment of cancer and in the conditioning therapy before bone marrow transplantation and before the allogenic transplantation of stem hematopoietic cells.[4,5] It leads to atrophy of the epithelium and to ulcerations that can be so severe to cause the interruption or the reduction of the treatment. Opportunistic infections can rise and parenteral nutrition of patients can become necessary. Few interventions of efficacy are available and there are still no universally accepted treatment protocols.

The corneal epithelium forms a protective barrier and serves as the main refractive element of the visual system. Corneal wounds can be caused by injuries or surgical interventions or pathologies such as keratitis and keratopathy. Healing is normally fast to reestablish epithelial barrier function.[6,7] Persistent corneal epithelial defects (CEDs) are associated with decreased production of tears or reduced corneal sensitivity and cause significant pain and visual impairment; the management of these conditions are problematic and healing with the standard protocols is often unattainable.[8]

In the past decade, increasing recognition of the relevance of painful diseases derived from chronic wounds and the lack of fully efficacious treatments have pressed further research on this topic.

WOUND HEALING

Wound healing is a complex, specific, biological process that involves the interactions between cells, extracellular matrix (ECM) components, and signaling compounds.[9] It progresses through a series of overlapping and independent phases that have been classified as hemostasis, inflammation,

proliferation, and remodeling.[1] Hereafter, the different stages involved in the healing of skin wounds will be briefly described. Analogous processes are involved in the healing of mucosal lesions.

The immediate responses to physical injury to the skin is bleeding and vasodilatation that permit the partial removal of debris and of potential pathogens and the arrival into injury site of plasma and immune cells.[10] Bleeding activates hemostasis, which plays a protective role; it is promoted by clotting factors with the aim of preventing blood from flowing out of the injured vessel. The clotting mechanism produces the coagulation of exudate with the formation of a clot, composed of cross-linked fibrin and of ECM proteins such as fibronectin, vitronectin, and trombospodin. The clot, upon drying, produces a scab that strengthens and supports the injured tissue by providing a barrier against microorganisms. Moreover, it acts as a matrix for invading cells and as a reservoir of GFs required during the different phases of the healing process.

The so-called inflammatory phase starts simultaneously to hemostasis; it occurs from within a few minutes to 24 hr and lasts for about three days.[1] Neutrophils transmigrate from local blood vessels and enter the wound site, followed by monocytes and lymphocytes. They produce a wide variety of proteinases and reactive oxygen species against microorganisms and remove by phagocytosis pathogen and debris not removed by bleeding.[11] Moreover, neutrophils secrete cytokines and GFs, molecules that play essential and differing roles in wound healing and initiate the proliferation phase of wound repair.[11] Cytokines such as tumor necrosis factor alpha (TNF-alpha), interleukin-1, -6, -8, and -10, granulocyte macrophage colony stimulating factor (GM-CSF), signal transducers and activators of transcription (STATs), and interferon-gamma (INF-gamma) act as signals essential for integrating and coordinating the immune response and for promoting keratinocyte reepithelization and granulation tissue formation.[1,9,10] During the inflammatory phase, platelets are activated, thus producing aggregates as a part of the clotting mechanism. Over the following 3–14 days, the proliferation phase occurs; it involves the initial repair processes of both epidermal and dermal layers. The entrance in the wound of fibroblasts and macrophages and the in-grown of capillary and lymphatic vessels begin the formation of a new dermal composite— that is, the granulation tissue.[9] In such a phase, GFs play a main role in stimulating angiogenesis, fibroblast proliferation, and protein deposition. A matrix of fibronectin, collagen, and hyaluronic acids is formed; it allows for myofibroblasts to contract the size of the wounds.[10] Simultaneously the reepithelization process occurs; it involves the migration of keratinocytes at the wound edge over the granulation tissue to form the new epidermis.

The last phase of wound healing is remodeling, also called maturation phase, during which a scar tissue is formed as the result of the synthesis of structural proteins, such as collagen, that remains elevated for 6–12 months.[9]

Vascular—Zwitterionic

The resolution of scar formation is due to the subsequent degradation of collagen by matrix metalloproteinases (MMPs) such as collagenase.

As previously cited, GFs are involved in all stages of the wound-healing process. Table 1 summarizes the functions of the main GFs in cutaneous wound repair.

A disruption in the event cascade at one or more stages of healing causes the formation of chronic wounds.

As for the healing of corneal lesions, they in most instances resurface, thanks to the migration of the remaining epithelial cells, which starts from the peripheral margins of the wound over the denuded surface of the cornea. This mechanism is dependent both on the interaction of the cells with the underlying substrate and on cell–cell adhesion.[12] However, when corneal ulcers are a consequence of conditions such as neurotrophic keratopathy or diabetic keratopathy, they are slower and more difficult to heal and often do not respond to conventional treatment regimens. Impaired reepithelialization is typical also of ophthalmic infections, which can be due to *acanthamoeba* in contact lens users, or to fungi especially in people with a suppressed immune system. A decrease in immune response can be responsible also for repeated attacks of herpes simplex keratitis.

Just after the damage, the cornea releases cytokines, GFs, proteases, and neuropeptides to restore its integrity.

Table 1 Functions of the main GFs in skin wound repair

Endogenous growth factors		
Family	**Members active in healing skin wounds**	**Functions in skin wound repair**
Platelet-derived growth factors (PDGFs)	PDGF-A, PDGF-B	Major player in wound healing
		Chemotactic effect for cells (neutrophils, monocytes, fibroblasts) migrating into the healing skin wounds
		Enhancer of fibroblast proliferation and of ECM production by these cells
		Stimulating effect on collagen matrix contraction
		Inducing effect of myofibroblast phenotype
Fibroblast growth factors (FGFs)	FGF1, FGF2, FGF5, FGF7, FGF10	Stimulatory effect on angiogenesis and cell migration
		Mitogenic effect for fibroblasts and keratinocytes
		Cytoprotective effect
Epidermal growth factors (EGFs)	EGF, transforming growth factor-α (TGF-α), heparin-binding EGF (HB-EGF)	Stimulatory effect on reepithelization and granulation tissue formation
		Preventing effect on excessive wound contraction
Vascular endothelial growth factors (VEGFs)	VEGF-A, VEGF-C, VEGF-D, placenta growth factor (PLGF)	Regulating effect on wound angiogenesis
Angiopoietins	Angiopoietin-1, Angiopoietin-2	Responsible for stabilization of blood vessels (angiopoietin-1)
		Responsible for vessel remodeling (angiopoietin-2)
Insulin-like growth factors (IGFs)	IGF-I, IGF-II	Stimulators of mitogenesis
Plasminogen-related growth factors	Hepatocyte growth factor (HGF/SF or PGRF-1), macrophage-stimulating protein (MSP) or hepatocyte growth factor-like protein (HGFL/SF2/PRGF2)	Stimulating effect on migration, proliferation, and matrix metalloproteinase production of keratinocytes, as well as on new blood vessel formation (HGF)
		Responsible for the stimulatory effect of wound exudate on macrophages, enhancer of macrophage-dependent wound debridement (MSP)
Nerve growth factor	NFG	Stimulatory effect on nerve ingrown, keratinocyte proliferation, and fibroblast migration
Transforming growth factor-b	TGF-beta	Chemo-attractive effect for neutrophils and macrophages
		Stimulatory effect on reepithelization, granulation tissue formation, angiogenesis, fibroblast proliferation, myofibroblast differentiation, and matrix deposition

Source: Data from Denis et al.[10] and Werner & Grose.[11]

In the beginning reepithelialization occurs, thanks to the stimulation of the proliferation, migration, and differentiation of the adjacent epithelium, which is initiated and controlled by soluble factors released from the epithelium itself, from keratocytes, and from the lachrymal glands. To avoid complications that could delay correct repairing process, in this phase it is of particular relevance that there is correct balance between stimulatory factors, such as fibronectin, tumor necrosis factor-alpha, colecystokinin gene-related peptide, platelet-derived growth factor, interleukin 6, and inhibitory factors, such as hepatocyte growth factor and tumor growth factor.

The factors released by the damaged epithelium affect the reparation of the stroma, where an increase of keratocytes apoptosis occurs, followed by an activation phase with migration and deposition of collagen fibrils. A correct repairing process of the stroma is of particular importance, since any defect would result in corneal scarring and loss of transparency.[8]

In detail, 12–24 hr after keratocyte apoptosis, the remaining keratocytes proliferate and migrate, and the stroma get populated by activated keratocytes and fibroblasts.[13] Also macrophages/monocytes, T cells, and polymorphonuclear cells infiltrate the stroma from limbal blood and tear film following the release of pro-inflammatory chemokines from the epithelium or from keratocytes responding to IL-1 and TNF-alpha, phagocyting necrotic debris. In this phase, also myofibroblasts can be found in the stroma, possibly deriving from keratocytes responding to transforming growth factor (TGF-beta). Their role is critical in ECM remodeling, thanks to the production of collagen, glycosaminoglycans, collagenases, gelatinases, and MMPs.[14]

The balance between GFs and different types of cells that participate in epithelial and stromal repair is crucial for the correct balance between cornea regeneration and fibrosis.[15]

Factors Impairing Wound Healing

The factors that inhibit or negatively influence wound healing can be divided into endogenous and esogenous factors. The endogenous factors comprise underlying pathologies that break off or slow down the wound-healing process, thus causing the occurrence of chronic wounds. Among these, there are venous insufficiencies that produce fluid transudation and fibrin cuffing of venules, arterial occlusion that determines hypossia and cell dysfunction,[16] and diabetes mellitus with consequent accumulation of advanced glycated endproducts (AGEs) in the tissues that results in hampering keratinocyte interactions with the provisional matrix proteins during reepithelialization.[17] Also, anemia delays wound healing, since an inadequate amount of oxygen reaches the wound site. A poor nutritional status that implies a defect in proteins, vitamins, and minerals is followed by an increase in healing time, jeopardizing the inflammatory phase and collagen synthesis.[1,18]

The exogenous factors include the presence of foreign bodies and bacterial infection.[3]

Foreign bodies can produce chronic inflammation that delays healing and can produce granuloma or abscess. Signs of bacterial overgrowth are change in the appearance of wound exudate that becomes purulent and characterized by an unpleasant odor, increased pain, fever, lymphangitis, and a rapid increase in the wound size.

Since wound environment is a good culture medium for microorganism proliferation, wound colonization is frequent and generally involves different microorganisms that can lead to infections. Among the most common bacteria responsible for wound infections are *Staphylococcus aureus*, *Pseudomonas aeruginosa*, β-*hemolytic streptococci*, and some *Proteus*, *Clostridium*, and *Coliform* species. Infections, besides resulting in prolonged healing times, can lead to cell inflammation (cellulitis) followed by bacteremia and septicemia that can be fatal.[1]

Even the uptake of drugs can affect the wound-healing process. For instance, glucocorticoids prejudice wound healing in both rats and humans, having an inhibitory effect on keratinocyte growth factor and impeding the inflammatory phase of such a process.[1]

A key factor in wound healing is wound exudate, whose function is to maintain the wound bed moist. It forms during inflammatory phase due to an increase in capillary permeability.[1]

Exudate is rich in GFs, nutrients, and leukocytes, which play important roles in wound healing. In the past, it was erroneously believed that dry wounds heal quicker than moist ones; it has been proved that a moisture balance is necessary for healing.

Chronic wounds are characterized by an excess of exudate which produces the maceration of the surrounding tissue. Such an exudate differs from that of acute wounds: it is characterized by higher concentrations of MMPs and polymorphonuclear elastase that are tissue-destructive proteinase enzymes. Therefore, chronic wound exudate is more "corrosive" than that of acute wounds. For this reason, an important characteristic of the modern wound dressings is the capability to absorb the excess of exudate to maintain skin moisture functional to healing.

POLYSACCHARIDES AS WOUND-HEALING AGENTS

Chitin and Chitosan

Chitin is a polysaccharide constituted by β(1-4)-linked *N*-acetyl-D-glucosamine residues; it is present in exoskeleton and in the cuticles of many invertebrates and in cell walls of green algae, yeasts, and fungi.[19] Chitin has a structure similar to cellulose: the only difference is that C-2 hydroxyl residues have been substituted by acetamide groups. It is insoluble in water, diluted acidic and basic

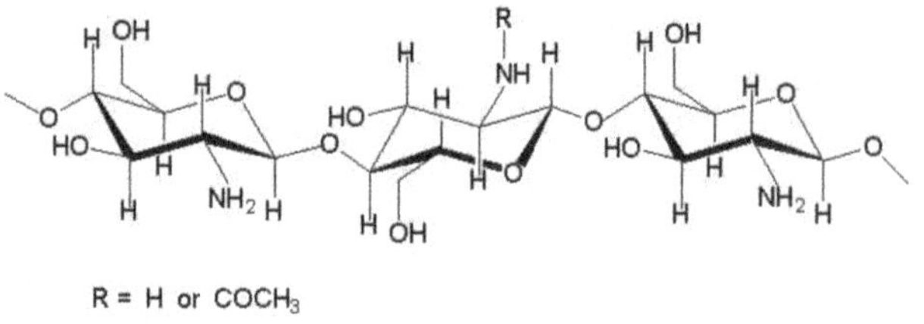

R = H or COCH₃

Fig. 1 Chemical structure of chitosan.

solutions and it is soluble in concentrated acids. In nature, chitin macromolecules with different molecular weights are present.

Chitosan is composed of *N*-acetyl glucosamine and glucosamine residues (Fig. 1). Grades characterized by different molecular weights and degree of acetylation are available on the market.

Chitin and chitosan are characterized by unique biological properties, such as biocompatibility, biodegradability, nontoxicity, antimicrobial, and hemostatic properties, that make them optimal candidates for wound healing.

Several chitin/chitosan-based dressings and bandages are commercially available. They have been recently reviewed by Muzzarelli.[9] They comprise Beschitin® Untika, Syvek-Patch® Marine Polymer Technologies, Chitipack® S Elisai Co. sponge-like chitin from squid, Chitodine® IMS, Tegasorb® 3M, Vulrisorb® Tesla-Pharma. Some of chitosan-based formulations (such as Chitoflex® HemCon, Chitoseal® Abbot) are specifically intended for the treatment of bleeding wounds. Various mechanisms are reported in literature for chitin and chitosan activity on wound healing. It is recognized that both chitin and chitosan act as chemo-attractants for neutropohils and stimulate granulation tissue formation or reepithelization. Evidences have been reported in the literature that chitin and chitosan are degraded by enzymes in the wound bed. Chitooligomers work as bricks in the synthesis of hyaluronan, which, in turn, promotes cell motility, adhesion, and proliferation and plays important roles in wound repair.[9] Chitosan promotes dermal regeneration, possesses a stimulatory effect on macrophages, and inhibits MMPs.[9] Moreover, chitosan is characterized by antimicrobial properties that are beneficial to healing of infected wounds.

Hereafter, a brief examination of such effects is reported.

Effects on polymorphonuclear neutrophils, macrophages, and fibroblasts

In a pioneer study, Peluso et al.[20] investigated *in vitro* the effect of chitosan on macrophage stimulation. In particular, the macrophage production of nitric oxide (NO), used against pathogens, was studied. The authors demonstrated the occurrence of an increased NO production due to *N*-acetyl glucosamine units of the polymer backbone. In a study conducted by Ueno et al.[21] macrophages were

stimulated with chitosan; subsequently, transforming growth factor-beta 1 (TGF-beta 1) and platelet-derived growth factor (PDGF) were assayed *in vitro*. The authors demonstrated that chitosan promoted the production of TGF-beta 1 and PDGF, which leads to an increase in ECM production. In another study, Ueno et al.[22] investigated *in vitro* the production of osteopontin (OPN), glycosylated phosphoprotein that promotes the attachment or spread of a variety of cell types, in human PMNs. They demonstrated that chitosan induced OPN production by polymorphonuclear neutrophils (PMNs) during the inflammation stage.

Okamoto et al.[23] studied the effect of chitin, chitosan, and their oligomers/monomers on macrophage migration. They demonstrated that chitin and chitosan oligomers, generated by chemo-enzymatic degradation in the wounds, significantly enhance the *in vitro* migratory activity of mouse peritoneal macrophages (PEMs).

As mentioned earlier, the activation of polymorphonuclear neutrophils and the phagocytosis of macrophages occurring in the wound-healing process aim at eliminating foreign bodies and bacteria. On the other hand, an excessive activation of such cells could lead to delayed healing due to prolongation of the inflammatory phase.[10] Some authors have demonstrated that chitosan membranes are able to reduce *in vitro* lysozyme release and reactive oxygen species production by PMNs.[24] Such results prove that chitosan is able to stimulate macrophages and PMNs with a not-excessive pro-inflammatory effect.[10]

The influence of chitin and chitosan on the proliferation of human dermal fibroblasts (HDFs) and keratinocytes was examined *in vitro* by Howling et al.[25] The degree of deacetylation affects chitosan stimulation effect: polymers with relatively high degrees of deacetylation strongly stimulated fibroblast proliferation. Such an effect required the presence of serum in the culture medium, suggesting that chitosan is able to interact with serum components, such as heparin and GFs, stabilizing and activating them. A shorter chain length 89% deacetylated chitosan hydrochloride was proved to increase proliferation rate by approximately 50% over the control and to inhibit human keratinocyte mitogenesis. Analogous results were obtained in another study, where the effects of different molecular weights of chitosan derivatives (carboxymethylated chitosan, CM-chitosan) on the growth and collagen secretion of normal skin fibroblasts and keloid fibroblasts were investigated *in vitro*.[26] CM-chitosan was proved to enhance significantly the proliferation of the normal skin fibroblast and to inhibit the proliferation of keloid fibroblast. Such an effect was more pronounced for lower-molecular-weight CM-chitosan.

Effects on coagulation process and angiogenesis

Different authors have demonstrated the capability of chitin and chitosan to promote hemostasis. In particular,

Fischer et al.[27] defined the mechanisms for the effect of poly-N-acetylglucosamine (pGlcNAc) containing fibers on platelet-mediated processes. They demonstrated that pGlcNAc fibers tightly bind plasma proteins and some platelet surface proteins, producing a platelet-dependent acceleration of fibrin gel formation. The contact between platelet integrins and pGlcNAc fibers, saturated by plasma proteins, accelerates fibrin clot formation. In another study, Fisher et al.[28] demonstrated that the interaction of chitin fibers with red blood cells (RBCs) determines the activation of the intrinsic coagulation cascade. This effect was related to the presence of phosphatidylserine on the outer layer of the surface membrane of nanofiber-bound RBCs.

More recently, Lord et al.[29] demonstrated that chitosan supports both platelet adhesion and activation, leading to thrombous formation. In particular, platelet adhesion was enhanced in the presence of absorbed plasma and ECM proteins and was mediated by $\alpha_{IIb}\beta_3$ integrins and P-selectin. The extent of activation was modulated by the presence of proteins, including perlecan and fibrinogen.

Pietramaggiori et al.[30] proved that pGlcNAc, in addition to its hemostatic properties, is able to accelerate wound closure in healing-impaired genetically diabetic mice. Granulation tissue showed higher levels of proliferation and vascularization after 1-hour treatment than the 24-hour and untreated groups.

Effects on matrix MMPs

Another effect recognized for chitosan is the capability of inhibiting MMPs, enzymes that degrade ECM components. As already mentioned in the "Factors Impairing Wound Healing" section, the exudate of chronic wounds is rich in such enzymes that have a "corrosive" effect. Then inhibition of MMPs is a primary therapeutic target for wound dressings.[9] Kim and Kim[31] studied the inhibitory effect of chitooligosaccharides on the activation and the expression of matrix metalloproteinase-2 (MMP2) in primary HDFs. Chitooligosaccharides characterized by 3–5 kDa MW showed the highest inhibitory effect on activity and protein expression of MMP2. This inhibition was caused by the decrease of the gene expression and transcriptional activity of MMP2. Other authors demonstrated, by means of atomic force microscopy, a direct molecular interaction between chitosan (MW 5×10^5 g mol^{-1}, degree of acetylation 30%) and MMP2 with the formation of a complex.[32] A high binding-specificity of chitosan to MMP2 was observed. This resulted in a noncompetitive inhibition of hydrolytic activity of MMP2. The authors supposed that the MMP-binding capacity of chitosan is at least partly responsible for the improved wound healing mediated by chitosan-based wound dressings.

Different papers have been published in the literature dealing with chitosan's inhibitory effect of MMPs that increase in the majority of malignant tumors and play an important role in the establishment of metastasis.[9]

Effects on microorganisms

In the past decade, the antibacterial properties of chitin, chitosan, and their derivatives have been an object of various studies.[33–39]

In particular, in our previous paper,[38] different chitosan grades and salts were compared for antimicrobial properties in view of their employment in the treatment of oral mucositis. In particular, two grades (high and low MW) of chitosan hydrochloride and chitosan ascorbate were considered. All the grades were characterized by a better activity against *Escherichia coli* than against *Staphilococcus aureus*. Such an activity was affected by pH, being higher at pH 5.0 than at pH 7.0, conceivably due to the poor polymer solubility at pH > 6.5. The highest antimicrobial activity was shown by the low-MW chitosan ascorbate solution.

In other studies, the antibacterial properties of oral thermally sensitive vehicles based on chitosan and its derivatives were investigated.[37,39] In particular, a trimethyl chitosan-based vehicle showed antimicrobial properties against *S. aureus*, *Streptococcus pyogenes* and *Streptococcus vestibularis*. It was characterized by higher microbicidal effect against the two streptococci than against *S. aureus*.[39]

Hyaluronic Acid

Hyaluronic acid (HA) is a high-molecular-weight linear glycosaminoglycan,[40] constituted of repetitive units of D-glucuronic acid and D-N-acetylglucosamine disaccharides (Fig. 2), firstly isolated by Meyer and Palmer in 1934.[41]

Also known as hyaluronan, HA is a naturally occurring polysaccharide found predominantly in the ECM, where concentrations are highest in rooster comb, human umbilical cord synovial fluids, connective tissue, and skin, followed by vitreous body.[42]

As the carboxyl groups (COO–) are completely ionized at physiological pH, it is very effective in absorbing water molecules to form a viscous gel, characterized by viscoelastic properties.

Concentration and mean molecular weight of HA determine its rheological properties. High viscosity at low shear rates and low viscosity at high shear rates are typical of pseudoplastic behaviour. These attributes are relevant for the structural function of HA in connective tissue, where it contributes to the turgor pressure,[42] as well as for the

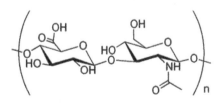

Fig. 2 Chemical structure of hyaluronic acid.

cushioning and lubricating effects in the aqueous humor of the eye and synovial fluid.[40]

The hygroscopic behavior of HA also facilitates the transfer of nutrients and ions, in addition to maintaining the extracellular space and tissue hydration during inflammation.[43]

The solubility of HA increases with longer-chain molecules, allowing the formation of a gel at low concentrations, thus conferring its hemostatic properties.[44]

Hyaluronic acid is characterized by a great number of biological functions modulating the integrity of tissues, and plays an important role in inflammation and wound healing.[44,45] These biological functions are contradictory depending on HA molecular weight and fragment size.[46]

High-molecular-weight HA grades, in their native form, maintain the structural integrity of tissue.[47] On the contrary, medium- and low-molecular-weight HA grades are involved in inflammation, since they stimulate gene expression in macrophages, endothelial cells, eosinophils, and selected epithelial cells.[48,49] Similarly, high-molecular-weight HA is antiangiogenic, whereas the degradation products stimulate endothelial cell proliferation, migration, and sprout formation.[50] The effects of HA are mediated by binding to specific cell surface receptors such as CD44, receptor for hyaluronic-acid-mediated motility (RHAMM), and intracellular adhesion molecule (ICAM-1 or CD54).[43,45] Hyaluronic acid is involved in all stages of wound healing[45] with different mechanisms, which hereafter will be summarized.

A day after injury, HA accumulates and binds to fibrinogen to form a temporary matrix that supports the fibroblasts and endothelial cells.[51] During the early inflammatory phase, HA stimulates the production of pro-inflammatory cytokines and facilitates adhesion of cytokine-activated lymphocytes to endothelium.[45]

Moreover, HA possesses free radical scavenger[52] and antioxidant[53] properties, which aid in moderating inflammation.

In particular, HA forms a viscous, pericellular meshwork around cells in a dose-dependent manner, thus restricting the movement of reactive oxygen species.[54] Spectroscopy studies regarding enzymatic digestion of HA indicated the presence of a double bond in the D-glucuronic acid unit,[55] which can form a complex with reactive oxygen species and reduce the toxicity of radicals. Exposing liposomal skin lipids to UV irradiation, HA and its fragments exerted antioxidative effects:[56] this supports the HA function as a radical scavenger.

During the proliferative phase of wound healing, HA stimulates the migration and proliferation of endothelial cells and the production of collagen (type 1 and 8), thus favoring angiogenesis.[55,57] As an integral part of the basal epithelial layer, HA promotes reepithelization by directly enhancing the migration of keratinocytes.[58] In addition, a persistently high level of HA is responsible for highly organized collagen deposition and scarless wound healing.[59,60]

Corneal cell migration in vitro is enhanced by the addition of hyaluronic acid, an effect that appears to be enhanced by the further addition of fibronectin or EGF.[61] Application of EGF to organotypic cultures results in the up-regulation of hyaluronan synthases (HASs) and the subsequent production of hyaluronic acid, which further results in increased motility and proliferation.[62] The EGF-dependent production of hyaluronic acid is associated with the deposition of hyaluronic acid around the cell and high levels of uptake, giving rise to the possibility that it may be acting in an autocrine or paracrine fashion.[63]

The remarkable properties and biocompatibility of HA have led to its use in a variety of clinical applications. There is, therefore, large evidence from scientific studies that HA might affect, mostly in a beneficial manner, several of the components of wound healing. With this in mind, HA has been used in vivo for a number of applications, resulting in some qualified success.

Several clinical studies have been performed with HA applied as cream or dressing.

Hereafter, some examples are cited.

In radiotherapy, severe skin reactions such as acute radioepithelitis, observed often at the beginning of the treatment, not only reduce the patients' quality of life but also their compliance. A randomized, double-blind, placebo-controlled study using HA cream 0.2% (Ialugen®) was carried out in 134 patients to investigate the prophylactic use of HA after radiotherapy treatment.[64] Twice-daily application of HA cream causes only a slight delay of the onset of acute skin reactions but significantly reduces their intensity compared with placebo cream. Additionally, the healing process using HA cream appears accelerated.

A pilot study evaluating the use of an HA-containing dressing (Hyalofill®) in the treatment of diabetic foot ulcers has been conducted.[65] The mean age of the patients was 60 years and 75% of wounds healed by secondary intention during the 20-week study. No epidermal grafting was used. All 36 patients first received surgical debridement for their diabetic foot wounds and then HA dressing until the wound bed approached 100% granulation tissue, followed with a moisture-retentive dressing until wound closure.

Many different hyaluronic acid-derived products have been developed for wound healing, mainly by Fidia Advanced Biopolymers (Padua, Italy) in association with the BRITE–Euram European Union Research Project. They are principally scaffolds that are cell culture/transfer devices. Since such topic is beyond the task of this entry, we only cite Laserskin® for the support of keratinocytes and HYAFF® materials for dermal regeneration.

Thanks to HA capability for promoting the healing process, many examples in literature refer to the use of hyaluronic acid in ophthalmic surgery[66] and in otolaryngology, namely in the middle ear and nasal cavity postsurgery.[67]

HA is an important component of both the vitreous and aqueous humor and intravitreal injection has been used extensively, for example, in traumatic perforation of the globe,[68] treating the flat anterior chamber following trab- eculectomy,[69,70] and after phacoemulsification.[71] In gel form, it has been used extensively to prevent desiccation of the cornea during a number of ophthalmic procedures.[66]

Many studies indicate that HA is biodegradable in the middle ear; with an average degradation of 4–12 weeks,[72–74] which is sufficient for healing prior to invoking an inflam- matory foreign body response.

Several intrinsic attributes of HA native isoform such as water solubility, rapid resorption, and short residence time in the tissue limit its clinical use.[75] Consequently, the ability to modify the chemical structure and alter the biode- gradability of HA to enable specific clinical needs is of great importance.

In this perspective, chemical modifications (e.g., cou- pling and crosslinking) are used to obtain more stable forms.[75] Coupling reactions involve alteration of specific functional groups of the polysaccharide (e.g., carboxyl, hydroxyl, and N-acetyl groups) by esterification, sulfida- tion, or amidation. Conversely, crosslinking is achieved by means of a cross-linker [such as poly(ethylene glycol) diacrylate (PEGDA)] stabilizing the hyaluronic acid chains. These materials can be produced as viscous hydrogels, hydrogel films, or lyophilized sponges.[76] Esterified hyaluronic acid has been profitably employed in tympanic membrane wound healing. Preparations based on such derivative [e.g., MeroGel (Medtronic Xomed, Inc.), EpiFilm (Xomed-Medtronic), EpiDisc otologic lamina (Xomed-Medtronic)] are retained longer than the unmodified isoform, due to their resistance to enzymatic degradation.[77]

Similarly to esterified HA, cross-linked hyaluronic acid derivatives [e.g., Carbylan-SX (Carbylan Biosurgery), Sepragel® (Genzyme Corp.), and Seprafilm® (Genzyme Corp.)] have improved mechanical properties and longer residence time compared with unmodified HA.[78,79]

Chondroitin Sulfate

Chondroitin sulfates (CHOS) are a group of glycosamino- glycans present in the ECM of all vertebrates made of sul- fated disaccharide units containing N-acetylgalactosamine (Fig. 3). The CHOS chains can be classified according to

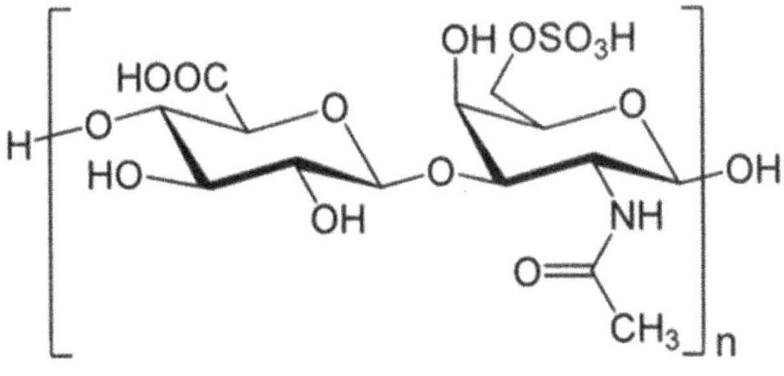

Fig. 3 Chemical structure of chondroitin sulfate.

the sulfate substitutions into different types, among which the most common are C-4-S and C-6-S.

Chondroitin sulfate is described as a biomaterial with favorable effects in healing of a variety of tissues. The lit- erature reports, for example, its ability to induce cell migra- tion and resistance to apoptosis of vascular smooth muscle cells (VSMCs).[80,81] On this basis, Charbonneau et al.[82] proposed a CHOS–EGF-based coating with antiapoptotic and pro-proliferative properties to counteract cell depletion caused by apoptosis and promote the healing of tissues surrounding vascular implants such as stent-grafts.

The wound-healing properties of a CHOS hydrogel have been demonstrated on wounds of the maxillary sinus mucosa, using New Zealand white rabbits' animal model. Although the average wound diameters in the treatment group were similar to the control group at 2 days, they were significantly lower at 4 days.[83] The healing was faster with the CHOS treatment, as revealed by histologic analysis. Moreover, the treated wounds displayed respiratory epithe- lium as opposed to the squamous epithelium exhibited on the untreated sides. A possible mechanism suggested to explain these promising results involves the role of the applied CHOS hydrogel as an ECM that can support fibroblasts for epithelial regeneration acting as a depot of the cytokines and GFs produced by the mucosa during the regeneration process.

Some other authors found that the addition of ECM components such as CHOS to titanium implants signifi- cantly enhanced bone remodeling and healing at early stages, and after 14 days caused newly formed bone covering a higher number of implants coated with CHOS with respect to the uncoated ones.[84]

Reyes et al.[85] demonstrated that a novel chondroitin sulfate–aldehyde adhesive was more effective than conven- tional sutures for sealing corneal incisions in an *ex vivo* study in rabbit eyes.

The positive effect of CHOS and of other negatively charged GAGs on regeneration can be also attributed to their capability of electrostatic interaction with positively charged GFs, such as basic fibroblast growth factor (bFGF), insulin-like growth factor, vascular endothelial growth factor, platelet-derived growth factor, and transforming growth factor-beta (TGF-beta). This electrostatic interac- tion has been indicated by the authors as a possible reason of the stabilization of GFs.[86,87]

This hypothesis is in line with the findings of Sandri et al.[7] about a sterile thermosensitive vehicle based on chondroitin sulfate sodium (CHOS) and hydroxypropyl- methyl cellulose (HPMC) and intended for healing of corneal ulcers. *In vitro* wound healing and proliferation test performed on rabbit corneal epithelial cells (RCE) showed that the formulation was effective in promoting cell growth.

A deep and systematic investigation of the effect of CHOS on tissue reparation and of mechanisms involved has been recently performed by Zou at al.[88] They demon- strated that CHOS promoted cell proliferation, possibly thanks to its ability to bind to FGFs, that stimulates fibroblasts proliferation. CHOS also affected a fundamental process in wound healing such as fibroblast adhesion.

Moreover, the authors demonstrated that the application of extraneous CHOSs modulated wound closure *in vitro* both in 2-D and in 3-D models and positively affected the contraction rate through suppression of the aberrant activation status of fibroblasts and of their contractile activity.

A further interesting finding of these authors concerns the importance of sulfate substitution for CHOSs effect in wound healing. In particular, the number and position of the sulfate groups are relevant for these interactions, so that C-4-S resulted less efficacious than C-6-S in the inhibition of cell–gel contraction. This can be related to the differences in binding of GFs and other signaling molecules to CHOSs. Although the mechanism of this interaction is not well understood, it seems clear that the sulfate groups specifically regulate the binding of signaling molecules with heparan sulfate.[88]

HEMODERIVATIVES IN WOUND HEALING

The hemoderivatives employed in the treatment of wounds are based on or derived from platelets. Platelets are specialized secretory cells that release, in response to activation, a large number of biologically active substances from intracellular alpha granules and in particular GFs. As already mentioned, GFs have a specific role in initiating and modulating tissue repair mechanisms, such as chemotaxis, cell proliferation, angiogenesis, ECM depositing, and remodeling. All the GFs present in the platelets have a role in the healing process because multiple signals are required to complete the regeneration process. This is why a single GF purified is not effective as much as the pool of GFs derived directly by platelets.

Among platelet derivatives, PRP, PG, and PL have been proved to be effective in wound healing.

Platelet-Rich Plasma

PRP consists of a limited volume of plasma enriched in platelets, and can be obtained starting from blood by centrifugation by means of apheresis procedures.

Autologous PRP is a potential wound-healing treatment because it contains fibrin and high concentrations of GFs.

There are different papers focused on the employment of PRP as healing enhancer for skin and corneal lesions.

Kushida et al.[89] investigated the *in vitro* effects of PRP and platelet-poor plasma on the proliferation and differentiation of skin fibroblasts into myofibroblasts and on wound contraction. The proliferation of dermal fibroblasts is an essential step in wound healing. Furthermore, wound contraction, which occurs subsequent to granulation tissue formation, is an important event that reduces the wound size for wound closure. For this reason, alpha-smooth

muscle actin (alfa SMA), which is present in fibroblasts of granulation tissue, known as myofibroblasts, and appears to be responsible for wound contraction, was quantified to put in evidence the contractile ability of fibroblasts during wound contraction. In this perspective, the authors used also an *in vitro* model based on floating collagen gel with incorporated fibroblasts to evaluate the decrease of collagen gel size due to the effect of myofibroblasts. The authors stated that PRP was able to promote proliferation of HDFs enhancing the expression of alpha SMA, as myofibroblast marker. PRP demonstrated more marked contraction in the collagen gel model than the platelet-poor plasma. These results suggested that PRP promoted proliferation, caused fibroblast differentiation into myofibroblasts, and promoted wound contraction, thus providing a potential therapeutic agent for skin wound healing.

Nakajima et al.[90] developed a freeze-dried PRP in an adsorbed form on a biodegradable polymer material (Polyglactin 910). Such system answers to the necessity for an immediate availability of autologous PRP samples. It was based on polymer filaments of PRP mesh, which was prepared by coating the polymer mesh with human fresh PRP and subsequent freeze-drying: these methods should allow incorporation of platelets and the preservation of the related GFs at high levels. A full-thickness skin defect model in a diabetic mouse demonstrated that the PRP mesh, although prepared from human blood, substantially facilitated angiogenesis, granulation tissue formation, and reepithelialization without inducing severe inflammation *in vivo*. The polymer matrix functioned as a bioactive material to facilitate tissue repair/regeneration.

Even there are some *in vitro* and *in vivo* preclinical data about the efficacy of PRP and a review about the treatment of chronic skin wounds by using PRP has been published in Cochrane Database.[91] This review evaluated the effectiveness and safety of PRP and included nine randomized clinical trials, with a total of 325 participants. The data analysis evidenced that there were no differences between the autologous PRP and the control groups in terms of healing, indicating insufficient evidence to support the routine use of autologous PRP as a treatment for chronic wounds. However, these results require confirmation in adequately powered, well-conducted, randomized, controlled trials.

As for the effects of PRP on cornea lesions, Kim et al.[92] evaluated PRP eyedrops in the treatment of persistent epithelial defects (PEDs). Autologous PRP and autologous serum (AS) were prepared from whole blood. The concentrations of transforming growth factor (TGF-beta1, TGF-beta2), epidermal growth factor (EGF), vitamin A, and fibronectin in the PRP and AS were analyzed and compared. The corneal epithelial healing efficacy of PRP was compared with that of AS in patients with PED induced by postinfectious inflammation.

The concentrations of TGF-beta1, TGF-beta2, EGF, vitamin A, and fibronectin in the PRP and AS were not statistically different. However, the concentrations of EGF in the PRP were significantly greater than in the AS. Autologous serum was used in 17 and PRP in 11 eyes of 28 patients. The healing rates of the corneal epithelia of the PRP-treated eyes were significantly higher than those treated with AS.

The PRP was effective in the treatment of PEDs. This may be attributable to its high concentration of platelet-contained GFs, most notably EGF. PRP could be an effective, novel treatment option for chronic ocular surface disease; however, an effective optimal concentration of platelets and GFs for ocular surface disease could not be determined in that study. This represents a drawback for the employment of GFs in the treatment of ocular disease, since higher levels of GFs could induce harmful effects: in particular, EGF proved to promote corneal neovascularization.

In the study of Kim et al.,[92] no evidence of corneal neovascularization in the PRP-treated eyes was pointed out, suggesting that the average concentration of EGF employed was not quite enough for the angiogenic effect. A further warning suggested by the authors was the possible complications resulting from the use of PRP due to pro-inflammatory cytokines (such as the soluble CD40 ligand).

Despite these limitations, the authors found that PRP had clinical effects in the treatment of PED comparable with those of AS.

Khaksar et al.[93] examined the effect of sub-conjunctival platelet-rich plasma (sPRP) in combination with topical acetylcysteine on corneal alkali burn ulcers in rabbits. PRP was obtained by rabbit plasma. Alkali wounds were inflicted on the central corneas of rabbits by applying a round filter paper, 6.0 mm in diameter, soaked in 1 M NaOH for 60 s. Only one eye in each rabbit was used.

A total of 20 rabbits were allocated into four groups of five animals each. Group 1 served as the control group. Group 2 received 3% N-acetylcysteine (NAC) topically three times daily for 2 weeks. The third group received only sPRP, whereas group 4 received sPRP with topical 3% NAC three times daily for 2 weeks. Clinical outcome was monitored by evaluation of epithelial defects, corneal opacity, duration of blepharospasm, corneal vascularisation, duration of ocular discharge, and wound area diameter measurement. After 3 weeks, eyes were enucleated and corneas were excised for histopathological analysis. Samples were assessed by evaluating the number of epithelial rows, stromal vascularization and inflammation, and stromal collagen arrangement.

Comparison between groups showed that group 3 had significantly shorter duration of blepharospasm than the control group. Additionally, group 3 had smaller mean defect area and greater wound healing. Histopathological investigation revealed significantly less inflammation and vascularization in the corneas in group 3; this group also had the best stromal collagen arrangement. The sPRP

seems to improve corneal epithelial burn healing. However, the acetylcysteine and sPRP combination may have a retarded healing effect as compared with PRP alone. The results of this study showed that the subconjunctival application of autologous PRP on corneal alkali burn ulcer is a simple and an economic treatment for ocular surface burns without undesirable side effects.

Platelet Gel

PG can be obtained starting from PRP by adding thrombin or calcium to form a three-dimensional and biocompatible fibrin scaffold (fibrin glue). In the literature, PG is also referred to as PRP gel.

Some authors[94] investigated the optimal composition of fibrin sealant preparations in vitro.

They compared the characteristics of a series of sealants from autologous components with those of a commercial glue (Bioseal Porcine Fibrin Sealent Kit). The concentrations of platelets as well as fibronectin in autologous fibrin glues were significantly higher than those in the commercial one. A dense platelet surface and fibrin net structures could be observed in the autologous samples, whereas there were only sparse fibrin nets without cellular components in Bioseal. Autologous and Bioseal fibrin sealants did not show significant difference in biochemical and mechanical properties, indicating that autologous PRP gel may find application in wound healing.

Marciel et al.[95] studied in vivo (animal model) the healing of deep second-degree burns (DSDb), which usually involve all epidermis layers, including the basal laminae, and heal with extensive areas of scarring. Four horses were placed into two groups G1 and G2: G1 received one PRP gel treatment (at day 0) and G2 two treatments of the same hermoderivative (at day 0 and day 3). Control groups (Gc1 and Gc2) were treated with saline solution, using the same schedule times of G1 and G2 groups. Epidermis dissections were made in the caudal regions of the horses and DSDb were made by infliction of hot iron. At day 40, G1 resembled intact tissue and G2 showed dense tissue. All groups had bacterial contamination but no infection. These results indicate that PRP gel accelerated repair, induced fibroses, and probably provided antibacterial activity in horses with DSDb. In particular, the hemoderivative increased the speed of repair of the ECM and its components in DSDb wounds in horses, but had a potential of fibroses formation.

In a pilot open study, Giuggioli et al.[96] evaluated the effect of PG in 12 patients affected by systemic sclerosis (SS) with skin ulcers resistant to conventional therapies from at least 6 months. PG was applied in the wound bed twice weekly for 2 weeks, then once a week for 12 weeks; in all cases, the ongoing treatments remained unchanged at the time of PG applications. Skin ulcers were evaluated at 0, 12, and 24 weeks; the patient's quality of life was also evaluated using the visual analogical scale (VAS) and the health assessment questionnaire (HAQ). During the 6-month follow-up, the skin ulcers consistently improved in 10/12 patients, with complete healing in 4. At the last evaluation, wound size significantly reduced from 23.4 ± 14.9 (SD) to 2.3 ± 2.2 (SD) cm^2. The patient's quality of life markedly improved: VAS significantly decreased from 87.08 ± 13.5 to 57.9 ± 12.6 and HAQ from 0.73 ± 0.43 to 0.57 ± 0.22. The authors suggested the usefulness of PG treatment in refractory scleroderma leg ulcers.

Other authors suggested the employment of PG for topical use in cutaneous ulcers of various ethiopathogenesis. In particular, Crovetti et al.[97] evaluated the application of PG once-weekly in skin ulcers of 24 patients affected by one of these pathologies: diabetes, arterious or venous insufficiency, neuropathy or post-traumatic diseases. Progressive reduction of the wound size, granulation tissue forming, wound bed detersion, regression, and absence of infective processes were considered for evaluating clinical response to hemotherapy. Only three patients could perform autologous withdrawal; in the others homologous hemocomponent was used. After a mean of ten applications, complete response was observed in nine patients, two were subjected to cutaneous graft, four stopped treatment, nine had partial response and continued the treatment. In each case, formation of granulation tissue increased following the first PG applications, while complete reepithelization was obtained later. Pain was reduced in every treated patient. The authors concluded that topical hemotherapy with PG may be considered as an adjuvant treatment of a multidisciplinary process, useful to enhance therapy of cutaneous ulcers.

Mazzucco et al.[98] evaluated the healing of dehiscent sternal wounds and necrotic skin ulcers by administering PG. Patient treated with PG were retrospectively compared with patients having similar lesions but undergoing similar treatments. The following clinical endpoints were considered: healing rate, hospital stay, and time required to have tissue regeneration adequate to undergo reconstructive plastic surgery. In patients with treated dehiscent sternal wounds, the healing rate and hospital stay were significantly reduced. Patients with treated necrotic skin ulcers required a notably shorter time to have surgery. In both situations, neither adverse reactions nor in situ recurrences were observed.

PG has been employed also in pediatric patients. Perotti et al.[99] reported a case study in which a decubitus ulcers in a newborn patient was successfully treated. The PG employed contained also white cells and was therefore named PLG (platelet leucocytes gel). The leucocytes in PLG (monocytes, lymphocytes, and neutrophils rich in myeloperoxidase) act as efficient antimicrobial agents. The healing process was evident as early as the fifth day of the treatment; this entailed a lower risk of infections and may even shorten the time spent in the hospital.

Platelet Lysate

PL is obtained by PRP by means of freeze–thawing cycles, which cause the lysis of the platelets and the release of their content.

Ranzato et al.[100] investigated the mechanisms underlying PL-induced wound healing using HaCaT keratinocytes, representing an *in vitro* model of proliferating and migrating keratinocytes. Cells were exposed to PL sample purified from whole blood. Cell metabolism and proliferation were determined using 3-(4,5-dimethylthiazol-2-yl)-5(3-carboxymethoxyphenyl)-2-(-4sulfophenyl)-2H-tetrazolium (MTS) and crystal violet assays, respectively; wound healing was evaluated by scratch wound assay and cell migration by transwell assay. Extracellular signal-regulated kinase (ERK) 1/2 and protein kinase p38 activations were studied using Western immunoblotting and intracellular Ca^{2+} dynamics by confocal imaging.

The *in vitro* wound closure rates showed a significant increase for cells exposed 6 and 24 hr to 20% PL. The cell migration assay showed a strong chemotactic effect toward PL. The intracellular Ca^{2+} chelator BAPTA-AM induced 100% inhibition of the PL effect on wound closure rate and cell migration, while kinase inhibitors (SB203580, PD98059, wortmannin, and LY294002) exerted variable % inhibition (50–30%) of wound closure as well as variable % inhibition (0–100%) of cell migration. The authors concluded that PL increased wound-healing rate by stimulating keratinocyte migration through a calcium- and mitogen-activated protein kinase p38-dependent mechanism.

Cipriani et al.[101] evaluated the biological effects of PL on human primary skin fibroblasts. PL demonstrated to be effective on the expression of various proteins, related principally to stress response, metabolism, and the cytoskeleton.

In a later work, El Backly et al.[102] reported that at physiologic concentrations, PL enhanced wound closure rates of NCTC 2544 human keratinocytes. The authors clearly evidenced the effect 6 hr after wounding. Moreover, PL induced a strong cell actin cytoskeletal reorganization that persisted up to 24 hr. The accelerated wound closure promoted by PL, in either presence or absence of serum, was associated with a high expression of the inflammatory cytokine interleukin-8. Moreover, 24 hr after PL treatment, confluent keratinocytes expressed low amounts of interleukin-8 and of the antimicrobial peptide neutrophil gelatinase-associated lipocalin, which dramatically increased under inflammatory conditions. These effects were associated with the activation of the inflammatory pathways, p38 mitogen-activated protein kinase, and NF-κB (nuclear factor kappa-light-chain-enhancer of activated B cells). The authors evidenced the trigger of inflammatory cascade and the antimicrobial role played by PL.

Geremicca et al.[8] studied the application of topically administered PL, obtained from PRP for the treatment of corneal ulcers caused by neurotrophic keratitis and of epithelial and stromal loss following physical or chemical trauma. The PL was administered in the form of eyedrops to patients who had not responded to conventional therapy and who were at risk of corneal scarring. The results were satisfactory in terms of both tissue regeneration and healing time. The clinical follow-up showed a clear reduction in regeneration time of the damaged epithelium and stabilization of the repair process. The epithelial defects disappeared completely in all the treated eyes within 6 to 32 days, depending on the type of lesion and the severity of the damage. The cornea reacts to the damage by releasing numerous substances, including cytokines, GFs, proteases, and neuropeptides, in order to restore its anatomical integrity. The use of PL was found to be effective in all cases characterized by epithelium loss, such as postherpetic corneal ulcers or ulcers occurring following trauma or exposure to caustic substances.

Pezzotta et al.[103] evaluated the employment of PL in the treatment of ocular GVHD (graft versus host disease), when the conventional therapies (represented by systemic immunosuppressive regimens and local therapies, mainly artificial tears and corticosteroids) give unsatisfactory results. The authors investigated the safety and efficacy of autologous PL rich in PDGFs in treating ocular GVHD unresponsive to standard medications. In particular, a total of 23 patients with refractory ocular GVHD (grades II–IV) unresponsive to standard therapy were treated with PL eyedrops four times/day for 6 months. Symptoms and signs (best visual acuity, Schirmer's test, and tear break up time (TBUT), evaluation of the anterior segment and fluorescein and lissamine staining) were always assessed by the same ophthalmologist. Patients were defined as "responders" when showing improvement for total complaints and at least one sign. After 30 days of treatment, 17 patients (73.9%) were classified as responders. The most improved symptom was photophobia (in 19 patients, 82.6%). TBUT improved in 20 patients (86.9%) and anterior segment score in 19 patients (82.6%). No serious adverse events occurred. PL eyedrops proved to be safe and effective in treating ocular GVHD and may represent a valid tool for the treatment from the early stages of the disease to avoid irreversible ocular damage.

FORMULATIONS BASED ON BIOPOLYMERS/ HEMODERIVATIVES ASSOCIATION

In the past years, one of the main research lines of our group has been devoted to the development of therapeutic platforms based on the association of biopolymers (chitosan, hyaluronic acid, and chondroitin sulfate) with PL for the treatment of epithelial (mucosal, cutaneous, and corneal) lesions. Such research was born as a joint project between the Department of Drug Sciences of the University of Pavia (Italy) and the Immunohaematology and Transfusion Service and Cell Therapy Unity of San Matteo University Hospital in Pavia. The project aimed at answering to the

unmet needs in the treatment of mucosal, ophthalmic, and skin lesions.[104]

The first platform developed was a semisolid formulation intended for the treatment of oral mucositis and corneal lesions.[104–106] It should be capable of assuring the maintenance of the PL bioactive molecules during the preparation, storage, and use and of allowing a prolonged permanence of such substances onto the diseased epithelia. Chitosan glutamate (CSG) salt was used as biopolymer, due to its well-known mucoadhesive properties.[38] The vehicle was mixed 1:1 with PL and stored at fridge temperature until used. The final formulation comes as single-dose preparation, to be dispensed in a number of doses tailored for the intended treatment. Figure 4 reports the preparation scheme of the formulation. The formulation content in PDGF AB growth factor, chosen as representative of those contained in PL, was not significantly different from the theoretical value, indicating that the developed vehicle allows the maintenance of growth factor activity. The formulation was subjected to *in vitro* cell proliferation and wound-healing tests effected on fibroblasts and RCE. Chitosan formulation was able to significantly enhance epithelial cell growth even after storage of up to 2 weeks (in-use conditions). Complete *in vitro* wound repair was achieved within 48 hr.

Recently, thermosensitive and mucoadhesive eyedrops based on chondroitin sulfate (CHOS) was developed to maintain and further prolong the contact of PL with corneal ulcers.[7]

In particular, a sterile vehicle based on CHOS and hydroxypropylmethyl cellulose (HPMC) was developed. It was intended for an extemporaneous loading of PL before the administration. In Fig. 5 the preparation scheme of the thermogelling formulation is reported.

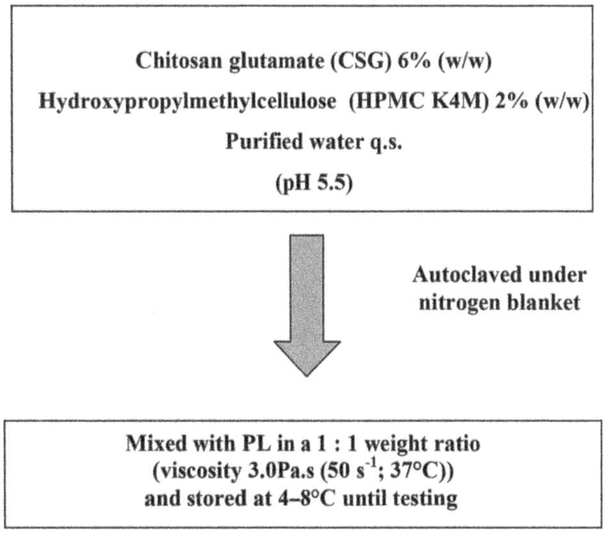

Chitosan glutamate (CSG) 6% (w/w)

Hydroxypropylmethylcellulose (HPMC K4M) 2% (w/w)

Purified water q.s.

(pH 5.5)

Autoclaved under nitrogen blanket

Mixed with PL in a 1 : 1 weight ratio
(viscosity 3.0Pa.s (50 s^{-1}; 37°C))
and stored at 4–8°C until testing

Fig. 4 Preparation scheme of PL/CSG-based formulation intended for the treatment of oral mucositis and corneal lesions.

Vascular—Zwitterionic

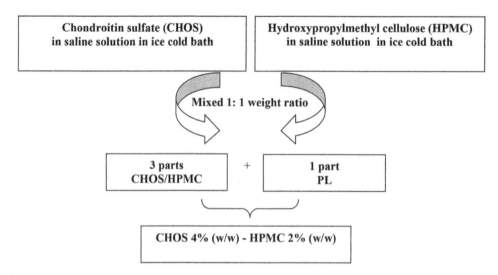

Fig. 5 Preparation scheme of CHOS/HPMC-based formulation intended for the delivery of PL in the treatment of corneal lesions.

The loaded formulation showed a quick in situ gelation at 32°C. The comparison between unloaded (blank) and loaded formulation evidenced that PL was able of lowering the sol–gel transition temperature of HPMC to the physiological ophthalmic range, probably due to the assistance of PL components to polymer chain dehydration. The formulation was also characterized by good mucoadhesive properties that allowed a high resistance toward eye removal mechanisms (lacrhrymation and blinking). *In vitro* cell proliferation and wound-healing tests (performed on RCE cell line) proved the occurrence of a synergic effect between CHOS and PL components.

PDGF AB assay during 15 days of storage at 4–8°C of loaded formulation and PL alone proved that the vehicle was characterized by stabilization properties toward the growth factor. In fact, while PL stored in the same conditions presented a significant decrease in PDGF AB content, the profile of the growth factor vs. time was close to 100% for the loaded formulation.

Recently, sponge-like dressings obtained by freeze-drying mixtures of a biopolymer [CSG or sodium hyaluronate (HA)] and PL were developed.[107] They were intended for PL delivery to chronic skin wounds. The dressings contained glycine as cryoprotectant agent and water as plasticizer. The addition of glycerophosphate to chitosan dressing (used to solubilize chitosan at pH close to neutrality) was also investigated. In Fig. 6 a photograph of a dressing based on CSG is reported.

The dressings were loaded with different amounts of PL. Schemes of dressing preparation are reported in Figs. 7 and 8. The simplicity of the technique employed for the preparation should make dressings easily available in a hospital pharmacy service, where autologous hemoderivative samples (derived for the same patient who needs therapy) could be used.

Fig. 6 Photograph of a CSG-based "sponge-like" dressing.

Depending on the composition, dressings showed different mechanical and hydration properties that make them suitable for wounds with different exudate amounts. In particular, the addition of 1% (w/w) glycine to the polymer solutions caused an improvement in the formulation mechanical resistance and 10% (w/w) water content increased formulation elongation properties. When placed in contact with pH 7.2 phosphate buffer, medium mimicking wound exudates, HA-based dressing gelified and dissolved in 12 min, whereas CSG-based formulation absorbed high buffer amounts and maintained its structure after 6 days.

The addition of glycerolphosphate to CSG formulation was responsible for an intermediate hydration behavior: dressing gelified in contact with buffer but with a lower

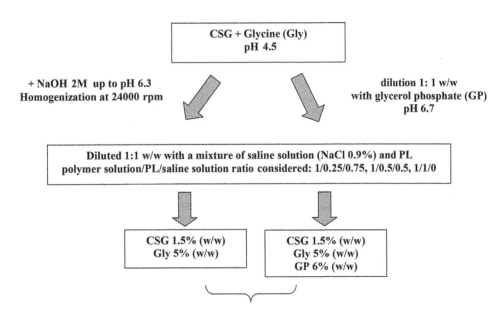

Fig. 7 Preparation scheme of CSG-based "sponge-like" dressings intended for the delivery of PL in the treatment of skin ulcers.

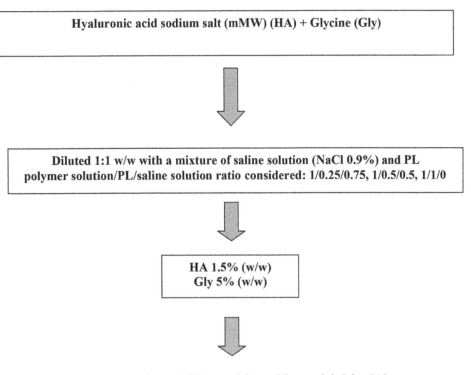

Fig. 8 Preparation scheme of HA-based "sponge-like" dressings intended for the delivery of PL in the treatment of skin ulcers.

hydration rate with respect to HA-based dressing (a complete gelification was obtained after 24 hr).

Dressings loaded with PL were characterized by % proliferation values on human fibroblast cells comparable to that of fresh PL, indicating that freeze-drying process and excipients employed did not disturb the activity of PL GFs. Such results were also confirmed by PDGF AB content.

Still intended for cutaneous administration, wound dressings based on solid lipid nanoparticles (SLNs), loaded with silver sulfadiazine (AgSD) were developed.[108]

Vascular—Zwitterionic

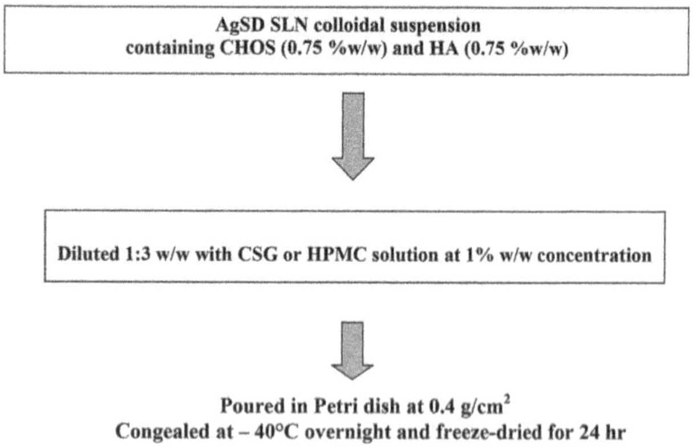

Fig. 9 Preparation scheme of dressings containing AgSD in SLNs to be extemporary loaded with PL for the treatment of skin ulcers.

The dressings were intended for extemporaneous loading with PL. SLNs were used to allow the association of PL with AgSD in the same formulation. This was necessary to avoid platelet protein degradation due to ion Ag⁺ and to prevent the cytotoxic effect of the drug on normal human dermal fibroblasts (NHDFs) used to test *in vitro* dressing proliferation properties.

AgSD-loaded SLNs were prepared by means of hot homogenization and ultrasound technique, by using glyceril behenate as lipid and poloxamer F68 as steric stabilizer and CHOS and HA in the aqueous phase, as schematically described in Fig. 9.

SLNs were loaded in wound dressings based on HPMC or CSG by mixing in a 1:3 weight ratio SLN dispersion with each polymer solution. These polymers were chosen to obtain a sponge matrix with elasticity and softness suitable to application on skin lesions. Dressings were prepared by freeze-drying. PL was loaded in the dressing by soaking.

SLNs were characterized by good biocompatibility on normal HDFs. The dressings were characterized by suitable mechanical properties. Moreover, the dressings based on CSG showed antimicrobial activity.

CONCLUDING REMARKS

It is generally recognized in the literature that the presence of biopolymers in formulations intended for wound repair greatly increases the success of the therapy. Recent findings have pointed out the peculiar action of platelet derivatives in promoting the healing of wounds, thanks to the presence of a pool of GFs and of other enabling substances.

For this reason, the use of platelet derivatives is more effective than the administration of a single purified growth factor. On the other hand, such hemoderivatives present the drawback to be characterized by variable amounts and type of bioactive substances. Therefore, they need a thorough characterization before use. Another drawback consists in

their poor stability on storage, which imposes their immediate use after preparation. Taking into account these problems, many attempts have been made to standardize the separation procedures of platelet derivatives from the other components of blood as well as to stabilize them by loading in suitable formulations. As far as stabilization is concerned, two different approaches can be pursued: one provides the set-up of suitable manufacturing processes such as freeze-drying, enabling the storage of the hemoderivatives in solid form; the other approach consists in the association of platelet derivatives with biopolymers, such as chondroitin sulfate. Such approaches can also be combined to further improve stabilization.

Moreover, the association of platelet derivatives with biopolymers, characterized by intrinsic bioactive properties, can determine a synergistic effect in the wound-healing process. The use of biopolymer platforms can also permit optimal biopharmaceutical properties such as prolonged release of bioactive substances and their longer permanence on the damaged tissue, thanks to the bioadhesion properties of polymers.

REFERENCES

1. Boateng, J.S.; Matthews, K.H.; Stevens, H.N.E.; Eccleston, G.M. Wound healing dressings and drug delivery systems: A review. J. Pharm. Sci. **2008**, *97* (8), 2892–2923.
2. Vasconcelos, A.; Cavaco–Paulo, A. Wound dressings for a proteolytic-rich environment. Appl. Microbiol. Biotechnol. **2011**, *90* (2), 445–460.
3. Gist, S.; Tio-Matos, I.; Falzgraf, S.; Cameron, S.; Beebe, M. Wound care in the geriatric client. J. Clin. Inter. Aging **2009**, *4*, 269–287.
4. Redding, S.W. Cancer therapy-related oral mucositis. J. Dent. Educ. **2005**, *69* (8), 919–929.
5. Sandri, G.; Bonferoni, M.C.; Ferrari, F.; Rossi, S.; Del Fante, C.; Perotti, C.; Gallanti, A.; Caramella, C. An in situ gelling buccal spray containing platelet lysate for the treatment of oral mucositis. Curr. Drug Discov. Technol. **2011**, *8* (3), 277–285.
6. Lu, L.; Reinach, P.S.; Kao, W.W. Corneal epithelial wound healing. Exp. Biol. Med. **2001**, *226* (7), 653–664.
7. Sandri, G.; Bonferoni, M.C.; Rossi, S.; Ferrari, F.; Mori, M.; Del Fante, C.; Perotti C.; Caramella, C. Thermosensitive eydrops containing platelet lysate for the treatment of corneal ulcers. Int. J. Pharm. **2012**, *426* (1–2), 1–6.
8. Geremicca, W.; Fonte, C.; Sisto Vecchio, S. Blood components for topical use in tissue regeneration: Evaluation of corneal lesions treated with platelet lysate and considerations on repair mechanisms. Blood Transfus. **2010**, *8* (2), 107–112.
9. Muzzarelli, R.A.A. Chitins and chitosans for the repair of wounded skin, nerve, cartilage and bone. Carbohydr. Polym. **2009**, *76* (2), 167–182.
10. Denis, T.G.; Dai, T.; Huang, Y.-Y.; Hambling, M.R. Wound-healing properties of chitosan and its use in wound dressings biopharmaceuticals. In *The Chitosan-Based Systems for Biopharmaceuticals: Delivery, Targeting and*

Polymer Therapeutics; Sarmento, B.; das Neves, J.; Eds.; John Wiley and Sons, Ltd.: Chichester, 2012; 137–158.

11. Werner, S.; Grose, R. Regulation of wound healing by growth factors and citokines. Physiol. Rev. **2003**, *83* (3), 835–870.

12. Suzuki, K; Saito, J; Yanai, R; Yamada, N; Chikama, T; Seki, K; Nishida, T. Cell-matrix and cell-cell interactions during corneal epithelial wound healing. Prog. Retin. Eye Res. **2003**, *22* (2), 113–133.

13. Fini, M.E. Keratocyte, and fibroblast phenotypes in the repairing cornea. Prog. Retin. Eye Res. **1999**, *18* (4), 529–551.

14. Dupps, W.J.; Wilson, S.E. Biomechanics and wound healing in the cornea. Exp. Eye Res. **2006**, *83* (4), 709–720.

15. Karamichos, D. Human corneal fibrosis: An *in vitro* model. Invest. Ophthalmol. Vis. Sci. **2010**, *51* (3), 1382–1388.

16. Clark, R.A.F.; Ghosh, K.; Tonnesen, M.G. Tissue engineering for cutaneous wounds. J. Invest. Dermatol. **2007**, *127* (5), 1018–1029.

17. Jacobsen, J.N.; Steffensen, B.; Aakkinen, L.; Krogfelt, K.A.; Larjava, H.S. Skin wound healing in diabetic b6 integrin-deficient mice. APMIS **2010**, *118* (10), 753–764.

18. Patel, G.K. The role of nutrition in the management of lower extremity wounds. Int. J. Low. Extrem. Wounds **2005**, *4* (1), 12–22.

19. Francesko, A.; Tzanov, T. Chitin. Chitosan and derivatives for wound healing and tissue engineering. Adv. Biochem. Eng. Biotechnol. **2011**, *125*, 1–27.

20. Peluso, G; Petillo, O.; Ranieri, M.; Santin, M.; Ambrosio, L.; Calabrò, D.; Avallone, B.; Balsamo, G. Chitosan- mediated stimulation of macrophage function. Biomaterials **1994**, *15* (15), 1215–1220.

21. Ueno, H; Nakamura, F.; Murakami, M.; Okumura, M.; Kadosawa, T.; Fujinaga, T. Evaluation effects of chitosan for the extracellular matrix production by fibroblasts and the growth factors production by macrophages. Biomaterials **2001**, *22* (15), 2125–2130.

22. Ueno, H.; Murakami, M.; Okumura, M.; Kadosawa, T.; Uede, T.; Fujinaga, T. Chitosan accelerates the production of osteopontin from polimorphonuclear leukocytes. Biomaterials **2001**, *22* (12), 1667–1673.

23. Okamoto, Y.; Inoue, A.; Miyatake, K.; Ogihara, K.; Shigemasa, Y.; Minami, S. Effects of chitin/chitosan and their oligomers/monomers on migration of macrophages. Macromol. Biosci. **2003**, *3* (10), 587–590.

24. Santos, T.C.; Marques, A.P.; Silva, S.S.; Oliveira, J.M.; Mano, J.F.; Castro, A.G.; Reis, R.L. *In vitro* evaluation of the behavior of polimorphonuclear neutrophilis in direct contact with chitosan-based membranes. J. Biotechnol. **2007**, *132* (2), 218–226.

25. Howling, G.I.; Dettmar, P.W.; Goddard, P.A.; Hampson, F.C.; Dornish, M.; Wood, E.J. The effect of chitin and chitosan on the proliferation of human skin fibroblasts and keratinocytes *in vitro*. Biomaterials **2001**, *22* (22), 2959–2966.

26. Chen, X.-G.; Wang, Z.; Liu, W.-S.; Park, H.-J. The effect of carboxymethyl-chitosan on proliferation and collagen secretion of normal and keloid skin fibroblasts. Biomaterials **2002**, *23* (23), 4609–4614.

27. Fischer, T.H.; Thatte, H.S.; Nichols, T.C.; Bender-Neal, D.E.; Bellinger, A.D.; Vournakisc, J.N. Synergistic platelet integrin signaling and factor XII activation in poly-*N*-acetyl glucosamine fiber-mediated hemostasis. Biomaterials **2005**, *26* (27), 5433–5443.

28. Fischer, T.H.; Valeri, C.R.; Smith, C.J.; Scull, C.M.; Merricks, E.P.; Nichols, T.C.; Demcheva, M; Vournakis, J.N. Non-classical processes in surface hemostasis: Mechanisms for the poly-*N*-acetyl glucosamine-induced alteration of red blood cell morphology and surface prothrombogenicity. Biomed. Mater. **2008**, *3* (1), 65–73.

29. Lord, M.S.; Cheng, B.; McCarthy, S.J.; Jung, M.; Whitelock, J.M. The modulation of platelet adhesion and activation by chitosan through plasma and extracellular matrix proteins. Biomaterials **2011**, *32* (28), 6655–6662.

30. Pietramaggiori, G.; Yang, H.J.; Scherer, S.S.; Kaipainen, A.; Chan, R.K.; Alperovich, M.; Newalder, J.; Demchea, M.; Vournakis, J.N.; Valeri, C.R.; Hechtman, H.B.; Orgil, D.P. Effects of poly-*N*-acetyl glucosamine (pGlcNAc) patch on wound healing in db/db mouse. J. Trauma **2008**, *64* (3), 803–808.

31. Kim, M.M.; Kim, S.K. Chitooligosaccharides inhibit activation and expression of matrix metalloproteinase-2 in human dermal fibroblasts. FEBS Lett. **2006**, *580* (11), 2661–2666.

32. Gorzelanny, C.; Poppelmann, B.; Strozyk, E.; Moerschbacher, B.M.; Schneider, S.W. Specific interaction between chitosan and matrix metalloprotease 2 decreases the invasive activity of human melanoma cells. Biomacromolecules **2007**, *8* (10), 3035–3040.

33. Helander, I.M.; Nurmiaho-Lassila, L.; Ahvenainen, R.; Rhoades, J.; Roller, S. Chitosan disrupts the barrier properties of the outer membrane of Gram-negative bacteria. Int. J. Food Microbiol. **2001**, *71* (2–3), 235–244.

34. Zheng, L.Y.; Zhu, J.F. Study on antimicrobial activity of chitosan with different molecular weights. Carbohydr. Polym. **2003**, *54* (4), 527–530.

35. Sun, L.; Du, Y.; Fan, L.; Chen, X.; Yang, J. Preparation, characterization and antimicrobial activity of quaternized carboxymethyl chitosan and application as pulp-cap. Polymers **2006**, *47* (6), 1796–1804.

36. Bonferoni, M.C.; Sandri, G.; Rossi, S.; Ferrari, F.; Caramella, C. Chitosan and its salts for mucosal and transmucosal delivery. Expert Opin. Drug Deliv. **2009**, *6* (9), 923–939.

37. Rossi, S.; Marciello, M.; Bonferoni, M.C.; Ferrari, F.; Sandri,.G.; Dacarro, C.; Grisoli, P.; Caramella, C. Thermally sensitive gel based on chitosan derivatives for the treatment of oral mucositis. Eur. J. Pharm. Biopharm. **2010**, *74* (2), 248–254.

38. Puccio, A.; Ferrari, F.; Rossi, S.; Bonferoni, M.C.; Sandri, G.; Dacarro, C.; Grisoli, P.; Caramella, C. Comparison of functional and biological properties of chitosan and hyaluronic acid, to be used for the treatment of mucositis in cancer patients. J. Drug. Deliv. Sci. Tech. **2011**, *21* (3), 241–247.

39. Rossi S.; Ferrari, F.; Bonferoni, M.C.; Sandri, G.; Faccendini, A.; Puccio, A.; Caramella, C. Comparison of poloxamer- and chitosan-based thermally sensitive gels for the treatment of vaginal mucositis. Drug Dev. Ind. Pharm. **2013**, *in press*.

40. Laurent, T.; Fraser, J. Hyaluronan. FASEB J. **1992**, *6* (7), 2397–2404.

41. Meyer, K.; Palmer, J. The polysaccharide of the vitreous humor. J. Biol. Chem. **1934**, *107*, 629–634.

42. Fraser, J.; Laurent, T.; Laurent, U. Hyaluronan: Its nature, distribution, functions and turnover. J. Intern. Med. **1997**, *242* (1), 27–33.

43. Weindl, G.; Schaller, M.; Schafer-Korting, M.; Korting, H.C. Hyaluronic acid in the treatment and prevention of skin diseases: Molecular biological, pharmaceutical and clinical aspects. Skin Pharmacol. Physiol. **2004**, *17* (5), 207–213.

44. Price, R.; Myers, S.; Leigh, I.; Navsaria, H. The role of hyaluronic acid in wound healing: Assessment of clinical evidence. Am. J. Clin. Dermatol. **2005**, *6* (6), 393–402.

45. Chen, W.; Abatangelo, G. Functions of hyaluronan in wound repair. Wound Repair Regen. **1999**, *7* (2), 79–89.

46. Stern, R.; Asari, A.; Sugahara, K. Hyaluronan fragments, an information-rich system. Eur. J. Cell. Biol. **2006**, *85* (8), 699–715.

47. Noble, P. Hyaluronan and its catabolic products in tissue injury and repair. Matrix Biol. **2002**, *21* (1), 25–29.

48. McKee, C.; Penno, M.; Cowman, M.; Burdick, M.D.; Strieter, R.M.; Bao, C.; Noble, P.W. Hyaluronan (HA) fragments induce chemokine gene expression in alveolar macrophages. The role of HA size and CD44. J. Clin. Invest. **1996**, *98* (10), 2403–2413.

49. Slevin, M.; Krupinski, J.; Kumar, S.; Gaffney, J. Angiogenic oligosaccharides of hyaluronan induce protein tyrosine kinase activity in endothelial cells and activate a cytoplasmic signal transduction pathway resulting in proliferation. Lab. Invest. **1998**, *78* (8), 987–1003.

50. Slevin, M.; Krupinski, J.; Gaffney, J.; Matou, S.; West, D.; Delisser, H.; Savani, R.C.; Kumar, S. Hyaluronan-mediated angiogenesis in vascular disease: Uncovering RHAMM and CD44 receptor signaling pathways. Matrix Biol. **2007**, *26* (1), 58–68.

51. Weigel, P.; Frost, S.; McGary, C.; LeBoeuf, R. The role of hyaluronic acid in inflammation and wound healing. Int. J. Tissue React. **1988**, *10* (6), 355–365.

52. Presti, D.; Scott, J. Hyaluronan mediated protective effect against cell damage caused by enzymatically produced hydroxyl (OH) radicals is dependent on hyaluronan molecular mass. Cell Biochem. Funct. **1994**, *12* (4), 281–288.

53. Trabucchi, E.; Pallotta, S.; Morini, M.; Corsi, F.; Franceschini, R.; Casiraghi, A.; Pravettoni, A.; Foschi, D.; Minghetti, P. Low molecular weight hyaluronic acid prevents oxygen free radical damage to granulation tissue during wound healing. Int. J. Tissue React. **2002**, *24* (2), 65–71.

54. Moseley, R.; Walker, M.; Waddington, R.J.; Chen, W.Y. Comparison of the antioxidant properties of wound dressing materials, carboxymethylcellulose, hyaluronan benzyl ester and hyaluronan, toward polymorphonuclear leukocyte derived reactive oxygen species. Biomaterials **2003**, *24* (9), 1549–1557.

55. Alkrad, J.A.; Mrestani, Y.; Stroehl, D.; Wartewig, S.; Neubert, R. Characterization of enzymatically digested hyaluronic acid using NMR, Raman,IR, and UV-Vis spectroscopies. J. Pharm. Biomed. Anal. **2003**, *31* (3), 545–550.

56. Trommer, H.; Wartewig, S.; Bottcher, R.; Poppl, A.; Hoentsch, J.; Ozegowski, J.H.; Neubert, R.H. The effects of hyaluronan and its fragments on lipid models exposed to UV irradiation. Int. J. Pharm. **2003**, *254* (2), 223–234.

57. Slevin, M.; Kumar, S.; Gaffney, J. Angiogenic oligosaccharides of hyaluronan induce multiple signalling pathways affecting vascular endothelial cell mitogenic and wound healing responses. J. Biol. Chem. **2002**, *277* (43), 41046–41059.

58. Gomes, J.A.; Amankwah, R.; Powell-Richards, A.; Dua, H. Sodium hyaluronate (hyaluronic acid) promotes migration of human corneal epithelial cells *in vitro*. Br. J. Ophthalmol. **2004**, *88* (6), 821–825.

59. West, D.C.; Shaw, D.M.; Lorenz, P.; Adzick, N.S.; Longaker, M.T. Fibrotic healing of adult and late gestation fetal wounds correlates with increased hyaluronidase activity and removal of hyaluronan. Int. J. Biochem. Cell Biol. **1997**, *29* (1), 201–210.

60. Mack, J.; Abramson, S.; Ben, Y.; Coffin, J.C.; Rothrock, J.K.; Maytin, E.V.; Hascall, V.C.; Largman, C.; Stelnicki, E.J. Hoxb13 knockout adult skin exhibits high levels of hyaluronan and enhanced wound healing. FASEB J. **2003**, *17* (10), 1352–1354.

61. Nishida, T.; Nakamura, M.; Mishima, H.; Otori, T. Hyaluronan stimulates corneal epithelial migration. Exp. Eye Res. **1991**, *53* (6), 753–758.

62. Pasonen-Seppanen, S.; Karvinen, S.; Törrönen, K.; Hyttinen, J.; Jokela, T.; Lammi, M.; Tammi, M.; Tammi, R. EGF upregulates, whereas TGF-β downregulates, the hyaluronan synthases Has2 and Has3 in organotypic keratinocyte cultures: Correlations with epidermal proliferation and differentiation. J. Invest. Dermatol. **2003**, *120* (6), 1038–1044.

63. Pienimaki, J.P.; Rilla, K.; Fulop, C.; Sironen, R.K.; Karvinen, S.; Pasonen, S.; Lammi, M.J.; Tammi, R.; Hascall, V.C.; Tammi, M.I. Epidermal growth factor activates hyaluronan synthase 2 in epidermal keratinocytes and increases pericellular and intracellular hyaluronan. J. Biol. Chem. **2001**, *276* (23), 20428–20435.

64. Liguori, V.; Guillemin, C.; Pesce, G.F.; Mirimanoff, R.O.; Bernier, J. Double-blind, randomized clinical study comparing hyaluronic acid cream to placebo in patients treated with radiotherapy. Radiother. Oncol. **1997**, *42* (2), 155–161.

65. Vazquez, J.R.; Short, B.; Findlow, A.H.; Nixon, B.P.; Boulton, A.J.; Armstrong, D.G. Outcomes of hyaluronan therapy in diabetic foot wounds. Diabetes Res. Clin. Pract. **2003**, *59* (2), 123–127.

66. Price, R.D.; Berry, M.G.; Navsaria, H.A. Hyaluronic acid: The scientific and clinical evidence. J. Plast. Reconstr. Aesthet. Surg. **2007**, *60* (10), 1110–1119.

67. Teh, B.M.; Shen, Y.; Friedland, P.L.; Atlas, M.D.; Marano, R.J. A review on the use of hyaluronic acid in tympanic membrane wound healing. Expert Opin. Biol. Ther. **2012**, *12* (1), 23–36.

68. Baykara, M.; Dogru, M.; Ozcetin, H.; Ertürk, H. Primary repair and intraocular lens implantation after perforating eye injury. J. Cataract Refract. Surg. **2002**, *28* (10), 1832–1835.

69. Hoffman, R.S.; Fine, I.H.; Packer, M. Stabilization of flat anterior chamber after trabeculectomy with Healon5. J. Cataract Refract. Surg. **2002**, *28* (4), 712–714.

70. Gutierrez-Ortiz, C.; Moreno-Lopez, M. Healon5 as a treatment option for recurrent flat anterior chamber after

trabeculectomy. J. Cataract Refract. Surg. **2003**, *29* (4), 635.

71. Sihota, R.; Saxena, R.; Agarwal, H.C. Intravitreal sodium hyaluronate and secondary glaucoma after complicated phacoemulsification. J. Cataract Refract. Surg. **2003**, *29* (6), 1226–1227.

72. Anniko, M.; Hellstrom, S.; Laurent, C. Reversible changes in inner ear function following hyaluronan application in the middle ear. Acta Otolaryngol. **1987**, *104* (Suppl.), 72–75.

73. Laurent, C.; Hellstrom, S.; Anniko, M. Inner ear effects of exogenous hyaluronan in the middle ear of the rat. Acta Otolaryngol. **1988**, *105* (3–4), 273–280.

74. Martini, A.; Rubini, R.; Ferretti, R; Govoni, E.; Schiavinato, A.; Magnavita, V.; Perbellini, A.; Fiori, M.G. Comparative ototoxic potential of hyaluronic acid and methylcellulose. Acta Otolaryngol. **1992**, *112* (2), 278–283.

75. Campoccia, D.; Doherty, P.; Radice, M.; Brun, P.; Abatangelo, G.; Williams, D.F. Semisynthetic resorbable materials from hyaluronan esterification. Biomaterials **1998**, *19* (23), 2101–2127.

76. Park, A.; Jackson, A.; Hunter, L.; McGill, L.; Simonsen, S.E.; Alder, S.C.; Shu, X.Z.; Prestwich, G. Cross-linked hydrogels for middle ear packing. Otol. Neurotol. **2006**, *27* (8), 1170–1175.

77. Martini, A.; Morra, B.; Aimoni, C.; Radice, M. Use of a hyaluronan-based biomembrane in the treatment of chronic cholesteatomatous otitis media. Am. J. Otol. **2000**, *21* (4), 468–473.

78. Balazs, E.; Bland, P.; Denlinger, J.; Goldman, A.L.; Larsen, N.E.; Leshchiner, A.; Morales, B. Matrix engineering. Blood Coagul. Fibrinolysis **1991**, *2* (1), 173–178.

79. Prestwich, G.; Shu, X.; Liu, Y.; Cai, S.; Walsh, J.F.; Hughes, C.W.; Ahmad, S.; Kirker, K.R.; Yu, B.; Orlandi, R.R.; Park, A.H.; Thibeault, S.L.; Duflo, S.; Smith, M.E. Injectable synthetic extracellular matrices for tissue engineering and repair. Adv. Exp. Med. Biol. **2006**, *585*, 125–133.

80. Laplante, P.; Raymond, M.A.; Gagnon, G.; Vigneault, N.; Sasseville, A.M.; Langier, Y.; Bernard, M.; Raymond, Y.; Hebert, M.J. Novel fibrogenic pathways are activated in response to endothelial apoptosis: implications in the pathophysiology of systemic sclerosis. J. Immunol. **2005**, *174* (9), 5740–5749.

81. Raymond, M.A.; Desormeaux, A.; Laplante, P.; Vigneault, N.; Filep, J.G.; Landry, K.; Pshezhetsky, A.V.; Hebert, M.J. Apoptosis of endothelial cells triggers a caspase-dependent anti-apoptotic paracrine loop active on VSMC. FASEB J. **2004**, *18* (6), 705–707.

82. Charbonneau, C.; Liberelle, B.; Hebert, M.J.; De Crescenzo, G.; Lerouge, S. Stimulation of cell growth and resistance to apoptosis in vascular smooth muscle cells on a chondroitin sulphate/epidermal growth factor coating. Biomaterials **2011**, *32* (6), 1591–1600.

83. Gilbert, M.E.; Kirker, K.R.;Gray, S.D.; Ward, P.D.; Szakacs, J.G.; Prestwich, G.D.; Orlandi, R.R, Chondroitin sulfate hydrogel and wound healing in rabbit maxillary sinus mucosa. Laryngoscope **2004**, *114* (8), 1406–1409.

84. Rammelt, S.; Illert, T.; Bierbaum, S.; Scharnweber, D.; Zwipp, H.; Schneiders, W. Coating of titanium implants with collagen, RGD peptide and chondroitin sulfate. Biomaterials **2006**, *27* (32), 5561–5571.

85. Reyes, J.M.; Herretes, S.; Pirouzmanesh, A.; Wang, D.A.; Elisseeff, J.H.; Jun, A.; McDonnell, P.J.; Chuck, R.S.; Behrens, A. A modified chondroitin sulfate aldehyde adhesive for sealing corneal incisions. Invest. Ophthalmol. Vis. Sci. **2005**, *46* (4), 1247–1250.

86. Park, Y.J.; Lee, Y.M.; Lee, J.Y.; Seol, Y.J.; Chung, C.P.; Lee, S.J. Controlled release of platelet derived growth factor-BB from chondroitin sulfate–chitosan sponge for guided bone regeneration. J. Control. Release **2000**, *67* (2–3), 385–394.

87. Park, J.S.; Yang, H.J.; Woo, D.G.; Yang, H.N.; Na, K.; Park, K.H. Chondrogenic differentiation of mesenchymal stem cells embedded in a scaffold by longterm release of TGF-beta 3 complexed with chondroitin sulfate. J. Biomed. Mater. Res. A **2010**, *92* (2), 806–816.

88. Zou, X.H.; Jiang, Y.Z.; Zhang, G.R.; Jin, H.M.; Hieu, N.T.M.; Ouyang, H.W. Specific interactions between human fibroblasts and particular chondroitin sulfate molecules for wound healing. Acta Biomater. **2009**, *5* (5), 1588–1595.

89. Kushida, S.; Kakudo, N.; Suzuki, K.; Kusumoto, K. Effects of platelet-rich plasma on proliferation and myofibroblastic differentiation in human dermal fibroblasts. Ann. Plast. Surg. **2013**, *71* (2), 219–224. doi:10.1097/SAP.0b013e31823cd7a4.

90. Nakajima,Y.; Kawase, T.; Kobayashi, M.; Okuda, K.; Wolff, L.F.; Yoshie, H. Bioactivity of freeze-dried platelet-rich plasma in an adsorbed form on a biodegradable polymer material. Platelets **2012**, *23* (8), 594–603.

91. Martinez-Zapata, M.J.; Martí-Carvajal, A.J.; Solà, I.; Expósito, J.A.; Bolíbar, I.; Rodríguez, L.; Garcia, J. Autologous platelet-rich plasma for treating chronic wounds. Cochrane Database Syst. Rev. **2012**, 10. Article ID: CD006899, doi:10.1002/14651858.CD006899.pub2.

92. Kim, K.M.; Shin, Y.-T.; Kim, H.Y. Effect of autologous platelet-rich plasma on persistent corneal epithelial defect after infectious keratitis. Jpn. J. Ophthalmol. **2012**, *56* (6), 544–550.

93. Khaksar, E.; Aldavood, S.J.; Abedi, G.R.; Sedaghat, R.; Nekoui, O.; Zamani-ahmadmahmudi, M. The effect of sub-conjunctival platelet-rich plasma in combination with topical acetylcysteine on corneal alkali burn ulcer in rabbits. Comp. Clin. Pathol. **2013**, *22* (1), 107–112.

94. Wu, X.; Ren, J.; Luan, J.; Yao, G.; Li, J. Biochemical, mechanical, and morphological properties of a completely autologous platelet-rich wound sealant. Blood Coagul. Fibrinolysis **2012**, *23* (4), 290–295.

95. Marciel, F.B.; DeRossi, R.; Modolo, T.J.C.; Pagliosa, R.C.; Leal, C.R.J.; Delben, A.A. Scanning electron microscopy and microbiological evaluation of equine burn wound repair after platelet-rich plasma gel treatment. Burns **2012**, *38* (7), 1058–1065.

96. Giuggioli, D.; Colaci, M.;·Manfredi, A.; Mariano, T.; Ferri, C. Platelet gel in the treatment of severe scleroderma skin ulcers. Rheumatol. Int. **2012**, *32* (9), 2929–2932.

97. Crovetti, G.; Martinelli, G.; Issi, M.; Barone, M.; Guizzardi, M.; Campanati, B.; Moroni, M.; Carabelli, A. Platelet gel for healing cutaneous chronic wounds. Transfus. Apher. Sci. **2004**, *30* (2), 145–151.

98. Mazzucco, L.; Medici, D.; Serra, M.; Panizza, R.; Rivara, G.; Orecchia, S.; Libener, R.; Cattana, E.; Levis, A.; Betta,

Vascular—Zwitterionic

P.G.; Borzini P. The use of autologous platelet gel to treat difficult-to-heal wounds: A pilot study. Transfusion **2004**, *44* (7), 1013–1018.

99. Perotti, G.; Stronati, M.; Figar, T.; Del Fante, C.; Scudeller, L.; Perotti, C. Allogeneic platelet leucocyte-gel to treat occipital decubitus ulcer in a neonate: A case report. Blood Transfus. **2012**, *10* (3), 387–389.

100. Ranzato, E.; Patrone, M.; Mazzucco, L.; Burlando, B. Platelet lysate stimulates wound repair of HaCaT keratinocytes. Br. J. Dermatol. **2008**, *159* (3), 537–545.

101. Cipriani, V.; Ranzato, E.; Balbo, V.; Mazzucco, L.; Cavaletto, M.; Patrone, M. Long-term effect of platelet lysate on primary fibroblasts highlighted with a proteomic approach. J. Tissue Eng. Regen. Med. **2009**, *3* (7), 531–538.

102. El Backly, R.; Ulivi, V.; Tonachini, L.; Cancedda, R.; Descalzi, F.; Mastrogiacomo, M. Platelet lysate induces *in vitro* wound healing of human keratinocytes associated with a strong proinflammatory response. Tissue Eng. A **2011**, *17* (13–14), 1787–1800.

103. Pezzotta, S.; Del Fante, C.; Scudeller, L.; Cervio, M.; Antoniazzi, E.R.; Perotti, C. Autologous platelet lysate for treatment of refractory ocular GVHD. Bone Marrow Transplant. **2012**, *47* (12), 1558–1563.

104. Caramella, C.M.; Sandri, G.; Rossi, S.; Mori, M.; Bonferoni, M.C.; Ferrari, F.; Del Fante, C.; Perotti, C. New therapeutic platforms for the treatment of epithelial and cutaneous lesions. Curr. Drug Deliv. **2013**, *10* (1), 18–31.

105. Caramella, C.; Bonferoni, M.C.; Rossi, S.; Sandri, G.; Ferrari, F.; Perotti, C.G.; Del Fante, C. Platelet Lysate and Bioadhesive Compositions Thereof for the Treatment of Mucositis. US Patent 0280952 A1, Nov 17, 2011.

106. Sandri, G.; Bonferoni, M.C; Rossi, S.; Ferrari, F.; Mori, M.; Del Fante, C.; Perotti, C.; Scudeller, L.; Caramella, C.; Platelet lysate formulations based on mucoadhesive polymers for the treatment of corneal lesions. J. Pharm. Pharmacol. **2011**, *63* (2), 189–198.

107. Rossi, S.; Faccendini, A.; Bonferoni, M.C.; Ferrari, F.; Sandri, G.; Del Fante, C.; Perotti, C.; Caramella, C.M. "Sponge-like" dressings based on biopolymers for the delivery of platelet lysate to skin chronic wounds. Int. J. Pharm. **2013**, *440* (2), 207–215.

108. Sandri, G.; Bonferoni, M.C.; D'Autilia, F.; Rossi, S.; Ferrari, F.; Grisoli, P.; Sorrenti M.; Catenacci, L.; Del Fante, C.; Perotti, C.; Caramella C. Wound dressings based on silver sulfadiazine SLN for tissue repairing. Eur. J. Pharm. Biopharm. **2013**, *84* (1), 84–90.

Zwitterionic Polymeric Materials

Vinod B. Damodaran
New Jersey Center for Biomaterials, Rutgers University, Piscataway, New Jersey, U.S.A.

Victoria Leszczak
Department of Mechanical Engineering, Colorado State University, Fort Collins, Colorado, U.S.A.

Melissa M. Reynolds
Department of Chemistry and School of Biomedical Engineering, Colorado State University, Fort Collins, Colorado, U.S.A.

Ketul C. Popat
Department of Mechanical Engineering and School of Biomedical Engineering, Colorado State University, Fort Collins, Colorado, U.S.A.

Abstract

Zwitterionic materials are an emerging class of biomaterials, characterized with their versatile structural features and excellent hemocompatibility. Consequently, this class of biomaterials is gaining significant importance in various biomedical applications including medical diagnostics, tissue engineering, and drug delivery. In this entry, the authors present a very descriptive summary of the development of this important class of materials, in terms of various synthetic approaches, bioefficacies, biomedical applications, and future prospectives.

INTRODUCTION

The most attractive molecular designs to improve the biocompatibility of various synthetic polymeric materials for biomedical applications include impregnating either biological molecules (biological approach) or biologically active components (biochemical approach) onto a polymer matrix. An extensive description of the biological approach to improve the biocompatibility by incorporating a biological molecule, such as nitric oxide (NO), is presented in detail by the authors in a separate entry.[1] A very attractive biochemical approach recently gaining both scientific as well as commercial interest to improve the biocompatibility is the use of zwitterionic polymers. Zwitterionic polymers are amphoteric materials that contain both opposite ionic species incorporated within the same pendant functional groups.[2] These materials are now being investigated for various biomedical utilities including anti-fouling applications, drug delivery, and biomedical imaging. This review summarizes the development of this very important class of biomaterial with a special emphasis on their biological evaluations and biomedical applicability.

ANTI-FOULING APPLICATIONS

Because of extremely low fouling properties and the capability to prevent protein adsorption, bacterial adhesion, and biofilm formation, zwitterionic materials have recently earned enormous importance in various biomedical applications including biomedical implants. Poly(ethylene glycol) (PEG) is also widely used for making protein-repellant surfaces,[3–6] however its susceptibility to oxidative damages can reduce their anti-fouling capabilities over long-term applications.[7] Moreover, while PEG and other anti-fouling materials (e.g., dextran, tetraglyme) achieved protein repellency via relatively weak hydrogen bonding hydration, the zwitterionic materials exerted a very strong hydration layer through electrostatic interactions and a strong dipole of zwitterions.[8–10] This results in "super-hydrophilicity" and ultimately leads to high protein resistivity. Consequently, the biomedical research field witnessed an increasing interest in developing zwitterionic materials as an efficient alternative to traditional non-fouling materials.

Polybetaines constitute a major class of zwitterionic materials, which have been extensively evaluated for non-fouling applications. These materials can be further classified into carboxybetaines (CB), sulfobetaines (SB), and phosphobetaines (PB), and a synopsis of the biomedical potential of these materials is given in the following sections.

Carboxybetaines

Compared to other zwitterionic polymers, carboxybetaines are the most attractive material because of their capability for ligand immobilization on a non-fouling background.

Concise Encyclopedia of Biomedical Polymers and Polymeric Biomaterials DOI: 10.1081/E-EBPPC-120050037

Vascular—Zwitterionic

Various ligands, including antibodies and proteins, can be conveniently modified onto carboxyl groups by amine conjugation chemistries without disturbing the overall non-fouling characteristics. Moreover, these groups are more hydrophilic and exhibit characteristic acid-base equilibria.[11] The very first synthetic polybetaine reported was a polycarboxybetaine, namely, poly(4-vinyl pyridine betaine) prepared through the hydrolysis of a polymeric quaternary ester salt of poly(4-vinylpyridine) and ethyl bromoacetate, by Ladenheim and Morawetz in 1957.[12] Even though these researchers observed a very high value of the effective dielectric constant in the swollen polymer coils caused by the presence of high concentration of zwitterionic groups in the polymer matrix, it took more than five decades to realize the potential applicability.

Jiang et al. prepared a highly protein-resistant surface by grafting poly(carboxybetaine methacrylate) (poly-CBMA) onto a gold surface via a surface-initiated atom transfer radical polymerization (ATRP).[13–15] The zwitterionic-modified surface prevented the nonspecific adsorption of a number of proteins with varying sizes and isoelectric points. Moreover, the researchers covalently immobilized monoclonal mouse antibody (mAb; anti-hCG) on the surface carboxyl groups of polyCBMA for specifically binding to human chorionic gonadotropin (hCG), while maintaining the nonspecific protein repellency.[13] This dual functional behavior of polyCBMA provided a very promising opportunity for making non-fouling surfaces for biosensors and diagnosis applications. Because of the structural similarity with the naturally occurring zwitterionic glycine betaine, polyCBMA also exhibited unique anticoagulation activities and high plasma protein adsorption resistivity compared to other zwitterionic analogues and PEGs.[16] As such, polyCBMA is an interesting polymer for making blood-compatible surfaces. Using polyCBMA polymer bushes with a thickness of 10–15 nm, Jiang et al. demonstrated a high level of protein resistance (adsorption 0.4 ± 0.9 ng/cm^2) against 100% human plasma compared to other non-fouling zwitterionic and PEG-modified surfaces having a comparable thickness.[16] A year later, the same researchers demonstrated that polyCBMA coatings can reduce the long-term biofilm formation of *Pseudomonas aeruginosa* by 95% up to 240 hr at 25°C and by 93% up to 64 hr at 37°C, and decreased *P. putida* biofilm accumulation by 95% up to 192 hr at 30°C.[17] This study also showed that the poly-CBMA surfaces retained the non-fouling activities over a range of temperatures from 25°C to 37°C (Fig. 1).

Extensive efforts are now ongoing to improve the physical and mechanical properties, stability, and uniformity of these non-fouling zwitterionic polyCBMA polymer surfaces and hydrogels by varying carboxybetaine methacrylate monomers as well as through crosslinking utilizing photopolymerization methods.[18–20] Additional efforts including the use of nitroxide-mediated free radical polymerization (NMFRP) in place of the commonly used ATRP technique

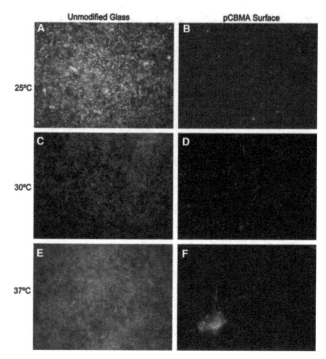

Fig. 1 Representative fluorescence microscopy graphs of biofilm formation on polyCBMA surfaces tested at different temperatures: (**A**) *P. aeruginosa* on unmodified glass for 48 hr at 25°C; (**B**) *P. aeruginosa* on polyCBMA surface for 240 hr at 25°C; (**C**) *P. putida* on unmodified glass for 96 hr at 30°C; (**D**) *P. putida* on polyCBMA for 192 hr at 30°C; (**E**) *P. aeruginosa* on unmodified glass for 15 hr at 37°C; (**F**) *P. aeruginosa* on poly-CBMA surface for 64 hr at 37°C.
Source: From Cheng et al.[17] ©, with permission from Elsevier.

in order to avoid the presence of any potentially cytotoxic catalyst contaminations are underway to improve the biocompatibility of polyCBMAs.[21]

Using a cationic precursor of polyCBMA, Cheng et al. prepared a switchable polymer surface coating with self-sterilizing and non-fouling capabilities.[22,23] The cationic precursor of polyCBMA killed more than 99.9% of *Escherichia coli* K 12 in 1 hr and switched to a zwitterionic non-fouling surface with a release of more than 98% of the dead bacterial cells upon hydrolysis (Fig. 2). Furthermore, the resulting non-fouling surface prevented any further protein attachment and biofilm formation offering a very promising material for making biomedical implants.

Very similar to the polyCBMA surfaces, functionalized poly(carboxybetaine acrylamide) (polyCBAA) surfaces also exhibited ultralow fouling properties in undiluted human plasma. Surfaces with nearly zero protein adsorption from complex undiluted human blood serum and plasma were achieved with polyCBAA coatings having ~20 nm thicknesses.[15,24,25] Furthermore, Vaisocherová et al. utilized polyCBAA coating immobilized with human monoclonal antibodies against activated leukocyte cell molecules (anti-ALCAM)[24] (Fig. 3) for detecting human

ALCAM antibodies in undiluted human blood plasma. The very high sensitivity and specificity achieved even in the presence of undiluted blood plasma provided a very unique strategy for preparing ultralow fouling biosensor platforms with selective biorecognition capabilities. Ultralow fouling polyCBAA polymer films with uniform and controlled thickness and high surface packing densities were achieved by Krause et al. using the surface-initiated photoinitiator-mediated polymerization (SI-PIMP) technique.[26] Anti-TSH (antibody to thyroid stimulating hormone) immobilization onto these polyCBAA polymer films,[26,27] while maintaining excellent post-functionalized non-fouling properties (through deactivation of unreacted surface groups back into the original non-fouling background), enabled a sensitive and selective detection of the corresponding antigen down to ~1 ng/mL directly from undiluted human plasma.

The non-fouling properties of CB groups were found to be greatly dependent on their spacer group distance between

the positive quaternary amine and the negative carboxyl groups, as well as environmental factors including pH and ionic strengths. For example, it has been found that in poly(CBAA)s containing CB groups with varying spacer groups (Fig. 4), the CB group with a long spacer showed less protein repellency than with shorter spacer groups at low ionic strengths and pH values. This observation is likely due to the higher hydrophobicity and higher pK_a.[28,29]

Sulfobetaine

Along with the poly(carboxybetaines) described previously, poly(sulfobetaines) constitute another attractive group of zwitterionic materials extensively evaluated for making non-fouling surfaces. Well-packed zwitterionic poly(sulfobetaine methacrylate) (polySBMA) brushes (Fig. 5) with a 7 nm thickness and prepared via surface-initiated ATRP were found to effectively reduce plasma protein adsorption from a platelet-poor plasma (PPP) solution over a wide range of ionic strengths (0.1 to 1 M), pH values (7.4 to 11), and temperature (22°C to 37°C).[30] Furthermore, a remarkable reduction in platelet activation and adhesion were also observed on the polySBMA-grafted surfaces, suggesting polySBMA as an effective and stable non-fouling surface coating material for various blood-contacting applications. Similarly prepared polySBMA surfaces were found to be almost completely resistant to the adsorption of a number of human proteins including human serum albumin (HSA), gamma globulin, fibrinogen, and lysozyme, even at low ionic strengths.[31–34] Additionally, polySBMA polymers exhibited anticoagulant activities in 100% human plasma and anti-hemolytic activities in RBC solutions, depending on the molecular weight of the polyzwitterions.[35]

In order to improve the hydrophilicity and protein repellency of polypropylene (PP) surfaces, high density polySBMA polymer brushes were successfully tethered onto the PP surfaces by Zhao et al., using UV-induced graft polymerization followed by the ATRP technique (Fig. 6).[36]

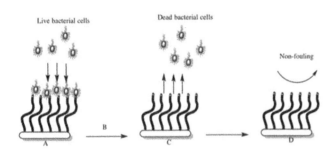

Fig. 2 Schematic representation of a surface that switches from an antibacterial surface to a non-fouling surface upon hydrolysis. (**A**) Antimicrobial cationic polyCBMA precursor effectively kills bacteria; (**B**) polyCBMA cationic precursor is converted into non-fouling zwitterionic polyCBMA upon hydrolysis; (**C**) dead bacteria remaining on the surface are released from non-fouling zwitterionic polyCBMA; (**D**) zwitterionic polyCBMA itself is highly resistant to bacterial adhesion.
Source: Adapted from Cheng et al.[22] ©, with permission from WILEY-VCH Verlag GmbH and Co. KGaA.

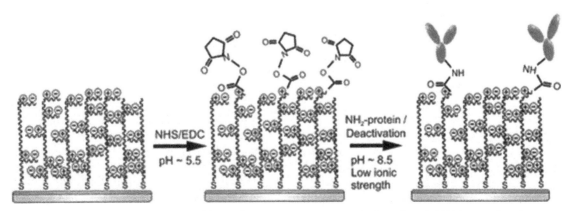

Fig. 3 Scheme of the surface activation, protein immobilization, and surface deactivation of a polyCBAA-coated surface.
Source: From Vaisocherová et al.[24] ©, with permission from American Chemical Society.

Vascular—Zwitterionic

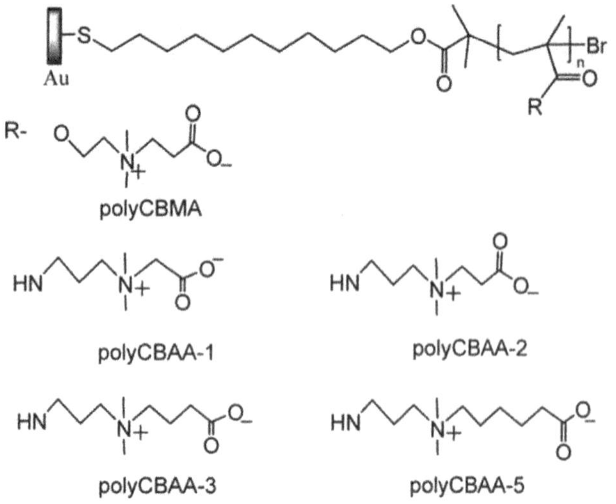

Fig. 4 Poly(CBAA)s containing CB groups with varying spacer lengths.
Source: From Zhang et al.[28] ©, with permission from American Chemical Society.

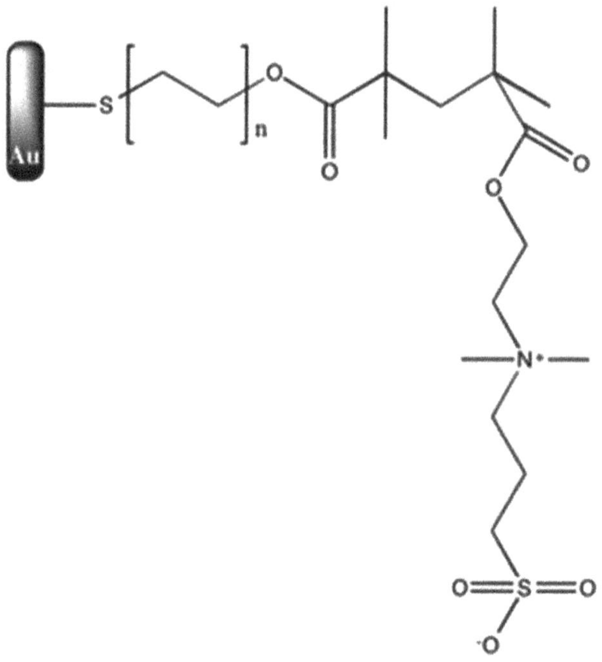

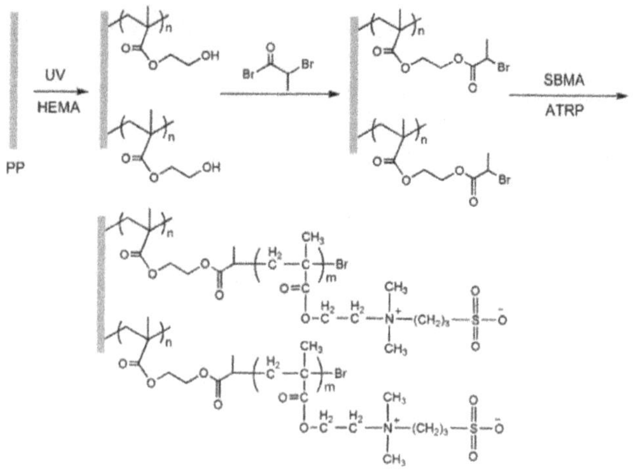

Fig. 6 Schematic representation for the UV-induced graft polymerization of HEMA and the surface-initiated ATRP of SBMA on PP surface.
Source: From Zhao et al.[36] ©, with permission from Elsevier.

Fig. 5 Structures of a polySBMA grafted surface.

Highly hydrophilic surfaces with water contact angle as low as 17.4° were obtained after surface grafting, and excellent anti-fouling properties resulted due to the introduction of zwitterionic groups on the polymer surface. In a similar way, Li et al. prepared a zwitterionic sulfobetaine-modified poly(vinylidene fluoride) (PVDF) by grafting a high-density zwitterionic polymer poly(3-(methacryloylamino)propyl-dimethyl-(3-sulfopropyl) ammonium hydroxide) (poly(MPDSAH)) using ATRP followed by cerium(Ce(IV))-induced graft copolymerization.[37]

Similar to polySBMA-grafted PP surfaces, the surface hydrophilicity of the sulfobetaine modified PVDF surfaces were also significantly enhanced (water contact angle 22.1°) and the amount of protein adsorption also decreased to zero.[37,38] The feasibility of extending the polySBMA surface grafting onto various surfaces including silicone and glass[39] and the excellent anti-fouling and blood compatibility capabilities demonstrated by these sulfobetaine-modified surfaces suggests that the method has a great potential in converting many generally hydrophobic surfaces to suit for various blood-contacting medical applications.

Recently, Chang et al. prepared a tunable and switchable thermoresponsive polymer coating using copolymers of zwitterionic sulfobetaine methacrylate (SBMA) and nonionic N-isopropyl acrylamide (NIPAAm) via homogeneous free radical copolymerization.[40] In aqueous solutions, the poly(SBMA-co-NIPAAm) copolymer presented unique and controllable lower and upper critical solution temperatures (LCST and UCST), conveniently altered by varying copolymer compositions, and solution concentration, polarity, and ionic strengths. Furthermore, the copolymers with a SBMA composition above 29 mol % demonstrated exceptionally low protein adsorption and high anticoagulant activities in normal human PPP solution. This suggests their use as a potential thermoresponsive biomaterial coating.

Thermally stable, well-defined poly(ethylene oxide) polymer brushes with terminal alkyl and sulfobetaine bristles (PECH-DMAPS) were reported by Kim et al.[41] Because of the characteristic formation of self-assembled multi-layer structures in films, the PECH-DMAPS polymer films exhibited uniform hydrophilic, zwitterionic sulfobetaine groups at the surface. Even though the polymer films were found to suppress bacterial adherence significantly with both Gram-positive (*Staphylococcus* epidermidis, *S. aureus*, and *Enterococcus faecalis*) and Gram-negative bacteria (*P. aeruginosa and E. coli*), the extent of bacterial adherence showed a dependence on the bacterial species and the content of sulfobetaine bristles in the polymer brushes. *P. aeruginosa* exhibited the highest level of adherence after 4 hr incubation; *E. coli* showed the lowest level of adherence, and the other bacteria exhibited intermediate adherence levels during the evaluation.

Similar to carboxybetaines, zwitterionic sulfobetaine polymer brushes were also explored for attaching antibodies to recognize specific microorganisms while maintaining the non-fouling background. Nguyen et al. immobilized anti-*Salmonella* antibodies on to polySBMA-grafted silicon nitride surfaces to evaluate the selective sensing properties of the modified surface in complex matrices (Fig. 7).[42] The antibody immobilized surfaces selectively detected *Salmonella* in complex media containing fibrinogen without any protein adsorption. Moreover, the polySBMA-grafted silicon surfaces demonstrated excellent protein-repellent capability (>99% repulsion) compared to the unmodified hydrophobic silicon nitride and other commonly used protein-repellent surfaces.[43]

Phosphobetaines

It has been demonstrated that the zwitterionic lipid phosphatidylcholine is a major constituent of the outer surface of the non-thrombogenic erythrocyte cell membranes.[44] Consequently, a number of approaches to prepare non-thrombogenic surfaces are initiated by incorporating various phosphobetaine (phosphorylcholines) groups or structural analogues. This approach is also termed "biomembrane mimicry" because of the structural similarity of the functional groups to the lipids that consist of cell membranes.[45]

The very first phosphatidylcholine-incorporated polymers were synthesized by Chapman et al.[46] and demonstrated a potential as hemocompatible materials for biological membranes to mimic the nonreactive cell surfaces.[47] Early approaches to prepare phosphobetaine-incorporated biomimetic surfaces were to include structural analogues of naturally occurring lipid dipalmitoyl phosphatidylcholine (DPPC), including diacetylenic phosphatidylcholine (DAPC)[48–50] and methacryloyloxyalkyl phosphorylcholines (MAPC) (Fig. 8),[51–55] and proved their ability to improve blood compatibility and to reduce protein adsorption. Similar to other polybetaines, the surface modification of various hydrophobic polymers with phosphobetaine groups also greatly enhanced the hydrophilicity of the resulting polymer surfaces.[56–59] However, the anti-bioadherent performances of the phosphobetaine derivatives were found to be markedly superior to that of other polybetaine analogues.[60]

Detailed biocompatibility evaluations of various MAPC copolymers in reducing plasma protein adsorptions, inflammatory responses, and cell adhesion were reported by Ishihara et al. in the early 2000s.[61–65] The researchers proposed that while exposing the MAPC copolymers to blood, there is a preferential adsorption of phospholipids onto the MAPC surfaces to the adsorption of proteins and adhesion of platelets to form a biomimetic surface.[66] The formation of such a biomimetic phospholipid layer on the surfaces enhances blood compatibility while reducing protein adsorption and platelet adhesion.

However, despite the excellent biocompatibility demonstrated by these materials, the majority of the first-generation poly(phosphobetaines) lacked the required mechanical properties in order to be suitable for various medical applications.[45,67] Conversely, because of the capability to retain a large amount of water with the zwitterionic head groups while preventing protein

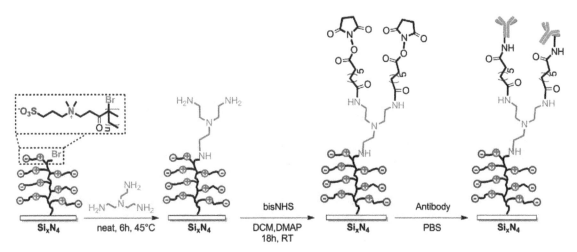

Fig. 7 Procedure for the attachment of anti-*Salmonella* antibodies to polySBMA-coated silicon nitride surfaces.
Source: From Nguyen et al.[42] ©, with permission from American Chemical Society.

adsorption, poly(phosphobetaines)-modified hydrogels improved lens wettability and enhanced surface properties compared to other conventional hydrogels used in contact lens applications.[68,69] Moreover, due to the enhanced hydration of the zwitterionic groups, these materials can improve the oxygen permeability, thereby ensuring optimum corneal health.[68] These characteristic advantages associated with the poly(phosphobetaine) zwitterionic materials led to the successful development of the Proclear™ family of contact lenses, prepared from poly(phosphobetaine)-hydroxyethylmethacrylate copolymers (poly(PB-*co*-HEMA)).[70]

Mechanically improved crosslinked poly(phosphobetaine) hydrogels were prepared by Goda et al. from 2-methacryloyloxyethyl phosphorylcholine (MPC) and a novel crosslinker, 2-(methacryloyloxy)ethyl-*N*-(2-methacryloyloxy)ethyl]phosphorylcholine (MMPC).[71] These hydrogels exhibited super-hydrophilicity, high oxygen and electrolyte permeability, and excellent resistivity to

protein adsorption, suggesting a promising biomaterial for ocular applications. Chu et al. prepared a soft contact lens biosensor (SCL-biosensor) for non-invasive in situ ocular biomonitoring of tear fluids using a polyMPC copolymer prepared with poly(dimethyl siloxane) (poly-DMS).[72] The polyMPC-incorporated SCL-biosensor showed an excellent relationship between the output current and glucose concentration sufficient for monitoring tear glucose levels when used on the eyes of rabbits, and confirmed to be useful in both static as well as dynamic states.

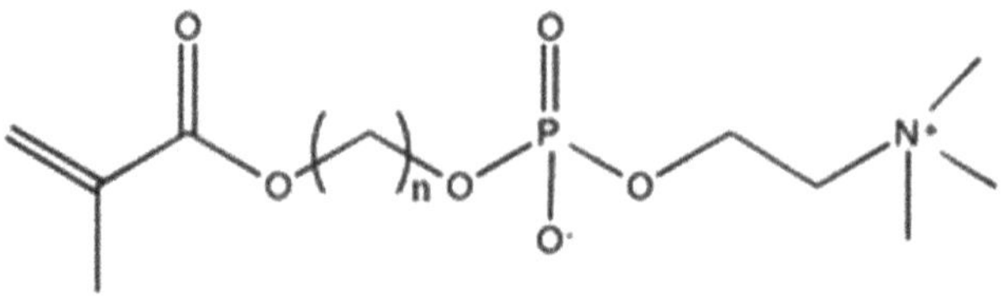

Recent advancements in the development of phosphobetaines also include improving the biocompatibility of various metallic surfaces through surface immobilization strategies. Using a covalent approach, Ye et al. immobilized poly(2-methacryloyloxyethyl phosphorylcholine-*co*-methacryl acid) (poly(MPC-*co*-MA)) onto a titanium alloy (TiAl$_6$V$_4$) surface, which had been functionalized with 3-aminopropyltriethoxysilane (APS).[73] The poly(MPC-*co*-MA)-modified TiAl$_6$V$_4$ surfaces (Ti-APS-PMA) displayed significant improvements in the blood compatibility, and substantial reduction in the acute platelet deposition, platelet activation, and fibrinogen adsorption compared to the control surfaces (Fig. 9). In a similar way, Kyomoto et al. grafted polyMPC onto the surface of a cobalt–chromium–molybdenum (Co-Cr-Mo) alloy using a 3-methacryloxypropyl trimethoxysilane (MPSi) intermediate layer and a photo-induced radical graft polymerization method, as a potential material for preparing orthopedic metal bearings for artificial hip joint system.[74] The grafted polyMPC layer provided superlubricity to the Co-Cr-Mo surface with

Fig. 8 Structure of MAPC derivatives: 2-methacryloyloxyethyl phosphorylcholine (MPC) (n = 2), 6-methacryloyloxyethyl phosphorylcholine (MHPC) (n = 6), and 10-methacryloyloxydecyl phosphorylcholine (MDPC) (n = 10).

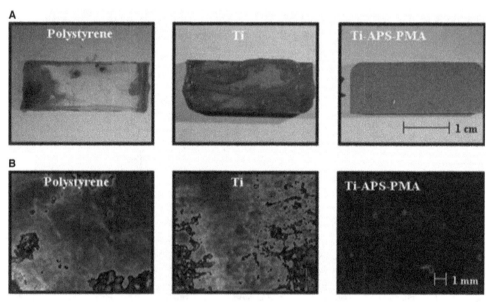

Fig. 9 (**A**) Macroscopic and (**B**) fluorescent micrograph images of unmodified and modified TiAl$_6$V$_4$ samples after contact with minimally anti-coagulated (1.5 U/mL heparin) ovine blood for 50 min at 37°C. Scale bars = 1 cm for (**A**) and 1 mm for (**B**).
Source: From Ye et al.[73] ©, with permission from Wiley Periodicals, Inc.

a friction coefficient as low as 0.006, similar to that of the natural cartilage interface. Furthermore, the polyMPC layer significantly enhanced the hydrophilicity of the Co-Cr-Mo-*g*-MPC surface (contact angle 20°) compared to the unmodified Co-Cr-Mo surface (contact angle 80°) and improved the non-fouling behavior. Similarly phosphobetaine-modified metallic surfaces including carbon, gold, stainless steel, and titanium were found to have greatly improved biocompatibility and reduced thrombogenicity compared to the unmodified forms.[75–81]

Similar to the Co-Cr-Mo-*g*-MPC surfaces discussed previously, polyMPC grafted onto the surface of cross-linked polyethylene (CLPE) is also shown to be a very promising material for making artificial hip joints.[82,83] Owing to the superlubricity exerted by polyMPC polymer, the polyPMC-grafted CLPE material exhibited an increased hydrophilicity and decreased friction torque. This resulted in a dramatic reduction in wear and ultimately provided a higher longevity of artificial hip joints.

DRUG DELIVERY

Recently, researchers focused on exploring the therapeutic capability of these important classes of biomedical materials as potent drug delivery vehicles. Cao et al. evaluated zwitterionic nanoparticles prepared from poly(carboxy-betaine)-poly(lactic-*co*-glycolic acid) block copolymer (PCB-PLGA) for the controlled and targeted delivery of docetaxel (Fig. 10).[84] The docetaxel-loaded PCB-PLGA nanoparticles showed a sustained release of the drug over 96 hr without any protein interferences in complex media. Similarly, zwitterionic polymer-lipid conjugates (1,2-diste aroyl-sn-glycero-3-phosphoethanolamine-poly(carboxy-betaine) (DSPE-PCB)) also resulted in greater stabilization of liposomes and provided extended blood circulating characteristics.[85] Wang et al. investigated hyperbranched poly(3-ethyl-3-(hydroxymethyl)oxetane)-poly(carboxy-betaine) (HBPO-PCB) micelles as a carrier for the site-specific delivery of doxorubicin (DOX).[86] Intracellular

uptake and *in vitro* evaluation of DOX-loaded HBPO-PCB micelles against hela cells showed a targeted cytotoxicity to inhibit the proliferation of cancer cells and exhibited an enhanced protein adsorption resistivity, suggesting a promising treatment option for increasing the site specificity of doxorubicin for the treatment of folate receptor (FR)-positive cancer.

Zwitterionic sulfobetaine copolymer functionalized mesoporous silica nanoparticles (MSN) were successfully evaluated by Sun et al. for temperature-responsive drug release applications.[87] Using rhodamine B (RhB) as a model drug, the researchers observed a thermo-triggered release of nearly 23% of the loaded RhB at 30°C and 80% at 55°C in saline solutions as a result of volume-phase transition of the sulfobetaine copolymer shells. Sulfobetaine-modified self-assembled nanoparticles loaded with paclitaxel exhibited extended and controlled release kinetics over 36 hr, and low cytotoxicity to osteoblast and hela cells compared to control samples.[88] Very recently, Zhai et al. demonstrated that a sulfobetaine-modified ε-polylysine copolymer cationic vector greatly enhanced the stability complex in serum circumstances, due to the stabilization of charges by the presence of sulfobetaines.[89] Moreover, the complex showed a 10-fold higher transfection efficiency than the non-sulfobetaine modified analogue, suggesting a very promising systemic gene vector for gene therapy.

Polyplexes prepared using phosphobetaine copolymers are currently gaining importance in preparing pH-responsive micellar vehicles for efficient and novel tumor-selective strategies. Licciardi et al. evaluated folic acid (FA)-functionalized polyMPC block polymers with 2-(dimethylamino)ethyl methacrylate (DMA) and 2-(diisopropylamino)ethyl methacrylate (DPA) as synthetic vectors for DNA condensation with very specific cell-targeting capabilities.[90] Similarly, prepared amphiphilic block copolymer micelles using polyPMC were found to very effective in the solubilization of highly hydrophobic anticancer drugs including paclitaxel.[91,92]

Madsen et al. evaluated a thermoresponsive amphiphilic diblock copolymer comprised of polyMPC and poly(2-hydroxypropyl methacrylate) (PHPMA) for the intracellular

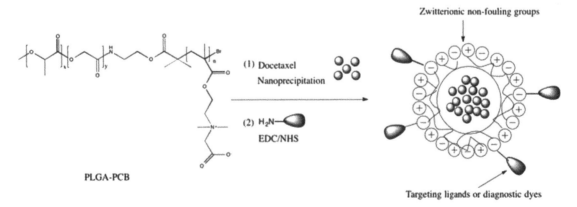

Fig. 10 Docetaxel-loaded PCB-PLGA nanoparticles for controlled and targeted drug delivery.
Source: Adapted from Cao et al.[84] ©, with permission from WILEY-VCH Verlag GmbH and Co. KGaA.

delivery of amphiphilic drugs.[93] The colloidal aggregates prepared from the copolymer efficiently solubilized a sample fluorescent dye, rhodamine B octadecyl ester, and provided a useful tool for enhancing the solubilization of similar drugs, thereby ensuring an efficient intracellular drug delivery. PolyMPC-camptothecin (polyMPC-CPT) conjugates were prepared by Chen et al. using an ATRP-click conjugation strategy with high drug loading capacity (up to 14% CPT) and excellent water solubility.[94] Drug release was achieved through ester cleavage and showed an enhanced induced cytotoxicity compared to the native CPT against different cancer cell lines, including breast (MCF7), ovarian (OVCAR-3), and colorectal (COLO 205) cancer cell lines.

In an interesting approach, Jia et al. prepared phosphobetaine-modified generation 5 poly(amido amine) (PAMAM) dendrimers (G5) via Michael addition reaction (Fig. 11).[95] The hydrophobic dendrimer interior was used for loading the anticancer drug Adriamycin (ADR). The biomimetic phosphobetaine-modified dendrimers showed less cytotoxicity and enhanced cancer cell recognition than the native PAMAM dendrimers. The drug conjugate showed a sustained release of ADR with effective inhibition of the growth of cancer cells. Site-specific delivery of the hydrophilic drug doxorubicin into HepG2 cancer cells was achieved by Liu et al. by incorporating the drug into biocompatible polymer vesicles prepared from α-cyclodextrins and double-hydrophilic poly(ethylene oxide)-b-polyMPC (PEO-b-polyMPC).[96] The doxorubicin polymer vesicles showed less cytotoxicity, enhanced drug efficiency, and an extended release over 50 hr than the parent drug.

MISCELLANEOUS APPLICATIONS

Even though the majority of the work reported to date describes the use of zwitterionic polymeric materials for non-fouling (protein-repellant) and drug delivery applications, a number of attempts are stated for other biomedical purposes including 3D hydroxyapatite mineralization,[97] bioimaging, and food processing.[98] 3D crosslinked hydrogel scaffolds prepared from zwitterionic

poly(sulfobetaine), polySBMA provided an efficient template for the nucleation and growth of hydroxyapatite (HA)-mineralization and osteointegration.[97] Enhanced cytocompatibility and appropriate swelling behavior under physiological conditions made these zwitterionic hydrogels exceptionally suited for potential skeletal tissue repair and regeneration applications.

Zwitterionic polyCBMA-modified multifunctional magnetic nanoparticles (MNPs) containing one polyCBMA chain and two 3,4-dihydroxyphenyl-L-alanine (DOPA) residue groups were developed by Zhang et al. for magnetic resonance imaging (MRI) applications.[99] Recently, the same researchers reported the development of degradable multifunctional polyCBMA nanogels encapsulated with a model drug (fluorescein isothiocyanate-dextran) and an MRI contrast agent (monodispersed Fe_3O_4 nanoparticles) with enhanced MRI performance and easy removal from the body after drug release.[100] Ultra-small sized contrast agents with very good superparamagnetic properties were reported by Kim et al. for MRI applications, using zwitterionic sulfobetaine (ASB-14, amidosulfobetaine-14)-coated superparamagnetic nanoparticles (SPIONs).[101] The SPIONs demonstrated exceptional water dispersibility and high *in vivo* stability, suggesting very promising MRI contrast agents for *in vivo* applications including tumor imaging and lymph node mapping.

CONCLUSIONS

Because of the unique and tunable structural properties and processabilities, zwitterionic materials gained a tremendous scientific interest and many are currently under evaluation for various biomedical applications, ranging from non-fouling surfaces to drug carriers. However, the exact potential of these biomaterials is still to be evaluated completely and many distinctive applications are waiting to be investigated. It is likely that we will see more use of zwitterionic materials in the future, specifically as interfaces for biomedical implants to modulate appropriate biological functions such as protein adsorption, blood coagulation, or cell adhesion.

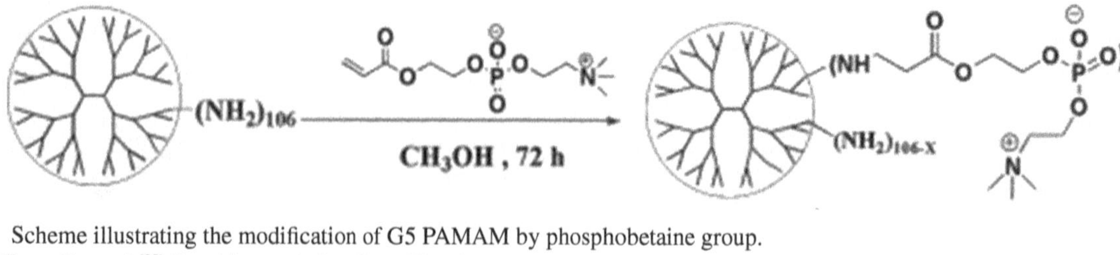

Fig. 11 Scheme illustrating the modification of G5 PAMAM by phosphobetaine group.
Source: From Jia et al.[95] ©, with permission from Elsevier.

REFERENCES

1. Damodaran, V.B.; Reynolds, M.M. Nitric oxide-releasing biomedical materials. In *Encyclopedia of Biomedical Polymers and Polymeric Biomaterials*; Mishra, M.K., Ed.; Taylor & Francis: USA, 2014, DOI: 10.1081/E-EBPP-120049951.

2. Lowe, A.B.; McCormick, C.L. Synthesis and solution properties of zwitterionic polymers. Chem. Rev. **2002**, *102* (11), 4177–4190.

3. Damodaran, V.B.; Fee, C.J.; Popat, K.C. Prediction of protein interaction behaviour with PEG-grafted matrices using X-ray photoelectron spectroscopy. Appl. Surf. Sci. **2010**, *256* (16), 4894–4901.

4. Gong, P.; Grainger, D. Nonfouling surfaces: A review of principles and applications for microarray capture assay designs. In *Methods in Molecular Biology: Microarrays*; Rampal, J., Ed.; Humana Press: USA, 2007; 59–92.

5. Hamming, L.M.; Messersmith, P.B. Fouling resistant biomimetic poly(ethylene glycol) based grafted polymer coatings. Mater. Matters **2008**, *3* (3), 52–54.

6. Damodaran, V.B.; Fee, C.J.; Popat, K.C. Modeling of PEG grafting and prediction of interfacial force profile using X-ray photoelectron spectroscopy. Surf. Interface Anal. **2012**, *44* (2), 144–149.

7. Ostuni, E.; Chapman, R.G.; Holmlin, R.E.; Takayama, S.; Whitesides, G.M. A survey of structure–property relationships of surfaces that resist the adsorption of protein. Langmuir **2001**, *17* (18), 5605–5620.

8. Chen, S.; Zheng, J.; Li, L.; Jiang, S. Strong resistance of phosphorylcholine self-assembled monolayers to protein adsorption: Insights into nonfouling properties of zwitterionic materials. J. Am. Chem. Soc., **2005**, *127* (41), 14473–14478.

9. Laughlin, R.G. Fundamentals of the zwitterionic hydrophilic group. Langmuir **1991**, *7* (5), 842–847.

10. Chen, S.; Li, L.; Zhao, C.; Zheng, J. Surface hydration: Principles and applications toward low-fouling/nonfouling biomaterials. Polymer **2010**, *51* (23), 5283–5293.

11. Weers, J.G.; Rathman, J.F.; Axe, F.U.; Crichlow, C.A.; Foland, L.D.; Scheuing, D.R.; Wiersema, R.J.; Zielske, A.G. Effect of the intramolecular charge separation distance on the solution properties of betaines and sulfobetaines. Langmuir **1991**, *7* (5), 854–867.

12. Ladenheim, H.; Morawetz, H. A new type of polyampholyte: Poly(4-vinyl pyridine betaine). J. Polym. Sci. **1957**, *26* (113), 251–254.

13. Zhang, Z.; Chen, S.; Jiang, S. Dual-functional biomimetic materials: nonfouling poly(carboxybetaine) with active functional groups for protein immobilization. Biomacromolecules **2006**, *7* (12), 3311–3315.

14. Ladd, J.; Zhang, Z.; Chen, S.; Hower, J.C.; Jiang, S. Zwitterionic polymers exhibiting high resistance to nonspecific protein adsorption from human serum and plasma. Biomacromolecules **2008**, *9* (5), 1357–1361.

15. Yang, W.; Xue, H.; Li, W.; Zhang, J.; Jiang, S. Pursuing "zero" protein adsorption of poly(carboxybetaine) from undiluted blood serum and plasma. Langmuir **2009**, *25* (19), 11911–11916.

16. Zhang, Z.; Zhang, M.; Chen, S.; Horbett, T.A.; Ratner, B.D.; Jiang, S. Blood compatibility of surfaces with superlow protein adsorption. Biomaterials **2008**, *29* (32), 4285–4291.

17. Cheng, G.; Li, G.; Xue, H.; Chen, S.; Bryers, J.D.; Jiang, S. Zwitterionic carboxybetaine polymer surfaces and their resistance to long-term biofilm formation. Biomaterials **2009**, *30* (28), 5234–5240.

18. Carr, L.R.; Krause, J.E.; Ella-Menye, J.-R.; Jiang, S. Single nonfouling hydrogels with mechanical and chemical functionality gradients. Biomaterials **2011**, *32* (33), 8456–8461.

19. Carr, L.R.; Xue, H.; Jiang, S. Functionalizable and nonfouling zwitterionic carboxybetaine hydrogels with a carboxybetaine dimethacrylate crosslinker. Biomaterials **2011**, *32* (4), 961–968.

20. Carr, L.R.; Zhou, Y.; Krause, J.E.; Xue, H.; Jiang, S. Uniform zwitterionic polymer hydrogels with a nonfouling and functionalizable crosslinker using photopolymerization. Biomaterials **2011**, *32* (29), 6893–6899.

21. Abraham, S.; Unsworth, L.D. Multi-functional initiator and poly(carboxybetaine methacrylamides) for building biocompatible surfaces using "nitroxide mediated free radical polymerization" strategies. J. Polym. Sci. Part A: Polym. Chem. **2011**, *49* (5), 1051–1060.

22. Cheng, G.; Xue, H.; Zhang, Z.; Chen, S.; Jiang S. A switchable biocompatible polymer surface with self-sterilizing and nonfouling capabilities. Angew. Chem. Int. Ed. **2008**, *47* (46), 8831–8834.

23. Cheng, G.; Xue, H.; Li, G.; Jiang, S. Integrated antimicrobial and nonfouling hydrogels to inhibit the growth of planktonic bacterial cells and keep the surface clean. Langmuir **2010**, *26* (13), 10425–10428.

24. Vaisocherová, H.; Yang, W.; Zhang, Z.; Cao, Z.; Cheng, G.; Piliarik, M.; Homola, J.; Jiang, S. Ultralow fouling and functionalizable surface chemistry based on a zwitterionic polymer enabling sensitive and specific protein detection in undiluted blood plasma. Anal. Chem. **2008**, *80* (20), 7894–7901.

25. Rodriguez-Emmenegger, C.; Brynda, E.; Riedel, T.; Houska, M.; Šubr, V.; Alles, A. B.; Hasan, E.; Gautrot, J.E.; Huck, W.T. Polymer brushes showing non-fouling in blood plasma challenge the currently accepted design of protein resistant surfaces. Macromol. Rapid Commun. **2011**, *32* (13), 952–957.

26. Krause, J.E.; Brault, N.D.; Li, Y.; Xue, H.; Zhou, Y.; Jiang, S. Photoiniferter-mediated polymerization of zwitterionic carboxybetaine monomers for low-fouling and functionalizable surface coatings. Macromolecules **2011**, *44* (23), 9213–9220.

27. Huang, C.-J.; Li, Y.; Jiang, S. Zwitterionic polymer-based platform with two-layer architecture for ultra low fouling and high protein loading. Anal. Chem. **2012**, *84* (7), 3440–3445.

28. Zhang, Z.; Vaisocherová, H.; Cheng, G.; Yang, W.; Xue, H.; Jiang, S. Nonfouling behavior of polycarboxybetaine-grafted surfaces: Structural and environmental effects. Biomacromolecules **2008**, *9* (10), 2686–2692.

29. Vaisocherová, H.; Zhang, Z.; Yang, W.; Cao, Z.; Cheng, G.; Taylor, A.D.; Piliarik, M.; Homola, J.; Jiang, S. Functionalizable surface platform with reduced nonspecific protein adsorption from full blood plasma—Material selection and protein immobilization optimization. Biosens. Bioelectron. **2009**, *24* (7), 1924–1930.

30. Chang, Y.; Liao, S.-C.; Higuchi, A.; Ruaan, R.-C.; Chu, C.-W.; Chen, W.-Y. A highly stable nonbiofouling surface

with well-packed grafted zwitterionic polysulfobetaine for plasma protein repulsion. Langmuir **2008**, *24* (10), 5453–5458.

31. Chang, Y.; Shu, S.-H.; Shih, Y.-J.; Chu, C.-W.; Ruaan, R.-C.; Chen, W.-Y. Hemocompatible mixed-charge copolymer brushes of pseudozwitterionic surfaces resistant to nonspecific plasma protein fouling. Langmuir **2009**, *26* (5), 3522–3530.

32. Yang, W.; Chen, S.; Cheng, G.; Vaisocherová, H.; Xue, H.; Li. W.; Zhang, J.; Film thickness dependence of protein adsorption from blood serum and plasma onto poly(sulfobetaine)-grafted surfaces. Langmuir **2008**, *24* (17), 9211–9214.

33. Li, G.; Cheng, G.; Xue, H.; Chen, S.; Zhang, F.; Jiang, S. Ultra low fouling zwitterionic polymers with a biomimetic adhesive group. Biomaterials **2008**, *29* (35), 4592–4597.

34. Li, L.; Marchant, R.E.; Dubnisheva, A.; Roy, S.; Fissell, W.H. Anti-biofouling sulfobetaine polymer thin films on silicon and silicon nanopore membranes. J. Biomater. Sci. Polym. Ed. **2011**, *22* (1–3), 91–106.

35. Shih, Y.-J.; Chang, Y. Tunable blood compatibility of polysulfobetaine from controllable molecular-weight dependence of zwitterionic nonfouling nature in aqueous solution. Langmuir **2010**, *26* (22), 17286–17294.

36. Zhao, Y.-H.; Wee, K.-H.; Bai, R. Highly hydrophilic and low-protein-fouling polypropylene membrane prepared by surface modification with sulfobetaine-based zwitterionic polymer through a combined surface polymerization method. J. Membr. Sci. **2010**, *362* (1), 326–333.

37. Li, Q.; Bi, Q.-Y.; Zhou, B.; Wang, X.-L. Zwitterionic sulfobetaine-grafted poly(vinylidene fluoride) membrane surface with stably anti-protein-fouling performance via a two-step surface polymerization. Appl. Surf. Sci. **2012**, *258* (10), 4707–4717.

38. Chang, Y.; Chang, W.-J.; Shih, Y.-J.; Wei, T.-C.; Hsiue, G.-H. Zwitterionic sulfobetaine-grafted poly(vinylidene fluoride) membrane with highly effective blood compatibility via atmospheric plasma-induced surface copolymerization. ACS Appl. Mater. Interfaces **2011**, *3* (4), 1228–1237.

39. Schön, P.; Kutnyanszky, E.; ten Donkelaar, B.; Santonicola, M.G.; Tecim, T.; Aldred, N.; Clare, A.S.; Vancso, G.J. Probing biofouling resistant polymer brush surfaces by atomic force microscopy based force spectroscopy. Colloids Surf. B Biointerfaces **2013**, *102*, 923–930.

40. Chang, Y.; Chen, W.-Y.; Yandi, W.; Shih, Y.-J.; Chu, W.-L.; Liu, Y.-L.; Chu, C.-W.; Ruaan, R.C.; Higuchi, A. Dual-Thermoresponsive Phase Behavior of Blood Compatible Zwitterionic Copolymers Containing Nonionic Poly(N-isopropyl acrylamide). Biomacromolecules **2009**, *10* (8), 2092–2100.

41. Kim, M.; Kim, J.C.; Rho, Y.; Jung, J.; Kwon, W.; Kim, H.; Ree, M. Bacterial adherence on self-assembled films of brush polymers bearing zwitterionic sulfobetaine moieties. J. Mater. Chem. **2012**, *22* (37), 19418–19428.

42. Nguyen, A.T.; Baggerman, J.; Paulusse, J.M.J.; Zuilhof, H.; van Rijn, C.J.M. Bioconjugation of protein-repellent zwitterionic polymer brushes grafted from silicon nitride. Langmuir **2011**, *28* (1), 604–610.

43. Nguyen, A.T.; Baggerman, J.; Paulusse, J.M.J.; van Rijn, C.J.M.; Zuilhof, H. Stable protein-repellent zwitterionic polymer brushes grafted from silicon nitride. Langmuir **2011**, *27* (6), 2587–2594.

44. Zwaal, R.F.A.; Comfurius, P.; Van Deenen, L.L. Membrane asymmetry and blood coagulation. Nature **1977**, *268* (5618), 358–360.

45. Lewis, A.L. Phosphorylcholine-based polymers and their use in the prevention of biofouling. Colloids Surf. B Biointerfaces **2000**, *18* (3–4), 261–275.

46. Johnston, D.S.; Sanghera, S.; Pons, M.; Chapman, D. Phospholipid polymers—synthesis and spectral characteristics. Biochim. Biophys. Acta (BBA) - Biomembranes **1980**, *602* (1), 57–69.

47. Hayward, J.A.; Chapman, D. Biomembrane surfaces as models for polymer design: The potential for haemocompatibility. Biomaterials **1984**, *5* (3), 135–142.

48. Hall, B.; Ie, E.; Bird, R.; Kojima, M.; Chapman, D. Biomembranes as models for polymer surfaces: V. Thrombelastographic studies of polymeric lipids and polyesters. Biomaterials **1989**, *10* (4), 219–224.

49. Hayward, J.A.; Durrani, A.A.; Shelton, C.J.; Lee, D.C.; Chapman, D. Biomembranes as models for polymer surfaces: III. Characterization of a phosphorylcholine surface covalenay bound to glass. Biomaterials **1986**, *7* (2), 126–131.

50. Pons, M.; Villaverde, C.; Chapman, D.A 13CNMR study of 10,12-tricosadiynoic acid and the corresponding phospholipid and phospholipid polymer. Biochim. Biophys. Acta (BBA) - Biomembranes **1983**, *730* (2), 306–312.

51. Kojima, M.; Ishihara, K.; Watanabe, A.; Nakabayashi, N. Interaction between phospholipids and biocompatible polymers containing a phosphorylcholine moiety. Biomaterials **1991**, *12* (2), 121–124.

52. Iwasaki, Y.; Kurita, K.; Ishihara, K.; Nakabayashi, N. Effect of methylene chain length in phospholipid moiety on blood compatibility of phospholipid polymers. J. Biomater. Sci. Polym. Ed. **1995**, *6* (5), 447–461.

53. Iwasaki, Y.; Kurita, K.; Ishihara, K.; Nakabayashi, N. Effect of reduced protein adsorption on platelet adhesion at the phospholipid polymer surfaces. J. Biomater. Sci. Polym. Ed. **1997**, *8* (2), 151–163.

54. Ishihara, K.; Fukumoto, K.; Iwasaki, Y.; Nakabayashi, N. Modification of polysulfone with phospholipid polymer for improvement of the blood compatibility. Part 1. Surface characterization. Biomaterials **1999**, *20* (17), 1545–1551.

55. Ishihara, K.; Ziats, N.P.; Tierney, B.P.; Nakabayashi, N.; Anderson, J.M. Protein adsorption from human plasma is reduced on phospholipid polymers. J. Biomed. Mater. Res. **1991**, *25* (11), 1397–1407.

56. van der Heiden, A.P.; Willems, G.M.; Lindhout, T.; Pijpers, A.P.; Koole, L.H. Adsorption of proteins onto poly(ether urethane) with a phosphorylcholine moiety and influence of preadsorbed phospholipid. J. Biomed. Mater. Res. **1998**, *40* (2), 195–203.

57. Ruiz, L.; Fine, E.; Makohliso, S.A.; Onard, D.; Johnston, D.S.; Textor, M.; Mathieu, H.J. Phosphorylcholine-containing polyurethanes for the control of protein adsorption and cell attachment via photoimmobilized laminin oligopeptides. J. Biomater. Sci. Polym. Ed. **1999**, *10* (9), 931–955.

58. Yung, L.-Y.L.; Cooper, S.L. Neutrophil adhesion on phosphorylcholine-containing polyurethanes. Biomaterials **1998**, *19* (1–3), 31–40.

59. Ishihara, K.; Hasegawa, T.; Watanabe, J.; Iwasaki, Y. Protein adsorption–resistant hollow fibers for blood purification. Artif. Organs **2002**, *26* (12), 1014–1019.

60. West, S.L.; Salvage, J.P.; Lobb, E.J.; Armes, S.P.; Billingham, N.C.; Lewis, A.L.; Hanlon, G.W.; Lloyd, A.W. The biocompatibility of crosslinkable copolymer coatings containing sulfobetaines and phosphobetaines. Biomaterials **2004**, *25* (7), 1195–1204.

61. Watanabe, J.; Ishihara, K. Phosphorylcholine and poly(D,L-lactic acid) containing copolymers as substrates for cell adhesion. Artif. Organs **2003**, *27* (3), 242–248.

62. Watanabe, J.; Eriguchi, T.; Ishihara, K. Cell adhesion and morphology in porous scaffold based on enantiomeric poly(lactic acid) graft-type phospholipid polymers. Biomacromolecules **2002**, *3* (6), 1375–1383.

63. Sawada, S.-I.; Sakaki, S.; Iwasaki, Y.; Nakabayashi, N.; Ishihara, K. Suppression of the inflammatory response from adherent cells on phospholipid polymers. J. Biomed. Mater. Res. Part A **2003**, *64A* (3), 411–416.

64. Iwasaki, Y.; Sawada, S.-I.; Ishihara, K.; Khang, G.; Lee, H.B. Reduction of surface-induced inflammatory reaction on PLGA/MPC polymer blend. Biomaterials **2002**, *23* (18), 3897–3903.

65. Iwasaki, Y.; Yamasaki, A.; Ishihara, K. Platelet compatible blood filtration fabrics using a phosphorylcholine polymer having high surface mobility. Biomaterials **2003**, *24* (20), 3599–3604.

66. Iwasaki, Y.; Ishihara, K. Phosphorylcholine-containing polymers for biomedical applications. Anal. Bioanal. Chem. **2005**, *381* (3), 534–546.

67. Nakaya, T.; Li, Y.-J. Phospholipid polymers. Prog. Polym. Sci. **1999**, *24* (1), 143–181.

68. Willis, S.L.; Court, J.L.; Redman, R.P.; Wang, J.-H.; Leppard, S.W.; O'Byrne, V.J.; Small, S.A.; Lewis, A.L.; Jones, S.A.; Stratford, P.W. A novel phosphorylcholine-coated contact lens for extended wear use. Biomaterials **2001**, *22* (24), 3261–3272.

69. Young, G.; Bowers, R.; Hall, B.; Port, M. Six month clinical evaluation of a biomimetic hydrogel contact lens. CLAO J. **1997**, *23* (4), 226–236.

70. Proclear PC technology, http://www2.coopervision.com/data/australia_new/documents/pc_technology_sheet.pdf (accessed March 2013).

71. Goda, T.; Matsuno, R.; Konno, T.; Takai, M.; Ishihara, K. Protein adsorption resistance and oxygen permeability of chemically crosslinked phospholipid polymer hydrogel for ophthalmologic biomaterials. J. Biomed. Mater. Res. Part B **2009**, *89B* (1), 184–190.

72. Chu, M.; Shirai, T.; Takahashi, D.; Arakawa, T.; Kudo, H.; Sano, K.; Sawada, S.-I.; Yano, K.; Iwasaki, Y.; Akiyoshi, K.; Mochizuki, M.; Mitsubayashi, K. Biomedical soft contact-lens sensor for in situ ocular biomonitoring of tear contents. Biomed. Microdevices **2011**, *13* (4), 603–611.

73. Ye, S.-H.; Johnson, C.A.; Woolley, J.R.; Snyder, T.A.; Gamble, L.J.; Wagner, W.R. Covalent surface modification of a titanium alloy with a phosphorylcholine-containing copolymer for reduced thrombogenicity in cardiovascular devices. J. Biomed. Mater. Res. Part A **2009**, *91A* (1), 18–28.

74. Kyomoto, M.; Moro, T.; Iwasaki, Y.; Miyaji, F.; Kawaguchi, H.; Takatori, Y.; Nakamura, K.; Ishihara, K. Superlubricious surface mimicking articular cartilage by grafting poly(2-methacryloyloxyethyl phosphorylcholine) on orthopaedic metal bearings. J. Biomed. Mater. Res. Part A **2009**, *91A* (3), 730–741.

75. Ye, S.-H.; Johnson, Jr, C.A.; Woolley, J.R.; Murata, H.; Gamble, L.J.; Ishihara, K.; Wagner, W.R. Simple surface modification of a titanium alloy with silanated zwitterionic phosphorylcholine or sulfobetaine modifiers to reduce thrombogenicity. Colloids Surf. B Biointerfaces **2010**, *79* (2), 357–364.

76. Lewis, A.L.; Tolhurst, L.A.; Stratford, P.W. Analysis of a phosphorylcholine-based polymer coating on a coronary stent pre- and post-implantation. Biomaterials **2002**, *23* (7), 1697–1706.

77. Zhu, L.; Jin, Q.; Xu, J.; Ji, J.; Shen, J. Poly(2-(methacryloyloxy) ethyl phosphorylcholine)-functionalized multi-walled carbon nanotubes: Preparation, characterization, solubility, and effects on blood coagulation. J. Appl. Polym. Sci. **2009**, *113* (1), 351–357.

78. Xu, F.-M.; Xu, J.-P.; Ji, J.; Shen, J.-C. A novel biomimetic polymer as amphiphilic surfactant for soluble and biocompatible carbon nanotubes (CNTs). Colloids Surf. B Biointerfaces **2008**, *67* (1), 67–72.

79. Collingwood, R.; Gibson, L.; Sedlik, S.; Virmani, R.; Carter, A.J. Stent-based delivery of ABT-578 via a phosphorylcholine surface coating reduces neointimal formation in the porcine coronary model. Catheter. Cardiovasc. Interv. **2005**, *65* (2), 227–232.

80. Yoshimoto, K.; Hirase, T.; Madsen, J.; Armes, S.P.; Nagasaki, Y. Non-fouling character of poly[2-(methacryloyloxy)ethyl phosphorylcholine]-modified gold surfaces fabricated by the 'grafting to' method: Comparison of its protein resistance with poly(ethylene glycol)-modified gold surfaces. Macromol. Rapid Commun. **2009**, *30* (24), 2136–2140.

81. Chen, X.; Lawrence, J.; Parelkar, S.; Emrick, T. Novel zwitterionic copolymers with dihydrolipoic acid: Synthesis and preparation of nonfouling nanorods. Macromolecules **2013**, *46* (1), 119–127.

82. Kyomoto, M.; Moro, T.; Konno, T.; Takadama, H.; Kawaguchi, H.; Takatori, Y.; Nakamura, K.; Yamawaki, N.; Ishihara, K. Effects of photo-induced graft polymerization of 2-methacryloyloxyethyl phosphorylcholine on physical properties of cross-linked polyethylene in artificial hip joints. J. Mater. Sci. Mater. Med. **2007**, *18* (9), 1809–1815.

83. Moro, T.; Kawaguchi, H.; Ishihara, K.; Kyomoto, M.; Karita, T.; Ito, H.; Nakamura, K.; Takatori, Y. Wear resistance of artificial hip joints with poly(2-methacryloyloxyethyl phosphorylcholine) grafted polyethylene: Comparisons with the effect of polyethylene cross-linking and ceramic femoral heads. Biomaterials **2009**, *30* (16), 2995–3001.

84. Cao, Z.; Yu, Q.; Xue, H.; Cheng, G.; Jiang, S. Nanoparticles for drug delivery prepared from amphiphilic PLGA zwitterionic block copolymers with sharp contrast in polarity between two blocks. Angew. Chem. Int. Ed. **2010**, *49* (22), 3771–3776.

85. Cao, Z.; Zhang, L.; Jiang, S. Superhydrophilic zwitterionic polymers stabilize liposomes. Langmuir **2012**, *28* (31), 11625–11632.

86. Wang, X.; Sun, X.; Jiang, G.; Wang, R.; Hu, R.; Xi, X.; Zhou, Y.; Wang, S.; Wang, T. Synthesis of biomimetic hyperbranched zwitterionic polymers as targeting drug delivery carriers. J. Appl. Polym. Sci. **2013**, *128* (5), 3289–3294.

87. Sun, J.-T.; Yu, Z.-Q.; Hong, C.-Y.; Pan, C.-Y. Biocompatible zwitterionic sulfobetaine copolymer-coated mesoporous silica nanoparticles for temperature-responsive drug release. Macromol. Rapid Commun. **2012**, *33* (9), 811–818.

88. Cao, J.; Xiu, K.-M.; Zhu, K.; Chen, Y.-W, Luo X.-L. Copolymer nanoparticles composed of sulfobetaine and poly(ε-caprolactone) as novel anticancer drug carriers. J. Biomed. Mater. Res. Part A **2012**, *100A* (8), 2079–2087.

89. Zhai, X.; Wang, W.; Wang, C.; Wang, Q.; Liu, W. PDMAEMA-b-polysulfobetaine brushes-modified?— Polylysine as a serum-resistant vector for highly efficient gene delivery. J. Mater. Chem. **2012**, *22* (44), 23576–23586.

90. Licciardi, M.; Tang, Y.; Billingham, N.C.; Armes, S.P.; Lewis, A.L. Synthesis of novel folic acid-functionalized biocompatible block copolymers by atom transfer radical polymerization for gene delivery and encapsulation of hydrophobic drugs. Biomacromolecules **2005**, *6* (2), 1085–1096.

91. Yusa, S.-I.; Fukuda, K.; Yamamoto, T.; Ishihara, K.; Morishima, Y. Synthesis of well-defined amphiphilic block copolymers having phospholipid polymer sequences as a novel biocompatible polymer micelle reagent. Biomacromolecules **2005**, *6* (2), 663–670.

92. Wada, M.; Jinno, H.; Ueda, M.; Ikeda, T.; Kitajima, M.; Konno, T.; Watanabe, J.; Ishihara, K. Efficacy of an MPC-BMA co-polymer as a nanotransporter for paclitaxel. Anticancer Res. **2007**, *27* (3B), 1431–1435.

93. Madsen, J.; Armes, S.P.; Bertal, K.; MacNeil, S.; Lewis, A.L. Preparation and aqueous solution properties of thermoresponsive biocompatible AB diblock copolymers. Biomacromolecules **2009**, *10* (7), 1875–1887.

94. Chen, X.; McRae, S.; Parelkar, S.; Emrick, T. Polymeric phosphorylcholine–camptothecin conjugates prepared by controlled free radical polymerization and click chemistry. Bioconjug. Chem. **2009**, *20* (12), 2331–2341.

95. Jia, L.; Xu, J.-P.; Wang, H.; Ji, J. Polyamidoamine dendrimers surface-engineered with biomimetic phosphorylcholine as potential drug delivery carriers. Colloids Surf. B Biointerfaces **2011**, *84* (1), 49–54.

96. Liu, G.; Jin, Q.; Liu, X.; Lv, L.; Chen, C.; Ji, J. Biocompatible vesicles based on PEO-b-PMPC/[small alpha]-cyclodextrin inclusion complexes for drug delivery. Soft Matter **2011**, *7* (2), 662–669.

97. Liu, P.; Song, J. Sulfobetaine as a zwitterionic mediator for 3D hydroxyapatite mineralization. Biomaterials **2013**, *34* (10), 2442–2454.

98. Mérian, T.; Goddard, J.M. Advances in Nonfouling Materials: Perspectives for the Food Industry. J. Agric. Food Chem. **2012**, *60* (12), 2943–2957.

99. Zhang, L.; Xue, H.; Gao, C.; Carr, L.; Wang, J.; Chu, B.; Jiang, S. Imaging and cell targeting characteristics of magnetic nanoparticles modified by a functionalizable zwitterionic polymer with adhesive 3,4-dihydroxyphenyl-l-alanine linkages. Biomaterials **2010**, *31* (25), 6582–6588.

100. Zhang, L.; Xue, H.; Cao, Z.; Keefe, A.; Wang, J.; Jiang, S. Multifunctional and degradable zwitterionic nanogels for targeted delivery, enhanced MR imaging, reduction-sensitive drug release, and renal clearance. Biomaterials **2011**, *32* (20), 4604–4608.

101. Kim, D.; Chae, M.K.; Joo, H.J.; Jeong, I.-H.; Cho, J.-H.; Lee, C. Facile preparation of zwitterion-stabilized superparamagnetic iron oxide nanoparticles (ZSPIONs) as an MR contrast agent for *in vivo* applications. Langmuir **2012**, *28* (25), 9634–9639.

Index

ABHS, *see* Albumin-based hydrogel sealant (ABHS)
Abrasion testing, 760, 761
Absorbable suture
  characteristics, 1527
  commercial structure, 1526–1527
  commercial suture materials, 1515–1518
  comparison, 1520, 1523
  degradation products, 1528
  mechanical properties, 1520–1521
  scanning electron images, 1515, 1519, 1527–1528
  tensile breaking force, 1527
Acid etching, 433
Actin gels, 682
Active pharmaceutical ingredients (APIs), 1076, 1465
Adenosine diphosphate (ADP), 1560–1561
Adhesives
  catechol derived, 1, 4–7
  cyanoacrylate glues, 2–3
  fibrin glue, 1–2
  Gecko-inspired, 9–11
  light-activated, 7–9
  nanostructured adhesives, 1
  polyethylene glycol-based glue, 1, 3
  polysaccharide adhesives, 1
Adipose-derived stem cells (ADSCs), 727, 1410
Adriamycin (ADR), 790
Adsorption theory
  bioadhesion, 113–114
  mucoadhesion, 963–964
Aerogels
  alumina, 41
  biopolymer-based, 42
  cellulosic (*see* Cellulosic aerogel)
  cobalt–molybdenum–sulfur, 42
  silica, 39–41
Aerosols, 308
Agar overlay tests, 131
Agarose, 681, 1218–1219, 1259
Agglutination techniques, 751
AGUs, *see* Anhydro-D-glucopyranose units (AGUs)
Albumin
  glue, 648–649
  small-molecule therapeutics delivery, 66
Albumin-based hydrogel sealant (ABHS), 650
Alcohol-resistant polyurethane, 1326
Aldehyde system
  albumin glue, 648–649
  aldehyde–gelatin film, 649
  gelatin and polysaccharides, 648
  glutaraldehyde, 649
Alginate, 681, 1259
Alginic acid
  applications, 1610
  manufacture, 1610

  structure, 1610
  wound care, 1609–1610
Alkaline phosphatase activity (ALP), 523, 525
Alkaline-catalyzed oxalkylation, 278
AM, *see* Amniotic membrane (AM)
Amalgam
  bonds, 436, 437
  restoration, 435, 436
  sealant, 435–437
Ambiphilic mucoadhesive polymers, 950
Amine-based MFC aerogels, 43
Amino acid *N*-carboxy anhydride (B-NCA), 589–590
(γ-Aminopropyl) triethoxysilane (APS) molecules, 589–590
Amniotic fluid-derived stem cells (AFSCs), 727
Amniotic membrane (AM), 1635
  corneal endothelium, 383–384
  corneal epithelium, 379
Amorphous solid dispersion system
  melt extrusion, 828
    bioavailability enhancement applications, 829
    controlled release products, 835, 836
    dissolution-enhanced products, 833–835
    miscibility, 831
    plasticizer, 829
    spring and parachute performance, 832
    surfactants, 831
  polymeric interactions
    anti-solvent process, 465
    basic techniques, 464
    cryogenic production, 465
    crystalline and glassy solid enthalpy and volume, 463
    dissolution rate, 468
    enteric compositions, 470, 471
    fluid bed coating, 464
    hot-melt extrusion, 466–468
    physicochemical properties, 462–463
    supercritical fluid technologies, 465
    tacrolimus solid dispersions, 469
    in vitro supersaturation profiles, 471
Amoxicillin-loaded electrospun nano-HAp/PLA (AMX), 497, 498
Amoxicillin mucoadhesive microspheres, 121
Amphiphilic diblock copolymer self-assembly, PNSs
  elongated micelles, 934
  micellar unimolecules and aggregates, 933
  micellization and gelation, 933
  spherical micelles, 933–934
  vesicles, 934
Amphiphilic hybrid block copolymers, 350–351
AMX, *see* Amoxicillin-loaded electrospun nano-HAp/PLA (AMX)
Amyl acetate (AA), 1551

AN69®, 868
Angiogenic process, 724, 1215
  angiogenic factor, 799–800
  ANSA, 801
  antibacterial polymers, 802
  fucoidan, 801–802
  HSPGs, 800–801
  interaction, 801
  mechanisms, 801
  pathological processes, 799
  therapeutic induction, 801–802
Angiography catheters, 1324–1325
Anhydro-D-glucopyranose units (AGUs), 271, 277
Anionic mucoadhesive polymers, 947, 948
Antibacterial polymers
  angiogenic process, 802
  biomedical applications, 802–803
  cationic biocides, 803
  cationic polymers, 804
  eukaryotic cells, 803–804
  phosphatidylethanolamine, 803
  requirements, 803
  strategies, 803
  surface functionalization, 803
Antibody
  biosensors, dentritic, 408
  small-molecule therapeutics delivery, 66
Antibody-targeted micelles, 1555–1556
Anticoagulants, 664
Anti-fouling substrates, 606–607
Antigen delivery, 143–144
Anti-infective biomaterials
  antimicrobial incorporation, 85
  bacterial adhesion (*see* Bacterial adhesion)
  DACRON® prostheses, 85
  Septopal®, 85
  silicone catheters, 85
  silver catheter, 85
  TEFLON® prostheses, 85
Anti-inflammatory biomaterials, 178
Antimicrobial polyurethanes
  co-extruded antimicrobial catheter, 1333
  fluorinated, 1335
  nosocomial infections, 1329
  polycarbonate-based, 1333
  polyethylene vinyl acetate, 1331
  silver and, 1330–1331
  thermoplastic medical-grade, 1334–1335
  wound dressings, 1333–1334
  zone of inhibition, 1331–1332
Antithrombogenicity *vs.* pseudoneointima, 665
Aprotinin, 2
Aptamers
  drug-encapsulated controlled-release, 1117–1118
  targeted drug delivery, 68–69
Arg-Gly-Asp (RGD) sequence, 344–345